ENGINEERED MATERIALS HANDBOOK®

Volume 3

ADHESIVES AND SEALANTS

Prepared under the direction of the
ASM International Handbook Committee

Hal F. Brinson, Technical Chairman

Cyril A. Dostal, Senior Editor
Mara S. Woods, Technical Editor
Alice W. Ronke, Assistant Editor
Scott D. Henry, Assistant Editor
Janice L. Daquila, Assistant Editor
Karen Lynn O'Keefe, Word Processing Specialist

Robert L. Stedfeld, Director of Reference Publications
Joseph R. Davis, Manager of Handbook Development
Penelope Allen, Manager of Handbook Production

Editorial Assistance
Lois A. Abel
Kelly Ferjutz
Heather F. Lampman
Penelope Thomas
Nikki D. Wheaton

**The Materials
Information Society**

First printing, December 1990

Engineered Materials Handbook is a collective effort involving hundreds of technical
specialists. It brings together in one book a wealth of information from world-wide sources to
help scientists, engineers, and technicians solve current and long-range problems.

Great care is taken in the compilation and production of this volume, but it should be made
clear that no warranties, express or implied, are given in connection with the accuracy or
completeness of this publication, and no responsibility can be taken for any claims that may
arise.

Nothing contained in the Engineered Materials Handbook shall be construed as a grant of any
right of manufacture, sale, use, or reproduction, in connection with any method, process,
apparatus, product, composition, or system, whether or not covered by letters patent,
copyright, or trademark, and nothing contained in the Engineered Materials Handbook shall be
construed as a defense against any alleged infringement of letters patent, copyright, or
trademark, or as a defense against liability for such infringement.

Comments, criticisms, and suggestions are invited, and should be forwarded to ASM
International.

Library of Congress Cataloging-in-Publication Data

ASM International

Engineered materials handbook.
Vol 3-: Hal F. Brinson, technical chairman.

Includes bibliographies and indexes.
Contents: v. 1. Composites—v. 2. Engineering plastics—v. 3. Adhesives and sealants.

1. Materials—Handbooks, manuals, etc. I. Reinhart, Theodore J.
II. Brinson, H.F. III. ASM
International. Handbook Committee.
TA403.E497 1987 620.1′1 87–19265
ISBN 0-87170-279-7 (v. 1)
SAN 204–7586
ISBN 0-87170-281-9

Printed in the United States of America

Foreword

The first two volumes of the *Engineered Materials Handbook* series, *Composites* and *Engineering Plastics*, were both devoted to materials based on synthetic polymer resins. Composites, as defined for this series, are polymer resin matrix materials (for the most part thermosets) reinforced with continuous fibers and engineered for high-performance properties that permit them to be used for primary structural applications in aerospace, transportation, and other high-tech industries. Engineering plastics are polymer resin-base materials (thermoplastics as well as thermosets) capable of being molded into load-bearing shapes, with high-performance properties that permit them to be used in the same manner as metals and ceramics.

This new volume is concerned with yet another category of polymeric materials: those used as engineering adhesives and sealants for joining and bonding a wide range of other materials, including metals, composites, engineering plastics, and ceramics. Another definition is in order here: Engineering adhesives and sealants, as defined for this series, are polymer resin-base materials used in the manufacturing industries for joining structural elements (or sealing joints between them) so that the joints themselves become load-bearing elements of the structural systems in which they are located.

Although the use of adhesives and sealants has been traced back to as early as 3000 B.C., only in relatively recent times has knowledge of the adhesion phenomena that enable this use begun to be recognized as a science in itself. As this body of scientific knowledge has grown, so has the use of engineering adhesives and sealants. With this growing use has come the need for a practical, comprehensive, basic reference on these materials, directed toward engineers in user companies, who in most cases, although perhaps experts in their product areas, are not adhesives and sealants experts.

Adhesives and Sealants is, to our knowledge, the first attempt, worldwide, to publish a book of this nature in the subject area. It could not, we believe, have been published a decade earlier. Our commitment to publish it at this time is evidence of our belief that it is a book whose time has come. We are pleased and very proud that this first attempt bears the ASM International imprint. Publication of this volume is a partial fulfillment of the society's continuing mission of leadership in gathering, processing, and disseminating technical information and in fostering the understanding and application of engineered materials and their research, design, reliable manufacture, use, and economic and social benefits.

We express our thanks and deep appreciation to technical chairman Hal F. Brinson and all the many professionals who gave freely of their time and expertise as section chairmen, authors, and reviewers for this volume. We also thank their organizations for supporting these contributions. The members of our Handbook Committee, likewise, are due our gratitude for their comments, suggestions, criticisms, and unstinting assistance with the peer review process for this volume. And also, we thank the members of our professional editorial staff for their commitment to holding this volume to the same high standards as its predecessors.

Stephen M. Copley
President
ASM International

Edward L. Langer
Managing Director
ASM International

Policy on Units of Measure

By a resolution of its Board of Trustees, ASM International has adopted the practice of publishing data in both metric and customary U.S. units of measure. In preparing this Handbook, the editors have attempted to present data primarily in metric units based on Système International d'Unités (SI), with secondary mention of the corresponding values in customary U.S. units. The decision to use SI as the primary system of units was based on the aforementioned resolution of the Board of Trustees, the widespread use of metric units throughout the world, and the expectation that the use of metric units in the United States will increase substantially during the anticipated lifetime of this Handbook.

For the most part, numerical engineering data in the text and in tables are presented in SI-based units with the customary U.S. equivalents in parentheses (text) or adjoining columns (tables). For example, pressure, stress, and strength are shown both in SI units, which are pascals (Pa) with a suitable prefix, and in customary U.S. units, which are pounds per square inch (psi). To save space, large values of psi have been converted to kips per square inch (ksi), where 1 ksi = 1000 psi. Some strictly scientific data are presented in SI units only.

To clarify some illustrations, only SI units are presented on artwork. References in the accompanying text to data in the illustrations are presented in both SI-based and customary U.S. units. On graphs and charts, grids correspond to SI-based units, which appear along the left and bottom axes; where appropriate, corresponding customary U.S. units appear along the top and right axes.

Both nanometers (nm) and angstrom units (Å) are used (1 nm = 10Å), the former to measure light wavelengths, and the latter as a unit of measure in x-ray crystallography. Data obtained according to standardized test methods for which the standard recommended a particular system of units are presented in the units of that system. Wherever feasible, equivalent units are also presented.

Conversions and rounding have been done in accordance with ASTM Standard E 380, with careful attention to the number of significant digits in the original data. For example, an annealing temperature of 1570 °F contains three significant digits. In this instance, the equivalent temperature would be given as 855 °C; the exact conversion to 854.44 °C would not be appropriate. For an invariant physical phenomenon that occurs at a precise temperature (such as the melting of pure silver), it would be appropriate to report the temperature as 961.93 °C (or 1763.5 °F). In many instances (especially in tables and data compilations), temperature values in °C and °F are alternatives rather than conversions.

The policy on units of measure in this Handbook contains several exceptions to strict conformance to ASTM E 380; in each instance, the exception has been made to improve the clarity of the Handbook. The most notable exception is the use of MPa $\sqrt{m}$ rather than $MN \cdot m^{-3/2}$ or $MPa \cdot m^{0.5}$ as the SI unit of measure for fracture toughness. Other examples of such exceptions are the use of "L" rather than "l" as the abbreviation for liter and the use of g/cm^3 rather than kg/m^3 as the unit of measure for density (mass per unit volume).

SI practice requires that only one virgule (diagonal) appear in units formed by combination of several basic units. Therefore, all of the units preceding the virgule are in the numerator and all units following the virgule are in the denominator of the expression; no parentheses are required to prevent ambiguity.

Preface

Adhesives and sealants have played an important role in the development of modern technology in the United States and elsewhere in the world. They are used extensively in such high-tech industries as aerospace, transportation, computers, telecommunications, and so on. In such uses, they are not just a means of joining structural elements or sealing joints between them, but are in fact load-bearing components of structural systems, in which their failure could cause the entire structure to fail or cease to perform its intended function. In these applications, adhesives and sealants must be included in the category of engineering materials, along with the structural metals, composites, engineering plastics, advanced ceramics, and so forth, that are normally categorized as such.

There has long been a need for a basic reference source on adhesives and sealants to which engineers and scientists can turn when dealing with the engineering design process and other complex technical issues associated with their use. This handbook is intended to serve this need. Its emphasis is on structural or engineering adhesives and sealants for manufacturing industry use, rather than on those materials intended for applications that do not fit this definition. There is no intent here to minimize the need for detailed, scientifically accurate information about these materials for other uses; in fact, we hope the volume will prove to be helpful to those involved with them, as well.

Although the employment of adhesives and sealants can be traced to ancient times, until recently the adhesion process has been regarded more as an art than a science. The last several decades have seen its gradual emergence as a scientific discipline in its own right. The Dutch-designed Fokker F27, an all-bonded turboprop aircraft that has been in service since the 1950s without a single adhesive bond failure, was an important milestone in this emergence. Similarly, the U.S. Air Force-funded PABST adhesives research program in the 1970s gave an order-of-magnitude boost to scientific understanding of the adhesion process. In these early years the Gordon Research Conference on Adhesion, as well as the establishment of the Adhesion Society and technical journals on adhesion, led to interaction in this technology of large numbers of industrial, government, and academic investigators from many scientific disciplines. During the 1980s government and industrial leaders, notably at the U.S. Office of Naval Research and the Adhesives and Sealants Council, supported the establishment of research and education programs at academic institutions, so that today there are several university centers or institutes devoted exclusively to adhesion science and numerous faculty members at many other universities who specialize in this area.

As the result of these and other similar activities in the last several decades, there has been a growing recognition of adhesion technology as a basic or fundamental science in its own right because it requires a knowledge of the interatomic forces that cause two dissimilar materials to adhere to each other. Further, this knowledge of microstructural events must be expressed in the macroscopic terms of engineering design parameters. The fundamental disciplines involved are:

- *Chemistry*, which provides information on the molecular nature of the materials to be joined, the chemistry of the adhesives and substrates, and the preparations (and/or primers) of the surfaces to be joined
- *Materials science*, which serves to connect the microscopic scale of chemistry with the macroscopic scale of solid mechanics through the examination of phenomena such as the loci of failure and deformation modes
- *Solid mechanics*, which involves stress and failure analysis associated with the design engineering process, including mathematical and experimental interpretation of mechanical properties

We have endeavored to ensure that each of these disciplines is well represented.

In recent years it has become clear that simply to include information from the different disciplines is not an adequate approach for properly addressing the cross-disciplinary nature of adhesion (including sealant) science. Rather, it is necessary to integrate these disciplines into a single science of adhesion—a singular entity instead of a collection of multiple sciences. Obviously, the merging of several sciences into a new one is difficult, at best. In this volume an attempt has been made to organize the information toward this end. We must let its readers judge the success of this endeavor.

The handbook has been organized with a view towards making it useful to both the novice and the experienced adhesion expert. Its first section covers fundamentals: a glossary of terms and definitions; overviews of adhesives and sealants technology, markets, and applications; specifications and standards applicable to both; and a review of general information sources. Sections that follow detail the major families of adhesives and sealant materials including their relation to function and use, surface considerations for both, and testing and analysis approaches for determining and utilizing their properties for design. Additional sections treat design use, environmental effects, manufacturing processes for using adhesives and sealants, and nondestructive inspection and quality control. A final section focuses on bonded structure repair of aircraft.

I am pleased to have the opportunity to thank all those who have contributed directly and indirectly to the making of this volume, including advisory committee members, section chairmen, article authors, technical reviewers, and so on. However, it is perhaps also appropriate here to acknowledge all of the efforts over the last four decades of those outstanding individuals who have, with dedication, perseverance, and foresight, brought our knowledge of adhesion into the realm of a rational science rather than that of a trial-and-error technology.

Hal F. Brinson, Technical Chairman
Professor of Mechanical Engineering,
University of Texas at San Antonio
Adjunct Institute Engineer,
Southwest Research Institute

Authors

Kent Adams
I.A.M., Inc.
R.D. Adams
University of Bristol
Sam C. Aker
Sam C. Aker, Inc.
Mark G. Allen
Georgia Institute of Technology
Dan Arnold
Boeing Military Airplane Company
Jamil Baghdachi
BASF Corporation
Yoseph Bar-Cohen
McDonnell Douglas Corporation
Willard D. Bascom
The University of Utah
Robert E. Batson
Dexus Research Inc.
C.E. Beck
Wright State University
M. Dale Beers
Loctite Corporation
Dean T. Behm
Ciba-Geigy Corporation
Thomas C. Bingham
Boeing Commercial Airplane Company
P.R. Borgmeier
The University of Utah
Hans J. Borstell
Grumman Aircraft Systems
W. Brockmann
Kaiserslautern University
William D. Brown
Dexter Adhesives and Structural
Materials Division
William O. Buckley
Ciba-Geigy Corporation
Edwin F. Bushman
Consultant
Lawrence D. Carbary
Dow Corning Corporation
Sarat Chandrasekharan
Ciba-Geigy Corporation
Francis H. Chang
General Dynamics Corporation
Harry K. Charles, Jr.
The Johns Hopkins University
Edwin C. Clark
Ciba-Geigy Corporation
H.M. Clearfield
IBM Corporation
John Comyn
John Comyn Ltd.
J.W. Connell
NASA Langley Research Center

Dennis J. Damico
Lord Corporation
Guy D. Davis
Martin Marietta Laboratories
Jerry W. Deaton
NASA Langley Research Center
William Devlin
Products Research and Chemicals
Corporation
James E. DeVries
Nordson Corporation
K.L. DeVries
The University of Utah
David A. Dillard
Virginia Polytechnic and State
University
John G. Dillard
Virginia Polytechnic and State
University
Kieran Drain
Ciba-Geigy Corporation
Richard K. Dropek
Hercules Aerospace Company
Lawrence T. Drzal
Michigan State University
David J. Dunn
Dexus Research Inc.
Michael G. Elias
Tremco, Inc.
John Gannon
Ciba-Geigy Corporation
Michelle M. Gauthier
Raytheon Company
Richard A. Giangiordano
Products Research & Chemical
Corporation
Garry Good
Battelle Memorial Institute
Victor A. Greenhut
Rutgers University
Donald J. Hagemaier
McDonnell Douglas Corporation
James Hagquist
H.B. Fuller Company
Wynford L. Harries
Old Dominion University
William F. Harrington
Uniroyal Adhesives & Sealants
Company, Inc.
L.J. Hart-Smith
McDonnell Douglas Corporation
Pedro J. Herrera-Franco
Michigan State University
J.J. Higgins
Exxon Chemical Company

Ray E. Horton
Boeing Commercial Airplane Company
F.C. Jagisch
Exxon Chemical Company
Jim Johnson
H.B. Fuller Company
Ronald W. Johnson
Boeing Military Airplane Company
Kristen T. Kern
Old Dominion University
Jerome M. Klosowski
Dow Corning Corporation
H. Kollek
Fraunhofer Institute for Material
Science
Cynthia L. Kreider
Polymerland Inc.
General Electric Company
Raymond B. Krieger, Jr.
American Cyanamid Company
M.H. Kuperman
United Airlines Maintenance Operations
Center
Arthur H. Landrock
U.S. Army Armament Research
Didier Lefebvre
AT&T Bell Laboratories
Kenneth M. Liechti
University of Texas at Austin
Sheila Ann T. Long
NASA Langley Research Center
Ajit K. Mal
University of California, Los Angeles
Gerald K. McKeegan
Dexter Adhesives and Structural
Materials Division
D.K. McNamara
Martin Marietta Laboratories
David P. Melody
Loctite International
Frederick K. Meyer
H.B. Fuller Company
Jack V. Morris
Polymeric Systems, Inc.
Mark Moulding
Thermal Equipment Corporation
S.H. Myhre
Northrop Corporation
John Nardone
Nardi & Associates
Tim A. Osswald
University of Wisconsin-Madison
D.E. Packham
The University of Bath

F.E. Penado
Hercules Aerospace Company
E.A. Peterson
Morton International, Inc.
John T. Quinlivan
Boeing Commercial Airplane Company
Russell Redman
Tremco, Inc.
John F. Regan
ChemRex Inc.
Theodore J. Reinhart
Air Force Materials Laboratory
Jeroen Rietveld
University of Wisconsin—Madison
Richard W. Roberts
General Dynamics Convair Division
Robert D. Rossi
National Starch and Chemical
Corporation
David Roylance
Massachusetts Institute of Technology
Erol Sancaktar
Clarkson University
David W. Schmueser
General Motors Research Laboratories

Jules E. Schoenberg
National Starch and Chemical
Corporation
Kevin J. Schroeder
Ciba-Geigy Corporation
Donald F. Sekits
Boeing Military Airplane Company
Louis H. Sharpe
Consultant
S.B. Smith (deceased)
Dow Corning Corporation
Loretta A. Sobieski
Dow Corning Corporation
Christer Sorensson
Products Research and Chemical
Corporation
Andrew Stevenson
Materials Engineering Research
Laboratory, Ltd.
Nak-Ho Sung
Tufts University
Sandra K.M. Swanson
H.B. Fuller Company
Ken M. Takahashi
AT&T Bell Laboratories

Todd Taricco
Thermal Equipment Corporation
James E. Thompson
Rhone-Poulenc, Inc.
Herb Turner
Nordson Corporation
Steven Tyler
Tremco, Inc.
Arthur M. Usmani
Boehringer Mannheim Corporation
James Van Twisk
McCann Manufacturing Co.
Alfred F. (Jerry) Waterland III
W.L. Gore & Associates, Inc.
Stanley E. Wentworth
U.S. Army Materials Technology
Laboratory
William G. Witte, Jr.
NASA Langley Research Center
Valerie Wheeler
Grumman Aircraft Systems
Denis J. Zalucha
S.C. Johnson Wax

Reviewers

R. Abbott
  Beech Aircraft Corporation
W. Aberth
  DAP Inc.
I.A. Abu-Isa
  General Motors Research Laboratories
K.W. Allen
  City University (London)
William F. Anspach
  Wright Research and Development
  Center
Richie J. Ashburn
  Silicones, Inc.
D.R. Baer
  Battelle Pacific Northwest Laboratories
Will Barbeau
  Barbeau Associates
Willard D. Bascom
  The University of Utah
J.P. Bates
  Bell Helicopter Textron Inc.
C. Bellanca
  Protective Treatments Inc.
Morris B. Berenbaum
  Allied-Signal, Inc.
C.W. Boeder
  3M Company
Ronald J. Botsco
  NDT Instruments
James A. Box
  Tremco, Inc.
Daniel R. Bradshaw
  Lord Corporation
D.M. Brewis
  Loughborough University of
  Technology
Brian Briddell
  Adco Products, Inc.
S. Brown
  U.S. Department of the Navy
D.R. Bryan
  LTV Aircraft Products Group
J. Burk
  Ciba-Geigy Corporation
D.L. Caldwell
  Dow Chemical U.S.A.
Paul Calvert
  The University of Arizona
Larry Carbary
  Dow Corning Corporation
Gil Carrilo
  Rohr Industries, Inc.
James M. Casey
  FLEXcon Company, Inc.

B.L. Chapman
  University of Washington
Sam Chase
  Loctite Corporation
Eric J.H. Chen
  E.I. Du Pont de Nemours & Company,
  Inc.
Lon Chen
  Bostik Inc.
Hun R. Choung
  Schnee-Morehead, Inc.
J.D. Clark
  Alcan International Ltd.
Howard M. Clearfield
  IBM Corporation
Rudy Cologna
  Consultant
John Comyn
  Consultant
Olly Cook
  LTV Missiles and Electronics Group
Andy Crocombe
  University of Surrey
A.W. Czanderna
  Solar Energy Research Institute
Dennis J. Damico
  Lord Corporation
T. Dando
  Ashland Chemical Company
I.M. Daniel
  Northwestern University
Rudolph D. Deanin
  University of Lowell
E. Demuts
  Wright Research and Development
  Center
Bart DeNeef
  De Havilland Aircraft Company of
  Canada
C.J. Derham
  Materials Engineering Research
  Laboratory Ltd.
K.L. DeVries
  The University of Utah
R.A. Dickie
  Ford Motor Company
K. Drain
  Ciba-Geigy Corporation
Lawrence T. Drzal
  Michigan State University
David J. Dunn
  Dexus Research Inc.
W.H. Emens
  McDonnell Douglas Corporation

Richard A. Eppler
  Eppler Associates
John Feeney
  Marketing Project Management
Henry R. Fenbert
  Boeing Commercial Airplane Company
William W. Feng
  Lawrence Livermore National
  Laboratory
Parker Finn
  The Tracking Station, Inc.
Dale Flackett
  Allied-Signal, Inc.
T.R. Flint
  Polymeric Systems, Inc.
Robert T. Foister
  General Motors Research Laboratories
B.W. Foster
  Texas Eastman Company
James C. Foster
  LTV Aircraft Products Group
Van R. Foster
  DAP Inc.
J.R. Fowler
  BREL Ltd.
Steve Gaarenstroom
  General Motors Research Laboratories
J. Goldstein
  Air Products and Chemicals, Inc.
Rudy Gonzalez
  Northrop Corporation
James A. Graham
  Lord Corporation
J. Grant
  The University of Dayton
George L. Grobe III
  Perkin-Elmer Corporation
Danton Gutierrez-Lemini
  Thiokol Corporation
D.J. Hagemaier
  McDonnell Douglas Corporation
Allegra Hakim
  U.S. Department of the Air Force
B.B. Hardman
  PCR Inc.
Alan Hardy
  Industrial chemist (retired)
Jerry Hartman
  Ferro Corporation
S.R. Hartshorn
  3M Company
Roy Hartz
  Uniroyal Goodrich Tire Company

William J. Havey
National Starch and Chemical
Corporation

Girard S. Haviland
Consultant

G.J. Helmstetter
National Starch and Chemical
Corporation

Edmund G. Henneke
Virginia Polytechnic Institute and State
University

Paul M. Hergenrother
National Aeronautics and Space
Administration

R.D. Hermansen
Hughes Aircraft Company

Sylvester Hill
Boeing Commercial Airplane Company

J.A. Hinkley
National Aeronautics and Space
Administration

Howard Holmquist
Boeing Commercial Airplane Company

Joseph W. Holubka
Ford Motor Company

Steven J. Hooper
The Wichita State University

William Houghton
Advanced Composites Laboratories

John F. Imbalzano
E.I. Du Pont de Nemours & Company,
Inc.

N.W. Jalufka
Hampton University

Steven Johnson
National Aeronautics and Space
Administration

John Johnston
Tesa Truck Inc.

Hooshang Jozavi
Conoco Inc.

Robert J. Jones
TRW, Inc.

J. Kardos
Washington University, St. Louis

Keith T. Kedward
McDonnell Douglas Technologies Inc.

Fred A. Keimel
Adhesives & Sealants Consultants

Albert A. Kilchesty
Pecora Corporation

Darrell A. Klemme
Black & Decker Inc.

F. Koblitz
AMP Inc.

H.W. Koehler
Hardman Inc.

Kathy Koszalka
Rohr Industries, Inc.

James A. Koutsky
University of Wisconsin

C. Kramer
Essex Specialty Products Inc.

David E. Kranbuehl
The College of William and Mary

Ronald J. Kuhbander
The University of Dayton

Frank Kury
Varian-Eimac

A.L. Landis
Lockheed Corporation

Scott F. Lange
Hercules Inc.

R.L. Lavallee
Ciba-Geigy Corporation

W.A. Lees
National Starch and Chemical
Corporation

J. Lincoln
U.S. Department of the Air Force

P. Locke
Graco Inc.

James Lomax
Rohm & Haas Company

Alfred C. Loos
Virginia Polytechnic Institute and State
University

A. Lustiger
AT&T Bell Laboratories

Kenn Luyk
Alcoa Laboratories

S. Mall
U.S. Air Force Institute of Technology

Art Marceau
Boeing Commercial Airplane Company

Paul Martakos
Atrium Medical Corporation

Shiro Matsuoka
AT&T Bell Laboratories

Bret Mayo
Battelle Memorial Institute

John E. McCarty
Consultant

Malcolm G. McLaren
Rutgers University

David McNamara
Martin Marietta Laboratories

Joey Mead
U.S. Army Materials Technology
Laboratory

L.E. Roy Meade
Lockheed Corporation

Ralph E. Meyer
Consultant

J.L. Miller
Lockheed Corporation

N.E. Millsap
Thiokol Corporation

J. Dean Minford
Consultant

Raja Mohan
Ashland Chemical Company

J.A. Moore
Rensselaer Polytechnic Institute

Thomas B. Moore
Thomas B. Moore Company, Inc.

F. Mopsik
National Institute of Standards and
Technology

G. Moran
Rohr Industries Inc.

Sheldon Mostovoy
Illinois Institute of Technology

Stephen J. Mumby
Biosym Technologies, Inc.

J.E. Myers
Simpson Gumpertz & Heger Inc.

L. Nativi
Loctite Corporation

Patricia Odegard
Honeywell Inc.

J.T. O'Connor
Loctite Corporation

Thomas F. O'Connor
Smith, Hinchman & Grylls Associates,
Inc.

W.J. van Ooij
Armco Inc.

Julian R. Panek
Consultant

Pat Patton
Loctite Corporation

H. Pearson
Lockheed Corporation

Dale L. Perry
The University of California—Berkeley

Don Pettit
Lockheed Corporation

Charles U. Pittman, Jr.
Mississippi State University

E. Plueddemann
Dow Corning Corporation

Joseph W. Prane
Consultant

R.L. Radecky
McDonnell Douglas Corporation

V. Reddy Raju
AT&T Bell Laboratories

C. Ramsey
Wright Research and Development
Center

M. Ratwani
Northrop Corporation

Theodore J. Reinhart
Air Force Materials Laboratory

Robert Rhein
U.S. Department of the Navy

James T. Rice
The University of Georgia

H. Ries
Norfolk State University

R.E. Robertson
The University of Michigan

Edward Rosenzweig
U.S. Department of the Navy

R.J. Ruch
Kent State University

E. Sancaktar
Clarkson University

T.M. Santosusso
Air Products and Chemicals, Inc.

Frederick F. Saremi
Arlon Products, Inc.

F.W. Scholz
The Boeing Company

F.H. Sexsmith
Lord Corporation

John J. Sheehan
  Loctite Corporation
Maynard A. Sherwin
  Union Carbide Corporation
A.R. Siebert
  The B.F. Goodrich Company
R. Simenz
  Lockheed Aeronautical Systems
  Company
Thomas J. Sliva
  DL Laboratories
Marvin B. Smith
  Flamemaster
Robert W. Smith
  Darworth Company
David R. Speth
  The Dow Chemical Company
Gerald Steele
  Ball State University
R.H. Stone
  Lockheed Corporation
Anne K. St. Clair
  National Aeronautics and Space
  Administration
Mark Stypczynksi
  Premier Industrial Corporation
Ephraim Suhir
  AT&T Bell Laboratories
Tim Sullivan
  AT&T Bell Laboratories
Ray Swann
  AMOTECH

L.M. Swope
  U.S. Department of the Air Force
Giuliana C. Tesoro
  Polytechnic University
R. Thomas
  Texas Research Institute, Inc.
Richard T. Thompson
  Loctite Corporation
J.S. Thornton
  TRI
J. Timberlake
  Union Carbide Chemicals and Plastics
  Company Inc.
Monika Tomasik
  Advanced Composite Products &
  Technology
Louis Toporcer
  Wacker Silicones Corporation
Aurora Turla
  Boeing Commercial Airplane Company
L.A. Tushaus
  3M Company
Fred Vaness
  Kenworth Truck Company
Robert Wallach
  National Starch and Chemical
  Corporation
David A. Wangsness
  3M Aerospace Materials Laboratory
Ben Ward
  Hoechst Celanese Corporation

Richard B. Warnock
  U.S. Department of the Air Force
W. Warrach
  Mobay Corporation
W. Waters
  Pecora Corporation
Karl Wefers
  Aluminum Company of America
R. Wegman
  Adhesion Associates
Dennis J. Werneke
  Honeywell Inc.
Steven White
  Beech Aircraft Corporation
James M. Whitney
  Wright Research and Development
  Center
Roy L. Whittaker
  Garlock Inc.
J. Wightman
  Virginia Polytechnic Institute and State
  University
Walter C. Wilhelm
  Lord Corporation
Gerald W. Yankie
  Techcon Systems Inc.
Denis J. Zalucha
  S.C. Johnson & Son, Inc.
Martel Zeldin
  Purdue University
Kevin Zilvar
  ALCOA/TRE Inc.

Contents

Section 8: Manufacturing Use
Chairman: Sam C. Aker, Sam C. Aker, Inc.

Section 9: Nondestructive Inspection and Quality Control
Chairman: Yoseph Bar-Cohen, Douglas Aircraft Company

Section 10: Bonded Structure Repair
*Chairman: Ray E. Horton, Boeing Commercial Airplane
Company*

Fundamentals of Adhesives and Sealants Technology

Chairman: Louis H. Sharpe, Consultant

Introduction

THE PERFORMANCE, and in many cases the very existence, of a wide range and variety of common products and structures in our modern society depend on adhesives and sealants. The many varieties of plywood would not exist without adhesives. Buildings utilizing curtain-wall construction would not exist without sealants. High-technology aircraft, particularly military types, could not achieve their present level of performance without adhesively bonded substructures. Multipane insulating glass windows, likewise, depend on adhesive sealants for their performance.

Between pressure-sensitive adhesive tapes, requiring only finger pressure to make them work, and the high-tech world of structural aerospace adhesives requiring a host of allied technologies for proper usage, lies a vast spectrum of adhesives of varying form and function. The same can be said of sealants, which cover the range from simple sealing tapes to the high-tech adhesive/sealant products used in multipane insulating windows.

Such a variety in materials alone is a serious problem to those who, realizing the advantages of using adhesive and sealant technology, wish to use the technology in their products. This, however, is not the only complication. Adhesives, for example, do not usually function well in a design that simply substitutes adhesive bonding for conventional mechanical fasteners. Therefore, prospective users must also be familiar with the "design rules" for using adhesives, which are, in general, different from those for conventional mechanical fastening. In addition, the initial performance and the durability of adhesively bonded structures in service are strongly dependent on how the parts to be joined are prepared and on the severity of the service environment. The structure and chemistry of the surface region of the parts to be joined and their response to service environments may well control performance. Therefore, prebonding treatments appropriate to service are a requirement for the application of the technology.

Finally, many adhesives and sealants must be cured, that is, converted from the relatively fluid, low molecular weight, non-load-bearing form in which they must be applied, to a solid, high molecular weight form able to bear sustained loads without creeping or otherwise failing. The most common way to accomplish this curing, in the case of adhesives, is by the application of heat and pressure. The required methods for applying heat and pressure are highly dependent on the geometry (and scale) of the parts to be joined and may have a large impact on manufacturability. Such issues are covered in more detail in the articles "Overview: Adhesives Technology" and "Adhesives and Adhesion."

What this means is that the application of adhesives bonding technology requires a careful integration of materials technology with (at least) design, manufacturability, serviceability, and cost considerations. Repairability may also be a factor to be considered, as noted below. In addition, if a bonded joint is a *part* of a larger structure, then the same considerations must be made in the context of the larger structure. That is, an overall *systems* approach is essential. In general, the same applies to the use of sealants.

For a variety of reasons, and despite carefully thought-out process controls, structures occasionally will fail short of their anticipated or designed service life. Because of this, the ability to do failure analysis—to determine what caused the failure and to know how to correct it—is an important added consideration.

Although it has often been said that adhesives and sealants are basically similar materials that perform many of the same functions, it will become clear upon reading this, and later, Sections of this Volume, that this is not, in general, a correct statement. In certain limited cases the functions of the two materials do, indeed, overlap. However, the fact that they are discussed in different parts of this (and other) Sections of this Volume and that different specifications (covering different properties) and test methods (covering different tests) apply to them, should make it clear that they are, in fact, different materials performing sometimes overlapping but, in general, different functions. Such issues are addressed in the articles "Adhesives Markets and Applications," "Sealants and Sealing Technology," and "Sealants Markets and Applications."

Section 1 of this Volume is an attempt to provide general background information on adhesives and sealants technology for engineering users who have limited familiarity with these materials. Its intent is to give the reader a feeling for the technologies. More detailed coverage of many of the subjects in Section 1 appears in other Sections.

A few final comments are in order. The information contained in this or, for that matter, any other handbook cannot make an instant expert of anyone who reads and digests it, except for cases of the simplest and least demanding applications of the technologies covered. It may be necessary to consider a large and complex set of interacting factors before a decision can be reached to use specific adhesives and sealants in a given assembly. The technologies are sophisticated, and technologically appropriate decisions require the input and advice of experts. In addition, all personnel responsible for the manufacture of structures using adhesive bonding should be thoroughly educated regarding its process requirements. Adhesive joints formed with state-of-the-art materials and processes will quickly become expensive junk in the hands of workers who are not well versed in the requirements of the proper practice of the technology.

The article "Guide to General Information Sources" describes the kind of information available to persons interested in the subject technologies at all levels, from fundamental research on adhesion to engineering applications of adhesives and sealants. In addition, a wealth of information on the practice of these technologies is usually available for the asking from the manufacturers of adhesives and sealants, through their technical service people or their distributors. One may expect, however, that the quality and extent of the guidance received from these and other experts will depend on the quality and extent of the information that is given them regarding a particular problem. The best policy, therefore, is to be free with information concerning *all* requirements.

Glossary of Terms

A

A. The symbol for a repeating unit in a polymer chain.

ABA copolymers. Block copolymers with three sequences, but only two domains.

A-basis. The "A" mechanical property value is the value above which at least 99% of the population of values is expected to fall, with a confidence of 95%. Also called A-allowable. See also *B-basis, S-basis,* and *typical-basis.*

abhesive. A material that resists adhesion. A film or coating applied to surfaces to prevent sticking, heat sealing, and so on, such as a parting agent or mold release agent.

ablation. A self-regulating heat and mass transfer process in which incident thermal energy is expended by sacrificial loss of material.

ablative plastic. A material that absorbs heat (with a low material loss and char rate) through a decomposition process (pyrolysis) that takes place at or near the surface exposed to the heat. This mechanism essentially provides thermal protection (insulation) of the subsurface materials and components by sacrificing the surface layer. Ablation is an exothermic process.

ABS. See *acrylonitrile-butadiene-styrene (ABS) resins.*

absolute humidity. The weight of water vapor present in a unit volume of air, such as grams per cubic foot, or grams per cubic meter. The amount of water vapor is also reported in terms of weight per unit weight of dry air, such as grams per pound of dry air, but this value differs from values calculated on a volume basis and should not be referred to as absolute humidity. It is designated as humidity ratio, specific humidity, or moisture content.

absolute viscosity. With respect to a fluid, the tangential force on unit area of either of two parallel planes at unit distance apart when the space between the planes is filled with the fluid in question and one of the planes moves with unit differential velocity in its own plane.

absorption. The penetration into the mass of one substance by another. The process whereby energy is dissipated within a specimen placed in a field of radiant energy. The capillary or cellular attraction of adherend surfaces to draw off the liquid adhesive film into the substrate.

AC. See *acetal (AC) copolymers, acetal (AC) homopolymers*, and *acetal (AC) resins.*

accelerated-life test. A method designed to approximate, in a short time, the deteriorating effect obtained under normal long-term service conditions. See also *artificial aging.*

accelerator. A material that, when mixed with a catalyst or a resin, speeds up the chemical reaction between the catalyst and the resin (usually in the polymerizing of resins or vulcanization of rubbers). Also called promoter.

acceptance test. A test, or series of tests, conducted by the procuring agency, or an agent thereof, upon receipt, to determine whether an individual lot of materials conforms to the purchase order or contract or to determine the degree of uniformity of the material supplied by the vendor, or both. Compare to *preproduction test* and *qualification test.* ASTM D 907.

acetal (AC) copolymers. A family of highly crystalline thermoplastics prepared by copolymerizing trioxane with small amounts of a comonomer that randomly distributes carbon-carbon bonds in the polymer chain. These bonds, as well as hydroxyethyl terminal units, give the acetal copolymers a high degree of thermal stability and resistance to strong alkaline environments.

acetal (AC) homopolymers. Highly crystalline linear polymers formed by polymerizing formaldehyde and capping it with acetate end groups.

acetal (AC) resins. Thermoplastics (polyformaldehyde and polyoxymethylene resins) produced by the addition polymerization of aldehydes by means of the carbonyl function, yielding unbranched polyoxymethylene chains of great length. The acetal resins, among the strongest and stiffest of all thermoplastics, are also characterized by good fatigue life, resilience, low moisture sensitivity, high solvent and chemical resistance, and good electrical properties. They may be processed by conventional injection molding and extrusion techniques and fabricated by welding methods used for other plastics.

acid-acceptor. A compound that acts as a stabilizer by chemically combining with acid that may be initially present in minute quantities in a plastic, or that may be formed by the decomposition of the resin.

acrylate resins. See *acrylic resins.*

acrylic plastic. A thermoplastic polymer made by the polymerization of esters of acrylic acid or its derivatives.

acrylic resins. Polymers of acrylic or methacrylic esters, sometimes modified with nonacrylic monomers such as the ABS group. The acrylates may be methyl, ethyl, butyl, or 2-ethylhexyl. Usual methacrylates are methyl, ethyl, butyl, laural, and stearyl. The resins may be in the form of molding powders or casting syrups and are noted for their exceptional clarity and optical properties. Acrylics are widely used in lighting fixtures because they are either slow burning or self-extinguishing, and do not produce harmful smoke or gases in the presence of flame.

acrylonitrile. A monomer with the structure (CH_2:CHCN). It is most useful in copolymers. Its copolymer with butadiene is nitrile rubber; acrylonitrile-butadiene copolymers with styrene (SAN) are tougher than polystyrene. Acrylonitrile is also used as a synthetic fiber and as a chemical intermediate.

acrylonitrile-butadiene-styrene (ABS) resins. A family of thermoplastics based on acrylonitrile, butadiene, and styrene, combined by a variety of methods involving polymerization, graft polymerization, physical mixing, and combinations thereof. The standard grades of ABS resins are rigid, hard, and tough, but not brittle, and possess good impact strength, heat resistance, low-temperature properties, chemical resistance, and electrical properties.

activation. The (usually) chemical process of making a surface more receptive to bonding with a coating or an encapsulating material.

activator. See *accelerator.*

addition polymerization. A chemical reaction in which simple molecules (monomers) are linked to each other to form long-chain molecules (polymers) by chain reaction.

additive. A substance added to another substance, usually to improve properties, such as plasticizers, initiators, light stabilizers, and flame retardants. See also *filler.*

adduct. A chemical addition product.

adhere. To cause two surfaces to be held together by adhesion. ASTM D 907.

adherend. A body held to another body by an adhesive. ASTM D 907. See also *substrate*.

adherend preparation. See *surface preparation*.

adhesion. The state in which two surfaces are held together by interfacial forces, which may consist of valence forces, interlocking action, or both. ASTM D 907. See also *mechanical adhesion* and *specific adhesion*.

adhesion promoter. A coating applied to a substrate, before it is coated with an adhesive, to improve the adhesion of the substrate. Also called primer.

adhesive. A substance capable of holding materials together by surface attachment. Adhesive is a general term and includes, among others, cement, glue, mucilage, and paste. These terms are loosely used interchangeably. Various descriptive adjectives are applied to the term adhesive to indicate certain characteristics: physical form, that is, liquid adhesive, tape adhesive; chemical type, that is, silicate adhesive, resin adhesive; materials bonded, that is, paper adhesive, metal-plastic adhesive; conditions of use, that is, hot-setting adhesive. ASTM D 907. See also *glue, gum, mucilage, paste, resin,* and *sizing*.

adhesive, anaerobic. See *anaerobic adhesive*.

adhesive assembly. A group of materials or parts, including adhesive, that are placed together for bonding or that have been bonded together. ASTM D 907. See also *assembly adhesive*.

adhesive, cold-setting. See *cold-setting adhesive*.

adhesive, contact. See *contact adhesive*.

adhesive dispersion. A two-phase system in which one phase is suspended in a liquid. ASTM D 907. Compare to *emulsion*.

adhesive failure. Rupture of an adhesive bond such that the separation appears to be at the adhesive-adherend interface. Sometimes termed failure in adhesion. Compare to *cohesive failure*. ASTM D 907.

adhesive film. A synthetic resin adhesive, with or without a film carrier fabric, usually of the thermosetting type, in the form of a thin film of resin, used under heat and pressure as an interleaf in the production of bonded structures.

adhesive, gap-filling. See *gap-filling adhesive*.

adhesive, heat-activated. See *heat-activated adhesive*.

adhesive, heat-sealing. See *heat-sealing adhesive*.

adhesive, hot-melt. See *hot-melt adhesive*.

adhesive, hot-setting. See *hot-setting adhesive*.

adhesive, intermediate-temperature-setting. See *intermediate-temperature-setting adhesive*.

adhesive joint. Location at which two adherends are held together with a layer of adhesive. ASTM D 907. See also *bond*.

adhesive, pressure-sensitive. See *pressure-sensitive adhesive*.

adhesive strength. The strength of the bond between an adhesive and an adherend.

adhesive, structural. See *structural adhesive*.

adhesive system. An integrated engineering process that analyzes the total environment of a potential bonded assembly to select the most suitable adhesive, application method, and dispensing equipment.

adiabatic. Occurring with no addition or loss of heat from the system under consideration.

admixture. The addition and homogeneous dispersion of discrete components, before cure.

adsorption. The adhesion of the molecules of gases, dissolved substances, or liquids in more or less concentrated form, to the surfaces of solids or liquids with which they are in contact. The concentration of a substance at a surface or interface of another substance.

advanced composites. Composite materials that are reinforced with continuous fibers having a modulus higher than that of fiberglass fibers. The term includes metal matrix and ceramic matrix composites, as well as carbon-carbon composites.

afterbake. See *postcure*.

aggregate. A hard, coarse material usually of mineral origin used with an epoxy binder (or other resin) in plastic tools. Also used in flooring or as a surface medium.

aggressive tack. Synonym for dry tack. ASTM D 907.

aging. The effect on materials of exposure to an environment for a prolonged interval of time. The process of exposing materials to an environment for a prolonged interval of time in order to predict in-service lifetime. Also called bursting strength.

air-bubble void. Air entrapment within a molded item or between the plies of reinforcement or within a bond line or encapsulated area; localized, noninterconnected, and spherical in shape.

alcohols. Characterized by the hydroxyl (—OH) group they contain, alcohols are valuable starting points for the manufacture of synthetic resins, synthetic rubbers, and plasticizers.

aldehydes. Volatile liquids with sharp, penetrating odors that are slightly less soluble in water than are corresponding alcohols.

aliphatic hydrocarbons. Saturated hydrocarbons having an open-chain structure, for example, gasoline and propane.

alkyd plastic. Thermoset plastic based on resins composed principally of polymeric

esters, in which the recurring ester groups are an integral part of the main polymer chain, and in which ester groups occur in most cross links that may be present between chains.

allophanate. Reactive product of an isocyanate and the hydrogen atoms in a urethane.

alloprene. Chlorinated rubber.

allotropy. The existence of a substance, especially an element, in two or more physical states (for example, crystals). See also *graphite*.

alloy. In plastics, a blend of polymers or copolymers with other polymers or elastomers under selected conditions; for example, styrene-acrylonitrile. Also called polymer blend.

allyl resins. A family of thermoset resins made by addition polymerization of compounds containing the group CH_2:CH—CH_2, such as esters of allyl alcohol and dibasic acids. They are available as monomers, partially polymerized prepolymers, or molding compounds. Members of the family are diallyl phthalate (DAP), diallyl isophthalate (DAIP), diallyl maleate (DAM), and diallyl chlorendate (DAC).

alpha (α) cellulose. A very pure cellulose prepared by special chemical treatment.

alpha (α) loss peak. In dynamic mechanical or dielectric measurement, the first peak in the damping curve below the melt, in order of decreasing temperature or increasing frequency.

alternating copolymer. A copolymer in which each repeating unit is joined to another repeating unit in the polymer chain (—A—B—A—B—).

alternating stress. A stress varying between two maximum values that are equal but of opposite signs, according to a law determined in terms of the time.

alternating stress amplitude. A test parameter of a dynamic fatigue test: one-half the algebraic difference between the maximum and minimum stress in one cycle.

amine adduct. Product of the reaction of an amine with a deficiency of a substance containing epoxy groups.

amino. Relating to or containing an —NH_2 or —NH group.

amino resins. Resins made by the polycondensation of a compound containing amino groups, such as urea or melamine, with an aldehyde, such as formaldehyde, or an aldehyde-yielding material. Melamine-formaldehyde and urea-formaldehyde resins are the most important family members. The resins can be dispersed in water to form colorless syrups. With appropriate catalysts, they can be cured at elevated temperatures.

amorphous plastic. A plastic that has no crystalline component, no known order or pattern of molecule distribution, and no sharp melting point.

amylaceous. Pertaining to, or of the nature of, starch; starchy. ASTM D 907.

anaerobic adhesive. An adhesive that cures only in the absence of air after being confined between assembled parts.

anchor pattern. A pattern made by blast cleaning abrasives on an adherend surface in preparation for adhesive application prior to bonding. Pattern is examined in profile.

anelastic deformation. Any portion of the total deformation of a body that occurs as a function of time when load is applied and which disappears completely after a period of time when the load is removed. In practice, the term also describes viscous deformation.

anhydride. A compound from which water has been extracted. An oxide of a metal (basic anhydride) or of a nonmetal (acid anhydride) that forms a base or an acid, respectively, when united with water.

aniline. An important organic base $(C_6H_5NH_2)$ made by reacting chlorobenzene with aqueous ammonia in the presence of a catalyst. It is used in the production of aniline formaldehyde resins and in the manufacture of certain rubber accelerators and antioxidants.

aniline-formaldehyde resins. Members of the aminoplastics family made by the condensation of formaldehyde and aniline in an acid solution. The resins are thermoplastic and are used to a limited extent in the production of molded and laminated insulating materials. Products made from these resins have high dielectric strength and good chemical resistance.

anisotropic. Exhibiting different properties when tested along axes in different directions. Not isotropic.

annealing. In plastics, heating to a temperature at which the molecules have significant mobility, permitting them to reorient to a configuration having less residual stress. In semicrystalline polymers, heating to a temperature at which retarded crystallization or recrystallization can occur.

antioxidant. A substance that, when added in small quantities to the resin during mixing, prevents its oxidative degradation and contributes to the maintenance of its properties.

antistatic agents. Agents that, when added to a molding material or applied to the surface of a molded object, make it less conductive, thus hindering the fixation of dust or the buildup of electrical charge.

aramid. A manufactured fiber in which the fiber-forming substance is a long-chain synthetic aromatic polyamide in which at least 85% of the amide linkages are directly attached to two aromatic rings.

arc resistance. Ability to withstand exposure to an electric voltage. The total time in seconds that an intermittent arc may play across a plastic surface without rendering the surface conductive.

aromatic. Unsaturated hydrocarbon with one or more benzene ring structures in the molecule.

aromatic polyester. A polyester derived from monomers in which all the hydroxyl and carboxyl groups are directly linked to aromatic nuclei.

artificial aging. The exposure of a plastic to conditions that accelerate the effects of time. Such conditions include heating, exposure to cold, flexing, application of electric field, exposure to chemicals, ultraviolet light radiation, and so forth. Typically, the conditions chosen for such testing reflect the conditions under which the plastic article will be used. Usually, the length of time the article is exposed to these test conditions is relatively short. Properties such as dimensional stability, mechanical fatigue, chemical resistance, stress cracking resistance, dielectric strength, and so forth, are evaluated in such testing. See also *aging.*

artificial weathering. The exposure of plastics to cyclic laboratory conditions, consisting of high and low temperatures, high and low relative humidities, and ultraviolet radiant energy, with or without direct water spray and moving air (wind), in an attempt to produce changes in the properties of the plastics similar to those observed after long-term continuous exposure outdoors. The laboratory exposure conditions are usually more intensified than those encountered in actual outdoor exposure in an attempt to achieve an accelerated effect. Also called accelerated aging.

ash content. Proportion of the solid residue remaining after a reinforcing substance has been incinerated (charred or intensely heated).

aspect ratio. The ratio of length to diameter of a reinforcing fiber.

assembly. See *adhesive assembly.*

assembly adhesive. An adhesive that can be used for bonding parts together, such as in the manufacture of a boat, an airplane, furniture, and the like. The term assembly adhesive is commonly used in the wood industry to distinguish such adhesives (formerly called joint glues) from those used in making plywood (sometimes called veneer glues). It describes adhesives used in fabricating finished structures or goods, or subassemblies thereof, as differentiated from adhesives used in the production of sheet materials for sale as such, for example, plywood or laminates. ASTM D 907.

assembly time. The time interval between the spreading of the adhesive on the adherend and the application of pressure or heat, or both, to the assembly. For assemblies involving multiple layers or parts, the assembly time begins with the spreading of the adhesive on the first adherend. ASTM D 907. See also *closed assembly time* and *open assembly time.*

A-stage. An early stage in the reaction of certain thermosetting resins in which the material is fusible and still soluble in certain liquids. Synonym for resol. ASTM D 907. Compare *B-stage* and *C-stage.*

atactic stereoisomerism. A chain of molecules in which the position of the side chains or side atoms is more or less random. See also *isotactic stereoisomerism* and *syndiotactic stereoisomerism.*

attenuation. The diminution of vibrations or energy over time or distance. The process of making thin and slender, as applied to the formation of fiber from molten glass.

autoclave. A closed vessel for conducting and completing either a chemical reaction under pressure and heat or other operation, such as cooling. Widely used for bonding and curing reinforced plastic laminates.

autoclave molding. A process in which, after lay-up, winding, or wrapping, an entire assembly is placed in a heated autoclave, usually at 340 to 1380 kPa (50 to 200 psi). Additional pressure permits higher density and improved removal of volatiles from the resin. Lay-up is usually vacuum bagged with a bleeder and release cloth.

average molecular weight. The molecular weight of the most typical chain in a given plastic; it is characteristic of neither the longest nor the shortest chain.

axial strain. The linear strain in a plane parallel to the longitudinal axis of the specimen.

B

B. The symbol for a repeating unit in a copolymer chain.

baffle. A device used to restrict or divert the passage of fluid through a pipeline or channel.

bagging. Applying an impermeable layer of film over an uncured part and sealing the edges so that a vacuum can be drawn.

bag molding. A method of molding or bonding involving the application of fluid pressure, usually by means of air, steam, water, or vacuum, to a flexible cover that, sometimes in conjunction with the rigid die, completely encloses the material to be bonded. ASTM D 907. Also called blanket molding.

Banbury. An apparatus for compounding materials. It is composed of a pair of contrarotating rotors that masticate the materials to form a homogeneous blend. This internal-type mixer produces excellent mixing.

Barcol hardness. A hardness value obtained by measuring the resistance to penetration of a sharp steel point under a spring load. The instrument, called the Barcol impressor, gives a direct reading on a 0 to 100 scale. The hardness value is often used as a measure of the degree of cure of a plastic.

barrier plastics. A general term applied to a group of lightweight, transparent, impact-resistant plastics, usually rigid copolymers of high acrylonitrile content. Barrier plastics are generally characterized by gas, aroma, and flavor barrier characteristics approaching those of metal and glass.

batch. The manufactured unit or a blend of two or more units of the same formulation and processing. ASTM D 907. Compare to *manufactured unit*.

B-basis. The "B" mechanical property value is the value above which at least 90% of the population of values is expected to fall, with a confidence of 95%. See also *A-basis, S-basis*, and *typical-basis*.

bend test. A test for ductility performed by bending or folding, usually by steadily applied forces, but in some instances by blows, to a specimen having a cross section substantially uniform over a length several times as great as the largest dimension of the cross section.

benzene ring. The basic structure of benzene, the most important aromatic chemical. It is an unsaturated, resonant, six-carbon ring having three double bonds. One or more of the six hydrogen atoms of benzene may be replaced by other atoms or groups.

beta (β) gage. A gage consisting of two facing elements, a β-ray-emitting source, and a β-ray detector. Also called beta-ray gage.

beta (β) loss peak. In dynamic mechanical or dielectric measurement, the second peak in the damping curve below the melt, in order of decreasing temperature or increasing frequency.

biaxial load. A loading condition in which a specimen is stressed in two directions in its plane. A loading condition of a pressure vessel under internal pressure and with unrestrained ends.

binder. A component of an adhesive composition that is primarily responsible for the adhesive forces which hold two bodies together. ASTM D 907. See also *extender* and *filler*.

bismaleimide (BMI). A type of polyimide that cures by an addition rather than a condensation reaction, thus avoiding problems with volatiles formation, and which is produced by a vinyl-type polymerization of a prepolymer terminated with two maleimide groups. Intermediate in temperature capability between epoxy and polyimide.

bleed. To give up color when in contact with water or a solvent. Undesirable movement of certain materials in a plastic, such as plasticizers in vinyl, to the surface of the finished article or into an adjacent material; also called migration.

bleeder cloth. A woven or nonwoven layer of material used in the manufacture of com-posite parts to allow the escape of excess gas and resin during cure. The bleeder cloth is removed after the curing process and is not part of the final composite.

bleeding. The removal of excess resin from a laminate during cure. The diffusion of color from a plastic part into the surrounding surface or part.

bleed-out. The spread of adhesive away from the bond area.

blister. An elevation of the surface of an adherend, the shape of which somewhat resembles a blister on the human skin. Its boundaries may be indefinitely outlined, and it may have burst and become flattened. A blister may be caused by insufficient adhesive; inadequate curing time, temperature, or pressure; or trapped air, water, or solvent vapor. ASTM D 907.

block copolymer. An essentially linear copolymer consisting of a small number of repeated sequences of polymeric segments of different chemical structure.

blocked curing agent. A curing agent or hardener rendered unreactive, which can be reactivated as desired by physical or chemical means. ASTM D 907. Compare to *hardener*.

blocking. An undesired adhesion between touching layers of a material, such as occurs under moderate pressure during storage or use. ASTM D 907.

bloom. A noncontinuous surface coating on plastic products that comes from ingredients such as plasticizers, lubricants, antistatic agents, and so on, which are incorporated into the plastic resin, or that occurs by atmospheric contamination. Bloom is the result of ingredients in the plastic coming out of solution and migrating to the surface.

blow hole. A void produced by the outgassing of trapped air during cure.

blushing. The condensation of atmospheric moisture at the bond line interface.

BMC. See *bulk molding compound*.

BMI. See *bismaleimide*.

body. The consistency of an adhesive, which is a function of viscosity, plasticity, and rheological factors.

bond. The union of materials by adhesives. To unite materials by means of an adhesive. Synonym for glue. ASTM D 907. See also *adhere*.

bond angle. The angle formed by the bonds of one atom to other atoms; for example, 109.5° for C—C bonds.

bond face. The part or surface of an adherend that serves as a substrate for an adhesive.

bond length. The average distance between the centers of two atoms; for example, 0.154 nm (1.54 Å) for C—C bonds.

bond line. The layer of adhesive that attaches two adherends. Synonym for glue line. ASTM D 907.

bond strength. The unit load applied to tension, compression, flexure, peel, im-pact, cleavage, or shear required to break an adhesive assembly with failure occurring in or near the plane of the bond. The term adherence is frequently used in place of bond strength. ASTM D 907. See also *adhesion, bond, dry strength, peel, or stripping, strength*, and *wet strength*.

branched polymer. In the molecular structure of polymers, a main chain with attached side chains, in contrast to a linear polymer. Two general types are recognized, short-chain and long-chain branching.

branching. The presence of molecular branches in a polymer. The generation of branch crystals during the crystallization of a polymer.

breaking extension. The elongation necessary to cause rupture of a test specimen. The tensile strain at the moment of rupture.

breaking factor. The breaking load divided by the original width of a test specimen.

breakout. Fiber separation or break on surface plies at drilled or machined edges.

B-stage. An intermediate stage in the reaction of certain thermosetting resins in which the material softens when heated and swells when in contact with certain liquids, but may not entirely fuse or dissolve. The resin in an uncured thermosetting adhesive, usually in this stage. Synonym for resitol. ASTM D 907. Compare to *A-stage* and *C-stage*.

bubble. A spherical, internal void or globule of air or other gas trapped within a plastic. See *void*.

buckling. A mode of failure generally characterized by an unstable lateral material deflection due to compressive action on the structural element involved.

built-up laminated wood. An assembly made by joining layers of lumber with mechanical fastenings so that the grain of all laminations is essentially parallel. ASTM D 907.

bulk adherend. With respect to interphase, the adherend, unaltered by the adhesive. ASTM D 907. Compare to *bulk adhesive*.

bulk adhesive. With respect to interphase, the adhesive, unaltered by the adherend. ASTM D 907. Compare to *bulk adherend*.

bulk modulus of elasticity. The ratio of the compressive or tensile force applied to a substance per unit surface area to the change in volume of the substance per unit volume. Also known as bulk modulus, compression modulus, hydrostatic modulus, modulus of compression, and modulus of volume elasticity.

bulk molding compound (BMC). Thermosetting resin combined with strand reinforcement, fillers, and so on, into a viscous compound for compression or injection molding.

burned. Showing evidence of thermal decomposition or charring through some discoloration, distortion, destruction, or conversion of the surface of the plastic, sometimes to a carbonaceous char.

burning rate. The tendency of plastic articles to burn at given temperatures.

butadiene. A gas (CH_2:CH·CH:CH_2), insoluble in water but soluble in alcohol and ether, obtained from the cracking of petroleum, from coal tar benzene, or from acetylene produced from coke and lime. It is widely used in the formation of copolymers with styrene, acrylonitrile, vinyl chloride, and other monomeric substances and imparts flexibility to the resultant moldings.

butadiene-styrene plastic. A synthetic resin derived from the copolymerization of butadiene gas and styrene liquids.

butt joint. A type of edge joint in which the edge faces of the two adherends are at right angles to the other faces of the adherends.

butylene plastics. Plastics based on resins made by the polymerization of butene or the copolymerization of butene with one or more unsaturated compounds, the butene being the greatest amount by weight.

C

calorimeter. An instrument capable of making absolute measurements of energy deposition (or absorbed dose) in a material by measuring its change in temperature and imparting the knowledge of the characteristics of its material construction.

caprolactam. A cyclic amide-type compound containing six carbon atoms. When the ring is opened, caprolactam is polymerizable into a nylon resin known as type 6 nylon, or polycaprolactam.

carbanion ion. Negatively charged organic compound (ion).

carbon. The element that provides the backbone for all organic polymers. Graphite is a crystalline form of carbon. Diamond is the densest crystalline form of carbon.

carbon black. A black pigment produced by the incomplete burning of natural gas or oil. It is widely used as a filler, particularly in the rubber industry. Because it possesses useful ultraviolet radiation protective properties, it is also much used in molding compounds intended for outside weathering applications.

carbon fiber. Fiber produced by the pyrolysis of organic precursor fibers, such as rayon, polyacrylonitrile (PAN), and pitch, in an inert environment. The term is often used interchangeably with the term graphite; however, carbon fibers and graphite fibers differ. The basic differences lie in the temperature at which the fibers are made and heat treated, and in the amount of elemental carbon produced. Carbon fibers typically are carbonized in the region of 1315 °C (2400 °F)

and assay at 93 to 95% carbon, while graphite fibers are graphitized at 1900 to 2480 °C (3450 to 4500 °F) and assay at more than 99% elemental carbon.

carbonium ion. Positively charged organic compound (ion).

carbonization. The process of pyrolyzation in an inert atmosphere at temperatures ranging from 800 to 1600 °C (1470 to 2910 °F) and higher, but usually at about 1315 °C (2400 °F). Range is influenced by the precursor, the processing of the individual manufacturer, and the properties desired.

casein. A protein material precipitated from skimmed milk by the action of either rennet or dilute acid. Rennet casein finds its main application in the manufacture of plastics. Acid casein is a raw material used in a number of industries, including the manufacture of adhesives.

casein adhesive. An aqueous colloidal dispersion of casein that may be prepared with or without heat; that may contain modifiers, inhibitors, and secondary binders to provide specific adhesive properties; and that includes a subclass, usually identified as casein glue, that is based on a dry blend of casein, lime, and sodium salts, mixed with water prepared without heat. ASTM D 907.

catalyst. A substance that markedly speeds up the cure of an adhesive when added in a minor quantity compared to the amounts of the primary reactants. ASTM D 907. See also *hardener*. Compare to *inhibitor*.

cation. A positively charged atom.

caul. A sheet of material employed singly or in pairs in the hot or cold pressing of assemblies being bonded. A caul is used to protect either the faces of the assembly or the press platens, or both, against marring and staining in order to prevent sticking, facilitate press loading, impart a desired surface texture or finish, and provide uniform pressure distribution. A caul may be made of any suitable material such as aluminum, stainless steel, hardboard, fiberboard, or plastic, the length and width dimensions generally being the same as those of the plates of the press where it is used. ASTM D 907.

cell. A single cavity formed by gaseous displacement in a plastic material. See also *cellular plastic*.

cellular adhesive. Synonym for foamed adhesive.

cellular plastic. A plastic with greatly decreased density due to the presence of numerous cells or bubbles dispersed throughout its mass. See also *cell, foamed plastics*, and *syntactic cellular plastics*.

cellulose acetate. An acetic acid ester of cellulose. It is obtained under rigidly controlled conditions, by the action of acetic acid and acetic anhydride on purified

cellulose usually obtained from cotton linters. All three available hydroxyl groups in each glucose unit of the cellulose can be acetylated, but in the material normally used for plastics it is usual practice to acetylate fully and then lower the acetyl value (expressed as acetic acid) to from 52 to 56% by partial hydrolysis. When compounded with suitable plasticizers, it gives a tough thermoplastic material.

cellulose acetate butyrate. An ester of cellulose made by the action of a mixture of acetic and butyric acids and their anhydrides on purified cellulose. It is used in the manufacture of plastics that are similar in general properties to cellulose acetate but are tougher and have better moisture resistance and dimensional stability.

cellulose ester. A derivative of cellulose in which the free hydroxyl groups attached to the cellulose chain have been replaced wholly or in part by acetic groups, for example, nitrate, acetate, or stearate groups. Esterification is effected by the use of a mixture of an acid with its anhydride in the presence of a catalyst, such as sulfuric acid. Mixed esters of cellulose, such as cellulose acetate butyrate are prepared by the use of mixed acids and mixed anhydrides. Esters and mixed esters, a wide range of which is known, differ in their compatibility with plasticizers, molding properties, and physical characteristics. These esters and mixed esters are used in the manufacture of thermoplastic molding compositions.

cellulose nitrate. A nitric acid ester of cellulose manufactured by the action of a mixture of sulfuric acid and nitric acid on cellulose, such as purified cotton linters. The type of cellulose nitrate used for celluloid manufacture usually contains 10.8 to 11.1% nitrogen. The latter figure is the nitrogen content of the dinitrate. Also called nitrocellulose.

cellulose propionate. An ester of cellulose made by the action of propionic acid and its anhydride on purified cellulose. It is used as the basis of a thermoplastic molding material.

cellulosic plastics. Plastics based on cellulose compounds, such as esters (cellulose acetate) and ethers (ethyl cellulose).

cement. See discussion under *adhesive*.

chain length. The length of the stretched linear macromolecule, most often expressed by the number of identical links.

chain transfer agent. A molecule from which an atom, such as hydrogen, may be readily abstracted by a free radical.

chalking. Dry, chalklike appearance of deposit on the surface of a plastic. See also *bloom*.

Charpy impact test. A test for shock loading in which a centrally notched sample bar is held at both ends and broken by striking the back face in the same plane as the notch.

chelate. Five- or six-membered ring formation based on the intramolecular attraction of hydrogen, oxygen, or nitrogen atoms.

chlorinated hydrocarbon. An organic compound having chlorine atoms in its chemical structure. Trichloroethylene, methyl chloroform, and methylene chloride are chlorinated hydrocarbon solvents; polyvinyl chloride is a plastic.

chlorofluorocarbon plastics. Plastics based on polymers made with monomers composed of chlorine, fluorine, and carbon only.

chlorofluorohydrocarbon plastics. Plastics based on polymers made with monomers composed of chlorine, fluorine, hydrogen, and carbon only.

chromatic aberration. In resinography, a defect in a lens or lens system resulting in different focal lengths for radiation of different wavelengths. The dispersive power of a single positive lens focuses light from the blue end of the spectrum at a shorter distance than from the red end. An image produced by such a lens shows color fringes around the border of the image.

chromatography. The separation, especially of closely related compounds, caused when a solution or mixture is allowed to seep through an absorbent (such as clay, gel, or paper), such that each compound becomes adsorbed in a separate, often colored, layer.

CIL flow test. A method of determining the rheology or flow properties of thermoplastic resins (developed by Canadian Industries Limited). In this test, the amount of molten resin forced through a specified-size orifice per unit time when a specified, variable force is applied gives a relative indication of the flow properties of various resins.

circuit board. A sheet of insulating material laminated to foil that is etched to produce a circuit pattern on one or both sides. Also called printed circuit board or printed wiring board.

CIS stereoisomer. A stereoisomer in which side chains or side atoms are arranged on the same side of a double bond present in a chain of atoms.

clean surface. A surface that is free of foreign material, both visible and invisible. A clean surface can be determined by the standard *water break test*. ASTM D 3932.

cleavage strength. The tensile load in terms of kgf/mm (lbf/in.) of width required to cause the separation of a test specimen 25 mm (1 in.) in length. ASTM D 1062.

closed assembly time. The time interval between completion of assembly of the parts for bonding and the application of pressure or heat, or both, to the assembly. ASTM D 907.

closed-cell cellular plastics. Cellular plastics in which almost all the cells are noninterconnecting.

coagulation. Precipitation of a polymer dispersed in a latex.

co-curing. The act of curing a composite laminate and simultaneously bonding it to some other prepared surface, or curing together an inner and outer tube of similar or dissimilar fiber-resin combinations after each tube has been wound or wrapped separately. See also *secondary bonding*.

coefficient of elasticity. The reciprocal of Young's modulus in a tension test. See also *compliance*.

coefficient of expansion. A measure of the change in length or volume of an object; specifically, a change measured by the increase in length or volume of an object per unit length or volume.

coefficient of friction. A measure of the resistance to sliding of one surface in contact with another surface.

coefficient of thermal expansion (CTE). The change in unit of length or volume accompanying a unit change of temperature.

cohesion. The state in which the particles of a single substance are held together by primary or secondary valence forces. As used in the adhesive field, the state in which the particles of the adhesive (or the adherend) are held together. ASTM D 907.

cohesive blocking. The blocking of two similar, potentially adhesive faces. ASTM D 1146.

cohesive failure. Rupture of an adhesive bond, such that the separation appears to be within the adhesive. ASTM D 907. Compare to *adhesive failure*.

cohesive strength. Intrinsic strength of an adhesive.

coin test. The use of a coin to tap a laminate at different sites to detect a change in sound, which would indicate the presence of a defect. A surprisingly accurate test in the hands of experienced personnel.

cold flow. See *creep*.

cold pressing. A bonding operation in which an assembly is subjected to pressure without the application of heat. ASTM D 907.

cold-setting adhesive. An adhesive that sets at temperatures below 20 °C (68 °F). ASTM D 907. See also *hot-setting adhesive*, *intermediate-temperature-setting adhesive*, and *room-temperature-setting adhesive*.

colligative properties. Properties based on the number of molecules present. Most important are certain solution properties used extensively in molecular weight characterization.

collimated. Rendered parallel.

collimator. In resinography, a device for controlling a beam of radiation such that its rays are as parallel as is possible.

colloidal. A state of suspension in a liquid medium in which extremely small particles are suspended and dispersed but not dissolved.

colophony. See *rosin*.

color concentrate. A measured amount of dye or pigment incorporated into a predetermined amount of plastic. The pigmented or colored plastic is then mixed into larger quantities of plastic material to be used for molding. This mixture is added to the bulk of plastic in measured quantity in order to produce a precise, predetermined color in the finished, molded articles.

compaction. In reinforced plastics and composites, the application of a temporary vacuum bag and vacuum to remove trapped air and compact the lay-up.

compatibility. The ability of two or more substances combined with one another to form a homogeneous composition having useful plastic properties, for example, the suitability of a sizing or finish for use with certain general resin types. Nonreactivity or negligible reactivity between materials in contact.

complex modulus. The ratio of stress to strain in which each is a vector that may be represented by a complex number. It may be measured in tension or flexure (E^*), compression (K^*), or shear (G^*).

complex Young's modulus. The vectorial sum of Young's modulus and the loss modulus. Analogous to the complex dielectric constant.

compliance. Tensile compliance is the reciprocal of Young's modulus. Shear compliance is the reciprocal of shear modulus. The term is also used in the evaluation of stiffness and deflection.

composite material. A combination of two or more materials (reinforcing elements, fillers, and composite matrix binder), differing in form or composition on a macroscale. The constituents retain their identities; that is, they do not dissolve or merge completely into one another although they act in concert. Normally, the components can be physically identified and exhibit an interface between one another.

compound. In reinforced plastics and composites, the intimate admixture of a polymer with other ingredients, such as fillers, softeners, plasticizers, reinforcements, catalysts, pigments, or dyes. A thermoset compound usually contains all the ingredients necessary for the finished product, while a thermoplastic compound may require the subsequent addition of pigments, blowing agents, and so forth.

compressive modulus. The ratio of compressive stress to compressive strain below the proportional limit. Theoretically equal to Young's modulus determined from tensile experiments.

compressive strength. The ability of a material to resist a force that tends to crush or buckle. The maximum compressive load sustained by a specimen divided by the original cross-sectional area of the specimen.

compressive stress. The normal stress caused by forces directed toward the plane on which they act.

condensation. A chemical reaction in which two or more molecules combine, with the resulting separation of water or some other simple substance; the process is called polycondensation if a polymer is formed. ASTM D 907. See also *polymerization*.

condensation resin. A resin formed by polycondensation, for example, the alkyd, phenol-aldehyde, and urea-formaldehyde resins.

conditioning. Subjecting a material to a prescribed environmental and/or stress history before testing.

conditioning time. See *joint-conditioning time*, *curing time*, and *setting time*.

conductivity. The reciprocal of volume resistivity. The electrical or thermal conductance of a unit cube of any material (conductivity per unit volume).

configurations. Related chemical structures produced by the cleavage and reforming of covalent bonds.

conformal coating. A coating that covers and exactly fits the shape of the coated object.

conformations. Different shapes of polymer molecules resulting from rotation about single covalent bonds in the polymer chain.

consistency. That property of a liquid adhesive by virtue of which it tends to resist deformation. Consistency is not a fundamental property but is composed of viscosity, plasticity, and other phenomena. ASTM D 907. See also *viscosity* and *viscosity coefficient*.

constituent. In general, an element of a larger grouping.

contact adhesive. An adhesive that is apparently dry to the touch and that will adhere to itself instantaneously upon contact. Synonym for contact bond adhesive and dry bond adhesive. ASTM D 907.

contact bond adhesive. Synonym for contact adhesive.

contact pressure resins. Liquid resins that thicken or polymerize upon heating and that require little or no pressure when used for bonding laminates.

contaminant. An impurity or foreign substance present in a material or environment that affects one or more properties of the material, particularly adhesion.

coordination catalysis. Ziegler type of catalysis.

copolymer. A long-chain molecule formed by the reaction of two or more dissimilar monomers. See also *polymer*.

copolymerization. See *polymerization*.

corona resistance. Resistance to an ionizing process: When an electric current passes through a conductor, it induces a surrounding electrostatic field. When voids exist in the insulation near the conductor, the high-voltage electrostatic field may ionize and rapidly accelerate some of the air molecules in the void. These ions can then collide with the other molecules and ionize them, thereby eating a hole in the insulation. Resistance to this process is called corona resistance.

corrosion resistance. The ability of a material to withstand contact with ambient natural factors or factors of a particular, artificially created atmosphere without degradation or change in properties. For metals, this could be pitting or rusting; for organic materials, it could be crazing.

cottoning. The formation of weblike filaments of adhesive between the applicator and substrate surface.

coupling agent. Any chemical designed to react with both the reinforcement and matrix phases of a composite material to form or promote a stronger bond at the interface.

coupon. Usually, a specimen for a specific test, such as a tensile coupon.

crack. An actual separation of material, visible on opposite surfaces of the part, and extending through the thickness. A fracture.

crack growth. Rate of propagation of a crack through a material due to a static or dynamic applied load.

cratering. Depressions on coated plastic surfaces caused by excess lubricant. Cratering results when paint is too thin and later ruptures, leaving pinholes and other voids. Use of less thinner in the coating can reduce or eliminate cratering, as can less lubricant on the part.

crazing. Fine cracks that may extend in a network on or under the surface of or through a layer of adhesive. ASTM D 907.

creep. The dimensional change with time of a material under load, following the initial instantaneous elastic or rapid deformation. Creep at room temperature is sometimes called cold flow. ASTM D 907.

creep recovery. The time-dependent decrease in strain in a solid following the removal of force.

creep-rupture strength. The stress that causes fracture in a creep test at a given time, in a specified constant environment. Sometimes referred to as the stress-rupture strength. In glass technology, the static fatigue strength.

critical damping. In dynamic mechanical measurement, that damping required for the borderline condition between oscillatory and nonoscillatory behavior.

cross laminate. A laminate in which some of the layers of material are oriented at right angles to the remaining layers with respect to the grain or strongest direction in tension. Balanced construction of the laminations above the centerline of the thickness of the laminate is normally assumed. ASTM D 907. Compare to *parallel laminate*.

cross linking. With thermosetting and certain thermoplastic polymers, the setting up of chemical links between the molecular chains. When extensive, as in most thermosetting resins, cross linking makes an infusible supermolecule of all the chains. In rubbers, the cross linking is just enough to join all molecules into a network.

crosswise direction. Crosswise refers to the cutting of specimens and to the application of load. For rods and tubes, crosswise is any direction perpendicular to the long axis. For other shapes or materials that are stronger in one direction than in another, crosswise is the direction that is weaker. For materials that are equally strong in both directions, crosswise is an arbitrarily designated direction at right angles to the lengthwise direction.

crystalline plastic. A material having an internal structure in which the atoms are arranged in an orderly three-dimensional configuration. More accurately referred to as a semicrystalline plastic because only a portion of the molecules are in crystalline form.

crystallinity. A regular arrangement of the atoms of a solid in space. In most polymers, including cellulose, this state is usually imperfectly achieved. The crystalline (ordered) regions are submicroscopic volumes in which there is some degree of regularity in the arrangement of the component molecules. In these regions there is sufficient geometric order to obtain definite x-ray diffraction patterns.

C-scan. The back-and-forth scanning of a specimen with ultrasonics. A nondestructive testing technique for finding voids, delaminations, defects in fiber distribution, and so forth.

C-stage. The final stage in the reaction of certain thermosetting resins in which the material is relatively insoluble and infusible. Certain thermosetting resins in a fully cured adhesive layer are in this stage. Synonym for resite. ASTM D 907. See also *A-stage* and *B-stage*.

cure. To change the physical properties of an adhesive by chemical reaction, which may be condensation, polymerization, or vulcanization; usually accomplished by the action of heat and catalyst, alone or in combination, with or without pressure. ASTM D 907. See also *dry* and *set*.

cure rate index. A measurement of length per unit time per lamp, at specific wavelengths, power per unit length, and focus.

cure stress. A residual internal stress produced during the curing cycle of, for example, composite structures. In this case, stresses originate when various components of a wet lay-up have different thermal coefficients of expansion.

curing agent. A catalytic or reactive agent that causes cross linking. Also called a *hardener*.

curing temperature. The temperature to which an adhesive or an assembly is subjected to cure the adhesive. The temperature attained by the adhesive in the process of curing (adhesive curing temperature) may differ from the temperature of the atmosphere surrounding the assembly (assembly curing temperature). ASTM D 907. See also *drying temperature* and *setting temperature*.

curing time. The period of time during which an assembly is subjected to heat or pressure, or both, to cure the adhesive. Further cure may take place after removal of the assembly from the conditions of heat or pressure, or both. ASTM D 907. See also *joint-conditioning time* and *setting time*.

cyanate resins. Thermosetting resins that are derived from bisphenols or polyphenols and are available as monomers, oligomers, blends, and solutions. Also known as cyanate esters, cyanic esters, and triazine resins.

cyanoacrylate. A thermoplastic monomer adhesive characterized by excellent polymerizing and bonding strength.

cyclic hydrocarbons. Cyclic or ring compounds. Benzene (C_6H_6) is a classic example.

D

damage tolerance. A design measure of crack growth rate. Cracks in damage-tolerant-designed structures are not permitted to grow to critical size during expected service life.

damping. The loss in energy, such as dissipated heat, that results when a material or material system is subjected to an oscillatory load or displacement.

dark reaction. The gelling of material in storage without light initiation.

debond. A deliberate separation of a bonded joint or interface, usually for repair or rework purposes. Also, an unbonded or nonadhered region; a separation at the fiber-matrix interface due to strain incompatibility. In the United Kingdom, the term often refers to accidental damage. See also *disbond* and *delamination*.

deflection temperature under load (DTUL). The temperature at which a simple cantilever beam deflects a given amount under load.

deformation under load. The dimensional change of a material under load for a specified time following the instantaneous elastic deformation caused by the initial application of the load. See also *creep*.

degassing. Opening and closing of a plastics mold to allow gases to escape during molding.

degradation. A deleterious change in the chemical structure, physical properties, or appearance of a plastic.

degrease. To remove oil and grease from adherend surfaces.

degree of polymerization. The number of structural units, or mers, in the average polymer molecule in a sample measure of molecular weight.

degree of saturation. The ratio of the weight of water vapor associated with a pound of dry air to the weight of water vapor associated with a pound of dry air saturated at the same temperature.

delamination. The separation of layers in a laminate because of failure of the adhesive, either in the adhesive itself or at the interface between the adhesive and the adherend. ASTM D 907.

deliquescence. The absorption of atmospheric water vapor by a crystalline solid until the crystal eventually dissolves into a saturated solution.

deposition. The process of applying a material to a base by means of vacuum, electrical, chemical, screening, or vapor methods, often with the assistance of a temperature and pressure container.

depth. In the case of a beam, the dimension parallel to the direction in which the load is applied.

depth dose. The variation of absorbed dose with distance from the incident surface of a material exposed to radiation. Depth-dose profiles give information about the distribution of absorbed energy in a specific material.

desiccant. A substance that can be used to dry materials because of its affinity for water.

design allowables. Statistically defined (by a test program) material property allowable strengths, usually referring to stress or strain. See also *A-basis*, *B-basis*, *S-basis*, and *typical-basis*.

desorption. A process in which an absorbed material is released from another material. Desorption is the reverse of absorption, adsorption, or both.

destaticization. A treatment of plastic materials that minimizes the effects of static electricity on the surface. It can be accomplished either by treating the surface with specific materials or by incorporating materials in the molding compound. Minimization of surface static electricity prevents dust and dirt from being attracted to and/or clinging to the surface of the article.

dew point. The temperature to which water vapor must be reduced to obtain saturation vapor pressure, that is, 100% relative humidity. As air is cooled, the amount of water vapor that it can hold decreases. If air is cooled sufficiently, the actual water vapor pressure becomes equal to the saturation water vapor pressure, and any further cooling normally results in the condensation of moisture.

diadic polyamide. Polyamide produced by the condensation of a diamine and a dicarboxylic acid.

dielectric. A nonconductor of electricity. The ability of a material to resist the flow of an electrical current.

dielectric constant. The ratio of the capacitance of an assembly of two electrodes separated solely by a plastic insulating material to its capacitance when the electrodes are separated by air.

dielectric curing. The curing of a synthetic thermosetting resin by the passage of an electric charge (produced from a high-frequency generator) through the resin.

dielectric heating. The heating of materials by dielectric loss in a high-frequency electrostatic field.

dielectric loss. A loss of energy evidenced by the rise in heat of a dielectric placed in an alternating electric field. It is usually observed as a frequency-dependent conductivity.

dielectric loss angle. The difference between 90° and the dielectric phase angle. Also called the dielectric phase difference.

dielectric loss factor. The product of the dielectric constant and the tangent of the dielectric loss angle for a material. Also called the dielectric loss index.

dielectric monitoring. Monitoring the cure of thermosets by tracking the changes in their electrical properties during material processing.

dielectric phase angle. The angular difference in phase between the sinusoidal alternating potential difference applied to a dielectric and the component of the resulting alternating current having the same period as the potential difference.

dielectric power factor. The cosine of the dielectric phase angle (or sine of the dielectric loss angle).

dielectric strength. The property of an insulating material that enables it to withstand electric stress. The average potential per unit thickness at which failure of the dielectric material occurs.

dielectrometry. Use of electrical techniques to measure the changes in loss factor (dissipation) and in capacitance during the cure of the resin in a laminate. Also called dielectric spectroscopy.

differential scanning calorimetry (DSC). Measurement of energy absorbed (endotherm) or produced (exotherm). May be applied to melting, crystallization, resin curing, loss of solvents, and other processes involving an energy change. May also be applied to processes involving a change in heat capacity, such as the glass transition.

differential thermal analysis (DTA). An experimental analysis technique in which a specimen and a control are heated simultaneously and the difference in their temperatures is monitored. The difference in temperatures provides information on relative heat capacities, the presence of solvents, changes in structure (that is, phase changes, such as the melting of one component in a resin system), and chemical reactions. See also *differential scanning calorimetry*.

diffusion. The movement of a material, such as a gas or liquid, in the body of a plastic. If the gas or liquid is absorbed on one side of a piece of plastic and given off on the other side, the phenomenon is called permeability. Diffusion and permeability are not due to holes or pores in the plastic but are caused and controlled by chemical mechanisms.

diluent. An ingredient added to an adhesive, usually to reduce the concentration of bonding materials. ASTM D 907. See also *extender* and *thinner*.

dimer. A substance (comprising molecules) formed from two molecules of a monomer.

diphenyl oxide resins. Thermosetting resins based on diphenyl oxide and possessing excellent handling properties and heat resistance.

disbond. An area within the bonded interface between two adherends in which an adhesion failure or separation has occurred. Also, colloquially, an area of separation between two laminae in the finished laminate (in this case, the term delamination is normally preferred). See also *debond*.

dispersion. Finely divided particles of a material in suspension in another substance. See also *adhesive dispersion*.

disproportionation. Termination by chain transfer between macroradicals to produce a saturated and an unsaturated polymer molecule.

doctor bar or blade. A scraper mechanism that regulates the amount of adhesive on spreader rolls or on the surface being coated. ASTM D 907.

doctor roll. A roller mechanism that is revolving at a different surface speed, or in an opposite direction, resulting in a wiping action for regulating the adhesive supplied to the spreader roll. ASTM D 907.

domain. A morphological term used in regard to noncrystalline systems, such as block copolymers, in which the chemically different sections of the chain separate, generating two or more amorphous phases.

dopant. A material added to a polymer to change a physical property.

dosimeter. A device for measuring radiation-induced signals that can be related to absorbed dose (or energy deposited) by radiation in materials and is calibrated in terms of the appropriate quantities and units. Also called dose meter.

double spread. The application of adhesive to both adherends of a joint. ASTM D 907.

dry. To change the physical state of an adhesive on an adherend by the loss of solvent constituents by evaporation or absorption, or both. ASTM D 907. See also *cure* and *set*.

dry bond adhesive. See *contact adhesive*.

dry-bulb temperature. The temperature of the air as indicated by an accurate thermometer, corrected for radiation if significant.

drying temperature. The temperature to which an adhesive on an adherend, an adhesive in an assembly, or the assembly itself is subjected to dry the adhesive. The temperature attained by the adhesive in the process of drying (adhesive drying temperature) may differ from the temperature of the atmosphere surrounding the assembly (assembly drying temperature). ASTM D 907. See also *curing temperature* and *setting temperature*.

drying time. The period of time during which an adhesive on an adherend or an assembly is allowed to dry with or without the application of heat or pressure, or both. ASTM D 907. See also *curing time*, *joint-conditioning time*, and *setting time*.

dry strength. The strength of an adhesive joint determined immediately after drying under specified conditions or after a period of conditioning in a standard laboratory atmosphere. ASTM D 907. Compare to *wet strength*.

dry tack. The property of certain adhesives, particularly nonvulcanizing rubber adhesives, to adhere on contact to themselves at a stage in the evaporation of volatile constituents, even though they seem dry to the touch. Synonym for aggressive tack. ASTM D 907.

DSC. See *differential scanning calorimetry*.

DTA. See *differential thermal analysis*.

DTUL. See *deflection temperature under load*.

ductility. The amount of plastic strain that a material can withstand before fracture. Also, the ability of a material to deform plastically before fracturing.

dynamic mechanical measurement. A technique in which either the modulus and/or damping of a substance under oscillatory load or displacement is measured as a function of temperature, frequency, or time, or a combination thereof.

dynamic modulus. The ratio of stress to strain under cyclic conditions (calculated from data obtained from either free or forced vibration tests, in shear, compression, or tension).

E

edge joint. A joint made by bonding the edge faces of two adherends.

elastic deformation. Any portion of the total deformation of a body that occurs immediately when load is applied and disappears immediately and completely when the load is removed.

elasticity. A materials property by virtue of which a material tends to recover its original size and shape after removal of a force causing deformation. See also *viscoelasticity*.

elastic limit. The greatest stress a material is capable of sustaining without permanent strain remaining after the complete release of the stress. A material is said to have passed its elastic limit when the load is sufficient to initiate plastic, or nonrecoverable, deformation.

elastic recovery. The fraction of a given deformation that behaves elastically. A perfectly elastic material has an elastic recovery of 1; a perfectly plastic material has an elastic recovery of 0.

elastic true strain (ϵ_e). Elastic component of the true strain.

elastomer. A macromolecular material which, at room temperature, is capable of recovering substantially in size and shape after removal of a deforming force. ASTM D 907.

electron spin resonance (ESR) spectroscopy. A form of spectroscopy similar to nuclear magnetic resonance, except that the species studied is an unpaired electron, not a magnetic nucleus.

elongation. Deformation caused by stretching. The fractional increase in length of a material stressed in tension. When expressed as a percentage of the original gage length, it is called percentage elongation.

elongation at break. Elongation recorded at the moment of rupture of a specimen, often expressed as a percentage of the original length.

emulsion. A two-phase liquid system in which small droplets of one liquid (the internal phase) are immiscible in, and are dispersed uniformly throughout, a second continuous liquid phase (the external phase). The internal phase is sometimes described as the dispense phase. ASTM D 907. See also *adhesive dispersion*.

emulsion polymerization. Polymerization of monomers dispersed in an aqueous emulsion.

encapsulated adhesive. An adhesive in which the particles or droplets of one of the reactive components are enclosed in a protective film (microcapsules) to prevent cure until the film is destroyed by suitable means.

encapsulation. The enclosure of an item in plastic, sometimes referring specifically to the enclosure of capacitors or circuit board modules. See also *potting*.

endurance limit. See *fatigue limit*.

energy loss. The energy per unit volume that is lost in each deformation cycle. Energy loss is the hysteresis loop area, calculated with reference to coordinate scales.

engineering adhesive. A bonding agent intended to join metal, plastics, wood, glass, ceramics, or rubber. The term differentiates such bonding agents from glues used to join paper and other nondurables.

engineering plastics. A general term covering all plastics, with or without fillers or reinforcements, that have mechanical, chemical, and thermal properties that render them suitable for use as construction materials, machine components, and chemical-processing-equipment components.

engineering stress. The stress calculated on the basis of the original dimensions of the specimen.

environment. The aggregate of all conditions (such as contamination, temperature, humidity, radiation, magnetic and electric fields, shock, and vibration) that externally influence the performance of an item.

environmental stress cracking (ESC). The susceptibility of a resin to cracking or crazing when in the presence of such chemicals as surface-active agents, or other environments.

epichlorohydrin. The basic epoxidizing resin intermediate in the production of epoxy resins. It contains an epoxy group and is highly reactive with polyhydric phenols such as bisphenol A.

epitaxy. The oriented growth of a crystalline substance on a substratum of the same or a different crystalline substance.

epoxide. Compound containing the oxirane structure, a three-member ring containing two carbon atoms and one oxygen atom. The most important members are ethylene oxide and propylene oxide.

epoxy plastic. A thermoset polymer containing one or more epoxide groups and curable by reaction with amines, alcohols, phenols, carboxylic acids, acid anhydrides, and mercaptans. An important matrix resin in reinforced composites and in structural adhesives.

epoxy resin. A viscous liquid or the brittle, solid-containing epoxide groups that can be cross linked into final form by means of a chemical reaction with a variety of setting agents used with or without heat.

equilibrium centrifugation. In resinography, a method for determining the distribution of molecular weights by spinning a solution of the specimen at a speed such that the molecules of the specimen are not removed from the solvent but are held at a point where the (centrifugal) force tending to remove them is balanced by the dispersive forces caused by thermal agitation.

ESC. See *environmental stress cracking.*

ESR. See *electron spin resonance spectroscopy.*

ester. The reaction product of an alcohol and an acid.

ethylene plastics. Plastics based on polymers of ethylene or copolymers of ethylene with other monomers, the ethylene being in greatest amount by mass.

eutectic. The composition within any system of two or more crystalline phases that melts completely at the minimum temperature. Also, the temperature at which such a composition melts.

eutectic bonding. The forming of a bond by bringing two solids to their lowest constant melting point at which the molten solids mix and harden upon cooling.

exotherm. The temperature-time curve of a chemical reaction or a phase change giving off heat, particularly the polymerization of casting resins. The amount of heat given off. The term has not been standardized with respect to sample size, ambient temperature, degree of mixing, and so forth.

expandable plastic. A plastic that can be made cellular by thermal, chemical, or mechanical means. Foam plastics such as expandable polystyrene and foam polyurethane are examples.

extender. A substance, generally having some adhesive action, added to an adhesive to reduce the amount of primary binder required per unit area. ASTM D 907. See also *binder, diluent, filler,* and *thinner.*

extensibility. The ability of a material to extend or elongate upon application of sufficient force, expressed as a percent of the original length.

extensometer. A mechanical or optical device for measuring linear strain due to mechanical stress.

extrusion. The process of compacting a plastic material into powder or granules in a uniform melt and forcing the melt through an orifice in a more or less continuous fashion to yield a desired shape. While held in the desired shape, the melt must be cooled to a solid state.

F

fadeometer. An apparatus for determining the resistance of resins and other materials to fading.

failure. See *adhesive failure* and *cohesive failure.*

fatigue. The failure or decay of mechanical properties after repeated applications of stress. Fatigue tests give information on the ability of a material to resist the development of cracks, which eventually bring about failure as a result of a large number of cycles.

fatigue ductility (D_f). The ability of a material to deform plastically before fracturing, determined from a constant strain amplitude, low-cycle fatigue test. Usually expressed in percent in direct analogy with elongation and reduction of area ductility measures.

fatigue ductility exponent (c). The slope of a log-log plot of plastic strain range and fatigue life.

fatigue life (N_f). The number of cycles of stress or strain of a specified character that a given specimen sustains before failure of a specified nature occurs.

fatigue limit. The stress level below which a material can be stressed cyclically for an infinite number of times without failure.

fatigue ratio. The ratio of fatigue strength to tensile strength. Mean stress and alternating stress must be stated.

fatigue strength. The maximum cyclical stress a material can withstand for a given number of cycles before failure occurs. The residual strength after being subjected to fatigue.

faying surface. The surfaces of materials in contact with each other and joined or about to be joined.

feathering. The tapering of an adherend on one side to form a wedge section, as used in a scarf joint.

FEP. See *fluorinated ethylene propylene.*

fiber. In general, filamentary materials. Often, the term fiber is used synonymously with filament. It is a general term for a filament with a finite length that is at least 100 times its diameter, which is typically 0.10 to 0.13 mm (0.004 to 0.005 in.). In most cases it is prepared by drawing from a molten bath, spinning, or depositing on a substrate. Fibers can be continuous or specific short lengths (discontinuous), normally no less than 3.2 mm (1/8 in.).

fiber content. The amount of fiber present in reinforced plastics and composites, usually expressed as a percentage volume fraction or weight fraction.

fiber count. The number of fibers per unit width of ply present in a specified section of a reinforced plastic or composite.

fiber diameter. The measurement (expressed in hundred thousandths) of the diameter of individual filaments.

fiber direction. The orientation or alignment of the longitudinal axis of the fiber with respect to a stated reference axis.

fiberglass. An individual filament made by drawing molten glass. A continuous filament is a glass fiber of great or indefinite length. A staple fiber is a glass fiber of relatively short length, generally less than 430 mm (17 in.), the length depending on the forming or spinning process used.

fiberglass reinforcement. The major material used to reinforce plastic. Available as mat, roving, fabric, and so forth, it is incorporated into both thermosets and thermoplastics.

fiber pattern. Visible fibers on the surface of laminates or molding. The thread size and weave of glass cloth.

fiber-reinforced plastic (FRP). In general, a plastic that is reinforced with cloth, mat, strands, or any other fiber form.

fiber show. Strands or bundles of fibers that are not covered by plastic because they are at or above the surface of a reinforced plastic or composite.

fiber wash. Splaying out of woven or non-woven fibers from the general reinforcement direction. Fibers are carried along with bleeding resin during cure.

fibrillation. Production of fiber from film.

filament. The smallest unit of a fibrous material. The basic units formed during drawing and spinning, which are gathered into strands of fiber for use as reinforcements. Filaments are usually of extreme length and very small diameter, usually less than 25 μm (1 mil). Normally, filaments are not used individually. Some textile filaments can function as a reinforcing yarn when they are of sufficient strength and flexibility.

filler. A relatively nonadhesive substance added to an adhesive to improve its working properties, permanence, strength, or other qualities. ASTM D 907. See also *binder* and *extender*.

filler sheet. A sheet of deformable or resilient material that, when placed between the assembly to be bonded and the pressure applicator, or when distributed within a stack of assemblies, aids in providing uniform application of pressure over the area to be bonded. ASTM D 907.

fillet. That portion of an adhesive that fills the corner or angle formed where two adherends are joined. ASTM D 907.

film. Sheeting having a nominal thickness not greater than 0.25 mm (0.010 in.).

film adhesive. A synthetic resin adhesive, usually of the thermosetting type, in the form of a thin, dry film of resin with or without a paper or glass carrier.

fines. Very small particles (usually under 200 mesh) accompanying larger grains, usually of molding powder.

fingerjoint assembly. A short portion of two boards joined at their ends by a fingerjoint obtained from a fingerjoint production line for testing, frequently referred to as an assembly. ASTM D 4688.

finish. A mixture of materials for treating glass or other fibers. It contains a coupling agent to improve the bond of resin to the fiber and usually includes a lubricant to prevent abrasion, as well as a binder to promote strand integrity. With graphite or other filaments, it may perform any or all of the above functions. To complete the secondary work on a molded part so that it is ready for use. Operations such as filling, deflashing, buffing, drilling, tapping, and degating are commonly called finishing operations.

first-degree blocking. An adherence between the surfaces under test of such degree that when the upper specimen is lifted, the lower specimen will cling thereto, but may be parted with no evidence of damage to either surface. ASTM D 1146.

first-order transition. A change of state associated with crystallization, melting, or a change in the crystal structure of a polymer.

fixture, or set, time. The shortest time required by an adhesive to develop handling strength such that test specimens can be removed from fixtures, unclamped, or handled without stressing the bond and thereby affecting bond strength. ASTM D 1144.

flame resistance. Ability of a material to extinguish flame once the source of heat is removed. See also *self-extinguishing resin*.

flame retardants. Certain chemicals that are used to reduce or eliminate the tendency of a resin to burn.

flame treating. A method of rendering inert thermoplastic objects receptive to inks, lacquers, paints, adhesives, and so forth, in which the object is bathed in an open flame to promote oxidation of the surface of the article.

flammability. Measure of the extent to which a material will support combustion.

flexibility. That property of a material by virtue of which it may be flexed or bowed repeatedly without undergoing rupture. ASTM D 3111.

flexibilizer. An additive that makes a finished plastic more flexible or tough. See also *plasticizer*.

flexural modulus. The ratio, within the elastic limit, of the applied stress on a test specimen in flexure to the corresponding strain in the outermost fibers of the specimen.

flexural strength. The unit resistance to the maximum load before failure by bending, usually expressed in force per unit area.

flocking. A method of coating by spraying finely dispersed powders or fibers.

flow. Movement of an adhesive during the bonding process, before the adhesive is set. ASTM D 907.

fluorinated ethylene propylene (FEP). A member of the fluorocarbon family of plastics that is a copolymer of tetrafluoroethylene and hexafluoropropylene, possessing most of the properties of polytetrafluoroethylene and having a melt viscosity low enough to permit conventional thermoplastic processing. Available in pellet form for molding and extrusion, and as dispersions for spray or dip-coating processes.

fluorocarbon plastics. Plastics based on polymers made with monomers composed of fluorine and carbon only.

fluorocarbons. A family of plastics that includes polytetrafluoroethylene, polychlorotrifluoroethylene, polyvinylidene, and fluorinated ethylene propylene. This family is characterized by good thermal and chemical resistance, nonadhesiveness, low dissipation factor, and low dielectric constant. A variety of forms are available, such as molding materials, extrusion materials, dispersions, film, and tape, depending on the particular fluorocarbon.

fluorohydrocarbon plastics. Plastics based on polymers made with monomers composed of fluorine, hydrogen, and carbon only.

fluoroplastics. Plastics based on polymers with monomers containing one or more atoms of fluorine or on copolymers of such monomers with other monomers, with the fluorine-containing monomer(s) being in greatest amount by mass.

flux. A plastics composition additive incorporated during processing to improve flow. Also, a state of fluidity.

foamed adhesive. An adhesive, the apparent density of which has been decreased substantially by the presence of numerous gaseous cells dispersed throughout its mass. Synonym for cellular adhesive. ASTM D 907.

foamed plastics. Resins in sponge form, flexible or rigid, with cells closed or interconnected and density over a range from that of the solid parent resin to 0.030 g/cm^3. The compressive strength of rigid foams is fair, making them useful as core materials for sandwich constructions. Also, chemical cellular plastics, the structures of which are produced by gases generated from the chemical interaction of their constituents. See also *expandable plastic*.

foaming agent. Chemicals added to plastics and rubbers that generate inert gases, such as nitrogen, upon heating, causing the resin to form a cellular structure.

folded chain. The conformation of a flexible polymer molecule when present in a crystal. The molecule exits and reenters the same crystal, frequently generating folds.

Fourier transform. An analytical method used automatically in advanced forms of spectroscopic analysis such as infrared and nuclear magnetic resonance spectroscopy.

fracture. The separation of a body. Defined both as rupture of the surface without complete separation of the laminate and as complete separation of a body because of external or internal forces.

fracture ductility. The true plastic strain of fracture.

fracture strength. The normal stress at the beginning of fracture. Calculated from the load at the beginning of fracture during a tension test and the original cross-sectional area of the specimen.

fracture toughness. A measure of the damage tolerance of a material containing initial flaws or cracks. Used in aircraft structural design and analysis.

free bend. The bend obtained by applying forces to the ends of a specimen without the application of force at the point of maximum bending.

free radical polymerization. A type of polymerization in which the propagating species is a long-chain free radical initiated by the introduction of free radicals from the thermal or photochemical decomposition of an initiator molecule.

free rotation. The rotation of atoms, particularly carbon atoms, about a single bond. Because the energy requirement is only a few kcal, the rotation is said to be free if sufficient thermal energy is available.

free vibration. A technique for performing dynamic mechanical measurements in which the sample is deformed, released, and allowed to oscillate freely at the natural resonant frequency of the system. Elastic modulus is calculated from the measured resonant frequency, and damping is calculated from the rate at which the amplitude of the oscillation decays.

friction, coefficient of. See *coefficient of friction*.

FRP. See *fiber-reinforced plastic*.

full-contour length. The length of a fully extended polymer chain.

functionality. The average number of reaction sites on an individual polymer chain.

furan resins. Dark-colored thermosetting resins available primarily as liquids ranging from low-viscosity polymers to thick, heavy syrups, which cure to highly cross-linked, brittle substances. Made primarily by polycondensation of furfuryl alcohol in the presence of strong acids, sometimes in combination with formaldehyde or furfuryldehyde.

furfural resin. A dark-colored synthetic resin of the thermosetting variety obtained by the condensation of furfural with phenol or its homologs. It is used in the manufacture of molding materials, adhesives, and impregnating varnishes. Properties include a high resistance to acids and alkalies.

fusion. In vinyl dispersions, the heating of a dispersion to produce a homogeneous mixture.

fuzz. The accumulation of short, broken filaments after passing glass strands, yarns, or rovings over a contact point. Often weighted and used as an inverse measure of abrasion resistance.

G

gage length. Length over which deformation is measured, for a tensile or compressive test specimen. The deformation over the gage length divided by the gage length determines strain.

gamma loss peak. In dynamic mechanical measurement, the third peak in the damping curve below the melt, in the order of decreasing temperature or increasing frequency.

gamma transition. See *glass transition*.

gap-filling adhesive. An adhesive, subject to reduced shrinkage upon setting, used as sealant.

gel. A semisolid system consisting of a network of solid aggregates in which liquid is held. Also, the initial jellylike solid phase that develops during the formation of a resin from a liquid. With respect to vinyl plastisols, a state between liquid and solid that occurs in the initial states of heating, or upon prolonged storage. In general, gels have very low strengths and do not flow like a liquid. They are soft, flexible, and may rupture under their own weight unless supported externally. In a cross-linked thermoplastic, gel is the fraction of polymeric material present in the network.

gelation. The point in a resin cure when the resin viscosity has increased to a point such that it barely moves when probed with a sharp instrument.

gelation time. That interval of time, in connection with the use of synthetic thermosetting resins, extending from the introduction of a catalyst into a liquid adhesive system to the start of gel formation. Also, the time under application of load for a resin to reach a solid state.

gel coat. A quick-setting resin applied to the surface of a mold and gelled before lay-up. The gel coat becomes an integral part of the finished laminate and is usually used to improve surface appearance and bonding.

gel permeation chromatography (GPC). A form of liquid chromatography in which the polymers are separated by their ability or inability to penetrate the material in the separation column.

gel point. The point at which a thermosetting system attains an infinite value of its average molecular weight. The viscosity at which a liquid begins to exhibit pseudoelastic properties. This stage may be conveniently observed from the inflection point on a viscosity-time plot.

gel time. The period of time from the initial mixing of the reactants of a liquid material composition to the point in time when gelation occurs, as defined by a specific test method.

glass cloth. Woven glass fiber material. See also *scrim*.

glass, percent by volume. The product of the specific gravity of a laminate and the percent glass by weight, divided by the specific gravity of the glass.

glass transition. The reversible change in an amorphous polymer or in amorphous regions of a partially crystalline polymer from, or to, a viscous or rubbery condition to, or from, a hard and relatively brittle one.

glass transition temperature (T_g). The approximate midpoint of the temperature range over which the glass transition takes place; glass fibers exhibit a phase change at approximately 955 °C (1750 °F), and carbon/graphite fibers exhibit the change at 2205 to 2760 °C (4000 to 5000 °F). The temperature at which increased molecular mobility results in significant changes in the properties of a cured resin system. Also, the inflection point on a plot of modulus versus temperature. The measured value of T_g depends considerably on the rate of temperature change in any experiment and on the frequency or rate of deformation in a mechanical test.

glue. Originally, a hard gelatin obtained from hides, tendons, cartilage, bones, and so on, of animals and also an adhesive prepared from this substance by heating with water. Through general use, the term is now synonymous with the terms bond and adhesive. ASTM D 907. See also *adhesive*, *gum*, *mucilage*, *paste*, *resin*, and *sizing*.

glue-laminated wood. An assembly made by bonding layers of veneer or lumber with an adhesive so that the grain of all laminations is essentially parallel. ASTM D 907.

glue line. Synonym for bond line. ASTM D 907.

glue line thickness. Thickness of layer of cured adhesive.

GPC. See *gel permeation chromatography*.

graft copolymers. A chain of one type of polymer to which side chains of a different type are attached or grafted.

granular structure. Nonuniform appearance of finished plastic material due to retention, or incomplete fusion, of particles of composition, either within the mass or on the surface.

graphite. A crystalline allotropic form of carbon.

graphite fiber. A fiber made from a pitch or polyacrylonitrile (PAN) precursor by an oxidation, carbonization, and graphitization process (which provides a graphitic structure). See also *carbon fiber*.

graphitization. The process of pyrolyzation in an inert atmosphere at temperatures in excess of 1925 °C (3500 °F), usually as high as 2480 °C (4500 °F), and sometimes as high as 9750 °C (5400 °F), converting carbon to its crystalline allotropic form. Temperature depends on precursor and properties desired.

green strength. The mechanical strength of material, that, even though cure is not complete, allows removal from the mold and handling without tearing or permanent distortion. Also called green grab or initial tack.

gum. Any of a class of colloidal substances exuded by or prepared from plants, sticky when moist, composed of complex carbohydrates and organic acids, which are soluble or swell in water. The term gum is sometimes used loosely to denote various materials that exhibit gummy characteristics under certain conditions, for example, gum balata, gum benzoin, and gum asphaltum. Gums are included by some in the category of natural resins. ASTM D 907. See also *adhesive*, *glue*, and *resin*.

H

habit. The characteristic mode of growth or occurrence of a crystal. Characteristic

assemblage of forms (free faces) at crystallization leading to a usual appearance.

hairline craze. Multiple fine surface separation cracks that exceed 6.4 mm (¼ in.) in length and do not penetrate in depth the equivalent of a full ply of reinforcement.

halocarbon plastics. Plastics based on resins made by the polymerization of monomers composed only of carbon and a halogen or halogens.

handling life. The out-of-refrigeration time during which a material retains its handleability. See also *pot life*.

handling strength. A low level of strength initially obtained by an adhesive that allows specimens to be handled, moved, or unclamped without causing disruption of the curing process or affecting bond strength. ASTM D 1144.

hardener. A substance or mixture of substances added to an adhesive to promote or control the curing reaction by taking part in it. Also, a substance added to control the degree of hardness of the cured film. See also *catalyst*. ASTM D 907. Compare to *blocked curing agent*.

hardness. The resistance to surface indentation, usually measured by the depth of penetration (or arbitrary units related to the depth of penetration) of a blunt point under a given load using a particular instrument according to a prescribed procedure. See also *Barcol hardness, Knoop hardness, Mohs hardness, Rockwell hardness*, and *Shore hardness*.

haze. Cloudy appearance under or on the surface of a plastic, not describable by the terms chalking or bloom.

HDPE. See *high-density polyethylene*.

head-to-head. On a polymer chain, a type of configuration in which the functional groups are on adjacent carbon atoms.

head-to-tail. On a polymer chain, a type of configuration in which the functional groups on adjacent polymers are as far apart as possible.

heat-activated adhesive. A dry adhesive film that is rendered tacky or fluid by the application of heat or heat and pressure to the assembly. ASTM D 907. Compare to *hot-melt adhesive*.

heat buildup. The rise in temperature in a part resulting from the dissipation of applied strain energy as heat or from applied mold cure heat. See also *hysteresis*.

heat-deflection temperature. The temperature at which a standard test bar deflects a specified amount under a stated load. Now called deflection temperature under load (DTUL).

heat-fail temperature. The temperature at which delamination occurs under static loading in shear. ASTM D 4498.

heat-sealing adhesive. A thermoplastic film adhesive that is melted between the adherend surfaces by heat application to one or both of the surfaces.

heat treating. Annealing, hardening, tempering, and other heat processes.

heterogeneous. In materials, dissimilar constituents that are separately identifiable. Consisting of regions of unlike properties separated by internal boundaries. It is noteworthy that not all nonhomogeneous materials are necessarily heterogeneous.

heterogeneous nucleation. In the crystallization of polymers, the growth of crystals on vessel surfaces, dust, or added nucleating agents.

hexa. An abbreviated form of hexamethylenetetramine, a source of reactive methylene for curing novolacs.

high-density polyethylene (HDPE). Generally, polyethylene ranging in density from about 0.94 to 0.96 g/cm^3 and over. While molecules in low-density polyethylene are branched and linked randomly, those in the higher-density polyethylenes are linked in longer chains with fewer side branches, resulting in more rigid material with greater strength, hardness, and chemical resistance and higher softening temperature.

high-frequency heating. The heating of materials by dielectric loss in a high-frequency electrostatic field. The material is exposed between electrodes and is heated quickly and uniformly by the absorption of energy from the electrical field.

high-impact polystyrene (HIPS). A thermoplastic resin produced from a styrene monomer and elastomers. It has good dimensional stability and low-temperature impact strength, high rigidity, and ease of processing.

high polymer. A macromolecular substance that, as indicated by the polymer by which it is identified, consists of molecules that are multiples of the low molecular unit and have a molecular weight of at least 20 000.

high-pressure laminates. Laminates molded and cured at pressures not lower than 6.9 MPa (1.0 ksi) and more commonly in the range of 8.3 to 13.8 MPa (1.2 to 2.0 ksi).

high-pressure molding. A molding or laminating process in which the pressure used is greater than 1400 kPa (200 psi) according to ASTM D 883, but commonly 7 MPa (1 ksi).

HIPS. See *high-impact polystyrene*.

homogeneous nucleation. In the crystallization of polymers, the primary nucleated species generated by the polymer molecules.

homologous. Belonging to or consisting of a series of organic compounds differentiated by the number of methylene groups (CH_2).

homopolymer. A polymer resulting from the polymerization of a single monomer.

honeycomb. Manufactured product of resin-impregnated sheet material (paper, glass fabric, and so on) or metal foil, formed into hexagonal-shape cells. Used as a core ma-

terial in composite sandwich constructions. See also *sandwich construction*.

hoop stress. The circumferential stress in a material of cylindrical form subjected to internal or external pressure.

hot-gas welding. A technique for joining thermoplastic materials (usually sheet) in which the materials are softened by a jet of hot air from a welding torch and joined together at the softened points. Generally, a thin rod of the same material is used to fill and consolidate the gap.

hot-melt adhesive. An adhesive that is applied in a molten state and forms a bond after cooling to a solid state. ASTM D 907. A bonding agent that achieves a solid state and resultant strength by cooling, in contrast to other adhesives, which achieve the solid state through evaporation of solvents or chemical cure. Also, a thermoplastic resin that functions as an adhesive when melted between substrates and then cooled. Compare to *heat-activated adhesive*.

hot-setting adhesive. An adhesive that requires a temperature at or above 100 °C (212 °F) to set it. ASTM D 907. Compare to *cold-setting adhesive, intermediate-temperature-setting adhesive*, and *room-temperature-setting adhesive*.

hot working. Controlled mechanical operations for shaping a product at temperatures above the recrystallization temperature.

humidity, absolute. See *absolute humidity*.

humidity ratio. In a mixture of water vapor and air, the mass of water vapor per unit mass of dry air.

humidity, relative. See *relative humidity*.

humidity, specific. See *specific humidity*.

hybrid. A composite laminate consisting of laminae of two or more composite material systems. A combination of two or more different fibers, such as carbon and glass or carbon and aramid, in a structure. Tapes, fabrics, and other forms may be combined; usually only the fibers differ.

hydraulic press. A press in which the molding force is created by the pressure exerted by a fluid.

hydrocarbon plastics. Plastics based on resins made by the polymerization of monomers composed of carbon and hydrogen only.

hydrolysis. Chemical decomposition of a substance involving the addition of water.

hydromechanical press. A press in which the molding forces are created partly by a mechanical system and partly by a hydraulic system.

hydrophilic. Capable of absorbing water. Easily wetted by water.

hydrophobic. Capable of repelling water. Poorly wetted by water.

hydroxyl group. A chemical group consisting of one hydrogen atom and one oxygen atom.

hygroscopic. Capable of attracting, absorbing, and retaining atmospheric moisture.

hygrothermal effect. Change in properties due to moisture absorption and temperature change.

hysteresis. Incomplete recovery of strain during unloading cycle due to energy consumption. This energy is converted from mechanical to frictional energy (heat).

hysteresis loop. In dynamic mechanical measurement, the closed curve representing successive stress-strain status of the material during a cycle deformation.

I

ignition loss. The difference in weight before and after burning. The burning off of binder or size, as with glass.

immiscible. With respect to two or more fluids, not mutually soluble; incapable of attaining homogeneity.

impact strength. The characteristic that gives a material the ability to withstand shock loading. The work done in fracturing a test specimen in a specified manner under shock loading.

impact test. The measure of the energy necessary to fracture a standard bar by an impulse load. See also *Izod impact test*, *reverse impact test*, and *Charpy impact test*.

impact value. The energy absorbed by a specimen of standard design when sheared by a single blow from a testing machine hammer. Expressed in J/m^2 or ft · $lbf/in.^2$. ASTM D 950.

inclusion. A physical and mechanical discontinuity occurring within a material or part, usually consisting of solid, encapsulated foreign material. Inclusions are often capable of transmitting some structural stresses and energy fields, but to a noticeably different degree than from the parent material. See also *void*.

indentation hardness. Hardness evaluated from measurements of area or indentation depth caused by pressing a specified indentation into the surface of a material with a specified force.

inert atmosphere. The use of a gas (usually nitrogen) that does not absorb or react with ultraviolet light in a curing chamber, in place of oxygen.

inert filler. A material added to a plastic to alter the end product properties through physical rather than chemical means.

infrared (IR). Pertaining to that part of the electromagnetic spectrum between the visible light range and the radar range. Radiant heat is in this range, and infrared heaters are frequently used in the thermoforming and curing of plastics and composites. Infrared analysis is used for the identification of polymer constituents.

infrared spectroscopy or spectrometry. A technique used to observe or plot the wavelengths in the electromagnetic spectrum lying beyond the red from about 750 nm (7500 Å) to a few millimeters.

inhibitor. A substance that slows chemical reaction. Inhibitors are sometimes used in certain types of adhesives to prolong storage or working life. Synonym for retarder. ASTM D 907. Compare to *catalyst* and *hardener*.

initial modulus. The slope of the initial straight portion of a stress-strain or load-elongation curve. See also *Young's modulus*.

initial strain. The strain produced in a specimen by given loading conditions before creep occurs.

initial stress. The stress produced by strain in a specimen before stress relaxation occurs. Also called instantaneous stress.

initiator. Sources of free radicals, often peroxides or azo compounds. They are used in free-radical polymerizations, for curing thermosetting resins, and as cross-linking agents for elastomers and cross-linked polyethylene.

inorganic pigments. Natural or synthetic metallic oxides, sulfides, and other salts that impart heat and light stability, weathering resistance, color, and migration resistance to plastics.

insert. An integral part of a plastic molding consisting of metal or other material that may be molded or pressed into position after the molding is completed.

instantaneous recovery. The decrease in strain occurring immediately upon unloading a specimen. A more reproducible value is obtained if the decrease in strain is measured after a given small increment of time (such as 1 min) following unloading. The value is expressed in the same units as strain, that is, the decrease in length divided by the gage length, usually in inches per inch. ASTM D 1780.

instantaneous strain. The strain occurring immediately upon loading a creep specimen. Value is measured after loading to obtain a more reproducible value. The value is expressed in the same units as strain, that is, the extension divided by the gage length, usually in inches per inch. ASTM D 1780.

insulation resistance. The electrical resistance between two conductors or systems of conductors separated only by insulating material. The ratio of the applied voltage to the total current between two electrodes in contact with a specified insulator. The electrical resistance of an insulating material to a direct voltage.

insulator. A material of such low electrical conductivity that the flow of current through it can usually be neglected. Similarly, a material of low thermal conductivity, such as that used to insulate structural shells.

integral skin foam. Urethane foam with a cellular core structure and a relatively smooth skin.

integral structure. A reinforced plastic structure in which several elements that would conventionally be fabricated separately and assembled by bonding or mechanical fasteners are instead laid up and cured as a single, complex, continuous structure. The term is sometimes applied more loosely to any structure not assembled by mechanical fasteners. All or some parts of the assembly may be co-cured.

interface. The boundary or surface between two different, physically distinguishable media. With fibers, the contact area between the fibers and the sizing or finish. In a laminate, the contact area between the reinforcement and the laminating resin.

interference fits. A joint or mating of two parts in which the male part has an external dimension larger than the internal dimension of the mating female part. Distension of the female by the male creates a stress, which supplies the bonding force for the joint.

interlaminar. Between two adjacent laminae, for example, an object (such as a void), an event (such as a fracture), or a potential field (such as shear stress).

interlaminar shear. Shearing force tending to produce a relative displacement between two laminae in a laminate along the plane of their interface.

intermediate-temperature-setting adhesive. An adhesive that sets in the temperature range from 31 to 99 °C (87 to 211 °F). Synonym for warm-setting adhesive. ASTM D 907. Compare to *cold-setting adhesive*, *hot-setting adhesive*, and *room-temperature-setting adhesive*.

interphase. The boundary region between a bulk resin or polymer and an adherend in which the polymer has a high degree of orientation to the adherend on a molecular basis. It plays a major role in the load transfer process between the bulk of the adhesive and the adherend or the fiber and the laminate matrix resin.

intralaminar. Within a single lamina, for example, an object (such as a void), an event (such as a fracture), or a potential field (such as a temperature gradient).

intrinsic viscosity. For a polymer, the limiting value of infinite dilution of the ratio of the specific viscosity of the polymer solution to its concentration in moles per liter.

introfaction. The change in fluidity and wetting properties of an impregnating material, produced by the addition of an introfier.

introfier. A chemical that converts a colloidal solution into a molecular one. See also *introfaction*.

ion exchange resins. Cross-linked polymers that form salts with ions from aqueous solutions.

ionomer resins. A new polymer that has ethylene as its major component, but that contains both covalent and ionic bonds. The polymer exhibits very strong interchain ionic forces. The anions hang from the hydrocarbon chain, and the cations are metallic, for example, sodium, potassium, or magnesium. These resins have many of the same features that polyethylene has, plus high transparency, tenacity, resilience, and increased resistance to oils, greases, and solvents. Fabrication is accomplished as it is with polyethylene.

irradiation. With plastics, bombardment with a variety of subatomic particles, usually alpha-, beta-, or gamma-rays. Used to initiate the polymerization and copolymerization of plastics and in some cases to bring about changes in the physical properties of a plastic.

isocyanate plastics. Plastics based on resins made by the condensation of organic isocyanates with other compounds. Generally reacted with polyols on a polyester or polyether backbone molecule, with the reactants being joined by the formation of the urethane linkage. See also *polyurethanes* and *urethane plastics*.

isomer. A compound, radical, ion, or nuclide that contains the same number of atoms of the same elements but differs in structural arrangement and properties. See also *stereoisomer*.

isotactic stereoisomerism. A type of polymeric molecular structure containing a sequence of regularly spaced asymmetric groups arranged in like configuration in a polymer chain. Isotactic (and syndiotactic) polymers are crystallizable.

isotropic. Having uniform properties in all directions. The measured properties of an isotropic material are independent of the axis of testing.

Izod impact test. A test to determine the resistance of a plastic material to a shock loading, in which a notched specimen bar is held at one end and broken by striking, and the energy absorbed is measured.

J

jig. A mechanism for holding a part and guiding the tool during machining or assembly operation.

joint. See *adhesive joint*, *lap joint*, *scarf joint*, and *starved joint*.

joint-aging time. Synonym for joint-conditioning time. ASTM D 907.

joint, butt. See *butt joint*.

joint-conditioning time. The time interval between the removal of the joint from the conditions of heat or pressure, or both, used to accomplish bonding and the attainment of approximately maximum bond strength. (Synonym for *joint-aging time*.) ASTM D 907. See also *curing time*, *drying time*, and *setting time*.

joint, edge. See *edge joint*.

joint, lap. See *lap joint*.

joint, scarf. See *scarf joint*.

joint, starved. See *starved joint*.

K

Kevlar. An organic polymer composed of aromatic polyamides (aramids) having a para-type orientation (parallel chain with bonds extending from each aromatic nucleus). A registered trademark of E.I. Du Pont de Nemours & Company, Inc. Often used as a reinforcing fiber.

Knoop hardness. Hardness that is measured by calibrated machines that force a rhomb-shape, pyramidal diamond indentor having specified edge angles under specified conditions into the surface of the test material; the long diagonal is measured after removal of the load. The microhardness Knoop tester uses a relatively small load to measure surface hardness.

L

ladder polymer. A polymer with two polymer chains cross linked at intervals.

lamella. The basic morphological unit of a crystalline polymer, usually ribbonlike or platelike in shape. Generally, about 10 nm (100 Å) thick, 1 μm (40 μin.) long, and 0.1 μm (4 μin.) wide, if ribbonlike.

lamellae. Plural of lamella.

lamellar thickness. A characteristic morphological parameter, usually estimated from x-ray studies or electron microscopy, that is usually 10 to 50 nm (100 to 500 Å). The average thickness of lamellae in a specimen.

laminate. To unite layers of material with adhesive. Also, a product made by bonding together two or more layers of material or materials. ASTM D 907. See also *cross laminate* and *parallel laminate*.

laminate ply. One fabric-resin or fiber-resin layer (of a product) that is bonded to adjacent layers in the curing process.

lamination. The process of preparing a laminate. Also, any layer in a laminate. ASTM D 907.

lap joint. A joint made by placing one adherend partly over another and bonding together the overlapped portions. ASTM D 907.

latent curing agent. A curing agent that produces long-term stability at room temperature but rapid cure at elevated temperatures.

latex. A stable dispersion of polymeric substance (natural or synthetic rubber or plastic) in an essentially aqueous medium. ASTM D 907.

latices. Plural of latex.

LCP. See *liquid crystal polymer*.

least count. In mechanical testing, the smallest change in indication that can customarily be determined and reported.

legging. The drawing of filaments or strings when adhesive-bonded substances are separated. ASTM D 907. Compare to *teeth*. See also *stringiness* and *webbing*.

let-go. An area in laminated glass over which an initial adhesion between interlayer and glass has been lost.

limited-coordination specification (or standard). A specification (or standard) that has not been fully coordinated and accepted by all interested parties. Limited-coordination specifications and standards are issued to cover the need for requirements unique to one particular department. This applies primarily to military agency documents.

linear expansion. The increase of a given dimension measured by the expansion of a specimen or component subject to a temperature gradient. See also *coefficient of thermal expansion*.

linear strain, tensile or compressive. The change per unit length (that is, percent deformation) due to an applied force in an original linear dimension.

liquid crystal polymer (LCP). A new thermoplastic polymer that contains primarily benzene rings in its backbone, is melt processible, and develops high orientation during molding with resultant improvement in tensile strength and high-temperature capability. The first commercial availability was as an aromatic polyamide. Available with or without fiber reinforcement.

liquid resin. An organic, polymeric liquid that becomes a solid when converted to its final state for use.

load. The force applied to the specimen at any given time. ASTM D 4027.

London dispersion forces. Weak intermolecular forces based on transient dipole-dipole interactions.

long-chain branching. A form of molecular branching found in addition polymers as a result of an internal transfer reaction. It primarily influences the melt-flow properties.

long period. A morphological parameter obtained from small-angle x-ray scattering. It is usually equated to the sum of the lamellar thickness and the amorphous thickness.

loss modulus. A quantitative measure of energy dissipation, defined as the ratio of stress 90° out of phase with oscillating strain to the magnitude of strain. The loss modulus may be measured in tension or flexure (E''), compression (K''), or shear (G''). See also *complex modulus*.

lot. A specific amount of material produced at one time using one process and constant conditions of manufacture and offered for sale as a unit quantity.

low-pressure laminates. In general, laminates molded and cured in the range of pressures from 2760 kPa (400 psi) down to and including pressure obtained by the mere contact of the plies.

low-pressure molding. The distribution of relatively uniform low pressure (1400 kPa, or 200 psi, or less) over a resin-bearing fibrous assembly of cellulose, glass, asbestos, or other material, with or without application of heat from an external source, to form a structure possessing specific physical properties.

lyotropic liquid crystal. A type of liquid crystalline polymer that can be processed only from solution.

M

macroscopy. Interpretation using only the naked eye, or a magnification no greater than 10×.

macrostrain. The mean strain over any finite gage length of measurement that is large in comparison with interatomic distances.

manufactured unit. A quantity of finished adhesive or finished adhesive component, processed at one time. The manufactured unit may be a batch or a part thereof. ASTM D 907. Compare to *batch*.

mass spectrometry (MS). An instrument that is capable of separating ionized molecules of different mass/charge ratios and measuring the respective ion currents. Used in the measurement of low molecular weights.

matrix. The part of an adhesive that surrounds or engulfs embedded filler or reinforcing particles and filaments. ASTM D 907.

maturing temperature. The temperature, as a function of time and bonding condition, that produces desired characteristics in bonded components. The term is specific for ceramic adhesives. ASTM D 907.

maximum elongation. The elongation at the time of fracture, including both elastic and plastic deformation of the tensile specimen. Applicable to rubber, plastic, and some metallic materials. Maximum elongation is also called ultimate elongation or break elongation.

mechanical adhesion. Adhesion between surfaces in which the adhesive holds the parts together by interlocking action. ASTM D 907.

mechanical hysteresis. The energy absorbed in a complete cycle of loading and unloading, including any stress cycle regardless of the mean stress or range of stress.

mechanically foamed plastic. A cellular plastic having a structure produced by physically incorporated gases.

mechanical properties. The properties of a material, such as compressive and tensile strengths and modulus, that are associated with elastic and inelastic reaction when force is applied. The individual relationship between stress and strain.

melamine plastics. Thermosetting plastics made from melamine and formaldehyde resins.

melt. A charge of molten plastic.

melt index. The amount, in grams, of a thermoplastic resin that can be forced through a 2.0955 mm (0.0825 in.) orifice when subjected to 20 N (2160 gf) in 10 min at 190 °C (375 °F).

melting point. The first-order transition in crystalline polymers. The fixed point between the solid and liquid phases of a material when approached from the solid phase under a pressure of 101.325 kPa (1 atm).

melt strength. The strength of a plastic while in the molten state.

mer. The repeating structural unit of any polymer.

metallic fiber. Manufactured fiber composed of metal, plastic-coated metal, metal-coated plastic, or a core completely covered by metal.

methyl methacrylate. A colorless, volatile liquid derived from acetone, cyanohydrin, methanol, and dilute sulfuric acid and used in the production of acrylic resins.

micro. In relation to reinforced plastics and composites, the properties of the constituents only, that is, matrix, reinforcements, and interface, and their effects on the properties of the composite.

microcracking. Cracks formed in composites when local thermal stresses exceed the strength of the matrix. Because most microcracks do not penetrate the reinforcing fibers, microcracks in a cross-plied tape laminate or in a laminate made from cloth prepreg are usually limited to the thickness of a single ply.

microdynamometer. An instrument for measuring mechanical force and observing the change in microscopic appearance of a small specimen.

microstrain. The strain over a gage length comparable to interatomic distances.

microstructure. A structure with heterogeneities that can be seen through a microscope.

modified acrylic. A thermoplastic polymer that has been altered to eliminate mixing, curing ovens, and odor and that cures rapidly at room temperature.

modifier. Any chemically inert ingredient added to an adhesive formulation to change its properties. ASTM D 907. Compare to *filler*, *plasticizer*, and *extender*.

modulus, initial. See *initial modulus*.

modulus of elasticity. The ratio of the stress to the strain produced in a material that is elastically deformed. If a tensile stress of 13.8 MPa (2.0 ksi) results in an elongation of 1%, the modulus of elasticity is 13.8 MPa (2.0 ksi) divided by 0.01, or 1380 MPa (200 ksi). Also called Young's modulus. See also *offset modulus* and *secant modulus*.

modulus, offset. See *offset modulus*.

modulus of resilience. The energy that can be absorbed per unit volume without creating a permanent distortion. Calculated by integrating the stress-strain curve from 0 to the elastic limit and dividing by the original volume of the specimen.

modulus of rigidity. The ratio of stress to strain within the elastic region for shear or torsional stress. Also called shear modulus or torsional modulus.

modulus of rupture (in bending). The maximum tensile or compressive stress value (whichever causes failure) in the extreme fiber of a beam loaded to failure in bending.

modulus of rupture (in torsion). The maximum shear stress in the extreme fiber of a member of circular cross section loaded to failure in torsion.

modulus, secant. See *secant modulus*.

modulus, tangent. See *tangent modulus*.

Mohs hardness. A measure of the scratch resistance of a material. The higher the number, the greater the scratch resistance (number 10 being termed diamond).

moisture absorption. The pickup of water vapor from air by a material, in reference to vapor withdrawn from the air only, as distinguished from water absorption, which is the gain in weight due to the absorption of water by immersion.

moisture content. The amount of moisture in a material determined under prescribed conditions and expressed as a percent of the mass of the moist specimen, that is, the mass of the dry substance plus the moisture.

moisture equilibrium. The condition reached by a sample when it no longer takes up moisture from, or gives up moisture to, the surrounding environment.

moisture regain. The moisture in a material determined under prescribed conditions and expressed as a percent of the weight of the moisture-free specimen. Moisture regain may result from either sorption or desorption and differs from moisture content only in the basis used for calculation.

moisture vapor transmission (MVT). A rate at which water vapor passes through a material at a specified temperature and relative humidity.

molecular mass. The sum of the atomic mass of all atoms in a molecule. In high polymers, because the molecular masses of individual molecules vary widely, they must be expressed as averages. The average molecular mass of polymers may be expressed as number-average molecular mass or mass-average molecular map. These averages are the first two moments of the molecular mass (or weight) distribution. Others in common use are (Z) and (Z + 1) averages, these being the third and fourth moments. Molecular mass measurement methods include osmotic pressure, light scattering, solution pressure, solution viscosity, and sedimentation equilibrium.

molecular weight. The sum of the atomic weights of all the atoms in a molecule. A measure of the chain length for the molecules that make up the polymer.

monochromator. A device for isolating radiation of one or nearly one wavelength from a beam of many wavelengths.

monomer. A single molecule that can react with like or unlike molecules to form a polymer. The smallest repeating structure of a polymer (mer). For addition polymers, this represents the original unpolymerized compound.

morphology. The overall physical form of the physical structure of a bulk polymer. Common units are lamellae, spherulites, and domains.

MS. See *mass spectrometry*.

mucilage. An adhesive prepared from a gum and water, and also in a more general sense, a liquid adhesive that has a low order of bonding strength. ASTM D 907. See also *adhesive, glue, paste,* and *sizing*.

multiple-layer adhesive. A dry-film adhesive, usually supported, with a different adhesive composition on each side; designed to bond dissimilar materials such as the core to face bond of a sandwich composite. ASTM D 907.

MVT. See *moisture vapor transmission*.

N

national standards laboratory. The laboratory that maintains the measurement standards of a nation, such as the National Institute of Standards and Technology in the United States.

NDE. See *nondestructive evaluation*.

NDI. See *nondestructive inspection*.

NDT. See *nondestructive testing*.

neat resin. Resin to which nothing (additives, reinforcements, and so on) has been added.

necking. The localized reduction in cross section that may occur in a material under tensile stress.

Newtonian fluid. A fluid in which the shearing rate is directly proportional to the applied torque.

nitrogen blanket. The use of nitrogen to produce an inert atmosphere.

NMR. See *nuclear magnetic resonance*.

nominal stress. The stress at a point calculated on the net cross section without taking into consideration the effect on stress of geometric discontinuities, such as holes, grooves, fillets, and so forth. The calculation is made using simple elastic theory.

nominal value. A value assigned for the purpose of a convenient designation. A nominal value exists in name only. In dimensions, it is often an average number with a tolerance in order to fit together with adjacent parts.

nondestructive evaluation (NDE). Broadly considered synonymous with nondestructive inspection (NDI). More specifically, the analysis of NDI findings to determine whether the material will be acceptable for its function.

nondestructive inspection (NDI). A process or procedure, such as ultrasonic or radiographic inspection, for determining the quality or characteristics of a material, part, or assembly without permanently altering the subject or its properties. Used to find internal anomalies in a structure without degrading its properties.

nondestructive testing (NDT). Broadly considered synonymous with nondestructive inspection (NDI).

nonhygroscopic. Lacking the property of absorbing and retaining an appreciable quantity of moisture (water vapor) from the air.

nonresonant forced and vibration technique. A technique for performing dynamic mechanical measurements in which the sample is oscillated mechanically at a fixed frequency. Storage modulus and damping are calculated from the applied strain and the resultant stress and shift in phase angle.

nonrigid plastic. For purposes of general classification, a plastic that has a modulus of elasticity either in flexure or in tension of not over 70 MPa (10 ksi) at 23 °C (70 °F) and 50% relative humidity.

normal stress. The stress component that is perpendicular to the plane on which the forces act.

notched specimen. A test specimen that has been deliberately cut or notched, usually in a V-shape, to induce and locate point of failure.

notch factor. Ratio of the resilience determined on a plain specimen to the resilience determined on a notched specimen.

notch sensitivity. The extent to which the sensitivity of a material to fracture is increased by the presence of a surface nonhomogeneity, such as a notch, a sudden change in cross section, a crack, or a scratch. Low notch sensitivity is usually associated with ductile materials, and high notch sensitivity is usually associated with brittle materials.

novolac (novolak). A phenol-aldehyde resin that, unless a source of methylene groups is added, remains permanently thermoplastic. Compare to *resinoid*. See also *thermoplastic*.

nuclear magnetic resonance (NMR). Relates to the radio-frequency-induced transitions between magnetic energy levels of atomic nuclei. An NMR instrument consists essentially of a magnet, radio frequency accelerator, sample holder sweep unit, and detector and is capable of producing an oscilloscope image or line recording of an NMR spectrum.

nucleating agent. A foreign substance, often crystalline, usually added to a crystallizable polymer to increase its rate of solidification during processing.

nylon. The generic name, by common usage, for all synthetic polyamides.

nylon plastics. Plastics based on a resin composed principally of a long-chain synthetic polymeric amide that has recurring amide groups as an integral part of the main polymer chain. Numerical designations (nylon 6, nylon 6/6, and so on) refer to the monomeric amides of which they are made. Characterized by great toughness and elasticity.

O

offset modulus. The ratio of the offset yield stress to the extension at the offset point.

offset yield strength. The stress at which the strain exceeds by a specific amount (the offset) an extension of the initial, approximately linear, proportional portion of the stress-strain curve. It is expressed as force per unit area.

olefin. A group of unsaturated hydrocarbons of the general formula C_nH_{2n} and named after the corresponding paraffins by the addition of -ene or sometimes -ylene to the root.

olefin plastics. Plastics based on polymers made by the polymerization of olefins or copolymerization of olefins with other monomers, the olefins being at least 50 mass %.

oligomer. A polymer consisting of only a few monomer units, for example, a dimer, trimer, tetramer, and so forth, or their mixtures.

one-component adhesive. An adhesive material incorporating a latent hardener or catalyst that is activated by heat.

open assembly time. The time interval between the spreading of the adhesive on the adherend and the completion of the assembly of the parts for bonding. ASTM D 907.

open-cell cellular plastic. Foamed or cellular plastic with cells that are generally interconnected.

organic. Originating in plant or animal life or composed of chemicals of hydrocarbon origin, either natural or synthetic.

organosol. A suspension of a finely divided resin in a volatile, organic liquid. The resin does not dissolve appreciably in the organic liquid at room temperature, but does at elevated temperatures. The liquid evaporates at an elevated temperature, and the residue upon cooling is a homogeneous plastic mass. Plasticizers may be dissolved in the volatile liquid.

orientation. The alignment of the crystalline structure in polymeric materials to produce a highly aligned structure. Orientation can be accomplished by cold drawing or stretching in fabrication. Orientation can generally be divided into two classes, uniaxial and biaxial.

oriented materials. Materials with molecules and/or macroconstituents that are aligned in a specific way. Oriented materials are anisotropic.

orthotropic. Having three mutually perpendicular planes of elastic symmetry.

outgassing. Devolatilization of plastics or applied coatings during exposure to vacuum in vacuum metallizing. Resulting parts show voids or thin spots in plating with reduced and spotty brilliance. Additional drying prior to metallizing is helpful, but outgassing is inherent to plastic materials and coating ingredients, including plasticizers and volatile components.

oven dry. The condition of a material that has been heated under prescribed conditions of temperature and humidity until there is no further significant change in its mass.

oxidation. In carbon/graphite fiber processing, the step of reacting the precursor polymer (rayon, polyacrylonitrile, or pitch) with oxygen, resulting in the stabilization of the structure for the hot stretching operation. In general, any chemical reaction in which electrons are transferred.

P

PA. See *polyamide*.

PAEK. See *polyaryletherketone*.

PAI. See *polyamide-imide resins*.

PAN. See *polyacrylonitrile*.

parabolic reflector. A reflector for ultraviolet curing that projects parallel light beams perpendicular to the assembly, but is not focused.

parallel laminate. A laminate in which all the layers of material are oriented approximately parallel with respect to the grain or to the strongest direction in tension. ASTM D 907. Compare to *cross laminate*.

PAS. See *polyaryl sulfone*.

paste. An adhesive composition having a characteristic plastic-type consistency, that is, a high order of yield value, such as that prepared by heating a mixture of starch and water and subsequently cooling the hydrolyzed product. ASTM D 907. Compare to *adhesive, glue, mucilage,* and *sizing*.

PBI. See *polybenzimidazole*.

PBT. See *polybutylene terephthalate*.

PC. See *polycarbonate*.

peel, or stripping, strength. The average load per unit width of bond line required to separate progressively one member from the other over the adhered surfaces at a separation angle of approximately 180° and a separation rate of 152 mm (6 in.) per minute. It is expressed in N/mm (lbf/in.) of width. ASTM D 903.

peel ply. A layer of open-weave material, usually fiberglass or heat-set nylon, applied directly to the surface of a prepreg lay-up. The peel ply is removed from the cured laminate immediately before bonding operations, leaving a clean, resin-rich surface that needs no further preparation for bonding, other than the application of a primer if one is required.

PEI. See *polyether-imide*.

penetration. The entering of an adhesive into an adherend. This property of a system is measured by the depth of penetration of the adhesive into the adherend. ASTM D 907.

permanence. The resistance of an adhesive bond to deteriorating influences. ASTM D 907.

permanent set. The deformation remaining after a specimen has been stressed a prescribed amount in tension, compression, or shear for a specified time period and released for a specified time period. For creep tests, the residual unrecoverable deformation after the load causing the creep has been removed for a substantial and specified period of time. Also, the increase in length, expressed as a percentage of the original length, by which an elastic material fails to return to its original length after being stressed for a standard period of time.

permeability. The passage or diffusion (or rate of passage) of a gas, vapor, liquid, or solid through a barrier without physically or chemically affecting it.

peroxy compounds. Compounds containing O—O linkage.

PESV. See *polyether sulfone*.

PET. See *polyethylene terephthalate*.

pH. The measure of the acidity or alkalinity of a substance, neutrality being at pH 7. Acid solutions are less than 7, and alkaline solutions are more than 7.

phase. A visibly separate, but not necessarily separable, portion of a system.

phase angle. The angle between a sinusoidally applied perturbation and the resultant sinusoidal reaction. Generally in reference to mechanical and dielectric processes.

phase change. The transition from one physical state to another, such as gas to liquid, liquid to solid, gas to solid, or vice versa.

phase separation. The formation of a second liquid portion from a previously homogeneous liquid over time.

phenolic resin. A thermosetting resin produced by the condensation of an aromatic alcohol with an aldehyde, particularly of phenol with formaldehyde. Used in high-temperature applications with various fillers and reinforcements. The resin in a single-stage compound, because of its reactive groups, is capable of further polymerization by the application of heat. In a two-stage compound, the resin is essentially not reactive at normal storage temperatures, but contains a reactive additive that causes further polymerization with the application of heat.

phenoxy resins. A high molecular weight thermoplastic polyester resin based on bisphenol A and epichlorohydrin. Developed in the United States, the material is available in grades suitable for injection molding and extrusion and for application as coatings and adhesives.

phenylsilane resins. Thermosetting copolymers of silicone and phenolic resins. Furnished in solution form.

phthalate esters. The most widely used group of plasticizers, produced by the direct action of alcohol on phthalic anhydride. The phthalates are generally characterized by moderate cost, good stability, and good all-around properties.

physical blowing agent. A gas, such as a fluorocarbon.

physical catalyst. Radiant energy capable of promoting or modifying a chemical reaction.

PI. See *polyimide*.

pick-up roll. A spreading device where the roll for picking up the adhesive runs in a reservoir of adhesive. ASTM D 907.

pitch. A high molecular weight material that is a residue from the destructive distillation of coal and petroleum products. Pitches are used as base materials for the manufacture of certain high-modulus carbon fibers.

planar. Lying essentially in a single plane.

plastic. A material that contains as an essential ingredient an organic polymer of large molecular weight; that is solid in its finished state; and that, at some stage in its manufacture or when it is being processed into finished articles, can be shaped by flow. Although materials such as rubber, textiles, adhesives, and paint may in some cases meet this definition, they are not considered plastics. The terms plastic, resin, and polymer are somewhat synonymous, but the terms resins and polymers most often denote the basic material as polymerized, while the term plastic encompasses compounds containing plasticizers, stabilizers, fillers, and other additives.

plastic deformation. Any portion of the total deformation of a body that occurs immediately when load is applied but that remains permanent when the load is removed.

plastic flow. Deformation under the action of a sustained hot or cold force. Flow of semisolids in the molding of plastics.

plastic foam. See *cellular plastic* (preferred term).

plasticity. A property of adhesives that allows the material to be deformed continuously and permanently without rupture upon the application of a force that exceeds the yield value of the material. ASTM D 907.

plasticizer. A material incorporated in an adhesive to increase its flexibility, workability, or distensibility. The addition of the plasticizer may cause a reduction in melt viscosity, lower the temperature of the second-order transition, or lower the elastic modulus of the solidified adhesive. ASTM D 907. Compare to *modifier*.

plastic memory. The tendency of a thermoplastic material that has been stretched while hot to return to its unstretched shape upon being reheated.

plastics, engineering. See *engineering plastics*.

plastic true strain (ε_p). Inelastic component of true strain.

plastigel. A plastisol exhibiting gellike flow properties. One having an effective yield value.

plastisols. Mixtures of vinyl resins and plasticizers that can be molded, cast, or converted to continuous films by the application of heat. Mixtures that contain volatile thinners as well are known as organosols.

plastometer. An instrument for determining the flow properties of a thermoplastic resin by forcing the molten resin through a die or orifice of specific size at a specified temperature and pressure.

ply. A single layer in a laminate. In general, fabrics or felts consisting of one or more layers. Yarn resulting from a twisting operation (for example, three-ply yarn consists of three strands of yarn twisted together). In filament winding, a ply is single pass (two plies forming one layer).

plywood. A cross-bonded assembly made of layers of veneer or veneer in combination with a lumber core or plies joined with an adhesive. Two types of plywood are recognized, veneer plywood and lumber core plywood. ASTM D 907.

PMMA. See *polymethyl methacrylate*.

PMR polyimides. A novel class of high-temperature-resistant polymers. PMR represents *in situ* polymerization of monomer reactants.

Poisson's ratio. The ratio of the change in lateral width per unit width to change in axial length per unit length caused by the axial stretching or stressing of a material. The ratio of transverse strain to the corresponding axial strain below the proportional limit.

polarimeter. An instrument for determining the amount of rotation of the direction of vibration of polarized light by the specimen; an instrument for determining the amount of polarization of light by the specimen or in the illuminating beam.

polariscope. An instrument consisting essentially of a pair of polars: the polarizer for polarizing the beam illuminating the specimen and the analyzer for analyzing the effect, if any, of the specimen on the polarized light. At least one of the polars should be rotatable for obtaining crossed and uncrossed polars. Either the polars should be simultaneously rotatable, or the specimen should be rotatable between polars.

polyacrylate. A thermoplastic resin made by the polymerization of an acrylic compound.

polyacrylonitrile (PAN). Used as a base material or precursor in the manufacture of certain carbon fibers.

polyamide (PA). A thermoplastic polymer in which the structural units are linked by amide or thio-amide groupings (repeated nitrogen and hydrogen groupings). Many polyamides are fiber forming.

polyamide-imide (PAI) resins. A family of polymers based on the combination of trimellitic anhydride with aromatic diamines. In the uncured form (ortho-amic acid) the polymers are soluble in polar organic solvents. The imide linkage is formed by heating, producing an infusible resin with thermal stability up to 290 °C (550 °F). These resins are used for laminates, prepregs, and electrical components.

polyamide plastic. See *nylon plastics*.

polyarylates. A family of aromatic polyesters derived from aromatic dicarboxylic acids and diphenols. These thermoplastics have an amorphous molecular structure and are tough, durable, and heat resistant. They have excellent dimensional and ultraviolet stability, electrical properties, flame retardance, and warp resistance.

polyaryletherketone (PAEK). A linear aromatic crystalline thermoplastic. A composite with a PAEK matrix may have a continuous-use temperature as high as 250 °C (480 °F). This definition is also germane to polyetheretherketone (PEEK), polyetherketone (PEK), and polyetherketoneketone (PEKK).

polyaryl sulfone (PAS). A thermoplastic resin consisting mainly of phenyl and biphenyl groups linked by thermally stable ether and sulfone groups. Its most outstanding property is resistance to high and low temperatures (from −240 to 260 °C, or −400 to 500 °F). It also has good impact resistance; resistance to chemicals, oils, and most solvents; and good electrical insulating properties. It can be processed by injection molding, extrusion, compression molding, ultrasonic welding, and machining.

polybenzimidazole (PBI). A thermoplastic resin that is strong and stable, has a high molecular weight, and contains recurring aromatic units. The resin is now mainly produced from the condensation of 3,3′, 4,4′-tetraaminobiphenyl (diaminobenzidine) and diphenyl isophthalate. It is used for fibers with resistance to high temperatures and flame.

polybutylenes. A group of polymers consisting of isotactic, stereoregular, highly crystalline polymers based on butene-1. Their properties are similar to those of polypropylene and linear polyethylene, with superior toughness, creep resistance, and flexibility.

polybutylene terephthalate (PBT). A member of the polyalkylene terephthalate family that is similar to polyethylene terephthalate in that it is derived from a polycondensate that in turn is derived from terephthalic acid, the diol component of which is butanediol rather than glycol, as in the case of polyethylene terephthalate. Properties include high strength, dimensional stability, low moisture absorption, good electrical characteristics, and resistance to heat and chemicals when suitably modified.

polycarbonate (PC). A thermoplastic polymer derived from the direct reaction between aromatic and aliphatic dihydroxy compounds with phosgene or by the ester exchange reaction with appropriate phosgene-derived precursors. It has the highest impact resistance of any transparent plastic.

polycondensation. See *condensation*.

polyester plastics. Plastics based on resins composed principally of polymeric esters, in which the recurring ester groups are an integral part of the main polymer chain, and in which ester groups occur in most cross links that may be present between chains.

polyesters, thermoplastic. See *thermoplastic polyesters*.

polyesters, thermosetting. See *thermosetting polyesters*.

polyesters, unsaturated. See *unsaturated polyesters*.

polyether-imide (PEI). An amorphous polymer with good thermal properties for a thermoplastic. Reported T_g of 215 °C (419 °F) and continuous-use temperature of about 170 °C (338 °F).

polyether sulfone (PESV). A high-temperature engineering thermoplastic consisting of repeating phenyl groups linked by thermally stable ether and sulfone groups. The material has good transparency and flame resistance and is one of the lowest-smoke-emitting materials available. Both polymer and reinforced grades are available in granule form for extrusion or injection molding.

polyethylene (PE). Thermoplastic materials composed by polymers of ethylene. They are normally translucent, tough, waxy solids that are unaffected by water and by a large range of chemicals. In common usage, these plastics have no less than 85% ethylene and no less than 95% total olefins.

polyethylene terephthalate (PET). A saturated, thermoplastic polyester resin made by condensing ethylene glycol and terephthalic acid and used for fibers, films, and injection-molded parts. It is extremely hard, wear resistant, dimensionally stable, and resistant to chemicals, and it has good dielectric properties. Also known as polyethylene glycol terephthalate.

polyimide (PI). A polymer produced by reacting an aromatic dianhydride with an aromatic diamine. It is a highly heat-resistant resin, ≥315 °C (600 °F). It is similar to a polyamide, differing only in the number of hydrogen molecules contained in the groupings. This polymer is suitable for use as a binder or adhesive and may be either a thermoplastic or a thermoset.

polyimides, thermoplastic. See *thermoplastic polyimides.*

polymer. A compound formed by the reaction of simple molecules having functional groups that permit their combination to proceed to high molecular weights under suitable conditions. Polymers may be formed by polymerization (addition polymer) or polycondensation (condensation polymer). When two or more monomers are involved, the product is called a copolymer. ASTM D 907. Compare to *monomer.*

polymerization. A chemical reaction in which the molecules of a monomer are linked together to form large molecules whose molecular weight is a multiple of that of the original substance. When two or more monomers are involved, the process is called copolymerization or heteropolymerization. ASTM D 907. See also *condensation.*

polymer matrix. The resin portion of a reinforced or filled plastic.

polymethyl methacrylate (PMMA). A thermoplastic polymer synthesized from methyl methacrylate. It is a transparent solid with exceptional optical properties. Available in the form of sheets, granules, solutions, and emulsions, it is used as facing material in certain composite constructions. See also *acrylic plastic.*

polymorphism. The ability of an element or compound capable of stable existence in different temperatures and pressure ranges to exist in two or more crystalline phases.

polyol. An alcohol having many hydroxyl groups. Also known as a polyhydric alcohol or polyalcohol. In cellular plastics usage, the term includes compounds containing alcoholic hydroxyl groups, such as polyethers, glycols, polyesters, and castor oil, which are used in urethane foams and other polyurethanes.

polyolefins. Plastics based on a polymer made with an olefin as essentially the sole monomer.

polyoxymethylenes (POM). Acetal plastics based on polymers in which oxymethylene is essentially the sole repeating structural unit in the chains. See also *acetal (AC) resins.*

polyphenylene oxides (PPO). Thermoplastic, linear, noncrystalline polyethers obtained by the oxidative polycondensation of 2,6-dimethylphenol in the presence of a copper-amine complex catalyst. These resins have a useful temperature range from less than −170 to 190 °C (−275 to 375 °F) with intermittent use possible up to 205 °C (400 °F), excellent electrical properties, unusual resistance to acids and bases, and processibility on conventional extrusion and injection molding equipment. Also known as polyphenylene ether (PPE).

polyphenylene sulfides (PPS). Crystalline polymers having a symmetrical, rigid backbone chain consisting of recurring para-substituted benzene rings and sulfur atoms. Known for excellent chemical resistance, thermal stability, and fire resistance. Their inertness to organic solvents, inorganic salts, and bases makes them corrosion resistant. Commercial engineering grades are always fiber reinforced.

polypropylene, reinforced. See *reinforced polypropylene.*

polypropylenes (PP). Tough, lightweight thermoplastics made by the polymerization of high-purity propylene gas in the presence of an organometallic catalyst at relatively low pressures and temperatures.

polystyrene, high-impact. See *high-impact polystyrene.*

polystyrenes (PS). Water-white homopolymer thermoplastics produced by the polymerization of styrene (vinyl benzene). Outstanding electrical properties, good thermal and dimensional stability, and staining resistance. Because this type of material is somewhat brittle, it is often copolymerized or blended with other materials to obtain desired properties.

polysulfide. A synthetic polymer containing sulfur and carbon linkages, produced from organic dihalides and sodium polysulfide. The material is elastomeric in nature; resistant to light, oil, and solvents; and impermeable to gases.

polysulfones (PSU). A family of sulfur-containing thermoplastics made by reacting bisphenol A and 4,4′-dichlorodiphenyl sulfone with potassium hydroxide in dimethyl sulfoxide at 130 to 140 °C (265 to 285 °F). The structure of the polymer is benzene rings or phenylene units linked by three different chemical groups, a sulfone group, an ether linkage, and an isopropylidene group. Polysulfones are characterized by high strength, the highest service temperature of all melt-processible thermoplastics, low creep, good electrical characteristics, transparency, self-extinguishing properties, and resistance to greases, many solvents, and chemicals. They may be processed by extrusion, injection molding, or blow molding.

polyterephthalate. A thermoplastic polyester in which the terephthalate group is a repeating structural unit in the chain, the terephthalate being greater in amount than other dicarboxylates that may be present.

polytetrafluoroethylenes (PTFE). Members of the fluorocarbon family of plastics made by the polymerization of tetrafluoroethylene. PTFE is characterized by its extreme inertness to chemicals, very high thermal stability, and low frictional properties. Among the applications for these materials are bearings, fuel hoses, gaskets and tapes, and coatings for metal and fabric.

polyurethanes (PUR). A large family of polymers with widely varying properties and uses. All of these polymers are based on the reaction product of an organic isocyanate with compounds containing a hydroxyl group. The reaction product of an isocyanate with an alcohol is called a urethan, according to rules of chemical nomenclature, but the terms urethane and polyurethane are more widely used in the plastics industry. Polyurethanes may be thermosetting or thermoplastic, rigid or soft and flexible, cellular or solid. The properties of any of these types may be tailored within wide limits to suit the desired application.

polyurethanes, thermoplastic. See *thermoplastic polyurethanes.*

polyvinyl acetals. Members of the family of vinyl plastics. Polyvinyl acetal is the general name for resins produced from a condensation of polyvinyl alcohol with an aldehyde. There are three main groups: polyvinyl acetal itself; polyvinyl butyral, and polyvinyl formal. Polyvinyl acetal resins are thermoplastics that can be processed by casting, extruding, molding, and coating, but their main uses are in adhesives, lacquers, coatings, and films.

polyvinyl acetate (PVAC). A thermoplastic material composed of polymers of vinyl acetate in the form of a colorless solid. It is obtainable in the form of granules, solutions, latices, and pastes and is used extensively in adhesives, for paper and fabric coatings, and in bases for inks and lacquers.

polyvinyl acetate emulsion adhesive. A latex adhesive in which the polymeric portion comprises polyvinyl acetate, copolymers based mainly on polyvinyl acetate, or a mixture of these and which may contain modifiers and secondary binders to provide specific properties. ASTM D 4317.

polyvinyl alcohol (PVAL). A thermoplastic material composed of polymers of the hypothetical vinyl alcohol. Usually a colorless solid, insoluble in most organic solvents and oils, but soluble in water when the content of hydroxy groups in the polymer is sufficiently high. The product is normally granular. It is obtained by the partial hydrolysis or by the complete hydrolysis of polyvinyl esters, usually by the complete hydrolysis of polyvinyl acetate. It is mainly used for adhesives and coatings.

polyvinyl butyral (PVB). A thermoplastic material derived from a polyvinyl ester in which some or all of the acid groups have been replaced by hydroxyl groups, and some or all of these hydroxyl groups have been replaced by butyral groups by reaction with butyraldehyde. It is a colorless, flexible, tough solid used primarily in interlayers for laminated safety glass.

polyvinyl carbazole. A thermoplastic resin, brown in color, obtained by reacting acetylene with carbazole. The resin has excellent electrical properties and good heat and chemical resistance. It is used as an impregnant for paper capacitors.

polyvinyl chloride (PVC). A thermoplastic material composed of polymers of vinyl chloride. It is a colorless solid with outstanding resistance to water, alcohols, and concentrated acids and alkalies. It is obtainable in the form of granules, solutions, latices, and pastes. Compounded with plasticizers, it yields a flexible material superior to rubber in aging properties. It is widely used for cable and wire coverings, in chemical plants, and in the manufacture of protective garments.

polyvinyl chloride acetate. A thermoplastic material composed of copolymers of vinyl chloride and vinyl acetate. It is a colorless solid with good resistance to water as well as concentrated acids and alkalies. It is obtainable in the form of granules, solutions, and emulsions. Compounded with plasticizers, it yields a flexible material superior to rubber in aging properties. It is widely used for cable and wire coverings, in chemical plants, and in the manufacture of protective garments.

polyvinyl formal (PVF). One of the groups of polyvinyl acetal resins made by the condensation of formaldehyde in the presence of polyvinyl alcohol. It is used mainly in combination with cresylic phenolics for wire coatings and impregnations, but can also be molded, extruded, or cast. It is resistant to greases and oils.

polyvinylidene chloride (PVDC). A thermoplastic material composed of polymers of vinylidene chloride (1,1-dichloroethylene). It is a white powder with a softening temperature of 185 to 200 °C (365 to 390 °F). The material is also supplied as a copolymer with acrylonitrile or vinyl chloride, giving products that range from the soft, flexible type to the rigid type. Also known as saran.

polyvinylidene fluoride (PVDF). This member of the fluorocarbon family of plastics is a homopolymer of vinylidene fluoride. It is supplied as powders and pellets for molding and extrusion and in solution form for casting. The resin has good tensile and compressive strength and high impact strength.

POM. See *polyoxymethylenes*.

porosity. A condition of trapped pockets of air, gas, or vacuum within a solid material. Usually expressed as a percentage of the total nonsolid volume to the total volume (solid plus nonsolid) or a unit quantity of material.

postcure. A treatment (normally involving heat) applied to an adhesive assembly following the initial cure, to modify specific properties. To expose an adhesive assembly to an additional cure, following the initial cure, for the purpose of modifying specific properties. ASTM D 907.

postforming. The forming, bending, or shaping of fully cured, C-staged thermoset laminates that have been heated to make them flexible. Upon cooling, the formed laminate retains the contours and shape of the mold over which it has been formed.

pot. To embed a component or assembly in liquid resin, using a shell, can, or case that remains an integral part of the product after the resin is cured.

pot life. The length of time that a catalyzed thermosetting resin system retains a viscosity low enough to be used in processing. Also called working life.

potting. Similar to encapsulating except that steps are taken to ensure complete penetration of all the voids in the object before the resin polymerizes.

power loss. The power, per unit volume, that is transformed into heat through hysteresis. It is the product of energy loss and frequency.

PP. See *polypropylenes*.

PPO. See *polyphenylene oxides*.

PPS. See *polyphenylene sulfides*.

prebond treatment. Synonym for surface preparation. ASTM D 907.

preconditioning. Any preliminary exposure of a material to specified atmospheric conditions for the purpose of favorably approaching equilibrium with a prescribed atmosphere.

precure. The full or partial setting of a synthetic resin or adhesive in a joint before the clamping operation is complete or before pressure is applied.

prefit. A process for checking the fit of mating detail parts in an assembly prior to adhesive bonding, to ensure proper bond lines. Mechanically fastened structures are sometimes prefitted to establish shimming requirements.

preheating. The heating of a compound before molding or casting, to facilitate the operation or reduce the molding cycle time.

prepolymer. A chemical intermediate with a molecular weight between that of the monomer or monomers and the final polymer or resin.

prepolymer molding. In the urethane foam field, a system in which a portion of the polyol is prereacted with the isocyanate to form a liquid prepolymer with a viscosity range suitable for pumping or metering. This component is supplied to end users with a second premixed blend of additional polyol, catalyst, blowing agent, and so forth. When the two components are subsequently mixed, foaming occurs.

preproduction test. A test or series of tests conducted by an adhesive manufacturer to determine conformity of an adhesive batch to established production standards, or by a fabricator to determine the quality of an adhesive before parts are produced, or by an adhesive specification custodian to determine conformance of an adhesive to the requirements of a specification not requiring qualification tests. ASTM D 907. Compare to *acceptance test* and *qualification test*.

pressure bag molding. A process for molding reinforced plastics in which a tailored, flexible bag is placed over the contact lay-up on the mold, sealed, and clamped in place. Fluid pressure, usually provided by compressed air or water, is placed against the bag, and the part is cured.

pressure intensifier. A layer of flexible material (usually a high-temperature rubber) used to ensure the application of sufficient pressure to a location, such as a radius, in a lay-up being cured.

pressure, saturation. See *saturation pressure*.

pressure-sensitive adhesive. A viscoelastic material which in solvent-free form remains permanently tacky. This material will adhere instantaneously to most solid surfaces with the application of very slight pressure. ASTM D 907.

primary nucleation. The mechanism by which crystallization is initiated, often by an added nucleation agent.

primary standard dosimetry system. A system that measures energy deposition directly without the need for conversion factors for interpretation of the radiation absorption process. Examples of such systems are calorimeters and Fricke dosimeters.

primer. A coating applied to a surface, prior to the application of an adhesive, to improve the performance of the bond. ASTM D 907. The coating can be a low-viscosity fluid that is typically a 10% solution of the adhesive in an organic solvent, which can wet out the adherend surface to leave a coating over which the adhesive can readily flow.

printed wiring board. A completely processed conductor pattern, usually formed on a stiff, flat base (laminated plastic). It serves as a means of electrical interconnection and physical attachment for printed circuits. Also called printed circuit board.

promoter. A chemical, itself a feeble catalyst, that greatly increases the activity of a given catalyst. See also *accelerator*.

proof. To test a component or system at its peak operating load or pressure.

proof pressure. The test pressure that pressurized components must sustain without detrimental deformation or damage. The proof pressure test is used to give evidence of satisfactory workmanship and material quality.

proportional limit. The greatest stress that a material is capable of sustaining without deviation from proportionality of stress and strain (Hooke's law). It is expressed in force per unit area. See also *elastic limit*.

propylene plastics. Plastics based on polymers of propylene or copolymers of propylene with other monomers, the propylene being the greatest amount by mass.

prototype. A model suitable for use in the complete evaluation of form, design, performance, and material processing.

PS. See *polystyrenes*.

PSU. See *polysulfones*.

PTFE. See *polytetrafluoroethylenes*.

PUR. See *polyurethanes*.

PVAC. See *polyvinyl acetate*.

PVAL. See *polyvinyl alcohol*.

PVB. See *polyvinyl butyral*.

PVC. See *polyvinyl chloride*.

PVDC. See *polyvinylidene chloride*.

PVDF. See *polyvinylidene fluoride*.

PVF. See *polyvinyl formal*.

Q

QPL. See *qualified products list*.

qualification test. A series of tests conducted by the procuring activity, or an agent thereof, to determine the conformance of materials, or materials system, to the requirements of a specification, normally resulting in a qualified products list under the specification. Generally, qualification under a specification requires a conformance to all tests in the specification, or it may be limited to conformance to a specific type or class, or both, under the specification. ASTM D 907. Compare to *acceptance test* and *preproduction test*.

qualified products list (QPL). A list of commercial products that have been pretested and found to meet the requirements of a specification, especially a government specification.

quality factor. The ratio of elastic modulus to loss modulus, measured in tension, compression, flexure, or shear. This is a nondimensional term and is the reciprocal of tan delta.

quasi-isotropic laminate. A laminate approximating isotropy by orientation of plies in several or more directions.

quench. In thermoplastics, a process of shock cooling thermoplastic materials from the molten state.

R

RA. See *reduction of area*.

radical. A very reactive chemical intermediate.

radio frequency (RF) preheating. A method of preheating molding materials to facilitate the molding operation and/or reduce molding cycle time. The frequencies most commonly used are those between 10 and 100 MHz.

radio frequency welding. A method of welding thermoplastics using a radio frequency field to apply the necessary heat. Also known as high-frequency welding.

reactive diluent. As used in epoxy formulations, a compound containing one or more epoxy groups that functions mainly to reduce the viscosity of the mixture.

recovery. The time-dependent portion of the decrease in strain following unloading of a specimen at the same constant temperature as the initial test. Recovery is equal to the total decrease in strain minus the instantaneous recovery. The recovery is expressed in the same units as *instantaneous recovery*. ASTM D 1780.

reduction of area (RA). The difference between the original cross-sectional area of a tension test specimen and the area of its smallest cross section. The reduction of area is usually expressed as a percentage of the original cross-sectional area of the specimen.

regular transmittance. Ratio of the light flux transmitted without diffusion to the flux incident.

reinforced plastics. Molded, formed, filament-wound, tape-wrapped, or shaped plastic parts consisting of resins to which reinforcing fibers, mats, fabrics, and so forth, have been added before the forming operation to provide some strength properties greatly superior to those of the base resin.

reinforced polypropylene. Polypropylene that is reinforced with mineral fillers, such as talc, mica, and calcium carbonate, as well as with glass and carbon fibers. The maximum concentration usually used is 50 wt%, although concentrates with higher levels of filler or reinforcement are available.

reinforcement. A strong material bonded into a matrix to improve its mechanical properties. Reinforcements are usually long fibers, chopped fibers, whiskers, particulates, and so forth. The term is not synonymous with filler.

relative humidity. The ratio of the actual pressure of existing water vapor to the maximum possible (saturation) pressure of water vapor in the atmosphere at the same temperature, expressed as a percentage.

relative rigidity. In dynamic mechanical measurements, the ratio of modulus at any temperature, frequency, or time to the modulus at a reference temperature, frequency, or time.

relative viscosity. For a polymer in solution, the ratio of the absolute viscosities of the solution (of stated concentration) and of the pure solvent at the same temperature.

relaxation time. The time required for a stress under a sustained constant strain to diminish by a stated fraction of its initial value.

relaxed stress. The initial stress minus the remaining stress at a given time during a stress-relaxation test.

release agent. A material that is applied in a thin film to the surface of a mold to keep the resin from bonding to the mold. Also called parting agent.

release film. An impermeable layer of film that does not bond to the resin being cured.

release paper. A sheet, serving as a protectant or carrier, or both, for an adhesive film or mass, which is easily removed from the film or mass prior to use. ASTM D 907.

reprocessed plastic. A thermoplastic prepared from melt-processed scrap or reject parts by a plastics processor, or from nonstandard or nonuniform virgin material. The term scrap does not necessarily connote feedstock that is less desirable or useable than the virgin material from which it may have been generated. Reprocessed plastic may or may not be reformulated by the addition of fillers, plasticizers, stabilizers, or pigments.

resenes. The constituents of rosin that cannot be saponified with alcoholic alkali, but that contain carbon, hydrogen, and oxygen in the molecule.

residual gas analysis (RGA). The study of residual gases in vacuum systems using mass spectrometry.

residual strain. The strain associated with residual stress.

residual stress. The stress existing in a body at rest, in equilibrium, at uniform temperature, and not subjected to external forces. Often caused by the forming and curing process.

resilience. The ratio of energy returned, on recovery from deformation, to the work input required to produce the deformation, usually expressed as a percentage. The ability to regain an original shape quickly after being strained or distorted.

resin. A solid, semisolid, or pseudosolid organic material that has an indefinite and often high molecular weight, exhibits a tendency to flow when subjected to stress, usually has a softening or melting range, and usually fractures conchoidally. Liquid resin, that is, an organic polymeric liquid which, when converted to its final state for use, becomes a resin. ASTM D 907. See also *adhesive*, *glue*, *gum*, and *rosin*.

resin content. The amount of resin in a laminate, expressed as a percentage of either total weight or total volume.

resin, liquid. See *liquid resin*.

resinography. The science of the morphology, structure, and related descriptive characteristics as correlated with the composition or conditions and with the properties or behavior of resins, polymers, plastics, and their products.

resinoid. Any of the class of thermosetting synthetic resins, either in their initial temporarily fusible state or in their final infusible state. ASTM D 907. See also *novolak* and *thermosetting*.

resin-rich area. Localized area filled with resin and lacking reinforcing material.

resin-starved area. Localized area of insufficient resin, usually identified by low gloss, dry spots, or fiber showing on the surface.

resin system. A mixture of resin and ingredients such as catalyst, initiator, and diluents required for the intended processing method and final product.

resite. Synonym for C-stage. ASTM D 907.

resitol. Synonym for B-stage. ASTM D 907.

resol (resole). Synonym for A-stage. ASTM D 907.

resol (resole) resin. Linear phenolic resin produced by alkaline condensate of phenol and formaldehyde.

resonant forced vibration technique. A technique for performing dynamic mechanical measurements, in which the sample is oscillated mechanically at the natural resonant frequency of the system. The amplitude of oscillation is maintained constant by the addition of makeup energy. Elastic modulus is calculated from the measured frequency. Damping is calculated from the additional energy required to maintain constant amplitude.

retarder. Synonym for inhibitor. ASTM D 907.

retrogradation. A change of starch pastes from low to high consistency upon aging. ASTM D 907.

reverse impact test. A test in which one side of a sheet of material is struck by a pendulum or falling object, and the reverse side is inspected for damage.

RGA. See *residual gas analysis*.

rheology. The study of the flow of materials, particularly the plastic flow of solids and the flow of non-Newtonian liquids. The science treating the deformation and flow of matter.

rigid plastics. For purposes of general classification, a plastic that has a modulus of elasticity either in flexure or in tension greater than 690 MPa (100 ksi) at 23 °C (70 °F) and 50% relative humidity.

rigid resin. A resin having a modulus high enough to be of practical importance, for example, 690 MPa (100 ksi) or greater.

Rockwell hardness. A value derived from the increase in depth of an impression as the load on an indentor is increased from a fixed minimum value to a higher value and then returned to the minimum value. Indentors for the Rockwell test include steel balls of several specific diameters and a diamond cone penetrator having an included angle of 120° with a spherical tip

having a radius of 0.2 mm (0.0070 in.). Rockwell hardness numbers are always quoted with a prefix representing the Rockwell scale corresponding to a given combination of load and indentor, for example, HRC 30.

room temperature. A temperature in the range of 20 to 30 °C (68 to 86 °F). The term room temperature is usually applied to an atmosphere of unspecified relative humidity.

room-temperature-curing adhesive. An adhesive that sets (to handling strength) within an hour at temperatures from 20 to 30 °C (68 to 86 °F) and later reaches full strength without heating.

room-temperature-setting adhesive. An adhesive that sets in the temperature range from 20 to 30 °C (68 to 86 °F), in accordance with the limits for standard room temperature specified in ASTM D 618 and ASTM D 907. Compare to *cold-setting adhesive*, *hot-setting adhesive*, and *intermediate-temperature-setting adhesive*.

room-temperature vulcanizing (RTV). Vulcanization or curing at room temperature by chemical reaction, particularly of silicones and other rubbers.

root-mean-square end-to-end distance. A measure of the average size of a coiled polymer molecule, usually determined by light scattering.

rosin. A resin obtained as a residue in the distillation of crude turpentine from the sap of the pine tree (gum resin) or from an extract of the stumps and other parts of the tree (wood rosin). ASTM D 907. Compare to *resin*.

row nucleation. The mechanism by which stress-induced crystallization is initiated, usually during fiber spinning or hot drawing.

rubbers. Cross-linked polymers having glass transition temperatures below room temperature that exhibit highly elastic deformation and have high elongation.

rupture. A cleavage or break resulting from physical stress. Work of rupture is the integral of the stress-strain curve between the origin and the point of rupture.

rupture strength. The true value of rupture strength is the stress of a material at failure based on the ruptured cross-sectional area itself.

S

sagging. Run-off or flow-off of adhesive from an adherend surface due to application of excess or low-viscosity material.

sandwich construction. A panel composed of a lightweight core material, such as cellular plastic, to which two relatively thin, dense, high-strength or high-stiffness faces, or skins, are adhered.

sandwich heating. A method of heating a thermoplastic sheet, prior to forming,

that consists of heating both sides of the sheet simultaneously.

satin. A type of finish having a satin or velvety appearance, specified for plastics or composites.

saturation pressure. The pressure, for a pure substance at any given temperature, at which vapor and liquid, or vapor and solid, coexist in stable equilibrium.

S-basis. The S-basis property allowable is the minimum value specified by the appropriate federal, military, Society of Automotive Engineers, American Society for Testing and Materials, or other recognized and approved specifications for the material.

SBS. See *short beam shear*.

scale. A condition in which resin plates or particles are on the surface of a pultrusion. Scales can often be readily removed, sometimes leaving surface voids or depressions.

scarf joint. A joint made by cutting away similar angular segments of two adherends and bonding the adherends with the cut areas fitted together. ASTM D 907.

schlieren. Regions of varying refraction in a transparent medium often caused by pressure or temperature differences and detectable especially by photographing the passage of a beam of light.

scrim. A low-cost reinforcing fabric made from continuous filament yarn in an open-mesh construction. Used in the processing of tape or other B-stage material to facilitate handling. Also used as a carrier of adhesive, for use in secondary bonding.

sealant. A material applied to a joint in paste or liquid form that hardens or cures in place, forming a seal against gas or liquid entry.

secant modulus. Idealized Young's modulus derived from a secant drawn between the origin and any point on a nonlinear stress-strain curve. On materials for which the modulus changes with stress, the secant modulus is the average of the 0 applied stress point and the maximum stress point being considered. See also *tangent modulus*.

secondary bonding. The joining together, by the process of adhesive bonding, of two or more already cured composite parts, during which the only chemical or thermal reaction occurring is the curing of the adhesive itself.

secondary crystallization. Slow crystallization process that occurs after the main solidification process is complete. Often associated with impure molecules.

secondary nucleation. The mechanism by which crystals grow.

secondary-standard dosimetry system. A system that measures energy deposition indirectly. It requires conversion factors to account for such considerations as geometry, dose rate, relative stopping power, incident energy spectrum, or other effects, in order to interpret the response of the system. Thus, it requires calibration against a primary dosimetry system or by means of a standard radiation source.

secondary structure. In aircraft and aerospace applications, a structure that is not critical to flight safety.

second-degree blocking. An adherence of such degree that when surfaces under tests are parted, one surface or the other will be found to be damaged. ASTM D 1146.

self-curing. See *self-vulcanizing*.

self-extinguishing resin. A resin formulation that will burn in the presence of a flame but will extinguish itself within a specified time after the flame is removed. The term is not universally accepted.

self-skinning foam. A urethane foam that produces a tough outer surface over a foam core upon curing.

self-vulcanizing. Pertaining to an adhesive that undergoes vulcanization without the application of heat. ASTM D 907. See also *vulcanization*.

semicrystalline. In plastics, materials that exhibit localized crystallinity. See also *crystalline plastic*.

semirigid plastic. For purposes of general classification, a plastic that has a modulus of elasticity either in flexure or in tension between 70 and 690 MPa (10 and 100 ksi) at 23 °C (70 °F) and 50% relative humidity.

separate-application adhesive. A term used to describe an adhesive consisting of two parts, one part being applied to one adherend and the other part to the other adherend and the two being brought together to form a joint. ASTM D 907.

set. To convert an adhesive into a fixed or hardened state by chemical or physical action, such as condensation, polymerization, oxidation, vulcanization, gelation, hydration, or evaporation of volatile constituents. ASTM D 907. See also *cure* and *dry*.

set (mechanical). Strain remaining after complete release of the load producing the deformation.

set (polymerization). To convert an adhesive into a fixed or hardened state by chemical or physical action, such as condensation, polymerization, oxidation, vulcanization, gelation, hydration, or evaporation of volatile constituents. See also *cure*.

setting temperature. The temperature to which an adhesive or an assembly is subjected to set the adhesive. The temperature attained by the adhesive in the process of setting (adhesive setting temperature) may differ from the temperature of the atmosphere surrounding the assembly (assembly setting temperature). ASTM D 907. See also *curing temperature* and *drying temperature*.

setting time. The period of time during which an assembly is subjected to heat or pressure, or both, to set the adhesive. ASTM D 907. See also *curing time, joint-conditioning time*, and *drying time*.

set up. To harden, as in the curing of a polymer resin.

shape factor. For an elastomeric slab loaded in compression, the ratio of the loaded area to the force-free area.

shear. An action or stress resulting from applied forces that causes or tends to cause two contiguous parts of a body to slide relative to each other in a direction parallel to their plane of contact. In interlaminar shear, the plane of contact is composed primarily of resin. See also *shear strength* and *shear stress*.

shear edge. The cutoff edge of the mold.

shear fracture. A mode of fracture in crystalline materials resulting from translation along slip planes that are preferentially oriented in the direction of the shearing stress.

shear modulus. The ratio of shearing stress to shearing strain within the proportional limit of the material.

shear rate. The overall velocity over the cross section of a channel with which molten polymer layers are gliding along each other or along the wall in laminar flow.

shear strain. The tangent of the angular change, caused by a force between two lines originally perpendicular to each other through a point in a body. ASTM D 4027.

shear strength. The maximum shear stress that a material is capable of sustaining. Shear strength is calculated from the maximum load during a shear or torsion test and is based on the original cross-sectional area of the specimen. See also *bond strength*.

shear stress. The component of stress tangent to the plane on which the forces act. The stress developing in a polymer melt when the layers in a cross section are gliding along each other or along the wall of the channel (in laminar flow). Shear stress is equal to force divided by the area sheared.

shelf life. The length of time a material, substance, product, or reagent can be stored under specified environmental conditions and continue to meet all applicable specification requirements and/or remain suitable for its intended function. Synonym for storage life.

shell tooling. A mold or bonding fixture consisting of a contoured surface shell supported by a substructure to provide dimensional stability.

Shore hardness. A measure of the resistance of material to indentation by a spring-loaded indentor. The higher the number, the greater the resistance. Normally used for rubber materials.

short beam shear (SBS). A flexural test of a specimen having a low test span-to-thick-

ness ratio, for example, 4:1, such that failure is primarily in shear.

short-chain branching. The dominant form of molecular branching in addition polymers, usually formed by a "backbiting" transfer reaction and resulting, primarily, in *n*-butyl side chains, but also other short-pendant groups such as methyl and amyl. Such branching results in reduced levels of crystallinity.

shortness. A qualitative term that describes an adhesive that does not string cotton, or otherwise form filaments or threads during application. ASTM D 907.

SI. See *silicones*.

silicones (SI). Plastics based on resins in which the main polymer chain consists of alternating silicon and oxygen atoms, with carbon-containing side groups. Derived from silica (sand) and methyl chlorides and furnished in different molecular weights, including liquids, solid resins, and elastomers.

single-lap specimen. In adhesive testing, a specimen made by bonding the overlapped edges of two sheets or strips of material, or by grooving a laminated assembly, as shown in ASTM D 2339 and D 3165. In testing, a single-lap specimen is usually loaded in tension at the ends. ASTM D 4896.

single spread. Application of adhesive to only one adherend of a joint. ASTM D 907.

size. Synonym for sizing. ASTM D 907.

sizing. The process of applying a material on a surface in order to fill pores and thus reduce the absorption of the subsequently applied adhesive or coating or to otherwise modify the surface properties of the substrate to improve the adhesion, and also, the material used for this purpose. Synonym for size. ASTM D 907. See also *primer*.

skin. The relatively dense material that sometimes forms on the surface of a cellular plastic or a sandwich construction.

slenderness ratio. The unsupported effective length of a uniform column divided by the least radius of gyration of the cross-sectional area.

slip. The relative collinear displacement of the adherends on either side of the adhesive layer in the direction of the applied load. ASTM D 4027.

slippage. The movement of adherends with respect to each other during the bonding process. ASTM D 907.

S-N diagram. A plot of stress (S) against the number of cycles to failure (N) in fatigue testing. A log scale is normally used for N. For S, a linear scale is often used, but sometimes a log scale is used here, too. Also, a representation of the number of alternating stress cycles a material can sustain without failure at various maximum stresses.

softening range. The range of temperatures within which a plastic changes from a rigid to a soft state. Actual values depend on the test method. Sometimes erroneously referred to as softening point.

solid-phase chemical dosimeter. An apparatus that measures radioactivity by using plastic, dyed plastic, or glass with an optical density, usually in the visible range, that changes when exposed to ionizing radiation. Examples currently in use include dyed polymethyl methacrylate (red perspex), undyed polyvinyl chloride, dyed polyamide (blue dye in a nylon matrix), and dyed polychlorostyrene (green dye in a chlorostyrene matrix). Solid-phase chemical dosimetry is generally considered to be a secondary-standard dosimetry system.

solids content. The percentage by weight of the nonvolatile matter in an adhesive. The actual percentage of the nonvolatile matter in an adhesive will vary according to the analytical procedure that is used. A standard test method must be used to obtain consistent results. ASTM D 907.

solvation. The process of swelling, gelling, or dissolving a resin by a solvent or plasticizer.

solvent adhesive. An adhesive having a volatile organic liquid as a vehicle. This term excludes water-base adhesives. ASTM D 907.

solvent-activated adhesive. A dry-film adhesive that is rendered tacky by the application of a solvent just prior to use. ASTM D 907.

specific adhesion. Adhesion between surfaces that are held together by valence forces of the same type as those that give rise to cohesion. ASTM D 907.

specific gravity. The density (mass per unit volume) of any material divided by that of water at a standard temperature.

specific heat. The quantity of heat required to raise the temperature of a unit mass of a substance 1° under specified conditions.

specific humidity. In a mixture of water vapor and air, the mass of water vapor per unit mass of moist air.

specific properties. Material properties divided by material density.

specific viscosity. The relative viscosity of a solution of known concentration of a polymer minus 1. It is usually determined for a low concentration of the polymer.

spectrometry. A method based on designation of the wavelengths within a particular portion of a range of radiation or absorptions, for example, ultraviolet emission, and absorption spectrometry.

spectroscopy. The study of spectra using an instrument for dispersing radiation for the visual observation of emission or absorption.

spiral-flow test. A method for determining the flow properties of a thermoplastic resin in which the resin flows along the path of a spiral cavity. The length of the material that flows into the cavity and its weight give a relative indication of the flow properties of the resin.

spread. The quantity of adhesive per unit joint area applied to an adherend, usually expressed in pounds of adhesive per thousand square feet of joint area. ASTM D 907. See also *double spread* and *single spread*.

spring constant. The number of pounds required to compress a spring or specimen 25 mm (1 in.) in a prescribed test procedure.

squeeze-out. Adhesive pressed out at the bond line due to pressure applied on the adherends. ASTM D 907.

stabilization. In carbon fiber forming, the process used to render the carbon fiber precursor infusible prior to carbonization.

stabilizers. Chemicals used in plastics formulation to help maintain physical and chemical properties during processing and service life. A specific type of stabilizer, known as an ultraviolet stabilizer, is designed to absorb ultraviolet rays and prevent them from attacking the plastic.

stacking sequence. For a laminate, the description of the orientations of the plies and their sequence.

staging. Heating a premixed resin system, such as in a prepreg, until the chemical reaction (curing) starts, but stopping the reaction before the gel point is reached. Staging is often used to reduce resin flow in subsequent press molding operations.

starved area. An area in a plastic part that has an insufficient amount of resin to wet out the reinforcement completely. This condition may be due to improper wetting, impregnation, or resin flow; excessive molding pressure; or incorrect bleeder cloth thickness.

starved joint. A joint that has an insufficient amount of adhesive to produce a satisfactory bond. This condition may result from too thin a spread to fill the gap between the adherends, excessive penetration of the adhesive into the adherend, too short an assembly time, or the use of excessive pressure. ASTM D 907.

static charge. The electric charge produced by the relative motion of a nonconducting material over a nonconducting plastic material. Charge separation is due to mechanical motion.

static fatigue. Failure of a part under continued static load. Analogous to creep-rupture failure in metals testing, but often the result of aging accelerated by stress.

static modulus. The ratio of stress to strain under static conditions. It is calculated from static stress-strain tests, in shear, compression, or tension. Expressed in force per unit area.

static stress. A stress in which the force is constant or slowly increasing with time, for example, test of failure without shock.

stereoisomer. An isomer in which atoms are linked in the same order but differ in their arrangement.

stereospecific plastics. Implies a specific or definite order of arrangement of molecules in space. This ordered regularity of the molecules in contrast to the branched or random arrangement found in other plastics permits close packing of the molecules and leads to high crystallinity (for example, in polypropylene).

stiffness. A measure of modulus. The relationship of load and deformation. The ratio between the applied stress and resulting strain. A term often used when the relationship of stress to strain does not conform to the definition of Young's modulus. See also *stress-strain*.

storage life. The period of time during which a packaged adhesive can be stored under specified temperature conditions and remain suitable for use. Synonym for shelf life. ASTM D 907. Compare to *working life*.

storage modulus. A quantitative measure of elastic properties, defined as the ratio of the stress, in-phase with strain, to the magnitude of the strain. The storage modulus may be measured in tension or flexure (E'), compression (K'), or shear (G').

strain. In tensile testing, the ratio of the elongation to the gage length of the test specimen, that is, the change in length per unit of original length. The term is also used in a broader sense to denote a dimensionless number that characterizes the change in dimensions of an object during a deformation or flow process. Strain is a nondimensional quantity, but is frequently expressed in inches per inch, meters per meter, or percent.

strain amplitude. The ratio of the maximum deformation, measured from the mean deformation to the free length of the unstrained test specimen. Strain amplitude is measured from 0 to peak on one side only.

strain, axial. See *axial strain*.

strain gage. Device used to measure strain in a stressed material based on the change in electrical resistance.

strain, initial. See *initial strain*.

strain relaxation. Reduction in internal strain over time. Molecular processes are similar to those that occur in creep, except that the body is constrained.

strain, residual. See *residual strain*.

strain, shear. See *shear strain*.

strain, transverse. See *transverse strain*.

strain, true. See *true strain*.

strength. See *bond strength*.

strength, compressive. See *compressive strength*.

strength, dry. See *dry strength*.

strength, flexural. See *flexural strength*.

strength, shear. See *shear strength*.

strength, tensile. See *tensile strength*.

strength, wet. See *wet strength*.

strength, yield. See *yield strength.*

stress. The internal force per unit area that resists a change in size or shape of a body. Expressed as force per unit area.

stress amplitude. The ratio of the maximum applied force, measured from the mean force to the cross-sectional area of the unstressed test specimen.

stress concentration. On a macromechanical level, the magnification of the level of an applied stress in the region of a notch, void, hole, or inclusion.

stress-concentration factor. The ratio of the maximum stress in the region of a stress concentrator, such as a hole, to the stress in a similarly strained area without a stress concentrator.

stress corrosion. Preferential attack of areas under stress in a corrosive environment, where such an environment alone would not have caused corrosion.

stress crack. External or internal cracks in a plastic caused by tensile stresses less than that of its short-time mechanical strength, frequently accelerated by the environment to which the plastic is exposed. The stresses that cause cracking may be present internally or externally or may be combinations of these stresses. See also *crazing.*

stress-cracking failure. The failure of a material by cracking or crazing some time after it has been placed under load. Time-to-failure may range from minutes to years. Causes include molded-in stresses, postfabrication shrinkage or warpage, and hostile environment.

stress-induced crystallization. The production of crystals in a polymer by the action of stress, usually in the form of an elongation. It occurs in fiber spinning and rubber elongation and is responsible for enhanced mechanical properties.

stress, initial. See *initial stress.*

stress, nominal. See *nominal stress.*

stress, normal. See *normal stress.*

stress relaxation. The decrease in stress under sustained, constant strain. Also called stress decay.

stress, relaxed. See *relaxed stress.*

stress, residual. See *residual stress.*

stress, shear. See *shear stress.*

stress-strain. Stiffness at a given strain.

stress-strain curve. Simultaneous readings of load and deformation, converted to stress and strain, plotted as ordinates (vertically) and abscissas (horizontally), respectively, to obtain a stress-strain diagram.

stress, tensile. See *tensile stress.*

stress, torsional. See *torsional stress.*

stress, true. See *true stress.*

stringiness. The property of an adhesive that results in the formation of filaments or threads when adhesive transfer surfaces are separated. Transfer surfaces may be rolls, picker plates, stencils, and so on. Compare to *teeth.* ASTM D 907. See also *legging* and *webbing.*

stripper. A chemical solvent or acid that can remove an adhesive bond.

structural adhesive. A bonding agent used for transferring required loads between adherends exposed to service environments typical for the structure involved. ASTM D 907.

structural bond. A bond that joins basic load-bearing parts of an assembly. The load may be either static or dynamic.

structural foams. Expanded plastic materials having integral skins and outstanding rigidity. Structural foams involve a variety of thermoplastic resins as well as urethanes.

styrene-acrylonitrile (SAN). A copolymer of about 70% styrene and 30% acrylonitrile, with higher strength, rigidity, and chemical resistance than can be attained with polystyrene alone. These copolymers may be blended with butadiene as a terpolymer or grafted onto the butadiene to make acrylonitrile-butadiene-styrene resins. They are transparent and have high heat-deflection properties, excellent gloss, chemical resistance, hardness, rigidity, dimensional stability, and load-bearing capability.

styrene-maleic anhydride (SMA). Copolymers made by the copolymerization of styrene and maleic anhydride, with higher heat resistance than the parent styrenic and acrylonitrile-butadiene-styrene families.

styrene-rubber plastics. Plastics based on styrene polymers and rubbers, the styrene polymers being the greatest amount by mass.

substrate. A material upon the surface of which an adhesive-containing substance is spread for any purpose, such as bonding or coating. A broader term than adherend. Compare to *adherend.* ASTM D 907.

surface preparation. A physical or chemical treatment, or both, of an adherend to render it suitable for adhesive joining. Synonym for prebond treatment. ASTM D 907.

surface tension. The force existing in a liquid-vapor-phase interface that tends to diminish the area of the interface. This force acts at each point on the interface in the plane tangent to that point.

surfactant. A compound that affects interfacial tensions between two liquids. It usually reduces surface tension.

symmetrical laminate. A laminate in which the stacking sequence of plies below its midplane is a mirror image of the stacking sequence above the midplane.

syndiotactic stereoisomerism. A polymer molecule in which side atoms or side groups alternate regularly on opposite sides of the chain.

syneresis. The exudation of small amounts of liquid by gels upon standing. ASTM D 907.

syntactic cellular plastics. Reinforced plastics made by mixing hollow microspheres of glass, epoxy, phenolic, and so forth, into fluid resins (with additives and curing agents) to form a moldable, curable, lightweight, fluid mass; as opposed to foamed plastic, in which the cells are formed by gas bubbles released in the liquid plastic by either chemical or mechanical action.

T

tack. The property of an adhesive that enables it to form a bond of measurable strength immediately after adhesive and adherend are brought into contact under low pressure. See also *dry tack, tack range,* and *tacky-dry.*

tack range. The period of time in which an adhesive will remain in the tacky-dry condition after application to an adherend, under specified conditions of temperature and humidity. ASTM D 907.

tacky-dry. The condition of an adhesive when the volatile constituents have evaporated or been absorbed sufficiently to leave it in a desired tacky state. ASTM D 907.

tan delta (tan δ). The ratio of the loss modulus to the storage modulus, measured in compression (K); tension or flexure (E); or shear (G). Also, the ratio of the out-of-phase components of the dielectric constant (that is, the loss) to the in-phase component of the dielectric constant (that is, the permittivity).

tangent modulus. The slope of the line at a predefined point on a static stress-strain curve, expressed in force per unit area per unit strain. This is the tangent modulus at that point in shear, tension, or compression. See also *secant modulus.*

teeth. The resultant surface irregularities or projections formed by the breaking of filaments or strings which may form when adhesive-bonded substrates are separated. ASTM D 907. Compare to *legging, stringiness,* and *webbing.*

telegraphing. In a laminate or other type of composite construction, a condition in which irregularities, imperfections, or patterns of an inner layer are visibly transmitted to the surface. Telegraphing is occasionally referred to as photographing. ASTM D 907.

telomer. A polymer composed of molecules having terminal groups incapable of reacting with additional monomers, under the conditions of the synthesis, to form larger polymer molecules of the same chemical type.

temperature, dry-bulb. See *dry-bulb temperature.*

tensile modulus. See *Young's modulus.*

tensile strength. The maximum load or force per unit cross-sectional area, within the gage length, of the specimen. The pulling stress required to break a given specimen.

tensile strength, ultimate. See *ultimate tensile strength.*

tensile stress. The normal stress caused by forces directed away from the plane on which they act.

terpolymer. A polymeric system that contains three monomeric units.

T_g. See *glass transition temperature.*

TGA. See *thermogravimetric analysis.*

thermal conductivity. The quantity of heat conducted per unit time through unit area of a slab of unit thickness having unit temperature difference between its faces.

thermal endurance. The time required at a selected temperature for a material or system of materials to deteriorate to some predetermined level of electrical, mechanical, or chemical performance under prescribed test conditions.

thermal expansion molding. A process in which elastomeric tooling details are constrained within a rigid frame to generate consolidation pressure by thermal expansion during the curing cycle of the autoclave molding process.

thermal stress cracking. The crazing and cracking of some thermoplastic resins from overexposure to elevated temperatures.

thermogravimetric analysis (TGA). The study of the change in mass of a material under various conditions of temperature and pressure.

thermoplastic. Capable of being repeatedly softened by heat and hardened by cooling. A material that will repeatedly soften when heated and harden when cooled. ASTM D 907. See also *novolak.*

thermoplastic fluoropolymers. See *fluoroplastics.*

thermoplastic polyesters. A class of thermoplastic polymers in which the repeating units are joined by ester groups. The two important types are polyethylene terephthalate (PET), which is widely used as film, fiber, and soda bottles; and polybutylene terephthalate (PBT), which is primarily a molding compound.

thermoplastic polyimides (TPI). Fully imidized, linear polymers with exceptionally good thermomechanical performance characteristics.

thermoplastic polyurethanes (TPUR). Linear (segmented) block copolymers. Such a copolymer consists of repeating groups of diisocyanate and short-chain diol, or chain extender, for a rigid block, and repeating groups of diisocyanate and long-chain diol, or polyol, for a flexible block. TPURs are commonly injection molded, blow molded, or extruded.

thermoset. Pertaining to the state of a resin in which it is relatively infusible. A material that will undergo or has undergone a chemical reaction by the action of heat, catalysts, ultraviolet light, and so on, leading to a relatively infusible state. ASTM D 907. See also *A-stage, B-stage, C-stage,* and *resinoid.*

thermosetting. Having the property of undergoing a chemical reaction by the action of heat, catalysts, ultraviolet light, and so on, leading to a relatively infusible state. ASTM D 907.

thermosetting polyesters. A class of resins produced by dissolving unsaturated, generally linear, alkyd resins in a vinyl-type active monomer such as styrene, methyl styrene, or diallyl phthalate.

thermotropic liquid crystal. A liquid crystalline polymer that can be processed using thermoforming techniques.

thin-layer chromatography (TLC). A microtype of chromatography in which a thin layer of special absorbent is applied to a glass plate, a drop of a solution of the material being investigated is applied to an edge, and that side of the plate is then dipped in an appropriate solvent. The solvent travels up the thin layer of absorbent, which selectively separates the molecules present in the material being investigated.

thinner. A volatile liquid added to an adhesive to modify the consistency or other properties. ASTM D 907. See also *diluent* and *extender.*

thixotropy. A property of adhesive systems to thin upon isothermal agitation and to thicken upon subsequent rest. ASTM D 907.

time profile. A plot of the modulus, damping, or both, of a material versus time.

TLC. See *thin-layer chromatography.*

tolerance. The guaranteed maximum deviation from the specified nominal value of a component characteristic at standard or stated environmental conditions.

torsion. Twisting stress.

torsional pendulum. A device for performing dynamic mechanical analysis, in which the sample is deformed torsionally and allowed to oscillate in free vibration. Modulus is determined by the frequency of the resultant oscillation, and damping is determined by the decreasing amplitude of the oscillation.

torsional stress. The shear stress on a transverse cross section caused by a twisting action.

toughness. A measure of the ability of a material to absorb energy. The actual work per unit volume or unit mass of material that is required to rupture it. Toughness is proportional to the area under the load-elongation curve from the origin to the breaking point.

T-peel strength. The average load per unit width of bond line required to produce progressive separation of two bonded, flexible adherends, under conditions designated in ASTM D 1876.

TPI. See *thermoplastic polyimides.*

TPUR. See *thermoplastic polyurethanes.*

transition, first-order. See *first-order transition.*

transition temperature. The temperature at which the properties of a material change. Depending on the material, the transition change may or may not be reversible.

trans stereoisomer. A stereoisomer in which atoms or groups of atoms are arranged on opposite sides of a chain of atoms.

transversely isotropic. In reference to a material, exhibiting a special case of orthotropy in which properties are identical in two orthotropic dimensions but not the third. Having identical properties in both transverse but not in the longitudinal direction.

transverse strain. The linear strain in a plane perpendicular to the loading axis of a specimen.

true strain. The natural logarithm of the ratio of gage length at the moment of observation to the original gage length for a body subjected to an axial force.

true stress. The stress along the axis calculated on the actual cross section at the time of observation instead of the original cross-sectional area. Applicable to tension and compression testing.

two-component adhesive. An adhesive supplied in two parts that are mixed before application. Such adhesives usually cure at room temperature.

typical-basis. The typical property value is an average value. No statistical assurance is associated with this basis.

U

UHMWPE. See *ultrahigh molecular weight polyethylene.*

ultimate elongation. The elongation at rupture.

ultimate tensile strength. The ultimate or final (highest) stress sustained by a specimen in a tension test. Rupture and ultimate stress may or may not be the same.

ultrahigh molecular weight polyethylene (UHMWPE). Those polyethylene resins having molecular weights in the 1.5 to 3.0 million range.

ultrasonic bonding. A method of joining using vibratory mechanical pressure at ultrasonic frequencies. Electrical energy is changed to ultrasonic vibrations by means of a magnetostrictive or a piezoelectric transducer. The ultrasonic vibrations generate frictional heat, melting the plastics and allowing them to join.

ultrasonic testing. A nondestructive test applied to materials for the purpose of locating internal flaws or structural discontinuities by the use of high-frequency reflection or attenuation (ultrasonic beam).

ultraviolet (UV). Zone of invisible radiations beyond the violet end of the spectrum of visible radiations. Because UV wavelengths are shorter than visible wavelengths, their photons have more energy, which initiates some chemical reactions and degrades most plastics, particularly aramids.

ultraviolet (UV) stabilizer. Any chemical compound that, when admixed with a resin, selectively absorbs UV rays.

unbond. An area within a bonded interface between two adherends in which the intended bonding action failed to take place, or an area in which two layers of prepreg in a cured component do not adhere. Also used to denote specific areas deliberately prevented from bonding in order to simulate a defective bond, such as in the generation of quality standards specimens.

undercure. An undesirable condition of a molded article resulting from the allowance of too little time and/or temperature or pressure for adequate hardening of the molding.

uniaxial load. A condition in which a material is stressed in only one direction along the axis or centerline of component parts.

uniform elongation. The elongation determined when the maximum load is reached. Applies to materials whose cross section decreases uniformly along the gage length up to maximum load.

unimeric. Pertaining to a single molecule that is not monomeric, oligomeric, or polymeric, such as saturated hydrocarbons.

unsaturated compounds. Any compound having more than one bond between two adjacent atoms, usually carbon atoms, and being capable of adding other atoms at that point to reduce it to a single bond.

unsaturated polyesters. The family of polyesters characterized by vinyl unsaturation in the polyester backbone, which enables subsequent hardening or curing by copolymerization with a reactive monomer in which the polyester constituent has been dissolved.

unsymmetric laminate. A laminate having an arbitrary stacking sequence without midplane symmetry.

urea-formaldehyde adhesive. An aqueous colloidal dispersion of urea-formaldehyde polymer which may contain modifiers and secondary binders to provide specific adhesive properties. Also, a type of adhesive based on a dry urea-formaldehyde polymer and water. A curing agent is commonly used with this type of adhesive. ASTM D 907.

urethane hybrids. Polymers that are formed by the reaction of two liquid components, an acrylesterol and a modified diphenylmethane-4,4'-diisocyanate (MDI). The acrylesterol is a hybrid of a urethane (nonalcohol) and an acrylic (unsaturated monoalcohol). The liquid-modified MDI contains two or more isocyanate groups that can react with the hydroxyl portion of the acrylesterol molecule.

urethane plastics. Plastics based on resins made by the condensation of organic isocyanates with compounds or resins that contain hydroxyl groups. The resin is furnished as two component liquid monomers or prepolymers that are mixed in the field immediately before application. A great variety of materials is available, depending on the monomers used in the prepolymers and polyols and the type of diisocyanate employed. Extremely abrasion and impact resistant. See also *isocyanate plastics* and *polyurethanes*.

UV. See *ultraviolet*.

V

vacuum bag. A flexible bag in which pressure may be applied to an assembly (inside the bag) by means of evacuation of the bag.

vapor. The gaseous form of substances that are normally in the solid or liquid state, and that can be changed to these states either by increasing the pressure or decreasing the temperature.

veil. An ultrathin mat similar to a surface mat, often composed of organic fibers as well as glass fibers.

vent cloth. A layer or layers of open-weave cloth used to provide a path for vacuum to "reach" the area over a laminate being cured, such that volatiles and air can be removed. Also causes the pressure differential that results in the application of pressure to the part being cured. Also called breather cloth.

venting. In autoclave curing of a part or assembly, turning off the vacuum source and venting the vacuum bag to the atmosphere. The pressure on the part is then the difference between pressure in the autoclave and atmospheric pressure. In injection molding, gases evolve from the melt and escape through vents machined in the barrel or mold.

Vicat softening point. The temperature at which a flat-ended needle of 1 mm^2 (0.0015 in.^{2}) circular or square cross section will penetrate a thermoplastic specimen to a depth of 1 mm (0.040 in.) under a specified load, using a uniform rate of temperature rise.

vinyl acetate plastics. Plastics based on polymers of vinyl acetate or copolymers of vinyl acetate with other monomers, the vinyl acetate being the greatest amount by mass.

vinyl chloride plastics. Plastics based on polymers of vinyl chloride or copolymers of vinyl chloride with other monomers, the vinyl chloride being the greatest amount by mass.

vinyl esters. A class of thermosetting resins containing esters of acrylic and/or methacrylic acids, many of which have been made from epoxy resin. Cure is accomplished, as with unsaturated polyesters, by copolymerization with other vinyl monomers, such as styrene.

vinylidene chloride plastics. Plastics based on polymer resins made by the polymerization of vinylidene chloride or by the copolymerization of vinylidene chloride with other unsaturated compounds, the vinylidene chloride being the greatest amount by weight.

virgin material. A plastic material in the form of pellets, granules, powder, flock, or liquid that has not been subjected to use or processing other than that required for its initial manufacture.

viscoelasticity. A property involving a combination of elastic and viscous behavior. A material having this property is considered to combine the features of a perfectly elastic solid and a perfect fluid. A phenomenon of time-dependent, in addition to elastic, deformation (or recovery) in response to load.

viscosity. The ratio of the shear stress existing between laminae of moving fluid and the rate of shear between these laminae. A fluid is said to exhibit Newtonian behavior when the rate of shear is proportional to the shear stress. A fluid is said to exhibit non-Newtonian behavior when an increase or decrease in the rate of shear is not accompanied by a proportional increase or decrease in the shear stress. ASTM D 907. Compare to *consistency*. See also *thixotropy*.

viscosity coefficient. The shearing stress tangentially applied that will induce a velocity gradient. A material has a viscosity of 0.1 Pa · s (1 P) when a shearing stress of 1 μN/mm (1 dyne/cm^2) produces a velocity gradient of (1 cm/s)/cm. ASTM D 907. See also *viscosity*.

viscous deformation. Any portion of the total deformation of a body that occurs as a function of time when load is applied but that remains permanent when the load is removed. Generally referred to as an elastic deformation.

void. Air or gas that has been trapped and cured into a laminate. Porosity is an aggregation of microvoids. Voids are essentially incapable of transmitting structural stresses or nonradiative energy fields.

void content. Volume percentage of voids, usually less than 1% in a properly cured composite. The experimental determination is indirect; that is, it is calculated from the measured density of a cured laminate and the "theoretical" density of the starting material.

volatile content. The percent of volatiles that is driven off as a vapor from a plastic or an impregnated reinforcement.

volatiles. Materials, such as water and alcohol, in a sizing or resin formulation, that are capable of being driven off as a vapor at room temperature or at slightly elevated temperature.

volume fraction. Fraction of a constituent material based on its volume.

volume resistance. The ratio of the direct voltage applied to two electrodes in contact with or embedded in a specimen to that portion of the current between them that is distributed through the volume of the specimen. Also, the electrical resistance between opposite faces of a 10 mm (0.40 in.) cube of insulating material. Also called specific insulation resistance.

volume resistivity. Ratio of the potential gradient parallel to the current in the material to the current density.

vulcanization. A chemical reaction in which the physical properties of a rubber are changed in the direction of decreased plastic flow, less surface tackiness, and increased tensile strength by reacting it with sulfur or other suitable agents. ASTM D 907. See also *self-vulcanizing*.

vulcanize. To subject to vulcanization. ASTM D 907.

W

warm-setting adhesive. Synonym for intermediate-temperature-setting adhesive. ASTM D 907.

warp. A significant variation from the original true, or plane, surface. ASTM D 907.

warpage. Dimensional distortion in a plastic object.

water absorption. The ratio of the weight of water absorbed by a material to the weight of the dry material.

water break test. Any surface that is chemically clean can be tested by the use of a drop of water, preferably distilled water. If the surface is clean, the water will break and spread; a contaminated surface will cause the water to bead. ASTM D 3932.

water-extended polyester. A casting formulation in which water is suspended in the polyester resin.

weathering. Exposure of plastics to the outdoor environment.

weathering, artificial. See *artificial weathering*.

webbing. Filaments or threads that sometimes form when adhesive transfer surfaces are separated. Transfer surfaces may be rolls, picker plates, stencils, and so on. ASTM D 907. See also *legging* and *stringiness*. Compare to *teeth*.

weeping. Slow leakage manifest by the appearance of water on a surface.

wet installation. A bolted joint in which sealant is applied to the head and shank of the fastener such that after assembly a seal is provided between the fastener and the elements being joined.

wet strength. The strength of an adhesive joint determined immediately after removal from a liquid in which it has been immersed under specified conditions of time, temperature, and pressure. The term is commonly used alone to designate strength after immersion in water. In latex adhesives the term is used to describe the joint strength when the adherends are brought together with the adhesives still in the wet state. ASTM D 907. Compare to *dry strength*.

wetting. The spreading, and sometimes absorption, of a fluid on or into a surface.

whisker. A short single-crystal fiber or filament used as a reinforcement in a matrix. Whisker diameters range from 1 to 25 μm (40 to 980 μin.), with aspect ratios between 100 and 15 000.

wicking. The flow of a liquid along a surface into a narrow space. This capillary action is caused by the attraction of the liquid molecules to each other and to the surface.

width. In the case of a beam, the shorter dimension perpendicular to the direction in which the load is applied.

wood failure. The rupturing of wood fibers in strength tests on bonded specimens, usually expressed as the percentage of the total area involved which shows such failure. ASTM D 907.

wood laminate. See *built-up laminated wood*, *glue-laminated wood*, *plywood*, and *laminate*.

wood veneer. A thin sheet of wood, generally within the thickness range from 0.3 to 6.3 mm (0.01 to 0.25 in.), to be used in a laminate. ASTM D 907.

work hardening. Increase in resistance to further deformation with continuing distortion. Hardening and strengthening of a metal or alloy caused by the strain energy absorbed from prior deformation.

working life. The period of time during which an adhesive, after mixing with catalyst, solvent, or other compounding ingredients, remains suitable for use. Synonym for pot life. ASTM D 907. Compare to *storage life*.

Y

yield point. The first stress in a material, less than the maximum attainable stress, at which the strain increases at a higher rate than does the stress. The point at which permanent deformation of a stressed specimen begins to take place. Only materials that exhibit yielding have a yield point.

yield point elongation. In materials that exhibit a yield point, the difference between the elongation at the completion and at the start of discontinuous yielding.

yield strength. The stress at the yield point. The stress at which a material exhibits a specified limiting deviation from the proportionality of stress to strain. The lowest stress at which a material undergoes plastic deformation. Below this stress, the material is elastic; above it, the material is viscous. Often defined as the stress needed to produce a specified amount of plastic deformation (usually a 0.2% change in length).

yield value. The stress (either normal or shear) at which a marked increase in deformation occurs without an increase in load. ASTM D 907.

Young's modulus. The ratio of normal stress to corresponding strain for tensile or compressive stress less than the proportional limit of the material. See also *modulus of elasticity*.

Z

zero time. The time at which the given loading and constraint conditions are initially obtained in creep and stress-relaxation tests, respectively.

Ziegler-Natta catalysts. Initially, a catalyst consisting of an alkylaluminum compound with a compound of the titanium group of the periodic table, a typical combination being triethylaluminum and either titanium tetrachloride or titanium trichloride. Subsequently, an enormous variety of such mixtures is used in polymerization to provide stereospecificity (isotactic or syndiotactic).

ACKNOWLEDGMENT

The assistance and advice of Will Barbeau, Barbeau Associates, and Robert S. Miller, Franklin Chemical Industries, in compiling this glossary are gratefully acknowledged.

SELECTED REFERENCES

- J. Frados, Ed., *Plastics Engineering Handbook*, 4th ed., Van Nostrand Reinhold, 1976
- Glossary of Terms, in *Engineering Plastics*, Vol 2, *Engineered Materials Handbook*, ASM INTERNATIONAL, 1988
- C.A. Harper, Ed., *Handbook of Plastics and Elastomers*, McGraw-Hill, 1975
- *High-Performance Adhesive Bonding*, 1st ed., Society of Manufacturing Engineers, 1983
- A.H. Landrock, *Adhesive Technology Handbook*, Noyes Publications, 1985
- T.A. Richardson, *Industrial Plastics: Theory and Application*, South-Western, 1983
- I. Skeist, Ed., *Handbook of Adhesives*, 2nd ed., Van Nostrand Reinhold, 1977
- "Standard Abbreviations of Terms Relating to Plastics," D 1600, *Annual Book of ASTM Standards*, American Society for Testing and Materials
- "Standard Definitions and Descriptions of Terms Relating to Dynamic Mechanical Measurements on Plastics," D 4092, *Annual Book of ASTM Standards*, American Society for Testing and Materials
- "Standard Definitions of Terms Relating to Conditioning," E 41, *Annual Book of ASTM Standards*, American Society for Testing and Materials

- "Standard Definitions of Terms Relating to Methods of Mechanical Testing," E 6, *Annual Book of ASTM Standards*, American Society for Testing and Materials
- "Standard Definitions of Terms Relating to Plastics," D 883, *Annual Book of ASTM Standards*, American Society for Testing and Materials
- "Standard Guide for Identification of Plastic Materials," D 4000, *Annual Book of ASTM Standards*, American Society for Testing and Materials
- "Standard Terminology of Adhesives," D 907, *Annual Book of ASTM Standards*, American Society for Testing and Materials
- L.R. Whittington, *Whittington's Dictionary of Plastics*, 2nd ed., Technomic, 1978

Overview: Adhesives Technology*

Louis H. Sharpe, Consultant

AN ADHESIVE, as defined by ASTM, is "a substance capable of holding materials together by surface attachment" (Ref 1). Although the definition is not explicitly stated as such, it could be expanded to say that an adhesive is a substance capable of holding materials together *in a functional manner* by surface attachment. This additional phrase is important because a substance may join two materials to form a joint capable of functional performance, yet may not be able to join two other materials at all in the sense that a joint would be so weak that one might say that it could be "wrenched apart by gravity." The same substance is clearly an adhesive in the first instance, but not in the second.

The capability of holding materials together is, therefore, not an intrinsic property of a substance but, rather, depends on the context in which that substance is used. Two important, basic facts about adhesive materials are that: a substance called an adhesive does not perform its function independent of a context of use and that an adhesive does not exist that will bond "anything to anything" with (implied) equal utility. The functions that adhesives serve are described below.

Mechanical Fastening. The major function of adhesives is for mechanical fastening. Because an adhesive can transmit loads from one member of a joint to another, it allows a more uniform stress distribution than is obtained using a mechanical fastener. Thus, adhesives often permit the fabrication of structures that are mechanically equivalent or superior to conventional assemblies and, furthermore, have cost and weight benefits. For example, adhesives can join thin metal sections to thick sections so that the full strength of the thin section is utilized. Conventional mechanical fastening or spot welding produces a structure whose strength is limited to that of the areas of the thin section that contact the fasteners or the welds. In addition, adhesives can produce joints with high strength, rigidity, and dimensional precision in the light metals, such as aluminum and magnesium, which may be weakened or distorted by welding.

Sealing and Insulating. Because the adhesive in a properly prepared joint provides full contact with mating surfaces, it forms a barrier to fluids that do not attack or soften it. An adhesive may also function as an electrical and/or thermal insulator in a joint. Its thermal insulating efficiency can be increased, if necessary, by foaming an adhesive with the appropriate cell structure in place.

On the other hand, electrical and thermal conductivity can be raised appreciably by adding metallic fillers. Oxide fillers, such as alumina, only increase thermal conductivity. Electrically conductive adhesives, filled with silver flake, are available with specific resistivities less than 50 times that of bulk silver.

Resisting Corrosion and Vibration. Adhesives can also prevent electrochemical corrosion in joints between dissimilar metals. They may also act as vibration dampers. The mechanical damping characteristics of an adhesive can be changed by formulation. However, changing such a property in an adhesive generally produces change in other properties of the joint, such as tensile or shear strength, elongation, or resistance to peel or cleavage.

Resisting Fatigue. A property somewhat related to the ability to damp vibration is resistance to fatigue. A properly selected adhesive can generally withstand repeated strains induced by cyclic loading without the propagation of failure-producing cracks.

Smoothing Contours. Adhesives usually do not change the contours of the parts they join. Unlike screws, rivets, or bolts, adhesives give little or no visible external evidence of their presence. They are used to join skins to airframes, and they permit the manufacture of airfoils, fuselages, stabilizers, and control surfaces that are smoother than similar conventionally joined structures and that consequently have better aerodynamic efficiency. These structures also have greater load-bearing capability and higher resistance to fatigue than conventionally joined structures. Helicopter rotor blades are now held together only with adhesives and are much more durable than their mechanically fastened predecessors.

Limitations. Adhesively bonded structures must be carefully designed and used under conditions that do not exceed the operational limitations of the adhesive. These limitations include certain types and magnitudes of stresses; whether the stresses are static or dynamic; and environmental factors such as temperature, humidity, and exposure to other vapors or liquids.

Probably all joints made with organic adhesives are adversely affected by the simultaneous application of moisture and mechanical stress. Some adhesives, especially those containing moisture-sensitive components, give particularly poor performance under these conditions. Joints made with such materials may not be capable of sustaining constant loads that produce more than 10% of the normal ultimate failing stress for only short periods. Joints made with other materials may sustain 50% of failing stress for extended periods of time.

The durability of an adhesive joint is influenced not only by adhesive properties but also by the surface preparation of the joined materials. If surface layers are weak and susceptible to moisture, the joint will be weak and susceptible to moisture. The advantages and disadvantages of adhesive joining are summarized in Table 1.

Adhesive Joints

The breaking strength of an adhesive joint is determined by the mechanical properties of the materials of the joint, the extent of interfacial contact (number, extent, type, and distribution of voids), the presence of internal stresses, the joint geometry, and the details of mechanical loading. Breaking strength is also influenced by the mechanical properties of any boundary layers (surface regions or interphases) that are present on the materials that make up the joint or that form in the adhesive during the manufacture of the joint. A uniform stress pattern in an adhesive joint is seldom produced by the application of an external force, and fracture initiates where local stress exceeds local strength. This is not acknowledged in the usual expression of breaking strength— an average determined by dividing breaking force by joint area. Local stresses, which may be many times the average stress and

*Adapted and updated from *Fastening and Joining*, reference issue, Penton Publishing, 1974

Table 1 Advantages and disadvantages of adhesive joining

Advantages	Disadvantages
Provides more uniform distribution of stress and larger stress-bearing area than conventional mechanical fasteners	Requires careful surface preparation of adherends
Joins thick or thin materials of any shape	Relatively long times are sometimes required for setting adhesive.
Joins any combination of similar or dissimilar materials	Limitation on upper service temperature is usually 175 °C (350 °F), but materials are available for limited use to 370 °C (700 °F).
Minimizes or prevents electrochemical corrosion between dissimilar metals	Heat and pressure may be required for assembly.
Resists fatigue and cyclic loads	Jigs and fixtures may be required for assembly.
Provides smooth contours	Rigid process control is usually necessary.
Seals joint, insulates (heat and electricity), and damps vibration	
Frequently faster and less expensive than conventional fastening	
Heat required to set adhesive is usually too low to affect strength of metal parts.	
Postassembly cleanup of parts is not difficult.	

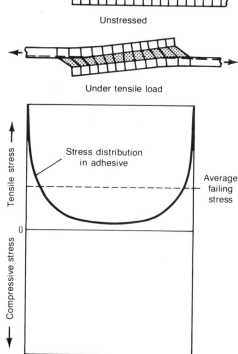

Fig. 1 Lap joint in tension

which determine the actual force that the joint can sustain, are hidden. The designer must compensate for, or minimize, their effects.

Stresses are also produced in a joint by the differential thermal expansion of the adhesive and adherends, either during setting or in service, and by the shrinkage of the adhesive when it changes from liquid to solid. An improperly designed or processed joint can contain so much internal stress that it has little or no ability to bear an external load. Such joints may fail during normal handling.

The Sections "Designing With Adhesives and Sealants" and "Testing and Analysis" in this Volume represent a resource for the types of considerations described below.

External Loads. Adhesive joints may be loaded in any of the three simple modes: tension, compression, or shear. All other loadings are combinations of these three modes. Because adhesive joints are composite structures of materials that usually have different mechanical properties, and because there are certain constraints in such joints, a simple external mode of loading usually produces complex, nonuniform stress. Eccentricities due to finite thickness of joints may produce stress modes that are totally absent in the external load. In addition, a small eccentricity in an external load may cause large stress concentrations in a joint.

Lap joints are probably the most commonly used adhesive test joints. When stressed in shear by tensile loading, they develop a nonuniform stress pattern along the lap (Fig. 1). Peak stresses, which may be several times the average failing stress, develop at the ends of the lap because of two factors, the differential strain induced between adhesive and adherends by the load, and the bending of the joint due to an eccentricity of loading that results from the thickness of the joint.

Eccentricities in loading, which are characteristic of simple lap joints, result in bending loads on the joint. If the bending moment is large enough, the adherends bend and load the ends of the joint in tension normal to the bond line.

Stress concentrations at the ends of the lap give these joints a characteristic behavior. Average breaking stress decreases with increasing lap length, but is independent of lap width. Actual breaking strength increases with overlap length, but reaches a limiting value as overlap becomes greater. Actual breaking strengths of simple lap joints are not proportional to overlap area, nor are the strengths of joints of equal area constructed of identical materials necessarily the same. The strengths of simple lap joints are proportional to the overlap area only at constant lap length, that is, only if the area is varied by changing lap width. Thus, a joint that is 25 mm (1 in.) wide by 125 mm (5 in.) long is weaker than a joint that is 125 mm (5 in.) wide by 25 mm (1 in.) long.

Modifications. Because the average breaking stress of a simple lap joint is determined by the maximum stresses at the ends of the lap, joint modifications that produce a more uniform stress distribution yield stronger joints. Four approaches are practical:

- Reduce the bending of the joint by stiffening the adherends
- Permit the joint to bend more easily by decreasing the thickness of the adherends. This, as in the first method, causes the adhesive to be less strained at the ends of the lap
- Modify the adhesive layer so that it can withstand greater strain under a given load by increasing the toughness and/or thickness of the adhesive
- Decrease the length of the overlap

A rivet or other mechanical fastener near the end of the lap can also moderate the effects of end stresses on the joint.

Preparation of Materials to Be Joined

Surface preparation of the adherends (materials to be joined) is necessary prior to the application of an adhesive. The treatment may range from a simple solvent wipe that is completed in seconds to a multistage cleaning and chemical treatment requiring 30 min or more.

The purpose of prebond treatments is to remove existing adventitious surface layers from the parts to be joined and to replace them with sound surface layers known to be suitable for the application. If a layer of grease or oil on an adherend is not removed, or if it is not absorbed and dispersed by the adhesive, it will affect the performance of the joint. Similarly, a thick or weak oxide layer on a metal adherend lowers joint strength and durability. It may be desirable to treat the surface, replacing an existing layer with a thinner and/or stronger oxide layer and/or one with different microroughness characteristics.

Metals. A solvent wipe or vapor degrease is a minimum prebond treatment. This treatment is also a prerequisite to more extensive surface preparation. The minimum treatment is usually satisfactory only for noncritical assemblies or those that will not be subjected to high stress or adverse environments.

Initial joint strength is not a sufficient indicator of the durability of a joint in service. Joints made from minimally prepared metal adherends may have the same initial breaking strengths as those made from similar adherends prepared with the most elaborate chemical treatments. Joints made with minimally prepared adherends, however, are markedly inferior in durability when exposed to adverse environments.

Chemical surface treatment for metals generally consists of immersing the degreased part in one or more aqueous solutions at room or somewhat elevated temperature, washing the part in water to remove the solution, and drying. Treatments may be strongly acidic and oxidizing, for example, a sulfuric acid/sodium dichromate solution for treating aluminum, or strongly alkaline, for example, the alkaline hypochlorite solution used for copper.

Ceramics and glasses are usually cleaned by immersing in a hot (60 °C, or 140 °F) glass-cleaning solution, consisting of an aqueous solution of sulfuric acid and sodium or potassium dichromate, and then rinsing with water and drying. This treatment ensures a more complete removal of organic materials, such as oils and greases, than is possible with solvent cleaning.

Thermoplastic polymers generally require surface treatments. A formulated polymer may contain migratory plasticizers, internal or external mold release agents, antioxidants, and so forth, any of which can cause weak boundary layers that complicate treatment.

Possible prebond treatments for highly crystalline polyolefins include flaming the surface, treating it in a "cold" plasma, permitting a corona discharge in air to impinge on the surface, and immersing the parts in glass-cleaning solution, as described above. Such treatments alter both the chemical and the mechanical nature of the surface region of a polymer.

The normal treatment for polytetrafluoroethylene fluorocarbon (Teflon) is immersion in a dispersion of sodium in an organic solvent, usually tetrahydrofuran, which apparently removes fluorine from the surface and produces a primarily carbonaceous surface layer.

Thermoset polymers, such as epoxies, phenolics, and styrene polyesters, behave somewhat better and are easier to bond than are the thermoplastics. Generally, all that is required is an abrasion of the surface, followed by a solvent wipe to remove debris, except in the case of silicones, which may require, for example, a "cold" plasma treatment. A state-of-the-art treatment for thermosets is the use of a scanning infrared or ultraviolet laser to ablate the surface, rather than using mechanical abrasion.

Other Materials. The Section "Surface Considerations" in this Volume comprises articles that focus on each of the materials identified above, as well as articles on electronic and biological materials, primers and coupling agents, and surface analysis techniques and applications.

Types of Adhesives

Adhesives, as already noted, do not perform as independent materials; rather, they are always part of a structure. They gain or lose effectiveness according to the environment, the bonded materials and structure, and the details of loading. Thus, tabular selection guides, which provide cookbook solutions to bonding problems, can, at best, eliminate grossly unsuitable materials from consideration; at worst, they can be misleading.

Thermosetting adhesives are available primarily in three forms: liquid, paste, and solid. The liquid is generally a free-flowing, one- or two-part system that may or may not contain solid fillers and/or colorants. Such systems may be so-called 100% solids systems, in which the contents are already solid or are completely reactable to a solid, or they may contain a nonreactive, volatile solvent or dispersant that is necessary for the existence of such a system or that imparts certain application characteristics. Occasionally, one component of a two-part system (usually the curing agent) is a solid, which requires that the two components be mixed at a higher temperature than the melting point of the solid.

A distinction is made between the liquid and paste forms (although both forms are fluids), because the flow behavior of the pastes is different from that of the unfilled or lightly filled liquids. Certain pastes are modified to be thixotropic (gel-like at rest, but fluid when agitated), so that an unrestrained mass of adhesive does not sag or flow unless forced to do so by mechanical action (for instance, spreading with a spatula) or so that it does not flow out of a vertical joint during the assembly and cure of a structure.

Thermosetting adhesives are also available in supported or unsupported films of various thicknesses. Films are flexible and can be cut or punched to shapes that conform to a particular joint. The ease of handling and lay-up of these materials and the lack of required cleanup required with liquids and pastes are their main advantages. The refrigeration of certain film-type adhesives increases storage life as is the case with all one-part materials. The refrigeration of film types that cure at room temperature is mandatory.

Thermoplastic adhesives are, in general, available in the same forms as are the thermosets, with the possible exception of pastes. The liquid form may be a solution and/or dispersion of the base thermoplastic polymer and other compounding ingredients in a volatile vehicle, or they may be 100% solids systems containing liquid monomer, prepolymer, and catalysts, to which is added an accelerator that induces polymerization to a solid, high molecular weight polymer.

Thermoplastic adhesives are also available in solid form (primarily, unsupported or supported film); in granules; and in lengths of extruded, flexible, cordlike materials wound into coils suitable for automatic application. The thermoplastic film adhesives are converted into the fluid form by either solvent or heat activation. Solvent activation is only suitable in applications in which one or both adherends permit the release of the solvent, for example, by diffusion. Heat activation is used when the adherends are not permeable and are able to withstand the required temperatures. In those compounds containing thermosetting resin, the heat cures the resin. These activation techniques are also used to join substrates, one or both of which are coated with an appropriate solvent-base adhesive that has been dried to a tack-free state.

Thermoplastic adhesives that do not contain any heat-reacting resin and which are heated to render them fluid for joining are called hot-melt adhesives. They are especially suitable for use with high-speed application machinery.

Other Types. Some adhesives have unique cure mechanisms. One-part anaerobic adhesives, which remain fluid in the presence of oxygen but cure to solids in its absence, have strength characteristics equivalent to those of the structural adhesives. One-part anaerobic adhesives cure at room temperature in several hours, but their cure can be accelerated to a matter of minutes at elevated temperatures by the use of certain primers/accelerators.

The cyanoacrylates cure in an extremely short period (sometimes 10 to 15 s) if confined in a thin film between close-fitting parts. The mechanism of cure is an anionic polymerization induced by the presence of a base. The thin film of moisture usually present on surfaces exposed to normal atmosphere is generally sufficient to harden these materials if they are squeezed to a thin film.

Adhesives that cure by exposure to ultraviolet light and others that cure by exposure to an electron beam are also available. Still others cure in the presence of moisture.

The Section "Adhesives Materials" in this Volume provides an overview of adhesive types and detailed descriptions of chemical families, advanced polymeric systems, and adhesive modifiers.

Selection of Adhesives

A number of factors upon which selection should be based are described below.

The type and state of an adhesive, its working properties, and the temperature and time required for its cure influence the choice of process for producing a structure. These factors, and others pertaining to the material and form of the structure, influence, specifically, the method of application and setting of the adhesive, as well as the assembly and fixturing of the structure.

Use. Whether a part to be assembled represents a production item or a prototype, or is still in a stage of development, may considerably influence adhesive selection. The cost of the adhesive may sometimes be cause for rejection in a particular production situation. Because the application of heat and/or pressure is not usually convenient or even available in the field, adhesives that are needed for field repair usually must perform without either. The Sections "Manufacturing Use" and "Bonded Structure Repair" in this Volume provide valuable information that relates to adhesives selection.

Application. Although the major function of adhesives is mechanical, that is, to fasten, sometimes they are also required to seal and insulate. As already stated, formulations that are good electrical and/or thermal conductors are also available. Further, adhesives prevent electrochemical corrosion in joints between dissimilar metals and resist vibration and fatigue. In addition, unlike mechanical fasteners, adhesives do not generally change the contours of the parts that they join. The reader is referred to the Section "Designing With Adhesives and Sealants" in this Volume for additional details.

Materials to Be Joined. Materials to be joined (adherends) influence the choice of adhesive with respect to their mechanical and physical properties, as well as the surface preparation they require prior to joining. Flexible materials should not be joined with a hard, brittle adhesive, nor should critical shapes be joined with a solvent-base adhesive that would tend to distort them. However, an adhesive that lightly attacks one or both of the materials to be joined may be used to reduce normally required surface preparation. The Section "Surface Considerations" in this Volume should be consulted.

Type of Joint. The performance of an adhesive in a joint is very much a function of the type of joint and the mode of loading. Joints that provide a more uniform stress distribution between members are generally stronger and more reliable than those that produce stress concentrations. For this reason, joints are better when loaded in shear and compression than in peel and cleavage. Tough adhesives respond more favorably and resist fracture better than hard, brittle materials under high strain conditions. It is highly desirable that joints be designed specifically for adhesive bonding. Direct conversion of a conventional, mechanically fastened joint to adhesive bonding is usually a poor idea. The Section "Designing With Adhesives and Sealants" consists of articles that are useful when considering the type of joint, as well as process limitations and mechanical requirements.

Process Limitations. Jigging and holding requirements of the parts during the setting of the adhesive can dictate the adhesive choice, as can the tolerance of fitted parts. For example, loose-fitting parts may require a gap-filling adhesive, whereas close-fitting parts may require a low-viscosity material. Limitations may be placed on the temperature and/or pressure that can be used because of temperature- or pressure-sensitive elements in the assembly to be joined.

Mechanical Requirements. In adhesive selection, the user must determine not only the nature and magnitude of the load to be imposed on the joint, but also whether the adhesive is required to hold temporarily or permanently. An adhesive may be required to hold parts in juxtaposition only long enough to be joined by other fastening methods or placed into another structure.

Load data should be obtained from test specimens that correlate in some known way with the actual joint to be used. Because this is seldom the case with data from the manufacturer, the user is sometimes forced to gather his own. Mechanical properties of the adhesive itself are seldom, if ever, tabulated; only the responses of standard joint specimens to various test procedures are tabulated.

Service Conditions. Maximum and minimum service temperatures influence adhesive selection because adhesives are temperature-sensitive materials. Although some available adhesives can withstand 370 °C (700 °F), a limitation of 65 to 95 °C (150 to 200 °F) is more usual for most types, even thermosetting adhesives, because strengths fall at temperatures considerably above this range. Low service temperatures, on the other hand, lead to the embrittlement of many adhesives and internal stressing of most joints.

Most, and probably all, organic adhesives are adversely affected by moisture, especially when in a stressed condition. However, performance under conditions of moisture varies considerably among adhesives. The effects of any fluids present, such as oils, solvents, or hydraulic fluids, must be considered. Design life under anticipated service conditions is also a factor in selection. See the Section "Environmental Effects" in this Volume for pertinent articles.

Influence on Associated Components. This factor assumes particular importance in closed systems, where parts may be subject to corrosion, gumming, or other damage by contact with either the adhesive or volatile substances that exude from it. Adhesives used with printed circuits and similar electrical applications must not corrode copper conductors or other associated components. Sometimes the adhesive mixing, application conditions, and curing conditions can influence its potential to corrode.

The Joining Process

Methods of Application. Common methods of applying liquid adhesives are by brush, spatula, trowel, dip, spray, roll, curtain, flow gun, and flow brush. The last five methods are particularly useful for the uniform application of adhesive on large, flat parts in high production rate situations.

Dip coating and spraying can be used for flat parts, but are especially suitable for contoured parts. A widely used method for applying liquid and thin-paste adhesive is brushing. Equipment is simple, waste is minimal, and limited areas of contoured shapes can be coated without masking. However, it is difficult to achieve uniform, high production rates, as well as uniform adhesive thickness, using this method.

If several coats of a solvent-base adhesive are to be applied, the most uniform thickness is obtained by applying each coat perpendicular to the preceding one. Proper drying times must be allowed between coats to avoid sagging, blistering, or lifting of earlier coats.

Film adhesives are relatively easy to apply. They are cut to size, placed in the area to be joined, and tacked in place at selected areas that are softened with a heat gun or soldering iron prior to assembly.

Another method of assembly is contact bonding. Certain materials, although dry and tack-free to the touch and unable to form a strong joint to many surfaces, will form a strong joint upon contact with a tack-free surface of the same or similar material. Such a joint is formed instantaneously and is completed by a brief application of pressure to ensure a good contact between the layers of adhesive.

Joints involving a thermosetting liquid or paste adhesive are made by a wet-assembly technique. Usually the parts must be held in position by fixtures or clamps until the fluid adhesive is cured. The same technique is used for polymerizing thermoplastic adhesives. Solvent-dispersed adhesives should be laid up wet only if solvent release is possible. Sometimes, to reduce the time required for solvent release after the joint is assembled, the wet-coated adhesive is allowed to dry partially before assembly.

Jigs and fixtures may be necessary for holding the parts together and for maintaining pressure on the adhesive during cure. The particular adhesive and the configuration of the assembly determine whether fixturing is necessary and its degree of complexity. Two flat sheets joined with a contact bond adhesive might not need any fixturing. A complex equipment cabinet or aircraft wing to be constructed to close dimensional tolerances with a high temperature curing adhesive may require many relatively complex fixtures. Often, thoughtful design of the product can markedly reduce the complexity of fixtures and simplify the maintenance of required dimensional tolerances.

During curing, most adhesives need follow-up pressure as a minimum. Spring clamps, or weights such as bags loaded with sand or shot, or inflatable pressure pads or tubes can be used. An alternate method, which provides uniform pressure on contoured surfaces, is to place the parts in a rubber blanket or bag and evacuate the bag. For flat parts, a hydraulically or pneumatically actuated platen press is useful.

Curing. The most important parameters relating to the cure of an adhesive are time and temperature. The time required for the process of curing an adhesive must be longer than the actual time required for cure so that time is allowed for the adhesive mass to reach cure temperature.

The type of heating apparatus also determines the processing time. For example, a forced-air recirculating oven transfers heat to an assembly more efficiently than a gravity-circulated oven of equivalent heat capacity.

If the parts of an assembly are of the same material and thickness, all bond lines should heat at equivalent rates. However, if thick and thin parts are being joined, or if the parts differ markedly in heat capacity or conductivity, not all bond lines will reach cure temperature simultaneously. The processing time for such assemblies must be adjusted so that the adhesive mass in the slowest-heating joint achieves proper cure.

The same consideration applies to applications in which a fixture is massive, compared with the portion of the assembly that it contacts, especially when certain film-type adhesives are used. A certain, but not necessarily critical, rate of temperature rise is necessary to enable the adhesive to become properly fluid and to wet the adherends before it sets. Too low a rate might cause the material to remain fluid for so long a period that it runs out of the joint. Too high a rate might cause the material to set to a solid state before it has had sufficient time to wet the adherends adequately.

Another factor to consider is the differential thermal expansion between components of an assembly, as well as the stress that may result from it. Differential expansion may also occur if the materials have different rates of heating or cooling. Such problems can often be alleviated by temperature staging the cure cycle, with the time at each stage being sufficient to ensure that all parts of the assembly reach thermal equilibrium. Cooling to ambient conditions is likewise staged.

During the construction of large or complex structures, it is often convenient to fabricate subassemblies and then bring them together to form the complete structure. The subassembly adhesive must, of course, be compatible with final assembly procedures and processes. The Section "Manufacturing Use" in this Volume expands on the various facets of the joining process.

Testing Adhesive Joints

Either destructive or nondestructive methods can be used to test adhesive joints or assemblies, but destructive testing is probably used more widely. The Sections "Testing and Analysis" and "Nondestructive Inspection and Quality Control" in this Volume augment the information summarized below.

Destructive Testing. Three types of specimens may be used for quality control by destructive testing: First, standard test specimens (such as lap-shear specimens or one of the many types of peels) that may have no exact mechanical counterpart in the finished assembly; second, specimens that are representative of or are exact mechanical analogs of one or more joints in the assembly; and third, finished assemblies.

The first type of test is widely used, primarily because of simplicity and low cost. Standard test specimens are processed, assembled, and cured with each assembly or lot of assemblies, so that they (and the assemblies) are subjected to the same set of conditions. These tested specimens give a measure of whether the bonding process is in control. They are part of the total information upon which decisions to accept or reject finished assemblies are based. The same procedures and comments apply to the second test method.

The third option is costly if the assembly is large, complex, or expensive. It is, however, the preferred method when economics permit. A statistically significant number of assemblies must be tested.

Nondestructive Testing. Present-day nondestructive testing (NDT) methods, while not adequate for determining the degree of success of the bonding operation, can indicate potential joint performance problems. They do not generally provide absolute indications of joint performance. Common NDT methods are described below.

Proof loading, which tests the ability of a structure to sustain the same type of load, but of greater magnitude, that it will bear in service is a useful form of NDT. Although the magnitude of the proof load determined during testing in the design or prototype stage is usually much higher than the anticipated service load, the proof load must not be so high that it causes irreversible or permanent deformation or damage to the structure.

Tapping, or sonic testing, operates on the principle that solid or supported areas of a joint, when tapped, emit a sound different from that of areas containing voids. The test can be performed by an experienced individual with a coin or, in a more objective way, by appropriate instrumentation.

Ultrasonic testing is the most commonly used instrumental technique for NDT. Although there are many variations of the technique, all require the buildup of an extensive data base that correlates instrumentally obtained parameters with the results of destructive tests on joints of identical materials, geometry, and processing. If the materials, processing, or conditions of destructive testing change, the correlations must be reestablished.

Visual inspection of the exterior of a joint is used to determine whether a joint contains an adequate amount of adhesive and whether the adhesive appearance and, with probing, the feel rule out a poor state of cure. A continuous bead of adhesive should be visible around the edge of the joint.

Cores may be taken from an assembly for testing in such a manner that the structural integrity of the assembly is not affected. Alternatively, small annular sections of a joint may be removed to produce disk-shape specimens that may be destructively tested in tension or shear. The resulting hole is then plugged with an appropriate material. This method is not, strictly speaking, nondestructive, because part of the bonded area is destroyed. However, if properly applied and used, this procedure should not measurably affect the structure.

Influence of Adhesives on Surroundings

When adhesives are confined with critical components in a closed container or when they contact fine wires or metal films whose dimensions and integrity are critical, the effect of the adhesive on these materials or components must be considered. The adhesive may contain volatiles, which, in an evacuated or sealed container and/or in the presence of heat, may evolve and condense on critical components and cause malfunctions. Examples of such occurrences are the fouling or pitting of gyroscope bearings and the creation of high-resistance contacts on sealed relays.

Another example of deleterious effects is the corrosion of some metals or alloys by particular adhesives. Epoxy adhesives with certain amine curing agents can be extremely corrosive to copper. This is especially true when adhesives bridge a pair of polarized conductors in the presence of moisture. Certain one-part silicone adhesives also exhibit this undesirable behavior.

Failure and Repair

Partial or total failure of one or more joints in an assembly can occur in handling, in proof testing, or in use. Assuming that a design has been proved adequate to withstand the rigors of handling, shipment, and use and that it has passed inspection, if it fails in use, the following points, among others, should be investigated:

- Was the structure subjected to stresses (environmental and/or mechanical) that exceeded design limitations?
- Was the joint failure cohesive and was it in the adhesive or in the adherend?
- Was the failure apparently in adhesion, indicating inadequate prebond treatment and/or the presence of a weak boundary layer?

An investigation of these and similar questions can uncover possible reasons for failure, determine whether the inadequacy may involve other joints in the structure, and determine whether the failure is limited to one particular structure or involves similar structures produced concurrently. Obtaining the answers may involve a complete and thorough examination and review of processing, application, assembly, and cure procedures. It is difficult, if not impossible, to give specific step-by-step instructions for the assessment and correction of failures in adhesive joints that cover all possible situations. Considerable reliance must be placed on the judgment of experienced personnel. The Section "Bonded Structure Repair" in this Volume provides practical guidelines.

Sometimes the repair of a failure must be accomplished on an assembly in the field. Such a repair can generally be carried out using either adhesives or conventional mechanical fasteners, or a combination of the two. Assuming that the joint was properly designed, it is preferable to repair it with an adhesive or adhesive-fastener combination, if the joint is accessible.

The repaired joint may not have the design strength of the original joint, because the repair adhesive will differ from the original adhesive, which was processed under controlled conditions. In addition, prebond treatments that are satisfactory for field use (assuming that such treatments can be implemented) do not, in general, produce the optimal surface preparation for adhesive joining.

REFERENCE

1. "Standard Terminology of Adhesives," D 907, *Annual Book of ASTM Standards*, American Society for Testing and Materials

Adhesives and Adhesion

Denis J. Zalucha, S.C. Johnson Wax

ADHESIVES, which have existed for thousands of years, have become much more important in the past few decades. The growing availability of a variety of new materials and significant advances in bonding technology have allowed people to routinely trust their lives and fortunes to adhesively joined constructions. The study and understanding of adhesives and the art and science of using them has never been more pertinent.

For any discussion to be productive and informative, it is important that the language be defined so that the words carry the same meaning to all the participants. In discussions of adhesives and adhesion, this can be critical because common words are often used to describe specific concepts that might be quite different from their casual meaning. Figure 1 shows an idealized adhesively bonded assembly. The terminology, as defined in ASTM D 907-89, has the following meaning:

- *Adherend*: A body held to another body by an adhesive
- *Substrate*: A material upon the surface of which an adhesive-containing substance is spread for any purpose, such as bonding or coating. A broader term than adherend
- *Primer*: A coating applied to a surface, prior to the application of an adhesive, to improve the performance of the bond
- *Adhesive*: A substance capable of holding materials together by surface attachment

The term "surface region" is discussed in detail in the section "Surface Parameters" in this article.

Basic Concepts

Adhesion

People generally have an intuitive, operational definition of adhesion, based on simple destructive tests. The difficulty with which the adhesively bonded assembly comes apart determines their opinion of the quality of adhesive used.

Such a definition often works well enough. However, there are constructions and situations in which "good" and "bad"

trade places. For example, masking tape can be put on a wall, painted over, and stripped off cleanly and easily; if the adhesive is considered at all, it is probably viewed as a "good" one because it came off easily. However, when recalling efforts to remove masking tape from something after it has been exposed to sun and weather for a few days, the tape is probably viewed as a bad adhesive because it is difficult to remove. A stubborn bumper sticker is another example of a strong but bad adhesive.

These situations illustrate both sides of the coin. In each case, what is being evaluated is much more than one adhesive or the interface between that one adhesive and one adherend. The bending, twisting, pulling, and stretching of an assembly tests the mechanical properties of the adhesive and of the substrates it joins. This allows a comparison of the cohesive strengths of the adhesive and adherends and the adhesive strengths of several interfaces. Ultimately, the behavior of the assembly is rated against the expectations of its behavior. The masking tape is expected to unroll easily but not so easily that the roll telescopes. It should adhere tightly to the wall for as long as needed and then should strip off easily and cleanly. The adhesive is a constant. What changes are the adherends that are combined with the adhesive, the magnitudes of the forces, the directions in which the forces are applied and the rates of deformation of the assemblies, and the conditions applied to the total composite.

In a practical sense, that which is called adhesion is not an intrinsic property of any adhesive or polymer, but rather the response of an adhesively bonded assembly to some destructive deformation. Perceptions of good and bad adhesion depend on from what the assembly is made, how it is put together and tested, and on the expectations of its response to the test conditions. The parts of the assembly can be spoken of separately—the adhesive, adherends, primers, surface treatments, and so on—but each must be considered as part of the whole because it is the relationship of the properties of each to those of the others that determines the performance of the assembly.

Although a huge number of adhesives are available today, each with its own characteristics, all share several important attributes:

- Through surface attachment only, they transfer and distribute load among the components of an assembly
- They must behave as a liquid, at some time in the course of bond formation, in order to wet the adherends
- They carry some continuous, sometimes variable, load throughout their lives
- They must work with the other components of the assembly to provide a durable product that offers resistance to degradation by elements of the environment in which it is used

Such attributes as load-carrying ability, wetting rate, durability, and lifetime, among others, can vary over several orders of magnitude, depending on the adhesive and on how the assembly is designed, constructed, and used.

The two attributes of adhesives listed above—ultimate load-bearing ability and

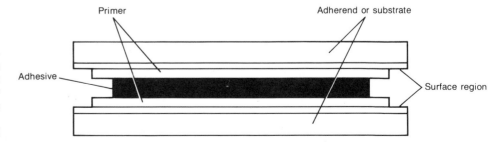

Fig. 1 Idealized adhesively bonded assembly

initial low viscosity to facilitate rapid wetting—help define which materials are usable as adhesives and how they can be used. In a broad definition, even solders or brazing alloys might be considered adhesives. Practically, though, only organic or silicone polymers are usually considered adhesives.

For rapid wetting and bond formation, the adhesive must be fluid, but in its final load-bearing state, it must be a high molecular weight solid. In the three ways to accomplish this, the adhesive is applied:

- In a molten state and solidified after the components of the joint are assembled
- As a solution or dispersion and the carrier liquid allowed to evaporate, leaving the high molecular weight polymer behind
- As a low-viscosity fluid containing reactive groups that undergo polymerization in the bond line to build the molecular weight sufficiently to carry a load

Merely having the adhesive in a liquid state is not sufficient to assure that it will wet and adhere to the substrate to which it is applied. Thermodynamic criteria for wetting and spreading have been investigated and a detailed review is given by Kaelble (Ref 1). Zisman and co-workers (Ref 2) have developed an empirical method for determining a critical surface tension of solid surfaces by extrapolating contact angles, the angle formed by a droplet of liquid on the surface where the solid, liquid, and air meet. If the surface tension of the liquid adhesive is less than the critical surface tension of the solid surface, the contact angle will be zero and the liquid will spread spontaneously and wet the surface. Surfaces are sometimes tested for the presence of oily contaminants by water break tests, where a clean (bondable) surface is indicated by the formation of a continuous film of clean water on it.

Time-Temperature Superposition

Pressure-sensitive adhesives (PSA), such as those used on tapes and labels, may appear not to fit any of the schemes above. However, they might actually be considered a special case of the first method, applied as a melt, where temperature has been replaced by time. Pressure-sensitive adhesives are high molecular weight, soft polymers having a high degree of chain segment mobility. The shear rates typically involved in applying adhesive-backed tapes by hand, for example, are low, and, on that time scale, the chain segments can move around and come into intimate contact with the adherend surface. In other words, the adhesive behaves as a liquid at deformation rates slower than the rate at which the segments can move (corresponding to high temperature). At high deformation rates or high peeling rates, analogous to lower temperature, the chains do not have time to

move or disentangle, and the adhesive behaves more like a solid. The apparent strength of the bond is a function of the rate at which the bond is tested to destruction, for example, by peeling.

The wetting behavior of pressure-sensitive adhesive is a specific example of the more general principle of time-temperature superposition. Williams, Landel, and Ferry (WLF) (Ref 3) have shown that under many conditions, for amorphous polymers less than about 100 °C (180 °F) above their glass transition temperatures (T_g), all mechanical and electrical relaxation times at some specific temperature can be related to their relaxation times at some chosen reference temperature. If T_g is chosen as the reference, and if a_T is the ratio of a particular relaxation time at temperature T to its value at T_g, then:

$$\log a_T \approx -17.44(T - T_g)/(51.6 + T - T_g) \quad \text{(Eq 1)}$$

This expression is commonly known as the WLF equation and describes this interchangeability of time and temperature. Heating a material is equivalent to waiting a long time; conversely, straining a polymer quickly is analogous to testing it at a lower temperature.

Bonding Mechanisms

Once the liquid adhesive has been applied to the adherends and intimate contact and wetting have been established, there are only three mechanisms that can be postulated to explain the adhesion of one material to another: mechanical interlocking, formation of direct chemical bonds across the interface, or electrostatic attraction.

Mechanical interlocking can certainly be a significant factor in adhesion. Ice on glass is a common example of adhesion by mechanical interlocking. Another is the mechanical adhesion of silver amalgam dental fillings to teeth. In general, though, mechanical interlocking is probably not a practical factor in most adhesive joining, although the roughness and porosity of a substrate and their effect(s) on stress distribution in the interfacial region can be a significant factor.

The formation of covalent chemical bonds across the interface is at the other extreme. This type of bonding would be expected to be the strongest and most durable. It does require that there be mutually reactive chemical groups tightly bound on the adherend surface and in the adhesive. There is evidence that such bonds do form and silane coupling agents are one example of using specific reactive groups to promote the formation of direct chemical bonds. The adhesion of solders to metals is probably enhanced by the formation of metallic bonds across the interface.

Electrostatic Attraction. By far the dominating adhesion mechanism, especially in

the absence of reactive groups, is the electrostatic attraction of the polar groups of the adhesive to the adherends and to each other. These are primarily dispersion forces (London forces) and forces arising from the interaction of permanent dipoles. These forces provide much of the attraction between the adhesive and adherend and contribute significantly to the cohesive strength of the adhesive polymer. Dispersion forces result from the interaction of *transient*, or virtual, dipoles present in all atoms, regardless of charge, and act over a very short range because the force decreases with the 6th or 7th power of the distance of separation. The attraction between *permanent* dipoles decreases with the 6th power of the distance of separation.

Because neither of these forces is significant beyond about 0.5 nm (5 Å), the need for intimate contact between adhesive and adherend becomes obvious. In a two-part article on the cohesive and adhesive strengths of polymers, H.F. Mark (Ref 4, 5) presents simple calculations of what hypothetical bond strengths might be expected from covalent bonds and polar forces. He proposes that even when there is a contribution from covalent bonding across the interface, the failure of polar bonds at lower load levels concentrates stress at the covalent bonds, causing these to fail also. Thus, the calculated mechanical strength of a partially covalent interface is still dominated by the polar forces.

Polar forces also play a role in the interdiffusion and entanglement of polymers. They provide a significant amount of the cohesive strength of the adhesive polymer. Where the adherend and adhesive can interdiffuse, the electrostatic forces provide the "friction" to prevent the polymers from simply sliding past one another and disentangling when the joint is loaded.

Practical Considerations

Classification of Adhesives

There are probably as many ways to classify adhesives as there are adhesives: by chemistry, use, application method, cure method, reactivity, market, rheology, lifetime, performance, origin, and others. Sometimes it is even difficult to determine what is an adhesive, depending on the use. Adhesives, sealants, and binders for composites have much in common.

The most common classifications involve a combination of origin and chemistry, such as animal and fish glues, vegetable glues, proteins, epoxy, urethane, cyanoacrylate, and so on. Skeist (Ref 6) describes 34 such classes; additional articles classify adhesives according to use, adherend, industry, or solvent. The lists are certainly not exhaustive. No author of the many handbooks available has succeeded in concocting a rational scheme for classifying all adhesives.

The broadest classifications are based on what needs to be done, and are then modified by how to accomplish it. This leads to three main adhesive classes: structural, laminating, and pressure sensitive. Pressure-sensitive adhesives are usually grouped separately because of their unique features, even when their use might place them in one of the other two classes. The main adhesive classes are:

- *Structural.* Loads can be large, sometimes several thousand pounds per square inch, and the bond area is small compared with the total surface area (automotive body sheet, aircraft assembly, furniture)
- *Laminating.* Loads are usually small and the bond covers a large fraction of the surface (plywood, multilayer packaging, labelling, decals)
- *Pressure-Sensitive.* Usage could be structural or laminating, but the easy application and rapid bond formation of the PSA is the main advantage. Usually, the loads are small, and often the bonds are temporary

Sometimes adhesives are classified according to performance. Remembering that other rankings and classes are possible, this is an example list of adhesives ranked by performance, with those considered high-performance first:

- *Hybrids*: Rubber-toughened epoxy, silicone, high-temperature polymers
- *Thermoset polymers*: Epoxy, phenolic, polyester, acrylic
- *Rubbery thermosets*: Polyether urethanes, polyester urethanes
- *Thermoplastics*: Polyamide, polyester
- *Elastomers*: Neoprene, nitrile rubber, styrene-butadiene block copolymers
- *Polyolefins*: Natural rubber, polybutadiene, polypropylene
- *Natural glues*: Casein, hide glue, fish glue, vegetable glue

Performance is defined as the ability of the adhesive to meet the needs of the product and expectations of the user. High-performance adhesives are able to carry larger loads and contribute to durable constructions that resist aggressive environments or are able to withstand some particular exposure extreme. These are the adhesives to which people trust their lives and fortunes. The silicone adhesive used to bond insulating tiles to the space shuttle can be considered high performance, although its bond strength is not necessarily high.

With all the choices possible, adhesive selection can become a bewildering task. Therefore, when selecting an adhesive for a particular application, it is usually easiest and most efficient to consider what the adhesive must do and to work from there.

This does not limit one to a particular adhesive chemistry or type and offers a better opportunity to select the adhesive most appropriate to the product desired and the assembly processes available. Again, it is important that the materials, processes, and their relationships be considered simultaneously. It is best to bring together, as early as possible, the user, potential adhesive suppliers, application equipment makers, and perhaps even the adherend suppliers to assure product success with few surprises.

Surface Parameters

Because adhesives must function by surface attachment only, the nature and condition of the adherend surfaces is critical to the success of any bonding venture. Surface usually refers to that portion of the adherend with which the adhesive interacts. For a freshly cleaved single crystal, this interaction region might be only one or two atomic layers. For a porous surface, such as wood or phosphoric acid anodized aluminum, a low-viscosity adhesive might reach a depth of several hundred nanometers or more. A rough surface provides more surface area (and the ability to modify stress distributions) than a smooth one of the same gross dimensions. In addition, the surface will usually have a chemical composition different from that of the bulk composition. In some cases, the surface composition has little in common with the bulk composition, owing either to adsorption of contaminants from the environment or segregation of bulk constituents at the surface. The outer layers can be loosely bound or tightly adhered and might themselves have high or low cohesive strength. Contaminants and segregated bulk components are not always homogeneously distributed in the surface region, and inhomogeneities can lead to potential corrosion sites, weak bond areas, and bond discontinuities that can concentrate stress.

The surface of metals, such as steel or aluminum alloys, might consist of several regions having no clearly defined boundaries. Moving outward from the bulk metal, there is likely to be a region that is still metallic, but has a chemical composition different from the bulk due to segregation of alloying elements or impurities. This is known as the segregation layer. Next, there is probably a layer of mixed oxide of the metals, followed by a hydroxide layer, and probably an adsorbed water layer. In addition, there will be contaminants adsorbed from the atmosphere, which might include compounds that contain sulfur, nitrogen, halogen, or others, depending on local air and rain pollutants and how the metal has been stored. There could also be processing aids, such as rolling oils, cutting lubricants, drawing compounds, and corrosion inhibitors. Finally, the mechanical working of the metal will probably have mixed all these

regions together, to some degree, into something of a nonhomogeneous surface frosting.

The situation with organic adherends such as wood or plastics is no better. Wood is inherently anisotropic and will vary widely in surface roughness, pH, porosity, and moisture content, even within a single sample or with time. The presence of sap, pitch, resins, or preservatives will also affect bondability.

Engineering plastics might be anisotropic as a result of the manufacturing or forming process, where some flow or "machine direction" is introduced. This is usually the case with fiber-reinforced sheets, such as those used in the automotive industry. Careful control of fiber orientation is sometimes used to enhance mechanical strength in a particular direction. Because components of the plastic can migrate to the surface, it is common to find low molecular weight polymers or oligomers, plasticizers, pigments, mold release agents, shrink control agents, and other processing aids, as well as adsorbed contaminants, in the surface region.

Plastic films will have a machine direction or some anisotropic polymer orientation if they are blown or otherwise stretched in the manufacturing process. These also can have low molecular weight fragments, oxidized polymer, plasticizer, processing aids, lubricants and slip aids, adsorbed water and organic contaminants, and other surprises in the surface region.

More so than with metals, the surface regions of plastics are dynamic, and thus continuously establish new equilibria internally and with their surroundings. In flexible, amorphous plastics, at temperatures above T_g, low molecular weight components are able to diffuse out of the bulk and into the surface region, while elements of the surroundings can diffuse into the plastic. A well-known example of the latter is the migration of plasticizer from vinyl film, causing the film to shrink and become brittle. If the film is part of a bonded construction, the plasticizer can migrate into the organic adhesive layer and change its mechanical properties. If the adhesive is an effective barrier to plasticizer migration, the plasticizer can build up in the interfacial region, causing bond failure.

The nature of a plastic surface can change rapidly in response to its surroundings. Even when the bulk of a material is in the glassy state, the surface region can be quite mobile because of the presence of low molecular weight polymer and contaminants. Polymers having polar and nonpolar regions in the chains can present different segments at the surface, depending on whether the surroundings are polar or not. Wiping a surface with an ionic solution will cause the polar groups to orient toward the surface, while the same treatment with a nonpolar solvent, such as hexane, can bring the non-polar aspect to the surface.

It is clear that for any adherend, the surface with which the adhesive must interact is an ill-defined, dynamic region whose nature depends on the bulk composition, history, and environment of the sample. The adhesive composition or morphology is not well defined in the interfacial region, either. Low molecular weight components can be drawn into porous materials, leaving higher molecular weight polymer and fillers behind. Compounds at the adherend surface can catalyze or inhibit polymerization. Solvents in the adhesive can extract components of plastics or cause the surface to swell or dissolve.

The interphase between the adherend and adhesive is complex, and its composition is usually unknown. However, there are ways to better control that interphase and to improve its reproducibility and reliability. Surface treatments and primers are valuable in providing more controlled bonding conditions.

The proper choice of surface treatments and primers is as important to the outcome as is the choice of adhesive and adherend. All must be considered simultaneously because success depends on their proper interaction. Surface treatments are operations that change the morphology or composition of the adherend surface and may involve adding, as well as removing or rearranging, material. Primers are usually thin, organic coatings applied to the surface (which might also have been treated); the primary reason for treating or priming is not necessarily to improve bond strength but rather to produce a controlled, reproducible, and durable surface in order to obtain a more predictable adhesive bond. A lower bond strength that remains constant and predictable over a long life is preferable to a high bond strength that decays rapidly.

Some of the more common surface treatments are:

- No treatment (low cost, poor reproducibility)
- Solvent wiping
- Vapor degreasing
- Mechanical abrasion
- Plasma treatment
- Etching or pickling
- Chemical conversion coating
- Phosphoric acid anodizing (high cost, good reproducibility)

As with adhesives, they are neither good nor bad, but it is important to select that which is most appropriate and cost effective. They are listed roughly in the order of the degree to which they produce a controlled surface.

The least expensive and least controlled method is to accept the present surface and make no attempt to change it. This is adequate in many constructions. Solvent wiping can remove many organic contaminants but requires ventilation and changing of the solvent as needed to avoid redepositing contaminants (the steady-state situation). Vapor degreasing avoids the redeposition problem but may be limited to small pieces. Mechanical abrasion, such as sanding or shotblasting, is effective where the substrate is sturdy enough. Abrasion usually leaves residue embedded in the surface, and the fresh surface of many materials can be quite reactive, changing rapidly on exposure to air or cleaning solvents.

Electrical discharge and plasma or corona treatment are often used to enhance the polarity and mechanical properties of the surface region and bondability of low surface energy plastics, such as polyethylene and polypropylene. Chemical stripping or acid etching removes the outer atomic layers and replaces them with a more controlled layer, but as with solvent cleaning, there is the danger of redeposition or poor stripping if the etchants are not kept clean and at the proper concentrations. Etching, followed by deliberate deposition of a controlled surface, can provide the most reproducible surface, but at the greatest expense. Chemical deposition of a new surface, as in chrome conversion coating or zinc phosphate treatment, has been used for many years where long life and durability of metal constructions are important. Phosphoric acid anodizing of aluminum for bonded aircraft assemblies has become the standard in that industry. It is also common to use several surface treatments on the same adherend to prepare the surface for adhesion.

Unlike the surface treatments above, primers always add a new, usually organic layer to the surface. They are used for the same reasons as surface treatments: to provide a durable and reproducible surface. Primers are frequently used to seal porous surfaces. Where the adhesive has desirable properties but poor adhesion to the substrate, primers can be used to provide a bridge between the adherend and the adhesive to obtain or improve adhesion. Primers can be used after or in conjunction with another surface treatment. That treatment might be needed to obtain adequate adhesion of the primer and durability of that interface, or the primer might serve to protect the new surface created by a previous treatment. Primers can sometimes take the place of a surface treatment. Instead of removing a weak layer, the primer can be designed to penetrate and bind weakly adhering material to provide a new, tightly anchored surface for the adhesive.

Classifying primers is as difficult as classifying adhesives and there are probably as many kinds. Selection of a primer is normally less of a problem than that of the adhesive because the adhesive supplier can offer a primer adhesive package or make primer recommendations.

Another important method of preparing surfaces is the use of coupling agents, particularly the silane coupling agents. These are organosilanes, which usually have three methoxy or ethoxy groups and one carbon chain attached to a silicon atom. Attached to the other end of the carbon chain is a group, such as epoxy or amine, that is capable of chemically reacting with groups in the adhesive. When applied in very thin films, the silicon-oxygen-carbon bonds react with moisture to release the alcohol and leave silanol (Si—OH) groups, which react with the surface or self-condense to form a continuous network having the reactive groups oriented into the adhesive. It is thus possible to form some chemical bonds through the interphase. Silane coupling agents have proven valuable in enhancing both strength and durability of bonded constructions.

Joint Design

Joint design plays the major role in using to advantage the ability of an adhesive to distribute loads over large areas. It is a general rule that adhesive bonds display the highest load-bearing capacity when the load is a shear force in the plane of the bond. Flaws and discontinuities in the bond line, including the edges, are regions of concentrated stress. Loads directed out of the plane of the bond tend to be concentrated in smaller areas so that the local stress can be several times the applied stress. Joint rotation or deformation under an applied stress can change the magnitude and direction of local stresses in the bond line. Adhesives display their lowest load-carrying capacity when subjected to peeling forces.

There is usually little choice in joint design for large-area laminates. Other constructions can present a wide array of possibilities for joint configuration and load distribution. There are several detailed discussions of joint design and behavior under load (Ref 7, 8) that can be consulted for specifics. The general rule in joint design is to maximize bond area and put as much as possible of the load in shear, while minimizing peel and cleavage forces.

An important part of any joint design effort is the test program. Initial tests might be done to determine whether any bond is formed with a particular choice of adhesive and other components. Once a system is chosen that provides acceptable initial adhesion, the test program should continue to provide useful information for predicting the life and performance of the construction. The best test program would include building the product, using it under worst-case conditions for its design life, and then determining how it holds up. Obviously, because this is impractical in most cases, test programs are developed to obtain useful predictive information as quickly and inexpensively as possible.

The fastest and least expensive programs are destructive tests, but these also have the lowest predictive value. There is probably no adhesive bond that is intended to be loaded only once and to fail the first time it is loaded, except for destructive-test specimens. Nearly all adhesive assemblies are subject to some continuous low level of stress, either from externally applied loads, such as the weight of the assembly itself, or from residual stresses from the adhesive solidification process. Therefore, creep in the assembly might be a major factor that is not uncovered in a rapid destructive test (recall WLF and Eq 1). In addition, the assembly might be subject to intermittent higher stresses and to cyclic stresses from vibrations.

Continuous low levels of stress can accelerate bond degradation by aggressive environments. Intermittent exposure to corrosive conditions can be more damaging than continuous exposure to the same conditions. For test results to have value in making predictions about the product, loading modes and environmental conditions should mimic, as far as possible, those expected in real life. Because lifelike tests tend to be more costly and time-consuming, their need should be determined by the cost and liability associated with the product.

Successful Use of Adhesives

To summarize the factors that lead to bonding success, it is important to remember that:

- Any bonded construction consists of at least two adherends and one adhesive and has two interphase regions
- Acceptable performance of the construction depends on the mechanical properties, durability, and reproducibility of the components and of the interphase regions and on proper joint design
- Performance of the interphase regions can be enhanced by the proper selection of adhesive, surface preparation, and assembly process

It is most important to remember that such things as adhesion, corrosion resistance, durability, bond strength, and other responses to a test or challenge are properties of the construction, rather than properties of the adhesive or any other individual component. Successful use of adhesives depends on taking account of all parts of the construction, from the beginning of the design process.

REFERENCES

1. D.H. Kaelble, *Physical Chemistry of Adhesion*, Wiley-Interscience, 1971
2. W.A. Zisman, *Adhesion and Cohesion*, P. Weiss, Ed., 1962
3. M.L. Williams, R.F. Landel, and J.D. Ferry, The Temperature Dependence of Relaxation Mechanisms in Amorphous Polymers and Other Glass-Forming Liquids, *J. Am. Chem. Soc.*, Vol 77, 1955, p 3701
4. H.F. Mark, Future Improvements in the Cohesive and Adhesive Strength of Polymers—Part I, *Adhes. Age*, Vol 22, No. 7, 1979, p 35
5. H.F. Mark, Future Improvements in the Cohesive and Adhesive Strength of Polymers—Part II, *Adhes. Age*, Vol 22, No. 9, 1979, p 45
6. I. Skeist, *Handbook of Adhesives*, 2nd ed., Van Nostrand Reinhold, 1977
7. R.D. Adams and W.C. Wake, *Structural Adhesive Joints in Engineering*, 1984
8. W.C. Wake, *Adhesion and the Formulation of Adhesives*, 2nd ed., Applied Science, 1982

SELECTED REFERENCES

- S.R. Hartshorn, Ed., *Structural Adhesives, Chemistry and Technology*, Plenum, 1986
- R.W. Hemingway, A.H. Conner, and S.J. Branham, Adhesives From Renewable Resources, *ACS Symposium Series 385*, American Chemical Society (Washington, D.C.), 1989
- A.J. Kinloch, *Adhesion and Adhesives Science and Technology*, Chapman and Hall, 1987
- H. Lee and K. Neville, *Handbook of Epoxy Resins*, McGraw-Hill, 1967
- L.-H. Lee, Ed., *Adhesive Chemistry, Developments and Trends*, Plenum, 1984
- G. Oertel, Ed., *Polyurethane Handbook*, Hanser, 1985
- J.R. Panek and J.P. Cook, *Construction Sealants and Adhesives*, 2nd ed., Wiley-Interscience, 1984
- R.L. Patrick, Ed., in *Theory*, Vol 1, *Treatise on Adhesion and Adhesives*, Marcel Dekker, 1967
- R.L. Patrick, Ed., in *Materials*, Vol 2, *Treatise on Adhesion and Adhesives*, Marcel Dekker, 1969
- R.L. Patrick, Ed., *Treatise on Adhesion and Adhesives*, Vol 3, Marcel Dekker, 1973
- D. Satas, Ed., *Handbook of Pressure-Sensitive Adhesive Technology*, 2nd ed., Van Nostrand Reinhold, 1989
- K. Tanabe, *Solid Acids and Bases, Their Catalytic Properties*, Academic Press, 1970

Adhesives Markets and Applications

James Hagquist, Frederick K. Meyer, and Sandra K.M. Swanson, H.B. Fuller Company

THE INDUSTRIES in which adhesives usage is either important or common, or both, are identified in this article, as are the applications specific to each industry.

Aircraft/Aerospace

In the aircraft/aerospace industry, adhesives have played a significant role, from the first hot-air balloons to the National Aeronautics and Space Administration (NASA) space shuttle. Of the specialty adhesives and sealants purchased in 1985, the aircraft/aerospace industry accounted for 7%, or $100 million of the total. Sales to this market were expected to reach $140 million in 1990, based on an annual average growth rate of 8%.

Structural adhesives account for the greatest market share of the adhesives used in aerospace applications and will be the focus of this discussion. Structural adhesive bonding has been used by the aircraft/aerospace industry since the 1940s. In the aerospace market, a distinction is made between primary and secondary structural applications. Joint failure in a primary structure will result in the loss of the aircraft, whereas failure in a secondary structure will result only in localized damage.

Structural adhesive bonding has primarily been developed for applications in defense. In 1975, the U.S. government sponsored a program known as the Primary Adhesively Bonded Structures Technology (PABST) to extend the use of adhesive bonding for aerospace and to optimize manufacturing. During the 1980s, structural adhesive bonding has been used on the F-16, F-18, and AV-8B by the military, on various helicopters and commercial transports, and on many business and commuter aircraft, such as the all-composite Beech Starship. Advanced development programs have also included structural adhesive bonding in their scope. The use of structural adhesive bonding produces structures that are lighter in weight, more resistant to fatigue, and that have improved aerodynamics compared to riveted structures.

The use of adhesives in this market is further described in the article "Aerospace Applications for Adhesives" in the Section "Designing With Adhesives and Sealants" in this Volume. In addition, the Section "Bonded Structure Repair" contains numerous articles that focus on advanced composite aircraft structures.

Structural adhesives used in aerospace applications can be further classified into the broad categories of structural film adhesives and structural liquid and paste adhesives.

Structural film adhesives applications include metal-to-metal and honeycomb bonding used for control surfaces; the empennage, or tail assembly of an airplane; skin panels; engines; and radomes. Miscellaneous secondary structural applications include fairings and flooring. Advanced composite-to-composite bonding using film adhesives is gaining in significance because of the rapid growth rate and use of these materials.

Resin system types used in structural film adhesives include epoxy-nitriles, epoxy-novolacs, epoxy-nylons, nitrile-phenolics, and polyimides, including bismaleimides. These films are manufactured by a hot-melt casting or solvent casting process to give an unsupported cast film or a film supported by a woven or nonwoven fabric. The manufacturing process of the polymer and/or film may be varied to obtain different degrees of tack (stickiness) and drape (handleability). Structural film adhesives are usually supplied with a polyfilm and/or release paper in tape, roll, or sheet form. A polyfilm is usually a polyethylene plastic that is approximately 125 μm (5 mils) thick and is available in rolls of various widths.

Most structural film adhesives are perishable and must be shipped and stored frozen or refrigerated in a moistureproof barrier package. The packaged material is allowed to attain room temperature before it is opened to prevent moisture condensation on the adhesive. The film is trimmed to size, one of the protective liners is removed, and the film is placed on the properly prepared adherend. Air pockets are squeezed out with a roller, the other protective liner (if present) is removed, and the part is assembled, clamped, and cured. The usual methods of curing are autoclave, vacuum or platen press, and oven vacuum bag. The Section "Manufacturing Processes" in *Composites*, Volume 1, *Engineered Materials Handbook*, published by ASM INTER-

NATIONAL in 1987, contains over 20 articles that cover aerospace requirements for curing, laminating, and tooling.

Structural liquid and paste adhesives are used for a variety of aerospace assembly operations, including liquid shimming and aircraft maintenance and repair. These adhesives are primarily one- and two-component epoxies.

One-component epoxies are often co-cured with film adhesives, and, like structural film adhesives, are usually refrigerated to prolong their shelf life. Two-component epoxy components are most often stored at room temperature and have a shelf life of 1 year. When mixed, most two-component epoxies will fully cure at ambient conditions, but can be accelerated with heat. Other liquid and paste adhesive types used in the aircraft/aerospace market include one-component and two-component urethanes, bismaleimides, acrylics, cyanoacrylates, polysulfides, anaerobics, silicones, and thermoplastic hot melts.

Nonstructural adhesives, such as neoprenes and nitriles, are used for the interior assembly of aircraft. Adhesive primers are also important to the market as a means to enhance the durability and strength of the adhesive bond.

Electrical/Electronics

The electrical/electronics market utilizes adhesives as insulative and conductive materials for a variety of applications. The various types include epoxies, polyamide hot melts, acrylics, polyimides, polyurethanes, silicones, and cyanoacrylates. Epoxies, because of their superior adhesion, purity, performance properties, and versatility, are the resins most commonly used.

Encapsulants and potting compounds used to protect an electrical device from adverse environmental and operative effects would best be classified as sealants. Adhesives represented only about 25% of the $150 million in adhesives and sealants used in the domestic electrical/electronics market in 1988.

Other articles in this Volume that provide details relevant to the use of adhesives in this market are "Surface Preparation of Electronic Materials" and "Electrical Properties."

Table 1 Approximate volume resistivities of various materials

Material	Volume resistivity, $\Omega \cdot cm$	Level of conductivity
Silver	1.6×10^{-6}	Conductive
Copper	1.8×10^{-6}	Conductive
Silver-filled epoxy	$1.0 \times 10^{-3} - 10^{-4}$	Conductive
Carbon-filled epoxy	5.0×10	Semiconductive
Silicon	1.0×10^5	Semiconductive
Glass	1.0×10^{10}	Insulative
Polyurethane	1.0×10^{14}	Insulative
Epoxy	1.0×10^{16}	Insulative
Polystyrene	1.0×10^{18}	Insulative

Table 2 Characteristics of important types of packaging adhesives

Types	Bonding speed	Adhesion	Typical applications
Dextrine	Slow to moderate	Paper	Case cartons, rigid boxes, bag paste laminating and mounting, envelopes
Starch conversion	Moderate	Paper to glass and treated polyethylene	Semiiceproof bottle labeling foil laminating
Polyvinyl and acrylic resin emulsions	Moderate to fast	Depends on formula; plastic, metal, wax- and ink-coated board	Like dextrine except where special properties are required. Also, heat seal, pressure sensitives, foil, plastic and foam laminates
Rubber and synthetic latexes	Moderate to extremely fast	Depends on formula, may bond a variety of materials	Foil laminates, polyethylene, bag seam pastes, untreated polyethylene labeling and laminating, pressure sensitives, surface printed adhesives
Casein	Moderate	Paper to glass	Fully iceproof bottle label adhesive
Animal glue	Slow to moderate	Paper, clay	Tightwrap, bookbinding
Hot melts	Very fast	Paper, metal coated board, plastics	Same as polyvinyl resin emulsions but at higher speeds and lower compression
Isocyanates (Polyurethanes)	Very fast	Metal, plastics, coated board, foils	Film laminating
Solvent cement	Fast	Metal, plastic films, coated board	Heat-seal coatings, laminating, bag sealing

Electrical properties include volume resistivity, surface resistivity, insulative resistance, dielectric constant, dissipation factor, dielectric strength, and arc resistance. Other properties important to the electrical performance and dependability of the adhesive include operating temperature, thermal conductivity, thermal expansion, compression strength, shock and impact resistance, and electrical conductivity. Table 1 gives the volume resistivity values for a variety of materials.

Insulative Adhesives. The most pertinent uses for electrically insulative, or nonconductive, adhesives are for battery construction, pressure-sensitive tapes, flexible printed wiring boards (PWBs) and other laminations, magnet bonding, television tube implosion proofing, and mica/mica paper insulation.

In battery construction, the hard rubber or polystyrene tops and bottoms are usually bonded with epoxies. Flexible PWBs, which are foil to film in nature, are laminated with the use of polyimide, acrylic, and polyester film adhesives. Magnet bonding for motors and speakers is usually achieved with free radical reactive adhesives.

Conductive and Semiconductive Adhesives. Conductive adhesives are generally very expensive because of the high loadings of silver, gold, gold-palladium, or other conductive filler. Electrically conductive adhesives applications include the attachment of lead wires, hybrid circuits, light-emitting diodes (LED), integrated circuits (dies), component-to-PWB bonding, and the bonding of ferroelectrical devices. Semiconductive adhesives are used where bonds require the dissipation of electrostatic charges, as in various computer parts, as well as in computer flooring.

Packaging

The packaging and converting of adhesives forms one of the largest sectors of the adhesives industry. In 1988, $365 million worth of packaging adhesives were used in the United States.

The types of adhesives used for packaging include dextrines, starch conversions, polyvinyl and acrylic resin emulsions, rubber latices, casein, silicates, animal glues, hot melts, isocyanates, and solvent cements. Table 2 provides the characteristics of each type.

Packaging products include packages converted from paper and paper boards, glass, metals, plastics, and a combination of two or more of these materials. The facets of the multidisciplined packaging industry include the packaging materials themselves; the converting of packaging materials; package fabrication, filling, and closing; packaging machinery; packaging testing; product properties; shipping, storage, and handling procedures; packaging economics; commercial art and design; marketing and advertising; and legal regulations. The choice of adhesive may affect each of these areas.

General adhesive applications include the manufacture of corrugated board, the lamination of paper and board, case and carton assembly, case and carton sealing, rigid setup box assembly, gummed tape and paper, envelopes, grocer mailer and specialty bags, cigarette manufacture, the lamination of films and foils, and the labeling of various adherends. Pressure-sensitive products are described in the section "Pressure-Sensitive Adhesives" in this article, as well as in the article "Types of Adhesives" in this Volume.

Appliance Market

The appliance market includes finished goods such as refrigerators, freezers, stoves, air conditioners, and small appliances. A variety of adhesives are used in the manufacture of these goods.

Probably the largest end-use of adhesives is the cabinet sealer or foam stop hot-melt adhesive, which is extruded along refrigerator cabinet seams first, to seal them from atmospheric moisture and, second, to act as a foam stop. This seal prevents the insulation foam from leaking out during the injection stage of assembly. Robots are often used to apply this adhesive.

Hot melts are also used for bonding freezer coils into place because the adhesive gains strength rapidly upon cooling. This eliminates time-consuming fixing operations and provides a temporary bond that holds the coils in place until the insulative foam has been injected into the housing. Another use for hot-melt adhesives is in the bonding of fiberglass insulation and sound-deadening material to the interior housing of air conditioners and dishwashers. Packaging-type hot melts are used widely as a means of packaging large appliances in containers that will protect them during shipment in truck and railcar.

Polyurethanes and epoxies have been used to bond metal to metal in housing assemblies and to bond plastic to plastic in handle assemblies and liner parts. Most of these plastic parts are polystyrene or acrylonitrile-butadiene-styrene (ABS).

Silicones have been used for bonding glass to metal in the construction of shelving and windows. There are also a number of smaller applications related to electronic potting in refrigerator ice makers and butter keepers.

Automotive Market

The automotive industry commonly uses thermoset adhesives such as epoxies, poly-

Table 3   Adhesive systems used in woodworking structural applications

Adhesive system	Assets	Liabilities	Application
Water-base.............	Low cost	Can raise grain	Wood veneering, cabinet components, film laminating
Hot-melt...............	Fastest, applied with gun	Surface bond, heat sensitive	Assembly and reinforcing, some wood veneering, cabinet construction
Solvent cement/weld.....	Easiest to use, contact bond	Lower strength, solvent problem	Plastic laminating, some wood veneering, some film laminating, cabinet components
Epoxies/acrylates........	High strength	Expensive, slow, possible odor and toxicity problem	Plastic laminating, some wood veneering, some film laminating, cabinet components

urethanes, anaerobics, cyanoacrylates, and acrylics, as well as thermoplastic hot melts, polyamides, water-base contact cements, and plastisols. The automotive adhesive market is constantly undergoing changes and advances in technology as automobile manufacturers work to decrease the weight of vehicles and improve the performance and integrity of their designs. The 1990 market size was expected to be $300 million.

Plastisol is probably the largest-volume adhesive/sealant used in the automotive industry because of its economical price and chemical resistance properties. It is a cross-linkable polyvinyl chloride (PVC) material, usually supplied in one-component form, that is applied to the surface to be sealed and is then heat cross-linked in an oven. The oven usually also cures the paint or metal primers used on the body, thereby serving a dual purpose. Common applications include sealing steel to steel or steel to sheet molding compound (SMC) for hood and deck and roofing panels, sealing welding seams for water repellency (such as in the gutters of the door), and minimizing air transmission to the inside of the vehicle. Plastisols are also used to form the gaskets for air filters and as sealants and bonding agents in hydraulic, oil, and fuel filters. The first corrosion-preventative undercoatings were made from plastisol and spray applied in the factory. Another large application is sound-deadening pads that are made from vinyl plastisol and applied to the wheel wells of automobiles.

Hot-melt adhesives are thermoplastic polymers that melt at elevated temperatures. They are applied as hot liquids to the substrate to be bonded, and upon cooling and solidification, the bond is formed. These polymers can be polyamides, polyesters, ethylene-vinyl acetate, polyurethane, or a variety of block copolymers and elastomers, such as butyl rubber, ethylene-propylene copolymer, styrene-butadiene, ethylene-butadiene, ethylene-butylene styrene, and styrene-isoprene.

Their applications are as body sealers, weather strip sealants, and taillight seal-

ants. The body sealers are applied when the body is being assembled, and their main function is to act as sealants against moisture and corrosive agents (such as salt). In the assembly of taillights, a sealant is extruded into a channel between the lens and the housing. The parts are then snapped together, and the sealant forms a moisture barrier to protect the bulb.

These adhesives are also used in automotive carpet bonding. The adhesive is extruded onto the floor of the vehicle, and the carpet is pressed into place. In some cases, the carpet is also stapled down. This product must be water and chemical resistant and must maintain its bond over a broad service temperature range.

Thermoset polyurethanes, acrylics, and epoxies are used for the structural bonding of metal to metal, SMC to metal, and SMC to SMC. This can include the bonding of hood, deck, and toolboxes (a feature of pickup trucks) and the bonding of body panels to the frame of the vehicle. A two-component polyurethane is used for most of the SMC bonding, but epoxies are being used as well.

In plastic bonding, such as taillights and headlights, two-component thermoset polyurethanes are used not only to seal the lenses but also to bond the parts structurally. These are very fast-setting systems that use robotics to apply them to substrates. Thermosets offer many advantages in that they usually are high-strength materials and, once reacted, will not "melt" or flow like a hot-melt product. They are available as one- or two-component products that must be mixed before using and are heat cured or cured with moisture from the air. Moisture-curing urethanes find their biggest use as sealants for windshields. The product is applied just before the glass is installed and is allowed to cure at ambient conditions with moisture from the atmosphere. This sequence efficiently seals the windshield from water and snow. Two-component urethanes have been tested as wheel well coatings and as antirust and anticorrosion undercoatings. The article "Automotive Applications for Adhesives" in this Volume

describes the market for this industry in more detail.

Woodworking

The largest use of structural adhesives in woodworking is in forest products, that is, in the production of particleboards, plywoods, hardboards, waferboards, fiberboards, composites, and laminated timbers. Formaldehyde complexes of phenol, urea, and resorcinol-phenol are used to make waterproof laminates.

The use of isocyanates has been investigated for many years. It is waterproof and bonds well to wood, perhaps too well, because it is difficult to demold from multi-platen presses. When this problem is solved, it could replace many of the formaldehyde complexes.

The production of furniture is a specialty area in the woodworking structural adhesives market. Each adhesive has advantages and disadvantages, and the selection and use of any adhesive in furniture construction is critical. Characteristics are given in Table 3.

Construction

New-home construction represents a very large market. The end-use products include gypsum board construction with water-base adhesives, particleboard or waferboard construction with various phenolic-type formaldehyde complexes, and sizing for fiberglass insulation using ethylene-vinyl acetate and modified water-base adhesives.

More than 50% of all products sold in construction are water based because the major substrate used is wood. Adhesives are used to dissipate or reduce the stress that bolts or nails can produce.

Adhesive and substrate sandwich panels (composite construction) are one of the largest end-uses of structural adhesives today. A panel can be made from two or more of the following materials: honeycomb paper, plastic, wood, sheet metal, masonry, and the adhesive. After the panels are constructed, they are tested to determine whether they meet requirements for appearance, strength, durability, insulation, acoustics, and permeability. The adhesive most commonly used in these panels is a solvent cement, an epoxy, or a polyvinyl acetate emulsion.

Insulated double-pane windows have become very important in saving energy in high-rise buildings. The most common materials used in the construction of these windows are polysulfides, silicones, thermoplastics, and polyurethanes.

Door construction is a modification of a standard sandwich panel, but must have better impact resistance and more flexibility. Floor coverings, such as vinyl, ceramic, and carpet, also use adhesives bonded to subfloors of concrete, plywood, or hardboard.

The article "Industrial Applications for Adhesives" in this Volume provides additional information.

Foundry

Phenolic resin binders are used in the foundry industry to hold sand particles together in the mold after they are deposited on a hot metal pattern plate. Excess sand and binder are removed, and the shell with the pattern is oven cured. However, a majority of foundry sand patterns are made in so-called green sand molds, which are uncured molds of sand, clay, and water. Each year a larger percentage of the molds are made with phenolics as binders because of reproducibility and dimensional accuracy. Hot melts and water-base adhesives have also been used by this industry.

Consumer

Consumer products represent a fast-growing market for adhesives. The common, general-purpose "white glue" is the most popular. Homemakers, homecrafters, and do-it-yourselfers love the water-washable, freeze/thaw-stable, mildew-resistant plasticized polyvinyl acetate homopolymers. However, over-the-counter adhesives, contact bonds (solvent cements), epoxies, pastes, hot-melt sticks, cyanoacrylates, and pressure-sensitive tapes and labels are now widely available.

Pressure-Sensitive Adhesives

The pressure-sensitive adhesive (PSA) market, which is very large, is expected to reach $320 million in 1990, with an annual growth rate of 11%. The industry can be divided into various market segments:

Packaging tapes

- Carton sealing
- Bundling
- Strapping

Industrial tapes

- Mounting and fastening
- Ductwork/heating, ventilation, and air conditioning
- Masking
- Pipe protection
- Electrical uses

Health care tapes

- Surgical and other medical uses
- Diaper closure

Consumer tapes

- Over-the-counter/do-it-yourself products

Adhesive products that are used in these applications are hot-melt coatings, water-base coating, and solvent-base adhesive systems. Examples of the types and complexity of these applications are described below.

Hot-melt adhesives are used extensively in the construction of diapers, specifically to bond the back sheet to the liner, to bond the leg elastic into place, to bind the absorbent material together so it does not bunch up when in use, to affix waist attachments, and to coat the adhesive tabs. These adhesives can be either sprayed or extruded in bead form, depending on the configuration of the substrates and the end-use.

Industrial tapes use solvent-base products, which are applied by reverse roll or fixed doctor blades. Solvent recovery systems are being added to production lines to ensure industry compliance with federal regulations.

In self-adhesive hot-melt applications, the hot melt is applied using an equipment system that is heated by diathermic oil. The system consists of a slot die and revolving bar, interconnection hose, and hot-melt supply gear pump electronically slaved to line speed in order to apply the set and targeted coat weight at all line speeds. This technology is not new and is well known.

In the PSA industry, adhesive performance is intricately influenced by application equipment. Equipment innovations are being accomplished rapidly as adhesive innovations are made. For example, the use of water-base PSAs in self-adhesive tape is a new area that is expected to grow in the coming years as solvents are phased out. The necessary equipment will also evolve. As market needs change, the evolution of the technology will adjust to meet these needs.

SELECTED REFERENCES

- *Adhes. Age*, Vol 29 (No. 11), Oct 1986, p 17
- P. Albericci, in *Durability of Structural Adhesives*, A.J. Kinloch, Ed., Applied Science Publishers, 1983
- "Basic Course on Adhesion and Adhesives," Seminar presented by H.B. Fuller Company, Philadelphia, Packaging Institute, Oct 1976
- J.C. Bolger, *Adhes. Age*, Dec 1980, p 15
- J.C. Bolger, in *Adhesives in Manufacturing*, G.L. Scheberger, Ed., Marcel Dekker, 1983
- *Chem. Week*, 15 March 1989, p 41
- G.H. Dietz, in *Building Construction Handbook*, F.S. Merritt, Ed., McGraw-Hill, 1975
- "European Tape and Label Conference Notes," International Marketing Inc., Exxon Chemical Company, Brussels, April 1989
- R.H. Gillespie, in *Adhesives for Wood*, Noyes Data Publication, 1984
- R.C. Griffin et al., *Principles of Package Development*, AVI Publishing, 1985
- S.R. Hartshorn, in *Structural Adhesives, Chemistry and Technology*, S.R. Hartshorn, Ed., Plenum Publishing, 1986
- I. Katz, *Adhesive Materials—Their Properties and Usage*, Foster Publishing, 1964
- S.M. Lee, in *Epoxy Resins, Chemistry and Technology*, C.A. May, Ed., Marcel Dekker, 1988
- G. Marra, Keynote address presented at Wood Adhesives, Research, Applications and Needs, Symposium on the Role of Adhesion and Adhesives in the Wood Product Industry, Madison, WI, 1980
- R.L. Patrick, in *Treatise on Adhesion and Adhesives*, Vol 4, R.L. Patrick, Ed., Marcel Dekker, 1976
- J. Rhude, in *Building Construction Handbook*, F.S. Merritt, Ed., McGraw-Hill, 1975
- C.L. Segal, Worldwide Markets for Advanced Fibers and Composites, in *Proceedings of the 21st International SAMPE Technical Conference*, Society for the Advancement of Materials and Process Engineering, Sept 1989
- Specialty Adhesives Will Top $2 Billion in Sales by 1990, *Adhes. Age*, Vol 29 (No. 11), Oct 1986
- J.H. Willis, in *Handbook of Adhesives*, I. Skeist, Ed., Van Nostrand Reinhold, 1977

Sealants and Sealant Technology

David J. Dunn, Dexus Research Inc.

SEALING has been defined as "the art and science of preventing leaks" (Ref 1). The historical development of sealing systems has sometimes been based more on art than on science. Some leakage from assembled joints was not only acceptable, it was considered a way of life. However, design engineers, architects, and sealant scientists have now begun to work together to build entire joint systems that have very low leak rates and retain their integrity for the life of the product or building. The impetus behind this effort has been the energy crisis of the 1970s and 1980s, the increase in environmental legislation and awareness, an ever-increasing focus on high quality, and a desire to eliminate the costs and hazards associated with gas, liquid, and vapor leaks.

Although adhesives and sealants are often formulated from the same types of polymeric materials, they are usually designed to have different properties. Adhesives typically have high tensile and shear strengths and are used to replace or enhance mechanical fasteners in load-bearing applications, that is, they are structural formulations, whereas sealants are usually lower-strength, flexible materials that are used as a barrier to the passage of liquids, vapors, gases, and solids. However, these are sweeping generalizations, because in non-load-bearing adhesive applications, a flexible "sealant-type" material is often the adhesive of choice, particularly when vibration resistance and impact strength are required or when materials with widely differing coefficients of thermal expansion need to be bonded.

In the potting and encapsulation of electronics components, modified adhesives are in fact sometimes the most effective sealants. In aircraft applications, sealants are used between faying surfaces (overlapping metal joints) to prevent both bimetallic corrosion and the ingress of moisture that could initiate corrosion.

In many applications, a material serves as both an adhesive and a sealant. A good example is the trend in the automotive industry toward replacing spot welding with adhesives in car body assembly, for example, in the bonding of the inner and outer panels of a car door. These structural "hem flange" adhesives, typically epoxies or vinyl plastisols, also play a major role as sealants in the prevention of water ingress into the joint (Fig. 1).

In modern high-rise building construction, structural silicone adhesive sealants are used both to seal and to bond glass windows. Materials typically regarded as adhesives also are often widely used as sealants. Anaerobic adhesives are a good case in point because their adhesive properties are an important reason behind their predominant use as sealants for threaded or flanged assemblies.

Sealants are used in every industrial segment, but their major markets are the construction, transportation, industrial manufacturing, and consumer do-it-yourself (DIY) industries. Numerous sealant applications do not require high-performance properties. This may account for the scarcity of specific coverage of sealant technology in the literature, represented by Ref 2 to 8. This Volume features many of the high-performance requirements of sealants in high-technology industries. The many federal, military, and industrial specifications and standards that now exist for sealants are described in the article "Specifications and Standards for Adhesives and Sealants" in this Volume.

Sealing Technology

This article addresses only those sealants that are applied as liquids, pastes, or extruded tapes, as distinguished from traditional preformed solid sealants such as cut or molded gaskets, O-rings, and shaft seals. There are numerous types of joints that require sealing (Fig. 2). All sealants, whether for construction joints, glazing, threaded assemblies, electrical encapsulation, or flange sealing, must perform three basic functions:

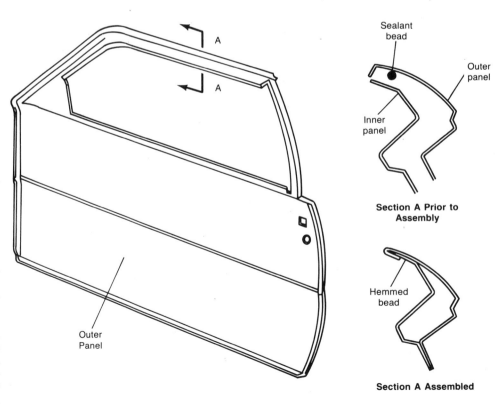

Fig. 1 Hem flange bonding of car door

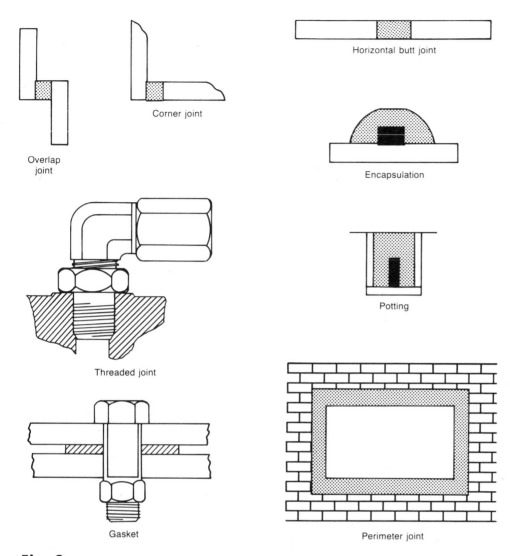

Fig. 2 Common sealed joints

(Labels within figure: Corner joint, Overlap joint, Threaded joint, Gasket, Horizontal butt joint, Encapsulation, Potting, Perimeter joint)

- Fill the space to create a seal
- Form an impervious barrier to fluid flow
- Maintain the seal in the operating environment

First, the sealant must flow and totally fill the space between the surfaces, and it is very important that it be able to conform to surface irregularities. Strong adhesion to the joint surfaces is also a necessary requirement for most sealant applications.

Second, the sealant must be totally impervious to fluid flow. Some older traditional sealing materials such as fiber sheet packings and cork are impermeable only when sufficiently compressed to close their inherent porosity. Most elastomers used in sealants are inherently impermeable to liquids, but care must be taken when gases or vapors are involved. For example, silicones are excellent sealants for liquid water but have high moisture vapor transmission rates.

Third, the sealant must maintain the seal throughout the life expectancy of the joint,

often under severe operating and environmental conditions. Exterior sealants in buildings must be able to accommodate large changes in joint width from temperature cycling, often over 50% in compression and tension, and must withstand the effects of heat, rain, and ultraviolet (UV) light.

Aircraft, automotive, and industrial sealed assemblies are often exposed to severe movements caused by vibration, mechanical strain, and changes in temperature, pressure, and velocity. The sealants in these systems must also be inert to the chemical effects of sealed liquids, which can encompass a wide range of materials, including aqueous solutions, hydraulic fluids, gasoline, organic chemicals, motor oil, jet fuel, liquid oxygen, strong acids, and alkalies. Aggressive liquids can cause the swelling, chemical degradation, and debonding of a sealant. Although a small amount of swelling can help to seal a joint, severe swelling is often the first symptom of the onset of chemical degradation.

Sealant Chemistries

Over the years, a wide range of chemical materials has been used for sealing. Modern demands for increased productivity, higher performance, and environmentally safer materials have led to a decline in the use of some of the older chemistries, particularly solvent-base sealants. Table 1 shows most of the different types of sealants, modes of cure, and major strengths and weaknesses. Selected features are described in this section of the article, whereas more detailed explanations of chemistry and cure mechanisms are provided in the Section "Sealant Materials" in this Volume.

Many key properties of sealants are determined not only by the molecular structure of the base polymer, but also by the selection of ingredients such as fillers, plasticizers, adhesion promoters, and stabilizers. Silicones are a good example for illustrating this point. If the base silicone polymers are fully cured and cross linked without additives, a soft polymer (consistency of cheese) with very low tensile strength is obtained. However, if several percent of a fumed silica filler are incorporated into the polymer, a tough rubbery polymer is formed, and a dramatic increase in strength is observed. The fumed silica, which often merely acts as a thickener or thixotrope when added to other types of sealants, is able to combine chemically with the silicone to provide significant reinforcement. Silicones and approximately a dozen other sealant chemistries are briefly described below.

Silicones were originally developed in the early 1940s (Ref 9, 10), but until recently have been used only for the most demanding sealing applications because of economics. The manufacturing and marketing of silicones is now practiced by many companies worldwide, including several that are totally backward integrated to the basic raw materials.

Many other companies purchase the intermediate silanes and silicone polymers and formulate specialty products for specific markets. Competitive pressures have forced prices down, allowing silicones to compete with lower-performance technologies and leading to increased use in many industries.

Silicone intermediates are manufactured from the reaction of elemental silicon (produced from silica) with organic alkyl halides in a process developed by the General Electric Company and called the Direct Process (Ref 10). The marriage of the organic and inorganic worlds yields unique materials that exhibit outstanding chemical and environmental stabilities. Silicones remain flexible at very low temperatures (to −60 °C, or −75 °F) and have exceptional thermal stability up to 250 °C (480 °F). They have unsurpassed oxidative, UV light, and weathering

Table 1 Sealant chemistries

Chemical family and curing type	Strengths(a)	Weaknesses
Silicones, one component (moisture-initiated condensation) and two components (condensation or addition)	Best weathering, highest flexibility, good adhesion, heat resistance	High MVTR, low depth of cure, slow curing(b)
Polyurethanes, one component (moisture-initiated condensation) and two components (condensation)	Good weathering, best adhesion, high flexibility	Weak UV resistance, weak heat resistance
Polysulfides, one component (moisture-initiated condensation) and two components (condensation)	Low MVTR, fuel resistant, good flexibility	Slow curing(b), low depth of cure
Acrylic latex (water evaporation)	Easy to use	High shrinkage, poor weathering, fair flexibility
Butyls (sulfur vulcanization)	Lowest MVTR, good flexibility	Fair weathering
Anaerobics (metal/peroxide initiated free radical)	Fast curing, chemical resistant, heat resistant	Brittleness, poor gap filling
Vinyl plastisols (heat fusion)	Good adhesion, low cost	Fair flexibility, fair weathering
Asphalts/coal tar resins (cooling oxidation)	Low cost, fuel resistant	Poor weathering
Polypropylene hot melts (cooling)	Low cost, expandable	Limited adhesion, fair flexibility

(a) MVTR, moisture vapor transmission rate. (b) Two-component version cures faster.

resistance, and good electrical insulative properties. Silicones shrink very little upon curing and are available as single-component moisture-cured versions (known as room temperature vulcanizing, or RTV, silicones) or as faster-curing two-component versions. High-elongation (≤1000%) construction sealants with the highest joint movement capability are formulated from silicones. Silicone use is also increasing as secondary sealants in insulated glass assembly. Lower prices have allowed RTV silicones to be marketed as high-performance caulks in the competitive consumer retail market. Silicone weaknesses include poor abrasion resistance, dirt pickup, limited depth of cure for RTVs, high water vapor transmission rates, and limited resistance to organic liquids.

RTV silicones have been used since 1971 for formed-in-place automotive gasketing in engine, transmission, and rear axle sealing. The weaknesses of early formulations, including poor oil resistance, poor adhesion, corrosion, and lack of flexibility, have recently been overcome (Ref 11). RTV silicones face strong competition from molded silicone gaskets used by U.S. manufacturers, but use in Japanese vehicles is increasing. The encapsulation or potting of electronic and electrical components is an important market for silicones, and materials that cure in less than 1 min under UV light have recently become commercially available.

A low-volume but very high-value application for specialty silicones is the use of fluorosilicones in the aerospace industry when resistance to both high temperatures and jet fuel is required. The high cost of these materials has prevented them from being used for automotive gasoline sealing applications.

Polyurethanes, often referred to simply as urethanes, were invented in Germany over 50 years ago (see Ref 12) but only recently have been formulated for demanding sealing applications. The chemistry of urethanes involves the condensation of polyester or polyether polyols with multifunctional aliphatic or aromatic diisocyanates. Two-component systems are common, as are one-component systems that cure on exposure to atmospheric moisture.

Urethane sealants are the leading sealants in the nonglazing segment of the construction market and possess most of the desirable properties displayed by silicones (good flexibility, low compression set, good weatherability, and very low shrinkage). Excellent primerless adhesion is obtained on most surfaces. Although joint movement capability (typically ±25%) is not as good as with silicones, it exceeds that of all other sealant types.

Urethanes are becoming the predominant adhesive sealant in the automotive industry for the installation of windshields and back lights, displacing the traditional butyl sealants. Systems use a one-component urethane in combination with a black-colored primer for UV protection and another primer to maximize adhesion. Urethanes have excellent adhesion to metals and automotive plastics, particularly to polyester sheet molding compounds. This capability makes them useful for bonding and sealing the new range of plastic and composite vehicle bodies. Modified hot-melt urethanes show great promise as automotive, construction, and industrial sealants.

Polysulfides, one of the earliest synthetic elastomers (Ref 13), owe their unique properties as sealants to the presence of sulfur atoms in the backbone of the polymer. They

are prepared by the condensation of terminal thiol groups in liquid prepolymers using oxidizing agents, usually peroxides. Polysulfides are manufactured exclusively by a few companies and are available as both one- and two-component sealants. They are flexible, giving them good joint movement capability, and are second only to butyls in terms of low moisture vapor transmission. As with butyls, this last feature allows their use as the primary sealant in insulated glass manufacture. Technological improvements are based on the use of liquid thiol-terminated polymers based on polyether urethane backbones (Ref 14) and polythioether backbones (Ref 15) to yield sealants with enhanced strength and thermal stability. Polysulfides have excellent resistance to jet fuel and are the sealant of choice for sealing integral fuel tanks in aircraft.

Acrylics are manufactured from copolymers of both methacrylate and acrylate monomers. The term vinyl acrylic is used to describe materials that also contain vinyl acetate units in the copolymer. Acrylics are available as solvent-base products and as latices. The latex products are moderate in cost, safe, have low odor, and are ideal for the DIY industry. They have moderate joint movement capability, fair moisture resistance, and are useful for interior caulking where excessive joint movement is not expected. Most so-called siliconized versions contain small amounts of silanes and are designed to give the perception of performance similar to that of silicone sealants.

Butyls is a term that encompasses products made from polyisobutylene, butyl rubber (copolymer of isobutylene and isoprene), and polybutenes (copolymers of 1-butene, 2-butene, and isobutylene). These one-part sealants are available as solvent-base caulks, deformable tapes, and hot-applied systems. Butyls have good flexibility and good adhesion without primers to most surfaces, and are resistant to UV light and water. Their extremely low moisture vapor transmission rates have made them a major primary sealant in the construction of insulated glass units. Hot- and cold-applied butyl sealants are used extensively in the automotive industry as sealers for back lights, as water-deflector films, and as wind noise baffles. Long used for automotive windshield glazing, butyls are rapidly being supplanted in this use by polyurethanes.

Anaerobics are often referred to as machinery adhesives, a term coined by the market leader (Ref 1). Anaerobics comprise filled or plasticized compositions based on multifunctional methacrylate or acrylate monomers. They cure when confined in narrow gaps where oxygen is excluded. Curing is accelerated by surfaces containing transition metals (usually iron and copper). The behavior of anaerobics can be considered exactly opposite to that of an alkyd paint. The paint is stored in a filled metal can to exclude oxygen and, when brushed on a surface, curing is initiated by the interaction of atmospheric oxygen with the paint driers. Anaerobics are stored in half-filled polyethylene containers to maximize oxygen diffusion and only cure when confined in an assembled joint.

Anaerobics have excellent chemical resistance to a wide range of industrial fluids and resistance to temperatures over 150 °C (300 °F). Although the products are best known as replacements for lock washers in the prevention of the vibration loosening of fasteners, the majority of uses are in industrial and automotive assembly and maintenance, for the sealing of threaded joints and the filling of pores in castings or powdered metal parts, or as formed-in-place gaskets. Weaknesses include the inability to fill large gaps without primers, an inherent brittleness, and, like most other organic sealants, the inability to function in very aggressive environments, such as liquid oxygen service.

Vinyl plastisols are dispersions of polyvinyl chloride (PVC) or PVC-polyvinyl acetate copolymers in plasticizers and are used widely in the automotive area as body seam sealants. The plastisols fuse and cure in the paint bake cycles. Weld-through versions are used as adhesive sealants in the bonding of body panels (Fig. 1). Solvent-base PVC pipe dopes or cements are used for bonding and sealing plastic plumbing joints.

Asphalts and coal tar resins are low-cost materials produced in oil refining and in the distillation of coal tar, respectively. They are used extensively as hot-applied highway and airfield joint sealants. Improved properties are obtained by blending with other resins and rubbers, for example, rubber-asphalt, coal tar-PVC, and coal tar-urethane. Naturally occurring asphaltic materials called Gilsonites are used in the auto industry as body seam sealers, but are being supplanted by plastisols.

Polypropylenes used as hot melts, particularly mechanically foamed versions, have established a strong position in appliance seal and gasket applications and are used as automotive and roofing weather seals.

Other Sealants and Sealing Aids. Low-cost and low-performance sealing caulks based on filled drying oils are still marketed in the DIY and factory glazing markets. Several other sealants based on rubbers are available, including styrene-butadiene, nitrile, chloroprene (neoprene), and styrene-butadiene-styrene thermoplastic elastomers. Sodium silicates and unsaturated polyesters are used for porosity sealing in castings. The sealing of threaded joints and gaskets in liquid oxygen systems cannot be accomplished with organic sealants because of the danger of combustion. Polytetrafluoroethylene (PTFE) tape is used for this

purpose, and special sealants based on gelled fluorocarbon oils and fluoroelastomers, often filled with PTFE, have also been developed. Epoxies are used for the encapsulation of electronic and electrical components. Typical formulations of sealants geared for certain industries have been published (Ref 16, 17).

For over 80 years, cut and molded gaskets have been treated with dressings to aid in sealing. Gasket dressings, such as mineral-filled mixtures of wood rosins, serve to hold gaskets in place on flanges during assembly and shipping; fill in substrate imperfections; and, through adhesion, assist in maintaining a seal under joint movement.

Properties

There are usually several possible sealants for any particular application. In 1990 the prices of sealants varied from about $0.8/kg ($0.4/lb) for the cheapest DIY and highway sealants to about $180/kg ($80/lb) for some high-performance aerospace sealants. The properties of sealants that are important for sealing are discussed below. The section "Design of Sealing Systems" in this article shows how these properties are combined with application requirements and mechanical designs of the joint to produce an optimum sealing system.

Cure Properties

Rate of Cure. The speed at which a material provides handling strength and then proceeds to full cure is critically important in some industries. In the construction industry, for example, slow cures imply that joint movement will take place during curing, which can sometimes lead to failure. However, it is important that materials not cure faster than can be handled in a production environment. In an automotive or industrial assembly, fast curing means productivity and is usually essential. Fast curing is also important in maintenance operations in which a piece of equipment or a vehicle must be back in service as soon as possible. Primers or two-component systems are used to increase the rates of cure.

Depth of cure is another important consideration for a sealant. One-component systems, such as silicones, urethanes, and polysulfides, which depend on the diffusion of atmospheric moisture into the sealant for curing, can take days or even weeks for deep section curing. Anaerobics, whose cure is initiated from metal surfaces, will not cure through gaps greater than about 0.5 mm (0.02 in.).

Shrinkage upon curing is a critical factor for any sealing system. Excessive shrinkage can lead to voids in a joint and induce stress in the sealant. Low-shrinkage products are those that have very high or 100% solids contents, including urethanes, silicones, polysulfides, plastisols, high-sol-

ids butyls, and anaerobics. Medium-shrinkage systems include hot-melt or hot-applied products such as asphalts, coal tars, and polypropylenes. High-shrinkage systems are all solvent- or aqueous-base sealants in which shrinkage is due to evaporation of the carrier. Some automotive butyl sealants are formulated with blowing agents and actually expand during curing to fill all voids. Foamed hot-melt sealants also expand after application.

Physical Properties

The vast majority of sealed joints experience some movement during service, and some flexibility in a sealant is essential. The only truly rigid sealants are anaerobics, because they tend to unitize an assembled joint and prevent it from moving. However, this lack of flexibility is sometimes considered a weakness in these products.

The properties that are important for a sealant are identified below. They should be measured after initial complete curing and then monitored under accelerated aging conditions.

Hardness is measured by the amount of penetration of a durometer in the cured sealant. Most sealants have Shore A durometer readings of 15 to 70. A change in hardness upon aging is an indication that further curing or degradation is taking place.

Modulus of elasticity is the ratio of the force (stress) required to elongate (strain) a sealant. In general, low- or medium-modulus sealants are able to accommodate much greater joint movement without putting a large stress on either the sealant or the substrate surfaces. The major exception to this generalization is a product that is used primarily as a structural adhesive, with sealing as the secondary benefit. Structural adhesives/sealants must have a high modulus (and, often, toughness) but must not creep under operating loads.

Tensile strength is needed to some degree to avoid cohesive failure under stress. High tensile strength is not necessary for most sealants, except in high-pressure fluid sealing applications.

Compressive strength and compression set are important parameters for sealing building and automotive joints, threaded fittings, and gasketing. Compressive strength is the maximum compressive stress that a material can withstand without breaking down or experiencing excessive extrusion. Compression set is the inability of a sealant to reexpand to its original dimensions after being compressed. It is usually expressed as the percentage loss of original thickness of a specimen after compression. High compression set is usually caused by further curing or degradative cross linking of the material and is very undesirable in a joint that can expand and contract. It is important, particularly for transportation and industrial sealants, that compression set be measured under actual conditions of use, for example, at high operating temperatures and in contact with the fluid to be sealed. In general, silicones and urethanes have the lowest compression sets, followed by polysulfides, butyls, and acrylics, although large variations in quality exist among suppliers.

Stress relaxation is a condition in which the stress decays as the strain (amount of elongation) remains constant. Some very low modulus sealants literally get pulled apart when held at quite low elongations.

Creep is the condition in which the strain (amount of elongation) increases as the stress remains constant. This property is very important for applications in which sealants must bear structural loads.

Thermal Properties

Heat and Cold Resistance. All elastomers of the type used in sealants get stiffer when cold and softer when hot. If a joint will experience low and high temperatures in service, test data should be generated to include cycling between the temperature extremes. The effect of heat aging on sealant properties should always be measured, particularly if continuous or intermittent exposure to high temperatures is expected. High temperatures usually soften a material initially and are often followed by degradation and embrittlement. Acrylics and butyls are not normally recommended for use in temperatures above 80 °C (175 °F). Urethanes and polysulfides are normally used below 100 °C (212 °F), but special high-temperature versions to 150 °C (300 °F) exist. Normal continuous service temperature for anaerobics is 150 °C (300 °F), but 200 °C (390 °F) versions are available. Most silicones are recommended for use to 200 °C (390 °F), but special reversion-stable versions have been tested and used successfully to over 300 °C (570 °F) in automotive gasketing applications. Typical data for the heat resistance of automotive under-hood elastomers (Ref 18) provide useful guidelines for selecting suitable sealants.

Chemical Properties

Fluid Resistance. Consideration must be given to the environment the sealant is likely to experience. For most aerospace, automotive, and industrial sealants, suppliers have generated compatibility charts for many different industrial fluids and chemicals. In general, highly cross-linked thermoset plastics like anaerobics have extremely good resistance to most chemicals. It should be borne in mind that, compared to the length of the seal (potential leak path), the area of exposure of the sealant to a chemical is very small in applications such as thread or porosity sealing. Thus, some degradation can be tolerated without loss of the seal. Elastomers are not as highly cross linked as anaerobics and can often be severely swollen, debonded, and degraded by certain fluids. Data for the resistance of elastomers to automotive fluids have been published (Ref 18).

It should be noted that resistance to one chemical does not necessarily transfer to nominally similar chemicals. For example, silicones can be formulated to be quite resistant to hot oils, but are highly swollen by gasoline. Water and moisture resistance is often very important in sealant applications in which weather exposure or water immersion is expected, for example, in construction, glazing, marine, electronics, and electrical applications. Urethanes are preferred for below-waterline applications in marine markets, whereas silicones are not recommended even though they are commonly used as seals in aquariums. Both silicones and urethanes allow diffusion of water vapor, whereas butyls and polysulfides have good water resistance, coupled with extremely low water vapor transmission rates.

Construction sealants are not designed to seal against chemicals, but can be exposed to them in use, for example, sealants subjected to road deicing chemicals or to chemicals used or stored nearby. Furthermore, many sealants are not exposed to aggressive chemicals in service, but are in manufacturing operations. Most electronics encapsulants experience only moderate temperatures and no chemical exposure in service, but commonly experience hot soldering operations and boiling solvent washes in production.

UV resistance is very important in all types of joints exposed to the sun, including those used in construction and automotive glazing. Some sealants, such as silicones, are inherently UV resistant, whereas others, such as urethanes, must be shielded or have UV absorbers and inhibitors added to the formulations.

Adhesion. Although the main function of a sealant is to fill a void completely, adhesion to the joint surfaces is critical for most sealing applications (there are exceptions, for example, a gasket that is held under compression). Loss of adhesion in a threaded or overlap joint can allow a fluid to wick into the sealant-substrate interface, leading to increased degradation of the sealant, or corrosion of the substrate, or leakage. It also can allow vibration loosening of a threaded joint, ultimately leading to failure.

In a construction joint, any loss of adhesion causes the sealant to move away from the surface, leaving a gap. Even in a gasket sealed by compression, adhesion can be a benefit. Loss of tension in such a system, such as occurs in the relaxation of fasteners, will ultimately lead to failure. Strong adhesion to the flanges postpones the onset of that failure until debonding occurs.

A primerless sealant that will strongly and permanently adhere to every surface under every possible condition is not considered realistic, especially when the enormous range of surfaces is considered, along with all the possible surface contaminants. The sealant chemist endeavors to make a sealant as versatile as possible and builds appropriate self-bonding adhesion into his sealant system by incorporating such features as oil-cutting chemicals and adhesion promoters.

Adhesion promoters are molecules that diffuse to the sealant-surface interface and behave like oil-in-water emulsifiers. Part of their molecular structure is very compatible with the sealant, and another part is very compatible with the surface; thus, they act as a bridge between the two. In some cases, solutions of these adhesion promoters are used as surface primers prior to sealant application. Reactive silanes have been found to be versatile adhesion promoters for many sealant and adhesive chemistries. The term primer can encompass these adhesion promoters, cure accelerators, corrosion-inhibiting primers, and cleaners. The importance of bond preparation and cleanliness cannot be overemphasized, although more than simple abrasion or solvent wiping is not practical in many industries. In critical applications, such as aircraft assembly, such preparation is routine. Corrosion-inhibiting primers, such as phosphoric acid anodizing, are used prior to application of adhesion promoter and sealant.

Electrical Properties

Most sealants are good insulators. In electronics and electrical applications that range from very low to very high voltage systems, special sealants with controlled dielectric constants, dielectric strengths, and dissipation factors are available. A few sealants are deliberately made electrically conductive for sealing and gasketing assemblies where radio frequency (RF) and electromagnetic interference (EMI) shielding are important.

Design of Sealing Systems

The design of a sealing system involves more than choosing the sealant with the right physical and chemical properties. The design of the component to be sealed, the performance expected of the sealant, the expected life of the finished product, and the practical and economic consequences of using a particular product must all be considered in selecting the best sealant.

Construction Joints. Klosowski (Ref 5) has published a checklist for sealant selection (Table 2) and has classified sealants as low, medium, and high range (see Table 3). There are numerous types of construction joints (for example, expansion and contraction joints, overlap joints, perimeter joints,

Table 2 Checklist of considerations for the selection of construction sealants

Required joint movement
Minimum joint width
Required strength
Chemical environment
In-service temperatures
Temperatures at time of application
Intensity of sun and weather in service
Longevity
General climate at application time
Materials cost, initial and lifetime
Installation cost
Other
   Fungicides
   Radiation resistance
   Insulating or conductive requirements
   Color
   Intrusion or abrasion resistance
   Cure rate
   Below-grade or continuous water immersion
   Accessibility of joint
   Priming
   Special cleaning requirements
   Dryness
   Other restrictions

Adapted from Ref 5

Table 3 Categorization of construction sealants

Range	Materials	Characteristics(a)
Low range	Vinyl latex, oil based, bituminous, polybutenes	JMC, ±3%; service life, 2–10 years; low cost
Medium range	Acrylic latex, solvent acrylics, neoprenes, butyls	JMC, ±5–±12.5%; service life, 5–15 years; medium cost
High range	Polysulfides, urethanes, silicones	JMC, ±12.5–+100/−50 (expansion/compression); service life, 10–50 years; high cost

(a) JMC, joint movement capability as percent of joint width

and butt joints). They can be horizontal, vertical, or inclined.

One of the fundamental design principles for a moving joint is to allow the flexible sealant to expand and contract. Thus, adhesion is required only to the two sides of the joint that will move. Figure 3 shows a corner joint with an installed release tape to prevent adhesion to the third side of the joint. The joint movement capability of a sealant is best exploited in a wide joint, rather than a narrow one. Joints that are too deep cause severe stresses and should be avoided or made shallower with a backing material.

An important consideration for any sealing operation is the ease of handling and applying the sealant. Single-component primerless sealants are easier to handle than two-component or primed systems. Hot-applied systems can be difficult to use, and products containing solvents may cause odor problems either during or after installation.

Table 2 is also very appropriate for similar joint systems in transportation and industrial applications. In the automotive industry, techniques learned from aircraft assembly are leading to the designing-in of sealants that contribute to the structural integrity and durability of the vehicle. Body panels are zinc coated prior to assembly. Sealants are placed in overlapping joints to prevent corrosion, and urethane adhesive sealants in windshield assembly enable vehicles to withstand rollover tests more successfully.

Threaded joints in fluid systems are considered semipermanent, being removed only for maintenance operations (Ref 1). Threaded fittings come with tapered or parallel threads or a combination of the two. The four common types of threads used for threaded fittings are shown in Fig. 4, and the spiral leak paths that occur when they are tightened are indicated. Regardless of thread design, fittings will leak unless some form of a sealant is used. PTFE tape has been used for many years to lubricate and seal assemblies, but it does not stop vibration loosening and is also banned in most hydraulic applications because of shredding problems that can cause system blockages. Elastomers, such as those used in confined O-rings, can be very effective but can suffer from sloppy assembly techniques. Latex sealants preapplied to the male threads are very effective for medium-pressure hydraulic, air, and water seals and can be assembled and reassembled four or five times. The oldest and least expensive way to plug spiral leak paths is to use drying or nondrying pipe dopes. They are often successful on small fittings where nonaggressive fluids and excessive vibration are not encountered. On large-diameter fittings, shrinkage upon drying can be a major problem.

Anaerobic sealants are perhaps the most versatile sealants for threaded fittings. Applied as thin liquids or creamy pastes, they lubricate the threads for assembly and cure to a hard, solvent-resistant sealant. They also act as a threadlocker to prevent vibration loosening. Because they develop strength by curing after they are in place, they are forgiving of tolerances, tool marks, and slight misalignments. Anaerobics are as effective as elastomeric O-rings at a fraction of the cost. There are versions that are preapplied on bolts in which the anaerobic is microencapsulated and released upon assembly, although a preapplied epoxy adhesive sealant is the market leader in this area.

Weaknesses of anaerobics are inherent brittleness, which sometimes causes failure and prevent them from being reused upon disassembly, and a tendency to stress crack plastic fittings. The limitations of gap curing must be observed. Pipes over 25 mm (1 in.) in diameter can have gaps that exceed 1 mm (0.04 in.) at the crest of the threads, extending cure times or giving marginal performance.

Gasketing. Although many benefits can be gained from using liquid sealants that cure to make formed-in-place gaskets, many failures due to badly designed systems have been observed over the years. Common failures include unsuccessful attempts to retrofit a liquid gasket to a flange system designed for a solid gasket, nonflat or bowed flanges, lack of control over joint gaps and gaps that far exceed specifications. Liquid gaskets are often applied by a robot on an automated production line. An obvious but often overlooked fact is that any gaps in the applied bead (caused by malfunctioning equipment or air bubbles in the sealant) are causes of instant leaks.

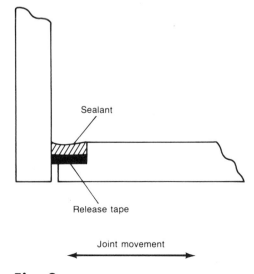

Fig. 3 Corner joint showing release tape

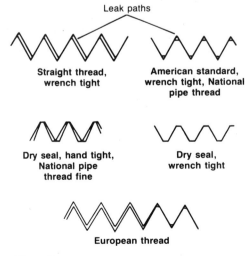

Fig. 4 Thread designs for threaded fittings

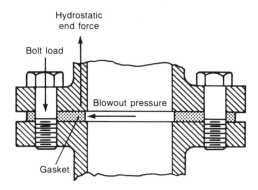

Fig. 5 Conventional compressed cut gasket

Some companies use visual identification systems to screen out misapplied gaskets prior to pressure testing.

The functioning of a formed-in-place gasket system is very different from that of a cut gasket. A conventional cut gasket (Fig. 5) maintains a seal by compression, induced by the tightening of the fasteners. It is essentially a compressed spring, comprising flanges, gasket, and fasteners. Any loss of tension in the system caused by stretching of the fasteners, vibration loosening, or creep or relaxation of the gasket will ultimately lead to failure. In contrast, a formed-in-place gasket does not cure until after assembly. It has metal-to-metal contact between the flanges, which, with the fasteners, carry. all of the tension. Even if some tension is lost, the adhesion of the sealant will often prevent the failure of the system. It is very important that flange systems be specifically designed for or modified for formed-in-place-gaskets.

Two types of sealants are commonly used, anaerobics and medium-modulus RTV silicones. Anaerobics require rigid flanges and small gaps and cure to a rigid system that is designed to prevent any joint movement. Medium-modulus RTV silicones are used in applications in which joint movement is expected. In this case, it is very important that the flange design allow a sufficient gap, primarily to permit atmospheric moisture to cure the silicone and secondarily to maximize the joint movement capability of the sealant. As with construction joints, two-sided adhesion is optimum (Fig. 6).

The use of formed-in-place gaskets is somewhat restricted by the limited range of usable sealant materials. Recently, the formation of preapplied conventional solid gaskets by curing a two-component or a UV-cured sealant on flanges has been commercialized.

Porosity Sealing. Most castings and powdered metal parts have microporosity. With castings, this porosity allows fluid leakage, and, with powdered metal parts, can cause severe plating problems. Sealing of the pores is carried out by impregnation with

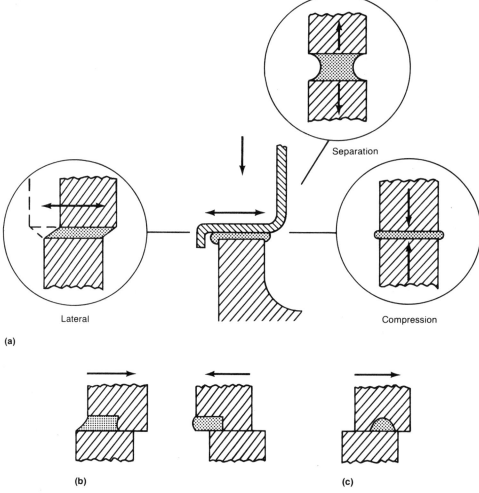

Fig. 6 Joint designs for formed-in-place gaskets. (a) Good design, two-sided adhesion. (b) Poor design, three-sided adhesion. (c) Poor design, four-sided adhesion

liquid sealants, using pressure or vacuum cycles, followed by curing (Ref 19). Three common types of impregnants are described below.

Water solutions of sodium silicates are the oldest products and are still used where temperature resistance over 260 °C (500 °F) is required. They suffer from very slow production and cure cycles and large shrinkage upon curing.

Styrene-base unsaturated polyesters are very effective for sealing a material with a large amount of porosity but produce hazardous vapor emissions and effluent after the rinsing of parts. They require heat curing for 1 to 2 hours at up to 140 °C (285 °F), which can produce resin bleedout.

Low-viscosity methacrylate resins are a much less hazardous alternative to the polyesters and are becoming the predominant sealant. They are available as products that are heat cured at 90 to 100 °C (195 to 212 °F) and as room-temperature-curing anaerobic versions. The latter are used for the most critical sealing operations where heat cannot be tolerated or where resin bleedout must be mitigated.

REFERENCES

1. G.S. Haviland, in *Machinery Adhesives for Locking, Retaining and Sealing*, Marcel Dekker, 1986
2. A. Damusis, *Sealants*, Reinhold, 1967
3. *Sealant Technology in Glazing Systems*, American Society for Testing and Materials, 1978
4. *Sealants: The Professionals Guide*, Sealants and Waterproofer's Institute, 1984
5. J.M. Klosowski, *Sealants in Construction*, Marcel Dekker, 1989
6. J.K. Panek and J.P. Cook, *Construction Sealants and Adhesives*, Wiley Interscience, 1984
7. R.B. Seymour, in *Encyclopedia of Chemical Technology*, Vol 20, Wiley-Interscience, 1982, p 549-558
8. R.E. Meyer, in *Encyclopedia of Polymer Science and Engineering*, Vol 15, Wiley-Interscience, 1989, p 131-145
9. H.W. Prost, *Silicones and Other Silicon Compounds*, Reinhold, 1949
10. E.G. Rochow, *An Introduction to the Chemistry of the Silicones*, John Wiley,

1946

11. M.D. Beers and D.M. Brassard, "3rd Generation Silicone Gasketing Technology," Paper presented at the 90th Society of Automotive Engineers Congress, Feb 1990

12. K.T. Frisch, in *Urethanes Technology*, Vol 4 (No. 1), March 1987, p 22-26

13. J.C. Patrick and N.M. Mnookin, British Patent 302,220, 1927

14. J. Hutt and H. Singh, U.S. Patent 3,923,748, 1975

15. J. Hutt, H. Singh, and M. Williams, U.S. Patent 4,366,307, 1982

16. E.W. Flick, *Adhesive and Sealant Compound Formulations*, Noyes Publications, 1984

17. E.W. Flick, *Adhesives, Sealants and Coatings for the Electronics Industry*, Noyes Publications, 1986

18. J.W. Horvath, *Rubber World*, Dec 1987, p 21

19. J.A. McNickle, *Die Cast. Eng.*, Jan/Feb 1985, p 68-69

Sealants Markets and Applications

Michael G. Elias, Russell Redman, and Steven Tyler, Tremco, Inc.

SEALANTS are formulated products used to fill discontinuities in a structure or assembly. A sealant fills joints, gaps, and cavities between two or more similar or dissimilar substrates, in contrast to an adhesive, which is primarily used to bond two materials together. Sealants create a working environment for the sealed assembly in which conditions can be isolated and controlled. Without sealants, these structures would not be as convenient or valuable to the user. Sealed assemblies must endure a wide range of thermal, environmental, chemical, and joint-movement conditions. Therefore, at a minimum, sealants must seal and perform within the service use conditions of the assembly.

Sealants are categorized in this article as caulks, sealants, or sealants adhesives. Caulks are materials used in applications where only minor or no elastomeric properties are needed. Sealants and sealants/adhesives are materials used in applications where elastomeric as well as structural strength properties are required and specified.

At this writing, the total market for sealants in the United States alone is valued at more than $1.5 billion. The main sealants markets are the construction/consumer, transportation, industrial, aerospace, appliance, and electronics market segments. The types of sealants used in each of these markets are identified in Table 1 and described below. In addition, the Section "Sealant Materials" in this Volume contains articles that fully describe every chemical family mentioned in this article.

Construction/Consumer Market

The key performance issues for the construction/consumer market are the adhesion, movement capability, and long-term durability of the sealant. In the construction industry, sealants must perform on a wide range of substrates, such as metal, glass, plastic, masonry, natural stone, and wood, and must provide a continuous seal to prevent wind, water, and other environment from entering the structure and causing cosmetic and structural damage.

A list of major exterior and interior construction sealant applications includes:

Exterior applications

- Horizontal and vertical control and expansion joints
- Perimeter joints
- Mullion and curtain wall panel joints
- Pointing of brick and stone masonry
- Horizontal paving or traffic-bearing joints
- Joints in water/chemical retainment structures
- Sealing of glass into window frames
- Internal seals in window framing systems
- Roof joints (parapets, reglets, flashings, pipes, ducts, vent openings)

Interior applications

- Horizontal and vertical control and expansion joints
- Perimeter joints
- Floor joints
- Duct, pipe, electrical penetration joints
- Fire stops
- Plaster and trim joints
- Sanitary fixture sealants

Sealant selection and specifications are based on a number of design and functional considerations, such as size and complexity, function of the structure, movement capability of the sealant, aesthetics, environment conditions (both exterior and interior), cost, and life expectancy. Table 2 lists general characteristics for the commonly used sealant types classified by movement capability.

The form and packaging of the sealant to be used for installation is a final consideration. It can be a wet sealant supplied in cartridge form, sausage packs, or in bulk units of 1.5 to 50 gal. Nonsag or self-leveling sealant forms are also available. These extruded materials may be tacky tapes or cured rubber gaskets.

The consumer market uses both exterior and interior sealants. Typical applications are for exterior seals around windows, doors, siding, and for roofing penetrations. For these applications, acrylic and silicone sealants and latex caulks are the predominant materials. Butyl caulks are widely used on aluminum and vinyl siding.

The commercial construction market uses high-performance exterior sealants, such as urethanes and silicones, on the wide range of substrates used in wall construction. Applications include control and expansion joints, perimeter joints, horizontal joints, pointing of masonry, and roofing joints. Glazing applications require that materials seal the glass into a frame using wet sealed systems, dry sealed systems, or wet/dry systems.

Wet systems include butyl tapes for low-rise buildings. Dry sealed systems use preformed vinyl, ethylene-propylene-diene monomer (EPDM) rubber, neoprene, and silicone gaskets to seal the glass in a window frame by compression. The wet/dry systems are combinations of sealants and compression glazing systems. These in-

Table 1 Sealant usage by generic sealant type

Generic type	Construction	Industrial	Transportation	Appliances	Aerospace
Asphalt, bitumen	X	X			
Oleoresins	X	X			
Butyl	X	X	X		
Hypalon	X				
EPDM	X	X	X		
Neoprene		X			
Styrene-butadiene		X			
Polyvinyl chloride		X			
Acrylic solution	X	X			
Acrylic emulsion	X	X			
Polyvinyl acetate	X	X			
Polysulfide	X	X	X	X	X
Polyurethane	X	X	X	X	X
Silicone	X	X	X	X	X
Fluoropolymer		X	X	X	X
Epoxies	X	X	X		
Intumescents	X				

Table 2 Sealant characteristics (wet seals)

Sealant type	Movement capability, %	Service life, years	Maximum joint width		Relative cost
			mm	in.	
Oil based	±1(a)	3–5	6.4	¼	Low
Butyls	±5(a)	5–10	12.7	½	Low
Latex	±7.5(a)	5–10	12.7	½	Low
Solvent acrylic	±12.5(b)	15–20	19	¾	Medium
Hypalon	±12.5(b)	10–20	6.4	¼	Medium
Neoprene	±12.5(b)	10–20	6.4	¼	Medium
Polysulfides	±25(c)	20		1⅓	High
Urethanes	±25–50(c)	20–30	>76	>3	High
Silicones	±25–50(c)	30	>76	>3	Very high

(a) Low movement. (b) Medium movement. (c) High movement

clude tapes, gaskets, and silicone sealants used in combination to form a weathertight seal on the glass. Acrylic, butyl, urethane, and silicone sealants are used for the internal seals in window and curtain wall systems. Silicone sealants are used for two- and four-sided structural glazed window systems, acting as a sealant/adhesive on two or four sides of the glass to seal and structurally retain it to the metal building framing.

A fast-emerging sealant application is use as a fire stop. Fire stops are needed in mechanical openings (including plumbing, electrical, air handling, and others) in walls and floors to prevent the passage of fire and smoke through a building. These fire stops must retain their integrity for a specified range of time and temperature, depending on construction code requirements.

Conventional fire stopping has typically included mortar and wood referred to as fire safe. Intumescent technology is based on the ability of the material in an opening to swell and fill the fire stop opening. These intumescent materials are provided in wraps, boards, and sealants. One-part, easily installed latex and solvent-base sealants that are highly filled with ceramic fibers are also used widely as fire stops.

Over the past ten years. new technologies have emerged in response to public concern for safety due to catastrophic fires. Silicone and urethane sealants are now available in a variety of forms, including nonsag and self-leveling sealants and foams. The sealants are typically backed by mineral wool to prevent fire penetration. Other materials used as fire stops include mechanical devices, pillows, and modified mortars.

Two types of sealant applications in the construction of metal buildings are metal-metal lap joints, which typically use butyl tapes or caulks, and typical construction joints, which use urethane, acrylic, latex, butyl, and silicone sealants, depending on the movement required and the metal building manufacturers guidelines.

Two common types of joints are used in highway construction: longitudinal (parallel to road) and transverse (expansion joints). These joints are sealed with a wide range of sealant types, including hot-pour and cold-pour modified asphalt or cold tar, preformed compression neoprene rubber seals, and very low-modulus silicone sealants.

Parking decks use one- and two-part urethane sealants for horizontal concrete expansion and control joints. The two-part urethane sealants are also available preformed in wide rubber strips, and sealed in place in the parking deck joints. Specialty applications for high-wear areas use a preformed neoprene rubber joint.

Suppliers. The major suppliers of sealants for the consumer market are Tremco, Inc.; DAP; General Electric Company, Silicone Products Division; and Macklanburg-Duncan Company. Major suppliers of sealants for the construction industry include Tremco, Inc.; Dow Corning Corporation; General Electric Company, Silicone Products Division; Mameco International; Sonneborn Division, Witco Chemical Corporation; Pecora Corporation; Sika Corporation; and Products Research & Chemical Corporation.

The automotive industry is the second largest consumer of sealants, after the construction industry. Sealants are used on cars, trucks, buses, and recreational vehicles. Important changes in the future will include the continued replacement of butyl tape by urethane sealant for windshield glazing, the demise of gilsonite dispersion as a body sealant, and the growth of butyl, styrene-butadiene rubber (SBR), and hot-applied, pumpable nitrile sealants in this same application. Hot-applied, pumpable grades are supplanting cold-applied, preformed SBR and nitrile-base mastics. Hot-melt polypropylenes are quickly making inroads in the sealing of water deflectors. As automakers continue to increase the use of plastics in vehicles, the consumption of sealants by the industry is expected to increase. The auto sealant market includes these market segments:

- Original equipment manufacturer (OEM) of cars, trucks, and buses
- Aftermarket (repair, repainting, body work) of cars, trucks, and buses, with both professional and do-it-yourself products
- OEM truck trailers
- Aftermarket truck trailers
- Recreational vehicles
- Other equipment, such as batteries and self-sealing tires

OEM cars, trucks, and buses consume a large volume of sealants for body sealing, windshield and backlight glazing, and miscellaneous end-uses, such as weather strips, air conditioning/heating components, and formed-in-place gaskets. Body sealants and glazing materials include plastisols, gilsonites, and preformed ethylene-vinyl acetate (EVA) hot melts, as well as preformed SBR, butyl, and nitrile mastics. Pumpable hot-applied butyls, SBRs, and nitriles, as well as hot-applied polyurethanes are also used. Gilsonite is expected to be eliminated and replaced by plastisols and hot-applied pumpable rubbers, the latter of which are projected to greatly increase in use from the 1990 volume.

OEM windshield glazing has been accomplished with urethane and with butyl tapes. At least two manufacturers are in the process of converting to polyurethane. The changeover is the result of stricter government regulations, which can be met more easily with sealants that provide high structural strength.

Other OEM applications consume a considerable amount of sealants. Weather stripping and taillights are sealed with pumpable hot-melt butyls, whereas air conditioning/heater components use pumpable EPDM. As mentioned above, water deflector sealing polypropylene hot melts are replacing preformed butyl mastics. Formed-in-place gaskets are the domain of silicone rubber. Anaerobics and unsaturated polyesters are employed as porosity sealants for engine blocks. They are also used to seal threaded pipe fittings, but the use of preapplied aqueous acrylics in place of anaerobics is increasing.

The professional aftermarket encompasses body shops, glazing shops, and service stations. As in OEM, sealants are used for body sealing, glazing, and formed-in-place gaskets. The body sealants used are cold-applied SBR mastics, nitriles, and preformed butyls. Although cellulose nitrate has been used to fill in the pinholes and scratches, it is being supplanted by magnesium silicate glazing putties and high-solids acrylic lacquers. Aftermarket glazing follows the lead of OEM in the conversion to polyurethanes, which now constitute one-third of the market, but are expected to account for nearly two-thirds in 1991. Formed-in-place gaskets are made of silicone, and rosin sealant is employed to seal preformed cork and rubber gaskets and threaded fittings.

Retail aftermarket sealants represent the do-it-yourself products available through hardware, home supply, and other outlets. Products include silicone formed-in-place gaskets, silicone sealants for glazing, rosins for gasket sealing, and materials for filling in surface imperfections.

Self-sealing tires use natural rubber and butyl sealants, which are applied to the inner tire surface to prevent the loss of air in the event of a puncture. Automobile storage batteries use sealants to bond and seal the lid to the casing body. Most automobile batteries now have polypropylene casings and are dielectrically sealed, and hence need no sealant. However, a relatively small number of hard rubber batteries are still used for heavy machinery, mining trucks, and other vehicles. These batteries use asphalt and epoxy sealants.

Truck trailer manufacturers use polyurethanes and neoprenes for exterior joints and silicones in the interior of the body. Small amounts of polysulfides are also used by at least one manufacturer. Aftermarket repair and maintenance of truck trailers use similar materials.

Recreational vehicles consume polybutene and butyl tapes, solvent-release butyl caulk, and acrylic latex sealant. These products are sold mainly to the OEM market. Tapes are used around door and window frames and for sealing gutter and drip rails.

Suppliers. The largest suppliers to the automotive industry include 3M, supplying a variety of products; Loctite Corporation, supplying silicones, anaerobics, and acrylics; Essex Chemical Corporation, supplying polyurethane windshield sealants; General Electric Company, supplying silicones for gaskets and glazing; Mortell Company, supplying plastisols; and Protective Treatments Inc., supplying butyl tape and plastisols. Others include SWS Silicones Corporation; Chemseco, Chemical Division (formerly Trenton Division); Chrysler Corporation; and Dow Corning Corporation.

Industrial

The industrial market consists of these segments: insulated glass, in-house glazing, and metal buildings. There are 2000 companies in the United States that manufacture insulated glass units. These units are commercialized at both the residential and nonresidential level. The larger, nonresidential manufacturers typically produce units with higher tolerances. Significantly longer warranties are often found here.

Conventional design in insulated glass construction includes various forms:

- Single seal, which requires a desiccant and a sealant between two units of glass
- Double seal, which requires a desiccant in combination with both a polyisobutylene (PIB) primary seal and a secondary seal
- Fused edge, which requires a monolithic layer of sealant material fused into a channel and used as a form

- Swiggle strip, which uses a patented PIB-base formulation of desiccant and mastic manufactured with a corrugated metal shim/spacer in the middle
- Triple-sided extrusion, which requires an extruded metal spacer that is preformed and encapsulated by desiccated mastic

The sealants used in these designs include dual-component polysulfides; polyurethanes with high solids contents and outstanding physical characteristics; thermoplastic hot-melt butyls; single- and dual-component silicones; and polyisobutylenes. Single-component sealants are typically more cost effective, but dual-component sealants have inherently faster cure rates and predictable adhesion. PIB sealants are mainly used as the primary seal in dual-seal construction.

In-house glazing requires both wet and dry applications that use sealants, tapes, or gasket extrusions. Wet-applied sealants can be either exposed or concealed. Exposed wet-applied sealants are typically silicones and acrylics. The medium-modulus structural silicones (both single- and dual-component), which are typically neutral, are used for cap beads, toe beads, heel beads, air seals, butt joints, and stopless applications.

Tapes are typically thermoplastics and butyls that can be used as bedding, end sealing, vision lites, panels of steel, and aluminum and porcelain surfaces. Extruded gaskets are used in dry glazing systems and are typically neoprene or EPDM. These extrusions have precise tolerances and are continually tested through production via a shadow graph or template. High-performance glazing systems typically combine both wet and dry components.

The metal building market uses acrylic and butyl sealants at window and door perimeters and in panel joints where specific performance criteria are required. Other industries also use these products and seam joints, either in a standing seam, such as on a metal roof, or at a roll line, such as on truck trailer panels. Butyl tapes or pumpable butyl sealants are easy to use and effective in lap joints on the roof panels and wall flashings of metal buildings. Concealed nondrying butyl tapes and sealants are used in this segment where a retained soft, uncured quality is required. These sealants serve as void fills at roll-forming lines and in lap joints.

Suppliers. Major suppliers of insulated glass are Tremco, Inc.; ADCO; H.B. Fuller Company; Bostik Division, Emhart Corporation; Morton Thiokol, Inc.; Delchem; Dow Corning Corporation; General Electric Company, Silicone Products Division; and Products Research & Chemical Corporation. Suppliers of in-house glazing sealants include Tremco, Inc.; Dow Corning Corporation; General Electric Company, Silicone

Products Division; and H.B. Fuller Company.

Suppliers of metal building tape are Chemseco; Schnee-Morehead; and Tremco, Inc.

Aerospace Industry

The aerospace industry represents a very large market. Its largest segment, aircraft, includes civil and military subsegments. Military aircraft production is growing faster than civil (general aviation) production.

Aerospace products other than aircraft include missiles, space systems, and support products and services. Increased defense expenditures for military aircraft, as well as healthy aircraft exports, will enable military sealant applications to increase annually. Although missiles represent a smaller market for sealants, sealant applications are expected to grow commensurate with increased spending on missile programs. The use of sealants in civil aircraft is expected to increase yearly because of a need for large transport aircraft. General aviation and civil helicopters represent smaller markets with little prospect of growth.

Integral fuel tanks and cabin pressure sealing are the two major applications for sealants in aircraft. Integral fuel tanks are the internal cavities, usually in the wing, that contain the fuel. Sealants are applied to joints, seams, channels, and fasteners to prevent leakage. The sealants are supplied in a range of viscosities to suit the application method. The methods used for applying sealants applications include filleting, for sealing sheet metal lap joints; the use of channel sealants, for injection between the two halves of a lap joint, which is inaccessible to filleting; and faying surface sealing on the contacting surfaces of a lap joint.

The aerospace industry primarily uses five sealant types: polysulfides, silicones, fluorosilicones, polyurethanes, and butyls. Polysulfides represent the majority of sealant usage and are principally used for sealing integral fuel tanks and for cabin pressure sealing, and, to a lesser degree, for aerodynamic smoothing and water tank sealing. Wide-body planes use more than 750 L (200 gal) of sealant per plane, including a significant quantity that is wasted in repackaging and dispensing. Smaller craft use about half that amount.

Two-component polysulfides are the most commonly used sealant because they have excellent resistance to the jet fuel stored in integral fuel tanks. They also have good aging and weathering properties. Their major drawbacks are sensitivity to high temperature and high weight per gallon (about 1.4 kg/L, or 12 lb/gal).

The polysulfides are usually cured with manganese dioxide or dichromates. Large aircraft manufacturers purchase the polysulfides in drums, mix the two parts, and repackage the product in cartridges. It is then frozen until use. Smaller users purchase polysulfides packaged in small kits.

Silicone sealants are also generally two-component formulations and are used where superior high-temperature resistance is a necessity, such as around engine firewalls. They lack jet fuel resistance and hence are not employed in fuel tanks or other areas where fuel may contact the sealant. They have excellent weathering resistance and are used in some window sealing and aerodynamic smoothing applications.

Fluorosilicones, like silicones, are used in high-temperature applications. However, fluorosilicones have jet fuel resistance, which silicones lack. Therefore, their use is reserved for higher-temperature areas that contact oil, jet fuel, or other solvents. Room-temperature vulcanizing (RTV) fluorosilicones are also used in and around the fuel tanks. Fluorosilicones of the noncuring type are used as channel sealants, primarily in military aircraft.

Polyurethanes have been under evaluation for cabin pressure sealing. One aircraft manufacturer has been testing polyurethane for several years, and limited commercial use is expected in the early 1990s. The major advantages of polyurethane are its formulation flexibility and its density, which is lower than that of polysulfide. Its major drawbacks are the necessity for a primer and concerns over isocyanates used in the shop area.

Butyls are primarily used as sealants for the vacuum bags that are necessary in the manufacture of the composite parts used in aircraft.

Suppliers. Products Research & Chemical Corporation is the largest overall supplier of aircraft sealants. In 1983 they acquired the Pro-Seal product line from Essex Chemical Corporation, which had been the second largest supplier. However, Dow Corning is the leading supplier of silicones and fluorosilicones. Other suppliers include Goal Chemical Sealants Corporation; Chem Seal Corporation of America; and General Electric Company, Silicone Products Division.

Appliances

Refrigerators and freezers have been consuming the bulk (85%) of the sealants used in appliances. Although highly filled, low-cost polypropylene hot melts are currently preferred for sealing refrigerator and freezer cabinets, they are expected to be displaced in about half of the units manufactured in 1991 by unfilled foamed polypropylene hot melts. The foaming of unfilled polypropylene hot melts reduces the amount of applied sealant, compared to the filled polypropylene sealant currently being used. However, the dollar value of polypropylene sealants is expected to increase yearly because of a slight growth in refrigerator production and because the unfilled polypropylene that is necessary for foaming costs more than filled polypropylene. In addition to polypropylene, asphalt, silicone, polybutene, and polyvinyl chloride (PVC) plastisol are also used in the appliance market.

As stated above, polypropylene hot melts are primarily used to seal cabinets in refrigerators and freezers. Small amounts are also used to seal off areas in air conditioners. Most polypropylene hot melt is modified with butyl rubber and is heavily filled, yielding a low-cost product. Foamed unfilled polypropylene hot melts cost twice as much per pound. These unfilled products are foamed to 50 to an additional 70% of their original volume, thereby giving a greater yield. The foamed product is also more readily applied by robots. It is anticipated that major appliance manufacturers will switch about half of their production to foamed hot melts by 1991.

Asphalt is still used as a cabinet sealer to a small degree, but it originally lost market share to polypropylene hot melts. It currently is used in air conditioners, as are polypropylene hot melts, to seal off areas from water penetration.

Polybutene caulk, supplied in putty, rope, or preformed gasket forms, is used to seal holes that are made in refrigerator and freezer cabinets for internal components, such as refrigeration lines. This nondrying, nonhardening caulk is applied manually. The increased automation of production lines will cause part of this market to be usurped by foamed hot melts because they can be applied automatically. However, a significant amount of polybutene caulk is used in areas where automated application is difficult.

Silicones are used in a wide variety of major and small appliances. The one-component RTV acetoxy-cured silicones predominate, although some alkoxy-cured silicone is used where the release of acetic acid upon curing would cause corrosion. Silicone is primarily used to seal interior gaps, for which it has U.S. Federal Drug Administration approval. Its clear or white color is desirable in applications that are visible to consumers.

The sealing of the bottom tray in microwave ovens is a growing market. Silicone is also used to seal the glass tops in cooking top ranges and the plumbing fixtures in dishwashers and washing machines, and as a gasket material for their transmissions. Silicone is well suited for many small appliances such as irons, coffeemakers, skillets, toaster ovens, and so on, because of its high-temperature properties and strong adhesion to dissimilar materials.

A small amount of PVC plastisol is currently used to seal components in refrigerators.

Suppliers. General Electric is the leading supplier to the industry because of its premier position in silicone production. Other suppliers include Dow Corning Corporation, silicone; H.B. Fuller Company, polypropylene hot melt and polybutene; Mortell Company, asphalt and polybutene; National Starch and Chemical Corporation, polypropylene hot melt; Findley Adhesives, Inc., polypropylene hot melt; DeVan Sealants, Inc., polybutene; Ohio Sealants Company, Inc., plastisol; and Inmont Corporation, polybutene.

Electronics

Sealants are used to protect electronic components that are used in aerospace, automotive, and appliance equipment. The primary sealant compounds used in these applications are acrylics, silicones, urethanes, polysulfides, epoxies, and unsaturated polyesters.

Liquid forms of these products are converted irreversibly to solids with no significant weight loss. This type of sealing utilizes techniques known as potting, encapsulating, and casting. These processing techniques are defined below.

Potting involves placing a device or component in a container, or pot, which is usually referred to as a can, if metal, or a shell, if plastic. The container is then filled with the potting material, which becomes the exterior wear surface of the component or device.

Encapsulating involves protecting an electronic or electrical device with a polymer-type material that forms a thick protective envelope (255 to 5080 μm, or 10 to 200 mil) around the device or component. This is normally accomplished by pouring or injecting the encapsulant into a cavity containing the component. Later, the cavity is removed. Although encapsulating in a cavity or shell is the predominant method for these types of sealing materials, it is by no means exclusive. A significant amount is accomplished by dip coating and by impregnation techniques.

Casting is the same technique as potting, except that the container is a temporary mold. The potting material becomes the exterior wear surface of the component or device.

Commercial demand sometimes requires automation of these processes. Automated dispensing machinery is available for integration into existing production lines.

Reasons for sealing electronic and electrical devices and components are:

- To protect the devices from the environment
- To allow the heat to dissipate
- To prevent the intrusion of moisture
- To prevent preset adjustments from being altered

- To shock proof the devices
- To mount the components
- To meet the requirements of a specification

Suppliers include Castall, Inc.; Conap Inc.; Emerson & Cuming; Epic Resin, Division of RTE Corporation; Hyson Division, The Dexter Corporation; 3M Company; Dow Corning Corporation; General Electric Company; McGhan Nusil Corporation; Isochem Products Company; Mereco Products, Division of Metachem Resins Corporation; Thermoset Plastics, Inc.

SELECTED REFERENCES

- P. Bugbee, "Principles of Fire Protection," Batterymarch, Nov 1978
- A. Damusis, Ed., *Sealants*, Van Nostrand Reinhold, 1967
- M. Elias, Silicone Sealant Technology, Markets Continue to Grow, *Adhes. Age*, Vol 8, May 1986
- "Low Rise Building Systems Manual," Metal Building Manufacturers Association, 1986
- J.R. Panek and J.P. Cook, *Construction Sealants and Adhesives*, 2nd ed., John Wiley & Sons, 1984
- J.R. Rundo, "Insulating Glass Evolution," Tremco, Inc., May 1988
- R. Salmon, *Encapsulation of Electronic Devices and Components*, Marcel Dekker, 1987
- I. Skeist, Ed., *Handbook of Adhesives*, 3rd ed., Van Nostrand Reinhold, 1989

Specifications and Standards for Adhesives and Sealants

John Nardone, Consultant, Nardi & Associates

STANDARDIZATION documents are the basis for the effective use of all materials. They are developed to ensure that the materials purchased are acceptable, that they were effectively processed, and/or that the final-product quality requirements are met. Recommended standardization practices serve a vital role in the successful utilization of adhesives and sealants.

Standardization documents originate within both industry and the U.S. government, each reflecting its interest in achieving high-quality products. Although the two types of standardization documents, that is, specifications and standards, are distinctly different, the terminology is sometimes used interchangeably, or the term standards is used broadly, as in much of this article.

Specifications are documents used in the procurement process that focus on raw materials and products. Commonly referred to as specs, they describe the technical requirements that are the basis for product acceptance and the procedures to be used in measuring the requirements. Specifications may also include guidance for shipping and storage, vital factors for many adhesives.

Standards are documents that establish requirements for "standardized" test methods, practices, or procedures connected with the procurement or performance evaluation of materials or products. Standards include a wide assortment of documents, including government handbooks.

Specifications relative to adhesives deal primarily with substances that hold materials together, whether the surface attachment is of a structural member of an airplane or the closure of a container. Hot-melt resins, cements, pastes, glues, and mucilages are included in the category of adhesive substances. On the other hand, sealants are used as gap-filling materials to seal joints and prevent the penetration of liquid or gaseous substances. Included in this category are caulking compounds, putties, elastomeric compounds, preformed gaskets, and sealing tapes. When an adhesive is capable of providing both adhesion and sealing functions, it may be designated as an adhesive/sealant, thereby indicating its dual role.

Numerous specifications and standards originate with manufacturers of adhesives and sealants and with those companies that apply these materials to products. For the most part, these are proprietary documents and are not available for consumer use. As a result, professional associations and societies, as well as the government, develop specifications and standards to ensure that the respective interests of industry and the government are considered.

Industry standards of interest to those using adhesives and sealants are issued by the following professional and trade organizations: the American Society for Testing and Materials (ASTM), the Society of Automotive Engineers (SAE), and the Technical Association of the Pulp and Paper Industry (TAPPI). A limited number of useful standards are published by other organizations, but these are mainly for special-use purposes.

U.S. government standards are primarily issued by the General Services Administration (GSA) and the military, officially referred to as the Department of Defense (DoD). Of secondary interest are those standards issued by the National Aeronautics and Space Administration (NASA) and the National Institute of Science and Technology (NIST), formerly the National Bureau of Standards (NBS). Within the Department of Defense, a program has been established that requires the military to use nongovernmental standards to the maximum possible extent (Ref 1). Also, the military requires coordination with industry associations to communicate DoD concerns, needs, and requirements. This practice eliminates the need for the development and maintenance of separate DoD specifications and standards. Government representatives participate in the review and preparation of industry standards and concur in their adoption for use in the acquisition of military material. Standards thus approved by the Department of Defense are designated DoD-adopted.

The following overview describes various organizations involved in adhesive and sealant specifications and standards and lists some of the important publications and available information.

Industry Specifications and Standards

ASTM Specifications and Standards. ASTM, 1916 Race Street, Philadelphia, PA 19103, was founded in 1898. It is a scientific and technical organization formed for the "development of standards on characteristics and performance of materials, products, systems, and services, and the promotion of related knowledge" and is the largest source of the world's voluntary consensus standards. ASTM standards for adhesives and sealants are issued by committees within ASTM, referred to as D14 and C24, respectively (Ref 2).

ASTM Committee D14 on adhesives has 11 subcommittees dedicated to the development of material specifications and methods for testing for adhesives and adhesive/sealant materials. The subcommittees also provide definitions of terms and recommended practices connected with research on adhesives and adhesion. ASTM Committee C24 on building seals and sealants has 20 subcommittees, whose interest is not only the development of material specifications and methods for testing, but in the promotion of knowledge and stimulation of research for the sealing of building joints and traffic decks considered part of a building complex.

The two ASTM committees are responsible for approximately 235 specifications and standards for adhesives and sealants. These and the other formally approved ASTM specifications, standard test methods, definitions, practices, guides, and classifications are published in the *Annual Book of ASTM Standards*, which consists of over 60 volumes. The set, issued annually, may be purchased directly from ASTM in hard copy or in microfiche form. Single copies, called separates, may also be purchased. The volumes of particular interest to those using adhesives and sealants are Volume 15.06, *Adhesives* (Committee D14) (Ref 3) and Volume 04.07, *Building Seals and Seal-*

Table 1 Typical ASTM standard specifications and test methods relevant to adhesives and sealants

Designation	Title
ASTM C 557	Specification for Adhesives for Fastening Gypsum Wallboard to Wood Framing
ASTM C 920	Specification for Elastomeric Joint Sealants
ASTM D 905	Test Method for Strength Properties of Adhesive Bonds in Shear by Compression Loading
ASTM D 950	Test Method for Impact Strength of Adhesive Bonds
ASTM D 1151	Test Method for Effect of Moisture and Temperature on Adhesive Bonds
ASTM D 2235	Specification for Solvent Cement for Acrylonitrile-Butadiene-Styrene (ABS) Plastic Pipe and Fittings
ASTM D 2559	Specification for Adhesives for Structural Laminated Wood Products for Use Under Exterior (Wet Use) Exposure Conditions
ASTM D 2851	Specification for Liquid Optical Adhesive
ASTM D 3933	Practice for Preparation of Aluminum Surfaces for Structural Adhesive Bonding (Phosphoric Acid Anodizing)
ASTM D 4317	Specification for Polyvinyl Acetate-Based Emulsion Adhesives
ASTM F 607	Test Method for Adhesion of Gasket Materials to Metal Surfaces

Table 2 Typical SAE specifications and test methods relevant to adhesives and sealants

Designation	Title
SAE AMS 1320	Decal Adhesive Remover
SAE AMS 3107	Primer, Adhesive, Corrosion Inhibiting for High Durability Structural Adhesive Bonding
SAE AMS 3374	Sealing Compound, One-Part Silicone, Aircraft Firewall
SAE AMS 3375	Adhesive/Sealant, Fluorosilicone Aromatic Fuel Resistant, One Part-Room Temperature Vulcanizing
SAE AMS 3376	Sealing Compound, Noncuring, Groove Injection, Temperature and Fuel Resistant
SAE AMS 3681	Adhesive, Electrically Conductive, Silver-Organic Base
SAE AMS 3686	Adhesive, Polyimide Resin, Film and Paste, High Temperature Resistant, 315 Degrees C or 600 Degrees F
SAE AMS 3695	Adhesive Film, Epoxy-Base for High Durability Structural Adhesive Bonding
SAE AMS 3704	Adhesive, Contact Chloroprene, Resin Modified
SAE ARP 1843	Surface Preparation for Structural Adhesive Bonding Titanium Alloy Parts
SAE ARP 4069	Aerospace Recommended Practice for Sealing Integral Fuel Tanks
SAE J 1523	Recommended Practice for Metal to Metal Overlap Shear Strength Test for Automotive Type Adhesives
SAE J 1525	Recommended Practice for Lap Shear for Automotive-Type Adhesives for Fiber Reinforced Plastics (FRP) Bonding

Table 3 Typical TAPPI technical information sheets and test methods relevant to adhesives and sealants

Designation	Title
TAPPI T 540	Determination of Polyethylene Adhesion to Non-Porous Substrates
TAPPI T 814	Peel and Shear of Hot Bonds at Elevated Temperature
TAPPI TI 0305-25-82	Buying, Storing, and Handling Adhesives
TAPPI UM 562	Adhesiveness of Gummed Paper Tape (McClaurin Test)
TAPPI UM 816-82	Glueability of Linerboard

ants; Fire Standards; Building Constructions (Committee C-24) (Ref 4). In addition, there is a subject index/alphabetical index (Volume 00.01) that is very helpful for locating specific documents. In certain instances, other ASTM subcommittees generate standards for adhesives or sealants. When this is the case, work is coordinated with ASTM Committees D14 or C24.

ASTM standards are preceded by the designation ASTM, followed by a letter designation (denoting the classification of material or product) and a number, as shown in Table 1. Reference to an ASTM document always uses this format. Adhesive and sealant specifications for materials acceptance, testing, methods of surface preparation, and product acceptance are provided.

SAE Aerospace Materials Specifications. The Society of Automotive Engineers, 400 Commonwealth Drive, Warrendale, PA 15096, publishes standards adopted by the aerospace industry as Aerospace Materials Specifications (AMS). This is a loose-leaf collection of more than 200 specifications for plastics, adhesives, elastomers, and related materials. The SAE Nonmetallics Committee has jurisdiction over these SAE-AMS specifications. The published listing of all SAE documents is contained in an AMS Index of Specifications. The military contributes to the development of these aerospace specifications by attending committee meetings. The Naval

Air Systems Command coordinates the adoption of AMS documents for the Department of Defense.

Typical SAE adhesive and sealant specifications and test methods are listed in Table 2. Approximately 30 in number, these standards are not as numerous as ASTM specifications, but cover a broad range of materials, primers, surface preparation practices, and methods of application and testing. All specifications are preceded by the SAE designation, followed by an AMS, ARP, or J designation, which generally classifies their use (see Table 2).

TAPPI Standards. The only other organization that promulgates a quantity of standards is the Technical Association of the Pulp and Paper Industry. The organization, located in Atlanta, GA, concerns itself with paper, pulp, and allied industries. TAPPI publishes standard test methods and technical information sheets that may be purchased directly from the organization. A sampling of the approximately 40 standards on adhesives it publishes is listed in Table 3.

The TAPPI standards, directed at the paper and box industry, cover materials acceptance, quality control, and the testing of paper materials and products. The stan-

dards deal with adhesive tapes, gummed paper, corrugated board, plastic film, release papers, laminated paper, and glue. The standards are preceded by the designation TAPPI, followed by the letter T, TI, or UM and a number, which generally categorize the subject area of the standard.

The National Aerospace and Defense Contractor's Accreditation Program (NADCAP) is a national third-party approval program actively supported by the Undersecretary of Defense for Acquisition. The charter for NADCAP states that it is to be nongovernment sponsored, industry supported, and government endorsed and used. The intent of this program is to avoid multiple qualification/approval efforts of subcontractor sources and material suppliers. One of the first NADCAP pilot programs was in the area of aerospace sealant materials.

International Organization for Standardization (ISO)

The ISO has an active program for the development of international standards for materials. International standards on adhesives and sealants are the responsibility of ISO Technical Committee 61 (ISO TC 61) on Plastics. The ISO, headquartered in Geneva, does not have a separate committee for adhesives and sealants. Fully accepted documents are designated International Standards (IS). Copies of ISO standards and lists of other industry standards are available from the American National Standards Institute (ANSI), 1430 Broadway, New York, NY 10018. ANSI provides the secretariat for many ISO committees in the United States.

Government Standards

Federal Specifications and Standards. Federal specifications are listed in the Index of Federal Specifications and Standards (FPMR 101-29.1), which is issued annually by the General Services Administration, Federal Supply Service (GSA-FSS), Superintendent of Documents, Government Printing Office, Washington, D.C. 20402. The index provides the means for classify-

Table 4 Typical federal specifications

Designation	Title
MMM-A-121	Adhesive, Bonding Synthetic Rubber to Steel
MMM-A-134	Adhesive, Epoxy Resin, Metal to Metal Structural Bonding
MMM-A-180C	Adhesive, Polyvinyl Acetate Emulsion
MMM-A-1931	Adhesive, Epoxy, Silver Filled, Conductive
SS-S-210A	Sealing Compound, Preformed Plastic, for Expansion Joints and Pipe Joints
TT-S-227B	Sealing Compound, Rubber Base, Two Component (For Calking, Sealing, and Glazing in Building Construction)
TT-S-1732	Sealing Compound, Pipe Joint and Thread, Lead-Free General Purpose
A-A-1556	Sealing Compound (Elastomeric Joint Sealants)
V-V-190	Sealing Compound, Dipcoating

Table 5 Typical military specifications

Designation	Title
MIL-A-83376	Adhesive Bonded Metal Faced Sandwich Structures, Acceptance Criteria
MIL-A-47040	Adhesive-Sealant, Silicone, RTV, High Temperature
MIL-A-46864	Adhesive, Epoxy, Modified, Flexible, Two Component
MIL-A-48611	Adhesive System, Epoxy-Elastomeric, for Glass to Metal
MIL-A-24179	Adhesive, Flexible Unicellular-Plastic Thermal Insulation
MIL-A-87135	Adhesive, Non Conductive, for Electronic Applications
MIL-C-2399	Cement, Liquid, Tent Patching
MIL-G-46030	Glue, Animal (Protective Colloid)
MIL-P-47279	Primer, Silicone Adhesive
MIL-S-12158	Sealing Compound, Noncuring, Polybutene
MIL-S-46897	Sealing Compound, Polyurethane Foam
MIL-S-11388	Sealing Material for Metal Container Seams
MIL-S-8802	Sealing Compound, Temperature Resistant, Integral Fuel Tanks and Fuel Cell Cavities, High Adhesion
MMM-A-132A	Adhesives, Heat Resistant, Airframe Structural, Metal to Metal
MIL-S-81733C	Sealing and Coating Compound, Corrosion Inhibitive
MIL-S-83430	Sealing Compound, Integral Fuel Tanks and Fuel Cell Cavities, Intermittent Use to 360 °F
MIL-S-29574(AS)	Sealing Compound, Polythioether, for Aircraft Structures, Fuel and High Temperature Resistant, Fast Curing at Ambient and Low Temperatures

Table 6 Military handbooks for adhesives

Handbook	Title
MIL-HDBK-337	Adhesive Bonded Aerospace Structure Repair
MIL-HDBK-691	Adhesive Bonding
MIL-HDBK-725	Adhesives—A Guide to Their Properties and Uses

ing all acquisition items by federal supply classification (FSC). To determine whether a specification is available, one should look under the FSC group and then scan the list for the adhesive or sealant of interest. Federal specifications are intended for use by all government organizations, including the Department of Defense.

The FSC category for adhesives is 8040, and more than 30 federal specifications are listed. The FSC category for sealants is 8030, with approximately 15 federal specifications listed. These specifications are listed by a unique alphanumeric designation; for example, an epoxy resin for metal-to-metal structural bonding is listed as MMM-A-134. The bulk of adhesive specifications use the MMM-A designation, while sealants use SS-S or other variations of letters. A list of typical federal specifications is given in Table 4.

Commercial item descriptions (CIDs) are also issued by the GSA-FSS. These are simplified forms of specifications and are generally one-page documents basically used to specify off-the-shelf items. These specifications carry an A-A designation, as seen in Table 4, and may apply to both adhesives and sealants.

Federal standards are mandatory for use by all federal agencies. These standards establish engineering and technical limitations and applications for items, materials, processes, methods, designs, and engineering practices. Although they are mainly used by government agencies, they are useful to industry as well. In many cases, they are also used by Department of Defense activities.

The DoD Index of Specifications and Standards (DoDISS) related standardization documents are the most useful sources of information used by the military services and the most widely used within the government. The DoDISS is available for a fee

from the Naval Publications and Forms Center (NPFC 106), 5801 Tabor Avenue, Philadelphia, PA 19120. It comprises an alphabetical (Part 1) and numerical (Part 2) listing of standards. The alphabetical listing reflects all active documents by nomenclature, cross-referenced to document number, document date, preparing activity, and date of publication. The numerical listing reflects all active documents, whether new, revised, amended, changed, or cancelled, in document-number sequence. A companion document to, but not part of, the DoDISS is the Federal Supply Classification Listing, which is a listing according to FSC group.

In addition to military specifications and standards, the DoDISS contains listings of military handbooks, federal specifications, commercial item descriptions, and the DoD-adopted industry standards, principally ASTM and SAE-AMS standards.

Military specifications are all preceded by the designation MIL, followed by the first letter of the title and a numeric designation. A typical listing of military specifications is given in Table 5. These specifications specify acceptance criteria for materials to be procured. Test methods for evaluating the materials for acceptance are mainly federal

test methods or DoD-adopted industry standards.

Whether searched by subject or number, the DoDISS provides the FSC number, preparing activity, user organizations within the government, date of latest issue, military service custodian, and appropriate code markings. The DoDISS also lists those industry and international standards recognized and adopted by the Department of Defense. For adhesives specifications listed in the Federal Supply Classification Listing, more than 30% are DoD-adopted industry standards.

A qualified products list (QPL) is another important standardization document. A QPL is a list of commercially available products that have been tested by, or under supervision of, the government and have been found to meet the requirements of a particular military or federal specification. Relatively few specifications have QPLs. Specification MIL-A-48611 has a QPL, as indicated in the DoDISS: The specification is preceded by the letter Q. To designate the QPL, the MIL-A is omitted and replaced by QPL. Thus, for this specific specification, the QPL is designated QPL-48611-6. QPLs are revised and updated more frequently than the basic specification, with the various version numbers tacked onto the QPL number.

Military standards establish engineering and technical requirements for processes, practices, and methods, as well as for material. These are designated by the form MIL-STD-XXX. Another type of military standard, designated by the form MS-XXXX, is specific to particular end items and is issued in sheet form. No military standard for adhesives or sealants is currently listed in the DoDISS.

The DoDISS also lists published military handbooks. These handbooks provide guidance information used in design, engineering, production, acquisition, and supply management operations in particular subject areas. Because a handbook is intended for guidance only, there are no mandatory provisions. The handbooks listed in the DoDISS for adhesives are given in Table 6. Like the DoDISS, the handbooks are also available through the NPFC.

The DoD also compiles a listing, called Federal Supply Classification Listing of DoD Standardization Documents. This is a cumulative listing of documents listed alphabetically within the Federal Supply Classification, for each FSC grouping. For adhesives and sealants, one should locate the 8040 or 8030 series to obtain a listing of all military, federal, and DoD-adopted industry standards.

The use of the DoDISS is mandatory for all military activities. This mandatory provision requires that it be used to identify standardization documents for the acquisition process. As with the federal standards, documents are subject to frequent revision, and this listing should be consulted to ensure applicability before procurement use. The DoDISS is available in hard copy and microfiche versions. The NPFC should be contacted for information and pricing.

Sources of Standards

Although the most direct source of industry-generated documents is the originators (ASTM, SAE, or TAPPI), a convenient means of obtaining all industry standards is from the microfilm and microfiche data files available from VSMF Data Control Services, a division of Information Handling Services, 15 Inverness Way East, P.O. Box 1154, Englewood, CO 80150.

The VSMF files are generally available from large companies or technical libraries. To procure the service would incur the expense of the microfilm cartridges or microfiche sheets, which are updated frequently, and a reader. However, the service does provide hard copy index guides, from which the user can locate all of the most recent standards. If the reader has reproduction capability, a copy can be made on the spot.

To obtain federal or military specifications, standards, and related documents, the best source is the NPFC. These documents are available free to government agencies, contractors, bidders, and so on. The most useful listing of available documents is the DoDISS. Federal specifications may be obtained from the Superintendent of Documents, U.S. Government Printing Office, Washington, DC 20402, but the NPFC is a more convenient source. It should be noted that DoD-adopted industry standards are available free from the NPFC.

ACKNOWLEDGMENT

The assistance of Mr. Arthur H. Landrock in the preparation of this overview is greatly appreciated. Mr. Landrock, a member of the U.S. Army Plastics Technical Evaluation Center (PLASTEC), is active in government specifications and standards work and in ASTM committees on plastics and adhesives.

REFERENCES

1. "Development and Use of Non-Government Specifications and Standards," Department of Defense Instruction 4120.20, Dec 1976
2. *ASTM 1989 Directory*, American Society for Testing and Materials
3. *Adhesives*, Volume 15.06, *Annual Book of ASTM Standards*, American Society for Testing and Materials
4. *Building Seals and Sealants; Fire Standards; Building Constructions*, Volume 04.07, *Annual Book of ASTM Standards*, American Society for Testing and Materials

Guide to General Information Sources

Arthur H. Landrock, Plastics Technical Evaluation Center (PLASTEC),
U.S. Army Armament Research, Development and Engineering Center

USEFUL SOURCES of information on adhesives and sealants are identified in this article. A brief discussion of literature from manufacturers is presented first, followed by a listing and discussion of 21 journals, trade magazines, and periodicals. Some publications are more focused on adhesives and sealants than others. Although the subject matter of a number of publications is primarily plastics, many have articles that describe adhesives and sealants, detail new products, and list meetings and conferences. The next section briefly discusses numerous books, handbooks, and monographs. Short courses, seminars, and conferences are then listed. The final sections describe the makeup of the ASTM committees on adhesives and sealants and the details of various adhesives data base programs. The article "Specifications and Standards for Adhesives and Sealants" in this Volume identifies both governmental and nongovernmental specifications and standards. This article should not be interpreted as representing the position of the U.S. Department of the Army or any other branch of the U.S. government.

Literature From Manufacturers

Much useful information is presented in the technical literature provided by raw materials suppliers and by manufacturers of adhesives, sealants, and engineering plastics. Readers who desire more information on products described in articles or advertisements in many journals and trade magazines can make their requests to manufacturers via the reader service cards supplied in the publications. Technical information of varying quality is then returned to the inquirer, and follow-up inquiries are often made by the manufacturers or suppliers. Other excellent and reliable sources of information on bonding the newer engineering plastics are the supplier bulletins, which usually include recommendations on adhesive bonding and solvent cementing.

Journals, Trade Magazines, and Periodicals

Adhesive Abstracts, published by Pergamon Press, Inc., Maxwell House, Fairview Park, Elmsford, NY 10523, (914) 592-7700, is published 12 times per year at a subscription price of $195. It is currently producing Volume 2 issues; back issues are available at a 25% reduction. A free issue is available on request.

The Adhesives & Sealants Newsletter is published monthly by Adhesives and Sealants Consultants, P.O. Box 72, Berkeley Heights, NJ 07922, (201) 464-3133. Editor-in-chief Fred A. Keimel is a well-known consultant and lecturer. Subscriptions are $165 for the first subscription. The newsletter supplies useful information. Features include announcements and reviews of technical meetings, courses, products, books, equipment, services, and supplies. Each issue has an editorial. Occasionally a supplement issue, *Information Resource Guide (IRG)*, is published. Its purpose is similar to that of this article, although its scope is much more comprehensive. This valuable publication is available only to subscribers to the newsletter.

Adhesives Age is a well-established monthly trade magazine published by Communication Channels, Inc., 6255 Barfield Rd., Atlanta, GA 30328. A 13th issue, *Adhesives Age Directory* (formerly the *Adhesives Red Book*), is published each May. The magazine emphasizes practical, rather than theoretical, topics. It frequently publishes issues devoted to special subject areas, such as hot-melt adhesives, pressure-sensitive adhesives, and sealants. The magazine also provides useful information under various regular departments such as adhesives and products, equipment, literature, calendar of events, and news of the industry from Washington, across the country, and abroad.

Advanced Materials & Processes is a monthly trade magazine (semimonthly in June and November) published by ASM INTERNATIONAL (formerly American Society for Metals), Materials Park, OH 44073 (216) 338-5151. It is intended for individuals involved in selecting, processing, and testing high-performance materials, including engineering plastics, metals, composites, adhesives, and sealants. Articles and departments cover material properties, design considerations, property-enhancing products, and equipment. The *Annual Guide to Engineering Materials* issue covers significant properties. Subscriptions are $48 per year. The publication is free to members of ASM INTERNATIONAL, a technical/engineering society that collects and disseminates information on the state-of-the-art selection, processing, and application of engineered materials.

Cassette is a new bulletin, published twice yearly by the Center for Adhesive and Sealant Science (CASS) at Virginia Polytechnic Institute and State University. The bulletin has news items, a list of technical meetings, and a calendar of events. The address of CASS is 2 Davidson Hall, Virginia Polytechnic Institute and State University, Blacksburg, VA 24061-0201, FAX 703-231-7826.

The Composites & Adhesives Newsletter is a 16-page publication published bimonthly by T/C Press, P.O. Box 36428, Los Angeles, CA 90036, (213) 938-6923.

The International Journal of Adhesion and Adhesives is a quarterly that publishes contributed papers and research reports. It also has a news section covering commercial and technological developments, conference reports, book reviews, and a calendar of events. Letters commenting on previously published material or presenting preliminary details of current work are encouraged. The journal is published by Butterworths Scientific Ltd., P.O. Box 63, Westbury House, Bury Street, Guildfords Surrey, GU2 5BH, UK. Subscriptions in the United States can be obtained through Expediters of the Printed Word Ltd., 527 Madison Avenue, New York, NY 10002, (212) 838-1304. The journal currently costs about $170 annually for U.S. delivery, but is of high quality. It draws together the many aspects of the science and technology of adhesive materials, from fundamental research and development work to industrial applications. Subject areas include interfacial interactions, surface chemistry, methods of testing, the accumulation of test data on physical and mechanical properties, environmental effects, new adhesive materials, sealants, the design of bonded joints, and manufacturing technology.

The Journal of Adhesion, in publication since 1969, is an excellent periodical that is of very broad interest to the large technical community concerned with the development of an understanding of the phenomenon of adhesion and its practical applications. This journal provides a forum for discussion of the basic and applied problems in adhesion. Papers are considered relevant if they contribute to the understanding of the response of systems of joined materials to mechanical or other disruptive influences. Experimental papers are required to incorporate theoretical background and theoretical papers must relate to practice. This periodical has two or three volumes of 4 issues each per year and currently costs $218 annually in the United States. The publisher is Gordon & Breach Science Publishers, Ltd., c/o STBS Ltd., 42 William IV Street, London WC2N, 4DE, UK. Subscriptions in the United States are available from Gordon & Breach Science Publishers, Inc., P.O. Box 786, Cooper Station, New York, NY 10276, (212) 206-8900. The editor-in-chief is Louis H. Sharpe, P.O. Box 3288, Hilton Head Island, South Carolina, (803) 671-4810.

The Journal of Adhesion Science and Technology is a new high-quality quarterly providing a forum for the basic aspects, theories, and mechanisms of adhesion. It particularly welcomes contributions covering the application of adhesion principles and theories in all areas of technology. The scope of the journal encompasses, but is not restricted to, the following areas of adhesion: theories of adhesion; surface energetics; fracture mechanics; the development and application of surface-sensitive methods for the purpose of studying adhesive phenomena; nondestructive methods for bond failure studies; techniques for the measurement of the adhesion of thin films, thick films, and coatings; the surface modification of polymers to improve their adhesion; mechanisms and studies of coupling agents; adhesive joints; wood bonding; structural adhesives; and adhesion aspects of composite materials, biomedical materials, paints and coatings (for example, in relation to corrosion control), microelectronics, lithographic materials, metallized plastics, rubbers/elastomers, and aerospace materials. The journal publishes review articles on the mechanisms or theories of adhesion and on areas of technological importance. In addition, original research pa-

pers or short communications on any of the above-mentioned topics are accepted. The journal also publishes a diary of international events and regularly highlights institutions that have concentrated activity on adhesion. Short reports on current research and letters to the editor are also published.

The journal is published by VSP-BV (formerly VNU Science Press BV), P.O. Box 346, 3700 A H Zeist, The Netherlands. The editors are K.L. Mittal, IBM U.S. Technical Education, Thornwood, NY, (914) 742-5747 and W.J. van Ooij, Colorado School of Mines, Golden, CO, (303) 273-3612. The subscription price for Volume 3 (1989), in eight issues, is approximately DM 700/US $382.

Materials Engineering is a monthly trade magazine with a mix of technical articles and news items that sometimes includes adhesives. It is published by Penton Publishing Company, 1100 Superior Avenue, Cleveland, OH 44114, (216) 696-7000. *Materials Selector*, issued each December, includes a brief tabular section on joining and sealing, including adhesive bonding. Reader service cards are provided for both the regular and *Materials Selector* issues. Subscriptions are free to qualified readers.

Modern Plastics is a well-established monthly trade magazine published by McGraw-Hill, Inc., 1221 Sixth Avenue, New York, NY 10020, (212) 512-6242. Subscriptions can be obtained by contacting *Modern Plastics*, P.O. Box 602, Hightstown, NJ 08520-9955, (609) 426-7118 or (800) 857-9402, extension 81. A regular section on new materials includes entries on adhesives and coatings. In mid-October of each year, *Modern Plastics Encyclopedia* is published in hardcover. The 1990 issue has more than 900 pages on all aspects of plastics technology. A discussion of adhesive bonding (for plastics) is usually given in a section on fabricating and finishing. Reader service cards are provided for the encyclopedia and for the magazine.

Packaging, formerly *Modern Packaging*, a monthly trade magazine for management, is published by Cahners Publishing Company, 1350 E. Touhy Avenue, P.O. Box 5080, Des Plaines, IL 60018, (312) 635-8800. Coverage includes all aspects of packaging and packaging materials, including some items on adhesives. Reader service cards are provided, and subscriptions are free to qualified readers.

Paper, Film and Foil Converter is published monthly by Maclean Hunter Publishing Company, 29 North Wacker Drive, Chicago, IL 60606, (312) 726-2802. Although it is a monthly trade magazine of the converting industry, it sometimes has useful information on adhesives used in converting. Reader service cards are provided. Subscriptions are free to qualified readers.

Plastics Compounding is published seven times a year by Resin Publications, Inc.,

a subsidiary of Edgell Communications Inc., P.O. Box 6350, Duluth, MN 55806-9933. The magazine is intended primarily for resin producers, formulators, and compounders. Occasional items on adhesives are included. Reader service cards are provided, and free subscriptions are available to qualified readers. Each June, *Plastics Compounding Redbook*, is published. This book, a comprehensive guide to the compounding of plastics materials, has three sections: feature editorial, product directories, and an alphabetical index of suppliers. The information on additives, modifiers, and fillers should be valuable to adhesives formulators. Reader service cards are provided. This guide is available for purchase as a separate item.

Plastics Design Forum is a bimonthly magazine intended for designers of plastics products and components. It is published by Edgell Communications Inc., 1 East First Street, Duluth, MN 55802, (218) 723-9200. Occasional items on adhesives are included. Free subscriptions are available to qualified readers, and reader service cards are provided.

Plastics Engineering, published monthly by the Society of Plastics Engineers, Inc. (SPE), 14 Fairfield Drive, Brookfield, CT 06804-0403, (203) 775-0471, is free to all society members. The articles generally cover practical aspects of plastics engineering. Information on new adhesives is often found in the "Materials" section.

Plastics Packaging is a relatively new bimonthly trade magazine published by Edgell Communications Inc., P.O. Box 6437, Duluth, MN 55806-9933. Information on adhesives is sometimes included. A subject index includes adhesives as an entry. Reader service cards are provided. Subscriptions are free to qualified readers.

Plastics Technology is a monthly trade magazine that calls itself "the magazine of plastics manufacturing" and occasionally includes items on adhesives. The publisher is Bill Communications, Inc., 633 Third Avenue, New York, NY 10017, (212) 986-4800. It emphasizes new developments. Reader service cards are included, and free subscriptions are available to qualified readers. A *Manufacturing Handbook and Buyers' Guide* is issued annually in mid-August. This Volume is of particular interest to adhesives formulators because of its large section on additives.

Plastics World is described as "the magazine for processors and designers." It is published monthly, with an extra issue in March, by Cahners Publishing Company, a Division of Reed Publishing USA, 275 Washington Street, Newton, MA 02158-1630. Each issue has a mix of business and technology news, as well as other information, all aimed at management. An annual *Plastics Directory* (March issue) has a useful list of additives and modifiers for adhesives. The magazine and directory both have reader service cards. Free subscriptions are available to qualified personnel.

SAMPE Journal is published bimonthly by the Society for the Advancement of Material and Process Engineering (SAMPE), a professional society covering all materials and processes. It can be ordered from the SAMPE International Business Office, 843 West Glentana, P.O. Box 2459, Covina, CA 91722, (818) 331-0616. The journal is issued at no charge to all SAMPE members and is available by subscription as well. A typical issue has six technical articles and nineteen features and departments, including news information, abstracts and selected titles, materials news, and product announcements. Occasional articles are published on adhesives. SAMPE was originally concerned with the aerospace field only, but later broadened its scope.

SAMPE Quarterly is also published by SAMPE (see above). It is available to SAMPE members at a reduced price or by subscription. As with *SAMPE Journal*, adhesives are occasionally included among the materials covered. A typical issue has ten articles. There is no advertising in the quarterly.

Books, Handbooks, and Monographs

Building Seals and Sealants; Fire Standards; Building Constructions, Annual Book of ASTM Standards, Vol 04.07, is a compilation of specifications, test methods, and practices detailing the property requirements for caulking and glazing compounds; structural, emulsion, hot-applied, and solvent-release sealants; and other building sealants, as well as information on how to measure sealant properties, such as adhesion, low-temperature flexibility, aging effects, extrudability, and slump. It is published each year by the American Society for Testing and Materials (ASTM), 1916 Race Street, Philadelphia, PA 19103. The 1989 issue of 1004 pages contains 186 standards.

Adhesives, Annual Book of ASTM Standards, Vol 15.06, is a compilation of specifications, practices, and test methods involving adhesives. The 1989 issue of 459 pages contains 10 specifications, 20 practices, and 75 test methods. It is available from ASTM (see above).

Military Handbook MIL-HDBK-691B, Adhesive Bonding, 12 March 1987, is a 439-page volume written for the Army Materials Technology Laboratory (MTL), Watertown, MA. It provides basic and fundamental information on adhesives and related bonding processes for the guidance of engineers and designers of military materiel. There are 17 chapters: "Introduction to Adhesive Bonding," "Joint Design," "Adherend Types," "Structural Adhesive Types and Selection Guidelines," "Adherend Surface Preparation," "The Conven-

tional Adhesive Bonding Process," "Solvent Cementing of Plastics (Solvent Welding)," "Recommended Adhesives for Specific Adherends," "Thermally and Electrically Conductive Adhesives," "Bonding of Honeycomb Structures," "Repair of Adhesive-Bonded Structures," "Weldbonding," "Test Methods," "Quality Control," "Environmental Effects (Durability)," "An Outline for a Bonding Process Specification," and "Specifications."

The 20-page table of contents is fully detailed. Copies of this guidance document are available from the Naval Publications and Forms Center, 5801 Tabor Avenue, Philadelphia, PA 19103, (215) 697-3321.

Military Handbook MIL-HDBK-337, Adhesive Bonded Aerospace Structure Repair, 1 December 1982, was prepared by the Air Force Materials Laboratory at Wright-Patterson Air Force Base, OH. It provides information on material and fabrication procedures for repairing adhesive-bonded aerospace structures and their parts. There are ten chapters: "Introduction," "Damage Assessment," "Repair Method Selection," "Materials and Processing," "Surface Preparation Procedures," "Repair Methods, Small," "Large Area Repairs," "Tools, Equipment and Facilities," "Tool Fabrication," and "Nondestructive Inspection." Copies of this handbook are available from the Naval Publications and Forms Center, 5801 Tabor Avenue, Philadelphia, PA 19103, (215) 697-3321.

Joining of Advanced Composites, Engineering Design Handbook, Pamphlet DARCOM-P 706-316, U.S. Army Materiel Development and Readiness Command, March 1979, is available to the general public for purchase from the National Technical Information Service (NTIS), U.S. Department of Commerce, Springfield, VA 22161, as publication ADA072362. This is an excellent, well-illustrated document with four chapters: "Introduction," "Composite Joint Design," "Design Data Requirements," and "Basic Bonded and Bolted Lap Joint Configurations and Design Variations." There is also a subject index.

Adhesives Technology Handbook, by A.H. Landrock, published by Noyes Publications, Park Ridge, NJ in 1985, is a comprehensive 44-page monograph on adhesives and sealants. The emphasis is on practical rather than theoretical aspects. An unusual feature is a lengthy 92-page chapter titled "Standard Test Methods, Practices and Specifications." The handbook also contains a detailed 16-page table of contents and 37 pages of indexes. Extensive references, many covering military publications, are given in each chapter.

High-Performance Adhesive Bonding, edited by G. DeFrayne, was published by the Society of Manufacturing Engineers (SME) in 1983. This 276-page book is concerned more with basic adhesive techno-

logy and state-of-the-art applications in manufacturing than with the chemistry and science of adhesives. One section of the book, however, is dedicated to the use of basic polymers and resins in adhesive formulations. The book is divided into four sections, each of which is a collection of five or six reprinted articles. The four are "Fundamentals," "Types of Adhesives," "Bonding Substrates," and "Industrial Applications." Also included are a listing of ASTM standard terms and an index.

Analysis and Testing of Adhesive Bonds, by G.P. Anderson, S.J. Bennett, and K.L. DeVries, was published by Academic Press, Inc., 111 Fifth Avenue, New York, NY 10003 in 1977. This 255-page book has six chapters: "Standard Adhesive Test Techniques," "Theory of Adhesive Fracture," "Adhesive Fracture Energy Tests," "Analytical Methods and Computer Techniques for Adhesive Bonding," "Chemical and Physical Aspects of Adhesive Fracture," and "Specific Applications and Aspects of Adhesive Fracture Mechanics." This book is for the advanced worker and is replete with complex equations.

Joining of Composite Materials, an ASTM special technical publication (STP 749), edited by K.T. Kedward, is the compiled proceedings of a symposium sponsored by ASTM Committee D-30 on High Modulus Fibers and Their Composites, Minneapolis, MN, 16 April 1980. Copies are available from ASTM, 1916 Race Street, Philadelphia, PA 19103. The book has four papers on adhesively bonded attachments.

Adhesively Bonded Joints: Testing, Analysis, and Design, by W.S. Johnson, is an ASTM special technical publication (STP 981) published in 1988 and available from ASTM (see above).

Adhesion and Adhesives, 2nd edition, edited by R. Houwink and G. Salomon, was published by Elsevier Publishing Company in two volumes. *Volume 1, Adhesives* (548 pages), published in 1965, is an introduction to adhesion, including adhesives and their use. *Volume 2, Applications* (590 pages), published in 1967, covers metal bonding and honeycomb panels for the aerospace industry.

Construction and Structural Adhesives and Sealants—An Industrial Guide, by E.W. Flick, Noyes Publications, Park Ridge, NJ, 1988, is a 754-page volume describing more than 1600 industrial construction and structural adhesives, sealants, and related products currently available for industrial, commercial, and consumer use. The book should be of value to industrial, technical, and managerial personnel involved in the specification and use of these products. It was compiled from information received from more than 100 industrial companies and other organizations. A trade name index is included. Products are presented by company, and the companies are listed alphabetically. The table of contents is organized in such a way that it also serves as a subject index.

The Handbook of Adhesive Raw Materials, by E.W. Flick, was published by Noyes Publications, Park Ridge, NJ in 1982. This 303-page handbook contains descriptions of hundreds of raw materials available to the adhesives industry. Only the most recent data has been compiled. For the most part, only trademarked raw materials are included. Common chemicals are virtually excluded. Most solvent-base raw materials are omitted.

Adhesives Technology—Developments Since 1979, edited by M. Gutcho, was published by Noyes Data Corporation in 1983. This 452-page book presents detailed descriptive information on U.S. patents issued between January 1980 and March 1982 relating to adhesives technology. A database publication, it provides information retrieved and made available from the U.S. patent literature. It supplies detailed technical information and can be used as a guide to the patent literature in this field. By indicating all the information that is significant and eliminating legal phraseology, the book presents an advanced commercially oriented review of recent developments in the field of adhesives technology.

Adhesives, Sealants, and Coatings for the Electronics Industry, by E.W. Flick, was published in 1986 by Noyes Publications, Park Ridge, NJ. This 197-page book presents technical data on more than 1600 contemporary adhesives, sealants, and coatings currently available to the electronics and related industries. The book is arranged alphabetically by company name and product category, tradenames, product number, main product features, and typical end-uses. Of particular interest is information on wash-away adhesives, conductive adhesives, and optoelectronic and semiconductor encapsulants.

Adhesive and Sealant Compound Formulations, 2nd edition, by E.W. Flick, was published by Noyes Publications, Park Ridge, NJ in 1984. This 366-page book has formulations for 444 new adhesives and sealants. It has information on neither solvent-base formulations nor on lead-containing raw materials. The book contains amounts and descriptions of raw materials, physical constants, and test results for formulations, key properties, and suggested modifications, along with the data source.

Adhesives for Structural Applications, edited by M.J. Bodnar, Picatinny Arsenal, *Journal of Applied Polymer Science*, Volume 6 (Number 20), March/April 1962, is the proceedings of a symposium held at Picatinny Arsenal, 27-28 September 1961. This paperbound volume of 17 papers represents the first adhesives symposium sponsored by Picatinny Arsenal.

Structural Adhesives Bonding, edited by M.J. Bodnar, Picatinny Arsenal, *Applied Polymer Symposia*, Number 3, 1966, published by Interscience Publishers, is the compiled proceedings of a symposium held at Stevens Institute of Technology, Hoboken, NJ, 14-16 September 1965. This paperbound volume contains 32 papers presented at the second adhesives symposium sponsored by Picatinny Arsenal.

Processing for Adhesive Bonded Structures, edited by M.J. Bodnar, Picatinny Arsenal, *Applied Polymer Symposia*, Number 19, 1972, published by Interscience Publishers, is the proceedings of a symposium held at Stevens Institute of Technology, Hoboken, NJ, 23 to 25 August 1972. This paperbound volume contains 35 papers presented at the symposium, which was sponsored by Picatinny Arsenal and Stevens Institute of Technology. It is the third Picatinny Arsenal adhesives symposium.

Durability of Adhesive Bonded Structures, edited by M.J. Bodnar, Picatinny Arsenal, *Applied Polymer Symposia*, Number 32, published by Wiley-Interscience, John Wiley & Sons, Inc., New York, NY, is the proceedings of a symposium held at Picatinny Arsenal, Dover, NJ, 27 to 29 October 1976. This paperbound volume has 29 papers presented at the symposium. It is the fourth adhesives symposium sponsored by Picatinny Arsenal.

Proceedings of the Joint Military/Government-Industry Symposium on Structural Adhesive Bonding, M.J. Bodnar and S.E. Wentworth, General Chairmen, was published in 1987 in a hardcover volume containing 41 papers presented at a symposium held at Picatinny Arsenal on 3 to 5 November 1987. Copies are available from the American Defense Preparedness Association, Rosslyn Center, Suite 900, 1700 North Moore Street, Arlington, VA 22209, (703) 522-1820. The symposium was the fifth adhesives symposium sponsored by Picatinny Arsenal.

Adhesives, Sealants, and Gaskets—A Survey, by R.B. Perkins and S.N. Glarum of Southern Research Institute, NASA SP-5066, 1967, is a useful, although not recent, Technology Utilization Division 60-page report on adhesives, sealants, and gaskets developed to operate in the extreme environment encountered in space work. Copies were originally available from the Superintendent of Documents, U.S. Government Printing Office, Washington, D.C.

Handbook of Adhesives, 3rd edition, edited by I. Skeist, was published by Van Nostrand Reinhold, 115 Fifth Avenue, New York, NY 10003, 1990. This massive book (864 pages) has information from 70 experts on 35 families of polymers and other components of adhesives and sealants. Coverage includes adhesive selection, joint design, surface preparation, mode of application, and end-use markets, including those of the construction, packaging, automotive, aircraft, textiles, tapes, abrasives,

electronics, and the home products industries. Specifications are thoroughly described in the chapters "Adhesive Selection and Screening Testing" and "Sealants and Caulks."

Cyanoacrylate Resins—The Instant Adhesives, edited by M. Lee and authored by the technical staff of Lee Pharmaceuticals, was published in 1981 by Pasadena Technology Press, 3533 East California Boulevard, Pasadena, CA 91107. This paperbound book of 241 pages is described as "a monograph of cyanoacrylate adhesives applications and technology." The book has 26 brief chapters, a bibliography of the chemical literature (1959 to 1980), one of the patent literature (1949 to 1980), and a chemical index.

Metal-to-Metal Adhesive Bonding, by S. Semerdjiev, was published by Business Books Ltd., London, in 1970. The book has nine chapters, including an introduction and the chapters "Basic Principles of Adhesives"; "Adhesives-Compositions Classification, Properties"; "Processes for Adhesive Bonding"; "Mechanical Properties and Performance of Adhesive Bonds"; "Testing and Inspection of Adhesive Bonds"; "Design of Adhesive Bonded Joints"; "Application of Adhesive Bonding of Metals"; and "Bonded Sandwich Structures." There are two appendixes, one on suppliers and manufacturers of specific adhesives and one on testing adhesives for metal and adhesive bonds: methods, standards, and specifications. An index is included.

Adhesives for the Composite Wood Panel Industry, by G.S. Koch, F. Klareich, and B. Exstrum, was published in 1987 by Noyes Data Corporation, Park Ridge, NJ. This 144-page book presents a market and technological analysis of current fossil-fuel-base adhesives for the composite wood panel industry. It identifies domestic and international research and development efforts in the area of bio-mass-derived alternative adhesives and assesses technical and economic factors that will influence the commercial success of these alternates. The book has five chapters, an appendix with tables of market data, references, and a glossary.

Adhesives for Wood—Research, Applications, and Needs, edited by R.H. Gillespie, Forest Products Laboratory, U.S. Department of Agriculture, was published by Noyes Publications, Park Ridge, NJ in 1984. This 250-page book surveys recent developments in and the current status of adhesives technology for wood.

Adhesive Bonding of Wood, by M.L. Selbo, is available as U.S. Forest Service Technical Bulletin 1512 (August 1975). It is available from the Superintendent of Documents, U.S. Government Printing Office, publication 001-000-03382.

Evaluating Wood Adhesives and Adhesive Bonds: Performance Requirements,

Bonding Variables, Bond Evaluation, Procedural Recommendations, prepared by the Forest Products Laboratory, U.S. Department of Agriculture, for the U.S. Department of Housing and Urban Development, February 1977. Available from the National Technical Information Service as PB 265646.

Wood Adhesives—Chemistry and Technology, Volume 1, edited by A. Pizzi, was published in 1983 by Marcel Dekker, Inc., New York, NY. Volume 2 was published in 1989 and has 11 chapters and 416 pages.

Home Construction Projects With Adhesives and Glues is a small, hardcover book published by Franklin Chemical Industries, Columbus, OH in 1983. This book was published in 1980 in almost identical form under the title *Adhesives and Glues—How to Choose and Use Them*. The revision is an excellent source, well illustrated with photographs and line drawings involving wood glues and their applications.

Joining Technologies for the 1990s—Welding, Brazing, Soldering, Mechanical, Explosive, Solid-State, Adhesive, edited by J.D. Buckley and B.A. Stein of the NASA Langley Research Center, National Aeronautics and Space Administration, was published by Noyes Data Corporation, Park Ridge, NJ in 1986. The book has six chapters and presents recent advances in joining technologies for the 1990s.

Advances in Adhesives: Applications, Materials and Safety, edited by D. Brewis and J. Comyn and published by T/C Press, P.O. Box 36428, Los Angeles, CA 90036 (1983), covers a number of useful topics and is extensively illustrated.

Structural Adhesives and Bonding is a compilation of papers presented at a special conference held March 1979 in El Segundo, CA. The publisher is T/C Press (see above). Topics covered include theoretical aspects, surface preparation and effects, adhesive materials and processes, factors influencing bond performance, testing considerations, applications, and case histories of structural adhesives for industrial and aerospace use.

Adhesives, Sealants and Primers, D.A.T.A. Digest, 5th edition, September 1989, is a softcover volume with 754 pages of properties and indexes covering more than 6000 products. There are also 261 introductory pages listing possible substrate combinations for the products listed and a general introduction to adhesives and adhesion. This useful volume is intended to serve as a starting point for designers, engineers, and other professionals who need to locate an adhesive, sealant, or primer suitable for a desired application. Tables of ranked properties cover flash point; glass transition temperature; lap-shear range, low; lap-shear range, high; service temperature range, low; service temperature range, high; and viscosity, Brookfield. The

publisher is International Plastics Selector, 8977 Activity Road, P.O. Box 26875, San Diego, CA 92126. The price is $150. Copies may be ordered by calling (800) 447-4666 (United States) or (619) 578-7600.

The International Adhesion Conference 1984 Proceedings. This is a softbound conference proceedings published by The Plastics and Rubber Institute, London, UK. The conference was held at the University of Nottingham, 12 to 14 September 1984. Twenty-five papers are included in the compilation, covering all aspects of adhesive bonding.

Adhesion—Fundamentals and Practice by the Ministry of Technology, UK, is the proceedings of an international conference on adhesion held at the University of Nottingham, UK, in 1966. The book of 308 pages is profusely illustrated. The publisher is Gordon & Breach Science Publishers, New York, NY. There are five parts, each consisting of an introduction, four or five technical papers, and a summary, followed by a general discussion at the end of the book. The five parts are "The Adherend Surface," "The Preparation of the Adherend Surface," "The Adhesive and Special Invitation Papers," "Joint Design," and "Methods of Test."

Adhesives Handbook 3rd edition, by J. Shields, SIRA Institute Ltd., UK, was published in 1984 by Butterworths, London, UK and is available from Butterworths, 10 Tower Office Park, Woburn, MA 01805. The book is extremely comprehensive, covering in considerable detail such subjects as joint design, adhesives selection, adhesives materials and properties, adhesives products directory, surface preparation, the bonding process, the physical testing of adhesives, adhesives tables, and adhesives and trade sources.

Adhesives in Modern Manufacturing, edited by E.J. Bruno, is a paperbound 183-page book published by the Society of Manufacturing Engineers, Dearborn, MI in 1970. It is an excellent, practical book with seven chapters, including one on case histories. The book has a glossary, an index, and an extensive bibliography. It is part of the SME Manufacturing Data Series.

Adhesive Bonding of Aluminum Alloys, edited by E.W. Thrall and R.W. Shannon, was published in 1985 by Marcel Dekker, Inc., New York, NY. This practical book is very well illustrated and has 16 chapters and an index. The chapters are as follows: "Introduction," "Chromic Acid Anodize Process Used in Europe," "Sealed Chromic Acid Anodize," "Phosphoric Acid Anodize," "Surface Analysis," "Adhesive Selection From the User's Viewpoint," "Epoxy Adhesives," "Elevated-Temperature-Resistant Adhesives," "Mechanical Properties of Adhesives," "Environmental-Durability Testing," "Chemical Analysis for Control," "Coatings," "Structural

Analysis of Adhesive-Bonded Joints," "Basic Principles for Tooling Design and Inspection," "Nondestructive Inspection," and "Adhesive-Bonded Aluminum Structure Repair."

Adhesive Bonding Aluminum is a 95-page booklet published in 1966 by Reynolds Metals Company, Richmond, VA. The book is a compilation of basic technology pertaining to the adhesive bonding of aluminum, with the emphasis on structural applications. Design information is supplemented by tables of adhesive properties and by detailed studies of typical applications.

Adhesive Bonding ALCOA Aluminum is a 106-page hardcover book providing useful information about the adhesive bonding of ALCOA aluminum. The book was published in 1967 by the Aluminum Company of America, Pittsburgh, PA. After an introduction, its chapters are: "Some Applications of Adhesive Bonded Joints," "Surface Preparation," "Adhesive Classifications," "The Design of an Adhesive-Bonded Joint," "Factors That Govern the Selection of an Adhesive," and "Safety Precautions." Also included are a glossary, a section of tables, and a bibliography.

Structural Adhesives—Chemistry and Technology, edited by S.R. Hartshorn, was published in 1986 by Plenum Publishing, New York, NY. The 505-page book is part of a series of topics in applied chemistry. The book has a detailed table of contents covering ten chapters, three appendixes, and an index. The chapters are: "Introduction"; "Fundamentals of Structural Adhesive Bonding"; "Phenolic Resins"; "Epoxy Structural Adhesives"; "Polyurethane Structural Adhesives"; "Cyanoacrylate Adhesives"; "High-Temperature Polymers and Adhesives"; "The Durability of Structural Adhesive Joints"; "Testing, Analysis, and Design of Structural Adhesive Joints"; and "Industrial Application Methods." The three appendixes are titled "Standard Definitions of Terms," "SI Units and Conversion Factors," and "ASTM Standards Relating to Adhesives." This book appears to be intended primarily for the advanced worker. It has considerable theory and is replete with equations and structural formulas.

Adhesion 1-Adhesion 13, edited by K.W. Allen, The City University, London, UK, is a hardcover series of conference proceedings of the annual conferences on Adhesion and Adhesives held at The City University, London, UK. Each volume presents an average of 10 to 12 well-illustrated and excellent papers on all aspects of adhesives technology. The publisher is Elsevier Science Publishers, Ltd. Volume 1 was published in 1977 and Volume 13 in 1989. The sole distributor in the United States and Canada is Elsevier Science Publishing Company, Inc., 52 Vanderbilt Avenue, New York, NY 10017.

The Handbook of Surface Preparation, by R.C. Snogren, is a 577-page book published in 1974 by Palmerton Publishing Company, New York, NY. This is an excellent source, covering surface preparation for a number of purposes, including adhesive bonding, although the format is somewhat confusing.

Surface Preparation Techniques for Adhesive Bonding, by R.F. Wegman, was published by Noyes Publications, Park Ridge, NJ in 1989. This 150-page handbook provides information on processing adherends prior to adhesive bonding. It presents processes for treating materials for adhesive bonding from the point of view of materials and process engineers. Where practical, the details of the processes are given in a process-specification format, including, for example, the controls for rinse waters, tank construction and processing and any known restrictions. Adherends covered include metals and alloys, metal and organic matrix composites, thermosets, and thermoplastics.

The Handbook of Pressure-Sensitive Adhesive Technology, 2nd Edition, edited by Donatas Satas, was published by Van Nostrand Reinhold Company, New York, NY in 1989. This excellent 960-page book with 39 chapters has information from experts from all over the world, in addition to more than 450 illustrations. It covers all aspects of pressure-sensitive adhesives. It is an updated revision of the first edition, published in 1982.

Thomas Register is a multivolume set (23 volumes in 1989) of hardcover volumes containing much useful information on adhesives, sealants, tapes, and so on, and can be obtained from Thomas Publishing Company, 1 Penn Plaza, New York, NY 10117, (212) 290-7277. It has a detailed listing of products and services, company profiles, and a catalog file (Thomcat). It contains much advertising, which can be informative.

Adhesive Bonding—Techniques and Applications, by Charles V. Cagle, was published in 1968 by McGraw-Hill, Inc., New York, NY. This is a book with a general approach "geared to the basic fundamentals pertinent to adhesive bonding." Considerable emphasis is placed on cleanliness, surface preparation, and training. The book has 351 pages and 12 chapters, including test methods and specifications. Also included are a list of manufacturers and adhesive suppliers and a glossary of terms associated with adhesive bonding.

Adhesives, Adherends, Adhesion, by Nicholas S. DeLollis, was published in 1980 and is the second edition of a book previously published by the same author in 1970 entitled *Adhesives for Metals: Theory and Technology*, Industrial Press, New York, NY. The 1980 revision, published by Robert E. Krieger Publishing Company, Inc., Hun-

tington, NY, has 341 pages and discusses all types of adherends in 17 chapters. There are three appendixes covering definitions, ASTM standards, and a U.S. Department of Defense index of specifications and standards. The book is well written, easy to read, and contains much useful material.

Adhesion Science and Technology Volume 9A and 9B are part of the American Chemical Society (ACS) Polymer Science and Technology Series. The editor is Lieng-Huang Lee, and the books were published in 1975 by Plenum Publishing Press, 233 Spring Street, New York, NY 10013, (212) 741-6680. The papers included are those given at the ACS Symposium on Science and Technology of Adhesion held in 1975 in Philadelphia, PA. The papers in both books cover interfacial phenomena and adhesion, synthetic polymers and adhesives, rubber adhesives and sealants, natural products and structural adhesives, growth and change in adhesives, the performance of adhesive joints, trends in adhesion research, and surface energetics of the printing process.

Adhesion and Bonding, edited by N.B. Bikales, is a paperbound book containing reprints of 13 articles published in the *Encyclopedia of Polymer Science and Technology* (first edition). The reprint book was published in 1971 by John Wiley & Sons, Inc., New York, NY. Papers reprinted were "General"; "Effect of Chemical Constitution"; "Theory of Adhesive Joints"; "Surface Properties"; "Adhesive Compositions"; "Glue, Animal and Fish"; "Applications"; "Evaluation"; "Analysis of Adhesives"; "Bonding"; "Dielectric Heating"; "Ultrasonic Fabrication"; and "Adherents." The reprint book has a number of illustrations and tables and an index.

Sealants in Construction, by J.M. Klosowski, was published by Marcel Dekker, Inc., New York, NY in 1989. There are six chapters: "Selecting Joint Sealants," "Types of Sealants," "Sealant Specification and Testing," "Sealant Applications," "Structural Glazing," and "Silicone Sealants." There is an index and list of references in this excellent technical treatise.

Industrial Applications of Adhesive Bonding, edited by M.M. Sadek of the College of Engineering and Petroleum, Kuwait University, Kuwait, was published by Elsevier Applied Science Publishers, London, UK and New York, NY in 1985. Copies are available from Elsevier Science Publishing Company, Inc., 52 Vanderbilt Avenue, New York, NY 10017. There are seven chapters: "Fabrication of Machine Tool Structures by Bonding," "Fabrication of Carbide Tipped End Mills by Epoxy Resin," "Testing of Bonded Joints," "Adhesives in the Automotive Industry," "Bonding Applications in the Japanese Machine Industry," "Industrial Adhesive Bonding in North America," and "Adhe-

sives for the Structural and Mechanical Engineers." The book has an index.

Adhesion and The Formulation of Adhesives, 2nd edition, by W.C. Wake, Applied Science Publishers Ltd., London, UK, 1982, is an excellent 332-page book by a recognized British authority. The book contains 14 chapters, as well as references, a bibliography, an author index, and a subject index. The chapters are "Introduction"; "Interactions Between Bonds"; "The Functions of Adhesives"; "The Properties of Surfaces and Interfaces"; "Theories of Adhesion and Adhesive Action"; "Joints, Rigid and Flexible"; "Joint Failure and Fracture Mechanics"; "Effect of Environment on Joint Performance"; "Materials Used in Adhesive Formulations"; "Adhesives in Common Use: Composition and Formulation"; "Structural Adhesives: Metal-to-Metal Adhesion"; "Sealants"; "Carbohydrates and Proteins as Adhesives"; "Hot-Melt Adhesives"; "Adhesion to Textiles"; and "Conclusion."

Treatise on Adhesion and Adhesives, edited by R.L. Patrick, is a set of six hardcover books published by Marcel Dekker, Inc., New York, NY and Basel, Switzerland. Each book contains an average of eight excellent chapters by outstanding authors. The six volumes, published from 1967 to 1989, cover all aspects of adhesive bonding.

The illustrations in each volume are excellent, and each volume is well indexed.

Developments in Adhesives 1, edited by W.C. Wake, was published by Applied Science Publishers Ltd., London, UK, in 1977. It contains 11 chapters, each by a different author. Seven chapters discuss the adhesives used in specific applications, and the remaining four discuss other adhesive subjects. The book has three indexes: a patent index, an author index, and a subject index.

Developments in Adhesives 2, edited by A.J. Kinloch, was published by Applied Science Publishers Ltd., London, UK, and Englewood, NJ in 1981. The book contains ten chapters, each by a different author. There are two indexes: an author index and a subject index.

Construction Sealants and Adhesives, 2nd Edition, by J.R. Panek and J.P. Cook, was published by Wiley-Interscience, John Wiley & Sons, Inc., New York, NY in 1984 and has 348 pages in 28 chapters. Architects, engineers, contractors, and students will find this book helpful in covering the latest technology for sealants, tapes, gaskets, and other materials. The book also provides specifications on construction sealants and adhesives.

Structural Adhesive Joints in Engineering, by R.D. Adams and W.C. Wake, was published in 1984 by Elsevier Applied Science Publishers, Ltd. Copies are available from Elsevier Science Publishing Company,

Inc., 52 Vanderbilt Avenue, New York, NY 10017. The intent of the authors was to include everything an engineer needs to know to be able to design and produce adhesively bonded joints that are required to carry significant loads. The advantages and disadvantages of bonding are given, together with a sufficient explanation of the mechanics and chemistry necessary to enable the designer to make a sound engineering judgment in any particular case. The eight chapters of the book are "Introduction," "The Nature and Magnitude of Stresses in Adhesive Joints," "Standard Mechanical Test Procedures," "The General Properties of Polymeric Adhesives," "Factors Influencing the Choice of Adhesives," "Surface Preparation," "Service Life," "Applications." Also included are references, an appendix (Standard American and UK Specifications and Adhesion Tests), an author index, and a subject index.

Machinery Adhesives for Locking, Retaining and Sealing, by G.S. Haviland, Loctite Corporation, was published by Marcel Dekker, Inc., New York, NY and Basel, Switzerland in 1986. Copies are available from Marcel Dekker, Inc., 270 Madison Avenue, New York, NY 10016. The objective of this book is to guide the designer, process engineer, or mechanic in selecting and using anaerobic machinery adhesives effectively. The book has eight chapters, a glossary, and an index. There are a number of illustrations and tables. Loctite is a leading manufacturer of anaerobic adhesives and sealants.

Handbook of Plastics and Elastomers, edited by C.A. Harper was published by McGraw-Hill, New York, NY, in 1975. Chapter 10, "Plastics and Elastomers as Adhesives," by E.M. Petrie, is an excellent source that concisely treats all aspects of adhesive bonding in 136 pages.

Wood and Wood Products Redbook, a 1989 wood glue reference guide, is available from Vance Publishing Company, Lincolnshire, IL.

Composites, Volume 1, *Engineered Materials Handbook*, published by ASM INTERNATIONAL in 1987, is the first volume in the series to which this *Adhesives and Sealants* volume belongs. Its 13 sections include topics that are useful in an adhesives context, such as resin properties; the analysis and design of composite materials and structures; adhesives selection, specifications, bonding surface preparation, and bonding cure considerations; dissimilar material separation; and faying surface sealing.

Adhesion and Adhesives, by A.J. Kinloch, was published in 1987 by Chapman and Hall, London and New York. It has eight chapters and 441 pages.

Adhesives from Renewable Resources, edited by R.W. Hemingway and A.H. Conner, was published by the American Chemical Society in 1989 as part of the ACS Symposium Series 385. It has 34 chapters and 510 pages.

Durability of Structural Adhesives, edited by A.J. Kinloch, was published by Applied Science Publishers Ltd., London, UK, in 1983. It has eight chapters and 360 pages.

Structural Adhesives: Developments in Resins and Primers, edited by A.J. Kinloch, was published by Elsevier Applied Science Publishers, Ltd. in 1986. It has eight chapters and 328 pages.

The Rauch Guide to the U.S. Adhesives and Sealants Industry, by Fred Keimel and Adhesives and Sealants Consultants, was published by Rauch Associates Inc., of Bridgewater, NJ in 1990. It contains data for 1988 and 1989 and projections to 1994, in 274 pages.

The Science of Adhesive Joints, 2nd Edition, by J.J. Bikerman was published in 1968 by Academic Press, New York and London.

Handbook of Epoxy Resins, by H. Lee and K. Neville, was reissued in 1982 by McGraw-Hill Book Company, New York. The book is considered the "bible" on epoxy resins, with a chapter on adhesives.

Short Courses, Seminars, and Conferences

Organizations that offer short courses or seminars in adhesives technology include the Center for Professional Development, P.O. Box H, East Brunswick, NJ 08816-0257, (201) 238-0600, and the Center for Adhesive and Sealant Science (CASS), Virginia Polytechnic Institute and State University, 2 Davidson Hall, Blacksburg, VA 24061-0201 (FAX 703-231-7826).

A two-day weekend short course on "Adhesion: Theory and Practice" is given by The Adhesion Society in connection with its annual meeting in February of each year. The course contains a half-day session on "Practical Applications of Adhesives." For information contact: Professor Gary R. Hamed, Polymer Science Department, The University of Akron, Akron, OH 44325; (216) 375-6831.

The Society for the Advancement of Material and Process Engineering (SAMPE) has two important conferences each year, both of which usually include papers on adhesive bonding technology. One conference is the National SAMPE Technical Conference, and the other is the SAMPE Symposium and Exhibition. Information on both conferences can be obtained from the SAMPE International Business Office, P.O. Box 2459, Covina, CA 91722. See also the SAMPE journals listed earlier in this article. Many of the trade and technical journals and magazines listed previously provide information on forthcoming short courses, seminars, and conferences. *The Adhesives and Sealants Newsletter* frequently critiques them.

ASM INTERNATIONAL also offers seminars, videotape courses, and home learning courses on a variety of topics in the realm of engineered materials, many of which give continuing education credits. In addition, the society sponsors or cosponsors expositions and conferences, such as *Materials Week* and the *World Materials Congress*. To ascertain the availability/scheduling of adhesives-related activities, the society can be contacted at (216) 338-5151, and assistance can be obtained from education or conference personnel.

ASTM Committees

Committee C-24 on Building Seals and Sealants is one of the committees of the American Society for Testing and Materials, 1916 Race Street, Philadelphia, PA 19103. The 1989 chairman was C.J. Parise, Smith, Hinchman & Grylis, Detroit, MI, (313) 964-3000, and the Staff Director at ASTM headquarters is Martha Kirkaldy, (215) 299-5531. Ms. Kirkaldy should be called for information on this committee and its meeting schedules.

Committee D-14 on Adhesives is another ASTM committee. The 1990 chairman is T.L. Wilkinson, Jr., Reynolds Metals Company, P.O. Box 27003, Williamsburg Road and Goddin Street, Richmond, VA 23261, (804) 788-7525, and the Staff Director at ASTM headquarters is Sharon L. Kauffman. Ms. Kauffman should be contacted for information on this committee and its meeting schedules, (215) 299-5599. The subcommittees are: D14.03, Research; D14.04, Terminology; D14.05, Editorial Review; D14.10, Working Properties; D14.20, Durability; D14.30, Wood Adhesives; D14.40, Adhesives for Plastics; D14.50, Special Adhesives; D14.70, Construction Adhesives; D14.80, Metal Bonding Adhesives; and D14.91, Long Range Planning.

Data Bases

PLASTEC Adhesives is a computer program that retrieves information on adhesives technology. Access is available by arrangement with the Plastics Technical Evaluation Center (PLASTEC) at the U.S. Army Armament Research, Development and Engineering Center, Building 355-N, Picatinny Arsenal, NJ 07806-5000, (201) 724-5859. W.S. DePiero can be contacted.

The technical data in the Adhesives Data Base has been abstracted from numerous adhesive bonding studies. Currently, the program contains laboratory data and supporting information describing adhesive materials, adherends, document sources, trade identifications, test methods, and surface preparations. A "lessons learned" section contains historical information describing known adhesive bonding problems, and a design and manufacturing section covers basic concepts such as joint design, materials selection, and manufacturing processes.

The PLASTEC Adhesives Data Base can be used by all federal agencies and by their contractors on a verified "need-to-know" basis. The Adhesives Data Base is available either by direct-access authorization or by individual inquiry through PLASTEC.

A subscription to the Adhesives Data Base, ensuring unlimited access to the data file, is available on a fee basis, renewable yearly. A fee schedule is available upon request. Direct access to the most current data with a telephone modem and compatible terminal allows immediate answers. Subscribers are supplied with a unique access code, a guide that describes the data base and how to use it, and a list of available specialists on call for access or operational assistance.

The Materials Information department of ASM INTERNATIONAL provides extensive coverage of the international literature on all aspects of engineered materials, including composites, plastics, adhesives, and sealants. This information is available monthly from *Engineered Materials Abstracts* (EMA) and its equivalent, the EMA data base. Materials Information also publishes *Polymers/Ceramics/Composites Alert* and its on-line equivalent, *Materials Business File*, which emphasizes the business aspects of the materials industry. Materials Information also has access to more than 300 other data bases.

Other computer-searchable data bases include those listed in Ref 1.

Standards and Specifications, prepared by the National Standards Association, Inc., is available on line through the Dialog Information Retrieval Service, Palo Alto, CA. It references over 113 000 U.S. and international documents, including standards from ASTM; American National Standards Institute (ANSI); the federal government; the U.S. Military; the Society of Automotive Engineers, Inc.; and others. In addition to giving information on the standard and its acceptance by ANSI and the U.S. Department of Defense, names of vendors of products and services conforming to the standard are provided. Hard copy of many of the standards is also available from the National Standards Association, Inc.

Standards Search, prepared by ASTM and SAE, is available on line through the Orbit Search Service, McLean, VA. It contains over 15 000 references (some with abstracts) to the standards included in the *ASTM Book of Standards* and in the *SAE Handbook, Aerospace Index*, and *Index of Aerospace Materials Specifications*.

Military and Federal Specifications and Standards, prepared by Information Handling Services, is available on line through BRS Information Technologies, Latham, NY. This cites more than 80 000 nonclassified U.S. Military and federal standards and specifications, joint Army-Navy specifications, Military Standard Drawings, and Qualified Product Lists.

Combined Industry Standards and Military Specifications is also prepared by Information Handling Services and available on line through BRS Information Technologies, Latham, NY. This source cites more than 150 000 government and industry standards. It covers some 50 U.S. and other national and international standards organizations, including ANSI, the International Standards Organization (ISO), and the National Bureau of Standards Voluntary Engineering Standards data base, which is based on the standards of some 400 societies. It also includes the military and related standards covered by *Military and Federal Specifications and Standards*.

REFERENCE

1. J.T. Rice, Adhesive Selection and Screening Testing, Appendix 1, *Handbook of Adhesives*, 3rd ed., I. Skeist, Ed., Van Nostrand Reinhold, 1990

Adhesives Materials

Chairman: Michelle M. Gauthier, Raytheon Company Missile Systems Division

Introduction

A LARGE NUMBER of adhesive materials are commercially available. Many adhesives do not function exclusively as bonding agents; they often perform additional functions such as sealing or vibration damping. Although some materials, such as silicones, can be classified as both adhesives and sealants, this Section is concerned with only the bonding function of the selected materials.

Adhesive materials can be classified in a variety of ways:

- Natural or synthetic polymer base
- Thermoplastics or thermosets
- Physical forms (one or multiple components, films, and so forth)
- Functional types (structural, hot melt, pressure sensitive, and so forth)
- Chemical families (epoxy, silicone, and so forth)

Synthetic thermoplastic and thermosetting adhesives are described in this Section. Because the emphasis of this Volume is on adhesives for engineering use, natural adhesives are omitted. Following an article on "Types of Adhesives" are subsections covering "Chemical Families," "Advanced Polymeric Systems," and "Adhesive Modifiers."

"Types of Adhesives" compares six functional types of adhesives, including structural, hot-melt, pressure-sensitive, organic solvent soluble, water-soluble, and ultraviolet/electron beam-cured adhesives. For each functional type, available chemical families are briefly described and compared, with some, such as acrylics, appearing under more than one functional type.

Selector tables are included for each functional type. They summarize the chemical families of adhesives that are applicable for bonding selected substrates. These tables are simplified to base selection of the adhesive solely on the adherend materials to be bonded, without taking into account other important factors such as the part design, the method of adhesive application and cure, the required end-use properties, and the environmental conditions to which the part will be subjected. Despite this simplification, these tables provide a starting point from which the design engineer can choose the many available adhesives.

The functional types of adhesive materials discussed in "Types of Adhesives" were selected based on their acceptance and use in major markets, such as transportation, aerospace, electrical/electronic, construction, and packaging.

The "Types of Adhesives" article and the "Chemical Families" subsection complement each other, and to an extent, overlap. (For example, epoxies are a chemical family, yet also function as structural adhesives. For this reason, they are discussed in both places in this Section as are thermoplastic and thermosetting adhesives and physical forms.)

To facilitate comparisons, articles in the "Chemical Families" subsection follow a common outline, which includes such information as chemistry, functional type, commercial forms, major markets and applications, costs, competing adhesives, uncured and cured properties, additives and modifiers, design considerations, and processing conditions. Adhesive materials included in the "Chemical Families" subsection were chosen on the basis of their large volume application in the previously stated end-use markets.

Articles in the "Advanced Polymeric Adhesives" subsection, covering polyimides, polyphenylquinoxalines, and polybenzimidazoles, follow the same common outline as those in the "Chemical Families" group. These materials were selected due to their increasing importance in applications such as aerospace, where conventional adhesives cannot withstand the high-temperature conditions.

A final subsection, "Adhesive Modifiers," contains information that complements each of the previous articles. Adhesive modifiers are found in all types of adhesives and in all chemical families. The article covering this topic describes five groups of materials, including electrically and thermally conductive fillers, extenders, adhesive promoters, tackifiers, and tougheners. In addition, it presents information on chemical families that contain these adhesive modifiers. The modifiers discussed are selected based on their widespread use in many adhesive formulations.

Types of Adhesives

Michelle M. Gauthier, Raytheon Company, Missile Systems Division

FIVE GROUPS of adhesives are discussed in this article:

- Structural
- Hot melt
- Pressure sensitive
- Water base
- Ultraviolet (UV) and electron beam (EB) cured

These five groups by no means include every available type of adhesive. Others, such as organic-solvent-base adhesives, are discussed briefly in this article.

Most of the adhesives in these groups are synthetic based, although natural adhesives, derived from vegetable and animal sources, are used in water-base and some pressure-sensitive adhesives in large quantities.

The characteristics of these five groups of adhesives are summarized in Table 1. Table 2 lists the advantages and limitations of the five groups.

Structural Adhesives

A structural adhesive is defined as a material used to transfer loads between adherends in service environments to which the assembly is typically exposed. Another definition of a structural adhesive is a material of proved reliability in engineering structural applications in which the bond can be stressed to a high proportion of its maximum failing load for long periods without failure (Ref 1). `

Most materials used in structural adhesives are thermosets. However, some thermoplastics, such as cyanoacrylates and anaerobics, are used. Thermoplastics harden rapidly, but have limited heat and creep resistance. The advantages of thermosets include insolubility and heat and creep resistance. Limitations for most include mixing requirements in which the monomer, resin, or prepolymer must be compounded with a hardener, activator, cross-linking agent, or catalyst in order to achieve cure. For most thermoset materials that are available fully compounded, heat must be applied to achieve a cure. Thermosets that cure by condensation polymerization, such as phenolics and polyimides, yield by-products as they cure. Addition polymerization, which does not result in the formation of by-products, is being used successfully with thermosets, including polyimides (Ref 2).

Characteristics

Structural adhesives can be classified as chemically reactive, evaporation or diffusion, hot-melt, delayed-tack, film, pressure-sensitive, or electrically and thermally conductive adhesives.

The first six were classified by the Society of Manufacturing Engineers. The last type was added because it does not completely meet the physical and chemical characteristics of any of the other types (Ref 3).

Chemically reactive adhesives can be subdivided into two groups: one-component systems, which include moisture cure and heat-activated cure categories, and two-component systems, which are subdivided into mix-in and no-mix systems. One-component formulations that cure by moisture from the surrounding air or by adsorbed moisture from the surface of a substrate include polyurethanes and silicones. Insufficient moisture in overlapping bond areas on nonporous substrates such as metals can result in insufficient cure.

Cyanoacrylates are also classified as adhesives that have a moisture cure. These thermoplastic materials form strong, rapidly setting bonds without the need for heat or a catalyst. In these materials, liquid monomer is converted to solid polymer in the presence of moisture.

A one-component heat-activated system usually consists of two components that are premixed. An advantage to the use of this

Table 1 Characteristics of various types of adhesives

Structural	Hot melt	Pressure sensitive	Water base	Ultraviolet/electron beam cured
Bonds can be stressed to a high proportion of maximum failure load under service environments	100% solid thermoplastics	Hold substrates together upon brief application of pressure at room temperature	Includes adhesives dissolved or dispersed (latex) in water	100% reactive liquids cured to solids
Most are thermosets.	Melt sharply to a low-viscosity liquid, which is applied to surface	Available as organic-solvent-base, water-base, or hot-melt systems	On porous substrates, water is absorbed or evaporated in order to bond.	One substrate must be transparent for UV cure, except when dual-curing adhesives are used (see below).
One- or two-component systems	Rapid setting, no cure	Some require extensive compounding (rubber base) to achieve tackiness, whereas others (polyacrylates) do not.	On nonporous substrates, water must be removed prior to bonding.	Some UV-curable formulations are dual curing; a second cure mechanism introduces heat or moisture or eliminates oxygen (anaerobics).
Room- or elevated-temperature cures	Melt viscosity is an important property.	Available supported (most) or unsupported on a substrate	Some are bonded following reactivation of dried adhesive film under heat and pressure.	In EB curing, density of material affects penetration.
Wide range of costs	Nonpressure sensitive and pressure sensitive	Primarily used in tapes and labels	Many are based on natural (vegetable or animal) adhesives.	UV/EB-curable formulations have laminating and PSA applications.
Various chemical families with varying strengths and flexibilities	Compounded with additives for tack and wettability		Nonpressure sensitive (most) or pressure sensitive applications	UV-curable formulations have laminating, PSA, and structural adhesive applications.

Table 2 Advantages and limitations of various types of adhesives

Structural	Hot melt	Pressure sensitive	Water base	Ultraviolet/electron beam cured
Advantages				
High strength	100% solids, no solvents	Labels and tapes have uniform thickness.	Low cost, nonflammable, nonhazardous solvent	Fast cure (some in 2 to 60 s)
Capable of resisting loads	Can bond impervious surfaces	Permanent tack at room temperature	Long shelf life	One-component liquid: no mixing, no solvents
Good elevated-temperature resistance (cross-linked)	Rapid bond formation	No activation required by heat, water, or solvents	Easy to apply	Heat-sensitive substrates can be bonded; cure is "cool."
Good solvent resistance	Good gap-filling capability	Cross-linking of some formulations possible	Good solvent resistance	Many are optically clear.
Good creep resistance	Rigid to flexible bonds	Soft or firm tapes and labels	Cross-linking of some formulations possible	High production rates
Some available in film form	Good barrier properties	Easy to apply	High molecular weight dispersions at high solids content with low viscosity	Good tensile strength
Limitations				
Two-component systems require careful proportioning and mixing.	Thermoplastics have limited elevated-temperature resistance.	Many are based on rubbers, requiring compounding.	Poor water resistance	Equipment expensive
Some have poor peel strength.	Poor creep resistance	Poor gap fillers	Slow drying	High material cost
Some are difficult to remove and repair.	Little penetration due to fast viscosity increase upon cooling	Limited heat resistance	Tendency to freeze	UV cures only through transparent materials (or secondary cure required).
Some require heat to cure.	Limited toughness at usable viscosities		Low strength under loads	Difficult curing on parts with complex shapes
Some yield by-products upon cure (condensation polymers).			Poor creep resistance	Many UV cures have poor weatherability because they continue to absorb UV rays.
			Limited heat resistance	
			Shrinkage of certain substrates in supported films and tapes	

type of system is the elimination of the metering and mixing requirements of two-component systems. However, special storage conditions (that is, 0 °C, or 32 °F, or below) are usually required, and these adhesives generally have a limited shelf life. Typical heat-activated systems include both liquids and films. Chemical families in this group include epoxies and epoxy-nylons, polyurethanes, polyimides, polybenzimidazoles, and phenolics, including epoxy-phenolics, nitrile-phenolics, polyvinyl formal-phenolics, and polyvinyl butyral-phenolics.

Anaerobics can also be classified as one-component adhesives that are chemically reactive. However, they are not included in either the moisture- or heat-activated cure categories because they are monomeric liquids that cure by free radical polymerization upon the elimination of oxygen. In general, anaerobics are based on methacrylates, acrylics, and acrylic-ester copolymers. Some structural adhesives are also based on urethane-acrylates (Ref 4, 5).

Two-component mix-in systems consist of two separate components that must be metered in the proper ratio, mixed, and then dispersed. Some will cure at room temperature, but heat is often applied to accelerate the cure and to achieve better bond quality. Chemical families in this group include epoxies, modified acrylics, polyurethanes, silicones, and phenolics.

Two-component no-mix systems consist of two separate components that do not require careful metering because no mixing is involved. Adhesive is applied to one surface, while an accelerator is applied to a second surface. The surfaces (substrates) are then joined. These adhesives will cure at both room and elevated temperatures. Modified acrylics are a chemical family that is included in this group.

The chemically reactive adhesives that are most widely used in structural adhesives include epoxies, polyurethanes, modified acrylics, cyanoacrylates, and anaerobics. These families are discussed in greater detail in the section "Structural Adhesive Chemical Families" in this article. Also included in that section are silicones, phenolics, and high-temperature adhesives, such as polyimides and polybenzimidazoles.

Evaporation or diffusion adhesives can be subdivided into materials that are based on either organic solvents or water. In solvent-base systems, the adhesive solution is coated on the porous substrates. Following solvent evaporation and/or absorption into the substrates, the surfaces are joined. Nonporous materials, such as metals, can also be bonded by this type of adhesive; however, heat and pressure are usually required to activate the adhesive. Chemical families included in this group are natural rubber, reclaimed rubber, synthetic rubbers, phenolics, polyurethanes, vinyls, acrylics, and naturally occurring materials.

The toxicity and flammability associated with the solvent limit the use of organic-solvent-base adhesives. These limitations do not apply to water-base adhesives, which comprise materials that are totally soluble (solution) or dispersive (latex) in water. A wide range of viscosities and solids contents are available. Limitations that do exist with water-base adhesives include slow setting rates, poor water resistance, and the possibility of freeze-coagulation of an unset water-containing formulation. Chemical families included in this group are natural rubber, reclaimed rubber, synthetic rubbers, vinyls, acrylics, and naturally occurring materials.

The section "Water-Base Adhesives" in this article provides more details. Although organic-solvent-base adhesives are not discussed elsewhere in this article, the chemical families in this group are, in general, the same as those discussed in the section "Water-Base Adhesives."

Hot-melt adhesives are 100% solid thermoplastics that are very loosely classified as structural adhesives because most will not withstand elevated-temperature loads without creep. High-performance hot melts, including polyamides and polyesters, will withstand limited loads. These adhesives are solid to a certain temperature, at which they will sharply melt to a liquid that is applied to a substrate. Cooling results in rapid setting. Chemical families in this group include ethylene-vinyl acetate copolymers, polyolefins, polyamides, polyesters, and thermoplastic elastomers.

The section "Hot-Melt Adhesives" in this article contains further details.

Delayed-tack adhesives remain tacky following heat activation and cooling. Tack remains over a period of time, ranging from minutes to days, and over a wide temperature range. Solid plasticizers used in the adhesive formulation contribute to nontackiness prior to heat activation. The use of solid plasticizers having different melting temperatures results in adhesives with different heat-activation temperatures. Typical solid plasticizers include dicyclohexyl phthalate, diphenyl phthalate, and ortho- or para-toluenesulfonamide. Chemical families used in making delayed tack adhesives include styrene-butadiene copolymers, polyvinyl acetates, polystyrene, and polyamides.

Film adhesives are applied quickly and easily and result in a uniform, controlled bond line thickness. Both two-sided and one-sided tapes and films are included in this category. These adhesives are either unsupported or are supported by various carriers, such as glass cloth, nylon, and paper. Some will cure at room temperature, but most require elevated temperature and nominal pressure. Although some film and tape adhesives are stored at room temperature, many require refrigerated storage. Film and tape adhesives usually consist of a high molecular weight backbone polymer for toughness, elongation, and peel strength; a low molecular weight cross-linking resin, such as an epoxy or a phenolic; and a curing agent for the cross-linking resin. Chemical families used in making film and tape adhesives include nylon-epoxies, elastomer-epoxies, nitrile-phenolics, vinyl-phenolics, epoxy-phenolics, and high-temperature-resistant aromatics, including polyimides and polybenzimidazoles (Ref 6). Chemical families used to make film and tape adhesives will be discussed in greater detail in the section "Structural Adhesive Chemical Families."

Vendors of both supported and unsupported film adhesives include AI Technology; Ablestik Laboratories; Adhesives Research, Inc.; BASF Corporation; Bostik Division, Emhart Corporation; Ciba-Geigy Corporation; Dow Chemical Company; Sheldahl; 3M Company; and Westinghouse Electric Corporation, Foretin Industries, Inc. (Ref 7).

Pressure-sensitive adhesives are capable of holding substrates together when they are brought into contact under brief pressure at room temperature. These adhesives are either unsupported or are supported by various carriers, including paper, cellophane, plastic films, cloth, and metal foil. Both single- and double-sided tapes and films are included. Most of these adhesives are based on rubbers compounded with various additives, including tackifiers. Chemical families in this group include natural rubber, styrene-butadiene rubber (SBR), reclaimed rubber, butyl rubber, nitrile rubber, polyacrylates, and polyvinylalkylethers. Rubber-base materials have poor aging characteristics, which can be overcome by the use of polyacrylates or polyvinylalkylethers (Ref 6). The section "Pressure-Sensitive Adhesives" in this article describes these materials in greater detail.

Conductive adhesives include both electrically and thermally conductive materials. Most adhesive fillers that result in electrical conductivity in a material also contribute to its thermal conductivity. However, thermally conductive adhesives that are electrically insulative are also available.

The majority of applications involve the use of silver, either in flake or powder form. This filler is preferred to gold because of its lower cost and its lower volume resistivity (that is, higher electrical

conductivity). The best silver-filled epoxies have a volume resistivity of 0.001 Ω · cm. With silver-filled adhesives, migration of silver to the surface under conditions of high humidity and direct current can occur. Neither gold- nor silver-coated copper fillers in various adhesives have this migration problem. Epoxies with silver fillers up to 85% by weight are available. However, adhesives with lower filler loadings generally have better strengths.

Other fillers that provide electrical conductivity include copper and aluminum, although oxide formation on the surfaces of these fillers can occur. This results in decreased conductivity because of decreased particle-to-particle contact.

In the past few years, conductive epoxies and polyimides have been used as die attach adhesives. Material requirements for die attach adhesives include:

- "Green" adhesive strength for preventing chip movement prior to cure
- Absence of resin bleed before or after cure
- Good elevated-temperature strength after a short cure period
- Thermal stability
- Absence of outgassing
- A minimum of ionic impurities

The chloride ion contents of polyimides are inherently lower than those of epoxies. However, these adhesives are more expensive, have lower bond strength, and are more difficult to cure. On the other hand, epoxies with low ionic impurities, including chlorides and sodium and potassium cations, have been available since 1981. These materials have a fast cure and excellent high-temperature strength, and a glass transition temperature that can be as high as 200 °C (392 °F).

Oxide fillers are nonelectrically conductive but are used to provide thermal conductivity. Alumina (aluminum oxide), the most commonly used filler, is fairly inexpensive. It can be added in high concentrations to epoxies and silicones without significantly increasing the uncured material viscosity. Minimum bond line thicknesses are desirable for thermally conductive adhesives because heat flow is proportional to the ratio of thermal conductivity to bond thickness. Alumina-filled epoxies contain filler in amounts as high as 75%, by weight, and have thermal conductivities ranging from 1.38 to 1.73 W/m · K (3.31 to 4.13 × 10^{-3} cal/cm · s · °C).

Other thermally conductive fillers include beryllia (beryllium oxide), which is high in cost and toxic, and boron nitride, which has a loading limitation of approximately 40%, by weight, in epoxies. Other inorganic oxides, including silica (silicon dioxide), are also used to provide thermal conductivity.

The chemical families that are used most often to provide electrical and/or thermal

conductivity include epoxies, polyurethanes, silicones, and polyimides. Epoxies are the most widely used (Ref 6, 8).

Vendors of conductive adhesives include AI Technology; Ablestik Laboratories; Amicon Corporation; Aremco Products, Inc.; Bacon Industries, Inc.; Chomerics, Inc.; Ciba-Geigy Corporation; Epoxy Technology, Inc.; Formulated Resins Inc.; Furane Products Company; Emerson & Cuming, Permagile Industries, Inc.; Technit; and Tra-Con, Inc. (Ref 7).

The selection of a particular structural adhesive for use in a specific application is dependent on certain factors:

- The substrates to be bonded, including their porosity and hardness and the absence or presence of surface coatings such as paint or plating
- End-use requirements, including stresses encountered during construction and in service, and the types of joints. (This is fully covered in the Section "Designing With Adhesives and Sealants" in this Volume.)
- Temperature requirements for cure and end-use
- Exposure conditions, including exposure to solvents, humidity, UV radiation, and outdoor weathering
- Flexibility requirements
- Aging stability
- Aesthetic requirements, including color and gloss
- Manufacturing conditions, including materials handling systems, processing equipment, and skill of personnel
- Cost, which should take into account waste, rejects, packaging, batch-to-batch reliability, and availability (Ref 4)

Markets and Applications

In 1987, the total estimated sales of adhesives and sealants was $4.2 billion. Of that amount, $1.45 billion, or 34.5%, was attributed to structural and specialty adhesives. The primary areas of application included automotive, aerospace, appliances, biomedical/dental construction, consumer electronics, fabric, furniture, general industrial, industrial machine, marine, and sports equipment applications. Of these areas, construction accounted for $600 million, or 41% of the total market.

It was predicted that from 1988 to 1992 the structural and specialty adhesive market would grow at a rate of 5% per year. Although the biomedical/dental area was a small market in 1987, accounting for $18 million, it is the fastest-growing market and is predicted to increase to $30 million by 1992. Adhesives in this area include cyanoacrylates, acrylics, modified epoxies, and inorganics. The electronics market is predicted to be the second fastest growing field by 1992.

Figure 1 shows the breakdown of structural and specialty adhesive sales by chem-

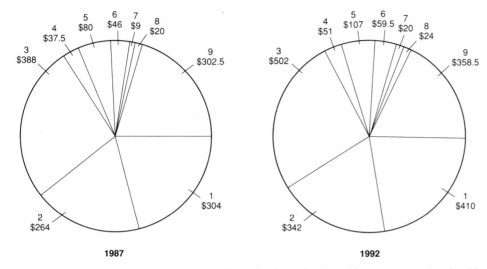

Fig. 1 Structural and specialty adhesives sales of formulated resins by chemical family in 1987 and predicted for 1992 (in million of dollars). (1) Epoxies. (2) Polyurethanes. (3) Modified acrylics. (4) Cyanoacrylates. (5) Anaerobics. (6) Silicones. (7) Polyimides. (8) Phenolics. (9) Other (polysulfide, and so on)

ical family in 1987 and that predicted for 1992. Sales of polyimides and cyanoacrylates are predicted to have the fastest growth (Ref 9).

Structural Adhesive Chemical Families

The most widely used structural adhesive groups are epoxies, polyurethanes, modified acrylics, cyanoacrylates, and anaerobics. Table 3 lists their advantages and limitations, and Table 4 compares typical properties. The sections below briefly discuss these and the chemical families of silicones, phenolics, and high-temperature adhesives.

Epoxies represent one of the most widely used structural adhesives. These chemically reactive systems include two-component room-temperature and two-component ele-vated-temperature curing, as well as one-component systems that usually require heat in order to cure.

Epoxy thermoset adhesives, available in many formulations, can be used to join most materials. These adhesives have good strength, will not yield volatiles during cure, and have low shrinkage. However, peel strength and flexibility are low, and epoxies are brittle.

A typical two-component system consists of a resin and a hardener, which are packaged separately. In a one-component system, the resin and hardener are packaged together. Other possible additives include accelerators, reactive diluents, plasticizers, resin modifiers, and fillers.

The most common epoxy resin is based on the diglycidyl ether of bisphenol A

(DGEBA). Typical hardeners (curing agents) include:

- Aliphatic polyamines, which cure at room temperature or at slightly elevated temperatures
- Fatty polyamides, which impart flexibility and are widely used
- Aromatic polyamines, which are solids
- Anhydrides, which require an elevated-temperature cure and yield adhesives with excellent thermal stability
- Boron trifluoride monoethylamine, which is a solid used in one-component epoxies

An exothermic reaction occurs upon the curing of epoxies. This reaction can be minimized by lowering the temperature of the mixed components in two-part systems, by limiting the batch size, and by using shallow mixing containers.

Epoxies are brittle, especially if cured with an anhydride. Therefore, thermoplastics and rubbery modifiers are often added to decrease the brittleness (Ref 1, 6).

Nylon-epoxies are among the best materials to use in film and tape adhesives. Nylon-epoxy, which first became available in 1960, has an approximate maximum service temperature of 138 °C (280 °F), compared to 177 °C (350 °F) for unmodified epoxies (Ref 10).

The main advantage of nylon-epoxies is increased flexibility and a huge increase in peel strength, compared to unmodified epoxy. Also, as indicated in Table 5, this very tough material has excellent tensile lap-shear strength. Fatigue and impact resistance are also good. Limitations include poor peel strength at low temperatures and poor creep resistance. Perhaps the most serious limitation is extremely poor moisture resistance in both the uncured and

Table 3 Advantages and limitations of the five most widely used chemically reactive structural adhesives

Epoxy	Polyurethane	Modified acrylic	Cyanoacrylate	Anaerobic
Advantages				
High strength	Varying cure times	Good flexibility	Rapid room-temperature cure	Rapid room-temperature cure
Good solvent resistance	Tough	Good peel and shear strengths	One component	Good solvent resistance
Good gap-filling capabilities	Excellent flexibility even at low temperatures	No mixing required	High tensile strengths	Good elevated-temperature resistance
Good elevated-temperature resistance	One or two component, room- or elevated-temperature cure	Will bond dirty (oily) surfaces	Long pot life	No mixing
Wide range of formulations	Moderate cost	Room-temperature cure	Good adhesion to metal	Indefinite pot life
Relatively low cost		Moderate cost	Dispenses easily from package	Nontoxic
				High strength on some substrates
				Moderate cost
Limitations				
Exothermic reaction	Both uncured and cured are moisture sensitive.	Low hot-temperature strength	High cost	Not recommended for permeable surfaces
Exact proportions needed for optimum properties	Poor elevated-temperature resistance	Slower cure than with anaerobics or cyanoacrylates	Poor durability on some surfaces	Will not cure in air as a wet fillet
Two-component formulations require exact measuring and mixing.	May revert with heat and moisture	Toxic	Limited solvent resistance	Limited gap cure
One-component formulations often require refrigerated storage and an elevated-temperature cure.	Short pot life	Flammable	Limited elevated-temperature resistance	
Short pot life (more waste)	Special mixing and dispensing equipment required	Odor	Bonds skin	
		Limited open time		
		Dispensing equipment required		

Table 4 Typical properties of the five most widely used chemically reactive structural adhesives

Property	Epoxy	Polyurethane	Modified acrylic	Cyanoacrylate	Anaerobic
Substrates bonded ...	Most	Most smooth, nonporous	Most smooth, nonporous	Most nonporous metals or plastics	Metals, glass, thermosets
Service temperature range, °C (°F)	−55 to 121 (−67 to 250)	−157 to 79 (−250 to 175)	−73 to 121 (−100 to 250)	−55 to 79 (−67 to 175)	−55 to 149 (−67 to 300)
Impact resistance	Poor	Excellent	Good	Poor	Fair
Tensile shear strength, MPa (ksi)	15.4 (2.20)	15.4 (2.20)	25.9 (3.70)	18.9 (2.70)	17.5 (2.50)
T-peel strength, N/m (lbf/in.)	<525 (3)	14 000 (80)	5250 (30)	<525 (3)	1750 (10)
Heat cure or mixing required	Yes	Yes	No	No	No
Solvent resistance ...	Excellent	Good	Good	Good	Excellent
Moisture resistance ..	Excellent	Fair	Good	Poor	Good
Gap limitation, mm (in.).	None	None	0.762 (0.030)	0.254 (0.010)	0.635 (0.025)
Odor.	Mild	Mild	Strong	Moderate	Mild
Toxicity	Moderate	Moderate	Moderate	Low	Low
Flammability.	Low	Low	High	Low	Low

cured materials. After exposure to 95% relative humidity for 2 months, a nylon-epoxy with an initial tensile lap-shear strength of 34 MPa (5 ksi) had a strength of only 6.8 MPa (1 ksi). Nylon-epoxies are not as durable as elastomer-epoxies or thermoplastic-modified epoxies. A typical application of nylon-epoxies is in laminates (Ref 1, 6).

Elastomer-epoxies generally contain nitrile rubber as the elastomeric component. This system is also referred to as a modified, or toughened, epoxy. One of the applications of widest use is in films and tapes.

Elastomer-epoxies cure at low pressures and low temperatures over a short time interval. This is achieved by adding a catalyst to the adhesive formulation.

As seen in Table 5, the bond strengths of elastomer-epoxies are lower than those of nylon-epoxies. However, the great advantage of elastomer-epoxies is in their subzero peel strengths, which do not decrease as fast as those of nylon-epoxies. In addition, the durability of elastomer-epoxies with regard to moisture resistance is better than that of nylon-epoxies but not as good as that of vinyl-phenolics or nitrile-phenolics. Limitations to the use of elastomer-epoxies include poor water immersion resistance and poor properties when exposed to marine conditions (Ref 1, 6).

There are many vendors of epoxies, including Bostik Division, Emhart Corporation; Ciba-Geigy Corporation; The Dexter Corporation; Emerson & Cuming; Furane Products Company; The BF Goodrich Company; Lord Corporation; Master Bond Inc.; Magnolia Plastics, Inc.; and National Starch and Chemical Corporation (Ref 11). The article "Epoxies" provides a more detailed discussion.

Polyurethanes. Chemically reactive polyurethanes include both one- and two-component systems. One-component systems are usually based on a polyether polyol reacted with a polyisocyanate, yielding an isocyanate-terminated polymer. A one-component system cures when exposed to moisture at room temperature. Two-component systems result from the reaction of low molecular weight polyols and isocyanates or from isocyanate-terminated prepolymers with either polyols or polyamines. Two-component systems cure at room and/or elevated temperatures (Ref 12). Both of these types of polyurethane adhesives are available in solvent-free and low-solvent systems.

One-component heat-curable urethanes are also available. In these formulations, free isocyanate groups are typically blocked by the addition of phenol. The prepolymer is then blended with the polyglycol curing agent and packaged. The mixture is stable until heated to elevated temperatures, whereupon the phenol is released and the isocyanate groups are regenerated. A rapid cure occurs when these groups come into contact with the polyglycol (Ref 6).

Polyurethanes will bond to most surfaces. They have an outstanding tensile lap-shear strength of 34.4 MPa (5 ksi) at −73.3 °C (−100 °F). In fact, polyurethanes have better low-temperature strength than any other adhesive, even epoxy-nylons. Good flexibility, abrasion resistance, and toughness are other advantages of polyurethanes. Limitations include sensitivity to moisture in both uncured and cured adhesives, toxicity of isocyanate, and poor tensile lap-shear strength at room temperature (12.4 MPa, or 1.8 ksi) (Ref 1, 6).

Polyurethane adhesive vendors include Emerson & Cuming; Furan Products Company; Transene Company Inc.; Formulated Resins Inc.; Applied Plastics Company, Inc.; Devcon Corporation; Lord Corporation; Ashland Chemical Company; and Mobay Chemical Corporation (Ref 11, 13). The article "Urethanes" in this Volume describes polyurethanes in more detail.

Modified acrylics, also referred to as second-generation acrylics or reactive acrylics, are composed of a modified acrylic adhesive and a surface activator. Typical modified acrylics are based on cross-linked polymethyl methacrylate grafted to vinyl-terminated nitrile rubber. Carboxy-terminated rubbers have also been used (Ref 12). Both one- and two-component systems are available.

Unlike epoxies, which cure by an ionic polymerization mechanism, modified acrylics cure by free radical addition. Therefore, careful proportioning of components is not required. In two-component systems, no mixing is required because the adhesive is applied to one substrate, the activator to the second substrate, and the substrates are joined. Handling strength is rapidly achieved in this fast-curing system.

Modified acrylics have good peel, impact, and tensile lap-shear strengths between −107 and 121 °C (−161 and 250 °F). High bond strengths are obtained with metals and plastics even if surfaces are oily or improperly cleaned. The cured adhesive exhibits little shrinkage. Resistance to high humidity is good, particularly when bonding plastic substrates. The limitations of modified acrylic adhesives include low hot-temperature strength, flammability, and the odor of the uncured acrylic adhesive (Ref 4, 6, 12). Vendors of modified acrylics include Lord Corporation and Hernon Manufacturing Inc. (Ref 11). The article "Acrylics" in this Volume provides a more detailed discussion of modified acrylic adhesives.

Cyanoacrylates are single-component liquids that cure when nucleophiles, such as water on the surface of a part, serve as initiation sites for polymerization. No by-products are released during cure. Factors that influence cure include the moisture on the surfaces to be bonded, relative humidity, pH, and bond line thickness. Problems may be encountered when bonding acidic

Table 5 Typical film adhesive properties

Material	High-pressure cure	Volatiles during cure	Tensile lap-shear strength		T-peel strength	
			MPa	ksi	kN/m	lbf/in.
Nylon-epoxy	No	No	39–49	5.5–7.2	14–22.8	80–130
Elastomer-epoxy	No	No	26–41	3.7–6.0	3.85–15.8	22–90
Nitrile-phenolic	Yes	Yes	21–31	3.0–4.5	2.63–10.5	15–60
Vinyl-phenolic	Yes	Yes	21–31	3.0–4.5	2.63–6.07	15–35
Epoxy-phenolic	Yes	Yes	14–22	2.0–3.2	1.05–2.10	6–12

(low pH) surfaces. Bond lines that are too thick may not cure in the center because water from substrate surfaces may be unable to penetrate completely.

Most cyanoacrylates are based on either methyl or ethyl cyanoacrylates. Methyl cyanoacrylates yield better strengths when bonding metals or other rigid surfaces. In addition, impact resistance is slightly better. Ethyl cyanoacrylates are stronger and more durable when bonding rubber or plastic substrates.

Cyanoacrylates will bond most substrates. Bond durability problems are encountered with silicones, polyolefins, and some fluoroelastomers, where extensive surface preparation is required, and also with glass (Ref 14).

In general, cyanoacrylates require no special curing equipment and have a fast cure, excellent tensile strength, and good shelf life. Limitations include high cost, poor peel strength, brittleness, limited elevated-temperature properties, and the capability to fill only small gaps (Ref 4, 15).

Cyanoacrylate vendors include Permabond International, Loctite Corporation, Master Bond Inc., Hernon Manufacturing Inc., 3M Company, Devcon Corporation, Lord Corporation, and Henkel Corporation (Ref 11, 15). The article "Cyanoacrylates" in this Volume discusses these materials in greater detail.

Anaerobics are single-component monomeric liquids that cure by free radical polymerization upon the elimination of oxygen. These materials are packaged in contact with air in order to maintain their monomeric form. The adhesives are available in machinery or structural grades. The former provides high cohesive strength in threaded-assembly bonding, and the latter provides high tensile and shear strengths in flat assemblies (Ref 6).

There are three types of surfaces to which anaerobic adhesives are applied: active, inactive, and inhibiting. Active surfaces include clean metals and thermosets and result in the fastest cure. Inactive surfaces include some metals and plastics, which result in a slow cure. Inhibiting surfaces include bright platings, chromates, oxides, and certain anodizes. Primers or heat must be used to achieve cure on inactive and inhibiting surfaces. It should be noted that some plastics and rubbers are attacked by anaerobics. Others, such as Teflon, acetal, polyolefins, nylon, and polyvinylidene chloride, can be bonded with anaerobics.

Average cure speeds range from fast (5 min to 2 h) to moderate (2 to 6 h) to slow (6 to 24 h) at room temperature in the absence of primers. Heat accelerates the cure (Ref 16).

In general, anaerobic adhesives have a service temperature limit of 149 °C (300 °F) and are moderately priced. Structural anaerobic adhesives have high strength and

good moisture, solvent, and temperature resistance. If air is present, it will attack certain thermoplastic and rubber surfaces and also prevent cure.

Vendors of anaerobics, which are generally acrylic based, include Devcon Corporation, Henkel Corporation, Hernon Manufacturing Inc., Loctite Corporation, Master Bond Inc., and Permabond International (Ref 11). These adhesives are discussed in more detail in the article "Anaerobics" in this Volume.

Silicones are available as both one- and two-component systems that cure to thermosetting solids. A one-component room-temperature-vulcanizing silicone cures at room temperature upon exposure to atmospheric moisture. Cures are either acidic or nonacidic in the presence of moisture. Adhesives that have an acidic cure have greater unprimed adhesion and a longer shelf life. However, the corrosion of metals due to the acid is a potential problem. Thin films of approximately 0.6 mm (0.024 in.) cure within 90 min, whereas thicker films of approximately 13 mm (0.512 in.) require 7 days to achieve full cure.

Two-component silicones do not require moisture in order to cure. During the addition polymerization of silicones, cure is achieved by catalytic action. Pot life setting time and cure time are dependent on the catalyst concentration. A system with 5% catalyst will typically have a 3-h pot life, a 22-h setting time, and a 7-day cure time at room temperature. Increasing the temperature accelerates the cure. Addition-polymerized silicones exhibit little shrinkage upon cure and have good high-temperature resistance. In the condensation polymerization of two-component silicones, by-products are released. These systems are less likely to be inhibited and can be used on a greater variety of substrates. However, reversion of polymerization is a potential problem (Ref 12).

In general, silicone adhesives have good peel strength over a service temperature range of −60 to 250 °C (−76 to 482 °F). Some have limited service to 371 °C (700 °F). Flexibility and impact resistance are good, as are moisture, hot water, oxidation, and weathering resistance. Typical lap-shear strengths are low in metal-to-metal bonds, with values ranging between 1.72 and 3.44 MPa (0.250 and 0.500 ksi). The cost of these adhesives is high.

Silicone adhesives are useful in bonding metals, glass, paper, plastics, and rubbers, including silicone, butyl rubber, and fluoroelastomers (Ref 1, 4, 6).

Vendors of silicone adhesives include Aremco Products, Inc.; Dayton Plastics, Inc.; Devcon Corporation; Dow Corning Corporation; General Electric Company; Foster Products Division, H.B. Fuller Company; Emerson & Cuming; Loctite Corporation; Polymeric Systems Inc.; and Ther-

moset Plastics, Inc. (Ref 7, 11). The article "Silicones" in this Volume discusses these adhesives in greater detail.

Phenolics or phenol-formaldehyde structural adhesives are chemically reactive systems that cure to thermosets. In one-component systems, meltable powders (resoles) are used as binders for particleboard or as alloys (including nitrile-phenolics, vinyl-phenolics, and epoxy-phenolics), which are used in the structural bonding of metals. In two-component systems, the resin and catalyst are mixed and then heated in order to cure. Both systems cure by a condensation reaction, yielding a by-product (Ref 6).

In general, phenolics are low-cost adhesives having good strength and resistance to biodegradation, hot water, and weathering. Elevated-temperature resistance is also good. Limitations include low impact strength and high shrinkage stresses, which lead to brittleness. Shelf life is limited, the adhesives are dark colored, and they can be corrosive (Ref 10).

Phenolics dominate the wood adhesives market, especially with regard to plywood. The structural adhesive bonding of metals, particularly with phenolic alloys, is another application (Ref 6).

Nitrile-phenolic alloys are composed of nitrile rubber and phenolic with additives. These systems are available in liquid or film form. Typical tensile lap-shear and T-peel strengths of nitrile-phenolic films are shown in Table 5. The advantages of this alloy include a maximum service temperature of 138 °C (280 °F), low cost, high bond strength, and excellent water, oil, biodegradation, and salt resistance. Disadvantages include poor to moderately poor low-temperature resistance and a high-pressure high-temperature cure. Typical applications include the bonding of metals, plastics, rubber, wood, glass, and ceramics. This structural adhesive is used in automobile brake-shoes and clutch disks and in aircraft applications (Ref 6, 12).

Vinyl-phenolics are alloys composed of polyvinyl formal (PVF)-phenolics or polyvinyl butyral (PVB)-phenolics. These additives are available as both liquids and films. Typical tensile lap-shear and T-peel strengths of vinyl-phenolics are found in Table 5. In general, vinyl-phenolics have a maximum service temperature of 82 °C (180 °F) and are equal to phenolic-nitriles in strength. These adhesives are better than epoxies in sandwich structures that require strength properties.

PVF-phenolics have good fatigue, weathering, fungi, salt, humidity, water, and oil resistance. Creep resistance is good in some formulations, but is poor in others above 90 °C (194 °F). These adhesives are used in metal-honeycomb and wood-metal applications.

PVB-phenolics also have good weathering, fungi, salt, humidity, water, and oil resistance, and creep and fatigue resistance are even better than they are in PVF-phenolics. However, PVB-phenolics lack the shear strength and toughness of PVF-phenolics. These adhesives are used to bond metal or reinforced-plastic facings to paper honeycomb, cork to rubber, and steel to rubber (Ref 6, 12).

Epoxy-phenolic adhesives are also available as liquids or films. This adhesive is one of the best for long-term use at 149 to 260 °C (300 to 500 °F). The minimum service temperature is usually −60 °C (−76 °F), although special formulations may be useful at temperatures as low as −260 °C (−436 °F). In general, epoxy-phenolics are relatively expensive. They have fairly good shear and tensile strengths over a wide temperature range. Table 5 gives ambient tensile lap-shear and T-peel strengths of films, and it can be seen that peel strength is poor. Impact resistance is also poor, whereas resistance to weathering, aging, aromatic fuels, glycols, hydrocarbon solvents, and water is good. Typical applications include the bonding of metals, glass, ceramics, phenolics, and honeycomb sandwich structures in aircraft applications (Ref 6, 12).

Vendors of phenolics and/or phenolic alloy adhesives include Adhesive Products Corporation; Aremco Products, Inc.; Borden, Inc., Chemical Division; Bostik Division, Emhart Corporation; H.B. Fuller Company; Hexcel Corporation; The Dexter Corporation; Key Polymer Corporation; Morton-Thiokol, Inc.; Narmco Materials, Inc.; and National Starch and Chemical Corporation (Ref 17). The article "Phenolics" in this Volume discusses these adhesives in more detail.

High-Temperature Adhesives. Structural adhesives having high-temperature resistance are based on synthetic organics having aromatic (benzene) and/or heterocyclic rings in the main structure. These chemical groups often include imidazoles and substituted imidazoles. These prepolymers have open-ring structures that close upon the application of heat. The condensation reaction leads to a highly cross-linked system. Adhesives are available as liquids and films.

High-temperature adhesives, including polyimides and polybenzimidazoles, are high-cost materials that are difficult to handle and require long cure times during which volatiles must be released. Polyimides are superior in long-term strength retention at elevated temperatures. At 260 °C (500 °F) in air, polyimides have higher bond strength than do epoxy-phenolics. However, strength retention following exposure to water is better in epoxy-phenolics. Polybenzimidazoles are stable in air for short-term exposures up to 288 °C (550 °F). Both polyimides and polybenzimidazoles are moisture sensitive. Applications of high-temperature adhesives are primarily in the aircraft and aerospace industries for bonding metals

and composites that will be exposed to temperatures of 260 °C (500 °F) or higher (Ref 1, 6).

Major Suppliers

Vendors of high-temperature adhesives include American Cyanamid Company, E.I. Du Pont de Nemours & Company, Inc., Furane Products Company, and Transene Company Inc. (Ref 11).

The articles "Polyimides," "Polyphenylquinoxalines," and "Polybenzimidazoles" in this Volume discuss high-temperature adhesives in greater detail.

Hot-Melt Adhesives

Hot-melt adhesives are 100% solids that, in the broadest sense, include all thermoplastics. Materials that are primarily used as hot-melt adhesives include ethylene and vinyl acetate copolymers (EVA), polyvinyl acetates (PVA), polyethylene (PE), amorphous polypropylene, block copolymers such as those based on styrene and elastomeric segments or other amide segments (that is, thermoplastic elastomers), polyamides, and polyesters. These are described more fully in the section "Hot-Melt Chemical Families" in this article.

In general, hot-melt adhesives are solid at temperatures below 79 °C (175 °F). Ideally, as the temperature is increased beyond this point, the material rapidly melts to a low-viscosity fluid that can be easily applied. Upon cooling, the adhesive sets rapidly. Because these adhesives are thermoplastics, the melting-resolidification process is repeatable with the addition and removal of the required amount of heat. Typical application temperatures of hot-melt adhesives are 149 to 188 °C (300 to 550 °F) (Ref 6, 18).

Characteristics

Hot-melt adhesives do not undergo a curing process because they simply cool from the melted state to form a solid film at the bond line (Ref 19). These adhesives are available in forms such as pellets, slabs, bars, slugs, and films that allow convenient handling by a variety of application equipment.

Because hot-melt adhesives must be in the molten state to be applied, melt viscosity is an important property. The limitations of the application equipment often influence the viscosity range that is selected for a hot-melt adhesive formulation.

Although hot-melt adhesives consist of 100% solids, they are rarely 100% polymers in composition. This is due to the limited adhesion of pure thermoplastics and their lack of molten properties, such as tack and wettability (Ref 5). The components of a hot-melt adhesive can be roughly divided into two categories, polymers and diluents. Typically, diluents are waxes, plasticizers,

tackifiers, stabilizers, extenders, and pigments. Functions of the diluents include:

- Lowering viscosity for ease of application
- Enhancing wettability
- Enhancing adhesive strength
- Increasing rigidity (extenders) or flexibility (plasticizers)

It is desirable to have a hot-melt adhesive with high strength. Yet, this property is generally accompanied by a high melt viscosity. Because a low melt viscosity is desirable for application purposes, hot-melt adhesive formulations must include diluents that decrease the polymer melt viscosity and, consequently, the overall adhesive strength.

Special dispensing equipment must be used to apply hot-melt adhesives. Typical application methods include the use of rollers, screw extruders, and squirting pumps or nozzle applicators (Ref 5).

Components. Polymers are generally of high molecular weight and provide strength and high viscosity to an adhesive. Waxes, such as paraffin, can function as both diluents and antiblocking agents. In addition, waxes promote surface wetting. Similarly, plasticizers such as phthalates, mineral oil, and glycolates provide both surface wetting and adhesive flexibility. Typical tackifiers include rosin, modified rosin, terpenes, modified terpenes, hydrocarbons, and chlorinated hydrocarbons. These materials provide tack and flexibility while promoting surface wetting and adhesion. Stabilizers such as hindered phenols help to maintain adhesive melt viscosity while functioning as antioxidants. Extenders such as talc, clay, and barites lower the cost of hot-melt adhesives and simultaneously help control melt flow (Ref 5).

Properties that are considered important are described below.

Melt viscosity is one of the most important properties of a hot-melt adhesive. In general, as the temperature of a polymer increases, its viscosity decreases. Therefore, in a hot-melt adhesive formulation, the melt temperature controls the viscosity, which greatly influences the extent of surface wetting. The temperature of the melt and the application equipment should be maintained as constant as possible.

The bond formation temperature is the minimum temperature below which surface wetting is inadequate. A hot-melt adhesive is applied at a running temperature, at which the viscosity is sufficient to wet surfaces properly. Running temperatures that are too high allow more time for surfaces to be wet properly. However, if low penetration is required and if the substrates are porous, this longer time may be detrimental and may result in a bond that contains insufficient adhesive. On the other hand, a running temperature that is too close to the

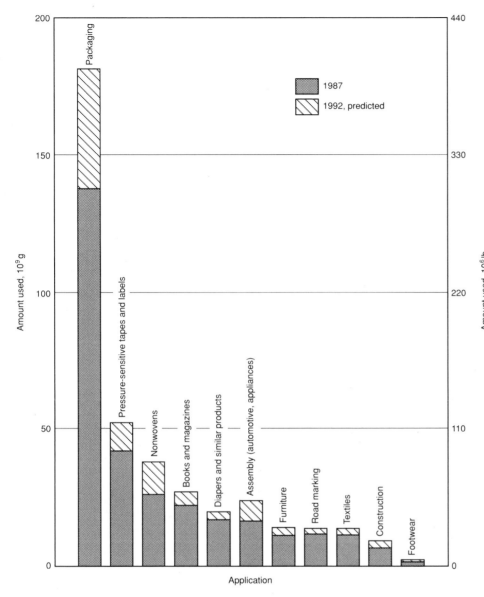

Fig. 2 Hot-melt adhesive amounts used in 1987 and predicted for 1992

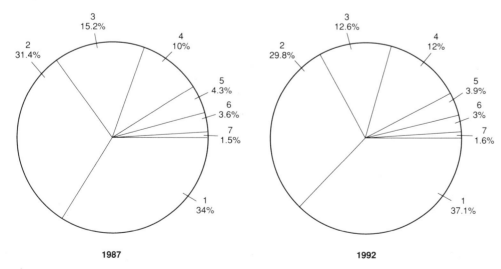

Fig. 3 Consumption of hot-melt adhesives by chemical family in 1987 and predicted for 1992. (1) Ethylene/vinyl acetate copolymer. (2) Polyethylene. (3) Polypropylene. (4) Block copolymers. (5) Polyamides. (6) Polyesters. (7) Other

bond formation temperature may result in a too-rapid adhesive bond solidification. As a result, overall properties such as strength may not be optimized.

The heat stability of the adhesive is another important property that is dependent on temperature. Running temperatures that are greater than the adhesive degradation temperature result in both charring and decreased overall properties. The addition of stabilizers to the adhesive formulation contributes to the stability of the material by hindering the acceleration of degradation due to the presence of oxygen.

The tack of the adhesive indicates the stickiness of the hot melt as it changes from a liquid to a solid state. This property affects the capability of substrates to be held together.

The range, or open, time is the lapse of time between the application of the hot melt to one substrate and the loss of its wetting ability on the second substrate due to solidification. This time duration typically ranges from fractions of seconds to minutes to infinity (in pressure-sensitive hot-melt adhesives).

End-use properties should also be considered. These include crystallinity, melt index, softening temperature, and mechanical properties.

The limitations of hot-melt adhesives are limited toughness at usable viscosities, low heat resistance, and poor creep resistance (Ref 7). All properties are affected by the polymer and diluents used in the formulation. The application conditions, including the amount of adhesive and the pressure applied to the bond line, also affect end-use properties. It should be noted that during the hot-melt adhesive-bonding operation, a minimum amount of pressure must be applied until the hot melt becomes solid or sufficient tack develops to hold the substrates in place (Ref 5).

Markets and Applications

In 1987, a total of 295.7×10^9 g (652×10^6 lb) of hot-melt adhesives were applied (excluding coatings and sealants). By 1992, the amount is expected to increase to 382.5×10^9 g (843×10^6 lb). Figure 2 shows the hot-melt adhesives applications in 1987 and those predicted for 1992. The consumption of hot-melt adhesives by chemical family in 1987 and that predicted for use in 1992 are found in Fig. 3 (Ref 20).

Hot-melt adhesives can be divided into two types, nonpressure sensitive and pressure sensitive. Nonpressure-sensitive adhesives include those used for direct bonding and heat sealing. In direct bonding, the hot-melt adhesive is applied to one surface, and the second surface is brought into contact with the adhesive while the second surface can still be sufficiently wet. In heat sealing, the hot-melt adhesive is applied to one surface and allowed to solidify. At a later time, this precoated surface is reheated and applied to the second surface.

Also included in the hot-melt, nonpressure-sensitive adhesive group are materials used for assistance bonding. In this application, the hot-melt adhesive bonds two surfaces quickly but temporarily. An additional adhesive, such as an epoxy, which often performs a structural bonding function, is then applied. By this means, the hot-melt adhesive provides a fast bond, and the structural adhesive provides additional heat resistance and load-bearing capability.

Pressure-sensitive hot-melt adhesives are tacky to the touch and can be bonded by the application of pressure alone. These adhesives have infinite open times and are bonded at room temperature. Typically, pressure-sensitive hot-melt adhesives are supplied as films on release-coated papers or in containers (Ref 5).

Hot-melt adhesives are used to bond all types of substrates, including metals, glass, plastics, ceramics, rubbers, and wood (Ref 4). Primary areas of application include packaging, book binding, assembly bonding (such as air filters and footwear), and industrial bonding, such as carpet tape and backings (Ref 5).

Hot-Melt Chemical Families

EVA copolymers and polyethylene are the lowest-cost hot-melt adhesives and are used in the greatest quantities. Many formulations are available. Typical properties of these adhesives include limited strength between approximately −34 and 49 °C (−30 and 120 °F), a limited upper service temperature of 60 to 80 °C (140 to 176 °F), low hot-melt viscosity, and creep under load with time (Ref 3, 6, 21). Most EVA-base hot-melt adhesives are used to bond paper, fabrics, wood, and some thermoplastics. Selected formulations are even used to bond glass and some metals, such as aluminum, copper, and brass. Vendors include The Dexter Corporation (polyolefin, EVA); Southeastern Adhesives Company (EVA); Bostik Division, Emhart Corporation (PE); Goodyear Rubber Products Corporation (polyterpene, EVA, PE); Arizona Chemical Company (EVA, polyterpene); and Bareco Division, Petrolite Corporation (EVA/PE) (Ref 11); as well as National Starch & Chemical Corporation (EVA); H.B. Fuller Company (EVA); Eastman Chemical Products Inc. (PE); and Findley Adhesives, Inc. (EVA, PE).

Polyvinyl acetal hot-melt adhesives are based on the reaction product of polyvinyl alcohol and an aldehyde. If the latter is formaldehyde, polyvinyl formal will result; if it is butyraldehyde, polyvinyl butyral will form. Both the formal- and butyral-base systems have fairly high strengths as hot-melt adhesives. Butyrals have better peel strength, are more flexible, and are widely used as an adhesive layer in laminated safe-

ty glass (Ref 6, 22). In addition to glass bonding, various polyvinyl acetal formulations are used to bond paper, leather, wood, and thermoplastics. Vendors include Monsanto Company; NL Industries, Inc., Industrial Chemicals Division; and National Starch and Chemical Corporation (Ref 11, 23).

Polybutene is saturated and olefinic in structure. Hot-melt adhesives based on polybutene are tough, partially crystalline, and exhibit slow crystallization rates, leading to long open times. Copolymers of butene that are 0.1 to 20 mol% ethylene result in softer and more flexible adhesives. In general, polybutene and its copolymers have low temperatures for recrystallization from the melt. This permits stress release in the adhesive bond, which may have been applied to cold surfaces. Polybutene and its olefinic copolymers have good bonding to nonpolar surfaces but poor compatibility with polar substances. These hot-melt adhesives have been used on rubbery substrates and are available as pressure-sensitive adhesives (Ref 24). The Chevron Chemical Company and Shell Chemical Company are suppliers of polybutene and its copolymer hot-melt adhesives.

Thermoplastic elastomers that are formulated into hot-melt adhesives include polyurethane, polyether-amide, and block terpolymers, such as styrene-butadiene-styrene; styrene-isoprene-styrene; and styrene-olefin-styrene, in which the olefin component is typically based on the saturated ethylene, propylene, and/or butylene polymer groups. The saturation in the midsegment of these terpolymers results in better UV and thermooxidative resistance than that found in unsaturated butadiene and isoprene midsegments.

In general, thermoplastic elastomer hot-melt adhesives are not as strong as polyesters. However, they exhibit good flexibility, high elongation, toughness, and vibration resistance. Polyurethanes and block terpolymers are used as hot-melt pressure-sensitive adhesives in tapes and labels (Ref 25, 26). Polyether-amide block copolymers exhibit good adhesion to glass. The high melting point and mechanical properties of the polyamide block and the elastomeric nature of the polyether block result in a material with good adhesion over a wide temperature range (Ref 27). Vendors of thermoplastic elastomer hot-melt adhesives include National Starch and Chemical Corporation; H.B. Fuller Company; and Findley Adhesives, Inc. (Ref 11).

Polyamides (nylons) and thermoplastic polyesters are sometimes classified as high-performance hot-melt adhesives. These adhesives have even greater strengths than olefinic and acetal-base polymers and exhibit adequate bonding from approximately −40 to 82 °C (−40 to 180 °F). Some formulations have upper service temperatures up

to 185 °C (365 °F) for applications that do not involve loads. One disadvantage to the higher service temperatures is the higher hot-melt application temperatures, which may require special equipment. Both polyamides and polyesters are sensitive to moisture absorption during application. This can result in hot-melt foaming, which leads to voids in the solidified adhesive and decreased strength. Nonetheless, polyamides exhibit good strength and flexibility. Polyesters are even stronger, but are more rigid and are usually highly crystalline. This results in a sharp melting temperature, which is desirable for high-speed bonding (Ref 3, 6, 21). Polyamide and polyester hot-melt adhesives are used to bond plastics, glass, wood, leather, foam, fabric, rubbers, and some metals, such as aluminum, copper, and brass (Ref 6). Polyamides are available from Bostik Division, Emhart Corporation; The Dexter Corporation; Henkel Corporation; Emser Industries; and General Mills Chemicals, Inc. (Ref 3, 11).

Foamable hot-melt adhesives have been commercially available since 1981. In these materials, either nitrogen or carbon dioxide is introduced into the hot melt, resulting in a 20 to 70% increase in adhesive volume. The foaming increases hot-melt spreading and open time. This method is usually used with polyethylene hot-melt adhesives that are applied to selected metal, plastic, paper, porous, and heat-sensitive substrates (Ref 3, 28).

Typical properties and applications of the various chemical families used for hot-melt adhesives are found in Table 6.

Major Suppliers

The suppliers of each hot-melt adhesive type are identified under the section "Hot-Melt Chemical Families" in this article.

Pressure-Sensitive Adhesives

Pressure-sensitive adhesives (PSAs) are capable of holding together objects when the surfaces are brought into contact under briefly applied pressure at room temperature (Ref 6, 18). The difference between contact adhesives and pressure-sensitive adhesives is that contact adhesives require no pressure to bond (Ref 11).

Materials used in PSAs must possess:

- Viscous properties to provide flow
- The capability to dissipate energy during adhesion
- Partial elastic behavior
- The tendency to resist excessive flow
- The ability to store bond rupture energy to provide peel and tack

In other words, PSA materials must be viscoelastic (Ref 29).

Pressure-sensitive adhesive materials include, in order of decreasing volume and

Table 6 Typical properties of hot-melt adhesives

Property	EVA/polyolefin homopolymers and copolymers	Polyvinyl acetate	Polyurethane	Polyamides	Polyamide copolymer	Aromatic polyamide
Brookfield viscosity, Pa · s	1–30	1.6–10	2	0.5–7.5	11	2.2
Viscosity test temperature, °C (°F)	204 (400)	121 (250)	104 (220)	204 (400)	230 (446)	204 (400)
Softening temperature, °C (°F)	99–139 (211–282)	. . .	. . .	93–154 (200–310)	. . .	129–140 (265–285)
Application temperature, °C (°F)	. . .	121–177 (250–350)	. . .	. . .	. . .	. . .
Service temperature range, °C (°F)	−34 to 80 (−30 to 176)	−1 to 120 (30 to 248)	. . .	−40 to 185 (−40 to 365)	. . .	. . .
Relative cost(a)	Lowest	Low to medium	Medium to high	High	High	High
Bonding substrates (Ref 10)	Paper, wood, selected thermoplastics, selected metals, selected glasses	Paper, wood, leather, glass, selected plastics, selected metals	Plastics	Wood, leather, selected plastics, selected metals	Selected metals, selected plastics	Selected metals, selected plastics
Applications (Ref 10)	Bookbinding, packaging, toys, automotive, furniture, electronics	Tray forming, packaging, binding, sealing cases and cartons, bottle labels, cans, jars	Laminates	Packaging, electronics, furniture, footwear	Packaging, electronics, binding	Electronics, packaging, binding

(a) Relative to other hot-melt adhesives

increasing price, natural rubber, styrene-butadiene rubber (SBR), reclaimed rubber, butyl rubber, butadiene-acrylonitrile rubber, thermoplastic elastomers, polyacrylates, polyvinylalkylethers, and silicones (Ref 6, 18).

Characteristics

Materials used as PSAs are usually available as solvent systems or hot melts. Substrates are coated with one of these two types of adhesives, and usually no cure of the material is involved upon its application (Ref 11). Adhesive-coated substrates, in their dry state, are permanently tacky at room temperature and do not require activation by water, solvents, or heat (Ref 3).

There are two major classes of PSAs: adhesives that are compounded to form PSAs and adhesives that are inherently pressure sensitive and require little or no compounding. Included in the former category are elastomers, and in the latter category are polyacrylates and polyvinylalkylethers (Ref 30).

Components. Pressure-sensitive adhesives are almost always supplied as a coating on a substrate. Both solvent-base and hot-melt-base adhesives are used. The latter components are discussed in more detail in the section "Hot-Melt Adhesives" in this article. The components of a solvent-base system are described below.

Adhesive. The coating weight of the adhesive is usually between 20 and 50 g/m² (0.066 and 0.164 oz/ft²). Materials are rubbers, acrylates, or silicones. Typical additives may include tackifiers, plasticizers, fillers, stabilizers and/or pigments.

Primers are applied at a coating weight of 2 to 5 g/m² (0.007 to 0.016 oz/ft²) from solvents or aqueous dispersions. Typical primers are based on nitrile rubber, chlorinated rubber, and acrylates.

Backing materials include paper; cellophane; plastic films, such as polyethylene, polypropylene, polyvinyl chloride, polyester, and polyimide; fabrics; and metallic foils.

Release coating materials, which are used optionally, are usually applied to the backing material if the PSA is to be rolled into tape. This reduces unwinding tension when the tape is used. These materials are applied in 1 to 5 g/m² (0.003 to 0.0165 oz/ft²) coating weights. Typically, acrylic acid esters of long-chain fatty alcohols, polyurethanes with long aliphatic chains, or cellulose ester are used. Release coating materials are not used in pressure-sensitive labels. In these products, the adhesive contacts both the backing material and a removable release material, usually paper (Ref 3, 6, 18).

Markets and Applications

Markets. Figure 4 shows the pressure-sensitive adhesive market in 1988. The largest area of market application was packaging. In 1987, sales of tapes amounted to $2.07 billion and sales of labels were $1.0 billion; these sales were predicted to increase to $2.46 billion and $1.21 billion, respectively, by 1992 (Ref 31).

Shipments of pressure-sensitive products are summarized in Fig. 5. These include single- and double-faced tapes, unprinted labels, base materials, and unspecified materials. Values were estimated for 1987 and predicted for 1992 (Ref 32).

PSAs are primarily used in tapes and labels, with tapes representing the largest area of application. PSA tapes have the following requirements:

- When the tape is unwound, the adhesive must remain on one side of the backing
- The adhesive must not split or transfer to the other side
- The unwinding force must be low

Release coating materials, which are usually applied to tape backings, influence unwinding force and the ability of the adhesive to remain on its backing material. In some cases, release coating interlayer materials may also be applied to the tape adhesive.

Applications. Tapes, the largest area of application of PSAs, can be classified by construction, function, application, or tex-

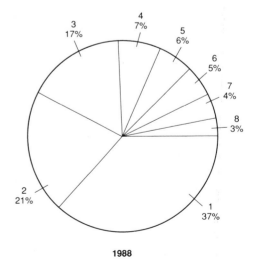

1988

Fig. 4 Pressure-sensitive adhesives market in 1988. (1) Packaging. (2) Hospital/first aid. (3) Office/graphic arts. (4) Construction. (5) Corrosion protection. (6) Automotive. (7) Electrical. (8) Other

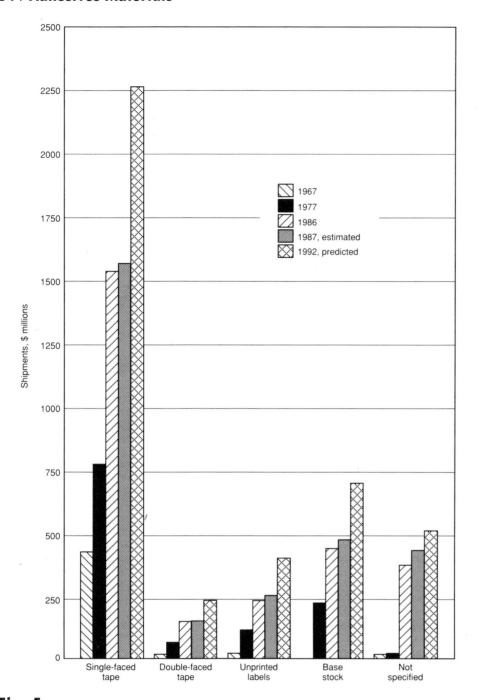

Fig. 5 Pressure-sensitive adhesives product shipments

dry. This material is then laminated to the label stock.

Pressure-Sensitive Chemical Families

The four primary families used as the adhesive component in PSAs are natural rubber, thermoplastic elastomers, polyacrylates, and silicones. The advantages and limitations of rubbers (that is, elastomers) in general, polyacrylates, and silicones are summarized in Table 7.

Table 8 summarizes typical PSA bonding substrates and applications by chemical family.

Natural rubber adhesives are usually applied as coatings made from organic solvents. The rubber itself is either a pale crepe grade for clear films or a darker grade. Natural rubber must be masticated to decrease its molecular weight and gel content; otherwise it is difficult to achieve tack and to solvent coat without using a large amount of solvent. Natural rubber must be compounded with tackifying resins and antioxidants. Typical tackifying resins include wood rosin and its derivatives, terpene resins, and petroleum-base resins. The tackifying resin, necessitated by the poor tack and adhesion of natural rubber, provides cohesive holding power, peel adhesion, and surface tack to the PSA. Other additives include plasticizers, which lower cohesive strength, reduce peel adhesion, and have a varying effect on tack; pigments, which provide color and can reduce the cost; and curing agents, which partially vulcanize the adhesive.

Natural rubber is cis-polyisoprene, which contains unsaturated double bonds susceptible to oxidative attack. Antioxidants are required to stabilize the PSA during its shelf life and while it is in service. Typical antioxidants include amines, which may stain; phenolics, which are nonstaining and are effective against UV radiation and sunlight; and dithiocarbamates, which are nonstaining and are effective against UV radiation and heat aging. Blends of antioxidants are often used. Natural rubber PSAs are used in masking tapes, double-sided tapes, electrical tapes, pressure-sensitive labels, and protective films (Ref 33).

Thermoplastic elastomers are block copolymers based on styrene-butadiene-styrene (SBS), styrene-isoprene-styrene (SIS), and styrene-olefin-styrene, with the olefin group being ethylene-butylene or ethylene-propylene. The saturated polyolefins provide better UV-radiation and oxidative resistance to the block copolymers than is found in the materials containing unsaturated butadiene or isoprene groups. Thermoplastic elastomers, unlike natural rubber, do not require mastication. In addition, thermoplastic elastomers are available as both solvent-coatable and hot-melt adhesives.

Thermoplastic elastomers are two-phase systems consisting of rigid, styrene-base

ture. The construction category includes fabric tapes, paper tapes, foil tapes, film tapes, nonwoven fabric tapes, reinforced tapes, foam tapes, two-faced tapes, and transfer tapes. The backing, rather than the adhesive, is the distinguishing feature of this tape classification.

The function category includes masking, holding, sealing, reinforcing, protecting, bundling, stenciling, splicing, identifying, insulating, packaging, and mounting. The application category includes hospital and first aid tapes, office and graphic art tapes, building industry tapes, packaging and sur-

face protection tapes, electrical tapes, automotive industry tapes, shoe industry tapes, appliance industry tapes, splicing tapes, and corrosion protection tapes. The texture category includes floor tiles, wall coverings, automobile wood-grained films, and decorative sheets (Ref 30).

Labels are the second largest application for PSAs. Specific characteristics that are important, and which differentiate labels from tapes, include backing material printability, flatness, ease of die cutting, and release paper properties. The adhesive is applied to the release paper and allowed to

Table 7 Advantages and limitations of rubber-, acrylate-, and silicone-base PSAs

Chemical family	Advantages	Limitations
Rubbers.................	Good flexibility High initial adhesion (better than acrylics) Ease of tackification (with additives) Lowest cost Good shear strength Good adherence to low- and high-energy surfaces Suitable for temporary or permanent holding	Low tack and adhesion (without additives) Poor aging, subject to yellowing Limited upper service temperature use Moderate service life
Acrylates................	Good UV resistance Good hydrolysis resistance (better than rubbers) Excellent adhesion buildup Good solvent resistance Good temperature use range (−45 to 121 °C, or −50 to 250 °F) Easier to apply (than rubbers) Good shear strength Good service life	Poor creep resistance (compared to rubbers) Fair initial adhesion Moderate cost (compared to rubbers, silicones)
Silicones	Excellent chemical and solvent resistance Wide temperature use range (−73 to 260 °C, or −100 to 500 °F) Good oxidation resistance Good adherence to low- and high-energy surfaces	Highest cost (compared to rubbers, acrylates) Lack of aggressive behavior

end blocks and a rubbery mid segment. The ratio of these two phases can be varied. If the block copolymers consist of approximately 15% styrene by weight, the material will have a lower modulus and tensile strength and a higher elongation than a similarly based elastomer having 30% styrene by weight. Additives in the adhesive formulation tend to associate with one phase or the other. For example, additives compatible with the rubbery phase aid in the development of tack, the processibility of the mid block, and adhesive softness. Aliphatic-olefin-derived resins, rosin esters, polyterpenes, and terpene-phenolic resins are examples of rubbery-phase-compatible additives. End-block-compatible resins increase the elastic modulus, stiffness, and tendency to draw. However, they may reduce the tack. Polyaromatic-base resins are compatible with the styrene end blocks.

Plasticizers are used in thermoplastic elastomers to decrease the hardness; elastic modulus; and melt, or solution, viscosity. Cohesive strength is thereby decreased, but tack is increased. The most effective thermoplastic elastomer plasticizers are completely insoluble in the end blocks and are completely miscible with the mid segment. Hydrocarbon oils are often used.

Other thermoplastic elastomer additives include fillers, pigments, and stabilizers. The latter are used in SBS and SIS elastomers to increase oxidative and UV-radiation resistance.

Thermoplastic elastomer PSAs are used as general-purpose tapes, high- and low-temperature tapes and labels, and weatherable tapes (Ref 34).

Polyacrylates having specific monomers in their composition are inherently pressure sensitive without compounding additives.

Therefore, the lower molecular weight materials used in compounded rubber-base adhesives are not present to migrate to the surface and interfere with bonding. Also, because polyacrylates are saturated in structure, they are resistant to UV and oxidative degradation. Polyacrylates used in PSAs are available in various forms: dispersed in organic solvents, in aqueous emulsions, as hot melts, or as 100% reactive solids.

Homopolymers are rarely used to make polyacrylate PSAs. Most are based on copolymers of acrylic acid ester with another monomer. Usually, acrylates and methacrylates having 4 to 17 carbon atoms are used as monomers. These include ethylhexyl acrylate, butylacrylate, ethylacrylate, and acrylic acid. Occasionally, vinyl acetate is used as a monomer to modify properties and decrease raw material cost.

The cross-linking of polyacrylates is accomplished using multifunctional monomers or monomers carrying reactive groups that are introduced into the polymer chain. A low degree of cross-linking is desirable because increased cross-linking results in decreased tack and peel adhesion. The advantages of cross-linking include improved shear and creep resistance, especially at elevated temperatures.

Although polyacrylates do not require compounding, selected materials are sometimes added to change properties. Tackifying resins and plasticizers improve tack and peel adhesion. Fillers and pigments are added occasionally (Ref 35). Polyacrylates are used in a variety of PSA applications, including tapes, labels, decals, trims, and moldings (Ref 11).

Silicone PSAs have a wider temperature use range than those materials described above. In addition, the adhesives have excellent chemical and solvent resistance and flexibility. Silicone PSAs are based on a gum and a resin. These materials are reacted following their dispersion in a solvent. Gums are either methyl-base or phenyl-modified silicones. Resins are also siloxane based. By adjusting the gum-to-resin ratio, a wide range of tack and peel adhesion properties can be obtained. A high gum content results in a PSA that is very tacky at room temperature, whereas a high resin content results in a material that is not very tacky at room temperature but becomes so upon the application of heat and pressure.

Methyl-base silicone PSAs have a wide range of viscosities (from 1 to 5 to 40 to 90 Pa · s) and other properties. Many different filler loadings and tape backings are used.

Phenyl-modified silicone PSAs have viscosities ranging from 50 to 100 Pa · s when low phenyl contents (6 mol%) are incorporated. Viscosities from 6 to 25 Pa · s are found in PSAs having high phenyl contents (12 mol%). In general, the phenyl-modified silicones have high peel strength and tack.

Table 8 Typical PSA bonding substrates and applications by chemical family

Property	Natural rubber	Thermoplastic elastomers	Acrylates	Silicones
Typical solids content, %(a).......	35–50	22–50	25–52	55–60
Bonding substrates....	Wood, metals, leather, paper, plastics, textiles	Wood, metals, leather, paper, glass, plastics, textiles	Glass, metals, plastics, paper	Elastomers, metals, plastics
Applications..........	Tapes General-purpose uses	Apparel, leather, footware, tapes, floor and wall coverings, furniture, fixtures	General industrial uses, tapes, labels, decals, trims, moldings	Tapes, coated films and fabrics, electronics

(a) Dispersed in solvents. Source: Ref 10

Table 9 Comparison of water-base and organic-solvent-base adhesives properties

Property	Water-base adhesives	Organic-solvent-base adhesives
Flammability	Low	High
Toxicity	Low	High
Cost	Low	Moderate
Dry time	Slow	Fast
Open time	Long	Varies
Bond strength development	Slow	Fast
Water resistance	Poor	Good
	Poor	
Cold resistance	(freezes)	Good
Electrical properties	Poor	Good

However, they are not completely compatible with methyl-base silicones. These adhesives are used in the specialty tape markets, particularly in aerospace, automotive, electrical, and appliance applications (Ref 36).

Major Suppliers

The largest suppliers of PSA tapes include 3M Company; Johnson & Johnson, Permacel; Nashua Corporation, Tuck Industries, Inc.; Mystik Tape Corporation; Shuford Mills, Inc.; Polyken Division, The Kendall Company; Arno Adhesive Tapes, Inc.; and Anchor Continental. The largest suppliers of PSA labels include Avery International; Morgan Adhesives Company; Dennison Manufacturing Company; Fitchburg Coated Products Inc., Division of Litton Industries Inc.; and Coated Products, Inc. (Ref 30).

Water-Base Adhesives

Adhesive materials that can be dissolved or dispersed in water are the basis of water-base adhesives. Many of these same materials are also used in organic-solvent-base adhesives (Ref 12).

In recent years, exposure to organic solvents has been increasily controlled by federal regulations. Therefore, it is advantageous to use water-base rather than organic-solvent-base adhesives. The use of water as a solvent results in lower cost, nonflammability, and lower toxicity. It also makes it possible to vent drying ovens directly into the atmosphere. In addition, many water-base adhesives can be substituted for organic-solvent-base adhesives with no modification to application equipment (Ref 6, 37).

Water-base adhesives have fairly good organic-solvent resistance. However, moisture resistance is usually poor, and adhesives are subject to freezing, which can affect properties (Ref 37). Table 9 compares several properties for water-base and organic-solvent-base adhesives.

Although there are many advantages to using water-base adhesives, there is still a large and important market for solvent-base adhesives. Water-base adhesives are usually unsuitable for hydrophobic surfaces, such as plastics, because of poor wettability. In addition, water shrinks some substrates, such as paper, textiles, and cellulosics, and is corrosive to selected metals, such as copper. Organic-solvent-base adhesives are suitable for application on hydrophobic surfaces and are compatible with most metal surfaces. However, organic solvents have a tendency to attack some plastic foams, whereas water-base adhesives may be suitable for this application (Ref 12).

A water-base adhesive formulation can be classified as a solution in which the polymer is totally soluble in water or alkaline water, or as a latex, which consists of a stable dispersion of polymer in an essentially aqueous medium (Ref 3). Drying of solution formulations involves solvent removal by the uniform diffusion of the solvent through the adhesive material mass. As drying progresses, viscosity increases, the diffusion rate decreases, and total solvent removal becomes difficult. On the other hand, in latex drying, the water moves freely around the adhesive material, rather than diffusing through it. The water then evaporates or is absorbed into a substrate. Drying usually occurs at a faster rate with latex dispersions (Ref 37).

Usually, the application of water-base adhesives to porous substrates does not necessitate the removal of the liquid. Rather, the absorption of the water into the substrate, as well as the process of evaporation, are the mechanisms for removal. On the other hand, when bonding water-base adhesives to nonporous substrates such as metals, the removal of the water is necessary before the bonding of parts. In many cases, bond formation depends on the reactivation of dried adhesive film under heat and pressure (Ref 12).

The performance properties of water-base adhesives can be varied by modifying formulations with various additives. A characteristic specific to water-base latices is that high molecular weight polymers can be dispersed at high solids content while maintaining a fairly low-viscosity liquid phase. These high molecular weight polymers lead to improved properties, including an increase in resistance to extreme service conditions. In addition, the cross-linking of many water-base adhesives is a possibility. This results in increased heat, moisture, and stress resistance (Ref 37).

Characteristics

Water-base adhesive solutions, in which the material is totally soluble in water, consist largely of natural adhesives. Materials that are soluble in water alone include animal glues, starch, dextrin, methylcellulose, and polyvinyl alcohol. Materials that are soluble in alkaline water include casein, rosin, carboxymethylcellulose, shellac, vinyl acetate, and acrylate copolymers containing carboxyl groups. Of all of these adhesive components, only polyvinyl alcohol, as well as vinyl acetate copolymers and acrylate copolymers containing carboxyl groups, can truly be classified as synthetic polymers.

Components. A water-base adhesive latex consists of a stable dispersion in an aqueous medium. A stable dispersion of two or more immiscible liquids held in suspension by small percentages of substances is an emulsion. Often the words latex and emulsion are used interchangeably to describe this type of water-base adhesive.

Latices can be classified into natural, synthetic, and artificial based latex. A natural latex is formed from natural rubber. A synthetic latex is based on an aqueous dispersion of polymers obtained by emulsion polymerization. Adhesive families in this category include neoprene, styrene-butadiene rubber, nitrile rubber, polyvinyl acetate, polyacrylates, and polymethacrylates. An artificial latex is made simply by dispersing the solid polymer. Included in this category are natural rosin and its derivatives, synthetic butyl rubber, and reclaimed rubber. When water-base adhesives are used to replace organic-solvent-base adhesives, a latex type is more likely to be used than a solution type (Ref 3).

When water-base adhesives are applied to a substrate, they usually do not undergo a curing reaction. They simply set, upon removal of the water. However, it should be noted that some materials (rubbers) can be vulcanized or cross linked by the application of heat or by activation provided by a catalyst.

Water-base adhesive families can be subdivided into materials used in making water-base adhesives only and materials used in making both water-base and organic-solvent-base adhesives. Included in the former classification are casein, dextrin, starch, animal glues, polyvinyl alcohol, sodium carboxy-methylcellulose, and sodium silicate. Included in the latter group are amino resins (urea and melamine formaldehydes), phenolic resins (phenol and resorcinol formaldehydes), polyacrylates, polyvinyl acetates, polyvinyl ethers, neoprene, nitrile rubber, SBR, butyl rubber, natural rubber, and reclaimed rubber (Ref 12).

Another classification of water-base adhesives arises from the natural or synthetic nature of the adhesive itself. Synthetics can be subdivided into early synthetics (developed during the years 1935 to 1960) and new synthetics (developed since 1960). The demand for natural water-base adhesives in 1990 was approximately the same as the demand for synthetics. However, by the mid-1990s the demand for synthetics is expected to outweigh that for the natural water-base adhesives. The gap between the

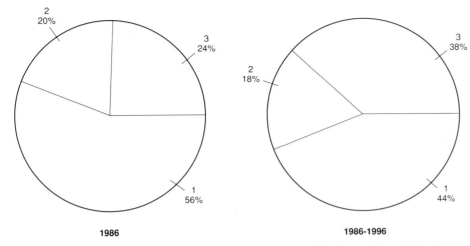

Fig. 6 Water-base adhesive families by demand in 1986 and by predicted growth from 1986 to 1996. (1) Natural adhesives. (2) Early synthetic adhesives. (3) New synthetic adhesives

two is predicted to widen in favor of synthetic water-base adhesives in the late 1990s and beyond because little growth is expected in the use of natural materials (Ref 38).

Water-base adhesives are available for either nonpressure-sensitive or pressure-sensitive applications. By far, the former comprises the majority of applications.

Properties. Dispersions consist of a continuous liquid phase into which particles of solid are suspended to form the dispersed phase. Surfactants are used to stabilize the system.

Solids content in water-base dispersions can be as high as 50% by volume. It is interesting to note that these dispersions still have good flow. In fact, even high concentrations of high molecular weight polymers in dispersions can flow well, whereas an identical concentration in an organic solvent might be highly viscous. Flow does not depend on the nature of the dispersed phase in these water-base adhesives.

In dispersions with solids content of 30% or less by volume, viscosity is unaffected by varying shear forces and is essentially the same as that of pure water. In dispersions containing a solids content greater than 30% by volume, the viscosity of the system increases significantly over that of pure water, with an increasing sensitivity to shear forces. Shear sensitivity includes pseudoplasticity behavior, in which an increase in shear forces decreases viscosity, and dilatancy behavior, in which an increase in shear forces further increases viscosity. The latter is often an indication of an unstable system.

In dispersions having high solids contents, small concentration changes can lead to significant viscosity changes. Overall, desirable properties of a dispersion include high molecular weight, high solids content, low viscosity, and Newtonian flow, in which the viscosity is unaffected by varying shear forces.

Factors that affect the rate of application of dispersions on substrates include (Ref 37):

- Required coating weight
- Solids content
- Specific heat of the adhesive
- Evaporation rate of the continuous phase
- Type of drying system used

Factors Affecting Bonding. As mentioned earlier, open time is the time after application of adhesive to one surface during which the adhesive remains sufficiently liquid for the second surface to be bonded. With water-base adhesives, the application of insufficient adhesive may result in partial setting before compression can be accomplished. On the other hand, the application of excessive amounts may result in squeeze-out and a longer solidification time. Factors that affect the open time of water-base adhesives include:

- Adhesive type and composition
- Solids content and viscosity of the liquid phase
- Application film thickness
- Substrate porosity and moisture content
- Temperature

Wet tack is the initial cohesive strength of the adhesive film prior to solidification. In dispersions, tack is dependent on the ability to increase rapidly both viscosity and solids content as the water is removed prior to coalescence of polymer particles. Wet tack is critical if compression time is short and if there is stress on the bond before the adhesive completely solidifies.

Ideally, compression should be maintained until the adhesive film cohesion is high enough to hold surfaces together against stresses that the bond may be subjected to when it is removed. If one or both surfaces are porous, an increase in compression will increase the rate at which the adhesive solidifies. The rate of absorption

of the continuous liquid phase into a porous substrate depends on:

- Molecular attraction, including polarity
- The structure of the porous substrate
- Pretreatments to which the porous substrate has been subjected
- The viscosity of the liquid

Finally, for high-speed bonding, solidification should take place as rapidly as possible. Factors that affect solidification include:

- Adhesive type
- Adhesive composition
- Compression

Solidification can be accelerated by applying heat to evaporate the water. The warm, dry adhesive film on one substrate is joined to a second substrate while the material is in a semiflowable state. In some cases, particularly when bonding nonporous substrates, the adhesive is allowed to dry completely on one substrate. Later it is heat activated in order to bond (Ref 37).

Markets and Applications

In 1930, the U.S. adhesives industry demand for water-base adhesives made up more than 90% of the entire market. By 1986, this demand had dropped significantly but still accounted for more than 60% of the adhesives market. From 1986 to 1996, the average water-base adhesives industry growth rate is predicted to be 3.9% per year.

In 1986, the market share of water-base adhesives used in PSA applications was 4.2%. Nonpressure-sensitive applications accounted for 95.8% of the market. In the period between 1986 and 1996, the market share of PSAs is expected to increase to 14%, with the nonpressure-sensitive adhesives market decreasing to 86%.

Figure 6 indicates the water-base adhesive families by demand in 1986 and by predicted incremental growth from 1986 to 1996. In 1986, demand for natural adhesives accounted for 56% of the total water-base market, and synthetics, 44%. This demand for each is expected to reverse by 1996 (naturals, 44%; synthetics, 56%).

Figure 7 indicates water-base adhesives applications by demand in 1986 and predicted demand by 1996. Rigid bonding applications include appliances, housewares, machinery, and electronics. Nonrigid bonding applications include fabrics, shoes, filters, and rugs. The largest application of water-base adhesives is in packaging; second, with a much smaller percentage, is construction. It is predicted that by 1996 these two application areas for water-base adhesives will continue to account for most of the demand (Ref 38).

Chemical Families

Natural adhesives are animal and vegetable based. Applications are usually limited to paper, paperboard, wood, and metal

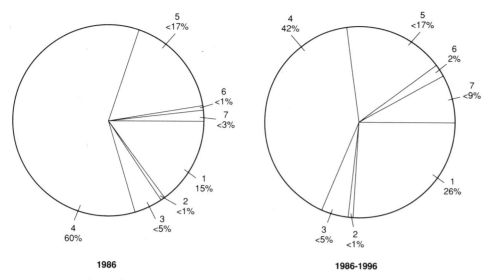

Fig. 7 Water-base adhesive applications by demand in 1986 and predicted growth by 1996. (1) Construction. (2) Transportation. (3) Rigid bonding. (4) Packaging. (5) Nonrigid bonding. (6) Consumer goods. (7) Tapes

foil. These adhesives have poor moisture and fungus resistance, good organic-solvent resistance, shear strengths between 0.034 and 6.9 MPa (0.5 and 1 ksi), and poor strength above 100 °C (212 °F) (Ref 3).

Casein is obtained by precipitation from milk protein. The dried powder is mixed with water prior to application and then hardened by the loss of water at room temperature, with some chemical conversion of protein to more insoluble calcium derivatives. This adhesive is unsuitable for outdoor use but is more resistant to temperature changes and moisture than other water-base adhesives. A casein adhesive tolerates dry heat to 70 °C (158 °F) and has good organic-solvent resistance. Although the adhesive is susceptible to biodegradation, chlorinated phenols can be used to inhibit this behavior. Typical applications include packaging (paper labels to glass), woodworking, and interior-grade plywood fabrication (Ref 3, 12).

Cellulosics include sodium carboxy-methylcellulose. This material is available either as a totally water-soluble solution or as a latex. The dried adhesive has poor moisture and biodegradation resistance but fair-to-good dry-heat resistance (Ref 12).

Rosin is derived from oleoresin, a product of pine trees. Plasticizers are sometimes added to decrease the brittleness of this material. Properties of the set adhesive include poor moisture and oxidation resistance, poor aging, and moderate bond strength. Rosin-based adhesives are used to bond paper (Ref 3).

A natural rubber latex usually consists of approximately 35% solids in a low-viscosity milky-white dispersion. The rubber is compounded with tackifiers and is usually treated with preservatives. Setting is accomplished by water evaporation. Vulcanization by heat or catalyst is also possible. Natural rubber-base adhesives have excellent flexibility; high initial tack; good water resis-

tance; poor solvent, oil, and oxidation resistance; and good resistance to biodegradation, if protected with preservatives. The adhesives also have poor resistance to outdoor exposure and become brittle at −34 °C (−29 °F). Unvulcanized rubber weakens at approximately 66 °C (151 °F), but vulcanization can improve this temperature to 93 °C (199 °F). Better water and solvent resistance are also found in vulcanized rubber. Typical bonding applications include paper, porous materials, footwear, upholstery, textiles, carpets, packaging, laminates (metal foil to paper, wood, or rubber) and PSAs (Ref 12).

Typical properties of natural rubber-base PSAs are summarized in Table 10. Terpene tackifiers are used extensively in natural rubber-base PSAs and are more effective than rosins. Compared to natural rubber organic-solvent-base PSAs, water-base PSAs have better shear adhesion because higher molecular weight rubber can be used. In solvent-base PSAs, natural rubber must be milled, which results in a large decrease in molecular weight (Ref 39).

Starch adhesives are prepared by heating the material in alkaline solution and then cooling to room temperature. Following application, the material sets after the removal of water. Tack develops rapidly at room temperature. The adhesive has poor resistance to moisture, elevated temperatures, weathering, and biodegradation. However, preservatives can be added to improve all of these deficiencies. Starch adhesives are suitable for indoor use only. Typical applications include paper cartons, bottle labeling, stationery, and indoor-grade plywood fabrication (Ref 12).

Dextrin results from the degradation of starch upon heating in the presence or absence of hydrolytic agents. Alkaline- or acidic-producing substances are usually used. Adhesives based on dextrin are used to bond paper and paperboard. Urea form-

Table 10 Water-base PSA properties

Water-base PSA/tackifier	Polyken tack(a) N	gf	90° quick stick(b) N	gf	180° peel adhesion(c) N/mm	lbf/in.	Shear adhesion(d), min	Latex type
Natural rubber latex/terpene (40/60)	>14.7	>1500	0.8	83	1.0	5.8	>3000	. . .
Natural rubber latex/rosin (40/60)	10.1	1031	0.50	51	0.76	4.4	105	. . .
SBR latex A/phthalate ester (40/60)	15.7	1600	0.39	40	0.76	4.4	1200	Hot latex
SBR latex D/low molecular weight polystyrene (60/40)	10.8	1100	0.31	32	0.48	2.8	>3000	Hot latex
SBR latex H/phthalate ester (40/60)	12.9	1320	0.20	20	1.0	5.8	100	Cold latex
SBR latex J/rosin (50/50)	10.8	1100	0.25	26	0.05	4.5	1620	Cold latex
Neoprene latex (noncarboxylated)/rosin (60/40)	. . .	. . .	0.54	55	1.36	7.8	1750	. . .
Neoprene latex (carboxylated)/rosin (60/40)	. . .	. . .	0.12	12	0.38	2.2	>4000	. . .
Acrylic latex E	13.1	1340	0.15	15	0.35	2	670	. . .
Acrylic latex G	10.2	1040	0.04	4	0.31	1.8	>3000	. . .
Acrylic latex G/rosin (50/50)	>14.7	>1500	0.42	43	0.90	5.2	>3000	. . .

(a) A mechanical probe tack test in which the top of a probe is brought into contact with a supported adhesive under low contact pressures for a short time and then pulled away at a fixed rate. The peak force of separation is measured. (b) A peel test in which a tape is placed on a flat, stainless steel test surface with no contact pressure beyond that of its own weight. The tape is peeled from the surface at a 90° angle at 300 mm/min (12 in./min), and the measured peel force is taken as the tack value. (c) Measures as force per unit of width. (d) On a specimen 13 × 25 mm (0.5 × 1.0 in.), using 9.8 N (1000 gf)

aldehyde is often added to increase water resistance (Ref 3).

A sodium silicate water solution consists of silicon dioxide (SiO_2) and sodium oxide (Na_2O) in a ratio that varies from 2 to 3.5. Because the water solution is viscous and has little tack, pressure must be used to hold substrates together while bonding. The dried adhesive is brittle and water sensitive. Aluminum salts can be added to the formulation to improve water resistance. Applications include the bonding of paper, corrugated boxes and cartons, wood, glass, porcelain, leather, and textiles (Ref 3, 12).

Animal glues are obtained from three sources: mammalian collagen, animal bones, and tannery wastes. Hide glues are stronger than bone glues and often exceed the strength of wood. These adhesives are primarily used to bond wood, leather, paper, and textiles. Use is limited to interior applications (Ref 3).

The early synthetic adhesives were developed between the years 1935 and 1960.

Polyvinyl alcohol is a water-soluble thermoplastic material that has limited application as an adhesive. When the water is removed, the material sets, forming a flexible, transparent bond. In general, polyvinyl alcohol adhesives are resistant to oils, solvents, and fungi. They are impermeable to most gases, but have poor moisture resistance and a fairly low upper service temperature of 66 °C (151 °F). This adhesive is used in packaging and in bonding porous materials, such as cork, leather, and paper. Polyvinyl alcohol is also used as a modifier in other aqueous systems to improve film-forming properties or promote adhesion (Ref 3).

Polyvinyl acetate and its copolymers are the most widely used resins in water dispersions. These adhesives are the basis for common, household white glues. Setting is accomplished by the removal of water due to evaporation or absorption into a substrate. These adhesives have poor weather, moisture, and solvent resistance but withstand grease, oil, and petroleum fuels. Additional properties are high initial tack, almost invisible bond lines, softening at approximately 45 °C (113 °F), good biodegradation resistance, poor resistance to creep under load, and low cost. Applications include bonding to paper, plastics, metal foil, cloth, and wood (Ref 3).

Urea formaldehyde is available as a spray-dried powder with an incorporated hardener. The system is reactivated by mixing with water. Setting is accomplished by condensation polymerization, yielding water as a by-product. This adhesive lacks durability and is unsuitable for extreme service conditions. However, it has good resistance to cold water and fungi. Urea formaldehyde adhesives can be compounded with other additives to increase durability. Phenol or resorcinol formaldehyde increase water, weather, and temperature resistance, but also increase the cost. Melamine formaldehyde is also used to increase water and temperature resistance. Overall, urea formaldehyde adhesives are nonstaining, light colored, and unsuitable for exterior or extreme temperature use. They are used for wood bonding, especially in plywood (Ref 12).

Melamine formaldehyde is also available as a powder that is reactivated by mixing with water or water-alcohol mixtures. Filled powders are also available. Setting is achieved by condensation polymerization, with water released as a by-product. This adhesive has excellent water resistance, good biodegradation resistance, and bond strengths that usually exceed wood strength when thin bond lines are used. Light-colored, nonstaining melamine formaldehyde adhesives are used for exterior, nonstructural wood bonding, including plywood (Ref 12).

Phenol formaldehyde is available as a spray-dried powder that is reactivated when mixed with water. Setting is accomplished following a condensation reaction, which releases water as a by-product. These adhesives have good resistance to weathering, boiling water, and fungi. However, bonds are brittle and are prone to fracture when subjected to vibration and impact. The alkaline nature of these adhesives can lead to corrosion on copper and brass. This adhesive is used to bond exterior-grade plywood (Ref 12).

Styrene-butadiene rubber is available as a latex that is usually compounded with tackifiers and plasticizers. Setting is accomplished when the water is removed by evaporation or absorption into the substrates. The tacky surfaces are then contact bonded. The properties of this adhesive include poor oil, solvent, and oxidation resistance; poor strength; and a tendency to creep. Water and heat resistance are better than in natural rubber. SBR adhesives are nonstaining and light-colored and have a useful temperature range between −40 and 71 °C (−40 and 160 °F). These low-cost adhesives are used to bond paper, textiles, plastic films, and metal foils (Ref 12).

Styrene-butadiene rubber is also used to form water-base PSAs. Although this rubber has some inherent tack, it is usually increased by the addition of tackifiers, such as rosin esters, terpenes, aromatic plasticizers, modified aromatics, and α-methylstyrene-vinyl toluene copolymers. In general, decreasing the molecular weight of SBR increases its tack. For example, in Table 10 the tack of a latex A rubber (lower molecular weight) is greater than that of a latex H rubber. A hot SBR latex is prepared at a higher polymerization temperature and yields a harder polymer of lower molecular weight than a cold SBR latex polymerization (Ref 39).

Neoprene (polychloroprene) is available as an aqueous latex emulsion compounded with additives such as tackifiers, antioxidants, wetting agents, and stabilizers. Setting is accomplished by the evaporation of water and the assembly of the tacky surfaces. Heat or solvent reactivation of dried films is also practiced, although it is not as successful as in other rubber-base adhesive systems. Neoprene adhesives have good water and biodegradation resistance. However, aromatic and chlorinated hydrocarbon resistance is poor. The service temperature range is approximately −50 to 95 °C (−58 to 203 °F). Soon after bonding, neoprene adhesives can withstand loads between 0.2 and 0.7 MPa (0.029 and 0.102 ksi). Neoprene is more effective than other rubbers in withstanding these continuous loads. Neoprene adhesives are used for bonding decorative plastic laminates, rubber, leather, linoleum, fabrics, wood, footwear, and thin sheets of aluminum to steel (Ref 12, 14).

Neoprene water-base materials are also used in PSAs. Both conventional, uncarboxylated neoprenes and carboxylated systems are used. As indicated in Table 10, carboxylation results in a significant increase in shear adhesion. Tackifiers include rosin esters, terpenes, and aromatic plasticizers (Ref 39).

Reclaimed rubber is based on devulcanized rubber and may be compounded with synthetic rubbers such as SBR. Most reclaimed rubber adhesives are also compounded with a variety of additives, including fillers, antioxidants, and tackifiers. Setting is accomplished following water evaporation or absorption into porous substrates. Most of these adhesives are black, although a few are gray or red-colored. Reclaimed rubber adhesives are comparable to natural rubber adhesives in properties, including heat, water, and solvent resistance. They are less resilient than natural rubber, but their adhesion to metals is better. Although reclaimed rubber adhesives develop strength rapidly, they are unsuitable for structural applications. Exposure to sunlight and heat above 70 °C (158 °F) has a deteriorating effect. These low-cost adhesives are suitable for applications requiring minimum strength, including use on wood, paper, fabric, leather, natural rubber, and some metals (Ref 12).

Nitrile rubber is also known as acrylonitrile-butadiene rubber (NBR), or Buna-N. Properties can be changed by varying the ratio of acrylonitrile to butadiene. In water-base nitrile rubber adhesives, the acrylonitrile content is usually greater than 25%. Nitrile rubber is usually compounded with tackifiers and, occasionally, vulcanizing agents. Setting occurs upon the evaporation of water and/or its absorption into porous substrates. Heat or solvent reactivation of dried films is also a possibility. These adhe-

Table 11   Typical water-base adhesive latex bonding substrates, applications, and other characteristics by chemical family

Characteristic	Natural rubber	SBR	Nitrile	Polyvinyl acetate	Neoprene	Ethylene-vinyl acetate	Acrylic
Typical solids content, %	60	55	64–65	50	50	34–47	37–65
Color	White	Cream (wet), tan (dry)	White	White	Natural	White	Clear to white
Bonding substrates	Textiles, various others	Textiles, metals	Plastics	Wood, plastics	Plastics, wood, leather, plaster	Plastics, wood	Plastics, textiles, foams
Applications	Textiles, various others	Textiles, lightweight metals	PSAs, tapes, labels	Woodworking, general purpose	General industrial, construction	Packaging, film bonding	Packaging, tapes, labels, textiles, film bonding

Source: Ref 11

sives have good oil and grease resistance, limited water resistance, and shear strengths in the range of 1.03 to 13.8 MPa (0.15 to 2.0 ksi). In general, nitrile rubber is the most versatile of the general-purpose water-base synthetic rubbers, but is less popular than neoprene. It does not bond well to natural rubber or butyl rubber. Typical applications include the bonding of plastics, neoprene, and nitrile rubber to metals and the bonding of fabrics, wood, and paper to glass and metals (Ref 3, 12).

Polyvinyl methyl ether adhesives are noncrystalline and colorless. These materials are soluble in water at temperatures less than 32 °C (90 °F), but will precipitate at higher temperatures. The adhesives, when set, have limited heat resistance and good UV stability and do not embrittle at low temperatures. Their primary application is in paper labels that will subsequently be remoistened. These labels will not curl and will adhere to glossy surfaces (Ref 12).

New synthetic adhesives were developed after the year 1960. Most are based on acrylics.

Acrylics that have a sufficient amount of carboxylation in their chemical structures are totally soluble in aqueous alkali. Carboxyl-containing copolymers are based on ethylene-acrylic acid and ethylene-methacrylic acid. Acrylate homopolymers are modified with unsaturated carboxylic acids. In general, carboxylated adhesives have enhanced metal adhesion and grease resistance, compared to noncarboxylated adhesives (Ref 6, 12).

Acrylic-base adhesives are inherently tacky because of the choice of monomers and molecular weights, and because of cross-linking. However, new tackifiable polymers that improve adhesion to low-energy surfaces, such as polyethylene and polypropylene, are available. In addition, unlike traditional acrylics, these tackified adhesives have a good balance of peel, shear, and tack properties (Ref 40).

Acrylic water-base adhesives have many applications as PSAs. Higher-alkyl acrylates, such as 2-ethylhexyl acrylate or *n*-butyl acrylate, are used with small amounts of polar monomers, such as acrylic acid,

vinyl acetate, acrylamide, maleic anhydride, or long-chain acrylamides. Systems are used with or without low levels of tackifiers. Typical tackifiers are based on aromatic copolymer resins or rosins. One advantage of aromatic copolymer tackifiers such as α-methylstyrene-styrene is that the white color of the acrylic itself is maintained.

Table 10 compares acrylic-base latex PSA properties to natural and synthetic rubber properties. In general, the lower molecular weight acrylic (latex E) resulted in better tack but poorer shear adhesion than the higher molecular weight adhesive (latex G). The addition of a rosin tackifier to the acrylic latex G resulted in improved tack, while maintaining the good shear adhesion of the acrylic itself (Ref 39).

Recent developments in acrylic latex adhesives include the formation of multicomponent systems by block or graft copolymerization with side-chain or end-group reactions through unsaturation or functional groups. Previously, these systems were available only as organic-solvent-base adhesives, but they are now being formulated for high-performance water-base adhesive applications. Multicomponent systems based on acrylated silicones, acrylated urethanes, and acrylated silicone urethanes have been formed. The addition of silicone resulted in adhesives with improved thermal stability, tensile strength, and resistance to oil and grease, UV rays, and solvents. The addition of urethane resulted in an increase in toughness, thermal stability, and resistance to solvents, abrasions, and UV rays. In particular, the silicone- and urethane-base acrylates had noticeably improved adhesion to ceramics, fiberglass, and metals (Ref 41). Table 11 summarizes typical water-base adhesive latex bonding substrates and applications by chemical family.

Major Suppliers

Various natural, water-base adhesives are available from Adhesive Products Corporation; H.B. Fuller Company; Borden, Inc., Chemical Division; and National Starch and Chemical Corporation (Ref 17). A variety of synthetic water-base adhesives

are available from Adhesives Research, Inc.; Bostik Division, Emhart Corporation; H.B. Fuller Company; Hernon Manufacturing Inc.; Monsanto Company; 3M Company; Morton Thiokol, Inc.; Polyvinyl Chemical Industries; and Uniroyal Plastics Division, Uniroyal, Inc. (Ref 11, 17).

UV/EB Cured Adhesives

The rapid conversion of specially formulated, 100% reactive liquids to solids can be accomplished using radiation curing. Potential energy sources include ultraviolet (UV), electron beam (EB), visible infrared (IR), and microwave sources. Cured materials are used as coatings, inks, adhesives, sealants, and potting compounds. This section is limited to a discussion of UV- and EB-cured adhesives.

The UV curing process typically involves the exposure of a reactive liquid that contains a photoinitiator to UV radiation at a wavelength between 200 and 400 nm. The liquid is rapidly converted to a solid, usually in less than 60 s. In the EB-curing process, electrons are artificially generated and accelerated to energies of less than 100 keV to greater than one billion keV. The amount of energy is controllable. Generally, 50 to 350 keV of electrons are used to cure adhesives having a 25 to 38 μm (1 to 15 mil) bond line. The reactive liquid in the EB process does not contain a photoinitiator (Ref 42). Because the main advantage of UV/EB-curable adhesives is rapid curing at room temperature, they can be used to bond heat-sensitive substrates, such as polyvinyl chloride. In addition, the rapid cure often eliminates the need to fixture parts and greatly increases production rates.

UV/EB-cured adhesives have been used to replace solvent-base adhesives because of the increasing cost of properly recovering and disposing of solvents. Most adhesives, being single-component materials, require no mixing and have little waste.

The cross-linked nature of UV/EB-cured adhesives results in good chemical, heat, and abrasion resistance; toughness; dimensional stability; and adhesion to many substrates. Unlike thermal curing, EB curing can be selective, and the depth of penetration can be controlled (Ref 43, 44).

There are several limitations to the use of UV/EB-cured adhesives. EB equipment is expensive, costing as much as $400 000 to $800 000, with accessories. UV equipment is less expensive, but the materials themselves are usually more costly because of the presence of photoinitiators (Ref 45). To cure adhesives properly, one substrate must be transparent to the UV radiation. However, the necessity of having a transparent substrate has been removed by the introduction of dual-curing adhesives. These adhesives are quickly set by a UV cure and are more fully cured by a second mechanism involving the introduction of heat or moisture or the elimination of oxygen (anaerobics). In EB-curable adhesives, the depth of EB penetration is limited by the density of the material, rather than its opacity (Ref 44).

Characteristics

The cure time of UV adhesives is usually less than 60 s and depends on:

- *Bond line joint thickness*. As the thickness increases, UV radiation loses its ability to penetrate totally, necessitating a second cure
- *Type of substrate*. A transparent substrate such as glass with a minimum gap may take as little as 5 s to cure. Opaque and darker substrates require longer cure times
- *Light intensity*. The more intense the UV light, the faster the cure (Ref 46)

The cure time of EB adhesives is comparable to that of UV adhesives. In general, EB radiation allows adhesive curing to be achieved at greater depths than that allowed with UV radiation. Electrons can pass through substrates that are opaque to UV light. In addition, the area of exposure and the depth of penetration can be controlled by the proper EB conditions (Ref 45).

The three types of UV/EB-curable adhesives are laminating adhesives, pressure-sensitive adhesives, and structural adhesives (UV only).

Laminating adhesives have the largest share of the UV/EB-curable adhesives market. Adhesives are applied and then cured in-line. The lack of solvent in these adhesives eliminates the drying step required when using solvent-base adhesives in laminates. The high degree of cross-linking in UV/EB-cured laminates contributes high bond strength, good heat resistance, and good chemical resistance. Substrates that are typically laminated include: film to paper, foil, fabric, glass, film, or wood; paper to foil, wood, or paper; wood to wood; and glass to glass or metal (Ref 44).

The PSA segment of the UV/EB-curable adhesives market is not nearly as large as that of the laminating adhesives segment. Because slight changes in exposure time greatly affect peel strength and tack, the formulation of these adhesives is critical. Other factors that must be considered include adhesive thickness, substrate type, and compatibility of the adhesive with the release coating.

Currently, there are three approaches used to make PSAs. The first is based on a UV-curable system employing conventional tackifiers blended with UV-reactive moieties. The process must be closely controlled for repeatable tack and peel strength. The second approach is based on an EB-curable system that is actually a hot-melt adhesive. EB-cross-linkable thermoplastic rubbers are usually used in these adhesives, resulting in hot-melt PSAs with better heat resistance than that of conventional elastomeric hot-melt adhesives. The final approach involves the development of inherently tacky oligomers for PSAs (Ref 44).

UV-curable structural adhesives require purer raw material grades. As a result, adhesive cost per pound is much higher than it is for laminating or PSA applications. The use of dual-curing systems allows opaque substrates to be cured.

The mechanical properties of UV-curable structural adhesives are dependent on polymer molecular weight and cross-linking density. These factors are related to the prepolymer size, degree of stiffness or flexibility, and functionality. The overall adhesive strength is affected by:

- Adequate UV transmission through the bond line
- Adhesive thickness
- UV intensity
- Postcuring (that is, by heat or exclusion of oxygen, as in anaerobics) (Ref 47)

Components of UV/EB-cured adhesives are described below.

Oligomers are reactive molecules that contribute adhesion, toughness, and flexibility to the overall properties of UV/EB adhesives. Typical reactive oligomers include acrylated epoxy resins and aromatic urethanes.

Monomeric diluents are low molecular weight molecules that are monofunctional, that is, containing only a single reactive group. These diluents reduce the viscosity of liquid oligomers. Some, such as methacrylates, increase the toughness and adhesion. Various types of acrylates are used as reactive monomers.

Cross-linking monomers are also found in UV/EB adhesive formulations. These low molecular weight materials contain two or more reactive groups in their molecular structures. Thus, these monomers provide cross-linking sites in the polymer chains. Typical cross-linking monomers include 1,3-butylene glycol dimethacrylate, tripropylene glycol diacrylate, and pentaerythritol tetracrylate.

Free radical initiators trigger the cross-linking reaction. In EB-cured adhesives, the electrons serve as free radical initiators for addition polymerization. Therefore, no chemical initiator additives are needed. In UV-cured adhesives, photoinitiators, which release free radicals when exposed to UV radiation, are required in order to initiate addition polymerization (Ref 45). The most recent UV- and EB-cure systems involve cationic polymerization mechanisms. In this method, special chemicals break down under UV radiation or EB radiation to produce active hydrogen ions. These ions break the bonds in cyclic monomers, resulting in ring opening and polymerization (Ref 48).

An adhesion promoter is another component that can be used in UV and EB adhesive formulations. This material bonds chemically with the substrate, becoming part of the adhesive bond. Its function is to provide good adhesion to substrates such as glass and certain metals.

A stabilizer is another component in UV adhesives that are formed by addition polymerization. This material is added to prevent premature polymerization by functioning as a free radical scavenger (Ref 42, 44, 45).

Markets and Applications

In 1988, $31 million in sales were reported worldwide for all UV-cured adhesives. In 1989 sales were predicted to rise 139%, to $74 million, by 1993 (Ref 49). Another report issued in 1987 estimated that materials for UV and EB systems will grow at a rate of 7% (Ref 50). Predicted areas of adhesive growth are fiber optics, metal finishing, release coatings, magnetic media, and package lamination (Ref 49).

Typical UV-curable adhesive applications include the electronics, automotive, medical, optics, packaging areas, and tapes and labels. EB-curable adhesives are used in magnetic tapes and floppy disks, where magnetic particles are bonded to films, as well as in packaging, tapes, and labels (Ref 43, 45, 46).

Chemical Families

Most UV/EB adhesives are based on an addition polymerization curing mechanism. Materials consist of acrylic acid esters of various forms or combinations of acrylates with aliphatic or aromatic epoxies, urethanes, polyesters, or polyethers (Ref 47, 48, 51). Table 12 compares typical properties of cured epoxy-acrylates and urethane-acrylates (Ref 51). Although the epoxy-base systems have higher tensile strengths, their elongations are less than those of the urethane-base systems. In addition, the urethane-base systems have better abrasion resistance.

Table 12 Comparison of typical properties of UV-curable epoxy-acrylates and urethane-acrylates

Adhesive(a)	Tensile strength		Elongation, %
	MPa	ksi	
Epoxy-acrylates			
DGEBA diacrylate	43.3	6.28	1.6
DGEBA ester diacrylate	73.9	10.7	7.5
Carboxy DGEBA diacrylate	39.4	5.72	2.4
Alkyd modified DGEBA diacrylate	55.8	8.09	7.0
Urethane-acrylates			
Urethane-aliphatic triacrylate	15.2	2.2	36
Urethane-aliphatic diacrylate	6.51	0.943	28
Urethane-aromatic multiacrylate	3.86	0.560	24.5

(a) DGEBA, diglycidyl ether of bisphenol A

UV/EB adhesives that undergo cationic polymerization are based on epoxies with reactive diluents and cyclic monomers. Producers of these new types of adhesives include Epolin Inc., Henkel Corporation, and Union Carbide Corporation. Recently, Allied-Signal Inc. reported a cationic-formulation based on vinyl ether and polyurethane oligomers (Ref 48). UV/EB pressure-sensitive adhesives are usually based on acrylics, synthetic rubbers, and/or silicones. Acrylics are the sole or major components (Ref 52). Table 13 compares selected UV-curable adhesives.

Major Suppliers

Suppliers of laminating adhesives include Sun Chemical Corporation; Morton Thiokol, Inc.; Lord Corporation; PPG Industries, Inc.; and Cavanagh Corporation. UV-curable PSAs are available from Beacon Chemical Company, Inc.; Master Bond Inc.; and Acheson Colloids Company. EB-curable PSAs are available from Shell Chemical Company and National Starch and Chemical Corporation. UV-curable structural adhesives are available from Loctite Corporation, Dymax Corporation, Emerson & Cuming, Norlan Products Inc., Hernon Manufacturing Inc., and Westinghouse Electric Corporation (Ref 44).

REFERENCES

1. A.H. Landrock, *Adhesives Technology Handbook*, Noyes Publications, 1985, p 15
2. B.C. Cope, "Advances in Adhesives," D. Brewis and J. Comyn, Ed., Warwick Publishing, 1983, p 42
3. A.H. Landrock, Chapter 5 in *Adhesives Technology Handbook*, Noyes Publications, 1985
4. *Mach. Des.*, Vol 55 (No. 11), 1987, p 36
5. T. Flanagan, Chapter 8 in *Handbook of Adhesive Bonding*, C.V. Cagle, Ed., McGraw-Hill, 1973
6. *Adhesive Bonding*, MIL-HDBK-691B, *Military Standardization Handbook*, U.S. Department of Defense, 1987, p 47-100
7. *Vendor Catalog Service—Vendor Name*, Inf. Handl. Services, Inc., Vol A, July-Aug 1989
8. J.C. Bolger and S.L. Morano, *Adhes. Age*, Vol 27 (No. 6), 1984, p 17
9. A. Barker, *Adhes. Age*, Vol 31 (No. 3), 1988, p 44
10. F. Keimel, "Adhesion Science and Technology," Paper presented at Seminar, The Center for Professional Advancement, East Brunswick, NJ, April 1989
11. J. Thuen, Ed., *Adhesives*, 4th ed., D.A.T.A. Inc. and the International Plastics Selector, Inc., 1986
12. J. Shields, *Adhesives Handbook*, 3rd ed., Butterworth, 1984, p 32-79
13. R.C. Kausen, *Mater. Des. Eng.*, Vol 60 (No. 3), 1964, p 108
14. B.D. Murray, *Adhesives*, 4th ed., D.A.T.A. Inc. and the International Plastics Selector, Inc., 1986, p A20-A21
15. C.F. Lewis, *Mater. Eng.*, Vol 105 (No. 10), 1988, p 57
16. M.B. Pearce, Jr., *Mater. Eng.*, Vol 90 (No. 7), 1973, p 50
17. A. Barker, *Adhes. Age*, Vol 30 (No. 6), 1987, p 95
18. E.J. Bruno, Ed., Chapter 1 in *Adhesives in Modern Manufacturing*, Society of Manufacturing Engineers, 1970
19. G.L. Schneberger, Chapter 1 in *Handbook of Adhesive Bonding*, C.V. Cagle, Ed., McGraw-Hill, 1973
20. M.M. Schlecter, *Adhes. Age*, Vol 30 (No. 12), 1987, p 18
21. D.R. Dreger, *Mach. Des.*, Vol 47 (No. 1), 1979, p 54
22. C.A.A. Rayner, in *Adhesion and Adhesives*, Vol 1, 2nd ed., R. Houwink and G. Salomon, Ed., Elsevier, 1965
23. E.W. Flick, *Adhesive and Sealant Compounds and Their Formulations*, Noyes Data Corporation, 1978
24. W.H. Korez, *Adhes. Age*, Vol 27 (No. 12), 1984, p 19
25. G. Holden and S. Chin, *Adhes. Age*, Vol 30 (No. 5), 1987, p 22
26. J.J. Bell and W.J. Robertson, SAE Paper 740261 presented at the Automotive Engineering Congress, Detroit, 1974
27. P. Borg and J. Boutillier, *Adhes. Age*, Vol 29 (No. 8), 1986, p 31
28. F.T. Hughes, *Adhes. Age*, Vol 25 (No. 9), 1982, p 25
29. S.W. Medina and F.V. DiStefano, *Adhes. Age*, Vol 32 (No. 2), 1989, p 18
30. D. Satas, Chapter 1 in *Handbook of Pressure Sensitive Adhesive Technology*, D. Satas, Ed., Van Nostrand-Reinhold, 1982
31. A. Barker, Ed., *Adhes. Age*, Vol 32 (No. 4), 1989, p 38
32. K. Ammons, *Adhes. Age*, Vol 31 (No. 9), 1988, p 26
33. G.L. Butler, Chapter 10 in *Handbook of Pressure Sensitive Adhesive Technology*, D. Satas, Ed., Van Nostrand-Reinhold, 1982
34. W.H. Korcz, D.J. St. Clair, E.E. Ewins, Jr., and D. deJager, Chapter 11 in *Handbook of Pressure Sensitive Adhesive Technology*, D. Satas, Ed., Van Nostrand-Reinhold, 1982
35. D. Satas, Chapter 13 in *Handbook of Pressure Sensitive Adhesive Technology*, D. Satas, Ed., Van Nostrand-Reinhold, 1982
36. D.F. Merrill, Chapter 15 in *Handbook of Pressure Sensitive Adhesive Technology*, D. Satas, Ed., Van Nostrand-

Table 13 Comparison of selected UV-curable adhesives

Property	Epoxy	Acrylic	Polyester	Methacrylates		Acrylics		Modified acrylates
Cure	UV	UV	UV	UV only	UV, anaerobic	UV, anaerobic	UV, anaerobic	UV, heat
Brookfield viscosity, Pa · s	2.5	0.55	6.5	0.105–6.0	0.500	2.25–20.0	0.300–2.0	7.0
Service temperature range, °C (°F)	−54 to 175 (−65 to 347)	−54 to 135 (−65 to 275)	−34 to 121 (−30 to 250)	−54 to 121 (−65 to 250)	−54 to 177 (−65 to 350)	−54 to 135 (−65 to 275)	−54 to 177 (−65 to 350)	−54 to 135 (−65 to 275)
Hardness	Shore D65	Barcol 70	···	Shore D63–D82	···	···	···	Barcol 75
Bonded substrates(a)	Glass	Glass, metals	Plastics, metals	Glass, metals, plastics	Glass, metals	Glass, metals	Metals	Glass, metals

(a) For a UV cure, one substrate must be transparent. Source: Ref 11

Reinhold, 1982

37. A.J. Kettleborough, Chapter 5 in *Advances in Adhesives*, D. Brewis and J. Comyn, Ed., Warwick Publishing, 1983
38. W. Broxterman, *Adhes. Age*, Vol 31 (No. 2), 1988, p 20
39. M.J. Jones, in *Adhesive Chemistry*, L.H. Lee, Ed., Plenum Press, 1984, p 693-724
40. F.V. DiStefano, S.W. Medina, and B.R. Vijayendran, *SAMPE J.*, Vol 24 (No. 4), 1988, p 27
41. R.R. Kelly and R.R. Alexander, *Adhes. Age*, Vol 31 (No. 2), 1988, p 26
42. *Radiation Curing—An Introduction to Coatings, Varnishes, Adhesives and Inks*, Education Committee of the Radiation Curing Division, The Association for Finishing Processes, Society of Manufacturing Engineers, 1984, p 3-20
43. A.H. Pincus, in *Proceedings of the Polymers, Laminations and Coatings Conference*, Technical Association of the Pulp and Paper Industry, 1986, p 207
44. A.H. Pincus, *Adhes. Age*, Vol 32 (No. 6), 1989, p 16
45. C.F. Lewis, *Mater. Eng.*, Vol 104 (No. 11), 1987, p 39
46. F. Marino, *Mach. Des.*, Vol 56 (No. 18), 1984, p 50
47. J. Woods, "Technical Paper FC85-414," The Association for Finishing Processes of the Society of Manufacturing Engineers, 1985
48. G.R. Smoluk, *Mod. Plast.*, Vol 65 (No. 6), 1988, p 250
49. A. Barker, Ed., *Adhes. Age*, Vol 32 (No. 9), 1989, p 47
50. A. Barker, Ed., *Adhes. Age*, Vol 31 (No. 6), 1988, p 48
51. W.J. Morris, "Technical Paper FC84-993," The Association for Finishing Processes of the Society of Manufacturing Engineers, 1984
52. K.C. Stueben, "Adhesive Chemistry—Developments and Trends," L.H. Lee, Ed., Plenum Press, 1984, p 319

Epoxies

Dean T. Behm and John Gannon, Ciba-Geigy Corporation

EPOXY ADHESIVES are perhaps the most versatile of the structural adhesives. Although generally characterized as being strong but brittle, they can be formulated to be more flexible without loss of tensile strength. Epoxy adhesives are able to bond a variety of substrates effectively and can be formulated to cure at either room temperature or elevated temperatures, under dry or wet conditions. Epoxy adhesives may not be as inexpensive as other adhesives described in this article, but they are quite competitive on a cost-performance basis.

Chemistry

Resin Chemistry

Epoxy resins are a group of thermosetting materials that possess the epoxy or oxirane group and are convertible into three-dimensional structures by a variety of curing reactions. The epoxy resins formed from bisphenol A and epichlorohydrin constitute the predominant types of resinous intermediates.

In the synthesis of epoxies, the liquid resins are derived by using an excess of epichlorohydrin with respect to bisphenol A, in the presence of sodium hydroxide, as shown in Fig. 1. The product is the crude diglycidyl ether of bisphenol A (DGEBA) possessing a viscosity of 11 to 15 Pa · s at 25 °C (77 °F).

Side reactions in the preparation form species other than epoxide, such as hydrolyzable chlorine, bound chlorine, and α-glycol, each of which is portrayed in Fig. 2. Moreover, the monomer content (DGEBA) is usually about 70 to 80% of the composition, with oligomeric species and side products (such as those shown above) representing the difference.

Solid epoxies can be prepared by chain extension of the liquid resin with bisphenol A in catalytic polymerizations using an excess of the former to provide epoxide terminal groups, as shown in Fig. 3.

The molecular weight or the degree of polymerization (n value) attained is a function of the stoichiometry employed in the chain extension reaction, with higher liquid resin-to-bisphenol-A ratios resulting in lower molecular weight products. The repeat unit of the polymer chain is characterized by the presence of a secondary hydroxyl group, which may be significant for adhesive properties. Increases in the molecular weight of solid resins result in higher viscosities, higher softening points, higher weights per epoxide contents, and higher values for hydroxyl contents, expressed as equivalents/gram (eg/g) of resin. Reactive diluents, that is, low molecular weight compounds containing one or more epoxy groups, are frequently added to liquid epoxide resins to lower viscosities, while solvents are employed to provide solutions of solid resins for application in the coatings, laminating, matrix, and adhesive areas where solvents can be removed during processing to the cured state. Prominent diluents are shown in Fig. 4, while the reduction in viscosity by various diluents is shown in Fig. 5.

In addition to the resins based on bisphenol A, structures derived from other backbones are used in the epoxy resin field, namely, cycloaliphatic resins, epoxidized phenol or epoxidized cresol novolacs, and other multifunctional species prepared by glycidylization with epichlorohydrin.

Cycloaliphatic resins encompass two varieties, one type being epoxies derived from peracid treatment of cycloolefins, as shown in Fig. 6(a), and the other type being glycidylization of cycloaliphatic acids with epichlorohydrin, as shown in Fig. 6(b).

Epoxidized phenol or cresol novolacs are prepared from phenol or cresol novolacs and epichlorohydrin with sodium hydroxide (Fig. 7) where n = 0.2 to >3 and R = H or CH_3, available from Dow Chemical Company and Ciba-Geigy Corporation.

Noteworthy among the other multifunctional species of epoxies are triglycidyl p-amino phenol (MY 0500, Ciba-Geigy Corporation), with a viscosity of 3 to 5 Pa · s at 25 °C (77 °F) (Fig. 8a) and N,N,N′,N′-tetraglycidyl-4,4′-diamino diphenyl methane (MY 720, Ciba-Geigy Corporation),

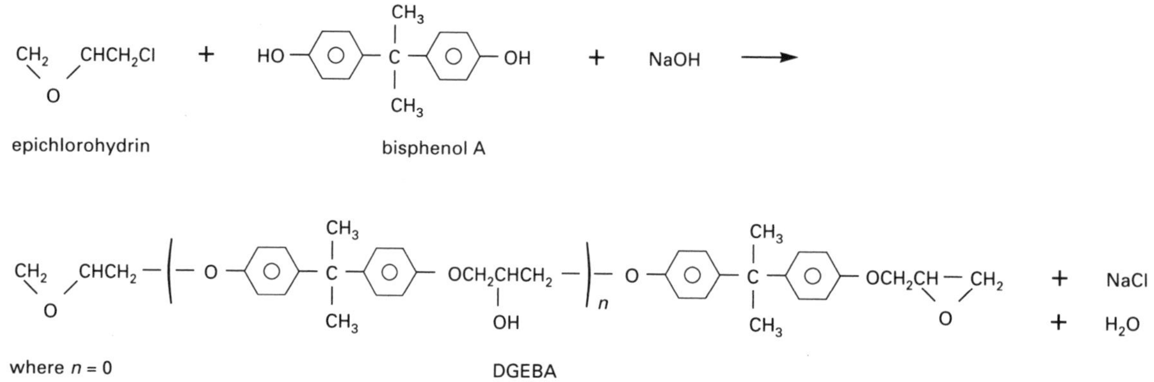

Fig. 1 Diglycidyl ether of bisphenol A (epoxy) synthesis

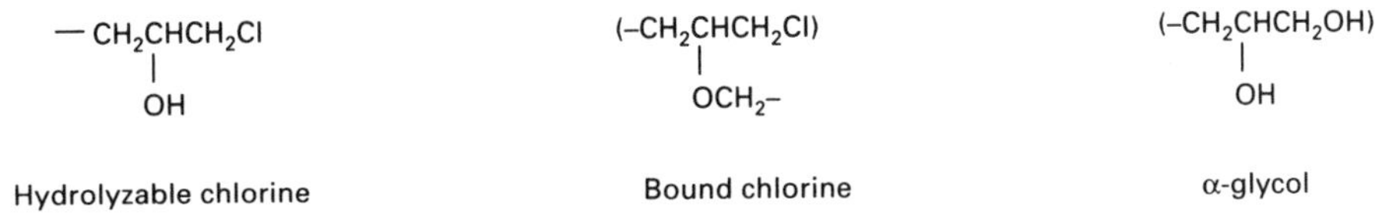

CH₂CHCH₂Cl with structure showing:

$$-CH_2CHCH_2Cl$$
$$|$$
$$OH$$

Hydrolyzable chlorine

$$(-CH_2CHCH_2Cl)$$
$$|$$
$$OCH_2-$$

Bound chlorine

$$(-CH_2CHCH_2OH)$$
$$|$$
$$OH$$

α-glycol

Fig. 2 Side reactions

with a viscosity less than 25 Pa · s at 50 °C (120 °F) (Fig. 8b). Epoxidation of bisphenol F (mixed isomers) with epichlorohydrin yields a liquid epoxide resin with a functionality of slightly greater than 2.0 and a viscosity of 4 to 6 Pa · s at 25 °C (77 °F), as shown in Fig. 9. Other epoxy resins comprise the diglycidyl esters of dimerized fatty acids, which are utilized to flexibilize aromatic-base epoxies, as shown in Fig. 10.

Cure Chemistry

Epoxies require curing to obtain useful properties, as noted above. The curing reactions can be performed with the terminal epoxy groups or the pendant hydroxyl groups or both.

The polyaddition of epoxy groups with amines or polyamides is a familiar curing reaction whereby a primary or secondary amine adds to the epoxy ring, forming a tertiary amine, as shown in Fig. 11. The formed hydroxyl groups accelerate the amine curing, and with excess epoxy present, the secondary hydroxyl groups can also add to the epoxy ring, as shown in Fig. 12. Aliphatic amines cure epoxies at room temperature, whereas aromatic amines require elevated temperatures. However, aromatic amine adducts of liquid epoxies can be accelerated to cure the resins at room temperature. Important polyamines for curing epoxies are identified in Fig. 13.

Polyamides are resinous hardeners with amide repeat units and terminal amine groups based on the polymerization of excess diethylene triamine and dimer acid (a C-36 difunctional carboxylic acid). As with polyaliphatic amines, curing of epoxies with polyamides is effected at room temperature, providing flexible, water-resistant thermosets. The polyamides, having varying viscosities and molecular weights, provide

lower exotherms and more latitude in stoichiometry than aliphatic polyamines.

Poly(amido amines) are similar to polyamides but are lower in viscosity because they are prepared from monofunctional fatty acids and polyamines. Performance characteristics are fairly similar to the polyamides.

Dicyandiamide (Dicy) is a solid latent hardener that reacts with both the epoxy terminal groups and the secondary hydroxyl groups formed on curing the resin. Being latent, it reacts with epoxies upon heating, but stops reacting upon the removal of heat. Its latent nature is due to its insolubility in the resin at ambient temperatures. Imidazoles constitute a class of epoxy hardeners that also react with epoxide groups and provide catalytic activity with high heat resistance. The mechanisms of the catalysis of epoxies by Dicy and imidazoles have been studied in detail in recent years (Ref 1, 2).

Anhydrides are another common reagent for curing epoxies, but acceleration is needed and elevated temperatures are usually employed to effect full conversion. The reaction of anhydrides with epoxies can be depicted as proceeding via a zwitterion, or dipolar ion, with tertiary amine catalysis, as shown in Fig. 14.

Epoxy-anhydride systems exhibit low viscosity, long pot life, and low exothermic heats of reaction, as well as low shrinkage when cured at elevated temperatures. A list of commercial anhydrides is shown in Fig. 15.

Accelerators for Epoxy Curing Reactions

Curing of primary and secondary amines is accelerated by hydroxyl compounds such

as phenols, alcohols, and water (hydrogen bond donors).

Anhydride cures are promoted by tertiary amines, for example, 2,4,6-tris(dimethylaminomethyl) phenol, N-benzyl-N,N-dimethylamine, and 1,4-diazabicyclo[2.2.2]octane (DABCO) and its salts. Other catalysts include quaternary phosphonium salts and imidazoles, as well as boron trifluoride or other Lewis acids. The use of stannous octoate promotes extensive polyetherification as well as polyesterification in curing reactions.

Homopolymerization of epoxies proceeds with both Lewis-base and Lewis-acid catalysis. As noted above, substituted imidazoles also catalyze homopolymerization. The mechanism of Lewis-base catalysis is illustrated in Fig. 16, where tertiary amines react with the methylene carbon of the epoxy group to form an intermediate zwitterion.

Chain propagation is shown proceeding by way of a polymeric anion to form a polyether. In practice, using Lewis-base catalysts leads to products with low heat distortion values because kinetic chain lengths are low, for example, six to eight epoxy groups polymerized per nitrogen group. In general, tertiary amines, such as N-benzyl-N,N-dimethylamine, are frequently employed as cure catalysts, as are Mannich bases, such as 2,4,6-tris(dimethylaminomethyl)phenol. Substituted imidazoles, like 2-ethyl 4-methyl imidazole, 2-methyl imidazole, and 1-methyl imidazole, have been utilized as homopolymerization catalysts while also functioning as hardeners for epoxy resins.

The use of Lewis acids, that is, cationic polymerization using boron trifluoride complexes, affords a controllable gelation rate and cure at moderate temperatures. Initiation of the homopolymerization has been attributed to the formation of an oxonium ion after solvation of an amine salt, as shown in Fig. 17. The curing is propagated by attack of other epoxy groups on the oxonium ion.

The monomethylamine complex of BF_3 is used frequently at a level of 2-4 parts per

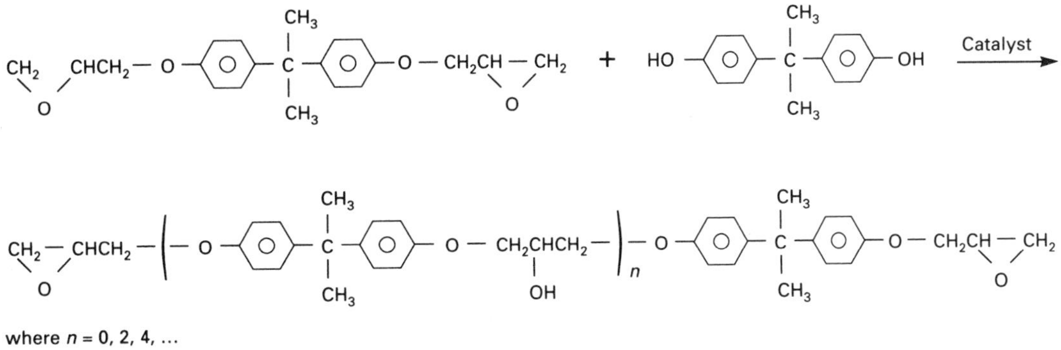

where n = 0, 2, 4, …

Fig. 3 Solid epoxy preparation

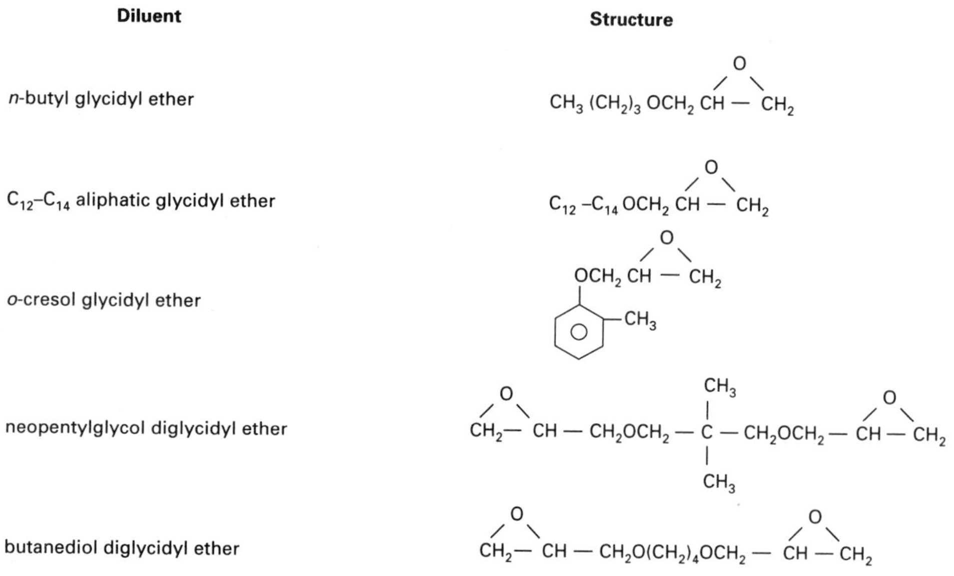

Diluent	Structure
n-butyl glycidyl ether	CH₃ (CH₂)₃ OCH₂ CH — CH₂
C₁₂–C₁₄ aliphatic glycidyl ether	C₁₂ –C₁₄ OCH₂ CH — CH₂
o-cresol glycidyl ether	
neopentylglycol diglycidyl ether	
butanediol diglycidyl ether	

Fig. 4 Diluents added to liquid epoxy resins to lower viscosity

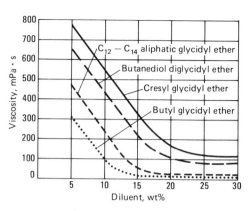

Fig. 5 Reduction of diglycidyl ether of bisphenol A viscosity by reactive diluents

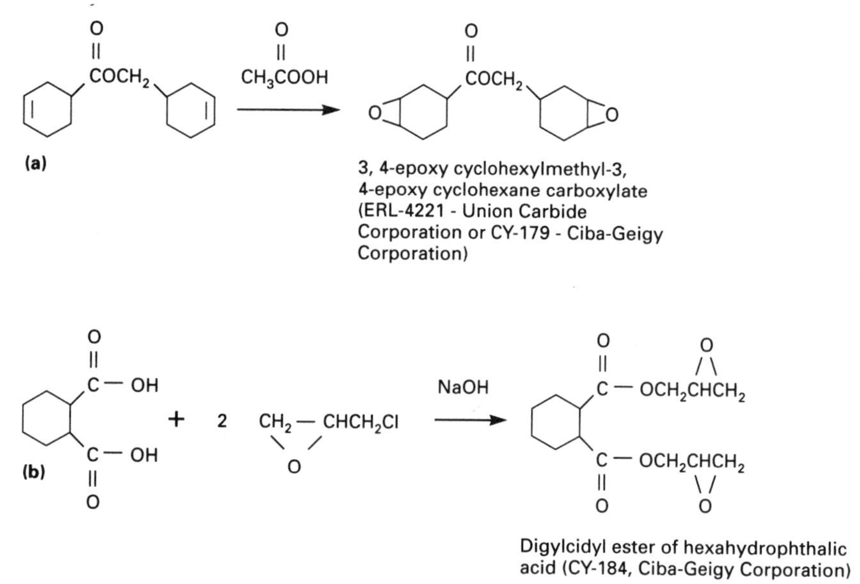

(a) 3, 4-epoxy cyclohexylmethyl-3, 4-epoxy cyclohexane carboxylate (ERL-4221 - Union Carbide Corporation or CY-179 - Ciba-Geigy Corporation)

(b) Digylcidyl ester of hexahydrophthalic acid (CY-184, Ciba-Geigy Corporation)

Fig. 6 Cycloaliphatic resins. (a) Epoxies derived from peracid treatment of cycloolefins (ERL-4221, Union Carbide Corporation, or CY-179, Ciba-Geigy Corporation). (b) Glycidylization of cycloaliphatic acids with epichlorohydrin

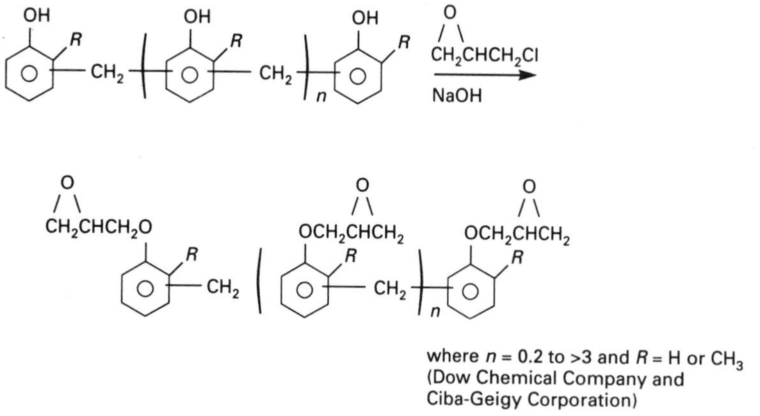

where *n* = 0.2 to >3 and *R* = H or CH₃ (Dow Chemical Company and Ciba-Geigy Corporation)

Fig. 7 Epoxidized phenol or cresol novolacs

hundred resin (phr) of epoxy resin at temperatures of about 100 °C (212 °F). The cured products are brittle without modification. Thiols, like alcohols or phenols, react with epoxies to form hydroxyl sulfides, as shown in Fig. 18. Catalysis by tertiary amines is very effective and reaction rates with thiols are faster than amine-epoxy reactions, as shown in Fig. 19.

The use of polyalkylene polysulfide polymers with mercaptan terminal groups (Thiokol liquid polymers) provides flexibility to room-temperature-cured epoxy-resin systems. The article "Polysulfides" in the Section "Adhesives Materials" in this Volume further describes various mercaptan-terminated formulations.

Alternate curing of epoxies can be accomplished by a latent hardener based on the reaction of phthalic anhydride with an excess of diethylene triamine (DETA). The product is a complex polyamide solid hardener that releases the aliphatic amine upon heating but whose latency with liquid epoxies is based on its physical nature. The fine dispersion of latent hardener (HT 939, Ciba-Geigy Corporation) in liquid epoxies is a commercially available system (HY 940, Ciba-Geigy Corporation). Modification of epoxies by other resinous curing agents affords flexibilization of the polymer chain by cocuring or phase separation of the modifying resin, leading to toughening of the epoxy network. Table 1 summarizes the salient effects noted in the modification procedure.

Functional Types

Structural. While capable of effectively bonding nonstructural substrates, epoxy adhesives also show excellent performance in structural bonding applications. Besides demonstrating good adhesion and mechanical properties in metal-to-metal bonding applications, epoxy adhesives are also effective in bonding such diverse substrates as plastics (thermosets and thermoplastics), concrete, and glass.

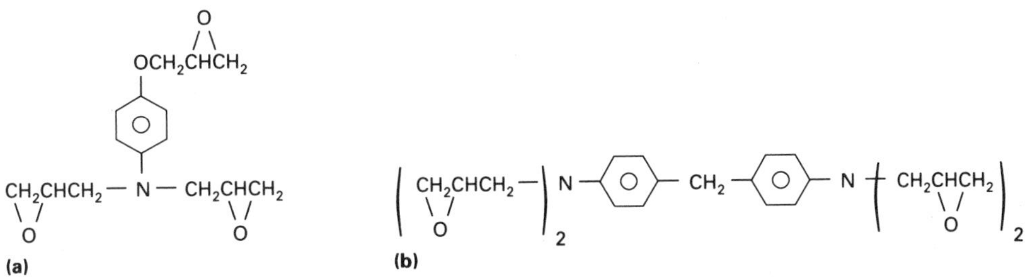

Fig. 8 Multifunctional species of epoxies. (a) Triglycidyl p-amino phenol. (b) N,N,N′,N′-tetraglycidyl-4,4′-diamino diphenyl methane

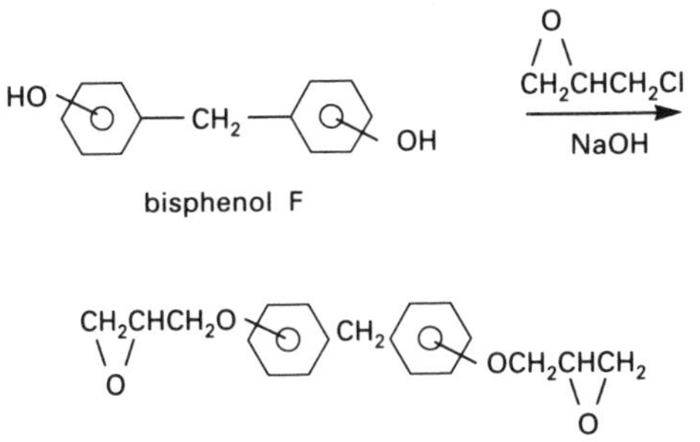

Fig. 9 Liquid epoxide resin based on epoxidation of bisphenol F with epichlorohydrin

Structural bonding with epoxy adhesives can range from applications in highway construction and automobile assembly to those in the aerospace and marine fields. Epoxy adhesives can often be formulated to meet the specific structural bonding needs of the end user.

Thermosetting Hot-Melts. Hot-melt adhesives are probably most often thought of as being nonreactive thermoplastics. However, thermosetting hot-melts are increasing in popularity in bonding applications. Similar to their thermoplastic counterparts, thermosetting hot-melts are applied by heating the adhesive and are cured to an infusible thermoset at an elevated temperature.

Thermosetting hot-melt epoxy adhesives rely on high molecular weight epoxy resins for the solid appearance of the uncured adhesive. Once cured, the thermosetting hot-melt demonstrates properties similar to other structural epoxy adhesives.

Commercial Forms

Liquids and Pastes. Epoxy adhesives can be formulated as either low-viscosity, unfilled liquids or higher-viscosity, filled pastes. These systems can be either one-component or two-component. The advantage of the two-component system is usually a longer shelf life, although after being mixed it has a limited pot life. The cure chemistry for both systems is discussed earlier in this article. Epoxy adhesives can also be found in a one-component powder form, but this is not as common as paste and liquids.

Films. Another form that epoxy adhesives are available in is film or film tape. These films can be used with or without a carrier. The carrier materials can range from glass cloths to thermoplastic films. The epoxy chemistry of films is often similar to that used in one-component pastes.

Syntactic Foams. While the majority of epoxy adhesives are supplied as liquids, pastes, or films, low-density syntactic foams are also available. The term syntactic refers to epoxy and glass microballoons. These one- or two-component epoxy systems typically exhibit good lap-shear and compressive strengths. Because of their low density, many are used in aerospace and marine applications.

Markets and Applications

Epoxy adhesives are versatile in that they can be used to bond nearly any substrate, and therefore have a large share of the consumer adhesives market. Epoxies in this market are often inexpensive, general-purpose structural adhesives for household applications.

In the general industrial market, epoxy adhesives are used with furniture, appliances, and recreational equipment. In addition, structural bonding applications in building or highway construction rely on the performance characteristics of epoxy adhesives.

The automotive market provides epoxy adhesives with a widely diverse range of uses. Besides subcomponent (engine and nonengine) assemblies, epoxy adhesives are used in metal-to-metal bonding of parts such as doors and hoods. With the advent of space-frame technology, the bonding of sheet molding compound (SMC) to a metal frame is necessary. Epoxy adhesives formulated to meet the requirements of this concept are currently used.

The aerospace and defense sectors also choose epoxy adhesives for a number of structural applications. Epoxies can be used in the assembly of missiles, composite repair, or bonding of aluminum skins to the aircraft body. Some epoxies can be used in aircraft engines, though this area often requires adhesive performance at temperatures above the upper limit for epoxy adhesive use (200 °C, or 390 °F). Aircraft interiors make use of epoxy adhesives, not only in paste or film form, but also as syntactic foams. These low-density epoxy

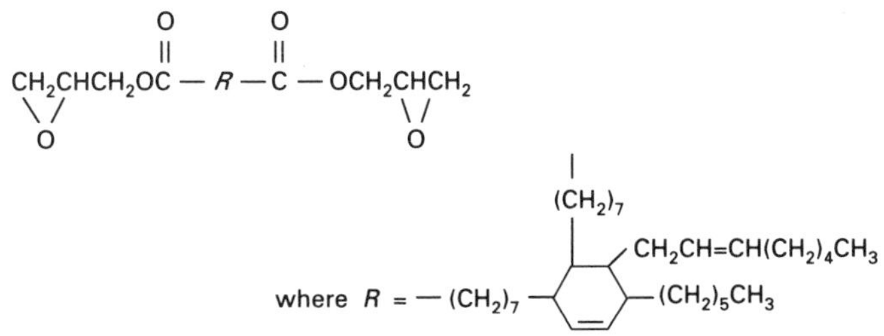

Fig. 10 Epoxy resin based on diglycidyl esters of dimerized fatty acid

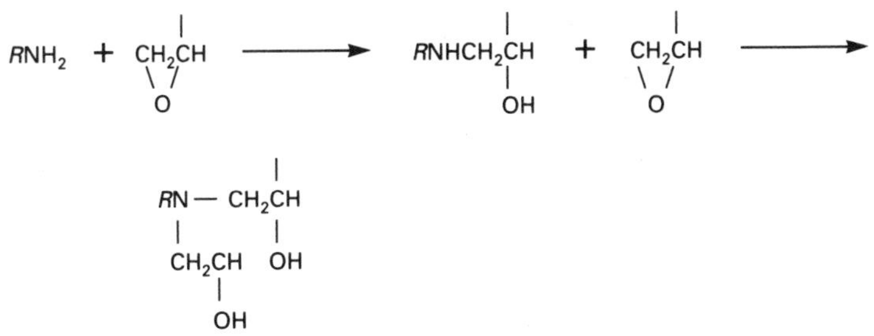

Fig. 11 Addition of primary or secondary amine to epoxy ring

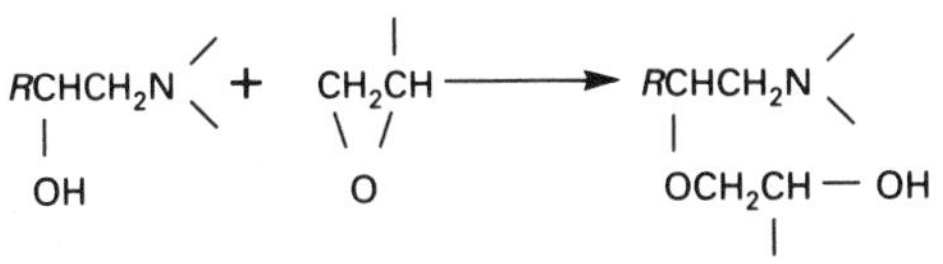

Fig. 12 Addition of secondary hydroxyl groups to epoxy ring

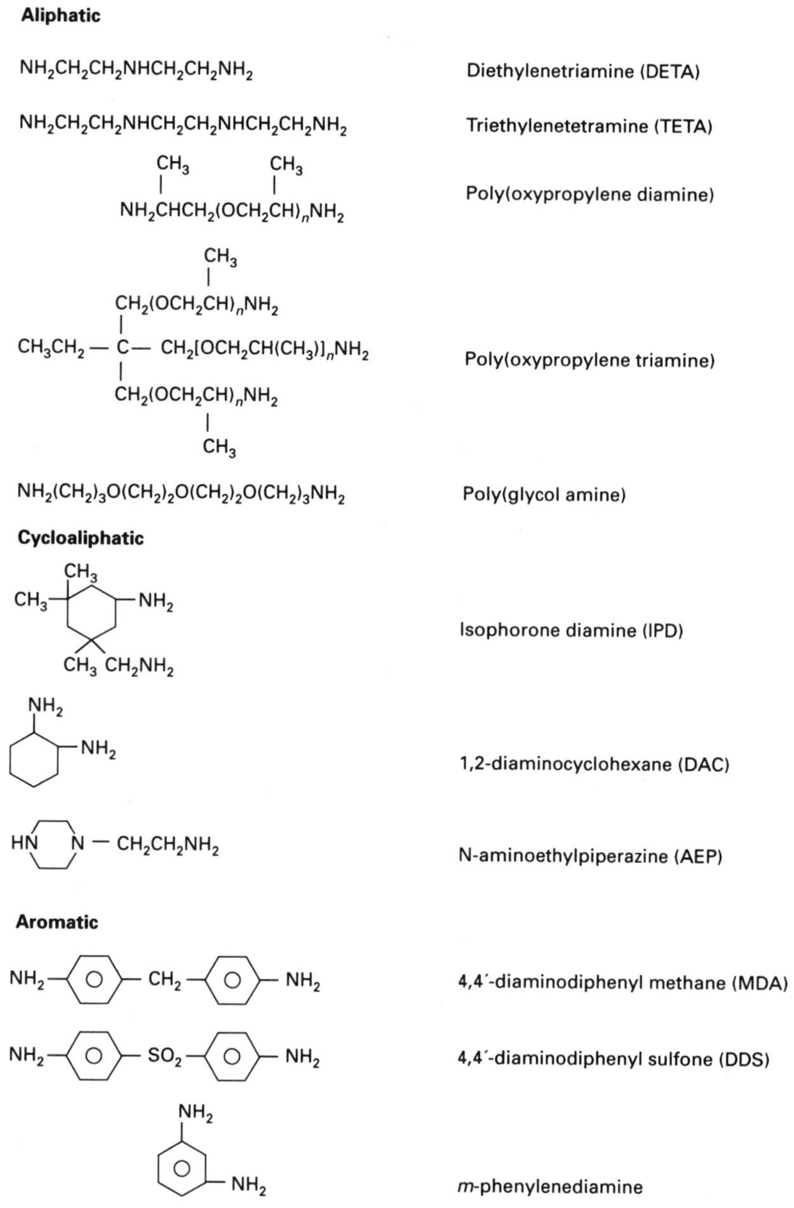

Aliphatic

$NH_2CH_2CH_2NHCH_2CH_2NH_2$ — Diethylenetriamine (DETA)

$NH_2CH_2CH_2NHCH_2CH_2NHCH_2CH_2NH_2$ — Triethylenetetramine (TETA)

Poly(oxypropylene diamine)

Poly(oxypropylene triamine)

$NH_2(CH_2)_3O(CH_2)_2O(CH_2)_2O(CH_2)_3NH_2$ — Poly(glycol amine)

Cycloaliphatic

Isophorone diamine (IPD)

1,2-diaminocyclohexane (DAC)

N-aminoethylpiperazine (AEP)

Aromatic

4,4'-diaminodiphenyl methane (MDA)

4,4'-diaminodiphenyl sulfone (DDS)

m-phenylenediamine

Fig. 13 Polyamines for curing epoxies

adhesives can be used in composite panel assembly or repair (see Fig. 20).

Major Suppliers

Companies supplying epoxy adhesives in the markets described above include 3M Company, Hardman, Inc., Lord Corporation, American Cyanamid Company, W.R. Grace & Company, H.B. Fuller Company, The Dexter Corporation, and Ciba-Geigy Corporation. Some companies specialize in certain market segments (for example, automotive, general industrial, consumer, aerospace, or defense).

Cost Factors. Epoxy adhesives are formulated and priced differently for each of the various market segments. For the automotive industry, one- or two-component epoxy adhesives can range from $4 to $15/kg ($2 to $7/lb), based on moderate quantities. In the general industry segment, prices can be slightly more ($7 to $18/kg, or $3 to $8/lb). Epoxy adhesives in the aerospace industry range from $22 to $66/kg ($10 to $30/lb).

Competing Adhesives

The primary benefit of phenolic adhesives over epoxy adhesives is better thermal resistance. However, phenolics require an elevated-temperature cure, while epoxies often need only a room-temperature cure. Phenolics also give off volatiles when cured.

Conversely, polyurethanes do not offer the thermal stability that epoxies do, but as a class possess better flexibility. The presence of water may interfere with the cure of polyurethanes, while the presence of water in epoxy adhesives may accelerate their cure.

Acrylics, cyanoacrylates, and anaerobics all offer fast cure and good strengths relative to epoxies, but they are not cost-competitive with epoxies. In addition, these materials are generally not thixotropic, so they have limited gap-filling capabilities.

Silicones possess thermal resistance characteristics similar to those offered by phenolics, compared to those of epoxies. However, silicones typically have low shear strengths. Silicones are most often used as sealants because the required mechanical strengths are often lower in sealing applications.

Thermoplastics also have low strength in structural applications. In addition, they have lower creep resistance than epoxies, and some solvent-base systems may have poorer chemical resistance.

Advanced polymeric systems such as polyimides and polybenzimidazoles offer extremely high service temperatures for organic compounds (350 °C, or 660 °F). These systems require a long cure for optimal properties and are expensive when compared to epoxies.

Properties

Physical. Epoxy adhesives can be formulated to meet many physical property requirements. Table 2 shows typical data for three two-component epoxy adhesives. These consist of a general-purpose adhesive, a fast-setting adhesive, and a high-performance adhesive.

Although epoxy adhesives are typically brittle, they can be modified, as shown by the elongation of the high-performance adhesive. The table also shows that their gel time and cure schedule can be manipulated for fast or slow cures.

In addition to two-component epoxy adhesives, typical physical properties for one-component epoxy adhesives are shown in Table 3. In this table, epoxy film tape is compared with thermosetting hot-melt and epoxy paste adhesives. Because the hardener is already incorporated in the resin in one-component adhesives, their shelf lives are considerably lower than two-component adhesives.

As seen in Table 4, the shelf life at room temperature for one-component epoxy adhesives is limited. In relation to two-component epoxy adhesives, the stability for each component is often 1 year or more.

Finally, Table 4 shows typical physical properties for epoxy-syntactic adhesives. Partly because of the high levels of glass balloons present as filler in the syntactic adhesives, they can be expected to possess good sag resistance.

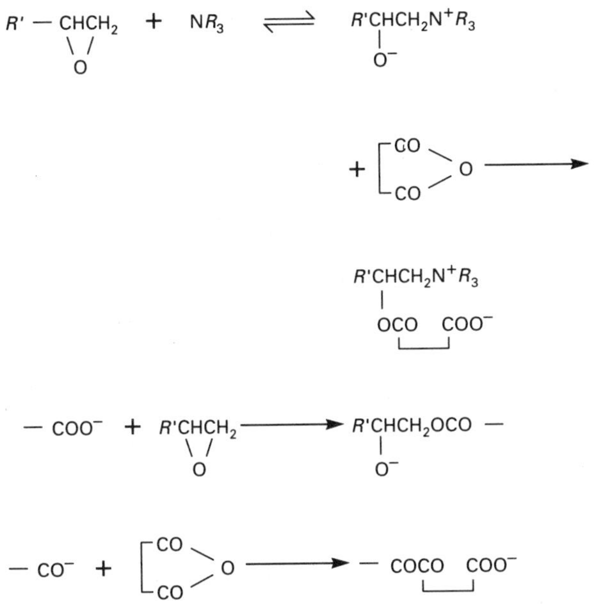

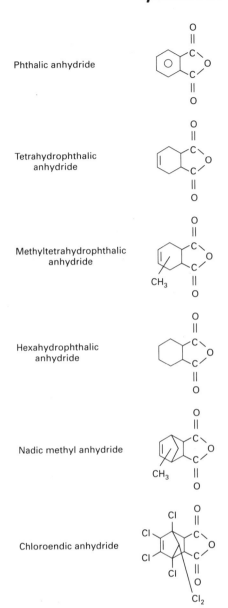

Fig. 14 Reaction of anhydrides with epoxies

Fig. 15 Commercial anhydride chemistry

Mechanical. As with the physical properties, a wide range of mechanical performance properties is available from epoxy adhesives. Table 5 shows typical mechanical properties of the same adhesives described in Table 3. As Table 6 shows, these three epoxy adhesives demonstrate good lap-shear strength to aluminum as well as to other substrates. Also evidenced is the good resistance to environmental exposure of epoxy adhesives.

The mechanical properties of the adhesives listed in Table 4 are shown in Table 6; again, initial lap-shear strength and some environmental data are listed. Because of their performance in the tensile and compression modes, syntactic adhesives are used. Data for both these tests are shown in Table 7, correlating with those in Table 4.

Additives and Modifiers

Fillers play an important role in adhesives performance. Besides the epoxy resin or the amine hardener, the filler often constitutes the greatest weight percentage used in an adhesives formulation.

The use of inorganic fillers (calcium carbonate, silica, and so on) lowers the cost of the adhesive. In addition, these fillers make the epoxy matrix less brittle and can contribute other physical properties (viscosity increase and thermal stability). The use of metallic fillers (iron, silver, copper, and aluminum) improves properties such as thermal conductivity, electrical conductivity, and system strength.

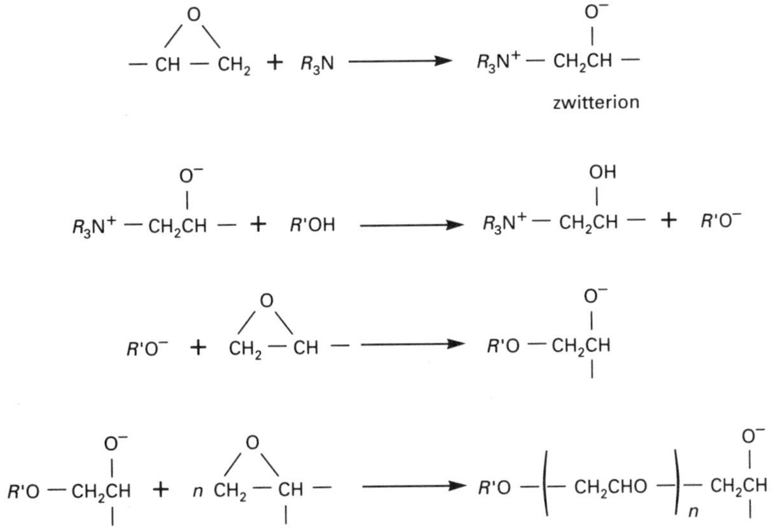

Thermoplastic fillers have also been incorporated into epoxide formulations, often to toughen the traditionally brittle epoxy adhesive. However, the cost of thermoplastic fillers is greater than inorganic fillers and may have a negative impact on overall cost.

In addition to unreactive thermoplastics used as toughening agents, epoxy adhesives can be modified with functional flexibilizers, as detailed in Table 2. Epoxidized polyurethanes can also be used as modifiers.

Colorants may be used in epoxy adhesive formulations. Pigments such as carbon black and titanium dioxide (white) may be used to color code resins and hardeners, and to let the end user know when mixing of the two components is complete (gray color).

Functional silanes are used as coupling agents to cohesively link the adhesive matrix with the substrate. They promote better

Fig. 16 Tertiary amines react with methylene carbon of epoxy group to form intermediate zwitterion

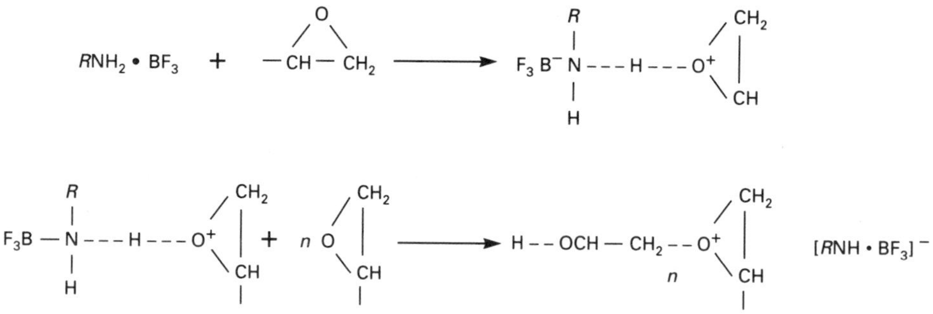

Fig. 17 Formation of oxonium ion after solvation of amine salt

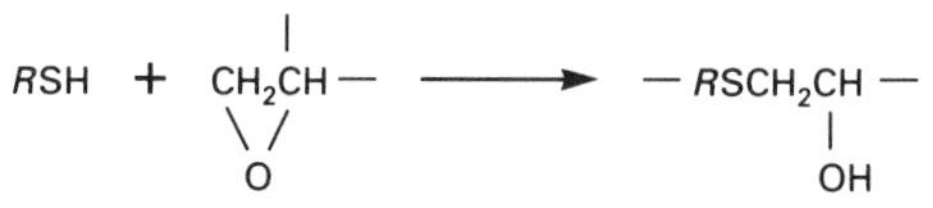

Fig. 18 Hydroxyl sulfide formed by thiols reacting with epoxies

overall strength and better adhesion to the substrate. Silanes can be added in bulk to the formulation or they can be bound to fillers that will react with the silanes (for example, silicas).

High-viscosity epoxy resins can also be "cut" with reactive diluents. These diluents are usually mono- or difunctional monomeric epoxides. Diluents can also serve as flexibilizers, though their performance is often inferior to the toughening agents already mentioned.

Design and Processing Parameters

Structural bonding applications for epoxy adhesives are numerous and varied. While it would be advantageous to be able to choose one adhesive for a variety of uses, the performance of the bonded part may not be optimal. Therefore, it is important to choose the adhesive appropriate to the application.

If a large part is to be bonded, or a large amount of adhesive is to be applied before joining substrates, a two-component adhesive with a short pot life would be inappropriate. To some extent automatic dispensing equipment can help by mixing the adhesive just prior to application.

The pot life of an epoxy adhesive can be as short as 5 min or as long as 90 min. The shelf life of each component in a two-component system is 6 months to 1 year or more. The shelf life of one-component adhesives ranges from 3 to 6 months. One-component epoxy adhesives can be used where large bond areas exist or long open times are necessary. However, the part must be placed in an oven to cure the adhesive and give the part its mechanical strength.

Some one-component epoxy adhesives must be frozen to prevent premature cure of the resin. Even at low temperatures, these

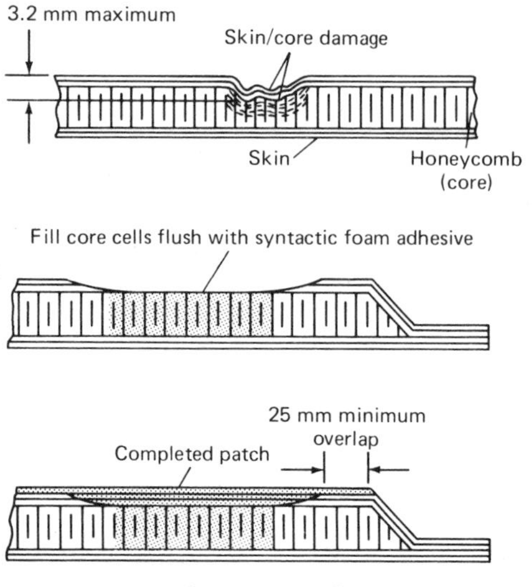

Table 1 Effects of resin modification

Flexibilizers	Concentration, %	Advantages	Disadvantages
Poly(propylene glycol) diglycidyl ether	10–60	Low viscosity Good flexibility Good abrasion resistance	Poor water resistance Fair impact resistance
Polyaminoamides	30–70	Good flexibility Good corrosion resistance	Fair chemical resistance
Liquid polysulfides	10–50	Excellent flexibility	Odor Poor heat resistance Tendency to cold flow
Aliphatic polyester adducts	10–30	Good water resistance Fair flexibility over a range of temperatures	High viscosity
Liquid butadiene-acrylonitrile copolymers	10–20	Good impact strength Good heat resistance	High viscosity

Table 2 Physical properties of two-component epoxy pastes

Property	General purpose	Fast setting	High performance
Color when mixed	Cream	Gray	Gray
Viscosity, Pa · s			
Resin	50	260	91
Hardener	35	160	103
Mixed	45	250	54
Mix ratio			
Weight/weight	100R/80H	100R/100H	100R/71H
Volume/volume	100R/100H	100R/100H	100R/100H
Specific gravity			
Resin	1.17	1.48	1.36
Hardener	0.92	1.44	0.97
Gel time	2 h	4 min	1 h
Elongation at break, %	. . .	. . .	11
Cure schedule	24 h/25 °C (77 °F) or 30 min/100 °C (212 °F)	4 h/25 °C (77 °F)	5 days/25 °C (77 °F) or 2 h/88 °C (190 °F)

R, resin; H, hardener

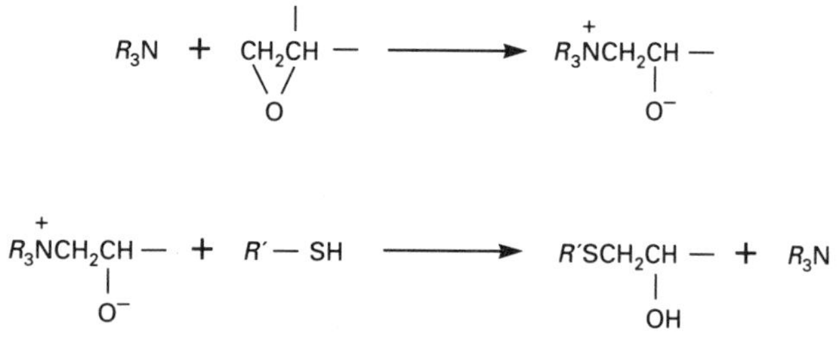

Fig. 19 Catalysis using tertiary amines

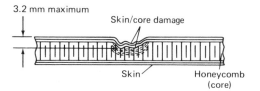

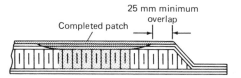

Fig. 20 Low-density epoxy adhesives used in composite panel assembly or repair

Table 3 Physical properties of one-component epoxy adhesives

Property	Epoxy-film tape	Thermosetting hot-melt	Epoxy paste
Appearance	Black film on carrier	Semirigid green solid	Red-brown paste
Viscosity	· · · ·	Extrude at 60–80 °C (140–175 °F)	300 Pa · s
Specific gravity	1.14	1.35	1.44
Elongation at break, % ..	· · · ·	1–2	7
Sag, mm (in.)	0 (2 mm, or 0.08 in., layer/200 °C, or 390 °F)	0 (6 mm, or 0.2 in., bead/171 °C, or 340 °F)	<3 (0.12) (6 mm, or 0.2 in., bead/171 °C, or 340 °F)
Cure schedule	24 h/25 °C (77 °F) or 30 min/100 °C (212 °F)	4 h/25 °C (77 °F)	5 days/25 °C (77 °F) or 2 h/88 °C (190 °F)
Shelf life	3 mo/25 °C (77 °F)	3 mo/25 °C (77 °F)	3 mo/25 °C (77 °F)

Table 4 Physical properties of epoxy syntactic adhesives

Property	One component	Two component
Appearance	White paste	Off-white paste
Viscosity		
Resin	· · ·	Paste
Hardener, Pa · s	· · ·	0.400
Mixed................	· · ·	Paste
Extrudability, 3.2 mm (0.125 in.) nozzle, 550 kPa (80 psi), cm³/min (in.³/min)	25 (150)	350 (21)
Specific gravity	1.0	0.7
Work life	8 h/25 °C (77 °F)	30 min/25 °C (77 °F)
Cure schedule	1 h/171 °C (340 °F)	24 h/25 °C (77 °F)
Shelf life...............	3–6 mo/−40 °C (−40 °F)	6 mo/25 °C (77 °F)

one-component systems have limited shelf lives of about 3 months.

As stated earlier, epoxies can bond a large number of substrates, though some result in more effective bonds than others. Epoxy adhesives are excellent candidates for bonding the substrates identified in Table 8.

Because most epoxies are brittle, the joint to be bonded should be designed to minimize a peel stress (Fig. 21). Most effective bonding for strength requires a bond line thickness less than 0.25 mm (0.010 in.).

When substrates are bonded and the part performs over a wide temperature range, it may be important to reduce stresses in the bond line by matching the coefficient of thermal expansion of the adhesive with that of the substrate.

Substrate preparation is as important in epoxy adhesive bonding as it is with all adhesives. Surfaces must be clean, abraded (or etched, in some cases), and usually primed. Glass and glass fibers are most often primed with silanes, while concrete, wood, and metals are treated with epoxy or phenolic primers. Primers are used not only to improve bonding and bond strengths, but also to improve environmental resistance of the joint. In metal bonding, primers are applied to cleaned surfaces to prevent reoxidation before the surface can be covered with adhesive.

The cure requirements of epoxy adhesives are largely dependent on the type of adhesive used. For example, two-component epoxy adhesives can cure at room temperature. An elevated-temperature postcure will often improve the properties of such an adhesive. Some two-component adhesives with high-temperature stability require a high-temperature cure for optimal performance. In a few cases, a heat cure should be avoided, as in the case of the fast-setting adhesive described in Table 2. For one-component (paste and film) and thermosetting hot-melt adhesives, an el-

Table 5 Mechanical properties of two-component epoxy pastes

Property	General purpose	Fast setting	High performance
Aluminum lap-shear strength, MPa (ksi)(a)			
At −60 °C (−75 °F)	20 (2.9)	10 (1.5)	29 (4.2)
At 25 °C (77 °F)	18 (2.6)	20 (2.9)	31 (4.5)
At 82 °C (180 °F)	<2 (0.3)	<8 (1.2)	18 (2.6)
At 121 °C (250 °F)	· · ·	· · ·	6.9 (1.0)
T-peel strength, N/mm (lbf/in.) ..	· · ·	· · ·	2 (11)
Thermal aging, MPa (ksi)	15 (2.2) (30 days/60 °C, or 140 °F)	21 (3.0) (30 days/60 °C, or 140 °F)	26 (3.8) (14 days/121 °C, or 250 °F)
Humidity aging, MPa (ksi)	12 (1.7) (40 °C, or 105 °F/92% RH)	7 (1.0) (54 °C, or 130 °F/95% RH)	20 (2.9) (54 °C, or 130 °F/95% RH)
Chemical resistance, MPa (ksi)			
Gasoline (90 days)	17 (2.4)	· · ·	· · ·
JP-4 (7 days)	· · ·	· · ·	34 (4.9)
Other substrate lap-shear strength, MPa (ksi), at 25 °C (77 °F)	23 (3.3) (copper)	· · ·	10 (1.5) (polyether-imide)

RH, relative humidity. (a) Aluminum lap-shear specimens tested according to ASTM D 1002, using chromic acid etched 2024-T3 aluminum

Table 6 Mechanical properties of one-component epoxy adhesives

Property	Epoxy film tape	Thermosetting hot melt	Epoxy paste
CRS lap-shear strength, MPa (ksi)(a)			
At −30 °C (−22 °F)	21 (3.0)	23 (3.3)	23 (3.3)
At 25 °C (77 °F)	16 (2.3)	20 (2.9)	22 (3.2)
At 82 °C (180 °F)	15 (2.2)	17 (2.5)	19 (2.8)
Salt spray endurance (500 h), MPa (ksi)	15 (2.2)	17 (2.5)	18 (2.6)

(a) CRS, cold-rolled steel (oily) lap-shear specimens tested according to ASTM D 1002

Table 7 Mechanical properties of epoxy syntactic adhesives

Property	One component	Two component
Aluminum lap-shear strength at 25 °C (77 °F), MPa (ksi)(a).............................	6.8 (0.99)	8.6 (1.2)
Compressive strength, MPa (ksi)		
At 25 °C (77 °F)	138 (20)	38 (5.5)
At 171 °C (340 °F)	69 (10)	· · ·

(a) Aluminum lap-shear specimens tested according to ASTM D 1002, using chromic acid etched 2024-T3 aluminum. Compressive test according to ASTM D 695

Table 8 Substrates for epoxy adhesive bonding

Metals
  Aluminum
  Chromium
  Magnesium
  Steel
  Titanium
  and so forth
Wood
Glass
  Fabric
  Plate
Ceramics
Polyesters
Rubber
Engineering plastics
Graphite
Concrete

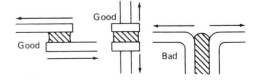

Fig. 21 Designs that minimize peel stress

evated temperature is necessary to cure the adhesive. Even with latent systems (Dicy), cures of 1.5 h or less at 170 °C (340 °F) are possible.

REFERENCES

1. S.A. Zahir, *Advances in Organic Coatings Science and Technology*, Vol IV, Technomic, 1982, p 83
2. F. Ricciardi, W. Romanchik, and M. Joullie, *J. Polym. Sci. A, Polym. Chem.*, Vol 21, 1983, p 1475

SELECTED REFERENCES

- D. Behm and E. Clark, High Performance Epoxy Adhesives, in *Proceedings of the 34th International SAMPE Symposium and Exhibition*, Society for the Advancement of Material and Process Engineering, May 1989
- H. Lee and K. Neville, Chapter 21, in *Handbook of Epoxy Resins*, McGraw-Hill, 1967
- C. May, Ed., Chapter 7, in *Epoxy Resins Chemistry and Technology*, 2nd ed., Marcel Dekker, 1988

Phenolics

Edwin F. Bushman, Consultant

PHENOLICS are condensation products of formaldehyde and phenol. Various phenols and/or aldehydes are sometimes substituted as reactants (Ref 1).

Phenolic resins were first formed in 1872 by A. Baeyer (Ref 2). The first phenolic resin patent was issued in 1899 to A. Smith (Ref 3). However, it was not until 1905 that L.H. Baekeland, who is considered to be the originator of the phenolic resin industry, entered the field (Ref 4). In 1910, Baekeland started the Bakelite Company, which became a division of the Union Carbide Company in 1939 (Ref 5).

Today, phenolics continue to dominate the wood adhesives (plywood) field and represent one of the largest volumes of any synthetic adhesive. They are also some of the lowest-costing adhesives. Phenolics can be formulated as water dispersions that can then penetrate wood cell structures, giving excellent formation of permanent bonds. Phenolics have also found applications in abrasives and foundry product lines. Other uses, which are outside the scope of this article, include phenolic molding compounds and coatings (Ref 6).

Chemistry

Numerous substituted phenols may be selected or combined to react with several aldehydes in the basic phenol formaldehyde reaction. Key conditions affecting type, processing, and application of the resin produced are the molar ratio of phenol to aldehyde; the presence of catalyst; reaction pH, temperature, and time conditions; and water presence.

Novolacs. Phenolics formed in two steps are called novolacs. A novolac is a thermoplastic phenol-formaldehyde resin formed with an excess of phenol.

The mole ratio of formaldehyde to phenol is always less than 1. The initial condensation reaction occurs when an excess of phenol is reacted with formaldehyde in the presence of an acid catalyst such as sulfuric acid. Figure 1 shows the structure of the phenol-terminated novolac, which has a typical degree of polymerization (n) of 5 to 6 (Ref 7). Novolac materials are relatively stable to molecular weight advancement because they contain no active methylol groups.

The two-step novolac resin is not heat reactive until a curing agent, such as hexamethylene tetramine (HMTA, or "hexa") is added. HMTA, the structure of which is indicated in Fig. 2, decomposes to form ammonia (NH_3) and formaldehyde. In the presence of the ammonia catalyst and possibly additional formaldehyde, a cured thermoset is formed.

Resols. Phenolics formed in one step are called resols. A resol is a thermosetting phenol-formaldehyde resin formed with an excess of formaldehyde.

Resols do contain active methylol groups and always tend to advance to a cured state. Resols are usually formed when an excess of formaldehyde is reacted with phenol in the presence of an alkaline catalyst such as sodium hydroxide or ammonium hydroxide. Figure 3 shows the structure of the methylol (or CH_2OH)-terminated resol, which initially has a degree of polymerization of 1 to 3 (Ref 7). The one-step resol resin is heat reactive and in most cases will cure to a thermoset structure when heated.

The resins based on parasubstituted phenols may be the one-step or two-step type, but are not capable of cross linking to a thermoset state. They find use as tackifiers in contact, pressure-sensitive, and hot-melt adhesives. The nonsubstituted phenolic resins are used as adhesives and in laminating applications.

Phenols and Related Monomers. Table 1 summarizes the physical properties of various phenols and related monomers. A range of substituted phenols, such as cresols, xylenols, butyl phenol, and so on, may be combined with phenol, in cross-linking reactions with formaldehyde, furfural, paraformaldehyde, or other aldehydes to produce resins with wide-ranging properties. Resorcinol differs from phenol by the addition of one hydroxyl group. The reactivity with aldehydes is greatly increased, so that room-temperature curing is possible without additional catalysts, although alkaline catalysts are sometimes used.

Formaldehyde and Related Monomers. Formaldehyde is produced by the dehydrogenation of methanol over a metal catalyst. U.S. commercial formaldehyde solutions contain 37% parts by weight (pbw) formal-

Fig. 1 Novolac prepolymer. n = 5 to 6

Fig. 2 Hexamethylene tetramine hardener used with novolac formulations

Fig. 3 Resol prepolymer. n = 1 to 3

dehyde in water and an additional 6 to 15% methanol to prevent precipitation. European formulas for formaldehyde (Formalin or Formol) are prepared by adding 40 g of formaldehyde to 100 cc of water at room temperature. Table 2 summarizes the physical properties of various aldehydes. Formaldehyde (HCHO) is the base member of the series of aliphatic aldehydes.

Furfural, another widely used aldehyde, is commercially derived from residues of corn cobs, bagasse, or rice hulls by means of sulfuric acid and steam distillation. Furfural, in combination with formaldehyde, is used to prepare resins for use in grinding and friction materials.

Furfural alcohol is derived from furfural by dehydrogenation. Furfural alcohol or phenol formaldehyde resin blends and acidic catalysts are used in preparing acid-resistant cements and mineral sand binding for

Table 1   Physical properties of various phenols and related monomers

Material	Phenol or monomer	Molecular weight	Melting point °C	°F	Boiling point °C	°F	pKₐ, at 25 °C (75 °F)(a)
Phenol	Hydroxybenzene	94.1	40.9	105.6	181.8	359.2	10.00
o-cresol	1-methyl-2-hydroxybenzene	108.1	30.9	87.6	191.0	375.8	10.33
m-cresol	1-methyl-3-hydroxybenzene	108.1	12.2	54.0	202.2	396.0	10.10
p-cresol	1-methyl-4-hydroxybenzene	108.1	34.7	94.5	201.9	395.4	10.28
p-tert butylphenol	1-tert-butyl-4-hydroxybenzene	150.2	98.4	209.1	239.7	463.5	10.25
p-tert octylphenol	1-tert-octyl-4-hydroxybenzene	206.3	85	185	290	554	. . .
p-nonylphenol	1-nonyl-4-hydroxybenzene	220.2	. . .	. . .	295	563	. . .
2,3-xylenol	1,2-dimethyl-3-hydroxybenzene	122.2	75.0	167.0	218.0	424.4	10.51
2,4-xylenol	1,3-dimethyl-4-hydroxybenzene	122.2	27.0	80.6	211.5	412.7	10.60
2,5-xylenol	1,4-dimethyl-2-hydroxybenzene	122.2	74.5	166.1	211.5	412.7	10.40
2,6-xylenol	1,3-dimethyl-2-hydroxybenzene	122.2	49.0	120.2	212.0	413.6	10.62
3,4-xylenol	1,2-dimethyl-4-hydroxybenzene	122.2	62.5	144.5	226.0	438.8	10.36
3,5-xylenol	1,3-dimethyl-5-hydroxybenzene	122.2	63.2	145.8	219.5	427.1	10.20
Resorcinol	1,3-dihydroxybenzene	110.1	110.8	231.4	281.0	537.8	. . .
Bisphenol A	2,2-bis(4-hydroxyphenyl)propane	228.3	157.3	315.1	. . .	. . .	. . .

(a) pKₐ is the negative logarithm (to base 10) of equilibrium constant, Kₐ, for reaction. pKₐ is used to express extent of disassociation or strength of weak acid. The weaker the electrolyte, the larger the pKₐ value; for example, H_2SO_4 (strong acid), pKₐ = 3; acetic acid (weak acid), pKₐ = 4.76; boric acid (very weak acid), pKₐ = 9.24. Source: Ref 1

foundry no-bake and hot-box core-making processes.

Phenolic Copolymers. Resorcinol formaldehyde, furfural alcohol and the amino resins, urea formaldehyde and melamine formaldehyde are frequently copolymerized or blended with the phenolic-type formaldehyde resins in industrial adhesive, laminating, rubber tire, bonding, sizing, fiber, paper, and coating formulations. Other phenolic copolymers include acrylic rubber phenolics, neoprene phenolics, nitrile phenolics, and silicone phenolics.

Epoxy-Novolac Base Phenolics. The bisphenol A-base epoxy resins are frequently limited in heat resistance for certain applications. Thus, epoxy-novolacs are used as modifiers.

Commercial Forms

Phenolics are commercially available in hundreds of formulations having various additives and modifiers. Typical formulations are discussed in the article ''Adhesive Modifiers'' in this Volume. In general, phenolics are available as novolacs or resols.

Novolacs are relatively stable and can be stored as liquids (solid novolac with solvent), powders, flakes, or lumps. Most are blended with a curing agent such as HMTA prior to packaging. Subsequent application of heat and pressure results in full cure.

Resols are available as solids or liquids. During the initial preparation of these materials, reaction temperatures must be carefully controlled in order to prevent premature curing. As with novolacs, heat and pressure are applied to cure fully. In addition to heat and pressure, catalysts may also be required to complete the curing process.

Dispersion resols are prepared in their hydroxymethyl form as water or solvent-base solutions. These materials must have a low molecular weight in order to have solution stability. Dispersion resols are usually prepared in two steps. An alkali or alkaline earth metal hydroxide catalyst is used to achieve molecular weight near the water insolubility point. Then, a polysaccharide gum is added as a protective colloid coating to control particle size and distribution. The dispersion has an average molecular weight of 800 to 46 000 g/mol (Ref 5).

Major Suppliers

Producers of phenolic resin adhesives, which include phenol formaldehyde, phenol formaldehyde/resorcinol formaldehyde, phenolics based on modified phenols and/or aldehydes, and phenolic/epoxy-novolacs, are many. Among the largest suppliers are Adhesive Products Corporation; Adhesives Research, Inc.; American Cyanamid Company, Polymer Products Division; Ashland Chemical Company; Borden, Inc., Packaging and Industrial Products; The Dexter Corporation, Adhesives and Structural Materials Division; Ferro Corporation, Composites Division; H.B. Fuller Company; Furane Products Company; Ciba-Geigy Corporation; BF Goodrich Company, Adhesives Division; Narmco Materials, Inc.; and 3M Company, Adhesives, Coating and Sealers Division.

Dow Chemical Company and Ciba-Geigy Corporation, suppliers of phenol and cresol-base epoxy-novolac resins, often purchase novolacs from other primary suppliers, such as Schenectady Chemicals, Inc.; Borden, Inc.; and others. The term modified epoxy labels many proprietary epoxy adhesives, prepregs, and film adhesives. It is often difficult to identify commercial epoxy-novolac material in proprietary epoxy adhesive formulations.

Suppliers of acrylic rubber phenolic adhesives include Adhesive Products Corporation; The Dexter Corporation, Adhesives and Structural Materials Division; H.B. Fuller Company; Industrial Adhesives Inc.; Pemco Adhesives, Inc., and Products Research and Chemical Corporation.

Neoprene phenolics and nitrile phenolics are produced by Armstrong World Industries Inc.; H.B. Fuller Company; National Starch and Chemical Corporation; PPG Industries, Inc.; Sherwin-Williams Company; 3M Company, Adhesives, Coating and Sealers Division; and Uniroyal Inc./Uniroyal Plastic Products, Inc. Additional suppliers of neoprene phenolics include Hernon Manufacturing Inc. and Industrial Adhesives Inc. Suppliers of nitrile phenolics also include: American Cyanamid Company; The Dexter Corporation, Adhesives and Structural Materials

Table 2   Physical properties of various aldehydes and related monomers

Type	Formula	Melting point °C	°F	Boiling point °C	°F
Formaldehyde	CH₂=O	−92	−134	−21	−6
Acetaldehyde	CH₃CH=O	−123	−189	20.8	69.4
Propionaldehyde	CH₃—CH₂—CH=O	−81	−114	48.8	119.8
n-butyraldehyde	CH₃(CH₂)₂—CH=O	−97	−143	74.7	166.5
Isobutyraldehyde	(CH₃)₂CH—CH=O	−66	−87	61	142
Glyoxal	O=CH—CH=O CH—CH H	15	59	50.4	122.7
Furfural	CH C—C=O ‖ ‖ ‖ \/ O	−31	−24	162	324

Source: Ref 1

Division; and BF Goodrich Company, Adhesives Division.

Suppliers of phenolic silicones include Adhesive Products Corporation; Fielco Chemical Corporation; Northern Products Inc.; Precision Adhesives, Inc.; and Research Sales, Inc. (Ref 8).

Major Markets and Applications

The percentage of annual phenolic output of 2.8 billion pounds that is allocated to finite adhesive end-use is difficult to pinpoint because phenolics have so many uses in which their adhesive properties are important to the end-use. Table 3 lists phenolic adhesive applications and consumption in 1989. This table does not include consumption data on phenolics used as contact or structural adhesives.

Wood, cellulosic fibers, synthetic fibers and thermoset phenolic and amino adhesive resins are routinely combined to make enormous quantities of plywood, particleboard, chipboard, wafer board, oriented strand board, hardboard and papers in the industrial countries. (The covers and pages of this book have such a source).

In 1979 softwood plywood output in the United States was 18.6 billion ft^2 (3/8 in. basis), hardwood plywood output was 1.14 billion ft^2, and particleboard output was 3.54 billion ft^2 (3/4 in. basis). The press equipment and throughput is impressive. Oriented strand board (OSB), a phenolic-wood sliver panel, is pressed in big sizes, as large as 2.6 × 9 m (8½ × 30 ft) untrimmed, or 24 × 8.5 m (8 × 28 ft) trimmed, with 10 mm (3/8 in.) and 19 mm (3/4 in.) thicknesses.

The strands are loaded and oriented in wire trays and then coated with 2 wt% resol phenolic in a spray-dry process (some mills use powdered novolacs). The strands are automatically loaded into the 12 press openings, pressed, and then automatically removed and trimmed.

The American National Standards Institute standard for mat-formed wood particleboard, ANSI A208.1, classifies particleboard by type of use; density (grade); and class, the voluntary standard for the industry. The National Particleboard Association standard limits formaldehyde to an emission limit of 0.3 parts per million (ppm) formaldehyde in air at a product loading of 0.13 ft^2 of board feet per cubic foot of room space. This is the same standard formaldehyde emission limit that is given in the U.S. Department of Housing and Urban Development 1984 regulation covering particleboard use in the construction of manufactured homes.

Youngquist and Rowell of the Forest Laboratory, U.S. Department of Agriculture, have been researching business and material opportunities for combining wood with nonwood materials. They have con-

Table 3 Phenolic adhesive use in 1989

Application	Consumption 10^6 kg	Consumption 10^6 lb
Coated and bonded abrasives	11	24
Wood, fibrous and granulated	122	269
Friction materials	18	40
Foundry and shell moldings	22	49
Insulation materials	235	518
Laminating	112	246
Plywood	738	1627
Total	**1258**	**2773**

Source: Ref 9

cluded that new composites can have performance characteristics that are superior to those of several components alone. In their example of wood fibers and glass fibers united by thermosetting phenol formaldehyde resin adhesive, several compositions and properties of pressed panels were evaluated.

The reduction of long curing times has been under study at facilities worldwide. Resin modification, catalysis, and steam preheating of chips or wafers have reduced cure rates. The radio frequency curing of veneer ply layup glue lines has been reported, as well as inductive heating techniques.

Phenolics are used in structural adhesives to bond wood, metals, plastics, and metal to paper. Typical adhesives are based on nitrile phenolics, neoprene phenolics, epoxy phenolics, and vinyl phenolics (Ref 6).

Resorcinols have long been used in structural wood bonding. Current increased interest in fire resistance, low smoke and toxic-gas emission and other environmental regulations, are favoring the increased use of resorcinol-modified phenolic resins as adhesives and matrix resins for structural composites.

In the early 1940s, cotton tire cord gave way to rayon, which was followed by nylon, polyester, glass, aramid fibers, and finally steel wire. Tire cord adhesion was enhanced for all these specialty tire reinforcements by resorcinol-formaldehyde bonding systems, in combination with vinyl pyridine copolymer latex.

Phenolics in the form of substituted novolacs or resols are used in combination with neoprene as contact adhesives. In the same way, nonsubstituted resols are used in combination with nitriles in the same contact adhesive applications. The function of the phenolic resin in these systems is primarily as a tackifier and adhesion promoter. Materials that are bonded with contact adhesives include leather, cloth, plastic, and rubbers. Contact adhesives are discussed in greater detail in the article "Silicones" in this Volume.

Powdered novolacs are usually used as binders to bond aluminum oxide and silicon carbide for applications as grinding

Table 4 Cost of phenolics in 1989

Material	Cost
Monomers	
Phenol	$0.42–0.44/lb(a)
Formaldehyde	$0.11–0.12/lb(a)
Phenolic adhesives	
Phenol formaldehyde resins and industrial resins	$0.90–1.40/lb
Phenolic film adhesive (low smoke, 1250 mm, or 50 in., wide)	$0.40–0.45/ft^2(b)
Other phenolics	
Phenol formaldehyde resol solution, 60% in alcohol (MIL-R-9299C approved)	$1.16/lb(c), or $1.12/lb(d), or $1.07/lb(a)
Phenol formaldehyde powders	$0.70–0.76/lb(a)
Phenol formaldehyde resin prepreg (1525 mm, or 60 in., wide, MIL-R-9299C approved)	$33.00/linear yard(e) or $27.00/linear yard(f)

(a) Cost per truckload. (b) For applications involving 2500 to 50 000 ft^2 total area. (c) Cost per 4 to 9 drums. (d) Cost per 70 drums. (e) For a 200 linear yard lot. (f) For a 1000 linear yard lot. Source: Ref 11

wheels. Phenolics used as binders in friction materials are based on novolacs, resols, novolac-resols, cresol resins, rubber-modified novolacs, and oil-modified novolacs. Most applications are in the automobile industry. These include brake liners, brake blocks, disk pads, clutch facings, and auto transmission blades. The materials are typically filled with asbestos, minerals, carbon, aramid, metal fibers, or a combination thereof.

Coated abrasives including sandpaper, sanding disks, and belts are usually bonded using resols. Abrasives that are typically applied to paper backings include diamond, silicon carbide, aluminum oxide, boron carbide, boron nitride, emery, quartz, and garnet.

Another application for phenolics is in binding sand used to produce foundry shell molds and cores. Novolacs available as powders, flakes, granules, and solvent- or water-base systems are typically used in shell molding.

Phenolics, usually in the form of resols, are used as binders in electrical and industrial laminates and in decorative laminates. Specific applications of the latter include furniture, decorative papers, and wood veneers (Ref 6).

Phenolics have also been used by the aerospace industry for both commercial and military applications. A typical application on the exterior of aircraft is the use of phenolics for finished skins. Although the volume of resin used is low, phenolic skins are a major capital expense. Inside the aircraft, a growing and large volume application is the use of phenolics for aircraft interiors.

Phenolics are also used in ablative areas (as ablative agents or materials) of rocket motor nozzles as well as for reentry vehi-

Table 5 Properties of a phenol formaldehyde resin adhesive solution for the production of particle board

Characteristic	Phenol resol, water solution
Dry solids, %	45 ± 1
Viscosity at 20 °C (68 °F); MPa · s........	130 ± 20
Specific gravity at 20 °C (68 °F)	1.17 g/cm³
Alkali content (solution), %..............	5
Water dilutability, ratio.................	∞ ratio
Shelf life at 20 °C (68 °F), weeks	3–4
B-time at 100 °C (212 °F), min(a)	20
Free formaldehyde, %...................	<0.1
Free phenol, %........................	<0.1

(a) B-time, time required to obtain B-stage (stage at which adhesive is partially cured) Source: Ref 13

Table 6 Rigidity of typical phenolic adhesives

Adhesive	Modulus of elasticity MPa	Modulus of elasticity ksi	Shear modulus of rigidity MPa	Shear modulus of rigidity ksi	Poisson's ratio
Modified phenolic	3450	500	1270	184	0.360
Epoxy phenolic	2725	395	1100	160	0.294
Vinyl phenolic	2240	325	805	117	0.385

Source: Ref 12

cles. In addition, phenolics are used in ballistic or armor applications. Carbon-carbon composites are also based on phenolics.

Cost Factors

The cost of unmodified phenolics is fairly low and very competitive with other adhesives. Copolymerization with neoprene, nitrile vinyl, or epoxy will increase the cost, but the high quality and general availability of phenolics make them the appropriate choice and even the standard for many applications (Ref 10). Table 4 summarizes the cost of phenolic monomers, adhesives, and related materials.

Properties

The largest application of phenolic adhesives is in wood product bonding. The phenolic contributes outdoor-weathering, moisture, heat, and microorganism resistance to the wood. Usually, bonded products are stronger and more durable than the wood substrates (Ref 12). Table 5 lists properties of a phenol formaldehyde resin adhesive solution for the production of particleboard.

In general, adhesives based upon phenolic or resorcinol resins have excellent resistance to moisture, oils, hydrocarbons, and solvents. Phenolics copolymerized with vinyls, neoprenes, nitriles, or epoxies also have excellent resistance to moisture, oil, hydrocarbons, and most solvents (Ref 14).

Phenolic adhesives have good mechanical properties including high tensile lap-shear strengths. Table 6 indicates that phenolics are moderately rigid. However, peel strengths are not very good, which is typical of most thermosets. Copolymerization with rubbery materials such as neoprene or nitrile improves peel properties. Phenolics also have good elevated-temperature resistance, electrical properties, and dimensional stability.

Table 7 summarizes typical properties of various phenolic adhesives that are used in structural applications. Table 8 compares tensile lap-shear strengths of phenolic and other high-temperature adhesives bonded to stainless steel.

Peel strength is increased by the copolymerization of phenolics with elastomers such as nitriles. Peel strengths of a nitrile phenolic adhesive are found in Table 9. These adhesives are very tough and durable. Even after 30 months at 120 °C (250 °F), the nitrile phenolic adhesive had a greater tensile lap-shear strength (12 MPa, or 1750 psi) than that of an epoxy phenolic (4.8 MPa, or 700 psi). After 30 months at 175 °C

Table 7 Typical properties of various structural phenolic adhesives

Property	Neoprene phenolic	Nitrile phenolic	Vinyl phenolic	Epoxy phenolic
Forms......................	Tapes	Tapes, pastes	Emulsions, tapes, separate components(a)	Tapes, pastes
Typical cure conditions	40–60 min at 150–230 °C (300–450 °F) with 0.17–3.5 MPa (25–500 psi) contact pressure	From 1–2 h at 150 °C (300 °F) to 8 min at 230 °C (425 °F) with 0.07–1.4 MPa (10–200 psi) contact pressure	Emulsions: 2–6 min at 60–80 °C (140–180 °F); contact pressure to 18 kg/linear cm (100 lb/linear in.) Nonemulsions: 15–30 min at 150 °C (300 °F); contact pressure to 0.70 MPa (100 psi)	40–60 min at 160–175 °C (325–350 °F) with 0.70 MPa (100 psi) contact pressure
Tensile lap-shear strength, MPa (ksi)(b)				
At −55 °C (−70 °F).........	34.5–41.4 (5–6)	22.1–32.7 (3.2–4.6)	15.2–20.5 (2.2–3.0)	17.2–20.5 (2.5–3.0)
At 25 °C (75 °F).............	17.2–24.1 (2.5–3.5)	22.1–32.7 (3.2–4.6)	15.2–20.5 (2.2–3.0)	17.2–20.5 (2.5–3.0)
At 80 °C (180 °F)...........	···	···	6.7–10.3 (1.0–1.5)	···
At 150 °C (300 °F)...........	···	10.3–17.2 (1.5–2.5)	···	10.3–12.4 (1.5–1.8)
Heat resistance, °C (°F)	80–95 (180–200)	260 (500) intermittent	80 (180)	>260 (500)
Applications	Metal-to-metal bonding; bonding of rubbers, ceramics, glasses, thermosets	Bonding of metals, plastics, rubbers, and frictional materials; bonding of honeycomb	Bonding of honeycomb to paper, metal, or fibrous glass; bonding foams or porous materials to metals	Metal-to-metal bonding; bonding of honeycomb

(a) Liquid phenolic plus powder polyvinyl formal. (b) Aluminum to aluminum. Source: Ref 15

Table 8 Tensile lap-shear strengths of high-temperature adhesives bonded to stainless steel

Adhesive	Tensile lap-shear strength At 25 °C (75 °F) MPa	At 25 °C (75 °F) ksi	At 65 °C (150 °F) MPa	At 65 °C (150 °F) ksi	At 120 °C (250 °F) MPa	At 120 °C (250 °F) ksi	At 175 °C (350 °F) MPa	At 175 °C (350 °F) ksi	At 205 °C (400 °F) MPa	At 205 °C (400 °F) ksi
Nitrile phenolic	34.5	5.0	31.0	4.5	24.8	3.6	13.1	1.9		0
Polybenzimidazole	20.7	3.0	19.7	2.85	19.0	2.75	17.9	2.6	16.5	2.4
Epoxy phenolic	19.7	2.85	18.3	2.65	15.9	2.3	13.8	2.0	12.1	1.75
Polyimide	17.9	2.6	16.2	2.35	14.5	2.1	12.0	1.75	10.3	1.5
Epoxy	15.9	2.3	12.8	1.85	9.7	1.4	6.2	0.9	1.0	0.15

Source: Ref 16

Table 9 Peel strengths of nitrile phenolic adhesive as a function of temperature

Temperature		Metal-to-metal climbing drum peel test		T-peel test	
°C	°F	kg/m	lb/in.	kg/m	lb/in.
−55	−70	400	22.5	70	4
25	75	1610	90	805	45
80	180	1305	73	450	25
120	250	1130	63	305	17

Source: Ref 17

Table 10 Tensile lap-shear strengths of glass-phenolic laminate bonded to 301 stainless steel with nitrile phenolic adhesive

	At 25 °C (75 °F)		At 120 °C (250 °F)		At −55 °C (−70 °F)	
Number of days	MPa	psi	MPa	psi	MPa	psi
0	23.8	3450	10.0	1450	23.1	3350
10	21.7	3150	5.9	850	22.1	3200
20	20.0	2900	4.1	600	20.1	2920
30	19.3	2800	4.0	580	19.7	2850
50	18.5	2680	3.1	450	19.3	2800
70	18.3	2650	2.6	380	19.2	2780
90	17.9	2600	2.1	300	18.6	2700

(a) Per MIL-E-5272. Source: Ref 16

Table 12 Formulations of extended, filled phenol formaldehyde resin adhesives for wood veneer plywood production

Material	Composition, pbw(a) 1	2	3
40–45% solids PF resin	100	100	100
300 mesh coconut (or walnut) shell flour	12	14	10
Industrial wheat flour	6	...	...
50% sodium hydroxide	2	...	...
Surfactant	0.1	0.1	0.1
Water	10	5	...
Total	130.1	120.1	110.1
Phenolic solids, %	31–35	33–37	36–41

(a) pbw, parts by weight. Source: Ref 18

Table 11 Several phenol formaldehyde resin adhesive formulations for hot pressing weather-resistant plywood

Formulation and processing(a)	Dry process	Wet process		
Phenol resin, pbw	100	100	100	100
Chalk, pbw	...	10	10	10
Coconut flour (300 mesh), pbw	5–10	8	8	8
Water, pbw	10–20	...	...	...
Hardener (paraformaldehyde), pbw	...	...	2	2
Accelerator (resorcinol resin), pbw	...	...	...	6
Glue service life, h	24	24	24	4
Press temperature, °C (°F)	130–150 (265–300)	130–140 (265–285)	120–125 (250–260)	100–110 (210–230)

(a) pbw, parts by weight. Source: Ref 13

Table 13 Typical neoprene phenolic contact adhesive formulation

Material	Composition, pbw
Neoprene	100
Antioxidant	2
Magnesium oxide	4
Zinc oxide	5
Phenolic resin	45
Magnesium oxide	4
Water	1
Toluene	115
Hexane	115
Ethyl acetate	115
Total	506

pbw, parts by weight. Source: Ref 6

Table 14 Typical nitrile phenolic structural adhesive formulation

Material	Composition, parts per hundred		
Nitrile rubber	100	100	100
Phenolic resin	50	80	100
Zinc oxide	5	5	5
Sulfur	1.5	1.5	1.5
Benzothiazole disulfide	1.5	1.5	1.5
Stearic acid	1.5	1.5	1.5

Source: Ref 17

(350 °F), the nitrile phenolic had a tensile lap-shear strength of 4.8 MPa (700 psi) while the epoxy phenolic had failed (Ref 17).

As indicated in Table 10, tensile lap-shear strengths of nitrile phenolic adhesives at −55 °C (−70 °F) and 25 °C (75 °F) are fairly constant regardless of exposure to humidity for up to 90 days. However, at 120 °C (250 °F), a decrease in strength occurs as a function of days of humidity exposure (Ref 17).

Additives and Modifiers

Table 11 lists several phenol formaldehyde resin adhesive formulations for hot pressing weather-resistant plywood (Ref 13). Table 12 summarizes formulations of extended, filled phenol formaldehyde resin adhesives for wood veneer plywood production. A typical neoprene phenolic contact adhesive formulation and several typical nitrile phenolic structural adhesive formulations are listed in Tables 13 and 14, respectively. Various additives and modifiers used in phenolic adhesive formulations are summarized in these tables.

REFERENCES

1. A. Knop and L. Pilato, Chapter 2 in *Phenolic Resins: Chemistry, Applications and Performance, Future Directions*, Springer-Verlag, 1985
2. A. Baeyer, *Ber. Dtsch. Chem. Ges.*, Vol 5, 1872, p 25
3. B.P. Barth, Chapter 23 in *Handbook of Adhesives*, 2nd ed., I. Skeist, Ed., Van Nostrand Reinhold, 1977
4. L.H. Baekeland, *J. Ind. Eng. Chem.*, Vol 1, 1909, p 149
5. *Mod. Plast.*, Vol 62, 1985, p 67
6. F.L. Tobiason, Chapter 17 in *Handbook of Adhesives*, 3rd ed., I. Skeist, Ed., Van Nostrand Reinhold, 1990
7. J.A. Brydson, Chapter 23 in *Plastics Materials*, 4th ed., Butterworth Scientific, 1982
8. A.B. Barker, Ed., *Adhes. Age*, Vol 33 (No. 6), 1990, p 86
9. *Mod. Plast.*, Vol 67 (No. 1), 1990
10. C.V. Cagle, Chapter 19 in *Handbook of Adhesive Bonding*, C.V. Cagle, Ed., McGraw-Hill, 1973
11. *Plast. News*, March 6, 1990
12. R.F. Blomquist, Chapter 17 in *Handbook of Adhesive Bonding*, C.V. Cagle, Ed., McGraw-Hill, 1973
13. A. Knop and L. Pilato, Chapter 11 in *Phenolic Resins: Chemistry, Applications and Performance, Future Directions*, Springer-Verlag, 1985
14. B. Fader, Chapter 13 in *Handbook of Adhesive Bonding*, C.V. Cagle, Ed., McGraw-Hill, 1973
15. W.H. Guttmann, *Concise Guide to Structural Adhesives*, Reinhold Publishing, 1961, p 7-9
16. C.V. Cagle, Chapter 14 in *Handbook of Adhesive Bonding*, C.V. Cagle, Ed., McGraw-Hill, 1973
17. N.J. DeLollis, Chapter 5 in *Adhesives for Metals—Theory and Technology*, Industrial Press, 1970
18. A. Pizzi, *Wood Adhesives: Chemistry and Technology*, Marcel Dekker, Inc., 1983

Urethanes

Cynthia L. Kreider, Polymerland Inc., a wholly owned subsidiary of General Electric Company

URETHANE-BASE ADHESIVES belong to a broad class of polymers that have in common interpolymeric links of carbamate esters known as urethanes. Polyurethanes are not made by polymerizing urethane units but, rather, are made by addition polymerization of a diisocyanate or polyisocyanate with a hydroxy(OH)-terminated polymer such as alcohol, polyol, polyester polyol, or polyether polyol.

The industrial synthesis of a simple urethane can be expressed as shown in Fig. 1. Polyurethane polymerization, accomplished by an addition reaction, requires at least difunctional monomers (Fig. 2).

Chemistry

Carbamate ester, or urethane, is the monoester/monamide of carbonic acid, as shown in Fig. 3. Carbonic acid can be formed in a solution of water and carbon dioxide. Carbonate esters (the building blocks of polycarbonate), carbamate esters (urethanes), and ureas are all derivatives of carbonic acid (Fig. 4).

Polyurethane adhesives are primarily synthesized using diphenylmethane-4,4'-di-isocyanate (MDI) or toluene diisocyanate (TDI) and polyols (hydroxyl-containing compounds). Isocyanates are synthesized by the reaction shown in Fig. 5, where R can be an aliphatic or aromatic group. Aromatic isocyanates are normally preferred for specialty polyurethane adhesive production because they are more reactive and more economical. An example of an aromatic isocyanate is shown in Fig. 6.

Polyurethane adhesive formulations use higher molecular weight and polymeric isocyanates as cross linkers. For example, TDI adducts polymeric MDI triphenylmethane-4,4',4"-triisocyanate and thiophosphoric acid-tris-(p-isocyanato-phenylester). These allow the attainment of three-dimensional cross-linked structures. Commonly used isocyanates for the cross-linking agents used in polyurethane adhesive formulations are listed in Table 1. Examples of polyols (hydroxy-functional polyesters and polyethers) are given in Fig. 7 and 8.

Handling. Both toluene diisocyanate (TDI) and diphenylmethane-4,4'-diisocyanate (MDI) should be handled carefully for several reasons. Because they react readily with water (see Fig. 9) as well as with alcohols, appropriate laboratory and production precautions should be taken. The threshold limit value for diisocyanates is 0.02 ppm, and the Occupational Safety and Health Administration has set a 20 ppb limit for an 8-h time-weighted average (TWA). Toluene diisocyanate, which has been studied extensively, has been determined to be a respiratory irritant and is a suspected carcinogen. Because of its reactivity with water, bioaccumulation is negligible. However, isocyanates are eye and skin irritants and may cause allergic reactions, asthma, bronchitis, emphysema, bronchopneumonia, and growth retardation. In polyurethanes, the isocyanate has been reacted to form the urethane group and it is therefore no longer present, unless polyisocyanates are added as cross-linking agents. Polyisocyanates have a much reduced volatility, compared to MDI and TDI, and can be safely handled when recommended precautions are observed.

Morphology and Reaction Intermediates. Polyurethane adhesives are generally composed of two phase-separated microdomains: soft domain and crystalline (hard domain). The viscoelastic behavior and toughness of these adhesives are attributed to the coexistence of these domains. As stated in Ref 1, "Hydrogen bonding is also responsible for some of the unique properties found in polyurethanes; however, it is currently felt that the effect of hydrogen bonding on structure should not be overstated. The good properties observed with polyurethanes appear to be due primarily to

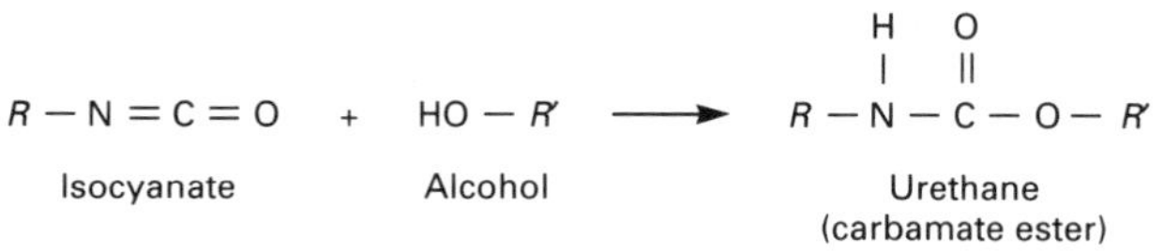

Fig. 1 Industrial synthesis of simple urethane

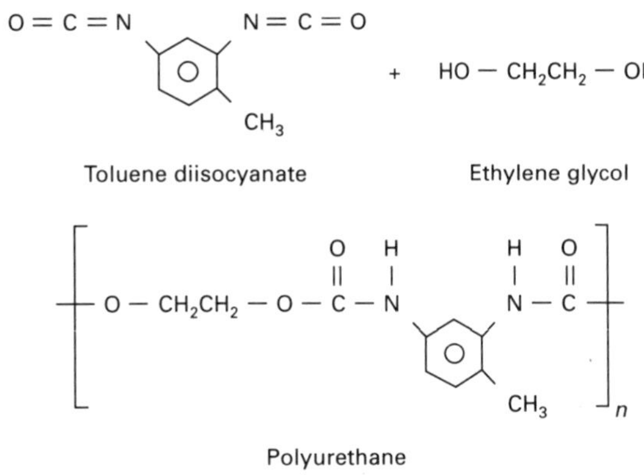

Fig. 2 Polyurethane polymerization by means of addition reaction using difunctional monomers

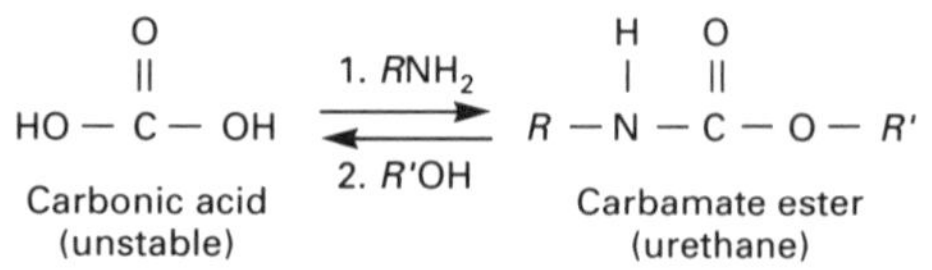

Fig. 3 Carbamate ester (urethane), derivative of carbonic acid

Fig. 4 Carbonic acid derivatives

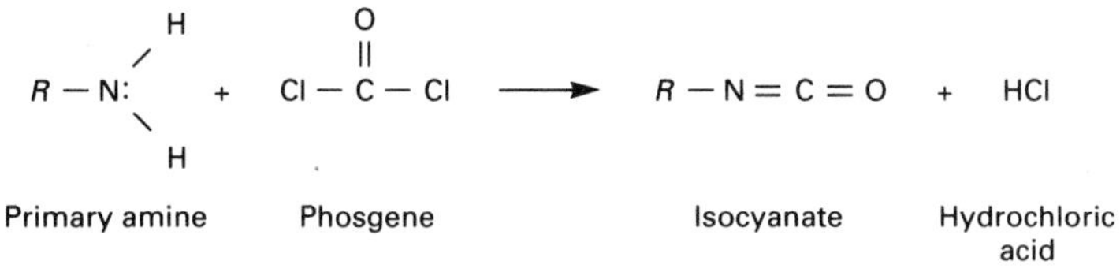

Fig. 5 Isocyanate synthesis

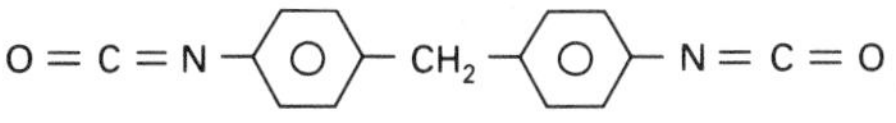

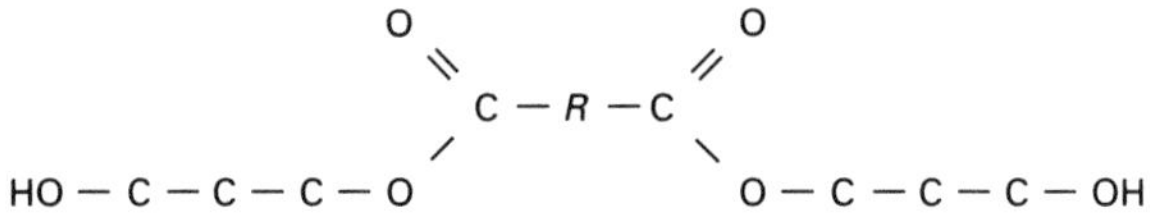

Fig. 7 Hydroxy-functional polyester polyol

Fig. 6 Diphenylmethane-4,4'-diisocyanate (MDI)

Fig. 8 Hydroxy-functional polyether polyol

formation of microdomains (that serve as tie-down points or pseudo crosslinking points)—and not to hydrogen bonding between hard and soft segments.'' The adhesion displayed by polyurethane adhesives may be attributable to the low surface energy of the adhesive.

The crystalline (hard) segments are formed by polyisocyanates and low molecular weight hydroxy compounds or small portions of hydroxy compounds. These hard segments of the polyurethane microstructure essentially ''cross link'' the amorphous (soft) segments. The amorphous domains are formed by hydroxy-terminated diols, that is, polyols based on polyesters or polyethers. These polyol portions are responsible for the elastic properties and the favorable low-temperature properties of polyurethanes. At elevated temperatures

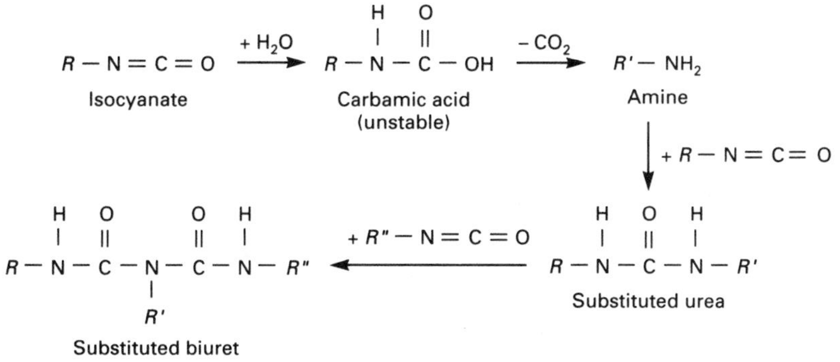

Fig. 9 Reaction of isocyanates with water

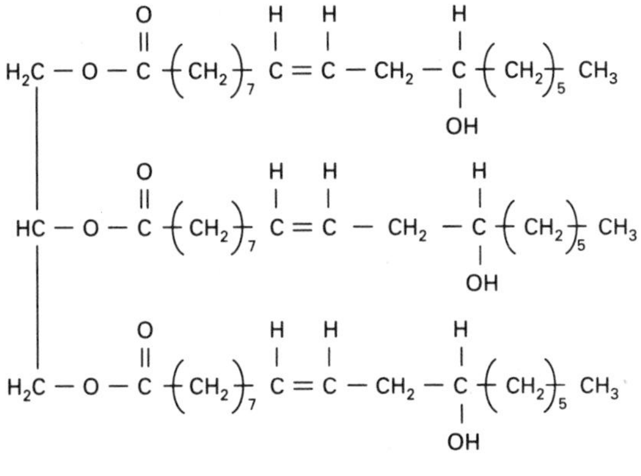

Fig. 10 Castor oil, a derivative of ricinoleic acid

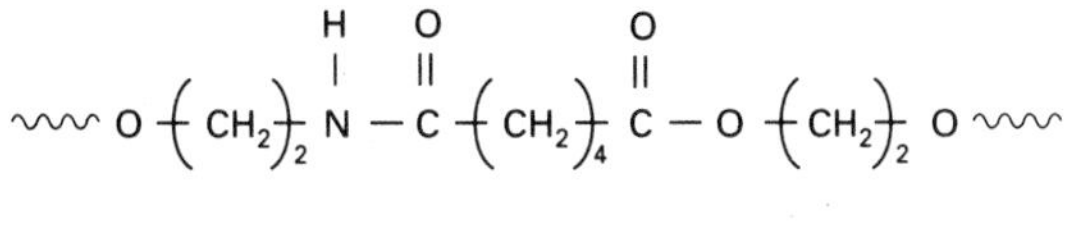

Fig. 11 Polyester amide

(above the glass transition temperature and the crystallization temperature), polyurethanes revert to one amorphous phase.

Polyols commonly used for the hydroxy portion of polyurethanes include polyester polyols (hydroxy-functional polyesters) and polyether polyols (hydroxy-functional polyethers). Polycarbonates, fatty alcohols, and castor oil (Fig. 10) are also used as polyols. Another type of polyol currently used in flexible adhesives is polyester amide (Fig. 11). The standard types of polyols used in urethane formulations are polyester polyols and polyether polyols.

Polyester polyol-based urethanes are noted for their good adhesion to substrates. When compared to polyether polyols, the polyester polyols yield urethanes with higher strength, modulus, and hardness values. Better resistance to oxidation at elevated temperatures is another characteristic of polyester polyol-based urethanes.

Polyester polyols are available in structures ranging from linear to highly branched. With a higher incidence of branching, more hydroxyl functionality is available for cross linking. The most commonly used polyester polyols are derived from aliphatic and aromatic dicarboxylic acids, and linear and branched alkanols. The advantages of low and high cross-link densities are described in the ''Properties'' section in this article.

Polyether polyol-base urethanes offer better low-temperature properties and hy-

Table 1 Commonly used polyisocyanate cross-linking agents

Cross-linking agent	Characteristics
Toluene diisocyanate-polyol adduct in ethyl acetate	75% solids
Polyol-modified diphenylmethane-4,4'-diisocyanate	Solvent free
Polymethylene polyphenylene isocyanate	Solvent free
Triphenylmethane-4,4',4''-triisocyanate in ethyl acetate ..	27% solids; provides the best adhesion to substrates (discoloring)
Tris-(*p*-isocyanato-phenyl)-thiophosphate in ethyl acetate	27% solids; provides the best adhesion to substrates (nondiscoloring)

The products listed here are normally used as cross-linking agents, whereas diphenylmethane-4,4'-diisocyanate and toluene diisocyanate are preferred building blocks in the production of polyurethane adhesives.

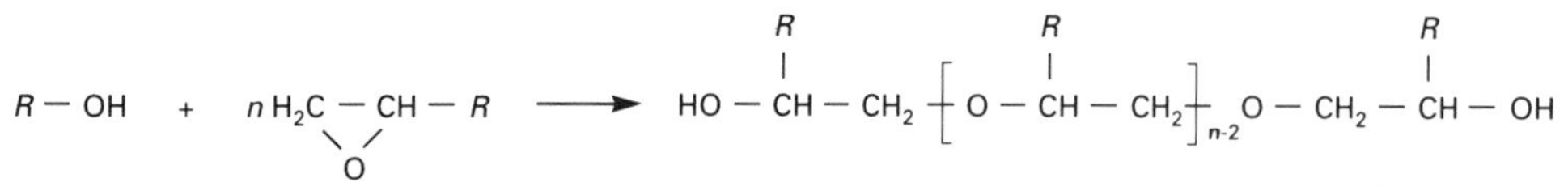

$$R-OH \quad + \quad nH_2C-CH-R \quad \longrightarrow \quad HO-CH-CH_2 \left[O-CH-CH_2 \right]_{n-2} O-CH_2-CH-OH$$

Fig. 12 Synthesis of polyether polyols, where $n = 10\text{-}80$ and $R = H$ or CH^3

drolysis resistance, lower cost, and easier processing than polyester polyol-base urethanes. They are particularly useful as solventless and high-solids-content reaction adhesives because of their low viscosity. Most polyether polyols are derivatives of polyethylene, polypropylene, and polytetramethylene oxide.

Because polyether polyols are low in viscocity, they are easier to use when formulating solventless and 100% solids polyurethane adhesives. Polyether polyol-based polyurethanes usually offer better hydrolytic resistance than do polyester polyol-based polyurethanes.

Because hydroxy groups on polyethers are secondary bonded, they are not as reactive as the primary bonded hydroxys in polyesters. (It should be noted, however, that primary hydroxy-terminated propylene glycols are commercially available.) Primary alcohols react three times faster than secondary ones, and tertiary alcohols react even slower. A simple example of the synthesis of polyether polyols is shown in Fig. 12.

Functional Types, Applications, and Cost

Two-component urethane adhesives are composed of a diisocyanate-terminated prepolymer as one component and a polyol and polyamine cross-linking agent and catalyst as the second component. These materials are particularly useful for solventless or low-solvent-content polyurethane adhesives. These two-component formulations are used to bond wood, plastics, foams, leather, and metals. Market areas include automotive, building and construction, and decorative applications. Specific examples include electrostatic flocking of plastics, textiles, or rubbers; lamination of building panels; rubber athletic flooring; and the bonding of filter elements in automotive body panels and windshields. This type of

polyurethane adhesive can function as a sealant and as an adhesive in structural and nonstructural applications.

One-component moisture-curing reaction adhesives are made with high molecular weight isocyanate urethanes called prepolymers. They are typically used in plastic film lamination and flexible packaging, for example, low-performance snack-food film adhesive and industrial applications such as panel laminations. Compatible substrates include polyesters, polyamides, cellophanes, and treated polyolefins. Figure 13 identifies the chemistry of this type of formulation.

Hydroxyl-terminated polyurethanes are used in solvent-base adhesives in one- or two-component formulations. Common solvents, which must be dry, include methyl ethyl ketone, acetone, ethyl acetate, tetrahydrofuran, and chlorinated hydrocarbons. Hydroxyl-terminated linear polyurethanes are synthesized from diisocyanates with an excess of high molecular weight polyester diols. The crystalline portion of this adhesive provides for its high cohesive strength. Above 50 °C (120 °F), hydroxyl-terminated polyurethanes become amorphous because the crystallization temperature (T_c) has been reached. However, a polyisocyanate may be added as a cross-linking agent. Cross linking with polyisocyanates serves to improve heat strength and solvent resistance at higher temperatures. The primary application area for this one-component adhesive is in shoe/boot manufacturing. Other areas include packaging film, vinyl repair, leather goods, and trim.

Two-component hydroxyl-terminated polyurethanes find applications in foam and fabric laminations in the shoe and textile markets, in laminated film packaging, electronic insulation, foam sponges, contour pattern flocking, and in laminations of polyvinyl chloride film with particle board for furniture. They are well suited for adhesion to polar substrates. Ingredients in certain

(a)

Polyurethane urea ionomer

(b)

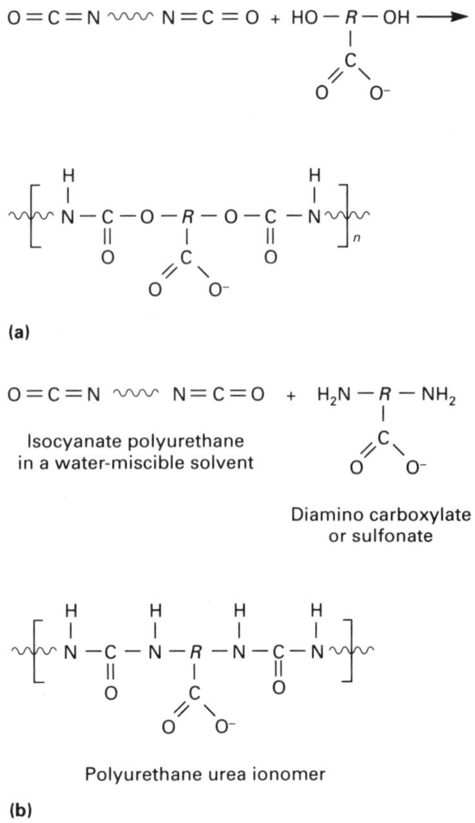

Fig. 14 Representative reactions of (a) water-base polyurethane dispersion and (b) polyurethane-urea dispersion

hydroxyl-terminated polyurethane systems can be sanctioned by the U.S. Federal Drug Administration for food contact.

Polyurethane-urea dispersions in water are under development to replace many solvent-base adhesives. A representative reaction is shown in Fig. 14(a). Polyurethane-urea dispersions are typically produced by the reaction shown in Fig. 14(b). An isocyanate-terminated prepolymer in acetone or another water-miscible solvent is chain extended with a diamine carboxylate or sulfonate to form a water-dispersible polyurethane-urea ionomer. The solution is then mixed in water and reacted, and the organic solvent is distilled out. The polyurethane dispersions can be blended with many water-base polymer dispersions

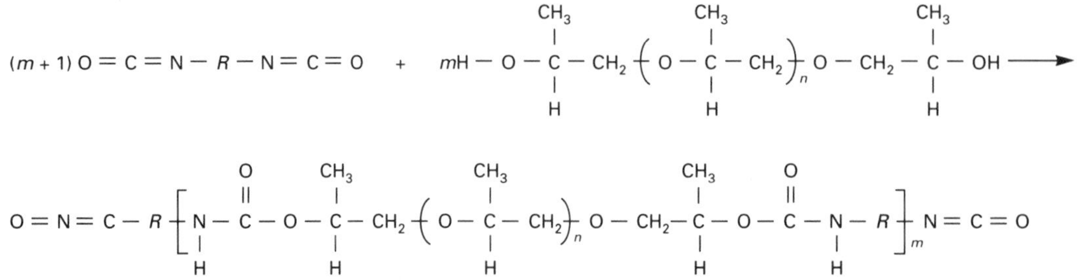

Fig. 13 One-component moisture-curing-reaction polyurethane adhesive based on high molecular weight isocyanates (prepolymers), where R = alkylene or arylene, $n = \sim 20$, and $m = \sim 3$. Source: Ref 2

and latices, such as polyvinyl acetate, eth-ylene-vinyl acetate copolymers, polyacryl-ics, and nitrile rubber. Polyurethane-urea dispersions display high adhesion levels and strength values at elevated tempera-tures. Applications include textile lamina-tions, textile-film laminations, flocking, shoe soles, automotive interior trim, and as an adhesion promoter in water-base coatings.

Applications. Polyurethane adhesives are used in the footwear, plastics pro-cessing, packaging, apparel, automotive, and building industries. In footwear, they are used to bond soles to the shoe uppers. In the apparel industry, fabric layers can be bonded into outer gear. In plastic pro-cessing, these adhesives are used to bond composite, plastic, metal, and ceramic substrates. They are also commonly used in transparent food packaging as the adhe-sive for flexible film. In the automotive industry, these adhesives are used for flocking, such as for glove compartments, and for adhering plastic sheeting, as in dash boards and doors. In electrostatic flocking, a good definition of contours can be achieved. Polyurethane structural adhesives/sealants can be used to bond automotive windshields, truck roofs, doors, lift gates, fenders, head lamps, rear lighting, supports, and grille openings. In the construction market, they are used to bond building panels to core materials and are suitable for adhering polyurethane foam and insulation.

Major suppliers of formulate polyure-thane adhesives are: Ashland Adhesives; Lord Corporation; Bostik Division, Emhart Corporation; 3M Corporation; Loctite Cor-poration; Bondwell; Uniroyal Inc.; H.B. Fuller Company; National Starch and Chemical Corporation; Synthetic Surfaces, Inc.; The B.F. Goodrich Company, Adhe-sives Systems Division; Pierce & Stevens (division of Pratt & Lambert); Imperial Ad-hesives and Chemicals; and Morton Thiokol, Inc.

Cost and Market Size. Although aliphat-ic urethanes are attractive because of their nonyellowing characteristics, they tend to cost more than aromatic urethanes. A poly-urethane dispersion is the most expensive type of polyurethane adhesive on a per-pound-of-polyurethane basis. Solvent-base adhesives are easier to process than one- or two-component reaction adhesives.

Sales of formulated polyurethane adhe-sives were $179 million in 1983 and $264 million in 1987 and are projected to reach $342 million in 1992, according to Business Communications Company, Inc. (report No. C-009U, ''Structural and Specialty Ad-hesives''). This growth reflects an average annual growth rate of 5.3% over the studied years. According to Frost & Sullivan Inc. (report No. A1953, ''Specialty Adhesive Market in the U.S.''), the polyurethane specialty adhesives market was $253 mil-lion.

Competing Adhesives

Epoxies, cyanoacrylates, anaerobics, phenolics, polysulfides, silicones, and acrylics compete with polyurethane adhe-sives. Of these, urethanes, epoxies, and acrylics are the adhesives of choice for structural applications. In general, polyure-thane adhesives have some advantages over competing adhesives:

- Highly versatile chemistry
- Flexibility (that is, suitable for bonding flexible films) or tough and rigid (for structural bonding)
- Ability to bond a variety of plastics (differ-ent types of plastics can be sandwiched)
- Ability to be formulated as one- or two-component systems
- Good low-temperature properties
- Good environmental resistance
- Good chemical, oil resistance

However, polyurethane adhesives have a few disadvantages:

- Moisture sensitivity during application
- Only average bond strength to metal if no primer is used (primer is therefore recom-mended for bonding metals)
- Requirement of precise mix ratio for cer-tain products
- Requirement of good mixing of compo-nents
- Maximum temperature of 150 to 163 °C (300 to 325 °F) for specially formulated polyurethanes

Polyurethane adhesives are often selected when bonding two different types of surfac-es, that is, plastic to ceramic, and so on.

Properties

Polyurethane adhesives can be tailored to achieve desired properties for a very wide variety of applications. It would thus be misleading to publish ''typical properties.'' Some generalizations, however, are de-scribed below. Bond strength can vary from soft (almost pressure sensitive) to high-strength structural types.

Higher cross-link density and/or more hard segments in an adhesive yields higher strength values and better resistance to chemicals and heat. Higher cross-link den-sity is usually achieved using an excess isocyanate component or through branching in the polymer backbone. Low cross-link density, achieved with linear polyols and an NCO/OH ratio close to 1, usually results in higher elasticity, higher peel strength, and better low-temperature properties. Com-pared to polyether-base urethanes, polyes-ter polyol-base urethanes are more oxida-tion resistant and are usually more resistant to oils and certain solvents, but are prone to hydrolytic degradation under extreme con-ditions.

Additives and Modifiers

All additives, modifiers, and solvents used in polyurethane adhesive formulations must be essentially free of water. In the presence of water, polyisocyanates will foam, and the gases that evolve will cause voids and result in poor adhesive strength. Solvents must have a water and alcohol content of less than 0.02%.

To ensure dryness of a polyurethane adhe-sive formulation, hydroscopics (drying agents) should be used. Triethyl orthofor-mate, calcium sulfate, tosylisocyanate, and molecular sieves, such as sodium aluminum silicate (Zeolites), at a 2 to 5% level are recommended. A carbodiimide is recom-mended in polyester polyol-base formulations to improve hydrolytic stability. To protect polyurethane adhesives from oxidation in bond lines, a phenolic antioxidant may be added at a 2% level. Biological attack can be prevented by adding an antimicrobial at a level of approximately 0.3%. Catalysts, such as tertiary amines, organometallic com-pounds such as dibutyltin dilaurate, and mer-cury catalysts, can be added to accelerate reaction with the isocyanate. For two-compo-nent formulations, it is customary to use additives with the polyol component. Addi-tives may include leveling agents (vicosity reducers), drying agents, antioxidants, fillers, colorants, and catalysts.

Polyurethane adhesive fillers include whit-ing, barytes, hydrated alumina, clays, quartz flour, and slate flour. These fillers help fill gaps and reduce shrinkage and the loss of adhesive during curing. They can also im-prove strength, reduce costs, and control viscosity. For porous substrates such as leather, where adhesive penetration into the substrate is not wanted, fumed silica filler may be added to thicken the adhesive. Filler is not usually used or is used at a low level in hydroxyl-terminated polyurethane adhesives. Colorants can be added to match the colors of substrates, to detect homogeneity in two-part adhesives, or to confirm that the adhesive has been applied.

Other polymers can be added to hydroxyl-terminated polyurethane adhesives for the purpose of customizing the adhesion, cohe-sive strength, solids content, viscosity, or other processing parameters. Compatibility with the adhesive system should be tested, particularly with respect to storage life. Table 2 lists the polymers that can be incorporated in polyurethane adhesive formulations.

Product Design Considerations

The choice between polyurethane adhe-sives with shorter and those with longer pot

Table 2 Polymers sometimes used in hydroxyl polyurethane adhesives

Nitrile rubber
Acetyl cellulose
Nitrocellulose
Alkyl phenolic resins
Terpene phenolic resins
Cyclohexanone, formaldehyde resins
Phthalate resins
Postchlorinated polyvinyl chloride(a)
Chlorinated rubber(a)
Chlorinated paraffins(a)
Vinyl chloride-vinyl acetate copolymer(a)

(a) Requires stabilization against hydrochloric acid evolution, or the polyester-base polyurethane will hydrolyze

lives becomes simpler when the necessary equipment is considered. Mixing and application equipment are available for two-component fast-reacting adhesives, whereas brushes, rollers, trowels, coating knives, spray guns, and a casting process can be used to apply slow-reacting adhesives.

In formulating solvent adhesives with highly volatile solvents such as acetone, care must be taken to balance the solvent blend properly. While the solvent volatilizes, it cools off the surface of the adhesive coat. The dew point can be reached on that substrate surface, and moisture may condense on the adhesive, preventing adequate adhesion with the second substrate.

Processing Parameters

The pot life of two-component systems can vary widely. If catalysts are employed, pot life can be shortened from hours to minutes or seconds. Towards the end of pot life, second-order kinetics are no longer applicable because microgels may form and later coalesce, which affects pot life measurements. In solvent-base systems, pot life can sometimes be extended by further dilution in appropriate solvents.

Storage Requirements. To achieve maximum shelf life, all ingredients used to formulate polyurethane adhesives must be pure and should have a maximum water and alcohol content of 0.02%. In addition, ingredients should be as pH neutral as possible. It is recommended that polyurethanes be stored in a dry, cool location, either a tightly sealed container (to avoid contamination) of tin or in a solvent-resistant polymer-lined container. Manufacturer guidelines for specific product storage requirements and periods should be followed.

Primer Use Requirements and Preparation of Surfaces. Roughening of surfaces is advised, particularly for shiny surfaces. Solvent washing is helpful to prepare clean substrate surfaces. Metals must be degreased or mechanically abraded. Some polyurethane adhesives are suitable for application to one substrate, whereas others require application to both substrates. Some suitable primers include diluted polyurethane adhesives, epoxies, silane coupling agents, polyisocyanates, and phosphoryl compounds. The latter three are best used as a component in the adhesive formulation.

Cure Requirements. Room-temperature-cured polyurethane adhesives will develop their bond strength under ambient conditions. For those formulated with heat-activated catalysts, brief exposure to infrared radiation or heating may be necessary. Bonds with moisture-cured polyurethane adhesives usually require aging for 1 week to achieve ultimate properties. However, some formulations are designed for excellent initial green strength and adhesion properties to allow substrates to be handled shortly after the adhesive is applied.

Solvent-base polyurethane adhesives require that the solvents evaporate. Moreover, the amount of water on the surface of the substrate (especially in moisture-curing systems) also greatly affects the speed of cure.

Recycling and Future Trends

In-plant recycling of solvent emissions is currently practiced for solvent-base adhesive production systems. Other attempts to recover solvents from the applied adhesive are impractical.

Alternatives to solvent adhesives are under full-scale development to meet legislative restrictions. Very-high-solids-content adhesives have been shown to be more successful thus far than water-base polyurethane adhesives. In some applications, solventless adhesives can be formulated to have performance characteristics similar to those of solvent-base adhesives and superior economics. In general, solventless systems take longer to cure, but do reach comparable ultimate properties.

ACKNOWLEDGMENT

The author wishes to express her appreciation to Wally Warrach and Bill Cibulas at Mobay Corporation for their assistance.

REFERENCES

1. B.H. Edwards, Polyurethane Structural Adhesives, in *Structural Adhesives: Chemistry and Technology*, S.R. Hartshorn, Ed., Plenum Publishing, 1986
2. M. Dollhausen, "Polyurethane Adhesives Based on Baycoll, Desmocoll, and Desmodur," Product Literature, Bayer AG
3. D.J. Zimmer and J.S. Morphy, The Solventless Approach to Adhesive Compliance, *TAPPI J.*, Dec 1988, p 123-126

SELECTED REFERENCES

● M. Dollhausen and W. Warrach, "A Review of Polyurethane Adhesives Technology," Paper presented at the Adhesive and Sealant Council, Philadelphia, Sept 1981; also, in *Adhes. Age*, Spring 1982
● K.C. Frisch, H.X. Xiao, and R.W. Czerwinski, Formulating Polyurethane Adhesives and Sealants, *Adhes. Age*, Sept 1988

Anaerobics

David P. Melody, Loctite International

ANAEROBICS form a subset of acrylic adhesives that has a fine balance of stability and reactivity. Their curing mechanism is triggered by active metals (for example, iron and copper) in the absence of oxygen. The term anaerobic refers to this feature of the hardening mechanism that requires the exclusion of oxygen, which occurs in closely mating engineering assemblies.

The first anaerobic formulations, devised in the early 1950s, were prepared by bubbling oxygen or air into dimethacrylate monomers. These formulations required continuously bubbling air to retain stability (Ref 1). Though promising, this form was impractical. The needed breakthrough came when it was found that a stable, yet reactive, one-part formulation could be prepared simply by mixing certain peroxy compounds, such as cumene hydroperoxide, into the methacrylate monomers. The commercialization of this invention led to the foundation of a leading producer and worldwide supplier of anaerobic products (Ref 2).

Chemistry

Anaerobic formulations still contain the methacrylate functional monomers and/or resins and free radical polymerization initiators found in the early formulations. However, as of 1990, a typical formulation comprises some or all of these components: monomers (and/or resins), initiators (of free radical polymerization), accelerators (for the initiation process), stabilizers, thickeners and other form modifiers, and performance modifiers (such as plasticizers and adhesion promoters). Each of these components is described below.

Monomer selection in anaerobic formulations is a major determinant of cured-adhesive properties. Most of the monomers used are methacrylates. Dimethacrylates of polyalkoxy glycols were, and are, useful bases of anaerobic formulations. Their formulas and those of other commonly used dimethacrylates are given in Fig. 1. The methacrylate-capped urethanes can vary from being relatively simple to being very sophisticated oligomers containing hard and flexible block moieties (Ref 3). Acrylate monomers are rarely used.

Formulations usually have a multifunctional methacrylate basis to facilitate rapid hardening and achieve a high level of oil and solvent resistance in the cured form. However, the use of diluent monomers, such as 2-hydroxypropylmethacrylate, is common. These monomers lower the viscosity of methacrylate- and acrylate-capped oligomers, control cross-link density, and enhance adhesion. Acrylic acid and methacrylic acid are occasionally used as copolymerizable adhesion promoters.

In contrast, monomers can be chosen to soften the cured anaerobic and make it flexible. To accomplish this, monomers such as 2-ethylhexyl methacrylate or lauryl methacrylate are used as diluents.

Initiators used in most anaerobic formulations are hydroperoxides. Cumene hydroperoxide is a frequently cited initiator. The hydroperoxide provides stability in anaerobic formulations, yet is suitably reactive (along with accelerators) when in contact with substrates and in the absence of oxygen. The distinctive feature of an anaerobic adhesive, in contrast to the larger set of "acrylic adhesives" is that it exists in one part, with a fine balance of stability and reactivity. The careful choice of initiator is just one element used to create that fine balance.

More complex "initiators" are cited in the literature (Ref 4), although the distinction between initiator and accelerator is

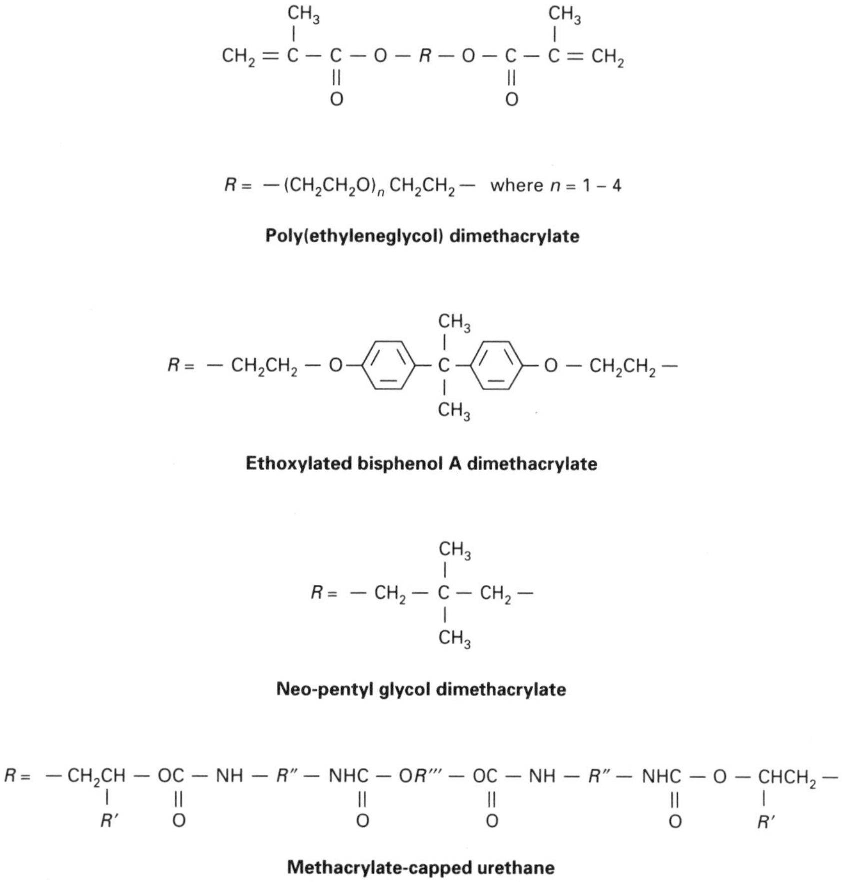

$R = -(CH_2CH_2O)_n CH_2CH_2-$ where $n = 1-4$

Poly(ethyleneglycol) dimethacrylate

$R = -CH_2CH_2-O-$⟨benzene⟩$-\overset{\overset{CH_3}{|}}{\underset{\underset{CH_3}{|}}{C}}-$⟨benzene⟩$-O-CH_2CH_2-$

Ethoxylated bisphenol A dimethacrylate

$R = -CH_2-\overset{\overset{CH_3}{|}}{\underset{\underset{CH_3}{|}}{C}}-CH_2-$

Neo-pentyl glycol dimethacrylate

$R = -CH_2CH-OC-NH-R''-NHC-OR'''-OC-NH-R''-NHC-O-CHCH_2-$

Methacrylate-capped urethane

Fig. 1 Dimethacrylates widely used in anaerobic adhesive formulations, where R' is —H or —CH$_3$, R'' is usually aryl, and R''' represents a wide variety of disubstituted and trisubstituted alkyl radicals

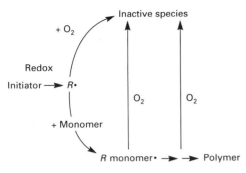

Fig. 2 The oxygen inhibition role in anaerobics (Ref 4)

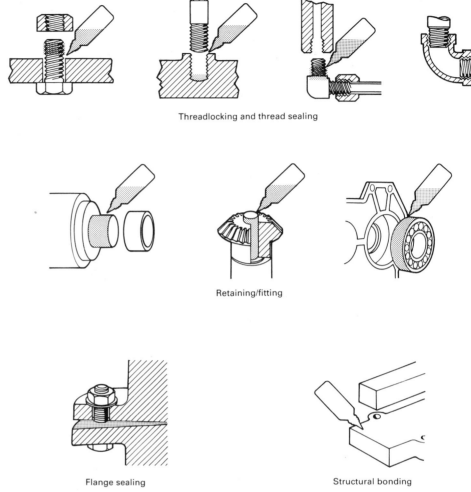

Threadlocking and thread sealing

Retaining/fitting

Flange sealing

Structural bonding

Fig. 3 Anaerobic adhesive applications

frequently difficult to establish. The potential to form peroxy derivatives of methacrylate monomers further complicates the mechanism (Ref 5).

Accelerators of anaerobically curing adhesives are necessary for rapid hardening, which provides fixturing. Many refinements of the technology have come from the work done in this area. Saccharin (benzoic sulfimide) and many compounds containing acidic-NH groupings (Ref 4) have proved to be accelerators and/or initiators, especially when used in combination with aromatic amines. Combinations such as saccharin with N,N-dimethyl-*p*-toluidine and saccharin with tetrahydroquinoline are widely cited as accelerators in the patent literature.

Stabilizers are critical to the formulation of useful anaerobic adhesive products. Phenolic or quinoidal stabilizers are often used. Chelators are used to trap traces of transition metals that could cause initiator decomposition with consequent preemptive curing.

Oxygen (in air) is the most important inhibitor because it gets added to reactive radicals, forming more stable species, as indicated in the reaction scheme of Fig. 2 (Ref 4). The inhibiting role of oxygen explains why most anaerobic products are packed in containers that are air permeable and partially filled, thereby maximizing shelf life.

Form modifiers include dyes, fluorescing agents, thickeners, thixotropes, and pigments. Dyes, pigments, and fluorescing agents are used to facilitate on-part inspection and, in some cases, to denote certain strength characteristics. Thickeners and thixotropes are used to create viscosity and flow characteristics appropriate to the target application. For example, water-thin anaerobic fluids are used to provide wicking into preassembled joints, whereas thixotropic anaerobic pastes are chosen for flange bonding and sealing. In all cases, great care must be taken to select purified and chemically compatible form modifiers to ensure that both stability and reactivity are unchanged in the adhesive.

Performance modifiers such as plasticizers and adhesion promoters may be includ-

ed. The former generate lower compressive strength, while the latter enhance adhesion. Other performance modifiers commonly described include lubricants, such as polytetrafluoroethylene (PTFE) powder, and fillers, such as mica. Heat-resistance enhancers (for example, imides containing reactive double bonds) are present in some products (Ref 6). The cure of these additives is thermally induced when the hardened anaerobic product is exposed to higher temperatures.

Compatible low/medium molecular weight elastomers have been incorporated in anaerobic formulations, resulting in enhanced toughness (Ref 7). The addition of photoinitiators to anaerobic formulations adds a further performance dimension. It allows convenient fixturing of joints by ultraviolet (UV) light curing of adhesive fillets.

To summarize, anaerobic formulation chemistry differs widely in composition, form, and cured properties, depending on the particular application. However, all are *one-part* adhesives that have a fine balance between package stability and reactivity in

contact with metal surfaces. This reactivity causes a redox-triggered hardening process (in the absence of oxygen). A heat and surface activator can be used to augment this curing mechanism.

Types and Forms

Functional Types. The chemistry of anaerobics allows the formulation of many functional types for specific applications, such as threadlocking adhesives, thread sealants, retaining adhesives, porosity sealants, flange sealants, structural adhesives, and UV-light-curing adhesives. Some of these application areas are depicted in Fig. 3, and each is described below.

Threadlocking adhesives were the first products produced using anaerobic technology. Contact with metal substrates favors anaerobic product cure and the high crosslink density attained provides excellent oil and solvent resistance. Careful formulation allows controlled variations in viscosity and locking strength.

Threadsealing anaerobic products, like threadsealing adhesives, can be formulated

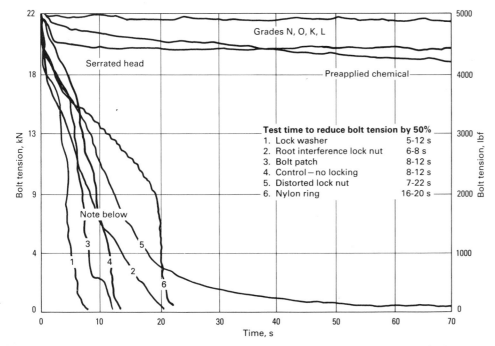

Fig. 4 Resistance of threaded assemblies to transverse shock using MIL-S-46163 anaerobics N, O, K, L versus other locking mechanisms. Bolts used were grade 5, 9.5 mm (3/8 in.) diam, 16 threads per inch. Source: Ref 8

- On-torque and the related torque tension, that is, the force that clamps the mated parts
- Break-loose torque, which is related to the choice of adhesive formulation and to the tightening torque used
- Prevailing (off) torque, that is, the resistance to removal after the joint has broken loose
- Vibration resistance, which is much better than for lock washers or other common mechanical techniques. The use of MIL-S-46163 grade (N, O, K, L) anaerobic products is compared with other locking mechanisms in terms of this property in Fig. 4

Many modern anaerobic adhesives have carefully controlled lubricity to lend precise clamp loading and are fairly tolerant of surface oils.

The benefits offered by properly formulated and applied threadlocking and threadsealing adhesives include:

- Simple and safe application
- Prevention of thread movement and consequent loosening
- Effective sealing and prevention of corrosion
- Lowering of weight and cost in assembly
- Tighter control of joint properties
- Controlled and easy disassembly (can be designed into the product, for example, in joints that need routine servicing)

Retaining or fitting cylindrical parts with anaerobic adhesives is now an established technology. Basically, anaerobic retaining adhesives bond slip-fitted components and augment the strength of interference-fitted assemblies. In the first case, components such as shafts, bushes, rotors, cylindrical liners, splines, and bearings are bonded or restored by applying these adhesives to the parts prior to slip fitting. Moderate-strength formulations can be used in applications (such as the mounting of bearings) requiring routine dismantling.

Anaerobic adhesives with carefully chosen rheology, lubricity, cure speed, and cured properties are particularly effective for augmenting "press" or "shrink" fits. Additional strength and performance are obtained without risking part damage by more extreme interference fitting geometries or more stress. This technology is exemplified in the assembly of an automotive differential ring using a light shrink fit along with a strong anaerobic adhesive (Fig. 5). Laboratory studies (Ref 9) on the fatigue resistance of such joints confirm the augmentation of performance by anaerobic adhesives (Table 1).

Porosity sealing is another application served by low-viscosity anaerobic formulations. Such sealants fall into two categories. In the first, the adhesive is applied topical-

with selected viscosity/rheology, curing speed, strength, solvent resistance, and other properties.

Retaining adhesives are formulated with anaerobic chemistry to bond cylindrically fitted metal parts, such as bearings into housings. Rheology, cured strength, and other properties can be designed to meet application needs.

Porosity sealants are thin anaerobic fluids that can be applied to porous metal substrate surfaces either topically or by a vacuum impregnation process (Ref 8).

Flange sealing anaerobic formulations are pastes that can be used to form seals in place between mating metal surfaces while allowing metal-to-metal contact. The rheology is chosen to facilitate application and maximize immediate physical sealing; subsequent cure develops the full sealing properties.

Structural adhesives based on anaerobic chemistry are best known for the bonding of small, rigid components because the cured products tend to be relatively hard and inflexible. In many cases, structural anaerobic adhesives are used in conjunction with surface activators laid down from solvent. This results in faster curing on a wide-ranging group of substrates.

UV-curing anaerobic adhesives combine the UV light and redox cure mechanisms in one product. While a UV-light-curing response can be built into most anaerobic formulations, it is used most widely in retaining and structural-bonding formulations. In such cases, the UV light induces cure and immobilizes any adhesive outside

the joint, thereby generating a rapid fixturing mechanism. The immobilization of the liquid also prevents the possible contamination of adjacent parts.

Commercial Forms. Anaerobic adhesives can be formulated in many physical fluid forms, from water-thin liquids to thixotropic pastes. The choice is largely determined by the application for the product and by the technique appropriate for applying the product to the substrates. For example, threadlocking and retaining adhesives are usually mobile fluids, whereas anaerobic adhesives for threadsealing and gasketing applications are generally pastes.

Markets and Applications

The markets for anaerobic adhesives include a wide range of assembly industries, such as the automotive, off-the-road vehicle, machine, and electromechanical-equipment construction markets. These adhesives also serve in maintenance and repair functions and are sold widely in the automotive repair and do-it-yourself markets. Applications for the functional types of anaerobic adhesives identified earlier are used in the various ways described below (Ref 8).

Threadlocking and threadsealing were the first applications found for anaerobic formulations. On threaded assemblies, anaerobics cure to fill the "inner space" with a highly cross-linked, chemically resistant, solid seal that prevents self-loosening. Anaerobic adhesives influence a number of important, measurable parameters in threaded assemblies:

Fig. 5 Automotive differential ring that uses a light shrink fit with strong anaerobic adhesives

Table 1   Values for slip fit and for press and shrink fits with and without anaerobic adhesive

Type of fit	Endurance limit(a)		Static shear strength	
	MPa	ksi	MPa	ksi
Slip fit, with adhesive....	7.2	1.1	17	2.5
Press fit				
Without adhesive......	9.9	1.4	19.7	2.9
With adhesive.........	21.5	3.1	42	6.1
Shrink fit				
Without adhesive......	20.4	3.0	29	4.2
With adhesive.........	20.2	2.5	53	7.7

(a) Endurance limit refers to a 50% probability of failure. For design purposes, an endurance limit equal to 30% of the static shear strength is used.
Note: Average shear stress calculated from moment. Tested on soft steel shaft, with cast iron hub, 25 mm (1 in.), in length, 25 mm (1 in.) pin diameter, and interference of 0.018–0.023 mm (0.7–0.9 mil)

ly. The fluid then wicks into pores and cures there because it contacts the metal and is away from air. An excellent example of this application is the sealing of welded wheel rims for use with tubeless tires.

In the second category, the adhesive is applied by a specialized process of vacuum impregnation suited for sealing components with fine pores, such as castings. In this process, a highly active anaerobic formulation in an immersion vessel is kept stable by refrigeration and continued air bubbling. The sealant is formulated with detergent ingredients to facilitate the removal of residual fluid by a water wash. This room-temperature-curing sealant gives particularly good properties in terms of sealing, part cleanliness, industrial hygiene, and machinability of the sealed components.

Flange sealing anaerobic formulations are widely used as form-in-place gaskets for relatively well-machined components in transmissions, engines, and pumps. Their features in such applications are:

- Provides ease of operation (screen printing, stencil, robotic tracing, or simple roller transfer)
- Eliminates the need for stocks of different gaskets
- Provides excellent thermal and solvent resistance
- Allows metal-to-metal contact, thereby eliminating the relaxation of tension in assemblies

Because early "plastic gasketing" adhesive compositions cured to rigid solids, they were best suited to applications in which joint movement between substrates of the same material was minimal. The development of formulations based on sophisticated urethane methacrylate resins now allows applications involving dissimilar substrates in which some movement can occur. This is particularly important in modern engines, which use both aluminum and cast iron components in light structures. Care is taken to minimize adhesion so that disassembly can be carried out easily.

Structural bonding with anaerobic adhesives is particularly effective for small and well-fitting metal, ceramic, and glass components. In these cases, formulations are used that maximize adhesion and introduce other characteristics required in face-to-face structural bonding (Ref 10). The basic anaerobic cure mechanism can be augmented by the use of heat or light or by the application of an activator applied from solvent or resin. Then, a wide variety of

cure systems, combined with low liquid viscosities, creates many application features:

- Precise, tidy applications
- Room-temperature cure, with removal or cure of adhesive fillets as desired
- Low shrinkage upon curing
- High tensile and shear strengths after cure, with adequate peel and cleavage strengths for relatively rigid components
- Fast fixturing (seconds) and rapid achievement of functional strength when used with surface activators

Particularly good UV-curing adhesives for glass and glass-metal bonding also have an anaerobic cure mechanism. Again, the most suitable applications involve small bonding areas. Low shrinkage and excellent clarity make these adhesives particularly useful when aesthetics are important, as in crystal glass bonding applications. One such adhesive is used successfully in the consumer adhesive market for the bonding and repair of glass items, with UV or visible sunlight being the agent of cure. Combinations of anaerobic and UV cure are widely used in the electrical industry for the bonding of stepping motors and for the precise bonding of video cassette recorder components.

Suppliers, Costs, and Competing Adhesives

Major Suppliers. Anaerobic technology was initiated by Loctite Corporation, which still has a strong market position in most countries. While many companies supply some anaerobic products, several provide an extensive range on a multiregional basis: Loctite Corporation, Henkel KGaA; Permabond International, a division of National Starch and Chemical Corporation; and Three Bond Corporation of Japan. Given the application specificity of many anaerobic products, technical advice from a major supplier should be obtained at a design stage, if possible.

Cost Factors. Anaerobic products are frequently formulated using expensive raw materials. Extensive and specific quality control systems are necessary to guarantee performance in these products. Although typical prices are high, they represent a good value because only small quantities are usually needed. In early 1990, typical prices ranged from $50 to 250 per kg ($23 to 115 per lb). Anaerobic adhesives frequently provide a unique reliability, rather than cost, benefit. However, where anaerobics replace mechanical alternatives, savings are usually 2 to 10 times the cost of the adhesive (per application).

Competitive Adhesives. Given the diverse nature of applications for anaerobic products and their unusual properties, physical fastening is the normal alternative.

Table 2   Adhesives that compete with anaerobics in specific applications

Application area	Competing adhesive	Advantages	Disadvantages
Porosity sealing	Sodium silicate solutions	Very low cost	Slow cure of water-soluble sealants
	Styrene-thinned polyesters	Low cost	Odor and toxicity
	Heat-accelerated methacrylates	Simple	Bleed-out, caused by heating
Flange sealing......	Room-temperature vulcanizing silicones	Good temperature resistance and gap filling	Tendency to shim
	Solvent-base resins	Low cost	Initially present solvent results in poor solvent resistance
Bonding..........	Epoxy adhesives	Lower cost, can gap fill	Two-part or one-part requiring heat cure
	Acrylic adhesives	Wider range of substrates	Odor and flammability

Table 3   Uncured properties of anaerobic adhesives

Property	Typical value range	Comment
Visual features	Colorless to highly pigmented	A fluorescing agent may be added to aid on-part inspection.
Viscosity	Water-thin liquid to viscous pastes	The rheology is chosen for the application involved.
Vapor pressure, Pa (psi)	10–30 (0.0015–0.0045)	The major contribution is usually minor solvent residue in the raw materials.
Density, g/cm³	1–1.3	The concentration of fillers determines the level of density.
Flash point, °C (°F)	80–150 (175–300)	. . .
Curing characteristics, response to redox/absence of air (O_2)		
Fixturing on steel, min	5–300	Heat, light, and activators can also be used.
Full cure on steel, h	6–100	

Table 4   Properties of cured anaerobic adhesives

Property	Typical values	Comment
Coefficient of thermal expansion, 10^{-4}/K	1	. . .
Thermal conductivity, W/m · K (Btu · in./h · ft² · °F)	0.1 (0.7)	. . .
Specific heat, kJ/kg · K (Btu/lb · °F)	0.3 (0.07)	. . .
Elongation at break, %	0.1–10	Most have low (1%) elongation to break
Adhesives strength		
Shear on steel substrates (MIL-R-46082), MPa (ksi)	1–40 (0.15–5.8)	
Tensile on steel substrates (ASTM D 1002), MPa (ksi)	1–30 (0.15–4.3)	Products are designed to have values appropriate to application needs.
Maximum off-torque resistance on unseated, threaded steel fasteners of 10 mm (0.4 in.) diam. (similar to MIL-S-46163), N · m (lbf · in.)	1–60 (9–530)	
Hot strength at room temperature, %		
At 100 °C (212 °F)	30–100	Some products contain B staging additives, which increase in strength upon heating.
At 150 °C (300 °F)	15–100	
Strength retained after aging for 1000 h, %		
At 100 °C (212° F)	80–100	Same as for hot strength
At 150 °C (300 °F)	20–100	

However, competing adhesive technologies are available for porosity sealing, flange sealing, and bonding (see Table 2).

Properties and Additives

Uncured properties differ greatly from cured properties in terms of viscosity and the rate at which they cure, but many other characteristics have fewer differences. A variety of parameters are presented in Table 3 with values typical of commercial anaerobic adhesives.

Cured properties are designed for specific applications. Therefore, products can have a wide range of both uncured and cured material properties. Cured material properties differ most in terms of strength and hardness (Table 4).

Additives and Modifiers. As already noted, anaerobic adhesives additives are categorized as form or performance modifiers. Form modifiers include dyes, pigments, fluorescing agents, thickeners (for example, polyacrylates), and thixotropic agents (for example, fumed silicas).

Performance modifiers include plasticizers (such as polyethylene glycol diesters), adhesion promoters (such as acrylic acid), lubricants (such as powder PTFE), toughening agents (such as acrylate-capped oligomers), photoinitiators (such as substituted acetophenones), compatible B-staging additives (such as imides containing polymerizable double bonds) (Ref 6), and fillers (such as mica). It should be noted that the choice of additive is limited and complicated by the sensitivity of anaerobic formulations to redox-active species.

Product Design and Processing Parameters

The design of adhesives products is highly application oriented. The choice of monomers is very important in determining the properties of the cured adhesives. High cross-link density is important for many applications in which thermal and solvent resistance are dominant needs and brittleness is not a problem.

Packaging is also a major factor. Partially filled, air-permeable packages (for example, low-density polyethylene bottles and tubes) successfully hold formulations of particularly fast-curing products. Obviously, it is necessary to choose light-screening packages for UV-light-sensitized versions of anaerobic adhesives.

Processing parameters with regard to shelf life and pot life; storage requirements; substrate, geometry, and cure requirements; use of primers and activators; and dispensing equipment are described below.

Shelf and pot life depend not only on the formulation, but also on the package used. As the word anaerobic implies, there is a strong dependence on access to the oxygen

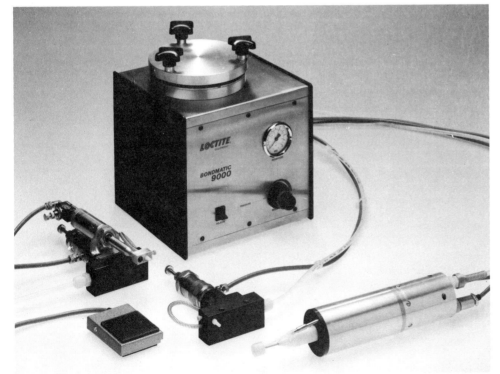

Fig. 6 Dispensing equipment for anaerobic adhesives

in air as a stabilizing factor. A package stability of months, afforded by large air-impermeable containers, becomes years in small, thin-walled low-density polyethylene bottles and tubes.

A unique advantage of anaerobic adhesives is the stability of the one-component formulation in air. Hence, on-part life (in air) is unlimited. Stability in dispensing equipment is good, too, if contact with redox-active metals (for example, copper and iron) is avoided. Properly formulated anaerobic adhesives self-harden before they stop functioning in a curing-adhesive fashion. Hence, shelf and pot lives are not limiting factors in the use of anaerobic adhesives.

Storage requirements can be qualified by the preceding comments on packaging and air exposure. Like all organic chemistry formulations, these adhesives should be stored in a cool and relatively dry place.

Substrate, geometry, and cure requirements are considered as a unit because anaerobic products rely on the substrate redox influence to bring about cure (in the absence of heat, light, or activators). Therefore, joint substrates and geometry require:

- Redox-active substrate materials (for example, containing accessible copper or iron)
- Freedom from excessive rust-inhibiting treatments
- Small gaps between substrates

- Bond areas large enough to cope with air inhibition around the edges
- Rigid substrates in joints designed to minimize peel and cleavage stresses in the assembly

Use of primers and activators is unnecessary on metal parts unless extremely fast cure (seconds to minutes) is required or unless the gaps between components are large. In these cases, surface treatments, which may be called primers or activators, can be used to extend the scope of the basic products in terms of cure speed and ability to cope with gaps between substrates. These primers or activators usually contain redox-active species in a solvent. However, new formulations containing methacrylate monomers suspend these active ingredients, which polymerize with the product, thus eliminating the need for solvent evaporation. True primers are occasionally used to enhance the performance of anaerobic adhesives in the bonding of glass or plated metal.

Dispensing equipment for anaerobic products has reached a very sophisticated level. Equipment ranges from attachments to plastic packages to facilitate hand dispensing to fully automatic systems that fit precisely into high-speed assembly lines. Figure 6 depicts a small selection of available dispensing equipment. Although anaerobic adhesives have a very good health-and-safety record, good practice suggests minimizing skin contact. The range of dis-

pensing equipment available from major suppliers ensures that this is possible.

REFERENCES

1. R.E. Burnett and B.W. Nordlander, U.S. Patent 2,628,178, 1953
2. B.D. Murray, M. Hauser, and J.R. Elliott, Anaerobic Adhesives, in *Handbook of Adhesives*, 2nd ed., I. Skeist, Ed., Reinhold, 1977
3. L.J. Baccei, U.S. Patent 4,309,526, 1982, and 4,295,909, 1981
4. C.W. Boeder, Anaerobic and Structural Acrylic Adhesives, in *Structural Adhesives—Chemistry and Technology*, S.R. Hartshorn, Ed., Plenum Publishing, 1986
5. W.A. Lees, C. Ford, D.J. Bennett, J.R. Swire, and P. Harding, U.S. Patent 3,795,641, 1974
6. B.M. Malofsky, U.S. Patent 3,988,299, 1976
7. T.R. Baldwin, D.J. Bennett, and W.A. Lees, U.S. Patent 4,138,449, 1979
8. G.S. Haviland, *Machinery Adhesives for Locking, Retaining and Sealing*, Marcel Dekker, 1986
9. M. Schwaiger and F. Schuch, Experimental Investigation on Bonded Cylindrical Joints, *Int. J. Veh. Des.*, 1985
10. F.R. Martin, Acrylic Based Adhesives, in *Structural Adhesives—Developments in Resins and Primers*, A.J. Kinloch, Ed., Elsevier, 1986

Acrylics

Dennis J. Damico, Industrial Adhesives Division, Lord Corporation

CURABLE ACRYLIC ADHESIVES can be defined as reactive cross-linking structural adhesives that cure by means of free radical initiation. They are based on methacrylate monomers and are generally toughened with elastomeric polymers. This type of adhesive is distinctly different from anaerobic, cyanoacrylate, and acrylic solution adhesives and emulsions.

Curable acrylic adhesives were first developed in West Germany in the late 1960s as an outgrowth of polymethyl methacrylate (PMMA) chemistry. Early formulations were simply low molecular weight solutions of PMMA dissolved in methyl methacrylate monomer. These simple systems, which could be cured by means of peroxide initiation, were used in the bonding of aluminum windows and doors. Since that time, a considerable amount of research has occurred, resulting in the emergence of the very sophisticated adhesive systems that are on the market today. A considerable amount of the most recent work has been conducted in the United States to develop what are now being referred to as high-performance acrylic adhesives.

This article will cover acrylic chemistry; the various generations of acrylic adhesives that have come into existence over the years; the advantages and limitations of these systems; the properties of adhesives, including lap-shear strengths; the uniqueness of these systems compared to other types of adhesives; and examples of performance capabilities.

Chemistry

Acrylic adhesives cure by addition polymerization reactions. The formation of free radicals initiates a very rapid chain reaction that results in the cure of the adhesive. This cure chemistry is significantly more rapid than a typical cure for condensation polymers such as those in epoxy and urethane adhesives. The cure profile of condensation (epoxy and urethane) polymers is compared to that of addition (acrylic) polymerization in Fig. 1.

The free radical reaction or addition polymerization used in acrylic adhesives offers the user of the adhesive certain advantages

over other types, that is, epoxies and urethanes. As can be seen from the cure curve, little polymerization is noticeable in the early stage of the reaction after the adhesive is mixed. This allows the user to position parts after they are mated without concern for disturbing the cure. When the actual cure is initiated, it proceeds very rapidly, with a clearly discernible end to the reaction process. This rapid achievement of final properties is different from the cure of urethanes and epoxies, which tend to get more and more viscous, with a more gradual achievement of final properties. The rapid-setting feature of acrylics allows the user to determine more clearly when parts are securely bonded and suitable for shipment. The fact that this rapid setting is possible at room temperature is unique to acrylics.

Furthermore, because of the availability of a large number of formulating tools and polymerization initiators, it has been possible for formulators to develop systems that have a whole spectrum of different cure rates. This allows manufacturing and design engineers to custom design specific adhesives to a variety of production schedules.

One disadvantage of acrylic adhesives is the tendency of the monomers used to achieve these desired properties to autopolymerize (that is, cure without the use of externally added accelerators or hardeners). Consequently, shelf stability can be a concern at temperatures as low as 45 °C (110 °F), a temperature at which some acrylic formulations will begin to cure, causing the product to harden before use. This problem is primarily historic in nature because recently introduced products have

much improved stability over early prototypes. Another concern with acrylic adhesives based on methyl methacrylate is the characteristic odor of the methyl methacrylate monomer, which, although not a health problem, can be objectionable to some users and may result in the use of other adhesive types for given applications.

Figure 2 shows the chemical formulations of methyl methacrylate and other related monomers typically used in acrylic adhesives. These monomers vary in volatility, with the lower molecular weight types having a strong odor and a low flash point. Higher molecular weight types offer less odor, higher flash points, and greater flexibility in the finished adhesives. Additionally, the lower molecular weight monomers tend to result in better adhesion and to be lower in overall cost. Whether lower molecular weight or higher molecular weight monomers or blends are used depends on which properties are of the most interest.

The differences among monomers can be appreciated by considering some specific examples. Methyl methacrylate, for instance, has a strong odor and is very volatile, with a flash point below 40 °C (100 °F). Butyl methacrylate and higher analogs of this monomer have decreasing amounts of odor and higher flash points. Adhesives formulated with butyl methacrylate, for example, are softer and have a lower glass transition temperature and less odor than their methyl methacrylate analogs.

As has been noted, a free radical must be generated in order for adhesive systems based on methacrylate monomers to cure. A typical means of generating a free radical is shown in Fig. 3. This common method of initiating the cure of acrylic adhesives is known as a redox (reduction and oxidation) reaction.

Redox reactions for acrylics typically involve an aromatic amine (shown in Fig. 3 as dimethyl aniline) in the resin portion of the adhesive interacting with a "mix-in" peroxide paste (Fig. 3). The result is the formation of a free radical that is capable of interacting with a double bond in methacrylated monomer in order to initiate a chain reaction.

Once begun, chain reactions ultimately lead to the formation of high molecular

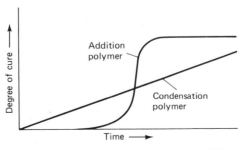

Fig. 1 Cure profile of condensation versus addition polymerization

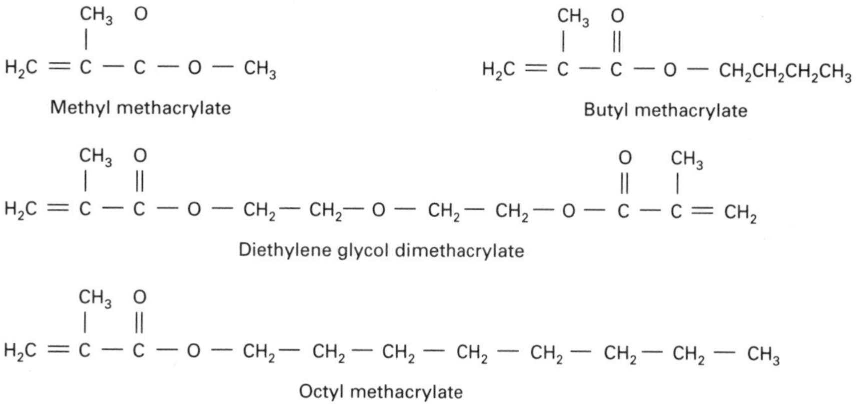

Fig. 2 Reactive species (monomers) for acrylics

Table 1 Other compounds for generating free radicals for acrylic adhesives

Compound	U.S. patent
Perester dibasic acid + metal ion ...	4,348,503 (Ref 2)
Saccharin salt + alpha hydroxy sulfone	4,081,308 (Ref 3)
Hydroperoxide + thiourea and metal salt(a)	4,331,795 (Ref 4)
N,N-dimethyl + saccharin p-toluidine	3,658,624 (Ref 5)

(a) For example, cobalt

weight polymers, provided that chain-terminating reactions do not interfere with the propagation step to any great extent. To explain further, free radicals, generally depicted as $R \cdot$ (where the dot represents an unpaired electron), are capable of interacting with free monomers to form a propagating chain with a free radical at one end. This can be depicted as $R-B-B-B-B-B \cdot$, where the initiating radical R is at one end of a polymer chain and the propagating reaction occurs through monomer B. This chain reaction continues as long as there are unreacted monomer molecules present and the reaction is not quenched. Quenching occurs when two free radicals interact to form a stable electron pair, terminating the reaction, or when excess stabilizers interfere with propagation by forming unreactive free radicals. The details of initiating and appropriately controlling chain reactions in acrylic adhesives have been carefully worked out by adhesive chemists to ensure that the final adhesive polymer is of high molecular weight and is useful as an adhesive.

Formulators use a number of different aromatic amines and peroxides to obtain a large variety of typical cure rates in fully formulated adhesives. Substituted aromatic amines have different reactivities with peroxide. Generally, substituted aromatic amines, such as dimethyl p-toluidine, are more reactive than unsubstituted parent compounds. The type of peroxide also affects the cure rate. Benzoyl peroxide is often chosen because it is reasonably stable at room temperature and can be used with the appropriate amine to obtain adhesives

that fully cure in minutes at room temperature.

Other compounds used for generating free radicals are shown in Fig. 4 and 5 and in Table 1. These mechanisms do not always involve the interaction of an amine compound and a peroxide, but all of them result in the formation of a free radical that is capable of interacting with monomeric species.

The tendency of acrylic monomers to polymerize spontaneously makes it necessary for adhesive formulators to include stabilizers in order to ensure good shelf stability. These stabilizers are generally quinone and hindered phenolic base compounds. They function by interacting with free radicals of oxygen or carbon to stop chain growth that might occur spontaneously, resulting in premature gelation. These materials are ultimately consumed when the preferred cure is initiated at the time of peroxide addition and a large flood of free radicals overwhelms the ability of the stabilizers to stop further polymerization.

In addition to the methacrylate monomers, initiators, and stabilizers that are typically found in formulations, another key ingredient in typical acrylic adhesives is the large variety of polymers used as modifiers (Table 2). These materials are high molecular weight polymers and/or oligomers and may or may not be substituted (that is, methacrylated). Typically, they are elastomeric materials capable of imparting flexibility and toughness to the finished adhesive. This added flexibility, in turn, increases the overall adhesive impact and peel strengths of bonded parts. The specific

choice of elastomeric modifier is an important element in the art of formulating and is the basis for much of the proprietary technology identified today.

Elastomeric modification is particularly important in systems based on methyl methacrylate, the most commonly used monomer. Methyl methacrylate monomer polymerizes into a high-modulus brittle material, which is not very useful as an adhesive. It can be generally assumed that any purchased, fully formulated acrylic adhesive on the market at this writing will contain some level of elastic modification and will probably be based on methyl methacrylate.

Curing Methods

The polymerization of acrylics by means of free radicals (chain reaction) allows the adhesives to be dispensed and cured using unique techniques that are not used with other adhesive types, for example, urethane and epoxy adhesives, which cure by a condensation polymerization reaction.

These different curing possibilities result because free radicals can propagate from monomer to monomer without the need for further peroxide formation or intimate mixing. Thus, the cure, even if begun at a very localized region, can propagate throughout the adhesive glue line for distances of up to 2.5 mm (100 mils) without the further influx of free radicals.

There are two distinctly unique methods of handling and curing acrylic adhesives. These are the "accelerator lacquer" cure and the "equal-mix," or "no-mix" ("honeymoon"), cure. In both these methods of curing acrylics, the adhesives do not need to be intimately mixed to obtain cure (see Fig. 6 and 7).

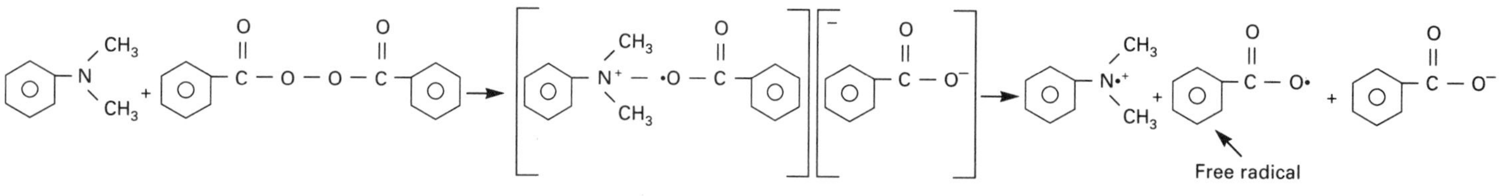

Fig. 3 Redox reactions for acrylics

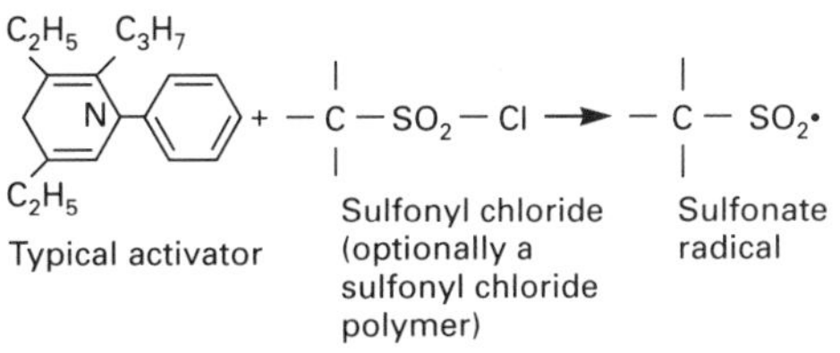

Fig. 4 structure

C_2H_5 C_3H_7

Typical activator + $-C-SO_2-Cl \rightarrow -C-SO_2\cdot$

Sulfonyl chloride (optionally a sulfonyl chloride polymer) Sulfonate radical

Fig. 4 Amine plus sulfonyl chloride initiation of the cure of acrylic adhesives (Ref 1)

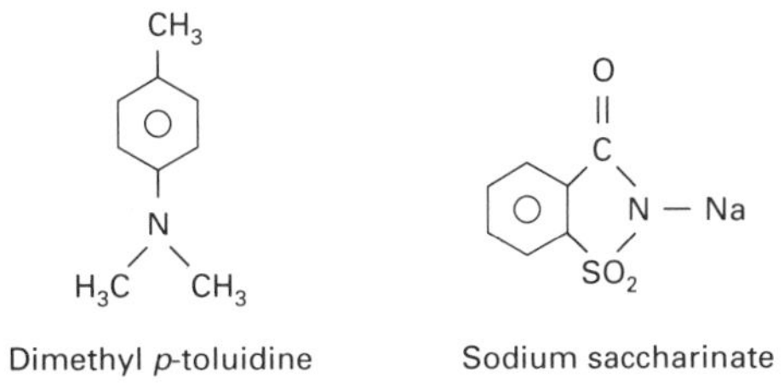

Fig. 5 structures

CH_3

Dimethyl *p*-toluidine

Sodium saccharinate

Fig. 5 Two-part no-mix system used to initiate acrylic adhesives cure

Table 2 Polymers used as modifiers

Polymer (oligomer)	U.S. patent
Chlorosulfonated polyethylene	3,890,407 (Ref 1)
Thermoplastic polyurethane........	3,994,764 (Ref 6)
Styrene-butadiene block copolymer ..	4,574,142 (Ref 7)
Carboxylated butadiene acrylonitrile	3,832,274 (Ref 8)
Polyvinyl methyl ether	4,223,115 (Ref 9)
Methacrylated urethane	3,873,640 (Ref 10)

Figure 6 depicts the use of a peroxide-containing accelerator lacquer (a solvent-base peroxide-containing lacquer film former) to cure the acrylic adhesive. When this accelerator lacquer is applied to one of the surfaces and subsequently dried, it represents a source of free radicals because of the presence of the peroxide. This peroxide remains unchanged until adhesive is applied on top of it, at which time amines in the adhesive interact with the peroxide-containing lacquer to dissociate the peroxide. The free radicals that are formed are then transported into the bulk of the adhesive and continue to propagate through the polymerizing methacrylate monomers to effect the final full cure.

The use of accelerator lacquers to cure acrylics is a very popular method because it allows the priming and storing of parts, the achievement of very rapid cures, and the use of inexpensive meter-mix-dispense equipment. As already noted, this cure technique cannot be used with condensation polymers such as epoxies and urethanes because they require full and intimate mixing of both components to achieve complete cures.

The other unique method of curing acrylics, the honeymoon, or no-mix, method (Fig. 7), also takes advantage of the ability of free radicals to propagate throughout the adhesive and complete the cure without the need for intimate mixing. In this approach, a bead of the A component can be applied to one surface, while a bead of the B component is applied to the other. As in the case of the accelerator lacquer, no reaction occurs until the two surfaces are mated. Once the two surfaces are joined, the rapid generation of free radicals occurs, followed by monomer polymerization and full adhesive cure. As with the accelerator lacquer, this technique cannot be used with other adhe-

sive types and requires the use of inexpensive dispensing equipment only.

Unlike many other adhesives currently on the market, a series of improved acrylic adhesives have appeared in the marketplace in rapid succession. These so-called generations of acrylic adhesives represent fundamental changes in the formulations and offer significant advantages to users. Their emergence, however, has caused some confusion among those trying to choose the best product for a given bonding application. In the descriptions below, each of the different acrylic types is discussed separately in order to clarify some of the confusion that still persists.

Functional Types and Properties

Conventional Acrylic Adhesives. Conventional, or first-generation, acrylic adhesives are the earliest examples of this technology. The formulation given in Table 3 represents the type of formulation that is cited in five U.S. patents: 3,994,764; 3,832,274; 3,873,640; 3,725,504; and 3,970,709. The formulation contains methacrylated monomers, a variety of polymers, aromatic amine initiator, and methacrylic acid. Formulations typical of that shown in Table 3 emerged in the early 1970s and found considerable utility in bonding thermoplastics. These first-generation products are still on the market and are still sold primarily for this purpose. They provide excellent gap-filling characteristics, a rapid cure, and an adhesion strength that is generally stronger than the substrates themselves.

Typical properties of first-generation adhesives are shown in Table 4. These adhesives have good adhesion to thermoplastics such as polystyrene, acrylonitrile-butadi-

ene-styrene (ABS), and polyvinyl chloride (PVC); to wood and rubber; and to steel and aluminum if the metals are clean. It can also be seen from Table 4 that adhesion to galvanized steel is lower than to other metals. Bonding to galvanized steel was a problem with first-generation acrylic adhesives. This problem has been overcome and is discussed in this article in the sections "Hybrid Epoxy-Acrylic Adhesives" and "Surface-Activated Acrylic (SAA) Adhesives."

It should be noted that if adhesives of the type shown in Table 3 are applied and cured using an accelerator lacquer, bond strengths equivalent to those achieved with mix-in peroxide pastes are obtained. In ABS bonding, for example, the literature reports substrate failures in the range of 2.9 kN (660 lbf) when either a mix-in accelerator or an accelerator lacquer is used.

Data have also been reported in company literature on the resistance of first-generation acrylic adhesives to aggressive environments, that is, alcohols and hot and room-temperature water. The patent literature also reports that when using mix-in accelerators, good bond strength retention is obtained for at least 35 days in these aggressive environments. The failure mode of substrate failure in plastic substrates, such as ABS, is also retained.

DH acrylic adhesives exemplify an emerging generation of acrylic adhesives. This type of material, cited in U.S. Patent 4,263,419 (Ref 13), represents a family of acrylics that first appeared in the early 1970s and were based on the carrier polymer chlorosulfonated polyethylene and acrylics.

These DH (DuPont Hypalon) adhesives had handling characteristics similar to those of other acrylics. They did differ, however, in the use of peroxides as part of the adhe-

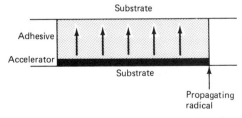

Fig. 6 Accelerator lacquer cure for acrylic adhesives

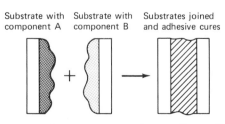

Fig. 7 Honeymoon, or no-mix, cure for acrylic adhesives

Table 3 Typical formulation of conventional, or first-generation, acrylic adhesives

Component	Parts by weight
Styrene-methyl methacrylate copolymer syrup................................	40
Methacrylic acid........................	9
Polymethyl methacrylate syrup...........	49
Diisopropanol *p*-toluidine	1.5
Toluhydroquinone	0.03

Source: Ref 8

Table 4 Properties of conventional, or first-generation, acrylic adhesives

U.S. Patent 3,994,764 (Ref 6)

Substrate	Shear strength	
	MPa	ksi
Steel/steel (clean)	26.2	3.80
Aluminum/aluminum	21.4	3.10
Wood/wood	12.4	>1.80 (wood failed)
Steel/glass	5.5	>0.80 (glass failed)
Steel/polyvinyl chloride	17.9	>2.60 (polyvinyl chloride failed)
Steel/glass-reinforced plastics	19.0	2.75
Steel/nylon	20.5	2.98
Steel/polystyrene	16.7	2.42
Steel/ acrylonitrile-butadiene-styrene	17.1	2.48
Steel/natural rubber	4.1	0.60 (rubber failed)
Wood/ acrylonitrile-butadiene-styrene	4.1	0.60
Hot dip galvanized steel/steel	2.9	0.42

Table 5 Typical high-performance acrylic adhesive formulation

Component	Parts by weight
Syrup, 25% polymer in methyl methacrylate (carboxylated butadiene acrylonitrile)	30–60
Methyl methacrylate	10–30
Methacrylic acid	1–5
Methacryloxyethyl phosphate	1–10
Mixed aromatic amines	1–3
Calcium molybdate/zinc phosphate blend	2–12

Source: Ref 14

amine accelerator, have limited availability, which interferes with the ability of manufacturers to supply product consistently.

High-Performance (HP) Acrylic Adhesives. The emergence of the DH acrylic adhesives was quickly followed by the introduction of so-called high-performance acrylic adhesives. This type of system, like the DH types, offered users the ability to bond through oily metals.

The appearance of the HP acrylic adhesives was marked by the publication of a number of patents, which specified the use of a variety of methacrylated and acrylated acid phosphate esters as key ingredients in the adhesive formulations. These special monomers were used to achieve the same amount of adhesion to unprepared metals as the DH systems, but without the need for an oily accelerator lacquer. Table 5 lists a typical HP acrylic adhesive formulation, obtained from the patent literature. The most notable of these patents is U.S. Patent 4,386,036 (Ref 14).

Adhesives of the type shown in Table 5 have bonded bronze, lead, nickel, magnesium, copper, aluminum, steel, and stainless steel. They do, however, continue to show weaknesses when zinc surfaces are bonded. The continued and increasing use in automobile panels of galvanized steel, which contains zinc on its surface, is the reason these adhesives are not well suited for the continually growing number of applications in automobile markets.

Table 6 gives adhesive lap-shear strengths when various substrates are bonded with high-performance acrylic adhesives. It can be seen that strengths obtained

sive itself and the use of amine as the mix-in hardener or accelerator lacquer.

Useful accelerator lacquers have been difficult to formulate with the DH technology because of the oily nature of the most popular amine (N-phenyl-3,5-diethyl, 2-propyl-1,4 dihydroxy pyridine). The oily accelerator lacquers that are generally available for curing DH acrylics are difficult to apply, and primed parts are difficult to handle.

At the time they were introduced, DH acrylic adhesives were state of the art because they were the first type of acrylic that could be successfully cured on unprepared metals. These systems had the ability to bond to dirty steel and aluminum surfaces and to bond to aluminum and steel surfaces that still had oils and drawing compounds on them. This ability marked a significant advance and established acrylics as a unique family of adhesives for bonding oily, unprepared metals.

Although DH acrylic adhesive systems enjoyed considerable popularity when they were first introduced, they have been replaced by other, more user-friendly types of adhesive formulations. In addition, some of the key raw materials, most notably the

when bonding to oily metals are at least as high, and often higher, than strengths obtained with clean surface bonds. When HP acrylics are used to bond oily metal and the bonded parts are subjected to aggressive environments (for example, condensing humidity, salt spray, water immersion, and gasoline immersion) for up to 1000 h, little or no reduction in overall bond strength is reported.

High-Impact (HI) Acrylic Adhesives. Another advance in acrylic adhesives is that of improved, low-temperature properties. Adhesives of this type are covered by U.S. Patent 4,769,419 (Ref 15) and others.

These patents describe novel adhesive structural compositions for metal bonding applications. The adhesives contain olefinic elastomers that are further reacted with monoisocyanates to eliminate any residual hydroxyl functionality. The olefinically modified elastomers can be used to formulate acrylics that are especially useful as room temperature curing adhesives that bond to oily metal and have good low-temperature performance.

The essential part of this generation of acrylics, as has been noted, is the use of the urethane-modified, olefinically terminated liquid elastomer. A typical polymer is shown in Fig. 8. Its advantage is that it is useful in many applications in which good adhesion and impact resistance at temperatures as low as −25 °C (−13 °F) are required (hence, high-impact, or HI, adhesives). Acrylic adhesives that existed prior to this advancement did not have acceptable performance in impact tests at these low temperatures. In some cases, significant embrittlement was observed at temperatures only as low as 0 °C (32 °F).

Other than this specific elastomer, cited in U.S. Patent 4,769,419, the remaining components of this particular family of acrylic adhesives are much like those previously described. Consequently, fully formulated adhesives can be expected to include a variety of fillers, other possible elastomer modifiers, typical acrylic monomers, metal adhesion promoting phosphorus compounds, and an initiator system most likely consisting of amines and peroxides.

Hybrid Epoxy-Acrylic Adhesives. Two other U.S. patents, 4,452,944 and 4,467,071 (Ref 16, 17), further expand HI acrylic adhesive technology. Both of these patents cite structural adhesives that have improved elevated-temperature performance characteristics.

The most significant of the two is U.S. Patent 4,467,071, which specifies an adhesive with improved heat resistance. This patent cites the use of large amounts of epoxy resin in conjunction with the previously described urethane-modified elastomer, to obtain both low-temperature properties and improved elevated-tempera-

Table 6 High-performance acrylic adhesive bonding of oily metals

Metal	Surface preparation	Shear strength	
		MPa	ksi
SAE 1010 cold-rolled steel	Oiled	44.1	6.39
	Solvent wiped	42.1	6.10
	Grit blasted	41.0	5.94
6061-T6 aluminum	Oiled	35.1	5.09
	Solvent wiped	33.6	4.87
	Grit blasted	34.7	5.03
2024-T3 aluminum-clad aluminum	Oiled	33.6	4.88
	Solvent wiped	32.5	4.72
5052-O aluminum	Oiled	14.8	2.14
	Solvent wiped	15.1	2.19

Note: Assemblies were made and tested according to ASTM D 1002 using 1.6 mm (0.063 in.) thick metal stock and a 0.25 mm (0.010 in.) thick glue line.

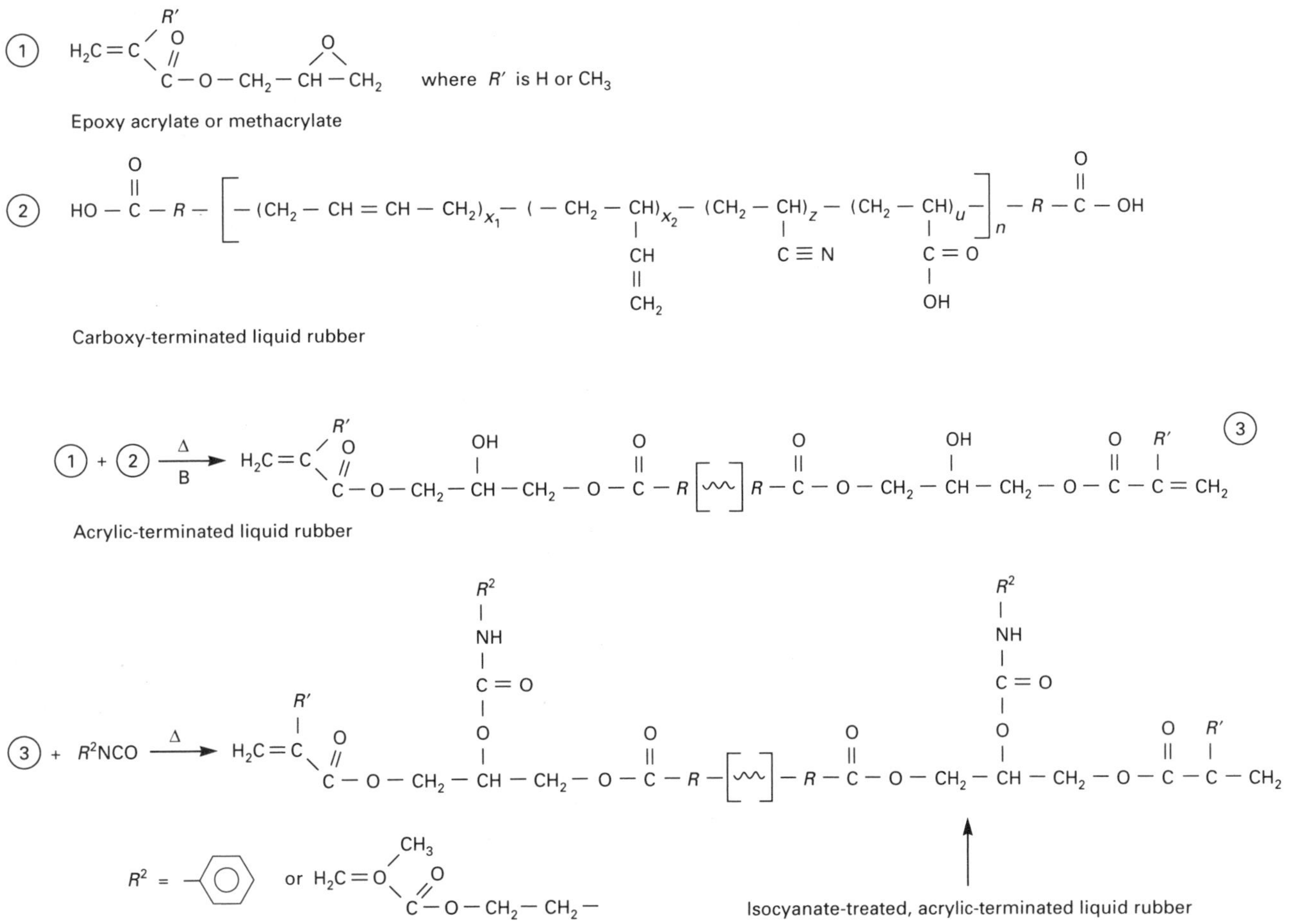

Fig. 8 Structurally generalized composition and process preparation of urethane-modified, olefinically terminated acrylic adhesive

ture properties. The specific combination of phosphorus-containing olefinically unsaturated monomers and epoxy resins of the bisphenol A type is what achieves elevated-temperature resistance. These adhesives are considered actual hybrids of epoxy and acrylic technology because of the significant levels of epoxy resin in the best-performing embodiments of this technology.

A typical formulation is shown in Table 7. In formulations cited in this patent, it is shown that improved lap-shear strength and resistance to temperatures up to 204 °C (400 °F) can be achieved when a ratio that ranges from 2:1 to 4:1 equivalents of epoxy to acid phosphate is used. With this optimum ratio, prolonged resistance of at least 30 min at 204 °C (400 °F) is possible. Adhesives that are prepared with this ratio will retain over 14 MPa (2 ksi) lap-shear strength after severe elevated-temperature postbakes.

It has been proposed that adhesives of the hybrid epoxy-acrylic type are very useful in automotive metal bonding applications in which resistance to high temperatures in paint bake ovens is required. The possibility of curing these systems at room temperature would also be an advantage. If room-temperature curing proved to be feasible,

expensive bake ovens and induction curing units would no longer be required.

Equal-Mix Acrylic Adhesives. Another special type of acrylic adhesives are the so-called equal-mix, or no-mix (honeymoon) type. This type of acrylic adhesive is unique among acrylics because polymerization is adequately achieved when the A component of the adhesive is put on one side of the substrate to be bonded, and the B component is put on one side of another substrate. When these two substrates, or halves, are joined, as shown in Fig. 7, enough free radicals are generated to initiate and complete the cure.

Table 7 Heat-resistant hybrid epoxy acrylic adhesive formulation
U.S. Patent 4,467,071 (Ref 17)

Component	Parts by weight
Polymer monomer syrup	20
Elastomer solution in methyl methacrylate	38
Methyl methacrylate	9
Inorganics	25
Methacryl phosphate	5
Stabilizers/initiators	3
Epoxy	Varies
Benzoyl peroxide	4

It is important to note that although most acrylic adhesives will cure to some extent using this method, certain initiators have been developed that are particularly well suited to this technique. One of the best examples of this technology is found in U.S. Patent 4,331,795 (Ref 4). This patent cites a cobalt salt cure accelerator used in one component of the adhesive and a hydroperoxide and aromatic amine used in the other component.

The formulations covered in this patent are typical of this type of acrylic adhesive. Reactive monomers, elastomers, fillers, and stabilizers are used. The uniqueness of this type of formulation is within the accelerator system, although the manner in which the initiator influences the propagating radical chain has not been clarified.

The equal-mix, or no-mix, technology has found extensive use in the electrical industry. It is a very useful tool for bonding magnets in electrical motors, where the fast cure and easy application technique are distinct advantages.

Surface-Activated Acrylic (SAA) Adhesives. The increased use of galvanized steel as a substrate in exterior automobile body panels has resulted in more attention to the

Table 8 Performance of surface-activated acrylic adhesives
Glue line 0.25 mm (10 mils) on G-90 galvanized steel

| Adhesive type | Lap-shear strength | | Failure mode |
	MPa	ksi	
Surface-activated acrylics	15.2	2.20	Cohesive and in metal
Cyanoacrylate	5.5	0.800	Adhesive
Anaerobic	1.48	0.215	Adhesive
Two-part urethane	7.79	1.13	Adhesive
Two-part epoxy	17.9	2.60	Adhesive and cohesive
Two-part acrylic	3.45	0.500	Adhesive and cohesive

Note: Parts allowed to cure 24 h at room temperature prior to testing

Table 9 Suppliers of acrylic adhesives

Supplier	Product	Form and type
Lord Corporation, Industrial Adhesives Division Erie, PA	Versilok	Pastes, all types
Loctite Corporation Hartford, CT	Depend	Modified pastes
Permabond International Englewood, NJ	Quickbond	Toughened acrylic
Devcon Corporation Danvers, MA	MVP	Acrylic adhesives
Hysol Industrial Products Pittsburg, CA	Engineering Adhesive	Modified types

bonding of this material. Galvanized steel helps automotive manufacturers realize the goal of building a vehicle with a body that will last at least 10 years without rusting. However, galvanized steel is difficult to weld and, even if welded successfully, has a tendency to corrode galvanically at the weld sites. In response to this problem, another class of acrylic adhesives emerged in the late 1980s.

SAA adhesives are specifically designed for use in bonding galvanized steel (U.S. Patents 4,855,001 and 4,703,089) (Ref 18, 19). These adhesives have the advantage of being truly one-part adhesives because they use the galvanized (zinc) surface as a catalyst to promote the curing reaction.

In a general sense, these systems might be considered anaerobic- or cyanoacrylate-type adhesives (both of which are discussed in separate articles in this Volume) because they are one-part systems that begin to cure when brought in contact with the surface to be bonded. However, when other performance and handling characteristics are considered, it is clear that they represent a completely different type of adhesive and are neither cyanoacrylates nor anaerobics.

Differences in handling are found primarily in their packaging. SAA adhesives can be packaged in large containers (for example, drums) without the risk of rapid polymerization and gelation. Anaerobics and cyanoacrylates, on the other hand, because of their composition and curing mechanism, cannot be packaged in conventional containers having a mass of more than a few pounds.

When the bond performance and cure on different substrates are compared, additional differences between SAA adhesives, cyanoacrylates, and anaerobics are found. Anaerobics will cure on essentially any substrate once air is excluded. SAA adhesives are specifically used on zinc and some related metals. Cyanoacrylate polymerization is initiated by basic catalysis and is essentially different compositionally. A comparison (of lap-shear strengths when bonding galvanized steel) of these different types to each other and to two-part epoxies, and acrylics and urethanes, is given in Table 8.

Because of their relatively recent emergence, SAA adhesives have not yet established a firm market niche in the area of galvanized steel bonding for automotive body panels. With time, the advantages they offer should result in market growth. Because they are one-component systems and are considerably lower in cost than cyanoacrylates and anaerobics, they may ultimately be used in areas in which anaerobics and cyanoacrylates have traditionally found application.

Photocurable acrylic adhesives represent a separate adhesives class that cure by exposure to photons in various forms. Of most interest are the ultraviolet (UV) wavelengths (below 400 nm), and, to some extent, the visible range (400 to 700 nm). This type of adhesive is used in bonding substrates that are transparent in the UV and visible regions, which is typical of many clear plastics and glass.

To achieve UV cure, a specific type of initiator, called a photoinitiator, must be used. This material is added to the formulation and results in the formation of free radicals when exposed to UV or visible light. Representative compounds that are initiated by UV light, particularly in the 200 to 300 nm wavelengths, are the benzoin ethers, benzophenones, and similar compounds. These compounds and others that can be activated in the visible region are shown in Fig. 9. When these photoinitiators

are exposed to a particular wavelength, they dissociate into smaller segments by a process known as scission, forming free radicals containing molecular fragments.

Compositionally, photocurable acrylics have formulations similar to those acrylics already described in this article, except that photoinitiators, rather than redox or other techniques, are used for the generation of free radicals. The only other significant difference is the use of an acrylate rather than a methacrylate monomer as the carrier. The acrylates are much more prone to UV initiation and propagation than the methacrylates and are consequently the better choice for formulators.

Company literature available from several adhesives suppliers lists UV-curable and visible light curing acrylic adhesives as part of their product package. Applications in the area of polycarbonate lens bonding exemplify the successful use of this type of acrylic adhesive.

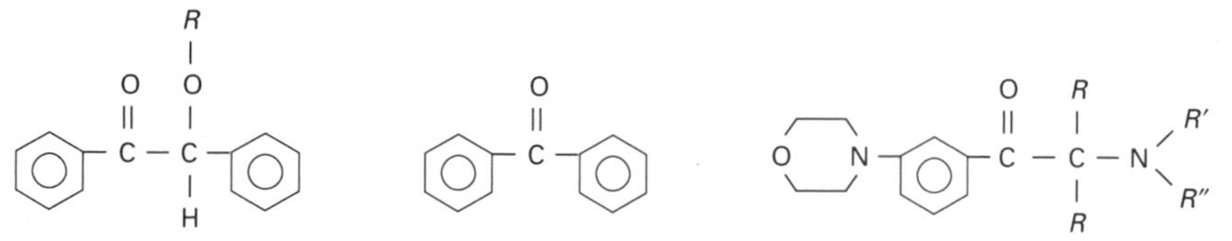

Major Suppliers

Acrylic adhesives are supplied by relatively few companies. The patent literature suggests that many companies share the same technology, and variations of it, under specific trade names. Table 9 lists companies and their products.

Cost Factors and Future Trends

At the beginning of the 1990s, worldwide sales for acrylic adhesives of the types discussed in this article range from 5 to 10 million dollars, making them a rather small segment of the overall industrial structural adhesive market, compared to epoxy and urethane. However, acrylic adhesives do fill certain market needs in the very-high-performance area, ensuring their continued use in certain applications for many decades.

The capability of acrylic adhesives to bond quickly to a variety of unprepared metals has been a significant factor for

Fig. 9 Typical photoinitiators used in photocurable acrylic adhesive formulations

certain electrical components. Bonding magnets for fractional horsepower motors is one example. Acrylics have also been very effective for bonding aluminum, leading to their use in construction (windows and doors), recreation (boats), and military use (aircraft repair and primary bonding). They also offer potential in the automotive area, where metal bonding is replacing welding, and primary-structure bonding is becoming a possibility. The experimental use of acrylics for bonding metal exterior structures is planned for the future.

Acrylics also offer the unique capability of bonding to plastics without the need for surface priming. They are also very good gap-filling materials, which makes their use preferable to solvent welding. Markets include recreational (boats, motorcycles, and sporting equipment), industrial (equipment enclosures and housings), and, to some extent, plastic appliance component bonding.

In the future, the various families of acrylic adhesives are expected to grow as increasing numbers of design engineers specify bonded parts and as specialty substrates that are bondable only with acrylic adhesives continue to emerge. The development of formulations with less odor, one-part systems, and hybrid types will also have a positive effect on overall growth.

REFERENCES

1. P.C. Briggs, Jr., Novel Adhesive Compositions; Chlorosulfonated Polyethylene or Sulfonyl Chlorides and Chlorinated Polyethylene, U.S. Patent 3,890,407, June 1975
2. A.G. Bachmann, Adhesive Composition; Quick-Setting, High Bonding Strength, U.S. Patent 4,348,503, Sept 1982
3. M.M. Skoultchi, Rapid Curing Two-Part Adhesives; Acrylic Monomer and Copper Saccharinate; Alpha-Hydroxy Sulphone or Alpha-Amino Sulfone as Activator, U.S. Patent 4,081,308, March 1978
4. I. Kishi, T. Nakano, and K. Ukita, Two Liquid-Type Adhesive Composition Comprising a Cobalt Salt Cure Accelerator in One Portion and a Hydroperoxide With an Aromatic Amine and/or a Pyridine Derivative Cure Accelerator in the Second Portion, U.S. Patent 4,331,795, May 1982
5. W.A. Lees, Bonding Method Employing a Two-Part Anaerobically Curing Adhesive Composition, U.S. Patent 3,658,624, April 1972
6. L.E. Wolinski, Adhesive Compositions; Polyurethanes, Free Radical Catalysts, Acrylic and Acid Monomers, U.S. Patent 3,994,764, Nov 1976
7. R.S. Charnock, Curable Adhesive Composition Toughened With Styrene-Butadiene Block Copolymer Rubbers, U.S. Patent 4,574,142, March 1986
8. W.J. Owston, Fast Curing Adhesives; Butadiene Elastomer, Acrylate, Methacrylic Acid, Reducing Agent, U.S. Patent 3,832,274, Aug 1974
9. T.H. Dawdy, E.C. Hornaman, F.H. Sexsmith, and D.J. Zalucha, Structural Adhesive Formulations; Addition Polymerizable System Containing Redox Couple Catalyst and Phosphorus Compound, U.S. Patent 4,223,115, Sept 1980
10. D.D. Howard and W.J. Owston, Adhesive Bonding of Polyvinyl Chloride and Other Synthetic Resins, U.S. Patent 3,873,640, March 1975
11. W.J. Owston, Fast Curing Polychloroprene Acrylic Adhesive, U.S. Patent 3,725,504, April 1973
12. W.J. Owston, Adhesive Formulations for Bonded Metal Assemblies With Resistance to Aggressive Environments, U.S. Patent 3,970,709, July 1976
13. H.G. Gilch and G. Piestert, Adhesive Composition, Addition Polymerizable Mixture With Activator, U.S. Patent 4,263,419, April 1981
14. H.J. Kleiner, Process for the Preparation of Vinylphosphonic Acid; Catalyzed Thermal Decomposition of 2-Acetoxyenthanephosphonic Dialkyl Ester, U.S. Patent 4,386,036, May 1983
15. T.H. Dawdy, Modified Structural Adhesives; Acrylated Conjugated Polymers Reacted With Isocyanates, U.S. Patent 4,769,419, Sept 1988
16. T.H. Dawdy, Structural Adhesive Formulations; for Bonding Metals and Plastics; Curable Unsaturated Urethane, Butadiene Polymer, Styrene, or Acrylate Polymer or Monomer and Phosphorous Compound, U.S. Patent 4,452,944, June 1984
17. T.H. Dawdy, Epoxy Modified Structural Adhesives Having Improved Heat Resistance; Unsaturated Organophosphorus Partial Ester, Redox Catalyst, U.S. Patent 4,467,071, Aug 1984
18. R.M. Bennett and D.J. Damico, Structural Adhesive Formulations and Bonding Method Employing Same, U.S. Patent 4,855,001, Aug 1989
19. D.J. Damico, Structural Adhesive Formulations; Heat Curable Addition Polymers, Polyurethanes; Bronsted Acid Catalysts, Reducible Metal Components, Sulfonyl Halides; Blends for Metal Bonding, U.S. Patent 4,703,089, Oct 1987

SELECTED REFERENCES

- A.G. Bachmann, Adhesive Composition; Acrylic Ester, Latent Catalyst, U.S. Patent 4,429,088, Jan 1984
- A.G. Bachmann, Adhesive Bonding Method; Acrylate Monomer, Elastomer Fillers, and Latent Catalyst System, U.S. Patent 4,432,829, Feb 1984
- A.G. Bachmann, "'Aerobic' Acrylic Adhesives (New Technology in Acrylic Adhesives)," Paper presented at the Adhesives '84 Conference (Cleveland), Society of Manufacturing Engineers, Sept 1984
- R.L. Bowen and H. Argentar, A Method for Determining the Optimum Peroxide-to-Amine Ratio for Self-Curing Resins, Appl. Polym. Sci., Vol 17, 1973
- W.H. Brendley, Jr., Fundamentals of Acrylics, Paint Varn. Prod., July 1973
- P.C. Briggs, Jr., Method of Adhesive Bonding; Chlorosulfonated Polyethylene or Mixture of Sulfonyl Chloride With Chlorinated Polyethylene in Polymerizable Vinyl Monomer, U.S. Patent 4,106,971, Aug 1978
- P.C. Briggs, Jr., Acrylic Monomers, U.S. Patent 4,112,013, Sept 1978
- Catalysts for the Polymerization of Acrylic Monomers, Rohm & Haas Company, June 1959
- D.J. Damico, Bonding Galvanized Steel With RT Curing Acrylics, Adhes. Age, Oct 1987
- T.H. Dawdy, E.C. Hornaman, F.H. Sexsmith, and D.J. Zalucha, Structural Adhesive Formulations; Addition Polymer, Reducing Agent, Oxidizer, U.S. Patent 4,293,665, Oct 1981
- L.R. Gatechair and D. Wostratzky, Photoinitiators: An Overview of Mechanisms and Applications, J. Radiat. Curing, July 1973
- W.A. Lees, The Science of Acrylic Adhesives, Br. Polym. J., Vol 11, June 1979
- W.A. Lees, Toughened Structural Adhesives and Their Uses, Int. J. Adhes. Adhes., Vol 1 (No. 5), July 1981
- G. Pasternack, Fundamental Aspects of Ultraviolet Light and Electron Beam Curing, J. Radiat. Curing, July 1982
- R.B. Seymour, Introduction to Polymer Chemistry, McGraw-Hill, 1971
- D.W. Wood and R.N. Lewis, Free Radical Initiators for Styrene Polymerization, Mod. Plast., July 1974
- J.E. Yeames, "A New Adhesive for Structural Bonding of Engineering Materials," Paper presented at the Adhesives '84 Conference (Cleveland), Society of Manufacturing Engineers, Sept 1984

Cyanoacrylates

Jules E. Schoenberg, National Starch and Chemical Corporation

CYANOACRYLATES are single-part solvent-free adhesives that form strong bonds on a wide variety of substrates. They harden at room temperature by anionic polymerization initiated either by adsorbed moisture or by alkaline sites on the surface.

These adhesives, developed in the early 1950s by the Tennessee Eastman Company, became commercially available in 1958. Since then, many companies worldwide have entered the cyanoacrylate field, some as chemical manufacturers, others as formulators and distributors.

Chemistry

Cyanoacrylate esters are 1,1-disubstituted ethylenes having the basic structure shown in Fig. 1. Their high reactivity is due to the ability of the electron-withdrawing ester and nitrile groups to share negative charges. Three resonance structures of the anion are shown in Fig. 2.

Commercially available 2-cyanoacrylate esters, their uses, and properties are listed in Table 1. Of the variety of monomers that can represent the radical group shown in Fig. 1, ethyl and methyl are the most commonly used because of their low price, high bond strengths, and availability.

The Manufacture of 2-Cyanoacrylate Esters. There are several known methods of synthesizing cyanoacrylate monomers. The most important industrial method uses the sequence:

1. Base-catalyzed Knoevenagel reaction of formaldehyde with a cyanoacetate ester to produce the 2-cyanoacrylate ester, which spontaneously polymerizes
2. Thermal depolymerization under acidic conditions to regenerate the monomer
3. Purification of the monomer by one or more distillations
4. Addition of stabilizer, thickeners, and other ingredients to produce a commercial adhesive

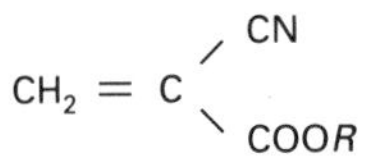

The first step, the reaction of formaldehyde with the cyanoacetate ester, is preferably carried out using an excess of the ester. The catalyst is usually a secondary amine, such as piperidine. Two likely pathways for this reaction are via the hydroxymethyl derivative and the Mannich base (Fig. 3).

The 2-cyanoacrylate esters are unstable in the presence of amines and undergo anionic polymerization. The excess cyanoacetate ester used in the Knoevenagel reaction acts as a chain transfer agent, reducing the molecular weight of the polymer and lowering the reaction viscosity (Ref 1). The two end groups are derived from the cyanoacetate ester (Fig. 4).

The second step is the depolymerization of the polymer by an unzipping mechanism. A cyanoacrylate polymer readily depolymerizes because the ceiling temperature (the temperature at which the monomer and polymer are in equilibrium) is low enough that other thermal reactions do not readily occur. The ceiling temperature is determined by the relative stabilities of the monomer and polymer. Structural features that either stabilize the monomer or lower the stability of the polymer reduce the ceiling temperature. In the case of cyanoacrylates, one feature that stabilizes the monomer is conjugation of the double bond. One feature that lowers the stability of the polymer is steric strain from the interactions of neighboring side groups.

The depolymerization step is carried out at high temperatures (150 to 190 °C, or 300 to 375 °F), and high vacuum (135 to 665 Pa, or 1 to 5 torr) in the presence of inhibitors that prevent repolymerization. The 2-cyanoacrylate ester distills over and is collected onto a nonvolatile acid anhydride or strong acid in the receiver and a free radical trap such as hydroquinone.

A by-product of the depolymerization is a 2,4-dicyanoglutarate (Fig. 5). In the third step, the monomer is separated from this material, as well as from excess polymerization inhibitors, by one or more distillations (Ref 2). The fourth step is described in the section ''Additives'' in this article.

Curing Mechanisms. Cyanoacrylates are capable of polymerizing by either an anionic or free radical mechanism. Anionic polymerization is much more facile and is the mechanism of bond formation in adhesives. Free radical polymerization can be a cause of instability during long-term storage, especially at elevated temperatures.

Almost any electron-donor compound, known as a nucleophile, can initiate the anionic polymerization of cyanoacrylates. Examples are water (possibly OH^-), alcohols (possibly OR^-), amines, phosphines, carboxylate anions, inorganic bases, halide ions, epoxides, mercaptans, and sulfoxides (Ref 3-6). Electron-rich olefins, such as styrene or vinyl ethers, react spontaneously with cyanoacrylates to produce alternating copolymers (Ref 7, 8).

The electrical charge on the propagating polymer depends on the initiator. Anionic initiators, such as OH^- and Br^-, produce

Fig. 1 Structure of cyanoacrylate esters

$$CH_2 = C \begin{array}{c} CN \\ \\ COOR \end{array}$$

Table 1 Commercially available 2-cyanoacrylate esters

Ester	Radical group, R	Uses and properties
Methyl	$-CH_3$	Strongest bonds on metals; good solvent and temperature resistance
Ethyl	$-CH_2CH_3$	General purpose; forms strong bonds on metals, plastics, and rubbers
Allyl	$-CH_2CH=CH_2$	Use temperatures above 100 °C (212 °F)
Butyl	$-CH_2CH_2CH_2CH_3$	Plastics and rubbers; improved flexibility; less irritating vapor than that of lower cyanoacrylates
2-methoxyethyl	$-CH_2CH_2OCH_3$	Low odor; improved flexibility; nonfogging
2-ethoxyethyl	$-CH_2CH_2OCH_2CH_3$	Same as 2-methoxyethyl
2-methoxy-1-methylethyl	$-CHCH_2OCH_3$ $\quad\quad\mid$ $\quad\quad CH_3$	Same as 2-methoxyethyl

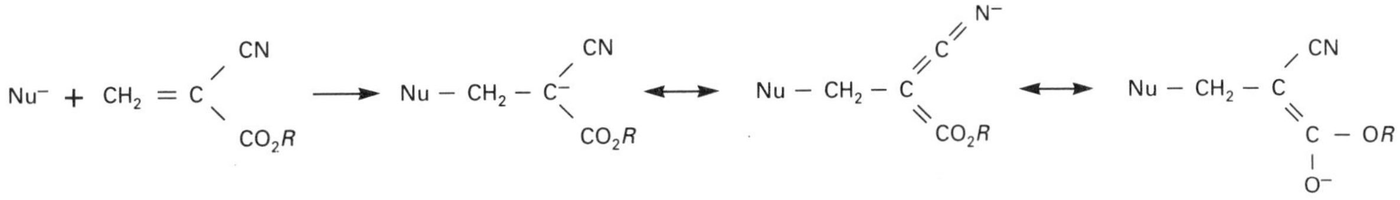

Fig. 2 Resonance structures of the anion of cyanoacrylate esters, where Nu^- refers to a negatively charged nucleophile

negatively charged polymers. Nonionic initiators, such as amines or phosphines, produce makrozwitterions with a negative charge on the growing end and a positive charge on the other end (Ref 9). This is shown in Fig. 6. The molecular weights of the polymers are very high, approaching 10^6 g/mol (Ref 9, 10).

The bonding action of cyanoacrylates depends on the presence of either alkaline (basic) sites or adsorbed water on the surface of the substrate. Glass has an alkaline surface and bonds very rapidly. Many rubbers and plastics contain additives, such as antioxidants and vulcanization accelerators, that can migrate to the surface and initiate polymerization. The bonding surface of a metal is an oxide or hydroxide. The basicity of the metal oxide or hydroxide increases with a decrease in the valence and an increase in the radius of the metal ion (Ref 11).

Surface Activators. Acidic and hydrophobic surfaces, which are poor at initiating cyanoacrylate polymerization, can be activated by the application of an accelerator. Surface activators are usually weak organic bases, such as aromatic tertiary amines, dissolved in a volatile solvent. They should be used sparingly. Too rapid a polymerization can lead to poor adhesion due to inadequate substrate wetting or to the formation of an inert skin that can prevent full cure. Accelerators can be applied from an aerosol dispenser, brush, or felt-tip pen. The aromatic amines retain their activity for several minutes after being applied. Most suppliers of cyanoacrylate adhesives sell surface activators.

Hydrolysis. Cyanoacrylates are susceptible to several types of hydrolytic reactions. These include hydrolysis of ester groups in both the monomer and polymer and hydrolysis along the backbone of the polymer (Fig. 7). Hydrolysis can have a pronounced effect on the properties of an adhesive. The 2-cyanoacrylate esters are more prone to

hydrolysis than most esters because of the presence of the electron-withdrawing nitrile group.

Ester hydrolysis of the monomer produces a relatively strong organic acid, 2-cyanoacrylic acid, which can retard the cure and alter the adhesive properties on various surfaces. As an example, old samples of cyanoacrylate adhesives that have undergone extensive hydrolysis are slow curing and produce weak bonds on rubbers.

Hydrolysis of the cured polymer is undesirable because it can lead to loss of bond strength and the formation of a toxic material, formaldehyde. This is one reason that cyanoacrylates have not received approval in medical applications from the U.S. Food and Drug Administration. As described in the section below, cyanoacrylate adhesives have been used in medical treatment in other countries.

Markets and Applications

A myriad of commercial uses exist for cyanoacrylate adhesives. Industries that are major markets include the automotive, beauty aid, consumer product, electrical and electronic, machinery, medical device, military, plumbing, and transportation industries. Typical commercial applications, categorized by bonding, fixturing, and potting functions, are listed in Table 2. There are two other applications of interest.

The first of these is the use of cyanoacrylates in many countries as medical adhesives to close surgical and other wounds. Butyl and higher ester cyanoacrylates are preferred because these polymers are flexible and have low tissue toxicity. They were used by the U.S. military during the Vietnam War, but have not received approval in the United States for the treatment of civilians.

Second, cyanoacrylates are used by law enforcement agencies to develop fingerprints (Ref 12). The vapors polymerize on fingerprints, forming visible lines. Methyl and ethyl 2-cyanoacrylates are preferred because of their relatively high vapor pressures.

Major Suppliers

The U.S. suppliers are listed in Table 3. Most of these companies provide literature for their products, which lists viscosities,

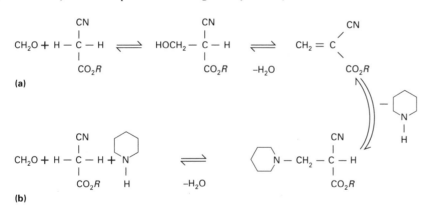

Fig. 3 Pathways for the Knoevenagel reaction in cyanoacrylate esters. (a) Via the hydroxymethyl derivative. (b) Via the Mannich base

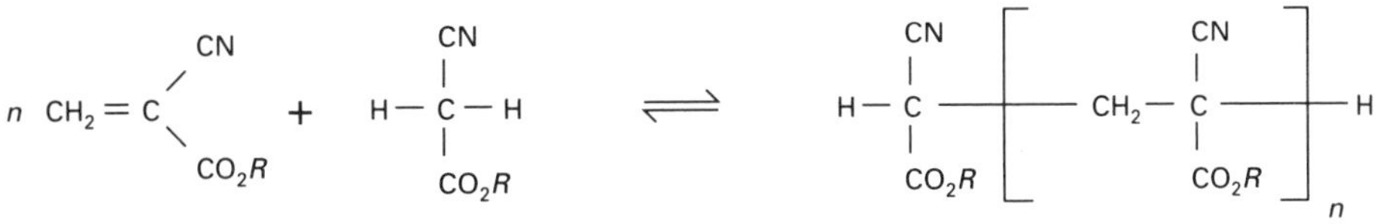

Fig. 4 Structure of the polymer based on cyanoacrylate esters

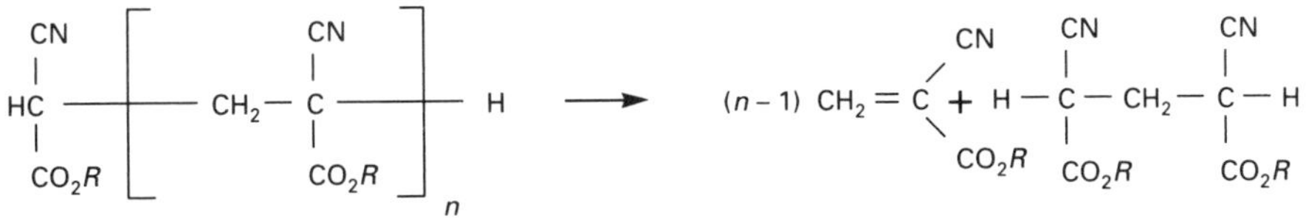

Fig. 5 Depolymerization of the polymer based on cyanoacrylate esters

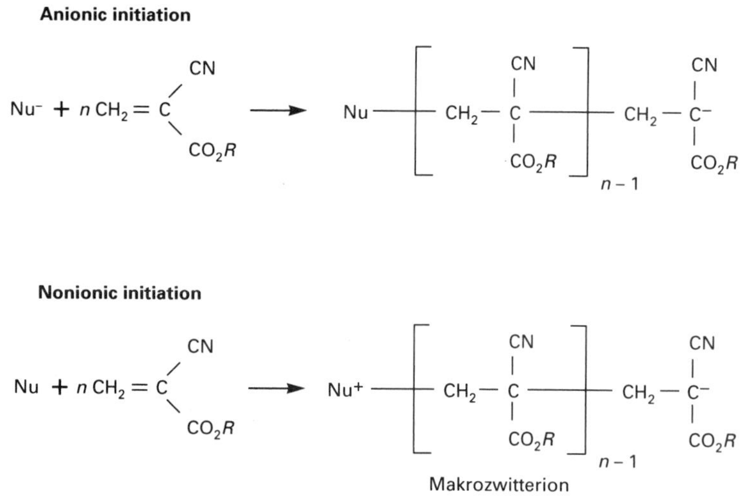

Anionic initiation

Nonionic initiation

Makrozwitterion

Fig. 6 Mechanisms of initiation in the polymerization of cyanoacrylate esters

Table 2 Typical commercial applications

Bonding	Fixturing (a)
Wiper blades	Jumper wires
Rubber bumpers	Spacers
Rubber feet	Filter caps
Strain gages	Shunt plates
Speaker cones	Heat sinks
Gears to shaft	Hardware to printed circuit
Video game cartridges	boards
Pushbuttons	Ferrite cores
Acrylic windows	Gaskets
Nameplates	Tennis racket parts
Dialysis membranes	
Catheters	
O-rings	**Potting**
Display panels	Transistors
Honing stones	Fiberglass molds
Gearshift indicators	Slotted screwheads

(a) Occurs when the assembly develops enough strength to be handled

cure rates, and bond strengths on various substrates.

Cost Factors

Although cyanoacrylate adhesives are relatively expensive on a weight basis, they are nonetheless considered economical by commercial users because:

- Very little is applied
- Automatic dispensing equipment can be used
- Bonding times are short
- Most substrates can be bonded
- Ovens are not needed because the adhesives cure at room temperature
- No mixing is required
- They do not contain solvents that have to be recovered or burned
- Expensive fixtures are not required

The 1989 price for methyl and ethyl 2-cyanoacrylates varied from about $130 to $175/kg ($60 to $80/lb, or $3.75 to $5/oz).

The higher esters cost $220/kg ($100/lb) or more. Projected 1989 sales in the United States were $168 million (Ref 14).

Properties

Cure rate, specifically, fast bonding, is one of the main reasons that cyanoacrylates are popular. Cure rates vary with the adhesive, substrate, surface treatment, and relative humidity. Fixturing times can vary from 2 s to 3 min, while full cure takes several minutes to 1 day. Moisture has a catalytic effect on the polymerization: The higher the relative humidity, the faster the cure. Relative cure rates for different surfaces are listed in Table 4.

The setting times of most commercial cyanoacrylates increase with increasing viscosity. This is due, in part, to slower spreading and wet-out on the surface, but also to higher amounts of inhibitors added by manufacturers to compensate for impurities in the thickener that can initiate poly-

merization. It also reflects the effects of larger gaps, which are usually present when a higher-viscosity type is chosen.

Bond strengths of cured cyanoacrylates on various surfaces are listed in Table 4. The actual values depend on such factors as monomer type, additives, surface treatment, temperature, and relative humidity. With plastics and rubbers, a bond is considered strong if the substrate rather than the joint fails.

Metals. Cyanoacrylates form strong bonds with most metals. Failure is usually adhesive (interfacial). Bond strengths decrease with the decreasing polarity of the 2-cyanoacrylate ester (that is, methyl > ethyl > butyl). Elongation to break increases with the decreasing polarity of the ester.

Plastics. Cyanoacrylates bond most plastics. In many cases, the bond strength is higher than the tensile strength of the substrate.

Polymers containing polar groups are easier to bond than polymers without polar groups. Polyolefins and other nonpolar

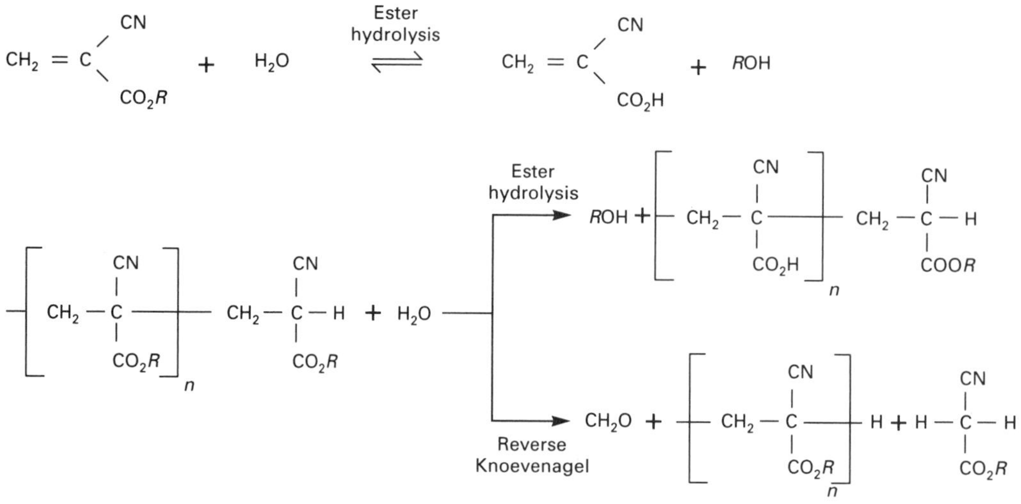

Fig. 7 Possible hydrolysis reactions in cyanoacrylates

Table 3 U.S. suppliers of cyanoacrylates

Alembic USA Mountain View, CA	Fel-Pro Inc. Skokie, IL	Oneida Electronic Manufacturing Co. Inc. Meadville, PA
Alteco USA, Inc. Torrance, CA	Firecat Technology Mountain View, CA	Pacer Technology Inc. Campbell, CA
Apple Adhesives Inc. Hollis, NY	Hernon Manufacturing Inc. Brooklyn, NY	Penn International Chemicals Mountain View, CA
Bostik Division Emhart Corp. Middleton, MA	Krazy Glue Inc. New York, NY	Permabond International Division of National Starch and Chemical Corp.
Cadillac Plastic & Chemical Co. Troy, MI	Kristy Wells Inc. New York, NY	Englewood, NJ
R.H. Carlson Co., Inc. Greenwich, CT	Loctite Corp. Newington, CT	Permatex Industrial Corp. Avon, CT
Ciba-Geigy Corp. Formulated Systems Group Madison Heights, MI	Lord Corp. Chemical Products Group Erie, PA	QCM Co. Kent, WA
Conap Inc. Olean, NY	Lu-Sol Corp. El Monte, CA	RTG Inc. Canton, MA
Dayton Division Whittaker Corp. West Alexandria, OH	Mavidon Corp. Palm City, FL	Super Glue Corp. Hollis, NY
DEVCON Corp. Danvers, MA	Morton Chemical Division Morton Thiokol, Inc. Chicago, IL	Three Bond of America, Inc. Torrance, CA 3M Co. Adhesives, Coatings and Sealers Division St. Paul, MN
The Dexter Corp., Adhesives and Structural Materials Division Pittsburg, CA	National Starch and Chemical Corp. Adhesives Bridgewater, NJ	VIGOR Co. Division of B. Jadow & Sons New York, NY

Source: Ref 13

Table 4 Cure rate and bond strength on selected substrates

Substrate	Cure rate	Bond strength
Metals		
Aluminum	Moderate	Moderate
Brass	Fast	Strong
Steel	Fast	Strong
Plastics		
Acrylonitrile-butadiene-styrene	Fast	Strong
Acrylic	Fast	Strong
Nylon	Fast	Moderate
Phenolic	Moderate	Strong
Polyacetal	Fast	Moderate
Polycarbonate	Moderate	Strong
Polyvinyl chloride	Fast	Strong
Rubbers		
Butyl	Very fast	Strong
Neoprene	Very fast	Strong
Natural rubber	Very fast	Strong
Nitrile rubber	Very fast	Strong
Miscellaneous		
Glass	Very fast	Strong
Wood	Slow	Strong

polymers require chemical treatment to impart surface polarity. Plasticized polymers pose a different problem. Plasticizer may migrate into the interface and weaken the bond.

Rubbers. Cyanoacrylates bond rapidly with most rubbers. Butt joints on freshly cut surfaces usually exceed the strength of the rubber. Hydrocarbon rubbers, such as an elastomeric polymer from ethylene, propylene, and a nonconjugated diene, as well as natural rubber, are harder to bond than rubbers containing polar groups. In addition, they require special grades of cyanoacrylates that have very low acidity. Silicone rubber has a very low surface energy and forms weak bonds.

Wood. Most hardwoods can be bonded with a fast-curing grade of cyanoacrylate. Soft woods are more difficult to bond because they tend to absorb adhesive. Because wood has an acidic surface, a surface activator is frequently used. Wood bonding is typically slow with conventional cyanoacrylates. Faster-curing wood-bonding cyanoacrylates contain additives, such as crown ethers, or polyethylene glycol derivatives, that greatly enhance the cure speed. These materials are described in the section "Miscellaneous Additives" in this article.

Glass and ceramics tend to have alkaline surfaces and to bond very rapidly. The bonds are strong initially, but weaken with time.

Heat Resistance. Most polymerized cyanoacrylates are thermoplastics that have poor heat resistance compared to thermoset adhesives such as epoxies. They are also susceptible to depolymerization at elevated temperatures. Poly(methyl 2-cyanoacrylate), which has the highest softening point of the cyanoacrylates, is a brittle polymer that can fail during temperature cycling because of differences between the coefficients of thermal expansion of the substrate and the adhesive.

The upper temperature limit for the sustained use of most cyanoacrylates is about 70 °C (160 °F). Bonds made from methyl and ethyl 2-cyanoacrylates can withstand short exposures to 100 °C (212 °F).

Allyl cyanoacrylate undergoes a secondary cure by thermal polymerization of the allyl groups (Ref 15). This produces a cross-linked polymer with better stability and heat resistance than the thermoplastic cyanoacrylates. The bond strength of the cross-linked polymer depends on the heat history. Tensile shear strengths at elevated temperatures increase with time as the allyl groups begin to polymerize and then may decrease because of further cross linking and embrittlement (Ref 16). It should be noted that the secondary cure can be so slow that the reduction in strength of the thermoplastic polymer often outpaces the development of strength that is due to the cross-linking reaction. If the bond must withstand any load during this period, the assembly will fail.

Moisture and Aging Resistance. Cyanoacrylate bonds usually undergo a loss in strength when immersed in water or exposed to high relative humidities (>90%). This occurs because moisture can migrate into the interface and form a weak boundary layer (Ref 17). Humidity resistance is excellent with plastics that have some solubility when in contact with the adhesive. A small amount of plastic may dissolve during the cure and eliminate the interface (Ref 18).

With respect to long-term aging, joints bonded with cyanoacrylates may undergo a gradual loss of strength after several years (Ref 19). In many cases, the changes are not large enough to cause bond failure. However, with some high-energy surfaces, water that diffuses into the interface can cause chemical changes in the substrate or adhesive. Bonds on glass deteriorate very rapidly, probably because glass has an alkaline surface that can catalyze the hydrolysis of the polycyanoacrylate. Surfaces that are easily corroded, such as steel, can oxidize and form a weak metal-to-metal oxide interface.

Solvent Resistance. Polymerized cyanoacrylates can be removed with certain polar aprotic solvents. Poly(methyl 2-cyanoacrylate), which has a solubility parameter of 14.7, requires a solvent such as dimethyl formamide, nitromethane, or propylene carbonate. The higher cyanoacrylates (solubility parameter of 11.2 for ethyl and 10.7 for butyl) are soluble in less polar

Table 5 ASTM tests suitable for cyanoacrylate adhesives

Test	Configuration	ASTM standard
Tensile strength		D 2095
Tensile shear strength		D 1002
T-peel		D 1876
Cleavage		D 1062
Impact strength		D 950
Viscosity	Capillary viscometer	D 445(a)

(a) Glass viscosimeter should be prerinsed with a 5% solution of phosphoric acid in acetone and then dried.

solvents, such as acetone, tetrahydrofuran, and others (Ref 20).

Test methods are fully described in the Section "Testing and Analysis" in this Volume. It should be noted that cyanoacrylates are sensitive to contamination and are subjected to a variety of quality control tests, including tests of purity, by the gas chromatography technique; water content by Karl Fischer titration; shelf life by accelerated aging; setting times on several surfaces; and both tensile strength and tensile shear strength on different substrates. The commonly used ASTM tests for determining adhesion properties are listed in Table 5.

Additives

Cyanoacrylate adhesives are rarely sold as pure monomers. They usually contain one or more of the following additives: inhibitors, thickeners, adhesion promoters, activators, plasticizers, fillers, and dyes.

Inhibitors are added to prevent polymerization during storage. Polymerization can occur in the liquid phase or on surfaces exposed to the vapor.

Protection from anionic polymerization is provided by chemical compounds that are electron acceptors, that is, protonic and Lewis acids. Some commonly used inhibitors are phosphoric acid, partial esters of phosphoric acid, sulfonic acids, and sulfur dioxide. Typical concentrations are 1 to 50 ppm. The key to manufacturing a commercially successful cyanoacrylate adhesive is to balance the acidity of the system so that the adhesive is stable during storage, yet remains fast curing.

Cyanoacrylate manufacturers also add inhibitors of free radical polymerization, usually phenolic types, such as hydroquinone, or t-butyl catechol. Typical levels are 50 to 1000 ppm.

Thickeners are added to control the rheology of the adhesive and to prevent the adhesive from being absorbed into the substrate or from running off the surface. The viscosities of commercial cyanoacrylates range from 0.002 Pa · s for the unthickened monomer to as high as 30 Pa · s for monomers thickened with organic polymers. Suitable thickeners include poly(alkyl acrylates), poly(alkyl methacrylates), poly(vinyl acetate), and poly(alkyl 2-cyanoacrylates). The higher-viscosity grades are required for bonding porous substrates and vertical surfaces and are necessary when large gaps exist between the two substrates.

Two disadvantages of adhesives based on the methyl and ethyl cyanoacrylates are their relatively low impact and peel strengths. The addition of elastomeric materials or rubber-toughened plastics can improve these properties (Ref 21-23). The addition of poly(methyl methacrylate), a glassy polymer commonly used as a thickener, has been found to increase impact strength and elongation-to-break values (Ref 24).

Thixotropic cyanoacrylates (those that break down to a liquid, when sheared) have been prepared by the addition of hydrophobic silicas, along with an organic thickener to prevent settling (Ref 25). These adhesives flow under high shear conditions, but are nonflowable after being applied. Hydrophobic silicas are manufactured by treating fumed silicas with organic silicon compounds containing groups that are reactive with the SiOH groups on the surface of the silica.

Miscellaneous Additives. Adhesion promoters, such as itaconic anhydride (Ref 26), acetic acid (Ref 27), and esters of gallic acid (Ref 28), improve bonding to metals. A possible mechanism for the improved adhesion is the displacement of adsorbed water molecules on the surface of the metal by the polar additive (Ref 29).

Certain chelating agents, such as crown ethers (Ref 30) and polymers of ethylene oxide (Ref 31), decrease the setting time without markedly reducing stability. Presumably, they function by complexing cations to produce bare anions.

Plasticizers are occasionally added to increase the flexibility of the cured adhesive and to improve impact resistance. Those most commonly used are C_1 through C_{10} esters of dicarboxylic acids, such as phthalic acid and sebacic acid. High concentrations of plasticizer reduce both the cure rate and bond strength. Lower bond strength has been used to advantage in temporary adhesives, which must be able to be broken easily (Ref 32).

Electrically conductive fillers, such as finely divided silver, have been used to produce electroconductive adhesives that have volume resistivities as low as 10^{-4} $\Omega \cdot$ cm. Metal powders have to be washed with dilute acid to prevent the spontaneous polymerization of the adhesive (Ref 33).

Design Considerations

Methyl and ethyl 2-cyanoacrylates form brittle films that are able to withstand shear, compressive, and tensile stresses better than peel or cleavage stresses. Designs that produce cleavage and peel forces should be avoided. Although tensile strengths are normally very high with butt joints, they should be avoided when both substrates are rigid because a slight misalignment can produce cleavage forces.

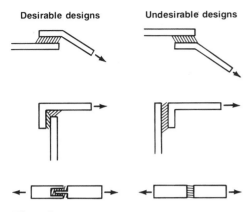

Desirable designs   **Undesirable designs**

Fig. 8   Joint designs for cyanoacrylate esters

Figure 8 shows both desirable and undesirable designs. Desirable designs distribute the load on the adhesive as shear and compressive stresses and minimize stresses outside the plane of the adhesive.

The lower molecular weight cyanoacrylates (that is, methyl and ethyl esters) have relatively poor impact strengths on rigid substrates. This property can be improved by bonding a strip of energy-absorbing rubber between the two rigid parts.

Processing Parameters

Surface Preparation. Cyanoacrylates give the best bonds on clean, dry, grease-free surfaces. In many cases, a simple solvent wipe is sufficient. To obtain the highest bond strengths, which in many cases exceed the internal strength of the substrate, a more elaborate treatment is necessary. The surface preparation of substrates is covered in the Section "Surface Considerations" in this Volume.

Methods of Application. Cyanoacrylate adhesives produce disappointing results when used incorrectly because polymerization initiation occurs only at the surface. Poorly mated parts may fail to bond because of incomplete polymerization. When used properly on smooth, well-matched surfaces, cyanoacrylates are among the most versatile adhesives.

Only the minimum quantity of material needed to coat the substrate should be applied. One drop per square inch is adequate for a smooth surface. Excessive amounts are not only wasteful but result in slower cures and weaker bonds.

The method of forming the bond depends on the cure rate on the particular substrate. Fast-curing adhesives with fixturing times of less than 5 s should not be spread because premature curing on that surface can result. A good technique is to apply a drop or more of the adhesive, assemble the part, and apply pressure for a few seconds until the bond sets. Slower-curing adhesives can be spread either with a plastic or metal rod or by quickly rubbing the pieces together.

In all cases, once polymerization has begun, the parts should not be repositioned because weak bonds will result.

One useful bonding technique is to preassemble the parts without any adhesive and apply a low-viscosity wicking-grade adhesive around the edges of the seam. The adhesive will penetrate into the joint by capillary action, making this method useful for bonding parts that are difficult to reach. The technique has also been used to bond balsa wood.

Porous surfaces or poorly fitting joints require thickened adhesives. Low-viscosity grades (<0.1 Pa · s) can fill gaps up to approximately 0.05 mm (2 mils); medium-viscosity grades (0.1 to 0.5 Pa · s) can fill those up to 0.25 mm (10 mils); and high-viscosity grades (>0.5 Pa · s), those up to 0.5 mm (20 mils).

A surface activator is required when there are acidic surfaces, porous surfaces, or large gaps, and so on. The activator is lightly applied to one surface, and the solvent is allowed to evaporate for a few seconds. The adhesive is applied to the other surface, and the parts are assembled. Initiation usually occurs within seconds.

Wire bonding requires another technique. The wire is positioned on the bonding surface, and a drop of a high-viscosity cyanoacrylate is applied. Accelerator is then misted onto the adhesive. A strong bond is formed within seconds. When the accelerator is applied before the adhesive, the bond is weaker and the wire can be removed.

One desirable feature of cyanoacrylates is the ease of application with automatic dispensing equipment. Several types of dispensers are available from cyanoacrylate adhesives suppliers.

Storage Stability. As mentioned earlier, cyanoacrylates are prone to both polymerization and hydrolysis. They should be stored in inert containers that have a low permeability to moisture. Typical packages include polyethylene bottles and tubes made of aluminum, tin, or polyethylene.

The stability of the adhesive is greatly improved by storing it at reduced temperatures, preferably in a freezer. The container should be allowed to warm to room temperature before it is opened to prevent moisture from condensing and penetrating the adhesive.

Most manufacturers claim that their products have shelf lives of 6 months to 1 year when stored at room temperature, and an almost indefinite life when stored in a freezer. Materials purchased in a retail market usually involve some risk. Even if manufactured properly, they may have been kept for long times at uncontrolled temperatures.

Health and Safety Considerations. Several hazards are associated with cyanoacrylate adhesives. One is polymerization on skin, which occurs almost immediately upon contact. If this occurs, the skin area should immediately be washed with cold water. The cured film will usually come off after several days, with an occasional scrub using soap and water. A skin softener can then be applied. In the case of eye contact, water should immediately be used to irrigate the eye, and prompt medical attention should be obtained. If skin is bonded, acetone or nail polish remover will soften the adhesive, which then can be gently pulled off. If solvents are not available, or if their flammability precludes their use, the harmless cured adhesive, which will be brittle, can be broken off with a peeling or kneading action or allowed to flake off by itself.

A large spill on the skin can cause tissue damage because of the exotherm heat of polymerization (ΔH = 59 kJ/mol, or 14 kcal/mol). This should be treated like a thermal burn.

Users of cyanoacrylate adhesives should wear safety glasses or goggles and polyethylene gloves. Rubber gloves are not recommended because they are easily bonded.

It is virtually impossible to swallow a cyanoacrylate because it would polymerize instantly in the mouth or throat. However, the inhalation of vapors can occur, and methyl, ethyl, and allyl 2-cyanoacrylates have harsh odors that are irritating to the eyes and respiratory system. The Occupational Safety and Health Administration (OSHA) recommends time-weighted average (TWA) concentrations of specific materials used during a normal 8 h workday and 40 h workweek, to which nearly all workers may be exposed without adverse effects. Because the TWA concentration recommended by OSHA for methyl cyanoacrylate is 2 ppm, the work area should be well ventilated.

Because cyanoacrylate esters are prone to hydrolysis, the toxicities of the resulting alcohols must be considered. Allyl alcohol has a high acute toxicity; 2-methoxyethanol and 2-ethoxyethanol have high reproductive toxicities (on both male and female test animals).

The accelerators used to activate surfaces are usually aromatic amines. Many are toxic chemicals and should be handled with caution. Some of the cleaning and debonding solvents are also toxic. Dimethyl formamide, which is commonly used, is readily absorbed through the skin.

Methyl and ethyl cyanoacrylate are combustible liquids with flash points of 75 °C (167 °F) and 83 °C (181 °F), respectively. Cyanoacrylate fires can be extinguished with water.

Large spills should not be wiped up. The area should be flooded with water to polymerize the adhesive and reduce vapor generation. The cured material can then be scraped up.

REFERENCES

1. J.M. Rooney, On the Mechanism of Oligomer Formation in Condensations

of Alkyl Cyanoacrylates With Formaldehyde, *Polym. J.*, Vol 13 (No. 10), 1981, p 975-978

2. G.F. Hawkins, Process for Manufacture of High Purity α-Cyanoacrylates, U.S. Patent 3,465,027, 1969

3. J.P. Cronin and D.C. Pepper, Zwitterionic Polymerization of Butyl Cyanoacrylate by Triphenylphosphine and Pyridine, *Makromol. Chem.*, Vol 189, 1988, p 85-102

4. E.F. Donnelly, D.S. Johnston, and D.C. Pepper, Ionic and Zwitterionic Polymerization of n-Alkyl 2-Cyanoacrylates, *J. Polym. Sci., Polym. Lett. Ed.*, Vol 15, 1977, p 399-405

5. I.C. Eromosele and D.C. Pepper, The Possibility of Oxonium/Carbenium Ion Initial Cations in the Zwitterionic Polymerization of Butyl 2-Cyanoacrylate by Cyclic Carbonates, *Makromol. Chem. Rapid Commun.*, Vol 7, 1986, p 531-532

6. T.H. Wicker and N.H. Shearer, Jr., Method of Adhesive Bonding, U.S. Patent 3,259,534, 1966

7. T.H. Wicker and N.H. Shearer, Jr., Adhesive Composition and Method of Bonding Using α-Cyanoacrylate Esters and Vinyl Aromatics, U.S. Patent 3,282,773, 1966

8. H.A.A. Rasoul and H.K. Hall, Jr., Cycloaddition and Polymerization Reactions of Methyl α-Cyanoacrylate With Electron-Rich Olefins, *J. Org. Chem.*, Vol 47, 1982, p 2080-2083

9. D.C. Pepper, Anionic and Zwitterionic Polymerization of α-Cyanoacrylates, *J. Polym. Sci., Sym.* 62, 1978, p 65-77

10. J. Guthrie, M.S. Otterburn, J.M. Rooney, and C.N. Tsang, The Determination of the Molecular Weight of Poly(ethyl 2-Cyanoacrylate) Adhesive, *Polym. Commun.*, Vol 25, 1984, p 318-319

11. G.A. Parks, The Isoelectric Points of Solid Oxides, Solid Hydroxides, and Aqueous Hydroxo Complex Systems, *Chem. Rev.*, Vol 65, 1965, p 177-198

12. J. Almog and A. Gabay, A Modified Super Glue Technique—The Use of Polycyanoacrylate for Fingerprint Development, *J. Forensic Sci.*, Vol 31 (No. 1), 1986, p 250-253

13. *Adhes. Age* (Directory, 21st ed.), Vol 32 (No. 6), 1989, p 33-75, 82

14. *Adhes. Age*, Vol 32 (No. 5), 1989, p 32

15. D.L. Kotzev, T.C. Ward, and D.W. Dwight, Assessment of the Adhesive Bond Properties of Allyl 2-Cyanoacrylate, *J. Appl. Polym. Sci.*, Vol 26, 1981, p 1941-1949

16. D.L. Kotzev, P.C. Novakov, and V.S. Kabaivanov, Synthesis and Properties of Some Alkenyl- and Alkinyl-2-Cyanoacrylates, *Angew. Makromol. Chem.*, Vol 92, 1980, p 41-52

17. K.F. Drain, J. Guthrie, C.L. Leung, F.R. Martin, and M.S. Otterburn, The Effect of Moisture on the Strength of Steel-Steel Cyanoacrylate Adhesive Bonds, *J. Adhes.*, Vol 17, 1984, p 71-82

18. K.F. Drain, J. Guthrie, C.L. Leung, F.R. Martin, and M.S. Otterburn, The Effect of Moisture on the Strength of Polycarbonate-Cyanoacrylate Adhesive Bonds, *Int. J. Adhes. Adhes.*, Vol 5 (No. 3), 1985, p 133-136

19. H.W. Coover, Jr. and J.M. McIntire, Cyanoacrylate Adhesives, in *Handbook of Adhesives*, 2nd ed., Section 34, I. Skeist, Ed., Van Nostrand Reinhold, 1977, p 569-580

20. E.F. Donnelly and D.C. Pepper, Solubilities, Viscosities and Unperturbed Dimensions of Poly(ethyl Cyanoacrylate)s and Poly(butyl Cyanoacrylate)s, *Makromol. Chem. Rapid Commun.*, Vol 2, 1981, p 439-442

21. C. Petrov, B. Serafimov, and D.L. Kotzev, Strength, Deformation and Relaxation of Joints Bonded With Modified Cyanoacrylate Adhesives, *Int. J. Adhes. Adhes.*, Vol 8 (No. 4), 1988, p 207-210

22. E.R. Gleave, Filled Cyanoacrylate Adhesive Compositions, U.S. Patent 4,102,945, 1978

23. J.T. O'Connor, Toughened Cyanoacrylates Containing Elastomeric Rubbers, U.S. Patent 4,440,910, 1984

24. C. Petrov, B. Serafimov, and D. Kotzev, Adhesive Bond Properties of Ethyl-2-Cyanoacrylate Modified With Poly(methylmethacrylate), *J. Adhes.*, Vol 25, 1988, p 245-253

25. A.E. Litke, Thixotropic Cyanoacrylate Compositions, U.S. Patent 4,477,607, 1984

26. E. Konig, Adhesive Compositions Containing a Cyanoacrylate and Itaconic Anhydride, U.S. Patent 3,948,794, 1976

27. J.E. Schoenberg and D.K. Ray-Chaudhuri, Adhesion Promoter for 2-Cyanoacrylate Adhesive Compositions, U.S. Patent 4,125,494, 1978

28. J.E. Schoenberg, 2-Cyanoacrylate Adhesive Compositions Having Enhanced Bond Strength, U.S. Patent 4,139,693, 1978

29. D.L. Kotzev, Z.Z. Denchev, and V.S. Kabaivanov, Adhesive Properties of Ethyl 2-Cyanoacrylate Containing Small Amounts of Acetic Acid as Adhesion Promoter, *Int. J. Adhes. Adhes.*, Vol 7 (No. 2), 1987, p 93-96

30. A. Motegi, E. Isowa, and K. Kimura, α-Cyanoacrylate-Type Adhesive Composition, U.S. Patent 4,171,416, 1979

31. W. Gruber and H. Bruhn, Cyanoacrylic Acid Ester Based Glues With a Content of a Diester of a Polyoxyalkylene Glycol, U.S. Patent 4,421,909, 1983

32. V.R. Allies and W.D. Zimmermann, Debondable Cyanoacrylate Adhesive Composition, British Patent 1,529,105, 1978

33. K.G. Chorbadjiev and D.L. Kotzev, The Effect of Fillers Upon the Properties of Electroconductive Cyanoacrylate Adhesives, *Int. J. Adhes. Adhes.*, Vol 8 (No. 3), 1988, p 143-146

Silicones

Loretta A. Sobieski, Dow Corning Corporation

COMMERCIAL SILICONE ADHESIVES are primarily available as pressure-sensitive adhesives (PSAs), although some can be cured with aminosilanes to perform as higher-strength laminating adhesives. They are applied as solvent dispersions that can be used as is or further diluted for roll coating, dipping, or brushing applications. Some converting companies have the technology to coat silicone PSAs on a release carrier and supply them as transfer films.

Chemistry

There are two types of cure systems used in silicone PSA products. One, a combination of condensation and radical polymerization, has been in use for more than 15 years. The other, addition cure, is relatively new in its use with silicone PSAs, but has been used extensively in silicone rubber and sealant products and release coatings. Although the resulting performance of the two types of products can be very similar, the chemistry dictates somewhat different processing considerations.

For example, in the traditional condensation cure, a reaction results from the condensation of silanol groups ($\equiv$SiOH) on silicone resin and/or polymer (Fig. 1). The reaction can be accelerated by the use of amines, metal complexes, and strong acids and bases. Generally, some interaction between the resin and polymer is required for the realization of optimum tack, adhesion, and cohesive strength (Ref 1). Because the resin and polymer are not mutually soluble in each other, the resulting network has a two-phase morphology consisting of a continuous polymer-rich phase and a discontinuous resin-rich phase (Ref 2). These are covalently bound together at the interface through siloxane linkages formed during condensation.

The ratio of resin to polymer has a significant effect on performance properties. A certain minimum amount of resin is required to tackify the polymer so that the mixture exhibits pressure-sensitive properties. Above that amount, the resin serves as a three-dimensional cross-linking site, resulting in increasing cross-link density. The adhesive becomes more elastic and displays higher adhesive and cohesive strength (Fig. 2 and 3). As the elastic component increases, however, the viscous component decreases, and lower tack values are obtained. This is consistent with traditional definitions of tack as a measure of viscous flow and bond formation under conditions of fast strain rates and short contact times (Ref 3).

Changes in both the resin and the polymer will affect the PSA performance. The level of silanol functionality on the resin and polymer can greatly affect tack, adhesion, and cohesive strength. With no functionality on either component, a cross-linked network cannot be established. The resulting adhesives are very tacky and exhibit only moderate peel strengths as a result of their very poor cohesive strength. Having silanol on the polymer only yields lower tack, but does give moderate peel and cohesive strength values. As the level of functionality on the resin is increased, cross-link density is also increased, and properties expected from a tighter network are achieved (Fig. 4). Tack and peel strength decrease only slightly, whereas a significant increase in cohesive strength is observed.

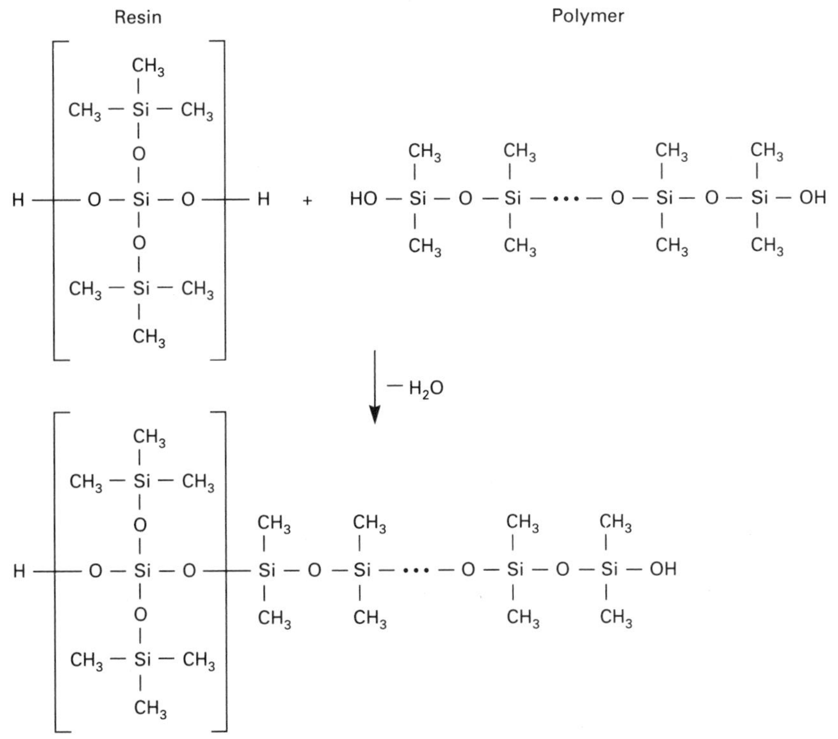

Fig. 1 Condensation reaction of silicone pressure-sensitive adhesives

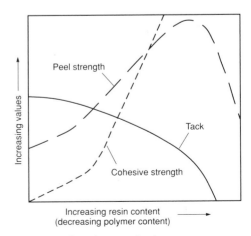

Fig. 2 Effect of resin and polymer content on performance properties

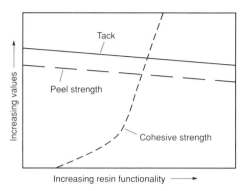

Fig. 3 Effect of resin functionality on performance properties

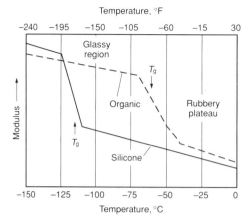

Fig. 4 Adhesive modulus versus temperature. T_g, glass transition temperature

Table 1 Solvent extraction of cured silicone PSAs

Benzoyl peroxide, %	Nonextractables, %	Extractables, %
1.0	34–38	62–66
2.0	48–52	48–52

The majority of silicone PSAs used are based on polydimethylsiloxane (PDMS). The substitution of diphenyl for some of the dimethyl groups changes the performance characteristics. For example, the incorporation of diphenyl groups raises the polymer glass transition temperature (T_g), making portions of the resin slightly more soluble in the polymer phase than is the case in PDMS polymers. The end result is that, at room temperature, a PSA based on polydiphenyl-PDMS copolymer may exhibit higher tack than a PSA similarly formulated with PDMS alone. Cohesive strength can also decrease somewhat.

For most applications, the level of cohesive strength obtained by condensation alone is not sufficient. Free radical polymerization through the methyl groups on both polymer and resin allows significant improvement in cohesive strength without greatly affecting tack or adhesion, as in the case of benzoyl peroxide (BPO):

$$C_6H_5C(O)O_2\ (O)CC_6H_5 \rightarrow 2C_6H_5C(O)O\cdot$$
$$C_6H_5C(O)O\cdot\ +\ \equiv SiCH_3 \rightarrow C_6H_5C(O)OH$$
$$+\ \equiv SiCH_2\cdot\ \equiv SiCH_2\cdot\ +\ \equiv SiCH_2\cdot$$
$$\rightarrow\ \equiv SiCH_2CH_2Si\equiv \qquad \text{(Eq 1)}$$

In addition, peroxide curing increases the resistance of silicone PSA to swelling caused by solvents and binds more of the unreacted resin (Table 1). It is also helpful in improving adhesion of the adhesive to a tape backing.

Organofunctional silanes, particularly aminofunctional silanes, can be used alone or in combination with peroxides to improve the cohesive strength by driving the condensation reaction further to completion. Because this reaction continues to progress until the adhesive becomes dry, the use of silanes is recommended only where materials will be bonded or laminated soon after solvent removal (Ref 4).

Silicone PSAs that cure by an addition reaction also use silicone resins and polymers, but their cross-link network is set up by the reaction:

$$\equiv SiCH=CH_2$$
$$+\ \equiv SiH\ \xrightarrow[\Delta]{Pt}\ \equiv SiCH_2CH_2Si\equiv \qquad \text{(Eq 2)}$$

In this type of network, the weight ratio and level of functionality of both components are also critical to obtaining the right balance of properties for a particular application.

Markets and Applications

The primary uses for silicone PSAs take advantage of their ability to wet low-energy surfaces, withstand high temperatures, and conform to irregular surfaces yet still be removed cleanly. They are usually used in the form of self-wound pressure-sensitive tapes. The single largest application is in the electronics market, specifically, the masking of printed circuit boards during the solder-stripping step and subsequent plating operations of the gold finger connectors. Usually supplied on a polyester backing, the silicone PSA must flow over edges and into corners to form a seal and thus protect other parts of the board from acidic stripping solutions. The silicone PSA tape can be removed cleanly from the board even after prolonged periods of contact. Other electronic applications include wave solder masking and component carrier tapes.

The electrical field is diverse in terms of specific uses, but general applications include wrapping and bundling of wires, wrapping transformers, and mounting wires and circuitry. Tape backings for these applications include polytetrafluoroethylene (PTFE), glass cloth, and polyimide. These applications utilize the ability of silicone to resist degradation at high temperatures while maintaining good electrical stability. They are the only PSAs that are rated for IEEE class H use, which requires a 180 °C (355 °F) continuous-use temperature.

Because silicone PSAs adhere well to low-energy surfaces, including other silicones, they are used as the adhesive for splicing tapes for silicone release liners, PTFE films, and PTFE-coated fabrics. Tapes made with backings of PTFE or glass coated with this material are used to wrap tension and processing rolls, as well as heat-seal bars, in the manufacture of plastic films, bags, and other containers because these backings provide a nonstick surface. Silicone rubber-backed tapes are used for high-temperature insulation and seals and for plasma spray masking in the repair of metal parts and equipment.

Major Suppliers

The major suppliers of uncured adhesives to the U.S. market are Dow Corning Corporation and General Electric Company. Transfer films are primarily supplied by Flexcon Company and Minnesota Mining & Manufacturing Company.

Cost Factors. Silicone PSAs are supplied at 55 to 60% solids levels at commercial prices ranging from $10 to 14/kg ($4.75 to 6.25/lb), with diphenyl silicone PSAs being in the higher price end of this range. This results in a cost of approximately $18 to 24/kg ($8 to 11/lb) solids.

Competing Adhesives

In pressure-sensitive applications, only high-performance acrylics approach the temperature resistance of silicones at both the high- and low-temperature regions. The T_gs of adhesives can be used to predict their suitability for pressure-sensitive applications at various temperatures. Below their T_gs, adhesives become stiff and brittle and cannot function as PSAs. For acrylic (organic) PSAs, this occurs over a temperature range of −65 to −40 °C (−85 to −40 °F) compared to approximately −120 °C (−185 °F) for a typical dimethyl silicone PSA (Fig. 4). Although acrylic PSAs have significantly higher peel strengths at 0 to 100 °C (32 to 212 °F), silicone PSAs will outperform them outside that range. In a study that compared the tensile static shear, lap-shear, and peel strengths of an acrylic and silicone PSA, the use temperature ranges were identified as −25 to 150 °C (−15 to 300 °F) for the acrylic and −75 to 250 °C (−105 to 480 °F) for the silicone (Ref 5). These values change according to the specific formulations used and the test regimen and should not be considered to be product specifications.

Properties

The uncured properties of some commonly used silicone PSAs are listed in Table 2. Typical performance properties of cured silicone PSAs are shown in Table 3.

Table 2 Silicone PSA physical properties

Product	Solvent	Solids, %	Viscosity, Pa · s	Comments
Dow Corning				
Q2-7406	Xylene	55	20–80	High shear strength
Q2-7566	Xylene	55	30–130	High temperature
X2-7656	Xylene	60	10–70	Low-temperature cure
280A	Xylene	55	30–80	High tack
282	Xylene	55	30–80	High adhesion
General Electric				
590	Toluene	60	2–16	Low viscosity
595/593	Xylene	55	40–90	General purpose
518	Xylene	55	50–100	Diphenyl, high temperature

Table 3 Typical silicone PSA performance properties

Material	Adhesion N/m	Adhesion ozf/in.	Tack N/mm²	Tack ozf/in.²
25 μm (1 mil) polyethylene terephthalate	382–491	35–45	0.11–0.14	249–328
50 μm (2 mil) polyethylene terephthalate	491–654	45–60	0.13–0.16	294–373
25 μm (1 mil) polyimide	273–382	25–35	0.10–0.13	226–305
50 μm (2 mil) aluminum	654–818	60–75	. . .	. . .
75 μm (3 mil) PTFE	382–491	35–45	0.07–0.14	158–316

Note: Cured with 2% benzoyl peroxide

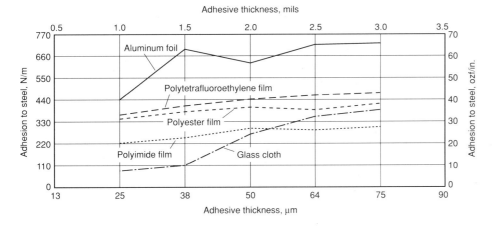

Fig. 5 Adhesion properties according to PSTC-1 versus adhesive thickness on various backings. For 25 to 50 μm (1 to 2 mil) thickness, cure is 1 min at 71 °C (160 °F) and then 2 min at 204 °C (400 °F). For 64 to 75 μm (2.5 to 3 mil) thickness, cure is 2 min at 71 °C (160 °F) and then 2 min at 204 °C (400 °F). Cured with 2% benzoyl peroxide

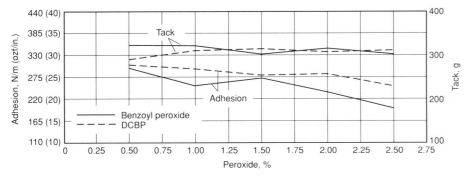

Fig. 6 Tack and adhesive peel strength according to PSTC-1 versus peroxide level and type. For 25 to 50 μm (1 to 2 mil) thickness, cure is 1 min at 71 °C (160 °F) and then 2 min at 204 °C (400 °F). For 64 to 75 μm (2.5 to 3 mil) thickness, cure is 2 min at 71 °C (160 °F) and then 2 min at 204 °C (400 °F). DCBP, 2,4-dichlorobenzoyl peroxide

In addition to the tape backing, which may affect testing, adhesive thickness influences adhesion and tack properties, as shown in Fig. 5. Tack, adhesion, and cohesive strength can be altered by changing the peroxide level and type. Examples of these effects are shown in Fig. 6.

Adhesion to some substrates, particularly those with low surface energies, increases over time, as shown in Table 4. The effect of temperature on peel strength is seen in Table 5. Electrical properties will vary with the particular adhesive, but typical values are shown in Table 6.

Additives and Modifiers

Fillers, pigments, or dyes are the usual additives used in silicone PSAs. Calcium carbonate can be used as an extender to reduce tack and adhesion. Treated calcium carbonates work best. Pigments are available as concentrated pastes in silicone fluid, but can also reduce tack. Dyes recommended for silicones are of the classes known as oil soluble or acetone soluble.

Silicone additives are available to modify the adhesion of silicone PSAs. Additive levels of 10% can be used to increase peel strength values slightly, but will simultaneously decrease tack.

Processing Parameters

Shelf Life and Pot Life. Uncatalyzed silicone PSAs have a shelf life of 6 to 12 months, but individual product data should be consulted for specifics. Once peroxide is added to peroxide-catalyzed silicone PSAs, the adhesive should be used within 24 h. It is not uncommon for catalyzed PSA to sit for 72 h because of scheduling factors or production interruptions, but lower performance may be observed. The peroxide half-life is shorter when in contact with solvents, thereby leaving less peroxide to do the cross linking as the catalyzed PSA ages. Tack and adhesion properties may appear to be normal, but a decrease in cohesive strength may be observed.

Platinum-catalyzed silicone PSAs also have a finite pot life once the platinum catalyst is mixed in. Although formulations may vary, it is recommended that they be used within 24 h. An increase in viscosity and the inability to obtain optimum tack and adhesion will be observed, as well as lower cohesive strength. Because contact with moisture should be avoided, drums should be sealed, or a drying tube should be placed over any opening.

Storage Requirements. There are no particular high- or low-temperature concerns for the storage of uncatalyzed silicone PSAs that would require special shipping considerations. However, because these materials are usually supplied in flammable solvents, they should be handled according-

Table 4 180° peel strength of silicone PSA on various substrates

	Immediate		2 days		7 days		14 days	
Substrate(a)	N/m	ozf/in.	N/m	ozf/in.	N/m	ozf/in.	N/m	ozf/in.
LDPE.................	190	17.5	220	20.0	245	22.5	285	26.0
HDPE	195	18.0	245	22.5	270	25.0	245	22.5
UHMWPE	150	14.0	190	17.5	180	16.5	205	19.0
PP....................	185	17.0	295	27.0	283	26.0	290	26.5
PVC..................	130	12.0	170	15.5	190	17.5	195	18.0
PS, white.............	230	21.0	300	27.5	270	25.0	300	27.5
PVDF.................	305	28.0	360	33.0	320	29.5	310	28.5
Polycarbonate	310	28.5	310	28.5	290	26.5	295	27.0
Acrylic...............	285	26.0	290	26.5	265	24.5	265	24.5
ABS, black	225	20.5	315	29.0	305	28.0	320	29.5
Polyurethane, 60 Shore D.	175	16.0	230	21.0	245	22.5	290	26.5
Nylon 101	295	27.0	315	29.0	290	26.5	305	28.0
Stainless steel........	300	27.5	315	29.0	305	28.0	365	33.5

Note: Based on 38 μm (1.5 mil) adhesive cured with 2.0% benzoyl peroxide on 25 μm (1 mil) polypyromelitimide film. (a) LDPE, low-density polyethylene; HDPE, high-density polyethylene; UHMWPE, ultrahigh molecular weight polyethylene; PP, polypropylene; PVC, polyvinyl chloride; PS, polystyrene; PVDF, polyvinylidene fluoride; ABS, acrylonitrile-butadiene-styrene

Table 5 Effect of temperature on peel strength

Temperature		180° peel strength	
°C	°F	N/m	ozf/in.
−73	−100	2109	190
−50	−60	1607	145
0	30	465	40
50	120	291	25
100	210	111	10
150	300	54	4.9
200	390	52	4.7
250	480	25	2.3

Note: Based on 25 μm (1 mil) polypyromelitimide coated with 38 μm (1.5 mil) adhesive, laminated to stainless steel

ly. Platinum-cured PSAs do have the potential of giving off flammable hydrogen gas during storage. However, in the absence of catalyst, this is unlikely. Platinum catalyst should be stored in opaque containers below 25 °C (77 °F). Peroxide catalyst should be stored at as low a temperature as possible and should be used fairly quickly to preserve a high level of activity. Storage at high temperatures could result in a very unstable and hazardous material. Recommendations should be sought from the supplier.

Substrate Requirements and Geometry. The curing of peroxide-catalyzed PSAs requires a temperature at which the peroxide has a reasonably short half-life. For the peroxides used with silicone PSAs, the lowest recommended cure temperature is 130 °C (265 °F). This becomes the major limiting factor for substrates used with silicone PSAs, which are typically applied to polyester, fluorocarbons, glass cloth, polyimide, and silicone rubber. These are all relatively high-temperature-resistant backings. They are also usually used because the applications require high-temperature resistance. The advent of platinum-catalyzed silicone PSAs reduces the temperature resistance requirements for backings. These PSAs can be cured at temperatures as low as 80 to 100 °C (175 to 212 °F), which allows the use of substrates such as polyolefins. Many additives used in the manufacture of films, however, can inhibit the cure of platinum-cata-

lyzed PSAs. Solvent resistance is a concern with all silicone PSAs, making it difficult to apply them directly to foams.

Industrial silicone PSAs do not release when cast onto silicone release liners, although some companies have developed proprietary release technology. These companies also sell silicone PSA transfer tapes. Many of these tapes are supplied with liners that are usually a form of crimped or embossed polyvinyl chloride or polyethylene. They work on the concept of reduced surface area of contact and can only be used for dry lamination to the PSA. Over time, the PSA will flow into the spaces, making it difficult to separate and leaving indentations in the PSA surface that reduce its initial tack. Other companies continue to pursue the development of transfer-coatable release technology, but no such product is available as of this writing.

Primer Requirements. In order to make superior self-wound tapes with silicone PSAs, it is necessary to apply a primer to the backing prior to adhesive application. Silicone release coatings are usually used for this purpose. They adhere well to the backings, and the PSA can both physically and chemically tie into the release coating layer upon cure. For peroxide-catalyzed PSAs, tin-catalyzed, condensation-cured silicone release coatings are used on polyester, polyimide, and fluorocarbon backings. Platinum-catalyzed addition-cured release coatings work best for platinum-

catalyzed silicone PSAs. Formulations can be obtained from PSA suppliers. For porous glass cloth, a precoat of either highly peroxide-loaded PSA or silane-catalyzed PSA can be applied to improve the bond between PSA and the glass cloth backing.

Cure Requirements. The peroxides most commonly used to cure silicone PSAs are 2,4-dichlorobenzoyl peroxide and benzoyl peroxide. The reaction in Eq 1 shows that the peroxides extract a hydrogen atom from a methyl group on silicone. Because the peroxides are nonspecific, they can extract a hydrogen atom from a xylene or toluene molecule as well. For this reason, it is important that as much solvent as possible be removed from the PSA prior to peroxide activation. If solvent is cured into the PSA, lower tack and adhesion properties will result.

Temperatures of 60 to 70 °C (140 to 160 °F) are recommended for the first oven zone in order to remove solvent, followed by temperatures of 130 to 200 °C (265 to 390 °F) for the remaining zones, depending on the backing substrate and the peroxide used. A cure of 1 min at 65 °C (150 °F) to effect solvent removal, followed by 2 min at 177 to 204 °C (350 to 400 °F), is used for adhesive that contains benzoyl peroxide. When 2,4-dichlorobenzoyl peroxide is used, the curing temperature may be reduced to as low as 132 °C (270 °F). Shorter cure times can be achieved with more efficient ovens. Inadequate cure of the peroxide can lead to blooming, observed as a whitish film or even crystalline shapes on the surface.

If equipment and the type of backing material permit the use of higher curing temperatures, the cure time may be shortened. Figures 7 and 8 show the effect of curing time and temperature on the adhesive peel strength of the material when the peroxide concentration is varied. The dashed lines in Fig. 7 indicate that the adhesive has not developed enough cohesive strength to prevent transfer. Higher cure temperatures develop the cohesive strength of the adhesive in less time than is the case at lower cure temperatures. The ultimate adhesive strength of the fully cured material is essentially the same, whether cured at higher or lower temperatures. The

Table 6 Electrical properties of cured silicone films

ASTM standard	Property(a)	Industrial adhesive A	Industrial adhesive B
D 149................	Dielectric strength, kV/mm (V/mil)(b)	60 (1500)	26 (653)
D 150................	Dielectric constant		
	At 10^2 Hz	2.93	2.93
	At 10^5 Hz	2.97	2.87
D 150................	Dissipation factor		
	At 10^2 Hz	0.005	0.004
	At 10^5 Hz	0.004	0.001
D 257................	Volume resistivity, Ω · cm	7.2×10^{15}	3.5×10^{14}

Note: These values are not intended for use in preparing specifications. (a) After 96 h at 23 °C (73 °F), 50% relative humidity. (b) Measured with 6.4 mm (¼ in.) diam electrodes on 50 μm (2 mil) film of adhesive cured on aluminum panel

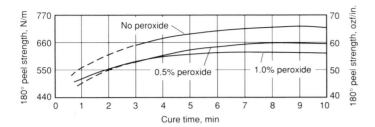

Fig. 7 Typical initial adhesive peel strengths of silicone PSA versus cure time and concentrations of benzoyl peroxide, at 150 °C (300 °F) using PSA tape with 50 μm (2 mil) thick aluminum foil backing. Cured adhesive coating weight, 62 g/m² (0.0014 oz/in.²). Percent benzoyl peroxide based on weight of adhesive solids

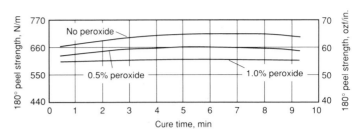

Fig. 8 Typical initial adhesive peel strengths of silicone PSA versus cure time and concentrations of benzoyl peroxide, at 200 °C (390 °F) using PSA tape with 50 μm (2 mil) thick aluminum foil backing. Cured adhesive coating weight, 62 g/m² (0.0014 oz/in.²). Percent benzoyl peroxide based on weight of adhesive solids

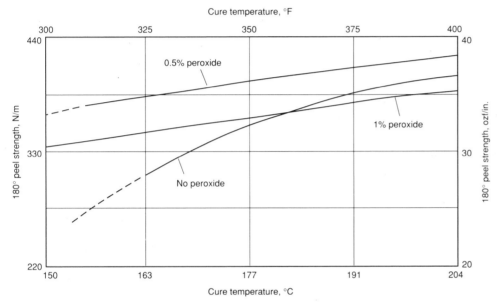

Fig. 9 Initial adhesive peel strength values of 50 μm (2 mil) thick primed PTFE tape using 280A (Dow Corning). Cured adhesive coating weight, 50 g/m² (0.0011 oz/in.²)

only difference is the time required to reach complete cure.

Figure 9 shows the interrelationship of peroxide concentration and cure temperature and the resulting effect upon adhesive peel strength. Higher levels of peroxide increase cross-link density and cohesive strength, but reduce adhesive strength.

Platinum-catalyzed silicone PSAs can be cured at temperatures as low as 80 °C (175 °F), although 100 °C (212 °F) is considered optimum. Unlike peroxide-cured PSAs, the solvent will not coreact. Therefore, gradient oven temperatures are not required. Undercure will result in lower tack, adhesion, and cohesive strength.

REFERENCES

1. L. Sobieski, Formulating Silicone Pressure Sensitive Adhesives for Application Performance, in *Technical Seminar Proceedings, Pressure Sensitive Tape Council*, May 1986
2. B. Copley, "The Thermal, Dynamic, Mechanical, and Morphological Properties of Silicone Adhesives," M.S. thesis, University of Minnesota, 1984
3. J. Schlademan, Tackifier Resins, in *Handbook of Pressure-Sensitive Adhesive Technology*, D. Satas, Ed., Van Nostrand Reinhold, 1982, p 353
4. D. Merrill, Silicone Pressure Sensitive Adhesives, in *Handbook of Pressure-Sensitive Adhesive Technology*, D. Satas, Ed., Van Nostrand Reinhold, 1982, p 346-347
5. T. Tangney, Sub-Ambient Pressure Sensitive Adhesive Applications: The Advantage of Silicones, in *Adhesives '87*, Publication AD87-530, Society of Manufacturing Engineers, 1987

Polysulfides

E.A. Peterson, Morton International, Inc.

LIQUID POLYSULFIDE POLYMERS are normally associated with high-performance sealants for the construction, marine, insulating glass, automotive, and aircraft industries. However, the lower molecular weight, low-viscosity liquid polymers are also widely used as elastomeric modifiers for the epoxy resins used in adhesives and coatings. Liquid polysulfide (LP) polymers have been used as the basis for sealants and adhesives since the 1940s. The technology behind the first liquid, room-temperature-curing elastomeric polymer evolved from an attempt in 1927 to synthesize antifreeze, the product of which was a gummy mass that would not dissolve. This article describes conventional polysulfides used in adhesives applications. Materials used as sealants are described in the article "Polysulfides" in the Section "Sealant Materials" in this Volume.

Chemistry and Forms

Liquid polysulfide polymers are formed by the reaction of bischloroethylformal and sodium polysulfide. Chemically, liquid polysulfides are polymers of bis(ethylene oxy) methane containing disulfide linkages terminated with mercaptan groups. The structure of the LP polymer is represented in Fig. 1, where the formal linkage provides flexibility and the disulfide linkage imparts excellent chemical and solvent resistance. Cross linking is provided by the incorporation of a trifunctional monomer, trichloropropane, into the polymer backbone during polymerization. The various LP polymers are available in molecular weights ranging from 1000 to 8000, and the degree of cross linking varies between 0.2 and 2%. These physical properties and others are featured in the "Properties" section in this article.

The liquid polysulfide polymers copolymerize with epoxy resins by an addition reaction between the mercaptan terminals and the epoxy group, as in Fig. 2. This reaction is the basis for formulating the flexible epoxy systems used in many varied application areas, such as adhesives, coatings, potting compounds, and plastic tooling.

A low molecular weight liquid polysulfide polymer with a corresponding low viscosity, such as LP-3, is used in epoxy systems. It acts as a reactive diluent as well as a flexibilizer, enabling greater ease in mixing and application. The addition of this polymer to the epoxy resin provides compounds with flexibility, high impact strength, and excellent chemical and solvent resistance, while maintaining the good adhesive properties of the epoxy.

Other mercaptan-terminated polymers are also available as modifiers for epoxy resins. One is the mercaptan-terminated low molecular weight polymer represented in Fig. 3.

Another polymer type is based on a polyoxypropylene polyol urethane backbone with no sulfur in it. A third type is a polythioether backbone with a monosulfide linkage in it. The structures of these two types are shown in Fig. 4 and 5, respectively.

The typical properties of these three mercaptan-terminated polymers are given in the section "Properties."

Concrete adhesives, that is, formulations used in concrete bonding applications, are also available. Two typical LP-epoxy concrete adhesive formulations are shown in Table 1. Their applications and properties are more fully described in the sections "Applications" and "Properties" in this article.

Applications

LP-epoxy resin systems are used in the construction, electrical, and transportation industries because of their unique combination of cured flexibility, adhesion to many substrates, and chemical resistance characteristics. They are typically used as adhesives for:

- Steel, aluminum, ceramics, wood, and glass
- Crack repair for concrete
- Patch repair for concrete and mortar
- Grouting compounds
- Lead-free automotive body solder

Concrete adhesive formulations are useful for bonding old to old or new to old concrete. Compositions of this type may also be used as a tack coat, thus improving adhesion considerably, especially when pouring wet concrete against cured concrete or steel. Typical uses of LP-epoxy concrete adhesives include:

- Raising floor levels
- Repairing badly degraded floors
- Patching old concrete
- Bonding preformed concrete structural units under dry or wet conditions (beams, piles, and so on)
- Bonding metal components, such as machinery to floors, under dry or wet conditions

Substrate Types and Preparation. Substrates that can be bonded with the LP-epoxy systems include concrete, steel, aluminum, wood, glass, carbon, ceramics, and some plastics. As with most adhesives, the substrate surface must be suitable before the adhesive is applied to achieve the best adhesion. Generally, loose corrosion, grease, oil, dirt, and excess moisture must be removed. Certain systems are available that are capable of being applied in underwater situations.

Major Suppliers

Liquid polysulfide polymers (LP, ELP-3, and ZL-1612) are manufactured by Morton International, Inc. Other polymers that are mercaptan-terminated, like the LP polymers, are produced by Products Research

$$HS(C_2H_4OCH_2OC_2H_4SS)_xC_2H_4OCH_2OC_2H_4SH$$

Fig. 1 Liquid polysulfide polymer. LP is a registered trademark of Morton International, Inc.

$$2R-CH-CH_2 + HS-R'-SH \longrightarrow -R-CHCH_2S-R'-S-CH_2-CH-R$$

Fig. 2 Copolymerization of LP polymer with epoxy resin. R is generally bisphenol A and R' is $-(C_2H_4OCH_2OC_2H_4SS)_xC_2H_4OCH_2OC_2H_4-$.

$$R[O(C_3H_6O)_n\,CH_2 - \overset{\diagup OH}{C}HCH_2SH]_3$$

Fig. 3 Capcure formula. *R* is an aliphatic hydrocarbon and *n* is 1 to 2.

$$-\!\!\!\!\!+\!CH_2CH(CH_3)O]_n\,\overset{O}{\overset{\|}{C}}NHC_6H_4CH_3NH\overset{O}{\overset{\|}{C}}O(CH_2)_3\,SC_2H_4OC_2H_4SH$$

Polyoxypropylene polyol urethane

Fig. 4 Permapol P-2 formula

and Chemical Corporation (Permapol P-2/P-3) and Diamond Shamrock Corporation (Capcure). The epoxies typically used are produced by Shell Oil Company (Epon) and Ciba-Geigy Corporation (Araldite).

Properties

The physical properties of a family of LP polymers are given in Table 2, while the effect on mechanical properties after adding LP-3 polymer to a typical unfilled epoxy adhesive formulation is shown in Fig. 6. The formulation uses a liquid epoxy resin of an epoxide equivalent-weight range of 185 to 192, with a viscosity of 11 to 15 Pa · s at 25 °C (77 °F). The amine curing agent was EH-330. The ratio of epoxy to amine was kept constant at 10 to 1 by weight. LP-3 polymer was added in varying amounts up to 50 parts by weight. Optimum room-temperature tensile shear is obtained by about 20 parts of LP-3 polymer, and flexural strength is improved in the range of 20 to 50 parts of polymer, with a peak around 35.

Figure 7 shows the effect of increasing the LP-epoxy ratio on impact resistance in a similar adhesion system. Note that at an LP-epoxy ratio of 0.5:1, it is less than 14 J (10.3 ft · lbf), while at a 1:1 ratio, it increases the impact value to approximately 95 J (70 ft · lbf) and eventually reaches values greater than 135 J (100 ft · lbf) at 1.5:1 and 2:1 ratios.

The tensile, flexural, and compressive shear strength values of LP-epoxy concrete adhesives are provided in Table 3. Different filler systems were used in each formulation, but both were based on an LP-epoxy ratio of 1:2 using an epoxy resin with 185 to 192 epoxide equivalent-weight range and a viscosity of 11 to 15 Pa · s.

The physical properties of the polymers described in the section "Other Mercaptan-Terminated Polymers" are given in Table 4. The properties of a new family of epoxy-terminated polysulfide polymers are discussed in the section "Epoxy-Terminated Polysulfides" in this article.

Temperature Performance. A more ductile adhesive also exhibits improved impact resistance over a wide temperature range. Figure 8 demonstrates the improved ductility over the range of −20 to 60 °C (−4 to 140 °F) as the result of incorporating only 35 parts per hundred resin (phr) LP-3 into the epoxy system. A DMTA trace shows that a lower glass transition temperature (54.5 °C, or 130 °F) for LP-modified epoxy, in contrast to 63.5 °C (146 °F) for straight epoxy material, also has been achieved and implies better low-temperature properties.

Adhesion to difficult-to-bond substrates is enhanced by the addition of LP-3 to an epoxy-based adhesive, as shown in Fig. 9. The lap-shear adhesion on steel with clean, rusty, oily, or wet surfaces is virtually doubled by the incorporation of 35 phr LP-3. This improved adhesion is a function of the better wetting characteristics imparted by the low viscosity of this polymer.

Chemical Resistance. Polysulfide-epoxy adhesives can be used in various chemical and solvent atmospheres, even under immersion conditions. Figure 10 gives the lap-shear data obtained after 14 days of complete immersion in 40 °C (104 °F) water. The LP-epoxy systems exhibit low permeability to water and water vapor, which gives adhesive products an improved water and corrosion resistance and allows their use in high-humidity environments. The LP-epoxy resin-cured adhesives display excellent resistance to a wide variety of oils, aromatic and aliphatic-type solvents, esters, ketones, and dilute acids and alkalis. However, tests of the adhesives under actual usage conditions should be conducted whenever possible. These adhesives cure with very low shrinkage, in contrast to unmodified epoxy resins, which may exhibit volume shrinkage as high as 6%.

Benefits of using elastomeric LP-modified epoxy adhesives are:

- Good wetting characteristics
- Adhesion to many substrates without priming
- Elastomeric ability, allowing them to withstand movement and vibration
- Reduced stress between dissimilar bonded materials
- Impact resistance, particularly at low temperatures
- Resistance to thermal shock/cycling
- Chemical resistance to oils and fuels

Table 1 LP-epoxy concrete adhesive formulations

	Formulation 1	Formulation 2
Part A, parts by weight		
Thiokol LP-3	100	100
Mortar white silica (HDS-100)	80	...
Hydrite 121	...	140
EH-330	20	20
Toluene	...	65
Part B, parts by weight		
Epoxy resin	200	200
Hydrite 121	...	105
Toluene	...	5
(Epoxide equivalent weight, 185–192; viscosity, 11–15 Pa · s at 25 °C, or 77 °F)		
Working and cure properties at 27 °C (80 °F) (50 g, or 1.75 oz, mass)		
Brush life (12 × 190 × 240 mm), h	0.2	1.4
Trowel life (12 × 190 × 240 mm), h	0.25	...
Set time (0.25 mm film), h	0.8	3.5
Tack free time (0.25 mm film), h	5.0	...
Cure time (0.25 mm, film), h	...	8–24
Mix ratio of A to B	100:100 by weight	100:100 by volume

Typical properties obtained with 0:1, 1:2, and 1:1 LP-epoxy ratios are given in Table 5.

Additives and Modifiers

Fillers such as calcium carbonate, graphite, milled glass fibers, silica, and talc can be added to LP-epoxy resin adhesives. Their addition modifies certain properties:

- Extends pot life
- Reduces the exotherm
- Changes pigmentation
- Improves shrinkage upon curing and dimensional stability at elevated temperatures
- Reduces cost of formulation
- Increases rigidity and impact strength

$$-\!\!\!\!\!+\!(O - C_2H_4 - S - C_2H_4)_x\!(O - \underset{CH_3}{\underset{|}{CH}} - CH_2 - S - CH_2 - CH_2)_y\!\!+_n O - CH_2 - CH_2 - S - R - S - R' - SH$$

Polythioether

Fig. 5 Permapol P-3 formula. *R* and *R′* are alkylene ethers of alkylene thioether radicals, the ratio of *x/y* is approximately 2:1, and *n* is greater than 8.

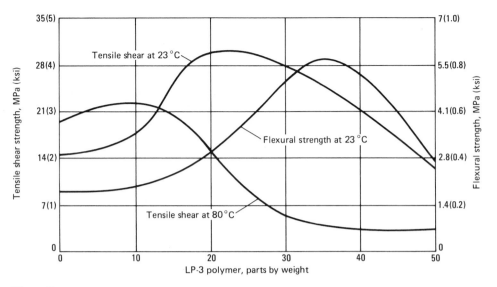

Fig. 6 Effect of polysulfide polymer addition on typical adhesive formulation cured for 1 h at 95 °C (205 °F)

Fig. 7 Effect of LP-epoxy ratio on impact resistance

The increase in impact resistance of three different fillers with increasing LP-3 content is shown in Fig. 11.

Solvents. Generally, the viscosity of the LP-epoxy mix is sufficiently low so that the use of solvents is not required, especially for conventional or airless spray or dispensing equipment. If solvents are necessary, ketones, esters, alcohols, and aromatic hydrocarbons are suitable.

Epoxy-Terminated Polysulfides

Although mercaptan-terminated polysulfide polymers are very effective modifiers in epoxy-base adhesives, the strong mercaptan odor is sometimes objectionable, especially for industrial plant application areas. Therefore, a new family of epoxy-terminated polysulfide polymers has become available, as represented by the formula in Fig. 12. To date, there are two members of this group. The first, ZL-1612, is an internally flexibilized resin whose physical properties are:

Physical state..................... Clear amber liquid
Color (Gardner) 3
Odor............................... Low (nonmercaptan)
Viscosity at 25 °C (77 °F), Pa · s... 40–50
Specific gravity................... 1.26
Epoxide equivalent weight
    (EEW) 310–350

Typical epoxy formulating methods can be utilized with this new polymer. The properties of this polymer, with the addition of a curative, are the same as would be achieved by a 1:2 LP-epoxy ratio (see Table 4).

The second polymer, ELP-3, is designed for use as a modifier for standard epoxy resins. Its physical properties are:

Physical state..................... Clear amber liquid
Color (Gardner)................... 9
Odor.............................. Low (nonmercaptan)
Viscosity at 25 °C (77 °F), Pa · s...0.8–2.5
Specific gravity................... 1.27
Epoxide equivalent weight
    (EEW) 575–625

Table 2 Physical properties of liquid polysulfide polymers

Specification requirements	LP-31	LP-2	LP-32	LP-12	LP-3	LP-33
Color MPQC-29-A, max.......	100	150	100	80	50	30
Viscosity, at 25 °C (77 °F), Pa · s	81.5–138	41–53	41–53	41–53	0.94–1.4	1.5–2.0
Moisture content, %	0.12–0.22	0.12–0.22	0.12–0.22	0.12–0.22	0.1 max	0.1 max
Mercaptan content, %........	1.0–1.5	1.5–2.0	1.5–2.0	1.5–2.0	5.9–7.7	5.0–6.5

General properties	LP-31	LP-2	LP-32	LP-12	LP-3	LP-33
Average molecular weight	8000	4000	4000	4000	1000	1000
Refractive index, D^n, 25 °C ± 0.15 °C (77 °F ± 0.27 °F)......	1.5728	· · ·	1.5689	· · ·	1.5649	· · ·
Pour point, °C (°F).............	10 (50)	7 (45)	7 (45)	7 (45)	−26 (−15)	−23 (−9)
Flash point, °C (°F)	225 (437)	208 (406)	212 (414)	208 (406)	174 (345)	186 (367)
Cross-linking agent content, %.....	0.5	2.0	0.5	0.2	2.0	0.5
Specific gravity at 25 °C (77 °F)..	1.30–1.32	1.25–1.29	1.28–1.30	1.28–1.30	1.26–1.28	1.26–1.28
Average viscosity at 4 °C (39 °F), Pa · s...........................	740	380	380	380	9	16.5
Average viscosity at 65 °C (150 °F), Pa · s.............................	14	6.5	6.5	6.5	0.15	0.21

Table 3 Properties of LP-epoxy concrete adhesives

	Tensile strength at 27 °C (80 °F)(a)				Flexural strength at 27 °C (80 °F)(a)				Compressive shear strength at 27 °C (80 °F)(a)			
	Formulation 1(b)		Formulation 2(b)		Formulation 1(b)		Formulation 2(b)		Formulation 1(b)		Formulation 2(b)	
Adhesive cure	MPa	ksi	MPa	ksi	MPa	ksi	MPa	ksi	MPa	ksi	MPa	ksi
Bonding old-to-old concrete(c)												
At 27 °C (80 °F) for 3 days	2.60	0.377	2.90	0.420	2.00	0.290	2.10	0.305	18.0	2.61	26.0	3.77
At 27 °C (80 °F) for 7 days	2.30	0.333	0.014	0.002	2.20	0.320	2.38	0.345	29.0	4.20	30.0	4.35
At 27 °C (80 °F) for 28 days	2.76	0.400	2.90	0.420	2.76	0.400	3.00	0.435	29.0	4.20	29.0	4.20
Bonding new-to-old concrete(c)												
At 27 °C (80 °F) for 3 days	2.60	0.377	2.70	0.390	2.38	0.345	1.69	0.245	· · ·	· · ·	· · ·	· · ·
At 27 °C (80 °F) for 7 days	2.90	0.420	3.10	0.450	2.38	0.345	2.69	0.390	· · ·	· · ·	· · ·	· · ·
At 27 °C (80 °F) for 28 days	3.21	0.465	3.50	0.507	2.59	0.375	2.76	0.400	· · ·	· · ·	· · ·	· · ·

(a) For 50 g (1.75 oz) mass. (b) Refer to Table 1 for breakdown of formulations. (c) Failures occurred in the concrete for all tests.

Table 4 Typical properties of other mercaptan-terminated polymers available as modifiers for epoxy resins

Property	Capcure	Permapol P-2	Permapol P-2	Permapol P-3
Mercaptan, %	3.3	2.5–3.3	1.4–1.7	2000(a)
Molecular weight	· · ·	1500	5500	5500
Viscosity at 25 °C (77 °F), Pa · s	15	2000	1400	700
Specific gravity	1.15	1.1	1.1	1.16
Color	1.0(b)	Straw	Water white to straw	Dark amber
Appearance	· · ·	Viscous liquid to semisolid	Viscous liquid	Liquid

(a) Mercaptan equivalent. (b) On Gardner-Hellige scale

Table 5 Typical properties obtained with various LP-3:epoxy ratios

Property	LP:EP ratio, parts by weight		
	0/1	1/2	1/1
Working, at 27 °C (80 °F)			
Viscosity, Pa · s	2.4	2.0	1.85
Pot life, min	25	17	20
Set time, min	35	19	20
Maximum temperature, °C (°F)	110 (230)	129 (265)	116 (240)
Mechanical, cured 7 days at 27 °C (80 °F)			
Tensile strength, MPa (ksi)	24 (3.5)	48 (6.9)	19 (2.75)
Elongation, %	0	10	30
Hardness, Shore D	80	78	69
Impact strength, J (ft · lbf)			
At 27 °C (80 °F)	0.68 (0.5)	3.8 (2.8)	97 (71.5)
At −37 °C (−35 °F)	0.68 (0.5)	0.68 (0.5)	2.0 (1.5)
Mechanical, cured 7 days at 27 °C (80 °F) and 70 h at 100 °C (212 °F)			
Tensile strength, MPa (ksi)	23 (3.4)	41 (6.0)	5.9 (0.85)
Elongation, %	0	12	65
Hardness, Shore D	80	76	35

Typically, ELP-3 would be used at polymer to epoxy resin ratios of 1:3-4. The physical properties of a 1:3 ELP-3:diglycidyl ether of bisphenol A (DGEBA) cured with a cycloaliphatic amine are given in Table 6, along with other formulations on which adhesives are based. Table 7 compares the lap-shear strength of unfilled systems to glass, fiber-reinforced plastic (FRP), plywood, and galvanized steel. In all cases, except for the galvanized steel, the strength of the adhesive exceeded that of the substrate.

One advantage of these ELP polymers over the mercaptan-terminated polysulfide polymers is the improved shelf stability when combined with an epoxy resin. This benefit allows one-component heat-curable adhesives to be formulated with room-temperature package stability. Lap-shear data of the one-component adhesive on oily galvanized steel are summarized in Table 8. Both of the polymers give improved strength compared with the straight epoxy system over the temperature range of −29 to 82 °C (−20 to 180 °F), as well as after high humidity or salt resistance.

These epoxy-terminated polysulfide polymers provide the same benefits in an epoxy-type adhesive as an LP polysulfide does:

- Flexibility
- Impact resistance
- Thermal shock resistance
- Adhesion to many varied substrates
- Chemical resistance
- Good electrical properties

In addition, these new polymers have:

- No mercaptan odor
- Good shelf stability in a one-component heat-curing adhesive

Processing Parameters

The LP-epoxy resin compounds are generally formulated as two-component systems to provide good shelf life and ease of handling. One component contains the liquid polysulfide and the curing agent, together defined as the curative, while the second component consists of the epoxy resin. After mixing the two components, the usable work life (pot life) depends on several factors:

- Temperature: the higher the temperature, the shorter the work life
- The reactivity of the epoxy resin
- The reactivity of the curative
- The level of LP-3 used
- The presence of diluents, such as fillers and/or solvents

Epoxy Resins. Liquid epoxy resins are most widely used with the LP-3 polymers. However, the epoxy resins that most benefit by modification with LP-3 include:

- Liquid unmodified bisphenol A resins
- Diluent-modified bisphenol A resins
- Liquid bisphenol A and bisphenol F blends

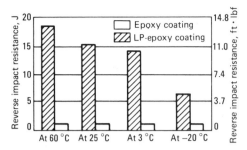

Fig. 8 Impact resistance over a wide temperature range for DGEBA epoxy resin-cycloaliphatic amine. LP-epoxy resin contained 35 phr of LP-3.

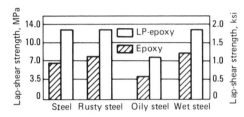

Fig. 9 Adhesion of DGEBA epoxy resin-cycloaliphatic amine to difficult substrates. LP-epoxy resin contained 35 phr of LP-3, and was room-temperature cured. Bond overlap was 15 × 25 mm (0.6 × 1.0 in.) and glue line was 0.2 mm (8 mils).

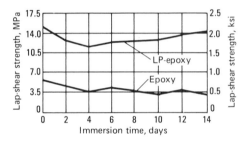

Fig. 10 Immersion in 40 °C (105 °F) water of aluminum substrate bonded with epoxy and an LP-epoxy based on 50 phr of LP-3. Adhesives cured for 0.5 h at 100 °C (212 °F)

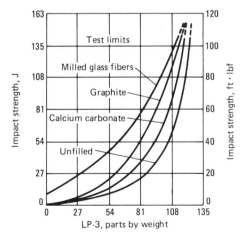

Fig. 11 Effect of fillers and LP-3 concentration on impact strength

$$CH_2-CH_2R - S(C_2H_4OCH_2OC_2H_4SS)_xC_2H_4OCH_2OC_2H_4 - SRCH_2 - CH_2$$
$$\diagdown O \diagup \qquad\qquad\qquad\qquad\qquad\qquad\qquad\qquad\qquad \diagdown O \diagup$$

Fig. 12 New family of epoxy-terminated polysulfide polymers

Table 7 Adhesion of three formulations to various substrates

Component(a)	Formulation(b) 1	Formulation(b) 2	Formulation(b) 3
Epoxy resin(c)	100	...	75
ZL-1612(d)	...	100	...
ELP-3(d)	...	...	25
Tertiary amine	10	6	8.5

Substrates	Lap-shear strength, MPa (ksi) 1	Lap-shear strength, MPa (ksi) 2	Lap-shear strength, MPa (ksi) 3
Glass/glass(e)	12.5 (1.81)	18.1 (2.63)	10.2 (1.48)
Fiber-reinforced plastic/fiber-reinforced plastic(e)	6.1 (0.89)	8.0 (1.16)	6.6 (0.96)
Plywood/plywood(e)	5.4 (0.78)	5.8 (0.84)	4.6 (0.67)
G-60 galvanized steel/G-60 galvanized steel	4.1 (0.60)	4.5 (0.65)	3.4 (0.49)

(a) Cured 7 days at 23 °C (73 °F). (b) In parts per weight. (c) Epoxide equivalent weight, 185–195. (d) Epoxy-terminated polysulfide. (e) Substrate failure

Table 8 Lap-shear strength on G-60 galvanized steel

Formulation(a)	Epoxy, parts by weight	ZL-1856-epoxy, parts by weight	ZL-1612, parts by weight
Epoxy resin(b)	100	60	...
ELP-3(c)	...	40	...
ZL-1612(c)	...	...	100
Dicyandiamide	10	7	6
3-phenyl-1,1-dimethyl urea	5	3.5	3

Tested at	Lap-shear strength after 15-min cure at 135 °C (276 °F), MPa (ksi) Epoxy	Lap-shear strength after 15-min cure at 135 °C (276 °F), MPa (ksi) ZL-1856-epoxy	Lap-shear strength after 15-min cure at 135 °C (276 °F), MPa (ksi) ZL-1612
−29 °C (−20 °F)	14.0 (2.03)	19.4 (2.81)	19.4 (2.82)
21 °C (70 °F)	12.3 (1.79)	18.8 (2.72)	19.8 (2.87)
>80 °C (180 °F)	12.8 (1.85)	13.7 (1.99)	15.4 (2.23)
21 °C (70 °F) after 500 h at 38 °C (100 °F), 95% relative humidity	9.10 (1.32)	14.5 (2.11)	15.0 (2.17)
21 °C (70 °F) after 500 h in salt spray	8.21 (1.19)	12.9 (1.87)	15.4 (2.24)

(a) In parts by weight. (b) Epoxide equivalent weight, 185–195. (c) Epoxy-terminated polysulfide polymer

- Solid bisphenol A resins
- Multifunctional epoxy resins
- Aliphatic diepoxide resins
- Polyamides
- Tertiary amines

Curatives. Ambient- or elevated-temperature cures can result from the choice of curative. The mercaptan-epoxy reaction is catalyzed by bases, particularly organic amines. Primary and secondary amines are preferred for better thermal stability and balance between low- and high-temperature performance. The optimum elevated-temperature performance is obtained with either an elevated-temperature cure or a room-temperature cure followed by a short postcure at high temperatures.

Suitable curatives for the LP-epoxy reaction include:

- Liquid aliphatic amines
- Liquid aliphatic amine adducts
- Solid amine adducts
- Liquid cycloaliphatic amines
- Liquid amide-amines
- Liquid aromatic amines

Ratio of Modifier to Epoxy Resin. Probably the most important processing factor is the polysulfide polymer to epoxy resin ratio. The most effective ratio varies between 1:2 to 2:1 of LP to epoxy. Typical properties of the unmodified epoxy and LP-modified epoxy with ratios of 1:2 and 1:1 were given in Table 5. An increase in percent elongation and impact strength are a result of the increase in the amount of LP polymer.

Viscosity Reduction. The low viscosity of the LP-3 polymer can result in a drastic reduction in the viscosity of the epoxy resin when the two materials are mixed. The effect of the reduced viscosity of two different epoxy resins is shown in Fig. 13. Thus, LP-3 can act as a reactive diluent as well as an elastomeric modifier for epoxy resins. This lower viscosity allows greater ease of mixing and application. It also provides a higher solids content for lower-solvent (solventless) systems.

Table 6 Typical properties of modified polysulfide

Component(a)	Formulation(b) 1	Formulation(b) 2	Formulation(b) 3
DGEBA epoxy resin(c) ..	100	...	75
ZL-1612(d)	...	100	...
LP-3(e)	50	...	...
ELP-3(d)	...	...	25
Tertiary amine	10	7	...
Cycloaliphatic amine ...	...	...	46

Property	Formulation 1	Formulation 2	Formulation 3
Tensile strength, MPa (ksi) 50 (7.23)	50 (7.23)	48 (6.98)	31 (4.53)
Elongation, %	3	4	10
Hardness, Shore D.....	82	82	74

(a) Cured 7 days at 22 °C (72 °F). (b) In parts per weight. (c) Epon 828. (d) Epoxy-terminated polysulfide. (e) Liquid polysulfide

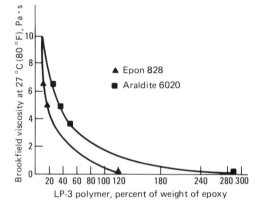

Fig. 13 Effect of LP-3 addition on viscosity of two typical epoxies

Diluents, which also affect pot life as well as the properties of LP-epoxy resin compounds, are discussed in the section "Additives" in this article.

ACKNOWLEDGMENT

Special appreciation is expressed to T. Rees and A. Wilford of Coventry, England, and to T. Fiorillo and J.R. Harding of Woodstock, IL, for data and assistance.

SELECTED REFERENCES

- "Capcure," Technical Bulletin, Diamond Shamrock Corporation
- "LP Polysulfide/Epoxy Systems," Technical Bulletin, Morton International, 1984
- E.A. Peterson, "Liquid Polysulfide Polymers for Adhesives and Sealants," Paper presented at ASC Short Course II, Adhesive and Sealant Council, April 1989
- M.J. Scherrer et al., "Epoxy Terminated Polysulfides and Polysulfide Water Dispersions: New Raw Materials for Adhesives and Sealants," Paper presented at ASC Spring Seminar, Adhesive and Sealant Council, March 1989

Elastomeric Adhesives

William F. Harrington, Uniroyal Adhesives & Sealants Company, Inc.

RUBBER-BASE ADHESIVES are a major element of the adhesives industry and are an intrinsic part of modern manufacturing environments. Roughly one-third of all adhesives used in the United States are made from synthetic or natural rubber in all the various forms that elastomeric adhesives can take. Approximately 400 adhesive manufacturers produce these compounds for resale to fabricators of other products.

The properties and characteristics of rubber-base adhesives that make these products the adhesive of choice for so many assembly operations include:

- *Versatility*. Elastomeric adhesives will provide adhesion to almost any surface used in manufacturing. The bond strength can vary from a temporary holding to a structural-support capability
- *Quick assembly*. Contact adhesives based on rubber materials develop virtually instantaneous bonds. The ability to hold two surfaces together when first contacted and before the adhesive develops its ultimate bonding properties when fully cured is known as green strength (Ref 1). The high green strength of rubber-base adhesives, particularly of polychloroprene/phenolic compounds, is vital in many applications because it allows handling immediately after assembly. Pressure-sensitive adhesives (PSA) also provide faster processing cycles
- *Flexibility*. The resilience of rubber-base adhesives helps absorb stress caused by the environment, which an assembled product must endure. Vibration, impact, shear, elongation, and peel are a few of the forces that elastomerics resist well
- *Variety of type*. Although there are five basic types of elastomers discussed in this article, many other polymer types exhibit rubberlike properties. Within each rubber class, several distinct grades of polymer are available, each of which can be compounded with a multitude of ingredients to alter properties
- *Economy*. On a per-gallon basis, elastomeric adhesives provide an inexpensive bond for most applications. Tooling is often low in cost, and the speed of assembly using elastomeric adhesives helps minimize labor costs

- *High peel strength*. The ability of rubber polymers to stretch and elongate, rather than shear apart, provides the strength necessary to hold components together when stressed by a peeling force
- *Variety of form*. Elastomeric adhesives can be supplied for assembly operations as solvent dispersions, water-borne compounds, hot melts, precast films, extruded tapes, or reinforced films and tapes. In addition, solvent and water-borne adhesives can be supplied as single-component or multicomponent reactive systems
- *Ease of modification*. Even with the thousands of rubber-base adhesives already available, a given product assembly or process may require a custom-tailored formulation. This is usually easily accomplished, once the proper elastomer for the application has been determined
- *Usability of thermoplastic and thermosetting polymers*. The chemical nature of most rubber polymers is such that for many applications no curing or crosslinking is necessary for the adhesive to perform its function. However, most rubber polymers can be cured or cross linked to provide higher strength, better heat resistance, or improved chemical resistance

While many polymers exhibit some of the characteristics of rubber and perform in much the same way in the course of bonding, five elastomers represent the majority of types available and most of the volume used. In certain polymeric configurations, acrylics and polyurethanes could be classified as elastomers, and in some cases, are. However, these materials are described separately in the articles "Acrylics" and "Urethanes" in this Section, "Adhesives Materials," of the Volume. (Acrylics and urethanes used in sealant applications are described in the Section "Sealant Materials" in this Volume.) Several cellulosic materials, such as cellulose acetate butyrate; several thermoplastic resins that are appropriately compounded, such as polyvinyl chloride and acrylonitrile-butadiene-styrene; and several vinyl acetate compounds (and their many derivatives) all possess many of the same characteristics as rubber-base adhesives, but these will not be covered in this article. Further, polysulfides

and silicones both exhibit properties similar to those of elastomeric compounds, but are described in separate articles in both the Section "Sealant Materials" and within this Section of the Volume.

This article describes a type of rubber from each of the following groups: natural, butyl, styrene-butadiene, acrylonitrile-butadiene, and polychloroprene. First, common properties are identified and described to facilitate an understanding of some terminology and concepts. Next, each type is defined in terms of its chemistry, the various final forms the adhesive can assume, some of the individual characteristics inherent to the rubber type, primary markets in which the adhesive type is used, processing properties, and additives/modifiers used as compounding ingredients.

It must be noted that a summary presentation such as this cannot provide much more than an overview because of the scope of the material being described. There are, quite literally, tens of thousands of rubber-base adhesives available for a great many manufacturing operations.

Common Properties

Differences in chemical structure and molecular weight will always provide distinctions between the elastomeric types. However, as a class of materials, rubber-base adhesives have a number of common characteristics described below. Other characteristics shared by elastomers are noted in the discussions of each elastomer type.

Fast strength development (green strength) is a major reason for choosing an elastomeric adhesive. Whether they are pressure sensitive or contact bond types, rubber adhesives develop strength faster than most other polymeric types. Figure 1 (Ref 2) shows the differences between several rubber polymers. In this case, the polymers were milled slightly before being dispersed in an appropriate solvent. The only other ingredient added was an antioxidant. The addition of resins and curatives would have provided significantly different results, such as faster strength buildup and higher peel strength. It should be pointed out that, although Fig. 1 shows a roughly similar starting point for peel strength for all

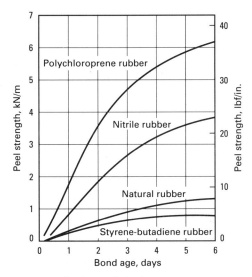

Fig. 1 Bond strength development at room temperature

elastomerics, natural rubber and styrene-butadiene rubber may provide higher initial peel strength values. This is usually the case in the first few minutes after mating coated surfaces. As time progresses, polychloroprene and nitrile rubber types develop much higher peel strengths, particularly after a few hours.

Viscosity is a measure of the thickness or consistency of a liquid. The viscosity of elastomeric adhesives somewhat restricts the method of application, particularly with solvent-dispersed compounds. Typically, solvent-borne rubber compounds require a viscosity range of 0.1 to 0.6 Pa · s to be sprayed. A similar viscosity range is suitable for curtain and dip coating. Brush application is most appropriate in the 1 to 5 Pa · s range, although products with viscosities as high as 60 Pa · s have been applied with a brush. Roll-coating compounds can have viscosities as low as 2 Pa · s but most often they are from 5 to 15 Pa · s. Trowelable and extrudable grades range from 50 to over 1000 Pa · s.

Although hot-melt adhesives require a viscosity range similar to that of solution types, water-borne adhesives generally have much more latitude in viscosity. Because most latexes have very low viscosities before compounding, virtually all are sprayable. Thickeners are often added to increase viscosity to make the adhesive suitable for other application methods. These thickeners usually provide a degree of thixotropy to the mixture, which means that even at relatively high viscosities (including over 10 Pa · s) many water-borne compounds can be sprayed. Dip and curtain coatings require viscosities in the 0.05 to 0.3 Pa · s range. The brush application method works readily with adhesives that have viscosities ranging from 1 to over 50 Pa · s, whereas 5 to 50 Pa · s is representative of roll-coatable latexes, and values above 250 Pa · s are appropriate for trowelable compounds.

Table 1 summarizes a number of properties and characteristics of rubber-base adhesives. Although this table is greatly simplified, it does cover the major differences between solvent-borne, water-borne, and hot-melt compounds.

Much attention has been directed to solvent-borne adhesives in recent years because the solvents are generally volatile, flammable, and toxic to some degree. Some solvents can enter into reactions with other air-borne contaminants and contribute to smog formation, stratosphere ozone depletion, and workplace exposure. The decision to use solvent-dispersed adhesives has become complicated by the necessity to comply with a variety of local, state, and federal rules and regulations. Although solvent recovery systems and afterburners can be effectively and economically attached to ventilation equipment, many factories are switching to water-borne, hot-melt, or 100% solids reactive systems, often at the expense of product performance or labor efficiency.

Natural Rubber

Natural rubber (NR) was well known and used by the natives of Central and South America before the arrival of Christopher Columbus. On his second voyage to the New World, Columbus discovered the use of rubber and introduced the product to Europe upon his return. Natural rubber, found in the sap of a number of plants and trees, of which *Hevea brasiliensis* is the most common source, remained an expensive curiosity until the latter part of the 18th century. The first patent for an adhesive compound, a naphtha dispersion of natural

Table 1 Summary of properties and characteristics of rubber-base adhesives

Property	Solvent-borne adhesives(a)	Water-borne adhesives	Hot-melt adhesives
Solids content, %	Usually 10–30%, with some in 40–50% range	Typically 40–60%	100%
Bonding methods	Contact, pressure sensitive, heat and solvent reactivated, wet bond	Same as solvent borne	Wet bond, heat activated, pressure sensitive
Drying rate	Usually fast; drying time adjustable by choice of solvents	Slow to dry; however, high solids content means less to dry.	No volatiles to evaporate
Open time	Wide range for contact adhesives; indefinite for pressure sensitive	Short range for contact types, unless compounded for open time; indefinite for pressure-sensitive types	Short range for thin film due to fast cool-down; indefinite for pressure sensitive
Surface wet-out	Excellent on most surfaces, even with minor contamination	Good on porous substrates, poor to fair on nonpolar surfaces like plastics	Good on most surfaces; may require heating the surface before application
Water resistance	Usually good to excellent	Fair to good	Excellent
Green strength	Excellent; allows early assembly handling	Fair; slow strength development	Fast bond development as bond line loses heat
Ultimate strength	Excellent, especially for curing types; highest film strength	Good to excellent; curing compounds provide high strength	Poor to fair (mostly noncuring types)
Application procedures	Wide range, including use of brush, spray, roll, extrusion, trowel	Wide range; may get misting with some spray techniques	Usually with brush, trowel, extrusion, or roll coat; may require preheating before applying
Temperature range	−55 to 150 °C (−65 to 300 °F) for end-use, freeze-thaw stable in liquid form	−40 to 120 °C (−40 to 250 °F) for end-use; requires temperature control to keep from freezing	−45 to 120 °C (−50 to 250 °F) for end-use; no storage or shipping problems; upper temperature range dependent on hot-melt type
Surface attack	May cause degradation of some plastics and paint	Causes shrinkage and wrinkling of fabric and paper; corrosive to metal	Usually no problem; heat may deform thin plastics
Flexibility	Excellent	Excellent	Excellent
Hazards	Requires vapor control to OSHA limits; many are flammable and may require emissions control	Usually no problem; some have trace levels of chemicals that require control of vapors	Few problems other than working with hot dispensing equipment
Cleanup	Solvent	Water when wet, solvent when dry	Solvent
Cost	Low-to-moderate cost per gallon, high cost per dry pound	Moderate cost per gallon, low cost per dry pound	Moderate cost per dry pound

(a) OSHA, Occupational Safety and Health Administration

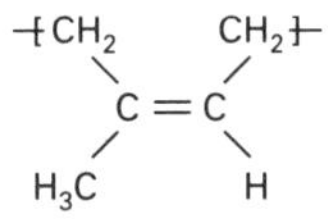

Fig. 2 Cis-1,4-polyisoprene chemical structure

rubber, was issued in 1791. This product was used for laminating and waterproofing textiles. Even today, a simple solvent solution of natural rubber is used as an art material cement and for the production of leather shoes.

As rubber became an important part of the industrial revolution and more uses were found for products manufactured with it, the adhesives used to fabricate these products also grew in importance. Manufacturers learned how to process rubber, compound it with other ingredients, and mix it with a variety of solvents to produce useful materials. The establishment of plantations in southeast Asia in the early 20th century provided a supply source for the growing automobile industry.

Chemistry. Natural rubber is harvested as a latex by tapping trees in much the same manner used to harvest maple syrup. Tree latex contains about 35% rubber solids, as well as small quantities of protein carbohydrates, resins, mineral salts, and fatty acids. The latex must immediately be ammoniated to avoid coagulation by these other ingredients and to prevent bacterial action. After collection, the latex is concentrated to 60% solids or higher if the product is to be shipped as latex; otherwise, the latex is coagulated, washed, dewatered, and pressed into bales for grading and shipment.

The chemical name for natural rubber is cis-1,4-polyisoprene, which has the structural formula shown in Fig. 2. Its molecular weight is typically about 1 million g/mol.

Forms. In solid form, natural rubber is graded according to the content of dirt remaining from the precipitation of latex at the plantation. The predominant grading system used since the mid-1960s has been the Standard Malaysian Rubber system, although some rubber is still available as smoked sheet, crepe, and so on. Chemically modified natural rubber is available, including Heveaplus rubber, a methyl methacrylate graft polymer, and an epoxidized natural rubber. Synthetic polyisoprene has been produced since the early 1960s, but exhibits lower tack and green strength in noncuring compounds. Depolymerized liquid natural rubber is available in several grades, but is not widely used. Also available are many grades of reclaimed rubber, produced from the grinding and heating of vulcanized rubber products such as tires, baby bottle nipples, and so on. Although reclaimed rubber offers some processing advantages to the compounder, its use has declined in recent

years because of the extensive use of blended polymers.

Natural rubber adhesives can be supplied to the end user in a variety of forms. Most often, the product is either a solvent dispersion or a latex for coating onto surfaces. Viscosity can range from very thin solutions for spray application to highly viscous mastics suitable for troweling. Natural rubber can be precoated onto fabric, paper, or film to provide pressure-sensitive tapes. This precoating can be accomplished using roll coaters to apply solvent-borne or water-borne compounds or using a friction calender to apply highly masticated, softened solid rubber to a surface. Depolymerized rubber can be applied directly as a 100% solids liquid polymer or as a solvent- or water-borne dispersion.

Properties. Generally, adhesives made from the various forms of natural rubber exhibit similar characteristics, although some properties are altered by the form or the presence of curatives. The properties of products based on rubber include:

- *Excellent tack.* Solvent solutions are very tacky even without resin modification. Latex compounds exhibit superior self-adhesion after coated surfaces have dried
- *A variety of bonding mechanisms.* Products can be used in contact, pressure-sensitive, or heat or solvent reactivating adhesive applications
- *Very good water and moisture resistance.* Solvent-borne adhesives have an advantage over latex, which contains surfactants. However, both are excellent for imparting moisture resistance to laminated products
- *High flexibility.* Any form of natural rubber possesses this characteristic
- *Wide range of substrates.* The inherent tackiness of natural rubber enables it to coat most nonpolar surfaces. A properly compounded solvent solution will stick to almost any surface. Polarity refers to electrostatic surface energy. The chemical nature of each substrate or adhesive inherently creates positive and negative charges at the interface, which attract each other. Many metals possess high polarity, whereas many plastics have low polarity.
- *High resilience and fatigue resistance.* Both properties are inherent to the natural rubber chemical structure
- *Brittleness with age.* The presence of a carbon-carbon double bond allows oxidative degradation to occur, even when vulcanized. Many compounds have antioxidant additives if aging is a key factor
- *Poor resistance to organic solvents and oils.* Although vulcanization reduces the rate of attack, it cannot prevent it
- *Low to moderate cost.* This factor depends on world market costs, degree of compounding, and form

- *Curability.* Although sulfur is used as the primary curative, requiring heat and pressure, isocyanates have also been used for room-temperature cures

Markets and Applications. Solvent-dispersed natural rubber adhesives can be either curing or noncuring. While the polymer is inherently tacky and can provide good building strength, tackifying resins are often added to improve attachment to polar surfaces. Because the solids levels of most solvent-based adhesives range from 5 to 35%, these adhesives are not the best product for gap filling. However, because natural rubber can be compounded as a pressure-sensitive adhesive with a high resin content, it is the foundation for many kinds of commercial tapes and surgical plasters. Many are used in leather footwear fabrication for temporary bonding and in rubber footwear as a curing laminating adhesive. Many natural rubber products, such as off-the-road tires, hoses, and belting, use a vulcanizing grade of solvent-dispersed natural rubber. These can be single- or multiple-component materials, with the curatives being added shortly before fabrication. Many retread and tire patch compounds are also solvent formulations of a vulcanizing type.

Reclaimed rubber and Heveaplus in solvents have widely divergent applications. Reclaimed rubber is most often used as a general-purpose, nonvulcanizing adhesive. It is extremely versatile in bonding such surfaces as insulation, polyethylene, felt, canvas, packaging materials, metal, wood, and cured rubber products. Heveaplus has a higher polarity than most natural rubber compounds and is often used as a primer for natural rubber solutions on such substrates as polyvinyl chloride (PVC) used for tapes and the synthetic upper materials used for shoes with crepe rubber soles.

Latex compounds, typically 40 to 80% solids, have been commonly used in paper, textile, and construction applications. One of the most popular applications for nonvulcanizing latex has been on self-sealing envelopes. Totally nontacky in the dried film, the natural rubber compound readily adheres to itself when the two surfaces are pressed together. Another application, albeit one that requires compounding for higher bulk, tack, and viscosity, is for bonding ceramic tile. Adhesives for synthetic resin floor tile are similar, but have less filler. Latex can also be used for bonding canvas and leather shoes and interior trim in some automotive applications and for food jar cork seals. Vulcanized latex is most often used for textile binding, particularly for tufted carpets.

Additives and Modifiers. Natural rubber adhesives accept a wide variety of other materials as compounding ingredients. Most mineral fillers and carbon black are

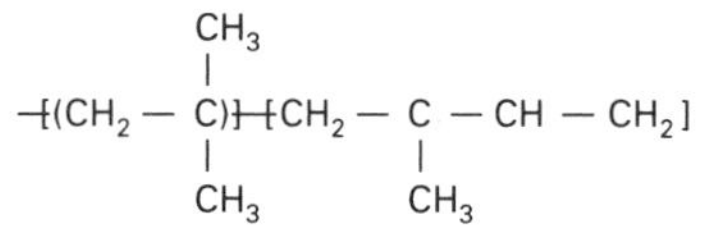

$$-[(CH_2 - \underset{\underset{CH_3}{|}}{\overset{\overset{CH_3}{|}}{C}})]-[CH_2 - \underset{\underset{CH_3}{|}}{C} - CH - CH_2]$$

Fig. 3 Butyl rubber chemical structure (isobutylene with a small percentage of isoprene)

easily incorporated into solvent-borne compounds, and, through the use of appropriate surfactants and wetting agents, into latex. Solvent-borne compounds also easily incorporate hydrocarbon, rosin, rosin ester, and coumarone-indene and terpene resins. Latex compounds require the emulsification of these resins before they are added. The same statements hold true for plasticizers and oils. Curatives, accelerators, and other dry ingredients may also be added, according to end-use requirements. In solvent systems, curatives are often packaged separately, while latex may be precured in the wet state without compromising properties, depending on the application.

Natural rubber adhesives can vary from a few lbf/in. width in peel strength in the case of unvulcanized pressure sensitive and temporary holding adhesives to a "substrate tear" level of bond strength in the case of vulcanized compounds used in hose, belting, and tire products. Nonvulcanizing adhesives typically withstand temperature ranges of −30 to 65 °C (−20 to 150 °F), whereas vulcanized adhesives can perform from −40 °C (−40 °F) to intermittent exposures above 150 °C (300 °F). Most natural rubber compounds are good electrical and thermal insulators, but individual adhesives can be formulated for conductivity, both electrical and thermal, and can be made nonflammable in the dry state.

Butyl Rubber

One of the early synthetic rubbers, butyl rubber is a polymerization product of isobutylene with a small percentage of isoprene to provide a degree of unsaturation to allow curing by several mechanisms. Although the amount of isoprene varies, depending on end properties desired, the degree of unsaturation is usually less than 2% of that found in natural rubber. Figure 3 depicts the chemical structure. The term saturation relates to the density of carbon-carbon double bonds in the polymer chain. If a polymer is fully saturated, it has no reactive sites (C=C) for cross-linking to occur.

Chemistry. Polyisobutylene (PIB) is a homopolymer of isobutylene with virtually no unsaturation in the polymer chain. Polyisobutylene can be both lower and higher in molecular weight than standard grades of butyl rubber. The lower molecular weight compounds, however, are most often used as modifiers of other compounds because of their permanent tackiness. An important

modification of the butyl polymer chain is halogenation, usually by chlorine or bromine, which increases functionality and provides a reactive site for alternate cure mechanisms. Additionally, a variety of partially cross-linked grades of butyl are available as are various depolymerized grades of butyl.

Forms. Butyl rubber adhesives are supplied to the end user in forms quite similar to those of natural rubber. Solvent-borne dispersions, water-borne emulsions, and a variety of pressure-sensitive adhesive (PSA) precoated films are available. Butyl and polyisobutylene can also be supplied as a major component of a hot melt or as a solid rubber extruded tape, or rope. Compounds can range from low-viscosity spray or brush materials to high-viscosity sealants and mastics.

Properties. Because of the extremely diverse nature of the polymer, general statements about butyl will be qualified by strong warnings about applicability. Its properties include:

- *Superior water and moisture resistance.* Butyl is by far the best elastomer for providing this characteristic
- *High initial tack.* The proper choice of butyl or polyisobutylene or mixtures thereof provides excellent building tack. Building tack is the capability for quick adhesion with low application pressure, but providing sufficient strength to continue manufacturing operations
- *Low air and gas permeability.* The amorphous, highly saturated nature of the polymer chain helps provide effective blockage to air and gases
- *Good resistance to vegetable and animal oils.* Although the product will swell slightly in some chemicals, overall resistance is good. Polyisobutylene PSAs are successful with high-oil-content foams and rubber
- *Good flexibility and impact resistance.* The amorphous nature of the polymer chain allows stretching and stress absorption
- *Wide substrate range.* Being pressure-sensitive adhesives, butyls and PIBs will stick to most other surfaces, including polyolefins
- *Relatively low strength.* Even when cured or cross-linked, butyls tend to creep under load
- *Excellent aging.* Their high degree of saturation resists oxidative attack
- *Heat resistance.* This variable characteristic depends on the content of PIB in the butyl rubber polymer. Generally, the higher the PIB content, the lower the heat resistance (and the higher the tack)
- *Variety of bonding mechanisms.* Butyl can be used as a contact adhesive, PSA, hot melt, or sealing tape
- *Low cost.* Butyl compounds can absorb high levels of fillers and softeners, which help reduce unit cost

- *Curability.* Butyl can be cured using sulfur, zinc oxide in halogenated functionalities, or isocyanates as the mechanism, or using a quinoid such as *p*-quinone dioxime

Markets and Applications. Butyl rubber adhesives are used predominately as solvent-borne dispersions, preformed tapes, and hot melts. A popular application is in the production of tapes and labels. This includes general-purpose tapes, pipe wrap tapes, surgical tape, electrical tapes, and others in which the polymer formulation is precoated onto a suitable film or reinforcement.

Hot-melt PSAs save the drying time necessitated by solvent dispersions, but require heating to melt temperatures and a subsequent cool-down cycle. Hot melts have also been used for a variety of sealing applications, for carton closing, on prefabricated metal buildings, in appliances, and very extensively as an insulated glass window sealant. Preformed sealing tapes are used widely in construction applications and are available in three grades: nonresilient, semiresilient, and resilient. Some preformed tapes are used for glazing applications in automotive glass, mobile home manufacture, rubber roof installation, and marine uses.

Self-curing butyl adhesives are required for laminating polyethylene film and for flocking adhesives. Most of the ethylene-propylene-diene monomer rubber roofs being installed on commercial and industrial buildings use a curing, solvent-dispersed, contact-grade butyl adhesive. Other curing butyl adhesives are used for hose, belting, and other fabricated rubber products. Some curing formulations are supplied in two parts that are mixed immediately prior to application.

Latex compounds can be used for PSAs that are precoated onto a film or fabric, for foil and paper laminations, and for PSA label production. They are also effective as a packaging adhesive and for bonding polyolefin films and fibers, when properly compounded.

Additives and Modifiers. Butyl rubber is similar to natural rubber in its ability to compound with other materials. The 100% solids hot melts and preformed tapes also exhibit this characteristic of being compatible with a wide array of fillers, resins, and other materials. Antioxidants are rarely needed because of the highly saturated nature of the polymer chain.

Bond strength with noncuring butyls is usually low, measuring less than that of natural rubber in PSA compounds, unless compounded with high-strength, high-polarity resins such as the phenolics. Generally, butyls are chosen for the capability to stick to almost anything, a high resistance to moisture, and the capability to age well.

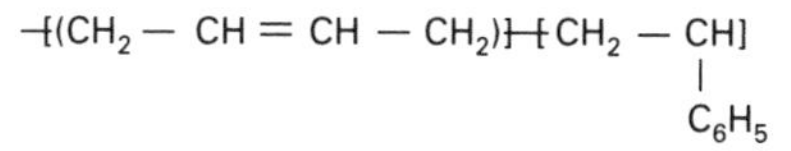

$$\text{--}(CH_2 - CH = CH - CH_2)\text{--}(CH_2 - CH]$$
$$| $$
$$C_6H_5$$

Fig. 4 Styrene-butadiene rubber chemical structure

The temperature range of butyl compounds is similar, in most respects, to that of natural rubber, whereas polyisobutylenes usually do not have as high an upper range for temperature resistance.

Styrene-Butadiene Rubber

Originally called the synthetic natural rubber, styrene-butadiene rubber (SBR) was developed as Buna-S in Germany in the late 1930s. Produced by the emulsion copolymerization of butadiene and styrene, SBR was the foundation for the growth of the synthetic rubber industry in the United States during World War II. When the Japanese cut off natural rubber supplies from Malaysian plantations, the United States production of SBR grew from minimal levels in 1941 to over 710 Mg (1.5×10^6 lb) in 1945.

More than 50 plants were built under the auspices of the government synthetic rubber program to produce SBR and other synthetic rubber compounds in order to meet the existing demand for the war effort and maintain supplies for other products.

After the war, when natural rubber again became available, the use of SBR slowed for many applications and the wartime plants were eventually sold to the rubber manufacturing firms that had run them during the war.

Although the basic advantage of SBR was low cost, it suffered from low building tack, cohesive strength, tensile strength, elongation, hot tear strength, and resilience, compared to natural rubber. However, by the 1950s, new polymerization and resin compounding procedures began to overcome these deficiencies, and SBR found its way into numerous adhesive formulas.

Chemistry. There are two basic emulsion polymerization processes, hot and cold, but hot polymerization predominates because it yields a lower molecular weight polymer and a broader molecular weight distribution. The lower molecular weight portion provides better quick-stick properties, whereas the higher molecular weight fraction provides a degree of shear strength. The SBR chemical structure is shown in Fig. 4.

In 1965, the first commercial grade of a new product, called a block copolymer, became available. Produced by solution polymerization rather than emulsion, these SBR materials possessed a more ordered chemical structure based on styrene-isoprene-styrene (SIS) or styrene-butadiene-styrene (SBS) and exhibited thermoplastic properties. The two-phase block copolymer had a continuous rubber phase (toughness and resilience) and a discontinuous plastic phase (solubility and thermoplasticity). Later developments included styrene-ethylene/butylene-styrene (S-EB-S) and styrene-ethylene/propylene-styrene (S-EP-S), both highly saturated compounds with polyolefin properties.

Emulsion-polymerized SBR is available in more than 100 grades, but only a few are readily utilized as a base for adhesives. Most variations are a result of differences in molecular weight or styrene content or depend on whether the polymer is linear or cross-linked. Solution polymerized block copolymers offer different grades by changing the molecular weight and the ratio of endblock to midblock. Carboxylated grades of emulsion polymers are also available as latex which can vary widely in properties depending on the density of the carboxyl groups.

Forms. Users can obtain SBR adhesives in a number of forms. Emulsion polymers are most often supplied as solvent-dispersed or water-borne emulsions. Often these forms are precoated onto paper or film and processed as pressure-sensitive labels and tapes. Solution-grade polymers are usually found in a solvent dispersion or as a hot-melt compound, either of which can be precoated onto various substrates as a PSA. Water-borne compounds of block copolymers can be made by emulsifying a solvent solution and stripping away the solvent. However, properties usually suffer because of the presence of the wetting agents and emulsifiers. Block copolymers can also be supplied as a powder that possesses hot-melt properties.

Properties. Emulsion- and solution-grade polymers sometimes have overlap of their properties, but often are widely different. Properties include:

- *Ease of dispersion.* Both emulsion- and solution-grade polymers provide low-viscosity solutions in aliphatic and aromatic solvents. This allows a low processing cost and yields high-solids compounds
- *Excellent water and moisture resistance.* The polymer is not affected by most aqueous chemicals
- *Good heat resistance.* The polymer is usually superior to natural rubber. Block copolymers can also provide good low-temperature properties
- *Poor tack.* Compounding with low molecular weight polymers, adding plasticizers, or mixing with a variety of tackifying resins is required
- *Excellent film flexibility.* Solution-grade polymers are often superior
- *Poor resistance to organic solvents and oils.* The polymer swells readily and absorbs oils. Vulcanizing improves this property

- *Wide range of substrates.* PSA and some solvent compounds will stick to almost any surface, including polyolefins
- *Fair aging.* The polymer is superior to natural rubber, but the carbon-carbon double bond is susceptible to oxidation, which causes the adhesive to harden and become brittle over time
- *Variety of bonding mechanisms.* The polymer is most often used as PSA, but it can be compounded for contact bonding or heat or solvent activation
- *Curability.* The polymer can usually be cured using sulfur, but other curing mechanisms are available to provide stronger bonds and higher temperature resistance
- *Low cost.* Emulsion polymers are inexpensive compared to most synthetic elastomers

Markets and Applications. Although block copolymers can be formulated as contact adhesives, heat reactivatable products, and curing cements, most are used as solvent-dispersed or hot-melt PSAs. The solids contents of the solvent-borne PSAs are often in the range of 40 to 50% and are easily roll or knife coated onto substrates or release paper. Polyolefin films and foams are often coated with a block copolymer for easy peel-and-stick application. These PSA-coated products, most of which are the hot-melt type, are used for construction applications, automotive padding and insulation, orthopedic devices, packaging, and book binding.

Emulsion polymers have extensive applications in both solvent and latex form, with curing and noncuring requirements. Many of the applications are similar to those of natural rubber, reflecting the similarities of the polymers. Solvent-borne products are used to bond paper, cardboard, wood, foam, plastic, and fabric, usually in a spray-grade mixture. A cross-linked SBR provides an excellent aerosol spray without cobwebbing. Sulfur-curable SBR cements are also widely used in tires.

The construction industry uses several types of solvent-dispersed SBR, including ceramic and tile bonding compounds. However, the most widely used type is a high-viscosity product for bonding flooring to joists and panels to studs, thereby reducing the amount of nailing and alleviating floor squeaks. SBR has long been used to produce a variety of tapes, such as masking and surgical, and labels. Miscellaneous applications include auto headliners, the bonding of expanded polystyrene to a variety of surfaces, and the bonding of rubber and fabric to metal.

The predominant applications of latex adhesives based on emulsion-polymerized SBR are to coat tire cord and tufted carpets, often in combination with natural rubber. SBR emulsions also find use as a binder for nonwoven goods, such as cloth wipes, diapers, and garment interliners. The wet combining and laminating of fabric, paper, leather, foil, films, and rubber are also common, although one of the substrates being wet bonded must be porous to allow moisture dissipation.

Additives and Modifiers. All the additives, fillers, modifiers, and other ingredients listed and discussed for natural rubber apply to SBR compounds. However, SBR can often be added to other compounds to provide specific properties and lower net cost. Temperature ranges and bond strengths for SBR adhesives are similar to those of natural rubber, with certain exceptions: Block copolymers provide higher tensile strength and elongation and often yield higher-strength PSAs; vulcanized or cross-linked versions exhibit higher cohesive strength when bonded to most surfaces and better oil and solvent resistance.

Acrylonitrile-Butadiene Rubber

More commonly known as nitrile or nitrile-butadiene rubber (NBR) or Buna-N, this product first became commercially available in about 1936. It is made by an emulsion polymerization process that combines acrylonitrile and butadiene. Adhesives normally use a polymer that contains 25% or more acrylonitrile, but the proportion can vary from 15 to 50%. Generally, the higher the acrylonitrile content, the greater the resistance to oil and the higher the polarity of the compound. NBR also becomes harder or more resinous with increasing acrylonitrile content (and therefore less elastomeric).

During World War II, NBR was an important part of the synthetic rubber program to replace critical supplies of natural rubber. The unique properties of high oil and plasticizer resistance, excellent heat resistance, and superior adhesion to metallic surfaces, combined with compatibility with a wide array of compounding ingredients, have contributed to the growing use of NBR adhesives since the end of World War II. Plasticizers are materials such as phthalate and medium-to-high molecular weight polyester fluids, which are added to rubber and plastic compounds to increase flexibility. Often, these materials migrate to the bond line interface.

Chemistry. NBR is produced by both emulsion and solution polymerization. Much of the commercially available rubber is polymerized at a relatively low temperature (5 °C, or 40 °F) and is therefore called cold rubber. The rubber most suitable for use in adhesives is produced by a hot polymerization process (25 to 50 °C, or 77 to 120 °F), which induces higher branching in the chemical structure. The chemical structure is shown in Fig. 5.

Solution-polymerized NBRs generally have lower-level acrylonitrile content, but

$$-(CH_2-CH=CH-CH_2)-(CH_2-CH)]$$
$$|$$
$$CN$$

Fig. 5 Acrylonitrile-butadiene rubber chemical structure

can be combined with monomers that possess a carboxyl group for higher reactivity. Most of these materials are low molecular weight liquid polymers. The high degree of functionality of solution-polymerized NBRs allows these products to react with many other reactive polymers, such as acrylics, epoxies, and polyurethanes.

Hot-emulsion-polymerized NBR is available in a number of grades from several suppliers, with progressive variations in acrylonitrile content and molecular weight. Another grading system is based on the requirement for milling the rubber prior to solvating it in the churn. Generally, milling always provides a lower-viscosity adhesive, even with soluble grades of NBR, and can improve the shelf stability of the final adhesive. Carboxylated polymers and methyl methacrylate graft polymers are also available.

Forms. Several NBR adhesive forms are available to the end user, reflecting the versatility of this polymer. Probably the most used form is the solvent-dispersed cement, which can be compounded for a wide range of viscosities. Water-borne NBR adhesives are growing in popularity for laminating applications but cannot match the film strength of the solvent-dispersed compounds. Pressure-sensitive versions can be precoated onto substrates and supplied in roll form. Heat- and solvent-reactivating cast films are popular for some manufacturing applications because they can be die cut or hand cut to match a component configuration. When liquid NBR is mixed with product thermoplastic polymers, a high-performance hot-melt product can be supplied. Liquid polymers can also be mixed into epoxies to provide 100% solids reactive adhesives.

Properties. There is such a diversity of NBR adhesives that listing general properties is difficult. The properties of these adhesives include:

- *Superior resistance to oils.* NBR has the highest resistance of any of the general elastomers and resists most greases and nonpolar solvents as well
- *High-temperature resistance.* NBR can easily perform at 150 to 175 °C (300 to 350 °F) if cured, with excursions to higher temperatures being possible
- *Structural strength.* When combined with phenolic resins or epoxies, shear strengths of over 21 MPa (3 ksi) are achievable on aluminum and other substrates
- *Low initial tack.* Unless tackifying resins are added, NBR requires high contact pressure to make bonds

- *Excellent aging properties.* Although this property is often due to the antioxidants added by the rubber manufacturer, it is aided by the polymer structure
- *Resistance to moisture and most chemicals.* The solvent, hot melt, and film grades (cast from solvent) are superior to the latex grade for this characteristic
- *Good flexibility and tensile strength.* These characteristics improve if the product is cross-linked
- *Curability.* NBR can often be cured using sulfur or sulfur-containing compounds, but isocyanates will also cross-link the polymer chain. Improved heat resistance and strength can be achieved by heating the bonded part, even without curatives
- *Variety of bonding mechanisms.* NBR can be used as a contact adhesive or PSA, or as a heat- or solvent-reactivating, hot-melt, or chemically reactive system
- *Wide range of substrates.* NBR has very strong bonds to high-polarity surfaces, such as steel and aluminum, but bonds poorly to low-polarity surfaces, such as polyethylene and natural rubber
- *Medium-to-high cost.* NBR is generally more expensive than comparable solids versions of other rubber polymers

Markets and Applications. Organic solutions of NBR are used throughout the industrial sector as general-purpose adhesives, especially where oil, plasticizers, and other organic fluids are encountered in service. Quite often, NBRs are used for bonding various kinds of gasketing (cork, fiber, foam, rubber, metal) to rigid superstructures, such as aircraft. NBRs are excellent for bonding plasticized PVC films to themselves and to metal, wood, glass, and masonry. However, staining can occur on light-colored vinyls if phenolic resin is part of the adhesive formulation.

Nitrile-phenolic adhesives have been used for automobile brake and clutch linings and for bonding abrasives to metal. Curing grades of NBR are used for nitrile hose and belt manufacture and for other kinds of rubber and plastic part assembly. Shoe sole attachment to leather and vinyl uppers is accomplished with NBR cements. A growing application over the past three decades has been the use of nitrile-phenolic compounds for the manufacture of printed wiring boards.

Water-borne versions of NBRs are used extensively for laminating, including the bonding of vinyl film and fabric-backed vinyl to fabrics and foams. Bonding NBR to fabrics is accomplished with latex compounds. NBR adhesives are also used to bond aluminum foil laminate to itself, composition boards, and wood, and to bond leather, canvas, and PVC.

Films cast from solution are often used to fabricate honeycomb structures for aircraft. However, these heat- and solvent-reactivating films can be used to bond virtually all of the substrates mentioned in previous paragraphs.

Additives and Modifiers. The fillers, resins, curatives, and other modifying ingredients previously mentioned in this article are all compatible with NBR. Being a highly polar polymer, NBR is often superior to other elastomers in terms of resin compatibility. The temperature resistance ranges of cured materials can range from more than 150 °C (300 °F) to as cold as −40 °C (−40 °F). Bond strength, as noted, can run well above 7 MPa (1 ksi) for many applications and can provide a structural bond to many substrates. Although tensile strength, temperature resistance, elongation, and resilience are improved with cross-linking, chemical resistance to organic compounds, acids, and alkalies remains excellent, even when the NBRs are not cured.

Polychloroprene Rubber

By far the most popular and versatile of the elastomers used for adhesives, polychloroprene rubber (CR) was developed following the pioneering work by Father Nieuwland of the University of Notre Dame. His experimental work with a copper chloride catalyst and the acetylene-enabled synthesis of divinylacetylene was done in the early 1920s. After a published paper reported his results, Father Nieuwland was joined in his research by DuPont scientists. Controlled reaction conditions and treatment with hydrochloric acid yielded the chloroprene monomer which, when polymerized, produced a rubberlike polymer. In 1932, this product was commercialized.

Although CR possessed many of the characteristics of natural rubber and had the advantage of higher polarity, it was not widely used for adhesives until World War II. Because it was a manufactured polymer, it was significantly more expensive than an agricultural product such as natural rubber. In addition, expensive aromatic solvents such as toluene were required for solvation, whereas natural rubber used low-cost naphtha solvents. Another problem was the fact that higher solids concentrations were needed with CR than with natural rubber to achieve a certain viscosity in a cement. This factor added to overall compound cost. Nonetheless, polychloroprene was chosen to replace natural rubber, which was in short supply during war time, in many adhesive formulas.

Chemistry. The original acetylene production process has been largely supplanted by the direct chlorination of butadiene. The chloroprene monomer, 2-chloro-1,3-butadiene, can exist in four different isomeric forms when polymerized. The predominant form is trans-1,4-polychloroprene, with the chemical structure shown in Fig. 6.

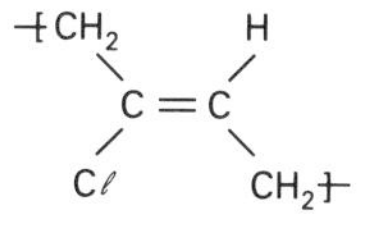

Polychloroprene rubbers come in more than two dozen grades of solid rubber for

Fig. 6 Trans-1,4-polychloroprene chemical structure

use in solvent formulations and more than a dozen emulsion grades. Molecular weight differences, the amount of polymer chain branching, and the rate of crystallization are factors that differentiate the grades. Crystallization is an ordering phenomena of polymeric chains that is reversible with exposure to heat or removal of the stresses that induced it. At temperatures over 50 °C (125 °F), noncuring CR adhesives begin losing cohesive strength, but regain strength upon cooling. The quick-grab and fast strength development of most of the CRs are the result of the high crystallization tendencies of this elastomer. The higher the crystallization rate, the faster the strength development. Liquid polymers are available, as well as carboxyl-terminated polymers, which allow rapid cohesive strength development.

Forms. Like NBR in particular and other rubber polymers in general, polychloroprene is available in a variety of forms. Approximately two-thirds of the CR adhesives presently marketed are solvent dispersions. Water-borne contact adhesives are growing in importance and are replacing many solvent-borne contact-grade compounds. Polychloroprene can be supplied as a pressure-sensitive or a hot-melt adhesive, but these forms are often replaced by lower-cost elastomers.

Properties. The general characteristics of CR adhesives are:

- *High green strength.* CR has rapid strength buildup, which speeds the assembly process
- *High ultimate strength.* Phenolic-modified CRs can provide shear strengths of over 7 MPa (1 ksi) on some surfaces
- *Resistance to moisture, chemicals, and oils.* The property retention of CR is good after continued exposure to a variety of conditions; oil resistance is not as good as in NBR, but better than in natural rubber
- *Wide range of substrates.* CR bonds to almost any high-polarity surface, as well as many low-polarity surfaces, including polyolefins
- *Excellent temperature resistance.* CR can easily range from −45 to 175 °C (−50 to 350 °F) with proper compounding
- *Good resilience.* This property is comparable to that of natural rubber
- *Excellent aging properties.* The chlorine atom in CR helps provide resistance to chain scission by oxidation
- *Medium-to-high cost.* CR is usually more expensive than SBR or natural rubber adhesives

- *Curability.* Several mechanisms can be used to cure CR, the most popular being a room-temperature cure with zinc oxide. Sulfur can be used, as can isocyanates and amines. Because of reactivity, many curing CR adhesives are supplied as two components to be mixed immediately prior to use

Markets and Applications. Contact cement solvent-borne CR adhesives are perhaps the most popular group of assembly adhesives used by industry. Properly compounded, they allow almost any two substrates to stick together. While solids contents can range from 5 to 90% in CR adhesives (or to 100% in solvent- and heat-reactivatable films and tapes), the most common range is from 15 to 30% and is used for spray applications. A high level of green strength allows early handling in laminating operations. The adhesives are used extensively in countertops and in panel fabrication for curtain walls, partitions, and recreational vehicle sidewalls.

Another major area for CR contact adhesives is in the manufacture of leather goods, particularly leather shoe sole bonding, as well as belt lamination and the production of other leather accessories. In the furniture industry, solvent-borne CR adhesives are widely used for a variety of assembly operations, such as bonding foam to foam, fabric to foam, foam to wood, and for bonding metal, fiberglass, and a variety of plastics. Many decorative laminates embody a rigid base of particleboard, plywood, and other composite woods and metals.

The transportation industry uses these contact adhesives for sponge insulation strips, vinyl landau top and padding, and a variety of trim bonding applications. However, when vinyl is a substrate, it is necessary to select a plasticizer that limits the effects of migration. A high-solids compound is often used to bond in place truck and trailer roofs, flooring, and scuff-resistant sidewalls. A similar application is the bonding of face panels to cores to create metal or wood doors. Another application is in the production of rubber goods, such as hose and belting, and the installation of synthetic elastomer roofing membranes. CR adhesives are also used to bond many types of plastics, including polyethylene, polycarbonate, and others.

Polychloroprene latex use is growing as a result of increasingly stringent air quality standards, although the film strength characteristics do not match those of solvent-borne products. Water-borne formulations are used to bond facing materials to fiberglass batts and insulation boards and for foil lamination, carpet installation, PVC floor tile mastics, and a variety of general-purpose contact bond applications. These contact applications include the bonding of foams, fabric, rubber, wood, and high-pressure laminates.

Additives and Modifiers. Like other elastomers, CR is compatible with and compounded with a wide array of fillers, softeners, curatives, and other ingredients. The most common ingredients used are the phenolics (to provide strength and heat resistance) and metallic oxides. Metallic oxides serve two general functions in CR adhesives. They act as scavengers for any hydrochloric acid or "free" chloride that may be generated during processing or aging. They also act as a curative to cause the cross-linking reaction to occur at room temperature. Room-temperature cures are commonly associated with this elastomer. The time factor for curing is variable, being dependent on both time and temperature.

Bond strength can vary from a temporary bond, such as that produced by a thermoplastic noncuring compound, to a substrate-tearing bond, when phenolic-modified curing products are used. Polychloroprene adhesives can be formulated to have very short open times for fast production operations or to retain contact bond characteristics for up to 24 h. Heat and solvent reactivation can be used to reinstate the contact bond properties of compounds that have "gone dead," that is, coated surfaces that no longer stick readily to themselves. Compounding to achieve special properties, such as color, electrical or thermal conductivity, or nonflammability, is easily accomplished with this highly versatile polymer.

Common Additives and Modifiers

Most of the compounding ingredients that can be used in elastomeric compounds have been mentioned in earlier sections. With solvent-borne adhesives, a variety of solvents can be chosen to accomplish differing objectives, such as controlling the drying rate, stabilizing the mixture, controlling viscosity, and dissolving important ingredients. Resins can be added to improve tack, surface wetting, heat resistance, bond strength, and oxidation resistance. Resins, or a combination of resins, can be chosen from the following: rosin, rosin esters, terpene resins, coumarone-indene, petroleum hydrocarbons, and both thermoplastic and thermosetting phenolics. Plasticizers and softeners reduce hardness, enhance tack, and reduce cost. A plasticizer must be compatible with other adhesive ingredients and with the surfaces to which it will be applied. Paraffinic oils, phthalate esters, and polybutenes are typical plasticizers.

Fillers are not often used in rubber adhesive compounds where strength is important because most inorganic ingredients adversely affect adhesive values. Fillers such as carbon black and titanium dioxide provide coloring to mixtures. Most fillers, however, are added for the purpose of lowering cost and increasing solution viscosity and body. Clay, calcium carbonate, various silicates, and carbon black can all serve those functions. In some cases, small amounts of fillers improve properties such as cohesive strength, enhance the characteristic of compatibility with certain substrates, and lower the coefficient of thermal expansion. Excess filler in a low-viscosity solution allows sediment to form because of specific gravity differences. This sediment can sometimes be difficult to remix into the solution.

Curatives for each rubber type have already been discussed. Other compounding ingredients can be extremely varied. Water-borne compounds usually require the addition of a protective colloid, a preservative, defoamers, wetting agents, and emulsifiers. These ingredients often contribute to the lower water resistance of films cast from water-borne adhesives in contrast to those cast from solvent-borne or hot-melt adhesives. Common to many kinds of rubber adhesives is the addition of a thixotrope for viscosity control, a variety of flame retardancy materials, products that contribute to electrical and thermal conductivity, antioxidants, ultraviolet-light protectants, and pigments for color matching.

REFERENCES

1. "The Importance of Green Strength in Adhesive Selection," Technical Sheet 472A, Synthetic Surfaces Inc.
2. D.G. Coe, "Neoprene Solvent Based Adhesives," Technical Bulletin ADH-100.1 (R1), E.I. Du Pont de Nemours & Company, Inc.

SELECTED REFERENCES

- C.V. Cagle, *Adhesive Bonding*, McGraw-Hill, 1968
- I. Skeist, Ed., *Handbook of Adhesives*, 2nd ed., Van Nostrand Reinhold, 1977
- A.H. Landrock, *Adhesives Technology Handbook*, Noyes Data Corporation, 1985

Polyimides

Robert D. Rossi, National Starch and Chemical Corporation

SINCE THE 1960s, lightweight, improved-performance polymeric materials have been replacing metal in environmentally harsh applications that require prolonged high-temperature stability and resistance to moist, humid environments; extreme abrasion; and radiation. The increased use of polymeric materials for harsh environments has been accompanied by an increasing need for adhesive systems that meet similar performance demands. Adhesives are needed to replace conventionally used metal fasteners or to effect speedy repair of in-service polymeric parts under less than ideal conditions.

Aromatic polyimides were introduced commercially in the early 1960s (Ref 1-4). Since then, they have been among the most heavily studied polymeric families for applications subjected to conditions of harsh environmental exposure. With more than 10 000 references in the literature since 1967, this family of thermally stable polymers continues to attract considerable interest for a variety of applications. This type of material has found applications as a high-temperature adhesive in a number of aerospace structural parts and electronics/microelectronics end-uses (Ref 5) because it has a continuous service temperature in the 250 to 350 °C (480 to 660 °F) range and excellent electrical and mechanical properties. Typical property values, primarily based on cured films, are given in Table 1, whereas values for specific polyimide adhesive products are given in the section "Properties" in this article.

A number of articles in the literature have reviewed all aspects of aromatic polyimides chemistry and processing (Ref 6-11). These sources are comprehensive and should be consulted when more detail is required. The scope of the present review, on the other hand, is restricted to polyimide materials and processes that are used to generate products of commercial significance.

Significant progress has been made throughout the 1980s to improve polyimide processibility and handling characteristics. This has resulted in the availability of a large number of commercial polyimide products in a variety of forms.

It should be noted that the only commercially significant class of polyimide structures is that based on aromatic monomeric subunits. Therefore, these are the only structures discussed in this article, and all references to polyimides assume this aromatic structure.

Chemistry

Speaking strictly from a chemical point of view, aromatic polyimides essentially fall into two categories: thermoplastics and thermosets. Thermoplastic polyimides are materials that do not undergo any cross-linking reaction and are able to be thermally reprocessed, albeit at elevated temperatures (Ref 12). Thermosetting polyimides actually undergo a chemical cross-linking reaction and hence can no longer be reprocessed through thermal means (Ref 13). Although the driving force behind the evolution in polyimide chemistry and technology has been a search for methods to make the material more processible, the resulting modifications and improvements relate only to processibility, not to any changes in the basic thermoplastic and thermoset chemical classifications.

Table 1 Typical polyimide properties

Property	Value
Electrical(a)	
Dielectric constant	2.6–4
Tan delta	<0.01
Breakdown voltage, kV/mm (MV/in.)	100 (2.5)
Volume resistivity, $\Omega \cdot cm$	10^{16}
Thermal conductivity, mW/m · K (mcal/cm · s · °C)	17 (0.040)
Mechanical(b)	
Tensile strength, MPa (ksi)	>140 (20)
Modulus of elasticity, GPa (10^6 psi)	3.4–8.3 (0.500–1.2)
Elongation, %	2–100
Compressive stress, MPa (ksi)	6.9 (1)

(a) Electrical property values are of cured film. (b) Mechanical properties are of a molded part, but film property values are about the same.

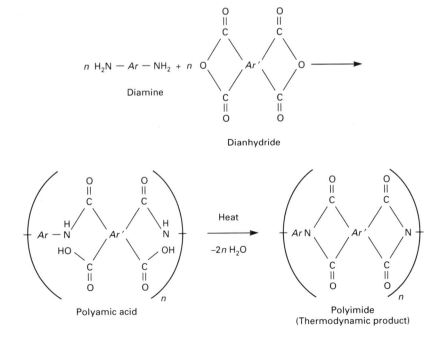

Fig. 1 Polyimide synthesis

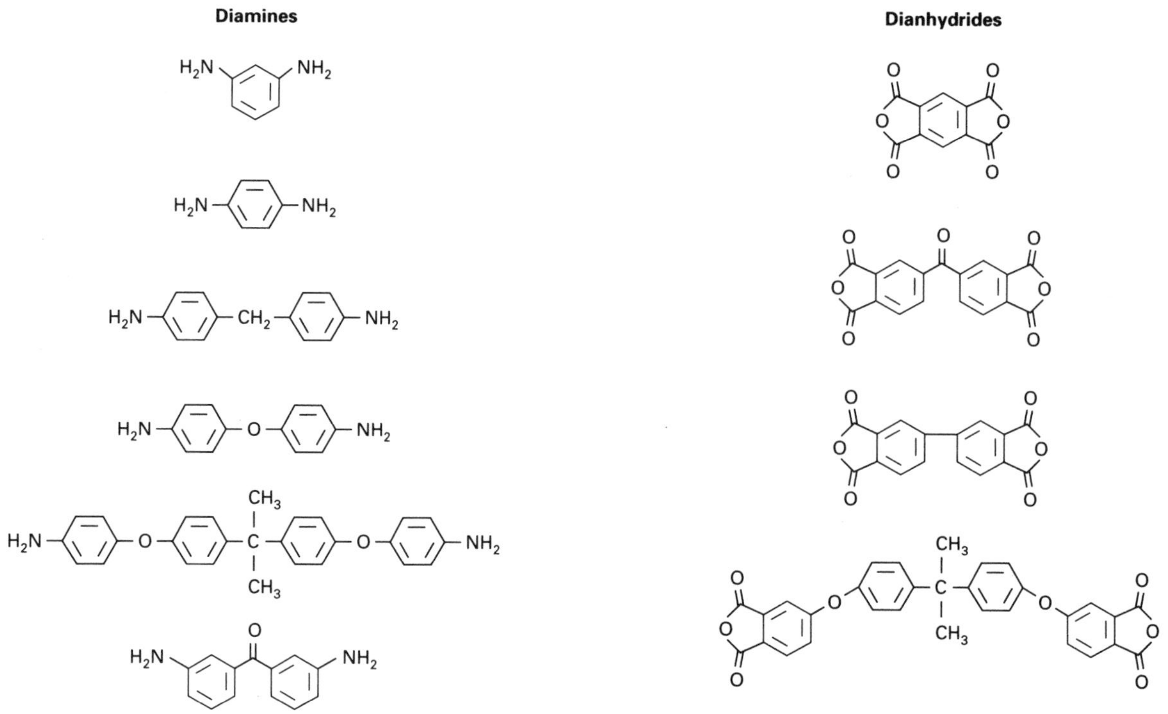

Diamines

Dianhydrides

Fig. 2 Aromatic monomers in commercial polyimides

Thermoplastic Polyimides

All thermoplastic polyimides (TPIs) are known as step growth (condensation) polymers because when the basic polyimide backbone unit is being prepared, monomers come together in a condensation reaction to form the polymer chain. This reaction is typically characterized by the loss of a small molecule by-product (either water or alcohol) during polymerization.

The most common method of polyimide synthesis is the reaction of an aromatic diamine with an aromatic dianhydride to form a polyamic acid initially and the subsequent

thermal ring closure to the polyimide structure (Fig. 1). Figure 2 depicts monomers used in commercially significant polyimides.

The initial reaction between the diamine and the dianhydride is extremely facile and, in most cases, is performed near or at room temperature. As with most chemical reactions, polyamic acid formation is a reversible reaction; in some cases, this reversibility is used to allow molecular weight equilibration to take place (Ref 13).

A modification of the scheme involves the formation of a polyamic ester during this initial reaction (Fig. 3). This process is

accomplished by using the half ester-acid chloride in place of the aromatic dianhydride (Ref 14).

In both cases, the cyclization to the polyimide structure is typically accomplished by a thermally induced intramolecular condensation. This reaction can be purely thermal, performed in a solvent system designed to remove the condensation by-product, or it can be catalyzed by an acidic species. The fully imidized form that is generated is referred to in the literature as preimidized.

By far, the most popular commercial form of thermoplastic polyimide is based on polyamic acid or polyamic ester because of their good solubility. Although somewhat dependent on the choice of monomers selected for the backbone, high molecular weight polyimides are usually found to be grossly insoluble in most common organic solvents. N-methyl pyrrolidone (NMP), a powerful, highly polar aprotic solvent, is successfully used to solvate polyimides, and, in many cases, is only effective at very dilute concentrations (typically <10%).

Monomers that have been found to improve the solubility of the preimidized form are those that break up the backbone structural regularity, causing the incorporation of so-called kinks in the polymer chain (Ref 15). Backbone regularity contributes to ordered chain stacking, resulting in insolubility. Disruption of this order allows solvents to penetrate and solvate the polymer chains.

Kinks are best achieved by substituting monomers that contain alkyl substituent groups for either the diamine or dianhy-

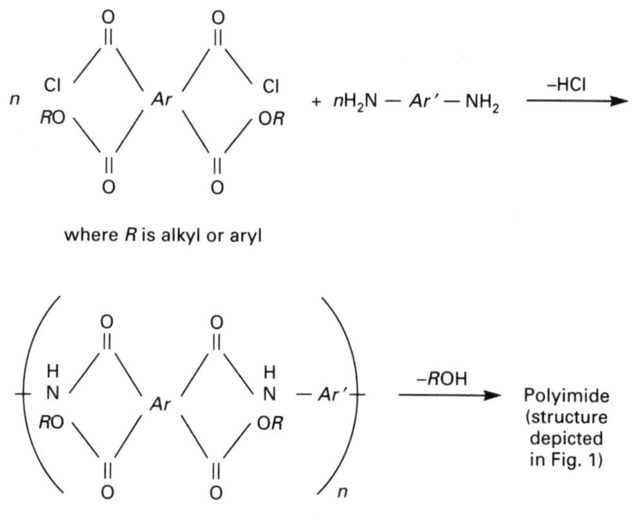

where R is alkyl or aryl

Fig. 3 Polyimide synthesis by means of polyamic ester formation

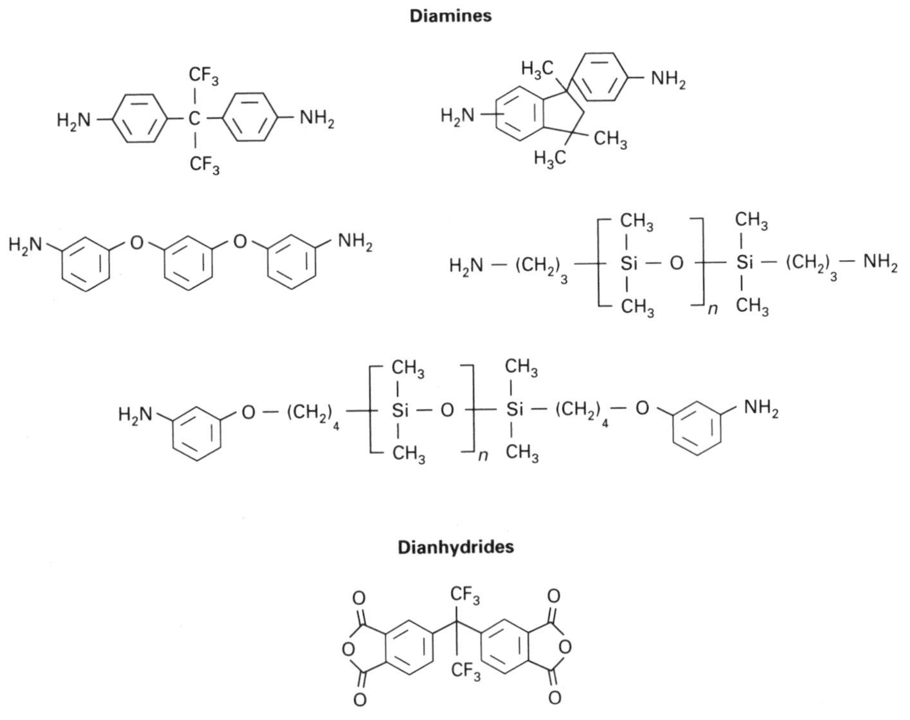

Diamines

Dianhydrides

Fig. 4 Monomers for improved solubility

dride. (Examples are shown in Fig. 4). Of this group, two structures have been used very successfully: monomers containing fluorine atoms and monomers containing the siloxane group. When incorporated in the polyimide backbone, these monomers dramatically improve the lipophilicity of the polymer, conferring solubility in even such nonpolar solvents as xylene and toluene. However, this improved solubility does have a disadvantage. Typically, these materials have glass transition temperatures (T_gs) that are lower than those of materials whose structures are based on the more regular monomer backbones (Ref 15). Lowering the T_g reduces the upper continuous-service temperature.

Thermosetting Polyimides

Unlike thermoplastic resins, thermosetting materials experience a chemical transformation during cure that renders the material thermally nonprocessible. This cross-linking reaction is typically accomplished using an organic functional group that undergoes a chain growth (addition) polymerization process (Ref 16).

Thermosetting polyimides, like thermoplastics, are prepared by combining an aromatic diamine with an aromatic dianhydride. However, unlike thermoplastics, thermosetting resins are purposely made as short molecular weight chains that are terminated at both ends with organic moieties that polymerize by means of an independent mechanism. This technology is referred to as the reactive oligomer approach (Ref 16). The chemistry is represented in Fig. 5.

In this approach, the oligomer molecular weight is controlled by the ratio of diamine to dianhydride. Figure 5 illustrates this technology using a molar excess of dianhydride. The capping group, which in the example is a monofunctional amine containing the reactive group X, does not participate in the polyimide condensation reaction, but does undergo cross-linking by an independent mechanism. Figure 5 shows just one way in which the reactive oligomer can be put together. It could also be prepared using the reverse ratio of diamine to dianhydride, in which case, end capping would be accomplished using a monofunctional anhydride containing the reactive group X.

The reactive oligomer approach to cross-linked thermosetting polyimides offers certain advantages over thermoplastic resins:

- Lower-viscosity solutions
- Better solvent resistance
- Higher glass transition temperature
- Less creep
- Low-temperature consolidation and densification prior to cure

The one disadvantage of these oligomers is reduced toughness. However, this disadvantage can be tempered somewhat by controlling the degree of polymerization (n) of the oligomer. Typically, the use of higher values of n results in improved toughness.

Three main categories of reactive end groups have commercial significance: acetylene-terminated oligomers, bisnadimides (BNIs), and the closely related bismaleimides (BMIs). The monofunctional reagents typically used to generate the reactive end groups are shown in Table 2.

An important consideration in reactive oligomer technology is the relationship between the T_g of the oligomer and the onset of the thermal polymerization of the end group. It is necessary, in order to generate a void-free, well-consolidated adhesive bond line or part, that the oligomer reach its T_g and flow before the end group polymerizes. Oligomer T_g is affected by its chain length and, more important, by the monomers employed. If cross-linking occurs prior to reaching the T_g, the resulting structural integrity of the polymer will be poor. Thus, the relationship between the T_g of

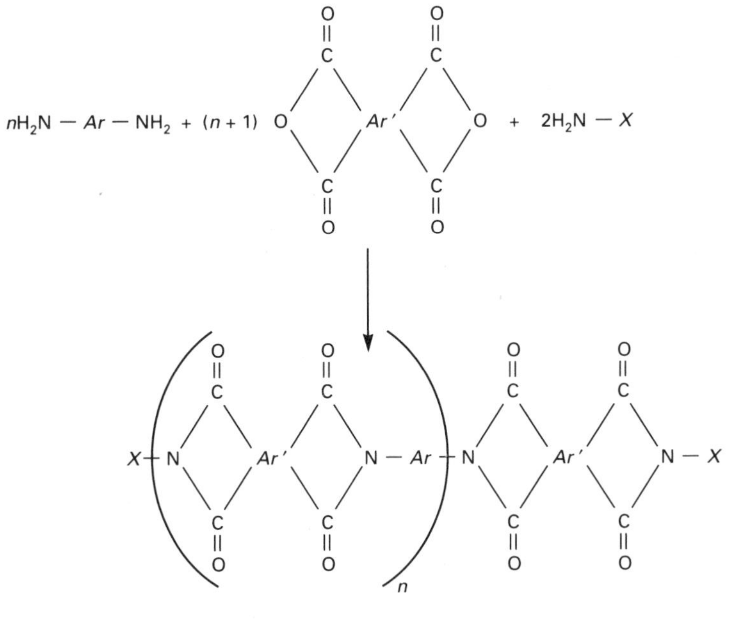

Fig. 5 Reactive oligomer approach

Table 2 Reactive oligomer end groups of thermoset polyimides

| Reagent | Type | Cure temperature | |
		°C	°F
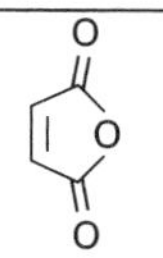	BMI.....................................	150–200	300–390
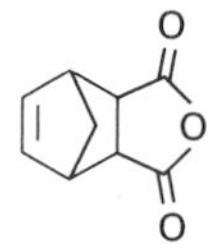	Bisnadimides.............................	275–300	525–570
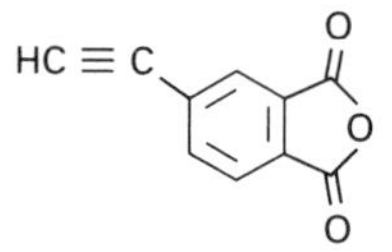	Acetylene terminated	250–275	480–525
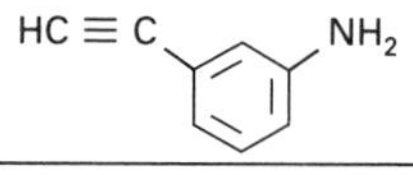	Acetylene terminated	250–275	480–525

the backbone oligomer and the onset polymerization temperature of the reactive end group places severe limitations on the oligomeric structure that can be commercially manufactured and used.

Bisnadimides and Bismaleimides. BNIs and BMIs were among the first reactive oligomers produced (Ref 17, 18). Once again, the driving force behind their development was improved processibility. Typically synthesized using the same type of step reaction described for thermoplastic polyimides, these materials are capped with nadic anhydride, in the case of BNIs, and with maleic anhydride, in the case of BMIs.

The initially formed amic acid can be dehydrated by either thermal or chemical means. The reaction is typically performed in polar aprotic solvents, for example, dimethyl formamide (DMF), dimethylacetamide (DMAC), or NMP (Ref 19). Chemical dehydration is performed using acetic anhydride/sodium acetate at moderate temperatures (<100 °C, or 212 °F) (Ref 20). However, as with most thermoplastic polyimides, the most processible form of BNI and BMI is the amic acid form.

The most popular BNI with commercial significance was developed by the Langley Research Center of the National Aeronau-

tics and Space Administration (NASA) and is therefore known as LARC-13 (Ref 21). Figure 6 shows the basic chemistry of the resin system. The nadic end group has been found to undergo a retro-Diels-Alder reaction at 250 to 270 °C (480 to 520 °F), releasing cyclopentadiene, which is rapidly consumed in a copolymerization reaction with the newly created maleimide-terminated oligomer (Fig. 7).

A discussion of BNIs would be incomplete without mention of the popular PMR-15 system, based on the *in-situ* polymerization of monomeric reactants with an M_n of 1500 g/mol. The bisnadimide PMR-15 is available as a monomer mix of methylene dianiline (MDA), benzophenone tetracarboxylic acid dimethyl ester (BTDE), and nadic ester (NE), as shown in Fig. 8 (Ref 22).

During thermal processing, the PMR-15 system, which is supplied in an alcohol solvent, undergoes *in-situ* polymerization of the polyimide. The monomers are present in stoichiometric proportions to give an *n* value (degree of oligomerization) of 2.087. Cross-linking occurs by means of the same mechanism described for Fig. 7. This mechanism accounts for the small number of volatiles encountered with *in-situ* polymerization.

The relatively high cure temperature for this nadic polymerization enables the use of a wide variety of monomer combinations. However, the only system that has been studied extensively to date is LARC-13.

The BMIs are closely related chemically to the BNIs. A number of BMI-terminated polyimide backbone systems have been produced and studied. However, because of the relatively low cure temperature of the maleimide group, practical limitations are quickly reached when considering BMIs with commercial impact. Therefore, the only BMI structure of significance is made from MDA (Fig. 9). Because this is a simple system (not oligomeric), the actual preparation can be performed in methylene chloride and other solvent systems that are less aggressive than those typically used in polyimide preparation. The thermally induced polymerization of BMIs has been studied

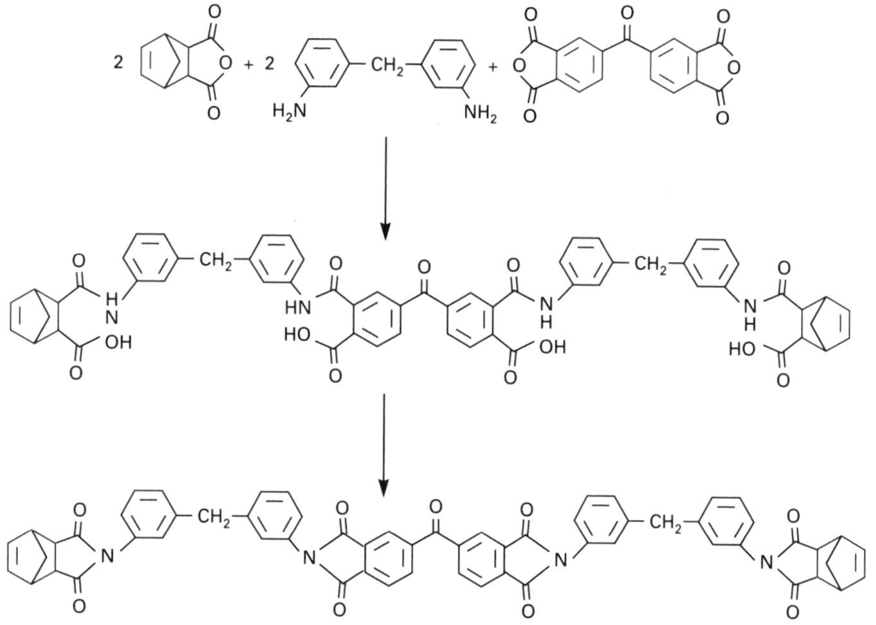

Fig. 6 LARC-13 chemistry

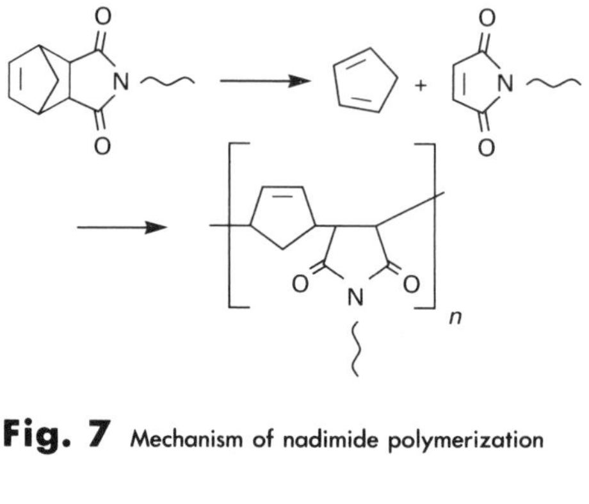

Fig. 7 Mechanism of nadimide polymerization

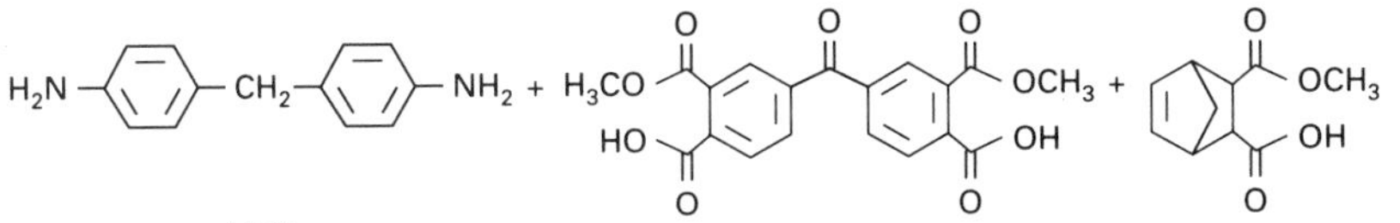

Fig. 8 PMR-15 monomer mix

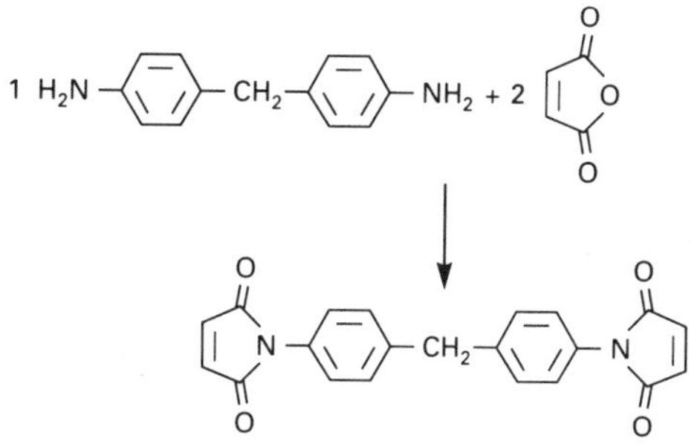

Fig. 9 Methylene dianiline bismaleimide synthesis

(Ref 23), and Fig. 10 shows the proposed mechanism.

Because the MDA-BMI system is not oligomeric, cross-link density is very high, and the resulting polymeric network is very brittle. This brittleness has been somewhat overcome by formulating these BMIs with other comonomers and with chain extenders, reactive diluents, and even elastomeric resins—the so-called rubber-toughened systems (Ref 24). An example of a toughened system using a comonomer is shown in Fig. 11, where the BMI is formulated with an allylic phenol. A mechanism for the copolymerization reaction is represented in Fig. 12. With this mechanism, three moles of maleimide resin can, in theory, coreact with each mole of allylic functionality present through two "ene" reactions and a Diels-Alder reaction. Alternatively, a mechanism that consumes two moles of maleimide for each allylic species has been proposed (Ref 25).

In both the BNI-capped and the BMI-capped oligomers, cross-linking leads to a structure that contains aliphatic linkages. Despite being highly cross-linked, these ma-

terials typically show lower thermal stability than materials in which cross-linking leads to aromatic structures. This is exemplified in the acetylene-terminated polyimide description below.

Acetylene-Terminated Polyimide Oligomers. The technology of using an acetylene group as the reactive moiety in a polyimide oligomer structure was introduced in 1974 (Ref 26). In particular, it was found that the aromatic ethynyl group undergoes a thermally induced polymerization, forming a thermally stable cross link.

The monofunctional reagent of commercial value containing the acetylene moiety is 3-aminophenylacetylene (APA):

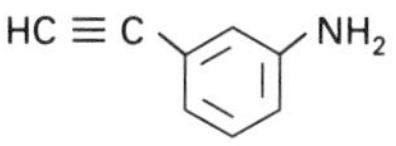

Acetylene-terminated oligomers are available commercially in the imidized and amic acid forms, and as a monomer mix. Figure 13 shows the basic chemistry for this resin system. Two important points should be made. First, the aromatic diamine monomer

of choice for this system is 1,3-bis(3-aminophenoxy)benzene (APB). This diamine helps maintain the T_g for the oligomer at a level well below the temperature of the acetylene polymerization. Again, the primary concern is consolidation before cure. APB is also critical to the maintenance of the superior solubility of this oligomer system.

The second point is that the ethynyl group in APA is meta with respect to the amino moiety. It was found that the use of the para isomer 4-aminophenylacetylene results in a side reaction that generates an acetophenone moiety during the thermal condensation reaction that is used to form the imide oligomer (Fig. 14) (Ref 27). This diversion of acetylene groups effectively results in a resin system that has a lower cross-link density than the acetylene in a meta position when fully cured and, hence, has inferior thermal and mechanical properties than expected.

Despite the relatively low molecular weight and lack of symmetry incorporated using the kinked diamine APB, these oligomers still showed very limited solubility when used in the fully imidized form. This led to the introduction of a unique form of imide chemistry, known as the isoimide (Ref 28). Isoimides (Fig. 15) are commercially available in various molecular weights, all of which are acetylene terminated.

The isoimide structure is prepared in a manner similar to that used for the normal imide, but is made under kinetically controlled conditions (Ref 29-31). This is accomplished by the chemical dehydration of the amic acid form, at low temperature, in a nonpolar solvent (Fig. 16). The isoimide successfully breaks up the regularity in the oligomer backbone, thereby allowing good solubility in solvents such as tetrahydrofuran (THF) and diglyme.

The mechanism of isoimide formation is shown in Fig. 17. The first step is the formation of the mixed anhydride from the amic acid, using trifluoroacetic anhydride as the dehydrating agent. Subsequent cyclization, by an addition-elimination reaction, followed by a rapid proton transfer,

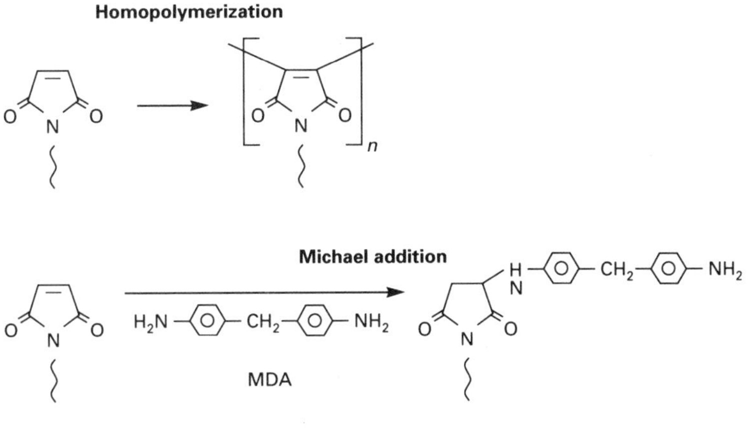

Homopolymerization

Michael addition

Fig. 10 Mechanism of maleimide polymerization

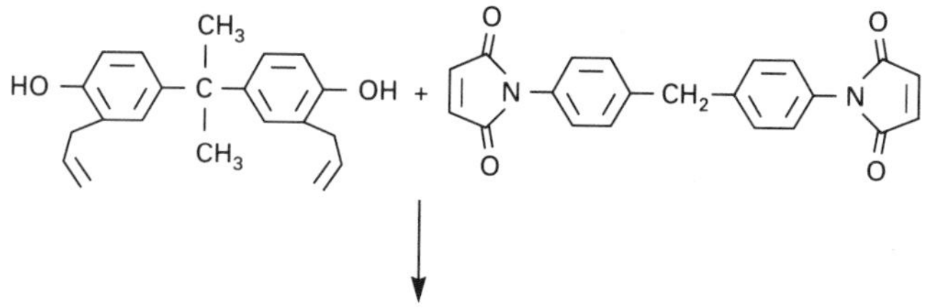

Fig. 11 Toughened BMI formulation

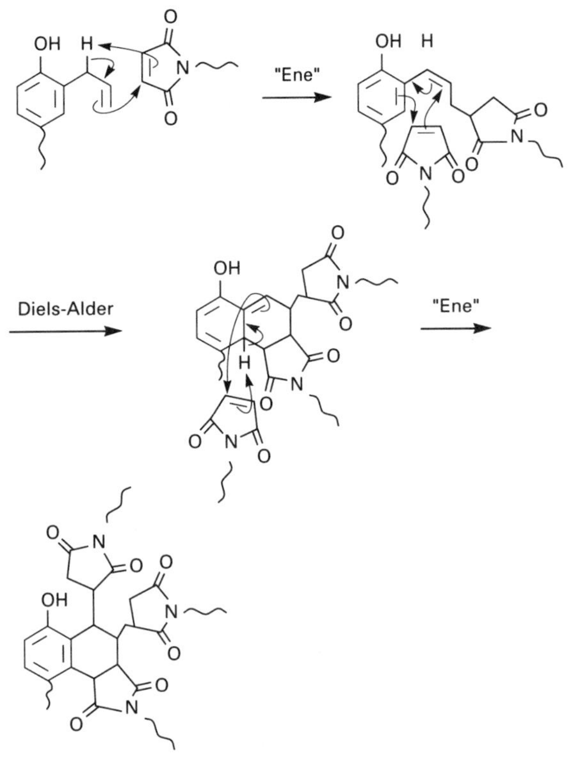

Fig. 12 Proposed mechanism of BMI/allylic phenol copolymerization

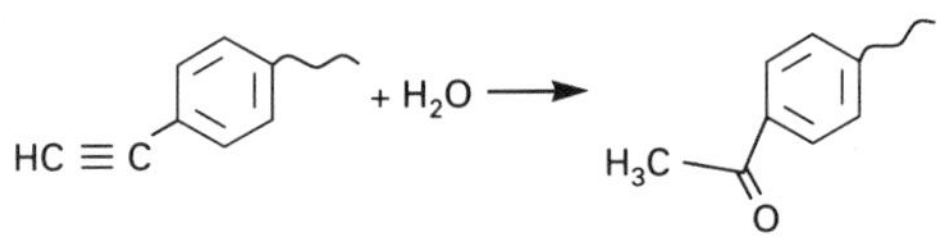

Fig. 14 Side reaction of 4-ethynylaniline

yields the isoimide structure, along with trifluoroacetic acid as the by-product.

This isoimide structure has been found to undergo rearrangement to the thermodynamically stable imide structure during thermal cure. This has been amply illustrated by two experiments. First, experiments were performed on a thin film sample of the $n = 1$ isoimide oligomer mounted in a controlled-temperature cell and continuously scanned by Fourier transform infrared spectroscopy (FTIR) as the cell was heated from room temperature to 350 °C (660 °F).

The data in Fig. 18 show that at higher temperatures, isoimide vibrations at the wave numbers at about 1840 and 1700 cm^{-1} are replaced by the more characteristic imide absorbance at 1730 cm^{-1}. Also worth noting in this IR experiment is the gradual disappearance of the H—C≡C vibration at about 3300 cm^{-1}, with increasing temperature. Complete disappearance is observed near 300 °C (570 °F).

Second, dynamic scanning calorimetry (DSC) experiments were conducted on an acetylene-terminated polyisoimide oligomer of $n = 1$, an acetylene-terminated normal polyimide oligomer of $n = 1$, and an aniline-terminated polyisoimide oligomer of $n = 1$. In this experiment, the large exothermic transition witnessed in the top thermogram in Fig. 19 is broken down into two components, the isoimide-to-imide rearrangement, shown in the bottom thermogram, and the cross-linking of the acetylene moiety, represented in the center thermogram.

Two independent studies have been performed to identify the chemistry of the

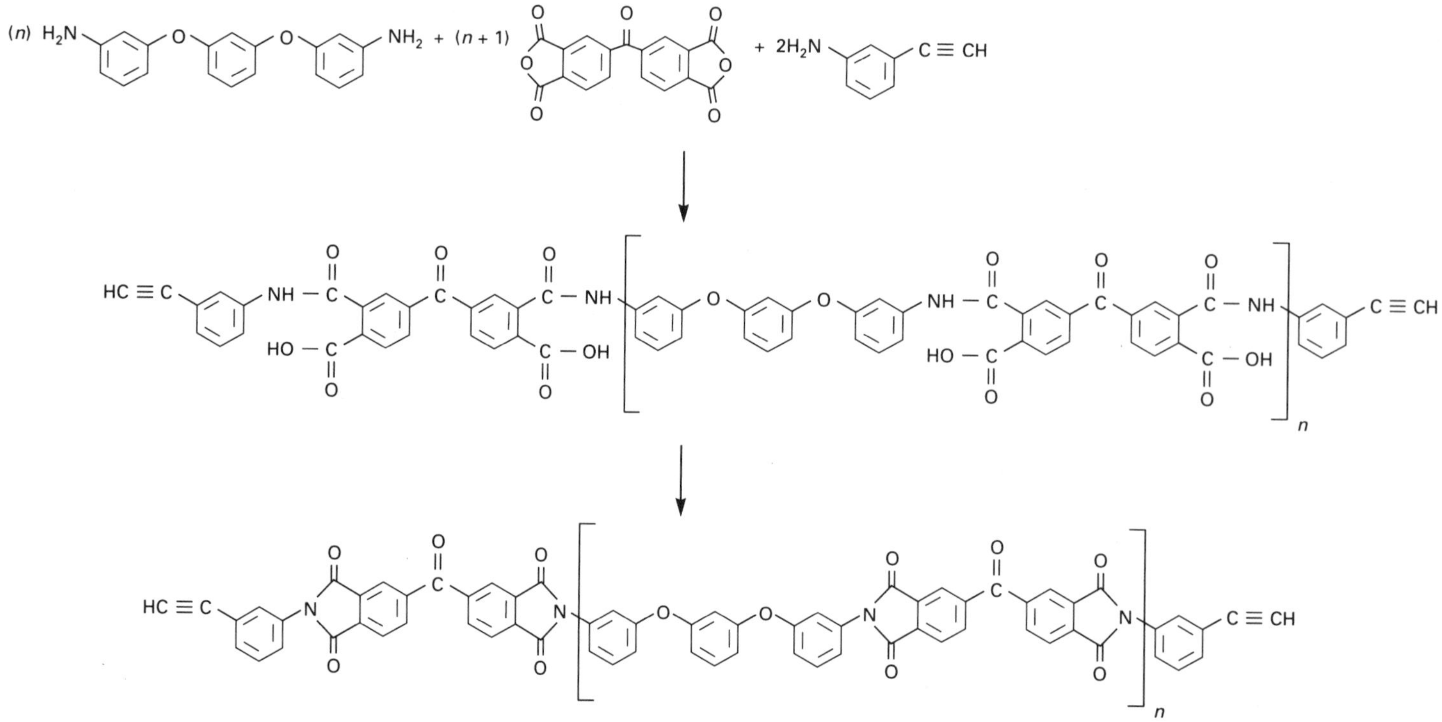

Fig. 13 Acetylene-terminated polyimide oligomer chemistry

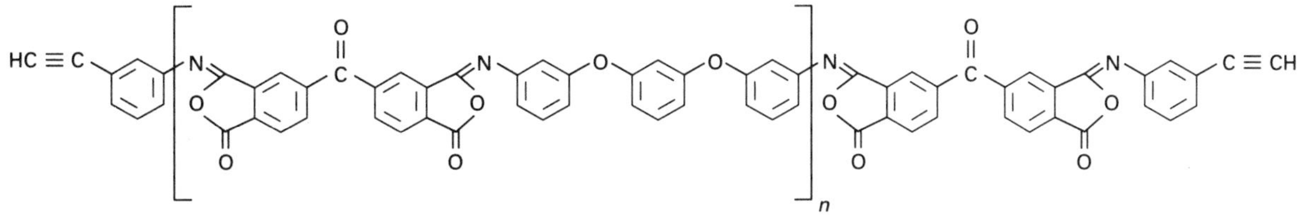

Fig. 15 Acetylene-terminated isoimide oligomer

acetylene cross-linking reaction. One that was reported by co-workers at Monsanto Company, after a series of natural abundance solid ^{13}C nuclear magnetic resonance (NMR) spectroscopy experiments, showed that about 30% of the structure formed during thermal cure consisted of benzenoid rings (Ref 32), which support a trimerization reaction mechanism (outlined in Fig. 20).

Another more recent and more elegant series of experiments was performed on an acetylene-capped imide oligomer prepared with independently labeled α and β ^{13}C on the acetylene of APA. Again, the cure was monitored in a solid ^{13}C NMR experiment (Ref 33). This study revealed that, in addition to benzenoid structures, naphthalenic units were detected, leading to the proposal

of the mechanism shown in Fig. 21. The initial reaction is a thermally catalyzed addition of one acetylene across another in a bimolecular Strauss (Ref 34) coupling reaction, followed by a thermal, probably through a free radical, Friedel-Crafts mechanism to form a naphthalene ring. Of course, this is only one mechanistic path (nonconcerted); others are possible. Reference 35 gives a general discussion of the Friedel-Crafts mechanism and related reactions. All evidence to date points to the generation of a substantial amount of aromatic structure during cure, which accounts for the excellent thermal properties of this resin system.

Miscellaneous Types. An additional category of cross-linkable polyimide precur-

sors is worth mentioning. These are the polyimides that cross-link upon exposure to ultraviolet (UV) radiation (Ref 36). This chemistry is shown in Fig. 22. These polyimides have gained commercial popularity mainly because of their use in microelectronic thin film dielectric applications (Ref 37). Although they do cross-link upon UV exposure, the subsequent thermal cure of the cross-linked polyamic esters generates a polyimide that is actually thermoplastic.

As Fig. 22 shows, the final cured polyimide does not contain an acrylate (or methacrylate) portion. This group is volatilized during thermal treatment. Consequently, the final cured polymer, in most instances, loses over half its original weight. Because this result has been problematic in some applications, a recently developed inherently photosensitive polyimide has been introduced (Ref 38). It is interesting that this chemistry (Fig. 23) requires no added photosensitizer to initiate the cross-linking reaction.

The mechanism of photocross-linking, although not yet completely understood, has been studied. Figure 24 represents current knowledge (Ref 39). Benzophenone and the ortho alkyl diamine are critical to the inherent photosensitivity, and it is believed that cross-linking actually occurs radically through these sites.

Debate exists concerning the final cured properties of photosensitive polyimides compared to those of their conventional counterparts (Ref 40). Some have reported inferior mechanical performance, speculating that this decrease is caused by the incomplete removal/volatilization of the acrylate (or methacrylate) portion during final cure.

It is sometimes necessary, depending on the substrate, to use an adhesion-promoting substance or an adhesion layer in order to improve adhesion to the polyimide (Ref 41-44). The adhesion of substrates such as silicon to polyimide is sometimes very difficult. In such cases, a thin (10^{-12} m, or 40 × 10^{-12} in.) coating of adhesion promoter, typically an aminosilane (3-aminopropyltrimethoxysilane), applied to the substrate surface improves adhesion dramatically. The action of this aminosilane is shown in Fig. 25. A general review of this activity is provided in Ref 45 to 47. Oxides on the silicon surface react to form a disiloxane linkage with the aminosilane, and it has

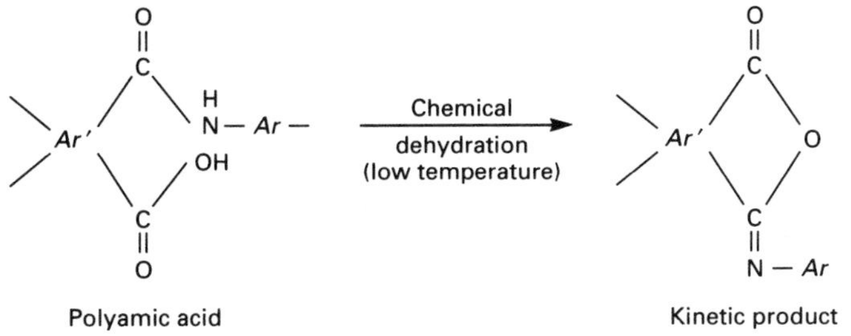

Fig. 16 Isoimide formation

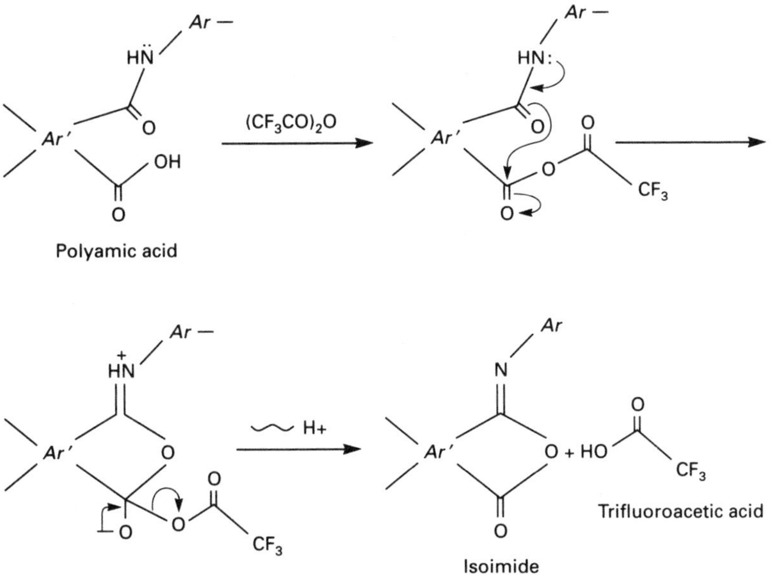

Fig. 17 Mechanism of isoimide formation

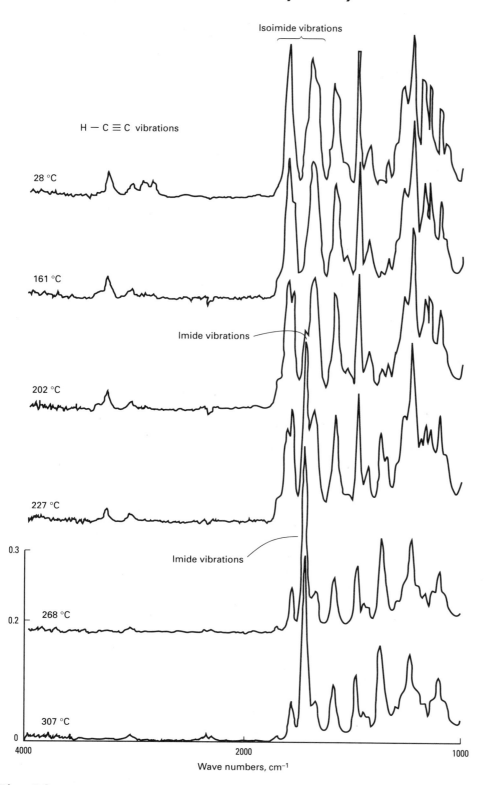

Fig. 18 FTIR study of Thermid IP-6001 in N_2

interaction of elemental chromium with polyimides occurs through a π-arene complex, rather than with the oxygen functionalities, which include the reactive carboxylic acid of a polyamic acid (Ref 48). It was also found that the tendency to form a π-arene complex increases with increasing electron density on the aromatic ring.

Forms and Suppliers

This section focuses on the most popular commercial forms of polyimide adhesives. Major suppliers include American Cyanamid Company, National Starch and Chemical Company, E.I. Du Pont de Nemours & Company, Inc., Rogers Corporation, Ferro Corporation, Monsanto Company, and Hercules Inc.

Lacquers, Varnishes, and Solutions. Polyamic acids, polyamic esters, and monomer mix prepolymers, supplied mainly in NMP or alcohol solvents, are typical constituents of polyimide lacquers and varnishes. Depending on the particular form employed, the percentage of solids in lacquers and varnishes can range from 13 to 70%. Typical application methods include dip, roll, spray, and brush coating.

Among the suppliers of lacquer, varnish, and solution forms of polyimide products are E.I. Du Pont de Nemours & Company, Inc., Rogers Corporation, Monsanto, and National Starch and Chemical Company. Du Pont Pyralin products, particularly in the polyamic acid form, use various combinations of aromatic diamines and dianhydrides in solvent systems that primarily comprise NMP.

Rogers Corporation Durimid thermoplastic polyimide offers a developmental product based on the NASA LARC TPI technology. Thermoplastic polyimide (TPI), the system based on 3,3'-diaminobenzophenone and BTDA, is available in solution form (Ref 49-51).

The Monsanto Skybond 700 series lacquer is a solution of the monomers BTDA diester/diacid and *m*-phenylene diamine. National Starch and Chemical Company offers two polyimide precursor products in lacquer form. One is the Thermid AL grade, which is a solution of APB, BTDA diester/diacid, and APA in ethanol. The other is a polyamic acid of these same monomers and is known as the Thermid LR series. Thermid belongs to the reactive oligomer class of polyimides and, as an oligomer, is supplied at various molecular weights. Thermid LR is offered at two weights: with an *n* of 1 and with an *n* of 30.

Another commercial reactive oligomer is PMR-15, developed by NASA Lewis Research Center and now marketed by Dexter Hysol. This, too, is a monomer mix composed of 5-norbornene-2,3-dicarboxylic

been proposed that this group then undergoes a transimidization with the polyimide or polyamic acid and chemically bonds the polyimide to the surface.

Chromium, nickel, or zinc is commonly used as an adhesion layer to help polyimide adhere to various metals such as copper (Ref 43). Poor adhesion of polyimides to

copper is due to the formation of an oxide between the copper and the polyimide. This oxide layer does not stick well to the copper. The formation of oxide is not unexpected because of the high processing temperature. The adhesion layer prevents oxide formation and greatly improves the adhesion. A recent study concludes that the

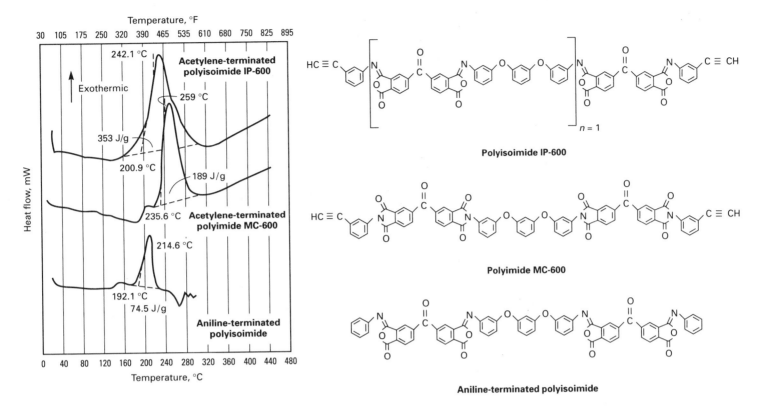

Fig. 19 DSC scan of different acetylene-terminated oligomers, taken at 10 °C/min (18 °F/min) in N₂

acid monoester with methylene dianiline and BTDA diester/diacid.

The solutions market has seen the participation of numerous competitors. E.I. Du Pont de Nemours & Company, Inc. and Hitachi produce primarily the polyamic acid form, including a low-stress product made from monomers containing biphenyl and/or terphenyl units that result in a rigid rod-type backbone. Hoechst Celanese Corporation supplies a preimidized material derived from fluorinated diamine and dianhydride monomers. General Electric Company has licensed its silicone polyimide technology to Hüls. National Starch and Chemical Company produces Thermid EL acetylene polyimide oligomer formulations. These products can be obtained in anywhere from 10 to more than 45% solids solutions, which are

typically applied by spin coating techniques to form very thin films of 25 μm (1 mil) or less. National Starch and Chemical Company also markets a poly(imide siloxane) based on technology originally developed by Atochem Inc. in North America (formerly M&T Chemicals, Inc.). These materials are mainly used as die attach adhesives or hybrid circuit encapsulants (Ref 52).

Die attach adhesive formulations deserve special mention. Polyimides have gained particular favor in this end-use because of the requirement for high ionic purity, which is not typically found with epoxy systems. The semiconductor integrated circuit (referred to as a die) is particularly sensitive to ionic contamination (notably chloride), and polyimides are commonly produced with less than 1 ppm chloride content.

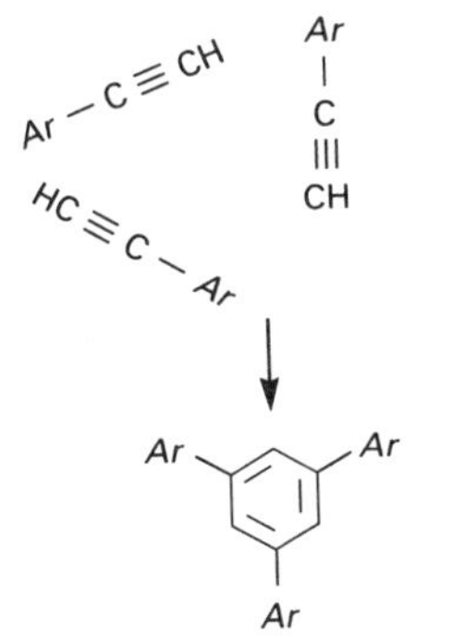

Polyimide-base die attach adhesives can be supplied as electrically conductive or nonconductive, depending on the specific requirement. In all cases, however, the adhesive is thermally conductive to allow heat flow from the integrated circuit die. Hence, these adhesives are typically formulated with precious metal fillers (for example, silver, gold, and sometimes nickel) to permit electrical conductivity or with ceramic fillers (for example, alumina or aluminum nitride) when good thermal conductivity alone is required (Ref 53, 54).

Major suppliers include Ablestik Laboratories, a wholly owned subsidiary of National Starch and Chemical Company, and Emerson & Cuming, a W.R. Grace & Company subsidiary. These products are usually supplied in paste form in a solvent (NMP, diglyme) or as a supported or unsupported film.

Films. The requirement for excellent interlayer adhesion, even in film applications, is vital. A film essentially acts as an adhesive that holds together entire structures, that is, metal and substrate, in addition to acting as an insulator. The most popular and well-known form of DuPont polyimide is Kapton film. Composed of pyromellitic dianhydride and oxydianiline, these materials are processed like the polyamic acid, cured, and supplied in ready-for-use film form of various thicknesses. Kapton is available as an uncoated or Teflon-coated film to provide superior moisture and chemical protection.

Fig. 20 Benzenoid cyclization (trimerization) reaction

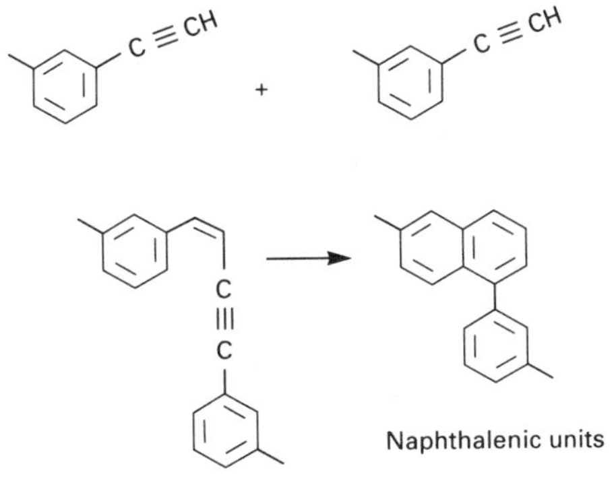

Fig. 21 Naphthalene cross-link formations (non-concerted mechanism)

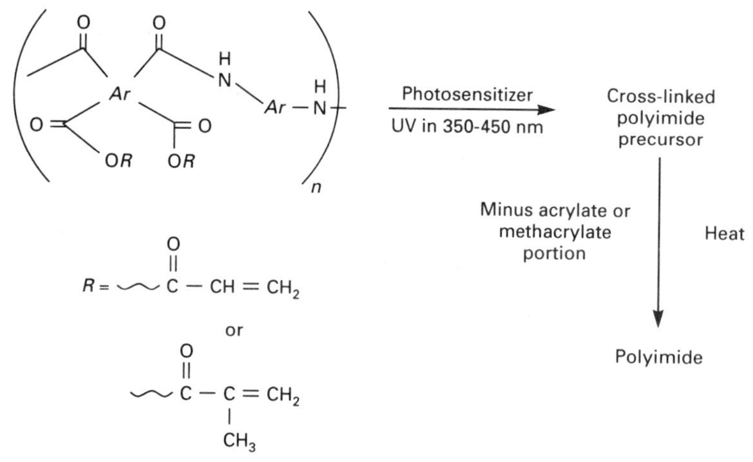

Fig. 22 Photosensitive polyimide chemistry

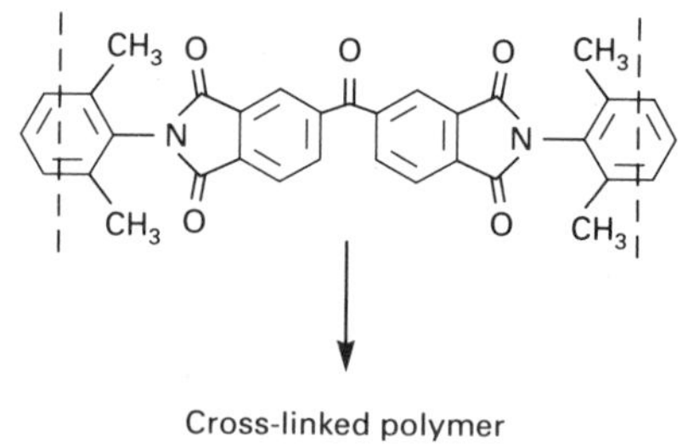

Fig. 23 Inherently photosensitive polyimide

ICI Americas Inc. and Allied-Signal Inc. are marketing in the United States a polyimide film manufactured in Japan by Ube Industries and Kanegafuchi Chemical Industry (Ref 55). Called Upilex and Apical, respectively, the suppliers claim these materials have properties and performance superior to those of Kapton film, including greater dimensional stability, lower moisture pickup, and improved UV and radiation resistance.

Rogers Corporation, under license from NASA, produces a film based on the LARC TPI technology. This material is characterized by easier processibility and improved mechanical performance than is the case with Kapton. The film, as is, can be used as an adhesive.

American Cyanamid Company has developed a supported (glass cloth) and unsupported film adhesive system based on modified PMR-15 technology (Ref 56). The FM series was designed for excellent adhesive strength retention after long exposure to extreme cold (−55 °C, or −65 °F) and elevated temperatures (370 °C, or 700 °F) in metal and composite applications. Ferro Corporation also markets an adhesive film based on the PMR-15 technology.

Cost Factors. Brief mention should be made of the polyimide marketplace. According to some references (Ref 55), high-technology applications for polyimides consume roughly 905 Mg (2×10^6 lb) annually. This translates into $170 million in sales per year. Polyimide adhesives can vary in price from about $22/kg ($10/lb) to $440/kg ($200/lb). Prices vary dramatically with chemistry, performance, and application, from $22/kg ($10/lb) for some molding compounds to $110/kg ($50/lb) for film to over $570/kg ($300/lb) for microelectronic end-uses. Polyimide molding compounds are not described in this article. Characteristics and property values are provided in the articles "Bulk Molding Compounds," "Injection Molding Compounds," and "Resin Transfer Molding Materials" in *Composites*, Vol 1, *Engineered Materials Handbook*, ASM INTERNATIONAL, 1987.

Markets and Applications

Major markets for polyimide adhesives include aircraft structural parts and repair, space vehicles, missiles, and electronic components.

The lacquer systems previously described have found application as high-temperature adhesives for bonding metal-to-metal, composites, and various other sandwich structures with continuous service temperature requirements of 290 °C (550 °F) or above. All have excellent chemical and radiation resistance.

Ready-for-use electronic-grade polyimide solutions have been available since the mid-1980s. These solutions of polyimide or polyimide precursors were produced in response to the increased use of ultrapure polyimides in the microelectronics industry for protective-coating applications (Ref 57). One of the largest potential applications of these pure polyimide solutions is as the dielectric interlayer in a new form of ultra-miniature printed wiring board for high-speed integrated circuits known as multichip modules (Ref 58-61). Polyimides appear to be the material of choice for this new application because of their excellent mechanical, thermal, and electrical properties when applied as a film.

A recently published, independent study was designed to choose the best dielectric polyimide for this multichip module application. The work was performed at the

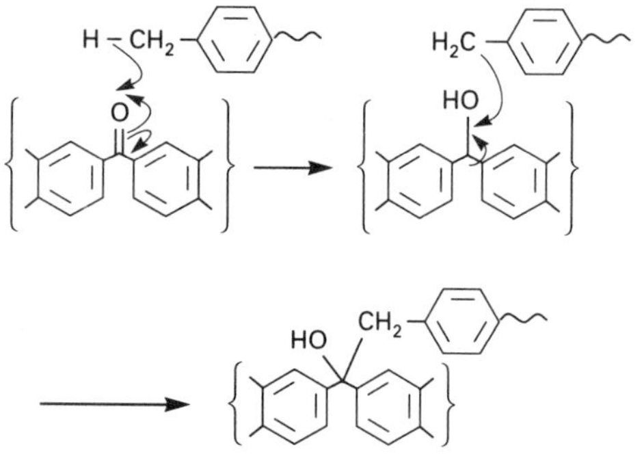

Fig. 24 Mechanism of inherently photosensitive cross-linking

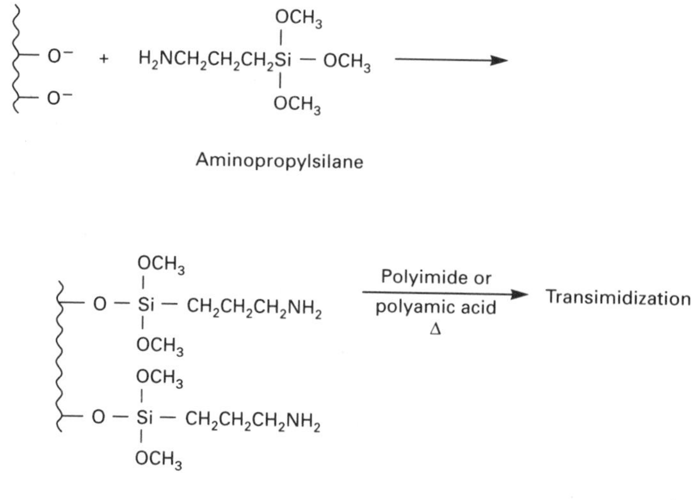

Aminopropylsilane

Fig. 25 Mechanism of aminosilane adhesion promoter

Table 3   MCNC study of electronic-grade polyimides for dielectric interlayer application

Scale 1 (best), 5 (worst)

Characteristics	Standard PI	Fluorinated PI	Silicone PI	Low-stress PI	Thermid
Solvent resistance	1	5	1	1	1
Thermal stability	2	3	3	1	2
Moisture absorption	5	3	3	2	4
Stress	5	5	5	1	5
Adhesion	1	2	2	5	2
Planarization	4	3	3	5	1
Etchability	1	1	5	1	1
Total (overall)	19	22	22	16	16

PI, polyimide. Source: Ref 60

Microelectronics Center of North Carolina (Ref 60). A number of processing characteristics were examined in this study, including adhesion to various inorganic substrates and metals that have potential utility in this important application. The results are presented in Table 3.

Although UV-curable polyimide and polyimide precursors are available from Ciba-Geigy Corporation; Toray Industries, Inc.; and, recently, E.I. Du Pont de Nemours & Company, Inc., they were not included in this study. However, these materials do represent important products for microelectronics applications because of the potential reduction in processing steps compared to conventional polyimide forms (four steps rather than nine) (Ref 42).

Other polyimide microelectronics applications include die attach and substrate attach adhesives (Ref 62), alpha-particle memory chip protection (Ref 54), integrated circuit dielectric films (Ref 63), hybrid circuit protective overcoat (Ref 64), and polymer thick film (Ref 65).

Die attach adhesive applications were described in the section "Forms and Suppliers" in this article.

Polyimide film has found wide commercial application. In many cases, films are used as adhesive laminates. Lamination is typically performed at elevated temperatures (>300 °C, or 570 °F) and high pressure (14 MPa, or 2 ksi).

Two key electronics/microelectronics applications are worth mentioning here. The first is the use of polyimide film as a flexible printed wiring board substrate material. These substrates have found use in everything from telephones to computer disk drives to any application involving flexible circuitry (Ref 66, 67). The second growing application is in a form of wire bonding for integrated circuit chips known as tape automated bonding (TAB) (Ref 68). With the increased complexity of modern integrated circuits, older wire bonding techniques are being replaced with thin copper traces constructed on polyimide film.

Competing Adhesives Systems and Future Trends

Many other polymeric systems have been or currently are under investigation as high-temperature adhesives. These include, but are not limited to, benzocyclobutenes (Ref 69, 70), polybenzimidazoles (Ref 71, 72), and polyquinoxalines (Ref 71, 72). Many of these systems have certain high-temperature advantages over polyimides but have problems of inadequate flow, volatile evolution, and the required high processing pressures and temperatures. The articles "Polyphenylquinoxalines" and "Polybenzimidazoles" in this Section of the Volume provide complete details on these materials.

It is not likely that future trends will involve new monomers and polymers because of the difficulty and cost of developing and bringing them to the marketplace. Instead, combinations of systems in semi-interpenetrating and fully interpenetrating polymer networks (IPNs) and other compatible blend forms (Ref 73-77) will be used to compensate for the deficiencies of some systems.

Properties

In addition to the discussion of property characteristics in the preceding sections of this article, typical physical, mechanical, and thermal properties of neat cured resins are compared in Table 4. The mechanical properties of a polyimide film are provided in Table 5.

Additives and Processing Parameters

Because of the high temperature and/or harsh environmental exposure requirements for polyimide-base adhesives, additives other than adhesion promoters are not typically incorporated. Wetting agents, colorants, bactericides, and antioxidants commonly used in other types of adhesives are not used with polyimides because they typically degrade the high-performance properties. The one exception is the use of toughening agents to counteract the brittleness of some of the BMI/BNI systems. With these systems, modifiers such as carboxy-terminated polybutadiene acrylonitrile rubber (CTBN) and silicones have been incorporated (Ref 78).

Preimidized polyimide adhesives in the resin form typically have an infinite shelf life. However, as a formulated lacquer or varnish, shelf life decreases dramatically, and the material (especially the amic acid forms) must be refrigerated or frozen for storage. Surface treatment prior to the application of the adhesive is very important and varies with the specimens to be bonded. A typical procedure for metal substrates (aluminum or titanium) is to remove organics by wiping specimens with a solvent such as acetone, toluene, or methyl ethyl ketone. After air drying, the specimens are soaked in an etchant (acidic or alkaline, depending on the substrate). The specimens are then thoroughly rinsed with distilled water and dried. At this point a surface primer may be used. This primer is typically a dilute (about 10%) solution of the polyimide adhesive to be used. Composite bonding is usually accomplished after surface cleaning and primer treatment, with no etchant step employed. The article "Curing Polyimide Resins" in *Composites*, Vol 1, *Engineered Materials Handbook*, ASM INTERNATIONAL, 1987, provides additional processing details.

REFERENCES

1. W.M. Edwards, U.S. Patent 3,179,614, 1965
2. W.M. Edwards, U.S Patent 3,179,364, 1965

Table 5   Mechanical properties of polyimide film

Property	Test temperature(a) At 23 °C (73 °F)	At 200 °C (390 °F)
Tensile strength, MPa (ksi)	170 (25)	120 (17)
Tensile modulus, GPa (10^6 psi)	2.96 (0.430)	1.79 (0.260)
Yield strength at 3% offset, MPa (ksi)	70 (10)	40 (6)
Elongation, %	70	90

(a) Test conducted on Kapton film

Table 4   Selected commercial polyimide properties

Property	Pyralin	Skybond	Thermid	PMR-15
Tensile strength, MPa (ksi)	130 (19)	70 (10)	140 (20)	55 (8.1)
Modulus of elasticity, GPa (10^6 psi)	4.1 (0.600)	4.1 (0.600)	2.8 (0.400)	3.2 (0.470)
Elongation, %	15	. . .	6–7	2
Glass transition temperature, °C (°F)	>300 (570)	>330 (625)	>200 (390)	>300 (570)
Coefficient of thermal expansion before T_g, 10^{-6}/K	40	35–40	38	28

3. E. Lavin, A.H. Markhart, and R.E. Kass, U.S. Patent 3,190,856, 1965

4. E. Lavin, A.H. Markhart, and R.E. Kass, U.S. Patent 3,347,808, 1967

5. L.K. English, Premium Performance From Polyimides, *Mater. Eng.*, Jan 1986

6. C.E. Sroog, *J. Polym. Sci., Macromol. Rev.*, Vol 11, 1976, p 161-208

7. M.I. Bessonov et al., Polyimides, Thermally Stable Polymers; L.V. Backinowsky and M.A. Chlenov, Trans., Consultants Bureau, 1987

8. A.L. Landis, in *Handbook of Thermoset Plastics*, S.H. Goodman, Ed., Noyes Publications, 1986, p 266-317

9. D.A. Scola and J.H. Vontell, *Chemtech*, 1989, p 112-121

10. D.A. Scola, in *Composites*, Vol 1, *Engineered Materials Handbook*, ASM INTERNATIONAL, p 78-89

11. I. Serlin et al., *Handbook of Adhesives*, 2nd ed., I. Skeist, Ed., Van Nostrand Reinhold, 1977, p 597-618

12. F. Rodriguez, *Principles of Polymer Systems*, McGraw-Hill, 1970, p 17

13. P.M. Cotts and W. Volksen, in ACS Symposium Series 242, American Chemical Society, 1984, p 227-237

14. M.I. Bessonov et al., Polyimides, Thermally Stable Polymers, L.V. Backinowsky and M.A. Chlenov, Trans., Consultants Bureau, 1987, p 5

15. C.J. Lee, *Proc. 34th Int. SAMPE Symp.*, Vol 34 (No. 1), 1989, p 929-940

16. P.M. Hergenrother, in *Reactive Oligomers*, F.W. Harris and H.J. Sinelli, Ed., ACS Symposium Series 282, American Chemical Society, 1985, p 1-29

17. H.R. Lubowitz, *ACS Polym. Prepr.*, Vol 12 (No. 1), 1971, p 329

18. H.R. Lubowitz, U.S. Patent 3,528,950, 1970

19. H. Stenzenberger, *Brit. Polym. J.*, Vol 20, 1988, p 383-396

20. I.K. Varma et al., *Polym. News*, Vol 12, 1987, p 294-306

21. P.M. Hergenrother, *Chemtech*, Aug 1984, p 496-502

22. D. Wilson, *Brit. Polym. J.*, Vol 20, 1988, p 405-416

23. K.N. Ninan, K. Krishnan, and J. Mathew, *J. Appl. Polym. Sci.*, Vol 32, 1986, p 6033-6042

24. S.J. Shaw and A.J. Kinloch, *Int. J. Adhes. Adhes.*, Vol 5 (No. 3), 1985, p 123

25. H. Stenzenberger et al., *Proc. 19th Int. SAMPE Tech. Conf.*, Vol 19, 1987, p 372

26. A.L. Landis, *ACS Polym. Prepr.*, Vol 15 (No. 2), 1974, p 537

27. P.M. Hergenrother, *Polym. Prepr.*, Vol 21 (No. 1), 1980, p 81

28. A.L. Landis and A.B. Naselow, *Proc. Nat. SAMPE Tech. Conf.*, 1982, Vol 14, p 236

29. W.I. Awad et al., *J. Iraqi Chem. Soc.*, Vol 2 (No. 2), 1977, p 5

30. J. Brown et al., *J. Am. Chem. Soc.*, Vol 88, 1966, p 4468

31. R.T. Otter et al., *J. Org. Chem.*, Vol 26, 1961, p 10

32. M.D. Sefcik et al., *Macromolecules*, Vol 12 (No. 3), 1979, p 423-425

33. W. Fleming, S. Swanson, and D. Hofer, American Chemical Society/Society of Applied Spectroscopy Pacific Conference on Chemistry and Spectroscopy (San Francisco, CA), 1985

34. L.F. Fieser and M. Fieser, *Advanced Organic Chemistry*, Reinhold, 1961, p 223-224

35. G.A. Olan, Ed., *Friedel-Crafts and Related Reactions*, Vol I and II, John Wiley, 1963

36. H. Ahne et al., in *Polyimides: Synthesis, Characterization and Applications*, Vol 2, K.L. Mittal, Ed., Plenum Publishing, 1984, p 905-918

37. P.G. Rickerl, J.G. Stephanie, and P. Slota, in *Proceedings of the 37th Electronic Components Conference*, Institute of Electrical and Electronics Engineers, 1987, p 220-225

38. J. Pfeifer, U.S. Patent 4,677,186, 1987

39. A.A. Lin et al., *Macromolecules*, Vol 21, 1988, p 1165-1169

40. M. Kojima et al., in *Proceedings of the 39th Electronic Components Conference*, Institute of Electrical and Electronics Engineers, 1989, p 920-924

41. A.D. Ulrich and W.G. Joslyn, *Proc. 34th Int. SAMPE Symp.*, Vol 34 (No. 1), 1989, p 127-138

42. R.J. Jenson and J.H. Lai, in *Polymers for Electronic Applications*, CRC Press, Inc., 1989, p 41

43. F. Faiycel et al., *J. Appl. Phys.*, Vol 65 (No. 5), 1989, p 1911-1917

44. L.P. Buchwalter, *J. Adhes. Sci. Technol.*, Vol 1 (No. 4), 1987, p 341-347

45. B. Arkles, *Chemtech*, Dec 1977, p 766-778

46. A.V. Patsis and S. Cheng, *J. Adhes.*, Vol 25, 1988, p 145-157

47. D. Suryanarayana and K. Mittal, *J. Appl. Polym. Sci.*, Vol 29, 1984, p 2039-2043

48. M. Nandi and A. Sen, *Chem. Mater.*, Vol 1, 1989, p 291-292

49. V.L. Bell, B.L. Stump, and H.J. Gager, *J. Polym. Sci., Chem. Ed.*, Vol 14, 1976, p 2275

50. V.L. Bell, U.S. Patent 4,094,862, 1978

51. C. Progar et al., U.S. Patent 4,065,345, 1977

52. A. Berger, U.S. Patent 4,395,527, 1983

53. W. Collins, *Connect. Technol.*, June 1987, p 35-38

54. E. Razon and Y. Tal, *Hybrid Circuit Technol.*, Dec 1989, p 17-20

55. A.S. Wood, Polyimides: Development Is Intense in This Little-Known Polymer Group, *Mod. Plast.*, May 1989

56. R.A. Pethrick, *Mater. Des.*, Vol 7 (No. 4), 1986, p 173

57. P. Burggraaf, Polyimides in Microelectronics, *Semicond. Int.*, March 1988

58. W.C. Shumay, Materials for High-Density Interconnection, *Adv. Mater. Process.*, Feb 1989

59. J.J. Licari, *Electron. Packag. Prod.*, Sept 1989, p 58-64

60. T.G. Tessier et al., in *Proceedings of the Electronic Components Conference*, Institute of Electrical and Electronics Engineers, May 1989, p 127-134

61. J.J.H. Reche, *Circuits Manuf.*, June 1989, p 46-51

62. J.C. Bolger, in *Polyimides: Synthesis, Characterization and Applications*, Vol 2, K.L. Mittal, Ed., Plenum Publishing, 1984, p 871-887

63. A.M. Wilson, in *Polyimides: Synthesis, Characterization and Applications*, Vol 2, K.L. Mittal, Ed., Plenum Publishing, 1984, p 715-733

64. C.J. Lee et al., in *Microelectronic Packaging Technology: Materials and Processes, Proceedings of the 2nd ASM INTERNATIONAL Electronic Materials and Processing Congress*, April 1989, p 359-367

65. T.K. Gupta, *Int. J. Hybrid Microelectron.*, Vol 12 (No. 3), 1989, p 139

66. K. Gilleo, 3-D Circuits—Have They Arrived?, *Electron. Packag. Prod.*, October 1989, p 112

67. L. Leonard, *Plast. Des. Forum.*, Sept-Oct 1989, p 17-22

68. P. Burggraaf, *Semicond. Int.*, June 1988, p 72-77

69. R.A. Kirchhoff, U.S. Patent 4,540,763, 1985

70. R.A. Kirchhoff et al., in *Proceedings of the 18th International SAMPE Technical Conference*, Society for the Advancement of Material and Process Engineering, 1986, p 476-489

71. C. Chih-Ching, *Polym. Sci. Technol.*, Vol 37, 1988, p 227

72. L.H. Lee, *Int. J. Adhes. Adhes.*, 1987, p 81

73. S. Choe et al., *Polym. Mater. Sci. Eng.*, Vol 56, 1987, p 827

74. E.C. Chenevey, European Patent Application 87302775.9, 1987

75. W. MacKnight, European Patent Application 87301367.6, 1987

76. G. Guerra et al., *Macromolecules*, Vol 21, 1988, p 231-234

77. F. Mercer, International Patent Application 688,794, 1985

78. A.K. St. Clair and T.L. St. Clair, *Proc. Nat. SAMPE Tech. Conf.*, Vol 12, 1986, p 721

Polyphenylquinoxalines

Stanley E. Wentworth, Polymer Research Branch, U.S. Army Materials Technology Laboratory

POLYPHENYLQUINOXALINES dif-
fer from nearly all other adhesive families
described in this Section of the Volume.
Although polyphenylquinoxalines are not
readily available in commercial adhesive
formulations, their inclusion in this Volume
reflects their great promise as adhesives for
the most harsh end-use conditions. This
promise derives not only from their outstanding thermal and thermooxidative stability, but also from their implied processibility based on their solubility in reasonable
solvents and thermoplastic nature. In a
comprehensive 1988 review of polyquinoxalines, including the polyphenylquinoxalines, it was stated that if a favorable
combination of price, processibility, and
performance is attained, quinoxaline polymers will find acceptance in industry, particularly in applications demanding high
chemical, thermal, and/or thermooxidative
stability (Ref 1).

Chemistry

Polyphenylquinoxalines constitute a family of linear heteroaromatic polymers containing the phenylquinoxaline unit in the
main chain. They are prepared by the step-
growth polycyclocondensation of bis(o-diamines) and bis(α-diketones), as shown in
the generic representation in Fig. 1. By the
appropriate variation of x, y, and/or Ar in
the monomer(s), where Ar is the aromatic
unit(s), the properties of the resultant polymer can be predictably varied over a very
wide range. For example, when x is H, y is
nil, and Ar is p-phenylene, the glass transition temperature is 372 °C (700 °F). If Ar is
p-oxydiphenylene, the glass transition temperature drops to 290 °C (555 °F). Similar
substitutions have led to polyphenylquinoxalines with glass transition temperatures
ranging from 207 to 390 °C (405 to 735 °F)
(Fig. 2).

As implied by this discussion of glass
transition temperatures, polyphenylquinoxalines are linear amorphous thermoplastics,
and therein lies both their advantage and
their principal weakness. Their non-
crosslinked nature permits their solubility in
several classes of organic solvents. However, their thermoplastic nature causes concern about elevated-temperature creep and
the substantial loss of properties above the
glass transition temperature.

Indeed, because they possess significant
thermooxidative stability above their glass
transition temperatures, a major goal of
polyphenylquinoxaline research has been
the reduction or elimination of this elevated-temperature plasticity. This has been
sought primarily through the synthetic modification of the tetraketone monomer in order to introduce latent thermal crosslink-
ability. Most often, this is achieved by
placing a thermally reactive group in the
para position of the pendant phenyl groups.
Groups that will not interfere in the poly-
mer-forming condensation reaction are selected. Thus, a high molecular weight linear
polymer is formed, on which processing
is performed. The finished end item is then
postcured to achieve crosslinking and
hence thermomechanical stabilization. Examples of thermally crosslinkable polyphenylquinoxalines are given in Fig. 3.
An alternative approach to crosslinking
employs an unsaturated crosslinking agent
that undergoes a Diels-Alder reaction with
a polyquinoxaline bearing pendant furyl
groups (Ref 14). To date, no crosslinkable polyquinoxaline has been commercialized.

In 1988, an alternative synthetic path
based on aromatic nucleophilic displacement was reported (Ref 15, 16). It is said to
offer a more facile structural modification
and potentially lower cost than the established approach.

Forms and Functional Types

Commercial Forms. At this writing, only
one polyphenylquinoxaline is commercially
available. It is the polymer depicted in Fig.
1 where Ar is p-phenylene, x is H, and y is
nil. It is available in 5, 10, or 15% solids
solutions in a 65% m-cresol-35% xylene
mixed solvent. While the principal intended
application for these products is in solid-
state electronic device fabrication, solutions of this sort are also suited to adhesive
bonding according to the process described
below.

Functional Types. Polyphenylquinoxaline adhesives could be considered primarily for structural applications in extremely
hot environments. Because they can be
used at temperatures that preclude the use
of aluminum, much of the literature focuses
on the bonding of titanium and advanced
composites (Ref 17-19).

Although hot-melt adhesives are not usually considered suitable for structural applications, the polyphenylquinoxalines are an
exception. They are processed in essentially the same manner as conventional hot
melt adhesives, but at temperatures hundreds of degrees higher. Thus, they have
very acceptable strength at ambient temperature and up to their glass transition temperature. Above that, thermoplasticity becomes a limiting factor.

No cure reaction occurs during bond formation. Rather, the polymer is raised above
its glass transition temperature, at which
point it becomes sufficiently mobile to flow
and wet the adherend surface. Subsequent
cooling immobilizes the polymer, and the
bond is formed.

In a typical bonding process (Ref 20), a
glass fabric carrier or scrim cloth is impregnated with a solution of the polyphenylquinoxaline, and the solvent is removed by
evaporation at elevated temperature. This

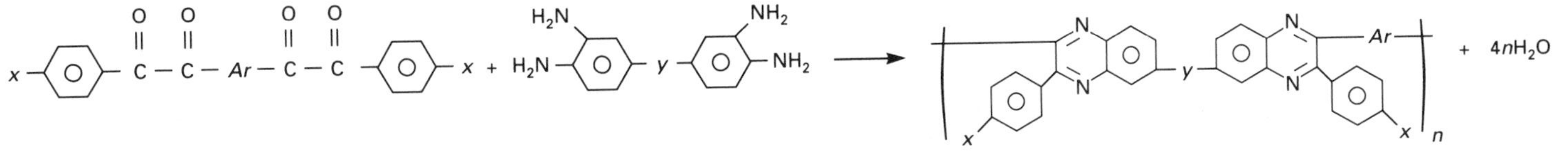

Fig. 1 General preparation of a polyphenylquinoxaline

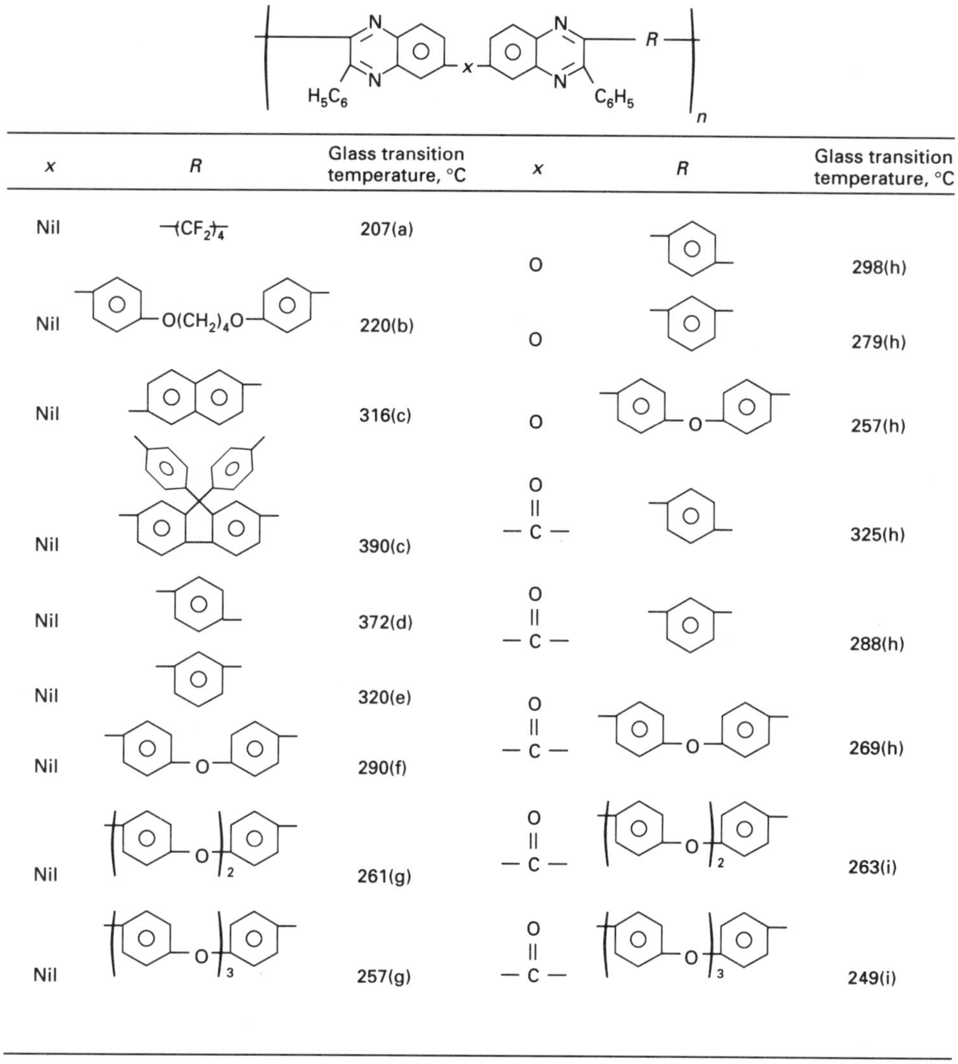

Fig. 2 Glass transition temperatures for selected polyphenylquinoxalines. (a) Ref 2. (b) Ref 3. (c) Ref 1. (d) Ref 4. (e) Ref 5. (f) Ref 6. (g) Ref 7. (h) Ref 8. (i) Ref 9

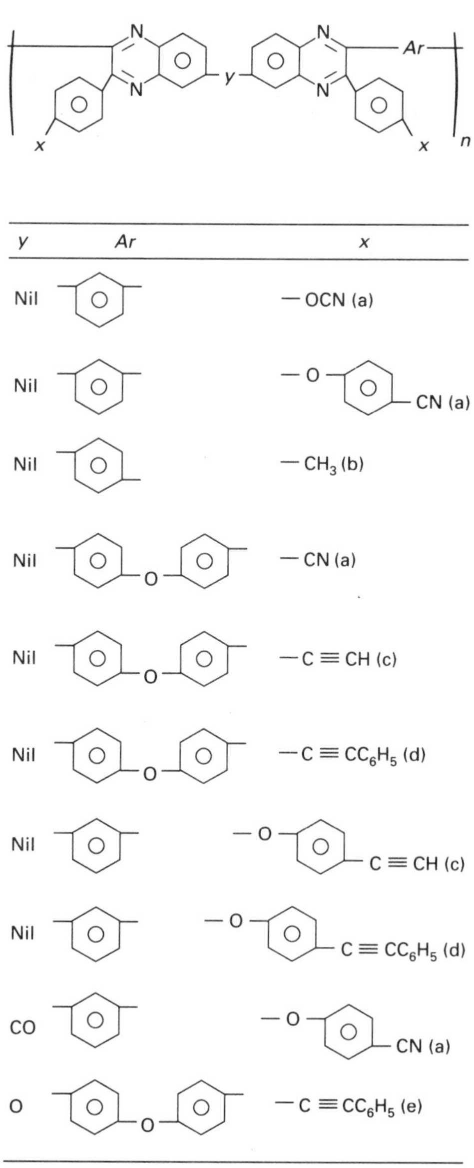

Fig. 3 Thermally crosslinkable polyphenylquinoxalines. (a) Ref 6. (b) Ref 10. (c) Ref 11. (d) Ref 12. (e) Ref 13

process is repeated until the desired film thickness (typically 250 to 375 μm, or 10 to 15 mils) is achieved. This film-coated cloth is then placed between the two adherends, the surfaces of which have been appropriately treated and primed with a very dilute solution of the polyphenylquinoxaline from which the solvent has been removed. This assembly is then vacuum bagged and subjected to an autoclave pressure of 1380 kPa (200 psi). The temperature is then brought to 370 °C (700 °F) at 3 °C/min (5.5 °F/min). After a 1-h hold at this temperature, the assembly is cooled to not less than 65 °C (150 °F), at which point the pressure is released and the part is debagged. This process will differ in details of processing temperatures and glass transition temperatures for particular polyphenylquinoxalines.

Markets and Applications

At this writing (early 1990), no major market exists for the polyphenylquinoxalines, which are still regarded as developmental. The product described above fills a low-volume requirement, at best. However, higher-volume applications are actively being sought. Examples include adhesives for extreme environments and matrix resins for advanced composites, especially for aerospace applications. The use of polyphenylquinoxalines as protective coatings for harsh environments, where not only high temperature but high humidity and/or corrosive or caustic conditions are encountered, is also an attractive possibility. The chemical process industry and geothermal energy industry come readily to mind. High-temperature wire insulation, another coatings application, has been shown to be a potentially valuable use of polyphenylquinoxalines (Ref 21).

Major Suppliers. To date, the only commercial supplier of polyphenylquinoxaline polymer products is Cemota, a French firm represented in the United States by IFP Enterprises. This firm supplies the solutions described in the section "Forms and Functional Types" in this article. These products are designated Syntorg IP 200-705, IP 200-710, and IP 200-715 for the 5, 10, and 15% solutions.

Other polyphenylquinoxalines can be purchased from custom synthesis houses, although the cost will reflect the low volume, bench-scale nature of such undertakings.

Cost Factors. Based on an approximate price of $990/kg ($450/lb) for the 15% solution, IP 200-715, the cost of this polyphenylquinoxaline on a solids basis is $6615/kg ($3000/lb) at this writing. In this price range, only very high value-added uses, such as electronic device fabrication, can be considered. The need to use only very small quantities in this application is an important factor.

While the price of this polymer is prohibitively high for most adhesive applications, the price for this polyphenylquinoxaline, which has great potential, can be reduced by one to two orders of magnitude if sufficient quantities are produced.

A lower-cost synthesis of the monomers has been sought, with most of the work being conducted on the bis(α-diketones) (Ref 22-24). The economics of the principal bis(*o*-diamine), 3,3'-diaminobenzidine, is controlled by the market for the polybenzimidazoles, for which it is also a monomer. A separate article, "Polybenzimidazoles," is provided in this Section of the Volume.

Reports have appeared describing lower-cost solvents for the preparation of polyphenylquinoxalines (Ref 25) and documenting the equivalence of polymers from lower-cost monomers and polymers from monomers prepared by the original method (Ref 26). A summary of cost-reduction efforts is provided in Ref 27.

To date, however, none of these potential cost savings has been realized. Polyphenylquinoxalines are very costly because they are not used in sufficient volume; they do not find application because they are very costly. The hope is that increasing use in electronics will eventually break this cycle, causing a downward spiral in price and making adhesive and other structural applications economically feasible.

Competing Adhesives

In the temperature range where polyphenylquinoxalines might be advantageously used, only the polyimides can be considered to be directly competitive. They have two very significant advantages over the polyphenylquinoxalines: They are readily available, and they are significantly less costly. Both of these factors reflect their substantially greater technological maturity and results in a much greater body of data on end-use applications and properties. The article "Polyimides" in this Section of the Volume provides complete details.

However, polyphenylquinoxalines have real and/or apparent advantages of their own, such as in processibility. Their thermoplastic nature and solubility suggest ease of processing, yet in practice this has not been demonstrated, at least for structural applications. Further, it is noted that some of the newer polyimides are thermoplastic and are as processible as the polyphenylquinoxalines.

However, the chemistry of the polyphenylquinoxalines provides the basis for one real advantage over the polyimides. They are totally inert to conditions that can cause hydrolytic degradation of polyimides (Ref 28). Although the degree to which polyimides hydrolyze varies tremendously with structure, physical form, and end-use environment and is not often a practical prob-

Table 1 Selected properties of commercially important polyphenylquinoxalines

Property	Polymer	
	Ar = *p*-phenylene	Ar = *p*-oxydiphenylene
Thermal		
Glass transition temperature, °C (°F)	372 (700) (Ref 4)	290 (555) (Ref 6)
Thermal stability(a), °C (°F)	560 (1040) (Ref 29)	530 (985) (Ref 1)
Thermooxidative stability(b), °C (°F)	500 (1040) (Ref 29)	480 (895) (Ref 1)
Coefficient of thermal expansion, 10^{-5}/K	5.5 (Ref 29)	. . .
Mechanical		
Tensile strength(c), MPa (ksi)	127.5 (18.5) (Ref 7)	117.2 (17.0) (Ref 13)
Tensile modulus(c), GPa (10^6 psi)	2.83 (0.410) (Ref 7)	2.76 (0.400) (Ref 13)
Elongation to break, %	7.1 (Ref 7)	23 (Ref 13)
Electrical (thin film)		
Dielectric constant(d)	2.7 (Ref 29)	. . .
Dissipation factor(d)	0.0005 (Ref 29)	. . .
Volume resistivity, $\Omega \cdot$ cm	3×10^{17} (Ref 29)	. . .
Dielectric strength, MV/mm (MV/in.)	0.2 (5.1) (Ref 29)	. . .
Other		
Density, g/cm^3	1.22 (Ref 29)	. . .

(a) Onset of rapid weight loss; in inert atmosphere. (b) Onset of rapid weight loss; in air. (c) Measured on thin films per ASTM D 882. (d) Measured at 1 kHz

lem, still the molecular structure of the polyphenylquinoxalines has the advantage of being devoid of a site for attack by water, which is present in all polyimides.

Density is another property in which the polyphenylquinoxalines may enjoy an advantage. The density of the commercially available product is 1.22 g/cm^3 (Ref 29), while the density of Kapton polyimide is greater than 1.4 g/cm^3 (Ref 30). The lower density of polyphenylquinoxalines can lead to a substantial weight savings on major aerospace components and systems, thereby contributing to overall system performance.

Of the numerous other heteroaromatic polymers that have been reported, only the polybenzimidazoles have received serious consideration as adhesives. Other classes of polymers, such as silicone resins and engineering thermoplastics, simply do not have the combination of thermal, thermooxidative, and thermomechanical properties to be considered competitive with polyphenylquinoxaline adhesives for use in extreme environments.

Uncured Properties

As already mentioned, polyphenylquinoxalines are directly obtained in the polymerization solvent as linear, high molecular weight thermoplastics. Hence, no cure reaction takes place during their processing after their isolation from solution. The one exception to this is the fabrication of graphite fiber reinforced composites by means of polymerization from monomeric reactants (Ref 31). This approach is probably not useful for adhesive bonding because of the difficulty in removing, through impervious adherends, the water produced during cure.

Because it has an influence on the bonding process, the viscosity of the polyphenylquinoxaline solution is analogous to an uncured property. Viscosity affects the ability of the solution to coat/wet the adherend, an important factor in bond quality. In turn, viscosity is directly controlled by the polymer concentration in solution and, to a lesser degree, by polymer molecular weight. The thickness of the polymer layer after solvent removal is directly related to polymer concentration. For most applications, optimum solution concentration can be quickly established by trial and error.

As with any solvent-borne adhesive, solvent removal is an extremely important issue. The most common solvent, *m*-cresol, alone or in combination with another solvent component, is extremely difficult to remove completely. This is due not only to the high boiling point of *m*-cresol (203 °C, or 400 °F) but also to the acid/base complex it forms with the basic nitrogen moiety in quinoxaline. Fortunately, the polyphenylquinoxalines can tolerate the elevated temperatures required for solvent removal, especially if vacuum is also used.

To avoid the use of *m*-cresol, polyphenylquinoxalines are sometimes used in a chlorinated hydrocarbon solution. However, the volatility of this solvent can cause other difficulties, principally the skinning over of the solution during application. Ultimately, the choice of solvent is governed by the specific requirements of the bonding process.

Cured Properties

Thermal, Physical, and Electrical. Although the polyphenylquinoxalines were first reported more than 20 years ago (Ref

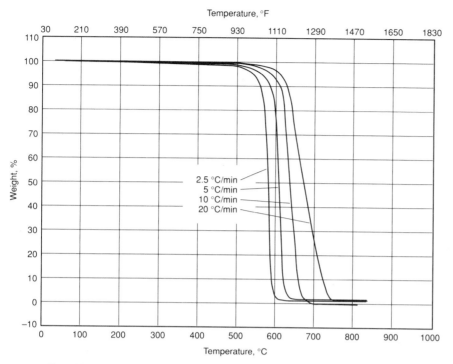

Fig. 4 Effect of heating rate on the onset of weight loss for the polyphenylquinoxaline containing *p*-phenylene, as determined by dynamic thermogravimetry in flowing air. Source: Ref 33

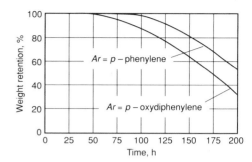

Fig. 5 Isothermal aging of polyphenylquinoxalines at 371 °C (700 °F) in air. Source: Ref 1

32), only two of them have ever approached the status of commercial availability. Because this is not likely to change, selected properties are given for these two polymers only. As discussed in the section "Chemistry" in this article and shown in Fig. 1, *x* is *H* and *y* is nil for both polymers, but *Ar* can be either *p*-phenylene or *p*-oxydiphenylene. The properties of *p*-phenylene and *p*-oxydiphenylene are presented in Table 1.

It is not possible to find a data source for these polymers that ensures that a given property was determined by the same method on an equivalent material, thereby allowing direct comparisons. Hence, these data should be regarded as representative of a range of values and not necessarily as intrinsic constants.

The data on thermooxidative stability is a case in point. The values reported were determined by dynamic thermogravimetric analysis in which weight is recorded as a function of temperature, which is increased at a constant rate. This rate of increase has an effect on the temperature at which rapid

weight loss begins: The faster the rate, the higher the onset temperature. Figure 4 shows this phenomenon (Ref 33).

Other factors also affect the thermooxidative stability temperature. The physical form of the sample (film versus powder) is especially important because of the major difference between the surface-to-volume ratios of the two. Molecular weight is important, too, partly because of the increased number of thermally labile end groups for low molecular weight material. A study detailing end group effects on the thermal stability of a polyphenylquinoxaline has been published (Ref 34).

Caution should also be exercised if these data are used to infer intermediate-to-long-term end-use performance. Thermooxidative stability, as derived from dynamic thermogravimetric analysis, is particularly misleading in this regard. The high stability temperatures indicated by this method are very much an artifact of the short residence time experienced by the polymer and should be used only for comparative screening purposes.

Much more useful thermooxidative stability data may be obtained by elevated-temperature isothermal-aging experiments in which weight is recorded as a function of time at a fixed temperature. Figure 5 shows the results of such an experiment for the polymers of Table 1. Here, the difference in thermooxidative stability is very evident, with the polymer containing *p*-oxydiphenylene beginning to lose weight much sooner and more rapidly than the *p*-phenylene analog. Even this method should be regarded only as a more realistic screening technique. Because only weight change is actually recorded, the effect on end-use properties, such as strength or stiffness, can only be inferred. Ultimately, it is necessary to test the effect of temperature on the properties of interest.

Adhesive. The use of polyphenylquinoxalines as adhesives for the bonding of advanced materials was described earlier in the section "Functional Types." Representative data are presented in the tables and figures in this section.

Table 2 gives the tensile shear strength of bonds to titanium using a polyphenylquinoxaline that contains *p*-oxydiphenylene. Several points emerge from these data. First, the ambient as-formed bond strength is more than acceptable for structural applications. Second, although the strength value drops approximately 25% after testing at 232 °C (450 °F), it is still acceptably high and is maintained at this temperature through very long aging periods. Finally, when its 290 °C (550 °F) glass transition temperature is exceeded, its strength at 316 °C (600 °F) drops catastrophically, illustrating the thermoplasticity problem previously discussed (Ref 16).

Another study that used this adhesive to bond titanium focused on the effect of adherend surface treatment on the long-term thermal stability of the bonds (Ref 35). The treatments selected were chromic acid anodization (CAA), phosphate-fluoride (PF) etch, and sodium hydroxide anodization (SHA). The results, shown in Fig. 6, show the importance of proper surface treatment to quality bonds and indicate that the SHA treatment is the most effective treatment. The results also demonstrate that useful

Table 2 Tensile shear strength of Ti-6Al-4V bonded with a polyphenylquinoxaline that incorporates *p*-oxydiphenylene

Test condition	Tensile shear strength(a)	
	MPa	ksi
At 25 °C (77 °F)	32.7	4.74
At 232 °C (450 °F) after 10 min at 232 °C (450 °F) in air	24.1	3.49
At 232 °C (450 °F) after 5000 h at 232 °C (450 °F) in air	23.9	3.47
At 232 °C (450 °F) after 8000 h at 232 °C (450 °F) in air	23.1	3.35
At 316 °C (600 °F) after 10 min at 316 °C (600 °F) in air	2.3	0.33

(a) According to ASTM D 1002. Source: Ref 17

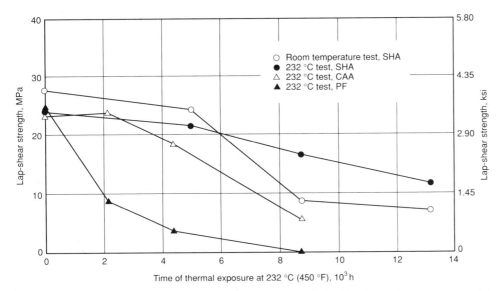

Fig. 6 Lap-shear strength test results after thermal exposure of Ti-6Al-4V with various surface treatments bonded with a polyphenylquinoxaline. SHA, sodium hydroxide anodization; CAA, chromic acid anodization; PF, phosphate-fluoride etch. Source: Ref 35

bonds to titanium can be achieved with this polymer. A detailed interpretation of the data and an excellent presentation of the bonding and testing procedures used are given in Ref 35.

The *m*-phenylene analog of the polymer shown in Table 1, where *Ar* is *p*-phenylene, has been evaluated as an adhesive for bonding titanium to titanium, titanium to a graphite-polyimide composite, and a graphite-polyimide composite to a graphite-polyimide composite (Ref 17). A number of surface treatments and bonding cycles were examined, and representative data are given in Table 3. The surprising retention of strength at 316 °C (600 °F) for the composite specimens compared to the titanium specimens is thought to be due to the partial softening of the composite matrix, resulting in less brittle behavior and better stress distribution. Other studies of a variety of polyphenylquinoxaline adhesives are available (Ref 3, 19, 20, 36-38).

Additives and Modifiers. Very little has been reported on additives and modifiers for polyphenylquinoxaline adhesives. Fillers such as aluminum and boron powder are sometimes used (Ref 1).

In a sense, residual *m*-cresol, which is not completely removed during the adhesive application process, acts as a plasticizer and assists in flow and adherend wetting during bond formation. However, this phenolic material has a rather noxious and corrosive nature and therefore is not desirable as a processing aid for most applications. On the other hand, the reverse situation has been reported, one in which a polyphenylquinoxaline is used as a toughening modifier for a brittle acetylene-terminated resin (Ref 39).

Product Design Considerations. Because of the developmental nature of polyphenylquinoxalines, there is essentially no experience with the design of end items that utilize these polymers as adhesives. As already noted, polyphenylquinoxalines are processed as very high-temperature hot-melt adhesives. This requires that the materials selected for a given system be able to withstand harsh process conditions.

Processing Parameters

Processing aspects have been discussed throughout this article, with the exception of the issues of shelf life, pot life, and storage conditions. Because of their exceptional thermal, thermooxidative, and hydrolytic stability, and because they are applied as adhesives in the fully cured state, polyphenylquinoxalines do not undergo deleterious changes or degradation during stor-

age. Solutions in *m*-cresol are stable indefinitely. The pot life of open *m*-cresol solutions, though not infinite, is normally long enough to accommodate most bonding operations. Solutions in lower boiling point chlorinated hydrocarbons should be kept covered to prevent skinning over and changes in concentration.

REFERENCES

1. P.M. Hergenrother, Polyquinoxalines, in *Encyclopedia of Polymer Science and Engineering*, Vol 13, 2nd ed., John Wiley & Sons, 1988, p 55
2. P.M. Hergenrother, Highly Fluorinated Phenyl-as-Triazine and Phenylquinoxaline Polymers, *Polym. Prepr.*, Vol 19 (No. 2), 1978, p 40
3. S.J. Havens, F.W. Harris, and P.M. Hergenrother, Polyphenylquinoxalines Containing Alkylenedioxy Groups, *J. Appl. Polym. Sci.*, Vol 32, 1986, p 5957
4. W. Wrasidlo, Transitions and Relaxations in Aromatic Polymers, *J. Polym. Sci.*, Part A-2, Vol 9, 1971, p 1603
5. P.M. Hergenrother and D.J. Progar, High Temperature Composite Bonding With PPQ, in *Proceedings of the 22nd National SAMPE Symposium*, Society for the Advancement of Material and Process Engineering, 1977, p 211
6. P.M. Hergenrother, Poly(phenyl-as-Triazines) and Poly(phenylquinoxalines). New and Cross-Linked Polymers, *Macromolecules*, Vol 7, 1974, p 575
7. P.M. Hergenrother, Paper presented at the 1984 International Chemistry Congress of Pacific Basin Societies, Honolulu, Dec 1984
8. P.M. Hergenrother, Poly(phenyl-as-Triazines) and Poly(phenylquinoxalines), New High Performance Polymers, in *Proceedings of the 19th National SAMPE Symposium*, Society for the Advancement of Material and Process Engineering, 1974, p 146
9. A.K. St. Clair and N.J. Johnston, Ether Polyphenylquinoxalines. II. Polymer Synthesis and Properties, *J. Polym. Sci.*, Polymer Chemistry ed., Vol 15, 1977, p 3009
10. S.E. Wentworth, Synthesis and Thermal Crosslinking of p-Tolyl Containing Polyquinoxalines, in *Chemistry and Properties of Crosslinked Polymers*, S.S. Labana, Ed., Academic Press, 1977, p 125
11. P.M. Hergenrother, Poly(phenylquinoxalines) Containing Ethynyl Groups, *Macromolecules*, Vol 14, 1981, p 891
12. P.M. Hergenrother, Poly(phenylquinoxalines) Containing Phenylethynyl Groups, *Macromolecules*, Vol 14, 1981, p 898
13. P.M. Hergenrother, unpublished research

Table 3 Tensile shear strength of bonds to several adherends using a polyphenylquinoxaline adhesive

Adherend	At room temperature		Tensile shear strength(a) At 288 °C (550 °F)		At 316 °C (600 °F)	
	MPa	ksi	MPa	ksi	MPa	ksi
Ti/Ti(b)	30.3	4.4	17.2	2.5	2.1	0.5
Ti/Gr-PI(c)	13.8	2.0	21.4	3.1	11.1	1.6
Gr-PI/Gr-PI	20.7	3.0	29.0	4.2	20.7	2.1

(a) According to ASTM D 1002. (b) Ti is Ti-6Al-4V. (c) Gr-PI is high tensile strength graphite fiber reinforced Skybond 710 polyimide.

14. R.J. Jones and M.K. O'Rell, Cross Linked Polyquinoxaline Polymers, U.S. Patent 3,904,584, 1975
15. J.L. Hedrick and J.W. Labadie, Synthesis of Poly(Arylene Ether Phenylquinoxalines), *Polym. Mater. Sci. Eng.*, Vol 59, 1988, p 42
16. J.W. Connell and P.M. Hergenrother, Synthesis of Polyphenylquinoxalines via Aromatic Nucleophilic Displacement, *Polym. Prepr.*, Vol 29 (No. 1), 1988, p 172
17. P.M. Hergenrother, High Temperature Organic Adhesives, *SAMPE Q.*, Vol 3, 1971, p 1
18. P.M. Hergenrother and D.J. Progar, High Temperature Composite Bonding With PPQ, *Adhes. Age*, Vol 20 (No. 12), 1977, p 38
19. C.L. Hendricks and S.G. Hill, Evaluation of High Temperature Structural Adhesives for Extended Service, *Adhesive Chemistry*, Vol 29, *Polymer Science and Technology*, Plenum Publishing, 1984, p 489
20. C.L. Hendricks, S.G. Hill, J.N. Hale, and W.G. Dumars, Evaluation of High Temperature Structural Adhesives for Extended Service—Phase V, NASA CR 178176, National Aeronautics and Space Administration, 1987; CA Vol 110, 155627
21. S.E. Wentworth, T.T. Rodzen, R.S. Yorgensen, and W.R. Diehl, Preliminary Evaluation of a Poly(phenylquinoxaline) as Thermally Resistant Wire Insulation, in *Proceedings of the 1988 IEEE International Symposium on Electrical Insulation*, Institute of Electrical and Electronics Engineers, 1988, p 279
22. S.E. Wentworth, Preparation of 1,4-Bis(phenylglyoxaloyl)benzene, U.S. Patent 3,829,497, 1974
23. R.H. Kratzer, K.L. Paciorek, and D.W. Karle, Modified Benzoin Condensation of Terephthalaldehyde With Benzaldehyde, *J. Org. Chem.*, Vol 41, 1976, p 2230
24. F.W. Harris and B.A. Reinhardt, Preparation of Tetraketones, U.S. Patent 4,082,806, 1978
25. S.E. Wentworth and D.J. Larsen, Low-Cost Solvents for the Preparation of Polyphenylquinoxalines, AMMRC TR 79-3, Army Materials and Mechanics Research Center, 1979; CA Vol 91, 124276
26. S.E. Wentworth and M.J. Humora, Evaluation of Polyphenylquinoxalines Derived From an Inexpensive Tetraketone Monomer, AMMRC TR 77-7, Army Materials and Mechanics Research Center, 1977; CA Vol 87, 118128
27. S.E. Wentworth, Recent Developments in Polyphenylquinoxaline Cost Reduction, in *Proceedings of the 11th National SAMPE Technical Conference*, Society for the Advancement of Material and Process Engineering, 1979, p 752
28. C.E. Sroog, A.L. Endrey, S.V. Abramo, C.E. Berr, W.M. Edwards, and K.L. Olivier, Aromatic Polypyromellitimides From Aromatic Polyamic Acids, *J. Polym Sci.*, Part A, Vol 3, 1965, p 1373
29. Syntorg IP 200, Technical Data Sheet 3, Cemota, 1988
30. S.A. Stern, Y. Mi, H. Yakamoto, and A.K. St. Clair, Structure/Permeability Relationship of Polyimide Membranes; Applications to the Separation of Gas Mixtures, *J. Polym. Sci., B, Polym. Phys.*, Vol 27, 1989, p 1887
31. T.T. Serafini, P. Delvigs, and R.D. Vannucci, In Situ Polymerization of Monomers for High Performance Poly(phenylquinoxaline)/Graphite Fiber Composites, *J. Appl. Polym. Sci.*, Vol 17, 1973, p 3235
32. P.M. Hergenrother and H.H. Levine, Phenyl-Substituted Polyquinoxalines, *J. Polym. Sci.*, Part A-1, Vol 5, 1967, p 1453
33. D.P. Macaione, unpublished research
34. J.M. Augl, Thermal Degradation of a Polyphenylquinoxaline in Air and Vacuum, *J. Polym. Sci.*, Part A-1, Vol 10, 1972, p 2403
35. D.J. Progar, Evaluation of Ti-6Al-4V Surface Treatments for Use With a Polyphenylquinoxaline Adhesive, NASA TM 89045, National Aeronautics and Space Administration, 1986; CA Vol 106, 106396
36. P.M. Hergenrother, Adhesive and Composite Evaluation of Acetylene-Terminated Phenylquinoxaline Resins, *Polym. Eng. Sci.*, Vol 21, 1981, p 1072
37. P.M. Hergenrother, Polyphenylquinoxalines Containing Pendant Phenylethynyl Groups: Preliminary Mechanical Properties, *J. Appl. Polym. Sci.*, Vol 28, 1983, p 355
38. P.M. Hergenrother, Quinoxaline/Phenylquinoxaline Copolymers, *J. Appl. Polym. Sci.*, Vol 21, 1977, p 2157
39. Y.P. Sachdeva and S.E. Wentworth, Toughening of AT-Resins With a Polyphenylquinoxaline, *J. Adhes.*, in press

Polybenzimidazoles

J.W. Connell, NASA Langley Research Center

POLYBENZIMIDAZOLES (PBIs) are a family of heterocyclic polymers that received extensive evaluation as adhesives and composite matrices in the 1960s. These polymers possess high thermal, thermooxidative, and chemical stability; good mechanical properties; and excellent flame resistance, making them high-performance/high-temperature materials that are attractive for use in harsh environments.

In the early 1960s the U.S. Air Force awarded contracts to several companies to develop methods for processing PBI and to determine its potential as a structural resin. Because high molecular weight PBI had poor melt flow characteristics, a prepolymer concept was developed. This approach involved preparing a low molecular weight, partially cyclized precursor to the high molecular weight PBI. This prepolymer had sufficient flow to achieve substrate wetting, but led to the evolution of large quantities of volatiles (that is, phenol and water) during curing and/or bonding cycles. In addition, the monomers used in the synthesis of PBI were expensive. These factors precluded the use and further development of PBI adhesives in a number of application areas.

This article will acquaint the reader with the chemistry and the physical and mechanical properties of various polybenzimidazoles that have displayed good adhesive properties. Those interested in a more comprehensive review of the synthesis; structure-property relationships; and neat resin, composite, and fiber mechanical properties of PBIs are directed to Ref 1 to 8. In addition, the Selected References in this article provide either overviews or the details of specific applications.

Chemistry

The preparation of PBIs was first reported in 1959 as the reaction of aliphatic dicarboxylic acids with aromatic bis(o-diamines) (Ref 9). Subsequently, wholly aromatic PBIs were prepared from the reaction of aromatic bis(o-diamines) and aromatic dicarboxylic acids or the phenyl ester derivatives (Fig. 1) (Ref 10). Since then many structurally different PBIs have been prepared and evaluated to varying degrees. It has been estimated that PBIs have been

prepared from more than 18 different bis(o-diamines) and hundreds of different diacids or derivatives thereof. Although many of the compositions offer some property advantages, the PBI in Fig. 1, poly[2,2'-(m-phenylene)-5,5'-bibenzimidazole], is representative of the structure of all commercial systems. The selection of this chemistry was made on the basis of the physical and mechanical properties it provided, the availability of starting materials, and the identification of suitable solvents for the preparation process. Other preparation methods that yield moderate and high molecular weight PBIs are described in the section "Processing Parameters" in this article.

Forms. Polybenzimidazole is commercially available in various forms: prepolymer, dry powder, solution, and molded parts.

Markets and Applications

Initially the primary application of PBI adhesive was for the bonding of various adherends. The test results of PBI used in this capacity are described in the section "Properties" in this article. PBI adhesives were also evaluated for bonding face sheet to honeycomb core and for use at elevated and cryogenic temperatures. The results of both investigations are also described in the "Properties" section. The future use of PBI as an adhesive will require applications in specialized areas where performance is critical. Developmental work to

address processing and cost issues is ongoing, although the cost of PBI is likely to be equivalent to or higher than that of other high-performance specialty adhesives.

Currently, the single largest use of PBI (in fiber form) is in flight suits, heat-protective apparel (that is, garments for fire fighters and race car drivers), and fire-resistant aircraft seats. PBI fibers have an excellent combination of properties, including nonflammability, high moisture recovery comfort, and textile processibility. Other forms (nonfiber) have been demonstrated in specialty applications, such as battery separators, printed circuit boards, fabricated structures, foams, and membranes and in separation technology (that is, microporous beads used to separate industrial and biological materials).

Major Suppliers

Suppliers currently include the PBI Products Division of the Hoechst Celanese Corporation and the Aerotherm Division of Acurex Corporation. The first PBI formulations on the market (late 1960s) were commercialized by the Whittaker Corporation, Narmco Materials Division.

Cost Factors. One PBI adhesive formulation is a prepolymer that sold in dry powder form for approximately $770/kg ($350/lb) in 1990.

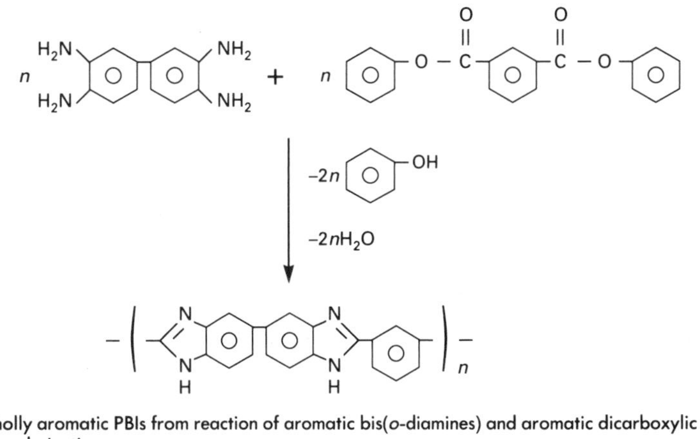

Fig. 1 Wholly aromatic PBIs from reaction of aromatic bis(o-diamines) and aromatic dicarboxylic acids or phenyl ester derivatives

Table 1 Titanium-to-titanium tensile shear strengths of PBI adhesive

Test condition(a)	Tensile shear strength(b)	
	MPa	ksi
At 23 °C (73 °F)	19.3	2.80
At 260 °C (500 °F)	18.9	2.75
At 371 °C (700 °F)	13.6	1.98
At 426 °C (800 °F)	10.3	1.50
At 260 °C (500 °F) after 100 h at 315 °C (600 °F)(c)	5.2	0.75
At 260 °C (500 °F) after 200 h at 315 °C (600 °F)(c)	3.4	0.50

(a) After 10 min residence at test temperature. (b) Bonding conditions: 1 h at 221 °C (430 °F), 1 h at 315 °C (600 °F) under 1.4 MPa (200 psi) load. Stabilized with unidentified inorganic arsenic compound. (c) Isothermal aging performed in air. Source: Ref 13

Table 2 Effect of substrate on tensile shear strength of PBI adhesive

Test condition(a)	Substrate	Tensile shear strength	
		MPa	ksi
At 23 °C (73 °F) after 1 h at 371 °C (700 °F)	Aluminum	9.6	1.40
At 23 °C (73 °F) after 1 h at 371 °C (700 °F)	17-7 PH stainless steel, annealed	13.1	1.90
At 23 °C (73 °F) after 5 h at 371 °C (700 °F)	Aluminum	10.3	1.50
At 23 °C (73 °F) after 5 h at 371 °C (700 °F)	17-7 PH stainless steel, annealed	5.5	0.80
At 23 °C (73 °F) after 10 h at 371 °C (700 °F)	Aluminum	8.9	1.30
At 23 °C (73 °F) after 10 h at 371 °C (700 °F)	17-7 PH stainless steel, annealed	0	0

PH, precipitation hardening. (a) Isothermal aging performed in air. Source: Ref 14

Table 3 Titanium-to-titanium tensile shear strengths of PBI adhesive

Test condition	Postcure(a)	Tensile shear strength(b)(c)	
		MPa	ksi
At 23 °C (73 °F)	Yes	16.9	2.45
At 23 °C (73 °F)	No	20.3	2.95
At 288 °C (550 °F) after 1 h at 288 °C (550 °F)	Yes	10.3	1.50
At 288 °C (550 °F) after 1 h at 288 °C (550 °F)	No	21.1	3.07
At 288 °C (550 °F) after 200 h at 288 °C (550 °F)	Yes	0.82	0.12
At 288 °C (550 °F) after 200 h at 288 °C (550 °F)	No	5.13	0.75

(a) PBI adhesive with 100 phr MD-105 aluminum powder subjected to postcure conditions of 1 h at 399 °C (750 °F) under vacuum. (b) Bonding conditions: 1 h each at 288 °C (550 °F) and 315 °C (600 °F) under 1.4 MPa (200 psi) of pressure. (c) AMS 4910 titanium, sandblasted; 112 E-glass, heat cleaned. Source: Ref 14

Properties

Polybenzimidazoles have good thermal stability, as measured by thermogravimetric analysis, and retention of mechanical properties at elevated temperatures. However, when subjected to isothermal aging at 371 °C (190 °F) in flowing air, a significant weight loss occurs after approximately 100 h. Polymers such as polyimide, polybenzoxazole, and polyquinoxaline retain approximately 80%, 65%, and 55% of their initial weights, respectively, after aging under these conditions (Ref 11). However, PBI does perform very well at elevated temperatures for short periods. In addition, antioxidants such as arsenic thioarsenate can significantly improve the isothermal aging performance of PBI, although the use of arsenic thioarsenate does raise a toxicity issue.

Unless otherwise stated, the adhesive properties described below are those obtained from the prepolymer of polybenzimidazole, the chemical structure of which is shown in Fig. 1. As already mentioned, early PBI adhesive applications were mainly the joining of various adherends. In 1965, an abundance of data on PBI adhesive properties became available (Ref 12-14). Using 17-7 PH steel adherends (acid etched), lap-shear specimens bonded at 1380, 690, and 69 kPa (200, 100 and 10 psi) showed no difference in strengths (Ref 11). The tensile shear strengths were about 24.1 MPa (3.5 ksi) at room temperature and about 15.8 MPa (2.3 ksi) at 371 °C (700 °F). The specimens were bonded by heating to 220 °C (430 °F) and holding for 1 h, then heated at 315 °C (600 °F) for 1 h, and postcured at 398 °C (750 °F) for 1 h under nitrogen. Titanium adherends (phosphate-fluoride treated) bonded under 690 kPa (10 psi) pressure using the time-temperature cycle described above gave tensile shear strengths of 11.9 MPa (1.73 ksi) at room temperature and 9.5 MPa (1.38 ksi) at 371 °C (700 °F). After isothermal aging at 371 °C (700 °F) for 50 h and testing at 371 °C (700 °F), it was found that essentially all strength was lost, regard-

less of the adherend. This was presumably due to adhesive degradation.

The work reported in Ref 13 is summarized in Table 1. The PBI adhesive shows good retention of strength up to 426 °C (800 °F). However, isothermal aging at 315 °C (600 °F) results in a significant loss of tensile shear strength. As noted in Table 1, these data were obtained on PBI stabilized with an unspecified amount of an unidentified inorganic arsenic compound. Elsewhere in the literature were reported tensile shear strengths as high as 26.6 MPa (3.86 ksi) when tested at room temperature on 17-7 PH annealed stainless steel (phosphate etched) (Ref 14). The PBI contained 100 parts per hundred (phr) of MD-105 aluminum powder. The bonding cure cycle consisted of 1 h at 343 °C (650 °F) under 1380 kPa (200 psi) pressure with a postcure of 6 h at 399 °C (750 °F) under vacuum. The effect of the substrate on tensile shear strength during isothermal aging is shown in Table 2. Although the PBI-bonding of aluminum adherends provided good strengths after aging 10 h at 371 °C (700 °F), the adhesive bonded to stainless steel lost all strength. Additional work showed that vapor-deposited aluminum on stainless steel adherends performed significantly better during isothermal aging than did uncoated stainless steel adherends (Ref 14).

It had been shown that inorganic arsenic compounds served to extend the useful lifetime of certain epoxies during isothermal aging. This approach was extended to PBI, and it was determined that 20% by weight of arsenic thioarsenate significantly improved the tensile shear strengths of PBI after isothermal aging (Ref 14). The bonding of titanium using PBI, as well as other adhesives, was not as successful as when other substrates, such as steel and aluminum,

were used. As a result, a variety of surface cleansers and treatments were evaluated. Some early results of titanium-to-titanium lap-shear strengths are presented in Table 3.

PBI adhesive formulations were also evaluated for bonding face sheet to honeycomb core. In this application, much larger bond areas are required than is the case for lap-shear specimens. Some results from sandwich column compression tests are presented in Table 4. The PBI adhesive was used to bond 17-7 PH stainless steel core to RH 950 stainless steel face sheet. The results of a test using an epoxy-phenolic adhesive are included for comparative purposes. The PBI-bonded sandwich showed good retention of skin strength at elevated temperatures. All specimens exhibited a buckling failure mode. The isothermal aging of PBI-bonded sandwich structures at 315 °C (600 °F) showed good retention of unaged skin strength up to about 50 h, after which adhesive failures began to predominate. The flatwise tensile strengths of the above-mentioned stainless steel sandwich configurations are presented in Table 5. Data from epoxy-phenolic adhesive are included for comparative purposes.

The mechanical behavior of PBI adhesive at cryogenic temperatures was also of interest for certain applications. Data for tensile shear strengths on aluminum and stainless steel substrates tested below 0 °C (32 °F) are presented in Table 6. Although Ref 16 presents PBI adhesive property data too voluminous to describe fully, the individually performed tests are briefly mentioned below.

In an extensive evaluation, fatigue, creep, environmental, T-peel, sandwich-peel, sandwich flatwise tension, panel shear, sandwich beam flexure, and sandwich edgewise compression tests were per-

Table 4 Sandwich column compression test for PBI and epoxy-phenolic adhesives

| Test condition(a) | PBI skin stress(b) | | Epoxy-phenolic skin stress | |
	GPa	10^6 psi	GPa	10^6 psi
At 23 °C (73 °F)...............	1.02	0.149	1.08	0.157
At 185 °C (365 °F)...............	1.33	0.192	0.94	0.136
At 260 °C (500 °F)...............	0.99	0.143	0.88	0.128
At 315 °C (600 °F)...............	1.16	0.169	0.89	0.129
At 371 °C (700 °F)...............	0.84	0.122	...	...
At 426 °C (800 °F)...............	0.59	0.854	...	...
At 482 °C (900 °F)...............	0.58	0.848	...	...
At 537 °C (1000 °F)...............	0.37	0.534	...	...

(a) Residence time of 10 min at test temperature. (b) Bonding conditions: 1 h each at 149, 204, 260, and 315 °C (300, 400, 500, and 600 °F) under 170 kPa (25 psi) of pressure. Source: Ref 14

Table 5 Flatwise tensile strength test for PBI and epoxy-phenolic adhesives

| Test condition(a) | PBI stress(b) | | Epoxy-phenolic stress | |
	MPa	ksi	MPa	ksi
At 23 °C (73 °F)	3.1	0.451	4.4	0.645
At 185 °C (365 °F)	3.3	0.482	2.5	0.375
At 260 °C (500 °F)	3.2	0.465	2.1	0.310
At 315 °C (600 °F)	3.4	0.501	1.7	0.256

(a) Residence time of 10 min at test temperature. (b) Bonding conditions: 1 h each at 149, 204, 260, and 315 °C (300, 400, 500, and 600 °F) under 170 kPa (25 psi) of pressure. Source: Ref 15

Table 6 Cryogenic tensile shear strengths for a PBI adhesive

| Test condition(a) | Substrate | Tensile shear strength | |
		MPa	ksi
At −240 °C (−400 °F)...............	15-7 PH Mo steel	31.7	4.60
At −240 °C (−400 °F)...............	2219-T81 aluminum	17.2	2.50
At −185 °C (−301 °F)...............	15-7 PH Mo steel	28.9	4.20
At −185 °C (−301 °F)...............	2219 T-81 aluminum	15.5	2.25

PH, precipitation hardening; Mo, molybdenum. (a) Residence time of 10 min at test temperature. Source: Ref 16

Table 7 Adhesive properties of PBI

| Test condition(a) | T-peel strength(b) | | Climbing drum peel strength(b) | |
	N/m	lbf/in.	kN/0.076 m	lbf/3 in.
At −240 °C (−400 °F)...............	770	4.4	2.3	12.6
At −185 °C (−301 °F)...............	718	4.1	...	...
At 23 °C (73 °F)...............	1225	7.0	1.5	8.5
At 149 °C (300 °F)...............	700	4.0	...	...
At 260 °C (500 °F)...............	858	4.9	3.0	16.9
At 371 °C (700 °F)...............	875	5.0	3.1	17.8
At 426 °C (800 °F)...............	700	4.0	...	...
At 482 °C (900 °F)...............	525	3.0	2.3	13.0

(a) Residence time of 10 min at test temperature. (b) Imidite 850 adhesive and 15-7 PH Mo stainless steel. Source: Ref 16

Table 8 Adhesive properties of poly(arylene ether imidazole)

| Test temperature(a) | Titanium-to-titanium tensile shear strength(b) | |
	MPa	ksi
At 23 °C (73 °F)...............	33.2	4.81
At 93 °C (200 °F)...............	26.2	3.80
At 177 °C (350 °F)...............	25.5	3.70
At 200 °C (390 °F)...............	21.0	3.05
At 23 °C (73 °F) after 500 h at 200 °C (390 °F)(c)...............	26.5	3.85
At 93 °C (200 °F) after 500 h at 200 °C (390 °F)(c)...............	25.9	3.75
At 177 °C (350 °F) after 500 h at 200 °C (390 °F)(c)...............	25.2	3.65
At 200 °C (390 °F) after 500 h at 200 °C (390 °F)(c)...............	23.9	3.47

(a) Polymer glass transition temperature was 243 °C (470 °F). (b) Bonding conditions: 1 h at 300 °C (570 °F) under 1380 kPa (200 psi) of pressure. PasaJell 107 surface treatment was used; failures were predominantly cohesive. (c) Isothermal aging performed in flowing air. Source: Ref 17

formed. The T-peel and climbing drum peel test results are presented in Table 7. The PBI adhesive was bonded to 15-7 PH Mo stainless steel under 1380 kPa (200 psi) of pressure followed by a 1 h postcure under nitrogen at 399 °C (750 °F). Flatwise tensile and panel shear tests of PBI bonded 15-7 PH Mo stainless steel at room temperature gave an ultimate strength of approximately 3450 kPa (500 psi) and a shear stress of approximately 2415 kPa (350 psi), respectively. The researchers concluded that the PBI adhesive exhibited good thermal stability for long exposures at moderate temperatures. However, long exposures at temperatures greater than 260 °C (500 °F) resulted in poor retention of unaged strengths. The adhesive provided good bond strengths up to 260 °C (500 °F) in most of the tests and offered very short-term strengths at temperatures greater than 315 °C (600 °F). In addition, no creep was observed during any of the tests.

The poly(arylene ether imidazoles) of the type shown in Fig. 2 are experimental materials that have shown good adhesive properties in preliminary evaluation (Ref 17). Representative titanium-to-titanium tensile shear strengths are presented in Table 8. The specimens were bonded by heating to 300 °C (570 °F) under 1380 kPa (200 psi) of pressure and held for 1 h. Although the glass transition temperature of the material is 243 °C (470 °F), the retention of tensile shear strength at 200 °C (390 °F) is excellent. In addition, after isothermal aging at 200 °C (390 °F) for 500 h in air, the retention of unaged strengths is very good. It should be noted that because isothermal aging is carried out on unstressed specimens, creep is not a factor. However, the isothermal aging of specimens under stress provides a more realistic test and could provide significantly different results.

Product Design Considerations

High-performance/high-temperature adhesives are required in many different applications in advanced aircraft, space vehicles, satellites, missiles, electronics, oil recovery systems, automotive industries, and domestic households. They can be used to join plastics, metals, ceramics, composites, and combinations thereof to each other or to like material. Other high-performance adhesives, used as insulators and coatings, are also needed where adherence to the substrate, rather than the joining of two components, is of prime importance.

There are many uses for advanced polymeric adhesives, and each application has a unique set of requirements. For example, a high-temperature adhesive for missile applications must perform for short periods of time, usually seconds, at very high temperatures (~760 °C, or 1400 °F). For advanced

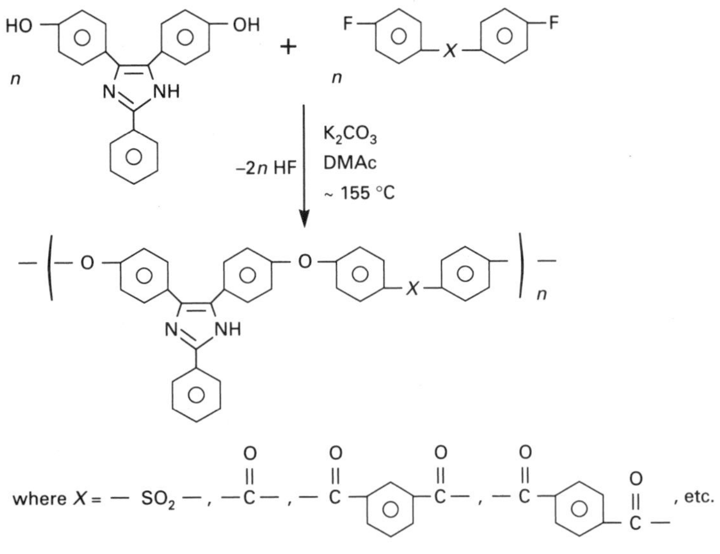

Fig. 2 Poly(arylene ether imidazoles). DMAc, N,N-dimethylacetamide. Source: Ref 17

aircraft applications, the adhesive must perform satisfactorily for tens of thousands of hours over the temperature range of −54 to 260 °C (−65 to 500 °F). In addition, the adhesive must resist the degradative effects of moisture, aircraft fuels, paint strippers, deicing fluids, and hydraulic fluids.

Another unique high-performance adhesive application is on orbital vehicles such as satellites and instrument platforms, where thermal cycling over the temperature range of −160 to 125 °C (−255 to 255 °F) may prevail. In this application, the adhesive must resist certain types of ionizing radiation, such as radiation from protons, electrons, and ultraviolet rays. Furthermore, it is desirable to have a good match between the coefficient of thermal expansion of the adhesive and that of the substrate to help prevent microcracking and debonding due to the effect of thermal cycling.

High-performance adhesives are also needed in the microelectronics industry. In addition to being stable up to 400 °C (750 °F) in an inert atmosphere, they must meet other very stringent requirements, including a low coefficient of thermal expansion, low moisture absorption, high strength and toughness, and high resistivity and stability in an electric field.

Another example in which high adhesion to the substrate is important is the nonstick interior and decorative exterior coatings on household cookware. The coating must adhere well to aluminum, be thermally stable, and resist hot oils, grease, and soaps. High-temperature adhesives are needed to bond polymeric films to copper foil or to the same kind of film and to bond high-performance composites to like material. A significant amount of research and development is currently under way to address the various needs and requirements for high-temperature adhesives.

High-performance/high-temperature adhesives are developed and formulated according to specific application. An adhesive used to join a honeycomb structure to face sheets has flow requirements that are different from one designed for metal-to-metal bonding. In honeycomb structures, the adhesive must flow sufficiently to form a fillet around the honeycomb cell under relatively mild pressures (such as 345 to 1380 kPa, or 50 to 200 psi, depending on the density of the honeycomb core). Adhesives with less flow can be used to join metal to metal because higher pressures can be tolerated. Various applications place different types of stress on the adhesive bond. Sandwich or metal-to-metal bonded structures must exhibit both shear and peel strengths, while an adhesive used to join brittle ceramic components is often designed to accommodate shear stresses.

The ideal high-temperature adhesive would have the following properties. As a tape, it would have tack and drape at room temperature, processibility under moderate conditions with no volatile evolution, compatibility with different adherends and surface treatments, low coefficient of thermal expansion, adequate mechanical properties under intended use conditions, high reproducibility in the fabrication of bonded components, repairability, and low cost. No such material exists today.

Processing Parameters

The commercial preparation of PBI is generally carried out in the melt (Ref 10). Polyphosphoric acid (Ref 18) and sulfolane or diphenylsulfone methods (Ref 19) are also used. The polyphosphoric acid method provides a homogeneous solution and offers the advantage of a moderate reaction temperature (~180 °C, or 355 °F). The use of inert molten heat transfer media such as sulfolane or diphenylsulfone also provides homogeneous mixtures and moderate reaction temperatures. Both methods are excellent for laboratory-scale preparations of high molecular weight PBI. However, there are disadvantages: Only low solid content (~5%) is attainable, and multistep isolation procedures must be used (that is, precipitation, washing, neutralizing, and solvent extraction).

Melt polymerization is the most widely used method for large-scale PBI preparation. In a representative procedure for the preparation of PBI (Ref 20), the high-purity monomers, such as 3,3′,4,4′-tetraaminobiphenyl and diphenyl isophthalate, are placed in a three-neck flask equipped with a mechanical stirrer, thermometer, moisture and/or phenol trap, condenser, and nitrogen inlet. All traces of oxygen are removed by alternate vacuum and purge cycles with purified nitrogen. A constant flow of nitrogen is maintained throughout the reaction. The mixture is heated with an oil bath, and the solids melt at about 150 °C (300 °F). Water and phenol begin to evolve, and the temperature is gradually increased to about 250 °C (480 °F). At this point the mixture is very viscous, and stirring becomes very difficult. Stirring is stopped, and the temperature is increased to about 290 °C (555 °F) and held for approximately 2 to 3 h before cooling. The foamy material, which has an inherent viscosity of 0.2 to 0.3 dL/g is removed and ground to a fine powder. The powder is heated in a nitrogen atmosphere to about 375 °C (705 °F), held for 3 h, and then cooled. A continuous flow of nitrogen is necessary to prevent oxidation and helps remove the phenol and water that evolves. The final polymer is a yellow-brown amorphous solid with an inherent viscosity of approximately 0.75 dL/g ($M_n \sim 18\,000$ to 20 000 g/mol). The procedure for producing adhesives is similar, except that the lower molecular weight prepolymer is the final product (that is, the reaction is stopped at an earlier stage).

Other less significant routes to the preparation of PBI are available. The alkoxide-catalyzed addition of bis(o-diamines) to aromatic dinitriles produces PBI of low molecular weight (Ref 21). A more successful means of yielding PBI involves the reaction of bis(o-diamines) with the bis(bisulfite) adduct of isophthalaldehyde (Ref 22). This method offers mild conditions and short reaction times and yields a polymer with moderate inherent viscosities (0.3 to 0.5 dL/g). A recently developed method for preparing PBI involves the reaction of aromatic bis(4-hydroxyphenylbenzimidazoles) with activated aromatic difluoro compounds, as shown in Fig. 3 (Ref 23). This

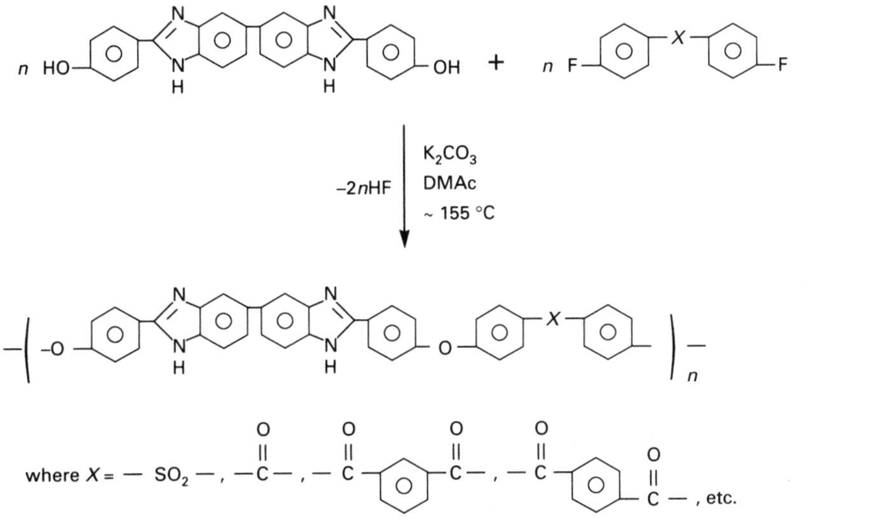

Fig. 3 Reaction of aromatic bis(4-hydroxyphenylbenzimidazoles) with activated aromatic difluoro compounds. Source: Ref 23

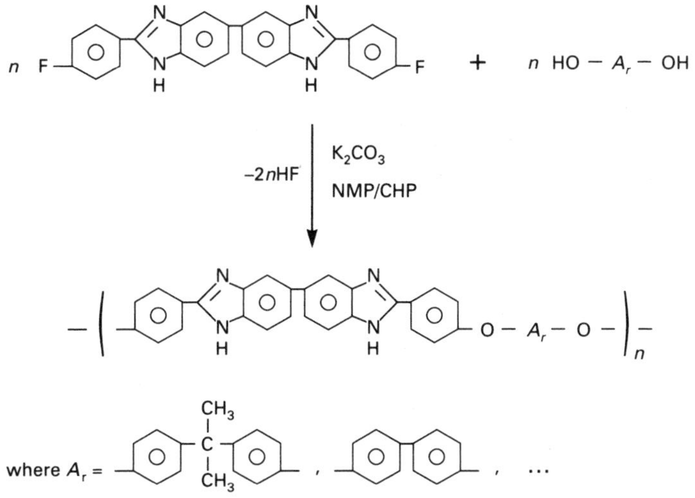

Fig. 4 Nucleophilic displacement based on bis(4-fluorophenylbenzimidazole) prepared from aromatic bis(o-diamines) and phenyl-4-fluorobenzoate. NMP/CHP, N-methyl pyrollidone/N-cyclohexyl-2-pyrollidone. Source: Ref 24

approach uses monomers in which the benzimidazole ring is already formed. The bis(phenolbenzimidazoles) are prepared by the reaction of aromatic bis(o-diamines) with phenyl-4-hydroxybenzoate. This method has yielded PBI of moderate molecular weight. Another procedure for achieving PBI, one involving nucleophilic displacement, is shown in Fig. 4 (Ref 24). The bis(4-fluorophenylbenzimidazole) is prepared from aromatic bis(o-diamines) and phenyl-4-fluorobenzoate. The fluoro groups are sufficiently activated for displacement to occur, yielding PBI of moderate molecular weight.

Aromatic polymers containing imidazole rings were first reported from the reaction of bis(phenyl-α-diketones) and dialdehydes in the presence of ammonia (Ref 25). However, this method produced polymers of low molecular weight due to side reactions. A more recent approach, which has produced a high molecular weight polymer, involves reacting bis(4-hydroxyphenyl)imidazoles with activated aromatic difluoro compounds in the presence of potassium carbonate (Fig. 2). The 4,5-bis(4-hydroxyphenyl)-2-phenylimidazole is prepared from 4,4'-dihydroxybenzil and benzaldehyde in the presence of ammonium acetate. This synthesis of poly(arylene ether imidazoles) has produced high molecular weight polymers, some of which have performed well as adhesives in preliminary evaluation.

REFERENCES

1. J.I. Jones, *Macromol. Sci. Rev. Macromol. Chem.*, Vol C2 (No. 2), 1968, p 343
2. A.H. Frazer, *High Temperature Resistant Polymers*, Interscience, 1968, p 138
3. V.V. Korshak, *Heat Resistant Polymers*, Keter Press, 1971, p 244 (in Hebrew)
4. H.H. Levine, *Encycl. Polym. Sci. Tech.*, Vol 11, 1969, p 188
5. A. Buckley, D.E. Stuetz, and G.A. Serad, *Encycl. Polym. Sci. Tech.*, Vol 11, 2nd ed., 1988, p 572
6. P.E. Cassidy, *Thermally Stable Polymers*, Marcel Dekker, 1980, p 163
7. E.J. Powers and G.A. Serad, in *High Performance Polymers: Their Origin and Development*, R.B. Seymour and G.S. Kirshenbaum, Ed., 1986, p 355
8. B.C. Ward, *SAMPE J.*, Vol 25 (No. 2), 1989, p 21
9. K.C. Brinker and I.M. Robinson, U.S. Patent 2,895,948, 1959
10. H.A. Vogel and C.S. Marvel, *J. Polym. Sci.*, Vol 50, 1961, p 511
11. P.M. Hergenrother, *SAMPE Q.*, Vol 3 (No. 1), 1971, p 1
12. J.R. Hill, *Soc. Aerosp. Mater. Proc. Eng. Ser.*, Vol 9, 1965, p V-2
13. R.B. Krieger, Jr. and R.E. Politi, *Soc. Aerosp. Mater. Proc. Eng. Ser.*, Vol 9, 1965, p V-3
14. H.H. Levine, *Soc. Aerosp. Mater. Proc. Eng. Ser.*, Vol 9, 1965, p V-1; S. Litvak, *Adhes. Age*, Vol 11 (No. 1), 1968, p 17; and Vol 11 (No. 2), 1968, p 24
15. S. Litvak, *Appl. Polym. Symp.*, Vol 3, 1966, p 279
16. T.J. Reinhart, Jr. and R. Hidde, *Appl. Polym. Symp.*, Vol 3, 1966, p 299
17. J.W. Connell and P.M. Hergenrother, *Polym. Mater. Sci. Eng. Proc.*, Vol 60, 1989, p 527
18. Y. Iwakura, K. Uno, and Y. Imai, *J. Polym. Sci.*, Vol 12, 1964, p 2605
19. F.L. Hedberg and C.S. Marvel, *J. Polym. Sci.*, Vol 12, 1974, p 1823
20. A.B. Conciatori and E.C. Chenevey, *Macromol. Syn.*, Vol 3, 1968, p 24
21. D.I. Packham, J.D. Davies, and H.M. Paisley, *Polymer*, Vol 10 (No. 12), 1969, p 923
22. J. Higgins and C.S. Marvel, *J. Polym. Sci., Part A-1*, Vol 8 (No. 1), 1970, p 171
23. P.M. Hergenrother, M.W. Beltz, and J.W. Connell, unpublished research, April 1989
24. J.L. Hedrick and J.W. Labadie, private communication, IBM Almaden Research Center, San Jose, CA, April 1989
25. V.B. Krieg and G. Manecke, *Makromol. Chem. Mater.*, Vol 108, 1967, p 210

SELECTED REFERENCES

- J.R. Hall and D.W. Levi, "Polybenzimidazoles: A Review," U.S. Army Plastics Technical Evaluation Center, 1966, p 1-48
- S. Litvak, "PBI Structural Adhesives," AFML-TR-65-426, Air Force Materials Laboratory, March 1966

- R. Reed and R. Hidde, *J. Polym. Sci., Part A-1*, 1963, p 1531
- W.D. Roper, "Spacecraft Adhesives for Long Life and Extreme Environments," Jet Propulsion Laboratory, Technical Report 32-1537, National Aeronautics and Space Administration, Aug 1971
- R.W. Vaughan and C.H. Sheppard, "Cryogenic/High Temperature Structural Adhesives," Contract Report NASA-16780, National Aeronautics and Space Administration, Jan 1974
- R.W. Vaughan and C.H. Sheppard, *Soc. Adv. Mater. Proc. Eng. Ser.*, Vol 19, 1974, p 7
- S.Y. Yoshino, M.A. Nadler, and D.H. Richter, *Int. Symp. Space Technol. Sci.*, Vol 7, 1967, p 199

Adhesive Modifiers

William O. Buckley and Kevin J. Schroeder, Ciba-Geigy Corporation

ADHESIVE RESINS and curing agents represent the basic chemistry and building blocks for the formulation of adhesives. However, these materials alone cannot satisfy the wide-ranging adhesive performance, processing, and economic profiles demanded in today's industrial marketplace. Therefore, in practice, adhesive formulators take advantage of many additional materials to modify adhesive properties in order to meet the specific end-use requirements of their customers. These adhesive modifiers are the topic of this article.

The first area of discussion is fillers and their major benefits and applications in adhesive technology. This is followed by a discussion of the considerations necessary for successful exploitation of these benefits.

Next, liquid materials that are used for improving adhesive properties are addressed. The typical materials used in formulated adhesives are identified, and their effects on final properties are illustrated.

Fillers for Adhesives

Fillers are an important class of materials employed for the modification of adhesive

properties. They typically represent the second highest percentage ingredient by weight in the formulated adhesive, after the base adhesive resin. Fillers are used for economic as well as performance reasons. The addition of fillers can lower the cost of the final adhesive product. Improvements in adhesive processing properties and end-use performance requirements are also realized by the incorporation of fillers into the resin component. Reduction of certain properties can also occur. These conditions are discussed in the section "Filler Considerations" in this article.

Typical properties of adhesive fillers are given in Table 1. Fillers, categorized by the specific contribution they provide to adhesive properties, are discussed in the following section. Only the most common adhesive fillers are discussed. More detailed information, including properties, suppliers, and costs, can be found in the references cited.

Benefits and Applications

Extenders and Dimensional Stabilizers. The major function of a filler in an adhesive is as an extender to reduce the raw material

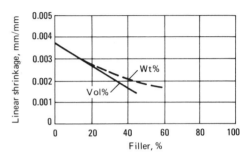

Fig. 1 Effect of filler on adhesive shrinkage during cure. Source: Ref 2

cost of the formulated product. Fillers used for this purpose are less expensive than the base adhesive resins typically employed and provide cost savings by replacing a portion of the resin. Additionally, the filler aids in reducing shrinkage during cure and lowers the thermal expansion coefficient from that of the resin alone. These benefits are shown in Fig. 1 and 2.

Calcium carbonate is the most widely used extender (Ref 4). Availability, as well as a balance of cost and performance properties, are the key reasons for this wide

Table 1 Selected properties of fillers

Property	Calcium carbonate	Kaolin	Dehydrated kaolin	Talc	Mica	Feldspar, nepheline
$K_E, \eta_{sp}/V_f$	4–5	6–10	6–10	6–10	6–10	4–6
Thermal conductivity, W/m · K	2.3	2.0	2.1	2.1	2.5	2.3
Specific heat, J/g · K (cal/g · °C)	0.860 (0.205)	0.90 (0.22)	0.84 (0.2)	0.87 (0.208)	0.86 (0.206)	0.88 (0.21)
Coefficient of thermal expansion, 10^{-6}/K	10	8	3–6	8	8	6.5
Young's modulus, GPa (10^6 psi)	255 (37)	196 (28)	196 (28)	196 (28)	294 (43)	294 (43)
Poisson's ratio	0.27	0.3	0.2	0.3	0.3	0.3
Hardness, Mohs	2.5–3	2	6–7	1	2.5–3	6
Absolute relative force	13	8	80	5	15	65
Dielectric constant	6.14	2.6	1.3	5.5–7.5	2–2.6	6
Density, g/cm³	2.71	2.58	2.5–2.63	2.8	2.82	2.6

Property	Wollastonite	Glass spheres	Glass balloons	Nutshell flour	Wood flour	Hydrous alumina	Asbestos	Coal	Barite	Silica
$K_E, \eta_{sp}/V_f$	4–6	2.5	2.5	5–6	6–10	6–10	10+	4–6	4–5	4–5
Thermal conductivity, W/m · K	2.5	0.7	0.008	0.6	0.3	0.08	2.1	0.25–0.33	2.5	2.9
Specific heat, J/g · K (cal/g · °C)	1.0 (0.24)	1.1 (0.27)	1.0 (0.24)	1.8 (0.43)	1.8 (0.42)	0.8 (0.19)	1.1 (0.26)	1.3 (0.3)	0.45 (0.11)	0.8 (0.19)
Coefficient of thermal expansion, 10^{-6}/K	6.5	8.6	8.8	5–50	5–50	4–5	0.3	5	10	10
Young's modulus, GPa (10^6 psi)	294 (43)	588 (85)	2 (0.28)	78 (0.01)	98 (14)	294 (43)	1420 (206)	196 (28)	294 (43)	294 (43)
Poisson's ratio	0.3	0.23	0.4	0.35	0.3	0.3	0.2	0.3	0.3	0.3
Hardness, Mohs	5–5.5	5	5	2	2	2	3–5	3	3–3.5	6.5–7
Absolute relative force	50	40	40	8	8	8	15–40	15	20	80
Dielectric constant	6	5	1.5	5	5	7	10	3	7.3	4.3
Density, g/cm³	2.9	2.48	0.15–0.3	1.3–1.35	0.5–0.7	2.4–2.42	2.4–2.6	1.47	4.4–4.5	2.65

Source: Ref 1

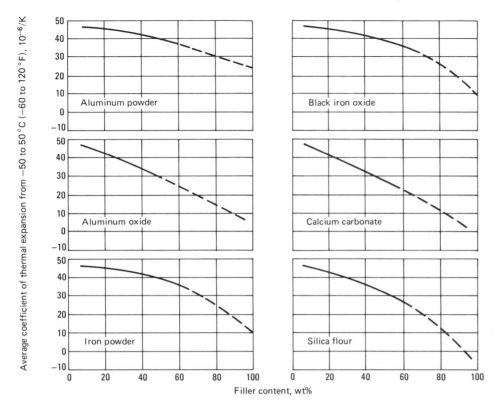

Fig. 2 Effect of various fillers on adhesive coefficient of thermal expansion. Source: Ref 3

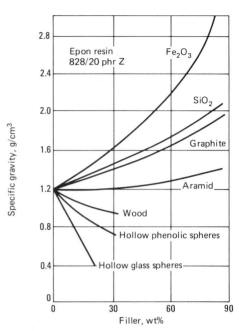

Fig. 3 Effect of fillers on adhesive specific gravity. Source: Ref 3

usage. This material is an abundant mineral that is mined from ore deposits throughout the world. Common forms include limestone, marble, calcite, chalk, and dolomite. It is also provided in various grades, manufactured by precipitation processes (Ref 1), and in many different particle sizes. To improve dispersion into the resin component, the filler is often coated with calcium stearate or stearic acid.

Silica is also often used as an extender in adhesive compositions. It, too, is an abundant mineral, found in a crystalline form as quartz and in an amorphous form as diatomaceous silica.

The crystalline structure of quartz varies, depending on the geographical region of the ore deposit. The quartz form of silica is very hard, which is a disadvantage to machinability. A more recent concern, over the respiratory problems associated with the inhalation of crystalline silica dust, has led to a greater use of the amorphous forms.

Diatomaceous silica deposits are the skeletal remains of prehistoric diatoms. This diatomite is soft and chalky and not nearly as abrasive as quartz. It is quite porous and has a higher surface area and oil adsorption than crystalline silica. Calcination reduces the surface area and moisture adsorption of diatomite, while retaining high oil adsorption (Ref 2).

Silicas in general are very acid resistant, except to hydrofluoric acid, but are not very resistant to strong bases. One benefit is that their surfaces can be modified by organofunctional silanes and titanates to improve compatibility with the resin matrix (Ref 5). In addition to the cost reduction advantage, silica provides improved dimensional stability and good electrical insulation properties in the cured adhesive (Ref 1).

Several types of amorphous silica are made synthetically. These types are differentiated by manufacturing process. Silica gels and precipitated silicas are manufactured by liquid processes. Fused silica is prepared by the high-temperature fusion of quartz. Fumed and electric arc silicas are a result of pyrogenic reactions (Ref 1).

Synthetic silicas also find application in adhesive formulations as extenders and for reinforcement. The effectiveness of each type is determined by the key properties of surface area, oil adsorption, moisture adsorption, particle size, porosity, and surface chemistry (Ref 1). Fumed silicas find their greatest benefit in rheology modification, which is discussed in the section "Rheology Control" in this article.

Kaolin, commonly called clay, is another naturally occurring mineral that is used as an extender in adhesives. Chemically, kaolin is hydrated aluminum silicate with a hexagonal platelike crystal structure. After being mined, kaolin is further processed by air-float, water-wash, or calcining operations. In addition to its function as a low-cost extender, it provides shrinkage con-

trol, dimensional stability, viscosity control, and pigmentation (Ref 6).

Feldspar and nepheline are minerals obtained from granite and similar formations. Both are anhydrous, alkali aluminosilicates. Their primary advantages in thermoset resins include improved chemical resistance and dimensional stability (Ref 1).

Wood flour, shell flour, and other cellulosic fillers are also used as adhesives extenders. Carbon black has also been used, but its reinforcement, electrical, and coloring properties are more often the reasons for its employment.

Hollow spheres are another class of fillers that have been utilized effectively to lower adhesive cost. In addition to reducing density, hollow spheres provide improved compressive and impact strengths because of a more uniform stress distribution than is found with particulate fillers. This class of low-density polymer formulators is referred to as syntactics. Low-density syntactic adhesives have found extensive use in the aerospace industry (Ref 7).

Hollow glass spheres or phenolic spheres are most commonly employed. The effect of these spheres and other fillers on specific gravity is shown in Fig. 3.

Solid microspheres of glass, ceramic, or other materials are also used, but without the same benefit of density reduction that attained with hollow spheres. The microspheres are also available with different surface treatments to improve adhesion with the resin.

Reinforcement. Improvements in the physical properties of adhesives, such as strength, modulus, impact resistance, and heat distortion temperature, have been re-

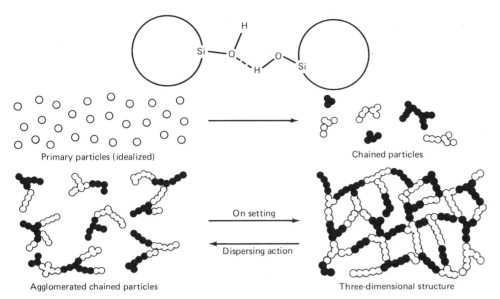

Fig. 4 Interaction between fumed-silica particles. Source: Ref 9

alized by the incorporation of reinforcing fillers. The reinforcement of adhesives has typically been accomplished through the use of miscellaneous flake and fibrous minerals and fillers. Various natural and synthetic fibers have also been used but to a limited extent because of their high contribution to adhesive viscosity. Particle and fiber packing concepts have been applied in an attempt to overcome this limitation (Ref 8).

Until recently, asbestos, or chrysotile, has been one of the most widely used reinforcing fillers in the adhesive industry because of its low cost and excellent reinforcement potential. However, this use has been severely hampered because of health effects associated with improper handling. Asbestosis and bronchial cancer can result upon overexposure. Therefore, other materials have been substituted, often at a sacrifice of cost and performance benefits.

Asbestos owes its excellent reinforcement properties to the acicular (needlelike) nature of its particles. It has been proposed that the reinforcing mechanism results from internal friction at the resin-asbestos interface (Ref 5). Asbestos also provides excellent environmental, chemical, and temperature resistance.

One material that has found use as an asbestos replacement is wollastonite, which is calcium metasilicate. It, also, is a fibrous mineral filler but has a lower aspect ratio than does asbestos. It is available in various particle sizes and aspect ratios. Surface-treated versions are also available to improve coupling with the matrix resins. Its additional benefits include low moisture adsorption and high loading capability (Ref 5). In addition, studies to date show no evidence that it causes lung disease (Ref 1).

Talc (hydrated magnesium silicate) is another mineral used for reinforcement. It owes its reinforcement properties to the platelike structure of its particles. Fibrous talcs are actually grades that contain asbestos fibers, along with the plate-type particles. In this form it poses the same health concerns as does asbestos.

The benefits of talc include very good stiffness and creep resistant properties at elevated temperatures. Talc also has good electrical insulating properties and chemical resistance (Ref 1). For a reinforcing filler, it disperses fairly well into the resin. Because talc is inexpensive, it also has a benefit as an extender.

Flake materials such as mica and flaked glass have been shown to have reinforcing properties in plastics, but are not frequently used in adhesive formulations. Organic fibers or discontinuous inorganic fibers such as microfibers and whiskers have also found use in composite technology but have not been utilized to a large extent in adhesive applications in part because of processing and cost considerations.

Processing Considerations

Rheology Control. Viscosity, thixotropy, and sag resistance are processing properties of adhesives that are controlled by the use of fillers. Specific requirements in each of these areas are dependent on the application. In general, reproducible and user-friendly performance is required. For instance, the adhesive must be capable of being easily dispensed and must not drip or run during the period between application and cure. Often several process steps are executed within this period, and they must not have a detrimental effect on the uncured material.

In other applications, it may be necessary for the adhesive to be self-leveling or to penetrate into inaccessible areas, both of which require a different rheology from that of the previous example. The primary rheological property affected by filler is viscosity, which is increased by the addition. Fibrous fillers cause a larger viscosity increase than particulate fillers.

Increased viscosity provides sag resistance and flow control of the adhesive. However, an excessive viscosity increase can yield a stiff paste that gives undesirable processing properties for many applications.

This has led to the use of fillers that provide a shear thinning effect; that is, viscosity decreases as shear rate increases, and vice versa. This not only allows easy pumping and dispensing of the adhesive, but also provides sag resistance once applied. In common practice, this property is referred to as thixotropy.

The most popular thixotrope is fumed silica, which is manufactured synthetically by the high-temperature hydrolysis of silicon tetrachloride, yielding agglomerations of fine, amorphous particles. The result is a powder that is light and fluffy in appearance and that, when added to typical adhesive resins, imparts thixotropy to the system (Ref 9).

The thixotropic effect is a result of the surface chemistry of the fumed-silica particles. Hydroxyl groups are present on some of the surface silicon atoms. These hydroxyl groups are capable of forming hydrogen bonds with other fumed-silica aggregates. At a sufficient silica concentration, this results in the formation of a three-dimensional network structure, which imparts thickening and thixotropy. When subjected to shear, the network breaks down, effecting a viscosity increase. When the shear stress is removed and the system is allowed to come to rest, the network is re-formed, with the associated viscosity increase (see Fig. 4).

In most systems, only a small percentage of fumed silica is necessary to promote thixotropy. The effect is most efficient in resins that are not hydrogen-bonding systems. In hydrogen-bonding resins, the resin competes with other silica aggregates for the hydrogen bonding sites, thereby disrupting network formation (Ref 9).

The use of various polar additives with hydroxyl groups, such as ethylene glycol, are often used to improve the thixotropic effect. They function by acting as bridges between surface hydroxyl groups, as shown in Fig. 5 (Ref 1, 9).

In addition to providing sag resistance, fumed silica also helps to stabilize viscosity as a function of temperature (see Fig. 6).

Untreated fumed silica is hydrophilic because of the hydroxyl groups on the surface. Hydrophobic grades are also manufactured in which the silica surface has been chemically modified to replace the silanol

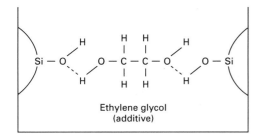

Fig. 5 Chemical compounds having short-chain polyfunctional molecules; hydrogen bonding with two or more fumed-silica aggregates. Source: Ref 9

groups with hydrophobic groups. For these grades, the network structure is built through mechanisms other than hydrogen bonding (Ref 10).

Hydrophobic surface treatment greatly improves moisture resistance. In many systems, the efficiency of the thixotropic action is enhanced (Fig. 7). Further, it has been reported that hydrophobic fumed silica provides better retention of sag-resistant properties upon aging compared to hydrophilic grades and other thixotropes (Fig. 8).

Other fillers are also capable of imparting thixotropy through the flocculation of particles induced by Van der Waals forces. These include talc, asbestos, modified bentonite, colloidal silica, and certain hydrated magnesium aluminum silicates (Ref 2). The use of titanium dioxide as an effective thixotrope has been reported (Ref 12). Organic

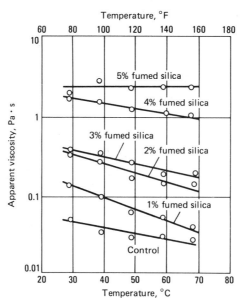

Fig. 6 Stabilization of viscosity of a paraffin oil with fumed silica. Source: Ref 9

fibers such as aramid have also been demonstrated to be useful thixotropes (Ref 13).

Thermal Conductivity Improvement. Fillers may also be added to improve thermal conductivity. The effect of various fillers on the thermal conductivity of cured adhesive is shown in Fig. 9. Metallic fillers such as aluminum and copper are most often used. The incorporation of metal fibers in

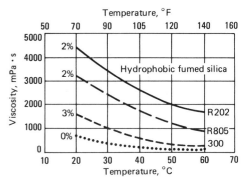

Fig. 7 Thickening effect of hydrophobic fumed silica in epoxy resin as a function of temperature. Resin: Araldite CY 179. Source: Ref 10

conjunction with metal powders has a synergistic benefit, as shown in Fig. 10 (Ref 1, 2). Oxides of aluminum, iron, beryllium, magnesium, and titanium have also been shown to improve thermal conductivity.

Electrical Conductivity Improvement. Cured adhesive resins alone are electrical insulators. Electrically conductive fillers have been used to formulate adhesives that are electrically conductive. These adhesives have been used extensively for die attach and other integrated circuit bonding applications (Ref 14).

Metal powders, flakes, or fibers are most often used. Silver is most widely used based on its high electrical conductivity. Gold is also used for applications in which resistance to chemical attack is desired (Ref 14). Aluminum, nickel, and copper are also used to some degree (Ref 2).

Carbon black also provides enhanced electrical conductivity to adhesive systems. These systems are used in the automotive industry to improve adhesive receptiveness to electrodeposited primers.

Electrical Property Enhancement. Various electrical properties of adhesives, such

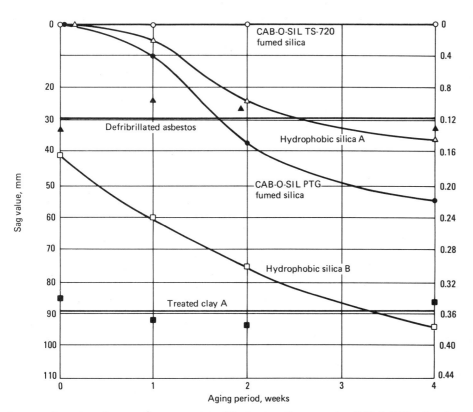

Fig. 8 Sag values of epoxy adhesives containing 3% of various thixotropes after 60 °C (140 °F) aging. Source: Ref 11

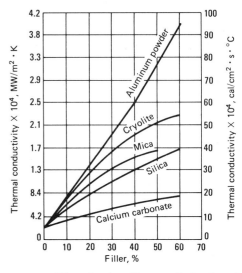

Fig. 9 Effect of various fillers on adhesive thermal conductivity. Source: Ref 2

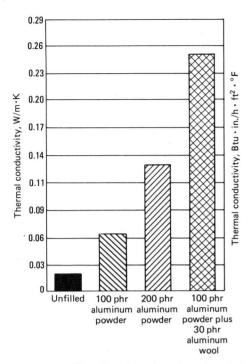

Fig. 10 Effect of metal powder and metal powder/metal fiber combinations on adhesive thermal conductivity. Source: Ref 2

as resistivity, dielectric constant, dissipation factor, and arc resistance, can be improved by incorporating nonconductive fillers into the resin. Special electrical-grade fillers are sold for this purpose. Metallic fillers have a detrimental effect on these properties. Volume and surface resistivities are modified by electrical-grade fillers such as silica, calcium carbonate, mica, and calcium silicate (Ref 2).

The dielectric constant and dissipation factor of adhesive are increased by fillers, including silica, hydrated alumina, mica, and zirconium silicate. Some oxides, however, will lower the dissipation factor (Ref 2). Dissipation factor is a measure of the amount of electrical energy dissipated when an electric current is applied. A high dissipation factor indicates a larger amount of energy dissipated. If it is too high, excessive heating will occur in the material. A low dissipation factor is therefore desired in electrical potting or insulating applications.

Fillers that heat up when exposed to electromagnetic radiation have also been investigated for accelerating adhesive cure. These include metal powders, metal oxides, barium titanate, and aluminosilicate (Ref 15, 16).

Flame Retardance and Smoke Suppression. Fillers in adhesives are used to impart flame retardance and smoke suppression, two very important applications. Increasing safety concerns, especially in the aerospace industry, have spurred significant growth in applying this technology to adhesives.

Flame-retardant fillers are additives that make the cured adhesive more difficult to ignite and limit flame propagation. These additives are not capable of completely preventing the polymer from burning in large fires, but they can slow the spread of fire.

Flame retardants operate through such mechanisms as:

- Heat absorption
- The liberation of nonflammable materials, which dilute the concentration of flammable material
- The formation of nonflammable char on the surface, which inhibits access to oxygen
- The generation of compounds that interfere with flame propagation
- Accelerating reactions that cause the adhesive to decompose and drip away from the flame (Ref 5)

Smoke suppression is also desired in these applications because many deaths in fires actually result from smoke inhalation. Some additives provide both flame retardance and smoke suppression properties.

The most popular flame retardant is antimony oxide, which is used as a synergist in conjunction with organic halogen compounds. When used in polyurethane systems, it has been found that the inclusion of zinc oxide gives improved performance (Ref 1).

Alumina trihydrate is another filler used for flame retardance, with the additional benefit of aiding smoke suppression. It provides flame retardance primarily by absorbing heat during endothermic dehydration (Ref 1). The smoke suppression performance is the result of the formation of char, rather than soot (Ref 17).

Flame retardance is also provided by organophosphorus compounds. They function by releasing nonvolatile acids during degradation, which increase char formation. Many of these compounds are liquid. Those that are solid include ammonium polyphosphate and ammonium orthophosphate (Ref 5).

Various borate compounds, such as zinc borate, boric acid, borax, barium metaborate, and ammonium fluoroborate are used in synergistic combinations with the organohalogen compounds to provide flame retardance. Often, these additives are used as cost-effective replacements for antimony oxide. Smoke reduction capability is also provided by the borate compounds (Ref 1, 5).

Flame retardance and smoke suppression are also exhibited by molybdenum compounds and certain tin compounds. Inorganic phosphorus compounds, such as red phosphorus, provide good flame retardance (Ref 1). Recently, surface-treated versions have become available, which satisfy handling concerns regarding flammability (Ref 18).

Pigmentation of adhesives is accomplished through the use of soluble organic dyes or insoluble pigments, with pigments being preferred. Inorganic pigments include titanium dioxide, aluminum powder, iron oxides, cadmiums, chrome yellows, and molybdate oranges. Organic pigments include carbon black, phthalocyanine blues and greens, and various reds and yellows (Ref 5).

Dispersion of the pigments is carried out by high-speed dispersion or three-roll milling to break down agglomerates. Prepared color dispersions in which the pigment is dispersed into a resin at high concentrations are available. A prepared dispersion is more easily blended into the final adhesive formulation.

Filler Considerations

Filler selection is based on the benefit desired from the filler and the end-use requirements of the adhesive. Typical fillers for various applications have already been discussed. Important filler considerations include availability, cost, surface chemistry, water adsorption, oil adsorption, density, coefficient of thermal expansion, and any other appropriate physical or chemical properties. Additional considerations should include the effect of the filler on adhesive properties such as viscosity, strength, modulus, adhesion performance, environmental durability, heat deflection temperature, creep resistance, machinability, and other pertinent physical or chemical properties.

Effect on Properties. The effect of fillers on several adhesive properties has already been discussed in relation to their applications and benefits. Their effect on the specific performance properties of the adhesive for the particular application should also be evaluated.

For an adhesive, the primary performance requirement is adhesion to the substrate. In general, fillers reduce shrinkage and coefficient of thermal expansion, which leads to reduced stresses in the bond line. This results in higher adhesive strengths.

To achieve a good adhesive bond, the adhesive must properly wet the surface of the substrate. Overloading the adhesive with filler or improper filler selection can result in an adhesive system that does not provide sufficient surface wetting.

The filler selection can also affect the chemical and moisture resistance of the adhesive. Fillers can often improve environmental resistance, with certain fillers providing better resistance to a particular environment than others. This effect is shown in Fig. 11, which compares the moisture resistance of unfilled adhesive resin to that of resin filled with various fillers. Additional improvements in environmental resistance can be achieved by using fillers that have been surface treated.

Fillers typically reduce the tensile, flexural, and compressive strengths of the

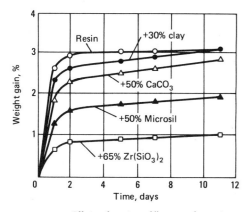

Fig. 11 Effect of various fillers on the moisture resistance of cured epoxy adhesive. Water absorption at 40 °C (105 °F). Source: Ref 2

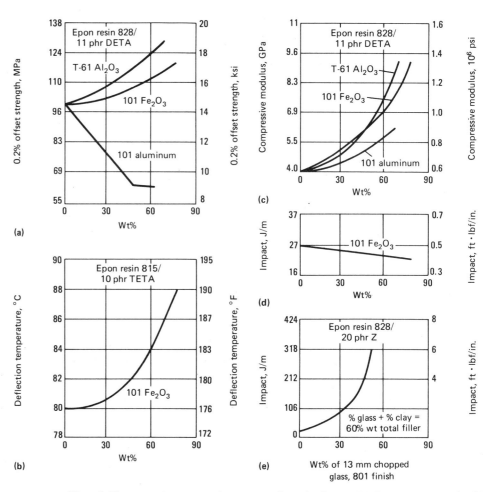

Fig. 12 Effect of fillers on various structural properties of cured adhesive. (a) Compressive strength. (b) Deflection temperature. (c) Compressive modulus. (d) Notched Izod impact strength. (e) Transfer molded specimens, notched Izod impact strength. DETA, tetrasodium ethylenediamine acetate; TETA, triethylene tetramine. Source: Ref 3

cured adhesive, but provide a modulus increase. As discussed above, reinforcing fillers can yield improvements in stiffness, heat deflection temperature, and impact resistance. The influence of fillers on various structural properties is shown in Fig. 12. The influence of fillers on the processing properties of the uncured adhesive, another important concern, was covered in the section "Rheology Control."

Filler Dispersion. In order to realize optimal properties when using fillers, it is necessary to ensure that the filler is adequately dispersed in the adhesive resin. Some fillers are more difficult to disperse because of their physical characteristics and surface chemistry.

Adsorbed water, which is present to some degree in most fillers, inhibits dispersion. Particle interactions resulting in aggregates of the flocculation of particles also impede efficient dispersion. Surface-modified fillers often present a solution.

Particle size and aspect ratio affect the degree of mixing required. Fibrous fillers

Table 2 Critical surface tensions of common plastics and metals

Materials	Critical surface tension	
	N/m	dynes/cm
Acetal	0.047	47
Acrylonitrile-butadiene-styrene	0.035	35
Cellulose	0.045	45
Epoxy	0.047	47
Fluoroethylene propylene	0.016	16
Polyamide	0.046	46
Polycarbonate	0.046	46
Polyethylene	0.031	31
Polyethylene terephthalate	0.043	43
Polyimide	0.040	40
Polymethylmethacrylate	0.039	39
Polyphenylene sulfide	0.038	38
Polystyrene	0.033	33
Polysulfone	0.041	41
Polytetrafluoroethylene	0.018	18
Polyvinyl chloride	0.039	39
Silicone	0.024	24
Aluminum	~0.5	~500
Copper	~1.0	~1,000

Source: Ref 20

are typically more difficult to disperse than are particulate fillers. In addition, high filler loadings make sufficient wetting of the filler more difficult because of the increased viscosities encountered. For these reasons, high shear mixing is required in these situations. Sometimes roll mill equipment is used to further assist dispersion.

Adhesion Promoters

A number of theories have been presented to explain adhesion. Adhesion promoters are used with adhesives in order to enhance the effects described by these theories. Most of the theories deal with the relative importance of a number of forces occurring at the interface of adhesive and adherent. These forces include Van der Waals attractions, hydrogen bonding, and chemical bonding. The latter force is the strongest, and Van der Waals attractions are the weakest of this group. However, physical adsorption (Van der Waals), although relatively weak, is important to the development of adhesion. This force influences the degree of physical entanglement and surface wet-

ting. The technical diversity of this subject is increased by the current methods of measurement. These are secondary in nature and, hence, reflect differences in the test methods themselves. Most adhesives incorporate a combination of additives, called adhesion promoters, or processes that exploit the chemical forces and physical phenomena in order to achieve maximum adhesion.

Surface Wetting. Interfacial wetting of the adhesive on the adherent must take place before the other physical and chemical forces can act. Without wetting, strong bonds are impossible. Wetting has been studied extensively from both the classical approach (Ref 3), and its effect on adhesion (Ref 19). The phenomenon of wetting may be described as a lateral spreading of the film and a penetration of the fluid adhesive into the rough morphology of the surface (Ref 19). To ensure adequate wetting, the adhesive should possess a surface tension no greater than the critical surface tension of the substrate. The critical surface tensions of common plastics and metals are shown in Table 2.

Table 3 Surface tensions of common adhesives and liquids

Material	Surface tension N/m	dynes/cm
Epoxy resin	0.047	47
Fluorinated epoxy resin(a)	0.033	33
Glycerol	0.063	63
Petroleum lubricating oil	0.029	29
Silicone oils	0.021	21
Water	0.073	73

(a) Experimental resin developed to wet low-energy surfaces (has low surface tension relative to most plastics). Source: Ref 21

Table 4 Surface tensions of solvents and other materials (in order of increasing surface tension)

Solvent	Surface tension N/m	dynes/cm	Temperature of test °C	°F
Hexane	0.0200	20.0	. . .	. . .
Isopropanol (anhydrous)	0.0208	20.8	25	77
Ethanol (anhydrous)	0.0221	22.1	25	77
Acetone	0.0226	22.6	25	77
n-propanol	0.0238	23.8	20	68
Isopropyl acetate	0.0245	24.5	25	77
Methyl ethyl ketone	0.0246	24.6	20	68
n-butanol	0.0246	24.6	20	68
1,1,1-trichloroethane	0.0251	25.1	20	68
Tributyl phosphate	0.0267	26.7	25	77
Ethylene glycol monobutyl ether	0.0274	27.4	. . .	. . .
n-butyl acetate	0.0276	27.6	27	80
Propylene glycol monomethyl ether	0.0277	27.7	. . .	. . .
Toluene	0.028	28.0	. . .	. . .
Xylene (meta)	0.028	28.0	. . .	. . .
Ethylene glycol monoethyl ether (2-ethoxy ethanol)	0.0282	28.2	. . .	. . .
Ethylene glycol monomethyl	0.0286	28.6	. . .	. . .
Dimethyl formamide	0.0373	37.3	20	68
Tricresyl phosphate	0.0407	40.7	. . .	. . .
Ethylene glycol	0.0484	48.4	20	68
Formamide	0.0582	58.2	20	68
Propylene glycol	0.0720	72.0	25	77
Water (distilled)	0.0728	72.8	20	68

Source: Ref 21

Additives that reduce the surface tension of adhesives are commonly used as adhesive modifiers. The goal in using these modifiers is to ensure complete wetting of the surface on the adherent so that the other adhesive forces may develop. Table 3 lists the surface tensions of common adhesives and liquids. The low surface tension exhibited by the fluorinated epoxy resin, silicone oils, and petroleum oil is an advantage in lowering the overall surface tension of the adhesive. The lubricating nature of the oils may seem to contradict the action of the adhesive, but in small amounts it may improve the adhesive strength by ensuring intimate contact with the substrate. Surface tension may also be lowered by the addition of solvents. Table 4 lists the surface tensions of some common solvents. Of course, solvent-base elastomeric adhesives are known to wet surfaces very efficiently, an occurrence due in part to this phenomenon.

Surface wetting may be inhibited by contamination or poorly prepared surfaces. For example, a metal surface with a high surface tension may be contaminated with oil used either in the processing of the metal or to protect the metal from oxidation (Ref 22). The end result is a contaminated metal surface that exhibits a low surface tension. Many additives that act to lower surface tension will also act to solubilize surface contaminants. Modifiers belonging to this group include many of the solvents and oils that are in this case acting as solvents. Another group of modifiers that serve this same function are the acrylics (Ref 23). They exhibit a high tolerance for the oils commonly known to contaminate surfaces. Acrylics, which are the main components of the anaerobic and acrylic families of adhesives, may be used to modify other adhesives classes. An advantage of the acrylics over solvent and oil modifiers is the presence of reactive end groups that may be used to react with the main resin component, permanently incorporating the modifier into the cured product.

The modifiers listed above all serve to improve the surface wetting of the adhesive. This allows other forces involved in adhesion to occur. The physical entanglement or mechanical interlocking aspect of adhesion has to do with the surface morphology and is examined with the surface pretreatment section in this Volume. Generally, a rough surface morphology enhances adhesive strength by increasing surface area and the degree of mechanical interlocking with the surface.

Polarity and Hydrogen Bonding. The hydrogen bonding of surface groups is a powerful adhesive force that may be influenced by additives to the adhesive systems. As with surface tension, hydrogen bonding is dependent on the surface of the adherent and the surface preparation method. The degree of polarity in an adhesive may also be influenced by additives, although this is normally determined by the functional groups on the backbone of the polymer. Two examples of systems that exhibit polarity and hydrogen bonding are neoprene adhesives and epoxy adhesives. Neoprene has chlorine groups, which contribute to the polarity of this polymer because chlorine is highly electronegative. Epoxies may have pendant hydroxyl functionality, which allows hydrogen bonding to occur.

Modifiers added to an adhesive system to increase either the electronegativity or the amount of hydrogen bonding present will increase its adhesive strength to many surfaces (Ref 24). Some examples of modifiers include 2-hydroxyethyl methacrylate, which increases hydrogen bonding capabilities; fluorinated compounds, which increase electronegativity; and compounds with acid functionality, such as methacrylic acid.

Phenolic resins are frequently used as modifiers to elastomeric adhesive systems and epoxies. The reason for their use is not necessarily to promote adhesion, although this secondary effect is realized because of increased hydroxyl functionality (Ref 14). Additives that promote this force are numerous, and many may not be thought of as true additives, although the effect on adhesive strength may be significant. Other examples of this effect are the use of polyester polyols in place of polyethers in urethanes and the use of phenoxy resins (polyhydroxy ethers), where the pendant functionality increases the polarity of the adhesive. The use of these types of modifiers must be balanced against additional moisture absorption characteristics that may occur with their use. Most of these additives contain two reactive groups, one to react with the polymer, and the other to interact with the surface.

Chemical bonding generally takes place through the use of coupling agents. These agents function as molecular bridges at the interface of two dissimilar substrates. Two classes of coupling agents are used extensively in adhesives: silanes and titanates.

Silanes, which are the most widely used, may be represented by the formula $YRSiX_3$ where X represents a hydrolyzable group, typically alkoxy, and Y is a functional organic group, such as amino, methacryloxy, epoxy, and so on. Typically R is a small aliphatic linkage that serves to attach the functional organic group to silicon (Si). Silanes are used with a number of adhesive families and may be used to link with the filler in the system, the fibers in tape systems, and the adherent, or bonding surface. An application guide based on adhesive and silane types is shown in Table 5. The reaction of silanes involves four steps, shown in Fig. 13. The X group (that is, the OMe

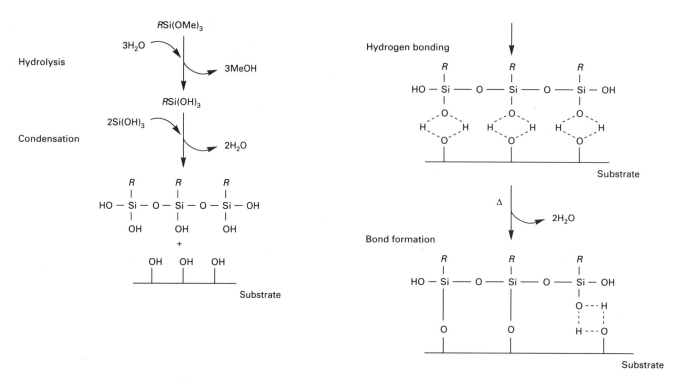

Fig. 13 Deposition of silane coupling agents

group) is involved in the reaction with the inorganic substrate. The R group is chosen from a wide variety of organic functionalities available to react with the polymer matrix (Ref 26). The completed coupling reaction results in the establishment of a covalent chemical bond. This type of bond accounts for an enhanced adhesion that is resistant to adverse environmental conditions and is up to 100% stronger than systems without silane. Typical addition levels used in formulations are less than 1%, which, in spite of the relatively high cost, adds relatively little to the formulation cost.

Titanates are also used as coupling agents. Although they have been used specifically to couple the polymer matrix to various fillers and fibers, adhesion to substrates is also improved. Improved bond strength to halocarbon substrates and im-

proved hydrolytic stability are claimed by the suppliers of titanates (Ref 27). A variety of organometallic coupling agents based on titanium and zirconium are available. The level in formulations is low for these compounds, 0.1 to 0.5 wt%, producing a dual function of superior processability and bonding (Ref 28). The cost of the organic titanates may be somewhat less than that of the silanes, depending on their chemical complexity.

Tackifiers

A tackifier is defined as an additive, usually a constituent of rubber-base and synthetic resin adhesives, intended to improve the stickiness of a cast adhesive film (Ref 29). A tackifier is therefore any compound that enhances this property. The various

chemical families used in adhesives use different types of tackifiers specific to the chemistry and processing of each family. This category of additive is therefore broad and specific to the chemical class of adhesive.

The elastomeric adhesives generally use resins to provide tack. The resins used for this purpose include C_5 and C_9 hydrocarbon resins, phenolics, liquid polybutenes, coumarone-indene resins, and a variety of oils, waxes, and other resin types. Included with tackifiers are plasticizers, which may act as either extenders or tackifiers to a polymer system. A list of common tackifiers and plasticizers used in one type of synthetic resin system is shown in Table 6 (Ref 30). Tackifiers are generally lower in cost than the elastomer component and are used to lower the overall cost of the formulated system. The benefits obtained, in addition to an increase in tack, may be improved heat resistance and improved processibility. Two examples that confer these additional benefits are thermoplastic elastomer systems, in which hydrocarbon resins are used to increase heat resistance, and neoprene systems, in which phenolic resins are used to improve processing and heat resistance. These synergistic effects may be found in many other elastomeric systems as well. The level of these additives may be a high percentage of the total formulation, and the effect on cost may be significant.

Urethanes may be classified as thermoplastics or thermosets, depending on the degree of cross linking present. The thermo-

Table 5 Silanes used as coupling agents

Adhesive type	Amino A-1100	Amino A-1110	Amino A-1120	Amino A-1130	Amino A-1106	Epoxy A-186	Epoxy A-187	Mercapto A-189	Methacryloxy A-174	Vinyl A-151	Vinyl A-172
Acrylic	●	●	●	●	WB	○	●	○	○		
Butyl	○						●		●		
Epoxy	●	●	●	●	WB	○	○	○			
Neoprene	○							●			
Nitrocellulose	●	○	○	○				●			
Nitrile	○							●			
Phenolic	○	○			WB	○	○	●			
Polyester	○	○					○	●	●	○	○
Polysulfide	○							●	●		
Polyurethane	○	●	○	○			●	●			
Vinyl	○	○	●	●	WB						
Hot melts						○	○		●	●	●

(a) ●, generally effective; ○, alternate. WB, water-borne systems. Source: Ref 25

Table 6 Examples of commercial resins and plasticizers used in rubber formulations

Material	Softening point °C	Softening point °F	Chemical base	Manufacturer
Rubber phase associating resins				
Super Sta-Tac	80, 100	176, 212		Reichhold
Quintone series	85, 100	185, 212		Nippon Zeon
Nevtac series	80, 100, 115, 130	176, 212, 240, 265		Neville
Piccotac 95-BHT series	92, 93	198, 199	Polymerized	Hercules
Escorez 2101	95	203	Mixed olefin	Exxon
Wingtack series	95–100	203–212		Goodyear
Escorez 1300 series	95, 100, 115	203, 212, 240		Exxon
Super Nevtac 99	99	210		Neville
Piccotac B	100	212		Hercules
Sta Tac/R	100	212		Reichhold
Hercotac AD	100, 115	212, 240		Hercules
Betaprene BC	100, 115	212, 240		Reichhold
Zonarez 7000 series	85, 100, 115, 125	185, 212, 240, 257		Arizona
Zonatac series	85, 100, 115	185, 212, 240		Arizona
Nirez 1000 series				Reichhold
Piccofyn A-100	100	212	Polyterpene	Hercules
Nirez V-2040	118	244		Reichhold
Piccolyte HM 110	110	230		Hercules
Piccolyte A..................	115, 135	240, 275		Hercules
Sylvatac series	69–107	156–225		Sylvachem
Super ester A series	75, 100, 115	167, 212, 240		Arakawa
Stabelite ester 10.............	80	176		Hercules
Foral 85.....................	82	180	Rosin ester	Hercules
Zonester series	83, 95	181, 203		Arizona
Foral 105....................	94	201		Hercules
Pentalyn H	104	219		Hercules
Escorez 5000 series	85, 105, 125	185, 220, 255		Exxon
Arkon P series...............	85, 125	188, 255	Hydrogenated	Arakawa
Regalrez series	17–125	63–255	Hydrocarbon	Hercules
Super Nirez 5000 series.......	100, 120	212, 248		Reichhold
Polystyrene phase associating resins				
Cumar series	10–130	50–265	Coumarone indene	Neville
Cumar LX-509...............	155	311		Neville
Picco 6000 series............	70–140	160–284		Hercules
Nevchem series..............	100–140	212–284	Hydrocarbon	Neville
LX 685 series................	98–155	208–311		Neville
Piccotex series...............	75–100, 120	167–212, 248	Alphamethyl	Hercules
Kristalex series	85–140	185–284	Styrene	Hercules
Amoco 18 series.............	100, 115, 145	212, 240, 293		Amoco
Piccolastic D-150.............	100	212	Polystyrene	Hercules
LX-1035	175	347		Neville
Rubber phase associating plasticizer				
Shellflex 371.................	...	...		Shell Chemical
Kaydol	...	...	Paraffinic/naphthenic	Witco
Tufflo 6000 series	...	...		Arco
Wingtack 10.................	...	...	Mixed olefin	Goodyear
Piccolyte S10................	...	...	Polyterpene	Hercules
Zonarez alpha 25.............	...	...		Arizona
Piccovar AP10...............	...	...	Alkylaryl resin	Hercules
Shellwax series	...	...	Paraffinic wax	Shell Chemical
Sun 5512	...	...		Sun
Shellmax series	...	...	Microcrystalline	Shell Chemical
Sun 5825	...	...		Sun

Source: Ref 30

plastics may be grouped with the elastomerics or they may be used as tackifiers in heat-activated systems because of their sharp softening points and their aggressive tack at elevated temperatures. The thermosetting urethanes are normally two-component systems with little tack present without a tackifier additive. Tackifiers for these systems are usually amines blended with the polyols in the hardener, which react immediately with the isocyanate functional groups in the resin component. This immediate reaction causes a rapid increase in molecular weight, which increases the cohesive strength of the polymer and therefore enhances the tack of the system. This type of tack is independent of the stickiness of a system. An increase in cohesive strength at the same level of stickiness will increase the energy necessary to pull substrate away from the adhesive. This may be considered as an overall increase in the tack of a system.

The other thermosetting adhesives generally follow the same trend as the urethane family. A higher molecular weight is used to develop the tack of the adhesive system. Epoxies generally use high molecular weight epoxies as additives to increase tack. The phenolics may employ higher molecular weight resins or higher softening point resins blended with low softening point resins to yield tacky semisolids at room temperature. The acrylics and anaerobics may utilize high molecular weight rubber or synthetic resins dissolved in the monomers to increase viscosity, molecular weight, and therefore tack. All of the adhesive systems may use solvents, including water, to disperse or dissolve the resin systems for processibility. The subsequent removal of solvent changes the physical state of the resin system from liquid to a tacky semisolid and, finally, to a solid. This process is used extensively, and in this sense the solvent may be considered a tackifier.

There are a number of processing methods that are used in conjunction with additives to promote tack in adhesive systems. The most common of these is the use of heat in conjunction with a thermoplastic additive. This is the basis of the hot-melt industry, but is also used in conjunction with other resin systems. A solid or semisolid adhesive is heated to the point at which a thermoplastic additive softens and provides tack to the system. The main resin component is then cured by standard methods, while the thermoplastic holds the parts together. The additives used for this method include the base components used in the hot-melt industry.

A second processing method is the use of acrylic components in a resin system, which are quickly reacted to high molecular weight by some processing means. The processing means may be ultraviolet light (UV) in conjunction with a UV initiator or the use of heat in conjunction with a peroxide. The result is a high molecular weight component in the formulation, which provides some cohesive strength until the main resin is cured.

Tougheners

The toughening of polymer systems has been an area of active research for many years. The main body of this work has involved the toughening of the highly crosslinked thermoset systems in the structural adhesive, composite, and electrical laminate industries. The chemical families involved in these industries include epoxies, phenolics, anaerobics, acrylics, cyanoacrylates, and the advanced polymeric systems. The other chemical families involved with

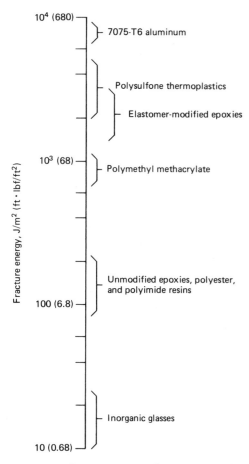

Fig. 14 Fracture energies of various materials. Source: Ref 31

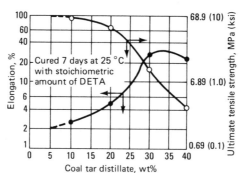

Fig. 15 Effect of coal tar distillate (plasticizer) content on tensile properties of Epon resin 834 (epoxy). Courtesy of Shell Chemical Company

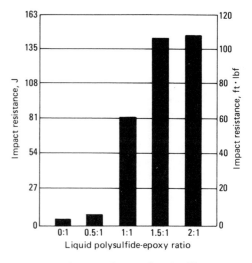

Fig. 16 Impact resistance of varying LP-epoxy ratios

adhesives, specifically the silicones, polysulfides, urethanes, and synthetic elastomers, do not require additives to achieve toughness because this property is inherent in the polymer systems. However, the highly cross-linked rigid systems all require the addition of some degree of toughness to the system to achieve useful performance properties.

Three types of toughening additives will be discussed based on the mechanisms of toughening. The mechanisms involved are plasticization, single-phase toughening, and two-phase toughening. The reason toughness is required in the systems described above is related to the inherent brittleness of the unmodified materials. A comparison of toughness based on fracture energy is shown in Fig. 14. This figure shows that epoxies, polyimides, and polyester resins are only slightly less brittle than the inorganic glasses.

Plasticization. For some time plasticizers have been added to resin systems belonging to all of the chemical families. This is not a preferred method, but before some of the newer methods came into use this was used extensively. The method is based on the softening effect that the plasticizer imparts to the polymer system. The advantages of this method include the low cost of the additives, improved processing, and synergistic effects that may occur (Ref 32). Dibutyl phthalate and other nonreactive diluents have been used to impart a degree of flexibility to cured epoxy systems at the expense of elevated temperature and chemical resistance properties. The increase in flexibility typical of this type of additive is shown as

percent elongation (Fig. 15). The disadvantages and limitations of this type of additive are a negative impact on the other performance properties, plasticizer migration, and incompatibility with the cured resin system (Ref 31).

Single-Phase Toughening. The generally positive effects of plasticizers can be achieved through the proper selection of a long-chain curing agent. This results in toughness induced by chain flexibility, rather than softening of the cured product. Because the toughener is reactive and actually becomes part of the resin structure, migration of the additive is not a problem, and incompatibility between resin and plasticizer may not exist. A number of additives may be included in this category. The main criterion for this category is long-chain difunctional material that, upon cure, becomes part of the thermoset matrix.

The use of this type of additive in acrylic systems is growing as improvements continue to be made on the second-generation acrylic systems. The first- and second-generation acrylics primarily use methyl methacrylate as the polymerizing monomer. Flexibility is increased through the use of higher molecular weight materials and long-chain difunctional materials, available through a number of sources. Use of these additives does not change the curing mechanism, and the resultant polymer incorporates these additives into the backbone. The result is a single-phase flexible system. The disadvantage of this approach is a reduction in the glass transition temperature (T_g) and, consequently, the heat resistance of the system. The cost of these additives may be slightly more than that of the lower molecular weight monomers, depending on the chemical complexity of the molecule. The level in formulations is dependent on the degree of flexibility required and the sacrifice in T_g that is acceptable.

Table 7 Typical epoxy formulations and their effect on properties

Highly flexible epoxy systems			
Araldite 6010, parts by weight	50	33	33
Araldite 508, parts by weight	50	67	67
n-aminoethylpiperazine, parts by weight	17	14	14
Cure schedule	7 days at 25 °C (77 °F)	7 days at 25 °C (77 °F)	Gel at 25 °C (77 °F) + 2 h at 100 °C (212 °F)
Property			
Tensile strength, MPa (ksi)	23 (3.34)	15 (2.23)	17 (2.51)
Elongation at failure, %	88	107	91
Modulus of elasticity, GPa (10^6 psi)	0.7 (0.1)	0.09 (0.013)	0.2 (0.029)
Hardness, Barcol	70	38	56
Dielectric constant at 25 °C (77 °F)			
At 60 Hz	4.82	5.19	6.05
At 10^5 Hz	4.05	4.03	4.36
Dissipation factor at 25 °C (77 °F)			
At 60 Hz	0.063	0.065	0.093
At 10^5 Hz	0.034	0.047	0.059
Volume resistivity at 25 °C (77 °F), $\Omega \cdot$ cm	9.5×10^{11}	9.7×10^{11}	8.4×10^{11}

Source: Ref 31

Table 8 Properties of various carboxyl-terminated butadiene acrylonitrile (CTBN) liquid polymers

Property	CTB	CTB	CTB	CTBN	CTBN	CTBN	CTBN X	CTBN X
Viscosity at 27 °C (80 °F), Pa · s.	40	40	33	55	125	625	155	265
Carboxyl, %	1.9	1.9	2.1	2.47	2.37	2.40	2.93	3.0
Molecular weight	4800	4800	· · ·	3500	3500	3500	3500	3500
Functionality	2.01	2.01	· · ·	1.9	1.85	1.85	2.3	2.3
Acrylonitrile, %	0	0	0	10	17	27	17	21.5
Solubility parameter	8.04	8.04	8.04	8.45	8.77	9.14	· · ·	· · ·
Heat loss, 2 h at 130 °C (265 °F), %	<2.0	<1.0	<1.0	<2.0	<2.0	<2.0	<2.0	<2.0
Specific gravity at 25 °C (77 °F), g/cm³	0.907	0.907	0.907	0.924	0.948	0.960	0.955	0.958

Epoxies offer a number of ways to flexibilize a system using the single-phase approach. Long-chain difunctional epoxy resins have been used for some time. The effect of typical epoxy formulation using Araldite 508 flexible liquid epoxy resin on polymer performance is shown in Table 7. Typical epoxy elongation without the use of Araldite 508 is in the 2% range.

The curing agents used with epoxies also offer the long-chain difunctional approach. Examples of this include the use of polysulfide polymers as curing agents for epoxy resins and the use of *n*-aminoethylpiperazine curing agents (Ref 33).

A graph representing improved impact resistance with the use of increasing amounts of liquid polysulfide (LP) hardener is shown in Fig. 16. Elongation properties also increase significantly.

Other curing agents that increase toughness appreciably are polyamide hardeners, available from a number of suppliers. The increase in toughness that these curing agents impart is less than that of the above examples, but it is still significant. The advantages of this approach are the high elongations made possible, the relatively low cost of the additives in comparison to two-phase systems, and a homogenous cured polymer system. The disadvantages are significant reductions in the T_g of the system due to a reduced cross-link density and reductions in solvent and chemical resistance in the resultant polymer.

Two-Phase Toughening. An increase in toughness due to the introduction of a second rubbery phase in thermoset materials is by

now widely known. Brittle resins can be toughened by dispersing a small amount of rubber or other elastomeric material into them as a discrete phase of microscopic particles. Such dispersed rubber particles impart apparent ductility and improve the crack resistance of a rigid resin matrix through their ability to absorb strain energy (Ref 35). The research work was pioneered at the B.F. Goodrich Company more than 30 years ago (Ref 36). The work that has received the most attention is on the carboxyl-terminated butadiene acrylonitrile (CTBN) liquid polymers.

Table 8 shows typical properties of CTBN liquid polymers (Ref 37). The utility of an approach based on a second rubbery phase is that the rubber with reactive end groups can be prereacted with the resin system, locking the rubber into the resin, which provides toughening and allows a retention of most of the heat-resistance properties.

More recently, thermoplastic additives have been used as tougheners in the very high performance market involving resin systems of the epoxy, polyimide, and bismaleimide classes (Ref 38-40). The result of this work has been the toughening of resin systems on a level that corresponds to some of the thermoplastics, along with a retention of the high-temperature properties of the neat resin systems.

Acrylic resins, beginning with the first-generation acrylic adhesives, used rubber to toughen the resin matrix (Ref 23). The second-generation acrylics and some of the

newer acrylic formulations probably make use of synthetic polymers or rubber that contains reactive end groups. Typically these formulations are improvements on the original patents, which used synthetic rubber (Ref 41). The cost of these formulations is relatively high, compared to the cost of epoxy and polyurethanes that offer the same general levels of performance. Ease of processing and short cure cycles are normally the reasons for choosing these systems. The degree of flexibility, as measured by peel strength, can be relatively high and probably falls between the epoxies, at the lower measured values, and the polyurethanes, which may have very high peel values.

Epoxy adhesives and other thermoset matrices may be toughened by a number of methods that use a second phase. These methods include the addition of various types of additives, including CTBN-ATBN (where ATBN is amine terminated butadiene acrylonitrile), unreactive rubber, chopped fiber, thermoplastics, fibrils, and crystalline segments (Ref 42).

The degree of toughening achieved is shown in Table 9, which compares a urethane toughening agent with a CTBN adduct. Fracture toughness improvements are shown in Fig. 17, which compares neat resin systems of varying equivalent weight to the same systems with a CTBN additive. Conclusions based on Fig. 17 are that the main source of toughness is intrinsic resin ductility and that shear band formation is

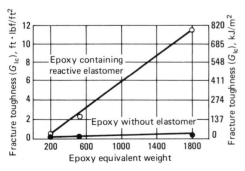

Fig. 17 Fracture toughness in epoxies containing systematic changes in cross-link density. Epoxy, DGEBA-DDA; elastomer, 1300 × 13 CTBN. Source: Ref 42

Table 9 Comparison of CTBN-epoxy and epoxy-urethane

Formulation components(a)	phr			
	A	B	C	D
DGEBAs liquid epoxy	100	77.5	50	· · ·
CTBN-DGEBA, EEW = 340	· · ·	37.5	· · ·	· · ·
Urethane/DGEBA, EEW = 235	· · ·	· · ·	50	100
Tabular alumina	40	40	40	40
CAB-O-SIL N70-TS	3.5	3.5	3.5	3.5
Dicyandiamide	6	6	6	6
3-phenyl-1,1-dimethyl urea	2	2	2	2
Property(b)				
Lap shear at room temperature, MPa (ksi)(c)	7.58 (1.1)	12.3 (1.78)	9.37 (1.36)	8.82 (1.28)
Lap shear at room temperature, MPa (ksi)(d)	· · ·	11.3 (1.6)	8.34 (1.2)	4.93 (0.72)
T-peel at room temperature, kN/m (kg/cm)(c)	2.45 (2.5)	6.8 (6.98)	3.8 (3.87)	4.0 (4.13)
T-peel at room temperature, kN/m (kg/cm)(d)	· · ·	7.2 (7.34)	3.0 (3.06)	1.2 (1.23)
Glass transition temperature, °C (°F)	125 (255)	113 (235)	113 (235)	105 (220)

(a) DGEBA, diglycidyl ether of bisphenol A; EEW, epoxide equivalent weight. (b) Tested on only drawing quality killed electrogalvanized (DQKEG) 16-20 E steel substrate; 25 × 100 × 0.76 mm (1 × 4 × 0.03 in.) specimen. (c) Specimen cured 1 h at 171 °C (340 °F). (d) Specimen cured 20 min at 204 °C (400 °F). Source: Ref 43

enhanced by the addition of a second-phase elastomer. The increase of fracture toughness is probably due to the second phase acting as a step to crack propagation. Large improvements may be seen by these comparisons. Table 8 shows that only a slight reduction in T_g is realized. On a cost-per-pound basis, the Hycar additives offer the best combination of properties improvement versus negative impact. The use of thermoplastics has primarily been in response to the needs of the aerospace industry. Compared to the Hycar additives, thermoplastics offer better high-temperature performance at the expense of ease of processing and ease of manufacture. The cost of these systems is generally much higher because of more difficult manufacturing and the use of proprietary, functionalized thermoplastics in the formulations.

REFERENCES

1. *Handbook of Fillers for Plastics*, H.S. Katz and J.V. Milewski, Ed., Van Nostrand Reinhold, 1987
2. H. Lee and K. Neville, *Handbook of Adhesive Resins*, McGraw-Hill, 1982
3. W.D. Harkins, *Physics of Surface Films*, Reinhold, 1952
4. "Reinforcements and Fillers for Plastics," Charles Kline and Company, 1980
5. *Additives for Plastics*, R.B. Seymour, Ed., Academic Press, 1978
6. T.G. Florea, "Kaolin, Hydrous and Anhydrous for Thermoplastic and Thermoset Resins," J.M. Huber Corporation
7. E.C. Clark, "Properties and Uses of Epoxy Syntatics," Technical Paper AD88-636, Society of Manufacturing Engineers, 1988
8. J.V. Milewski, "Identification of Maximum Packing Conditions in the Bimodal Packing of Fibers and Spheres," Paper presented at the 29th Annual Technical Conference, Reinforced Plastics Division, The Society of the Plastics Industry, Inc., 1974
9. "Cab-O-Sil Fumed Silica Properties and Functions," Technical Bulletin, Cabot Corporation, June 1987
10. "Aerosil for Solvent-Free Epoxy Resins," Technical Bulletin 27, Degussa Corporation, Nov 1986
11. "Cab-O-Sil TS-720 Hydrophobic Fumed Silica," Technical Bulletin, Cabot Corporation, April 1988
12. F. Ivan, Romanian Patent RO 76,642, CA 98, 176955, 1983
13. Japanese Patent, Fuji Electric Company, Ltd., Kokai Tokkyo Koho 58,19,324, CA 99, 54560, 1983
14. L.S. Buchoff, in *Handbook of Adhesives*, 3rd ed., I. Skeist, Ed., Van Nostrand Reinhold, 1989
15. Japanese Patent, Aisin Kako Company, Ltd., Kokai Tokkyo Koho 58,74,773, CA 99, 177166, 1983
16. Technical Literature, Hellerbond Division, Alfred E. Leatherman Company, Inc.
17. D.F. Lawson, E.L. Kay, and D.T. Roberts, Mechanism of Smoke Inhibition by Hydrated Fillers, *Rubber Chem. Technol.*, Vol 48 (No. 1), 1975
18. "Amgard Flame Retardants," Technical Literature, Albright and Wilson Americas Company
19. D.H. Kaelble, in *Treatise on Adhesion and Adhesives*, R.L. Patrick, Ed., Marcel Dekker, 1967, p 169
20. E.M. Petrie, Adhesively Bonding Plastics: Meeting an Industry Challenge, *Adhes. Age*, 15 May 1989, p 6
21. J.W. Prane, Some Insight Into Why Adhesives Adhere, *Adhes. Age*, June 1989, p 52
22. R.T. Foister and K.J. Schroeder, Adhesive Bonding to Galvanized Steel, *J. Adhes.*, Vol 24, 1987, p 259-278
23. A.H. Landrock, *Adhesives Technology Handbook*, Noyes Publications, 1985, p 136
24. "Acrylics as Adhesion Promoters," Technical Bulletin, Rohm Tech
25. "Organofunctional Silanes—A Profile," Technical Brochure, Union Carbide Corporation, 1983
26. Silanes Coupling Agents, Polymer and Sealant Additives, Technical Brochure, Petrach Systems, Inc., 1984
27. S.J. Monte, Titanates, in *Modern Plastics Encyclopedia*, McGraw-Hill, 1989, p 177
28. S.J. Monte and G. Sugerman, "Enhanced Bonding of Fiber Reinforcements to Thermoset Resins," Paper presented at the Water-Borne and Higher-Solids Coatings Symposium, New Orleans, University of Southern Missippi, 1988
29. J. Shields, *Adhesives Handbook*, 2nd ed., Newnes-Butterworths, 1976
30. "Kraton Thermoplastic Rubber," Technical Brochure, Shell Chemical Company, 1983
31. C.A. May, Epoxy Resins Chemistry and Technology, 2nd ed., Marcel Dekker, 1988, p 552
32. Dynamit Nobel, Netherlands Patent Application 6,602,522
33. Araldite 508, Technical Brochure, Ciba-Geigy Corporation
34. LP Polysulfide/Epoxy Systems, Technical Brochure, Morton Thiokol, Inc., 1984
35. "Toughen Epoxy Resins With Hycar RLP," Technical Brochure RLP-2, B.F. Goodrich Company
36. R.S. Drake and A.R. Siebert, *SAMPE Q.*, Vol 6 (No. 4), 1975, p 1
37. "Hycar Reactive Liquid Polymers," Technical Brochure RLP-1, B.F. Goodrich Company, 1983
38. H.G. Recker, Highly Damage Tolerant Carbon Fiber Epoxy Composites for Primary Aircraft Structural Applications, *SAMPE Q.*, Oct 1989
39. M.T. Blair, and P.A. Steiner, The Toughening Effects of PBI in a BMI Matrix Resin, in *Proceedings of the 33rd International SAMPE Symposium*, Society for the Advancement of Material and Process Engineering, March 1988, p 524
40. G.R. Almen *et al.*, Semi-IPN Matrix Systems for Composite Aircraft Primary Structures, in *Proceedings of the 33rd International SAMPE Symposium*, Society for the Advancement of Material and Process Engineering, March 1988, p 979
41. Hypalon Synthetic Rubber, Technical Brochure, E.I. Du Pont de Nemours & Company, Inc., 1984
42. N.J. Johnston, "Resin Property—Composite Property Relationships," NASA Langley Research Center, National Aeronautics and Space Administration
43. "Adhesive Systems Using Hycar Reactive Liquid Polymers," Technical Brochure, B.F. Goodrich Company, March 1988

Sealant Materials

Co-Chairmen: J.M. Klosowski and S.B. Smith (deceased), Dow Corning Corporation

Introduction

THE PRIMARY FUNCTION of sealant materials is to seal holes and gaps between other engineering materials. A secondary function performed by selected commercial sealants is to provide, in addition to a seal, the primary, or only, bonding mechanism between substrates. Such materials are often referred to as adhesive sealants. Although in actual use virtually all sealing applications require some level of adhesion in order to function successfully as a seal, no effort will be made here to differentiate between sealants and adhesive sealants. The articles in this Section deal primarily with sealing, and with bonding only as it relates to overall seal quality.

Caulks and putties will receive minimal treatment, as these materials are encountered more in aesthetic enhancement than in true sealing functions. True sealant materials have the capability to undergo considerable substrate movement and still perform the basic sealing function. Caulks and putties, on the other hand, are limited to areas with minimal substrate movement.

The section begins with a general overview of families of sealants, their common characteristics, current applications, and current markets, along with a brief comparison of material costs and an assessment of cost effectiveness.

The "Chemical Families" subsection contains articles covering the major polymeric families, with a brief review of their polymeric chemistry and synthetic routes and cure chemistries used within the generic family. The chemistry of additives and modifiers is described only if it is applicable to that polymeric family. These articles follow a common outline, to facilitate cross comparisons.

Application capabilities are delineated, and special attention is given to highly suitable as well as proscribed applications. Performance data feature typical mechanical, electrical, and handling properties. Special environmental durability test data are included, with typical data for high- and low-temperature environments, water and oil immersion, long-term weathering, thermal endurance, electrical endurance, and any pertinent combinations. Product design considerations are discussed briefly, with information on commercial forms, major suppliers, markets and applications within markets, cost factors, and competitive materials or systems.

The polymeric families selected as article subjects for this Section were chosen on the basis of their current breadth of use in transportation, construction, aerospace, electronics, industrial maintenance, and consumer markets, as well as on the uniqueness of each family in providing critical solutions to typical engineering sealant problems. In sum, this Section is intended to provide the reader rapid access to current state-of-the art information concerning the major sealant materials.

Types of Sealants

Lawrence D. Carbary, Dow Corning Corporation

SEALANT TYPES and their general characteristics are the focus of this article. Chemical families, performance capabilities, relevant specifications, and physical properties are described. This overview primarily uses the building construction industry as the context in which sealant types are discussed and compared. The other articles in this Section of the Volume describe other industries in which particular sealants may be used.

The building construction industry relies heavily on high-performance sealants, but most contractors and architects agree that improper sealant selection and usage can cause problems. Many architects and contractors do not adequately select and install the products that are applied. Sealants in industrial, electronic, and automotive applications are generally much better researched by the user. If a user understands the application, selects two or more products, and tests them for that application, a more successful product selection is likely. The money saved on lower-performance products is wasted if a contractor must be hired to remove and replace the initial installation, or if the product fails and its replacement becomes necessary. Generally, the high-performance sealants will perform in low-performance applications, whereas the opposite is not true. Often the anticipated performance of a sealant will be less than that actually attained. Therefore, it may be prudent to start with a higher-performance product than that which is anticipated as being necessary.

Performance Characteristics

Among the several ways to classify sealants, depending on the application are: high-temperature stability versus no stability at high temperatures; low-temperature flexibility versus brittleness at low temperature; and fuel resistant (no weight or volume change or swelling when exposed to fuel) versus dissolution or decomposition in fuels. However, the most common way to classify building industry sealants is by their ability to handle movement.

Low-performance sealants typically have movement capacities of 0 to 5% of the joint width. These sealants are extremely low cost and were the first type to be introduced, in the early 1900s. These products, which can be considered crack fillers only, include oil-base, resin-base, polyvinyl acetate-base, and bituminous-base caulks that are highly filled. They can experience severe hardening and shrinkage after installation, and although they may perform as a crack filler, they serve this function only until they shrink enough to pull away from the joint walls. They are used in the consumer market and have a very limited useful service life in that capacity.

Medium-performance sealants have movement capacities of 5 to 12% of the joint width and have a longer service life than low-performance sealants. Butyl, latex acrylic, and solvent acrylic sealants are included in this group.

The sealants that are solvent- or water-release-curing systems will have considerable shrinkage: 10 to 30% after installation. Some of these sealants use either a silicone fluid to facilitate gun-dispensible application and minimize shrinkage or a silane coupling agent to promote adhesion. However, these siliconized products are generally very similar in nature to their nonsiliconized parent product. A siliconized butyl is a butyl and a siliconized acrylic is an acrylic. Neither is a silicone. The acrylic sealants and the skinning butyls are used in residential construction because they perform well for several years on common substrates if used within their limits and are readily available at most hardware stores. Their price is generally between that of low-performance and high-performance sealants.

High-performance sealants have movement capacities greater than ±12.5% of the joint width, even up to or greater than ±25%. Sealants in this class are chemically curing elastomers. The families include urethanes, the best of the acrylics, polyether-modified silicones, polysulfides, and silicones. In building construction, these sealants are usually intended for use in exterior high-movement applications. Some of the silicones are also used as structural adhesives to attach stone, metal, or glass to the side of a building.

High-performance sealants are used as below-grade sealants, in reservoirs and canals, as well as in sealing joints between concrete slabs in a roadway. They are typically the most expensive sealants, and very few are found in common hardware stores. Many of the high-performance sealants have a short shelf life, which prohibits their use in the consumer market, but have sufficient quality to be used in most industrial applications. These sealants must be purchased from a construction or industrial distributor or directly from the manufacturer, who can identify points of purchase.

Specifications and Physical Properties

ASTM C 962, "Standard Guide for Use of Elastomeric Joint Sealants," provides an excellent source of information regarding joint design for the building industry, the importance of two-sided adhesion, recommended joint widths for various movement capabilities, types of sealants appropriate for various substrates, and a table of the coefficients of linear thermal expansion of common building construction substrates. This document can assist in determining the proper width of sealant joints for a particular building construction application, as well as provide joint design and installation guidelines. This specification is being revised to include medium-performance sealants. A useful guide to proper application procedures is available from the Sealant, Waterproofers, and Restoration Institute (SWRI) in the form of an application manual and a videotape.

ASTM C 920, "Standard Specification for Elastomeric Joint Sealants," is a comprehensive specification for high-performance sealants for the construction industry. This specification describes the minimum acceptable properties required to meet the specification, as well as movement capabilities.

The Japanese Industrial Standard JIS A 5757 is also a comprehensive specification for high-performance construction industry sealants. The International Standards Organization is working toward the goal of establishing global sealant standards. The European community has standards for each country. European sealant manufacturers may make various products, but one may be sold only in Germany and another only in France. The unification of Europe in 1992 should bring a common European standard. Global standards may not be in place until the early part of the next century.

There are many other standards unique to industries or applications other than building construction, such as standards for highways, runways, fire seals, oven seals, and electronic seals. The article "Specifications and Standards for Adhesives and Sealants" in this Volume discusses these topics in more depth. Several physical properties that are universal to most sealants are described below.

Hardness. The most typical hardness scale for sealants is the Shore A scale, which ranges from 1 to 100. No resistance is represented by 0, and 100 represents total resistance. A durometer is used to measure hardness based on the amount of penetration of a blunt probe into the sealant. Most sealants range from 10 to 60 in Shore A durometer values. The Shore 00 scale is used to measure extremely soft sealants. A reading of 100 on Shore 00 correlates to a reading of 10 on a Shore A scale. A pencil eraser would have a Shore A reading of about 70, and a clear silicone purchased from a hardware store would have a Shore A of about 30. ASTM C 661, "Standard Test Method for Indentation Hardness of Elastomeric-Type Sealants by Means of a Durometer," is the method most often used to determine hardness.

Modulus. There are many kinds of modulus, such as tangent modulus, secant modulus, and Young's modulus, but the most common form of modulus used in the building industry is described here. It is the measure of stress at a given strain of the sealant. Typical sealants exhibit a nonlinear stress-strain diagram when they are pulled to destruction. Two methods of measuring tensile strength are used when testing sealants. ASTM D 412, "Test Methods for Rubber Properties in Tension," measures the tensile strength and elongation at break for a slab of rubber. In this test, a slab of rubber is cut into a dumbbell shape and is pulled apart. When comparing sealants using this test, one can easily identify a higher-modulus (more stiff) product from a lower-modulus (less stiff) product. However, that is all that can be determined. Although this stress-strain relationship is very important from a theoretical point of view, most sealant users need a more practical value.

The best way to determine field performance and stress-strain relationships of a joint using an elastomeric sealant is to make a joint in accordance with ASTM C 719, "Standard Test Method for Adhesion and Cohesion of Elastomeric Joint Sealants Under Cyclic Movement (Hockman Cycle)." A 12.5 × 12.5 × 50 mm (½ × ½ × 2 in.) sealant joint is placed between plates that measure 25 × 75 mm (1 × 3 in.) and is

pulled in tension. Pull rates used by manufacturers vary from 12.5 to 50 mm (0.5 to 2 in.) per minute. With the kinds of stress-strain curves that result, an engineer can easily estimate the forces being transferred through an elongated sealant joint.

All sealants exhibit a change in modulus with temperature. Silicones have the least change with temperature whereas the acrylics and PVA sealants exhibit a much larger change. Sealants will raise in modulus with a drop in temperature. As the temperature drops, joint widths increase due to panel shrinkage. A low modulus sealant product with a minimum modulus change in temperature is desirable for an exterior application because the sealant must extend relatively easily in cold weather for a long-term performance. Sealants that exhibit large modulus changes with temperature will generally fail in an exterior application within one or two winters because the joints will open in the cold and the product will become hard and brittle, causing a failure.

Movement Capability. In the United States, movement is determined by ASTM C 719, "Standard Test Method for Adhesion and Cohesion of Elastomeric Joint Sealants Under Cyclic Movement." Other countries use similar methods with variations. ASTM C 719 is probably the most severe general performance test to which building construction sealants are subjected. This test involves the cycling of sealant joints at a movement rate of 3.2 mm (⅛ in.) per hour after various exposures to water and hot and cold temperatures. The test method can be modified to provide results for nearly any range of movement. Typical building industry sealant specifications recognize only +12.5% and +25% joint movement capabilities. The ASTM C 719 test method allows the testing of any movement range, and some manufacturers make claims for such movements as +40/−25%, ±50%, and +100/−50%. The rule of thumb is that the lower the modulus, the higher the movement rating.

Cure Time. Each sealant will have a different cure time stated by the manufacturer. Users may differ with the manufacturer because certain values and times given by the manufacturer do not represent the users needs. Very few if any manufacturers will state a deep section cure rate in millimeters per day or week. This is an important consideration for people waiting for sealant cure to perform tests. Another important property rarely stated is tooling or working time. This is the time a mechanic can work the sealant into a joint before the sealant starts to skin or solidify. Almost always a tack-free time is stated. ASTM C 679, "Standard Test Method for Tack-Free Time of Elastomeric-type Joint Sealants," is the test method to determine this. It will give the time when a sealant will be dry to the touch of polyethylene. This time

is considerably longer than the tooling time.

Tear Strength. Tear strength is measured by ASTM D 624. This is a general measure of the toughness or abrasion resistance. Organic sealants will typically have a high tear strength value where silicones will not. One can get a general idea of the tear strength of a sealant by scraping the sealant with a fingernail. High-tear sealants are less prone to physical abuse by people and birds when compared to low-tear sealants. High-tear sealants will typically be found in traffic and sidewalk joints. Low-tear sealants can be successfully used in traffic joints, however, the joint must be recessed below the pavement so that the sealant is not abraded away by traffic.

Major Sealant Types

Oil-Base Caulks. As mentioned previously, these types of materials are low-performance, low-cost products that primarily comprise oils and fillers. They are used as crack fillers only and are not suited for exterior applications where some degree of movement is anticipated.

A typical formulation for an oil-base caulk would be (Ref 1) 25 to 30% linseed or soy oil, 6 to 12% fibrous filler, 40 to 60% calcium carbonate filler, 0.05 to 3% pigment, 0 to 2% gelling agents, and 0 to 1% catalyst.

The oils dry and or oxidize because of the cross linking of the conjugated double bonds in the oil in the presence of catalyst. These types can generally be painted over in 24 h and are typically used by consumers who buy a tube of caulk based on price only. Although their short life span makes them a poor investment in many applications, there is a market for them nonetheless.

Latex acrylic sealants, as the name implies, cure by the evaporation of water. These materials, first introduced in the late 1960s, are based on latex acrylic polymers designed by Rohm & Haas Company (Ref 2). The products have grown in popularity because of some excellent handling and durability characteristics. The advantages of these medium-performance materials include exterior durability, ease of application, water cleanability, low toxicity, low cost, low odor, and no flammability. In addition they can be applied to damp surfaces and have generally good adhesion to glass and ceramic. They are used by contractors in the residential construction market because they are easy to apply and can be painted over, after a brief cure, with latex paint.

The disadvantages of the product are that the application temperature must be above 5 °C (40 °F), shrinkage does occur, rain washout is possible, and tremendous modulus changes with temperature change can take place. Most latex caulks are medium-performance grades, with a movement capacity of 5 to 12.5% of joint width. Recently, Rohm & Haas has shown that a latex formulation can withstand ±25% movement (Ref 2). As with many other sealant types, surface priming is often not needed. The ease of cleanup contributes to the popularity of latex products in the do-it-yourself consumer market. However, the limited availability of primers to this market is a problem.

A current specification to which these materials are subjected is ASTM C 834, "Standard Specification for Latex Sealing Compounds." As the materials continue to improve, they may be included under ASTM C 920. The future for latex products is bright, regardless of whether their base is acrylic or not. Government regulations are becoming more strict in regards to chemical cure into the air. The manufacturer who produces a durable and high-performance latex caulk might have an advantage in the late 1990s and thereafter.

Polyvinyl acetate caulks have been available for a number of years. Typically a low-performance consumer product, they are based on the same technology used for white glues. These materials lack flexibility and are formulated with plasticizers to obtain flexibility.

Advantages of this low-performance material include ease of application, water cleanup, paintability, and good adhesion characteristics. Disadvantages are low movement, high modulus, shrinkage, and tremendous modulus change with temperature change.

Solvent acrylics have the durability found in latex acrylics and are well known for their unprimed adhesion. Curing is accomplished by the release of solvent, rather than water. Consequently, their odor, toxicity, and flammability levels are not low, as in their latex counterparts. Because of odor their chief applications tend to be for outdoor use. Solvent acrylics will tend to harden to a 45 to 60 Shore A durometer after a year and will become extremely stiff in cold weather.

The product has a moderate cost and medium performance. The solvent within the system allows the sealant to gain adhesion through dirt and grease, whereas other sealants require extremely clean surfaces. These materials have good exterior durability, second only to that of silicones. Disadvantages are that they require heating in the cartridge for cold weather gun dispensibility and have offensive odors during cure. Their joint movement capability is low and the material has tremendous modulus change with temperature change.

Butyl sealants described here are in the form of solvent-release sealants, deformable preformed tapes, hot applied sealants, and polyisobutylene sealants.

Solvent-release butyl sealants that meet ASTM C 1085, "Standard Specification for Butyl Rubber Based Solvent Release Sealants," have medium performance capabilities, with a maximum 10% joint movement rating. These materials are moderate in cost and were the premium sealants of the 1940s and 1950s. Adhesion characteristics are a benefit. The disadvantages include shrinkage, staining of adjacent surfaces, bubbling on hot surfaces, poor recovery from extension, and a 5 to 15 year service life.

Preformed butyl tapes are used extensively in the glazing industry. They are 100% solids and do not exhibit shrinkage. Used as a press-in-place application between glass and aluminum, they must have 20 to 30% compression. Their service life is about 10 years, after which the glazing system is resealed with a high-performance sealant.

Hot-applied butyl sealants are used in the insulated glass (IG) industry. The sealant bonds two pieces of glass and keeps moisture vapor from entering the unit. The material is applied at 150 to 204 °C (300 to 400 °F) and is rigid at room temperature.

Polyisobutylene is used in two forms. One is a noncuring mastic sealant used in splice joints in curtain wall applications. This form is self healing and is used in internal high-movement applications where that property is desirable. However, staining of the exterior of the building is possible because water runs over the sealant and out the weep holes in the glazing system. This sealant is typically 25% solvent, and shrinkage can be a problem if insufficient sealant is used.

Polyisobutylene is also sold as a 100% solids material extruded into 3 mm (⅛ in.) diam beads that are then used as the primary moisture seal in a dual-sealed IG unit construction. This application as a primary seal is typically used in conjunction with a secondary seal of silicone, urethane, or polysulfide.

Polysulfides, which entered the construction market in the early 1950s, were the first high-performance sealants in this market segment. Polysulfide sealants come in three forms, one-part, two-part manually mixed, and two-part mechanically meter-mixed formulations. Their many applications are described in the article "Polysulfides" that follows. This overview addresses only construction industry applications.

Two-part manually mixed polysulfide sealants are excellent high-performance sealants. Their advantages include good extensibility (±25%), good recovery, paintability, and good cure rate. They have reasonable adhesion, durability, good weatherability, and solvent resistance. Some formulations are used in reservoirs and canals for long-term water-immersion applications.

Disadvantages include the required mixing, occasional failure after 10 years due to

compression set, the need for primers or adhesion promoters with certain surfaces, and surface cracking due to ultraviolet (UV) radiation exposure.

Two-part mechanically meter-mixed polysulfide sealants are used extensively as IG sealants and, in the aerospace industry, as fuel-resistant sealants. Polysulfide IG sealants do not require a primary seal for some applications because the polysulfide has a low moisture vapor transmission rate.

One-part polysulfide sealants have the same advantages as the two-part sealants, and they do not require mixing. Because of their low moisture vapor transmission rates, they are extremely slow to cure. This can result in a deformed sealant due to joint movement before cure. The one-part sealants are as susceptible to compression set failure as are the others, and have an offensive odor during cure.

Polysulfides are moderate to high in price, as one would expect for high-performance sealants. Their use in the United States is on the decline, but polysulfides are the premium products in Europe, particularly in IG applications. European sealant formulators use higher polymer-to-filler ratios than do U.S. formulators, resulting in a higher-performance, more durable product.

Variations of these sealants, such as those that have mercaptan end groups in their chemistry, may cause some modifications in this market data. Mercaptan-terminated polymers are further described in Ref 3.

Urethanes also come in three forms: one-part; two-part manually mixed, field pigmentable; and two-part mechanically meter-mixed sealants. These sealants are high-performance sealants with excellent adhesion, durability, and life expectancy. The two-part formulations have a fast cure rate, whereas the one-part formulation has a slow cure rate. The two-part meter-mixed sealants have an excellent moisture vapor transmission rate and can be used in IG applications without a primary polyisobutylene seal. The urethanes are paintable, and new formulations have lower modulus and higher movement capabilities than previous products. These sealants have high tear strength and are used in traffic applications.

The disadvantages of these sealants are that they lose adhesion to glass upon UV exposure, harden with age, crack and chalk because of UV exposure, and have significant modulus change with temperature change. The upper-temperature limit of 80 °C (180 °F) can be a disadvantage, for example, on the south wall of a dark facade in Phoenix, AZ. At 80 °C (180 °F), the cured sealant can revert into liquid and bleed into the facade. These sealants should not be used beyond their capabilities and should be protected from climate extremes.

Although the sealants do crack and chalk from UV exposure, this effect can be a

Table 1 Summary of sealant properties

	Type of sealant			
	Low performance	Medium performance		
Characteristic	Oil base, one part	Latex (acrylic), one part	Butyl skinning, one part	Acrylic, solvent-release
Maximum recommended joint movement, percent of joint width	±3	±5, ±12.5	±7.5	±10 to 12.5
Life expectancy, years(a)	2–10	2–10	5–15	5–20
Service temperature range, °C (°F)	−29 to 66 (−20 to 150)	−20 to 82 (−20 to 180)	−40 to 82 (−40 to 180)	−29 to 82 (−20 to 180)
Recommended application temperature range, °C (°F)(b)	4–50 (40–120)	4–50 (40–120)	4–50 (40–120)	4–80 (40–180)
Cure time to a tack-free condition, h(c)	6	0.5–1	24	36
Cure time to specified performance, days(c)	Continues	5	Continues	14
Shrinkage, %	5	20	20	10–15
Hardness, new (1–6 mo), at 25 °C (75 °F), Shore A scale	· · ·	15–40	10–30	10–25
Hardness, old (5 years) at 25 °C (75 °F), Shore A scale	· · ·	30–45	30–50	30–55
Resistance to extension at low temperature	Low to moderate	Moderate to high	Moderate to high	High
Primer required for sealant bond to				
Masonry	No	Yes	No	No
Metal	No	Sometimes	No	No
Glass	No	No	No	No
Applicable specifications				
United States	TT-C-00593b	ASTM C 834, TT-S-00230	TT-S-001657	TT-S-00230
Canada	CAN/CGSB 19.6-N87		CGSB 19-GP-14M	CGSB 19-GP-5M

	Type of sealant					
	High performance					
	Polysulfide		Urethane		Silicone	
Characteristic	One part	Two part	One part	Two part	One part	Two part
Maximum recommended joint movement, percent of joint width	±25	±25	±25	±25	±25 to +100/−50	±12.5 to ±50
Life expectancy, years(a)	10–20	10–20	10–20	10–20	10–50	10–50
Service temperature range, °C (°F)	−40 to 82 (−40 to 180)	−51 to 82 (−60 to 180)	−40 to 82 (−40 to 180)	−32 to 82 (−25 to 180)	−54 to 200 (−65 to 400)	−54 to 200 (−65 to 400)
Recommended application temperature range, °C (°F)(b)	4–50 (40–120)	4–50 (40–120)	4–50 (40–120)	4–80 (40–180)	−30 to 70 (−20 to 160)	−30 to 70 (−20 to 160)
Cure time to a tack-free condition, h(c)	24(c)	36–48(c)	12–36	24	1–3	0.5–2
Cure time to specified performance, days (c)	30–45	7	8–21	3–5	5–14	0.25–3
Shrinkage, %	8–12	0–10	0–5	0–5	0–5	0–5
Hardness, new (1–6 mo), at 25 °C (75 °F), Shore A scale	20–40	20–45	20–45	10–45	15–40	15–40
Hardness, old (5 years) at 25 °C (75 °F), Shore A scale	30–55	20–55	30–55	20–60	15–40	15–50
Resistance to extension at low temperature	Low to high	Low to moderate	Low to high	Low to high	Low	Low
Primer required for sealant bond to						
Masonry	Yes	Yes	Yes	Yes	No	Yes
Metal	Yes	Yes	No	No	. . .	No
Glass	No	No	No	No	No	No
Applicable specifications						
United States	ASTM C-920, TT-C-00230C	ASTM C 920, TT-S-00227E	ASTM C 920, TT-S-00230C	ASTM C 920, TT-S-00227C	ASTM C 920, TT-S-00230C TT-S-001543A	ASTM C 920, TT-S-001543A
Canada	CAN 2-19.13M		CAN 2-19.13M		CAN/CGSB 19.18	TT-S-00227E

Note: Data from manufacturer data sheets; U.S. made sealants are generally considered. (a) Affected by conditions of exposure. (b) Some sealants may require heating in low-temperature environments. (c) Affected by temperature and humidity. Source: Adapted from Ref 1

self-cleaning feature when used in exterior applications. The color of the joint will remain intact, with the exception of lighter colors, which will tend to yellow upon UV exposure.

These products are moderately high in price, as one would expect of high-performance sealants with good physical properties, durability, and a 10 to 20 year service life. Urethane formulations are further described in Ref 4.

Modified silicones represented a new class of sealants for the early 1980s. They were developed in Japan to solve some of the bleeding and staining problems that occurred with highly plasticized Japanese silicone sealants. It would be more appropri-ate to call them modified urethane sealants because they use a polyether backbone, like urethanes, but are cross linked with silane cross linkers, like silicone sealants. The silicone or silane typically represents less than 5% of the total composition. The sealant produced has high-performance capabilities with many of the same characteristics as urethane, but the one-part systems, like a one-part silicone, cure much faster than a one-part urethane.

As this sealant gains more acceptance in the world marketplace and is continuously improved in weatherability, it will take a share from both the urethane and the silicone market. Although it does have the limitation of an 80 °C (180 °F) maximum temperature and shows surface cracks with UV radiation exposure, it is still in its infancy.

Silicone sealants are clearly the highest-performance sealants in the world market today because of their high-temperature stability, low-temperature flexibility, resistance to weathering, capability as structural adhesives, quick cure, and excellent adhesion and because they have the largest choice of modulus. However, their distinct disadvantages include dirt pickup, low tear strength, and nonpaintability. Silicones usually command the highest prices because they offer the highest performance characteristics and weathering abilities, properties unsurpassed by any other sealant.

Most silicone sealants cure by reaction with moisture in the air, but some cure in confinement. Silicones come in one package (ready to use) and two-part manual or mechanical meter-mix forms. For some applications, such as fire-rated seals, a foam seal is available. The products generally take moisture from the air and react to become an elastomer, giving off a by-product.

Silicone sealants have major markets in structural glazing, construction, automotive, consumer do-it-yourself, highway, aerospace, fuel containment, appliance, industrial, and specialty areas. Most markets are based on one or more of the special properties listed earlier, but the longevity of the sealant in a particular application is also very important.

Government regulations are more closely scrutinizing the by-products given off by sealants. Water-base silicone caulks have been developed because their water cleanup, ease of application, low toxicity, and low volatility are desirable features. These new products are true silicones and not siliconized acrylics as was mentioned earlier. The cured rubber properties of these products show excellent weatherability and durability found in a true silicone.

The various sealants described in this article are compared in Table 1, which can be used as a guide in the process of product selection.

REFERENCES

1. J.M. Klosowski, *Sealants in Construction*, Marcel Dekker, 1989
2. J. Lomax, "Acrylic Polymer Caulks and Sealants," Paper presented at the Sealant Short Course, Adhesives and Sealants Council, Jan 1990
3. E.A. Peterson, "Formulating Sealants With Mercaptan Terminated Polymers," Paper presented at the Sealant Short Course, Adhesives and Sealants Council, Jan 1990
4. J.F. Regan, "Urethane Formulations," Paper presented at the Sealant Short Course, Adhesives and Sealants Council, Jan 1990

Polysulfides

Richard A. Giangiordano and Christer Sorensson, Products Research & Chemical Corporation

POLYSULFIDE SEALANTS are described in this article in terms of their characteristics, product forms, applications, advantages, limitations, and other parameters. The article "Polysulfides" in the Section "Adhesives Materials" in this Volume offers corresponding descriptions in an adhesives context.

Chemistry

A polysulfide liquid polymer is synthesized by the reaction of aqueous sodium polysulfide with bis(2-chloroethyl)formal. Cross linking can be built in by including a predetermined amount of 1,2,3-trichloropropane. A mixture of sodium disulfide and bis(2-chloroethyl)formal and 1,2,3-trichloropropane is reacted at 100 °C (212 °F), producing a mixture of chain lengths in which sulfur is present as —C—S—S—C— or —C—S—S—S—C—; some chains have even higher sulfur contents. Polysulfide terminals (—S—S—S) are produced at the ends of the chain. A reaction with sodium sulfite reduces most of the polysulfide groups to disulfides and the terminal groups to thiols (—SH).

Conventional mercaptan-terminated liquid polymers (HS—R—SH) are available in molecular weights ranging from 1000 to 7000, with cross-linking densities ranging from 0.05 mol% to 2.0 mol%. A typical polysulfide structure is shown in Fig. 1. When polymerized by mixing with an oxidizing agent, water is separated out, with the hydrogen coming from the mercaptan groups of two polymer molecules, and the oxygen being supplied by the oxidizing agent, as shown in Fig. 2. The molecules are joined at the sulfurs.

Manganese dioxide, lead dioxide, calcium peroxide, and dichromates are among the preferred oxidation agents. Manganese dioxide activated with an alkaline wash is the most common curing agent used today, except where light colors are required. Zinc peroxide or calcium peroxide can be used in

a two-part system to produce a white or cream-colored sealant. Polysulfides can also be reacted with epoxy resins in an addition curing system. A typical reaction is shown in Fig. 3.

Sources of conventional polysulfide polymer include Morton International (Moss Point, MS) and Toray Thiokol in Japan. An additional source of polysulfide polymer is Chemiewerke, located in East Germany.

Modified Polysulfide. A recent advancement has been made by chemically modifying conventional polysulfide polymers in a one-step reaction. This new class of polymers exhibits lower viscosity and improved compatibility with plasticizers, pigments, and fillers. They therefore offer greater latitude in the formulation of high-performance sealants, adhesives, and coatings with lower volatile content.

The products formulated from these polymers show improved thermal and chemical resistance, lower permeability to inert gases and fuel vapors, and superior physical properties, when compared to sealants, adhesives, and coatings based on conventional polysulfide polymers. The chemical modification reaction and the resultant chemical structure are shown in Fig. 4.

Other Mercaptan-Terminated Polymers. Polysulfide polymer chemistry was advanced in 1974 when a polyoxypropylene urethane backbone with mercaptan terminals was synthesized. This development greatly simplified the control of molecular weight and minimized the presence of impurities from side reactions. Although the backbone is significantly different from conventional polysulfide, the cross-linking chemistry is the same, and the cured product is a polysulfide rubber. A typical structure is shown in Fig. 5.

Polythioether Polymer. A new polymer that contains sulfur in the form of thioether is also available. Unlike conventional polysulfide polymer, it does not contain S—S or formal linkages in the polymer structure. These bonds are the weakest

$$-\!(CH_2CH_2-O-CH_2-O-CH_2-CH_2-S-S)_n$$

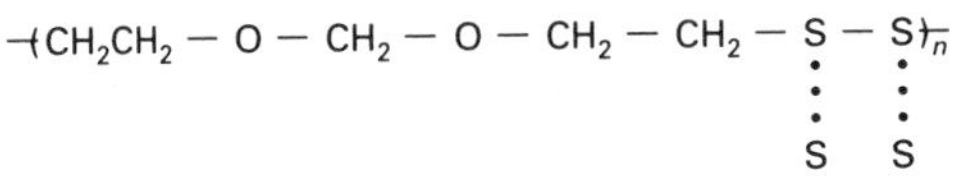

Fig. 1 Conventional polysulfide structure

links in conventional polysulfides. The polythioether polymer is highly resistant to fuel and organic fluid, and has improved thermal stability, compared to conventional polysulfide.

Polythioether polymers may be terminated with a variety of end groups. In addition to the common polysulfide terminal of mercaptan, the options include hydroxy, silyl, olefin, or nonreactive end groups. The unique ability to terminate this polymer in a variety of ways allows the formulator access to several curing system chemistries. This avails a range of polysulfide polymers with unique properties for special high-performance applications. The chemical structure of the polythioether polymer is compared to conventional polysulfide polymers in Fig. 6.

Forms

Functional Types. Polysulfide sealants are used in the aerospace and defense industries as well as the construction and industrial areas. Compounded products are significantly different because various curing systems and application properties are designed into the products to meet the needs of the specified industry.

In the aerospace industry, jet fuel resistance is a common requirement. Therefore, all sealants compounded for this application must meet certain specifications for fuel resistance. In the construction industry, polysulfide sealants must be compounded to yield a low-modulus sealant that is capable of accommodating the movement that occurs in an expansion joint.

By way of example, ASTM TTS00230 is a specification that suits many of the require-

$$HS-[R-S]_{\overline{n/2}}\,S-(R')-S-[S-R]_{\overline{n/2}}\,SH + HS-[R-S]_{\overline{n/2}}\,S-(R')-S-[S-R]-SH \xrightarrow{[O]} HS-R-S-S-R-SH + H_2O$$

Fig. 2 Polymerization of conventional mercaptan-terminated polysulfide

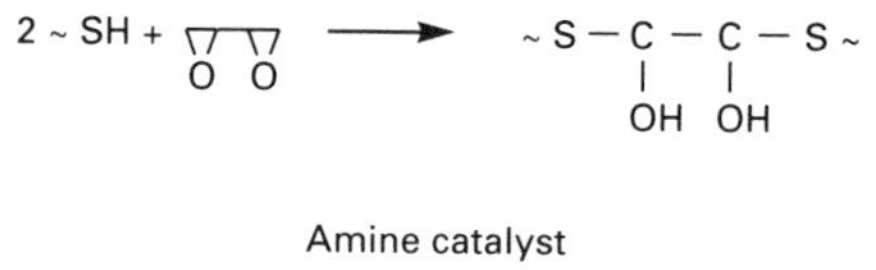

Amine catalyst

Fig. 3 Addition curing mechanism for mercaptan-terminated polymers

ments for a sealant used in high-performance construction areas. The specification includes a cycling portion in which samples of the sealant bonded to various substrates, including concrete, aluminum, and glass, are extended and compressed at various temperatures. The sealant must be elastic in order to accommodate this movement without raising a high load on the adhesive bond. Sealants with high-modulus properties can, in fact, tear the concrete apart during the extension cycle. Low-modulus properties can be achieved in several ways. Selection of a polymer with lower cross-link density, the use of efficient and effective plasticizers, and compounding with pigments that have less of a reinforcing effect, are all suitable means.

Polysulfide sealants are used extensively in the production of insulating glass units worldwide. This application requires that the compounded sealant possess excellent barrier properties, both to moisture and to inert gases. Improvement in the barrier properties of a sealant begins with the polymer selection. Various polysulfide polymers have differing permeability rates, which can be improved (lowered) by using plasticizers, as well as carefully selected fillers.

Commercial Forms. Polysulfide sealants are available in various viscosity and thixotropy ranges, from flow-type liquids to heavy-type nonsagging pastes. These sealants are available in multicomponent kits (0.004, 0.02, and 0.2 m^3, or 1, 5, and 55 gal), as well as standard caulking cartridges, and two-part injection cartridge kits. All kits are available from the sealant producer or through a distribution network.

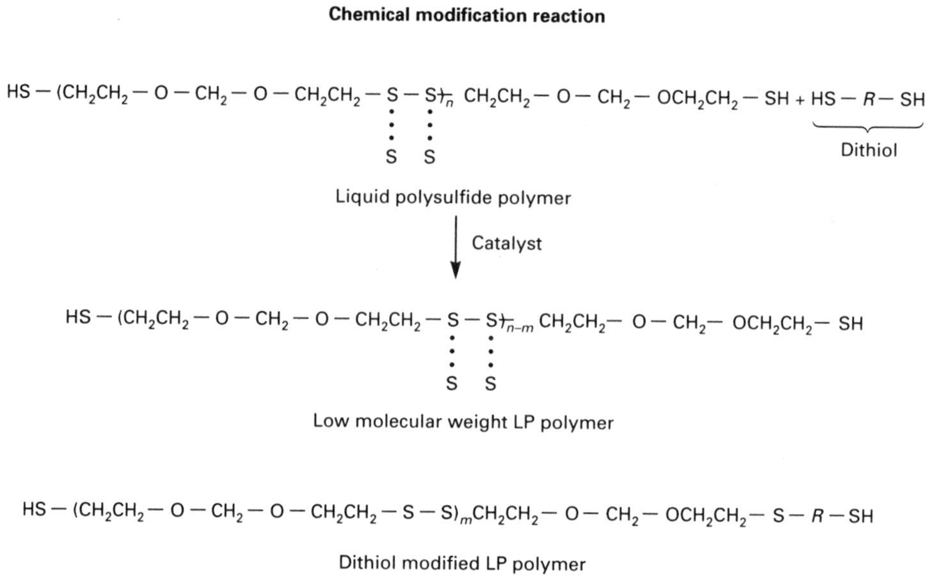

Markets and Applications

Aerospace and Defense

Polysulfide sealants are used in a large number of specialized aerospace applications, including fuel tank sealants, pressure sealants, corrosion-inhibiting sealants, windshield sealants, aerodynamic smoothing compounds, quick-repair sealants, and, more recently, electrically conductive sealants.

Aircraft sealants are typically available in a wide range of pot lives (10 min to 336 h) and in three main consistency classes: brushable (Class A), fillet or gun grade (Class B), and faying-surface grade (Class C). Some sealants are also available in a sprayable version.

Chemical modification reaction

Liquid polysulfide polymer

Catalyst

Low molecular weight LP polymer

Dithiol modified LP polymer

Fig. 4 Modified polysulfide

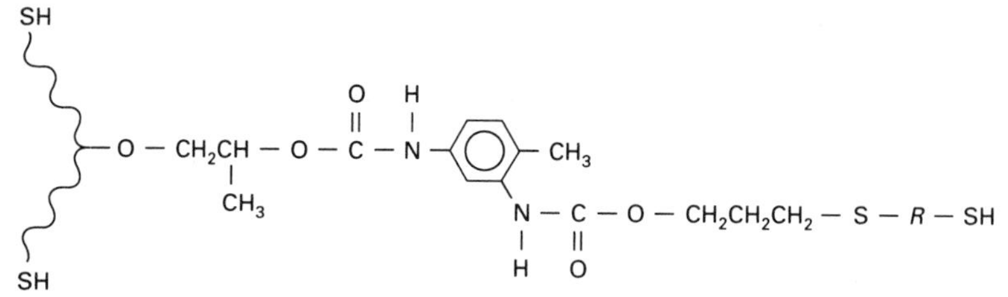

Fig. 5 Mercaptan-terminated polyoxypropylene urethane

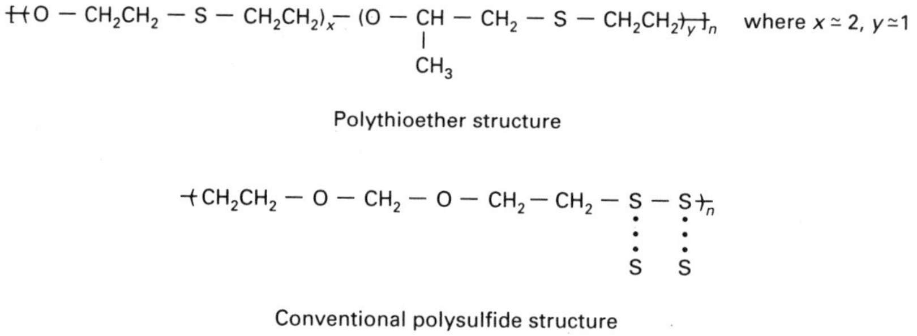

Polythioether structure

Conventional polysulfide structure

Fig. 6 Polythioether and conventional polysulfide structure comparison

The use of sealants in the aerospace industry is governed by a multitude of military, industrial, and customer specifications. Some of the major specifications are shown in Tables 1 and 2.

Construction

Polysulfide sealants were introduced to the construction industry in the 1950s as the first high-performance sealants. Because of their elastomeric and adhesive properties, they replace mastic materials, such as butyls and oil-base caulking putties, in high-performance expansion joints.

Originally, polysulfide sealants for this industry had two components. Later, one-component, high-performance sealants were introduced. In North America today, polysulfides have lost market share in the construction industry to other synthetic elastomeric products, such as silicone and polyurethane. However, polysulfide sealants are still used in some specialty construction and civil engineering applications that require chemical resistance. In Europe and Asia, polysulfides still maintain a significant share of the construction sealant market.

Industrial

The primary industrial use of polysulfide sealants is in the manufacture of insulating

Table 1   Major aerospace sealant specifications

Specification	Product governed by specification
MIL-S-8516	Sealing electrical components against air, dust, and moisture
MIL-S-8802	Fuel tank sealant resistant to aviation gasoline, jet fuel, and hydraulic fluids, temperature range of −55 to 135 °C (−65 to 275 °F)
MIL-S-81733	Corrosion-inhibitive, fuel-resistant sealing compound
MIL-S-83430	Fuel tank sealant resistant to jet fuel, temperature range of −55 to 180 °C (−65 to 360 °F)
MIL-S-29574	Polythioether sealing compound, fuel and high temperature resistant, fast curing at ambient and low temperatures
AMS 3266	Two-part polythioether sealing compound, elastomeric, electrically conductive
AMS 3267	Low adhesion, corrosion-inhibiting sealant, for access doors
AMS 3268	High-temperature resistant, corrosion-inhibiting sealant
BMS 5-26	Integral fuel tank sealant
BMS 5-95	Pressure, environmental, and fuel cavity sealant, corrosion inhibitive

Table 2   Major construction sealant specifications

Specification	Product governed by specification
Federal Specification SS-S-200E	Airport runway sealant
Federal Specification TT-S-00227E	Two-part elastomeric joint sealant
Federal Specification TT-S-00230C	One-part elastomeric joint sealant
ASTM C 920	Elastomeric joint sealant

glass units. Sealed dual-pane glass units are produced around the world to improve the energy efficiency of windows. Polysulfides maintain the largest market share in this industry.

In North America, insulating glass sealants based on a mercaptan-terminated polyoxypropylene urethane have gained a significant market share, replacing many conventional polysulfide-base insulating glass sealants. The superior ultraviolet and water resistance of the newer mercaptan-terminated polymer, as well as its ability to retain inert gas such as argon, has allowed the insulating glass industry to upgrade its performance requirements. The technology whereby argon gas is used in conjunction with low-emissivity glass to significantly improve the thermal properties of a window has only recently been introduced to North America, although it currently holds a small market share. It is believed that by the mid 1990s, high-performance windows will be the standard of the industry.

Several important specifications are used to qualify the performance of insulating glass units (ASTM E 774 is used in the U.S.; 12 GP-8, in Canada; and DIN 1286, in Germany). Artificial aging tests monitor the performance of the sealants through various temperature and humidity extremes. The DIN 1286 Part 2 also measures the ability of an insulating glass unit to retain inert gas, such as argon. This is the only gas-retention specification for insulating glass units that is active in the world.

Major Suppliers

Aerospace and Defense Market. The major suppliers in North America are ChemSeal, Division of Flamemaster, Sun

Valley, CA; and Products Research & Chemical Corporation, Glendale, CA. In Europe, the major suppliers include Chem-Seal (through distributors); Le Joint Francais and 3-M, in France; and Products Research & Chemical, UK, Limited, Newcastle upon Tyne, England. In Asia, the primary suppliers are ICI, Melbourne, Australia; and Yokohama Rubber Company, Tokyo, Japan.

Industrial and Construction Market. The major U.S. suppliers are:

- ChemRex, Minneapolis, MN
- Morton International, Chicago, IL
- Pecora, Philadelphia, PA
- Products Research & Chemical Corporation, Glendale, CA
- Tremco, Cleveland, OH

In Europe, major suppliers are:

- Chemetal, Frankfurt, West Germany
- Evode, Stafford, England
- Expandite, England
- Kommerling, Pirmasens, West Germany
- Le Joint Francais, Bezons, France
- Products Research & Chemical, UK, Limited, Newcastle upon Tyne, England
- Servicized, Slough, England
- Teroson, Heidelberg, West Germany

Major suppliers of polysulfide sealants in Asia include Yokohama Rubber Company, Tokyo, Japan; and ICI, Melbourne, Australia.

Cost Factors

Depending on the particular application and the degree of performance needed, selling prices of aerospace polysulfide sealants, adhesives, and coatings vary from $11/L ($40/gal) to several hundred dollars per gallon. Prices vary according to kit size and the performance parameters required.

Sealants for the standard construction market range in price from $4 to 7/L ($15 to 25/gal). Certain specialty sealants are priced higher.

The selling price of sealants for industrial applications varies greatly, depending on performance requirements, purchase volume, package size, and so forth. Therefore, sealant selling prices can range from $4/L ($15/gal) to more than $26/L ($100/gal).

Competing Materials and Systems

Aerospace. Because of the uniqueness of its backbone and the sulfur groups contained therein, polysulfide polymers have fuel- and chemical-resistance properties that offer performance advantages. Although not directly a competing system for normal riveted airframe manufacture, the use of adhesives to bond fuel tank structures eliminates the use of polysulfide faying-surface sealants. However, this type of airframe still requires a polysulfide sealant to impart fuel resistance to the fuel tank structure.

Conventional polysulfide sealants are somewhat limited in regard to long-term, high-temperature performance. Sealants based on polythioether polymers that eliminate the disulfide link in the backbone exhibit improved long-term, high-temperature performance, making them suitable for high-performance military aircraft applications.

Fluorosilicone-base sealants have not been, and probably will not be, used as extensively as polysulfide-base sealants in fuel and nonfuel applications, for a number of reasons:

- As of 1990, sealants that use fluorosilicones are more costly than those using polysulfides
- The availability of different product classes is not as broad for fluorosilicones as it is for polysulfide sealants, which are available for brush, fillet, faying surfaces, and spray applications. In addition, a wide variety of worklife versions are offered to suit various environmental conditions and production/engineering requirements
- Fluorosilicone sealants provide only very moderate adhesion, compared to polysulfides (1.75 kN/m versus 3.50 kN/m, or 10 lbf/in. versus 20 lbf/in.)
- Like silicone, fluorosilicone tends to contaminate adjacent surface areas, thereby increasing the possibility of causing bond failures in these areas
- Fluorosilicone is outperformed by polysulfide in terms of fuel resistance
- Fluorosilicone sealants have application limitations because they require moisture to activate the curing reaction and therefore cannot be applied to faying surfaces
- The reactivity to moisture of one-part fluorosilicone shortens its shelf life (6 mo), compared to polysulfide (9 mo)

Construction. Because the fuel and chemical resistance of polysulfides are not necessary in most construction applications, other one- and two-component elastomeric materials, such as polyurethanes, acrylics, and silicones, have replaced many of the polysulfide sealants used in the construction industry.

Table 3 Property data comparison

	AMS 3357 requirements (fluorosilicone)	MIL-S-83430 requirements (polysulfide)	PR-1770 B-2(a)	CS-3204 B-2(b)
Nonvolatile content (minimum), %	92	97	97.8	93
Flow, mm (in.)	1.3–13 (0.05–0.5)	2.5–19.0 (0.1–0.75)	2.5 (0.10)	10 (0.40)
Application life at 2 h (minimum), g/min (oz/min)	N/A	15 (0.5)	17 (0.6)	N/A
Tack-free life (maximum), h	1.3	24	<24	24
Hardness (Shore A)	35	35	53	50
Tensile strength (dry), MPa (ksi)				
Jet fuel reference fluid immersion	400	250	420	300
14 days, 60 °C (140 °F)	180	200	270	N/A
Elongation (dry), %				
JRF immersion	200	250	320	350
14 days, 60 °C (140 °F)	90	200	420	N/A
Peel strength, kN/m (lbf/in.)				
7 day, immersion, 60 °C (140 °F)	1.75 (10)	3.50 (20)	9.6 (55)	7.0 (40)
70 day, immersion, 60 °C (140 °F)	0.88 (5)	3.50 (20)	9.6 (55)	7.0 (40)

(a) Polysulfide sealant of Products Research and Chemical Corporation, where B-2 designates a heavy-type product with a minimum 2-h application life. (b) Polysulfide sealant of ChemSeal

Table 4 Compounding ingredients

Chemical ingredient	Function
Fillers	Usually various forms of a calcium carbonate, but also may be carbon black, glass or phenolic bubbles (for light weight), metallic oxides, vermiculite, and others. Fillers affect viscosity, hardness, block flow, extrusion rate, thixotropy, elongation, cohesive strength, and other properties of the system.
Accelerators and retarders	Affect the work life and cure rate of the formulation. A common retardant is stearic acid. Accelerators are commonly amine compounds.
Adhesion promoters	Influence the ability to adhere to many surfaces. Adhesion additives include such compounds as epoxies, phenolics, and silanes.
Plasticizers	Affect modulus, elongation, and other properties. They are used as vehicles for the catalyst. Excessive use, or poor selection, can compromise environmental resistance, reduce hardness, and reduce strength. Typical plasticizers used include low volatility phthalates and chlorinated paraffins.
Pigments	Coloring agents are often required by the customer for a particular end use, such as silver color for aerodynamic smoothing, black for certain trim areas, and other distinctive colors to identify special properties. Metallic oxides, carbon black, and other materials are used.
Thixotropic agents	Several types are used to produce nonslumping characteristics while permitting good extrusion rates.

Industrial. Low-performance materials used in insulating glass (IG) production, such as hot-melt butyl sealants, have reduced somewhat the market share once enjoyed by polysulfides because they are less expensive. However, polysulfide sealants still maintain the largest market share worldwide in this industry.

High-performance silicone sealants have been used for approximately 10 years in IG production. However, the poor moisture vapor transmission and barrier properties of silicone sealants necessitate the inclusion of a barrier when constructing an IG unit that uses a silicone exterior seal. The normal barrier used is polyisobutylene. This construction is commonly referred to as a dual-sealed construction. The polyisobutylene acts as the water vapor barrier in this unit construction. Silicones also do not possess the necessary barrier properties for inert gases. Argon gas is commonly used in the insulating glass unit construction in order to improve its thermal performance. Polyisobutylene is a mastic material that does not chemically react with the glass. It cannot be relied upon as an inert gas barrier; therefore, silicones are not used in units that are manufactured containing an inert gas such as argon.

Silicone dual-sealed units are used exclusively in structural glazing applications because of the excellent ultraviolet resistance of silicone. This design incorporates the sealed insulating glass units as a structural member of the building. Because it is bonded directly to the building, no metal mullions or sashes are used to contain the unit.

Application Parameters

Aerospace. High-performance polysulfide sealants supplied to this industry are two-component systems that require mixing. Several types and sizes of containers are available. The most common vehicle is the two-component kit, in which a premeasured amount of base and catalyst are packaged separately and mixed on site. In larger applications, the kit may be a 0.2 m³ (55 gal) drum and a 0.02 m³ (5 gal) pail of catalyst, which are dispersed via a two-component mixing system.

A unique package available to aerospace manufacturers is a frozen cartridge supplied with one-part premixed sealants. The user brings the frozen cartridge to room temperature and applies the sealant. Mixing on site is thus eliminated.

Construction and Industrial. The construction industry generally uses polysulfide sealants in two-component kits (0.006 m³, or 1.5 gal), and in one-component, standard-size caulking cartridges. The two-part packages are premeasured for mixing on the job site. A half-inch drill with a mixing paddle is generally employed to mix the two components at the time of application.

In an industrial insulating glass application, materials are generally supplied in a 0.2 m³ (55 gal) drum and a 0.02 m³ (5 gal) pail. A metering, mixing, and dispensing system is used to apply the sealant.

Properties

Polysulfide sealants possess extremely good fuel resistance, and therefore can be used in many environments where other polymer systems are not suitable. In addition, their resistance to hydrocarbon fluids is considered excellent. Table 3 provides a variety of property values.

Additives and Modifiers

Polysulfide sealants are compounded products that contain many ingredients, all of which are selected to impart specific physical properties. Table 4 indicates the general class of materials used in the compounding of high-performance polysulfide sealants.

Processing Parameters

Shelf and Pot Lives. In the aerospace and defense industries, the typical shelf life for two-component polysulfide sealants can be 6, 9, or 12 months, depending on sealant type and specification. Pot lives vary from extremely short, in the case of quick repair sealants (10 to 15 min), to extremely long (up to 336 h), in the case of faying-surface sealants that are intended for large assemblies. Typical storage requirements are for locations where temperatures are maintained above 0 °C (32 °F) but below 25 °C (80 °F).

Premixed and frozen polysulfide sealants can be stored at −40 °C (−40 °F) or below for months without significant change. Industry practice, however, involves a 21- to 30-day shelf life, after which some companies retest and extend shelf life, as indicated.

Substrate Requirements and Primers. Most aerospace sealants are formulated to adhere, without the use of primers, to normal aircraft construction materials, such as aluminum alloys, titanium, and stainless steel. Most metals are initially passivated in some way. However, surface conditioners or adhesion promoters (actually, very thin primers) are used extensively in aircraft manufacture for additional corrosion resistance, as well as adhesion promotion. The surface conditioners or adhesion promoters are often dyed (red, green, yellow, or blue) to provide a visible film for inspection.

Quick-repair sealants are usually used with primers. If the surface area to be resealed has been cleaned without damage to the original primer, an adhesion promoter is applied and resealing is accomplished. If, on the other hand, the surface of the primer has been penetrated, the primer is removed, the aluminum surface is treated with chromate and sodium phosphate, and resealing is accomplished, using an adhesion promoter before applying the sealant.

Butyls

J.J. Higgins and F.C. Jagisch, Exxon Chemical Company

BUTYL RUBBER and polyisobutylene are elastomeric polymers used as primary binders, tackifiers, and modifiers in a wide variety of adhesive and sealant compounds. These rubbers are blended with the conventional fillers, pigments, resins, plasticizers, and stabilizers used in other sealing compounds and are formulated to give specific end-use properties. The principal difference between these polymer types is that butyl is a copolymer of isobutylene with a minor amount of isoprene, whereas polyisobutylene is a homopolymer.

Chemistry

The development of polyisobutylene (PIB) can be traced to the early 1870s, when very low molecular weight homopolymer was produced at room temperature, catalyzed by boron trifluoride (BF_3). In the 1920s and 1930s, more useful, higher molecular weight polyisobutylenes were made using lower reaction temperatures. In 1937, two Exxon scientists successfully copolymerized isobutylene with small amounts of isoprene. The reaction required very low temperatures (~ -95 °C, or -140 °F), pure feed stocks, Friedel-Craft-type catalysts, and alkyl halide solvents for the catalyst and feedstock. The isoprene introduces unsaturation, thereby permitting the copolymer (butyl rubber), to be vulcanized. Later developments included the introduction of small amounts (1 to 3%) of halogen, as either chlorine or bromine, into the polymer to further enhance reactivity and vulcanization (Ref 1).

Polymer Properties. As shown in Fig. 1, the hydrocarbon backbone of the butyl polymers is relatively long and straight. This regular structure, with few double bonds or reactive sites, renders butyl very stable and quite inert to the effects of weathering, age, and heat. It has good resistance to vegetable and animal oils and to attack by chemicals. Being an all-hydrocarbon material, the butyl polymer has very low water absorption and is soluble in typical hydrocarbon solvents. The many side groups attached to the polymer chain produce a high degree of damping or shock absorption. Because these side groups are not large in size and are regularly spaced,

they do not interfere with the orientation of the rubber molecule, and instead contribute to its very low air and moisture permeability and outstanding barrier properties.

Polymers of isobutylene have very little tendency to crystallize and depend on molecular entanglement or cross linking, rather than on crystallinity, for their strength. The completely amorphous character of these polymers gives an internal mobility that imparts flexibility, permanent tack, and damping. Although their tack is high, because these polymers are hydrocarbons with no or very little polarity, their chemical attraction to many surfaces is weak. Therefore, they are often mixed with resins and other materials that impart some polar character to the blend.

Isobutylenes also have a glass transition temperature of about -60 °C (-75 °F), which assures flexibility even at low temperatures. Except for halobutyl, their nonpolar and low unsaturation characteristics also provide excellent electrical insulation properties, which are reinforced by their low water-absorption property.

The physical, chemical, and compounding properties of the butyl polymers will vary with molecular weight, degree of un-

saturation, minor constituents in the polymer, and, in certain instances, chemical modification. With proper allowance for gross differences in molecular weights, butyl and polyisobutylene can be considered as one family and can be used interchangeably in formulating activities unless some degree of vulcanization is required (Ref 2).

Sealant Formulating. Sealant manufacturers typically purchase various butyl polymers to mix and blend with other compounding ingredients and then form or package these compounded sealants. Because of the tough, high molecular weight nature of the butyl polymers, relatively expensive high-shear mixing equipment is needed to accomplish these processing steps including Banbury internal mixers, rubber mills, sigma-blade kneaders, and so on. Other specialized equipment for the manufacture of preformed tape sealants can involve the use of ram- or auger-fed extruders.

Butyl rubber or its modified versions are the prime binders used in most sealant applications. The very high molecular weight PIB polymers are used only in low additive levels to impart cohesive strength. The low molecular weight, semisolid PIB

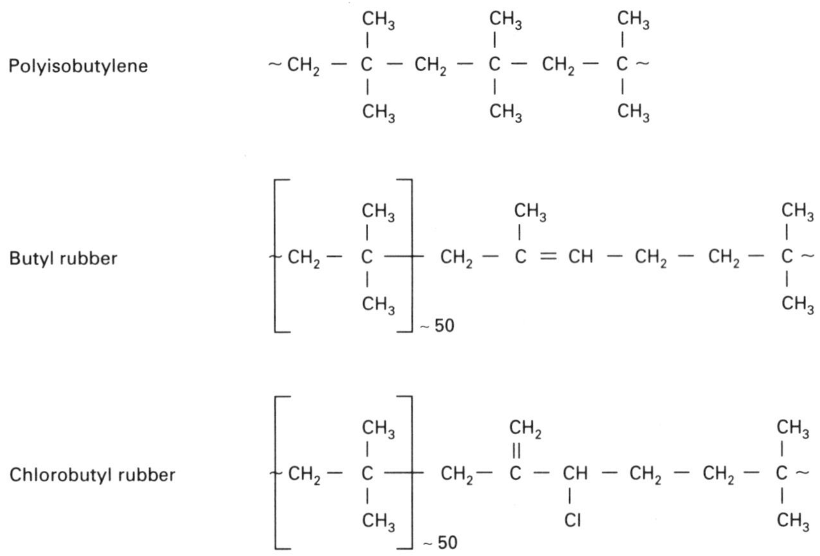

Fig. 1 Butyl rubber molecules

Table 1 Typical molecular weights of selected isobutylene polymers

Polymer grade(a)	Viscosity average molecular weight (Flory)
Vistanex(b) LM-MS	45 000
Vistanex LM-MH	55 000
Exxon Butyl 065	350 000
Exxon Chlorobutyl 1066	400 000
Exxon Butyl 268	450 000
Vistanex L-100	1 250 000
Vistanex L-140	2 110 000

(a) All are products of Exxon Chemical Company. (b) Polyisobutylene homopolymer

grades are also used mainly as additives because they can lower viscosity; provide tack, surface wetting, and adhesion, particularly to difficult substrates; and aid weathering and aging properties. These polymers can also serve as the prime binders in certain sealant uses, such as barrier tape or bead seals in insulated glass manufacture, as well as in certain electrical sealants and void fillers.

Polymer Grades and Suppliers. Typical molecular weights of selected isobutylene polymers are given in Table 1. Other butyl, halobutyl, and polyisobutylene grades are also available from Exxon Chemical Company. Polysar Ltd in Canada also produces a line of butyl polymers for sealant usage, including both chlorobutyl and bromobutyl grades. In addition, the company offers terpolymer XL grades, which can serve as partially cross-linked rubbers in selected sealant applications requiring high strength and resiliency (Ref 3).

Badische Anilin and Soda Fabrike (BASF) produces a full line of polyisobutylene polymers, including low, medium, and high molecular weight materials under the Oppanol trade name.

Partially cross-linked butyl rubbers made in a mixing, blending, and reacting operation by Hardman, Inc., are also offered to sealant manufacturers under the Kalar trade name. Hardman also produces a low molecular weight, semiliquid, reactable butyl as part of the Kalene polymer series (Ref 4).

Solutions of butyl rubber and various blends, cutbacks, and partially cured grades are also offered by several companies, including ADCO Products and A-Line Products.

Forms, Applications, and Suppliers

Sealing Tapes

Preformed tapes are the largest and most varied form of butyl sealant. They consist of 100% solids materials and are supplied in a variety of widths and thicknesses, typically in the form of a roll on a core, with the layers separated by release paper. These tapes often contain a string or a cured rubber or metal insert core to prevent stretching or "drawdown" in the course of

use. In special cases, this core can serve as a built-in spacer. The tapes exhibit some degree of pressure-sensitive adhesion and are relatively soft for ease of installation. They are almost always used in compression between nonporous substrates.

The tape form allows clean, quick, and accurate installation. Neither special mixing nor special primers are required prior to use. Butyl tapes also have a nearly indefinite shelf life.

Tapes form a functional seal immediately, without having to wait for a chemical cure or solvent evaporation. Because of their softness and compressibility, sealing tapes assure continuous, void-free seals, despite variations in the tolerances of the joints or materials to be sealed. However, because of their form, dimensions, and the need to be used in compression, tapes cannot be readily used on rough or irregular surfaces, such as concrete, stone, and so forth.

Generally, tape sealants can be categorized as nonresilient, semiresilient, or resilient, depending on their elastic properties. Highly extensible, resilient sealants are usually based on partially cured butyl or chlorobutyl rubber. In this case, the cure capability of butyl is utilized to yield a sealant with increased strength and recovery from deformation properties. When properly formulated, this type of tape can readily handle movements of ±25% (Ref 5).

Semiresilient tapes are elastic in nature and typically contain high levels of regular butyl or PIB polymers. Nonresilient tapes are extremely soft products and usually contain butyl rubber that has been highly extended with polybutene plasticizers and fillers. These softer tapes are generally designed for lower joint-movement applications in the ±7.5% range (Ref 6).

Construction. One of the most common uses for sealing tapes is in glazing windows and wall panels in modern high-rise commercial buildings. Figure 2 illustrates two typical units constructed using sealing tapes. As shown in Fig. 2(a), tapes can be used to seal both the outside and the inside of a window. Because it has an initial thickness that is slightly greater than the seal channel width, when the tape is compressed, it is held firmly against the glass by the interior sash. Figure 2(b) depicts the tape being used to seal the outside of a window, while a gun-dispensed caulk is employed on the inside. Various types of rubber-base, gun-dispensable sealants are often used in combination with tapes. However, in the types of units shown, tapes have a distinct advantage because they can always be applied from the inside.

A simple procedure used when glazing with sealing tape, as in Fig. 2(a), would be:

- Wipe clean the surface of the window frame to which the tape will be applied with a solvent

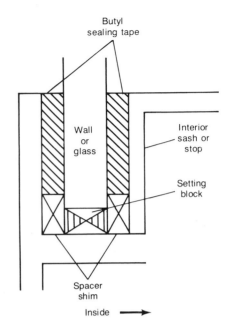

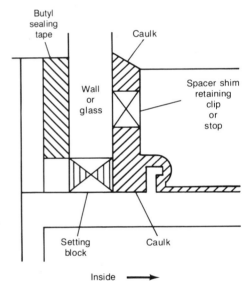

Fig. 2 Typical architectural tape installations

- Apply tape by unrolling it onto cleaned surface or applying precut strips with the release liner still on one side
- Use hand pressure to adhere most tapes to frame surface
- Butt together and splice adjacent ends of tape with hand pressure sufficient to form a continuous seal
- Peel release liner from tape, and set the window in on rubber setting blocks
- Apply tape used to seal the interior of the window in a similar manner
- Once tapes are in place, compress them against the window when the interior sash or molding is snapped or screwed into place

The techniques used to install sealing tapes will vary somewhat among particular appli-

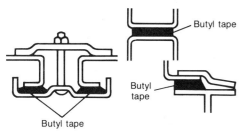

Fig. 3 Typical constructions for prefabrication and appliance sealing

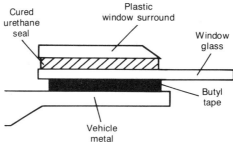

Fig. 4 Automotive fixed window installation

cations, but the broad principles are the same.

The architectural glazing described above requires tapes that are soft and compressible enough to be installed under any condition. However, once they have been installed, these tapes should exhibit excellent elastic recovery from compression and extension so that the seal will remain intact when the glass and window frame are subject to wind pressures and temperature differences (Ref 6-8). Squeeze-out or flow of the tape can occur because of glass deflection under these conditions, and this flow must be countered by appropriate formulating techniques, including partial vulcanization. A well-formulated, well-installed tape should have a life expectancy of greater than 20 years.

Prefabricated Construction. When tapes are used, an immediate leakproof seal can be made as soon as assembly is completed. This capability makes tapes ideal sealants for factory assembly and prefabricated construction.

Industrial buildings, silos, truck and trailer bodies, and mobile homes are all examples of tape sealant applications. Factory glazing of various preformed window units would also be in this category. These tapes tend to be permanently flexible, soft, tacky, and easily workable compounds that are either nonresilient or semiresilient in nature. These seals are quite effective for 10 years and sometimes longer, depending on temperature and movement conditions.

Appliance sealing tapes are very similar in nature and properties to those described above. This type of sealing tape is used in refrigerators, freezers, washers, dryers, and air conditioners for the purposes of sealing, damping, vibration absorbing, and sound deadening. These tapes also tend to be soft, easily workable, and nonresilient in nature. Figure 3 illustrates typical constructions for sealing both prefabricated units and appliances.

Automotive Applications. For a number of years, butyl tapes were used as primary windshield and back light sealants in the original equipment manufacturing (OEM) of automotive products. Although robotic application techniques and integral glass-bonding requirements displaced tape in

OEM use, tape systems continue to be very widely used as replacement sealants in glass and body shops. This application requires a good balance between high shear strength and low compression, with little recovery or spring back. A partially cured butyl with a fairly high level of plasticizer and filler is usually necessary. Current automotive OEM production continues to use butyl tape, but in the less demanding assembly of fixed window units, such as that shown in Fig. 4.

Aside from glass sealing, preformed butyl tapes are used in a number of other joints and seams during automotive construction. This includes foam-in-place expandable seals and void fillers, as well as butyl thermoplastic compounds that can flow and seal following bake-oven heating. Soft butyl preformed or die-cut gaskets are also used to seal a variety of other automobile-body openings.

Hot-Applied Sealants

Butyl sealants that are 100% solids can also be formulated so that applied heat will melt or soften the compound during sealant installation. Usually, high levels of thermoplastic materials such as tackifier resins are added to the compound to provide these hot-melt properties. These sealants will typically possess all of the normal butyl rubber properties, but will also have slightly higher adhesion levels and greater cohesive strength. There will be some loss in elongation and flexibility and also some debits in low-temperature properties (Ref 9).

Butyl thermoplastic sealants can be supplied in bulk form. Equipment is available to completely melt these compounds and pump them into place as hot liquids. Melt viscosity is quite high, and temperatures of approximately 175 °C (~350 °F) are not uncommon. Additional plasticizer is usually added to these bulk compounds to further reduce melt viscosity. Irregular shapes and forms can be sealed using this approach, and the compound will quickly cool and reset to give a finished seal (Ref 10). This pumping technique is designed for in-plant application, and the Pyles Thermo-O-Flow System is one of the more widely used hot-melt butyl applicators. Major applica-

tions involve insulated glass and automotive sealing.

Melt sealants in pellet, rod, or bulk form are also available, in which the prime binders are polyolefin plastics such as ethylene vinyl acetate, polyethylene, and amorphous polypropylene. Low levels (5 to 10%) of butyl rubber are used as additives to impart flexibility, tack, and adhesion-promoting properties. Low-molecular-weight PIB is widely used in this regard.

Another hot-melt approach involves the use of the conventional tape form, but with higher levels of resins and thermoplastic materials in the formulation. Here the tape can be heated and softened using any number of techniques, including ovens, hot-air jets, heat lamps, resistance wires, and so forth. The heated tape can then be more easily compressed and installed and will show superior surface-adhesion properties. Melt or flow-in-place automotive tape was previously mentioned, and the use of heat-softened tape is even more common in insulated glass applications (Ref 11).

Insulated Glass. A unique tape sealant system for insulated glass applications is a thermoplastic butyl-base compound that contains an integral, fluted, aluminum spacer strip with its own desiccant for moisture absorption. This heat-softened tape thus enables the insulated-glass manufacturer to use one product to do the job of several components (Ref 12).

Butyl rubber is an ideal polymer for insulated-glass sealing because of its extremely low moisture-vapor transmission, coupled with its excellent aging properties. Glass units with long-term guarantees can be produced when butyl compounds serve as the primary seal.

Figure 5 shows the integral spacer strip tape, as well as one approach to a pumpable hot-melt-applied butyl glass sealant. In addition, a dual-sealed unit is depicted, in which a low molecular weight PIB primary barrier seal is backed up by a curable silicone or polysulfide sealant that provides strength and integrity to the window unit (Ref 13).

Gun-Dispensable and Pumpable Sealants

Butyl compounds can be cut back with organic solvents and used in gun-dispensable and pumpable mastics, sealants, and caulks. Because of the tough, highly elastic nature of the butyl binder, fairly large amounts of filler and permanent plasticizer such as polybutene are commonly used in these formulations. A good quality, gun-dispensable butyl caulking compound will typically contain only about 15 wt% of butyl polymer.

Relatively large amounts of solvent are also used in these systems to achieve the necessary soft, smooth consistency required in gun-dispensable sealants. These

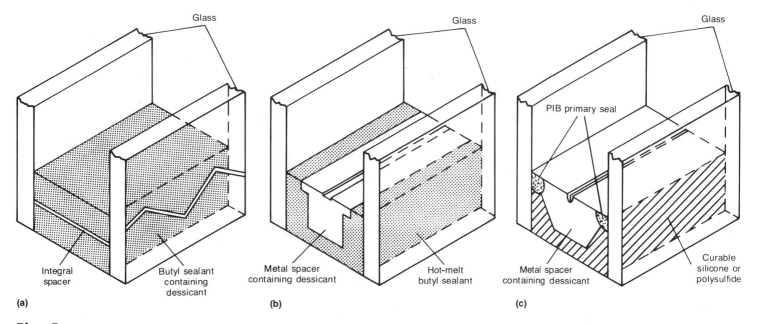

Fig. 5 Insulated glass sealing. (a) Integral spacer strip (Tremco). (b) Three-sided hot melt butyl extrusion. (c) Dual-sealed unit

compounds cure or set strictly through solvent loss. When put in place, the volatile solvent evaporates into the atmosphere or migrates into porous substrates, leaving the tough, rubbery sealant compound.

Unlike conventional caulks and putties, butyl caulks age very well and maintain their softness and flexibility for extended periods. However, because of the low polymer level and the uncured nature of the sealant, only moderate joint movement can be tolerated. The solvent evaporation will also cause some shrinkage. Butyl caulks are typically nonreactive and nonstaining, will set up quickly, and can be readily painted over. Federal Specification TT-S-001657 also covers a sealant-compound, single-component, butyl rubber-based, solvent-release type (for building and other types of construction). Properties of butyl-base caulks are given in Table 2.

Butyl caulks are used in a variety of new construction and home repair activities. They have a large over-the-counter market, being used in glazing and bedding, in some marine applications, and as secondary sealants in construction applications with tapes or curable compounds. They can also serve as acoustical sealants in interior applications (Ref 14, 15).

Specialty Sealants

Butyl polymers are covered under a broad range of Food and Drug Administration regulations. A number of compounds have been developed for food-contact and food-handling applications. Gun-dispensable butyl compounds are used to seal some can ends and are used in certain cap and closure sealant systems. These usually involve contact with fatty foods, where butyl shows an exceptionally good resistance to degradation.

The use of single-ply, cured-rubber sheet roofing that is based on ethylene-propylene elastomers has grown substantially since 1980. In conjunction with these systems, both gun-dispensable and tape butyl sealants have been developed to serve in flashing, patching, seaming, and bonding applications.

The long-term aging and environmental resistance properties of butyl rubber make it ideal as a sealant for precast concrete piping used in a number of drainage and sewer applications. Certain federal regulations also call for the use of high hydrocarbon-content butyl sealants.

Because of its saturated-hydrocarbon, nonpolar nature and closely packed regular structure, butyl polymers have excellent electrical insulating properties. This makes butyl an ideal base for compounds used in splicing, sealing, and void-filling during cable construction, in junction boxes, and in a number of field-applied systems.

Table 2 Typical properties of a butyl-base caulk

Property	Description
Solids	75 to 90%
Usual colors	White and aluminum most common, others available
Shelf life	1 to 2 y
Primer	Usually none required
Back-up materials	For joints more than 13 mm- (0.5 in.-) deep. Almost any compressible material that is not oil- or asphalt-impregnated; for example, butyl tape, rubber and plastic foams, sponges, and so forth
Curing type	Evaporation of vehicle
Curing characteristics	Dries from surface inward by solvent evaporation. Those compounds containing a small amount of "drying" material give a skinning-type surface.
Tack-free time	0.5 to 6 h
Application temperature	5 to 50 °C (40 to 120 °F)
Service temperature	−30 to 95 °C (−20 to 205 °F)
Aged hardness	15 to 30, Shore A
Effect of weathering	Good resistance to ozone, oxidation, and moisture. Slight stiffening but remains permanently flexible
Life expectancy	10 to 20 y
Adhesion	Good
Joint movement	±5 to 10%
Common uses	Fills stable joints and joints with some movement; bedding material; glazing applications; general sealant for wood, metal, masonry, and glass
Advantages	Relatively low cost, easy workability, minor cleanup problems, elastic compounds with permanent flexibility, and good weathering properties
Limitations	Noncuring systems; therefore, strength and elasticity are not as high as with curing rubber-based compounds. Possibly some shrinkage with low solids-content materials

Table 3 Typical tape properties

Property	Architectural tape	Automotive tape	Appliance tape
Compression deflection, applied pressure to compress to ½ height at 2 in./min, kPa (psi)	160 (23)	260 (38)	80 (12)
Durometer, Shore A	17	20	10
Penetrometer, needle penetration for 2 N (200 gf) total load for 5 s, mm (in.)	9.1 (0.36)	8.9 (0.35)	13.5 (0.54)
Shear modulus, applied pressure to shear tape at 2 in./min, kPa (psi)	28 (4)	70 (10)	14 (2)
Recovery, % original height regained 24 h after compression	85	12	5

Properties and Testing

The suitability of a preformed tape for a given sealing job depends on the properties of the particular butyl compound used. Specialized testing tries to simulate conditions that tapes may encounter during installation and use (Ref 16).

The degree of softness or hardness of a tape is related to the ease with which it can be compressed during installation and its ability to conform to variations in the surfaces being sealed. Tests used to indicate softness would include the Shore A or 00 Durometer, needle/cone penetration, and compression/deflection measurements.

Sealing tapes generally need not possess a great deal of strength because most applications require that the tape function solely as a sealant. In applications that do require some support function, a shear strength test is usually conducted, and either a shear-modulus value is calculated or a yield-strength value is determined.

The ability to deflect and recover when members being sealed are stressed is one of the desirable features of tapes. The elastic properties of sealing tapes are indicated by testing for recovery from elongation and compression, and by determining the rate of stress relaxation.

Sealing tapes also must possess some level of pressure sensitivity and have sufficient adhesion to their substrates, such that a typical tensile adhesion test will result in a cohesive failure.

Once installed, sealing tapes are exposed to a number of environmental conditions under which they are expected to perform. Aside from original properties, these sealants can be tested after exposure to temperature, ultraviolet light, water or salt spray, outdoor exposure, or any other unique end-use environment.

Typical properties of tapes designed for architectural, automotive, and appliance applications are shown in Table 3.

Additives

Pigments and fillers contribute to the overall performance of the sealant and allow rubber compounds to be more easily processed and extruded. Carbon black, talcs, and silicas can enhance strength properties, while clays and calcium carbonates are mainly extenders and diluents. Pigment levels per 100 vol% of polymer can range from 50 vol% in certain tapes to 200 vol% in some gun-dispensable caulks.

The main tackifiers and plasticizers used in butyl sealants are the very low molecular weight polybutene polymers. These can also serve as secondary binders. Mineral oils are also used as plasticizers. Tackifier resins are used as adhesion promoters and as thermoplastic processing aids in hot-flow sealant compounds.

A variety of stabilizers, processing aids, adhesion promoters, and other ingredients can also be blended into these compounds, allowing a wide variety of butyl sealants to be formulated and tailored for specific end uses (Ref 17, 18).

Sealant Suppliers

Because sealants based on the butyl polymers are supplied in such a wide variety of types and forms, from tapes to melts to gun-dispensable solvent cutbacks, there are a large number of manufacturers serving this market. Most major sealant producers will have a line of butyl-base products in their sealant portfolio. Naturally, there will be some specialization as companies serve a particular market such as construction, industrial, or automotive, or serve major subsections and specialties such as insulated glass or prefabricated construction or electrical seals. A number of companies will concentrate on the extruded, preformed tape type butyl sealant, while others are involved strictly with gun-dispensable caulks supplied in bulk or cartridge form.

Regardless of the specific sealant supplier involved, technology advances in terms of new equipment for dispensing sealants and new techniques for formulating a greater diversity of butyl sealants have extended the range of applications in which butyl can be used. Therefore, more sealing applications can now benefit from the outstanding aging, weathering, and barrier properties of butyl.

REFERENCES

1. E.N. Kresge, R.H. Schatz, and H.-C. Wang, Isobutylene Polymers, in *Encyclopedia of Polymer Science and Engineering*, 2nd ed., Vol 8, John Wiley & Sons, 1987
2. J.V. Fusco and P. Hous, Butyl and Halobutyl Rubbers, in *Introduction to Rubber Technology*, M. Morton, Ed., Van Nostrand Reinhold, 1987
3. D.A. Paterson, XL Butyl Rubber Improves Preformed Sealant Tape, *Adhes. Age.*, Vol 12 (No. 8), Aug 1969, p 25
4. "Kalar," Technical Data Sheet 12382, and "Kalene," Technical Data Sheet 32983 and 4483, Hardman Inc., 1983
5. A.J. Berejka and A. Lagani, U.S. Patent 3,597,377, 31 Aug 1971
6. A.J. Berejka, Sealing Tapes, in *Sealants*, A. Damusis, Ed., Van Nostrand Reinhold, 1967
7. Architectural Tape Sealants, *U.S. Glass, Metal and Glazing*, Protective Treatments Inc., Dec 1966
8. C.L. Bellanca, Preformed Butyl Tapes in Glazing Applications, in *Science and Technology of Glazing Systems*, STP 1054, C.J. Parise, Ed., American Society for Testing and Materials, 1989
9. A.J. Berejka and J.J. Higgins, Broadened Horizons for Butyl Sealants, *Adhes. Age*, Vol 16 (No. 12), Dec 1973
10. J.E. Callan, Cross-Linked Butyl Hot Melt Sealants, *J. Adhes. Sealant Counc.*, Vol 3 (No. 1), 1974
11. J.J. Higgins, A.J. Berejka, and L. Spenadel, U.S. Patent 3,654,005, 4 April 1972
12. T.W. Greenlee, U.S. Patent 4,431,691, 14 Feb 1984
13. D. Quade, New Technologies in Insulating Glass, *Glass*, Vol 36, March 1986
14. J.J. Higgins, Butyl and Related Solvent Release Sealants, in *Sealants*, A. Damusis, Ed., Van Nostrand Reinhold, 1967
15. J. Panek and J.P. Cook, *Construction Sealants and Adhesives*, 2nd ed., John Wiley & Sons, 1986
16. A.J. Berejka, "Test Methods for Architectural Tape Sealants," Report EPL-6804-2747, June 1968
17. J.J. Higgins, F.C. Jagisch, and N.E. Stucker, Butyl Rubber and Polyisobutylene, in *Handbook of Adhesives*, I. Skeist, Ed., Van Nostrand Reinhold, 1990
18. F.C. Jagisch, Polyisobutylene Polymers in Sealants, *Adhes. Age*, Vol 21 (No. 11), Nov 1978, p 47

Urethanes

John F. Regan, ChemRex Inc.

A MAJOR COMPONENT of urethane sealants is diisocyanate or poly(isocyanate), which can be aromatic or aliphatic in nature. Another major component of urethane sealants is a dihydroxyl or polyhydroxyl compound that could have a polyester or polyether backbone linkage. Other active hydrogen compounds, such as mercaptans and amines, are also used extensively as curatives or as chain extenders. A wide variety of these raw materials is available. They differ from one another in functionality, backbone structure, molecular weight, and hence, properties of specific end-use applications.

Chemistry

The polyether hydroxy terminated compounds are widely accepted in the sealant industry because the price is lower than that of polyesters and because it is possible to design a wide spectrum of physical properties for this polymer. As shown in Table 1, polyethers are prepared by the base-catalyzed addition of alkylene oxides to difunctional or polyfunctional alcohols or amines. Depending on the functional nature of the active hydrogen species, di-, tri-, or tetrahydroxyl materials can be prepared, the choice of which has great impact on physical and chemical properties. The polymerization proceeds by the successive attack of an anion upon the monomeric alkylene oxide. Chain termination occurs by the combination of the polymer anion with a proton. Propylene oxide is perhaps the most widely used monomer and leads primarily to the formation of secondary alcohol groups. However, a certain number of primary alcohol groups are also formed.

Table 2 lists isocyanates that are commonly used in the manufacture of sealants, including both aliphatic and aromatic types. Some of these isocyanates, such as toluene diisocyanate, are more volatile, and some are more reactive, such as 4,4'-diphenyl methane diisocyanate (MDI). In general, the reactivity of the aliphatic isocyanates, such as hydrogenated MDI and isophorone diisocyanates, are much slower in reactivity than the corresponding aromatic isocyanates. Certain shelf life considerations must be considered with specific isocyanates, particularly MDI. Isocyanate fumes or skin contact can represent serious hazards. Necessary precautions should be taken according to manufacturer recommendations.

The formation of both urethane and urea is shown in Fig. 1. The primary linkage created from the reaction of an alcohol with an isocyanate is a urethane linkage. In a one-component system, the reaction of the excess isocyanate with water leads to an amine, which then reacts with the terminal isocyanate group in the polymer to form a disubstituted urea. This is the basic curing mechanism for a one-part urethane moisture-curing system. In a two-part system, it is quite usual to have a polyol-rich component as one unit that is then reacted with a second unit, an isocyanate-rich component, at a nominal one-to-one equivalent ratio that quickly results in a urethane rubber because of the stoichiometric nature of the chemistry.

In other instances, it is possible to use an amine as one unit and the isocyanate as the other, resulting in the formation of a urea as the stoichiometric reaction product. In this case, polyurea is the primary linkage.

Table 2   Isocyanate used in urethane sealant manufacture

Toluene 2,4-diisocyanate (TDI)
Toluene 2,6-diisocyanate
4,4'-diphenyl methane diisocyanate (MDI)
Isophorone diisocyanate
Hydrogenated TDI
Hydrogenated MDI
1,6-hexamethylene diisocyanate
Methylene bis 4-cyclohexyl isocyanate

$$R - \overset{..}{N} - C \equiv O: \leftrightarrow R - \overset{..}{N} = C = \overset{..}{O} \leftrightarrow R - N \equiv C - \overset{..}{O}:$$

(−)	(+)		(+)	(−)
minor			major	

Thus, a one-part system requires the moisture cure of excess isocyanate, leading to the formation of an amine, with the resultant formation of a urea linkage. One-part systems require a nominal stoichiometric ratio of 2:1 of the isocyanate to hydroxyl group for stability. Two-part systems are reacted at a stoichiometric level, which is a nominal 1:1 equivalent reaction, so that the chemistry that yields the finished product occurs relatively fast.

Figure 2 identifies two other types of reactions that can occur with urethanes. In the presence of excess isocyanate, a reaction of the isocyanate with a urethane linkage to form an allophanate is possible. In

$$RNCO + R'OH \longrightarrow RNHCOOR'$$

Formation of urethane

$$RNCO + H_2O \longrightarrow [RNHCOOH]$$
$$\downarrow$$
$$RNH_2 + CO_2$$
$$\downarrow + RNCO \quad \text{(Carbamic acid)}$$
$$2RNCO + H_2O \longrightarrow RNHCONHR + CO_2$$

(Disubstituted urea)

Formation of urea

Fig. 1 Formation of urethane and urea

Table 1   Steps in the manufacture of polyols

Chemistry
Organic oxide + active hydrogen
Compound + catalyst

Manufacture

1. Organic oxides
   Ethylene oxide
   Propylene oxide   } Base-catalyzed { Active hydrogen species
   1,2-butylene oxide   addition   Glycol (diol)
   1,4-butylene oxide   Glycerine (triol)
                                              Ethylene diamine (tetrol)
2. Catalysts added to compound formed in Step 1
3. Chain termination
4. Primary and secondary hydroxyl formation

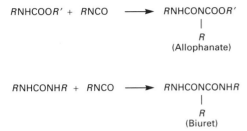

Fig. 2 Cross-linking reactions that yield allophanate and biuret

addition, if urea linkages are present, the reaction of the isocyanate with the urea to form biuret linkages is possible. The chemistry is controlled by temperature, stoichiometric ratio, mixing techniques, type of catalyst system involved, and, to a certain extent, some of the additives to the system.

A wide variety of catalysts is available for use in urethane chemistry. Tin catalysts are quite common and they promote not only the reaction of the isocyanate with the polyols, but also the reaction with water. The presence of tertiary amines with a tin catalyst is synergistic for this same chemistry.

The use of salts such as bismuth tends to promote the chemistry of the isocyanate with the polyol, rather than the chemistry of the isocyanate with water. Selected reactivities can be used advantageously. The chemistry can be carried out by the use of a catalyst at room temperature, with the resulting exotherm in the range of −4 to 15 °C (25 to 60 °F), depending on the nature of the reactants. However, it is possible, in the absence of a catalyst, to heat the material to approximately 70 to 77 °C (160 to 170 °F) and carry out the chemistry without any catalyst.

Functional Types

Polyurethane sealants, characterized by their common urethane linkage, have varying molecular weights and varying hard and soft segments to suit most applications. In addition, end groups can be modified, particularly in two-part systems. End groups can be hydroxyl, amine, epoxide, mercaptan, isocyanate, and others. The choice of end group is best left to the particular formulator to suit the specific properties to be established in the sealant. Like other polymer types, urethanes can be made either weather resistant or nonresistant. Other characteristics designed for end-use application can be the prevention of heat loss and the penetration by water, dust, or air. Urethanes can accommodate expansion, compression, vibration, and other hostile environmental effects, such as ultraviolet (UV) radiation and heat.

Commercial Forms

The modern urethane sealant is composed of pigmented or unpigmented synthetic elastomeric material, which, in its noncured state, is pourable or gun dispensable. The consistency of urethane sealants in the noncured state can be adjusted to meet the requirements of application, such as self-leveling or nonsagging characteristics on vertical joints. As the sealant cures, it becomes an elastomeric material with specific properties tailored to the end-use requirement. Material is normally available in cartridges of several sizes, including pouches, pails that range from 4 to 20 L (1 to 5 gal), drums of 200 L (55 gal), and in bulk form. Two-part systems are available with ratios ranging from 1:1 to 10:1 and in several pail sizes, ranging from 6 to 23 L (1.5 to 6 gal).

Major Markets and Applications

The urethanes have proved to possess most, if not all, of the qualities desired in current sealant applications. They can be tailored to specific end-use requirements, offering excellent aging resistance, elongation, and recovery characteristics. Polyurethanes are tough and abrasion resistant. Their high movement capability allows them to be used in wide joints. The cure rate of a two-component urethane system can be adjusted from very rapid to very slow. Most importantly, urethanes can withstand as much as 20 years of service. These capabilities have led to a number of marketplace opportunities, which are described below.

In the construction industry there are two major types of joints that require sealant. Vertical joints are necessary in curtain walls and perimeter caulking. Horizontal joints are placed in floors, sidewalks, parking lots, and swimming pools. Most of the sealant volume used is in vertical joints. Both one- and two-part systems are used.

In addition to the nonsag and weathering characteristics needed for vertical joints in the applications described above, abrasion resistance and toughness are required of horizontal joints because they are used in areas of considerable traffic. The advantages of selected polyurethanes in these applications are their good resistance to contamination by asphalt, salt, and other chemicals.

Polyurethanes are also used as expansion joint and paving joint sealants. Expansion joints are used on certain parking areas and decks and require a joint sealant that can expand as necessary to accommodate temperature cycle changes. The requirements in this application are weatherability, traffic-bearing ability, abrasion resistance, and ability to withstand expansion/contraction due to hot and cold temperatures.

Urethane sealants are also applied as liquid membranes or sheet goods between slabs of concrete to prevent water from leaking and running through critical spots.

The requirements for waterproofing membranes are good adhesion to concrete and good moisture permeability resistance. Minimal weathering capability is required and strength features are not essential to the performance of the membrane. Parking deck membranes require good adhesion and abrasion, as well as weathering, solvent, and water resistance.

Consumer applications represent a fairly large market for sealants. However, the share held by urethanes in this market is relatively low, because most of the over-the-counter sales are products of acrylic, butyl, and silicone. The polyurethanes can offer the advantage of general usage in that they can be used to repair windows, tub perimeters, gutters, sidewalks, and automobiles.

Automotive applications comprise an extremely large market for urethane sealants. The major automotive application is as a sealant and adhesive for passenger car windshields and taillights. The polyurethane used to seal windshields is actually considered part of the construction material that prevents roof collapse during rollovers. A primer is normally required to bond the urethane windshield sealant to metal. The principal properties required are good adhesion, weatherability, and flexibility. Other automotive applications include the opera windows in luxury models, T-top roofs, and drip rail steel molding. Automotive aftermarket products also represent a large market for urethane sealants.

Major Suppliers and Cost Factors

In alphabetical order, the major suppliers of urethane sealants are ChemRex Inc.; Mameco International; Pecora Chemical Corporation; Products Research and Chemical Corporation; Sika Corporation; and Tremco, Inc.

Cost factors are a rather complex issue affected by the amount and type of isocyanate used; the amount and type of polyol used; the amount of polymer in the sealant system; specialty additives, such as UV-radiation stabilizers, antioxidants, and adhesion promoters; packaging and processing costs; and the amount of scrap involved in manufacture. The price of the raw materials can be influenced by the size of the business and its ability to negotiate good price contracts based on a sufficiently high volume. Polyether polyols are less expensive than polyester polyols. Aromatic isocyanates cost less than aliphatic isocyanates. The incorporation of 0.25% of a silane adhesion promoter can add $0.15/L ($0.50/gal) to the cost of the material. The high cost of hazardous waste disposal also enters the pricing equation. Thus, a system that incorporates rework and improved manufacturing capabilities has a cost advantage. Continuous systems are less expensive than batch systems.

Given these considerations, the best estimate of the cost of urethane sealants at this writing is in the range of $1.30 to 2.40/kg ($0.60 to 1.10/lb). This range represents raw material cost only. Most raw materials used in a sealant formulation would cost roughly $1.65 to 1.85/kg ($0.75 to 0.85/lb).

Competing Materials and Systems

In the field of high-performance joint sealants, urethanes compete with silicones and polysulfides. In the waterproofing membrane area, the liquid-applied systems are being replaced by solid sheet goods and panel-type materials, which appear to be quite effective.

Application Parameters

Sealants are normally applied by roller, spray, brush, or trowel, or through a nozzle, depending on the viscosity and thixotropy of the urethane, as well as the application method desired by the particular applicator. Normally, most urethanes can be applied at temperatures as low as 5 °C (40 °F) and as high as 50 °C (120 °F). Lower and higher application temperatures have been observed. The low-temperature limits are the ability to move the material into the joint or onto the surface of application, the ability for adhesion to occur, and the ability to keep moisture away from the substrate. The upper-temperature limits are the capability for sag control and the product pot life. In certain two-part urethane systems, application temperatures as high as 70 °C (160 °F) have been observed.

Urethanes are suitable for continuous use at the quite low temperature of −40 °C (−40 °F) and upper temperature of 80 °C (180 °F), although higher temperatures on a continuous basis are suitable for specific polymers. It has been observed that under certain circumstances, temperature changes, holes in backer rods, and the incorporation of air during mixing can lead to bubbling of the sealant in a joint. In addition, on applied liquid membranes, very fast skinning under conditions of high heat and wind can lead to entrapment of the solvent, resulting in blistering of the coating.

Substrates should be dry for application. One of the most common causes of bubbling in urethane sealants and membranes is residual moisture in the substrate, whether it be concrete or wood. Wood, glass, aluminum, and concrete represent traditional substrates with established characteristics. With concrete, density is a consideration as is the type of surface treatment used on aluminum substrates. Both factors can limit adhesion. An oil coat on aluminum can also minimize good adhesion. Thus, moisture,

dirt, and all other foreign matter must be removed from surfaces before the application of a urethane sealant.

Joint failure is normally the result of excessive or premature movement of the substrate. The application of sealant into a joint should be monitored carefully, not only to ensure the appropriateness of the sealant for the specific type of joint, but also to preclude excessive joint movement both before and after the sealant has cured through. These sealants are designed for specific limits of movement (normally, less than ±25%) due to temperature changes. Some urethane sealants can tolerate movement of ±50%.

Movement may occur within the first several days of application, particularly in lightweight applications. Sealant failure can be likely. Lightweight applications normally require the use of a two-part urethane sealant because of the speed with which the sealant can adhere and develop good physical properties.

Joint design is a critical feature of good urethane sealant performance. Fixed limits of depth should be considered. A backer rod is commonly used to control sealant depth when two-surface bonding, rather than three, is required. A backer rod prevents the sealant from sliding deep into the joint, not only controlling joint depth, but also preventing air from migrating to the back side of the sealant.

Sealant flammability is becoming increasingly important in certain applications. Future requirements for the limitation of flame spread rate and smoke density may exist. Special additives, such as metal oxides, hydroxides, carbonates, and chlorinated hydrocarbons, can be used to minimize the flammability of urethane sealants. Halogen-containing polyols are also available.

Properties Data

Typical characteristics of urethane sealants are presented in Table 3.

Additives and Modifiers

Additives and modifiers are used in urethane sealants to achieve cost effectiveness and to enhance specific performance properties. Specific fibers, as well as carbon black in very high levels, can be used to promote slump resistance. Calcium carbonate can be used as a filler to decrease cost and enhance certain sag control characteristics. Specific types of clay (kaolin) that are hydrous and calcined (aluminosilicate), talc (magnesium silicate), and both fumed and precipitated silica are all used as fillers.

The choice of filler type is made by the formulator, as governed by the end-use performance requirements. For example, carbon black would not be used in a normal urethane sealant unless a black color is

Table 3 Typical characteristics of urethane sealant

Characteristic	Value or comment
Tensile strength, MPa (ksi) ...	1.4–5.5 (0.2–0.8)
Elongation (ultimate), %	200–1200
Hardness, Shore A	25–60
Weatherometer rating	No elastomeric property change after 3000 h
Low-temperature flexibility, °C (°F)	−40 (−40)
Service temperature range, °C (°F)	−40 to 80 (−40 to 180)
Expected life, years	≤20
Dynamic movement, %	±25
Adhesion peel strength, kN/m (lbf/in.)	3.5–8.8 (20–50)
Cure time, days	1–7
Durability	Excellent
Shrinkage, %	0–2
Gun dispensability at 5 °C (40 °F)	Excellent
Water immersion	Yes(a)
Sag resistance	Excellent
Objectionable odor	No
Primer required	No(a)
Water permeability, [kg/Pa · s · in.)] × 10^{-12} (perm · in., 23 °C)	0.15–1.2 (0.1–0.8)

(a) Primer is required for water immersion conditions.

Table 4 Blocking agents

Phenol
m-nitrophenol
p-chlorophenol
Catechol
Phloroglucinol
Hydrogen cyanide
Ethyl malonate
Acetyl acetone
Ethyl acetoacetate
Caprolactam
B-thionaphthalene
Isooctylphenol
4-hydroxybiphenyl
a-pyrrolidone
Ketimine
Oxazolidine

desired. Fibers may give a textured appearance. Calcium carbonate may introduce some moisture sensitivity to the product. Fumed silica may be too costly to be used for sag control enhancement.

Plasticizers commonly used in urethane sealants are phthalates, chlorinated hydrocarbons, and aliphatic hydrocarbons. These help to lower both modulus and Shore A hardness values, while enhancing gun dispensability. Extenders, which primarily lower cost by extending the product and provide resistance to moisture permeability, include coal tar and asphalt. A small amount of solvent can also be used in urethane sealants to decrease cost as well as to lower viscosity in order to promote processing. In addition, the interaction of solvents may play a role in the formation of thixotropic materials.

Table 4 lists blocking agents, which assist in the stabilization of polyurethane systems. In some cases, blocking agents help promote the cure rate and tack-free time of the polyurethane systems.

Table 5 UV stabilizers and antioxidants

Secondary amines
Phosphites
Phenolics
Bisphenolics
Hindered amines
Benzophenone
Benzotriazole

Table 5 lists the functional materials used as UV stabilizers and as antioxidants. Antioxidants and UV stabilizers must be chosen carefully because some cause color changes and some can interact with other additives in the urethane sealant. Because these materials represent a significant cost factor, a study of the necessary level of addition is quite important. The combination of UV stabilizers, such as the hindered amines, with the antioxidants, such as phosphites or phenols, can offer a good synergistic mix. Long-term studies of the benefits are warranted.

Figure 3 shows the chemical formulas of silane adhesion promoters. As a class of chemical compounds, the silanes appear to offer the best system for the adhesion of urethane sealants, particularly to glass and aluminum. The type and amount of silane needed should be system specific. The order in which specialty additives are added becomes significant in good formulation technology.

Table 6 lists those materials used in sag control which is one of the more important physical aspects of achieving good performance of a urethane sealant in vertical applications. Combinations of these materials are widely used, and it is up to the individual formulator to study the different sag-control agents in terms of performance, processibility, long-term shelf stability, and cost.

Product Design Considerations

Product design considerations are closely tied to end-use performance requirements. It is often difficult to obtain proper information on the performance requirements of a urethane sealant. Some properties that need to be considered are not, and others that do not need to be considered are. It is important to assign the proper level of importance to each property.

Design considerations include the determination of the nature of the substrate, the necessary degree of adhesion, the anticipated amount of joint movement, and the time lapse between application and joint movement as well as whether the material will be used outside or inside. Inside use still requires the use of an antioxidant. Long-term usage also requires the use of an antioxidant. Outside use requires a UV stabilizer, primarily to stabilize color and

$CH_2 = CHSi(OC_2H_5)_3$

Vinyltriethoxysilane

$CH_2 = CHSi(OCH_2CH_2OCH_3)_3$

Vinyl-tris-(beta-methoxyethoxy) silane

$CH_2 = C - C - OCH_2CH_2CH_2Si(OCH_3)_3$ with CH_3 and O

gamma-methacryloxypropyltrimethoxysilane

O S $CH_2CH_2Si(OCH_3)_3$

beta-(3,4-epoxycyclohexyl)-ethyltrimethoxysilane

$CH_2 = CHCH_2OCH_2CH_2CH_2Si(OCH_3)_3$ with O

gamma-glycidoxypropyltrimethoxysilane

$CH_2 = CHSi(OOCCH_3)_3$

Vinyltriacetoxysilane

$HSCH_2CH_2CH_2Si(OCH_3)_3$

gamma-mercaptopropyltrimethoxysilane

$NH_2CH_2CH_2CH_2Si(OC_2H_5)_3$

gamma-aminopropyltriethoxysilane

$NH_2CH_2CH_2NHCH_2CH_2CH_2Si(OCH_3)_3$

N-beta-(aminoethyl)-gamma-aminopropyltrimethoxysilane

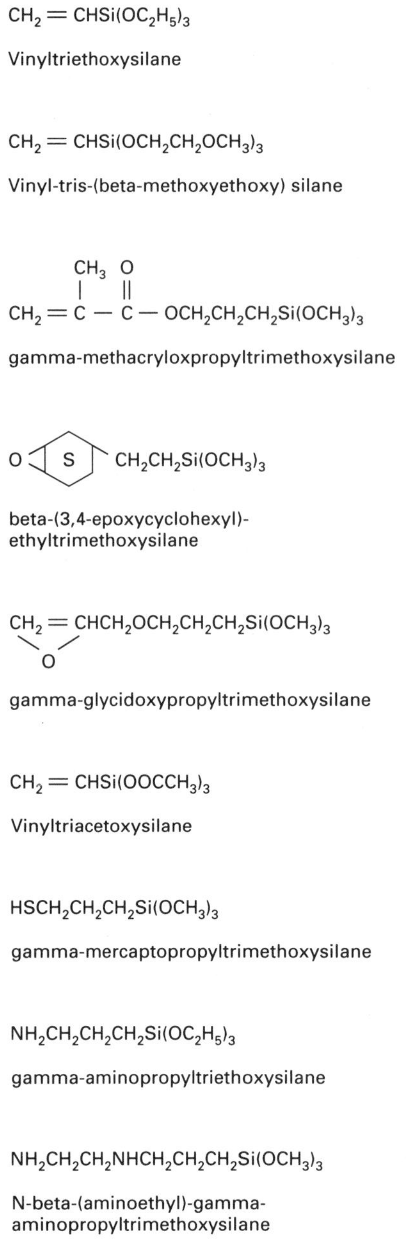

Fig. 3 Silane adhesion promoters

minimize chalking and crazing. There may be other considerations, such as the weathering capability of the polymer backbone structure.

The requirement for solvent resistance, such as jet fuel resistance on an airplane runway, may be another design consideration. Weathering, solvent resistance, and abrasion resistance are also important to sealant membrane applications on parking decks. In the case of jet fuel resistance, a material must have minimal volume and weight change when submerged in jet fuel. This can be accomplished with the proper use of cross-linking systems to yield a cross-link density that sufficiently resists jet fuel. In the case of the parking deck membrane, a polyester backbone with an aliphatic isocyanate is a requirement because

Table 6 Sag control materials

Clays
Talcs
Fumed silica
Polyvinyl chloride gel
Surfactants and polar additives
Combinations of the above

the combination of abrasion resistance and solvent resistance is very difficult to attain. In the case of water immersion, primers are normally recommended for all substrates. A higher cross-link density in the polymer backbone to achieve long-term resistance to water is also recommended. For applications that are under water and exposed to sunlight, additional sunlight protection may be required.

If significant joint movement capability is going to be required, an important design consideration is to use a relatively low-modulus product capable of fairly high (25% minimum) expansion and contraction and to employ other good performance criteria, such as those described in federal specification TT-S-00230C. If the joint must withstand movement shortly after sealant application, then a two-part system is preferred over a one-part system.

If the sealant must be subjected to painting, a more rapid skinning product may be needed, using very little solvent to ensure good paintability in a short period of time and to preclude cracking.

In addition to the above end-use considerations, it is important to treat urethane polymers as a two-phase morphology. Not only hydrogen bonding, but the relative volume percentages of the hard (isocyanate) and soft (polyester or polyether) segments of the polymer backbone must be given serious consideration. The polymer is also a factor in terms of tensile capabilities, modulus, hardness, oxidative stability, hydrolytic stability, abrasion resistance, elongation, elasticity, and recovery characteristics. To summarize, design considerations are extensive and complex and require good insight into the whole area of chemistry and the use of additives.

Processing Parameters

Shelf life is an important consideration for the one-part urethane systems and for the activator component of two-part systems. Higher isocyanate ratios in a one-part system and the absence of any moisture greatly promote good shelf life performance of the product. In addition, system pH tends to play a significant role. The pot life of the product depends particularly on temperature in the two-part system. At higher temperatures, the pot life tends to diminish. Thus, some consideration must be given to the impact of the catalyst on the chemistry as affected by temperature: Normally, the control of the catalyst level can significantly influence pot life.

In terms of storage requirements, the best stability and long-term storage life is achieved when the material is stored in dry containers, whether they be cartridges, pouches, or drums, and when the temperature is kept below 40 °C (100 °F).

For all applications, it is necessary that the substrate be clean and dry to maximize the long-term performance of the sealant, particularly its adhesion. Certain substrates may require special treatments to ensure good adhesion. For old surfaces, it is often necessary to clean, abrade, or remove old coatings. Primers are normally recommended for any underwater end-use, or when special substrates are used, or when the application will receive a high level of wear.

The available cure chemistries are thermal, moisture, and chemical. Moisture cure is used most predominantly in one-part urethane systems. Chemical cures, used in two-part systems, can be polyols, amines, or some other reactive hydrogen species that reacts with the isocyanate. The one-part system can also be chemical curing if certain materials, such as oxazolidines, are used. First, these react with water to produce an amine and hydroxyl group, which then can complete the chemical cure by reaction with the isocyanate.

SELECTED REFERENCES

- O. Bayer, Polyurethanes, *Mod. Plast.*, Vol 24, June 1947, p 149-152, 250, 262
- P.F. Bruins, *Polyurethane Technology*, Wiley-Interscience, 1969
- D.J. David and H.B. Stanley, *Polyurethanes: Analytical Chemistry of the Polyurethanes, Part III*, Wiley-Interscience, 1969
- B.A. Dombrown, *Polyurethanes*, 2nd ed., Plastics Application Series, Reinhold, 1965
- E.N. Doyle, *The Development and Use of Polyurethane Products*, McGraw-Hill, 1971
- K.N. Edwards, "Urethane Chemistry and Applications," Paper presented at the ACS Symposium Series 172, Las Vegas, American Chemical Society, Aug 1980
- R.M. Evans, in *Advances in Urethanes—Adhesives and Sealants*, Technomic, 1985
- R.M. Evans and R.B. Greene, Urethane Sealants, in *Building Seals and Sealants*, STP 606, American Society for Testing and Materials, 1976, p 112-133
- K.C. Frisch, in *Advances in Urethanes—Elastomers and Coatings*, Technomic, 1985
- N.G. Gaylord, Ed., *Polyethers, Part 1. Polyalkylene Oxides and Other Polyethers*, Wiley-Interscience, 1963
- C.E. Leyes, R.E. Weber, and R.E. Jones, Urethane, in *Modern Plastics Encyclopedia*, Nopco Chemical Company, 1967, p 349
- "Properties and Essential Information for Safe Handling of Toluene Diisocyanate," Chemical Safety Data Sheet SD-73, Manufacturing Chemists Association, 1971
- D.V. Rosato and J.R. Lawrence, Ed., *Plastics Industry Safety Handbook*, Cahners Books, 1973, p 156-158
- J.H. Saunders and K.C. Frisch, *Polyurethanes: Chemistry and Technology, Part 1, Chemistry*, Wiley-Interscience, 1962
- J.H. Saunders and K.C. Frisch, *Polyurethanes: Chemistry and Technology, Part II, Technology*, Wiley-Interscience, 1964
- I. Skeist, *Polyurethanes II*, Skeist Laboratories, Inc., Oct 1983

Solvent Acrylics

Jamil Baghdachi, BASF Corporation, Coatings and Colorants Division

SOLVENT-BASE ACRYLIC SEAL-ANTS have been used in glazing and small joint industrial sealing since the 1960s. First introduced to the construction industry about 1960, these materials are now used mainly in building construction and are not particularly well suited for use by the home-owner or small contractor.

These single-component, gun-grade seal-ants, like aqueous acrylic sealants, are char-acterized by the typical properties of acry-lates, that is, outstanding resistance to aging and good adhesion to many substrates. In addition, solvent-base acrylic sealants are capable of adhering to wet, oily, and dusty surfaces. Therefore, a careful and costly surface pretreatment is not necessary. Fur-thermore, these low-recovery elastomeric sealants will self heal in the event of cohe-sive failure (Ref 1).

Chemistry

The acrylics now used for sealants are tailor-made copolymers of acrylic and/or methacrylic acid esters and other mono-mers (Ref 2, 3). Specifically, acrylic seal-ants are primarily based on ethyl, butyl and 2-ethyl hexyl acrylate monomers, as well as small quantities of methyl methacrylate, acrylic and/or methacrylic acids, and other specialty acrylic monomers. Examples are shown in Fig. 1.

Frequently, the acrylic monomers are co-polymerized with other vinyl monomers, such as vinyl acetate, vinyl chloride, sty-rene, and so on. The high reactivity of vinyl groups permits the synthesis of linear poly-mers of very high molecular weight. By mixing several monomers, the desired prop-erties, such as resistance to ultraviolet (UV) radiation, resistance to chemicals, elastici-ty, and hardness, can be achieved. Suitable polymers are those that are linear and ther-moplastic and those that can be rendered adequately elastic by cold vulcanization or oxidation or by the effect of alkaline sub-stances, such as caustic soda, cement, or lime.

The physical properties of these products are influenced by the alpha substituent of the acid (hydrogen atom or methyl group) and by the length of the alcohol side-chain modification (Ref 4). The ester side-chain group also effects properties significantly. As the ester side chain becomes longer, the tensile strength of the polymer is decreased and its elongation increases (Ref 5, 6).

Most commercial processes are free rad-ical type addition reactions conducted at elevated temperatures in the presence of an initiator. The characteristic properties of these polymers are greatly influenced by the conditions of polymerization. The variation of catalyst level, reaction time, tempera-ture, and monomer concentration make it possible to adjust the molecular weight and ultimately the physical properties of the polymer.

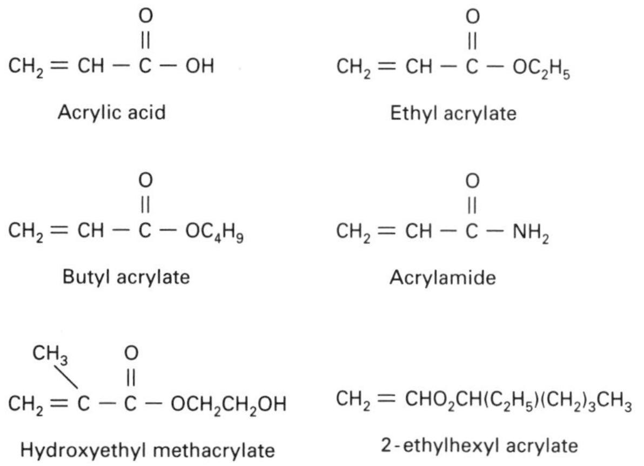

Formulation

Solvent acrylic sealants are compounded from base polymers in solvents such as xylene (xylol). The acrylic sealant contains approximately 40 to 50% polymer. Many of the same fillers, such as calcium carbonate and talc, that are used with other elastomers can be successfully incorporated into the acrylics. A typical formulation of a solvent acrylic sealant is shown in Table 1.

Other ingredients include pine oil, used as a pigment dispersant and penetrating agent for oil or grease on the joint surface; ethyl-ene glycol, used as a bridging agent for fumed silica; and a solvent, used for viscos-ity reduction (Ref 7, 8).

Properties

Solvent acrylics can be produced in a wide variety of architectural colors, and the color stability of these sealants is very good. These sealants are cured by solvent release, and they are generally 80 to 85% solids. The relatively high solvent content can cause considerable shrinkage, which is moderated somewhat by the slow solvent release of the system, resulting in sufficient stress relief to overcome joint distortion. The surface of the sealant remains tacky for a day or so, depending on the environmen-tal temperature, and the complete cure of the average-size sealant bead may take from 2 to 5 weeks.

Solvent acrylics are considered nontoxic and nonallergenic. However, because they

Fig. 1 Monomers used in solvent acrylic formulations

Table 1   Typical formulation of solvent acrylic sealant

Material	Composition, %
Acrylic solution polymer (86% nonvolatiles)..	56
Calcium carbonate	24
Talc	5
Xylene...................................	9
Ethylene glycol	2
Pine oil	1
Fumed silica.............................	3
**Total...................................	100**

Table 2 Typical properties of solvent acrylic sealant

Elongation at break, %	200–250
Tensile strength, MPa (ksi)	1.4–2.1 (0.200–0.300)
Hardness, Shore "A"	20–25
Peel adhesion, N/mm (lbf/in.)	2.1–2.6 (12–15)

Table 3 Solvent acrylic sealant characteristics

Advantages	Disadvantages
Requires no priming	Relatively poor low-temperature elasticity
Good elongation and excellent adhesion	Can only be used in small joints
One-component system	Poor recovery and shear strength
Excellent ultraviolet-ray resistance	Slow cure speed
Good moisture and chemical resistance	Relatively poor flexibility
Remains elastic indefinitely	Strong odor
Highly resistant to digging or gouging	Poor water immersion resistance
Good color stability and available in many colors	Remains slightly tacky when pressure is applied and picks up dirt
Good weathering and durability	High cold flow
Self healing and nonstaining to most materials	

contain solvent, they should be handled with a normal amount of care.

Odor is a distinct problem. Not only is the odor somewhat offensive, it can contaminate food products if the sealant is used in food storage areas. Adequate ventilation is mandatory when using the sealant for interior applications.

Solvent acrylic sealants have good resistance to a fairly broad range of chemical solvents after the material has completely cured. Resistance to water is good, but not outstanding. Therefore, these materials should not be used for joints that are immersed in water, such as in swimming pools. These sealants also show good resistance to salt spray. In addition, their UV resistance is excellent, which makes them very good glazing sealants. Solvent acrylics are formulated to meet the requirements of federal specification TT-S-00230. They age and weather very well, retain colors, and show very little chalking or surface crazing.

Although these sealants are rubbery materials with low recovery, they are not true elastomers. The stress relaxation is so rapid that it is virtually impossible to plot a modulus for this type of sealant. The specimen in the testing machine will stretch with no change in load. When failure does occur, it is almost invariably a cohesive failure, with about 200 to 250% elongation (Table 2).

Tensile adhesion values are not reported because adhesive failure is almost unknown in a tensile test. The peel adhesion strength of solvent acrylics is about 2.1 to 2.6 kN/m (12 to 15 lbf/in.), which is quite good. This combination of properties gives a sealant with excellent adhesion but poor recovery characteristics. Therefore, this sealant is best suited for small joints with limited joint movement. When a cohesive failure occurs, the two fractured portions of the sealant can be pushed back together, whereupon the material will knit and continue to function.

Solvent acrylic sealants, when fully cured, have a Shore A hardness of 20 to 25. This value increases by about 15 to 20 points as the temperature drops below −18 °C (0 °F). Table 3 lists the advantages and disadvantages of the solvent acrylic sealant.

Applications

The excellent adhesion of solvent acrylics makes them one of the most versatile types of sealants for building construction work. They can be used for glazing, control joints, glass-to-mullion joints, and the pointing of brick and stone masonry. They require no priming and can be successfully used in joints that have had minimal or no cleaning. Solvent acrylics do not require the "hospital clean" joints demanded by some of the more elastic sealants.

Because of their high adhesion and low recovery, these materials can be used for corner beads, reglets, and oddly shaped joint openings that many of the other elastomers cannot tolerate. This versatility and capability for tolerating abuse are two of the prime reasons these materials are selected by architects. The architect can select just one sealant for a building, to function in exterior panel joints, to act as heel filler beads for glazing, and to seal the perimeters of air conditioning ducts.

Solvent acrylics are also used in those areas where low joint movement is expected, for example, joints between window frames and concrete, for which the butyl compounds are unsuitable because of their high shrinkage and inadequate elongation, and for which the more elastomer compounds are too expensive. Solvent acrylic sealants are also used to glaze insulating glass panels into place.

Major suppliers of solvent-free and solvent-containing base polymers include BASF Corporation, Rohm and Haas Company, and B.F. Goodrich Company. Solvent acrylic sealants are available from local compounders in caulking cartridges or in bulk.

REFERENCES

1. J.P. Cook, *Construction Sealants and Adhesives*, John Wiley & Sons, 1970
2. U.S. Patent 3,551,374, BASF Corporation
3. Belgium Patent 689,663, BASF Corporation
4. W.H. Brendley, "Fundamentals of Acrylic Polymers, Paint and Varnish Production," Rohm and Haas Company, July 1973
5. J.S. Amstock, *Adhes. Age*, Feb 1969, p 18
6. W.J. Reid, *Adhes. Age*, April 1970, p 21
7. "Acryloid CS-1" and "Room Temperature Gunnable Sealants Made With Acryloid CS-1," RC-34, Rohm and Haas Company, 1967
8. F.L. Miller, Acrylic—Sealants—Long Service Life, *Glass Dig.*, Vol 7, Dec 1971

Water-Base Acrylics and Polyvinyl Acetates

Jim Johnson, H.B. Fuller Company

CAULKS AND SEALANTS made from emulsion polymers first became available commercially in the 1950s and were soon followed by latex paints. Since 1960, latex use has grown significantly to the point that it now constitutes a large portion of the total sealant market. The vast majority of applications is in the residential construction market, where they are used for general sealing purposes. Some of the early latex caulks experienced problems with aging, cracking, and freeze-thaw stability, but improvements in formulation have now made them very acceptable products for many applications.

Recently, new latex formulations have been developed for more demanding applications in which more joint movement is experienced, but latex product use is still very limited in these areas. Sealant failures are very expensive to fix, especially in relation to the cost of premium sealants, compared to new higher-performance latex sealants. For this reason, there can be reluctance on the part of contractors to try these new products.

Chemistry and Formulation Constituents

Latex caulks and sealants are highly formulated emulsions that contain a base polymer from which most of the adhesion and elastomeric properties are derived. Base polymers are generally of three types: polyvinyl acetate homopolymers, vinyl acetate and acrylic or ethylene copolymers, and acrylic copolymers.

All three of these polymer classes are produced by a free radical emulsion polymerization process. Typical monomers that are incorporated include vinyl acetate, ethyl acrylate, butyl acrylate, methyl methacrylate, acrylonitrile, and acrylic and methacrylic acid. These are copolymerized in water in the presence of surfactants to make stable, small-particle-size dispersions. These polymers are characterized as clear, tough, high molecular weight plastics. They can be brittle or flexible, depending on monomer type and ratio. More information

will be presented later on the differences between them.

In some emulsions, specialty surfactants and monomers are used to enhance certain properties of the final sealant. For example, a polymerizable surfactant can be used if improved water resistance is desired. All latex polymers require some level of surfactant to stabilize the microscopically sized polymer particles. If a polymerizable surfactant is used, it is bound to the polymer backbone and cannot migrate to the adhesive-substrate interface and detract from water resistance.

The base polymer formulation includes additives that give desired end-use properties. A typical formulation for a latex caulk is given in Table 1. Proper formulation is essential to obtain a caulk with acceptable final properties. Formulation components are described below.

Polymer Latex Base. The caulk or sealant derives most of its adhesion and elastomeric properties from the polymer base. The least-expensive and lowest-performing grades are made from vinyl acetate homopolymer, which is very brittle and has poor aging/weathering resistance. Vinyl acetate ages poorly because of the relative ease of hydrolysis of the pendant acetate groups. Cleavage of these groups occurs much more readily than it does in acrylic polymers in exterior environments. A high

Table 1 Typical formulation for an acrylic latex caulk

Raw material	Parts by weight
Polymer latex (55–60% solids)	38.0
Nonionic surfactant	0.7
Pigment dispersant	0.8
Plasticizer	9.0
Mineral spirits	1.5
Ethylene glycol	1.0
Calcium carbonate	48.0
Titanium dioxide	1.0

Physical properties

Solids, %	83–85
pH	6.0–8.0
Weight/volume, g/cm^3 (lb/gal)	1.3–1.6 (11–13)
Volume shrinkage, %	25–30%

level of plasticizer is required to attain any flexibility. Caulks designed for interior applications only, in which little or no joint movement will occur, should be made from polyvinyl acetate. Acrylic copolymers are used for premium-grade latex caulks. Their main advantages include improved weathering/aging resistance and better adhesion to most substrates, such as wood, some metals, and glass. Vinyl acetate and acrylic and vinyl acetate and ethylene copolymers are used as a compromise for medium-grade formulations. As one would expect, property values fall between those of the other two classes, although some raw material cost savings do result.

Nonionic Surfactant. A nonionic surfactant is added to improve mechanical stability, help disperse pigment and fillers, and impart freeze/thaw and shelf stability to the formulation. A widely used type is polyethylene oxide nonyl phenol with very hydrophilic characteristics (high hydrophile-lipophile ratio).

Pigment Dispersant. As the name implies, a pigment dispersant incorporates the filler/pigment into the latex emulsion. Without it the filler would "steal" surfactant from the polymer latex, and coagulation of the polymer would result. Phosphate salts are very good dispersants in many formulas, but they can hydrolyze during long-term storage. For this reason, a secondary dispersant is added.

Plasticizers are used to increase the percent elongation and flexibility, improve cold crack resistance, and add solids content to the formulation to minimize shrinkage. They can also contribute to improved adhesion and affect flow characteristics. Some plasticizers have migration problems, however, which lead to paint glossing, dirt pickup, and loss of flexibility. Many people are of the opinion that polymeric plasticizers are an excellent choice because their higher molecular weight contributes to superior aging properties (less migration). Some caulks are designed to be internally plasticized, which means that the polymer base is sufficiently low in glass transition temperature (the lower the value, the softer and more flexible the polymer) to make a plasticizer unnecessary. Internally plasticized polymer emulsions must be high in solids content because an increase in solids is not possible through plasticizer addition. This formulating approach has not gained wide acceptance because of the other favorable contributions imparted by the plasticizer.

Solvents. A small amount of mineral spirits is added to improve the "tooling" of the formulation. The mineral spirits prevent immediate skinning over of the caulk and allow it to maintain a wet edge. Ethylene or propylene glycol promote freeze/thaw stability in the formulation. This property is very important because in many applications the caulk is applied and exposed to freezing temperatures before it is fully cured.

Fillers. Most often, calcium carbonate is used as the filler, or extender. Talc or clay is also used occasionally. Fillers give the formulation higher solids content, which ensures less shrinkage, and a masticlike consistency, which confers nonslump properties. Rheology (flow) and, to a limited extent, elastic properties, are affected by the particle size and oil absorption characteristics of the selected extender.

Caulks and sealants can be characterized by their pigment-to-binder ratio. The amount of extender solids divided by the amount of polymer base solids defines this ratio. Typically, a ratio of 2 to 3:1 is used in formulations. Using too much extender results in a decrease in adhesion and aging properties, such as crack resistance. Lower levels of extender improve adhesion, but can result in excessive volume shrinkage, slump or sag, or cohesive cracking.

Miscellaneous Additives. A widely used additive in latex caulks is a glycidyl or organoamino functionalized silane adhesion promoter. Its use in very small quantities (0.03 to 0.1% of total formulation) improves adhesion to glass after extended water immersion. For some applications, this can be an advantage. Unfortunately, the extent of its contribution to improved properties has been overshadowed by the product labeling abuse resulting from its use. On many labeled tubes, the word silicone or siliconized has been used in conjunction with latex caulk, implying that these products may perform like premium silicone sealants, which is not always the case. It is true, however, that the addition of this compound improves wet adhesion characteristics to a limited number of substrates and improves peel adhesion values after extended water soaks.

Other optional additives include antimildew agents (especially for tub and tile sealants), preservatives, emulsion stabilizers, and various colored pigment dispersions for custom coloring. One advantage of latex sealants is that they are available in a wide variety of colors and they are also paintable

after a short drying time. Many caulks contain a cellulosic or polyacrylic acid copolymer thickener. Clay is used to add "body," and sometimes associative thickeners are used to adjust gun dispensability. In fact, many more additives than can be described here are available to the formulator.

Forms and Applications

All of the latex caulks and sealants are fairly similar in their commercial forms. They are high-viscosity mastics of a gun-dispensable grade. These products are available in various-size tubes for use with caulking guns, in squeeze tubes (for tub and tile applications), and even in pressurized tubes for do-it-yourselfers who might not have a caulking gun. For industrial use, caulks and sealants can be purchased in drums and 20 L (5 gal) pails. Special pumping equipment is available from the manufacturer to handle the requirements for this type of product. All of these caulks and sealants are one-component systems, requiring no mixing or special preparation.

Among all of the caulk and sealant classes, latex caulks and sealants are often the easiest and most trouble-free to use, if they are used only in acceptable applications. They give off no irritating odors, unlike many other sealants, and clean up with water. For these reasons they have gained widespread acceptance for residential and light construction use.

Some typical uses for latex caulks are:

- Perimeter sealing of windows, doors, vents, louvers, and so on
- Filling of gaps between dissimilar substrates such as concrete blocks and wood framing
- Sealing cracks in concrete blocks or brick masonry
- Sealing metal buildings
- General construction sealing
- Use as back bedding for wood, aluminum, or vinyl-clad windows
- Use as glazing compound for window repair
- Sealing interior decks, pavements, curtain wall structures

It is significant that all these uses involve applications in "nonworking" joints, which are defined as those in which total joint movement is expected (or designed) to be less than 15% (see Fig. 1). Total joint movement must be minimized for most caulking applications. Higher expansion and contraction occurs when large structures, such as curtain walls or concrete highway sections, are sealed together. Temperature extremes and other factors that occur with large structures cause the excessive joint movement that leads to premature failure, either cohesive or adhesive, of sealed joints. An improperly designed joint can also contribute greatly to premature failure,

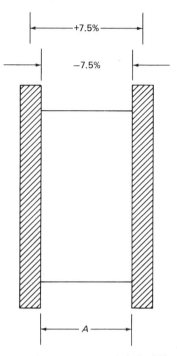

Fig. 1 Total joint movement, which should be limited to ±7.5% for most latex caulks. *A* is nominal width of sealant.

but this subject will not be discussed in this article.

Thus, the primary function of latex caulks and sealants is to prevent air infiltration and to bridge gaps between various substrates that do not move appreciably. Even the suggested total joint movement of 15% is excessive for some of the low-grade caulks. The combination of poor aging resulting from brittleness and even moderate movement will cause adhesion failure over a very short service life.

Latex caulks should not be used for any applications where extended water immersion is possible, although intermittent exposure to water is permissible. The premium acrylic latex grades can handle more water contact without losing adhesion than less expensive grades. When glass is one of the substrates, a sealant containing a silane adhesion promoter should be used. Certain new high-performance all-acrylic latex sealants should be considered for some applications in which more joint movement (25 to 50%) is expected. Recent development work has resulted in greatly improved properties, but field testing of these types has not been extensive.

Major Suppliers and Costs. Manufacturers of latex caulks and sealants include DAP, Inc.; Red Devil Inc.; Macklanburg-Duncan Company; Macco Adhesives; Sherwin Williams Company; H.B. Fuller Company; and others. Rohm & Haas Company and Union Carbide Corporation are two major suppliers of the acrylic and vinyl-acrylic polymer emulsions for these products. In 1990, latex caulks were priced from $1.00 to $3.00 for a 0.4 L (1/10th gal) cartridge, depending on quality as well as markup by distributors and retail outlets.

Application Parameters

The guidelines below are suggested for the use of latex caulks and sealants (Ref 1).

Surface Preparation. Substrates that are to be sealed should be as clean as possible. Oily residue should be removed, as well as peeled paint, old caulk, and other material that interferes with adhesion. Prior to use on plastics and metals, with or without coatings, the sealant supplier should be contacted for recommendations on achieving adequate adhesion. One advantage of latex sealants is that even if the substrate is damp, good adhesion will still result.

Application Conditions. Ideally, the temperature should be between 5 and 40 °C (40 and 100 °F). Low-temperature application causes problems with freezing, diminishes substrate adhesion, and creates difficulty in gun dispensibility. One difficulty arises when millwork (especially metal and vinyl clad millwork) is sealed with a caulk (for example, in window manufacture using back bedding) and the finished glazed units are stored in an unheated warehouse (<0 °C, or 32 °F). The latex caulk will freeze and immediately stop curing. If the window units are installed in freezing weather, the caulk will be expected to perform in a frozen, uncured condition. However, it will be very rigid if it is frozen and will not have the flexibility that the cured caulk would be expected to have, leading to glass breakage from expansion and contraction of the glass-framework interface. It is important that caulks be freeze/thaw stable so that when freezing does occur prior to cure, they will return to a noncoagulated form.

In warm or hot weather, less time is available for tooling of the caulk. Because the consistency (viscosity) of most caulks will degrade in hot weather, caulks should be formulated to be nonslumping under these conditions. A standard test that checks this property will be discussed in the section "Testing Methodology" in this article.

Latex caulks should not be used in exterior applications if rain is expected, or during or after a rainfall. Washout will almost certainly occur unless at least 4 to 8 h elapses after application, especially before a hard-driving rainfall. Washout problems have been one of the main disadvantages of using latex caulks and sealants, although many manufacturers have had few or no complaints in this area at the present time.

Use of Primers. Rarely is a primer used for a latex caulk application. However, if one is required for use with unique substrates, either the manufacturer should be consulted before proceeding or the adhe-

Table 2 ASTM C 834 requirements for latex sealing compounds

Test	Specification	Requirements
C 731	Extrudability	2 g/s (0.07 oz/s), min
C 732	Artificial weathering	
	Wash-out	None, after weathering
	Slump	None, after weathering
	Cracking	None, after weathering
	Discoloration	As acceptable to purchaser
	Adhesion loss	25% max
C 733	Volume shrinkage	30% max
C 734	Low-temperature flexibility	No cracking through to substrate
C 736	Recovery	75% min
C 736	Adhesion loss	25% max
D 2204	Slump	4 mm (0.16 in.) max
D 2203	Stain index	3 max
D 2377	Tack-free time	No material adhering to plastic strip

sion should first be tested on a sample piece of the substrate. After a bead is dispensed with a gun, it should be allowed to cure for a minimum of 48 h and, ideally, for 7 days. Some amount of force should be required to remove the bead. Other more elaborate tests for checking adhesion are described in the section "Properties" in this article.

Backup Materials. If the joint to be sealed is quite large, a backup material should be used to prevent the excessive use of sealant. In the case of an open-ended cavity, backup material helps ensure good contact between the sealant and the substrate surface. A backup material with a round cross-section will optimize the cured shape of the sealant, resulting in good adhesion in expansion and contraction.

If the gap to be sealed is a three-sided joint, a bond breaker should be used. This is a tape or material that prevents adhesion to the surface underneath it and optimizes adhesion in extension and compression of the joint by eliminating stress fractures.

The use of these guidelines will ensure a successful application and minimize any potential for problems.

Properties

Most latex caulk and sealant suppliers offer products that conform to ASTM C 834 requirements (Ref 2). The vast majority of the acrylic and vinyl-acrylic grades obtained from reputable suppliers will meet these criteria. Those caulks based on vinyl acetate homopolymers are not likely to meet this specification because of the embrittlement of the base polymer that occurs upon aging. Some will not meet the requirements even before aging because of their limited flexibility.

The extension-recovery and adhesion test is probably the most difficult aspect of ASTM C 834, but does not require a 7-day water soak prior to testing and also does not involve extension at freezing temperatures, like the more difficult durability and cyclic testing in ASTM C 920 (Ref 3) and TT-S-00230C (Ref 4).

Over the last few years, sealant suppliers have been working to develop latex sealants

that conform to TT-S-00230C, Type II, Class A or B (25 to 50% total joint movement) or ASTM C 920 requirements. This effort has met with some success, although very few products are available. The three major challenges in developing products to meet the TT-S-00230C specification are described below.

Test 4.3.5 requires that weight loss be less than 10% of original weight. This is practically impossible to achieve with a latex system because it necessitates more than a 90% total solids content, which is extremely difficult to achieve while maintaining good flow, mechanical stability, and so on. Those suppliers who claim to satisfy TT-S-00230C performance levels take exception to this test. The typical weight loss of latex caulks is 15 to 20%; volume shrinkage is 25 to 30%.

Test 4.3.9 is a durability test in which adhesion to various substrates is maintained after a 7-day water soak and after extension at freezing temperatures. It is difficult to obtain good water resistance with a latex system because of the components required to stabilize the system. It is also difficult to attain the minimal adhesion and elongation requirements to survive 25 to 50% extension at freezing temperatures.

Test 4.3.10 is a 180° peel adhesion test of at least 875 N/m (5 lbf/in.) with less than 25% adhesive failure. Most latex caulks are relatively strong, cohesively, because of the high molecular weights of the polymer bases used in their formulations (compared to those of solution acrylics). For this reason, cohesive failure is much less likely to occur. This test requires a 7-day water soak as well, which causes problems for many formulations. The development of low molecular weight polymer bases that allow cohesive failure at lower adhesion values would be a significant contribution to the improvement of latex properties. There are, in fact, several newer latex products that have met the requirements of tests 4.3.9 and 4.3.10.

Testing Methodology

There are several specifications that apply to latex caulk and sealant testing. The

Table 3   Federal specification TT-S-00230C, sealing compound type II, Class B

Test	Specification	Requirements
3.3.	Stability	6 months min at <26 °C (80 °F)
3.4	Toxicity	No adverse effects when used as intended
4.3.2.2.	Type II (nonsag)	Maximum sag of 4.8 mm (³⁄₁₆ in.) in vertical displacement
4.3.3	Extrusion rate	<45 s under test conditions
4.3.4	Hardness property at standard conditions	After 21 days of aging, Shore A hardness of 15–50 (35–50 range for walkways, and so on)
4.3.5	Weight loss, cracking and chalking after heat aging	Weight loss less than 10% of original weight; no cracking or chalking after heat treatment
4.3.6	Tack-free time	Compound must cure to tack-free condition in <72 h
4.3.7	Staining	No stain on white cement mortar base
4.3.9	Durability	With mortar, glass, aluminum, or other specified substrate, the total loss in bond area and cohesion area and/or equivalent amount of sealant deformation among the three specimens tested shall be no more than 25% of area.
4.3.10	Adhesion in peel	Minimum peel strength of 875 N/m (5 lbf/in.) with no more than 25% adhesive failure

Note: Type II, nonsagging; class B, 25% total joint movement (class A allows 50% total joint movement)

many factors in the process of formulating a good caulk or sealant involve a balancing of product properties. Tests have been developed that use standard procedures to check most or all of these formulation properties. In addition to ASTM C 834, ASTM C 920, and TT-S-00230C (Ref 2-4), is AAMA specification 802.3 (Ref 5). Tables 2 and 3 describe the test requirements of ASTM C 834 and TT-S-00230C, respectively.

All of the specifications cited above share similar testing requirements, which encompass the needs of most applications. If the tested compounds satisfy the specification requirements, they will normally work well in most approved applications, although there is no substitute for successful field history. Most of these specifications include tests such as those described below.

An extrudability test measures flow characteristics of a compound under normal and adverse use conditions. A basic requirement is that the compound must be gun dispensable after freeze thaw cycling and heat aging. This type of aging tests not only the product, but also the container in which it is packaged, which should be tightly sealed.

Artificial Weathering and Aging. All sealing compounds are expected to perform after extended aging in exterior environments. These tests attempt to simulate exterior aging conditions using heat, ultraviolet (UV) light, and water exposure. Typically, the test samples are checked for visual appearance, adhesion loss, cohesive cracking, washout, discoloration, dirt pickup, mildew, and so forth. Artificial weathering units are also used when one or more of these conditions apply at any one time. Other tests use southern Florida exposure for intense heat and UV light for multiple months.

Durability is a type of testing that closely resembles artificial weathering and aging, but relates more to extensibility and compressibility under adverse conditions (joint movement capability). This type of testing is critical because it often differentiates between compounds that are capable of withstanding 50% or more total joint movement (high level), those that can withstand 25 to 50% total joint movement (moderate level), and those that can withstand less than 25% total joint movement (low level). For example, the durability test in Ref 4 involves the extension/compression of heat-aged samples after a 7-day water soak and the extension of samples at very cold temperatures (−25 °C, or −15 °F). A loss of more than 25% in area from adhesion or cohesive cracking indicates a failed compound.

Adhesion. Most of the specifications under discussion use 180° peel strength tests to check adhesion to various substrates. A layer of sealant is placed on the substrate, and a fabric is embedded in the sealant. After the sealant is cured, the force required to remove a 25 mm (1 in.) long strip is recorded, along with the percentage of cohesive failure (see Fig. 2). Any substrate of interest can be used. The Architectural Aluminum Manufacturers Association (AAMA) peel adhesion test is unique in that it requires the presence of a standard amount of oil on an aluminum substrate to simulate levels of oil residue in a production environment.

Slump/Sag. Compounds must stay in place on vertical installations prior to and after cure. This capability is tested by putting a prescribed amount of uncured sealant on a vertical surface or channel, elevating the temperature, and observing slump or sag properties.

Staining. Latex compounds should be formulated such that minimal or no staining occurs on substrates. This is tested by the application of sealant to either a sample

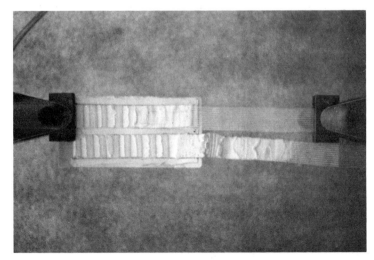

Fig. 2 The 180° peel adhesion test on anodized aluminum as in test 4.3.10 of TT-S-00230C. The sealant is cut down to the substrate when it begins to fail at the cloth-sealant interface.

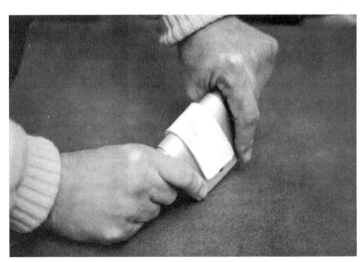

Fig. 3 Low-temperature flexibility test as in ASTM C 834. Artificially aged sealant is bent over a 25 mm (1 in.) mandrel at −17 ± 1 °C (0 ± 2 °F). Sample is then checked for both loss of adhesion and cracking.

piece of substrate or standard filter paper. Observations are made for the migration of an oil or caulk vehicle after a set period of time. Very few latex caulks have difficulty passing this test.

Volume Shrinkage. Compounds should have minimal volume shrinkage (less than 30%). Excessive shrinkage results in cracking or the possible loss of adhesion in certain types of joints. Testing is accomplished by a volumetric displacement measurement before and after cure.

Low-Temperature Flexibility. Compounds of slabs of prescribed size are artificially aged using either an oven or weatherometer. After curing and aging, the samples are conditioned at low temperature (typically -25 or -18 °C, or -15 or 0 °F) and bent over a 13 or 25 mm (½ or 1 in.) mandrel (see Fig. 3). Samples are checked for both loss of adhesion and cracking.

REFERENCES

1. "Standard Practices for Use of Latex Sealing Compounds," C 790, *Annual Book of ASTM Standards*, American Society for Testing and Materials
2. "Standard Specification for Latex Sealing Compounds," C 834, *Annual Book of ASTM Standards*, American Society for Testing and Materials
3. "Standard Specification for Elastomeric Joint Sealants," C 920, *Annual Book of ASTM Standards*, American Society for Testing and Materials
4. "Sealing Compound: Elastomeric Type, Single Component," Interim Federal Specification TT-S-00230C, National Bureau of Standards
5. "Type I Back Bedding Glazing Compounds for Use with Architectural Aluminum," AAMA Specification 802.3, Architectural Aluminum Manufacturers Association

SELECTED REFERENCES

- "Voluntary Specifications for Sealants," Architectural Aluminum Manufacturers Association, 1976
- A. Nelson, "UCAR Latexes for Water-based Caulks and Sealants," Union Carbide Corporation, 1974
- J.R. Panek, *Construction Sealants and Adhesives*, 2nd ed., John Wiley & Sons, 1984

Silicones

M. Dale Beers, Loctite Corporation
Jerome M. Klosowski, Dow Corning Corporation

SILICONES encompass a wide range of materials, including fluids, resins, heat vulcanizing elastomeric gum stocks, specialty chemicals, and room-temperature vulcanizing (RTV) liquid rubbers. Their commercial importance is based on their unique combination of properties. These property characteristics are the result of scientific endeavors to combine some of the most stable behavior attributes of the inorganic world with the highly utilizable aspects of organic substances. Silicone rubber, which maintains flexibility at extremely low temperatures and stability at high temperatures, illustrates this quite well.

The subject of this article is confined to the product family of liquid RTV elastomeric silicone sealants. Silicone adhesives are described in the Section "Adhesives Materials" in this Volume.

Chemistry

Polymers. The most common ingredient of silicone sealants is a polymer known as a silanol-terminated polydimethylsiloxane, as shown in Fig. 1. Silanol is an OH group attached to silicon. Several manufacturing steps are involved in obtaining this type of polymer. The first step consists of the reaction of the appropriate organic halide, or halocarbon, which is usually methyl chloride, with finely divided silicon metal in the presence of a copper catalyst to yield a mixture of materials known as chlorosilanes (Ref 1), as shown in Fig. 2. The predominant product is dimethyldichlorosilane. The reaction is carefully controlled at elevated temperatures using a pressurized fluid bed reactor.

The reaction products shown in Fig. 2 are separated through distillation. The dimethyldichlorosilane is then reacted with a large excess of water. This hydrolysis gives a mixture of cyclic and linear silanol-terminated polydimethylsiloxanes, as shown in Fig. 3.

The hydrolyzate mixture can be polymerized by a number of different catalysts, or the cyclic polydimethylsiloxanes can be removed by vacuum distillation and polymerized separately to form silano terminated polymers. The molecular weight of the polymer is dependent on the amount of moisture (chain stopper, giving the terminal silanol groups) retained in the system. Either basic or acidic polymerization catalysts may be used. Some examples of the basic catalysts are the hydroxides of lithium (Ref 2), sodium (Ref 2), potassium (Ref 2), cesium (Ref 3), and potassium amide (Ref 4).

Some of the acid catalysts are sulfuric acid (Ref 5), phosphoric acid (Ref 6), ethylsulfuric acid (Ref 7), pyrophosphoric acid (Ref 8), nitric acid (Ref 8), selenic acid (Ref 9), boric acid (Ref 10, 11), activated fuller's earth (Ref 11, 12), and chlorosulfonic acid (Ref 13).

For more specialty-type sealants, various synthetic routes allow some of the pendant methyl groups on the polydimethylsiloxane polymer to be replaced with other organic groups to produce various product characteristics. For example, aerospace-type sealants may contain approximately 5 mol% diphenylsiloxy groups to lower the brittle point of the sealant from −65 to −110 °C (−85 to −165 °F). Sealants requiring very low solvent-swell features for immersion in jet aviation kerosene or ordinary gasoline are based on polymers that have a 1,1,1-trifluoropropyl group on each silicon atom in the polymer chain. These types of changes will be discussed in more detail in the section "Applications and Suppliers" in this article.

Fillers. Another important ingredient present in RTV silicone sealants is filler. Fillers are selected to serve a variety of functions in elastomeric silicones. Frequently, high-surface-area fumed silicas are used to reinforce the relatively weak cured polymer network. High-surface-area carbon blacks are also useful for increasing the strength of the cured sealant. For example, the cured tensile strength of the polydimethylsiloxane polymer can be increased from the typical 275 to 415 kPa (40 to 60 psi) to as high as 8.3 MPa (1200 psi) using these types of high-surface-area fillers.

A variety of semireinforcing and nonreinforcing fillers are used to improve physical properties slightly or to act as extenders in order to decrease the cost of the sealant. Some examples are ground calcium carbonates, ground quartz, iron oxide, diatomaceous earth, zinc oxide, and various types of clays. Low-density fillers like plastic or glass microballoons can be used to produce low-density compounds. A large number of thermally stable inorganic pigments are used to provide the required colors.

Curing Systems. Another necessary ingredient of a silicone sealant is a reactive cross-linking system that will cause a transformation from the liquid or pastelike product composition to an elastomeric silicone rubber. The majority of these cross-linking

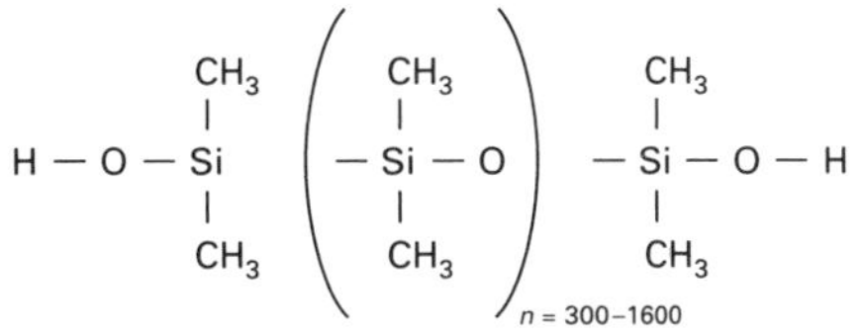

Fig. 1 Silanol-terminated polydimethysiloxane

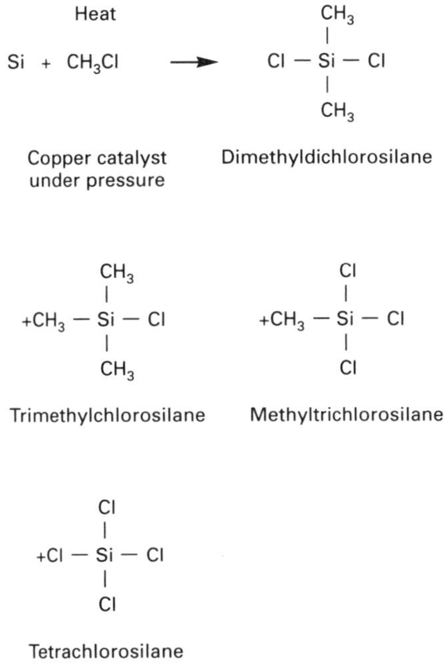

Fig. 2 Chlorosilane reaction products

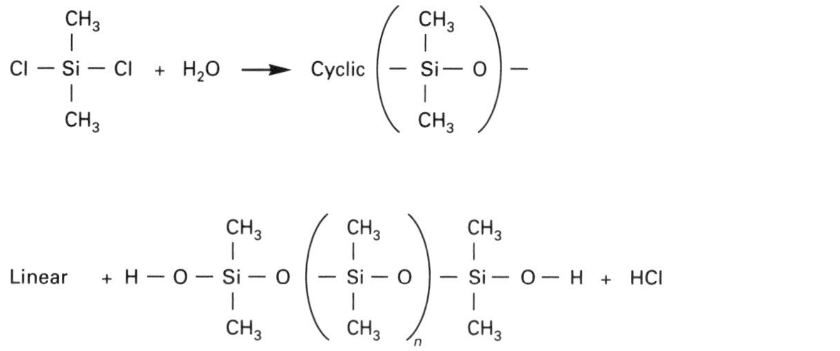

$$CH_3 - \underset{\underset{CH_3}{|}}{\overset{\overset{CH_3}{|}}{Si}} \ \ Cl - Si - Cl \ + \ H_2O \longrightarrow Cyclic \left(- \underset{\underset{CH_3}{|}}{\overset{\overset{CH_3}{|}}{Si}} - O \right) -$$

$$Linear \ + \ H - O - \underset{\underset{CH_3}{|}}{\overset{\overset{CH_3}{|}}{Si}} - O \left(- \underset{\underset{CH_3}{|}}{\overset{\overset{CH_3}{|}}{Si}} - O \right)_n - \underset{\underset{CH_3}{|}}{\overset{\overset{CH_3}{|}}{Si}} - O - H \ + \ HCl$$

Fig. 3 Hydrolysis of dimethyldichlorosilane to yield cyclic and low molecular weight linear silanol-terminated polydimethylsiloxanes

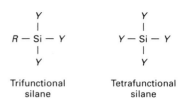

Multifunctional silanes

$$R - \underset{\underset{Y}{|}}{\overset{\overset{Y}{|}}{Si}} - Y \qquad Y - \underset{\underset{Y}{|}}{\overset{\overset{Y}{|}}{Si}} - Y$$

Trifunctional silane   Tetrafunctional silane

Fig. 4 Moisture-reactive polyfunctional silanes used as cross-linking agents

agents are moisture-reactive polyfunctional silanes of the type shown in Fig. 4. To accelerate the rate of cure, an organometallic compound known as a condensation catalyst is also used. Examples of these catalysts are identified below.

In the silanes shown in Fig. 4, the Y groups are readily hydrolyzable, low molecular weight organic radicals that are released during the cross-linking reaction. These by-products, which diffuse out of the RTV upon contact with moisture, include alcohols (Ref 14-18), ketoximes (Ref 19), carboxylic acids (Ref 20-23), amides (Ref 24-25), ketones (Ref 26, 27), and amines (Ref 28). Because the cure is dependent on the diffusion of moisture into the sealant and the diffusion of the by-product out of the sealant, the sealant cures from the top down, which is a relatively slow process.

Some of the cross-linking agents that may be used are depicted in Fig. 5. The condensation catalyst may be a tin soap (Ref 14), such as stannous octoate and dibutyltindilaurate, or an organotitanate (Ref 15-17).

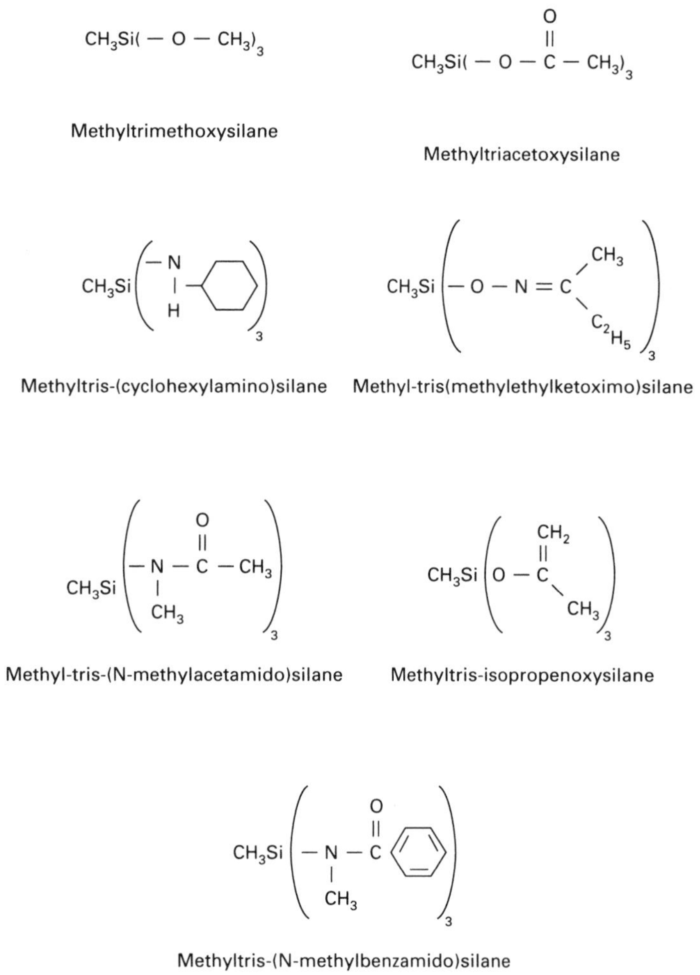

Most of the curing chemistry referred to above typically applies to one-component silicone sealants. They are packaged in collapsible metal tubes, plastic caulking cartridges, 5 gal pails, or 55 gal drums. However, two-component sealant products are also available. The way in which the above-mentioned ingredients are incorporated into the final product is dependent on whether the desired RTV sealant is to be packaged in one or two units. The condensation catalyst is most frequently packaged as one of the two units, while the polymer, filler, and cross-linking agent are combined to form the other unit. In some cases, the condensation catalyst and cross-linking agent are both incorporated in an inactive carrier fluid or paste, but the silanol-terminated reactive polymer-filler composition is kept apart in another package.

The decision about whether the cross-linking agent is incorporated with the polymer and filler mixture depends on the tendency of the specific organosilane to react with particular ingredients in the absence of a condensation catalyst. Curing takes place when the two units are mixed together. Two-component RTV silicones can be designed to cure in deep sections at room temperature in a matter of minutes or hours, without being dependent on atmospheric moisture. The vulcanization process is shown in Fig. 6.

As already noted, the most common type of RTV silicone sealant is a one-component silanol condensation curing product. It contains the various types of fillers described above for strength, flow control, and other properties, in combination with a moisture-reactive polyfunctional organosilane and usually a condensation catalyst. The one-component sealant compositions must be kept completely anhydrous to prevent curing from occurring in the package. During manufacture in the absence of atmospheric moisture, the reactive cross-linking agent terminates the silanol functional polymer, as shown in Fig. 7.

The R shown in Fig. 7 is usually methyl, vinyl, phenyl, or trifluoropropyl. Upon extrusion from its container and exposure to atmospheric moisture, the sealant vulcanizes to an elastomeric silicone, releasing one

$$CH_3Si(-O-CH_3)_3$$

Methyltrimethoxysilane

$$CH_3Si\left(-O-\overset{\overset{O}{\|}}{C}-CH_3\right)_3$$

Methyltriacetoxysilane

$$CH_3Si\left(-\underset{\underset{H}{|}}{N}-\bigcirc\right)_3$$

Methyltris-(cyclohexylamino)silane

$$CH_3Si\left(-O-N=C\overset{\diagup CH_3}{\diagdown C_2H_5}\right)_3$$

Methyl-tris(methylethylketoximo)silane

$$CH_3Si\left(-\underset{\underset{CH_3}{|}}{N}-\overset{\overset{O}{\|}}{C}-CH_3\right)_3$$

Methyl-tris-(N-methylacetamido)silane

$$CH_3Si\left(O-\overset{\overset{CH_2}{\|}}{C}\diagdown_{CH_3}\right)_3$$

Methyltris-isopropenoxysilane

$$CH_3Si\left(-\underset{\underset{CH_3}{|}}{N}-\overset{\overset{O}{\|}}{C}-\bigcirc\right)_3$$

Methyltris-(N-methylbenzamido)silane

Fig. 5 RTV silicone cross-linking agents

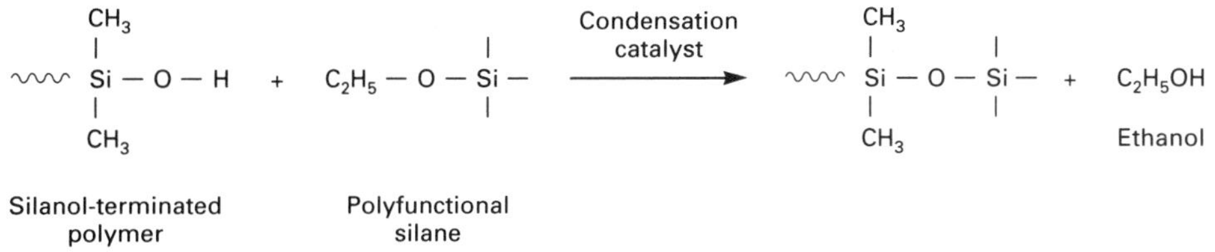

Fig. 6 Condensation curing process of two-component RTV silicone sealants

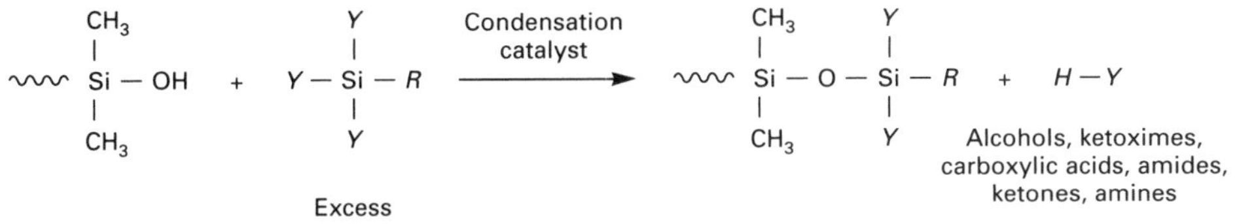

Fig. 7 Termination of silanol functional polymer by reactive cross-linking agent

or more of the cure by-products already described. The moisture-catalyzed vulcanization is shown in Fig. 8. The expected mechanism, of course, is Y group hydrolysis to form a silanol, which then reacts with another Y group to give a cross link.

Another type of silicone sealant that is utilized to a much lesser degree involves a product class of materials known as addition-curing liquid silicone elastomers. Like the RTV condensation-curing materials, they have pourable to thixotropic pastelike rheological properties. They differ in that they do not require atmospheric moisture to cure and generally employ heat to obtain rapid vulcanization. Special proprietary adhesion promoters are included in the product compositions to give them adhesive characteristics.

More generally, this type of cure chemistry is used to design molding products

known for their excellent release characteristics. They can be either one- or two-component types of materials. In the case of one-component compositions, an inhibitor is added to complex with a platinum catalyst, which is responsible for initiating the vulcanization reaction of silicon hydride functional groups with vinyl moieties on the silicone polymer. Elevated temperatures are required to release the platinum from the inhibitor to effect the cure to a silicone elastomer. This cross-linking process is shown in Fig. 9.

The platinum catalyst can be in a variety of forms, ranging from platinum-vinyl silicone complexes to chloroplatinic acid, H_2PtCl_6. The platinum can be present in quantities as low as 5 ppm and up to about 60 ppm. In the case of a two-component composition, the platinum catalyst, as well as the vinyl functional polymer and fillers represent one unit,

while the silicon hydride counterpart represents the other. The two components are mixed together in the proper ratio when the sealing operation is ready to begin.

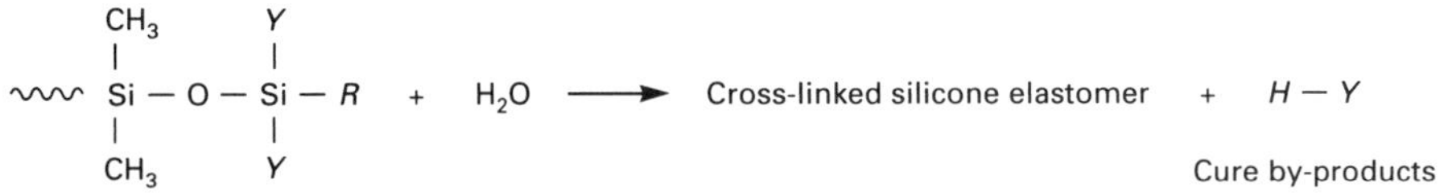

Forms and Properties

Because of their unique product properties, silicone sealants have gained a segment in the consumer, construction, and general industrial markets. Described below are the various product types in conjunction with their related properties.

Flow and Viscosity. One of the many attractive characteristics of silicone sealants is that they are fairly easy to use because of their relatively low viscosities and controllable flow properties. Their viscosities are affected by changes in temperature to a much lesser extent than are those of other organic-type sealant products. For example, silicone construction sealant will remain extrudable at temperatures as low as -40 °C (-40 °F) (Ref 29).

In both one- and two-component systems, a wide range of rheological properties such as pastelike and nonsagging properties, is available, as well as pourable and self-leveling characteristics. The viscosities of the pastelike nonsagging sealants are usually expressed in terms of the extrusion rate out of a 3.2 mm (⅛ in.) orifice at an air pressure of 620 kPa (90 psi). Extrusion rates range from approximately 100 to 1000 g/min (3.5 to 35 oz/min). Almost all silicone sealants are solventless and have minimal shrinkage upon curing.

Thermal Properties. Silicone products exhibit the highest thermal stability characteristics of the available elastomeric sealants. They are considered to occupy an intermediate position between organic polymers and silicate materials such as glasses or enamels. Industrial grades of silicone sealants can be heated in air for 1 year or more at temperatures between 160 and 205 °C (325 and 400 °F) without appreciable change in their physical properties. At temperatures between 205 and 345 °C (400 and 650 °F), the exposure period ranges from a matter of hours to several months.

This stability is due to the siloxane structure of the polymer backbone, as well as the thermal stabilizers and fillers used in them. Table 1 demonstrates the thermal stability of a typical extremely high-temperature RTV product. These extremely high-temperature characteristics have been used both in the consumer and industrial markets and will be discussed in more detail in the section "Applications and Suppliers" in this article.

Weathering. The superior resistance of silicone sealants to ultraviolet (UV) radiation and ozone, compared to organic types of elastomeric sealants, makes them ideal for use in the construction industry. This outstanding stability is derived from the silicon-oxygen linkage in the polymer chain, as

Fig. 8 Moisture-catalyzed vulcanization of one-component RTV silicone sealants

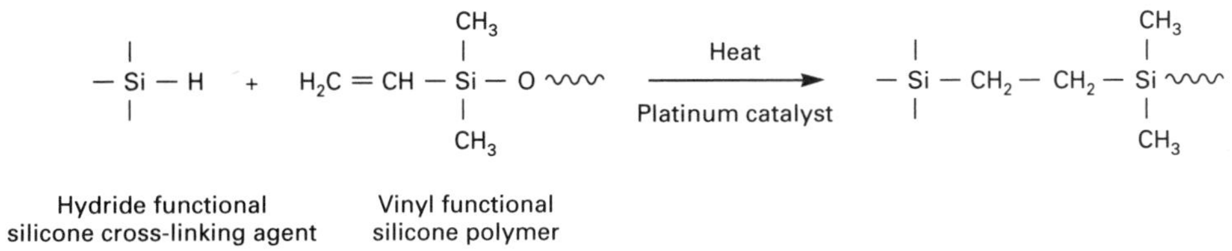

Fig. 9 Addition-curing process of platinum-catalyzed silicone sealants

Table 1 Characteristic thermal aging properties of an extremely high-temperature RTV silicone sealant

Property	Original properties	After 168 h at 250 °C (480 °F)	After 168 h at 315 °C (600 °F)	After 336 h at 315 °C (600 °F)
Shore A, hardness	33	28	45	65
Tensile strength, MPa (psi)	2.4 (350)	2.7 (390)	2.9 (420)	3.3 (475)
Elongation, %	400	540	300	180

Table 2 Physical properties of the early two-component RTV silicone sealants

Property	Value
Shore A, hardness	15–70
Tensile strength, MPa (psi)	0.7–6.2 (100–900)
Elongation, %	80–200

Table 3 Typical strength and elongation properties of RTV silicone adhesive sealants

Property	Low strength	Medium strength	High strength
Shore A, hardness	10–30	20–70	25–60
Tensile strength, MPa (psi)	0.7–1.7 (100–250)	1.7–4.1 (250–600)	4.1–8.3 (600–1200)
Elongation, %	100–1200	80–700	300–1000
Tear strength, kN/m (lbf/in.)	1.75–5.3 (10–30)	3.5–14 (20–80)	14–22/44 (80–125/250)

well as to the absence of any double-bond unsaturation. Many organic polymers contain some degree of unsaturation and are therefore subject to attack by ozone and UV radiation. This behavior is evidenced by surface cracking of the organic rubber after long-term weather exposure. For this reason, RTV silicone is frequently used where long-term weather exposure is expected.

Ultraviolet Radiation Stability. Silicones do not absorb UV radiation and for this reason they do not degrade in sunlight. This important feature is the basis for their position as the premier glazing sealant and as the only sealant allowed in structural glazing applications where the sun shines through glass on the bonded interface.

Strength and Elongation. The first RTV silicone sealants marketed in the late 1950s were of the two-component variety. They were relatively weak materials compared to the heat-vulcanizable silicone rubber stock available at that time. The typical cured properties of these early two-component RTV materials are listed in Table 2.

Since the early 1950s, significant progress has been made in enhancing both the strength and elongation properties of one- and two-component RTV silicone products. Tensile strengths as high as 8.3 MPa (1200 psi) and elongations as high as 1600% are now available in these products. The characteristic high tensile strength materials are generally used in the aerospace industry, whereas the extremely high elongation products are used in the construction industry, where low-modulus sealants are required to maintain long-term adhesion to concrete in building and highway expansion joints. The characteristic strength and elongation properties available in the large variety of products manufactured by the silicone industry today are given in Table 3.

Electrical Insulation. Silicone elastomers have long been known for their excellent electrical insulation properties. This is mainly due to the low polarity of the silicone polymers in these compositions. They

Table 4 Typical RTV silicone rubber electrical properties

Property	Value
Volume resistivity, $\Omega \cdot$ cm	3×10^{15}
Dielectric strength, 0.075 m (3 in.) thick specimen, kV/mm (V/mil)	20 (500)
Dielectric constant at 10 Hz	2.8
Dissipation factor at 60 Hz	0.0026

have extremely high resistance to high-voltage ionization, and their corona resistance approaches that of mica.

The main advantage of RTV silicone rubber adhesive sealants is that they retain their good electrical properties, such as dielectric strength, power factor, and insulation resistance, at elevated temperatures at which other typical organic insulating materials become useless. Their electrical properties are listed in Table 4.

Curing Characteristics. The curing rates of one- and two-component RTV products vary widely, depending on the desired application. The faster-curing materials are generally found in industrial and consumer applications, whereas the slower-reacting products are directed toward the construction market, where longer sealant surface finishing times are important.

Because one-component systems depend on the diffusion of atmospheric moisture into the sealant for their cure, they are not recommended for applications requiring deep-section vulcanization. Instead, they are usually restricted to those areas where the needed curing thickness in one direction does not exceed 13 mm (0.5 in.). The skinning time of the sealant can vary anywhere from minutes to more than 1 h, depending on the product design requirements for the intended application. Higher temperatures and humidities will accelerate the cure, whereas the converse is true at lower temperatures and humidities.

In general, the faster-curing one-part sealants will vulcanize through a 3.2 mm (⅛ in.) thickness from a single surface in 8 to 24 h at 25 °C (77 °C) and 50% relative humidity. Slower-curing products will require 24 to 72 h.

In the case of two-component systems, the cure rate is dependent on the level and type of curing catalyst used. The moisture necessary to vulcanize a two-component RTV completely is already present in one of the two packaged component units. Deep-section cure is readily obtained in a properly designed product. The cure rate can be adjusted to take place in a matter of minutes or hours. Stannous octoate and dibutyltin-dilaurate are among the most common catalysts employed. The former is the most reactive.

Applications and Suppliers

Product applications of silicone adhesive sealants are described in this article by three categories: construction, industrial, and consumer sealants.

Construction

Silicone sealants are used in the construction industry because of their superior long-term weathering resistance (Ref 30). This is due to the extreme stability of the silicone polymer when subjected to UV radiation and ozone oxidation. In addition, silicone sealants can be easily applied from caulking cartridges, pails, or drums with automated pumping equipment attached to dispensing lines and collapsible tubes. The surface tooling times of the silicone sealants are designed by the manufacturer to meet various application requirements. They can be either self-leveling in nature for use on horizontal on-grade joints or nonsag for vertical joints.

The five principal application areas for silicone construction sealants are use in building and highway expansion joints, structural glazing in which the cured silicone sealant becomes part of the overall load-bearing design, general weatherproofing glazing on porous and nonporous materials, sanitary mildew-resistant sealant used mainly for sealing kitchen and bathroom fixtures, and flame-retardant fire-stop sealants for sealing around pipes, electrical conduits, ducts, and electrical wiring within building walls. Each of these product categories is discussed below, along with manufacturers and general purchase prices as of 1990.

Building and Highway Expansion Joints. The building and highway expansion joint market is a fast growing part of this industry. The silicone sealants designed for building and highway expansion joints are low modulus in nature (Ref 31), that is, they exhibit high elongation and low tensile strength properties so that there is very low stress applied to the sealant-joint bonding interface. This low stress is an absolute necessity because the surface of the joint is traditionally composed of concrete having relatively weak tensile strength characteristics. If the modulus is too high, the surface of the concrete will be pulled off, causing the joint to fail. Also, the adhesion at the interface must be adequate in order to prevent adhesive loss of the seal. Adhesion is obtained through the use of proprietary adhesion promoters in the composition or by coating the surface with a silicone resin primer before the sealant is applied. To

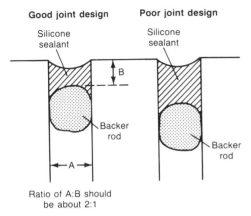

Ratio of A:B should
be about 2:1

Fig. 10 Recommended concrete expansion joint design using low-modulus silicone construction sealant

Table 5 Typical physical properties of silicone sealant for concrete building and highway expansion joints

Property	Value
Shore A, hardness	10–25
Tensile strength, MPa (psi)	0.7–2.0 (100–290)
Elongation, %	650–1600
Joint movement capability, %	±50 to 100/−50

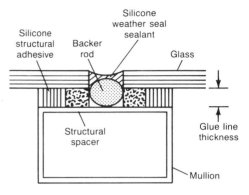

Fig. 11 Typical structural glazing design

obtain the proper degree of adhesion, the surfaces are cleaned with a mechanical surface abrading device such as a wire brush. The recommended joint design is shown in Fig. 10. It should be noted that the backer rod in Fig. 10 is composed of foamed polyethylene, from which the silicone easily releases, thereby allowing the sealant to operate with the lowest possible stress on the bonding interface.

The low-modulus construction silicone sealants represent the lowest modulus sealants in the industry. Because building movements should put very little stress on the joint walls, this type of sealant is the most appropriate for most building systems. These sealants are critical to building systems with weak surfaces.

Another important feature of a building or highway silicone expansion joint sealant is that its cure by-products must be neutral, or basic, in nature. If the sealant gives off an acidic cure by-product, such as acetic acid, a chemical reaction will occur, by which the concrete leaves behind a weak salt deposit interlayer that will lead to joint failure. The proper neutral-curing sealants will give off amides, oximes, alcohols, or amines as cure by-products. Properties of silicone sealants used in this market are given in Table 5.

Structural Glazing. Silicone construction sealants designed for structural glazing applications are of the medium- to high-modulus range and are substantially higher in modulus than some of the very low-modulus sealants used in building or highway expansion joints (Ref 32). This medium- to higher-strength sealant, which has excellent bonding properties to nonporous substrates such as glass, aluminum, and various types of coated metals, is necessary because it becomes an integral part of the structural support design. This type of silicone sealant is utilized in a glazing system where it bonds the glass to the structural framing of a building. Dynamic wind loads are then transferred from the glass to the perimeter support by the cured elastomeric

sealant. However, they cannot be so stiff as to not allow movement, because thermal movement is an important part of this sealing application. Only silicone sealants are allowed in structural glazing applications.

Both one- and two-component RTV silicone products are used in structural glazing applications. The two-component neutral-curing product is used when faster cure is needed for assembly line production, such as the in-shop glazing of structurally glazed curtain wall modules. Because the two-component product does not depend on the diffusion of atmospheric moisture into the sealant for its cure but instead cures through the entire cross-sectional area of the applied sealant, the manufactured assemblies can be handled much sooner without jeopardizing the bonding interface. Typically, the two-component version will change from its pastelike consistency to an elastomer in 4 to 8 h, whereas a one-component RTV silicone will take several days.

Each silicone construction sealant manufacturer will review structural glazing joint designs and make recommendations for the best utilization of specific sealant products. A typical structural glazing joint design is shown in Fig. 11. Table 6 lists the typical physical properties of silicone structural glazing products.

General Glazing for Weatherproofing. A large number of silicone construction sealants are used for general exterior glazing for weatherproofing. Some products have been designed solely for use on nonporous surfaces, and others can be used on both porous and nonporous surfaces. Examples of porous surfaces are concrete and other types of masonry.

Generally, the acetic acid releasing products are confined to use on nonporous materials such as glass, plastics, aluminum, and various types of coated metals. The neutral-curing silicone sealants are usually suggested for use on both types of surfaces.

Most of these products exhibit primerless adhesion characteristics. As is the case with all elastomeric silicone sealants, these products do not become embrittled with age and can be expected to perform for 20 years or more on properly prepared surfaces. Table 7 lists the typical physical properties of general silicone glazing sealants.

Sanitary Mildew-Resistant Sealants. The silicone industry supplies a number of mildew-resistant sealants to be used on nonporous surfaces such as around ceramic tile, showers, tubs, sinks, and plumbing fixtures where conditions of high humidity and temperatures exist. Such elastomeric silicones retain their elasticity and will not crack with age like some of their organic sealant counterparts. These sanitary mildew-resistant silicone sealants can be based on either acetic-acid- or neutral-curing, byproduct releasing, cross-linking systems. Table 8 lists the typical physical properties of sanitary mildew-resistant sealants. The chemical resistance and resistance to attack by harsh cleaning agents combined with longevity are the chief reasons behind the wide use of silicones in this market. Although the typical silicone does not serve as a nutrient and support mildew growth, the mildew-resistant sealants typically contain a fungicide.

Flame-Retardant Fire Stop Sealants. Flame-retardant silicone RTV sealing materials are used to seal wall openings through which electrical wiring, conduits, ducts, and pipes are run. They serve to stop the passage of fire, smoke, and water between rooms. Both one-component and two-component RTV sealants have been designed for this purpose. The latter consists of two

Table 6 Typical physical properties of silicone structural glazing sealants

Property	Value
Shore A, hardness	25–45
Tensile strength, MPa (psi)	1.9–3.3 (270–485)
Elongation, %	300–500
Joint movement capability, %	±25

Table 7 Typical physical properties of silicone glazing sealants

Property	Value
Shore A, hardness	15–40
Tensile strength, MPa (psi)	1.0–2.8 (150–400)
Elongation, %	300–800
Joint movement capability, %	±25

Table 8 Typical physical properties of sanitary mildew-resistant sealants

Property	Value
Shore A, hardness	25–40
Tensile strength, MPa (psi)	0.97–2.8 (140–400)
Elongation, %	300–500
Joint movement capability, %	±25%

Table 9 Typical properties of flame-retardant, fire-stop, one-component silicone sealing products

Property	Value
Shore A, hardness	25–40
Tensile strength, MPa (psi)	1.7–2.8 (250–400)
Elongation, %	300–500
Joint movement capability, %	±25%
Oxygen index	32–40

Table 10 Typical properties of a liquid two-component RTV silicone foam sealer

Property	Value
Density, g/cm^3	0.19–0.32
Oxygen index	28–36
Consistency	Pourable
Uncured product viscosity of two components, Pa · s	6–10

low-viscosity liquid components that are typically combined by automatic mixing and dispensing equipment during application, whereupon a foaming reaction commences immediately. The oxygen index for these materials typically ranges from about 32 to 40. The sealants are usually used in applications in which the fire penetration rating ranges from 1 to 3 h. Tables 9 and 10 describe the properties of these two sealing systems.

Suppliers. The principal manufacturers of silicone construction sealants in the United States are Dow Corning Corporation; the General Electric Company, Silicone Products Division; Wacker Silicones Corporation; Mobay Chemical Corporation, Silicone Division; Rhone-Poulenc Inc.; and Tremco Inc. The silicone construction sealants are considered to be among the higher-cost premium-type products. This market grows in spite of the initial cost of these products because of their longevity and durability, as well as their cost effectiveness.

Costs. The total utilization costs depend on the total volume of product purchased, as well as on container size. The glazing and concrete expansion joint product classes run from about $8 to 13/L ($30 to 50/gal). The more expensive flame-retardant fire stop products range from about $16 to 26/L ($60 to 100/gal).

Industrial Silicones

The industrial RTV silicone adhesive sealants consist of myriad products. However, for the purposes of this discussion, they will be broken into four basic product groups: automotive, aerospace, electronics, and general industrial.

Automotive. Aside from some specialized sealant applications that will be discussed later, the largest use of RTV silicone is for automotive gasketing, both at the

original equipment manufacturer (OEM) level and in garage repair situations.

The use of RTV silicone formed-in-place gasketing began in the early 1970s. Several product changes have been made since that time. The first generation of silicone formed-in-place gasketing involved products that evolved acetic acid as a cure by-product. This gave rise to corrosion concerns in enclosed environments in which the acetic acid had no immediate escape route available. Typically these materials were used to seal valve covers, oil pans, intake manifolds, engine side seals, and thermostat housings. The corrosion concerns were eliminated with the introduction of neutral-curing silicone sealants. The corrosion concern was then followed by the need for products that had high joint movement capability combined with excellent oil resistance. All of these product improvement requirements have been addressed by the silicone industry.

The latest improved joint movement technology and oil resistance properties are shown in Fig. 12 and Table 11. The oil resistance of the sealant has been optimized to bring about the stabilization of the silicone rubber against the acidic thermal decomposition by-products of the engine oil. The current engine sealing needs are being met by both molded silicone rubber and the RTV silicone sealants.

Aerospace. In the aerospace industry, high-strength adhesive sealants are used for adhering heat-cured silicone rubber gaskets to metal substrates around the doors and windows of aircraft. Virtually every U.S. missile or satellite contains silicone sealant because of its long life and resistance to temperature extremes. Special very low-volatile-content two-component RTV sealants have been used as window sealants on space vehicles to prevent the condensed

vapor fogging of visual observation ports. The low outgassing features of these types of products are given in Table 12. RTV silicones are also used as coating materials in the aerospace industry to protect mechanical assemblies from moisture, chemicals, and temperature extremes. As in the automotive industry, a number of silicone sealants are used in electronic assemblies; this application is discussed below.

The electronics industry uses both one- and two-component RTV silicones for sealing and encapsulation. They consist of materials that give off noncorrosive cure by-products, such as alcohols, or addition-curing silicone materials that have no cure by-products. There are also UV-radiation curable sealants that are ideal when rapid assembly line production procedures are used. These products serve to protect the electronic components, such as electrical connectors, switches, coils, and chip boards, from dust, chemicals, and moisture.

These materials can range from low-viscosity conformal-type coatings to thixotropic nonsag-type pastes. Again, silicones are used because of their outstanding elevated-temperature stability, excellent electrical insulation properties, and low-temperature flexibility characteristics. The high-temperature stability and low-temperature flexibility properties are very important to electronic aerospace applications. As was pointed out earlier, the basic silicone polymer system present in these sealant products will remain flexible at temperatures as low as −65 to −109 °C (−85 to −165 °F). Depending on product selection and curing temperatures, this family of silicone sealants can be made to cure to an elastomer in a matter of minutes or just a few hours. The cured physical properties are very similar to those described previously for various RTV silicone sealant products.

General Industrial Use. A multitude of silicone sealant products are available for general industrial use. Here, too, they are selected because of their stability over wide temperature ranges, chemical stability, extreme weather resistance, bonding capabilities on a wide range of substrates, excellent electrical insulation properties, and ease of use.

Aside from general production use, many silicone sealants are employed in maintenance and repair applications. These uses include sealing, gasketing, bonding, insula-

Table 11 Comparison of the oil resistance properties of current low-modulus, high joint movement RTV automotive gasketing product versus older technology

Property	Initial properties		Properties after 14 days at 150 °C (300 °F) in 5W-30 engine oil	
	Old	Current	Old	Current
Shore A, hardness	35	34	3	22
Tensile strength, MPa (psi)	3.0 (430)	1.7 (240)	0.8 (114)	1.2 (170)
Elongation, %	530	450	690	550
Volume swell, %	· · ·	· · ·	28	28

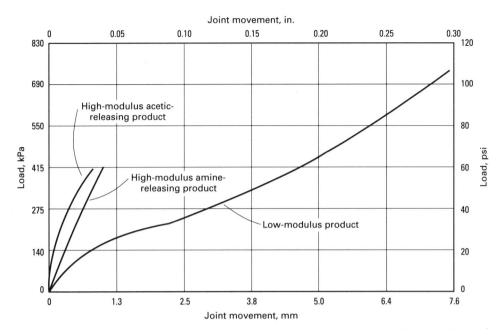

Fig. 12 Joint movement behavior of a low-modulus product compared to some high-modulus materials. Based on cast iron and cast aluminum lap-shear specimens 25 × 9.5 mm (1 × 0.375 in.), with a shear gap of 1.4 mm (0.054 in.). Cure time was 14 days.

tion, encapsulating, and coating applications. Again, their performance properties are well described in earlier tables.

Suppliers. The primary manufacturers of industrial silicones in the United States are Dow Corning Corporation; General Electric Company, Silicone Products Division; Loctite Corporation; Wacker Silicones Corporation; Rhone-Poulenc Inc., Silicone Division; and Mobay Corporation, Silicone Division.

The costs of industrial silicone products depend greatly on how specialized the required manufacturing technology is to produce them. For example, the extremely low volatility requirements for some aerospace products dictate such low-volume slow production procedures that the final product purchasing prices are driven to the $660 to 1100/kg ($300 to 500/lb) level. Specialty electronic sealant materials usually range from about $22 to 66/kg ($10 to 30/lb). General industrial silicone sealing products sell for approximately $9 to 33/kg ($4 to 15/lb).

Consumer Silicones

One-component RTV sealants are used in many do-it-yourself consumer applications. They are based on acetic acid- and neutral-curing products and a relatively new group of water-dispersed silicones. Various sealants are sold as bathtub caulks, windshield sealants, high-temperature instant-gasketing sealants, regular instant-gasketing sealants, general automotive sealants, sealants specifically for use on metals, glass sealants, hobby glues, general household glues and sealants, and paintable all-weather silicone caulk. These materials possess self-bonding adhesion properties to glass, metal, ceramics, fabrics, wood, leather, plastics, and paper. They exhibit the typical properties of elastomeric silicones, such as permanent flexibility, high-temperature stability, good electrical insulation properties, resistance to chemicals, and excellent weathering properties.

The water-dispersed silicones warrant further discussion. They are typically dispersions of precured or pregelled polymers in combination with fillers and other additives typical of latex acrylic systems. These materials have properties similar to conventional silicones, but the advantages of water clean-up and little, if any, dirt pick-up or staining. These materials will represent an ever-increasing part of the consumer silicone market.

Suppliers. The main consumer sealant manufacturers are Dow Corning Corporation; General Electric Company, Silicone Products Division; Loctite Corporation; Wacker Silicones Corporation; Rhone-Poulenc Inc., Silicone Division; and DAP.

Costs. The consumer product costs range from approximately $0.096 to 0.17/cm³ ($2.85 to 5.00/fl oz) in a tube and $0.135 to 0.237/305 cm³ ($4 to 7/10.3 fl oz) in a caulking cartridge.

REFERENCES

1. E.G. Rochow, U.S. Patent 2,380,995, Aug 1945
2. J.F. Hyde, U.S. Patent 2,490,357, Dec 1949
3. D.T. Hurd and R.C. Osthoff, U.S. Patent 2,737,506, March 1956
4. D.T. Hurd, R.C. Osthoff, and M.L. Corrin, *J. Am. Chem. Soc.*, Vol 76, 1954, p 249
5. W.I. Patnode and D.F. Wilcock, *J. Am. Chem. Soc.*, Vol 68, 1946, p 358
6. E.L. Warrick and R.R. McGregor, U.S. Patent 2,435,147, Jan 1948
7. K.A. Andrianov, S.I. Dzhenchelskaya, and Y.K. Petrashkov, *Sov. Plast. Plast. Massy*, No. 3, 1960, p 20
8. D.T. Hurd, *J. Am. Chem. Soc.*, Vol 77 (No. 2), 1955, p 988
9. N. Reiso and K.K. Yushi, Japanese Patent 3,738, Jan 1955
10. E.L. Warrick and R.R. McGregor, U.S. Patent 2,431,878, Dec 1947
11. E.C. Britton, H.C. White, and C.L. Moyle, U.S. Patent 2,460,805, Feb 1949
12. H. Knopf, A. Beerwald, and G. Brinkmann, West German Patent 957,662, Feb 1954
13. J. Marsden and G.F. Roedel, U.S. Patent 2,469,883, May 1949
14. C.A. Berridge, U.S. Patent 2,843,555, July 1958
15. S. Nitzche and M. Wick, U.S. Patent 3,127,363, Mar 1964; West German application, Aug 1955
16. D.R. Weyenberg, U.S. Patent 3,334,067, Aug 1967
17. S.O. Smith and S.B. Hamilton, U.S. Patent 3,689,454, Sept 1972
18. M.A. White, M.D. Beers, G.M. Lucas, R.A. Smith, and R.T. Swiger, U.S. Patent 4,395,526, July 1983
19. E. Sweet, U.S. Patent 3,189,576, June 1965
20. L. Ceyzeriat, U.S. Patent 3,133,891, May 1964, French application, July 1957
21. L.B. Brunner, U.S. Patent 3,032,532, May 1962
22. M.D. Beers, U.S. Patent 3,382,205, May 1968
23. T.A. Kulpa, U.S. Patent 3,296,161, Jan 1967
24. D. Golitz, K. Damm, R. Muller, and W. Noll, U.S. Patent 3,417,047, Dec 1968; West German application, Feb 1964
25. L. Toporcer, U.S. Patent 3,776,933, Dec 1973
26. T. Takago, T. Sato, and A. Hisashi, U.S. Patent 3,819,563, June 1974; Jap-

Table 12 Properties of a typical low-volatiles, low-outgassing RTV silicone sealant

Temperature			Vacuum		
°C	°F	Time, h	Pa	torr	Weight loss, (%)(a)
120	250 168		784 × 10⁻⁸	5.9 × 10⁻⁸	0.117
205	400 168		226 × 10⁻⁷	1.7 × 10⁻⁷	0.31

(a) Weight loss as a function of time and temperature

anese application, July 1972

27. T. Takago, U.S. Patent 4,180,642, Dec 1979; Japanese application, June 1977

28. S. Nitzche and M. Wick, U.S. Patent 3,032,528, May 1962; West German application, Feb 1959

29. J.M. Klosowski and G.A.L. Gant, "The Chemistry of Silicone Room Temperature Vulcanizing Sealants," ACS Symposium Series 113, Plastic Mortar, Sealants and Caulking Compounds, R.B. Seymour, Ed., American Chemical Society, 1979, p 1257

30. J.M. Klosowski, *Sealants in Construction*, Marcel Dekker, 1989, p 265

31. S. Spell and J.M. Klosowski, Silicone Sealants for Use in Concrete Construction, AM. Conc. Inst., SP-70, Vol 1, p 217

32. J.M. Klosowski, *Sealants in Construction*, Marcel Dekker, 1989, p 215

Fluorocarbons

Alfred F. (Jerry) Waterland III, W.L. Gore & Associates, Inc.

FLUOROCARBON AND FLUORO-ELASTOMER sealants offer the designer and user a wide range of tough physical properties, chemical inertness, and thermal stability. The unique chemistry of these materials allows for the unequaled combination of almost universal chemical compatibility, a wide operating temperature range, and good strength characteristics.

Although fluorocarbon and fluoroelastomer sealants have similar chemistries, their major applications differ significantly. Each of these two sealant types is discussed separately in this article.

Fluorocarbons

Two fluorocarbon polymers are predominantly used as sealant materials, polytetrafluoroethylene (PTFE) and polychlorotrifluoroethylene (PCTFE). These materials are used in static sealant applications in a cut-and-place form and as a solid form-in-place, cord material. They are also used in some low-speed dynamic applications, such as compression packings and seals.

Chemistry

Polytetrafluoroethylene, which is probably the widest known and most versatile fluorocarbon resin, was invented in 1938 by a research scientist conducting experiments with an unusual gas, tetrafluoroethylene (TFE). Like ethylene (C_2H_4), the TFE molecule also contains two carbon atoms. However, rather than these being surrounded by hydrogen atoms, the carbon atoms have four fluorine atoms surrounding and linked to them with very strong bonds. At the conclusion of one particular experiment, it was discovered that one of the cylinders previously con-

taining the TFE gas was empty, except for a white, solid material at the bottom. This material was the product of the polymerization, or joining together, of the TFE molecules to form the plastic PTFE, better known as Teflon (a DuPont trade name), which is shown as:

$$\sim\ \begin{array}{ccc} F & & F \\ | & & | \\ C & - & C \\ | & & | \\ F & & F \end{array}\ \sim$$

Because of its novel combination of properties, this new material would soon revolutionize many industries, including the sealants industry.

The polymerization of TFE can be controlled to produce an extremely high (10^6 to 10^7) molecular weight material that is white in appearance. This high molecular weight, combined with the chemical bonding of carbon and fluorine (one of the strongest chemical bonds), gives PTFE its superior mechanical and chemical properties. However, also as a result of its high molecular weight, the melt viscosity of PTFE is much too high for it to be processed directly into its end-use shapes by typical melt and form type processing. Thus, depending on the molecular weight produced, the PTFE must be further processed using methods uncommon to other melt-processible plastics.

Forms

The common commercial polymerization processes for PTFE are controlled to produce three distinct and separate PTFE polymers. These polymers are identical chemically yet different in physical form. A coarse granular PTFE resin is produced through the method

of suspension polymerization. PTFE fine powder or paste is produced through emulsion polymerization. A PTFE aqueous dispersion or a milky, liquid PTFE is formed through dispersion polymerization. Table 1 shows various end-use applications for the three types of PTFE resins.

Applications

One of the major applications for granular PTFE is industrial gaskets, packings, and seals. The nature and properties of these sealant products are described below.

PTFE sealant products are primarily produced from the granular type resin using a modification of the powder sintering technique used in metallurgy. Finely ground granular resin particles are compressed into a mold to form a solid cylinder called a billet. The billet is then heated above its melting point to sinter, or fuse, the resin particles into a solid mass. The sintered billet is then carefully cooled to room temperature.

This cylinder is then used in a variety of secondary operations to make useful PTFE products. For example, if sheet-type gaskets are to be manufactured, this solid PTFE billet is skived (much like planing a piece of wood) into sheets. Sheets may have a variety of thicknesses, ranging from 0.8 to 6.4 mm ($\frac{1}{32}$ to $\frac{1}{4}$ in.). Gaskets are then cut from these sheets. Dynamic seals or O-rings are machined to shape from made-to-order sized and sintered tubes and bars. PTFE compression packings made from thin, narrow PTFE ribbons are formed in a fashion similar to that used for the PTFE sheet gaskets, except that a very thin, continuous slice is taken off the sintered billet. This thin sheeting is then slit into narrow ribbons to be braided into compression packing materials.

The petrochemical, food, beverage, pharmaceutical, and chemical pulping industries are major consumers of PTFE sealants. The vast array of aggressive chemicals and solvents used and produced in the chemical processing and paper making industries mandate a sealant material that does not degrade during use. PTFE sealants used for flanges, pumps, reaction vessels, valves, and so on, can provide this protection reliably throughout the very wide temperature

Table 1 Polytetrafluoroethylene resin types and applications

Type	Description	Primary uses
Granular	Free-flowing powder or granulated powder	Gaskets, packing, seals, sheet, and molded shapes for mechanical applications; tapes for chemical, mechanical, electrical, and nonadhesive applications
Fine powder	Agglomerated powder	Thin-walled tubing, over-braided hose, spaghetti tubing, thread-sealant tape, pipe liners, porous structures, gaskets, seals
Dispersion	Aqueous dispersion	Impregnation coating, packing, film

Table 2 Typical chemicals with which PTFE resins are compatible

Abietic acid	Cyclohexane	Hydrochloric acid	Phosphoric acid
Acetic acid	Cyclohexanone	Hydrofluoric acid	Phosphorus pentachloride
Acetic anhydride	Dibutyl phthalate	Hydrogen peroxide	Phthalic acid
Acetone	Dibutyl sebacate	Lead	Pinene
Acetophenone	Diethyl carbonate	Magnesium chloride	Piperidene
Acrylic anhydride	Dimethyl ether	Mercury	Polyacrylonitrile
Allyl acetate	Dimethylformamide	Methyl ethyl ketone	Potassium acetate
Allyl methacrylate	Diisobutyl adipate	Methacrylic acid	Potassium hydroxide
Aluminum chloride	Dimethylhydrazine,	Methanol	Potassium permanganate
Ammonia, liquid	unsymmetrical	Methyl methacrylate	Pyridine
Ammonium chloride	Dioxane	Naphthalene	Soap, detergents
Aniline	Ethyl acetate	Naphthois	Sodium hydroxide
Benzonitrile	Ethyl alcohol	Nitric acid	Sodium peroxide
Benzoyl chloride	Ethyl ether	Nitrobenzene	Solvents, aliphatic and
Benzyl alcohol	Ethylene bromide	2-nitro-butanol	aromatic(a)
Borax	Ethylene glycol	Nitromethane	Stannous chloride
Boric acid	Ferric chloride	Nitrogen tetroxide	Sulfur
Bromine	Ferric phosphate	2-nitro-2-methyl-	Sulfuric acid
n-butylamine	Fluoronaphthalene	1-propanol	Tetrabromoethane
Butyl acetate	Fluoronitrobenzene	n-octadecyl alcohol	Tetrachloroethylene
Butyl methacrylate	Formaldehyde	Oils, animal and	Trichloroacetic acid
Calcium chloride	Formic acid	vegetable	Trichloroethylene
Carbon disulfide	Futane	Ozone	Tricresyl phosphate
Cetane	Gasoline	Perchloroethylene	Triethanolamine
Chlorine	Hexachloroethane	Phenol	Vinyl methacrylate
Chloroform	Hexane	Phosphoric acid	Water
Chlorosulfonic acid	Hydrazine	Pentachlorobenzamide	Xylene
Chromic acid		Perfluoroxylene	Zinc chloride

Note: The absence of a specific chemical does not mean that it is incompatible with PTFE resins. (a) Some halogenated solvents may cause moderate swelling.

Table 3 Typical end user prices for common sheet-type gaskets from 1.6 mm (^{1}/16 in.) thick sheets

Asbestos/beater addition		Premium nonasbestos		Conventional mechanical-grade PTFE	
$/m²	$/ft²	$/m²	$/ft²	$/m²	$/ft²
48	4.5	54	5	120	11

range of −240 to more than 260 °C (−400 to >500 °F).

It is the basic structure of the PTFE polymer that allows such high-temperature capabilities, chemical inertness, and other unique properties. The fluorine atoms that surround the carbon chain literally serve as a protective glove effectively shielding the carbon atoms from attack. This fluorine sheath, with its extremely strong chemical bond linking the fluorine atoms to the carbon atoms, enables PTFE to resist chemical attack even at elevated temperatures (Table 2).

The food, beverage, and pharmaceutical industries also find wide use for PTFE sealants because of their chemical inertness and, more important, their noncontaminating properties, which are not only keenly desired, but which are required by federal law. Other sealant materials contain mixtures of organic and inorganic substances that either leach out into the process stream or are slowly degraded by exposure and temperature, resulting in process contamination.

PTFE gaskets are extremely well suited for use in these industries. The minimum design seating stress (the Y factor in the American Society of Mechanical Engineers boiler and pressure vessel code calculations) for PTFE gaskets is much lower than it is for other types of gasketing materials, such as compressed asbestos and nonasbestos. As a result, PTFE gaskets perform very well and are being used widely as a replacement for asbestos gaskets. Flanges and bolts that were previously designed for use with asbestos gaskets will produce higher compressive loads than is actually required to effect a seal with a PTFE gasket.

Despite their chemical inertness, high-temperature capabilities, and high-performance physical properties, the widespread use of PTFE sealants is somewhat limited because of economics and one key mechanical restriction, cold flow.

Processing Parameters

The manufacturing of PTFE resins is a very sensitive, energy-intensive operation that results in an expensive raw material for gaskets relative to other sealant materials. By the time the PTFE resin is processed into finished-product sheet gaskets, the end user price is roughly 2.5 times that of a premium grade, compressed, nonasbestos sheet gasket. The high cost of the PTFE raw resins, $11 to 18/kg ($5 to 8/lb), will often restrict its use to applications that require its specific performance abilities. Table 3 compares typical end user sheet-type gasket prices.

The federal government has scheduled a mandatory reduction in the use of asbestos-containing materials. Because asbestos gaskets still represented a sizable portion of the world gasket market in 1990, there has been a focused movement toward the development of nonasbestos replacement gasket materials. As more and better nonasbestos materials are developed, mostly with higher-priced raw materials, the cost of sheet gaskets in general will rise. The price differential between conventional PTFE sheet gaskets and the nonasbestos alternatives will be considerably less than that between PTFE and asbestos gaskets.

Cost aside, there still exist mechanical and physical problems with conventional-grade PTFE sheet gasket materials that will continue to restrict its widespread use. The primary property that makes PTFE an ideal slippery solid lubricant, that is, its low coefficient of friction, plays a large part in its weakest mechanical characteristic when used as a gasket, that is, creep and cold flow.

For a gasket to provide a tight seal in a flanged joint for an indefinite, extended period of time, it must be able to withstand and reach equilibrium with the enormous compressive forces that are exerted against it by the tightening of the bolts. Although all sealant materials (with the exception of metal gaskets) encounter some difficulties withstanding the effects of the compressive force indefinitely, conventional-grade PTFE gaskets typically require periodic re-torquing of the flange bolts to maintain seal integrity when under load for an extended period of time. The cause of this retightening requirement for long-term performance is the inherent structure of the PTFE polymer itself.

The chemical bonds securing the fluorine atoms to the carbon atoms are strong and durable, giving a very stable PTFE molecule, but the attractive forces between the polymer chains, one to another and side by side, are not nearly as strong. As a result, the slipperiness of the PTFE chains allows them to slide across one another in an effort to achieve equilibrium under a large compressive force. Elevated temperatures accelerate this process. Consequently, conventional PTFE sealants are limited to a maximum upper service temperature of 260 °C (500 °F), although their actual melt temperature is much higher.

A scanning electron microscope (SEM) photomicrograph of conventional mechanical-grade PTFE gasketing is shown in Fig. 1. It is obvious that there exists no mechanical substructure to prevent or retard the material flow when under load.

If the mechanical considerations of conventional mechanical-grade PTFE gaskets are understood, this material can appropriately be used for low-pressure applications in which excessive compressive stresses can be avoided, even with temperatures up to 260 °C (500 °F). The flange bolts should still be checked periodically and retorqued to compensate for any gasket creep.

Fig. 1 SEM photomicrograph of conventional mechanical-grade PTFE. 340×

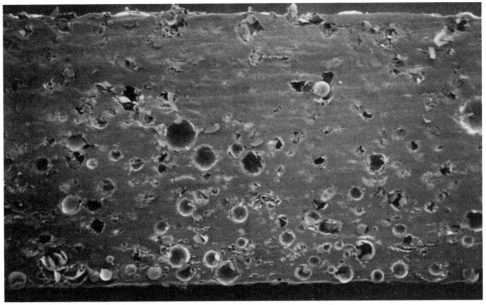

Fig. 2 SEM photomicrograph of Blue Gylon. 80×

Additives and Modifiers

Significant strides have been made in the production of PTFE gaskets that effectively eliminate the long-term creep and cold flow problems of conventional mechanical-grade PTFE. The most common alteration involves either the addition of high-modulus fillers to the PTFE prior to forming it into sheets or the sandwiching of conventional PTFE around an expanded or perforated metal core.

In the first approach, high-modulus fillers are used to form a bridge through the gasket such that the bulk of the compressive load applied to the gasket is carried by the filler itself and not by the PTFE. The result is a gasket material that retains much of the chemical inertness of conventional mechanical-grade (unfilled) PTFE, yet can better withstand long-term exposure to compressive forces without suffering from excessive creep and cold flow. However, because the fillers themselves do not share the range of inertness and chemical compatibility that the PTFE has, some restrictions in chemical compatibility will result.

Garlock Mechanical Packing Division of Colt Industries manufactures a range of filled PTFE gasket materials and sells them under the trade name Gylon. These various types of Gylon materials together offer the entire range of fluid compatibility that is found in conventional mechanical-grade, unfilled PTFE. Figure 2 shows the high-modulus filler beads interspersed throughout the structure of a Gylon product.

In the PTFE sandwich construction, the resistance to creep and cold flow is obtained by the presence of the perforated core, which retards the flow of the PTFE by giving it something to attach to. Typical core materials are stainless steel and Inconel. However, as with the filled materials, chemical compatibility of these materials is limited by the metal core.

Another means of reducing the degree of creep and cold flow in PTFE gaskets is available, with the added benefit of maintaining the universal chemical inertness of PTFE. This is accomplished by introducing an internal structure to the PTFE material. Expanded PTFE, produced by W.L. Gore & Associates, Inc. and sold under the trade name GORE-TEX, is a highly microfibrous PTFE material that is 100% PTFE. As shown in Fig. 3, this material has a defined oriented fibril substructure, which is essentially a fiber reinforcement that imparts significant structural integrity to the PTFE. Thus, the chain network of PTFE fibrils aids in resisting excessive creep and cold flow under compression, and does this without altering the chemical compatibility of the PTFE.

Markets and Applications

The movement away from asbestos gaskets, coupled with the acute environmental need to protect against leaks and chemical spills in industrial facilities, has created the need for better, more dependable gasketing systems. As a result, the market for PTFE gasket materials is growing rapidly. The U.S. fluorocarbon gasket market (that is, PTFE) will have grown from relative non-existence in 1977 to a projected $75 million market by 1993. No other nonmetallic gasket material market is expanding as rapidly.

Fluorocarbon PTFE sealants are widely used as compression packing materials within the same industries previously listed for PTFE gaskets. Compression-type packings are used to suppress leakage around shafts, rods, and valve stems in dynamic sealing applications. The most commonly used PTFE packing is a 100% PTFE, white fiber packing, often with a PTFE dispersion coating applied to the surface. This type of packing is most appropriate for use in dynamic fluid handling equipment, which requires high resistance to constant exposure to aggressive chemicals or high-purity protection from contamination. The chemical inertness and nonleaching characteristics of the PTFE valve or pump packing provide trouble-free service without contamination, chemical attack, or degradation of the sealing material.

One limitation of pure PTFE compression packings, when used as a pump packing, is their poor thermal conductivity, which is

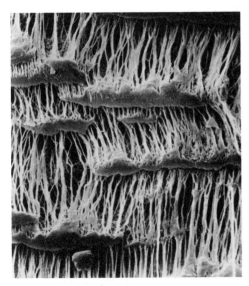

Fig. 3 SEM photomicrograph of GORE-TEX expanded PTFE. This material is dramatically different from other forms of PTFE. 340×

Table 4   Typical end user prices for common PTFE sheet-type gaskets from 1.6 mm (1/16 in.) thick sheets

Conventional mechanical-grade PTFE		Filled PTFE		Stainless steel reinforced PTFE		Expanded PTFE	
$/m²	$/ft²	$/m²	$/ft²	$/m²	$/ft²	$/m²	$/ft²
120	11	280	26	840	78	495	46

Table 5   Typical service and recommended performance range of various sheet-type gasket materials

Performance parameter	Asbestos	Nonasbestos	Conventional mechanical-grade PTFE	Filled PTFE	Expanded PTFE
Chemical compatibility, pH	3–9	4–8	0–14	0–14(a)	0–14
Temperature limits, °C (°F)	400 (750)	315 (600)	260 (500)	260 (500)	315 (600)
Pressure limits, MPa (ksi)	10.3 (1.5)	0.62 (0.09)	0.55 (0.08)	0.55 (0.08)	20.5 (3.0)
Intended service	General fluid service; check with manufacturer for limitations	General fluid service; check with manufacturer for limitations	All chemical service	Varied chemical service; check with manufacturer for limitations	All chemical service

(a) With a range of filled PTFE sheets

much lower than that of other commonly used compression packing materials.

Despite the low coefficient of friction (that of wet ice) of the PTFE packing, a significant amount of heat is nonetheless generated by the rubbing action of the pump shaft against the packing as the shaft is turning. If the heat that develops is not transferred quickly enough through the packing to the gland casing, damage to the pump shaft or sleeve will result, causing premature pump failure. As the packing becomes hotter, it expands, and as it attempts to expand, it presses tighter against the rotating shaft, thereby aggravating the problem. Once the packing temperature approaches 260 °C (500 °F), the enormous thermally induced compressive stress creates a fairly rapid creep or flow of the packing. This results in excessive leakage around the packing and eventual packing failure.

In light of these mechanical constraints, the use of pure PTFE compression packings is typically limited to valve-packing applications under 260 °C (500 °F), where rotational friction is negligible, and in low-pressure, slow-shaft-speed pumps (preferably with gland water cooling) where the compressive stress on the packing and rotational friction are within acceptable limits.

PTFE compression packings can also be made using expanded PTFE as previously described. The expanded-PTFE structure easily accommodates a wide range of fillers that can be used to facilitate heat transfer and other desirable characteristics.

The U.S. compression packing market in 1989 had shipments in excess of $90 million, with shipments of $125 million projected by 1993. It is estimated that a third of this

market is composed of PTFE or PTFE blend packings. As the industry moves away from asbestos-containing compression packings, the PTFE packing market share is expected to grow.

Properties

Expanded PTFE offered as a form-in-place, cord-type sealant provides the same chemical inertness, high-temperature usage, and resistance to creep and cold flow that the sheet-type expanded PTFE product provides, but at a significantly lower cost. Expanded PTFE form-in-place gasket material is analogous to a form-in-place room-temperature vulcanizing (RTV) type sealant. Because of the low density of expanded PTFE ($\sim$0.6 g/cm^3, compared to 2.2 g/cm^3 for conventional, mechanical-grade PTFE) and its flexibility, gaskets can be formed by simply adhering a bead of this material around the perimeter of the sealing surface of the flange. Furthermore, significant cost savings are realized with gaskets formed by expanded PTFE cords because there is little if any scrap associated with its use, whereas gaskets cut from sheet stock can produce anywhere from 20 to 60% unusable scrap. All of the modified forms of PTFE gasketing previously listed help to reduce creep and cold flow in PTFE gaskets to varying degrees. As a result, modified fluoropolymer gaskets are used in many applications requiring high-performance, specific properties in which high gasket cost is more than offset by the higher cost of gasket failure. Table 4 provides typical end user prices for common forms of PTFE sheet-type gaskets. Table 5 provides typical service and recommended performance ranges of various sheet-type gasket materials.

Fluoroelastomers

The second major category of fluorocarbon sealants is fluorocarbon elastomers, better known as fluoroelastomers. Fluoroelastomers have chemical compatibility features similar to those of PTFE fluorocarbon sealants, but because they are elastomeric by construction, they have been formulated to produce excellent resiliency or resistance to compression set properties that are essentially nonexistent in conventional fluorocarbon sealants. Although a small portion of the fluoroelastomers market is in static gasketing, either molded or cut from sheets, fluoroelastomers were developed for and have their greatest use in high-temperature, dynamic-type sealing applications, where their resistance to chemical attack and low compression set are absolutely critical. Fluoroelastomers such as Viton, Aflas, and Kalrez are predominantly used in these types of challenging environments.

Chemistry

Fluoroelastomers were developed in the 1950s to meet demands created by military and jet aircraft applications for specialty elastomers with improved heat, chemical, and solvent resistance. Fluoroelastomers, or fluorine-containing elastomers, are copolymers that usually contain two or more of the following monomers: hexafluoropropylene (HFP), vinylidene fluoride (VF$_2$), and tetrafluoroethylene (TFE). The thermal resistance, chemical resistance, and low-temperature flexibility of fluoroelastomers can all be adjusted to suit the needs of particular applications by varying the ratio of these monomers. For instance, the first commercial fluoroelastomer developed was a vinylidene fluoride-hexafluoropropylene copolymer. The chemical resistance was then further improved by creating a terpolymer based on vinylidene fluoride, hexafluoropropylene, and tetrafluoroethylene. Several types of commercially available fluoroelastomers, including a brief description of their use properties, are:

- Aflas, manufactured by Asahi Glass Company, Ltd. In structure it is a tetrafluoroethylene-propylene copolymer. Properties include high-temperature resistance to 230 °C (450 °F) and broad chemical resistance (acids, bases, sour oil, oxidizing agents, methanol, methyl ethyl ketone, and so on)
- Viton, manufactured by DuPont. Its structure is vinylidene fluoride and hexafluoropropylene copolymer. Properties include good thermal and chemical resistance
- Fluorel, manufactured by 3M. Its structure and properties are essentially the same as those of Viton

- Kalrez, manufactured by DuPont. Its structure is perfluoromethyl vinyl ether and TFE copolymer. Properties include outstanding high-temperature and chemical resistance, both of which are comparable to that of PTFE

Markets and Applications

Fluoroelastomers are used extensively in the automotive and aerospace industries in particular because of the long service life and the requirements for high-temperature use and fluid compatibility in the hostile confines of jet and automobile engines. For example, fluoroelastomer seals are replacing conventional, mechanically energized, fluorocarbon lip seals in jet engines and provide a service life up to 14 times longer. Additionally, automakers are using fluoroelastomer valve stem seals that better resist the chemical attack of aggressive engine oil and fuel additives that help today's higher temperature and higher rpm engines. The effective positive sealing of these valve stems and long service life are critical factors in reducing oil consumption and phosphorus buildup in these engines.

Fluoroelastomer sealants can be molded directly into their required end-use form using conventional rubber processing equipment and techniques such as injection or transfer molding and extrusion.

Major Suppliers and Costs

The largest producers of fluoroelastomers are DuPont and 3M. Fluoroelastomer usage is currently growing at a rate of approximately 8% per year, and growth is expected to continue. Worldwide demand for fluoroelastomers in 1992 is projected to be more than 8.5×10^6 kg (19×10^6 lb), despite 1990 prices that ranged from $22 to 190/kg ($10 to 85/lb). About 70% of all the fluoroelastomers produced are used in the aerospace and automotive industries as O-rings, seals, gaskets, and packing materials.

Properties

The aggressive fuel additives and some alternative fuels used in jet and automobile engines cause even traditionally fuel-resistant elastomers to swell excessively. Ethanol or methanol concentrations of just 15 to 20% will permeate typical elastomers and cause this undesirable swelling. On the other hand, the resistance to chemical attack and the low permeability of fluoroelastomers allows them to perform much more favorably in contact with these fuels and additives.

Fluoroelastomers also outperform other elastomers in oil field down-hole sealing applications, where operating temperatures can exceed 260 °C (500 °F) with pressures in excess of 70 MPa (10 ksi). The presence of methane, hydrogen sulfide, carbon dioxide, and steam can very quickly destroy typical elastomers. The use of fluoroelastomers increases the reliability and durability of oil field equipment.

Fluoroelastomers are the material of choice in virtually all solid rocket space booster joints. This application typically uses this material in a pressure-activated O-ring configuration and is the last barrier to prevent leakage of hot-combustion gases.

All commercialized fluoroelastomer sealants exhibit sound mechanical properties at elevated temperatures, but at low temperatures, resilience decreases. Although fluorosilicone elastomers are not fluorocarbon elastomers, they retain their resilience and remain flexible at temperatures as low as −60 °C (−75 °F). The resilience of the fluoroelastomer Kalrez decreases at low temperatures and is therefore not recommended for use at temperatures below −30 °C (−20 °F). Fluorosilicone sealants, however, are appropriate for chemical resistance applications and yet maintain their flexibility and resiliency at lower temperatures. Thus, while a host of fluoroelastomer sealants may be suitable for aggressive, high-temperature applications, the lowest-temperature applications are most reliably served by fluorosilicone elastomers, if they are chemically compatible with the fluid being sealed.

Modern combinations of fluorocarbon materials have caused revolutionary changes in the sealant industry. Engines, equipment, and processes run and operate reliably because of the performance properties of fluorocarbon sealants. For long-term durability in static or dynamic sealant applications, there is often no alternative to the use of fluorocarbon sealants.

SELECTED REFERENCES

- R.A. Brullo, Seal of Reliability, *Des. News*, 23 Nov 1987
- R.B. Bush, Fluorosilicone Elastomers: Exceptional Performers for Severe Environments, *Mech. Eng.*, July 1987
- L.K. English, First Family in Engineering Resins, *Mech. Eng.*, Jan 1988
- M.M. Lynn, Monomers, Fluorine Level Affect Fluoroelastomer Usage, *Rubber Plast. News*, 27 July 1987
- A.C. Mayer III, "Gaskets and Seals," Business Research Report B99, The Freedom Group, Inc.
- P.A. Schweitzer, *Handbook of Corrosion Resistant Piping*, Industrial Press, 1969
- Versatile Fluoroelastomers are High in Performance, Growth and Price, *Elastomerics*, July 1988

Polyether Silicones

Jamil Baghdachi, BASF Corporation, Coatings & Colorants Division

POLYETHER SILICONE sealants, adhesives, and caulks are important industrial, architectural, and engineering compounds. Common applications involve the functions of bonding, sealing, waterproofing, and weatherproofing (Ref 1, 2).

The chemical family of polyether silicones encompasses a large variety of materials, ranging from resins and room-temperature moisture-curable rubbers to heat-vulcanizable and radiation-curable elastomeric gum stocks. Polyether silicone is a common name applied to silicon- or silane-modified polyethers in their cured state; they contain Si—O—Si bonds. The resin or sealant in the wet state is properly referred to as polyether silicon. This chemical family represents a unique inorganic and organic hybrid that has the combined properties of silicones and organic polyethers. This article describes just one member of the family of polyether silicones: room-temperature-curing, silane-modified polyurethane or polyether sealants.

Polyether silicons are moisture curable and form an elastomeric product similar to room-temperature-vulcanizing (RTV) silicone sealants. Their outstanding adhesion to a wide variety of surfaces and excellent weathering properties make them ideal for use in a large number of industrial and construction applications.

Chemistry

Silane-modified polyethers currently consist of a polyoxyalkylene backbone with alkoxysilane pendant or terminal groups. The modification can be achieved in many ways; the examples described in this section are the most practical and likeliest to be used in the known sealant formulations.

In the procedure shown in Fig. 1, low molecular weight polyether polyols, such as polyoxypropylene polyols, are extended with diisocyanates to form a polyurethane-containing polyether backbone. The procedure and related steps are described in Ref 3 to 8 and can be formulated into one- or two-part moisture-curable sealants.

Upon exposure to atmospheric moisture, the alkoxy units of silane-capped resin hydrolyze to form silicone bonds:

$$\sim\!\!\sim\!\!\sim \; Si - O - Si \; \sim\!\!\sim\!\!\sim$$

These silicone bonds connect the polyether-polyurethane chains, resulting in elastomeric materials with the structure:

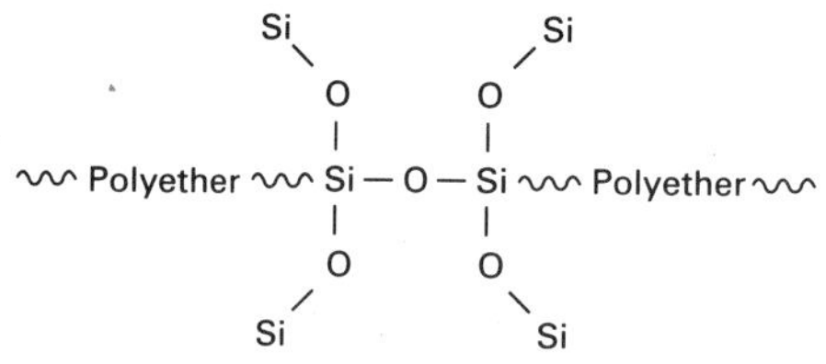

Another route for producing silicone-containing polymers with a pendant silyl group is shown in Fig. 2 and described in Ref 9 to 15.

In addition to the procedures described above, polyether silicon can be prepared as shown in Fig. 3. According to this procedure, a high molecular weight polyether polyol is first reacted with allyl ether, followed by hydrosilation to give a polyether that is end-capped with methyldimethoxysilane groups (Ref 16-19). In the presence of an appropriate catalyst and moisture, the methoxysilicon groups hydrolyze to silanol and liberate a small amount of methanol. Further reaction of the silanol with either another silanol or a methoxysilicon group gives the silicon-oxygen-silicon linkage, affording cross linking of the resin, as shown in Fig. 4.

Compounding

The cured, unfilled polyether silicone is a soft, rubbery material that has relatively low elongation, tear, modulus, and tensile strength properties. Both its physical and mechanical properties can be improved dramatically when properly formulated for use as a sealant.

Polyether silicon sealants can be formulated as one-part or two-part products that contain nearly 100% solids, of which about 40% is resin. These high-performance systems have a joint movement capability of 25 to 50%. Polyether silicons might contain some solvents, but in very small amounts (1 to 5%). Because of their low solvent content, they hardly shrink upon solvent volatilization and do not harden with age.

Modulus, flexibility, and elasticity are controlled by the presence of nonreactive modifiers, fillers, and other additives. Typical polyether silicon sealants are filled with carbon black, clay, calcium carbonate, aluminum silicate, titanium dioxide, and other components. Adhesion promoters, diluents, plasticizers, processing aids, catalysts, thixotropic agents, and miscellaneous other materials are also added to impart desired properties.

The two-component sealant systems tend to cure more rapidly without a need for maintaining optimum cure conditions. The second component is usually a paste that contains water, a filler such as clay or

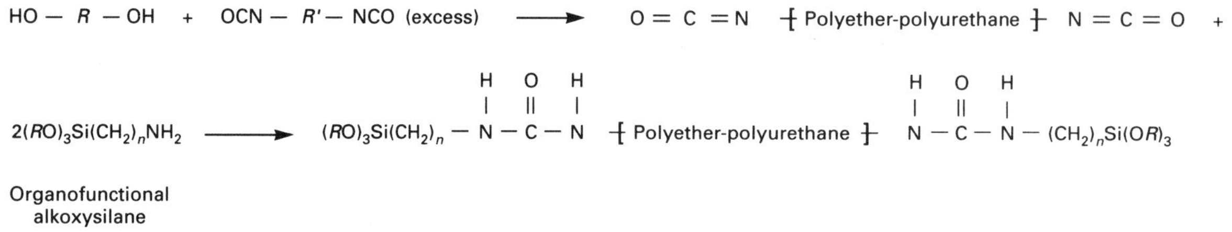

Fig. 1 Polyurethane-containing polyether based on low molecular weight polyols extended with diisocyanates

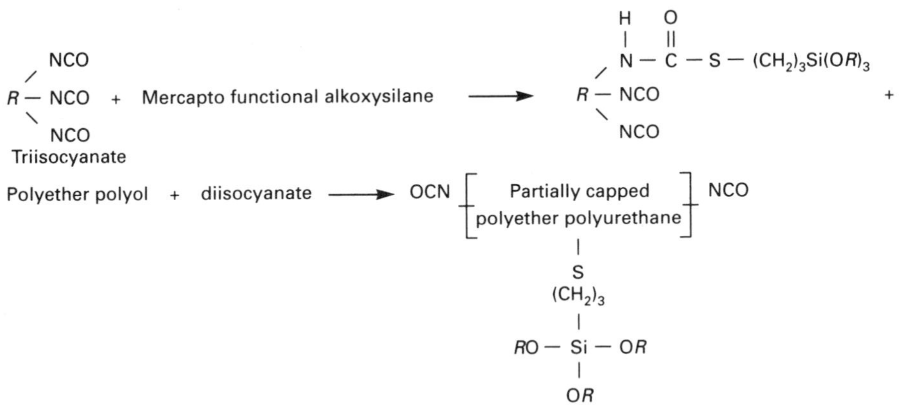

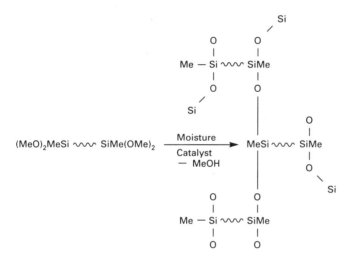

Fig. 2 Polyurethane-polyether with pendant silyl group

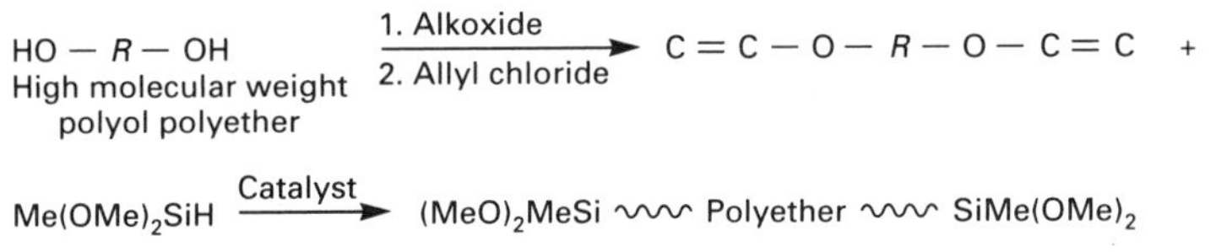

Fig. 3 High molecular weight polyether polyol with methyldimethoxysilane groups

Fig. 4 Cross linking of product of Fig. 3

Table 1 Typical polyether-polyurethane silicon sealant formulation

Component	phr
Polyether silicon resin	100
Carbon black	25
Clay	45
Plasticizer	20
Aromatic solvent	10
γ-methacryloxypropyl trimethoxysilane	1

phr, parts per hundred parts of resin by weight

Table 2 Typical sealant formulation

Component	phr
Silmod 20A	60
Silmod 300	40
Plasticizers	50
Calcium carbonate	120
Titanium dioxide	20
Thixotropic agent	3
Antioxidant	1
Ultraviolet absorber	1
Dehydrating agent	2
Adhesion promoter	3
Dibutyl dilurate	2
Lauryl amine	0.5

phr, parts per hundred parts of resin by weight

calcium carbonate, and other optical ingredients. Depending on the formulation, polyether silicons can be used as sealant adhesives to bond and/or seal a variety of substrates, such as steel, glass, aluminum, painted and unpainted surfaces, concrete, plastics, and elastomers.

Table 1 lists the components in a typical formulation of a moisture-curing polyether-polyurethane silicon that is based on resin prepared according to the route shown in Fig. 2. This one-part formulation is suitable for such applications as bonding windshield glass to the body of a motor vehicle.

Table 2 gives a formulation suitable for construction applications, based on resin prepared as shown in Fig. 3 (Ref 17, 18). Unlike the two-part systems, this formulation is kept in a completely anhydrous condition to prevent curing from taking place within the package. This composition, as mixed by the manufacturer, is stable until it is applied and exposed to atmospheric moisture.

Product Types and Properties

Polyether silicon sealants are usually formulated to a high viscosity and an over-90% solids content. Specifically compounded systems are labeled as fast curing and/or primerless sealants. Primerless systems are primerless only to specific surfaces, such as glass, concrete, and electrocoated surfaces. Usually, a surface conditioner or primer is required if other substrates are to be used.

The versatility of these sealant systems and the range of properties derived from them have yielded a vast array of products from various manufacturers.

Cure Properties. The curing rates of one- and two-part polyether silicons vary widely. The sealant product is usually designed to meet the cure rate required for a specific end use. The faster-curing materials are generally used for industrial and consumer applications, whereas the slower-vulcanizing products are tailored for the construction market, where longer tooling times are frequently desired.

Conventional one-part systems depend on the diffusion of atmospheric moisture into the sealant for their cure. Therefore, they are not recommended for applications requiring deep section vulcanization. However, significant advances have been made to improve the cure speed of one-part systems. A new polyether-polyurethane silane-modified sealant, based on the resin prepared according to Fig. 1, tends to cure quite rapidly, approaching the cure rate of a two-component system. It also appears to be autocatalytic in nature; that is, it requires only a small amount of moisture to initiate the cure mechanism. Once the curing reaction begins, it continues until a complete state of cure has been reached.

Because the sealant described above does not require constant exposure to moisture, it can be used as a gap-filling adhesive and also serves in applications requiring deep section vulcanization. The skinning time of the sealant can vary anywhere from several minutes to more than 1 h, depending on the formulation and on the temperature and humidity conditions of the environment. Higher temperatures and humidities will

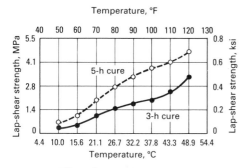

Fig. 5 Effect of cure temperature on cure rate of lap-shear specimen cured at 25 °C (77 °F)

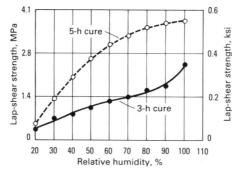

Fig. 6 Effect of environmental humidity on cure rate of lap-shear specimen cured at 25 °C (77 °F)

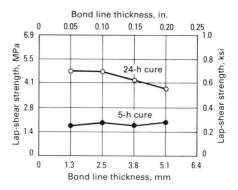

Fig. 7 Bond line thickness versus lap-shear strength

accelerate the cure. The converse is true at lower temperatures and humidities.

The cure characteristics of a typical sealant of this type are shown in Fig. 5 and 6. Under these conditions, a value of 2.8 MPa (0.4 ksi) normally indicates a 90% cured sealant. An open bead of this sealant will cure more quickly. One-part sealant curing data are summarized in Table 3.

At a normal humidity level, the one-part sealant cures on the surface (tack free) within minutes of application and to a depth of 12.5 mm (0.5 in.) or more within 24 h. Good cure and tack-free surfaces are obtained relatively quickly at temperatures as low as 10 °C (50 °F). The relatively fast cure rate makes this sealant an ideal candidate for bonding substrates that require a thin bond line thickness (0.25 to 2.5 mm, or 0.01

to 0.1 in.) (Fig. 7). Thus, bonded joints can attain handling strength within minutes, depending on temperature and humidity.

Curing data for a construction-grade one-part sealant are summarized in Table 4, and curing rate is given in Fig. 8. A lower relative humidity level will slow the cure, but the ultimate properties should be the same when cured.

Two-part construction grade sealants are less dependent on humidity and will cure to their ultimate properties within 7 days over a temperature range from 5 to 70 °C (41 to 158 °F). The data are summarized in Table 5, and the cure performance is given in Fig. 9.

Strength Properties. The strength of a polyether silicone elastomer is designed to

be application-specific. For example, low-modulus (low tensile strength combined with high elongation) types of sealants are required in concrete expansion joints to prevent failure at the weak substrate interface. High-strength adhesive sealants are used in automotive and aerospace applications. Presently, the medium-strength seal-

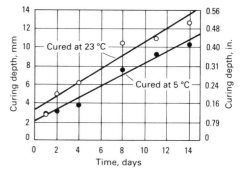

Fig. 8 Curing depth versus time for one-part sealants at two temperatures

Table 3   Curing depth of one-part sealant at two different temperatures

Cure	Time, h					
	1	2	4	8	16	24
Depth, mm (in.)						
At 10 °C (50 °F), 50% RH	0.4 (1/64)	0.8 (1/32)	1.6 (1/16)	3.2 (1/8)	4.8 (3/16)	6.4 (1/4)
At 23 °C (73 °F), 50% RH	1.6 (1/16)	3.2 (1/8)	4.8 (3/16)	6.4 (1/4)	9.5 (3/8)	12.7 (1/2)

Based on FC-2000 sealant, BASF Corporation, Coatings & Colorants Division. RH, relative humidity

Table 4   Curing depth of one-part Silmod sealant at two different temperatures and relative humidity of 50 to 60%

Cure	Time, days					
	1	2	4	8	11	14
Depth, mm (in.)						
At 5 °C (40 °F)	2.9 (0.12)	3.1 (0.12)	3.8 (0.15)	7.5 (0.30)	9.2 (0.37)	10.3 (0.41)
At 23 °C (73 °F)	2.8 (0.11)	5.0 (0.20)	6.1 (0.24)	10.5 (0.42)	11.0 (0.44)	12.7 (0.51)

Based on Silmod, product of Union Carbide Corporation

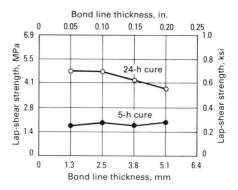

Fig. 9 Tensile strength of two-part sealants after 7 days as a function of curing temperature

Table 5   Curing rates for two-part sealants under varying conditions

Curing conditions		Time, days	Modulus at 150% elongation		Tensile strength at break		Elongation at break, %
Temperature °C	°F		kPa	psi	kPa	psi	
−10	14	7	69	10	294	43	740
		21	167	24	578	84	610
5	41	3	127	18	559	81	710
		7	176	25	559	81	600
23	73	3	186	27	657	95	610
		7	176	26	549	80	600
50	122	3	186	27	676	98	630
		7	167	24	559	81	560
70	158	3	147	21	519	75	620
		7	157	23	549	80	620

Table 6   Typical mechanical properties of one- and two-part polyether silicon sealants

Property	Value
Shore A hardness	15–50
Tensile strength, kPa (psi).........	196–980 (28–142)
Elongation, %	100–800
Tear strength, kPa (psi)	98–490 (14–71)
Modulus at 100%, kPa (psi)	98–980 (14–142)

Table 7 Physical properties of one-part industrial-grade sealant

Property	Value
Shore A hardness	50–65
Tensile strength, MPa (ksi)	6.9–10.3 (1.0–1.5)
Elongation, %	200–300
Tear strength, N/mm (lbf/in.)	35–53 (200–300)
Modulus at 100%, MPa (ksi)	1.7–2.8 (0.25–0.40)

Based on FC-2000 BASF Corporation, Coatings & Colorants Division

Table 8 Adhesion of one-part sealant to several substrates

Substrate	Lap-shear adhesion strength(a) MPa	ksi
Glass	6.9	1.0
Aluminum	3.4	0.5
Cold-rolled steel	3.4	0.5
Painted surface(b)	6.9	1.0
Polyurethane rubber	2.1	0.3
Electrocoated surface	8.3	1.2

Values based on bond line thickness of 6.4 mm (¼ in.). (a) Cohesive failure. (b) Primer required

Table 9 Adhesion of one-part Silmod sealants to several substrates

Substrate	Adhesive strength N/mm	lbf/in.
Aluminum	5.0	28.1
Stainless steel	10.0	56.2
Granite	5.0	28.1
Mortar	10.3	57.9
Plywood	10.3	57.9
Polyvinyl chloride plate	10.0	56.2

Using 180° peel test on 10 mm (0.4 in.) wide specimen. The aluminum, stainless steel, and granite substrates were washed with acetone before application.

ants are utilized in the greatest variety of applications.

The mechanical properties of sealants can be dramatically varied by formulation. The range of mechanical properties of construction-grade sealant is given in Table 6.

Products exhibiting tensile strengths as high as 10.3 MPa (1.5 ksi), accompanied by tear strengths of 43.8 kN/m (250 lbf/in.) and good elongations, have been developed. The properties of a typical industrial-grade polyether silicon are summarized in Table 7.

Adhesion characteristics of polyether silicone sealants vary greatly from product to product. The curing of polyether silicons produces siloxane bonds similar to those of a silicone sealant, but the main portion of the polymer is still polyether. As a result, the adhesion and surface properties are similar to urethane sealants, which are also mostly polyether. Therefore, paintability characteristics are similar to those of urethane sealants. However, the adhesion characteristics of polyether silicons are superior to those of urethane sealants because of the adhesion-promoting mechanism obtained by the silicon modification.

Industrial-grade polyether-polyurethane silicone sealants have excellent adhesion to a variety of prepared and unprepared surfaces. Manufacturer product information data can best determine the adhesion performance limitation. The data in Table 8 show that the sealant adheres well to painted surfaces, electrocoated steel, aluminum, cold-rolled steel, glass, and rubbers.

The adhesion of construction-grade sealants to a variety of surfaces is shown in Table 9. The good adhesion of these products is maintained even after weathering, and separation (adhesive failure) from substrates after years of use does not readily occur. Good adhesion prevents cracks from forming between the sealant and the building and reduces moisture penetration and air flow.

Weathering and durability are two of the important features that distinguish silicone-modified sealants from conventional polyether-polyurethane sealants. Both industrial- and construction-grade sealants must maintain their initial strength values and bond integrity under severe conditions. The joints and seals are expected to function when exposed to wide temperature ranges and to fuel, oil, high humidity, vibration, impact, ultraviolet (UV) radiation, and dust. The data in Table 10 show that there is little change in the adhesion properties under accelerated-weathering studies.

Construction-grade polyether silicons show similar trends. Figure 10 shows that the change in tensile strength values with time is small.

Accelerated-weathering testing of a one-part polyether silicon is also shown in Table 11, whereas another set of tests of the durability of silicon-modified sealants is summarized in Table 12. Both one- and two-part sealants show only minor changes in physical properties after being aged at 90 °C (195 °F) for up to 29 weeks.

Applications and Processing Parameters

Silicon-modified moisture-curing polyether-polyurethane sealants are important to the automotive industry for their ability to adhere glass automotive parts, such as windshield, rear and side windows, and the like to metal chassis and parts. The adhesive must be strong enough to retain the windshield in place under roll-over conditions and crash situations, a requirement of the Federal Motor Vehicle Safety Standards. In addition, a strongly bonded windshield and rear window provide additional support to the vehicle roof. Polyether silicons are considered to be the best adhesive sealants available for these purposes because windshields are assembled in the trim shop, where heat cure is not possible.

Table 10 Accelerated-weathering testing of one-part sealant

Test condition	Metal-glass lap-shear samples MPa	ksi
3 days at room temperature	6.9	1.00
7 days at 100% RH, 38 °C (100 °F)	6.6	0.950
14 days at 80 °C (175 °F)	6.9	1.00
14 day water immersion	6.6	0.950
1000-h weatherometer(a)	6.8	0.980
2000-h weatherometer(a)	6.9	1.00
2 years Florida exposure	6.4	0.925
2 years Arizona exposure	6.1	0.890

Based on PL 3000 Sealant, BASF Corporation, Coatings & Colorants Division. (a) ASTM G 53 test apparatus: 24 h continuous run; 8 h ultraviolet radiation at 82 °C (180 °F), 4 h condensation at 70 °C (158 °F)

Automotive adhesives must also satisfy a number of requirements that are somewhat independent of joint performance. These products must be usable under conditions that include high production rates with short, unvarying times for each operation; low tolerance for health and safety hazards; and cure times, pressures, temperatures, and so on, that are somewhat variable. Once these factors and requirements have been met and adhesive bonding has been performed, the joint must perform under very severe conditions throughout the life of the vehicle. No other mass-produced product of comparable complexity is expected to function under such a wide range of conditions.

To improve and enhance the adhesion of painted surfaces, as in the case of windshield bonding, a primer is conventionally used. Primers are normally brushed on or sprayed over the painted surface and allowed to air dry before a sealant bead is applied. Recently, a new primerless adhesive sealant has become available (Ref 20). It eliminates the need for a primer and reduces the interfaces that the sealant encounters. As a result, bonded joints are more durable, and both the process and the application of the sealant are more convenient and economical.

Open time of the adhesive is a critical parameter in industrial applications. The adhesive should have enough time, before it sets, to allow careful orientation of the parts and/or to tolerate occasional interruptions in the assembly lines. The open times of a

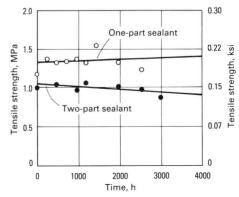

Fig. 10 Accelerated-weathering data for one- and two-part sealants

Table 11 Accelerated-weathering testing

Time, h	Modulus at 150% elongation		Tensile strength at break		Elongation at break, %
	kPa	psi	MPa	ksi	
0........	764	111	1.15	0.166	326
240......	735	107	1.33	0.193	290
480......	843	122	1.28	0.186	257
720......	755	109	1.31	0.190	273
960......	843	122	1.33	0.193	257
1200.....	813	118	1.29	0.188	247
1440.....	804	117	1.52	0.220	308
1980.....	794	115	1.31	0.190	273
2520.....	764	111	1.25	0.176	257

Based on one-part Silmod sealant

Table 12 Heat resistance at 90 °C (195 °F)

Time, weeks	Modulus at 150% elongation		Tensile strength at break		Elongation at break, %
	kPa	psi	MPa	ksi	
One-part Silmod sealant					
0.......	774	112	1.18	0.170	328
2.......	696	101	1.08	0.156	373
7.......	647	94	1.0	0.145	328
13.......	598	87	0.90	0.131	332
17.......	657	95	0.92	0.134	318
23.......	588	85	0.78	0.114	273
26.......	510	74	0.74	0.108	320
29.......	500	72	1.13	0.163	250
Two-part Silmod sealant					
0.......	274	40	0.92	0.134	745
1.......	274	40	1.25	0.182	880
7.......	304	44	1.11	0.161	805
13.......	274	40	1.04	0.151	800
23.......	274	40	0.92	0.134	785
29.......	265	38	0.87	0.126	748

Sunshine weatherometer, black panel temperature of 63 °C (145 °F), water sprinkle for 12 min during every 120 min

fast-curing sealant under various temperatures and humidities are shown in Fig. 11.

The data show that when the sealant adhesive reaches the end of its open time, and when enough skin has formed on the surface to prevent the sealant from wetting the second substrate, the joints usually show unacceptable adhesive failure.

Handling strength is another important characteristic that a good sealant should possess. This processing parameter is especially critical in industrial applications and in assembly plants where manufactured articles rock or vibrate on the conveyer belt.

Therefore, a fast-curing sealant that does not totally depend on the moisture from the environment is especially attractive.

The commercial one-part silicon-modified polyethers are stable and have a good shelf life, which ranges from 3 to 12 months when stored unopened and kept in moder-

ate temperature and humidity conditions. As mixed by the manufacturer, the one-part component sealant is stable under anhydrous conditions. Currently, it is widely believed that under completely anhydrous conditions and rather low temperatures, the shelf life can be extended beyond the 3-to-12-month period.

Limitations

Lingering areas of weakness exist with any new technology. The most notable weakness of polyether silicones, compared to silicone sealants, is their sensitivity to high temperature. The bonds do not perform well above 150 °C (300 °F). However, this is not considered a serious weakness, because environmental temperatures range from −40 to 65 °C (−40 to 150 °F), which is well below this critical temperature.

Suppliers and Competing Materials

Silicon-modified polyether sealants are currently supplied in the United States by BASF Corporation, Coatings & Colorants Division, and by Essex Specialty Chemicals. Commercially available Silmod polyether silicone resin, produced by

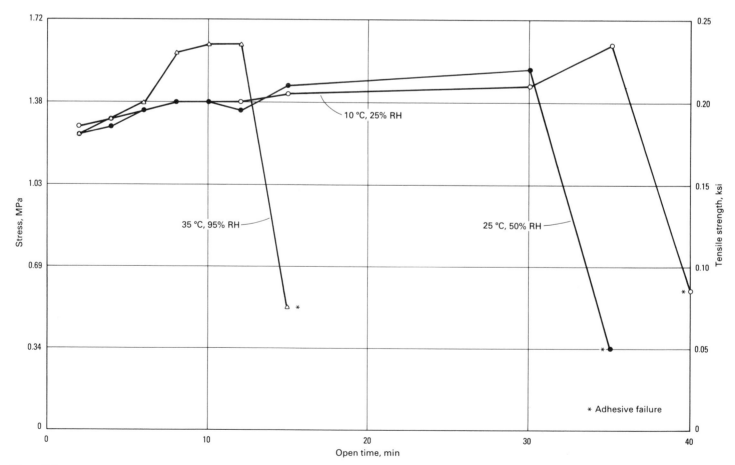

Fig. 11 Open times of fast-curing primerless sealant

Union Carbide Corporation, is the base resin for construction-grade polyether silicones.

Because of their versatility, ease of processing, and varied end uses, a direct comparison of polyether silicons with existing materials such as latex and solution acrylics and polyurethanes is not possible. In some respects silicone sealants are a close match for silicon-modified polyethers, but in general they appear to be more costly than polyether-silicon sealants.

REFERENCES

1. W.J. Reid, What's Ahead for Adhesives and Sealants, *Adhes. Age*, April 1970, p 21
2. J.P. Cook, *Construction Sealants and Adhesives*, Wiley-Interscience, 1970
3. J. Seiter, U.S. Patent 3,627,722, Dec 1971
4. B. Brose, *et al.*, U.S. Patent 3,632,557, Jan 1972
5. M. Barron, *et al.*, U.S. Patent 4,067,844, Jan 1978
6. E. Chu, *et al.*, U.S. Patent 3,711,445, Jan 1973
7. E.R. Bryant, *et al.*, U.S. Patent 3,979,344, Sept 1976
8. E.R. Bryant, *et al.*, U.S. Patent 4,222,925, Sept 1980
9. S.D. Rizk, U.S. Patent 4,687,533, Aug 1987
10. S.D. Rizk, *et al.*, U.S. Patent 4,652,012, Aug 1985
11. S.D. Rizk, *et al.*, U.S. Patent 4,345,053, Aug 1982
12. R.M. Evans, *et al.*, U.S. Patent 3,372,083, March 1968
13. R.L. McKellar, U.S. Patent 3,502,704, March 1970
14. E. Chu, *et al.*, U.S. Patent 3,711,445, Jan 1973
15. K. Asai, *et al.*, U.S. Patent 3,886,226, May 1975
16. K. Isayama, U.S. Patent 3,971,751, July 1976
17. J.F. Timberlake, "Silmod™ Silicon Modified Polyethers for Moisture Cured Sealants," Technical Publication, Union Carbide Corporation
18. J. Takase, U.S. Patent 4,444,974, April 1984
19. T. Mita, *et al.*, U.S. Patent 4,507,469, March 1985
20. J. Baghdachi, *Adhes. Age*, Dec 1989, p 28-31

Surface Considerations

Chairman: Guy D. Davis, Martin Marietta Laboratories

Introduction

PROPER ADHEREND SURFACE PREPARATION is essential for a good adhesive bond. In the case of a structural bond, the best adhesive in the world is useless unless the stresses can be transferred from one adherend to the adhesive and on to the other adherend. There are three requirements for a good surface preparation. First, it must allow the adhesive or primer to wet the adherend surface. Second, it must enable the formation of bonds across the adherend-adhesive/primer interface. These bonds can be chemical (covalent, acid base, van der Waals, hydrogen, and so on), mechanical (physical interlocking), or diffusional; the relative importance of the different types depends on the particular system and the environmental conditions to which the bond is exposed. Third, the adherend and its interface with the adhesive/primer must be stable over the lifetime of the bonded structure.

The above requirements can be met by many different means, and these means can be specific to the material, to the environment, or to the application. The need for a surface preparation specific to a particular adherend material is fairly obvious. For example, a chemical treatment for aluminum would not be appropriate for a plastic, for a ceramic, or even for titanium or steel. As can be seen in the article "Surface Preparation of Metals" in this Section, even different alloys of steel can require different chemical treatments. On the other hand, mechanical treatments are less material specific; for example, abrasion of one form or another (grit blasting, sanding, and so on) can be used under certain conditions for metals, plastics, composites, coatings, wood, and other materials.

The environmental-specific dependency of surface treatments is less obvious at first, but it can also be important from a practical aspect. For example, an aluminum-epoxy interface is sensitive to moisture-induced degradation in two ways. First, because aluminum oxide-epoxy chemical bonds are readily disrupted by moisture, an aluminum-epoxy adhesive bond that relies solely on chemical bonds (such as that formed with a microscopically smooth adherend) will fail when water reaches the interface. Second, the aluminum oxide formed during many pretreatments is susceptible to hydration. The morphological and volume changes caused by hydration induce high stresses at the bond line and promote failure at the hydroxide-metal interface. For a bond in an environment that will never be exposed to moisture, this sensitivity can be ignored and simple treatments that are not resistant to attack can be used. Similarly, in a high-temperature environment, titanium surface oxides will dissolve into the substrate, providing an easy path for crack propagation. At moderate temperatures, the rate at which this occurs is extremely slow, so that oxide-forming treatments can be suitable; at high temperatures, nonoxide preparations must be used.

Finally, surface treatments are application specific because not all applications require the same interfacial properties. A bond to a sealant, a paint, or other coating is not subject to the same stresses as a structural bond. Consequently, the surface preparation for a coating may not need to be as complicated as that for a surface meant to bear structural loads. For a coating, degreasing and abrading to roughen the surface can give excellent results; however, the same treatment might not be adequate for a structural bond.

The first article in this Section, "Surface Analysis Techniques and Applications," describes methods to characterize the chemistry and physics of the adherend surface and its interaction with the adhesive or primer. The second article, "Primers and Coupling Agents," discusses how these materials help meet the requirements of good surface preparation by improving wetting, forming strong chemical bonds across the interface, and stabilizing the surface by inhibiting corrosion and other forms of deterioration. The remaining articles describe and discuss specific surface treatments for a variety of materials. In some cases, there is an inherent overlap between classes of materials; for example, composites can be considered a subset of plastics and polymerics. In all cases, the appropriate surface treatment depends on the specific application; the organization of this Section should facilitate selection of a particular preparation.

Surface Analysis Techniques and Applications

Guy D. Davis, Martin Marietta Laboratories

SURFACE AND INTERFACE SCIENCE is a multidisciplinary study of the physics, chemistry, materials science, and engineering of the nearly two-dimensional boundaries between two phases of matter. A material and its environment (ranging from a liquid to a vacuum) or two discrete constituents of a solid give rise to the surfaces and interfaces most frequently investigated. The analysis of the properties of surfaces is becoming increasingly important in studies of a wide range of processes, including adhesive bonding and failure analysis. The properties of a thin film or of a surface can be very different from those of the bulk material because of the asymmetry introduced by the boundary. These differences are caused by differences between "bulk" and "surface" atomic bonds, as well as by the formation of a reacted layer (for example, an oxide or hydroxide film), segregation of alloying components, incorporation of electrolytic ions, adsorption of molecules from gases and liquids, or other phenomena.

Many of the chemical and physical properties of a material change dramatically from the first monolayer of atoms on the surface to the second, whereas others change only over hundreds of nanometers. Accordingly, the definition of the thickness of a surface depends on the properties being measured and can range from the outer electronic cloud to the first atomic or molecular monolayer (0.1 to 0.3 nm, or 0.004 to 0.012 μin., for an atomic monolayer) to the detection depth of the common analytical techniques (1 to 2 nm, or 0.04 to 0.08 μin.) to the thickness of a thin film (>100 nm, or 4 μin.). Unless otherwise indicated, the surface is assumed here to include the volume investigated by the surface-sensitive techniques.

It is the properties of the surface that control the way a material interacts with its environment. Thus, surface properties govern such macroscopic phenomena as adhesion, corrosion, wear, wettability, catalysis, and biocompatibility. Similarly, the properties of internal interfaces can govern such phenomena as the strength of materials and

their electrical and optical properties. In adhesive bonding, the chemical properties of an adherend surface govern the interaction of the adherend with a primer or adhesive, as well as the long-term resistance of the bond to environmental degradation (Ref 1-4). On the other hand, the morphology of the adherend controls the degree of physical interlocking with the primer or adhesive such that a microscopically rough surface can provide a stronger (and more durable) bond than a smooth surface (Ref 3, 5).

Although the sophisticated study of surfaces is a relatively new field of investigation, the importance of surfaces has long been recognized, even during the Middle Ages (Ref 6):

> When a plate of gold shall be bonded with a plate of silver or joined thereto, it is necessary to beware of three things, of dust, of wind and of moisture: for if any come between the gold and silver they may not be joined together...
>
> —*De Proprietatibus Rerum*
> (The Properties of Things) 1250 A.D.

This realization of the significance of surfaces, however, has not implied a ready ability to investigate and understand surfaces, especially compared to bulk material:

> God made solids, but surfaces were the work of the devil.
>
> —Wolfgang Pauli

Various techniques have been developed to study the surfaces and interfaces of materials. Several of these are classified in Table 1 according to the incoming and outgoing particles used to probe the surface. Each of these techniques provides different information about the composition, structure, chemical bonding, or electronic properties of the material. Specifically, these techniques can provide qualitative and quantitative analyses and indicate chemical bonding or they can measure the distribution of material, elements, or, in a few cases, even individual atoms as a function of lateral depth or displacement along the surface. Some techniques are extremely

surface sensitive, whereas others are best applied to thin films.

Most of these techniques are not commonly used for adhesive bonding and other applied problems because of sample constraints, the equipment needed, the information obtainable, the maturity of the technique, or the analysis required. However, the usefulness of surface analysis has been well established for basic and applied research, as well as for quality control and failure analysis.

In this article, several techniques commonly used to measure surface composition, chemistry, or morphology are discussed. Some detailed aspects of the analysis of x-ray photoelectron spectroscopy (XPS) and Auger electron spectroscopy/ scanning Auger microprobe (AES/SAM) data are then examined, followed by an illustration of the application of surface analysis to the study of adhesive bonding. For further information on these and other techniques, the reader is referred to the several reviews of the subject matter (Ref 1, 6-35).

Surface-Sensitive Techniques

The ideal technique for surface chemical analysis should provide elemental identification and a quantitative analysis of the first monolayer of the surface and identify the type of bonding present at the surface. In addition, the measurement process should not alter the surface; it should be capable of probing only a small surface area to resolve regions of inhomogeneity; it should have near-uniform high sensitivity for all elements; it should be suitable for any sample of interest; it should be amenable to rapid and simple measurement and analysis; and it should provide, or at least permit, a profile into the near-surface (<1 μm, or 0.04 mil) region of the sample. No one technique satisfies all of these criteria. In some cases, it is not necessary that all criteria be met, then one analytical technique may be able to answer all the questions raised. In other cases, the surface will have to be examined

Table 1   Surface-sensitive and near-surface-sensitive techniques

	Outgoing particle(a)			
Incoming particle	Electrons	Photons	Ions	Neutrals
Electrons	AES/SAM EELS LEED RHEED SEM TEM STM	EDS WDS XES	ESD	. . .
Photons	XPS (ESCA) XAES UPS	EXAFS SEXAFS XANES FTIR	. . .	. . .
Ions	IAES INS	PIXE	SIMS ISS RBS	SNMS(a) SALI(a)
Neutrals	. . .	. . .	FIM(b) FAB	. . .

AES, Auger electron spectroscopy; EDS, energy dispersive spectroscopy; EELS, electron energy loss spectroscopy; ESCA, electron spectroscopy for chemical analysis; ESD, electron-stimulated desorption; EXAFS, extended X-ray absorption fine structure; FAB, fast ion bombardment; FIM, field ion microscopy; FTIR, Fourier transform infrared spectroscopy; IAES, ion-induced Auger electron spectroscopy; INS, ion neutralization spectroscopy; ISS, ion-scattering spectroscopy; LEED, low energy electron diffraction; PIXE, proton-induced X-ray emission; RBS, Rutherford backscattering spectroscopy; RHEED, reflection high-energy electron diffraction; SALI, surface analysis with laser ionization; SAM, scanning Auger microprobe; SEM, scanning electron microscopy; SEXAFS, surface extended x-ray absorption fine structure; SIMS, secondary ion mass spectrometry; SNMS, secondary neutral mass spectroscopy; STM, scanning tunnelling microscopy; TEM, transmission electron microscopy; UPS, ultraviolet photoelectron spectroscopy; WDS, wavelength dispersive spectroscopy; XAES, x-ray-induced Auger electron spectroscopy; XANES, x-ray absorption near-edge spectroscopy; XES, x-ray emission spectroscopy; XPS, x-ray photoelectron spectroscopy. (a) Neutral; are positioned for detection. (b) Requires electric field or incoming neutral

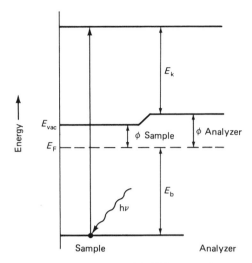

Fig. 1   Representation of the XPS process. An x-ray photon of energy, $h\nu$, is absorbed by an atom, which emits a photoelectron with kinetic energy, E_k, from a core level of binding energy, E_B.

by more than one method, thus leading to a more complete understanding of a problem.

X-Ray Photoelectron Spectroscopy (XPS)

XPS is arguably the most useful of the surface-sensitive techniques employed to investigate problems associated with adhesive bonding. It can provide a very good qualitative and quantitative elemental analysis of a surface and a reasonably good determination of the chemical bonding. It involves the bombardment of the sample surface with monoenergetic x-rays in ultrahigh vacuum (UHV) ($<10^{-6}$ Pa; 1 atmosphere = 1.01×10^5 Pa). As the photons (of energy, $h\nu$) travel through the material, some are absorbed by atoms, with photo-

electrons being emitted to conserve energy (Fig. 1). If the emissions are close to the surface and in its general direction, the photoelectrons can escape into the vacuum with little loss of energy and be detected. The conservation of energy requires that:

$$E_k = h\nu - E_B - \Phi_{analyzer} \qquad \text{(Eq 1)}$$

where E_k is the measured kinetic energy of the photoelectron relative to the vacuum level, E_B is the energy relative to the Fermi level that binds the electron to the atom, and $\Phi_{analyzer}$ is the work function of the electron energy analyzer. An XPS spectrum (Fig. 2a), the number of electrons detected per unit energy, $N(E)$, versus binding energy, consists of a series of peaks superimposed on a sloping or stepped background.

The energies of the peaks allow elemental identification and yield chemical-state information of the surface (Fig. 2b). The heights or areas of the peaks permit a quantification of the elements present (except for hydrogen, which has no characteristic binding energy). Compilations of spectra or core-level binding energies (Ref 9, 16, 21, 23, 36), including a National Institute of Standards and Technology (NIST) computerized data base (Ref 36), are available to allow chemical-state analysis. Spectral interpretation has been discussed in detail by Briggs and Riviere (Ref 37).

The surface sensitivity of XPS (and AES) is due to the high cross section of inelastic collisions between the photoelectrons (with typical kinetic energies of 200 to 1200 eV) and other electrons of the sample. As a result, only photoelectrons originating very

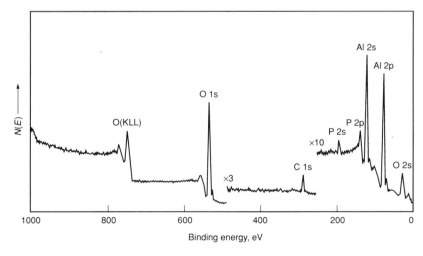

(a)

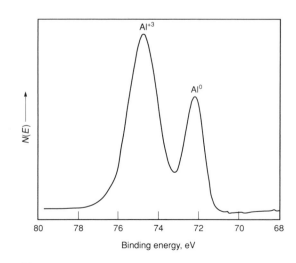

(b)

Fig. 2   (a) XPS survey spectrum of a phosphoric acid anodized (PAA) adherend surface. (b) High-resolution Al 2p spectrum of a native oxide on aluminum showing metallic and oxidized Al contributions

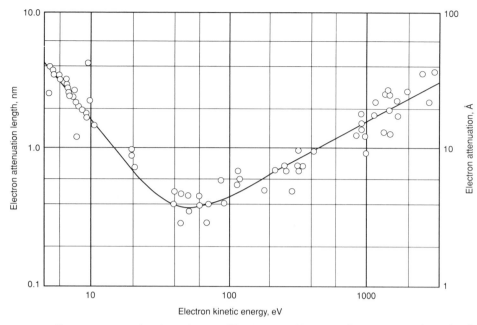

Fig. 3 Electron attenuation length as a function of kinetic energy. Maximum surface sensitivity is obtained at the minimum of the curve. Source: Ref 38, 39

near the surface have a high probability of being emitted into the vacuum without loss of energy. (Those electrons that have lost energy form the sloping background of the spectrum.) The signal, I_0, from any depth, d, is exponentially attenuated so that its value at the surface, I, is:

$$I = I_0 e^{[-d/(\lambda \cos\theta)]} \qquad \text{(Eq 2)}$$

where θ is the angle that the detected photoelectrons make with respect to the surface normal and λ is the attenuation length, which varies with electron kinetic energy (Ref 38, 39), as shown in Fig. 3. Although there is variation from material to material, as reflected by the scatter around the "universal" curve in the figure (Ref 39), the attenuation lengths slowly increase with electron kinetic energies above 100 eV. This variation allows information on the elemental depth distribution to be determined (see the section "Multiple Anodes/Different Transitions" in this article).

The ability to quantify XPS data is one of the major advantages of the technique. Not only is it good at detecting relative changes in the surface composition as a function of different surface treatments, it can also provide a good absolute concentration for all elements except hydrogen. Additionally, the elemental detection limits generally vary less than an order of magnitude and average 0.1 at.% with data acquisition times of several minutes. The quantitative analysis procedure and some of the possible pitfalls of data interpretation are discussed in the section "Quantification" in this article.

The other major advantage of XPS is the chemical state information it provides. This is discussed in the section "XPS Spectra",

but, in brief, photoelectron peaks are commonly shifted in energy depending on the charge localized on the atom so that the chemical state of the element can often be determined. For example, aluminum exhibits a 2.0 to 2.5 eV shift of its 2p line between the metallic and +3 state of Al_2O_3 (Fig. 2b).

A final advantage is that the surface composition of nearly all vacuum-compatible samples can be measured by XPS with little, if any, chemical changes being caused for most classes of samples. However, damage may occur in sensitive materials as a result of x-ray or electron bombardment or radiated heat. The electrons can originate from an electron gun used to neutralize the positive charge left on the surface during the photoemission process or from the window of the x-ray source. In some instruments, radiant heat from the anode can raise the temperature of the specimen. This electron and/or heat radiation may result in desorption of surface species or induce other changes in the near-surface region. Additionally, the requirement of UHV can prevent some polymers from being placed in the spectrometer, especially those with high levels of retained solvent or other volatiles. Such specimens can sometimes be examined by inserting them into a separately pumped introduction chamber and allowing the volatiles to be pumped away or by cooling to liquid nitrogen temperatures to minimize evaporation. However, it must be realized that in the first case the surface being measured is not identical with that initially prepared.

The major disadvantages to XPS are poor lateral resolution, a relatively weak signal, and the possible nonuniqueness of its chemical shift information (see the section

"Chemical State Information"). Recent advances in instrumentation have improved the first two problems, but XPS will always be inherently less capable than AES in these characteristics because of the difficulty of building intense x-ray sources (relative to electron sources) and of focusing x-rays (relative to charged particles, in general). Small-spot XPS spectrometers allow signals to be restricted to areas of the order of 10 to 100 μm (0.4 to 4 μin.) in diameter, compared to traditional analysis areas of several mm^2.

The advantages and limitations to XPS are given in Table 2. The table also compares XPS to the other chemical probes discussed below.

Auger Electron Spectroscopy/Scanning Auger Microscopy (AES/SAM)

When a vacancy in a core electron energy level of an atom is formed, the excited ion may become more stable by the well-known x-ray fluorescence process or by the Auger process, which involves electron emission. Figure 4 illustrates the process, in which the energy of the ion is reduced when an electron from a higher level fills the core hole. To conserve energy, a second electron is emitted, leaving the atom doubly ionized. The measured kinetic energy, $E_k(WXY)$, relative to the vacuum level, of an Auger electron from transition WXY is approximated by:

$$E_k(WXY) = E_B(W) - E_B(X) - E_B(Y) - \Phi_{analyzer} \qquad \text{(Eq 3)}$$

where $E_B(W)$ is the binding energy of an electron in core level W (which had the initial core hole); $E_B(X)$ and $E_B(Y)$ are the corresponding quantities for the "down" and "up" electrons in levels X and Y; and $\Phi_{analyzer}$ is the work function of the electron analyzer. One may say, more accurately, that $E_B(X)$ and $E_B(Y)$ are the binding energies of electrons in levels X and Y in the presence of a hole in level W. The energy of an Auger electron is determined only by the binding energies of the electrons involved in the transition and is therefore independent of the energy of the ionizing radiation.

Because an Auger electron requires two electrons above the lowest level, the elements hydrogen, helium, and gaseous lithium exhibit no Auger transition. Each of the remaining elements possesses a unique spectrum with one or more Auger peaks, and compilations of spectra (Ref 41-43) or energies (Ref 44, 45) are available, as are detailed instructions on peak identification (Ref 37, 46). The intensity of the transition provides a means of quantitative analysis, and, in some cases, the line shape gives chemical information (as described in the sections "Quantification" and "AES Spectra" in this article).

Table 2 Advantages and limitations of XPS, AES, SIMS, and SNMS

Technique	Advantages	Disadvantages
X-ray photoelectron spectroscopy (XPS)	Is sensitive to 2–10 monolayers; metals < oxides << polymers	Has relatively poor lateral resolution and poor imaging capability
	Can detect about 10^{-3} at.%	Requires deconvolution techniques (for example: surface monolayer)
	Is quantitative to ~10 to 20% without standards	Has adequate data acquisition rates for compositional depth profiles; inferior to other methods
	Is especially useful for chemical shifts from the same element in different compounds	
	Is the least destructive of all techniques (x-ray excitation)	
	Has a sensitivity range within a factor of 10 for most atomic numbers	
	Has minimal sample charging	
Auger electron spectroscopy (AES)	Is sensitive to 2–10 monolayers	May alter surface composition from ESD
	Can detect about 10^{-3} at.%	May have severe charging problems
	Is outstanding for compositional depth profiles	Will form carbon from polymers (electron beam cracking)
	Is quantitative to ±10% with standards	Has a slow rate of element mapping
	Has superb lateral resolution (20–50 nm, or 0.6–2.0 min beam)	
	Is fastest of the four methods	
	Has superb imaging and lateral mapping capabilities	
	Is useful for chemical shifts for some elements	
Secondary ion mass spectrometry (SIMS)	Is sensitive down to 1–2 monolayers	Requires sample destruction
	Can detect 10 ppm or less	Is quantitative with difficulty, at best
	Is superb for compositional depth profiles	Has varying elemental sensitivity
	Can detect isotopes	Has complex spectra
	Can detect hydrogen and deuterium present	May have chemical state changes from ion bombardment
	Can acquire data rapidly	
	Has lateral imaging capability	
Secondary neutral mass spectrometries (SNMS)	Has the first six advantages of SIMS	Requires sample destruction
	Is quantitative with modest use of standards (a major improvement over SIMS)	May have chemical state changes from ion bombardment

Source: Ref 40

The initial core hole can be created by an electron, x-ray, or energetic ion. As a result, Auger transitions appear in XPS spectra, along with the photoemission peaks (Fig. 2) and are used in some forms of data analysis (see the section "XPS Spectra"). However, most AES and all SAM measurements are obtained with electron excitation to allow for high signal strength and small beam size.

Traditionally, Auger spectra (Fig. 5a), obtained using phase detection techniques,

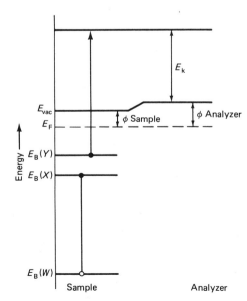

Fig. 4 The AES process in which a core hole in the *W* level is filled by an electron from the *X* level, causing an electron from the *Y* level to be emitted

were taken as d*EN(E)*/d*E* versus *E* to accentuate the small, sharp Auger peaks that are superimposed on a large, slowly varying background of inelastically scattered electrons. However, reductions in the electron beam size and total beam current have caused decreases in the signal strength, usually necessitating the use of pulse counting or beam brightness modulation techniques to acquire high spatial resolution data. Although this mode of acquisition generates an *EN(E)* spectrum (Fig. 5b), the data are frequently differentiated digitally prior to analysis.

The primary advantage of AES/SAM is its lateral resolution and imaging of the surface (Ref 32). Instruments can generate electron beams with widths as small as 50 nm (2 μin.) during SAM analysis, a resolution that is three orders of magnitude better than that possible with small-area XPS spectrometers. This capability can be used to measure the composition of a single point of interest, to detect variations in composition along a line (Fig. 6) (Ref 7), or to map the distribution of elements on the surface (Fig. 7) (Ref 7).

An almost equally valuable capability of AES/SAM is the ease with which sputter-depth profiles can be obtained (see the section "Depth Distribution Information" in this article). In this procedure, the surface is slowly sputtered or milled away by inert ion bombardment, with AES monitoring the composition as a function of time. Although this can also be done with small-spot XPS, AES is more commonly used, first, because of its smaller analysis area,

which permits a smaller sputtered area and, hence, a faster sputter rate, and, second, because of its faster data acquisition rate.

The quantification of AES/SAM data is more difficult than the quantification of XPS. These difficulties, which are discussed below, primarily result from the complex Auger line shape and from data analysis in the d*EN(E)*/d*E* mode, which inherently produces a poorer signal-to-noise ratio than the *EN(E)* mode.

In addition, the electron beam of AES/SAM heats the sample and desorbs surface species much more readily than does an x-ray beam. The electron bombardment will also cause more problems with insulating samples such as polymers and oxides. This is manifested in two ways: The sample, especially a polymer sample, can be damaged so that the data are not representative of the original surface; and the sample is likely to become charged because the number of electrons absorbed can be greater or less than those that emitted. In the worst case, no usable spectrum is obtained; otherwise, spectral features will shift in kinetic energy, but still be identifiable. Possible ways to reduce charging include lowering the beam current, changing the beam voltage, orienting the sample to obtain a near glancing incident beam, bombarding with low-energy positive ions, heating the sample, or coating the sample with a conductive coating (but leaving a small uncoated area for analysis) (Ref 47, 48). Different samples frequently require different methods, and the choice is often made by trial and error.

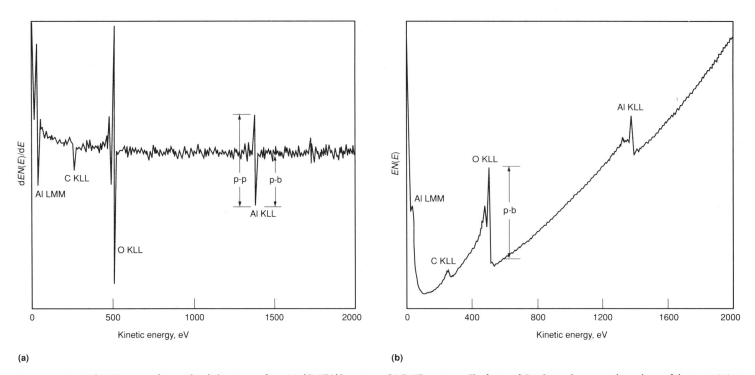

Fig. 5 Typical AES spectra of an oxidized aluminum surface. (a) d*EN(E)/dE* spectrum. (b) *EN(E)* spectrum. The factor of *E* is due to the energy dependence of the transmission function of the analyzer. The usual means for measuring peak intensity, peak-to-peak height (p-p), and peak-to-background height (p-b) are indicated.

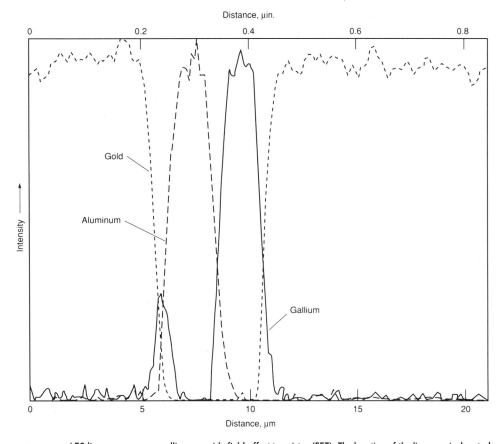

Fig. 6 AES line scan across a gallium arsenide field-effect transistor (FET). The location of the line scan is denoted by the horizontal line in the SEM micrograph of Fig. 7.

Secondary Ion Mass Spectrometry/Secondary Neutral Mass Spectrometry (SIMS/SNMS)

SIMS involves the detection of ions ejected from a surface by momentum transfer induced by an incident ion beam (Fig. 8). Again, Table 2 lists its advantages and limitations. It is capable of detecting all elements (including hydrogen), allows isotope discrimination, and has very good sensitivity for some elements (as low as a few parts per billion in favorable cases) (Ref 50). The surface sensitivity results because the ejected ions originate in the top few layers of the solid.

The number $N^+(A)$ of positive ions of isotope A detected by the analyzer can be written:

$$N^+(A) = \alpha^+(A)n(A)SIT \qquad \text{(Eq 4)}$$

where $\alpha^+(A)$ is the ratio of positive secondary ions to the total number of sputtered A atoms, $n(A)$ is the number of A atoms, S is the total elemental sputtering yield, I is the incident ion current, and T is the transmission of the analyzer. An analogous expression exists for negative ions.

Quantitative analysis can be very difficult with SIMS because $\alpha^+(A)$ is very material dependent, with the fraction of positive metal ions varying by up to five orders of magnitude, depending on whether the surface is metallic or oxidized (Ref 21). In some cases, this dependence can be used to advantage by flooding the surface with oxygen to enhance the ion yield and reduce the variation in signal that otherwise occurs

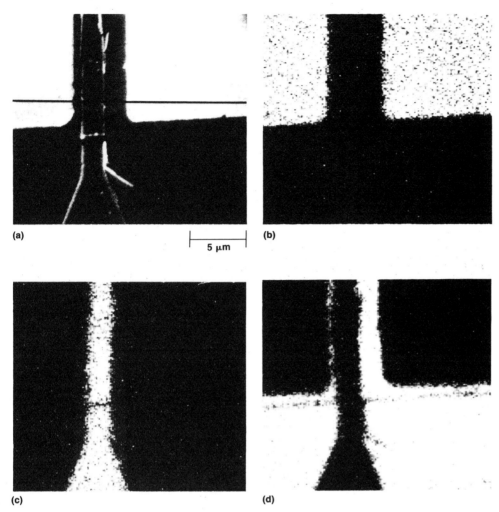

(a)

5 μm

(b)

(c)

(d)

Fig. 7 (a) SEM micrograph and (b) gold, (c) aluminum, and (d) gallium AES maps of a gallium arsenide field-effect transistor

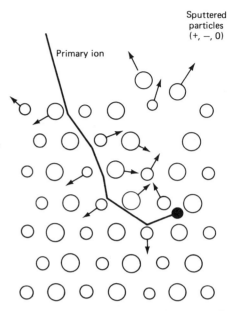

Fig. 8 The SIMS process. The incident ion transfers momentum to atoms of the sample. Some of these gain sufficient momentum normal to and away from the surface to be ejected from the sample. Most of these ejected atoms are neutral, but some are ionized either positively or negatively. Source: Ref 49

at the oxide-metal interface. In other cases, matrix effects are inevitable. As a result, quantitative analysis is generally feasible only in limited cases, such as for dopant profiles in ion-implanted semiconductors, where the matrix is constant although the impurity concentration may vary by several orders of magnitude.

SNMS reduces such matrix effects by detecting sputtered neutrals (Ref 50). Because sputtered atoms are much more likely to be neutrals than charged particles, variations in the ion yield have very little effect on the number of neutrals. For detection, the neutrals are postionized by electrons (Ref 51), plasma (Ref 52, 53), or a laser (Ref 54-57). Accordingly, an equation similar to Eq 4 applies, with α^+ being the fraction of postionized neutrals. Although this value is dependent on the means of postionization and the element being measured, it is independent of the sample, and thus instruments can be calibrated for quantification. Elemental detection limits for SNMS are relatively uniform and average in the parts per million range (Ref 21). Depending on the

specific element and material, these values can be greater or less than the corresponding values in SIMS.

By its very nature—analysis of sputtered material—SIMS and SNMS are destructive. However, the amount of damage varies dramatically with the ion current density. In static SIMS/SNMS (Ref 8, 30, 58), the ion beam current density is kept very low (1 nA/cm², or 6.45 nA/in.²) so that the lifetime of a surface monolayer can be several hours to several days. The focus of static SIMS is generally on the chemistry of the surface, as revealed by ionized clusters of atoms ejected from the sample, as opposed to detection of the depth distribution of trace impurities. Surface sensitivity for static SIMS is governed by the escape depth of secondary ions and is typically about 0.4 nm (0.016 μin.) (Ref 24). Lateral resolution is usually similar to that of XPS.

Dynamic SIMS/SNMS, in contrast, uses a high ion beam current density for a fast sputter rate (Ref 49). Consequently, it is more concerned with depth distributions through thin films than with surface compo-

sitions. Its forte is the low-level detection of impurities. The spatial resolution is comparable to that of many older SAM instruments, whereas the depth resolution is primarily governed by atomic mixing induced by the ion beam and depends on the ion energy (Ref 24, 45, 59). Other factors that can control the depth resolution include surface roughening (described in the section "Ion Sputtering" in this article) and crater wall effects (Ref 45, 59). Crater wall effects are important if ions that originate from the walls of the sputter crater are detected, because this material originates at a shallower depth than that of current interest (at the bottom of the crater). Such ions are customarily excluded from detection by electronic gating, or by a lens to exclude the ions from the extraction optics.

As with AES and any other bombardment with charged particles, charging of insulating samples can be a problem (Ref 30). Usually charging is overcome by flooding the surface with low-energy electrons, but it can also be minimized by replacing the primary ion beam with one of neutral atoms, a technique referred to as fast atom bombardment (FAB).

High-Resolution Scanning Electron Microscopy (SEM)

In addition to the surface composition and chemical bonding measured by the techniques described above, it is often critical to characterize the surface morphology of an adherend. For macrorough adherend surfaces (defined as containing features on the order of microns), any SEM and stan-

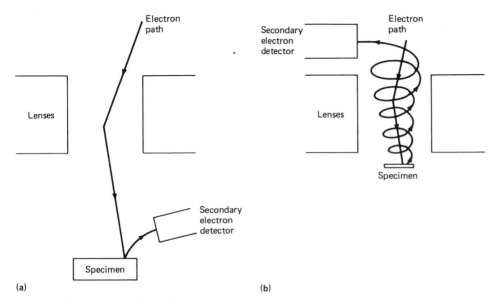

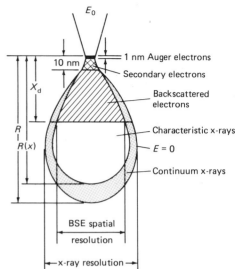

Fig. 9 (a) A conventional SEM. (b) A STEM operating in the SEM mode. In a STEM, the sample is placed closer to the lens, allowing a smaller probe diameter; additionally, the lens filters the emitted electrons so that only secondary electrons are detected. Source: Ref 61

Fig. 10 The spread of an electron beam with depth into a specimen. BSE, backscattered electron

dard sample preparation technique will suffice. However, adherend bonding properties are usually governed by microroughness (defined as containing features on the order of tens of nanometers) (Ref 3, 5, 60). Such morphologies are not observable in a normal SEM, but require the ultrahigh resolution available only on scanning transmission electron microscopes (STEMs) or on SEMs that use the same design principles.

In a STEM, the specimen to be studied is placed within the objective lens (Fig. 9). Because of its short focal length, this lens minimizes spherical aberrations and permits a smaller probe diameter (Ref 22). It also acts as a momentum filter, blocking backscattered electrons and permitting detection of only the true secondary electrons with energies approximately equal to or less than 50 eV that originate in the first 10 nm (0.4 μin.) of the sample, a depth at which there is minimal spreading of the incident beam. Consequently, the secondary electrons are detected from a volume having a projected area that is only slightly larger than the probe diameter. In contrast, in conventional SEMs, backscattered electrons are detected from a micron-size volume (Fig. 10) (Ref 62). On the other hand, in both types of instruments, x-rays are emitted and detected (using either EDS or WDS) from an even larger volume.

Employing the ultrahigh resolution of a STEM also requires proper sample preparation. Surfaces that are not electrically conducting require a metal coating that bleeds away the charge of the incident beam. It is important to choose a coating that does not mask the true adherend morphology. Early investigators coated their surfaces with aluminum and reported a popcornlike mor-

phology for Forest Products Laboratories (FPL)-etched aluminum; however, the transmission electron microscopy (TEM) of stripped oxides indicated a porous morphology (Ref 3). This discrepancy was resolved by using platinum coatings approximately equal to 5 nm (0.2 μin.) in thickness, which are sufficient to dissipate charge, but do not introduce structural artifacts at typical magnifications required to determine microroughness (Ref 5). The morphologies obtained in this manner are identical to those obtained by TEM.

Even with this ultrahigh resolution, the interpretation of three-dimensional surface features can be difficult with conventional micrographs. However, a stereo pair can be obtained by taking micrographs of the same area at slightly different angles (typically ±7°) so that a three-dimensional view can be realized by examining the pair under a stereo viewer. Several of these micrographs are given in the article "Surface Preparation of Metals" in this Volume.

Data Analysis

The following discussion of data analysis incorporates quantification, chemical state information, depth distribution information, and surface behavior diagrams.

Quantification

Quantification is possibly the most important chemical information derived from surface analysis, aside from the initial elemental identification. It can be achieved from XPS and AES/SAM survey spectra showing the entire spectral range at low resolution (Fig. 2a and 5) or, in the case of XPS, from high-resolution spectra of individual peaks (Fig. 2b). Quantification is usually obtained

by using sensitivity factors, $S_{Z,j}$, which are functions of the emission cross section for a given peak, j, and the incident radiation, the attenuation length of the measured electrons, the transmission of the analyzer, and the detection efficiency of the electron detector (Ref 7, 63, 64). The depth-weighted average concentration (C_Z) over the detection volume is calculated by:

$$C_Z = \frac{I_{Z,j}/S_{Z,j}}{\Sigma_Z I_{Z,j}/S_{Z,j}} \qquad \text{(Eq 5)}$$

Several compilations of these sensitivity factors for XPS (Ref 16, 65-72) and AES (Ref 41-43) are available and have been critically reviewed by Seah (Ref 73). These values were obtained from measurements on standard reference compounds. They provide good indications of relative changes in a series of spectra and good first approximations to absolute quantification (±10 to 20% for XPS, ±20% for AES). Errors occurring in the analysis from the choice of sensitivity factors will be systematic, and there will be internal consistency within a set of measurements. In many cases, this will be sufficient to reveal chemical differences and trends. In others, more accurate analysis affords an improved understanding of surface reactions (Ref 2, 74-77).

For improved quantification, it is important to understand the conditions under which a given set of sensitivity factors are derived (for example, how the spectra were taken and how the data were analyzed) before presuming applicability to one's own measurements (Ref 78). Many of the parameters influencing signal strength are dependent on the spectrometer and the particular operating conditions used. The signal intensities obtained in round robins involving the analysis of identical samples in different laboratories using a variety of instruments and specified operating conditions showed considerable scatter (Ref 79, 80). In another, more controlled instrument comparison, spectra were taken in the same laboratory on three different spectrometers (Ref 81). Again, the relative intensities of different peaks were found to differ, due most likely to differences in detection efficiency (Ref 73, 81). Such results demonstrate that measurements on one instrument by one operator do not necessarily give the same results as those on another instrument by another operator.

To obtain the most reliable measurements of surface compositions, it is necessary to derive sensitivity factors from measurements of standard reference materials on a given instrument under fixed operating conditions. These will allow excellent reproducibility and accuracy (typically ±5% for XPS, ±10% for AES) (Ref 2, 74). Nonetheless, several caveats need to be emphasized. The first is to know the surface composition of your standard, which is not necessarily the same as the bulk composition because of extraneous contamination, oxidation or hydration, surface segregation of one or more components, and so forth. Consequently, it is advisable to confirm the derived sensitivity factor with measurements on another known standard of different composition. The second caveat is to keep the operating conditions fixed or to derive separate sensitivity factors for each set of conditions because they can vary with resolution (Ref 63, 76), primary electron beam voltage (Ref 41), sample placement (Ref 80), and other parameters.

The choice of signal to be used in quantification differs for XPS and AES/SAM, and these shall be considered separately. For XPS, the area under the largest photoemission peak of a given element is the most commonly used signal. For the most part, area measurements are made according to one of two standard background subtraction schemes, which usually yield only small differences in the area (Ref 63). However, the choice of background can become important with peaks that exhibit complex shapes. For these cases, integral background (Ref 82) (in which the value of the background is assumed to be proportional to the fraction of the peak area at higher kinetic energy) is preferred over a straight-line background (Ref 58). Other more accurate approximations of the background also exist (Ref 64, 83, 84), but are not widely used.

The situation for AES/SAM is less straightforward because there is no ideal choice of signal. This is due in part to the practice of analyzing $dEN(E)/dE$ spectra and in part to the more complicated line shapes of AES transitions. The peak-to-peak (p-p) height of a $dEN(E)/dE$ transition is the most common measure of signal intensity, although the negative peak-to-background (p-b) height can also be used (Fig. 5a) (Ref 64, 85). Although these procedures give reasonably good qualitative analyses, they are inherently less reliable than those of XPS for several reasons. First, differentiation is a noise-inducing process. Because a height measurement on a $dEN(E)/dE$ spectrum is essentially a second derivative of an area measurement on an $EN(E)$ spectrum, the former has a poorer signal-to-noise ratio. Second, a peak-to-peak height measurement, especially one performed on a computer, involves only two data points (the maximum and minimum). It ignores the vast majority of the data and is subject to errors introduced by noise spikes. Third, and additionally, the peak-to-peak height measurement is not invariant, but is dependent on energy resolution (Ref 63, 64), sample positioning, and other instrumental parameters, as well as changes in the line shape of the Auger transition (Ref 41, 63, 64).

Changes in the Auger line shape can occur with changes in the chemical state of an element (Fig. 11) (Ref 86, 87). Thus, the line shape can sometimes be used as a probe of the local chemical environment (Ref 1, 77, 86, 87). However, the changes complicate quantification and can require different sensitivity factors for different chemical states. Additionally, the peak-to-peak height from an element present in two or more chemical states can actually decrease even though the elemental concentration has increased if a negative peak for one species interferes with a positive peak for another.

The use of $EN(E)$ spectra avoids some of these difficulties. As with XPS spectra, the peak height or the peak area can be measured from $EN(E)$ spectra. With peak area, the biggest problem is the question of background. Because the complex shape of the Auger peak makes contributions to the background difficult to determine (Ref 88), there is no good, generally accepted method of background removal. With peak height determination, the background is not as great a problem. The general procedure is to measure from the high kinetic edge of the peak to its maximum (Fig. 5b) (Ref 43, 89). Although these results will be affected somewhat by changes in line shape, they will be reasonably reliable unless two or more peaks overlap.

This section has focused on the potential problems of quantifying XPS and AES measurements. This has been done more to alert the reader to possible pitfalls, not to suggest that quantification is hopeless or even difficult. Indeed, much valuable and elegant work has been done using quantification from XPS and AES results. In most cases, the accuracy obtained is reasonably good and the internal consistency is sufficient to understand or solve a problem. The concerns discussed above become important primarily when high accuracy is needed or when results are compared to work performed by other groups.

Chemical State Information

XPS Spectra. Another important aspect of XPS is its ability to provide chemical state information. This information is conveyed by small shifts in binding energy (typically up to 5 eV) of the photoelectron peaks. As electronic charge is transferred from (to) an atom during chemical bond formation, the screening it provides of the positively charged nucleus decreases (increases) and the binding energies of the core electrons increase (decrease) in response. Tabulations exist of the chemical shifts of the principal line of an element for many compounds (Ref 10, 16, 36, 90) to assist in the analysis. In many cases, the shifts are large enough to determine the concentrations of the different chemical states separately (for example, Fig. 2b).

In order to compare peak energies to tabulated values, it is necessary to have an energy reference to adjust for any surface potential that develops on the surface of the sample as a result of the escape of photoelectrons. The most common method is to reference to the adventitious carbon present on all surfaces not cleaned *in situ* (Ref 91, 92), with the C 1s binding energy being most frequently assigned the value 284.8 eV (Ref 36, 90). Other reference schemes include the use of internal standards, the deposition of an overlayer, physical mixtures, and low-energy incident electrons. An internal standard assumes that the binding energy of one component of a system is invariant. It is particularly useful in the study of polymers (Ref 92). The deposition of a standard, such as gold or an organic compound, provides a peak of known energy as long as there is no interaction with the substrate and no thickness dependence or differential charging (Ref 92). However, the deposition may mask minor components of the surface. Mixtures of powdered samples are also occasionally used for referencing, but care must be taken to ensure intimate contact of the powders and the absence of differential charging. The key to charge compensation is consistency. When making comparisons between different sets of measurements, one needs to know the method of calibrating the energies and the charge referencing in order to make any necessary adjustments before the identification of chemical species.

When there is the possibility of species having similar chemical shifts, additional information in the spectrum can sometimes help. A particular example relevant to adhesive bonding is the identification of silicon-containing compounds as either silicates, which may be fillers in an adhesive, or siloxanes, such as silicones, which can lead to bond failure in the case of adherends contaminated with a mold-release agent. To eliminate the effects of charging, which shift all peaks an equal amount, and to avoid reference to the C 1s signal, which would be the sum of signals from the adhesive and from adventitious hydrocarbons, the binding energy difference between the O 1s and the Si 2p peaks can be obtained. Then, by comparing the results from the unknown with those of standards (428.5 eV for silicates and 429.7 for siloxanes), the source of the silicon may be determined (Ref 1, 93).

In many cases, the Auger peaks present in the XPS spectrum also provide chemical state information and can supplement energy shifts from the photoelectron peaks. The modified Auger parameter, α', is defined as (Ref 94):

$$\alpha' = E_B(P) - E_B(A) + h\nu \qquad \text{(Eq 6)}$$

where $E_B(P)$ and $E_B(A)$ are the (apparent) binding energies of the principal photoelectron peak and the Auger transition, respectively, and $h\nu$ is the energy of the x radiation. The modified Auger parameter is independent of the x-ray energy and sample charging; values for many compounds have been tabulated to assist in chemical state assignments (Ref 16, 36, 90, 93, 94). With these the identification of many more compounds is possible than is the case with photoelectron chemical shift results alone.

AES Spectra. The use of AES spectra to obtain chemical state information is less advanced than the use of XPS spectra. In part, this is because AES spectra are customarily acquired with less energy resolution than XPS spectra. Thus, most of the energy shifts reflected by the modified Auger parameter are not detectable. However, some energy shifts are readily detectable (Fig. 11) (Ref 41), for example, the oxides of elements in the third row of the periodic table. In some cases, the line shape of a transition can also be used for chemical state identification. Most commonly, the line shape serves as a fingerprint of a compound (Ref 87, 95), but line shapes of transitions involving valence band electrons can also be used to obtain a species-specific density of valence states (Ref 77, 86). In either type of analysis, the line shape can be a sensitive measure of the local chemical environment. However, because there is no large collected data base of Auger line shapes, the analyst is restricted to a few simple cases, such as carbon (Ref 87) or the third-row oxides (Ref 41), or to situations in which the data base is generated in-house or through the individual literature.

Depth Distribution Information

Frequently, the distribution of elements in the near-surface region is important. Simple distributions (Fig. 12) include a random mixture, a surface enrichment of one species, an overlayer, a buried layer, or lateral inhomogeneities (Ref 14). With lateral inhomogeneities, it is a simple matter of whether or not

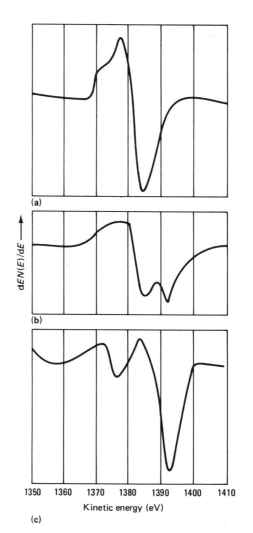

Fig. 11 Al KLL Auger transition from oxidized aluminum. (a) Oxide region. (b) Interfacial region. (c) Metallic region. One may note the change in both line shape and peak energy from the oxide to the metal. The changes can provide chemical state information, but complicate elemental quantification. Source: Ref 7

the instrument has the required lateral resolution. Features larger than the analysis area can be examined by point spectra, line scans, and maps (Ref 32). Depth distribution information is available by a variety of methods, depending on the probed depth. Several of these methods are discussed below.

Ion Sputtering. The most common means of obtaining a depth profile is ion sputtering combined with AES/SAM, although small-area XPS can also be used. A beam of ionized inert gases, generally Ar^+, erodes away the sample while AES/SAM or XPS measures the composition of the newly exposed surface. A rastering of the ion beam provides a flat crater, making the detection area representative of a single depth. Sputtering is a simple, routine process that is most useful in probing depths from 2 nm to 2 μm (0.08 μin. to 0.08 mil) and has led to the understanding of many

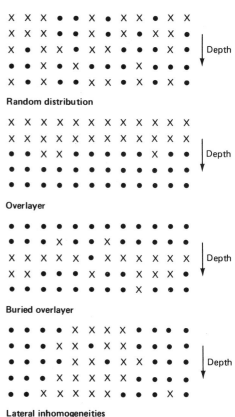

Fig. 12 Possible distribution of elements in the near-surface region. Source: Ref 14

problems. As with any procedure, it has its limitations, which are discussed to acquaint the reader with the possible pitfalls.

Because the sputtering process is destructive, the profile is a one-time event. It can be repeated only on a fresh area of sample (if it is large enough) or on a new sample. In most cases, this is not a concern, but it can be a problem if there is an instrument malfunction or if the sample has unknown constituents whose signals are not properly monitored.

Although the composition is measured as a function of sputtering time, it is usually preferred that it be a function of depth. This conversion is normally achieved using a sputter or erosion rate derived from Ta_2O_5 or SiO_2 films of known thickness. However, sputtering rates differ from material to material because of differences in sputter yields and change as a function of depth if the composition of a sample changes. Consequently, a precise determination of the depth scale is not normally available (Ref 96). The situation is even more complicated if the overlayer is porous and not dense, such as oxidized aluminum and titanium adherend surfaces (Ref 97, 98). For these types of samples, a comparison of sputter depth profiles and SEM micrographs shows that the former measures the total mass of the film and not its physical dimensions. However, because the two techniques com-

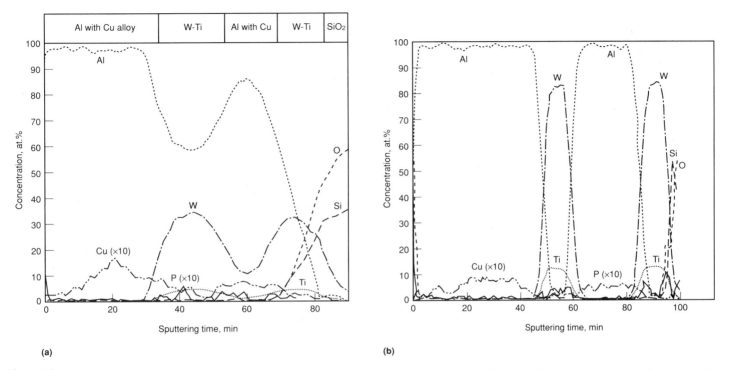

Fig. 13 AES sputter-depth profiles of a microelectronic bonding pad. (a) Obtained with stationary specimens. (b) Obtained with rotating specimens. Source: Ref 103

plement each other, a measure of the porosity of the film can be obtained.

Depth resolution decreases with increasing sputtering time or depth into the sample. This degradation is one of the limiting factors in the depth to which profiles can be usefully made. (The other is the time required.) Several factors, which have been reviewed in Ref 96, serve to reduce the resolution. These include original surface roughness and sample inhomogeneities, ion beam nonuniformities, atomic mixing and enhanced diffusion, and sputter-induced roughness. The result of these effects is a constant depth resolution at shallow depths and a decreased resolution at greater depths that is proportional to depth.

One way to improve the depth resolution is to use two ion guns (Ref 99). A more refined and less costly way is to rotate the sample during the sputtering procedure (Ref 100, 101). This rotation averages the incident ion beam orientation over all azimuthal angles and minimizes the effect of differently oriented grains in the material (Ref 102). As a result, sputter-induced roughness is minimized, and initial roughness can even be reduced. This improved depth resolution is illustrated in Fig. 13 (Ref 103), which shows sputter-depth profiles of an integrated circuit bonding pad with and without specimen rotation. With rotation, each interface is sharp, and each layer is distinct; without rotation, the interfaces are very broad, and the layers are poorly defined so that one could erroneously conclude that significant interdiffusion has occurred. If rotation occurs only during ion bombard-

ment and if the sample is returned to the same orientation at the end of each cycle, analysis can be performed at multiple locations away from the axis of rotation. Otherwise, analysis must be done at the center of rotation.

Although the composition of a layer is determined by measuring the sputtered surface, this surface does not necessarily have the same composition as the layer before ion bombardment. This difference may be due to displacement mixing, radiation-enhanced diffusion, preferential sputtering, or sputter-induced reduction (Ref 104). The first two of these affect depth resolution because they cause a physical broadening of the interface that is an artifact of the ion bombardment procedure. Preferential sputtering of multicomponent samples is a result of different atomic sputtering yields and depends on both the chemical bonding within the material and the relative masses of the incident ion and the atoms of the sample (Ref 105). The surface composition is altered so that, at equilibrium, the sputtered material has the same composition as the initial local layer. Consequently, the concentration of the component with the lowest sputter yield will be enhanced at the sputtered surface. The sputter-induced reduction of oxides is a special case of preferential sputtering resulting in chemical changes. TiO_2, MoO_3, and Ta_2O_5 are examples of the reduction of the oxide (Fig. 14) (Ref 7), whereas Al_2O_3, MgO, and SiO_2 are not (Ref 96). The net result of these effects is that the measured composition of a layer may not accurately indicate the true film

composition (Ref 104-106), but changes observed during sputtering (after equilibrium is reached) most likely reflect actual compositional changes.

Ball cratering is an alternative method of depth profiling with small probe beams (for example, AES or SIMS) that is suited for thicker regions (1 to 100 μm, or 0.04 to 4 mils) (Ref 107-111). In this procedure, a rotating steel ball or cylinder of radius R is coated with diamond paste and used to grind a shallow spherical or cylindrical crater into a sample. The crater edge is lightly ion sputtered to remove contamination, and Auger line scans or separate point spectra are taken along the edge in the direction normal to the rotation. Assuming that x, the distance from the crater edge, is much less than R, the depth (z) can be approximated by (Ref 110):

$$z = \frac{x(D - x)}{(2R)} \qquad \text{(Eq 7)}$$

where D is the diameter of the crater. Although this method is destructive, it has the advantage over ion bombardment that the measurement along the crater wall can be repeated. Similar measurements can also be performed on a sputtering crater edge, but there is no easy means to convert the lateral distance to depth.

The depth resolution of ball cratering is dependent on the lateral resolution of the SAM or SIMS and on any material smearing that may occur during polishing. Aside from small changes due to the spherical geometry, the resolution is constant with depth

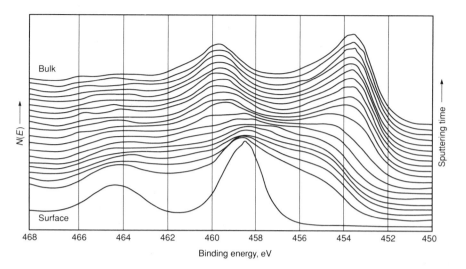

Fig. 14 XPS spectra of the Ti 2p peak from oxidized titanium as a function of sputter time. The oxide film is TiO$_2$ (at 458.5 eV), but ion bombardment causes a reduction to a lower oxidation state. Metallic titanium from the substrate (at 453.3 eV) begins to appear midway in the sputtering.

(Ref 110). It is then likely to give better resolution than ion sputtering at large depths, but not as good resolution at shallow depths.

Depth profiles of an engine bearing obtained by ball cratering and ion bombardment can be compared in Fig. 15. The effects of degraded depth resolution are clearly seen with the nickel diffusion barrier. The ball crater profile shows a sharp, nearly pure nickel layer, whereas the sputter-depth profile shows a broader, apparently diffused nickel layer. The depth resolutions for the two cases were estimated at 0.7 and 1.3 μm (0.03 and 0.05 mil), respectively (Ref 111).

Variable Take-Off Angle. The remaining techniques discussed in this article for obtaining depth distribution information are nondestructive, but are capable of probing only relatively shallow depths of a few nanometers into the sample. The first of these uses a variable take-off angle. Two or more measurements are made with different values for the take-off angle of the electrons (θ from Eq 2). By using a near-glancing take-off angle, the exponential attenuation factor of Eq 2 is increased so that the measurement is more surface sensitive. The most common application is in the comparison of measurements of a photoelectron peak that exhibits a splitting, denoting two chemical states. The component whose fraction of the total increases in the near-glancing spectrum has a greater relative concentration on the surface than does the other.

A similar comparison can also be made between the concentrations of different elements. The procedure is limited to depths of less than approximately 5 nm (0.2 μm), so that the bulk component can be detected. Although profiles of the distribution of elements or chemical states have been obtained from measurements at several different take-off angles (Ref 112, 113), a particular form of the profile must be assumed beforehand. When an abrupt, uniform overlayer is assumed, a single measurement is sufficient, and Eq 2 can be integrated over the respective ranges of the thin film and bulk components:

$$\frac{I_{film}}{I_{bulk}} = \frac{c_{film} \int_{z=0}^{D} e^{[-z/(\lambda\cos\theta)]}dz}{c_{bulk} \int_{z=D}^{\infty} e^{[-z/(\lambda\cos\theta)]}dz} \qquad (Eq\ 8)$$

which gives:

$$\frac{I_{film}}{I_{bulk}} = \frac{c_{film}}{c_{bulk}}(e^{[D/(\lambda\cos\theta)]} - 1) \qquad (Eq\ 9)$$

which can be inverted to find D. In these expressions, I_{film} and I_{bulk} are the intensities of the signals from species originating in the thin film and the bulk, respectively (for example, for Al$_2$O$_3$ on Al, Al^{3+}, and Al^0); c_{film} and c_{bulk} are the concentrations (number densities) of these species in the film and bulk, respectively (0.4 and 1 for the same example); D is the thickness of the thin film, λ is the electron attenuation length; and θ is the take-off angle. In this simple analysis, the species representative of the film (bulk) is assumed not to be present in the bulk (film); c_{film} and c_{bulk} are assumed to be constant with depth in the

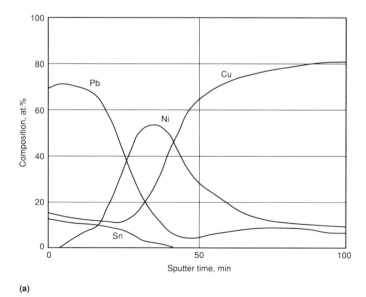

(a)

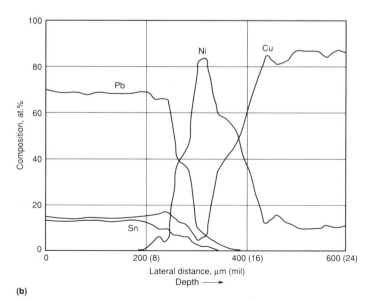

(b)

Fig. 15 Depth profiles of a multilayer engine bearing. (a) Obtained by sputtering. (b) Obtained by ball cratering. Source: Ref 105

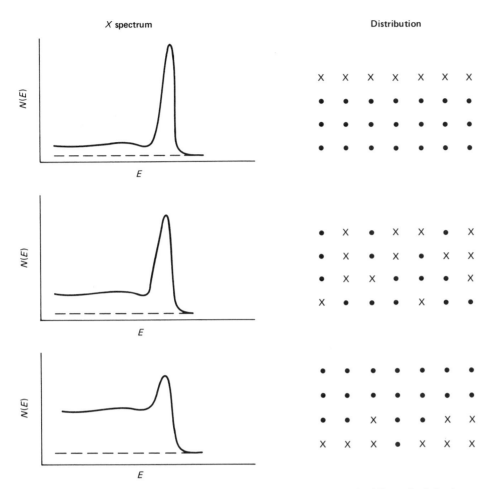

Fig. 16 Plots showing increase in the background for XPS spectra of element *X* for different depth distributions. Source: Ref 114

film and bulk regions, respectively; and λ is assumed to be the same for both signals in both regions.

Multiple Anodes/Different Transitions. Another way to change the surface sensitivity is to alter the attenuation length by varying the electron kinetic energy. This can be done for both AES and XPS by using two different transitions or peaks. However, this requires either separate sensitivity factors for the two peaks, or, at the least, their intensity ratio from a homogeneous specimen as measured by the same spectrometer.

An alternative means of achieving XPS photoelectrons with different attenuation lengths is to vary the energy of the incident x-rays (Ref 7). As with the different transitions, separate sets of sensitivity factors are needed for the different x-ray lines. Because many high energy x-ray lines are broad, have complex shapes, or have strong satellite lines, separate chemically shifted components may be more difficult to resolve and observe.

Both the multiple-anode and different-transition approaches are nondestructive and provide information only in a depth range from the surface of about 5 nm (0.2

μm). Although depth profiles are unobtainable, except in the most elementary cases, the procedures are simple, and a qualitative sense of the composition as a function of depth is readily available from the spectra.

Inelastic Scattering Ratio. A third means of obtaining depth distribution information nondestructively from shallow depths is to compare the ratio of the area of an XPS, or an $N(E)$ AES, peak to the increase in background due to the inelastically scattered electrons (Ref 114). This approach is based on the principle that electrons generated on the surface have a relatively small probability of being inelastically scattered (and therefore cause little increase in the background), whereas photoelectrons from several atomic layers into the sample have a much higher probability of losing energy before being detected (and therefore cause a greater background increase). This is shown in Fig. 16 and is also seen in Fig. 2, which shows little increase in background for the carbon signal that originates as an adventitious overlayer, and a greater increase for the oxygen and aluminum signals that originate throughout the oxide layer. Tougaard and Ignatiev showed that the ratio of peak area to background rise is nearly indepen-

dent of analyzer resolution and element for a variety of metals (Ref 114).

Because different distributions can result in identical ratios (for example, a buried thin film at a particular depth will give equivalent results as a homogeneous distribution), the procedure does not generate a unique distribution profile. Additionally, it cannot readily give information on the distribution of an element that is present in two chemical states dispersed differently because their background contributions mix too completely. Nonetheless, this procedure has the advantages of being available from spectra without additional data acquisition and being easy to perform. It is a simple test to determine whether an element is localized on the very surface.

Comparison With Less Surface-Sensitive Techniques. The final method of determining depth distribution information is to compare the surface-sensitive measurements with those from less surface-sensitive, that is, thin-film, techniques, such as energy-dispersive spectroscopy (EDS), wavelength dispersive spectroscopy (WDS), Rutherford backscattering spectroscopy (RBS), or Fourier transform infrared spectroscopy (FTIR), or even with those from purely bulk composition techniques. Such a comparison will tell whether the constituent in question is localized on the surface, but will provide neither a distribution of constituents nor the thickness of an overlayer unless the overlayer is a significant fraction of the sample or detection volume for the thin-film techniques. These comparisons are useful in thick-film analysis or when the bulk or thin-film analysis is used as a screening procedure before surface analysis.

Surface Behavior Diagrams (SBDs)

Surface behavior diagrams are a graphic means for the display and analysis of quantitative XPS and AES/SAM data (Ref 115). They can be very useful in the detailed investigation of surface and interfacial interactions. The SBDs resemble ternary or quaternary phase diagrams in that they represent a surface composition as a weighted sum of three or four basis compounds. However, they differ from phase diagrams in that surface compositional information rather than bulk phase information is represented and changes in the surface composition during nonequilibrium reactions can be traced as a function of reaction time, solution concentration, applied potential, depth into the sample, or other parameters of interest.

By comparing the path along which the composition evolves to those predicted by proposed models, the mechanisms involved during a reaction at the surface or at an interface can be established. Often, insights into the surface chemistry that would be

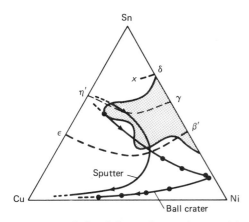

Fig. 17 Surface behavior diagram (expressed in weight percent) of the depth profiles shown in Fig. 15. Source: Ref 111

less apparent by other means can be obtained. SBDs have been used to study the evolution of the surface composition under a variety of conditions: those relevant to adhesive bonding include the hydration of aluminum oxide surfaces (Ref 2-4, 69, 115, 116) and the adsorption of hydration inhibitors from aqueous solutions (Ref 2, 4, 115, 116).

An example of the use of SBDs to display data graphically is given in Fig. 17, which shows the same depth profiles as are in Fig. 15 (neglecting the Pb contribution). The SBD clearly shows the enhanced depth resolution of the ball crater profiles and also indicates the presence of poor score-resistant compositions (indicated by the shaded regions), which increase with diffusion during use. These detrimental compositions are not apparent from the traditional means of display (Fig. 15). Based in part on these measurements and analysis, the bearing composition was changed to provide better score resistance.

Examples of additional studies will be presented in the sections "Hydration of Phosphoric-Acid-Anodized Aluminum" and "Adsorption of Hydration Inhibitors" in this article to illustrate the applications of surface analysis and SBDs to the investigation of adhesive bonding issues.

Construction of an SBD requires good quantitative measurements of the surface composition. For a compound SBD, the analyst chooses a set of basis compounds with which the surface composition can be expressed as a linear combination (Ref 109). Normally these compounds are selected with prior knowledge or expectations of the surface. In the example of the hydration of phosphoric-acid-anodized aluminum, the basis compounds chosen were Al_2O_3, H_2O, and $AlPO_4$ (Ref 69). The compositions of the possible hydration products, AlOOH and $Al(OH)_3$, could then be represented as $Al_2O_3 + H_2O$ and $Al_2O_3 + 3H_2O$, respectively. The surface composition, in the form of molar concentrations of the basis com-

pounds, is plotted on an equilateral triangle or a tetrahedron with each vertex representing the composition of a basis compound. (At this point, an improperly selected set of basis compounds will be evident because some of the compositions will fall outside of the SBD.) The evolution of the surface composition is then traced as a function of some variable of interest, either as a parameter or as a separate axis normal to the plane of a ternary SBD (Ref 109).

Applications to Adhesive Bonding

Surface analysis has several roles in adhesive bonding: routine quality control, failure analysis, process development, and detailed scientific investigation. An example of its use in quality control and failure analysis, the identification of silicon-containing contamination, was presented earlier. This section describes failure analysis in general and gives examples of scientific investigations in which surface analysis was crucial to an understanding of the particular problem. These examples demonstrate the synergism possible in a detailed, multitechnique approach to a problem where no one technique (XPS, AES, or SEM) would give a complete understanding. They also illustrate the usefulness of specialized data analysis such as SBDs.

Failure Analysis

In adhesive bonding, failure analysis is a critically important aspect of both manufacturing and scientific investigations. Identification of the locus of failure of a manufactured product or a test structure is necessary to establish the cause of the failure and to either recommend a remedy or understand the mechanisms of crack initiation and propagation and identify the weakest link in the structure.

Bond failure, either initially or after environmental exposure, occurs in one of four modes: cohesive, adhesive, interfacial, or mixed (Fig. 18). Cohesive failure can occur within the adhesive system, in the oxide film, within a composite member, or in the metal itself (unlikely in most applications of adhesive bonding). A cohesive locus of failure indicates that the ultimate performance has been reached for a given system; improved performance can only be achieved by redesigning the joint or by replacing the weakest component with a stronger or more durable one. On the other hand, an adhesive failure between the adhesive or primer and the adherend indicates that further bond performance can be obtained by strengthening that interface. Similarly, an interfacial failure (for example, one between the oxide film and the metallic substrate) suggests that improvements in performance can be made. In both of these cases and also when there is mixed-mode failure, the means for

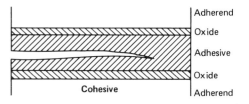

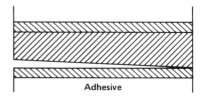

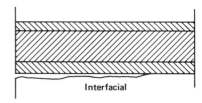

Fig. 18 Possible loci of failure of an adhesively bonded system

increasing the performance may be indicated by the failure analysis. This could include removing contamination or changing the adherend pretreatment procedure.

The first step in failure analysis is visual inspection. In a few cases, such as a failure through the middle of an adhesive, this examination is sufficient to identify the locus of failure. However, in many situations, this is not the case. For example, a failure may visually appear interfacial or adhesive, but crack propagation may have occurred within one of the bond components close to the interface. Similarly, contamination can not generally be detected visually unless it is gross.

A problem that potentially can prevent failure analysis is lack of sample integrity. Postfailure contamination, such as human handling or packaging in unclean containers or wrappings, can irreparably alter the surface chemistry. Unfortunately, this is common in a manufacturing environment where samples can be passed from hand to hand before they are submitted for surface analysis. Another type of loss of surface integrity is a change occurring during testing, for example, corrosion of a metal surface after crack propagation. In this case, although a failure may occur at the metal-oxide interface, it will appear to have occurred cohesively in the oxide. Only an awareness of the potential reactions and a more detailed failure analysis allow the determination of the actual locus of failure (Ref 3).

Examination of the failed surfaces should be made initially with both electron micros-

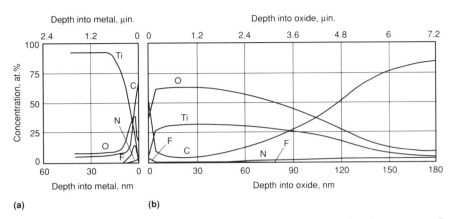

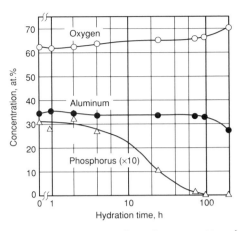

Fig. 19 Sputter depth profiles of the (a) metal and (b) oxide sides of a propagated crack in a titanium adhesive system. The depth profiles allow the locus of failure to be fixed at or near the oxide-metal interface. Source: Ref 117

Fig. 20 Variation of the surface composition of PAA samples exposed to 100% relative humidity at 50 °C (120 °F). Source: Ref 115

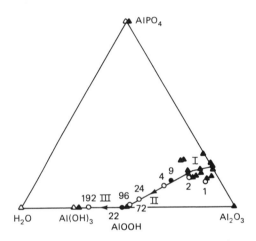

Fig. 21 Al$_2$O$_3$-AlPO$_4$-H$_2$O SBD showing the three steps of the hydration of PAA oxides in 100% relative humidity at 50 °C (120 °F), shown by ◯, and 60 °C (140 °F), shown by ●. Calculated compositions are represented by △; compositions of standards and of PAA samples prior to hydration are shown by ▲; ◯ and ● represent compositions of samples at various stages of hydration. The numbers beside the points correspond to the exposure time, in hours, to the humid environment. The open circles represent the same data shown in Fig. 20.

copy (Ref 1, 3, 22) and either XPS or AES/SAM, depending on the instruments available, the spatial resolution required, and the nature of the surface. In these investigations, it is important to realize that SEM specimens are routinely coated with a conductive film which will mask the surface chemistry so that no useful information can be obtained from surface analysis. Consequently, separate, but supposedly identical, samples must undergo SEM and surface analysis or a sample must undergo surface analysis first and then subsequent microscopic examination.

Depth profiles of one or both sides of the failure might also be helpful or even necessary. Figure 19 shows an example of an interfacial failure between the oxide and the titanium adherend. Because the freshly exposed metal surface oxidized immediately, the initial surface spectra could not distinguish between the interfacial failure and a failure entirely within the oxide (Ref 117). Depth profiles might also be needed to identify the specific locus of failure in the analysis of complex structures, such as those composed of many layers (Ref 118).

For the interpretation of failure analysis results, it is often necessary to have either spectra of the various components of the bond and of possible contaminants or knowledge of their chemical composition. This allows a unique element or chemical state to be chosen as a signature, or fingerprint, of each component. (A corresponding morphology fingerprint may also be possible from SEM micrographs.) For the metal and its oxide, obvious signatures are Al^0 and Al^{3+}, for example. Selecting a fingerprint for adhesives and primers is less straightforward. Many times the primer has SrCrO$_4$ or a similar compound for corrosion inhibition, and it can serve as identification. Fillers in adhesives and primers can also be used as fingerprints. In some cases, differences between possible components may be more subtle. The example of siloxanes and silicates and the use of the binding energy

difference between the O 1s and Si 2p lines have already been discussed.

Once the locus of failure is determined, the cause of failure can generally be deduced from the chemistry or morphology. Some of the possible causes of failure of aluminum adhesive bonds include contamination of the surface prior to bonding (Ref 1), poor surface preparation resulting in a smooth oxide film (Ref 3), hydration of the aluminum oxide surface (Ref 2-4, 74), incompatibility of a surface treatment with the adhesive (Ref 2, 4), disengagement (either partial or complete) of the adhesive from the porous oxide (Ref 2, 119, 120), and elastoplastic fracture of the adhesive (Ref 4, 120). For the failure analysis of a manufactured system, the identification of the cause of failure allows remedies to be made that will correct the problem or, at least, prevent it from happening again. For the failure analysis of a test structure, the identification should lead to a greater understanding of the processes and reactions involved.

Hydration of Phosphoric-Acid-Anodized Aluminum

Phosphoric acid anodization (PAA) is a common method of surface preparation of aluminum adherends prior to adhesive bonding for aerospace applications, as discussed in the article "Surface Preparation of Metals" in this Volume. PAA surfaces are microscopically rough (Ref 5) and are more resistant to hydration than are aluminum surfaces prepared by other treatments (Ref 74, 121). The microroughness allows mechanical interlocking with the primer or adhesive, which results in a strong adhesive bond, while the environmental stability of the oxide (together with its porosity) results in excellent bond durability in hot, humid environments.

To understand the reason that PAA surfaces resist hydration, fresh PAA surfaces and surfaces at various stages of hydration have been examined with XPS and high-

resolution SEM (Ref 1, 3, 4, 74, 121). The surface compositions as a function of time in a humidity chamber are presented in Fig. 20. In the course of the experiment, the phosphorus level decreases to zero, whereas the oxygen and aluminum levels change only slightly. Although little insight into the hydration chemistry can be derived from this representation, the Al$_2$O$_3$-H$_2$O-AlPO$_4$ SBD of Fig. 21 (Ref 74, 115) provides a clearer description of the initial surface composition and the hydration process. The fresh surfaces (solid triangles) are clustered around line I, which corresponds to the equivalent of approximately one molecular layer of phosphate on top of Al$_2$O$_3$.

Hydration is seen to be a three-step process. The first step (line I) can occur prior to insertion of the sample into the humidity chamber and involves the adsorption of

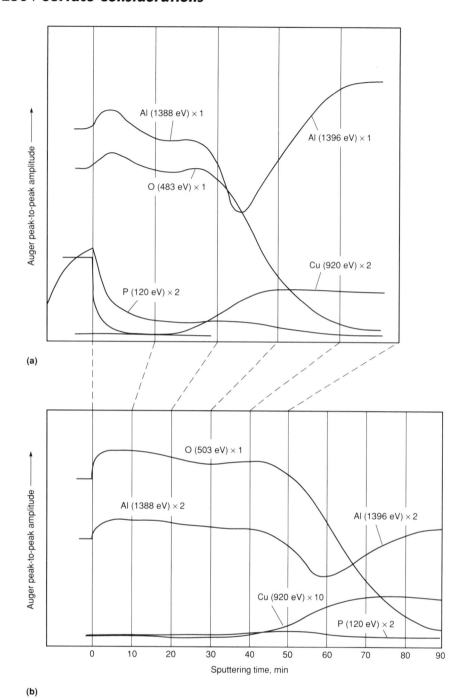

(a)

(b)

Fig. 22 AES sputter-depth profiles of PAA samples. (a) Before hydration. (b) After hydration. Note the small amount of phosphorus present in the hydrated sample. Source: Ref 74

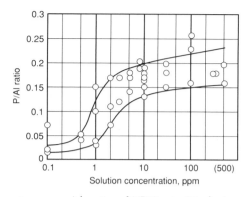

Fig. 23 Adsorption of NTMP onto FPL aluminum surfaces as measured by XPS as a function of NTMP solution concentration. Samples were immersed for 30 min. Most of the measurements are also plotted on Fig. 24. A monolayer of NTMP corresponds to a P/Al ratio of ~0.2. Source: Ref 124

would have been buried below the hydration product so that AES sputter depth profiles should have allowed the distinction (Fig. 22). These profiles clearly show that very little of the initial phosphorus was present in the hydrated layer. Thus, the limiting factor in the hydration resistance is the dissolution of the inhibiting phosphate.

Adsorption of Hydration Inhibitors

Based, in part, on the results of the PAA study, the use of several organic hydration inhibitors on PAA and Forest Products Laboratory (FPL) (Ref 122) sodium dichromate/sulfuric acid etched surfaces was investigated (Ref 1, 2, 4, 123, 124). One of the inhibitors that proved to be very successful was nitrilotris methylene phosphonic acid, $N[CH_2P(O)(OH)_2]_3$. It provides hydration resistance for FPL surfaces in a manner similar to anodization in phosphoric acid; that is, a protective molecular layer forms that does not allow the aluminum oxide substrate to hydrate until the inhibitor complex dissolves.

Treatment with nitrilotris methylene phosphoric acid (NTMP) consists of immersing the adherend into a dilute aqueous solution (generally 100 to 300 ppm) of inhibitor and rinsing. For an understanding of the processes involved during the adsorption of NTMP, FPL surfaces were examined by XPS following treatment in solutions with concentrations ranging from 0.1 to 500 ppm. The adsorption isotherm in Fig. 23 indicates that little adsorption occurred until the solution concentration reached 1 to 10 ppm. Saturation coverage of approximately one molecular layer was then observed at higher concentrations. The resulting SBDs provide a more complete description of the adsorption process.

Much of the data of Fig. 23 are plotted on the Al_2O_3-NTMP-H_2O SBD of Fig. 24. Figure 25 shows the initial FPL surface to be

water from ambient air. This adsorption involves no change in morphology and is reversible. More extensive hydration occurs once the sample is exposed to high humidity (line II). During this stage, dramatic changes in surface morphology are observed as the surface is converted to boehmite, AlOOH. Later, further hydration ensues with the growth of bayerite, $Al(OH)_3$, crystallites (line III). Because adhesive bonds fail during the boehmite-forming step (Ref 3), this stage was examined more closely. A comparison with various

possible mechanisms showed that two processes were compatible with the direct evolution of the surface composition to boehmite: the growth of boehmite on top of the original surface, and the slow dissolution of the surface phosphate followed by rapid hydration of the newly exposed amorphous aluminum oxide (Ref 74). Because the dissolution process is much slower than the hydration process, no differentiation between the two cases with the SBD alone was observed. However, in the first case, the original phosphorus-containing surface

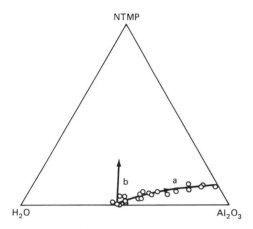

Fig. 24 Al_2O_3-NTMP-H_2O SBD showing a, the surface composition of FPL surfaces after 30-min immersion in aqueous solutions of NTMP at concentrations ranging from 0.1 to 500 ppm, with increases going from left to right, and b, the hypothetical path representing no displacement of water. Source: Ref 2

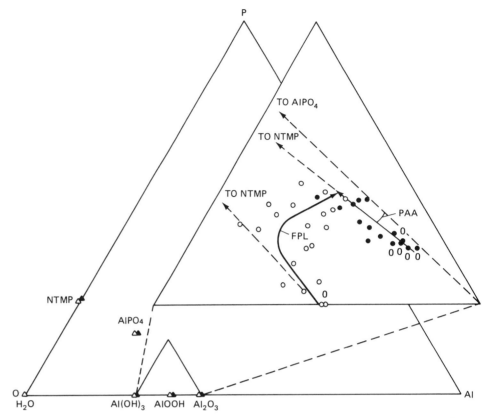

Fig. 25 Aluminum-phosphorus-oxygen SBD, showing the surface composition of FPL, indicated with ○, and PAA surfaces after immersion in NTMP solutions at various concentrations, with ●. The FPL data are the same as in Fig. 24. Calculated compositions are shown as △; the measured compositions of standards are shown by ▲. Compositions denoted by ○ represent surfaces not immersed in NTMP solutions. Source: Ref 116

Al_2O_3 with a sizable quantity of adsorbed water that is displaced by the NTMP. (An adsorption process that did not involve the displacement of water, for example, if the initial surface were AlOOH and not Al_2O_3 with adsorbed water, would follow line b toward the NTMP vertex.) Furthermore, upon treatment of the adherend in 100 to 500 ppm concentration solutions, the NTMP displaced all the water initially present and formed a molecular layer on the Al_2O_3 surface. However, adsorption is a two-step process, as indicated by the curvature of path a, which is more pronounced in the aluminum-phosphorus-oxygen SBD of Fig. 25. At low concentrations, the NTMP competes less successfully with water for the available adsorption sites; as a consequence, only one inhibitor-oxide bond is formed, with the resulting release of a molecule of water. At the higher solution concentrations, all three methylene-phosphonic-acid legs of NTMP form bonds to the oxide, thereby displacing more water. This displacement of water is so efficient that the surfaces immersed in the highest-concentration solutions exhibit no residual water.

For comparison, the adsorption characteristics of NTMP on PAA surfaces are also shown in Fig. 25. Because the initial PAA surface had very little adsorbed water, adsorption was a simple addition reaction, with the surface composition evolving along a line drawn to the NTMP composition point. Nonetheless, the saturation coverage by phosphorus-containing groups was very similar to that of NTMP on FPL. The ratio of nitrogen concentrations on the NTMP/PAA and the NTMP/FPL surfaces (1.5 at.% and 2.5 to 3.0 at.%) was compared to the corresponding ratio of additional phosphorus (1:2) to differentiate between NTMP adsorption on unoccupied active sites only and NTMP displacement of the phosphate

incorporated during anodization. Because the two ratios were equal, the possibility of a significant displacement of the original phosphate was eliminated; that is, NTMP adsorbed only on unoccupied active sites. Regardless of the details of the adsorption process, however, NTMP increases the hydration resistance of the PAA surface and further improves bond performance (Ref 120).

ACKNOWLEDGMENT

The author would like to thank J.S. Ahearn, C.R. Anderson, H.M. Clearfield, G.O. Cote, K.A. Olver, and J.D. Venables for numerous valuable discussions.

REFERENCES

1. G.D. Davis and J.D. Venables, in *Durability of Structural Adhesives*, A.J. Kinloch, Ed., Applied Science Publishers, 1983
2. G.D. Davis, J.S. Ahearn, L.J. Matienzo, and J.D. Venables, *J. Mater. Sci.*, Vol 20, 1985, p 975
3. J.D. Venables, *J. Mater. Sci.*, Vol 19, 1984, p 2431
4. J.S. Ahearn and G.D. Davis, *J. Adhes.*, Vol 28, 1989, p 75
5. J.D. Venables, D.K. McNamara, J.M. Chen, T.S. Sun, and R.L. Hopping, *Appl. Surf. Sci.*, Vol 3, 1979, p 88
6. J.M. Walls, Ed., *Methods of Surface Analysis*, Cambridge University Press, 1989
7. G.D. Davis, in *Adhesive Bonding*, L.H. Lee, Ed., Plenum Publishing, 1990
8. A. Benninghoven, *Surf. Sci.*, Vol 53, 1975, p 596
9. D. Briggs and M.P. Seah, Ed., *Practical Surface Analysis*, John Wiley & Sons, 1983
10. T.A. Carlson, *Photoelectron and Auger Spectroscopy*, Plenum Publishing, 1974
11. C.C. Chang, *J. Vac. Sci. Technol.*, Vol 18, 1981, p 276
12. P.H. Holloway, *Appl. Surf. Sci.*, Vol 26, 1986, p 550
13. J.L. Koenig, *Acc. Chem. Res.*, Vol 14, 1981, p 171
14. C.J. Powell, *Appl. Surf. Sci.*, Vol 4, 1980, p 492
15. R.S. Swingle and W.M. Riggs, *CRC Crit. Rev. Anal. Chem.*, Vol 5, 1975, p 262
16. C.D. Wagner, W.M. Riggs, L.E. Davis, J.F. Moulder, and G.E. Muilenberg, *Handbook of X-ray Photoelectron Spectroscopy*, Perkin-Elmer, 1979
17. J.T. Grant, *Appl. Surf. Sci.*, Vol 13, 1982, p 35

18. H.W. Werner and R.P.H. Garten, *Rep. Prog. Phys.*, Vol 47, 1984, p 221

19. A.W. Czanderna, Ed., *Methods of Surface Analysis*, Elsevier, 1975

20. L. D'Esposito and J.L. Koenig, *Fourier Transform Infrared Spectroscopy*, Vol 1, Academic Press, 1978

21. L.C. Feldman and J.W. Mayer, *Fundamentals of Surface and Thin Film Analysis*, North-Holland, 1986

22. J.L. Goldstein and H. Yahowitz, Ed., *Practical Scanning Electron Microscopy*, Plenum Publishing, 1975

23. K. Siegbahn, C. Nordling, A. Fahlman, R. Norberg, K. Hamrin, J. Hedman, G. Johansson, T. Bergmark, S.E. Karlsson, J. Lindgren, and B. Lindberg, *Nova Acta Reg. Soc. Sci. Upsal. Serv. IV*, Vol 20, 1967; Technical Report AFML-TR-68-189, U.S. Air Force Materials Laboratory, Oct 1968

24. S. Hofmann, *Surf. Interface Anal.*, Vol 9, 1986, p 3

25. W.L. Baun, in *Adhesion Measurements of Thin Films, Thick Films, and Bulk Coatings*, STP 640, K.L. Mittal, Ed., American Society for Testing and Materials, 1978, p 41

26. W.L. Baun, in *Industrial Applications of Surface Analysis*, L.A. Casper, Ed., American Chemical Society, 1982, p 121

27. W.J. van Ooij, A. Kleinhesselink, and S.R. Leyenaar, *Surf. Sci.*, Vol 89, 1970, p 165

28. P.J.K. Paterson and W. Broughton, *Int. J. Adhes. Adhes.*, Vol 1, 1981, p 181

29. J.F. Watts and J.E. Castle, *J. Mater. Sci.*, Vol 19, 1984, p 2259

30. D. Briggs, *Surf. Interface Anal.*, Vol 9, 1986, p 391

31. D. Briggs, *J. Adhes.*, Vol 21, 1987, p 343

32. I.F. Ferguson, *Auger Microprobe Analysis*, Adam Hilger, 1989

33. A. Benninghoven, F.G. Rüdenauer, and H.W. Werner, *Secondary Ion Mass Spectrometry, Basic Concepts, Instrumental Aspects, Applications, and Trends*, John Wiley & Sons, 1987

34. R.G. Wilson, F.A. Stevie, and C.W. Magee, *Secondary Ion Mass Spectrometry, A Practical Handbook for Depth Profiling and Bulk Impurity Analysis*, John Wiley & Sons, 1989

35. A.W. Czanderna and D.M. Hercules, Ed., *Ion Spectroscopies for Surface Analysis*, Plenum Publishing, 1990

36. C.D. Wagner and D.M. Bickham, *NIST X-ray Photoelectron Spectroscopy Database*, Version 1.0, Standard Reference Database 20, National Institute of Standards and Technology, 1989

37. D. Briggs and J.C. Riviere, in *Practical Surface Analysis*, D. Briggs and M.P. Seah, Ed., John Wiley & Sons, 1983, p 87

38. J. Tejeda, M. Cardona, N.J. Shevchik, D.W. Langer, and E. Schonherr, *Phys. Status Solidi (b)*, Vol 58, 1973, p 189

39. M.P. Seah and W.A. Dench, *Surf. Interface Anal.*, Vol 1, 1979, p 2

40. A.W. Czanderna, Overview of Ion Spectroscopies for Surface Compositional Analysis, in *Ion Spectroscopies for Surface Analysis*, A.W. Czanderna and D.M. Hercules, Ed., Plenum Publishing, 1990

41. L.E. Davis, N.C. MacDonald, P.W. Palmberg, G.E. Riach, and R.E. Weber, *Handbook of Auger Electron Spectroscopy*, Perkin-Elmer, 1976

42. G.E. McGuire, *Auger Electron Spectroscopy Reference Manual*, Plenum Publishing, 1979

43. T. Sekine, Y. Nagasawa, M. Kudoh, Y. Sakai, A.S. Parkes, J.D. Geller, A. Mogami, and K. Hirata, *Handbook of Auger Electron Spectroscopy*, Japan Electron Optics Laboratory Company, 1982

44. W.A. Coghlan and R.E. Clausing, *At. Data*, Vol 5, 1973, p 317

45. C.D. Wagner, in *Practical Surface Analysis*, D. Briggs and M.P. Seah, Ed., John Wiley & Sons, 1983, p 521

46. "Standard Practice for Elemental Identification by Auger Electron Spectroscopy," E 827, *Annual Book of ASTM Standards*, American Society for Testing and Materials; *Surf. Interface Anal.*, Vol 5, 1983, p 266

47. W. Wei, *J. Vac. Sci. Technol. A*, Vol 6, 1988, p 2576

48. S. Ichimura, H.E. Bauer, H. Seiler, and S. Hofmann, *Surf. Interface Anal.*, Vol 14, 1989, p 250

49. D.E. Sykes, in *Methods of Surface Analysis*, J.M. Walls, Ed., Cambridge University Press, 1989, p 216

50. W. Reuter, in *Secondary Ion Mass Spectrometry: SIMS V*, A. Benninghoven, R.J. Colton, D.S. Simons, and H.W. Werner, Ed., Springer-Verlag, 1986, p 94

51. O. Ganschow, in *Secondary Ion Mass Spectrometry: SIMS V*, A. Benninghoven, R.J. Colton, D.S. Simons, and H.W. Werner, Ed., Springer-Verlag, 1986, p 79

52. H. Oechsner, in *Thin Film and Depth Profile Analysis*, H. Oechsner, Ed., Springer-Verlag, 1984, p 63

53. H. Oechsner, in *Secondary Ion Mass Spectrometry: SIMS V*, A. Benninghoven, R.J. Colton, D.S. Simons, and H.W. Werner, Ed., Springer-Verlag, 1986, p 70

54. C.H. Becker and K.T. Gillen, *Anal. Chem.*, Vol 56, 1984, p 1671

55. C.H. Becker and K.T. Gillen, in *Secondary Ion Mass Spectrometry: SIMS V*, A. Benninghoven, R.J. Colton, D.S. Simons, and H.W. Werner, Ed., Springer-Verlag, 1986, p 85

56. J.B. Pallix, C.H. Becker, and K.T. Gillen, *Appl. Surf. Sci.*, Vol 32, 1988, p 1

57. U. Schühle, J.B. Pallix, and C.H. Becker, *J. Vac. Sci. Technol. A*, Vol 6, 1988, p 936

58. J.C. Vickerman, in *Methods of Surface Analysis*, J.M. Walls, Ed., Cambridge University Press, 1989, p 169

59. C.W. Magee and R.E. Honig, *Surf. Interface Anal.*, Vol 4, 1982, p 35

60. P.F.A. Bijlmer, *J. Adhes.*, Vol 5, 1973, p 319

61. J.P. Wightman, Virginia Polytechnic Institute and State University, private communication

62. H.M. Clearfield, D.K. McNamara, and G.D. Davis, in *Adhesive Bonding*, L.H. Lee, Ed., Plenum Publishing, 1990

63. M.P. Seah, in *Practical Surface Analysis*, D. Briggs and M.P. Seah, Ed., John Wiley & Sons, 1983, p 181

64. J.T. Grant, *Surf. Interface Anal.*, Vol 14, 1989, p 271

65. C.D. Wagner, L.E. Davis, M.V. Zeller, J.A. Taylor, R.H. Raymond, and L.H. Gale, *Surf. Interface Anal.*, Vol 3, 1981, p 211

66. C.D. Wagner, in *Practical Surface Analysis*, D. Briggs and M.P. Seah, Ed., John Wiley & Sons, 1983, p 511

67. C.D. Wagner, *Anal. Chem.*, Vol 49, 1977, p 1282

68. C.D. Wagner, *Anal. Chem.*, Vol 44, 1972, p 1050

69. V.I. Nefedov, N.P. Sergushin, I.M. Band, and M.B. Trzhakovskaya, *J. Electron Spectrosc. Relat. Phenom.*, Vol 2, 1973, p 383

70. V.I. Nefedov, N.P. Sergushin, Y.V. Salyn, I.M. Band, and M.B. Trzhakovskaya, *J. Electron Spectrosc. Relat. Phenom.*, Vol 7, 1975, p 175

71. S. Evans, R.G. Pritchard, and J.M. Thomas, *J. Phys. C, Solid State Phys.*, 1977, p 2483

72. S. Evans, R.G. Pritchard, and J.M. Thomas, *J. Electron Spectrosc. Relat. Phenom.*, Vol 14, 1978, p 341

73. M.P. Seah, *Surf. Interface Anal.*, Vol 9, 1986, p 85

74. G.D. Davis, T.S. Sun, J.S. Ahearn, and J.D. Venables, *J. Mater. Sci.*, Vol 17, 1982, p 1807

75. G.D. Davis, W.C. Moshier, J.S. Ahearn, H.F. Hough, and G.O. Cote, *J. Vac. Sci. Technol. A*, Vol 5, 1987, p 1152

76. W.C. Moshier, G.D. Davis, J.S. Ahearn, and H.F. Hough, *J. Electrochem. Soc.*, Vol 134, 1987, p 2677

77. G.D. Davis, D.E. Savage, and M.G. Lagally, *J. Electron Spectrosc. Relat. Phenom.*, Vol 23, 1981, p 25

78. C.G.H. Walker, D.C. Peacock, M. Pratton, and M.M. El Gomati, *Surf. Interface Anal.*, Vol 11, 1988, p 266

79. C.J. Powell, N.E. Erikson, and T.E. Madey, *J. Electron Spectrosc. Relat. Phenom.*, Vol 17, 1979, p 361

80. C.J. Powell, N.E. Erikson, and T.E. Madey, *J. Electron Spectrosc. Relat. Phenom.*, Vol 25, 1982, p 87

81. D.R. Baer and M.T. Thomas, *J. Vac. Sci. Technol. A*, Vol 4, 1986, p 1545

82. D.A. Shirley, *Phys. Rev. B, Condens. Matter*, Vol 5, 1972, p 4709

83. S. Tougaard and P. Sigmund, *Phys. Rev. B, Condens. Mater.*, Vol 25, 1982, p 4452

84. S. Tougaard, *Surf. Interface Anal.*, Vol 11, 1988, p 453

85. M.T. Anthony and M.P. Seah, *J. Electron Spectrosc. Relat. Phenom.*, Vol 32, 1983, p 73

86. D.E. Ramaker, in *Springer Series in Chemical Physics*, ISISS Proceedings, R. Vanselow, Ed., Springer-Verlag, 1982

87. T.W. Haas, J.T. Grant, and G.J. Dooley, III, *J. Appl. Phys.*, Vol 43, 1972, p 1853

88. G.D. Davis and M.G. Lagally, *J. Vac. Sci. Technol.*, Vol 18, 1981, p 727

89. A. Mogami, *Surf. Interface Anal.*, Vol 7, 1985, p 241

90. C.D. Wagner, in *Practical Surface Analysis*, D. Briggs and M.P. Seah, Ed., John Wiley & Sons, 1983, p 477

91. P. Swift, *Surf. Interface Anal.*, Vol 4, 1982, p 47

92. P. Swift, D. Shuttleworth, and M.P. Seah, in *Practical Surface Analysis*, D. Briggs and M.P. Seah, Ed., John Wiley & Sons, 1983, p 437

93. C.D. Wagner, D.E. Passoja, H.F. Hillery, T.G. Kinisky, H.A. Six, W.T. Jansen, and J.A. Taylor, *J. Vac. Sci. Technol.*, Vol 21, 1982, p 933

94. C.D. Wagner, L.H. Gale, and R.H. Raymond, *Anal. Chem.*, Vol 51, 1979, p 466

95. G.D. Davis, M. Natan, and K.A. Anderson, *Appl. Surf. Sci.*, Vol 15, 1983, p 321

96. S. Hofmann, in *Practical Surface Analysis*, D. Briggs and M.P. Seah, Ed., John Wiley & Sons, 1983, p 141

97. J.S. Ahearn, T.S. Sun, C. Froede, J.D. Venables, and R. Hopping, *SAMPE Q.*, Vol 12, 1980, p 39

98. T.S. Sun, D.K. McNamara, J.S. Ahearn, J.M. Chen, B. Ditchek, and J.D. Venables, *Appl. Surf. Sci.*, Vol 5, 1980, p 406

99. D.E. Sykes, D.D. Hall, R.E. Thurstans, and J.M. Walls, *Appl. Surf. Sci.*, Vol 5, 1980, p 103

100. A. Zalar, *Surf. Interface Anal.*, Vol 9, 1986, p 41

101. J.D. Geller and N. Veisfeld, *Surf. Interface Anal.*, Vol 14, 1989, p 95

102. R. Smith, T.P. Valkering, and J.M. Walls, *Philos. Mag. A*, Vol 44, 1981, p 879

103. J.D. Geller, Geller Microanalytical Laboratory, private communication

104. N.Q. Lam, *Surf. Interface Anal.*, Vol 12, 1988, p 65

105. R. Kelly, *Surf. Sci.*, Vol 100, 1980, p 85

106. H.M. Clearfield, G.O. Cote, K.A. Olver, D.K. Shaffer, and J.S. Ahearn, *Surf. Interface Anal.*, Vol 11, 1988, p 347

107. V. Thompson, H.E. Hintermann, and L. Chollet, *Surf. Technol.*, Vol 8, 1979, p 421

108. J.M. Walls, D.D. Hall, and D.E. Sykes, *Surf. Interface Anal.*, Vol 1, 1979, p 204

109. I.K. Brown, D.D. Hall, and J.M. Walls, *Vacuum*, Vol 31, 1981, p 625

110. C. Lea and M.P. Seah, *Thin Solid Films*, Vol 75, 1981, p 67

111. J.C. Bierlein, S.W. Gaarenstroom, R.A. Waldo, and A.C. Ottolini, *J. Vac. Sci. Technol. A*, Vol 2, 1984, p 1102

112. T.D. Bussing and P.H. Holloway, *J. Vac. Sci. Technol. A*, Vol 3, 1985, p 1973

113. M Pijolat and G. Hollinger, *Surf. Sci.*, Vol 105, 1981, p 114

114. S. Tougaard and A. Ignatiev, *Surf. Sci.*, Vol 129, 1983, p 355

115. G.D. Davis, *Surf. Interface Anal.*, Vol 9, 1986, p 421

116. G.D. Davis, J.S. Ahearn, and J.D. Venables, *J. Vac. Sci. Technol. A*, Vol 2, 1984, p 763

117. H.M. Clearfield, D.K. Shaffer, J.S. Ahearn, and J.D. Venables, *J. Adhes.*, Vol 23, 1987, p 83

118. G.D. Davis, H.M. Clearfield, W.C. Moshier, and G.O. Cote, *Surf. Interface Anal.*, Vol 11, 1988, p 359

119. J.S. Ahearn and G.D. Davis, in *Adhesion '87*, Plastics and Rubber Institute, 1987, p 291

120. D.A. Hardwick, J.S. Ahearn, A. Desai, and J.D. Venables, *J. Mater. Sci.*, Vol 21, 1986, p 179

121. G.S. Kabayaski and D.J. Donnelly, Report D6-41517, The Boeing Company, Feb 1974

122. H.W. Eichner and W.E. Schowalter, Report 1813, Forest Products Laboratory, 1950

123. J.S. Ahearn, G.D. Davis, T.S. Sun, and J.D. Venables, in *Adhesion Aspects of Polymer Coatings*, K.K. Mittal, Ed., Plenum Publishing, 1983, p 281

124. D.A. Hardwick, J.S. Ahearn, and J.D. Venables, *J. Mater. Sci.*, Vol 19, 1984, p 223

Primers and Coupling Agents

Willard D. Bascom, The University of Utah

IN NEARLY ALL adhesive bonding operations, some type of coating is applied to the adherend surface prior to the application of the adhesive. These precoats generally fall into two categories, primers and coupling agents. The function of a primer is viewed as an adhesion promoter and protective coating, whereas a coupling agent serves only to create chemical bonds between adhesive and adherend. However, the distinction between primer and coupling agent is not always clear. A primer may, by accident or design, act as a coupling agent. Conversely, a coupling agent may serve as a primer. Quite frequently, a precoat will include both.

This article presents a general discussion of primers and coupling agents and gives several examples of their use. Much of the discussion relates to the bonding of inorganic adherends, metals, glasses, and ceramics, where the need for primers and coupling agents is almost axiomatic. The use of precoats in the bonding of polymers and polymer composites is not well documented.

Primers

Virtually all adhesive suppliers recommend a primer for their paste or film adhesives when they are used to bond metals or other high-surface-energy adherends such as glasses and ceramics. These primers have at least three functions.

First, primers protect the adherend from contamination or changes in adherend surface chemical constitution between the cleaning or surface treatment operation and the bonding operation. These adherends invariably receive a pretreatment, which may simply be a surface cleaning to remove loose oxide and/or organic contamination, or which may create a specific surface condition, for example, an oxide that resists corrosion and possesses a special microporosity into which the primer penetrates to create mechanical interlocking, or "keying."

A second function of a primer is to penetrate surface roughness and oxide microporosity. Paste and, certainly, film adhesives have high viscosities, which prevent them from penetrating into the roughness that normally exists as a result of machining or the microporosity deliberately generated

on the surface. A primer, on the other hand, is a low-viscosity fluid that is typically a 10% solution of the adhesive in an organic solvent, which can wet out the adherend surface. This leaves a coating over which the adhesive can readily flow and attain intimate contact. Bishopp et al. (Ref 1) examined aluminum-epoxy bonds with and without a primer by sectioning across the bond line with an ultramicrotome and viewing the sections using transmission electron microscopy. In the absence of a primer, trapped air was found in the adherend surface roughness that had been created by various methods ranging from light abrasion to acid-dichromate etching. Much of this porosity was eliminated when the adherends were precoated with a primer. Finally, primer solutions may dissolve low levels of organic contamination that otherwise would remain at the adhesive-adherend interface as a weak boundary layer.

Not only the adhesive, but the primer solution as well may contain wetting agents, flow control agents, elastomeric toughening additives, and quite frequently, corrosion inhibitors such as zinc and strontium chromate and other inorganic chromate salts. The actual composition of a commercial primer is proprietary information and in some cases may be tailored to a specific application.

Ecological concerns are forcing a change from organic solvent-base primers to water-base primers. In the past, a solvent was simply allowed to evaporate into the atmosphere. This practice is no longer acceptable, especially because adhesives are being used increasingly for large-scale production and for large structures, for example, automotive and aerospace construction, resulting in large volumes of solvent waste. Recovery and recycling of the solvents are not cost effective.

Water-base primers are being developed that are essentially aqueous emulsions of the adhesive and the other ingredients that make up a primer formulation. Ideally, the emulsifying agent is a nonionic surfactant because it remains behind in the dried film. Cationic or anionic surfactants are hygroscopic and, left in the primer film, would attract water and result in the subsequent degradation of bond strength.

The U.S. Air Force has undertaken a number of development efforts to improve primers for the structural bonding of aluminum and titanium (Ref 2). These include the development of water-soluble polymers for water-base primer systems, the electrodeposition of primers, and more effective corrosion-inhibiting additives.

There are a number of water-soluble polymers, notably the phenol/resorcinol-formaldehyde novolac polymers, and there are even some water-soluble epoxy compositions that may be potential primer resins. Obviously, they must be compatible with the adhesive and with the other ingredients in the primer system.

The electrodeposition of primers is a well-established technology employed for priming automotive parts prior to painting. The use of this technology for adhesive bonding requires the development of electroprimer formulations suitable for adhesives. The process involves an aqueous suspension of an electrophoretic (charged) organic polymer that is attracted to the charged (direct current) metal adherend. This process has advantages over spraying or dip coating. It is adaptable to automation, is pollution free, and offers uniform film distribution, controlled film thickness, and rapid film application.

The more effective inorganic chromate inhibitors, such as zinc chromate and strontium chromate, have low solubilities in water. Consequently, their incorporation into a water-base primer may be ineffective. In the Air Force program, efforts have been made to develop water-soluble primers based on dichromate salts of organic nitrogen compounds. These materials are inherently unstable and readily decompose by oxidation in air. However, guanidine dichromate has been reported to be as effective as zinc chromate (Ref 1).

Foister et al. (Ref 3) have presented a detailed investigation of electrodeposited primers on steel for structural adhesive bonding. Their studies reveal the complex chemistry of the primer film formation and interaction of the primer with the adhesive. In the case of one adhesive composition, the amine curing agent degraded the primer, but after the amine was replaced by an imidazole catalyst, there was no degrada-

Table 1 Typical silane coupling agents

Organofunctional group	Chemical structure
Vinyl	$CH_2{=}CHSi(OCH_3)_3$
Chloropropyl	$ClCH_2CH_2CH_2Si(OCH_3)_3$
Epoxy	$CH_2\overset{O}{\overbrace{CHCH_2}}OCH_2CH_2CH_2Si(OCH_3)_3$
Methacrylate	$CH_2{=}\underset{CH_3}{C}{-}COOCH_2CH_2CH_2Si(OCH_3)_3$
Primary amine	$H_2NCH_2CH_2CH_2Si(OC_2H_5)_3$
Diamine	$H_2NCH_2CH_2NHCH_2CH_2CH_2Si(OCH_3)_3$
Methyl	$CH_3Si(OCH_3)_3$

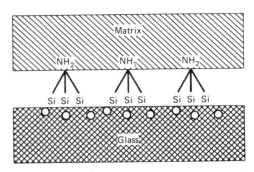

Fig. 1 Coupling of a silane between adherend (glass) and adhesive (matrix). Exposure occurred in water at 60 °C (140 °F).

tion of the primer, and excellent, moisture-resistant bonds were formed. The result was far better than bonding to cleaned but unprimed steel.

Primerless Bonding. Ideally, it would be more cost effective if primers could be eliminated from the adhesive bonding process. In principle, this can be done if two criteria are met. First, the adhesive must be applied immediately after the surface treatment of the adherend, with no time allowed for recontamination of the surface or loss of the chemical or morphological characteristics imparted by the surface treatment. Second, the surface treatment must result in an adhesive-adherend boundary that is resistant to moisture. These conditions might be realized in a high-production-rate, automated-processing scenario and by including a corrosion inhibitor in the adhesive itself.

Primers are rarely used in the bonding of polymers or polymer matrix composites. Conventional practice is to abrade the surface slightly and then rinse or wipe with a solvent to remove dust and oily contamination. Kinloch and Taig, for example, describe the adhesive bonding of thermoplastic matrix composites (Ref 4), whereas Petrie (Ref 5) reviews the bonding of plastics, including surface treatments.

Except in the aerospace industry, there is a trend that not only departs from the use of primers, but eliminates surface cleaning steps as well. In the automotive industry, adhesive bonding is frequently performed directly on oily steel surfaces (Ref 6). The surface must be free of corrosion products, and any thick layers of oil are reduced or removed by solvent wiping. In addition, the adhesive must be cured at temperatures of 121 °C (250 °F) or greater. Siebert *et al.* (Ref 7) found that the addition of a carboxy-terminated butadiene acrylonitrile liquid rubber to a conventional dicyandiamide-cured bisphenol A epoxy adhesive improved adhesion to oily steel by as much as 65% in both lap-shear and peel tests.

The bonding of sheet molding compound (SMC) parts to metals and other SMC parts does not involve primers. The parts are wiped clean using a solvent such as methylene chloride, and urethane adhesives are applied. Recently, it has been found that a two-part epoxy composition containing amine-terminated butadiene acrylonitrile gives good SMC-metal and SMC-SMC bonds without the need for a solvent wipe (Ref 8).

There is a growing trend toward the development of adhesive bonding systems that do not require primers or even the removal of low-level contamination (see, for example, Ref 9). Certain industries, notably aircraft manufacturing, continue to rely on special surface treatments and primers including coupling agents, although coupling agents, as such, are not used by the aerospace industry in high-strength metal-to-metal adhesive bonding. In general, there is usually a delay between surface treatment and the application of the adhesive. In such a case, a primer coating is essential. For metals with low corrosion resistance, a corrosion inhibitor must also be included in the primer.

Coupling Agents

The term coupling agents initially referred to surface coatings used to improve the moisture resistance of glass fiber reinforced polymer composites. Because the resin matrix (epoxy, polyester) is hygroscopic, the bond between fiber and matrix is easily degraded when exposed to water, especially under hot-wet conditions.

This problem was solved by applying silane coupling agents to the glass fiber. Chemically, coupling agents have the structure:

$$R'{-}Si(OR)_3 \qquad \text{(Eq 1)}$$

where R is an alkyl group, for example, $C_2H_5{-}$, and R' is a functional group designed to be reactive with the adhesive. Several commercially available silanes are listed in Table 1. The concept of coupling envisions the alkoxy group (OR) being hydrolyzed to silanols ($SiOH$), which then chemically bond or hydrogen bond to the glass surface.

$$R'{-}Si(OR)_3 + 3H_2O \rightarrow R'{-}Si(OH)_3 + 3ROH$$

$$R'{-}Si(OH)_3 + HO{-}Si_s \rightarrow R'{-}Si{-}O{-}Si_s + H_2O$$
$$\underset{(OH)_2}{} \qquad \text{(Eq 2)}$$

or

$$R'{-}Si(OH)_3 + HO{-}Si_s \rightarrow R'{-}SiOH{-}{-}{-}OH{-}Si_s$$
$$\underset{(OH)_2}{}$$

$$+ H_2O \qquad \text{(Eq 3)}$$

The organofunctional group (R') chemically reacts with the adhesive (Fig. 1). Although this simplistic view has been shown to be unrealistic, no satisfactory alternative explanation of how these agents work has been offered, despite more than two decades of research.

There are three major arguments against the coupling theory, the first being the low probability of a silane molecule bonding chemically with all three silanol groups attached to the surface. The steric constraints required for this conformation of the silane molecule are too severe, considering the random distribution of silanol groups ($OH{-}Si_s$) on the surface. The second argument is that the trialkoxysilanes readily polymerize to form three-dimensional networks, especially in the presence of acidic or basic catalysts. In fact, they are usually applied to surfaces in the presence of weak acids. The polymerization of a widely used coupling agent, aminopropyltriethoxy silane, $NH_2CH_2CH_2CH_2Si(OCH_2CH_3)$ is autocatalytic because of the high basicity of the amine group. Third, although the silanes are effective adhesion promoters on metals as well as on glasses, ${-}Si{-}O{-}Me{-}$ bonds are not hydrolytically stable.

It has been recognized for some time that the silanes form strongly adsorbed polysiloxane films on glass, metal (even gold), and ceramic surfaces (Ref 10). The mechanical and chemical conformation of these films and their thickness, cross-link density, and reactivity have been shown to be highly dependent on the application conditions. The most critical factors appear to be solution concentration (Ref 11), solution pH (Ref 12), and drying time and temperature (Ref 12). Also, the acid-base character of the adherend surface may influence the polysiloxane film structure (Ref 12).

It was suggested in the early 1970s that the adsorbed polysiloxane film is an open network that allows interdiffusion of the adhesive molecules to form an interpene-

Table 2 Tensile lap-shear and flatwise tension strengths

Test condition	Tensile lap-shear		Flatwise tension	
	MPa	ksi	MPa	ksi
At 24 °C (75 °F), dry				
Without primer 15.4		2.23	6.03	0.875
With primer 15.6		2.26	5.87	0.850
At 24 °C (75 °F), after 30 days in saltwater spray at 38 °C (100 °F)				
Without primer 19.5		2.83	3.66	0.530
With primer 22.3		3.23	4.28	0.620
At 24 °C (75 °F), after 60 days in 95–100% relative humidity at 49 °C (120 °F)				
Without primer 16.7		2.42	4.66	0.675
With primer 17.8		2.58	4.28	0.690

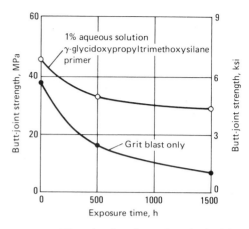

Fig. 2 Effect of a silane "primer" on the durability of an epoxy-mild steel joint

trating network (Ref 10). Subsequent work, much of which has been reviewed by Boerio (Ref 12), has provided circumstantial evidence, and in some cases direct evidence, supporting this hypothesis. It is not difficult to envision the penetration of relatively small molecules, that is, epoxy monomers, into the polysiloxane film. However, the successful use of silane coatings to improve the adhesion of high molecular weight thermoplastics to a variety of substrates taxes this concept of interpenetration. One would not expect these long polymer molecules to diffuse significantly into the polysiloxane layer.

A study by Sung et al. (Ref 13) demonstrated that interpenetration of a polysiloxane film by a thermoplastic is not only possible, but can be related to adhesive bond strength. In this study, the effect of aminopropylsilane coatings on the adhesion of polyethylene to aluminum oxide (sapphire) was investigated. The adhesive joints were prepared by hot pressing the polyethylene to the alumina at 149 °C (300 °F). It was found that the peel strength of the bond depended on the concentration of the silane and the extent of drying prior to bond formation. At the optimum concentration, about 2%, the investigators found a loss in bond strength for drying temperatures greater than 110 °C (230 °F). Using scanning electron microscopy, they found that failure was primarily cohesive in the polyethylene for drying temperatures of 25 °C (77 °F) in vacuum for 1 h. Heating at higher temperatures and longer drying times resulted in failure at the silane-polyethylene boundary. Presumably, the polyethylene interpenetrates into a relatively open polysiloxane network formed at low temperatures, but excessive drying results in a more highly cross-linked network into which the polyethylene cannot penetrate.

Gettings et al. (Ref 14) examined the effect of glycidoxypropyltrimethoxysilane on the bond strength between an epoxy and iron. Without any silane coating, the dry strength was high, and failure occurred through the adhesive, albeit near the adhesive-adherend boundary. When tested after exposure to water, the bond strength was seriously degraded, and failure occurred at the interface. Application of the silane significantly improved the wet bond strength, and failure was through the silane layer. The effect of the silane film was attributed to the resistance of the silane-iron interface to attack by water.

The detailed mechanism by which the silanes effect adhesion and adhesive bond durability is still illusive. There is little doubt that they form polysiloxane films on the adherend surface and that the adhesive may penetrate these films to form interpenetrating polymer networks. However, the properties of these interphase IPNs have not been determined, and it is not obvious why they should strengthen the interphase region or affect the resistance of the adherend to attack by water. There is substantial evidence that the silanes form chemical and hydrogen bonds to the surface. Much of this evidence is based on chemical bonding to metal surfaces (Ref 15). However, considering the hydrolytic instability of siloxane bonds to metals, the importance of such bonds to adhesive durability against moisture is problematic.

The rate of water diffusion to the bond line is thought to be a critical factor in adhesive bond durability, as discussed in the work by Gledhill et al. (Ref 16). It is tempting to speculate that the primary role of the silanes is to develop an interphase morphology that significantly reduces the rate of water diffusion into the bond line.

Other coupling agents have been developed, notably the organo-titanates and the zircoaluminates (Ref 17). Like the silanes, they have an organofunctional group that presumably reacts with the adhesive and a hydrolyzable functionality that is displaced during application to the adherend, allowing the titanium or zirconium ion to form metal oxide bonds to the adherend surface. There do not appear to have been any in-depth investigations of these agents comparable to the studies on silane adhesion promoters.

Schmidt and Bell (Ref 18-20) have developed a series of mercaptoester coupling agents for bonding to steel. The chemical structure of these agents is:

$$SH—CH_2—CO—O—$$
$$[CH—CH_2—(CH_2—CH_2)_n—]_m \qquad \text{(Eq 4)}$$

Both the mercapto functionality (SH—) and the carbonyl group (CO—) are believed to bond to the steel surface; the SH— forms chemical bonds, and the CO— goes through dipole-dipole interaction. Both are considered important to increasing the dry and wet strengths of epoxy adhesives to steel, compared to uncoated steel.

In addition to the work described in this section, particularly Ref 12, the reader should consider the works listed in the Selected References.

Examples

The tables and graphs covered below illustrate the effect of primers and coupling agents on adhesive bond strength and durability.

Primers. In Table 2, the lap-shear and flatwise tension strengths of a commercial adhesive applied to aluminum with and without primer are compared (Ref 21). Published data, particularly under severe field conditions, are difficult to obtain. Table 2 does not fully reflect the effectiveness of primers.

Coupling Agents. Table 3 shows the effect of silane coupling agents on the wet strength of a polyester-glass cloth laminate tested in three-point flexure (Ref 22). The nonreactive ethyl silane (second entry in Table 3) actually degrades the composite dry strength and has little effect on wet strength. The vinyl silane (third entry), which can react with the polyester, increases the dry strength and maintains the wet strength to a level comparable to that of the untreated composite tested dry. Figure 2 shows the effect of an epoxy-terminated silane on the durability of an epoxy-steel adhesive bond (Ref 23).

Combined Effect of Primer and Coupling Agent. In a recent study of the adhesive bonding of steel bridge components, the effect of bond durability using primer and coupling agent, separately and in combination, was determined (Ref 24,

Table 3 Effect of silane coupling agents on the flexural strength of polyester-glass cloth laminate

| Silane | Flexural strength | | | |
| | Dry | | Wet | |
	MPa	ksi	MPa	ksi
None....................................	420	61.0	158	23.0
CH₃CH₂Si(OCH₂CH₃)₃..................	300	43.5	205	29.7
CH₂ CHSi(OCH₂CH₃)₃..................	476	69.0	420	61.0

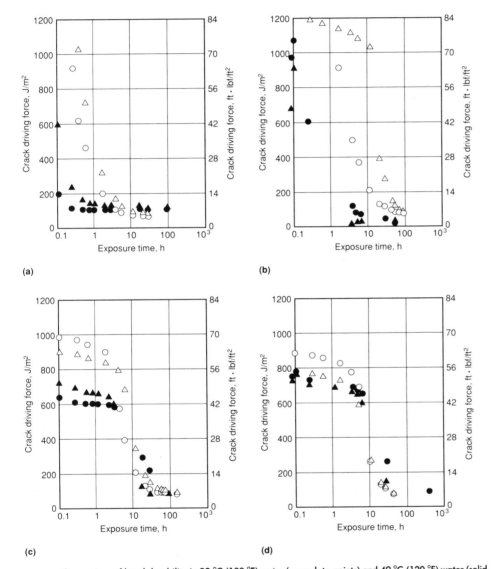

(a)

(b)

(c)

(d)

Fig. 3 Comparison of bond durability in 38 °C (100 °F) water (open data points) and 49 °C (120 °F) water (solid data points). (a) On a surface primed with the primer formulation recommended by the adhesive manufacturer. (b) On a surface precoated with an epoxy-terminated alkoxysilane. (c) On an unprimed surface but with 1% of the silane in the adhesive. (d) On a surface treated with the recommended primer and 1% of the silane in the adhesive. Circles and triangles represent data from different specimens.

25). Figure 3 gives the crack driving forces (applied strain energy release rate, G), measured using a constant displacement wedge specimen based on four different scenarios. The exposure time at which the crack driving force decreases precipitously is a measure of the adhesive bond resistance to hot-wet conditions. Bonds made using the manufacturer-recommended primer showed the least resistance to water (Fig. 3a),

whereas a silane coating on the adherends gave some improvement (Fig. 3b). The use of the silane in the adhesive was most effective and also gave the most consistent results (Fig. 3c and d).

REFERENCES

1. J.A. Bishopp, E.K. Sim, G.E. Thompson, and G.C. Wood, The Adhesively Bonded Aluminum Joint: The Effect of Pretreatment on Durability, *J. Adhes.*, Vol 26, 1988, p 237
2. T.J. Reinhart, Novel Concepts for Priming Metallic Adherends for Structural Adhesive Bonding, in *Adhesion 2*, K.W. Allen, Ed., *Appl. Sci. Publ.*, 1978, p 87
3. R.T. Foister, R.K. Gray, and P.A. Madsen, Structural Adhesive Bonds to Primers Electrodeposited on Steel, *J. Adhes.*, Vol 24, 1989, p 17
4. A.J. Kinloch and C.M. Taig, The Adhesive Bonding of Thermoplastic Composites, *J. Adhes.*, Vol 21, 1987, p 291
5. E.M. Petrie, Adhesively Bonding Plastics: Meeting an Industry Challenge, *Adhes. Age*, Vol 32, 1989, p 6
6. R.F. Wegman, *Surface Preparation Techniques for Adhesive Bonding*, Noyes Publications, 1989, p 68
7. A.R. Siebert, L.L. Tolle, and R.S. Drake, CTBN-Modified Epoxies Work in Poor Bonding Conditions, *Adhes. Age*, Vol 29, 1986, p 19
8. D.N. Shah, Rubber Modified Epoxy Adhesive Compositions, U.S. Patent 4,803,232, Feb 1989
9. Cost-Effective SMC Bonding Through Supplier Teamwork, *Adhes. Age*, Vol 31, 1988, p 20
10. W.D. Bascom, Structure of Silane Adhesion Promoter Films on Glass and Metal Surfaces, *Macromolecules*, Vol 5, 1972, p 792
11. R.L. Patrick, Chapter 4 in *Treatise on Adhesion and Adhesives*, Vol 1, R.L. Patrick, Ed., Marcel Dekker, 1973
12. F.J. Boerio, Coupling Agents as Adhesion Promoters in Adhesive Bonding, Chapter 7 in *Treatise on Adhesion and Adhesives*, Vol 6, R.L. Patrick, Ed., Marcel Dekker, 1989
13. N.H. Sung, A. Kaul, I. Chin, and C.S.P. Sung, *Polym. Eng. Sci.*, Vol 22, 1982, p 637
14. M. Gettings, F.S. Baker, and A.J. Kinloch, *J. Appl. Polym. Sci.*, Vol 21, 1977, p 2375
15. M. Gettings and A.J. Kinloch, Surface Characterization and Adhesive Bonding of Stainless Steel, *Surf. Interface Anal.*, Vol 1, 1979, p 189
16. R.A. Gledhill, A.J. Kinloch, and S.J. Shaw, A Model for Predicting Joint Durability, *J. Adhes.*, Vol 11, 1980, p 3
17. E. Galli, Update: Surface Modification, *Plast. Compd.*, March/April, 1986, p 40
18. R.G. Schmidt, J.P. Bell, and A. Garton, Chemical Interactions Between Mercaptoester Coupling Agents and Steel, *J. Adhes.*, Vol 27, 1989, p 127
19. R.G. Schmidt and J.P. Bell, Investigation of Steel/Epoxy Adhesion Durability Using Polymeric Coupling Agents.

II. Factors Effecting Adhesion Durability, *J. Adhes.*, Vol 25, 1988, p 85

20. R.G. Schmidt and J.P. Bell, Investigation of Steel/Epoxy Adhesion Durability Using Polymeric Coupling Agents, *J. Adhes.*, Vol 27, 1989, p 135

21. Product Literature, Polymer Products Division, American Cyanamid Company

22. S. Sterman and J.G. Marsden, Silane Coupling Agents, *Ind. Eng. Chem.*, Vol 58, 1966, p 33

23. A.J. Kinloch, Predicting and Increasing the Durability of Structural Adhesive Joints, in *Adhesion 3*, K.W. Allen, Ed., 1978, p 1

24. E.J. Ripling, P.B. Crosley, W.D. Bascom, and E. Munley, The Use of Adhesives to Replace Welded Connections in Bridges, in *Proceedings of the 20th SAMPE International Technical Conference*, Session C2, Society for the Advancement of Material and Process Engineering, Sept 1988

25. E.J. Ripling, P.B. Crosley, and W.D. Bascom, "Use of Adhesives to Replace Welded Connections in Bridges," Final Report, Contract DTFH61-87-C-00029, U.S. Federal Highway Administration, 1989

SELECTED REFERENCES

● W.D. Bascom, Fiber Sizing, in *Composites*, Vol 1, *Engineered Materials Handbook*, ASM INTERNATIONAL, 1987, p 122-124

● E.P. Plueddemann, *Silane Coupling Agents*, Plenum Press, 1982

● *Silanes, Surfaces, and Interfaces*, D.E. Leyden, Ed., Gordon and Breach Science Publishers, 1986

Surface Preparation of Metals

H.M. Clearfield, IBM Corporation, Thomas J. Watson Research Center
D.K. McNamara and Guy D. Davis, Martin Marietta Laboratories

SURFACE PREPARATION is essential for the successful implementation of adhesive bonding technology. Both the initial bond strength and the subsequent bond durability are critically dependent on the interaction between the adhesive (and/or primer) and a pretreated adherend surface. For metals, surface preparation involves both the removal of weak boundary layers or layers that are chemically incompatible with the adhesive and the formation of stable, adherent layers that are mechanically and chemically compatible with the adhesive. This article discusses the surface preparation of metal adherends, concentrating on those used most often for structural applications, that is, aluminum, titanium, copper, and steel. Because of the intense interest in bonding these metals, there has been considerable research in the correlation of surface treatment with bond performance.

Only the most common surface preparations are described for each of these metals. This is not intended to be a comprehensive review. References 1 and 2 are examples of more lengthy treatments of surface preparations. This article correlates microscopic properties, that is, surface morphology and chemistry, with bond durability, which is the most difficult property to maximize. That is, in addition to providing process procedures, surface characterization and bond durability results are also presented in order to provide enough information to allow the engineer to choose the appropriate surface preparation for a particular application.

Background. A general conclusion of many adhesive bonding studies conducted within the last 15 years is that bond durability is significantly enhanced when the adherend surface contains a significant degree of microscopic roughness (Ref 3, 4). Here the term microscopic implies features on the order of tens of nanometers. This definition is arbitrary, but has proved suitable in practice. For purposes of distinction, macroscopic roughness is defined as features on the order of micrometers. The surface treatments producing the greatest durability exhibit microroughness, and durability is increased even further when the

adherend surface is chemically compatible with the adhesive.

The approach taken to surface preparation depends on the type of metal used. Although it may be possible to prepare atomically clean aluminum and titanium in the laboratory, it is not practical to do so on large structures. Thus, an oxide is usually formed intentionally. This is accomplished by etching or anodization, most often in an acidic medium and less often in an alkaline medium. The resulting oxides are porous, tenaciously adherent to the metal, and, in some instances, chemically tailored for the adhesive. While reactive metals are generally prepared this way, the approach does not work for steel. Because iron oxide forms rather slowly and is loosely adherent, the most common method of preparing steel surfaces is by grit blasting, which results only in macroroughness. However, as will be shown, it is possible to deposit a microrough, adherent coating on various steel surfaces. The surface preparation of copper adherends is also described.

Analysis. The advent of surface analysis techniques has made possible the correlation of surface morphology and chemistry with bond durability. High-resolution scanning electron microscopy (SEM) and x-ray photoelectron spectroscopy (XPS) have been the most useful and have been used in a wide variety of applications, encompassing fundamental studies, failure analysis (production and experimental), and raw materials qualification. All are discussed in more detail in the article "Surface Analysis Techniques and Applications" in this Volume, which should make the utility of surface analysis clear.

Aluminum Adherends

Aluminum surfaces are prepared for adhesive bonding for aerospace applications either by etching or anodization in acid solutions. For less stringent strength and durability requirements, mechanical abrasion is adequate (Ref 5). Commonly used preparations result in microrough adherend morphologies, which have been shown to yield the best overall bond durability. Four of these surface preparations, the Forest

Products Laboratory (FPL) etching procedure, P2 etching procedure, phosphoric acid anodization (PAA), and chromic acid anodization (CAA), are described below.

The FPL and other chromic-sulfuric acid etching procedures are the oldest surface pretreatments for aluminum adherends (Ref 6), with the exception of simple degreasing or mechanical abrasion. Variations of chromate-containing solutions, for example, $NaHSO_4$ instead of H_2SO_4, are used for low-stress applications. In addition to being used as a complete adherend pretreatment, FPL is also frequently used as the first step in other pretreatments, such as PAA and CAA.

The P2 etch, a recently developed process, avoids the use of toxic chromates, but still provides the complex oxide surface morphology that is crucial to a mechanically interlocked interface and strong bonding. This will be discussed further in the section "Oxide Morphology and Chemistry" under "Aluminum Adherends" in this article. Ferric sulfate is used as an oxidizer in place of sodium dichromate (Ref 7). The P2 solution produces an oxide morphology very similar to that seen on chromic-sulfuric acid etch surfaces over a broad range of time-temperature solution concentration conditions (Ref 8). Mechanical testing confirms that P2-prepared surfaces are equivalent to FPL-prepared specimens. Thus, the P2 solution appears to have great promise as a less hazardous replacement for the chromic-sulfuric acid etches.

Phosphoric acid anodization was developed by The Boeing Company in the late 1960s and early 1970s to improve the performance of bonded primary structures (Ref 9). Bonds formed with PAA-treated adherends exhibit durability during exposure to humid environments that is superior to those formed with FPL-treated adherends, especially when epoxy adhesives are used. In addition, PAA bonds are less sensitive than FPL bonds to processing variables such as rinse-water chemistry and time before rinsing. As a result, the PAA procedure has become the treatment of choice in the United States for critical applications.

Chromic acid anodization (Ref 10) is widely used to improve the corrosion resis-

Table 1 Optimized Forest Products Laboratory etching procedure

Step(a)	Agent
1. Solvent degrease	1,1,1-trichloroethane (typically vapor)
2. Alkaline deoxidize.....	Turco 4215S
3. Rinse, 5 min, RT	Water
4. Etch, 65 °C (150 °F) ...	FPL solution 60 g/L (8 oz/gal) $Na_2Cr_2O_7 \cdot 2H_2O$, 173 g/L (23 oz/gal) 96% H_2SO_4, 1.9 g/L (0.25 oz/gal) 2024 Al, bal water
5. Rinse	Deionized water
6. Oven dry	· · ·

(a) RT, room temperature

Table 2 Some of the most commonly used chromic-sulfuric acid etching procedures

Process	Etch bath composition, wt% $Na_2Cr_2O_7 \cdot 2H_2O$ (a)	H_2SO_4	H_2O	Temperature °C	°F	Time, min
FPL etch 2.5		24.3	73.2	68	155	15–30
UK defense standard 03-2/1 (1970) 6.4		23.3	70.3	60–65	140–150	30
DIN 53 281 .... 7.5		27.5	65.0	60	140	30
Alcoa 3........................ 3.2		16.5	80.3	82	180	5
FPL optimized 5.0		26.7	68.3	65	150	10
FPL-RT 6.4		23.4	70.2	22	72	240
P2 .... 15.0 Fe SO_4		37.0	48.0	60–70	140–160	8–15

(a) Except as noted. Source: Ref 13

tance of aluminum surfaces, for example, for window frames and other architectural applications. Similarly, it was thought that the use of a good protective coating on the aluminum would protect the metal interface and thereby increase the bond durability of the joint. Although CAA is not as popular as FPL and PAA treatments in the United States, it has been extensively developed and is widely used for aerospace applications in Europe (Ref 11, 12).

Processing

The optimized FPL etching procedure is shown in detail in Table 1. Other chromic-sulfuric acid etches and the P2 etch are summarized in Table 2 (Ref 13).

The initial degreasing removes gross organic contamination from the surface, whereas the alkaline cleaning removes some of the oxide coating formed during the aluminum heat treatment and rolling. The remaining oxide is dissolved in the etching solution (Ref 14). The addition of the aluminum alloy seed optimizes the FPL solution by releasing copper into the solution, resulting in better properties in bonded systems that use clad adherends; that is, clad adherends etched in a copper-free FPL solution do not develop the characteristic FPL morphology (Ref 15, 16).

Although there are some differences in detail among the anodization processes commonly used for aluminum adherend preparation, these processes share many common features. As the processing steps in Table 3 show, the anodization step for PAA is performed at a constant voltage, whereas for CAA, voltage is applied to the adherend in a gradual or stepwise fashion until the maximum voltage is reached, and then the voltage is held for roughly 30 min. Both anodization processes are more costly and time-consuming than the FPL process but are frequently justified by the increase in bond durability.

Oxide Morphology and Chemistry

The FPL oxide morphology is shown in the high-resolution SEM stereo micrograph and isometric drawing in Fig. 1 (Ref 19). The oxide consists of a network of shallow pores and protrusions or whiskers on top of a thin barrier layer. This microroughness provides a means for mechanical interlocking between the adhesive and the oxide surface that is critical for the durability of epoxy-bonded structures (Ref 3). Buffered chromate-containing solutions are somewhat less aggressive as etchants and generally result in shallower pores without the whiskers.

Chemically, the FPL film is amorphous Al_2O_3, with varying quantities of adsorbed water (Ref 20) that can be removed by heating, vacuum exposure, or adsorption with certain organic hydration inhibitors. The heat treatment of magnesium-containing alloys, following the growth of an FPL oxide, can result in an outdiffusion of magnesium and the formation of MgO, but does not change the surface morphology. Small amounts of magnesium in the FPL oxide may be beneficial to bond durability (Ref 21), although studies have suggested that

Table 3 Processing steps for PAA and CAA treatments

Step	PAA	CAA MIL-SPEC (Ref 10)	Bell Helicopter (Ref 17)	Fokker-DIN (Ref 18)
Pretreatment.....	FPL	FPL	FPL	Chromic-sulfuric acid etch
Anodization	100 g/L (13 oz/gal) H_3PO_4, 20–25 °C (68–77 °F), 10 V, 25 min	40–50 g/L (5.3–6.6 oz/gal) CrO_3, 32–38 °C (90–100 °F), 8 V/min to 40 V, hold 55 min	60–100 g/L (8–13 oz/gal) CrO_3, 33–37 °C (91–99 °F), 3–5 V/min to 40 V, hold 30–35 min	50 g/L (6.6 oz/gal) CrO_3, 38–42 °C (100–108 °F), 5 V/min to 40 V, hold 20 min, raise to 50 V, hold for 10 min
Rinse	Deionized water, 20 °C (68 °F), air dry	Rinse, 1–5 min, 20 °C (68 °F) oven dry, $T < 65$ °C (150 °F)	Rinse, 20–25 °C (68–77 °F) seal, 75–125 ppm CrO_3, 82–85 °C (180–185 °F), 7–9 min air dry	Rinse, 5 min, 20–25 °C (68–77 °F) air dry, $T < 60$ °C (140 °F)
Process controls A/m² (A/ft²), minimum ...· · ·		10 (0.9)	10 (0.9)	· · ·
Weight, g/m² (oz/ft²) ......· · ·		2.2 (0.007)	3.2 (0.10)	Coating weight
pH ....· · ·		0.4–0.6	<0.8; free acid, 30 g/L (4 oz/gal)	Free acid, 30–50 g/L (4–6.6 oz/gal)

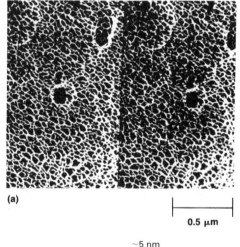

(a)

0.5 μm

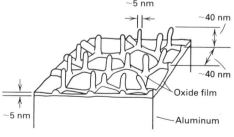

(b)

Fig. 1 FPL 2024 aluminum surface. (a) High-resolution stereo electron micrograph. (b) Isometric drawing

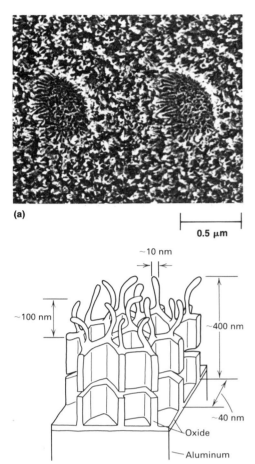

(a)

0.5 μm

(b)

Fig. 2 PAA 2024 aluminum surface. (a) Stereo micrograph. (b) Isometric drawing

magnesium in other oxides may have a deleterious effect (Ref 22).

The PAA oxide morphology has an even greater degree of microroughness than the FPL surface, as shown in Fig. 2 (Ref 19). The oxide consists of a well-developed network of pores on top of a barrier layer. Whiskers protrude from the pores away from the high-strength alloy adherend. The total oxide thickness is approximately 400 nm (16 μin.). (Purer aluminum alloys have taller cells, but no whiskers.) This degree of microroughness can provide more mechanical interlocking than FPL-treated adherends for improved bond strength and durability (Ref 3, 9, 23), but this increase occurs in interlocking only if the polymeric primer or adhesive penetrates the pores of the oxide. Cross-sectional SEM views of PAA coated with typical epoxy primers show this type of penetration (Ref 3).

Chemically, the PAA film is amorphous Al_2O_3 (Ref 3, 24-27), with the equivalent of a monolayer of phosphate incorporated onto the surface (Ref 26, 28-33). Alloying constituents of the adherend are not generally found in the as-anodized oxide (Ref 26). Depending on storage conditions, some water can adsorb on the surface, but is readily

removed by heating or storing in a dehydrating environment (for example, a desiccator or a vacuum).

The inherent CAA oxide (that is, without prior FPL etching or other pretreatment) is a dense structure of tall columns (Fig. 3) (Ref 19). The outer surface can be quite smooth, with fine pores running through most of the oxide layer at the junction of the column walls. The total oxide thickness is 1 to 2 μm (0.04 to 0.08 mil), which is much greater than that provided by other surface treatments. The barrier layer at the bottom of the columns is also relatively thick, because of the high anodization voltages used. For example, transmission electron microscope (TEM) photographs of microtomed cross sections of similar oxides show barrier layers 40 nm (1.6 μin.) thick at the bottom of the pores (Ref 34, 35). The CAA oxide is Al_2O_3; the upper portion is amorphous, and there are indications that the lower part is crystalline (Ref 12, 26, 27, 33). In contrast to the phosphate incorporated in the PAA oxide, little chromium is incorporated into the CAA oxide (Ref 26, 36).

The denser oxide formed by CAA treatments offers less interlocking potential than do PAA oxides. Nonetheless, primers have been shown to penetrate the pores (Ref 12). In addition, because the morphology of the outer surface of the oxide is strongly dependent on the process steps used just prior to the CAA treatment, the type of pretreatment can be chosen to enhance the bonding. Figure 4 shows the surface morphologies of three surfaces given different pretreatments prior to chromic-acid anodization. The sample that was FPL-etched and then anodized (Fig. 4a) has a typical FPL cell-and-whisker morphology. The surface that was PAA treated before the CAA anodization (Fig. 4b) has a deeper cell structure, typical of PAA. Finally, the surface that had a smooth tartaric acid-anodized (TAA) oxide before CAA treatment retained a smooth surface (Fig. 4c), although there is evidence of structure underneath the defects in the surface film. These results suggest that the growth of the oxide is due to the diffusion of oxygen through the oxide to the metal interface.

The morphology of CAA oxides has also been altered by varying the processing conditions, including using higher-temperature anodizing solutions and postanodizing phosphoric acid etches (Ref 31, 37). Csanady *et al.* (Ref 35) reported that the aluminum-clad surfaces that performed poorly in durability testing due to a smooth oxide morphology could be improved by treatment modifications that roughen the surface. In addition, Brockmann and co-workers (Ref 12, 38) have indicated that adherends treated with the dual-voltage CAA process form more durable bonds than those treated with the single-voltage process. One advantage of the CAA process is

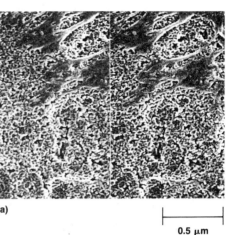

(a)

0.5 μm

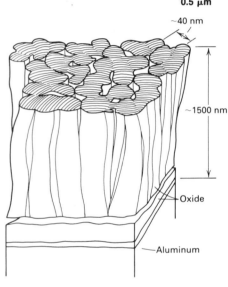

(b)

Fig. 3 CAA 2024 aluminum surface. (a) Stereo micrograph. (b) Isometric drawing

that the oxide is less friable, that is, less susceptible to mechanical damage, than the PAA oxide (Ref 12, 39).

Durability

Nearly all failures in aluminum adhesive joints in aircraft have been initiated by moisture (Ref 12). Consequently, the projected long-term durability of adhesively bonded systems is most often determined by accelerated testing in a saturated aqueous environment. In one common test, a double-cantilever beam geometry (ASTM D 3672, the wedge-crack propagation test) is used, as shown in Fig. 5. A crack is initiated at one end of the test specimen, and its propagation in the wet environment is recorded as a function of time. Failure analysis is then performed to provide insight into the mechanism of crack propagation. Because water is present at the crack tip while it is under stress, the wedge test is more severe than a lap-shear or stress-durability test, in which moisture must diffuse into the bond line from the edges of the specimen.

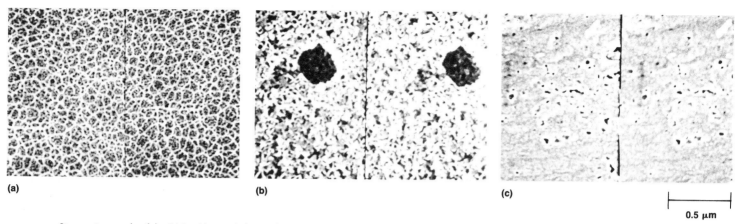

Fig. 4 Stereo micrographs of the CAA oxide morphology with variations in pretreatment prior to anodization. The morphology reflects the pretreatment that was used rather than any inherent CAA morphology. (a) FPL. (b) PAA. (c) TAA

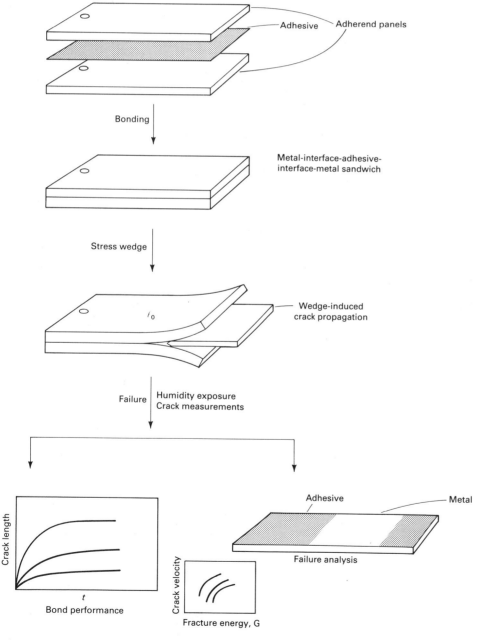

Fig. 5 Diagram of the wedge-crack propagation test for bond durability

Another benefit of the wedge-crack propagation test is its amenability to a fracture mechanics analysis. Thus, it provides the design engineer with an environment-sensitive ultimate strain-energy release rate that can be used when designing a bonded component or system.

As mentioned earlier, studies of the durability of epoxy-aluminum bonded systems have shown that bond performance is a function of the mechanical interaction between the oxide and the adhesive and of the resistance of the oxide surface to hydration (Ref 3, 40-42). For many microrough hydration-stable aluminum adherend surfaces, crack propagation during the wedge test is relatively slow and occurs within the adhesive; in these cases, the mechanical interlocking holds the bond intact even if the chemical bonds (between adhesive and oxide) are broken. Others have suggested that the microroughness allows the strain energy at the crack tip to be distributed over a larger area, resulting in a more durable bond (Ref 12). An example of the gain in durability realized from microroughness is shown in Fig. 6, where wedge test results are shown for FPL-, PAA-, and $CrO_3/NaHSO_4$-etched adherends.

When the test specimens are subjected to moisture, the oxide surface adsorbs water from the adhesive and subsequently hydrates. This hydration at the monolayer level is believed to be responsible for the initial bond failure, that is, the disruption of the adhesive-oxide bond. The now-exposed oxide continues to hydrate, forming the oxyhydroxide boehmite, or $AlOOH$. This phase is weakly bound to the aluminum substrate, making bond "healing" or reformation impossible. More importantly, the volume change accompanying this transformation leads to further stress at the bond line and continued crack propagation. Continued exposure causes the boehmite to hydrate further into bayerite, that is, $Al(OH)_3$.

Because crack propagation occurs as a consequence of hydration, bond durability

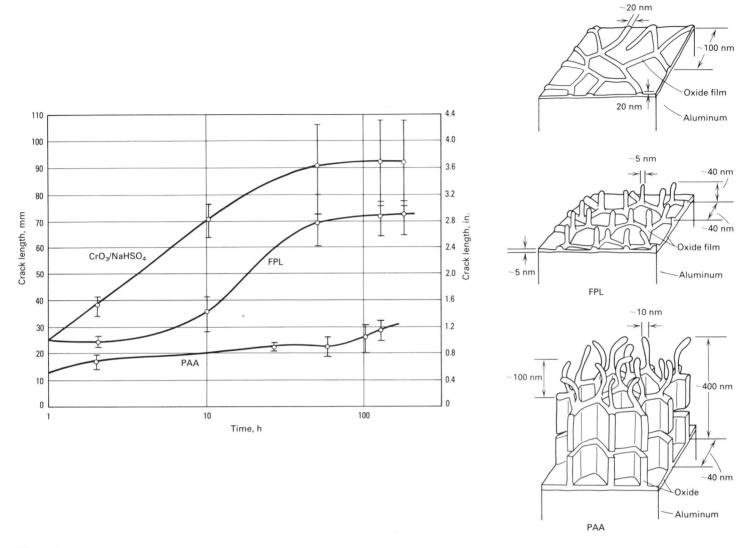

Fig. 6 Wedge-crack propagation test results (crack length versus time) for 2024 aluminum adherends treated with CrO₃/NaHSO₄, FPL, and PAA processes

can be improved by decreasing the rate of water diffusion through the adhesive or along the bond line and/or increasing the hydration resistance of the oxide (Ref 3, 12, 20, 28, 33, 43-45). Both concepts are demonstrated in Fig. 7 using typical crack-propagation data (Ref 20, 46-49) for FPL

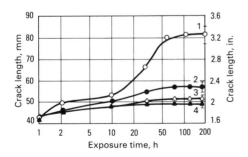

Fig. 7 Wedge-crack propagation test results (crack length versus time) for FPL-treated aluminum surfaces: (1) FPL with FM 123 (moisture wicking); (2) FPL with FM 300 (moisture resistant); (3) FPL with an organic hydration inhibitor and FM 123; and (4) FPL with BR 127 primer and FM 300. Sources: Ref 15, 18, 23, 48

adherends with (1) a water-wicking adhesive (a worst-case scenario), (2) a water-resistant adhesive, (3) an organic hydration inhibitor with a water-wicking adhesive, and (4) a corrosion-inhibiting epoxy primer with a water-resistant adhesive.

For water-wicking and water-resistant adhesives (as-prepared FPL adherends), crack propagation occurs at the oxide-metal interface because of hydration. In the third case, where an organic hydration inhibitor is applied to the adherend prior to bonding, any absorbed water is removed, leaving a single molecular layer of inhibitor on the oxide. This inhibitor monolayer dramatically improves the hydration resistance of the oxide so that, when used with a chemically compatible adhesive, it increases the durability of the bonded structure to such a degree that it approaches that of a PAA-treated structure. In some cases, hydration is sufficiently slowed that it is no longer the determining factor in crack propagation. Rather, the adhesive slowly disengages from the oxide because of moisture-induced

plasticization in the polymer (Ref 20). In the fourth case, in which a chromate-containing epoxy primer is applied to the FPL adherend prior to bonding, the primer slows the transport of water to the oxide and sufficiently reduces the hydration rate of the oxide so that crack propagation occurs in the adhesive. This is a common practice in bonded aerospace structures.

In cases in which surface treatment produces a smooth oxide (intentionally or unintentionally), bond performance is controlled by chemical bonds across the oxide-epoxy interface. This situation can arise, for example, when an FPL adherend is rinsed in fluorine-contaminated water or is exposed to fluorine vapor (Ref 50). The oxide-epoxy bonds are relatively weak and are readily disrupted by moisture (Ref 51, 52). As a result, bond failure is rapid upon exposure to humid conditions, and the crack propagates along the adhesive-oxide interface. In cases in which a smooth oxide is formed intentionally, coupling agents such as silanes can be used to improve durability.

Bond durability has also been improved using inorganic primers (Ref 53). Such agents have been found to improve the durability of bonds formed with smooth adherends, but no more than a microrough morphology does.

Bonds made with PAA-treated adherends exhibit greater inherent durability than those made with FPL adherends (Ref 23, 46) because of both the better hydration resistance of PAA surfaces (Ref 28) and the more developed microroughness and subsequent interlocking between the oxide and the adhesive (Ref 20). Although the hydration of PAA surfaces is slower than that of FPL surfaces, the former occurs by a three-step process: reversible adsorption of water, slow dissolution of the surface phosphate followed by rapid hydration of the freshly exposed Al_2O_3 to boehmite (AlOOH), and further hydration of the boehmite to bayerite. As with FPL adherends, it is the second step (hydration to boehmite) that causes crack propagation. The initial stability of the PAA surface results from the very thin layer of phosphate ions that is incorporated in the oxide during anodization. FPL oxides with an adsorbed layer of phosphonate-containing hydration inhibitor derive their hydration resistance from a similar mechanism. In all cases, the bonding degradation processes can be retarded by the use of moisture-resistant adhesives or primers.

CAA oxides protect the metal surface from hydration because of their inherent thickness; the important factor for bond durability is the stability of the outer oxide structure when water diffuses through the bond line to the polymer-oxide interface. Because hydration rates are controlled by the thickness of the barrier layer (which is directly proportional to the anodizing voltage), typical CAA processes yield oxides that are more resistant to hydration than are FPL surfaces. Direct comparisons of the normal surface treatments based on durability tests show that CAA adherends can perform as well as PAA adherends (Ref 11, 12, 54, 55).

Production Considerations

Although the procedures already described are generally straightforward, there are many ways in which improper handling of the surfaces, both during and after processing, can destroy their bonding potential. The extremely fine-scale FPL oxide morphology, for example, can easily be destroyed without leaving a trace visible to any but the highest-resolution electron microscopes. Because anodized oxide layers are much thicker, they are detectable by easily applied optical means (observation with crossed polarizers; see for example, Ref 56). Although optical indications do not guarantee good bond performance, they can be used to ensure the correct performance

of chemical processing steps. This inspectability often gives anodizing processes an advantage in practice. It is true, however, that the bonding potential of the anodized oxides often lies in the detailed nature of the outermost oxide surface, and not simply in the presence of an oxide. Careful process and solution controls are essential for all surface treatments if reliable bonds are to be made.

Aluminum oxide surfaces are particularly sensitive to halogen ions. Low concentrations of the F^- and Cl^- in the etch/anodizing solutions markedly reduce the height and roughness of the oxide morphology produced, leading to poor bond performance, especially in adverse environments (Ref 57). In addition, F^--contaminated rinse water (fluoridated tap water) can deposit a sufficiently concentrated ion residue on the fresh aluminum oxide surface so as to react with ambient moisture and form an acid that attacks the oxide structures (Ref 19). The use of distilled water and solution monitoring is necessary for reliable adherend surface preparation.

Because of the nature of the processing, some solution drag-out is inevitable during rinsing operations; the pH of the rinse bath in the vicinity of the aluminum surfaces can frequently be quite low. If low-pH rinse water is allowed to dry on the oxide, the surface may become sufficiently acidic to react with the amine curing agents that typically are in epoxy primers, leading to an undercured epoxy at the oxide interface (Ref 57, 58). Similar surfaces bonded with a partially cured (B-stage) film adhesive do not suffer bonding problems, presumably because the curing agent in these films is not sufficiently mobile to diffuse to the interface and be neutralized. Because the acidic residue could not be detected on the surface either by electron microscopy or XPS, a two-step rinse after etching/anodizing is recommended.

The greater pore depth of anodizing treatments generally improves their resistance to degradation by these chemical effects, but all pretreatments are subject to handling contamination. The extremely fine pores on all well-prepared bonding surfaces are perfect traps for oils found on virtually all surfaces, such as fingers, nylon gloves, kraft paper, silicone rubber, and Teflon (Ref 59). Contaminant layers that fill the pores and cover the surface will prevent the crucial mechanical interlocking from forming and lead to low-strength bonds. Surface contact can also damage the oxide asperities, but the relative roughness of the surfaces ensures that such damage is limited to relatively small areas. The most significant problem is contamination, and the only prevention is to avoid all contact with areas that are to be bonded. Once the surfaces have been coated with primer or adhesive, they are much less susceptible to damage.

Usually, badly contaminated primer layers can be successfully cleaned by solvent wiping. However, contact of a bare oxide surface with solvent is as detrimental as other forms of contamination.

One particularly detrimental contaminant is any form of silicone, which can disrupt normal bond formation even at low levels. It has generally been found that 2 to 3% silicon on a surface in the form of silicone, as detected by XPS, is sufficient to cause disruption. The 2 to 3% silicon corresponds to a coverage of 50 to 100% of a surface with a single molecule layer of silicone molecules. Silicone is very difficult to remove from any surface, for example, silicone-contaminated primer surfaces generally cannot be cleaned by solvent wiping. Thus, all bonding operations should be carried out in a silicone-free environment. Often this is difficult because silicones or their analogues are present in almost all mold-release agents and in many machining fluids. Experience has indicated that even mold-release agents claiming to be nonsilicone materials usually have a siliconelike moiety that is equally effective at disrupting bonds. Care must be taken to ensure that surfaces to be bonded never contact any surface treated with a mold-release agent.

Titanium Adherends

Titanium alloys are particularly attractive for aerospace structures because of their high strength-to-weight ratio. In addition, they retain their mechanical properties at high temperatures and can therefore be used as components in structures where operating temperatures up to 370 °C (700 °F) (and possibly beyond) are expected. One such alloy, Ti-6Al-4V, is used widely throughout the aerospace industry in bonded structures. The desire to bond to this alloy adhesively and use the resulting structure at elevated temperatures has played a key role in the development of high-temperature adhesives (those curing at temperatures up to 400 °C, or 750 °F) (Ref 60, 61). In this section, surface preparations for Ti-6Al-4V and the durability of the subsequent adhesive bonds are described.

According to an informal survey, grit blasting is perhaps the most common method for the surface preparation of Ti-6Al-4V. However, it will be shown that bond durability is poor for grit-blasted adherends, and thus it is not recommended as a surface preparation, especially for high-stress applications.

As with aluminum, durable surface preparations for Ti-6Al-4V can be achieved by forming oxides in anodizing and/or etching solutions. Typically, anodization results in the best bond durability for titanium alloys, primarily because of the microrough surface morphology that results from the treatment. The most commonly used process (Ref 3, 60, 62-66) is CAA (but in a different form

Table 4 Anodization procedures for Ti-6Al-4V

	CAA (Ref 65)	SHA (Ref 68)	NaTESi (Ref 71)
1. Degrease	Methyl ethyl ketone	Methyl ethyl ketone	Alkaline
2. Pickling			
15 vol% of 70% HNO_3, min	10	10	. . .
3 vol% of 49% HF, at RT, s	30	30	. . .
3. Rinse, deionized water, 25 °C (77 °F), min	1–5	1–5	1–5
4. Anodization	50 g/L (6.6 oz/gal) CrO_3, 1 g/L (0.13 oz/gal) NH_4HF_2, 10 V, 20 min, 20–25 °C (68–77 °F)	5M NaOH, 10 V, 20–30 min, 20–30 °C (68–86 °F)	7.50 mol/L NaOH, 0.33 mol/L Na-tartrate, 0.10 mol/L ethylene diamine tetraacetic acid, 10 V, 15 min, 30 °C (86 °F)
5. Rinse, deionized water, min	5–20	5–20	5–20
6. Air dry, °C (°F)	25–60 (77–140)	25–60 (77–140)	25–60 (77–140)

Table 5 Processing steps for the AP surface treatment

Step	Comment
1. Solvent degrease	Wipe with methyl ethyl ketone
2. Alkaline degrease/deoxidation	. . .
3. Rinse	Deionized water, 5–20 min, 20–25 °C (68–77 °F)
4. Etching	0.5M NaOH + 0.2 M H_2O_2, 20 min, 65 °C (150 °F)
5. Rinse	Deionized water, 5–20 min, 20–25 °C (68–77 °F)
6. Air dry	. . .

Source: Ref 73

Table 6 Processing steps for the plasma-sprayed surface treatment

Step	Comment
1. Solvent degrease	Methyl ethyl ketone, wipe or vapor
2. Grit blast	Al_2O_3, 120 mesh
3. Substrate preheating	Plasma flame, one pass
4. Plasma spray, Ti-6Al-4V	
Power	500 A, 60 arc V
Gas pressure, kPa (psi)	
Primary (Ar)	690 (100)
Secondary (H_2)	345 (50)
Gas flow, m^3/min (ft^3/min)	
Primary	2.3 (80)
Secondary	0.42 (15)
Spray distance, mm (in.)	100–150 (4–6)
Coating thickness, mm (in.)	0.05 (0.002)
5. Air cool	. . .

than that used for aluminum adherends) (Ref 67). Recently, a treatment using sodium hydroxide anodization (SHA) has gained interest because it yields bond durability in laboratory studies comparable to (Ref 68) or exceeding (Ref 69, 70) that of CAA. More recently, the NaTESi process, a variation of the SHA process in which a tartrate titanium-complexing agent is used, has been investigated (Ref 71). The solution is also effective as an etchant. An alkaline peroxide (AP) solution used as an etching treatment also results in a microrough surface and has been shown to provide good bond durability (Ref 72-74). However, its use has declined recently because of the instability of one of its components, H_2O_2, at the bath temperature of 65 °C (150 °F). The AP solution has been used in anodization, but with no gain in durability over that achieved with AP etching (Ref 73).

In early studies, the durability of Ti-6Al-4V adherends was determined as a function of surface preparation for ten adherend preparations and several adhesives (Ref 62, 74, 75). The section "Durability" under "Titanium Adherends" in this article provides additional details. The general result was the same as that found for aluminum-epoxy bonded systems: Surface preparations that produce oxides with no roughness (macro- or micro-) yield the poorest bond durability (class I adherend); those that produce a large degree of macroroughness, with little or no microroughness, fall into an intermediate class of moderate-to-good durability (class II adherend); and those preparations that produce oxides with significant microroughness lead to the best durability (class III). This improvement in performance with microroughness is due in part to capillary forces that cause the polymer to displace air and completely fill the pores. The morphology and bond durability of class III adherends are described below.

Surface preparations for Ti-6Al-4V that result in oxides have a limited range of applicability because the oxide is dissolved in the alloy at high temperatures (for example, 200 °C, or 390 °F) (Ref 76). Thus, for high-temperature applications, a surface

treatment that is a microrough Ti-6Al-4V coating deposited by plasma spraying was developed. In plasma spraying, an electric arc is struck between a cathode and an anode (nozzle), both of which are water cooled. A working gas, typically a mixture of argon and hydrogen (or nitrogen and hydrogen), is forced through the arc, creating a plasma. A powder of the material to be deposited (in this case, Ti-6Al-4V) is injected into the plasma either within or downstream from the nozzle, and the plasma softens or melts the powder in flight. The molten particles, traveling at supersonic velocities, strike the substrate and "splat cool." The result is a metallic coating that is both macro- and microrough.

Processing

The steps used to provide an anodized surface on titanium adherends are similar to those used for aluminum adherends. Initially, the adherend must be degreased to remove organic contaminants. Degreasing is followed by an acid etch to remove the oxide scale. The anodization (if used) is done at constant voltage, and the adherend is rinsed and dried. (See Table 4 for details.) Grit blasting is often included in the process, although studies suggest that it provides no gain in durability. The steps for producing the AP surface shown in Table 5 are similar. The processing steps for plasma spraying are somewhat different because the process involves a physical coating deposition rather than a chemical one. The steps are summarized in Table 6 (Ref 76).

Oxide Morphology and Chemistry

The oxide formed by the CAA process is shown in Fig. 8(a) and (b). At low magnifications, there is very little roughness. However, a two-level morphology is evident. This is typical of CAA-treated Ti-6Al-4V and is due to the two-phase nature of the alloy, that is, large grains of the aluminum-rich α phase with an intergranular, vanadium-rich β phase. At higher magnifications, a honeycomblike structure consisting of amorphous TiO_2 is evident (Ref 60, 63-67,

74, 76). The cell diameters are on the order of 30 to 40 nm (1.2 to 1.6 μin.) with wall thicknesses of about 5 to 10 nm (0.2 to 0.4 μin.). The thickness of the oxide depends on the concentration of the solution and the anodization voltage used. Typically, a 5% CrO_3 solution (10 V, 20 min at room temperature) is used, as outlined in Table 4. This results in an oxide that is 120 to 130 nm (4.8 to 5.2 μin.) thick. Cross-sectional transmission electron micrographs reveal that the honeycomb structure (~10 nm, or 0.4 μin. thickness) is on top of a thin, dense barrier layer (20 to 30 nm, or 0.8 to 1.2 μin.) (Ref 59).

One characteristic of CAA oxides is that they contain a significant amount of fluorine. Sputter-depth profiles obtained by Auger electron spectroscopy (Ref 60, 65, 74, 76) show a buildup of fluorine in the barrier layer. Although adsorbed fluorine is considered detrimental to aluminum adhesive bonds, as discussed earlier in the section "Production Considerations" (Ref 19), CAA titanium oxides exhibit remarkable bond durability, probably because the fluorine is incorporated into the oxide, rather than adsorbed onto it, and does not attack the oxide morphology.

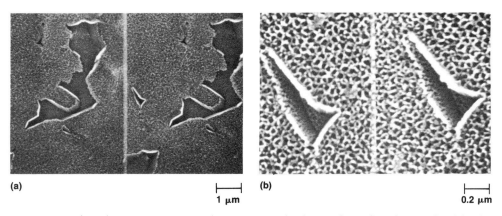

(a)

|— 1 μm —|

(b)

|— 0.2 μm —|

Fig. 8 High-resolution, stereo-scanning electron micrographs showing the oxide surface produced by the anodization of Ti-6Al-4V adherends in a 5% chromic-acid solution. (a) Low magnification. (b) High magnification

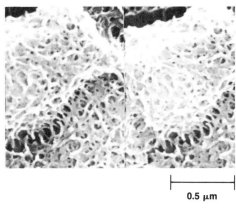

|— 0.5 μm —|

Fig. 10 Oxide surface produced by etching Ti-6Al-4V adherends in an alkaline peroxide solution

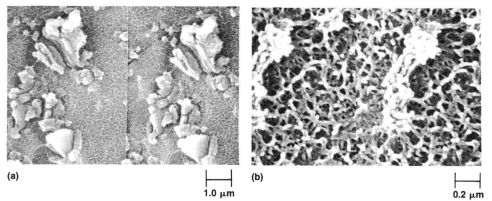

(a)

|— 1.0 μm —|

(b)

|— 0.2 μm —|

Fig. 9 Scanning electron micrographs of the oxide surface produced by the anodization of Ti-6Al-4V adherends in a 5.0 M NaOH solution. (a) Low magnification. (b) High magnification

The inclusion of the SHA oxide as a class III titanium adherend is rather arbitrary. As Fig. 9 shows, macroroughness is evident even at low magnifications (Ref 68, 70, 76, 77). In this case, the Ti-6Al-4V was anodized in a 5.0M NaOH solution at 10 V. The resulting morphology consisted of mountainlike features 1 to 2 μm (0.04 to 0.08 mil) high (and wide), with flat areas between. Higher magnification revealed that the flat areas contained some microroughness, but not to the same degree as with CAA oxide. (Filbey *et al.* (Ref 70) reported the SHA morphology is identical to that of CAA in the flat areas, with substantial microroughness; the authors have not observed this, although some variations do occur with processing conditions.) The sputter-depth profiling of the flat areas in a scanning Auger microprobe (SAM) determined that the SHA oxide was approximately 40 to 50 nm (1.6 to 2 μin.) thick (Ref 77). The composition of the mountainous features was not determined. In the flat areas, the oxide consisted of amorphous TiO$_2$.

In addition, CAA surfaces are more acidic than SHA surfaces after anodization (Ref 70). This is a potentially significant factor in oxide-adhesive chemical interactions, that

is, acid-base reactions at the adhesive-oxide interface (Ref 78).

In the earlier studies, the AP treatment was the only etchant that yielded a class III Ti-6Al-4V adherend surface (Ref 72-74). The high-magnification micrograph in Fig. 10 (Ref 74) clearly shows the microroughness and porosity of the surface. The pore size appears to be somewhat larger than that of the CAA surface. The oxide consists of TiO$_2$, the crystallinity of which has been disputed (Ref 73, 74). The thickness of the oxide varies from 60 to 130 nm (2.4 to 5.2 μin.), depending on the bath temperature and immersion time. One striking difference between the AP oxide and either the CAA or SHA oxides was the buildup of carbon at the oxide-metal interface (Ref 74); the significance of this was not investigated. The AP solution has also been used as an anodizing treatment (Ref 73, 79), producing a microrough morphology similar to that produced by etching with AP solutions.

The NaTESi solution (Table 4) has also been used as both an etchant and an anodizing treatment (Ref 71). As with the AP solution, the resultant morphology is similar for both. It is predominantly macrorough,

with a small amount of microroughness. Both oxides are mixed phase, that is, they consist of TiO$_2$ (rutile) and hexagonal TiO, as determined by TEM. The main difference between the etched and anodized surfaces is in the oxide chemistry. The NaTESi etch procedure includes a posttreatment in HNO$_3$, which removes complexing agent residues and leaves the surface with a higher concentration of OH groups.

The morphology of the plasma-sprayed (PS) surface is different from any of the oxide surfaces previously studied, as can be seen in Fig. 11 (Ref 76). At the highest magnification (Fig. 11b), the microroughness is more random than that produced by any chemical process. The pores are quite deep, and many knoblike protrusions can be seen. These protrusions result from the splat cooling and rapid solidification of the Ti-6Al-4V droplets as they strike the substrate. At lower magnifications (Fig. 11a), some macroscopic inclusions are evident. According to SAM investigations, these are Ti-6Al-4V, presumably from droplets that cooled and solidified prior to striking the substrate.

Durability

Moderate Environments. In the earlier studies mentioned above, the durability of Ti-6Al-4V bonded systems was tested in both wedge (tensile mode) (Ref 62) and lap-shear (shear mode, under stress) (Ref 75) configurations. The conditions for these studies were 60 °C (140 °F) and 95% relative humidity. Epoxy-base adhesives were used in all cases. Representative data are shown in Fig. 12(a) and (b). Brief descriptions of the surface treatments and resultant oxides are given in Table 7. For both test geometries, the class III adherends yielded clearly superior bond durability, although some of the class II adherends performed almost as well. (SHA was not included in these tests; see explanation below.) The failure mode in the class III bonded systems was predominantly cohesive, that is, entirely within the adhesive layer. From this, it was generally

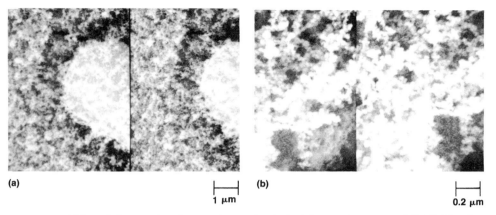

Table 7   Classification of surface preparation for titanium alloys

Process	Oxide thickness, nm	μin.	Classification
Phosphate fluoride	20	0.8	I
Modified phosphate			
fluoride	8	0.3	I
Dapcotreat	6	0.2	II
PasaJell 107			
Dry hone	20	0.8	II
Liquid hone	20	0.8	II
Turco 5578 etch	17.5	0.7	II
CAA 5% solution			
5 V	40 (a)	1.6	III
10 V	80	3.2	III
Alkaline peroxide	45–135	1.8–5.4	III

(a) This value is in conflict with those reported in later studies.

Fig. 11   High-resolution scanning electron micrographs obtained from a plasma-sprayed Ti-6Al-4V surface exhibiting random microscopic roughness. (a) Low magnification. (b) High magnification

concluded that microroughness is essential for the durability of epoxy-titanium bonds over the limited range of environments used in these particular studies. Other factors are important in more severe environments or for bonded systems formed with polyimides or some of the newer, high-temperature-stable adhesives and titanium adherends. For instance, the failure mode in lap-shear specimens (bonded with a thermoplastic polyimide, LARC TPI) of CAA Ti-6Al-4V changed from cohesive to adhesive after 5000 h of aging at 232 °C (450 °F) (Ref 60, 61, 80). In some cases, it appeared that the adhesives pulled away from the CAA oxide, whereas in other cases, the oxide itself failed (Ref 81). Oxide failure in CAA adherends that were exposed to high temperatures prior to bonding with epoxy has been observed and is discussed below.

Because SHA is a relatively new preparation, SHA adherends were not included in the early studies. In wedge tests conducted in a humidity chamber over a temperature range of 55 to 80 °C (130 to 175 °F) and 95% relative humidity, epoxy-SHA bonded systems have shown durability comparable to that of epoxy-CAA systems (Ref 68, 70). In one of these studies (Ref 69), the authors concluded that the SHA adherend was more durable than the CAA adherend. However, the crack-propagation results showed only a 2.5 mm (0.1 in.) difference between the two after 150 h of exposure. This difference is within the range of error in typical wedge tests (Ref 20) and can be considered negligible.

Matz investigated the durability of Na-TESi-prepared adherends with six adhesive-primer combinations using wedge-crack propagation, climbing drum peel, and lap-shear tests (Ref 71). In general, small gains in joint strength and durability were realized for anodized surfaces (over etched ones). The major differences, however, were in the mode of failure. Specimens with etched surfaces tended to exhibit adhesive failure between the primer and the adherend, whereas those with anodized surfaces

exhibited cohesive failures in the adhesive or the primer. Representative data from the wedge crack-propagation tests are shown in Fig. 13. Matz did not compare the durability of NaTESi adherends and CAA and SHA adherends, but his results suggest that they are comparable (assuming chemical compatibility with the adhesive).

It is not surprising that class III Ti-6Al-4V adhesively bonded systems fail within the adhesive under the reported test conditions. Work has shown that the bare, that is, unbonded, titanium oxides are stable under such conditions (Ref 64, 65, 74) and that crack propagation can be attributed to strain energy that is released by the water-saturated adhesive. However, the oxide is not stable under more severe conditions, as has been shown for CAA titanium oxides.

High-temperature, aqueous environments cause the CAA oxide morphology to change (Ref 64, 65). After immersion in boiling water for 24 h, the surface develops crystallites, and the honeycomb structure starts to disappear. This is in marked contrast to FPL-treated aluminum adherends, in which the oxide is completely hydrated typically within 5 min of immersion in boiling water. After 72 h of exposure, the honeycomb structure is totally disrupted, and the surface is completely covered with crystallites (Fig. 14). It should be noted that the direct exposure of bare adherends to aqueous environments such as described is a more severe test than the exposure of actual bonded structures to similar environments. If such drastic changes in morphology occurred in an actual bond line, they would probably cause a loss of bond strength or bond failure. Such a test is meant only to identify an environment that causes adherend degradation and to compare the stability of titanium adherends with that of aluminum adherends.

The durability of CAA-, SHA-, and PS-treated adherends (bonded with epoxy and cured at 175 °C, or 350 °F) in boiling water, using the wedge-crack propagation test, has been compared (Ref 76). Included in this

test were several panels that were prepared by grit blasting alone. The results are shown in Fig. 15. The CAA, SHA, and PS adherends performed identically, within process variability. The larger error in the CAA data was attributed to scatter in the initial crack length. For all three, the final crack lengths are nearly identical. The grit-blasted adherends showed much poorer performance than the other three: Most debonded completely within 24 h, and those that survived the test could be separated by hand.

Although the CAA, SHA, or PS adherends exhibited nearly identical crack propagation values, the loci of failure were different. For CAA adherends, crack propagation was cohesive, that is, entirely within the adhesive. However, for the SHA and PS adherends, the locus of failure changed during the test. Initially, it was through the adhesive, but, during exposure, it switched almost immediately to propagation between the adhesive and the adherend. The mode of failure for grit-blasted specimens was identical to that for the SHA and PS adherends.

The similarity in locus of failure between PS and grit-blasted panels is significant. The two surfaces are chemically identical, but the morphologies are vastly different. The difference in crack propagation can most likely be attributed to a mechanical interlocking between adhesive and adherend and the reduction of stress per unit area at the crack tip in the presence of microroughness; the coarser morphology of the grit-blasted surface results in less capillary force during the wetting and curing processes so that trapped air at the bottom of the pores forms areas of stress concentration.

The differences in failure mode between the CAA and SHA specimens can be attributed to chemical effects. Filbey *et al.* determined that the CAA surface had a residual pH of less than 6 and that the SHA surface had a residual pH greater than 8 (Ref 70). This suggests an acid-base interaction between the adhesive and the oxide, as proposed by Fowkes (Ref 78). This is entirely feasible because curing agents for epoxy

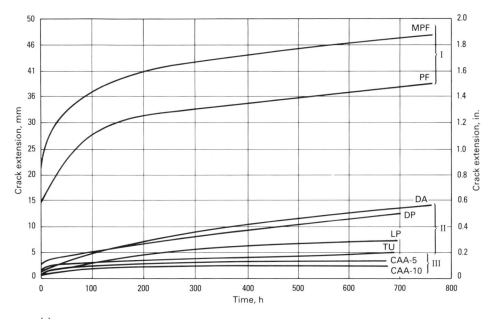

(a)

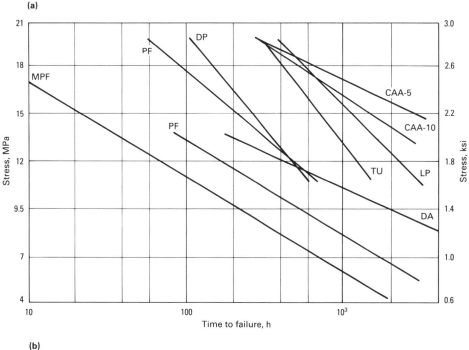

(b)

Fig. 12 Typical bond durability data for Ti-6Al-4V adherends bonded with epoxy adhesives at 60 °C (140 °F) and 100% relative humidity. (a) Crack propagation versus time for the wedge-crack propagation test (Ref 62). (b) Applied stress versus time to failure for the lap-shear geometry (Ref 75). PF, phosphate fluoride; MPF, modified phosphate fluoride; DP, PasaJell 109 dry hone; LP, PasaJell 107, liquid hone; CAA 5% solution, 5; CAA 5% solution, 10; TU, Turco 5578 etch; DA, Dapcotreat; PF, MPF

adhesives contain basic functionalities that can migrate to an acidic surface during the cure cycle (Ref 58, 82). The wedge-crack propagation test results (Ref 76) support this acid-base interaction, as does additional work by Filbey et al. on Ti-6Al-4V adherends treated by various preparations and coated with inorganic primers (Ref 83). They found that durability improved for class I adherends with the inorganic primer, but decreased for CAA adherends. Thus, in the preparation of titanium adherends for service in normal environmental conditions, the

chemical compatibility between the adhesive and the substrate surface, as well as the mechanical interaction, must be considered.

High-Temperature Environments. As mentioned earlier, oxide surface preparations are limited to service environments with temperatures less than about 200 °C (390 °F). Epoxy adhesives are generally used for these applications. For higher service temperatures, polyimides are often used. Their cure temperatures are from 330 to 400 °C (625 to 750 °F). In this section, the issues relating to long-term durability at

high temperatures are described, starting with an explanation of the temperature dependence of the adherend morphology.

Unlike aqueous environments, dry, high-temperature environments have little or no effect on CAA morphology. When bare adherends exposed to air (330 °C, or 625 °F) for 1200 h and to vacuum (400 °C, or 750 °F) for 165 h were studied, it was found that the CAA specimens retained their morphology. However, bonds formed after such exposures failed under minimal force (Ref 65, 76). In these cases, the failures occurred at the oxide-metal interface, caused by the dissolution of the oxide into the base metal at the exposure temperature (Ref 76). Because oxygen is highly soluble in the metal (up to 25 at.%), dissolution of the oxide (or diffusion of oxygen away from the oxide) could cause defects such as voids or microcracks to occur at the oxide-metal interface. Any stress applied to the adherend after exposure would then be concentrated in small areas at the interface. Under these conditions, very little tensile force would be needed to pull the oxide away from the metal. This phenomenon was observed for an adherend exposed in vacuum at 400 °C (750 °F) for as little as 1 h (Ref 76, 84). The minimum bond strength is obtained at the onset of this process. Some strength can be recovered by dissolving the oxide completely, although this is not recommended.

SHA adherends that were exposed to dry, high-temperature environments did not exhibit as dramatic a loss in bond strength over similar exposure times (Ref 76, 77, 84). However, SHA-treated lap-shear specimens (bonded with a polyimide adhesive) that were aged at 232 °C (450 °F) exhibited dramatic losses in bond strength at long exposure times (>1 year) (Ref 61). The specimens from the study in Ref 61 were obtained for failure analysis. Analysis of the debonded surfaces showed that failure occurred within the oxide, results similar to those observed for CAA adherends. This suggests similar fracture mechanisms, but different kinetics, for both SHA and CAA adherends.

PS adherends have also been exposed in vacuum at temperatures up to 800 °C (1470 °F) (1-h exposure at each temperature) (Ref 76). No changes were observed at temperatures of 650 °C (1200 °F) and below. Above 650 °C (1200 °F), changes in the bulk microstructure, and hence the surface morphology, were observed. At all temperatures, the PS coating remained adherent.

To determine whether these effects are observed in actual bonded structures, wedge-crack propagation tests on PS and CAA specimens bonded with a polyimide adhesive (Durimid 100, Rogers Corporation), were conducted. The bonding cycle included a 1-h hold (under pressure) at 400 °C (750 °F). One set of panels was held for 10 h at 400 °C (750 °F) (labeled CAA-10).

Results are shown in Fig. 16. It is immediately obvious that the initial crack length for the PS specimens was 20 to 30 mm (0.8 to 1.2 in.) less than those for the CAA specimens. PS and CAA-1 specimens exhibited roughly 5 to 10 mm (0.2 to 0.4 in.) of crack propagation.

The most significant difference between the PS and CAA specimens was in the mode of crack propagation. For the PS specimens, propagation occurred entirely within the adhesive. However, for the CAA specimens, over 90% of the crack propagation occurred at the oxide-metal interface. Using elementary fracture mechanics, it is estimated that the PS adherends can withstand strain-energy release rates that are at least five times greater than those of the CAA specimens. The difference in performance is limited only by the adhesive itself. Most likely, the PS adherends are capable of sustaining much higher strain-energy release rates.

Copper Adherends

Copper is not generally regarded as a structural metal. Indeed, the largest use of it in bonded applications is in the microelectronics industry. However, copper alloys are used in a variety of marine applications in which high stresses occur. Again, common practice is to prepare the surface by mechanical abrasion. Although this produces adequate initial bond strength, the resultant durability is poor (Ref 1). Several etch treatments have been developed for copper (and copper alloys such as brass and bronze) that are suitable for long-term structural applications. Two of these, an Ebonol C etch and an alkaline-chlorite etch (Ref 1, 85, 86), result in black oxide surfaces that have microrough morphologies. The processing steps are similar to those used for titanium and aluminum adherends.

Processing

The processing steps for the Ebonol C and the alkaline-chlorite etches are shown in Table 8. These etches are used for alloys containing at least 95% Cu. An etching procedure for other Cu alloys, based on a ferric chloride-nitric acid solution, is also shown in Table 8. The rationale for each step was discussed in the sections describing aluminum and titanium adherends.

Chemistry and Morphology

Both the Ebonol C and the alkaline-chlorite procedures result in a matte-black oxide, presumably named for the color of the film. The black color immediately suggests that the oxide is CuO, and Packham and co-workers have verified this by XPS (Ref 85, 87). The thickness of the oxide, according to Packham, increases linearly with etching time for both procedures (Evans

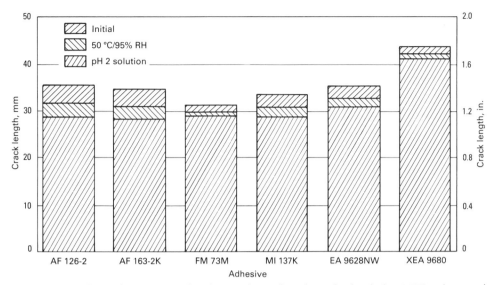

Fig. 13 Wedge-crack propagation data for Ti-6Al-4V adherends anodized with the NaTESi solution and bonded with various epoxy adhesives. Specimens were first exposed to 95% relative humidity (RH) at 50 °C (120 °F) for 75 min. Then they were immersed in a Na$_3$PO$_4$ solution adjusted to pH 2 at 50 °C (120 °F) for 24 h. Source: Ref 71

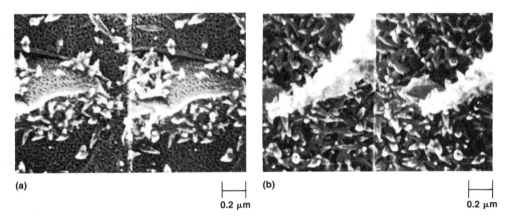

Fig. 14 CAA oxide (Ti-6Al-4V) after immersion in water at 100 °C (212 °F). (a) For 24 h. (b) For 72 h. The development of crystallites is concurrent with the disappearance of the honeycomb morphology.

and Packham used the Ebonol C etch at 90 °C, or 190 °F) (Ref 86). They found, however, that the initial bond strength to polyethylene did not increase significantly beyond the first 10 min of etching. The resulting oxide thicknesses were 750 nm (30 μin.) for the Ebonol C procedure and 150 nm (6 μin.) for the alkaline-chlorite procedure.

As mentioned above, both procedures result in microrough (or microfibrous) morphologies. After 10 min, the Ebonol C etch produces a whiskerlike morphology, with local clusters emanating from a single point. Thus, the appearance is actually floral. For similar conditions, the alkaline-chlorite solution produces a more needlelike or nodular surface.

Bonding

Evans and Packham studied the bonding of polyethylene to the Ebonol C and alkaline-chlorite etched surfaces (Ref 86). Using 180° peel tests, they determined the strength

as a function of etch conditions and subsequent posttreatments. The respective values after 10-min etching were 2.6 and 1.4 N/mm (15 and 8 lbf/in.) for the Ebonol C and alkaline-chlorite etches. No explanation was given for the difference, nor was there any attempt to identify the locus of failure. To investigate the effects of surface morphology versus surface chemistry, Evans and Packham "damaged" the oxides by buffing, which resulted in a relatively smooth surface. Peel strengths dropped approximately 75% for both. In the same study, they reduced the alkaline-chlorite oxide to a metallic surface (but maintained the morphology) by cathodic polarization in NaOH solutions. In this case, the peel strength dropped only approximately 30%, and Evans and Packham concluded that the surface morphology was the dominant factor.

In a later study, Hine *et al.* conducted fracture studies of joints formed with chemically polished and alkaline-chlorite-treated

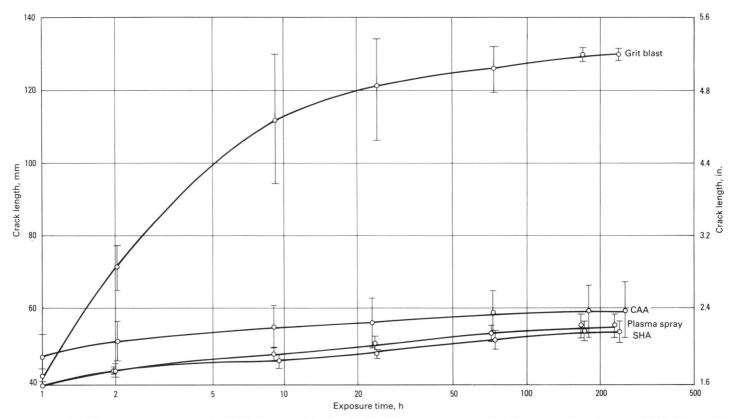

Fig. 15 Crack length versus exposure time for Ti-6Al-4V adherends bonded with FM 300M epoxy and immersed in boiling water. The performance of CAA, SHA, and PS adherends is comparable.

copper adherends and an unmodified epoxy adhesive (Ref 87). They found that the fracture toughness in both single-edge-notched (SEN) and tapered-double-cantilever-beam (TDCB) geometries increased with the microroughness produced by the etch—by a factor of 6.5 for SEN and 2.5 for TDCB. Additionally, the failure mode was interfacial for the polished adherends, but cohesive (in the epoxy) for the alkaline-chlorite-etched adherends. When a rubber toughening agent was added to the epoxy, the results changed somewhat. Comparable

fracture toughnesses were obtained for all TDCB specimens, but the alkaline-chlorite-etched joints exhibited 15 times the toughness of the polished joints. Hine *et al.* also observed that the copper oxide inhibited the cure of the modified epoxy. The normal, micron-size rubber particles did not exist. Rather, a wormlike microstructure developed, consisting of phase-separated rubber and modified epoxy with less-than-normal rubber content. Dissolution of the copper oxides during cure was believed responsible for this (Ref 87).

Steel Adherends

Despite intense commercial interest in the bonding of steel structures and considerable research efforts dedicated to the achievement of durable, structural bonds to steel surfaces, at this writing no general-purpose pretreatments for steel substrates have been developed. Unlike aluminum and titanium, iron does not form coherent, adherent oxides, making it difficult to create a stable film with the fine microroughness needed for good adhesion. Grit blasting is commonly used as a pretreatment. Although it is adequate in some applications, as discussed below, it is not suitable for cases in which the bonded structure is exposed to severe environments.

Table 8 Etching procedures for copper adherends

Step	Ebonol C	Ferric chloride/ alkaline chlorite	Nitric acid
1. Degreasing	Trichloroethane (vapor), 87–90 °C (190–194 °F) as needed	Trichloroethane	Trichloroethane (vapor), 87–90 °C (190–194 °F) as needed
2. Cleaning	10 vol% HNO₃ (70%) 90 vol% H₂O, 30 s, 25 °C (77 °F)	11 vol% HNO₃, 68 vol% H₃PO₄, 11 vol% acetic anhydride, 10 vol% H₂O, 4 min, 25 °C (77 °F)	. . .
3. Rinse	H₂O, running transfer to etchant	Distilled H₂O, then acetone	. . .
4. Etch	Ebonol C, 6.8 kg (240 oz) H₂O to make 3.8 L (1 gal), 2 min, 36 °C (95 °F)	30 g/L (4 oz/gal) NaClO₂ 100 g/L (13 oz/gal) Na₃PO₄ 50 g/L (6.6 oz/gal) NaOH, 15 min, 95 °C (205 °F)	6 wt% FeCl (42%) 12 wt% HNO₃ 82 wt% deionized H₂O, 2 min, 25 °C, (77 °F)
5. Rinse	H₂O, 2–3 min, 10–15 °C (50–60 °F)	Deionized H₂O, 25 °C (77 °F), then acetone	H₂O, 3–5 min, 10–15 °C (50–60 °F)
6. Dry	Air or dry N₂	Air	Air or dry N₂

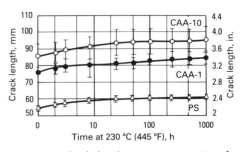

Fig. 16 Crack length versus exposure time for CAA- and PS-treated Ti-6Al-4V adherends bonded with LARC TPI polyimide and aged in laboratory ambient at 230 °C (445 °F). CAA-treated adherends exhibited oxide failures.

Table 9 Processing steps for two acid etches used on steel adherends

Step	Phosphoric acid-alcohol	Nitric-phosphoric acid
1. Grit blast	Al_2O_3, 150 mesh, 170–240 kPa (25–35 psi)	. . .
2. Etch	33 vol% H_3PO_4 (85%), 67 vol% ethanol, 10–12 min, 55–60 °C (130–140 °F)	30 vol% H_3PO_4 (85%), 5 vol% HNO_3, 64.99 vol% deionized H_2O, 0.01 vol% surfactant, 5–7 min, 20–25 °C (68–77 °F)
3. Rinse	Deionized H_2O, 3–7 min, 20–25 °C (68–77 °F)	Deionized H_2O, 3–8 min, 20–25 °C (68–77 °F)
4. Dry	Forced air, 60–65 °C (140–150 °F)	Forced air, 60–65 °C (140–150 °F)

Source: Ref 1

⊢—————⊣
10 μm

Fig. 17 Phosphoric-acid-etched A606 mild steel surface, showing heavy smut formation

Additionally, the variety and complexity of steel microstructures greatly complicate the task of developing a universal surface treatment. Although various cleaning treatments and chemical etchants have been used for both low-carbon and stainless steels over many years, none has been widely adopted or shown to be superior to grit blasting (Ref 88). There have been encouraging results recently for stainless steels, but an effective etch or anodization process for low-carbon steels has not been developed. The best approach for preparing low-carbon steels appears to be the deposition of a more durable and bondable coating using, for example, a conversion coating treatment. The remainder of the article describes surface treatments for steels, including bonding results for mass-production applications in which no surface preparation is used.

Cleaning as a Pretreatment for Steel

As with aluminum and titanium, the most critical test for bonded steel joints is durability in hostile (that is, humid) environments. Results from a study by Pocius *et al.* illustrate the difficulty of forming durable bonds on steels (Ref 89). These experiments compared solvent-cleaned (smooth) 1010 cold-rolled steel surfaces with FPL aluminum (microrough) substrates. Although the dry lap-shear strengths were not markedly different, stressed lap-shear joints of steel adherends that were exposed to a humid environment failed in less than 30 days, whereas the aluminum joints lasted for more than 3000 days.

Virgin alumina grit-blasted steel surfaces that were tested at Martin Marietta Laboratories showed poor durability as well. Wedge-test specimens exhibited rapid crack growth during the humidity-chamber exposure (Ref 90). Electron micrographs of opposing failure surfaces taken in the crack-growth region showed purely interfacial failure. The metal surface had not been corroded, nor was there visual evidence of moisture attack on the polymer. Gledhill and Kinloch (Ref 52) investigated joints immersed in water and found a similar displacement of adhesive-to-metal bonds. They concluded, "Thermodynamics indi-

cate that if only secondary forces are acting across the interface, water will virtually always desorb an organic adhesive from a metal oxide surface." This suggests that simply cleaning the adherend surface is not an adequate treatment for structural bonding applications.

Further work has been done in an effort to optimize mechanical cleaning. Gilibert and Verchery tried different types of grinding and grit blasting and found that grit blasting outperformed fine or coarse grinding (Ref 91). They concluded that the best results were obtained when the grit size was matched to the size of the filler material in the adhesive. Watts and Castle also investigated a variety of mechanical cleaning treatments and judged grit blasting to be the best (Ref 92). They concluded that improved durability was strictly related to the increased surface area of the grit-blasted surfaces.

A number of novel techniques have been attempted to provide (atomically) clean steel surfaces for optimal chemical bonding, but generally there has been no improvement over grit blasting. Ultraviolet light/ ozone exposures have been used very successfully to provide extremely clean surfaces for semiconductor processing. However, when this technique was applied to oily steel, the cleaned surfaces showed no improvement in bond strength over control specimens (Ref 93). Glow-discharge etching, another method of providing very clean semiconductor surfaces, was used to prepare steels for an anticorrosion, polymer-film-coating process (Ref 94). Two other novel means of cleaning steel surfaces, which were investigated by Ruhsland (Ref 95), incorporate cleaning components into the adhesive itself. Although all the treatments improved adhesion to dirty surfaces, they were no better than wiping with a solvent before bonding.

Chemical Etches

Low-Carbon Steels. Although chemical treatments for cleaning steel surfaces have been widely used for many years, there are no general chemical treatments designed to prepare steels for adhesive bonding. Much research has been done to find such a treatment, but has achieved only limited

success. In one case, a superheated steam treatment was used following etching in dilute HCl to induce a microfibrous topology similar to those of aluminum and titanium bonding surfaces (Ref 96). However, bonded surfaces treated in this way failed at the oxide-metal interface and at lower stress levels than would be expected for normally prepared, smoother steels.

One of the main problems with chemical-etch treatments of low-carbon steels is the formation of a loosely adhering layer of iron oxide "smut" on the adherend surface. The oxide is difficult to remove before bonding, but easily pulls away with the adhesive when the joint is stressed. Figure 17 shows an electron micrograph of an A606 steel surface after etching in phosphoric acid. Despite vigorous rinsing, the steel surface is mostly obscured by Fe_3O_4 smut. Russell *et al.* attempted to suppress the formation of the smut by using acid etches in alcohol solutions (Ref 97, 98). Although they demonstrated that such solutions were reasonably effective, the treatments were no better than grit blasting in stressed durability testing. The processing steps for Russell's process, that is, phosphoric acid-alcohol etch, are listed in Table 9, along with the processing steps for another common steel etch, a nitric acid/phosphoric acid mixture. In some applications, a treatment that is "as-good-as-grit-blasting" is still acceptable; Soviet researchers have indicated that a chemical-etch/corrosion-inhibitor treatment can reduce the cost of surface preparation to 15% of that for grit blasting (Ref 99).

Another major problem with chemical etches is that the resulting morphology is a function of the metallurgy of the substrate. Figures 18 and 19 show electron micrographs of clean A606 and A514 steel surfaces after a phosphoric acid etch. Although A606 and A514 substrates are chemically similar to high-strength low-alloy steels, their etched surfaces offer different

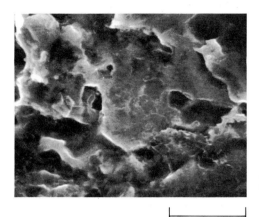

Fig. 18 Phosphoric-acid-etched A606 steel surface showing a smut-free, smooth-walled, crevice morphology

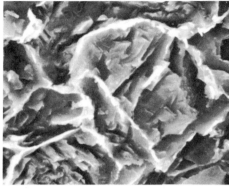

Fig. 19 Phosphoric-acid-etched A514 steel surface showing extensive etching along martensite boundaries. The heat treatment of this alloy differs from that of the A606 alloy in Fig. 18.

bonding/interlocking potentials because different heat treatments are used to develop their microstructures. In both, the acid primarily attacks grain boundaries, etching out surface grains. However, the A606 surface has smooth-sided, dimpled morphological features similar to grit-blasted surfaces, whereas the A514 surface has small crevices that can allow adhesive penetration. Etched A514 performs much better than either grit-blasted A514 or etched A606 surfaces in wedge-crack propagation tests.

The adherend metallurgy also indirectly determines the degree of smut buildup because the rate of smut formation is proportional to the etch rate. For instance, the etch rate of A606 is four times greater than that of A514 because of differences in grain size. As a result, a different etch treatment is required for each. In general, the roles of alloy and heat-treatment differences have not been widely investigated in bonding research on steel; details of the substrate metallurgy are usually not reported.

The high etch rates for standard, cold-rolled steels cause smut buildup that is difficult to eliminate. Simply reducing the

activity of the etch solution is frequently not possible because many steel surfaces do not react at all to weak etching solutions. Using surface analysis techniques, Leroy has shown that standard processing can cause graphite and other carbonaceous species to coat the steel (Ref 100). A similar effect has also been observed with "drawing quality steel" (Ref 101). These layers are relatively impervious to mineral acids and must be undercut to be removed. Research has shown that carefully controlled phosphoric acid solutions can be used to clean steel (Ref 102). However, because these solutions do not consistently provide an improved bonding morphology, they offer no performance enhancement over normal, more easily monitored techniques, such as grit blasting.

Stainless Steels. Although there is considerable evidence that chemical surface treatments improve the substrate bondability of stainless steels, there is no general agreement on which treatment is the best. One etchant commonly used with stainless steels is an HNO_3-HF mixture (Ref 103-

106); another is chromic acid. Various solutions based on sulfuric acid have also been beneficial (Ref 107-109), and several researchers have investigated a wide variety of solutions based on acids with dichromates added, with and without anodization (Ref 110-113). The treatment judged to be the best by one researcher did not receive the same ranking from another. All of the treatments used a highly concentrated, very active solution that attacked grain boundaries and the chromium-poor regions around chromium carbide particles. One common finding by researchers who used peel tests (Ref 106, 109-111) was that surface roughness correlated with bond performance. Pocius *et al.* (Ref 89), for example, noted that peel performance could be correlated to the microscopic surface roughness seen in a scanning electron microscope. However, the type of substrate alloy and heat treatment were different in many cases. The treatment that worked best for any one researcher may have been related to the precise metallurgy of the samples investigated. For example, Hirko (Ref 113) reported that three different treatments, a nitric acid anodization, a sulfuric acid/dichromate etch, and a sulfuric acid/dichromate anodization, provided essentially equivalent wedge-crack propagation test performances.

Conversion Coating Treatments

A better approach to the formation of durable bonded joints with steels is to use a conversion coating process to deposit a rough, corrosion-resistant layer on the adherend. To achieve such a coating, zinc and iron phosphate solutions are used to precipitate crystallites onto the steels. This frequently provides good bonding morphology (as shown in Fig. 20) (Ref 114). Research at the West German Academy of Science showed that iron and zinc phosphate surfaces had 15% lower bonding strengths when tested under dry conditions, but typically had 100% better values than a ground steel surface after the bonds had been soaked in water (Ref 115). Work done at Martin Marietta Corporation, using carefully controlled zinc-phosphate treatments, supports this finding. Trawinski *et al.* demonstrated lap-shear strengths equivalent to those on grit-blasted steel substrates (Ref 116) but greatly improved wedge-crack propagation test performances (Ref 90). However, to achieve this performance, the crystallites in the coating must be small. Although the same technique can work well for aluminum alloys (Ref 102), it is more advantageous for steel, because steel substrates corrode more readily, even under polymer coatings.

Typical processing steps for conversion coating treatments are given in Table 10. The solutions themselves are proprietary to the manufacturers. Because the results of-

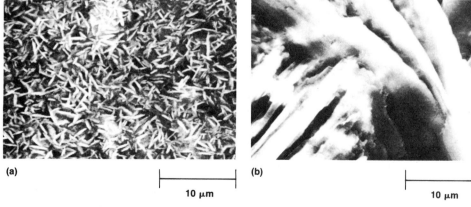

(a)   (b)

Fig. 20 Electron micrographs of conversion coatings. (a) Grain-refined microcrystalline conversion coating. (b) Conventional zinc phosphate conversion coating

Table 10 Typical processing steps for conversion coatings deposited onto steel adherends

Step	Method or agent
1. Precleaning..........	Grit blast, grinding or machining
2. Cleaning	Alkaline, 1–5 min, 50–75 °C (120–165 °F)
3. Rinse	Water, 1 min, 20–25 °C (68–77 °F)
4. Grain-refining rinse ..	Titanium/salt water, 0.25–1 min, 20–25 °C (68–77 °F)
5. Conversion coating...	Immersion 1–2 min, 40–85 °C (105–185 °F)
6. Rinse	Chromate/water, 0.25–1 min, 20–75 °C (68–165 °F)

(a)
⊢—⊣ 10 μm

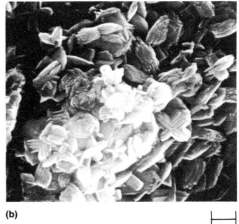
(b)
⊢—⊣ 10 μm

Fig. 21 Electron micrographs of microcrystalline conversion coatings on two steel adherends, showing different morphologies for the same process. (a) A514. (b) A606

ten depend on the adherend metallurgy, only an outline of the process is given.

Recent efforts to improve the corrosion resistance of thinner conversion coatings have led to the development of grain refiners and initiators that allow small-grained zinc-phosphate coatings to be readily deposited (Ref 117-121). If the grain size is not controlled, the coating becomes so thick that one phosphate crystal precipitates onto another, and a weak interface is introduced (Fig. 20). This may be the reason that in earlier studies (Ref 122), heat-curing epoxies were not found to be compatible with conversion coatings. Thermal stress between the thick conversion coating and the steel substrate may have caused fracture in the coating. Gan *et al.* investigated the relationship between the surface condition before phosphatization and the strength of the coating and found that the phosphate layer had failed for all coatings tested (Ref 123). Electron micrographs of the coatings showed large phosphate crystals. In a subsequent study, they demonstrated improved performance using a finer-grained coating, although failures still occurred within the coating layers (Ref 124).

As with chemical etches, the development of optimal conversion coatings requires an assessment of the microstructure of the steel. Correlations have been found between the microstructure of the substrate material and the nature of the phosphate films formed. Aloru *et al.* demonstrated that the type of phosphate crystal formed varies with the orientation of the underlying steel crystal lattice (Ref 125). Figure 21 shows the different phosphate crystal morphologies that formed on two heat-treated surfaces. The fine-flake structure formed on the tempered martensite surface promotes adhesion more effectively than the knobby protrusions formed on the cold-rolled steel.

No Surface Preparation

In many high-volume production applications (that is, the automotive and appliance industries), the elaborate surface preparation of steel adherends is undesirable or impossible. Thus, there has been wide-spread interest in bonding directly to steel coil surfaces that contain various protective oils (Ref 126-131). Debski *et al.* proposed that epoxy adhesives, particularly those curing at high temperatures, could form suitable bonds to oily steel surfaces by two mechanisms: first, thermodynamic displacement of the oil from the steel surface and, second, absorption of the oil into the bulk adhesive (Ref 126, 127). The relative importance of these two mechanisms depends on the polarity of the oil and the surface area-to-volume ratio of the adhesive (which can be affected by adherend surface roughness).

Charnock (Ref 130) determined the bond strengths of joints made from steel substrates coated with a variety of oils and waxes (~60 g/m², or 0.20 g/ft²) and joined with acrylic adhesive. Lap-shear strengths up to 15 MPa (2.2 ksi) were obtained with room-temperature curing. Very little degradation was seen after 1000 h of exposure at 40 °C (105 °F) and 95% relative humidity. However, the shear strength dropped by roughly half when the joints were subjected to a 30-min bake at 200 °C (390 °F). Although the bond strengths reported by Charnock do not approach those reported for aluminum, titanium, or copper adherends, they are adequate for the intended application.

REFERENCES

1. R.F. Wegman, *Surface Preparation Techniques for Adhesive Bonding*, Noyes Publications, 1989
2. *The Handbook of Adhesive Bonding*, C.V. Cagle, Ed., McGraw-Hill, 1973
3. J.D. Venables, *J. Mater. Sci.*, Vol 19, 1984, p 2431
4. A.J. Kinloch, *J. Mater. Sci.*, Vol 17, 1982, p 617
5. J.D. Minford, *Adhes. Age*, Vol 17, 1974, p 24
6. H.W. Eichner and W.E. Schowalter, Report 1813, Forest Products Laboratory, 1950
7. N.L. Rodgers, in *Proceedings of the 13th National SAMPE Technical Conference*, Society for the Advancement of Material and Process Engineering, 1981, p 640
8. A. Desai, J.S. Ahearn, and D.K. McNamara, "Cleanliness of External Tank Surfaces," Technical Report MML TR 85-65, Martin Marietta Laboratories, 1985
9. G.S. Kabayashi and D.J. Donnelly, Report D6-41517, The Boeing Company, 1974
10. MIL-A-8625C, Military Specification
11. W. Brockman and O.D. Hennemann, in *Proceedings of the 11th SAMPE Technical Conference*, Society for the Advancement of Material and Process Engineering, Nov 1979, p 804-816
12. W. Brockman, O.D. Hennemann, H. Kollek, and C. Matz, *Int. J. Adhes. Adhes.*, Vol 6, 1986, p 115
13. I. Olefjord and L. Kozma, *Mater. Sci. Technol.*, Vol 3, 1987, p 860
14. T.S. Sun, J.M. Chen, J.D. Venables, and R. Hopping, *Appl. Surf. Sci.*, Vol 1, 1978, p 202
15. A.V. Pocius, in *Adhesion Aspects of Polymeric Coatings*, K.L. Mittal, Ed., Plenum Publishing, 1983, p 173
16. A.V. Pocius, T.H. Wilson, Jr., S.H. Lunquist, and S. Sugii, in *Progress in Advanced Materials and Processes: Durability, Reliability, and Quality Control*, G. Bartelds and R.J. Schliekilmann, Ed., Elsevier, 1985, p 71
17. Process Specification 4352, rev J, Bell Helicopter Textron, Inc., June 1980
18. Process Specification TH 6.7851, Fokker VFW B.V., Aug 1978
19. J.D. Venables, D.K. McNamara, J.M. Chen, T.S. Sun, and R.L. Hopping, *Appl. Surf. Sci.*, Vol 3, 1979, p 88

20. G.D. Davis, J.S. Ahearn, L.J. Matienzo, and J.D. Venables, *J. Mater. Sci.*, Vol 20, 1985, p 975

21. J.M. Chen, T.S. Sun, J.D. Venables, D.K. McNamara, and R. Hopping, in *Proceedings of the 24th National SAMPE Symposium*, Society for the Advancement of Material and Process Engineering, 1979, p 1188

22. A.J. Kinloch, H.E. Bishop, and N.R. Smart, *J. Adhes.*, Vol 14, 1982, p 105

23. D.A. Hardwick, J.S. Ahearn, A. Desai, and J.D. Venables, *J. Mater. Sci.*, Vol 21, 1986, p 179

24. S.M. El-Mashri, R.G. Jones, and A.J. Forty, *Phil. Mag. A*, Vol 48, 1983, p 665

25. R.H. Olsen, in *Proceedings of the 11th National SAMPE Technical Conference*, Society for the Advancement of Material and Process Engineering, 1979, p 770

26. H.E. Franz, *Z. Werkstofftech.*, Vol 14, 1983, p 164, 290

27. H. Kollek, *Int. J. Adhes. Adhes.*, Vol 5, 1985, p 75

28. G.D. Davis, T.S. Sun, J.S. Ahearn, and J.D. Venables, *J. Mater. Sci.*, Vol 17, 1982, p 1807

29. A.J. Kinloch and N.R. Smart, *J. Adhes.*, Vol 12, 1981, p 23

30. J.S. Solomon and D.E. Hamlin, *Appl. Surf. Sci.*, Vol 4, 1980, p 307

31. P. Poole and J.F. Watts, *Int. J. Adhes. Adhes.*, Vol 5, 1985, p 33

32. A.J. Kinloch, H.E. Bishop, and N.R. Smart, *J. Adhes.*, Vol 14, 1982, p 105

33. M.F. Abd Rabbo, J.A. Richardson, and G.C. Woods, *Corros. Sci.*, Vol 16, 1976, p 689

34. V. Singh, P. Rama, G.J. Cocks, and D.M.R. Taplin, *J. Mater. Sci. Lett.*, Vol 14, 1979, p 745

35. A. Csanady, I. Imre-Baan, E. Lichtenberger-Bajza, E. Szontagh, and F. Domolki, *J. Mater. Sci.*, Vol 15, 1980, p 2761

36. A.E. Yaniv, N. Fin, H. Dodiuk, and I.E. Klein, *Appl. Surf. Sci.*, Vol 20, 1985, p 538

37. D.J. Arrowsmith and A.W. Clifford, *Int. J. Adhes. Adhes.*, Vol 5, 1985, p 40

38. W. Brockmann and O.D. Hennemann, in *Adhesive Joints*, K.L. Mittal, Ed., Plenum Publishing, 1984, p 469

39. K.W. Allen and M.G. Stevens, *J. Adhes.*, Vol 14, 1982, p 137

40. J.S. Ahearn, T.S. Sun, C. Froede, J.D. Venables, and R. Hopping, *SAMPE Q.*, Vol 12, 1980, p 39

41. T.P. Hoar and N.J. Mott, *J. Phys. Chem. Solids*, Vol 9, 1959, p 97

42. J.S. Ahearn and G.D. Davis, in *Proceedings of Adhesion '87*, 1987, p 291

43. W. Brockmann, AGARD Lecture Series 102, Advisory Group for Aerospace Research and Development, North Atlantic Treaty Organization, 1979

44. S. Naviroj, J.L. Koenig, and H. Ishida, *J. Adhes.*, Vol 18, 1985, p 93

45. P. Walker, *J. Coat. Technol.*, Vol 52, 1980, p 29

46. J.S. Ahearn, G.D. Davis, T.S. Sun, and J.D. Venables, in *Adhesion Aspects of Polymeric Coatings*, K.L. Mittal, Ed., Plenum Publishing, 1983, p 281

47. J.D. Venables, M.E. Tadros, and B.M. Ditchek, U.S. Patent 4,308,079, 1981

48. D.A. Hardwick, J.S. Ahearn, and J.D. Venables, *J. Mater. Sci.*, Vol 19, 1984, p 223

49. D.K. Shaffer, H.M. Clearfield, and J.S. Ahearn, Technical Report MML TR 86-76(c), Martin Marietta Laboratories, N00014-85-C-0804, Office of Naval Research, 1986

50. J.M. Chen, T.S. Sun, J.D. Venables, and R. Hopping, in *Proceedings of the 22nd National SAMPE Symposium*, 1977, p 25

51. A.J. Kinloch, *J. Adhes.*, Vol 10, 1979, p 193

52. R.A. Gledhill and A.J. Kinloch, *J. Adhes.*, Vol 6, 1974, p 315

53. R.A. Pike, *Int. J. Adhes. Adhes.*, Vol 5, 1985, p 3; Vol 6, 1986, p 21

54. D.M. Brewis, in *Durability of Structural Adhesives*, Applied Science Publishers, 1983, p 215-254

55. D.J. Arrowsmith and Q.W. Clifford, *Int. J. Adhes. Adhes.*, Vol 5, 1985, p 40

56. "Aerospace Recommended Practice," ARP 1524, Society of Automotive Engineers, 1978

57. D. Trawinski, S. Kodali, R. Curley, D.K. McNamara, and J.D. Venables, in *Proceedings of the 14th National SAMPE Technical Conference*, Society for the Advancement of Material and Process Engineering, 1983

58. D.K. McNamara, L.J. Matienzo, J.D. Venables, J. Hattayer, and S.P. Kodali, in *Proceedings of the 28th National SAMPE Symposium*, Society for the Advancement of Material and Process Engineering, 1983

59. D.K. McNamara, J.D. Venables, T.S. Sun, J.M. Chen, and R.L. Hopping, in *Proceedings of the 11th National Technical Conference*, Society for the Advancement of Material and Process Engineering, 1979

60. S.G. Hill, P.D. Peters, and C.L. Hendricks, NASA contractor report 177936, Contract NAS1-15605, National Aeronautics and Space Administration, 1985

61. D.J. Progar and T.L. St. Clair, *Int. J. Adhes. Adhes.*, Vol 6, 1986, p 25; D.J. Progar, NASA Technical Memorandum 87738, National Aeronautics and Space Administration, 1986

62. S.R. Brown, in *Proceedings of the 27th National SAMPE Symposium*, Society for the Advancement of Material and Process Engineering, 1982, p 363

63. B.M. Ditchek, T.I. Morgenthaler, T.S. Sun, and R.L. Hopping, in *Proceedings of the 25th National SAMPE Symposium*, Society for the Advancement of Material and Process Engineering, 1980, p 909

64. M. Natan, J.D. Venables, and K.R. Breen, in *Proceedings of the 27th National SAMPE Symposium*, Society for the Advancement of Material and Process Engineering, 1982, p 178

65. H.M. Clearfield, D.K. Shaffer, and J.S. Ahearn, in *Proceedings of the 18th National SAMPE Technical Conference*, Society for the Advancement of Material and Process Engineering, 1986, p 921; H.M. Clearfield, D.K. Shaffer, J.S. Ahearn, and J.D. Venables, *J. Adhes.*, Vol 23, 1987, p 83

66. M. Assefpour-Dezfuly, C. Vlachos, and E.H. Andrews, *J. Mater. Sci.*, Vol 19, 1984, p 3626

67. Y. Moji and J.A. Marceau, U.S. Patent 3959091, 1976

68. A.C. Kennedy, R. Kohler, and P. Poole, *Int. J. Adhes. Adhes.*, Vol 3, 1983, p 133

69. D.J. Progar, NASA Technical Memorandum 89045, National Aeronautics and Space Administration, 1986

70. J.A. Filbey, J.P. Wightman, and D.J. Progar, *J. Adhes.*, Vol 20, 1987, p 283

71. C. Matz, *Int. J. Adhes. Adhes.*, Vol 8, 1988, p 17

72. K.W. Allen, H.S. Alsalim, and W.C. Wake, *J. Adhes.*, Vol 6, 1974, p 153

73. J.L. Cotter and A. Mahoon, *Int. J. Adhes. Adhes.*, Vol 2, 1982, p 47

74. B.M. Ditchek, K.R. Breen, and J.D. Venables, Technical Report MML TR 80-17c, Martin Marietta Laboratories, 1980; M. Natan, K.R. Breen, and J.D. Venables, Technical Report MML TR 81-42c, Martin Marietta Laboratories, 1981; both submitted under Contract N00019-80-C-0508, Naval Air Systems Command

75. R.F. Wegman and D.W. Levi, in *Proceedings of the 27th National SAMPE Symposium*, Society for the Advancement of Material and Process Engineering, 1982, p 440

76. H.M. Clearfield, D.K. Shaffer, S.L. VanDoren, and J.S. Ahearn, *J. Adhes.*, Vol 29, 1989, p 81

77. H.M. Clearfield, G.O. Cote, K.A. Olver, D.K. Shaffer, and J.S. Ahearn, *Surf. Interface Anal.*, Vol 11, 1988, p 347

78. F.M. Fowkes, *J. Adhes. Sci. Technol.*, Vol 1, 1987, p 7

79. A. Mahoon, in *Proceedings of the 27th National SAMPE Symposium*, Society for the Advancement of Material and

Process Engineering, 1982, p 150

80. P.D. Peters, E.A. Ledbury, C.L. Hendricks, and A.G. Miller, in *Proceedings of the 27th National SAMPE Symposium*, Society for the Advancement of Material and Process Engineering, 1982, p 940

81. T.L. St. Clair, P.M. Hergenrother, and D.J. Progar, private communication, 1987

82. R.G. Dillingham and F.J. Boerio, *J. Adhes.*, Vol 24, 1987, p 315

83. J.A. Filbey and J.P. Wightman, *J. Adhes.*, Vol 28, 1989, p 23, 35

84. D.K. Shaffer, H.M. Clearfield, C.P. Blankenship, Jr., and J.S. Ahearn, in *Proceedings of the 19th SAMPE Technical Conference*, Society for the Advancement of Material and Process Engineering, 1987, p 291

85. D.E. Packham, in *Adhesive Joints: Formation, Characteristics, and Testing*, K. Mittal, Ed., Plenum Publishing, 1984

86. J.R.G. Evans and D.E. Packham, *J. Adhes.*, Vol 10, 1979, p 39

87. P.J. Hine, S. El Muddarris, and D.E. Packham, *J. Adhes. Sci. Technol.*, Vol 1, 1987, p 69

88. A.J. Duke, *J. Iron Steel Inst.*, Vol 207, 1969, p 570

89. A.V. Pocius, D.A. Wangness, C.J. Almer, and A.G. McKown, in *Proceedings of the Adhesion Society Conference*, 1984

90. D.L. Trawinski, D.K. McNamara, and J.D. Venables, *SAMPE Q.*, Vol 15, 1984, p 6

91. Y. Gilibert and G. Verchery, in *Adhesive Joints: Formation, Characteristics, and Testing*, K. Mittal, Ed., Plenum Publishing, 1984, p 69

92. J.F. Watts and J.E. Castle, *J. Mater. Sci.*, Vol 19, 1984, p 2259

93. E.H. Andrews and N.E. King, *J. Mater. Sci.*, Vol 11, 1976, p 2004

94. V.G. Zadorozhnyi, D.M. Rafalovich, and I.L. Roikh, *Elecktronnaya Obrab. Mater.*, Vol 1, 1977, p 43

95. K. Ruhsland, in *Adhesive Joints: Formation, Characteristics, and Testing*, K. Mittal, Ed., Plenum Publishing, 1984, p 257

96. P.J. Hine, D.E. Packham, and S. El Muddarris, in *Proceedings of the International Adhesion Conference*, Plastics and Rubber Institute, 1984, p 10.1-10.4

97. W. Russell, R. Rosty, and D.W. Levi, Technical Report ARSCD-TR-82005, U.S. Army Research and Development Center, June 1982

98. W.J. Russell, R. Rosty, K.M. Adelson, M.J. Bodnar, R.F. Wegman, E.A. Garnis, and D.W. Levi, Technical Report ARSCD-TR-83014, U.S. Army Research and Development Center, Oct 1983

99. A.K. Mindyuk, B.N. Lavrishin, and A.M. Krutsan, *Sov. Mater. Sci.*, Vol 19, 1983, p 560

100. V. Leroy, *Mater. Sci. Eng.*, Vol 42, 1980, p 289

101. N.N.C. Hsu, J.S. Noland, J.S. Brinen, and A.W. Graham, in *Conference Proceedings of the Seventh Annual Adhesion Society Meeting*, 1984

102. E.A. Podoba, S.P. Kodali, R.C. Curley, D.K. McNamara, and J.D. Venables, *Appl. Surf. Sci.*, Vol 9, 1981, p 359

103. N.L. Bottrell, *Sheet Metal Ind.*, Vol 42, (No. 461), 1965, p 667

104. A.W. Henderson, in *Aspects of Adhesion*, University of London Press, 1965, p 33-46

105. H.L. David, *Met. Deform.*, Vol 26, 1974, p 45

106. H.L. David, *Met. Deform.*, Vol 19, 1973, p 27

107. N.L. Rodgers, in *Proceedings of the Conference on Specialized Cleaning, Finishing and Coating Processes*, American Society for Metals, Feb 1980, p 63-77

108. K.W. Allen, *J. Adhes.*, Vol 8, 1977, p 183

109. L.G. Look, in *Proceedings of the Conference on Technology Transfer*, Society for the Advancement of Material and Process Engineering, 1981, p 150

110. A.V. Pocius, C.J. Almer, R.D. Wald, T.H. Wilson, and B.E. Davidian, *SAMPE J.*, Vol 20, 1984, p 11

111. T. Smith, *J. Adhes.*, Vol 17, 1984, p 1

112. R.P. Haak and T. Smith, *Int. J. Adhes. Adhes.*, Vol 1, 1983, p 5

113. A.G. Hirko, in *Proceedings of the Conference on Hi-Tech Review 1984*, Society for the Advancement of Material and Process Engineering, April 1984

114. D.L. Trawinski, in *Proceedings of the Conference on Advancing Technology in Materials and Processes*, Society for the Advancement of Material and Process Engineering, 1985, p 1065-1072

115. C. Bishof, A. Bauer, R. Kapelle, and W. Possant, *Int. J. Adhes. Adhes.*, Vol 5, 1985, p 97

116. D.L. Trawinski and R.D. Miles, in *Proceedings of the Conference on Hi-Tech Review 1984*, Society for the Advancement of Material and Process Engineering, 1984, p 633

117. H.E. Chandler, *Met. Prog.*, Vol 121, 1982, p 38

118. T. Sugama, in *Proceedings of the Conference on Structural Adhesives Needs Through the 1990s*, U.S. Army Research Office, 1985, p 11

119. F.J. Spaeth, in *Proceedings of the Conference on Finishing '83*, Society of Manufacturing Engineers, Feb 1983, p T-1 to T-4

120. M. Kronstein, Paper 800444, in *Proceedings of the SAE Congress*, Society of Automotive Engineers, 1980

121. E. Janssen, *Mater. Sci. Eng.*, Vol 42, 1980, p 309

122. M.C. Ross, R.F. Wegman, M.J. Bodner, and W.C. Tanner, *SAMPE J.*, Vol 12 (No. 5), 1976, p 4

123. L. Gan, H.W. Ong, and T.L. Tan, *J. Adhes.*, Vol 16, 1984, p 233

124. H.W. Ong, L. Gan, and T.L. Tan, *J. Adhes.*, Vol 20, 1986, p 117

125. S. Aloru, P. Bartoli, M. Leoni, and M. Memmi, Paper 19 in *Proceedings of the Conference of the International Deep Drawing Research Group*, Hoogoven Ijmiuden BV, 1985

126. M. Debski, M.E.R. Shanahan, and J. Schultz, *Int. J. Adhes. Adhes.*, Vol 6, 1986, p 145

127. M. Debski, M.E.R. Shanahan, and J. Schultz, *Int. J. Adhes. Adhes.*, Vol 6, 1986, p 150

128. N.N.C. Hsu, J.S. Noland, J.S. Brinen, and A.W. Graham, in *Proceedings of the 7th Annual Meeting of the Adhesion Society*, 1984

129. P. Commercon and J. Wightman, *J. Adhes.*, Vol 22, 1987, p 13

130. R.S. Charnock, *Int. J. Adhes. Adhes.*, Vol 5, 1985, p 201

131. J.R. Arnold, in *Proceedings of the Automotive and Corrosion Prevention Conference*, Society of Automotive Engineers, 1986, p 21

Surface Preparation of Plastics*

SURFACE PREPARATION of plastics before adhesive bonding is critical to bond reliability and integrity. The selection of surface preparation methods requires careful evaluation and analysis. Factors influencing the decision include cost, size of the components, availability of equipment, the type of adhesive being used, the material(s) being bonded, joint design, the bonding process chosen, and the availability of adequate process control.

Proper surface preparation ensures that joint failure will occur within the adhesive, not at the adhesive-adherend interface. In this type of fracture, described as cohesive, a layer of adhesive remains on both adherends. When fracture occurs at the adhesive/adherend interface, failure is said to be adhesive. Joint failure modes can be described in terms of the percentage of failure that is adhesive or the percentage that is cohesive. A specimen with 100% cohesive failure would be ideal in terms of surface preparation.

More information on test methods and the characterization of failures of adhesive bonds is available in the Section "Testing and Analysis" in this Volume. The present article describes the steps necessary to prepare the surfaces of all types of polymeric materials for adhesive bonding. The surface preparation of other classes of materials, that is, metals, ceramics and glasses, electronic materials, and biological materials, is discussed in detail in the other articles in this Section of the Volume.

Cleaning

Regardless of the materials, processes, and process controls used, contaminated adherends are not amenable to adhesive bonding. After cleaning to remove obvious surface contamination, chemical or physical treatments can be used to ensure the complete removal of dirt, oils, and other contaminants and to produce a surface ready for bonding. These include solvent cleaning, intermediate cleaning, and chemical treatment. The effects of these treatments on adhesive bond strength in various polymers is shown in Table 1.

Solvent Cleaning

Solvent cleaning does not chemically or physically alter the surface being cleaned. Solvent cleaning methods include wiping, immersion, spraying, vapor degreasing, and ultrasonic scrubbing. *Caution*: Regardless of the solvent cleaning method chosen, safety is an important consideration. Many organic solvents carry with them warnings regarding toxicity, flammability, incompatibility with other chemical agents, and proper use of equipment. All written safety procedures and manufacturer instructions should be strictly followed.

Wiping is adequate for many surface preparation treatments. It is convenient and versatile; however, possible disadvantages include incomplete removal of contamination and recontamination of a freshly cleaned surface with unclean solvents. For these reasons, only fresh solvents should be used, and materials used for wiping should be clean and never immersed in the solvent.

Immersion is often adequate for removing light soil and is useful for relatively small parts. For more heavily contaminated components, scrubbing by agitation, brushing, wiping, or ultrasonic means may be necessary. After immersion, the parts must be rinsed. Rinsing is the critical step in immersion cleaning.

Spraying has the advantage of the scrubbing effect produced by the impingement of the solvent particles on the adherend surface. Loosened contaminants are washed away, and only clean solvent is added to the surface.

Vapor degreasing uses hot solvent vapors, which condense on the component being cleaned to dissolve and remove contaminants from the surface. It is effective in removing oils, greases, and waxes, as well as other contaminants. The choice of solvent is important in vapor degreasing. Solvents must be nonflammable, nonexplosive, and nonreactive during use; they must be capable of dissolving surface contaminants; they must have low boiling points, low heats of vaporization, and high vapor density to reduce losses; and they must conform to local pollution control ordinances. Commonly used vapor degreasing solvents include trichloroethylene, perchloroethylene, methyl chloroform, and methylene chloride.

Ultrasonic scrubbing uses high-frequency sound waves to produce agitation of the solvent bath and cavitation (repeated formation and collapse of tiny bubbles) on the surface of the component being cleaned. It quickly removes many contaminants, including particulates, strongly adherent soils, and insoluble materials. Ultrasonic scrubbing is often followed by a solvent rinse to remove residues and loosened particulates completely. This procedure need not employ organic solvents; it is often used with detergents and surfactants. The only restrictions are that the cleaning rinse not attack equipment or the component being cleaned.

Intermediate Cleaning

Intermediate cleaning includes any operation that removes surface contaminants by physical, mechanical, or chemical means without chemically altering the adherend surface. Solvent cleaning should always precede intermediate cleaning. Examples of intermediate cleaning procedures include alkaline or detergent cleaning, grit blasting, wire brushing, sanding, and abrasive scrubbing. More information on intermediate cleaning treatments for specific plastics is available in the sections "Preparation of Thermosetting Materials" and "Preparation of Thermoplastic Materials" in this article.

Table 1   Effect of various surface preparation treatments on lap-shear strength of adhesively bonded polymers

| Polymer | Relative bond strength(a) | | | |
	Control	Plasma	Abrasion	Chemical
High-density polyethylene	1.0	12.1	. . .	5.1
Polypropylene	1.0	221	. . .	649
Valox 310 polyester (General Electric Company)	1.0	18.9	2.9	1.0
Silicone rubber	1.0	>20	4.7	. . .

(a) Results normalized to the control for each material. Source: Branson International Plasma Corporation

*Adapted from *Adhesive Bonding*, MIL-HDBK-691B, *Military Standardization Handbook*, U.S. Department of Defense, March 1987

Table 2 Results of contact-angle testing of plastics after various surface treatments

Material	Contact angle, degree			
	Control	Plasma	Flame	Chemical
Polypropylene	87	22	87	60
Polyester	71	18	. . .	75
Polyvinyl chloride	90	35	. . .	79
Polycarbonate	75	33	. . .	76
Silicone rubber	93	17	. . .	. . .
High-density polyethylene	87	42	38	54

Source: MIL-HDBK-691B

Chemical Treatment

Chemical treatment, as the name implies, involves a change in the chemical nature of the adherend surface to improve its adhesion characteristics. Common chemical treatments include etching, priming, and plasma treatment. These are described briefly below; more detailed information is available in the later sections "Preparation of Thermosetting Materials" and "Preparation of Thermoplastic Materials."

Etching promotes adhesion and wettability of polymer adherend surfaces. Depending on the adherend material and the surface effects desired, a variety of etchants can be used. These are outlined in the sections "Preparation of Thermosetting Materials" and "Preparation of Thermoplastic Materials" in this article.

Primers are usually dilute adhesive solutions in an organic solvent. They are used to enhance wetting, protect the clean adherend surface, improve adhesive properties, serve as a barrier coat to prevent unwanted reactions between adhesive and adherend, and hold adhesive or adherend in place during assembly. The use of primers also results in joints with higher reliability and durability.

Plasma treatment is an effective means of preparing the surface of many polymers for bonding. The polymer surface is exposed to a gas plasma generated by glow discharge. This causes atoms to be expelled from the polymer surface, resulting in the formation of a strong, wettable skin with good adhesion characteristics. Adhesive bonds on plasma-treated surfaces are usually from two to four times stronger than those produced with untreated surfaces. Plasma treatment is a rapid way to prepare the surfaces of nearly any polymeric material, including polyester, polyethersulfone, polyaryl sulfone, polyphenylene sulfide, fluoropolymers, and nylons.

Evaluating Preparation Effectiveness

Tests are available to measure the effectiveness of surface preparation treatments both before and after bonding. Prebonding tests are designed to check the wettability of the prepared adherend surface. These tests, including the water-break test and the more quantitative contact-angle test, are briefly described here.

The water-break test is based on the premise that a clean, well-prepared surface will hold a continuous film of water rather than isolated droplets. The surface to be tested is covered with distilled water, then allowed to drain for about 30 s. A continuous water film indicates adequate preparation, whereas a break in the film is taken as evidence of a soiled or contaminated area on the surface. The surface should be recleaned until a water-break-free condition is obtained in testing.

The contact-angle test measures wettability by determining the contact angle between the prepared adherend surface and a drop of reference fluid, often distilled water. Large contact angles indicate poor wettability, whereas smaller angles show increasing wettability. Results of contact-angle tests on various polymers after different surface preparation treatments are shown in Table 2.

Postbonding Tests. After bonding, the joint may be destructively tested by pulling to failure. This allows the measurement of properties such as shear strength and peel strength, as well as the characterization of the joint failure mode. The Section "Testing and Analysis" in this Volume contains more information on the destructive testing of adhesively bonded joints.

Effect of Surface Exposure Time. It has been observed that the time between surface preparation and bonding, known as surface exposure time (SET), can affect the strength of adhesively bonded joints in lap-shear and peel tests. In general, increasing the SET results in reduced critical joint strength, especially at above normal temperatures and relative humidities. However, if prepared surfaces are protected and relative humidity is maintained at about 50%, up to 30 days may elapse between surface preparation and bonding without serious losses in joint integrity.

Preparation of Thermosetting Materials

Thermosetting plastics become rigid (cure) upon exposure to sufficient heat. Most thermosetting plastics are relatively easy to bond. Adhesive bonding is usually the only practical nonmechanical joining method for thermosetting plastics, because the materials are not soluble in solvent cements.

Many parts molded from thermosetting materials have a mold release agent on the surface that must be removed before adhesive bonding can be accomplished. These agents can usually be removed by washing or wiping with detergent or solvent, followed by light sanding to break the surface glaze. A final solvent wipe also is frequently used.

Solvents used for thermosetting materials include acetone, toluene, trichloroethylene, methyl ethyl ketone (MEK), low-boiling petroleum ether, and isopropanol. Surface abrasion can be accomplished using fine sandpaper, carborundum or alumina abrasives, metal wools, or steel shot.

The following discussion includes very brief descriptions of the properties of thermosetting plastics. Detailed information on the processing, properties, and applications of these materials is available in Volume 2 of *Engineered Materials Handbook*. Thermoset matrix composites, that is, thermosetting materials reinforced with whiskers or continuous fibers of glass, graphite, aramid, or other materials, are discussed in detail in Volume 1 of *Engineered Materials Handbook*.

Epoxy resins are among the most commonly used thermosetting materials. They have a range of desirable physical and mechanical properties, including moderate to high strength, good dimensional stability and creep resistance, and temperature stability over a range of −75 to 425 °C (−100 to 800 °F). Epoxy composites are available with numerous continuous and discontinuous reinforcing phases. The preparation of composite materials for adhesive bonding is discussed in the article "Surface Preparation of Composites" in this Volume.

A common surface preparation treatment for epoxy resins is to:

- Wipe with any of the solvents listed above
- Abrade by the methods listed above
- Wipe with a clean cloth to remove grit
- Repeat the solvent treatment

A silane-type primer diluted with methanol or ethanol may then be applied, although this step is optional.

Phenolics and Polyesters. Phenolics combine high strength, good electrical properties, and good dimensional stability. Polyesters have good resistance to oils and solvents. Both materials are frequently bonded. The surface preparation treatment recommended for epoxies is also suggested for phenolics and polyesters.

Polyimides have exceptional thermal stability, with some materials able to withstand temperature excursions to 480 °C (900 °F). The recommended treatment for parts fabricated from Vespel polyimide from DuPont is:

- Solvent clean in trichlorofluoroethylene (Freon TF), perchloroethylene, or trichloroethylene to remove surface contaminants
- Wet or dry abrasive blast to enable application of a uniform thickness of adhesive
- Repeat solvent cleaning to remove grit
- Dry

Polyimide components can also be prepared by etching according to the following procedure: degrease in acetone, etch for 1 min in 5% sodium hydroxide solution at a temperature of 60 to 90 °C (140 to 195 °F), rinse in cold water, and air dry.

Silicone resins are used as thermal and electrical insulation compounds, paint vehicles, molding compounds, laminates, impregnating varnishes, encapsulating materials, and baked-on release agents. They can be prepared for bonding by wiping with acetone or other organic solvent; sand, grit, or vapor blasting; wiping with a clean cloth to remove grit and particles; and repeating the solvent treatment.

Thermosetting polyurethanes can be obtained in a variety of different densities and forms, including sheet and molding and casting resins. They are excellent for cryogenic applications and have good chemical resistance, skid resistance, and electrical properties. Surface preparation usually involves wiping with acetone or MEK, abrading, wiping to remove particles and grit, repeating the solvent wipe, drying, and applying a urethane or silane primer.

Diallyl phthalate has exceptional electrical insulating properties, high-temperature stability, and good resistance to most chemicals and moisture. It is available with a variety of fillers, including glass and minerals. A recommended surface preparation is:

- Wipe with one of the organic solvents recommended for thermosetting plastics
- Abrade surface by sand, grit, or vapor-blasting, or by using steel wool
- Wipe with a clean, dry cloth to remove grit and particles
- Repeat solvent wiping

Urea-formaldehyde resins are hard, rigid, and abrasion resistant. They have superior electrical properties, excellent dimensional stability, high impact strength, and good solvent and creep resistance. The surface preparation procedure given earlier for epoxies may be used, or the following procedure may be followed: scrub with an abrasive detergent; rinse first with tap water, then with deionized water; dry; sand; wipe with isopropanol; dry; and prime or bond.

Preparation of Thermoplastic Materials

Unlike thermosetting resins, thermoplastic materials will repeatedly soften upon

heating and harden when cooled. This behavior is a result of the linear macromolecular structure of thermoplastics.

Also, thermoplastic surfaces, unlike those of thermosetting materials, usually require physical or chemical modification to achieve acceptable bonding. This is especially true of crystalline thermoplastics such as polyolefins, linear polyesters, and fluoropolymers. Methods used to improve the wettability and adhesion of thermoplastic surfaces include:

- Oxidation by chemical means or flame treatment
- Roughening of the surface by electrical discharge
- Gas plasma treatment (see the section "Plasma Treatment" in this article)

The following discussion includes very brief descriptions of some thermoplastics. Detailed information on the processing, properties, and applications of these materials is available in Volume 2 of *Engineered Materials Handbook*. Thermoplastic composites are discussed in detail in Volume 1 of *Engineered Materials Handbook*.

Nylons are semicrystalline thermoplastics that are not solvent sensitive; therefore, relatively strong solvents are used for surface cleaning. A wide range of nylon materials, with a corresponding breadth of properties, is available.

Solvents for nylon include acetone, MEK, perchloroethylene, trichloroethylene, and Freon TIC. Detergents also may be used instead of organic solvents.

A common method for the surface preparation of nylon is:

- Wash with acetone
- Dry
- Hand sand with 120-grit abrasive cloth
- Remove sanding dust with a short-bristled stiff brush

Bonding should be completed as soon as possible after preparation of nylon parts, which should be slightly warmed before bonding. Nitrile-phenolic, resorcinol-formaldehyde, vinyl-phenolic, or silane primers should be used. Epoxy adhesives can be used for nylon-to-metal bonding. Because most nylon materials readily absorb moisture, relative humidity should be closely controlled throughout all phases of preparation and bonding.

Acetal Copolymer. This material, available as Celcon from Hoechst Celanese Corporation, is a highly crystalline copolymer with excellent resistance to solvents and chemicals. Because of this resistance, acid etching treatments may be used for surface preparation. Chromic acid or hydrochloric acid treatments, followed by the application of a thin coat of specially developed primer, are suggested.

The procedure for chromic acid etching is as follows: wipe parts with acetone; air dry;

etch surfaces for 10 to 15 s in a solution of (by weight) 400 parts sulfuric acid, 11 parts potassium dichromate, and 44 parts deionized water; flush with tap water; rinse with deionized water; and oven dry at 60 °C (140 °F).

Concentrated hydrochloric acid at room temperature removes approximately 0.008 to 0.013 mm (0.0003 to 0.0005 in.) of Celcon surface each minute of exposure. Parts should be thoroughly rinsed and dried at room temperature for 4 h after exposure, and heat-formed or machined Celcon components should be stress relieved before etching.

Acetal homopolymer (Delrin from DuPont) is a highly crystalline material with excellent solvent resistance. The following procedure is recommended for chromic acid etching:

- Wipe with acetone or MEK
- Immerse 10 to 20 s at room temperature in a solution (by weight) of 15 parts potassium dichromate, 24 parts tap water, and 300 parts concentrated sulfuric acid
- Rinse in running tap water for a minimum of 3 min
- Rinse in distilled or deionized water
- Dry in air-circulating oven at 40 °C (100 °F) for 1 h

DuPont has also developed a patented surface preparation process termed Satinizing, which results in strong adhesion to the Delrin surface. In this procedure the Delrin parts are dipped for 10 to 30 s in a mixture of (by weight) 0.5% pyrogenic silica, 3% 1,4 dioxane, 0.3% paratoluene sulfonic acid, and 96.2% perchloroethylene at 80 to 120 °C (175 to 250 °F). *Caution*: Although mixing these components should pose no particular problem, formaldehyde fumes are released during and after dipping. The bath area should be well ventilated to ensure the removal of these fumes.

After dipping, the parts are transferred to an oven at 40 to 120 °C (100 to 250 °F) to induce rapid etching of the surface. Etch depth is controlled by oven temperature and exposure time; at 120 °C (250 °F), adequate etching should take no longer than 1 min. The oven area should be ventilated to remove any residual formaldehyde fumes. The parts are then sprayed with hot (70 to 80 °C, or 160 to 175 °F) water to stop the etching and are subsequently air dried.

Acrylonitrile-butadiene-styrene (ABS) parts may be prepared by abrasion or by chromic acid etching. The abrasive treatment involves sanding with medium-grit sandpaper, wiping to remove dust, oven drying at 70 °C (160 °F) for 2 h, and priming.

The chromic acid etching procedure begins with degreasing in acetone. This is followed by etching for 20 min in a solution of (by weight) 26 parts concentrated sulfuric acid, 3 parts potassium dichromate, and 11 parts water. Parts are rinsed in tap water and distilled water and are then dried in warm air.

Cellulosics are solvent cemented unless they are to be bonded with dissimilar materials. In this case conventional adhesives must be used, and the recommended surface preparation steps are:

- Solvent degrease in methanol or isopropanol
- Grit or vapor blast, or abrade with 220-grit emery cloth
- Repeat solvent degreasing
- Heat 1 h at 95 °C (200 °F) and apply adhesive while hot

Fluoropolymers generally have good resistance to chemicals and tend to be difficult to prepare for adhesive bonding because of their inertness. Fluoropolymers that are all prepared for bonding by an identical treatment include:

- Ethylene-chlorotrifluoroethylene (E-CTFE) copolymer (Halar, Allied Chemical Company)
- Fluorinated ethylene propylene (FEP) copolymer (Teflon FEP, DuPont)
- Perfluoroalkoxy (PFA) resins
- Polyfluorotrichloroethylene (PCTFE)
- Polytetrafluoroethylene (PTFE)

Surface preparation for these materials consists of wiping with acetone; treating with commercially available sodium naphthalene solutions (use extreme caution and follow all manufacturer recommendations); removing from solution with metal tongs; washing with acetone and then with distilled or ionized water; and drying in an air-circulating oven at 40 °C (100 °F) for 1 h.

Polyvinylidene fluoride (PVDF) is a high molecular weight homopolymer (Kynar from Pennwalt Corporation) with good resistance to many solvents. It is partially soluble in strongly polar solvents such as ketones and esters. Recommended surface preparation is to clean in methanol or isopropanol; abrade by sanding, scouring, or abrasive blasting; and repeat the solvent cleaning. Gas plasma treatment also is effective for this material.

Ionomers (Surlyn resins from DuPont) do not develop high-strength adhesive bonds. When bonding these materials, the suggested surface preparation technique is:

- Solvent degrease in acetone or MEK
- Grit blast with 180-grit alumina abrasive, vapor blast, or abrade with 100-grit emery cloth
- Repeat solvent degreasing step

Polyphenylene Oxide (PPO) Materials. Noryl polystyrene-modified PPO resins from General Electric have excellent dimensional stability over a wide temperature range. They are usually joined by solvent cementing. After cleaning with isopropanol or a commercial detergent, Noryl PPO parts

may be prepared by sanding or by chromic acid etching in a series of steps:

- Etch 1 min at 80 °C (175 °F) in a solution of 375 g concentrated sulfuric acid, 18.5 g potassium dichromate, and 30 g water
- Rinse in distilled water
- Dry

Studies of unmodified PPO materials indicate that the chromic acid solution may become ineffective for these materials after one use. This situation can be avoided by vacuum blasting and then wiping with acetone.

Polysulfone (Udel), Polyarylate (Ardel), and Polyaryl Sulfone (Astrel) Resins. These materials, from Union Carbide Corporation (Udel and Ardel) and 3M Company (Astrel), have high service temperatures and good resistance to a variety of chemicals. They can be prepared by sandblasting, solvent washing, or acid etching. The recommended steps for acid etching are:

- Ultrasonic clean in an alkaline solution
- Wash with cold water
- Bathe 15 min at 65 to 70 °C (150 to 160 °F) in a solution (by weight) of 3.4% sodium dichromate and 96.6% concentrated sulfuric acid
- Wash with cold water
- Dry at 65 °C (150 °F) in an air-circulating oven

Polycarbonate. Solvent cementing is recommended for bonding this resin to itself or to materials soluble in the same solvents. Adhesive bonding must be used for bonding to dissimilar plastics, metal, or wood. Recommended solvents for cleaning are methanol, isopropanol, petroleum ether, heptane, and white kerosene. Ketones, toluene, trichloroethylene, benzol, and a number of other solvents such as paint thinners cause crazing or cracking.

After cleaning, possible surface preparations include flame treatment and abrasion with fine-grit cloth or sandpaper. In flame treatment, the part is passed through the oxidizing portion of a propane flame. Treatment is complete when all surfaces are polished to a high gloss, and the part must be cooled 5 to 10 min before bonding.

Polyesters. Thermoplastic polyesters, including polyethylene terephthalate (PET) and polybutylene terephthalate (PBT), are available from a variety of manufacturers. PET is used primarily in films, whereas PBT is used for molding. Both materials have good general resistance to solvents and chemicals.

One surface preparation treatment for Valox thermoplastic polyester from General Electric is to abrade lightly with 240-grit sandpaper, then degrease with toluene or trichloroethylene. Plasma treatment, however, results in bonds 3 to 4 times stronger than those obtained with this procedure.

Celanex polyester from Hoechst Celanese Corporation may be prepared by

abrading with sandpaper and degreasing by wiping with a solvent such as acetone. Tenite thermoplastic polyester from Eastman Chemical Products Inc. can be prepared by dipping for 10 to 15 s in methylene chloride before bonding.

Polyetheretherketone (PEEK) from ICI Americas Inc. is a high-performance material developed primarily as a coating and insulation material for wiring. Isopropanol, toluene, and trichloroethylene can be used to clean PEEK surfaces, which can then be roughened to enhance adhesion. Degreasing should be repeated after abrasive treatment. Bond strength can be substantially improved (one study indicated ~20% increase) by the oxidizing flame treatment, described in the section "Polycarbonate" in this article.

Polyethylene (PE). Adhesives cannot link chemically or mechanically with untreated polyethylene surfaces, and because polyethylene is relatively inert in most solvents, neither can solvent cementing be used. A number of surface preparation methods, including flame, chemical, electronic, and primer treatments, are in use (Table 3). A chromic acid etching method is convenient for complex parts:

- Wipe with acetone, MEK, or xylene
- Immerse for 60 to 90 min (room temperature) or 30 to 60 s (70 °C, or 160 °F) in a solution (by weight) of 15 parts potassium dichromate, 24 parts tap water, and 300 parts concentrated sulfuric acid
- Rinse in running tap water for at least 3 min
- Dry in an air-circulating oven at 40 °C (100 °F) for 1 h

The oxidizing flame treatment, described in the section "Polycarbonate," provides stronger bonds than does chromic acid etching. After flame treatment, the surface should be scoured with soap and water, rinsed in distilled water, and dried at room temperature. Close control is required during heating to prevent warpage and distortion of parts. Gas plasma treatment is also effective for polyethylene and is recommended for geometrically complex parts and components that require very strong bonds.

Polystyrene and styrene acrylonitrile (SAN) parts are prepared using identical methods. These methods range from simple abrasion and removal of dust particles to etching with a sodium dichromate/sulfuric acid solution. The steps for the latter are:

- Degrease in methanol or isopropanol
- Immerse for 3 to 4 min at 100 to 105 °C (212 to 220 °F) in a solution (by weight) of 90 parts concentrated sulfuric acid and 10 parts sodium dichromate
- Rinse thoroughly in distilled water
- Dry below 50 °C (120 °F)

Table 3 Effect of various surface treatments on lap-shear strength of adhesively bonded high-density polyethylene

Control MPa (psi)	Flame MPa (psi)	Abrasion MPa (psi)	Acid(b) MPa (psi)	Acid(c) MPa (psi)	Acid(d) MPa (psi)	Plasma(e) MPa (psi)	Plasma(f) MPa (psi)
— Lap-shear strength after indicated treatment(a)							
Bonded to polyamide-modified epoxy							
0.32 (46)	3.31 (480)	1.34 (195)	3.28 (475)	3.44 (499)	3.43 (497)	3.20 (464)	3.21 (466)
Bonded to styrene-nonsaturated polyester							
0.59 (85)	2.99 (434)	1.21 (175)	1.91 (277)	2.46 (357)	2.72 (394)	1.31 (190)	1.81 (263)
Bonded to nitrile-rubber phenolic							
0.30 (44)	0.95 (138)	0.39 (56)	0.66 (96)	0.76 (110)	0.77 (112)	. . .	. . .

(a) Single-lap shear specimens were prepared from 100 × 25 × 25 mm (4 × 1 × 1 in.) polyethylene coupons. Overlaps of 12.7 mm (0.5 in.) were used. After bonding, all specimens were cured for 72 h at room temperature, postcured for 4 h at 40 °C (100 °F), and air conditioned at room temperature and 50% relative humidity for at least 24 h before testing. (b) Oven dried at 90 °C (195 °F) after treatment. (c) Wiped and air dried at room temperature after treatment. (d) Acetone-dried after treatment. (e) Helium, 30 s exposure. (f) Oxygen, 30 s exposure. Source: MIL-HDBK-691B

For selective etching, a thixotropic paste made with 3 parts concentrated sulfuric acid, 1 part powdered potassium, and enough fused silica earth to obtain the desired consistency can be applied to mating surfaces. The parts are then heated to 80 °C (180 °F) for 3 to 4 min before the paste is removed from the surface by rinsing with distilled water.

Polyether sulfone (PESV) is available as Victrex resin from ICI Americas Inc. It can be bonded to itself using solvent cementing or ultrasonic welding techniques. Parts made from PES can be cleaned using ethanol, methanol, isopropanol, or low-boiling petroleum ether. Solvents that should not be used are acetone, MEK, perchloroethylene, tetrahydrofuran, toluene, and methylene chloride.

Polymethyl methacrylate (PMMA) is available as Plexiglas (Rohm & Haas Company) and Lucite (DuPont). Solvent cementing is the usual method for bonding PMMA to itself. Surface preparation for bonding to a dissimilar material can be accomplished by wiping with methanol, acetone, MEK, trichloroethylene, isopropanol, or detergent; abrading with fine (180- to 400-grit) sandpaper or abrasive blast; wiping with a clean, dry cloth to remove particles; and repeating the solvent wipe.

Polyphenylene Sulfide (PPS). A crystalline thermoplastic sold as Ryton from Phillips Petroleum Company, PPS has a high melting point, good thermal stability, and excellent chemical resistance. It is almost completely inert below about 190 to 205 °C (375 to 400 °F). A suggested surface preparation method is to solvent degrease in acetone, sandblast, and repeat the degreasing step. Alternatively, surfaces may be wiped with ethanol-soaked paper, sanded with 120-grit sandpaper, and cleaned with a stiff-bristled brush.

Polyetherimide (Ultem from General Electric Company, Plastics Operations) is an amorphous thermoplastic with high heat-distortion temperature and good resistance to a range of chemicals and solvents. Only partially halogenated hydrocarbons (for example, methylene chloride and trichloroethane) will dissolve Ultem; therefore, these compounds should not be used for cleaning.

Polypropylene (PP) is similar to polyethylene, and chromic acid etching methods are generally similar to those described in the section "Polyethylene" in this article. Polypropylene components, however, require longer immersion, that is, 1 to 2 min at 70 °C (160 °F), as opposed to 30 to 60 s at that temperature for polyethylene parts. The oxidizing flame and gas plasma treatments described in the section "Polyethylene" are also applicable.

SELECTED REFERENCES

- M.J. Bodnar, Ed., *Processing for Adhesives Bonded Structures*, Interscience, 1972
- J.R. Hall, Activated Gas Plasma Surface Treatment of Polymers for Adhesive Bonding, *J. Appl. Polym. Sci.*, Vol 13, 1969, p 2085-2096
- A.H. Landrock, "Processing Handbook on Surface Preparation for Adhesive Bonding," Picatinny Arsenal Technical Report 4883, 1975
- A.H. Landrock, *Adhesives Technology Handbook*, Noyes Publications, 1985
- E.M. Petrie, Plastics and Elastomers as Adhesives, in *Handbook of Plastics and Elastomers*, C.A. Harper, Ed., McGraw-Hill, 1975
- J. Shields, *Adhesives Handbook*, 3rd ed., Butterworths, 1984
- I. Skeist, Ed., *Handbook of Adhesives*, 2nd ed., Van Nostrand Reinhold, 1977
- R.C. Snogren, *Handbook of Surface Preparation*, Palmerton Publishing, 1974

Surface Preparation of Composites

John G. Dillard, Department of Chemistry, Virginia Polytechnic Institute and State University

THE REQUIREMENTS for adhesive bonding related to composite materials necessitate that two aspects be addressed. One is the bonding of fibers or fillers to the matrix in the composite, and the other is the bonding of the fabricated composite to other adherends, including metals, plastics, and other composites.

The adhesive bonding of fibers or fillers to the matrix resin usually requires a unique surface treatment of the reinforcing material. The bonding of fabricated composites to other adherends also may require a surface treatment or at least an engineering of the adherend surfaces so that adherends are easily bonded using the adhesive selected for joining. This article first describes the surface treatment of fibers, including glass, polyaramid, and carbon/graphite, in order to promote adhesive bonding of the fiber to the matrix resin. This extensive discussion is followed by a description of the treatment of composite and plastic (thermoplastic and thermoset) material surfaces in preparation for adhesive bonding. The practical aspects of the latter topic are described in detail in the article "Surface Preparation of Composites for Adhesive-Bonded Repair" in the Section "Bonded Structure Repair" in this Volume. That Section includes other articles that also relate to composite structures. Although thermal welding and solvent cementing represent other methods for joining plastics or composites, they are not within the scope of this article, and the reader is referred to Ref 1 and 2.

Fiber-Matrix Adhesion

Parameters that affect the adhesion of glass, polyaramid, and carbon/graphite fibers to various matrices are described below.

Glass

Glass fiber reinforced composites have been used as body materials for automobiles (Ref 3, 4), for marine applications (Ref 5), and in airplane construction (Ref 6). Glass fiber reinforcement may be combined with thermoplastic or thermoset polymers (Ref 7). In the preparation of glass fibers (Ref 7, 8), attention is paid to the handleability of the fibers, to the introduction of

coatings to ensure bonding with the resin matrix, and to the techniques or procedures that involve the fiber in the fabrication process (Ref 9). Components of the fiber finish, or sizing, may include lubricants, coupling agents/adhesion promoters, film formers, wetting agents, and binders to satisfy the requirements for incorporating glass fiber into the polymer matrix (Ref 10, 11). Significant effort has been devoted to understanding the role of the coupling agent (Ref 12-22) in relation to the performance of glass fiber reinforced composites. The properties and preparation characteristics of glass-filled (short glass fibers) thermoplastics are discussed in Ref 8, but a detailed discussion of adhesive interaction between fiber and resin is not presented.

The principal goal in treating the glass fiber is to permit its bonding to the resin matrix. The compound(s) used to treat the glass fiber are coupling agents. Historically, silane compounds have been developed as effective coupling agents (Ref 12-14). Plueddemann (Ref 12, 15) has indicated that there is no "best" primer/coupling agent for the adhesion of filler or fiber with a given polymer. It is necessary to optimize the primer compatibility and degree of condensation with the properties of the resin. However, once a silane-containing primer has been developed for a given resin, the primer should be effective for bonding with glass, ceramic compounds, and most metals (Ref 12-15).

Organosilicon compounds have been prepared to contain functional groups that adhere strongly to glass and also interact with the matrix resin. Silane coupling agents with the general formula X-Si$(OR)_3$ have been employed most extensively for the modification of glass surfaces (Ref 12, 14, 15). The X group is selected to be compatible with the polymer matrix, and the R group is usually methyl or ethyl so that hydrolysis can occur to obtain the silanetriol. The treatment of glass fibers utilizes the —OH functional groups on the glass surface, which react by the elimination of water with —OH groups of the silanetriol. Thus, as a result of the silane derivatization, the polar glass surface can be altered chemically to exhibit the functionality of the —X group. The wide variety of —X groups

available (Ref 12, 14) permits alteration of the glass surface chemistry so that bonding with almost any adhesive is possible. An important advantage of the use of silane coupling agents is the enhancement of durability and performance (Ref 12-17).

Although the concept of modifying the glass surface to be compatible with a given resin is easily understood, a number of factors influence the structure of the silane layer and thus the chemical and physical characteristics of the resulting composite (Ref 16). Among these factors are the structure of the silane in solution, which is dependent on the pH of the solution, the preparation time, and the temperature; the chemical nature of the —X group in the silane; the postcure conditions, especially the drying conditions; the topography of the glass fiber; and the chemical composition of the reinforcement surface (Ref 16). The interaction of the coupling agent with the surface leads to —Si—O— surface bonds and also cross linking of the silane to produce polysiloxanes attached to the filler or fiber surface.

Treatment of glass fibers may result in the deposition of multilayers of coupling agent on the surface (Ref 16). Plueddemann (Ref 12-14) has commented that adhesion to fillers may require the deposition of up to 100 monolayer equivalents of the coupling agent on the surface. Interaction of the —X group of the silane with the matrix resin may occur via chemical processes or by interdiffusion and the formation of an interpolymer network (Ref 12, 14, 16). The recognized necessity for interpolymer network formation is based on the knowledge that certain coupling agents are effective in enhancing the interaction between matrix and fiber or filler, where no chemical reaction between the constituents takes place (Ref 12-16).

Plueddemann (Ref 12, 14) has adopted rapid screening tests to evaluate potential resin-curative combinations to yield durable bonds for glass with thermoset and thermoplastic materials (Ref 14, 15). The procedure is to treat commercial microscope slides by dipping the slide in 0.5% aqueous primer solution or by wiping the slide with a 10 to 25% solution of the prehydrolyzed silane in alcohol. Thermoset or thermoplastic materials are then adhered to the slide.

Table 1 Silane coupling agents, X-Si(OR)$_3$, wet adhesion

X functional group	Chemical structure	Thermoset matrix resin and cure conditions(a)						
		Polyester with benzoyl peroxide at 130 °C (265 °F)	Polyester with MEK peroxide at 70 °C (160 °F)	Epoxy with diethylene triamine at room temperature	Epoxy with amide at room temperature	Epoxy with anhydride at 150 °C (300 °F)	Epoxy with m-phenylene diamine at 150 °C (300 °F)	Epoxy with dicyanate at 175 °C (350 °F)
Chloropropyl	Cl(CH$_2$)$_3$-Si(OCH$_3$)$_3$	NR	NR	VG	G	G	E	VG
Epoxy	CH$_2$—CH—CH$_2$—O—(CH$_2$)$_3$Si(OCH$_3$)$_3$	G	NR	E	VG	G	VG	VG
Methacrylate	CH$_2$=CH—C—O—(CH$_2$)$_3$—Si(OCH$_3$)$_3$ CH$_3$ O	VG	VG	VG	NR	G	VG	VG
Primary amine	NH$_2$—(CH$_2$)$_3$—Si(OC$_2$H$_5$)$_3$	NR	NR	NR	NR	NR	NR	G
Ethylenediamine	NH$_2$(CH$_2$)$_2$NH(CH$_2$)$_3$Si(OCH$_3$)$_3$	NR	G	NR	G	NR	G	VG
Mercapto	HS(CH$_2$)$_3$Si(OCH$_3$)$_3$	G	G	VG	G	NR	G	VG
Cationic styryl	CH$_2$=CH—C$_6$H$_4$CH$_2$NH(CH$_2$)$_2$NH(CH$_2$)$_3$Si(OCH$_3$)$_3$·HCl	G	E	G	G	G	VG	G
Mixed phenyl silane	(OCH$_3$)$_3$/NH$_2$(CH$_2$)$_2$NH(CH$_2$)$_3$Si(OCH$_3$)$_3$	NR	NR	VG	NR	G	VG	VG
Modified melamine resin	CH$_2$—CH—CH$_2$-O—(CH$_2$)$_3$Si(OCH$_3$)$_3$ CH$_3$	G	G	VG	E	VG	NR	G
Aluminum alkoxide	Al(O—CH—CH$_2$CH$_3$)$_3$	...	...	G	...	...	NR	G
H$_2$O test limits		48 h, boil	...	7 day, 70 °C (160 °F)	...	7 day, boil	...	...

(a) MEK, methyl ethyl ketone; NR, not recommended; G, good (significant improvement compared to the control, perhaps 100-fold improvement in time to failure); VG, very good (up to 1000-fold improvement in performance compared to the control); E, excellent (most outstanding performance of a series of treatments). Source: Ref 14

The specimen peel strength test or other strength tests are carried out to evaluate bonding effectiveness. Specimens are also soaked in room-temperature or elevated-temperature water to determine bond durability. Based on the debonding results, a scale of effectiveness is established.

The screening results from Ref 14 are summarized in Table 1 for thermosets and in Table 2 for thermoplastics. A general observation is that it is possible to select a single silane primer or combination of primers to enhance glass-matrix bonding and durability. The effectiveness of the primer is attributed to the formation of hydrophobic cross-linked siloxanes at the glass-primer interface and to interpenetration and cross linking of the polymer within the primer-polymer interphase. Enhanced performance in water-exposure tests is also related to the equilibrium established (Ref 15, 18) in the hydrolysis and re-formation of siloxane bonds involving silane coupling agents.

An understanding of the interaction of the coupling agent with the glass surface and with the polymer matrix has been derived primarily from extensive infrared (IR) spectroscopic studies (Ref 16, 17, 19-22). A review of the important findings from the IR studies has recently been presented (Ref 19). A significant conclusion is that at the silane-glass interface, chemical bonding occurs but no penetration of silane into the glass takes place. On the other hand, at the silane-matrix (epoxy) interface, silane is dispersed in the resin. Koenig and Emadipour (Ref 17) have identified several regions in glass-epoxy composites:

- Glass fiber
- Glass-silane interface. Covalent bonds involve interconnecting regions of Si—O—Si bonds
- Chemisorbed region, which contains higher silane oligomers exhibiting cross linking; chemisorbed species are extractable by hot water
- Physisorbed region, which contains low silane oligomers. There is limited bonding among the molecules. Physisorbed species are extractable by cold water
- Silane-epoxy region, which has epoxy-low silane oligomers. Covalent bonding involves a transition region to the silane-epoxy interphase
- Silane-epoxy interphase, which has a three-dimensional cross-linked network. This strong, rigid, interpenetrating network is characterized by many layers
- Epoxy resin, which has characteristics that depend on composition and treatment or preparation of the resin

In the study of mechanical performance in the glass-silane-epoxy system, single fi-

Table 2 Silane coupling agents, X-Si(OR)$_3$, wet adhesion

X functional group	Chemical structure	Thermoplastic matrix resin and fuse temperature, °C (°F)(a)						
		PVC 180 (355)	Urethane 175 (350)	HDPE 240 (465)	HDPP 250 (480)	Nylon 275 (525)	Polycarbonate 275 (525)	PEEK 400 (750)
Vinyl	CH$_2$=CH-Si(OCH$_3$)$_3$	NR	NR	G	NR	NR	NR	NR
Chloropropyl	Cl(CH$_2$)$_3$-Si(OCH$_3$)$_3$	NR	G	G	G	G	G	NR
Epoxy	CH$_2$—CH-CH$_2$-O—(CH$_2$)$_3$Si(OCH$_3$)$_3$	NR	VG	NR	NR	G	G	NR
Methacrylate	CH$_2$=CH —C—O—(CH$_2$)$_3$—Si(OCH$_3$)$_3$ CH$_3$ O	NR	NR	G	NR	G	G	NR
Primary amine	NH$_2$—(CH$_2$)$_3$—Si(OC$_2$H$_5$)$_3$	G	NR	NR	NR	G	G	NR
Ethylenediamine	NH$_2$(CH$_2$)$_2$NH(CH$_2$)$_3$Si(OCH$_3$)$_3$	G	NR	NR	NR	G	G	NR
Mercapto	HS(CH$_2$)$_3$Si(OCH$_3$)$_3$	NR	VG	NR	NR	G	NR	NR
Cationic styryl	CH$_2$=CH—C$_6$H$_4$CH$_2$NH(CH$_2$)$_2$NH(CH$_2$)$_3$Si(OCH$_3$)$_3$·HCl	VG	VG	VG	VG	VG	G	G
Mixed phenyl silane	(OCH$_3$)$_3$/NH$_2$(CH$_2$)$_2$NH(CH$_2$)$_3$Si(OCH$_3$)$_3$	VG	VG	NR	NR	VG	VG	G
Modified melamine resin	CH$_2$—CH—CH$_2$—O—(CH$_2$)$_3$Si(OCH$_3$)$_3$	G	E	NR	NR	VG	VG	VG
Azide (composition not specified)		NR	NR	G	VG	NR	...	...

(a) PVC, polyvinyl chloride plastisol; HDPP, high-density polypropylene; PEEK, polyetheretherketone; HDPE, high-density polyethylene; NR, not recommended; G, good (significant improvement compared to the control, perhaps 100-fold improvement in time to failure); VG, very good (up to 1000-fold improvement in performance compared to the control); E, excellent (most outstanding performance of a series of treatments). Source: Ref 14

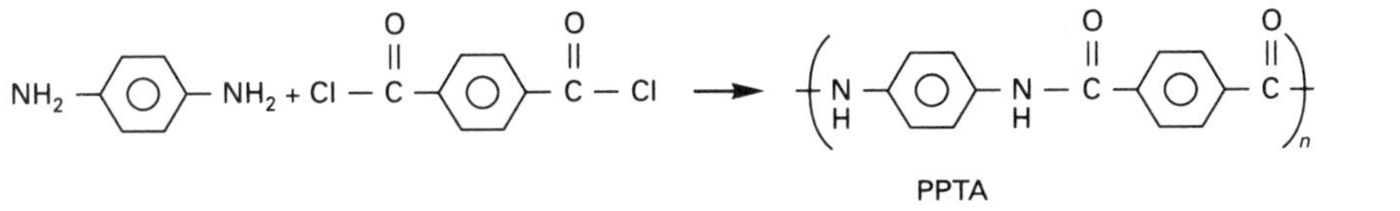

Fig. 1 Chemical structure of poly(p-phenylene terephthalamide) (Kevlar)

ber interfacial bond strengths were measured (Ref 17). The measured shear stress was at a maximum value following treatment with 0.5% aqueous (acidified) N-2-aminoethyl-3-aminopropyl trimethoxy silane (AAPS), resulting in a 70% improvement over an untreated fiber. An initial decrease in the interfacial bond strength upon hydrothermal treatment was attributed to hydrolysis of the oligomeric silane region. A less drastic decrease occurred at longer times, presumably because significant cross linking exists near the glass surface. A consequence of outer surface silane removal, prior to the formation of composites, was the enhancement of interfacial shear strength to optimum values. Recent evidence from differential scanning calorimetry, IR spectroscopy, and solid-state ^{13}C nuclear magnetic resonance (NMR) has demonstrated that penetration of the epoxy resin into the hydrolyzed silane took place (Ref 22).

Enhancement of fiber-thermoplastic matrix interaction via the formation of ionomer bonds with silanes has been developed (Ref 23). The interaction is promoted as a result of the bonding of acid-functional silanes on the glass substrate to acid groups (carboxylic) on the polymer in the *presence of metal ions*, namely zinc or sodium. The discovery of greater bond strength is particularly important for the fabrication of composites under high shear conditions, for example, injection molding or extrusion. Greater adhesion in water at room temperature was found for high-density polyethylene and polypropylene using Zn^{2+} modified primer. Good adhesion to glass was found, using Zn^{2+} ionomers, for nylon, poly(ethylene-vinyl alcohol), styrene-butadiene block copolymer, rigid polyethylene, acrylonitrile-butadiene-styrene (ABS), carboxylated polyethylene, and urethane. Compression molded laminates of fiberglass cloth-nylon 6,6 prepared using Zn^{2+} acid silane primer showed excellent flexure strength retention after exposure to boiling water. It is important to the performance of the thermoplastic-primer-glass system that the film former in the primer be compatible with the matrix polymer.

Additional studies have addressed the mechanical performance of glass-resin systems, specifically investigating interfacial bonding with silanes. Single-fiber pull-out tests of poly(butylene terephthalate) bonded to E-glass have been reported (Ref 24). The role of morphology (especially the on-

set of crystallization at the filler-matrix interface) in failure behavior of glass-filled polypropylene has been investigated (Ref 25). New coupling agents built around octahedrally coordinated silicon and containing a phthalocyanine ring have exhibited the potential for forming strong, hydrolytically stable bonds between glass and polymers (Ref 26, 27). Improved mechanical properties for a composite system composed of a fluoropolymer coreacted with epoxy resin and diamino-diphenyl sulfone-glass cloth were reported (Ref 28).

Recent studies have evaluated dynamic-mechanical properties of silane coatings for glass-polyester composites (Ref 29-31). The findings from the study were: desirable coupling agents interact with the resin via interpenetration and chemical bonding; "poor" coupling agents yield rigid boundary layers; the most sensitive indicators of efficient stress transfer for glass-polyester composites are a low E'' modulus and glass transition temperature (T_g) (nature of the boundary layer); and strong joints result from a synergistic effect of chemical bonding and a soft boundary layer. The existence of chemical bonds between the polyester matrix and glass fiber alone is insufficient to obtain a strong polymer-glass bond.

Additional studies of silane coupling agents (Ref 32) have used various techniques to demonstrate that the epoxy structure is modified in the interface region, probably as a result of amine catalyst on the glass surface. In another investigation (Ref 33), a variation in the amount of coating on "epoxy-compatible" glass fibers was found. Surface analytical techniques were used to evaluate the distribution of components on the glass and to indicate that the coupling agents had reacted with other components in the coating mixture.

Polyaramid

The preparation and properties of polyaramid-type fibers have been summarized by Chiao and Chiao (Ref 34). A condensation polymer produced by the reaction of p-phenylene diamine with terephthaloyl chloride to yield poly(p-phenylene terephthalamide) (PPTA), is shown in Fig. 1.

Extended chains exist in the fiber axis direction, with hydrogen bonding in the transverse direction (Ref 35, 36), forming a radially arranged pleated sheet structure. Microvoids (5 × 25 nm, or 0.2 × 1 µin.) arranged along the fiber axis have been observed by transmission electron micros-

copy (TEM) (Ref 37). The core structure is covered with a 0.1 to 1.0 µm (4 to 40 µin.) thick skin structure that is assumed to be less crystalline than the fiber core. The skin structure potentially allows the penetration of small molecules into the fiber structure. However, Penn *et al.* (Ref 38) have commented that solvent processes intended to introduce sizing reagents into the fiber reside at the fiber surface, and that no interpenetration of sizing and polyaramid fiber occurs.

To enhance interfacial adhesion, it is necessary to increase intermolecular forces across the interface. The ideal solution is to incorporate strongly adhering chemical functional groups on the fiber that can also form strong bonds with the resin. Among the methods that have been used to improve fiber-matrix adhesion are chemical treatments, including coupling agents, coatings, grafting, and a variety of plasma treatments. In some studies it was reasoned that polyaramid-type fibers could be hydrolyzed in acids or bases to form amine sites (Ref 39) that would bond favorably with resin matrices, particularly epoxy resins. The active amine sites were reacted butanediol diglycidyl ether to create epoxy pendant groups on the fiber surface. Upon bonding the modified fibers to epoxy resin and testing via single-fiber adhesion tests, no improvement in adhesion was found. It was reported that cohesive failure was caused by overtreatment of the fibers. Thus any advantage gained by improved adhesion was offset by mechanical damage resulting from the harsh chemical treatment of fibers. In another study (Ref 40), the production of carboxylate groups by mild hydrolysis of polyaramid fibers was accomplished. Although tensile properties were not measured, it was suggested that the treatment altered the fiber surface chemically without destruction of the skin structure and that such alterations potentially could improve fiber-matrix adhesion.

Polyaramid fiber surfaces were modified by bromination followed by ammonolysis, as well as by nitration followed by reduction, to create amine functional groups (Ref 41, 42). A correlation of surface amine group concentration with tensile properties was demonstrated. Fiber strength remained unchanged for amine surface concentrations up to 0.86 per 100 $Å^2$, but significant strength loss was found for fibers with 6.31 amine groups per 100 $Å^2$. Interlaminar shear strength tests on composites prepared using an epoxy resin and fabric containing 0.86 amine groups per 100 $Å^2$ produced via the nitration/reduction process exhibited about a 52% improvement in strength over composites using nontreated fabric. Such studies demonstrate that the judicious selection of reagents or reaction conditions can lead to a beneficial surface modification of fibers without loss of tensile strength.

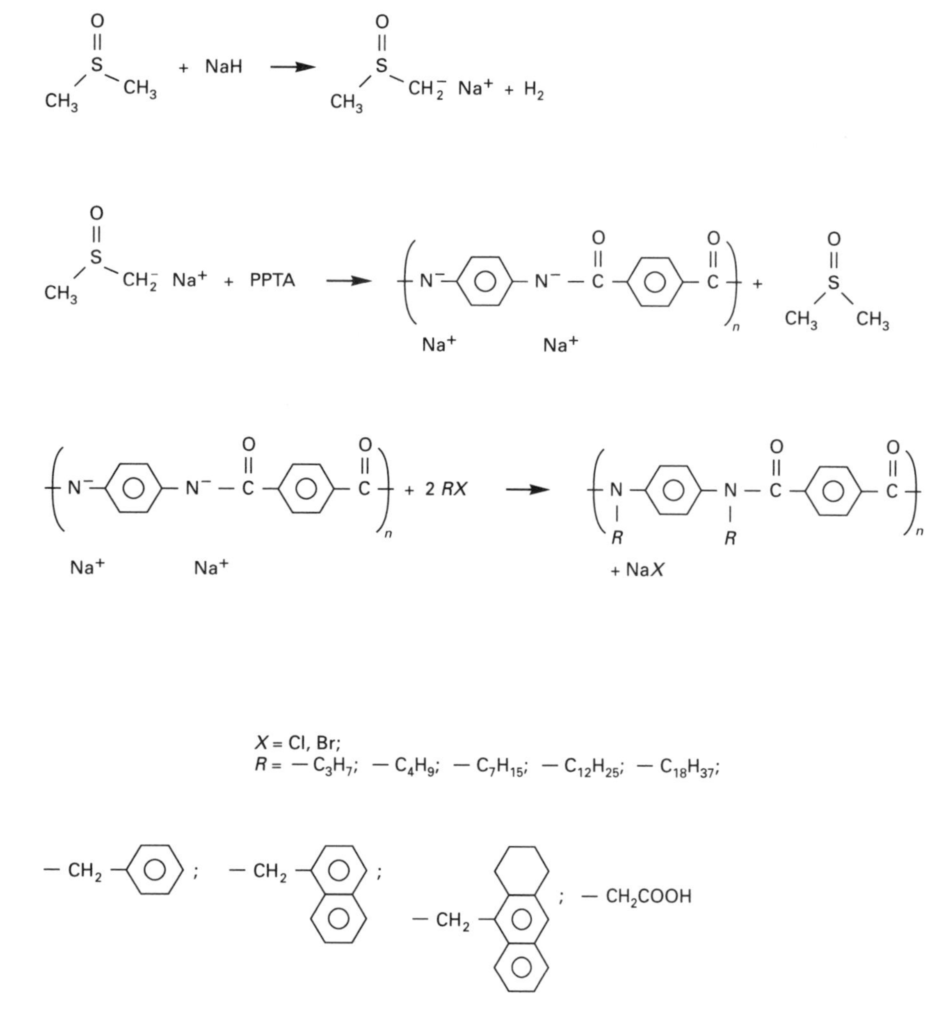

Fig. 2 Derivatization of Kevlar

Table 3 Interfacial bond strength of single-filament aramid-epoxy

| Surface sizing(a) | Bond strength | |
	MPa	ksi
None	32.7±7.9	4.74±1.1
None(b)	31.1±6.8	4.51±0.98
Aromatic sulfonamide	35.7±8.9	5.18±1.29
Aromatic polyimide	26.5±6.8	3.84±0.98
Polyphenyl quinoxaline	28.0±8.6	4.06±1.25
Amidoamine	23.2±6.4	3.36±0.92
Titanate	24.9±5.8	3.61±0.84
Azide	24.5±12.7	3.55±1.84
Silicone	7.91±3.3	1.15±0.48

(a) Unless specified, matrix was N,N,N',N'-tetraglycidyl methylene dianiline-4,4' diaminodiphenyl sulfone. (b) Matrix was diglycidyl ether of bisphenol A/aliphatic amine. Source: Ref 38

Recent efforts have focused on the development of specific reagents that will bond to the amide nitrogen and at the same time provide extended chain functional groups that can interact with the resin (Ref 43-45). In the processes, referred to as metalation reactions, PPTA amide hydrogens are ionized using sodium methylsulfinylcarbanion in dimethylsulfoxide (DMSO), according to the reaction shown. The subsequent reaction of ionized PPTA with halogenated reagents leads to the formation of N-substituted PPTA compounds. The functionality of the N-substituent can be selected or formulated to enhance adhesive bonding with a given resin (Fig. 2).

The procedure for the modification of PPTA fibers incorporating the octadecyl group is summarized as (Ref 45): PPTA fibers (1 g, or 0.04 oz) were added to a 100 mL (3.4 oz) solution of $CH_3(S{=}O)CH_2^-$ Na^+ in DMSO at 30 °C (85 °F) followed by reaction for 1 h; 3.1 g (0.11 oz) of n-octadecyl bromide were added, and the reaction was continued for 2 h. The modified fibers were filtered and then washed alternately with water and acetone. N-modified fibers were subsequently dispersed in xylene.

Composites containing polyethylene or ionomer resin and surface-modified PPTA fibers were prepared, and their mechanical properties were measured. In the preparation of the composite (Ref 45), 0.5 g (0.018 oz) of surface-modified PPTA dispersed in xylene was combined with 9.5 g (0.34 oz) polyethylene in xylene at 120 °C (250 °F). Composites were precipitated by the addition of the xylene suspension to large quantities of methanol. The composites were filtered, washed with methanol, and dried for 10 h at 80 °C (175 °F) in a vacuum oven. Discontinuous fiber reinforced composites (reinforcement volume fraction 0.022) made with random in-plane fiber orientation were prepared by compression at 190 °C (375 °F). Some difficulty was encountered in dispersing unmodified fibers into the matrix. The molded specimens were approximately 2 mm (0.08 in.) thick. Tensile tests were conducted at room temperature and at 80 °C (175 °F), and results indicated that the mechanical properties of surface-modified PPTA composites prepared using polyeth-

ylene or ionomer are more desirable than composites prepared from nonmodified fibers (Ref 44, 45).

Penn and coworkers (Ref 38) have studied the effect of sizing agents on interfacial shear strength using a single-fiber pull-out test. Sized fibers were bonded to epoxy beads for the tests. Sizings, selected to adhere to polyaramid, were aromatic sulfonamide, aromatic polyimide, polyphenylene quinoxaline, amidoamine, silicone, epoxy azide, and organic titanates. The preparation of sized fibers included passing the fiber through a 1 to 3 wt% sizing solution (usually with methylene chloride as solvent) to add 1 to 2 wt% to the fiber, and then heating the fiber in an oven to remove solvent and also to polymerize the sizing reagent into an insoluble coating. Surface energetic evaluation via contact angle measurements indicated that the application of sizing did not make the fiber and cured epoxy equivalent. Interfacial bond strengths shown in Table 3 also reveal that the application of these sizings to polyaramid did not improve the shear strength of the composite.

In a more recent publication (Ref 46), plasma treatment combined with subsequent chemical treatment was used to enhance adhesive bonding with epoxy. The procedure for modifying the fibers was:

- *Fiber wash.* Wash fiber in 0.2 M Na_2HPO_4 (pH 9.1) at reflux 24 h. Rinse thoroughly in distilled water; wash in spectroscopic grade acetone. Vacuum dry at 80 °C (175 °F) and store in a vacuum desiccator (in the dark)
- *Plasma treatment.* Expose to monomethyl amine plasma 2, 5, or 15 min; use radio frequency of 27.1 MHz and gas pressure of 25 Pa (0.2 torr). (Treatment optimized to minimize surface etching or to prevent loss of fiber strength). Store in the dark. Desiccator not needed. Wash fiber as described above
- *Pendant group attachment.* React fibers with 1,6-diisocyanatohexane as a neat liquid 48 h at room temperature in the presence of 1 drop dibutyl tin dilaurate

Table 4 Tensile properties of treated aramid fibers subjected to bromination/ammonolysis

Sample	NH$_2$/100 Å^2	Bromination conditions(a)	Ammonolysis conditions	Tenacity, grams/denier
Control	⋯	⋯	⋯	35.9
Brominated	⋯	2.87% NBS/DMAC/BPO 85 °C (185 °F)/2 h	⋯	36.8
Brominated/ammonolysis	0.47	⋯	EDA/NaNH$_2$/CuCl reflux/118 °C (245 °F)/1 h	23.8
Brominated	⋯	1.5% NBS/DMAC/BPO 85–90 °C (185–195 °F)/2 h	⋯	40.5
Brominated/ammonolysis	6.31	⋯	EDA/DMSO/NaNH$_2$/CuCl 100–120 °C (212–250 °F)/40 min	16.0
Brominated	⋯	1.0% NBS/DMAC/BPO-CHP(a) 100 °C (212 °F)/1 h, 150 °C (300 °F)/1 h	⋯	31.4
Brominated/ammonolysis	0.86	⋯	EDA/CCl$_4$/NaNH$_2$/CuCl 40–60 °C (105–140 °F)/10 min	32.5

(a) CHP, cumyl hydroperoxide. Source: Ref 42

Table 5 Tensile properties of treated aramid fibers subjected to nitration/reduction

Sample	NH$_2$/100 Å^2	Nitration conditions(a)	Reduction conditions(b)	Tenacity, grams/denier
Control	0.05	⋯	⋯	35.9
Nitration	⋯	OAc:HOAc:HNO$_3$, 100:28:28, RT/1 h	⋯	34.9
Nitration/reduction	0.48	⋯	Cocomplex/NaBH$_4$, RT/7 h	32.6
Nitration scheme 1	⋯	OAc:HOAc:HNO$_3$:H$_2$SO$_4$, 100:37:25:1.25, RT/2 h, 40 °C (105 °F)/15 min	⋯	37.2
Nitration/reduction	0.59	⋯	Cocomplex/NaBH$_4$, RT/24 h, 100 °C (212 °F)/2 h	32.4
Nitration	⋯	OAc:HOAc:HNO$_3$:H$_2$SO$_4$, 100:100:25:0.5, RT/0.5 h, 40 °C (105 °F)/0.5 h	⋯	19.7
Nitration/reduction	0.56	⋯	Nitration scheme 1 above, SBH/THF, 0.7:100, RT/24 h, reflux/1 h	39.0
Nitration scheme 2	⋯	OAc:HOAc:HNO$_3$:H$_2$SO$_4$, 100:37:32:1.6, 5–10 °C (40–50 °F)/1 h, 15 °C (60 °F)/3 h	⋯	27.1
Nitration/reduction	0.54	⋯	SBH/THF, 7:100, RT/14 h, reflux 6 h	23.2
Nitration	⋯	Nitration from nitration scheme 2, RT/3 h	⋯	38.7

(a) RT, room temperature. (b) SBH/THF, sulfurated borohydride in tetrahydrofuran. Source: Ref 42

Table 6 T-peel strengths of treated aramid fabric-Epon 828 epoxy resin

Composite	NH$_2$/100 Å^2 fabric	T-peel strength N/m	T-peel strength lbf/in.
Control(a)	⋯	105	0.6
N/R(a)	0.62	210	1.2
B/A(a)	3.12	230	1.3
Control(b)	⋯	160	0.9
N/R(b)	0.65	195	1.1
B/A(b)	1.64	245	1.4

(a) Composite of 25 wt% fabric; cure using *m*-phenylene diamine.
(b) Composite of 35 wt% fabric; cure using xylylene diamine.
Source: Ref 42

- *Nitration*: React fiber in acetic anhydride, glacial acetic acid, and fuming nitric acid at 10 to 15 °C (50 to 60 °F)
- *Reduction*: Place nitrated fibers in 50 vol% aqueous ethanol, buffered to pH 6.7 to 7.0 with KH$_2$PO$_4$ and K$_2$HPO$_4$; reduce with NaBH$_4$ in the presence of Co(bipy)$_3$(ClO$_4$)$_3$ (catalyst)

Details for the modification of polyaramid fabric are given (Ref 42) via bromination and ammonolysis or nitration and reduction processes. The chemical approach is similar to that used for single fibers, but the details were altered to optimize the process for fabric. Composites were prepared using Epon 828 epoxy resin.

The results for single-fiber tests that are shown in Tables 4, 5, and 6 indicate that fiber tensile properties are not significantly altered. On the other hand, the incorporation of amine sites in the fabric leads to improved peel strength and apparent interlaminar shear strength.

The effect of cure conditions (Ref 47) on the interfacial strength properties has been reported. The level of adhesion produced by —NO$_2$ or —NH$_2$ incorporation was comparable. Improved adhesion with —NH$_2$ modified fibers is expected by means of covalent bonding with epoxy groups. However, such bonding is not possible for NO$_2$-modified fibers. The identification of a specific cause for such improved fiber-resin adhesion is difficult.

Other investigators have studied the effect of silicone fluid, polyester and polyether resins (Ref 48), and a polyurethane varnish (Ref 49). Fibers coated with these materials gave similar tensile strengths, but the effect on toughness was dissimilar.

The surface modification of aramid fibers by treatment in various gaseous microwave plasmas has been reported (Ref 50). The properties of polyaramid fibers and a multifilament cloth treated in a nonreacting gas plasma, a nonpolymerizable gaseous plasma, polymerizable reactive gas plasmas, and in an activation plasma followed by exposure to a polymerizable vapor were compared (Ref 50). Variables in the plasma study were a gas pressure of 25 to 135 Pa (0.2 to 1.0 torr), microwave power of 100 to

(catalyst). Conduct two Soxhlet extractions to remove unreacted —NCO, with methylene chloride and then acetone used as solvents. Store fibers in the dark to eliminate potential photodegradation

Wu and Tesoro (Ref 42) and Penn *et al.* (Ref 47) have studied polyaramid fibers modified to contain amine surface groups by bromination followed by ammonolysis, as well as by nitration followed by reduction. Improved peel strengths and interlaminar shear strengths in epoxy laminates were attributed to the formation of covalent bonds in aramid-epoxy composites. The surface modification process was:

- *Cleaning*: Wash fiber in CCl$_4$ (reflux) 2 h; wash with distilled water, dry 110 °C (230 °F), 2 h
- *Bromination*: React fiber in a dimethylacetamide (DMAC) solution of N-bromosuccinimide (NBS) containing benzoyl peroxide (BPO) and N,N′-dimethylbenzylamine
- *Ammonolysis*: Reflux brominated fiber in ethylenediamine containing sodium amide and trace cuprous chloride

Table 7 Bond strength performance of plasma-treated aramid laminates (representative data)

Plasma	Effectiveness ratio, r(a)
Argon	1.10
NH$_3$	1.33
Air	1.40–1.83
N$_2$	1.83
Allylamine	0.81
Propane epoxy	1.34
Hexamethyl disiloxane	2.20

(a) Ratio of ultimate peel strength (plasma treated)/ultimate peel strength (control). Source: Ref 50

700 W, and exposure time of 5 to 60 s. Plasma-treated polyaramid materials were immediately reacted with triazine resin upon removal from the treatment apparatus. Two-ply laminates were prepared for peel tests to measure laminate bond strengths. The results, summarized in Table 7, clearly demonstrate that each of the plasma treatments enhanced the bond strength in polyaramid-triazine composite laminate materials. This result was found even when the ultimate tensile strength of fibers was decreased by the plasma treatment. It was reasoned that the plasma treatment enhanced cohesive bonding among individual fibers.

The modification of polyaramid fibers by means of a hydrosol/coupling agent scheme has been described (Ref 51). In this process, called the SNS process, a hydrosol of tin is prepared by dissolving SnCl$_4$ and adding SnCl$_2$ in water at room temperature (Ref 52). The deposition of a mixed valence hydrous tin oxide occurs upon interaction with the fibers. The application of a silane coupling agent is carried out using a 1 wt% aqueous solution of an amino-triethoxy silane. It is reported that such coatings are corrosion resistant and thermally stable (to 150 °C, or 300 °F). The preparation of test specimens using thermoplastic resin is accomplished by solvent casting and drawdown techniques. For thermoset resins, the fiber and resin were mixed, a curing agent was added, the sample was cast, and the resin was cured. Compared to nontreated fibers, improved adhesion was found for systems using three different epoxy systems (Epon 828-DTA-BDMA, Epon 828-NMA-BDMA, and DEN 438-BPA-BDMA), a polystyrene system, and a polysiloxane res-

in, DC 648. Enhanced adhesion was found for tests under either dry or wet conditions. The mechanism for improved adhesion was not discussed.

In-situ bulk polymerization of styrene on polyaramid using divinylbenzene as the cross-linking agent has been described (Ref 53). Polyaramid fiber was swollen in dimethylformamide (DMF) (10 min, 135 to 140 °C, or 275 to 285 °F) and then introduced into the reaction mixture given in Table 8. Following surface modification, coated fibers were heated at 75 °C (165 °F) for 24 h. Polystyrene was injected into the aligned fibers using a special mold. The results demonstrated that coating the fibers increased the tensile strength of the composite, with *in-situ* polymerization producing an increase of up to 25%. Additional improvement is achieved by cross linking the polystyrene.

The adhesion of polyaramid to rubber has been reported (Ref 54). The principal finding was that adhesion of fiber-to-rubber is due to hydrogen bonding.

It is apparent that chemical surface modification of polyaramid fibers enhances adhesion, but because of the inertness of the fibers, fairly complex chemical processes are required to effect modification. Treatment in various gaseous plasmas also alters fiber surface chemistry and improves adhesion.

Carbon/Graphite

The preparation of carbon fibers and the treatment of fiber surfaces for inclusion in carbon fiber-resin matrix systems has been described in detail (Ref 55). The effects of surface treatments on the mechanical properties of fibers and composites were given in Ref 55, and patents (1974 to 1977) for the surface treatment of carbon fibers were cited. A specific surface treatment and/or the sizing reagents can be used.

Common sizing agents are organic-based components, including epoxy, polyimide, and polyvinyl alcohol. Water is also used as a sizing agent. Typically, the weight of sizing applied to the fiber is of the order of 0.5 to 7 wt% (Ref 55).

An extensive study of chemical surface treatments of carbon fibers and their effect on bonding with epoxy adhesives was conducted (Ref 56). The study examined a thermal oxidative treatment followed by the

addition of a polymeric coating and by selected wet chemical oxidation processes. Flame treatment (1900 °C, or 3450 °F) was selected as a preprocessing step to accomplish the oxidation of fibers, decomposition of contaminants, and/or removal of volatile, low molecular weight constituents before polymer application. Each process is expected to enhance fiber-polymer interaction by chemical bonding, greater fiber-matrix contact, and increased porosity. The polymeric coatings studied were polyhydroxyl ether and polyphenylene oxide. A general result was that the shear strength properties of graphite-epoxy composites prepared with polyvinyl acetate or water-sized fibers and treated with thermoplastic resins improved by a factor of two or more. No reduction in fiber tensile strength was found.

The incorporation of carbonyl groups on carbon fibers can be accomplished by means of controlled aqueous (or aqueous dioxane) oxidation using IO_4^- (periodate) and a mixture of OsO_4 and $NaIO_4$ (Ref 56). The effects of wet chemical treatments of graphite fibers included the formation of carbonyl functionality, enhancement of epoxy-graphite adhesion (scanning electron microscopy, or SEM, data), minimal fiber degradation, improvement of composite interlaminar shear strength, the appearance of a grainy fiber surface containing crystallites, and pores that separate the crystalline regions.

A discussion of surface treatments for increasing interlaminar shear strength (Ref 57) emphasized both the oxidation of fibers with gases and processes in the liquid phase. Oxidation in air or oxygen resulted in the production of phenolic or carboxyl groups on the fiber surface. Significant increases in interlaminar shear strength for carbon fiber-epoxy composites were found. Similar changes in fiber surface properties were noted following oxidation in CO_2, NO_2, and O_3. Liquid phase oxidants included nitric acid (HNO_3), sodium hypochlorite (NaOCl), and potassium permanganate ($KMnO_4$). The surface properties of the fibers were dependent on the liquid reagent used in the oxidation. Nitric acid oxidation produced predominantly carbonyl groups as established from electron spectroscopy for chemical analysis/x-ray photoelectron spectroscopy (ESCA/XPS) studies. Reaction of type I fibers with permanganate led to the

Table 8 Aramid-polystyrene composites: preparation constituents and tensile strength

Polystyrene		Styrene, 10^{-6} m^3	Benzene, 10^{-6} m^3	Benzoyl peroxide		Divinyl benzene		Change in fiber, wt%(a)	Tensile strength	
g	oz			g	oz	g	oz		MPa	ksi
...	...	...	...	...	...	...	...	...	43.2	6.27
10.0	0.35	...	100.0	...	...	...	...	45.3	46.7	6.77
10.0	0.35	50.0	...	...	...	...	...	54.5	47.2	6.85
10.0	0.35	50.0	...	4.0	0.14	...	...	66.1	51.2	7.43
5.0	0.18	50.0	...	4.0	0.14	0.4	0.014	44.6	56.9	8.25
2.0	0.07	50.0	...	4.0	0.14	0.4	0.014	25.8	56.8	8.23

(a) 75 °C (165 °F)/24 h. Source: Ref 53

Table 9 Acid group results on anodized carbon fibers

Fiber	Base equivalents of neutralized NaOH, μeq/g	NaOC₂H₅, μeq/g	Tensile strength GPa	10⁶ psi
AC (Type II fibers), nontreated	7	10	2.16	0.31
AC, HNO₃	14	20	2.10	0.30
AC, NaOH	16	17	1.97	0.29
AG (Type I fibers), nontreated	3	2	2.25	0.33
AG, HNO₃	205	216	<0.8	0.12
AG, NaOH	16	17	1.81	0.26

Source: Ref 57

Table 10 Interlaminar shear strength for treated carbon-epoxy composites

Treatment	Interlaminar shear strength MPa	ksi
Commercial process		
At 37.5 vol%	54.3	7.88
At 42.0 vol%	59.0	8.56
At 46.0 vol%	64.6	9.37
At 51.5 vol%	71.5	10.37
Electrocoated		
At 38.0 vol%	48.2	6.99
At 45.0 vol%	48.3	7.01
At 49.5 vol%	52.6	7.63
At 54.5 vol%	70.1	10.2

Source: Ref 60

Table 11 Izod impact strength for treated carbon-epoxy composites

Treatment	Izod impact strength kJ/m²	ft · lbf/ft²
Commercial process		
At 38.0 vol%	35.6	2420
At 43.0 vol%	43.5	3000
At 46.5 vol%	53.2	3670
At 52.0 vol%	63.5	4380
Electrocoated		
At 39.0 vol%	57.1	3940
At 46.5 vol%	71.8	4955
At 47.0 vol%	63.7	4395
At 57.5 vol%	85.2	5880

Source: Ref 60

formation of graphite oxide and acid functional groups (—COOH). ESCA/XPS indicated a significant formation of graphite oxide, compared to acid groups.

Anodic oxidation in basic solutions yielded CO_2 and degradation products (polycarboxylic acids), the latter being stable in the electrolyte solution. By contrast, degradation products (including carboxylic acids and phenolic groups, that resulted from anodization in HNO_3 remained on the fiber surface. The interlaminar shear strength of composites containing fiber with degradation products on the surface was less than that for composites prepared from equivalent fibers where degradation products were removed. The loss of adhesion was attributed to a layer of weak interfacial adhesion in the region containing degradation products.

The distribution of strong acid groups, that is, hydroxylic and carboxylic, was evaluated by titration with NaOH and $NaOC_2H_5$, respectively. The results shown in Table 9, with the exception of type I fibers anodized in nitric acid, reveal near-equivalent concentrations of weak surface groups. The tensile strength data indicate that graphite oxide formation (AG-HNO_3) seriously weakens the cohesion of graphite in the fibers. The effect of surface acid groups on the interlaminar shear strength of carbon fiber-epoxy composites is revealed in the mode of failure and in the correlation of interlaminar shear strength with the number of weak acid groups, that is, those neutralized by $NaOC_2H_5$. The presence of strong acid surface groups generally results in a tensile-mode failure.

The surface chemistry of carbon fibers prepared by anodization in selected electrolytes was investigated with the goal of enhancing adhesion to thermoplastic matrices. Fibers were anodized in H_2SO_4, NaOH, $(NH_4)HCO_3$, $(NH_4)_2SO_4$, and H_2O (Ref 58). Surface analysis results (ESCA/XPS) for derivatized (Ref 59) as-received AS-4 fibers or Celion 6000 fibers indicated the presence of amine groups on the AS-4 fibers and amine and carboxyl groups on the Celion 6000 fibers. The anodization of carbon fibers increased the surface oxygen content to values from 17 to 30 at.%, while increases in nitrogen were found for anodizations in the ammonium salts. The tensile

strength of the fibers anodized in sulfuric acid or in sodium hydroxide decreased by about 20%, a result indicating too severe a treatment of the fibers. Fast atom bombardment mass spectra of the fibers demonstrated that fibers anodized in sulfuric acid had more aromatic character than fibers anodized in sodium hydroxide.

The electrodeposition of polymer components onto carbon fibers was studied in the effort to enhance fiber-resin adhesion (Ref 60-63). In the approach used, a polymer was selected to provide an optimal interface or a flexible interlayer between fiber and matrix. Improved shear strength was attained by providing strong bonds between matrix and fiber, and increased toughness was achieved by energy absorption through deformation and reduction in stress concentration in the interface. Electrodeposition of butadiene-co-maleic anhydride (BMA) (Ref 60) was expected to enhance the properties of carbon fiber-epoxy composites by the interaction and chemical bonding of hydrolyzed anhydride groups (—COOH) with the epoxy. Improved toughness and the arrest of crack growth were expected from the flexible butadiene segments of the electrodeposited polymer.

The mechanical testing of graphite-epoxy composites prepared from electrodeposited fibers and a commercially treated fiber (Fortafil 5T) were carried out, with the results summarized in Tables 10 and 11. It is apparent that the interlaminar shear strength is greater for the composite prepared from the commercially treated fiber when compared at equal volume fractions. SEM examination of failed surfaces revealed failure between the interface and matrix as a result of weak bonding between the epoxy matrix and the copolymer. It was argued that failure at the copolymer-matrix interface occurs as a result of loss of —COOH functional groups by means of a Kolbe reaction process, permitting radical formation and cross linking of the interphase polymer. On the other hand, failure at the copolymer interphase leads to superior impact strength for composites prepared from polymer-coated carbon fibers (see Table 10).

The electrodeposition of different polymers onto the graphite fiber resulted in improved mechanical properties in graph-

ite-epoxy composites (Ref 61). The electrodeposition of poly(ethylene-co-acrylic acid) (EAA) and poly(methyl vinyl ether-co-maleic anhydride) (MVEMA) onto graphite fibers resulted in a negligible decrease in strength for EAA-coated fibers, but did reduce strength for MVEMA-treated fibers. Interpretation of interlaminar shear strength test data was complicated by the fact that pure shear failure was not observed for composites containing electrodeposited fibers. Mixed-mode failure was attributed to poor fabrication and lack of matrix penetration between carbon fibers. Results of interfacial shear strength tests in single-fiber composites (Table 12) revealed that electrocoated layers were better able to transfer stress than were commercially treated fibers. Improvement in interfacial shear stress properties was due to the interpenetration of and chemical reactions between epoxy and carboxyl groups. Absorption of energy by deformation, crack blunting, and crack deflection led to improved impact strength.

The application of a so-called top layer of vinyl monomer on electrocoated coatings was suggested to improve matrix-interface adhesion (Ref 64). The electrocopolymerization of methyl acrylate, acrylamide, and glycidyl acrylate with acrylonitrile on

Table 12 Interfacial shear stress of electrocoated carbon fibers

Coating/treatment	Mean interfacial shear stress	
	MPa	ksi
None	26.6	3.86
EAA	43.2	6.27
MVEMA-I	46.6	6.76
MVEMA-II	40.4	5.86
HNO₃	47.6	6.90
Commercial treatment	36.2	5.25

EAA, ethylene-acrylic acid; MVEMA, methyl vinyl ether-co-maleic anhydride. Source: Ref 61

Table 13 Interfacial shear strength and surface atomic oxygen concentrations, based on ESCA/XPS analysis

Fiber designation	Interfacial shear strength		Oxygen, at. %
	MPa	ksi	
AU(a)	24.1	3.50	9.0
AS(b)	74.0	10.7	20.0
AS(b)(c)	71.7	10.4	17.5
AS(b)(d)	68.0	9.86	7.0
AS(b)(e)	65.8	9.54	3.0
HMU(a)	13.9	2.02	5.0
HMS(b)	20.3	2.94	9.0
HMS(b)(c)	19.8	2.87	3.0

(a) Nontreated. (b) Commercially treated (specified only to be compatible with epoxy resins). (c) Thermally treated at 300 °C (570 °F). (d) Thermally treated at 600 °C (1110 °F). (e) Thermally treated at 750 °C (1380 °F) and then reduced in hydrogen. Source: Ref 65

Table 14 Interfacial shear strength of graphite fibers-Epon 828

Fiber and surface finish	Interfacial shear strength	
	MPa	ksi
AU-1, Hercules type A-1, untreated	44.9	6.51
AU-1C, Hercules type A-1, untreated but coated with DER 330 epoxy	58.0	8.41
AS-1, Hercules type A-1, surface treated	74.5	10.8
AS-1C, Hercules type A-1, surface treated and coated with DER 330 epoxy	93.8	13.6
AU-4, Hercules type A-4, untreated	37.3	5.41
AU-4C, Hercules type A-4, untreated but coated with DER 330 epoxy	42.1	6.11
AS-4, Hercules type A-4, surface treated	68.3	9.91
AS-4C, Hercules type A-4, surface treated and coated with DER 330 epoxy	81.4	11.8

Source: Ref 66

graphite fibers produced relatively uniform coatings and allowed the variation of properties by controlling the monomer ratio in the electrolyte bath. Random copolymers of acrylonitrile and methyl acrylate can be produced (Ref 64). Mechanical behavior (lower interlaminar shear strength and greater impact strength) of the carbon fibers was similar to the results reported earlier (Ref 61).

The importance of additional surface treatments in enhancing graphite fiber-epoxy matrix interactions was studied for two types of fibers (Ref 65). A comparison of the properties of fibers with no surface treatment, fibers with surface treatment, and fibers with selected thermal treatments revealed a two-part mechanism for enhanced performance involving the removal of weak outer fiber layer (defect laden) and the addition of surface chemical groups. Measured interfacial shear strengths, along with surface oxygen atomic concentrations based on ESCA/XPS data, are summarized in Table 13. The principal correlation is that higher surface oxygen concentration is associated with greater interfacial shear strength. The exception to this general observation is the high strength yet low oxygen content of the hydrogen-reduced AS fiber. Regarding the treatment of this fiber, it was postulated that removal of the weak boundary layer by means of the thermal treatment did not alter the ability of the surface to sustain high shear loads.

The epoxy finish layer applied to surface-treated graphite fibers affects the properties of the composite in which the fiber is used. Studies of fiber-matrix adhesion characteristics for untreated, untreated coated, surface-treated, and surface-treated coated fibers have elucidated the role of the finish (Ref 66). In each case the coated fiber contained an epoxy finish (Dow Chemical DER-330, a diglycidyl ether of bisphenol A). Surface analysis measurements (ESCA/XPS) indicated that the AS fiber was covered by the finish layer. Interfacial shear strength results (Table 14) showed that the epoxy coating led to an increase in strength, both for the untreated fibers and for the surface-treated fibers. In addition, the failure mode changed from interfacial to ma-

trix, comparing uncoated with coated fibers. It was reasoned that such a change occurs as a result of the formation of a brittle interface layer between the matrix and the fiber. The finish interphase is believed to exhibit higher modulus along with reduced fracture toughness. These two factors promote better stress transfer. However, lower fracture toughness is associated with the failure mode change.

In a study (Ref 67) in which the effect of surface treatments and an epoxy "finish" on hygrothermal conditioning was investigated, interfacial shear strength was reduced for exposure at 20 °C (70 °F) under moisture-saturated conditions. Plasticization of the matrix (not chemical changes) was found to be responsible for the reduction in strength. The presence of an epoxy finish was effective in altering the level of interfacial shear strength reduction. Increases in the temperature of hygrothermal exposure caused the reduction in interfacial shear strength, with the extent of reduction being greater for the fiber that had no finish. The reduction in strength is irreversible in that drying the fiber does not restore strength.

In an attempt to improve fiber durability upon exposure to moisture, plasma cleaning, combined with plasma polymerization, was used to alter the surface energies of AS4 graphite fibers (Ref 68). First, fibers were etched in an oxygen or hydrogen plasma. The effect of this pretreatment was to remove a portion of the sizing and to produce free radicals. The fibers were then coated with plasma-polymerized monomers. A post-plasma-treatment procedure was followed in which the fibers were retained in the plasma chamber under partial vacuum and exposed to monomer vapor for a specified period of time in an effort to reduce the possibility of reactions with atmospheric components upon removal of fibers from the treatment chamber. Plasma treatments produced fibers for which the dispersive component of surface

energy was approximately equal and unchanged, compared to as-received untreated fiber. Alterations in the polar component were correlated with elemental surface analysis (ESCA/XPS) (Table 15). Interlaminar shear tests of fiber-epoxy composites in a dry atmosphere revealed improvements in strength following fiber treatment in oxygen or methyl methacrylate plasmas. Interlaminar shear strengths in moist conditions showed approximately equivalent values for composites prepared from treated and untreated fibers and were independent of fiber surface energy. It was reasoned that failure in wet conditions occurred in the epoxy.

A recent study of a continuous cold oxygen plasma treatment of carbon fibers indicated an increase in the interlaminar shear strength of carbon fiber-epoxy composites (Ref 69). Characterization of the treated fibers using IR and ESCA/XPS methods suggested the formation of —C—OH, —C(O)—, and COOH groups. Fiber wetting and surface roughness increased, and striations were found following the cold plasma treatment. Fiber strength was not reduced significantly as a result of the plasma treatment.

The importance of the chemical nature of sizing has been studied for short carbon fiber-polypropylene composites (Ref 70). Fiber sizings included epoxy, nylon (polyamide), and polycarbonate for the preparation of composites with polypropylene and composites with blends of propylene-acrylic acid graft copolymer (2%). The tensile strength of fibers in polypropylene increased with increasing fiber fraction (0.05 to 0.16) and varied with fiber sizing (at fiber fraction = 0.16) in the manner where polycarbonate, at 68 MPa (10 ksi), is stronger than both polyamide and epoxy, at 45 MPa (6.5 ksi). For a matrix system with a blend of polypropylene-2% acrylic acid matrix

Table 15 Treated graphite fiber (AS4) surface energies and surface analysis results

Treatment	Dispersive component		Polar component		At. %	
	N/m	dynes/cm	N/m	dynes/cm	Carbon	Oxygen
Untreated	0.025	25.0	0.028	28.1	76	20
Oxygen etch	0.022	22.1	0.020	20.3	23	17
Methyl methacrylate coated	0.023	23.0	0.022	22.1	. . .	. . .
Bromobenzene coated	0.026	26.0	0.003	3.2	84	6
Styrene coated	0.027	27.0	0.005	5.3	91	9
Ethylene coated	. . .	. . .	. . .	. . .	69	31
Oxygen etch sputtered	. . .	. . .	. . .	. . .	90	10
Hydrogen etch	. . .	. . .	. . .	. . .	86	14

Source: Ref 68

system, the resulting variation in tensile strength (at 0.18 fiber loading) with fiber sizing was polyamide at 95 MPa (14 ksi), polycarbonate at 80 MPa (12 ksi), and epoxy at 53 MPa (7.7 ksi). It was suggested that the enhanced tensile properties of the polypropylene-acrylic acid matrix were related to greater hydrogen bonding between the sizing and matrix resin functional groups. Greater matrix-fiber adhesion for polypropylene was attributed to greater matching of the polymer and sizing dielectric constants.

Improvement in toughness characteristics was achieved by incorporating thermoplastic fibers and fiber coatings into the graphite-epoxy matrix (Ref 71). Fibers were coated with either polyvinyl alcohol or polysulfone. The inclusion of thermoplastic fibers influenced the interlaminar load distribution, while the coating permitted the control of fiber-matrix debonding and fiber pull-out and decreased fiber-flaw communication. The studies demonstrate that the fiber coating increased tensile strength, notched fracture toughness, and low-velocity impact toughness.

Studies to characterize the acid-base properties of carbon fibers have been carried out to evaluate the use of such information in the selection or development of fiber surface treatments (Ref 72, 73). The acid-base properties of untreated, oxidized-treated, and sizing-treated (coated) fibers were measured using inverse gas chromatography (Ref 72). It was found that the untreated fiber had acid (Lewis) character and practically no base (donor) chemistry. The oxidized fiber exhibited a strong acid character, and the sized fiber was "amphoteric," showing a strong acid and high base character. A correlation between the interfacial shear resistance and the specific interaction parameter was found. One interpretation of the correlation was that interfacial adhesion results principally from acid-base interactions between fiber and matrix. Other factors, such as interdiffusion and co-cross-linking effects, were not evaluated.

Inverse gas chromatographic studies of unsized fibers demonstrated the nonpolar nature of these fibers (Ref 73). Corresponding studies of commercial-size epoxy-polyester carbon fibers exhibited better solvation properties than epoxy-carbon fibers. These experiments emphasized the value of the method used to classify fiber coatings and the potential value of screening fiber properties in the selection of favorable resin matrices.

To better understand the chemical nature of carbon fiber surfaces, analytical methods exhibiting surface sensitivity have been employed. Hammer and Drzal (Ref 74) studied untreated and treated A-type and HM-type fibers using ESCA/XPS. The treatment of fibers resulted in an increase in surface oxygen and nitrogen content for A-type fibers. An increase in oxygen content was expected because an oxidation "treatment" was used. It was suggested that nitrogen could arise if the fibers were oxidized in nitric acid, although the measured N 1s binding energy was indicative of —CN or C—NH$_2$ functional groups. For HM fibers, only oxygen and carbon were detected, and the concentration of oxygen was greater for the treated material. No other surface elements were detected. Thermal treatments in vacuum led to lower oxygen concentrations, presumably by the evolution of CO, CO$_2$, and H$_2$O.

Carbon fibers subjected to cleaning in trichloroethylene, followed by oxidation in air, chromic acid, nitric acid, or sodium hypochlorite (Ref 75) were characterized by ESCA/XPS. The treated fibers were also used to prepare graphite-epoxy composites. The fiber treatment conditions were: air oxidation, 400 °C, (750 °F), 1 h; chromic acid, reflux, 15 min; nitric acid, reflux, 3 h; and sodium hypochlorite, 45 °C (115 °F), 24 h. A detailed examination of the C 1s and O 1s photopeaks indicated the presence of carbonyl (carboxyl, ester) and —COR (alcohol, ester) groups. Treatment of the fibers resulted in at least a fourfold increase in the O:C ratio, with the increase varying with the treatment in the manner where nitric acid > hypochlorite > chromic acid > air ≫ as received. Interestingly, the surface treatments did not have a large effect on the mechanical properties of the fiber. On the other hand, the failure mode in graphite fiber-epoxy composites changed from failure in the composite for nontreated fibers to failure in the adhesive layer for treated fibers.

The effect of electrochemical treatments on the surface chemistry of carbon fibers has been published by Sherwood and co-workers (Ref 76-80). The total surface oxide increased when fibers were electrochemically treated in sulfuric acid or ammonium bicarbonate. The relative surface concentrations of acid- and ester carbon-containing groups was not influenced by electrode potential. Carbonyl, or —C(O)—, groups were most abundant at low potentials, while —C—O—R functionality increased with increasing potential (Ref 76).

The surface analysis of industrially oxidized carbon fibers (agent/method not revealed) demonstrated an increase in C/O and C/N surface functional groups with the increasing level of treatment (Ref 77). The carbon-oxygen functionality was principally —C(O)—, although —C—O— groups were present. The carbon-nitrogen groups appeared to have amine or amide functionality.

The surface carbon-oxygen functional groups for fibers electrochemically treated in nitric acid were principally carbonyl types, with contributions from carbonate and ester-type groups (Ref 78). The nature of the surface oxide was dependent on the duration of the treatment and the concentration of the electrolyte. The intensity of the carbonyl functionality was severely dependent on the magnitude of the polarization potential. The intensity of oxidized carbon species decreased with time during x-ray bombardment in the ESCA/XPS analysis. Nitrogen was not detected on the fiber surface.

The anodic oxidation of carbon fibers in sodium hydroxide (Ref 79) produced —OH and —CO$_2$R groups. The alcohol functional group was not detected for fibers oxidized in nitric acid. Oxidation in sodium nitrate yielded C 1s spectra similar to those found for oxidations in nitric acid. The variation in carbon-oxygen functionality was studied as a function of pH for anodization in sodium nitrate. The results reveal that the carboxyl group concentration increased with increasing pH, while —C(O)— groups were present at low pH, and —C—O—H functionality dominated at high pH (Table 16).

Electrochemical oxidation of carbon fibers in ammonium salts (Ref 80) led to a variety of carbon-oxygen functional groups. Oxidation above 1.5 V in 10% ammonium nitrate produced keto/enol functionality and carboxyl/ester groups, with the keto/enol groups being dominant. Increasing the electrolyte concentration, from 10^{-4} to 1 M, resulted in an increase in the carbon-oxygen functionality and an increase in the oxygen-to-carbon ratio from 0.31 at 10^{-4} to 0.53 at 1 M. Nitrogen incorporation as amine, and possibly —CN and/or pyridine groups, was

Table 16 ESCA/XPS analysis data for anodic oxidation of carbon fibers

Electrolyte	Photopeak chemical assignment	Relative area(a)
Nitric acid	—C(O)—	0.651
	—CO_2R (carboxylic acid/ester)	0.191
	Satellite/plasmon	0.037
Sodium nitrate, pH 0.9	—C(O)—	0.673
	—CO_2R (carboxylic acid/ester)	0.220
	Satellite/plasmon	0.074
Sodium nitrate, pH 7.0	—C(O)—	0.660
	—CO_2R (carboxylic acid/ester)	0.253
	Satellite/plasmon	0.079
Sodium nitrate, pH 11.9	—C—OH	0.543
	—CO_2R (carboxylic acid/ester)	0.292
	Satellite/plasmon	0.040
Sodium hydroxide	—C—OH	0.095
	—CO_2R (carboxylic acid/ester)	0.112
	Satellite/plasmon	0.100

(a) Relative to graphitic peak. Source: Ref 79

found. The presence of amine groups could be beneficial in the effort to increase the mechanical properties in carbon fiber composites.

Adhesive Bonding of Polymeric Materials

Successful adhesive bonding in polymeric materials or composites is dependent on the nature of the adherend surfaces. Factors that influence the adhesion of polymers to polymers or composites to composites include wetting of the adherend, cleanliness of the substrate, chemical and physical properties of the adherends, and joint design. Procedures for preparing adherend surfaces for adhesive bonding are described below (Ref 81).

First, the adherend surface is either abraded or abraded and then degreased. Degreased is a term that is a carryover from the surface treatment of metals and may not have the same meaning for the treatment of polymers and composites. In a general sense, degreasing is a surface treatment with a solvent, usually organic based, to remove undesirable surface contaminants and/or mold release agents. Second, the surfaces of the adherends should be treated by a specific process that is optimized for joining the respective substrates with the adhesive of interest. Third, the test specimens should be adhesively bonded, and the appropriate cure cycle for the adhesive/adherend system should be initiated.

Because of the variety and potential combinations of adherend-adhesive systems, it is not reasonable to discuss all available procedures. Therefore, selected general treatments are indicated in Table 17 for each class of polymeric materials. Specific procedures are given in various literature sources (Ref 81-84). It should be recognized that the surfaces of composite materials are usually composed of polymeric coatings, and it is to this surface that the adhesive may be applied. Thus, surface preparation methods or procedures that are applicable to polymers may also be successful in bonding composites containing the respective polymer. For polymeric/composite materials, the surface energy (Ref 85) or acid-base properties (Ref 86) have been used as guidelines for determining what procedures, either physical or chemical, may be required to enhance adhesive bonding. Methods used to alter the adherend surface include solvent cleaning, abrasion, chemical modification, gaseous plasma, flame, and laser ablation treatments (Ref 81).

Selected procedures for the treatment of polymer/composite adherends deserve special comment because of the difficulty of bonding these materials. Surface characterization studies have been carried out in an attempt to better understand the relationship between surface chemistry, topography, and bond performance.

Fluoropolymers, such as polytetrafluoroethylene (PTFE) are attractive materials because of chemical inertness, insulating properties, and thermal stability. Methods and procedures used to alter fluoropolymer surfaces have included chemical processes, ion and electron bombardment treatment, and glow discharge plasma exposure (Ref 87 and those references cited within it). Reactions of PTFE in Na/NH_3 and subsequent analysis by ESCA/XPS and IR spectroscopy reveal the formation of —C=C— (ethylenic) and —C≡C— (acetylenic) unsaturation, as well as the formation of —C(O)— functional groups. Treatment of PTFE in benzoin dianion, $K_2[Ph(CO)—(CO)—Ph]$, yields a material that readily reacts with Cl_2 or Br_2. The presence of aromatic rings and acetylenic groups was also confirmed by IR studies. The introduction of —COOH, —CHO, and —COH functional groups to PTFE can be done using organolithium or Grignard reagents. Reactions of polychlorotrifluoroethylene (PCTFE) with methyllithium result in the loss of chlorine and fluorine and the introduction of oxygen (ESCA/XPS data) as carbonyl groups. Lithium amalgam modifies the fluorocarbon surface to yield a graphitelike structure. The process is believed to involve the reaction of Li^+ with carbon radicals to yield LiC_x species, which subsequently reacts with water to give LiOH and CH groups. Increased lap-shear strength was found for PTFE specimens (aluminum-epoxy-PTFE-epoxy-aluminum) that had been electrochemically reduced in the solvent DMF, which contained tetrabutylammonium tetrafluoroborate, or $(t\text{-but})_4N^+BF_4^-$.

Exposure of fluoropolymers to glow-discharge plasmas produces near surface alterations in chemical and physical properties. Sputter etching in argon by radio frequency (RF) glow discharge produces cones and —C—O groups. Bond strengths using an epoxy adhesive were greater than the corresponding strengths for sodium-etched adherends. Exposure of PTFE to ionizing radiation in the presence of oxygen results in the formation of acid fluoride functionality. Ionizing radiation degrades the polymer by means of chain scission. Chemical changes taking place during RF sputtering include cross-linking reactions and the generation of —C(O)— functional groups, with the formation of —C(O)— groups being dominant for longer exposure times.

The irradiation of fluoropolymers with ions, electrons, or energetic neutrals results in topographical changes in the surface. Materials that were produced by irradiation in methyl acrylate vapor and subsequently hydrolyzed exhibited improved adhesion with epoxy adhesives compared to the behavior of sodium-etched specimens. The bombardment of PTFE with nitrogen or nitrous oxide ions results in the reduction of carbon where fluorine is evolved and then replaced by nitrogen. ESCA/XPS results indicate the formation of —NCF— groups from —CF_2— functionality.

The dramatic influence of curing agent on binding of poly(vinylidene fluoride) or PVF_2, was recently studied (Ref 88). By the appropriate selection of the amine curing agent for the epoxy resin, it was possible to enhance the adhesive bonding of untreated PVF_2. Modification of PVF_2 by the amine curing agent was achieved by selecting slow-reacting amines, which permit a modification reaction with the polymer and curing of the epoxy. The PVF_2 surface modification reaction was most effective at 70 °C (160 °F) with diethylaminopropylamine for the bonded system aluminum-epoxy-PVF_2-epoxy-aluminum. Based on PVF_2 color changes and IR spectral evidence it was demonstrated that the chemical process was dehydrofluorination, followed by cross linking (gelation). This finding indicates that the selection of the appropriate curing agent may promote the formation of strong adhesive bonds without prior adherend surface modification.

The modification of polyethylene (PE), polypropylene (PP), and other polymer surfaces has been investigated extensively (Ref 89-101). Early investigations (Ref 89) reported that chemical processes using ammonium peroxydisulfate enhanced the bonding of polyolefin films. A review of the effects of various treatments on the adhesive bonding of PE and PP has been completed, and comments on how such treatments alter the surface chemistry have been given (Ref 90). The surface reaction processes cited included chlorine exposure

Table 17 Surface treatments for polymeric surfaces

Adherend	Degrease treatment (solvents)	Surface treatment	Potential adhesive
Acetal (copolymer)	Ketone	Abrasion, dichromate/sulfuric acid, plasma	Epoxy, polyester (—NCO cured), cyanoacrylate
Homopolymer............	Ketone	Abrasion, dichromate/sulfuric acid	Epoxy, nitrile, and nitrile-phenolic
Acrylonitrile-butadiene-styrene	Ketone	Abrasion, dichromate/sulfuric acid	Epoxy, urethane, thermosetting acrylic, nitrile-phenolic, cyanoacrylic
Cellulose, cellulose-acetate, cellulose-acetate-butyrate, cellulose nitrate	Alcohol	Abrasion, warm to 100 °C (212 °F), 1 h, bond warm	Epoxy, urethane, polyester (—NCO cured), nitrile-phenolic, cyanoacrylate
Fluorocarbons	Chlorinated alcohol or ketone	Sodium-ammonia, sodium-naphthalene/THF, potassium acetate/295 °C (565 °F), electrical discharge Gaseous plasma	Epoxy, urethane
Polyamide (nylon)	Ketone	Abrasion, resorcinol-formaldehyde, adhesive primer	Phenol-formaldehyde, epoxy, nitrile phenolic, cyanoacrylate
Polycarbonate............	Alcohol	Abrasion	Epoxy, urethane, cyanoacrylate
Polyethylene, polypropylene..........	Ketone	Flame, dichromate/sulfuric acid, electric discharge, gaseous plasma, titanate primer	Epoxy, nitrile-phenolic
Polyethylene terephthalate (PET-Mylar)	Ketone	Solvent treat, NaOH solution, plasma, titanate primer	Polyester (—NCO cured), epoxy, urethane
Polyimide................	Ketone	Abrasion, sodium hydroxide, plasma	Epoxy
Polymethylmethacrylate, Methacrylatebutadiene ...	Ketone or alcohol	Abrasion	Epoxy, urethane, cyanoacrylate, thermosetting acrylic
Polymethylpentene	Ketone	Flame, dichromate/sulfuric acid	Epoxy, nitrile-phenolic
Polyphenylene oxide.......	Alcohol	Abrasion, dichromate/sulfuric acid	Epoxy, urethane, thermosetting acrylic
Polyphenylene sulfide......	Ketone, chlorinated solvents	Abrasion	Epoxy, urethane
Polystyrene..............	Alcohol	Abrasion	Urethane, epoxy, unsaturated polyester, cyanoacrylate
Polysulfone	Alcohol	Abrasion	Urethane, epoxy
Polyvinyl chloride, polyvinyl fluoride........	Ketone, chlorinated solvents	Abrasion, solvent treat	Rigid PVC, epoxy, urethane, cyanoacrylate, thermosetting acrylic, flexible PVC, nitrile-phenolic
Thermoplastic polyester....	Ketone	Abrasion, ketone wash	Epoxy, thermosetting acrylic, urethane, nitrile-phenolic
Thermoset plastic	Ketone	Abrasion	Epoxy, thermosetting acrylic, urethane

Source: Ref 81

plus UV irradiation, UV irradiation only, chromic acid (dichromate/sulfuric acid), hot chlorinated solvents, flame, corona discharge treatments, and cross linking by activated species of inert gases (CASING). However, the principal method used with films is corona discharge; for other materials (particularly bottles), flame is used; and for complex shapes, chromic acid (dichromate/sulfuric acid) is used. The primary concerns regarding adhesion in polyolefins are the presence or absence of a weak boundary layer, the importance of wetting, the role of the specific surface treatment, and the chemical and morphological changes resulting from a given treatment. Consideration of experimental results for treated polyolefins including surface en-ergy measurements, surface chemical characterization (ESCA/XPS), and adhesive tests (measured shear or peel strengths) have led Brewis and Briggs (Ref 90) to propose a model for adhesion to polyolefins:

- As a result of dispersion forces, perfect wetting of the polyolefin surface would result in high levels of adhesion
- Because high levels are not achieved, wetting must be incomplete or inadequate
- Surface treatments result in chemical modifications that increase the degree of wetting and/or produce functional groups that lead to interactions that are stronger than dispersion forces

Experimental evidence indicates that the surface treatments (except CASING, perhaps) that oxidize the polyolefin surface give rise to polar functional groups. Chromic acid etching produces —C—OH, —C(O)—, —O—C=O, and —SO$_3$H groups on PP and low-density PE (LDPE). A correlation of the adhesive strength and the degree of oxidation (number of carbon-oxygen groups; chemical nature unspecified) was noted. High levels of oxidation were found for flame-treated LDPE. However, the oxidation was confined to a layer estimated to be from 4 to 9 nm (0.16 to 0.36 μin.) thick on the polymer surface. Air oxidation resulted in the formation of —C(O)— groups. The ESCA/XPS analysis of corona-treated polyolefins indicated the probable presence of alcohol, ketone, and carboxylic acid functional groups. Derivatization studies permitted a more detailed identification of the functional groups and allowed a better understanding of the adhesion mechanism in treated polyolefins.

Modification of PE surfaces using activated gaseous species was reported in 1967 (Ref 91). Low-power RF was used to excite helium to remove the weak boundary layer on PE and to cross link the material. It was argued that cross linking increased the cohesive strength of PE, leading to improved tensile shear strength for aluminum-epoxy-treated polymer-epoxy-aluminum samples. Other excitation sources, such as microwave excitation and corona discharge treatments, were cited as additional ways to produce excited species, with the potential for surface modifications.

The surface chemical characterization of polyethylene terephthalate (PET) (Ref 92) indicated little change in the CH$_n$, —O—CH$_2$—, or —COO functional group intensities following plasma treatment in N$_2$ or Ar. Nitrogen or argon treatment of other polymers, including poly(oxymethylene), —(CH$_2$—O—)$_n$—, cellulose acetate, polyacrylonitrile, Nylon 6, and PTFE resulted in the incorporation of oxygen functionality, which is usually associated with carbonyl carbon. Nitrogen functionality following the nitrogen treatment was attributed to amide-type groups.

The electrical discharge treatment of polymers was also studied, and the surfaces were characterized by surface-sensitive analytical methods (Ref 93). The discharge treatment yields phenolic-OH and carboxylic acid groups, thereby allowing enhanced self-adhesion by hydrogen bonding. Chain scission was also found. Adhesion behavior was examined by measuring the thermal activation required to produce good seals (that is, self-adhesion) at a given fixed pressure and temperature. Treated PET bonded to untreated PET resulted in better seals than treated PET bonded to treated PET.

Low molecular weight oxidized material produced on corona-treated PP (Ref 94) was

Table 18 The effect of derivatization on automobile adhesion combined with surface analysis

Treatment(a)	C 1s (3 × 10⁴)(b)	O 1s (10⁴)(b)	F 1s (10⁴)(b)	Ti 2p₃/₂ (3 × 10³)(b)	Peel strength N/25 mm	gf/in.
DT	20.4	12.8	. . .	. . .	2.45	250
DT-PFPH	20.0	7.5	12.3	. . .	0	0
DT-PFPH-TAA	18.7	11.2	9.7	15.2	2.36	241
DT-TAA	19.8	12.9	. . .	15.3	3.82	390
DT-TAA-PFPH	18.8	15.0	9.1	11.6	1.76	180

(a) DT, discharge treated. (b) Counts, full scale. Refer to Fig. 3 for structures of PFPH and TAA. Source: Ref 101

found to enhance the adhesion of surface species, especially inks. It was noted that while the formation of a weak boundary layer is often associated with such low molecular oxidized material, these surface materials are soluble in polar liquids and may therefore be incorporated into the surface coating, while also being strongly bound to the oxidized polypropylene substrate. (In the case cited, a polyamide ink was investigated.)

Adhesion between LDPE and cellulose (paper laminate) was improved with oxygen or ammonia treatment of the LDPE (Ref 95). Although plasma treatment of the paper laminate was also examined, it was less effective than the treatment of the polymer. ESCA/XPS analysis of oxygen-treated materials indicated increased oxygen concentration in the form of hydroperoxide and carboxyl groups. The ammonia treatment produced an LDPE surface containing approximately 6 at.% nitrogen, but nitrogen was not detected in untreated LDPE. Plasma treatment of the paper surface also improved adhesion, principally by the formation of polar nitrogen- and oxygen-containing groups on the fiber surfaces.

Corona-treated PE films exhibited better self-adhesion than nontreated films (Ref 96, 97). Earlier workers had reported the formation of carbonyl groups on corona-treated PE (Ref 98). Peel strengths at a 180° angle were compared for corona-treated and chemically treated fibers, leading to the suggestion that enhanced adhesion resulted from hydrogen bonding by means of enol-keto interactions on adjacent PE surfaces.

Details of the changes in surface chemistry were examined using ESCA/XPS (Ref 99), but the large range of chemical functional groups prevented unambiguous determination of the functionalities present. Subsequent studies (Ref 100, 101) in which surface analysis of derivatized-treated PE was accomplished with ESCA/XPS demonstrated that the necessary keto-enol transformation occurred and that self-adhesion took place by enolic-carbonyl hydrogen bonding. Curve resolution of C 1s spectra of corona-treated PE suggested the presence of —C—O—R, —CHO, or —C(O)—, and carboxyl groups. However, the analysis following functional group derivatization permitted an evaluation of functional group content per original surface —CH₂— group. The results are shown in Fig. 3, along with the reaction employed to identify the respective groups. The effect of derivatization on autoadhesion combined with surface analysis is given in Table 18. The results in Table 18 demonstrate that the elimination of enolizable carbonyl groups prevents adhesion. The enhancement of adhesion by TAA (Fig. 3), above that due to the discharge, was attributed to cross linking of —OH groups by the titanium complex. The results also indicate that the ability to derivatize enolic- and alcoholic-OH groups separately permits independent control over these groups for specific interaction.

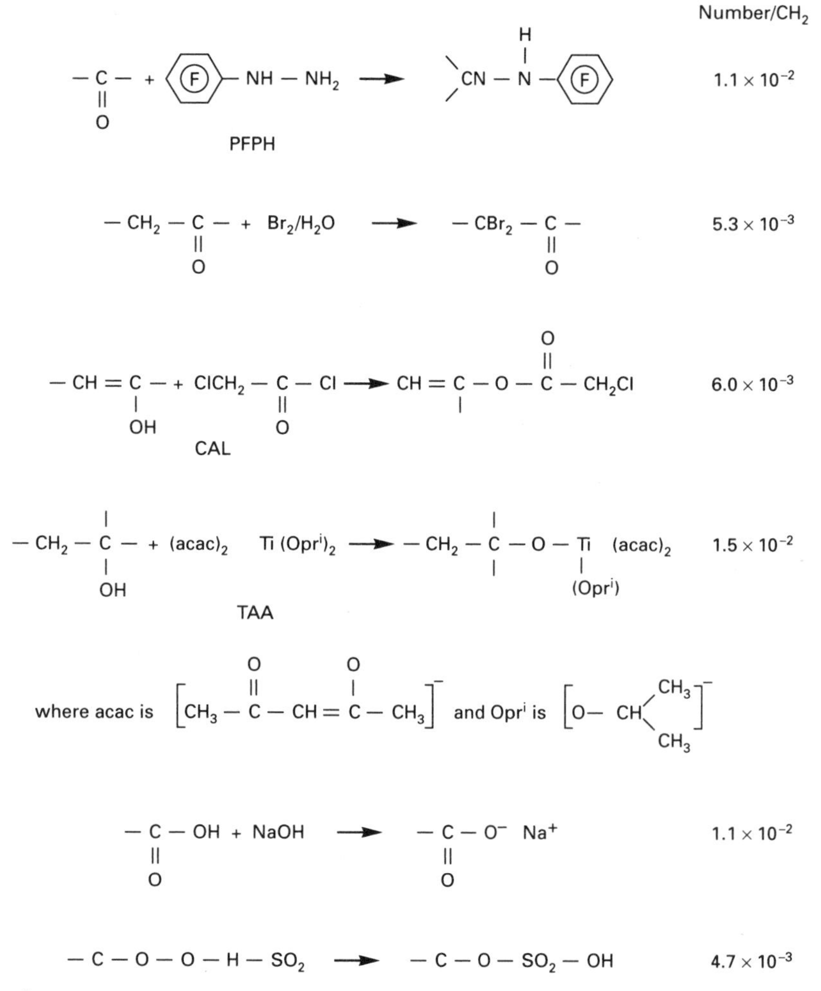

Bonding of Composites to Composites

It was suggested earlier in this article that methods used for adhesive bonding of polymers may be employed for composite bonding. Many aspects of composite bonding (joining) have been discussed at the annual Advanced Composites Conferences sponsored by ASM INTERNATIONAL and the Engineering Society of Detroit (Ref 102-104). Generally, composite surface treatments consist of abrasion and solvent cleaning. However, with growing attention on more efficient production and the use of automation, the demand for no surface preparation and no hazardous solvent or other pretreatment procedures for composites is increasing (Ref 105). A discussion of selected studies is presented to illustrate the variety of approaches that have been employed to enhance the adhesive bonding of composites.

A survey of the potential for plasma pretreatments for bonding a variety of polymers/composites using epoxy or urethane adhesives has demonstrated the success of plasma treatments (Ref 106). Poly-

Fig. 3 Derivatization scheme for treated polymer surfaces. Source: Ref 101

Table 19 Lap-shear strength versus adherend pretreatment

Material	Treatment	Lap-shear strength of epoxy			Lap-shear strength of urethane		
		MPa	ksi	Failure mode	MPa	ksi	Failure mode
Valox 310 polyester resin ..	None	3.6	0.52	(a)	1.3	0.19	(a)
	Oxygen	11.3	1.64	(b)	1.5	0.22	(a)
	Ammonia	9.8	1.42	(b)	6.6	0.96	(a)
Noryl 731 polyphenylene ether	None	4.3	0.62	(a)	1.5	0.22	(a)
	Oxygen	10.2	1.48	(a)	13.0	1.89	(b)
	Ammonia	12.4	1.80	(b)	6.6	0.96	(a)
Durel polyaryl ester	None	1.7	0.25	(a)	0.8	0.12	(a)
	Oxygen	14.9	2.16	(b)	6.1	0.88	(a)
	Ammonia	13.7	1.99	(b)	2.3	0.33	(c)
Vectra A625 liquid crystal polymer...............	None	7.2	1.04	(a)	0.9	0.13	(a)
	Oxygen	11.0	1.60	(b)	6.7	0.97	(b)
	Ammonia	8.5	1.23	(b)	7.2	1.04	(b)
Ultem 1000 polyetherimide........	Control	1.3	0.19	(a)	. . .	. . .	
	Oxygen	13.4	1.94	(b)	. . .	. . .	
	Ammonia	14.2	2.06	(b)	13.3	1.93	(b)
Lexan 121 polycarbonate ..	Control	11.7	1.70	(a)	3.7	0.54	(a)
	Oxygen	15.5	2.25	(b)	7.8	1.13	(c)
	Ammonia	4.7	0.68	(a)	3.2	0.46	(c)
Delrin 503 polyacetal	Control	1.1	0.16	(a)	0.2	0.03	(a)
	Oxygen	4.5	0.65	(a)	9.3	1.35	(a)
	Ammonia	3.9	0.57	(a)	2.7	0.39	(a)

(a) Adhesive failure. (b) Cohesive failure of adherend. (c) Cohesive failure within the adhesive. Source: Ref 106

meric materials that had no pretreatment or that received plasma treatment in oxygen or ammonia were bonded, and the lap-shear strengths and failure modes were determined. The results are collected in Table 19. Treatment in either oxygen or ammonia led to greater lap-shear strength for all materials, with the exception of polycarbonate treated in an ammonia plasma. For the most effective treatment for each adherend, lap-shear strength usually increased by at least a factor of two (Ref 106). Studies of the surface functionalities present before and after plasma treatment indicated that sizing components, which impaired adhesion, were removed. The results in Table 19 do not indicate a preferred treatment for these engineering materials. It appears that the treatment procedure must be evaluated and optimized for the adherend-adhesive combination of interest.

The surface treatment of adherends for the adhesive bonding of thermoplastic-base fiber-composites using epoxy or acrylic adhesives has been investigated (Ref 107). A double-cantilever-beam specimen was used to evaluate adhesive fracture energy. Surface treatments included:

- Abrasion, in which composite material was abraded with alumina, wiped with methyl ethyl ketone, and dried
- Acid etch, using a potassium permanganate 1% solution in 5:2:2 sulfuric acid: ortho phosphoric acid:distilled water
- Corona discharge, in which the composite was abraded as defined above, and treated in corona using 15 to 20 kHz and 0.1 to 0.9 kW

The composites studied were:

- 60 vol% unidirectional carbon-polyetheretherketone (PEEK)
- 55 vol% unidirectional carbon-amorphous polyamide (PA) copolymer
- 55 vol% woven carbon-PA (same as above, but with 11 plies of woven fiber)
- 60 vol% unidirectional aramid-PA
- 60 vol% woven aramid-PA (same as above, but with 9 plies of woven fiber)
- 62 vol% woven carbon-polyether imide (PEI)
- 53 vol% unidirectional carbon-polyphenylene sulfide (PPS)
- 60 vol% unidirectional carbon-polyimide (PI)
- 60 vol% woven carbon-PI (same as above, but with 6 plies of woven fiber)
- 34 vol% unidirectional carbon-epoxy

The three adhesives used in the study were two-part rubber-toughened epoxy paste, rubber-toughened epoxy film, and two-part acrylic paste (from three different manufacturers, using different cure procedures).

For comparison purposes, hot-melt adhesive films composed of the same chemical type as the thermoplastic matrix were investigated. These materials were based on PEEK, PA, and PPS. The significant findings provide guidelines for the selection of an effective adherend surface treatment and identification of the method for adhesive application:

- A simple abrasion/solvent treatment is adequate to bond freshly prepared thermoset-base composites using epoxy or acrylic adhesives
- A simple abrasion/solvent treatment followed by bonding using a hot-melt adhe-

sive film of the same chemical composition as the thermoplastic resin matrix leads to relatively high adhesive failure energies
- Failure for abraded and solvent-cleaned thermoplastic composites bonded with acrylic- or epoxy-type adhesives occurred along the adhesive-composite interface and exhibited low adhesive fracture energies (G_c)
- A corona discharge treatment of the thermoplastic composites was effective in increasing adhesive-thermoplastic intrinsic adhesion. The G_c values increased with the level of corona treatment until a plateau was reached

These studies emphasize the need to select appropriate adhesives and surface pretreatments, but also point out the need to recognize the possible influence of composite properties (transverse stress) on bonded-joint performance.

The effect of surface contaminants and abrasive pretreatment on the adhesion bonding of carbon fiber reinforced epoxy composites (CFRP) was studied (Ref 108). Variations in laminate surface contamination were obtained by laying up the prepreg on several different materials, that is, a commercial glass cloth coated with polymeric fluorinated hydrocarbon, a satin weave cloth impregnated with PTFE, aluminum sheet coated with a polymeric fluorinated hydrocarbon, and aluminum sheet coated with a silicone-containing solution. Composites were abraded with silicon carbide grit (240 or 300) paper, with abrasive cloth (Scotchbrite), and by blasting with dry alumina grit (280 grade). Lap-shear samples were prepared using selected epoxy adhesives. ESCA/XPS elemental compositions for the cleaved composite and the matrix were equivalent, and no contaminating elements were detected. Surface analysis results for prepared samples contained information indicating the presence of —C—C—, —C—H, —C—O, —C(O)—, and C—F functional groups. The presence of carbon-silicon groups was inferred from the knowledge of possible silicone structures. A general surface analysis result was that the concentrations of contaminant functional groups and fluorine- and silicon-containing species decreased with the increasing severity of the abrasion treatment. Similarly, it was found that bonded surfaces contaminated with release agents exhibited low-strength joints with failure cleanly at the adhesive-composite interface. The significance of the study was that the composite must be treated such that virtually all traces of release agent are removed.

Contamination from fluorine-containing constituents has also been detected on graphite fiber reinforced polyimide composites (Ref 109, 110) by ESCA/XPS characterization. Surface treatments were carried out after washing the composite specimens in methanol. The individual surface treatments included 600 SiC grit hand sand, 180 SiC grit hand sand, 120 alumina grit blast, Boeing (100 alumina) grit blast, glass bead blast, ethanolic KOH soak, sulfuric acid soak, hydrazine hydrate soak, and Maxwell Laboratory Flashblast treatment.

Treated composites were primed with BR 34B-18 and then bonded with FM 34B-18 (a supported, aluminum-filled polyimide film) in a lap-shear configuration. Specimens were also aged at various temperatures to evaluate bond degradation. The published results showed no apparent effect of surface pretreatment on lap-shear strength. However, a small decrease in strength with increasing test temperature or aging temperature was noted. The failure modes were adhesive failure for nontreated samples, cohesive (within the adhesive) or composite failure for mechanically treated specimens, and cohesive or adhesive failure for chemically treated composites. Surface analysis data for unaged control fracture samples revealed low fluorine content that was independent of surface treatment. Corresponding thermally aged samples had increased fluorine content which, as a rule, was proportional to the fluorine content on the prebonded sample surfaces.

The treatment of fluorine-contaminated carbon fiber-polyimide matrix composites in an oxygen plasma resulted in the formation of carbon-oxygen functional groups (Ref 110). The surface treatments that were studied for comparison included methanol wash, methanol wash and sulfuric acid soak, grit blast and methanol wash, methanol wash and plasma treatment, and no pretreatment (as received). Surface-sensitive analytical methods such as ESCA/XPS, ion scattering spectroscopy (ISS), and photoacoustic-Fourier transform IR spectroscopy (PAS-FTIR) were used to evaluate fluorine loss and chemical surface changes. The results demonstrated that grit blasting, as well as argon or oxygen plasma treatments, removed most of the fluorine contamination. The plasma treatment also resulted in the formation of various carbon-oxygen polar groups on the surface. Such groups are expected to enhance adhesive bonding. Topographical examination of the oxygen-plasma-treated composite indicated a roughening of the surface and etching (removal) of material. Methanol washing, sulfuric acid soaking, and nitrogen plasma treatment methods were not effective in the removal of fluorine from the composite surface.

Various mechanical, physical, and chemical methods were used to pretreat carbon-fiber-reinforced plastic (epoxy) laminates (Ref 111). Surface abrasion was achieved with Scotchbrite and SiC or corundum grinding papers. Pretreatments also included corona discharge treatments and chemical treatments with solvents and etching agents. For bonded specimens exposed to a hot/wet environment, the mechanically abraded specimens exhibited the best performance. It was noted that optimization of the pretreatments and bonding parameters was still needed for CFRP materials.

Low-pressure plasma and corona treatment of high-performance thermoplastics, polyether-imide, a liquid crystalline polymer, and, for comparison, polypropylene, were reported (Ref 112). Bonding of the treated materials with a two-component epoxy, a one-component epoxy filled with metal powder, and a two-component urethane in lap-shear specimens was accomplished. Surface energies for all materials increased significantly following corona or plasma treatment. The lap-shear test results showed that a correlation with surface energy occurred for polypropylene, but not for polyether-imide or the liquid crystal polymer. The pretreatment was valuable for PEI only if the filled one-component epoxy adhesive was used. Improvement in lap-shear strength was found for treated LCP bonded with urethane. Because no surface analysis results were reported, it is not possible to correlate failure behavior directly with surface chemistry.

The importance of sheet molding compounds (SMC) and/or composites in the transportation industry is well recognized (Ref 113). The increased use of adhesives and SMC/composites in automobile production has prompted studies to aid in understanding the surface of SMC (Ref 114-118). Cheever (Ref 114) has provided a model of SMC, including the nature of the surface, derived from IR-spectroscopic, SEM, and contact angle measurements. The surface, which is energetically inhomogeneous, was represented as a capillary layer consisting of 2 to 3 μm (0.08 to 0.12 mil) deep pores and 1 μm (0.04 mil) diam calcium carbonate particles. This outer layer is about 25 μm (1 mil) thick. Below the outer layer, a substrate approximately 100 μm (4 mils) thick contains resin-calcium carbonate aggregates. The bulk material is composed of glass fiber-resin and $CaCO_3$-resin domains. Preparation of the surface for adhesive bonding may involve abrasion (Scotchbrite) and solvent washing (usually methylene chloride) (Ref 116). Surface characterization measurements (Ref 115) confirmed the chemically heterogeneous nature of the SMC surface and indicated that light abrasion removed the outer surface layer, exposing substrate resin and calcium carbonate. Thus, abrasion alters the surface chemistry of SMC prior to adhesive bonding.

Procedures to ablate the surface by IR laser treatment were developed to remove or eliminate surface bonding difficulties due to surface contaminants and internal or external mold release agents (Ref 116). The method is rapid, can be adapted to robotic control, and does not involve potentially hazardous solvents. Laser ablation could be controlled in varying degrees. Light treatment involves the removal of contaminants, but does not destroy the resin. Medium treatment ablates the resin-rich surface, exposing resin-encapsulated filler. Medium-heavy treatment exposes filler with resin removed below the filler, yet does not produce loose filler on the surface. Heavy treatment severely chars the surface, ablating surface resin and generating loose filler on the surface. Lap-shear tests of specimens prepared using a two-part epoxy adhesive revealed a lower green strength for laser-ablated samples. However, strengths following paint cure cycling, 60 °C (140 °F) water soaking for 1 week, or testing at 82 °C (180 °F) were comparable to strengths measured for Scotchbrite-treated SMC. The failure mode was judged to be acceptable for the samples prepared using the laser ablation treatment.

The adhesive bonding of lap-shear specimens using SMC was enhanced for coupons treated in reducing, stoichiometric, or oxidizing flames (Ref 117). Treatment of SMC in oxygen, dry air, nitrogen, and argon RF plasmas (Ref 118) also led to improved adhesive strengths for lap-shear test samples. Samples were bonded using a two-part urethane adhesive. Surface characterization measurements revealed oxidation of the surface following the flame treatment with the formation of —C—O—, CHO, —C(O)—, and CO_2R functional groups (Ref 118). Alterations in the surface following plasma treatment included significant oxidation of the carbon constituents in the oxygen plasma, particularly —C(O)— (carbonyl) and —CO_2R (carboxyl) groups. Carbonate functionality was evident primarily by means of plasma etching. SEM studies confirmed that a roughened surface was produced by etching of the SMC surface, and PAS-FTIR measurements provided additional evidence of carbonate formation. The extent of surface chemical oxidation/etching in the plasma treatment was dependent on the gas used and varied in the manner where oxygen > dry air > nitrogen > argon.

Adhesive bonding studies of a composite prepared by structural reaction injection molding emphasized the effect of mold release agents (Ref 119). Surface preparations for bonding with a two-part urethane adhesive required surface sanding of the composite with 100-grit paper to remove mold release agents. Following the treatment, lap-shear strengths measured at 22 or 82 °C (70 or 180 °F) were greater by more than a factor of two, and the failure mode also changed. Arimax, which was prepared using soap or wax mold release agents, was bonded using a two-component hybrid adhesive (primerless). Lap-shear strengths at 22 °C (70 °F) for all samples showed virtually no improvement as a result of sanding. Failure occurred by delamination. Tests conducted at 82 °C (180 °F) gave mixed results, and it was suggested that the variability was due to the uniformity and concentration of mold release agent on the test surfaces before adhesive bonding.

The influence of mold release agents and/or release cloth components on the surface chemistry of composites requires much additional study. The control of such agents at bonding surfaces is important if consistent and reproducible bonding of composites is to be realized.

ACKNOWLEDGMENT

The preparation of this manuscript was supported in part through the National Science Foundation-Science and Technology Center program at Virginia Polytechnic Institute and State University under grant DMR-8809714.

REFERENCES

1. E.M. Petrie, Adhesively Bonding Plastics: Meeting an Industry Challenge, *Adhes. Age*, 15 May 1989, p 6-13

2. W.V. Titow, Solvent Welding of Plastics, in *Adhesion 2*, K.W. Allen, Ed., Applied Science Publishers, 1978, p 181-196

3. M. Martin and J.F. Dockum, Composites in Land Transportation, in *Handbook of Composites*, G. Lubin, Ed., Van Nostrand Reinhold, 1982, p 679-698

4. D. McCosh, Car Wars, *Pop. Sci.*, 1989, Vol 234, p 115-119, 151

5. W.R. Graner, Marine Applications, in *Handbook of Composites*, G. Lubin, Ed., Van Nostrand Reinhold, 1982, p 699-721

6. G. Lubin and S.J. Dastin, Aerospace Applications of Composites, in *Handbook of Composites*, G. Lubin, Ed., Van Nostrand Reinhold, 1982, p 722-743

7. F.R. Jones, Interfacial Aspects of Glass Fibre Reinforced Plastics, in *Interfacial Phenomena in Composite Materials '89*, F.R. Jones, Ed., Butterworths, 1989, p 25-32

8. A.M. Shibley, Glass-Filled Thermoplastics, in *Handbook of Composites*, G. Lubin, Ed., Van Nostrand Reinhold, 1982, p 115-135

9. C.E. Knox, Fiberglass Reinforcement, in *Handbook of Composites*, G. Lubin, Ed., Van Nostrand Reinhold, 1982, p 136-159

10. D. Santrach, Industrial Applications and Properties of Short Glass Fiber-Reinforced Plastics, *Polym. Compos.*, Vol 3, 1982, p 239-244

11. K.L. Lowenstein, *The Manufacturing Technology of Continuous Glass Fibers*, Elsevier, 1973

12. E.P. Plueddemann, *Silane Coupling Agents*, Plenum Publishing, 1982

13. E.P. Plueddemann, Silane Compounds for Silylating Surfaces, in *Silanes, Surfaces, and Interfaces*, D.E. Leyden, Ed., Gordon & Breach, 1986, p 1-24

14. E.P. Plueddemann, Coupling Agents, in *International Encyclopedia of Composites, Vol. 1.*, S.M. Lee, Ed., VCH Publishers, 1990, p 507-511

15. E.P. Plueddemann, Silane Primers for Epoxy Adhesives, *J. Adhes. Sci. Technol.*, Vol 2, 1988, p 179-188

16. H. Ishida, A Review of Recent Progress in the Studies of Molecular and Microstructure of Coupling Agents and Their Functions in Composites, Coatings and Adhesive Joints, *Polym. Compos.*, Vol 5, 1984, p 101-123

17. J.L. Koenig and H. Emadipour, Mechanical Characterization of the Interfacial Strength of Glass-Reinforced Composites, *Polym. Compos.*, Vol 6, 1985, p 142-150

18. E.R. Pohl and F.D. Osterholtz, Kinetics and Mechanism of Condensation of Alkylsilanols in Aqueous Solution, in *Silanes, Surfaces, and Interfaces*, D.E. Leyden, Ed., Gordon & Breach, 1986, p 481-500

19. J. Comyn, Silane Coupling Agents, in *Structural Adhesives: Developments in Resins and Primers*, A.J. Kinloch, Ed., Elsevier, 1986, p 269-312

20. H. Ishida and J.L. Koenig, A Fourier-Transform Infrared Spectroscopic Study of the Hydrolytic Stability of Silane Coupling Agents in E-Glass Fibers, *J. Polym. Sci., Polym. Phys. Ed.*, Vol 18, 1980, p 1931-1943

21. J.L. Koenig, FTIR Studies of Interfaces, in *Silanes, Surfaces, and Interfaces*, D.E. Leyden, Ed., Gordon & Breach, 1986, p 43-57

22. K.-P. Hoh, H. Ishida, and J.L. Koenig, Spectroscopic Studies of the Gradient in the Silane Coupling Agent/Matrix Interface in Fiberglass-Reinforced Epoxy, *Polym. Compos.*, Vol 9, 1988, p 151-157

23. E.P. Plueddemann, Composites Having Ionomer Bonds With Silanes at the Interface, *J. Adhes. Sci. Technol.*, Vol 3, 1989, p 131-139

24. K.P. McAlea and G.J. Besio, Adhesion Between Polybutylene Terephthalate and E-Glass Measured With a Microdebond Technique, *Polym. Compos.*, Vol 9, 1988, p 285-290

25. J. Denault and T. Vu-Khanh, Role of Morphology and Coupling Agent in Fracture Performance of Glass-Filled Polypropylene, *Polym. Compos.*, Vol 9, 1988, p 360-367

26. D.F. Sounik and M.E. Kenney, New Potential Silane Coupling Agents, *Polym. Compos.*, Vol 6, 1985, p 151-155

27. H. Ishida, J.L. Koenig, B. Asumoto, and M.E. Kenney, Application of UV Resonance Raman Spectroscopy to the Detection of Monolayers of Silane Coupling Agent on Glass Surfaces, *Polym. Compos.*, Vol 2, 1981, p 75-80

28. R.W. Rosser, T.S. Chen, and M. Taylor, Modification of Epoxy-Reinforced Glass-Cloth Composites With a Perfluorinated Alkyl Ether Elastomer, *Polym. Compos.*, Vol 5, 1984, p 198-201

29. Y. Eckstein, Role of Silanes in Adhesion. Part I. Dynamic Mechanical Properties of Silane Coatings on Glass Fibers, *J. Adhes. Sci. Technol.*, Vol 2, 1988, p 339-348

30. Y. Eckstein, Role of Silanes in Adhesion. Part II. Dynamic Mechanical Properties of Silane-Treated Glass Fiber/Polyester Composites, *J. Adhes. Sci. Technol.*, Vol 3, 1989, p 337-355

31. D. Pawson and F.R. Jones, The Role of Oligomeric Silanes on the Interfacial Shear Strength of AR Glass-Vinyl Ester Composites, in *Interfacial Phenomena in Composite Materials '89*, F.R. Jones, Ed., Butterworths, 1989, p 188-192

32. A.T. DiBenedetto, G. Haddad, C. Schilling, and F. Osterholtz, Evaluation of Silane Coupling Agents for Glass Fibers in Composite Materials, in *Interfacial Phenomena in Composite Materials '89*, F.R. Jones, Ed., Butterworths, 1989, p 181-187

33. J.L. Thomason, Characterisation of Fibre Surfaces and the Interphase in Fibre-Reinforced Polymer Composites, in *Interfacial Phenomena in Composite Materials '89*, F.R. Jones, Ed., Butterworths, 1989, p 171-180

34. C.C. Chiao and T.T. Chiao, Aramid Fibers and Composites, in *Handbook of Composites*, G. Lubin, Ed., Van Nostrand Reinhold, 1982, p 272-317

35. L.S. Penn and F. Larsen, Physiochemical Properties of Kevlar 49 Fiber, *J. Appl. Polym. Sci.*, Vol 23, 1979, p 59-73

36. M.G. Northolt, X-Ray Diffraction Study of Poly(p-Phenylene Terephthalamide) Fibers, *Eur. Polym. J.*, Vol 10, 1974, p 799-804

37. M.G. Dobb, D.J. Johnson, A. Majeed, and B.P. Saville, Microvoids in Aramid-Type Fibrous Polymers, *Polymer*, Vol 20, 1979, p 1284-1288

38. L.S. Penn, F.A. Bystry, and H.J. Marchionni, Relation of Interfacial Adhesion in Kevlar/Epoxy Systems to Surface Characterization and Composite Performance, *Polym. Compos.*, Vol 4, 1983, p 26-31

39. T.S. Keller, A.S. Hoffman, B.D. Rat-

ner, and B.J. McElroy, Chemical Modification of Kevlar Surfaces for Improved Adhesion to Epoxy Resin Matrices: I. Surface Characterization, in *Physiochemical Aspects of Polymer Surfaces. Volume 2*, K.L. Mittal, Ed., Plenum Publishing, 1983, p 861-879

40. E.G. Chatzi, S.L. Tidrick, and J.L. Koenig, Infrared Studies of Kevlar, *Polym. Prepr.*, Vol 28, 1987, p 13-15

41. M. Breznick, J. Banbaji, H. Guttmann, and G. Marom, Surface Treatment Technique for Aramid Fibres, *Polym. Commun.*, Vol 28, 1987, p 55-56

42. Y. Wu and G.C. Tesoro, Chemical Modification of Kevlar Fiber Surfaces and of Model Diamides, *J. Appl. Polym. Sci.*, Vol 31, 1986, p 1041-1059

43. M. Takayanagi and T. Katayose, N-Substituted Poly(p-Phenylene Terephthalamide), *J. Polym. Sci., Polym. Chem. Ed.*, Vol 19, 1981, p 1133-1145

44. M. Takayanagi, Polymer Composites of Rigid and Flexible Molecules, *Pure Appl. Chem.*, Vol 55, 1983, p 819-832

45. M. Takayanagi, T. Kajiyama, and T. Katayose, Surface-Modified Kevlar Fiber-Reinforced Polyethylene and Ionomer, *J. Appl. Polym. Sci.*, Vol 27, 1982, p 3903-3917

46. L.S. Penn, T.J. Byerley, and T.K. Liao, The Study of Reactive Functional Groups in Adhesive Bonding at the Aramid-Epoxy Interface, *J. Adhes.*, Vol 23, 1987, p 163-185

47. L.S. Penn, G.C. Tesoro, and H.X. Zhou, Some Effects of Surface-Controlled Reactions of Kevlar 29 on the Interface in Epoxy Composites, *Polym. Compos.*, Vol 9, 1989, p 184-191

48. Y.W. Mai, Controlled Interfacial Bonding on the Strength and Fracture Toughness of Kevlar and Carbon Fibre Composites, *J. Mater. Sci. Lett.*, Vol 2, 1983, p 723-725

49. Y.W. Mai and F. Castino, Fracture Toughness in Kevlar-Epoxy Composites With Controlled Interfacial Bonding, *J. Mater. Sci.*, Vol 19, 1984, p 1638-1655

50. M.R. Wertheimer and H.P. Schreiber, Surface Property Modification of Aromatic Polyamides by Microwave Plasmas, *J. Appl. Polym. Sci.*, Vol 26, 1981, p 2087-2096

51. J.E. Sohn, Improved Matrix-Filler Adhesion, *J. Adhes.*, Vol 19, 1985, p 15-27

52. J.T. Kenney, W.P. Townsend, and J.A. Emerson, Tin and Iron Hydrous Oxide Deposits on Polyethylene, Teflon, and Paraffin, *J. Colloid Interface Sci.*, Vol 42, 1973, p 589-596

53. H. Kaufer and E.M. Abdel-Bary, Synthetic Organic Fiber-Reinforced Thermoplastics. I. Enhancement of the Ad-

hesion of Poly(ethylene terephthalate) and Kevlar Fibers to Polystyrene, *Colloid Polymer Sci.*, Vol 260, 1982, p 788-793

54. Y. Iyengar, Adhesion of Kevlar Aramid Cords to Rubber, *J. Appl. Polym. Sci.*, Vol 22, 1978, p 801-812

55. D.M. Riggs, R.J. Shuford, and R.W. Lewis, Graphite Fibers and Composites, in *Handbook of Composites*, G. Lubin, Ed., Van Nostrand Reinhold, 1982, p 196-271

56. R.J. Dauksys, Graphite Fiber Treatments Which Affect Fiber Surface Morphology and Epoxy Bonding Characteristics, *J. Adhes.*, Vol 5, 1973, p 211-244

57. P. Ehrburger and J.B. Donnet, Surface Treatment of Carbon Fibres for Resin Matrices, in *Handbook of Composites, Vol 1, Strong Fibres*, W. Watt and B.V. Perov, Ed., Elsevier, 1985, p 577-603

58. T.A. DeVilbiss and J.P. Wightman, Surface Characteristics of Carbon Fibers, in *Composite Interfaces*, H. Ishida and J.L. Koenig, Ed., Elsevier, 1986, p 307-316

59. D.S. Everhart and C.N. Reilley, Chemical Derivatization in Electron Spectroscopy for Chemical Analysis of Surface Functional Groups Introduced on Low Density Polyethylene Film, *Anal. Chem.*, Vol 53, 1981, p 665-676

60. R.V. Subramanian and A.S. Crasto, Electrodeposition of a Polymer Interphase in Carbon-Fiber Composites, *Polym. Compos.*, Vol 7, 1986, p 201-218

61. A.S. Crasto, S.H. Own, and R.V. Subramanian, The Influence of the Interphase on Composite Properties: Poly(Ethylene-Co-Acrylic Acid) and Poly(Methyl Vinyl Ether-Co-Maleic Anhydride) Electrodeposited on Graphite Fibers, *Polym. Compos.*, Vol 9, 1988, p 78-92

62. R.V. Subramanian and J.J. Jakubowski, Electropolymerization on Graphite Fibers, *Polym. Eng. Sci.*, Vol 18, 1978, p 590-600

63. R.V. Subramanian, Electrochemical Polymerization and Deposition on Carbon Fibers, *Pure Appl. Chem.*, Vol 52, 1980, p 1929-1937

64. J.P. Bell, J. Chang, H.W. Rhee, and R. Joseph, Application of Ductile Polymeric Coatings Onto Graphite Fibers, *Polym. Compos.*, Vol 8, 1987, p 46-52

65. L.T. Drzal, M.J. Rich, and P.F. Lloyd, Adhesion of Graphite Fibers to Epoxy Matrices: I. The Role of Fiber Surface Treatment, *J. Adhes.*, Vol 16, 1982, p 1-30

66. L.T. Drzal, M.J. Rich, M.F. Koenig, and P.F. Lloyd, Adhesion of Graphite Fibers to Epoxy Matrices: II. The Ef-

fect of Fiber Finish, *J. Adhes.*, Vol 16, 1983, p 133-152

67. L.T. Drzal, M.J. Rich, and M.F. Koenig, Adhesion of Graphite Fibers to Epoxy Matrices: III. The Effect of Hygrothermal Exposure, *J. Adhes.*, Vol 18, 1985, p 49-72

68. A. Benatar and T.G. Gutowski, Effects of Moisture on Interface Modified Graphite Epoxy Composites, *Polym. Compos.*, Vol 7, 1986, p 84-90

69. S. Mujin, H. Baorong, W. Yisheng, T. Yin, H. Weiqiu, and D. Youxian, The Surface of Carbon Fibers Continuously Treated by Cold Plasma, *Compos. Sci. Technol.*, Vol 34, 1989, p 353-364

70. R.A. Weiss, Strength Improvement of Short Graphite Fiber-Reinforced Polypropylene, *Polym. Compos.*, Vol 2, 1981, p 89-94

71. J.H. Williams and P.N. Kousiounelos, Thermoplastic Fibre Coatings Enhance Composite Strength and Toughness, *Fibre Sci. Technol.*, Vol 11, 1978, p 83-88

72. J. Schultz, L. Lavielle, and C. Martin, The Role of the Interface in Carbon Fibre-Epoxy Composites, *J. Adhes.*, Vol 23, 1987, p 45-60

73. M.-F. Grenier-Loustalot, Y. Borthomeiu, and P. Grenier, Surfaces of Carbon Fibres Characterized by Inverse Gas Chromatography, *Surf. Interface Anal.*, Vol 14, 1989, p 187-193

74. G.E. Hammer and L.T. Drzal, Graphite Fiber Surface Analysis by X-Ray Photoelectron Spectroscopy and Polar/Dispersive Free Energy Analysis, *Appl. Surf. Sci.*, Vol 4, 1980, p 340-355

75. D.M. Brewis, J. Comyn, J.R. Fowler, D. Briggs, and V.A. Gibson, Surface Treatments of Carbon Fibres Studied by X-Ray Photoelectron Spectroscopy, *Fibre Sci. Technol.*, Vol 12, 1979, p 41-52

76. A. Proctor and P.M.A. Sherwood, X-Ray Photoelectron Spectroscopic Studies of Carbon Fibre Surfaces. II. The Effect of Electrochemical Treatment, *Carbon*, Vol 21, 1983, p 53-59

77. A. Proctor and P.M.A. Sherwood, X-Ray Photoelectron Spectroscopic Studies of Carbon Fibre Surfaces. III. Industrially Treated Fibres and the Effect of Heat and Exposure to Oxygen, *Surf. Interface Anal.*, Vol 4, 1982, p 212-219

78. C. Kozlowski and P.M.A. Sherwood, X-Ray Photoelectron Spectroscopic Studies of Carbon-Fibre Surfaces. Part 4. The Effect of Electrochemical Treatment in Nitric Acid, *J. Chem. Soc., Faraday Trans. I.*, Vol 80, 1984, p 2099-2107

79. C. Kozlowski and P.M.A. Sherwood, X-Ray Photoelectron Spectroscopic Studies of Carbon-Fibre Surfaces. Part

5. The Effect of pH on Surface Oxidation, *J. Chem. Soc., Faraday Trans. I.*, Vol 81, 1985, p 2745-2756

80. C. Kozlowski and P.M.A. Sherwood, X-Ray Photoelectron Spectroscopic Studies of Carbon Fibre Surfaces. VII. Electrochemical Treatment in Ammonium Salt Electrolytes, *Carbon*, Vol 24, 1986, p 357-363

81. E.M. Petrie, Adhesively Bonding Plastics: Meeting an Industry Challenge, *Adhes. Age*, 15 May 1989, p 6-13

82. A.H. Landrock, *Adhesives Technology Handbook*, Noyes Publishers, 1985

83. J. Shields, *Adhesives Handbook*, Butterworths, 1984

84. C.E. Chastain and N. Berry, Surface Preparation, in *International Plastics Section, Digest Adhesives, Sealants, and Primers*, 5th ed., D.A.T.A. Business Publishers, 1989

85. A.J. Kinloch, *Adhesion and Adhesives: Science and Technology*, Chapman and Hall, 1987

86. F.M. Fowkes, D.O. Tischler, J.A. Wolfe, L.A. Lannigan, C.M. Ademu-John, and M.J. Halliwell, Acid-Base Complexes of Polymers, *J. Polym. Sci., Polym. Chem. Ed.*, Vol 22, 1984, p 547-566

87. L.M. Siperko and R.R. Thomas, Chemical and Physical Modification of Fluoropolymer Surfaces for Adhesion Enhancement: A Review, *J. Adhes. Sci. Technol.*, Vol 3, 1989, p 157-173

88. H. Schonhorn and J.P. Luongo, Adhesive Bonding of Polyvinylidene Fluoride: Effect of Curing Agent in PNF_2 Surface Modification, *J. Adhes. Sci. Technol.*, Vol 3, 1989, p 277-290

89. C.E.M. Morris, Adhesive Bonding in Polypropylene, *J. Appl. Polym. Sci.*, Vol 15, 1971, p 501-505

90. D.M. Brewis and D. Briggs, Adhesion to Polyethylene and Polypropylene, *Polymer*, Vol 22, 1981, p 7-16

91. H. Schonhorn and R.H. Hansen, Surface Treatment of Polymers for Adhesive Bonding, *J. Appl. Polym. Sci.*, Vol 11, 1967, p 1461-1474

92. H. Yasuda, H.C. Marsh, S. Brandt, and C.N. Reilley, ESCA Study of Polymer Surfaces Treated by Plasma, *J. Polym. Sci., Polym. Chem. Ed.*, Vol 15, 1977, p 991-1019

93. D. Briggs, D.G. Rance, C.R. Kendall, and A.R. Blythe, Surface Modification of Poly(ethylene terephthalate) by Electrical Discharge Treatment, *Polymer*, Vol 21, 1980, p 895-900

94. M. Strobel, C. Dunatov, J.M. Strobel, C.S. Lyons, S.J. Perron, and M.C. Morgen, Low-Molecular-Weight Materials on Corona-Treated Polypropylene, *J. Adhes. Sci. Technol.*, Vol 3, 1989, p 321-335

95. B. Westerlind, A. Larsson, and M. Rigdahl, Determination of the Degree of Adhesion in Plasma-Treated Polyethylene/Paper Laminates, *Int. J. Adhes. Adhes.*, Vol 7, 1987, p 141-146

96. C.Y. Kim and D.A.I. Goring, Surface Morphology of Polyethylene After Treatment in a Corona Discharge, *J. Appl. Polym. Sci.*, Vol 15, 1971, p 1357-1364

97. D.K. Owens, Mechanism of Corona-Induced Self-Adhesion of Polyethylene Film, *J. Appl. Polym. Sci.*, Vol 19, 1975, p 265-271

98. G.D. Cooper and M. Prober, The Action of Oxygen Corona and of Ozone on Polyethylene, *J. Polym. Sci.*, Vol 44, 1960, p 397-409

99. A.R. Blythe, D. Briggs, C.R. Kendall, D.G. Rance, and V.J.I. Zichy, Surface Modification of Polyethylene by Electrical Discharge Treatment and the Mechanism of Autoadhesion, *Polymer*, Vol 19, 1978, p 1273-1278

100. D. Briggs and C.R. Kendall, Chemical Basis of Adhesion to Electrical Discharge Treated Polyethylene, *Polymer*, Vol 20, 1979, p 1053-1054

101. D. Briggs and C.R. Kendall, Derivatization of Discharge-Treated LDPE: An Extension of XPS Analysis and a Probe of Specific Interactions in Adhesion, *Int. J. Adhes. Adhes.*, Vol 2, 1982, p 13-17

102. P. Beardmore and C.F. Johnson, in *Advanced Composites: The Latest Developments*, Proceedings of the Second Annual Conference on Advanced Composites, ASM INTERNATIONAL, 1986

103. R.H. Sjoberg and E.J. Lesniak, in *Advanced Composites III: Expanding the Technology*, Proceedings of the Third Annual Conference on Advanced Composites, ASM INTERNATIONAL, 1987

104. J.E. Hill, in *How to Apply Advanced Composites Technology*, Proceedings of the Fourth Annual Conference on Advanced Composites, ASM INTERNATIONAL, 1988

105. G.Y. Yee, An Overview of Structural Adhesive Bonding Needs for Composites in the Automobile Industry, in *Advanced Composites: The Latest Developments*, Proceedings of the Second Annual Conference on Advanced Composites, ASM INTERNATIONAL, 1986, p 217-219

106. S.L. Kaplan, P.W. Rose, and D.A. Frazer, Gas Plasma and the Treatment of Plastics, in *How to Apply Advanced Composites Technology*, Proceedings of the Fourth Annual Conference on Advanced Composites, ASM INTERNATIONAL, 1988, p 193-198

107. G.K.A. Kodokian and A.J. Kinloch, The Adhesive Fracture Energy of Bonded Thermoplastic Fibre-Composites, *J. Adhes.*, Vol 29, 1989, p 193-218

108. B.M. Parker and R.M. Waghorne, Surface Pretreatment of Carbon Fibre-Reinforced Composites for Adhesive Bonding, *Composites*, Vol 9, 1982, p 280-288

109. T.A. DeVilbiss, D.L. Messick, D.J. Progar, and J.P. Wightman, SEM/XPS Analysis of Fractured Adhesively Bonded Graphite Fibre-Reinforced Polyimide Composites, *Composites*, Vol 16, 1985, p 207-219

110. D.J.D. Moyer and J.P. Wightman, Characterization of Surface Pretreatments of Carbon Fiber-Polyimide Matrix Composites, *Surf. Interface Anal.*, Vol 14, 1989, p 496-504

111. C.W. Matz, Adhesion on Carbon-Fibre-Reinforced Plastics, *J. Adhes.*, Vol 22, 1987, p 61-65

112. H. Gleich, R.M. Criens, H.G. Mosle, and U. Leute, The Influence of Plasma Treatment on the Surface Properties of High-Performance Thermoplastics, *Int. J. Adhes. Adhes.*, Vol 9, 1989, p 88-94

113. P.R. Young, Thermoset Matched Die Molding, in *Handbook of Composites*, G. Lubin, Ed., Van Nostrand Reinhold, 1982, p 391-448

114. G.D. Cheever, Role of Surface Morphology and Composition in the Painting of Sheet Molding Compound (SMC) Plastics, *J. Coat. Technol.*, Vol 50, 1978, p 36-49

115. J.G. Dillard, C. Burtoff, and T. Buhler, Surface Chemical Study of Sheet Molded Composite (SMC) as Related to Adhesion, *J. Adhes.*, Vol 25, 1988, p 203-227

116. J. Newbould and K.J. Schroeder, Infrared Laser Surface Treatment of Sheet Molding Compound (SMC), in *Advanced Composites III: Expanding the Technology*, Proceedings of the Third Annual Conference on Advanced Composites, ASM INTERNATIONAL, 1987, p 365-386

117. J.G. Dillard, T.F. Cromer, C.E. Burtoff, A.J. Cosentino, R.L. Cline, and G.M. MacIver, Surface Properties and Adhesion of Flame Treated Sheet Molded Composite (SMC), *J. Adhes.*, Vol 26, 1988, p 181-198

118. J.G. Dillard and I. Spinu, Plasma Treatment of Composites for Adhesive Bonding, in *How to Apply Advanced Composites Technology*, Proceedings of the Fourth Annual Conference on Advanced Composites, ASM INTERNATIONAL, 1988, p 199-208

119. T.M. Liang, T.A. Tufts, and G.E. Kimes, Using Structural Adhesives to Bond Composites Made by Structural Reaction Injection Molding (SRIM), in *How to Apply Advanced Composites Technology*, Proceedings of the Fourth Annual Conference on Advanced Composites, ASM INTERNATIONAL, 1988, p 217-223

Surface Considerations for Joining Ceramics and Glasses

Victor A. Greenhut, Department of Ceramics, Rutgers University

THE JOINING of ceramics and glasses to other ceramic and glass parts and to metallic structures requires an understanding of the features that distinguish ceramics from metals and polymers. Ceramics and glasses usually show strong covalent/ionic bonding and are typically based on various combinations of silicates, aluminosilicates, metal oxides, carbides, and nitrides. Following conventional definitions, in this article glasses are considered to be amorphous materials based on these systems, whereas ceramics can be single crystals, polycrystalline materials, or combinations of crystalline and amorphous phases.

Thermodynamically, solid glasses are supercooled liquids similar to their ceramic counterparts in terms of various engineering properties. Ceramics often contain substantial glass as a continuous or quasi-continuous phase containing crystalline ceramic grains. Such a structure is shown both by traditional clay-base ceramics and by engineering ceramics, such as high aluminas. So-called glass-ceramics are glasses that devitrify during a thermal hold, nucleating and growing a crystalline phase or phases.

Ceramics and glasses typically show chemical inertness relative to metals and polymers. They have low electrical and thermal conductivity and are usually transparent to a broad spectrum of electromagnetic radiation. Mechanically, ceramics have very high elastic modulus (stiffness) and compressive strength properties, coupled with the relatively low fracture toughness of brittle materials. The latter property makes ceramics susceptible to intrinsic flaws, such as pores and large grains, and extrinsic flaws, such as grinding marks and chips. Such flaws can dominate the mechanical strength behavior of a ceramic or glass. Ceramics have high melting points, and properties are maintained at elevated temperatures. They also have relatively low thermal expansion characteristics, compared to other classes of materials. Ceramics exhibit low friction and wear when subjected to moving contact. Glasses show a quasi-continuous variation in properties with the adjustment of glass chemistry,

whereas ceramics show more complex composite interactions that are often not fully understood and that are dependent on chemistry and microstructure.

Those unfamiliar with ceramics are often frustrated in attempts to bond or seal the material, particularly with metals or polymers. The designation system for ceramics is not precise. For example, the term alumina is applied to a wide range of compositions and microstructures with varied properties and bonding behaviors. Specifications are elusive and difficult to make precise because the chemical and physical formulation of the ceramic is often dependent on raw-material sources, forming method, and available furnaces. Glasses are more predictable on the basis of chemistry, but frequently, users do not recognize that a wide range of chemistries and properties are available within a class of glasses. Although one glass may bond well to a metal, another with related chemistry may not.

Ceramics and glasses, while chemically resistant, are not inert. Often, water from surface preparation or the atmosphere forms an adherent surface film or reaction layer. Organic solvents may also bond to the ceramic or glass. These residues may interfere with bonding or sealing. Alternatively, when a ceramic or glass is used as a protective seal, it must be selected for use in a particular chemical environment. For example, some "solder" glasses used in low-temperature electronic joining and sealing may be considered water soluble.

Recent advances in monolithic and composite ceramics have substantially improved upon the brittle characteristic, in terms of both a reduction of intrinsic defects and damage tolerance. However, flaws that are introduced either during surface preparation or the bonding process can lead to failure at or near the bond interface as a result of the brittle nature of the ceramic-metal or ceramic-ceramic bond. This feature, as well as differences in properties between classes of materials, must be taken into account in order to achieve the successful joining or sealing of ceramics and glasses.

Various important considerations for bonding ceramics and glasses are covered in the section "Fundamentals of Bonding," below. The technology is approached chiefly by covering ceramic/glass-metal bonding and sealing. This is because many of the considerations involved are common to all ceramic bonding, although some special considerations are necessary because of the differences between ceramic and glass bonding and properties. Subsequent sections in this article consider glass-to-metal, ceramic-to-metal, glass-to-glass, and glass-to-ceramic bonding and sealing. Greater attention is given to the first two areas because they represent technologies of wider engineering interest (the latter two technologies largely being the province of ceramic production). The final section of this article describes the use of polymeric materials in bonds and seals.

Fundamentals of Bonding

When a metal and a ceramic are joined, frequently one is a liquid phase and the other is a solid. Vapor deposition or solid-solid joining may also be used. There is intimate physical contact, and bonding may occur by surface or interfacial interaction, depending on the specific chemical nature of the materials. Solution and reduction/oxidation interactions may proceed, which promote good adherent bonds. In technologies involving a liquid phase, it is a usual requirement that wetting by the liquid phase occur. In many technologies, the liquid phase is obvious as bulk melting occurs, but other cases may be misleading with regard to this requirement because the liquid phase is either a thin layer or a microscopic constituent. Often, interactions between the materials themselves and with the environment will alter the wetting response with time.

Stresses between joined components may also significantly affect the bond, either deteriorating bond quality, causing components to delaminate from seals, or resulting in interfacial failure. Such stresses may arise because of differences in the elastic

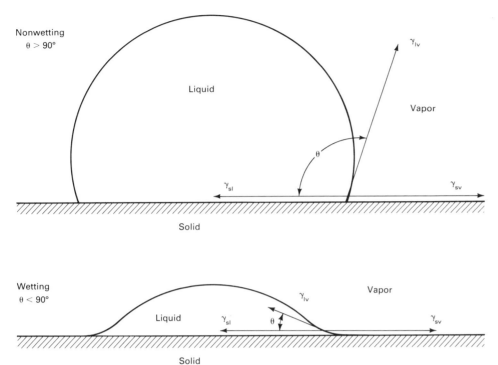

Fig. 1 Sessile drop configuration

Nonwetting
$\theta > 90°$

Liquid

γ_{lv}

Vapor

θ

γ_{sl} γ_{sv}

Solid

Wetting
$\theta < 90°$

Liquid γ_{sl} γ_{lv} Vapor γ_{sv}

θ

Solid

modulus under an applied or residual stress. Differences in thermal expansion or phase transformations can give rise to stresses that exceed the bond strength. Thermal cycling may introduce cyclic fatigue due to this expansion difference.

Wetting

Melting the metal or softening a glassy material is common when bonding ceramics, glasses, and/or metals. Wetting is a usual requirement to provide a quality bond or seal (Ref 1, 2). To be capable of joining or coating, the liquid must flow or spread across the surface or joining interface. This condition is important for several reasons. Wetting ensures good contact between materials, with the liquid phase filling asperities, thereby providing a mechanical interlock. Intimate contact also avoids discontinuous bonding flaws at the interface, which can act as failure origins. In addition, the lower interfacial energy associated with a wetting system implies a more adherent bond upon solidification.

Figure 1 shows a cross section of the sessile drop method for determining the wetting response of a liquid on a flat solid. Alternative methods involve capillary liquid rise, wicking in a porous solid, and force produced between phases (Ref 2-4). The sessile drop is perhaps most universally applicable and convenient to set up. Because a liquid metal or flowing glass will be used at elevated temperature, the liquid drop and substrate are best set in a tube or other appropriate furnace. End plates with fused-quartz ports permit a controlled at-

mosphere and the ability to illuminate and view the drop(s).

The contact angle (θ) is measured with a telescopic protractor or may be recorded for later measurement with a short telephoto single-lens reflex (SLR) camera or a video recorder. An angle greater than 90° correlates to nonwetting behavior, whereas an angle less than 90° correlates to wetting behavior. For an angle of 0°, the liquid flows across the surface in a "spreading" condition. The wetting angle provides valuable engineering information because as the system interacts, the wetting response can change with time and temperature.

If the flow behavior of a viscous glass wetting a solid is understood, the system can be timed and controlled. The effect of surface preparation, as well as changes in the vapor, liquid, or solid, can be determined in order to promote a bond. Although it is termed the sessile drop, in industrial practice the drop is rarely allowed to come to equilibrium as it reacts with films and impurities on the surface or gas that is has absorbed, or as bulk interaction proceeds. The wetting response indicates the degree to which the liquid spreads or infiltrates the solid. Usually, a good wetting response implies that an adherent bond is feasible, whereas nonwetting rarely corresponds to an adequate bond. Other factors, such as the stress state in the joint and interfacial defects, also affect the quality of the bond.

The interfacial energies indicated in Fig. 1 must be balanced once equilibrium is achieved. Under such conditions, the interfacial energy can be estimated or derived

(Ref 3-6). When extrapolated to room temperature, this provides an estimate of the theoretical bond strength for the system. The actual bond strengths observed are often many orders lower than such estimates, a condition that can be attributed to interfacial defects and the brittle nature of the interface.

In joining ceramics and metals, a transition from metal oxide to metallic layers may be used to promote wetting. It is common for ceramics and metals to show poor wetting without some engineering artifice. A moly-manganese paste (see "The sintered metal powder process [SMP]" section) may be applied to a ceramic to provide a metallizing layer. Such pastes, upon firing, can result in a ceramic-and-glass based structure that bonds to the ceramic and a metallic surface that can be joined to metals by conventional brazing technologies (Ref 7). It is common to preoxidize the metal in glass-to-metal joining and sealing (Ref 3, 5, 8). The oxide layer adheres to the metal and may lower the interfacial energy of the surface, thereby promoting wetting by providing a covalent/ionic bond that is more compatible with the glass.

It is quite common to employ a forming gas, which tends to oxidize the metallic component or partially reduce the ceramic or glass at the surface. Such surface modification can promote wetting and significantly enhance bonding. Indeed, atmosphere control is often vital to successful bonding between metals and ceramics. All of the above techniques suggest that a transition between metallic and covalent/ionic bonding is an important consideration for successful wetting and adherence.

Engineering to attain wetting is less of a concern in the joining or sealing of ceramics and glasses with each other. Generally, oxide glasses and ceramics wet each other readily so that a glass with sufficient fluidity at fabrication temperature will successfully flow across an oxide ceramic and glaze the surface. The glass phase in the ceramic or a glass frit allows microscopic wetting and joining. Because it is difficult to obtain good wetting between many oxide and nonoxide ceramics and/or glasses, a transition or a mixed microstructure is usually needed.

Bonding and Adherence Mechanisms

Convenient categories for considering and promoting bonding of ceramics and metals are physical, mechanical, and chemical bonding. Generally, all three bonding mechanisms are found in well-bonded systems, and it is difficult to separate the contribution of each. Enhancing the mechanisms when preparing a material for bonding and when joining is important in order to engineer a strong bond.

Physical bonding is the effect that occurs when two perfectly flat surfaces are brought together to atomic interaction distances. Local atomic rearrangement results in adhesion. The energy difference between the specific surface energy of one material and that of another is the work of adhesion. The work of adhesion can yield a theoretical breaking stress similar to the strength of the metal or ceramic (Ref 4, 9). The actual failure stress is often several orders of magnitude smaller because of flaws in the bond. Physical bonding provides a guideline to the selection of systems that will bond well.

Mechanical bonding refers to the interlocking microstructure of rough surfaces to provide tensile strength and, in the case of shear, frictional strengthening. During the bonding process, liquid metal or glass can flow into cavities and asperities, a ductile metal can conform to a rough ceramic surface, or a vapor can deposit in surface asperities. The surface may be roughened by acid or base chemical attack, electrolytic corrosion, scouring, grinding, or grit/sand blasting to enhance mechanical bonding. In grit blasting, passing the apparatus in multiple directions at an inclined angle helps produce undercuts that enhance the bond. The effectiveness of different surface roughening methods is dependent on their optimum application, as well as on the specific method and materials used. Chemical interaction between the materials being bonded may also lead to mechanical bonding. In addition, the increase in surface area of the rough surface may provide an increase in physical bonding.

The degree of mechanical bonding is often modeled on the interface ratio or undercut density. The former is the ratio of actual bonded surface to that of a flat, smooth interface and is often taken on a cross-sectional sample. The undercut density measures the number of reentrant features on a cross section per linear distance. An increase in either number usually enhances the bond, but undercut density tends to correlate better with tensile adherence. Either measure may correlate with shear failure strength.

Chemical Bonding. Chemical reactions may change the nature of the bond interface, providing a transition zone by dissolution, interdiffusion, redox process, and precipitation. As the bond between ceramics and metals is fundamentally different, a theory of bonding first developed for glass-to-metal bonds may have general applicability. The model hypothesis is that the glass (ceramic, more generally) at the interface must be saturated with the lowest-oxidation-state oxide of the metal being bonded (Ref 5). There is an idealized molecular monolayer of metal oxide at the interface providing a continuity of chemical activity (bond energy) and structure between metal and ceramic. Incorporation of the metallic species may occur either by diffusion or reaction during bonding or by prior pro-

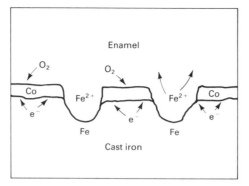

Fig. 2 Electrolytic effect between enamel and iron

cessing. The incorporation of metallic species in the ceramic or glass can promote bonding. This is often termed the chemical, or oxide, layer theory.

In glass-to-metal bonding, such as enameling, the electrolytic theory may operate, as shown in Fig. 2. Adherence-promoting metal oxides dissolved in the enamel are reduced during firing to a metallic state by the base metal. Alternatively, a flash of the metal may be deposited on the base metal. An example is the use of cobalt, either in the base coat or as a flash in the porcelain enameling of cast iron and steel. A local galvanic cell is formed, which strongly corrodes the base metal. The metal that goes into solution provides chemical bonding through the continuity of the activity discussed above. The rough surface with anchoring etch pits provides mechanical adhesion. Supersaturated metal may be precipitated during this process or upon cooling, resulting in dendrites that extend from the base metal, which provides a further mechanical anchor into the glass. This is termed the dendritic theory.

Other chemical reactions may produce a transition structure that varies from metallic to ceramic structure. Often metals are preoxidized to give an oxide that will be wet by and react with glass or glass within a ceramic, saturating the interfacial region. An adherent oxide on the base metal may provide a ceramic surface for bonding to a glass. A physical mixture of ceramic and metal, such as that in the sintered metal powder process (to be discussed) may provide a transition structure between the metal and the ceramic or glass. Interfacial phases that form may have properties that are different from those of either metal or ceramic (Ref 1-9).

In practice, complex combinations of the various mechanisms actually occur. As indicated above, porcelain enameling often shows mechanical bonding by means of etch pits and dendrites, as well as chemical bonding by the enrichment of the glass close to the interface in metallic species. Thus, features of physical, mechanical, and chemical adhesion can all be identified in prac-

tice, but their individual contributions may be difficult to determine.

Stresses in Ceramic-Metal Bonds

Thermally Induced Stresses. The thermal expansion of metals is usually considerably different from that of ceramics, as can be seen in Fig. 3. Ceramics or glasses may also show a mismatch of thermal expansion. If two materials with different expansions are constrained by a bond, the difference in thermal expansion can result in a residual stress. Such stresses add to the residual stresses that result from rapid cooling, thermal gradients, and phase changes. These residual stresses in themselves or in combination with applied stresses may result in bond failure.

In a flat-sheet geometry, the greater shrinkage of a metal upon cooling causes net compressive stress in the ceramic and tension in the metal. This is usually a favorable situation because the ceramic displays relatively high compressive strength, while the metal has similar tensile and compressive values. Precautions must be taken not to develop interfacial stresses that exceed the bond strength. In flat geometries, enamels are selected to provide a compressive stress in the glass. A similar condition is sought in glazing of a ceramic body. The compressive stress can enhance fracture strength and wear resistance, provided that critical failure stress values are not exceeded. Several approaches can be taken to limit or control the level of thermally induced stresses.

The thermal expansion of the ceramic and metal may be chosen to match or produce desirable stress distributions. An examination of Fig. 3 shows that tungsten and molybdenum have a close match with several ceramic and glass compositions. Therefore, they can be used in applications in which the high-temperature oxidation resistance of these metals is not a problem. Platinum has often been used as the primary bond metal with alumina because of the good expansion match. For cost reasons, other metals are bonded to this intermediate platinum layer. The alloy ASTM F15 (Fe, 28 to 29% Ni, 17 to 18% Co), often called Kovar (Carpenter Technologies), has frequent application in bonding to borosilicate glass. Such alloys show a particularly good match with the glass below the glass strain point upon cooling. These alloys are ductile and usually do not embrittle. They can be soldered, brazed, or welded. Glass chemistry can be adjusted to provide a reasonable expansion match with metal, such as, for example, the use of lead silicate and other glass compositions for enameling steels.

A "Housekeeper seal" is another technique for minimizing stresses due to differential thermal expansion. A thin piece of metal is used so that its thin dimension, low yield strength, and ductility allow it to fol-

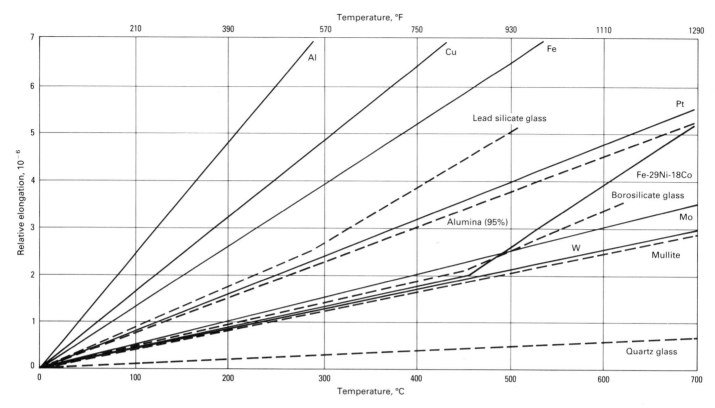

Fig. 3 Thermal expansion values of typical ceramics, glasses, and metals

low the thermal expansion characteristic of ceramic. The metal deforms to accommodate the difference in expansion (Ref 3, 8). This technique is often used in electrical feed-throughs, such as the conductor that passes through the envelope of high-power incandescent lamps.

Continuum mechanics can be used to calculate the residual stresses that result from expansion differences (Ref 3, 8, 10). Such calculations can be used in the iterative design of a part or the selection of applied loads that do not exceed the critical stress for bond failure.

A compressive seal is one example. In this type of seal, the ceramic and interface are placed in compression, while the metal is put in tension, taking advantage of the mechanical properties of each. Such compressive seals may be used to attain a purely mechanical bond between metal and ceramic. This can be done by shrink fitting the materials together, using the difference in thermal expansion during cooling to develop an interference fit. An example is the bonding of concentric cylinder geometries in which the metal is the outer component and shrinks on the ceramic during cooling because the metal shows higher contraction. This can produce a strong, tight fit and, if the metal deforms slightly, a hermetic seal. Caution must be used not to exceed the critical tensile failure stress for any component of the stress, either in the ceramic or the interface.

Elastic Modulus Effect. Interfacial or material failure may result because of the difference between the elastic modulus of a ceramic and that of a metal, as shown in Fig. 4. Typically, ceramics show greater stiffness than do metals. If the ceramic must follow the deformation of the metal because the metal is the larger member in an assembly, a relatively low stress in the metal (σ_m) will result in a high stress (σ_c) in the ceramic at the ceramic-metal interface. This is because the strain must be compatible ($\epsilon_{m,c}$) across the interface. Even if the ceramic is much stronger than the metal, as in the case shown, the ceramic fracture stress may be exceeded. Failure in the ceramic or at the bond interface will result.

Glass-to-Metal Joining and Sealing

There are two general categories of glass-to-metal joining: glass-to-metal seals and enamels. A clear and consistent distinction between the two areas is not made in practice. The categories may be distinguished by their application, the base metal employed, or the technological details of the methods. In terms of application, glass-to-metal seals are primarily used in electronics applications to provide hermeticity and electrical isolation. Examples are hermetic lamp and device sealing, the isolation of electrical feed-throughs, and packaging and vacuum tube technologies. Enamels tend to

be used in structural applications to provide corrosion resistance and decoration. The primary geometry is that of a coating or sealant. Examples are the coating of household appliances, bathroom fixtures, and cookware.

Enamels are most commonly applied to low-carbon steels and cast irons. This distinguishes them from glass-to-metal seals, which are usually made on nonferrous metals, iron-nickel-cobalt ASTM F15 alloy, and stainless steel. Often, glass coatings on titanium and copper for decorative applications

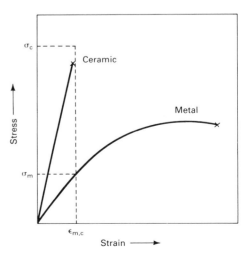

Fig. 4 Modulus effect in stressing ceramic-metal bond

such as jewelry are referred to as enamels. The latter use shows the difficulty in applying a concise designation system.

Glass-to-metal seals and enamels can also be characterized by the way in which the hydrogen, carbon, or nitrogen based gases that evolve during joining are treated. These elements may either be dissolved in the base metal or exist as hydrocarbons on the surface of the metal. At elevated temperature, these elements combine with oxygen, forming gas bubbles in the molten glass or the metal-glass interface. This is because the glass is relatively impervious to the gases evolved from the metal.

In glass-to-metal sealing, precautions are taken to clean the base metal of hydrocarbon residues scrupulously. The metal is then typically baked out or decarburized to remove dissolved species in the metal, as well as to further clean the surface. This process may be combined with a preoxidation step before joining. Higher levels of carbon or hydrogen are permitted in enameling, and in many applications no special treatment is used to remove these elements before fabrication. Additives, such as clay that contains organic material, are included in the enamel formulation to produce a fine, distributed bubble structure that acts as a getter for the evolved gases. This avoids large, objectionable bubbles in the glass, which could expose the base metal to the environment, and prevents bubbles from forming at the glass-metal interface, which would weaken the bond.

Glass-to-Metal Seals

In glass-to-metal sealing, the materials are usually selected both for required engineering properties and for thermal expansion coefficients that match reasonably over the temperature range of fabrication and use. After the forming and metal-cutting operations, the metal is usually cleaned chemically in caustic and washed. It may then undergo acid pickling, neutralization, and washing, followed by thorough degreasing. A plating or metal flash may be introduced to promote bonding, in which case the sequence above is repeated. Usually the metal is then decarburized and/or hydrogen baked, both to clean the surface and to diffuse out species that produce gas. Other products may also need to be removed, such as nitrogen in the case of titanium and its alloys.

Frequently, the metal then undergoes a controlled-oxidation step. This creates a tightly adherent layer of oxide, which must remain adherent during the formation of the glass-to-metal bond. In some cases, the adherence-promoting oxide is produced in the same step as the glass sealing, by appropriate atmosphere control and controlled viscous flow of the glass. The adherence oxide is usually wet by the glass and then dissolves partially or wholly in the glass,

providing a chemical glass-to-metal bond (Ref 5). References 3, 8, and 11 can be consulted for reviews of this subject.

Tungsten, molybdenum, titanium, tantalum, stainless steels, nickel, nickel-iron, ASTM F15-type, platinum, copper, gold, copper-clad nickel, and iron-nickel Dumet alloys have been used in common engineering applications. These metals have been coupled with a large range of borosilicate and lead silicate glasses. Aluminosilicate, lithium, barium, and soda-lime-silica glasses have been used to a lesser extent. Borosilicate glasses are most commonly used in glass-to-metal sealing, but lead silicate glasses are more common for bonding to copper, copper-clad alloys, and iron-chromium alloys such as stainless steels. Barium is being substituted for lead-base glasses in production applications where lead vapor may represent a potential hazard.

The glass is usually applied as a frit, paste, bead, or solid preform. It may be heated to the required viscosity range by electrical, radio frequency, or flame heating, either with the glass already applied to the preoxidized metal surface or presoftened and then pressed into contact with the metal. Graphite is frequently used to align or press parts together because it shows little bonding with the glass. Top-offs may be used to guide glass flow or prevent its flow into unwanted areas. It is common for the joining process to be conducted in a protective or special forming gas atmosphere to promote wetting and adhesion. Either rapid cooling is avoided or an annealing step is interposed to avoid the development of residual stresses due to rapid cooling.

ASTM F15 alloy (Fe-29Ni-17Co), termed Kovar, is a common glass-sealing alloy because it provides very good expansion matching with borosilicate glasses. These products can also be oxidized to provide a controlled, highly adherent oxide that is wet well by the glass and creates a strong bond. Oxidation along near-surface grain boundaries of the alloy gives mechanical anchoring, even without ostensible surface roughness. ASTM F15 alloy is used in microelectronic packaging, electrical feed-throughs, high-frequency devices, windows and ports, and other diverse applications. It provides a common and convenient example of the procedure for glass-to-metal sealing.

After forming, the Kovar alloy is alkaline degreased, water washed, chemically cleaned, water washed again, and Freon degreased with careful attention to the quality of the final degreasing steps. If a shaped glass preform is used or if it has protective sizing (starch), it may be washed and cleaned with hydrofluoric acid followed by water washes and possibly Freon or organic rinses. The metal is then subjected to a bake out and preoxidation treatment. It is rare, but possibly beneficial, for the glass to be

baked out. The Kovar is decarburized in hydrogen, hydrogen/nitrogen, or a dissociated ammonia atmosphere with a controlled water level at 900 to 1100 °C (1650 to 2010 °F). The bake out temperature should be higher than the final sealing temperature. A typical cycle is 10 to 15 min at 1100 °C (2010 °F) in a 1 vol% water, 10 vol% hydrogen/nitrogen atmosphere. Some manufacturers repeat the procedure. The alloy should not be stored at this or subsequent steps because oxidation, contamination, and adsorbed water may interfere with the quality of bonding.

The ASTM F15 alloy is then oxidized in a controlled way. Because Kovar is sufficiently tolerant, the alloy can be oxidized in air and natural gas, but bond quality will vary chiefly because of variations in the relative humidity and water content of the gas. Pure oxygen gives a thick, nonadherent, flaky oxide that is unsatisfactory for bonding. When produced in pure oxygen, the oxide does not penetrate along grain boundaries providing anchoring points. A high-quality oxide structure with suitable thickness and grain-boundary penetration can be accomplished with a 10-min treatment at 800 to 1050 °C (1470 to 1920 °F) in a gas composed of 1 vol% water and 0.4 vol% hydrogen/nitrogen. There are procedures in which the oxidation and sealing steps are combined.

The preoxidized metal is now ready for glass sealing. The glass frit, bead, paste, or preform is applied. The atmosphere may be like that of the preoxidation step, or natural gas may be used with somewhat greater variability in adherence. Sealing is done at about 1000 °C (1830 °F). Glass viscosity, wetting response, part geometry, part loading, and part geometry determine the time and temperature used in the sealing step (Ref 12).

Enamels

Enamels, that is, fused, vitreous, superficial coatings on metals, were used in prehistoric times. A technology developed during the Egyptian epochs, when enamels were applied to gold, silver, copper, and bronze jewelry. The application of enamels to iron and steel provided a means for sealing a metal surface with a corrosion-resistant coating and for decorating it also. The first method, still practiced, involved dusting red-hot metal with a dry powder, which softened and fused to provide a coating. Repeated application was employed to build up a continuous coating of sufficient thickness. Improvements to the process, such as improved surface preparation of the metal, the application of a liquid suspension of frit prior to firing, improved and controlled glass compositions, and, more recently, electrostatic spraying have made enamels a convenient protective seal and

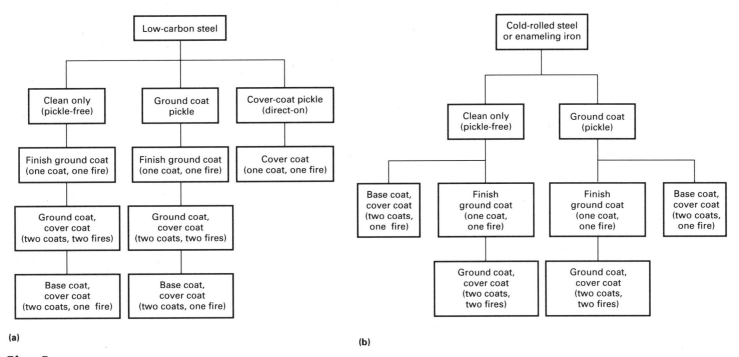

Fig. 5 Flow diagram for enameling. (a) Low-carbon steels. (b) Cold-rolled steel or enameling (cast) iron

decoration for consumer products and industrial production.

Preparation of Metal for Enameling. Figure 5(a) outlines the enameling process for low-carbon steel, whereas Fig. 5(b) depicts the process for cold-rolled steel and enameling (cast) iron. The first step is the preparation of the base metal, which requires a suitably roughened surface free from dirt, oxides, rust, machining oils, and grease. Surface roughening may be produced by acid pickling or grit blasting, which may also serve to clean the surface. The precise sequence of preparation steps depends on the choice of metal and enamel, method of application, and firing procedure.

Precious metals require no pretreatment if their surfaces are clean. Copper, bronze, and brass are cleaned and then given a 2 to 4% nitric acid pickle, followed by a clean-water rinse. Stainless steel and high-temperature alloys are usually prepared by grit or sand blasting. Aluminum alloys require special surface treatment, such as chromatizing and preoxidation. Cast iron and heavy-gage steels are often grit blasted before following the procedures given below.

The general steps for the metal preparation of steels, enameling irons, and cold-rolled steel, as well as several other alloys, are described below (Ref 13).

Annealing. The metal to be enameled is heated to a temperature sufficient to remove organic surface impurities that may act as a source of gases, which become trapped in the enamel or enamel-metal interface during firing. Such bubbles or blisters may weaken the coating, cause discoloration of the coating, or expose base

metal. The annealing procedure may also be used to decarburize a steel surface for similar reasons. Stress relief and heat treatment may also be accomplished if a low-temperature enamel is to be applied in subsequent steps.

Grit or Sand Blasting. Heavy scale, rust, and dirt are removed. The surface is roughened to provide for mechanical bonding. The best practice involves multiple passes in different directions to ensure full surface cleaning and roughening. Frequently, the grit blasting is done with normal impingement, but a superior surface roughness with anchoring undercuts can be accomplished by using an angle that is inclined 30 to 45° from the surface.

Solvent Cleaning. A suitable alkaline wash, often in caustic soda at 40 to 60 °C (105 to 140 °F) is employed to remove residues of oils, greases, soaps, and rust preventatives from the surface.

Acid pickling is done to remove metal oxidation products and scale. It also roughens the surface, promoting mechanical adhesion. Uninhibited, 6 to 9 wt% sulfuric acid at 65 to 75 °C (150 to 165 °F) is most commonly used for reasons of cost and suitability. Hydrochloric acid (10 to 12 wt%) may also be used at room temperature. Phosphoric acid and acid salts are used as well. After pickling, the metal is rinsed in cold, running water. The water is often acidified (pH 3 to 3.5) to prevent the hydrolysis of iron salts.

Nickel Flash. After the cold wash, nickel is frequently deposited on the surface. The nickel helps to provide controlled oxidation of the base metal, thereby preventing the

occurrence of defects in the enamel. It promotes the electrolytic mechanism of corrosion, thereby providing surface pits for mechanical bonding as well as increased metal oxide in the enamel near the metal interface for chemical bonding in accordance with the oxide layer theory (Ref 14-16). Aqueous nickel solutions are used in proportions ranging from 0.25 to 3 wt% nickel sulfate at 70 to 80 °C (160 to 175 °F) for 5 to 15 min. The pH should be between 2.6 and 3.2. Nickel concentration, temperature, and pH must be closely controlled for consistent results. The nickel coat ranges from 0.27 to 1.6 g/m^2 (0.008 to 0.05 oz/yd^2).

Neutralization. The pH of the metal surface is increased as acid and undesirable surface chemical residues are neutralized. This also provides rust protection if the metal is held prior to enameling. Typically, alkaline soda ash or borax solutions are used with concentrations equivalent to 0.02 to 0.3 wt% Na_2O. Residues of these solutions may have a slight fluxing effect during enameling.

Ground Coats. Typical porcelain enamel ground coats for sheet steel and enameling iron are alkali-metal aluminoborosilicate glasses containing adherence-promoting oxides of cobalt, nickel, or copper. The primary purpose of the ground coat is to provide satisfactory adherence to the base metal. To this end, the ground coat is designed to promote mechanical and chemical bonding by the complex interactions discussed previously under the section "Bonding and Adherence Mechanisms." The glass composition is selected to minimize the thermal expansion mismatch. Where the

Table 1 Chemical oxide composition of ground coats for steel and cast iron, parts by weight

Component	I	II	III	IV	V
KNa_2O	17.8	18.5	21.0	18.0	19.4
B_2O_3	16.0	15.2	11.8	14.1	13.3
Al_2O_3	7.7	8.8	8.7	5.4	8.0
SiO_2	51.1	52.1	44.4	44.1	56.6
CaF_2	5.5	3.8	8.7	6.8	
BaO.........			2.0		
CoO	0.5	0.4	0.4		
Mn_2O_3	0.9	1.2	0.9		
NiO..........	0.5		0.2		
CuO			0.7		
CaO.........			1.2		
Sb_2O_3				4.8	1.4
ZrO_2				3.7	6.2

Ground coats I, II, and III are typical blue ground coats, whereas coats IV and V are white ground coats for use with white enamel cover coats.

form and function permit, a slight compressive stress in the bond interface and enamel is selected.

The ground coat often contains a small amount of clay with organic material. As the clay decomposes, small bubbles are produced in the enamel. This fine bubble structure functions as a sink for gases that evolve from the base metal during firing, thereby preventing interfacial gas flaws, surface roughness, and exposure of the base metal. The bubble structure, as well as crystalline phases, also serves to opacify the enamel, partially hiding the base metal for decorative purposes.

The ground coat also provides a smooth surface to which the cover coat can attach. In addition, it contributes to the chemical resistance, thermal resistance, and decorative functions of the enamel system. For some steels, adequate adherence may be achieved without a ground coat, but this usually involves special surface treatments.

Ground coats are usually composed of a mixture of ground-glass powders or frits, as well as other additives, such as those used to promote a bubble structure. Each component frit may contribute to the successful adherent bond with the base metal. One frit may fuse at a low temperature, sealing and protecting the surface of the metal from atmosphere during the enameling process or promoting glass-to-metal bonding at early stages. Another may provide sag resistance as the molten glass is heated. Yet another may provide color or decorative function, obscuring defects that would otherwise be apparent in the cover coat. Table 1 gives typical ground coat compositions for application to steels.

The ground coat may be applied to heated metal as a dry powder, but more often is applied to the metal at lower temperatures by dipping, spraying, or electrostatic spraying of a slurry. The slurry is dried and then fired at temperatures ranging from 780 to 870 °C (1435 to 1600 °F).

Cover coats are designed to provide protection or resistance to particular environ-

ments, often coupled with appearance requirements, such as color, opacity, and surface texture. Enamel cover coats are similar in composition to ceramic glazes and are intended to provide resistance to liquid and atmospheric corrosion, abrasion, and thermal shock. Typically, titania-base cover coats are used for acid resistance, whereas zirconia-base cover coats provide alkali resistance.

The cover coat has a softening temperature that is somewhat lower than the ground coat so that the ground coat does not flow and the bubble structure does not coalesce during cover-coat firing. The thermal expansion of the cover coat is matched to that of the ground coat and base metal. A slightly lower expansion value is usually selected so that modest compression in the final enamel protects the surface from crack opening. Excessive compressive stress can result in "shivering" or "pop-off" of segments of enamel, while too great a tensile stress can lead to crazing or, in extreme cases, to "fish-scaling."

Cobalt in the ground coat provides a dark blue color that is commonly hidden by opacifying the cover coat. The cover coat develops fine crystals with a different index of refraction from that of the surrounding glass matrix. This promotes diffuse scattering of light and opacity. Titania and zirconia which have a large difference between their refractive indexes and that of the glass matrix and also have a tendency to crystallize out of the glass, are common opacifiers. Colorants may be added, and, in the case of a white cover coat, slight bleed-through from the undercoat may contribute to a blue-white appearance.

The cover coat may range from 0.08 to 0.7 mm (3 to 28 mils) in thickness, depending on application. The frit is typically brushed or sprayed on, dried, and fired. The cover coat may be applied directly as a dry powder and fused to the hot metal and ground coat. It is also common to apply the ground and cover coats and then dry and fire both layers in a single firing step, known as the two-coat, one-fire process. Table 2 gives typical cover-coat compositions used for steels.

Ceramic-to-Metal Joining and Sealing

Ceramics and metals can be joined with fasteners or adhesives, but when a durable, hermetic join capable of extended service in a variety of environments and temperatures is required, a direct bond between metal and ceramic is desirable. Such bonds require a preparation step that provides a transition structure from ceramic covalent/ionic bonding to the metallic bond. This ensures wetting and chemical adherence between ceramic and metallic parts. Often these methods are characterized by the way

Table 2 Chemical oxide composition of cover coats used for steel and cast iron, parts by weight

Component	I	II	III	IV	V
KNa_2O	37.1	21.8	32.0	37.6	32.4
B_2O_3	11.8	13.4	9.1	11.8	14.9
PbO.........	10.4	13.4	16.0		
Al_2O_3	6.4	4.6	5.4	6.5	5.3
SiO_2.........	39.0	2.9	24.0	44.2	14.9
CaF_2	6.2	5.6	7.0	6.2	4.9
CaO.........	3.9	6.7	2.3	4.0	10.5
ZnO........		7.5	7.9	5.0	9.2
Sb_2O_3	3.7		4.2	3.7	
ZrO_2........		4.7			5.9
BaO........		12.0	8.0		
MgO					5.4

in which bonding or sealing materials are applied, that is, liquid-phase, solid-phase, or vapor-phase joining. Two broad categories are described below: ceramic metallization and joining and ceramic coatings on metals. For each category, the techniques are listed in the approximate order of use frequency.

Ceramic Metallization and Joining

Comprehensive reviews of ceramic metallization and joining can be found in Ref 3, 11, and 17 to 19. Surface preparation of the ceramic is important. Often the surface chemistry and structure of a fired ceramic are quite different from the bulk material. The machining, grit blasting, polishing, or chemical etching of the ceramic surface may substantially decrease or increase the adhesion of a metallizing layer. If the ceramic surface is processed or handled after firing, cleaning and removal of adsorbed water or hydrocarbons may be needed. Chemical cleaning, acid etching, and grit blasting may be preliminary steps. For ultimate cleanliness, the ceramic is baked briefly at elevated temperature (800 to 1500 °C, or 1470 to 2730 °F). As-fired ceramic may need no cleaning to produce a satisfactory bond.

The sintered metal powder process (SMPP) is also referred to as the moly-manganese paste process. It uses a paste composed of molybdenum and manganese metals, together with their oxides, suspended in an organic vehicle. Compositions containing tungsten, iron, tantalum, rhenium, and titanium with their oxides and in combination with molybdenum and manganese can also be used. Some moly-manganese pastes contain neither molybdenum nor manganese. A typical composition is: 60 g (2.1 oz) molybdenum powder (0.2 um, or 0.008 mil); 15 g (0.5 oz) manganese powder (0.2 um, or 0.008 mil); 0.2 g (0.007 oz) nitrocellulose; and 30 mL butyl acetate.

The paste is applied by brushing, spraying, applying with a roller, silk screening, dipping, and using transfer tape. The particle-size organic binder and vehicle are adjusted to provide a suitable rheology for the particular application method. A layer thickness of about 20 μm (0.8 mil) is typically applied to a clean, as-fired, machined or ground ceramic surface. The higher glassy phase content near the surface of many as-fired ceramics has been found to enhance bonding.

The SMPP has been used sucessfully to metallize a wide variety of oxide, carbide, and nitride ceramics and ceramic composites. Ceramics containing little or no glassy silicate phase require the incorporation of such material into the paste composition. For example, to bond to 99% alumina or sapphire, the composition above may be mixed with an equal amount of powder composed of $13Al_2O_3$-$52MnO$-$35SiO_2$ or $41Al_2O_3$-$54CaO$-$5MgO$.

After application of the paste, drying is performed in air, often using heat lamps. Firing is done in a hydrogen atmosphere (dew point of approximately 25 °C, or 77 °F) at 1200 to 1600 °C (2190 to 2910 °F) for about 0.5 h. Titanium and manganese additions or surface vapor deposited coatings may enhance bonding and lower the firing temperature a few hundred degrees.

The structure of the coating provides a transition from a mixed glassy and crystalline structure, which "migrates" from the ceramic and the paste to the ceramic interface, to a metallic structure formed by hydrogen reduction at the surface of the metallization (Ref 1, 2, 7, 20). The structure is that of a liquid-phase sintered product consisting of a glassy phase with metallic and oxide ceramic particles. Near the ceramic interface, the particles are generally more ceramic in nature, whereas at the surface, metal particles predominate. This gives a bridging structure between ceramic and metallic bonding.

It is difficult to braze this metallized surface directly because most brazes do not wet the metallization. It is more common to either electroless or electroplate nickel onto the surface to a layer thickness of at least 5 to 8 μm (0.2 to 0.3 mil). Copper or other metals can also be plated. In some cases, a metal powder is applied and allowed to diffusion bond to the surface in a reducing atmosphere. The nickel-coated surface can be brazed using conventional metal joining/brazing technologies. In electronics applications, the nickel or copper coating may be a conductor or contact or may be soldered or wire bonded.

The SMPP requires a number of steps and unit operations. The procedures are relatively easy to engineer and can yield reliable, consistent results. The bond is among the strongest achieved. It is the most common metallization and ceramic-to-metal joining method in applications other than microcircuitry. It has the advantage of allowing the final joining of a metal to the metallized ceramic by conventional brazing

methods. This makes it amenable to mass production and the joining of large or complex shapes and structures.

The metal powder-glass frit method is used for the metallization of hybrid microelectronic circuitry and, to some extent, is used for ceramic-to-metal joins. In the precious metal powder-glass frit method, a noble metal alloy based on platinum, palladium, gold, silver, and/or iridium in finely divided form is mixed in a glass frit and suspended in an organic binder and vehicle. The pattern is usually silk screen printed onto an electronic substrate material, such as alumina, or a multilayer capacitor material, such as barium titanate. One advantage of the precious metal systems is that they permit air firing.

The glass composition is selected to provide low viscosity for good flow at the desired bonding temperature. A bonding glassy phase forms at the ceramic-metal interface to provide adherence. The surface is metal rich and may be used directly as a conductor, or may be plated, soldered, or brazed, without further surface preparation. The transition from a glass-bonded ceramic structure at the ceramic interface to a metallic structure at the surface is akin to the SMPP and may also be regarded as a liquid-phase bonding technology.

The as-fired surface of the ceramic may show better adhesion than a polished or etched surface. This is either because there is more glassy phase at the ceramic surface or because of differences in the grain microstructure. Active metals may be included.

A clean, dry substrate is used. An as-fired alumina substrate may show better adhesion because of the glass content or microstructure at the surface. The incorporation of more active metals in the composition (titanium, zirconium, nickel, copper) may enhance the formation of mixed-metal oxides at the ceramic-metal interface. These, as well as glassy devitrifying phases, may improve adherence (Ref 21). The incorporation of more active metals may require firing under inert or reducing conditions. Rare earth elements may be vapor deposited on the surface first in order to enhance adhesion.

Copper and nickel can be used in much the same way at a lower material cost than the precious metals. The glass frit must be a composition that is stable at low oxygen partial pressures (Al_2O_3, B_2O_3, CaO, MgO, SiO_2, and ZnO) because the metal particles must be fired in a nitrogen atmosphere with a low oxygen partial pressure (10 ppm). Controlled oxidation of the metal is required in order to ensure wetting of the metal particles by the glass (Ref 21, 22).

The active metal process involves the use of an active metal constituent such as titanium, zirconium, or hafnium in a braze metal composition. This causes conventional brazing alloys to wet and adhere to a

ceramic. The active metal may be alloyed with copper or nickel in order to reduce its reactivity. Brazing is usually performed at carefully controlled temperatures from 800 to 1200 °C (1470 to 2190 °F) in a vacuum or inert atmosphere. The braze metal may be applied as a powder, wire, or shim directly to the ceramic. Because the metal part must be attached in the metallizing operation, joining by the active metal process has geometric and size limitations. If a hydride of the active metal is applied as a separate layer on the ceramic, a more conventional braze can be used (Ref 23).

Active metal is transported in the liquid braze to the ceramic interface and interacts, forming a transition microstructure (Ref 24). Wetting can be so enhanced by the use of active metal that great care must be taken to prevent braze metal from flowing into unwanted areas. Bonds are among the strongest observed for ceramic-to-metal joining. The active metal cannot be used at elevated temperature, because the reactions with the ceramic will continue and adherence will deteriorate. For the same reason, process control is critical to obtaining reproducible, high-quality bonds. The application of a thin layer of active metal by vapor-phase deposition techniques has been found to improve bonding. Recently, new compositions have been developed to bond nonoxide structural ceramics for elevated-temperature use (Ref 25).

The gas-metal eutectic (direct) method, also called direct bonding, is being employed increasingly. It can be used to bond oxide ceramics such as aluminum oxide, silica, beryllia, zirconia, and spinels, as well as nonoxide ceramics, such as aluminum nitride, silicon nitride, and silicon carbide, to such metals as copper, nickel, iron, molybdenum, and their alloys (Ref 26-28). The metal can quickly be bonded directly to the ceramic without intermediary material, with little or no pressure, and without melting the metal.

Several metals show a low-melting eutectic with their oxide (sometimes sulfide or silicide), which is an advantage in direct bonding. For example, at low partial pressures of oxygen, copper has a eutectic with copper oxide below the melting point of copper. Copper can be preoxidized and bonded in a vacuum or inert atmosphere or alternatively oxidized while bonding in a low partial pressure of oxygen. The metal/metal oxide melts, while the bulk metal remains solid. The process requires careful atmosphere and temperature control to avoid melting the bulk-metal shape. The eutectic melt wets and bonds to the ceramic.

It has been demonstrated that in the case of copper bonded to alumina, a reaction takes place to form a further interfacial bonding phase, copper aluminate. This is a low-melting eutectic in the tertiary system and shows that a further bonding reaction may be important (Ref 28). The minor element composition of the ceramic has also been shown to have major effects on bond strength (Ref 28).

Pressed Diffusion Joining. Joining by solid-phase diffusion and reaction may be employed to bond metals and ceramics. To obtain very high strength bonding requires careful surface preparation, elevated temperature, long times, and high pressures. Such bonds are termed pressure diffusion, ram, or crunch joins. This approach can be particularly suitable for low yield stress metals such as copper. Copper can be pressed to conform to the ceramic, or the ceramic can be forced into a cavity in the metal. This can give good contact between the metal and ceramic, with a residual stress at their interface.

The parts must be as clean as possible before contact. They are held with a high external force applied normal to the interface at as high a temperature as possible, but above half the melting temperature of the metal (Kelvin). The gas can have a profound effect on joining, in terms of both trace impurities and different levels of effectiveness depending on the inert gas chosen. A variety of metals and ceramics can be coupled by this method (Ref 29, 30).

Vapor-Phase Metallizing. A thin metallic film can be applied by various deposition methods from the vapor phase. Such methods include evaporation, sputtering, flame spraying, reactive sputtering, chemical vapor deposition (CVD), plasma spraying, and ion implantation. Frequently, an active metal, such as titanium, zirconium, and chromium, is deposited to promote bonding, followed by noble or base metals, such as gold, platinum, palladium, nickel, and copper. Final layers are usually oxidation resistant metals. Platinum shows similar thermal expansion, good ductility, and excellent adhesion to alumina ceramics and is frequently used directly.

Often, a very tenacious bond is developed with thin-film vapor-phase applications for metals that bond poorly as bulk material. Rather different bonding may occur when active atoms or ions are deposited, compared to liquid-phase or solid-phase bonding. Contact can occur atom by atom, providing superior physical bonding, even in systems that do not wet. Additionally, the mechanical behavior of thin films differs from that of bulk material, which may also account for the tenacity of many bonds.

Usually, metallizations applied by vapor-phase methods are quite thin, although layers several micrometers thick can be developed. Plasma spraying and CVD techniques are most often used to build up thicker structures. After deposition, layers can be built up by conventional solder and braze methods, or, alternatively, electroforming can be used for joining to metal parts once vapor-phase deposition has been complet-

ed. High-temperature brazing is typically done in hydrogen, under vacuum, or with a cracked ammonia atmosphere. Ceramics and glasses that are not tolerant of brazing temperatures can be soldered at lower temperatures. Strong, hermetic seals can be made on glasses, as well as on single-phase, polyphase, and single-crystal ceramics.

Vapor-Phase Ceramic Coatings. A wide variety of methods is available for applying ceramic films to metals, including thermal spraying, plasma deposition, radio frequency (RF) and reactive sputtering, CVD, and ion implantation. Such techniques have been reviewed, particularly with a view of protective, thermal, and friction coatings. These coatings can be used alone (the techniques are reviewed in the section "Ceramic Films and Coatings on Metals") or, occasionally, as the base upon which a bulk ceramic is bonded. In such applications, a strong bond is required between the base metal and the primary ceramic coating. This can best be achieved by vapor techniques having higher energy inputs, such as plasma spraying and reactive and/or RF sputtering. These techniques provide a rather intimate and potentially strong bond to the metal, which may then function as a joining phase for successive layers or as a bulk join. Thermal mismatches must be taken into account for successful application. Even when the application is made on a small area, the high-energy input can affect bond quality.

The nonmetallic fusion process is similar to the technologies of glass sealing and enameling. A glass frit is heated to its working range and is used to bond a metal and ceramic component. This liquid-phase method has technological requirements similar to those for glass-to-metal bonding. Bonding to the ceramic is usually straightforward because the glasses generally wet and bond well. Thermal expansion, residual stresses, and the formation of undesirable ceramic phases need to be considered.

The glass layer is purposely kept thin. It must suffice for bonding to the metal but should be incorporated into the ceramic structure to the greatest extent possible. A residual, thick glass layer may be a zone of structural weakness. The incorporation of the glass into the ceramic structure may also allow the joint to be used at temperatures higher than the softening or fabrication temperature of the bond.

An alternative is the use of a bonding glass that devitrifies to form a glass-ceramic. This has the advantage of greater strength, a wider range of thermal expansions for matching purposes, and better chemical durability than sealing glasses. An example is a lithium-phosphate-nucleated lithium-silicate (glass-ceramic), which matches and seals well to Inconel and Hasteloy alloys (Ref 25).

A typical application of conventional nonmetallic fusion is the use of manganese aluminosilicate glasses, which may be ap-

plied to the joint interface as a preform of suspension. The assembly is heated in inert gas at 1200 to 1500 °C (2190 to 2730 °F) for about 15 min and then cooled rapidly (100 °C/min, or 180 °F/min), so that the glass does not devitrify. A high fabrication temperature and rapid cooling can lead to flaws that result in interfacial failure. The frit composition is selected to match expansion and provide wetting and adhesion to both the metal and ceramic. The bond can be heated much higher than is possible with an SMPP bond, on the order of 888 °C (1630 °F). Nonmetallic fusion has been used successfully to bond high-purity alumina and sapphire.

Liquid-Phase Metallizing. A salt solution of a precious metal (platinum, gold, silver, palladium) is applied to the ceramic and then thermally reduced. A thin conductive coating results, which can be soldered, brazed, or electroplated directly. Reactive metals can be included in the composition in order to improve adherence. This method is particularly useful for putting electrodes on ceramics for electrical characterization and for very low volume applications.

Electroforming. After a ceramic part has been metallized by vapor-phase or liquid-phase techniques to provide a thin, conductive layer, electroforming can be used to avoid high-temperature brazing. The metallized ceramic and metal (often copper) are pressed together, and the assembly is placed in an electroplating bath for some days. Electrodeposition in the gap bonds the metal and metallized ceramic. Large, complex parts can be joined this way, avoiding thermal expansion mismatch stresses. The method can be coupled with pressure diffusion joining techniques, but this no longer avoids elevated temperature or thermally induced stresses upon cooling.

The graded-powder process is frequently closely related to either the metal powder-glass frit process or the SMPP. A powder compact is made in which the composition varies in layers from ceramic-rich to metal-rich. This mechanically produced variation in the preform permits a more ceramiclike bond material at the ceramic interface and greater metal content for joining to the metal. The composition, properties, and structure are graded between the ceramic and metal. The powder is fired at a temperature similar to those used in the SMPP or metal powder-glass frit process upon which it is based. Because this method results in bonds only slightly stronger than those of other methods and because the preform increases costs, it has seen limited use.

Ceramic Films and Coatings on Metals

Ceramic films and coatings on metals can perform a variety of functions, as listed in Table 3. Tribological performance is a com-

Table 3 Uses of typical ceramic films and coatings

Material	Use
Al_2O_3, B_4C, Cr_3C_2, CrB_2, $CrSi_2$, Cr_3Si_2, DLC(a), Mo_2C, $MoSi_2$, SiC, TiB_2, TiC, TiN, WC	Wear reduction
MoS_2, BN, BaF_2–CaF_2	Friction reduction
Cr_2O_3, Al_2O_3, Si_3N_4, SiO_2	Corrosion reduction
Ca_2Si_4, $MgAl_2O_4$, MgO, ZrO_2 (stabilized with Mg or Ca)	Thermal protection
In_2O_3–SnO_2	Electrical conductivity
GaAs, Si	Semiconductors
SiO_2	Electrical insulation
$Bi_4Ti_3O_{12}$	Ferroelectricity
AlN	Electromechanical use
BaF_2–ZnS, CeO_2, CdS, CuO–Cu_2O, Ge–ZnS, SnO_2	Selective optical transmission and reflectivity
SiO_2	Optical wave guides
GaAs, InSb	Optical processing (electrooptic, and so on)
SiO_2, SnO_2, ZrO_2	Sensors

(a) DLC, diamondlike carbon

Table 4 Categorization of ceramic coating techniques

Atomic deposition	Particulate deposition	Bulk coatings	Surface modification
Chemical vapor environment	Fusion coatings	Diffusion	Chemical (liquid) oxidation
Chemical vapor deposition	Sol-gel	Diffusion bonding	
Reduction	Thermal spraying	Hot isostatic pressing	Chemical (vapor)
Decomposition	Plasma spraying	Wetting processes	Thermal
Plasma enhanced	Low-pressure plasma spraying	Dipping	Plasma
Plasma environment	Laser-assisted plasma spraying	Enameling	Ion implantation
Sputter deposition	Electric arc spraying	Spraying	Sputtering
Diode			
Triode			
Reactive			
Evaporation			
Direct			
Activated reactive			
Vacuum environment			
Vacuum evaporation			
Ion implantation			

mon application area. Low friction and wear, particularly for elevated-temperature applications, favor the use of ceramic coatings (Ref 31-33). Ceramics can provide corrosion resistance to strong chemicals, even at elevated temperature, as well as thermal protection because of their oxidation resistance. Nonoxide ceramics are useful in protective applications for which water may cause slow crack growth (stress-corrosion cracking) in other ceramic coatings (Ref 34). The minimization of stresses in the ceramic-to-metal bond are particularly important, especially when high-temperature thermal cycling occurs (Ref 35). Thin films of ceramic are vital to such modern electronics applications as semiconductors, thin-film resistors, photothermal and photoelectric converters, insulator layers, and high-temperature superconductors (Ref 36).

Ceramic coatings can be applied by the four general categories of methods given in Table 4. These categories are sometimes further subdivided into those that directly deposit a ceramic compound, that is, physical deposition; those that chemically react precursors, that is, chemical deposition; or those that react metallic species with gases, that is, reactive deposition (Ref 37, 38).

The chemical vapor deposition of ceramics most often involves the thermal decomposition or chemical reduction of organometallic compounds. In some cases, halides or simpler organic compounds are used. A variety of ceramic coatings, including oxides, carbides, and nitrides, can be applied. Table 5 gives a partial list of ceramic coatings and their gas reactions for CVD. Such coatings can be applied to bulky, complex shapes at high deposition rates and are amenable to high-volume production.

Coatings tend to have elongated grains normal to the deposition direction, but the grain size and composition can be controlled by the reactive gas concentrations. A gradual change in composition and structure can be accomplished by gradually altering the chemistry of the gas flows.

Multiple CVD coatings in various combinations are deposited on cutting tools to improve performance. Alumina and titanium carbide can provide abrasion resistance, whereas titanium nitride yields low cutting friction. All three provide a thermal or diffusion barrier, while alumina provides oxidation resistance. Combinations of various layers of these and other ceramics can increase the cutting speeds of metal matrix cemented carbides by three times and their lifetime by two to ten times (Ref 39).

Proper coating by chemical vapor deposition requires critical design of the reactor vessel and careful attention to gas flows within the reactor and around individual parts. The parts must be heated uniformly to rather high temperatures. This may limit applicability to particular metals. The part must be tolerant of the gases and their reaction products and of a low-pressure environment (Ref 40).

Evaporation. Ceramics can be evaporated in a high vacuum and deposited on a metal substrate. The ceramic or ceramic precursor is heated either directly or with an electron beam to a temperature that exceeds or is close to its melting point. The substrate can be cooled, but is heated by the deposition process.

In direct evaporation, the ceramic itself is evaporated with or without dissociation of the compound. Some ceramic coatings that can be directly evaporated are: MgF_2, B_2O_3, CaF_2, SiO, GeO, and SnO. Most other ceramics dissociate and can be deposited with cation- or anion-deficient structures or chemistries different from the starting material.

Reactive evaporation involves the interaction of a reactive gas with an evaporated metal or low-valence compound (suboxide) to deposit a ceramic. Examples of compounds deposited by reactive evaporation are Al_2O_3, Y_2O_3, TiO_2, In_2O_3, SnO_2, BeO, SiO_2, TiC, ZrC, NbC, Ta_2C, VC, W_2C, HfC, TiN, Ti_2N, MoN, Cu_xS, and $Cu_xMo_6S_8$ (Ref 41).

Sputtering transfers a high kinetic energy ion to the metal surface, which has been produced by ion bombardment of the ceramic or ceramic precursor. The metal surface can first receive ultimate cleaning in the sputtering system by ion bombardment of the substrate itself. The ability to clean and the high energy of the impinging material usually give a more adherent bond than the preceding methods. For ceramics, it is more usual to use RF rather than dc sputtering, and reactive techniques are frequently required to produce the ceramic coating. RF techniques permit direct sputtering by eliminating charge buildup. They also allow a lower pressure and therefore purer films. Diode, triode, and magnetron sputtering techniques can be employed.

Plasma spraying is well suited to the application of high-melting-point materials such as ceramics. The inert nature of the heating source permits the deposition of oxide and nonoxide ceramics. Because the heat source is quite efficient and imparts high energy to the coating material, a tenacious bond can result. The coated material can be kept either relatively cool or heated so that thermally induced interfacial stresses can be controlled. Low-pressure plasma spraying permits the spraying of reactive materials and of materials such as carbides, which react in air. High coating rates and the capability to build up relatively thick layers are advantages of this process.

The deposited ceramic coating can be quite porous, and this may limit application where environmental protection is required. Sintering the material, vitrifying, or impregnating with a molten glass can be used to seal the porosity. Porosity can be reduced by lowering the velocity and hence the

Table 5 Chemical vapor deposition of ceramics

Materials	Reactive gas	Temperature °C	Temperature °F
Nitrides			
BN	$BCl_3 + NH_3$	1000–2000	1830–3630
	Thermal decomposition $B_3N_3H_3Cl_3$	1000–2000	1830–3630
HfN	$HfCl_x + N_2 + H_2$	950–1300	1740–2370
Si_3N_4	$SiH_4 + NH_3$	950–1050	1740–1920
	$SiCl_4 + NH_3$	1000–1500	1830–2730
TaN	$TaCl_5 + N_2 + H_2$	2100–2300	3810–4170
TiN	$TiCl_4 + N_2 + H_2$	650–1700	1200–3090
VN	$VCl_2 + N_2 + H_2$	1100–1300	2010–2370
ZrN	$ZrCl_4 + N_2 + H_2$	2000–2500	3630–4530
Oxides			
Al_2O_3	$AlCl_3 + CO_2 + H_2$	800–1300	1470–2370
SiO_2	$SiH_4 + O_2$	300–450	570–840
	Thermal decomposition $Si(OC_2H_5)_4$	800–1000	1470–1830
Silicon oxide nitride	$SiH_4 + H_2 + CO_2 + NH_3$	900–1000	1650–1830
SnO_2	$SnCl_4 + H_2O$	. . .	. . .
TiO_2	$TiCl_4 + O_2$ + hydrocarbon (flame)	. . .	. . .
Silicides			
V_3Si	$SiCl_4 + VCl_4 + H_2$	. . .	. . .
MoSi	$SiCl_2$ + Mo (substrate)	800–1100	1470–2010
Borides			
AlB_2	$AlCl_3 + BCl_3$	≈1000	≈1830
HfB_x	$HfCl_4 + BX_3$ (X = Br, Cl)	1900–2700	3450–4890
SiB_x	$SiCl_4 + BCl_3$	1000–1300	1830–2370
TiB_2	$TiCl_4 + BX_3$ (X = Br, Cl)	1000–1300	1830–2370
VB_2	$VCl_4 + BX_3$ (X = Br, Cl)	1900–2300	3450–4170
ZrB_2	$ZrCl_4 + BBr_3$	1700–2500	3090–4530
Carbides			
B_4C	$BCl_3 + CO + H_2$	1200–1800	2190–3270
	$B_2H_6 + CH_4$	. . .	. . .
	Thermal decomposition $(CH_3)_3B$	≈550	≈1020
Cr_7C_3	$CrCl_2 + H_2$	≈1000	≈1830
Cr_3C_2	$Cr(CO)_5 + H_2$	300–650	570–1200
HfC	$HfCl_4 + H_2 + C_6H_5CH_3$	2100–2500	3810–4530
	$HfCl_4 + H_2 + CH_4$	1000–1300	1830–2370
Mo_2C	$Mo(CO)_6$	350–475	660–890
	Mo + hydrocarbon	1200–1800	2190–3270
SiC	$SiCl_4 + C_6H_5CH_3$	1500–1800	2730–3270
	$MeSiCl_3 + H_2$	≈1000	≈1830
TiC	$TiC_4 + H_2 + CH_4$	980–1400	1800–2550
W_2C	Thermal decomposition $W(CO)_6$	300–500	570–930
	$WF_6 + C_6H_6 + H_2$	400–900	750–1650
VC	$VCl_2 + H_2$	≈1000	≈1830

deposition rate. Care must be taken to produce a uniform coating with compositional consistency.

Sol-Gel Coatings. In sol-gel processes, a sol, which is a stable dispersion of hydroxides or hydrous oxides, is applied to a substrate metal. The sol may be applied by dipping, pouring, screening, or spinning. After application, it is catalyzed to form a gel. The coating is then dried. The coating has a microporous structure that may be used to advantage. Alternatively, the applied coating is fired or infiltrated to achieve an impermeable structure. Pore size and distribution can be controlled by the initial deposition process, as well as by drying and densification steps. Bonds to metals that may be difficult to accomplish can be made using other techniques (Ref 42).

Sol-gel techniques allow a high degree of chemical homogeneity and purity. Densification can be done at low temperatures, thereby minimizing thermal stresses. Multi-component systems are possible with mono-scale microstructures. The coating may show high shrinkage during drying and firing due to the presence of considerable solvent and porosity. This often limits the thickness of coating that can be accomplished.

Ion Implantation. An ionized, gaseous species such as oxygen, nitrogen, carbon dioxide, or argon bombards the metal with high energy. The species penetrates and interacts with the metal. The depth of penetration and the interactions depend on the type and energy of the ion and of the base material. The interaction of the bombarding ions with the target material results in oxide, carbide, and nitride coatings. The energetic nature of the process ensures a strongly bonded coating. The interaction of the bombarding species within surface layers of the base material can yield properties quite different from those of other coating techniques (Ref 43).

Nitrogen ion implantation of a nitride on titanium-aluminum-vanadium alloys can improve the wear characteristics by a factor of ten. A thin film of ceramic nitride, carbide, or boride formed by sputtering can be improved in adhesion and hardness by subsequent ion implantation with inert gas ions (Ref 44).

Ceramic and Glass Bonding

Glazes are glass coatings applied to ceramic bodies. Common glazes are suspensions of clay and glass frit in water. Additives may be included to disperse the slip properly and obtain appropriate rheology. The glaze is applied to a green, unfired ceramic or to a bisque, prefired ceramic by brushing, spraying, or dipping. Spraying is favored for continuous coating. Dipping or painting may be required to cover internal surfaces adequately. Generally, no special surface preparation is required if the ceramic surface is clean.

Oxide ceramics are usually wet by oxide glasses. Special consideration may be necessary to coat nonoxide ceramics. Binders added to the ceramic or glaze composition may aid in the wetting process. In some cases they function irreversibly, with a larger contact angle for the advancing rather than the receding liquid. In this case, once a surface is wetted, it remains wetted. This is important because cracks in the unfired glaze may not completely rewet the surface upon fusion. In many cases, some of the base ceramic dissolves into the glaze composition, producing an interlocking structure.

Typically, the thermal expansion difference between glaze and enamel is chosen to put the glaze into compression. A typical value for a ceramic ware at room temperature is about 70 MPa (10 ksi). To accomplish this, the expansion of the glaze must be lower than that of the ceramic body. A typical expansion behavior and the stresses that develop are shown in Fig. 6. Increasing the firing temperature or time, increasing the amount of alkaline earth relative to alkaline content in the composition, and increasing low-expansion glass formers can reduce glaze expansion. The stresses that develop in the glaze also depend on the relative thicknesses of body and glaze, as well as the overall design. An excessive tensile stress, either upon cooling or reheating during use, will cause the enamel to craze. Excessive compression can cause portions of glaze to shiver off. Table 6 gives typical glaze compositions.

Glazes provide opacity, texture, and color. Opacity is provided by developing crystals (during firing) that have an index of refraction different from that of the glass matrix. Diffuse scattering results in opacity. Some common opacifiers are given in Table 7. Most glazes are vitreous in character

Table 6 Typical basic glaze compositions

	Firing temperature		$\begin{bmatrix} Na \\ K \end{bmatrix}_2 O$	$\begin{bmatrix} Mg \\ Ca \end{bmatrix} O$	Composition, %(a)				
Description	°C	°F			PbO	Al$_2$O$_3$	B$_2$O$_3$	SiO$_2$	Other
Leadless raw porcelain glaze									
Glossy	1250	2280	0.3	0.7	⋯	0.4	⋯	4.0	⋯
Matte	1250	2280	0.3	0.7	⋯	0.6	⋯	3.0	⋯
High-temperature glaze, glossy	1465	2670	0.3	0.7	⋯	1.1	⋯	14.7	⋯
Bristol glaze, glossy	1200	2190	0.35	0.35	⋯	0.55	⋯	3.30	0.3 ZnO
Aventurine glaze (crystals precipitate)	1125	2060	1.0	⋯	⋯	0.15	1.25	7.0	0.75 Fe$_2$O$_3$
Lead containing fritted glaze, glossy	1080	1975	0.33	0.33	0.33	0.13	0.53	1.73	⋯
	930	1705	0.17	0.22	0.65	0.12	0.13	1.84	⋯
	1210	2210	0.05	0.50	0.45	0.27	0.32	2.70	⋯
Raw lead glaze									
Glossy	1100	2010	0.1	0.3	0.6	0.2	⋯	1.6	⋯
Matte	1100	2010	0.1	0.35	0.55	0.35	⋯	1.5	⋯
Opaque	1100	2010	0.1	0.2	0.7	0.2	⋯	2.0	0.33 SnO$_2$

(a) Mole fraction, with RO + R$_2$O = 1.00

because of the glassy nature of the glaze. Crystals can be allowed to develop to provide a matte finish or, with the correct submicron size, an opalescent luster. Color is provided by the presence of either transition metal oxides or colloidal colorants.

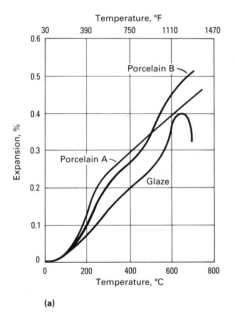

(a)

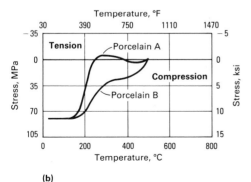

(b)

Fig. 6 (a) Thermal expansion and (b) stress development of a glaze for two clay ceramic (porcelain) bodies

Glass-Glass and Ceramic-Ceramic Joining. Oxide glasses are relatively easily joined as long as surfaces are clean or pristine. The good mutual wetting response and solubility of oxide glasses are responsible. Adsorbed water or organics must be baked out before the glass is allowed to fuse. Attention must be paid to relative expansion so that excessive stresses are not developed. This possibility can be alleviated either by compositional control or by grading the glass composition and hence expansion across a join. A glass frit can be applied to the interface, which, upon heating, will fuse glass pieces. The faceplate and funnel of a television picture tube are joined in this way. A high-lead, low fusion temperature solder glass is applied to the interface, which, upon heating, fuses and joins the glasses without softening the glass components. In all processes, care must be exercised to avoid weakening flaws.

The joining of ceramics can be accomplished in a similar fashion. A glass-base frit or slurry can be applied to a clean interface, and, upon softening, it will wet and join the ceramic pieces. Glass will penetrate the pores of a ceramic and may interact chemically, providing mechanical interlocking, and, in many cases, making the glass frit "disappear." Often the composition of the glass bonding layer is that of a glaze composition. This approach can be used for both fired and green ceramics.

Green ceramics can be joined directly. Moist pieces can be pushed together. In some cases, the surfaces are wetted, in others, a small amount of the ceramic slip is applied to the surface. Upon firing, either the ceramic parts will sinter together or a glassy phase will form continuously across the bond.

Organic Bonding

This section provides a brief discussion of the use of adhesives for joining ceramics and glasses with metals and with each other. When adhesives are used to join ceram-

ics with other ceramics or with metals, most of the issues already discussed such as surface roughness, wetting, modulus mismatch, and so on, hold true. Certainly, the use of coupling agents can substantially improve the quality of bond to a ceramic. However, the appropriateness of adhesive bonds between these systems is limited. Usually, the ceramic is involved in the design to provide environmental resistance, thermal protection, hermeticity, wear resistance, and so forth. In such circumstances, the adhesives frequently will not be adequate for the intended end-use conditions, although there are specific numerous applications for which an adhesive bond will prove to be adequate.

Surface Preparation of Ceramics and Glasses. In order to prepare a ceramic or glass surface properly for joining, the bonding material and ceramic must be selected with care. The mode of loading and potential bond failure must also be considered. Generally, a roughened ceramic will provide a superior bond in shear, and an interlocking keyhole structure will provide a strong bond in tension or shear. The roughness can be provided by the porous structure of the ceramic itself, by diamond grinding the surface, or by grit blasting at a relatively high energy input. Chemical roughening is possible, but will often involve the use of strong acids or bases unless a relatively soluble grain-boundary phase exists in the ceramic. Removal of the solution used for chemical attack may be difficult. This surface must be wet by the adhesive in order to penetrate the roughness or surface pores of the ceramic.

In some applications the surface may remain smooth and a satisfactory bond may be achieved, nonetheless. This is particularly true of ornamental attachments to glass, for which the major forces are applied in a tearing mode. In this case little advantage may be provided by a rough surface. Contact adhesives also benefit from a smooth surface, which can activate the bonding reaction. Contact adhesives may prove to

Table 7 Typical opacifying glazes

Characteristic	Fluoride opal glass	Zircon glaze, cone 11	Tin oxide glaze, cone 11	Zircon glaze, cone 06
Glass composition, %				
SiO_2	71.7	73.1	61.1	43.0
Al_2O_3	5.3	14.6	12.1	3.1
Na_2O	4.7	0.8	1.0	4.2
K_2O	3.5	3.2	5.2	1.6
B_2O_3	6.4			8.5
CaO		2.0	12.1	2.5
ZnO	6.4		8.4	
PbO	2.0			37.2
MgO		6.3		
TiO_2				
Opacifier composition, wt%	7.7 NaF 8.7 CaF_2	13.6 $ZrSiO_4$	4.1 SnO_2	14.0 $ZrSiO_4$
Glaze composition, vol%	88.2 glass 5.9 NaF 5.9 CaF_2	92.8 glass 7.2 $ZrSiO_4$	98.4 glass 1.6 SnO_2	90.8 glass 9.2 $ZrSiO_4$
Refractive index				
Glass	1.5	1.5	1.5	1.6
Opacifiers	NaF = 1.3 CaF_2 = 1.4	$ZrSiO_4$ = 2.0	SnO_2 = 2.0	$ZrSiO_4$ = 2.0

be difficult to bond to a porous ceramic because inadequate contact pressure is developed and adhesive is forced into the pores. Filling the surface pores before bonding with a contact adhesive may be desirable. One technique that can be used when appropriate is to wet the porous ceramic surface lightly with water prior to the application of the adhesive. The water fills the pores and prevents adhesive from penetrating. If the captured water can escape through open porosity, or if it will not affect performance, this simple expedient can be used. Other penetrating liquids could be used in a similar manner.

The state of a ceramic or glass surface can also have a profound influence on bond quality. Naturally, the surface must be free of dirt, contamination, and loosely adhering particles. Ceramics and glasses may appear nominally clean, yet be difficult to bond to because of their surface condition. Most ceramics and glasses bond readily with water and many organic solvents. They may appear clean, yet possess a surface film that interferes with adhesive bonding. Atmospheric humidity or washing may be the source of water, while the cleaning of machining oils may introduce solvents.

One approach for water removal is to use organic solvents such as alcohols and acetone. The organic residue may not interfere with the bonding adhesive. Alternatively, water may "clean" the surface of a ceramic that has organic on it. When water is not adequate, a hydrogen peroxide treatment may aid in final organic removal.

When its use is possible, the as-fired ceramic or glass surface may provide the most successful bonding. After firing the ceramic, bonding should be done immediately. A pristine surface can also be achieved by reheating the ceramic to high temperature in a vacuum or controlled gas environment. Bonds may also be enhanced by the treatment of the ceramic surface with a material that enhances the bonding or coupling of the ceramic or glass and the adhesive bonding material. A number of proprietary treatments are available. Such materials are often used to develop a reasonable bond strength between ceramic-glass reinforcements and polymer matrices in polymer matrix composites.

REFERENCES

1. J.T. Klomp, Interfacial Reactions Between Metals and Oxides During Sealing, *Bull. Am. Ceram. Soc.*, Vol 59, 1980, p 794-800
2. M.E. Twentyman, High Temperature Metallizing, *J. Mater. Sci.*, Vol 10, 1975, p 765-766
3. W.H. Kohl, *Handbook of Materials and Techniques for Vacuum Devices*, Reinhold, 1967
4. J.R. Tinklepaugh and W.B. Crandall, Ed., *Cermets*, Reinhold, 1960
5. J.A. Pask and R.M. Fulrath, Fundamentals of Glass-to-Metal Bonding: VIII. Nature of Wetting and Adherence, *J. Am. Ceram. Soc.*, Vol 45, 1962, p 592-596
6. F. Bashforth and S.C. Adams, *An Attempt to Test Theories of Capillarity*, Cambridge University Press, 1883
7. A.G. Pincus, Mechanism of Ceramic-to-Metal Adherence, Adherence of Molybdenum to Alumina Ceramics, *Ceram. Age*, Vol 63, 1954, p 16-20, 30-32
8. J.D. Partridge, *Glass-to-Metal Seals*, Society of Glass Technology, 1949
9. D. Tabor, Interaction Between Surfaces: Adhesion and Friction, in *Surface Physics of Materials*, Vol 11, Academic Press, 1975, p 475-529
10. S.S. Rekhson, Annealing of Glass-to-Metal and Glass-to-Ceramic Seals, *Glass Technol.*, Vol 20, 1979, p 27-35
11. H.E. Pattee, R.E. Evans, and R.E. Monroe, Joining Ceramics and Graphite to Other Materials, WASH SP-5052, National Aeronautics and Space Administration, 1968
12. J.J. Schmidt and J.L. Carter, Using Nitrogen Based Atmosphere for Glass-to-Metal Sealing, *Met. Prog.*, Vol 128, 1985, p 29-34
13. A.I. Andrews, *Enamels: The Preparation, Application and Properties of Vitreous Enamels*, Twin City Printing, 1945
14. M. Borom and J. Pask, Role of Adherence Oxides in the Development of Chemical Bonding at Glass-Metal Interfaces, *J. Am. Ceram. Soc.*, Vol 49, 1966, p 1-6
15. B.W. King and W.H. Duckworth, Nature of Adherence of Porcelain Enamels to Metals, *J. Am. Ceram. Soc.*, Vol 42, 1959, p 504-525
16. L.S. O'Bannon, "The Adherence of Enamels, Glass, and Ceramic Coatings to Metals—A Review," Battelle Memorial Institute, 1959
17. G.R. Van Houten, A Survey of Ceramic-to-Metal Bonding, *Ceram. Bull.*, Vol 38, 1959, p 301-307
18. G.P. Chu, Some Aspects on Glass-Ceramics to Metal Sealing, in *Proceedings of the Tenth International Congress on Glass*, Ceramic Society of Japan, 1967, p 94-102
19. C.I. Helgesson, *Ceramic-to-Metal Bonding*, Boston Technical Publishers, 1968
20. S.S. Cole and G. Sommer, Glass Migration Mechanism of Ceramic to Metal Seal Adherence, *J. Am. Ceram. Soc.*, Vol 44, 1961, p 265-271
21. R.G. Loasby, N. Davey, and H. Barlow, Enhanced Property Thick Film Pastes, *Solid State Technol.*, Vol 15 (No. 5), 1972, p 46-72
22. M. Monneraye, Les Encres Sengraphiables en Microélectronique Hybride—les Matériaux et Leur Comportement, *Acta Electron.*, Vol 21, 1978, p 263-281
23. C.S. Pearsall, New Brazing Methods for Joining Non-Metallic Material to Metals, *Mater. Methods*, Vol 30, 1949, p 61-62
24. J.E. McDonald and A. Eberhart, Adhesion in Metal Oxide Systems, *Trans. Metall. Soc. AIME*, Vol 233, 1964, p 512-517
25. R.E. Loehman, Interfacial Reactions in Ceramic-Metal Systems, *Bull. Am. Ceram. Soc.*, Vol 68, 1989, p 891-896
26. J.F. Burgess, C.A. Neugebauer, and G. Flanagan, The Direct Bonding of Metals to Ceramics by the Gas-Metal Eutectic Method, *J. Electrochem. Soc. Solid-State Sci. Technol.*, May 1975
27. J.T. Klomp and P.J. Vrugt, Interfaces Between Metals and Ceramics in Surfaces and Interfaces, *Mater. Sci. Res.*,

Vol 14, 1980, p 97-105
28. J.E. Holowczak, V.A. Greenhut, and D.J. Shanefield, Effect of Alumina Composition on Interfacial Chemistry and Strength of Direct Bonded Copper-Alumina, *Cer. Eng. Sci. Proc.*, Vol 10, 1989, p 1283-1294
29. W. Dawihl and E. Klinger, "Mechanical and Thermal Properties of Welded Joints Between Alumina and Metals, *Ber. Dtsch. Keram. Ges.*, Vol 46, 1969, p 12-18
30. H.J. de Bruin, A.F. Moodie, and C.E. Warble, Reaction Welding of Ceramics Using Transition Metal Intermediates, *J. Aust. Ceram. Soc.*, Vol 7 (No. 2), 1971, p 57-58
31. J.B. Wachtmann and W.H. Rhodes, Ceramics—Advanced Forms Find Essential Uses as an Enabling Technology, *Science*, Vol 6, May 1985, p 6-18
32. J.B. Wachtmann and R.A. Haber, Ceramic Films and Coatings, *Proc. Chem. Eng. Prog.*, Vol 82 (No. 1), 1986, p 39-46

33. W.A. Brainard, "The Friction and Wear Properties of Sputtered Hard Refractory Compounds," NASA Technical Memorandum TM-78 895, National Aeronautics and Space Administration, 1978
34. E. Lang, Ed., *Coatings for High Temperature Applications*, Applied Science Publishers, 1983
35. I. Kvernes and P. Fartum, Use of Corrosion Resistant Plasma Sprayed Coatings in Diesel Engines, *Thin Solid Films*, Vol 2, 1978, p 119-128
36. W.A. Pliskin, Comparison of Properties of Dielectric Films Deposited by Various Methods, *J. Vac. Sci. Technol.*, Vol 14, 1977, p 1064-1081
37. R.F. Bunshah, *Deposition Technologies of Films and Coatings*, Noyes Press, 1982
38. D.M. Mattox, Commercial Applications of Overlay Coating Techniques, *Thin Solid Films*, Vol 84, 1981, p 361-365
39. B.M. Kramer, Requirements for Wear-

Resistant Coatings, *Thin Solid Films*, Vol 2, 1980, p 117-128
40. W.A. Bryant, Review: The Fundamentals of Chemical Vapor Deposition, *J. Mater. Sci.*, Vol 12, 1977, p 1285-1306
41. R.F. Bunshah and A.C. Raghuram, Activated Reactive Evaporation Process for High Rate Deposition of Compounds, *J. Vac. Sci. Technol.*, Vol 9, 1972, p 1385-1389
42. C.J. Brinker and S.P. Mukherjee, Comparison of Sol-Gel Derived Thin Films With Monoliths in a Multicomponent Silicate Glass System, *Thin Solid Films*, Vol 77, 1981, p 141-148
43. T.S. Picraux and P.S. Peercey, Ion Implanation of Surfaces, *Sci. Am.*, Vol 252 (No. 3), 1985, p 102-110
44. B.H. Kear, J.W. Mayer, J.M. Poate, and P.R. Strutt, Surface Treatments Using Laser, Electron and Ion Beam Processing Methods, *Metall. Treatises*, 1981, p 321-342
45. W.W. Coffeen, *Ceram. Ind.*, Vol 70, May 1958, p 77

Testing and Analysis

Co-Chairmen: David A. Dillard, Virginia Polytechnic Institute & State University
Didier Lefebvre, AT&T Bell Laboratories

Introduction

TESTS ON MONOLITHIC MATERIALS have been conducted for decades and have been developed into well-established techniques for measuring constitutive and fracture and strength properties. Although many of these test techniques have been successfully applied to measure bulk properties on neat adhesive coupons, there are questions concerning how these bulk properties correspond with the properties of thin adhesive layers within bonded joints. Some of the questions arise because any test on a bonded joint actually represents a test on a *system* rather than on a (single) *material*. The systems may consist of the adherends, the adhesive, perhaps primers and oxide layers, and finally the interphase regions that encompass the comingled zones between the distinct phases. Even if the perfect test geometry were available in which all stresses were absolutely uniform and totally identified, the test results obtained would be difficult to interpret because they would represent system properties rather than material properties. Small changes to any component that makes up the system could significantly alter the adhesive performance. These ideas suggest great care be used in producing and reporting the details associated with any given test specimen. Results should always be reported as, and considered to be, system properties rather than material properties.

Further complications also arise with the nonideal test geometries used in practice. Careful analysis of even the best test specimens reveals that there are complicated, nonuniform, three-dimensional stress distributions. Most adhesive joints also exhibit singular stress fields at certain locations within the bond. Rather than the simple geometries and stress distributions found in monolithic coupons, every bonded joint is a *structure*. The results obtained will represent the adhesive system performance when used in the specific structure of the given test geometry. This evokes statements such as "Single-lap joint strengths can only be used to predict strengths of other identical lap joints." Although this statement is not completely true, it does underscore the difficulty in attempting to use the structural performance of a test specimen to predict the behavior of a bonded joint for a practical structure. Concepts such as average shear strength are highly dependent on the (specimen) structure, and only through very careful analysis can these breaking strengths be even rudimentarily employed as design allowables. Fracture tests are able to provide a more meaningful approach to design because they introduce consistent singularities into the test geometries. Three-dimensional effects, mode mix ratios, and debond locus again can introduce complications into the test results and may necessitate careful analysis to establish design allowables and methodologies.

As one selects test techniques to meet specific needs, it must be understood that bonded joint test results represent the performance of a material system in a test geometry structure. Failure to recognize this fact can lead to serious errors in applying laboratory data to engineering designs. Despite these difficulties, accurate designs are possible through careful analysis of test geometry results. The title of this Section serves to suggest that testing and analysis of adhesive joint structures must always go hand-in-hand.

In the article "Measuring Constitutive Properties," tests for determining bulk moduli for static and time-dependent behavior are discussed. Relatively simple tests on neat coupons avoid the system and structural problems but should be used with care, since they may give properties that are different than those developed *in situ*. A brief discussion of correlation between bulk and *in situ* properties is given. This chapter also discusses several *in situ* techniques for determining constitutive behavior at the macro and micro levels. Rheological properties during wetting and cure can play a significant role in the quality of bond produced. Techniques for determining this behavior and applying it to manufacturing process optimization are mentioned.

The article "Failure Strength Tests and Their Limitations" presents an overview of several commonly used tests to measure a breaking strength of *in situ* adhesive systems. These approaches are fraught with difficulties in interpretation because each technique represents a bonded structure with a complex state of stress, including singularities at the bond termini. To illustrate this problem, a discussion of the general stress distributions within the ubiquitous single-lap joint is given. Further complications arising from spew and singular stress fields are also presented. Although sometimes considered as fracture tests, peel tests strengths are often complicated by large deformations and plastic yielding. Because of these difficulties, several peel tests have also been included in this Section. The Section summarizes the difficulties associated with obtaining meaningful design allowables, and recommends that these tests be used primarily as quality control tests rather than tests for engineering properties.

In the article "Fracture Testing," the advantages of fracture tests over failure strength tests are detailed. The basic philosophy for a fracture-mechanics-based approach to testing and design is presented, along with (nondestructive) techniques for measuring debond growth in test geometries. Descriptions and typical results for several commonly used test geometries are presented. Techniques needed for data interpretation and analysis are presented to assist with implementation. Complications arising from mode mix, debond locus, and three-dimensional effects are also briefly discussed.

In many applications, loads are imposed over long periods of time and may be cycled due to hygrothermal or mechanical input. The article "Static and Dynamic Fatigue Testing" describes some of the debond mechanisms that occur under these loading conditions. Test techniques for performing debond rate tests for these two loading modes are discussed. Typical results are given, and special techniques for monitoring debond growth are discussed. Methods used to interpret these data and establish design criteria are also presented.

The article "Special Tests for Membranes and Miniature Components" addresses the increasingly important adhesion issues for membranes, thin coatings, and miniature devices, including those from the microelectronics industry. The test techniques presented are not as well developed as some of the more popular geometries discussed in the article "Fracture Testing," but they are especially well suited to these applications. Several membrane techniques are presented and assessed. Techniques for microfabricating test specimens for the electronics industry are presented, along with test methods and typical results. Although developed for the microelectronics industry, these geometries are appropriate for other small-scale devices.

Sealants and elastomeric adhesives require specialized techniques for testing and analysis because of the large deformations and nearly incompressible behavior. These complications have limited the analysis of relevant test geometries and have resulted in some rather crude tests. Many applications for these materials tend to be displacement controlled rather than load controlled. An approach for measuring fracture properties based on the butt specimen, and several recent fracture approaches are discussed. Concepts of failure by cavitation, especially relevant to elastomeric materials, are discussed. Tests for measuring bond strength for tire cord are discussed, and quality control test techniques are also given.

Fiber- and particle-reinforced materials are rapidly replacing more conventional materials for many applications. These unique material systems represent very important adhesive problems where the adhesive (matrix) typically makes up more than 40% of the material. The article "Composite Fiber/Matrix Bond Tests" discusses some of the difficulty in determining bond strengths between reinforcement and matrix. A variety of techniques used to estimate the quality of the interface are discussed, along with the relative advantages and limitations of each.

In "Microstructural Analysis," typical techniques used for microstructural analysis of polymers and surfaces are briefly described. Examples of microstructure in semicrystalline polymers and cross-linked polymers are given, along with an explanation of their relationship with processing variables and macroscopic properties. Particular emphasis is given to structures specific to boundary regions. Defects and heterogeneities commonly encountered in adhesive bonds are shown, and a description of their role in the failure process is given.

In the article "Thermal Properties and Temperature Effects," two major themes are discussed: adhesive durability and adhesive processing. The topic of thermal effects and durability includes: the role of thermal stresses and characterization techniques used to determine the coefficient of thermal expansion, and the change in temperature from the anchoring temperature, the techniques used to characterize thermally induced degradation (TGA, thermal fatigue, thermal spiking, and so on), and the use of temperature in accelerated life prediction. The topic of processing covers the techniques customarily used to measure properties needed to optimize the processing and final properties of thermosets and thermoplastic adhesives.

The article "Electrical Properties" reviews the electrical properties and relevant tests that are critical in cable insulation, device encapsulation, and IC protection where both adhesive performance and electrical performance are of concern. As with mechanical properties, it is emphasized that bulk polymer properties may not be directly usable to predict the performance of the bond (system) or device (structure). In particular, conduction mechanisms or electrochemical processes specific to the boundary regions are discussed, along with the effects on the electric field of changing the geometry and field strength. Test methods capable of separating bulk properties from interfacial properties are discussed. Characterization methods for electrochemical degradation (anodic or cathodic) are also outlined, because of their importance in durability predictions.

Although a wide variety of appropriate test geometries exists for measuring adhesive system properties, many applications are difficult to translate into preexisting test geometries. In cases where the adherends are very brittle, difficult to machine, or available only in certain dimensions and shapes, it becomes necessary to adapt techniques for special applications. The process of slicing up a prototype into test specimen-size coupons and running them through a test machine may provide quality control type data, but it seldom results in meaningful quantitative information unless careful analysis accompanies the experiments. Different gripping conditions, constraint effects, and loading mode may all lead to errors in interpretation. In the article "Evaluating Test Geometries," some proved techniques and newer approaches to evaluating the stress distributions in such custom joints are discussed.

In summary, this Section emphasizes the necessity of considering test results as the performance of an adhesive material system in a test specimen structure. The inappropriateness of carelessly using failure strength tests to obtain engineering properties has been established. The necessity of analysis of test geometries to interpret experimental results is concluded.

Measuring Constitutive Properties

Tim A. Osswald and Jeroen Rietveld, University of Wisconsin—Madison

ADHESIVES FOR ENGINEERING APPLICATIONS are typically derived from low molecular weight materials, which are then polymerized during a curing step to form the adhesive joint (Ref 1-4). Prior to curing, the resin is in a low-viscosity fluid state in order to maximize the initial wetting of the adherend surfaces. As the resin polymerizes, the adhesive converts from a fluid to a solid state, with an intermediate highly viscous, elastic gel state. Once solidified, the resin exhibits its adhesive character by mechanical and/or specific adhesion. Depending on the chemistry of the curing reaction, these resins form either network or linear macromolecular systems. The glass transition temperatures (T_g) of adhesives based on either linear or network systems are high enough that the adhesives typically exhibit high strengths and moduli even at elevated temperatures

Epoxy- and acrylic-base adhesives, phenolic-resorcinols, polyurethanes, and polyimides are all examples of polymeric materials that fall into the category of structural adhesives. The acrylic-base adhesives include cyanoacrylates, anaerobics, and modified acrylics. Many of these resins are toughened by the addition of rubbers in order to make them less brittle, while others are reinforced by the addition of rigid particulates. To ensure that the filled adhesive is homogeneous, the additives should be finely dispersed throughout the matrix.

In addition to the five groups of structural adhesives already mentioned, several other chemically different polymeric resins are used in adhesive systems, such as ureaformaldehydes, silicones, phenolics, and plastisols. These adhesive resins are also either linear or network polymeric systems. Because of the macromolecular nature of adhesives, many of their constitutive physical, thermal, rheological, and mechanical properties can be evaluated according to the testing procedures developed for thermosetting and thermoplastic polymers. Furthermore, the results of these tests can then be interpreted in terms of the behavior of polymeric systems in general.

This article considers several testing procedures that can be used to measure the static, transient, and dynamic mechanical properties, as well as the rheological properties of engineering adhesives. First discussed are common methods of testing bulk specimens, beginning with static testing, used to measure the modulus of elasticity, E; rigidity, G, and Poisson's ratio, v. Next discussed is long-term mechanical behavior in creep and stress relaxation experiments, which determine the time-dependent creep compliances, $D(t)$ or $J(t)$, and the time-dependent relaxation moduli, $E(t)$ or $G(t)$. This is followed by a discussion of the dynamic mechanical analysis of polymeric adhesives to measure their storage, E' (or G'), and loss, E'' (or G''), moduli, and their loss tangent, tan δ. Information on the microstructure and T_g of adhesives, especially in terms of their effect on mechanical properties, is also presented. Finally, several standard test methods for measuring the mechanical (sometimes referred to as *in situ* mechanical) and rheological behavior of polymeric adhesives are described.

Static Testing

Static tests of bulk specimens are the most widely conducted mechanical measuring experiments in the engineering world. This is true, in part, because historically they are the oldest tests used to measure material properties and also because the tensile member and the torsion bar, commonly used bulk specimens, are often seen both in structures and in machines. When testing adhesives, the testing of bulk specimens is easier than the testing of films because much larger deformations can be attained, which, in turn, are much easier to measure.

Many polymers and liquid adhesives are easily cast into bulk specimens without the need for subsequent machining. However, with thermosetting materials, the exothermic reaction can lead to overheating when casting specimens that are too thick. One must also watch out for air bubbles, which because of the occasional high viscosity of the material, remain trapped inside the specimen. In addition, high-temperature gradients during curing lead to residual stresses inside the specimen, which makes results from such specimens subject to question.

In the tensile test, a slowly increasing load is applied to a material specimen, and the resulting load-deflection curve is recorded. A typical stress-strain curve for a polymer is shown in Fig. 1. For polymeric materials this test is described by ATSM standard D 638. The information gained from the tensile test includes such fundamentally important parameters as yield strength, modulus of elasticity, E, ultimate strength, percent elongation, and percent reduction of area. Less common parameters, such as modulus of toughness, can also be taken from the tensile test results. The modulus of elasticity is the slope of the elastic portion of the tensile stress-strain curve in Fig. 1.

The stress-strain relationship of polymeric materials may vary substantially with temperature and strain rate. Additionally, plastics do not have a sharply determined yield point. Because plastics have a higher Poisson's ratio than do metals, necking in the tensile specimen occurs more frequently. Hence, if stress were calculated in terms of instantaneous cross-sectional area, the yield point would become an indistinct section on the stress-strain curve.

The torsion test is used to determine the modulus of rigidity or shear modulus, G, of a material. Figure 1 shows a typical stress-strain curve of a polymer under pure shear. In this test a torque, τ, is applied on a specimen that deflects an amount θ radians. To measure shear properties of low-modulus adhesives, the ASTM D 3983 and D 4027

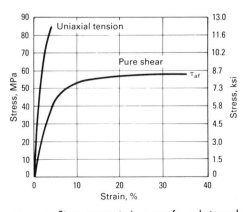

Fig. 1 Stress versus strain curves for a hot-cured rubber-toughened epoxy adhesive, at 23 °C (73 °F) and an average strain rate of 7.5×10^{-4}/s, in uniaxial tension and pure shear. Source: Ref 2

Table 1   Adhesive material properties used in theoretical stress analysis and failure predictions of high-tensile steel-carbon fiber composite

Property	Value
Young's modulus (E_a), GPa (10^6 psi)	3.05 (0.44)
Tensile fracture stress (σ_{af}), MPa (ksi)	82 (11.9)
Tensile fracture strain (ϵ_{af}), %	4.75
Shear modulus (G_a), GPa (10^6 psi)	1.13 (0.16)
Shear yield stress (τ_{ay}), MPa (ksi)	54 (7.8)
Shear (plastic) fracture (γ_{af}), %	35
Poisson's ratio (ν_a)	0.35

Note: Based on single-part rubber-toughened epoxy. Source: Ref 2

Table 2   Comparison of modulus values for different epoxy particulate composites

	Modulus of elasticity			
	At −70 °C (−95 °F)		At 50 °C (120 °F)	
Composite	GPa	10^6 psi	GPa	10^6 psi
Epoxy	4.1	0.59	2.8	0.41
Epoxy-glass(a)	4.3	0.62	3.1	0.45
Epoxy-rubber(b)	3.3	0.48	2.2	0.32
Hybrid (a)(b)	4.1	0.59	2.6	0.38
Hybrid (using silane-treated glass particles)(a)(b)	4.3	0.62	2.5	0.36

(a) Glass volume fraction, 0.1. (b) Rubber, 15 parts per hundred resin (phr). Source: Ref 5

tests are available. The shear stress, τ, and strain, γ, are calculated using the relationships:

$$\tau = \frac{T * r}{J} \quad \text{(Eq 1)}$$

$$\gamma = \frac{r * \theta}{L} \quad \text{(Eq 2)}$$

where r is the radius, J is the polar moment of inertia, and L is the length of the specimen. The modulus of rigidity is the slope of the elastic portion or the shear stress-strain curve in Fig. 1. Poisson's ratio, ν, can now be determined by:

$$\nu = \frac{E}{2G} - 1 \quad \text{(Eq 3)}$$

and the bulk modulus, K, can be calculated using:

$$K = \frac{E}{3(1 - 2\nu)} \quad \text{(Eq 4)}$$

Bulk modulus is defined as the ratio between hydrostatic stress, where $\sigma_{xx} = \sigma_{yy} = \sigma_{zz}$, and volumetric strain. The volumetric strain is defined as volume change divided by initial volume. Nearly incompressible materials, such as rubber, have a Poisson's ratio of about 0.5 and a very high bulk modulus. Representative values for the mechanical properties of a single-part rubber-toughened epoxy adhesive are listed in Table 1.

Because engineering adhesives may be subjected to a range of temperatures and because many adhesives are actually filled systems, it is of interest to show how the mechanical properties can be affected by temperature and fillers. Table 2 and Fig. 2 and 3 clearly demonstrate the effect that filler content and temperature have on the mechanical properties of a material.

Transient Testing

Because the stress-strain behavior of polymeric materials often requires inclusion of a time dependence, a polymeric material exhibits viscoelastic properties rather than merely elastic properties (Ref 6-15). For a viscoelastic material, the relationship between stress and strain can be generally written as:

$$\sigma = f(\epsilon, t) \quad \text{(Eq 5)}$$

To simplify the use of Eq 5, the behavior of the material could be assumed to be that of a linear viscoelastic material and therefore could be given by:

$$\sigma(t) = f(t)\epsilon_0 \quad \text{(Eq 6)}$$

The time-dependent function in Eq 6 may be viewed as the modulus of the material, but a modulus that now depends on the time frame of measurement rather than a time-independent modulus of constant value. The viscoelastic nature of polymers affects not only the modulus of the material, but may strongly influence other mechanical properties as well.

Usually, materials exhibit nearly linear behavior only over limited and/or certain ranges of stress, strain, and time. Linearity of the viscoelastic functions may also be influenced by the temperature at which the material happens to be. If conditions are such that relatively large strains will be encountered, a nonlinear relationship will probably be required to describe the viscoelastic material mathematically. Equation 5 gave the general form of a relationship between stress and strain for a nonlinear time-dependent system. Likewise, relationships may be found that can directly predict the strain instead of the stress. Such an expression will be of the form:

$$\epsilon = g(\sigma, t) \quad \text{(Eq 7)}$$

As an approximation it may be assumed that Eq 7 can be rewritten as:

$$\epsilon = g_1(\sigma)g_2(t) \quad \text{(Eq 8)}$$

by factoring it into two independent functions of stress and time. Equations of this form are analytically convenient, and although they may not provide an entirely accurate description of the actual deformational behavior of polymeric materials, they do offer a good first approximation for nonlinear behavior. Time-dependent viscoelastic properties can be examined experimentally by creep and stress-relaxation tests.

Creep tests measure the time-dependent deformation of a material while it is held under a constant applied load at a given temperature (Ref 6-11, 13-15). Creep data are generally collected over a time period starting from about 10 s and extending up to 10 years. The complete testing method is described by ASTM D 2990. Tests are typ-

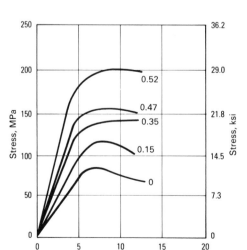

Fig. 2 Stress-strain curves for particle-reinforced epoxy resins determined in uniaxial compression. Different volume fractions (indicated) of a silica-filled epoxy resin. Source: Ref 5

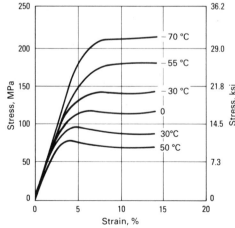

Fig. 3 Stress-strain curves for particle-reinforced epoxy resins determined in uniaxial compression and dependent on the temperature of the behavior of a glass particle reinforced epoxy resin. Source: Ref 5

ically conducted on simple-shape bars under uniaxial tension, for which the stress is applied as a step excitation. The exact dimensional change of a plastic part over the course of time when subjected to a constant load also depends on the state of stress itself. Actual stress states on an in-service plastic article are probably a multiaxial combination of tensile, shearing, and/or compressive stresses. The following discussion, however, is confined to uniaxially loaded samples in tension, which is a standard configuration during a creep experiment. It should be mentioned that there are ways to take into account the action of a combined stress, such as by using the concept of an effective stress or by conducting multiaxial creep experiments.

An objective of a creep test may be to define a time- and stress-dependent function for the overall strain, ϵ, or γ, of a particular polymeric material. At rather low levels of strain, a time-dependent function for the compliance may be specifically derived. If the test sample is measured in tension, then at low strains the following equation is obtainable from a creep test:

$$\epsilon(t) = D(t)\sigma_0 \qquad \text{(Eq 9)}$$

Similarly, for a test sample in shear, the following equation results at low strains for a creep test:

$$\gamma(t) = J(t)\tau_0 \qquad \text{(Eq 10)}$$

The total strain response of a typical viscoelastic material subjected to a constant load is shown in Fig. 4. Included in this diagram are the instantaneous elastic and plastic contributions to the total strain, ϵ_{tot}, which can be written as:

$$\epsilon_{tot} = \epsilon_{el} + \epsilon_{pl} + \epsilon_{cr} \qquad \text{(Eq 11)}$$

The elastic strain, ϵ_{el}, is a recoverable deformation, while the plastic strain, ϵ_{pl}, is an irreversible contribution to the total deformation. The creep component, ϵ_{cr}, can be further broken down into three stages:

$$\epsilon_{cr} = \epsilon_p + \epsilon_s + \epsilon_t \qquad \text{(Eq 12)}$$

The first stage, referred to as primary creep, ϵ_p, occurs at a decreasing creep rate with time. The second stage, referred to as secondary creep, ϵ_s, occurs at a constant rate of creep. Finally, tertiary creep, ϵ_t, occurs at an increasing creep rate with time for the last creep stage and usually ends in rupture of the material. Of the three creep stages, only primary creep is a recoverable deformation for polymeric materials. It is important to realize that a relatively low-magnitude load applied to a polymeric material causes the material to deform gradually over time, eventually culminating in creep-rupture. How much deformation occurs, and how long it takes to develop, depends on the viscoelastic properties of the material. It is also beneficial to have at least a qualitative knowledge of secondary and ter-

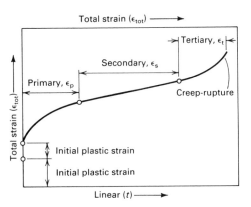

Fig. 4 Idealized constant stress creep curve for a constant temperature. Source: Ref 8

tiary creep so that primary and secondary creep experimental data or mathematical expressions are not erroneously extrapolated.

With the exception of tertiary creep, a linear relationship that contains all the deformations found in Eq 11 and 12 is:

$$\epsilon = A\sigma + B\sigma[1 - \exp(-Ct)] + D\sigma t \qquad \text{(Eq 13)}$$

Also, for nonlinear behavior, a common approach is to include higher than first-order stress terms, leading to:

$$\epsilon = A\sigma + B\sigma^n[1 - \exp(-Ct)] + D\sigma^n t \qquad \text{(Eq 14)}$$

where A, B, C, D, and n are material-dependent coefficients. These expressions allow for the instantaneous recovery of the elastic strain and the gradual recovery of the primary creep strain when the load is removed. In Eq 13 and 14, the amount of plastic strain is assumed to be dependent on both the speed of loading of the material and the duration of load application, such that the amount of plastic strain is only a function of the load, $\epsilon_p = f(\sigma)$.

Several commonly used nonlinear expressions exist to describe the initial portion of the creep deformation alone, and these include Findley's empirical equation:

$$\epsilon_{cr} = \epsilon_0 + mt^n \qquad \text{(Eq 15)}$$

as well as Nutting's empirical relationship:

$$\epsilon_{cr} = k\sigma^a t^b \qquad \text{(Eq 16)}$$

In Eq 15 and 16, ϵ_0 and m are functions of stress for a given material, while n, k, a, and b are material constants. A temperature-dependent function can also be explicitly included. Equation 8 can, in this instance, be generally expanded as:

$$\epsilon = g_1(\sigma)g_2(t)g_3(T) \qquad \text{(Eq 17)}$$

An empirical equation along these lines that describes the initial creep behavior of polymeric materials may be given by (Ref 16):

$$\epsilon_{cr} = k\sigma^a t^b \exp\left(\frac{-E_a}{RT}\right) \qquad \text{(Eq 18)}$$

where an Arrhenius type of temperature dependence has been included. In Eq 18, E_a is an activation energy, R is the gas constant, and T is the absolute temperature.

A typical family of creep curves is shown in Fig. 5(a). Usually, such data will appear on a graph with a logarithmic time scale instead of the linear time scale shown in Fig. 4. It should be apparent from the data in Fig. 5 that when considering large intervals of the strain, stress, and time variables, especially at higher values of these variables, nonlinear viscoelastic behavior can be observed. Creep curves can also be replotted as stress versus time on what is called an isometric graph (Fig. 5b), or as stress versus strain on what is called an isochronous graph (Fig. 5c). The isometric graph may be used as an indication of the stress-relaxation behavior of plastic materials at constant levels of strain. An isochronous graph such as Fig. 5(c) is an indicator of the stress-strain behavior of polymeric materials at any segment in time during a wide time span. The more familiar short-term stress-strain tests are essentially obtained for the $t \approx 0$ time frame. Any initial linear portion of isochronous curves measures the time-dependent "creep modulus" of linear viscoelastic materials, while the nonlinearity of these curves will be due to nonlinear viscoelastic behavior. Strictly speaking, the "creep modulus" is only a convenient definition; fundamentally, it has no formal basis in the theory of viscoelasticity.

The deformational behavior in the secondary creep region may be better examined by plotting creep data on log-log coordinates as strain rate versus stress. Creep data that are presented in terms of the creep rate as a function of time rather than in terms of the total strain as a function of time will remove any time-independent terms from Eq 11. Certain potential errors in the vertical position of the creep curve would thus be eliminated.

Because of the viscoelastic nature of polymeric materials, their deformational response to an applied stress turns out to be very dependent on temperature as well as time. The exact temperature dependence will, in turn, be influenced by the thermal properties of the polymer itself, which are categorically different for thermoplastics and thermosets. For some conditions, the time and temperature dependence of the relationship between stress and strain may be superposable and therefore essentially equivalent. Such time-temperature equivalency can be used to evaluate the deformational behavior of a polymeric material in very long time frames without always having to resort to experimental methods that can last up to several years.

The mechanical behavior of many polymers, particularly amorphous ones, may be extrapolated to much longer times by conducting the measurements over a wide

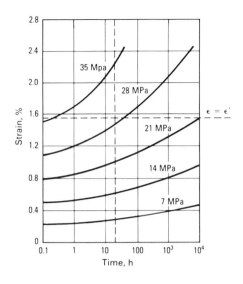

(a)

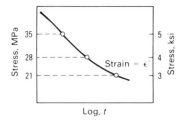

(b)

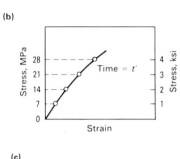

(c)

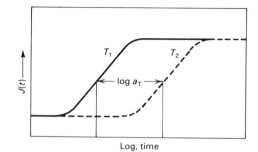

Fig. 6 Diagram showing the simplest form of time-temperature equivalence for compliance, $J(t)$. Source: Ref 15

Fig. 5 (a) Typical family of creep curves taken at a constant temperature but different stresses. (b) Isometric curve constructed from family of creep curves at strain, ϵ'. (c) Isochronous curve constructed from family of creep curves at time, t'. Source: Ref 17

range of temperatures but within a reasonable time frame. The results may then be manipulated by shifting the data relative to a constant reference temperature, thereby going beyond the experimental time frame. An example of the shifting procedure for two creep compliance curves of an idealized polymer at two different temperatures is shown in Fig. 6. The horizontal distance between the two curves along the logarithmic time axis defines a shift factor, a_T, and can be obtained from:

$$\log a_T = \log[t(T)] - \log[t(T_R)] \qquad \text{(Eq 19)}$$

where T_R is an arbitrarily chosen reference temperature. If the shifted curves are plotted with respect to a constant reference temperature, the time axis should be labeled t/a_T instead of t.

Stress relaxation tests measure the decay of an applied load within a material while it is held at a constant deformation and at a given temperature (Ref 6-11, 13-15). These tests are not run as commonly as are creep tests. The complete testing method is described in ASTM D 2991. Tests are typically conducted on simple-shape bars under uniaxial tension, for which the strain is applied as a step excitation. For a stress-relaxation test, the objective may be to obtain the time- and strain-dependence of a function describing the measured stress, σ, or τ, for a particular material. At rather low levels of strain, a time-dependent function for the modulus may be specifically derived. If the test sample is measured in tension, then at low strains, the following equation is obtainable from a stress-relaxation test:

$$\sigma(t) = E(t)\epsilon_0 \qquad \text{(Eq 20)}$$

Similarly, for a test sample in shear, the following equation results at low strains for a stress-relaxation test:

$$\tau(t) = G(t)\gamma_0 \qquad \text{(Eq 21)}$$

With regard to the relationship between a modulus (E or G) and a compliance (D or J), it is important to note that although for a perfectly elastic solid J will equal $1/G$, the same cannot be said for a viscoelastic material; that is, $J(t)$ does not simply equal $1/G(t)$. This distinction is sometimes overlooked. The formal relationship between the stress-relaxation modulus and the creep compliance is given by:

$$\int_0^t G(\tau)J(t-\tau)d\tau = t \qquad \text{(Eq 22)}$$

In addition, the equation:

$$E(t) = 2G(t)(1 + \nu) \qquad \text{(Eq 23)}$$

taken from the elasticity relationship between the tensile modulus and the shear modulus applies to viscoelastic materials only if they are at equilibrium, that is, if ν is approximately equal to a constant, which is not to be confused with a system at steady state. In Eq 23, ν is a time-dependent Poisson's ratio for the polymeric material. For a viscoelastic material, the more general relationship between the tensile and shear moduli is given by:

$$E(t) = 2[G(t) + \int_{-\infty}^t \nu(\tau)G(t-\tau)\,d\tau] \qquad \text{(Eq 24)}$$

Dynamic Mechanical Testing

Dynamic mechanical testing of bulk specimens can also be used to evaluate a material for its viscoelastic or time-dependent properties (Ref 7, 10-12, 15). Dynamic experiments provide data over an effective time interval of 10^{-8} s to 10^3 s. Measurements may be made with the sample in shear or in tension, and the sample may be subjected to a periodically varying stress or a periodically varying strain. The periodic input is usually sinusoidal and can be imposed at a variety of frequencies and amplitudes. Dynamic measurements are conducted at a constant temperature, while the frequency of the cyclic input is varied; or equivalently, the temperature is varied, while a constant frequency is maintained. The frequency ω (rad/s) in a dynamic test qualitatively relates to the time varied in a transient experiment through $\omega = 1/t$. Depending on the frequency range of interest, the experimental analysis employs free vibration, resonance vibration, or wave propagation methods. The amplitude of the periodic input is kept relatively low to ensure that the behavior of the polymer is that of a linear viscoelastic material and to avoid the possibility that at high strain or stress amplitudes the polymer sample fatigues or ruptures as higher frequencies are encountered.

Analogous to the creep and stress-relaxation experiments, which determine the time- and temperature-dependent functions, are the dynamic mechanical experiments, which define the frequency- and temperature-dependent functions characteristic of viscoelastic materials such as polymers. Such testing methods can be used to measure values that are called the storage and loss moduli (or compliances) for a polymeric material. The storage modulus relates to the amount of recoverable energy stored within a deformed material (solidlike behavior), whereas the loss modulus relates to the amount of unrecoverable energy lost within a deformed material (fluidlike behavior).

Storage and Loss Moduli. If a low-amplitude, sinusoidally varying tensile strain with frequency, ω, in rad/s, given by:

$$\epsilon = \epsilon_0 \sin \omega t \qquad \text{(Eq 25)}$$

is applied to a viscoelastic material, the stress response of the material will be at the same frequency, but will lead the strain by some phase angle, δ, and will be given by:

$$\sigma = \sigma_0 \sin(\omega t + \delta) \qquad \text{(Eq 26)}$$

A perfectly elastic material would have the resulting stress in phase with the applied strain, whereas a purely viscous material would have 90° phase difference between the stress and the strain. By defining a complex stress as:

$$\sigma^* = \sigma_0 \exp(i\omega t + \delta) \qquad \text{(Eq 27)}$$

Eq 25 and 26 can be rewritten using complex variable notation as:

$$\frac{\sigma^*}{\epsilon^*} = E^* = \left(\frac{\sigma_0}{\epsilon_0}\right)\exp(i\delta) = \left(\frac{\sigma_0}{\epsilon_0}\right)\cos\delta + i\left(\frac{\sigma_0}{\epsilon_0}\right)\sin\delta \qquad \text{(Eq 28)}$$

or, in its shortened form, as:

$$E^* = E' + iE'' \qquad \text{(Eq 29)}$$

The real part of the complex dynamic modulus, E^*, is called the storage modulus, E', and the imaginary part is called the loss modulus, E''. Respectively, E' and E'' are the in-phase (elastic) and out-of-phase (viscous) components of the complex modulus. The ratio of the energy dissipated per cycle to the potential energy stored per cycle is called the loss tangent, defined as:

$$\tan\delta = \frac{E''}{E'} \qquad \text{(Eq 30)}$$

Similar expressions can be defined for a sinusoidally varying shear strain so that a complex shear modulus, $G^* = G' + iG''$, is obtained.

As previously mentioned, the analysis may also be carried out in terms of a sinusoidally varying applied stress instead of a strain. For a sinusoidal stress or strain input, the experimental results, however, are not independent of one another. Definite relationships exist between the storage or loss modulus and the storage or loss compliance, which, for a shearing deformation, are given, respectively, by:

$$G' = \frac{1/J'}{1 + \tan^2\delta} \qquad \text{(Eq 31)}$$

and

$$G'' = \frac{(1/J'')\tan^2\delta}{1 + \tan^2\delta} \qquad \text{(Eq 32)}$$

From Eq 31 and 32 it should be apparent that $J^* = 1/G^* = J' - iJ''$.

Typical plots of the storage modulus, loss modulus, and loss tangent as a function of frequency for an amorphous polymer are shown in Fig. 7. These properties measure the capacity of a material for storing energy (Fig. 7a) and for dissipating energy (Fig. 7b) and the internal friction (damping) of a material (Fig. 7c). The energy is dissipated as heat, and the amount of heat generated within the material per cycle is given by:

$$Q = \pi E'' \epsilon_0^2 \qquad \text{(Eq 33)}$$

for a sinusoidally varying tensile deformation with maximum strain amplitude, ϵ_0. A high damping material smooths out undesirable vibration and, in so doing, increases its internal temperature. This ability to dissipate mechanical energy as well as store it is indicative of a behavior between that of a viscous fluid and that of an elastic solid, that is, viscoelastic behavior.

The experiment results from dynamic mechanical tests can be used to assess the elastic versus the viscous nature of a polymeric material at a given frequency and temperature. By using the principle of time and temperature superposability, dynamic tests can be run at different temperatures in order to judge the change in the viscoelastic behavior of a material for experimentally inaccessible frequency ranges, or vice versa. An example of the shifting procedure for two loss tangent peaks of an idealized polymer at two different temperatures is shown in Fig. 8. The horizontal distance between the two curves along the logarithmic frequency axis defines a shift factor, a_T, and can be obtained from the following expression:

$$\log a_T = \log[\omega(T)] - \log[\omega(T_R)] \qquad \text{(Eq 34)}$$

where T_R is an arbitrarily chosen reference temperature. If the shifted curves are plotted with respect to a constant reference temperature, the frequency axis should be labeled ωa_T instead of ω.

The nonconstant modulus of a polymeric material as determined by dynamic mechanical experiments may give essentially the same result as the moduli obtained from creep or stress-relaxation tests. This equivalency is shown in Fig. 9. Appropriate empirical equations can be used to relate the time-dependent modulus to the frequency-dependent storage and loss moduli. Other empirical relationships exist for calculating the loss tangent from transient data. Fourier transform methods may also be used to invert one type of data into the other type, that is, transient to dynamic and dynamic to transient.

Torsion Pendulum and Torsional Braid Analysis (TBA). Several instruments are available for characterizing the dynamic mechanical properties of polymeric materials (Ref 10, 12, 15). The frequency and temperature range accessible with such instrumentation depends on the particular operational method of each instrument. Free-vibration and resonance methods are quite accurate and relatively simple to use over a wide temperature range, but the effective frequency is set by the rigidity of the sample. Forced-vibration nonresonance methods are more complex, but the test frequency can be independently adjusted. Typical instrumentation includes the rheo- or autovibron, the dynamic mechanical analyzer, the torsion pendulum, torsional braid analysis, and the vibrating reed. A brief discussion on the use of just two of these

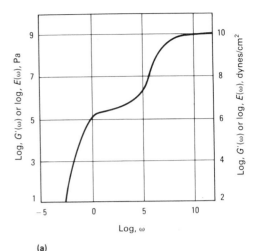

(a)

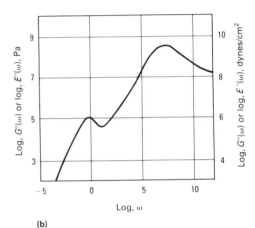

(b)

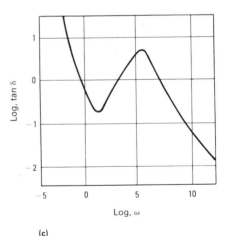

(c)

Fig. 7 Modulus versus frequency for amorphous plastic. (a) Storage modulus. (b) Loss modulus. (c) Loss tangent. Source: Ref 10

methods to obtain dynamic mechanical properties is found below.

The torsion pendulum and TBA methods are used to measure the dynamic shear modulus and mechanical damping of a material. In the former method, the sample typically has the shape of a long cylindrical rod or a long rectangular bar/film, while in the latter method, the sample is usually

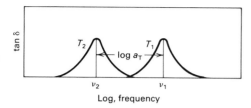

Fig. 8 The simplest form of time-temperature equivalence for loss factor tan δ. Source: Ref 15

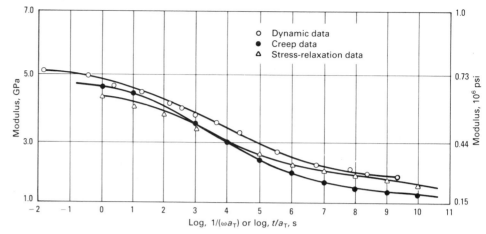

Fig. 9 Adjusted curves illustrating the equivalency of various experimental methods. Source: Ref 12

coated on a braided substrate consisting of long glass fibers. Torsional braid analysis offers the advantage that a reactive sample, such as a precured adhesive resin, can be monitored as the coated resin polymerizes on the glass fiber substrate. With either method, specimens are vertically mounted with one end securely clamped and the other attached to an inertia disk. The mounted specimen is often contained in an environment controlled by a thermostat.

During an experimental run, the inertia disk is set in motion, and the specimen is subjected to a twisting oscillation. The amplitude of the damped oscillation decreases with time while the natural frequency, f, of the oscillation remains constant. The storage modulus, G', can be calculated from:

$$G' = KI(2\pi f)^2[1 + (\Delta/2\pi)^2] \qquad \text{(Eq 35)}$$

where K is a geometric constant for the sample, I is the moment of inertia (obtained in a separate calibration experiment), f is the natural frequency of the oscillation in hertz (Hz), and Δ is the logarithmic decrement. The logarithmic decrement is given by $\Delta = \ln (A_i/A_{i+1})$ where A_i is the amplitude of the ith oscillation, and it is related to the loss tangent, tan δ, by $\Delta = \pi \tan \delta$. For small damping, Eq 35 reduces to:

$$G' = KI(2\pi f)^2 \qquad \text{(Eq 36)}$$

Therefore, the loss modulus is given by:

$$G'' = (KI\Delta\pi)(2f)^2 \qquad \text{(Eq 37)}$$

From Eq 36 it can be seen that a relative modulus instead of an absolute modulus can be measured even if K and I are not known, because G' will be proportional to f^2. To summarize, the storage modulus G' is obtained from a measurement of the natural frequency of the oscillation, whereas the loss tangent, tan δ, and the loss modulus, G'', are obtained by measuring the amplitude damping of the oscillation.

Microstructure

The static, transient, and dynamic mechanical properties of adhesives are modified by material-dependent factors. These factors are indicative of the chemical, physical, and/or compositional nature of the polymer itself. Of course, it is still quite possible to describe the mechanical behavior of a material without any knowledge of its internal detail, by experimentally evaluating the parameters of the preceding stress-strain relationships. Nevertheless, it is important to be aware of the manner in which these factors affect the properties of a polymeric material, and thus it is helpful to categorize adhesives into thermoplastic and thermosetting, and filled and unfilled groups. Furthermore, if the polymer is approached from a molecular level, it becomes easier to anticipate the influence of material-dependent factors on the mechanical behavior. The polymer molecules within adhesives are made of many thousands of smaller molecules linked together by strong covalent bonds. In the case of thermoplastic materials, the individual polymer molecules are not interconnected, and these macromolecules are potentially capable of being quite mobile relative to one another. On the other hand, thermosetting materials consist of macromolecules that are extensively interconnected, forming a network system of covalently linked macromolecules. In such a macromolecular network, the individual mobility of polymer chains is severely restricted. It makes sense, therefore, that the fluidlike response of a thermoplastic material is inherently greater than that of a thermosetting material.

The glass transition temperature, T_g, is one of the more important thermal properties of polymeric materials. At temperatures below the T_g, the polymer chains exhibit very little mobility and are essentially frozen in a glassy state, whereas at temperatures above T_g, the polymer chain mobility is substantially increased. For many polymeric materials, the retention of the strength and modulus of the polymer at elevated temperatures is limited by the T_g of the polymer. Bismaleimide resins are a class of structural adhesives that exhibit excellent high-temperature stability due to their relatively high T_g, which are typically in excess of 250 °C (480 °F) (Ref 18).

The cross-link density of a thermosetting resin has a significant effect on the physical, thermal, and mechanical properties of the cured adhesive (Ref 1, 2, 4, 5, 18-22). As the molecular weight between cross links is decreased, the T_g and the modulus of the polymeric material increases. For example, the curing conditions of a diglycidyl ether of bisphenol A epoxy resin has the following effect on the T_g of the adhesive (Ref 23): a cure cycle of 40 h at 100 °C (212 °F) leads to a T_g of 122 °C (250 °F), whereas a cure cycle of 3 h at 200 °C (390 °F) leads to a T_g of 161 °C (320 °F). The final cross-link density can also influence the adhesive properties of the resin. For highly cross-linked polymeric systems, depending on the extent of cross linking, the relevance of T_g may be less significant because the molecular mobility of the polymer chains becomes restricted by the presence of cross links. A continuous increase in the temperature for a very highly cross-linked material often leads to thermal degradation of the polymer before it can leave the glassy state.

Dynamic mechanical tests are very useful for identifying the important transition temperatures of polymeric materials (Ref 7, 10-12, 15). In particular, the loss tangent and loss modulus contain information about the material at a molecular level. Polymers are in a hard, glassy state below the T_g. A temperature increase may cause the material to become more flexible or rubbery. In the vicinity of this transition temperature, the polymer usually exhibits a noticeable drop in storage modulus and displays a maxima in loss modulus and loss tangent. This behavior is shown in Fig. 10 for a rubber-modified epoxy system. The T_gs of the filler and the matrix show up as separate maxima on the log decrement curves. Also, the relative position of the loss peak for the adhesive matrix is affected by the isothermal cure conditions, whereas the relative position of the loss peak for the rubber filler is not affected.

Other transitions that are more subtle than the glassy-to-rubbery transition exist

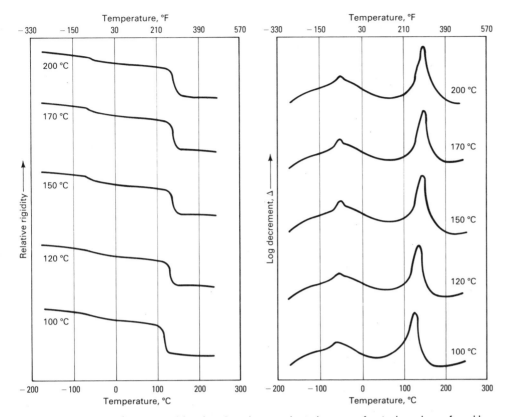

Fig. 10 Representative torsional braid analysis thermomechanical spectra after isothermal cure for rubber-modified system. Temperatures in graphs are isothermal cure temperatures. Source: Ref 19

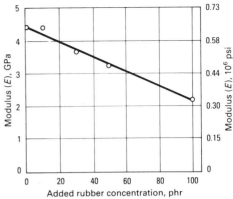

Fig. 11 Effect of rubber addition on the modulus (at 23 °C, or 73 °F) of a rubber-toughened polybismaleimide. Source: Ref 23

mechanical properties of an adhesive can be modified, often by adding fine rubber particles and/or rigid particulate fillers (Ref 2, 5, 23). Assuming that there is good interfacial adhesion between the matrix and the additive, the basic premise is that the additives will impart some of their properties to the polymer matrix. Figures 2 and 3 show the influence of rigid particulate fillers on the stress-strain behavior of epoxy adhesives. The effect of rubber concentrations on the elastic modulus of a rubber-toughened polybismaleimide adhesive is shown in Fig. 11.

Standard Test Methods

The testing and mechanical behavior of various standard joint designs are discussed in this section. The *in situ* mechanical property data obtained from standard test specimens is comparable to data on mechanical properties obtained from bulk specimens if the stress state is simple, no stress concentrations are present, and failure occurs within the adhesive and not near or at the interface (Ref 2). The butt joint test and the napkin ring test are two *in situ* methods by which the more difficult testing of bulk specimens can perhaps be avoided. If there are discrepancies between the results from *in situ* and bulk specimen testing procedures, it could be argued that the mechanical behavior of the thin adhesive layers in the butt joint test and the napkin ring test may be more representative of real joint behavior. See Table 3 for mechanical properties obtained by such methods.

Butt Joint Test. The cylindrical butt joint shown in Fig. 12 is commonly used to test thin glue line specimens loaded in torsion, tension, or compression (Ref 3, 4). The test is used to measure the moduli of rigidity and elasticity and Poisson's ratio. If the adherend is much stiffer than the adhesive, the adhesive experiences a radial constraint by the adherend. For a butt joint axially loaded in tension or compression and assuming the contraction of the adherend to be 0, the

as well, and these are also indicative of changes in mobility at the molecular level that may in turn be exhibited as changes in the macroscopic properties of the polymer. Therefore, the viscoelastic character of the polymer above and below these transition temperatures can be quite different, and the location of these transition temperatures relative to the end-use temperature may be important in determining the long-term performance of an adhesive under load. If the experiment is run over a wide temperature range and a constant frequency of about 1 Hz is maintained, the

dynamically determined transition temperatures of the polymer roughly coincide with dilatometrically or calorimetrically determined transition temperatures. However, it is again important to remember that because of the time-temperature superposability principle for viscoelastic materials, a polymer may appear to exhibit either glassy or rubbery behavior that is wholly dependent on the particular time frame of measurement.

Filled Adhesives. Polymers are often filled with various additives in order to improve particular properties. Likewise, the

Table 3 Shear properties of some structural adhesives at room temperature and at a moderate rate of test

Adhesive(a)	Modulus of adhesive		Yield stress		Fracture stress		Fracture strain, %
	GPa	10⁶ psi	MPa	ksi	MPa	ksi	
FM 73(b)(c)	0.58	0.084	31	4.5	37	5.4	61
FM 73(c)	0.56	0.081	30	4.4	38	5.5	140
REDUX 775RN(d)	1.23	0.178	50	7.3	55	8.0	120
BSL 312(e)	0.67	0.097	38	5.5	42	6.1	75
Hot-cured (single-part) toughened epoxy	1.13	0.164	54	7.8	58	8.4	35
Cold-cured (two-part) toughened epoxy	0.51	0.074	28	4.1	35	5.1	105
Cold-cured (two-part) toughened acrylic	0.16	0.023	8	1.2	34	4.9	~150

(a) All properties measured using the thick-adherend shear test specimen, except where otherwise indicated. (b) Napkin ring shear specimen was used. (c) FM 73, a toughened epoxy film adhesive (hot cured, 60 min/125 °C, or 255 °F) from American Cyanamid Company. (d) REDUX 775RN, a vinyl phenolic film adhesive (hot cured, 30 min/150 °C, or 300 °F) from Ciba-Geigy Corporation. (e) BSL 312, a modified epoxy film adhesive (hot cured, 30 min/120 °C, or 250 °F) from Ciba-Geigy Corporation. Source: Ref 2

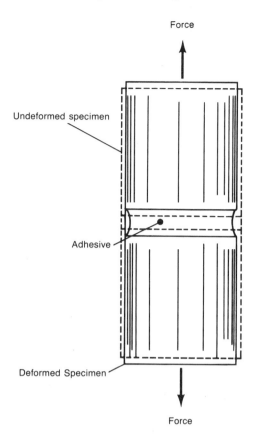

Fig. 12 A butt joint loaded in tension

ratio of an apparent modulus, E^*, to the true modulus of elasticity is given as a function of Poisson's ratio of the adhesive:

$$\frac{E^*}{E} = \frac{(1 - \nu)}{(1 + \nu)(1 - 2\nu)} \qquad \text{(Eq 38)}$$

From Eq 4 and 38 we obtain:

$$\nu = \frac{2G - E^*}{2(G - E^*)} \qquad \text{(Eq 39)}$$

The butt joint can also be loaded in torsion to determine the modulus of rigidity, which is given by:

$$G = \frac{2b}{\pi R^4} \left(\frac{T}{\theta} \right) \qquad \text{(Eq 40)}$$

where b is glue line thickness. Once G and ν have been measured, Eq 3 can be used to determine the modulus of elasticity of the adhesive.

The napkin ring test is very similar to the butt joint test (Ref 24). It consists of two thin-walled cylinders bonded together and is commonly used to measure the modulus of rigidity. This type of arrangement reduces the amount of shear stress variation across the adhesive. The modulus of rigidity is given by:

$$G = \frac{2b}{\pi(R_o^4 - R_i^4)} \left(\frac{T}{\theta} \right) \qquad \text{(Eq 41)}$$

where R_o and R_i are the outer and inner radii of the cylindrical specimen.

Measuring Surface Cleanliness. One important aspect of adhesives is to determine the degree of surface cleanliness. Krieger and Wilson (Ref 25) developed a technique that measures surface cleanliness by the success with which indium adheres onto another solid surface. The Section "Surface Considerations" in this Volume contains articles that describe the surface preparation of a variety of adherend materials, as well as analysis techniques.

Measuring Rheological Properties

In order to be able to relate stress to the strain rate of deformation in uncured polymers or adhesives, it is important to understand their rheological behavior. The most commonly used rheological material function is viscosity, which, in arbitrary steady shear flow, is defined as:

$$\eta = \frac{\tau_{ij}}{\dot{\gamma}} \qquad \text{(Eq 42)}$$

In most industrial applications, polymers exhibit a non-Newtonian shear thinning behavior, as shown in Fig. 13. Here, for practical ranges of processing deformation rates, the viscosity can be approximated by the power law model of Ostwald and de Waele:

$$\eta = m\dot{\gamma}^{n-1} \qquad \text{(Eq 43)}$$

where m and n are constants characteristic of each individual polymer or adhesive. Other important mechanical functions are the primary normal stress coefficient, ψ_1, and the secondary normal stress coefficient, ψ_2, which are defined as:

$$\psi_1 = \frac{\tau_{11} - \tau_{22}}{\dot{\gamma}^2} \qquad \text{(Eq 44)}$$

$$\psi_2 = \frac{\tau_{22} - \tau_{33}}{\dot{\gamma}^2} \qquad \text{(Eq 45)}$$

Both stress coefficients are functions of the strain rate of deformation, the primary being much larger than the secondary. It should be noted that for Newtonian fluids, ψ_1 and ψ_2 are both 0 and η is a constant. The material functions η, ψ_1, and ψ_2 completely determine the state of stress of a material in a steady shear flow, and they can be measured using various experimental setups, two of which are discussed below.

The capillary viscometer is the most widely used instrument for measuring viscosity. The technique is based on the steady laminar flow through a circular capillary or tube. The capillary rheometer is depicted in Fig. 14. In this apparatus the polymer melt is forced through a heated tube, usually at a constant flow rate, Q. From tube flow analysis, the shear rate at the wall, $\dot{\gamma}_w$, can easily be calculated:

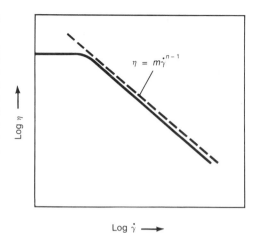

Fig. 13 Viscosity versus strain rate curve

$$\dot{\gamma}_w = \frac{1}{\tau_w^2} \frac{d}{d\tau_w} \left(\tau_w^3 \frac{\tau_w^3 Q}{\pi R^3} \right) \qquad \text{(Eq 46)}$$

where shear stress at the wall of the capillary, τ_w, can be related to the tube pressure drop:

$$\tau_w = \frac{R\Delta P}{2L} \qquad \text{(Eq 47)}$$

Once the shear rate and the shear stress at the wall are known, the viscosity can easily be calculated:

$$\eta(\dot{\gamma}_w) = \frac{\tau_w}{\dot{\gamma}_w} \qquad \text{(Eq 48)}$$

Although the expressions in Eq 46 to 48 are convenient for calculating the viscosity of polymer melts or adhesives, they do require numerical differentiation of the data. Additionally, one should be aware that there are end effects in the measurements and that no information on normal stress differences is obtained with this experiment. The end effects are corrected by taking measurements at different (L/D) ratios but at the same wall shear rates (Ref 26).

For free-flowing Newtonian liquids, either ASTM D 1084 (Saybolt) or D 2556 can be used. For thixotropic materials, ASTM D 2556 (Brookfield) is commonly used.

The cone-and-plate viscometer is often used when measuring the viscosity of non-flowing materials and the primary and secondary normal stress coefficient functions as a function of shear rate and temperature (Ref 26). The geometry of a cone-and-plate viscometer is shown in Fig. 15. Because the angle θ_0 is very small, typically less than 5°, the shear rate can be considered constant and is given by:

$$\dot{\gamma}_{\theta\phi} = \frac{\Omega}{\theta_0} \qquad \text{(Eq 49)}$$

where Ω is the angular velocity of the cone. The shear stress can also be considered constant and can be related to measured torque, T:

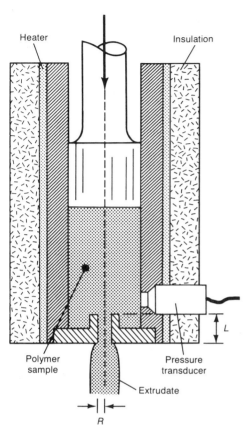

Fig. 14 Capillary viscometer

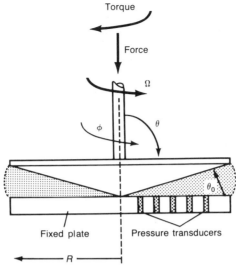

Fig. 15 Cone-and-plate viscometer

$$\tau_{\theta\phi} = \frac{3T}{2\pi R^3} \qquad \text{(Eq 50)}$$

The viscosity function can now be obtained from:

$$\eta(\dot{\gamma}_{\theta\phi}) = \frac{\tau_{\theta\phi}}{\dot{\gamma}_{\theta\phi}} \qquad \text{(Eq 51)}$$

The primary normal stress coefficient function can be calculated by measuring the force, F, required to maintain the cone in place and is given by:

$$\psi_1 = \frac{2F}{\pi R^2 \dot{\gamma}^2} \qquad \text{(Eq 52)}$$

Although it is also possible to determine the secondary stress coefficient function from the normal stress distribution across the plate, it is very difficult to obtain accurate data.

Rheological Properties During Cure. Thermosets may cure either by addition polymerization, as is the case with unsaturated polyesters, or by condensation polymerization, as with urethanes. One common way of measuring the kinetics of a reacting material is to use a differential scanning calorimeter (DSC). A DSC can control the temperature in a sample and at the same time can measure the rate of heat into or out of the sample. The curing reaction can be traced this way because it is frequently exothermic; that is, each time a

linkage is formed, a certain amount of chemical energy is converted into heat. The total heat per unit mass released from a sample is Q_t, and the degree of cure can be defined as:

$$C^* = \frac{Q}{Q_t} \qquad \text{(Eq 53)}$$

and the rate of heat liberated can be defined as:

$$\dot{Q} = Q_t \left(\frac{dC^*}{dt} \right) \qquad \text{(Eq 54)}$$

Several empirical models exist that define dC^*/dt (Ref 27). For example, for polyester resin, a model that represents the curing reaction fairly well could look like:

$$\frac{dC^*}{dt} = K_0 e^{-E/RT} (1 - C^*)^n C^{*n} \qquad \text{(Eq 55)}$$

where the constants K_0, E, and n are adjusted to fit experimental data. Previously, we discussed how the rheological properties of adhesives are a function of temperature and strain rate. This is not sufficient to describe reacting flows because the molecular weight of the material increases as the reaction proceeds and the viscosity increases with molecular weight. To better model the flow and processing of reacting materials, their viscosity must also be a function of degree of cure. As an example, polyurethanes, which seem to be Newtonian, obey an empirical constitutive equation:

$$\eta = \eta_0 e^{-E/RT} \left(\frac{C_g^*}{C_g^* - C^*} \right)^{b+C^*} \qquad \text{(Eq 56)}$$

where C_g^* represents the degree of cure at which gelation occurs and η_0, E, and b are adjusted to fit the experimental data. One can see that with this model the viscosity goes to infinity when C^* reaches C_g^*. For processing, the significance of the gel

point, C_g^*, is that no flow occurs after it is reached.

REFERENCES

1. W.A. Lees, *Adhesives in Engineering Design*, Springer-Verlag, 1984
2. A.J. Kinloch, *Adhesion and Adhesives: Science and Technology*, Chapman and Hall, 1987
3. J.J. Bikerman, *The Science of Adhesive Joints*, 2nd ed., Academic Press, 1968
4. K.W. Allen, *Adhesion 1*, Applied Science Publishers, 1977
5. R.J. Young, in *Structural Adhesives: Developments in Resins and Primers*, A.J. Kinloch, Ed., Elsevier, 1986
6. E. Passaglia and J.R. Knox, in *Engineering Design for Plastics*, E. Baer, Ed., Reinhold, 1964
7. L.E. Nielsen, *Mechanical Properties of Polymers and Composites*, Vol 1, Marcel Dekker, 1974
8. R.L. Thorkildsen, in *Engineering Design for Plastics*, E. Baer, Ed., Reinhold, 1964
9. R.E. Chambers, *Structural Plastics Design Manual*, ASCE Manuals and Reports on Engineering Practice 63, American Society of Civil Engineers, 1984
10. J.D. Ferry, *Viscoelastic Properties of Polymers*, 3rd ed., John Wiley & Sons, 1980
11. J.J. Aklonis, W.J. MacKnight, and M. Shen, *Introduction to Polymer Viscoelasticity*, Wiley-Interscience, 1972
12. T. Murayama, *Dynamic Mechanical Analysis of Polymeric Material*, Elsevier, 1978
13. M.W. Darlington and S. Turner, in *Creep of Engineering Materials*, C.D. Pomeroy, Ed., Mechanical Engineering Publications, 1978
14. W.N. Findley, J.S. Lai, and K. Onaran, *Creep and Relaxation of Nonlinear Viscoelastic Materials*, North-Holland, 1976
15. I.M. Ward, *Mechanical Properties of Solid Polymers*, 2nd ed., Wiley-Interscience, 1983
16. G.G. Trantina, *Polym. Eng. Sci.*, Vol 26, 1986, p 776
17. R.J. Crawford, *Plastics Engineering*, Pergamon, 1987
18. H. Stenzenberger, in *Structural Adhesives: Developments in Resins and Primers*, A.J. Kinloch, Ed., Elsevier, 1986
19. J.K. Gillham, in *Structural Adhesives: Developments in Resins and Primers*, A.J. Kinloch, Ed., Elsevier, 1986
20. F.R. Martin, in *Structural Adhesives: Developments in Resins and Primers*, A.J. Kinloch, Ed., Elsevier, 1986
21. E.W. Garnish, in *Structural Adhesives: Developments in Resins and Primers*, A.J. Kinloch, Ed., Elsevier, 1986
22. G.C. Stevens, in *Structural Adhesives:*

Developments in Resins and Primers, A.J. Kinloch, Ed., Elsevier, 1986

23. A.J. Kinloch, in *Structural Adhesives: Developments in Resins and Primers,* A.J. Kinloch, Ed., Elsevier, 1986

24. N.A. de Bruyne, in *Adhesion and Adhesives,* N.A. de Bruyne and R. Houwink, Ed., Elsevier, 1951

25. G.L. Krieger and G.J. Wilson, *Mater. Res. Stand.,* Vol 5, 1965, p 341

26. R.B. Bird, R.C. Armstrong, and O. Hassager, *Dynamics of Polymeric Liquids,* 2nd ed., John Wiley & Sons, 1987

27. M.R. Kamal and S. Sourour, *Polym. Eng. Sci.,* Vol 13, 1973, p 59

Failure Strength Tests and Their Limitations

R.D. Adams, University of Bristol (England)

THE SUCCESSFUL USE of adhesives in load-bearing engineering structures relies critically on a correctly balanced interaction between design, testing, and experience. First, it is necessary at the design stage to know what loads are likely to occur in practice and the lifetime pattern over which these loads will act. For instance, different designs are needed to withstand impact, fatigue, creep, environmental damage, or short-term on-off loading. Knowing the lifetime loading pattern, the designer can initially introduce allowables (safety factors) to reduce the problem to one of short-term loading. For instance, it is possible to design and test a joint for short-term loading (with the load increasing monotonically to failure over, for example, a minute) and to factor this by, for example, three or five times if fatigue loads are anticipated. Experience, rather than an understanding of the fundamental mechanics, is usually the guide regarding the factor to be applied. In short, the development of a successful design methodology can be achieved only by underpinning it with a sensible testing program. Knowing how to interpret the word sensible is the key.

During a development program, many of the components produced are tested to destruction. It is obviously impossible to do this to an entire production run. A proportion of the product chosen in accordance with some statistical analysis, is therefore selected for testing. The intensity of the testing will depend on the number of items being produced (10 or 10 000), their cost, and how critical it would be if failure were to occur.

Often it is not necessary to test a whole component or structure in order to ascertain the likely behavior under load. For instance, if aluminum sheet, which is made to recognized standards, is bonded into a structure, it is usually only the quality of the adhesive bond that needs to be assessed. Sometimes this can be done nondestructively on the component, but there are severe uncertainties about the success of most, if not all, current nondestructive tests for adhesive bonds (Ref 1). This is because,

while it is often possible to determine whether or not adhesive is present in a bond line, it is impossible to be certain that it is adequately attached to the substrates. The solution is therefore to test to destruction some joint coupons that are typical of those in the production process and that have gone through the same time-temperature-pressure profile as the actual component. The surfaces of the test coupons should be prepared in the same manner as those of the production components. These coupons will normally be of a standard form in accordance with industry, national, or international standards.

It is the goal of this article to describe some of these standard tests, to give a mathematical model where possible, and then to interpret the modes of failure. In this way, the results of these standard tests can be seen objectively, and the modes of failure as well as the loads at which they occur, can be correctly interpreted.

The most frequently used standard test is the single-lap joint shown in Fig. 1(a). Other versions of this joint exist, such as the double-lap joint, shown in Fig. 1(b), and various derivatives of these. In addition, butt joints (Fig. 1c), both in tension and torsion, are sometimes used to obtain an indication of material behavior, as are various peel joints, one of which is shown in Fig. 1(d). This article concentrates on the lap joints, rather than on the mechanics of all these joints, but appropriate references are provided to relevant texts. A fuller treatment of the subject can be found in Ref 2.

Lap Joints and Linear Elastic Algebraic Solutions

The average shear stress (τ_m) for the single-lap joint shown in Fig. 1(a) is given by:

$$\tau_m = P/bl \qquad \text{(Eq 1)}$$

where P is the applied load, b is the joint width, and l is the joint length. Many designers analyzing stresses find this simple equation, which is the definition of adhesive

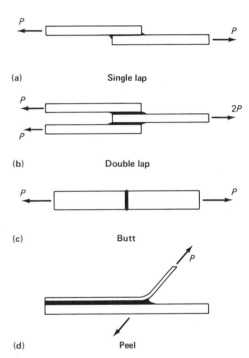

Fig. 1 Some typical bonded joints used in testing

shear strength used in standard tests such as ASTM D 1002, to be sufficient. The equation is, of course, rather simplistic and does not take into account the flexibility of the adhesive and the adherends.

Volkersen (Ref 3) tried to analyze the stresses in riveted panels, but could deal only with the case of an infinite number of tiny rivets, which effectively created a continuum, for which he developed his well-known shear lag equations. The continuum is, of course, identical to the case of an adhesive layer. Volkersen assumed that the adhesive deformed only in shear and that the adherends deformed only in tension. The equation he developed for the shear stress distribution at any position x along the length of a single-lap joint is:

$$\bar{\tau} = \frac{\omega \cosh \omega X}{2 \sinh \omega/2} + \left(\frac{\psi - 1}{\psi + 1} \right) \frac{\omega \sinh \omega X}{2 \cosh \omega/2}$$

$$\text{(Eq 2)}$$

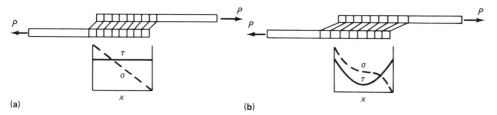

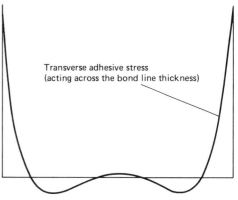

Fig. 2 Exaggerated deformations in loaded single-lap joint showing the adhesive shear stress, τ, and the adherend tensile stress, σ. (a) With rigid adherends. (b) With elastic adherends

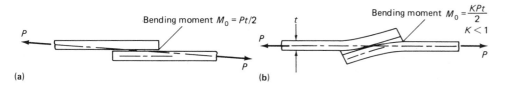

Fig. 3 The Goland and Reissner bending moment factor. (a) Undeformed joint. (b) Deformed joint

Fig. 4 Transverse (peel) stresses in a single-lap joint according to Goland and Reissner

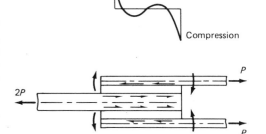

Fig. 5 Bending moments induced in the outer adherends of a double-lap joint, together with adhesive peel stresses

where $\omega^2 = (1 + \psi)\phi$, $\psi = t_1/t_2$, $\phi = Gl^2/(Et_1t_3)$, $X = x/l$, and $-\frac{1}{2} \leq X \leq \frac{1}{2}$. In addition, G = shear modulus of the adhesive, E = Young's modulus of the adherends, l = length of the bonded region, t_1 and t_2 = adherend thicknesses, and t_3 = adhesive thickness. For $\bar{\tau} = \tau_x/\tau_m$, τ_m = average applied shear stress, as defined in Eq 1.

This stress distribution is shown in Fig. 2(b) and should be compared with the constant distribution of shear stress from Eq 1 given in Fig. 2(a). In Fig. 2(a), the adherends are considered to be rigid, while the actual distribution in Fig. 2(b) depends on the joint dimensions and the moduli of the adherends and adhesive. In practice, the adherend tensile stress decreases progressively from a maximum at the loaded end to 0 at the unloaded end. The rate of change of adherend stress depends mainly on the stiffness of the adherends, as shown by Volkersen's equations. The shear strain (and stress) in the adhesive layer is at a maximum at each end and at a minimum in the middle. If the adherends are not of equal stiffness, the adhesive shear and adherend tensile stresses will be asymmetric.

But Volkersen ignored the fact that the directions of the two forces in Fig. 2 are not colinear, which creates a bending moment applied to the joint in addition to the in-plane tension. The adherends bend, and the rotation alters the direction of the load line in the region of the overlap in such a way that the joint displacements are no longer directly proportional to the applied load. Goland and Reissner (Ref 4) took this effect into account by using a bending moment factor (k), which relates the bending moment on the adherend at the end of the overlap (M_0) to the in-plane loading, by the relationship:

$$M_0 = kPt/2$$

where t is the adherend thickness (the thickness of the adhesive layer was neglected)

(see Fig. 3). If the load on the joint is very small, no rotation of the overlap takes place; thus, $M_0 = Pt/2$, and $k = 1.0$. As the load is increased, the overlap rotates, bringing the line of action of the load closer to the centerline of the adherends, thereby reducing the value of the bending moment factor. Goland and Reissner gave a similar shear stress distribution to that of Volkersen, but also gave the transverse (peel) stresses σ_y, in the adhesive layer as:

$$\sigma_y = \frac{\sigma t^2}{C^2 R_3}\left[\left(R_2\,\lambda^2\,\frac{k}{2} - \lambda k'\cosh\lambda\cos\lambda\right)\right.$$

$$\cosh\frac{\lambda x}{C}\cos\frac{\lambda x}{C}$$

$$+ \left(R_1\lambda^2\,\frac{k}{2} - \lambda k'\sinh\lambda\sin\lambda\right)$$

$$\left.\sinh\frac{\lambda x}{C}\sin\frac{\lambda x}{C}\right]$$

where σ = the mean tensile stress in the adherends
t = adherend thickness
x = position along glue line
$C = l/2$
$\lambda = C/t(6E_3t/Et_3)^{1/4}$
$k' = k(C/t)\,[3(1 - \nu^2)\,\sigma/E]^{1/2}$
$R_1 = \cosh\lambda\sin\lambda + \sinh\lambda\cos\lambda$
$R_2 = \sinh\lambda\cos\lambda - \cosh\lambda\sin\lambda$
$R_3 = (\sinh 2\lambda + \sin 2\lambda)/2$
ν = Poisson's ratio
The shear and peel stresses were both assumed to be uniform across the adhesive thickness, and it can also be seen from Fig. 4 that the maximum values of peel stress occur at the ends of the overlap.

A double-lap joint (shown in Fig. 1b) is essentially two single laps, back to back. This arrangement eliminates gross joint rotation because there is no net bending moment, but there is internal bending, as shown in Fig. 5.

Renton and Vinson (Ref 5) and Allman (Ref 6) subsequently produced analyses in which the adherends have been modeled to account for bending, shear, and normal stress. They also accounted for the 0 adhesive shear stress at the overlap ends where the adhesive has a free surface (law of complementary shears). Allman also allowed for a linear variation of the peel stress across the adhesive thickness, although the adhesive shear stress remains constant through the thickness.

Adams and Peppiatt (Ref 7) considered the three-dimensional stress state and showed the existence of shear stresses in the adhesive layer and direct stresses in the adherends acting across the width of the joint, these stresses being caused by Poisson's ratio strains in the adherends. They also allowed for *adherend* shear strains because, for many practical joints, Goland and Reissner's criteria for neglecting adherend shear strains are not applicable. This is particularly so with composite materials such as carbon fiber reinforced plastics (CFRP) because of the low in-plane shear modulus, G_{12}.

All the closed-form analyses show that the maximum adhesive stresses (shear and peel) always occur near the ends of the bond line. Unfortunately, they all predict that the stresses will decrease as the bond

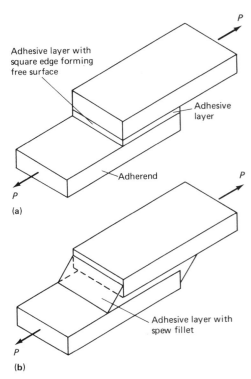

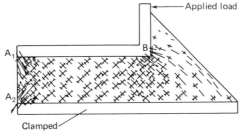

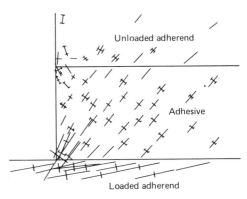

Fig. 6 Lap joints. (a) Adhesive layer with square edge. (b) Adhesive layer with fillet

Fig. 7 Principal stress pattern for silicone rubber model showing end effects. The lines represent principal stresses in magnitude and direction: A bar at the end of a line indicates compression. (This convention also is used in Fig. 8 and 9.)

Fig. 8 Finite-element prediction of the principal stress pattern at the end of a square-edge adhesive layer

line thickness increases. This is not the case in practice because bond line thickness only weakly affects lap joint strength (Ref 8). Also, it has been assumed that the adhesive layer ends in a square edge, as shown in Fig. 6(a). However, even a rectangular plate with shear loading on its two opposite sides experiences high tensile and compressive stresses at its corners because of the requirement that the direct and shear stresses acting on the free surface (edges) must be 0. Thus, in an adhesive layer with a square edge, similar tensile and compressive stresses must occur in the corners of this layer because of the free surface. However, practical adhesive joints rarely have a square edge. Instead, they are formed with a fillet (see Fig. 6b), which is squeezed out under pressure while the joint is being manufactured.

Photoelastic stress analysis has shown that the position and magnitude of the maximum stress depend on the edge shape. While the algebraic solutions correctly predict that the highest stresses are near the ends of the joint, they are unable to take into account the influence of the fillet. However, it is in just these regions of maximum stress where failure is bound to occur that the assumed boundary conditions of the algebraic theories are the least representative of reality. Some alternative must therefore be found, and this is best realized by obtaining a numerical solution to the stress problem that can allow for realistic geometry at the ends of the bond line. With modern computers, the finite-element method (FEM) is the preferred approach.

Lap Joints and FEM (Elastic Case)

The finite element method is now a well-established means for mathematically modeling stress (and many other) problems. Its advantage lies in the fact that the stresses in a body of almost any geometrical shape under load can be determined. The method is therefore capable of being used for analyzing an adhesive joint with a spew fillet. Figure 7 shows the principal stress pattern obtained by finite-element analysis applied to a silicone rubber model joint (Ref 8). The length and direction of the lines represent the magnitude of direction of the principal stresses at the centroid of each element. A bar at the end of a line implies a compressive (negative) principal stress. The presence of a fillet clearly causes the stress pattern to differ significantly from the pattern at the end with no fillet. At the points A1 and A2, high tensile and compressive stresses are shown, while the region away from the end is in pure shear, as shown by the equal and opposite principal stresses in the elements. The stresses in the fillet are predominantly tensile, with a maximum stress concentration at this end being at the sharp corner, B.

Figure 8 shows the stress pattern at the end of a square-edge adhesive layer in a typical aluminum-to-aluminum lap joint bonded with a structural epoxy adhesive. The highest tensile stress exists at the corner of the adhesive adjacent to the loaded adherend and represents a stress concentration of at least 10 times the average applied shear stress.

The influence of a fillet on the stress pattern is shown in Fig. 9, which is at the tension end of a double-lap joint. Even though only a very small triangular fillet, 0.5 mm (2 mils) high, was used, the stress system is very different from that of Fig. 8. Also, it can be seen that the adhesive at the ends of the adhesive layer and in the spew fillet is essentially subjected to a tensile load at about 45° to the axis of loading. The highest stresses occur near the corner of the unloaded adherend because the 90° corner introduces a stress-concentrating effect.

Because the maximum stress occurs within the fillet and not at or near the adhesive surface, it is unlikely that the approximation to the spew shape by the triangular fillet has a significant effect on the stress distribution.

Observation of the failure of aluminum-to-aluminum lap joints bonded with typical structural adhesives shows that cracks are formed approximately at right angles to the directions of the maximum principal stresses predicted by the elastic finite-element analysis. In general, these cracks run close to the corners of the adherends, as shown in Fig. 9. Thus, it can be proposed that failure in a lap joint is initiated by the high tensile stresses in the adhesive at the ends of the joint. Cohesive failure of the adhesive occurs in this manner in normal, well-bonded joints. Under further loading, the initial crack in the fillet is turned to run along (or close to) the adhesive-adherend interface. It meets a similar crack running in the opposite direction, resulting in the familiar fracture path shown in Fig. 10.

The essential advance afforded by finite-element stress analysis is that it allows for realistic boundaries (such as the fillet) and can readily accommodate variations in stress across the thickness. More detailed

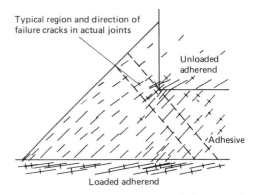

Fig. 9 Finite-element prediction of the principal stress pattern at the end of an adhesive layer with a 0.5 mm (0.02 in.) fillet

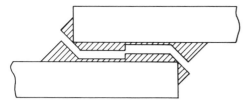

Fig. 10 Failure surfaces of a single-lap joint

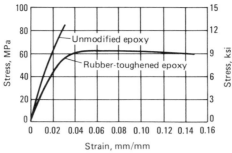

Fig. 11 Bulk uniaxial tensile stress-strain curves for an epoxy and a rubber-modified epoxy adhesive

Lap Joints and Plasticity

Complicated mathematics is required if the elastic stress situation in a single-lap joint is to be determined algebraically. However, modern structural adhesives can develop large plastic strains to failure. Thus, it is necessary to consider what happens to the stress and strain distribution when the adhesive can yield. Further, these new adhesives are so strong that the adherends, too, may be caused to yield.

Two opposite effects occur when the adherends yield. On the one hand, increased differential straining of the adherends causes the adhesive stresses to be increased, thus leading to premature failure. On the other hand, if the adherends are stressed to yield, they will more easily rotate under the effect of the noncolinear applied loads, causing the Goland and Reissner joint factor (k) to decrease more than if the adherends remained elastic, thus reducing the stresses. It is therefore necessary to investigate both adhesive plasticity and adherend plasticity, using either continuum mechanics or numerical (finite-element) techniques.

Figure 11 shows the tensile stress-strain behavior of two structural epoxy adhesives. One is an unmodified epoxy that is almost elastic to failure, whereas the other is toughened by a dispersion addition of rubber (rubber toughened), giving a lower strength but a considerably enhanced failure strain.

The main advocate of continuum mechanics is L.J. Hart-Smith, who produced an enormous amount of work on this subject on the Primary Adhesively Bonded Structure Technology (PABST) program under contract to the U.S. Air Force Flight Dynamics Laboratory. This method is a development of the shear lag analysis of Volkersen and the two theories of Goland and Reissner. While Hart-Smith has published many papers on the analysis of joints, the reader is referred to Ref 9 in particular.

The first difficulty is deciding how to characterize the adhesive. Hart-Smith chose an elastic-phase model such that the total area under the stress-strain curve was equal to that under the true stress-strain curve. He developed computer programs for analyzing various joints, both double-lap and single-lap, with equal or dissimilar adherends, for parallel, step, scarf, and double-strap configurations. Similar programs are available from the Engineering Sciences Data Unit (ESDU) for elastic and elastic-plastic calculations. The ESDU programs were developed from the work of P. Grant of British Aerospace and, again, are based on the work by Volkersen and by Goland and Reissner. In the EDSU/Grant method, the adhesive is modeled with the following conditions: If $\gamma < \gamma_e$, then $\tau = \gamma G_e$; but if $\gamma > \gamma_e$, then $\tau = \tau_e + [(\alpha\beta)/(\alpha + \beta)]$, where $\alpha = \gamma G_e - \tau_e$ and $\beta = \tau_{max} - \tau_e$. In these expressions, γ is the adhesive shear strain, γ_e is the shear strain at the onset of yield, τ is the shear stress, τ_{max} is the limiting shear stress, τ_e is the shear stress at the onset of yield, and G_e is the elastic shear modulus.

Hart-Smith equated the yield stress and the failure stress (elastic, perfectly plastic behavior) and said that failure occurs when the adhesive reaches its limiting plastic *shear strain*. It should be noted that the shear strain distribution is not simply a multiple of the low-load case because the plastic shear stress in the adhesive layer causes a distortion to the elastic shear lag theory.

Hart-Smith showed the effect of lengthening the overlap in a double-lap joint that had been loaded so that plastic deformation of the adhesive had occurred. There remained a large, low-stress region in the middle of the joint. Far from accepting this as structurally inefficient, the PABST/Hart-Smith design philosophy claims that it is essential in order to overcome the effects of creep at the ends if low-cycle creep/fatigue loading is encountered.

As mentioned already, the basic Hart-Smith approach entails neglecting the normal or peel stresses acting across the glue line. However, in practice, these may be the main contributors to joint failure, even in double-lap joints. Hart-Smith has recognized this possibility and, in one of his analyses, combined elastic peel stresses with plastic shear stresses. It is argued that, for sufficiently thin adherends, the peeling stresses are not important.

When nonlinear material properties are treated by a closed-form analysis such as Hart-Smith's, the limitation is: How tractable is a realistic mathematical model of the stress-strain curve within an algebraic solution? With the finite-element techniques, the limit becomes that of computing power. The high elastic stress and strain gradients at the ends of the adhesive layer need to be accommodated by using several eight-node quadrilateral elements across the thickness. It is then necessary to define yield (of the adhesive usually, but sometimes also of the adherend) and then to adopt a suitable failure criterion.

The yield behavior of many polymers, including epoxy resins, is dependent on both the hydrostatic (dilatational) and shear (deviatoric) stress components. Thus, there is a difference between the yield stresses in uniaxial tension and those in compression. For epoxy resins, ratios of compressive to tensile yield stresses of the order of 1.3 have been obtained by various authors. This behavior can be incorporated into the finite-element analysis by assuming a paraboloidal yield criterion of the form:

$$(\sigma_1 - \sigma_2)^2 + (\sigma_2 - \sigma_3)^2 + (\sigma_3 - \sigma_1)^2 + 2(Y_C - Y_T)(\sigma_1 + \sigma_2 + \sigma_3) = 2 Y_C Y_T$$

where σ_1, σ_2, and σ_3 are a combination of principal stresses causing yield, and Y_C and Y_T are the uniaxial compressive and tensile yield stresses (neglecting signs). This type of yield criterion applies to many amorphous polymers over a wide range of stress states. It should be noted that when $Y_C = Y_T$, the paraboloidal yield criterion reduces to the more familiar von Mises cylindrical criterion.

Recently, Harris and Adams (Ref 10) have produced results for lap joints that allow for the rotation of the adherends (large-displacement analysis) and the significant plastic deformation of both the adherends and the adhesive. References 11 and 12 are works by other authors who have thoroughly studied the yielding of adhesives and whose approach is influenced by, but remains slightly different from, the approach described in Ref 10.

The effects of adherend yielding were investigated by modeling the adherends with elastoplastic properties. The adhesive maximum principal stress distributions are shown in Fig. 12 for a case where the adherend properties correspond to a relatively low-strength aluminum alloy (a 0.2% proof stress of 110 MPa, or 16 ksi) and the adhesive is linearly elastic. Under the action of tension and bending, the adherends begin to yield at an applied load of approximately 1.5 kN (340 lbf). By 3 kN (675 lbf), the adherend plastic deformation has had two effects on the adhesive stresses. First, it has led to a reduction in the peak stress

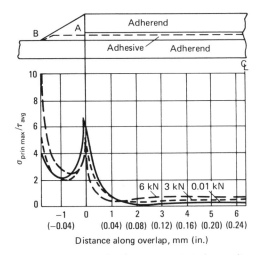

Fig. 12 Normalized maximum principal stress distributions along the adhesive layer with adherend yielding, at various applied loads (elastic adhesive)

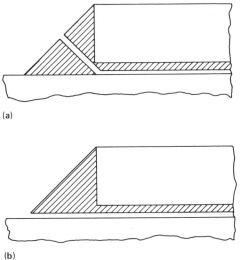

Fig. 13 Modes of fracture in lap joints with adhesive and adherend yielding. (a) Type I. (b) Type II

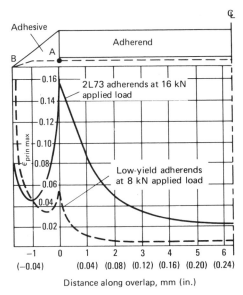

Fig. 14 The effect of adherend yielding on the maximum principal adhesive strain distribution for high-yield (2L73) and low-yield aluminum alloy adherends. Adhesive is a rubber-modified epoxy.

concentration at the end of the overlap, at point A, over and above that for the elastic case, as a result of the enhancement of the joint rotation. Second, the concentration of adhesive stress at the ends of the fillet, point B, has increased, because of the maximum adherend deformation occurring adjacent to this point, producing a localized increase in the differential shear effect. At an applied load of 6 kN (1350 lbf), further plastic deformation in the adherends has taken place, and the peak at point A is further reduced but, more significantly, the peak at B has dramatically increased: The stresses at this point are now the highest in the adhesive. It may be concluded, therefore, that when adherend plastic deformation takes place, the joint strength will be reduced and, at the same time, failure will no longer initiate from point A, but from point B. The two types of failure observed in practice, designated as types I and II, are shown in Fig. 13. Type II is indicative of adherend plastic deformation.

The other extreme of behavior occurs when a high-strength, low-ductility adherend is combined with a ductile adhesive. Figure 14 shows the computed principal strains in a carboxyl-terminated butadiene-acrylonitrile adhesive between two 2L73 high-strength aluminum adherends. This combination gave a joint strength of 16 kN (3600 lbf), but when ductile (low-yield-strength) adherends were used with the same adhesive, the joint strength was reduced by half, to 8 kN (1800 lbf).

The important point to be grasped here is that it is essential when doing lap-shear tests that the correct grade of adherend be used. If a low-yield adherend is substituted for a high-yield material, *the adhesive will appear to be weaker* on the basis of Eq 1 ($\tau = P/bl$), which is the normal ASTM 1002 test (Ref 2).

There is a complete range of adhesive behavior between the two extremes given here, in which the failure criterion may not be based on a single parameter, such as maximum principal stress or strain. Further investigation into failure criteria for adhesives is required in order to determine the limits to the strength of simple lap joints, and of more complex bonded structures under a variety of loads. It has been found that failure in ductile adhesives best correlates with the tensile strain to failure in a bulk specimen, whereas in brittle adhesives (less than 3% strain to failure), tensile stress gives the best correlation. Although adhesives in a single-lap joint appear to be loaded in shear, they do not fail in shear. It is strongly contended that few, if any, structural adhesives fail in shear but that they fail in tension when the principal stresses or strains reach some limiting value.

Local Stress Concentrations and Fracture Mechanics

A sharp corner or crack in theory causes an infinite stress (or strain) concentration, often referred to mathematically as a singularity. Because it is impossible in reality to have an infinite stress concentration, the science of fracture mechanics has evolved to explain why such infinite stresses and strains do not exist or, alternatively, if they do exist, why structures do not collapse under very light loads.

With bonded joints, it is necessary to be able to predict the strength of an uncracked adhesive. If there is no crack present, fracture mechanics requires that either an artificial or theoretical one must be introduced, which clouds the issue. Alternatively, as the author believes, fracture mechanics cannot be used. Also, if the adhesive becomes

plastic, the size of the so-called plastic zone in elastoplastic fracture mechanics quickly exceeds the adhesive thickness, such that the restraining effect of the adherends becomes crucial. Thus, the fracture mechanics approach, especially with ductile adhesives in thin bond lines, must be seen as intellectually suspect. Although fracture mechanics has been used by some authors in adhesive bond studies, there is little or no evidence of a joint having been designed on this basis.

In practice, the sharp corners at the ends of a lap joint are always rounded slightly during manufacture, such as by abrasion or by etching during surface treatment. Also, the adhesive is not linearly elastic to failure, but can yield. Finally, the author has observed experimentally that, despite the theoretical stress concentration at the corner of the unloaded adherend, the crack leading to failure does not always cut across the corner, but is usually about the same distance from it as the glue line thickness (for example, 0.2 mm, or 8 mils). This implies that whatever condition causes failure, it is not at the actual corner, but some distance from it. The influence of the geometry of the corners in adhesive joints has been studied by Adams and Harris (Ref 13). They showed that rounding the corner removed the singularity and produced a uniform stress field in this region because of the restraining effect of the relatively rigid adherend. When plastic energy density in the adhesive was analyzed, it was shown that the maximum value was generally away from the corner, thereby explaining why failure initiated away from the corner and not at it. Figure 15 shows results for a rigid corner with various degrees of rounding.

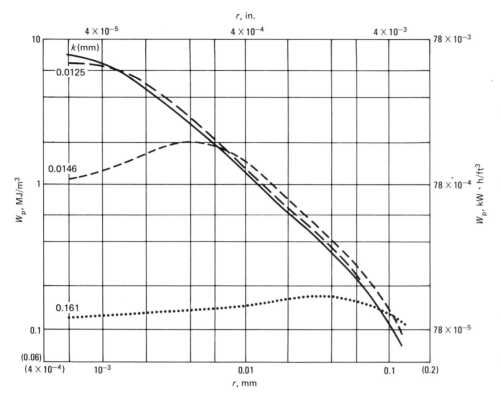

Fig. 15 Variation of the adhesive plastic energy density distribution with corner rounding for a rubber-toughened epoxy adhesive

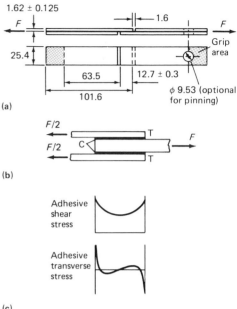

Fig. 16 (a) Double-lap joint as specified in ASTM D 3165. (b) True double-lap joint. (c) Stresses in a true double-lap joint. All units given in mm

The value of the parameter k in Fig. 15 is of the same order as the effective radius of the corner, and it should be noted that 0.01 mm (0.4 mil) is about as small a radius as can reasonably be created.

Using their mathematical model, Adams and Harris were able to predict the strengths of various aluminum-to-aluminum joints, bonded with rubber-toughened epoxy (see Fig. 11). These predictions are compared with their experimental results in Table 1, where it can be seen that excellent agreement has been obtained. Maximum adhesive principal strains were used as the failure criteria.

The essence of these results is that the shape of the ends of the joint is important in determining the measured lap-shear strength. In particular, there is a 50% increase in strength when a square-edge joint has a fillet added. Put another way, the variability in fillets inherent in joint preparation can cause scatter in the results, while the true adhesive properties are unchanged.

Table 1 Comparison of joint strength predictions with experimental values

	Predicted		Experimental (median)	
Joint type	kN	lbf	kN	lbf
Square edged	14.4	3240	15.9	3580
Filleted...............	21.2	4770	20.4	4590
Filleted and radiused...	25.3	5690	24.5	5510

Double-Lap Joints

Various forms of double-lap joints are used in adhesive testing. The classical form, shown in Fig. 1(b), is essentially two single laps laid back to back, as noted earlier.

A common form of the lap joint test that is used industrially is ASTM D 3165. The test configuration is shown in Fig. 16(a). Two sheets are bonded together, and the specimens are cut from the assembly. With film adhesives and a press, reasonably uniform bond line thickness can be achieved. However, liquid or paste adhesives tend to get squeezed out, leading to variable bond lines, and it is difficult with sheets to avoid entrapped air and to vent volatiles. The sheets are cut to form the test area. The main problem is to define the cut: too deep a cut weakens the adherends (especially laminates), and too shallow a cut causes the adherend to carry some load, and unknown but not insignificant damage is done to the adhesive in the highly stressed region at the joint ends. The ASTM D 3165 specimen is not a true double-lap joint, but is really a single-lap one with its adherends stiffened outside the test section.

In the true double-lap joint, shown in Fig. 16(b), together with its shear and peel stress distributions (Fig. 16c), it is difficult to make two bond lines with the same thickness. The great time and care needed to prepare good specimens is not rewarded by improved results.

Finally, the so-called thick-adherend test (ASTM D 3983) is a development of the test shown in Fig. 16(a). Differential straining is reduced by using thick adherends, and these are also stiff in bending. Unfortunately, the adhesive is in a far-from-uniform stress state, as is popularly believed, and there exist substantial transverse peel loads (see Renton in Ref 2 and Ref 14). In effect, the thick-adherend test was designed for low-modulus adhesives, but it has been used for structural epoxies (and similar high-modulus adhesives), thereby giving misleading results.

Butt Joints

Various tests have been developed that purport to give the mechanical properties of adhesives by using butt joints in tension or shear (it is almost impossible to fracture such joints in compression). ASTM has a variety of tests in this category (Ref 2), and the principles of these tests are briefly reviewed below.

Because some adhesives have a strong exothermic reaction upon curing, these can be made only in the thin film form, where the adherends can act as a heat sink. Thus, butt joints provide an apparently convenient, direct, and simple means of stressing these adhesives. In the elastic region, a simple, closed-form analysis is sufficient, although this is not as easy to apply when yield occurs. On the face of it, the stress distribution is simple. However, the problem of end effects remains. If joints are to be loaded to failure and if the failure stress is to mean anything, it must be the true

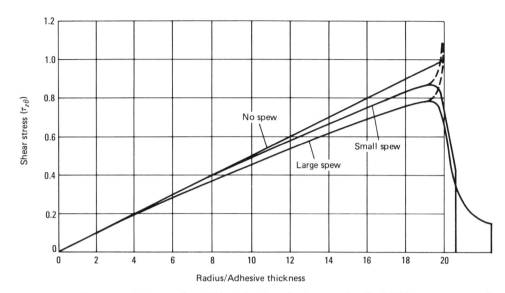

Fig. 17 Shear stress distributions for solid butt joints in torsion (aspect ratio: 40). Solid lines are $\tau_{z\theta}$ at $z = 0$, and dashed lines are $\tau_{z\theta}$ at $z = t/2$. Arbitrary units shown for shear stress

stress and not a convenient but misleading approximation made by taking the stress as the load divided by the area.

ASTM E 229 specifies a test in which an annular ring of adhesive is sheared between two thin cylinders, leading to the term napkin ring test. This form of specimen minimizes the variations of shear stress in the adhesive and has been used by various authors for measuring the shear strength of adhesives. The shear-stress distribution calculated from simple elasticity theory is:

$$\tau_{z\theta} = \frac{2Tr}{\pi(r_o^4 - r_i^4)}$$

where $\tau_{z\theta}$ is the shear stress at a radius (r) caused by an applied torque (T), while r_i and r_o are the inner and outer radii of the annulus. When bonding the specimen, it is usual for some of the adhesive to be squeezed out to form a spew fillet, and this may be expected to modify the simple, theoretical stress distribution. Over much of the joint, the fillet reduces the stresses slightly. Unfortunately, at the outer and inner circumferences, adjacent to the substrate, there is a large stress concentration (Ref 15). Removing the fillet minimizes the stress concentration, and this is recommended in the ASTM E 229 test, albeit for other reasons. Adams and Coppendale (Ref 16) estimated in their investigation that the fillet gave a stress increase of about 1.8 times at the edge, compared with the situation with no fillet. It is therefore important with napkin ring tests (and similar shear tests) to remove the fillet before testing. Otherwise, misleading strength results will be given.

Adams and Coppendale (Ref 16) recommend the use of a solid (that is, not annular) butt joint. This is easier to make, and only the outer fillet need be removed, which is

fairly easy to do. A simple graphic correction of the torque-twist curve can be used to obtain true shear stress versus shear strain, even when the adhesive yields. Figure 17 shows the variation of shear stress with radius for a solid butt joint in torsion.

Butt joints for testing adhesives in tension also are usually designed with a circular cross section to facilitate manufacture and to maintain symmetry. In this case, it would appear to make little difference whether they are annular or solid. In a butt joint subjected to a tensile load, the adhesive is restrained in the radial and circumferential directions by the adherends. In the absence of this restraint, the adhesive would tend to contract radially with respect to the adherends due to its much lower modulus. The presence of the adherends has the effect of inducing radial and circumferential stresses in the adhesive, thereby increasing the stiffness of the joint. The simplest analysis makes the assumption that the radial and circumferential strains in the adherend and the adhesive are 0, in which case the radial and circumferential stresses are given by:

$$\sigma_r = \sigma_\theta = \left(\frac{\nu}{1 - \nu}\right)\sigma_z$$

where ν is Poisson's ratio of the adhesive and σ_z is the nominal applied axial stress. The apparent Young's modulus (E'—) (defined as the applied axial stress divided by the axial strain) is given by:

$$E'— = \frac{\sigma_z}{\epsilon_z} = \frac{E(1 - \nu)}{(1 + \nu)(1 - 2\nu)}$$

where E is Young's modulus of the adhesive. Taking the adherend strains into account by assuming that the radial strain in the adhesive is equal to the Poisson's ratio strain in the adherends gives:

$$\epsilon_r = \epsilon_\theta = -\left(\frac{\nu_a}{E_a}\right)\sigma_z$$

where ν_a and E_a are the Poisson's ratio and Young's modulus of the adherends. The radial and circumferential stresses now become:

$$\sigma_r = \sigma_\theta = \left(\nu - \frac{E\nu_a}{E_a}\right)\left(\frac{\sigma_z}{1 - \nu}\right)$$

These simple analyses ignore the requirement that the radial stress must be 0 at the free boundary of the adhesive, which implies that shear stresses must exist in the adhesive layer.

Using the same finite-element techniques as for torsion, Adams *et al.* (Ref 15) analyzed butt joints in tension to elucidate the various radial, circumferential, and axial stress distributions. Figure 18 shows their results for the stresses in a butt joint without a fillet. This shows major stress concentrations near the outer edges. It is also in a region of high, complex stresses. The presence of a fillet tends to reduce this stress concentration, in contrast to the case in shear discussed above.

Further work by Adams and Coppendale (Ref 16) considered what happens when the adhesive yields. They used a pressure-dependent yield criterion (that is, one in which the von Mises deviatoric-stress-causing yield increases linearly with hydrostatic pressure). They postulated that, under compressive loads, the adhesive in a butt joint should not yield, but if yielding does begin to occur, then because Poisson's ratio will increase towards 0.5, the yield zone will be very restricted. Tensile butt joints appear to provide a simple means for establishing the direct stress-strain and tensile strength properties of an adhesive. However, it is difficult to measure the small deflections involved (compensating also for adherend strains) (Ref 2), and the large stress concentrations at the periphery of the joint imply that any strength values thus obtained are likely to be meaningless. Other problems, such as avoiding bending, only compound the difficulties. It is therefore strongly recommended that axial butt joint tests for adhesive properties be avoided.

Peel Tests

Various forms of peel tests are used to assess the performance of structural adhesives. In effect, this form of test deliberately stresses the adhesive in a very small region, subjecting it to a large tensile stress, although there is usually a complex stress situation present. The stress situation is not easily assessed, and peel is normally used to compare adhesives rather than to measure their properties.

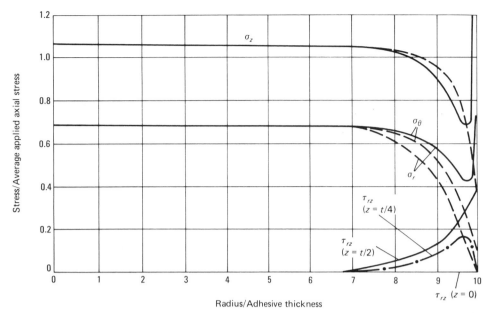

Fig. 18 Stress distributions for solid butt joint in tension (no spew fillet, aspect ratio: 20). Solid lines are $z = t/2$, and dashed lines are $z = 0$.

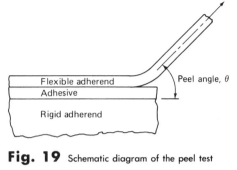

Fig. 19 Schematic diagram of the peel test

The basis of the peel test is shown in Fig. 19. A thin substrate, such as aluminum alloy, is bonded to a thick substrate that is itself clamped (or bonded) to a rigid support. A force (P) is applied at some angle (θ) and the strip is peeled off. In testing, care has to be taken to keep the angle (θ) constant, which is not easy.

Crocombe and Adams (Ref 17) analyzed this test, allowing for both the adhesive and the strip to become plastic. They showed that the main stresses in the adhesive would drive a crack toward the interface with the flexible adherend. The actual transverse stress at the point where fracture was expected to occur was a function of the peel angle, as shown in Fig. 20. It should be noted that these are average stresses across the adhesive thickness and that much larger values are found near the flexible adherend.

Figure 21 shows various forms of peel test. The most commonly used is the T-peel test (effectively, a 90° peel test), as specified by ASTM D 1876 (Fig. 21a). It has the

advantage that the two ends can be gripped in the jaws of a testing machine and pulled without complications, other than the necessity to ensure that the unfractured portion remains at 90° to the pulling direction. The results are usually expressed as load divided by the strip width (N/mm, or lbf/in.). In Fig. 21(b), the strip is bent backwards 180° and peeled as shown. This test is useful only for very thin materials.

The climbing drum test (Fig. 21c) is used for peeling the face plates from honeycomb sandwich constructions and is specified in ASTM D 1781. The floating roller arrangement shown in Fig. 21(d) is described in ASTM D 3167. In theory, the arc at which the skin is peeled away in both the climbing drum and the floating roller tests is controlled by the roller geometry, but in practice the conformity is not as certain.

On the whole, the floating roller test gives the most reputable results, the climbing drum test is preferred for honeycomb, and the T-peel test is the easiest to carry out.

As a further consideration of peel tests, it is necessary to look at the test often referred to as the Boeing wedge test, which is now specified in ASTM D 3762 (Fig. 22). The stresses are generated by forcing the wedge between the two bonded plates, forming peel (cleavage) stresses across the bond line. This test is not normally used for

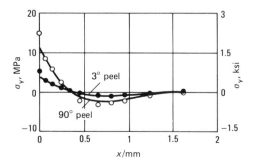

Fig. 20 Variation of adhesive transverse direct stress σ_y with distance x from the bond end of a peel test

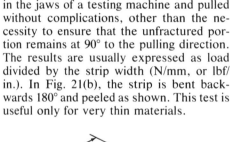

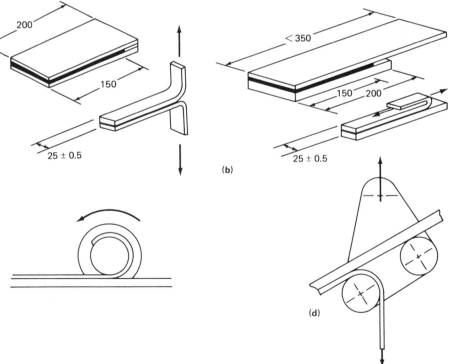

Fig. 21 Various forms of the peel test. (a) 90° T peel. (b) 180° peel. (c) Climbing drum. (d) Floating roller. All dimensions given in mm

Fig. 22 Boeing wedge test

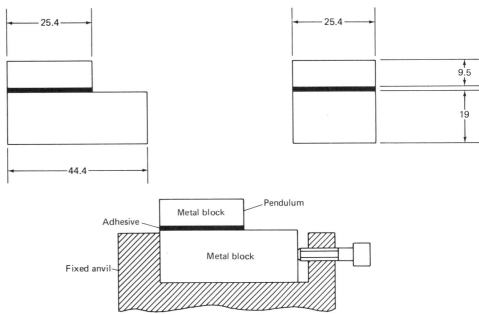

Fig. 23 ASTM standard impact test (ASTM D 950). All dimensions given in mm

quantitative or short-term testing, but is a cheap and sensitive way of investigating environmental effects and differences in surface preparation. The test yields information on crack growth rate and can indicate whether the fracture is between the adhesive and the substrate or wholly within the adhesive.

Dynamic Tests. Normal "static" tests usually take about a minute to fracture. However, sometimes designers need information on creep, fatigue, and impact loading. This testing may be considered "dynamic" because the load is applied such that its rate of application (slow, quick, or cyclic) is an important test parameter.

A variety of ASTM creep tests are available, such as D 1780, D 2293, D 2294, D 2918, and D 2919. These are mainly lap joint tests. In fatigue, ASTM D 3160 specifies a lap-shear specimen similar to that recommended for single-lap tests (D 1002).

Of recent interest is the ability of an adhesive to withstand shock loads, especially in bonded constructions for road vehicles. It is well known that most polymers have rate-dependent mechanical properties. Increasing the strain rate usually increases the modulus and fracture strength and decreases the fracture strain. The principal standard test of impact is ASTM D 950, shown in Fig. 23. The bonded metal block is struck by a pendulum traveling at 3.5 m/s (11.5 ft/s), and the energy absorbed is measured as in a standard Izod or Charpy pendulum test, with a correction being made for the kinetic energy imparted to the block. This is not a very scientific test, and ASTM recommends that it be used only for comparison purposes. Indeed, the energy absorbed depends on the stiffness to ground of the lower (fixed) anvil.

Table 2 Mechanical properties of carbon fiber reinforced plastic adherends

Property	Value
Longitudinal Young's modulus (E_1), GPa (10^6 psi)	135 (19.6)
Transverse Young's modulus (E_2), GPa (10^6 psi)	7 (1.0)
Interlaminar (longitudinal) shear modulus (G_{12}), GPa (10^6 psi)	4.5 (0.65)
Longitudinal tensile strength (σ_1), MPa (ksi)	1550 (225)
Transverse tensile strength (σ_2), MPa (ksi)	50 (7.25)
Interlaminar shear strength (τ_{12}), MPa (ksi)	110 (16.0)
Longitudinal Poisson's ratio (ν_{12})	0.3
Transverse Poisson's ratio (ν_{23})	0.3

In effect, there is no satisfactory standard method for impact testing. It is possible to obtain properly instrumented impact testers that can be customized to a given application. The results must be interpreted carefully, and a good dynamicist should be consulted. Harris and Adams (Ref 18) measured the static and impact properties of a variety of adhesives and used finite-element analysis to predict the corresponding static and impact strengths of lap joints made according to ASTM D 1002. They obtained excellent agreement between the theoretical and experimental results, thereby vindicating the claim that a properly instrumented impact test can give sensible answers.

Problems Associated With Testing Composites

Advanced composite materials, such as carbon/graphite or glass fiber reinforced plastics, are normally bonded rather than riveted or bolted. There is therefore a need to test bonded composites during the development phase, as well as during production. The essential feature of aligned composites is that the material is anisotropic locally and, even for laminates with fibers in various directions in different layers, there is often a degree of global anisotropy. Worse, unsymmetric or antisymmetric layering can lead to coupling between stretching, bending, and twisting. The adhesive bond layer may therefore be subjected to unexpected stresses, which may lead to premature failure.

Because metals are mostly isotropic, small stresses through the thickness or across the width of an adherend are rarely a cause for concern. Although composites are strong and stiff in the fiber direction, however, they are weak and compliant transverse to the fiber direction. These transverse properties are of the same order as those of the matrix. Table 2 gives some typical properties of unidirectional composites. From this table, it can be seen that if a bonded joint in which one or more of the adherends is a composite experiences transverse loading, there is a strong likelihood of the composite failing transversely in tension before the adhesion fails. Such failures are likely to be wrongly interpreted as adhesive failures, whereas in fact they are *joint assembly* failures. The author has recently reviewed the stresses in joints involving composite adherends (Ref 19). In this work, it was shown that premature transverse fracture of the composite may lead to a fivefold reduction in joint strength, compared with a sample of the same dimensions and materials but in which due attention was paid to the design of the joint ends.

Recommendations

It can be seen from the discussion of the mechanics of various forms of joint that the interpretation of results requires care. It is an old axiom in materials and structural testing that the test method should not drive the results, but should be a passive means of correctly simulating the loading that will be experienced in practice. In many cases, this will be a desired, but unachievable aim. Most testing will be carried out on single specimens that are easy to test. The important point is that the numbers obtained from these tests are mainly comparative and do not give direct material properties. As a simple example, no one should believe that the lap-shear test will give the true shear strength of an adhesive. Equally, the tensile butt joint does not define the true strength (or modulus) of a joint.

If researchers and designers need true strength and modulus values, great care must be taken in carrying out the tests and interpreting the results, because most of the tests involving adhesive joints do not give directly interpretable values. It is suggested that the best test for determining the mechanical properties of an adhesive is to make specimens from bulk samples, ensuring that these reach the same state of cure that the investigator assumes to exist. For some adhesives, exothermic reactions upon curing may cause a runaway temperature increase, leading to overheating and damage. In addition, in most bulk specimens, the careful avoidance of air bubbles is important.

In the marketplace, one often hears the Latin phrase *caveat emptor*, buyer beware. For the testing of adhesives, perhaps *caveat investigator* would apply!

REFERENCES

1. R.D. Adams and P. Cawley, A Review of Defect Types and Nondestructive Testing Techniques for Composites and Bonded Joints, Vol 21, *NDT Int.*, 1988, p 208-222
2. R.D. Adams and W.C. Wake, *Structural Adhesive Joints in Engineering*, Applied Science Publishers, 1986
3. O. Volkersen, Die Nietkraftverteilung in Zugbeanspruchten mit Konstanten Laschenquerschritten, *Luftfahrtforschung*, Vol 15, 1938, p 41-47
4. M. Goland and E. Reissner, Stresses in Cemented Joints, *J. Appl. Mech. (Trans. ASME)*, Vol 66, 1944, p A17-A27
5. W.J. Renton and J.R. Vinson, The Efficient Design of Adhesive Bonded Joints, *J. Adhes.*, Vol 7, 1975, p 175-193
6. D.J. Allman, A Theory for the Elastic Stresses in Adhesive-Bonded Lap Joints, *Q. J. Mech. Appl. Math.*, Vol 30, 1977, p 415-436
7. R.D. Adams and N.A. Peppiatt, Effect of Poisson's Ratio Strains in Adherends on Stresses of an Idealised Lap Joint, *J. Strain Anal.*, Vol 8, 1973, p 134-139
8. R.D. Adams and N.A. Peppiatt, Stress Analysis of Adhesive-Bonded Lap Joints, *J. Strain Anal.*, Vol 9, 1974, p 185-196
9. L.J. Hart-Smith, Stress Analysis: A Continuum Mechanics Approach, in *Developments in Adhesives*, Vol 2, A.J. Kinloch, Ed., Applied Science Publishers, 1981
10. J.A. Harris and R.D. Adams, Strength Prediction of Bonded Single Lap Joints by Non-linear Finite Element Methods, *Int. J. Adhes. Adhes.*, Vol 4, 1984, p 65-78
11. P. Czarnocki and K. Piekarski, *Int. J. Adhes. Adhes.*, Vol 6, 1986, p 93
12. P. Czarnocki and K. Piekarski, Yielding of Adhesives, *J. Mater. Sci.*, 1986
13. R.D. Adams and J.A. Harris, The Influence of Local Geometry on the Strength of Adhesive Joints, *Int. J. Adhes. Adhes.*, Vol 7, 1987, p 69-80
14. W.J. Renton, The Symmetric Lap-Shear Test—What Good Is It?, *Exp. Mech.*, Vol 33, 1976, p 409-415
15. R.D. Adams, J. Coppendale, and N.A. Peppiatt, Stress Analysis of Axisymmetric Butt Joints Loaded in Torsion and Tension, *J. Strain Anal.*, Vol 13, 1978, p 1-10
16. R.D. Adams and J. Coppendale, The Elastic Moduli of Structural Adhesives, in *Adhesion 1*, K.W. Allen, Ed., Applied Science Publishers, 1977, p 1-17
17. A.D. Crocombe and R.D. Adams, Peel Analysis Using the Finite Element Method, *J. Adhes.*, Vol 12, 1981, p 127-139
18. A.J. Harris and R.D. Adams, An Assessment of the Impact Performance of Bonded Joints for Use in High Energy Absorbing Structures, in *Proc. Inst. Mech. Eng.*, Vol 199, No. C2, 1985, p 121-131
19. R.D. Adams, Strength Predictions for Lap Joints, Especially With Composite Adherends—A Review, *J. Adhes.*, Vol 30, 1989, p 219-242

Fracture Testing and Failure Analysis

Kenneth M. Liechti, Engineering Mechanics Research Laboratory, Department of Aerospace Engineering
and Engineering Mechanics, University of Texas at Austin

STRUCTURES that are designed for stress levels below the yield or fatigue strengths of their constituent materials have long been observed to experience material failure by cracking. In many cases, the failure could be traced to stress concentrations, preexisting flaws, or some previously generated localized damage. These observations led to the conclusion that structures could fail at relatively low levels of applied load due to the presence of cracks. Consequently, the art of fracture mechanics was developed for determining the resistance of materials to the growth of cracks, as well as for designing damage-tolerant structures.

Because the bondline of adhesively bonded joints may contain stress concentrations, manufacturing flaws, and defects, fracture mechanics has been applied to the failure analysis of adhesively bonded joints. In adhesively bonded joints, cracks can develop in the adhesive (cohesive failure) or along the interface between the adhesive and an adherend (adhesive failure). This article will provide background information (in the section "Concepts" below) and then discuss specific details of specimen geometries and test procedures.

Concepts

Most of the current fracture mechanics practice in the testing and analysis of adhesives and the design of adhesively bonded joints is limited to linear elastic fracture mechanics methods. The development of the background material presented here will therefore be similarly constrained. Historically, fracture mechanics developed around various criteria for crack initiation and propagation. The two concepts that will be examined here are based on stress distributions around cracks and energy balances.

Crack Tip Stress Analysis. If an asymptotic analysis is conducted to determine the stress distribution in the region surrounding a crack tip, it can be shown (Ref 1) that the stress state is given, with reference to Fig. 1, as:

$$\begin{Bmatrix} \sigma_{11} \\ \sigma_{22} \\ \sigma_{12} \end{Bmatrix} = \frac{K_I}{\sqrt{2\pi r}} \cos \theta/2 \begin{Bmatrix} 1 - \sin \theta/2 \sin 3\theta/2 \\ \sin \theta/2 \cos 3\theta/2 \\ 1 + \sin \theta/2 \sin 3\theta/2 \end{Bmatrix}$$

$$+ \frac{K_{II}}{\sqrt{2\pi r}} \begin{Bmatrix} -\sin \theta/2(2 + \cos \theta/2 \cos 3\theta/2) \\ \cos \theta/2(1 - \sin \theta/2 \sin 3\theta/2) \\ \sin \theta/2 \cos \theta/2 \cos 3\theta/2 \end{Bmatrix}$$
(Eq 1)

The solution was obtained for a planar body under the action of in-plane loads and by enforcing the stress-free condition of the crack faces ($\theta = \pm\pi$). The stress-free condition and the sharp crack lead to the square root singular and angular terms in the solution, which are the same for all cracks. The coefficients K_I and K_{II} are known as the mode I and II stress-intensity factors, respectively. They reflect symmetric (tensile opening) and asymmetric (in-plane shear sliding) components, respectively; the fracture mode depends on the nature of the applied load (Fig. 2). The stress-intensity factors in Eq 1 are undetermined from the asymptotic analysis. They depend on the globally applied loads and the complete geometry of the configuration, and thus require a separate stress analysis. However, the important consequence of the asymptotic analysis is that the stress-intensity factor distinguishes the local crack tip stress distribution from one configuration to another. Alternatively, it can be said that when two cracked configurations have the same stress-intensity factor, the stress distributions are the same in the vicinity of the crack tips.

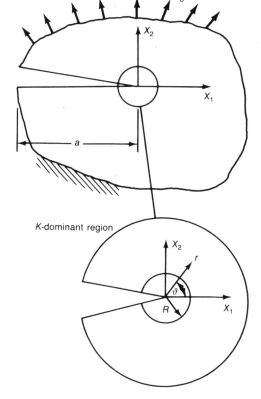

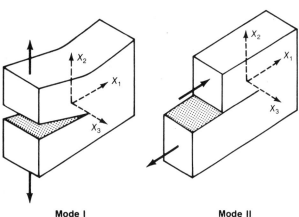

Fig. 1 Crack tip coordinate system

Fig. 2 Fracture modes

The mode I and II stress-intensity factors are defined as:

$$K_I = \lim_{r \to 0} \{(2\pi r)^{1/2} \sigma_{22}|_{\theta=0}\} \qquad \text{(Eq 2)}$$

and

$$K_{II} = \lim_{r \to 0} \{(2\pi r)^{1/2} \sigma_{12}|_{\theta=0}\} \qquad \text{(Eq 3)}$$

and are generally a function of the applied load (σ) and crack length (a):

$$K = K(\sigma, a) \qquad \text{(Eq 4)}$$

The in-plane displacement components around the crack tip are given by:

$$\begin{Bmatrix} u_1 \\ u_2 \end{Bmatrix} = \frac{K_I}{2\mu} \sqrt{\frac{r}{2\pi}} \begin{Bmatrix} \cos \theta/2(\kappa - 1 + 2 \sin^2 \theta/2) \\ \sin \theta/2(\kappa + 1 - 2 \cos^2 \theta/2) \end{Bmatrix}$$
$$+ \frac{K_{II}}{2\mu} \sqrt{\frac{r}{2\pi}} \begin{Bmatrix} \sin \theta/2(\kappa + 1 + 2 \cos^2 \theta/2) \\ -\cos \theta/2(\kappa - 1 - 2 \sin^2 \theta/2) \end{Bmatrix}$$
$$\text{(Eq 5)}$$

where μ is the shear modulus, $\kappa = 3 - 4\nu$ is the plane strain, $\kappa = (3 - \nu)/(1 + \nu)$ is the plane stress, and ν is the Poisson's ratio of the material.

For antiplane shear, the mode III stress-intensity factor is introduced to obtain near-tip stresses and displacements:

$$\begin{Bmatrix} \sigma_{31} \\ \sigma_{32} \end{Bmatrix} = \frac{K_{III}}{\sqrt{2\pi r}} \begin{Bmatrix} -\sin \theta/2 \\ \cos \theta/2 \end{Bmatrix} \qquad \text{(Eq 6)}$$

$$u_3 = \frac{2K_{III}}{\mu} \sqrt{\frac{r}{2\pi}} \sin \theta/2 \qquad \text{(Eq 7)}$$

$$K_{III} = \lim_{r \to 0} \{(2\pi r)^{1/2} \sigma_{32}|_{\theta=0}\} \qquad \text{(Eq 8)}$$

Again, it is the stress-intensity factor that distinguishes the crack tip stress distribution from one loading and crack length to another.

A natural consequence of the asymptotic solutions given above is that the stress-intensity factor can be used as a crack initiation criterion. In other words, if a crack initiation experiment had been conducted on a given material with measurements of the applied load and crack length so that the associated critical value of the stress-intensity factor could be determined, then crack initiation in a component made of the same material, but having a different geometry and/or loading, would be expected to occur at the same critical value of the stress-intensity factor that was noted in the first experiment. The crack initiation criterion can be formally expressed as:

$$K(\sigma, a) = K_c \qquad \text{(Eq 9)}$$

where K_c is the critical value of the total stress-intensity factor K and serves as a measure of the resistance of the material to crack initiation. The coefficient K_c is also known as the fracture toughness of the material.

For the purposes of this discussion K_c is a "generic" fracture toughness under any mode or combination of fracture modes. Mixed-mode fracture in adhesively bonded joints is an active research area and will be discussed in more detail later in this article.

The elasticity solutions for the stresses (Eq 1 and 6) suggest that the stresses at the crack tip are infinite. In practice, the stresses exceed the yield strength of the material, and zones of damaged and inelastically deformed material surround the crack tip. The damage can take the form of voids, crazes, and so on, whereas the inelastic deformations can be due to plasticity, viscoelasticity, or viscoplasticity. The extent of the plastic zone (R) under mode I conditions can be obtained by considering σ_{22} ahead of the crack ($\theta = 0$) and equating it to the tensile yield strength of the material (σ_y). Equation 1 indicates that:

$$R = \frac{1}{2\pi} \left(\frac{K_I}{\sigma_y}\right)^2 \qquad \text{(Eq 10)}$$

Within this region, the linear elastic solution is invalid. However, if the region $r < R$ is much smaller than the region over which the asymptotic elastic solution dominates, then the stress-intensity factor can still be used to characterize the stress distribution that controls crack initiation.

There are a number of situations where crack growth can occur slowly at stress-intensity factor values below the toughness of the material (K_c), which marks the onset of catastrophic rapid growth. This subcritical growth arises under cyclic (fatigue) or static loadings, when the time-dependent behavior of the material is important; it can also arise from environmental effects. Under these conditions, resistance to crack growth is characterized by correlations of crack growth rates with the stress-intensity factor as shown schematically in Fig. 3. Alternatively, it can be said that for fatigue growth:

$$\frac{da}{dN} = f(K) \qquad \text{(Eq 11)}$$

represents the resistance to fatigue crack propagation, where N is the number of cycles. This material property [the specific form of the function $f(K)$] can then be used to predict fatigue crack growth by integrating Eq 11 so that:

$$N = \int_{a_0}^{a} \frac{da}{f[K(\sigma, a)]} \qquad \text{(Eq 12)}$$

where the stress-intensity factor in Eq 12 is obtained from a stress analysis of the fatigue crack path in the component of interest; that is, $K(\sigma, a)$ is known. The initial crack length (a_0) may correspond to detectable flaw sizes or some other convenient

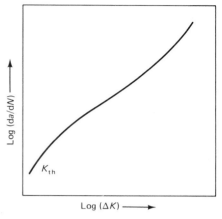

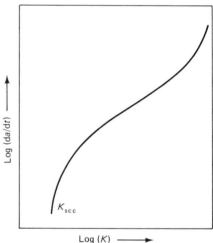

Fig. 3 Representations of the resistance to subcritical crack growth

scale. In the case of time-dependent growth due to viscoelastic effects, environmental effects, or a combination of the two, this correlation can be made:

$$\frac{da}{dt} = g(K) \qquad \text{(Eq 13)}$$

Equation 13 can be integrated to predict the crack growth history so that inspection frequencies and the probable lifetime of the component can be established.

The procedures outlined above can be classified as a finite life design approach. However, crack growth rates (particularly for fatigue crack growth) may be so high that it is better to preclude altogether the further growth of any preexisting flaw. This can be done by noting the threshold values shown in Fig. 3, where K_{th} and K_{scc} are the fatigue and environmentally assisted thresholds, respectively. For stress-intensity factors below these values no crack growth will occur. A more conservative design approach is therefore to consider the smallest preexisting flaw that can be detected in a structure and make sure that the stress-intensity factor associated with the maximum load is lower than K_{th} or K_{scc}.

A large variety of methods exist for determining the stress-intensity factor associated with a particular configuration (see Ref 2). When finite-element methods are used for the stress analysis of cracked components, stress-intensity factors can be extracted by examining the displacement solution near the crack and making use of Eq 5 and 7. Some finite-element codes make use of so-called hybrid elements that contain fracture parameters as degrees of freedom and therefore yield them directly as part of the solution without further postsolution processing. Many other codes use energy principles and the relationship between the stress-intensity factor and the energy-release-rate parameter.

Energy Concepts. The analysis of crack initiation in brittle materials was initially approached from an energy balance viewpoint (Ref 3). It was postulated that during an increment of crack extension (da) there can be no change in the total energy (E) of the cracked body. The total energy E was viewed as being composed of the potential energy of deformation (II) and the surface energy (S). Therefore, during crack extension:

$$dE = dII + dS = 0 \qquad \text{(Eq 14)}$$

The rate of change of potential energy with respect to crack extension (da) in a planar component of thickness (b) is defined as the energy-release rate (G):

$$G = \frac{-dII}{bda} \qquad \text{(Eq 15)}$$

If the surface-energy density is denoted by γ, then $dS = 2\gamma bda$ for the two increments of fracture surfaces formed during crack extension. Equations 14 and 15 then combine to yield:

$$G = 2\gamma \qquad \text{(Eq 16)}$$

as the criterion for crack extension in a brittle solid. Because G is derivable from a potential function, it is often referred to as a crack driving force. Therefore, Eq 16 represents the balance that is achieved at the point of crack initiation between the energy provided by the loaded component and the energy required for the creation of new surface or the fracture resistance. The fracture resistance is a characteristic of the material, whereas the energy-release rate depends on the loading and geometry of the crack component.

Perhaps the simplest and most common method of determining the energy-release rate is to consider the change in component compliance as a crack grows in it. With reference to Fig. 4, it can be seen that a cracked adhesive joint is being subjected to a constant force P, and Δ is the associated displacement through which P does work. The potential energy of the component is the difference between the strain energy (U) and the work done by the force:

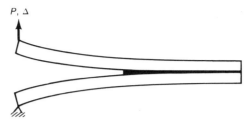

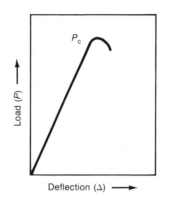

Fig. 4 Representation for energy release from cracked adhesive joint

$$II = U - P\Delta \qquad \text{(Eq 17)}$$

For a linearly elastic material, the strain energy $U = \frac{1}{2}P\Delta$, and Eq 15 and 17 yield:

$$G = \frac{P}{2b} \left(\frac{\partial \Delta}{\partial a} \right)_P \qquad \text{(Eq 18)}$$

The compliance (C) of a linearly elastic component is given by $C = \Delta/P$, and the energy-release rate in Eq 18 becomes:

$$G = \frac{P^2}{2b} \frac{dC}{da} \qquad \text{(Eq 19)}$$

The same expression can be obtained for a fixed-grip (displacement) loading and a compliant loading device. This simple result is quite powerful because it can be used to determine energy-release rates directly from load-displacement records of fracture toughness tests without the need for any further stress analysis. Furthermore, in the event that laminated beams are used for fracture tests, the energy-release-rate solutions can be obtained from relatively simple beam theory analyses.

Because the stress-intensity factor and energy-release rate approaches both apply to linearly elastic components, they can be related (Ref 1):

$$G = \frac{1}{E'} (K_I^2 + K_{II}^2) + \frac{1}{2\mu} K_{III}^2 \qquad \text{(Eq 20)}$$

where $E' = E/(1 - \nu^2)$ for plane strain and $E' = E$ for plane stress. Therefore, for a mode I crack at initiation:

$$G_c = K_c^2/E' \qquad \text{(Eq 21)}$$

When G_c is computed from K_c, as indicated in Eq 21, it is generally several orders of magnitude higher than the surface energy

2γ. Only for very brittle materials is the equality $G_c = 2\gamma$ preserved. In the tougher materials, most of the energy released at crack initiation is dissipated in inelastic deformation near the crack tip. Because of the equivalence between K and G noted above, the energy-release rate can also be used to characterize fatigue and environmentally assisted crack growth in a manner similar to that depicted in Fig. 3.

Adhesive Fracture Property Specimens

Over the years, quite a variety of specimen geometries have been developed for determining the fracture resistance of adhesives to cohesive or adhesive failure. The main reason for this variety is that adhesives are used for joining a wide range of materials under equally diverse loading conditions. Furthermore, because the adhesive layer is sometimes constrained between relatively stiff and tough materials, crack growth is constrained to be coplanar in mixed-mode loading. In homogeneous monolithic components it is often sufficient to determine the mode I fracture resistance because crack paths are usually free to orient themselves normal to the mode I load; however, the fracture resistance of adhesives *in situ* must be determined for all three modes and various mode combinations. This condition has contributed greatly to the variety of specimens available. The specimen geometries considered in this section are used mainly for structural adhesives with relatively stiff adherends. The fracture testing of adhesively bonded elastomers as well as that of sealants and thin films is discussed in subsequent sections.

Specimen Geometries. The number and variety of fracture property specimens that have been proposed for obtaining mode I fracture resistance of monolithic materials is small compared with the number of specimens with pure and mixed-mode capabilities that have been proposed for determining the fracture properties of adhesives and interfaces. Nevertheless, the specimens that have been devised over the years can be grouped into three categories: laminated-beam specimens, blister specimens, and prismatic specimens. The laminated-beam specimens are simplest and provide a surprisingly wide range of fracture mode mixtures. The blister specimens are useful for thin films or for examining environmental effects on crack growth. The prismatic specimens can be useful for compact geometries and for using biaxial loading to obtain a wide and easily varied range of fracture mode mixtures.

Laminated-Beam Specimens. The specimens in this class consist of two adherends that are not necessarily of the same material or height. The out-of-plane thickness is generally of the same order as the specimen

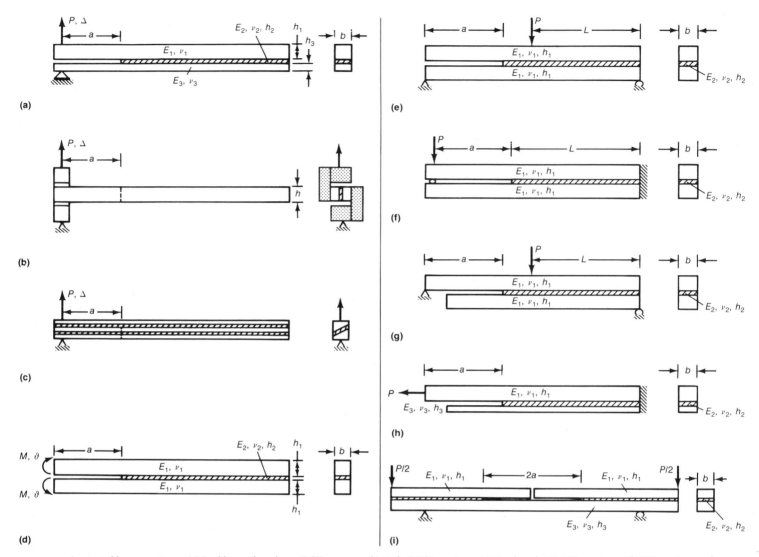

Fig. 5 Laminated-beam specimens. (a) Double-cantilever-beam (DCB) specimen. (b) Mode III DCB specimen. (c) Mixed-mode I-III DCB specimen. (d) DCB specimen under pure bending. (e) End-notched-flexure specimen. (f) End-notched-cantilever specimen. (g) Mixed-mode-flexure specimen. (h) Cracked-lap-shear specimen. (i) Four-point-bend specimen

height, which is much less than the length of the specimen. The adhesive height (h_2) is usually much less than the adherend heights (h_1 and h_3) (Fig. 5a). The specimens are subjected to shear loads that are applied to the bond and crack planes at different lengthwise locations and orientations to obtain various mode mixes.

The simplest example, the double-cantilever-beam (DCB) specimen, was first introduced for fracture toughness testing of adhesives by Ripling and Mostovoy (Ref 4). When the load is applied perpendicular to the crack plane, the DCB can be used for mode I fracture toughness tests if the adherends are made of the same material and have the same thickness. When loaded by a fixed displacement (Δ) the configuration is known as the wedge test and is useful for examining environmentally assisted or time-dependent crack growth. The DCB specimen can also be used to obtain mode I/mode II mixtures (Ref 5) if the adherends have different bending stiffnesses (Fig. 5a).

If the load is instead applied parallel to the crack plane (Fig. 5b), mode III cracking can be obtained (Ref 6). Mode I/mode III mixtures have been examined by orienting the crack plane (Fig. 5c) at some angle $\alpha (0 < \alpha < 90)$ to the direction of the applied load (Ref 7). To simplify data reduction and provide a specimen with an energy-release rate and a stress-intensity factor independent of crack length, the adherend heights h_1 and h_3 can be tapered (Ref 4). Alternatively, the fixed adherend heights can be retained, but a concentrated moment (M) (Fig. 5d) instead of the force P (Fig. 5a to 5c) can be applied (Ref 8). The laminated-beam geometry can also be used to determine mode II fracture resistance simply by applying a three-point bending load (Fig. 5e) instead of an end load. The concept was first applied to interlaminar delamination of composite materials (Ref 9, 10). The specimen has been dubbed the end-notched-flexure (ENF) specimen and has been used to determine the mode II fracture

properties of structural adhesives (Ref 4, 11-13).

Alternatively, one end of the laminated beam can be clamped, and a shear load (Fig. 5f) can be applied at the other end (Ref 14, 15). If the three-point bending is retained but is reacted by only one of the cracked arms of the specimen, then a mode I/mode II ratio of 57% in opening to total-energy-release rates can be obtained. The specimen (Fig. 5g) is known as the mixed-mode-flexure (MMF) specimen (Ref 10, 11, 13). The mode mix can presumably be altered by using adherends of different bending stiffnesses.

More mode II-dominant mixtures are provided by the cracked-lap-shear (CLS) specimen. It has essentially the same geometry as the MMF specimen, but is loaded axially (Fig. 5h). Generally, G_I/G_T ratios ranging from 20 to 35% can be obtained by changing the ratio of the upper- and lower-adherend bending stiffnesses. Although difficulties (due mainly to interpretation of boundary

Table 1   Specimen compliances

Specimen	Figure	Compliance ($C = \Delta/P$)
Double-cantilever beam (shear load)	5a,b	$C = \dfrac{8a^3}{E_1 bh_1^3} + \dfrac{b(1 + v_1)a}{E_1 bh_1}$ ($E_1 = E_3$, $h_1 = h_3$)
Double-cantilever beam (bending moment)	5d	$C = \dfrac{24a}{Ebh^3}$
End-notched flexure	5e	$C = \dfrac{2L^3 + 3a^2}{8E_1 bh_1^3} + \dfrac{(1 + v_1)(1.2L + 0.9a)}{2E_1 bh_1}$
Mixed-mode flexure	5g	$C = \dfrac{2L^3 + 7a^3}{8E_1 bh_1^3} + \dfrac{(1 + v_1)(0.6L + 0.45a)}{2E_1 bh_1}$
Cracked lap shear (tension)(a)	5h	$C = \dfrac{a}{b}\left(\dfrac{1}{E_3 h_3} - \dfrac{1}{E_1 h_1 + E_3 h_3}\right)$
Cracked lap shear (bending)	· · ·	$C = \dfrac{a}{Eb}\left(\dfrac{1}{h_3^3} - \dfrac{1}{(h_1 + h_3)^3}\right)$ ($E_1 = E_3 = E$)

(a) Infinitely long specimen

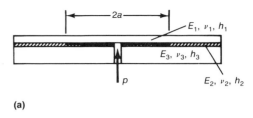

(a)

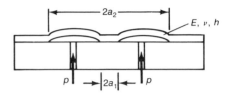

(b)

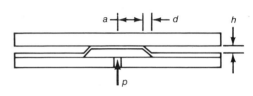

(c)

Fig. 6 Blister specimens. (a) Pressurized blister specimen. (b) Island blister specimen. (c) Constrained blister specimen

conditions) have been noted (Ref 16, 17) in using the specimen in its originally proposed form (Ref 18), it has been used extensively (Ref 19-22) in the simpler manner noted in Fig. 5(h). The specimen has been the subject of an ASTM Round Robin (Ref 23), which further emphasized the need to account for geometrical nonlinearities (Ref 17, 24) when stress analyzing the specimen. Specimen compliances are given in Table 1 for the specimens depicted in Fig. 5 (a to h). The CLS specimen can also be subjected to pure bending (Ref 18), in which case more mode I-dominated mixtures can be obtained. Pure bending can be achieved in the four-point-bend configuration (Fig. 5i), which provides an energy-release rate that is independent of crack length (Ref 25).

Blister specimens have central cracks, and the load is usually introduced through one of the adherends. The load is often applied by pressure loading, which results in a circular crack front, an axisymmetric geometry that is free of three-dimensional (edge) effects. It is ideal for examining crack initiation and propagation where three-dimensional effects are to be avoided, most notably for environmentally assisted crack growth (Ref 26). The pressure loading makes the specimen useful for delaminating thin films.

The specimen geometry is shown in Fig. 6(a). It may be a simple bimaterial geometry where a film is directly bonded to a substrate, or it can consist of two adherends joined by an adhesive layer. In either case, the load is usually introduced through one of the layers. In the original configuration (Ref 27), the load was applied by a punch that was driven through a hole in the substrate. The idea of a pressurized loading was introduced by Williams and co-workers (Ref 28-32), who also gave a fracture mechanics interpretation to the failure pro-

cesses in the specimen. For relatively strong adhesion, the delamination may branch and cause a shear-out of the more flexible adherend. A number of innovations have recently been introduced that address this problem. The first is the island blister (Ref 33), which is shown in Fig. 6(b). The delaminated region in this specimen is annular and reduces the tension in the delaminating layer for the same crack-front loading or stress-intensity factor in the original blister specimen. The second recent innovation is the constrained blister test (Ref 34, 35), in which a rigid constraint is placed at a fixed height above the delaminating layer in order to limit its deflections (Fig. 6c). An added feature of this configuration is that the energy-release rate is independent of crack length (Ref 34, 35).

Prismatic specimens have not been as widely used as other configurations, perhaps because of the greater adherend heights that are involved and, in some cases, their more-complex shape. Nevertheless, they have certain advantages that occasionally override these factors. In some cases, the advantage is a compact geome-

try, which minimizes the use of adhesive. In other cases, prismatic specimens subjected to biaxial loadings can provide a wide range of mode mixes from a single specimen.

A number of investigators (Ref 36-38) have adapted the compact tension specimen that is routinely used for mode I fracture toughness testing of homogeneous materials to adhesive bond layers (Fig. 7a). Similarly, compact mode II specimens have been devised (Ref 4, 36, 37), although care must be taken to minimize bending effects that introduce a mode I component. As long as there is an adhesive layer of finite thickness, there will always be a bending effect. A specimen proposed by Arcan offers the possibility of minimizing bending effects in mode II testing. Although originally developed for obtaining the shear deformation

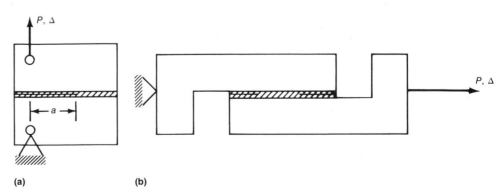

(a)   (b)

Fig. 7 Prismatic specimens. (a) Compact tension specimen. (b) Compact shear specimen

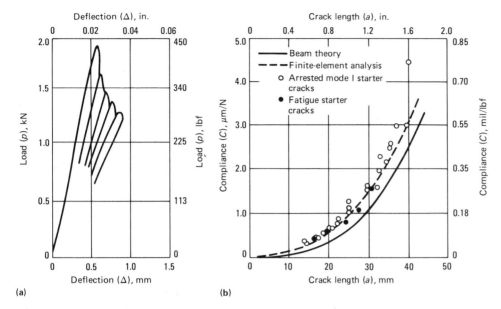

Fig. 8 Crack initiation in a DCB specimen. (a) Load versus deflection. (b) Compliance versus crack length

(Ref 39) and fracture (Ref 40) characteristics of bulk homogeneous materials, the specimen has recently been considered for characterizing the shear deformation of adhesive layers (Ref 41, 42) and could presumably be extended to the fracture of adhesive layers. Furthermore, the specimen loading is not restricted to shear, but can also be oriented to obtain mode I and intermediate mode mixes, in the manner suggested for homogeneous materials (Ref 43).

Scarf specimens have also been used (Ref 44, 45) to obtain mixed-mode fracture properties of adhesive layers. The use of biaxial loading rather than geometry changes for controlling mode mixes was first suggested by Ripling *et al.* (Ref 46); such an approach uses an independently loaded mixed-mode (ILMM) specimen. This specimen has recently been used with some modification by Sancaktar *et al.* (Ref 47) for determining mixed-mode fracture envelopes for adhesive layers of various thicknesses, both with and without scrim cloth.

Fracture Analysis of Specimens

To obtain the fracture toughness or resistance of a particular adhesive, it is necessary to associate the value of a fracture parameter with crack initiation or crack velocity. This can be done in a number of ways, of which the most direct is the compliance method, based on Eq 19. However, the method does not yield any information on the fracture mode mix; this information must be obtained by supplementary stress analyses, which can also be used to check fracture parameter values derived from the compliance method. For laminated-beam and blister specimens, analytical expressions for specimen compliance can be derived that can be compared directly with measured compliances. In the absence of measured compliances, the analytical expressions can be substituted into Eq 19 to obtain the dependence of the energy-release rate on load level and crack length. The compact prismatic specimens require finite-element analyses, and the solutions from these analyses can be used to extract fracture parameter values and check measured compliances. The details of these approaches and their associated data reduction are given in the following sections.

The direct compliance method relies totally on measured quantities to extract energy-release-rate values and does not rely on any intermediate stress analysis steps. It can be used with any specimen as long as the specimen response is elastic. It is the only method for extracting fracture parameters from data without any accompanying stress analysis. The disadvantages of the method are that it involves taking derivatives of measured quantities, and no information on fracture mode mixtures can be extracted.

The method depends on measurements of the load (P) and associated displacement (Δ) applied to the specimen. A DCB specimen with aluminum adherends and FM 300 adhesive (Ref 13) can serve as an example. The load-deflection record (Fig. 8a) was obtained for a number of crack lengths as cracks were initiated (loaded) and arrested (unloaded). As the cracks get longer, the stiffness of the specimen (load versus deflection slope) decreases. The compliance $C = \Delta/P$ is the inverse of the stiffness and is plotted as a function of crack length (Fig. 8b). The energy-release rate for any particular load level can then be obtained by fitting a curve to the compliance data and differentiating it in order to apply Eq 19. The fracture toughness of the material in mode I is obtained from:

$$G_{Ic} = \frac{P_c^2}{2b} \frac{dC}{da} \qquad \text{(Eq 22)}$$

where P_c is the load level at which the crack initiated. The value of P_c can be determined from direct observation of the crack and synchronization with the load-deflection record. It can also be determined from the departure from linearity of the load-deflection record. The mode mix in this case is

Table 2 Energy-release-rate solutions

Specimen	Figure	Energy-release rate
Four-point bend	5i	$G = \dfrac{P^2 l^2}{8 E_3 b} \left(\dfrac{1}{I_3} - \dfrac{E_1}{E_3 I_c} \right)$
		$I_c = \dfrac{b}{12} \left[h_1^3 + h_3^3 + \dfrac{3 E_1 h_1 h_3 (h_1 + h_3)^2}{(E_3 h_1 + E_1 h_3)} \right]$
Mode III DCB	5b	$G = \dfrac{3 P \Delta}{2ba}$
Wedge test		$G = \dfrac{3 E_1 h_1^3 \Delta^2}{4 a^4} \quad (E_1 = E_3, h_1 = h_3)$
Blister specimen	6a	$G = \dfrac{3(1 - v^2) p^2 a^4}{32 E h^3}$
Island blister	6b	$G = \dfrac{p^2 a_1^2}{32 \sigma_0 h} \left(\dfrac{\beta^2 - 1}{\ln\beta} - 2 \right)^2, \ \beta = \dfrac{a_2}{a_1}$
Constrained blister	6c	$G = phg$
		$g = \left(1 - \dfrac{d}{2a} \right) + \left(\dfrac{d}{3a} - \dfrac{1}{2} \right) \dfrac{\partial d}{\partial a}$

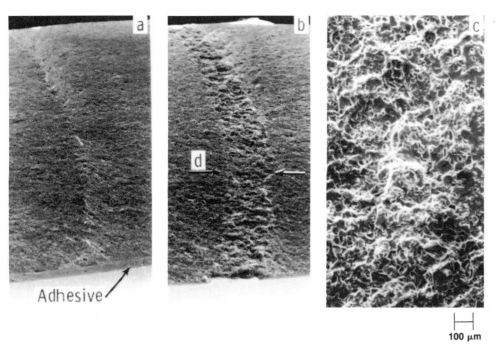

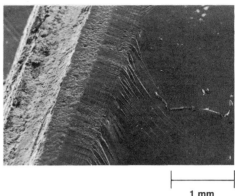

Fig. 10(a) Brittle unmodified epoxy undergoing mode I fracture. Source: Ref 52

Fig. 9 Fractographic features associated with mode I crack initiation. Crack growth direction, left to right. (a) Adhesive thickness, 0.875 mm (0.035 in.). The ductile band is well contained within the adhesive layer. (b) Adhesive thickness, 0.18 mm (0.0072 in.). The ductile band is extended lengthwise due to interaction between the damage zone and the adherend; this extension is denoted as d. (c) Enlarged view of the central portion of the ductile band in (b). Source: Ref 51

obvious from the symmetry of the specimen geometry and loading, but it is not a product of the compliance method itself.

The Hybrid Compliance Method. The measured compliances that are used in the direct method outlined above are subject to some scatter. This can lead to complications when taking the derivative dC/da. As indicated above, one solution to the scatter problem is to fit a smooth curve to the measured compliances. Another approach, which is also an excellent consistency check on the measured compliances, is to obtain a compliance prediction from some form of stress analysis. For the laminated-beam (Fig. 5) and blister (Fig. 6) specimens, beam or plate theory analyses are often sufficient. Most compact specimens that do not conform to the assumptions of these theories are usually analyzed by finite-element methods. In any case, the compliance can be extracted from the stress analyses, compared with measured values, and used with critical-load measurements to extract the toughness. This hybrid method of using compliances obtained from analysis and measured critical loads is useful when displacement measurements are not possible. However, every effort should be made to

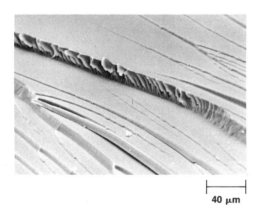

Fig. 10(b) Very brittle unmodified epoxy undergoing mode I fracture. Source: Ref 52

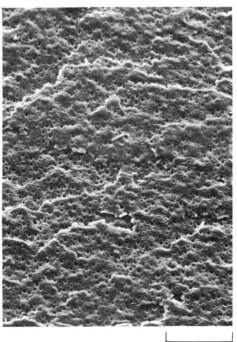

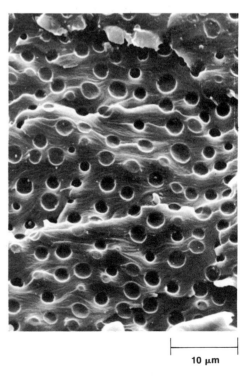

Fig. 10(c) Tough rubber-modified epoxy undergoing mode I fracture. Source: Ref 52

|⊢————⊣|
1 mm

Fig. 11(a) Effect of temperature on the fractography of an unmodified epoxy undergoing mode I fracture. Source: Ref 52

obtain displacement measurements so that consistency checks can be made.

Continuing with the example depicted in Fig. 8, a beam theory analysis (Ref 13) that included the effect of shear yielded the compliance:

$$C = \frac{8a^3}{E_1 bh_1^3} + \frac{6(1 + v_1)a}{E_1 bh_1} \qquad \text{(Eq 23)}$$

The beam theory compliance is plotted as the solid line in Fig. 8(b). A linearly elastic, small-strain, small-rotation, finite-element analysis of the same geometry but including the adhesive layer was also conducted. The predicted compliance from this analysis is shown as the dashed line in Fig. 8(b), and it is in close agreement with the measured compliances. The beam theory analysis predicted a stiffer response because it was assumed that the cracked beam halves were cantilevered at the crack tip, whereas rotation of the adherends actually takes place beyond the crack tip. A correction can be made for this effect (Ref 48). The finite-element analysis confirmed that the energy-release rates were purely mode I. The procedures outlined above can be conducted for any specimen as long as the variation of its compliance with crack length is known. Table 1 gives the compliance relations for a number of specimens.

Energy-Release-Rate Solutions. There are a number of situations where the energy-release rate is obtained without first determining the compliance. Analytical energy-release-rate solutions for particular geometries and loadings can often be obtained from energy arguments (Ref 1). In other cases they can be obtained from a postprocessing algorithm of a finite-element code. In the former case, the energy-release rate at crack initiation or corresponding to a particular crack growth rate can be obtained by substituting the appropriate load level and crack length into the analytical energy-release-rate solution. Table 2 contains a listing of energy-release-rate solutions for a number of specimens. In case a finite-element analysis has been used to obtain an

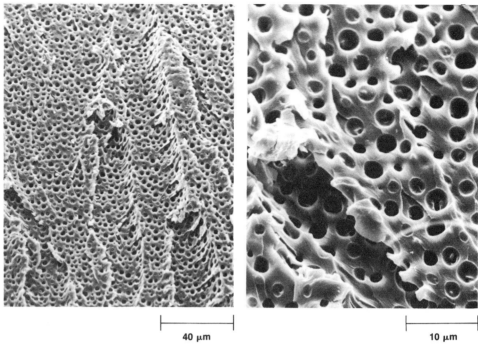

|⊢————⊣|
40 μm

|⊢————⊣|
10 μm

Fig. 11(b) Effect of temperature on the fractography of a modified epoxy undergoing mode I fracture. Source: Ref 52

energy-release-rate solution, the energy-release rate (G_c) corresponding to a critical load level (P_c) can be scaled as:

$$G_c = \left(\frac{P_c}{P}\right)^2 G \qquad \text{(Eq 24)}$$

where P and G are the load level and corresponding energy-release rate used in (P) and obtained from (G) the finite-element analysis. In other words, as long as the specimen response is linear, there is no need to conduct a finite-element analysis for each load level. When using this method, it is again only necessary to measure the load that is applied to the specimen to determine the value of P_c.

Crack Length Measurements. A number of the specimens considered in this article have energy-release rates or stress-intensity factors that depend strongly on the crack length. The importance of obtaining accurate crack length measurements has provided some impetus for the development of so-called iso-G specimens, which have energy-release rates that are independent of crack length. However, if fatigue, environmental, or time-dependent crack growth is being considered, then crack length measurements must still be made in order to determine the dependence of crack growth rates on energy-release rates.

The most common method of measuring crack length involves mounting a low-power traversing microscope on the loading device in order to detect the crack tip. The detection task is made easier by putting the specimen under load and coating the specimen with a thin, brittle coating such as

liquid paper. Nevertheless, it can still be difficult to determine the crack tip location in this way, particularly under mode II conditions. If a toughness specimen is being precracked under fatigue loading, or if fatigue crack growth is being examined, it is easier to detect the crack tip location during the loading if the frequency is low enough. Crack length measurements can often be made by examining the fracture surface after a test has been conducted.

When adherends are thin, as is often the case with specimens made of fiber-reinforced composites, it is possible to detect the crack-front location by coating the top of the adherend with a photoelastic coating and viewing the coating through crossed polars. Because the specimen is viewed in planar view, the crack-front geometry can be examined for any deviations from linearity. Ultrasonic methods are also useful for wide specimens.

The methods described above are quite labor intensive, which can make them unacceptable for long-term crack propagation tests that involve cyclic or static loadings. In such cases, it may be desirable to automate the crack length measurement process. Visual detection methods can be automated using digital image analysis techniques that provide feedback signals to stepper-motor-actuated positioners (Ref 49).

Another method involves measuring specimen compliance and extracting the crack length from prior calibration with a similar specimen having known crack lengths; this method is particularly suited to cyclic loading and automated data acquisi-

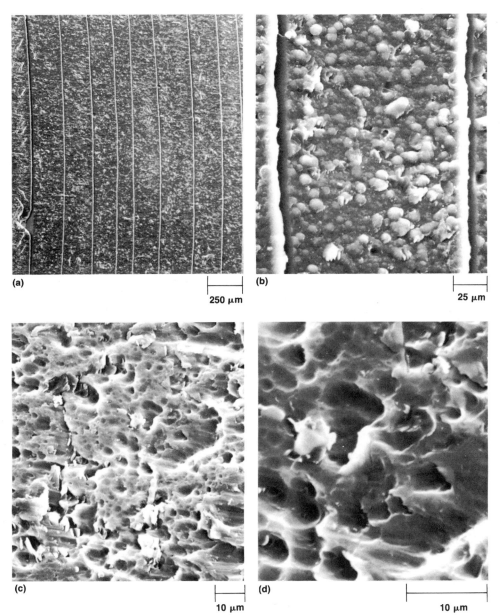

(a)

250 μm

(b)

25 μm

(c)

10 μm

(d)

10 μm

Fig. 12 Microcracking in an unscrimmed brittle adhesive layer under mode II fracture. (a) Fracture surface. Adhesive thickness, 41 μm. (b) Enlarged view of (a). (c) Adherend side of fracture surface. Adherend thickness, 9 μm. (d) Enlarged view of (c). Source: Ref 6

and depth of field can be obtained from transmission electron microscopy, but the technique entails the development of replicas, and the production of replicas is time-consuming and may produce artifacts.

The main factors that influence the appearance of the fracture surface are the toughness of the material; the loading direction, history, and rate; the adhesive layer thickness; and the environment. Therefore, a fracture surface may reveal not only the initial cause of the fracture but also the subsequent sequence of events. The fractographic features described in the following sections have been arranged according to fracture mode mix.

Mode I Fractographic Features. The fracture surfaces of brittle adhesives usually appear relatively smooth and featureless, even when examined under the scanning electron microscope. This is because the fracture involves a pure cleavage of the material without any associated inelastic deformation mechanisms, which are additional sources of energy dissipation over and above the energy required to create new surface. In moderately brittle materials, some inelastic deformation may be associated with the slow, stable initiation of a crack, but as the crack accelerates, the loading rates that a material point experiences as it is approached by the crack are much higher, and inelastic deformation is inhibited.

Because of this deformation behavior, the crack initiation and propagation sequence very often leaves two flat regions separated by a rough region (Fig. 9). The roughness is due to the crack blunting, void formation, and material breakdown that are associated with the initiation phase. The increased width of the ridge in Fig. 9(b) relative to that in Fig. 9(a) is due to interactions between the damage zone and the adherend that arise when the adhesive layer is thin. In thin adhesive layers, damage is confined in the height dimension and therefore grows in the length dimension.

The magnified view (Fig. 9c) of the rough region in Fig. 9(b) reveals the formation of ridges caused by void growth and coalescence. The material in the ridges has therefore been highly drawn and torn. The ridges are the edges of dimples roughly 100 μm (4 mils) in diameter. The size of the dimples reflects the amount of plastic dissipation associated with crack initiation (Ref 51). The smaller voids are bubbles that were formed during the relatively low-pressure cure of these particular samples. The band of roughness in Fig. 9 (a and b) is convex in the direction of crack growth because edge effects usually result in crack fronts that have a thumbnail shape. In cases where crack growth follows a slip-stick pattern of growth and arrest, many flat and rough regions may exist along the fracture surfaces.

tion systems. If any of the compliance relations listed in Table 1 are valid, then they can also be used to extract crack length. Electronic crack gages are often used on monolithic fracture specimens to automate crack length measurements. The gages consist of a conductive foil bonded to the specimen edges to which a potential is applied. The resistance of the foil changes as the crack proceeds. A similar concept has been used to determine crack initiation in an adhesive joint under impact (Ref 50), although in this case a conductive film was sprayed across the anticipated crack path.

Fractography. Much information about crack growth mechanisms can be obtained by examining the appearance of the fracture surfaces once a specimen or component has completely failed. The fracture surface features may provide a check on the crack tip stress state, particularly in shear tests, where bending or tensile loads can sometimes be inadvertently introduced. As indicated earlier, the fracture surface can be used to determine the length of cracks prior to a fracture toughness test if the features associated with initiation (blunting) and propagation are different. In some cases, it may be sufficient to make the observations under the optical microscope. However, if very fine features are to be discerned, then an electron microscope (usually a scanning electron microscope, SEM) is used because of its higher resolution and better depth-of-field characteristics. Improved resolution

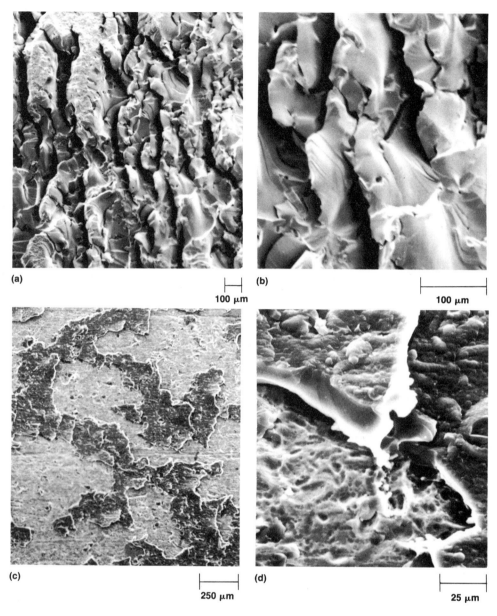

(a)

100 μm

(b)

100 μm

(c)

250 μm

(d)

25 μm

Fig. 13 Microcracking in an unscrimmed tough adhesive layer under mode II fracture. (a) Adhesive thickness, 0.30 mm (0.012 in.). (b) Enlarged view of (a). (c) Adhesive thickness, 11 μm. (d) Enlarged view of (c). Source: Ref 6

growth. The voids scatter light and give rise to the stress whitening effect.

Temperature and load rate also affect the toughness of adhesives and the appearance of the fracture surfaces. Increases in temperature and/or decreases in load rate generally toughen the material because inelastic deformation occurs more easily. The increase in inelastic deformation due to increased temperature (Ref 52) appears in the form of enhanced river line features in the unmodified epoxy (Fig. 11a) and deeper ridges and voids in the modified epoxy (Fig. 11b). The fact that the river lines in Fig. 11(a) curve outward toward the specimen edges is due to three-dimensional (edge) effects. In the unmodified epoxy, the specimen thickness was reduced by as much as 20% as the crack propagated through the specimen, suggesting that shear yielding was occurring.

Mode II Fractographic Features. In homogeneous materials it is very difficult to propagate a coplanar crack under mode II loading. The crack will usually branch and grow under mode I conditions. When an adhesive layer is constrained by stiffer adherends, there is an interplay among the relative toughnesses of the adhesive, adherend, and interface, and the thickness of the adhesive layer (Ref 6). The fractographic features are consequently dictated by similar considerations.

If the adhesive is relatively brittle and thick, the high stresses ahead of an initially coplanar cohesive crack cause microcracks to form at 45° to the plane of the bond. As the load is increased, the microcracks grow until they reach the interface, where they may link along or near the interface or by tearing of the interconnecting ligaments. The mode of linkage probably depends on the toughness of the interface and the spacing of the microcracks.

An example of linkage close to the interface is shown in Fig. 12. The lower magnification (Fig. 12a) reveals a series of lines across the specimen width that are convex in the direction of crack growth (left to right); these lines are the mouths of microcracks that have intersected the near-interface region, which accounts for the bulk of the fracture surface. The higher-magnification view (Fig. 12b) shows that the right-hand microcrack faces have relatively sharp branches into the near-interface region, whereas the material near the left microcrack faces has been highly stretched as the connecting crack approached. These observations are consistent with the overall crack growth direction and branching of the microcracks. The adherend side of the fracture surface (Fig. 12c) reveals ovalized cavities that are reminiscent of shear dimpling in metals. There is also a hackle pattern (Fig. 12d), where adhesive-filled holes fractured (Ref 12).

The microcrack spacing increases for tougher adhesives (Fig. 13). However, the

Another characteristic of crack initiation in relatively brittle adhesives is the presence of river lines that mark the intersection of different levels (planes) that a crack path can take. The difference in the elevation of the fracture planes is often caused by the way that the crack is introduced. For example, Fig. 10(a) shows a crack initiated by a saw cut and razor blade in an unmodified (brittle) epoxy. When the critical load level was applied, the minor imperfections in the starter crack produced by the razor blade insertion caused cleavage on planes of slightly differing elevations. The intersections of these planes give rise to lines that are known as river lines.

For very brittle materials, the only place that any plastic shear yielding can occur is

along the river lines. As indicated in Fig. 10(b), the major crack planes are very smooth; only the river lines, or cliff faces, are rough, indicating that two major crack planes coalesced by shear yielding. The rough features on the cliff faces are generally known as hackles. For tougher, rubber-modified epoxies (Fig. 10c), the fracture surfaces are a lighter color than the bulk material, indicating the possibility of stress whitening (Ref 52). Higher magnification reveals the presence of ridges and circular depressions on the fracture surface. The circular depressions are associated with the location of the rubber particles that are used to modify the epoxy. The cavitation of the rubber particles and particle-matrix debonding give rise to substantial void

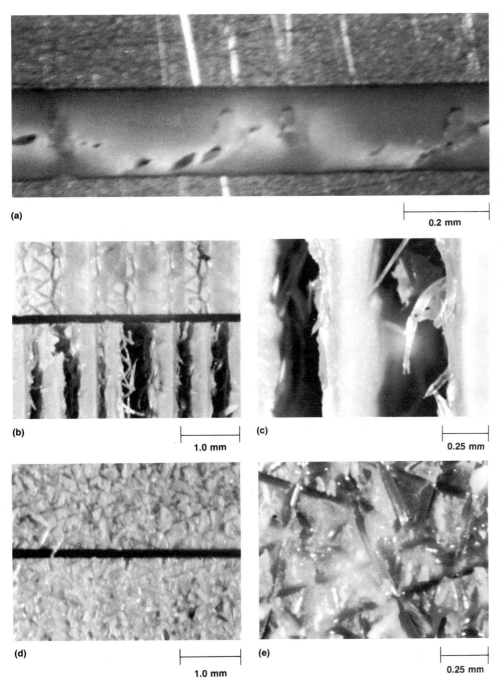

Fig. 14 Shear microcracking in a scrimmed adhesive layer under mode II fracture. (a) Development of microcracks ahead of a major crack. Crack growth direction, left to right. (b) Development of microcracks in uncracked adhesive layers subjected to pure shear. (c) Enlarged view of (b). (d) Development of microcracks in uncracked adhesive layers under bond-normal loading. (e) Enlarged view of (d). Source: Ref 13, 42

layers subjected to pure shear (Fig. 14 b and c). The scrim fibers again act as stress concentrators, initiating microcracking (Ref 42) that is linked by ligament tearing rather than interfacial linking. The introduction of any bond-normal loading dramatically flattens the fracture surfaces (Fig. 14 d and e).

Mode III Fractographic Features. Microcracking also occurs in brittle adhesive layers undergoing mode III fracture (Fig. 15). However, the microcracks leave an arrowhead pattern on the fracture surface that points in the direction of crack growth (right to left, Fig. 15 a and b). The microcracks are therefore parallel to the crack direction in the center of the specimen and perpendicular near the specimen edge. At the edge, the microcracks are again oriented at 45° to the long axis of the beam. As the thickness of the adhesive layer decreases (Fig. 15 c and d), the microcrack density increases and the degree of plastic deformation associated with crack propagation increases; this increase is indicated by the numerous stress-whitened regions which run parallel to the microcracks. The fracture surface of a thick, tough adhesive layer (Fig. 16 a and b) still retains the arrowhead markings but is much rougher between microcracks, as was the case in mode II. The roughness again decreases with decreasing adhesive thickness (Fig. 16 c and d) with more evidence of plastic flow.

In tests on even tougher adhesives such as thermoplastic polyetheretherketone (PEEK), microcracking did not occur (Ref 6). Instead, the cohesive starter crack branched to the interface and the remaining crack growth was interfacial. However, the interfacial cracking did not occur for very thin adhesive layers. As can be seen from Fig. 17, there was a lot of distortion and tearing with stress whitening. The mode II fracture surface (Fig. 17 a and b) is made up of ovalized dimples emanating from voids, whereas the mode III fracture surface (Fig. 17 c and d) has highly drawn triangular features associated with voids.

REFERENCES

1. M.F. Kanninen and C.H. Popelar, *Advanced Fracture Mechanics*, Oxford University Press, 1986
2. H. Tada, P.C. Paris, and G.R. Irwin, "The Stress Analysis of Cracks Handbook," Del Research Corporation, 1973
3. A.A. Griffith, The Phenomena of Rupture and Flow in Solids, *Philos. Trans. R. Soc. (London) A*, Vol A221, 1920, p 163-198
4. E.J. Ripling, S. Mostovoy, and H.T. Corten, Fracture Mechanics: A Tool for Evaluating Structural Adhesives, *J. Adhes.*, Vol 3, 1971, p 107-123
5. P.D. Mangalgiri, W.S. Johnson, and R.A. Everett, Jr., Effect of Adherend Thickness and Mixed-Mode Loading on

microcracks are harder to discern because of the increase in plastic deformation between them. The region between microcracks is much rougher (Fig. 13 a and b) than was the case for the more-brittle adhesive. However, as the adhesive thickness is reduced (Fig. 13 c and d), the microcrack spacing decreases and the fracture surface roughness between microcracks decreases, giving rise to a lower toughness (Ref 6).

Scrim cloth in the adhesive enhances microcracking and makes it more likely that linkage will be removed from the interface, giving rise to much rougher fracture surfaces. Figure 14(a) reveals the development of microcracks ahead of a major crack under mode II loading (Ref 13). Microcracking development under shear is not limited to the presence of major cracks. Microcracks can also develop in uncracked adhesive

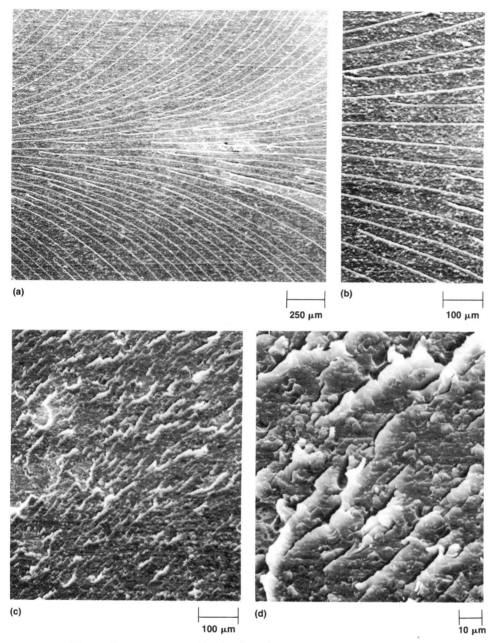

Fig. 15 Microcracking in an unscrimmed brittle adhesive layer under mode III fracture. (a) Adhesive thickness, 23 μm. Crack growth direction, right to left. (b) Enlarged view of (a). (c) Adhesive thickness, 3.3 μm. (d) Enlarged view of (c). Source: Ref 6

Debond Growth in Adhesively Bonded Composite Joints, *J. Adhes.*, Vol 23, 1987, p 263-288

6. H. Chai, Shear Fracture, *Int. J. Fract.*, Vol 37, 1988, p 137-159

7. E.J. Ripling, P.B. Crosley, and W.S. Johnson, A Comparison of Pure Mode I and Mixed-Mode I-III Cracking of an Adhesive Containing an Open Knit Cloth Carrier, in *Adhesively Bonded Joints: Testing, Analysis, and Design*, W.S. Johnson, Ed., STP 981, American Society for Testing and Materials, 1988, p 163-182

8. P.W. Osborne, Steady State Fracture Device, *Rev. Sci. Instrum.*, Vol 37, 1966, p 664

9. A.J. Russell and K.N. Street, Factors Affecting the Interlaminar Fracture Energy of Graphite/Epoxy Laminates, in *Progress in the Science and Engineering of Composites*, T. Hayashi, K. Kawata, and S. Unekawa, Ed., Pergamon Publishing, 1982, p 279-286

10. A.J. Russell and K.N. Street, Moisture and Temperature Effects on the Mixed-Mode Delamination Fracture of Unidirectional Graphite/Epoxy, in *Delamination and Debonding of Materials*, W.S. Johnson, Ed., STP 876, American So-

ciety for Testing and Materials, 1985, p 349-370

11. S. Mall and N.K. Kochar, Criterion for Mixed Mode Fracture in Composite Bonded Joints, *Proc. Inst. Mech. Eng.*, Vol 6, 1986, p 71-76

12. H. Chai and S. Mall, Design Aspects of the End-Notch Adhesive Joint Specimen, *Int. J. Fract.*, Vol 36, 1988, p R3-R8

13. K.M. Liechti and T. Freda, On the Use of Laminated Beams for the Determination of Pure and Mixed-Mode Fracture Properties of Structural Adhesives, *J. Adhes.*, Vol 28, 1989, p 145-169

14. A.J. Russell and K.N. Street, The Effect of Matrix Toughness on Delamination: Static and Fatigue Fracture of Unidirectional Graphite/Epoxy in Toughened Composites, in *Toughened Composites*, N.J. Johnston, Ed., STP 937, American Society for Testing and Materials, 1987, p 275-294

15. E. Moussiaux, H.F. Brinson, and A.H. Cardon, "Bending of a Bonded Beam as a Test Method for Adhesive Properties," Report CAS/ESM-87-2, Virginia Polytechnic Institute and State University, Center for Adhesion Science, 1987

16. R.A. Everett, Jr. and W.S. Johnson, Repeatability of Mixed-Mode Adhesive Bonding, in *Delamination and Debonding of Materials*, W.S. Johnson, Ed., STP 876, American Society for Testing and Materials, 1985, p 267-281

17. C. Lin and K.M. Liechti, Similarity Concepts in the Fatigue Fracture of Adhesively Bonded Joints, *J. Adhes.*, Vol 21, 1987, p 1-24

18. T.R. Brussat and S.T. Chiu, Fatigue Crack Growth of Bondline Cracks in Structural Bonded Joints, *J. Eng. Mater. Technol. (Trans. ASME)*, Vol 100, 1978, p 39-45

19. S. Mall, W.S. Johnson, and R.A. Everett, Jr., Cyclic Debonding of Adhesively Bonded Composites, in *Adhesive Joints: Their Formation Characteristics and Testing*, K.L. Mittal, Ed., Plenum Publishing, 1984, p 639-658

20. W.S. Johnson and S. Mall, A Fracture Mechanics Approach for Designing Adhesively Bonded Joints, in *Delamination and Debonding of Materials*, W.S. Johnson, Ed., STP 876, American Society for Testing and Materials, 1985, p 189-199

21. S. Mall, M.A. Rezaizadeh, and R. Gurumurthy, Interaction of Mixed-Mode Loading on Cyclic Debonding in Adhesively Bonded Composite Joints, *J. Eng. Mater. Technol. (Trans. ASME)*, Vol 109, 1987, p 17-21

22. S. Mall and K.T. Yun, Effect of Adhesive Ductility on Cyclic Debond Mechanisms in Composite to Composite Bonded Joints, *J. Adhes.*, Vol 23, 1987, p 215-231

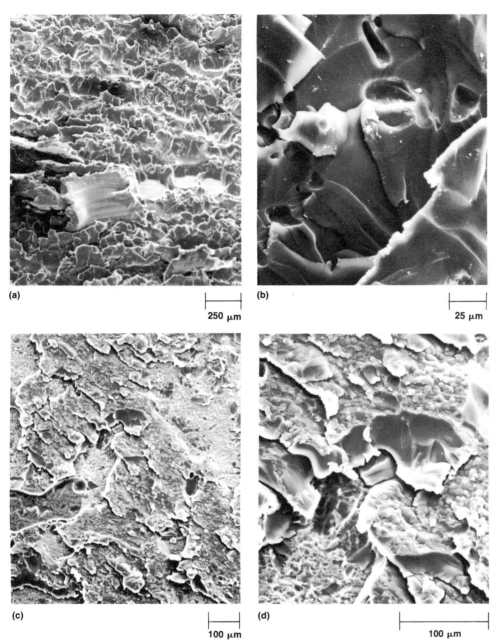

Fig. 16 Microcracking in an unscrimmed tough adhesive layer under mode III fracture. (a) Adhesive thickness, 0.10 mm (0.004 in.). (b) Enlarged view of (a). (c) Adhesive thickness, 20 μm. (d) Enlarged view of (c). Source: Ref 6

23. W.S. Johnson, Stress Analysis of the Cracked Lap Shear Specimen: An ASTM Round Robin, *ASTM J. Test. Eval.*, Vol 15, 1987, p 303-324

24. B. Dattaguru, R.A. Everett, Jr., J. Whitcomb, and W.S. Johnson, Geometrically Nonlinear Analysis of Adhesively Bonded Joints, *J. Eng. Mater. Technol. (Trans. ASME)*, Vol 106, 1984, p 59-65

25. P.G. Charalambides, J. Lund, A.G. Evans, and R.M. McMeeking, A Test Specimen for Determining the Fracture Resistance of Bimaterial Interfaces, *J. Appl. Mech. (Trans. ASME)*, Vol 56,

1989, p 77-82

26. D.A. Dillard, K.M. Liechti, D.R. LeFebvre, C. Lin, and J.S. Thornton, Development of Alternate Techniques for Measuring the Fracture Toughness of Rubber to Metal Bonds in Harsh Environments, in *Adhesively Bonded Joints: Testing, Analysis, and Design*, W.S. Johnson, Ed., STP 981, American Society for Testing and Materials, 1988, p 83-97

27. H. Dannenberg, Measurements of Adhesion, *J. Appl. Polym. Sci.*, Vol 5, 1961, p 125-134

28. M.L. Williams, The Continuum Inter-

pretation for Fracture and Adhesion, *J. Appl. Polym. Sci.*, Vol 13, 1969, p 29-40

29. M.L. Williams, The Fracture Threshold for an Adhesive Interlayer, *J. Appl. Polym. Sci.*, Vol 14, 1970, p 1121-1126

30. J.D. Burton and W.B. Jones, Theoretical and Experimental Treatment of Fracture in an Adhesive Interlayer, *Trans. Soc. Rheol.*, Vol 15, 1971, p 39-50

31. S.J. Bennett, K.L. DeVries, and M.L. Williams, Adhesive Fracture Mechanics, *Int. J. Fract.*, Vol 10, 1974, p 33-43

32. G.P. Anderson, K.L. DeVries, and M.L. Williams, Mixed Mode Stress Field Effect in Adhesive Fracture, *Int. J. Fract.*, Vol 10, 1974, p 565-583

33. M.G. Allen, P. Nagarkar, and S.D. Senturia, Aspects of Adhesion Measurement of Thin Polyimide Films, *Proceedings of the 5th Annual Conference on Polyimides*, in press

34. Y.-S. Chang, Y.-H. Lai, and D.A. Dillard, The Constrained Blister—a Nearly Constant Strain Energy Release Rate Test for Adhesives, *J. Adhes.*, Vol 27, 1989, p 197-211

35. M.J. Napolitano, A. Chudnovsky, and A. Moet, The Constrained Blister Test for the Energy of Interfacial Adhesion, *J. Adhes. Sci. Technol.*, Vol 2, 1988, p 311-314

36. Y.W. Mai and A.S. Vipond, On the Opening and Edge-Sliding Fracture Toughness of Aluminum-Epoxy Adhesive Joints, *J. Mater. Sci.*, Vol 13, 1978, p 2280-2284

37. A. Anandarajah and A.E. Vardy, Mode I and II Fracture of Adhesive Joints, *J. Strain Anal.*, Vol 19, 1984, p 173-183

38. C.O. Arah, D.K. McNamara, H.M. Hand, and M.F. Mecklenburg, Techniques for Screening Adhesives for Structural Applications, *J. Adhes. Sci. Technol.*, Vol 3, 1989, p 261-269

39. N. Goldenberg, M. Arcan, M. Nicolau, and E. Nicolau, On the Most Suitable Specimen Shape for Testing Shear Strength of Plastics, in STP 247, American Society for Testing and Materials, 1958, p 115-121

40. L. Banks-Sills and M. Arcan, A Compact Mode II Fracture Specimen, in *Fracture Mechanics: Seventeenth Volume*, STP 905, American Society for Testing and Materials, 1986, p 347-363

41. V. Weissenberg and M. Arcan, A Uniform Pure Shear Testing Specimen for Adhesive Characterization, in *Adhesively Bonded Joints: Testing, Analysis, and Design*, STP 981, American Society for Testing and Materials, 1988, p 28-38

42. K.M. Liechti and T. Hayashi, On the Uniformity of Stresses in Some Adhesive Deformation Species, *J. Adhes.*,

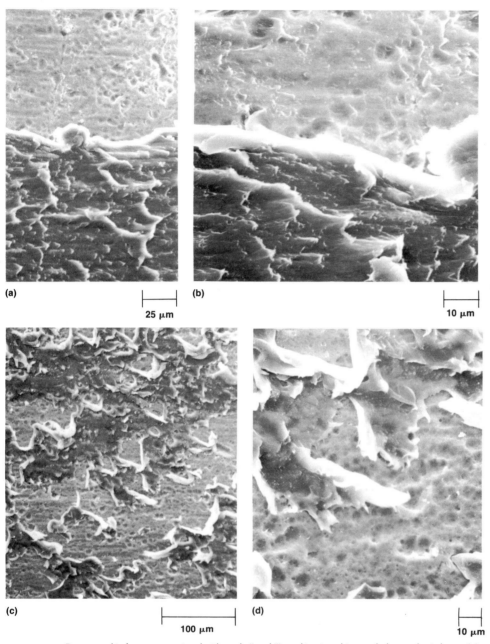

Fig. 17 Fractographic features associated with mode II and III cracking in a thin, tough thermoplastic layer. (a) Mode II. Adhesive thickness, 3 μm. (b) Enlarged view of (a). (c) Mode III. Adhesive thickness, 2 μm. (d) Enlarged view of (c). Source: Ref 6

Vol 29, 1989, p 167-191

43. L. Banks-Sills, M. Arcan, and Y. Bortman, A Mixed-Mode Fracture Specimen for Mode II Dominant Deformation, *Eng. Fract. Mech.*, Vol 20, 1984, p 145-157

44. G.C. Trantina, Combined Mode Crack Extension in Adhesive Joints, *J. Compos. Mater.*, Vol 6, 1972, p 371-385

45. W.D. Bascom, R.L. Cottington, and C.O. Timmons, Fracture Reliability of Structural Adhesives, *J. Appl. Polym. Sci.*, Vol 32, 1977, p 165-188

46. E.J. Ripling, S. Mostovoy, and R.L. Patrick, Application of Fracture Mechanics to Adhesive Joints, in STP 360, American Society for Testing and Materials, 1963, p 5-14

47. E. Sancaktar, H. Jozavi, J. Baldwin, and J. Tang, Elastoplastic Fracture Behavior of Structural Adhesives Under Monotonic Loading, *J. Adhes.*, Vol 23, 1987, p 233

48. S. Mostovoy, P.B. Crosley, and E.J. Ripling, Use of Crack Loaded Specimens for Measuring Plane Strain Fracture Toughness, *J. Mater.*, Vol 2, 1967, p 661-681

49. D. Knight, "QRMSII Questar Remose Measurement System," Specifications and user manual, Questar Corporation, 1985

50. A.J. Kinloch and G.A. Kodokian, The Impact Resistance of Structural Adhesive Joints, *J. Adhes.*, Vol 24, 1987, p 109-126

51. H. Chai, Bond Thickness Effect in Adhesive Joints and Its Significance for Mode I Interlaminar Fracture of Composites, in *Composite Materials: Testing and Design (Seventh Conference)*, J.M. Whitney, Ed., STP 893, American Society for Testing and Materials, 1986, p 209-231

52. A.J. Kinloch, S.J. Shaw, and D.A. Tod, Deformation and Fracture Behavior of a Rubber-Toughened Epoxy: I. Microstructure and Fracture Studies, *Polymer*, Vol 24, 1983, p 1341-1363

Static and Dynamic Fatigue Testing

Erol Sancaktar, Clarkson University

ADHESIVELY BONDED JOINTS may exhibit fatigue failure during service because of sustained or cyclic exposure to loads and hostile environments. Cyclic or transient exposure of adhesive joints usually results in deterioration and possible failure of bonds under dynamic fatigue conditions. The topic of dynamic fatigue is addressed briefly in this article, and more fully in the article "Fatigue and Fracture Mechanics" in this Volume. Bond deterioration and failure under sustained exposure is usually called static fatigue and is governed mostly by the laws of viscoelasticity and viscoplasticity, as well as the kinetic theories of chemistry as they apply to the polymeric adhesive matrix and interphase and possibly to the adherend. Obviously, viscoelasticity may also play an important role under cyclic exposure conditions, especially when the mean level of the applied cyclic load is high or when the adhesive joint is exposed to high-temperature and/or high-humidity environments. Under such conditions the applied stresses or strains may undergo cyclic relaxation or cyclic creep, respectively, as shown in Fig. 1.

In this article the three basic components of an adhesively bonded joint, namely the bulk adhesive, the interphase, and the adherend are identified along with the main factors affecting their durability. It is noted that determination of threshold (stress or strain) values (below which failures are not expected to occur during the design life) become the critical issue from the durability point of view when designing adhesively bonded joints. Obviously, experimentation is necessary to obtain such threshold values. For this purpose, a general test program may encompass three different categories of specimens. The first category involves less expensive, simple specimens such as the single-lap specimen and the thin adherend wedge test specimen, which are used primarily for screening (comparison) purposes. The second category includes more expensive and difficult to machine specimens such as the independently loaded mixed-mode specimen, or Arcan and Iosepescu specimens to obtain more accurate quantitative design data. In the third category are the actual structural prototype components tested in limited numbers to verify theoretical predictions.

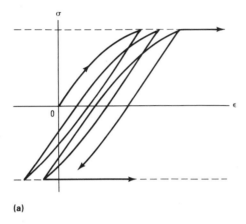

(a)

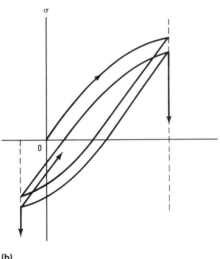

(b)

Fig. 1 (a) Creep and (b) relaxation under cyclic loading

In order to be able to make accurate durability predictions based on experimental data, phenomenological description of the interaction among the environment, adherend, interphase, and the adhesive is needed. In those cases, such as the single-lap geometry, that involve large deviations from what is actually being measured (that is, pure shear behavior), it is necessary to have an estimate of the discrepancy involved.

This article reviews phenomenological models to describe creep and delayed failure behavior of adhesively bonded joints including the effects of such viscoelastic behavior on (mixed-mode) fracture conditions. Phenomenological description of moisture ingression in adhesive joints is also included. The relationship between specimen geometry and test results is considered in terms of analyzing typical experimental results. For example, stress concentration factors are calculated for single-lap joints for which shear stress threshold values are reported. Finally, durability of sealants is also briefly discussed.

Factors Affecting Joint Durability. The study of fatigue phenomena should include the following factors because they are all potential factors in deteriorating adhesively bonded structures:

- The stress and strain levels and their modes of application with regard to time and joint geometry
- Exposure to thermal environments, including high- and low-temperature environments
- Exposure to water
- Exposure to chemicals
- Exposure to radiation

When all these factors are present, the combined degradation effect can be expressed with the use of a combined degradation function (Φ) in the form:

$$\Phi = F(-Kt^b/a_M a_T a_C a_R) \qquad \text{(Eq 1)}$$

where K = a system constant, t = exposure time, b = a time exponent, a_M, a_T, a_C, a_R = time shift factors relating to respective mechanical, thermal, chemical (including exposure to water), and radiation effects. Note that the use of time shift factors a_i can be equivalent to an effective increase in exposure time depending on the adversity of the environmental conditions in comparison to design or test conditions. In Eq 1, the function Φ is a decreasing function that is usually asymptotic in form. Such functions multiply global joint strength and performance parameters such as ultimate strength, rigidity, and durability parameters, such as time to failure. Obviously, the problem of bond fatigue is an issue of durability. In those cases in which an asymptotic value does exist for the degradation function (Φ), it represents an important engineering design parame-

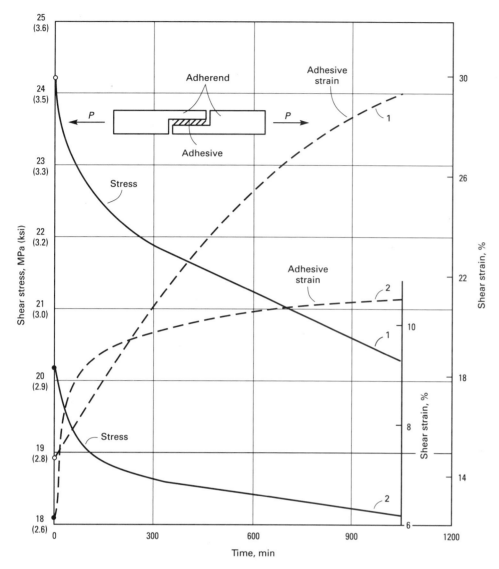

Fig. 2 Simultaneous occurrence of adhesive creep and adherend relaxation in aluminum symmetric single-lap specimens bonded with Metlbond 1113 adhesive (constant deformation was maintained at the end of adherends)

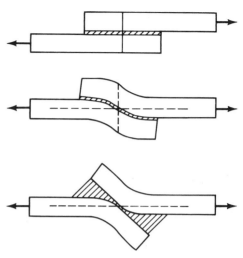

Fig. 3 Bending of the overlap area due to load eccentricity in single-lap joints

ter: the threshold value for environmental degradation.

The Components of an Adhesive Joint. In studying the durability of adhesively bonded joints under exposure to the above degradation parameters, one needs to consider three entities:

- The bulk of adhesive, which often exhibits viscoelastic constitutive behavior
- The interphase, which exists between the adhesive and the adherend and has properties different from those of the bulk adhesive because of the action of the mechanical and chemical adhesion process, adherend surface treatment, and surface topography
- The adherend, which may act viscoelastically and can react to chemical and thermal environments

The nature of load transfer between the substrates by means of the adhesive and the interphase is quite complex, especially when any or all of the three constituents, adhesive, interphase, and adherend, exhibit viscoelastic behavior. For example, Sancaktar and Brinson (Ref 1) showed that when aluminum adherends are bonded with Metlbond 1113 modified epoxy adhesive (with carrier cloth) in symmetrical single-lap geometry, the joint can exhibit relaxation and creep behaviors simultaneously, depending on the location of measurement. Figure 2 shows this behavior. When constant strain is maintained at the ends of aluminum adherends, the joint relaxes while the elastic recovery of the adherends exerts tensile forces on the adhesive, which strains as shown in Fig. 2. Despite the 31/1 ratio between the adherend/adhesive stiffnesses, considerable strain transfer occurs between the two materials, causing the adhesive to creep under fixed displacement boundary conditions for the adherends.

In assessing the long-term durability of adhesive joints for structural applications, it is necessary to test the joints under conditions simulating actual design load conditions. For most engineering applications, this means stressed-durability testing of shear specimens. Some applications, however, involve direct loading of the adhesive layer in the perpendicular (cleavage) direction. Furthermore, because almost all adhesive joining applications involve geometric discontinuities at the ends of the bond line, cleavage (peel) stresses are created at those locations as required by stress and moment equilibrium conditions. An example of peel action created by eccentricity in the shear loading path of single-lap joint geometries is shown in Fig. 3. The severity of these peel stresses depends on the joint geometry and the relative magnitudes of the material properties. The possible presence of voids and disbonds that may develop into cracks and the presence of the interphase, both of which may be weaker under cleavage loading than under shear loading, require additional stressed-durability testing of the adhesive joints in the cleavage mode.

The Single-Lap Geometry. The stress nonuniformities resulting from the geometric discontinuities in adhesive joints make it difficult to interpret the experimental results obtained. When material properties in particular are needed, the use of specimens such as the standard single-lap geometries (ASTM D 1002), which assume that the shear stress is uniform along the bond line and is equal to the applied load divided by the bond area, should be considered approximate at best. The ASTM D 1002 standard geometry not only is eccentric in loading path, but also suggests the use of thin adherends to reduce this eccentricity (1.6 mm, or 0.064 in. thick adherends). With the development of high-strength structural adhesives that can bear shear stresses in excess of 55 MPa (8 ksi) (Metlbond 1113),

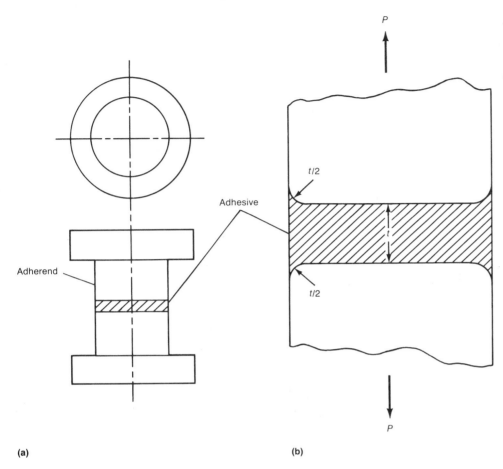

Fig. 4 (a) Butt (tensile) joint geometry. (b) Improvement by rounding the adherend edges

out-of-plane bending and even the plastic deformation of metal adherends of this thickness are possible because of the loading eccentricity.

The Tensile (Butt) Joint Geometry. Direct tensile stressed-durability testing of adhesive joints without starter cracks can be performed using any one of several ASTM standards: D 897 (Fig. 4), D 1344, and D 2095. All these specimens, however, possess stress singularities at the overlap edges. Consequently, so-called singularity participation needs to be quantified (Ref 2) for the range of test configuration cited. It was reported that rounding the adherend butt edges to a diameter equal to the thickness of the adhesive (Fig. 4) renders the specimen "singularity-free" (Ref 2). It should be not-

ed, however, that this procedure increases the adhesive thickness locally, and nonlinear stress effects may result, especially in the presence of high-temperature and high-humidity environments.

A crack in a bond line can be stressed in three different modes, denoted as modes I, II, and III (Fig. 5). Superposition of these three modes constitutes the general case of crack loading. Mode I, known as the opening or cleavage mode, requires the least amount of energy for crack propagation in homogeneous materials. This may or may not be the case for adhesively bonded joints because a crack propagating along a joint is most often constrained to run between the adherends regardless of the applied loading orientation. In fact, in practice almost all

joints fail in a mixed-mode fashion. Most practical adhesive applications involve only mode I and mode II because of the peel and in-plane shear that arise in the use of lap joint geometries. In such geometries, adhesive cracks are exposed to both tension and shear, resulting in mixed-mode cracking.

The Boeing Wedge Test. The simplest method of stressed-durability fracture testing under mode I condition is the Boeing wedge test (Ref 3) (Fig. 6), which is also a standard test method (ASTM D 3762). This test uses self-stressing in mode I for inexpensive screening purposes. The self-stressing is induced by the insertion of a wedge, which creates the initial crack (a) (Fig. 6), and as the crack propagates, the effective cleavage load (P) on it decreases because:

$$P = \delta/C \qquad (Eq\ 2)$$

where δ is the crack opening displacement (COD), which is equal to the thickness of the wedge (Fig. 6), and C is the compliance. For double-cantilever-beam (DCB) configurations such as this one, the compliance is given by (Ref 4):

$$C = 2[(a + a_0)^3 + h^2 a]/(3E\ I) \qquad (Eq\ 3)$$

where a_0 = the empirical rotation factor due to additional deflection at the "built-in" end of the beam, h = beam height (3.2 mm, or 0.125 in. in Fig. 6), E = substrate modulus, $I = Bh^3/12$ (B is 13 mm, or 0.5 in., in Fig. 6).

The rotation factor (a_0) of Eq 3 can be obtained by using the compliance calibration method originally described by Stone (Ref 5). With this procedure a specimen is loaded monotonically to a specific level while the displacement at the point of loading is measured. A COD gage is used for

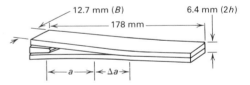

(a)

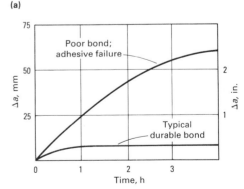

(b)

Fig. 6 (a) The Boeing wedge test specimen and (b) its crack propagation behavior at 49 °C (120 °F) and 100% relative humidity. The B and h values relate to Eq 3. *a*, distance from load point to initial crack tip; Δ*a*, growth during exposure. Source: Ref 3

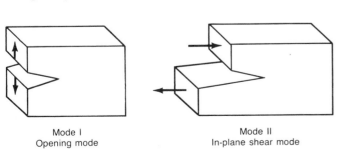

Mode I
Opening mode

Mode II
In-plane shear mode

Mode III
Tearing mode

Fig. 5 Modes of loading

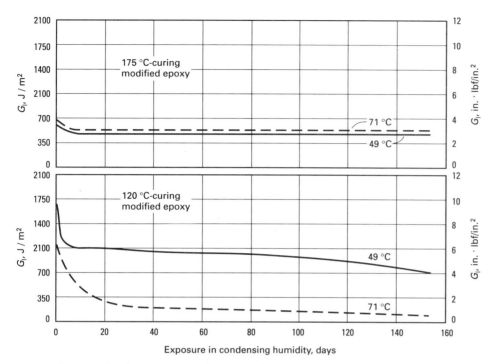

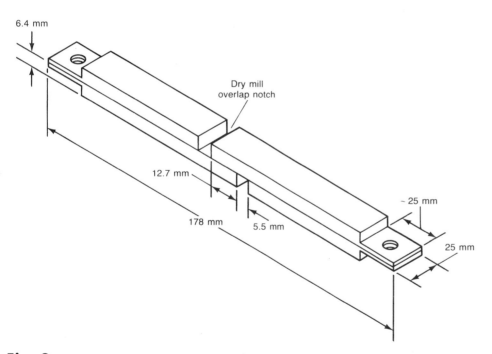

Fig. 7 Variation of mode I strain-energy release rate with test temperature for two different (cure temperature) adhesives tested with thick-adherend DCB specimens (2024-T3, bare, phosphoric-acid anodized). Source: Ref 3

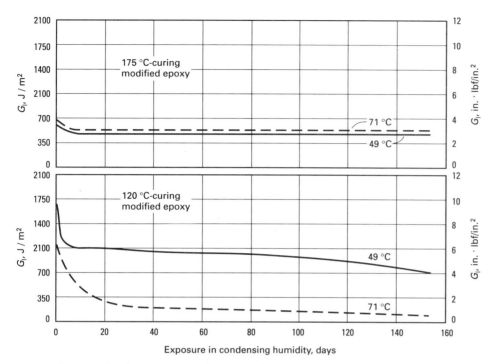

Fig. 8 Thick-adherend (symmetric) lap-shear specimen. Source: Ref 3

this purpose to be able to measure relatively small displacements accurately. This measurement procedure is repeated several times, while increasing the crack length slightly each time. Different values for the rotation factor (a_0) are then submitted into Eq 3 until a good fit is made with the experimental compliance data (Ref 6). A value of $a_0 = 0.6h$ has been used in the literature for wedge test specimens having

the beam dimensions 25 × 150 × 3.8 mm (1 × 6 × 0.15 in.) (Ref 7).

The decrease in the effective load provides a self-arrest capability for the wedge test, which enables the establishment of the threshold level in terms of the mode I load (P) or the strain-energy release rate in mode I, which is a function of the load as given by:

$$G = (P^2/2B)(\partial C/\partial a) \qquad \text{(Eq 4)}$$

Substitution of Eq 2 and 3 into Eq 4 results in:

$$G_I = \delta^2 Eh^3 [3(a + a_0)^2 + h^2]/16[(a + a_0)^3 + ah^2]^2 \qquad \text{(Eq 5)}$$

Again, δ = the wedge thickness in a wedge test. Examination of Eq 4 and 5 reveals that the G_I decreases with increases in a polynomial function of the crack length (a) because they are inversely related. In other words, as the crack progresses, the G_I decreases nonlinearly at a decreasing rate. This decrease in the strain-energy release rate continues until sufficient energy is no longer available to create new crack surfaces in the constant-input energy system of the wedge test specimen. Based on the elastic theory, which governed the derivation of Eq 5, the Griffith criterion describes this threshold level with:

$$G_I = W_S \qquad \text{(Eq 6)}$$

where W_S is the energy necessary to create a surface of unit dimension. Figure 7 shows the presence of such thresholds for a 175 °C (347 °F)-curing modified-epoxy system and a 120 °C (248 °F)-curing system exposed to 49 °C (120.2 °F) and 70 °C (159.8 °F) temperature environments with condensing humidity.

The obvious shortcoming of Eq 5 is its exclusion of adhesive properties, which are likely to be viscoplastic. Furthermore, the calculation for the compliance function of Eq 3 for the DCB geometry is based solely on elastic theory and does not take into account any plastic deformation of the adherend beams. However, the relatively thin adherend beams shown in Fig. 6 and also recommended by the ASTM D 3762 procedure are prone to permanent deformation and may lead to serious errors in the calculation of G_I values.

Thick-adherend specimen usage represents a partial solution to the problem of severe stress concentrations in thin-adherend single-lap specimens and the problem of plastic deformation and the possible presence of state-of-plane stress in the wedge test specimens. In lap geometries, load eccentricity can be avoided with the use of the thick-adherend symmetric lap specimen shown in Fig. 8. This geometry, however, also has cleavage action at the overlap ends because shear stresses are 0 at the ends of the bond line and this creates an unbalanced shear moment over the overlap area, which tends to load it in mode I. This cleavage action can cause the in-plane bending of the overlap area during tensile-shear loading of the joint, as shown in Fig. 9. For mode I stressed-durability testing, again the use of thick adherends, as shown in Fig. 10, is recommended to obtain more accurate results. Note that the total specimen thickness in Fig. 10 is 25 mm (1 in.), while it is 6.35 mm (¼ in.) for the wedge test specimen shown in Fig. 6. It should also be noted that

Fig. 9 Bending of the overlap area in symmetric single-lap specimens during loading

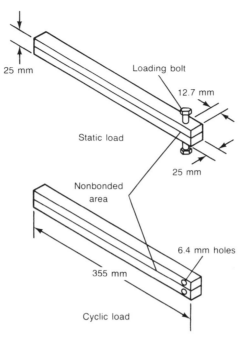

Fig. 10 Thick-adherend double-cantilever-beam specimen. Source: Ref 3

the DCB type cleavage peel test specimen prescribed by ASTM D 3807 has 12.7 mm (½ in.) total specimen thickness.

A General Test Program. Based on the above discussions, it is clear that one would use different test geometries, methods, and analyses depending on the purpose.

For comparison and preliminary screening purposes, test specimens of the thin-adherend single-lap (ASTM D 1002) and thin-adherend (Boeing) wedge (ASTM D 3762) geometries can be used to obtain qualitative results. They are inexpensive and are easy to machine and bond.

To obtain more accurate quantitative results for design use, thick-adherend specimens, which need to be machined to symmetric configuration for lap-shear testing, should be used. However, these specimens are more expensive to fabricate and are larger. Consequently, they require larger

environmental-exposure facilities. In addition, the loads involved to test them may be higher and may require more sophisticated methods of application and monitoring. In this case, "the quantitative aspects do outweigh the cost disadvantages once the preliminary screening has been accomplished" (Ref 3).

Testing of structural configurations in limited numbers is performed as the final stage in a testing and design program for verification purposes. This testing procedure is obviously costly and provides statistically limited data. It does, however, serve the important purpose of identifying any configuration or design fault that may limit the durability of the integral structure (Ref 3).

Analysis of Test Results and Joint Specimen-Test Result Relationship

This section of the article describes viscoelasticity-viscoplasticity considerations, examples of the assessment of temperature-dependent delayed failure behavior of bonded single-lap joints, and the determination of stress-concentration factors for single-lap specimens.

Viscoelasticity-Viscoplasticity Considerations

Creep and Delayed Failure. Creep behavior of structural adhesives is an important design consideration, especially if high levels of stress, temperature, and humidity are present. Mechanical properties of polymeric materials are affected by creep, par-

ticularly at elevated temperatures and humidity. Creep is time- and temperature-dependent inelastic deformation and is usually measured in a test environment of constant stress and constant temperature. Designers are usually interested in two aspects of creep: creep at low stress levels over a long period of time, in which creep strains are described as a function of time for constant stress and temperature values; and creep at high stress levels resulting in delayed failures. Each of these aspects is described below.

Creep at Low Stress Levels. There are a number of empirical and theoretical methods to express creep strains as functions of time. A complete empirical method for predicting creep curves has been described by Larke and Parker (Ref 8). They use mathematical equations for primary, secondary, and tertiary tensile creep regions to use existing creep data for the prediction of other creep curves. Their method also enables estimation of strain and time values when the primary creep merges into the secondary stage and the secondary creep merges into the tertiary stage.

Theoretical equations describing creep strains as functions of time and stress are based on linear or nonlinear viscoelastic constitutive equations. Each category is described below.

Linear constitutive equations include:

- The Kelvin model, which has the constitutive equation (to apply to pure shear):

$$\dot{\gamma} + (G\gamma)/\mu = \tau/\mu \qquad \text{(Eq 7)}$$

where the dot in this equation and others that follow designates time differentiation, and G and μ refer to the shear elastic modulus and viscosity coefficient, respectively, provides the creep equation:

$$\gamma = (\tau_0/G)\{1 - \exp[-(Gt)/\mu]\} \qquad \text{(Eq 8)}$$

where τ_0 is the level of constant shear stress. Equation 8 represents delayed elasticity in creep as the creep strain γ approaches the value (τ_0/G) at infinite time.

- The Maxwell model, which has the constitutive equation:

$$\dot{\gamma} = (\dot{\tau}/G) + (\tau/\mu) \qquad \text{(Eq 9)}$$

provides the creep equation:

$$\gamma = (\tau_0/\mu)[t + (\mu/G)] \qquad \text{(Eq 10)}$$

The Maxwell model is usually suitable for the characterization of noncross-linked polymers because their deformation process may occur in the form of flow under long time or slow rate loading.

- The modified Bingham model (Ref 9) (Fig. 11), which has the constitutive equation:

$$\gamma = \tau/G \ (\tau \le \theta)$$
$$\dot{\gamma} = (\dot{\tau}/G) + [(\tau - \theta)/\mu] \ (\theta < \tau < \tau_{ult}) \qquad \text{(Eq 11)}$$

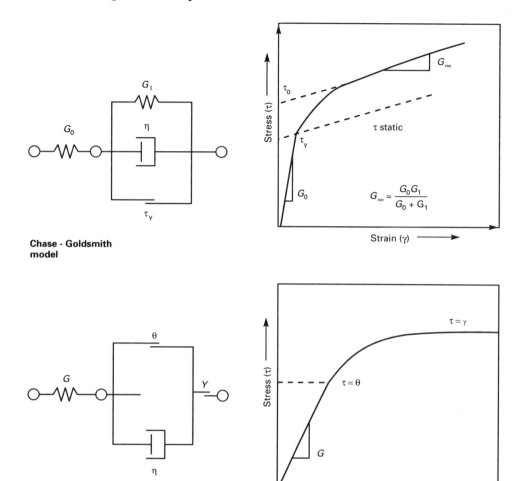

Chase - Goldsmith model

$$G_\infty = \frac{G_0 G_1}{G_0 + G_1}$$

Modified Bingham model

Fig. 11 Two viscoelastic overstress models used to describe adhesive behavior

where θ and τ_{ult} are elastic limit and ultimate shear stresses, respectively, provides the creep equation:

$$\gamma = [(\tau_0 - \theta)t/\mu] + (\tau_0/G) \quad (\tau_0 > \theta) \qquad \text{(Eq 12)}$$

This model allows the inclusion of plastic stress via τ_{ult} and was used by Sancaktar to describe the constant strain rate shear stress-strain behavior of two different cross-linked (epoxy) thermosetting adhesives in the bonded form (Ref 10).

- The Chase-Goldsmith model (Ref 11) (Fig. 11), which has the constitutive equations:

$$\gamma = \tau/G_0 \quad (\tau \leq \tau_s)$$
$$\dot{\gamma} = (\dot{\tau}/G_0) + [(\tau - \tau_s)/\mu'] \quad (\tau > \tau_s) \qquad \text{(Eq 13)}$$

where $\tau_s = (G_0(\theta + G_1\gamma)/(G_0 + G_1)$ and $\mu' = G_0\mu/(G_0 + G_1)$, will provide the creep equation:

$$\gamma = (\theta/G_1)[\exp(-G_1 t/\mu) - 1] + (\tau_0/G_\infty)$$
$$[1 - \exp(-G_1 t/\mu)] + (\tau_0/G_0)\exp(-G_1 t/\mu) \qquad \text{(Eq 14)}$$

where $G_\infty = (G_0 G_1)/(G_0 + G_1)$. The Chase-Goldsmith model was applied by Sancaktar *et al.* to describe the creep be-

havior of LARC-3 thermoplastic adhesive in the bonded lap-shear mode (Ref 12).

In the nonlinear constitutive equation, the nonlinear power-law response in creep is represented by:

$$\gamma = \tau_0\{D_0 + D_1[t \exp(\psi\tau_0)]^n\} \qquad \text{(Eq 15)}$$

where D_0 is the instantaneous compliance and D_1, n, and ψ are material parameters that characterize the nonlinear time- and stress-dependent behavior. Equation 15 is described on the basis of Schapery's nonlinear constitutive equation given in Ref 13. It was applied by Sancaktar *et al.* in creep characterization of bulk thermoplastic resins (Ref 14).

The well-established time-temperature superposition principle (TTSP) can be used in conjunction with the aforementioned creep relations (Eq 8, 10, 12, 14, and 15) to extend the creep data obtained at higher temperatures to values at longer times in the same strain range. As long as no structural changes occur in the material at high temperatures, TTSP can be used for amorphous

and semicrystalline thermoplastics and thermosets for extrapolation purposes. The application of the TTSP can even be extended into the nonlinear viscoelastic region (Ref 15). In order to apply this principle, one should first construct a "master curve," in which the complete compliance-time behavior is plotted at a constant temperature. For the "master curve" a reference temperature T_0 is chosen. The relation for the compliance observed at any time (t) and temperature (T_0) in terms of the experimentally observed compliance values at different temperatures (T) is given as (Ref 16):

$$D(T_0,t) = [\rho(T)T/\rho(T_0)T_0]D(T,t/a_T) \qquad \text{(Eq 16)}$$

where $D(T,t) = \gamma(T,t)/\tau$ and $\rho(T)$ is the temperature-dependent density. The term a_T that appears in Eq 16 is a time shift factor, which can be obtained with the application of the well-known Williams-Landel-Ferry equation.

Creep at High Stress Levels. Two equations, one by Crochet and one by Zhurkov, are discussed below. In relation to Crochet's delayed failure equation, Naghdi *et al.* (Ref 17) defined a yield surface as a function of stress (σ_{ij}), plastic strain (ϵ^P_{ij}), and work hardening due to the path history (κ_{ij}) and time history of loading-deformation (χ_{ij}) so that:

$$f = f(\sigma_{ij}, \epsilon^P_{ij}, \kappa_{ij}, \chi_{ij}) = 0 \qquad \text{(Eq 17)}$$

The general yield condition given by Eq 17 was later simplified (Ref 17, 18) with the use of a von Mises yield criterion to the form:

$$f = (1/2)S_{ij}S_{ij} - k^2 = 0 \quad [k = k(\chi, \kappa)] \qquad \text{(Eq 18)}$$

where S_{ij} is the deviatoric stress tensor.

On the basis of the theory of elastic perfectly plastic solids, Crochet neglected the work-hardening term and assumed k to be a function of χ only (Ref 18). He described k as a monotonically decreasing function of χ and asserted that $k(\chi)$ has a lower bound for large values of χ. This behavior was graphically shown by Crochet (Ref 18). The functional behavior of k described above was later interpreted mathematically by Brinson (Ref 9) in the form:

$$Y(t) = A' + B' \exp(- C'\chi) \qquad \text{(Eq 19)}$$

where $Y(t)$ is the time-dependent maximum stress and A', B', and C' are material parameters. The specific functional form of χ was described by Crochet as:

$$\chi = [(\epsilon^V_{ij} - \epsilon^E_{ij})(\epsilon^V_{ij} - \epsilon^E_{ij})]^{1/2} \qquad \text{(Eq 20)}$$

where the superscripts V and E refer to the viscoelastic and elastic strains, respectively. Equation 20 was later interpreted by Sancaktar (Ref 12) to apply to the lap-shear condition with:

$$\tau = A + B \exp(- C\chi_s) \qquad \text{(Eq 21)}$$

$$\chi_s = \gamma^V_{12} - \gamma^E_{12} \qquad \text{(Eq 22)}$$

and γ_{12} refers to the familiar "engineering" shear strain. Apparently, an expression for χ_s can be obtained by subtracting the elastic shear strain $\gamma^E = \tau/G$ from the proposed creep equation (such as Eq 8, 10, 12, 14, and 15) to obtain χ_s. This can be done because delayed failure occurs only in those elements loaded up into the viscoelastic region.

Following the procedure described above, the use of the Maxwell model creep relation (Eq 10) to evaluate the χ_s (Eq 22) first and subsequent substitution in Eq 21 results in:

$$\tau = A + B \exp(-C\tau t/\mu) \qquad (Eq\ 23)$$

Following the same procedure with the modified Bingham model, we get:

$$\tau = A + B \exp(-C(\tau - \theta)t/\mu) \qquad (Eq\ 24)$$

Similar application with the Kelvin model results in:

$$\tau = A + B \exp[C(\tau/G) \exp(-Gt/\mu)] \qquad (Eq\ 25)$$

The asymptotic value for Eq 23 and 24 is the material constant A, whereas for Eq 25 it is $A + B$. Apparently, Eq 25 describes a much faster approach in time of the rupture stresses to their asymptotic values. Other constitutive models such as the Chase-Goldsmith model and the nonlinear model described can be applied similarly to result in other forms of the delayed-failure relation (Eq 21).

The asymptotic values A of Eq 23 and 24 and $A + B$ of Eq 25 are important design parameters, because they represent the maximum safe stress values below which creep failures are not expected to occur. Sancaktar proposed to express the asymptotic values of Crochet's Eq 21 (such as the values A and/or $A + B$) as a function of the applied temperatures to represent experimental data in an empirical fashion (Ref 19, 20, 21). The application of Crochet's equation to high-temperature creep data is justified by the applicability of the time-temperature superposition principle to the creep equations (Eq 8, 10, 12, 14, and 15).

The second equation to be described here, Zhurkov's temperature-dependent creep-rupture equation, is based on the theory of chemical reaction rates, which employs an exponential relation between time and applied stress. The theory of rate processes assumes that the rate at which the number of unbroken (polymeric) molecular bonds N diminishes is proportional to the number of unbroken bonds, that is:

$$dN/dt = -\zeta N \qquad (Eq\ 26)$$

The rate factor ξ is assumed to be proportional to the probability of bond breakage due to thermal fluctuation and tensile force with the equation:

$$\zeta = (1/t_0) \exp\{-[U_0 - F(\sigma)]/KT\} \qquad (Eq\ 27)$$

where K is Boltzmann's constant, T is the absolute temperature, U_0 is the activation energy (interatomic bond energy for polymers), t_0 is a constant on the order of the molecular oscillation period of 10^{-12} seconds, and $F(\sigma)$ is a function of the applied stress. The form of $F(\sigma)$ can be arbitrarily chosen as $\lambda\sigma$, where the material constant (λ) alludes to the free energy density functional of the stress history for viscoelastic solids. On this basis, Zhurkov et al. (Ref 22) determined that the time-to-rupture (t_r) of fibers of strips subjected to uniaxial stress (σ) was given by:

$$t_r = t_0 \exp[(U_0 - \lambda\sigma)/KT] \qquad (Eq\ 28)$$

Because Eq 28 permits failure at vanishing stress, it was modified by Slonimskii et al. (Ref 23) to the form:

$$t_r = t_0 \exp[(U_0/RT) - \alpha\sigma] \qquad (Eq\ 29)$$

where U_0 is a constant activation energy per mole (associated as an energy barrier term), R is the universal gas constant, and α is a constant. It should be noted that a theoretical relation similar in form to Eq 28 was reported by Bueche in 1958.

This author believes that the limited number of material constants available in Eq 29 for modeling complex multiphase epoxy materials, some of which may also contain carrier mats, severely limits its predictive ability, especially when the adhesive is in the bonded lap-shear mode. Indeed, experiments performed by Sancaktar et al. (Ref 20, 21) on the creep-rupture behavior of three different structural adhesives in the bonded (single) lap-shear mode confirmed this assertion. The use of Crochet's equation provides much more latitude in fitting creep-rupture data because of the availability of many different viscoelastic models, including the nonlinear ones. Even when the application of a viscoelastic model to the constant strain rate or when the application of creep stress-strain data of an adhesive is not possible (or not known), Crochet's equation can still be applied in an empirical fashion to represent the temperature-dependent creep-rupture data. Data on the room- and elevated-temperature creep behavior of three different structural adhesives in the bonded (single) lap-shear form will be presented in the following section to illustrate the applicability of Crochet's equation for describing temperature-dependent delayed-failure behavior.

As mentioned earlier, the single-lap joint geometry induces severe cleavage and shear stress concentrations at the overlap edges because of the eccentricity of the applied load and unbalanced shear load moments on the bond line. Furthermore, if the adhesive material exhibits nonlinear viscoelastic constitutive behavior, it is possible that the creep behavior of the adhesive will be enhanced at higher stress levels, as described by Eq 15. In fact, Brinson et al. reported that such stress-enhanced material nonlinearities can be present even at low stress levels, except that "they may not become apparent for very long times" (Ref 24). Obviously, when stress-enhanced creep is present, the overlap edges of a single-lap joint will creep faster than the rest of the bond line, and, consequently, creep failures will initiate at those locations. Therefore, it is desirable to know the magnitude of stress concentrations at the overlap edges in order to be able to interpret any delayed failure data obtained with the use of single-lap joint geometry. A procedure to obtain such stress concentration values based on the analysis of Goland and Reissner (Ref 25) will also be presented in this article following the presentation of the experimental data.

Examples of the Assessment of Temperature-Dependent Delayed-Failure Behavior of Bonded Single-Lap Joints

In order to assess the applicability of Crochet's delayed-failure equation and its empirical extension to include the temperature effects, data obtained by Sancaktar et al. (Ref 20, 21) on the room- and elevated-temperature creep-rupture behavior of three different structural adhesives (in the bonded lap-shear mode) will be presented. The model adhesives are: FM 73, FM 300, and thermoplastic polyimidesulfone.

Experimental Procedures. The model adhesives were used to bond two 25 mm (1 in.) wide, 1.27 mm (0.05 in.) thick titanium alloy strips that were exact duplicates. The resulting single-lap specimens were in accordance with ASTM D 1002 specifications. The overlap length was 12.7 mm (0.5 in.), and the adhesive thicknesses varied from approximately 0.127 mm (0.005 in.) to 0.229 mm (0.009 in.). All test specimens were prepared at NASA Langley Research Center (LRC). The preparation of thermoplastic polyimidesulfone adhesive and the processing of the test specimens have been described in detail by St. Clair (Ref 26). Thermoplastic polyimidesulfone is a high-temperature adhesive with thermoplastic properties and solvent resistance. It is a novel adhesive developed at NASA LRC.

The adhesive FM 300 is a modified-epoxy adhesive film supplied with a tricot knit carrier cloth, which it is claimed offers a good mixture of structural and handling properties (Ref 27). Its product information brochure reports a service range of -55 to $150\ °C$ (-67 to $300\ °F$), with excellent moisture and corrosion resistance. The adhesive FM 73 is a modified-epoxy adhesive film with a polyester knit fabric that the manufacturer claims offers optimum physical properties (Ref 27). Its product information brochure reports a service range of -55 to $120\ °C$ (-67 to $248\ °F$), with high resistance to moisture. The bonding procedures for FM 73 and FM 300 adhesives are described in detail in their product information brochures (Ref 27).

The shear stress in the lap joints was assumed to be uniform and was calculated by dividing the load applied to the adherends by the overlap area. The shear strains were defined as bond line deformation divided by the average bond line thickness. For this reason, precision measurement of bond line thicknesses and bond line lengths were made with the aid of a microscope under a magnification of 40×. A high-precision, air-cooled clip-on gage was used for shear strain measurements (see Fig. 12b).

Two notches were milled 25.1 mm (0.99 in.) apart on the overlap side of the single-lap specimens so that the clip-on gages could be attached in them. The notches were 2.54 mm (0.1 in.) deep and 0.254 mm (0.01 in.) wide. Stress concentrations due to the notches in the adherends were estimated to be negligible. Any adherend deformation that occurred within the measuring range of the clip-on gage was calculated and subtracted from the experimental data to obtain the adhesive strains.

Groups of 24 to 28 single-lap specimens bonded with the model adhesives were provided by NASA LRC. The mechanical testing of these single-lap specimens was performed either on a Tinius-Olsen Universal testing machine or on a Model 1331 Instron Servohydraulic testing machine, depending on machine availability. Room- and high-temperature creep tests were performed with an initial crosshead rate of 7.6 mm/min (0.3 in./min). Each adhesive was tested in creep at a minimum of four different stress levels above the elastic limit shear stress (found from both room- and high-temperature constant strain rate tests) and at three or four different temperature levels. The FM 73 adhesive creep tests were performed at temperature levels of 21 °C (70 °F), 54 °C (130 °F), and 82 °C (180 °F). The FM 300 adhesive creep tests were performed at temperature levels of 21 °C (70 °F), 82 °C (180 °F), 121 °C (250 °F), and 148 °C (300 °F). The thermoplastic polyimidesulfone adhesive creep tests were performed at temperature levels of 21 °C (70 °F), 121 °C (250 °F), 177 °C (350 °F), and 232 °C (450 °F). The maximum test temperature for each adhesive was determined on the basis of product information available in the literature (Ref 26, 27). The high-temperature creep tests were performed with the aid of an environmental chamber, which was attached to the testing machine.

Experimental Results and Discussion. Experimental results revealed that the constant strain rate shear stress-strain behavior of both FM 73 and FM 300 adhesives could be represented with the use of the modified Bingham model (Ref 10). Room-temperature creep tests on FM 73 adhesive showed significant viscoelastic-plastic effects in the

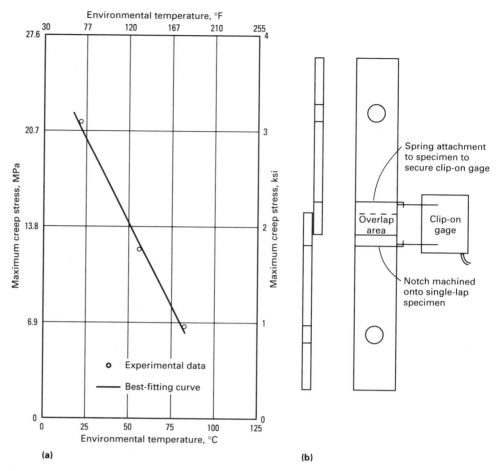

(a)

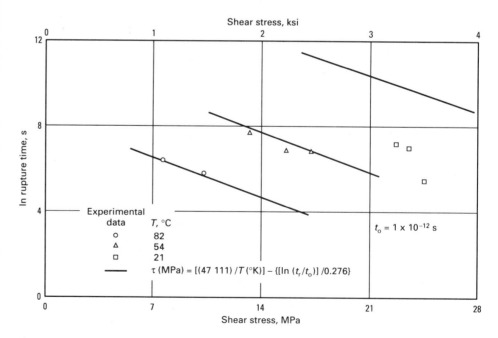

(b)

Fig. 12 (a) Variation of maximum (safe) creep stress with environmental temperature for FM 73 adhesive. (b) Gage used for shear strain measurements

Fig. 13 Creep-rupture data and comparison with Zhurkov's equation for FM 73 adhesive

tertiary region where high levels of creep stress were applied. Higher magnitudes of creep strains were reached at elevated temperatures. The sudden plastic flow to failure

(the tertiary region) observed at room temperature was not evident at elevated temperatures. Creep-rupture data for FM 73 are shown in Fig. 13 and 14 and Table 1. Figure

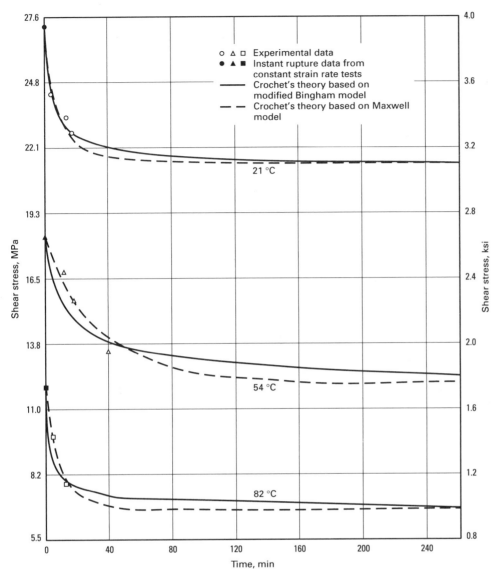

Fig. 14 Creep-rupture data and comparison with Crochet's equation for FM 73 adhesive. The asymptotic values are well defined experimentally by no failure at long times.

Table 1   Material constants obtained on the basis of Crochet's equation for FM 73 creep-rupture data

A		B		θ		K, × 10⁻⁶		C, × 10⁻⁶		T	
MPa	ksi	MPa	ksi	MPa	ksi	1/KPa · min	1/psi · min	1/KPa · min	1/psi · min	°C	°F
21.4	3.097	5.7	0.826	20.7	2.999	40.0	276	4.5	31	21	70
12.1	1.750	6.9	0.960	11.9	1.727	17.4	120	2.3	16	54	130
6.8	0.982	5.3	0.768	6.7	0.975	94.7	653	13.5	93	82	180

Note: See Fig. 14.

13 shows the comparison of Zhurkov's (modified) equation (Eq 29) with the experimental data. As may be observed, data and theory show good agreement at elevated temperatures. The room-temperature data, however, do not show agreement with theory. The constants U_0 and α used to fit Zhurkov's equation to the creep-rupture data are also shown in Fig. 13.

Crochet's equation (Eq 21) is shown in Fig. 14. It can be seen that the behavior is a decaying exponential one with respect to time. The plots reach an asymptotic stress level, below which creep failures are not expected to occur. The highest points in Fig. 14 (shown as solid black symbols) represent ultimate shear stress values obtained from the highest rate shear stress-strain curves (Ref 10). Figure 14 reveals that the delayed-failure behavior of FM 73 can be predicted accurately by employing either the Maxwell or modified Bingham models. The solid and dashed lines in Fig. 14 represent Crochet's relations (Eq 23 and 24)

based on the Maxwell and modified Bingham models, respectively. The elastic limit stresses (θ) in Eq 24 are determined from the constant strain rate stress-strain data (Ref 10) by extrapolation. The appropriate constants used in fitting Crochet's equation to FM 73 creep-rupture data are shown in Table 1.

Figure 12(a) shows the variation of the maximum (safe) level of creep stress values (below which creep failures are not expected to occur) with environmental temperature. The maximum creep stress values represent the asymptotic stress levels (that is, material constants A in Table 1) obtained by the application of Crochet's equation. Apparently, the relation between the asymptotic shear stress level and temperature is a linear one for the range of data shown. Figure 12(b) shows the specimen geometry and the method of strain measurement used in obtaining this creep data.

Room- and elevated-temperature creep testing of FM 300 adhesive revealed the presence of tertiary creep regions. As expected, higher levels of creep strain were reached at elevated temperatures. Comparison of experimental creep-rupture data with Zhurkov's relation (Eq 29) resulted in poor agreement. Comparison of the creep-rupture data with Crochet's equation based on the Maxwell and modified Bingham models (Eq 23 and 24) is shown in Fig. 15. The appropriate constants used in fitting Crochet's equations to FM 300 creep-rupture data are shown in Table 2. The use of the modified Bingham model results in a slightly better fit. Figure 16 shows variation of the maximum (safe) level of creep stress values (that is, material constants A in Table 2) with environmental temperature. Apparently, the relation between the asymptotic stress level and temperature is a nonlinear one, for the range shown.

Experimental results revealed that thermoplastic polyimidesulfone exhibits weak rate dependence. For this reason, a nonlinear elastic powder function relation of the form:

$$\tau = P\gamma^m \qquad \text{(Eq 30)}$$

where P and m are material constants, was fitted to the constant strain rate stress-strain data (Ref 10). Room-temperature creep data on thermoplastic polyimidesulfone revealed lower secondary creep rates in comparison to the other thermosetting adhesives studied. The presence of tertiary creep regions was evident in both room- and elevated-temperature creep data. The creep behavior of polyimidesulfone was affected considerably by the interfacial (adhesive-adherend and adhesive-fiber) fracture processes, especially at elevated temperatures. Visual examination of postfailure surfaces revealed that the extent of interfacial separation increased with increasing environmental temperature, as shown in Fig. 17. This

Table 2 Material constants obtained on the basis of Crochet's equation for FM 300 creep-rupture data

A		B		θ		K, × 10⁻⁶		C, × 10⁻⁶		T	
MPa	ksi	MPa	ksi	MPa	ksi	1/KPa · min	1/psi · min	1/KPa · min	1/psi · min	°C	°F
25.3	3.675	5.0	0.725	20.5	2.974	34.7	239	12.2	84	21	70
19.3	2.800	11.0	1.600	16.8	2.441	20.0	138	3.48	24	82	180
14.3	2.075	10.9	1.577	13.8	2.005	79.6	549	16.2	112	121	250
9.3	1.350	10.9	1.579	9.3	1.343	115	796	24.1	166	149	300

Note: See Fig. 15.

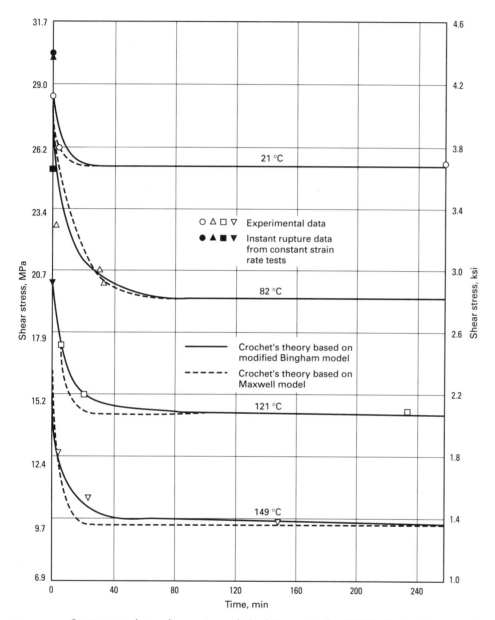

Fig. 15 Creep-rupture data and comparison with Crochet's equation for FM 300 adhesive. The asymptotic values are well defined experimentally by no failure at long times.

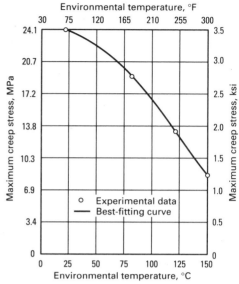

Fig. 16 Variation of maximum (safe) creep stress with environmental temperature for FM 300 adhesive

Fig. 17 Typical fracture surfaces of titanium-polyimidesulfone specimens exhibiting the effects of increasing temperature. Environmental temperatures were (left to right) 21 °C (70 °F), 120 °C (250 °F), 175 °C (350 °F), and 230 °C (450 °F).

was attributed to inadequate surface treatment of the carrier cloth and adherends. Consequently, the interfacial bonds deteriorated quickly at elevated temperatures. Such interfacial separations resulted in failures without appreciable adhesive deformation. Hence, lower levels of creep strains were observed at comparable stress levels when the test temperatures were increased.

Comparison of Zhurkov's equation with the experimental data showed poor agreement. The author attributes this partially to the interfacial fracture processes described above. The discrepancy between the data and theory was greater at the room-temperature and 232 °C (450 °F) conditions.

Figure 18 and Table 3 show comparison of creep-rupture data with Crochet's equa-

tion based on the Maxwell model (Eq 23). Because constant strain rate stress-strain behavior of the polyimidesulfone adhesive was determined to be approximately nonlinear elastic (that is, a viscoelastic model could not be fitted to the experimental data) and because the creep behavior could not be characterized with any viscoelastic model, the simplest model that yielded the best-fitting form of Crochet's equation was used. On this basis, Crochet's equation provided a better fit to the experimental data than did Zhurkov's equation. The appropriate constants used in fitting Crochet's equation (based on the Maxwell mode) to the creep-rupture data are shown in Table 3.

Figure 19 shows variation of the maximum (safe) level of creep stress values (below which creep failures are not expected to occur) with environmental temperature. The maximum creep stress values represent the asymptotic shear stress levels (that is, material constants A in Table 3) obtained with the application of Crochet's

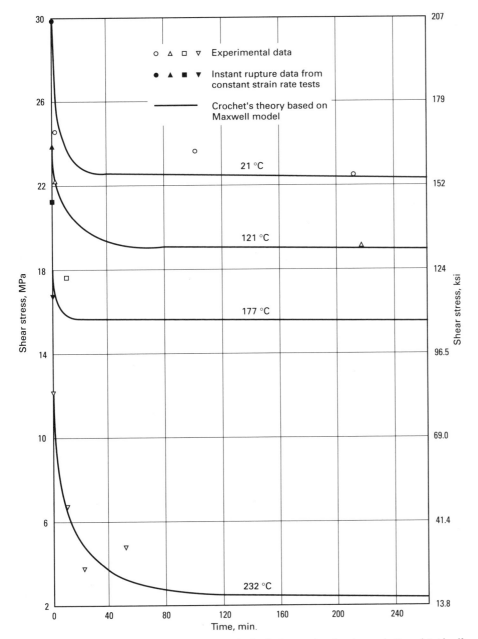

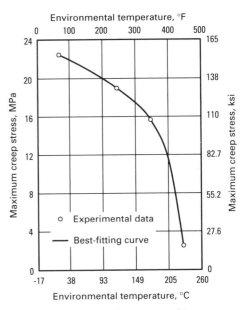

Fig. 19 Variation of maximum (safe) creep stress with environmental temperature for thermoplastic polyimidesulfone adhesive

Fig. 18 Creep-rupture data and comparison with Crochet's equation for thermoplastic polyimidesulfone adhesive. The asymptotic values are well defined experimentally by no failure at long times.

Table 3 Material constants obtained on the basis of Crochet's equation for thermoplastic polyimidesulfone

A		B		$C, \times 10^{-6}$		T	
MPa	ksi	MPa	ksi	1/KPa · min	1/psi · min	°C	°F
22.5	3.267	7.3	1.054	6.77	46.7	21	70
19.1	2.765	4.7	0.685	3.97	27.4	121	250
15.7	2.270	7.4	1.069	22.9	157.7	177	350
2.6	0.370	14.1	2.052	15.9	109.4	232	450

Note: See Fig. 18.

equation. A sharp decrease in the level of maximum creep stress is observed for temperatures above 177 °C (305 °F).

Based on the single-lap-shear experimental data from the three adhesives studied, we can conclude that Crochet's equation provides a better description of the temperature-dependent creep-rupture data than does Zhurkov's equation. Zhurkov's equation, in general, provided poor fit for the room-temperature creep-rupture data. Furthermore, when interfacial separations were present (as is common in adhesively bonded joints), the fit was poor at all temperatures. Additional discussions on the behavior of the interphase will be presented later in this article. It should also be noted that crack propagations originating from stress concentration sites (see next paragraph) and inherent flaws in the bond line may also be the dominant cause of joint failure. Consequently, discussions on the fracture behavior of adhesive joints, including viscoelastic fracture, will also be presented later in this paper. More information on adhesive joint fracture can be found in the article "Fatigue and Fracture Mechanics" in this Volume. A comparison of the mechanical properties of FM 73, FM 300, and thermoplastic polyimidesulfone adhesives at comparable testing conditions is shown in Table 4.

Determination of Stress-Concentration Factors for Single-Lap Specimens

The quantity reported above and in many other publications as the lap-shear strength is the breaking load per unit bond area. However, failure initiation in materials is a localized phenomenon that is dependent on the maximum stresses at a point reaching some critical value. In single-lap specimens, the maximum stress is reached at the overlap edges.

Volkersen (Ref 28) analyzed the stress distribution in single-lap joint geometry under a load due to stretching of the adherends while ignoring the tearing stresses at the free ends. He considered the case in which two identical elastic adherends of uniform thickness (t) and Young's moduli (E) were bonded over a length (2C) by an adhesive of thickness (m), which was assumed to be-

Table 4 Mechanical properties of FM 73, FM 300, and thermoplastic polyimidesulfone adhesives in the bonded single-lap mode at comparable testing conditions

Property and test condition	Adhesive		
	FM 73	FM 300	Thermoplastic polyimidesulfone
At 21 °C (70 °F), elastic shear modulus, MPa (ksi)	87.6 (12.7)	136.1 (19.7)	173.0 (25.1)
At 21 °C (70 °F)			
Under constant strain rate conditions, %/s	0.75	1.94	1.35
Maximum shear stress, MPa (psi)	23.9 (3469)	29.7 (4300)	27.2 (3947)
Maximum shear strain, %	58.5	31.2	59.6
At 121 °C (250 °F)			
Under constant strain rate conditions, %/s	2.36	384	59.1
Maximum shear stress, MPa (psi)	3.45 (500)	25.2 (3652)	23.8 (3450)
Maximum shear strain, %	107	72.8	37.5
At room temperature			
Creep strain, %	85.25	32.01	56.76
At comparable creep stresses, MPa (psi)	24.1 (3499)	24.4 (3535)	23.6 (3423)
At 82 °C (180 °F)			
Maximum creep stress, MPa (psi)	7.6 (1104)	20.2 (2925)	. . .
Maximum creep strain, %	130	65	. . .
At 121 °C (250 °F)			
Maximum creep stress, MPa (psi)	. . .	15.2 (2200)	21.5 (3117)
Maximum creep strain, %	. . .	125	37.76
At 21 °C (70 °F), maximum (safe) level of creep stress, MPa (ksi)	21.4 (3.097)	25.3 (3.675)	22.5 (3.267)
At 82 °C (180 °F), maximum (safe) level of creep stress, MPa (ksi)	6.8 (0.982)	19.3 (2.800)	. . .
At 121 °C (250 °F), maximum (safe) level of creep stress, MPa (ksi)	. . .	14.3 (2.075)	19.1 (2.765)

Table 5 Overlap edge stress concentration values for single-lap joints bonded with various structural adhesives

Adhesive	Specimen	n_V	n_{GR}	n_τ	n_σ
FM 73	SJ-9-4	1.8828	2.4552	2.7458	3.9752
	SJ-10-2	2.1493	2.8000	3.1404	4.5404
FM 300	2078-3	2.0283	2.6549	2.9721	4.3081
	2083-3	2.0017	2.6264	2.9404	4.2626
	2078-2	1.9937	2.6126	2.9242	4.2376
	2083-4	1.8189	2.4305	2.7164	3.9295
Thermoplastic polyimidesulfone	2027T-1	2.3917	3.0312	3.4023	4.9476
	2029T-3	2.3684	3.0296	3.3992	4.9408

Note: For FM 73 and FM 300 adhesives, E_a = 2.324 GPa (0.337 × 10⁶ psi) (Ref 29) and ν = 0.40 (Ref 29). For thermoplastic polyimidesulfone, E_a = 4.958 GPa (0.719 × 10⁶ psi) (Ref 26) and ν = 0.38 (Ref 26).

have like an elastic solid with shear modulus (G_a). The bonded faces of the adherends were assumed to be parallel so that the thickness m was considered constant.

Volkersen's analysis showed that when the bonded members are in pure tension, the shear stress in the adhesive layer is a maximum at each end of the overlap. He compared the maximum shearing stress at the end of the overlap with the mean stress to evaluate the stress concentration factor (n_V):

$$n_V = \zeta \coth(\zeta) \qquad (Eq\ 31)$$

where

$$\zeta = [(2C^2G_a)/(Etm)]^{1/2} \qquad (Eq\ 32)$$

Goland and Reissner (Ref 25), in their analysis of the single-lap joint geometry, showed the existence of high stress concentrations near each end of the overlap region, in the form of peak normal (tearing or cleavage) and shear adhesive stresses. The magnitudes of these stresses depend on:

- The flexibility of the adhesive
- The length of overlap
- Young's moduli of the adherends
- The thickness of the adherends
- The thickness of the adhesive layer

Goland and Reissner categorized adhesively bonded joints in two approximate groups to simplify their analysis.

In the first approximation, the work of the normal (σ_0) and shear (τ_0) adhesive stresses were neglected in comparison to the work of the adherend stresses (σ_y and τ_{xy}), which were assumed to be continuous across the adhesive layer. Adhesive layers were considered inflexible for joints satisfying the conditions:

$$(m/E) << (t/E)\ \text{or}\ (m/G) << (t/G) \qquad (Eq\ 33)$$

In the second application, the work of the adherend stresses was ignored in comparison to the work of adhesive stresses (σ_y and τ_{xy}) for joints satisfying the conditions:

$$(m/E) >> (t/E)\ \text{or}\ (m/G) >> (t/G) \qquad (Eq\ 34)$$

Such joints were considered to have flexible adhesive layers. The single-lap joints that were used for the experimental program reported above can be considered to have flexible adhesive layers. Based on the peak shear stresses given by the Goland and Reissner analysis, the stress concentration factor (n_{GR}) for flexible joints can be calculated as:

$$n_{GR} = (1/4)[(BC/t)(1 + 3k) \coth (BC/t) + 3(1 - k)] \qquad (Eq\ 35)$$

where

$$B = (8G_at/Em)^{1/2} \qquad (Eq\ 36)$$

$$k = [\cosh(UC)]/\{\cosh(UC) + 2[2\sinh(UC)]^{1/2}\} \qquad (Eq\ 37)$$

$$UC = [3(1 - \nu^2)/2]^{1/2}(C/t)[F/(tWE)]^{1/2} \qquad (Eq\ 38)$$

and

$$Y = (6E_at/Em)^{1/4} \qquad (Eq\ 39)$$

In addition, ν = Poisson's ratio for the adherends (0.3), F = load applied to the adherends, E_a = elastic modulus for the adhesive, m = adhesive layer thickness, and W = width of the adherend. It should be noted that this approach again neglects the presence of tearing stresses in the adhesive layer. A more accurate analysis should include the effect of the tearing stresses as described below.

Because both normal (($\sigma_0)_{max}$) and shear (($\tau_0)_{max}$) stresses exist at the overlap edge, principal (normal) (σ_1) and maximum shear (τ_{max}) stresses need to be calculated for the assessment of the failure condition. These stresses can be calculated with the use of:

$$\sigma_1 = (\sigma_0)_{max}/2 + \{[(\sigma_0)_{max}^2/2]^2 + [(\tau_0)_{max}]^2\}^{1/2} \qquad (Eq\ 40)$$

and

$$\tau_{max} = \{[(\sigma_0)_{max}^2/2]^2 + [(\tau_0)_{max}]^2\}^{1/2} \qquad (Eq\ 41)$$

where ($\tau_0)_{max}$ is given by Eq 35 and ($\sigma_0)_{max}$ is determined by the method of Goland and Reissner to be:

$$(\sigma_0)_{max} = (F/tW)(C/t)^2\{(L^2k/2)\ [\sinh(2L) - \sin(2L)]/[\sinh(2L) + \sin(2L)] + Lk'[\cosh(2L) + \cos(2L)]/[\sinh(2L) + \sin(2L)]\} \qquad (Eq\ 42)$$

where

$$L = (YC/t) \qquad (Eq\ 43)$$

$$k' = (V_0CW/Ft) \qquad (Eq\ 44)$$

and

$$V_0 = (kF/W)[3F(1 - \nu^2)/(tEW)]^{1/2} \qquad (Eq\ 45)$$

The stress-concentration factors n_τ and n_σ can now be calculated by dividing τ_{max} and σ_1, respectively, by the average shear stress (τ = F/A).

The four methods above were applied by Sancaktar et al. in calculating the stress-concentration factor (n) for single-lap joints

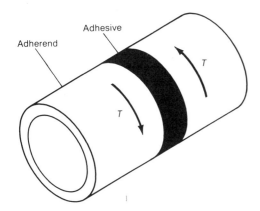

Fig. 20 Napkin ring test geometry

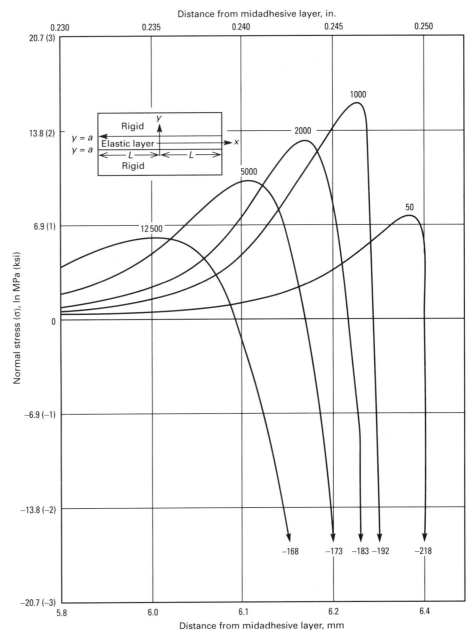

Fig. 21 Normal stress (σ) at the interface between 1.52 mm (0.06 in.) thick adhesive and adherends for moisture diffusion at 50 to 12 500 s. Arrows indicate values of σ at the exposed outer edge. Source: Ref 35

bonded with FM 73, FM 300, and thermoplastic polyimidesulfone adhesives (Ref 21), yielding the results shown in Table 5. As a first approximation, the stress values given in Fig. 14, 15, 16, 18, and 19 should be multiplied by one of these stress-concentration factors to approximate the actual failure stress at the overlap edges. Because the structural adhesives discussed here have bulk strength values of up to 62 MPa (9 ksi), the stress-concentration values in the range of 2 to 3 given in Table 4 should be considered reasonable. It should also be noted that these stress-concentration values are calculated on the basis of linear elastic theory and even though they represent stress ratios, their applicability in the presence of stress- or temperature-enhanced nonlinearities will be severely diminished.

Torsional Test Geometries. The shear stress-strain behavior of adhesives can also be effectively determined using torsional joint geometries. The napkin ring test geometry shown in Fig. 20 is made of hollow adherend tubes and offers the most uniform state of stress among the torsional specimens. Additional information on bonded torsional test specimens can be obtained from ASTM standards D 3658 and E 229 and from Ref 30 and 31. Grant recently reported an improved version of the napkin ring geometry with the use of a new cone-and-plate geometry (Ref 32). Grant claims that this new geometry has a more uniform shear state than does the standard napkin ring specimen.

Effects of Moisture Ingression on Joint Performance

Moisture ingression can affect a joint through two mechanisms, deterioration of the adhesive bulk, and deterioration of the interface. On the bulk of the adhesive, the effects of moisture manifest themselves in mechanical dilatation (swelling) and plasticization. In cross-linked polymers, the extent of swelling is related to the concentration of cross links by the relation (Ref 33):

$$-[\ln(1 - v_2) + v_2 + x_1 v_2^2] =$$
$$(V_1/v_2'M_c)[1 - (2M_c/M)][v_2^{1/2} - (v_2/2)] \quad (Eq\ 46)$$

where v_2 = volume fraction of polymer in swollen gel, also inverse of swelling ratio; V_1 = molar volume of solvent; v_2' = specific volume of polymer; x_1 = a solvent-polymer interaction term; M = molecular weight of polymer before cross linking; M_c = molecular weight of polymer chain between points of cross linking.

Wong (Ref 34) claimed that additional cross linking can take place in epoxy materials during a water sorption process. He attributed this to the plasticization of the epoxy network, which leads to higher segmental mobility and, consequently, increased diffusion of chain segments with unreacted epoxide and amino groups. Increases in molecular weight and limiting viscosity numbers have also been observed in alkylcyanoacrylate adhesives humid aged in tetrahydrofuran at 30 °C (86 °F) for 28 days. This finding, in light of Eq 46, reveals that the swelling stresses due to moisture ingression are a nonlinear function of time.

Weitsmann (Ref 35) determined the swelling stresses induced in Epoxide 3501 adhesive bonding rigid adherends with a 12.7 mm (0.5 in.) overlap. For this purpose he used variational principles and Fickian diffusion. His results show compressive

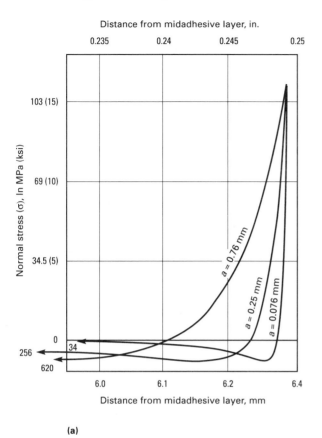

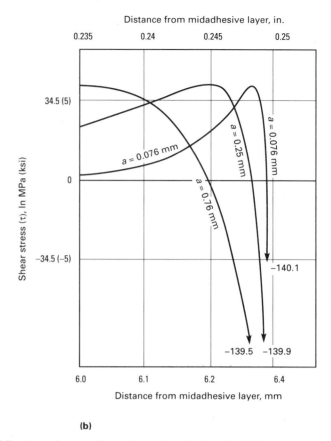

Fig. 22 (a) Normal stress, σ, at the interface between adhesive and adherend for the fully saturated case and for various indicated values of half-thickness, a. Arrows indicate values at the middle of the overlap. (b) Shear stress, τ, at the interface between adhesive and adherend for the fully saturated case and for various values of half-thickness, a. Arrows indicate values at the edge. Source: Ref 35

stresses at the overlap edges changing to tensile values along the interface during the diffusion process. The whole stress curve shifts inwards along the interface in time, while the tensile stress peaks decrease (Fig. 21). The interlaminar normal and shear stresses for the fully saturated case are shown in Fig. 22. It should be noted that in the fully saturated case, the edge peak stresses are tensile (Fig. 22a). It should also be noted that Weitsmann used the fully elastic adhesive material assumption in this study with swelling strains (for times where $0 < t < 12\,500$ s) given by the relation:

$$\epsilon(x) = 0.03 \text{ erfc } \{[1 - (x/L)]/2(k_w t/L^2)^{1/2}\} \quad (Eq\ 47)$$

where the swelling due to moisture saturation was assumed to be $\epsilon_0 = 3\%$ and $k_w =$ coefficient of diffusion ($k_w = 1.3 \times 10^{-6}$ mm²/s, or 2×10^{-9} in.²/s), and $2L =$ overlap length.

It is well known that the mechanical properties of polymer-base adhesives can be degraded by the absorption of moisture from humid environments. For example, Schmidt and Bell (Ref 36) report 66% and 70% reductions in tensile modulus and yield stresses, respectively, for 0.127 mm (5 mil) thick Epon 1001/V115 films that absorbed 3 to 3.5% water by weight as a result of immersion at 57 °C (134.6 °F) (Fig. 23).

Effect of Stress and Free Volume on Sorption Behavior. In most cases, bonded structures are subjected to static or dynamic stresses while they are exposed to environmental conditions. The effect of stress on the sorption and diffusion of organic vapors in polymers has been the topic of many previous investigations. For example, Treloar (Ref 37) found that uniaxial stress applied to cellulose filaments caused an increase in their equilibrium water content. Fahmy and Hurt correlated the changes in diffusion coefficient for 5208 epoxy resin in water to a uniaxial stress using the concept of free volume change due to an applied stress (Ref 34). They argued that the application of tensile stress resulted in an increase in the diffusion coefficient. Wong (Ref 34) reported that the sorption behavior in polymers is directly related to their free volume. He also reported that, in polymers subjected to water diffusion, the equilibrium water content and the sorption rate increase with increasing tensile stress. Wong also suggested that the resistance to creep deformation of an epoxy resin is significantly lowered upon absorption of water.

Recent work at Virginia Polytechnic Institute and State University (Ref 38-40) developed a model for the diffusion of moisture in adhesive joints. The model

developed is based on free volume theories and relates the diffusion coefficient for small penetrant molecules in polymers to temperature, strain, and penetrant concentration (Ref 38). In this work Lafebvre et al. assumed that the transport kinetics is governed by the constant redistribution of the free volume, caused by the segmental motions of the polymer chains. The stress dependence of solubility was predicted using the Hildebrand theory. A non-Fickian driving force arising from differential swelling was also included in their governing equations. Lefebvre et al. reported that externally applied stresses, residual stresses, and swelling stresses affect diffusion rate.

In the first part of their study (Ref 38), the authors concluded that the diffusion boundary value problem is highly nonlinear and is coupled with the mechanical response of the polymer. Their theoretical predictions were validated experimentally using a modified sorption technique (Ref 39). For this purpose, diffusivity measurements were performed under tensile, compressive, and pure shear stress states using polyimide films. The state of tensile stress was applied both uniaxially and biaxially. The concentration dependence and temperature dependence of the diffusion coefficient based on their model were also validated using literature data.

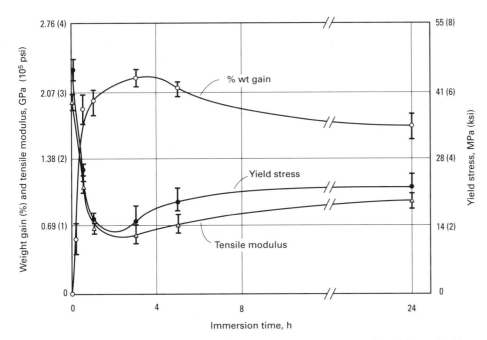

Fig. 23 The effect of 57 °C (135 °F) water immersion on percent weight gain, tensile modulus, and yield stress of Epon 1001/V115 epoxy resin system in bulk film form. Source: Ref 36

The researchers also incorporated their theoretical model into a two-dimensional finite-element code with viscoelastic capability and showed that the main accelerating effects in the diffusion rate of moisture through a bond line can be traced to the strain and concentration dependence of the diffusion coefficient (Ref 40). In their conclusions, they stated that "the bulk properties (also called volumetric properties) of the adhesive play a central role in the coupling mechanisms between the diffusion behavior and viscoelastic response" (Ref 40).

The Effect of Stress Whitening on Fluid Ingression. Many polymers used for bonding and protection functions contain secondary phases for the purposes of increasing the toughness, matching the coefficient of thermal expansion to the substrate, modifying the elastic modulus, increasing the strength, modifying the electrical and magnetic properties, improving the handling and manufacturing conditions and thickness control, and so on. The secondary phase, in these cases, may be a dispersed phase, such as randomly dispersed particles, fibers, and so forth, or a continuous and orderly structure, such as a glass carrier cloth.

In either case, loading of the polymer leads to stress concentrations at the interphase regions between the polymer matrix and the secondary phase. Such stress concentrations may result in the debonding of the polymer from the secondary phase and may also lead to the plastic deformation of the polymer in these regions in the form of cavitation, shear banding (to connect the cavitated regions), and microcracking. Obviously, such localized failure will lead to an increase in the free volume within the polymer structure. The presence of macroscopic discontinuities such as cracks (as in the wedge test), overlap edges, and so on, facilitates this internal failure mechanism locally. Consequently, the described microscopic failure process often remains localized without causing catastrophic failure but may deteriorate the whole polymer structure by acting as a path of moisture diffusion. Whether stress-whitened zones act as moisture diffusion paths has not yet been proved conclusively. However, the fact that they absorb more water than the virgin material and material stressed below stress whitening was shown in 1990 by Sancaktar et al. (Ref 41), as illustrated in Fig. 24. The model adhesive used in this case was EA 9628NW modified-epoxy film containing a nonwoven nylon mat. Detailed information on stress whitening in structural adhesives is given in the article "Fatigue and Fracture Mechanics" in this Volume.

In an elastoplastic analysis of brittle epoxy adhesives, several investigators (Ref 6, 42, 43) used Irwin's concept of plastic deformation zone at the tip of a crack (Ref 44) to resolve the issue of stress singularity at that location. The approximate diameter for this zone, calculated on the basis of measured mode I strain-energy release rate (G_{Ic}) and bulk tensile properties (tensile strength, Young's modulus, and Poisson's ratio), was also suggested as the optimum adhesive thickness for bonded joints (Ref 6, 42). This suggestion was offered to ensure that the maximum G_{Ic} value is obtained in bonded specimens, as is the case in bulk samples.

As confirmed by visual and microscopic evidence, it is this author's belief that this plastic deformation zone at the tip of a crack encompasses the stress-whitened zone. An exact correlation between the plastic zone diameter and the size and shape of the stress-whitened zone is not yet available. Because Irwin's analysis is based on small-scale yielding assumption and does not accurately describe the shape of the plastic deformation zone, the presence of a circular deformation zone should be considered hypothetical only. In fact, mathematical functions describing the exact shape of crack tip plastic deformation zones for monolithic samples are available in the literature (Ref 45, 46).

Diffusion Mechanism of Water: Regions of Different Densities. If two different regions in the path of diffusion—a less-dense stress-whitened region and a dense, or nonstress-whitened region—are assumed, then the effective diffusion coefficient (D_{eff}) in the path of diffusion becomes:

$$D_{eff} = D_1 F^1 (C) + D_2 [1 - F^1 (C)] \qquad \text{(Eq 48)}$$

where the coefficients D_1 and D_2 depend on the density of the regions in the absorption path, with D_1 representing the less-dense regions. The value of F^1 depends on the concentration of the absorbed water molecules, with $F^1(C)$ approximately 1 at low concentrations and $F^1(C)$ less than 1 at high concentrations, where the contribution from the dense regions becomes significant.

Broutman and Wong originally proposed Eq 48 to describe the effects of free volume distribution produced spontaneously by density fluctuations created by molecular motions (Ref 34). This author's interpretation is that density fluctuation applies to the case of stress-whitened (less dense) versus nonstress whitened (more dense) regions based on the free volume concept.

Lefebvre et al. (Ref 38) proposed the following expression for the diffusion coefficient (D):

$$D = (D_0/T_0)T \exp \{(B^D/f_0)[\alpha(T - T_0) + \epsilon_{kk}^f + \gamma C^N]/[f_0 + \alpha(T - T_0) + \epsilon_{kk}^f + \gamma C^N]\} \qquad \text{(Eq 49)}$$

where T = temperature, B^D = a numerical parameter related to the minimum hole size for the jump of a small penetrant molecule, f = fractional free volume, α = coefficient of thermal expansion of the free volume, ϵ_{kk}^f = volume dilatation of the free volume due to external loads, and γ = coefficient of swelling. The subscript 0, on any variable, refers to a reference temperature (T_0).

Experimental data can be used to assess the effects of the presence of stress whitening on the relevant parameters of Eq 49. Subsequently, Eq 49 can be expressed in the form of Eq 48 to allow for spatial dependence of the diffusion coefficient due to stress whitening.

Path of Moisture Diffusion. Theoretical prediction for the path of diffusion and a spatial concentration profile can be done on

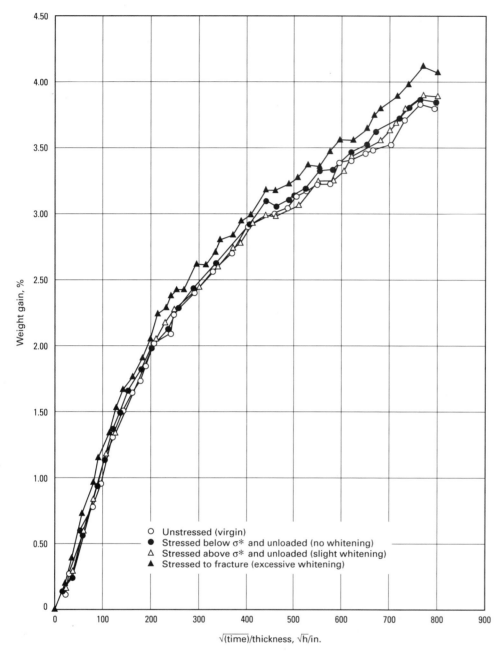

Fig. 24 Reduced absorption plots exhibiting the effects of stress whitening on water retention in a modified-epoxy adhesive in bulk form (Dexter-Hysol EA 9628NW). σ* = stress whitening stress

Legend in figure:
○ Unstressed (virgin)
● Stressed below σ* and unloaded (no whitening)
△ Stressed above σ* and unloaded (slight whitening)
▲ Stressed to fracture (excessive whitening)

Y-axis: Weight gain, %
X-axis: √(time)/thickness, √h/in.

the basis of the model proposed by Lefebvre *et al.* (Ref 38):

$$(\partial C'/\partial t) = \nabla^* [D(\epsilon_{kk}^f C) (\nabla^* C' - C' \nabla^* \epsilon_{kk}^f)] \tag{Eq 50}$$

where the normalized concentration (C') is defined by (Ref 38):

$$C' = C/[S_0(1 + \epsilon_{kk}^f)P_p] \tag{Eq 51}$$

with S_0 = solubility in the absence of strain and P_p = partial pressure of the sorbent in the vapor phase.

Equation 50 can be solved numerically for given spatial boundary conditions.

Diffusion to Interface. In addition to the diffusion mechanisms described above, fluids can reach the adhesive-adherend interface by means of the so-called wicking process, which involves transport along the interface. There is disagreement in the literature on the dominant path for water diffusion in adhesive joints. Comyn (Ref 47) suggests that "fiber-matrix interface is not an active path for water ingress." Ko and Wightman (Ref 48) in their study with Udel P-1700 polysulfone thermoplastic adhesive used to bond aluminum-lithium substrates, determined that "water molecules diffuse into the polysulfone and thence to the interface." In this work Wightman *et al.* used a scanning transmission electron microscope (STEM) to study the surface morphology and x-ray photoelectron spectroscopy (XPS) to determine surface composition.

Drain *et al.* (Ref 49), in their study of steel-steel cyanoacrylate bonds, concluded that "the most likely method of ingress into the bond would be via wicking mechanism along the substrate/polymer interface." They claim that water vapor diffuses through the metal oxide-adhesive interface, causing the formation of ferric oxide and, thus, weakening the interfacial oxide layer. Other general mechanisms that weaken adhesive-adherend interphases as a result of water ingression are the displacement of the adhesive by water molecules at the bonding surface and hydrolytic degradation of the adhesive material. Of course, hydrolytic degradation of the bulk adhesive can occur anywhere in the diffusion path, as shown in Fig. 23.

Piggott *et al.* (Ref 50) studied single production glass fibers, embedded in a polyester resin and immersed in water at temperatures up to 75 °C (167 °F) for various lengths of time. Their results showed that water at elevated temperatures degrades the interphase, but if dried for more than 2 h, "the loss is recoverable at 140 °F (60 °C), and may perhaps be at 167 °F (75 °C) as well" (Ref 50). They used autoradiographic tests in which tritiated water was allowed to diffuse through a two-fiber composite slice 2 mm (0.079 in.) thick and exposed a photographic film bonded to the other side. These tests revealed that water diffuses preferentially along the interphase.

Dodiuk *et al.* studied the effect of moisture absorption into the solid-film adhesive material in the uncured form (Ref 51). They concluded that moisture ingression into the uncured, supported epoxy film adhesives FM 73 and FM 300K, Redux 319 supported epoxy film, and AF-126 adhesives resulted in inferior lap-shear strengths when these adhesives were subsequently cured to bond 2024-T351 bare aluminum adherends coated with BR-127 primer. When the absorbed moisture content was below a 0.3% level, predrying of the adhesive under vacuum (405 to 675 Pa, or 0.12 to 0.2 in. Hg) for 3 to 4 h at room temperature to remove the absorbed moisture helped restore the subsequent lap-shear strength values to expected levels.

Viscoelasticity Considerations for DCB Fracture Testing

Creep Behavior in Bulk. The bulk tensile and shear creep compliances for an adhesive material can be characterized using standard finite exponential series based on a linear or nonlinear mechanical (spring-dashpot) model. Such a model can be chosen based on the constant strain rate behavior or it can be directly fitted with creep data. It can also be an overstressed model, as described earlier in this article. Consequently, the creep compliance can have the form:

$$D(t,T,C,\sigma) = D_0(T,C,\sigma) + \Sigma D_i(T,C,\sigma)$$
$$[1 - e_i^{-ta}(T,C,\sigma)][1 - e_i^{-ta(t,c,\sigma)}] \quad \text{(Eq 52)}$$

which will allow time (*t*)-temperature (*T*)-moisture (*C*)-stress (σ) shifts.

Creep-rupture (delayed-failure) behavior of the adhesive material can be described using Crochet's equation along with this compliance, as described earlier.

Relaxation Behavior in Bulk. The bulk tensile and shear relaxation compliances for the adhesive material can be characterized similarly using standard finite exponential series based on a linear or nonlinear mechanical (overstress) model. The relaxation modulus, therefore, can have the form:

$$E(t,T,C,\sigma) = E_0(T,C,\sigma) + \Sigma E_n(T,C,\sigma)\, e_n^{-ta(T,C,\sigma)} \quad \text{(Eq 53)}$$

which will allow time-temperature-moisture-stress shifts.

The separate characterizations of bulk tension and shear modes is suggested because of the possibility that the model adhesive may have different viscoelastic behavior in tension and shear. Obviously, such different constitutive behaviors in tension and shear will also affect the mixed-mode fracture criteria. In fact, it is not unusual for polymeric materials to have different mechanical behavior in tension and shear, especially at high strain and temperature levels. The earlier work of Sancaktar on viscoelastic characterization of Metlbond epoxy adhesive illustrated such behavior (Ref 1).

It should also be noted that other constitutive models in differential and integral forms, for adhesive materials, can be found in the literature (Ref 13, 24, 52, 53).

Determination of Modes I and II Strain-Energy Release Rate Values for Viscoelastic Adhesives. The determination of both G_I and G_{II} based on viscoelastic theory is described below.

Determination of G_I. The G_I value can be determined by taking into consideration the viscoelastic deformation of the adhesive layer in a DCB specimen:

$$G_I = G_I{}' + G_I{}'' + G_I{}''' \quad \text{(Eq 54)}$$

where G_I = energy release rate of the unbonded portion of the cantilever, $G_I{}''$ = energy release rate of the bonded portion of the beam on a viscoelastic (adhesive layer) foundation, and $G_I{}'''$ = energy release rate of the viscoelastic adhesive layer. It should be noted that adhesive deformation is taken into consideration in Eq 54.

Similar elastic analysis (Ref 54) of DCB geometry, which treats the adherends as finite beams that are partly free and partly supported by an elastic foundation and the adhesive bond as a thin strip under prescribed displacement, results in the expression:

$$G_I = G_I^l\,(\phi^2 + \alpha\psi^2) \quad \text{(Eq 55)}$$

or the mode I strain-energy release rate. In Eq 55, the term G_I^l is the familiar strength-of-materials formulation of G_I for bending deflection, as given by Eq 5, less the shear deflection term. The terms φ, α, and ψ are defined as:

$$\phi = [(\sinh^2\beta L + \sin^2\beta L)/(\sinh^2\beta L - \sin^2\beta L)]$$
$$+ (1/\beta a)[(\sinh\beta L\cosh\beta L - \sin\beta L\cos\beta L)/$$
$$(\sinh^2\beta L - \sin^2\beta L)] \quad \text{(Eq 56)}$$

$$\psi = \sin\beta L\sinh\beta L - (1/\beta a)[(\zeta - 1)/2]$$
$$\sin\beta L\cosh\beta L + \phi\cos\beta L\cosh\beta L$$
$$-(1/\beta a)\,[(\zeta + 1)/2]\cos\beta L\sinh\beta L \quad \text{(Eq 57)}$$

Here

$$\zeta = [(\sinh^2\beta L + \sin^2\beta L)/(\sinh^2\beta L - \sin^2\beta L)]$$
$$+ 2\,\beta a[(\sinh\beta L\cosh\beta L + \sin\beta L\cos\beta L)/$$
$$(\sinh^2\beta L - \sin^2\beta L)] \quad \text{(Eq 58)}$$

where L is the width of the uncracked specimen (that is, specimen width minus crack length, *a*). Also,

$$\alpha = (1 - v_2)/(1 + v_2)\,(1 - 2\,v_2) \quad \text{(Eq 59)}$$

where v_2 is the Poisson's ratio of the adhesive. The term β is a parameter of dimension (length)$^{-1}$ and is given by:

$$\beta = [(3E_2/E_1)/th^3]^{1/4} \quad \text{(Eq 60)}$$

where E_1 and E_2 are the elastic moduli for the adherend and adhesive, respectively; h is the beam height; and $2t$ is the adhesive thickness. Examination of Eq 55 through 60 reveals that when softer adhesives with lower E_2 values are used, larger deformations, and, consequently, larger G_I values result. This conclusion corroborates the findings of Hunston *et al.* (Ref 43), which showed higher G_{Ic} values for a carboxy-terminated butadiene acrylonitrile (CTBN)-epoxy system at higher temperatures. On the other hand, if adhesive deformation includes plastic strain, then, obviously, the energy utilized in this irreversible process is no longer available to propagate the crack.

Based on these considerations, the results from the viscoelastic testing of the bulk adhesive material can be used in quasi-static numerical analysis with Eq 55 through 60 to determine time-temperature-water uptake-stress-dependent G_I values. This type of analysis allows the exclusion of work done in the plastic deformation of the adhesive in calculation of G_I values. Actual experimentation information on the extent of plastic deformation in the adhesive layer can be obtained by loading-unloading tests on the bulk material. Furthermore, the effects of delayed elasticity can, similarly, be accounted for. It should be noted that a numerical solution to the time-dependent G_I problem obtained using a compliance or relaxation function would not allow the exclusion of plastic work. However, if nec-

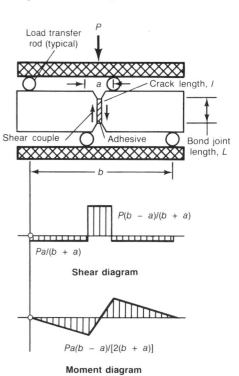

Fig. 25 Adaptation of Iosipescu specimen for testing adhesives in the bonded shear mode. Also shown are simplified shear and moment diagrams for the bonded specimen. Source: Ref 55

essary, this type of analysis can also be performed and compared with the results from quasi-static analysis.

Determination of G_{II}. The G_{II} can be determined by taking into consideration the viscoelastic deformation of the adhesive layer. For this purpose an analytical method similar to the one described above can be used. A quasi-static analysis based on an elastic G_{II} equation (Ref 6) can be performed to determine time-temperature-moisture-stress-dependent G_{II} values.

To summarize, the calculation of G_I and G_{II} values can be done on the basis of an accurate energy balance, which will include irreversible losses (plastic strain) and delayed elasticity (time effects). Consequently, the exact amount of elastic energy that contributes to the creation of new fracture surfaces can be determined. This allows determination of an accurate failure criterion based on the experimental data.

Other Test Specimen Geometries for Accurate Quantitative Testing

The Iosipescu specimen shown in Fig. 25 is basically a beam under four-point bending that contains a machined notch between the shear couples to provide a "close-to-uniform" state of stress. Jonath reported that in this specimen "the shear stress exceeds its average value by less than 3% in magnitude for less than 5% of the bond length" (Ref 55). He also reported that the variation

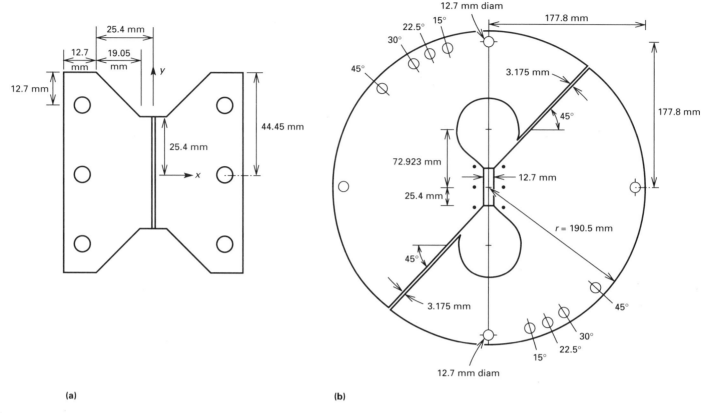

Fig. 26 (a) The stiff-adherend Arcan specimen. (b) The associated grips. Source: Ref 57

of peel stresses is more than that of the shear stress, but their magnitudes are two orders less than that of the shear stress.

The Arcan specimen (Ref 56), a modified form of which is shown in Fig. 26 (Ref 57), allows mixed-mode testing of the adhesive layer with different orientations of loading. Liechti *et al*. report that the "uniformity coefficient," which is the bond line length fraction over which the stresses are within 10% of the average shear and/or normal stresses, can be as high as 99% for this specimen (Ref 57).

Additional test geometries include cracked lap-shear specimen (Ref 58-60) and bonded cantilever beam (Ref 61).

Sancaktar recently used the independently loaded mixed-mode specimen (ILMMS) for fatigue and fracture testing of structural adhesives (Ref 6, 62-66). The ILMMS geometry allows independent application of sequential (transient) loading, load hold, and unloading in modes I and II directions. Detailed information on this ILMM specimen is given in the article "Fatigue and Fracture Mechanics" in this Volume. In interpreting the ILMMS data, modes I and II strain-energy release rate values G_I and G_{II} can be calculated by taking into consideration the viscoelastic deformation of the adhesive layer, as described earlier. Stress relaxations at the crack tip and the effects of adhesive creep on crack opening and shear-ing displacements can be taken into consideration in determining the fracture criteria.

Determination of Mixed-Mode Fracture Criteria

Several theoretical approaches for determining mixed-mode fracture (failure) criteria are described below.

Energy Balance Criterion. The G_I and G_{II} values that result in catastrophic failure can be calculated using the viscoelastic approach described previously. The energy balance criterion:

$$G_I + G_{II} = \text{constant} \qquad \text{(Eq 61)}$$

can then be applied at different rate-time-temperature-water uptake conditions.

Strain-Energy Density Criterion. For adhesive joints, Sih (Ref 67) proposed application of the strain-energy density criterion, rather than the energy release rate concept. With this criterion, unstable fracture is assumed to occur when the strain energy stored in an element ahead of the crack reaches a critical value, defined by:

$$S_{cr} = [(1 - v)(1 - 2v)/2E] \, K_{Ic}^2 \qquad \text{(Eq 62)}$$

For an adhesive joint under mixed-mode plane crack extension:

$$S_{cr} = a_{11}K_I^2 + 2a_{12} \, K_I K_{II} + a_{22}K_{II}^2 \qquad \text{(Eq 63)}$$

where a_{ij} are material constants. It should be noted that (Ref 67):

$$K_I = F[(t/a), (E_2/E_1), (v_2/v_1)] \, \sigma \, \sqrt{a} \qquad \text{(Eq 64)}$$

and

$$K_{II} = G[(t/a), (E_2/E_1), (v_2/v_1)] \, \tau \, \sqrt{a} \qquad \text{(Eq 65)}$$

Based on this criterion, therefore, not only do relaxations in σ and τ need to be taken into account, but changes in functions F and G with rate-time-temperature-water uptake should also be considered because they are dependent on the adhesive properties. This, again, can be done with the use of a quasi-static analysis based on the viscoelastic material.

Maximum Principal Stress and Crack Opening Displacement Criteria. In applying the maximum principal stress criteria, the relaxation of stresses at the crack tip should be taken into consideration based on the relaxation functions and elastic residual stresses determined during the viscoelastic bulk characterization of the adhesive. Stress relaxation in both tensile and shear stresses in modes I and II (respectively) can be considered separately because of the possibility of different mechanical behaviors in these two modes at high strain, temperature, and water uptake levels. Such behavior should also be monitored by means of displacement transducers placed

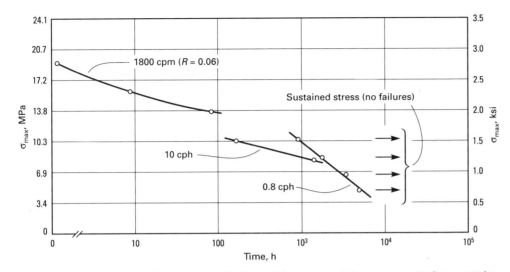

Fig. 27 The effect of cyclic frequency on the fatigue behavior under high temperature (60 °C, or 140 °F), condensing humidity conditions. cpm, cycles per minute; cph, cycles per hour. Source: Ref 68

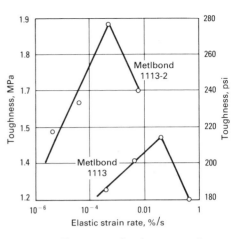

Fig. 28 The area under the stress-strain curve (toughness) as a function of test rate for Metlbond 1113 and 1113-2 modified structural adhesives with and without carrier cloth. The presence of the scrim cloth lowers the toughness.

in both mode I and II directions. This will allow the determination of any shift in principal stress direction due to different tensile and shear behaviors.

Data obtained with displacement gages can also be used to arrive at a failure criterion based on crack tip displacements. Creep behavior of the adhesive layer can be taken into account in this criterion with the use of creep compliances and transient creep strains determined during bulk tensile and shear testing of the adhesive.

In using the maximum principal stress criterion expressed in terms of stress-intensity factors, as suggested by Sancaktar (Ref 6, 64, 66):

$$(K_I/K_{Ic}) + (K_{II}/K_{IIc})^2 = 1 \qquad \text{(Eq 66)}$$

the dependence of K_I and K_{II} on adhesive properties as suggested by Sih in Eq 64 and 65 can be taken into consideration in fitting Eq 66 to experimental data.

Dynamic Fatigue

Although a detailed discussion on dynamic fatigue is provided in the article "Fatigue and Fracture Mechanics" in this Volume, it is appropriate to provide some discussion here on the viscoelastic effects on dynamic fatigue, including the effect of environmental conditions. Fatigue testing was performed using thick-adherend single-lap-shear specimens with 2024-T3 (anodized) aluminum adherends bonded with an unspecified 121 °C (250 °F) cure system (Ref 68, 69). Figure 27 shows their results for the effect of cyclic frequency on fatigue strength and fatigue life.

This author believes that strong viscoelastic effects are evident in Fig. 27, where slow cyclic loading allows dominant creep process due to the mean level of the applied cyclic load ($R = P_{min}/P_{max} = 0.6$).

Obviously, the cyclic nature of the load itself causes material degradation because sustained loading at comparable loads does not result in failure whereas 0.8 cycles per hour does (Fig. 27). However, the viscoelastic effects are presented because fast cyclic loading results in high-strength failures in shorter times, while slow cyclic loading results in low-strength failures at longer times. It is possible that the viscoelastic effects are exasperated because of the presence of a high-temperature, condensing humidity environment. A similar rate-dependent increase in bulk adhesive toughness was shown earlier for Metlbond 1113 and 1113-2 modified-epoxy resins with and without carrier cloth, respectively, by Sancaktar *et al.* as shown in Fig. 28 (Ref 70).

Based on Fig. 27, the effects of viscous heating seem to be negligible because higher frequencies do not result in notable reductions in fatigue strength or life. This observation is in agreement with other fatigue experiments, which revealed a negligible temperature rise in the bond line for specimens bonded with modified structural adhesives (see the article "Fatigue and Fracture Mechanics").

Figure 29 shows the effects of moisture ingression and environmental temperature on fatigue behavior. An increase in the environmental temperature results in the reduction of fatigue strength and life. It appears that the effect of moisture ingression is coupled with the thermal effect. At room temperature, wet specimens exhibit higher strength and longer fatigue life. At 60 °C (140 °F), however, the effect is reversed because the higher temperature possibly accelerates the diffusion and hydrolytic degradation by water and its plasticizing effect may exasperate the high-temperature creep process in the adhesive.

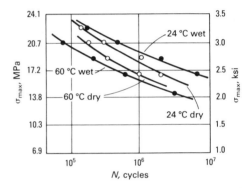

Fig. 29 The effects of high temperature and humid environments on the fatigue behavior of adhesively bonded joints cycled at 30 Hz. Source: Ref 68

Behavior of the Interphase

At the beginning of this article, it was stated that when any or all of the components of a bonded joint—the adhesive, the interphase, and the adherend—exhibit viscoelastic behavior, considerable strain transfer may take place between these components. Sancaktar *et al.* recently performed a nonlinear viscoelastic analysis of the fiber-matrix interphase that illustrated this behavior (Ref 71, 72).

Figure 30 shows the variation of the maximum shear stress that exists of an embedded (bonded) fiber with time and material properties. In this case, both the adhesive matrix and the interphase were assumed to have nonlinear viscoelastic behavior as described by Eq 15, but were assigned different material properties. The substrate (fiber) was assumed to have elastic behavior. Figure 30 reveals the possibility of transition from time-dependent stress reductions at small creep compliance values (D_{1m}) for the matrix (that is, matrix approaching elastic behavior) to time-dependent increases in interfacial peak shear stresses at high D_{1m}

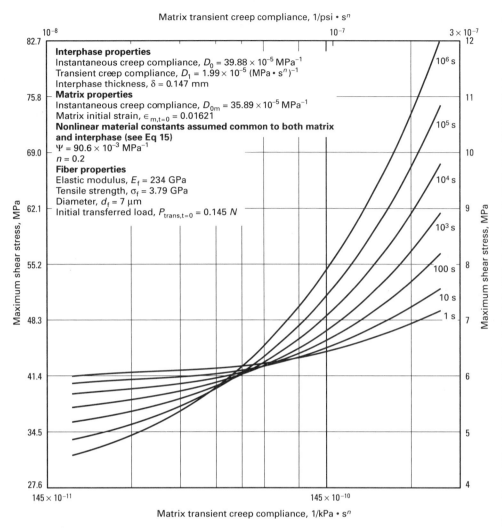

Matrix transient creep compliance, 1/psi · s^n

Interphase properties
Instantaneous creep compliance, $D_0 = 39.88 \times 10^{-5}$ MPa^{-1}
Transient creep compliance, $D_1 = 1.99 \times 10^{-5}$ (MPa · s^n)$^{-1}$
Interphase thickness, $\delta = 0.147$ mm
Matrix properties
Instantaneous creep compliance, $D_{0m} = 35.89 \times 10^{-5}$ MPa^{-1}
Matrix initial strain, $\epsilon_{m,t=0} = 0.01621$
Nonlinear material constants assumed common to both matrix and interphase (see Eq 15)
$\Psi = 90.6 \times 10^{-3}$ MPa^{-1}
$n = 0.2$
Fiber properties
Elastic modulus, $E_f = 234$ GPa
Tensile strength, $\sigma_f = 3.79$ GPa
Diameter, $d_f = 7$ μm
Initial transferred load, $P_{trans,t=0} = 0.145$ N

Matrix transient creep compliance, 1/kPa · s^n

Fig. 30 The variation of maximum interfacial shear stress at the fiber ends, with time and matrix transient creep compliance. The results are based on nonlinear viscoelastic analysis of a cylindrical interphase zone in a fiber fragmentation test specimen.

values. This transition depends on the comparative values of the creep compliances for the matrix and the interphase with stress reductions occurring when the compliance for the matrix is smaller than that for the interphase. Obviously, such a transition can occur during the life of a bonded joint because the viscoelastic properties of the adhesive matrix and/or the interphase may change preferentially because of environmental changes such as moisture diffusion and exposure to thermal environments.

Sancaktar *et al.* also performed single fiber fragmentation tests to assess the effects of high temperature and loading rate on the interfacial shear stress (Ref 73). These experiments involve embedding a single fiber in the adhesive matrix coupon, which is loaded in tension subsequent to cure. During the tensile loading of the coupon, the load is transferred to the fiber by means of interfacial shear stress that equilibrates the fiber tensile stress, which causes it to fragment until its length is not long

enough to transfer sufficient interfacial shear to break it any further. When reached, such fiber length is called the critical fiber length and it is inversely proportional to the magnitude of the interfacial shear stress. Figures 31 and 32 show the effects of loading rate on the interfacial shear stress with the use of critical fragment length for sized (surface-treated) and unsized fibers, respectively. Obviously, the interfacial strength is reduced with slower loading rates for both sized and unsized fibers. Similarly, increases in the environmental temperature result in reductions in the interfacial strength (Fig. 33).

Durability of Sealants

A methodology for determining the durability of sealants has been discussed in detail by Sandberg and Albers (Ref 74). They suggested the use of sealant strain capacity as the critical parameter for gaging sealant durability. They defined an allowable design strain value as:

ϵ_a = 5% exclusion limit for basic strain
$\quad \times$ Durability factors $\times$ State of strain factor
$\quad \times$ Geometric shape factor $\times$ Substrate
$\quad$ factor $\times$ Safety factor   (Eq 67)

In Eq 67, the basic strain value is measured using symmetric single-lap or rectangular butt joints for shear and tension cases, respectively, with the loads applied monotonically. The durability factors are determined to take into account the effects of exposure to water, heat, cold, chemicals, ultraviolet radiation, fatigue, and creep loading. Accelerated and/or long-term tests are performed to measure these values. The state-of-strain factor is used to account for the differences in shear and tensile behaviors of the sealant joint. The geometric-shape factor is used to account for any possible variations in strain capacity due to the geometric differences between the test specimens and the actual sealant joint. Different adhesion characteristics of the sealant on various adherends is accounted for with the substrate factor. For the specific case of a solvent-base acrylic tension joint on aluminum in a building wall, Sandberg and Albers reported the following values for some of the above-mentioned parameters:

Basic tension strain = 176%
State-of-strain factor = 1.00
2500 cycle fatigue factor = 0.32
Heat aging = 1.00
Cold flexibility at −30 °C , or −22 °F = 0.49
Ultraviolet radiation = 1.00
Compression set = 0.52
Geometric-shape factor (width/thickness) = 1.00
Safety factor = 0.67
Allowable strain (calculated using the above factors and Eq 67) = 6.3%

Recommendations

In testing for the durability of adhesives and adhesively bonded joints for design purposes, the determination of threshold values becomes the critical issue. Threshold values can be analytically modeled and experimentally verified using the methods described in this article and the article "Fatigue and Fracture Mechanics." In order to be able to make accurate durability predictions based on experimental data, two important issues have to be addressed. First is the accurate phenomenological description of the interaction between the environment, adherend, interphase, and adhesive. Second is the accurate measurement of the phenomena to minimize the discrepancy between what needs to be measured and what is actually measured, or at least an accurate assessment of such a discrepancy when it is unavoidable.

Unfortunately, however, in adhesive joint design the process of minimizing the discrepancy between the measured quantities and what needs to be measured be-

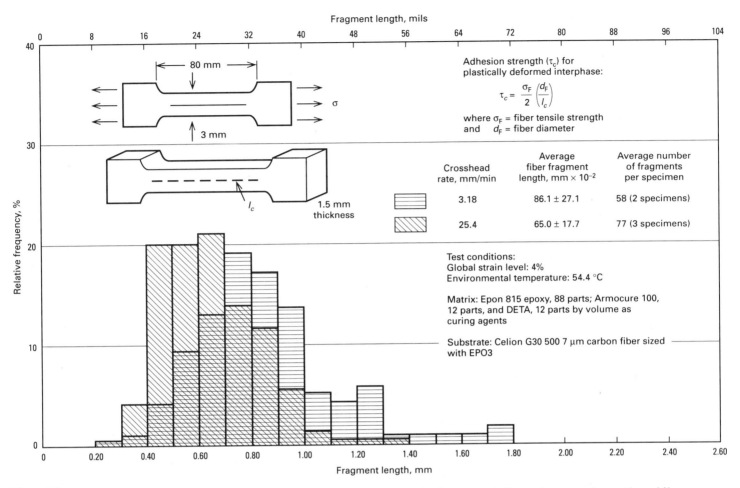

Fig. 31 The effect of crosshead rate on fiber-matrix adhesion as indicated by fragment size distribution in single fiber tension test specimens with sized fibers

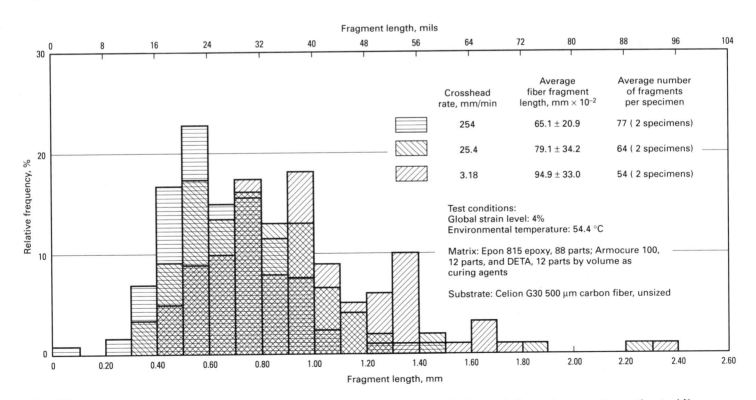

Fig. 32 The effect of crosshead rate on fiber-matrix adhesion as indicated by fragment size distribution in single fiber tension test specimens with unsized fibers

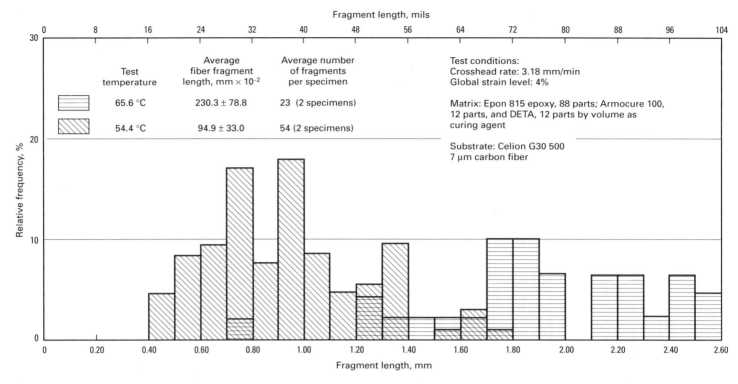

Fig. 33 The effect of test temperature on fiber-matrix adhesion as indicated by fragment size distribution in single fiber tension test specimens with unsized fibers

comes a lengthy and expensive task that may be prohibitive for many designers if economizing becomes important. For this reason, a general test program is suggested to encompass the economical as well as the technological and scientific considerations, based on a procedure in which less expensive, simple specimens, such as the single-lap specimen or thin-adherend wedge test specimen, are used for screening (comparison) purposes because they do not provide accurate quantitative data for design calculations; the more expensive, difficult-to-machine specimens, such as the ILMMS, Arcan, and Iosipescu specimens, are used selectively to obtain more accurate quantitative design data; and actual structural components are tested to very limited numbers to verify theoretical predictions.

ACKNOWLEDGMENT

The research work reported in this article in the sections "Examples of the Assessment of Temperature-Dependent Delayed-Failure Behavior of Bonded Single-Lap Joints" and "Determination of Stress Concentration Factors for Single-Lap Specimens" was supported by NASA under grant NAG-1-284. Other works partially referred to in this paper were supported by the Grumman Corporation (fiber-matrix interphase), NASA grant NGR-47-004-090 (rate effects on structural adhesives, Dr. H.F. Brinson, principal investigator), and Clarkson University (effects of stress whitening on moisture diffusion, and others).

Thanks are also due to numerous authors and publishers who allowed their works to be referenced and reproduced in this paper.

REFERENCES

1. E. Sancaktar and H.F. Brinson, The Viscoelastic Shear Behavior of a Structural Adhesive, in *Adhesion and Adsorption of Polymers*, L.H. Lee, Ed., Vol 12-A, Polymer Science and Technology Series, Plenum Publishing, 1980, p 279-300
2. M. Okajima and G.B. Sinclair, Towards Improved Testing of Adhesive Tensile Strengths, in *Proceedings of the XIIIth Southeastern Conference on Theoretical and Applied Mechanics*, Vol 1, College of Engineering, University of South Carolina, 1986, p 367-371
3. J.C. McMillan, Durability Test Methods for Aerospace Bonding, in *Developments in Adhesives*, 2, A.J. Kinloch, Ed., Applied Science Publishers, 1981, p 243-278
4. S. Mostovoy, P.B. Crosley, and E.J. Ripling, Use of Crack-Line Loaded Specimens for Measuring Plane Strain Fracture Toughness, *J. Mater.*, Vol 2, 1967, p 661-681
5. S.F. Stone, R.A. Westmann, and M.E. Fourney, "Analytical and Experimental Studies in Adhesive Mechanics," Report UCLA-ENG-7556, University of California, Los Angeles, July 1975
6. E. Sancaktar, H. Jozavi, J. Baldwin, and J. Tang, "Elastoplastic Fracture Behavior of Structural Adhesives Under Monotonic Loading," *J. Adhes.*, Vol 23, 1987, p 233-262
7. J.A. Filbey and J.P. Wightman, Factors Affecting the Durability of Ti-6Al-4V/Epoxy Bonds, *J. Adhes.*, Vol 28, 1989, p 1-22
8. E.C. Larke and R.J. Parker, *Advances in Creep Design*, John Wiley & Sons, 1971
9. H.F. Brinson, *Deformation and Fracture of High Polymers*, H.H. Kausch, J.A. Hassell, and R.I. Jaffee, Ed., Plenum Publishing, 1974
10. E. Sancaktar, S.C. Schenck, and S. Padgilwar, Material Characterization of Structural Adhesives in the Lap Shear Mode, Part I: The Effects of Rate, *Ind. Eng. Chem., Prod. Res. Dev.*, Vol 23, 1984, p 426-434
11. K.W. Chase and W. Goldsmith, Mechanical and Optical Characterization of an Anelastic Polymer at Large Strain Rates and Strain, *Exp. Mech.*, Vol 14, 1974, p 10-18
12. E. Sancaktar and S. Padgilwar, The Effects of Inherent Flaws on the Time and Rate Dependent Failure of Adhesively Bonded Joints, *J. Mech. Des.*, Vol 104, 1982, p 643-650
13. R.A. Schapery, On the Characterization of Nonlinear Viscoelastic Materials, *J. Poly. Eng. Sci.*, Vol 9, 1969, p 295-310
14. N.A. Cenci and E. Sancaktar, "Geometrical and Material Nonlinearities in Single Lap Joints Bonded With Ther-

moplastic Adhesives,'' Report 095, Clarkson College of Technology, 1984

15. M.W. Darlington and S. Turner, *Creep of Engineering Materials*, Mechanical Engineering Publications Ltd., 1978

16. J.J. Ahlonis, W.J. MacKnight, and M. Shen, *Introduction to Polymer Viscoelasticity*, Wiley-Interscience, 1972

17. P.M. Naghdi and S.A. Murch, On the Mechanical Behavior of Viscoelastic-Plastic Solids, *J. Appl. Mech.*, Vol 30, 1963, p 321-328

18. M.J. Crochet, Symmetric Deformations of Viscoelastic-Plaster Cylinders, *J. Appl. Mech.*, Vol 33, 1966, p 327-334

19. E. Sancaktar, Material Characterization of Structural Adhesives in the Lap Shear Mode, *Int. J. Adhes. Adhes.*, Vol 5, 1985, p 66-68

20. E. Sancaktar and S.C. Schenck, Material Characterization of Structural Adhesives in the Lap Shear Mode, Part II: Temperature Dependent Delayed Failure, *Ind. Eng. Chem., Prod. Res. Dev.*, Vol 24, 1985, p 257-263

21. S.C. Schenck and E. Sancaktar, "Material Characterization of Structural Adhesives in the Lap Shear Mode," NASA Contractor Report 172237, National Technical Information Service, 1983

22. S.N. Zhurkov, Kinetic Concept of the Strength of Solids, *Int. J. Fract. Mech.*, Vol 1, 1965, p 311-323

23. G.L. Slonimskii, A.A. Askadskii, and V.V. Kazanteseva, Mechanism of Rupture of Solid Polymers, *Polym. Mech.*, Vol 13, 1977, p 647-651

24. M.E. Tuttle and H.F. Brinson, "Accelerated Viscoelastic Characterization of T300/5208 Graphite-Epoxy Laminates," Report VPI-E-84-9, Virginia Polytechnic Institute, 1984

25. M. Goland and E. Reissner, The Stresses in Cemented Joints, *J. Appl. Mech.*, Vol 77, 1944, p 17-27

26. T.L. St. Clair and D.A. Yamaki, "A Thermoplastic Polyimidesulfone," NASA Technical Memorandum 84574, National Technical Information Service, 1982

27. FM 73 and FM 300 Adhesive Film Product Information Brochures, American Cyanamid Company

28. L. Greenwood, The Strength of a Lap Joint, *Aspects Adhes.*, Vol 5, 1969, p 40-50

29. S. Mall, W.S. Johnson, and R.A. Everett, Jr., "Cyclic Debonding of Adhesively Bonded Composites," Technical Memorandum 84577, National Aeronautics and Space Administration, 1982

30. G. Dolev and O. Ishai, Mechanical Characterization of Adhesive Layer in-Situ and as Bulk Material, *J. Adhes.*, Vol 12, 1981, p 283-294

31. W.T. McCarvill and J.P. Bell, Torsional Test Method for Adhesive Joints, *J. Adhes.*, Vol 6, 1974, p 185-193

32. J.W. Grant and J.N. Cooper, Cone-and-Plate Shear Stress Adhesive Test, *J. Adhes.*, Vol 21, 1987, p 155-167

33. L.E. Nielsen, *Mechanical Properties of Polymers*, Reinhold, 1962

34. T.C. Wong, "Moisture Diffusion Mechanisms in Epoxy Resins," Ph.D. thesis, Illinois Institute of Technology, 1982

35. Y.J. Weitsman, Stresses in Adhesive Joints Due to Moisture and Temperature, *J. Compos. Mater.*, Vol 11, 1977, p 378-394

36. R.G. Schmidt and J.P. Bell, Investigation of Steel/Epoxy Adhesion Durability Using Polymeric Coupling Agents. II. Factors Affecting Adhesion Durability, *J. Adhes.*, Vol 25, 1988, p 85-107

37. L.R.G. Treloar, The Absorption of Water by Cellulose and Its Dependence on Applied Stress, *Trans. Faraday Soc.*, Vol 48, 1952, p 567-576

38. D.R. Lefebvre, D.A. Dillard, and T.C. Ward, A Model for the Diffusion of Moisture in Adhesive Joints. Part I: Equations Governing Diffusion, *J. Adhes.*, Vol 27, 1989, p 1-18

39. D.R. Lefebvre, D.A. Dillard, and H.F. Brinson, A Model for the Diffusion of Moisture in Adhesive Joints. Part II: Experimental, *J. Adhes.*, Vol 27, 1989, p 19-40

40. S. Roy, D.R. Lefebvre, D.A. Dillard, and J.N. Reddy, A Model for the Diffusion of Moisture in Adhesive Joints. Part III: Numerical Simulations, *J. Adhes.*, Vol 27, 1989, p 41-62

41. E. Sancaktar and D.R. Beachtle, The Effect of Stress Whitening on Moisture Diffusion in Thermosetting Polymers, Report MIE-207, Clarkson University, 1990

42. W.D. Bascom, R.L. Cottington, and C.D. Timmons, Fracture Reliability of Structural Adhesives, *J. Appl. Polym. Sci.*, Vol 32, 1977, p 165-188

43. D.L. Hunston, A.J. Kinloch, and S.S. Wang, Characterization of the Fracture Behavior of Adhesive Joints, in *Adhesive Joints*, Plenum Publishing, 1984, p 789-807

44. G.R. Irwin, Plastic Zone Near a Crack and Fracture Toughness, in *Proceedings of the Seventh Sagamore Conference*, U.S. Army, 1960, p IV-63, IV-78

45. D. Broek, *Elementary Engineering Fracture Mechanics*, 3rd ed., Noordhoff International Publishing, 1974

46. J.W. Dally and R.J. Sanford, Strain Gage Methods for Measuring the Opening Mode Stress Intensity Factor, K_I, in *Proceedings of the 1985 Spring Conference on Experimental Mechanics*, Society for Experimental Mechanics, 1985, p 851-860

47. J. Comyn, The Relationship Between Joint Durability and Water Diffusion, in *Developments in Adhesives, 2*, A.J.

Kinloch, Ed., Applied Science Publishers, 1981, p 279-313

48. C.U. Ko and J.P. Wightman, Experimental Analysis of Moisture Intrusion Into the Al/Li-Polysulfone Interface, *J. Adhes.*, Vol 25, 1988, p 23-29

49. K.F. Drain, J. Guthrie, C.L. Leung, F.R. Martin, and M.S. Otterburn, The Effect of Moisture on the Strength of Steel-Steel Cyanoacrylate Adhesive Bonds, *J. Adhes.*, Vol 17, 1984, p 71-82

50. M.R. Piggott and P.S. Chua, The Effects of Water on the Glass Fibre Polymer Interface, in *Composites and Other New Materials for PVP: Design and Analysis Considerations*, G.E.O. Widera, Ed., PVP, Vol 174, American Society of Mechanical Engineers, 1989, p 143-146

51. H. Dodiuk, L. Drori, and J. Miller, The Effect of Moisture in Epoxy Film Adhesives on Their Performance: I. Lap Shear Strength, *J. Adhes.*, Vol 17, 1984, p 33-44

52. E. Sancaktar, Constitutive Behavior and Testing of Structural Adhesives, *Appl. Mech. Rev.*, Vol 40, 1987, p 1393-1402

53. H.F. Brinson, Durability (Lifetime) Predictions for Adhesively Bonded Structures, in *Mechanical Behaviour of Adhesive Joints*, G. Verchery and A.H. Cardon, Ed., Editions Pluralis, 1987, p 3-25

54. M.B. Ouezdou, A. Chudnovsky, and A. Moet, Re-evaluation of Adhesive Fracture Energy, *J. Adhes.*, Vol 25, 1988, p 169-183

55. A.D. Jonath, New Approach to the Understanding of Adhesive Interface Phenomena and Bond Strength, in *Adhesion and Adsorption of Polymers*, L.H. Lee, Ed., Plenum Publishing, 1980, p 175-198

56. V. Weissberg and M. Arcan, A Uniform Pure Shear Testing Specimen for Adhesive Characterization, *Adhesively Bonded Joints: Testing, Analysis, and Design*, STP 981, American Society for Testing and Materials, 1988

57. K.M. Liechti and T. Hayashi, On the Uniformity of Stresses in Some Adhesive Deformation Specimens, *J. Adhes.*, Vol 29, 1989, p 167-191

58. W.S. Johnson and S. Mall, "A Fracture Mechanics Approach for Designing Adhesively Bonded Joints," NASA Technical Memorandum 85694, National Technical Information Service, 1983

59. S. Mall, W.S. Johnson, and R.A. Everett, Jr., Cyclic Debonding of Adhesively Bonded Composites, *Adhesive Joints*, K.L. Mittal, Ed., Plenum Publishing, 1984, p 639-658

60. S. Mall and W.S. Johnson, Characterization of Mode I and Mixed-Mode Failure of Adhesive Bonds Between Composite Adherends, in STP 893,

American Society for Testing and Materials, 1986, p 322-334

61. E. Moussiaux, A.H. Cardon, and H.F. Brinson, Bending of a Bonded Beam as a Test Method for Adhesive Properties, in *Mechanical Behaviour of Adhesive Joints*, G. Verchery and A.H. Cardon, Ed., Editions Pluralis, 1987, p 163-174

62. E. Sancaktar, H. Jozavi, and J.R. Baldwin, Elastoplastic Fracture Behavior of Structural Adhesives Under Monotonic Loading, in *Proceedings of the XIIIth Southeastern Conference on Theoretical and Applied Mechanics*, Vol 2, College of Engineering, University of South Carolina, 1986, p 752-758

63. E. Sancaktar and J.R. Baldwin, Mixed-Mode Fracture in Adhesively Bonded Joints Under Monotonic Loading, in *Proceedings of the 1986 SEM Spring Conference on Experimental Mechanics*, Society for Experimental Mechanics, 1986, p 529-536

64. E. Sancaktar, Elastoplastic Fracture Behavior of Structural Adhesives, in *Mechanical Behaviour of Adhesive Joints*, G. Verchery and A.H. Cardon, Ed., Editions Pluralis, 1987, p 351-362

65. E. Sancaktar and J. Tang, Mixed-Mode Fracture in Adhesively Bonded Joints Under Dynamic Loading, in *PRI Proceedings of the IIIrd International Conference on Adhesion*, Plastics and Rubber Institute, 1987, p 18-1, 18-7

66. E. Sancaktar, Elastoplastic Fracture Behavior of Structural Adhesives Under Monotonic and Dynamic Loading, in *Proceedings of the Joint Military/Government-Industry Symposium on Structural Adhesive Bonding*, American Defense Preparedness Association, 1987, p 120-139

67. G.C. Sih, Fracture Mechanics of Adhesive Joints, *Polym. Eng. Sci.*, Vol 20, 1970, p 977-981

68. J.A. Marceau, J.C. McMillan, and W.M. Scardino, Cyclic Testing of Adhesive Bonds, in *Proceedings of the 22nd National SAMPE Symposium*, Society for the Advancement of Material and Process Engineering, 1977

69. J. Romanko, Behavior of Adhesively Bonded Joints Under Cyclic Loading, Paper presented at the NATO-AGARD Lecture Series 102, *Bonded Joints and Preparation for Bonding*, Oslo, Norway and The Hague, The Netherlands, North Atlantic Treaty Organization-Advisory Group for Aerospace Research and Development, 1979

70. D.W. Dwight, E. Sancaktar, and H.F. Brinson, Failure Characterization of a Structural Adhesive, in *Adhesion and Adsorption of Polymers*, L.H. Lee, Ed., Plenum Publishing, 1980, p 141-164

71. E. Sancaktar and P. Zhang, Viscoelastic Modeling of the Fiber-Matrix Interphase in Composite Materials, Report MIE-187, Clarkson University, 1989

72. E. Sancaktar and P. Zhang, Nonlinear Viscoelastic Modeling of the Fiber-Matrix Interphase in Composite Materials, *J. Mech. Des. Trans. ASME*, 1990

73. E. Sancaktar, A. Turgut, and F. Guo, "The Effects of Test Temperature, Loading Rate, Fiber Sizing, and Global Strain Level on the Fiber/Matrix Interfacial Strength," MAE Report, Clarkson University, 1990

74. L.B. Sandberg and M.P. Albers, Methodology for Determining the Durability of Sealants, *J. Adhes.*, Vol 15, 1982, p 27-47

Special Tests for Membranes and Miniature Components

Mark G. Allen, School of Electrical Engineering, Georgia Institute of Technology

ADHESION TESTS that are particularly suited for measuring the adhesion of membranes and miniature components, especially microelectronics, are described in this article. Guidelines that indicate which tests are most appropriate for particular applications are presented, along with equations that translate observed quantities (such as peel strength, pressure, or current) into a debond energy characteristic of an adhered layer/substrate system. Other adhesion tests, such as those described in the other articles in this Section of the Volume, may also be useful in particular applications. Because the development of new adhesion tests for microelectronic materials is a subject of great current interest in both the adhesion and the microelectronics communities, the current literature may also prove to be valuable in the selection of an adhesion test for a particular membrane or miniature component application.

Membrane Tests

One of the most popular adhesion tests performed on membranes is the blister test, or one of its many modifications. The blister test, first introduced by Dannenberg in 1961 (Ref 1), has recently been the subject of much study on the measurement of the adhesion of coatings of relatively flexible materials adhered to relatively rigid substrates.

In this test, a suspended membrane of an adhered layer is pressurized through a hole in the substrate by a fluid (liquid or gas) until the adhered layer begins to peel from the substrate. The blister test is shown in Fig. 1. The quality of the adhesion of the adhered layer is determined by measuring the pressure necessary to initiate peel (known as the critical, or blow-off, pressure) and relating this pressure to a debond energy (a quantity characteristic of the adhesive strength) through the mechanical properties of the adhered layer and the blister geometry.

The debond energy can also be measured by directly measuring the pressure-volume (PV) work done by the pressurizing fluid on the blister as the adhered layer peels. This work has two components, a reversible component (as long as the adhered layer is not plastically deformed) that is due to elastic stretching of the adhered layer, and an irreversible component that is due to detaching the adhered layer from the substrate. By measuring the elastic component independently (for example, by a postpeel inflation of the blister), the work necessary to debond the adhered layer from the substrate can be calculated from a simple subtraction (Ref 1). Both methods have been used to measure the debond energy of adhered layers peeled in the blister test.

Fabrication of Samples

The fabrication of samples for the blister test can be accomplished in various ways. If the adherend is a tapelike or other relatively solid material, it can be pressed over a hole in the substrate to create a suspended membrane. If the adherend is a spin-cast or spray-cast material (such as a paint), several fabrication methods are available. The hole in the substrate can be sealed with a thin foil, which adheres only weakly to the substrate. The paint is then applied over the substrate and foil. When pressurized through the hole, the foil peels easily from the substrate until it reaches the area in which the paint film itself is adhered to the substrate. Alternatively, the adhering layer can be cast onto the solid substrate and a hole drilled or etched from the back to expose a suspended membrane of the material (Ref 2). Another method is to fill the hole in the substrate temporarily with a plug of polytetrafluoroethylene or other nonadhering material while the adhering layer is cast. Once the layer has dried, the plug can be removed from the back, leaving behind a suspended membrane of material. Any other method that results in a region of adhered layer suspended over a hole in the substrate can also be used. For example, microfabrication techniques have been employed to make both ordinary and specialized blister geometries for use in measuring the debond energies of materials of importance in microelectronics. This is discussed in greater detail in the next section.

Relation of Measured Critical Pressure to Debond Energy

In order to relate the observed quantities in any adhesion experiment (such as critical pressure) to a material or system property describing adhesion (such as debond energy), models of the mechanical behavior of the system must be developed. Such models for the classical blister test are described below for two cases: adhered layers under zero in-plane stress, and adhered layers under nonzero in-plane tensile stress.

Griffith Energy Balance. The relationship between the debond energy and the critical pressure can be determined using an energy balance argument. Consider a pressurized blister of film adhered to a substrate, as shown in Fig. 1. Using a Griffith argument (Ref 3), during any virtual increment in crack area, the total energy of the peeling system must be constant. Differentially, this can be expressed as:

$$\delta E = \delta \Pi + \delta S = 0 \qquad \text{(Eq 1)}$$

where Π is the potential energy of deformation of the blister and S is the energy that forms a new surface. For the adhesion tests to be described here, other dissipative processes occurring during peel, such as plastic deformation, are lumped into the term dS. With this definition, a rearrangement of Eq 1 yields:

$$\delta \pi = -\gamma_a \delta A \qquad \text{(Eq 2)}$$

where γ_a is defined as the debond energy and δA is an increment of crack area. Thus, if expressions for the potential energy of deformation of the blister can be obtained, the debond energy can be found from the critical pressure by applying Eq 2. It should be noted that, in general, the relationship between the debond energy and the critical pressure will involve the mechanical properties of the film, as well as the geometry of the test site.

Adhered Layers Under Zero In-Plane Stress. The experimental measurement of the critical pressure to determine the debond energy is usually much easier than the PV method described earlier. However, care must be taken to ensure that the anal-

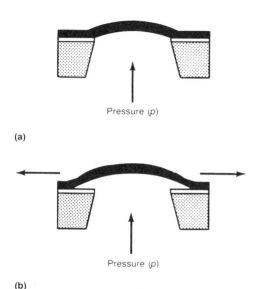

Fig. 1 The blister test. (a) A suspended membrane of film is deflecting as a function of applied pressure. (b) As the pressure is increased to the critical pressure, the film begins to peel radially outward from the substrate.

ysis applicable to the particular blister geometry is used, or erroneous results will be obtained. It should also be noted that the following arguments do not explicitly take into account the effect of different mode loadings on the measured debond energy. The calculation of the various mode loadings for the blister test is treated in Ref 4.

Gent (Ref 5) has identified three regimes for the blister test, which must be analyzed differently when relating the critical pressure to the film debond energy. Let the dimensionless parameter λ be defined as the ratio of the blister diameter to the adhering film thickness, that is:

$$\lambda = \frac{2a}{t} \qquad \text{(Eq 3)}$$

where a is the blister radius and t is the adhered film thickness. The three regimes can then be identified as follows:

$$\lambda \ll 1 \quad \text{(Regime I)}$$

$$\lambda - 1 \quad \text{(Regime II)}$$

$$\lambda \gg 1 \quad \text{(Regime III)}$$

These three regimes actually correspond to three different loadings of the film caused by the different geometries. In regime I, pressurization of the adhered layer results mainly in the creation of high-stress regions around the edge of the blister diameter, with little or no deformation of the film itself. In regime II, pressurization of the adhered layer results in a bending deformation of the adhesive layer. In this case, the adhered layer can be considered to be a plate with a built-in edge constraint undergoing a bend-

ing deflection. The effect of tensile stretching can be ignored as long as the maximum deflection at the center of the blister (caused by applied pressure) is also small compared with the blister thickness. Finally, in regime III, pressurization of the adhered layer results mainly in its tensile deformation. In this case, the adhered layer can be considered to be a membrane with almost no resistance to bending. Only blisters fabricated in portions of regime II (moderately large λ) and regime III (large λ) can be considered membranes. Membrane blisters containing mixtures of bending and stretching (that is, the transitions between regimes II and III) are treated in the section "Calculation of Debond Energy Relations."

By determining these potential energy expressions for the three regimes described above, the following relationships between critical pressure and debond energy can be obtained (Ref 4, 5):

$$\gamma = \frac{3\, a\, p_c^2}{2\, \pi\, E} \qquad \text{(Regime I)} \qquad \text{(Eq 4)}$$

$$\gamma = \frac{9\, a^4\, p_c^2}{128\, E\, t^3} \qquad \text{(Regime II)} \qquad \text{(Eq 5)}$$

$$\gamma = \left(\frac{a^4\, p_c^4}{17.4\, E\, t} \right)^{(1/3)} \qquad \text{(Regime III)} \qquad \text{(Eq 6)}$$

where λ is the debond energy of the adhered layer (the quantity that is directly related to the quality of the adhesion of the film); a is the blister radius prior to the initiation of peel; p is the critical pressure, or pressure necessary to initiate peel; E is the Young's modulus of the adhered layer; and t is the thickness of the adhered layer. The units of debond energy are work (or energy) per unit area; debond energies are commonly reported in J/m². References 4 and 5 should be consulted for restrictions and assumptions in the derivation of Eq 4 to 6. More general expressions (Eq 7, 8) will reduce to Eq 4 to 6 under the appropriate assumptions.

Adhered Layers Under Nonzero In-Plane Stress. Described below are the causes and implications of in-plane stress, the calculation of debond energy relations, and application of the peel criterion for stressed adhered layers.

Causes and Implications of In-Plane Stress. In-plane stress may arise from a variety of causes. For example, the cure of an adhered layer on a substrate may result in shrinkage and corresponding residual in-plane tensile stress. Alternatively, the thermal expansion coefficient mismatch between adhered layer and substrate may give rise to either residual in-plane compressive stresses or tensile stresses. If the adhered layers are under in-plane stress, the above relationships (regimes I to III) are not valid. Films under sufficient compressive in-plane

stress may blister and spontaneously deadhere from their substrates as a means of relieving this stress. Cases such as these have been treated by Hutchinson (Ref 6). When the film is under compressive stress, a membrane of film usually cannot be formed without the occurrence of buckling. Therefore, blister or other membrane tests are not usually used; but indentor methods, described in the section "Indentation Techniques" in this article, are more commonly employed. However, the blister test is directly applicable to the measurement of films with residual tensile stress. In this case, the suspended membrane of film is more like a stretched drumhead. As one might expect, however, differing relations between debond energy and critical pressure exist for these films.

Calculation of Debond Energy Relations. The case of in-plane tensile stressed blisters in regimes II and III, as well as a mixture of the two, has been treated in Ref 7. The potential energy of the blister used in Eq 2 is related to both the blister load-deflection behavior (how the blister deforms in response to an applied load) and the blister geometry. Therefore, in order to apply Eq 2, various blister geometries are assumed and the load-deflection behavior of each is calculated. The blister geometries considered are a square of side length $2a$ and a circle of radius a. The adhered layers to be debonded can be treated as membranes (no resistance to bending) or as plates, and are characterized by a Young's modulus (E), an in-plane Poisson's ratio (v), and an in-plane residual tensile stress (σ_0). It should be noted that the transition from plate to membrane behavior is analogous to the transition from regime II to regime III as described in the previous section. Details of the load-deflection calculation can be found in Ref 8. It is found that the load-deflection behavior of all three cases (square membrane, circular membrane, and clamped circular plate) can be described by the equation:

$$p = k_1 d^3 + (k_2 + k_3)\, d \qquad \text{(Eq 7)}$$

where p is the applied (uniform) pressure, d is the deflection of the blister at its geometric center, and k_1, k_2, and k_3 are functions of the geometry of the test site and of the blister regime (plate or membrane). This equation is the general load-deflection relation that will be used in the analysis of the blister potential energy. The three values of k in Eq 7 as a function of the Poisson's ratio of the adhered layer are given in Table 1. See also Fig. 2.

For the case of a lateral load uniformly distributed over a thin plate or membrane (ratio of film thickness, t, to blister size, a, $\leqslant 1$), the potential energy of the deformed adherend layer can be calculated and substituted into Eq 2. In this analysis, it is assumed that the substrate is rigid and the

Table 1 Parameters for the load-deflection relation (Eq 7)

$k_1(r)$	k_2	k_3
Square membrane		
$[\pi^6/128(1-\nu^2)]\{5/16-[4(5-3\nu)^2/9\pi^2(9-\nu)+64(1+\nu)]\}(Et/a^4)$ 0		$(3.04\,\sigma_0 t)/a^2$
Clamped circular plate		
$[3/(1-\nu^2)](1.221-7.848\times10^{-3}\,(6-\nu)(1+11\nu)-8.965\times10^{-3}$		
$(23-41\nu)(1.98-\nu))(Et/a^4)$ .... $5.68(Et^3/a^4)$		$(4\,\sigma_0 t)/a^2$
Circular membrane		
$[8/3\,(1-\nu)](Et/a^4)$ 0		$(4\,\sigma_0 t/a^2)$

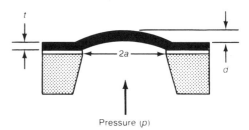

Fig. 2 A suspended membrane undergoing a deflection (d) at its center in response to an applied pressure (p)

film is elastic, so that increasing crack size does not change the energy stored in the residual in-plane stress. Unlike the blistering of an adhered layer under compression, described earlier in the section "Causes and Implications of In-Plane Stress," if the blister is under residual tensile stress, increasing crack size a does not change the energy stored in the residual in-plane stress. This is because as long as the edges of the film are attached to the substrate, relaxation of the residual stress cannot occur. Therefore, although the residual stress and strain affect the peel criterion through the load-deflection behavior, the energy balance need take into account only the elastic strain energy stored in the blister. The assumption is invalid if the blister substrate is not infinitely rigid. For example, it is known that films on a thin silicon substrate can cause the substrate to curve. A common technique to determine the magnitude of the residual stress in a film is to measure this curvature. In a structure such as this, blister peeling can allow some substrate relaxation, causing the strain energy due to the residual stress to decrease as the blister size increases, effectively lowering the critical peel pressure.

Application of Eq 2 yields the following relationship between debond energy and critical deflection (that is, deflection at the center of the blister at the point that peel initiates) for films under residual in-plane tensile stress:

$$\gamma_a\left[\frac{5\,c_1 a}{2}\left(\frac{\Delta}{a^3}\right)^4+\frac{3\,c_2}{a}\left(\frac{\Delta}{a^3}\right)^2\right.$$
$$\left.+2c_3 a\left(\frac{\Delta}{a^3}\right)^2\right]\frac{da}{dA} \qquad \text{(Eq 8)}$$

where da/dA is the incremental dependence of blister size on blister area, Δ is the blister volume, and c_1, c_2, and c_3 are constants that depend on the mechanical properties of the

film. Table 2 gives the values of the various parameters in Eq 8 for three blister geometries. By substitution of the appropriate values from Table 2 into Eq 8 and simultaneous solution of Eq 8 with the corresponding load-deflection relation (Eq 7), a value for γ_a can be determined as a function of the critical debond pressure p_c.

It should be noted that the constants depend on both the modulus and the residual stress of the film. Therefore, with the exception of some limiting cases discussed in the following section, the experimental measurement of p_c cannot be related to γ_a unless the mechanical properties of the film have been accurately determined.

Application of the Peel Criterion. Under certain conditions, Eq 8 reduces to important special cases that have been previously reported. For example, in the case of a clamped circular plate with 0 residual stress undergoing small deflections ($c_1 = c_3 = 0$, regime II), Eq 7 and 8 become identical to that of Williams (Ref 9):

$$\gamma_a = 0.5\,p_c d_c \qquad \text{(Eq 9)}$$

where d_c is the deflection at the center of the plate at pressure p_c. Alternatively, for the case of a circular membrane undergoing large deflections with zero residual stress ($c_2 = c_3 = 0$, regime III), Eq 7 and 8 become:

$$\gamma_a = 0.625\,p_c d_c \qquad \text{(Eq 10)}$$

This result is very close to the result obtained in Eq 6. Although Eq 6 and Eq 10 appear to be different, the substitution of Eq 7 into Eq 10 to eliminate d_c yields Eq 6. A similar substitution will show the equivalence of Eq 9 and 5.

In order to illustrate the effect of the residual stress, a circular membrane undergoing large deflections (regime III) that is also under varying degrees of in-plane tensile stresses will be considered. In this case, the substitution of the appropriate parameters from Table 1 into Eq 7 yields:

$$p_c = 3.56\,\frac{Et}{a^4}d_c^3 + 4\,\frac{\sigma_0 t}{a^2}d_c \qquad \text{(Eq 11)}$$

(A Poisson's ratio of 0.25 has been assumed in Eq 11.) A corresponding substitution of parameters into Eq 8 yields:

$$\gamma_a = 2.22Et\,(d_c/a)^4 + 2.00\sigma_0 t\,(d_c/a)^2 \qquad \text{(Eq 12)}$$

Equation 12 is a relationship between the debond energy and the critical deflection. The relation between the debond energy and the critical pressure can be obtained by the simultaneous solution of Eq 11 and Eq 12. The results of this solution are presented in Fig. 3, a logarithmic plot of the debond energy (normalized by the film modulus and thickness) versus critical pressure (normalized by blister size, film modulus, and film thickness) for various residual stresses. As can be seen, for zero residual stress, a slope of 4/3 is obtained, agreeing with Eq 6. As the residual stress is increased, the adhesive energy corresponding to a given critical pressure decreases; this is due to energy expended in deflecting the film against the residual stress. At sufficiently high stress, the slope of the critical pressure relation becomes equal to 2. This corresponds to a linear load-deflection relation (with the cubic term in Eq 11 being negligible) and thus leads to the same functional form as a blister undergoing small deflections (where the load-deflection relation is also linear) (Ref 9). Finally, at large pressures, the effect of the residual stress on the critical pressure becomes small because the load-deflection relation is dominated by stretching against the modulus (with the linear term in Eq 11 being negligible).

Other Membrane Adhesion Tests

In addition to the classical blister test already described and analyzed, a variety of new tests involving suspended membranes have been developed to meet specific applications. Two of the tests described below, the constrained blister test and the island blister test, are used for thin, well-adhered systems. One of the potential problems with the application of the blister test to well-adhered layers in regimes II and III is that the layers tend to be tensile-strength limited. In other words, in the attempt to remove the film from the substrate in a controlled

Table 2 Parameters for the debond energy relation (Eq 8)

Parameter	Square membrane	Clamped circular plate	Circular membrane
c_1	(a)	(a)	(a)
c_2	0	$5.44\,Et^3$	0
c_3	$1.88\,t\sigma_0$	$3.82\,t\sigma_0$	$2.55\,t\sigma_0$
Δ	$16a^2\,d/\pi^2$	$a^2\,d\pi/3$	$a^2 d\pi/2$
da/dA	$1/2a$(b)	$1/2\pi a$	$1/2\pi a$

(a) c_1 is calculated from k_1 using the relation $c_1 = k_1\,(a^3/\Delta)^3\,ad^3$. (b) Assumes incrementally symmetric peel

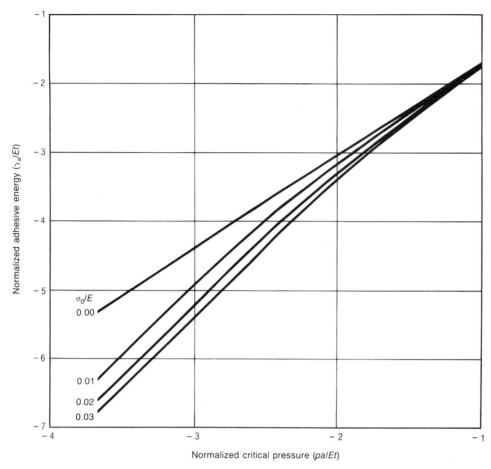

Fig. 3 A log-log plot of the normalized debond energy as a function of normalized pressure parameterized by the in-plane stress in the film. Source: Ref 7

Constrained Blister Test. One membrane test that seeks to overcome this tensile strength limit is known as the constrained-blister test (CBT) (Ref 10, 11). Shown in Fig. 4, the CBT approach to increasing the tensile strength limit of the film is to place a flat plate over the growing blister, thereby limiting the maximum strain that the film experiences. In this way, much higher critical pressures (corresponding to higher debond energies) can be measured before the tensile strength limit of the film is exceeded (that is, before the film tears).

Unlike the tests described previously, the CBT is usually performed in such a way that the blister growth as a function of time can be observed. This is accomplished by using a transparent constraining plate over the blister and observing the blister growth at constant pressure. The growth rate will depend not only on the debond energy, but also on the magnitude of the dissipative processes during peel. For example, a very brittle film would be expected to have a large growth rate once the critical pressure was reached, while a relatively ductile film would be expected to have a relatively lower growth rate. Higher pressures also result in an increase in the growth rate. Applying the Griffith energy balance as for the classical blister, the following relationship for debond energy as a function of applied pressure has been proposed (Ref 10):

$$\ln\left(\frac{A(t)}{A_0}\right) = \left(\frac{\beta\, p^2 h t}{\gamma - ph}\right) \quad \text{(Eq 13)}$$

where $A(t)$ is the blister area at time t, A_0 is the initial blister area (at $t = 0$), h is the spacing of the constraining plate above the film, p is the applied pressure, γ is the debond energy, and β is a phenomenological coefficient that describes the dissipative

manner, the film tears before it peels. Almost every effort to solve this problem has involved strengthening the film in some way. For example, in the standard 90° peel test, often a very thick layer of polymer is built up so that it can be peeled without

tearing. The problem in this approach is that the measured peel strength may change drastically as the film thickness increases. Thus, it is desirable to investigate the feasibility of tests that can be adapted to thin films without modifying them in any way.

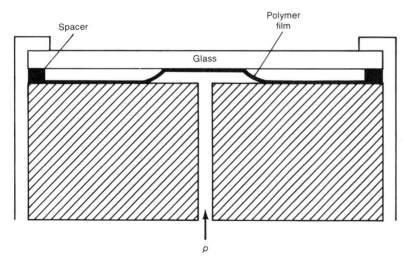

(a)

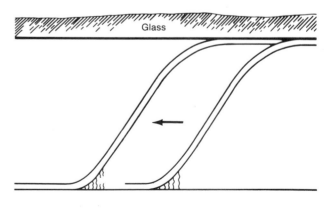

(b)

Fig. 4 The constrained blister test. (a) Test apparatus. (b) Detail of film debonding from the substrate

processes occurring during peel. Although there is some confusion about the exact interpretation of the coefficient β, Eq 13 has been used successfully to fit data taken in the constrained blister test. By taking area data as a function of time for several different spacer heights, the debond energy and the dissipative coefficient can be independently determined. This method of analysis neglects the possibility that changes in the spacer height may also result in a change in the mode mix, potentially affecting the measured debond energy.

If the dissipative processes can be neglected ($\beta = 0$), or alternatively, are included in the debond energy (γ) (as with other membranes described here), then Eq 13 can be written as:

$$\gamma = ph \qquad \text{(Eq 14)}$$

Various corrections to the above equations have been proposed. For example, a correction factor recommended in Ref 11 results in the equation:

$$\gamma = phq \qquad \text{(Eq 15)}$$

where q, a dimensionless quantity ranging from 0.8 to 1.0, is a geometric correction for the slope of the adhered layer (Fig. 4). This slope is assumed to be vertical between the substrate and the constraining plate in Eq 13 and 14. Finite-element models have also been carried out in an attempt to improve the above relationships (Ref 12).

The constrained blister test has been successfully demonstrated on various adhesive tapes, and the functional forms of the above equations have been verified. However, there is still some uncertainty about the exact meaning of the dissipative coefficient (β). Attempts to apply the constrained blister test to thin solvent-cast films (in order to overcome their tensile strength limit) have as yet met with limited success only. One speculation is that although the maximum strain in these films is limited and higher debonding pressures can be applied, these high pressures cause failure to initiate around defects (for example, dust particles or pinholes) in the solvent-cast film.

The island blister test is another geometric modification of the standard blister adhesion test. This modification consists of a suspended membrane of film with an island of substrate at its center. The membrane support and the island are secured to a rigid plate, and the film is pressurized, which peels the film off the island. A model for this annular peel indicates that even for systems of good adhesion, peel can be initiated at low enough pressures to prevent film failure by making the center island sufficiently small. Thus, the tensile strength limit of thin, well-adhered films can be overcome. In principle, this test, like the tests described above, could be applied to many different types of paints and coatings. As yet, however, the island blister test has

been confined to microelectronic applications. It is therefore described more fully in the subsection "Island Blister Test" under the section "Microelectronics Tests" in this article. The applicability of this test to thin, well-adhered non-microelectronic films is currently under study.

Peninsula Blister Test. One of the characteristics inherent to island blister geometry is the increase in the strain-energy release rate as peel progresses. In other words, an island blister held at constant pressure will result in a debond with an increasing peel rate as a function of time. One test that has been proposed to address this issue is a modification of the island blister test, called the peninsula blister test (Ref 13).

In this test, a rectangular suspended membrane is fabricated over a hole in a substrate that has a "peninsula," or strip of substrate projecting from one end underneath the film. The film is adhered to this peninsula, which will become the site of the debond test. The film is then pressurized from the back, as in the conventional, constrained, and island blister tests, and the critical pressure necessary to sustain peel along the peninsula is measured. This pressure can then be related to the debond energy, as is done for the other blisters described above. This technique is capable of measuring very large debond energies at relatively low pressures by making the ratio of peninsula width to film width sufficiently small; thus, it may also be applicable to thin, well-adhered systems. In addition, in certain regimes, the peel rate is no longer a function of adhered width and therefore does not change as peel progresses. One disadvantage mentioned in Ref 13 is that the peninsula blister is no longer axisymmetric; this potentially reduces its utility for environmental-exposure testing. Because this technique is relatively new, interested readers are referred to Ref 13 for the equations relating observed critical pressures to film debond energies.

Inverted Blister Test. Another test that has been proposed for measuring the debond energy of thin, well-adhered films is known as the inverted blister test (Ref 14). In this test, the thin film is adhered to a substrate that is reasonably compliant. The film-substrate system is then inverted (that is, turned upside down), and the film is bonded to a second, rigid substrate. At this point, the system looks like a sandwich, with the original, slightly compliant substrate on top, the film in the middle, and the new, rigid substrate on the bottom. A hole is made in the rigid substrate and film from the back, as in the standard blister test. Pressure is then applied to the hole, and the original, compliant substrate is peeled from the film, as in the standard blister test. Because the compliant substrate is presumably more resistant to rupture than is the

original, thin film, higher pressures can be applied to the film-substrate interface (and therefore larger debond energies can be measured) than in the noninverted configuration. The inverted blister test is analyzed using the same equations as those used in the standard blister test, except that the mechanical properties of the compliant substrate are used instead of the mechanical properties of the film.

Tests Driven by In-Plane Stresses

As described in the section "Adhered Layers Under Zero In-Plane Stress," films under compressive stress can spontaneously delaminate from substrates. Films under tensile stress can also delaminate if a flaw is induced in the film (see the section "Cutout Tests"). Finally, even if no stress is present in the film, in-plane stress can be induced by a normal force on a film coupling through the Poisson's ratio of the film. This is the basis of indentation techniques for adhesion measurement.

Indentation Techniques. In-plane stresses can be induced in films by the use of indentation techniques (Fig. 5) (Ref 15-17). In one popular method, a thin film is coated onto a substrate, and a ball of well-defined radius is pressed with a known and controllable force onto the film substrate. The coating underneath the ball is thus deformed, and a complicated stress state consisting in part of shear is induced in the coating-substrate interface. At a sufficiently high applied force, these stresses cause failure at the film-substrate interface. The applied load at which this failure is observed can then be related to the shear strength of the film-substrate bond. This technique is particularly applicable for the very thin films often encountered in microelectronics, which usually cannot be peeled from a substrate.

Cutout Tests. As mentioned previously, films under residual tensile stress can delaminate from the edges or they can delaminate if a flaw is introduced into the film. For example, consider a film under residual stress adhering to a substrate. If the thickness of the film is increased, the total energy stored due to the film residual stress will increase. As the film thickness is increased, the energy adhering the film to the substrate (that is, the debond energy) will be exceeded, and spontaneous film delamination will occur. The application of an energy balance indicates that delamination will occur when:

$$\gamma = \frac{t_{cr} \, \sigma_0^2 \, (1 - \nu)}{E} \qquad \text{(Eq 16)}$$

where γ is the film debond energy, σ_0 is the residual stress in the film, E is the Young's modulus of the film, and ν is the Poisson's ratio of the film. Thus, the debond energy of

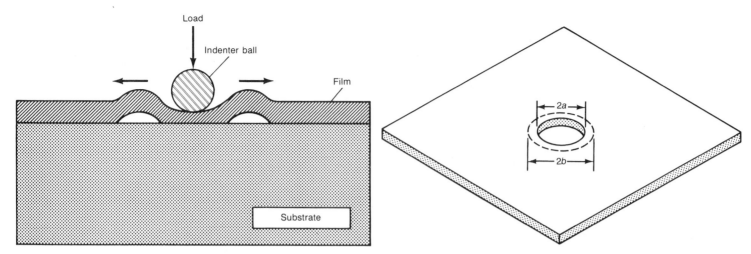

Fig. 5 The indentation test. A normal force causes a lateral displacement of the film and initiates debonding.

Fig. 6 The cutout test. Delamination of a stressed film on a rigid substrate due to the presence of a circular cut

the film can be measured by measuring the film critical thickness.

In practice, varying the thickness of the films is often difficult to do. In order to determine the debond energy, a series of films of increasing thickness would have to be prepared until the film spontaneously delaminated. Thus, much effort would be necessary to prepare samples that would give a null result.

A better method has been proposed by Farris (Ref 18). The film shown in Fig. 6 is under residual tensile stress and has had a circular cut of radius a inscribed in its center. If the thickness of the film is above the critical thickness, the central region of the film will spontaneously debond. The debond will also propagate circularly into the outer region of the film. It is expected that the debond will arrest at some radius b that is greater than the initial cut a. It should be noted that the film in this case does not peel completely to the edge of the sample because only radial stresses are relieved by debonding. By measuring the radius of the debond after peel has stopped, the debond energy of the film can be calculated from the following equation:

$$\gamma = \frac{t\sigma_0^2}{E}\left[\frac{a^2(1-\nu^2)}{b^2(1-\nu)+a^2(1+\nu)}\right] \quad \text{(Eq 17)}$$

where t is the film thickness, a is the radius of the initial cut, b is the radius of the final debond, and the other parameters are the same as those noted in Eq 16.

It should also be noted that the calculation of the debond energy requires knowledge not only of the Young's modulus of the material, but also of the material residual stress. In some instances, such as stress buildup due to a mismatch of the coefficients of thermal expansion, it may be possible to calculate the film stress. Otherwise, it must be measured independently on the same substrate if the debond energy is to be calculated. Alterna-

tively, if the debond energy is known or is measured by an independent method, this test can be used to determine the residual stress in the adhered layer.

Microelectronics Tests

The importance of thin film technology in the fabrication of integrated circuits (ICs) has become well established since the early 1960s. Films that have been used regularly include metals (especially aluminum), silicon, silicon nitride, silicon dioxide, and, more recently, polymers. Polymers are used as photoresists, radiation masks, dielectrics, and interlevel insulators. In many of these applications, the accurate determination of the mechanical properties and adhesion of these films is essential in order to ensure device performance and reliability. For example, the loss of adhesion of a photoresist pattern to a substrate during processing can lead to pattern distortion and loss. Similarly, the loss of interlayer adhesion between a dielectric film and a metal signal line in an IC multilevel interconnect scheme could lead to corrosion of the line and, ultimately, failure of the device. There are many such examples that illustrate the need for a reliable, quantitative, and reproducible adhesion test for these polymeric films, as well as for the more common inorganic films used in microelectronics.

A desired feature of adhesion tests in microelectronics is that they be performed *in situ*, that is, on as-deposited films, to ensure that the film under test most closely resembles a film in an actual-device application. This section describes adhesion tests that have been used to measure, *in situ*, the adhesion of films used in microelectronics devices.

Peel Tests

A very widely used test, both in microelectronics and in other applications, is the peel test (Fig. 7). A conceptually simple

test, the peel test involves peeling the film off of the substrate at a specified angle and rate (known as the peel angle and peel rate, respectively). The peel test therefore allows control over the mechanics of debonding (including estimation of viscoelastic and other rate effects), which is not available in many of the previously described tests (Ref 19, 20). Contact to the film in order to apply the peel force can be made either directly or by attaching a secondary substrate to the film and applying the peel force to this substrate. The peel force is measured using a tensile testing machine. A rolling platform or other device should also be employed to maintain the peel angle at a constant value.

In Fig. 7, if the strip undergoing peel has a width w and a tensile force F, the peel force per unit width is simply F/w. A common data-reporting procedure is to measure the peel force of strips of various widths and plot the measured force as a function of

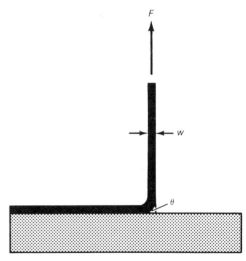

Fig. 7 The peel test, in which the film is peeled from the substrate at a constant, specified rate by the force F at an angle θ.

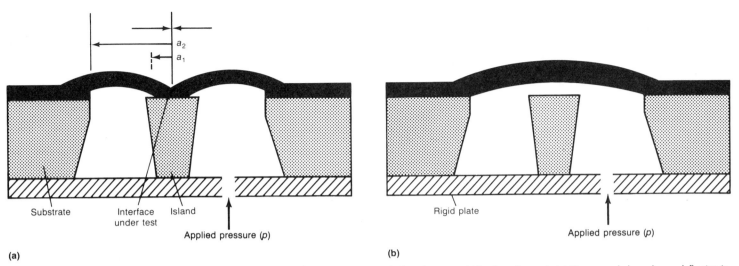

Fig. 8 The island blister test. Pressure is applied through a hole in a rigid plate to which the substrate and island are fastened. (a) The suspended membrane deflecting in response to the applied pressure. At the critical pressure, the film begins to peel off the center island. (b) Geometry of structure after peel has been completed

strip width. The slope of the best-fit line through these data is then taken as a measure of the peel strength of the material.

The units of peel strength (force per unit width) are the same as those of debond energy (work per unit area) because the units of work are force times distance. Occasionally, results are reported in mass per unit width; it is then necessary to multiply by the acceleration due to gravity in order to convert between peel strength and debond energy. It must be remembered, however, that because debond energies are representative not only of the interface but also of the type of loading (and therefore of the individual test), a direct comparison of debond energies measured in one test with those measured in another may not be possible.

The main advantage of the peel test in microelectronics applications is the relative ease of preparation of samples. For this reason, it is widely used (Ref 21-23). Once again, however, thin, well-adhered films may be impossible to peel without tearing. One solution to this problem is to make and peel much thicker films. However, it has been shown that plastic and viscoelastic effects can dramatically alter the measured peel strength of a thick versus a thin film, even if they are identical materials on identical substrates. The use of careful modeling (Ref 19) can help overcome this difficulty and attempt to relate the peel strength to a debond energy that is more characteristic of the interface.

Island Blister Test

The difficulties involved in the peeling of thin well-adhered films were discussed in the section Constrained Blister Test. In that test, large pressures could be applied to thin films without exceeding their tensile strength limit because the maximum strain was limited by a constraining plate. An alternative method is to limit the maximum strain by keeping the critical pressure low. One geometry that can accomplish this is known as the island blister (Ref 7, 24). The island blister is a modification of the standard blister site in that the suspended membrane of film has an island of substrate still attached at its center. The island and the substrate are both fastened to a rigid plate, and pressure is applied, as in the standard blister test (Fig. 8). Film peeling will now occur only off the center island. The pressure necessary to initiate peel can be made low, compared to the tensile strength of the film, simply by making the center island sufficiently small. Thus, the tensile strength limit of the film can be overcome geometrically. It should be noted that the same argument applies to the previously described peninsula blister test. This argument is discussed briefly below.

The Griffith criterion suggests that if a geometry can be found in which, for an increment of crack extension (δa), the area of created surface (δA) can be decreased without simultaneously decreasing the potential energy ($\delta \pi$), larger values of γ_a may be measured at the same load. For simple blisters, this derivative da/dA is inversely proportional to the membrane size. Decreasing the membrane size fails because the potential energy will also decrease. However, consider an annular island structure of outer radius (a_2) and inner radius (a_1) as shown in Fig. 8. The blistering will now occur only off the center island. The potential energy stored in this deformed geometry prior to peel is a function of the difference $a_2 - a_1$, which changes only slightly during peel. The peel circumference (crack area), however, is proportional only to a_1. Thus, a large geometric advantage can be obtained by decreasing a_1 (small crack area) while keeping $a_2 - a_1$ large (large potential energy).

These arguments can be confirmed mathematically by applying Eq 2. For the case of the load-deflection behavior of the film dominated by its residual stress (that is, the induced stresses due to stretching deformations are much lower than the residual in-plane stresses in the film), the relationship between the critical pressure and the debond energy can be calculated to be (Ref 25):

$$\gamma_a = \frac{p_c^2 a_1^2}{32\sigma_0 t} \left[\frac{(\beta^2 - 1)}{\ln \beta} - 2 \right]^2 \qquad \text{(Eq 18)}$$

where β is defined as the geometric ratio:

$$\beta = \frac{a_2}{a_1} \qquad \text{(Eq 19)}$$

Thus, as the outer radius of the film grows large compared to the inner radius, the term β also grows large, while as the outer radius of the film shrinks toward the inner radius, the term β approaches unity. It can also be shown that if the debond energy of the adhered layer to both the outer and inner (island) regions is equal, the critical pressure necessary to propagate a debond inward is always less than the critical pressure necessary to propagate a debond outward. Thus, peeling will always occur inward "off the island." The mode mix of this test should be relatively constant until the adhered radius a_1 approaches the thickness of the film.

Equation 18 suggests one method for overcoming the tensile strength limit of thin, well-adhered films. If the film under test has a particular debond energy, thickness, and stress, and there is a maximum pressure that the film can withstand, the critical, or peel, pressure can always be made lower than the maximum pressure simply by making the quantity β sufficiently large. In other words, by making the island sufficiently small with respect to the sus-

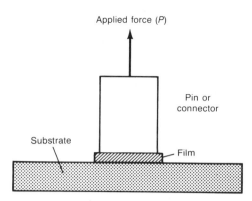

Fig. 9 The direct pull-off test. The interface under test is the one between the substrate and the film.

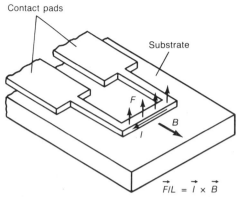

Fig. 10 The electromagnetic tensile test. The film experiences a normal force *F* due to the interaction of the magnetic field *B* and the current *I*. *L* = length of U where lateral force is generated.

pended membrane of film, peel even of thin, well-adhered films can be initiated.

Electromagnetic Tensile Test

The simplest type of tensile adhesion test is the direct pull-off test (Fig. 9). It involves connecting (for example, by cement or solder) a pin or connector to the film, clamping the substrate, and applying a force on the pin normal to the interface. The force is increased until failure occurs, and a value for the debond energy is calculated from this critical force. Although conceptually simple, this test suffers from a host of complications, such as the difficulty of uniform attachment to a thin film, the creation of residual film stresses due to the curing of the film cement, and the necessity of good alignment in order to achieve uniform loading (peel stresses caused by an inclination of the force *P* will cause debonding at much lower forces than if the loading is completely normal). Despite these and other drawbacks, a number of authors have employed this method (Ref 26).

An elegant extension of this test, especially applicable to microelectronic materials, has been developed (Ref 27, 28). In this test, termed the electromagnetic tensile test, metallic or other conducting films are deposited on a polymeric or other insulating surface. The film is then patterned into a rectangular U form using lithographic techniques (Fig. 10). The entire specimen is placed in a lateral magnetic field, and current pulses are passed through the metal film. Because of the interaction of the current and the magnetic field, a force on the film normal to the surface of the polymer is generated in one portion of the U (Fig. 10). By increasing the current until the film detaches, a value for the detachment force of the metal film to the polymer can be determined.

The adhesive strength can be determined from the observed variables using the equation:

$$P = \frac{BI}{w} \tag{Eq 20}$$

where *B* is the applied magnetic field, *I* is the magnitude of the applied current, *w* is the width of the strip (Fig. 10), and *P* is the force per unit area generated by the interaction of the current and the applied magnetic field.

The maximum current that can be passed through the line is limited by Joule heating of the line. Current is therefore usually applied in pulses in order to circumvent this problem. The maximum current and pulse duration are dictated by the maximum allowable temperature rise in the metal line. The estimated temperature rise due to the pulsed current is given by:

$$\Delta T = \frac{\tau I^2 r}{c d A^2} \tag{Eq 21}$$

where *τ* is the pulse duration in seconds, *A* is the cross-sectional area of the line, *d* is the density of the line material, and *c* is the heat capacity of the line material. Typical numbers used in this test for the adhesion of aluminum films to polyimide were a width of 10 μm (0.4 mil), a current of 8 A, a pulse duration of 1 μs, and a temperature rise of approximately 3 °C (5 °F).

The debond force per unit area is measured by observing the thin lines under a microscope and applying a pulse of current. The current is increased, and the procedure is repeated until separation of the line from the substrate is observed. It should be noted that the appropriate quantity taken to be a measure of the bond strength is a force per unit area (that is, a pressure), as opposed to an energy per unit area (for example, for the blister or peel tests described above).

REFERENCES

1. H. Dannenberg, Measurement of Adhesion by a Blister Method, *J. Appl. Polym. Sci.*, Vol 5, 1961, p 125
2. J.A. Hinkley, A Blister Test for Adhesion of Polymer Films to SiO$_2$, *J. Adhes.*, Vol 16, 1983, p 115
3. M.F. Kanninen and C.H. Poplar, Chapter 3 in *Advanced Fracture Mechanics*, Oxford University Press, 1985
4. G.P. Anderson, S.J. Bennett, and K.L. DeVries, Chapter 3 in *Analysis and Testing of Adhesive Bonds*, Academic Press, 1977
5. A.N. Gent and L. Lewandowski, Critical Blow-Off Pressures for Adhering Layers, *J. Appl. Polym. Sci.*, Vol 33, 1987, p 1567-1577
6. A.G. Evans and J.W. Hutchinson, On the Spalling and Delamination of Compressed Films, *Int. J. Solids Struct.*, Vol 20, 1984, p 455-466
7. M.G. Allen and S.D. Senturia, Analysis of Critical Debonding Pressures of Stressed Thin Films in the Blister Test, *J. Adhes.*, Vol 25, 1988, p 303-315
8. M.G. Allen, "Measurement of Mechanical Properties and Adhesion of Thin Polyimide Films," M.S. thesis, Massachusetts Institute of Technology, 1986
9. M.L. Williams, The Continuum Interpretation for Fracture and Adhesion, *J. Appl. Polym. Sci.*, Vol 13, 1969, p 29
10. M.J. Napolitano, A. Chudnovsky, and A. Moet, The Constrained Blister Test: A New Approach to Interfacial Adhesion Measurements, *Proceedings of the American Chemical Society Division of Polymeric Materials: Science and Technology*, Vol 57, 1987, p 755-759
11. Y.S. Chang, Y.H. Lai, and D.A. Dillard, The Constrained Blister—A Nearly Constant Strain Energy Release Rate Test for Adhesives, *J. Adhes.*, Vol 27, 1989, p 197-211
12. Y.H. Lai and D.A. Dillard, Numerical Analysis of the Constrained Blister Test, *J. Adhes.*, in review
13. D.A. Dillard and Y. Bao, The Peninsula Blister Test: A High and Constant Strain Energy Release Rate Fracture Specimen for Adhesives, *J. Adhes.*, in review
14. M. Fernando and A.J. Kinloch, Department of Mechanical Engineering, Imperial College (London), private communication
15. A.J. Perry, *Thin Solid Films*, Vol 107, 1983, p 167
16. J.E. Ritter, L. Rosenfeld, M.R. Lin, and T.J. Lardner, Interfacial Shear Strength of Thin Polymeric Coatings on Glass, in *Thin Films: Stresses and Mechanical Properties*, J.C. Bravman, Ed., Materials Research Society, 1989, p 117-122
17. T.P. Weihs, S. Hong, J.C. Bravman, and W.D. Nix, *J. Mater. Res.*, Vol 5, 1988, p 931
18. R.J. Farris and C.L. Bauer, A Self-Delamination Method of Measuring the Surface Energy of Adhesion of Coatings, *J. Adhes.*, Vol 26, 1988, p 293-300
19. K. Kim and J. Kim, Elasto-Plastic Analysis of the Peel Test for Thin Film Adhesion, *Trans. ASME*, Vol WA/

EEP-3, 1986

20. H. Fukuda and K. Kawata, On The Mechanics of Peeling, in *Resins for Aerospace*, C.A. May, Ed., Symposium Series 132, American Chemical Society, 1980, p 70

21. D. Suryanarayana and K.L. Mittal, Effect of pH of Silane Solution on the Adhesion of Polyimide to a Silica Substrate, *J. Appl. Polym. Sci.*, Vol 29, 1984, p 2039

22. R. Narechania, J. Bruce, and S. Fridmann, Polyimide Adhesion—Mechanical and Surface Analytical Characterization, *J. Electrochem. Soc.*, Vol 132, 1985

23. L.B. Rothman, Properties of Thin Polyimide Films, *J. Electrochem. Soc.*, Vol 127, 1981, p 2216

24. M.G. Allen and S.D. Senturia, Application of the Island Blister Test for Thin Film Adhesion Measurement, *J. Adhes.*, Vol 29, 1989, p 219-231

25. M.G. Allen, "Measurement of Adhesion and Mechanical Properties of Thin Films Using Microfabricated Structures," Ph.D. thesis, Massachusetts Institute of Technology, 1986

26. K.L. Mittal, Adhesion Measurement of Thin Films, *Electrocomponent Sci. Technol.*, Vol 3, 1976, p 21

27. S. Krongelb, in *Adhesion Measurements of Thin Films, Thick Films, and Bulk Coatings*, K.L. Mittal, Ed., STP 107, American Society for Testing and Materials, 1978

28. B.P. Baranski and J.H. Nevin, Adhesion of Aluminum Films on Polyimide by the Electromagnetic Tensile Test, *J. Electron. Mater.*, Vol 16, 1987, p 39

Special Tests for Sealants and Elastomeric/Foam Adhesives

Didier R. Lefebvre, AT&T Bell Laboratories

SEALANTS, ELASTOMERIC ADHE-SIVES, and foam adhesives can be categorized for convenience as soft adhesives, with respect to their mechanical behavior. Likewise, this article refers to structures that contain soft adhesives as soft joints. Intuitively, a soft material is defined as one having low tensile and shear moduli and low hardness; that is, it can undergo large deformations at small loadings. Of more significance to material testing and characterization are three other features often associated with soft materials: nonlinear behavior, energy dissipation due to viscoelastic behavior, and high Poisson's ratio.

First, soft materials tend to exhibit strong deviations from linearity in their stress-strain curves. In the case of elastomers and foams, a stiffening is typically observed at high strains. This kind of nonlinearity is referred to as a material (physical) nonlinearity and should not be confused with the geometric (structural) type. Geometrically, nonlinear behavior is seen when strains are so large that the relationship between strain and displacements is no longer linear. Thus, large deformations necessitate the use of strain definitions, such as the Lagrangian strain, that include terms of higher order than the commonly used engineering strain. It is not unusual to simultaneously observe the two kinds of nonlinearities in soft materials undergoing large deformations.

Secondly, soft materials tend to respond to mechanical perturbation in a time-dependent manner often called viscoelastic behavior. This means that the stress-strain curve is rate dependent and shows a hysteresis loop upon unloading. In viscoelastic materials, mechanical energy is converted into heat as a result of dissipative mechanisms at the molecular level. The area of the hysteresis loop is a measure of the dissipated heat.

Lastly, soft materials, such as rubbers, tend to deform with minimal volume change. This property gives rise to highly triaxial, highly nonuniform stress distribution in loaded joints.

It is quite clear that the above properties impose additional requirements on testing techniques that necessitate mapping the nature and magnitude of the stresses. For this reason, it is advantageous to test soft joints with a method that does not necessitate a detailed knowledge of stress distributions and stress singularities and is capable of coping with energy dissipation. This can be achieved by using the energy balance approach of fracture mechanics. Note that linear fracture mechanics (LFM) is invalid when materials do not behave in a linear-elastic manner away from the crack tip. There are special cases, however, where tests can be designed for evaluation by LFM or by the energy balance approach, interchangeably. The principles used to select one approach over another are discussed next.

Fracture Mechanics for Testing Soft Joints

The basic tenet of fracture mechanics is that the strength of structures is governed by the initiation and propagation of inner flaws. Flaws can be introduced by a damage mechanism during service or may be inherent to the morphology of the material used. An inherent flaw in an elastomer can be an imperfectly bonded inorganic filler particle, for example. With the help of fracture mechanics, the resistance of a material to flaw propagation is characterized by a parameter known as the fracture energy (G_c) or, alternatively, the fracture toughness (K_c). Because these parameters are true material properties, independent of the geometry of the cracked body, they can be used to predict ultimate strength for a given flaw size or critical flaw size for a given loading. Fracture energy and fracture toughness are interrelated as long as the behavior of the material is linear-elastic away from the crack region. Because G_c is derived from the energy balance approach, it is the parameter of choice here for characterizing the fracture resistance of soft joints.

The energy balance approach was first used by Griffith (Ref 1) and Orowan (Ref 2) for brittle materials. According to this description, flaw propagation will occur whenever the stored elastic energy released per unit area of virtual crack is sufficient to overcome the cleavage energy of the material. It may be written:

$$\frac{\partial W}{\partial A} - \frac{\partial U}{\partial A} \geq 2\gamma \qquad \text{(Eq 1)}$$

where W is work done by the external force, U is stored elastic energy, A is crack area, and γ is surface free energy.

In joints containing highly deformable materials, the above criterion must be modified for two reasons: first, some mechanical energy may be converted to heat as a result of viscoelastic dissipation within the adherends, and second, the energy required to cause crack growth is far greater than twice the surface free energy because of dissipative mechanisms in the crack tip region. Thus, the criterion for crack growth becomes:

$$\frac{\partial W}{\partial A} - \frac{\partial U}{\partial A} - \frac{\partial V}{\partial A} \geq G_c \qquad \text{(Eq 2)}$$

where V is the energy dissipated from the crack tip and G_c is fracture energy. The left side of the expression in Eq 2 is the energy release rate, or the driving force for debond, and is often denoted G. (As the driving force reflecting the actual state of stress, G should not be confused with the material property G_c, describing the separation energy.) The dissipation term is significant when a soft adherend undergoes large deformation. This situation is encountered either when a soft adherend is peeled away from a rigid substrate or in a blister specimen with a thin elastomeric membrane. The dissipation term can be evaluated directly by extracting the energy loss from the hysteresis loop of the stress-strain curve (Ref 3-5). Alternatively, the viscoelastic constitutive behavior of the soft material can be determined by creep or relaxation studies, and then V can be calculated by subtracting the calculated stored energy from the external work. Another more rigorous approach to this problem is the J-contour integral method (Ref 6, 7).

When possible, it is convenient to use fracture specimens that behave in a linear-elastic

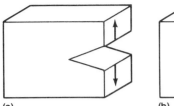 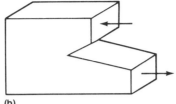 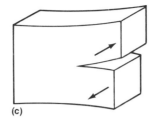

Fig. 1 Modes of loading of a crack. (a) Cleavage or tensile opening, mode I. (b) In-plane shear, mode II. (c) Antiplane shear, mode III. Source: Ref 8

manner, so the term $\partial V/\partial A$ can be neglected. In such cases, the expression in Eq 2 can be shown (Ref 8) to be equivalent to:

$$\frac{F^2}{2}\frac{\partial C}{\partial A} \geq G_c \qquad \text{(Eq 3)}$$

where F is the external load and C is the compliance of the structure (displacement/load). The expression in Eq 3 is valid whenever the soft material is highly constrained between rigid adherends, as in the case of the double cantilever beam specimen and the butt joint specimen. When applicable, the expression in Eq 3 is more convenient than that in Eq 2 because the compliance is easy to obtain experimentally by monitoring the stiffness of the structure. If the joint structure behaves in a nonlinear, nonelastic manner, the expression in Eq 3 is not applicable and the rate of energy dissipation must be evaluated. Failure to do so will yield an apparent fracture energy of $G_c + \partial V/\partial A$, which is not a true measure of adhesive fracture toughness.

The fracture energy G_c contains a dissipative component, even in the most brittle adhesives (Ref 8):

$$G_c = G_0 + \psi \qquad \text{(Eq 4)}$$

where G_0 is the energy required for rupturing secondary and primary bonds (intrinsic fracture energy), and ψ is the energy dissipated in viscoelastic and plastic deformations. In spite of the dissipative contributions, G_c remains an intrinsic material property as long as the dissipative mechanisms are confined to the crack tip region. As discussed in the next section, ψ is a function of rate and temperature.

In the sections in this article that describe commonly used test geometries, the concept of fracture mode is used. Figure 1 describes the three fundamental modes of fracture, denoted I, II, and III. In general, fracture energy is the result of three components, G_{Ic}, G_{IIc}, and G_{IIIc}, which correspond to the three modes of loading depicted in Fig. 1. Mode I (cleavage) is usually the predominant loading mode in fracture tests.

Effect of Rate and Temperature

The intrinsic fracture energy G_0 in Eq 4 is a measure of the various binding forces acting across the fracture plane, such as

intermolecular, intramolecular, or interatomic interactions. Prior to attaining the breakage point of these primary or secondary bonds, a stressed region must develop ahead of the crack tip and extend far beyond the scale of the ruptured bonds. As the bonds are broken and the crack moves further, the region will experience stress relaxation. This stress-strain cycle experienced by the region bordering the path of fracture gives rise to the viscous dissipation, ψ, defined earlier. Because the dimension of the region undergoing the viscoelastic hysteresis is quite large compared to the dimension of molecular bonds, ψ is typically orders of magnitude higher than G_0. Furthermore, because binding forces must be nonzero for a stress field to develop, G_0 and ψ are expected to be interrelated. Based on intuitive arguments, some authors (Ref 9-13) have proposed this relationship:

$$\psi = G_0 f(\dot{a}, T, \epsilon) \qquad \text{(Eq 5)}$$

where f is a function of crack growth rate ($\dot{a}$), temperature (T), and strain (ϵ). Combining Eq 4 and 5 yields:

$$G_c = G_0 [1 + f(\dot{a}, T, \epsilon)] \qquad \text{(Eq 6)}$$

Thus, function f provides a description of the rate and temperature dependence of G_c.

Both rate and temperature variations allow the experimentalist to adjust the ratio of the debond time scale to the time scale of the segmental motions involved in the dissipative processes. In rubbery materials, the relationship between the time scale of molecular relaxations and temperature is described by the Williams-Landel-Ferry (WLF) equation (Ref 14). In this case, rate and temperature may be related:

$$\ln\left(\frac{\dot{a}_{T_0}}{\dot{a}_T}\right) = \frac{-C_1 (T-T_0)}{C_2 + T - T_0} \qquad \text{(Eq 7)}$$

where T is temperature, T_0 is reference temperature, $\dot{a}_{T_0}$ is the debond rate at T_0, and $\dot{a}_T$ is the debond rate at T. Both C_1 and C_2 are constants determined by the choice of T_0. It is often convenient to select the glass transition temperature as the reference temperature. With Eq 7, it is possible to construct a single master curve from G_c versus $\dot{a}$ data collected at various temperatures. This master curve can span a much wider rate range than otherwise would be accessible with a single temperature. If the

material's behavior is not well described by the WLF equation, a master curve still may be obtained empirically by shifting the various curve segments graphically.

In general, the higher the debond rate or the lower the temperature, the further the material is from thermodynamic equilibrium. Therefore, a high rate or low temperature will each tend to increase the area of the stress-strain hysteresis loop experienced by the volume elements in the path of the crack. The higher energy dissipation, in turn, results in a higher fracture energy. Conversely, a lower rate or higher temperature will cause a decrease in G_c. In the limit of an infinitely slow debond rate, the viscoelastic function f in Eq 6 goes to zero and G_c approaches the intrinsic fracture energy G_0. The effect of rate and temperature is illustrated in Fig. 2. The various curves represent the fracture energy versus debond rate for different substrates adhesively bonded with cross-linked styrene-butadiene rubber. Each is a master curve obtained from data collected over a wide range of temperatures and shifted with the WLF equation. As expected from Eq 6, changing the substrate causes a change in G_0, but the curves remain roughly parallel, because the dissipative component of fracture energy is mainly a characteristic of the soft adherend (invariant here).

So far, it has been implicitly assumed that the failure mechanism is invariant over the entire range of debond rates and temperatures. This is not always the case, however, as Fig. 3 shows. The curve represents G_c versus the debond rate obtained by peeling a cloth-backed layer of uncross-linked styrene-butadiene rubber from a polyethylene terephthalate (PET) substrate. Again, the plot is a master curve obtained from points collected over a wide range of temperatures. An abrupt drop in G_c is observed as the locus of failure shifts from cohesive within the uncross-linked rubber adhesive to interfacial between the adhesive and the substrate. The second maximum in G_c is associated with unstable crack growth, often referred to as stick-slip debond. The stick-slip phenomenon is due to the response of the material going from rubbery to glassy at some critical debond rate, causing a sharp decrease in energy dissipation.

Peel Tests

Experimental. Peel tests are widely used for evaluating the relative strength of adhesives and for comparing surface treatments. They usually consist of peeling away a soft adherend from a rigid substrate at a constant angle, as depicted in Fig. 4. The peel angle can be varied to obtain various combinations of mode I and mode II loadings.

The floating roller peel setup shown in Fig. 5 offers a convenient way to maintain a constant angle of peel as well as a constant curvature in the peeled strip, while gripping

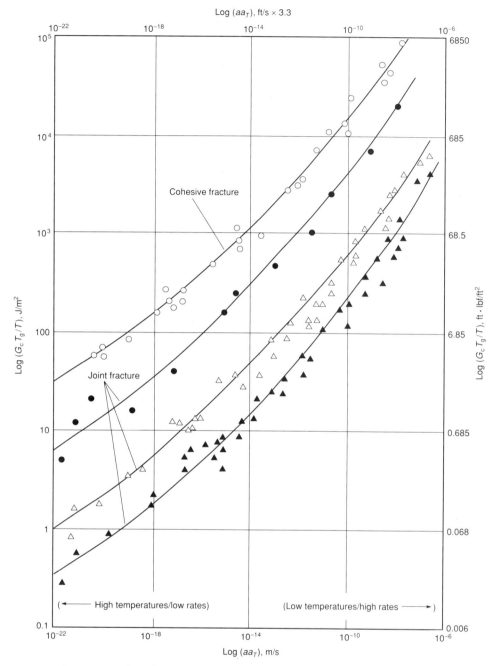

Log (aa_T), ft/s × 3.3

Cohesive fracture

Joint fracture

(← High temperatures/low rates)

(Low temperatures/high rates →)

Log (aa_T), m/s

Fig. 2 Master curves for adhesive fracture energy, G_c, versus effective rate of debond for cross-linked styrene-butadiene rubber (SBR) adhering to various substrates. The points for each curve represent data from various test rates and temperatures. Reference temperature = glass transition temperature (T_g), 233 K (−40 °C). ○, G_c for cohesive fracture of cross-linked SBR; ●, G_c for fracture of cross-linked SBR/etched fluorinated ethylene-propylene copolymer joint; △, G_c for fracture of cross-linked SBR/treated fluorinated ethylene-propylene copolymer joint; ▲, G_c for fracture of cross-linked SBR/polyethylene terephthalate joint. Source: Ref 8, 10

the specimen with a conventional tensile tester. Alternatively, angle control may be achieved by installing a stage capable of translating the substrate in concert with the crosshead. A number of variations on the peel test have been reported. Figure 6 shows a symmetrical arrangement designed to adjust the degree of bending of the peeled strip by varying the distance, D, between the symmetrically positioned substrates. Figure 7 shows the T-peel test arrangement used for studying adhesion between soft adherends. With very

soft materials it is often advantageous to reinforce the soft strip with a backing. The backing should be both rigid (in stretching) and flexible (in bending) to avoid excessive stretching and bending of the peeled strip near the crack tip region. Large stretching and sharp bending are undesirable because they are a source of viscoelastic dissipation and/or plastic deformation away from the crack tip region.

Peel specimens are usually tested by applying a given displacement rate and recording

the peel force versus time. A proper mechanical analysis then converts the obtained data to a peel energy versus peel rate curve. Standard peel tests at constant rate may be modified by inserting a spring between the crosshead and the peeling strip (Ref 8, 18) to decouple the peel rate from the crosshead rate. This soft machine arrangement can eliminate stick-slip behavior and reveal the value of G_c at zero rate (Ref 18).

Fracture Mechanics. The analysis of peel specimens can be quite involved for several reasons:

- The peeled material typically has nonlinear, viscoelastic behavior and sometimes experiences plastic deformation
- The peeled strip often undergoes large stretching and bending
- A change in peel rate may drastically alter the mix of fracture modes at the crack tip

Extreme caution should be used, therefore, when converting peel loads to fracture energies. Unfortunately, closed-form solutions are generally unavailable for problems of this complexity, and few finite-element programs can simultaneously handle nonlinear viscoelastic behavior, large stretching, and large bending. However, there are a few relatively simple cases of elastic behavior for which a closed-form solution can be obtained for the left side of Eq 2. The analyses by Anderson *et al.* (Ref 16) and Lindley (Ref 19) are suggested reading.

In general, high peel angles are not desirable because they tend to increase energy dissipation in the highly bent region away from the crack tip, making it difficult to analyze (Ref 20). The problem may be alleviated by using rollers, as shown in Fig. 5, or by using an elastic backing. From fracture or stress analyses that assumed linear elastic behavior (Ref 20), it was found that the peel force is inversely proportional to (1 − cos θ), where θ is the peel angle defined in Fig. 4. Experimentally, this means that angle θ should not be too small, in order to avoid excessive deformation or failure in the peeled strip.

Defined previously was a region bordering the crack plane, which undergoes a stress-strain cycle as the crack tip approaches and then recedes. It follows that if the size of this region is limited by the thickness of the strip, then the fracture energy will be thickness dependent (Ref 20). Conversely, if the region undergoing dissipation is smaller than the thickness of the adherend, then G_c is not thickness dependent (all other parameters remaining equal).

Blister Tests

Experimental. In its most general form, the blister specimen consists of a layer of soft material completely adhering to a rigid substrate, except over a central portion where loading is applied (Ref 8, 16, 21). The

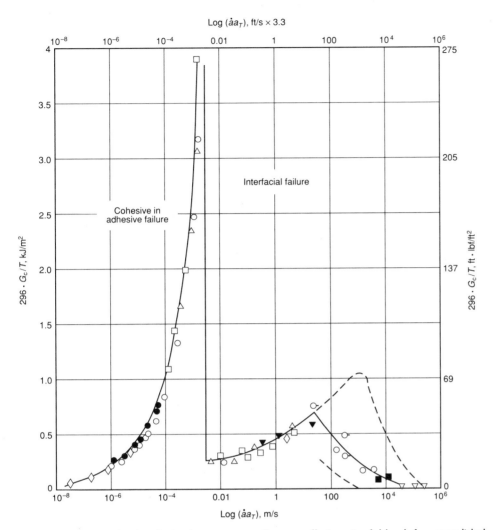

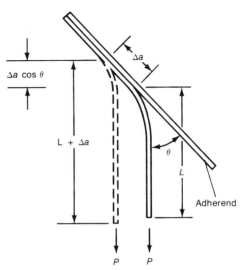

Fig. 4 Peel test specimen. Source: Ref 16

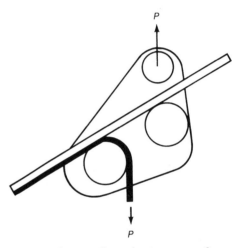

Fig. 3 Master curve for the adhesion fracture energy, G_c, versus effective rate of debond, for uncross-linked SBR/polyethylene terephthalate joint. The different symbols represent different test temperatures from −35 to 60 °C (−31 to 140 °F). Broken curves denote the extreme values when unstable, stick-slip peeling occurred. Reference temperature = 296 K (23 °C). Source: Ref 8, 15

Fig. 5 Floating roller peel resistance test. Source: Ref 16

most common blister geometries have a circular unbonded region at the center. When the unbonded region is pressurized by injection of a compressed fluid, the soft adherend lifts off and forms a blisterlike cavity. The thickness of the soft material ranges from very thin (membrane) to very thick (semi-infinite medium), as illustrated in Fig. 8. The geometry depicted in Fig. 9, known as the cylindrical peel specimen, is used in special cases where premature shear failure in the soft adherend would occur in the absence of lateral constraint.

When a critical pressure (P_c) or, equivalently, a critical energy-release rate (G_c) is attained, the disc-shape flaw propagates radially. If one layer is transparent, crack initiation and propagation may be monitored optically. For opaque materials, alternate techniques based on pressure monitoring have been proposed (Ref 16). Caution must be used when testing large blister specimens. Catastrophic failure may occur at constant pressure because the energy-release rate in-

creases rapidly with debond radius. Therefore, it is recommended to avoid designs that store too much elastic energy prior to reaching G_c, and to limit the flow rate of the pressurizing fluid with a bleed valve.

A simplified version of the blister concept is the strip blister specimen (Ref 21) shown in Fig. 10. This linear configuration of the blister specimen is usually loaded by inserting a dowel into the initial flaw. Flaw size can be measured by using ultrasonic or optical techniques, or simply by using calipers (Ref 21).

Axisymmetric Blister Fracture Mechanics. When only elastic behavior in the soft layer is considered, simple closed-form solutions have been derived for the limiting cases when layers obey plate theory or when layers are infinitely thick. For intermediate thicknesses, numerical analysis must be used (Ref 16). Figure 11 shows a comparison between numerical solutions and closed-form solutions. This plot is given in nondimensional form and can be used to obtain a solution to the left side of the

expression in Eq 3 graphically, whenever closed-form solutions are unavailable. Note that the numerical curve coincides with the closed-form solutions only at low or high values of h/a (where h is plate thickness and a is blister radius). In the low h/a condition, it is preferable to use thick-plate theory over thin-plate theory. This is because shear deformations in the bent plate are not negligible, unless $4h^2/[(1 − v)a^2]$ is very small (Ref 21), where v is the Poisson's ratio of the soft layer. Reference 16 has more details on the mechanical analysis of the axisymmetric blister specimen.

In general, thin membranes should be avoided because deformations rapidly become too large as the blister grows, resulting in a combined bending and stretching deformation that is difficult to analyze (Ref 16). This problem may be alleviated by constraining the membrane with a rigid plate or by reinforcing the membrane with an elastic backing.

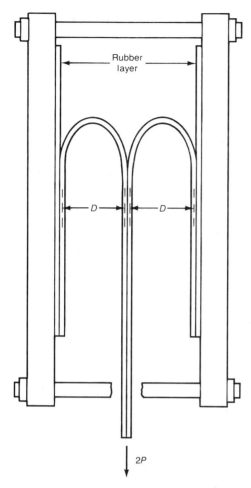

Fig. 6 Symmetric peel test arrangement. Source: Ref 17

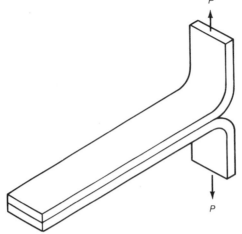

Fig. 7 T-peel test specimen. Source: Ref 16

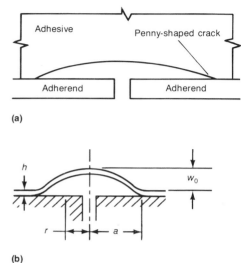

Fig. 8 Blister specimens. (a) Debond at the interface of a rigid plate and a semi-infinite (very thick) medium. (b) Blister specimen with low thickness-to-radius ratio. Source: Ref 16

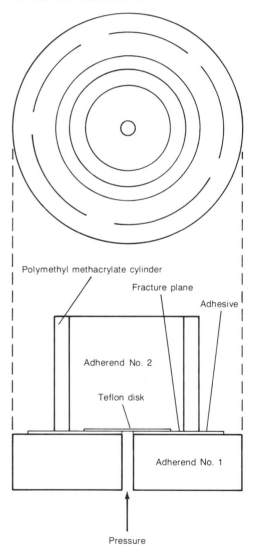

Fig. 9 Cylindrical peel specimen for testing soft adhesives prone to cohesive shear failure. Source: Ref 16

Attention should be given to the change in mode mixture as debond progresses radially. It has been shown that the loading tends to change from mainly mode I to a combination of mode I and II as h/a becomes smaller (Ref 21). However, mode I remains predominant as long as h/a is greater than 4 (Ref 8).

Strip Blister Fracture Mechanics. If only elastic behavior is considered in the soft layer, the strip blister specimen can be analyzed with the closed-form solution for a clamped beam with a concentrated load (Ref 21). As with the axisymmetric blister, shear deformations in the soft material should be taken into account, except in the case of very thin strips (Ref 21). The mode II component of fracture tends to increase with debond distance in a more pronounced way than in the axisymmetric geometry (Ref 21). More details on the analysis of the strip blister can be found in Ref 21.

Double Cantilever Beam Tests

Experimental. A schematic of the double cantilever beam (DCB) specimen (Fig. 12) is shown fitted with an extensometer for mea-

suring compliance. The specimen consists of a sandwiched layer of soft adhesive material bonded between more rigid adherends. Because the soft material is constrained between two more rigid substrates, large deformations and nonlinear, nonelastic behavior away from the crack tip are avoided. If the beams behave elastically, the energy-release rate may be obtained experimentally by measuring compliance, C, versus crack distance, a.

When a DCB specimen such as in Fig. 12 is subjected to a constant opening load, crack growth tends to be unstable because the energy-release rate increases with the square of crack distance. However, when loaded with a constant displacement, the DCB produces an energy-release rate that decays with the fourth power of crack distance, leading to rapid crack arrest. Such behavior is observed in the wedge test where a displacement is maintained using a shim or wedge inserted at the extremity.

To avoid rapidly changing energy-release rates, the DCB specimen can be tapered as shown in Fig. 13 (Ref 8). In this case the compliance varies linearly with crack distance, which, according to the expression in Eq 3, means G is independent of crack distance. Mathematical descriptions of the contours used to achieve $\partial C/\partial a$ = constant are given in Ref 8.

DCB specimens are usually loaded perpendicular to the crack plane with two equal forces acting in opposite directions. This type of loading yields pure mode I at the crack tip. A mode II component may be introduced in a variety of ways: by clamping the specimen and applying two unequal forces normal to the center line; by clamping the specimen and applying forces at an angle; or by using beams of unequal thicknesses. Pure mode II may be obtained by clamping the unflawed extremity and by applying two normal forces of equal magnitude in the *same* direction.

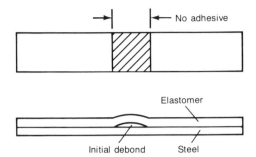

Fig. 10 Strip blister specimen. Source: Ref 21

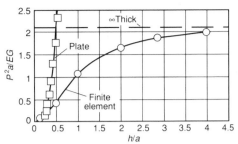

Fig. 11 Fracture behavior of the blister specimen. Comparison of numerical solutions with closed-form solutions. P = blister pressure, a = debond radius, E = modulus of the soft adherend, G = energy-release rate, and h = thickness of the soft layer. Source: Ref 16

Fracture Mechanics. Compared with the peel and blister specimens, the DCB geometry greatly reduces viscoelastic dissipation in the soft material for three reasons (Ref 22): the volume of soft material is minimal; the strain levels are smaller, thereby minimizing nonlinear effects; and the soft layer is highly constrained, forcing it to deform in the bulk, rather than shear, mode. (The bulk behavior of polymers tends to be substantially less dissipative than shear behavior.)

In the ideal case of a sandwiched layer with zero compliance and elastic beams, simple solutions based on elementary beam theory are available. References 8, 16, and 23 supply more details. Unfortunately, these elementary solutions are restricted to thin and rigid adhesives between elastic adherends (Ref 8, 22). With softer adhesives, the compliance of the sandwiched layer is not negligible, causing the measured compliance of the specimen to exceed that predicted by elementary beam theory. This deviation is illustrated in Fig. 14. In this case, a more accurate closed-form solution for G can be obtained by treating each beam as being on an elastic foundation. The resulting improvement in the solution is illustrated in Fig. 14. Irrespective of the type of adhesive layer used, it is highly recommended that the specimen always be calibrated by measuring compliance versus crack dimension. This method not only provides correct G values with minimal effort,

but can also be used to monitor crack distance during a test by conducting compliance measurements periodically (provided that the loads used never exceed P_c).

When designing a DCB specimen, the selection of beam thicknesses that can provide the desired G values, while remaining in the small-deflection envelope, is recommended. This requirement is based on the fact that most analytical solutions, including those with elastic foundation corrections, are valid for moderate deflections only, and that large beam deflections tend to introduce a beam foreshortening error. Finally, the remaining bonded length should always be large enough to ensure that the boundary conditions used in the analysis remain valid (Ref 22).

Cone Tests

In its most general form, the cone test specimen consists of a conical plug bonded to a conical hole, as shown in Fig. 15 (Ref 16). This geometry is ideal for introducing various amounts of mode I, II, and III components at the crack tip. For example, by applying a tensile load along the specimen axis and varying the cone angle, θ, the mode I to mode II ratio can be varied. When desired, a mode III component can be introduced by applying a torsional load about

the longitudinal axis (Ref 16). A cylindrical configuration ($\theta = 0$) with an initial circumferential debond will yield pure mode I when loaded in tension or pure mode III when loaded in torsion. In the limit of $\theta = 90°$, the specimen simply becomes a butt joint between two cylinders.

The fracture mechanics of the cone test for elastomeric adhesives sandwiched between two rigid conical adherends has been studied using finite-element analysis (Ref 16). It was found that for $\theta = 5, 45,$ or $90°$, the energy-release rate is maximal halfway between the edges, while for $\theta = 0°$, the energy-release rate is maximal at the edges. These predictions were confirmed experimentally by studying the locus of debond initiation for various cone angles.

Tensile and Shear Tests

The butt joint specimen and the lap-shear specimen are commonly used to characterize adhesive strength. Typically, no macroscopic flaws are introduced and a strength-of-material approach is used in the analysis: the applied load at failure is simply converted to an average tensile strength or average shear strength by dividing load by the area of the bonded region. Although this simple

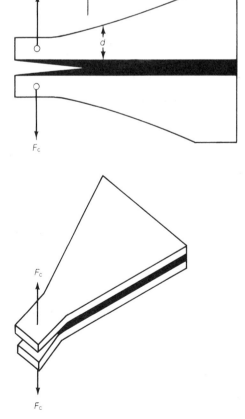

Fig. 13 Tapered double cantilever beam specimens for constant G fracture tests. Source: Ref 8

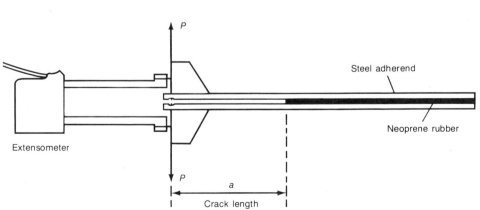

Fig. 12 Double cantilever beam specimen fitted with an extensometer for compliance measurements. Source: Ref 22

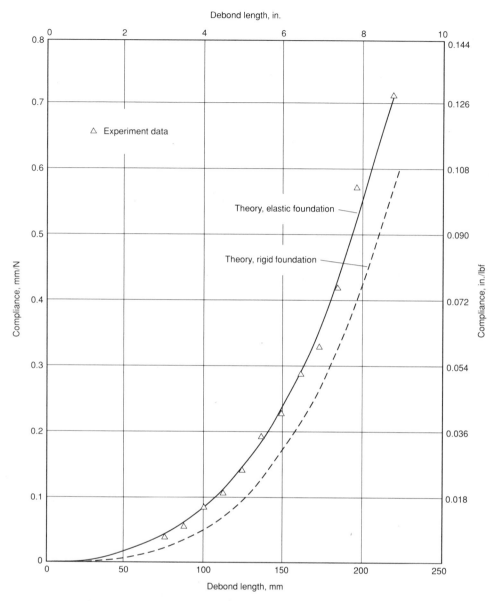

Fig. 14 Double cantilever beam specimen. Experimental and predicted compliance versus debond distance. Source: Ref 22

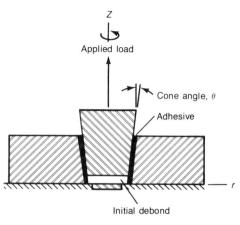

Fig. 15 Cone test specimen. Source: Ref 16

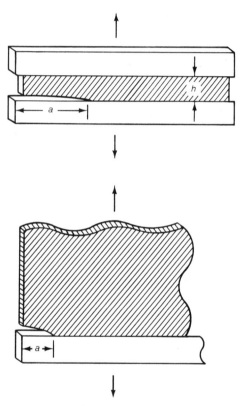

Fig. 16 Tensile tests with (a) thin adhesive layers and (b) thick adhesive layers. Source: Ref 23

approach can be useful for qualitative studies, the obtained failure parameters are useless for design purposes. Stress analysis of the butt joint and of the lap-shear specimen have demonstrated that several components of the stress tensor are usually present simultaneously and that the distribution of these stresses is highly nonuniform (Ref 8, 16). Because of the complex stress states, the load at failure cannot be readily converted into a strength-of-material type of criterion, such as the maximum principal strain or the maximum octahedral stress. Furthermore, because of the absence of a well-defined crack, the load at failure cannot be converted to a fracture energy or a fracture toughness. Thus, in the conventional approach, the tensile strength or shear strength reflects the presence of unknown flaws whose size may vary among speci-

mens. For this reason, it is suggested that a better way to use tensile or shear tests is to introduce known flaws and to use an analysis based on fracture mechanics.

Fracture Tensile Test. The tensile geometries depicted in Fig. 16 were analyzed by Gent *et al.* (Ref 23) in the case of small strains and linearly elastic adhesives. A relationship between failure load and fracture energy was determined in the two limiting cases of large and small crack dimension-to-thickness (a/h) ratios. With a thin adhesive layer (Fig. 16a), the failure load was found to be insensitive to crack dimension and inversely proportional to the square root of bond thickness. Conversely, with a thick adhesive layer (Fig. 16b), the failure load was found to be insensitive to adhesive thickness and inversely proportional to the square root of crack length.

Thus, bond thickness is the critical dimension for thin adhesive layers, while crack distance is the critical dimension for thick adhesive layers. In both cases, the failure load decreases with the critical dimension. An empirical relation that spans the two limiting cases was also given (Ref 23).

Fracture Shear Test. The same two limiting cases were studied for a shear loading, assuming an incompressible, linearly elastic adhesive between rigid adherends (Fig. 17a, 17b). The relationships among failure load, fracture energy, bond thickness, and crack size are analogous to those found for the tensile test (Ref 23).

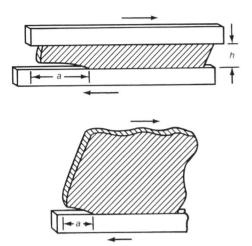

Fig. 17 Shear tests with (a) thin adhesive layers and (b) thick adhesive layers. Source: Ref 23

Combined Mechanical/Environmental Testing

Two general approaches are available for assessing the effect of environment on the mechanical properties of a bonded joint. In the first, unloaded fracture specimens are exposed to the environment for a prescribed length of time, and then are tested destructively to determine their residual strength. The lowering of fracture strength from the initial value is taken as a measure of environmental degradation (Ref 8). By varying exposure time and temperature, it is sometimes possible to extract the activation energy of the rate-limiting mechanism of the

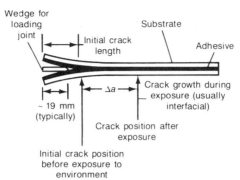

Fig. 18 Wedge test specimen for durability studies. Source: Ref 8

degradation process (Ref 8). In the second approach, the specimens are simultaneously subjected to load and environmental attack.

One of the most severe tests is the combination of mode I loading and environmental exposure. In the case of soft joints, this can be achieved by using simple self-loading devices such as the wedge specimen (Fig. 18) or the strip blister specimen loaded with a dowel (Fig. 19). The advantage of these methods is that specimens can be made in large numbers because they are easy and inexpensive to fabricate. Because no costly testing machine is tied up, another advantage is that a large number of specimens can be tested simultaneously, under a wide variety of conditions and for long periods of time.

Both the wedge and strip blister specimens are loaded with a constant displacement, producing an energy-release rate that

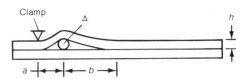

Fig. 19 Strip blister specimen loaded with a dowel, for durability studies. The clamp serves to confine debond to one side of the dowel only. Source: Ref 21

decays rapidly with debond distance until crack arrest occurs. Because debond distance for a given opening is an indirect measure of G, the final debond distance is often used as a measure of bond quality. In principle, the rate of decay of the energy-release rate with debond distance can be reduced by replacing the wedge or the dowel by a jig of lower stiffness. This concept is illustrated in Fig. 20 (Ref 21) with a spring-loaded DCB specimen immersed in a harsh liquid.

Harsh liquids/vapors diffuse through adhesives and sealants, causing a degradation of adhesion by a variety of chemical and physical mechanisms (Ref 8). Depending on the applied load, the debond tip may coincide with the advancing weakened bond, or it may lag behind. If the mechanical loading is high enough, debond can proceed faster than the diffusion process. Conversely, in the absence of applied or residual load, diffusion of the environment still occurs, although adherend separation may not occur. At intermediate load levels, the debond tip coincides with the advancing weakened region; stress opens the weakened region, allowing convection rather than diffusion to be the predominant mode of transport. Because of this complex interplay between the mechanical loading and the kinetics of degradation, the combined effect of load and environment is best described by plotting log (debond rate) versus log G. This representation commonly used in fatigue studies covers the entire spectrum of debond mechanisms and may be useful to define an endurance limit, G_{scc}. The endurance limit is defined as the upper bound of the G range where no fracture occurs. An example of log ($\dot{a}$) versus log G plot is given in Fig. 21.

The above procedure for studying durability is difficult to implement when the environment diffuses rapidly from the exposed edges of the specimens toward the centerline. Sealants can be used but are not always effective because they may affect the result of the experiment or may obstruct the view of the crack. The axisymmetric blister test offers an elegant solution to this problem if the environment is used as the pressurizing fluid. In this case, diffusion *must* occur perpendicular to the crack front.

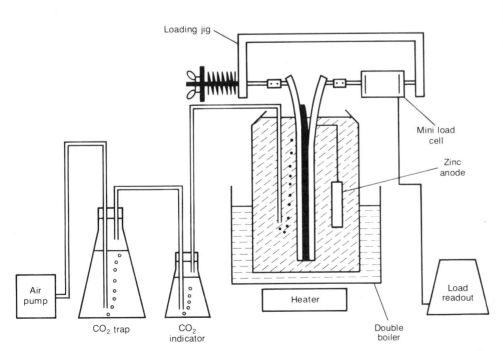

Fig. 20 Spring-loaded double cantilever beam specimen undergoing cathodic delamination in seawater. The spring lowers the rate of decay of G with debond distance. Source: Ref 21

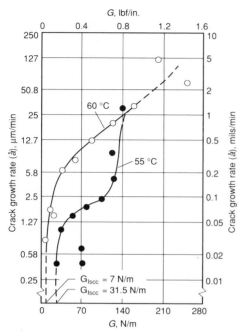

Fig. 21 Debond rate versus energy-release rate for an epoxy-to-aluminum bond exposed to water at 55 °C (130 °F) and 60 °C (140 °F). G_{Ic} = fracture toughness, mode I. The variations in the shapes of the curve with temperature illustrate the complex interplay between mechanical loading and the kinetics of degradation. Source: Ref 24

REFERENCES

1. A.A. Griffith, VI, The Phenomena of Rupture and Flow in Solids, *Philos. Trans. Soc.*, Vol A221, 1920, p 163
2. E. Orowan, Fracture and Strength of Solids, *Rep. Prog. Phys.*, Vol 12, 1948, p 185
3. A. Ahagon, A.N. Gent, H.J. Kim, and Y. Kumagi, Fracture Energy of Elastomers in Mode I (Cleavage) and Mode III (Lateral Shear), *Rubber Chem. Technol.*, Vol 48, 1975, p 896
4. A. Kadir and A.G. Thomas, Tearing of Unvulcanized Natural Rubber, *J. Polym. Sci., Polym. Phys. Ed.*, Vol 22, 1984, p 1623
5. A.J. Kinloch and A.D. Tod, A New Approach to Crack Growth in Rubbery Composite Propellants, *Propel. Explos. Pyro.*, Vol 9, 1984, p 48
6. A.J. Kinloch and R.J. Young, *Fracture Behavior of Polymers*, Applied Science Publishers, London, 1983
7. J.G. Williams, *Fracture Mechanics of Polymers*, Ellis Horwood, Chichester, 1984
8. A.J. Kinloch, *Adhesion and Adhesives, Science and Technology*, Chapman and Hall, 1987
9. A.N. Gent and A.J. Kinloch, Adhesion of Viscoelastic Materials to Rigid Substrates, III, Energy Criterion of Failure, *J. Polym. Sci.*, Part A2, Vol 9, 1971, p 659
10. E.H. Andrews and A.J. Kinloch, Mechanics of Adhesive Failure, *Proc. R. Soc.*, Vol A 332, 1973, p 385
11. E.H. Andrews and A.J. Kinloch, Mechanics of Adhesive Failure, *Proc. R. Soc.*, Vol A 332, 1973, p 401
12. E.H. Andrews and A.J. Kinloch, Mechanics of Elastomeric Adhesion, *J. Polym. Sci.*, Symp. No. 46, 1974, p 1
13. A.N. Gent and J. Schultz, Effect of Wetting Liquids on the Strength of Adhesion of Viscoelastic Materials, *J. Adhes.*, Vol 3, 1972, p 281
14. M.L. Williams, R.F. Landel, and J.D. Ferry, The Temperature Dependence of Relaxation in Amorphous Polymers and Other Glass-Forming Liquids, *J. Am. Chem. Soc.*, Vol 77, 1955, p 3701
15. A.N. Gent and R.P. Petrich, Adhesion of Viscoelastic Materials to Rigid Substrates, *Proc. R. Soc.*, Vol A 310, 1969, p 433
16. G.P. Anderson, S.J. Bennett, and K.L. DeVries, *Analysis and Testing of Adhesive Bonds*, Academic Press, 1977
17. A.N. Gent and G.R. Hamed, Peel Mechanics for an Elastic-Plastic Adherend, *J. Appl. Polym. Sci.*, Vol 21, 1977, p 2817
18. E.H. Andrews, A.T. Khan, and A.H. Majid, Adhesion to Skin: Part 1, Peel Tests With Hard and Soft Machines, *J. Mater. Sci.*, Vol 20, 1985, p 3621
19. P.B. Lindley, Ozone Attack at a Rubber-Metal Bond, *J. Inst. Rubber Ind.*, Vol 5 (No. 6), 1971, p 243
20. A.N. Gent and G.R. Hamed, Peel Mechanics of Adhesive Joints, *Polym. Eng. Sci.*, Vol 17 (No. 7), 1977, p 462
21. D.A. Dillard, K.M. Liechti, D.R. Lefebvre, C. Lin, J.S. Thornton, and H.F. Brinson, in *Adhesively Bonded Joints: Testing, Analysis and Design*, STP 981, W.S. Johnson, Ed., American Society for Testing and Materials, 1988, p 83
22. D.R. Lefebvre and D.A. Dillard, The Development of a Modified Cantilever Beam Specimen for Measuring the Fracture Energy of Rubber-to-Metal Bonds, *Exp. Mech.*, Vol 28 (No. 1), 1988, p 38
23. A.N. Gent, Fracture Mechanics of Adhesive Bonds, *Rubber Chem. Technol.*, Vol 47 (No. 1), 1974, p 202
24. E.J. Ripling and S. Mostovoy, The Fracture Toughness and Stress-Corrosion Cracking Characteristics of an Anhydride-Hardened Epoxy Adhesive, *J. Appl. Polym. Sci.*, Vol 15, 1971, p 641

Composite Fiber-Matrix Bond Tests

Lawrence T. Drzal and Pedro J. Herrera-Franco, Michigan State University

EXTENSIVE RESEARCH EFFORTS over the past decade have been devoted to the development of fiber-reinforced composites of higher mechanical capability and structural reliability. A fundamental understanding and knowledge of the stress distribution induced by the applied load is necessary in order to utilize these materials effectively. Because the stiffness of the reinforcing fibers in a composite material is typically much higher than that of the matrix, differences in displacements of the two constituents will arise due to the applied loads. It is also known that the mechanism of load transfer at the fiber-matrix interphase plays a major role in the mechanical and physical properties of the composite.

The concept of interphase has been described as the region existing between bulk fiber and bulk matrix (Ref 1). The complexity of this interphase can best be visualized from a schematic model (Fig. 1), which shows the different characteristics.

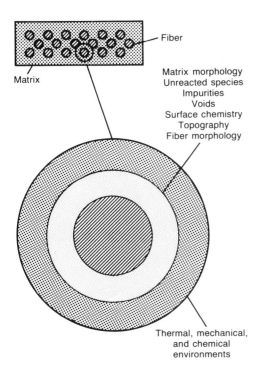

Fiber

Matrix

Matrix morphology
Unreacted species
Impurities
Voids
Surface chemistry
Topography
Fiber morphology

Thermal, mechanical,
and chemical
environments

Fig. 1 Characteristics of the interphase in a composite material

Research efforts to develop structure-property relationships for fiber-matrix interfacial behavior either have been based entirely on mechanics, with some assumptions being made about the level of fiber-matrix adhesion in the composite, or have taken a surface chemistry approach in which the interface was assumed to be the only factor of importance in controlling the final properties of the composite. Neither effort has had much success.

It has also been shown (Ref 1-5) that in coupling the mechanics and the surface chemistry approaches, the structure and composition of the fiber-matrix interphase, as well as the state of stress, are very important to a thorough understanding of the composite static and dynamic final properties.

Commercial reinforcing fibers are usually surface treated to enhance their adhesion to the binding resin (Ref 6). Thus, tools are required to characterize the interfacial bond strength, the interpretation of which may further be complicated by various contributions, including chemical, physical, and mechanical effects.

It would be desirable to have a fiber-matrix adhesion technique in which measurements are made on fibers in a high-fiber-volume fraction composite that has been subjected to the same processing or environmental exposures encountered either during manufacturing and fabrication or while in service. Factors such as fiber orientation in chopped fiber systems, the presence of fillers, interfiber spacing, moisture and solvent sorption, fatigue and thermal exposures, and so on, could then be properly evaluated for their effect on composite properties.

The experimental methods of characterizing the fiber-matrix interphase during recent years have relied on the use of single fiber-matrix adhesion and failure modes. The oldest technique is the fiber pull-out method (Ref 7), which was developed in the early stages of composite research when fibers were much larger and easier to handle. The variations in detail that have evolved mostly involve the matrix portion. The fiber may be pulled out of the matrix, which can be a block of resin, a disk, or a droplet (Ref 8). Current trends include the

use of very small droplets in order to reduce the difficulties of preparing thin disks of resin and to reduce the variability in exit geometry.

In a second method, the fiber may be totally encapsulated in a matrix coupon. A tensile load is applied to the coupon, and the interfacial shear stress transfer mechanism is relied upon to transfer tensile forces to the encapsulated fiber through the interphase (Ref 9, 10). The fiber tensile strength (σ_f) is exceeded, and the fiber fractures inside the matrix. This process is repeated, producing shorter and shorter fragments, until the remaining fragment lengths are no longer sufficient in size to produce further fracture through this stress-transfer mechanism, and a shear lag analysis is completed on the fragments in order to calculate the interfacial shear strength.

More recently, an *in-situ* measurement technique has been proposed for measuring the fiber interfacial shear strength (Ref 11). It involves the preparation of a polished, cut surface of a composite in which the fibers are oriented perpendicular to the surface. A small probe is placed on an individual fiber, and the force and displacements are monitored in small steps to the point at which the fiber detaches from the matrix. In this case the interfacial shear strength is dependent on the local properties of the fiber and matrix as well as on geometric parameters.

These three methods rely on the evaluation of a single fiber to provide information about the level of fiber-matrix adhesion or the interfacial failure mode. Other less well-known methods are available for providing fiber-matrix adhesion information.

Outwater and Murphy (Ref 12) used a single fiber aligned axially in a rectangular prism of matrix. A small hole was drilled in the center of the specimen through the fiber. The prism was placed under a compressive load, and the propagation of an interfacial crack was followed with increasing load. The fracture energy of the interface could be calculated from this data through the use of the expression:

$$G_{II} = \left\{ (\epsilon_r E_f)^2 - \left[4\tau \left(\frac{x}{a} \right) \right]^2 \right\} \left(\frac{a}{8E_f} \right) \quad \text{(Eq 1)}$$

where ϵ_r is the strain in the resin, E_f is the tensile modulus of the fiber, τ is the frictional shear stress, x is the length of the debond, and a is the fiber diameter.

Narkis (Ref 6) proposes using a single-fiber specimen in which a single fiber is embedded along the centerline in the neutral plane of a uniform cross-sectional beam. The beam is placed in nonuniform bending according to an elliptical bending geometry. This causes the shear stress to build up from one end of the fiber according to the gradient of curvature of the specimen. Careful observation of the fiber allows location of the point at which the fiber fails because of a maximum shear stress criterion. The stress along the fiber can be calculated according to:

$$\tau(x) = Q(3E_mB/A^4b) \, [(B^2/A^2) - 1]$$

$$\{(x^2/A^2) \, [(B^2/A^2) - 1] + 1\}^{-5/2} \qquad \text{(Eq 2)}$$

where E_m is the matrix tensile modulus, b is the beam width, Q is the first moment of transformed cross-sectional area, and A and B are constants from the equation of the ellipse.

Other less direct methods have also been reported for measuring the interaction between fiber and matrix. Ko *et al.* (Ref 13) have examined a carbon fiber-epoxy system in which the interfacial properties have been varied by the use of dynamic mechanical analysis (DMA). They reported a change in the tan δ peak attributable to changes in the fiber-matrix interface. Chua (Ref 14) also measured a shift in the loss factor for glass-polyester systems that corresponded to changes in the condition of the fiber-matrix interface. Perret (Ref 15) measured both the loss factor and the change in the shear modulus with increasing displacement and detected a change in composite properties with a change in the fiber-matrix interface.

For systems in which the level of fiber-matrix adhesion is very low, ultrasonic methods may be useful. Yuhas *et al.* (Ref 16) have used ultrasonic wave attenuation to establish correlations with short-beam shear data. This method was useful for poorly bonded systems, but was not sensitive to well-bonded interfaces. Recently, localized heating has been coupled with acoustic emission events to detect interfacial debonding (Ref 17). Correlations were made between acoustic parameters and interfacial strength.

The ability to measure fiber-matrix interfacial properties, such as modulus of the interphase material, will be necessary for predictive models that couple interfacial properties to composite properties. Further development of interphase-sensitive techniques will be forthcoming.

Numerous other techniques have been developed for measuring shear properties in fiber-reinforced-composite laminates. The most commonly used test methods for in-plane shear characterization are the $(\pm 45)_s$

tension test and the Iosipescu test. To determine the interlaminar shear strength, the short-beam shear test is more frequently used.

This article reviews state-of-the-art test techniques, focusing on the methods of assessing the fiber-matrix mechanical interactions, and addresses the theoretical analyses upon which these methods are based.

Fiber Pull-Out and Microdrop Technique

The pull-out experiment, which is believed to possess some of the characteristics of fiber pull-out in composites, consists of a fiber or filament embedded in a matrix block or thin disk of known geometry. A steadily increasing force is applied to the free end of the fiber in order to pull it out of the matrix (Ref 7). The strength of the fiber-matrix interface can be calculated to a first approximation by balancing the tensile and shear stress applied on the fiber, obtaining a simple relationship of the form $\tau = (\sigma_f/2)(d/l_c)$, where it is assumed that the shear stress is uniformly distributed along the embedded length (Ref 7, 9).

Several theoretical models have been proposed to determine the shear stresses developed during the pull-out. Greszczuk (Ref 18) considered the case of an elastic matrix in which the shear stress distribution was no longer uniform and the load transferred between the fiber and matrix did not change uniformly along the fiber. He showed that the distribution of stresses and forces depends on the properties of the elastic matrix. Lawrence (Ref 19) further developed Greszczuk's theory by including the effect of friction. Takaku and Arridge (Ref 20) considered the effect of the embedded length on the debonding stress and the pull-out stress and also the effect of Poisson's contraction on the variation of pull-out stresses. Gray (Ref 21) reviewed and applied the previously mentioned theories (Ref 19, 20) to calculate the maximum shear stress when the fiber is pulled out from the elastic matrix. He concluded that the mixture of adhesional bonding and friction resistance that occurs in a pull-out test specimen depends on the length of the embedded fiber. Adhesional bonding increases with embedded fiber length, whereas the frictional resistance to pull-out due to friction decreases. Laws (Ref 22) calculated the load-displacement curve of a pull-out test based on Lawrence's theory (Ref 19). He also calculated the crack spacing and strength of an aligned short-fiber composite and the effect of the interfacial and frictional bonds on pull-out.

More recently, Banbaji (Ref 23, 24) presented a theoretical model that considered the effect of normal transverse stresses on the pull-out force. He first analyzed the case in which the normal stress is constant,

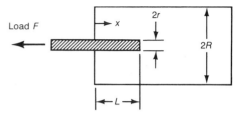

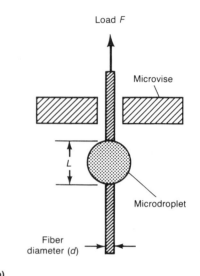

Fig. 2 Fiber pull-out experiment showing geometric parameters. (a) Piggott. (b) Miller

and then the case in which the stresses depend on the way the tensile force changes during an actual test. He applied the results to a polypropylene-cement system.

In a series of papers, Chua and Piggott (Ref 25-28) re-examined the pull-out process and showed that it is governed by at least five different variables: interfacial pressure (p), friction coefficient (μ) along the debonded length, work of fracture of the interface (G_i), the embedded fiber length (L), and the free fiber length (l_f).

Stress-Strain Relationships. Based on previous work by Greszczuk and Lawrence (Ref 18, 19), Chua and Piggott (Ref 25) developed a relationship for the tensile stress within the fiber at any point along the embedded length:

$$\sigma_f = \sigma_{fe} \sinh \, [n(L - x)/r]/\sinh(ns) \qquad \text{(Eq 3)}$$

where $s = L/r$ and σ_{fe} is the fiber stress at the polymer surface. The geometric terms are defined in Fig. 2.

$$n^2 = E_m/E_f(1 + \nu_m)\ln(R/r) \qquad \text{(Eq 4)}$$

The term $\ln(R/r)$ in the denominator results from the integration of shear stresses from the interface to the periphery of the matrix volume.

The shear stress at the interface can be calculated from the equilibrium of forces exerted on a differential fiber element, of length dx, to give the well-known equation:

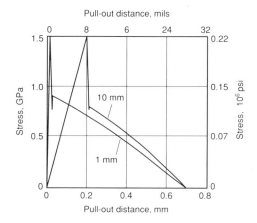

Fig. 3 Typical force-displacement plot of a pull-out experiment

$$\tau_i = (r/2) \frac{d\sigma_f}{dx} \qquad \text{(Eq 5)}$$

which, using Eq 3, results in:

$$\tau_i = n\sigma_{fe} \cosh[n(L - x)/r]/2\sinh(ns) \qquad \text{(Eq 6)}$$

It should be pointed out that no bonding is considered across the end of the fiber.

As stated by Piggott (Ref 29), there are three possible routes to failure during a pull-out test:

First, failure may occur when the maximum shear stress reaches the interface strength (τ_{iu}), which has a maximum absolute value at $x = 0$, that is, at the polymer surface. The debonding force is then $F_d = \pi r^2 \sigma_{fe}$. From Eq 6 is obtained:

$$F_d = (1/n)2\pi r^2 \tau_{iu} \tanh(ns) \qquad \text{(Eq 7)}$$

Second, failure may occur upon yielding at the interface at some value of stress τ_{iy}, in which case a constant shear stress distribution can be assumed along the embedded fiber length, as long as work-hardening effects are negligible, and thus:

$$F_d = 2\pi r L \tau_{iy} \qquad \text{(Eq 8)}$$

Finally, failure may occur if the interface fractures with work of fracture (G_i) per unit area of interface. The source of the required fracture surface energy is the strain energy stored within the specimen components (Ref 26-37).

In their work, Chua and Piggott (Ref 25, 29) considered only the extensional strain energy stored in the embedded fiber length (U_L) and the shear strain energy in the matrix immediately surrounding the fiber (U_m) as:

$$U_L + U_m = \frac{1}{(2nE)_f} \pi r^3 \sigma_{fe}^2 \cosh(ns) \qquad \text{(Eq 9)}$$

where n is given by Eq 4 and $s = L/r$.

Equating the total strain energy to $2\pi r L G_c$, where G_c is the unit fracture energy at the interface, the debonding load is found to be:

$$P_d = 2\pi r [E_f G_c r (ns) \tanh(ns)]^{1/2} \qquad \text{(Eq 10)}$$

Penn and Lee (Ref 30) considered the existence of an initial microcrack of length a at the fiber-matrix interface. They also considered the effect of the strain energy contributed by the free fiber length to the crack propagation process.

Using an energy balance, they arrived at the following expression for the debonding force of the microdrop:

$$P_d = \frac{2\pi r (r G_c E_f)^{1/2}}{[1 + csch^2(ns)]} \qquad \text{(Eq 11)}$$

Friction During Pull-Out. After debonding, friction at the interface has to be overcome (Ref 20, 22, 24, 27) in order for pull-out to proceed. Friction at the interface is due to the normal compressive stresses that are caused by the pressure p_0 acting on the fiber from the matrix. Such stresses ($\tau_i = \mu p_0$) (μ is the coefficient of friction) arise from resin shrinkage during curing of the specimen and from the difference of coefficients of thermal expansion of matrix and reinforcing fiber (Ref 23, 27, 38).

After failure of adhesion, the interfacial shear stress ($\tau_i = \mu p$) increases with increasing pull-out distance (Ref 39), but the normal stress acting from the matrix will produce on the fiber a reduction in cross-sectional area, due to Poisson's effect, resulting in a reduction of interfacial normal stresses.

In a typical force-displacement plot (Fig. 3) describing the pull-out process, the first peak is attributed to debonding and frictional resistance to slipping, whereas subsequent lower peaks are attributed to friction and a stick-slip mechanism, giving rise to a serrated portion of the curve (Ref 2, 20, 23, 24). Because of relaxation at both the free and embedded lengths of the fiber, the slope of a curve (τ_i) against pull-out distance gives only an approximate value of the interfacial shear stress (μp_0) where p_0 is the pressure exerted by the matrix shrinkage at the moment the fiber emerges from the polymer.

The experimental value for shear stress obtained from the slope of the pull-out curve is related to the true value of $\tau_i = \mu p_0$ by Eq 12:

$$\tau_i = \tau \exp/[1 + 2\tau \exp(1+2L)/E_f r] \qquad \text{(Eq 12)}$$

If, in addition to the curing stresses, an external pressure p_e is applied, the shear stress increases as:

$$\tau_i = \mu(p_e + p_0) \qquad \text{(Eq 13)}$$

The value of μ can be determined from experimental results by plotting τ_i against p_e

and evaluating the slope of the curve. The intersection of the curve with the vertical axis should give μp_0. If no external pressure is applied, μ can be estimated from the curve of the pull-out force as a function of the pull-out distance:

$$F_A = \pi d^2 E_f (1 + v_m)/v_f E_m (1 - e^{\eta y/L}/e^\eta) \qquad \text{(Eq 14)}$$

where v_f is the fiber Poisson's ratio and $\eta = 2\mu v E_m L/E_f r(1 + v_m)$.

The effect of the external pressure applied to the specimen is more noticeable on the pull-out force at short embedded lengths (Ref 23, 40).

It can be seen in Table 1 that the effect of increasing pressure is increasing pull-out forces. However, this increment is also affected by the decreasing thickness of the fiber or filament due to both the pressure and lateral Poisson's contractions that are produced upon application of the initial tension, which results in the decrease in the coefficient of friction.

Experimental Procedure. The following experimental procedure can be used to measure the interfacial shear strength by means of the microdrop technique. It was used to test the adhesion strength of AS4 carbon fibers and ultrahigh molecular weight polyethylene (UHMW-PE) (Allied-Signal Inc. SPECTRA-1000) to epoxy resin Epon 828 (diglycidyl ether of bisphenol A, Shell Chemical Company) with 14.5 wt% of *m*-phenylenediamine (mPDA) curing agent (Aldrich Chemicals).

First, the ends of 100 mm (4 in.) long single-carbon AS4 fibers are taped to parallel sides of a specially built frame using double-stick tape. After mixing and degassing the resin, the microdrops are applied to the fibers using a syringe and needle. A small drop of the thermoset resin is made to flow to the needle tip and is allowed to come in contact with a fiber. After retraction of the needle tip, some of the resin remains, forming a microdrop around the fiber. The microdrop size ranges from 80 to 200 μm (3 to 8 mils) in diameter. Next, the microdrops are allowed to cure at room temperature for 24 h, and are then postcured for 2 h at 75 °C (165 °F) and for 2 h at 125 °C (255 °F). It has been found that immediate curing at elevated temperature causes the curing agent to diffuse out of the droplet (Ref 39, 41). After curing, the fiber is affixed to an aluminum plate and kept in a dessicator to await testing.

Gaur *et al.* (Ref 42) described a method for measuring the interfacial shear strength

Table 1 Effect of pressure on the maximum pull-out forces

Pressure		Embedded length		Pull-out force		Shear strength	
MPa	ksi	mm	in.	N	lbf	MPa	ksi
0	0.	6.4	0.25	37.4	8.42	0.8	0.12
8.8	1.28	4.9	0.19	53.8	12.1	1.0	0.15
17.6	2.55	4.2	0.16	41.2	9.27	1.2	0.17
35.2	5.10	4.3	0.17	62.0	13.9	1.2	0.17

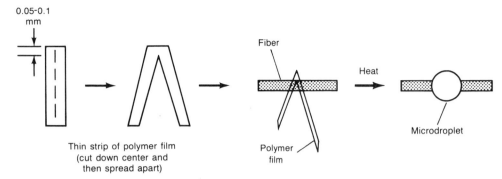

Fig. 4 Procedure for forming thermoplasic resin microdrops. Source: Ref 42

in various fiber-thermoplastic systems. In particular, they measured the interfacial shear strength of carbon and aramid fibers embedded in four thermoplastic resins: polyetheretherketone (PEEK), polyphenylene sulfide (PPS), polycarbonate (PC), and polybutylene terephthalate (PBT). The microdrops are formed using a small, thin piece of film (about 2 to 30 mm, or 0.08 to 1.2 in.).

A longitudinal cut is made on the film piece, along nearly its entire length, to form two strips joined at one end for a distance of 50 to 100 μm (2 to 4 mils). The strips are suspended on the horizontal fibers already affixed to a holding frame, and the thermoplastic is melted on the fibers. Upon melting, nearly uniform-sized droplets are obtained. Their lengths are controlled by the film thickness. This procedure is shown in Fig. 4.

In the microbond or microdrop experiment, a small aluminum plate is attached to a 50 or 250 g (2 or 9 oz) load cell. The droplet is gripped with micrometer blades, which are brought together until they nearly touch the fiber. The blade micrometer, mounted on a translation stage, is made to move away from the load cell and become parallel to the longitudinal axis of the fiber, at a speed of 0.11 mm/min (4 mils/min). The position and speed of the translation stage are controlled remotely using a motorized actuator. The translation of the stage causes the microdrop to be sheared off of the fiber surface. The force required to debond the microdrop is recorded using a pen plotter. Figure 5 shows a photograph of the droplet pull-off test apparatus. The entire apparatus is mounted on an optical microscope stage for measurement of the embedded length and fiber diameter. The procedure outlined here is similar to those described by other investigators (Ref 42-48).

Advantages and Limitations. One of the major advantages of the pull-out and microdrop techniques is that the value of force at the moment of debonding can be measured. Also, these techniques can be used for almost any fiber-matrix combination. On the other hand, there are serious inherent limitations to the pull-out technique, especially when using very fine reinforcing fibers with diameters ranging from 5 to 50 μm (0.2 to 2 mils). Because the debonding force is a function of the embedded length, the maximum embedded length is of the range of 0.05 to 1.0 mm (0.002 to 0.04 in.). Longer embedded lengths cause fiber fracture. It is extremely difficult to keep the length to such short values, and the handling of the specimens is very delicate. Furthermore, the meniscus that is formed on the fiber by the resin makes the measurement of the embedded length very inaccurate. For microdroplet specimens, the small size makes the failure process difficult to observe. Most important, the state of stress in the droplet can vary both with size and with variations in the location of points of contact between the blades and the microdrop. Furthermore, it has been shown by Rao and Drzal (Ref 39) that the mechanical properties of the microdrop vary with size because of variations in the concentration of the curing agent.

For a given fiber-matrix combination, a relatively large scatter in the test data is obtained from the microdrop or the pull-out tests. Such a wide distribution of shear strengths has been attributed mainly to testing parameters such as position of the microdrop in the loading fixture, faulty measurement of fiber diameters, and so on (Ref 8, 45). In addition, variations in the chemical, physical, or morphological nature of the fiber along its length will affect the results of interfacial shear strength measurements, which only consider very small sections (Ref 8, 30, 42).

Single-Embedded-Fiber Technique

The single-embedded-fiber tensile specimen was originally used by Kelly and Tyson (Ref 9), who observed a multiple fiber fracturing phenomenon in a system consisting of a low concentration of brittle fibers embedded in a copper matrix, upon application of a tensile force.

As shown in Fig. 6, the fiber axial stress that is introduced by the interfacial shear stress produced on planes parallel to the fiber axis rises from the ends until the fiber fracture strength (σ_f) is reached. At that point, the fiber fractures at some site where the fiber stress is at a maximum, depending on fiber defect distribution and probability. Continued application of stress to the specimen will result in the repetition of this fragmentation process until all remaining fiber lengths become so short that the shear stress transfer along their lengths can no

Fig. 5 Apparatus of the microdrop technique

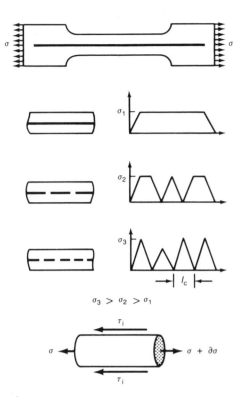

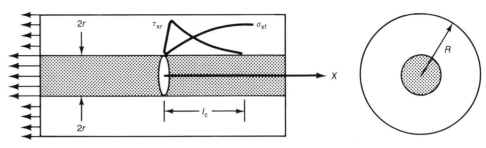

Fig. 7 Micromechanical model of a single-embedded fiber. Source: Ref 54

Fig. 6 Single-fiber fragmentation test specimen

longer build up enough tensile stresses to cause any further failures with increasing strain. This maximum final fragmentation length of the fiber is referred to as the critical length (l_c). If the matrix surrounding the fiber is a perfectly plastic matrix, the shear stress at the interface can be considered to be constant along the short fiber critical length. An average shear strength (τ) can be determined from a simple balance of force that results in:

$$\tau = \frac{\sigma_f}{2}\left(\frac{d}{l_c}\right) \qquad \text{(Eq 15)}$$

where d is the fiber diameter for a circular fiber cross section. Because the fiber-matrix interface is placed under shear, the calculated value of τ should be an excellent estimator of the interface shear strength (Ref 10).

The distribution of stress around discontinuous fibers in composites has been studied by a number of researchers. Theoretical analyses have been performed by Cox (Ref 49) and Rosen (Ref 50). In these models only fiber axial stress distribution and the fiber-matrix interfacial shear stress distribution are determined. Amirbayat and Hearle (Ref 51) studied the effect of different levels of adhesion on the stress distribution, that is, no bond, no adhesion, perfect bond, and the intermediate case of limited friction. They also considered the inhibition of slippage by frictional forces resulting from interfacial pressure due to Poisson's lateral contractions of the matrix, but did not consider the shrinkage of the matrix during curing.

An interesting model was proposed by Theocaris (Ref 52). He incorporated an interphase which he named a mesophase, which constitutes a boundary layer between the main phases of the composite. From a physical basis, a continuous and smooth transition of the properties from one phase to the other is assumed. Because the mechanical properties of this region also contribute to the composite properties, the determination of the local mechanical modulus is important. Dynamic mechanical analysis is used to identify the mesophase properties, primarily the glass transition temperature (T_g), through changes in the loss modulus peak.

Lhotellier and Brinson (Ref 53) developed a mathematical model that includes the mechanical properties of the interphase, the stress concentration near fiber breaks, and the elastoplastic behavior of both the matrix and the interphase.

Analysis. Whitney and Drzal presented an analytical model to predict the stresses in a system consisting of a broken fiber surrounded by an unbounded matrix (Ref 54). The model (see Fig. 7) is based on the superposition of the solutions to two axisymmetric problems, an exact far-field solution and an approximate transient solution. The approximate solution is based on the knowledge of the basic stress distribution near the end of the broken fiber, represented by a decaying exponential function multiplied by a polynomial. Equilibrium equations and the boundary conditions of classical theory of elasticity are exactly satisfied throughout the fiber and matrix, while the compatibility of displacements is only partially satisfied. The far-field solution away from the broken fiber end satisfies all the equations of elasticity. The model also includes the effects of expansional hygric and thermal strains and considers orthotropic fibers of the transversely isotropic class.

A relationship is obtained for the axial normal stress (σ_x) in the fiber:

$$\sigma_x = [1 - (4.75\bar{x} + 1)e^{-4.75\bar{x}}]C_1\epsilon_0 \quad (r \geq R \geq 0) \qquad \text{(Eq 16)}$$

where $\bar{x} = x/l_c$, ϵ_0 is the far-field strain, and C_1 is a constant dependent on material properties, expansional strains, and the far-field axial strain. It can be noticed that σ_x is

independent of the fiber radius. The critical length l_c is defined such that the axial stress recovers 95% of its far-field value, that is:

$$\sigma_x(l_c) = 0.95C_1\epsilon_0 \quad (r \geq R \geq 0) \qquad \text{(Eq 17)}$$

The interfacial shear stress is given by the expression:

$$\tau_{xR}(\bar{x}, r) = -4.75\mu C_1\epsilon_0\bar{x}e^{-4.75\bar{x}} \qquad \text{(Eq 18)}$$

$$\mu = [G_m/(E_{1f} - 4\nu_{12f}G_m)]^{1/2} \qquad \text{(Eq 19)}$$

E_{1f} denotes the axial elastic modulus of the fiber, whereas ν_{12f} is the longitudinal Poisson's ratio of the fiber, determined by measuring the radial contraction under an axial tensile load in the fiber axis direction, and G_m denotes the matrix shear modulus. It should be noted that the negative sign in the expression for the shear stress is introduced to be consistent with the definition of an interfacial shear stress in classical theory of elasticity. The radial stress at the interface is given by:

$$\sigma_r(\bar{x}, r) = [C_2 + \mu^2 C_1(4.75\bar{x} - 1)e^{-4.75\bar{x}}]\epsilon_0 \qquad \text{(Eq 20)}$$

The constant C_2, like C_1, is dependent on material properties, expansional strains, and the far-field strain. Numerical results are normalized by σ_0, which represents the far-field fiber stress in the absence of expansional strains. In particular:

$$\sigma_0 = C_3\epsilon_0 \qquad \text{(Eq 21)}$$

The constants C_1, C_2, and C_3 are given by:

$$C_1 = E_{1f}(1 - \bar{\epsilon}_{1f})/\epsilon_0 + 4K_fG_m\nu_{12f}/(K_f + G_m)$$
$$\{\nu_{12f} - \nu_m + [(1+\nu_m)\bar{\epsilon}_m - \bar{\epsilon}_{2f} - \nu_{12f}\bar{\epsilon}_{1f}]/\epsilon_0\} \qquad \text{(Eq 22)}$$

$$C_2 = 2K_fG_m/(K_f + G_m)$$
$$\{\nu_{12f} - \nu_m + [(1 + \nu_m)\bar{\epsilon}_m - \bar{\epsilon}_{2f} - \nu_{12f}\bar{\epsilon}_{1f}]/\epsilon_0\} \qquad \text{(Eq 23)}$$

$$C_3 = E_{1f} + [4K_f\nu_{12f}G_m(\nu_{12f} - \nu_m)]/(K_f + G_m) \qquad \text{(Eq 24)}$$

and

$$K_f = E_m/2[(2 - E_{2f}/2G_{2f} - 2\nu_{12f}^2E_{2f}/E_{1f})] \qquad \text{(Eq 25)}$$

where E_{2f} is the radial elastic modulus of the fiber and K_f is the plane-strain bulk modulus of the fiber.

A numerical example is presented for a single-fiber composite of AS4 fiber-Epon

Table 2 Material properties

Property	Epon 828	AS-4
E_1, GPa (10^6 psi)	3.8 (0.55)	124 (18)
E_2, GPa (10^6 psi)	3.8 (0.55)	21 (3)
ν_{12}	0.35	0.25
G_{23}, GPa (10^6 psi)	1.4 (0.20)	8.3 (1.2)
α_1, 10^{-6}/K	68	−0.11

828 epoxy matrix. The material properties are given in Table 2.

The specimens were cured at 75 °C (167 °F) and postcured at 125 °C (257 °F). The difference between room temperature, 21 °C (70 °F) and the postcure temperature yields $\Delta T = 104$ °C (187 °F), which is the worst case for thermal residual stresses. Because it is most likely that some residual stresses will be relieved during cool-down from the postcure temperature, in this example the value $\Delta T = 75$ °C (135 °F) was chosen.

Figures 8, 9, and 10 show plots of σ_{xf}/σ_0, τ_{rx}/σ_0, and σ_r/σ_0, where the far-field stress was used to normalize out the numerical results. The axial fiber stress and the interfacial shear stress are relatively insensitive to thermal strains, but the radial strain is quite sensitive to thermal strains.

It should also be noted that the peak value of interfacial shear predicted by this model occurs at a small distance from the broken end of the fiber. This was also noticed by other researchers (Ref 55-57).

Determination of the Critical Length. Most high-performance engineering fibers have strengths that vary considerably along their length because of flaws introduced through handling or manufacturing or because of intrinsic anomalies of the material (Ref 58, 59).

Assuming an average value for σ_f, the fiber still can be considered to have randomly spaced flaws or point defects that introduce a slight variation in strength, which, depending on its value, may or may not prevent the fiber from fracturing once

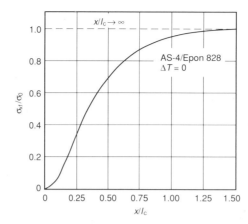

Fig. 8 Distribution of axial stresses along fiber length. Source: Ref 54

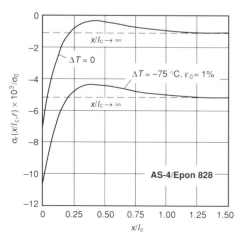

Fig. 10 Distribution of radial stresses along fiber length. Source: Ref 54

the average σ_f is reached. When the built-up stress in the fiber approaches the true value of σ_f, the fiber will instantaneously and repeatedly break into shorter and shorter pieces until the remaining fragments are not long enough for the linear stress to build up from either end to exceed the fiber strength anywhere. This final length is usually referred to as the critical length (l_c). Any fragment with a length slightly exceeding l_c will break in two, yielding, at the conclusion of the experiment, a random distribution of fragment lengths between $l_c/2$ and l_c.

Because of the random nature of this problem, the distribution of fiber fragment length should be characterized. Drzal et al. (Ref 4, 10) used a two-parameter Weibull distribution to achieve such a characterization. Then, using the arithmetic mean fragment length, that is, the original unbroken length divided by the number of fragments, and an average value for σ_f at the critical length (l_c) extrapolated from simple tension tests, they obtained expressions for the mean interfacial shear strength:

$$\bar{\tau} = \frac{\sigma_f}{2\beta} \Gamma(1 - 1/\alpha) \qquad \text{(Eq 26)}$$

$$\text{Var}(\tau) =$$
$$\frac{\sigma_f^2}{4\beta^2} [\Gamma(1 - 2/\alpha) - \Gamma^2(1 - 1/\alpha)] \qquad \text{(Eq 27)}$$

where β and α are the maximum likely estimates of the scale and shape parameters, respectively, and Γ is the gamma function.

Several other investigators have used similar techniques to characterize the critical length. Bascom and Jensen (Ref 60) used an approach similar to that of Drzal and co-workers. Wimolkiatisak et al. (Ref 61) found that the fragmentation length data fitted both the Gaussian and Weibull distributions equally well. Fraser et al. (Ref 62) developed a computer model to simulate the stochastic fracture process and, together with the shear lag analysis, described the shear transmission across the interface.

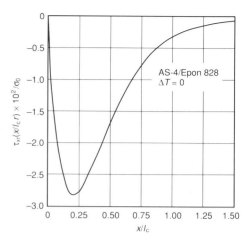

Fig. 9 Distribution of shear stresses along fiber length. Source: Ref 54

Rather surprisingly, it has not been widely used, probably because of its required computational intensity.

Netravali et al. and Henstenburg and Phoenix (Ref 58, 59) used a Montecarlo simulation of a Poisson/Weibull model for the fiber strength and flaw occurrence to calculate an effective interfacial shear strength (τ_i) using the relationship:

$$\tau_i = K \frac{\sigma_f}{2} d/\bar{l} \qquad \text{(Eq 28)}$$

where K is a correction factor to be determined from the Weibull/Poisson model and $\bar{l}$ is the mean fiber fragment length. In their analysis, Netravali et al. (Ref 58) and Henstenburg and Phoenix (Ref 59) suggested that the correction factor K has a value equal to 0.801, which is considered quite high for brittle fibers. They found that a value of 0.889 is more appropriate for the correction factor K in such a case.

Because the measured fiber fragment lengths are distributed between $l_c/2$ and l_c, an average fragmentation length can be approximated as $\bar{l} = 0.75l_c$, and the given correction factors yield errors of 7 and 19% respectively, with regard to the average value of 0.75.

Experimentally, it is very difficult to measure the strength of individual fibers at such short lengths, and most analyses extrapolate mean strength and strength distribution data obtained from longer gage lengths. Asloum et al. (Ref 63) studied the dependence of the strength of high-strength carbon fibers on gage length by means of the Weibull model. They showed that the mathematical form of the estimator chosen and the sample size, when larger than about 20, do not influence the results of the analysis. Also, they found that neither the three-parameter nor the two-parameter Weibull distribution is appropriate for describing the critical length dependence on gage length of the fiber during testing. Furthermore, it was shown that a linear logarithmic dependence of strength on gage length is the most accu-

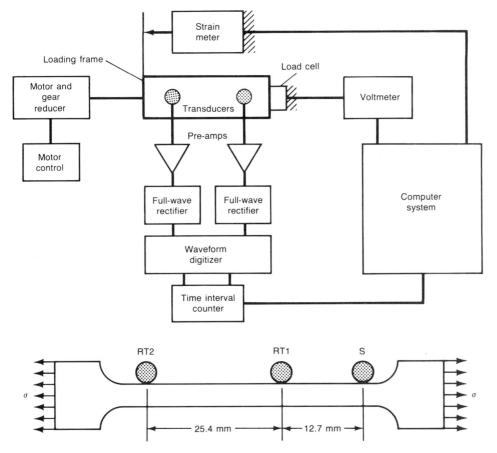

Fig. 11 Configuration of the experimental apparatus used in the acoustic emission technique. Source: Ref 67

rate, simple method for extrapolating the fiber tensile strength at short lengths.

There are, as can be seen, different models that have been used to describe the distribution of the fiber fragment lengths on the single-fiber fragmentation test. In spite of slight differences between the statistical treatments of the experimental data, they should prove to be more useful if they are combined with a more accurate theoretical model of the mechanical events that occur in this technique.

Other studies have reported the influence of mechanical properties of the matrix and fiber on the critical aspect ratio and, consequently, on the stress transfer in single-fiber composites. The effect of adhesion as affected by the surface treatment on the fiber, the ratio of elastic moduli of fiber and matrix, and temperature on the critical aspect ratio was analyzed experimentally by Asloum (Ref 64). Termonia (Ref 65) used a finite-difference approach to show that the critical length for efficient stress transfer to the fiber is a function of the ratio between elastic moduli of the fiber and matrix. In his model he also considered the dependence of the critical length on the adhesion by including an adhesion factor. A decrease in adhesion is seen to increase the critical length, particularly when the adhesion factor becomes less than 30%.

Folkes and Wong (Ref 66), in their study of adhesion between fiber and matrix of thermoplastic composites, noticed that the formation of transcrystalline morphology around glass fibers in polypropylene has an effect on the critical fiber length, probably through the change in local interphase modulus.

Measurement of the Critical Fiber Fragment Length. One shortcoming of the simple fiber fragmentation technique is that the positions of fiber breaks must be measured in order to calculate the fiber fragment lengths. Although a large statistical sampling of the interface occurs with the fragmentation test (for some matrix combinations, the number of breaks is of the order of 50 to 100 in a gage length of 20 mm, or 0.8 in.), this process is tedious and time consuming, because each test coupon must be analyzed individually.

Traditionally, the fiber failure positions have been measured optically using a microscope equipped with a calibrated filar eyepiece, leaving room for errors due to the inherent limitations of light microscopes. The optical method also requires that the matrix be transparent and have a strain to failure at least three times that of the fiber.

An acoustic emission technique (AE) has been developed recently for the measurement of fiber fragment length distributions

(Ref 67). This technique is based on the fact that the speed of wave propagation is a function of the specimen material itself. It may also be influenced, however, by geometric parameters, dominant frequency of the emitted ultrasonic signal, and, more important, the deformation of the material. The experimental configuration of the AE apparatus is shown in Fig. 11. A simple algorithm—incorporating the average wave speed in the epoxy (or any other matrix), the distance between the two receiving transducers, the offset distance between one specific receiving transducer and the fixed grip, time intervals, and the corresponding strains—was used to obtain the location of the fiber breaks, the fiber fragment lengths, and fiber aspect ratios.

Figure 12 compares the aspect ratios measured by optical and acoustic emission methods for glass fibers in two different epoxy blends. Some discrepancies were obtained between the acoustic emission and optical techniques for low aspect ratios in the brittle epoxy blend. This may have been because one of the AE probes used had a diameter of 1 mm (0.04 in.). Table 3 shows interfacial shear strength values obtained from optical and acoustic emission techniques. Slight differences could be attributed to small variations in the scale parameter for the aspect ratios, which, in turn, reflect differences in the observed variability due to differences in the shape parameter. Better agreement was obtained for the interfacial shear strength for the flexible resin blend using both acoustical and optical techniques. One advantage of the acoustic measurement technique is that it does not require a transparent matrix and therefore can be used in matrices such as metal and many polymeric materials.

An alternative technique for measuring fiber fragment length has been developed by Waterbury and Drzal (Ref 68). It uses a special software package called "Fibertrack," together with a computer-interfaced translation stage. While the coupon is translated at a constant velocity, the operator presses a "mouse" button as each break passes a set of cross hairs. The time intervals between breaks are stored in RAM memory and converted in displaced distances by entering the overall distance traveled. These distances are written to a magnetic disk and combined with fiber strength and diameter data to produce Weibull distributions and gamma function calculations to find the interfacial shear strength.

This technique thereby makes use of the human ability to discriminate events and to identify breakpoints visually while freeing the operator from much of the drudgery associated with the manual operation of the test. The entire process of test coupon mounting, loading, fiber fragment length measurement, data storage, and interfacial shear calculation requires approximately 6

Table 3 Aspect ratios, mean fragment lengths, and interfacial shear strength values obtained by optical and acoustic emission techniques

Epoxy blend and measurement method	Fragment aspect ratio			Mean fiber diameter		Mean fragment length		Interfacial shear strength	
	Mean	Shape parameter	Scale parameter	mm	in.	mm	in.	MPa	ksi
1. Optical	42.89	3.92	47.37			0.97	0.0381	27.53	3.993
Acoustic emission	42.93	2.08	48.47	22.5	0.89	0.97	0.0381	27.50	3.988
2. Optical	40.44	3.45	44.98			0.86	0.0339	29.46	4.273
Acoustic emission	44.88	2.44	50.62	21.33	0.85	0.96	0.0378	26.30	3.814

min per fiber, which represents a considerable improvement over the current state of the art.

Photoelastic Evaluation of the Interface. Some common polymeric matrices, when undeformed, can be considered optically isotropic. However, when subjected to stresses, whether due to externally applied loads or thermally induced stresses from differential shrinkage during sample cool-down from cure temperature, the material becomes optically anisotropic (birefringent). If the resin is sufficiently transparent, it can be studied with polarized light (Ref 1, 2-4, 10, 69). It would be beneficial, for a better understanding of fiber-matrix interactions, to study the stress birefringence adjacent to the fibers, before, during,

and after application of load in a single-embedded-fiber test.

Drzal *et al.* (Ref 1, 2, 4, 10) observed qualitative differences in the stress pattern resulting from interface changes when working on graphite fibers and epoxy matrices. Normal chromatic cycles were observed in the epoxy at low strains, but they disappeared at higher load levels. Isochromatic fringes were not observed under high strain conditions present at the fiber fragment ends. Instead, a light, bulbous region was observed.

Figure 13 shows a series of photoelastic stress patterns with increasing strain. The photoelastic data were generated with untreated fibers (AU) in an epoxy matrix. At low strains after a fiber break, extensive

resin birefringence can be seen around the fiber ends. With increasing strain, this birefringence activity rapidly extends down the fragment away from the break.

Dynamic observation of this process suggests that the stresses proceed along the fragment by a stick-slip mechanism. That is, the stresses build up and then appear to release and move ahead an incremental amount repetitively. For this particular fiber combination, each fiber fragment fails interfacially at low levels of strain. With increasing load, the fiber fragments interact with the matrix only through weak frictional forces, resulting in a very low interfacial shear strength.

Surface-treated fibers (AS), which are mechanically identical to the AU fibers,

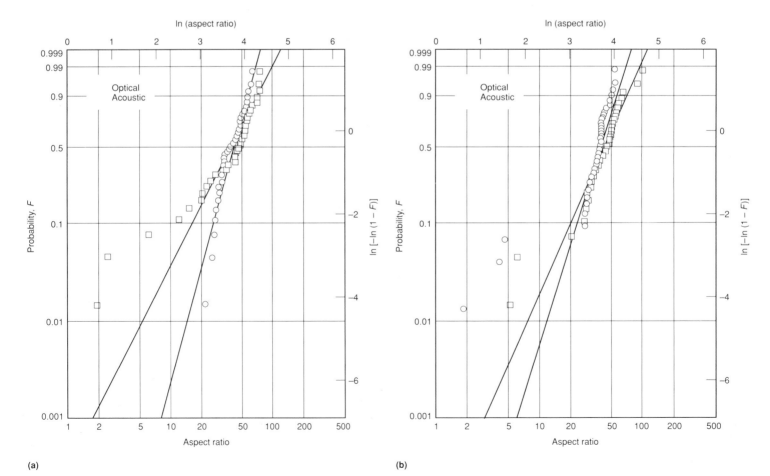

Fig. 12 Comparison of aspect ratios measured with acoustic emission and optical techniques. (a) For glass fiber-stiff epoxy system. (b) For glass fiber-soft epoxy system. Source: Ref 67

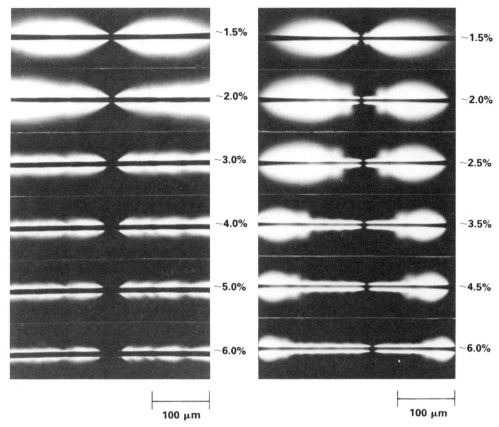

Fig. 13 Photoelastic patterns obtained for AU fibers in an epoxy matrix. Source: Ref 1

100 μm

Fig. 14 Photoelastic patterns obtained for AS fibers in an epoxy matrix. Source: Ref 1

100 μm

~1.5%
~2.0%
~3.0%
~4.0%
~5.0%
~6.0%

~1.5%
~2.0%
~2.5%
~3.5%
~4.5%
~6.0%

behave, under polarized light, in a completely different manner from the untreated fibers. Figure 14 shows that at the fiber break, the stress builds up at the ends of the fiber. However, at higher strain levels, a narrow, very intense region of photoelastic activity remains around the fiber, while the initial bulbous region moves away from the fiber ends toward the center of the fiber fragment. The interfacial shear strength that was measured from this fiber-matrix system was more than three times greater than that obtained for the AU fiber, indicating a higher degree of fiber-matrix interaction. Subsequent examination of ultramicrotomed sections of each fiber-matrix interphase and comparison with the photoelastic photographs suggest that the narrow intense area that remains around the fiber is a region where an interfacial crack has passed, whereas the bulbous region at the tip of the photoelastically active zone appears to be the plastically deformed region due to the moving crack tip. It is thus evident that photoelastic observations before, during, and after application of loads to a single-fiber-composite coupon could help to elucidate the fiber-matrix interactions, as well as to judge the effect of surface chemistry and morphology on the properties of the fiber-matrix interface.

Experimental Apparatus and Procedure. Thermoset test coupons for the sin-

gle-fiber fragmentation test can be fabricated by a casting method with the aid of a silicone room-temperature-vulcanizing (RTV) 664 eight-cavity mold. Standard ASTM 64 mm (2.5 in.) dog bone specimen cavities with a 3.175 mm wide by 1.5875 mm thick by 25.4 mm (⅛ × ⅟₁₆ × 1 in.) gage section are molded into a 76.2 × 203.2 × 12.7 mm (3 × 8 × 0.5 in.) silicone piece. Sprue slots are molded in the center of each dog bone to a depth of 0.7938 mm (⅟₃₂ in.) and through the end of the silicone piece.

Single fibers approximately 150 mm (6 in.) in length are selected by hand from a fiber bundle. Single filaments are carefully separated from the fiber tow without touching the fibers, except at the ends. Once selected, a filament is mounted in the mold and held in place with a small amount of rubber cement at the end of the sprue. The rubber cement is not in contact with the cavity, which contains the grip sections, nor the gage length section in the mold.

After setting for a minimum of 4 h under controlled temperature and humidity conditions to permit the rubber cement to dry, the resin is added with the aid of a disposable pipette. The long, narrow tip should be removed so that the resin can readily enter and exit the pipette chamber. Air bubbles are avoided by first degassing the silicone mold and the resin in a vacuum chamber before filling the mold cavities. The assem-

bly is transferred to an oven where the curing cycle is completed. After cooling down to room temperature, the mold can be curled away from the specimens parallel to the fiber to prevent fiber damage. The test samples can then be stored in a dessicator until ready for analysis. Prior to testing, the coupons should be inspected for defects, and any defective ones should be discarded.

The single-embedded-fiber specimens are then tested to failure in uniaxial tension, fracturing the fiber repeatedly within the matrix until critical length is achieved (that is, fracture ceases). A microstraining machine capable of applying enough load to the tensile coupon (Fig. 15) is fitted to the microscope stage so that the x-y stage controls manipulate the jig position. This allows the operator to assess the fracture process along the entire gage length of the coupon.

A transmitted-light polarizing microscope should be configured such that there is one polarizer below and one above the test coupon when the loading jig is in place. The embedded fiber is located in the center of the polymeric coupon and is therefore difficult to observe at high magnification with standard objectives. The microscope should be equipped with a long-distance 20× or greater objective lens. The fiber diameter is measured using a calibrated filar eyepiece. The fiber fragment length measurement procedure described earlier is then applied for data collection and analysis.

Microdebonding/Microindentation Technique

Another approach for the characterization of the interfacial shear strength was first proposed by Mandell and co-workers (Ref 11, 70-73). In this technique, single fibers perpendicular to a cut and polished surface of a regular high-fiber-volume fraction composite are compressively loaded to produce debonding and/or fiber slippage (Ref 71).

In contrast to conventional methods, which use a model system to provide information on fiber-matrix adhesion, this microindentation technique is an *in-situ* interface test for real composites and has the advantage of reflecting actual processing conditions. It can allow determination of the interface strength due to fatigue or environmental exposure or possibly monitor interface properties of parts in service (Ref 73).

Analysis. Interfacial shear strength is calculated through the use of a finite-element analysis of the stressed area for each fiber under examination. The microindentation test is run on individual selected fibers on a polished cross section. An individual fiber in a composite is surrounded by neighboring fibers located at various distances and distributed in a variety of arrangements, which range from a hexagonal array to random and dispersed configurations. The diameter of the tested fiber and the distance to the nearest

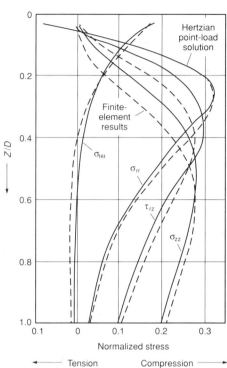

Fig. 15 Photograph of loading jig used in the single-fiber fragmentation test

Fig. 16 Interfacial stress components along an imaginary interface normal to the free surface in a homogeneous isotropic material. D, fiber diameter. Source: Ref 73

neighbor fiber are recorded for each test on a micrograph, and a simplified axisymmetric finite-element model (FEM) is used. This model includes the fiber, surrounding matrix, and average composite properties beyond the matrix (Ref 73). It was shown that the maximum shear stress along the interface is insensitive to probe stiffness as long as the contact area does not approach the interface.

Figure 16 shows results for a case in which the fiber and matrix are considered to have the same mechanical properties. The finite-element results are compared with those obtained from an analytical solution of the Hertz contact problem of a point load on a half-space with imaginary boundaries. Good agreement was found between the FEM and the analytical solution in spite of slight differences in loading conditions. Figures 17 and 18 show results for Nicalon (SiC) fibers in an aluminosili-

cate glass matrix and for untreated high modulus carbon fibers in a borosilicate glass matrix system, respectively. The stress distribution for the Nicalon fibers is very similar to that for the isotropic homogeneous case because of the low E_f/E_m ratio. It can also be noticed that the anisotropy of the carbon fibers affects the stress distribution by spreading out the shear stress transfer along the length of the fiber.

Figure 19 shows interfacial shear stresses for a carbon fiber-epoxy system. The low modulus of the epoxy matrix produces a similar effect on the stress distribution, as in the case of isotropic fibers.

Calculations of interfacial shear strength assume either a maximum interfacial shear stress criterion or a maximum radial tensile stress criterion, ignoring other stress components and residual stresses in both cases. The interfacial shear strength (τ_i) is calculated from:

$$\tau_i = \sigma_{fd}\,(\tau_{max}/\bar{\sigma}_f)_{FEM} \qquad \text{(Eq 29)}$$

where σ_{fd} is the average compressive stress applied to the fiber end at debonding and $(\tau_{max}/\bar{\sigma}_f)_{FEM}$ is the ratio of the maximum shear stress to applied stress. Figure 20 shows normalized interfacial shear stress ($\tau_{max}/\bar{\sigma}_f$) as a function of G_m/E_f (matrix shear modulus/fiber axial elastic modulus) for a variety of materials.

Calculation of the interfacial tensile strength σ_r/i for tensile radial stress at the surface is given by:

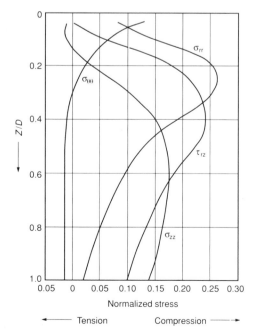

Fig. 17 Interfacial stress distribution for Nicalon (SiC) fibers in an aluminosilicate glass matrix obtained using the microindentation technique. Source: Ref 73

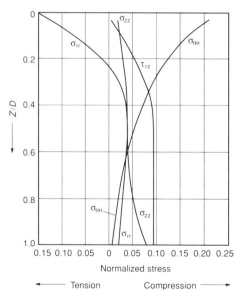

Fig. 18 Interfacial stress distribution for untreated high modulus carbon fibers in a borosilicate glass matrix obtained using the microindentation technique. Source: Ref 73

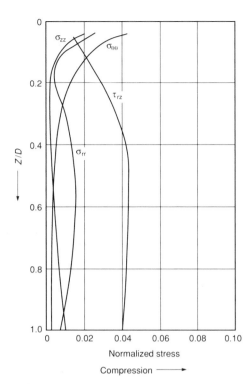

Fig. 19 Interfacial stress distribution for a carbon fiber-epoxy system obtained using the microindentation technique. Source: Ref 73

$$\sigma_i^r = \sigma_{fd} (\sigma_{max}^r / \bar\sigma_f)_{FEM} \qquad \text{(Eq 30)}$$

where $(\sigma_{max}^r / \bar\sigma_f)_{FEM}$ is the ratio of the maximum radial tensile stress at the surface to the applied fiber pressure resulting from the finite-element analysis. In all cases, the FEM results are calculated for the ratio of matrix layer thickness to fiber diameter of 0.40.

It is evident that further refinements are required to treat the free-surface problem adequately and that the accuracy of stress in this area is still uncertain. Also, the effect of thermal and elastic residual stresses should be included because they should be very significant near the free surface.

Slight variations to this technique have been reported by Netravali *et al.* (Ref 74) in studies of interfacial shear strength between E-glass fibers and an epoxy matrix. To avoid any cyclic loading on the specimen and allow for visual detection of fiber debonding, a test procedure was developed to monitor the load and depth of indentation continuously as a characteristic change in the load-depth curve. In addition, acoustic emissions generated by various events are monitored, and the rate of loading and specimen geometry can be adjusted to simulate different situations that can occur in actual service.

Thin samples approximately 4 to 10 fiber diameters thick were used, and the assumption of constant shear stress along the fibers was made. The shear stress was calculated from the relation:

$$\tau_a = F/\pi dt \qquad \text{(Eq 31)}$$

where F is the load value, d is the diameter of the fiber, and t is the thickness of the specimen. Differential shrinkage between the epoxy and the fiber after elevated-temperature curing will generate a hydrostatic pressure, which is given by:

$$P = \alpha_m E_m \Delta T (1 + \nu_m) \qquad \text{(Eq 32)}$$

where α_m is the linear coefficient of thermal expansion (CTE) of the matrix material, E_m is the matrix Young's modulus, ΔT is the difference between the curing and room temperatures, and ν_m is Poisson's ratio of the matrix. It should be noticed that the linear CTE of the fiber is considered to be negligible and also that its elastic modulus is several times greater than that of the matrix. As a consequence, the lateral expansion of the fiber at the point of application of the force is also considered negligible.

Because of its thickness, the specimen bends upon application of the indentation force, resulting in radial compression at the top and tension at the bottom. The resulting stress distribution is calculated from the elastic analysis of the theory of plates, according to:

$$\sigma_r = (6d/t)(F/2\pi dt)\{1/2 + [\rho^2/(d^2 - \rho^2)]\ln(\rho/d)\} \qquad \text{(Eq 33)}$$

where σ_r is the radial stress at the top surface, d is the fiber diameter, t is the specimen thickness, and ρ is the inside radius of the brass annulus on which the specimen is rigidly mounted.

Marshall *et al.* (Ref 75-77) reported on the use of a nanohardness apparatus to determine interfacial mechanical properties in fiber-reinforced ceramic composite materials. They calculated the interfacial sliding friction using the relationship:

$$\tau_F = \frac{\sigma_f^2}{4E_f}(r/L) \qquad \text{(Eq 34)}$$

where σ_f is the pressure applied to the fiber, r is the radius of the fiber, L is the distance the fiber has been displaced into the matrix, and E_f is the elastic modulus of the fiber.

The Dow Chemical Company has developed a microdebonding indentation system (Ref 78) that overcomes some deficiencies of the method described by Mandell and co-workers. This fully automated instrument is designed to be used outside the research environment. It is based on a Zeiss optical microscope. A diamond-tipped stylus mounted on the objective holder of the microscope is used to compress single fibers into the specimen. Initiation of fiber debonding is sensed by a load cell attached to the sample holder. Other components include a precision-controlled motorized stage with three degrees of freedom (linear motion in three orthogonal axes), a television camera and monitor, and an IBM PC-AT-compatible computer.

The specimens are prepared following standard metallographic techniques, ensuring that the fibers of the composite are always perpendicular to the surface of the specimen holder. The force required to debond the fiber is input to an algorithm that calculates the interfacial shear strength. In the commercial system the test data are reduced using a closed-form algorithm derived from Mandell's finite-element analysis using the method of least squares and resulting in an expression that is a function of σ, the axial stress in the fiber at debond; G_m, the shear modulus of the matrix; E_f, the axial tensile modulus of the fiber; T_m, the distance from the tested fiber to the nearest adjacent fiber; and d, the diameter of the tested fiber.

The ($\pm$45)$_s$ Tensile Test

Based on expressions developed by Petit (Ref 79) for the shear stress-strain results of a uniaxial tensile test for a ($\pm$45) symmetrical laminate using plate theory, Rosen (Ref 80) showed that the expressions for the in-plane shear stress-strain curve could be obtained from the longitudinal and transverse stress-strain curves:

$$\tau_{12} = \bar\sigma_x / 2 \qquad \text{(Eq 35)}$$

$$\gamma_{12} = (\epsilon_x^0 - \epsilon_y^0) \qquad \text{(Eq 36)}$$

where $\bar\sigma_x$ is the applied tensile stress and ϵ_x^0 and ϵ_y^0 are the longitudinal and transverse strains. Caution should be exercised regarding the interpretation of the ultimate stress and strain results from this test, because the laminate is under a state of combined stress rather than pure shear (Ref 81), and the presence of normal stress components would be expected to have a deleterious effect on ultimate shear strength. Figure 21 shows the geometry of the specimen used in this test. Details of the test are presented in ASTM D 3518.

The Iosipescu Shear Test Specimen

A state of pure shear is achieved within the test section of the Iosipescu shear test specimen by applying two counteracting moments produced by two force couples (Ref 82). From simple statics considerations, the shearing force acting at the center of the cross section is equal to the applied force as measured from the loading device. As stated by Iosipescu, there is no stress concentration at the sides of the notches of the specimen (see Fig. 21 for specifications), because the normal stresses are parallel to the sides at that point of the specimen. Consequently, the shear stress is obtained by simply dividing the value of the applied force by the net cross-sectional area between the two notch tips.

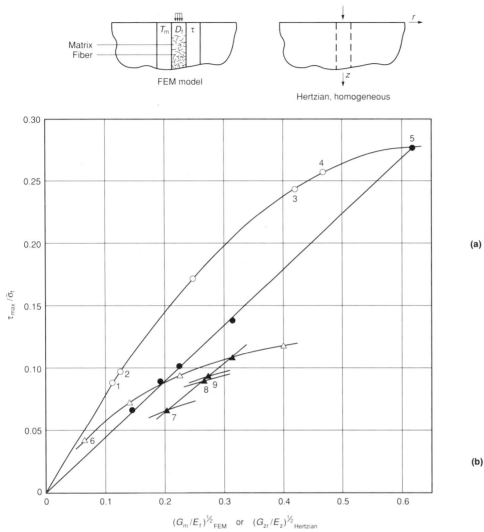

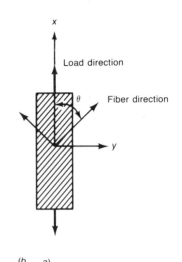

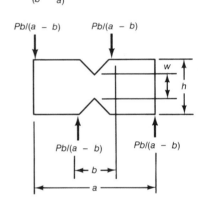

Fig. 20 Maximum normalized interfacial shear stresses for various hypothetical material systems and trends from Hertzian solutions for transversely isotropic properties referenced to an imaginary interface position. ○, FEM model, isotropic fibers; ●, Hertzian with $G_{zr} = 14$ MPa (2 ksi), E_z varying; △, FEM with $E_z = 276$ GPa (40×10^6 psi), matrix varying; ▲, FEM with $G_m = 28$ GPa (4×10^6 psi), carbon fiber E_z varying. (1) S-glass-epoxy, (2) E-glass-epoxy, (3) Nicalon-1723 glass, (4) Nicalon-BMAS, (5) Hertzian isotropic, (6) carbon-epoxy, (7) P100 aluminum, (8) P55 aluminum, (9) Untreated high modulus borosilicate. Source: Ref 73

The Short-Beam Shear Test

The short-beam shear method is used to estimate interlaminar shear strength only. It is based on elementary beam theory and involves a three-point flexure specimen with the span-to-width ratio (L/h) chosen to produce interlaminar shear failure. The shear distribution along the thickness of the specimen is a parabolic function symmetrical about the neutral axis, where it attains a maximum value at the upper surface and a value of 0 at the lower surface. When used in conjunction with laminated materials, the interlaminar shear stress will be parabolic within each layer, but a discontinuity in slope will occur at the ply interfaces. As a result, the maximum value of the shear stress will not necessarily occur at the center of the beam. Thus, caution should be exercised when interpreting the results of

composite materials that cannot be treated as homogeneous (Ref 81). The interlaminar shear stress is given by:

$$\tau_m = 0.75 \, \frac{W}{A} \qquad \text{(Eq 37)}$$

where τ_m is the maximum value of τ_{13}, W is the applied load, and A is the cross section of the specimen. Details of this test are documented in ASTM D 2344. The geometry of this specimen is given in Fig. 21.

Comparison of Test Methods

In order to provide an objective comparison between the six methods used to measure the interfacial shear strength, an AS-4 carbon fiber-Epon 828 mPDA (epoxy) resin system was selected. Table 4 summarizes test results obtained from the different techniques.

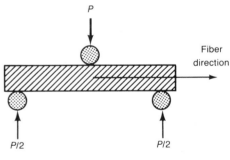

Fig. 21 Schematic representation of specimens for different shear test methods in laminates. (a) (± 45)$_s$ tension shear test. (b) Iosipescu shear test. (c) Short-beam interlaminar shear test

If the first two model techniques are compared, it can be seen that the microdrop technique yields lower results. Several reasons can be mentioned to explain this difference. First, bulk properties of the microdrops are slightly lower than those of the single-fiber technique matrix. It has been indicated by Rao and Drzal (Ref 39) that significant diffusion of the curing agent out of the droplet can occur if a regular curing cycle is used immediately following the formation of the microdrops on the fiber. In

Table 4 Summary of shear test results

Test	AS-4/Epon 828 mPDA	
	MPa	ksi
Single-embedded-fiber test	68.3	9.9
Microdrop technique..............	50.3	7.3
Microindentation technique........	65.8	9.5
$(\pm45)_{3s}$ tension test	72.2	10.4
Iosipescu shear test	95.6	13.8
Short-beam shear test.............	84.0	12.2

order to achieve a full cure, a modified curing cycle needs to be used.

Other studies have not reported this problem in their work with carbon fiber-epoxy matrix systems, but different curing agents (Ref 8, 43, 45) were used in these studies. Also, it has been shown (Ref 25-29, 42) that the mechanical events that occur in the pull-off technique are different than those occurring in the single-fiber technique. In the microdrop technique, the total fracture energy is contributed by the strain energy stored in the free-and-embedded fiber length, as well as in the matrix immediately surrounding the fiber. In the single-fiber fragmentation method, the total energy is contributed by the embedded fiber and the surrounding matrix. The free fiber length should be made as short as possible during the experiment in order to obtain accurate results. The small size of the droplet may also influence the interfacial shear stress due to local variation in the adhesion properties along the fiber (Ref 36).

Microindentation system results seem to agree well with the results obtained from the single-fiber fragmentation method, although further study is planned to assess the effect of factors such as free surface, residual stresses, neighboring fibers, polishing damage to the free surface, and so on.

As reported by Madhukar and Drzal (Ref 83), the interfacial shear strength values obtained from a (±45) tension test agree well with those obtained from the single-embedded-fiber technique. They strongly recommend caution when working with fiber-matrix systems with low adhesion levels, because of scissoring effects observed in composites having poor strength.

The Iosipescu shear test yields slightly higher values of interfacial shear strength than those obtained with the other techniques. A linear relationship is observed between values obtained from the short-beam shear test and the single-fiber test. However, the differences in the nature and assumptions of these techniques make it obvious that the shear-strength values should yield slightly different results.

It is clear that none of these techniques offers a complete and unambiguous method for measuring the interfacial shear strength between fiber and matrix, but, for a proper interpretation of test results, the model methods should provide a clearer idea of the level of adhesion and the failure mode between fiber and matrix. The tests performed in laminates are recommended only when the level of adhesion between fiber and matrix is known; otherwise, they should be complemented with the single-fiber model techniques.

REFERENCES

1. L.T. Drzal, M.J. Rich, and P.F. Lloyd, Adhesion of Graphite Fibers to Epoxy Matrices: I. The Role of Fiber Surface Treatment, *J. Adhes.*, Vol 16, 1982, p 1-30
2. L.T. Drzal, M.J. Rich, M. Koenig, and P.F. Lloyd, Adhesion of Graphite Fibers to Epoxy Matrices: II. The Effect of Fiber Finish, *J. Adhes.*, Vol 16, 1983, p 133-152
3. L.T. Drzal, M.J. Rich, and M. Koenig, Adhesion of Graphite Fibers to Epoxy Matrices: III. The Effect of Hygrothermal Exposure, *J. Adhes.*, Vol 18, 1985, p 49-72
4. L.T. Drzal, Composite Interphase Characterization, *SAMPE J.*, Vol 19, 1983, p 7-13
5. L.T. Drzal, Effect of Graphite Fiber-Epoxy Adhesion on Composite Fracture Behavior, in *Proceedings of the 2nd U.S./Japan ASTM Conference on Composite Materials*, American Society for Testing and Materials, 1985
6. M. Narkis, E.J.H. Chen, and R.B. Pipes, Review of Methods for Characterization of Interfacial Fiber-Matrix Interactions, *Polym. Compos.*, Vol 9 (No. 4), Aug 1988, p 245-251
7. L.J. Broutman, Measurement of the Fiber-Polymer Matrix Interfacial Strength, in *Interfaces in Composites*, STP 452, American Society for Testing and Materials, 1969, p 27-41
8. B. Miller, P. Muri, and L. Rebenfeld, A Microbond Method for Determination of the Shear Strength of a Fiber/Resin Interface, *Compos. Sci. Technol.*, Vol 28, 1987, p 17-32
9. A. Kelly and W.R. Tyson, Tensile Properties of Fiber-Reinforced Metals: Copper/Tungsten and Copper/Molybdenum, *J. Mech. Phys. Solids*, Vol 13, 1965, p 329-350
10. L.T. Drzal, M.J. Rich, J.D. Camping, and W.J. Park, Interfacial Shear Strength and Failure Mechanisms in Graphite Fiber Composites, Paper 20-C, 35th Annual Technical Conference, Reinforced Plastics/Composites Institute, The Society of the Plastics Industry, 1980
11. J.F. Mandell, J.-H. Chen, and F.J. McGarry, "A Microdebonding Test for In-Situ Fiber-Matrix Bond and Moisture Effects," Research Report R80-1, Department of Materials Science and Engineering, Massachusetts Institute of Technology, Feb 1980
12. J.O. Outwater and M.C. Murphy, The Influences of Environment and Glass Finishes on the Fracture Energy of Glass-Epoxy Joints, Paper 16-D, in *Proceedings of the 24th Annual Technical Conference*, The Society of the Plastics Industry, 1969
13. Y.S. Ko, W.C. Forsman, and T.S. Dziemianowicz, Carbon Fiber-Reinforced Composites: Effect of Fiber Surface on Polymer Properties, *Polym. Eng. Sci.*, Vol 22, Sept 1982, p 805-814
14. P.S. Chua, Characterization of the Interfacial Adhesion Using Tan Delta, *SAMPE J.*, Vol 18 (No. 3), April 1987, p 10-15
15. P. Perret, J.F. Gerard, and B. Chabert, A New Method to Study the Fiber-Matrix Interface in Unidirectional Composites: Application for Carbon Fiber-Epoxy Composites, *Polym. Test.*, Vol 7, 1987, p 405-418
16. D.E. Yuhas, B.P. Dolgin, C.L. Vorres, H. Nguyen, and A. Schriver, Ultrasonic Methods for Characterization of Interfacial Adhesion in Spectra Composites, in *Interfaces in Polymer, Ceramic and Metal Matrix Composites*, H. Ishida, Ed., Elsevier, 1988, p 595-609
17. W.L. Wu, "Thermal Technique for Determining the Interface and/or Interply Strength in Polymeric Composites," Polymers Division Preprint, National Institute of Standards and Technology, 1989
18. L.B. Greszczuk, Theoretical Studies of the Mechanics of the Fiber-Matrix Interface in Composites, in *Interfaces in Composites*, STP 452, American Society for Testing and Materials, 1969, p 42-58
19. P. Lawrence, Some Theoretical Considerations of Fibre Pull-Out From an Elastic Matrix, *J. Mater. Sci.*, Vol 7, 1972, p 1-6
20. A. Takaku and R.G.C. Arridge, The Effect of Interfacial Radial and Shear Stress on Fibre Pull-Out in Composite Materials, *J. Phys. D, Appl. Phys.*, Vol 6, 1973, p 2038-2047
21. R.J. Gray, Analysis of the Effect of Embedded Fibre Length on the Fibre Debonding and Pull-Out From an Elastic Matrix, *J. Mater. Sci.*, Vol 19, 1984, p 861-870
22. V. Laws, Micromechanical Aspects of the Fibre-Cement Bond, *Composites*, April 1982, p 145-151
23. J. Banbaji, On a More Generalized Theory of the Pull-Out Test From an Elastic Matrix. Part I, Theoretical Considerations, *Compos. Sci. Technol.*, Vol 32, 1988, p 183-193
24. J. Banbaji, On a More Generalized Theory of the Pull-Out Test From an Elastic Matrix. Part II, Application to Polypro-

pylene-Cement System, *Compos. Sci. Technol.*, Vol 32, 1988, p 195-207

25. P.S. Chua and M.R. Piggott, The Glass Fibre-Polymer Interface: I, Theoretical Considerations for Single Fibre Pull-Out Tests, *Compos. Sci. Technol.*, Vol 22, 1985, p 33-42

26. P.S. Chua and M.R. Piggott, The Glass Fibre-Polymer Interface: II, Work of Fracture and Shear Stresses, *Compos. Sci. Technol.*, Vol 22, 1985, p 107-119

27. P.S. Chua and M.R. Piggott, The Glass Fibre-Polymer Interface: III, Pressure and Coefficient of Friction, *Compos. Sci. Technol.*, Vol 22, 1985, p 185-196

28. P.S. Chua and M.R. Piggott, The Glass Fibre-Polymer Interface: IV, Controlled Shrinkage Polymers, *Compos. Sci. Technol.*, Vol 22, 1985, p 245-258

29. M.R. Piggott, Debonding and Friction at Fibre-Polymer Interfaces: I, Criteria for Failure and Sliding, *Compos. Sci. Technol.*, Vol 30, 1987, p 295-306

30. L.S. Penn and S.M. Lee, Interpretation of Experimental Results in the Single Pull-Out Filament Test, *J. Compos. Technol. Res.*, Vol 11 (No. 1), Spring 1989, p 23-30

31. J.K. Morrison, S.P. Shah, and Y.S. Jenq, Analysis of Fiber Debonding and Pull-Out in Composites, *J. Eng. Mech.*, Vol 114 (No. 2), Feb 1986, p 277-294

32. L.S. Sigl and A.G. Evans, Effects of Residual Stress and Frictional Sliding on Cracking and Pull-Out in Brittle Matrix Composites, *Mech. Mater.*, Vol 8, 1989, p 1-12

33. H. Stang and S.P. Shah, Failure of Fibre Reinforced Composites by Pull-Out Fracture, *J. Mater. Sci.*, Vol 21, 1986, p 953-957

34. C. Atkinson, J. Avila, E. Betz, and R.E. Smelser, The Rod Pull Out Problem, Theory and Experiment, *J. Mech. Phys. Solids*, Vol 30 (No. 3), 1982, p 97-120

35. J. Bowling and G.W. Groves, The Debonding and Pull-Out of Ductile Wires From a Brittle Matrix, *J. Mater. Sci.*, Vol 14, 1979, p 431-442

36. A. Kelly, Interface Effects and the Work of Fracture of a Fibrous Composite, *Proc. R. Soc. (London) A*, Vol 319, 1970, p 95-116

37. K. Kawata and N. Takeda, Analysis of Shear Tests of Tapered Lap Adhesive Joints and Fiber Pull-Out Tests Based Upon Energy Balance Concept, *Trans. JSCM*, Vol 4 (No. 1), 1978, p 23-30

38. J. Ostrowski, G.T. Will, and M.R. Piggott, Poisson's Stresses in Fibre Composites. I: Analysis, *J. Strain Anal.*, Vol 19 (No. 1), 1984, p 43-49

39. V. Rao and L.T. Drzal, unpublished results, Michigan State University, 1989

40. M.R. Piggott, A. Sanadi, P.S. Chua, and D. Andison, Mechanical Interac-

tions in the Interfacial Region of Fibre Reinforced Thermosets, in *Composite Interfaces*, Proceedings of the First International Conference on the Composite Interface, May 1986, p 109-121

41. A. Ozzello, D.S. Grummon, L.T. Drzal, J. Kalantar, I.H. Loh, and R.A. Moody, Interfacial Shear Strength of Ion Beam Modified UHMW-PE Fibers in Epoxy Matrix Composites, *Proc. Mater. Res. Soc. Symp.*, Vol 153, 1989, p 217-222

42. U. Gaur, G. Besio, and B. Miller, Measuring Fiber/Matrix Adhesion in Thermoplastic Composites, *Plast. Eng.*, Oct 1989, p 43-45

43. L.S. Penn and S.M. Lee, Interpretation of the Force Trace for Kevlar/Epoxy Single Filament Pull-Out Tests, *Fibre Sci. Technol.*, Vol 17, 1982, pp 91-97

44. U. Gaur and B. Miller, Microbond Method for Determination of Shear Strength of a Fiber/Resin Interface: Evaluation of Experimental Parameters, *Compos. Sci. Technol.*, Vol 34, 1989, p 35-51

45. K.P. McAlea and G.J. Besio, Adhesion Between Polybutylene Terephthalate and E-Glass Measured With a Microdebond Technique, *Polym. Compos.*, Vol 9 (No. 4), Aug 1988, p 285-290

46. L.S. Penn, G.C. Tesoro, and H.X. Zhou, Some Effects of Surface-Controlled Reactions of Kevlar 29 on the Interface in Epoxy Composites, *Polym. Compos.*, Vol 9 (No. 3), June 1988, p 184-191

47. R.A. Latour Jr., J. Black, and B. Miller, Fatigue Behavior Characterization of the Fibre-Matrix Interface, *J. Mater. Sci.*, Vol 24, 1989, p 3616-3620

48. K.P. McAlea and G.J. Besio, Adhesion Kinetics of Polybutylene to E-Glass Fibers, *J. Mater. Sci. Lett.*, Vol 7, 1988, p 141-143

49. H.L. Cox, The Elasticity and Strength of Paper and Other Fibrous Materials, *Brit. J. Appl. Phys.*, Vol 3 (No. 1), March 1952, p 72-79

50. B.W. Rosen, Mechanics of Composite Strengthening, Chapter 3 in *Fiber Composite Materials*, American Society for Metals, 1964, p 37-75

51. J. Amirbayat and J.W.S. Hearle, Properties of Unit Composites as Determined by the Properties of the Interface. Part I: Mechanism of Matrix-Fibre Load Transfer, *Fiber Sci. Technol.*, Vol 2 (No. 2), 1969, p 123-141

52. P.S. Theocaris, *The Mesophase Concept in Composites*, Springer-Verlag, 1987

53. F.C. Lhotellier and H.F. Brinson, Matrix-Fiber Stress Transfer in Composite Materials: Elasto-Plastic Model With an Interphase Layer, *Compos. Struct.*, Vol 10, 1988, p 281-301

54. J.M. Whitney and L.T. Drzal, Axisym-

metric Stress Distribution Around an Isolated Fiber Fragment, in *Toughened Composites*, STP 937, American Society for Testing and Materials, 1987, p 179-196

55. A.S. Carrara and F.J. McGarry, Matrix and Interface Stresses in a Discontinuous Fiber Composite Model, *J. Compos. Mater.*, Vol 2 (No. 2), April 1968, p 222-243

56. A.K. Soh, On the Determination of Interfacial Stresses in a Composite Material, *Strain*, Vol 21 (No. 4), Nov 1985, p 163-172

57. S.L. Pu and M.A. Sadowski, Strain Gradient Effects on Force Transfer Between Embedded Microflakes, *J. Compos. Mater.*, Vol 2 (No. 4), April 1968, p 138-151

58. A.N. Netravali, R.B. Henstenburg, S.L. Phoenix, and P. Schwartz, Interfacial Shear Strength Studies Using the Single-Filament Composite Test. I: Experiments on Graphite Fibers in Epoxy, *Polym. Compos.*, Vol 10 (No. 4), Aug 1989, p 226-241

59. R.B. Henstenburg and S.L. Phoenix, Interfacial Shear Strength Studies Using the Single-Filament-Composite Test. Part II: A Probability Model and Monte Carlo Simulation, *Polym. Compos.*, Vol 10 (No. 5), Dec 1989

60. W.D. Bascom and R.M. Jensen, Stress Transfer in Single Fiber Resin Tensile Tests, *J. Adhes.*, Vol 19, 1986, p 219-239

61. A.S. Wimolkiatisak and J.P. Bell, Interfacial Shear Strength and Failure Modes of Interphase-Modified Graphite-Epoxy Composites, *Polym. Compos.*, Vol 10 (No. 3), June 1989, p 162-172

62. W.A. Fraser, F.H. Ancker, A.T. Dibenedetto, and B. Elbirli, Evaluation of Surface Treatments for Fibers in Composite Materials, *Polym. Compos.*, Vol 4 (No. 4), Oct 1983, p 238-248

63. El M. Asloum, J.B. Donnet, G. Guilpain, M. Nardin, and J. Schultz, On the Determination of the Tensile Strength of Carbon Fibers at Short Lengths, *J. Mater. Sci.*, Vol 24, 1989, p 3504-3510

64. El M. Asloum, M. Nardin, and J. Schultz, Stress Transfer in Single-Fiber Composites: Effect of Adhesion, Elastic Modulus of Fibre and Matrix and Polymer Chain Mobility, *J. Mater. Sci.*, Vol 24, 1989, p 1835-1844

65. Y. Termonia, Theoretical Study of the Stress Transfer in Single Fibre Composites, *J. Mater. Sci.*, Vol 22, 1987, p 504-508

66. M.J. Folkes and W.K. Wong, Determination of Interfacial Shear Strength in Fibre-Reinforced Thermoplastic Composites, *Polymer*, Vol 28, July 1987, p 1309-1314

67. A.N. Netravali, L.T.T. Topoleski,

W.H. Sachse, and S.L. Phoenix, An Acoustic Emission Technique for Measuring Fiber Fragment Length Distributions in the Single-Fiber-Composite Test, *Compos. Sci. Technol.*, Vol 35, 1989, p 13-29

68. M.C. Waterbury and L.T. Drzal, accepted for publication in *J. Compos. Technol. Res.*

69. K.H.G. Ashbee and E. Ashbee, Photoelastic Study of Epoxy Resin/Graphite Fiber Load Transfer, *J. Compos. Mater.*, Vol 22, July 1988, p 602-615

70. J.F. Mandell, J.H. Chen, and F.J. McGarry, Paper 26D, in *Proceedings, 35th Conference on R.P./Composites Institute*, Society of the Plastics Industry, 1980

71. D.H. Grande, "Microdebonding Test for Measuring Shear Strength of the Fiber/Matrix Interface in Composite Materials," M.S. thesis, Massachusetts Institute of Technology, June 1983

72. J.F. Mandell, D.H. Grande, T.H. Tsiang, and F.J. McGarry, A Modified Microdebonding Test for Direct In-Situ Fiber/Matrix Bond Strength Determination in Fiber Composites, Research Report R84-3, Massachusetts Institute of Technology, Dec 1984

73. D.H. Grande, J.F. Mandell, and K.C.C. Hong, Fibre-Matrix Bond Strength Studies of Glass, Ceramic and Metal Matrix Composites, *J. Mater. Sci.*, Vol 23, 1988, p 311-328

74. A.N. Netravali, D. Stone, S. Ruoff, and T.T.T. Topoleski, Continuous Microindentor Push Through Technique for Measuring Interfacial Shear Strength of Fiber Composites, *Compos. Sci. Tech.*, Vol 34, 1989, p 289-303

75. D.B. Marshall, "An Indentation Method for Measuring Matrix-Fiber Frictional Stresses in Ceramic Composites," American Ceramics Society, Dec 1984, p c-259 to c-260

76. D.B. Marshall and A.G. Evans, Failure Mechanisms in Ceramic Fiber/Ceramic-Matrix Composites, *J. Am. Ceram. Soc.*, Vol 68 (No. 5), May 1985, p 225-231

77. D.B. Marshall and W.C. Oliver, Measurement of Interfacial Mechanical Properties in Fiber Reinforced Ceramic Composites, *J. Am. Ceram. Soc.*, Vol 70 (No. 8), Aug 1987, p 542-548

78. M.K. Tse, U.S. Patent 4,662,228, 1987

79. P.H. Petit, A Simplified Method of Determining the Inplane Shear Stress-Strain Response of Unidirectional Composites, STP 460, American Society for Testing and Materials, 1969, p 63

80. B.W. Rosen, A Simple Procedure for Experimental Determination of the Longitudinal Shear Modulus of Unidirectional Composites, *J. Compos. Mater.*, Vol 6, 1972, p 552-554

81. J.M. Whitney, I.M. Daniel, and R.B. Pipes, Experimental Mechanics of Fiber Reinforced Composite Materials, Monograph 4, Society for Experimental Stress Analysis, 1982

82. D.E. Walrath and D.F. Adams, Analysis of the Stress State in an Iosipescu Shear Test Specimen, Department Report UWME-DR-301-102-1, Composite Materials Research Group, Department of Mechanical Engineering, University of Wyoming, Laramie, June 1983

83. M.S. Madhukar and L.T. Drzal, Fiber-Matrix Adhesion and Its Effect on Composite Mechanical Properties. I. Inplane and Interlaminar Shear Behavior of Graphite/Epoxy Composites, submitted for publication in *J. Compos. Mater.*

Microstructural Analysis

D.E. Packham, The University of Bath (England)

THE GOAL OF THE SCIENCE OF MATERIALS is to account for the properties of materials in terms of their structure. An understanding of microstructure is an integral part of this endeavour. This general principle applies in particular to the performance of adhesive joints.

This article considers the relevant aspects of the microstructure of the component parts of an adhesive joint. These include the morphology, cross-link density, and phase structure of the adhesive, as well as the topography and nature of the surface layers of the substrate. The need to regard an adhesive joint as a "system" or composite, that is, as more than just the sum of its components, is emphasized. Ways in which the microstructure of the adhesive can be modified as a result of interaction with the substrate are described in the section "Influence of Substrate on Adhesive Microstructure" in this article.

Many of the influences will be rationalized by means of Eq 1, which shows the wide range of factors that, in principle, contribute to joint toughness. Changes in the microstructure in any part of the joint have the potential for affecting the overall properties.

It is hoped that sufficient background is provided the engineer wishing to exploit the advantages of adhesives so that, in general terms, the relationship between microstructure and performance can be appreciated. Attention is directed to those aspects of microstructure over which the practitioner can exert an influence, and thus prepare joints such that pitfalls can be avoided and the full potential of adhesives technology can be realized.

An adhesive joint, even the simplest, is best regarded as a system in which its properties reflect the properties of its different parts: the substrate or substrates, the adhesive, the interfaces, and particularly the interaction among these parts.

Fracture Energy. This point may be illustrated by considering the fracture energy per unit area (P) of some hypothetical joint. The energy (P), supplied to break the joint in a particular way, will be given by the sum of the various energy-absorbing processes involved in the fracture process. This is essentially an illustration of the first law of thermodynamics and can be written as:

$$P = W + \psi_{v/e} + \psi_{plast} + \psi_{sub} + \ldots \quad \text{(Eq 1)}$$

The right-hand side of the equation represents the different energy dissipation processes involved in the fracture. The term W represents the energy required to create the new surfaces. Thus, it will be work of adhesion (W_a) for adhesive failure, work of cohesion (W_c) for cohesive failure, and some combination of the two for mixed failure. Because P is defined in terms of the nominal unit area of a hypothetical flat surface, any increase in the roughness of the fracture surface will lead to greater actual fracture area, thus increasing P.

The ψ terms refer to other energy loss mechanisms. In most practical joints, the surface energy term (W) is several orders of magnitude lower than the most significant ψ terms.

The bulk mechanical properties of the adhesive influence the fracture energy. As the adhesive is stressed and relaxes during fracture, viscoelastic energy losses ($\psi_{v/e}$) may occur. Their magnitude depends on the temperature and deformation rate. This contributes to a rate and temperature dependence of P.

Where ductile adhesives are involved, significant plastic work (ψ_{plast}) may occur during fracture. This, again, is sensitive to rate and temperature.

In some joints, the substrate absorbs work during fracture. The term ψ_{sub} is included in Eq 1 to draw attention to this. A simple example would be the permanent deformation of an aluminum strip during the shear testing of some strong joints.

Not only does the fracture energy reflect the bulk properties of the materials it comprises, the influence of the interface is pervasive as well. This is perhaps obvious when failure is adhesive, that is, at the interface. The value of W_a then depends on the interfacial energies. It may be less obvious that the "strength" of the interface dictates the load the joint will bear and thus the extent to which the bulk materials, adhesive and substrate, are stressed. This, in turn, influences the magnitude of the energy dissipated viscoelastically or plastically. In some mechanically simple joints, the magnitude of the viscoelastic losses have been shown to be proportional to the surface energy term (Ref 1).

The nature of the substrate surface can influence the joint strength in other ways. It may react chemically with the adhesive, altering its properties, perhaps near the interface, or through the full thickness. Changes in surface roughness can act physically by altering the wetting characteristics or by changing the stress field in the interfacial regions.

From these considerations it can be seen that a series of joints between the same adhesive and substrate, which differ only in the substrate surface treatment, may break at different loads and yet all show failure within the adhesive.

The list of energy dissipation modes in Eq 1 is not exhaustive: Each joint has to be considered a system, and the important modes assessed. As already indicated, a useful feature of the equation is that it shows that a change in the properties of any part of the system may influence the fracture energy and practical toughness of a joint.

The Microstructure. Earlier articles in this Section of the Volume, "Testing and Analysis," are concerned with the bulk properties of adhesive materials and with formal methods for measuring the "strength" of adhesive joints. This article is concerned with those microstructural aspects of a joint that influence its performance. Some microstructural effects are concerned with the bulk adhesive polymer, and some with the surface of the substrate. The discussion above in relation to Eq 1 showed how each of these can affect the performance of the joint.

The relevant microstructural features in a particular joint depend on the materials involved. Some adhesives contain filler particles or fibers, which greatly affect their mechanical properties; these have been considered in the article "Composite Fiber-Matrix Bond Tests" in this Volume. Some polymers exhibit crystallinity, making both the extent of crystallinity and the morphology present relevant to the discussion.

Many adhesives are thermosets. They cross link by reactions carried out *in situ*. Commonly, the engineer can exert control

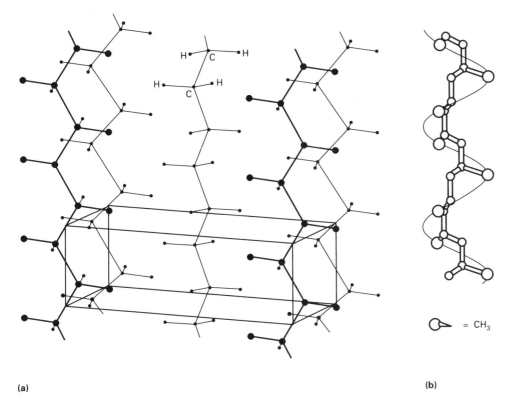

(a)

(b)

= CH₃

Fig. 1 Unit cells of (a) polyethylene and (b) polypropylene

over the cross-link density and, thus, over the joint properties.

Some polymeric adhesives are heterogeneous, showing a two-phase structure. The toughened adhesives, where a rubbery phase is dispersed in a brittle matrix, are especially important. The properties of two-phase adhesives depend on the disposition of the phases.

Defects such as voids occur in adhesive joints. Usually, they occur inadvertently and result in a weakening of the joint. Sometimes, however, they can have a beneficial influence, in which case they can be produced deliberately.

The microstructure of the substrate can be influenced by pretreatment. Changes in roughness are easily brought about, for example, by abrasion. However, more subtle techniques that produce porous oxide surfaces or microfibrous needles may have more dramatic effects. In some joints, the substrate influences the microstructure of

the polymer in the interfacial regions by affecting its morphology, cross-link density, phase structure, or void content.

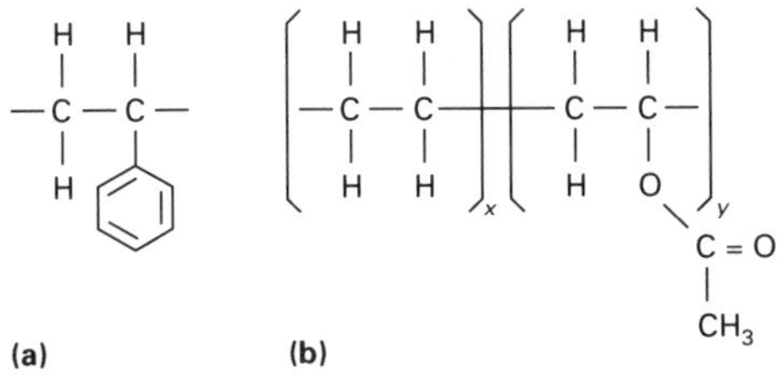

In this article, then, the microstructure of adhesive joints is discussed with emphasis on three aspects: its relationship to process variables, its characterization, and its influence on macroscopic properties of the joint.

Microstructure of the Adhesive

This section describes crystallinity, crystalline morphology, adhesion, cross-linked adhesives, phase structure, and voids.

Crystallinity

The scientific use of the term crystal is reserved for materials that have "long-range order." This means that on the molecular scale there is a regular structure in the sense that a unit cell of atoms or molecules is repeated indefinitely throughout the crystal. This is recognized experimentally by the sharp x-ray diffraction pattern that a crystalline material gives. Analysis of the diffraction pattern enables the size of the unit cell and the arrangement of atoms in it to be deduced.

It is relatively easy, in principle, to imagine how the spherical atoms or ions of a metal or ionic compound may be arranged in such a regular manner. The close packing of spheres of equal radius, for example, gives rise to the face-centered cubic struc-

ture shown by such metals as copper and aluminum.

The unit cell of crystalline polymers shows the polymer chains aligned in characteristic ways. As an example, Fig. 1(a) shows the unit cell of polyethylene $(CH_2—CH_2)_n$, in which the carbon backbone takes a planar configuration. When a space-filling model is built, it can be seen that the chains fit closely together: CH_2 protuberances on one chain fit into the hollow caused by the 109° 28¹ C—C—C bond angle on an adjacent chain.

Polypropylene has a molecular structure similar to that of polyethylene, except that there is a methyl group on every other carbon atom: $(CH_2—CHCH_3)_n$. In order to accommodate the bulky methyl group within the unit cell, the carbon backbone is twisted helically (Fig. 1b).

Extent of Crystallinity. Not all polymers show evidence of crystallinity in their x-ray diffraction patterns. Many polymers have no long-range order and are amorphous. For those polymers that do exhibit crystallinity, the diffraction patterns are less sharp than those from more conventional crystalline materials. Their crystallinity is partial, that is, they exhibit properties of both crystalline and amorphous materials.

The percentage crystallinity of such materials can be measured by various means. In general terms, the extent of crystallinity can be related to the molecular structure by considering the ease with which it could be regularly arranged in a unit cell. Thus, chain irregularities, bulky and highly polar side groups, long- and short-chain branching, and cross linking all tend to reduce crystallinity, often to zero.

High-density polyethylene has relatively few side groups and thus a high crystallinity, such as 70 or 80%, depending on the grade. Low-density polyethylene has many short and some long side chains. Its crystallinity is about 45%. Commercial polystyrene (Fig. 2a), with its bulky phenyl side group, is amorphous, as are most cross-linked adhesives.

Ethylene-vinyl acetate copolymers (Fig. 2b) are widely used in hot-melt adhesive compositions. As the vinyl acetate content increases, crystallinity decreases, and, above an approximate 50% level of vinyl acetate, the materials are amorphous (Fig. 3).

Table 1 gives approximate crystallinities of a number of polymers that form the basis of adhesive systems. It shows in broad terms the correlation of crystallinity and structure described above.

Nucleation and Growth. The crystallinities given in Table 1 are only approximate because they vary from grade to grade when features such as chain branching are altered. They also vary according to the thermal history of the sample. This may enable the engineer to exert some control

(a)   **(b)**

Fig. 2 Molecular structure of (a) polystyrene and (b) ethylene-vinyl acetate copolymer

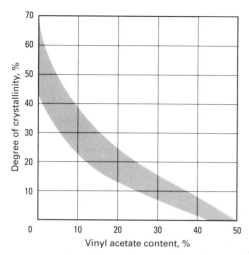

Fig. 3 Crystallinity and composition of ethylene-vinyl acetate copolymers. Source: Ref 2

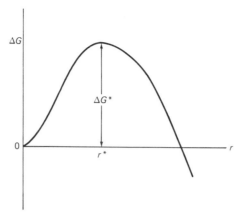

Fig. 4 Representation of the change in Gibbs free energy (G) with the radius of a nucleus of a crystal in a melt below the melting point

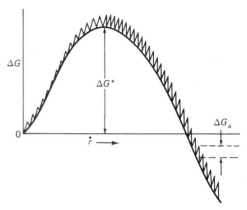

Fig. 5 Same as Fig. 4, but showing the difference between activation energies for critical nucleus formation (ΔG^*) and the transfer from melt to crystal (ΔG_a). Source: Ref 3

over crystallinity by the method of forming the adhesive joint. It may also lead to changes in crystallinity during service life. An understanding of these effects requires some appreciation of the theory of formation of a crystalline solid from the liquid, that is, of the theory of nucleation and growth.

Nucleation. The thermodynamic definition of melting point is the temperature at which the liquid and solid (crystal) have the same free energy. Below the melting point, the crystal has lower free energy and is therefore more stable. The lower the temperature, the greater the relative stability of the solid.

For the solid to form, however, some growth units (for example, atoms, for simplicity) have to come together to form a tiny fragment of the crystal (an embryo), which has to grow into a nucleus and then into a crystal. Like any other surface, the interface between the embryo (or nucleus or crystal) and the melt has an excess energy (the surface energy) associated with it. Because of its minute size, the embryo has a

very high surface area-to-volume ratio and is therefore unstable.

In classical nucleation theory (Ref 3), which addresses this question, the relationship between overall free energy and nucleus size is easily derived. Figure 4 shows the relationship schematically. For small nuclei, the surface energy term dominates. Above the critical radius (r^*), growth is favored, and the nucleus soon becomes stable. However, to achieve this, the critical activation energy (ΔG^*) has to be available.

The classical theory, for a spherical nucleus, states:

$$r^* = \frac{2\gamma}{\Delta T \cdot \Delta S_v} \qquad \text{(Eq 2)}$$

and

$$\Delta G^* = \frac{16}{3} \cdot \frac{\pi\gamma^3}{(\Delta T)^2 \Delta S_v^2} \qquad \text{(Eq 3)}$$

where γ is the surface energy, ΔS_v is the entropy difference per unit volume between crystal and melt, and ΔT is the undercooling, that is, the difference between the melt temperature and the melting point.

It is clear from these equations that some undercooling is essential and that, from the thermodynamic point of view, the greater the undercooling, the easier will be nucleation: both r^* and ΔG^* fall with ΔT.

Classical nucleation theory has been adapted to apply specifically to the crystallization of polymers (Ref 4), and the equations, such as Eq 2 and 3, have been modified to apply to the lamellar morphology of polymers. For the qualitative understanding of the effects of process variables and service conditions that are needed by the engineer using crystalline adhesives, such detail is not necessary.

Rate of Nucleation. An understanding of the effect of temperature on nucleation rate is important. Classical theory (Ref 3) treats this as being proportional to the product of the number of nucleation sites available and the rate at which atoms surmount the activation energy barrier (ΔG_a) between melt and crystal. The first term in the product is proportional to the fraction of atoms having sufficient energy (ΔG^*) to enable a stable nucleus to form. From the Maxwell distribution, this is proportional to $\exp(-\Delta G^*/kT)$, where k is Boltzmann's constant and T is absolute temperature. The second term in the product is given by a similar exponential term involving ΔG_a. Thus the rate of nucleation I is given by:

$$I = A \exp\left(-\frac{\Delta G_a}{kT} - \frac{\Delta G^*}{kT}\right) \qquad \text{(Eq 4)}$$

which, incorporating Eq 3, can be written:

$$I = A \exp\left[-\frac{\Delta G_a}{kT} - \frac{16\pi\gamma^3}{3(\Delta T)^2 \Delta S_v^2 kT}\right] \qquad \text{(Eq 5)}$$

where A is a constant.

The difference between the two activation energies in these equations may be appreciated from the representation in Fig. 5.

Equation 5 shows that, at low undercooling, the nucleation rate is principally determined by the difficulty of forming a critical

Table 1 Crystalline melting points and approximate crystallinities of some polymers of importance in adhesive application

Polymer and structure	Approximate crystallinity, %	Melting point °C	Melting point °F
High-density polyethylene —CH₂—CH₂—	80	137	279
Isotactic polypropylene —CH₂—CH— with CH₃	70	165	329
Polyethylene terephthalate —O—CH₂CH₂OOC—⟨benzene⟩—CO—	55	265	509
Polyvinyl alcohol —CH₂—CH— with OH	55	Decomposes	
Nylon 6 —NH(CH₂)₅CO—	50	225	437
Low-density polyethylene as high-density polyethylene, but with many more branches	40	120	248

Note: As explained in the text, the crystallinity of a polymer sample depends greatly on its detailed structure and thermal history, and thus the values shown should be viewed as only roughly indicative of the upper limit for commercial materials.

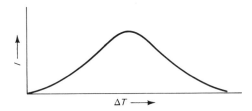

Fig. 6 Variation of rate of nucleation (*I*) with undercooling (Δ*T*) according to Eq 5

nucleus, whereas at high undercooling, the rate is determined by thermal energy available to surmount the Δ*G*_a barrier.

Figure 6 gives a graphic representation of Eq 5. Thermodynamic considerations (compare to Eq 2 and 3) suggested that nucleation became easier as undercooling increased, and kinetic factors show that there will be a maximum rate of nucleation at some finite degree of undercooling. This has many practical consequences, especially for polymer systems.

Heterogeneous Nucleation. The difficulty of nucleation discussed above may be reduced if the crystals nucleate on a second solid phase. Such heterogeneous nucleation may occur on particles adventitiously present in the melt (dust, catalyst residues, additive particles, and so on) or on the side of the vessel. It is considered that in almost all practical cases polymers nucleate heterogeneously, rather than homogeneously (Ref 4). The effect is to increase the nucleation rate, but the broad qualitative effects, which were discussed above, still apply.

Crystal Growth. Once nucleation has occurred, whether homogeneously or heterogeneously, growth tends to be an easier process. As the melt temperature falls, the free energy difference between liquid and solid increases, with the result that growth is favored. On the other hand, the thermal energy decreases, making activation energy barriers more difficult to overcome. This applies both to the diffusion of molecules to the crystal surface and to the barrier (Δ*G*_a) to be crossed from liquid to crystal.

Thus, theories of crystal growth predict a finite degree of undercooling at which the growth rate is greatest. The shape of the curve given by the classical theory is shown in Fig. 7. For low degrees of undercooling,

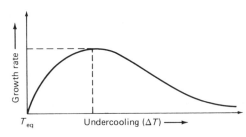

Fig. 7 Variation of crystal growth rate (*I*) with undercooling (Δ*T*) according to classical theories. Source: Ref 3

Fig. 8 Optical micrograph of a spherulitic structure taken between crossed polars

the growth rate is proportional to the undercooling.

For polymers, it is recognized that competition between growth and secondary nucleation can be important. Secondary nucleation (the nucleation of a new layer on an existing crystal surface) becomes relatively easier at lower temperatures. This leads to more branching and a finer structure when growth occurs under these conditions.

Process Variables and Crystallinity. The principles of nucleation and growth summarized in Fig. 6 and 7 enable the effects of thermal history on crystallinity to be understood.

At low degrees of undercooling, the nucleation rate is low. From the few nuclei that do form, relatively large crystals or crystalline entities grow until they impinge on one another. Different grades, or even different batches, of the same polymer may have different concentrations of heterogeneous nucleation sites and therefore a different density of nuclei. It may be possible to increase the nucleation rate deliberately by adding nucleating agents. At higher degrees of undercooling, nucleating rates will be higher, thus, crystal size will be smaller.

In adhesives technology it is much more common for crystallization to occur as an adhesive cools to ambient temperature, rather than when it is held isothermally. A low cooling rate tends to give higher overall crystallinity. With a higher cooling rate, the temperatures of maximum nucleation and growth may be passed long before either process is anywhere near completion.

In some polymers, the activation energies are so high that even at the temperature of maximum nucleation rate, no nuclei form within a practical time scale. Such a polymer is always amorphous.

In less extreme cases, the maximum rate may be great enough for nucleation to occur when cooling is slow, but not when it is fast. Thus, some polymers, such as the linear polyester polyethylene terephthalate used in some hot-melt adhesives, are crystalline unless quenched rapidly from the melt.

In many polymers, the absence of polar and bulky side groups makes the activation energies sufficiently low to ensure that

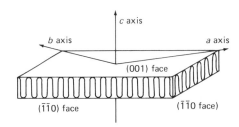

Fig. 9 Main features of polymer lamellar single crystals

some crystallization takes place at all practical cooling rates. The percentage crystallinity may be altered by a few percent by changing the cooling rate.

If a crystalline adhesive experiences elevated temperatures in service, the nucleation and growth, kinetically inhibited at lower temperatures, may recommence. A change of crystallinity, and thus of properties with time, may be expected. Not only is the percentage crystallinity itself likely to increase, the crystalline morphology will probably alter as well.

Crystalline Morphology

Polymer molecules are much more complex building blocks than the atoms of metals or the ions of ceramics, and, as a consequence, polymer morphology is a complex subject that is only recently becoming well understood (Ref 4, 5).

Spherulites are one of the most easily observed manifestations of crystallinity in polymers. These are entities of spherical symmetry that can often be seen when a thin section of a crystalline polymer is observed between crossed polars in a microscope. Under these conditions they show a characteristic Maltese cross pattern (Fig. 8). They vary in size from less than 1 μm (0.04 in.) in diameter to a few millimeters, in extreme cases. As a melt cools, nuclei form and the spherulites grow radially from the central nucleus, maintaining spherical symmetry until they impinge on their neighbors. Despite this radial growth direction, diffraction experiments show that the direction of the polymer chains is tangential.

Polymer Single Crystals. Many features of the morphology of melt-crystallized polymers have been elucidated by analogy with the structure of crystals obtained from solution. From dilute solutions of many polymers, single crystals can be obtained. A typical lamellar structure of such a crystal is shown in Fig. 9. Because the crystal grows in the *a* and *b* directions, its dimensions in these directions may extend to μm's, but the thickness in the *c* direction is limited, typically to about 110 Å (0.44 μin.). The *c* direction is the direction of the carbon chain, but polymer chains may be as long as 10 000 Å (40 μin.), and therefore chain folding must occur, and the molecule must reenter the crystal.

The Structure of Spherulites. It is thought that the structure of a nucleus from which a spherulite grows in the melt is similar to that of a lamellar single crystal. The single crystallike nucleus grows laterally, but with successive branching of the lamellae. The effect of this branching is to fill the whole volume around the original nucleus with fibrils of growing lamellae. Once the volume around the nucleus has been filled, a spherical symmetry is achieved, and the fibrils continue to grow radially.

It must be remembered that these spherulites may comprise the whole of the solid polymer, both the crystalline and the often extensive amorphous material. The relative dispositions of the crystalline and amorphous material is the subject of some uncertainty. Some amorphous material occurs in the interfibrillar and interspherulitic regions. The fibrils themselves, although consisting essentially of bundles of chain-folded lamellae with the chain axis tangential, are far from perfectly regular. Chain branches, chain ends, and many irregularities and defects of alignment mean that a significant proportion of the amorphous content is within the fibrils (see Fig. 10).

Grains and Spherulites. There is a superficial similarity between spherulites in a semicrystalline polymer and the grains in a metal or ceramic. From what has been said above, several fundamental differences are obvious. A spherulite contains the substantial amorphous content of the polymer. The detailed structure of the spherulite, including the disposition of the amorphous content, will vary according to the history of the sample. The crystal orientation within a single grain is everywhere the same, whereas in a spherulite it varies from place to place because the *c* direction is always tangential.

Other Morphologies. Spherulites are not always seen in melt-crystallized polymer. For instance, with some polyethylenes, the nucleation density is so great that a birefringent mass of unresolved spherulitic embryos appears. Where the polymer flows before solidifying (for instance, in injection molding), row nucleation may occur, giving rise to "linear spherulites." The influence of surfaces on morphology is discussed below.

Upon cold drawing, the spherulitic structure is destroyed, as the lamellae are reoriented with the chain axis in the draw direction. Unless elaborate precautions are taken (Ref 6), extensive chain folding persists. Much interesting work has been done (Ref 4, 7, 8) to attempt to crystallize polymers with extended chain morphologies, but these methods are not relevant to the processing of adhesives.

Process Variables and Morphology. The thermal history of a crystalline adhesive may affect the morphology, as well as the crystallinity. Large spherulites tend to grow at elevated temperature and low cooling

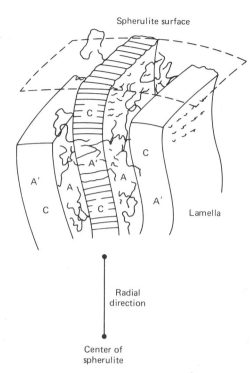

Fig. 10 Possible structure of a spherulite. Source: Ref 12

rates, whereas small spherulites grow at lower temperatures and fast cooling rates. Annealing may lead to the enlargement of spherulites.

It is not only the size of the spherulite that is affected by cooling rate and annealing; its quality may also change. The extent of branching, the fineness of the fibrils, and the degree of perfection of the lamellae are among the features involved.

Characterization of Crystallinity and Morphology (Ref 9, 10). The most fundamental measure of crystallinity is x-ray diffraction. A graph of the intensity of scattered x-rays against angle shows maxima associated with the crystallinity. The ratio of the area associated with crystallinity to that of the broad amorphous scattering gives the value of the crystallinity.

Differential scanning calorimetry enables enthalpy changes, such as heat of fusion, to be measured. By measuring the magnitude of the melting endotherm as a polymer is heated through its melting region, an estimate of crystallinity can be obtained, if a value for the heat of fusion of a completely crystalline sample is known. If this is not experimentally accessible, calibration by x-ray diffraction can be used.

A simple, secondary method of crystallinity measurement relies on closer packing, and thus higher density, of the crystalline material. Once densities of pure crystalline and amorphous materials are measured or deduced, the assumption of a linear relationship between density and crystallinity enables the crystallinity of any other sample to be obtained from its density.

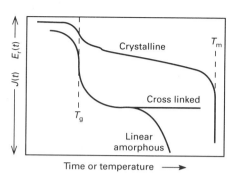

Fig. 11 Variation of stress relaxation modulus, $E_r(t)$, and creep compliance, $J(t)$, with temperature (isochronous) or time (isothermal) for linear amorphous, cross-linked amorphous, and crystalline forms of the same polymer

The infrared spectrum of some polymers shows separate absorption peaks attributable to crystalline and amorphous material. When calibrated by another method, the ratio of absorbances allows the crystallinity to be measured. For example, the ratio of the absorbances at 730 cm^{-1} and 1300 cm^{-1} in polyethylene gives the percentage crystallinity.

The morphology may be visually characterized by microscopy. A solid polymer may be microtomed into thin sections, typically 10 to 20 μm (0.4 to 0.8 mil) thick, with the sections then observed in an optical polarizing microscope, giving an overall impression of the morphology. The orientation may be assessed by measurement of birefringence by standard techniques.

It is more difficult to use reflection microscopy. The surfaces have to be etched, and extensive practical experience is needed to obtain good results. Finer detail is obtained by transmission and scanning electron microscopy, where roughly similar techniques can be used.

It is unlikely that a detailed study of the crystallinity or morphology of adhesive materials will often be needed in the context of their practical use. Many commercial adhesives represent complex mixtures of many components in addition to the crystalline polymer. This necessarily complicates, and sometimes invalidates, the application of many of the techniques described above.

Crystallinity, Morphology, and Adhesion

The emphasis in the preceding sections has been on the influence that process variables and service conditions may have in altering polymer morphology and crystallinity. These alterations, in turn, affect the mechanical properties which, as was shown in the discussion of Eq 1, help to determine the fracture energy of the adhesive joint. Some account of the effect of changes in crystallinity and morphology on mechanical properties is relevant to the use of adhesives.

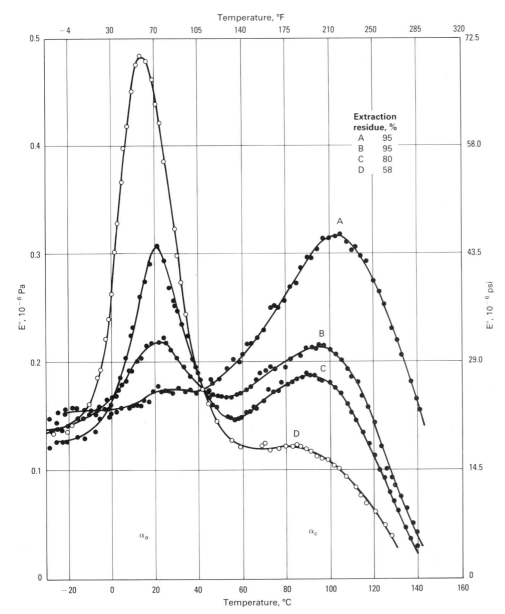

Temperature, °F

Extraction residue, %
A 95
B 95
C 80
D 58

Fig. 12(a) Effect of crystallinity on viscoelastic loss properties of polypropylene. Loss modulus for drawn samples of atactic content increasing A, B, C, D. Source: Ref 12

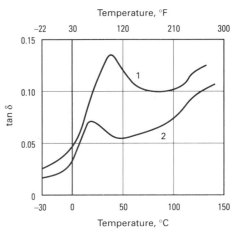

Fig. 12(b) Loss tangent in the region of the amorphous loss peak α_a for sample 1 quenched to -78 °C (-108 °F) and sample 2 annealed to 100 °C (212 °F) for 30 min. Source: Ref 13

tallinity decreases with vinyl acetate content. The consequent effect on tensile properties is shown in Fig. 13.

Crystallinity is certainly among the factors that influence the adhesion of these polymers. Some of the complexity can be seen in Fig. 14, which gives peel loads for four different EVAs applied as hot-melt coatings to oxidized copper (Ref 14, 15). Table 2 gives some properties of these polymers, and an interpretation of the results is described below.

The decrease in peel load in the sequence A-B-D is associated with the increase in crystallinity, specifically reducing the elongation at break (Fig. 13). The overall effect of this reduced ductility is a reduction in the strain energy density of the material, despite an increase in modulus and in yield strength. In terms of Eq 1, the plastic energy term (ψ_{plast}) is reduced.

Although polymer C has a higher vinyl acetate content than does D, its crystallinity is almost the same (Table 2). This is probably because there is less branching in the polyethylene sequences. Polymer C also

The polymer chain segments incorporated in the crystal lattice do not have the freedom of movement of those in amorphous regions. Elastic modulus and yield strength then increase with crystallinity. The restraining influence on chain mobility gives increased relaxation times and consequently slower creep and stress relaxation values, which are important in some applications of adhesive bonds. The appropriate viscoelastic functions for crystalline and amorphous polymers are shown in contrast in Fig. 11.

Viscoelastic loss is one of the explicit terms in Eq 1 contributing to the fracture energy of an adhesive bond. In some bonds this contribution is so significant that the rate and temperature dependence of adhesion correlates directly with peaks of loss

tangent (tan δ) or of loss modulus (E'') (Ref 11).

In a crystalline polymer, some loss peaks are associated with crystalline regions, and some with amorphous regions. In Fig. 12(a) the loss spectra of samples of polypropylene of differing isotactic content are compared (the higher the isotactic content, the higher the crystallinity). The α_c peak increases, but the α_a decreases in magnitude as crystallinity increases (Ref 12). Figure 12(b) shows the effect of quenching and annealing on the α_a peak (Ref 13). Thus, the effect of crystallinity changes on adhesion are complex, and are sensitive to the rate of test and temperature.

Ethylene-vinyl acetate (EVA) copolymers are widely used in hot-melt adhesive formulations. As Fig. 3 shows, their crys-

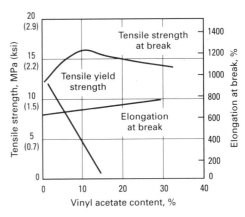

Fig. 13 Effect of vinyl acetate content on the tensile properties of ethylene-vinyl acetate polymers of constant melt flow index. Source: Ref 2

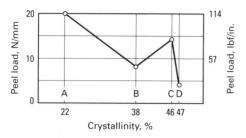

Fig. 14 Peel strength of fabric-backed ethylene-vinyl acetate copolymers versus the effect of crystallinity. Source: Ref 14, 15

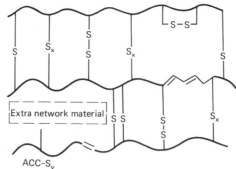

Fig. 15 Principal structural features of sulfur-vulcanized natural rubber

has a lower melt index and a higher molecular weight. This makes interpretation of the adhesion in terms of these molecular properties very difficult. Whatever the precise influence of microstructure on properties here, the correlation of strain energy density (Table 2) with peel load (Fig. 14) remains good when values for C are included.

Cooling Rate. There are various accounts in the literature of the rapid cooling of polyolefin and polyester hot-melt adhesives providing enhanced adhesion (Ref 15). Typical examples are low-density polyethylene coatings on oxidized copper and steel, for which the peel strengths increased upon quenching from 1.4 to 8.2 N/mm (8.0 to 46.7 lbf/in.) and from 2.4 to 6.3 N/mm (13.7 to 35.9 lbf/in.), respectively. Quenching lowers crystallinity a modest amount (about 2% in this example), but produces much smaller spherulites. The former gives a moderately enhanced yield stress, whereas the latter gives an enormous increase in ductility (Table 3). The result is that the quenched polymer has a much higher fracture energy, which, by means of ψ_{plast}, is reflected in the higher peel strengths.

Cross-Linked Adhesives

Many of the most widely used adhesives are cross-linked polymers. Epoxies, phenolics, unsaturated polyesters, and all vulcanized elastomeric adhesives fall within this category. These "set" by a chemical reaction, usually between two distinct components, for example, a resin and a curing agent. Heating is often involved. The cross-linking reactions are complex for most of the adhesives. A number of competing reactions occur simultaneously, and the extent of each reaction and the balance be-

tween them depends on the experimental conditions. The technologist himself is usually responsible for effecting the cure, and often for mixing the components as well. There is wide scope for variation, both deliberate and inadvertent.

The Cross-Linked Network. The concept of the ideal network is described in mathematical terms by Flory (Ref 17). In simple qualitative terms, it comprises a network of covalently linked molecules extending to the limits of the specimen. Such a network is uniform in the sense that the linkages are similar in type and density throughout the sample. The network forms by the reaction of low molecular weight molecules until, at the gel point, an insoluble three-dimensional network is formed.

Practical networks can deviate from the ideal in many respects. Network defects, such as loose chain ends, intramolecular loops, and entanglements, may form. Uneven cure can result from temperature gradients during cure or from incomplete mixing of the reagents, leading to local gelation and the resultant formation of microgel inhomogeneities, which may then be connected by a more loosely cross-linked network.

Rubbers. Many types of rubbery adhesives are employed. Some of these cure by the sulfur-base vulcanization technology that is widely used in the rubber processing industry. The complex reactions involved have been extensively studied, especially for natural rubber, but many of the details remain obscure (Ref 18). Figure 15 shows the more important aspects of such a cross-linked network.

Different cure formulations and reaction conditions give a different balance between

the structural features and, thus, different properties. For example, a conventional curing system uses a relatively high sulfur-to-accelerator ratio and gives a preponderance of polysulfidic cross links, but a majority of monosulfide links result from an "efficient vulcanizing" (EV) system with a low sulfur-to-accelerator ratio (Ref 19). Because polysulfide cross links are thermally labile, an EV rubber has better thermal aging properties.

The most important molecular property affected by changing vulcanizing conditions is the overall cross-link density. The statistical theory of rubber elasticity (Ref 17) gives the following expression for the deformation of a rubber produced by a tensile force (f):

$$\frac{f}{A_0} = \frac{\rho RTf'X}{2}(\lambda - 1/\lambda^2) \qquad \text{(Eq 6)}$$

where A_0 is the original cross-sectional area, ρ is the density, R is the gas constant, T is the absolute temperature, f' is a constant depending on the number of chains emanating from each cross link (commonly, $f' = 4$), and λ is the extension ratio. Further, X is the cross-link density (in moles of cross link per unit mass of rubber), and therefore Eq 6 predicts that the constant $\rho RTf'X/2$, which in terms of conventional elasticity is the shear modulus, is directly proportional to the cross-link density.

Equation 6 has to be refined to take network defects into account, but, with this reservation, it should apply to any network polymer in the rubbery temperature range (Ref 17).

The increase in modulus with cross-link density is accompanied by a decrease in extension to break. The effect on tensile strength is complex, but for some natural rubbers, ethylene-propylene rubber, and styrene-butadiene rubber (SBR), it has been found to pass through a maximum (Ref 20). At constant cross-link density, the strength depends on the dominant type of cross link. It increases in the order of carbon-carbon, monosulfide, disulfide, polysulfide (see Fig. 15). The intrinsic failure energy (analogous to W in Eq 1), like tensile strength, shows a maximum at moderate cross-link densities (Ref 20).

These changes in cross linking, with the consequent effects on mechanical properties, may occur as a result of changes in cure conditions (temperature or composi-

Table 2 Properties of four ethylene-vinyl acetate copolymers

Property	Designation			
	A	B	C	D
Melt flow index, g/10 min (oz/10 min)	7.2 (0.25)	7.2 (0.25)	3.1 (0.11)	6.2 (0.22)
Amount of vinyl acetate, wt%	27.5	17.5	12.0	9.8
Crystallinity, %	22	38	46	47
Strain energy density, MJ/m³ (kWh/in.³)	55 (250)	44 (200)	49 (223)	38 (173)

Source: Ref 14, 15

Table 3 Properties of polyethylene cooled from the melt at different rates

Property	Cooled slowly	Quenched in iced water
Crystallinity, %	41.7	39.6
Yield stress, MPa (ksi)	10.0 (1.45)	8.2 (1.19)
Elongation at break, %	230	600

Source: Ref 16

Fig. 16 Fracture surface of adhesive bond showing the two-phase structure of the rubber-toughened epoxy adhesive. (A void is located near center of the micrograph.)

tion) either locally or throughout the sample. They will produce corresponding changes in adhesion.

Rubber bonds to fabric play a crucial part in the pneumatic tire and in many other industrial products. Significantly higher peel strengths have been found with conventional cure than with EV systems. For the same rubber bonded to nylon fabric, Wootton found peel strengths of about 20 N/mm and 10 N/mm (114 and 57 lbf/in.), respectively (Ref 21). Peel strength passed through a maximum as cure time was increased. It is interesting to note that the maximum was well beyond the "optimum cure" as judged by the conventional criteria of the rubber technologist.

Glasses. Some of the important structural adhesives, such as epoxies and phenolics, are cross-linked polymers used below their glass transition temperatures (T_gs). Many of the broad principles discussed for rubbers should apply to glasses, all differences having been considered, but the difficulty of characterizing the glassy material, particularly with the high cross-link densities usually involved, makes it difficult to obtain a quantitative understanding of the structure-property relationships (Ref 22).

A number of authors have produced electron microscopic evidence for a "nodular" feature, approximately 100 nm (4 μin.) in size, in glassy polymers such as epoxies (Ref 20, 22). Although some interesting correlations have been reported, it is difficult to relate their size or density with confidence to the adhesive composition, cure details, or mechanical properties. Some authors regard the features as inadvertent artifacts of specimen preparation.

Characterization. Equation 6 gives the basis of an important method of measuring the cross-link density of a rubbery material from the elastic modulus. For reliable results, refinement of the equation and careful experimental technique are necessary (Ref 23). The method may be applied to glassy

resins by making the measurements above their T_gs (Ref 20).

A complementary method makes use of the swelling of rubber in suitable organic liquids. The higher the cross-link density, the less liquid is absorbed. The application of thermodynamics to the situation gives the Flory-Rehner equation:

$$\ln(1 - V_r) + V_r + \chi V_r^2 + \rho V_m V_r^{1/3}(f'X/2) = 0 \quad \text{(Eq 7)}$$

where V_m is the molar volume of the liquid density ρ; V_r is the volume fraction of swollen rubber; and χ is the polymer-liquid interaction parameter. The terms f' and X were defined in the context of Eq 6.

The interaction parameter has to be obtained by calibration of the method with a sample of known cross-link density. Although values are available for most types of rubber (Ref 24), the precise values alter with the cure conditions used; thus, for accurate work, they should be measured for the type of sample of interest (Ref 25).

Phase Structure

Many polymer blends and copolymers show a two-phase structure. Of particular interest in the context of adhesion are the rubber-toughened structural adhesives. These typically consist of a relatively brittle matrix in which are dispersed rubbery spheres a few microns in diameter (Fig. 16). Some bonding between the two phases is required. The principal effect on the properties is to increase the fracture energy markedly.

Well-known examples of plastics toughened in this way include high-impact polystyrene and acrylonitrile-butadiene-styrene (ABS) (Ref 26). Toughened epoxies (Ref 20, 27, 28) and acrylics (Ref 29) are important as adhesives. Because they have cross-linked matrices, both the network and the phase structure are susceptible to manipulation by altering the process variables.

Optimum Structure. For effective toughening, a dispersion of rubber particles of micron size is required. The usual diameter range of 0.5 to 5 μm (0.02 to 0.2 mil) is thought to be the optimum. Without some chemical bonding between the phases, toughening is greatly reduced. The proportion of rubber phase should be high, as long as it is consistent with obtaining the correct microstructure. In practice, the upper limit of volume fraction is 0.2 to 0.3. Above this level, phase inversion occurs, giving resin particles in a rubbery matrix and disastrous properties.

Phase Separation and Gelation. These rubber-toughened adhesives are made by dissolving the reactive rubber as a liquid in the resin containing curing agent. In order to obtain the required bond properties, phase separation, gelation, and, ideally, vit-

rification have to occur in order at appropriate times.

The free energy of mixing (ΔG_m) for a polymer system can be expressed in terms of the enthalpy (ΔH_m) and entropy (ΔS_m) of mixing:

$$\Delta G_m = \Delta H_m - T\Delta S_m \quad \text{(Eq 8)}$$

With reasonable assumptions about the enthalpy and entropy (Ref 30), a more explicit relationship may be expressed:

$$\Delta G_m = V_m(\delta_1 - \delta_2)^2\phi_1\phi_2 + RT(n_1 \ln \phi_1 + n \ln \phi_2) \quad \text{(Eq 9)}$$

where V_m is molar volume, n_1 and n_2 are the number of moles of resin and rubber, ϕ_1 and ϕ_2 are the volume fractions of resin and rubber, R is the gas constant, T is absolute temperature, and δ_1 and δ_2 are the solubility parameters of resin and rubber. The solubility parameter is defined as the root of the cohesive energy density, that is, the ratio of the molar energy of vaporization to the molar volume:

$$\delta = \left(\frac{E}{V_m}\right)^{1/2} \quad \text{(Eq 10)}$$

If ΔG_m is negative, the two components will be soluble, whereas if it is positive, they will be insoluble. The enthalpy term in Eq 9 is positive, unless the solubility parameters are identical, in which case it is 0. The entropy term is negative, but is not very large in magnitude. Thus, for the system to be a solution, the solubility parameters have to be close in value. In practice, a difference of about 1 (J/cm³)^{1/2} [0.49 (cal/cm³)^{1/2}] is sought for the resin and the rubber. This difference is small enough for them to form a solution when mixed.

As the polymerization proceeds, the molar volume increases, and eventually the magnitude of the enthalpy term becomes greater than that of the entropy term, and thus the rubber becomes insoluble. Particles of rubber then have to nucleate and grow in the matrix of the still-liquid resin (see the section "Nucleation and Growth" in this article).

Once the rubber particles have grown to an appropriate size, the polymerization of the resin reaches the gel point and the system gels. As the resin is still above the T_g, further reaction and some further growth of the rubber particles will occur, although much more slowly. Eventually, the degree of cross linking may reach the stage where the resin vitrifies, that is, becomes a glass at the reaction temperature.

In order to obtain the correctly toughened adhesive, a formidable range of thermodynamic criteria have to be met and certain kinetic factors controlled within acceptable limits, that is, initial solubility of the rubber, subsequent insolubility of the rubber, rate of nucleation, rate of growth, and formation of the gel. Because each of these is temper-

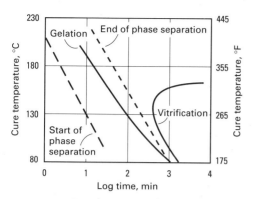

Fig. 17 Typical time-temperature transformation diagram showing phase separation and gelation for a rubber-toughened epoxy adhesive

ature dependent to a different extent, the scope for error is large.

Gillham (Ref 28) has represented the situation on a time-temperature transformation diagram (Fig. 17). This shows the effect of cure temperature on the critical stages of the process and shows how some of the process variables might be expected to affect the adhesive. He uses a specially developed combination of dynamic mechanical analysis and cloud point measurement to obtain the data needed (see the article "Thermal Properties and Temperature Effects" in this Volume).

Thus, the microstructure, properties, and consequent adhesion of rubber-toughened adhesives are critically dependent on the correct process variables. The recipes of the manufacturers will be based on some kind of optimization experiments. The practitioner should be aware of the likely effects of deviating from them. The success of these adhesives can be monitored by examination of the microstructure in a scanning electron microscope and by measuring the strength of trial joints.

Voids

Voids can easily occur in an adhesive as a result of air entrapment during preparation or failure to displace air from the substrate surface. Many adhesives have a high viscosity in the liquid stage, which makes the removal of air bubbles difficult, even when a vacuum is applied to assist the process. Moisture, either in the adhesive or on the substrate surface, is a further source of voids when the system is heated. Voids and crazes may develop during the lifetime of adhesive bonds, as a result of service stress or impact.

The presence of voids would normally be regarded as undesirable. They reduce the quantity of adhesive available to bear the stress, and, even more serious, they act as stress concentrators leading to premature failure. On the other hand, the crack sensitivity of adhesives can vary enormously from material to material. The presence of voids in

ductile materials is normally less serious than in brittle ones, and occasionally voids have even been shown to be beneficial.

Brittle Adhesives. The idea that brittle materials contain intrinsic flaws, the size of which determine their strength, plays a central role in fracture mechanics. Because this subject was developed extensively in the article "Fracture Testing and Failure Analysis," it is only briefly referred to here.

A void in a stressed adhesive will act to concentrate the stress to an extent that depends on the size and shape of the void and its orientation with respect to the stress. As an illustration, the stress σ_t at the tip of an elliptical void in an elastic lamina subject to a normal stress σ_0 at right angles to the major axis of the ellipse is given by (Ref 20):

$$\sigma_t = \sigma_0 \left(1 + 2\sqrt{a/r}\right) \qquad \text{(Eq 11)}$$

where $2a$ is the major axis of the ellipse and r is the radius of curvature of the tip of the crack. It is clear that long, sharp cracks can give rise to very high stress concentrations.

A complementary approach originally developed by Griffith and modified by Orowan leads to Eq 12, relating failure stress σ_f to crack length a:

$$\sigma_f = \left(\frac{EG_c}{\pi a}\right)^{1/2} \qquad \text{(Eq 12)}$$

where E is the modulus of the adhesive and G_c is its fracture energy. The relationship between strength and flaw size predicted by Eq 12 has been demonstrated for a number of brittle polymers for both natural and artificial flaws (Ref 20).

Crazes (Ref 20) are thought to play an important part in the failure of many brittle polymers, especially thermoplastics. Crazing involves cavitation of the material leading to a region of void that is spanned by microfibrils of polymer orientated in the stress direction. Flaws such as cracks and air bubbles are considered to be one source of the stress concentration necessary for craze initiation. Such flaws could lead to craze formation in an adhesive joint subject to service stresses and minor accidental damage. Once initiated, crazes may grow under the influence of stress to form a crack and eventually cause failure of the joint.

Detection of Voids. Voids have a potentially serious effect on the strength of an adhesive joint. Care in the preparation of the adhesive (to avoid entrapment of air and dust particles), in the pretreatment of the substrate (to ensure optimum wetting characteristics), and in the application and cure of the adhesive (to enable spreading to occur before the adhesive gels) will reduce the probability of the formation of voids. In many production processes, nondestructive tests are performed to detect bonds with serious voids. In critical structures, such as in aircraft, bonded regions are examined periodically during their service life.

A range of ultrasonic techniques, as well as x-radiography and thermography, have the potential for detecting voids in adhesive bonds. Some are suitable only for the research laboratory at present, but others have been used for many years on the factory floor. Details can be found in the Section "Nondestructive Inspection and Quality Control" in this Volume.

The size of detectable defect varies according to the method and the circumstances, but it is rarely smaller than a few square millimeters. Thus, gross defects such as might be associated with the misapplication of adhesive can be detected, but small voids, which the discussions above show to be very serious, cannot.

When a bond failure is examined in the optical or electron microscope, it is sometimes possible to identify voids or other defects at the site of initiation of the fracture. It is likely that the void seen in Fig. 16 acted in this way.

Beneficial Effects of Voids. The potentially deleterious consequences of voids in an adhesive have been emphasized in the discussion above. The actual seriousness of a void depends on the type of adhesive involved. A ductile material with crack blunting capability will be able to tolerate voids that would be catastrophic in a brittle polymer. There is even evidence that in some circumstances, voids can actually toughen a ductile adhesive by involving larger volumes of material in plastic deformation during fracture.

Adam, Evans, and Packham (Ref 31, 32) studied the effect of the deliberate introduction of spherical voids into a polyethylene applied as a hot melt to several metal substrates. The voiding was achieved by incorporating a blowing agent into the layer of polymer adjacent to the interface. Some of the results are shown in Table 4, where considerable increase in peel strength is shown when the blowing agent was employed. Electron microscopical examination of the failure surfaces showed extensive plastic deformation of the polymer around the sites of the voids.

It appears that in this example the concentration of stress around the void initiates yielding in the polymer during peeling, leading to enhanced energy dissipation by plastic deformation compared to Eq 1.

Table 4 Effect of adding a blowing agent to polyethylene on the adhesion to copper

Pretreatment of copper	Use of blowing agent	Peel strength N/mm	lbf/in.
Polishing	No	0.1	0.57
	Yes	1.8	10.3
Oxidizing (microfibrous surface)	No	1.5	8.6
	Yes	5.8	33.1

Source: Ref 31

Independent of the work of Adam *et al.*, foamed hot-melt adhesives have been developed (Ref 33). The types available include polyethylene, polypropylene, ethylene-vinyl acetate, polyester, and polyamide.

Microstructure of the Substrate

Adhesives are wide in scope, and the substrates to which they are applied are also very diverse in nature and correspondingly varied in microstructure. For instance, much of the discussion of adhesive microstructure above would be relevant to a polymer substrate, especially if it were flexible so that ψ_{sub} (Eq 1) contributed to the joint toughness. In order to keep the length of the discussion within acceptable bounds, the emphasis here is on rigid substrates that are primarily, but not exclusively, metallic in nature.

Engineering Surfaces

It is possible to conduct scientific experiments on pristine surfaces of pure compounds in ultrahigh vacuums, but in almost all cases adhesives are used with practical materials in a mundane environment. Such engineering surfaces are never smooth or clean.

Surface Roughness Measurement (Ref 34). The complex property of surface roughness may be investigated by a number of techniques. In a profilometer, a fine needle is drawn over the surface and is supposed to follow the features of the topography. From its vertical displacement, various parameters may be calculated to represent surface roughness, and a profile graph may be drawn. With microprocessor control, the needle can make a series of traverses, and a three-dimensional map of the surface can be drawn.

A commonly quoted parameter is the roughness average (also known as the center line average), which is defined as:

$$R_A = \frac{1}{l} \int_0^l |y| \, dx \qquad \text{(Eq 13)}$$

where l is the length traversed in the x direction and y is the vertical displacement. R_A gives a number that partially characterizes the roughness, but when used alone, it can be very misleading. Figure 18 shows five different surface profiles, all with the same roughness average (Ref 35).

The topography of relatively smooth surfaces can be studied by optical interference methods. An optical flat is placed on the surface, and interference patterns give information about surface roughness.

Microscopy provides a powerful means of obtaining an overall qualitative impression of the features of a surface. In order to get an adequate depth of field, it is desirable to use a stereo microscope. An optical

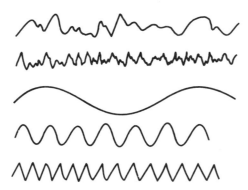

Fig. 18 Different profile graphs having the same roughness average (R_a). Source: Ref 35

microscope is simple to use and can give valuable information about coarser features, but the scanning electron microscope has much greater magnification (in favorable circumstances, a resolution of 10 nm, or 0.4 μin., may be obtained). It will often be used in this type of work at low magnification, making use of its great depth of field. The broad conclusion from these and other methods of characterization of surface roughness is that as the molecular scale is approached, engineering surfaces are always rough, often grossly so.

Surface Structure. The microscopic examination of sections through surfaces and the use of analytical techniques such as those described in the article "Surface Analysis Techniques and Applications" in this Volume enable the study of the structure and chemical nature of surfaces. The conclusion of these studies is that the surfaces of engineering materials never have the same structure or composition as the bulk.

A typical surface of a metal is shown in Fig. 19. The subsurface is worked, and the grains are fractured and partially reoriented. Nearer the surface there may be a very heavily worked layer, perhaps incorporating process oils and grease. For all metals, except for a few noble ones such as platinum, a surface oxide forms within frac-

tions of a second of exposure to air. The thickness may range from a few angstroms to several microns. The oxide will usually be partially hydrated and perhaps carbonated. On its surface will be layers of chemically and physically adsorbed water.

This simplified description is complicated in practice by further contaminants, which depend on the alloy and its history. For example, the surface of pickled mild steel strip (Ref 36) might contain a carbonaceous deposit, chlorine from the etch, calcium from the rinsing water, sulfur and nitrogen from a corrosion inhibitor, and copper (a minor component of the alloy concentrated in the surface).

It is not only metal surfaces that are different in composition from the bulk. Silicate glasses have silanol groups on the surface, over which are adsorbed multilayers of water. The difficulty of achieving solubility in a polymer was mentioned above in connection with the discussion of Eq 9. A further consequence is that additives and low molecular weight fractions tend to migrate to the surface during molding and, more slowly, during the service life of the material. Thus, it can be seen that the surfaces of engineering materials are neither smooth nor clean.

Surface Microstructure and Adhesion

The chemical nature and the physical nature of a surface are crucial in adhesion. It is often difficult to separate the effects of the two. The chemical structure, which influences the reactivity of the substrate to the adhesive, its surface energy, and, in turn, its fundamental wetting characteristics, also affects the strength and stability of surface layers such as oxides.

It is obvious that a weak and friable surface layer provides a poor substrate for adhesion. Apart from this, it is sometimes in the substrate surface layer that fracture of an adhesive bond occurs. In bonding rubber to metal for components such as automobile suspension units and engine mounts, oxide

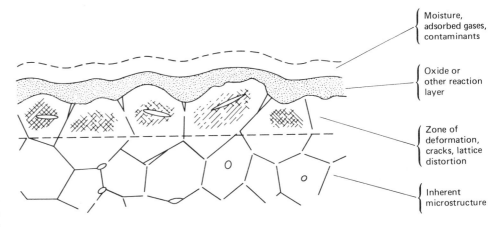

Fig. 19 Typical surface structure of a metal. Source: Ref 34

Moisture, adsorbed gases, contaminants

Oxide or other reaction layer

Zone of deformation, cracks, lattice distortion

Inherent microstructure

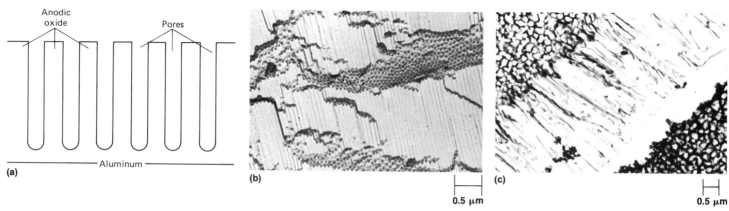

Fig. 20 Structure of porous anodic film on aluminum. (a) Idealized. (b) Sulfuric acid-formed film (section). (c) Phosphoric acid-formed film showing pore openings (top left and bottom right) and section (central band)

failure noted in quality control experiments is taken as an indication that the adhesive has been correctly applied and the vulcanization is satisfactory. Hine *et al.* (Ref 37) found that a particular oxide structure on steel was suitable for strong bonding with a polyethylene hot melt, but became the site of fracture when a toughened epoxy adhesive was used.

It is sometimes found that surface layers that are satisfactory when a bond is formed are unstable in service conditions. The articles in the Section "Environmental Effects" in this Volume provide detailed information. Thus, many metal oxides will be unsatisfactory at low pH levels. Some change the degree of hydration in a humid environment and weaken. Noland found that the oxide formed on aluminum by sulfochromating was subject to such a change; thus the environmental stability of the bonds was unsatisfactory (Ref 38).

Surface pretreatments, such as those described in the Section "Surface Considerations" of this Volume which play such a vital role in adhesive practive, have evolved, usually by trial and error, to remove weak layers and to produce surface structures with adequate strength and stability. Many failures in practice are caused by producing an unsuitable substrate microstructure because an inappropriate pretreatment is used or an appropriate one is misapplied.

Roughness. Many surface treatments that improve adhesion have the effect of roughening the surface involved. It is simplistic to always attribute the improvement to roughening *per se*. Treatments, such as abrasion and etching, also remove weak boundary layers, increase the reactivity of the surface, and improve its wetting behavior.

There are theoretical reasons to ask whether roughening improves adhesion. Calculations of the theoretical strength of adhesive bonds suggest that levels of adhesion greater than those found practically should be achieved from van der Waals interactions between a plane substrate and an adhesive that has fully wetted it (Ref 39). The discrepancy is accounted for in such terms as incomplete wetting and the presence of voids and other defects.

For any particular solid and liquid surface energy, a rough surface is more difficult to wet than a smooth one. At equilibrium, pores may not be filled, and air may be trapped. The degree of difficulty depends on the details of the roughness produced (Ref 40). Reentrant cavities, for example, are particularly difficult to wet. A further problem with rough surfaces is that wetting may be further inhibited because the adhesive sets before equilibrium wetting is attained. Once the adhesive has set, voids of trapped air and surface asperities may produce stress concentrations that reduce the strength of a brittle adhesive. Thus, roughening *per se* is not necessarily of benefit to adhesion.

Mechanical Keying. From the previous discussion, it might seem that the commonsense view that mechanical interlocking was important in adhesion was completely wrong. Despite what was said, there are well-documented examples of better adhesion or of different adhesion mechanisms to rough surfaces, some of which are described below.

Anchor Coats and Metallized Plastics (Ref 40). There are many examples in the patent literature of loose sintered ceramic or metal being an effective substrate for polytetrafluoroethylene. This polymer, which is notoriously difficult to bond, penetrates into the voids and pores of the sintered coating.

Another commercial process where it seems clear that mechanical keying plays a part is in the metal plating of plastics such as ABS and polypropylene. The first stage involves the electroless deposition of a thin metal layer. To get good adhesion of this layer, the polymers are etched with a strongly oxidizing acid. This produces a very rough surface. The ABS has a two-phase structure rather like that of the rub-

ber-toughened adhesives discussed earlier. The acid etches out the rubber particles in the surface, leaving spherical pits into which the metal penetrates. There is evidence that the chemical nature of the surface that is produced, as well as the topography, is important to good adhesion in these examples.

Anodized Surfaces. When aluminum is anodized in aggressive electrolytes, such as sulfuric, phosphoric, and chromic acids, a thick, porous oxide is produced. Anodizing has been of increasing interest since the late 1960s as a method of pretreatment (Ref 41), especially for the aerospace industry. Phosphoric acid anodizing forms the basis of at least one aircraft specification.

The classical description of the porous anodic coating is of a regular hexagonal array of cylindrical pores penetrating normal to the surface and almost to the base metal (Fig. 20a). Typical structures of films formed in sulfuric acid and phosphoric acid are shown in Fig. 20(b) and (c). The pore diameter depends on the electrolyte and the anodizing conditions. Typical values are 200 Å (0.8 μin.) and 400 Å (1.6 μin.) for films produced in sulfuric and phosphoric acids respectively. The thickness varies according to the current density and anodizing time and it can be of the order of 1 μm (0.04 mil) or more.

Adhesion to Anodized Aluminum. Using polyethylene as a model hot-melt adhesive, it has been shown that good adhesion can be obtained to anodized surfaces, and that both polyethylene and low-viscosity epoxy resin penetrate for considerable distances into the pores (Ref 32). The attraction of anodized surfaces to the aerospace industry is a result of good reproducibility, good initial adhesion, and, especially, good environmental stability (Ref 42, 43).

Anodized Titanium. Titanium alloys are also of interest to the aerospace industry. Recent work on the effect of various surface treatments on the durability of epoxy-titanium bonds has pointed to the importance of porous anodizing in sodium hydroxide or

Fig. 21 Microfibrous surfaces on (a) copper, (b) steel, and (c) zinc. Source: Ref 46

chromic acid. Bonds formed after these pretreatments showed good durability in hot, wet environments (Ref 44, 45).

Microfibrous Surfaces. The anodizing of copper in a sodium hydroxide solution produces a topography consisting of a forest of black copper (II) oxide needles. Similar microfibrous surfaces can be prepared by chemical oxidation in an alkaline chlorite solution. Microfibrous surfaces differing in the detailed structure can be produced on steel by oxidation and on zinc by electrodeposition (Fig. 21) (Ref 32, 37, 46). Polyethylene and ethylene-vinyl acetate hot-melt adhesives and epoxy thermosetting adhesives have been applied to these substrates in studies on the effect of topography on adhesion. Good adhesion has been found: The microfibrous surfaces enable a greater part of the potential toughness of the adhesive to be exploited. In terms of Eq 1, they enhance the plastic dissipation (ψ_{plast}). Thus, the fracture energy for a toughened epoxy bonded to flat zinc was 700 J/m^2 (48 ft · lbf/ft^2), but to a dendritic zinc surface, was 2550 J/m^2 (175 ft · lbf/ft^2).

Mechanism of Adhesion. A number of examples of adhesion have been discussed in which roughening of the surfaces, pro-

ducing microporous or microfibrous surfaces, gives good, and sometimes enhanced, adhesion. In bonds where the failure mode has been investigated, it is usually found that there is extensive plastic deformation of the polymer. The implication is that the surface microstructure of the substrate alters the stress field in the interfacial regions, initiating plastic deformation in the adhesive. This effect is particularly great when a ductile adhesive is involved, but useful increases in joint toughness have been found with zinc dendrites and brittle, untoughened epoxy resin (Ref 46).

Influence of Substrate on Adhesive Microstructure

The microstructure of the adhesive and the substrate have already been considered, and the effects on joint strength were emphasized. The adhesive joint has to be considered as a system, not just as the sum of the various parts. The adhesive sets in contact with the substrate, and its properties and structure are not independent of the substrate. This final section of the article describes some ways in which the substrate

can influence the microstructure of the adhesive.

Morphology

In some joints involving crystalline polymers, such as polyethylene, a nonspherulitic structure, termed a transcrystalline layer (TCL), can be observed at the substrate surface. There has been speculation about whether this is a function of the surface energy of the substrate and whether its presence leads to higher adhesion.

The influence of TCLs in adhesion have been studied using polyethylene as a paradigm (Ref 47). A microscope was used with a hot stage capable of establishing thermal gradients. The cooling of the polymer in the presence of a metal surface was observed. Sometimes the polymer nucleated in the bulk, and spherulites grew to the metal surface without the formation of a TCL. More commonly, surface nucleation, leading to a TCL, did occur. The TCL grew until it impinged on the bulk-nucleated spherulitic polymer (Fig. 22). A TCL could form even when the metal temperature was higher than that of the melt, indicating that the surface acted as a source of heterogeneous nuclei. The several different surface

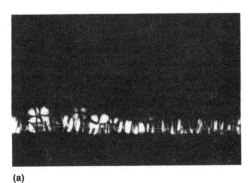

(a)

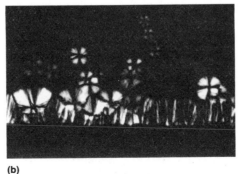

(b)

(c)

Fig. 22 Successive stages in the solidification of polyethylene from the melt, showing the formation of a transcrystalline layer at a copper surface (bottom of micrographs) and bulk-nucleated spherulites. Source: Ref 47

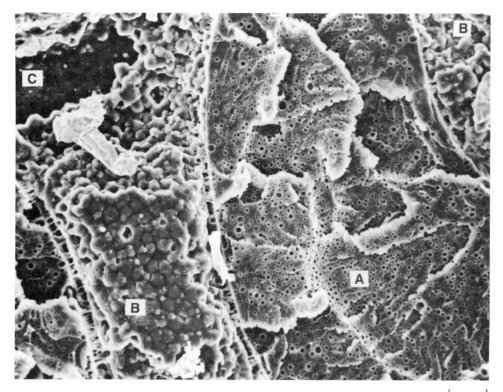

Fig. 23 Fracture surface of a bond between copper and rubber-toughened epoxy resin showing gross phase separation in the resin. A, soft translucent phase; B, hard white phase; C, copper substrate. Source: Ref 51

treatments of the metal investigated did not appear to influence the ease of formation of the TCL.

It seems clear that the origin of the layer is similar to that of "columnar" regions recognized in metallurgical samples. These form at the surfaces of ingots as a result of two factors. The crucible surface provides a site for heterogeneous nucleation. Because heat is lost through the crucible walls, this is the region of greatest undercooling.

The TCL provides a clear example of the influence of the substrate on adhesive morphology. It might be thought that the boundary between the TCL and spherulitic structure would be weak because of a concentration of low molecular weight fractions. Fracture studies suggest that the failure path passes through the TCL, not along this boundary, and an influence of the TCL on measured adhesion has not been demonstrated.

Cross-Link Density

In principle, it is likely that a substrate surface will have an influence on the cross-link density of the adhesive in its vicinity. In many joints, when the adhesive is setting, the substrate is likely to be at a different temperature from that of the bulk adhesive. This will influence the rates of the various reactions occurring in the adhesive. Stevens (Ref 22) has commented on the different cure kinetics observed in thin curing samples adjacent to interfaces. Because bulk cross-link density measurement on thermoset adhesives is often difficult, it is even more difficult to detect changes in cross-link density in a small interfacial region.

A well-documented example of the influence of substrate on the adjacent microstructure is provided by studies of the bonding of rubber to brass or copper. It has been shown that the sulfur of the curing system reacts with copper in the alloy, leading to the formation of a nonstoichiometric copper sulfide that links in with the sulfur cross-link network of the rubber (Ref 48, 49).

In other work, the presence of different alloys has been shown by thermal analysis to alter the cure kinetics of nitrile-butadiene rubber, with the implication that the cross-link structure at the interface will be different from that in the bulk, and moreover will depend upon the alloy involved (Ref 50). These differences appear to be reflected in the different levels of adhesion of the rubber molded against the different alloys.

It is well known (Ref 32) that metals such as steel and copper exert a catalytic effect on the oxidation of polyolefins and that this profoundly influences their adhesion. The oxidation of these polymers follows a complex series of free radical reactions, which leads to some cross linking as well as to the incorporation of oxygen-containing groups. Here then is a further example of substrate influence on the cross linking adjacent to it.

Phase Structure Influence

In this article, emphasis has been placed on the importance of obtaining the appropriate phase structure in rubber-toughened adhesives. To achieve this, a delicate balance of chemical and physical changes has to occur at the correct rates. The adhesion of rubber-toughened epoxy to copper provides an interesting example of the upset of this balance because of the substrate, with consequent poor adhesion (Ref 51). Gross phase separation occurred, producing two intermingled phases in domains approximately 100 μm (4 mils) wide and a millimeter or more in length (Fig. 23).

Copper is a transition metal with potential reactivity towards resin, hardener, and rubber in this adhesive. It is not surprising that some reaction took place, resulting in a change in solubility or reaction rate, which led to a different phase structure. It would be strange if copper proved to be the only substrate to affect the phase structure of such adhesives in the critical interfacial region.

REFERENCES

1. A.N. Gent and G.R. Hamed, *Plast. Rubber Mater. Appl.*, Vol 3 (No. 1), 1978, p 17
2. G.W. Gilby, in *Developments in Rubber Technology*, Vol 3, A. Whelan and K.S. Lee, Ed., Applied Science Publishers, 1982
3. M.E. Fine, *Phase Transformations in Condensed Systems*, Macmillan, 1964
4. D.C. Bassett, *Principles of Polymer Morphology*, Cambridge University Press, 1981
5. A.S. Vaughan and D.C. Bassett, in *Comprehensive Polymer Science*, Vol 2, G. Allen and J.C. Bevington, Ed., Pergamon Press, 1989
6. I.M. Ward, in *Advances in Oriented Polymers*, Vol I, I.M. Ward, Ed., Applied Science Publishers, 1982
7. P.J. Barham and A. Keller, *J. Mater. Sci.*, Vol 20, 1985, p 2281
8. S. Nomura, in *Comprehensive Polymer Science*, Vol 2, G. Allen and J.C. Bevington, Ed., Pergamon Press, 1989
9. J.P. Runt, in *Encyclopedia of Polymer Science and Engineering*, Vol 4, H.F. Mark *et al.*, Ed., John Wiley & Sons, 1986, p 482
10. A.E. Woodward, *Atlas of Polymer Morphology*, Hanser, 1989
11. A.N. Gent and R.B. Petrick, *Proc. R. Soc. (London) A*, Vol 310, 1969, p 433
12. M. Takayanagi, *Mem. Fac. Eng. Kyushu Univ.*, Vol 20, 1963, p 43
13. S. Minami, S. Uemura, and M. Takay-

anagi, *Rep. Prog. Polym. Phys. Japan*, Vol 7, 1964, p 37

14. T.A. Hatzinikolaou and D.E. Packham, *Adhesion*, Vol 9, 1984, p 31
15. T.A. Hatzinikolaou, Ph.D. thesis, University of Bath, 1985
16. J.R.G. Evans and D.E. Packham, *Int. J. Adhes. Adhes.*, Vol 1, 1981, p 149
17. P.J. Flory, *Principles of Polymer Chemistry*, Cornell University Press, 1953
18. N.J. Morrison and M. Porter, in *Principles of Polymer Morphology*, Cambridge University Press, 1981
19. B.G. Crowther, P.M. Lewis, and C. Metherell, in *Natural Rubber Science and Technology*, A.D. Roberts, Ed., Oxford University Press, 1988
20. A.J. Kinloch and R.J. Young, *Fracture Behaviour of Polymers*, Applied Science Publishers, 1983
21. D.B. Wootton, *Adhesion*, Vol 9, 1984, p 1
22. G.C. Stevens, in *Structural Adhesives*, A.J. Kinloch, Ed., Applied Science Publishers, 1986
23. L.R.G. Treloar, *Physics of Rubber Elasticity*, 3rd ed., Clarendon Press, 1975
24. J. Bandrup and E.H. Immergut, *Polymer Handbook*, 3rd ed., John Wiley & Sons, 1989
25. C.G. Moore and B.R. Trego, *J. Appl. Polym. Sci.*, Vol 8, 1964, p 1957

26. C.B. Bucknall, *Toughened Plastics*, Applied Science Publishers, 1977
27. A.J. Kinloch, in *Structural Adhesives*, A.J. Kinloch, Ed., Applied Science Publishers, 1986, p 127
28. J.K. Gillham, in *Structural Adhesives*, A.J. Kinloch, Ed., Applied Science Publishers, 1986, p 1
29. F.R. Martin, in *Structural Adhesives*, A.J. Kinloch, Ed., Applied Science Publishers, 1986, p 29
30. R.A. Orwoll, in *Encyclopedia of Polymer Science and Engineering*, Vol 15, H.F. Mark *et al.*, Ed., John Wiley & Sons, 1986, p 380
31. T. Adam, J.R.G. Evans, and D.E. Packham, *J. Adhes.*, Vol 10, 1980, p 279
32. D.E. Packham, in *Developments in Adhesions*, Vol 2, A.J. Kinloch, Ed., Applied Science Publishers, 1981
33. W.H. Cobbs, *Adhes. Age*, Vol 22, 1979, p 31
34. D.F. Moore, *Principles and Applications of Tribology*, Pergamon Press, 1975
35. B. Pugh, *Friction and Wear*, Newnes, 1973
36. V. Leroy, *Mater. Sci. Eng.*, Vol 42, 1980, p 289
37. P.J. Hine, S. El Muddarris, and D.E. Packham, *J. Adhes.*, Vol 17, 1984, p 207
38. J.S. Noland, in *Adhesion Science and Technology*, L.H. Lee, Ed., Plenum Publishing, 1975

39. J.R. Huntsberger, in *Treatise on Adhesion and Adhesives*, Vol 1, R.L. Patrick, Ed., Marcel Dekker, 1967
40. D.E. Packham, in *Adhesion Aspects of Polymeric Coatings*, K.L. Mittal, Ed., Plenum Publishing, 1983
41. K. Bright, B.W. Malpass, and D.E. Packham, *Nature*, Vol 223, 1969, p 1360
42. A.J. Kinloch, *Adhesion and Adhesives*, Chapman and Hall, 1987
43. A.C. Moloney, in *Surface Analysis and Pretreatment of Plastics and Metals*, D.M. Brewis, Ed., Applied Science Publishers, 1982
44. J.A. Filbey and J.P. Wightman, *Adhesion*, Vol 12, 1988, p 17
45. J.A. Filbey and J.P. Wightman, *J. Adhes.*, Vol 28, 1989, p 1
46. D.E. Packham, *Int. J. Adhes. Adhes.*, Vol 6, 1986, p 225
47. P.T. Reynolds and D.E. Packham, Paper presented at the 14th European Physics Conference, Vilafranca del Penedes, Spain, 1982
48. W.J. van Ooij, *Rubber Chem. Technol.*, Vol 51 (No. 1), 1978, p 52
49. G. Haemers, *Adhesion*, Vol 4, 1980, p 175
50. D.E. Packham, internal report, University of Bath, 1989
51. P.J. Hine, S. El Muddarris, and D.E. Packham, *J. Adhes. Sci. Technol.*, Vol 1, 1987, p 69

Thermal Properties and Temperature Effects

David Roylance, Massachusetts Institute of Technology

TEMPERATURE is a dominant variable affecting the properties of polymeric adhesives. It acts in several distinct ways, one of which is that increases in temperature increase the internal energy of the polymer, causing the individual atoms of the material to oscillate around positions of greater interatomic separation. This leads to an expansion of the specimen at large, which in turn generates a number of effects of engineering significance.

Increases in temperature also act to increase the importance of entropic, as opposed to energetic, effects, as is evident in the familiar expression for Helmholtz free energy:

$$dG = dH - T\,dS$$

where G is free energy, H is internal energy, T is temperature, and S is entropy. The material is thermodynamically driven to seek the state of lowest free energy. As the temperature is increased, the material will tend to increase its entropy (disorder) rather than reduce its internal energy.

It should be noted that a moderate increase in temperature can produce a dramatic increase in the rates of many physical and chemical processes. Because many important polymer properties are time dependent, the dependence of rate on temperature is of strong engineering significance.

This article explains these thermal effects in some detail, attempting to show how temperature-dependent molecular processes generate important consequences for the macroscopic engineering properties of the material. Polymer-specific effects are described at length.

Thermal Expansion Effects

Temperature has the same volume-expanding effect on most polymers that it does on metallic or ceramic materials. Increases in temperature increase the interatomic separation distances in essentially all materials, due to the well-known anharmonicity of the bond energy function. This, in turn, produces thermal expansion of the solid at large, an expansion that is usually proportional to the temperature increase:

$$\frac{\Delta V}{V} = \alpha_v \Delta T$$

where V is the volume and α_v is the coefficient of volumetric thermal expansion.

This thermal expansion produces a number of considerations in materials engineering, for adhesives as for any material. The expansion is often large enough to be of geometrical significance, and parts machined to close tolerances must consider the effect of thermal expansion. In addition to this obvious effect, structures experiencing nonuniform temperature distributions may fracture by thermal shock. If hotter material tries to expand but is constrained by cooler material, the cooler material is pulled into tension. If the stress is sufficient to cause fracture, the material fails. A common example of this is plunging a hot drinking glass into cold water. Surface flaws that are always present on such a glass are placed in tension, and unstable crack growth ensues. Temperature variations can also be important in stress analysis. All finite-element stress analysis codes include a material constitutive law relating the stresses to the strains. When temperature variations exist, the analyst must choose codes that adjust the strains for thermal expansion, or the code results will be erroneous.

A number of experimental means are available for determining the expansion coefficient in polymers, perhaps the most popular being thermomechanical analysis (TMA). TMA is one of a number of methods included in the general term thermal analysis, all of which measure various materials responses to controlled temperatures or time-dependent temperature programs. The TMA module of a thermal analysis system is a small indentor-type probe that is pressed with a small force against the surface of a specimen. When the material undergoes a thermal transition affecting its hardness, notably the polymer glass-rubber transition, the indentation of the probe changes. The TMA module records this change, and its position as a function of temperature is a convenient measure of transition temperature. However, because the probe position is also affected by thermal expansion, even when the material does not undergo a transition, it can be used directly to measure the thermal expansion coefficient.

In addition to the familiar effects listed above, polymeric adhesives are dramatically influenced by temperature in a number of ways not usually appreciable in metals and ceramics, and the remainder of this article deals primarily with these polymer-specific effects.

One of these polymer-specific effects relates to thermal expansion, and it is appropriate to mention it here. Normal thermal expansion is an internal energy effect: The increase in interatomic spacing is predicted directly by the shape of the bond energy function. However, when a rubber band is stretched under the influence of a fixed weight and the temperature is increased, the rubber band is observed to shrink, rather than elongate. This is sometimes claimed, erroneously, to indicate a negative coefficient of thermal expansion. Unstretched rubber expands when heated because of the increase in internal energy, as is the case in essentially all materials. The contraction of a stretched rubber band is a dramatic illustration of an entropic, as opposed to energetic, effect. When the temperature is increased, the importance of the $T\,dS$ term in the Helmholtz free energy relation is also increased. The Brownian motion of the molecular structure is increased, and the material tries with increased vigor to recover its random (maximum entropy) unstretched conformation.

Time-Temperature Relations in Polymers

The response of a polymeric material to a thermal or mechanical stimulus is often markedly time-dependent. Upon the application of a small constant stress, for instance, a polymer will exhibit an instantaneous elastic deformation, just as a metallic or ceramic material would. However, if the molecular mobility is sufficient, as is often the case even at room temperature, a polymer will thereafter continue to deform in the direction of the applied load. This process, called creep, is widely observed in polymers and arises as the polymer experiences molecular conformational change due to rotation about main-chain covalent bonds. This conformational change can be modeled as a thermally activated rate process, and the mathematics of rate processes are very useful in rationalizing thermal effects in polymeric adhesives.

One important function of materials testing is that of determining or assuring the service lifetime of a product using that material. Of course, it is often not possible to wait in real time to measure durability directly, and some means of estimating durability from analytical predictions or short-term testing is needed. Many physical and chemical processes proceed at rates that vary strongly with the temperature, and this provides a means of accelerated testing, which has become a cornerstone of polymer engineering. If relaxation occurs too slowly at room temperature for convenient measurement, the experimenter can simply raise the temperature to speed things up. There is often a quantitative relationship between time and temperature, and some approaches to modeling this relationship will be outlined below.

Concept of Relaxation Time. As described earlier, a polymer subjected to a mechanical or thermal excitation will begin a rate process in which it moves gradually to a new equilibrium conformation. There are a number of useful mathematical strategems for modeling the kinetics of this process (for example, spring-dashpot mechanical analogies, integral formulations of linear superposition), but an interesting alternative approach is simply to adopt the reasonable assumption that the rate at which the polymer approaches equilibrium is proportional to the interval between it and equilibrium at any instant (Ref 1). (This is equivalent to a first-order rate process.) Letting the deviation from equilibrium be denoted as α, we can write:

$$\frac{d\alpha}{dt} = -k\alpha \qquad \text{(Eq 1)}$$

where the rate constant (k) is the constant of proportionality. For instance, in a creep process, α would be the final equilibrium value of strain minus the instantaneous value of strain: $\alpha(t) = \epsilon_r - \epsilon(t)$. Solving Eq 1:

$$\alpha = \alpha_0 \exp(-kt)$$

where α_0 is the initial value of α, for example the final equilibrium strain minus the initial strain in a creep process. At that particular time when $t = 1/k$, α will have been reduced to 1/e of its original value, α_0.

This value of time is a characteristic of the process and is often defined as the relaxation time:

$$\tau = \frac{1}{k}$$

The relaxation time varies strongly with temperature and is perhaps the most important of the various ways by which temperature influences polymer response. The end points of the time-dependent process are also temperature dependent. For instance, as the temperature is increased, the equilibrium creep strain tends to increase also. However, this is a relatively minor effect, which is more or less linear in temperature, in contrast to the very strong and usually exponential dependence of *rate* on temperature. The sections to follow describe some of the means by which the influence of temperature on relaxation time can be modeled.

Most polymer processes involve a wide range of molecular responses and possess kinetics that are too complex to be modeled by a single relaxation time. When more accurate models are desired, it is often possible to extend the concept to a distribution of relaxation times. If $f(\tau)$ is the fraction of molecular responses having a particular relaxation time, τ, then:

$$\alpha = \alpha_0 \int_0^\infty f(\tau)\exp(-t/\tau)d\tau$$

More detailed accounts of the continuous-distribution approach to polymer response and descriptions of how these distributions can be extracted from experimental data are available in the classic texts by McCrum (Ref 1) and Ferry (Ref 2).

Eyring Formulation. In the Eyring (or Arrhenius) view of rate processes, τ varies exponentially with temperature:

$$\tau = \tau_0 \exp\frac{U}{RT} \qquad \text{(Eq 2)}$$

where τ_0 is a constant, U is an apparent activation energy for the process, and $R = 8.31$ J/mol · K (2 Btu/lb · mol · °R) is the gas constant. This relation is often inconvenient for computation because in this exponential form the numbers may exceed computer capacity as the temperature is varied over a wide range. Taking logarithms:

$$\log \tau = \log \tau_0 + \frac{U}{2.303RT} \qquad \text{(Eq 3)}$$

If the relaxation time at a convenient reference temperature, T_{ref}, is known, and it is desirable to compute it for another temperature, T, then Eq 3 is evaluated at the reference temperature:

$$\log \tau_{ref} = \log \tau_0 + \frac{U}{2.303RT_{ref}} \qquad \text{(Eq 4)}$$

Subtracting Eq 4 from Eq 3:

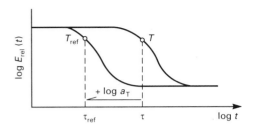

Fig. 1 Definition of the shift factor

$$\log \tau - \log \tau_{ref} \equiv \log a_T = \frac{U}{2.303R}\left(\frac{1}{T} - \frac{1}{T_{ref}}\right) \qquad \text{(Eq 5)}$$

where $\log a_T$ is a "time-temperature shift factor," which gives the shift in logarithmic time of a relaxation in terms of a reference value at T_{ref}. If relaxation curves are measured at two different temperatures, as shown, then $\log a_T$ is the horizontal displacement along the log t axis, as shown in Fig. 1.

We also have:

$$\log a_T = \log\left(\frac{\tau}{\tau_{ref}}\right)$$

$$a_T = \left(\frac{\tau}{\tau_{ref}}\right)$$

so a_T can be found by ratio rather than horizontal shifting.

A series of viscoelastic data, for example, relaxation data, taken over a range of temperatures can be converted to a single master curve by means of this horizontal shifting (Fig. 2).

A particular curve is chosen as a reference, and then the other curves are shifted horizontally to obtain a single curve spanning a wide range of log time. Curves representing data obtained at temperatures lower than the reference temperature appear at longer times, to the right of the reference curve, they so will have to shift left. This is a *positive* shift, as defined by the shift factor in Eq 5. Each curve produces its own value of a_T, so that a_T becomes a tabulated function of temperature. The master curve is valid only at the reference temperature, but it can be used at other temperatures by shifting it by the appropriate value of $\log a_T$.

If the kinetic treatment used to develop Eq 5 is in fact applicable, then a plot of log a_T versus $1/T$ should produce a straight line whose slope is $U/2.303R$, as shown in Fig. 3.

The reader will have noted that several restrictive assumptions have been made in the foregoing discussion, principally that the physical process can be captured with a single-characteristic time model, and that the characteristic time varies in a simple exponential way with the temperature. Because neither assumption is true for all

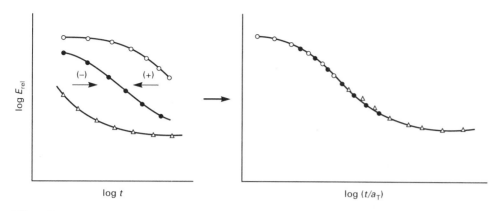

Fig. 2 Construction of the master curve

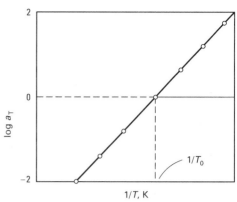

Fig. 3 Arrhenius plot for the computation of activation energy. $U/2.303R$ where $R = 8.314$ J/mol · K (2 Btu/lb · mol · °R)

materials or all time-temperature regimes, caution is necessary. An alternative approach, often preferable when the temperature is above the glass transition temperature (T_g), is presented below.

Williams-Landel-Ferry Equation. While the above kinetic treatment is usually applicable to secondary polymer transitions, the glass-rubber primary transition appears to be governed by other principles. For this transition, the Williams-Landel-Ferry (WLF) equation is often preferred:

$$\log a_T = \frac{-C_1(T - T_{ref})}{C_2 + (T - T_{ref})} \quad \text{(Eq 6)}$$

where C_1 and C_2 are arbitrary material constants whose values depend on the choice of reference temperature T_{ref}. It has been noted that if T_{ref} is chosen to be T_g, then C_1 and C_2 assume "universal" values applicable to a wide range of polymers:

$$\log a_T = \frac{-17.44(T - T_g)}{51.6 + (T - T_g)} \quad \text{(Eq 7)}$$

which is valid in the range near T_g. It should be noted that some caution must be exercised when using WLF equations in automatic computation because there is a singularity at $T = T_g - 51.6$, and values of log a_T become meaningless.

The original WLF paper took pains to state this approach as an empirical one, but it is often viewed in terms of free-volume concepts. To sketch this concept, we take the fractional free volume (f) available for molecular motions to be governed by simple thermal expansion:

$$f = f_g + \alpha(T - T_g)$$

where f_g is the volume available at the T_g, and α is the coefficient of volumetric thermal expansion above the T_g. The relaxation time (τ) is now taken to vary exponentially with the free volume:

$$\tau = \tau_0 \exp\left(\frac{\beta}{f}\right)$$

We now proceed as with the Eyring approach, obtaining the shift factor by comparing the relaxation time at an arbitrary temperature with that at a reference temperature. Taking $T_{ref} = T_g$:

$$\ln a_T = \ln \tau - \ln \tau_g$$
$$= \beta\left[\frac{1}{f_g + \alpha(T - T_g)} - \frac{1}{f_g}\right]$$
$$= \frac{-(\beta/f_g)(T - T_g)}{(f_g/\alpha) + (T - T_g)}$$

This is clearly of the same form as Eq 7, where $(f_g/\alpha) = 51.6$. Because $\alpha \approx 5 \times 10^{-4}$/K for most amorphous polymers, this implies that $f_g \approx 0.025$ for all amorphous polymers as well. In this simple view, the glass temperature is simply that temperature above which the free volume is sufficient for large-scale cooperative chain motion, with that critical free volume having a universal value of 2.5%.

Physical Aging

The free-volume approach to the WLF equation, described above, may be difficult to justify rigorously. However, it is useful as an aid to intuition, and another application of this approach is an important phenomenon termed physical aging. The topic has been studied in depth by Struik (Ref 3).

Briefly stated, the Struik concept begins by noting that as a polymer is cooled at a finite rate through its glass transition, kinetic hindrances will keep the polymer from attaining its equilibrium density; an "excess" free volume will be frozen into the polymer. This excess volume is unstable, and as the material ages, it will move gradually toward its equilibrium density at a rate determined by the molecular mobility. It is this gradual densification, and the property changes caused by it, that make up the phenomenon of physical aging.

The excess free volume is not necessarily deleterious. It provides a capacity for molecular motion, which may impart increased toughness to the material. As the material densifies, it may therefore become more brittle, and this aging-induced embrittlement is perhaps the most serious conse-

quence of the physical aging phenomenon. Embrittlement can also be induced in such systems by thermal annealing, which is counterintuitive to materials engineers accustomed to using annealing treatments to relieve internal residual stress or high levels of work hardening. If the anneal is near, but not above, the T_g, all that may be accomplished is the rapid elimination of ductility-enhancing free volume.

The excess volume also enables more rapid molecular motions than are possible at the equilibrium density. This sets up a dog-chasing-its-own-tail process: Excess free volume is unstable and is lost at a rate related to molecular mobility, however, because molecular mobility decreases as free volume is lost, the rate of physical aging decreases as the process continues.

Finally, free volume can be induced by a number of factors in addition to the thermal effects described above. The hydrostatic component of mechanical stress influences volume, as does the infusion of a compatible fluid agent. Organic liquids or vapors can serve such a function, as can water in the case of polymers containing polar groups that are compatible with water molecules. Such fluids can promote toughness by enhancing molecular mobility. If the fluid is lost with time by outdiffusion, the polymer will appear to behave much as it would during thermal physical aging. This also indicates the possibility of synergism between multiple environmental factors: When an adhesive is exposed simultaneously to, for example, both humidity and elevated temperature, the effect may be greater than the sum of the effect induced by either acting alone, because the increased free volume induced by one exacerbates the influence of the other.

Processing Thermoset Adhesives

Many materials processing operations involve a rendering of the material into a fluid state, so that it can flow to a desired shape or establish intimate physical contact with adherends. The fluidization step may involve heating, as in the casting of metals and injection molding of thermoplastic polymers. In contrast, thermosets are typically processed at molecular weights low enough to achieve reasonable fluidity without substantial heating. Once the desired shaping or encapsulation has been achieved, the resin is rendered solid by a chemical cross-linking (or occasionally chain extension) reaction.

As in all processing operations, the materials engineer attempts to select materials and processing parameters that will develop a microstructure having the optimal properties for the application at hand. In adhesives technology, this may involve increasing the temperature enough to lower the viscosity so that suitable flow may take place, while maintaining the hydrostatic pressure in the resin above the vapor pressure of water or other dissolved diluents. Heating is then continued in order to increase the cross-link density in the resin. The desired final cross-link density is a balance between having a high glass temperature and sufficient toughness.

Thermoset processing involves a large number of simultaneous and interacting phenomena, notably transient and coupled heat and mass transfer. This makes an empirical approach to process optimization difficult. The following sections will attempt to outline the nature of these cure processes and show how relatively simple models can be used to quantify them.

Cross-Linking Reactions in Thermoset Processing

Physical Consequences of Cross Linking. Upon the initial heating of an uncured thermoset, the viscosity falls, just as with most fluids. However, chemical reaction eventually overshadows this thermal thinning, and the viscosity rises as the mobility of the resin is increasingly restricted by intermolecular cross linking. When enough cross linking has occurred to cause incipient gelation (some molecules having become large enough to span the entire specimen), the viscosity becomes indefinitely large. Technically, the resin has now passed from a liquid to a solid state and will no longer permit unrestricted deformations. Many processing operations, such as molding or void suppression, must be completed before gelation occurs.

The resin T_g also rises as the curing reaction proceeds, because cross linking serves to reduce molecular mobility for a given temperature. If the T_g rises enough to equal the cure temperature, the resin by definition will have vitrified. The reaction then becomes largely diffusion limited, with a substantial reduction in its rate.

The work of Gillham (Ref 4, 5) provides a convenient means of visualizing the transformations that take place during the cure of a thermosetting resin. Gillham used torsion-

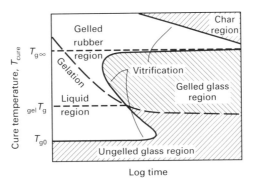

Fig. 4 Time-temperature-transformation diagram for isothermal cure. Source: Ref 4

al braid analysis and other methods to measure the times to gelation and vitrification at various isothermal cure temperatures. Figure 4, termed a time-temperature-transformation (TTT) diagram by Gillham, illustrates the major features of such data.

The TTT diagram shows that the time to gelation, which is a point of fixed conversion calculable from statistical theory, increases as the cure temperature is lowered. This is expected for a thermally activated rate process. The S-shape vitrification curve reflects the competition between the usual tendency of rising temperature to soften the material and the role of increasing reaction rate to accelerate the curing reaction and harden the material by cross linking. The temperature at which even the completely uncured resin vitrifies is T_{g0}, and $T_{g\infty}$ is the temperature of the fully cured resin. The value of $T_{g\infty}$ is of considerable practical importance, as it is the softening temperature of the fully cured adhesive, and thus limits its useful temperature range for most applications. The vitrification curve obviously must remain between these limiting temperatures.

The TTT diagram is very useful for visualizing what would otherwise be a confusing interplay of phenomena, notably the dual role played by temperature. The diagram is also valuable as a processing aid. The temperature at which the gelation and vitrification lines cross ($_{gel}T_g$) indicates the upper temperature at which the resin can be stored. Once the resin has gelled, that is, has developed a three-dimensional network that prevents unbounded deformation, the material cannot be further processed. At temperatures less than the $_{gel}T_g$, vitrification occurs before gelation. Because vitrification quenches the cross-linking reaction, gelation is avoided and the material remains processible.

Thermal degradation can also take place during cure, due to a variety of chemical mechanisms that include oxidation or chain scission. These mechanisms often produce chromophoric chemical species that result in the polymer exhibiting discoloration in addition to a degradation of physical and mechan-

ical properties. This degradation takes place at a rate that decreases exponentially with temperature, just as the cross-linking reaction does. Accordingly, the time to excessive degradation can be plotted on the TTT diagram along with the gelation and vitrification curves, and this line is denoted "char" in Fig. 4. The curing program should obviously be set to avoid the time-temperature region above the char line.

Example: Chemical Reaction Mechanism. Thermosetting resins can achieve chemical cross linking by a variety of mechanisms, the only requirement being that functionality exist that can produce covalent bonding between either the resin entities themselves or between the resin and a cross-linking agent (a hardener). A functionality of at least two in one constituent and three in the other will be needed to form three-dimensional networks. Labana has provided a review of the various types of resins and hardeners in current use (Ref 6). Rather than repeat such a list here, it may be sufficient to present a single example that illustrates two of the cross-linking mechanisms found in many systems.

A bismaleimide resin based on short, linear, polymerizable monomers containing the imide group (Ref 7) was developed to overcome the problems associated with off-gas generation in conventional polyimides. At elevated temperatures, the preimidized segments polymerize at the end groups without producing any volatiles (Ref 8). The resin can be cross linked by mixing 2.5 moles of the 4,4'-bis(maleimide diphenyl methane) with 1 mole of 4,4'-diamino diphenyl methane. The reaction may proceed by homopolymerization at the maleimide double bonds, or by the addition of the diamine at the maleimide double bonds, as shown in Fig. 5. During cure, both reactions can occur simultaneously. The amine addition yields a chain extension rather than cross linking. Control of the relative rates of chain extension and cross linking, which can be achieved by adjusting the cure cycle, permits a measure of control over the toughness and other properties of the cured adhesive.

Kinetics of Cure

A Simple Rate Model. In stoichiometric mixtures of resin and hardener, it is often possible to model the kinetics of thermoset cure by relatively simple n^{th} order rate expressions of the form:

$$\frac{d\alpha}{dt} = Z \exp\left(\frac{E}{RT}\right)(1 - \alpha)^n \qquad \text{(Eq 8)}$$

where now α is the extent of reaction, Z is a frequency factor, E is the activation energy, R is the gas constant, T is the absolute temperature, and n is the reaction order. The material-dependent parameters in this expression (Z, E, and n) must be determined by

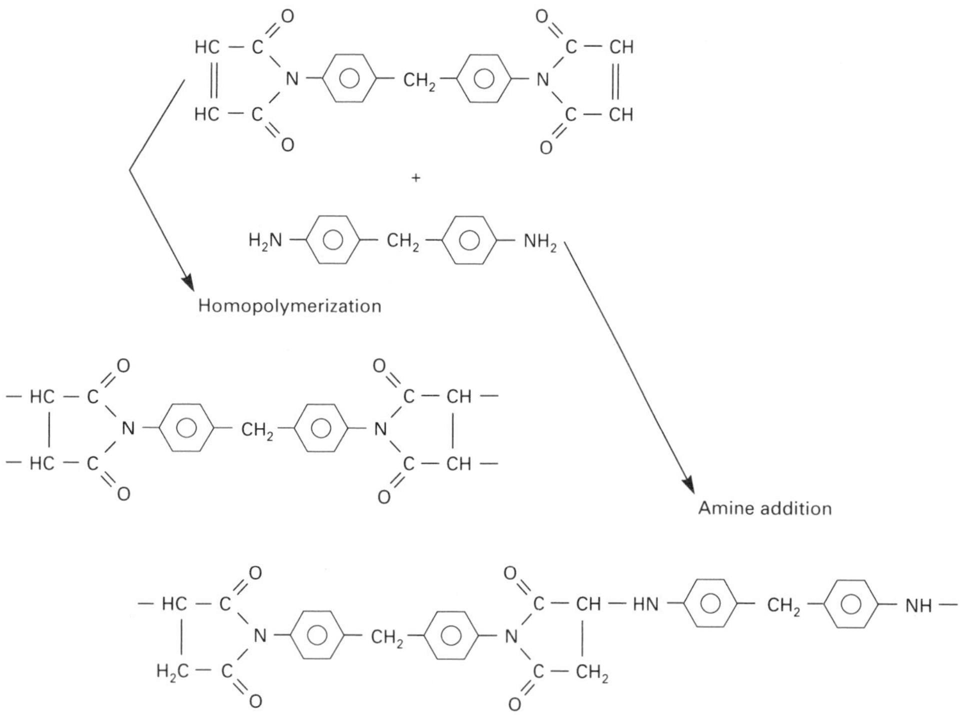

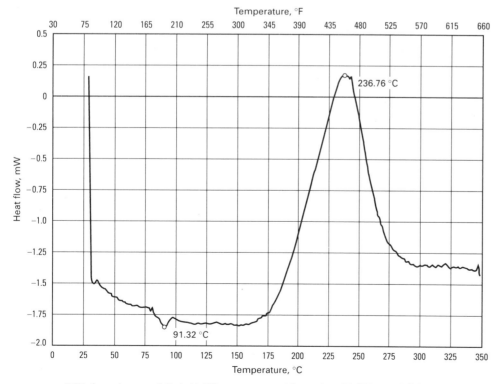

Fig. 5 Cure chemistry of bismaleimide resin Kerimid 601. Source: Ref 9

experimentation, with the ability to monitor the reaction as a function of time and temperature. With these $\alpha(t, T)$ data in hand, numerical methods must be used to fit Eq 8 to the data by a suitable choice of parameters. An example of this procedure is given below.

Kinetic Thermal Analysis. Differential scanning calorimetry (DSC) can be used to measure the rate of heat absorption or evolution by a specimen as the temperature is raised at a constant rate. In these scanning experiments, the data are recorded on a

thermal curve as a trace of the heat flow (or rate of heat evolution (dH/dt) given as a function of the temperature (T). Positive deviation from the baseline in the thermal curve indicates either an increase in the specific heat of the specimen or the occurrence of an exothermic reaction in the specimen. The thermoset curing reaction is usually exothermic and can be observed easily using DSC.

The use of DSC in the preliminary kinetic analysis of thermoset cure can be illustrated using a recent study of the bismaleimide resin discussed earlier (Ref 10). The material used in this research was supplied in the form of B-stage prepreg, with an unspecified quantity of glass fiber in the reactive polyimide resin. All of the specimens were taken from the same sheet of prepreg to avoid the possibility of inconsistency between lots. The prepreg was stored in a sealed bag at -18 °C (0 °F) until the specimens were prepared (to prevent exposure of the material to heat and moisture), and specimens were not prepared until the time of each individual run. Using the earlier work by Pappalardo (Ref 11) as a guideline, the DSC was set to heat the specimen from room temperature to 350 °C (660 °F) to ensure that the entire reaction exotherm would be recorded. A typical thermal curve is shown in Fig. 6.

Thermal curves of the sort shown in Fig. 6 are valuable for providing an intuitive guide to the range of temperatures and energies expected in curing reactions. In addition, the thermal curves can provide an experimental means for determining the numerical parameters in hypothesized kinetic models. This model-fitting procedure can yield various analytical means for studying and optimizing the fabrication of complex items containing the reactive polymer. Several procedures have been proposed during the past several decades to perform model fitting using thermal analysis data (Ref 12-16). In 1986, Dhar provided a useful review that includes a numerical program for comparing the results of several methods (Ref 17).

To illustrate the general nature of the process, the nth order kinetic model of Eq 8 is assumed, and the method of Freeman and Carroll (Ref 18, 19) is employed to fit the model of the thermal curve. This method is one of the most convenient available, in that only a single scanning DSC experiment is needed to obtain all of the kinetic parameters. However, one might expect this convenience to be accomplished by a loss of accuracy, and a number of workers feel this to be true (Ref 20). Further, the assumption of thermally activated n^{th} order kinetics might be incorrect as well. The bismaleimide resin cure is thought to consist of simultaneous addition and cross-linking reactions, as shown in Fig. 5, and the combination of these reactions might not be

Fig. 6 DSC thermal curve of Kerimid 601 prepreg material tested at 10 °C/min (18 °F/min) in nitrogen atmosphere (14.1 mg, or 0.0005 oz., specimen mass, including glass fibers)

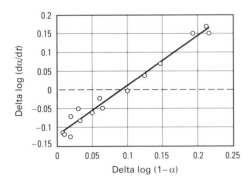

Fig. 7 Freeman-Carroll plot corresponding to thermal curve of Fig. 6

Table 1 Numerical parameters obtained from three plots of separate DSC experiments

Test	1	2	3
Reaction order	1.35	1.40	1.31
Activation energy, kJ/mol (kcal/mol)	116.4 (27.8)	115.9 (27.7)	112.6 (26.9)
Log, Z, 1/s	11.2	10.6	10.2
Correlation coefficient	0.97	0.94	0.95

describable with a single n^{th} order model. Finally, Sichina (Ref 21) notes that curing reactions that exhibit autocatalyzing effects may not obey the simple kinetic scheme used here.

Although the above concerns should be kept in mind, the results shown below are useful for illustrating a general approach. Of course, a more exhaustive study would be required to assess fully the validity of the underlying assumptions. The extent of reaction α and the rate of reaction $d\alpha/dt$ can be measured directly from the DSC thermogram, assuming that each reaction event liberates the same quantity of heat:

$$\frac{d\alpha}{dt} = \frac{1}{H_0} \cdot \frac{dH}{dt}$$

$$\alpha = \int \frac{d\alpha}{dt}\, dt$$

Here dH/dt, the rate of heat evolution, is just the ordinate of the DSC thermogram, and H_0 is the area between the DSC baseline and the ordinate over the range of temperatures comprising the reaction. H_0 is the total heat evolution of the cure reaction; in this case this quantity is not that of the resin alone, because the specimen also contains an unmeasured quantity of glass fibers.

The Freeman-Carroll approach linearizes Eq 8 by taking logarithms:

$$\log \frac{d\alpha}{dt} = \log Z + \frac{E}{2.303RT} + n \log(1-\alpha) \quad \text{(Eq 9)}$$

This equation is then evaluated at various temperatures spaced at equal increments of $1/T$, and the corresponding experimental values of α and $d\alpha/dt$ are used in a difference equation obtained from Eq 9:

$$\Delta \log \frac{d\alpha}{dt} = \Delta \log Z + \frac{E}{2.303R}\Delta\left(\frac{1}{T}\right)$$
$$+ n\Delta\log(1-\alpha)$$

This equation predicts that a plot of the differences of the logs of the rates of reaction at various temperatures, spaced at equal increments of reciprocal temperature, versus the corresponding differ-

ences of logs of the extent of reaction at those temperatures should be a straight line whose slope is the reaction order n and whose intercept can be used to compute the activation energy E. (Of course, $\Delta \log Z = 0$, because Z is a constant.) Once E and n are known, Z can be found by evaluating Eq 9 at any selected temperature. This procedure can be performed using spreadsheet software on personal microcomputers. Figure 7 shows a typical Freeman-Carroll plot for the material identified in Fig. 6.

Three such plots were made in this study, each from a separate DSC experiment. (Each experiment, however, used specimens from the same batch of prepreg.) The analysis appears reproducible, as seen in Table 1.

Improved Model for Constant Heating Rates. In addition to the general restrictions mentioned earlier regarding the simplistic nature of the above analysis, the scanning thermal experiment may at some time produce sufficient cross linking so that the resin T_g catches up to the cure temperature. From that time on, the resin may be reacting in a diffusion-controlled manner, with the T_g keeping a few degrees ahead of the steadily increasing cure temperature. Gillham (Ref 22) reports that this effect can be modeled satisfactorily by adopting a relation between T_g and the extent of reaction α of the form:

$$\frac{T_g - T_{g0}}{T_g} = \frac{(C_1 - C_2)\alpha}{1 - (1 - C_2)\alpha} \quad \text{(Eq 10)}$$

When the glass temperature T_g as given by Eq 10 is greater than the current cure temperature, the reaction rate equation (Eq 8) is then replaced by one, attributable to Kaeble (Ref 23), that modifies the preexponential constant Z:

$$\ln Z(T) = \ln Z(T_g) + \frac{C_1(T - T_g)}{C_2 + (T - T_g)} \quad \text{(Eq 11)}$$

The $Z(T_g)$ is the frequency factor at T_g and is assumed to have the same value as in the rubbery or liquid state. The second term is a WLF-type expression, which attempts to account for the reduced free volume and molecular mobility at temperatures below the T_g. Equation 11 clearly requires more computation than the simple expression of Eq 8. However, this is not a serious drawback in light of today's inexpensive yet very powerful computers.

A General Process Model

Governing Equations. During the processing of thermoset resins in actual applications, the kinetics of the chemical curing reaction are complicated by the nonuniform nature of the process. An overall process model must combine the curing kinetic model with the equation for conductive transient heat transfer in order to obtain the spatial and temporal variation of temperature and extent of reaction during the process.

Numerical approaches are suggested because of the complexity of these processes, and the author has developed a general finite-element code for this purpose. This code, which has been implemented on PC-compatible microcomputers (and which is available from the author), is able to solve simultaneously the conservation equations for the transport of momentum, thermal energy, and reactive species for a variety of two-dimensional geometries and a number of problem types (Ref 24-26). As developed in standard texts in transport theory (for example, Ref 27), these equations can be written:

$$\rho \left[\frac{\partial u}{\partial t} + u\nabla u \right] = -\nabla p + \nabla(\eta \nabla u) \quad \text{(Eq 12)}$$

$$\rho c \left[\frac{\partial T}{\partial t} + u\nabla T \right] = Q + \nabla(k\nabla T) \quad \text{(Eq 13)}$$

$$\left[\frac{\partial C}{\partial t} + u\nabla C \right] = R + \nabla(D\nabla C) \quad \text{(Eq 14)}$$

Here u, T, and C are fluid velocity (a vector), temperature, and the concentration of reactive species; these are the principal variables in the formulation. Other parameters are density (ρ), pressure (p), viscosity (η), specific heat (c), thermal conductivity (k), and species diffusivity (D). The ∇ operator is defined as $\nabla = (\partial/\partial x, \partial/\partial y)$. The similarity of these equations is evident and leads to considerable efficiency in the coding of their numerical solution. In each case, the time rate of change of the transported variable (u, T, or C) is balanced by the convective or flow transport term, for example, $u\nabla T$; the diffusive transport, for example, $\nabla(k\nabla T)$; and a generation term, for example, Q.

In conventional closed-form analysis, one generally seeks to simplify the governing equations by dropping those terms that are zero or those whose numerical magnitudes are small relative to the others, and then proceeding with a mathematical solution. In contrast, this finite-element code is written to contain all of the terms, and the particularization to specific problems is done entirely by the selection of appropriate numerical parameters in the input data set. In this present problem, the momentum equation is not needed, and all terms containing the flow velocity are dropped. This is accomplished by setting appropriate flags in the data set.

The units must be given special attention in these equations, particularly because materials properties obtained from various handbooks or experimental tests are usually reported in units that must be converted to obtain consistency in the above equations. The governing equations are volumetric rate equations. For instance, the heat generation rate (Q) is energy per unit volume per unit time, such as $N \cdot m/m^3 \cdot s$. The user must select units for all parameters so that each term in the energy equation has these same units.

Q and R are generation terms for heat and chemical species, respectively, whereas the pressure gradient (∇p) plays an analogous role for momentum generation. Heat generation arises from viscous dissipation and reaction heating:

$$Q = \tau : \dot{\gamma} + R(\Delta H) \qquad \text{(Eq 15)}$$

where τ and $\dot{\gamma}$ are the deviatoric components of the stress and strain rate, R is the rate of chemical reaction, and ΔH is the heat of reaction. R, in turn, is given by a kinetic chemical equation. In this model, an mth order Arrhenius expression has been implemented:

$$R = k_0 \exp\left(\frac{-E^\dagger}{R_g T}\right) C^m \qquad \text{(Eq 16)}$$

where k is a preexponential constant, $E^\dagger$ is an activation energy, and R_g is the gas constant.

Viscosity (η) is a strong function of the temperature and the shear rate for many fluids, and the flow code has been written to include a Carreau power-law formulation for shear thinning and an Arrhenius expression for thermal thinning. The formal equation is:

$$\eta = \eta_0 \exp\left(\frac{-E_\eta}{R_g T}\right) [1 + (\lambda \dot{\gamma})^2]^{(n-1)/2} \qquad \text{(Eq 17)}$$

where η_0 is the 0 shear viscosity limit, E_η is an activation energy for thermal thinning, λ is a shape parameter, and n is the power-law exponent. This formulation is admittedly not suitable for all cases, such as liquids exhibiting strong elastic effects, but it is commonly used in much of the literature for viscous flow rheology.

The boundary conditions for engineering problems usually include some surfaces on which values of the problem unknowns are specified, for instance, points of known temperature or initial species concentra-

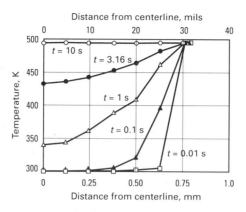

Fig. 8 Predicted temperature profiles at various times for cure program in which outer surfaces of wiring board are suddenly brought to 491 K (425 °F). t, time

tion. Some other surfaces may have constraints on the gradients of these variables, as on convective thermal boundaries when the rate of heat transport by convection away from the surface must match the rate of conductive transport to the surface from within the body. Such a temperature constraint might be written:

$$h(T - T_a) = -k\nabla T \cdot n \text{ on } \Gamma_h \qquad \text{(Eq 18)}$$

where h is the convective heat transfer coefficient, T_a is the ambient temperature, and n is the unit normal to the convective boundary Γ_h.

Numerical Results. As an example of this approach, consider the curing of a multilayer printed wiring board consisting of symmetric outer layers of copper-clad laminate and one central copper-clad layer, bonded by bismaleimide adhesive prepreg material. Noting that the thermal and reactive species gradients are much larger in the board through-thickness direction than in the plane of the board, a single strip of elements can be placed to run from the board centerline to the outer edge. The time-temperature program of the press cycle is imposed on the topmost nodes as a boundary condition, and the code is operated in a time-stepping fashion to compute the internal temperatures and degrees of reaction as the cure proceeds.

Figure 8 shows the temperature profiles that are predicted within the board at various times when the board is placed in a heated mold and the outer surfaces are suddenly brought to the curing temperature of 220 °C (425 °F). The temperatures equilibrate in times on the order of 10^1, and in this time the reaction proceeds to completion as well. The processing time is obviously minimized by such a cure cycle, but the thermal strains produced by the nonuniform temperatures at early on may be problematic in terms of board warpage and interlayer delamination.

Figure 9 shows the results of heating more gradually at 6 °C/min (10 °F/min) until

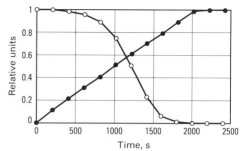

Fig. 9 Normalized temperature (ascending line) and fraction of unreacted resin (descending curve) for cure program consisting of heating at 6 °C/min (10 °F/min) to 220 °C (491 K, or 425 °F), followed by holding at 220 °C (425 °F)

220 °C (425 °F) is reached and then holding at that temperature. In this case the heating rate is slow enough, compared to the thermal equilibration time, to keep the board at an essentially uniform temperature and degree of reaction during the process, and the reaction proceeds to completion during the heat-up phase of the cure cycle. The high-temperature hold following the heat-up phase is unnecessary with regard to chemical reaction in the adhesive.

REFERENCES

1. N.G. McCrum, *Anelastic and Dielectric Effects in Polymeric Solids*, John Wiley & Sons, 1967
2. J.D. Ferry, *Viscoelastic Properties of Polymers*, John Wiley & Sons, 1961
3. L.C.E. Struik, Physical Aging in Plastics and Other Glassy Materials, *Polym. Eng. Sci.*, Vol 17, 1977, p 165-173
4. J.B. Enns and J.K. Gillham, Time Temperature Transformation (TTT) Cure Diagram: Modeling the Cure Behavior of Thermosets, *J. Appl. Polym. Sci.*, Vol 28, 1983, p 2567-2591
5. J.K. Gillham, Curing, in *Encyclopedia of Polymer Science and Engineering*, Vol 4, John Wiley & Sons, 1986, p 519-524
6. S.S. Labana, Crosslinking, in *Encyclopedia of Polymer Science and Engineering*, Vol 4, John Wiley & Sons, 1986, p 349-395
7. F.P. Darmory, "Kerimid 601 Polyimide Resins for Multilayer Printed Wiring Boards," Rhodia Inc.
8. D. Kumar, G.M. Fohlen, and J.A. Parker, High-Temperature Resins Based on Aromatic Amine-Terminated Bisaspartimides, *J. Polym. Sci., Polymer Chemistry Edition*, Vol 21, 1983, p 245-267
9. C. DiGiulio, M. Gautier, and B. Jasse, Fourier Transform Infrared Spectroscopic Characterization of Aromatic Bismaleimide Resin Cure States, *J. Appl. Polym. Sci.*, Vol 29, 1984, p 1771-1779

10. R. Fullerton, D. Roylance, A. Acton, and R. Allred, Cure Analysis of Printed Wiring Boards Containing Reactive Adhesive Layers, *Polym. Eng. Sci.*, Vol 28, 1988, p 372-376

11. L.T. Pappalardo, DSC Evaluation of Epoxy and Polyimide-Impregnated Laminates (Prepregs), *J. Appl. Sci.*, Vol 21, 1977, p 809-820

12. H.J. Borchardt and F. Daniels, The Application of Differential Thermal Analysis to the Study of Reaction Kinetics, *J. Am. Chem. Soc.*, Vol 79, 1957, p 41-46

13. E.S. Watson, M.T. O'Neill, J. Justin, and N. Brenner, The Analysis of a Temperature Controlled Scanning Calorimeter, *Anal. Chem.*, Vol 36, 1964, p 1238-1245

14. R.A. Fava, Differential Scanning Calorimetry of Epoxy Resins, *Polymer*, Vol 9, 1968, p 137-151

15. O.R. Abolafia, Application of Differential Scanning Calorimetry to Epoxy Curing Studies, *Proc. SPE Annu. Tech. Conf.*, Vol 15, 1969, p 610-616

16. A.A. Duswalt, The Practice of Obtaining Kinetic Data by Differential Scanning Calorimetry, *Thermochim. Acta*, Vol 8, 1974, p 57-68

17. P.S. Dhar, A Comparative Study of Different Methods for the Analysis of TGA Curves, *Comput. Chem.*, Vol 10, 1986, p 293-297

18. E.S. Freeman and B. Carroll, The Application of Thermoanalytical Techniques to Reaction Kinetics, *J. Phys. Chem.*, Vol 54, 1957, p 394-397

19. L. Reich and S.S. Stivala, *Elements of Polymer Degradation*, McGraw-Hill, 1971

20. Presentation at the Meeting of the New England Thermal Forum (Newton, MA), April 1987

21. W.J. Sichina, "Considerations in Modeling of Kinetics by Thermal Analysis," Publication E-81774, E.I. Du Pont de Nemours & Company, Inc.

22. J.K. Gillham, Modeling Reaction Kinetics of an Amine-Cured Epoxy System at Constant Heating Rates From Isothermal Kinetic Data, *Polym. Mater. Sci. Eng.*, Vol 57, 1987, p 87-91

23. D.H. Kaeble, Chapter 4 in *Computer-Aided Design of Polymers and Composites*, Marcel Dekker, 1985

24. D.K. Roylance, Use of "Penalty" Finite Elements in Analysis of Polymer Melt Processing, *Polym. Eng. Sci.*, Vol 20, 1980, p 1029-1034

25. D.K. Roylance, Finite Element Modeling of Nonisothermal Polymer Flows, in *Computer Applications in Applied Polymer Science*, American Chemical Society Symposium Series, Number 197, American Chemical Society, 1982, p 265-276

26. C. Douglas and D.K. Roylance, Chemorheology of Reactive Systems: Finite Element Analysis, in *Chemorheology of Thermosetting Polymers*, American Chemical Society Symposium Series, Number 227, American Chemical Society, 1983, p 251-262

27. R.B. Bird, W.E. Stewart, and E.N. Lightfoot, *Transport Phenomena*, John Wiley & Sons, 1960

Electrical Properties

Ken M. Takahashi, AT&T Bell Laboratories

POLYMERIC MATERIALS are used in a wide variety of electrical applications, as insulators, dielectrics, encapsulants, adhesives, coatings, and resists. This article outlines, in three major sections, the experimental techniques for measuring the electrical properties that determine material performance in these applications. The first section covers the measurement of dielectric permittivity and loss factor over the broad frequency range that characterizes dipolar relaxations. The second section discusses resistivity measurements with mention of conduction mechanisms and nonohmic effects. The third section describes common breakdown processes and techniques that are used to characterize them.

Dielectric Properties

Dielectric permittivity and loss factor are key properties of polymers used in electronic applications as substrates, coatings, encapsulants, and dielectric films. To maintain the speed and integrity of electronic signals, it is necessary to select polymers in which both properties are low and insensitive to temperature, frequency, and humidity. More generally, the dielectric response of polymers provides a wealth of information about their molecular relaxation processes and mechanical properties (Ref 1).

Several techniques for measuring dielectric permittivity and loss factor are summarized in the sections below. As shown in Fig. 1, a variety of experimental techniques is available for measuring the dielectric properties of the 15 or more orders of magnitude of frequency over which important dielectric relaxations may take place. The following discussion divides the techniques into two broad categories. The first category contains techniques used for low ($<10^8$ Hz) frequencies, in which the wavelength of the electric field is large compared to sample dimensions. This category includes time-domain and alternating current (ac) bridge techniques, which employ lumped parameter treatments of circuit response to the applied electric field. At higher frequencies, electric field varies with both time and position within the sample. The sample must then be considered to be a distributed parameter system, and alternate techniques, such as transmission line and resonant-cavity measurements, are used. These techniques can be used to measure dielectric properties at frequencies up to about 10^{11} Hz, beyond which small cell size and large signal attenuation make them impractical.

Definitions

The temporal representation of a sinusoidal voltage:

$$V_0 \cos(\omega t + \phi_1) = Re \, (V_0 e^{j\phi_1} e^{j\omega t}) \qquad \text{(Eq 1)}$$

has an amplitude (V_0) and phase angle (ϕ_1) that are contained in the complex voltage amplitude $V(\omega)$:

$$V(\omega) = V_0 e^{j\phi_1} = V' + jV'' \qquad \text{(Eq 2)}$$

where $V' = V_0 \cos(\phi_1)$ and $V'' = V_0 \sin(\phi_1)$, and $j = (-1)^{1/2}$. When V is imposed across an arbitrary sample, the result is a phase-shifted complex current response of the form:

$$I(\omega) = I_0 e^{j\phi_2} = I' + jI'' \qquad \text{(Eq 3)}$$

The sample response is determined by its complex impedance:

$$Z(\omega) = Z' + jZ'' = \frac{V(\omega)}{I(\omega)} \qquad \text{(Eq 4)}$$

or, equivalently, its admittance:

$$Y(\omega) = \frac{1}{Z(\omega)} \qquad \text{(Eq 5)}$$

Typically, the response of dielectric samples can be modeled with an equivalent circuit consisting of a network of capacitors and resistors. The impedance of a capacitor is $Z_C = (j\omega C)^{-1}$, whereas the impedance of resistive elements is simply $Z_R = R$. We now define the complex relative permittivity:

$$\epsilon^* = \epsilon' - j\epsilon'' = \frac{C^*}{\epsilon_0 K_b} = \frac{Y}{j\omega \epsilon_0 K_b} \qquad \text{(Eq 6)}$$

where $\epsilon_0 = 8.854$ pF/m is the permittivity of free space (rationalized absolute practical units). The cell constant K_b (in meters, m) is a function of electrode and sample geometry. For parallel plate cells, K_b is nominally the electrode area divided by the sample thickness. The dielectric constant of the material is ϵ', and ϵ'' is the loss factor. Both are frequency dependent.

When voltage (V_0) is applied between two electrodes in vacuum, the electrodes acquire a charge of $\pm Q_v$. The vacuum capacitance is then defined as $C_v = Q_v/V = \epsilon_0 K_b$. When the space between electrodes contains a dielectric material, the electrode charges are increased by the polarization of the material, such that $C/C_v = Q/Q_v = \epsilon'$ defines the dielectric constant of the material.

The material polarization contains at least three contributions. The first two are the electronic and atomic polarization, which can be considered instantaneous for our purposes. The contribution of these fast polarizations to the dielectric constant is termed ϵ_∞ and is measured at frequencies high enough to freeze dipole relaxations. The third polarization contribution involves the orientation of permanent dipoles, and is of primary interest in dielectric studies. Dipole relaxations may require the cooperative motion of attached and surrounding polymer chain segments and can take place on time scales corresponding to the entire range of frequencies represented on the axis of Fig. 1. The dielectric constant ϵ' has a maximum value at frequencies that are sufficiently low to allow all dielectric polarizations to respond to the applied field, and therefore it contains the contributions from all dipole relaxations. Its value in the low-frequency limit is the static dielectric constant (ϵ_s). In Fig. 2, for example, the temperature-dependent dielectric response of poly(vinyl acetate) (Ref 1) is shown. The values of the dielectric constant (ϵ') demonstrate both low-frequency (ϵ_s) and high-frequency (ϵ_∞) limits.

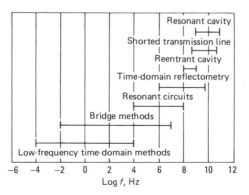

Fig. 1 Dielectric techniques

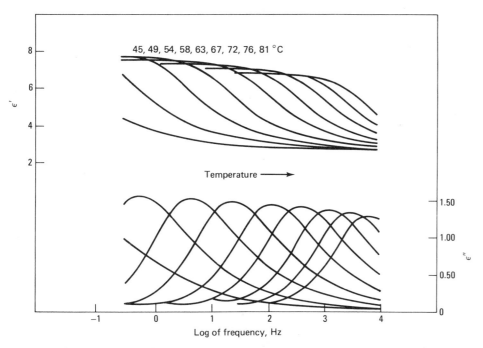

Fig. 2   Dielectric spectra for poly(vinyl acetate). Upper curves represent the dielectric constant (ϵ'), and lower curves the loss factor (ϵ''). Source: Ref 1

Dissipative processes in the sample, such as the dipole relaxations described above, are characterized by the loss tangent or dissipation factor $D = \tan \delta = \epsilon''/\epsilon'$. Equivalently, the quality $Q = 1/D$ or the loss angle δ is sometimes reported. In addition to dissipation by dipole relaxations, often ϵ'' contains a conduction contribution $\epsilon''_{cond} = -1/\omega\rho_b\epsilon_0$. Loss factors have maxima at frequencies that are characteristic of the relaxation processes. The loss peaks in Fig. 2 show the temperature dependence of the dipole relaxations, and thus of the fundamental molecular relaxation process. The frequency, amplitude, and shape of loss peaks are the primary descriptors of dipole relaxations.

When the wavelength of the electromagnetic field is comparable to sample dimensions, spatial variation of the instantaneous field within the sample must be considered. A transverse electromagnetic (TEM) wave traveling in the x direction has electric and magnetic fields that vary with time and position as (Ref 2):

$$E = E_0 \exp(j\omega t - \gamma x)$$
$$H = H_0 \exp(j\omega t - \gamma x) \qquad \text{(Eq 7)}$$

and are related by the Maxwell equation:

$$\nabla \times H = j\omega\epsilon^* \epsilon_0 E \qquad \text{(Eq 8)}$$

where the propagation factor

$$\gamma = j\omega(\epsilon_0\epsilon^* \mu)^{1/2} = \alpha - j\beta \qquad \text{(Eq 9)}$$

and μ is the magnetic permeability of the sample (equal to μ_0, the permeability of free space, for most polymers), α is the attenuation factor of the material, and β is the phase constant. In distributed parameter measurements, the complex permittivity is generally determined by comparing wave propagation in a waveguide or cavity containing a section of sample to propagation in the empty waveguide.

Low-Frequency Measurements

Most dielectric measurements in the low-frequency ($<10^8$ Hz) region are made using time-domain, ac bridge, impedance meter, and resonant-circuit methods. The section "Sample Geometries" describes sample and electrode designs that are suitable for these measurements. Low-frequency time-domain techniques, which are preferred for measurements at frequencies less than 1 Hz, are presented in the section "Time-Domain Measurements." The ac bridge and vector impedance meter methods are outlined in the section "Bridge Methods." Resonant circuits, which are useful for measuring dielectric properties in the upper-frequency range of lumped parameter measurements, are described in the section "Resonant Circuits."

Sample Geometries. Four commonly used sample and electrode geometries are shown in Fig. 3(a to d). All consist of sample films or cylinders sandwiched by parallel electrodes. The guarded configuration (Fig. 3a) is the most useful because it minimizes fringe fields and eliminates the ability of surface conduction to create the illusion of bulk conduction or increased electrode area. Typically, the guard electrode is grounded and electrode 2 is held at virtual ground. Current through electrode 1 includes fringing field and surface conduction currents; calculation of sample impedance is based on current in electrode 2. The cells in Fig. 3(b to d) are more easily fabri-

cated and are used in circuits that are not designed for guarded measurements. The ring cell (Fig. 3e) is not used for dielectric measurements. Electrode materials are discussed below.

Several stray capacitances interfere with dielectric measurements and must be accounted for, either by experimental design or by calibration of the measuring system. These include fringing, or edge effects, as well as capacitances of leads and electrodes to ground and to each other. In the lower-frequency ranges, cable shielding effectively eliminates lead-lead and lead-electrode capacitances. Capacitances to ground vary with the details of the lead and electrode position with respect to the ground and are measured. Cell constants and edge effect corrections are summarized in Table 1. The edge corrections are empirical formulas based on the perimeter and thickness of the sample. Fringe corrections for a wider variety of electrode shapes and dimensions can be found in ASTM D 150 (Ref 3).

Electrode Materials. Electrodes must provide intimate contact with the sample, have resistances that are low compared to sample resistance, be stable to oxidation or corrosion in the test environment, and not react with or change the properties of the sample. The common electrode materials are described below.

Metal plates can be used for resistance measurements if intimate contact between the electrodes and samples can be ensured. Metal surfaces upon which the polymer samples are cast or coated perform well, but contact between a sample film and a flat metal plate held in contact by pressure alone is of questionable continuity. If the samples are sufficiently deformable, pressures of 140 to 700 kPa (20 to 100 psi) have been found to provide adequate electrical contact (ASTM D 257, in Ref 3).

In spite of the risk of poor contact and air gap formation, metal foil is sometimes used as an electrode material. Contact can be ensured by applying thin films, such as those of silicone liquids, between the sample and the foil. If the film is sufficiently thin and has low dielectric loss, the series film impedance may be tolerable.

Evaporated or sputtered metal films are generally excellent electrode materials. Complex electrode shapes and guard rings are easily fabricated by depositing the electrodes through masks. Vacuum-deposited aluminum electrodes may oxidize when in contact with very conductive or corrosive materials, or if measurements are made in an aggressive environment. Gold films are perhaps the most reliable, although sputtered gold has been known to penetrate thin (of the order of μm) polymer samples, causing cell short circuits.

Conductive paints and adhesives are generally polymer composites containing high volume fractions of metal particles. These

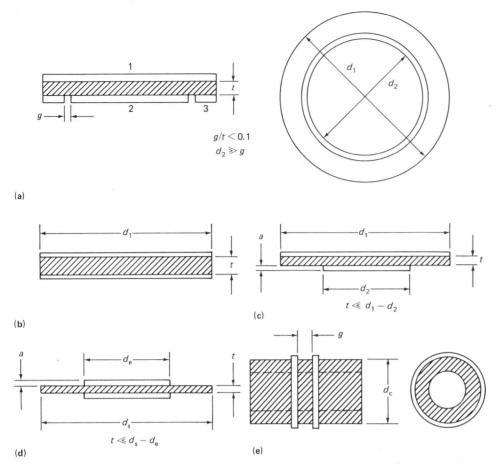

$$g/t < 0.1$$
$$d_2 \gg g$$

(a)

(b)

(c)

$$t \ll d_1 - d_2$$

(d)

$$t \ll d_s - d_e$$

(e)

Fig. 3 Sample geometries. (a) Disks with guard electrode. (b) Equal disk electrodes. (c) Unequal disk electrodes. (d) Equal disk electrodes smaller than sample. (e) Cylindrical sample with ring electrodes

Table 1 Geometric and edge correction factors for disk-shape dielectric samples

Sample type(a)	K_b'(b)	A(b)	B(b)
Guarded	$\dfrac{\pi}{4t}(d_2 + g)^2$	0	0
Equal electrodes	πd^2	0	$d_1\,(0.027 - 0.00792 \ln t)$
Unequal electrodes	$\pi\dfrac{d_2^2}{4t}$	$0.013\,d_2$	$d_2\,(0.0383 - 0.0105 \ln t)$
Equal electrodes smaller than sample	$\dfrac{\pi d_e^2}{4t}$	$0.0060\,d_e$	$d_e\,(0.021 - 0.00792 \ln t)$

$$\epsilon' = \frac{C_{\text{meas}}(\text{pF}) - B}{0.008854\,K_b' + A}$$

Note: Corrections are based on thin electrodes ($a \ll t$). (a) See Fig. 3 for specimen schematics. (b) Empirical fringe correction factors A and B are dimensionally incorrect, and therefore capacitance must be measured in pF and lengths in mm (after ASTM D 150)

composites can make good electrodes as long as they have much lower impedance than the samples and as long as the low molecular weight species that are present in the organic matrix before hardening (solvent, for example) do not swell or otherwise affect the sample properties.

Time-domain measurements involve monitoring the transient charge across a sample in response to an applied voltage waveform, most commonly a voltage step. The response contains information about relaxations taking place on a continuum of time scales and can be Laplace transformed

into the frequency domain using the superposition principle.

The major advantage of time-domain measurements is speed. An entire dielectric relaxation spectrum can be measured in a single experiment on a time scale on the order of one cycle at the lowest desired frequency. For this reason, time-domain techniques are best suited to the measurement of low-frequency relaxation spectra, which can be very time consuming when performed using sequential steady-state frequency-domain techniques. Nevertheless, a variety of time-domain measurements have been used to

measure dielectric relaxation spectra at frequencies up to 15 GHz (Ref 4, 5).

In this section, discussion is limited to measurements made in the audio to subaudio range, where a lumped parameter treatment is valid. Using the following approach detailed by Mopsik (Ref 6), modern instrumentation and digital data acquisition allow the technique to provide high accuracy (0.1%) and high sensitivity (tan $\delta = 10^{-5}$) over the frequency range from 10^{-4} to 10^4 Hz.

Figure 4 shows the voltage step and charge response across a dielectric sample. The transient capacitance is:

$$C(t) = \frac{Q(t)}{V_0} \qquad \text{(Eq 10)}$$

and can be transformed into the frequency domain according to:

$$C^*(\omega) = C'(\omega) - C''(\omega) = \int_0^\infty \frac{dC(t)}{dt} e^{-j\omega t}\,dt \qquad \text{(Eq 11)}$$

The step charge accumulated before the first sampling at $t = t_1$ contains primarily the fast electronic and atomic polarization contributions to C^*:

$$C_0 = \int_0^{t_1} \frac{dC(t)}{dt} e^{-j\omega t}\,dt \approx \frac{Q_0}{V_0} \qquad \text{(Eq 12)}$$

To enhance charge measurement sensitivity, the instantaneous charge (Q_0) is compensated by simultaneously imposing a reverse voltage step on a lossless reference capacitor, shown in Fig. 5. The components of complex capacitance $C^*(\omega)$ are then given by:

$$C'(\omega) = \int_{t_1}^\infty \frac{dC(t)}{dt} \cos \omega t\,dt + C_0 + C_{\text{ref}} \qquad \text{(Eq 13)}$$

$$C''(\omega) = \int_{t_1}^\infty \frac{dC(t)}{dt} \sin \omega t\,dt$$

In practice, data are collected in constant t or log t increments over the time range from t_1 to t_2 represented in Fig. 4(b). When sampling digitally, t_1 is comparable to the minimum sample period ($\Delta t_{\min}$), and the maximum frequency that can be resolved by the experiment is limited to $\omega_{\max} = {\sim}1/\Delta t_{\min}$. The lowest accessible frequency is controlled by the duration of the experiment: $\omega_{\min} \sim 1/t_2$. A variety of methods for performing the numerical integrations of the data have been used, including the commonly available fast Fourier transform (FFT). In a more accurate and computationally efficient method (Ref 6), a cubic spline representation of the $Q(t)$ data measured in constant log t increments is integrated from $t = 0$ to $t = \infty$. The $Q(t)$ data are extrapolated to $t = 0$ for the high-frequency response and to $t = \infty$ using an assumed decay function. The result is an accurate and sensitive measurement of complex permit-

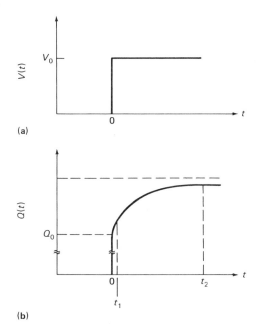

(a)

(b)

Fig. 4 Time-domain experiment. (a) Voltage step. (b) Transient charge response

tivity over the maximum theoretical limits set by t_1 and t_2.

Bridge Methods. Alternating current bridges are the complex impedance analogs of the Wheatstone bridge and are used to measure permittivity and dissipation in dielectric materials. The traditional bridges have fairly simple equivalent circuits and can provide accurate measurements of capacitances as low as 10^{-7} pF and dissipation factors as low as 10^{-5}. Most bridges are based on the basic Harris (10^{-2} to 10^2 Hz) (Ref 7), Schering (10^1 to 10^6 Hz) (Ref 8), inductive ratio arm (10^1 to 10^5 Hz) (Ref 9) and parallel T-network (5×10^5 to 3×10^7 Hz) (Ref 2) bridges. As described in detail by Vaughan (Ref 10), numerous circuit modifications have resulted in bridges that provide accurate dielectric data throughout the frequency range from 10^{-2} to 10^7 Hz.

Because manual bridge balancing is time consuming, integrated circuits and digital data acquisition have led to the development of several instruments that allow the automated measurement of dielectric properties. Some are true bridges, such as the automated transformer arm bridges described by Hedvig (Ref 11) that use voltage feedback from phase-sensitive lock-in detectors to obtain balance, while admittance meters or vector voltmeters use direct phase sensitive detection to compare sample current response to input voltage. As reviewed in Ref 12, several microprocessor-controlled automation schemes are used to provide fast and convenient dielectric measurements in the frequency range where ac bridge techniques are employed.

Resonant Circuits. The application of resonant circuits to measure dielectric properties is similar to that of ac bridges, except

that the circuits are tuned to the resonance point rather than to the zero of the circuit. Resonant circuits are useful for measuring dielectric properties in the frequency range from 10^4 to 10^8 Hz, where series impedances and shunt admittances interfere with bridge measurements. The techniques described here use two-terminal measurements. Stray admittances are eliminated by comparing results obtained by tuning circuits to resonance both with and without the sample in place between micrometer-driven parallel plates.

In the capacitance variation method of Hartshorn and Ward (Ref 13), resonance is obtained by tuning the single variable capacitor (C_v) in Fig. 6(a) in the presence of the sample. Then the sample is removed from between the micrometer-driven parallel plates, and the interelectrode gap is reduced until the sample capacitance is duplicated and resonance is reestablished. The ϵ' value of the sample is calculated based on the electrode geometry and the ratio of electrode gaps with and without the sample: ϵ' is the ratio of electrode gaps t_{air}/t_{sample}. The dissipation factor is then obtained from the shape of the resonance curve (Fig. 6b):

$$\tan \delta = \frac{(\Delta C_i - \Delta C_0)}{2 C_s (q-1)^{1/2}} \quad \text{(Eq 14)}$$

where ΔC_v is the capacitance range about the resonance point within which the square of measured voltage exceeds the fraction $1/q$ of its resonance value (V_r^2). The peak width in the absence of sample is ΔC_0.

The Q-meter method involves tuning the circuit shown in Fig. 7 to resonance by adjusting the variable capacitor (C). The inductance (L) of the coil is chosen to allow the circuit to resonate in the frequency range of interest. Circuit quality $Q = V/IR$ is measured at resonance without sample (subscript 1) and with sample in place (subscript 2). Thus sample quality is given by:

$$Q_s = \frac{Q_1 \, Q_2 (C_1 - C_2)}{C_1 (Q_1 - Q_2)} \quad \text{(Eq 15)}$$

Sample capacitance is the difference $C_1 - C_2$.

Radio Frequency and Microwave Measurements

In the high-frequency ranges, finite ratios of sample size to field wavelength place additional importance on sample dimensions. Measurement apparatus tends to be specialized for narrow frequency bands. In this section, three high-frequency techniques are presented. Time-domain reflectometry covers the widest frequency band, bridging the frequency range between lumped parameter and transmission line measurements. The shorted transmission line and resonant-cavity techniques allow

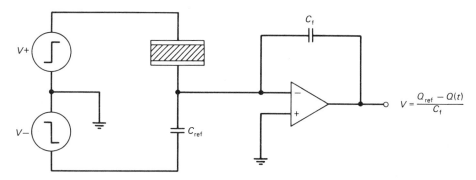

Fig. 5 Time-domain measurement circuit

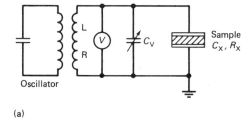

(a)

(b)

Fig. 6 (a) Hartshorn-Ward measurement circuit. (b) Resonance curve

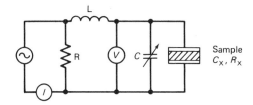

Fig. 7 Q-meter measurement circuit

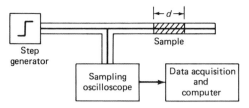

Fig. 8 Time-domain reflectometer

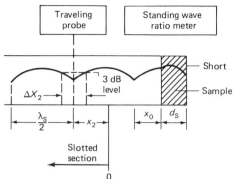

Fig. 9 Slotted waveguide (after ASTM D 2520)

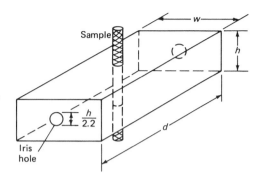

Fig. 10 Resonant cavity with cylindrical sample (after ASTM D 2520)

measurement of fast dielectric relaxations at frequencies up to 10^{11} Hz.

Time-domain reflectometry (TDR) is perhaps the most useful of the high-frequency time-domain techniques reviewed by van Gemert (Ref 4). It can be used to measure permittivity in the frequency range from 10^6 to 5×10^9 Hz. Because of its broad frequency range, TDR is more useful than the reentrant-cavity technique described by Works *et al.* (Ref 14) (Fig. 1).

In TDR measurements, a fast voltage step is applied to a coaxial cable that is air filled except for a section containing the material to be studied (Fig. 8). When the signal reaches the material-filled section of the cable, the impedance discontinuity results in a reflected signal. Sample length d is chosen so that only the first reflection from the loading air-sample interface is detected during the measurement. The incident (amplitude V_0) and reflected [$V_r(t)$] signals are detected at the oscilloscope sampler. Fourier transforms of the voltage signals, integrated numerically as in the earlier section "Time-Domain Measurement," are then used to calculate the complex permittivity of the material:

$$\epsilon^*(\omega) = \left[\frac{V_0(\omega) - V_r(\omega)}{V_0(\omega) + V_r(\omega)} \right]^2 \quad \text{(Eq 16)}$$

Details of the TDR method are presented in Ref 4 and 15. As in low-frequency time-domain measurements, the high-frequency limit of the measurement is limited by the slew rate of the voltage pulse and the oscilloscope sampling rate. The low-frequency limit is determined by the time scale over which voltages are integrated. The technique is not sensitive to low-loss materials; perhaps the best sensitivity of TDR to dissipation is tan δ ≈ 10^{-3} (Ref 16).

Slotted Transmission Lines. In this technique, a micrometer-controlled power (square-law voltage) probe is used to determine the location and width (x and Δx_i) (Fig. 9) of standing wave nodes in a transmission line that is terminated with a short. Node width Δx_i is the distance between the two points near the node where the probe reads twice the power that is registered at the node point. When a short length (d_s) of sample that completely fills the cross section of the transmission line is placed flush against the short, node location and width are changed in a manner that is character-

istic of the complex propagation constant (γ) and thus of ε*. The technique is capable of providing ε' to ±1% and tan δ to 2×10^{-4} at frequencies up to 50 GHz (Ref 2).

For a shorted transmission line with no wall losses, the propagation constant is given by:

$$\gamma = 2\pi(\lambda_c^{-2} - \epsilon^*\lambda_0^{-2})^{1/2} \quad \text{(Eq 17)}$$

where λ_c is the cutoff wavelength for the mode of interest based on cross-sectional dimensions of the waveguide, and $\lambda_0 = 2\pi c/\omega$ is the wavelength of the signal in free space. The propagation constant in the presence of sample (γ₂) is defined by:

$$\frac{\tanh(\gamma_2 d_s)}{\gamma_2 d_s} = \frac{\lambda_{gh}(1 - j \, r \tan \phi)}{2\pi j d_s(r - j \tan \phi)} \quad \text{(Eq 18)}$$

where

$$\lambda_{gh} = (\lambda_0^{-2} - \lambda_c^{-2})^{-1/2} \quad \text{(Eq 19)}$$

The voltage standing wave ratio r is:

$$r = \frac{\lambda_{gs}}{\pi\Delta x} \quad \text{(Eq 20)}$$

The measured wavelength in the slotted section of the guide is λ_{gs}, whereas Δx is the difference in node widths with (denoted by subscript 2) and without (denoted by subscript 1) sample in the waveguide:

$$\Delta x = \Delta x_2 - \Delta x_1 \quad \text{(Eq 21)}$$

and the phase distance φ is:

$$\phi = 2\pi \left(\frac{N}{2} - \frac{d_s}{\lambda_{gh}} \pm \frac{x_2 - x_1}{\lambda_{gs}} \right) \quad \text{(Eq 22)}$$

The guided wavelength in the sample section λ_{gh} may be different from that in the slotted section λ_{gs} if the sample section is heated to study temperature effects; λ_{gh} is corrected for thermal expansion of the waveguide in the heated sample section. Equations 17 and 18 are solved numerically to obtain the complex permittivity of the sample. Mathematical and experimental details may be found in ASTM D 2520 (Ref 3).

Resonant Cavities. When a small sample is placed in a resonant cavity, the resonant frequency and quality (Q) of the cavity are perturbed. Comparison of these values with

those of the empty cavity are sufficient to calculate the ε* of the sample. Resonant cavities can have very high Q factors when empty and therefore have the potential for measuring low-loss materials and are the preferred technique for measurements above 5 GHz (Ref 2). Resonant frequency is controlled by the sample size. In perturbation measurements, the sample is small compared to the wavelength; this places an upper limit on the frequency in the 100 GHz (millimeter wavelength) range.

Several cavity and sample shapes have been used (Ref 2, 3, 11). The present discussion is limited to the rectangular cavities described in ASTM D 2520 (see Fig. 10), in which small samples of well-defined shape are placed at a location where the electric field is a maximum. Iris holes in the shorted ends of the cavity are used to drive and monitor the power intensity in the cavity. Calculations are for a horizontal transverse electromagnetic wave, which resonates in the empty cavity at a frequency of:

$$f_0 = 15 \left[\left(\frac{1}{w} \right)^2 + \left(\frac{N}{d} \right)^2 \right]^{1/2} \quad \text{(Eq 23)}$$

where N is the (odd) number of half-waves at resonance in the cavity; w and d are internal dimensions, in centimeters, of the cavity; and f_0 is the resonant frequency in GHz. Dissipation factor D is measured by varying the frequency to find the frequencies, both above ($f_{\alpha+}$) and below ($f_{\alpha-}$) resonance, at which power intensity is an arbitrary α dB below the resonant intensity. D is given by:

$$D = \frac{(f_{\alpha+} - f_{\alpha-})}{Bf_0} \quad \text{(Eq 24)}$$

where frequencies are in GHz and B is defined as $(10^{\alpha/10} - 1)^{1/2}$. For the cylindrical sample shown in Fig. 10:

$$\epsilon' = \frac{w \, h \, d \, (f_2 - f_1)}{2\pi r^2 \, h \, f_2} \quad \text{(Eq 25)}$$

and r is the cylinder radius. Calculations are also given in ASTM D 2520 for bar-, strip-, sheet-, and sphere-shape samples.

Conduction

Coatings and encapsulants used for the mechanical protection of metals are often required for electrical isolation as well. The material conductivity and the identity of its charge carriers can control the electrochemical processes at the metal surface that lead to corrosion, blistering, or cathodic delamination. At the other extreme, adhesives may be required to behave as solid electrolytes for batteries, or as metallic conductors to allow adhesive interconnection between circuit boards and packaged devices or modules. In either case, the conduction properties are critical. In this section of the article, techniques for measuring electron and ion transport in polymers are discussed.

The principal parameter describing conduction in polymers is bulk conductivity (σ_b). Polymers have a very wide range of conductivities, from metallic, for example, p-doped poly(p-phenylenes), $\sigma_b \sim 10^5$ S/m, to highly insulating, for example, polystyrene, $\sigma_b \sim 10^{-16}$ S/m. The sections "Definitions," "Direct Current Measurements" and "Transient Techniques" focus on techniques that are used for measuring ohmic conductivities of polymers. However, Ohm's law is seldom a complete descriptor of conduction in polymers. The section "Deviations From Ohm's Law" outlines some of the common interfacial and nonlinear effects. In most cases, the understanding of transport mechanisms, or even of charge carrier identity, is tentative. However, there are several techniques that can be used to clarify the charge transfer process. The section "Conduction Mechanisms" briefly describes some of these techniques.

Definition

The definition of conductivity is based on Ohm's Law, which states that current density i (A/m^2) is proportional to the applied potential gradient $\nabla \phi$ or electric field E (V/m).

$$i = -\sigma_b \nabla \phi = \sigma_b E \qquad \text{(Eq 26)}$$

where σ_b (S/m) is the bulk conductivity of the material. Bulk conductivity is related to the mobility (u_i), m^2/mole $\cdot$ J $\cdot$ s and concentration (c_i), mole/m^3 of all charge carriers by:

$$\sigma_b = F^2 \sum_i z_i^2 u_i c_i \qquad \text{(Eq 27)}$$

where z_i is the charge of carrier i and $F =$ 9648 C/equiv is Faraday's constant. When conduction is ohmic, the potential distribution in the test specimen is determined only by the test electrode geometry, and the conductivity (or resistivity, $\rho_b = 1/\sigma_b$ can be determined by measuring the resistance, R_b (Ω), between test electrodes:

$$R_b = \frac{V}{I} \qquad \text{(Eq 28)}$$

$$\sigma_b = \frac{1}{R_b K_b} \qquad \text{(Eq 29)}$$

where V is the voltage applied between test electrodes, I is the steady current passing through the sample, and K_b (in meters) is a cell constant having the same meaning as that used in the section "Dielectric Properties."

Polymer surfaces, near-surface films, or interfaces with hydrophilic substrates can be quite conductive, particularly in the presence of surface contamination or adsorbed moisture. Surface conduction is termed ohmic if it obeys the two-dimensional analog of Eq. 26:

$$i_s = -\sigma_s \nabla \phi \qquad \text{(Eq 30)}$$

where the surface conductivity $\sigma_s = (\rho_s, \Omega \cdot$ m/m)$^{-1}$.

Direct Current Measurements

Direct current experiments are the simplest and most direct way to measure resistivity. General comments about experimental apparatus and humidity control are provided in the sections below, followed by descriptions of measurements made using the parallel plate and four-point probe electrode configurations.

General Experimental Considerations. Instrumentation considerations and humidity effects are described below.

Instrumentation. A common arrangement for measuring the dc conductivity of polymers is the voltmeter-ammeter circuit shown in Fig. 11(a). Voltage is applied across the sample between the high electrode (1) and the measuring electrode (2), which is usually held at ground potential. The optional guard electrode (3), which is held at the same potential as the measuring electrode, minimizes fringe fields and interference from surface conduction effects.

There are many variations on the circuitry used to measure current and voltage, including a variety of commercial meters that are capable of precisely measuring resistances up to 10^{16} Ω. High-resistance measurements are limited by the ammeter sensitivity and voltmeter input impedance and are susceptible to electromagnetic interference. All cables should be coaxial, and the sample itself should be shielded in a metal enclosure. For very resistive samples, currents may take some time to stabilize after application of the voltage. The decay time constant is R_bC, where C is the total capacitance of the sample, the measurement circuitry, and the leads. When current decay is observed, it is important to consider the possibility of interfacial polarization, described in the sections "Interfacial Effects" and "Diffusion."

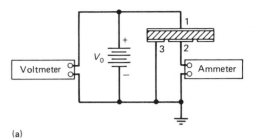

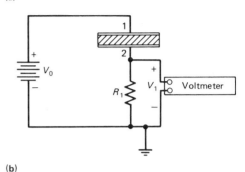

Fig. 11 (a) Voltmeter-ammeter circuit. (b) Current divider circuit

An alternative circuit commonly used for high-resistance measurements is the voltage divider circuit shown in Fig. 11(b). In this measurement, only the voltage V_1 across a high precision resistor R_1 must be measured. Sample resistance is given by:

$$R_b = R_1 \left[\frac{V_0 - V_1}{V_1} \right] \qquad \text{(Eq 31)}$$

Generally, input impedance of the voltmeter must be much higher than the circuit resistances. However, if the input resistance of the meter is known precisely, sample resistances that are higher than R_{input} can be measured:

$$R_b = \frac{R_1 R_{input}}{R_1 + R_{input}} \left[\frac{V_0 - V_1}{V_1} \right] \qquad \text{(Eq 32)}$$

The majority of resistance measurements use one of the two circuits described above, or commercial direct-reading resistance meters of varying range, precision, and circuit complexity, rather than the classical galvanometer and Wheatstone bridge methods detailed in ASTM D 257 (Ref 3).

Humidity Effects. Polymer conductivity can be sensitive to ambient humidity, with a several-orders-of-magnitude conductivity change not uncommon for polymers that contain ionic charge carriers. Surface conduction is also very sensitive to humidity (see Fig. 12). Typically, water adsorbs or condenses into films on hydrophilic or contaminated sample surfaces above a critical relative humidity, leading to greatly enhanced ionic surface conduction. Where humidity effects are pronounced, sample preparation, storage, and measurement must be done in controlled environments.

Even when the loss of insulation resistance is not severe enough to cause short-

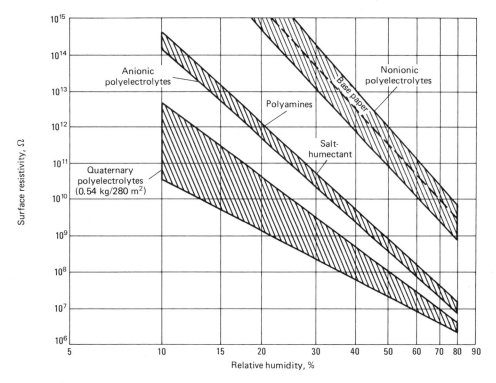

Fig. 12 Surface resistivity of several polyelectrolytes versus relative humidity. Source: Ref 17

Table 2   Cell constants for resistivity specimens

Type	K_B	K_S
Guarded disk.........	$\dfrac{\pi(d_2 + g)^2}{4t}$	$\dfrac{\pi(d_2 + g)}{g}$
Equal disks..............	$\dfrac{\pi d^2}{4t}$	
Unequal disks......	$\dfrac{\pi(d_2 + 0.883t)^2}{4t}$	
Equal disks smaller than sample.........	$\dfrac{\pi(d_e + 0.442t)^2}{t}$	
Tube or rod		$\dfrac{\pi d_c}{g}$

For highly resistive materials, very thin films can be used, keeping in mind the possibility that near-surface properties of the material may differ significantly from the bulk properties.

The cells of Fig. 3(a) (electrodes 2 and 3) and Fig. 3(e) are used for surface conduction measurements. Surface conductivity is very sensitive to sample preparation and can rarely be measured without some interference from bulk conduction. For these reasons, σ_s is not usually considered a material property. Electrode materials are generally the same as those described for dielectric measurements in the earlier section "Electrode Materials."

Four-point probes are most useful for measuring the resistivity of semiconducting materials. They consist of two electrode pairs, one for imposing a current through the sample, and one for sensing voltage drop. Their major advantage is that interfacial effects such as contact rectification and charge carrier injection at the current-providing electrodes do not influence bulk conduction in the vicinity of the voltage-sensing electrodes.

The standard four-point probe configuration is shown in Fig. 13. Sharp electrode tips contact the samples at four colinear points. Because electric fields are highly nonuniform in these measurements, bulk conductivity must be linear, homogeneous, and isotropic, and surface conduction must not be present. Conductivity is defined by Eq 28 and 29, where I is the current applied between outer electrodes 1 and 4 and V is the voltage measured between inner electrodes 2 and 3. In each case, the sample is assumed to be very large compared to probe dimensions in the plane of four-point contact. Cell constants have been calculated for several sample shapes. For semi-infinite samples (all dimensions being large compared to s_i), K_b is given by (Ref 24):

$$K_b = 2\pi \left[\frac{1}{s_1} + \frac{1}{s_3} - \frac{1}{s_1 + s_2} - \frac{1}{s_2 + s_3} \right]^{-1} \quad \text{(Eq 33)}$$

For equally spaced probes ($s_i = s$) on thin sheet samples,

term device failure, moisture-enhanced conductivity is still the most important factor in the long-term environmental failure of polymer encapsulants and coatings that are used in electronic applications. A number of long-term failure mechanisms have been identified, including corrosion (reviewed by Schnable and co-workers in Ref 18 and Sinclair in Ref 19), electrolytic metal migration and dendrite formation (reviewed by Steppan *et al.* in Ref 20), and conductive anodic filament growth (Ref 21, 22).

A number of accelerated tests are used routinely to screen organic materials under consideration for use in electronic applications. Most common is the temperature-humidity-bias (THB) test. A large dc bias is placed between electrode lines that are in contact with a polymer sample in a humid environment (typically, 85 °C, or 185 °F, at 85% relative humidity). To pass the test, leakage current must remain below a certain threshold, and electrode continuity must be maintained. The THB test duration ranges from 24 to more than 1000 h. Testing times are sometimes reduced by controlling relative humidity at temperatures above 100 °C (212 °F) in an autoclave. Such tests are known as highly accelerated stress tests (HAST), or "pressure-cooker" tests. THB and HAST tests are rigorous; materials that pass them rarely fail in the field. However, it should be noted that failure mechanisms observed in "accelerated" tests are not necessarily ones that take place in operating environments (Ref 23). Therefore, it may not be meaningful to extrapolate accelerat-ed failure results to predict reliability in expected operating conditions.

Parallel-Plate Measurements. Common electrode configurations are shown in Fig. 3. Cell constants and dimensional constraints are listed in Table 2. Figure 3(a) is a guarded configuration and as such is the most useful for making bulk measurements when surface conduction is suspected. The guard electrode protects the measurement electrode from surface currents and, alternatively, allows surface conductivity to be measured. Further, electric fields are nearly uniform (conduction is one-dimensional) in the guarded cells, an important consideration when nonlinear (high-field) or anisotropic conduction is studied.

The cell of Fig. 3(b) is an example of an unguarded parallel-plate cell with insulating edges. For thin film samples, the insulator is air. For thicker samples, it consists of an insulating cylindrical holder fabricated from a machinable insulating ceramic, which is filled with sample and capped by the electrodes. This geometry is useful for measurement of conductivity in materials that are not readily formed into uniform films. As in the guarded cells, the electric field is one-dimensional. Because the cell is not guarded against interfacial conduction, and because sample dimensions cannot be controlled down to extremely small thickness, the ceramic cell is not well suited to measurements on highly resistive materials.

The cells of Fig. 3(c) and (d) are suitable for polymers that can be solvent cast, spin coated, or extruded into uniform thin films.

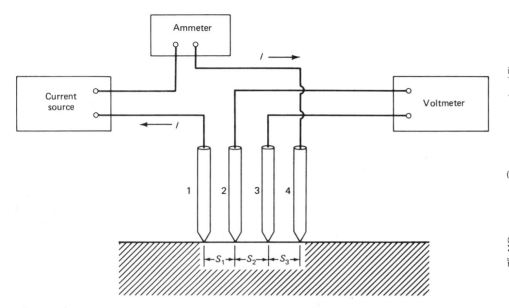

Fig. 13 Four-point probe

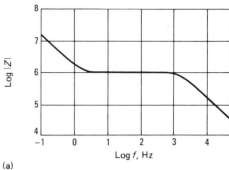

(a)

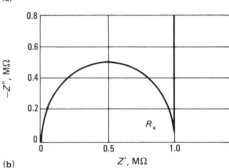

(b)

Fig. 14 (a) Bode plot of blocked sample impedance spectrum. (b) Cole-Cole plot of blocked sample impedance spectrum

$$K_b = \frac{t\pi}{\ln 2} F_1\left[\frac{t}{s}\right] \qquad \text{(Eq 34)}$$

where t is sample thickness and $F_1(t/s) \approx 1$ for $t/s < 0.625$. For thicker samples, $F_1(t/s)$ has been tabulated by Smits (Ref 25). The K_b values have also been calculated for thick sheets and thin sheets on conductive substrates (Ref 24), for probe positions near sample edges (Ref 24, 26), and for thin circular and rectangular samples (Ref 25).

Transient Techniques

When the voltage across measurement samples is varied as a function of time, the current response depends on, among other things, the sample resistivity. A variety of transient measurements have been used to follow charge transfer processes in polymers and at polymer-electrode interfaces. Of these, voltage rate-of-change measurements are the simplest, and ac impedance measurements have proved to be the most versatile.

Voltage rate-of-change measurement is based on the simple model that sample response to an applied voltage is determined only by bulk resistivity and a known frequency-independent bulk dielectric constant; that is, the sample responds as a capacitor and resistor in parallel. The high electrode of an unguarded parallel plate sample is charged to a constant voltage and then disconnected from the voltage source. The voltage of the high electrode is monitored as a function of time as it discharges through the sample to the ground electrode. The transient voltage is given by:

$$V = V_0\, e^{-(t/\tau)} \qquad \text{(Eq 35)}$$

where V_0 is the initial charging voltage, and $\tau = R_b C = \rho_b \epsilon' \epsilon_0$ is the relaxation time constant. Thus, the time rate of decay can

be obtained from the slope of the plot of ln V versus t.

Alternating Current Impedance Measurements. The strength of ac impedance measurements is that they are sensitive to all the conduction and displacement processes that contribute to sample impedance and that they offer the possibility of distinguishing these processes based on their mutual interactions and differing frequency dependences. This section introduces some of the basic concepts of impedance plot analysis and equivalent circuit modeling. For more comprehensive discussions of the use of impedance measurements to study bulk conduction and interfacial electrochemistry, the reader is referred to reviews by Sluyters-Rehbach and Sluyters (Ref 27) and Walter (Ref 28).

Impedances of capacitive and resistive elements have already been defined. Typically, samples can be modeled by a capacitor and resistor in parallel. However, if the sample is an ionic conductor, impedance response may be complicated by a diffusional, or Warburg, impedance. The Warburg impedance for parallel-plate electrodes is (Ref 27):

$$Z_W = \frac{\sigma_W}{\omega^{1/2}}\left[\frac{\sinh(t^*) + \sin(t^*)}{\cosh(t^*) + \cos(t^*)} \right. \qquad \text{(Eq 36)}$$
$$\left. - j\,\frac{\sinh(t^*) - \sin(t^*)}{\cosh(t^*) + \cos(t^*)}\right]$$

where $t^* = t(\omega/2D_i)^{1/2}$ is the dimensionless interelectrode gap and D_i is the mean ionic diffusivity. For relatively high frequencies ($t^* \gg 1$), diffusion layers reside near electrode surfaces, and the Warburg impedance for any electrode geometry follows the form:

$$Z_W = \frac{\sigma_W}{\omega^{1/2}}(1 - j) \qquad \text{(Eq 37)}$$

In this case, the Warburg impedance reflects interfacial concentration and polarization and acts in series with ionic, but not electronic, conduction.

Conduction, dielectric displacement, and diffusion processes are distinguished by their different frequency dependences. For simple systems, these differences result in characteristic forms for Bode ($\log|Z|$ versus $\log f$) and Cole-Cole (Z'' versus Z') impedance plots. Figures 14(a) and (b), for example, show impedance behavior that might be expected for a sample with an interfacial blocking capacitance, based on the equivalent circuit model of Fig. 15. At both low and high frequencies, the Bode plot (Fig. 14a) shows the sample response to be primarily capacitive, with $|Z| \approx Z'' \sim \omega^{-1}$. At intermediate frequencies, the impedance is dominated by sample resistance ($|Z| \approx Z' = R_x$) and is therefore frequency independent. Resistance can be obtained from the impedance magnitude in this frequency region, $|Z| \approx Z'$, or from the real intercept of the Cole-Cole plot (Fig. 14b). Many of the simple impedance plot shapes and corresponding equivalence circuits are presented in Ref 27 and 29.

Ideally, the correct equivalent circuit and corresponding parameter values can be inferred from the impedance response. Often, however, sample response is more complex than that of the sample described above. As an example, a model that has been used to describe conduction between two electrode lines in an epoxy-SiO_2 interface (Ref 23) in the absence of high humidity is shown in Fig. 16. For this experiment, a thin film of

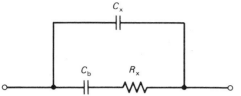

Fig. 15 Blocked sample equivalent circuit

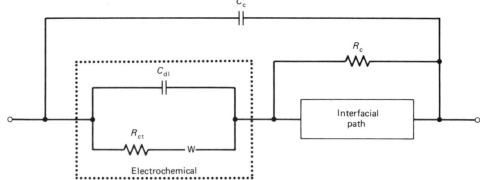

Fig. 16 Model of conduction in an epoxy-SiO₂ interface. Source: Ref 23

epoxy was deposited onto a silicon chip that contained a pattern of microfabricated aluminum lines. Line length was 10^4 times the gap between lines; therefore, the measurement was highly surface sensitive. Further, a porous gold film was deposited on top of the epoxy to allow an independent measurement of bulk impedance response. The epoxy was ionically conductive, and therefore an electrochemical reaction modeled by the Randles equivalent circuit (the dotted box in Fig. 16) was used to describe the interface. Further, the sample response included an epoxy-SiO₂ interfacial conduction process, the model of which is not shown here. Interpretation of the interfacial impedance allowed conclusions to be drawn about the effects of humidity on adhesion in the epoxy-SiO₂ system. As in this example, when sample behavior is nontrivial, it is the task of the experimentalist to select and verify the appropriate equivalent circuit model. The model will consist of circuit elements in series and in parallel. The impedances of two elements in series add:

$$Z_{series} = Z_1 + Z_2 \qquad \text{(Eq 38)}$$

while the admittances of parallel elements add:

$$\frac{1}{Z_{parallel}} = \frac{1}{Z_1} + \frac{1}{Z_2} \qquad \text{(Eq 39)}$$

Based on these relationships, the impedance response of the assumed equivalent circuit model is calculated. Circuit parameters are fit to the data using numerical regression.

The major drawback of equivalent circuit modeling is that the models are not unique. Therefore, the fact that a model is capable of closely describing impedance data is only weak evidence that the model correctly describes the physical conduction process. It is the (often difficult) task of the experimentalist to ascertain whether the physical significance ascribed to model parameters is correct. Conclusions based on circuit elements used to model interfacial impedances, for example, will be unfounded if the corresponding fitted parameter is actually controlled by a bulk process. In this example, the response of the measured parameter to variation of electrode area or sample thickness would confirm that it represents a bulk process. More generally, the validation of complex models usually requires independent measurement of at least some of the physical processes that contribute to sample impedance.

Deviations From Ohm's Law

Although σ_b, as defined by Ohm's Law (Eq 26), is the parameter most commonly used to characterize conduction, a number of factors often contribute to nonohmic behavior. These include high-field effects, interfacial impedances, diffusion, and conduction anisotropy, described below.

Nonlinear Bulk Conduction. Bulk conductivity in many polymers becomes field-enhanced when the applied electric field is sufficiently high. The high-field conduction mechanisms reviewed in Ref 30 and 31 include, for example, the Poole-Frenkel and hopping mechanisms, which allow superlinear conduction. Field-dependent conductivities can be identified and studied in cells that ensure a uniform field in the measurement region of the sample.

Interfacial Effects. Several interfacial effects can lead to electrode-dependent interfacial impedances. These include the formation of blocking contacts, when the charge carriers are electrons or holes, and interfacial high-field effects. When a metal electrode is placed in contact with a semiconducting sample whose charge carriers are electrons, for example, a blocking contact forms if the work function of the metal exceeds that of the material, $\phi_m > \phi_s$. In this case, a positive space charge region (or depletion region) forms in the sample near the interface. The interface is rectifying, with a Schottky potential barrier to electron transfer from the metal to the sample under reverse bias. When the electric field is sufficiently high, nonlinear electrode-limited conduction mechanisms take place, including tunneling and Schottky injection (Ref 32). Techniques that avoid interference from interfacial effects include the four-point probe measurements, xerographic discharge measurements, and ac measurements, all of which are described in this article.

Diffusion. When charge carriers are ionic, charge transport is driven not only by electric fields but also by concentration gradients:

$$i = -F^2 \nabla \phi \sum_i z_i^2 u_i c_i - F \sum_i z_i D_i \nabla c_i \qquad \text{(Eq 40)}$$

where the ionic diffusivities are related to their mobilities by the Nernst-Einstein relationship $D_i = RTu_i$. Apparent sample resistance increases with time after a voltage is imposed as concentration gradients develop. At early times, diffusion layers are thin compared to sample dimensions (see the section "Alternating Current Impedance Measurements"), and the concentration polarization acts in series with sample resistance. However, in low frequency or dc experiments, Eq 40 couples diffusion and conduction throughout the sample. In general, steady-state conduction of ionic charge carriers is not ohmic; thus carrier mobilities can only be inferred from conduction data by simultaneously solving for potential and concentration profiles through the sample. Further, the identity of charge carriers is generally not known. For these reasons, ionic conductivity and diffusivities are best measured by ac techniques, which minimize concentration polarization and decouple diffusion effects from bulk conduction.

Conduction Anisotropy. Several organic solids exhibit metallic conduction, with conductivities exceeding 10^4 S/m (Ref 33). Nearly all of them conduct along a strongly favored molecular direction. When the molecules are aligned by crystallization or flow, the bulk conductivity becomes highly anisotropic. Conduction anisotropies ($\sigma_\perp/\sigma_\parallel$) of 16 have been observed in weakly aligned polyacetylene (Ref 34), and anisotropy on the order of 10^3 has been observed in single crystals of tetrathiofulvalenium tetracyanoquinodimethane (TTF-TCNQ).

Similarly, when polymers are filled with aligned, conductive carbon fibers to create effective conductivities in the metallic range, conduction anisotropies as high as 10^4 have been observed (Ref 35). In these systems, conductivity is a tensor rather than the scalar quantity indicated by Eq 27,

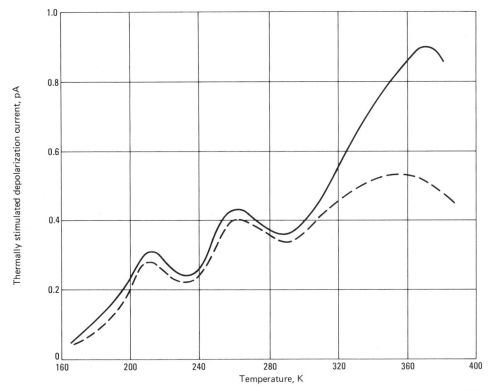

Fig. 17 Thermally stimulated depolarization currents for cross-linked polyethylene polarized at 333 K and 500 V for 30 min. The solid line represents a sample aged for several hours with about 90 kV ac. Source: Ref 40

and cell constants such as those listed in Table 2 generally do not apply. It is therefore essential to perform conductivity measurements in cell geometries that ensure truly one-dimensional electric fields. In particular, care must be taken when measuring conductivity perpendicular to the preferred direction, because conduction in the favored direction may increase the apparent electrode area. Suitable geometries include the parallel-plate cells of Fig. 3(a) and (b).

Conduction Mechanisms

To understand conduction mechanisms, the identity, concentration, and mobility of the dominant charge carriers must be known. Some of the techniques that are used for measuring these variables are described briefly in the sections below.

Photoconduction. Exposure to electromagnetic radiation causes charge carriers to be generated in many polymers that are doped with charge transfer complexes. The study of the resulting photoconduction (Ref 36-38) has proved to be useful for measuring carrier mobility and charge, trap depth, and bulk conductivity of the polymers. The principal techniques used are the xerographic-discharge and time-of-flight techniques. Both techniques can be applied to dark conduction studies, but are most commonly used to study conduction in photoconductive materials.

Xerographic Discharge. In a xerographic discharge experiment, the free surface of a grounded sample film is charged to several kilovolts using a corona discharge. Charge carriers are generated by a flash of light at the appropriate wavelength, and the surface potential decay is monitored with a capacitive probe. The charge carrier mobility can be obtained from the $V(t)$ characteristic. In this experiment, the need for a transparent top electrode is avoided, and space charge effects and photoinjection from electrodes are not considered to be important, as they might be in time-of-flight measurements. In a modification of this technique, the top surface of a weakly or nonphotosensitive sample is coated with a thin film of photosensitive material that generates carriers with high quantum efficiency. With this modification, the transport of selected carriers can be studied.

Time-of-Flight Measurements. In time-of-flight measurements, the sample is sandwiched between two biased electrodes, at least one of which is transparent. A sheet of charge is generated near the transparent electrode by a flash of light, and the current through the sample is monitored as a function of time. If the sheet charge is small compared to the capacitive charge displacement of the measurement cell, and the dark conductivity of the sample is also sufficiently small, the current will have a constant value that is proportional to the carrier mobility until the photogenerated charge is depleted.

Hall Effect. When a semiconducting sample carrying a current (I) is subjected to an orthogonal magnetic field (H), a voltage across the third orthogonal dimension of the sample is generated. This Hall voltage (V_H) indicates electronic conduction and is proportional to both I and H, with a proportionality constant called the Hall constant (R_H). As outlined in Ref 39, R_H provides information about the sign, concentration, and mobility of the charge carriers.

Thermally stimulated depolarization (TSD) measurements are used to study thermal transitions in dipole and chain segment relaxations, as well as charge trapping processes. The experiment involves polarizing the parallel-plate sample by applying an electric field at a temperature (typically just above the glass transition temperature) that is sufficiently high to allow molecular dipoles and trapped charges to redistribute in response to the field. Then the sample is quenched to liquid nitrogen temperature, still under the influence of the field. Finally, the field is removed, and the frozen-in polarization relaxes as temperature is slowly increased, providing current peaks as the sample temperature passes through molecular relaxation and charge-trapping transition temperatures.

Figure 17 shows three TSD peaks measured on a sample of cross-linked polyethylene used as the insulator for a high-voltage cable (Ref 40). In this case, the differences in peak height have been attributed to a change in the supermolecular structure of the partially crystalline material (Ref 41). After exposure to a high-voltage ac signal for several hours, the high-temperature TSD peak increased dramatically, while the two lower-temperature peaks remained essentially unchanged. Dipole relaxations can be identified by dielectric loss measurements (Ref 1, 11), and the remaining characteristic trap relaxations reveal information about electron trap densities and energies. Detailed descriptions of TSD measurements can be found in Ref 41 and 42.

Breakdown

Dielectric breakdown is a primary cause of failure of coatings used to provide insulation in high-voltage applications. In the breakdown process, mobile electrons acquire energy by acceleration in an electric field and then transfer the energy to other mobile or trapped carriers, or to the lattice by dissipative interactions. The breakdown strength of a material is the minimum electric field (F_B), above which the ability of the material to insulate is unstable. Intrinsic breakdown strength is controlled by a wide variety of structural and environmental variables that affect the ability of a material to acquire and dissipate electronic energy.

Because experimental isolation of the intrinsic breakdown process is problematic, intrinsic breakdown mechanisms are not

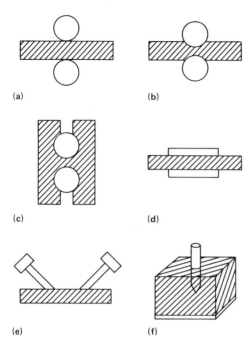

Fig. 18 Sample geometries for breakdown measurements. (a), (b), and (c) Spherical or cylindrical electrodes. (d) Flat-plate electrodes. (e) Tungsten rod assembly for dry arc measurements. (f) Needle-plate geometry

completely understood (Ref 43). The most useful practical tests of breakdown strength are therefore those in which sample preparation and geometry approximate the intended application of the insulating polymer. Both short-term breakdown mechanisms and long-term or aging mechanisms are discussed here. Reviews of breakdown mechanisms can be found in Ref 43 to 45. For details of standard experimental methods, the reader is referred to Ref 43, 46, and 47.

Electrode Geometries for Bulk Measurements

Six electrode and sample geometries used for making breakdown measurements are shown in Fig. 18. The first three use rounded (cylindrical or spherical) electrodes to eliminate the divergent electric fields at electrode edges and the possibility of subsequent flashover (see the section "Water Treeing") that can occur in parallel-plate measurements (Fig. 18d) (Ref 46). The needle-plate geometry is used for treeing experiments (see the section below).

Short-Term Breakdown

The time lag between the onset of voltage stress and breakdown is varied. Short-time measurements are used to characterize breakdown processes that do not depend on the chemical or mechanical degradation of the insulator, or on the absorption of impurities. Such processes include thermal and electromechanical breakdown and surface flashover.

Thermal Breakdown. Joule heating or dielectric dissipation can lead to the thermal and electrical instability of polymers. The experimental literature indicates that for most nonpolar polymers and many polar polymers, breakdown strength is constant, or increases slightly with temperature up to a transition temperature (T_c). Above T_c, F_B falls off rather sharply with temperature (Ref 45). Further, electrical (ionic) conductivity increases with temperature, whereas thermal conductivity decreases with temperature. Therefore, if the applied field causes sufficient sample heating, sample temperature runs away, leading to thermal breakdown.

Heat generation depends on the amplitude and frequency of voltage stress, while heat dissipation depends on sample and electrode geometry. The simplest measurement is that of impulse thermal breakdown, in which breakdown takes place before significant thermal conduction can occur. In general, a description of thermal breakdown requires the solution of coupled heat and charge transfer equations for each electrode geometry, with temperature-dependent parameters. O'Dwyer (Ref 44) discusses the experimental and mathematical problems.

Electromechanical Breakdown. The electrostatic attractive force between oppositely polarized electrodes may be sufficient to cause thermoplastic polymers to flow to creep to the point that the decreased electrode gap causes the interelectrode field to exceed the breakdown threshold. As reviewed by Fischer (Ref 43), there is sufficient experimental evidence that this electromechanical breakdown does take place in some materials and geometries, although the rheological models that have been applied are not sufficiently rigorous to predict breakdown conditions quantitatively.

Flashover (Surface) Breakdown. Flashover is the occurrence of surface breakdown (both electrical and chemical) initiated by arcing in the surrounding medium or corona discharge near the electrode-sample contact edge. Usually, flashover occurs when thin surface layers in the arc region become conductive. Flashover can be reversible if it is caused by surface melting, for example, or irreversible if it results from decomposition of the sample surfaces. Flashover can be a problem in both bulk breakdown experiments and in practical applications.

ASTM D 495 (dry arc) and D 2275 (surface corona) standard tests (Ref 3) are commonly used to characterize the flashover resistance of a treated surface. In dry arc measurements, short-duration dc arcs of increasing severity are imposed between strip or rod electrodes (Fig. 18e) until flashover takes place; the current and duration of the electric arc required to cause flashover provide a basis for comparison with other surfaces. In surface corona measurements, two interfacial electrodes are sufficiently polarized to obtain corona formation near the electrode-sample edges. Surface corona resistance is measured as time to breakdown. In both dry arc and surface corona measurements, careful attention to surface preparation and relative humidity are required to obtain reproducible results.

In bulk breakdown measurements, the chance of flashover is minimized by using electrodes with rounded edges to avoid divergent fields near sample surfaces, and by constructing samples with long surface paths between electrodes.

Aging

Polymer degradation by long-term environmental exposure, mechanical stress, and sustained voltage stress is known to cause a number of gradual breakdown processes. In general, it is difficult to predict service lifetimes on the basis of accelerated breakdown tests without actual field data to back up the results; thus, the following accelerated aging tests should be treated as qualitative failure predictors.

Discharge in Voids. Discharge in gas-filled voids is similar to surface corona discharge. It adds thermal and chemical energy to the void surfaces, leading to damage of the polymer and, with time, complete sample breakdown. Void discharges are most destructive when the material is subjected to ac stresses, and failure is accelerated by increased frequency. Some success in characterizing void discharges has been achieved (Ref 44). A void may be completely stable to discharge when the voltage across it is below a characteristic inception voltage. In addition, the characteristic discharge magnitude has been used to describe discharge power in voids. However, descriptions of the degradation processes that lead to complete breakdown are lacking. Further, the size, shape, gas composition, and gas-polymer interfacial properties are generally difficult to control or characterize, and discharges measured in artificial voids poorly reproduce the natural discharge processes. As a result, standard measurement techniques have not been developed.

Electrical Treeing. Electrical trees are branched structures of tubular filaments that originate from points of divergent electric field, such as the tips of conductive inclusions of water-filled voids. The filaments are regions of bulk decomposition of the polymer into gaseous fragments that form structures ranging in appearance from sparsely branched to bushy. Trees grow at relatively lower applied fields than those required to cause short-term breakdown, and trees cause breakdown in materials that would otherwise have high dielectric strength.

Experimental techniques for measuring the initiation times and propagation rates of electrical trees have been reviewed by Mason (Ref 47). All measurements are made by embedding two electrodes, at least one of which is a sharp needle (see, for example, Fig. 18f), into the polymer prior to voltage exposure. Sample geometry, particularly sharpness of the needle tip, influences treeing. Although the maximum electric field is easily calculated from the radius of curvature of the needle tip (ASTM D 3756 in Ref 3), there is evidence that conduction and space charge moderate the electric field strength near the electrode tip (Ref 45). Therefore, tip sharpness beyond a certain point may not be important. Direct current or relatively low ac voltages are applied to the sample, either with increasing intensity or for extended periods. Initiation and growth are detected electrically or optically by the characteristic spurts of localized breakdown and charge flow. The most commonly measured characteristics are tree inception time at a given voltage, the voltage at which trees initiate in half of the samples after a given time, and the time to breakdown. It is commonly found that breakdown time varies with applied field by an inverse power law ($t \sim F^{-n}$) where n can be as high as 20. The power law provides a useful means for extrapolating insulator service lifetimes from (relatively) short-term breakdown data.

Water Treeing. Water trees, which are similar in appearance to electrical trees, form in the simultaneous presence of electrical stress (usually ac) and water or other polar liquid. Water treeing is a common problem in cable insulators such as the cross-linked polyethylene used in undersea cables. Unlike electrical trees, water filaments are water-filled cracks rather than gas-filled tubes. When water trees are dried out, they become invisible, but quickly refill upon water exposure by osmotic or capillary forces. Water tree inception times are short compared to electric tree times, even at lower electric fields. However, water tree growth is relatively slow and is not accompanied by measurable electrical discharges. Proposed mechanisms of water tree growth are reviewed by Fischer (Ref 43). All involve the development of high mechanical stress at branch tips in the presence of electrical field, usually as a result of water decomposition at crack tips, where the electric field diverges.

Water tree inception and propagation can be measured using similar test geometries that are similar to those used in electrical tree measurements. However, samples must be exposed to the voltage stress and an aqueous environment at the same time. Thus, the ground electrode is immersed in the aqueous solution. As pointed out by Noto (Ref 48), most water trees of practical importance are of the "bow-tie" type, which form at both ends of slender natural inclusions or water-filled (osmotic) voids.

As in dry-void measurements, the understanding of water-tree-induced breakdown is not sufficient to simulate or characterize such inclusions properly.

Tracking. When hygroscopic contaminants such as dust accumulate on an insulator surface, sufficiently high relative humidities can cause water condensation and the subsequent formation of surface conduction paths. If highly biased electrodes lie on the same surface, unsteady flashover will occur, accompanied by resistive heating and local drying of the surface film. After a time, surface decomposition will result in the growth of conductive carbonaceous tracks that ultimately bridge the electrodes and form a short.

Tracking resistance is commonly measured using the ASTM D 2303 (Ref 3) method. Electrode geometry is similar to that used in the flashover measurement (Fig. 18e). However, the electrode is inclined and is subjected to a steady flow of 0.1% NH_4Cl solution, which can be modified to simulate expected contaminants. A 60 Hz voltage stress of stepwise increasing amplitude is imposed on the sample until tracking initiates. Tracking resistance is usually characterized by the minimum tracking voltage, as well as by the time required for tracks to reach a specified length.

REFERENCES

1. S. Matsuoka, Dielectric Relaxation, in *Encyclopedia of Polymer Science and Engineering*, Vol 5, 2nd ed., John Wiley & Sons, 1986
2. R. Bartnikas, Alternating-Current Loss and Permittivity Measurements, in *Engineering Dielectrics*, Vol IIB, STP 926, R. Bartnikas, Ed., American Society for Testing and Materials, 1987
3. *Electrical Insulation and Electronics*, Vol 10.02, *Annual Book of ASTM Standards*, American Society for Testing and Materials
4. M.J.C. van Gemert, High-Frequency Time-Domain Methods in Dielectric Spectroscopy, *Phillips Res. Rep.*, Vol 28, 1973, p 530
5. A. Suggett, Time Domain Methods, in *Dielectric and Related Molecular Processes*, Vol 1, M. Davies, Ed., The Chemical Society, 1972
6. F.I. Mopsik, Precision Time-Domain Dielectric Spectrometer, *Rev. Sci. Instrum.*, Vol 55 (No. 1), 1984, p 79
7. W.P. Harris, "Operators Procedure Manual for the Harris Ultra-Low Frequency Impedance Bridge," Report 9627, National Bureau of Standards, 1968
8. D.J. Scheiber, An Ultra-Low Frequency Bridge for Dielectric Measurements, *J. Res. Natl. Bur. Stand.*, Vol 65C, 1961, p 23
9. R.H. Cole and P.M. Gross, A Wide Range Capacitance-Conductance Bridge, *Rev. Sci. Instrum.*, Vol 20 (No. 4), 1949, p 252
10. W.E. Vaughan, Experimental Methods, in *Dielectric Properties and Molecular Behavior*, Van Nostrand Reinhold, T.M. Sugden, Ed., 1969
11. P. Hedvig, *Dielectric Spectroscopy of Polymers*, John Wiley & Sons, 1977, p 156-160
12. P. Hedvig, Dielectric Relaxation Phenomena: Experimental Aspects, *IEEE Trans. Electr. Insul.*, Vol EI-19 (No. 5), 1984, p 371
13. L. Hartshorn and W.H. Ward, The Measurements of the Permittivity and Power Factor of Dielectrics at Frequencies from 10^4 to 10^8 Cycles per Second, *J. Inst. Electr. Eng. (London)*, Vol 79, 1936, p 597
14. C.N. Works, T.W. Dakin, and F.W. Boggs, A Resonant Cavity Method for Measurement of Dielectric Properties at Ultrahigh Frequencies, *Trans. Am. Inst. Electr. Eng.*, Vol 63, 1944, p 1092
15. R.H. Cole, Time Domain Reflectometry, *Annu. Rev. Phys. Chem.*, Vol 28, 1977, p 283
16. D. Chahine and T.K. Bose, Drift Reduction of the Incident Signal in Time Domain Reflectometry, *Rev. Sci. Instrum.*, Vol 54 (No. 9), 1983, p 1243
17. M.F. Hoover and H.E. Carr, *TAPPI J.*, Vol 51, 1968, p 552
18. G.L. Schnable, R.B. Comizzoli, W. Kern, and L.K. White, A Survey of Corrosion Failure Mechanisms in Microelectronic Devices, *RCA Rev.*, Vol 40, 1979, p 417
19. J.D. Sinclair, Corrosion of Electronics: The Role of Ionic Substances, *J. Electrochem. Soc.*, Vol 153 (No. 3), 1987, p 175
20. J.J. Steppan, J.A. Roth, L.C. Hall, D.A. Jeannotte, and S.P. Carbone, A Review of Corrosion Failure Mechanisms During Accelerated Tests, *J. Electrochem. Soc.*, Vol 134 (No. 1), 1987, p 175
21. D.J. Lando, J.P. Mitchell, and T.L. Welsher, Conductive Anodic Filaments in Reinforced Dielectrics: Formation and Prevention, in *Proceedings of the 17th Annual Reliability Physics Conference*, Institute of Electrical and Electronics Engineers, 1979, p 51
22. J.P. Mitchell and T.L. Welsher, Conductive Anodic Filament Growth in Printed Circuit Materials, *Proc. Printed Circuit World Conv. II*, Vol 1, 1981, p 80
23. K.M. Takahashi, AC Impedance Measurements of Moisture in Interfaces Between Epoxy and Oxidized Silicon, *J. Appl. Phys.*, Vol 67 (No. 7), 1990, p 3419
24. L.B. Valdes, Resistivity Measurements on Germanium for Transistors, *Proc.*

IRE, Vol 42, 1954, p 420

25. F.M. Smits, Measurement of Sheet Resistivities With the Four-Point Probe, *Bell System Tech. J.*, 1958, p 711

26. A. Uhlir, Jr., The Potentials of Infinite Systems of Sources and Numerical Solutions of Problems in Semiconductor Engineering, *Bell System Tech. J.*, 1955, p 105

27. M. Sluyters-Rehbach and J.H. Sluyters, Sine Wave Methods in the Study of Electrode Processes, in *Electroanalytical Chemistry*, Vol 4, A.J. Bard, Ed., Marcel Dekker, 1970

28. G.W. Walter, A Review of Impedance Plot Methods Used for Corrosion Performance Analysis of Painted Metals, *Corros. Sci.*, Vol 26 (No. 9), 1986, p 681

29. K.S. Cole and R.H. Cole, Dispersion and Absorption in Dielectrics: I. Alternating Current Characteristics, *J. Chem. Phys.*, Vol 9, 1941, p 341

30. M. Ieda, Electrical Conduction and Carrier Traps in Polymeric Materials, *IEEE Trans. Electr. Insul.*, Vol EI-19 (No. 3), 1984, p 162

31. K.C. Kao and W. Hwang, *Electrical Transport in Solids, With Particular Reference to Organic Semiconductors*, Pergamon Press, 1981

32. D.M. Taylor and T.J. Lewis, Electrical Conduction in Polyethylene Terephthalate and Polyethylene Films, *J. Phys. D, Appl. Phys.*, Vol 4, 1971, p 1346

33. M.R. Bryce, Organic Conductors, *Chem. Britain*, Vol 24 (No. 18), 1988, p 781

34. C.K. Chiang, A.J. Heeger, and A.G. MacDiarmid, Synthesis, Structure, and Electrical Properties of Doped Polyacetylenes, *Ber. Bunsenges, Phys. Chem.*, Vol 83, 1979, p 407

35. F. Carmona and A. El Amarti, Anisotropic Electrical Conductivity in Heterogeneous Solids With Cylindrical Conductive Inclusions, *Phys. Rev. B*, Vol 35 (No. 7), 1987, p 3284

36. J. Mort and G. Pfister, Photoelectronic Properties of Photoconducting Polymers, in *Electronic Properties of Polymers*, J. Mort and G. Pfister, Ed., John Wiley & Sons, 1982

37. J. Ulanski, J.K. Jeszka, and M. Kryszewski, Electrical Properties of Polymers Doped With Charge Transfer Complexed Forming Additives, *Polym.-Plast. Technol. Eng.*, Vol 17 (No. 2), 1981, p 139

38. J.M. Pearson and S.R. Turner, The Nature and Applications of Charge-Transfer Phenomena in Polymers and Related Systems, in *Molecular Association*, Vol 2, R. Foster, Ed., Academic Press, 1979

39. F. Gutmann and L.E. Lyons, *Organic Semiconductors*, John Wiley & Sons, 1967

40. S.M. Gubanski and B. Masurek, "Aging Studies of a 20 kV Polyethylene Cable Insulation in an Increased Electric Field," Paper presented at the European Symposium on Polymeric Materials, Lyon, France, 1987

41. R.J. Fleming, Thermally-Stimulated Conductivity and Luminescence in Organic Polymers, *IEEE Trans. Electr. Insul.*, Vol 24 (No. 3), 1989, p 523

42. *Engineering Dielectrics, Vol IIB: Electrical Properties of Solid Insulating Materials: Measuring Techniques*, STP 926, R. Bartnikas, Ed., American Society for Testing and Materials, 1987

43. P. Fischer, Dielectric Breakdown Phenomena in Polymers, in *Electrical Properties of Polymers*, D.A. Seanor, Ed., Academic Press, 1982

44. J.J. O'Dwyer, *The Theory of Electrical Conduction and Breakdown in Solid Dielectrics*, Clarendon Press, 1973

45. M. Ieda, Dielectric Breakdown Process of Polymers, *IEEE Trans. Electr. Insul.*, Vol EI-15 (No. 3), 1980, p 206

46. K.N. Mathes, Electrical Properties, in *Methods of Experimental Physics*, Vol 16C, Academic Press, 1980

47. J.H. Mason, Assessing the Resistance of Polymers to Electrical Treeing, *IEE Proc.*, Part 1, Vol 128 (No. 3), 1981

48. F. Noto, Research on Water Treeing in Polymeric Insulating Materials, *IEEE Trans. Elect. Insul.*, Vol EI-15 (No. 3), 1980, p 251

Evaluating Test Geometries

David W. Schmueser, Engineering Mechanics Department, General Motors Research Laboratories

STRUCTURAL ADHESIVE JOINT DESIGN is frequently dependent on extensive experiments that are needed to assess the mechanical and physical properties of different adhesive systems. Adhesives tests can be used for various reasons, including checking the effectiveness of different surface preparations, comparing the mechanical properties of a group of adhesives, and generating parameters that can be used to predict the actual performance of a bonded joint. The first two of these can be classified as qualitative tests, in which a variety of test methods could be applied. Quantitative information, required for the third reason, is more difficult to obtain.

The most commonly used qualitative test methods for assessing adhesive joint performance are those issued by the American Society for Testing and Materials (ASTM). Several ASTM standard tests are listed in Table 1, along with schematics of the particular test geometry. With the exception of a few of these test methods that measure basic engineering properties of the adhesive (the napkin ring test, for example), the usefulness of most of the test methods is that they allow comparative tests to be conducted because they are standardized from one laboratory to another. However, individual applications of adhesive joints often require novel test techniques and geometries for quantitative joint assessment. For cases in which the adherends are very brittle or are difficult to machine, it becomes necessary to develop specialized geometries and test procedures. This article is devoted to a discussion of these specialized procedures.

Stress analysis of adhesive joints requires a knowledge of the basic engineering properties of the adhesive. Typically, the main properties required are Young's, or tensile, modulus (E); shear modulus (G); and yield stresses and fracture strains for uniaxial tension and pure shear. Most of the common tests listed in Table 1 do not enable these properties to be determined because of the complex state of stress that is induced in the adhesive layer by the specimen geometry. Consequently, two different approaches have been adopted to overcome this problem: The first is to measure the properties by preparing bulk adhesive specimens, and the second is to measure these properties by using specially designed joint geometries with a thin film bond line.

The preparation of bulk adhesive test samples has been an approach taken by many experimentalists (for example, Ref 1, 2). Tensile stress versus strain properties of the adhesive are obtained by testing specimens that are cast and machined into desired shapes, such as a cylindrical hourglass (Ref 2). The bulk adhesive samples are usually carefully polished to avoid premature failure due to surface scratches. One problem with many brittle structural adhesives is that the tensile stress-strain response and corresponding fracture strain are particularly sensitive to grip conditions, surface marks, and air bubbles.

The second approach to determining basic adhesive engineering properties is to employ specific geometries that test, *in situ*, the adhesive in a thin film form. For these ideal geometries, the stress state should be simple, no stress concentrations should be present, and adequate surface treatment should ensure that the failure mode is cohesive in the adhesive layer.

The tensile joint geometries listed in Table 1 do not achieve these objectives. The single-lap-shear specimen, for example, is easy to prepare and test, and the lap-shear strength, the maximum load divided by the bond area, is taken to be the strength of the adhesive. However, the lack of complete tensile and bending rigidity in the specimen adherends gives rise to differential bend and shear in the joint which in turn produces peeling stresses and nonuniform shear stresses in the adhesive layer. Even if the adherends were absolutely rigid, nonuniform shear stresses would still result from stress concentrations formed at the bimaterial corner by the adhesive termination and the adherend. Thus, while the single-lap-shear specimen is useful for ranking adhesive formulations or surface preparations, the lap-shear strengths are not really the strength of the adhesive and cannot be used as a joint design parameter.

The designs of some of the additional specimens listed in Table 1 represent attempts to obtain more uniform stress states based on modifications to the single-lap-shear specimen. The double-lap-shear joint reduces the severity of the bending deformations through its more symmetric geometry. Nonetheless, bimaterial corners still give rise to nonuniform stress states.

The thick adherend single-lap-shear specimen is another modified single-lap-shear design. Although the thick adherends greatly reduce peeling, stress concentrations are still present toward the adhesive edges.

Finally, concluding the consideration of tension-loaded shear specimens, Arcan has recently proposed a stiff tensional shear specimen (Ref 3) based on a specimen geometry that had previously been developed for homogeneous and composite materials (Ref 4). When comparing adhesive shear stresses in the stiff tensional specimen to those in the thick-adherend single-lap joint, it was determined that shear stress concentrations were reduced by 30% in the stiff tensional specimen.

The application of shear loads to a thin film specimen is another type of *in situ* test. The most commonly used geometry in this regard is the napkin ring specimen (Table 1), which approaches the ideal of an adhesive in a state of uniform shear stress (Ref 5). However, the ratio of wall thickness to specimen diameter should be kept small to reduce the influence that misalignment might have on bond line displacements. A cone-and-plate specimen (Ref 6), commonly used for viscous fluid tests, has also been proposed as an alternative torsional thin film specimen. Based on the assumption that the adherends are rigid, it was shown that the shear stress distribution was more uniform than that of the napkin ring specimen. However, for typical structural adhesives, the rigidity assumption is not valid.

Another consideration in designing adhesively bonded joints is the possibility of crack growth, either within the adhesive or at or near the interface between the adhesive and the adherend. This crack growth may be catastrophic when the fracture toughness of the adhesive or adhesive-adherend interface has been exceeded. Because the mechanics of fracture is involved, it is useful to express the resistance to crack growth in terms of some fracture parameter that reaches a critical value for catastrophic growth. Crack paths in bonded joints are frequently constrained by stiffer and tough-

Table 1 Standard test methods for adhesive joints

Joint geometry	ASTM standard	Description	Joint geometry	ASTM standard	Description
Tensile loaded butt joints	D 897	For substrates in block form	Torsional shear joints	D 3658	Specifically for ultraviolet-light-cured glass-to-metal joints
	D 2094 and D 2095	Specifically for bar- and rod-shape substrates		D 229	Uses a napkin ring test to determine shear modulus and shear strength of structural adhesives
	D 429	Rubber-to-metal bonding			
	D 816	Specifically for rubbery adhesives	Cleavage joints	D 1062	Metal-to-metal joints
	D 1344	Cross-lap specimen specifically for glass substrates			
Tension loaded lap joints	D 1002	Basic metal-to-metal single-lap joint for single-lap-shear strength			
	D 2557	Basic low-temperature specimen		D 3807	For engineering plastics substrates
	D 3163	Joint for rigid plastic substrates			
	D 2295	Basic elevated-temperature specimen		D 3433	Contoured-cantilever-beam specimen for determining G_{1c}
	D 3164	Single-lap joint in which a small rectangle of plastic is sandwiched between metal substrates	Peel joints	D 3167	Floating-roller test
	D 3528	Basic metal-to-metal double-lap joint giving double-lap-shear strength		D 903	180° peel test
	D 3165	Metal-to-metal laminate test for large bond areas (also can be used with plastic substrates)			
	D 906	For adhesives used in plywood laminate construction		D 1876	T-peel test
	D 2339	For wooden two-ply laminate substrates			
	D 3983	Thick substrates used in a single-lap joint; shear modulus and strength of adhesive determined		D 1781	Climbing drum peel for sandwiches
				D 429	For rubber-to-metal bonding
				D 2558	For shoe-soling materials
Compressively loaded lap joints	D 905	Specifically intended for wooden substrates	Impact resistance	D 950	Block shear specimen
	D 4027	Rail-constrained specimen for measuring shear modulus and strength			

er adherends. Therefore, crack growth is frequently of a mixed-mode nature, involving the opening (mode I) and forward shearing (mode II) components. The question of an appropriate choice of mixed-mode fracture parameter for joint design has received a considerable amount of attention recently. Finite-element analysis can be applied to determine mode mix for a given joint geometry. However, some difficulties have been encountered in applying this computational technique. First, the magnitude of the fracture parameters are sensitive to the number of elements used to model the joint. Secondly, the assumption of self-similar crack growth, coplanar in the middle of the adhe-sive thickness, is typically made. This assumption of coplanar growth is not valid for pure mode II crack growth.

In addition to finite-element analysis, a great deal of work has been focused on developing specialized test specimens that could be used to evaluate fracture behavior over the full range of mode mixes. The double cantilever beam specimen gives pure peel (mode I) debond modes (Ref 7-10), while the end-notched flexure specimen yields pure forward shear (mode II) behavior (Ref 11, 12). Intermediate-mode mixes can be obtained by using the cracked lap-shear specimen (Ref 13, 14) to yield peel behavior ranging from 20 to 35%.

Finally, an important feature of any transportation vehicle structure is its behavior under impact loading conditions. Adhesive bonding provides an alternative joining technique for bonding lightweight structures composed of aluminum and/or composite materials. Adhesives, being polymeric materials, have properties that are, in general, rate dependent. Therefore, there is a need to give structural designers some confidence that bonded joints designed for normal operating conditions will maintain their structural integrity under impact conditions. The limited work that has been completed in this area has consisted of applying specialized test procedures and

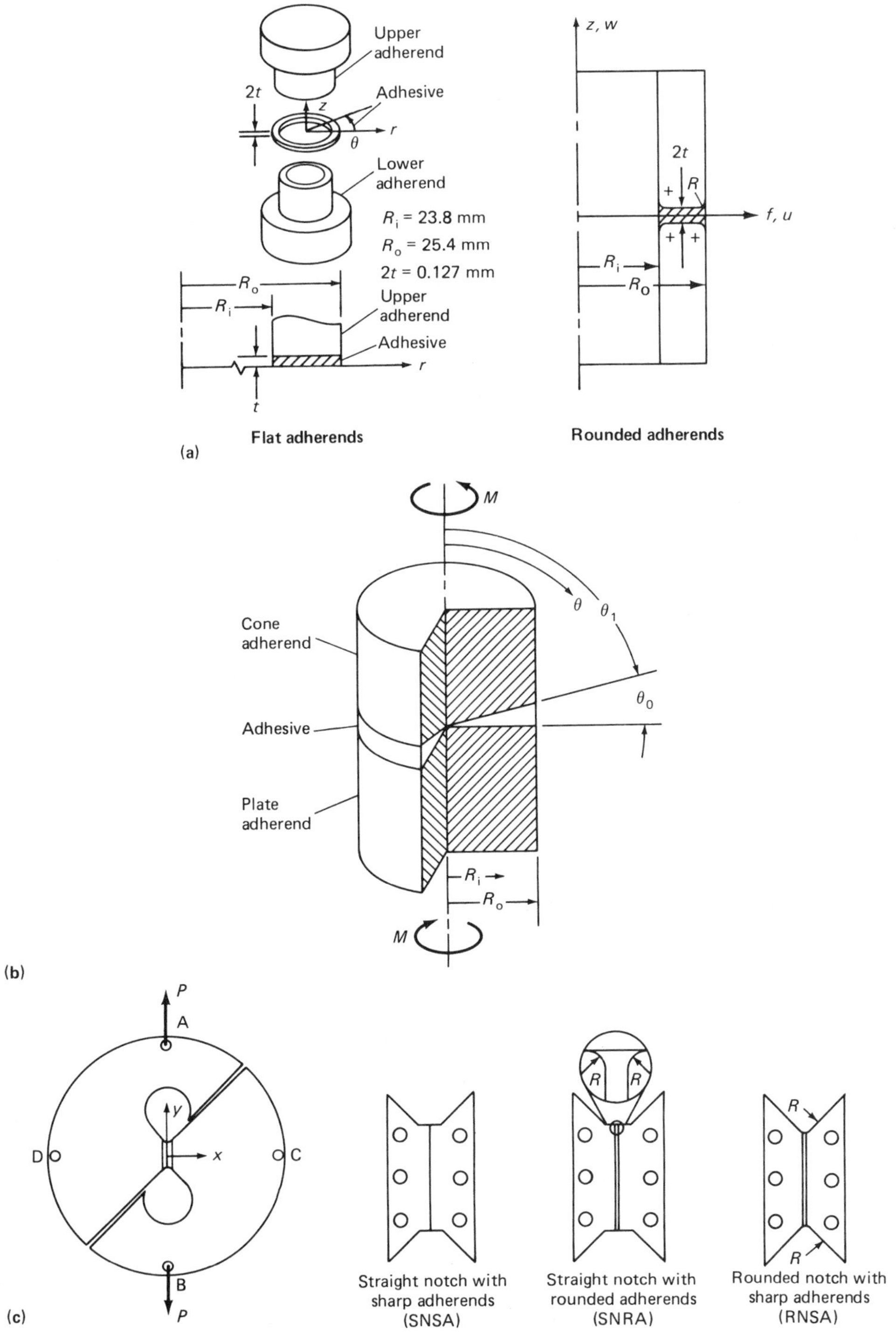

Fig. 1 Specimen geometries for *in situ* testing. (a) Napkin ring. (b) Cone and plate. (c) Stiff adherend. Source: Ref 19

ing, the evaluation of mixed-mode fracture behavior of adhesives and surface preparations, and the characterization of the performance of adhesives under impact load conditions. The following section presents appropriate philosophies for designing and implementing test techniques for those applications. A discussion of experimental and computational tools to assess the performance of several of the specialized tests concludes the evaluation of test geometries.

Approaches for Designing Specialized Test Techniques

This section specifically describes *in situ* testing of bonded joints, the characterization of mixed-mode bond line fracture, and the impact performance of bonded joints.

In Situ Testing of Bonded Joints

When applying *in situ* tests to determine the engineering properties of an adhesive, the specific joint geometries should allow the application of external loads that produce simple stress states, produce uniform stress states by minimizing the effects of stress concentrations, and ensure cohesive failure of the specimen. Several of these issues are discussed below.

Significant Specimen Loading Modes. The simplicity of the stress state in a specimen can be assessed by determining stress distributions that result from shear, bond-normal, and thermal loads.

Shear. For the case of uniaxial tension loads, the single-lap-shear joint does not achieve a simple stress state with a lack of stress concentration. Differential shear and bending of the joint produce a combination of shear and peel stress within the adhesive. Stress concentrations also are significant at the bimaterial corner where the adhesive terminates at the adherend. For this reason, shear stresses are frequently achieved by applying torsion loads. The stiff-adherend specimen described below is a special case in that significant amounts of shear deformation are achieved by applying tension loads through S-shape grips.

Bond-normal loading is an important consideration because slight misalignment of a specific specimen geometry can result in displacements across the adherends. Another reason for the importance of this loading mode is that, as bonded joint analyses become more sophisticated, multiaxial yielding of the adhesive will have to be modeled. Therefore, there will be a need for specimen geometries that provide uniform stress distributions under bond-normal loads for the establishment of multiaxial failure criteria.

Thermal loading results from cool-down of the adhesive following cure. The resulting thermal residual stress state can be quite severe, particularly for the high-temperature-curing (371 °C, or 700 °F) structural

joint geometries to characterize the impact strength of joints. Harris and Adams (Ref 15) used an instrumented impact test configuration for the single-lap-shear joint to compare strength at high rates of loading to static strength values. Kinloch and Kodokian (Ref 16, 17) employed a single-edged notched specimen for bulk and *in situ* tests to characterize linear elastic fracture toughness values for mode I fracture behavior under impact loads. Test results showed

impact toughness values for rubber-modified epoxy adhesives to be significantly greater than that of unmodified epoxy adhesives. The differences arose mainly from the energy-dissipating mechanisms at the crack tip that involve rate-dependent shear deformations in the epoxy.

This introductory section has identified three applications that require specialized test geometries: the determination of adhesive material properties through *in situ* test-

adhesives that currently are under development.

Specimens for In Situ Testing. When shear, bond-normal, and thermal loads are applied to an *in situ* specimen, there is the possibility of stress concentrations developing at bimaterial corners formed by adhesive termination points. Various modifications to specimen geometries, such as rounding or beveling adherend edges, can be considered in order to improve the uniformity of adhesive layer stresses.

The napkin ring geometry is shown in Fig. 1(a). It consists of two adherend tubes butt joined by an adhesive ring. When a torque load (*T*), is applied to the specimen, the shear stress (τ), at a radial distance (*r*), from the z axis is:

$$\tau = \frac{2Tr}{\pi(R_o^4 - R_i^4)} \qquad \text{(Eq 1)}$$

Using Eq 1 and the dimensions shown in Fig. 1(a), the shear stress can be shown to be within 3% of a mean value throughout the adhesive ring. However, work completed by Hein and Erdogan (Ref 18) suggests that significant stress concentrations near the adhesive edges corresponding to the inner and outer radii occur for bond-normal loads. Thus, as shown in Fig. 1(a), rounding the adherend edges is one approach to reducing stress concentrations. As described below in performance comparisons of the *in situ* specimens, beveling the adherend geometries is another method for stress-concentration reduction.

The cone-and-plate specimen geometry, shown in Fig. 1(b), consists of solid lower and upper circular adherends having flat and conical surfaces, respectively, which define the adhesive volume. For the performance comparisons discussed below, the outer radius R_o was taken to be 12.7 mm (0.5 in.), while the cone angle was 3°. As discussed in Ref 6, typical adherend materials only approximate the assumption of rigid adherends that must be made with this specimen to produce uniform shear stresses. For structural adhesives, the rigidity assumption is not valid, and shear stresses are accompanied by highly compressive stresses near the specimen center. Therefore, a modification that can be applied to this specimen is a hole drilled down the center axis.

The stiff-adherend specimen, shown in Fig. 1(c), is a modification of a specimen that was employed for the shear testing of homogeneous materials (Ref 3). The specimens are attached to S-shape grips in the central region. A tension load is applied to these grips parallel to the bond line to produce shear stress in the adhesive. Because the right-corner bond line terminations for this specimen are sources of significant stress concentrations, the adherend edges can be rounded.

The baseline specimen geometry (Fig. 1c) has a straight notch with sharp adherends (SNSA). As described in Ref 19, the adherends can be rounded to a value equal to the adhesive thickness (*R* = 2*t*). This configuration can be termed straight notch with rounded adherends (SNRA). Another modification considered in Ref 19 was a rounded notch with sharp adherends (RNSA). The SNRA and RNSA modifications are shown in Fig. 1(c).

Comparison of Specimen Performance. The performance of *in situ* specimens was compared in Ref 19 by conducting stress analysis using the finite-element method. The adherends and adhesive are assumed to be isotropic, linearly elastic materials. The adherends considered in Ref 19 were aluminum with a Young's modulus of 68.9 GPa (10×10^6 psi), Poisson's ratio of 0.33, and a coefficient of thermal expansion (CTE) equal to 54×10^{-6}/K. The corresponding adhesive properties were 2.3 GPa (0.33×10^6 psi) for Young's modulus, 0.4 for Poisson's ratio, and 167×10^{-6}/K as a CTE. Shear loads were applied to the specimens by either torque or tension-shear loads. The bond-normal loads were approximated by imposing a displacement equal to 25.4 μm (1 mil) normal to the bond line. Thermal loads were represented by a cool-down to 21 °C (70 °F) from a stress-free cure temperature of 121 °C (250 °F).

Stress distributions at the middle of the adhesive layer can be normalized by dividing local stresses by an average shear stress (τ_0) or normal stress (σ_0). In order to compare specimen performance, a uniformity coefficient can be defined as the fraction of the length of the bond line over which the stresses are within ±10% of τ_0 or σ_0.

Uniformity coefficients computed in Ref 20 for each of the three *in situ* specimens are compared in Fig. 2. The percentages for the coefficients are grouped according to load type. As shown in the bar graphs, the uniformity coefficients are usually significantly reduced by attempting to reduce stress concentrations at bimaterial corners. Therefore, the most promising specimen is determined not only by the highest uniformity coefficient, but also by the degree of stress concentration. For example, for the modified napkin ring results, the flat heads had the greatest uniformity coefficient under bond-normal or shear loads. However, stress concentrations were eliminated from the rounded adherend specimens. Therefore, even though its uniformity coefficient was relatively low, it is a more desirable specimen because stress concentration reduction has a more significant influence on bond line strength.

Of the test geometries examined, the core-and-plate geometry was the least promising in that it had the lowest stress uniformity. The napkin ring and stiff-adherend geometries differed in residual stress states with the latter specimen providing the most uniform state. Within the family of

Table 2 Mode-mix ratios for adhesive fracture specimens

Specimen type	Mode mix (G_I/G_t)
Double cantilever beam	1.00
End-notched flexure	0.00
Mixed-mode flexure	0.57
Cracked lap shear	0.28

modifications made to the stiff adherend geometry, the specimen with a straight notch and rounded adherends (SNRA) was the optimum configuration providing reasonable uniformity without stress concentrations for all load cases. Thus, of the three geometries considered for *in situ* testing, the stiff-adherend specimen is the most promising for optimum performance. However, these butterfly-shaped specimens require significantly more fabrication steps than the other specimens.

Characterization of Mixed-Mode Bond Line Fracture

A primary consideration in designing structural, adhesively bonded joints is the possibility of crack growth, either within the adhesive or at a bond line interface. The crack or debond growth can be catastrophic if the fracture toughness of the adhesive or interface is exceeded. Therefore, because debond growth is characteristic of bond failure, it is desirable to express the resistance to fracture in terms of a debond parameter that reaches a critical value for catastrophic growth. One such parameter is the strain-energy release rate (*G*), the amount of energy dissipated per unit amount of crack extension.

Adhesives are used for joining a wide variety of materials under equally diverse loading conditions. Because adhesive layers are usually constrained between relatively stiff and tough materials, crack growth is often constrained to be coplanar in spite of the mixed-mode nature of the loading. Whereas in homogeneous, monolithic materials it is usually sufficient to characterize fracture resistance through the mode I component of fracture resistance because crack paths can orient themselves appropriately, the fracture behavior of adhesives tested *in situ* must be determined for mode I and mode II behavior and various combinations thereof.

The following presents a discussion of similar specimen geometries that can be applied under slightly different loading conditions to determine pure and mixed-mode fracture resistance of adhesives.

Specimens for Fracture Characterization. The specimens considered for this review are the double-cantilever-beam (DCB), end-notched-flexure (ENF), mixed-mode-flexure (MMF), and the cracked-lap-shear (CLS) specimens. The geometries are shown in Fig. 3 and are typical of applications for aluminum adherends. The adherend thick-

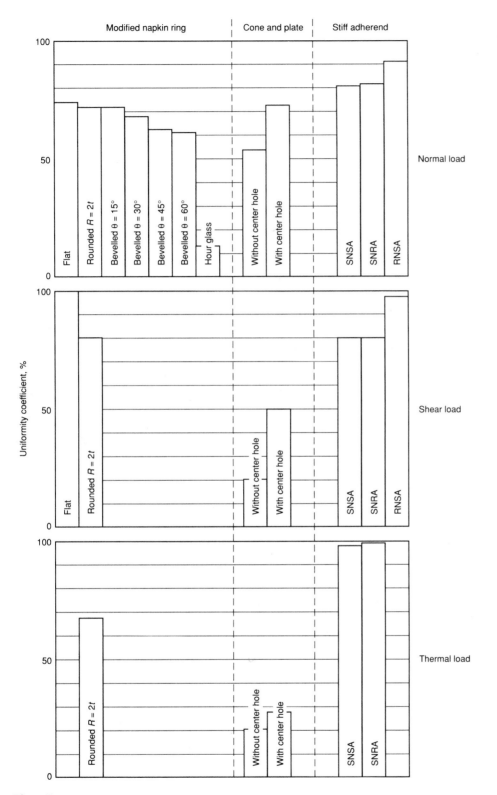

Fig. 2 Uniformity coefficients for *in situ* specimens. Source: Ref 20

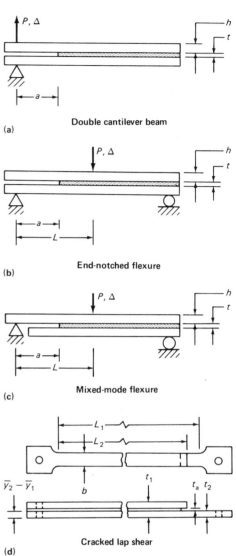

Fig. 3 Specimens for pure and mixed-mode debond characterization. Source: Ref 22, 25

nesses for the specimens are selected to ensure that yielding of the adherend does not occur prior to crack growth in the adhesive. The mode-mix ratios (G_I/G_t) for the specimens are listed in Table 2.

The DCB specimen is a mode I specimen that is loaded in tension, as shown in Fig. 3(a). U-joints can be used to ensure moment-free load transfer to the specimen. The specimen has been standardized according to ASTM D 3433.

The ENF specimen is a mode II specimen that is loaded in three-point bending. The ENF specimen is usually supported by roll-er pins (Fig. 3b) to allow unconstrained movement of the support points. The first use of the ENF specimen was for measuring fracture toughness of unidirectional composite materials (Ref 11).

The MMF specimen provides a 57% mode-mix ratio. The load configuration, shown in Fig. 3(c), is similar to that of the ENF specimen. The MMF specimen was initially applied to investigate environmental effects on fracture toughness (Ref 21).

The CLS specimen provides a 28% mode-mix ratio for specimens with equal-thickness adherends. Unlike the ENF and MMF specimens, the CLS specimen is loaded in tension, as shown in Fig. 3(d). The specimen has been the subject of a recent ASTM round robin analysis (Ref 22), which emphasized the need to take into account geometrical nonlinearities when applying finite-element techniques to compute debond parameters.

Methods for Computing Mixed-Mode Debond Parameters. Critical values for

Table 3 Summary of averaged toughnesses

Method of data reduction	Mean G_{Ic} kJ/m²	Mean G_{Ic} ft · lbf/ft²	CV, %	Mean $G_{(I,II)c}$ kJ/m²	Mean $G_{(I,II)c}$ ft · lbf/ft²	CV, %	Mean G_{IIc} kJ/m²	Mean G_{IIc} ft · lbf/ft²	CV, %
Measured compliance	1.120	76.7	14	1.103	75.4	16	1.593	109.1	42
Measured compliance(a)	1.015	69.5	25	0.963	66.0	11	1.785	122.3	45
Beam theory	0.822	56.3	17	1.208	82.7	35	1.978	135.5	66
Beam theory(a)	0.858	58.8	12	1.120	76.7	31	2.048	140.3	28
Finite element	0.980	67.1	16	1.295	88.7	14	2.258	154.7	57
Finite element(a)	0.910	62.3	17	1.190	81.5	12	2.223	152.3	21

(a) Precracks are grown by fatigue.

strain-energy release rates can be determined from measured specimen compliance, beam theory, and finite-element analysis. For the measured compliance and beam theory methods, the total strain-energy release rate (G_t), for a specific load (P), can be obtained from the rate of change of compliance (c), with crack length (a), through the expression:

$$G_t = \frac{P^2 \partial C}{2b \; \partial a} \qquad \text{(Eq 2)}$$

where b is the specimen width. Brief descriptions of the three methods are provided below.

Compliance Measurements. For each of the above specimens, a load can be applied under constant displacement rate until the onset of debond growth. When the load displacement ceases, the load will decrease and the crack will grow a small distance before arresting. After measuring the new crack length, the specimen can be reloaded under displacement control, and the crack growth process repeated. The inverse of the slope of each load-displacement curve is the specimen compliance. The critical loads for crack propagation correspond to load values that deviate from the linear portion of the curves. A third-degree polynomial can be used to allow the dC/da term to be determined in Eq 2.

Beam Theory. When beam theory is applied to each of the above specimens, expressions for the compliance as a function of crack length can be determined. The expressions for the strain-energy release rates of each specimen are given in Eq 3 to 6 for the DCB, ENF, MMF, and CLS specimens, respectively.

$$G_{Ic} = \frac{P_c^2}{2b} \left(\frac{24a^2}{Ebh^3} + \frac{3}{Gbh} \right) \qquad \text{(Eq 3)}$$

$$G_{IIc} = \frac{P_c^2}{2b} \left(\frac{9a^2}{8Ebh^3} + \frac{0.9}{4Gbh} \right) \qquad \text{(Eq 4)}$$

$$G_{tc} = \frac{P_c^2}{2b} \left(\frac{21a^2}{8Ebh^3} + \frac{0.45}{Gbh} \right) \qquad \text{(Eq 5)}$$

$$G_{tc} = \frac{P_c^2}{2b} \left(\frac{1}{E_2 A_2} - \frac{1}{E_1 A_1} \right) \qquad \text{(Eq 6)}$$

For the expressions in Eq 3 to 6, E and G are the tensile and shear moduli, respectively, of the adherends. The beam theory analyses for the equations do not take into account geometric nonlinearities or large deflections. However, shear deflections are included for the DCB, ENF, and MMF specimens because of their relatively short length. In Eq 6 for the CLS specimen, $(EA)_2$ is the tensile rigidity of the strap, and $(EA)_1$ is the tensile rigidity of the total specimen (strap plus lap). Details of the analyses for Eq 3 to 6 are given in Ref 7 and 23.

Finite-Element Techniques. A third method used to determine critical debond parameters is the finite-element method. Two-dimensional finite-element models for various crack lengths located at the middle of the adhesive bond line can be applied to the specimens using eight-node isoparametric elements. At the crack tip, these elements are collapsed into triangular quadralaterals with quarter-point nodes (Ref 24). A linear, static analysis can be completed for each debond length model for an experimentally applied load. Critical strain-energy release rates can then be determined from computed crack opening displacements, as summarized by:

$$G_I = \frac{\pi E_a (\Delta v)^2}{32(1 - v_a^2)r} \qquad \text{(Eq 7)}$$

$$G_{II} = \frac{\pi E_a (\Delta u)^2}{32(1 - v_a^2)r} \qquad \text{(Eq 8)}$$

where E_a is Young's modulus of the adhesive, v_a is Poisson's ratio of the adhesive, Δv is the displacement normal to the crack plane between adjacent nodes on opposite crack flanks, Δu is the crack plane displacement between adjacent nodes on opposite crack flanks, and r is the distance from the nodes to the crack tip.

Because, for linear fracture mechanics, critical strain-energy release rates are proportional to the square of the critical load, the critical strain-energy release rates from finite-element analyses can be expressed as:

$$G_{(I,II)c} = G_{(I,II)f} \left(\frac{P_c}{P_f} \right)^2 \qquad \text{(Eq 9)}$$

where $G_{(I,II)f}$ is the strain-energy release rate component associated with the load P_f used in the finite-element analysis.

Summarizing the three methods discussed above for determining debond parameters, the compliance method relies totally on measured quantities and does not rely on results from analytical stress analysis. It can be applied to any specimen as long as the specimen exhibits elastic behavior. The disadvantage of the compliance method is that it involves taking derivatives of measured quantities that usually exhibit scatter. When displacement measurements are not possible, the beam theory method of using compliances from stress analysis and measured critical loads is useful. However, every effort should be made to use displacement data for consistency checks for compliances derived from beam theory. Finally, when information relating to fracture mode mix is needed, finite-element methods must

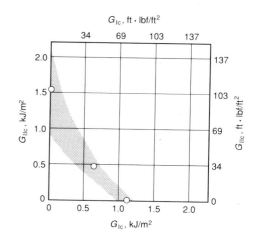

(a)

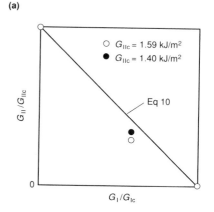

(b)

Fig. 4 Fracture toughness envelopes for FM 300 adhesive. (a) Absolute. (b) Normalized. Source: Ref 25

be applied as the compliance and beam theory methods can provide no information on debond fracture parameter mode mix.

Comparisons of mixed-mode specimen response for the DCB, MMF, and ENF specimens are summarized in Table 3 and Fig. 4. The CLS geometry is not considered here because of the large deformation analyses that are needed to compute the strain-energy release rates for the specimen accurately (Ref 22). Mixed-mode results for the CLS geometry are discussed in a following section directed at specialized applications of finite-element techniques. The results summarized in Table 3 and Fig. 4 are based on studies reported in Ref 25 for aluminum adherends bonded with FM 300, a relatively tough adhesive. The specimen geometries are those shown in Fig. 3. Fracture toughness values were obtained either from arrested starter cracks or from precracks extended by fatigue loads (Table 3).

Tests results for the DCB, MMF, and ENF specimens are summarized in the fracture envelope shown in Fig. 4(a). The toughness values determined from compliance measurements are used to plot the envelope because they are direct and can be applied by a wider population of investigators. The values show the mode I and mixed-mode toughnesses to be much the same. However, the mean mode II toughness value is 42% greater than the mean mode I value, but with greater scatter. As shown in Fig. 4(b), the criterion:

$$\frac{G_I}{G_{Ic}} + \frac{G_{II}}{G_{IIc}} = 1 \qquad \text{(Eq 10)}$$

holds for the FM 300 adhesive where $G_{Ic} \neq G_{IIc}$. However, the more simple criterion:

$$G_I + G_{II} = G_{tc} \qquad \text{(Eq 11)}$$

could also be valid if the scatter in the data could be reduced.

It is also important to note that in Table 3 the average toughness values obtained from beam theory and finite-element analyses show increasing toughness with an increasing mode II component. The measured compliances, on the other hand, gave rise to $G_{(I,II)c}$ values that were slightly lower than the G_{Ic} values. However, the mixed-mode toughness based on compliance was notably higher than the mode I toughness based on beam theory and finite-element computations. Furthermore, the mode II toughnesses obtained from the latter two methods were greater than corresponding values from compliance measurements. As a result, the fracture envelope follows Eq 10 more closely, although it is clear from the results given in Table 3 that shifts in toughness component values can be expected when different methods are used for their computation. However, the results also indicate that this shifting may not be in the same direction for all mode mixes, which

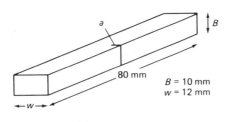

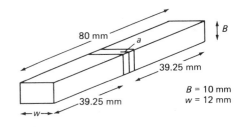

Fig. 5 SENB three-point bend specimens for impact testing. Source: Ref 16

then affects the fracture envelope shape and choice of mixed-mode fracture parameter (Eq 10 and 11).

In conclusion, G_c values can be most directly obtained using the beam theory method, provided a limited number of displacement data are available to check the consistency of the results. If results for G_c mode-mix values are needed, the finite-element method should be applied.

Impact Performance of Bonded Joints

The performance of structural adhesives and adhesive joints under impact loads is an important consideration in the design of air and surface transportation structures. The following describes some work on the development of an instrumented impact test for characterizing the G_{Ic} fracture properties of bulk and *in situ* specimens (Ref 16, 17).

Specimen Geometry and Test Procedure. Bulk specimens in the form of single-edged, notched, three-point bend (SENB) geometries were developed by Kinloch and Kodokian (Ref 16). The specimens are prepared by casting sheets of epoxy adhesives and machining the sheets into bars with the dimensions shown in Fig. 5(a). Cracks are then inserted into the bars by using a narrow milling cutter or by machining a notch and then inserting a razor blade into the notch to propagate a sharp crack. Cracks of various lengths can be inserted using these techniques. The section "Data Reduction Techniques" in this article describes how the impact resistance is determined from the parameters measured by the test equipment.

Adhesive joint specimens can be prepared using the SENB geometry, as shown in Fig. 5(b). The end faces of the aluminum alloy bars should be surface treated with chromic acid for 20 min, rinsed with water, and dried prior to bonding. As described in Ref 16, the bars can be bonded using a rubber mold with a gap of 1.5 mm (0.06 in.) between the bars. Cracks can be machined into the center of the adhesive using a slender flywheel cutter with a radius of 12 μm (0.5 mil).

A crack initiation gage can be applied to one surface of the specimen so that the onset of crack growth and the associated

bar impact force level can be accurately determined. The gage design developed by Beguelin *et al.* (Ref 26) is appropriate for this application. For this design, the face of the specimen of width W and perpendicular to the face on which the crack has been inserted is drilled to accept two small pins. The pins are inserted about 10 mm (0.4 in.) from the crack face, on either side of the crack. A graphite-reinforced polymeric-base film is then sprayed completely across the width of the specimen to a distance of from 1.5 to 2.0 mm (0.06 to 0.08 in.) on either side of the crack. A thin line of conductive silver paint is then applied completely across the specimen width. The gage is completed by painting, on either side of the crack, a silver line to the pin that has been mounted on that half of the specimen.

A common method used for impact testing is for a pendulum striker or hammer to impact on a supported SENB specimen. The velocity with which the striker impacts the specimen can be varied by changing the angle from which the striker is released. For the experiments described in Ref 16, SENB specimens were placed on the shoulders of a vice of a pendulum impact machine to give a span of 48 mm (1.9 in.). The specimens were struck by a striker on the opposite face of that containing the inserted crack. A strain-gage transducer was mounted in the tup of the pendulum, and a record of the tup force versus time and the resistance of the crack initiation gage versus time were digitized and recorded using a microcomputer-based data acquisition system.

Data Reduction Techniques. The fracture energy (G_{Ic}) for the SENB specimen is given from a linear-elastic fracture mechanics analysis by:

$$G_{Ic} = \frac{F_c^2 \partial C}{2B \partial a} \qquad \text{(Eq 12)}$$

where F_c is the force at the onset of crack growth, B is the specimen thickness, and C is the compliance of the test specimen, given by the displacement/load ratio. As derived in Ref 17, a dimensionless geometric factor, Φ, can be introduced such that:

$$\Phi = \frac{C}{\partial C / \partial(a/W)} \qquad \text{(Eq 13)}$$

Table 4 Comparison of slow-rate and impact test results for rubber-modified adhesive

Test temperature		Fracture energy, slow rate (G_{Ic})				Fracture energy, impact (G_{Ic})			
		Bulk		Joint		Bulk		Joint	
°C	°F	kJ/m²	ft · lbf/ft²	kJ/m²	ft · lbf/ft²	kJ/m²	ft · lbf/ft²	kJ/m²	ft · lbf/ft²
−40	−40	0.75	51.4	· · ·	· · ·	1.08	74.0	1.17	80.1
−20	−4	1.10	75.4	· · ·	· · ·	1.18	80.8	1.18	80.8
0	30	1.36	93.2	· · ·	· · ·	1.17	80.1	1.19	81.5
20	70	1.88	128.8	1.89	129.5	1.38	94.5	1.31	89.7
46	105	3.76	257.6	· · ·	· · ·	1.40	95.9	1.39	95.2
60	140	6.29	430.9	· · ·	· · ·	2.18	149.3	2.10	143.9

Source: Ref 16

where W is the specimen width. Using this factor, it may be shown that:

$$G_{Ic} = \frac{U_c}{BW\Phi} \qquad \text{(Eq 14)}$$

where U_c is the stored elastic strain-energy at the onset of crack growth. The value of Φ may be evaluated either from measuring the compliance as a function of crack length or from published tables (Ref 27) of the value of Φ as a function of a/W and L/W, where L is the span of the test specimen between the supported points.

In order to determine true material property values for impact fracture toughness and fracture energy from the above equations, the force that is needed is that acting in the specimen, and the energy needed is that gained by the specimen. However, a problem that arises is that the measured force values are not necessarily the same as the forces acting on the specimen.

At relatively slow impact velocities, differences between the measured values and those acting within the specimen may be negligible. However, differences between measured and required values may become significant as the velocity of the striker is increased. Dynamic effects that contribute to these differences arise from the relatively high contact stiffness of the striker-specimen interface compared to that of the specimen and from the loss and reestablishment of contact between the specimen and the striker during a test. An important consequence of these dynamic effects is that severe oscillations in the measured force-time response may occur.

Williams and Adams (Ref 28) have developed a lumped mass-linear spring model, which quantitatively takes these dynamic effects into account. An important aspect of the model is that it takes into account the contact stiffness between the specimen and striker and that this factor controls the dynamics of the test system. The model predicts that the true impact fracture energy (G_{Ic}^d), corrected for dynamic effects, is given by:

$$G_{Ic}^d = \frac{U_c}{BW\Phi} F(a,k_1,k_2,m) \qquad \text{(Eq 15)}$$

The correction term, $F(a,k_1,k_2,m)$, is a function of the contact stiffness, (k_1), the SENB spring stiffness, (k_2), and the effective specimen mass (m). A more detailed description of the dynamic model is given in Ref 28.

Specimen Response for Low and High Loading Rates. Values of the fracture energy (G_{Ic}) for a rubber-modified epoxy adhesive tested at different temperatures (Ref 16) are given in Table 4. The slow-rate test results were conducted using an Instron tensile testing machine at a displacement rate of 1.7×10^{-5} m/s (5.6×10^{-5} ft/s), giving time to failure (t_f) values ranging from 30 to 300 s. The impact tests were conducted at a striker velocity of 0.5 m/s (1.6 ft/s), giving t_f values of between 400 and 1000 μs. The G_{Ic} values from the impact test were computed from Eq 14, because t_f values for the striker velocity were determined to be greater than values influenced by dynamic effects (Ref 16). All of the joints failed by cohesive fracture in the adhesive layer.

Several interesting trends are evident from the data shown in Table 4. First of all, there is good agreement between the values of G_{Ic} for the bulk and adhesive joint specimen. Second, at the lower temperature values, there is only a small difference between values from the slow-rate tests and those from the impact tests. Third, at temperatures below 0 °C (32 °F), the values of G_{Ic} from the impact tests are lower than corresponding values from the slow-rate tests. The differences between the values for the slow-rate and impact tests increase with increasing test temperature.

Test results for the SENB bulk specimen, which illustrate dynamic effects (Ref 17), are shown in Fig. 6. Discrete measured values of the fracture energy (G_{Ic}) are plotted against time to failure (t_f) for the same rubber-modified epoxy adhesive described above. The solid lines in Fig. 6 represent theoretical values for G_{Ic} determined from Eq 15. Bounding values for the dynamic model stiffness were determined by computing specimen stiffnesses at the maximum values of a employed in the tests ($a = 3.0$ mm, or 0.12 in.) and by computing corresponding stiffnesses for the minimum a values ($a = 0.0$ mm). As shown in Fig. 6, as the impact velocity of the striker is increased, the measured G_{Ic} versus t_f relation shows an increasing number of oscillations. The theoretical curves for G_{Ic} are in good agreement with the experimental values. The observed increases in G_{Ic} at the higher impact velocities appear to arise from dynamic effects and, therefore, do not represent inherent properties of the material. The experimental data points suggest that true material properties of G_{Ic} are not strongly dependent on values of t_f greater than 100 μs.

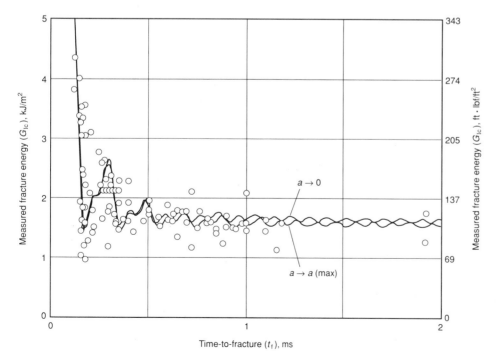

Fig. 6 Measured fracture energy (G_{Ic}) versus time-to-failure (t_f) for a rubber-modified epoxy polymer adhesive. Source: Ref 17

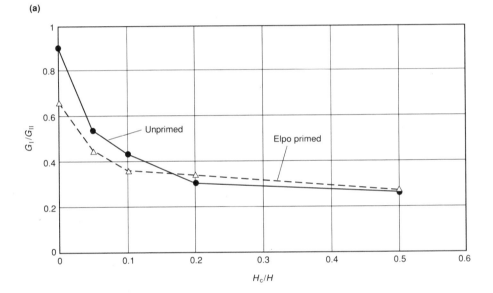

(a)

(b)

Fig. 7 (a) Mesh sensitivity results and (b) variation of G_I/G_{II} through the thickness of the CLS joint. Source: Ref 31

Tools for Assessing Specimen Performance

Finite-Element Methods. As described above, the finite-element method can be applied to compute fracture mechanics parameters that characterize bond strength at an interface by adhesion energy per unit of bond area. These parameters provide a unifying link between bonding mechanical loads, bond geometry, adherend and adhesive material properties, and interface conditions. The finite-element method has also been applied to characterize the stress concentrations that occur at free edges in joints (Ref 29, 30). Described below are two specialized applications of the finite-element method for assessing bonded-joint performance. The first examines the effect of bond line thickness on the mixed-mode response of CLS joints bonded with primed and unprimed adherend surfaces. The second applies finite-element analysis to determine the effects of friction on the computation of strain-energy release rates for the ENF specimen.

The CLS specimen represents a simple structural joint that exhibits both peel and shear response when subjected to in-plane loads. A recently completed study (Ref 31) investigated the mixed-mode response of CLS joints bonded with unprimed and primed steel adherend surfaces. Three different bond line thicknesses (0.3, 0.8, and 1.3 mm, or 0.012, 0.032, and 0.052 in.) were evaluated for the joints. Strain-energy release rates were computed based on a singular finite-element formulation (Ref 32), which was incorporated into plane-strain, two-dimensional analyses. The analyses took into account geometric nonlinearities.

A typical finite-element model for the CLS joint consisted of 1212 eight-node isoparametric elements in regions remote from the debond tip, and eight six-node triangular elements of a variable singularity type at the

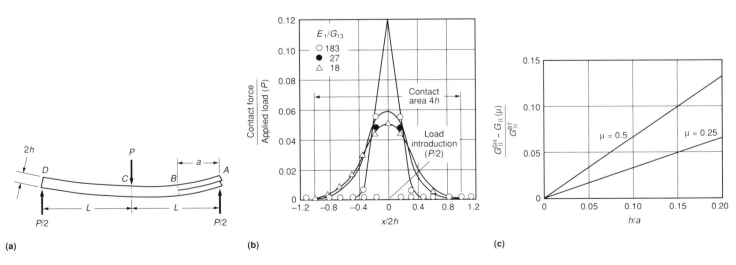

(a)

(b)

(c)

Fig. 8 (a) ENF specimen parameters. (b) Contact normal force distribution. (c) Reduction of G_{II} due to friction for the ENF joint. Source: Ref 34

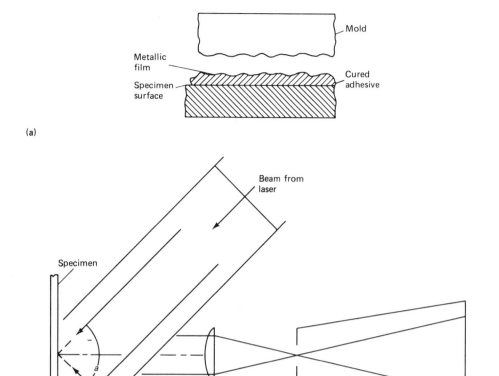

(a)

(b)

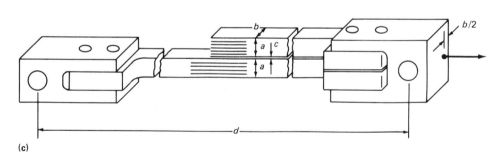

(c)

Fig. 9 Reflectance grating technique and optical arrangement for Moire interferometry. (a) Formation of reflective grating. (b) Optical arrangement. (c) CLS specimen geometry, where a = 5 mm (0.197 in.), b = 3.78 mm (0.149 in.), c = 0.21 mm (0.0083 in.), and d = 250 mm (10 in.). Source: Ref 35

debond front. The singular element is conformable with adjacent eight-node isoparametric elements. Five elements with a nonuniform through-the-thickness distribution were used to model the adherend thickness. The number of elements used to model the bond line thickness was determined from a mesh sensitivity study, which is summarized in Fig. 7(a). As shown, the computed values for G_t, the total strain-energy release rate, are relatively insensitive to the number of elements. However, the ratio of peel to shear (G_I/G_{II}) release rate components was very sensitive to the number of through-the-thickness elements. As a result of the sensitivity study, a nonuniform mesh consisting of 17 elements was used to model the bond line thickness.

Experimental values for load versus debond extension of the CLS joints were input to the finite-element models for computing the debond parameters. The CLS specimens failed in an adhesive mode either at the adhesive-steel interface (unprimed joint) or at the primer-steel interface (primed joint). The failure location was independent of the bond line thickness. The computational results of the study showed that the debond always grew at a through-the-thickness location that had the greatest G_I/G_{II} ratio. This important result is shown in Fig. 7(b), where G_I/G_{II} is plotted versus H/H_c, the ratio of debond location from the adherend-bond line interface to the total bond line thickness. The plot shows a significant gradient in G_I/G_{II} behavior as the

debond location is moved from the middle of the bond line (H_c/H = 0.5) to the bond line interface (H_c/H = 0.0). It is also important to note that the primed joint has significantly lower G_I/G_{II} ratios than the unprimed joint at the adherend-bond line interface. This result directly correlates with increases in joint strength of the primed joints compared to the unprimed joints. Thus, finite-element analysis can be an important tool for investigating debond behavior at varied locations through the bond line thickness.

The ENF debond specimen is essentially a three-point flexure specimen with a through-width debond area for measuring mode II fracture toughness, as shown in Fig. 8(a). When this specimen is used with continuous-fiber composite material adherends, interlaminar shear deformations and friction between the debond faces can influence the strain-energy release rate values. A study has been completed that applies finite-element analyses to examine the influence of friction arising from normal contact forces (Ref 33). The following expressions derived from beam theory were used in Ref 33 to assess the effects of friction and transverse shear deformations.

$$G_{II}^{SH} = G_{II}^{BT} \left[1 + 0.2 \left(\frac{E_1}{G_{13}} \right) \left(\frac{h}{a} \right)^2 \right] \quad \text{(Eq 16)}$$

$$G_{II}^{BT} = \frac{9a^2P^2}{(16E_1w^2h^3)} \quad \text{(Eq 17)}$$

In these expressions, the superscript SH represents the inclusion of interlaminar shear deformation quantified by E_1/G_{13}, where E_1 is the flexural modulus and G_{13} is the interlaminar shear modulus. Superscript BT denotes beam theory expressions, which do not take into account interlaminar shear deformation. Geometric and load parameters are defined in Fig. 8(a).

A potential energy-absorbing mechanism in the ENF specimen is the friction between the crack faces in the region above support pin A in Fig. 8(a). As load is transferred from the lower beam to the upper beam, crack growth causes sliding of the crack surfaces. Friction between the crack surfaces opposes this sliding and consequently is an energy-absorbing mechanism.

The determination of the contact force distribution over the crack faces is a difficult task because of localized loading, coupled with the anisotropic material properties of the adherends. To overcome this difficulty, the contact force was investigated numerically in Ref 34 with the finite-element code ADINA. Coulomb's sliding friction was modeled for two friction coefficients (μ = 0.25 and μ = 0.5) applied to a variety of crack lengths and adherend thicknesses. Nonlinear truss elements were used to study the normal contact force distribution. In Fig. 8(b), the normal contact force distribution as a function of E_1/G_{13} is pre-

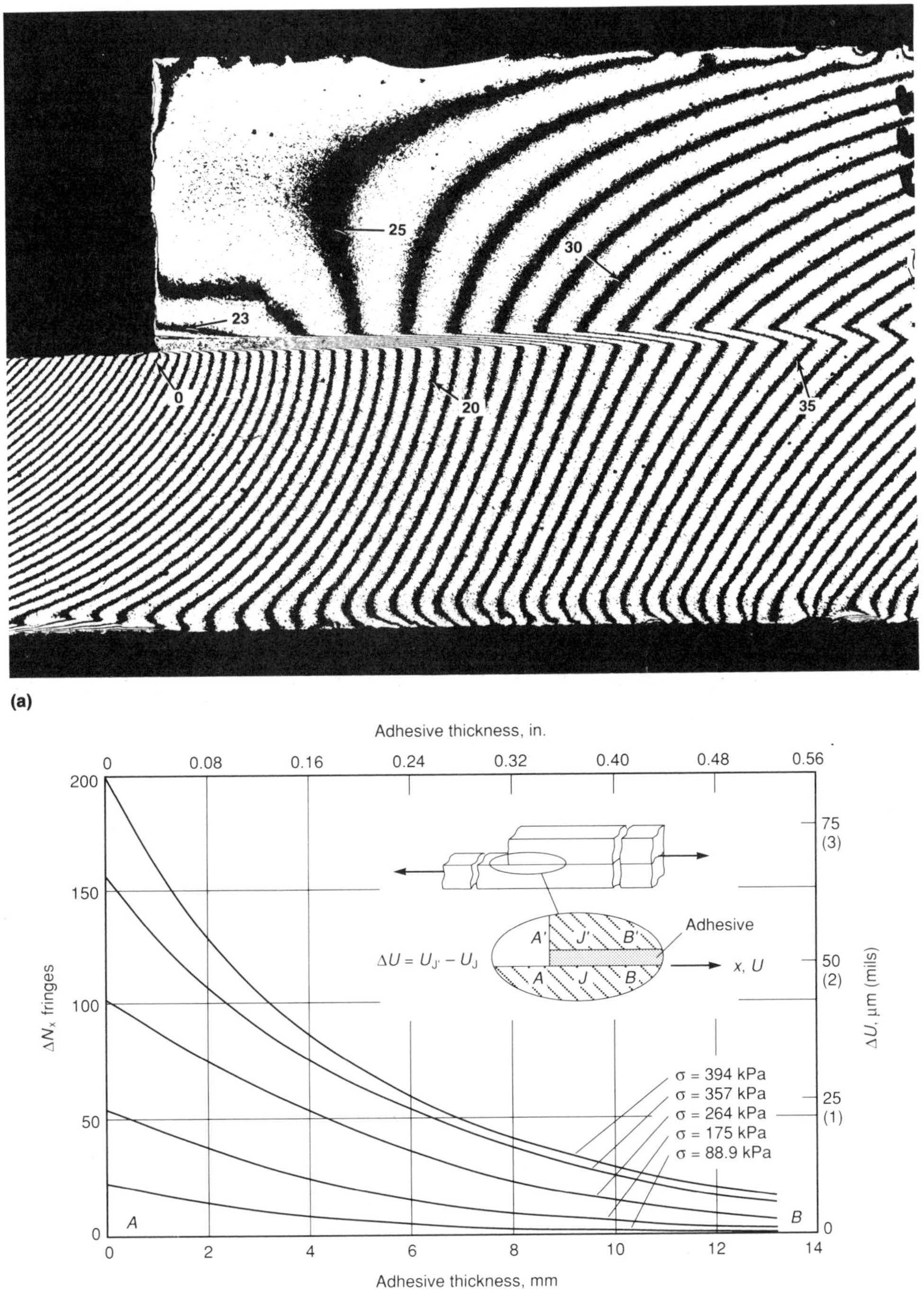

(a)

(b)

Fig. 10 (a) U displacement field (σ = 89 MPa, or 13 ksi). (b) Δu across the adhesive thickness. Source: Ref 35

results so that frictional effects can be minimized through suitable sizing of the ENF specimen.

Post-Moire Interferometry. Moire interferometry is a whole-field optical method using coherent laser light to measure in-plane displacements. The method consists of applying a crossed-line diffraction grating onto a specimen surface using a special mold, as shown in Fig. 9(a). The result is a thin (0.025 mm, or 1 mil) reflective phase grating, which deforms, together with the specimen surface. The specimen grating is monitored with the optical arrangement shown in Fig. 9(b). Two beams of light incident at angles α and $-\alpha$ form a virtual reference grating. Its frequency is $f = (2 \sin \alpha)/\lambda$, where λ is the wavelength of the light. This reference grating interacts with the specimen grating to form a two-beam interference pattern recorded by the camera. The orientation of the CLS specimen shown in Fig. 9(c) should be noted. When the reference lines are perpendicular to the x axis, the pattern is a contour map of displacements governed by the relationship:

$$U = \frac{1}{f} N_x \qquad \text{(Eq 19)}$$

where U is the in-plane x component of displacement at any point, and N_x is the fringe order at that point in the fringe pattern. In the work by Post *et al.* (Ref 35) applied to the CLS specimen, the cross-line grating was applied in the x-y plane with lines perpendicular to the x and y axes. When the reference grating lines were perpendicular to the y axis, in-plane displacements (V) were obtained from fringe orders (N_y) by:

$$V = \frac{1}{f} N_y \qquad \text{(Eq 20)}$$

The CLS specimen shown in Fig. 9(c) was analyzed in Ref 35 using the Moire interferometry technique above. The frequency (f) was 2400 lines/mm (60 lines/mil). An argon laser was used at wavelength 0.5 μm (20 μin.) and 100 mW power. The specimen was loaded in five steps such that plastic deformation in the aluminum adherends initiated at the third step. The U displacement fields were photographed 7 min after the load application. The specimen exhibited a minimum amount of creep and relaxation. Fringe patterns and order numbers corresponding to the first load step (σ = 89 MPa, or 13 ksi) are shown in Fig. 10(a). Figure 10(b) shows the longitudinal displacements (Δu) across the adhesive thickness as a function of the distance x from the joint discontinuity. The displacement attenuated gradually with x. At x = 13 mm (0.5 in.), Δu had diminished to approximately 10% of its peak value. Corresponding results for transverse displacement are presented in Ref 35.

sented, which shows that the contact area between the debonded regions is less than $4h$ (two total adherend thicknesses) in length and is centered about the point of load application. Based on this result, the following nondimensional strain-energy release rate parameter was proposed (Ref 34) to quantify the influence of friction on the reduction of the mode II strain-energy release rate:

$$\frac{G_{II}^{SH}(\mu = 0) - G_{II}(\mu)}{G_{II}^{BT}} = 4\mu(h/a)/3 \qquad \text{(Eq 18)}$$

This parameter is plotted versus h/a in Fig. 8(c). For reasonable values of μ and E_1/G_{13}, the error in G_{II} is only 2 to 5%, according to this analysis. To summarize, finite-element analyses (Ref 34) show that Eq 18 provides a conservative upper bound on numerical

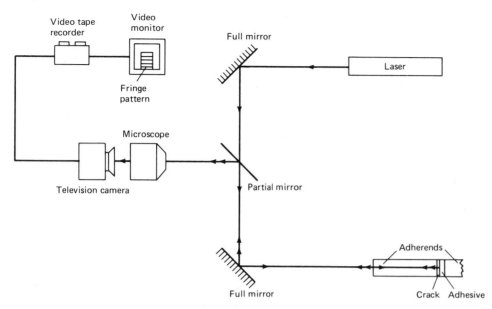

Fig. 11 Crack profile measurement system. Source: Ref 36

This application illustrates the usefulness of the Moire interferometry technique for measuring in-plane displacements of adhesive joints. Its high spatial resolution allows accurate whole-field analyses of joints.

Liechti-COD Techniques. Crack-opening interferometry is an optical technique that has been developed by Liechti and Knauss (Ref 36, 37) to measure directly the normal deformations in the crack-front regions of debonded specimens. Crack-face separations

can be resolved to within at least half a wavelength of the viewing light at any point in the field of view for sufficiently small profile gradients. A schematic of the crack-profile measurement system is shown in Fig. 11. A laser light source is used to monitor the adherend displacements. Mirror arrangements channel the incident laser beam to the specimen, and beams are reflected at the crack faces into a microscope. A closed-circuit television camera is mounted on the microscope to monitor changing fringe patterns, which are also recorded on video tape. The microscope is mounted on an orthogonal micrometer transverse mechanism, which allows tracking of the crack motion.

When the coherent and monochromatic beams reflected from the crack face are combined, an interface pattern is formed that consists of light and dark fringes corresponding to loci of constructive and destructive interference. At a given extinction fringe (m) counted from the crack tip, the crack-face separation distance, h, is given by:

$$h = \frac{m}{2}\lambda, \quad m = 1, 2, \ldots \qquad \text{(Eq 21)}$$

For sufficiently small angles between the crack faces, fringes can be observed with the naked eye. Higher fringe densities produced by larger angles must be resolved with a microscope. With regard to the interpretation of crack-opening interferograms, two issues are discussed in detail in Ref 36: crack-front location, and fringe resolution and crack-profile slope limits. In sum, the data presented in Ref 36 show that while the resolution of the method is insufficient to resolve fully the high-deformation gradients for low-adhesive moduli that are nearly incompressible, the restriction is not severe, and the method has been successfully applied to the measurement of crack profiles in model adhesive joints.

A loading device was constructed to apply orthogonal boundary conditions independently to the model joints, within 0.16 μm (6.4 μin.), ensuring complete compatibility with the crack-opening interference method described above. Crack-opening displacement results from adherend displacements imposed either normal or parallel to a bond line are described next.

The model specimens used in Ref 37 and shown in Fig. 12 are composed of glass adherends and use a polyurethane elastomer as the adhesive. A sequence of video recordings of interface separation resulting from applied normal displacements is shown in Fig. 13(a). Clearly, the crack does not extend completely across the specimen width, but has a fingerlike shape. The full three-dimensional shape of the crack could be generated by a sequence of scans in the width direction. However, because of the great amount of labor required, the results

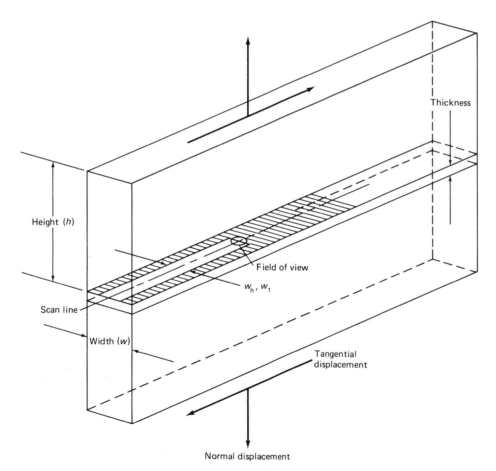

Fig. 12 Model specimen and debond geometry. Source: Ref 37

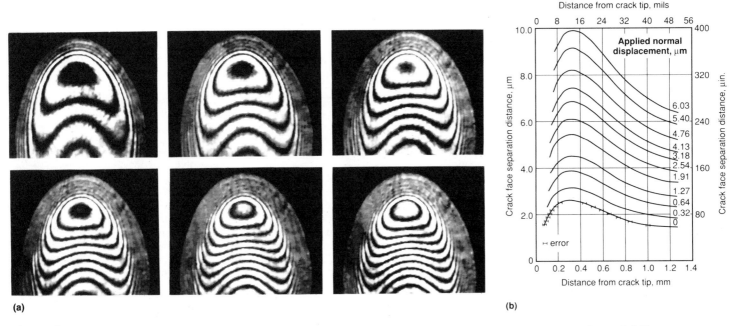

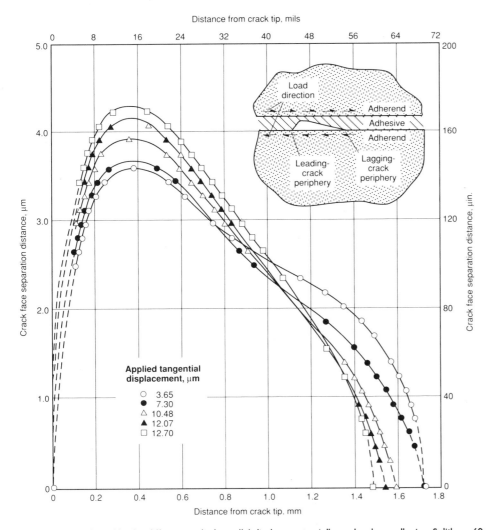

Fig. 13 Applied normal displacements. (a) Interface fringe patterns. (b) Crack profiles. Adherends: glass; adhesive: Solithane 60:40. Source: Ref 37

Fig. 14 Crack profiles for different applied parallel displacements. Adherends: glass; adhesive: Solithane 60:40. Source: Ref 37

reported in Ref 37 were based on a single scan along the crack center. A sequence of crack profiles for increasing normal displacements corresponding to Fig. 13(a) are shown in Fig. 13(b). The separation of the crack faces for no applied displacement indicates the presence of residual stresses. The resulting high-deformation gradients near the crack tip cannot be resolved fully with the present technique. Therefore, the profiles in Fig. 13(b) are incomplete near the crack tip. However, in the resolved profile regions, the consistency of the technique is good.

Crack profiles resulting from applied displacements parallel to the bond line are shown in Fig. 14. The sequence of imposed parallel displacements produces normal crack-flank displacements. Furthermore, the crack develops a finite length, with a maximum in crack-face separation occurring at the leading-crack periphery, as defined in the insert of Fig. 14. The separation decreases with a much smaller profile gradient towards the other, lagging-crack periphery. The steepening of the crack profile at the leading crack front, as well as the smooth crack closure near the lagging end of the crack, have been theoretically predicted by Comninou and Schmueser (Ref 38) for a crack at a bimaterial interface. Thus, optical interferometry has been shown to be a useful tool for examining the validity of criteria for the interfacial debonding of joints.

Dillard-Kynar Sensors. There currently exists a void in the area of experimental adhesive joint testing for measuring peel stresses outside of laboratory environments. In an effort to fill this void, stress sensors made from Kynar piezoelectric film

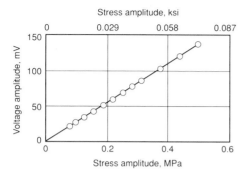

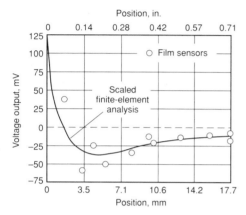

Fig. 15 (a) Calibration results for a butt joint; 52 μm copper-nickel film, plexiglas spacers, 3.2 × 3.2 mm sensor. (b) Peel stress results for a lap-shear joint with the Kynar sensor. Source: Ref 40

are currently under development (Ref 39). Kynar piezoelectric film consists of a thin sheet of poly(vinylidene) fluoride (PVDF), which has been stretched laterally in one direction. A thin layer of metal alloys is then vapor deposited on both surfaces of the film. The PVDF film is then transversely poled in a high electrical film at elevated temperature.

When subjected to a normal strain, a transient open-circuit voltage is created between the two metallization surfaces. The transient nature of the produced voltage indicates that the technique is most useful in cyclic or impact-loading situations. The inherent nature of the film presents some limitations to its use as an embedded stress sensor. One limitation is that PVDF loses its piezoelectric properties at about 70 °C (160 °F). This property loss will seriously limit the widespread use of the sensors in adhesive joints because the film cannot be used with high-temperature-cure adhesives. A second limitation is that achieving a good bond to the film is difficult. It appears that a weak interphase between the PVDF and an adhesive significantly lowers the bond strength of adhesive bonds containing imbedded stress sensors.

In spite of these limitations, calibration studies (Ref 40) using epoxy butt joints that correlate the voltage output of the film to dynamic stress have been performed. The butt joint configuration was chosen because normal stresses in the bond center vary by no more than 1.5% from the average normal stress in the joint. The calibration results shown in Fig. 15(a) indicate a linear relationship between the amplitude of dynamic stress and the measured voltage. The output voltages were independent of testing frequencies above 5 Hz.

Using the calibration values obtained from the butt joint, peel stresses in a single-lap-shear joint were measured (Ref 40). As shown in Fig. 15(b), the resulting voltage measurements have the same trends as do finite-element solutions for peel stresses. However, further calibration work is needed for the graphite-epoxy adherends used for this study. Further work is also needed to examine the effects of three-dimensional stress components on the response of the sensor.

REFERENCES

1. H.C. Schjeldrup and W.B Jones, Mechanical Behavior of Cast Adhesive Film, *Adhes. Age*, Vol 21 (No. 2), 1978, p 35-38
2. R.D. Adams, R.W. Atkins, J.A. Harris, and A.J. Kinloch, Stress Analysis and Failure Properties of Carbon-Fiber-Reinforced-Plastic/Steel Double-Lap Joints, *J. Adhes.*, Vol 20, 1986, p 29-53
3. V. Weissberg and M. Arcan, A Uniform Pure Shear Specimen for Adhesive Characterization, in *Adhesively Bonded Joints: Testing, Analysis, and Design*, STP 981, American Society for Testing and Materials, 1988
4. M. Arcan, Z. Hashin, and A. Voloshin, A Method to Produce Uniform Plane-Stress States With Application to Fiber-Reinforced Material, *Exp. Mech.*, Vol 18, 1978, p 141-146
5. W.T. McCarvill and J.P. Bell, Torsional Test Method for Adhesive Joints, *J. Adhes.*, Vol 6, 1974, p 185-193
6. J.W. Grant and J.N. Cooper, Cone-and-Plate Shear Stress Adhesive Test, *J. Adhes.*, Vol 6, 1974, p 155-167
7. E.J. Ripling, S. Mostovy, and H.T. Corten, Fracture Mechanics: A Tool for Evaluating Structural Adhesives, *J. Adhes.*, Vol 3, 1971, p 107-123
8. T.R. Brussat, S.T. Chiu, and S. Mostovy, "Fracture Mechanics for Structural Adhesive Bonds," AFML TR 77-163, U.S. Air Force Materials Laboratory, 1977
9. G.P. Anderson, S.J. Bennett, and K.L. DeVries, *Analysis and Testing of Adhesive Bonds*, Academic Press, 1977
10. A.J. Kinloch, *Adhesion and Adhesives: Science and Technology*, Chapman and Hall, 1987
11. A.J. Russell and K.N. Street, Factors Affecting the Interlaminar Fracture Energy of Graphite/Epoxy Laminates, in *Progress in Science and Engineering of Composites*, Vol ICCM-IV, 1982, p 279-286
12. G.S. Giore, Fracture Toughness of Unidirectional Fiber-Reinforced Composites in Mode II, *Eng. Fract. Mech.*, Vol 20, 1984, p 11-21
13. G.C. Trantina, Combined Mode Crack Extension in Adhesive Joints, *J. Compos. Mater.*, Vol 6, 1972, p 371-385
14. J.A. Brussat and S.T. Chiu, Fatigue Crack Growth of Bondline Cracks in Structural Bonded Joints, *J. Eng. Mater. Technol.*, Vol 100, 1978, p 39-45
15. J.A. Harris and R.D. Adams, An Assessment of Impact Performance of Bonded Joints for Use in High Energy Absorbing Structures, *Proc. Inst. Mech. Eng.*, Vol 199, 1985, p 121-131
16. A.J. Kinloch and G.H. Kodokian, The Impact Resistance of Structural Adhesive Joints, *J. Adhes.*, Vol 24, 1987, p 109-126
17. A.J. Kinloch, G.H. Kodokian, and M.B. Jamarani, Impact Properties of Epoxy Polymers, *J. Mater. Sci.*, Vol 22, 1987, p 4111-4120
18. V.L. Hein and F. Erdogan, Stress Singularities in a Two-Material Wedge, *Int. J. Fract. Mech.*, Vol 7, 1971, p 317-330
19. T. Hayashi, "Pure Shear Testing of Adhesives In-Situ," EMRL Report 87/7, Engineering Mechanics Research Laboratory, University of Texas, 1987
20. K.M. Liechti and T. Hayashi, On the Uniformity of Stresses in Some Adhesive Deformation Specimens, in *Advances in Adhesively Bonded Joints*, Publication MD-6, American Society of Mechanical Engineers, 1988, p 123-133
21. A.J. Russell and K.N. Street, Moisture and Temperature Effects on the Mixed-Mode Delamination Fracture of Unidirectional Graphite/Epoxy, in *Delamination and Debonding of Materials*, STP 876, American Society for Testing and Materials, 1985, p 349-370
22. W.S. Johnson, Stress Analysis of the Cracked-Lap-Shear Specimen: An ASTM Round-Robin, *J. Test. Eval.*, Vol 15 (No. 6), Nov 1987, p 303-324
23. T. Freda, "The Use of Laminated Beams for the Determination of Pure and Mixed-Mode Fracture Properties of Adhesives," EMRL Report 87/5, Engineering Mechanics Research Laboratory, University of Texas, 1987
24. L. Banks-Sills and D. Sherman, Comparison of Methods for Calculating Stress Intensity Factors With Quarter-Point Elements, *Int. J. Fract.*, Vol 32, 1986, p 127-140
25. K.M. Liechti and T. Freda, On the Use of Laminated Beams for the Determina-

tion of Pure and Mixed-Mode Properties of Structural Adhesives, in *Advances in Adhesively Bonded Joints*, Publication MD-6, American Society of Mechanical Engineers, 1988, p 65-77

26. P. Beguelin, B. Stalder, and H. Kausch, A Simple Velocity Gage for Measuring Crack Growth, *Int. J. Fract.*, Vol 22, 1983, p R47-R54

27. J.G. Williams, *Fracture Mechanics of Polymers*, Ellis Horwood, 1984

28. J.G. Williams and G.C. Adams, The Analysis of Instrumented Impact Tests Using a Mass-Spring Model, *Int. J. Fract.*, Vol 33, 1987, p 209-222

29. R.D. Adams and W.C. Wake, *Structural Adhesive Joints in Engineering*, Elsevier, 1984

30. D.W. Schmueser and N.L. Johnson, Stress Analysis of Adhesively Bonded Electroprimed Steel Lap Shear Joints, *J. Adhes.*, Vol 24, 1987, p 47-64

31. D.W. Schmueser and N.L. Johnson, Effect of Bondline Thickness on Mixed-Mode Debonding of Adhesive Joints to Electroprimed Steel, *J. Adhes.*, in press

32. R.E. Smelser, Evaluation of Stress Intensity Factors for Bimaterial Bodies Using Numerical Crack Flank Displacement Data, *Int. J. Fract.*, Vol 15, 1979, p 135-143

33. L.A. Carlsson, J.W. Gillespie, and R.B. Pipes, On the Analysis and Design of the End Notched Flexure (ENF) Specimen for Mode II Testing, *J. Compos. Mater.*, Vol 20, Nov 1986, p 594-604

34. J.W. Gillespie, L.A. Carlsson, and R.B. Pipes, Finite Element Analysis of the End Notched Flexure Specimen for Measuring Mode II Fracture Toughness, *Compos. Sci. Technol.*, Vol 27, 1986, p 177-197

35. D. Post, R. Czarnek, J. Wood, D. Joh, and S. Lubowinski, Deformation and Strains in Adhesive Joints by Moire Interferometry, Research Report VPI-E-84-25, Virginia Polytechnic Institute and State University, July 1984

36. K.M. Liechti and W.G. Knauss, Crack Propagation at Material Interfaces I: Experimental Technique to Determine Crack Profiles, *Exp. Mech.*, Vol 22, July 1982, p 262-269

37. K.M. Liechti and W.G. Knauss, Crack Propagation at Material Interfaces II: Experiments on Mixed-Mode Interaction, *Exp. Mech.*, Vol 22, Oct 1982, p 383-391

38. M. Comninou and D. Schmueser, The Interface Crack in a Combined Tension-Compression and Shear Field, *J. Appl. Mech.*, Vol 46, 1979, p 345-348

39. D.A. Dillard, G.L. Anderson, and D.D. Davis, Jr., A Preliminary Study of the Use of Kynar Piezoelectric Film to Measure Peel Stresses in Adhesive Joints, *J. Adhes.*, Vol 29, 1989, p 245-255

40. G.L. Anderson and D.A. Dillard, The Use of KYNAR Piezoelectric Film to Measure Normal and Peel Stresses in Adhesive Joints, in *Proceedings of the 13th Annual Meeting of the Adhesion Society*, Feb 1990

SECTION 6

Designing With Adhesives and Sealants

Co-Chairmen: John T. Quinlivan, Boeing Commercial Airplane Company
Dan Arnold, Boeing Military Airplane Company

Introduction

THERE ALWAYS EXIST many ways to design an adhesive joint. The critical issues facing the design engineer include consideration of the basic function, size, and dimensions of the joint, the adhesive and materials to be joined, and the proposed means of tooling and manufacturing the joint.

In adhesive bonding, the issue often is to design for bonding; not simply to substitute bonding for other means of fabrication. Adhesive bonded joints tend to be strong in shear but weak in peel. Consequently, the design of an adhesive joint will often maximize the adhesive bond shear area and minimize the tensile or peel forces applied to the adhesive. In addition, the load intensity will influence the details of the bonded joint. The key point is that design practice for adhesive bonding departs significantly from conventional, mechanical joining methods. The article on adhesive bonding analysis and design reviews the static strength factors important in adhesive bonding design.

The article on numerical analysis methods summarizes the state of the art in adhesive bond analysis. Both closed-form and finite-element methods are presented. Design details often render closed-form techniques inadequate for the analysis of complex bonded structures. Fortunately, numerical techniques are available, which, when coupled with the number-crunching capabilities of today's computers, provide useful information to the analyst. The state of stress or strain at the local detail, as well as the large bonded assembly level, are addressed.

One of the major technical advantages of adhesive bonding is the dramatic improvement in fatigue life compared with more conventional joining methods. Many applications of adhesive bonding can be traced to the need to reduce or to eliminate the fatigue problems associated with other designs. The nature of fatigue is given considerable attention in the article that reviews fatigue design considerations and the nature of fatigue in adhesive bonded joints.

Damage tolerance, the ability of a structure to tolerate a reasonable level of damage or defects without catastrophic failure, is a necessary requirement in the design of many structures. Adhesive bonding provides one means of achieving this requirement with a high degree of structural efficiency. The inherent differences between the damage tolerance characteristics of conventional metal and bonded structure designs must be recognized early and accounted for in the design process.

Adhesive bonding has its own unique defects that can occur during the manufacturing cycle or be encountered in service. The article on effects of defects reviews the types of defects that commonly occur and the types of destructive and nondestructive testing employed to discover and remedy them. Failure to consider these issues is one of the most common errors committed by design engineers when producing bonded joints.

Design proof testing is the critical experimental verification of analytical prediction. In adhesive bonding it is often used to validate the production method and process also. The type of testing used to validate the design will depend upon the application but can range from simple coupon tests to full-scale evaluations. Regardless of the approach taken, the critical load conditions, specimen size, and service environment are critical issues that must be addressed when determining the test plan. This chapter describes the role of design proof testing in assessing adhesively bonded structures and qualifying them for in-service use.

Sealants are not often considered with adhesives, but adhesives often make excellent sealants. The design considerations article points out the design features unique to sealants. Several sealant designs are reviewed to illustrate the principles. Assemblies involving materials with widely differing coefficients of thermal expansion, the effective joint seals as in aircraft fuel tanks or automotive environmental seals, and vibration isolators are some of the more common applications. The materials discussed in this article generally are low-moduli, castable elastomers that possess limited value as structural adhesive bonding agents but have significant importance in uses as environmental seals, vibration isolation designs, and in the joining of unlike materials.

The success of any technology is its application to useful products. A series of articles reviews applications of adhesives and sealants. The intent is to provide the design engineer with a view of the history of adhesive bonding and sealant design developments and experience. Current production design and manufacturing practices are discussed with extensive use of graphs, tables, drawings, and photographs.

Adhesive Bonding Design and Analysis

Raymond B. Krieger, Jr., Engineered Materials Department, American Cyanamid Company

THE SAFE AND EFFECTIVE DESIGN of bonded structure is fundamentally dependent on both a valid stress analysis of the adhesive in the bond and reliable tests that determine *allowable* stresses in the adhesive. Safe design is achieved by comparing the calculated stresses with the allowable stresses, thereby predicting the performance of the bond. Effective design is the most lightweight, economical configuration consistent with safety. This is achieved through the stress analysis itself because it contains parameters relating to the size and geometry of the adherends. These sizes and geometries can be selected for minimum weight and cost, consistent with safe adhesive stress levels as determined by the analysis.

For stress analysis, a model specimen is presented that embodies the fundamental stress distribution found in the adhesive in typical bonded joints. This analysis is shown to be valid when the adherends are in tension, compression, or shear. It also shows that the indispensable property of the adhesive, for stress analysis, is stiffness in shear.

The older tests of lap-shear and peel strengths are discussed to show the problems of these methods in providing allowable stresses and stiffness data. Fracture mechanics tests are also discussed regarding their value in predicting bond performance in service. The shear stiffness of the adhesive is obtained by a special extensometer used on a thick-adherend lap-shear specimen. For nonlinear work, a second extensometer is used on the model joint. Tests to obtain allowable stresses are described using these two extensometers.

Finally, a series of design geometries based on the linear stress analysis parameters is presented. These geometries show, by example and discussion, how stress analysis is used to provide designs with maximum efficiency at minimum weight.

Model Joint for Stress Analysis

It is desirable to establish a simple configuration of a bonded joint for study and stress analysis. Such a joint should represent a real-life structure in a typical and fundamental way. The so-called lap-shear specimen does not serve this goal well because it is too short to develop the full, efficient, shear stress distribution in the adhesive. The configuration selected for discussion is the skin-doubler specimen described in Ref 1 and shown in Fig. 1. The bonded joint is that between an infinitely long skin and a doubler that is also infinitely long. The skin is loaded in tension, and the adhesive transfers tension load into the doubler.

The shear stress distribution in the adhesive is shown in Fig. 1. There is a peak at the doubler tip that fades away to 0 stress at some point along the joint. At this point, because the tension stress, or strain, in the doubler equals that in the skin, no further load transfer is needed. The equations shown in Fig. 1 are for the case when stiffness parameters are linear, the skin gage equals the doubler gage, and the glue line thickness is constant. The equations shown are developed in Ref 1 and 2.

The formulas shown in Fig. 1 give the maximum adhesive shear stress (at the doubler tip) and the distance required for the adhesive stress to fade to 0. These two values are seen to be driven by the stiffness of the glue line. If the shear modulus (G) is reduced, the peak adhesive shear stress is reduced, with no change in load on the structure. Thus, the adhesive, as a fastener, exhibits very unique behavior. It will *vary* its peak stress in a joint with changes in its stiffness (from temperature, fluid absorption, and glue line thickness) without any change in the load on the structure. The distance to 0 adhesive shear is also dependent on adhesive stiffness change.

A concept that is fundamental to an understanding of the behavior of an adhesive is that the adhesive must be considered a fastener. The usual material properties of tension, compression, flexure, and modulus do not serve well. The indispensable property is stiffness in the shear mode. The stiffness of the glue line relative to the stiffness of the adherends determines the stress distribution in the adhesive.

Figure 1 shows the skin-doubler specimen in tension. A less familiar situation is

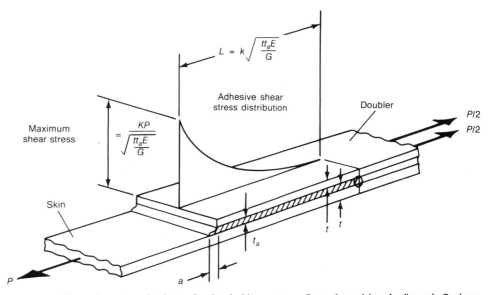

$$L = k\sqrt{\frac{tt_aE}{G}}$$

Adhesive shear stress distribution

Doubler

$P/2$
$P/2$

Maximum shear stress $= \dfrac{KP}{\sqrt{\dfrac{tt_aE}{G}}}$

Skin

t_a

t t

P

a

Fig. 1 Adhesive shear stress distribution for skin-doubler specimen. E, tensile modulus of adherend; G, shear modulus of adhesive

Fig. 2 Adhesive shear distribution when adherends are in shear

one in which the specimen is in shear, as in Fig. 2. This condition occurs in aircraft wings or helicopter blades when torsion loads and/or shear flows due to bending are present. Figure 2 (shear mode) shows the glue line deformation to be quite similar to that in Fig. 1 (tension mode), except that the directions of the two strains are at 90° to each other. It is apparent that the shear curve in Fig. 2 will fade to 0, as in Fig. 1. The formulas for shear stress peak and distance to 0 stress are identical, except, of course, that the metal shear modulus (G_m) must be substituted for the metal tension modulus (E).

Perhaps the most important thing about this concept is that the shear modulus is generally one-third that of the tension modulus. Because these moduli are in the denominator, a reduction to one-third means a stress peak increase in the ratio of the square root of 1 divided by the square root of $\frac{1}{3}$ (~1.73). In other words, if the shear load per unit width is the same as the tension load per unit width, the adhesive stress will be 73% higher in the shear mode. It is apparent that adhesive stresses from torsion or shear due to bending cannot be taken lightly.

Inspection of the formulas in Fig. 1 and 2 shows clearly that the adhesive shear stiffness (shear modulus, G) is the indispensable property for calculating the shear stress in the adhesive. Because this modulus represents the shear stress divided by the shear strain, it is imperative to have a test where both adhesive stress and strain can be accurately measured. As will be seen, lap-shear and peel tests do not produce accurate stress values. Accurate, economical measurement of shear strain requires an extraordinary extensometer.

Lap-Shear, Peel, and Crack Propagation Tests

Lap-shear, peel, and crack propagation tests do not relate well to stress analysis.

Their usefulness is unquestioned for receiving inspection, process control, process variable evaluation, estimating the useful service temperature range of an adhesive, and, in a limited way, miscellaneous evaluation of service problems. The salient feature of each of these tests is that the result is limited to finding the ultimate strength of the test specimen. Nothing is learned about basic properties of the adhesive, such as shear strength or shear deformation. This is the reason that these tests do not provide the information needed for stress analysis. The remainder of this article discusses the pros and cons of these tests in some detail to establish their true value.

The lap-shear test is described in Federal Specification MMM-A-132. It also has many variations in the details of construction and in the variety of materials used as adherends. Essentially, two pieces of 25 mm (1 in.) wide material are bonded together using a 13 mm (0.5 in.) overlap. The ends are pulled in order to fail the adhesive in the shear mode. The most common material might be 2024-T3 aluminum of 1.6 mm (0.064 in.) thickness.

The failure is expressed as a stress that represents the failing load divided by the bond area of 320 mm² (0.5 in.²). Its simplicity is matched by its economy, which accounts in large measure for its popularity.

Perhaps the most telling concept for the test is that it does not describe a basic property of the adhesive, but merely the strength of a specific joint made with the adhesive. This means that the information obtained does not relate to stresses in an actual design.

Figure 3 illustrates reasons for this limitation. First, the drawing shows the loaded deformation of the adherends, using an exaggerated scale. It is seen that the metal undergoes severe bending at the beginning of the joint. Aluminum reaches its yield point here at a relatively low load of approximately 680 kg (1500 lb), as evidenced by permanent set in failed specimens. This means that the metal strains are grossly unequal over the length of the joint. In addition, the stiffness of the adhesive is not linear with load, that is, at about two-thirds of ultimate strength a rapid drop in stiffness takes place, followed by a large strain-to-failure value. This property causes the true shear stress distribution to begin with high sharp peaks at the ends of the test area. At high stresses, these peaks fade away. This peak reduction is caused by a loss of adhesive stiffness at the ends, which, in turn, causes additional load to go to the central, stiffer path.

Figure 3 shows four stress distributions, in a qualitative way, as the load increases from 1P to 4P (near ultimate strength). Thus, it can be seen that the shear stress is never the simple load divided by area but actually has many distributions during the

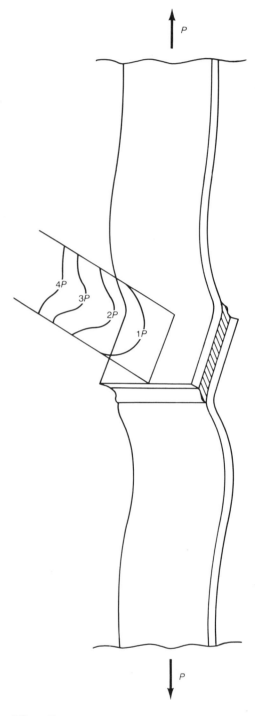

Fig. 3 Deformations of a lap-shear specimen

test. It is apparent that this characteristic makes the lap-shear test questionable for tests involving creep or fatigue. Over the years many lap-shear tests have been run in an attempt to characterize an adhesive performance in hostile environments.

The stiffness of the adhesive changes with temperature and/or fluid absorption, adding to the problem of knowing the true stresses in testing. Also, there is a very serious time problem in exposure of lap-

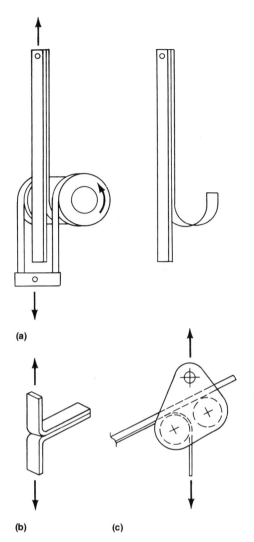

Fig. 4 Peel test methods. (a) Climbing drum. (b) "T" peel. (c) Bell

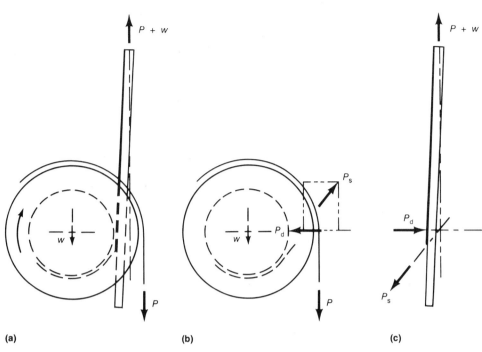

Fig. 5 Climbing drum peel test, with forces due to drum weight

shear coupons to fluids. For water exposure, the time for the test area to reach equilibrium, or even saturation, can be six weeks or longer. Hence, any test conducted before equilibrium or saturation involves testing partly dry and partly wet adhesive, as well as a wide range of degradation at the interface. Nevertheless, actual service environment must be addressed. Environmentally conditioned shear test coupons are often used to establish design allowables.

The metal-to-metal peel test has been popular since the inception of structural bonding. The results obtained have value in process control and in receiving inspection for a specific adhesive. Designers concerned about toughness have persistently demanded high peel strength adhesives. Peel strength has always been a part of adhesive performance specifications and thus has played a dominant role in adhesive formulation.

Figure 4 describes three popular peel tests, the climbing drum peel test (Fig. 4a), the "T" peel test (Fig. 4b), and the Bell peel test (Fig. 4c). Each gives different numerical results but all have essentially the same faults and virtues. Generally, a peel test is viewed by designers and materials engineers as a measure of toughness, albeit in a subjective, even emotional, way. The test results have little use in the stress analysis of a bond.

It is supposed that accumulated good service performance means that the adhesive peel strength value must be maintained. The fault has been that designers and materials engineers tend to demand ever higher peel values with any program allowing a new adhesive selection. The adhesive suppliers have met these demands, to the best of their ability, because competitive success has required it.

The fault really lies in the fact that high peel strength has been a trade-off for high-temperature performance and environmental resistance (particularly water exposure). The subjectivity of the test is perhaps most clearly shown by the fact that various adhesives that are successful in service have a comparative peel strength ratio as high as 9:1.

Figure 5 shows the climbing drum peel test apparatus. A thin strip of metal is peeled from a thick strip by winding the thin strip around a drum. Torque is applied to the drum by pulling down on straps wrapped around the drum. The thin strip of metal is wound on the drum at a smaller radius than are the straps. This difference in radius (moment arm) allows the straps to place a large torque on the drum than on the thin strip. This causes the drum to roll upward, thus failing the bond by peeling the thin strip from the thicker one. The force involved to do this is combined with the two radii, to arrive at a torque value that describes the peel strength.

A further indication of the subjectivity of the test lies in the improper use of a correction factor because of the drum weight. It was reasoned that the load used simply to raise the drum did not contribute to the failing of the bond. Therefore, this load is deducted from the peel strength. This load is found by running the test with a thin metal strip or a strip of canvas only, rather than with a bonded specimen. Careful analysis shows this procedure to be invalid, because the force due to the weight of the drum is actually carried on the bond. Figure 5 presents this analysis. Figure 5(a) shows the drum and specimen in equilibrium with the downward strap load (P) and drum weight (w) equal to the upward force on the specimen.

Next, Fig. 5(b) shows the drum as a free body with the weight of the drum w and the downward pull of the straps P being reacted by the tension load in the thin strip of the test specimen P_s, and a sidewise force P_d on the drum caused by the specimen pushing on the drum. Both P_s and P_d are vectors whose resultant force is $P + w$, or the upward force on the specimen.

To complete the analysis, Fig. 5(c) shows the test specimen as a free body. It is in equilibrium, with the upward force $P + w$ being balanced by the same two vectors as before, P_s and P_d. Inspection of Fig. 5(c) shows that the force P_s, which is caused by the weight of the drum, is clearly loading the glue line of the test specimen. This

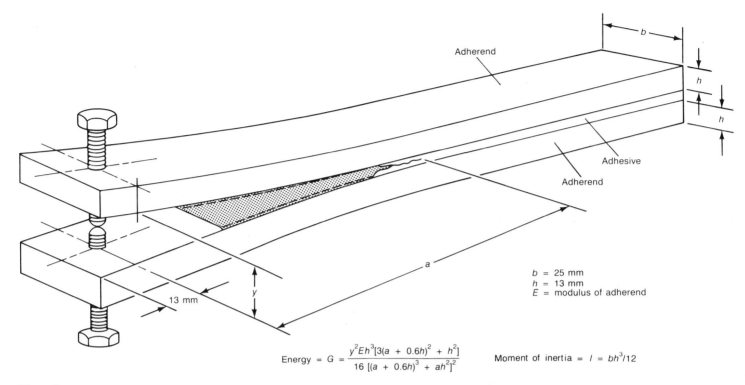

$$\text{Energy} = G = \frac{y^2 E h^3 [3(a + 0.6h)^2 + h^2]}{16 [(a + 0.6h)^3 + ah^2]^2}$$

$b = 25 \text{ mm}$
$h = 13 \text{ mm}$
$E = \text{modulus of adherend}$

$$\text{Moment of inertia} = I = bh^3/12$$

Fig. 6 Double-cantilever-beam (DCB) fracture mechanics test

means that no reduction of peel strength should be made because of the weight of the drum. These reductions have been on the order of 1.75 kN · m (10 lbf/in.) and, if not taken for a popular high-temperature adhesive, would increase its peel strength from 1.58 to 3.33 kN/m (9 to 9 + 10 = 19 lbf/in.), a value perhaps more pleasing to designers.

Peel testing has nevertheless contributed to the success of structural bonding. This article has dwelt on the negative side in order to better understand the true nature of the test. In this way, it may be put in proper perspective and thus contribute to the development of ever more useful, durable, adhesive systems.

Crack Propagation Tests. In the early 1970s, fracture mechanics principles were enthusiastically applied to the testing of structural adhesives. Rather than testing crack propagation in a bulk, or neat, mass of adhesive, proponents combined aluminum with the specimen, as shown in Fig. 6. In earlier times, such an arrangement was called a cleavage test, and the adhesive was rated by the force at the screws that was necessary to fracture the bond.

In the fracture mechanics approach, the adhesive is rated by a, the distance from the screws to the crack tip, together with the stiffness parameters of the metal in the a region. This energy, or strain energy release rate, is designated G, and its formula is shown in Fig. 6.

Serious doubts are raised by this approach. First, the science of fracture mechanics is not based on a construction; it is based on a monometallic (or monoplastic) specimen with a crack in it that is being driven by force, causing stress concentration at the crack tip. Further, the theory states that for the crack to stop moving, there must be a plastic zone at the crack tip, while the rest of the metal is strained inside its elastic limit. The size of the plastic zone can be calculated using values of G, the tensile modulus, and the yield stress of the material.

When this is computed for a typical adhesive, the plastic zone is huge, compared to the glue line thickness (see Fig. 7). This clearly indicates that the test does not meet the requirement of a plastic zone at the crack tip; that zone must be surrounded by an elastic continuum.

Another problem is that the specimen is blind to the stiffness of the adhesive. The formula in Fig. 6 does not contain terms for glue line thickness or the tension stress-strain properties of the adhesive. This is unacceptable because different specimens showing identical energy values (G) at critical or arrest points or under environmental exposure can show a variation in tensile stress that exceeds 100%.

This has been determined by stress analysis of the metal beam that forms one-half of the specimen. Careful measurement of glue line thickness distribution gives the bending pattern of the beam, and because its stiffness is known, the adhesive forces (tension and compression) can be calculated. The deformation patterns force a stiff glue line to undergo much higher tensile forces than the softer glue line experiences, yet the crack propagation formulas show them to be under identical energy levels (a = constant, G = constant).

This phenomenon could be lethal in any environmental study with different adhesives because heat or moisture will soften different adhesives to different degrees. Studies of interfacial variables could be compromised because different adhesives could produce different stresses for the same value of G.

The exposure of precracked specimens to environment also has a disquieting feature, in that the glue line has already been strained beyond yield, in fact, almost to ultimate, regardless of whatever lower loads it may subsequently experience. Because the degradation rate of such a glue line may be much faster than that of an unstrained one, it would seem that the data would be applicable solely to a bond on an aircraft that had previously been loaded to within a hair of ultimate.

To demonstrate this effect, two identical specimens were prepared. The first was loaded and cracked to G_{arrest}, and the distance a was measured. The second was not loaded, but the bond was removed with a thin jeweler's saw the same distance a as in the first specimen. (To eliminate any possible metal stiffness variable from loss of metal to the saw, the same saw was used on the first specimen, after it had been cracked, to clear out the crack over the distance a. The glue lines were 125 μm, or 5 mils, and the saw width was 355 μm, or 14 mils.)

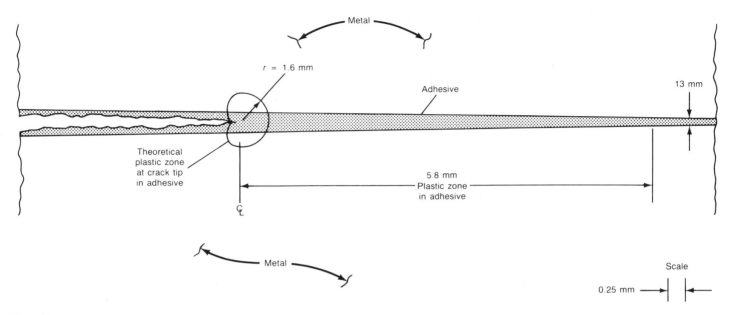

Fig. 7 Adhesive crack tip geometry in fracture mechanics specimen

Both specimens were then reloaded to G_{critical} while recording the loading force versus tension strain at the crack tip. The damage caused by the previous straining in the first specimen was reflected by its stiffness, which was half that of the second specimen.

In sum, the fracture mechanics approach appears to be subject to misleading variables. From the viewpoint of stress analysis, it appears that establishing calculated values for G in an actual structure would be cumbersome, uncertain, and inaccurate where adhesive stiffness patterns are unknown. These problems may rule out fracture mechanics as a means of predicting service performance or structural bonds.

Testing for Adhesive Shear Stiffness

The need for adhesive shear stiffness data is apparent. Not only is it necessary for all stress analysis, it must be known for all aircraft environments because it changes with temperature and fluid absorption. The early test methods, such as torsion tubes, proved to be too costly. The expense of sufficient data for statistical confidence has generally been prohibitive. To find a solution, a simple specimen was selected and an extraordinarily sensitive extensometer was developed.

The first requirement was to obtain a specimen in which the accurate shear stress could be essentially uniform and accurately known for all load levels up to ultimate. Consideration of the standard lap-shear coupon in Fig. 3 shows that this requirement is defeated by a lack of stiffness in the adherends. Accordingly, the specimen selected was the so-called thick adherend specimen shown in Fig. 8.

This is a lap-shear specimen with an overlap of 9.52 mm (0.375 in.) and with 9.52 mm (0.375 in.) adherends. For economy and uniformity of glue line thickness, the specimens are made by bonding together two thick plates, which are then cut into 25 mm (1 in.) wide strips and notched on each side to form the lap-shear geometry. The thickness of the metal and the shorter overlap reduce the adherend strains to an acceptable minimum.

The KGR-1 extensometer was developed (Ref 3) to measure the shear movement of the glue line. This instrument is shown in Fig. 9. The movement of the glue line is made to move the core in a linear variable differential transformer (LVDT). The voltage change is fed to a recorder, which produces a curve of load versus glue line movement. Two instruments are used in order to eliminate errors from unsymmetrical specimens. They are held on the specimen by being attached to each other by springs. These springs cause the steel points of the extensometers to grip the specimen and thus support each other.

Figure 10 presents a typical shear stress-strain curve obtained with a KGR-1 extensometer. The recorder produces a curve of specimen load versus total specimen deformation in shear. For any point on the curve, the shear stress is found by dividing the load by the bond area.

The adhesive shear deformation is found by simply subtracting the metal deformation from the total deformation signal. (Reference 3 presents the method and validity of obtaining the specimen metal deformation.) The shear strain is the deformation divided by the glue line thickness. The shear modulus G is found (for the initial linear portion of the curve) by dividing the shear stress by the shear strain.

Three basic points are presented to define the curve adequately: linear limit (LL), knee (KN), and ultimate strength (UL). The

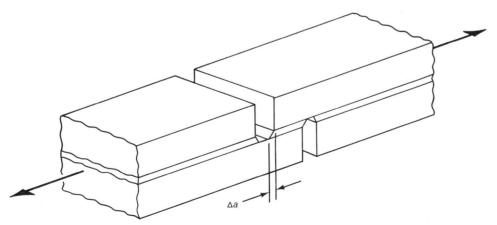

Fig. 8 Thick adherend lap-shear specimen

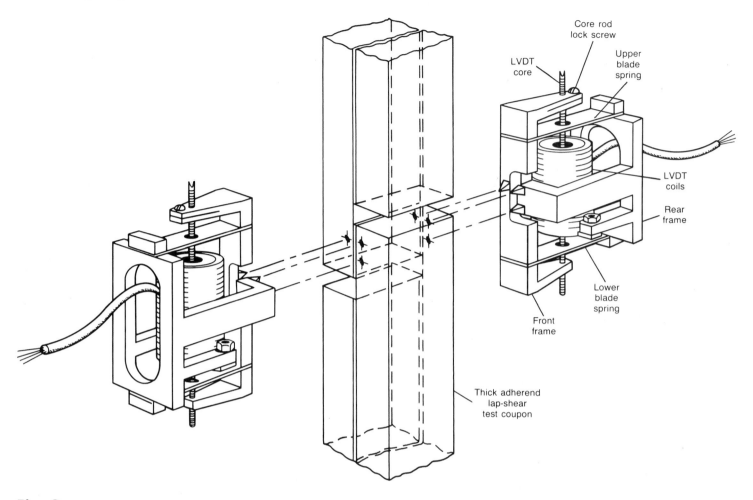

Fig. 9 The KGR-1 extensometer. LVDT, linear variable differential transformer

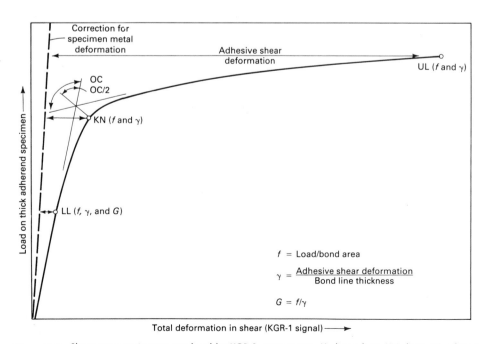

f = Load/bond area

$$\gamma = \frac{\text{Adhesive shear deformation}}{\text{Bond line thickness}}$$

$G = f/\gamma$

Fig. 10 Shear stress-strain curve produced by KGR-1 extensometer. LL, linear limit; KN, knee; UL, ultimate strength; f in units of MPa (ksi); γ in units of mm/mm; and G in units of Pa (psi)

rationale for simplifying the full curve to these three points is described below.

Linear limit is arbitrarily chosen as that point at which the curve departs from a straight line through the origin. The coordinates of this point establish the shear modulus G for use in stress analysis in the linear range. Because modulus is the prime usefulness of the LL point, it does not matter if personal judgment produces a scatter in the stress values as long as the point is taken on the curve. If the curve has no straight-line portion, the coordinates of LL are presented as 0 stress and 0 strain.

Knee is the point found by bisecting the angle between the initial tangent and another tangent drawn to the curve beyond the knee region (Fig. 10). The worth of this technique can be judged from the following considerations: First, the purpose, no more and no less, is to establish a region, not a point, where a rapid drop in stiffness occurs. Second, this region has been found, so far, to be possibly decisive for repeated loads. For loads above the knee region, there appears to be a loss of useful fatigue life. This suggests that irreversible mechanical damage has occurred and that it may reduce environmental durability. Repeated loads, quite near to KN, have shown a reduction, for each single cycle, of LL stress and shear modulus. This seems to indicate that useful fatigue life exists below KN, but above LL (see Ref 4). Next, it is rational to assume a range, not a point, for fatigue life, that is, higher stress with a lower number of cycles, and vice versa.

Finally, this is considered to mean that the yield point concept for metals is not applicable. Metals can be taken beyond yield (forming parts by permanent set), yet retain stiffness, fatigue life, and environmental durability. This cannot be safely assumed for structural adhesives. In view of these considerations, it is deemed fruitless to attempt to define a precise point for any value between LL and KN.

Ultimate strength is simply the point at which ultimate failure occurs. It corresponds to that load which the structure must endure just once in its service life.

Skin-Doubler Analysis Verification

With the advent of accurate adhesive strain data by use of the KGR-1 extensometer, it became possible to verify the skin-doubler formulas in Fig. 1. To do this, another extensometer was needed to measure the adhesive shear strain at the doubler tip. This measured value could then be compared to the calculated value.

The second extensometer is identified as KGR-2 (see Fig. 11). As before, the adhesive shear movement (Δ_a) at the doubler tip is made to move the core of an LVDT. The voltage change is fed to a recorder, which plots a curve of load versus adhesive strain at the doubler tip.

The results of tests run on a skin-doubler specimen are shown in Fig. 12. In addition to measuring the adhesive strain at the doubler tip, scribe lines were made at intervals across the glue line. These lines are offset by the adhesive shear movement. Figure 12 shows these offset values as they describe the shear stress distribution. Reference 2 describes this work in detail. As can be seen in Fig. 12, the agreement between calculated and measured adhesive shear strain is very convincing.

Skin-Doubler Adhesive Strain in the Nonlinear Range

Up to this point, the stress analysis has been in the linear range of the adhesive shear stress-strain curve. This range is important and possibly decisive in design because it is here that durability in creep and fatigue comes into play. For analysis above limit load and on to ultimate failure, the nonlinear portion of the stress-strain curve must be used. The mathematics for this is necessarily complex, but the KGR-2 can

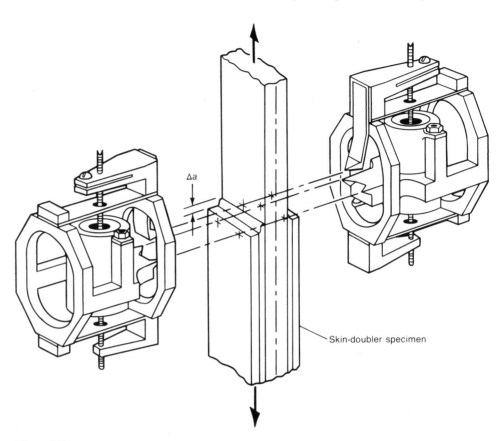

Fig. 11 The KGR-2 extensometer

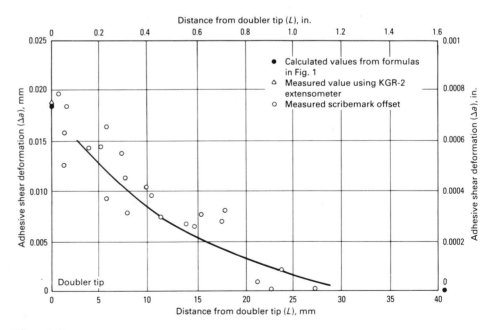

Fig. 12 Skin-doubler shear stress distribution, calculated and measured data

supplement the work for simple joints. This is illustrated below.

Shear Stress-Strain Data. Figure 13 presents data for three adhesives tested on the thick adherend specimen and on the skin-doubler specimen. In order to show relationships conveniently, the two sets of data are plotted using identical adhesive strain scales. Strain is shown on the horizontal axes. The two sets of data are set one above the other with their strain scales exactly aligned. Given a point on one curve, its mate on the other can quickly be found by traveling on the strain line until it inter-

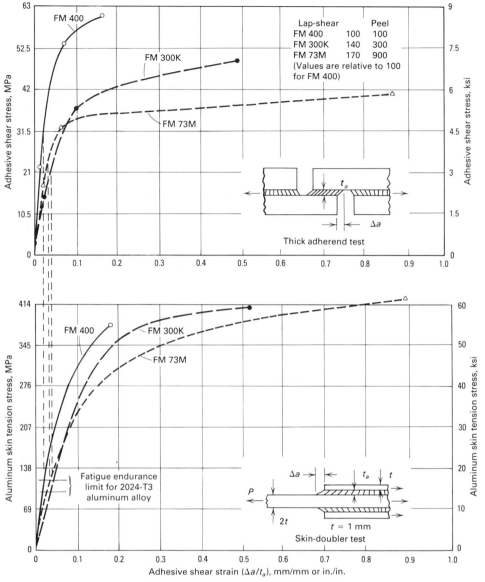

Fig. 13 Shear stress-strain data at room temperature for three adhesives on thick adherend and skin-doubler specimens

Table 1   Conventional test comparisons

Adhesive	Metal-to-metal peel ball method	Lap-shear MMM-A-132
FM 400	100	100
FM 300K	300	140
FM 73M	900	170

Note: Values are relative to 100 for FM 400.

100% relative humidity than at room temperature, dry.

Returning to Fig. 13 and the skin-doubler tests, durability can be addressed, as can the ability of the bond to survive repetitive loading near limit load on the aircraft. Limit load corresponds to skin metal yield point, for example, 276 MPa (40 ksi). Here, the adhesive strains are FM 400 at 0.08 mm/mm and FM 300K at 0.15 mm/mm. Following these strain lines upwards into the graph of the thick adherend, it can be seen that FM 400 and FM 300K are working near their knee points, and FM 73M is slightly beyond.

It may be concluded from this, first, that the drop from curve FM 400 to curve FM 300K is not a disaster. The reduction in knee is counteracted by the reduction in modulus, so the value of adhesive stress remains at the knee.

The second conclusion is that the drop from curve FM 400 to curve FM 73M is more serious because the knee reduction dominates the modulus reduction. The adhesive stress has gone slightly beyond the knee.

Third, it can be seen that by this method the designer gains access to decisive information that is far superior to data from peel and lap-shear tests. If prudence dictates, the designer can run repetitive load tests on skin-doubler specimens, reduce the doubler gage, and increase the skin gage.

Finally, the data pleasantly reaffirms that bonded attachment is superior to rivets because of the skin area lost to their holes.

Next, the ability of these adhesives to outlive the skin metal in fatigue can be examined. If the metal endurance limit is assumed to be between 100 and 120 MPa (14 and 17 ksi), the adhesive works at strains between 0.01 and 0.04 mm/mm, or in./in. Following these strain lines upwards into the graph of the thick adherend, it can be seen that all three adhesives are working below their knee stresses. This indicates that the attachment has a realistic potential for outliving the metal. Again, it is impressive to realize that the metal is represented at its full capability, not compromised by rivet holes.

It appears objectively that proper stress analysis exists for bonded joints. In a more subjective way, this foundation will lead to even more improvement in testing. Specifically, improved testing to produce allow-

sects the pertinent curve. For example, in the thick adherend test, the adhesive FM 400 has its knee at a strain of 0.08 mm/mm, or in./in. Moving down to the skin-doubler curve on this strain line, the skin metal stress is found to be 276 MPa (40 ksi).

In this way, real-life conditions for skin-doubler bonds can be quickly considered because the adhesive strain (stress) is readily found for any skin metal stress. Three adhesives were chosen because they represent actual data for three major differences in properties. First, there is a difference in shear modulus (the slope of the initial portion of the curve). Adhesive FM 400 is approximately three times stiffer than FM 300K or FM 73M. Second, there is a difference in knee value. Adhesive FM 300K has a higher knee than adhesive FM 73M, with the moduli being essentially the same. Third is a difference in elongation, defined as

strain at the point of ultimate failure. In addition to these stress-strain curves, conventional test comparisons, on the basis of a value of 100 for adhesive FM 400, are given in Table 1. Thick adherend test data suggest that FM 400 is superior.

The confusion is dispelled by the real-life skin-doubler data, which show (for ultimate strength) that adhesive FM 73M is superior because it can work the skin metal to the highest stress before adhesive failure. However, the advantage of FM 73M is by no means as great as lap-shear and peel values would suggest. There is a strong implication that environment (such as hot-wet), causing a drop similar to the change in values of FM 400 to those of FM 300K would not cause a loss in ultimate joint strength. In fact, this has been demonstrated with FM 73M adhesive. It is actually stronger on the skin-doubler specimen at 60 °C (140 °F) with

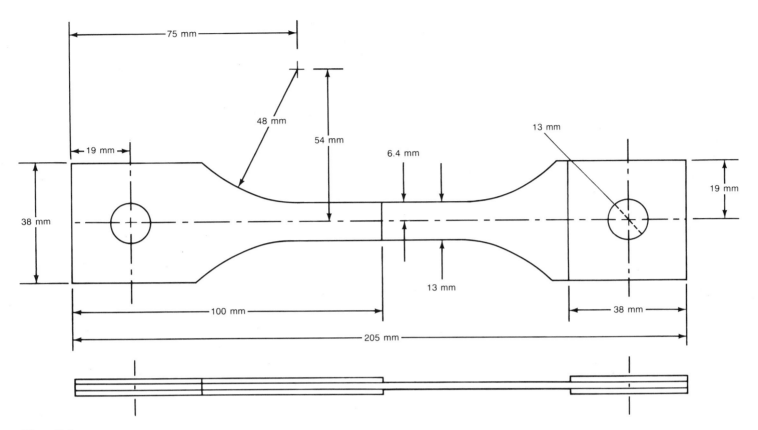

Fig. 14 Specimen for the fatigue testing of structural adhesives

able stresses for fatigue and creep are definable and can be perfected in the near future.

Tests for Allowable Stresses for Fatigue and Creep

Much work has already been done in testing adhesives for fatigue and for creep. Unfortunately, this work has largely been a matter of testing the adhesive either in lap-shear or in thick adherend shear. The drawback to these methods, as has been shown, is that they do not represent real-life situations. This is especially true of the skin-doubler specimen. Thus the results obtained have little relation to the actual structure and no value as allowable stresses. With the skin-doubler specimen under load, real-life conditions exist in which the adhesive acts as a fastener. For example, if shear strain in the thick adherend specimen is taken into the knee region, fatigue life is nil: Failure will occur in only a few load cycles. In the skin-doubler specimen, however, many load cycles that work the adhesive strain beyond the knee can be applied at the doubler tip without load failure (Ref 4). This is because the adhesive is less stiff at its knee strain and some of the load is transferred elsewhere in the joint.

In the case of creep, a similar phenomenon exists. The thick adherend specimen

was loaded at two strain levels near its knee, and creep strain was measured for a 15-min period. The lower strain level showed an accumulated creep of 0.024 mm/mm, or in./in., whereas the higher level showed 0.133 mm/mm, or in./in.

Next, two tests were run in which the higher of these strains was placed on the adhesive at the doubler tip. The first strain showed creep of 0.00833 mm/mm, whereas the other showed 0.0125 mm/mm, over a 15-min period. These are an order of magnitude lower than the thick adherend values. The lower values are due to a load redistribution in the adhesive with much of the joint remaining at strains well below the knee (in the elastic range).

Upon the removal of the load, the strain at the doubler tip returned to 0. In other words, no set or hysteresis was present. This is because the aluminum was in the elastic range (as well as most of the glue line) and able to force the adhesive back to 0 strain.

Another difficulty in understanding a real-life bonded construction is the presence of secondary and even tertiary strains at the crucial point, the doubler tip. There are tension strains (normal to the plane of the glue line) and extra shear strains at the sides of the tip where the skin becomes narrower because of Poisson's ratio. These strains (stresses) cannot be neglected because they most certainly will be

decisive regarding total strain for failure in fatigue and creep.

There are those who have confidence that these strains can be calculated accurately. It is doubtful that a sufficiently high level of accuracy can be obtained to produce safe allowable stresses that will allow analysts to predict the performance of a design. The production of such fatigue and creep allowables is best done by the empirical testing of skin-doubler specimens. Many recoil intellectually at empirical methods. The key response to this is that we are dealing with a fastener, not a material, and this fact produces a telling precedent. The fatigue and creep abilities of rivets and bolts are not obtained by testing aluminum and steel materials. Rather, they are obtained empirically by testing pertinent constructions using the rivets and bolts themselves. This comparison may underline the merit of empirical testing with skin-doubler specimens to obtain fatigue allowables for adhesives.

Fatigue Testing

Figure 14 shows the fatigue test specimen. The edge view shows a single center plate (the skin) with shorter plates (the doublers) bonded to each end. One set of doubler plates is quite short and does not enter into the test except to provide extra bearing strength at the bolt. The other set of doubler plates continues to the center of the specimen. It will be seen that under load (P), the center

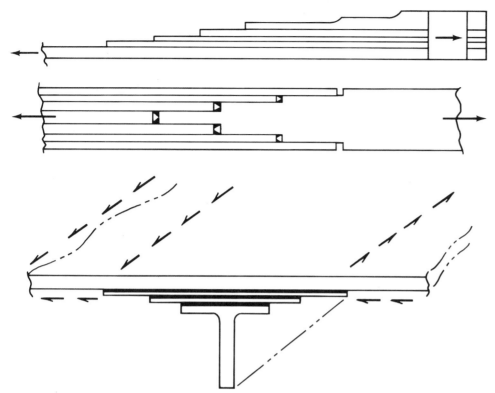

Fig. 15 Cross sections of typical design with multiple doublers

plate is made to strain (stretch) out from between the doubler plates, exactly as on the skin-doubler specimens (Ref 1).

The test site is at the doubler tip. The adhesive stress is maximum here, and the failure therefore originates at this point. It is important to note that adhesive failure will not necessarily be catastrophic. The instant of failure is when the first fracture or disbond appears at the doubler tip. This failure can be difficult to detect, because it may not be audible or visible and there will not be a drop in load on the specimen.

The current technique consists of placing a small strain gage on the doubler, as near to the tip as is practical. When load is applied to the skin, the adhesive works in shear to pass part of it to the doubler. The strain gage readily senses the doubler strain because the adhesive loading is maximum at the doubler tip.

At the first damage to the bond, the doubler tip is unloaded, and the strain gage signal is drastically reduced. Quantitative readings are not needed. The drop in signal strength is made to activate a relay, the fatigue machine load is shut down, and cycles to failure are recorded.

Fatigue Specimen Design. The basic problem is to ensure a failure in the adhesive rather than in the metal in tension or in bearing at the bolt. The allowable stress in bearing for the 13 mm (0.5 in.) diam bolt in 2024T3 aluminum is taken as 48 MPa (7 ksi). If the doubler gage is equal to the skin gage, the skin stress will be twice the bolt bearing

stress, or 96 MPa (14 ksi). These stresses are low enough to avoid fatigue in the metal.

A skin gage (and doubler gage) for 1.6 mm (0.064 in.) is not always high enough to load the adhesive near its knee stress and cause failure of the adhesive. For gages at or below 1.6 mm (0.064 in.), the allowable specimen load for fatigue in the metal may be too low to cause the adhesive to fail. This suggests that many adhesives may be capable of failing the metal (causing the metal to fail in tension just in front of the doubler tip) in any reasonable gage, except possibly when tested in a hostile environment. The maximum cohesive stress for the adhesive would seem to occur in cold or cold-wet conditions. Cold-wet or hot-wet conditions may also be critical for the bond at the metal interface.

Room-temperature testing has shown that FM 400 can fail the metal in fatigue at skin and doubler gages of 1.6 mm (0.064 in.). Designs have been made for fixtures to test in environments of cold-wet and hot-wet conditions. Briefly, these tests are accomplished by attaching the specimen to the inside bottom of a small cup. The outside of the cup bottom is attached to the fatigue machine, as is the opposite end of the specimen. To achieve the requisite environment, hot or cold fluids are pumped through the cup while the test is in progress.

Creep Testing

Testing adhesives for creep stress allowables appears to represent a different problem from testing thick adherend specimens

to obtain compliance curves and to document nonlinear viscoelastic behavior. If it is assumed that a skin-doubler specimen is strained so that the adhesive at the doubler tip is strained beyond linearity, two concerns arise: Will the adhesive be destroyed near the tip, and will the joint exhibit permanent set when unloaded.

A fundamental difference between the thick adherend specimen and the skin-doubler specimen is this: Upon unloading, the thick adherend specimen no longer puts any load on the glue line, whereas upon unloading, the skin-doubler specimen can put considerable force on the glue line. In the latter case, the elastic energy in the metal and in the adhesive try to restore the geometry to the original condition, that is, one in which there is no adhesive strain at the doubler tip. The important concept is that the skin-doubler specimen has great potential, in no small part because of the ability to include the effects of hostile environment.

Design of Adhesive Joints

As stated earlier, the basic principle for the design of adhesive joints is the achievement of the safe transfer of load between the parts of the structure in the most effective way. The term safe means not exceeding the strength and durability limits of the adhesive and the structure. Efficient means with minimum weight and cost. A safe joint can be assured only by proper stress analysis. This means calculating the stresses in the adhesive and adherends, and then comparing these calculated values with allowable (safe) values in order to predict the performance of the joint. Efficiency is achieved by working closely below the allowable stresses and by simplifying design complexities to save weight and cost.

The designer can learn very little from simple drawings or bonded blocks in shear, tension, cleavage, or peel. There is also limited value in simply reviewing a long series of randomly selected examples of bonded joints. A more useful approach is to establish a fundamental design procedure that can be used to solve virtually any load transfer problem.

Toward this end, this article has presented fundamental joint geometries and developed stress analysis methods. These will enable the designer to create safe, efficient bonded joints.

The joint geometries are those shown in Fig. 1 and 2, while the formulas shown in these figures provide the chief tool in the design of the optimum joint. The salient variables are glue line thickness and distance to 0 shear. The ways that these variables are used in design are described below.

Effect of Glue Line Thickness. Increasing the glue line thickness reduces the peak adhesive stress and increases the distance

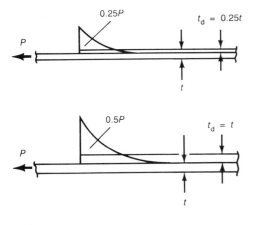

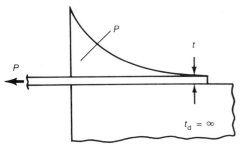

Fig. 16 Adhesive shear stress distribution versus doubler thickness

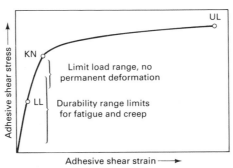

Fig. 17 Allowable stress ranges on shear stress-strain curve

(L) to 0 shear, with no change in load on the joint. Therefore thick glue lines are desirable up to the point of serious weight penalty. Before this point is reached, the glue line at the doubler tip can be increased locally by cutting away the doubler. This cut must be on the adhesive side of the doubler and need not taper, an expensive, accurately machined taper.

Effect of Skin Thickness. The significant point regarding skin thickness is that the thicker it is, the higher the load that must be carried on the bond. This is because the skin, in an efficient structure will be thicker only if it carries a higher load. It must be recognized that in Fig. 1 there is only one glue line available to unload the skin.

If the skin is thick enough to overload the glue line, the solution is to reduce the doubler gage and add more doublers, each stepped back from the other, as in Fig. 15. This arrangement provides more than one glue line to carry the load.

Effect of Doubler Thickness. Changes in doubler thickness affect the load on the adhesive because the doubler must eventually work at the same strain as does the skin. Thus, a thin doubler will receive a smaller percentage of load. Reference 5 shows that this is accomplished by variations of skin gage divided by doubler gage. For skin and doubler of equal gages, $k = 0.71$. For 0 doubler thickness, $k = 0$. The other extreme is when the doubler is infinitely thick, that is, when the skin is bonded to a very thick block or mass. In this case the skin/doubler thickness ratio becomes 0

and $k = 4$. The distance (L) to 0 stress is also affected. For 0 doubler thickness, $k = 0$ and for a doubler of infinite thickness, $k = 4$. Figure 16 shows these relationships.

Effect of Adherend Stiffness. Inspection of the formulas in Fig. 1 shows that the shear stress peak reduces as adherend stiffness increases. For example, changing adherends from aluminum ($E = 10 \times 10^6$ psi) to steel ($E = 30 \times 10^6$ psi) reduces the peak by the ratio of the square roots of these moduli. This amounts to a drop of 73%. The reason for this is that increasing the adherend stiffness effectively amounts to reducing the adhesive stiffness. Thus, the load is transferred over a longer distance (L) and at a lower stress level.

Effect of Distance to 0 Shear. In many designs the need for compromise can result in short overlaps of the bonded joint. With the formulas of Fig. 1 and 2, an adequate length of overlap can be calculated, and, if necessary, the design can be corrected.

Nonlinear Stress Analysis. The formulas in Fig. 1 and 2, as already noted, are for the linear condition of all stiffness parameters. Briefly, this means that glue line thickness is constant; glue lines are either dry or 100% saturated with fluid, that is, the value of G is constant over the distance L; and E and G are constant, that is, the stress levels in adherend and adhesive are in the straight-line portions of their stress-strain curves. This linear range is generally at load levels where durability is the design criterion, that is, repeated loads (fatigue) and sustained loads (creep). For analysis at higher loads, the difficulty is that E and G are reduced as stress levels rise. These reductions offset each other, but do not necessarily occur at the same rates.

These conditions are analyzed by bilinear techniques and finite-element approaches, both of which are beyond the scope of this article. It is proposed that these methods are complex and somewhat compromised by inevitable assumptions made during stress analysis. In order to ensure adequate safety margins, designs are often tested to failure. In a general way, this can be done with skin-doubler specimens while monitor-

ing the peak stress (strain) with a KGR-2 extensometer, as in Ref 2 and 5.

Allowable Stresses. The primary source for allowable stresses is the adhesive shear stress-strain curve, described in Ref 3. This information can be obtained over an environmental range of temperature from -55 to 260 °C (-67 to 500 °F) with the adhesive dry or saturated with water. Other fluids can also be used. These data are invaluable in that they give adhesive stiffness (shear modulus), which varies with temperature and/or fluid saturation.

Figure 17 presents a general diagram of the shear stress-strain curve. It has been established that the adhesive suffers permanent damage in the stress region of KN. Also, quite good elastic performance in the region of LL, that is, linear response over as many as 80 repeat loadings with no hysteresis has been shown. This indicates that for fatigue and creep, operating stresses should not greatly exceed LL. As discussed earlier, the ultimate stress levels are described by UL (Fig. 17).

It is recognized that fatigue life may well exist above LL for specific numbers of cycles. Proper curves (S-N) for strength versus the number of cycles will be developed from skin-doubler specimens adapted for fatigue testing. It is important to recognize that acceptable stress levels and margins of safety are the responsibility of the designer and the stress analyst, because such work inevitably requires assumptions and judgment beyond the scope of test data.

In conclusion, the basic concept for the adhesive joint is that the critical adhesive stresses are stiffness driven. For the same adherends and load, the adhesive stress will increase with a stiffer adhesive and decrease with a softer adhesive. For the same adhesive and load, the adhesive stress will increase with a stiffer adherend. The adhesive is best considered a fastener whose stiffness, rather than strength, determines its performance. This is also true beyond the linear region, where ultimate strength of the joint depends not on adhesive strength, but on adhesive elongation (strain) at its ultimate stress.

Proper joint design cannot be achieved by educated guessing or by unrelated test data. Correct stress analysis is indispensable as a design guide. Such analysis is now feasible because adhesive stiffness properties are available at a low cost commensurate with a data population large enough for statistical confidence.

REFERENCES

1. R.B. Krieger, Evaluating Structural Adhesives Under Sustained Load in Hostile Environment, in *Proceedings of SAMPE Conference*, Society for the Advancement of Material and Process Engineering, Oct 1973

2. R.B. Krieger, "Shear Stress-Strain Properties of Structural Adhesives in Hostile Environments," American Cyanamid Company, Oct 1976; *J. Appl. Poly. Sci.*, 1977, p 321-339

3. R.B. Krieger, "Stiffness Characteristics of Structural Adhesives for Stress Analysis in Hostile Environment," American Cyanamid Company, Oct 1975

4. R.B. Krieger, Fatigue Testing of Structural Adhesives, in *Adhesion 4*, K.W. Allen, Ed., Applied Science Publishers, 1979-1980

5. R.B. Krieger, "Stress Analysis of Metal-to-Metal Bonds in Hostile Environment," American Cyanamid Company, Oct 1977; *Adhes. Age*, Vol 21 (No. 6), 1978

Rating and Comparing Structural Adhesives: A New Method

L.J. Hart-Smith, Douglas Aircraft Company, McDonnell Douglas Corporation

A PROPOSED new method of rating and comparing structural adhesives under shear loading and a new specimen for accomplishing the task are described in this article. The recommended change is to switch from the ASTM D 1002 standard lap-shear coupon, in which the adhesive is forced to fail under artificial conditions not at all representative of real structural joints, to a different kind of coupon in which the adherends, rather than the adhesive, can fail. The merit of each adhesive is assessed according to the strength of an adherend it can break with standardized test coupons of varying thicknesses; the stronger the adhesive, the stronger (and thicker) the members it can break.

The new test specimen, called the inverse skin-doubler coupon, is a geometric inversion of the conventional double-lap-shear joint, but is inherently free of all parasitic peel stress effects associated with that joint. This specimen permits valid comparisons to be made between different adhesives and provides a strength measurement that is directly applicable in design. The basis of this concept is that two adhesives can be considered equivalent if both are just strong enough to break identical adherends outside the joint, even if the individual adhesive properties are not all identical. Likewise, one adhesive may be rated twice as strong as another if it can break an adherend of the same material that is twice as thick as the other adhesive can break.

This concept is a departure from past practices in which the intent was to fail the adhesive rather than the adherends. However, the objective in designing adhesively bonded structural joints should be to ensure that the adhesive layer will never fail. Therefore, the proposed change in the philosophy of testing would make the testing more compatible with proper structural design practice.

It should be noted that this elimination of adverse peel stresses from the test coupons implies that they also be removed from the design, or be reduced to a level of insignificance, using the tapering techniques described later in this article or some equivalent stepped runout. Abrupt thickness discontinuities are known to cause a loss in effective adhesive shear strength as well as fatigue cracks in metal structures or delaminations in composite structures.

The task of designing structural bonded joints has long since progressed beyond reliance on adhesive lap-shear data. The proportions of adhesively bonded joints of both simple and complex design can now be established in terms of nonlinear stress-strain curves measured to failure under pure shear loads on thick-adherend specimens. Stepped-lap joints designed this way hold the wings on the F-18 and the tails on nearly all current fighter aircraft.

Although rational design methods are now available (even though not all adhesives have yet been suitably characterized) the task of selecting adhesives remains more an art than a science. The test specimen proposed in this article offers the greatest benefits in this area.

Proposed Test Specimen

The proposed specimen is shown in Fig. 1. Two modes of failure are possible: The continuous outer adherends will break if the adhesive is sufficiently strong, or the discontinuous inner member will disbond if it is stronger than the adhesive. The effective shear strength of the adhesive layer is established by performing such tests for various thicknesses (and, hence, strengths) of adherends. The tests should be repeated throughout the range of environments anticipated in service (temperature, humidity, and so forth).

Each outer member should be at least half as strong as the inner member; the strength is then controlled by the central member. The load transferred through the adhesive layers is proportional to the strength of the inner member. Any load associated with excess strength in the continuous members bypasses the adhesive layers. It is convenient to make each adherend the same thickness and to compensate analytically for the associated minor effect from the stiffness imbalance.

The strength of the adhesive layers in the inverse skin-doubler specimen shown in Fig. 1 is independent of overlap when the overlap is long. The load transfer through the adhesive occurs in narrow bands adjacent to each end of the doubler, and the adhesive shear stresses and strains are precisely the same as would occur in a structural double-lap or double-strap joint. The adhesive at each discontinuity in a member is not affected by the presence or absence of any other remote discontinuities. This similarity between the adhesive behaviors in doublers and joints is shown in Fig. 2.

Once an adhesive has been shown to be stronger than an adherend of a specified

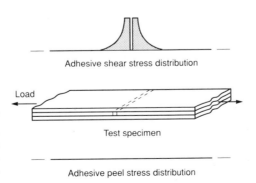

Fig. 1 Inverse skin-doubler specimen

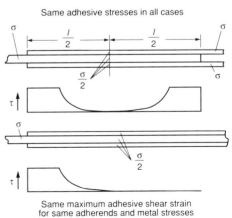

Fig. 2 Similarity of bonded stresses in joints and doublers

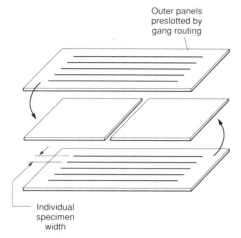

Fig. 3 Manufacture of inverse skin-doubler specimens

thickness for a certain range of environments, the adhesive will inevitably be stronger still than any thinner members of the same material. The use of such an adhesive for thicker members would require the more complex stepped-lap joint to develop more bond strength. In such a multistep joint, adhesive shear load is transferred at both ends of each step instead of only at the outer ends of the overlap, as is the case with uniformly thick (unstepped) adherends.

The only drawback to the inverse skin-doubler specimen is that the failure of the adhesive might not be apparent by visual inspection. An external clip-on extensometer should be installed to bridge the gap between the inner members. Alternatively, a crack-opening displacement transducer could be inserted into the gap. Failure of the adhesive would register as a deviation from the linearity of the load-elongation signal prior to its occurrence. Of course, it would not be difficult to detect the failure of the outer plates that would always occur directly over that gap. The manufacture of such test coupons would be simple and straightforward. All details would be made from sheets of a single thickness. The only critical step would be to ensure that any burrs left by cutting the inner sheets be removed prior to bonding; the anodizing or etching process used in surface preparation would assist in their elimination. Any tendency to trap air bubbles or thick adhesive layers in the middle of the specimen could be avoided by perforating the cover sheets or by partially preslotting them, as shown in Fig. 3, in much the same manner that panels of multiple lap-shear coupons are manufactured today.

In applying this concept, the adhesive would not be tested on a large number of adherends with very similar thicknesses. Rather, it would be qualified against one of several standard thicknesses. In the case of aluminum alloys, it is suggested that the standard thicknesses be some of the cus-

tomary sheet gages of 7075-T6. Each of the standard thicknesses adopted would define a class of structural adhesives.

Prior experience with chemically similar adhesives would be used to reduce the number of tests needed to be performed in each case. This stepwise approach could give rise to discontinuities in the adhesive characterization as a function of service temperature. However, because the adhesive would require characterization by its complete stress-strain curve measured on the thick-adherend test or its equivalent, precise calculations for design would still be possible. Further, there is obviously no objection to rating any adhesive against a nonstandard sheet thickness to substantiate some specific design of an adhesively bonded structure.

This proposed inverse skin-doubler specimen is easy to build and has no critical tolerances. It is also easy to test and, being inherently free of induced peel stresses, can rate adhesives under the same kind of pure-shear loading to which properly designed structural bonded joints are subjected. This rating is far more useful in design than the so-called strength determined by the standard lap-shear test. Instead of characterizing the shear behavior of the adhesive, that test is influenced more by the actual peel stress concentrations that a good designer does his best to eliminate.

To enable comparisons, a description of the lap-shear test coupon, its limitations, and the manner in which structural bonded joints are actually designed are provided in the next section of this article.

Current Lap-Shear Test Specimen

The most widely used adhesive test specimen is undoubtedly the single-lap-shear coupon, ASTM D 1002. The data obtained with this coupon, when used in conjunction with the Boeing wedge-crack test to verify the adequacy of the surface preparation, are quite useful for quality control and incoming receiving and inspection. However, this lap-shear test is an extremely poor way to compare different adhesives, and no way has been found to use the average shear strength thus measured for design. Despite these severe limitations, this lap-shear test is still the basis of most military specifications, perhaps because it is such an easy test to perform and because there has been no agreement on an alternative specimen.

The limitations of the current lap-shear specimen (Fig. 4) should be explained so that the merits of the new test specimen can be understood. The basic problem with the standard lap-shear test is that the average "shear strength" that is determined is not a unique material property characterizing the particular adhesive. Instead, it is a rather vague quantity that depends strongly on the

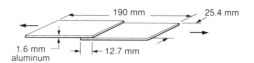

Fig. 4 Standard lap-shear bond test

geometry of the joint being tested. For instance, if the overlap were doubled and all other variables were left unaltered, the strength of the joint would *not* be doubled. Likewise, if the joint geometry and adhesive were not altered and the adherend material were changed (to steel, for example), the apparent strength of the adhesive would be changed dramatically. Even a simple change in adherend thickness is known to affect the so-called lap-shear strength, as demonstrated by Wegman and Devine (Ref 1). Such a test result cannot form the basis of rational design process.

Experimental data of Wegman and Devine have been interpreted analytically by the author in an earlier work (Ref 2) and are reproduced in Fig. 5 and 6. As indicated in Fig. 5, the stronger and stiffer adherend materials appear to enhance the measured "strength" of the adhesive. Similarly, Fig. 6 shows that doubling the adherend thickness can appear to double the adhesive "strength" for long-overlap joints, because failure of the adhesive is preceded by yielding of the adherend for both thicknesses shown. For all but very short overlaps, Fig. 6 reveals a distinctly nonlinear joint strength that cannot be explained under any circumstances as the product of a universal bond shear-strength "allowable" and the bonded area.

The pioneering analysis of this single-lap joint was performed in 1944 by Goland and Reissner (Ref 3). Since then, there have been many further analyses, including one by the author (Ref 4), that have refined their work by taking into account adhesive plasticity, by improving the estimation of the boundary conditions, and so on. There is no perfect analysis yet, and most of the differences between the modern analyses are of the type in which one detail has been covered thoroughly, but at the expense of accuracy in some other detail. Such differences are not germane to the present discussion.

It should be noted, however, that Adams and Wake (Ref 5) found significant and often overlooked errors in the derivations by Goland and Reissner. Not only is there an error in their formula for the peel stresses, the induced bending moment in the adherends at the ends of the overlap is overestimated for all but the limiting cases of 0 load or 0 overlap. In every other case, their derivations underestimate the joint strength because that bending moment becomes the boundary condition establishing the peak adhesive stresses at the ends of the

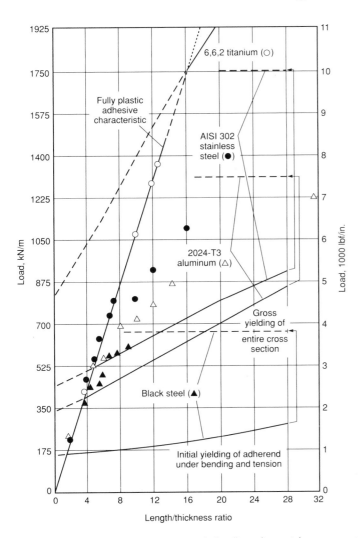

Fig. 5 Influence of 3.18 mm (0.125 in.) thick adherend material on apparent shear strength of adhesives

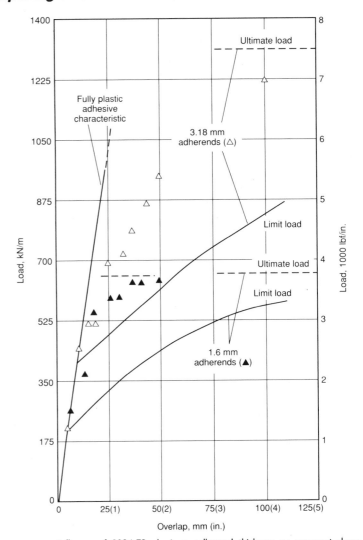

Fig. 6 Influence of 2024-T3 aluminum adherend thickness on apparent shear strength of adhesives

overlap. The same error is also implicit in many finite-element analyses of such joints.

There is complete agreement that the adhesive shear stress (or strain) for the standard lap-shear test specimen varies considerably along the length of the bond and is much higher near the ends of the overlap than at the middle. There is also complete agreement that, because of the eccentricity in load path, there are severe peel stresses induced in the adhesive and that these peak at the ends of the overlap. In addition, there is irrefutable evidence that the best of the adhesives used to bond aircraft structures are so strong that the adherends in this test coupon usually yield prior to failure of the adhesive and that they remain permanently bent after the specimen has failed. These effects are shown in Fig. 7. Adhesives are characterized by the ASTM D 1002 lap-shear test procedure using 7075-T6 aluminum alloy even when the actual structure is made from 2024-T3 aluminum alloy because the higher yield strength of 7075-T6 results in a higher apparent adhesive shear strength.

In designing adhesively bonded structure, the prime task is to maximize its efficiency. This is done, first, by using double-lap joints or very long single-lap joints to minimize the adherend bending and, second, by tapering the ends of the overlaps to reduce the induced peel stresses and thereby permit the adhesive to develop its full shear strength. Accordingly, the adhesive in a properly designed structural joint, in which the proportions are established so that the adhesive will never fail, does not behave at all in the same manner as the adhesive in a short-overlap test coupon. These differences in behavior, which are discussed in Ref 2, explain why the standard lap-shear test is inherently unsuitable as a source of design data for adhesives and should not be used to compare two different adhesives that might have different shear and peel strengths. Conversely, the combination of shear and peel stresses in a single test coupon makes it quite economical for quality assurance work.

In fact, for the overlap-to-thickness ratio of 8 to 1 used for the ASTM D 1002 lap-shear coupon made from 1.6 mm (0.063 in.)

thick aluminum adherends, the main cause of failure of the adhesive is usually the induced peel stresses, with only a minor contribution from the applied shear stresses in the adhesive. This is indicated in Fig. 8, which depicts single-lap joints that have been analyzed by considering each possible failure mode in turn, isolated from the other two. No shear-dominated failure occurs anywhere near the geometry used for the test coupon. Figure 8 also indicates that quite high structural efficiencies can be obtained with single-lap joints, provided that the adherend thickness is limited and the overlap is made long. An overlap-to-thickness ratio of 80 to 1, 10 times the ratio in the test coupon, was used in the skin splices of the adhesively bonded fuselage program known as PABST (Ref 6). Only by means of a rational, nonlinear structural analysis and the actual adhesive stress-strain curves in shear is it possible to relate those bonded skin splices and the short-overlap test coupons. They cannot be magically correlated in terms of any fictitious universal "allowable" adhesive shear strength.

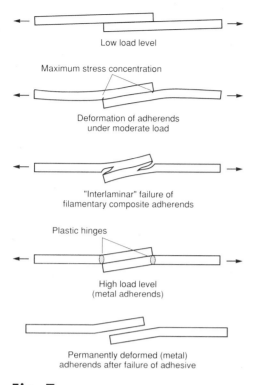

Fig. 7 Failure of single-lap bonded joints

Rational Design of Adhesively Bonded Structural Joints

The benefits of the new test specimen may be better appreciated after an explanation of the design of double-lap or double-strap adhesively bonded splices is given. Such designs are exceedingly simple. The integrated strength of the adhesive bond is independent of the total overlap for all but impractically short overlaps. Therefore, the overlap is made so long that the adhesive in the middle of the overlap cannot creep under the worst circumstances. Any shorter overlap could permit failure by creep-rupture, and any longer overlap could not add to the strength or durability of the bonded joint. These considerations are explained in Fig. 9. For a double-shear joint, typical overlaps for bonding aluminum aircraft structures in subsonic transports are 30 times the skin thickness (Ref 6).

While the design process is extremely simple, more is involved in the stress analysis to ensure that the integrated shear strength of the adhesive is at least 50% greater than the strength of the adherends outside the joint. It should be noted that the strength of the bond *must* exceed the strength of the adherends, even if the design ultimate load was actually far less. If the adhesive bond were ever designed to be weaker than the adherends outside the joint, there could be no tolerance to damage in either the bond or the adjacent adher-

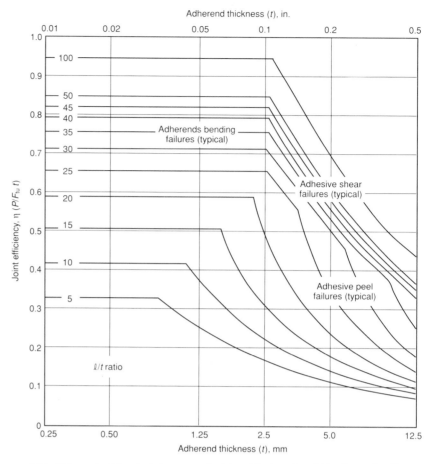

Fig. 8 Joint efficiency chart for single-lap composite joints. Numbers 5 to 100 refer to the ℓ/t ratio. Brittle adhesive used with $(0°/45°/-45°/90°)$ high tensile strength graphite-epoxy adherends at room temperature

ends; the adhesive would act as a weak-link fuse (Ref 7).

The analysis of the joint is based on stress-strain curves for the adhesive under shear loading, as shown in Fig. 10. Such curves can be obtained by using the KGR-I extensometer, developed by Krieger (Ref

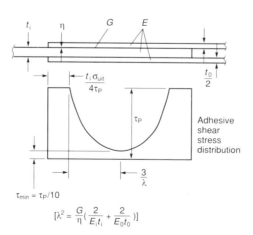

Fig. 9 Design of double-lap bonded joints. The plastic zones must be long enough for ultimate load, the elastic trough must be wide enough to prevent creep at the middle, and the specimen must be checked for adequate strength.

8). The installation of a pair of these extensometers on a thick-adherend test coupon is shown in Fig. 11.

The analysis of such double-lap or double-strap joints using methods developed by the author (Ref 9) leads to two conclusions. First, there is always an upper limit on the thickness of adherends that can be safely bonded using this simple joint; thicker adherends require the more complex stepped-lap joint. Second, the ends of the adherends must be tapered, as shown in Fig. 12, if the adhesive is to realize its full potential shear

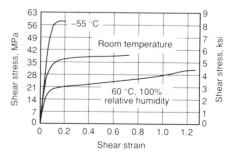

Fig. 10 FM 73 adhesive stress-strain curves in shear

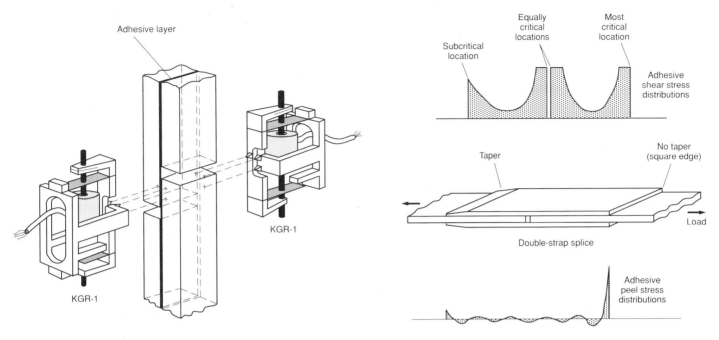

Fig. 13 Adhesive stresses in tapered and untapered bonded joints

strength and avoid a premature failure due to induced peel stresses.

For bonding typical aluminum aircraft structures, the taper should be down to a tip thickness of no more than 0.76 mm (0.030 in.), with a slope of 1 in 10. The maximum sheet thickness for safe design is about 3.18 mm (0.125 in.), although the best adhesives could actually break plates more than 4.78 mm (0.188 in.) thick.

The pioneering elastic analyses of the double-lap bonded joint are those of Volkersen (Ref 10) and de Bruyne (Ref 11). While their absolute strength predictions would differ from those given by the nonlinear theory in Ref 9, the characteristics are the same. All of the load is transferred in narrow zones adjacent to the ends of the members.

The inherent characteristics of long-overlap double-lap or double-strap splices would seem to make them excellent for comparing

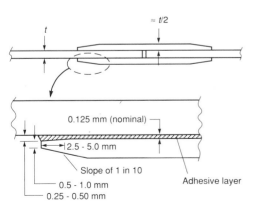

Fig. 12 Tapering of edge of splice plates to relieve induced adhesive peel stresses

different adhesives. The joint strength is completely insensitive to the precise value of the long overlap, and the thickness of the sheet that can be bonded without failing the adhesive is very sensitive to the shear strength of the adhesive (or, more correctly, to the shear strain energy). The stronger the adhesive, the thicker the sheets that can be spliced together. Moreover, because such a test actually matches real structure, the test result becomes directly applicable design data that need no further interpretation whatsoever.

Unfortunately, the double-lap and double-strap joints have one characteristic that renders them unsuitable as test specimens: They are not inherently free of all induced peel stresses at the ends of the overlap. That is why the inverse skin doubler may be preferable. It is apparent from the right side of Fig. 13 that high peel stresses develop at the end of the untapered splice plate. These would prevent the adhesive from developing its full shear strength in this type of joint. Accordingly, such joints should be tapered in the manner shown at the left of Fig. 13 in order to render the remaining peel stresses insignificant. In such a case, the strength of the adhesive is then limited by the conditions that prevail where the central members butt together and any normal stresses that might develop in the adhesive would be small and compressive instead of large and tensile.

A comparison of Figures 1 and 13 shows that the conditions that exist in the middle of each specimen are exactly the same, provided that the overlap is sufficiently long. Thus, the inverse skin-doubler specimen does accurately model the critical lo-

cation in a properly designed double-lap joint.

While the adherends in the inverse skin-doubler test coupon are not tapered, the outer extremities of the splice plates in structural joints and doublers must be tapered in accordance with Fig. 12 if such joints are to attain as high a strength as would be predicted from tests on the inverse skin-doubler specimen. Fortunately, the structural double-lap joint strength is not adversely affected by any excessive tapering, as explained in Fig. 14. This insensitivity can easily be understood in terms of Fig. 2. The total load transferred at both ends of the joints in Fig. 14 must inevitably be identical, even if the adhesive shear stress distributions are not. Compatibility of deformations for such long-overlap stiffness-balanced joints requires that precisely half the load be transferred at each end of the joint; otherwise, the strains in the adherends would not be uniform throughout the interior of the joint.

The skin-doubler specimen, shown in Fig. 15, looks just like the double-strap specimen, except that the central member is continuous. Further, the skin-doubler specimen develops high induced peel stresses at the ends of the doublers. In structural applications, such doublers might need to be tapered in the manner shown in Fig. 12. Because of this, the skin-doubler specimen would not make a satisfactory test coupon for rating adhesives; the result would depend more on the precision of the tapering than on the properties of the adhesive being tested.

While the inverse skin-doubler specimen would overestimate the shear strength of a

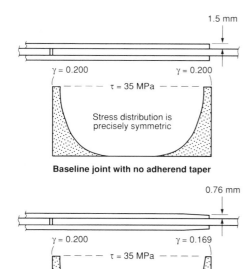

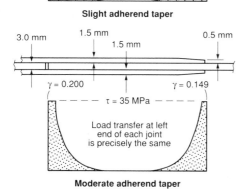

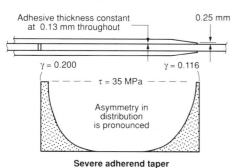

Fig. 14 Insensitivity of adhesive bonded joint strength to modifications at one end of joint only. Adhesive strain at right end of joint decreases with more taper.

bonded joint in which the outer splice plates were neither tapered nor laminated and stepped, the basic skin-doubler specimen with thick square-cut splice ends would lead one to underestimate the shear strength of the adhesive and unnecessarily limit the use of adhesive bonding to very thin structures.

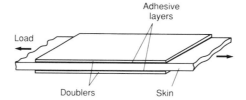

Fig. 15 Skin-doubler adhesive test specimen

The compromise of laminating thick splice plates out of a series of layers each sufficiently thin to minimize induced peel stresses and stepping the ends back progressively, as in a practical design, would demonstrate the true shear strength of an adhesive but would be more costly than using the inverse skin-doubler specimen advocated here.

On the other hand, in conjunction with the KGR-II extensometer, the skin-doubler specimen makes an excellent coupon for investigating the durability of adhesive bonds without fearing a sudden, catastrophic failure, as could occur with a bonded joint rather than a doubler (see the section "Adhesive Bonding Design and Analysis" in this article).

Unlike lap-shear strength which exhibits a strong sensitivity to the temperature under which the test is performed, the inverse skin-doubler specimens will show very little variation in strength below the upper service temperature limit, which is in turn set by the glass transition temperature of the adhesive. The reason for this insensitivity is that the load transferred through long-overlap bonded joints is defined by the adhesive strain energy in shear and not by any individual property, such as the maximum shear strength. The area under each of the stress-strain curves in Fig. 10 is quite similar. The adhesive becomes stronger and more brittle at low temperatures and weaker but more ductile at high temperatures. Consequently, the use of the proposed test specimen would remove some of the confusion that the lap-shear test coupon has caused about the effects of temperature below the glass transition point on the strength of structural bonded joints.

While the discussion of this new specimen has focused on adhesively bonded aircraft structures, the same principle is also applicable to the different adherend materials and adhesives that are used in other industries. The specimen overlap will probably need to be adjusted to suit the material, particularly in the case of wood, which can be glued in much greater thickness than is possible for the stronger aluminum alloys. However, in every case

it should be possible to establish limits on the thickness of what can safely be bonded and what should not be bonded with any given adhesive. More important, there will be no ambiguity about the meaning of the test result, as there is with the current lap-shear test coupon.

REFERENCES

1. R.F. Wegman and A.T. Devine, Adherend Mechanical Properties Limit Adhesive Bond Strengths, Paper 1-4-2, in *SAMPE 14th National Symposium Proceedings*, Cocoa Beach, FL, Society for the Advancement of Material and Process Engineering, Nov 1968
2. L.J. Hart-Smith, "Differences Between Adhesive Behavior in Test Coupons and Structural Joints," Douglas Aircraft Company Paper 7066, presented to ASTM Adhesives Committee D-14 Meeting, Phoenix, American Society for Testing and Materials, March 1981
3. M. Goland and E. Reissner, The Stresses in Cemented Joints, *J. Appl. Mech.*, Vol 11, 1944, p A17-A27
4. L.J. Hart-Smith, "Adhesive-Bonded Single-Lap Joints," NASA CR-112236, National Aeronautics and Space Administration, Jan 1973
5. R.D. Adams and W.C. Wake, *Structural Adhesive Joints in Engineering*, Elsevier, 1984, p 23-27
6. L.J. Hart-Smith, Adhesive Bonding of Aircraft Primary Structures, *SAE Trans.*, 801209, 1980
7. L.J. Hart-Smith, Design and Analysis of Bonded Repairs for Metal Aircraft Structures, in *Bonded Repair of Aircraft Structures*, A.A. Baker and R. Jones, Ed., Martinus Nijhoff, 1988, p 31-47
8. R.B. Krieger, Jr., Analyzing Joint Stresses Using an Extensometer, *Adhes. Age*, Vol 28 (No. 11), Oct 1985, p 26-28
9. L.J. Hart-Smith, "Adhesive-Bonded Double-Lap Joints," NASA CR-112235, National Aeronautics and Space Administration, Jan 1973
10. O. Volkersen, Die Nietkraftverteilung in zugbeanspruchten Nietverbindungen mit Konstanten Laschenquerschnitten, *Luftfahrtforschung*, Vol 15, 1938, p 41-47; also, in "The Distribution of Forces on the Rivets in Stretched Joints With Constant Strap Cross Sections," J. Helledoren, Trans., Durand Reprinting Committee, California Institute of Technology
11. N.A. de Bruyne, The Strength of Glued Joints, *Aircr. Eng.*, Vol 16, 1940, p 115-118, 140

Numerical Design and Analysis

F.E. Penado and Richard K. Dropek, Hercules Aerospace Company

NUMERICAL ANALYSIS TECHNIQUES such as the finite-element (FE) method provide powerful tools in the design of efficient bonded joints. This is particularly true when one considers that the complicated geometry and material dissimilarity in the joint region pose mathematical difficulties that render exact solutions too complicated for most practical joints. Some of these difficulties include hard-to-satisfy boundary conditions in the joint region and the existence of points of singularity (theoretically infinite stresses) at the interface between the adherend and the adhesive.

Numerical methods, on the other hand, are easily adaptable to complex loadings, geometries, material properties, and boundary conditions. In spite of their versatility, however, numerical solutions are only approximations to the actual solution. The degree of correlation between these approximations and the actual solution depends on the nature of the problem being analyzed. Thus, the designer/analyst must understand the limitations of the numerical techniques being used and also the nature of the stresses and failure mechanisms in bonded joints.

This article outlines the basic considerations that must be kept in mind for the successful numerical design and analysis of adhesively bonded joints. Emphasis is given to the FE method because it is the most powerful and commonly available numerical technique. The solution methodologies presented are based on a balance between theory and practice.

Analysis Methods

The analysis methods that the engineer has available for the design of bonded joints can be divided into two groups: closed form and numerical (approximate). Each method has advantages and disadvantages, but the solution of problems with complex geometries and boundary conditions becomes possible only through numerical approximations. Numerical methods are treated in detail in this article; however, closed-form solutions are also included because they elicit characteristics of bonded joint behavior important in failure considerations that otherwise would go undetected. One of

these characteristics is the existence of stress singularities at the interface between the adhesive and the adherend. Closed-form solutions also provide a means of checking numerical results for accuracy and convergence. Following is a discussion of the applicability of each of these methods.

Closed-Form Methods. Exact elasticity solutions that satisfy the equations of equilibrium and compatibility, along with the prescribed boundary conditions, are possible for bonded dissimilar bodies. For example, a rigorous elasticity solution for the case of a cylindrical inclusion bonded to a plate of different material properties was derived in Ref 1. It was shown, however, that there are mathematical difficulties encountered in solving this kind of problem. It was also shown that stress singularities exist at the intersection of the inclusion and the free surface of the plate. Other authors have also shown the existence of singularities when joining dissimilar materials (Ref 2-5). The concept of stress singularities is shown in Fig. 1 for the single-lap joint. In

the immediate vicinity of the point of singularity (for example, as $r \to 0$):

$$\sigma_{ij} \sim \frac{1}{r^{\alpha}} \qquad \text{(Eq 1)}$$

where σ_{ij} are the stresses, r is the distance from the point of singularity (see Fig. 1), and α is a positive real number, $0 < \alpha < 1$, that represents the strength of the singularity. Note that α depends on the local geometry and material properties but is independent of the global geometry or loading. More details about the asymptotic character of the stresses near the singularity point are given in the Appendix. The singular behavior exhibited by unflawed, bonded dissimilar materials is similar to that at the tip of a crack in a homogeneous material, except that in the latter case $\alpha = 0.5$ (square root singularity).

In view of the mathematical difficulties posed by exact elasticity solutions, many researchers have attempted to make simplifying assumptions to obtain the state of stress in certain joints of practical interest. A joint that has received considerable atten-

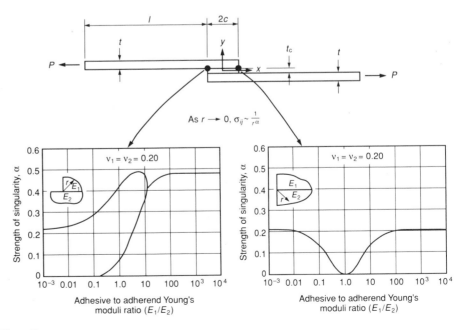

Fig. 1 Strength of singularities at the interface corners of the single-lap joint. Source: Ref 4

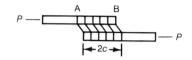

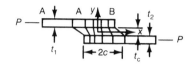

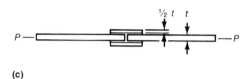

Fig. 2 Schematic representation of the joint deformations under the rigid adherend and Volkersen's assumption. (a) Rigid adherend assumption. (b) Volkersen's assumption. (c) Double-lap joint

tion is the single-lap joint shown at the top of Fig. 1. This joint is considered below in order to illustrate the limitations and difficulties associated with the closed-form analysis of bonded joints.

The simplest single-lap joint solution considers the adherends to be rigid and the adhesive to deform only in shear (Fig. 2a). This assumption is justifiable on the basis that in many practical joints the elastic moduli of the adherends are much larger than that of the adhesive. Then, the shear stress in the adhesive (τ) is constant and given by:

$$\tau = \frac{P}{2c} \tag{Eq 2}$$

where P and $2c$ are, respectively, the applied force per unit width and overlap length.

A refinement of the rigid-adherend assumption was provided by Volkersen (Ref 6). Volkersen took into account the adherend deformation by assuming that the adherends deform in tension only, whereas the adhesive deforms in shear. This is known as differential shear or shear lag (see Fig. 2b). This assumption leads to the following expression for the shear stress in the adhesive (see Fig. 2b for notation):

$$\tau = \frac{P\Omega}{2} \left[\frac{\cosh \Omega x}{\sinh \Omega c} + \left(\frac{t_1/t_2 - 1}{t_1/t_2 + 1} \right) \frac{\sinh \Omega x}{\cosh \Omega c} \right] \tag{Eq 3}$$

where $\Omega = \{[1 + (t_1/t_2)]G_c/Et_1t_c\}^{1/2}$, $G_c =$ adhesive shear modulus, and $E =$ adherend Young's modulus.

However, the analysis of Volkersen does not consider two additional effects that occur in the single-lap joint: bending induced by the eccentricity in the load path, and the

free-surface effects at the ends of the bond line (where the shear stresses vanish). Thus, the solution presented by Volkersen approximates only the shear stress distribution in joints that exhibit no net structural bending moment, such as the double-lap joint (see Fig. 2c). It will be shown later that, in addition to shear stresses, normal stresses also exist in the double-lap joint (see Fig. 20a).

An examination of the effect of load path eccentricity on the single-lap joint was made by Goland and Reissner (Ref 7). The effect of bending is shown schematically in Fig. 3. Their analysis considered two limiting cases: (a) the adhesive layer is so thin that its effect on the flexibility of the joint may be neglected, and (b) the joint flexibility is mainly due to that of the adhesive layer. Case (b) is the one that is of practical interest. Their solution to (b) assumed that the two adherends were plates under cylindrical bending, while the adhesive layer was analogous to a system of springs (constant stresses across the adhesive layer). Their analysis, which violated the condition of zero shear stresses at the overlap ends, showed that significant normal stresses also occur in addition to the shear stresses at the ends of the adhesive layer. Their solution for the normal and shear stresses in the adhesive is (see Fig. 1 for notation):

$$\sigma = \frac{Pt}{c^2\Delta} \left[\left(R_2\lambda^2 \frac{K}{2} - \lambda K' \cosh \lambda \cos \lambda \right) \right.$$

$$\cosh \frac{\lambda x}{c} \cos \frac{\lambda x}{c} + \left(R_1 \lambda^2 \frac{K}{2} \right.$$

$$\left. - \lambda K' \sinh \lambda \sin \lambda \right) \sinh \frac{\lambda x}{c} \sin \frac{\lambda x}{c} \right] \tag{Eq 4}$$

$$\tau = -\frac{P}{8c} \left[\frac{\beta c}{t} (1 + 3K) \frac{\cosh(\beta x/t)}{\sinh(\beta c/t)} + 3(1 - K) \right] \tag{Eq 5}$$

where $P =$ applied force/unit width; $\beta^2 = 8G_ct/Et_c$; $\gamma^4 = 6E_ct/Et_c$; $\lambda = \gamma(c/t)$; $R_1 = \cosh \lambda \sin \lambda + \sinh \lambda \cos \lambda$; $R_2 = \sinh \lambda \cos \lambda - \cosh \lambda \sin \lambda$; $\Delta = \frac{1}{2}(\sinh 2\lambda + \sin 2\lambda)$; $K' = (Kc/t)[3(1 - \nu^2)(P/Et)]^{1/2}$; $u_2 = \{[3(1 - \nu^2)/2](P/Et^3)\}^{1/2}$; $K = [\cosh(u_2c)]/[\cosh(u_2c) + 2\sqrt{2}\sinh(u_2c)]$; $E_c, G_c =$ Young's modulus, shear modulus of adhesive; and $E, \nu =$ Young's modulus, Poisson's ratio of adherend.

It should be noted that the solution for the normal stresses, Eq 4, is written incorrectly as presented in Ref 7. The correct form, used in Eq 4, is given in Ref 8 to 10. Other extensions of the work of Goland and Reissner to consider nonidentical or anisotropic adherends, zero shear at the overlap ends, and thickness effects are found in Ref 11 to 14. All of these solutions, however, assume either constant or linearly varying stresses across the adhesive thickness, thus forcing the singularities shown in Fig. 1 to disappear.

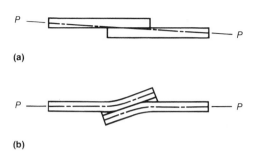

Fig. 3 Schematic representation of bending deformations in a single-lap joint. (a) Undeformed joint. (b) Deformed joint. Source: Ref 37

Figures 4(a) and 4(b) show, respectively, the shear and normal stress comparisons of the rigid-adherend, Volkersen, and Goland and Reissner solutions, with FE models of the single- and double-lap joints for $P = 1.78$ kN (400 lbf), $l = 63.50$ mm (2.5 in.), $t = 3.18$ mm (0.13 in.), $2c = 12.70$ mm (0.5 in.), and $t_c = 0.254$ mm (10 mils). Aluminum adherends with an epoxy adhesive were used. In Fig. 4(a), it can be seen that Volkersen's solution for the shear stresses approximates the finite-element solution for the double-lap joint, whereas that of Goland and Reissner approximates the single-lap joint. Goland and Reissner's solution predicts that the maximum shear stress is at the ends of the overlap, whereas the FE model shows that this stress tends toward zero at the ends, the maximum being a small distance from the ends. The rigid-adherend solution for the shear stresses is seen to be in error, particularly at the overlap ends where it significantly underpredicts the stress concentrations. In Fig. 4(b), large normal stress concentrations can also be seen at the ends of the overlap of the single-lap joint. Details about the FE models used are given below.

Numerical Methods. Closed-form methods are limited because only the simplest geometries and boundary conditions can be accommodated. For more complex situations, approximate numerical solutions become necessary. Following is a brief description of the two most commonly used numerical methods: finite difference (FD) and finite element.

The FD approach evolved by replacing the governing partial differential equations (PDEs) of the continuum with *pointwise* approximations (for example, the PDEs are substituted by their FD approximations at discrete points; the solution is improved as more points are used). This method allows the accommodation of more complex boundary conditions than are possible with closed-form solutions. However, problems are encountered with curved or irregular boundary conditions (see Fig. 5a). This method, however, has been found useful in the solution of two-dimensional heat transfer and fluid mechanics problems, as well as solid mechanics problems with boundaries parallel to the coordinate axes.

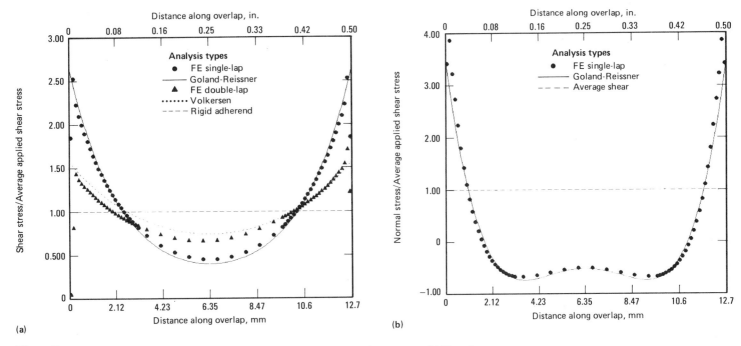

Fig. 4 Comparison of closed-form solutions with finite-element (FE) results. (a) Shear stresses. (b) Normal stresses

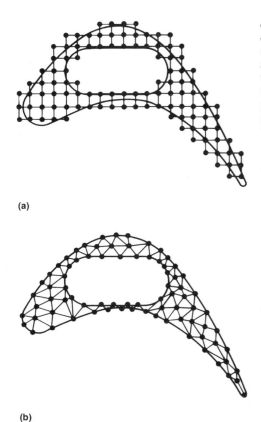

(a)

(b)

Fig. 5 Finite-difference and finite-element discretizations of a turbine blade profile. (a) Typical finite-difference model. (b) Typical finite-element model using triangular elements. Source: Ref 17

The FE method, on the other hand, utilizes a *piecewise* approximation to the governing PDEs. The region of interest is divided into an assemblage of smaller subregions or elements interconnected by nodes on the element boundaries. Continuous piecewise smooth functions over the element region are used to approximate the unknown quantities. These functions, known as interpolation functions, are typically selected as polynomials because they are easy to integrate and differentiate. The foundation of the method is that the global response of the structure is the summed response of the individual elements. By using an integral formulation a system of algebraic equations in the unknown nodal quantities is generated. A typical FE mesh is shown in Fig. 5(b). Because the elements can be arranged in a variety of ways to conform to the boundaries of the structure, the FE method is attractive for the solution of problems with very complex geometries.

Another feature that makes the FE method well suited for the analysis of bonded joints is the ability to interface elements with different material properties. The solution can be refined as much as needed by increasing the number of elements, but limitations are imposed by the storage capacity of the computer being used. The reader is referred to the literature for a complete description of the theory behind this method (Ref 15-19).

Numerous FE analysis codes are commercially available, such as NASTRAN, ADINA, ANSYS, and so forth. These codes provide built-in elements for special applications, for example, truss and beam elements, two-dimensional solid plane stress or plane strain elements, plate and shell elements, axisymmetric shell or solid elements, three-dimensional solid or "brick" elements, and crack-tip elements. Capabilities include static or dynamic analysis, material and geometric nonlinear formulations, isotropic or anisotropic materials, and so on. In addition, automatic mesh (element) generators are available that include postprocessing capabilities. Table 1 lists the capabilities of seven representative commercial codes that are available (Ref 15).

The FE analysis results for the single- and double-lap joints shown in Fig. 4 were obtained using four-node quadrilateral plane strain elements and the MSC NASTRAN code. The mesh geometries are shown in Fig. 6(a) and 6(b). For the double-lap joint, advantage was taken of the double symmetry that exists and only a one-quarter model was necessary, as in Fig. 6(b). The meshes in Fig. 6 and the plots in Fig. 4 for the FE results were created using, respectively, the PATRAN mesh generator and postprocessor. Note in Fig. 6 that it was possible to take advantage of the closed-form solutions of Volkersen and Goland and Reissner to refine the mesh near the ends of the joint, where steep stress gradients were expected.

The FE results in Fig. 4 were obtained at the center row of elements in the adhesive. Three elements were used across the adhesive thickness, as seen in Fig. 6. The center row of elements provides the best approximation to the assumption of Volkersen and Goland and Reissner of constant stresses across the adhesive thickness. In order to demonstrate the through-the-thickness variation of stresses in the single-lap joint, results for the top and bottom rows of elements are compared with the center row in Fig. 7(a) and 7(b). Looking at the top row of elements, both the normal and shear

Table 1 Numerical codes available for structural analysis

Attribute	ABAQUS FE	ADINA FE	ANSYS FE	BOSOR FD	MSC NASTRAN FE	STAGS FE	PATRAN-II composites D&A modules FE
Interactive capability (mesh generator, material input, FE input)............	Poor	Good (ADINA-IN)	Fair	Good (interactive BOSOR version)	Fair (FEMGEN)	Poor	Excellent
Typically used mesh generator........	Internal code (fair to poor)	PATRAN, ADINA-IN	PATRAN, Code internal	Code internal	PATRAN	Code internal	PATRAN
Local lamina lay-up input	Yes (shell only)	None	Yes (triangular shell)	Yes	Yes (shell only)	Yes	Yes
Local-to-global transformation........	Yes	None	Yes	Yes	Yes	Yes	Yes
Global-to-local lamina stress or strain recovery	Yes (shell only)	None	Limited	Limited	Yes (shell only)	Yes, stresses	Yes
Analysis types							
Static............................	Yes	Yes	Yes	Yes	Yes	Yes	Yes
Dynamic	Modal, transient	Modal, transient	Modal, transient	Modal	Modal, transient	Modal, transient	No
Collapse	Yes	Yes	Yes	No	Yes	Yes	No
Bifurcation	Yes	Yes	Yes	Yes	Yes	Yes	No
Geometry nonlinearity	Yes	Yes	Yes	Yes	Yes	Yes	No
Hygrothermal	Yes	No	Yes	Yes	Yes	Yes	Yes
Orthotropic element types							
Three-dimensional plates...........	No	No	Yes	No	Yes	Yes	No
Two-dimensional axisymmetric shell...	Yes	Yes	Yes	Yes	No	No	No
Three-dimensional shell...........	Yes	Yes	Yes	No	Yes	Yes	No
Two-dimensional solid	Yes	Yes	Yes	No	Yes	Yes	No
Two-dimensional axisymmetric solid...	Yes	Yes	Yes	No	Yes	No	Yes
Three-dimensional solid...........	Yes	Yes	Yes	No	Yes	No	Yes
Code support	Hibbitt, Carlson & Solenson, Inc.	ADINA Engineering	Swanson Analysis Systems Inc.	David Bushnell	MacNeal-Schwendler Corporation	COSMIC	PDA

Source: Ref 15

stresses have a higher peak on the left end than at the same location for the center row. At the right end, top row, the peaks are lower than at the left end. A similar situation is seen to occur at the bottom row of bond-line elements, except that the left and right end behavior is reversed due to anti-symmetry. Away from the ends, the thickness variation disappears.

This behavior can be easily understood when one considers the singularities of stress at the ends of the joint shown in Fig. 1. For the moduli of elasticity considered, the strength of the singularity on the top left corner of the joint is seen to be larger than at the top right corner ($\alpha_{\text{left}} \approx 0.27 > \alpha_{\text{right}} \approx 0.17$). Thus, on the top row (with similar but antisymmetric behavior on the bottom row) at both corners the stresses are expected to theoretically become infinite, with the left corner approaching large values faster than the right corner. As the FE mesh is further refined near these corner points, increasingly larger values for the stresses would be obtained. This singular behavior would obviously create problems when considering failure criteria based on maximum stresses. Methods to circumvent this apparent difficulty are discussed in the section "Failure Criteria" in this article.

The preceding information illustrates the power of the FE method, as well as the need for an understanding of closed-form solutions of bonded joints before FE solutions are attempted. For many applications where a closed-form solution is not feasible, the FE method provides the only alternative. However, there are additional aspects of bonded joint failure and numerical mod-

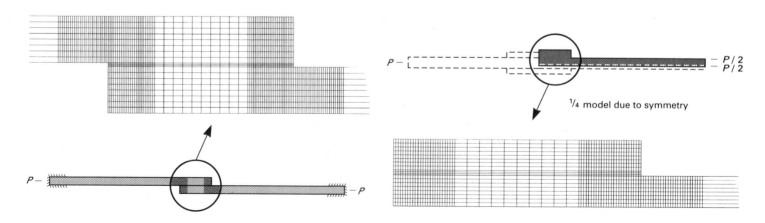

(a)   (b)

Fig. 6 Mesh discretizations used in the FE models of the single-lap and double-lap joints. (a) Single-lap joint. (b) Double-lap joint

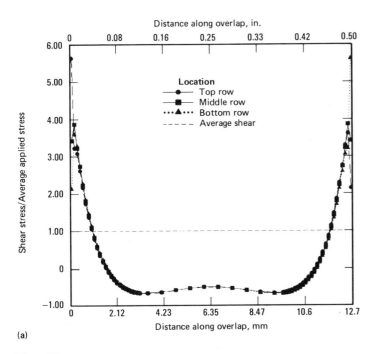

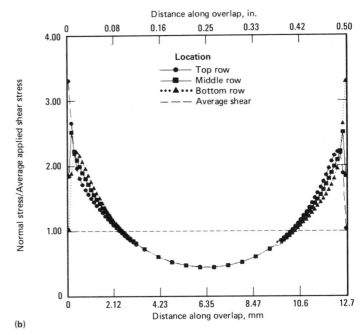

Fig. 7 Through-the-thickness variation of stresses in the bond line of the single-lap joint analyzed using the FE method. (a) Normal stresses. (b) Shear stresses

eling that the engineer must keep in mind to obtain accurate FE stress and failure predictions. These aspects are discussed in the following sections.

Failure Criteria

Calculating the state of stress in a joint is only part of the design process. The engineer also needs to predict failure loads and margins of safety. This is accomplished by the application of a suitable failure criterion. Several methods to predict adhesive joint strengths are discussed below, followed by a failure-prediction approach using the FE method. The advantages and limitations of each method are presented. In general, which method is best depends on the application at hand.

The average stress method simply considers the average shear or normal stress in the bond (P/A) and compares it to the average stress allowable from test specimens. It will produce adequate predictions of bond strength if test coupons have the same loading, geometry, material properties, and are under the same temperature and moisture conditions as the structural part being designed. Thus, a precise test specimen analogue is required for strength predictions. This method is also useful to obtain a qualitative strength comparison between several different adhesives tested using the same type of coupon (for example, lap-shear strength).

The engineer should be aware, however, that deviations between the structural part and the test coupons may lead to large errors, up to a factor of three or more. An example taken from Ref 20 is shown in Fig. 8, which shows the load at failure for butt-

tension specimens that differ only by the thickness of the adhesive. The average stress method in this case is unable to account for the variation in strength due to bond line thickness.

The maximum stress method, which is widely used in industry, compares the maximum stress in the bond (normal, shear, or von Mises) with the corresponding allowables from tests. It improves on the average stress method by considering the variation of stresses in the adhesive along the bond line and across the thickness, but more analysis effort is required. One of its main advantages is that, in principle, the test specimens do not need to be of identical geometry to the part being analyzed.

However, limitations occur due to the inability of the method to account for the points of stress singularity that were shown earlier to occur when joining dissimilar ma-

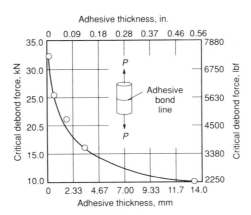

Fig. 8 Effect of the adhesive thickness on the debond load of EA 934 adhesive. Source: Ref 20

terials (see Fig. 1). For example, if the stresses become infinite, the predicted load at failure is zero. From a practical point of view, infinite stresses are not obtained because FE results will produce an averaged value over the size of the element, thus bounding the stresses. However, there will be a dependency of the FE results on element size, with the value of the stresses increasing as the size of the element is made smaller near the singularity point. A method to determine the element size near the singularity is discussed below in the subsection "A Practical Joint Failure Criterion."

The maximum stress method has been found useful in predicting failure initiation of a joint. For example, experimental results with brittle adhesives have shown crack initiation at the predicted location. However, there are cases when a joint can carry substantially more load after failure initiation. This will occur in brittle adhesives where a crack initiates but is arrested at a later stage, or in ductile adhesives where plasticity occurs and an elastoplastic analysis is necessary. These topics are discussed in the next two subsections.

Fracture Mechanics Methods. The presence of stress singularities in the bond region (see Fig. 1) indicates that fracture mechanics methods may be suitable to investigate bond failures. In fact, several researchers have applied fracture mechanics methods to study the failure of bonded joints. One approach is applicable to the extension of existing cracks of known size (Ref 21-25). This method is useful when the location and size of cracks or flaws in the adhesive are known from experimental data.

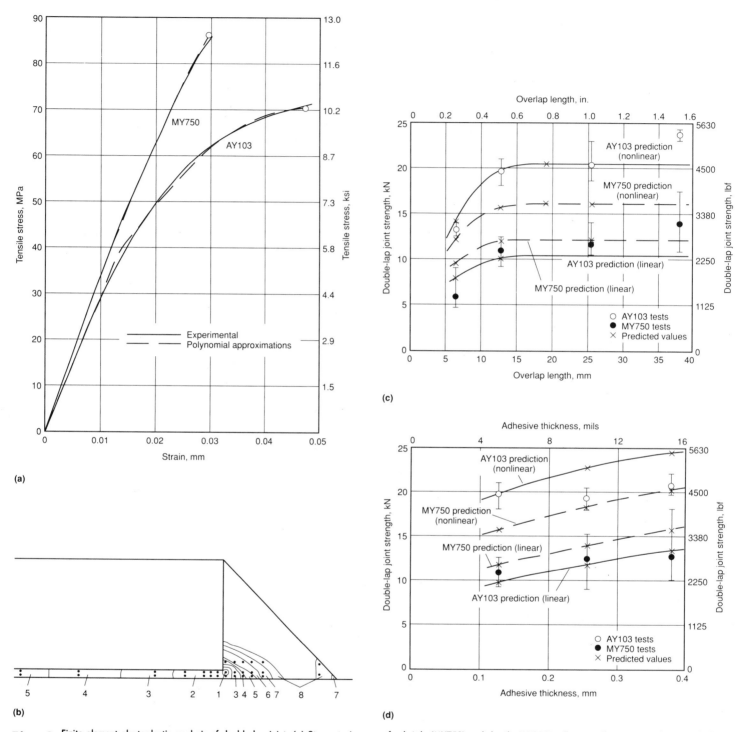

Fig. 9 Finite-element elastoplastic analysis of double-lap joint. (a) Stress-strain curves for brittle (MY750) and ductile (AY103) adhesives. (b) Incremental zones of plastic deformation for the ductile adhesive (AY103). (c) Comparison of failure predictions with experimental data for various overlap lengths. (d) Comparison of failure predictions with experimental data for an overlap of 12.7 mm (0.5 in.) and varying adhesive thicknesses. Source: Ref 37

A different fracture mechanics approach to predicting the strength of unflawed bonded joints considers the propagation of inherent flaws or imperfections, even if macroscopically the material appears to be unflawed. These inherent flaws are those that exist in all bonds due to voids, microscopic inclusions, debonds, and so forth. Griffith (Ref 26, 27), in developing his theory of rupture of brittle materials in the 1920s, assumed that all materials have microscopic cracks or flaws that produce high stress concentrations. Failure, then, is produced by the propagation of these flaws. Based on the original concept of Griffith, Wang and Crossman (Ref 28, 29) presented an energy-release-rate method to study delamination and transverse cracks in graphite-epoxy laminates. They assumed the existence of a macroscopic crack distribution in the composite. Then, the ultimate strength of the uncracked composite was calculated from the worst flaw size in that distribution. A similar approach to predict tensile bond strength based on an inherent flaw and fracture mechanics was described recently by Anderson *et al.* (Ref 30, 31).

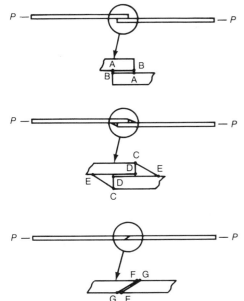

Fig. 10 Locations where stress singularities may occur in three typical joint configurations (points A-G). The strongest singularities are at points A, D, and F (sites of failure initiation).

Table 2 Typical environmental effects to be considered

Space	Earth
Radiation	Moisture
Ultraviolet	Temperature
Electromagnetic	Chemical
Other	Creep
Temperature distribution	Thermal fatigue
Atomic oxygen	Mechanical fatigue
Creep	
Thermal fatigue	

adhesive joints. They found that rounding the corner reduces the theoretically infinite stresses and increases the joint strength.

Due to the complexities encountered in the use of fracture mechanics to predict bond strength, this continues to be an area of active research.

Elastoplastic Analysis. The fracture mechanics criteria considered above assume that failure in the adhesive occurs with little or no plasticity. However, many adhesives currently used in the industry exhibit large plastic strains to failure. An example is rubber-modified epoxies. For very strong adhesives, adherend plasticity is also possible. In such cases, a solution that includes the material nonlinearities becomes necessary. Although fracture mechanics can, in principle, be extended to consider plasticity effects at the crack tip or singularity point, a more popular approach with bonded joints is based on elastoplastic analysis. The FE method is well suited for such problems. However, the iterative nature of the solution in such cases requires considerably more computational time and effort. An example is presented below to illustrate this procedure.

Figure 9 shows the results of an FE elastoplastic analysis of the double-lap joint taken from Ref 37. Two types of adhesives were used, brittle (MY750) and ductile (AY103). Figure 9(a) shows the stress-strain curves for both adhesives, along with the third-order polynomial approximations used in the computer model. The ratio of the compressive to tensile yield stress for the AY103 and MY750 adhesives was experimentally determined from bulk specimens to be, respectively, 1.27 and 1.14.

Figure 9(b) shows the incremental zones of plastic deformation for the ductile adhesive. The yield surfaces were calculated using a paraboloidal yield criterion of the form:

$$(\sigma_1 - \sigma_2)^2 + (\sigma_2 - \sigma_3)^2 + (\sigma_3 - \sigma_1)^2 + 2(|\sigma_c| - \sigma_t)$$

$$(\sigma_1 + \sigma_2 + \sigma_3) = 2|\sigma_c|\sigma_t \quad \text{(Eq 6)}$$

where σ_1, σ_2, and σ_3 are the principal stresses and σ_c and σ_t are, respectively, the compressive and tensile yield stresses.

Figure 9(c) shows a comparison of the failure predictions with experimental data for the two adhesives for various overlap

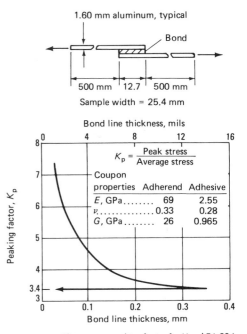

Fig. 11 Shear stress peaking factor for Hysol EA 934 adhesive, single-lap shear test specimen

lengths. Figure 9(d) shows the comparison with experimental data for an overlap of 12.7 mm (0.5 in.) and varying adhesive thicknesses. A maximum strain failure criterion was used to predict failure. The strain at failure was determined from uniaxial bulk specimens. The main characteristic of the results in Fig. 9(c) and 9(d) is that the linear analysis for the ductile adhesive leads to very conservative joint strength predictions, indicating the need for an elastoplastic analysis in such cases.

In general, a qualitative agreement between experiments and predictions is observed, where the ductile adhesive joints are stronger than the brittle adhesive joints. Even though the stress singularities have disappeared due to plastic flow, mesh refinement will have an effect on the solution because strain singularities are still present at the points of singularity.

A Practical Joint Failure Criterion. As indicated above, questions about mesh refinement and failure prediction arise due to the existence of stress and strain singularities at the ends of the bond. A finite-element methodology for bond strength prediction that accounts for the presence of singularities and minimizes the inconsistencies between the test specimens and the structural part is described below:

1. Determine the elastic moduli of the adhesive (for example, Young's modulus and Poisson's ratio) from uniaxial bulk specimens. If the adhesive exhibits nonlinear behavior, uniaxial stress-strain curves will also be needed from bulk specimens. These properties will be used

They assumed that failure loads can be predicted from the inherent flaw size (determined experimentally) and an appropriate energy balance based on energy-release rate. They considerd Mode I failure only (for example, tension-dominated failures) and obtained very accurate bond-strength predictions (within 5% in many cases) in butt-tension, blister, and single-lap-shear specimens. Both the failure loads and debond initiation locations for adhesives of various thicknesses were predicted using this theory. However, combined (mixed-mode) loading, which is very common in most joints, was not considered.

Finally, an approach proposed by Gradin and Groth (Ref 32-34) considers crack initiation using a fracture criterion based on the singularities that appear at the interface between bonded dissimilar materials. Their criterion is formulated in terms of a singular intensity factor (Q) that is analogous to the stress-intensity factor for the case when cracks are present. Their method assumes that no cracks are present in the vicinity of the material-induced singularity. However, in some cases, stress-intensity factors for the initial bond geometry may not control debond initiation because inherent flaws (for example, manufacturing-induced voids) that are larger than the entire singularity region could exist in the high-stress area (Ref 35). In addition, in many practical joints the sharp corners of the adherends are slightly rounded in the manufacturing process (for example, as a result of surface treatments such as abrasion or etching). Adams and Harris (Ref 36) studied the influence of the geometry of the corners in

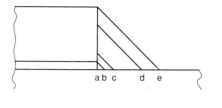

Distance from center of overlap, in.

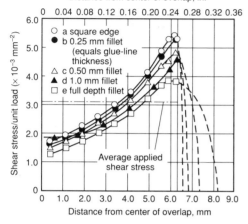

Fig. 12 Influence of the fillet size on the shear stress distribution at the tension end of a double-lap joint. Source: Ref 37

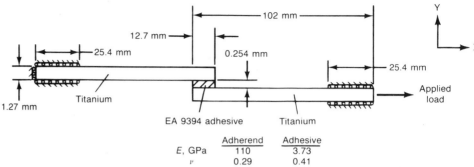

Fig. 13 Schematic of titanium adherend lap-shear coupon showing the boundary conditions used in the FE model

	Adherend	Adhesive
E, GPa	110	3.73
ν	0.29	0.41

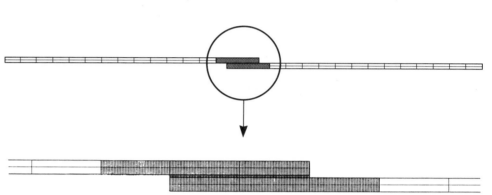

Fig. 14 Finite-element grid for the titanium adherend lap-shear coupon

in the FE code input in steps 3 and 4 below

2. Determine failure loads from simple analogue joint specimens (for example, lap-shear and butt-tension). The important factor to keep in mind here is that the strength of the singularity (α in Equation 1) at the site of failure initiation in the test specimens must be the same as in the structural joint to be analyzed (the significance of this requirement is explained in step 4 and in Example 1). Figure 10 shows the locations where stress singularities may occur in three typical joint configurations. The sites of failure initiation, which in this case correspond to the strongest singularities, are points A, D, and F. Note that only the local geometry (the angle of intersection between the adhesive and the adherend at interface corners) and material properties are required to ensure the same strength of singularity (see Fig. 1). For example, if the structural joint has fillets at the ends of the bond, use fillets in the test specimens. If a scarf joint is to be designed and analyzed, use scarf joint specimens with the same scarf angle. The reader is referred to Ref 4 and 5 for details on the effect of the angle of intersection and material properties on the strength of the singularity. Failure will normally initiate at highly stressed points where singularities are present

3. Determine stress or strain allowables from the failure loads determined in the above test specimens using FE analysis. For linear (brittle) adhesives, determine the maximum normal or von Mises stress at failure. For nonlinear (ductile) adhesives, use a nonlinear FE analysis to determine the maximum normal strain at failure. In either case, the element size in the regions where steep gradients are expected, such as bond termini and failure initiation sites, should be at least one-third of the adhesive thickness. Note that in the FE analysis infinite stresses are not obtained because the results correspond to an average over the size of the element

4. Conduct the appropriate FE analysis (linear or nonlinear) on the structure being considered to obtain the maximum normal or von Mises stress (for linear adhesives), or the maximum normal strain (for nonlinear adhesives) under the applied loads. Compare this maximum stress or strain with the corresponding allowable stress or strain in step 3 to determine margins of safety or predicted failure loads. In this step, it is very important that the element size near the site of failure initiation be the same as in step 3, so that the stresses are averaged over the same distance. Therefore, similarity between the test specimens and the structural part is achieved by having the same strength of singularity and the same element size at the failure point

Using the above procedure will ensure more accurate bond strength predictions. Further considerations on the FE modeling itself are given in the next section. Joint analyses that illustrate this methodology are given in the "Examples and Test Correlation" section.

Design Considerations in FE Modeling

In this section certain parameters that must be kept in mind to ensure successful FE modeling of the structure being designed and analyzed are discussed.

Boundary condition selection is one of the fundamental problems in the formulation of a representative mathematical model of a structure. It also requires insight into the stiffness and response of adjacent attaching structures. The analyst can employ one of the following modeling techniques to successfully establish appropriate boundary conditions:

1. Terminate the boundary edges of the model far enough away from the adhesive joint area so as not to influence deflection, strain, and stress response in the joint because of local boundary effects. The analyst must prescribe the translation and rotation constraints at the boundary. Sometimes this is easily done when symmetry conditions or a

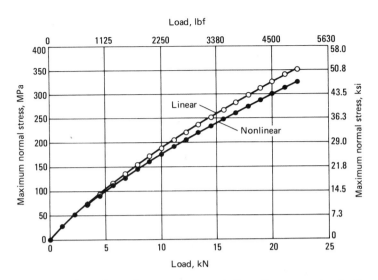

Fig. 15 Finite-element linear versus geometrically nonlinear comparison for the maximum normal stresses in the titanium lap-shear coupon

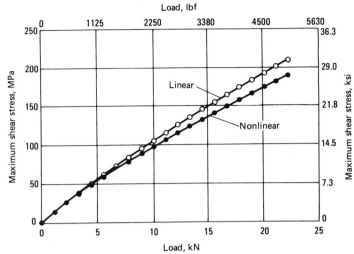

Fig. 16 Finite-element linear versus geometrically nonlinear comparison for the maximum shear stresses in the titanium lap-shear coupon

"rigid" boundary exists. Often, however, these constraints may only represent the boundary conditions in a global sense and locally generate incorrect tractions because of the artificial stiffnesses imposed at the boundary. Thus, it is usually prudent that the boundary be sufficiently far away from the region of interest to preclude overlap of the joint stress concentrations with the local boundary stresses. Generally, if the length from the joint region to the boundary is 10 times the adherend thickness, there will be no stress impingement from the boundary conditions. Invoking Saint-Venant's principle, such boundary stresses should dissipate two to three thicknesses from the end

2. A second approach in achieving realistic boundary conditions is to represent the adjacent structure using a coarse model. This approach is useful when deflections and/or rotations are expected at the boundary because of an elastic support. For example, if a detailed joint model was constructed using solid finite elements, the adjacent structure could be represented using beam elements at the boundary. The analyst must still provide sufficient "distance" between the element transition area (beam-to-solid in this example) and the detailed joint elements so as to prevent boundary stress oscillations from impinging on the joint stresses. This approach provides a more accurate representation of the boundary stiffnesses and removes the more stringent rotation and translation constraints away from the detailed joint area. However, some stress oscillations are likely to remain at the element transition area

3. A third boundary condition approach may consist of using superelements to more accurately represent the interface

stiffness. Superelements are generally used when the interfacing structures are extremely complex. Attachment nodes are specified in the superelement definition at locations where the detailed joint model would interface. The superelement stiffness matrix need only be solved once and the resulting stiffnesses assigned to each attach node. The detailed joint model boundary elements are then defined. Transformation constraints that link the deformations of the superelement attach nodes and joint boundary

elements are prescribed. Numerous parametric analyses may then be run on the joint model as long as the boundary elements of the joint, which are constrained to the superelement nodes, are retained. Subsequent FE runs need only solve for the joint stiffness matrix

Material properties to be considered for the joint analysis must include all environmental effects to which the structure would be exposed in its end-use condition. Specifically, environmental factors that affect the stiffness and strength of the adhesive and adherends must be considered. Some of the more important environmental effects are

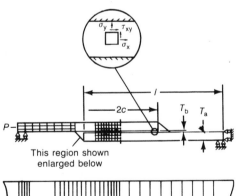

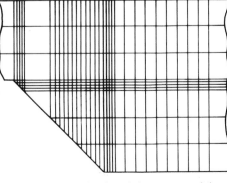

Fig. 17 Example of good element size and shape selection in the FE model of a single-lap joint with a fillet. Source: Ref 48

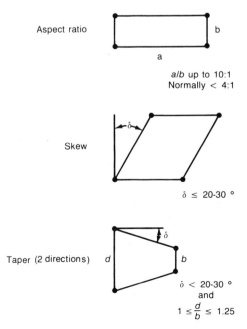

Fig. 18 Reasonable limits for element shapes. Source: Ref 49

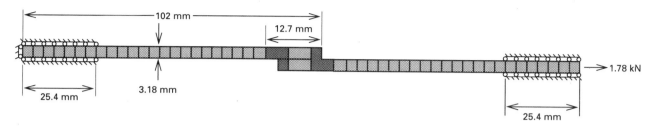

Joint Geometry and Boundary Conditions Used in Analyses

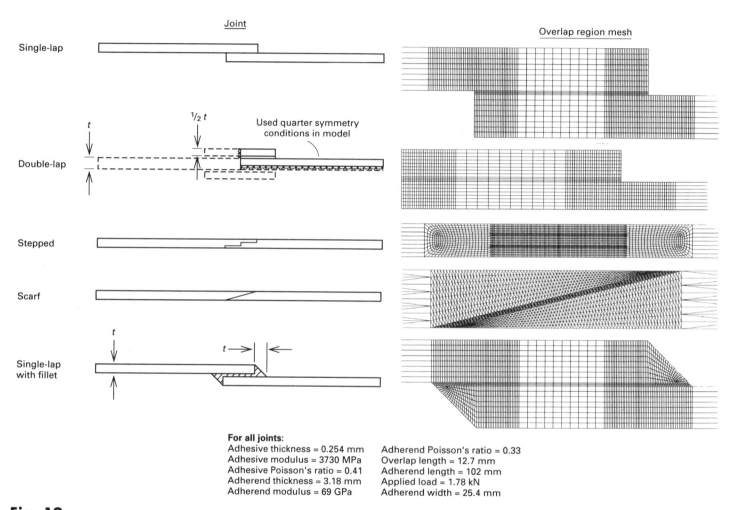

For all joints:
Adhesive thickness = 0.254 mm   Adherend Poisson's ratio = 0.33
Adhesive modulus = 3730 MPa   Overlap length = 12.7 mm
Adhesive Poisson's ratio = 0.41   Adherend length = 102 mm
Adherend thickness = 3.18 mm   Applied load = 1.78 kN
Adherend modulus = 69 GPa   Adherend width = 25.4 mm

Fig. 19 Joint geometry, material properties, boundary conditions, and mesh discretizations used in the FE analysis of the four typical joint geometries

listed in Table 2, which includes some of the more critical space and earth environmental conditions.

Characterization tests must be conducted on the materials to assess the effect of each of the environmental conditions on stiffness and strength. Detailed characterization tests are discussed in the Section "Testing and Analysis" in this Volume. These characterization tests may take the form of typical bulk adhesive tests (for example, where the adhesive is tested as a tensile dogbone) or *in-situ* tests.

In-situ testing attempts to characterize the adhesive as a thin bond line of material. The manufacturing and testing of *in-situ* coupons requires nonstandard equipment and methods. Specialized test coupons are fabricated to measure the moduli of the bond line as load is applied using small-deflection transducers. Fortunately, it has been shown (Ref 37, 38) that bulk adhesive tests are adequate for characterization of the material properties with differences between *in-situ* and bulk coupon moduli of less than 10%. This eliminates the need for the more costly and time-consuming *in-situ* testing.

The stress-strain properties must also be understood prior to designing and analyzing the joint. If the adhesive exhibits a high degree of nonlinearity, it may drastically impact the resulting stress profile and peak stresses in the joint, as shown in Ref 37 and 39. Analysis of this effect requires that the finite-element code be able to handle nonlinear material property input. Numerous computer codes are able to provide this feature, including ABAQUS, ADINA, and NASTRAN. Note that if long-term joint creep is of concern (as, for example, joints fabricated from low-modulus, high-elongation polymers), a viscoelastic analysis would need to be conducted. Fortunately, many structural adhesives have a high modulus (>690 MPa, or 100 ksi), possess a fairly

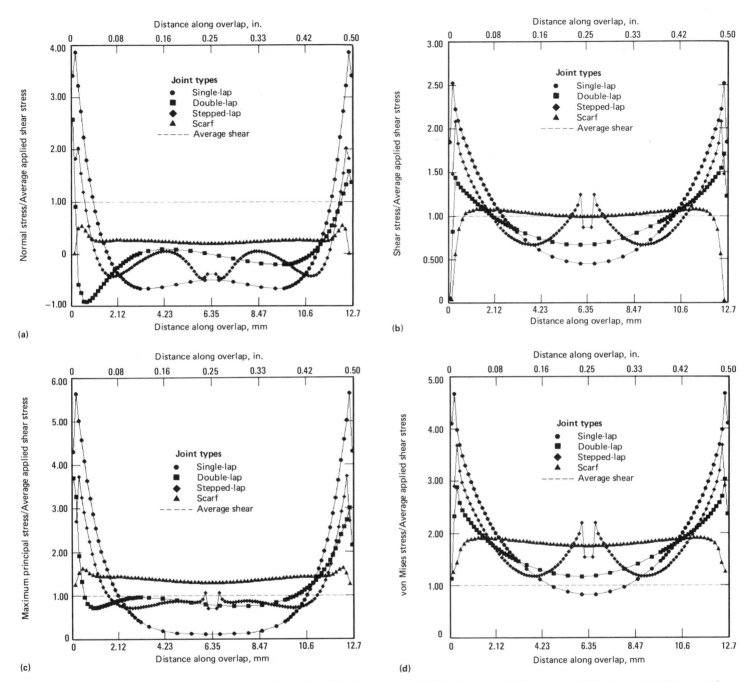

Fig. 20 FE results for the stresses in the adhesive along the overlap of the four joint types. (a) Normal stresses. (b) Shear stresses. (c) Maximum principal stresses. (d) von Mises or equivalent stresses

linear stress-strain curve to failure, and exhibit low creep. Joints fabricated with these adhesives do not require nonlinear or viscoelastic material analyses.

It is important to note that an adhesive that exhibits linear behavior at ambient conditions may show marked nonlinear response when subjected to elevated moisture and temperature environments. A typical adhesive characterization test matrix would, at a minimum, include temperature and moisture environmental test conditions that would establish if material nonlinearity exists under these conditions. Thus, the adhesive must be modeled at its

thermal and moisture extremes using the appropriate moduli and strength allowables. The adhesive may be subjected to significant stresses, particularly at the boundaries, when it is subjected to large moisture or thermal gradients (see, for example, Ref 40 and 41). These stresses are caused by the mismatch of the coefficients of moisture expansion (CME) or thermal expansion (CTE) between the adherends and the adhesive. Thus, CTE and CME values must be input into the model for all materials.

The thermal- and moisture-induced strains are then superimposed with the me-

chanical strains to obtain the total strain state. This is commonly expressed as:

$$\{\epsilon_i\}_{total} = [s_{ij}] \{\sigma_j\} + \{\epsilon_i\}_{thermal} +$$

$$\{\epsilon_i\}_{moisture} \quad i, j = 1, 2, \ldots, 6 \quad \text{(Eq 7)}$$

where $\{\epsilon_i\}$ are strain components, $[s_{ij}]$ is the compliance matrix, and $\{\sigma_j\}$ are the stresses. The term $[s_{ij}] \{\sigma_j\}$ contains the mechanical strain components.

The material properties of the adherend may require special attention if it is made of a composite material, such as a fiber-reinforced plastic (FRP). These materials are generally composed of two-dimensional, or

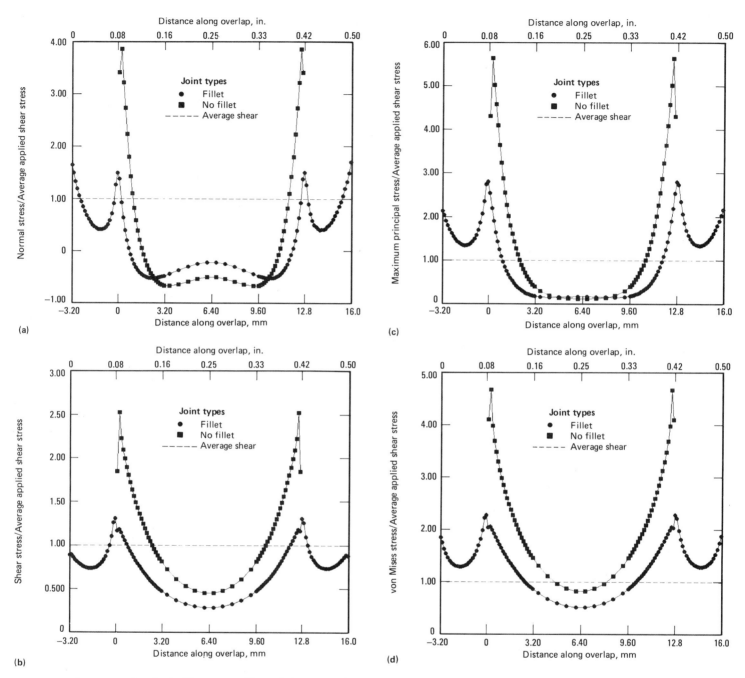

Fig. 21 Stress comparison for the fillet versus no-fillet single-lap joint conditions. (a) Normal stresses. (b) Shear stresses. (c) Maximum principal stresses. (d) von Mises or equivalent stresses

occasionally, three-dimensional fiber-oriented laminates. The two-dimensional laminates, because of a lack of reinforcement in the thickness direction, are of particular concern because they exhibit a propensity to delaminate between the plies adjacent to the adhesive layer during joint loading.

If possible, each group of plies in a composite adherend should be modeled as a separate layer of elements so as to discern the interlaminar stresses during load application. Such ply-by-ply models are generally possible for two-dimensional plane

stress, plane strain, or axisymmetric analyses. If each ply is modeled as a separate layer, as in a two-dimensional model, then the material properties for that specific lamina are used as the element material properties. However, it is difficult to model laminate plies in a three-dimensional model because the number of elements and resulting degrees of freedom become quite large. Consequently, approximations are necessary, and fewer elements through the thickness must be modeled for the three-dimensional case with groups of plies lumped together to form the stiffness of each three-

dimensional element. The element properties are obtained by laminating the individual plies together using a suitable lamination theory and assuming that the resulting element is homogeneous and anisotropic. *Composites*, Volume 1 of *Engineered Materials Handbook* provides detailed methods for obtaining both lamina and laminate properties. The resulting stresses are for the homogeneous element, and individual ply stresses are not obtained unless some sort of postprocessing is conducted.

In order to obtain three-dimensional element properties and recover the individual

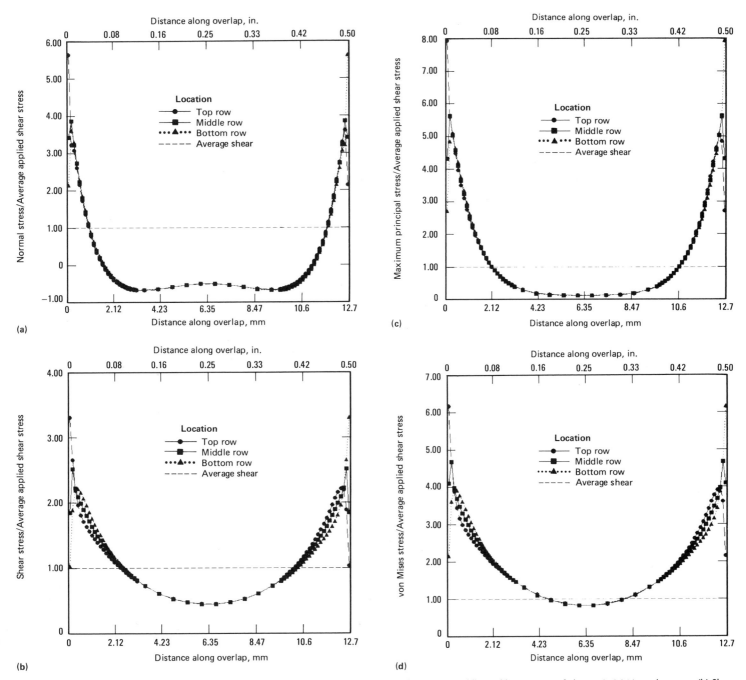

Fig. 22 Stresses in the bond line of the single-lap joint at different locations across the thickness (top, middle, and bottom rows of elements). (a) Normal stresses. (b) Shear stresses. (c) Maximum principal stresses. (d) von Mises or equivalent stresses

ply and interlaminar stresses, special pre- and postprocessors have been developed (Ref 42, for example). The preprocessor must store the lamina information for each element and laminate the plies to form the element stiffness matrix. After model solution, the postprocessor solves for the element ply stresses using, as a boundary condition, the resulting deformation field from the FE model.

Occasionally, resin plies are inserted between the lamina of composite adherends in order to enhance toughness. References 43 to 45 discuss how the use of adhesive inter-

leaf layers may reduce adherend interlaminar shear stresses. These interleaved layers may also be included in a ply-by-ply FE model to evaluate their effectiveness in reducing the shear and normal stress gradients.

Geometric Considerations. Although it may appear obvious, it is important to accurately model the adherend and adhesive geometries. Adherend and adhesive thicknesses and bond length are of prime importance in the determination of the stresses in the joint. Figure 11 shows the effect of adhesive thickness on the peak shear stress

of the single-lap joint. It can be seen that there is a significant variation depending on the thickness of the bond. The joint stiffness is determined by the adherend and adhesive thicknesses, bond length, and material properties. Thus, the thickness and bond length play a key roll in determining the stress distribution in the joint. Simplifying assumptions, such as neglecting the effect of the adhesive thickness, may lead to incorrect results.

Adhesive fillets, adherend tapers, and other joint details must also be modeled because they will significantly alter the local

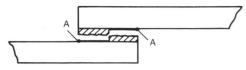

Fig. 23 Diagram of failure surfaces of single-lap joint. A, failure initiation sites

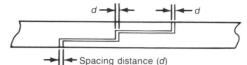

Fig. 24 Definition of the spacing distance (d) in the stepped-lap joint

stress gradients at the ends of the joint. Generally, square corner adhesive bond features do not occur during the fabrication of the joint. Rather, fillets are the rule and should be modeled as a matter of course. Figure 12 shows the results from a parametric analysis of four different sizes of a 45° fillet in a double-lap joint. The analysis showed a reduction in bond line shear stresses depending on the size of the fillet used, the no-fillet peak stress being 1.4 times the full-fillet peak shear stress. This example underscores the importance of modeling such features.

Similarly, stresses may be locally reduced by tapering adherend thickness at the end of a joint. This taper feature should be modeled so as to accurately include the stress gradients at the bond termination. The effect of fillets will be further discussed in the section "Finite-Element Results of Common Joint Geometries," and an exam-

ple is provided in the "Examples and Test Correlation" section.

The analyst must also be careful as to the type of approximation used to model the joint in a global sense. For example, tubular joints should not, in general, be modeled as flat plates. The results in Ref 46 show that tubular joints only approach flat-plate stress conditions for large tube radius to joint thickness ratios (on the order of $R/t \geq 100$). These results show that as the radius-to-thickness ratio increases, the curvature stiffening effect is reduced and the tubular joint stresses approach the flat-plate solutions.

Geometric nonlinear conditions may exist in many common joint designs (for example, single-lap joints). The geometric nonlinearity is generally a result of local moment loading at the joint caused by offset load application at the adherends. Figure 3

shows a typical single-lap-shear joint and the resulting rotation that occurs during loading. The effect of this large degree of rotation can be evaluated by a geometric nonlinear analysis.

The geometric nonlinear analysis is based on developing an element stiffness matrix composed of the standard small-deflection stiffness matrix and a geometric stiffness matrix. References 18, 19 and 47 discuss the formulation of the element geometric stiffness matrix and recommended solution procedures. The fundamental characteristic of the element geometric stiffness term is that it is a function of the current deformed geometry.

One common solution approach is to apply load to the FE model in a series of incremental load steps. The geometric stiffness matrix is then recalculated for each incremental load step, and the resulting incremental displacement is summed with the displacements of previous load steps. The solution continues until the total applied load is achieved. Solutions of this type are discussed in detail by Zienkiewicz (Ref 18) and involve consideration of large displacements and structural instability (if loaded in compression). The case of compression loading in the joint may produce collapse conditions that must be accommodated in the nonlinear solution algorithm. Zienkiewicz also shows how the geometric nonlinear problem may be readily extended to fully dynamic situations, and he gives several illustrative examples.

As a rule, the analyst should first conduct a linear analysis of the problem, study the results, and determine where local moments, compressive loads, and so forth may combine to indicate regions where geometric nonlinearity must be considered. Only when the analyst fully understands the basic linear model should a geometric nonlinear analysis be considered.

A geometric nonlinear analysis for the single-lap joint is presented in Fig. 13 to 16. Figure 13 shows the specimen geometry, material properties, and boundary conditions used. Isoparametric, quadrilateral, and generalized plane-strain elements were used in the model. The finite-element grid details are shown in Fig. 14. Three elements were used across the adhesive thickness. The peak normal and shear stresses are shown in Fig. 15 and 16 (these peak stresses occur at locations A in Fig. 10). Experimental results indicate that the typical failure load for this joint is about 11.1 kN (2475 lbf). At this load level, the differences between the linear and nonlinear results are 6.7% for the peak normal stress and 8.0% for the peak shear stress. Larger differences are possible with thinner adherends and a thicker bond line.

Element Shape and Element Selection. Sufficient element refinement must be provided in the FE grid so that areas of high strain gradient may be accurately modeled.

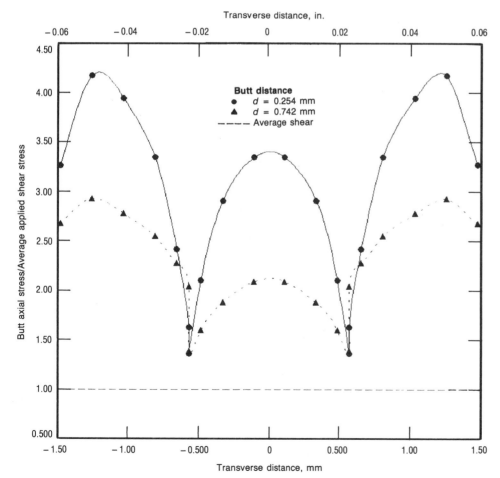

Fig. 25 Effect of spacing distance on the butt axial stresses in the stepped-lap joint

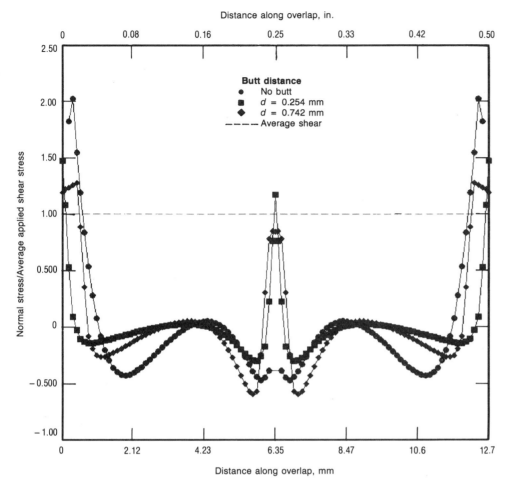

Fig. 26 Effect of spacing distance on the normal stresses in the stepped-lap joint

This requires some engineering judgment and foresight, and existing closed-form solutions for simple geometries may serve as an aid. Typical locations that require element refinement are those areas where geometry or material changes occur.

Figure 17 shows a grid of a single-lap joint and a detail of the local grid refinement in the area of an adhesive fillet (Ref 48). The grid shows four elements through the thickness to obtain the stress gradients within the adhesive layer. Refinement is also provided in the fillet area to obtain the stress gradients due to the material discontinuity between the adherend and adhesive. It is recommended that at least three elements be used through the adhesive thickness because most element strain functions (used in the element formulation) seldom match the actual strain distribution within the adhesive layer.

Element shape is also important in that highly skewed and large aspect ratio elements should, in general, be avoided. In particular, if isoparametric quadrilateral elements are used, numerical difficulties may result in the element integration scheme if the elements are highly skewed. Certain highly skewed quadrilateral element shapes

may even lead to negative element stiffness matrices. Such regions in the finite-element model should be remeshed. In addition, the use of different element types less susceptible to skewness should be considered as, for example, triangular or four-triangle quadrilateral elements.

Figure 18 shows typical guidelines for element geometry. Large aspect ratio elements (elements having large length-to-width ratios), should be avoided because they provide an inaccurate representation of the structural stiffness. A maximum element aspect ratio of 10:1 should be considered as a guideline. Elements should not be skewed more than 30°. Tapered elements should have side angles of less than 30° and d/b (top-to-bottom) length ratios of less than 1.25.

The selected type of element used in the analysis may have a significant impact upon the final result. It is suggested that four-node isoparametric quadrilateral elements be used in plane-stress, plane-strain, or axisymmetric analyses, whereas eight-node solid elements be used in three-dimensional adhesive joint analyses. Triangular elements may also be utilized for complex geometries. An example of this is the scarf joint (see Fig. 19), where the use of quadri-

lateral elements would lead to considerably more meshing effort to avoid highly skewed geometries.

Special elements may also be formulated to analyze adhesive joints, as shown in Ref 50 to 52. In Ref 50 and 51, special elements were formulated to account for across-the-width stresses and bending stress-concentration effects in single-lap-shear joints. This makes possible the use of fewer elements through the thickness of the joints (in the case of the bending load element) since the rotation terms are included in the displacement formulation. Typical quadrilateral elements would require four or more elements through the thickness to achieve an equivalent approximation. In Ref 52, two- and four-noded finite elements, which isolated various elasticity control parameters, were formulated. These control parameters were selected to simulate various closed-form solution results. These examples illustrate that the analyst may develop unique elements to study structural features of key interest.

Finite-Element Results of Common Joint Geometries

The typical joint geometries shown in Fig. 19 have been modeled using the FE method. The selected joint geometries include single-lap, double-lap, stepped, and scarf. These specific joints were selected to illustrate numerical modeling methods as well as to demonstrate the key attributes of the four distinct joint types.

Figure 19 shows the overall grid of the single-lap-shear coupon specimen. A similar grid refinement was used for all of the other joint models. The single-lap-shear joint also shows the boundary conditions, applied loading, and dimensions that were used on all of the joints. All joints used identical adherend thicknesses and lengths. The adhesive bond line in all joints was 0.254 mm (10 mils) thick with three elements used through the thickness. The adherend overlap was 12.7 mm (0.5 in.). The location where the joints differ is in the adhesive bond region. The detailed finite-element grid at the overlap region is also shown in Fig. 19 to highlight the different joining configurations. Identical material properties were used in all cases.

Joint Stress Comparison. Figures 20(a) to 20(d) show a comparison of the results for normal, shear, maximum principal, and von Mises stresses, respectively. The stresses were plotted using the middle row of elements in the adhesive (through-the-thickness effects will be discussed later in this section). Each stress component is normalized to the average applied shear stress at the joint bond line. The average shear stress is defined as the component of the applied load in the bond line direction divided by the area of adhesive sandwiched

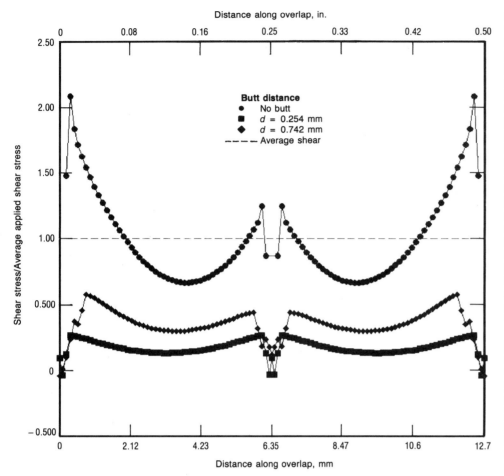

Fig. 27 Effect of spacing distance on the shear stresses in the stepped-lap joint

between adherends. It should be noted that the stepped-lap joint was modeled without adhesive at the butt region on the end of each step. This forces the joint to carry the load in shear and not in normal tension at the butt region. This modeling refinement allows for a more realistic comparison of the joint designs. The effect of bonding the butt regions is discussed in more detail later in the section "Stepped-Lap Joint Design Details."

Figure 20(a) indicates that the single-lap, double-lap, and stepped-lap joints have significant normal stress concentrations at the overlap ends. The single-lap joint has the highest component of normal tension, which is 1.50, 1.91, and 7.02 times that of the double-lap, stepped-lap, and scarf joints, respectively. The scarf joint has the lowest normal stress and exhibits the most uniform stress state across the adhesive bond, while the other joints distribute the load less efficiently.

Figure 20(b) shows the shear stress along the adhesive. It is immediately obvious that the scarf geometry provides the most efficient joint for carrying shear load (while minimizing normal stress). The scarf joint shear stress indicates a peak of only 1.08 times the average stress. In comparison, the double-lap, stepped-lap, and single-lap joints

show shear stress peaking of 1.71, 2.08, and 2.52 times the average stress, respectively. This peaking effect is sometimes called shear lag and is used to address joint efficiency in shear. Because the scarf joint may be difficult to fabricate for certain structures, the stepped-lap joint has been used as an approximation. The results presented suggest that additional steps would have to be added to the stepped-lap model to approach the scarf joint efficiency because the stress concentrations at the edges vary significantly between the two types of joint.

Figures 20(c) and 20(d) are presented here as potential failure criteria. Both maximum principal stress and von Mises stress have been suggested as failure criteria for brittle and ductile adhesives, respectively. As expected, the scarf joint shows the most uniform stress state for both maximum principal and von Mises stresses. The single-lap, double-lap, and stepped-lap joints show significant stresses at the adherend edge discontinuities, with the single-lap joint showing the highest peaks.

Effects of Adhesive Fillet. The advantages of an adhesive fillet were discussed in some detail in the section "Design Considerations in FE Modeling." An analysis of a

single-lap joint has been conducted with and without a fillet to illustrate the effect on the stresses at the overlap ends. Figures 21(a) to 21(d) show the results of the normal, shear, maximum principal, and von Mises stresses with and without a 45° adhesive fillet. The plots compare the middle row of adhesive elements. All four plots show a drastic reduction in the discontinuity peak stresses at the adherend termination due to the fillet. The fillet peak normal and shear stresses are reduced, respectively, to 44 and 52% of the no-fillet normal and shear stresses. The maximum principal and von Mises stresses show similar improvement in stress reduction. Similar percentage improvements were also observed for the peak stresses in the top and bottom rows of elements. However, the actual increase in the load-carrying capacity of the no-fillet versus fillet conditions also depends on the strength of the stress singularities. This is discussed in more detail in Example 1 of the "Examples and Test Correlation" section.

Through-the-Thickness Adhesive Stresses. Through-the-thickness effects were studied using the single-lap joint. Three elements were used through the thickness. The importance of using three elements is seen in the plots shown in Fig. 22(a) to 22(d). Figures 22(a) and 22(b) show through-the-thickness stress variations near the overlap ends for normal and shear stresses. The top row of elements shows peak stresses on the left side of the adherend-adhesive interface, whereas the bottom row of elements exhibits peak stresses on the right side interface. After the first 0.05 mm (2 mils), the normal stresses are constant through the thickness, whereas shear stresses do not converge until approximately 2.0 mm (0.08 in.) into the joint. More through-the-thickness variation is seen in the shear stress than in normal stress. These stress peaks at the interface corners were discussed earlier in the section "Analysis Methods" and were shown to be caused by stress singularities. The maximum peak normal stress is 5.64 times the average shear stress, while the middle row of elements has a maximum peak normal stress of 3.86. Similarly, the maximum peak shear stress is 3.31 times the average shear stress, while the middle row of elements has a peak shear stress of 2.52. As indicated earlier, the magnitude of these stress peaks is dependent on the element size.

As one might expect, the maximum principal stress plot follows the normal stress plot characteristics, while the von Mises plot follows the trends (not the magnitude) of the shear stress plot. Following either maximum principal or von Mises stresses, one would expect failure to begin in the top row left side and bottom row right side simultaneously. Figure 23 shows a typical single-lap-shear specimen failure where this indeed does happen (Ref 14). Without ade-

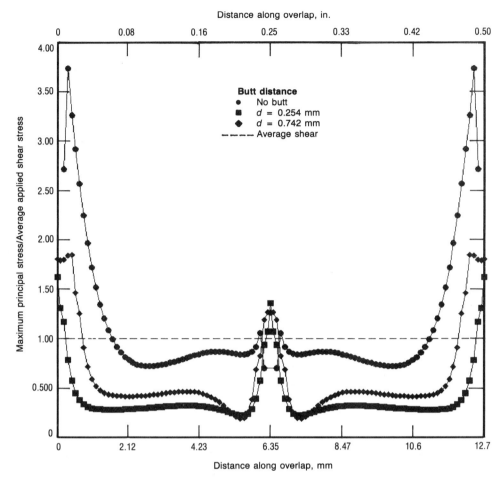

Distance along overlap, in.

Fig. 28 Effect of spacing distance on the maximum principal stresses in the stepped-lap joint

quate elements through the thickness, this type of detailed failure information would not be recovered.

Stepped-Lap Joint Design Details. The stepped-lap joint has several design details that are unique. Specifically, if the butt regions at overlap ends are bonded, the joint has large axial stresses that develop at these locations. Figure 24 defines the butt-joint distance (d). The axial stresses are plotted in Fig. 25 for d values of 0.254 mm (10 mils) and 0.742 mm (30 mils). The plot shows a 30% reduction in butt axial stresses when d is increased from 0.254 mm (10 mils) to 0.742 mm (30 mils). To optimize the stepped-lap joint, both the adhesive modulus and the butt-joint thickness (d) would need to be varied. For the stepped-lap joint modeled herein, the modulus of the resin is too high, causing normal stresses to exceed the adhesive allowable at the butt-joint terminations. A modulus on the order of 690 MPa (100 ksi) and a thickness (d) ranging from 1 to 2 mm (0.04 to 0.08 in.) should be considered.

Figures 26 to 29 show the normal, shear, maximum principal, and von Mises stresses for the two different spacing distances (d) and for the case in which the adhesive at the butt joint is totally eliminated. Figure 27 shows that the shear stresses are signifi-

cantly reduced when the butt region is allowed to carry axial load. However, this shear stress decrease is at the expense of high butt axial stresses, which may exceed adhesive allowables. A balanced design would take into account how much the butt distance (d) needs to be varied to reduce both the axial and shear stresses.

Examples and Test Correlation

Three examples are presented of various joint geometries to illustrate the analysis methods and present failure predictions. These specific adhesive joint examples were selected because test data were readily available for failure criteria correlation. In addition, these examples illustrate the usefulness of the finite-element approach in identifying the peak stresses in the areas of material discontinuity.

Example 1: Single-Lap-Shear Specimen With and Without an Adhesive Fillet

This example provides both an analytical and experimental comparison of the load-carrying capacity of a single-lap joint with

and without fillets at ambient conditions. Several fillet sizes are considered.

Structural Schematic. Figure 19 shows the single-lap-shear joint under consideration. The figure shows the specimen dimensions, boundary conditions, and the material properties used in the finite-element model. The specimen has a 12.7 mm (0.5 in.) nominal adherend overlap with a 0.254 mm (10 mil) bond line thickness. Aluminum adherends 3.18 mm (0.13 in.) thick were used in the specimen. EA 9394 adhesive, manufactured by Hysol Division, The Dexter Corporation, was used (see Fig. 19 for properties). The specimen width was 25.4 mm (1 in.). Figure 30 shows the fillet geometries used in the study, both the manufactured (curved) geometry and the FE model (straight line) approximations. (It was later found, by running an FE model having the curved fillet geometry, that in all cases the error in the peak von Mises and principal stresses in the straight line approximations was less than 3%, except for the 12.7 mm (0.5 in.) radius fillet, where the error was less than 7%.) The study included no fillet, half fillet, 45° full fillet, long taper (elongated) fillet, and large fillet conditions, as shown in Fig. 30.

Finite-Element Selection. The FE method was used to analyze this problem using the MSC NASTRAN code with four-noded isoparametric quadrilateral plane-strain elements. This element assumes a linear strain function in the element, thus providing an accurate approximation to bending.

Simplifying Assumptions and Analytical Approach. This problem was modeled assuming plane strain. Loads were applied at the end of the specimen uniformly across the end nodes (see Fig. 19). A linear material analysis was conducted because the EA 9394 epoxy is a relatively stiff material and the stress-strain curves are fairly linear. Geometric nonlinearity effects were neglected.

Element Mesh Discretization and Boundary Conditions. The finite-element grid is shown in Fig. 19. Three elements were used through the thickness to obtain enough grid refinement for through-the-thickness effects. Because the objective of the analysis was to simulate conditions in the load frame, the boundary conditions were selected to represent the test machine loading grips. Ends of the metal adherends were constrained from movement in the y-direction for a distance of 25.4 mm (1 in.), but allowed to deform axially. The left end nodes were constrained from movement in the x-direction. A load of 1.78 kN (400 lbf) was applied to the right end.

Key Results. Figures 31(a) and 31(b) show, respectively, the normalized maximum principal and von Mises (equivalent) stresses along the top row of elements for the four fillet conditions. Note in Fig. 31(a) and 31(b) that the peak principal and von

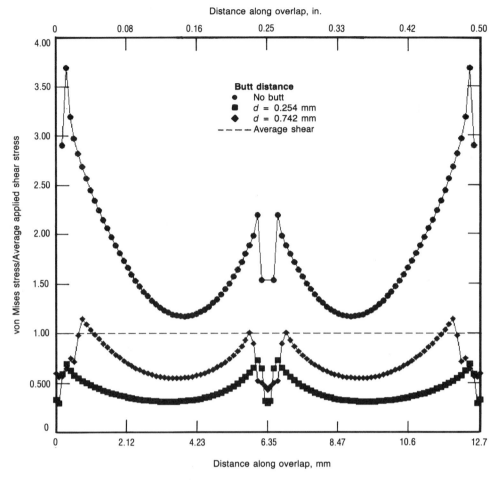

Fig. 29 Effect of spacing distance on the von Mises or equivalent stresses in the stepped-lap joint

Mises stresses always occur at the 12.7 mm location (which corresponds to point D in Fig. 10), except for the large fillet condition, where the peak stresses occur at the −6.35 mm location (end of fillet). In the latter case, small cracks were observed to initiate in the test specimens at this location at a load well below failure. These cracks did not propagate and did not contribute to the overall failure of the joint. Therefore, for all the joints analyzed, the critical location for failure is always at 12.7 mm (point D in Fig. 10, where the strongest singularity occurs).

Comparing the peak von Mises stress for the no-fillet condition from Fig. 22(d) (top row of elements) with Fig. 31(b) for the 45° full-fillet condition shows that the no-fillet peak is 2.25 times the fillet case. From this

the analyst might imply that the experimental failure load for the full-fillet specimens should also be 2.25 times higher than the no-fillet specimens. As discussed below, this assumption is incorrect because of the different singularity conditions between the two cases.

Comparison to Test Data and Conclusions. The single-lap joint specimen described herein was tested according to Fig. 30. Ten tests each were conducted with five different fillet conditions. For simplicity, the no-fillet versus 45° full-fillet condition is considered first. The test specimen radius that most closely matches the 45° full fillet is the 6.35 mm (0.25 in.) radius fillet test case. The average load at failure for the 6.35 mm (0.25 in.) radius fillet was 8.11 kN (1823 lbf).

The no-fillet condition delivered 5.92 kN (1331 lbf) (see Fig. 30). Thus, the ratio of fillet to no-fillet condition is 1.37. The analysis, however, suggests that a factor of 2.25 should be the difference. One explanation for the reason the test values are lower than the factor of 2.25 suggested by the model lies in the difference in the strength of singularities between the fillet and no-fillet conditions.

For the no-fillet condition, the strongest singularity from Fig. 1 is $\alpha = 0.27$ and occurs at points A in Fig. 10. For the fillet condition, it is $\alpha = 0.34$ (from Ref 5) and occurs at points D in Fig. 10 (note that for all of the cases with a fillet, the strength and location of the strongest singularity is the same). Thus, the stress gradients near the singularity points are different for the fillet and no-fillet conditions. Even though the same sized element was used in all FE models, the averaging over the element is different because of the difference in stress gradients.

Figures 32 and 33 illustrate this point. Figure 32 shows the predicted versus experimental comparison when the no-fillet specimen is used as a base to predict the failure loads of the specimens with fillets. Both the maximum principal stress and von Mises were used as failure criteria. The predicted failure loads for the specimens with fillets were calculated by using the following relationship:

$$\text{Predicted failure load} = \frac{(\text{FE peak stress})_{base}}{(\text{FE peak stress})_{structure}}$$

$$\times\ (\text{Experimental failure load})_{base}$$

For example, the von Mises predicted failure load for the 3.18 mm fillet radius in Fig. 32 was obtained as follows:

$$10.48\ \text{kN} = \frac{34.05\ \text{MPa}}{19.24\ \text{MPa}} \times 5.92\ \text{kN}$$

It can be seen in Fig. 32 that there is poor correlation between test results and predictions because the strength of singularity at the failure initiation point for the no-fillet and fillet conditions is different.

Figure 33, on the other hand, shows good agreement between predicted and experimental values (within 17%). The predicted values in Fig. 33 were calculated in a similar manner to those in Fig. 32, except that the 6.35 mm radius fillet specimen, instead of the no-fillet specimen, was used as the base. Thus, the strength of singularity at the point of failure initiation for the base was the same as in the predicted cases, resulting in good correlation between experimental and predicted failure load values.

This example shows the importance of determining the allowables from test specimens that have the same strength of singularity at the point of failure initiation as the structure being analyzed.

Table 3 Stress results and predictions for joint with aluminum end fitting

FE model applied force		FE model peak von Mises stress		Single-lap-shear coupon average failure load		Single-lap shear coupon von Mises allowable		Predicted failure load(a)		Tested failure load	
kN	lbf	MPa	ksi	kN	lbf	MPa	ksi	kN	10³ lbf	kN	10³ lbf
17.5	3930	23.4	3.4	6.88	1550	101.2	14.7	75.7	17.0	62.8	14.1

(a) Predicted load= von Mises allowable/von Mises peak stress in model × Applied FE force

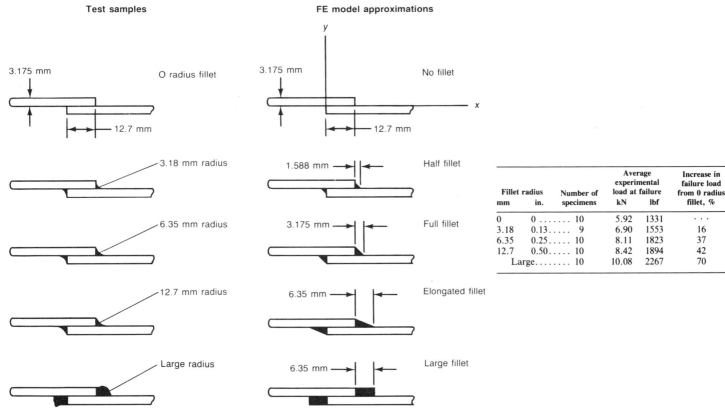

Fig. 30 Fillet geometries used in both the test specimens and the finite-element approximations. Failure loads for the test specimens are also shown.

| Fillet radius | | Number of | Average experimental load at failure | | Increase in failure load from 0 radius |
mm	in.	specimens	kN	lbf	fillet, %
0	0	10	5.92	1331	. . .
3.18	0.13	9	6.90	1553	16
6.35	0.25	10	8.11	1823	37
12.7	0.50	10	8.42	1894	42
	Large	10	10.08	2267	70

Example 2: Small-Diameter Graphite-Epoxy Tube With a Bonded Aluminum End Fitting

This particular problem was designed to explore potential threaded stud joints for small-diameter tubes. The application was geared to space structures where numerous small-diameter tubes would be used for truss networks.

Structural Schematic and Laminate Definition. Figure 34(a) shows the axisymmetric truss tube joint under study. The metal fitting is 6061-T6 aluminum. The composite laminate in the tube is $(40/0_3/-40/0)_s$. The fiber and resin used are IM6/3501-6. The lamina thicknesses are shown in Fig. 34(a), along with the laminate moduli. The

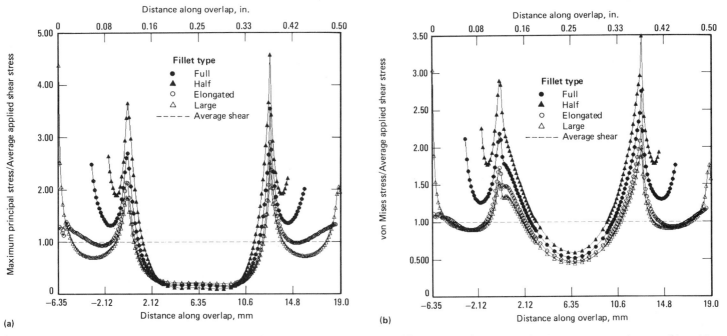

Fig. 31 Stress plots from the finite-element results at the top row of elements for the various fillet geometries shown in Fig. 30. (a) Maximum principal stresses. (b) von Mises or equivalent stresses

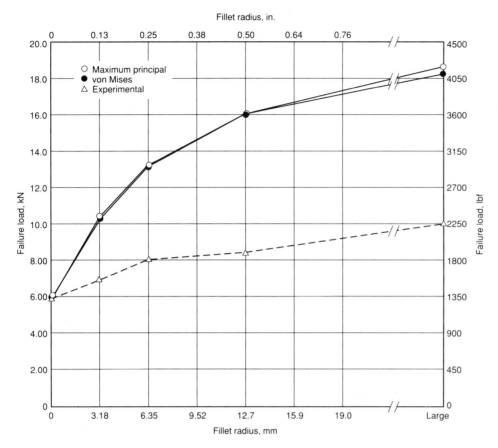

				Peak stresses(a) from FE analysis				Predicted failure load from FE analysis			
Fillet radius		Experimental load at failure		FE approximation	von Mises		Principal		von Mises		Maximum principal	
mm	in.	kN	lbf		MPa	ksi	MPa	ksi	MPa	ksi	kN	lbf
No fillet		5.92	1331	No fillet	34.05	4.94	43.80	6.35	5.92 (base)	1331	5.92 (base)	1331
3.18	.0.125	6.90	1551	Half fillet	19.24	2.79	25.30	3.67	10.48	2356	10.25	2304
6.35	.0.25	8.11	1823	Full fillet	15.17	2.20	19.57	2.84	13.29	2988	13.25	2979
12.7	.0.5	8.42	1894	Long fillet	12.52	1.82	16.03	2.33	16.10	3620	16.18	3636
Large	.10.08		2267	Large fillet	11.03	1.60	13.91	2.02	18.28	4107	18.64	4191

(a) At 1.78 kN (400 lbf) applied FE load

Fig. 32 Predicted versus experimental trend comparison using the no-fillet case as a base. Note that the trend is not predicted well because of the difference in the strength of singularities between the no-fillet and fillet conditions.

fiber volume fraction in the tube laminate is 62%. The nominal bond thickness is 0.254 mm (10 mils), and the bond length is 31.8 mm (1.3 in.).

Finite-Element Selection. A four-noded isoparametric quadrilateral axisymmetric element was used in the model. This is a similar element to that used in the previous example. A linear strain function is again used in the formulation of the element.

Simplifying Assumptions and Analytical Approach. The joint was modeled using axisymmetric elements. Load was uniformly applied to the end of the threaded stud. The composite was modeled as a uniform, anisotropic, homogeneous solid rather than as separate groups of plies. This approach is less accurate when predicting composite laminate interlaminar failure but generally provides a reasonable approximation. Only one element through the bond line thickness

was modeled. One element through the bond is not generally recommended. It was used in this model to maintain simplicity for a quick-look study.

Element Mesh Discretization and Boundary Conditions. Figure 34(b) shows the boundary conditions and loads used in the FE model. The model was axially constrained at the tube end of the model. Note that this axial boundary constraint is more than 10 tube thicknesses from the joint bond region so that boundary effects were minimized.

Test Correlation and Conclusions. Table 3 lists the predicted and the actual test load values. Failure was predicted to occur in the adhesive bond at location A in Fig. 34(b). The peak von Mises (equivalent) stress was used as the failure criterion. A von Mises allowable stress value was obtained from FE analysis conducted on

failed single-lap-shear test coupons without a fillet. Table 3 lists the single-lap-shear allowable for the von Mises stress (101.2 MPa, or 14.7 ksi). The predicted load at failure was determined to be 75.7 kN (17×10^3 lbf). This was calculated by multiplying the FE applied load by the ratio of the allowable von Mises stress divided by the von Mises peak stress in the joint (see Table 3). Table 3 shows that the predicted failure load is within 20% of the tested failure load of 62.8 kN (14×10^3 lbf). This is a reasonable answer, considering that the model contained several simplifying assumptions for a quick-look study.

Example 3: Joint Analysis With Metallic Fitting and Graphite-Epoxy Composite Tube

The problem presented is to determine the critical stresses in the bond region of the joint that are due to the mechanical and hygrothermal (nonmechanical) loads.

Structural Schematic and Laminate Definition. Figure 35(a) shows the axisymmetric bonded joint under study. The metal fitting was 7075-T73 aluminum. The composite laminate was made of the following lay-up inside diameter (ID) to outside diameter (OD):

	−45°	$0_2°$	90°	$0_2°$	+45°
Material	HMS-4	P75	HMS-4	P-75	HMS-4
Thickness, μm (mils) ..	76 (3)	305 (12)	76 (3)	305 (12)	76 (3)

A fiber volume fraction of 62% was used. An even-weave glass cloth was used between the metal fitting and the composite tube to reduce stress discontinuities and provide corrosion protection. The glass cloth was 0.50 mm (0.02 in.) thick. EA 934 adhesive was used to bond the metal and composite together.

Selected Numerical Method. The FE method was used to analyze this problem. Isoparametric, four-noded quadrilateral axisymmetric elements were used. This element type assumes a linear strain function.

Simplifying Assumptions and Analytical Approach. The problem was modeled using axisymmetric elements in an axisymmetric-loading FE code. Loads were applied using terms of a Fourier series. Load terms were summed to produce the various load states on the joint. For example, the moment load (*M*) was applied using the first term of the Fourier series, while the constant axial load (*F*) was applied using the zeroth term of the series. Resulting strains and stresses were summed using the principle of superposition because deflections were small.

Each fiber-reinforced plastic (FRP) composite layer of the structure was modeled as a discrete layer of finite elements so as to obtain accurate interlaminar normal and shear stresses. Testing has shown these stresses to govern failure in composite materials for many lap-joint designs.

Element (Mesh) Discretization and Boundary Conditions. Figure 35(b) shows the boundary conditions and loads used to model the structure. The model was axially and circumferentially constrained at the end of the metallic fitting. Radial growth was unrestrained. Axial and moment mechanical loads were applied at the FRP composite end of the model. A steady-state change (ΔT = −93.3 °C, or −136 °F) in thermal environment was also imposed on the joint. The discretized configuration is shown in Fig. 35(c).

Key Results. Maximum critical stresses occurred in the tube laminate and not the bonding adhesive. Figure 36 shows the interlaminar shear stress for the joint at the outermost composite-ply interface of the tube. The peak stress occurs at the termination of the composite tube, as one might expect. The allowable stresses in the shear stress mode stated in the figure show that interlaminar shear drives the design. Using a maximum stress failure criterion, the margin of safety in the design under this load condition is $MS = 0.08$. It should be noted that the shear stress was primarily due to the cold condition, not the applied mechanical loading.

Comparison to Test Data and Conclusions. The joint described above had glass cloth bonded at room temperature after tube cure and used a scalloped aluminum fitting. Tube failure, not adhesive failure, was predicted, and in fact, actually occurred during tests.

Earlier joint designs used a co-cured glass cloth and straight tapered (nonscalloped) fitting. Analyses of the earlier joint predicted outer-tube ply shear failure due to thermal cool-down to occur at ΔT = −120 °C (−184 °F). Acoustic events and inspection after cool-down indicated thermal cracking at $\Delta T \approx$ −110 °C (−166 °F). Sections cut from thermally cycled joints showed cracking initiating at the peak shear location point A of Fig. 36. Thus, model predictions showed good agreement with the experimental data. It should be noted that co-curing of the glass cloth added 6.9 MPa (1 ksi) shear stress to the outer ply, helping to precipitate the thermal failure.

ACKNOWLEDGMENT

The authors would like to thank DeVor R. Taylor for providing analytical support for the geometrically nonlinear analysis of the single-lap joint and Vicki L. Pearson for producing many of the figures using CATIA.

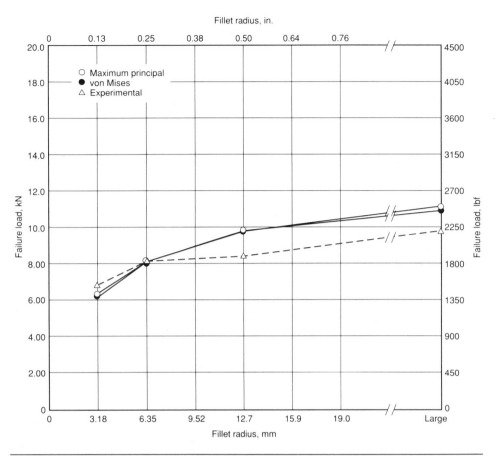

Fillet radius		Experimental load at failure		FE approximation	Peak stresses(a) from FE analysis				Predicted failure load from FE analysis			
					von Mises		Principal		von Mises		Maximum principal	
mm	in.	kN	lbf		MPa	ksi	MPa	ksi	kN	lbf	kN	lbf
3.18	0.125	6.90	1551	Half fillet	19.24	2.79	25.30	3.67	6.39	1437	6.27	1410
6.35	0.25	8.11	1823	Full fillet	15.17	2.20	19.57	2.84	8.11 (base)	1823	8.11 (base)	1823
12.7	0.5	8.42	1894	Long fillet	12.52	1.82	16.03	2.33	9.83	2209	9.90	2225
Large		10.08	2267	Large fillet	11.03	1.60	13.91	2.02	11.15	2505	11.41	2565

(a) At 1.78 kN (400 lbf) applied FE load

Fig. 33 Predicted versus experimental trend comparison using the full-fillet case as a base. Trend is predicted well because all the fillet types have the same strength of singularity.

Appendix: Character of Stress Singularities in Bonded Dissimilar Materials

The asymptotic solution for the stresses in the immediate vicinity of the singularity point (for example, small values of r in Fig. 37a and 37b) is, for isotropic materials:

$$\sigma_{ij}^{(s)} = r^{-\delta}\, f_{ij}^{(s)}(\phi, \delta) + \ldots \qquad \text{(Eq 8)}$$

where the superscript s = 1, 2 indicates material 1 or 2; r and ϕ are polar coordinates defined in Fig. 37; $\sigma_{ij}^{(s)}$ are the components of the stress tensor; δ is an exponent that depends on the local geometry (for example, the angles ϕ_1 and ϕ_2 in Fig. 37a and 37b) and material properties, but is independent of the global geometry or loading; and $f_{ij}^{(s)}$ are functions, independent of r, that depend on the angle ϕ, the

parameter δ, and the global geometry and loading.

Equation 8 is applicable to the two interface corner cases shown in Fig. 37(a) and 37(b). The difference between these two cases lies in the boundary conditions at ϕ_1 and ϕ_2. Practical situations where the case illustrated by Fig. 37(a) is applicable include points A, B, C, E, F, and G in Fig. 10, whereas an example of the application of Fig. 37(b) is given by point D in Fig. 10.

In general, δ is a complex exponent occurring in complex conjugate pairs, $\delta = \alpha \pm i\beta$. Because the strain energy must be finite as the origin is approached, $\alpha < 1$. Furthermore, a singular state of stress prevails if α

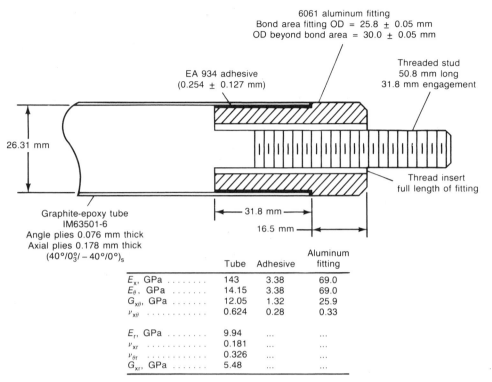

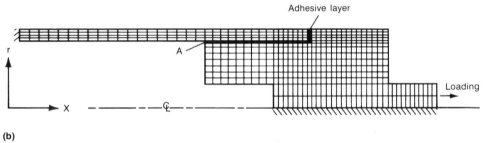

Fig. 34 Graphite-epoxy tube with aluminum end fitting. (a) Schematic showing dimensions. (b) Axisymmetric finite-element grid and boundary conditions

> 0. Expanding Eq 8 in terms of the real and imaginary parts of δ and its complex conjugate, and taking the appropriate linear combinations of terms to produce real stresses, gives the following expression:

$$\sigma_{ij}^{(s)} = r^{-\alpha}\ g(r,\beta)\ f_{ij}^{(s)}(\phi,\delta)\ + \ldots \qquad \text{(Eq 9)}$$

where $g(r, \beta)$ is of the form $\cos[\beta \cdot \ln(r)]$ or $\sin[\beta \cdot \ln(r)]$.

The factor $g(r, \beta)$ in Eq 9 gives rise to stress oscillations, which become pronounced close to the origin. The characteristics of this oscillatory behavior are discussed in Ref 53. When $\beta = 0$ (real δ), the oscillatory behavior disappears. It can be seen from Eq 9 that, as the origin is approached (for example, as $r \to 0$),

$$\sigma_{ij}^{(s)} \sim r^{-\alpha} \qquad \text{(Eq 10)}$$

since in this case the factor $r^{-\alpha}\ g(r, \beta)$ in Eq 9 is dominated by $r^{-\alpha}$.

For anisotropic materials, the asymptotic form of the stresses is similar to Eq 8 and 9 in most cases. For certain combinations of material properties, however, singular terms of $r^{-\alpha}\ \ln(r)$, $r^{-\alpha}\ [\ln(r)]^2$, $r^{-\alpha}\ [\ln(r)]^3$, and so forth (logarithmic singularities) may also exist. A thorough discussion of the nature of singularities in bonded dissimilar anisotropic materials is given in Ref 54 to 56. It should be noted that logarithmic singularities may also occur in isotropic materials (Ref 57).

REFERENCES

1. F.E. Penado and E.S. Folias, The Three-Dimensional Stress Field Around a Cylindrical Inclusion in a Plate of Arbitrary Thickness, *Int. J. Fract.*, Vol 39, 1989, p 129-146
2. D.B. Bogy, Edge Bonded Dissimilar Orthogonal Elastic Wedges Under Normal and Shear Loading, *J. Appl. Mech.*, Vol 35, 1968, p 460-466
3. D.B. Bogy, Two Edge-Bonded Elastic Wedges of Different Materials and Wedge Angles Under Surface Tractions, *J. Appl. Mech.*, Vol 38, 1971, p 377-386
4. V.L. Hein and F. Erdogan, Stress Singularities in a Two-Material Wedge, *Int. J. Fract. Mech.*, Vol 7, 1971, p 317-330
5. D.B. Bogy and K.C. Wang, Stress Singularities at Interface Corners in Bonded Dissimilar Isotropic Elastic Materials, *Int. J. Solids Struct.*, Vol 7, 1971, p 993-1005
6. O. Volkersen, Die Nietkraft Verteilung in Zugbeanspruchten Nietverbindungen mit Constanten Laschenquerschnitten, *Luftfahrtforschung*, Vol 15, 1938, p 41-47, (in German)
7. M. Goland and E. Reissner, The Stresses in Cemented Joints, *J. Appl. Mech. (Trans. ASME)*, Vol 11, 1944, p A17-A27
8. R.D. Adams and N.A. Peppiatt, Stress Analysis of Adhesive-Bonded Lap Joints, *J. Strain Anal.*, Vol 9, 1974, p 185-196
9. I. Sneddon, The Distribution on Stress in Adhesive Joints, in *Adhesives*, D. Eley, Ed., Oxford University Press, 1961
10. E.W. Kuenzi and G.H. Stevens, "Determination of the Mechanical Properties of Adhesives for Use in the Design of Bonded Joints," Report FPL-011, U.S. Forests Products Laboratory, 1963
11. L.J. Hart-Smith, "Adhesive-Bonded Single-Lap Joints," NASA Report CR-112236, Douglas Aircraft Company, Jan 1973
12. T. Wah, Stress Distribution in a Bonded Anisotropic Lap Joint, *J. Eng. Mater. Technol. (Trans. ASME)*, Vol 95, 1973, p 174-181
13. D. Chen and S. Cheng, An Analysis of Adhesive-Bonded Single-Lap Joints, *J. Appl. Mech.*, Vol 50, 1983, p 109-115
14. I.U. Ojalvo and H.L. Eidinoff, Bond Thickness Effects Upon Stresses in Single-Lap Adhesive Joints, *AIAA J.*, Vol 16, 1978, p 204-211
15. R.K. Dropek, Numerical Design and Analysis of Structures, in *Engineering Materials Handbook*, Vol 1, ASM INTERNATIONAL, 1987, p 463-478
16. L.J. Segerlind, *Applied Finite Element Analysis*, John Wiley & Sons, 1984
17. K.H. Huebner and E.A. Thornton, *The Finite Element Method for Engineers*, John Wiley & Sons, 1982
18. O.C. Zienkiewicz, *The Finite Element Method*, McGraw-Hill, 1977
19. K.J. Bathe, *Finite Element Procedures in Engineering Analysis*, Prentice-Hall, 1982
20. G.P. Anderson and K.L. DeVries, Predicting Strength of Adhesive Joints

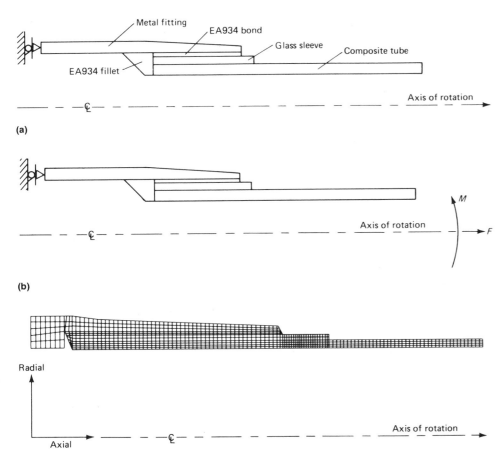

Fig. 35 Axisymmetric bonded joint under study. (a) Typical joint cross section. (b) Boundary conditions. (c) Axisymmetric finite-element grid. Source: Ref 15

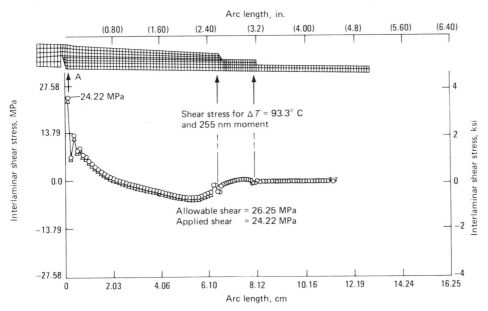

Fig. 36 Finite-element results for the interlaminar shear stresses at the outermost composite ply interface. Source: Ref 15

From Test Results, *Int. J. Fract.*, Vol 39, 1989, p 191-200

21. B.M. Malyshev and R.L. Salganik, The Strength of Adhesive Joints Using the Theory of Cracks, *Int. J. Fract. Mech.*, Vol 1, 1965, p 114-128

22. W.G. Knauss, Fracture Mechanics and the Time Dependent Strength of Adhesive Joints, *J. Compos. Mater.*, Vol 5, April 1971, p 176-192

23. G.G. Trantina, Fracture Mechanics Approach to Adhesive Joints, *J. Compos. Mater.*, Vol 6, April 1972, p 192-207

24. G.G. Trantina, Combined Mode Crack Extension in Adhesive Joints, *J. Compos. Mater.*, Vol 6, 1972, p 371-385

25. W.S. Johnson and S. Mall, A Fracture Mechanics Approach for Designing Adhesively Bonded Joints, in *Delamination and Debonding of Materials*, STP 876, W.S. Johnson, Ed., American Society for Testing and Materials, 1985, p 189-199

26. A.A. Griffith, The Phenomena of Rupture and Flow in Solids, *Philos. Trans. R. Soc. (London)*, Vol A221, 1921, p 163-197

27. A.A. Griffith, The Theory of Rupture, in *Proceedings of the First International Congress of Applied Mechanics*, Biezeno and Burgers, Ed., 1924, p 55-63

28. A.S.D. Wang and F.W. Crossman, Initiation and Growth of Transverse Cracks and Edge Delaminations in Composite Laminates: Part 1—An Energy Method, *J. Compos. Mater. Suppl.*, Vol 14 (No. 1), 1980, p 71-108

29. A.S.D. Wang and F.W. Crossman, "Fracture Mechanics of Sublaminate Cracks," *Air Force Office of Scientific Research Technical Report*, Contract F-49620-79-C-0206, 1982

30. G.P. Anderson and K.L. DeVries, Predicting Bond Strength, *J. Adhes.*, Vol 23, 1987, p 289-302

31. G.P. Anderson, D.H. Brinton, K.J. Ninow, and K.L. DeVries, A Fracture Mechanics Approach to Predicting Bond Strength, in Paper presented at the ASME Winter Annual Meeting (Chicago, IL), 28 Nov-2 Dec 1988

32. P.A. Gradin, A Fracture Criterion for Edge-Bonded Bimaterial Bodies, *J. Compos. Mater.*, Vol 16, 1982, p 448-456

33. P.A. Gradin and H.L. Groth, A Fracture Criterion for Adhesive Joints in Terms of Material-Induced Singularities, in *Proceedings of the 3rd International Conference on Numerical Methods in Fracture Mechanics*, Pineridge Press, 1984, p 711-720

34. H.L. Groth, Stress Singularities and Fracture at Interface Corners in Bonded Joints, *Int. J. Adhes. Adhes.*, Vol 8 (No. 2), April 1988, p 107-113

35. G.P. Anderson and K.L. DeVries, Analysis of Standard Bond-Strength Tests, in *Treatise on Adhesion and Adhesives*, R.L. Patrick, Ed., Marcel Dekker Inc., 1988, p 55-121

36. R.D. Adams and J.A. Harris, The Influence of Local Geometry on the Strength of Adhesive Joints, *Int. J. Adhes. Adhes.*, Vol 7, 1987, p 69-80

37. R.D. Adams and W.C. Wake, *Structural Adhesive Joints in Engineering*, Elsevier, 1984

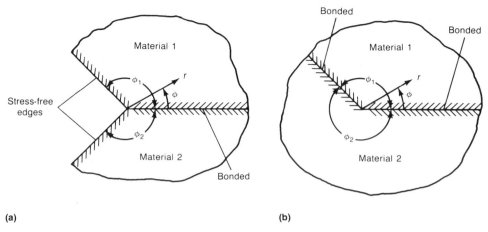

Fig. 37 Two types of interface corners in bonded dissimilar materials. (a) Notched wedges. (b) Fully bonded wedges

38. D. Jangblad, P. Gradin, and T. Stenstrom, Determination and Verification of Elastic Parameters for Adhesives, in *Adhesively Bonded Joints: Testing, Analysis, and Design*, STP 981, W.W. Johnson, Ed., American Society for Testing and Materials, 1988, p 54-68

39. K.J. Lee, J.W. Small, and V. Miller, The Non-Linear Analysis of Bonded Joints, in *Proceedings of the 29th International SAMPE Technical Conference*, Oct 1987, p 76-86

40. Y. Weitsman, Stresses in Adhesive Joints Due to Moisture and Temperature, *J. Compos. Mater.*, Vol 11, Oct 1977, p 378-394

41. Y. Weitsman, Residual Thermal Stresses in a Symmetric Double-Lap Joint, *J. Therm. Stresses*, Vol 3, 1980, p 521-535

42. A. Alexander, Implications of Advanced 3-D Analytical Methods, in Paper presented at the 13th Annual Mechanics of Composites Review (Bal Harbour, FL), 2-3 Nov 1988

43. R.B. Krieger, Jr., Stress Analysis Concepts for Adhesive Bonding of Aircraft Primary Structure, in *Adhesively Bonded Joints: Testing, Analysis, and Design*, STP 981, W.S. Johnson, Ed., American Society for Testing and Materials, 1988, p 264-275

44. R.B. Krieger, Jr., An Adhesive Interleaf to Reduce Stress Concentrations Between Plys of Structural Composites, *SAMPE J.*, July/Aug 1987, p 30-32

45. R.E. Politi, Factors Affecting the Performance of Composite Bonded Structures, in *Proceedings of the 19th International SAMPE Technical Conference*, Society for the Advancement of Material and Process Engineering, Oct 1987, p 36-45

46. J.L. Lubkin and E. Reissner, Stress Distribution and Design Data for Adhesive Lap Joints Between Circular Tubes, *J. Appl. Mech. (Trans. ASME)*, Vol 78, Aug 1956, p 1213-1221

47. H.C. Martin and G.F. Carey, *Introduction to Finite Element Analysis, Theory and Application*, McGraw-Hill, 1973, p 151-183

48. A.D. Crocombe and R.D. Adams, Influence of the Spew Fillet and Other Parameters on the Stress Distribution in the Single Lap Joint, *J. Adhes.*, Vol 13, 1981, p 141-155

49. T.L. Rose, "Practical Finite Element Modeling and Techniques Using MSC/NASTRAN," Seminar Notes, NA/1/000/PMSN, MacNeal-Schwendler Corporation, Sept 1989, p 1-15

50. S. Amijima, T. Fujii, and A. Yoshida, Two Dimensional Stress Analysis on Adhesive Bonded Joints, in *Proceedings of The 20th Japan Congress on Materials Research*, The Society of Materials Science, 1977, p 275-281

51. S. Amijima, T. Fujii, Y. Wadayama, and W. Kishimoto, On the Strength of Single Lap Joints Under Bending Load (Effect of Lap Length and Taper Angle), in *Proceedings of The 26th Japan Congress on Materials Research*, The Society of Materials Science, 1983, p 227-232

52. W.C. Carpenter and R. Barsoum, Two Finite Elements for Modeling the Adhesive in Bonded Configurations, *J. Adhes.*, Vol 30, 1989, p 25-46

53. G.P. Anderson, S.J. Bennet, and K.L. DeVries, *Analysis and Testing of Adhesive Bonds*, Academic Press, 1977

54. T.C.T. Ting and S.C. Chou, Edge Singularities in Anisotropic Composites, *Int. J. Solids Struct.*, Vol 17, 1981, p 1057-1068

55. R.I. Zwiers, T.C.T. Ting, and R.L. Spilker, On the Logarithmic Singularity of Free-Edge Stress in Laminated Composites Under Uniform Extension, *J. Appl. Mech.*, Vol 49, 1982, p 561-569

56. S.S. Wang and I. Choi, Boundary Layer Effects in Composite Laminates: Part 1—Free-Edge Stress Singularities, *J. Appl. Mech.*, Vol 49, 1982, p 541-548

57. J.P. Dempsey and G.B. Sinclair, On the Stress Singularities in the Plane Elasticity of the Composite Wedge, *J. Elasticity*, Vol 9, 1979, p 373–391

Fatigue and Fracture Mechanics

Erol Sancaktar, Department of Mechanical and Aeronautical Engineering, Clarkson University

THE SUBSTANTIALLY INCREASED use of adhesively bonded joints since 1960 includes many applications that subject the bonded structure to cyclic (fatigue) loadings, as shown in Fig. 1. Although adhesive bonding for fatigue loading is primarily used in the aerospace industry, many other applications in modern engineering practice also require consideration of cyclic loading. Such applications include adhesive bonding in automotive, marine, home appliance, machinery, medical, dental, construction, toy, and other industries. Obviously, the loading environments, such as environmental temperature and humidity; the level, nature, and frequency of the loads; the type of adhesives used; the geometry and materials of the structure to which the adhesive is applied; and the cure methods and constraints are all variable and at least as numerous as the number of industries cited above.

It is possible to find information that was published as early as 1957 on the fatigue strength of adhesively bonded joints (Ref 1), but most of the earlier works utilized a methodology and extrapolation from the science of metal fatigue. Only since the late 1970s have researchers paid special attention to the fatigue of adhesively bonded joints using modern stress analysis methods such as finite- and boundary-element methods, microstructure characterization methods, and the nonlinear and time-dependent constitutive modeling methods required for polymeric and composite materials.

The reported knowledge in the aircraft industry regarding the fatigue strength superiority of adhesively bonded joints to riveted connections dates back to the early 1960s (Ref 2). In 1962, Argyris reported that the fatigue strength-to-ultimate strength ratio (obtained under monotonic loading conditions) for adhesively bonded aluminum plates that were fatigued for 10^5 cycles was 63% higher than for riveted counterparts, based on an increase in the ratio going up from 0.3 for the latter to 0.49 for the former (Ref 2).

One important source of fatigue failure in aircraft is sonic fatigue. During certain aircraft operational modes, combined thermal and acoustic conditions subject the fuselage, wing, and empennage structures to sonic and thermal fatigue (Ref 3). The locations and magnitudes of this type of loading depend on the vehicle/engine configuration and emanate from engine air intake duct turbulence, jet engine exhaust, and thermal loads created by aerodynamic heating (Ref 3). Noise reflections or heat flow due to air flow during takeoff (or lift-off) may significantly increase the acoustic and thermal-load intensities. Consequently, it is thought that sonic fatigue should govern the design of advanced composite aircraft structures. For example, analytical studies for vertical or short take-off and landing (V/STOL) aircraft resulted in predictions of 153 to 167 dB overall sound pressure levels for the side and bottom fuselage skins at lift-off, decreasing to about 7 dB at an altitude of one wing semispan (Ref 3). The temperature at which these sonic environments exist for the lightweight fuselage panels has been estimated to be about 121 °C (250 °F) during lift-off (Ref 3).

Fatigue Fundamentals. The basic diagram that has been in use for the fatigue design of steels and other metals and alloys is the stress amplitude (ratio) versus fatigue life (cycles) plot, often referred to as the *S-N* plot. These plots are obtained by using the fatigue data from identical specimens tested under identical and nontransient cyclic conditions. For steels and some other materials, the *S-N* curve has a characteristic bend with an asymptotic stress amplitude value, which results in a horizontal line on a graph, where the log(S) ratio is vertical and the log(N) is horizontal for a million or more cycles of fatigue loading. The strength corresponding to this bend is called the endurance limit, or the fatigue limit. The log(S) versus log(N) curve for nonferrous metals and alloys never becomes horizontal; hence, these materials do not have an endurance limit. Even for those materials that exhibit an endurance limit, the intended life of the structural design that will be subjected to cyclic loading may be less than would be needed to reach the endurance limit. In such cases the concept of fatigue strength is used. The fatigue strength corresponds to the stress amplitude that causes failure in an arbitrary number of cycles. Consequently, the corresponding number of cycles should always be given with the values of fatigue strength. *S-N* data are also plotted in the form of a strain amplitude versus fatigue life diagram. For example, a random fatigue curve for bonded skin-stiff-

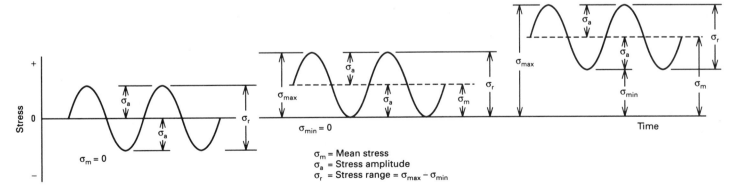

σ_m = Mean stress
σ_a = Stress amplitude
σ_r = Stress range = $\sigma_{max} - \sigma_{min}$

Fig. 1 Cyclic stress modes based on the mean stress level

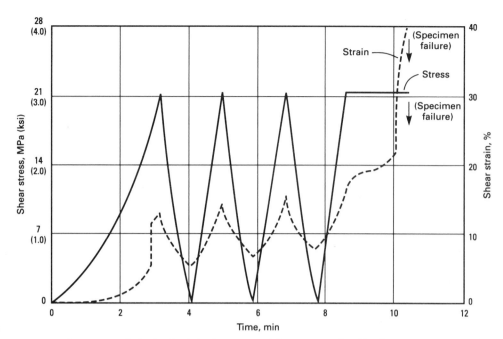

Fig. 2 Symmetric single-lap loading/unloading creep-to-failure response of Metlbond 1113 (Narmco Materials, Inc.)

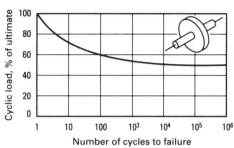

Fig. 3 Fatigue performance of a typical toughened anaerobic adhesive (Ref 4)

ener joints predicts approximately 444 μm/m root mean square (rms) strain for failure at 10^7 cycles of sonic fatigue. In this case a nomograph is used to relate the applied overall sound pressure levels to overall rms strain levels in different (composite) structures (Ref 3).

For short-life (low-cycle) fatigue tests that last less than 10^5 cycles to failure, strain control is preferred to avoid runaway strains that result from cyclic creep of the material. An example of such cyclic creep in aluminum symmetric single-lap specimens that are bonded using toughened epoxy adhesive with carrier cloth is shown in repeated stress loading in Fig. 2.

The above-mentioned fatigue data by Argyris presented in the form of an *S-N* curve does not exhibit an endurance limit (Ref 2). Other data that contain an endurance limit for adhesively bonded joints do exist, however, and an example is presented in Fig. 3 (Ref 4). In this figure, the endurance limit for a (typical) toughened anaerobic adhesive appears to be 50% of the ultimate shear strength for the adhesive tested in the rod pull-out configuration shown. The geometry of a bonded joint has a substantial influence on the resulting stress peaks and, consequently, on the location and initiation of fatigue failures.

The presence of an endurance limit in the *S-N* curves of some adhesively bonded joints imply the presence of a threshold for nonpropagating cracks in the adhesive layer. In fact, recent work (Ref 5, 6) has revealed the presence of two thresholds when the crack propagation rate (da/dN) is plotted against a strain-energy release rate range (ΔG) or stress-intensity range (ΔK)

parameter. These thresholds are the onset threshold level separating the nonpropagating region (at low ΔK values) from the "steady" crack propagation region and the catastrophic threshold level for the fast crack propagation that results in immediate failure of the structure. Figure 4 shows these threshold levels, along with a size scale for molecular and macromolecular features present in adhesive layers and adhesive-adherend interphases.

So far, this discussion has highlighted the important issues of fatigue analysis for adhesively bonded joints. The remainder of the article contains detailed discussions of each of these issues, that is, the effects of joint and loading geometry and the resulting state of stress with its influence on the fracture and crack propagation behavior; the material behavior of the adhesive and the adherend, including the effects of the environment; the mode of cyclic loading, including cyclic frequency; the micro- and macrostructures of the adhesive material; and the effects of these preceding issues on the determination of failure criteria for onset and catastrophic threshold levels.

Joint and Loading Geometry Effects

Effect of Joint Geometry. Nonuniform stress distribution in adhesively bonded joints has been studied and reported in the literature since 1909 (Ref 7). The classical paper by Goland and Reissner based on elastic analysis (Ref 8) and the later works by Hart-Smith based on elastic-plastic adhesive material assumption (Ref 9) describe the stress distributions in single-lap joint

geometry, which reveal high peel and shear stresses at the overlap edges. These stresses at the overlap edges subject any pre-cracks or cracks to mixed-mode loading. Results of these analyses also reveal that the magnitudes of adhesive stresses are affected similarly by adhesive thickness (especially at the overlap ends) and overlap length. When the values of these parameters are increased, the stress levels are decreased. The effects of these geometric parameters on the adhesive stresses are also functions of the location on the overlap length, the strongest effect being on the overlap edge where stress concentrations exist. For example, Wang performed fatigue tests on 2024-T3 aluminum alloy adherend single-lap joints bonded with FM-47 adhesive (Ref 10). The thickness of the adherends was 1.6 mm (0.06 in.). Two different overlap lengths were used: 12.5 and 9.5 mm (0.50 and 0.37 in.). At 10^6 cycles, a single-lap joint with 12.5 mm (0.50 in.) overlap length failed at 47.2 MPa (6.8 ksi), while the one with the 9.5 mm (0.37 in.) overlap had a fatigue strength of 41.4 MPa (6.0 ksi). Specimens with corresponding overlap lengths had monotonic loading strengths of 251 MPa and 229 MPa (36.4 and 33.2 ksi), respectively.

During the last decade or two, considerable work has been done on maximizing the load-carrying capacity of adhesively bonded joints by reducing the stress peaks and making the stress distributions more uniform. For example, the use of prebent adherends has been shown to increase the load-carrying capacity of single-lap joints (Ref 11), while the use of tapered overlap edges has been shown to improve single- and double-lap joints (Ref 12), as well as tubular butt joints with flanges (Ref 13). Using rounded butt edges improves torsional and butt joints (Ref 2). Attempts to develop finite-element codes for the unified and optimum design of adhesive joints have been reported in the literature (Ref 14, 15). References 2 and 15 also focus attention on the adhesive fillet geometry at the overlap edges.

Using an elastic shear lag analysis, Szepe (Ref 7) expressed what he called the average ultimate shear stress (τ_{ult} = load/

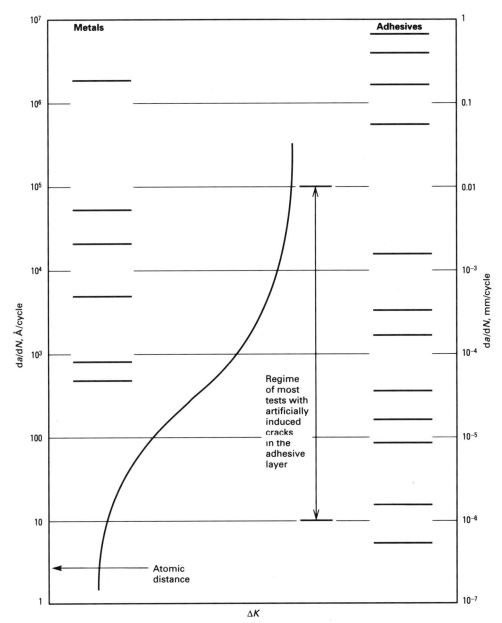

Fig. 4 Different crack growth rate regimes in adhesive layers and comparison with molecular and macromolecular features

adhesive area) for the adhesive layer of a double-lap joint by:

$$\tau_{ult} = \tau_0 \left\{ 1 + \left[\left(\frac{kL^2}{3E} \right) \left(t_2^{-1} - (2t_1)^{-1} \right) \right] \right\} \quad \text{(Eq 1)}$$

where τ_0 is the shear strength of the adhesive (may be interpreted as the bulk shear strength); k is the spring constant of the adhesive layer, which is equal to the adhesive shear modulus (G)/adhesive layer thickness (η); L is the overlap length; E is the adherend elastic modulus; $2t_2$ is the thickness of the main plate; and t_1 is the thickness of individual cover plates.

With this method Szepe suggested that the fatigue strength of a double-lap adhesive joint with any geometry can be calculated when the average ultimate shear stress of two specimens with different lengths of overlap (or bulk material properties of the adhesive) and the S-N diagram for any overlap length are known. If the spring value, G/η, for the adhesive layer is not constant, another S-N diagram, preferably for a different adhesive thickness, is necessary. This methodology is shown in Fig. 5.

Use of Fracture Mechanics. The fatigue life prediction method described above does not account for the presence of any cracks in the adhesive layer and, consequently, does not utilize the methods of fracture mechanics. Previous investigations by Sancaktar *et al.* (Ref 16-20) on toughened structural adhesives with and without a carrier (Metlbond 1113 and 1113-2 solid film adhe-

sives, respectively) showed that the bonded adhesive behavior was predominantly affected by fracture processes. Figure 6 shows the variability of the shear stress-strain behavior in the bonded form for similar joints tested under monotonic loading. Prior to testing, a neutron radiography examination of the joints shown in Fig. 6 revealed that the variations in the stress-strain behavior were a result of inherent flaws (voids) in the bond line. Figure 7 shows a fracture surface with its corresponding neutron radiograph. This figure reveals that a large initial void shown on the radiograph (Fig. 7a) can also be detected on the photograph of a fracture surface that failed under monotonic loading (Fig. 7b). Because a fracture mechanics approach, such as the Griffith theory (Ref 21), predicts that larger void sizes may result in failures under smaller fracture stresses when cracks initiate from them, the presence of flaws such as the ones shown may easily result in catastrophic joint failures.

Inherent flaws cause the bond line to fail in a brittle manner. Figure 8 shows a scanning electron micrograph (and enlarged detail) of a joint with a brittle fracture surface. Obviously, the brittle fracture was promoted by the inherent flaws. The presence of a carrier cloth also affects the fracture pattern. A previous investigation on the adhesive (Ref 22) reported that slow crack propagation (usually a result of ductile fracture) occurred along a plane of fiber, while fast crack propagation (usually a result of brittle fracture) was confined to planes without fiber.

Because fracture processes exert the dominant effects on the behavior of adhesively bonded joints, it is necessary to determine accurate relations for the fracture stresses. If the adhesive behaves in a brittle manner in the bonded form (Fig. 8), the use of a linear elastic fracture mechanics (LEFM) approach can be considered appropriate for characterizing the fracture stresses. Extensions of this approach using the elastoplastic fracture mechanics principles of Irwin (Ref 23) and others (Ref 24-26) and extensions using geometrically nonlinear analysis (Ref 27) have been reported in the literature.

The flaws in a bond line may develop into cracks when subjected to loads. The energy required to propagate a crack is called fracture energy. In isotropic solids, fracture energy is a function of the stress state. This is the reason there is a higher likelihood of the development and propagation of cracks at the stress-concentration areas, such as the overlap edges, of adhesive joints. The stress-intensity factor (K) is often used as a measure of the stress in the vicinity of a crack tip. For an isotropic material, the stress-intensity factor can be expressed as (Ref 21):

$$K = \alpha P (a)^{1/2} \quad \text{(Eq 2)}$$

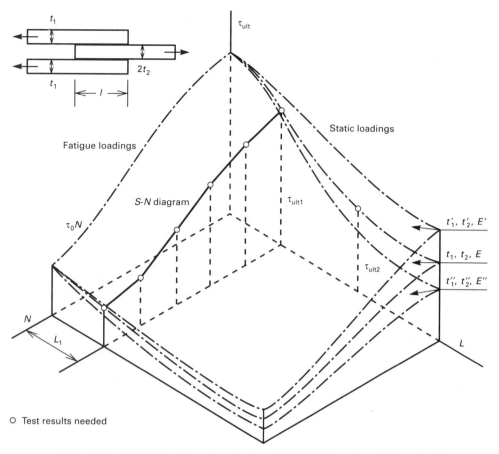

Fig. 5 Schematic showing that the fatigue strength of any double-lap joint can be calculated when the average ultimate shear stress of two specimens with different lengths of overlap and the S-N diagram for any length of overlap are known (Ref 7)

where α is a function determined by the stress state and geometry, P is the applied load, and a is the flaw size. In order to use Eq 2 to predict the fracture stress, one needs to know the size of the largest flaw and to measure K experimentally.

The strain-energy release rate (G) can also be used to determine the fracture stress. The expression defining the strain-energy release rate is given by (Ref 21):

$$G = \left(\frac{P^2}{2B}\right)\left(\frac{\partial C}{\partial a}\right) \qquad \text{(Eq 3)}$$

where B is the material thickness in the vicinity of the crack and $\partial C/\partial a$ is the change in compliance (C) of the structure with crack length (a). The advantage of using the concept of the strain-energy release rate and Eq 3 is the fact that it is independent of the crack tip stress distribution because $\partial C/\partial a$ is a global property of the specimen. The compliance (C) is defined as:

$$C = \frac{\delta}{P} \qquad \text{(Eq 4)}$$

where δ is the displacement. Because the displacement is a function of the loading mode, the strain-energy release rate must

be determined for each of the three principal modes of failure: opening mode (G_I), in-plane shear (G_{II}), and torsional shear (G_{III}).

The variation of the stress-intensity factor (K) with crack opening displacement (or overall deformation of the member) is analogous to a stress-strain curve. For an elastic material, K increases essentially linearly (in some cases, deviations from linearity may occur because of bending of the parts of the structure transmitting load to the crack tip) with displacement and stress (σ) up to a critical event similar to yielding in a standard uniaxial stress-strain test. The occurrence of this event is represented by the fracture toughness (K_c); the notch or the flaw suddenly begins to grow, and complete fracture occurs. The strain-energy release rate corresponding to this event is G_c, and it is called the critical strain-energy release rate, or fracture energy. When the fracture toughness (K_c) is plotted using the vertical axis versus the thickness of the structure, the curve exhibits a horizontal asymptotic value for K_c. This value is commonly known as the plane-strain fracture toughness. Both plane-strain K_c and G_c values are considered material properties.

It is possible to relate strain-energy release rates in the three modes of loading to stress-intensity factors (Ref 21, 23) using the relations:

$$G_I = K_I^2 (1 - \nu^2)/E \qquad \text{(Eq 5)}$$

$$G_{II} = K_{II}^2 (1 - \nu^2)/E \qquad \text{(Eq 6)}$$

$$G_{III} = K_{III}^2 (1 + \nu)/E \qquad \text{(Eq 7)}$$

where ν is the Poisson's ratio.

Although the cleavage or tensile opening mode (mode I) is the one that most readily results in the failure of monolithic, isotropic materials, this is not the case for adhesively bonded joints because crack propagation is constrained within the adhesive layer. Therefore, when designing adhesively bonded joints, one needs to consider mixed-mode fracture, and not assume that G_{Ic} alone is critical. In adhesive bonding applications, the majority of the cases involve only G_I and G_{II} due to peel and shear stresses, respectively, with the use of some form of lap-shear geometry. In such geometries, adhesive cracks are exposed to both tension and shear, resulting in mixed-mode cracking, with a mixture of modes I and II.

The Adhesive Advantage in Fracture. The U.S. Air Force airplane damage tolerance requirements (MIL-A-83444) specify the initial flaw size and the critical flaw size in an airplane structure. The initial flaw must not grow to the critical size after a defined period of service use and fail if a specified monotonic (residual strength) load is applied subsequently. Under these guidelines two different categories of structures are specified. The "slow crack growth" structure must assume an initial flaw a 6.35 mm (0.25 in.) long surface flaw or a 1.27 mm (0.05 in.) long bolt hole corner flaw. A "fail-safe multiload path independent" structure can assume an initial surface flaw of 2.54 mm (0.1 in.) in length or a bolt hole corner flaw of 0.51 mm (0.02 in.) in length (Ref 28). It was reported that adhesively bonded laminated structures are ideal for fail-safe structure designs because they provide independent load paths and crack arrest features (Ref 28). Furthermore, bolt holes in the primary structures are eliminated. This is why the stress levels in adhesively bonded multilaminate structures can be higher than those in monolithic (metal) structures without reducing damage tolerance capability, while the weight of the structure is considerably reduced. This type of improvement has been demonstrated for both aluminum (Ref 28, 29) and titanium laminates (Ref 30). In the case of the titanium laminates, the fracture toughness (K_c) is reported to be 39% higher and the through-the-thickness crack growth rate to be 20% slower than in corresponding monolithic metal plates. For the surface-cracked laminates, the damage tolerance life is reported to range from 6 to more than 15 times the

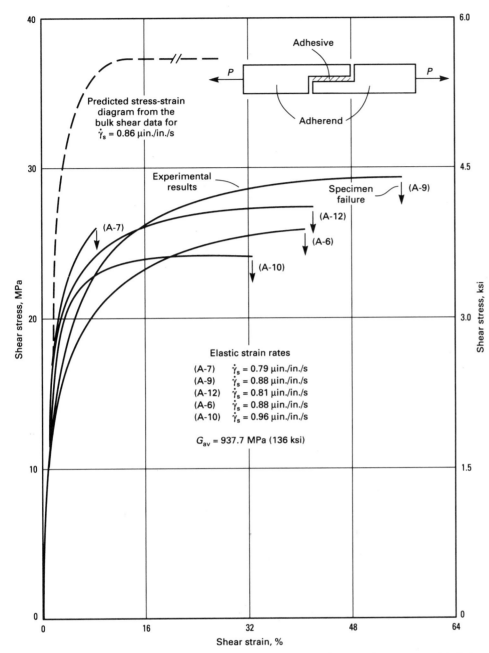

Fig. 6 Symmetric single-lap adhesive stress-strain response of Metlbond 1113 and comparison with bulk shear prediction. The A-12 curve corresponds to the specimen in Fig. 7.

- For specimens with 0° and 45° plies at the interface, the fatigue damage initiated as cyclic debonding. For specimens containing 90° plies at the interface, ply cracking occurred in the 90° lamina
- For specimens with 0° plies at the interface, cyclic debonding occurred in the adhesive layer. For specimens with 45° and 90° plies at the interface, however, intraply damage occurred first in the adherend that was directly loaded (strap adherend) as a result of cyclic loading. When the fatigue damage reached the first 0° ply, it continued as delamination
- For specimens with 0° and 45° plies at the interface, damage initiation at the adhesive layer occurred at stress levels 13 and 71% higher for adhesives without a carrier (EC 3445) and with a carrier (FM 300), respectively, in comparison to specimens with 90° plies at the interface
- The total strain-energy release rate threshold values were very close for specimens with 0° and 45° interface plies because damage occurred at the adhesive layer

Consequently, Johnson concluded that the use of 90° plies at the interface should be avoided to prevent delamination in the adherend.

Improvements in the fracture behavior of structures through the use of polymer resin matrix composite materials are known in industries other than aerospace. For example, composite crack arrestors are excellent candidates for the replacement of their metal counterparts for gas pipeline applications (Ref 32).

Optimized Joint Geometry for Fatigue Resistance. Methods of improving the load-carrying capacity of adhesively bonded joints (under monotonic loading) have already been discussed. For example, Sancaktar *et al.* (Ref 11) showed that the load-carrying capacity of single-lap joints with inflexible adhesive layers (Ref 8) were increased by as much as 71% when prebent adherends were used (Fig. 9 and 10). Similarly, the prebending of adherends improves the fatigue behavior of single-lap joints because of reductions in the overlap edge stress-intensity factors at given joint loads. An example of this improvement was reported by Johnson (Ref 33) and is illustrated in Fig. 11.

In another example of joint geometry modification to improve fatigue behavior, Johnson *et al.* reported that the no-growth threshold nondamaging stress range was increased by almost 50% when the adherend taper angle was changed from 90° (regular specimens) to 5° (Ref 34) (Fig. 12). It should also be noted that Adams *et al.* reported a three-fold increase in the monotonic strength of adhesively bonded double-lap joints when the adherend tapering was combined with the use of an adhesive fillet

life of the monolithic specimen (Ref 30). In this case, experiments were performed using compact tension (for through-the-thickness) and part-through-surface (for surface-cracked) specimens.

Johnson (Ref 29) attributes the higher toughness in adhesively laminated plates to the plane-stress failure of the individual plies, in comparison to the plane-strain failure of a monolithic plate of the same thickness. The advantages of laminates containing thin layers of adhesive are claimed to be higher fracture toughness, slower crack growth in the adhesive material, lower stress-intensity factors in cracked plies due to load transfer to an uncracked ply, and the resistance of adhesive material to the propagation of cracks to adjacent uncracked plies (Ref 29). The same work also revealed a significant reduction in the fatigue growth of impact damage when the aluminum substrate was prestrained.

The constant-amplitude cyclic loading behavior of graphite-epoxy composite specimens with different adherend stacking sequences was studied by Johnson *et al.* using cracked lap-shear specimens (see insert in Fig. 12), which induce mixed-mode loading on the adhesive material at the overlap edge (Ref 31). The study resulted in the following conclusions:

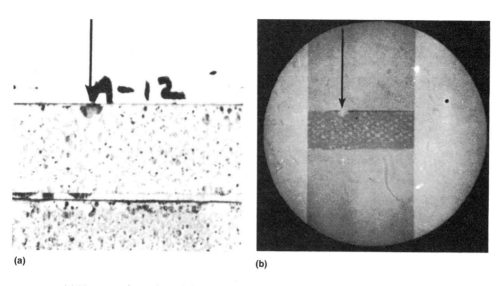

(a)

(b)

Fig. 7 (a) Neutron radiography and (b) postfracture surface photographs of a symmetric single-lap specimen. The large inherent void (arrows) is also visible on the postfracture surface.

at the edge of the overlap (Fig. 13) (Ref 15). However, data on the effect of this type of geometric modification on joint fatigue behavior are not available at this time.

Stress Singularity Methods. The use of asymptotic singular fields in the fracture analysis of adhesive joints and micromechanics of adhesive failure is a serviceable tool because adhesively bonded joints almost always contain geometric discontinuities. A recent work by Barsoum (Ref 35) in this area included a study of singularities at the interfaces of nonlinear materials, 0° interface cracks, and square-corner singu-

larities. His work revealed that the singularity is oscillatory when cracks are inclined at angles of less than Φ_c, with this angle taking on values between 22.5° and 30°, and, in a manner similar to interfacial cracks, when Φ is the angle of inclination of the crack to the interface. His results also indicated that the oscillatory exponent for an aluminum adherend is stronger than that of a graphite-epoxy adherend, indicating a larger mixture of modes I and II for the former. Barsoum suggests that the degree of mixture of modes can be considered a rationale for the resulting

scatter in the monotonic G_{Ic} and G_{IIc} data available in the literature.

Another recent work that used the stress singularity approach for evaluating adhesive strength was by Hattori *et al.* (Ref 36). The researchers used the boundary-element method to calculate stress distributions along with stress singularity parameters to express stress fields near bonding or contact edges. The method used allowed estimation of both delamination occurrence and propagation. Delaminations were observed using scanning acoustic tomography with 0.1 mm (0.004 in.) pitch.

Crack Propagation Laws. As in the monotonic load fracture, the principle of energy conservation must be satisfied during fatigue loading. Consequently, for unit plate thickness we have:

$$(d/da) (F + Q) = (d/da) (U + D + S) \qquad \text{(Eq 8)}$$

where a is the crack length per unit crack width, F is the work done by external forces, Q is heat added to the crack tip, U is elastic strain energy, D is energy dissipated, and S is energy needed to create fracture surfaces. This equation is a mathematical statement of the first law of thermodynamics, as well as a generalization of the Griffith's energy balance approach to fracture.

During the cyclic loading of viscoelastic materials, there is a phase difference between the time-dependent stress and strain functions, and heat is produced because of viscous dissipation. Consequently, in a cyclic test the inherent viscoelastic nature of the adhesive becomes more effective in its

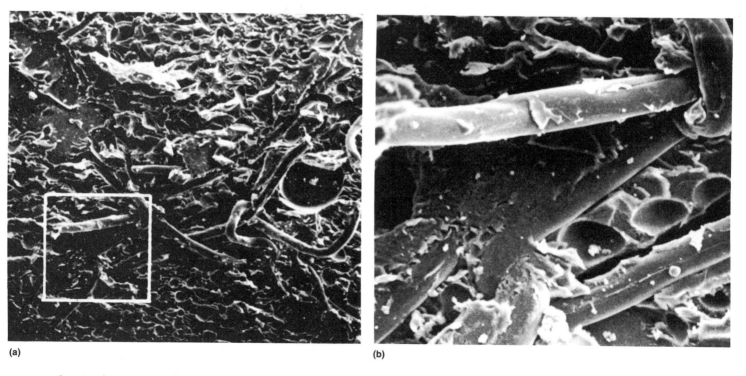

(a)

(b)

Fig. 8 Scanning electron micrographs of a Metlbond 1113 single-lap specimen with brittle fracture surface. (a) 4.1 mm² (6.3 × 10⁻³ in.²). (b) 0.041 mm² (6.3 × 10⁻⁵ in.²)

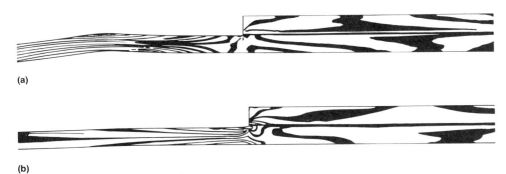

(a)

(b)

Fig. 9 Isochromatic fringe patterns at 223 N (50 lbf) tensile load for (a) 15° and (b) 0° prebent adherends

fracture behavior because of the long duration of load application. Therefore, during lengthy fatigue loading, heat dissipation caused by viscous action and creep deformation caused by the mean level of the sinusoidally varying load become nonnegligible factors. Thus, the D term of Eq 8, representing the irreversible work output, is no longer negligible, as opposed to cases of fast crack propagation occurring under monotonic loading. Also, because of the viscoelastic effects cited, the formation of microcracks, voids (pores), and stretched fibrils is facilitated. Therefore, void coalescence and fibril rupture become more likely at lower loads in comparison to the likelihood of fracture of the adhesive under monotonic loads.

During multiaxial (or uniaxial) cyclic loading, K_I and K_{II} values increase as the crack propagates in a stable fashion. Consequently, a fracture criterion based on monotonically obtained values is expected to be valid only at the onset of catastrophic crack propagation (at best), when K_I and K_{II} approach K_{Ic} and K_{IIc} values, respectively. In fact, in most cases K_{Ic} and K_{IIc} values obtained from monotonic experiments do not provide a good prediction of catastrophic threshold levels because crack tip geometries may change because of heat dissipation and plastic deformation, which result in the crazing, microcracking, cavitation, voiding, and fibrilation of the adhesive material during cyclic loading as mentioned above. These microstructural deformation and failure mechanisms lower the fatigue resistance of the material and become more pronounced as the stress amplitude and duration are increased. Therefore, catastrophic crack propagations that follow stable crack growth are expected to be much

more random, compared to the catastrophic crack propagations induced by the monotonic loading of a starter crack. Furthermore, the former occur at lower K or G values.

For the reasons cited, empirical formulas are usually adopted to fit the data obtained in cyclic-load dominant fracture tests and are used as the governing model for crack propagation prediction.

In fatigue loading cases, even when the loading level is relatively low compared to the loading level applied for catastrophic crack propagation under monotonic loading, stresses around the crack tip are still beyond the yield stress of the adhesive in the whole cycle or in part of the cycle because of the crack tip singularity; thus, crack propagation, even though not catastrophic, can still occur in small spurts with the applied cyclic load. Intuitively, it is easy to understand that the crack growth rate depends on P_{Imax}, P_{Imin}, and P_{II} because it is induced by crack opening displacements (COD). Therefore, in terms of elastic energies, the crack growth rate should depend on parameters such as strain-energy release rate range ($\Delta G = G_{max} - G_{min}$), total strain-energy release rate (G_{Tmax}), crack length (a), material properties, and specimen geometry, as well as environmental effects such as temperature (T). This relationship can be shown as:

$$da/dN = f(\Delta G, G_{Tmax}, E, T, \ldots) \qquad \text{(Eq 9)}$$

Among these parameters, the strain-energy release rate range (ΔG) due to the dynamic part of the loading plays the most important role in fatigue crack propagation, again, because it is proportional to the COD.

Typically, when fatigue crack growth rate (da/dN) is plotted (on log-log paper) against strain-energy release rate range (ΔG) for mode I loading, a multimodal curve is obtained with three different crack growth rate regions. The first region is identified by a threshold strain-energy release rate range for nonpropagation cracks where the ΔK level is too small to induce an appreciable amount of crack propagation. Furthermore, in this range, crack tip plastic deformation may be sufficient to blunt any crack propagation. The second region is expected to be governed by a power-law relationship between crack growth rate and strain-energy release rate. The third begins with the threshold for catastrophic crack propagation (Fig. 4). This article refers to the threshold at which the crack starts growing as the onset threshold, whereas the threshold at which catastrophic failure occurs is called the catastrophic threshold.

For adhesively bonded joints, a widely used crack growth model for the second region described above is given by:

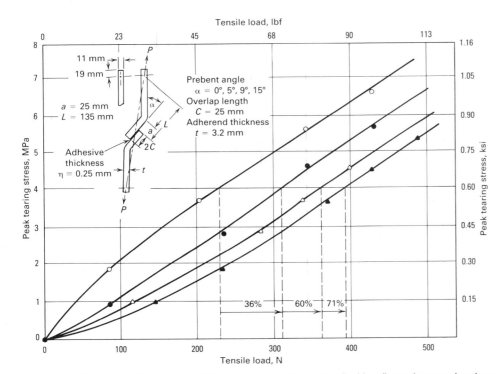

Fig. 10 Effect of tensile load on peak tearing stress for joints with inflexible adhesive layers and prebent adherends

$$da/dN = c(\Delta G)^n \qquad \text{(Eq 10)}$$

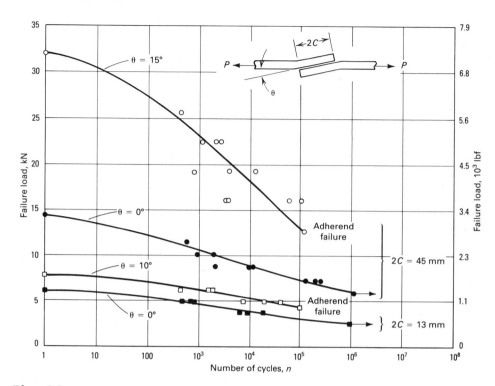

Fig. 11 Effect of prebend angle and overlap length on the fatigue strength of single-lap joints (Ref 34)

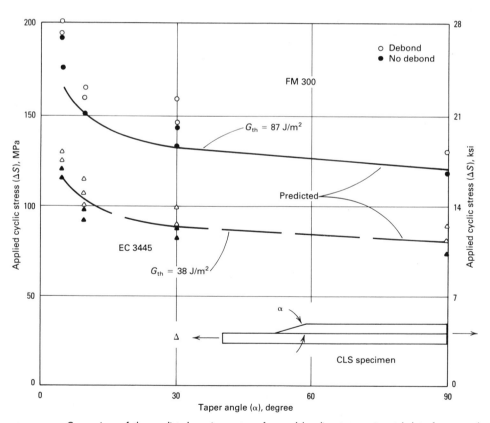

Fig. 12 Comparison of the predicted maximum stress for no debonding to experimental data for tapered cracked lap-shear (CLS) specimens. Open symbols represent debonding. Solid symbols represent no debonding (Ref 34).

where c and n are material constants obtained by "curve fitting" the experimental fatigue data to Eq 10. This relation, however, does not include any G_{II} term and hence is not expected to correlate mixed-mode crack propagation data efficiently, even though the terms c and n can be expressed as functions of G_{II}.

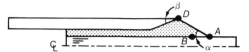

Fig. 13 Double-lap joint with inside taper and adhesive fillet (Ref 15)

Jablonski (Ref 37) used Eq 10 to describe the crack growth rate in aluminum tapered double cantilever beam (TDCB) specimens bonded with two different high-temperature structural adhesives: AF-163 (3M Company), which is rubber-filled 121 °C (250 °F) curing solid film adhesive with nylon scrim cloth support, and EA-9649 (Hysol Division, The Dexter Corporation), an aluminum-, asbestos-, and rubber-filled solid film adhesive with a 177 °C (350 °F) curing temperature. The specimens were tested under load control using a sine wave at 40 Hz and a load ratio of $R = P_{min}/P_{max} = 0.1$. His results revealed that the bond line thickness influences the crack growth rate, da/dN. In fact, the da/dN values for 0.5 mm (0.02 in.) thick bonds were an order of magnitude higher than those of the 0.25 mm (0.01 in.) thick bond lines. Jablonski also predicted that the onset threshold values for the thicker bond samples (0.5 mm, or 0.02 in.) will be lower than for the thinner bond samples (0.25 mm, or 0.01 in.) because the cracks propagate in thicker samples at ΔG values below the onset threshold for the thinner bonds.

Crack Closure. Jablonski (Ref 37) attributes the increased crack growth rate associated with thicker bonds to the phenomenon of crack closure. In this phenomenon, the stresses at a crack tip reach high values upon loading, causing yielding in a zone called the crack tip plastic zone. For deformable materials, the crack propagation process includes plastic energy that is due to this crack tip plastic deformation. Furthermore, the mathematical formulation of the stress field based on perfectly elastic material behavior suffers from singularity at the crack tip. In order to resolve this problem, Irwin assumed elastic-plastic material behavior and stated that the failure stress at the crack tip is equal to the uniaxial yield stress (σ_y) acting over a region called the plastic zone. Irwin calculated the size (diameter) of the plastic deformation zone by considering an infinite plate containing a single crack under the action of tensile in-plane loading perpendicular to the crack (Ref 23). The diameter (r_p) of the plastic (deformation) zone was thus shown to be:

$$r_p = \frac{K_{Ic}^2}{M \pi \sigma_y^2} \qquad \text{(Eq 11)}$$

where M is the plastic constraint factor. For adhesively bonded joints, the size and shape of this zone have been related to the bond thickness (Ref 24, 25, 38). Hunston *et al.* reported that when the thickness of the bond is large, the plastic zone volume and the fracture behavior of the bond line are similar to that in the bulk adhesive (Ref 38). However, as the bond thickness decreases, the zone size at failure stretches farther down the bond line and the fracture energy increases. They claimed that with further decreases in the bond line, approximately when the height of the deformation zone equals the bond thickness, the bond fracture energy begins to decrease. They also reported that, under monotonic loading, bulk samples and very thick bonds exhibit subcritical crack growth prior to failure. Thinner bonds, however, show little evidence of such crack growth.

Under cyclic loading, this plastic zone is stretched and enlarged by elastically deformed material during the loading cycle. When the crack is unloaded, the plastically deformed part keeps its size, but the elastically deformed part tends to resume its original size and position, so that a compressive stress is exerted on the plastic zone. From stretch to compression, there must be a load level at which the stress on the plastic zone is 0. This load level is called the crack closure load (P_{cc}) level. In the load range of 0 to P_{cc}, no crack propagation is expected because the crack is kept closed.

To account for the crack closure load, the term ΔG should be written as:

$$\Delta G = \Delta G_{Ieff} \qquad \text{(Eq 12)}$$

where ΔG_{Ieff} is the effective strain-energy release rate range in mode I and can be expressed as:

$$\Delta G_{Ieff} = (\tfrac{1}{2}B)\,(\partial C/\partial a)$$
$$(P_{Imax}^2 - P_{Imin}^2) \text{ for } P_{cc} < P_{Imin} \qquad \text{(Eq 13)}$$

or

$$\Delta G_{Ieff} = (\tfrac{1}{2}B)\,(\partial C/\partial a)\,(P_{Imax}^2 - P_{cc}^2) \text{ for } P_{cc}$$
$$> P_{Imin} \qquad \text{(Eq 14)}$$

where P_{cc} is the crack closure load.

For double cantilever beam configurations, the term $(\partial c/\partial a)$ in Eq 13 and 14 can be calculated using the formula (Ref 39):

$$C = 2[(a + a_0)^3 + h^2a]/(3EI) \qquad \text{(Eq 15)}$$

where a_0 is the empirical rotation factor due to additional deflection at the "built-in" end of the beam, h is the beam height, E is the substrate modulus, and $I = Bh^3/12$.

Based on the mechanism of crack closure explained above, it follows that when the cyclic load is applied in the range of zero to P_{cc}, the crack tip will not open and crack propagation will not occur. This theoretically verifies the existence of a threshold in the first region. However, in practice, the threshold value is usually higher than the

one calculated according to the mechanism of crack closure because of global geometric effects.

Correlation coefficients are often used to assess the "goodness of fit" of the assumed crack growth criterion to the experimental data. In his work, Jablonski determined that for the adhesive used with the carrier, the correlation coefficients were much higher when ΔG_{eff} values were used in Eq 10 (c coefficient = 0.706 for ΔG and c coefficient = 0.941 for ΔG_{eff}). For the adhesive used without the carrier, no such effect of the bond thickness could be observed because of scatter in the data.

The mode of fluctuating load (that is, reversed stress, repeated stress, and so on) indicated by $R = P_{min}/P_{max}$ can conceivably have an effect on the crack closure behavior of adhesively bonded joints. Detailed investigations on this effect, however, are not reported in the literature.

Mixed-Mode Cyclic Loading. It has already been mentioned that crack propagation is constrained within the adhesive layer in adhesively bonded joints, regardless of the loading orientation. Because of this condition, mixed-mode fracture needs to be considered. Most practical adhesive applications involve only G_I and G_{II} because of the peel and in-plane shear stresses that arise in the use of lap joint geometries. In these geometries, adhesive cracks are exposed to both tension and shear, resulting in mixed-mode cracking. In such cases, the total strain-energy release rate (G_T) is given by:

$$G_T = G_I + G_{II} \qquad \text{(Eq 16)}$$

Johnson and Mangalgiri (Ref 40) predicted that the fracture of brittle resins with low G_{Ic} values are controlled by the G_{Ic} component. In tough resins, however, the fracture process is controlled by the G_T (Ref 40). The authors related the higher dilatational capability and the presence of a larger free volume in modified epoxies (due to rubber toughening or extended chains), plasticized epoxies, and epoxies at higher temperatures, to higher G_{Ic} values. They claimed that only a limited amount of G_{IIc} improvement results from the modification of the epoxy because shear deformation does not require volume dilatation. However, the G_I versus G_{II} data for the brittle (unmodified) systems 5208 and 3501-6; for the toughened systems Hx205, F-185, EC 3445, and FM 300; and for semicrystalline thermoplastic polyetheretherketone (PEEK) (Fig. 14) indicate an increase in G_{IIc} values with increasing G_{Ic} values when different adhesives are compared. It should be noted that these values were obtained under monotonic loading conditions. They also reported that the G_I/G_{II} ratio varies with matrix modulus, which in turn varies with moisture and temperature. This variation was not large below the 100 °C (212 °F) test temper-

ature for the 3501-6 unmodified epoxy system they studied.

Johnson *et al.* used the cracked lap-shear (CLS), specimen geometry extensively to study the mixed-mode fatigue fracture behavior of structural adhesives (Ref 34, 41, 42). The CLS specimen is a modified form of single-lap geometry, with which both peel and shear stresses are obtained at the tip of a crack propagating from the overlap edge. Based on experiments that used EC-3445 (without a carrier) and FM 300 (with a carrier) under conditions of 10 Hz and an R value of 0.1, it was concluded that the G_T correlated better than either G_I or G_{II} in the equation where:

$$da/dN = c(G)^n \qquad \text{(Eq 17)}$$

Mall and Johnson concluded that "G_T appears to be the governing parameter for cohesive debond growth under static and fatigue loadings." They also reported that the n values of Eq 17 ranged from 4 to 4.5 and are much higher than the corresponding values for aluminum and steel alloys, which range from 1.5 to 3 (Ref 41). They suggested, therefore, that adhesive debond growth is more sensitive to possible errors in design loads and that a safe-life design criterion that avoids debond initiation (that is, onset threshold) should be used. Some of their other significant findings were:

- The onset threshold values under fatigue loading were an order of magnitude below those of their monotonic (static) counterparts (Ref 42)
- Debond initiation and growth can occur in the absence of peel stresses (Ref 34)
- Thicker adherends (in double cantilever beam specimens) result in slower crack growth rates and higher onset threshold values (Ref 43). It is interesting to note that Hart-Smith's prediction of higher shear strength (for the same bond area) with thicker or stiffer adherends (Ref 9) corroborates this finding

Case Study

A more efficient way to describe mixed-mode crack propagation data is to make da/dN an explicit function of G_{Tmax} and ΔG, as given by:

$$da/dN = c\, G_{Tmax}^m\, (\Delta G)^n \qquad \text{(Eq 18)}$$

The environmental effects are not included in Eq 10, 17, and 18, but can be incorporated through the G values as in Eq 19, which will be discussed later in this article. Equation 18 was recently used by Sancaktar *et al.* (Ref 5, 6, 26) for describing the crack growth behavior of Metlbond 1113 (with a carrier) and 1113-2 (without a carrier) structural adhesives. Sancaktar used the independently loaded mixed-mode specimen (ILMMS) geometry to study the fracture

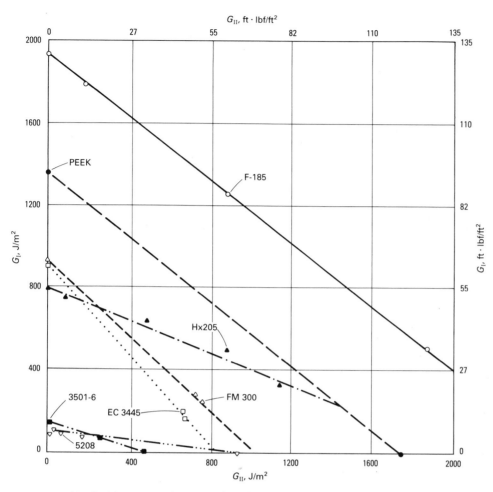

Fig. 14 Mixed-mode monotonic fracture toughness for various brittle and tough structural adhesives (Ref 40). 5208, Narmco; 3501-6, Hercules; Hx205, Hexcel Corporation, chain extended; F-185, Hexcel Corporation, chain extended, rubber modified; EC 3445, 3-M Company, rubber modified; FM 300, American Cyanamid; PEEK, Imperial Chemical Industries

behavior of structural adhesives under monotonic and cyclic loadings.

There are three basic specimen geometries for the mixed-mode testing of adhesives in the bonded form: cracked lap-shear, scarf joint (Ref 44) and the ILMMS geometries. Among these three types, the only specimen geometry that allows the independent measurement of P_I and P_{II} (and consequently of G_I and G_{II}) is the ILMMS. The other two require finite-element analysis to calculate G_I and G_{II} values from a load applied in only one direction. In order to test different P_I to P_{II} ratios, a hydraulic actuator force is applied statically in mode II, and the specimen is loaded monotonically or cyclically in mode I until failure. Subsequently, the static force is applied in mode I while the specimen is loaded monotonically or cyclically in mode II until failure.

The ILMMS geometry was originally proposed by Ripling *et al*. (Ref 45). The original dimensions and geometry proposed in this work were altered substantially by Sancaktar *et al*. on the basis of experimental stress analysis through the use of reflection and

transmission photoelasticity and finite-element analysis in order to minimize any interference between the two modes (that is, cleavage loads resulting from shear loads, and so on) and to reduce the effects of stress discontinuities. This new geometry allows the use of a small hydraulic actuator inside the specimen.

Initially, an attempt was made to develop a single, modified ILMMS geometry that could be used for both loading cases: monotonic loading in opening, static in shear, and monotonic in shear, static in opening. This first attempt proved unworkable because the two sets of circular loading holes and the larger size of the square static load cut out that was needed to fit the hydraulic piston in either of two loading positions reduced the section modulus of the specimen so much that yielding invariably occurred in the steel when high shear loads were applied. This problem led to the development of two separate specimen geometries. Geometry A, shown at left in Fig. 15, is used for static loading in mode II, and geometry B (at right) is used for static loading in mode I. Examination of Fig. 15

reveals the presence of frictional resistance affecting the loads applied orthogonal to the static loads. These frictional loads were measured experimentally and subtracted from the applied (monotonic or cyclic) load. During the experimental program, all of the cyclic load-dominant mixed-mode fracture tests were performed on an Instron 1331 servo-hydraulic testing machine. The static shear load was applied in the mode II direction through a piston, and the cyclic load was applied in the mode I direction using sine wave load control with 10 Hz frequency and a load ratio where $R = 0.1$. These load ratio and frequency values allowed accelerated testing of the samples in the cyclic mode.

Each level of the static shear load used during the experimental program was coupled by a number of cyclic opening loads in the mode I direction. Five static load levels (between 0 and 17.8 kN, or 0 and 4000 lbf) were applied in the mode II direction. For each static load level, the onset threshold strain-energy release rate range value (G_{th}) was determined first. The G_{th} value was defined such that the specimen should fail in 10^6 cycles under the lowest P_{Imax} level. If the specimen did not fail in 10^6 cycles, the P_{Imax} level was increased by 4.45 kN (1000 lbf) at one time until the onset threshold was reached. After attaining the onset threshold, the P_{Imax} was increased by 4.45 kN (1000 lbf) at one time until the specimen failed in a very short time. This procedure was repeated for five different static load levels for both Metlbond 1113-2 and Metlbond 1113 adhesives.

Dynamic Measurement of Rolling Friction Force. It was found that if a piston was directly inserted in the specimen to apply the static shear load, as in the case of monotonic load-dominant fracture, the fatigue life of the specimen did not correspond to the applied cyclic opening load, even when the sliding friction force was deleted from the cyclic opening load. This is because the direct contact between the piston and the specimen surfaces increased the stiffness of the system considerably. To overcome this difficulty, a pin was placed to roll on two blades that were already lubricated with polytetrafluoroethylene to reduce the rolling friction force between the surfaces of the piston and the specimen.

The procedure for measuring the rolling friction force is to first apply a certain level of cyclic load P_{Imax_1} at a no-shear load and record the corresponding COD amplitude. Second, apply a certain level of static shear load P_{II}, after which the amplitude of COD will decrease because of the increase of rolling friction force. Third, increase the cyclic opening load to P_{Imax_2} until the amplitude of COD returns to its original level. Finally, the rolling friction force, P_f, can be calculated as $P_f = P_{Imax_2} - P_{Imax_1}$.

Measurement of Crack Closure Load by COD Method. Sancaktar *et al*. used a sim-

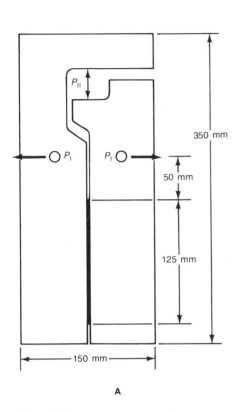

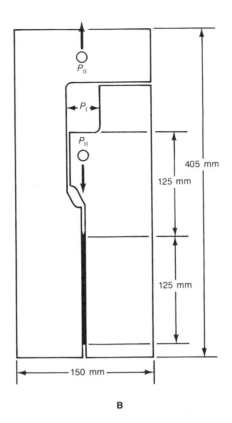

A

B

Fig. 15 Independently loaded mixed-mode specimen geometries A and B

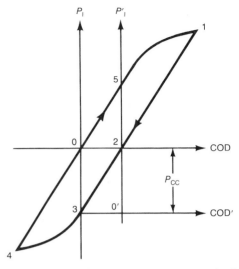

Fig. 16 The measurement of crack closure load based on the hysteresis phenomenon

ple method based on the hysteresis curve of P_I versus COD to measure the crack closure load. This method, depicted in Fig. 16, is based on elastic-plastic adhesive material behavior assumption.

The measured results revealed that the crack closure load is relatively small compared to the $P_{I min}$ applied.

This is attributed to the relatively brittle material behavior of the adhesive and to the plane-strain conditions of the specimen, resulting in a small plastic deformation zone. The relation between the adhesive layer thickness, the crack tip plastic deformation zone size, and the crack closure load was discussed earlier in the section "Crack Closure."

Establishment of Plane-Strain Conditions. Researchers, including this author (Ref 24-26), have established that tough adhesives, that is, adhesives that develop a zone of plastic deformation with a radius, r_{yc}, at the tip of an advancing crack, have an optimum bond thickness, n, where G_{Ic} is a maximum. This optimum thickness was found to be just slightly greater than $2r_{yc}$.

Using bulk tensile and fracture toughness tests, Sancaktar *et al.* (Ref 25) determined that for the cure conditions used, Metlbond 1113-2 adhesive (without carrier cloth) has a value of $r_{yc} = 0.219$ mm (0.00863 in.). This value agreed fairly well with results from experiments run on ILMM specimens. For specimen widths of both 6.4 and 12.7 mm

(0.25 and 0.50 in.), G_{Ic} is maximized at $n = 6.3$ mm (0.025 in.). This corresponds to a plane-strain condition at the crack front. For Metlbond 1113 adhesive (with carrier cloth) the G_{Ic} versus bond thickness behavior reveals that stable condition is achieved at $n = 0.18$ mm (0.007 in.), much below that without the carrier cloth. This demonstrates the stabilizing effect of carrier cloth on G_{Ic} values. It also should be noted that this optimum bond thickness value agrees very well with the previously calculated $2r_{yc}$ value for 1113 obtained with bulk specimens under identical cure conditions (Ref 25).

As far as the effects of specimen thickness (denoted by B in Eq 3, 15) are concerned, the experimental evidence indicates that for 1113-2, G_{Ic} is maximized when the adherend thickness (that is, crack front length) is equal to 6.4 mm (0.25 in.). As the specimen thickness increases to 13 mm (0.50 in.), a state of plane-strain is reached and the value for G_{Ic} then remains constant, even with further increases in thickness.

The results are similar for the adhesive with the carrier cloth (Metlbond 1113). In this case, the G_{Ic} value maximizes at 3.2 mm (0.125 in.) and achieves a plane-strain condition at 9.5 mm (0.375 in.)

The maximum G_{Ic} values calculated in these experiments for both adhesives closely match those calculated by Sancaktar *et al.* earlier (Ref 25), using bulk edge-crack compact tension specimens and by O'Con-

ner (Ref 22), using tapered double cantilever beam specimens, where both used the same adhesives.

Results

Saw Tooth Phenomenon on the Fracture Surface. In the cyclic load-dominant mixed-mode fracture tests, it was found that the main crack in the adhesive bond usually propagated near an adhesive-adherend interface. It was also found that many small cracks with similar angles branched from the main crack and approached the same interface, resulting in a sawtooth appearance of the fracture surfaces. One hypothetical explanation of this phenomenon is offered below based on the maximum principal stress criterion.

An investigation by Wang shows that the stress field very close to the crack tip in the double cantilever beam type of adhesively bonded specimen is similar to that in monolithic specimens (Ref 46). Therefore, with the understanding that the analysis is not very accurate because of the constraints imposed by the rigid adherends, the results obtained for monolithic specimens are introduced here. If a monolithic plate containing a central crack is loaded in combined mode I and mode II directions, the relation between the angle (θ_m) corresponding to the maximum principal stress direction and the ratio of the stress-intensity factors K_I and K_{II} in modes I and II can be plotted as shown in Fig. 17(a) (Ref 21). It should be noted that the $\theta_m = -70.53°$ value calculated theoretically for the condition $K_I = 0$ varies by 26% from the $\theta_m = -56°$ value determined experimentally using monolithic plates. Also shown in Fig. 17(a) are the experimental data for mixed-mode fracture (solid points 1 through 7), which reveal a variation from 0° to approximately 37° in the direction of crack propagation as the mag-

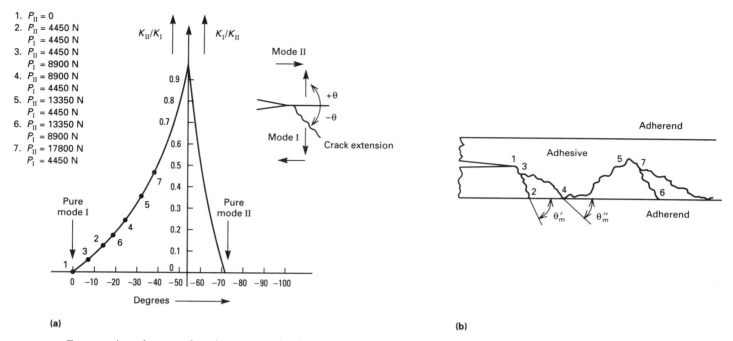

Fig. 17 The sawtooth postfracture surface phenomenon analysis based on the maximum principal stress criterion. (a) Crack extension angle predicted by the maximum principal stress criterion. (b) The sawtooth crack propagation on the fracture surfaces

nitude of P_{II} (K_{II}) is increased. Although what is shown in Fig. 17(a) is only a theoretical prediction based on monolithic plate theory, it has been confirmed qualitatively by visual observations on the postfracture surfaces of bonded samples.

As shown in Fig. 17(b), for bonded specimens under mixed-mode loading, the crack should start propagating at an angle approximately equal to what is predicted in Fig. 17(a). The crack branch from point 1 to point 2 is created in this fashion, but as soon as the crack reaches the interface at point 2, it has to change the propagation direction because of the constraint of the adherend. It is assumed that point 3 is weaker than point 2 because of stress concentration. Consequently, a new crack will initiate at point 3 and propagate toward point 4 in a direction θ_m'', close to the θ_m' forming the sawtooth configuration described by points 2, 3, and 4. After the crack reaches point 4 on the interface, it might propagate along the interface for a while and then return to the adhesive bond at point 5 because of defects such as flaws and voids inside the bond. When the procedure that occurred at point 1 repeats at point 5, another sawtooth is formed.

Figure 17(b) shows a main crack between points 1, 3, 4, 5, and 7 and branched cracks at points 3 and 2 and at 7 and 6. The branched cracks have angles close to the corresponding θ_m. With different combinations of P_I and P_{II} or different ratios of K_I and K_{II}, the branch cracks have different angles (θ_m), as shown in Fig. 17(a). The possibility of crack branching is cited here because of consistent observations of crack tip crazing, cavitation, and microcracking

mechanisms, which occur in bulk samples. Furthermore, the presence of voids and flaws in adhesive bonds is a well-documented phenomenon. However, it should also be noted that a finite-element analysis and the related experimental work on mixed-mode crack propagation (in CLS specimens) under cyclic loading revealed that G_I was maximized when the crack propagated near an interface (Ref 40). On the other hand, this analysis did not account for the presence of any microcracks that might propagate simultaneously or immediately after the propagation of the main crack. It should further be noted that the discussion presented here on the prediction of crack propagation direction is based on the maximum principal stress criterion. The relationship between the adhesive material behavior and the corresponding fracture criterion will be discussed later in the sections "Ductile Versus Brittle Behavior" and "Fatigue Fracture Criteria, Threshold Levels" in this article.

Experimentally, the crack propagation mechanism described above was verified visually by observing the post-fracture surfaces of the specimens. Under pure opening mode, the failure surfaces exhibited uniform adhesive distribution with relatively smooth surface texture. Under pure shear or high shear loading conditions, however, the adhesive post-fracture surface has a jagged, curled up condition. The amount of "peel up," or surface roughness, found on mixed-mode failure surfaces was proportional to the P_{II} to P_I ratio. This condition supports the hypothesis associated with the maximum principal stress failure criterion, in that adhesive joints under a global biaxial

load actually fail in mode I locally except that with higher P_{II} to P_I ratios, toughened and brittle adhesives will produce a larger amount of fractured surface area. A possible method for quantifying adhesive deformation and fracture processes by using a Talysurf profilometer on the post-fracture surfaces has been described in Ref 18.

Crack Propagation Under Mixed-Mode Cyclic Loading. To determine the effect of static shear load on the cyclic dominant mixed-mode fracture, five different static shear load levels were applied to the ILMM specimens. Each level of static shear load was coupled with a number of cyclic opening load levels. A computer program subroutine ZXSSQ from the software package IMSL available in the UNIX system at Clarkson University was used to determine the coefficients in Eq 10 and 18 by fitting the experimental data. ZXSSQ is a finite-difference Levenberg-Marquardt routine for solving nonlinear least-square problems.

Because the first model (Eq 10) does not contain the strain-energy release rate G_{II} due to the static shear load P_{II}, it should be fitted with the steady state (da/dN) versus G values corresponding to a variety of P_Is for each P_{II} level. Therefore, it governs crack growth rate (da/dN) versus G for one level of P_{II} coupled with different P_I levels. On the other hand, the second model (Eq 18) contains one more parameter, $G_{Tmax} = G_{Imax} + G_{II}$, which involves the strain-energy release rates due to P_{Imax} and P_{II}. Hence, it can be fitted to all (da/dN) versus G_{Tmax} and G values corresponding to different P_Is for all five P_{II} levels. Consequently, it governs the crack growth rate under different P_{Imax} and P_{II} combinations.

Table 1 Material parameters for the crack growth model $da/dN = c(\Delta G_I)^n$ where ΔG_I is in units of in. $\cdot$ lbf/in.^{2}

P_{II}	Adhesive without carrier cloth		Adhesive with carrier cloth	
	c	n	c	n
0..	6.71×10^{-5}	4.2381	5.08×10^{-4}	2.36
At 4.4 kN (1000 lbf)...................	2.44×10^{-4}	3.507	6.15×10^{-7}	6.785
At 8.9 kN (2000 lbf)...................	3.49×10^{-4}	3.457	1.95×10^{-5}	4.357
At 13.3 kN (3000 lbf).................	1.14×10^{-3}	2.612	2.41×10^{-5}	4.2689
At 17.8 kN (4000 lbf).................	5.68×10^{-4}	3.3672	9.123×10^{-4}	2.365

(a) Metlbond 1113-2. (b) Metlbond 1113

Obviously, the first model requires five different values for the parameters c and n to describe the crack growth rates for five different P_{II} levels coupled with a variety of P_Is. The second model, however, requires only one set of the parameters c, m, and n to describe all the data and therefore may be considered more efficient for data fitting and representation purposes. Obviously, because the sums of the square of errors for the two models are expected to be different, efficiency in data representation is not necessarily expected to provide accuracy also.

It was found that, for most specimens, steady-state crack propagation prevailed when crack length was less than 76.2 mm (3 in.) (Fig. 15), and catastrophic failures occurred when the crack propagated to or beyond this length. Therefore, the 76.2 mm (3 in.) crack length was taken as the threshold level for catastrophic crack propagation.

The material parameters c and n of Eq 10 for the model with a carrier cloth and without a carrier cloth are shown in Table 1. The parameters c, m, and n for the second model (that is, Eq 18) are 4.77×10^{-5}, 1.2533, and 3.6812, respectively, in the case of Metlbond 1113-2, and 3.65×10^{-5}, 0.77173, and 3.19886, in the case of Metlbond 1113 if the G values are represented in units of in. $\cdot$ lbf/in.^{2}. For the adhesive without the carrier cloth, the first model provides a better fit because the average error with the first model ($\sim 2.32 \times 10^{-2}$) is about half of that for the second model. However, for the adhesive with the carrier cloth, the average error is about the same for both the first and second models (2.48×10^{-2} and 2.68×10^{-2}, respectively).

Material Behavior

The material behavior of the adhesive and its effects on the fatigue and fracture behavior of the bonded joint are described in this section. Discussions on the effects of environmental conditions on the material behavior of adhesives are included, and the viscoelastic and microstructural behavior of adhesives are considered in detail.

Viscoelasticity Considerations

Previous investigations by Sancaktar *et al.* of the high-temperature constant strain rate and creep behavior of FM 73, FM 300 (American Cyanamid Company) and thermoplastic polyimide sulfone (National Aeronautics and Space Administration) structural adhesives revealed drastic reductions in strength properties at high temperatures (Ref 47, 48, 49). Changes in thermal environments from 21 to 149 °C (70 to 300 °F) resulted in reductions in tensile strength by as much as 87% and increases in tensile strains by as much as 100%. Among the three adhesives studied, the thermoplastic polyimide sulfone, which had the best strength retention, still exhibited a reduction in ultimate shear strength of approximately 44% in the bonded form from its room-temperature strength value. The creep behavior of the same adhesive was also affected considerably by the environmental temperatures, with faster creep rates resulting in delayed failures at lower shear levels. At high testing temperatures (232 °C, or 450 °F) interfacial separation between the carrier cloth and the adhesive material was also observed with the polyimide sulfone adhesive (Ref 48). In fatigue loading, the viscoelastic behavior of the adherends should also be taken into account because of the possibility of energy dissipation there.

Heat Dissipation and the Effect of Cyclic Frequency. Knauss has reported that in glassy polymers subjected to cyclic loading, the heat buildup in the crack tip region cannot be neglected (Ref 50). He predicted that glassy polymers, as well as filled elastomers, can soften extensively because of the energy dissipated during cyclic loading. He predicted further that the cycle frequency should affect the crack propagation rate as the glass transition temperature (T_g) of the polymer is approached. However, recent investigations on bonded structural adhesives using cyclic frequencies of 40 Hz (Ref 37) to 66.7 Hz (Ref 51) revealed that the temperature rise in the specimens was negligible. For example, Jablonski (Ref 37) reports a temperature increase of about 0.02 °C (0.04 °F) at the crack tip for tapered double cantilever beam specimens bonded with EA 9649 adhesive without a carrier.

Application of Time-Temperature Superposition to Fracture Energy. Recent work by Hunston *et al.* (Ref 52) showed that it is possible to apply time-temperature su-

perposition techniques to help characterize the fracture behavior of epoxy adhesives using a relation of the form:

$$G = G_s + Zt_f^r e^{-\Delta Er/RT} \qquad \text{(Eq 19)}$$

where G_s is the minimum value of G at low temperatures and high speeds and Z, r, and ΔE characterize the magnitude, rate dependence, and temperature dependence of G. The parameter t_f represents the time necessary for the load to increase from 0 to the value required for the onset of crack growth.

Because of their polymeric structure, most adhesives exhibit some degree of viscoelastic response. Constitutive modeling of viscoelastic structural adhesives has been discussed in detail by Sancaktar (Ref 53).

An analysis of the fatigue crack growth in viscoelastic materials with possible application to tread cracking in auto tires has been given by Williams (Ref 54). For this purpose, Williams obtained a criterion for spherical flaw instability and predicted a growth rate cycle similar to that seen in metals, where a steady-state growth rate per cycle is achieved at long times.

In many structural adhesives, the addition of secondary phases, such as rubber tougheners, carrier cloth, and metal powders, usually suppresses viscoelastic behavior when another failure mechanism becomes dominant (especially in the presence of cracks). Examples of such failure mechanisms imposed by the presence of a secondary phase are plastic deformation in the form of stress whitening and interfacial separation.

Material Microstructure Considerations

Most polymers used for bonding and protection functions contain secondary phases for the purposes of increasing toughness or strength, matching the coefficient of thermal expansion to the substrate, modifying the elastic modulus or the electrical and magnetic properties, improving the handling and manufacturing conditions, and thickness control, among others. In these cases, the secondary phase may either be a dispersed phase, such as randomly dispersed particles or fibers, or a continuous and orderly structure, such as a glass carrier cloth. In either case, loading applied to the polymer leads to stress concentrations at the interphase regions between the polymer matrix and the secondary phase. Such stress concentrations may result in the debonding of the polymer from the secondary phase and may also lead to the plastic deformation of the polymer in these regions in the form of cavitation, shear banding (to connect the cavitated regions), and microcracking. The presence of macroscopic discontinuities, such as drilled holes, facilitates this internal failure mechanism locally.

The Stress-Whitening Phenomenon.
Previous investigations by Sancaktar *et al.*
on Metlbond 1113 and 1113-2 rubber-toughened epoxy adhesives (with and without a carrier cloth, respectively) revealed that stress whitening takes place prior to complete failure. This observation was made on fracture samples tested for bulk tensile properties (Ref 55), bulk shear properties (Ref 56), bulk fracture energy properties (Ref 57), and bonded properties (Ref 25). Similar observations on the same model adhesive were reported earlier by Brinson *et al.* (Ref 58, 59). They observed that the stress-whitening pattern was very similar to the crazes that formed prior to the maximum stress during a tensile test on polystyrene (Ref 60). Brinson *et al.* reported that the "crazes" they observed occurred at various angles to the loading direction. These angles ranged from approximately 45 to 90° with respect to the loading axis. A considerable number of these angles were similar (Ref 59) to the angle (54.1° to the loading axes) at which slip bands occur in many metals in uniaxial tension (Ref 61). Brinson *et al.* reported that the crazing process they observed in Metlbond adhesives was evidence of local damage or of a failure mechanism that occurred well in advance of gross fracture. They also stated that the crazed areas of the model adhesives were not true cracks, but areas of lower density.

In microscopic scale, stress whitening in rubber-modified epoxies is caused by cavitation, voiding, and rupture of the rubbery phase due to the high triaxial stress state that exists at the dispersed rubber particles and their close proximity. This local failure process is followed by coalescence of the formed cavities through shear banding and usually results in gross failure. Stress whitening is an energy-dissipative mechanism due to the creation of new surfaces during the cavitation process. For this reason, the phase separation and/or dispersion of the rubbery phase (which can be monitored in the cure process) serves to toughen the brittle epoxy matrix. It should be noted that although cavitation occurs as a result of a triaxial state of stress in a microscopic scale, the global bulk stress-whitening stress (σ^*) value that corresponds to the occurrence of local failure events in the specimen is defined and measured as a macroscopic uniaxial stress on a bulk specimen.

Yee *et al.* (Ref 62, 63) conducted mechanical and microscopy studies on the toughening mechanisms in rubber-modified epoxies. They performed microscopic investigations on bulk tensile and three-point-bend single-edge-notched specimens in order to study the nature of the stress whitening of a number of epoxies with varying amounts of rubber content. They found that in the absence of rubber, no stress

whitening was observed during the deformation and fracture of their epoxy resins. They also found that as the rubber content increased, the size of the crack tip whitening, as well as the fracture energy (G_{Ic}), increased. Through scanning electron microscopy of the whitened zone, they found that rubber particles and the matrix had cavitated and that the size of these cavities was larger than the diameters of the undeformed rubber particles as determined by transmission electron microscopy. They also reported that the stress-whitening effect visible to the naked eye is simply due to the scattering of light from these cavitated regions. They maintained that the energy dissipation, crack tip blunting capability, and/or toughening mechanisms of these materials is facilitated by the fracture of the rubber phase at the walls of the cavities in stress-whitened zones.

Kinloch (Ref 64) reported that one of the most successful methods of increasing the toughness of thermosetting adhesives is to incorporate a second phase of dispersed rubbery particles into the cross-linked polymer. He states that the deformation processes that dissipate energy in the vicinity of a crack tip in rubber-toughened epoxies are, first, cavitation in the rubber or at the rubber/matrix interface and, second, plastic shear yielding that is localized in the bulk of the epoxy matrix. Kinloch concludes that it is this localized cavitation process that gives rise to the stress whitening that often accompanies crack growth. Figure 18(a) shows the toughening mechanism described by Kinloch, and Fig. 18(b) shows the time-temperature-transformation (TTT) isothermal cure diagram for the phase-separating epoxy system. The TTT diagram reveals that cure history controls the degree of dispersion and/or phase separation of the rubber-epoxy system, which most likely results in varying degrees of stress whitening.

Visual and microscopic evidence strongly suggests that the plastic deformation zone at the tip of a crack encompasses a stress-whitened zone. An exact correlation between the plastic zone diameter and the size and shape of the stress-whitened zone is not yet available. Because Irwin's analysis (Ref 23) is based on a small-scale yielding assumption and does not accurately describe the shape of the plastic deformation zone, the presence of a circular deformation zone should be considered hypothetical only. In fact, mathematical functions describing the exact shape of crack-tip plastic deformation zones for monolithic samples are available in the literature (Ref 65, 66).

During previous bulk fracture experiments on Metlbond adhesives (Ref 67, 68), crack tip stress whitening on the specimen surfaces perpendicular to the fracture plane was observed to occur prior to rapid crack growth. It also was observed that as the

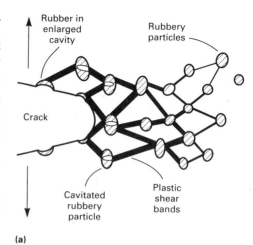

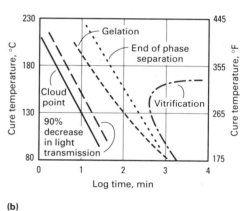

Fig. 18 Analysis of the rubber-toughening mechanism in epoxy systems. (a) The toughening mechanism. (b) Time-temperature-transformation isothermal cure for the phase-separating diglycidyl ether of bisphenol A epoxy system (Ref 64)

load increased, a stress-whitened zone developed at the crack tip up to a certain extent, beyond which the specimen failed rapidly. The specimens that had a carrier cloth exhibited larger areas of crack tip stress-whitening zones (CTSWZ). It also was observed that the specimens with a higher degree of CTSWZ had rougher fracture surfaces, indicating a more stable crack growth, whereas those exhibiting little or no CTSWZ (low cure temperature) had very smooth fracture surfaces. The presence of voids in the highly stressed regions of the adhesive also seemed to increase the degree of stress whitening there.

The formation of voids during the thermosetting cure process of structural adhesives was discussed in some detail by Bikerman (Ref 69). He maintained that the flaws form in the adhesive layers of bonded joints (and, consequently, in composite materials) as a result of the cure process for a number of reasons: physical changes resulting in unfavorable stress concentrations in the form of voids due to shrinkage effects associated with the curing of thermosetting adhesives; physical and chemical changes resulting in weak boundary layers between the adhesive and adherend and also in the formation of weak spots in the bulk of the adhesive film; and trapped air bubbles initially present at the interface of adhesive-adherend or adhesive-adhesive layers. It should be noted that some voids may also be present in individual layers of a solid film adhesive prior to the final cure of a multi-layered specimen.

Modeling of the Stress-Whitening Phenomenon. Brinson et al. (Ref 58, 59) reported that the adhesive properties for the rubber-toughened Metlbond adhesives were different before and after stress whitening due to changes in material behavior and that bilinear equations were therefore needed to describe this behavior. Specifically, they determined that the Poisson's ratio decreased considerably once the stress whitening had occurred. This reduction was as much as 42% (when the Poisson's ratio decreased from 0.388 to 0.226). When they performed loading-unloading experiments on the Metlbond adhesives, they found that beyond the elastic limit stress, unloading resulted in higher levels of plastic strains as the initial loading stresses were increased. They determined that the most suitable model that would fit the stress-strain response of the model adhesives with prediction of the stress-whitening stress level σ^* was a modified version of the Ramberg-Osgood equation (Ref 70), which they called the bilinear RAMOD-2 equation. The bilinear behavior was obtained when log ϵ_p was plotted against log $(\sigma-\theta)$. The plastic strain (ϵ_p) was assumed to be a function of the overstress above the elastic limit stress (θ), and the stress levels defining the intersection point for the bilinear behavior were found to occur slightly below the stress-whitening stress values. Thus, the bilinear RAMOD-2 has the form:

$$\epsilon = \frac{\sigma}{E} \qquad\qquad 0 \leq \sigma \leq \theta$$

$$\epsilon = \left(\frac{\sigma}{E}\right) + K_1 (\sigma - \theta)^{n_1} \quad \theta \leq \sigma \leq \sigma^* \quad \text{(Eq 20)}$$

$$\epsilon = \left(\frac{\sigma}{E}\right) + K_2 (\sigma - \theta)^{n_2} \quad \sigma^* \leq \sigma \leq \sigma_{max}$$

where K_1, n_1, K_2, and n_2 are the material constants determined empirically from a bilinear log-log plot of ϵ_p versus $(\sigma-\theta)$. In the bilinear RAMOD-2 model, σ^* is defined as the stress at the intersection of the bilinear plots of log ϵ_p versus log $(\sigma-\theta)$.

Subsequent research by Sancaktar et al. (Ref 67, 68) on the same phenomenon confirmed the findings of Brinson et al. Furthermore, it was also shown that the material behavior represented by the parameter $1/n_1$ is similar to that represented by the tensile strength (σ_y) of the bulk material (Ref 67, 68). In other words, a higher $1/n_1$ would indicate a higher extent of cure, less

shrinkage stresses, less degradation, less softening, and consequently more stable material behavior prior to stress whitening. The material behavior represented by the parameter $1/n_2$, however, is similar to that represented by the fracture toughness (K_{1c}) of the bulk material. Therefore, $1/n_2$ may be related to the softening after stress whitening up to failure. Higher $1/n_2$ values would indicate higher resistance to complete failure and resistance to the propagation of microcracks that eventually result in total failure.

This author believes that the use of σ^* rather than σ_y in Eq 11 is more appropriate for modified structural adhesives. The reason for this suggestion is that σ^* represents local failure in the bulk material more closely than does σ_y. Furthermore, the material behavior after σ^* has been reached (locally), represented by the parameter $1/n_2$, is of great importance in relating local failures to catastrophic failures in flawed structures, as explained above.

Effect of Cure Conditions. Sancaktar et al. also showed that the cure conditions of temperature and time considerably affect the Ramberg-Osgood exponent n_1 and the stress-whitening parameters σ^* and $1/n_2$ (Ref 67, 68). The stress-whitening stress (σ^*), the inverse Ramberg-Osgood exponent $(1/n_1)$, and the tensile strength (σ_y) all exhibit similar, bell-shaped increasing/decreasing behavior with respect to cure temperature, with peaks in the 99 to 110 °C (210 to 230 °F) range, which coincides well with the T_g for the model adhesive.

The material behavior represented by the parameter $1/n_2$ is shown to be similar to that represented by the fracture toughness (K_{1c}) of the adhesive material. Both $1/n_2$ and K_{1c} exhibit decreasing behavior values when the cure temperature is increased.

Notch Blunting. Woan et al. claimed that any possible improvements in fatigue behavior that may be obtained by using rubber tougheners may depend on whether any stress concentrations (notches) are already present in the structure (Ref 71). Their experiments showed that unnotched fatigue samples made of rubber-toughened (high-impact) polystyrene, with a high volume of rubber phase (approximately 30%) grafted to the matrix, have less fatigue resistance than the unmodified resin. These results were obtained regardless of whether the cyclic loading mode was fluctuating with nonzero mean stress or reversed (that is, zero mean stress). They cited another reference that reported lower fatigue resistance for a glassy polymer and nitrile elastomer blend than for its unmodified version. Woan et al. claimed that "in rubber modified polymers subject to alternating loads, craze and crack initiation occur sooner than in unmodified polymers and appear to have a greater influence on fatigue life than a possibly reduced fatigue crack propagation

rate due to presence of the dispersed rubber phase" (Ref 71). They also reported that the resistance to fatigue fracture of glassy polymers increases substantially when molecular weight is increased.

Effect of Metal Powder Addition. In a 1989 investigation by Davies and Kilik, 75 μm (3 mil) mesh size copper and aluminum powders were added to a toughened, single-part epoxy (Permabond ESP 110) to improve thermal conductivity in metal cutting tool applications (Ref 51). The concentrations of these powder additives were 1 vol% (copper) and 5 vol% (aluminum). The fatigue samples were tested at 66.7 Hz, and no appreciable increase of specimen temperature was observed during experimentation. Davies et al. reported that "for all conditions the higest value of cycles to failure was obtained with 1 vol% copper powder addition and the lowest values with 5 vol% aluminum powder addition. Additions of 1 vol% of either powder gave higher values than the unfilled adhesive, whereas a 5 vol% addition of either powder generally gave lower values than the unfilled adhesive, except for copper powder additions at high stress levels" (Ref 51). They attributed the shorter fatigue life in aluminum-filled fatigue samples to the stress-raising effect of the irregular-shape aluminum powder in the adhesive. They also claimed a crack-stopping effect occurs by the addition of metal powder at low volumetric concentrations.

Effect of the Carrier Cloth. The inclusion of a synthetic carrier cloth (most often a solid film) in a structural adhesive obviously constitutes a secondary phase. The recent work by Sancaktar et al. presented in this article in the section "Case Study" revealed that the adhesive with the carrier cloth (Metlbond 1113) has slower crack propagation than the adhesive without the carrier cloth (Metlbond 1113-2). This conclusion was based on the comparison of the parameters c, m, and n of Eq 18. Results reported in the open literature on the effects of carrier cloth on the cyclic behavior of bonded joints seem to vary. For example, comparing FM 300 adhesive (with a carrier) to EC 3445 (without a carrier), Johnson and Mall reported that the adhesive with the carrier has superior debond resistance (Ref 34). A later paper by Mall et al., however, reported the opposite result (Ref 72), but attributed the observed inferior debond resistance to weak bonds between the mat carrier and the adhesive.

Jablonski's results in a comparison of AF 163 (with a carrier) and EA 9649 (without a carrier) revealed faster crack propagation in the adhesive with a carrier (Ref 37). Fractographic studies by Jablonski showed that debonding at the scrim-adhesive interface occurred progressively by fatigue cycling. At the scrim-adhesive interface, Jablonski observed tear ridges parallel to the scrim fibers, confirming progressive debonding.

Table 2 Onset threshold values for Metlbond 1113-2

P_{Imax} kN	10^3 lbf	P_{II} kN	10^3 lbf	G_{Tmax} kJ/m²	in. · lbf/in.²	G_{Imax} kJ/m²	in. · lbf/in.²	G_{II} kJ/m²	in. · lbf/in.²	ΔG_I kJ/m²	in. · lbf/in.²	N_c cycle
6.23	1.40	0	0	402.9	2.301	402.9	2.301	0	0	398.9	2.278	40 142
6.23	1.40	4.45	1.00	405.7	2.317	402.9	2.301	2.751	0.016	398.9	2.278	181 022
5.79	1.30	8.90	2.00	385.5	2.047	374.4	1.984	11.07	0.063	343.9	1.964	817 273
4.90	1.10	13.4	3.00	273.7	1.563	248.7	1.420	25.04	0.143	246.2	1.41	84 874
3.56	0.80	17.8	4.00	175.9	1.00	131.5	0.751	44.31	0.253	130.3	0.74	462 256

Table 3 Onset threshold values for Metlbond 1113

P_{Imax} kN	10^3 lbf	P_{II} kN	10^3 lbf	G_{Tmax} kJ/m²	in. · lbf/in.²	G_{Imax} kJ/m²	in. · lbf/in.²	G_{II} kJ/m²	in. · lbf/in.²	ΔG_I kJ/m²	in. · lbf/in.²	N_c cycle
6.23	1.4	0	0	402.95	2.301	402.95	2.301	0	0	398.9	2.278	569 002
6.23	1.4	4.45	1.00	405.7	2.317	402.95	2.301	2.751	0.016	398.9	2.278	511 813
6.23	1.4	8.90	2.00	414.0	2.364	402.95	2.301	11.06	0.063	398.9	2.278	426 616
5.79	1.3	13.35	3.00	372.3	2.126	347.44	1.984	24.87	0.142	343.9	1.964	897 389
5.79	1.3	17.80	4.00	391.7	2.237	347.44	1.984	44.31	0.253	343.9	1.964	277 469

The cracks, which primarily occurred at fiber bend locations of the scrim, were a few microns deep and appeared to have formed as a result of bending. Jablonski also observed surface fatigue cracks in the scrim cloth attached to both fracture surfaces, indicating that the scrim itself carried fatigue loading. These results provide some clarification regarding variations observed in the effects of carrier cloth in fatigue behavior. It is known that adhesives that contain a carrier strain less at comparable stress levels (Ref 16, 17, 58, 59). Consequently, crack opening displacements are expected to be smaller in the presence of a scrim carrier cloth. Therefore, based on the crack opening displacement criterion, which states that crack extension takes place when the material at the crack tip reaches a maximum permissible strain, crack extensions are hindered at comparable stress levels when a carrier is present. In such cases the presence of a carrier increases the fatigue resistance of the bond. However, when carrier/adhesive debonding takes place in a progressive manner because of weak interphases, such improvements may be surpassed by the weakening of the bond during cyclic loading.

Ductile Versus Brittle Behavior. Another important material feature that affects the fatigue behavior of bonded joints is the duc-tility of the adhesive. Mall *et al.* reported that "the relative fatigue resistance and threshold value of cyclic debond growth in terms of (normalized w.r.t.) its static fracture strength is higher in the brittle adhesive than its counterpart in the ductile adhesive" (Ref 72). The dominant effect, however, is the mode of failure, that is, shear versus tensile, which determines the amount of energy that can be absorbed and the type of microdamage produced during the loading-unloading cycle. For example, Luckyram and Vardy reported that under cyclic loading, an adhesive (one part, heat curing) that fails mainly by shear yielding (no crazing) has higher crack initiation and propagation behavior than a two-part cold-cure adhesive that fails in the brittle mode with crazing at the crack tip (Ref 73). This is true in spite of the much greater fracture toughness (under monotonic loading) of the one-part adhesive compared to the two-part adhesive. They also reported that neither adhesive exhibits any dependence on cycle frequency within the range of 0.5 to 5.0 Hz.

Fatigue Fracture Criteria, Threshold Levels

The case study discussed earlier is further described here. The onset threshold values for the Metlbond 1113-2 and 1113 adhesives are shown in Tables 2 and 3, respectively. Examination of these tables reveals that for both adhesives, the G_{Imax} values at onset threshold decrease with increasing static shear load values in mode II. Tables 4 and 5 show the catastrophic threshold values for 1113-2 and 1113 adhesives, respectively. Again, an increase in mode II loads results in a general decrease in ΔG_I and the number of cycles at catastrophic crack propagation (N_c) values.

A comparison of P_I and P_{II} values corresponding to catastrophic crack propagation under cyclic loading and the same values under monotonic loading (Ref 25) revealed that catastrophic crack propagation occurs under lower P_I-P_{II} load combinations when the specimen is subjected to cyclic loading. In other words, P_{Ic} and P_{IIc} values are reduced as a result of cyclic load application.

It should also be noted that for monotonic mixed-mode fracture of the model adhesives, the maximum principal stress criterion provided a better correlation for predicting failure. This criterion predicts failure in mode I under the action of the maximum principal stress because a biaxial state of stress is applied globally. The mode I in such cases will be at an angle to the applied P_I load, with this angle being determined by the applied P_I-to-P_{II} ratio. Obviously, such an inclined crack will have to connect with similar ones or it will simply continue propagating in the P_{II} direction (possibly close to or at the interface), resulting in catastrophic failure. The likelihood of this type of failure is high, especially for brittle adhesives.

Sancaktar proposed the following simplified approach to describe this type of failure: If one takes a biaxial stress element well ahead of the crack tip where the stresses are defined and uniform, the elementary theory defines the maximum principal stress as:

$$\left(\frac{\sigma}{2}\right) + \left[\left(\frac{\sigma^2}{4}\right) + \tau^2\right]^{1/2} \qquad \text{(Eq 21)}$$

Table 4 Catastrophic threshold values for Metlbond 1113-2

P_{Imax} kN	10^3 lbf	P_{II} kN	10^3 lbf	ΔG_I kJ/m²	in. · lbf/in.²	N_c cycle
10.24	2.30	0	0	1576.6	9.0032	1618
9.35	2.10	4.45	1.00	944.7	5.3945	6388
9.35	2.10	8.90	2.00	944.7	5.3945	4536
9.35	2.10	13.4	3.00	944.7	5.3945	3692
9.35	2.10	17.8	4.00	944.7	5.3945	3623

Table 5 Catastrophic threshold values for Metlbond 1113

P_{Imax} kN	10^3 lbf	P_{II} kN	10^3 lbf	ΔG_I kJ/m²	in. · lbf/in.²	N_c cycle
10.68	2.40	0	0	1233.9	7.046	9530
10.24	2.30	4.45	1.00	1313.1	7.498	1486
10.24	2.30	8.90	2.00	985.4	5.625	2741
10.24	2.30	17.8	4.00	1076.6	6.15	1271

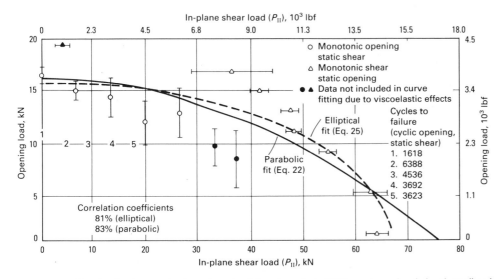

Fig. 19 Opening load versus in-plane shear load for plane-strain ILMM specimens bonded with Metlbond 1113-2 (without carrier cloth) and loaded under monotonic and cyclic loading conditions

Assuming that loads and stresses are linearly related, substitution into Eq 21 yields:

$$\Psi A P_I + B^2 P_{II}^2 = \Psi^2 \qquad \text{(Eq 22)}$$

Obviously, Eq 22 should contain at least an additional P_I^2 term resulting from the contribution of shear loading to the opening direction. However, this contribution is neglected, as it is assumed to be small.

It should also be noted that Eq 22 can be rewritten in terms of the stress-intensity factors as:

$$\left(\frac{K_I}{K_{Ic}}\right) + \left(\frac{K_{II}}{K_{IIc}}\right)^2 = 1 \qquad \text{(Eq 23)}$$

Another fracture criterion that may apply to the failure of adhesively bonded joints under monotonic loading is the energy balance criterion given by:

$$K_I^2 + K_{II}^2 = \text{constant} \qquad \text{(Eq 24)}$$

where the locus of failure is a quarter circle. However, previous experimentation by other researchers (Ref 40) showed

that an elliptical condition is more realistic, that is:

$$\left(\frac{P_I}{P_{Ic}}\right)^2 + \left(\frac{P_{II}}{P_{IIc}}\right)^2 = 1 \qquad \text{(Eq 25)}$$

One may note that Eq 22 describes a parabolic failure condition, whereas Eq 25 describes an elliptical one.

These criteria represent the upper limit for catastrophic threshold values under cyclic loading. Experimental data available in the literature, including the data obtained by Sancaktar et al., indicate substantial reductions in failure loads as a result of cyclic load application. Figures 19 and 20 show the monotonic mixed-mode fracture data (in terms of mode I and II loads) for the adhesives Metlbond 1113-2 (without a carrier) and 1113 (with a carrier), respectively, along with catastrophic threshold load levels for comparison purposes.

Cumulative Damage Under Varying Frequencies. The discussion thus far has revealed that during cyclic loading irreversible damage accumulates in the adhesive layer. For transient (nonuniform) cyclic histories, the cumulative damage model can be used to describe such damage accumulation. The simplest way to account for cumulative damage would be to define a damage fraction (D) as the fraction of life used during a particular transient cyclic event. Consequently, failure would be expected to occur when the sum of damage fractions equals unity:

$$\Sigma D_i \geq 1 \qquad \text{(Eq 26)}$$

A nonlinear damage relation between the damage fraction (D) and the cycle ratio (n_i/N_i), where n_i and N_i are the cycles applied and life expectancy at a stress (or stress amplitude) level (σ_i) would be defined as:

$$D = \left(\frac{n_i}{N_i}\right)^{F(\sigma)} \qquad \text{(Eq 27)}$$

where the exponent is a function of the stress (amplitude) level and is considered to have values between 0 and 1. When the exponent is unity, the method becomes identical to the famous Miner's rule (Ref 74). It should be noted that the method described accounts for both sequence and stress (amplitude)-level effects.

It is also possible that transient loading may accelerate the fatigue process, with the amount of damage induced being a strong function of the extent of damage already inflicted upon the sample. In such cases a rate factor may be necessary to multiply Eq 27. Obviously, such a rate factor would have to be a function of the damage accumulated (ΣD_i).

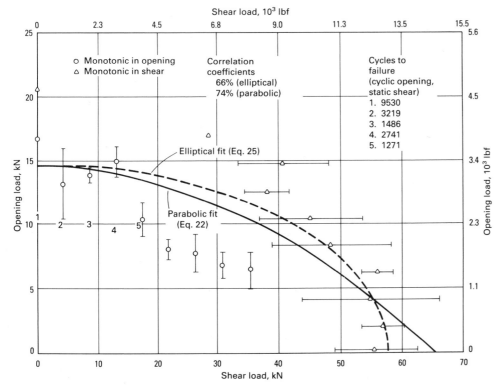

Fig. 20 Opening load versus in-plane shear load for plane-strain ILMM specimens bonded with Metlbond 1113 (with carrier cloth) and loaded under monotonic and cyclic loading conditions

Experiment Synopsis

Sancaktar's fatigue experiments with the toughened-epoxy adhesives with carrier cloth bonding (Metlbond 1113) and without carrier cloth bonding (Metlbond 1113-2) ILMM specimens led to certain conclusions:

- The strain-energy release rate range (ΔG) due to the dynamic part of the sinusoidally fluctuating load plays the most important role in cyclic load dominant mixed-mode fracture. This is shown by a comparison of the material parameters m and n, which are the exponents of G_{Tmax} and ΔG, respectively, in Eq 18. For Metlbond 1113-2, the m and n values were 1.2533 and 3.6812, and for Metlbond 1113, they were 0.7717 and 3.1989, respectively. It is obvious that the n values are more than three times the m values
- The adhesive with the carrier cloth has slower crack propagation than the adhesive without the carrier cloth. This is revealed by the smaller m and n values for the 1113 adhesive
- The G_{Imax} and ΔG values at onset and catastrophic threshold levels decrease with increasing static shear load values in mode II
- The shear loads affect the crack propagation directions within the adhesive bonds
- An increase in the mean level of the sinusoidally fluctuating loads results in shorter fatigue life
- With cyclic loading, catastrophic crack propagation occurs under lower P_{I} and P_{II} load combinations than is the case under fast monotonic loading (Fig. 19 and 20)

Life Prediction Methodology. Based on the discussions presented in this article, it is apparent that the life prediction methodology for adhesively bonded joints should include the definition of loads and environment; joint and structure geometry; and adherend and adhesive material properties, including the cure properties and interphase properties and its quality. With this input, stress analysis should be performed to obtain a through-the-thickness, point-by-point description of stresses and displacements. Information obtained in this manner can then be used in conjunction with the (experimental) information on the mode of fatigue and/or fracture failure to ensure structural integrity for the component that is to be joined by adhesive bonding (Ref 75).

ACKNOWLEDGMENT

Dr. E. Sancaktar's research work reported in the section "Case Study" in this article was supported by the National Science Foundation under Grant CEE-8317428. His other works referred to in this paper were supported by the National Science Foundation Grant CME-8007251 (the effects of cure conditions), National Aeronautical and Space Administration (NASA) Grant NAG-1-284 (viscoelastic adhesive characterization), Olin Corporation Trust Grant (prebent adherends), NASA Grant NGR-47-004-090 (rate effects on structural adhesives, Dr. H.F. Brinson, principal investigator), and Clarkson University. Thanks are also due to the numerous authors and publishers who allowed their works to be referenced and reproduced in this paper.

REFERENCES

1. N.N.S. Chen, P.I.F. Niem, and R.C. Lee, Fatigue Behaviour of Adhesive Bonded Joints, *J. Adhes.*, Vol 21, 1987, p 115-128
2. R.D. Adams and W.C. Wake, *Structural Adhesive Joints in Engineering*, Elsevier, 1984
3. M.J. Jacobson, "Sonic Fatigue of Advanced Composite Panels in Thermal Environments," AIAA-81-1699, American Institute of Aeronautics and Astronautics, 1981, p 1-9
4. W.A. Lees, *Adhesives in Engineering Design*, Springer-Verlag, 1984
5. E. Sancaktar and J. Tang, Mixed-Mode Fracture in Adhesively Bonded Joints Under Dynamic Loading, in *Proceedings of Adhesion '87*, Plastics and Rubber Institute, 1987, p 18-1, 18-7
6. E. Sancaktar, Elastoplastic Fracture Behavior of Structural Adhesives, in *Proceedings of Fifth International Joint Government-Industry Symposium on Structural Adhesive Bonding*, American Defense Preparedness Association, 1987, p 120-139
7. F. Szepe, Strength of Adhesive-Bonded Lap Joints With Respect to Change of Temperature and Fatigue, in *Proceedings of Second SESA International Congress on Experimental Mechanics*, Society for Experimental Stress Analysis, 1965, p 478-484
8. M. Goland and E. Reissner, The Stresses in Cemented Joints, *J. Appl. Mech.*, Vol 11, 1944, p A17-A27
9. L.J. Hart-Smith, Stress Analysis: A Continuum Mechanics Approach, *Developments in Adhesives—2*, A.J. Kinloch, Ed., Applied Science Publishers, 1981, p 1-44
10. D.Y. Wang, Influences of Stress Distribution on Fatigue Strength of Adhesive-Bonded Joints, *Exp. Mech.*, Vol 4, 1964, p 173-181
11. E. Sancaktar and P. Lawry, A Photoelastic Study of the Stress Distribution in Adhesively Bonded Joints With Prebent Adherends, *J. Adhes.*, Vol 11, 1980, p 233-241
12. L.J. Hart-Smith and B.L. Bunin, "Selection of Taper Angles for Doubles, Splices, and Thickness Transition in Fibrous Composite Structures," Douglas Paper 7299, McDonnell Douglas Corporation, 1983
13. A.M. Woolfrey, G.C. Eckold, and J.C. McCarthy, "Analysis of Adhesive Joints in G.R.P. Vessels and Pipework, in *Proceedings of Adhesives, Sealants and Encapsulants Conference*, Network Events Ltd., Vol 2, 1985, p 48-63
14. A.D. Crocombe and A. Tutarch, A Unified Approach to Adhesive Joint Analysis, in *Proceedings of Adhesives, Sealants and Encapsulants Conference*, Network Events Ltd., Vol 2, 1985, p 18-32
15. R.D. Adams, R.W. Atkins, J.A. Harris, and A.J. Kinloch, Stress Analysis and Failure Properties of Carbon-Fibre-Reinforced-Plastic/Steel Double-Lap Joints, *J. Adhes.*, Vol 20, 1986, p 29-53
16. E. Sancaktar and H.F. Brinson, "The Viscoelastic Shear Behavior of a Structural Adhesive," Report VPI-E-79-33, Virginia Polytechnic Institute, 1979
17. E. Sancaktar and H.F. Brinson, The Viscoelastic Shear Behavior of a Structural Adhesive, *Adhesion and Adsorption of Polymers*, L.H. Lee, Ed., Polymer Science and Technology Series 12-A, Plenum Publishing, 1980, p 279-300
18. D.W. Dwight, E. Sancaktar, and H.F. Brinson, Failure Characterization of a Structural Adhesive, *Adhesion and Adsorption of Polymers*. L.H. Lee, Ed., Polymer Science and Technology Series 12-A, Plenum Publishing, 1980, p 142-164
19. E. Sancaktar and S. Padgilwar, The Effects of Inherent Flaws on the Time and Rate Dependent Failure of Adhesively Bonded Joints, *Mech. Des.*, (*Trans. ASME*), Vol 104, 1982, p 643-650
20. E. Sancaktar, Non-destructive Examination of Adhesive Bonds With Neutron Radiography, *Int. J. Adhes. Adhes.*, Vol 1, 1981, p 329-330
21. D. Broek, Elementary Engineering Fracture Mechanics, 3rd ed., Martinus Nijhoff Publishers, 1983
22. D.G. O'Conner, "Factors Affecting the Fracture Energy for a Structural Adhesive," Master of Science Thesis, Virginia Polytechnic Institute, 1979
23. G.R. Irwin, Plastic Zone Near a Crack and Fracture Toughness, in *Proceedings of Seventh Sagamore Conference*, U.S. Army, 1960, p IV-63, IV-78
24. W.D. Bascom, R.L. Cottington, and C.O. Timmons, Fracture Reliability of Structural Adhesives, *J. Appl. Poly. Sci.*, Vol 32, 1977, p 165-188
25. E. Sancaktar, H. Jozavi, J. Baldwin, and J. Tang, Elastoplastic Fracture Behavior of Structural Adhesives Under Monotonic Loading, *J. Adhes.*, Vol 23, 1987, p 233-262
26. E. Sancaktar, Elastoplastic Fracture Behavior of Structural Adhesives, in *Mechanical Behavior of Adhesive*

Joints, G. Verchery and A.H. Cardon, Ed., Editions Pluralis, 1987, p 351-362

27. B. Dattaguru, R.A. Everett, Jr., J.D. Whitcomb, and W.S. Johnson, Geometrically Nonlinear Analysis of Adhesively Bonded Joints, *J. Eng. Mater. Technol. (Trans. ASME)*, Vol 106, 1984, p 59-65

28. W.S. Johnson, W.C. Rister, and T. Spamer, Spectrum Crack Growth in Adhesively Bonded Structure, *J. Eng. Mater. Technol. (Trans. ASME)*, Vol 100, 1978, p 57-63

29. W.S. Johnson, "Impact and Residual Fatigue Behavior of ARALL and AS6/5245 Composite Materials," NASA Technical Memorandum 89013, National Aeronautics and Space Administration, National Technical Information Service, 1986

30. W.S. Johnson, Damage Tolerance Evaluation of Adhesively Laminated Titanium, *J. Eng. Mater. Technol. (Trans. ASME)*, Vol 105, 1983, p 182-187

31. W.S. Johnson and S. Mall, Influence of Interface Ply Orientation on Fatigue Damage of Adhesively Bonded Composite Joints, *ASTM Composite Technology Review*, American Society for Testing and Materials, Spring 1986, p 3-7

32. H. Jozavi, C.W. Dupuis, and E. Sancaktar, Investigation of Fracture Behavior of a Composite Crack Arrestor, *J. Compos. Mater.*, Vol 22, 1988, p 427-446

33. W.S. Johnson, "Overview of NASA Programs Directed Toward Adhesive Joint Service Life Prediction," DOD/NASA Workshop on Adhesive Joint Service Life Prediction, National Aeronautics and Space Administration, 1981

34. W.S. Johnson and S. Mall, "A Fracture Mechanics Approach for Designing Adhesively Bonded Joints," NASA Technical Memorandum 85694, National Technical Information Service, 1983

35. R.S. Barsoum, Asymptotic Fields in Adhesive Fracture, *J. Adhes.*, Vol 29, 1989, p 149-166

36. T. Hattori, S. Sakata, and T. Watanabe, A Stress Singularity Parameter Approach for Evaluating Adhesive and Fretting Strength, in *Advances in Adhesively Bonded Joints*, MD-V6, American Society of Mechanical Engineers, 1988, p 43-50

37. D.A. Jablonski, Fatigue Crack Growth in Structural Adhesives, *J. Adhes.*, Vol 11, 1980, p 125-143

38. D.L. Hunston, A.J. Kinloch, and S.S. Wang, Micromechanics of Fracture in Structural Adhesive Bonds, *J. Adhes.*, Vol 28, 1989, p 103-114

39. S. Mostovoy, P.B. Crosley, and E.J. Ripling, Use of Crack-Line Loaded Specimens for Measuring Plane Strain Fracture Toughness, *J. Mater.*, Vol 2, 1967, p 661-681

40. W.S. Johnson and P.D. Mangalgiri, "Influence of the Resin on Interlaminar Mixed-Mode Fracture," NASA Technical Memorandum 87571, National Technical Information Service, 1985

41. S. Mall, W.S. Johnson, and R.A. Everett, Jr., Cyclic Debonding of Adhesively Bonded Composites, in *Adhesive Joints*, K.L. Mittal, Ed., Plenum Publishing, 1984, p 639-658

42. S. Mall and W.S. Johnson, Characterization of Mode I and Mixed-Mode Failure of Adhesive Bonds Between Composite Adherends, in STP 893, American Society for Testing and Materials, 1986, p 322-334

43. P.D. Mangalgiri, W.S. Johnson, and R.A. Everett, Jr., Effect of Adherend Thickness and Mixed Mode Loading on Debond Growth in Adhesively Bonded Composite Joints, *J. Adhes.*, Vol 23, 1987, p 263-288

44. A.J. Kinloch and S.J. Shaw, A Fracture Mechanics Approach to the Failure of Structural Joints, in *Developments in Adhesives—2*, A.J. Kinloch, Ed., Applied Science Publishers, 1981, p 83-124

45. E.J. Ripling, S. Mostovoy, and R.L. Patrick, Application of Fracture Mechanics to Adhesive Joints, in STP 360, American Society for Testing and Materials, 1963, p 5-19

46. S.S. Wang, J.F. Mandell, and F.J. McGarry, An Analysis of the Crack Tip Stress Field in DCB Adhesive Fracture Specimen, *Int. J. Fract.*, Vol 14, 1978, p 39-58

47. E. Sancaktar, S.C. Schenck, and S. Padgilwar, Material Characterization of Structural Adhesives in the Lap Shear Mode, Part I: The Effects of Rate, in *Industrial and Engineering Chemistry—Product Research and Development*, Vol 23, American Chemical Society, 1984, p 426-434

48. E. Sancaktar and S.C. Schenck, Material Characterization of Structural Adhesives in the Lap Shear Mode, Part II: Temperature Dependent Delayed Failure, in *Industrial and Engineering Chemistry—Product Research and Development*, Vol 24, American Chemical Society, 1985, p 257-263

49. E. Sancaktar, Material Characterization of Structural Adhesives in the Lap Shear Mode, *Int. J. Adhes. Adhes.*, Vol 5, 1985, p 66-68

50. W.G. Knauss, The Mechanics of Polymer Fracture, *Appl. Mech. Rev.*, Vol 26, 1973, p 1-17

51. R. Kilik and R. Davies, Mechanical Properties of Adhesive Filled with Metal Powders, *Int. J. Adhes. Adhes.*, Vol 9, 1989, p 224-228

52. D.L. Hunston, A.J. Kinloch, S.J. Shaw, and S.S. Wang, Characterization of the Fracture Behavior of Adhesively Bonded Joints, in *Adhesive Joints*, K.L. Mittal, Ed., Plenum Publishing, 1984, p 789-807

53. E. Sancaktar, Constitutive Behavior and Testing of Structural Adhesives, *Appl. Mech. Rev.*, Vol 40, 1987, p 1393-1402

54. M.L. Williams, Fatigue-Fracture Growth in Linearly Viscoelastic Material, *J. Appl. Phys.*, Vol 38, 1967, p 4476-4488

55. E. Sancaktar, H. Jozavi, and R.M. Klein, The Effects of Cure Temperature and Time on the Bulk Tensile Properties of a Structural Adhesive, *J. Adhes.*, Vol 15, 1983, p 241-264

56. E. Sancaktar, Bulk Shear Testing of Solid Film Adhesives With Symmetric Rail Shear Test Specimen, *Exp. Tech.*, Vol 8, 1984, p 27-30

57. H. Jozavi and E. Sancaktar, The Effects of Cure Temperature and Time on the Bulk Fracture Energy of a Structural Adhesive, *J. Adhes.*, Vol 18, 1985, p 25-48

58. H.F. Brinson, M.P. Renieri, and C.T. Herakovich, Rate and Time Dependent Failure of Structural Adhesives, in STP 593, American Society for Testing and Materials, 1975, p 177-199

59. M.P. Renieri, C.T. Herakovich, and H.F. Brinson, "Rate and Time Dependent Behavior of Structural Adhesives," Report VPI-E-76-7, Virginia Polytechnic Institute and State University, 1976

60. H.H. Kausch, *Polymer Fracture*, Springer-Verlag, 1987

61. R. Hill, *The Mathematical Theory of Plasticity*, Oxford University Press, 1956

62. A.F. Yee and R.A. Pearson, Toughening Mechanisms in Elastomer-Modified Epoxies Part I. Mechanical Studies, *J. Mater. Sci.*, Vol 21, 1986, p 2462-2474

63. R.A. Pearson and A.F. Yee, Toughening Mechanisms in Elastomer-Modified Epoxies Part II. Microscopy Studies, *J. Mater. Sci.*, Vol 21, 1986, p 2475-2488

64. A.J. Kinloch, Multiphase Thermosetting Adhesives, in *Abstracts*, Eighth Annual Meeting, Adhesion Society, 1985, p 6a

65. D. Broek, Elementary Engineering Fracture Mechanics, 3rd ed. (Noordhoff International Publishing, Leyden, Netherlands), 1974

66. J.W. Dally and R.J. Sanford, Strain Gage Methods for Measuring the Opening Mode Stress Intensity Factor, K_I, in *Proceedings of the 1985 Spring Conference on Experimental Mechanics*, Society for Experimental Mechanics, 1985, p 851-860

67. H. Jozavi and E. Sancaktar, The Effects of Cure Temperature and Time on the Stress Whitening Behavior of Struc-

tural Adhesives. Part I. Analysis of Bulk Tensile Data, *J. Adhes.*, Vol 27, 1989, p 143-157

68. H. Jozavi and E. Sancaktar, The Effects of Cure Temperature and Time on the Stress Whitening Behavior of Structural Adhesives. Part II. Analysis of Fractographic Data, *J. Adhes.*, Vol 27, 1989, p 159-174

69. J.J. Bikerman, *The Science of Adhesive Joints*, Academic Press, 1961

70. W. Ramberg and W.R. Osgood, Description of Stress-Strain Curves by Three Parameters, NACA TN 902, National Advisory Committee for Aeronautics, 1943

71. D.J. Woan, M. Habibullah, and J.A. Sauer, Fatigue Characteristics of High Impact Polystyrene, *Polymer*, Vol 22, 1981, p 699-701

72. S. Mall and K.T. Yun, Effect of Adhesive Ductility on Cyclic Debond Mechanism in Composite-To-Composite Bonded Joints, *J. Adhes.*, Vol 23, 1987, p 215-231

73. J. Luckyram and A.E. Vardy, Fatigue Performance of Two Structural Adhesives, *J. Adhes.*, Vol 26, 1988, p 273-291

74. J.A. Bannantine, J.J. Comer, and J.L. Handrock, *Fundamentals of Metal Fatigue Analysis*, Prentice-Hall, 1990

75. J. Romanko, K.M. Liechti, and W.G. Knauss, Life Prediction Methodology for Adhesively Bonded Joints, in *Adhesive Joints*, K.L. Mittal, Ed., Plenum Publishing, 1984, p 567-586

Flaw Detection (Effects of Defects)

Francis H. Chang, General Dynamics/Fort Worth Division

ADHESIVELY BONDED STRUCTURES used in the aircraft industry consist mainly of metallic or composite substrates bonded either to each other or to some substructure by one or several adhesive layers. In a typical adhesively bonded aircraft wing structure, the composite or metallic wing skins may consist of several thin layers, or laminae of fiber-reinforced plastic or metals, bonded together. The bonded skin layers can be bonded to spars and/or ribs to form a wing structure. Spaces between the skin layers and the spars/ribs in a wing structure can be used for fuel storage; therefore, the spars/ribs represent the only structure-stiffening elements.

In flaps, ailerons, and other control surfaces, metallic or nonmetallic honeycomb structures can be placed between the skin layers to serve as stiffeners. In composite structures, the laminates and substructure may be co-cured, and the bonding of the substrate and substructure can be considered the same as the bonding of the composite laminate.

Two types of bonded structures are shown in Fig. 1. The substrate-to-substrate bonding (Fig. 1a) can be extended to a multilayered structure by adding on other layers. In the adhesively bonded honeycomb structure (Fig. 1b), bonding between the honeycomb core and the skin is achieved mainly by the interfacial bonds between the honeycomb core/adhesive, and the skin/adhesive. In a more complicated structure, a substructure may replace the lower skin shown in Fig. 1(b).

Adhesive Defects

The most common types of defects that occur at interfaces A and C in Fig. 1 are voids and foreign objects. A void is a total lack of adhesion caused by noncontact at the interface. Voids can be generated during fabrication if a large volume of trapped gas evolves from the adhesive during cure. They also can be formed by a nonuniform application of pressure (vacuum) on the substrates and/or by a variety of shimming effects at the bond line.

Foreign objects at the interfaces isolate the substrate and adhesive and prevent bonding between them. The degree of dis-

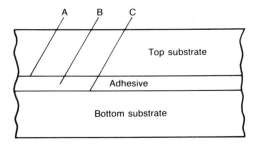

(a)

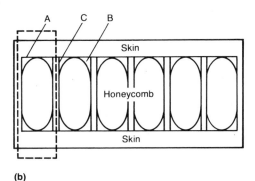

(b)

Fig. 1 Categories of adhesively bonded structures. (a) Multilayered structure. (b) Honeycomb structure. A, B, and C refer to interfaces, which are described in the section "Categories of Adhesive Defects" in this article.

bond depends on the effects that the material properties of the foreign objects have on promoting a bond with the substrate. The similarities of the foreign objects with the adhesive, notably in acoustic impedance and opacity to x-ray and neutron radiation, are important factors in the selection of an appropriate nondestructive inspection (NDI) method to detect these inclusions. Well-bonded, cohesively failed surfaces of two single-lap-shear specimens are shown in Fig. 2(a), and interfaces with varying degrees of disbonds in three specimens that failed adhesively are shown in Fig. 2(b).

Another type of bonding defect that occurs at the A and C interfaces is the weak bond. With this defect, there is no void at the interface separating the substrate and adhesive, which are in intimate contact with each other. Rather, there is little or no adhesion to bind the two surfaces together.

A weak bond can be formed when the substrate surface is not properly prepared; that is, when it is not primed or not properly etched (in the case of metallic substrates), or the inadvertent presence of release agents or chemical agents prohibits bonding in both metallic and composite substrates. Another possible cause of a weak bond is insufficient bonding pressure, which prevents the substrate and adhesive surfaces from forming a strong mechanical link. An insufficient curing temperature can also cause a weak bond. In this case, the adhesive lacks the proper environment to achieve the chemical reactions conducive to the formation of a strong bond.

At present, no reliable method exists for detecting a weak bond after the structure is fabricated. Thus, stringent in-process control is required before and during the fabrication of an adhesively bonded structure.

Bonding Sequence. A typical adhesive bonding sequence that can prevent a weak bond condition involves these steps:

1. Prefit all detail parts to ensure adequate mating
2. Disassemble prefit structure
3. Forward metallic detail parts for surface treatment
4. Degrease aluminum core details
5. Apply adhesive
6. Assemble in bond form
7. Move to autoclave or blanket press
8. Cure
9. Conduct postbond cleanup
10. Perform secondary operations, such as fastener installation, potting, sealing, and so on
11. Conduct a leak check
12. Inspect

Surface Preparation. Primary surface treatments for aluminum alloys start with a pretreatment solvent wipe, followed by a vapor degrease with trichlorethylene or 1,1,1-trichloroethane. Precleaning is accomplished using a nitric acid solution followed by a rinse, and final cleaning uses a sulfuric acid or sodium dichromate solution followed by a rinse. The surface is then visually inspected. For multiple-stage bonds and repairs, an acid cleaner is usually used in the cleaning process.

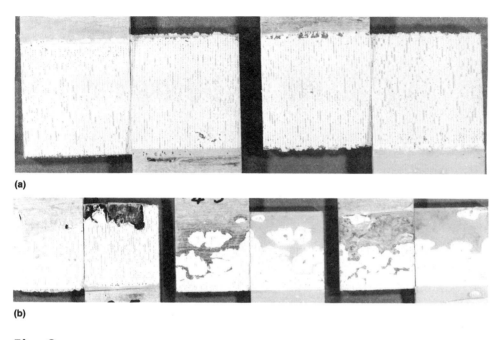

(a)

(b)

Fig. 2 Interfaces of (a) good bonds and (b) disbonds

For titanium alloys, a grit blast with 120 to 320 mesh aluminum oxide should be applied, followed by a degreasing procedure. The alloys are then immersed in Pasa Jell 107M solution for the next cleaning step, followed by a rinse. They are dried at a maximum temperature of 70 °C (160 °F) and visually inspected.

Processing Environment. In the assembly area, a controlled environment for primary adhesive bonding operations is maintained. This involves dust-free conditions, temperature control to a maximum 27 °C (80 °F), humidity control to a maximum relative humidity of 55%, and a "white-glove" operation. Certain procedures can be done outside of the controlled area, such as splicing of honeycomb core details, laminating operations in connection with the adhesive, minor repairs, and filling the end closures of bonded assemblies with potting compound. However, these procedures should still be subjected to "white-glove" operation and cleanliness precautions, and they should be conducted by specially trained personnel.

Storage Temperature. Certain maximum out-time restrictions are placed on adhesives to prevent deterioration in their quality before use. For example, the time prior to application at room temperature for one adhesive primer is 7 days, whereas for other primers it may be 14 days. For certain adhesives, the out-time is 3 days. After aluminum alloys are cleaned, the maximum time until an adhesive primer is applied is 120 hours, whereas the maximum time after primer application until adhesive application can be as long as 30 days. After the adhesive application and before the assembly is cured, the maximum time allowed for

Reliabond 398 and AF-147/FM 300, for example, is 7 and 4 days, respectively. The heat-up-rate limits during cure are within the range of 2 to 6 °C per minute (3 to 10 °F per minute). Cure pressure can vary between 170 and 345 kPa (25 and 50 psi).

Bonding Quality Control. Quality control methods for adhesively bonded structures include adhesive overlays, nondestructive inspection, visual and dimensional inspections, leak tests by water submersion at 77 °C (170 °F) for 1 to 2 min, surface treatment checks by daily peel tests, and test tabs. For adhesive film, each roll is checked by mechanical and physical tests to ensure that it meets procurement specification requirements.

Test tabs are made by using the same materials and procedures, under the same conditions, as those for the actual structure. Destructive tests are conducted on these test tabs to ensure the mechanical strength of the specimens. Mechanical test specimen configurations are shown in Fig. 3. If the mechanical strength meets the required strength specification, it is inferred that other structures will have the required strength when NDI methods are applied to ensure that no voids or inclusions exist in the bond line. A semidestructive means of inspection has been investigated by Chuang *et al.* to test the bond strength of an adhesively bonded structure and to screen weak bonds (Ref 1).

Categories of Adhesive Defects. Bond characteristics and bonding defects in adhesively bonded structures can be discussed in terms of interfaces A, B, and C (Fig. 1). Interface A separates the top substrate from the adhesive itself. In Fig. 1(a), interface C

separates the adhesive itself from the bottom substrate, whereas in Fig. 1(b) it separates the adhesive from the honeycomb wall. Bonding generally refers to interfaces between the substrate and the adhesive; however, in the case of metallic substrates, adhesion actually occurs between a thin oxide layer (1 μm, or 0.04 mil, thick) and the adhesive. The interface between the substrate itself and the oxide layer is usually considered too microscopic in size to be of practical consideration. However, the formation and nature of the oxide layer itself are considered to be of prime importance to the bond.

Interface B refers to the adhesive material itself. The ability of the adhesive to withstand external loading depends on its modulus of elasticity, fracture toughness, and other factors. Subsequent to fabrication of the structure, the adhesive may be vulnerable to environmental degradation, such as moisture ingression and extreme temperature effects. Defects caused by improper processing can produce porosities in the bond line that degrade the mechanical strength of the adhesive. Porosities can be readily detected using various NDI methods. Bond line thickness, which can affect the mechanical strength of the bond (Ref 2), can also be measured quite accurately by the ultrasonic NDI method. Initial cohesive bond strength, as a collective representation of factors affecting the integrity of the bond itself either before or after the structure has entered into service, can be measured by the Fokker bond tester (Ref 3).

Effects of Defects

The effects of defects in adhesively bonded structures have been studied over the years in many industrial application areas in an effort to establish accept/reject criteria. Studies in the past were conducted mainly on coupon-type specimens to correlate NDI parameters with adhesive bond strength. These studies showed, in general, that most of the parameters that are measurable by NDI methods can in fact be correlated with the ultimate failure loads of the specimens. However, the oversimplified configurations of the specimens used in these studies produced uncertainty in the scale-up of the test results (Ref 4). Several studies were conducted on larger-scale and even full-scale flight articles (Ref 2). Results of these studies showed that adhesively bonded structures have a high tolerance to bonding defects, including delaminations.

Mechanically Induced Artificial Delamination. Delamination is the predominant bonding defect. Single- or double-lap-shear specimens are used to study the effects of induced or simulated delaminations. In single-lap-shear specimens, the load application is off-axis and tends to produce a large peeling effect at the bond.

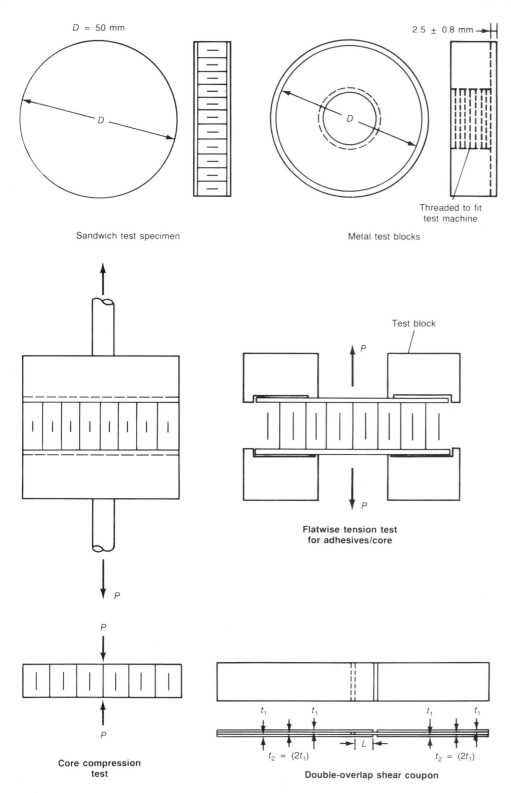

Fig. 3 Mechanical test specimen configurations for quality control

dominant effect, causing most of the strength reduction to occur in the last 25% of the overlap joint.

Single-lap-shear specimens were made from a bonded plate by milling 1.6 mm (1/16 in.) grooves 25 mm (1 in.) apart on opposite sides of the plates. The plates were then cut to 25 mm (1 in.) widths. The aluminum plates were made of 2024 aluminum, and the adhesive used was RB-7116. The aluminum plates were cleaned, etched (Forest Products Laboratory, or FPL, process), and primed according to standard bonding procedures. The finished plates were 205 × 205 mm (8 × 8 in.) to provide seven 25 mm (1 in.) wide specimens from each plate. Mechanically induced delaminations were inflicted on the specimens by placing them in a vise between steel blocks; a specified length of the overlap was left protruding from the vise. The protruding overlap was delaminated by placing a narrow chisel in the saw cut and tapping with a hammer. Delaminations started at the saw cut and progressed to the edge of the clamped area. Specimens were produced with delaminations covering 25, 50, and 75% of the overlap, with good control of the delamination length.

Ultrasonic through-transmission inspection was conducted before and after mechanical delamination inducement on the specimens to verify the delaminated areas. Figure 4 shows the C-scan results of several specimens. The effect of delaminations on shear strength reduction is shown in Fig. 5. The strength reduction corresponds to the results of the stress analysis. As the overlap was reduced by increasing the delaminated area, the ultimate shear strength reduced slowly until a 50% reduction was reached. As the overlap continued to decrease beyond the 50% point, the average shear stress increase became the dominant effect, causing most of the strength reduction to occur in the last 25%. Because delaminations do not act as a stress raiser, the study concluded that discrete delaminations are not as detrimental to the adhesive bond strength as other parameters in single-lap-shear specimens (Ref 5).

Bond Line Porosities Produced by Shimming. Effects of defects tests were conducted on single-lap-shear adhesively bonded specimens using FM 73 adhesive under a Wright Research and Development Center Flight Dynamic and Materials Laboratories program titled Primary Adhesively Bonded Structure Technology (PABST) in 1977. This program is discussed in Ref 2, as well as in the article "End-Product Nondestructive Evaluation of Adhesive-Bonded Metal Joints" in this Volume. Results obtained on 100, 80, and 50% bonded areas in 25 mm (1 in.) wide specimens showed roughly the same ratio in shear strength reduction.

In the PABST program, single-lap-shear specimens with varying degrees of porosi-

One study considered the leading edge of the overlap as a stress concentrator (Ref 5). Stress analysis showed that the stress concentration at the leading edge is reduced by shortening the overlap. A stress analysis was conducted for single-lap-shear specimens carrying a constant load with the overlap length shortened in steps of 25%. Shortening the overlap reduced the bending moment on the lap. Although the average shear stress was increased, the maximum peel stress remained almost constant. As the overlap approached zero, the increase in the average shear stress became the

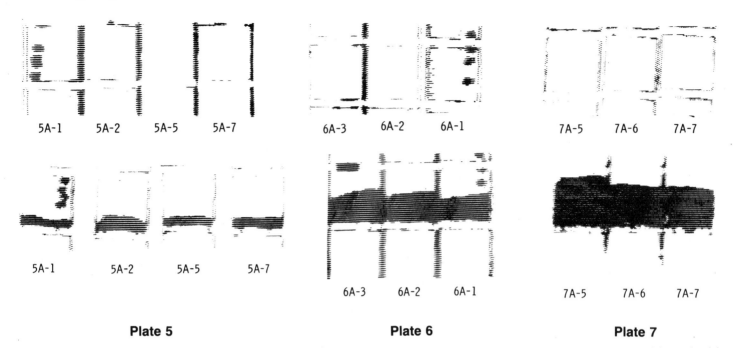

| Plate 5 | Plate 6 | Plate 7 |

Fig. 4 Ultrasonic C-scan images of RB-7116/2024 single-lap-shear specimens before (top) and after (bottom) defect inducement. Plate 5 specimens had defects induced that covered approximately 25% of the overlap. Plate 6 specimens had defects induced that covered approximately 50% of the overlap. Plate 7 had defects induced that covered approximately 75% of the overlap. Source: Ref 5

ties were made by using spacer shims and multiple adhesive layers to vary the adhesive layer thickness. Neutron radiography (NR) inspection was conducted on the specimens to verify the degree of porosities. Representative NR inspection images are presented in Fig. 6. Figure 7 presents the lap-shear strength versus adhesive thickness for both porous and nonporous adhesives, as determined by NR inspection. An 80% strength level is reached at approximately 0.33 mm (0.013 in.) for a thick porous adhesive and at 0.48 mm (0.019 in.) for a multiple-layer nonporous adhesive with an adherend thickness of 1.6 mm (0.063 in.).

Bonding Defects From Surface Treatments. A recently conducted study on the strength of contaminated bonded lap-shear joints (Ref 6) used test specimens of graphite-epoxy substrates joined in single-lap-shear joints with EC 3448, a 120 °C (250 °F) curing paste adhesive. Each specimen had a width of 100 mm (4 in.), a total length of 190 mm (7.5 in.), and an overlap area of 13 mm (0.5 in.). Contaminated areas were introduced into the overlap area; these contaminations occupied 19, 25, and 31% of the 100 mm (4 in.) wide bonded area. The size of the contamination was scaled to simulate the size that could occur in a typical full-scale production bonded joint. Sixteen different contamination types were introduced to the test specimens and to duplicate specimens (four each); there were four control specimens. The specimens were then destructively tested to measure their ultimate shear strength.

Tensile test results for the specimens are presented in Fig. 8 with descriptions of the contaminants. Peak stress was calculated using the entire bond area as if there was no contamination. The specified acceptable bond strength for composites with EC 3448 adhesive was 17.2 MPa (2.5 ksi). Results presented in Fig. 8 indicate that marker pencil, release agent, masking tape residue, and both 25 and 31% voids caused failures below the specified acceptance level.

The study concluded that bonded joints in composites are basically damage tolerant. Joint strength effects should be investigated more fully to take advantage of damage tolerance design.

Incipient Defects in Bonded Structures. A study of honeycomb sandwich structure, including a bonded joint for local transfer, was conducted (Ref 7) using a control surface made of F16H581 fiberglass epoxy and epoxy edge components, an aluminum sub-

structure, and FM 300 and FM 37 structural adhesives. Representative specimens were prepared that included artificial flaws typical of a bonded honeycomb structure. These specimens and the control surface structure were bonded in an autoclave at 345 kPa (50 psi) and a soaking temperature of 175 °C (350 °F) for 2 h. For quantitative calibration of the neutron radiography evaluation method, aluminum-to-aluminum single-lap-shear reference specimens 25 mm (1 in.) wide × 198 mm (7.8 in.) long were prepared. In these specimens, different shim thicknesses ranging from 0.20 to 0.71 mm (0.008 to 0.028 in.) were used to produce bond-line thicknesses representing various bond quality.

The shim thickness in a specimen determines the quality of the bond because a thicker bond line, beyond a certain limit, produces a large amount of voids in the bond line. The neutron radiography inspection method is based on the principle that the neutron beam can penetrate through a metallic substrate but will be attenuated or absorbed by materials such as adhesives that contain hydrogen atoms. Bond line layers containing large amounts of voids between metallic substrates are observable in the NR images. Figure 9 shows destructive test results for lap-shear strength in terms of MPa (psi) for test specimens with various shim thicknesses; these results are compared with the findings of an NR inspection of the specimens. The bond quality indicated by the void content from the NR inspection has a positive correlation with the bond quality indicated by the shear strength results.

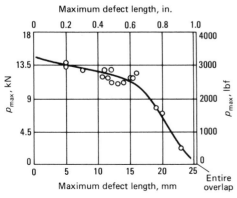

Fig. 5 Failure load as a function of defect length for delaminated single-lap-shear specimens. Source: Ref 5

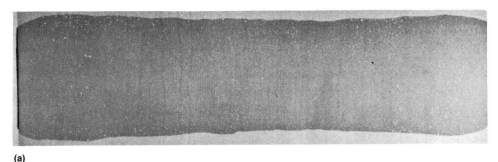

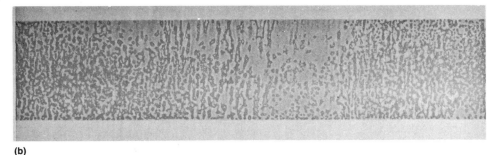

(a)

(b)

(c)

(d)

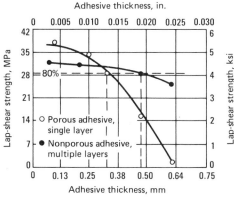

Fig. 7 Lap-shear strength versus adhesive thickness for porous and nonporous FM 73 adhesive. Source: Ref 2

Fig. 6 Positive prints of neutron radiographs of typical shimmed test panels. Approximate test panel width, 38 mm (1½ in.). Facing sheets are 1.0 to 1.25 mm (0.040 to 0.050 in.) thick 2024-T3 aluminum. (a) Shim thickness, 0.20 mm (0.008 in.); actual thickness, 0.36 mm (0.014 in.). (b) Shim thickness, 0.30 mm (0.012 in.); actual thickness, 0.51 mm (0.020 in.). (c) Shim thickness, 0.40 mm (0.016 in.); actual thickness, 0.53 mm (0.021 in.). (d) Shim thickness, 0.50 mm (0.020 in.); actual thickness, 0.64 mm (0.025 in.). Courtesy of Donald Hagemaier, Douglas Aircraft Company

Several control surfaces were fabricated and subjected to NDI both before and after nominal loading to 150% of the design load. Residual displacement (RD) measurements of several key points on the control surfaces were taken during loading to evaluate the stiffness of the control surface structure. The honeycomb bonded area on the full-size control surface was divided into 20 squares with sizes inversely proportional to the stress level in that area. Five articles that were judged to be unacceptable based on the NDI and RD evaluation were loaded to failure to determine their ultimate strength. A summary of the test results for these five articles is presented in Table 1.

In Table 1, column 1 identifies the five rejected articles and column 2 lists the grading of the bond quality judged by the Fokker bond tester before loading to 150% of the design load. The grading was based on the number of squares on each article with unacceptable bond tester readings. After loading to 150% of the design load, NDI measurements were repeated on each article. The increase in the number of squares with unacceptable bond tester readings after loading is listed in column 3. Column 4 gives strength ratings based on the results in column 3. The total void area measured by neutron radiography is listed in column 5. No significant difference in void area was observed before and after nominal loading. A grading for the stiffness of each structure based on the maximum residual displacement after nominal loading is listed in column 6. Grading according to the ultimate load to failure is listed in column 7. The variation between the ultimate load sustained by the strongest and the weakest bond is 50% of the nominal load.

The data in Table 1 indicate that neither the Fokker bond tester grading done before loading nor that done after loading to 150% of the design load (columns 2 and 4, respectively) correlates with the ultimate load grading. The void area measurements (column 5) and the gradings for residual displacement under nominal loading (column 6) also do not correlate with those for ultimate load. The only column that correlates with that for ultimate load is column 4, which contains gradings for the difference in Fokker bond tester results after nominal loading.

The data in Table 1 reveal several pertinent points about the effect of defects on bond strength. First, Fokker bond tester measurements may reveal cohesive or adhesive bond anomalies on individual locations. However, the collective effects of these anomalies do not affect the ultimate load to any significant extent. Even after the nominal loading, these anomalies show no correlation with ultimate load data. Second, void areas in the bond as measured by neutron radiography do not significantly affect the ultimate load. In Table 1, the total void area in the worst case was 195 mm^2 (0.3 in.^{2}), an insignificant percentage of the total bonded area. Third, stiffness and strain measurements at key points in a

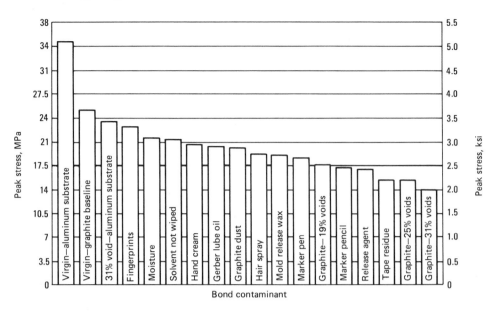

Fig. 8 Contaminated bond lap-shear test results. Source: Ref 8

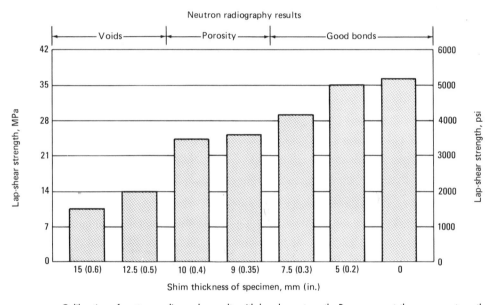

Fig. 9 Calibration of neutron radiography results with lap-shear strength. Bars represent the average strength of single-lap-shear joint specimens. Source: Ref 7

Table 1 Grading of nondestructive and destructive test results

Grading from I (best) to V (worst)

| (1) Control surface | Nondestructive results | | | | | Destructive results | |
| | Fokker bond tester | | | Neutron radiography | | | |
	(2) Grading of control surface strength	(3) Number of defective squares(a)	(4) Grading of strength according to column 3	(5) Voided area mm²	in.²	(6) Grading according to nominal load(b)	(7) Grading according to ultimate load
A IV		1	I	140	0.22	II	I
B III		3	II	160	0.25	I	II
C I		6	III	140	0.22	V	III
D II		6	IV	50	0.08	IV	IV
E V		6	V	180	0.28	III	V

(a) Number of additional defective squares after nominal loading. (b) Grading for the stiffness of the structure based on the maximum residual displacement after nominal loading. Source: Ref 7

structure after nominal loading also have no correlation with the ultimate load results.

In summary, it appears that NDI measurements alone cannot reveal the bond strength of an adhesively bonded structure. Neutron radiography can measure the void area in the bond line, and in the case of coupon-type specimens, these measurements show a positive correlation with bond strength. However, because of the small percentage of voids or porosities in an entire structure, these defects have an insignificant effect on the bond strength. This conclusion appears to substantiate the theory that porosities and small voids in the bond line have little effect on the strength of an adhesively bonded structure (Ref 4).

NDI Methods

Different NDI methods for detecting bond defects are discussed briefly here, as well as in the Section "Nondestructive Inspection and Quality Control" in this Volume. Generally, voids and inclusions at the interfaces and porosities in the bond line can be detected by ultrasonic methods. The x-ray radiographic method can detect large voids, inclusions, and porosities if there is sufficient difference in x-ray absorption between the normal and defective areas. As already noted, neutron radiography is most effective in penetrating through a metallic substrate to detect the absence of hydrogen atoms contained in an adhesive. Infrared thermography can detect large areas of debonding and inclusions if their thermal conductivity values are different from those of substrate and adhesive system. No reliable NDI method exists to date for the detection of a weak bond at the substrate-adhesive interface. A semidestructive screening method has been investigated.

Ultrasonic methods are perhaps the most widely used inspection methods for bonded structures. The through-transmission mode of operation is the most practical technique to use on multilayered laminates and assembled structures (Ref 8). This technique utilizes two transducers, one on each side of the part. One transducer serves as a transmitter, and the other serves as a receiver. The transmitting transducer sends a pulsed ultrasonic beam into the part; the beam traverses through the part to the opposite side, where it is picked up by the receiving transducer. Any discontinuities created by disbonds in the path of the sound beam cause a reduction in the intensity of the sound beam reaching the receiving transducer. The variations of sound energy transmitted through the part are monitored on the CRT screen of the ultrasonic transmission equipment and recorded on a C-scan recording device. The major limitation of the through-transmission technique is the requirement to have access to both sides of the part.

Table 2 Results of flat-panel tests using the Shurtronics harmonic bond tester

Panel number	Void detected by Standard probe	Void detected by Miniature probe(a)	Void measurable by Standard probe	Void measurable by Miniature probe(a)	Response	Skin thickness mm	Skin thickness in.	Core thickness mm	Core thickness in.
1	Yes	Yes	Yes	Yes	Down scale	2.5/2.3	0.100/0.090	22.9	0.900
2	Yes	Yes	Yes	Yes	Down scale	3.2/2.8	0.125/0.110	25.4	1.00
3	Yes	Yes	Yes	Yes	Up scale	2.0/1.8	0.080/0.070	20.3	0.800
4	Yes	Yes	Yes	Yes	Up scale	0.6/2.2	0.025/0.032	19.1	0.750
5	Yes	Yes	Yes	Yes	Up scale	0.5/0.3	0.020/0.012	10.2	0.400
6	Yes	Yes	Yes	Yes	Up scale	0.2/1.8	0.080/0.070	12.7	0.500
7	Yes	Yes	Yes	Yes	Up scale	0.6/1.0	0.020/0.040	12.7	0.500
8	Yes	Yes	Yes	Yes	Up scale	3.2/2.8	0.125/0.110	12.7	0.500
9	Yes	Yes	Yes	Yes	Up scale	1.5/0.5	0.060/0.020	12.7	0.500
10	Yes	Yes	Yes	Yes	Up scale	0.8/1.0	0.030/0.040	12.7	0.500

(a) The miniature probe has more gain and will detect voids measurable in top-skin thicknesses from 0.3–4.0 mm (0.012–0.160 in.). Source: Ref 14

For some assembled structures that have only one accessible side, the pulse-echo mode of operation can sometimes be used (Ref 9). This technique uses a single transducer that serves as both transmitter and receiver, alternating between pulses. The ultrasonic beam traverses through the part and reflects off the bottom of the structure, causing it to traverse back through the part to the transducer, which now serves as the receiver. If there is a discontinuity anywhere in its path, the ultrasonic beam will not be able to reach the bottom of the structure, and the back-surface signal will be absent from the CRT display on the ultrasonic unit. The pulse-echo operation is generally limited to simple bonded structures; in multilayered structures, the signals reflected from the bond lines tend to excessively attenuate the beam intensity. This attenuation prevents the beam from reaching the back surface, even in the case of a good bond.

The through-transmission method can be applied to honeycomb structures for the detection of fairly large debonds. The core walls of honeycomb structures act as sound-conducting media. If there is a disbond between the skin and the core, the ultrasound cannot traverse from the skin to the core, and the discontinuity will cause a loss of signal at the receiving transducer.

The pulse-echo method can also detect a disbond by monitoring the signal reflected from the back surface of the honeycomb structure. However, the ultrasound will have to travel round-trip before reaching the transducer. For thick honeycomb structures and highly attenuated honeycomb materials, a round-trip attenuation may diminish the reflected signal to an undistinguishable point and render the technique unusable.

The pulse-echo method has been significantly improved in the area of signal pattern recognition as a flaw criteria discriminant (Ref 10). The ultrasonic signals are digitized and stored in a massive computer memory to permit detailed comparison of the time domain and a reference signal obtained from an unflawed area. The digitized ultrasonic waveforms can also be transformed into the frequency or other domains to extract bond line features such as porosity density and bond line thickness. Ultrasonic spectroscopy has been a useful tool for characterizing bond lines in the research laboratory (Ref 11, 12).

Resonance methods for inspecting adhesive bonds use the principles of both electromagnetics and acoustics (Ref 13). In the Shurtronics harmonic bond tester, the probe coil induces an alternating eddy current with an attendant alternating magnetic field (Ref 14). Interacting forces between the magnetic field and the induced eddy current cause the panel being tested to vibrate. The probe contains high-sensitivity microphone coaxial with the coil; this microphone measures the acoustic vibration response of the panel test. Well-bonded structures produce dampened vibrations. Bonded structures of poor quality and structures in disbond conditions produce undamped vibrations.

The Shurtronics harmonic bond tester can detect both amplitude and phase characteristics in the acoustic response of the structure under test. Because no couplant is required in its use, there is no possibility of surface contamination. Curved and flat surfaces can be easily scanned with an inspection speed of up to 0.5 m²/min (5 ft²/min). The limitation on applicable skin thicknesses is a major disadvantage of this bond tester. Another disadvantage is that the probe has a tendency to cancel out the response when it is over the center of a near-side void in test panels with skin thicknesses over 2.5 mm (0.1 in.). The detection of far-side voids requires verification by using a reference part that is an exact duplicate of the test part.

Results of test specimen inspection using this bond tester are summarized in Table 2; results obtained for the reference panels are given in Table 3. In the flat-panel tests, a standard probe was used with the switch set to the amplitude measurement position. The optimum stand-off proved to be from approximately flush to 0.25 mm (0.010 in.), as determined by the meter response. At these settings, any variation in core thickness did not affect the response from near-surface voids within the thin-skin range from 0.30 to 2.25 mm (0.012 to 0.089 in.). However, for thick skins (≧2.29 mm, or 0.090 in.), the void response was variable. Test results indicated that the variation in the response of near-surface voids may be a function of core thickness. For this reason, the use of the Shurtronics bond tester is limited to the thin-skin range.

The tap hammer inspection method of locating disbonds is based on sound produced when a well-bonded part is struck

Table 3 Results of reference-panel tests using the Shurtronics harmonic bond tester

Skin thickness mm	Skin thickness in.	Void detected	Void measurable	Response
0.51	0.020	Yes	Yes	Up scale
1.01	0.040	Yes	Yes	Up scale
1.52	0.060	Yes	Yes	Up scale
2.03	0.080	Yes	Yes	Up scale
2.54	0.100	Yes	Yes	Down scale
3.05	0.120	Yes	Yes	Down scale
3.56	0.140	Yes	Yes	Down scale
4.06	0.160	No	No	None
4.57	0.180	No	No	None
5.08	0.200	No	No	None
5.59	0.220	No	No	None
6.10	0.240	No	No	None
6.60	0.260	No	No	None
7.11	0.280	No	No	None

Note: Core thickness in each case was 25.4 mm (1 in.). Source: Ref 14

Table 4   Results of flat-panel tap hammer inspections

Panel number	Skin thickness mm	in.	Void detection
1	2.54	0.100	Yes—very little change in sound
	2.29	0.090	Yes—very little change in sound
2	3.18	0.125	Yes—very little change in sound
	2.79	0.110	Yes—very little change in sound
3	2.03	0.080	Yes
	1.78	0.070	Yes
4	0.64	0.025	Yes
	0.81	0.032	Yes
5	0.51	0.020	Yes
	0.30	0.012	Yes
6	2.03	0.080	Yes
	1.78	0.070	Yes
7	0.51	0.020	Yes
	1.01	0.040	Yes
8	3.18	0.125	Yes—very little change in sound
	2.79	0.110	Yes—very little change in sound
9	1.52	0.060	Yes
	0.51	0.020	Yes
10	0.76	0.030	Yes
	1.01	0.040	Yes

Source: Ref 14

and that produced when a disbonded part is struck (Ref 14). The experience of the inspector in discerning between the sounds is the factor that determines the success of this method. Generally, a void will produce a sound that is less sharp or clear than that produced by a void-free area. Responses from voids can be described as dull or dead; voids are sometimes indicated by rattling sounds. Results for tap hammer inspections of test specimens and reference panels are given in Table 4. This method is reasonably accurate for panels with skin thicknesses of less than 2.0 mm (0.080 in.). It provides, at best, a qualitative inspection of bonded structures with flaw areas that are very difficult to estimate.

The Fokker bond tester is a resonance-type ultrasonic tester with a piezoelectric transducer coupled to the part surface by an ultrasonic couplant. The couplant is driven by a continuous-wave module at various frequencies (Ref 15). The frequency response and resonance amplitude of the bonded structure under the probe area are measured as the driving frequency is swept through several modes of vibration range. Stone (Ref 16) and Guyott *et al.* (Ref 17) examined the analytical background of the Fokker bond tester pertaining to cohesive bond strength. They concluded that changes in cohesive properties in the bond line can be revealed by the tester, but only if the parameters that affect the tester measurements are fully understood.

Radiographic Methods. The conventional x-ray radiographic method places an x-ray source on one side of a part and a film that is sensitive to x-rays on the opposite side. An image of the part on the film is created showing the x-ray attenuation in various lateral areas on the part. In general, the x-ray attenuation of inclusions and large voids is sufficiently different from that of

the remainder of the part to clearly distinguish them on the x-ray record. For honeycomb structures, water entrapped in the core and corrosion on the inner surfaces of the metallic skins are also detectable. Conventional x-ray equipment is quite common and widely used in the industry.

An electronic recording device can also be used instead of x-ray films to produce a real-time image of the part in real-time radiography. The focal point of the x-ray source can be made very small (2 to 5 mm, or 0.08 to 0.20 in.) in a microfocus x-ray to improve image resolution and enable enlargement of the x-ray image; these microfocus techniques can be used in the real-time radiography mode. Conventional and microfocus real-time radiography are becoming popular in the aerospace industry.

To obtain the x-ray image of a cross section of a part, x-ray computer tomography (CT) can be used. The x-ray CT method rotates either the source or the detectors to measure the x-ray attenuation through the part at different angles. The data are stored in a high-power computer. The attenuation at various angles is then analyzed and reconstructed to produce an image of the part at that cross section. The advantage of x-ray CT is that it can show the attenuation characteristics at a specific cross section, such as the bond line. Porosities, disbonds, and inclusions at the bond lines can be detected by this method. The chief disadvantage of the technique is its high initial cost.

In an alternative method, a thermal neutron source can be used in place of the x-ray source. The unique characteristic of neutron beams is their ability to readily penetrate through metals; however, the beams are heavily attenuated by elements containing hydrogen atoms. Adhesives are generally polymer compounds and thus offer high

attenuation to neutron beams. In an adhesively bonded structure with metallic substrates, the adhesive in the bond line is the major attenuator of the neutron beam. Any disbonds in the bond line will allow the neutron beam to traverse the structure and show up in the radiographic image. Water-filled core cells in honeycomb structures also show up distinctly in the record.

Thermographic methods measure the surface temperature of a test part. Vibrothermography (Ref 18, 19) is an active method for monitoring the surface temperature of a part as it is being cyclically stressed. An adhesive debond in a bonded structure will be indicated by an increase in local temperature on a thermal measuring device, which is generally an infrared-sensitive measurement instrument. Alternatively, an external heat or cooling source can be used to raise or lower the transient temperature on the part surface. As the transient increase or decrease in temperature on the local area returns to the ambient, it is monitored by a temperature-sensitive device. Disbonds or inclusions in the bond line in the adhesively bonded structure will show up as hot or cool spots in the thermal image. Voids as small as 25 mm (1 in.) square can reportedly be detected below an aluminum substrate that is 0.25 mm (0.01 in.) thick (Ref 20).

Comparison of NDI Methods. In a study described in Ref 14, flaw detection capabilities of the through-transmission NDI method were compared with those of the Shurtronics harmonics bond tester and the tap hammer technique. Adhesively bonded structures with metal skins and honeycomb cores were used to conduct the tests.

Test specimens were prepared using 2024-T6 aluminum skin with a thickness ranging from 0.30 to 3.2 mm (0.012 to 0.125 in.); an aluminum honeycomb core with a thickness ranging from 10 to 25 mm (0.4 to 1.0 in.); and FMS 1013 type 1B adhesive. The specimen dimension was 100×230 mm (4×9 in.), and flaws on both sides of the panel were created by mechanically depressing a 25 mm (1.0 in.) diam circular area to a depth of 0.5 mm (0.020 in.) in the metallic core. The honeycomb core size was 3.2 mm (⅛ in.). Figure 10 depicts the specimen configuration.

After the detail parts were prepared and prefitted, they were cleaned by a phosphoric acid anodizing process. The adhesive was applied to one of the skins so that the adhesive layer was facing up. The honeycomb with the mechanically depressed circular holes and a dummy skin on top was assembled with the first skin and cured in a blanket press for 1 h at 125 °C (260 °F) using 310 kPa (45 psi) pressure. Following the cure, the dummy skin was removed, and a layer of adhesive was applied to the second skin to be bonded. The adhesive layer on the second skin, like that on the first, was facing up to prevent the adhesive from flowing into the

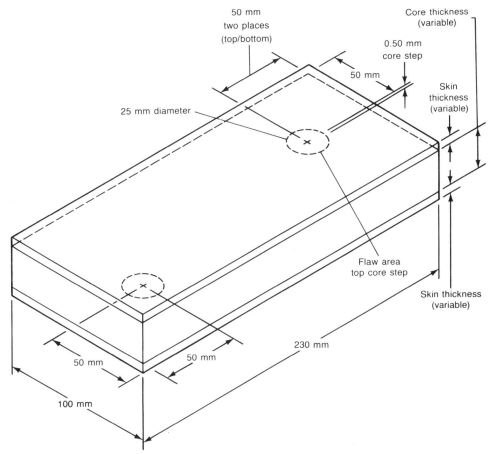

Fig. 10 Typical honeycomb 2024-T6 Al nonstepped test panels bonded with FMS 1013 type 1B adhesive. Source: Ref 14

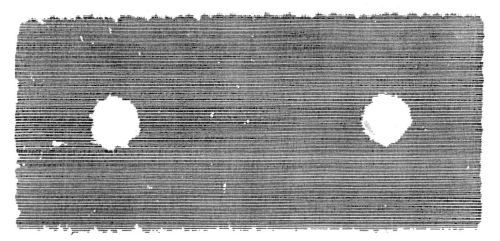

Fig. 11 Typical ultrasonic through-transmission C-scan recording. Source: Ref 14

Fig. 12 Through-transmission C-scan of a multi-layered bonded structure

The test panels were inspected by the immersion method using a reflectoscope with a pulser/receiver. The flat-faced transmitting and receiving transducers were 6.4 mm (¼ in.) in diameter with a frequency of 2.25 MHz. A typical C-scan recording obtained using the through-transmission inspection method is shown in Fig. 11. Implants located in multiple bond lines can be detected by this inspection method. Figure 12 shows the C-scan recording of a ten-ply adhesively bonded panel with metal substrates.

Weak Bond Inspection

The nondestructive inspection of adhesive joints and advanced composite structures has been successful in detecting disbonds, voids, delaminations, and foreign inclusions of a size that would affect the service life of a component. However, there is another type of bond defect that does not produce a void but still results in weak bond strength. This condition is created by improperly cleaned surfaces or a defective adhesive system, and it cannot be detected by even state-of-the-art NDI methods.

For the production inspection of aircraft structures, the inability of current NDI procedures to detect weak bonds is remedied by subjecting structures to proof loading as an acceptance inspection criterion (Ref 21). Proof tests are expensive and require extensive facilities. A more cost-effective method is needed to guarantee the joint integrity of adhesive bonds and the quality of composite laminates.

Another issue is that aircraft may suffer service damage or environmental degradation not anticipated in the damage envelope prescribed in the proof-loading acceptance test. The need for an NDI method to screen the bond strength in adhesive joints and

depressed core area. After the second skin was attached, the assembly was cured under the same conditions as those for the first skin. Both skins were then cleaned and the bond integrity was verified using the through-transmission inspection method.

A reference panel was fabricated as a calibration standard for the NDI methods to be evaluated. The panel was a honeycomb sandwich structure with a 2024-T6 aluminum skin, an aluminum honeycomb core, and FMS 1013 type 1B adhesive. The top skin was step-machined to provide a variation in skin thickness from 0.50 to 7.1 mm (0.020 to 0.280 in.). A 25 mm (1 in.) wide depression with a depth of 1 mm (0.040 in.) was machined across the length of the panel 13 mm (0.5 in.) from both ends.

composite laminates is especially urgent for field and depot inspection.

The problem of weak bond detection was investigated in several independent research and development programs and in a subsequent contractual research and development program sponsored by the Federal Aviation Administration (FAA) (Ref 1). An approach was developed that used stress waves generated by a high-power ultrasonic unit to impinge on the adhesive bond line, thereby disrupting weak bonds. The stress waves were of sufficient amplitude to disrupt marginal bonds so that they could be detected by conventional ultrasonic techniques. Bonds with acceptable strength levels were not affected by the stress waves.

A portable high-power ultrasonic bond strength screening system was constructed. Metal-to-metal, composite-to-composite, composite-to-metal, and composite-to-honeycomb-core adhesively bonded structures were fabricated and evaluated by the system. The power level of the ultrasound required to disrupt weak bonds without damaging good bonds was determined for each type of bonded structure. The effects of different types of defects in the bonded joints were studied by performing mechanical testing on the different types of specimens.

Test Specimens. Three types of specimens were made from 255 × 610 mm (10 × 24 in.) and 510 × 510 mm (20 × 20 in.) panels. Metal-to-metal bonded specimens made from 3.2 mm (0.125 in.) thick 2024-T3 aluminum sheets bonded by high-temperature-curing adhesives (EA 9649R or AF 147) represented type I. Type II were composite-to-metal and composite-to-composite specimens bonded by EA 9649R adhesive; both 10-ply and 20-ply graphite-epoxy composites were used to fabricate type II specimens. Honeycomb-core-reinforced beam specimens with graphite-epoxy skins and/or aluminum skins represented type III. Polyamide paper and aluminum honeycomb cores were used for the three types of specimens.

Each type of specimen consisted of two groups, one with strong bonds and the other with weak bonds. Strong-bond specimens were fabricated using normal bonding procedures, that is, proper surface preparation and proper cure cycles. Specimens with weak-bond conditions were prepared using improper surface treatments, bad adhesive, or improper cure processes. Improper surface treatments included the use of unetched surfaces, oil application or a release agent, water soaking, and exposure to high temperature and humidity.

Test Methods. To investigate the correlation between the input power level of high-power ultrasound and the bond strength of the specimens, compressive-shear or flatwise tensile tests were performed for each group of specimens. The compressive-shear test was used to evaluate the bond

strength of type I specimens. These specimens were originally fabricated as 255 × 610 mm (10 × 24 in.) panels and machined into 25 × 255 mm (1 × 10 in.) strips. In each group, some of the strip specimens were used in the high-power ultrasound test, and others were further machined into 13 × 25 mm (0.5 × 1 in.) compression specimens (Fig. 13a).

A total of about 2000 compression shear specimens of the type I configuration with various bonding conditions were tested. The average failure loads of type I specimens bonded with EA9649R, AF 147, and FM 300K adhesives are plotted in Fig. 14. Results for specimens with various bonding defects are shown alongside those for the control specimens, which were bonded using normal procedures without adverse treatment. Specimens with unetched bonding surfaces suffered a 10% loss in bond strength, and contamination of the bonding surfaces with oil or release agents produced a minimum 60% loss in bond strength. Some of the bonded panels that were prepared with oil contamination on the bonding surfaces were debonded during machining. The bond strength of these specimens was therefore estimated to be less than the lowest value observed in the compressive-shear test for type I specimens, that is, 14 MPa (2 ksi).

A flatwise tensile test was conducted to obtain the bond strength of graphite-epoxy-to-graphite-epoxy specimens. In all cases, the tensile failure was attributed to composite laminate failure instead of adhesive bond failure, even in the weak-bond specimens.

Figure 3 shows a typical flatwise tension specimen and test block configuration. The top and bottom surfaces of the test specimens were first adhesively mounted on the metallic test blocks, which were then threaded into the tensile test machine. The ultimate tensile strength was obtained by dividing the load-to-failure results by the surface area of the specimen. A high-temperature adhesive was used to bond the specimens on the flatwise test blocks. The block assembly was cured at a temperature of 175 °C (350 °F). Because the weak-bond specimens were undercured, they experienced additional curing during the curing of the specimen-test block assembly. Results of the tests, therefore, do not represent the true bond strength of the undercured graphite-epoxy-to-graphite-epoxy specimens.

The bond strength of type III honeycomb-core-reinforced sandwich specimens bonded with EA 9649R adhesive was determined by the flatwise tensile tests. Three or more 50 mm (2 in.) diam circular plugs were cut from each 100 × 255 mm (4 × 10 in.) specimen for testing. A few 25 × 255 mm (1 × 10 in.) strip specimens that were not suitable for the flatwise tensile test were evaluated in the sandwich beam test. The effect of defective adhesive and improper

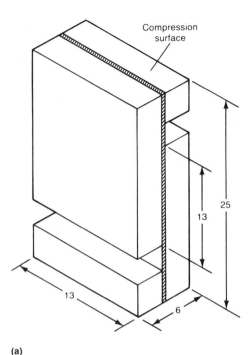

(a)

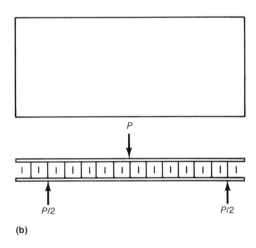

(b)

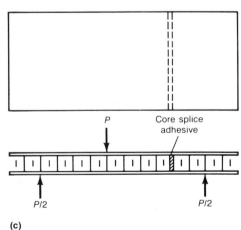

(c)

Fig. 13 Specimen configurations for a weak-bond screening system study. (a) Compression shear specimen. All dimensions in millimeters. (b) Simple sandwich shear beam for testing honeycomb core. (c) Simple sandwich shear beam for testing core splice adhesive. Source: Ref 1

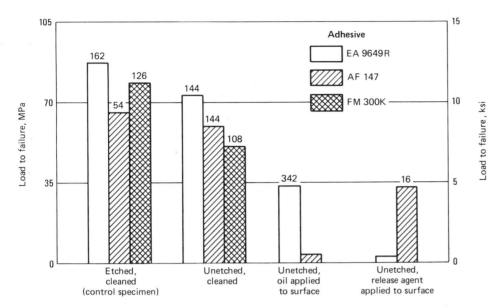

Fig. 14 Test results of type I (Al-Al) specimens. The numbers above the bars refer to the number of specimens tested.

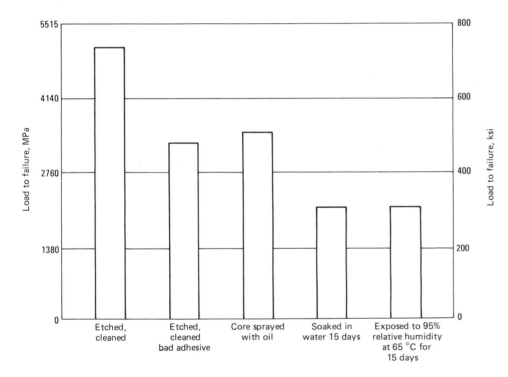

Fig. 15 Test results of type III specimens with EA 9649R adhesive

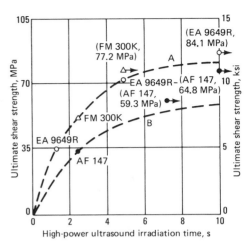

Fig. 16 Ultimate shear strength versus high-power ultrasound irradiation time for Al-Al specimens. Source: Ref 1

treatment of the bonded joints for specimens bonded with EA 9649R adhesive is shown in Fig. 15 in terms of the failure loads of the type III specimens. The use of bad adhesive and contamination of the core surface produced a 30% loss in bond strength. The environmental effect of a water soak and a combination of high humidity and temperature was a 60% loss in bond strength.

Test Results. A meaningful correlation between ultimate shear strength and high-power ultrasound irradiation time to induce

debond was achieved only for type I specimens (Fig. 16). The values in parentheses in Fig. 16 represent the ultimate shear strength of the specimens that could not be debonded after being irradiated for the amount of time indicated in the abscissa. Therefore, these values represent the high limit for the estimated correlation curves.

The dashed curves A and B are the estimated correlation curves for specimens bonded with EA 9649R and AF 147 adhesives, respectively. The data for specimens bonded with FM 300K are similar to those

for EA 9649R. Although they represent rough estimates, such correlation curves are necessary to ensure that the power setting for screening weak bonds in a specimen will not affect the good bonds.

REFERENCES

1. S.Y. Chuang, F.H. Chang, and J.R. Bell, "Adhesive Bond and Composite Strength Screening System Development Study," DOT/FAA/CT-84/2, U.S. Federal Aviation Administration, April 1984

2. D.J. Hagemaier, Adhesive-Bonded Joints, in *Nondestructive Evaluation and Quality Control*, Vol 17, 9th ed., *Metals Handbook*, ASM INTERNATIONAL, 1989, p 610-640

3. J.L. Caarls, R.J. Querido, and A.G. Julier, "Principles and Applications of the Fokker Bond Tester," Paper presented at the Air Transport Association of America Nondestructive Testing Forum, Sept 1985

4. L.J. Hart-Smith, "Effects of Flaws and Porosity on Strength of Adhesive-Bonded Joints," Report MDC J4699, McDonnell Douglas Corporation, Report Nov 1981

5. P.L. Flynn, "NDE of Adhesively Bonded Metal Laminates," Report ERR-FW-1777, General Dynamics, Dec 1976

6. R. Wong and R. Abbott, "Adhesively Bonded Composite Structures: Process Control, Non Destructive Inspection and Strength Effects," SAE Report 891044, Presented at the General Aviation Aircraft Meeting, April 1989

7. E. Segal and S. Kenig, Acceptance Criteria for Nondestructive Evaluation of Adhesively Bonded Structures, *Mater. Eval.*, Vol 47 (No. 8), Aug 1989, p 921-927

8. E. Segal and J.L. Rose, Nondestructive Testing Technique for Adhesive Bond

Joints, in *Research Techniques in Nondestructive Testing*, Vol IV, Academic Press, 1980

9. H.G. Tattersall, The Ultrasonic Pulse-Echo Technique as Applied to Adhesion Testing, *J. Phys. D, Appl. Phys.*, Vol 6, 1973

10. A.N. Mucciardi and R.K. Elsley, "Characterization of Defects in Adhesive Bonds by Adaptive Learning Networks," Paper presented at the Advanced Research Projects Agency/Air Force Materials Laboratory Conference (La Jolla, CA), 1978

11. F.H. Chang, P.L. Flynn, D.E. Gordon, and J.R. Bell, Principles and Application of Ultrasonic Spectroscopy in NDE of Adhesive Bonds, *IEEE Trans. Sonics Ultrason.*, Vol 23 (No. 5), 1976

12. E.A. Lloyd, Developments in Ultrasonic Spectroscopy in *Ultrasonics International Conference Proceedings*, IPC Guildford, 1975

13. Y.V. Lange, Application of the Ultrasonic Resonance Method to Nondestructive Evaluation of Adhesive Bonding, Translated from *Defektoskopiya*, No. 2, March-April 1974

14. G.L. Marsh, "Evaluation of the Shurtronics and Sondicator Bond Testers to Determine Inspection Capabilities," Report FZM-6222, General Dynamics, Dec 1973

15. R.J. Schliekelmann, Non-Destructive Testing of Adhesive Bonded Metal-to-Metal Joints, *Non-Destr. Test.*, June 1972

16. D.E.W. Stone, "Non-Destructive Methods of Characterising the Strength of Adhesive-Bonded Joints—A Review," Technical Report 86058, Royal Aircraft Establishment, Sept 1986

17. C.C.H. Guyott, P. Cawley, and R.D. Adams, The Non-Destructive Testing of Adhesively Bonded Structure: A Review, *J. Adhes.*, Vol 20, 1986

18. E.W. Bielss, "Infrared Inspection of Bonded Panels," Report FMR67-1429A, General Dynamics, Dec 1967

19. C.J. Pye and R.D. Adams, *NDT Int.*, Vol 17, 1984

20. R.J. Schliekelmann, Bonded Joints and Preparation for Bonding, Chapter 8 in *AGARD Lecture Series 102*, 1979

21. J.W. Goodman, J.W. Lincoln, and C.L. Petrin, "Certification of Composite Structures for USAF Aircraft," Paper presented at the Aerospace Industries Association of America Aircraft Systems and Technology Meeting (Dayton, OH), Aug 1981

SELECTED REFERENCES

● R. Abbott, "Design and Certification of the All-Composite Airframe," SAE paper 892210, Presented at the Aerospace Technology Conference, Sept 1989

● G.A. Alers and R.K. Elsley, "Ultrasonic Measurement of Adhesive Bond Strength," Report SC595.32SA, Project III, Unit A, Task 1, Rockwell Science Center, 1977

● G.A. Alers, R.K. Elsley, J.M. Richardson, and K. Fertig, "Ultrasonic Measurement of Interfacial Properties in Completed Adhesive Bonds," Paper presented at the Advanced Research Projects Agency/Air Force Materials Laboratory Conference (La Jolla, CA), July 1978

● G.A. Alers, P.L. Flynn, and M.J. Buckley, Ultrasonic Techniques for Measuring the Strength of Adhesive Bonds, *Mater. Eval.*, Vol 35, April 1977

● J.B. Beal, Ultrasonic Emission Detector Evaluation of the Strength of Bonded Materials, *NASA Tech. Memo.*, NASA TMX-53602, National Aeronautics and Space Administration, 1966

● A.W. Bethune, Durability of Bonded Aluminum Structure, *SAMPE J.*, July/Aug/Sept 1975

● F.H. Chang, R.A. Kline, and J.R. Bell, "Ultrasonic Evaluation of Adhesive Bond Strength Using Spectroscopic Techniques," Paper presented at the Advanced Research Projects Agency/Air Force Materials Laboratory (La Jolla, CA), July 1978

● G.J. Curtis, Acoustic Emission Energy Relates to Bond Strength, *Non-Destr. Test.*, Oct 1975

● W.A. Dukes and A.J. Kinlock, Non-Destructive Testing of Bonded Joints—An Adhesion Science Viewpoint, *Non-Destr. Test.*, Dec 1974

● H. Kwun and D.G. Alcazar, "Evaluation of Bond Testing Equipment for Inspection of Army Advanced Composite Airframe Structures," Report MTL TR 88-28, U.S. Army Materials Technology Laboratory, Oct 1988

● G.A. Matzkanin, C.G. Gardner, and G.L. Burkhardt "Improved Inspection for Bonded C-5A Structure," Report 15-1311, Southwest Research Institute, 1974

● P.A. Meyer and J.L. Rose, Ultrasonic Determination of Bond Strength Due to Surface Preparation Variation in an Aluminum-to-Aluminum Adhesive Bond System, *J. Adhes.*, Vol 8 (No. 6), 1976

● T.H. Norriss, Non-Destructive Testing of Bonded Joints—The Control of Adhesive Bonding in the Production of Primary Aircraft Structures, *Non-Destr. Test.*, Dec 1974

● J.L. Rose and P.A. Meyer, Ultrasonic Procedure for Predicting Adhesive Bond Strength, *Mater. Eval.*, Vol 31 (No. 6), 1973

● M.C. Ross, R.F. Wegman, M.J. Bodnar, and W.C. Tanner, Effect of Surface Exposure Time on Bonding of Ferrous Metals, *SAMPE J.*, July/Aug/Sept 1975

● R.J. Schliekelmann, "Holographic Interference as a Means for Quality Determination of Adhesive Bonded Metal Joints," Paper presented at the Internationaal Congrescentrum Rai (Amsterdam), Aug-Sept 1972

● T. Smith, "Surface Contamination: NDE Mapping and Effects on Bond Strength," Paper presented at the Advanced Research Projects Agency/Air Force Materials Laboratory (La Jolla, CA), July 1978

● G. Walters and D.M. Halom, "Destructive Test Results of 16T7303-A811 Rudder Assembly," Report 16PR7087, General Dynamics, Nov 1987

● D.O. Thompson, R.B. Thompson, and G.A. Alers, Nondestructive Measurement of Adhesive Bond Strength in Honeycomb Panels, *Mater. Eval.*, April 1974

● J.R. Zurbrick, "Nondestructive Test Technique Development Based on the Quantitative Prediction of Bond Adhesive Strength," Report AUSD-0331-70-RR, Avco Government Products Group, July 1970

● J.R. Zurbrick, E.A. Proudfoot, and C.H. Hastings, "Nondestructive Test Technique Development for Evaluation of Bonded Materials," Report AUSD-0494-71-CR, Avco Government Products Group, Nov 1971

Design Proof Testing

Ronald W. Johnson, Boeing Military Airplanes, The Boeing Company

ADHESIVE BONDING for aircraft components has been used extensively since aircraft designers discovered that bonded joints were structurally more efficient than mechanically fastened joints. One of the earliest and best known uses of bonding was for the Mosquito bomber, built in Great Britain from 1941 to 1945 (Ref 1). In recent times, adhesive bonding for composite and metallic components has been extensively used for both secondary and primary structures.

Examples of composite bonded secondary structures are the components shown in Fig. 1. Examples of composite bonded primary structures are the NASA/ACEE horizontal stabilizers (Ref 2) for the aircraft described in Ref 3 and 4. Examples of metallic bonded primary structures are the wing and fuselage stiffeners of the aircraft described in Ref 5 and the empennage components of the aircraft depicted in Fig. 2 (Ref 6).

Another example of adhesive bonding for primary structures was a developmental metal bond program (Ref 7) conducted for the U.S. Air Force to evaluate the benefits of bonding metal components for fuselage structures. In the development of all these structures, the extensive testing of coupons, subcomponents, components, and full-scale parts was performed to produce data to guide the design work, to establish allowables, and to provide test evidence that the final design had met all criteria.

This article discusses testing rationale and environmental considerations, as well as coupon, subcomponent, component, and full-scale testing. It also discusses the test approach for validation of bonded aircraft structure. The same approach is applicable for any structural development program where strength and/or stiffness margins of safety are critical.

Testing Rationale

For large aircraft components, such as the fuselage, wings, and empennage sections, complete test simulation of all critical load conditions is extremely expensive. Therefore, new designs are normally substantiated by analysis and supported by test evidence. Thus the first step that must be taken in defining a test program is to recognize that all test data must support the final design. The recommended testing rationale is to use the building block approach, shown in Fig. 3. In this process, the test results from each step in the building blocks are used in the following step so that eventually the full-scale test is achieved. Using this procedure, design surprises and "show stoppers" are kept to a minimum. The composite horizontal

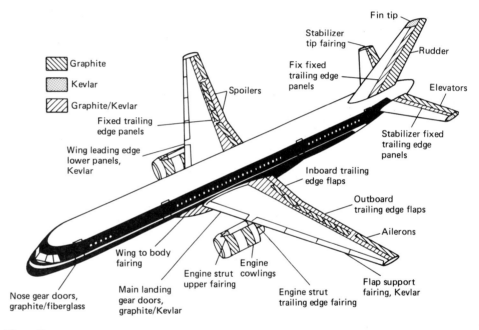

Fig. 1 Secondary structural composite components on the Boeing 757

Fig. 2 Boeing YC-14

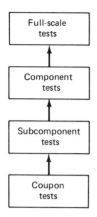

Fig. 3 Building block approach

stabilizer described in Ref 2 was developed using the building block test program approach. This approach was also used in the course of defining a development program for the application of advanced composites for the fuselage structures of large transports (Ref 8). The main objectives to be achieved at each building block level are described below.

Coupon tests are developed to produce test data for material screening and process control and to develop design allowables for margin-of-safety calculations. Coupons represent relatively small and flat specimens at the lowest level of complexity of a complete structure.

Subcomponent tests produce test data early in the developmental process to provide guidance for critical design details. These tests produce data in the form of allowables for complex details that contain critical load paths.

Component tests are used to validate design concepts early in the development cycle before committing these components to final design. They also produce test data as part of the final substantiation process.

Full-scale tests are typically static and cyclic in order to demonstrate strength, stiffness, durability (fatigue), and damage tolerance (safety). The latter may not require a full-scale test. Full-scale tests are designed to address particular questions about the structure:

- Is the internal load distribution obtained by analysis the same as that obtained in the test?
- Have there been any errors or omissions in design, manufacture, or quality assurance measures?
- Are there any unexpected deflections that impose functional constraints?
- Are there any deflections that significantly alter the load paths and increase the stress-strain level of a structural element (that is, are there large displacement effects)?

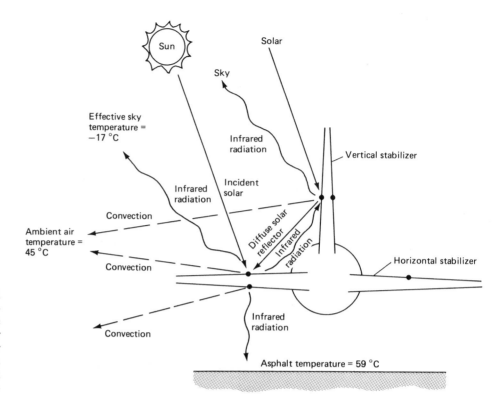

Fig. 4 Boeing 737 horizontal stabilizer thermal model

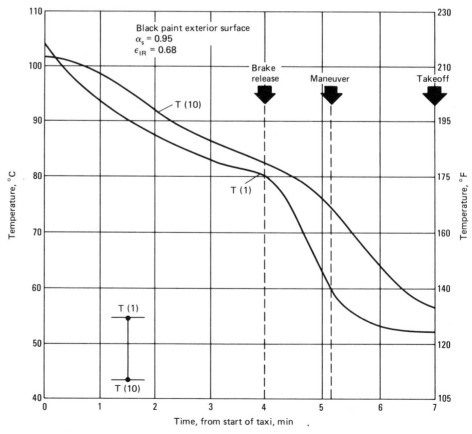

Fig. 5 Transient thermal response. T(1), thermal analysis node on skin surface. T(10), thermal analysis node on the skin stiffener inner chord

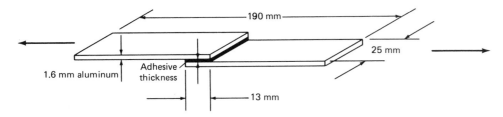

Fig. 6 Standard ASTM D 1002 lap-shear bond test

- Does the analytic assessment of the durability of composites, metals, and the hybrid structure, particularly in the interface area, correspond to the appropriate test data?
- Does the analytic damage tolerance evaluation of the structure match the test data?

Each of these building blocks is further described in a separate section within this article.

Environmental Considerations

All adhesives are influenced by temperature and moisture; the influence of these environmental conditions must be taken into account in the analyses. Environmental testing is normally performed at the coupon and subcomponent level because the relatively small sizes of the specimens can make testing at these levels economical.

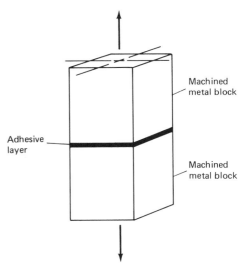

Fig. 7 Flatwise tension test

One of the major considerations for environmental testing is to determine the test environment to be imposed on the test structure. The most direct and conservative approach is to assume that all areas of the aircraft are thermally soaked at the environmental extremes. If this approach results in excessive weight penalties, then thermal analyses should be performed to produce a more realistic distribution of thermal environments in the individual areas of the structure to minimize the associated weight penalties.

To develop the environmental analyses, heat transfer models of the structure and the variations in the external environment must be defined. This requirement is essential for composite structures because the relatively poor heat transfer properties of composite components, in general, result in higher thermal gradients and temperatures than would exist for comparable metallic structures.

During the development of the horizontal stabilizer described in Ref 2, a thermal analysis of the external skin and stiffeners was performed to establish the maximum design temperature. The thermal model, shown in Fig. 4, contained the influence of the vertical fin, the runway, and the sky. Several different paint color combinations for the vertical fin and stabilizer were evaluated, and the results showed that the external skin temperature was strongly influenced by color combinations. The maximum skin temperature for the stabilizer was calculated using white paint on the vertical fin and black on the stabilizer because of the high solar reflection from the white fin surface and the high absorption capability of the black stabilizer surface. The external skin and stiffener inner flange temperatures for a thermal soak condition, using black paint on the horizontal fin and white paint on the vertical fin, for a 0 wind velocity day,

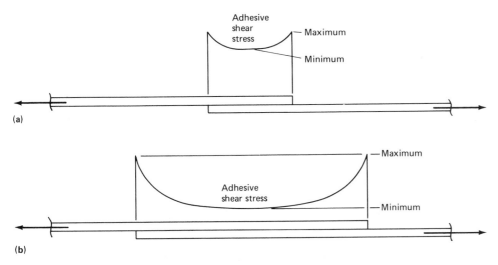

Fig. 8 Comparison of adhesive shear stress versus lap length. (a) Short lap length. (b) Long lap length

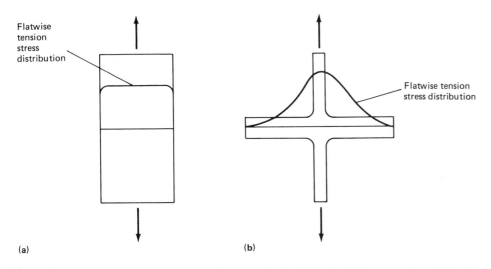

Fig. 9 Comparison of flatwise tension stress distributions. (a) Bonded blocks. (b) Back-to-back bonded T sections

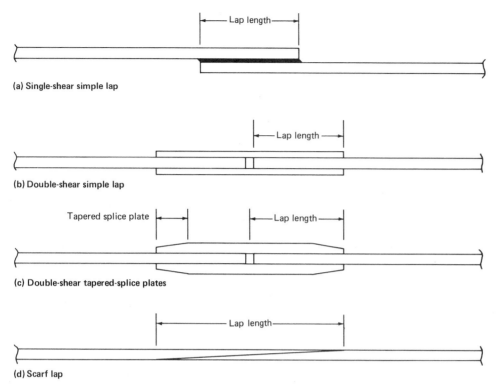

(a) Single-shear simple lap

(b) Double-shear simple lap

(c) Double-shear tapered-splice plates

(d) Scarf lap

Fig. 10 Examples of design-specific joint configurations

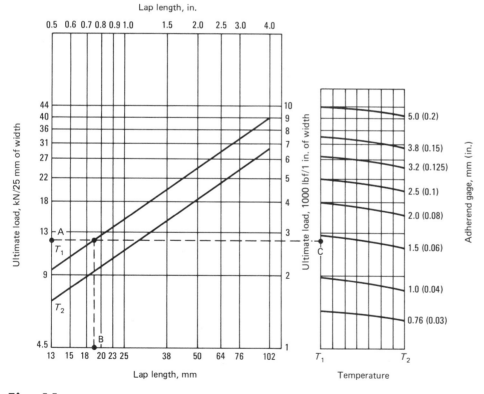

Fig. 11 Examples of bonded-aluminum lap-shear allowables

is shown in Fig. 5. This figure also shows how the temperatures are reduced after a 7-min taxi and takeoff. The temperature used as the design point was the tempera-ture of the inner flange of the skin stiffener 1.2 min after takeoff.

In addition to thermal analyses, moisture analyses need to be performed for compos-ite materials because strength properties are measurably influenced by moisture. An ex-ample of a moisture analysis is presented in Ref 2. The consideration of aircraft usage must be an integral part of the thermal and moisture analysis. A conservative approach would be to assume that the aircraft will spend its entire lifetime at the temperature and moisture extremes. This analysis ap-proach will usually result in excessive weight penalties. Therefore, consideration should be given to performing a basing study to develop a more realistic environ-ment for the lifetime of the aircraft. An example of a thermal, moisture, and basing study is presented in Ref 9.

Because the environmental analysis pro-cedure will have a very strong influence on structural sizing, the certifying agency must be informed of the planned procedure, and a consensus on the validity of the procedure should be obtained early in the design cycle.

Coupon Testing

Coupon testing can be divided into sever-al categories, that is, material screening, process control, and design allowables. Ma-terial screening coupon testing is performed to obtain the relative strengths of adhesive and surface preparation combinations at room temperature and at environmentally extreme conditions to select the best bond system for a particular design situation. Once the adhesive and surface preparation method have been selected, continuous pro-cess control testing is performed during fabrication of all components as a check on the consistency of the adhesive and of sur-face preparation. Allowables testing is per-formed at the coupon level to establish a statistical base of strength data for margin-of-safety calculations for design-specific de-tails.

The standard ASTM D 1002 test coupon, shown in Fig. 6, represents the universal standard test for adhesive shear strength. This specimen is an effective and inexpen-sive quality assurance coupon that is used to verify cure processing and to determine the relative shear strengths and environ-mental sensitivity of various adhesives for material screening. The flatwise tension specimen, shown in Fig. 7, is also used to obtain baseline material properties. It should be emphasized that the strength data obtained from either one of these specimen configurations should not be used for de-sign-specific details.

In the case of the lap-shear specimen (Fig. 6), the strength for different lap lengths cannot be ratioed up by the lap length because of the highly nonlinear shear stress versus lap length comparison curves, as shown in Fig. 8. In the case of the flatwise tension specimen, the vastly differ-ent stress distributions between a typical flatwise-tension structural detail and the

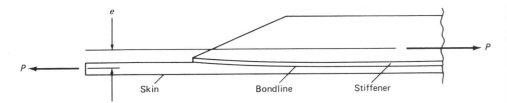

Fig. 12 Bonded-stiffener runout detail

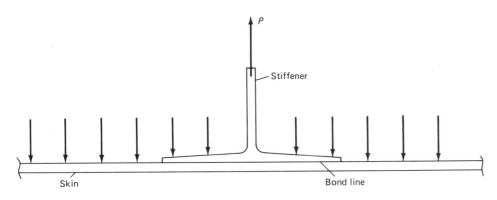

Fig. 13 Pressure restraint test detail

Fig. 14 Honeycomb panel attachment detail

simple flatwise-tension blocks is shown in Fig. 9. A more detailed discussion of the differences between simple lap-shear specimens and structural joints is presented in Ref 10.

Because a large number of variables exist, such as lap length, lap geometry, and adherend materials, and because a large number of specimens are required to develop the statistical base for allowables, design-specific details need to be minimized. Examples of lap splice geometries that can be considered for designs are shown in Fig. 10. Of the configurations shown, the single-shear simple lap (Fig. 10a) is the most common and the easiest to manufacture. However, this joint is asymmetric, and the adherends, when under load, bend to line up the load, resulting in high adhesive peel stresses that develop at the ends of the adherends.

The next most common joint is the double-shear simple lap (Fig. 10b). This joint is more efficient than the single-shear splice because adherend rotations and high peel stresses are reduced because of the symmetric load transfer. The double-shear tapered-slice plate joint (Fig. 10c) is more efficient than the double-shear simple lap because the tapered splice plates reduce the

peel stresses at the end of the lap. Tapering the ends of the adherends in the single-shear lap joint configuration would also improve joint efficiency. A detailed discussion of lap splice shear stresses and peel stresses is presented in Ref 11.

The scarf joint (Fig. 10d) is the most structurally efficient of all the configurations because the adherends stay lined up under load and peel stresses are minimized because of the tapered geometry. However, this joint is extremely expensive to manufacture over long lengths, and it is seldom used except for bonded repairs where external aerodynamic smoothness is a major design requirement.

Allowables for joint strength versus lap length and adherend thickness need to be obtained for all joint configurations considered for design. Environmental design parameters also need to be considered in the test matrix. An example of a design curve for bonded aluminum adherends that relates ultimate load capacity to lap length and adherend thickness for specified design temperatures is shown in Fig. 11. Given a specific design load in Fig. 11 (point A) and a specific temperature (T_1 for example) and dropping straight down at this intersection, a required lap length can be found (point B).

Point C will give the nominal thickness of adherence for a specific alloy.

Subcomponent Testing

The general classification of subcomponent testing is those tests performed on

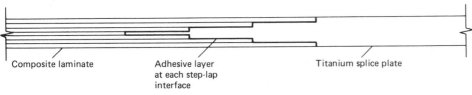

Fig. 15 Composite-to-titanium step-lap bonded joint

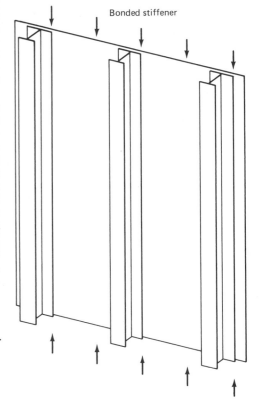

Fig. 16 Short-column crippling test specimen

Fig. 17 YC-14 honeycomb panel shear test

Fig. 18 YC-14 honeycomb panel stability test

design details that are critical to the strength, stiffness, fatigue performance, and damage tolerance capability of the airframe being developed. Early subcomponent testing is done at the preliminary design stage to guide the structural designer in determining whether new structural configurations will perform their desired function. Subcomponent testing is also performed on detail design parts to provide test substantiation for certification to the regulatory agencies. In many cases, subcomponent testing provides the data to support margin-of-safety calculations directly for details that contain very complex load paths. For these cases, several replicate tests have to be performed to provide a statistical confidence level.

Figures 12 to 19 exemplify various subcomponent tests. Figure 12 depicts a tension test of a bonded stiffener runout detail. The main consideration in this test is that

under a tension load, the eccentric load path causes tension stresses to exist at the end of the stiffener. Design parameters to be considered are the thickness ratios of the bonded parts, the rate of taper of the stiffener outstanding leg, and the adhesive properties. In many designs, antipeel fasteners are used to prevent this failure mode.

Figure 13 shows a pressure restraint test specimen in which internal pressure, due either to fuselage pressure or wing fuel pressure, is restrained by a bonded stiffener. The main design parameters in this test are the adhesive properties and the thickness and taper geometry of the stiffener flange bonded to the skin. Analysis methods for the design of stiffeners bonded to composite laminated details are found in Ref 12 to 14. Examples of pressure restraint tests are found in Ref 15.

Figure 14 is a photograph of a honeycomb panel attachment detail (Ref 6) in which the

load carried by the inner face sheet is sheared by the honeycomb core into the outer face. In a compression design case, the joint will be stability critical because of the eccentric load path. The main design parameters to be considered are the shear stiffness and strength of the core, the taper rate of the outer face sheet, the thickness ratio of the two face sheets, and the adhesive properties.

Figure 15 depicts a step-lap bonded transition between a fibrous composite laminate and a metal splice detail. The significant parameters to be considered in this specimen are the lengths of the overlaps and the number and orientation of the composite plies that are bonded to each step. Analysis procedures for the design of composite-to-titanium step-lap bonded joints are found in Ref 16.

Figure 16 shows a short column crippling specimen in which the skin is designed to buckle below compressive ultimate load

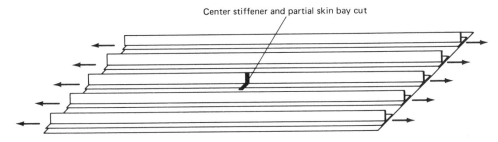

Center stiffener and partial skin bay cut

Fig. 19 Bonded stiffener tension damage tolerance panel

Fig. 20 Boeing 737 graphite composite horizontal stabilizer stub box test. Source: Ref 2

and the stiffeners carry the compressive column load to ultimate. The design concerns are the crippling stability of the stiffener element and the bond integrity between the stiffener and the sheet in the presence of the buckled skin due to compressive load.

Figure 17 is a photograph of a honeycomb shear panel test (Ref 6). Here the design concerns are the bond integrity between the skin and core and the core shear stiffness. For a skin-stiffener design, similar testing would be performed to evaluate the bond integrity between the stiffener and the skin in the presence of diagonal shear buckles. An example of a skin-stiffener shear panel test is found in Ref 15. It should be noted that skins may be stiffened by either honeycomb or discrete stiffeners and that their attachment to the skin may be either mechanical or adhesive bond.

Figure 18 is a photograph of a bonded honeycomb panel stability test (Ref 6) in which the design concern is the buckling stability of the panel sections between the support elements. In the case of multirib wing and empennage designs, the support points would be the ribs and spars. In the case of fuselage designs, the support points would be the body frames. For skin-stiffener designs, similar testing would be performed to evaluate the column stability of the stiffener sections between the support elements.

Figure 19 shows a bonded stiffener tension damage tolerance panel in which the design concern is the bond integrity between the skin and stiffener when the panel cut load is redistributed into the adjacent stiffeners. If the adhesive between the skin and adjacent stiffeners cannot assume the additional load caused by the cut, then the stiffeners will disbond from the skin and structural integrity will be lost. When designing cut stiffener damage tolerance panels, multiple skin and stiffener bays need to be included to ensure that the stresses in the cut region are not influenced by the panel free edges; that is, a sufficient number of uncut stiffeners must be provided between the cut stiffener and the panel free edge. A detailed description and test results of graphite composite subcomponent testing are given in Ref 2 and 15.

Component Testing

Component testing is normally referred to as the testing of complete, full-scale sections of the flight vehicle. Component tests are usually performed early in the design cycle to evaluate critical or highly loaded regions. Component test specimens must contain complete structural details to ensure that all load paths of the structure to be designed are fully represented. If a component is a piece of a large continuous structure, such as a wing, fuselage, or empennage section, the influence of the continuous structure on the test component must be fully represented to ensure that the load distribution within the test component is valid; that is, the load introduction at the boundaries of the component must realistically simulate the actual loads in the complete structure at these points. Whenever possible, results of finite-element structural analyses of the complete structure should be compared with the component test results to ensure that stiffnesses and test loading are properly represented.

Environmental tests should be considered at the component level rather than at the full-scale level for economic reasons. Damage tolerance testing should also be considered at the component level because component test configurations will contain all the alternative load paths for load redistribution, a necessary consideration in damage tolerance assessment.

Examples of component tests are shown in Fig. 20 to 24. Figure 20 is a photograph of an aircraft graphite composite horizontal stabilizer stub box test. In this test, the box section was mounted to a rigid strongback at the stabilizer lug attachment points. Finite-element analyses were performed to obtain representative stiffnesses of the existing aluminum center section. These calculated stiffnesses were then incorporated into the lug attachment points. The air loads and inertia loads that occurred within the box area were applied by bonded load pads. The outboard box loads were introduced by an aluminum transition section. The primary test objective was to evaluate the static strength adequacy of the design in the highly loaded and complex stabilizer to center section attachment area.

This test was performed early in the design cycle to ensure that critical composite design details such as the lug attachments, spars, and skins would perform as designed. Damage tolerance testing was also performed on this component. Because the design did not significantly change from preliminary to final form, the damage tolerance test results obtained from this component test were considered valid for, and were used as part of, the final certification.

Figure 21 shows a graphite composite outboard elevator test setup from a composite elevator development program (Ref 17). In this test setup, the elevator box section was supported at the hinge attachment points, and the air loads and inertia loads were introduced by bonded load pads. The primary test objective was to verify the torsional stiffness of the box structure. Finite-element models of the complete composite elevator and the test setup were developed, and the calculation of stiffnesses from these models and the test results were compared to the existing aluminum elevator to ensure that the composite elevator would provide the same flight and flutter characteristics.

Fig. 21 Boeing 727 graphite composite outboard elevator test

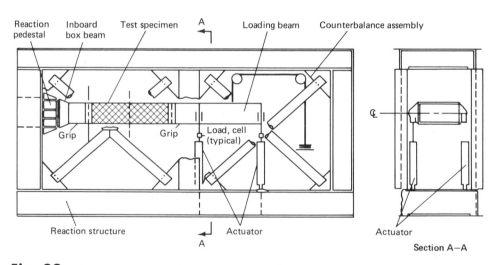

Fig. 22 Multirib wing box test fixture

Figure 22 shows the test fixture that was used to test the multirib box structure in a development program to determine the damage tolerance of composites (Ref 9). In this setup, the 2.5 m (8 ft) long test box was loaded as a cantilever beam. Shear and bending moments were introduced through a loading box beam at one end of the test box, while the other end was attached to the reaction pedestal through a grip transition box. Figure 23 shows the test setup. The primary test objective was to evaluate the damage tolerance of the wing box component. The test box contained two spars, five stringers, and four rib bays, which provided all the major load paths to ensure adequate load redistribution for damage tolerance investigations.

Figure 24 depicts a test setup for an aircraft aft fuselage. The test setup contains the production vertical fin, the horizontal stabilizer attachment fittings, and the production aft body, including the pressure bulkhead. The test article was cantilevered off a rigid strongback through a fuselage transition section. The structure was loaded by applying loads to the vertical fin and the horizontal stabilizer

Fig. 23 Multirib wing box in the test fixture

Fig. 24 Boeing 757 aft fuselage test

attachment points. The fuselage was pressurized to simulate high-altitude flight. The primary purpose of this test was to validate fatigue durability and damage tolerance. Because this article contained all production components, the test results were used as test evidence of fatigue durability and damage tolerance for certification.

Full-Scale Tests

The full-scale static test is the most important test in qualifying new structural designs. The parameters to consider when developing the basic requirements for this test are the type of test article, the type and number of load conditions, the usage environment to be simulated, the load level, and the type and quantity of data to be obtained. The ability of the test data to meet certification requirements must be inherent in each of these static test requirements.

The full-scale test enables the most realistic simulation of internal load distribution and the stress-strain level of each structural element. Because this test is the best representation of the true structural performance of the system, it must be a significant part of the certification or qualification process. Whether the test article is a major component or an entire airframe, it is considered representative of the production structure because it is assumed to have been fabricated according to production drawings using specification-controlled materials on production tooling, and to have followed fabrication and assembly specifications, and to have been inspected according to production quality control requirements.

The first step in planning the full-scale test is the selection of the load cases to be applied in the test. The selection of the critical load conditions must start with a review of the analysis of the structure, because the selection of the most critical load conditions for the structure is based on this analysis.

Minimum safety margins are usually not the sole consideration, although they are the main one, in the selection of critical conditions. Additional considerations may include:

- Stability-critical structures that have low but not minimum margins of safety
- Combined loading conditions that are difficult to analyze
- Major structural joints or intersections that are difficult to analyze

Once the critical load conditions have been selected, the means of load application must be established to facilitate the most cost-effective method of simulating the real flight and ground loads application. The detail method of loading a structure needs careful consideration. Selection of a load application method must take into account the

Fig. 25 Boeing 727 composite elevator full-scale test

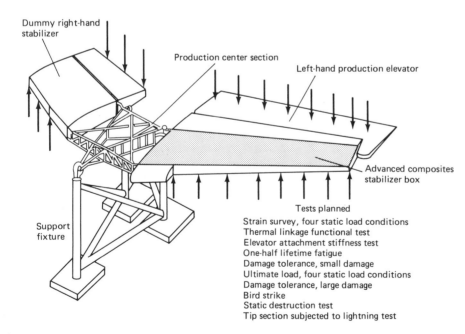

Dummy right-hand stabilizer

Production center section

Left-hand production elevator

Advanced composites stabilizer box

Support fixture

Tests planned
Strain survey, four static load conditions
Thermal linkage functional test
Elevator attachment stiffness test
One-half lifetime fatigue
Damage tolerance, small damage
Ultimate load, four static load conditions
Damage tolerance, large damage
Bird strike
Static destruction test
Tip section subjected to lightning test

Fig. 26 Boeing 737 composite stabilizer full-scale ground test setup

engineer's desire for the truest simulation of the load distribution, as well as the cost of the test setup. Loading methods include the use of formers that contact the structure on the exterior where the surface is supported by the substructure; direct fastening to the substructure by penetrating the surface panels, and bonding or mechanically fastening to the surface panels.

Formers usually provide a more concentrated loading and do not give as good a representation of the distributed air load as do the other methods. This method usually has a less complex loading setup and is therefore often the least costly.

Direct attachment to the substructure is usually similar in loading simulation and cost to the former method. However, for both composite and metal structures, care must be taken in the method of attachment to the substructure. If the attachment can be made at the same fastener locations that are used for attaching the surface panels to the substructure elements, the procedure is relatively easy. However, if special access holes are needed, this method is less acceptable for both metals and composites. The effects of holes are particularly bad for composites because of their stress-concentration sensitivity.

Direct surface attachment usually offers a better chance of uniform load representation, although this closer representation of the real vehicle structural load usually involves a more complex test setup. The application of the load directly to a composite surface must be done more carefully than to a metal surface because of the greater sensitivity to stress concentration and greater through-the-thickness weakness of the composite.

Once the load application method and test setup have been selected, the instrumentation required to collect data (for submittal as part of the qualification base for the structure) must be defined. The data display must:

- Ensure correct application and introduction of the load
- Allow monitoring of the testing in real time and protect the test article from being incorrectly loaded
- Validate that the loads applied produce not only the correct loading, but also the correct deflected shape
- Validate that the correct interaction of the applied load is being made in the correct ratio and magnitude for combined load conditions
- Provide a means to terminate loading automatically when loading errors are detected, thereby precluding a human reaction to the error and avoiding loss of the test article
- Validate that the internal strain distribution is as predicted by analysis

The types of instrumentation that are often applied in full-scale testing are strain gages; deflection measurement indicators; stress coats; photostress, moiré fringe, and acoustic emission detectors; and accelerometers. All data are electrically recorded, and computers are used to enhance the on-line presented critical locations data.

A discussion of full-scale testing relative to composite structures is presented in Ref 18. Examples of full-scale tests are shown in Fig. 25 to 27. Figure 25 shows the test setup for an advanced composite elevator (Ref

17). In this test, the elevator hinge fittings were fixed to a series of support points that could be moved to simulate the horizontal stabilizer deflected shape. The air loads and inertia loads were introduced into the elevator through pads bonded to the surface, as shown in Fig. 25. For each load case, the elevator hinge attachment supports were displaced to develop the stabilizer deflected shape, and the elevator was rotated to the angle setting corresponding to the load case. The air loads and inertia loads were then applied through the bonded load pads. Strain gage surveys were obtained to provide test evidence data for certification of the composite elevator.

Figures 26 and 27 show the test setup and the full-scale test of a composite horizontal stabilizer (Ref 2). As shown in the test setup (Fig. 26), the full-scale test stabilizer was supported on a production aluminum center section. The air loads and inertia loads on the test stabilizer were balanced by a comparable set of loads applied to a dummy box section on the other side of the center section. Static load strain surveys, durability, and damage tolerance tests were performed on this test article to provide test evidence for certification of the composite horizontal stabilizer.

Fig. 27 Boeing 737 composite stabilizer full-scale test

REFERENCES

1. "DeHavilland, D H 98, Mosquito," Jane's All the World Aircraft, 1948
2. R.B. Aniversario, S.T. Harvey, J.E. McCarty, J.T. Parson, D.C. Peterson, L.D. Pritchett, R.D. Wilson, and E.R. Wogulis, "Design, Ancillary Testing, Analysis, and Fabrication Data for the Advanced Composite Stabilizer for Boeing 737 Aircraft," Volume II, Final Report, NASA CR-166011, The Boeing Company, Dec 1982
3. "Bell/Boeing V-22," Jane's All the World Aircraft, 1988
4. "Navy A-6 Intruder, Grumman/Boeing," Jane's All the World Aircraft, 1988
5. "Fokker F-27 and 28," Jane's All the World Aircraft, 1964
6. "Boeing YC-14 Advanced Transport Prototype," Jane's All the World Aircraft, 1976
7. E.W. Thrall, Jr., et al., "Primary Adhesively Bonded Structure Technology, (PABST), Phase II; Detail Design," Technical Report AFFDL-TR-77-135, Douglas Aircraft Company, Aug 1977
8. R.W. Johnson, L.W. Thomson, and R.D. Wilson, "Study on Utilization of Advanced Composites in Fuselage Structures of Large Transports," NASA CR 172406, Boeing Commercial Airplanes, Feb 1985
9. R.E. Horton, R.S. Whitehead, et al., "Damage Tolerance of Composites," Vol 1, AFWAL-TR-87-3030, U.S. Air Force Wright Aeronautical Laboratories, July 1988
10. L.J. Hart-Smith, "Differences Between Adhesive Behavior in Test Coupons and Structural Joints," Paper presented to American Society for Testing and Materials, Adhesives Committee D-14 Meeting, Phoenix, March 1981
11. L.J. Hart-Smith, "Adhesive Layer Thickness and Porosity Criteria for Bonded Joints," Report AFWAL-TR-82-4172, U.S. Air Force Wright Aeronautical Laboratories, Dec 1982
12. H.C. Tsai, "Solution Method for Stiffener-Skin Separation in Composite Tension Field Panel," Report NADC 82171-60, Naval Air Development Center, Oct 1982
13. H.C. Tsai, "Approximate Solution for Skin-Stiffener Separation Including Effect of Interfacial Shear Stiffness," Report NADC 83131-66, Naval Air Development Center, Nov 1983
14. S.B. Biggers and J.T.S. Wang, "Skin/Stiffener Interface Stresses in Composite Stiffened Panels," NASA CR 172261, Lockheed-Georgia Company, Jan 1984
15. P.J. Smith, L.W. Thomson, and R.D. Wilson, "Development of Pressure Containment and Damage Tolerance

Technology for Composite Fuselage Structure in Large Transport Aircraft," NASA CR 3996, The Boeing Company, Aug 1986

16. L.J. Hart-Smith, "Analysis and Design of Advanced Composite Bonded Joints," NASA CR-2218, Douglas Aircraft Company, Jan 1973

17. D.V. Chovil, S.T. Harvey, J.E. McCarty, O.E. Desper, E.S. Jamison, and H. Syder, "Advanced Composite Elevator for Boeing 727 Aircraft," Vol I, Technical Summary, NASA CR-3290, The Boeing Company, 1981

18. J.E. McCarty, Full-Scale Testing, in *Composites*, Vol 1, *Engineered Materials Handbook*, ASM INTERNATIONAL, 1987

Design Considerations Unique to Sealants

James E. Thompson, Rhone-Poulenc Inc.

SEALANTS are unique materials used to separate liquid or gaseous phases. They are generally used between sections of a rigid or semirigid container for one or both phases. The places at which these sections come together are called joints. Because some movement is typically experienced between the sections, the sealant must often have some movement capabilities.

Expressed in terminology similar to that of adhesive technology, a joint consists of two substrates of the same or different material, with a layer of sealant between them. However, the possible orientations of the substrates with respect to one another are much broader with a sealant than with an adhesive. Sealants are used to seal joints in liquid containers, ranging from fish tanks to fuel tanks, or in chemical processing equipment, where they function to hold liquid in or keep air out. Sealants are also used in joints in construction materials, such as curtain wall panels, to separate indoor air from outdoor air for environmental control and weatherproofing. Additional uses include various electronics, aerospace, and marine applications (Ref 1-3).

Sealants, as opposed to seals or gaskets, are generally in a viscous, that is, liquid or paste, state when they are applied to the joint. They then may be cured or transformed to an elastic or plastic state by various means. The sealant material properties may in fact approach those of a preformed gasket in final usage. Whereas gaskets and seals function by being compressed between the substrates, a sealant is generally expected to provide a seal by adhering to the substrates; in certain cases, however, such as some cured-in-place gasketing applications, the sealant may be cured before joint assembly. An important special case is that of preformed sealant tapes, which might be considered an intermediate form between gaskets and sealants.

Sealants, in contrast to adhesives, generally function to accommodate movement of the substrates with respect to one another, rather than to hold the substrates together. The sealant is thus expected to allow, rather than restrict, movement. The primary function of a sealant is to seal, but it usually possesses a certain adhesive character as well. If it also has sufficient physical strength, it can be used partially or entirely as a structural element. Such is the case in modern structural sealant glazing, which relies entirely on the adhesive properties and physical strength of a structural-quality sealant to hold the glass in place in the face of extreme environmental stresses and in the absence of mechanical fasteners.

Most useful sealant materials must possess a combination of properties. The design considerations for sealants share some features of those for adhesives and those for gaskets, but are fundamentally different from both.

High-performance sealants that cure to a highly elastomeric state, adhere tenaciously to substrates, and have optimum performance that is highly dependent on joint design are now available. The principles of joint design outlined in this article, while applicable to sealants in general, are specific to these modern high-performance sealants.

General Design Considerations

Often when structures requiring joints are designed, sealant joint design and installation workmanship are an afterthought. Thus it is not surprising that sealants in joints often fail, resulting in rather expensive sealant replacement and possibly joint redesign. While joint design is unique for each application, some general principles are useful.

Stresses on a Joint. The stresses on a joint may be resolved into shear stress and extension/compression stresses, which are caused by movement of the substrates due to thermal expansion and contraction; environmental forces, such as wind loads and traffic loads; or other forces. These forces deform the sealant, and this deformation is called strain. Sealants must be able to accommodate a certain amount of movement; this property of a sealant is called its movement capability (Ref 4).

Figure 1 shows a simple butt joint and the extension/compression strain experienced by the sealant. A sealant with a movement capability of ±50% can undergo a strain of 50% in extension or compression. It is obvious that a joint with a greater distance between the substrates can accommodate a greater amount of movement. For example, a sealant with a movement capability of ±50% can accommodate 5 mm (0.2 in.) of movement in a 10 mm wide (0.4 in.) joint, whereas it will accommodate only 2.5 mm (0.1 in.) of movement in a 5 mm (0.2 in.) joint. These relationships are not exactly linear, but most designers treat them as such for the sake of simplicity.

The strain caused by shear forces on a joint is shown in Fig. 2. Using a simple Pythagorean relationship (Ref 5), the strain on a joint undergoing shear deformation can be adequately approximated for most joint applications (Fig. 3).

Materials Considerations. Good joint design is dependent on the properties of the sealant, on the characteristics of the materials forming the joint (substrates), and on the design configuration of the joint. Assuming that the general sealant properties

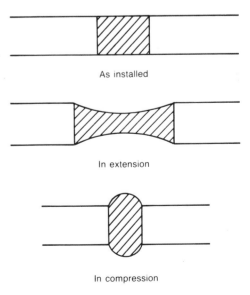

As installed

In extension

In compression

Fig. 1 Simple butt joint

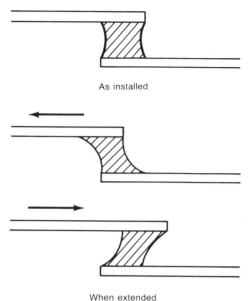

Fig. 2 Simple lap joint

(labels within figure: As installed; When extended)

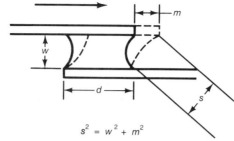

$$s^2 = w^2 + m^2$$

Fig. 3 Strain on a lap joint

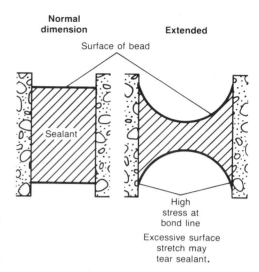

Fig. 4 Bead thickness and stress

(labels within figure: Normal dimension; Extended; Surface of bead; Sealant; High stress at bond line; Excessive surface stretch may tear sealant. Low stress at bond line; Surface stretch is not excessive if a thin bead is used.)

are suitable for the application, for example, chemical and environmental resistance, high- and low-temperature performance, electrical properties, and gas impermeability, there are certain other sealant and substrate material characteristics that must be considered by the joint designer:

- Adhesion of the sealant to the substrate
- Cohesive strength of the sealant
- Cohesive strength of the substrate
- Movement capability of the sealant

Adhesion of the sealant to the substrate is an inherent property of the sealant-substrate interface and may be enhanced by the use of a primer. Most manufacturers provide information on available primers, and their recommendations should be followed.

The cohesive strength of the sealant, that is, its resistance to tearing internally, is an obvious consideration. It will determine the minimum cross section necessary for the sealant in the joint.

An often neglected, but critical, factor in joint design is the cohesive strength of the substrate. A substrate that does not have sufficient strength to support the stresses that develop at the sealant-substrate interface will fail internally. This situation occurs most often with composite materials, such as concrete or laminates, or natural materials, such as sandstone, when used in conjunction with high-modulus sealants. A special case is that of a coated substrate, where the coating may not have sufficient adhesion to the substrate. Pretesting of the sealant on a sample of the actual substrate to be used is always recommended because the sealant itself may, through the release of solvent or by-product, affect the substrate.

If the sealant has sufficient adhesion to the substrate, the most important sealant

property is its movement capability, which is an intuitively obvious concept, yet is difficult to determine from basic sealant properties. It is best determined through functional testing, but some general guidelines can be stated. The movement capability of a sealant is a function of its elongation at failure (ultimate elongation) and its modulus (stress at a given strain, most often reported as 100% modulus, or stress when the sealant is elongated 100%, where the lower the modulus value, the higher the joint movement capability). Elastic recovery, that is, the ability of the sealant to return to its original shape after being stretched (or compressed), also plays a major role in movement capability. If the joint design allows the sealant to be repeatedly stressed beyond its elastic recovery limit, the joint will ultimately fail.

For the joint itself, the important design considerations are:

- Sufficient sealant-substrate interface to provide adequate adhesion
- Maximum surface area-to-volume ratio to reduce strain
- Sufficient cross-sectional area to avoid cohesive failure
- Accessibility
- Accommodation of sealant cure requirements

Because sealants may fail either cohesively or adhesively, both the cohesive strength of the sealant itself and the adhesion of the sealant to the substrate are important.

A sealant that is too weak (has too-low a cohesive strength) may adhere to the substrate, but will not afford a satisfactory seal because it will fail in its bulk. This situation is most common with noncuring sealants, but may occur with any sealant if the joint design is not correct. The strength of the sealant in the joint is obviously dependent on its cross-sectional area. The larger the cross section, the more resistant to cohesive failure the joint will be. If cohesive strength were the only strength consideration, joints with a high cross-sectional area would be preferred. In practice, however, joints usually fail adhesively.

The adhesive strength of the joint is a function of the adhesive interface between the sealant and the substrate. Although it

might seem that a large interface would be better, in practice the adhesive interface rarely fails catastrophically. Rather, it begins to fail in a peel mode at the points of highest stress. Failure of the adhesive interface of a joint occurs when the local stress exceeds the adhesive strength. Thus, it is the stress at the bond line that must be minimized.

The stress at the bond line is a function of the strain (or elongation) of the free surface of the sealant. Because most high-performance sealants are very elastic in their behavior, the value of Poisson's ratio for these materials is approximately 0.5 and their volume remains constant as they are elongated. This important concept is shown in Fig. 4. A thinner sealant joint produces less stress at the bond line and accommodates more joint movement than a thicker bead.

Two other considerations that can easily be overlooked when designing a joint are accessibility (it must be possible to apply the sealant to the joint easily, as well as to clean and prepare/prime substrates) and accommodation of the cure requirements of the sealant. Many sealants cure by absorption of moisture, and many release solvents or volatile by-products, which must be allowed to evaporate freely. Therefore, the joint design must provide sufficient free surface area for these processes to occur.

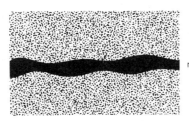

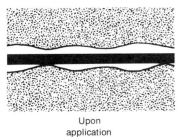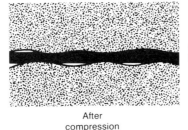

Formed-in-place gasket fills most voids and surface irregularities

Preformed gaskets may not fill voids or scratched or nicked components

Fig. 5 FIPG conformability to mating surfaces

A final consideration in joint design is the sealant application conditions. Because a joint is designed to accommodate a certain amount of movement, it will be at a certain point in its movement cycle when the sealant is applied. If the sealant is applied and cured when the joint is at an extreme point of movement, the sealant may not provide the movement capability originally calculated.

Applications

In actual practice, the uses of elastomeric sealants may be broadly grouped into thin bond line and thick bond line joint applications. The former include many liquid containment applications, such as automotive valve and differential cover sealing, which usually involve the parts being held together with mechanical fasteners. The latter include weather-sealing and glazing applications, in which the parts are fixed to a stationary grid or superstructure. Although a variety of possible joint types and designs can be listed, it is far more informative to illustrate several application types to demonstrate the design principles.

Form-in-Place Gasketing. Early joint sealing compounds for automotive gasketing applications, developed at the beginning of this century, were applied to cut gaskets made of cork, paper, mineral composition, or soft metals and gave some adhesive character to the gasket-substrate interface. These compounds were made from natural oils and plasticizers and were usually called gasket dressings. In the 1960s, high-performance materials, that is, one-part moisture-curing silicone sealants, were first used as replacements for the gaskets themselves. This gasketing technique came to be known as form-in-place gasketing (FIPG) and has been in use ever since, with various modifications, in both the automotive original equipment market (OEM) and aftermarket (Ref 6). The same technology and joint designs are used for environmental seals of electrical enclosures and various tank sealing applications.

Very similar applications exist for anaerobic sealants (Ref 7), but because of their rigidity and cure characteristics, anaerobics are limited to joints with a tight fit and little movement. Anaerobics are ideal for certain pipe flange applications and press-fit joints, such as shaft seal collars that intimately contact metal (necessary to catalyze the curing reaction) and exclude air (also necessary for curing).

In FIPG applications, the sealant is applied to the substrates, and the parts are joined while the sealant is still wet. This technique provides several advantages, foremost of which is the conformability of the wet sealant to surface irregularities (Fig. 5). Whereas a dry, preformed gasket will ride on the high spots of the surface, a semiliquid sealant will fill the entire space between the surfaces to be sealed, and thus may lower machining costs. Other advantages over preformed gasket technology include inventory reduction, accommodation to part configuration changes, possible material savings, and production volume adaptability, from low-volume hand applications to medium-volume semiautomated equipment to high-volume fully automated equipment. FIPG technology is best suited to joints that are infrequently disassembled; generally the sealant must be replaced for reassembly.

Because a sealant is often expected to function in a joint that has dissimilar substrates, such as aluminum and steel, and in environments with large temperature gradients, such as in automobile engines, some movement capability is required, and thus joint thickness must be controlled. Several methods of controlling joint thickness are in general use.

The most positive method of controlling joint thickness is by the use of spacers. Spacers may be integral to one or both of the mating parts, as shown in Fig. 6, or they may be added as washers during the construction of the joint. A less desirable method is to torque the fasteners to less than the amount of torque required to squeeze out the sealant completely from the joint, in which case the sealant bead size and rheological characteristics become critical. In all cases, some method of locking the fasteners must also be considered.

Because the joint is sealed without compression of the cured sealant, it is the adhesion of the sealant to the substrates that allows the joint to remain leak-free. Some squeeze-out of the sealant is desirable in these configurations (Fig. 6) because the cured bead of excess sealant will assist the primary seal and, in case of pressure build-up in the interior of the sealed container, may relieve the adhesive interface of some stress by acting as an O-ring. As with other sealant applications in which the joined parts are held in place with fasteners, the primary motion is in the lateral direction, producing shear forces on the sealant.

A generally accepted joint design is one 13 mm (0.5 in.) or more in width and 1.0 mm (0.040 in.) or more in gap size when a typical silicone sealant is used. Obviously, cleanliness of the substrate, cure time before pressure testing, rheological properties of the sealant, and bead size and positioning will contribute to the effectiveness of the joint sealant. Testing should be conducted before the joint design is finalized.

In these applications, some alternate joint designs are obvious. A groove configuration (Fig. 7a), while controlling sealant thickness, will negate the joint movement capability of the sealant because the material in the groove has no room to flex and elongate (Fig. 7b). The same is true in a tensile stress mode because the material at the exterior corners of the groove will experience extreme stress, resulting in an adhesive failure in a peel mode (Fig. 7c). Although this configuration has been used in some automotive joint designs in Europe, it has proved to be less than satisfactory for general use and requires an accelerated system that can cure in the absence of air or outside moisture.

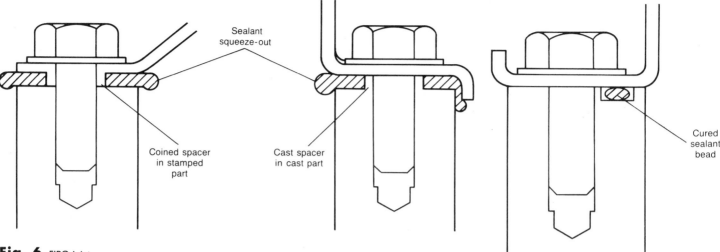

Fig. 6 FIPG joint

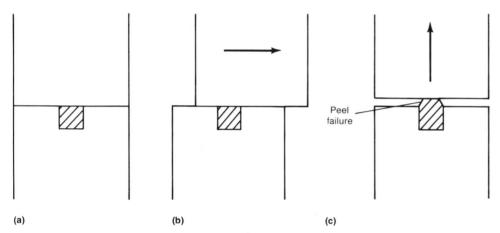

Fig. 7 FIPG groove configuration

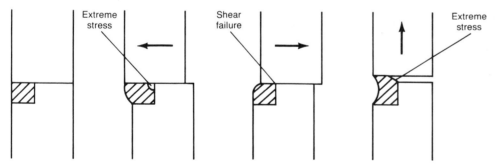

Fig. 8 FIPG step configuration

Fig. 9 CIPG joint

A step configuration of the joint will result in similar types of failure (Fig. 8). Joint configurations similar to the step joint configuration are used with thicker joints, but bond breakers are recommended to reduce strain at the enclosed face of the joint.

Cure-in-Place Gasketing. Parts using form-in-place gasketing may also be assembled after the sealant has cured. These systems have come into use as an outgrowth of the original FIPG technology and are referred to as cured-in-place gasketing (CIPG), or cured FIPG. A different type of joint design is preferred for these applications.

Because CIPG systems form a seal by compression of the cured sealant bead, adhesion to both surfaces of the joined parts is not necessary. This design is particularly suitable for joints requiring frequent disassembly, including many automotive parts, such as thermostat housings, and many electrical enclosures.

Because there is a limit to the compressibility of the cured sealant bead, some method of limiting the compression must be used. The same methods used to ensure a minimum thickness of the wet form-in-place sealant bead may be used, but unlike the wet gasket method, the preferred method is to provide a groove, or step, in which the sealant can rest (Fig. 9). In a typical automotive application using silicone sealant, the groove allows a 40 to 70% squeeze. For example, in a joint with a 1.6 mm (1/16 in.) diam bead, a satisfactory groove would be 0.8 mm (1/32 in.) deep (50% squeeze) and 4.8 mm (3/16 in.) wide.

The cured-in-place joint design requires application of the sealant enough in advance of assembly to allow sufficient curing of the sealant bead. Although this is a disadvantage in some cases, in most cases the sealant may be applied several days in advance (for instance, by a fabricator who subsequently ships the parts to an assembly plant to be used days or weeks after sealant application). An additional advantage is that no sealant is present on the final assembly line, and the problems associated with handling a reactive substance are eliminated. Because the CIPG requires a precise bead size, most applications are made by automated equipment, which easily accommodates two-component sealant systems. Other accelerated curing systems are also available for this application, including heat-accelerated systems and ultraviolet-light-accelerated systems.

Weather Sealing. Early construction techniques for large buildings used heavy masonry walls. Joints were generally packed with natural fiber, bituminous compounds, or nonflexible mortar. The massive construction minimized temperature change within the building material and resulting expansion and contraction. Leakage of

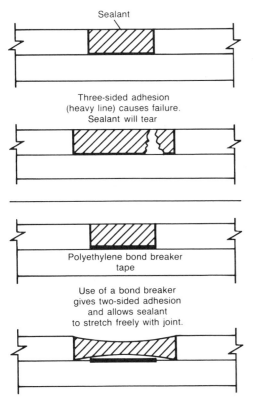

Fig. 10 Bond breaker

these joints was often masked by the ability of the construction to absorb rain without admitting water into the interior.

Curtain wall construction, pioneered after World War II and now widely used in commercial construction, uses a relatively thin skin of panels. These panels change temperature quickly and as a result undergo rapid and extreme expansion and contraction, placing a high demand on the joint sealant. Additionally, the thin outside skin, often made of glass, makes any leakage of the joint obvious.

Joints of this type require highly elastic and weather-resistant sealants with excellent adhesion to the substrates. Often, sili-

cone, polyurethane, and polysulfide sealants are used. Although joint design may be partially dependent on sealant type and substrate, some general design considerations are useful (Ref 8-10).

As noted in the section "General Design Considerations" in this article, the sealant in a simple butt joint must have a certain extension capability, which involves both the elongation properties of the bulk-cured sealant and its adhesion to the substrates. The volume of sealant in a joint, as well as the surface area bonded to the panel, will remain constant, so that an extended sealant bead will "neck down," as shown in Fig. 4.

Failure of the adhesion interface of a joint occurs when the local stress exceeds the adhesive strength. To minimize stress on the bond line, a thinner bead is generally desirable (Fig. 4). Because the sealant may also fail cohesively, that is, by tearing internally, and because this mode of failure is more likely to occur with a smaller cross section, some minimum cross section is necessary. Most sealant manufacturers require a minimum sealant design depth of 3.2 mm (0.125 in.), though in practice a minimum depth of 6.4 mm (0.25 in.) is more typical. Generally, a sealant depth of half the joint width is recommended. Another technique to minimize stress in a joint is to tool a concave surface on the sealant bead. This increases the surface area while decreasing the volume.

The required joint width is determined by the movement capability of the sealant, as well as by other considerations, such as tolerances of construction. For example, with a sealant having a movement capability of ±50%, the designed joint width should be at least twice the expected joint movement, plus tolerances. For a sealant with a joint movement capability of ±5%, the designed joint width should be 20 times the expected joint movement plus tolerances. For joints experiencing shear stresses, the strain on the joint can be approximated by using a

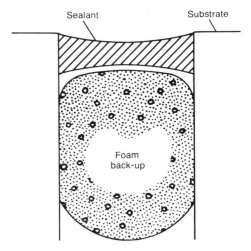

Fig. 11 Typical joint detail

Pythagorean relationship (Fig. 3). Sealants intended for these applications are rated with respect to joint movement capability by the manufacturers, though current building practice is to assume a movement capability of ±25% or less.

In practical use, the sealant bead must be backed by something. This requirement introduces the concept of a bond breaker (Fig. 10). Three-sided adhesion will accelerate failure of the sealant in the joint by concentrating the stress at the interior corners of the sealant bead. By using a bond breaker, such as polyethylene tape to which the sealant will not adhere or a foamed shape that will not restrict sealant movement, the sealant bead can stretch freely with joint movement.

A typical finished weather-sealing joint is shown in Fig. 11. The backup material is usually an extruded polyethylene or polyurethane foam rod with a round cross section. This allows sealants to deform freely,

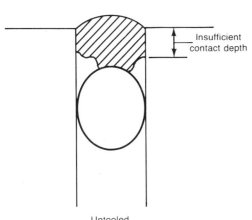

Fig. 12 Joint tooling

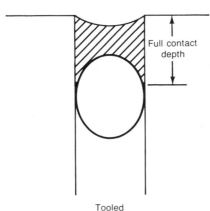

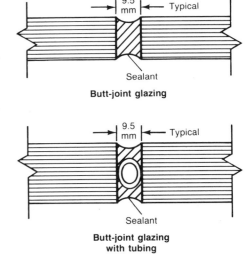

Fig. 13 Butt-joint glazing

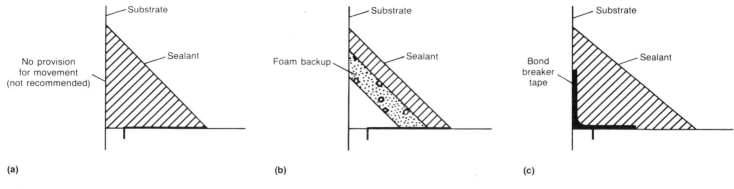

(a)   (b)   (c)

Fig. 14 Fillet joints

as with a bond breaker. The placement of this backer rod determines the thickness of the finished sealant bead. It is important to tool the sealant in the joint to provide good adhesion contact and to fill the joint opening completely with sealant (Fig. 12).

Butt glazing joints (Fig. 13) require a different type of application technique. If no tubing is used in the interior of the joint, one side of the joint is taped along its entire length, after which the sealant is applied and tooled from the opposite side. When the tooled surface has skinned over, the tape is removed and that side is tooled. An alternate method of constructing this type of joint is to apply and tool the sealant from both sides simultaneously, but this requires experienced installers and is not recommended because of the possibility of air entrapment.

Fillet Sealant Joints. Commonly used in all types of construction and many industrial applications, the fillet joint (Fig. 14a) is prone to failure if not properly designed (Ref 11). The same basic design principles apply to this joint that apply to others. The adhesion of the sealant to the substrate must be maximized and the sealant bead must be allowed to stretch freely. A pre-

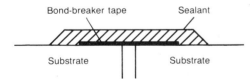

Fig. 15 Band-aid joint

ferred design for a fillet joint is shown in Fig. 14(b). A practical problem with this design is the unavailability of backing materials with a triangular cross section. An alternative design is shown in Fig. 14(c), where a bond-breaking tape is applied directly to the substrate. The elongation of the sealant in relationship to the joint movement can be determined from simple geometric relationships. The points of maximum stress on the adhesive interface will depend on the actual sealant cross section and the movement of the substrate relative to each other, but will generally be directly related to the highest strain (highest elongation) of the sealant surface.

Band-Aid Joints. A type of joint developed for use in special applications or in the repair of improperly designed joints is the band-aid joint (Fig. 15). The objective of the band-aid joint is to provide sufficient free stretch of the sealant to allow expansion if the dimensions of the space between the substrates are not adequate. The ability of this type of joint to accommodate substrate movement will generally be determined by the minimum distance between the points bonded to the substrate, because that is usually where the highest sealant strain will occur. The joint should be designed with this in mind.

REFERENCES

1. A. Damusis, Ed., *Sealants*, Van Nostrand Reinhold, 1967
2. J.W. Prane, Sealants and Caulks, in *Handbook of Adhesives*, I. Skeist, Ed., Van Nostrand Reinhold, 1977
3. J.P. Cook, *Construction Sealants and Adhesives*, Wiley-Interscience, 1970
4. T.F. O'Connor, Ed., *Building Sealants: Materials, Properties and Performance*, STP 1069, American Society for Testing and Materials, 1990
5. F.W. Shishler and J.M. Klosowski, Sealant Stresses in Tension and Shear, in *Building Sealants: Materials, Properties and Performance*, STP 1069, American Society for Testing and Materials, 1990
6. "Joint Design for Formed-in-Place Gasketing," CDS-4128, General Electric Company, 1980
7. J. Tokarski, "Formed-in-Place Gaskets: Concept vs. Reality," Paper presented at Gaskets: A Symposium, Warrendale, PA, Society of Automotive Engineers Inc., 1977
8. "Joint Sealers," "One Part Silicone Building Sealants," Rhone-Poulenc Inc.; "Rhodorsil 3B," "Rhodorsil 5C," "Rhodorsil 6B," "Rhodorsil 70," "Rhodorsil 90," Specification Data, Construction Specifications Institute, 1987-1989
9. "Rhodorsil Silicone Sealants Data Sheet, Rhodorsil 4B," Rhone-Poulenc Inc., 1989
10. "Sealant Manual," Flat Glass Marketing Association, 1983
11. J.C. Myers, Behavior of Fillet Sealant Joints, in *Building Sealants: Materials, Properties and Performance*, STP 1069, American Society for Testing and Materials, 1990

Automotive Applications for Adhesives

Kieran Drain and Sarat Chandrasekharan, Ciba-Geigy Corporation

ADHESIVES BONDING in automobile assembly has increased substantially in recent years as adhesives have displaced traditional methods of fastening. Adhesives have a very long history of use in automobile assembly, although the rapid growth of their applicability has resulted in increased awareness in this industry. Adhesives have been used in decorative and trim applications for more than 40 years. Their earliest use in structural bonding was in bonding brake linings (1949) (Ref 1). This article summarizes adhesive application areas, focusing on current production, design, and manufacturing practices.

Adhesives bonding offers many advantages over mechanical fastening (Ref 2):

- Uniform stress distribution over the entire bond line
- Design flexibility
- The joining of substrates unsuitable for conventional fastening techniques
- Improved production line efficiency
- Weight reduction
- Improved corrosion resistance
- Improved bond quality in inaccessible joints

All of these advantages contribute to increased competitiveness in the marketplace, which supports the further increase in adhesive use.

A major advantage, design flexibility, relates to the use of new manufacturing methods and new materials or substrates. Although new materials will not have as large an impact as was originally envisaged in the mid-1980s, a trend toward new materials that require new methods of fastening will still exist, as shown in Table 1.

Possibly the most significant trend in vehicle design is improved corrosion resistance, with a corresponding increase in service life, made possible by the replacement of welding with adhesives bonding. This further supports the growth of adhesives use.

The increased acceptance of adhesives by automotive engineers has resulted from the response of adhesive suppliers to the limitations of early adhesives by producing new rapid-curing and higher-performance materials (Ref 2). Materials are now available that rectify earlier problems of adhesives application, curing, and performance of the cured product. The increased acceptance has led engineers to design for adhesives rather than use adhesives in joints designed for welding or other means of fastening. This, more than any other factor, will drive adhesive growth.

Applications for adhesives and sealants include many areas. This article describes the application areas of brake shoe bonding, decorative trim and fabric, subcomponents, gasketing and threadlocking, body assembly, structural bonding, sound absorption, and repair.

A diverse category of general bonding that has developed includes such applications as hemmed flange bonding of closure panels and roof bow, roof rail, and drainage ditch bonding, among others (see Fig. 1). In all of these applications, a high degree of automation has been employed for adhesive dispensing and curing, thereby providing significant productivity gains.

The acceptance of adhesives continues to improve in the noncritical areas described above. True structural bonding applications have been slow to develop. Major research programs to produce fully bonded vehicles are underway at a number of the major automobile manufacturers. The industry can gain confidence from the historical examples of brake shoe bonding and, more recently, windshield bonding, which play an important role in the structural integrity of the vehicle.

Brake Shoe Bonding

Brake shoe bonding is a long-established application. The same technology is used for disk brake pads, clutch facings, and transmission bands. The dominant technology is that of phenolic resin adhesives. These adhesives are the adhesive of choice because they have exceptionally high resistance to the shear loads experienced between the lining and the shoe, or backing plate. In addition, they retain their properties at the high service temperatures of the application. By replacing riveting, they allow greater economy by reducing lining wastage and minimizing the risk of damage to metal contact surfaces.

Table 1 Change in material composition of North American produced passenger car

Material	1990		1995	
	kg	lb	kg	lb
Total steel...........	691	1520	600	1320
Total cast iron	172	380	136	300
Total plastic	100	220	136	275
Total aluminum	66	145	77	170

Source: Ref 3

The adhesive for brake shoe bonding is usually provided as a one-component solvent-base system. Application by the extrusion of a thixotropic paste or roller coating of a low-viscosity liquid is most common. Curing the adhesive requires a combination of temperature (150 °C, or 300 °F) and pressure (700 kPa, or 100 psi). A certain temperature is needed to promote the reaction, whereas pressure is required to prevent volatile products of the reaction from causing blistering and bubbling in the bond line. The process is very laborious, as shown in Fig. 2. Newer methods of assembly have not been pursued for reasons that include the risk factor associated with making a change to a successful, critical application, as well as the higher costs associated with more process-friendly adhesive types.

Decorative Trim and Fabric Adhesive

Adhesives have long been used for noncritical decorative applications. Typical automobile decorative applications include (Ref 5):

Interior

- Trim panel fabric
- Door panel fabric
- Ceiling fabric
- Carpet adhesive
- Weather stripping

Exterior

- Body side molding
- Wood grain decals
- Stripping decals: weather stripping, vinyl roof, mirror to metal frame

These applications, while important to aesthetics, play no part in vehicle structural

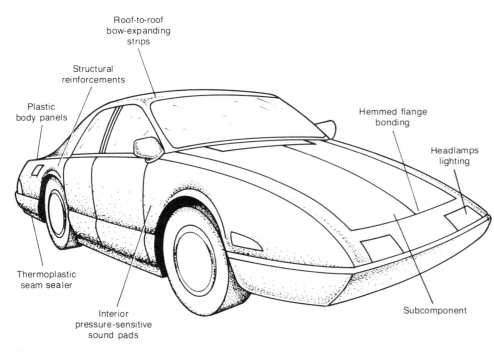

Fig. 1 Automotive applications for adhesives

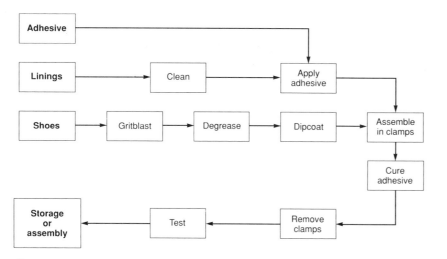

Fig. 2 Manufacturing process for brake shoe linings. Source: Ref 4

integrity (Ref 5). The absence of a significant design load permits the use of lower-strength and lower-cost adhesive types. This application area was once dominated by solvent-base elastomeric adhesives. Environmental concerns have resulted in their replacement, at least in part, by latex, hot-melt, and pressure-sensitive tape products.

Solvent cements of neoprene or styrene butadiene were for a long time the workhorse of this industry segment, particularly for the attachment of vinyl roofs. Today, acrylate pressure-sensitive adhesives are being used in their place. These adhesives offer the low-temperature flexibility and durability necessary in this application. While water-base pressure-sensitive adhesives are a non-toxic alternative to solvent cements, their

long drying times are often unacceptable. An alternative approach is the use of thermoplastic hot-melt adhesives. These materials have been used historically for carpet installation, but now find use in weather strip attachment and as headliner adhesives. Other interior trim applications include door panels, dash panels, and pack trays.

Another significant advance has been the introduction of pressure-sensitive tapes to provide increased holding power in exterior trim and weather strip applications. The tapes usually comprise an elastomeric foam core double faced with poly (butyl acrylate) adhesive (Ref 6). Both tapes and thermoplastic hot melts have the advantage of developing immediate adhesion. In the case of tapes, peel strength builds with time as a

result of a diffusion process (Ref 7). A further advantage of tape adhesives is ease of application, with no need for spraying or hot-melt dispensing equipment, or for the use of hazardous cleanup solvents. However, application is by hand.

Subcomponents

Automotive subcomponents continue to represent one of the areas of major growth in adhesives use. Factors such as extremely high production rates and minimal impact on vehicle structural integrity have contributed to increased adhesives use.

One of the most significant subcomponent applications is lighting, both primary and secondary lighting, in particular the bonding of lenses to reflectors. Conventional glass lenses are being replaced by polycarbonate and acrylic materials. Similarly, new thermoplastic materials, specifically filled polycarbonate, nylon, and polyester, are being used for reflectors. These new materials allow design engineers to develop aerodynamic shapes to fit new body styles. These substrate changes have resulted in a decreased use of heat-cured epoxy adhesives and vibration welding and an increased use of room-temperature-vulcanizing silicones, two-part urethanes, hot-melt thermoplastics (dominant in secondary lighting applications), and, to a limited degree, ultraviolet (UV)-curable acrylates. Key performance requirements in all cases are stress distribution (low elastic modulus), durability, chemical and thermal stability, and outgassing resistance (Ref 8). Two-part urethanes show promise of becoming the dominant adhesive type, despite the necessity for two-component mixing, offering adhesion to a wide range of substrates without the need for the metal clips used with silicones and some hot-melt systems.

A more recent area of adhesives growth has been that of automotive electronics. Specific applications include (Ref 9):

Power train electronics

- Engine control modules
- Transmission controls
- Electronic spark timing controls

Body electronics

- Instrument panel display electronics

Chassis electronics

- Suspension and ride controls
- Electronic brake and traction controls
- Security systems
- Electronic power-assist steering systems

Hybrid electronics

- Ignition modules
- Pressure sensors
- Power switching

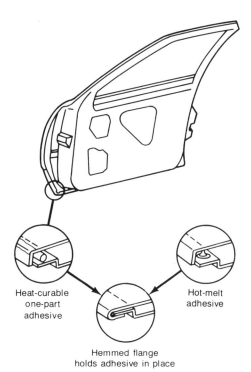

Fig. 3 Automotive hemmed flange bonding

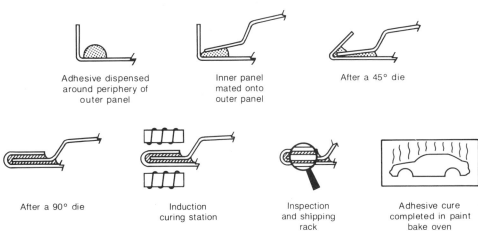

Fig. 4 Adhesive bonding/curing process for automotive closure components

Entertainment and comfort electronics

- Heater and air controls
- Music system speakers

Conventional adhesive applications, such as bonding motor magnets and fixturing wire harnesses, are now in the minority. More often, the adhesive is required to perform additionally as a barrier to the environment, that is, in potting or encapsulating applications. Preferred adhesive types in use are epoxy resins, urethanes, silicones, and acrylates. The epoxy systems being used are highly filled to reduce the coefficient of thermal expansion and may be formulated to dissipate heat from active devices. Urethanes and silicones compensate for the differential expansion and contraction experienced in complex assemblies by their low elastic modulus. Acrylates have seen limited use in shallow potting applications because of the productivity benefits of UV curing. In all cases, ionic purity is a requirement, as is low moisture uptake, so that insulative properties are retained upon environmental exposure (Ref 9).

Other subcomponents in which adhesives are used include mirror assemblies (silicones) and window assembly attachments (acrylates and epoxies). Exterior mirrors are bonded to an adjustable back support using a silicone adhesive and no mechanical fasteners (Ref 1). Conventional window mechanism hardware is attached to door window glass with mechanical fasteners. A lower-cost system is possible when adhesively bonded brackets are used because they reduce the number of parts and the assembly time. Adhesives use is expected to increase in radiator, heater, and air conditioner assemblies.

Gasketing and Threadlocking

One of the smallest automotive applications for adhesives, but one of the most critical, is the area of assemblies that lock or seal, thread, or slip-fit or mate (gaskets). Anaerobic adhesives are a specialty class of materials developed for these applications. Adhesives of this class are one-component products that cure rapidly by a combined action in which cure is initiated by the metal substrate and accelerated by the exclusion of oxygen as parts are mated. In use, these adhesives allow perfect metal-to-metal fit, while correcting for poor tolerances and producing an effective seal (gasket) or full fastener torque (threaded assembly). Their use is dictated where vibration loosening is a problem, for example, in oil pan screws. In many gasketing applications, for example, engine gasketing, the service temperature is beyond the limitations of acrylate adhesives. The material of choice for such applications is RTV silicone (Ref 10).

Body Assembly

Adhesives are now used extensively in body assembly operations. The major application is for the bonding of hemmed flanges in doors, engine hoods, truck lids, and tailgates (Fig. 3). These bonds are partly load bearing, and the adhesive is used in conjunction with widely spaced spot welds, which stabilize the joint while the adhesive cures. Other body assembly applications include the bonding of stiffener panels to roofs, hoods, and trunk lids, as well as the bonding of roof rails, drainage ditches, and wheel house flanges.

In addition to providing a superior joint, adhesive bonding allows the use of more cost-effective manufacturing methods. Subassemblies comprising large, highly contoured outer panels and internal stiffeners can be assembled in a single operation (Ref 11).

Although vinyl adhesives once dominated this segment, more demanding corrosion specifications have resulted in the acceptance of epoxy adhesives, particularly in hemmed flange applications. However, U.S. and foreign manufacturers who have changed to substrates of mild steel can use vinyl adhesives (induction cured) for hemmed flange bonding and sealing. Other adhesives used in hemmed flange bonding are, in most cases, single-component systems. This is possible because the oven cycles from the paint operations are used to cure the adhesive.

Conventional oven curing adhesives are increasingly being replaced by induction precurable or pregelling adhesives (Fig. 4). Induction curing uses the principle of electromagnetic induction to cause resistive heating in metal substrates. The heat generated is then conducted into the adhesive, initiating or accelerating cure (Ref 2). Induction curing allows for significant weld reduction and a consequent improvement in the surface quality of exterior panels (Ref 12).

Induction curing is used to provide dimensional stability while parts are transported prior to painting and adhesive curing. In addition, induction pregelling improves adhesive wash-out resistance. This is not a major issue in hemmed flange joints, but is critical in other, more exposed joints. In these cases, wash-out resistance may be achieved using epoxy adhesives presented in a hot-melt or film tape delivery form (Ref 2). More recent developments in body assembly adhesives include the development of reactive plastisol adhesives and sprayable epoxy systems (100% solid). Reactive plastisol (epoxy and vinyl interpenetrating networks) attempt to combine the

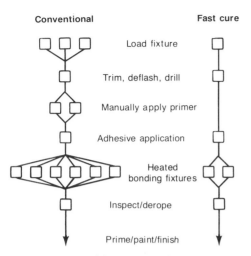

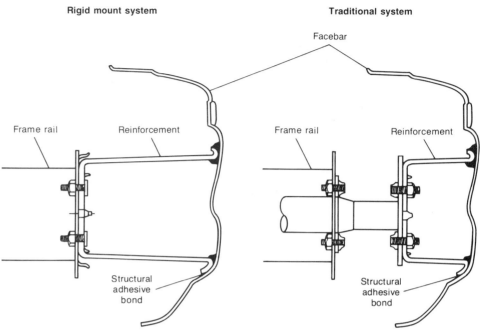

Fig. 5 Impact of fast-cure primerless adhesives on sheet molding compound bonding

Fig. 6 Structural bonding of bumper systems

excellent impact resistance of vinyls with epoxy durability (Ref 13).

Steel remains the dominant material used in body panel construction. While few vehicles have totally plastic bodies, many contain some plastic components, most commonly tailgates and hoods. Most plastic panels are not exposed to the electrodeposition primer process (E-coat) and, lacking this opportunity for thermal curing, use two-component adhesive systems. Both epoxy and urethane adhesive systems are currently being used. Urethanes are used whenever possible because of their lower raw material costs. The use of an epoxy adhesive is dictated when a paint process would expose the panel assembly to temperatures that exceed the heat distortion point of the urethane adhesive. Feasibility has been established by a number of automobile manufacturers for using induction curing to bond plastics with epoxy adhesives and to use high-frequency curing or microwave curing to cure urethane adhesives (Ref 14).

The most significant advance in plastics bonding has been the development of primerless systems, allowing the elimination of environmentally unacceptable solvents (Ref 15). This, coupled with the increase in cure speed, has the effect of streamlining assembly operations (Fig. 5) (Ref 15, 16). The increased use of both metal and plastic panels in one vehicle, as well as the extension of the plastic-clad space-frame concept, will dictate the need for new adhesives with performance tailored to differential substrate bonding.

Structural Bonding

The use of adhesives in the applications described thus far has had a significant impact on vehicle styling and total vehicle cost. The ultimate goal of a fully bonded body shell has yet to be achieved. Adhesives are showing promising results in laboratory and prototype vehicle tests for this application, but other than isolated examples, true structural bonding in automobile assembly is limited at present to bumper bonding, windshield bonding, and metal reinforcement.

Adhesively bonded bumpers are being used in small vehicles. In one instance, the bumper comprises a Xenoy facia (polycarbonate and polybutylene terephthalate blend) adhesively bonded to a Xenoy reinforcement beam mounted directly to the frame of the vehicle (Fig. 6, rigid-mount system). Replacing a traditional collision control system that includes a foam absorber with this design can reduce weight by as much as 20% and cost by 10% and still offer 8 km/h (5 mph) impact performance (Ref 17). Energy is absorbed by the plastic structure upon impact, and shock absorbers or foam are not required.

The adhesive of choice in bonding the thermoplastic described above is a two-part urethane system. Good adhesive performance has also been observed with certain epoxy and acrylate adhesives. In addition to providing high structural strength and easier component assembly, adhesive bonding reduces stress in the joint that would otherwise compromise bumper performance. A further advantage of adhesive bonding is increased design flexibility.

Windshield installation adhesives have evolved from their initial role of sealing against water leakage to a significant role in ensuring torsional rigidity in the vehicle (contributing 10 to 50% of vehicle torsional rigidity). The application of Federal Motor Vehicle Safety Standards (FMVSS) to windshields in 1970 forced the automotive industry to adopt structural bonding for windshield installation, ending the use of polysulfide sealants for this application (Ref 18). Moisture-curable polyurethane adhesives are now used worldwide for windshield and back light (rear window) installation. The added strength provided by the adhesive has allowed a 3.5 to 4.5 kg (8 to 10 lb) metal weight reduction from A pillars and headers in some models, while still adding an extra margin of safety against the roof crush requirements of FMVSS 216 (Ref 19).

Urethane adhesives used in windshield bonding are predominantly single component, while some two-component systems are used for repair installations. Although the single-component adhesive is convenient, it must be used in conjunction with a series of primers. These primers are of two types, that is, silane primers to provide durability in bonding to both glass and painted metal, and blackout primer applied to the glass to extend adhesive service life by screening out UV radiation that would otherwise cause photochemical degradation of the (aromatic) urethane polymer. Because cure is achieved through a reaction with water, full properties are slow to develop, being governed by the rate of diffusion of water through the adhesive bead. To compensate for seasonal variations in plant humidity, a two-component system has been developed in which the second component is a water-base paste. This system is necessary when production rates demand faster cure speed and when flush glass mounting restricts moisture diffusion.

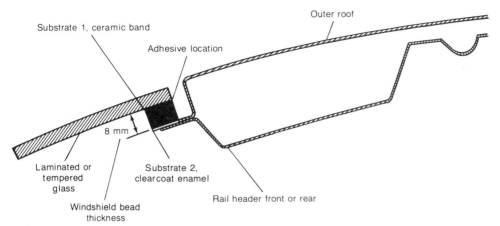

Fig. 7 Typical glass installation joint

The need for the blackout primer has diminished with the introduction of a band of ceramic frit on the windshield to protect the adhesive from UV degradation (Ref 20). The rubber dam used to prevent primer and adhesive flow-out onto appearance areas of the glass can often be eliminated because of the development of high-viscosity "pumpable tape" adhesives (Ref 19). The developmental efforts of adhesive suppliers are largely focused on the elimination of adhesion-promoting primers and toxic solvents (used in both adhesives and primers). Recent reports in the industry suggest the availability of fast-setting, one-part, moisture-curable systems with a cure rate equivalent to those of accelerated systems (water-injected paste). The greatest challenge is to create an adhesive that has universal adhesion to the many possible paint systems that can be used.

Figure 7 shows a typical joint design, in which the adhesive bonds to both the ceramic frit layer on the glass and the painted pinch weld area of the car body. New developments that may affect this design include the introduction of antilacerative automobile glass or the reaction injection molded modular glass concept.

Localized metal reinforcement has been achieved through the use of structural adhesive patches. These patches are die-cut pieces of epoxy prepreg, that is, glass fiber mat impregnated with epoxy resins. The adhesive is formulated to be pressure sensitive so that it can be applied to oily or contaminated metal surfaces. Curing is achieved in assembly plant paint bake ovens. As with single-component epoxy paste adhesives, the patches must be stored frozen; they have a limited shelf life at plant temperatures. Epoxy patches provide high flexural strength, eliminating the need for more bulky metal reinforcements to provide localized reinforcement in quarter panels, roofs, trunk lids, and other closure panels (Ref 1).

Structural bonding in adhesive body shell construction remains at the concept vehicle stage. The absence of accepted accelerated test methods to predict in-service performance accurately may mean that it will be many years before structural adhesive bonding becomes a reality for major model production. The accepted representative test specimen for characterizing adhesive performance is the box beam. This section is typical of many structural members in a body shell, including cross members, side rails, door pillars, roof rails, and so on. The testing of adhesively bonded box beams over a range of temperatures has shown encouraging results, but it also highlights the need for tougher adhesives (Ref 21).

A good correlation of box beam tests to simple impact tests has been reported, in addition to recent reports in the literature of adhesives with exciting impact resistance/temperature profiles (Ref 22). Significant research is being carried out by adhesives suppliers and automobile manufacturers to characterize more effectively the performance of structural adhesives. New test methods have been developed, most specifically in relation to adhesive impact resistance (Ref 23). Adhesive properties in the bond line have been correlated with bulk polymer properties. Such characterization of adhesives is providing data for finite-element analysis in support of joint design for structural bonding (Ref 24).

The significant advantages of structural bonding are (Ref 21):

- Improved strength/stiffness
- Improved noise, vibration, and harshness (NVH) and aerodynamics
- Reduced corrosion
- Elimination of spot weld metal finishing operations
- Reduction in overall manufacturing costs
- Increased design freedom

These advantages make the fully bonded vehicle a major goal for the industry. Perhaps the most significant impact of adhesives bonding will be the increased freedom in design, with joint accessibility no longer being an issue. However, if the issues of adhesive impact, fatigue, and durability were successfully resolved, a large number of processing issues would have to be addressed. Current manufacturing practices in body stamping, assembly, and painting do not favor adhesive bonding. Stamped metal parts are heavily contaminated with mill oils, drawing compounds, press lubricants, and anticorrosive chemicals. Manufacturing tolerances vary between 0 and 3 mm (0 and 0.12 in.). In addition to these issues, the use of adhesives will require training in the safe handling of new chemicals.

Problems related to adhesive sag in curing, adhesive displacement during power wash cycles, and the deroping of adhesives following cure (cutting away of the cured "rope") will also have to be addressed. Unlike welding, adhesive bonding does not allow visual confirmation of bond quality. Liability considerations dictate the need to establish on-line nondestructive testing if adhesive bonding is to replace welding. All of these issues will offset some of the cost advantages of adhesive bonding and may delay the introduction of this technology.

Sound Absorption

The use of adhesive bonding in body shell construction offers the potential for significant noise reduction. As of 1990, the automotive industry commonly uses elastomeric noise control materials to reduce vehicle interior noise. These elastomeric materials, which are significantly different from all the adhesive types described above, have a low modulus and provide good damping characteristics. The frequency range of interest is 20 Hz to 20 kHz, and the dominant noise sources are the drivetrain, tires/suspension system, and wind. It is possible for noise to travel through the air (by means of the climate control system, for example), pillars, and floor rails, or through solid components such as windows and closure panels. Another possible noise path is vibrational transmission from the source to a surface that can, in turn, radiate this energy as noise, for example, transmission from engine to floor panels (Ref 25). Because each vehicle will have its own unique noise signal, sound absorption treatments must be devised for every new model.

Adhesive damping materials are distinct from barriers (for example, dashboards) or absorptive materials (for example, polyurethane foam) used to control vehicle interior noise. Damping materials reduce the kinetic energy of vibration by transforming it into heat and dissipating it, rather than simply by reflecting and absorbing it. The ability of a polymeric damper to attenuate sound is related to material stiffness and thus is temperature dependent. Viscoelastic dampers find use in roof, door, and floor pan applications, reducing wind, road, and engine noise. Although dampers provide effec-

Fig. 8 Adhesive repair of crash damage

tive vibration control, the control of vibration noise by dampers is limited to the control of resonant frequencies in the acoustic element of the vibration. The performance of viscoelastic dampers can often be enhanced by the use of spacer layers or constrained layers (Ref 25). The preferred delivery form for viscoelastic dampers is as extruded, die-cut, pressure-sensitive preforms. They may be considered "chemical parts" and are applied to their target area by hand.

Accepted methods for measuring damping performance are the Geiger plate test (SAE J-671) and the complex modulus test (Oberst Bar, ASTM E 756). These methods measure decay rate and loss factor, respectively. Recently published work suggests that vibration damping may be measured by dynamic mechanical spectroscopy, suggesting that efficiency is related to the loss factor peak width (Ref 26). It may also be possible to correlate dynamic mechanical analysis data with sound damping. The ultimate goal of this work would be using mechanical spectroscopy to design better damper systems.

Repair

The use of adhesives to repair crash damage in automobiles is being investigated at several automobile insurance company research centers, with limited commercial activity already underway. Escalating vehicle cost is forcing insurance companies to consider repairing vehicles that previously would have been written off. The technique of bonding replacement body panels to crash-damaged automobiles was originally developed by General Motors Corporation (Opel) in West Germany. The adhesive bonding of replacement body panels offers certain advantages over welding. Specifically, bonding distributes stress evenly over the total repair area, and it reduces refinishing costs. The major advantage may be the reduced time required for the repair because there is no need to remove flammable or thermally degradable materials from the work area, for example, from seating, gas tanks, or interior trim (Ref 27). Significant research is needed before adhesives bonding can be considered for the repair of structural members.

Adhesives used for this application are two-component epoxy systems. The absence of a paint bake cycle in aftermarket operations prevents the use of one-component systems. This factor contributes to reduced adhesive performance when compared to high-performance single-component epoxies, but is offset by the ability to clean and abrade the substrate surface thoroughly. Adhesives are available that have been formulated in convenient dual-chamber cartridges for accurate mixing (Ref 28). Figure 8 shows the type of repairs that are currently possible.

REFERENCES

1. Y.F. Chang, M.F. Milewski, and K.P. Tremonti, in *Adhesives Technology for Automotive Engineering Applications, Proceedings of the SME Conference,* Society of Manufacturing Engineers, Oct 1987, AD87-527
2. K.J. Schroeder and K.F. Drain, Paper presented at the Adhesives Technology for Automotive Engineering Applications Conference, Dearborn, MI, Society of Manufacturing Engineers, Nov 1989
3. *Automot. Eng.,* Vol 97 (No. 2), 1989, p 149
4. CIBA-GEIGY "Adhesives for Brake and Clutch Bonding," Instruction Manual RTB1F, Ciba-Geigy Corporation, 1983
5. P.G. Phillips and W.E. Broxterman, *Adhes. Age,* No. 8, 1989, p 40
6. *Automot. Eng.,* Vol 96 (No. 2), 1988, p 6
7. P.G. McGonlgal, *J. Adhes.,* Vol 26, 1988, p 265
8. V. Gheyara, Paper presented at the Adhesives Technology for Automotive Engineering Applications Conference, Dearborn, MI, Society of Manufacturing Engineers, Nov 1989
9. "Electronics for Automotive Applications," Publication 600, Delco Electronics Division, General Motors Corporation
10. G.L. Schneberger, in *Handbook of Adhesives,* 2nd ed., I. Skeist, Ed., Van Nostrand Reinhold, 1977, p 782
11. B. Scheidle, *CIBA-GEIGY Resin Aspects,* Vol 21, 1989, p 8
12. Y.F. Chang *et al., Adhes. Age,* Vol 29 (No. 2), 1986, p 27
13. I. Namiki, *J. Appl. Polym. Sci.,* Vol 20, 1976, p 1933
14. R.C. Rains, *Mater. Res. Symp. Proc.,* Vol 124, 1988, p 323
15. K. Wongkamolsesh, in *Proceedings of the 43rd Annual SPI Conference,* The Society of The Plastics Industry, Feb 1988, p 24E
16. D.N. Shah, in *Proceedings of the SME Autocom 88 Conference,* Society of Manufacturing Engineers, May 1988, p AD88-232
17. V. Wigotsky, *Plast. Eng.,* Sept 1989, p 23
18. M.T. Cocozzoli, "Sealants and Structural Adhesives," Fisher Body Division, General Motors Corporation, March 1982
19. M.A. Sermolins, Paper presented at the Adhesives Technology for Automotive Engineering Application Conference, Dearborn, MI, Society of Manufacturing Engineers, Oct 1987
20. M.D. Kirby, Technical Paper 870308, SAE Technical Paper Series, Society of Automotive Engineers Inc., 1987
21. I.N. Moody, P.A. Fay, and G.D. Suthurst, Chapter 7 in *Adhesion II,* K.W. Allen, Ed., Elsevier, 1987, p 97
22. W.F. Marwick and J.H. Powell, in *Proceedings of the '88 SAE Conference,* Society of Automotive Engineers Inc., Oct 1988
23. A.J. Kinloch and G.A. Kodokian, *J. Adhes.,* Vol 24, 1987, p 109
24. M. Fischer and M. Pasquier, *Constr. Build. Mater.,* Vol 3 (No. 1), 1989, p 31
25. "Application of Noise Control Materials in the Automotive Industry," Paper presented at SAE Noise Vibration Conference, Traverse City, MI, Society of Automotive Engineers Inc., May 1989
26. D. Klempner, *Recent Developments in Polyurethane I.P.N.'s,* Technomic, K.C. Frisch, Ed., 1988, p 50
27. C.S. Adderley, *Mater. Des.,* Vol 9 (No. 5), 1988, p 289
28. *Auto Body Repair Manual,* Publication 28326e, Ciba-Geigy Corporation

Aerospace Applications for Adhesives

Edwin C. Clark, Ciba-Geigy Corporation, Furane Aerospace Products

ADHESIVES USAGE in aerospace applications can be traced back to the infancy of the industry. In the early years, many primary structural materials, such as wood, thread and fabric, were bonded with adhesives. As technology advanced and the industry grew, aluminum became the structural material of choice, and metal fasteners were predominately used in aircraft assembly.

Metal fasteners had certain advantages over adhesives in terms of providing consistent, reproducible performance, simple manufacturing techniques, and higher load-bearing capability. From the 1940s through the 1960s, these advantages sufficiently met customer performance needs. However, with the approach of the 1970s, the aerospace industry experienced a demand for higher performance, improved fuel economy, and more cost-effective manufacturing, which highlighted the deficiencies associated with metal fasteners and renewed an interest in adhesive bonding.

During the 1970s, the driving force for increased usage of adhesives in this industry came from the military sector. This focus on adhesive bonding was driven by new advanced aircraft designs, requiring lighter-weight materials that could be used in construction of highly contoured parts. The joining of these lightweight structures required the use of technologies that allowed improved dynamic load-bearing capabilities and adaptability to more demanding manufacturing techniques resulting in smoother, more aerodynamic surfaces. All these criteria could be satisfied using adhesives.

Military use of adhesives in the early 1970s could be described as a compromise between meeting advanced design requirements and maintaining reliable, adhesively bonded structures. The primary concern during this time was to determine if, when, and how an adhesive would fail. It was not uncommon to experience either uniform adhesive failure throughout an application or an isolated instance where one bond failed and the others maintained their integrity. To address this issue, the U.S. Air Force funded the Primary Adhesively Bonded Structure Technology (PABST) (Ref 1) program in 1975. Its primary objective was to achieve significant improvements in cost, weight, integrity, and durability of primary fuselage structures. This project considered several adhesive bonding factors to determine how to design and implement adhesive technology with a high level of reliability.

The PABST program involved the analysis of surface preparation, surface contamination, joint design, bond line uniformity, peel strength, and environmental exposure of the adhesive prior to bonding. It was determined that the best surface preparation for aluminum was a phosphoric acid anodizing process. To maintain this surface for bonding, no contact was allowed prior to assembly. Testing showed that when an adhesive was uniformly applied over the bonded surface, no point stresses developed and the reliability of the assembly improved. By proper joint design, that is, increased bond area or the utilization of available joint designs (Fig. 1), the maximum loading of a part could be directed away from the bond line toward other load-bearing components. When an adhesive was exposed to a humid environment prior to bonding, bond strength dropped dramatically. This observation is quantified in Fig. 2 and highlights the need for proper storage of adhesives. Results like these helped to create an understanding of how adhesives could be reliably used. With ongoing research in adhesives technology and the availability of ever-increasing amounts of historical applications data, there continues to be a constant movement toward the acceptance of adhesives in aerospace applications.

It must be realized that the demands placed on adhesives are as varied as the environments they encounter and functions they perform. With this in mind, this article will attempt to highlight the major classifications of adhesives used in the aerospace industry, the forms in which they are available, how they are applied, examples of their use, and major attributes in each application.

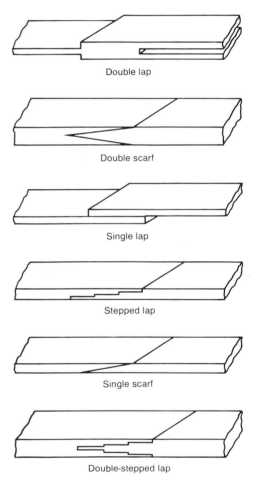

Fig. 1 Joint designs for adhesive bonding. Source: Ref 2

Double lap

Double scarf

Single lap

Stepped lap

Single scarf

Double-stepped lap

A number of classes of adhesives are used in the aerospace industry today. Table 1 lists those most frequently encountered. Of the classes listed, epoxy-base adhesives dominate the industry. This is due to their ease of use, high-temperature capability (205 °C, or 400 °F), good solvent resistance, and their ability to formulate a wide range of performance characteristics. Other notable adhesives include urethanes, selected primarily for high-peel-strength applications and their ability to

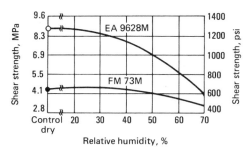

Fig. 2 Effect of moisture in uncured adhesive film. Lap-shear specimens were subjected to 120 °C (250 °F) after a 10-min soak at temperature and then dried for 24 h at 40 °C (100 °F) and 10% relative humidity; specimens were preconditioned for 10 days at 25 °C (75 °F) at specified humidity. Source: Ref 3

Table 1 Classes of adhesives used in aerospace

Class	Application
Acrylic	Plastics bonding
Bismaleimide . . .	High-temperature, metal, or composite bonding
Cyanoacrylate . .	Plastics
Epoxy	Metal, composite, and plastics bonding
Phenolic	Fabrication of honeycomb core
Polyimides	High-temperature, metal, or composite bonding
Silicone	Plastics
Urethane	Plastics, composite

Table 2 Adhesive product forms used in aerospace

Two-component liquids and pastes
One-component paste (usually frozen, heat activated)
Film adhesive (with or without textile carrier)
One-component hot melts
One- and two-component pressure sensitives

Table 3 Aerospace adhesive applications

Honeycomb core construction
Honeycomb sandwich construction
Secondary honeycomb sandwich bonding
Noise-suppression systems for engine nacelles
Bonding honeycomb to primary surfaces
Bonded aileron
F-15 composite speedbrake
Adhesive bonding of aircraft canopies and windshields
Missile radome-bonded joint
Electrically conductive adhesive for lightning strike application
Laminate repair
Honeycomb structure damage
Repair of adhesive disbonds

bond plastics, and polyimides, used in high-temperature environments (330 °C, or 625 °F). Although these adhesives are encountered quite frequently within the industry, it should become very clear, as applications are described, that each of the adhesives classes listed in Table 1 has individually unique characteristics.

As critical as the chemical nature of the adhesive is to performance, the product forms of the adhesive will dictate how and where it can be used in aircraft manufacturing. Table 2 lists the product forms that are most commonly found. Two-component adhesives can be packaged in several different configurations, such as simple individual containers for hand or meter mixing, double-barrel syringes, two-component tubes, and plastic bags with a barrier that separates the two components. The materials in two-component packages range from low-viscosity liquids to very thick pastes. One-component adhesives come in individual containers, extrudable tubes, and films. The one-component adhesives generally require either cold storage to extend shelf life, and/or high temperatures to cure them (>82 °C, or 180 °F). The product forms listed in Table 2 are made available to meet application needs that vary from hand application to meter mix spraying. Table 3 lists specific adhesive applications, each of which is discussed in this article. For reference, Fig. 3 shows the major components of a commercial aircraft.

Honeycomb Core Construction

The primary objective in designing and building an aircraft is to use construction materials that are strong and lightweight. Honeycomb core is an important material used to achieve this combination of properties. As shown in Fig. 4, honeycomb core consists of thin sheets of material that are bonded in such a way that creates open-cell areas between adjoining sheets, resulting in a material that is low in density per unit volume.

Two basic manufacturing methods used to make honeycomb core are the expansion and the corrugated methods. In the expansion method, thin sheets of substrate are cut to size, an adhesive is printed onto the sheets, and the sheets are assembled into a block. The block is then placed into a heated press where the adhesive is cured. After curing, the block is then pulled apart (expanded), which separates the sheets and creates the individual core cells. This process is shown in Fig. 5. During core manufacturing, it is important that the adhesive be printed accurately so that cell structure is uniform. When the core substrate is aluminum alloy, it must first be cleaned and a corrosion-resistant

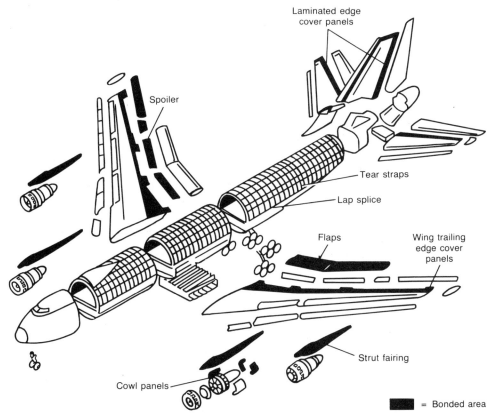

Fig. 3 Bonded areas on commercial aircraft. Source: Ref 4

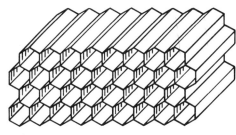

Fig. 4 Hexagonal cell honeycomb. Source: Ref 5

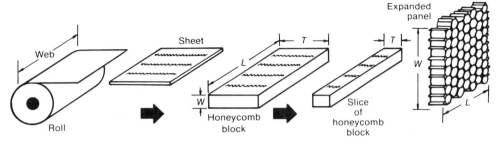

Fig. 5 Expansion method of honeycomb core fabrication. Source: Ref 5

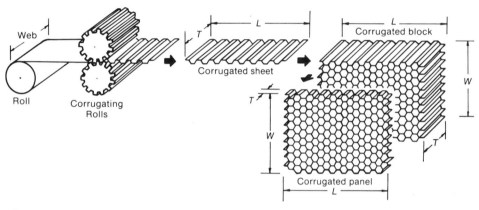

Fig. 6 Corrugation method of honeycomb core fabrication. Source: Ref 5

treatment applied before bonding takes place. This is done to prevent corrosion of the core during service.

The corrugated method of manufacturing honeycomb core is primarily used when higher-density or higher-temperature cores are needed. In this process, the substrate comes in sheet form, which passes through corrugated rollers to preform the core walls.

Adhesive is applied to the surfaces to be bonded, and the sheets are stacked. The stacks are then placed in an oven to cure the adhesive. This process is shown in Fig. 6. In both the expansion and corrugated methods, the adhesives used must meet the requirements for temperature, vibrational, and environmental resistance. Typically, phenolic or epoxy adhesives are used.

Honeycomb Sandwich Construction

One of the main uses of honeycomb core in the aerospace industry is in the construction of sandwich panels. The basic design of a sandwich panel is shown in Fig. 7. There are two common ways of making a sandwich panel based on the required sandwich structure configuration. When a flat sandwich panel is desired, the core and face sheets are cut to size, an adhesive is applied between them, and the assembly is placed in a heated press. After the adhesive is cured, the sandwich panel can be cut to the size needed.

A second method is used when contoured parts are needed and the application does not justify the cost of a mold. This method uses a contoured tool to configure the part and a vacuum bag to consolidate the face sheet, core, and adhesive. Figure 8 shows the steps involved in this process. Before the components of the sandwich structure are placed on the tool for consolidation, it is frequently necessary to preshape the face sheets and honeycomb core. When face sheets are made of composite material, they can be vacuum bagged and cured on the tool prior to bonding to the honeycomb. This preshaping is done when a minimum amount of part distortion is required.

There are two common ways to preshape honeycomb core. For metallic and high-density nonmetallic core, preshaping is done by machining. For lighter-weight core, the core is placed in a tool and sufficient pressure is applied to actually crush the core walls until the desired shape is achieved. To assemble, one of the face sheets is placed on the tool, adhesive is applied, and the core is placed on the adhesive. A second layer of adhesive is then applied to the top of the core and the second face sheet is positioned on top of the adhesive. This sandwich is then vacuum bagged to consolidate the components. The assembly can be left at room temperature under vacuum if the adhesive can be cured under this condition. If heat is required, the assembly can be transferred to an oven or autoclave.

Throughout this discussion of honeycomb construction, the form of the adhesive has not been identified. The most frequently used adhesive form in this application is epoxy film. This adhesive may be applied as shown in Fig. 7, where the film is simply placed between the face sheet and core, pressure and heat are applied, and the adhesive is cured.

A second method for applying adhesive to honeycomb core is called reticulation. It is used when density is a critical factor and the use of excess adhesive is not acceptable. Reticulation simply involves applying the adhesive to the ends of the core walls without covering the open cell ends. This is

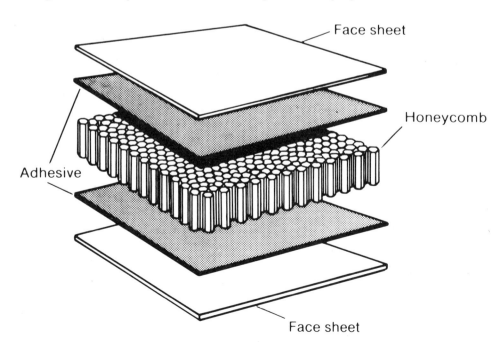

Fig. 7 Bonded honeycomb sandwich assembly. Source: Ref 5

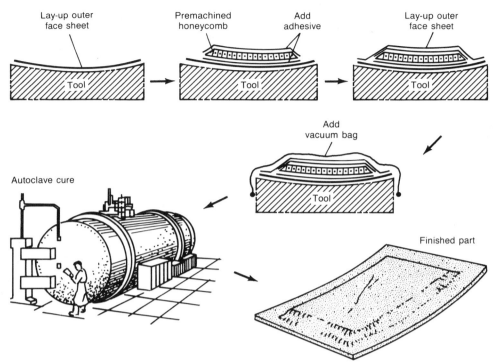

Fig. 8 Contoured honeycomb sandwich manufacturing technique. Source: Ref 6

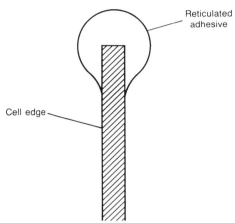

Fig. 9 Deposition of reticulated adhesive on honeycomb cell edges. Source: Ref 7

shown in Fig. 9. Two methods can be used to reticulate an adhesive onto honeycomb core. The first is to roller coat a medium-viscosity liquid adhesive that will migrate to the cell edge. The second method involves placing a film adhesive over the core and heating the film until it becomes fluid. When the adhesive begins to flow, it will migrate to the cell edge, leaving the cell open. From this point, a face sheet is placed over the core, and pressure and heat are applied until full cure is achieved. The resulting honeycomb structure is much lighter than one made with a film adhesive.

Secondary Honeycomb Sandwich Bonding

There are several techniques currently used in the aerospace industry to bond honeycomb sandwich panels together. The first and easiest method is to bond the face sheets of each panel to each other. Al-

though simple, this method typically does not allow for detailed contouring of the part. A second method is best described as a tongue and groove method. This involves cutting a channel into one side of the panel, applying adhesive into the channel, then sliding the end of a second panel into the channel. The resulting assembly provides for the connection of panels at or near a 90° angle. Figure 10 shows this technique. In both cases, the adhesive used is typically an epoxy, formulated to meet design bonding strengths, flame resistance (if used in aircraft interiors), and consistency for flow control.

In the two sandwich joining methods previously described, the adhesive had the primary responsibility for joining the panels

together. One of the most frequently encountered secondary applications for adhesives in panel construction is in fastener core potting. This technique is shown in Fig. 11. Fastener core potting involves drilling a hole into one side of the honeycomb, placing a fastener insert into the hole, and then forcing adhesive through the top and down around the insert, to set it in place. The fastener insert has two holes in the potting shield. The adhesive is dispensed through a nozzle into one of these holes. As the chamber fills up, adhesive starts to come out of the other hole, signaling the completion of the adhesive application. Once cured, this fastener can be used to bolt panel assemblies together. This type of application can be found throughout aircraft interiors, including all galley assemblies, overhead storage bins, and floor panels.

The adhesives used in this application are often required to be low in density to support the resulting low-density structure. This low-density adhesive, referred to as a syntactic, is characterized by adhesive and compressive strength. Table 4 lists typical properties of syntactics. In addition to the properties shown, syntactics must perform well under humid conditions and in areas where fluid resistance is critical.

Noise-Suppression Systems for Engine Nacelles

Since 1970, there have been increasing requirements for commercial aircraft manufacturers to reduce the noise levels of takeoff and landing. One method used to reduce the noise level is by the acoustical modification of engine nacelles. This modification involves lining the inlet and exhaust areas of the nacelles with sound-absorbing materials that are constructed by bonding honeycomb core to solid and microporous face sheets.

One example of this type of noise-suppression system is based on the concept of restrictive air movement controlled by the

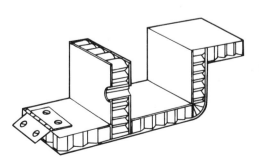

Fig. 10 Tongue and groove bonding of honeycomb sandwich panels. Source: Ref 8

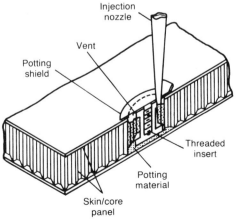

Fig. 11 Potting of fastener inserts. Source: Ref 9

Table 4 Typical physical properties of reacted epoxy syntactics

Property	Value
Density, g/cm³	0.35–1.0
Compressive strength, MPa (ksi)	14–140 (2–20)
At 25 °C (77 °F)	14 (2)
At 177 °C (350 °F)	14–70 (2–10)
Tensile lap-shear strength, MPa (ksi)	7–17 (1–2.5)
Glass transition temperature (a)	65–177 (150–350)
Coefficient of thermal expansion, 10^{-5}/K(a)	8
Shore D hardness	40–90
Dielectric constant, at 60 Hz	3.0–3.5
Volume resistivity, at 25 °C (77 °F) $\Omega \cdot$ cm	1.2×10^{13} to 8.7×10^{13}
Shrinkage, %	0.1–0.5
Other properties	Flame retarded Sound dampening Thermal insulation

(a) Determined by dynamic scanning calorimetry. Source: Ref 10

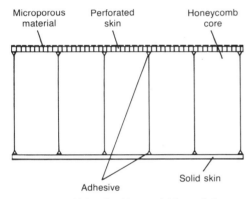

Fig. 12 Noise-absorbing sandwich panel. Source: Ref 11

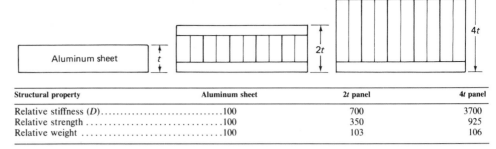

Structural property	Aluminum sheet	2t panel	4t panel
Relative stiffness (D)	100	700	3700
Relative strength	100	350	925
Relative weight	100	103	106

Fig. 13 Structural efficiency of honeycomb sandwich panels relative to that of aluminum sheet. Source: Ref 5

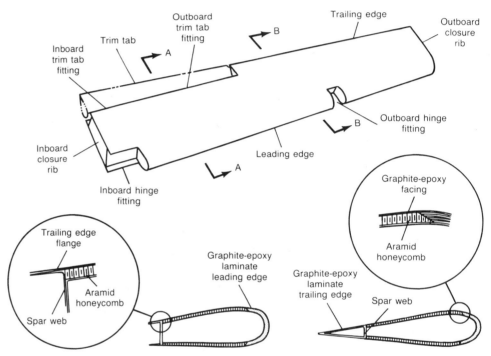

Fig. 14 SF 340 aileron configuration. Source: Ref 12

presence of thousands of small baffle chambers. To implement this concept, the design shown in Fig. 12 was developed. The design requires an adhesive to be placed on a perforated skin. A layer of microporous fabric is placed on top of the adhesive and the assembly is cured. A reticulating adhesive is then placed on the honeycomb core and the perforated skin/microporous fabric assembly is placed on the reticulating adhesive. This assembly is then vacuum bagged and cured in an autoclave. The resulting structure provides openings in the perforated skin/microporous fabric so that sound vibration may enter the cell chamber. Once in the chamber, restrictions in air movement will dampen the noise and ultimately result in reduced noise emissions.

Several critical factors are involved in the design and construction of this type of noise-suppression material. First, in order for the system to work, there must be holes through which air can travel to get to the cell chambers. This requires an adhesive(s) that flows sufficiently to bond the core to perforated skin and the skin to microporous fabric without filling the pore openings. Second, the adhesive(s) used must perform well through a temperature range of 80 to 155 °C (176 to 312

°F), must be resistant to fluids, and must withstand a high vibrational environment. By meeting these requirements, the noise-suppression concept described here has become the standard in the industry.

Bonding Honeycomb to Primary Surfaces

Honeycomb sandwich structures provide a source of construction material that is lightweight and strong. Figure 13 shows a comparison between a sheet of aluminum and two sandwich structures with aluminum face sheets of half the thickness of the aluminum sheet. It becomes quite apparent from this comparison that by utilizing the sandwich design, higher levels of stiffness and strength can be obtained with minimal impact on weight. It is due to the performance of this design that adhesives are increasingly being used to bond primary

structures. Two examples that highlight this concept are described below.

Bonded Aileron. The aileron configuration used in SF 340 aircraft is shown in Fig. 14. The part was constructed on a two-piece, hollow aluminum mandrel (Fig. 15). After release coats were applied, the inner graphite-epoxy plies were draped over the main mandrel and connected at the trailing edge mandrel. At this point, additional plies of graphite-epoxy were added to build up hinge areas. A layer of epoxy adhesive film was then placed over the graphite-epoxy plies. A tungsten-powder-filled adhesive was applied to the leading edge, and another layer of film adhesive was laid down. Preformed core was then placed on top of the film adhesive and the final layers of graphite-epoxy were placed. The assembly was then vacuum bagged and cured at 176.6 °C at 62 Pa (350 °F at 90 psi) for 2 h. Testing of this structure showed no discernible damage at 150% of design load.

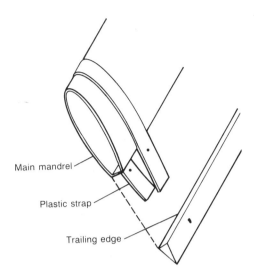

Fig. 15 Mandrel assembly on SF 340 aileron. Source: Ref 13

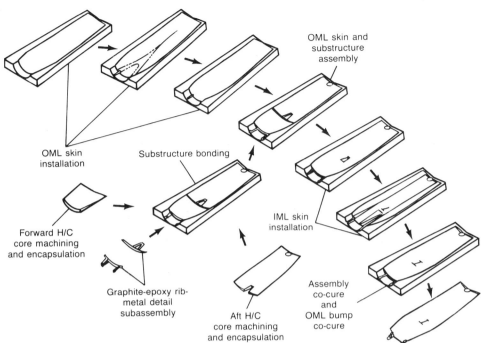

Fig. 16 F-15 composite speedbrake. OML, outer moldline; IML, inner mold line; HC, honeycomb core. Source: Ref 14

Critical factors important in this type of application are that the adhesive has a similar cure schedule to the composite material and the adhesive acts as a buffer between the composite material and the core so that shrinkage during cure does not damage the core.

F-15 Composite Speedbrake. The F-15 composite speedbrake was the first high-production-rate graphite-epoxy structure used in a U.S. Air Force weapons system. Its construction is based on graphite-epoxy skins co-cured over encapsulated honeycomb core. Figure 16 shows the manufacturing sequence for the speedbrake.

Fabrication of the speedbrake starts with the machining of the honeycomb core. Actuator ribs are foam bonded to the core during one of the co-curing processes. Titanium ribs are then bonded to the core with a high-performance room-temperature-curing epoxy paste adhesive. Once all the subassemblies are in place, the graphite-epoxy skins are positioned and the assembly is vacuum bagged and cured. Afterward, secondary machining is performed for fastener adhesive attachment.

The main attributes of the adhesive used in the speedbrake assembly were high performance in terms of temperature, strength, and vibration. These adhesives had the secondary benefit of availability in forms that met the manufacturing need. The foaming adhesive was able to fill voids where matched surfaces were not always available. The paste adhesive could be easily applied by hand and had the benefit of curing at room temperature.

Adhesive Bonding of Aircraft Canopies and Windshields

Aircraft canopies and windshields are fabricated from laminates of glass, acrylic, or polycarbonate. Traditionally, a thermo-

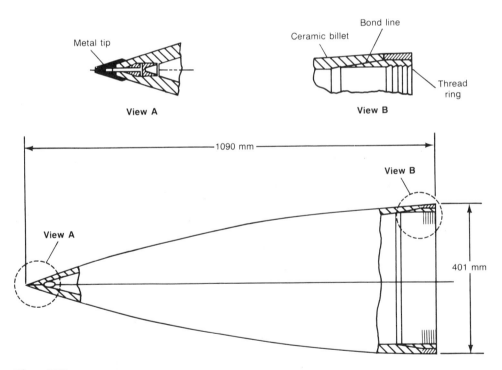

Fig. 17 Radome configuration. Source: Ref 16

plastic is used as the interlayer for these laminates. Recently, silicone adhesives have been identified as viable candidates to replace these thermoplastics (Ref 15). The benefits of silicone adhesives are long-term optical clarity in temperatures ranging from 54 to 177 °C (65 to 350 °F), flexibility during thermal cycling, and ability to compensate for differential rates of expansion between substrates.

The silicone adhesive comes in either film or liquid form and is applied between sheets of substrate for bonding. The surfaces should be cleaned with hexane or methanol and primed with a silane coupling agent. When assembling the substrate, it is important that the purity of the adhesive be maintained and that adhesive film thickness remain constant by optimizing cure schedules. In silicones, the higher the temperature and

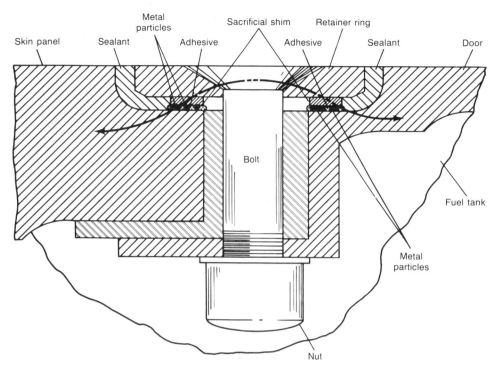

Fig. 18 Bonding application in lightning strike design. Source: Ref 17

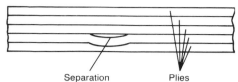

Fig. 19 Composite ply delamination

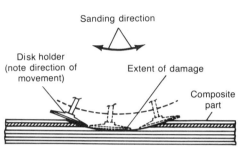

Fig. 20 Removal of damaged composite material

component adhesive is mixed and applied to both bonded surfaces with a serrated trowel. The two substrates are then joined and fixtured in place until the adhesive is cured.

Lightning Strike Applications

One of the environmental challenges facing an aircraft in flight is to withstand being struck by lightning. During a lightning strike, the electrical current enters and passes through the conductive components of the aircraft, ultimately exiting from another point more in line with the final destination of the lightning bolt. If the current passing through the plane encounters a region that is nonconductive, sparking and damage to the substrate will occur. To prevent damage from a lightning strike, it is important that conductive paths be avail-

longer the cure, the better the optical clarity.

Missile Radome-Bonded Joint

Missile bonding applications represent one of the greatest challenges facing adhesives used in the aerospace industry. The physical configuration of a radome is shown in Fig. 17. The substrates for this application are ceramic and filament-wound aramid-epoxy. The adhesive used to bond these two very dissimilar substrates is an epoxy. The criteria that the adhesive must meet includes room-temperature cure, long-term environmental performance at temperatures ranging from 34 to 54 °C (30 to 130 °F), and high-temperature performance associated with rapid heating.

The rapid heating element of these requirements is specifically noteworthy because the bond line temperature during launch of the missile can range from the cold temperature of storage to more than 200 °C (392 °F) in less than 2 min. Most room-temperature-curing epoxy adhesives have not reacted sufficiently at room temperature to provide the needed level of performance at 200 °C. It is this requirement alone that is very difficult to meet.

The bonding of the ceramic nose cone to the aramid-epoxy ring starts with the application of an adhesive primer to the bonding surfaces. Once the primer is dried, the two-

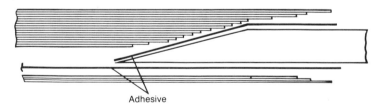

Fig. 21 Repair of damaged composite

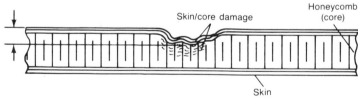

Fig. 22 Damaged honeycomb sandwich panel

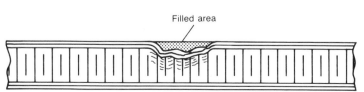

Fig. 23 Repair of minor impact damage

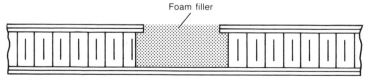

Fig. 24 Replacement of honeycomb with syntactic foam

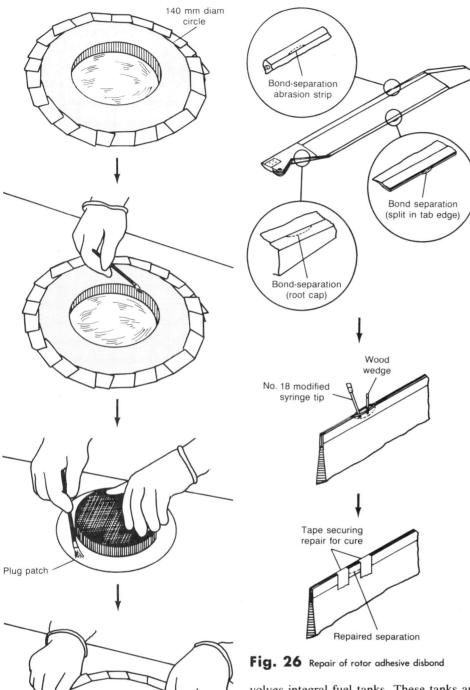

Fig. 25 Repair of honeycomb core using core splicing adhesive

Fig. 26 Repair of rotor adhesive disbond

able to allow electrical currents to pass through and out of the aircraft.

One of the areas of major concern when considering lightning strike protection in-volves integral fuel tanks. These tanks are part of the wing assembly and are closely associated with the outer wing surface. To compound their exposure to lightning strikes, fuel tank access panels in the skins are bolted. These bolts tend to attract lightning. To protect the fuel tank from sparks associated with a lightning strike, conductive adhesives are used, providing pathways for the escape of electrical current.

The design used to provide lightning strike protection for an integral fuel tank is shown in Fig. 18. The areas of importance in this figure include the bolt, nut, door, skin panel, retainer ring, sacrificial shim, sealant, fuel tank, metal particles, and adhesive. The adhesive in this case is a metal-particle-filled epoxy resin. It is used to bond the sacrificial shim to the door. When lightning strikes the bolt, the current can travel through it to the retainer ring, then through the sacrificial shim, and then through the adhesive into the door and out into other parts of the aircraft. By providing this pathway, no sparking will occur.

Repair Applications

Three areas where aerospace repairs can be performed are at the original manufacturing facility, at a repair depot, and in the field (Ref 18). At the manufacturing site, many of the adhesives used for repair are the same ones used to manufacture the original parts. This facility typically has the equipment resources to remanufacture a damaged component. At a repair depot, the repairs performed require more skill and a variety of alternate products to perform the job correctly. Without the original fixtures and tools located at the manufacturing facility, the repair depot is often forced to use different techniques or materials to repair the structures. Field repair involves the repair and maintenance of aerospace structures at sites other than the original manufacturing site or repair depot. Field repair is the most demanding because of requirements for short turnaround times, maximum equipment utilization, need for simplicity and to complement the limited facilities and trained personnel available. Three repair techniques in which the use of adhesives is critical are described below.

Laminate Repair. Damage to composite laminates typically involves either delamination between plies and/or destruction of material due to impact. Figure 19 shows a delamination between plies in a composite structure. The current field repair technique for repairing this type of damage is to remove the damaged material, either by cutting or sanding, as shown in Fig. 20. An adhesive film is then applied to the surface to be repaired. Impregnated fabric is placed on top of the adhesive in the same ply orientation as the original composite (Fig. 21). The assembly is then vacuum bagged and cured at elevated temperature. Through the use of a tapered bond, ply alignment, and proper adhesive and resin selection, over 60% of the unflawed strength may be achieved. With design allowables in the range of 40 to 60%, this type of repair can often return the part to its original capability. The function of the adhesive in this application is to provide additional bonding between the damaged surface and the new impregnated fabric.

Honeycomb structure damage generally takes on the appearance of delamination between the laminate and honeycomb, or deformation of the structure because of impact. Figure 22 shows an example of a damaged honeycomb structure. There are three basic repair techniques based on the level of damage incurred. The first method is to simply fill in the dented area with a body filler or syntactic type material (Fig. 23). The second method involves the removal of the sandwich skin and honeycomb from the damaged area. After removal, the hole is filled with a structural syntactic foam and a new skin is constructed over it (Fig. 24). The third method is to remove the damaged skin and honeycomb core. A new piece of honeycomb is cut to replace the material that was removed. The exterior surfaces of the remaining core and the replacement core are then coated with a splicing adhesive and the new core is set into place (Fig. 25). Again, a new skin is applied with the adhesive being cured either before or in combination with the skin. The advantage of this method of repair is that it can frequently achieve original strength levels with minimal weight gains.

Repair of Adhesive Disbonds. Adhesive disbonds generally occur due to environmental attack on the bond line, improper use of adhesive, poor surface preparation, or when design strength levels are exceeded. Figure 26 shows a repair sequence for the disbond of a rotor blade laminate. The procedure here is to open the disbond so that an adhesive can be applied. The bonded surfaces are then pretreated by either solvent wipe, sandblasting, and/or application of an adhesion promoter. With a narrow area for application, this repair would require a very low viscosity adhesive. To control its application, the adhesive is generally dispensed through a syringe. The two surfaces are then brought together through some mechanical means and the assembly is cured at a temperature appropriate for the laminate. Most cure schedules are below 100 °C (212 °F), so that absorbed moisture in the laminate is not forced out where it can affect the adhesive during cure. The result-

ing repaired structure can then be placed back into service.

REFERENCES

1. "Primary Adhesively Bonded Structure Technology (PABST)," General Material Property Data, AFFDL TR 77 107, 1977
2. N. Allbee, Adhesives for Structural Bonding, *Adv. Compos.*, Nov/Dec 1989, p 44
3. E.W. Thrall, Jr., Prospects for Bonding Primary Aircraft Structure in the 80s, in *Proceedings of the 25th National SAMPE Symposium and Exhibition*, 6-8 May 1980, p 726
4. R.E. Politi, Structural Adhesives in the Aerospace Industry, in *Handbook of Adhesives*, 3rd ed., 1990, p 717
5. R. Corden, Honeycomb Structure, in *Composites*, Vol 1, *Engineered Materials Handbook*, ASM INTERNATIONAL, 1987, p 721
6. J.C. McMillan, Advanced Composites in New Boeing Commercial Aircraft, Kevlar in Aircraft, in *Summary of the Technical Symposium IV*, Section 9, 1983, p 1-8
7. S. Quick, The Art of Reticulation and Applicable Adhesives, in *Proceedings of the 18th International SAMPE Technical Conference*, Vol 18, 1986, p 657
8. "Bringing Aircraft Technology Down to Earth," Report 28123/e, Ciba-Geigy Corporation
9. E.C. Clark, Properties and Uses of Epoxy Syntactics, in *Adhesives 88 Conference Papers*, Society of Manufacturing Engineers, 11-13 Oct 1988, p 636-5
10. E.C. Clark, Properties and Uses of Epoxy Syntactics, in *Adhesives 88 Conference Papers*, Society of Manufacturing Engineers, 11-13 Oct 1988, p 636-6
11. F.J. Riel and P.M. Rose, Adhesive Bonded Noise Suppression Structures for Commercial and Military Aircraft, *SAMPE Q.*, Oct 1984, p 48
12. P. Oliva, SF 340 Airfoil Structure: A Unique Approach, in *Proceedings of the 17th National SAMPE Technical Conference*, Vol 17, Society for the Advancement of Material and Process Engineering, 1985, p 116
13. P. Oliva, SF 340 Airfoil Structure: A Unique Approach, in *Proceedings of the 17th National SAMPE Technical Conference*, Vol 17, Society for the Advancement of Material and Process Engineering, 1985, p 119
14. R.B. Kollmansberger and M.N. Botkin, F 15 Composite Speedbrake in Production, in *Proceedings of the 8th National SAMPE Conference*, Vol 8, Society for the Advancement of Material and Process Engineering, 1976, p 147
15. M. Baile *et al.*, Optically Transparent Silicone Adhesives, *Rubber World*, Oct 1988, p 34-37
16. D. Peterson and D. Boyce, Development of an Adhesive System for a Tactical Missile Radome Joint, in *Proceedings of the 5th International Joint Military/Government Industry Symposium on Structural Adhesive Bonding*, U.S. Army Armament Research, Development, and Engineering Center, 3-5 Nov 1987, p 311
17. M. Billias and E. Borders, Electrically Conductive Structural Adhesive, U.S. Patent 4,428,867, 31 Jan 1984
18. E.C. Clark and K.D. Cressy, Field Repair Compounds for Thermoset and Thermoplastic Composites, in *Proceedings of the 32nd International SAMPE Symposium*, Society for the Advancement of Material and Process Engineering, 6-9 April 1987, p 271-278

Industrial Applications for Adhesives

Robert E. Batson, Dexus Research Inc.

THE MAJOR INDUSTRIAL SECTORS of the U.S. economy have been undergoing dramatic change. Specifically, the heavy metal production and fabrication industries centered in the "rust belt" are being replaced by the light manufacturing industries of electronics, medical disposables, consumer products, sporting goods, and others. The dominance of the automotive industry has been supplanted by the electronics industry as the largest domestic industry in terms of gross national product.

The light manufacturing industries use light alloy metals, plastics, and ceramic materials in their manufacturing processes where the traditional steel assembly processes of welding or use of mechanical threaded fasteners and interference press fits are no longer viable assembly methods. Adhesives of all types are proving to be among the most cost-effective assembly methods used today. They can provide stronger, more durable joints and be designed with weight-saving components in order to provide greater reliability in all types of environments.

Industrial adhesive applications fall into several major application categories. Some basic areas in which industrial adhesives are used by original equipment manufacturers (OEM) in terms of dollar volume (Ref 1) are:

- Paper, labels, and packaging adhesives, 26%
- Wood and furniture, 11.4%
- Textiles and fiber bonding, 6.5%
- Construction adhesives/sealants for the construction trades, 31.4%
- Structural adhesives for transportation applications, 8%
- Nonindustrial, 3.3%
- Structural or specialty adhesives for specific markets, 13.4% (electronic and electrical, 4.5%; medical, sporting goods, and aerospace, 3.5%; and metal assembly, 5.4%)

Packaging adhesives represent the majority of adhesives by volume in the industrial market, with hot-melt and solvent-base adhesives being the two primary types. They are followed by wood glues for the furniture industry and flexible latex and silicone adhesives for the construction industry. Engineering and specialty adhesives represent only a small portion of total adhesive consumption in terms of volume, but they use high-cost raw materials. Engineering and specialty adhesives are of prime importance in the high-tech, industrial sectors of electronics, medical, electrical, specialty construction, and sporting goods areas, and are the focus of this article.

Adhesives Technologies

Several hundred companies currently produce adhesives in the United States, offering more than 25 000 different adhesive formulations, but 95% of all engineering (load-carrying) bonding that occurs in OEM applications uses six adhesive families, which cover the majority of application requirements (Ref 2). These six families of engineering adhesives are:

- *Epoxies*: High strength, good temperature and solvent resistance
- *Urethanes*: Flexible; good peel, shock, and fatigue resistance
- *Acrylics*: Versatile; bonds oily parts; fast cure, good overall properties
- *Anaerobics*: Metal bonding of threads, cylindrical shapes
- *Cyanoacrylates*: Fastest bonding; plastics and rubbers; limited temperature and moisture resistance
- *Silicones*: Flexible; good outdoor weathering, good sealing/bonding

Each of these families can be further divided by physical properties, cure system, and/or application orientation for a wide range of practical products. The Section "Adhesives Materials" in this Volume contains articles describing each of these families, as well as others. "Brittle epoxies" and "flexible urethanes" are no longer valid concepts because toughened, flexible epoxies and harder urethane adhesives are currently available. New formulations of adhesives, primers (adhesion promoters), and activators (to increase curing rate) are constantly being developed to keep pace with new plastic and coating formulations, electronic component requirements, and end user process requirements.

Other types of adhesives, such as solvent-base cements that are used either as plastics "welding" adhesives (Table 1) or as contact cements, are very useful in industrial applications, but are facing increased legislated restrictions and higher costs. Many manufacturing facilities have curtailed new applications usage of solvent-base adhesives and will replace these products with 100% solids, reactive adhesives in the near future (Ref 3). Starches, polyesters, phenolics, and polyvinyl-base adhesives are also used in industrial applications, but have processing or performance considerations that limit their use.

For reactive adhesives, such as the methacrylates and urethane acrylates, ultraviolet (UV) wave energy curing is used as an alternative or supplementary curing system to thermal, chemical, or room-temperature curing systems.

Ultraviolet-curing adhesives are increasing in popularity in the high-tech industrial

Table 1 Examples of solvent cements

Plastic	Solvents
Polystyrene	Toluene, methylene chloride, trichloroethylene, xylene, 1,1,1-trichloroethane
Cellulosics	Ethyl acetate, acetone, methyl ethyl ketone, cellosolve acetate
Polyvinyl chloride	Tetrahydrofuran, cyclohexane, cyclohexanone
Polycarbonate	Methylene chloride, ethylene dichloride
Acrylonitrile-butadiene-styrene	Tetrahydrofuran, methyl ethyl ketone, methyl isobutyl ketone, methylene chloride
Acrylics	Methylene chloride, ethylene dichloride
Nylon	Aqueous phenol, alcoholic resorcinol, calcium chloride in ethanol
Polysulfone	Methylene chloride

sectors of electronics and medical disposables because of their low-temperature, ultrafast (≦20 s) polymerization of many acrylated formulations of silicones, urethanes, and methacrylates. UV cure refers to the cure system rather than a chemical type and is used with one or more transparent or translucent substrates through which the UV energy can be transmitted to the bond line (Ref 4).

From an application standpoint, the adhesive selection process should take into account an adhesive type that has had widespread use, although only for nonstructural bonding applications. This type is represented by hot-melt adhesives, which are thermoplastic sticks or chips that are heated to a liquid application state, applied to the surface, and then naturally cooled to fixture the components and form a joint. New technology involving reactive urethanes that are "piggy-backed" onto the hot melt as a carrier is becoming commercially available for a new generation of "structural" hot melts. The urethane adhesive can moisture cure over time to form a tough structural bond, while the hot-melt constituent can fixture the components together quickly. For the large-area bonding of office partition panels, composite materials, and metal to plastic, this type of adhesive appears to be more cost effective than are traditional two-component urethanes or contact adhesives.

Systematic Approach to Adhesive/Application Selection

Three different processing phases must be considered when selecting an industrial adhesive for a specific application. These phases are uncured, or liquid; curing, or transition; and cured, or solid. Cured adhesive performance is the most important concern because it affects reliability. In many cases, several types of adhesives may meet the performance criteria satisfactorily, in which case other product features may become the important selection factors.

Table 2 shows the uncured, curing, and cured properties that are important to every industrial application. All these physical properties must be evaluated before the final selection is made. In many cases, safety considerations may prohibit the use of an adhesive that has a flash point below 40 °C (100 °F) or a toxic ingredient, regardless of its cured performance and cost-effectiveness. This is particularly true of bench-type assembly operations in the electronics or medical industries, rather than factory floor usage.

Table 3 shows the implications for each processing phase in terms of direct and indirect processing costs in the actual application (that is, labor has to be weighed against automated application equipment,

capital costs, and parts handling options). When all performance considerations have been satisfied, the selection of an adhesive may be between several adhesive systems (a thermal-curing epoxy, two-component urethane, or UV-curing methacrylate).

The interaction of the three processing phases is also shown in Table 3, in terms of the critical processing factors of reproducibility in application, productivity in processing, and reliability in field service performance. These three areas must be evaluated before selecting an adhesive in order to obtain the real "applied" cost, rather than an adhesive material cost alone. For high-volume industrial applications, end users should consider processing time and capital costs, as well as material costs.

In many cases, end users look only at the direct material cost of an adhesive and use this as a basis of comparison. End users who use this method fail to realize that the true value of an adhesive is the applied, cured cost per unit component, not the direct material cost. This cost includes the direct material and labor costs, in addition to the indirect costs for fixtures and equipment, floor space, curing energy, work-in-progress inventory, and rework or scrap allowances related to bonding. By including all costs incurred in the bonding process, the true cost of the adhesive application can

be determined, which allows end users to justify the introduction of or replacement with a new adhesive.

Although many of the applications presented in this article appear to be "naturals" for adhesives, the adhesives were designed into the application at the earliest possible point, not used as a "fast fix" on the production line to correct an inherent design or assembly problem. "Fixes" are, at best, used for a transition period until a redesign is available and seldom realize the full benefits offered by an engineered adhesive system. Many bad application experiences can be traced to the use of adhesives as "fixes" without consideration for such factors as joint design, surface preparation, and application equipment.

Adhesives Applications

The remainder of the article discusses the adhesive applications in the electronics, electrical, medical, sporting goods, and construction industries.

Electronics

The electronics market comprises several different industries, the predominant ones being telecommunications, consumer, computer, defense, industrial process controls,

Table 2 Major properties of adhesives

Uncured (liquid)	Curing (transition)	Cured (solid)
Viscosity	Ultraviolet energy	Strength, static
Color	Thermal cure	Strength, dynamic
One or two components	Room-temperature cure	Environmental resistance
Fluorescence	Stress inducement	Electrical
Toxicity	Shrinkage	Thermal
Flash point		Toughness
Shelf life		
Pot life (if applicable)		
Wetting		
Solvent content		

Table 3 Major cost and performance considerations

Processing phase	Consideration
Uncured (liquid)	Application method
	Adhesive application
	Packaging
	Safety
	Manual, semiautomated, or automated application
	Reproducibility
	Right amount in the right place at the right time, every time
Curing (transition)	Curing method
	Capital expenditure
	Work-in-progress inventory
	Floor space
	Fixtures
	Rate of cure
	Fixture/full cure
	Productivity
	On-line (throughput) versus off-line (batch processing)
Cured (solid)	Performance
	Specifications
	Warranty
	Field service
	Reliability
	In-service durability to meet product life

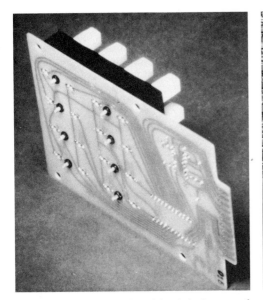

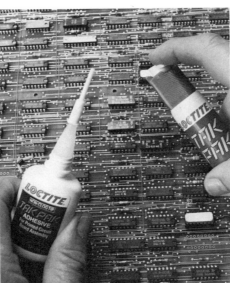

Fig. 1 PWB with keyboard bonded. Courtesy of Loctite Corporation

Fig. 2 Cyanoacrylate adhesive wire tacking of ECO wires to PWB. Courtesy of Loctite Corporation

Fig. 3 UV-curing adhesive for wire tacking with special hand-held dispenser and curing wand. Courtesy of Dymax Corporation

and automotive electronics (Ref 5). All of these industries use the basic building block of electronics, that is, the printed wiring board (PWB), but each has its own board designs, components, and processing methods.

A typical PWB is a multilaminate reinforced-plastic board that has a plastic protective coating (permanent solder mask). This represents one prime substrate for adhesives used in electronics, the other substrate being wire insulation, electronic devices, or mechanical components (Fig. 1). Wires, devices, and mechanical switches are made or coated with plastic for dielectric insulation of the electrical conductive metals. The majority of electronic applications involve bonding thermoplastics, ceramics, or aluminum substrates. Key adhesive selection criteria are cure speed, dielectric properties, ease of application, and thermal stress performance.

The principal adhesive applications for PWBs are in surface-mount technology (SMT), thermally or electrically conductive applications, engineering change order (ECO) wire tacking, and general device bonding (strain relief of solder connections). Off-board applications involve die attach (using high-purity polyimide resins) and component subassembly fabrication prior to PWB assembly (Ref 6). The adhesives for microelectronics use in die attach and hybrid construction have specific properties and performance requirements (Ref 7). SMT adhesives can be electrically or thermally conductive and/or dielectric in nature and yet be used to mount small parts. The quantities used in each application are relatively small, measured in 10^{-2} or 10^{-3} g, but the number of applications is measured in the millions. This results in small adhesive package sizes (1 to 10 g, or 0.35 to 35

oz) of specialized formulations at very high comparative prices.

Wire-Tacking Adhesives. In the computer industry, the wire tacking of ECO wires to the PWB is critical because of the rapid obsolescence of circuitry design and the need to update existing assemblies constantly. Between 10 and 100 ECO wires must commonly be secured to the surface of the PWB and must not break when the board is inserted or removed from equipment. At first, hot-melts, thermoplastics, room-temperature vulcanizing (RTV) silicones, two-part epoxies, and various latex or solvent cements were used to bond the wires down. Wire hold-down loops were also employed. These required that the operator feed individual wires through the hold-down device. With a large number of changes, this became very expensive.

Standard off-the-shelf cyanoacrylates were also used with a brush-on accelerator to fixture the wire to the board in seconds, where bonds were made every 50 to 75 mm (2 to 3 in.) along the wire. This product combination was effective in terms of speed and processing, but had marginal performance because of bond popping (adhesive failure to PWB) and sloppy accelerator application. Adhesive companies developed specially formulated cyanoacrylates and spray-type accelerators, which could achieve total cure in 3 to 5 s with maximum strength and minimum bond stressing. Cyanoacrylates, being rapid-room-temperature curing, low-viscosity, nontoxic adhesives, proved to be ideal for this application and quickly became the industry standard, replacing other adhesives, mechanical wire ties, and clips (Fig. 2).

Cyanoacrylate adhesives could be used with polyvinyl chloride, polytetrafluoroethylene, and other insulation types by bridg-

ing the wire to form a "captive adhesive wire tie," allowing one product to be used for all insulation types. The products were made easier to apply by dispensing directly from the small package (squeeze bottle), followed by an overspray of accelerator to the exposed adhesive (two-step system). Cyanoacrylate product development has kept pace with PWB material improvements toward smoother protective board coatings (dry-film permanent solder masks). Elastomeric polymers are added to the formulation to increase adhesion during wave solder or burn-in conditioning of the PWB, when thermal stressing may cause bond popping.

UV-curing adhesives and a special focused light source are also used to achieve wire tracking (Fig. 3). A special syringe applicator and wand light allows the operator to control the application and cure the applied product with the UV light source with one hand. The bonds cure in 1 to 3 s and form a tough, resilient bond to the PWB and the wire, with little odor and without the wasted product that characterizes oversprays (Ref 8).

SMT Adhesives. For decades, the electronics industry designed circuits based on leaded components. These components were electrical devices with wire leads protruding from the body of the device for attachment to the board via through holes, which allowed the leads to be bent or crimped to the bottom of the board before being wave soldered (Fig. 4). The crimping of leads kept the components firmly in place during processing, and all components were attached on one side of the board to undergo bottom-side soldering.

The advent of SMT technology quickly changed all processing standards. This technology involved capped devices that had no leads and that could be mounted to the top and bottom surfaces of the board (Fig. 5). The only logical method was to bond the surface-mount devices (SMDs) to the board, usually the bottom, and then insert leaded components through the top surface in the normal way. This clearly is a transitional bonding method based on hybrid boards of SMDs and

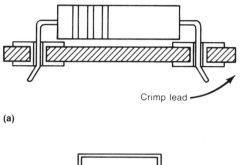

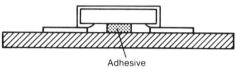

Crimp lead

(a)

(b)

Fig. 4 Leaded components versus SMD device mounting to PWB. (a) Leaded component mounted by means of through holes, crimped, and ready for wave solder. (b) SMD attached to PWB with adhesive and ready for soldering step

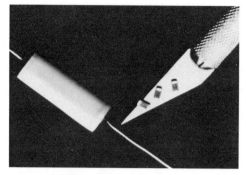

Fig. 5 SMD capacitors compared to leaded capacitors

leaded components. Boards with all SMD components will probably use tacky solder paste to hold the SMDs in place at their solder connections rather than using adhesives at the center point.

In practice, SMDs are placed on small beads or dots of uncured adhesive and cured in place by thermal or UV energy. This holds the SMDs firmly in place during subsequent processing operations, such as leaded device insertion or cleaning, until they can be soldered to the board solder pads. SMT adhesives serve two functions: They act as a processing aid (for example, holding a part temporarily until it can be permanently attached with the soldering operation), and they provide stress relief to solder connections during service life to prevent premature failure of the electrical connections (Ref 9).

Three chemical types of adhesives are commonly used for SMT bonding: acrylics,

urethane-acrylates, and epoxies. Each has specific application benefits related to cure method, speed of cure, and tear strength. Other adhesive systems that have found limited use are silicones, thermoplastic hot melts, and pressure-sensitive adhesives (Ref 10). The acrylics were early developments and have generally been supplied by Japanese companies to support Japanese placement equipment, such as computer-programmable pick-and-place machines that accurately locate the small SMDs on the solder pads at high rates of speed. Urethane-acrylates and epoxies have been supplied by U.S. adhesive companies specializing in the electronics market. They offer more sophisticated cure systems, shelf stability, and improved performance, and represent second-generation SMT adhesives.

Ideally, these adhesive formulations need to be the single-component type to avoid pot life problems, viscosity changes, air entrapment due to mixing, and other problems associated with two-component systems. An SMT adhesive must be capable of being dispensed reproducibly by automatic equipment and must have some type of identity that can be machine sensed (visible or fluorescent dye). It must be repairable

while also being strong enough to hold the component reliably in place, curable in less than 2 min at temperatures that are safe for the PWB, sag resistant in order to retain its shape, but able to flow and wet the surfaces when assembled. In most cases, it must be nonconductive electrically, noncorrosive, nontoxic, and have acceptable odor and flammability ratings. In short, it must be a unique adhesive incorporating many specific properties not found in generic adhesives. Successful SMT adhesives demand prices up to $1/g ($28/oz) in small package sizes. A large assembly plant will use $1 million of product annually.

SMT adhesive formulations are also based on their application method. These methods include screen printing (squeezing the adhesive through accurately defined patterns in a screen for high-production applications), pin transfer (multipin grids, which convey patterns of adhesive drops to the PWB), and syringe application (single shots of adhesive delivered by a pressure-controlled syringe) as shown in Fig. 6 (Ref 11). Epoxy, acrylic, and urethane-acrylate chemistries can be formulated by rheology to be screenable (low, creamy viscosity), pin transferable (liquid viscosity), or syringe applicable (high viscosity with thixotropy). The most popular method is the syringe method employing 3 to 10 mL (0.18 to 0.60 in.^{3}) syringes controlled electropneumatically for a moderate production of many different types of PWBs (Fig. 7).

One main consideration when selecting an SMT adhesive is the cure method desired, with epoxies and acrylics being thermally cured and urethane-acrylates being cured by heat, UV wave energy, activators, or anaerobically (absence of air). The most popular method is thermal curing using infrared (IR) thermal energy provided by a short in-line heating tunnel. UV curing is often used at the beginning of the curing cycle to give fast fixturing to the SMDs

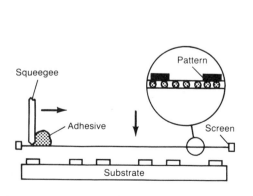

(a)

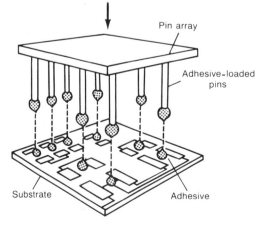

(b)

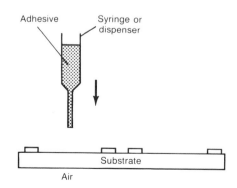

(c)

Fig. 6 SMT adhesive application methods. (a) Screen printing. (b) Pin transfer. (c) Syringe

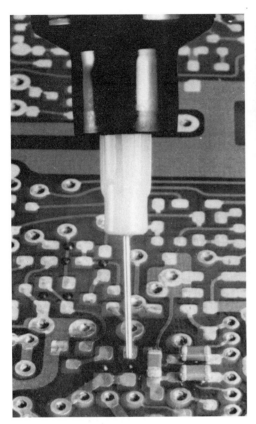

Fig. 7 SMT adhesive applied by syringe to PWB. Courtesy of Dymax Corporation

Fig. 8 Transistors bonded to heat sink rail using a thermally conductive, two-part methacrylate-type adhesive and surface activator. Courtesy of Loctite Corporation

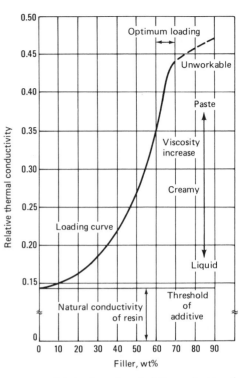

Fig. 9 Relative thermal conductivity as a function of filler level (aluminum oxide powder in an anaerobic resin)

and prevent off-placement movement, such as skewing, tombstoning, and sliding. A UV cure is followed by an IR thermal cure.

One question that arises with regard to SMT adhesives is that if adhesives must be used, why not use an electrically conductive adhesive to replace the solder for electrical connections? These adhesives would eliminate fluxes, flux cleaning steps, and provide faster processing with significant material cost savings. The answer is that conductive adhesives cannot fulfill all the requirements of solder with respect to conductivity, metal (silver) migration from the bond, and long-term durability. Many adhesive companies believe they are close to overcoming these problems and providing a solder replacement; in fact, in many low-volume, low-power applications, conductive adhesives are being used for that purpose (Ref 12, 13).

Thermally Conductive Adhesives. Miniaturization of electronic circuitry may result in problems of heat buildup, which can cause premature failure of the electronic components if their maximum operating temperature is exceeded. Electronic packaging engineers and designers have incorporated heat sinks, which are small, metallic, finned radiators that use conduction and convection to carry the heat passively from the power device (heat generator) to the air. The efficiency of heat transfer depends a

great deal on the interface of the heat generator and the heat sink.

Traditionally, a transistor, diode, or other power device has had a metal case (used for heat conduction and mounting holes) fastened to the heat sink with mechanical fasteners, such as pop-rivets or threaded fasteners. A characteristic problem is that a maximum of 20% of the metal surfaces are in contact, regardless of finish or fit, while air occupies the remaining 80% of the joint (air insulates the surfaces and prevents heat transfer).

One popular way to solve this problem has been to use a conductive medium, such as silicone grease and/or mica washers, to displace the air and provide the necessary thermal path. However, thermal greases are dust collectors, have a tendency to migrate, and can flow out of the joint at higher temperatures. Other methods have involved compressible pads of silicones or composite material that is soft enough to be an effective interface between the two surfaces. These methods all require an inventory of mechanical fasteners to mount the parts to the heat sink and the PWB, as well as an inventory of shaped pads or washers that must be inserted at the assembly operation and then fastened. (See Table 4 for costs related to heat sink assembly methods.)

A more efficient method is the use of thermally conductive adhesives to provide both the thermally conductive path and the attachment method (Ref 14). Metallic (electrically conductive) or nonmetallic (electrically insulative) powders are blended into the adhesive formulation to make high-viscosity, pasty adhesives (Fig. 8, 9). The most popular conductive systems are formulated with epoxies, silicones, anaerobic acrylates, and methacrylates. "Spacer" parti-

cles also can be added to adhesives to provide a definitive gap between the two substrates to ensure electrical isolation. This often provides a quantitative measure of dielectric strength in the bond line.

Despite the advantages of reduced inventory of parts, increased thermal transfer efficiency, and reduced labor and parts costs, adhesives have some disadvantages that should be evaluated. Two-component epoxies and silicones require metering and mixing. In the case of thermally and electrically conductive adhesives, it is essential that the bond line be free of air bubbles entrapped during mixing (effective conductive pathways). Vacuum degassing of the mixed adhesives prior to application is inconvenient and requires additional equipment and time. Pot life (working time after mixing) can be short with two-component systems, reducing the application time to 30 min or less before the adhesive is too thick to apply.

Packaging two-component epoxies in small so-called dual pouches with a removable separator between the hardener and resin is one way to overcome the meter/mix/air problems of electronic-grade adhesives. These packages have premeasured chemicals that can be mixed in the pouch without entrapping significant amounts of air. Application is direct from the pouch to the substrate to avoid mess or contamination. Using dual product dispensers for materials with a 1:1 to 1:10 ratio mix and static mixer

Fig. 10 TO3 power device bonded to finned heat sink with high-strength, thermally conductive methacrylate adhesive. Courtesy of Loctite Corporation

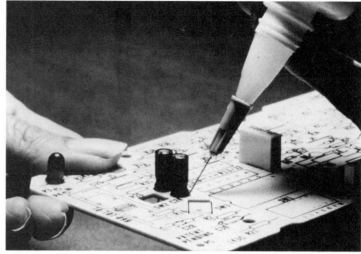

Fig. 11 Mounting of large components to PWB surface with cyanoacrylate adhesive for strain relief of solder connections

tubes is another way to meter/mix filled resins and hardeners without air entrapment if their viscosities are low enough.

Other thermally conductive adhesive systems are single component and thermally curing. They are degassed at the supplier plant and are ready to use from the package. These packages can be refrigerated to maintain their stability for 3 to 6 months and allow curing at lower temperatures than normally needed for single-component products. The inconvenience of refrigeration is offset by the improved thermal conductivity and reproducibility.

Another adhesive family used in thermally conductive products is the two-part (requires no metering/mixing) anaerobic/aerobic system consisting of a heavily filled resin and a preapplied surface activator. Anaerobic (cures in the absence of air) and aerobic (cures in the presence of air) systems are convenient to use because the liquid activator can be applied to one (or

both) surface(s), and the adhesive to the other. Cure commences when the surfaces are brought into contact. Fixture strength is obtained within 3 to 7 min, and full cure, within 2 h (Ref 9) (Fig. 10).

Thermally curable conductive tapes and films are also available in strip and precut forms, which offer adequate levels of adhesive strength without the handling problems of paste materials. Films come on a plastic release carrier, with 105 to 120 °C (220 to 250 °F) cures in 15 min (Ref 8), and are available in a variety of thicknesses, 205 μm (8 mils) and up. Double-sided adhesive tapes are pressure-sensitive adhesives of a Kapton film core with 25 to 50 μm (1 to 2 mils) thick acrylic adhesive on each side.

Electrically conductive adhesives are based on epoxy, silicone, or acrylate chemical technology, with the most popular commercial products being one- and two-part epoxies. Typical fillers used to gain electrical conductivity are silver, nickel, and car-

bon, although pure gold may be used in special applications involving microelectronics. Carbon- and nickel-filled products with lower electrical conductivity than silver are used in the grounding of circuits, electrical static discharge protection, electromagnetic interference/radio frequency interference (EMI/RFI) protective coatings, gaskets, and adhesives.

Silver, being the most conductive of the popular fillers, is normally used for most products. It can achieve the highest level of conductivity while being cost effective. Many developmental products have been made in order to reduce the amount of silver (for example, the 60 to 70% filler level) used in obtaining high electrical conductivity (5 to 0.7 m$\Omega \cdot$ cm range) by the use of silver-plated microspheres, rods, and other shapes. The theory is that most of the current is carried on the surface of the filler rather than through the matrix and that physical contact between particles is what establishes the conductive pathways between substrates. Higher percentages of fillers result in greater frequency of contacts between particles and smaller spacing between filler particles. The same effect could be achieved by placing conductive media on surfaces of insulative particles and using them at high filler levels. This has not been totally accepted, and the standard in the industry remains high purity, silver flake at high filler levels for maximum conductivity, both thermal and electrical.

A new technology involving UV-curing electrically conductive adhesives has recently been introduced. At present this product line is still in the introduction stage and is being evaluated by several large companies for future use. The concept is based on special chemical groups of oligomers that can be blended together to form urethane, epoxy, and silicone combinations that offer unique properties. Surface cure

Table 4 Cost impact of heat sink assembly methods

Manufacturing and service costs	Solid insulator and heat sink grease (mica, silicone)	Soft gel pad or film	Two-component filled epoxy	Heat-curing filled epoxy	High-strength filled anaerobic	Repairable filled anaerobic
Inventory requirements	Moderate to high	Moderate to high	Low	Low	Low	Low
Hand labor input (automatic assembly not practical)	High	High	Moderate	Low	Low	Low
Mixing and measuring requirements	None	None	High	None	None	None
Curing time required	None	None	High	Moderate	Low	Low
Fixturing required for cure cycle	None	None	High	High	Low	Low
Product migration and dust pickup problems	High	None	Low	Moderate	Low	Low
Capital equipment required	None	None	None	High	None	None
Disassembly problems	Low to moderate	Low to moderate	High (not repairable)	High (not repairable)	High (not repairable)	None

and fast fixturing are provided by UV energy, while the bond line under the component is cured by a secondary system that depends on moisture, air, heat, or an activator. As of 1990, this is the only UV-curing electrically conductive product on the market, but it could be successful if it can be silk screened onto PWBs and then cured at ambient temperatures with UV energy.

Tape adhesives offering varying levels of conductivity ($\leq 0.1 \, \Omega/\text{in.}^2$) are also available but are used in applications where convenience outweighs the necessity to minimize electrical resistance. Typical tape applications are for splicing flat cables, mounting backup batteries and leaded chip carriers to PWBs, and ceramic hybrid substrate assembly. In certain instances, acrylic adhesives containing metal particles can experience anisotropic conductivity. That is, the adhesive allows an electrical path in one direction, in this case, the thickness, but does not conduct in the lengthwise or widthwise direction of the adhesive tape, which has a dielectric strength of 500 V dc.

Significant applications for electrically conductive adhesives include fabricating hybrid circuits for transistors, integrated circuits, and the surface-mount devices already mentioned. Electrically conductive adhesive materials are epoxy-base formulations of a liquid or preform nature. The preforms are glass supported, tack-free adhesive films of 50 to 75 μm (2 to 3 mils) thick and are die cut to the size and shape required prior to bonding. In some cases, silver foil is used with the adhesive. The adhesive is applied to both sides of the foil to make a sandwiched tape, instead of using silver fillers within the adhesive (Ref 7). Preforms are always thermal-cure formulations that involve moderate temperatures (typically, 120 °C, or 250 °F, for 1 to 2 h). Preforms are attractive for bonding substrates to the package because they control the bond line thickness, uniformity, and degree of flow of the adhesive during thermal cure.

The three bonding areas of microelectronics are substrate to package, chip to substrate (the most critical bonding application), and lip sealing, which involves hermetically sealing the package together, as well as bonding. Normally, lip sealing is done with preforms because adhesive migration will occur if excessive material is applied and leakage will occur if insufficient material is applied.

General Component Bonding. Many electronic components must be sealed against moisture or other environmental agents for optimum reliability. A major advantage of adhesives (with the exception of silicone, which allows moisture and gas permeation through the bond) is that they provide a seal and join parts permanently if they are applied in a continuous manner over the entire joint area. In small electronic components, gaskets or other sealing devices are unnecessary if appropriate adhesives are used. These adhesives can be gap filling, relatively flexible, and impervious to harmful environmental agents.

Adhesives that are in this category are the toughened, higher-viscosity cyanoacrylates; thermoplastic hot melts; toughened epoxies; and methacrylates. As already mentioned, silicones are excellent sealants/adhesives and are one of the most cost-effective gap-filling adhesives, but they do allow moisture vapor and other gases to permeate the bond line with time.

Toughened cyanoacrylates, which are single-component adhesives, offer a good compromise of ease of application, moisture cure, fast fixturing, and good durability. The addition of an elastomeric monomer to a previously "brittle" type adhesive has provided a wide range of new uses for toughened products. The same chemical concept has occurred with toughened epoxies (high performance, flexible grades), where elastomer additives have been added to improve low-temperature performance and increase flexibility.

Acrylate adhesives of the methacrylate, anaerobic, and acrylated-urethane families are also used in many electronic applications with a dual-cure or tricure system. These products are normally single-component formulations that can be used with chemical surface activators as a two-step bonding program. Recently, suppliers of all these products have incorporated UV-curing technology for ultrafast fixturing and surface curing, and thermal cure initiators for those areas not exposed to UV light. These cure combinations allow parts to be fixtured together more quickly, while allowing full cure during subsequent processing steps (involving heat, if required).

Mounting relatively large electronic components to PWBs, such as axial and radial electrolytic capacitors, resistors, potentiometers, relays, transformers, switches, connectors, and others, by fastening methods other than solder connections is necessary to provide strain relief to the electrical connections (Fig. 11). Stresses caused by thermal expansion and mechanical forces caused by handling can lead to connector failure if the connector is used as the primary holding force. Originally, mechanical fasteners were used to mount components, but with the high cost of labor and parts, cyanoacrylate and UV-curing adhesives have been found to be more cost-effective (less than ½ ¢ per application in terms of material cost).

Potentiometers and other adjustable electrical components used on PWBs require adjustment to balance the circuit and then must be locked in place. Cyanoacrylates and UV-curing adhesives provide this locking function of the adjustable element by bonding it to the housing of the device, preventing any further movement. Many end users use this technique for two rea-

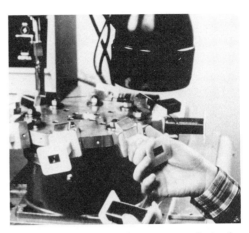

Fig. 12 UV-curing coil termination adhesive for wire-winding operation. Courtesy of Loctite Corporation

sons: to maintain original settings from vibrational movement, and to prevent or indicate unauthorized tampering with the device. Similarly, fasteners such as screws and rivets on access ports for cabinetry can be partially covered with adhesive to indicate whether the fasteners have been tampered with, which would negate the warranty.

Many small, plated, threaded fasteners are used in electronic and electrical assemblies that are subject to high-frequency vibration, thermal stressing, and improper tightening. Threaded fasteners depend on frictional forces on the thread surfaces for natural holding power. However, this is difficult to achieve in practice because of their small size and low seating torques. Anaerobic or cyanoacrylate adhesives offer a removal bond between the nut and screw that will hold the parts together and yet allow minor adjustment if required.

Electrical

Electrical applications for adhesives are different from electronic applications in that they generally involve significant load-carrying requirements. Several electrical industries lend themselves to adhesive use, such as in rotating electrical devices, motors, alternators and generators, transformers, solenoids, and the processing controls of instruments, relays, and switches. Most of these devices are electromechanical assemblies involving movement due to electrical and magnetic force fields.

Laminated components, such as transformers, solenoid coils, rotors, and stators, must be assembled from thin laminates and held in place mechanically or adhesively. Transformers and solenoids produce a low-frequency vibration in the laminates, resulting in an audible humming noise and minor loss of energy. High-quality transformers use very low viscosity, wicking adhesives to penetrate and bond together individual laminates, preventing noise generation. These adhesives can be dip applied in vac-

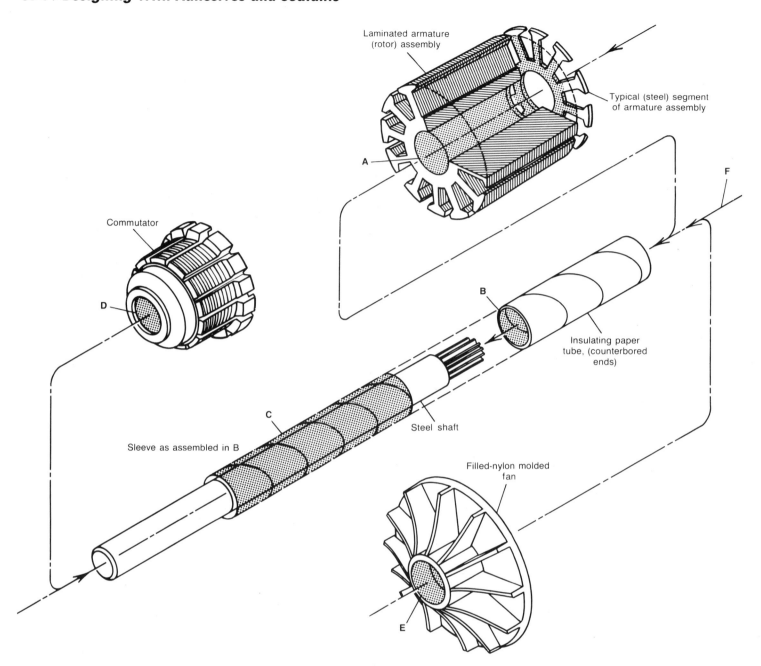

Fig. 13 Adhesive applications for motor armature assembly

uum or pressure tanks or applied to the laminate edges.

In the production of transformer and solenoid coils, in which wire is wound at high speed onto shaped bobbins, a process problem occurs when securing the cut wire (coil termination) after winding in order to remove the coil from the winder station. The end of the wire must be quickly and reliably secured to the other windings without damaging the wire insulation. Tapes, rubber bands, waxes, and other fast-tacking methods have been used with varying degrees of success but often with a great deal of scrap coil. Cyanoacrylates were first used with the overspray accelerator to give fast wire

tacks in 1 s or less. Although product performance was satisfactory, coil winding machine users found it very difficult to automate the termination process with reliability. UV-curing adhesives were then tried and found to be ideal for the application, because they are one-component, low-odor adhesives that can be cured in less than 3 s to form a tack-free, strong bond (Fig. 12).

Many electrical devices that use coils need to provide additional security to the wires as they pass out of the body of the device to the electrical connectors, which are usually pins. UV-curing or epoxy adhesives are used to form a strain relief coating

over the fine wires to bond them securely to the plastic form. This prevents overstressing the wire wrap or solder joint that forms the electrical connection. UV-curing adhesives work much faster.

One of the most dramatic electrical applications for adhesives was in the production of double-insulated motors for the power tools industry. One manufacturer redesigned its basic power tool motors in 1969 to bond and provide insulation between armature components. The goal was to increase safety and productivity, while keeping overall assembly costs comparable to those of noninsulated motors. The new design was that of a 100% adhesive-bonded power tool

Fig. 14 Ferrite magnet bonding for fractional horsepower motor using aerobic adhesive and surface activator. Courtesy of Dymax Corporation

Fig. 15 Speaker assembly bonded to plastic speakerphone housing

Fig. 16 Head lamp bonding using polyurethane two-part adhesive with plastic lens and housing. Courtesy of Ciba-Geigy Corporation

motor to replace welding, press-fitting, injection-molding, and other mechanical joining techniques.

Figure 13 shows the component parts and sequence of assembly operations for building a double-insulated motor. The adhesive applications involve various steps: (A) bond together the individual laminates to make up the rotor assembly (spray the inner bore of laminate stack with low-viscosity adhesive to wick into the stack, followed by induction cure); (B) bond the insulating, impregnated kraft-paper tube to the shaft (coat the bore of insulating tube and outside diameter of shaft with medium-viscosity urethane-acrylate retaining adhesive, position, and then thermal cure); (C) bond the laminate stack to the insulating kraft-paper tube (coat the bore of laminate stack with retaining adhesive, which should be the same adhesive as in steps B, C, and D, assemble, and heat cure); (D) bond the commutator to the insulating tube (heat commutator, coat the bore with retaining adhesive, assemble, let cure); (E) assemble wind armature to the shaft; and (F) assemble the cooling fan impeller to the shaft.

The adhesives used in this application were anaerobic grades of retaining compounds. These compounds are very useful for bonding metal mechanical components together as a single component and for bonding nonmetallic substrates using heat or surface activators. The resulting assembly in Fig. 13 was stronger, could be automatically processed, reduced scrap losses, and improved overall rotor quality.

The production of most small, fractional-horsepower motors involves the bonding of magnet sections to the metal or plastic case that forms the stator section of the motor (as opposed to the rotor assembly, discussed above). These types of motors are widely used in automotive uses (blower, windshield wiper, power windows), timer motors (appliances, industrial timers), and

many other applications requiring high production manufacturing methods.

Methacrylate and urethane-acrylate adhesives are the two most popular types of adhesives for motor magnet bonding because of their capability to be automated and rapidly cured by heat or activator. Automated assembly machines are able to dispense these adhesive gels onto vertical surfaces as the magnets are held in their fixtures. The fixtures locate and press the magnets into position against the inner case wall, where cure is achieved in 30 to 60 s (Fig. 14). Methacrylates and urethane-acrylates are toughened adhesives that bond to a variety of plated-metal, plastic, and painted surfaces without requiring primers or extensive surface cleaning. The adhesive cushions the brittle ferrites against vibration, impact, and thermal stresses; increases motor life; and reduces scrap loss due to ferrite breakage.

Magnet bonding is also important in loud speaker manufacturing, where the Alnico magnets are bonded to the plated steel housing. Adhesives similar to those used in motor magnet bonding are used with surface activators to achieve 15 to 30 s fixturing and 1 h total cure. The bonds are much stronger than the ferrite substrates because of the surface area of the bond and the brittle nature of the ferrite. Drop tests result in 100% ferrite failure, rather than joint failure. In most cases involving magnets, the geometry and bond area of the joint provide a wide margin of safety for adhesive survival over that of the substrates.

Speaker voice cones, dust covers, gaskets, and many other components are assembled with cyanoacrylate and methacrylate adhesives. These components have no inherent strength and need to be securely bonded to the speaker frame (basket) in order to work properly. Speakers for telecommunications are generally bonded into their end product as the most effective means of creating a vibration-free joint line

that is accurate in reproducing sound without hum or distortion (Fig. 15).

Lens bonding for industrial and automotive lighting is almost standard in the industry today. Plastic (polycarbonate) or glass lenses are bonded to plastic, steel, or metal parabolic reflectors to effect a watertight seal and permanent assembly. New bonded plastic designs are replacing older glass lens-to-glass reflector and earlier ultrasonic welded plastic lens designs. Adhesives have proved to be the most cost-effective method of lamp assembly and allow greater design latitude and material selection.

In the automotive industry, composite head lamps that are smaller in size and aerodynamically designed also allow automatic assembly methods with higher quality standards (Fig. 16). One of the major problems in bonded lens assemblies is the wide variety in coefficients of thermal expansion (CTE) between glass, polycarbonate, steel, and thermoset polyester. This CTE difference can create high stress levels on the bond during normal service. Adhesives used for bonding these different materials have to be specially formulated to ensure plastic adhesion, overall strength, flexibility, and gap filling. Polyurethane two-component adhesives, toughened epoxies, and methacrylates are the most popular adhesives for lens bonding applications, which, in one automotive supply plant, cover 20 different head lamp and fog lamp designs.

Medical

The manufacture of medical durable (life support) equipment, medical disposable items, and sensing/monitoring/reporting devices are the principal application areas in the medical industry. Adhesives must pass U.S. Pharmacopoeia (USP) class VI tests to prove that they are compatible with human tissue and blood. Both the medical durable and disposable industrial sectors are based on plastic, glass, and stainless steel substrates. These substrates require unique ad-

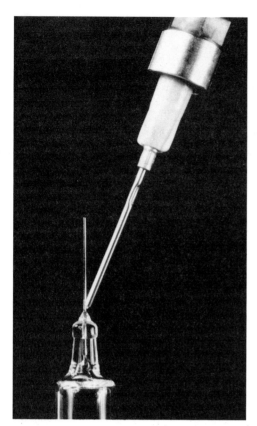

Fig. 17 Hypodermic cannula bonding using a UV-curing adhesive

Fig. 18 Industrial grade needles bonded with UV-curing adhesive. Courtesy of Loctite Corporation

Fig. 19 Bonding of wear plate to golf club head with a toughened cyanoacrylate. Courtesy of Loctite Corporation

hesive properties to withstand sterilization processing (steam autoclaving, gamma radiation, or ethylene oxide exposure) and have a 0 defect quality level.

The plastics used in the medical durable business are some of the most difficult to bond: polyether-imide, polyether sulfone, polysulfone, polycarbonate, polyphthalate carbonate, thermoplastic polyesters, and polyphenylene ether (Ref 15). Adhesives and other joining methods used with these plastics and other inert, noncorroding substrates are often required to provide a seal as well as a structurally strong bond. Fusion joining techniques based on ultrasonic welding or electromagnetic welding have been used when possible. When adhesives have been used, silicones, UV-curing urethane-acrylates, UV-curing methacrylates, and cyanoacrylates have generally been the most popular because they are single-component products that have class VI approval and bond well to plastic in short cure times.

Medical disposable products typically use polyvinyl chloride (PVC) (both flexible and rigid), polycarbonate, polypropylene, acrylonitrile-butadiene-styrene (ABS), silicone rubber, and stainless steel. Typical medical disposable devices are catheters (all types), pressure-monitoring lines, esophogeal stethoscopes, tubing adaptors, cannulas, hypodermic needles, and metering valves.

PVC is the most popular plastic for medical applications involving tubing, whereas polycarbonate and ABS are used for molded rigid tube connectors and device housings. Polycarbonate and stainless steel are normally used for hypodermic needle components. Stainless steel is also used for maximum-strength applications. Adhesives for this market include solvent-base types for use with PVC (normally, cyclohexonone cements with PVC fillers), cyanoacrylates, UV-curing and RTV silicones, and UV-curing methacrylates.

Syringe bonding is a high-volume use of UV-curing adhesives because millions of needles can be produced each year on automated machines. Because bonding reliability is very critical, pull-out strength and sealing are measured on a routine basis as process quality specifications. UV-curing adhesives of low viscosity are typically used with a small drop of adhesive that is applied automatically to the assembled needle and a transparent or translucent hub (Fig. 17).

The hub is made of polypropylene and therefore does not allow maximum strength development because of the low adhesion forces on untreated polypropylene surfaces. In many designs, an annular cavity ring is molded into the bore of the hub. The adhesive then flows into this ring and cures by UV energy (8 s cure), creating a mechanical, captive interlocking design that is bonded firmly to the outer diameter of the stainless steel needle passing through the hub. This method has largely replaced the mechanical crimping of the stainless steel tube in the hub or the use of heat-curing epoxies.

Opaque grades of PVC used as hubs for needle assembly can be bonded with low-viscosity grades of cyanoacrylates or special dual-cure grades of UV-curing adhesives in medical applications. Two-part methacrylates with a chemical activator can be used on industrial needles for internal bore curing (Fig. 18).

Sporting Goods

Golf clubs, tennis rackets, skis, jogging shoes, wind surfer boards, sailplanes, archery products, snowboards, skis, and the majority of other sporting products are manufactured today using adhesives to reduce weight, reduce finishing costs, and improve durability. The relatively large bond areas and use of dissimilar substrates make bonding the best way to provide positive holding power of components that are subjected to high localized stresses.

Most sporting goods are made with reinforced plastic composites (fiberglass or aramid) and lightweight metal components (tubing or structural sections of aluminum, stainless steel, and alloys) that are hand assembled using adhesives. Golf club manufacture is a good example of adhesive bonding using these types of substrates. The steel shafts are assembled to the wood or plastic composite heads with a high-strength, toughened, two-part epoxy that is heat cured. Performance for the joint is critical. The environment to which the clubs are subjected can have temperature extremes of −30 to 95 °C (−20 to 200 °F) (as in automobiles in summer), as well as high-impact torque forces. Shaft-head joints can be heat cured at temperatures as low as 65 °C (150 °F) in 35 min. These parameters have replaced 105 °C (225 °F) cures for 24 h, without loss of performance level.

Another bonding application is the metal strike plate on the face of a wood or plastic club head. It is bonded with contact adhe-

Fig. 20 Ski laminate construction using adhesives. Courtesy of Ciba-Geigy Corporation

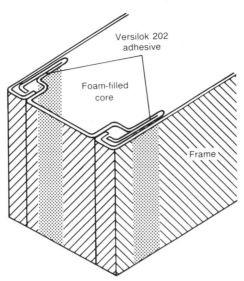

Fig. 21 Residential door bonding cross section of the application bonded with a two-part methacrylate using surface activator and room-temperature cure

Fig. 22 Robotic tracer head for door bonding application. Courtesy of Lord Corporation

sives or urethane adhesives and is further secured by screws. The bottom wear plate on a driver is bonded to the head with a toughened cyanoacrylate adhesive. Although elastomer-toughened adhesives cure a bit more slowly than the standard product and therefore require increased time during the processing operation, they provide greatly improved durability (Fig. 19).

A modern multilaminate archery bow is made from composite plastics in shapes designed by computers to give increased power and accuracy. Stress values of 125 MPa (18 ksi) may be attained on the wood-fiberglass laminate interface, and a shear stress value of over 21 MPa (3 ksi) on the adhesive bond line.

One of the problems encountered in the use of adhesives for laminate bonding is the necessity to find a system that cures rapidly in the laminating process and provides uniform stress distribution and high heat distortion temperatures. To avoid lengthy oven cures and expensive fixturing, radio frequency (RF) curing has been used with hydraulic laminating press fixtures to give cycle times of 30 min or less for sets of 5 to 10 laminates per cycle. The key requirement of adhesive bonding, particularly laminate bonding, is even heat and pressure distribution at the bond line.

Skis also use laminate construction of ten or more plies to build up the desired strength, flexibility, and exterior surface to provide outstanding impact, abrasion, and durability characteristics (Fig. 20). Surfboards are another good application for adhesive bonding. Their lightweight foam-filled thermoformed hulls (ABS thermoplastic sheet) and plastic (vinyl) components require even stress distribution and sealing of the bond line. Methacrylate adhesives using surface activators have been found to be the most cost-effective adhesive system from the standpoint of material, process time, and capital equipment costs. The hull halves are bonded together on one side first and cured. Then, adhesive is applied on the

other side before the foam-filling and final bonding operations. Vinyl plastic bumpers, reinforcing components, and sealing strips are bonded in place during the last steps in the operation. All use activators and methacrylate adhesives.

Construction

Construction adhesives are often dual-function products, being sealants as well as adhesives. Many adhesives/sealants used in the manufacture of glazed windows, buildings, and doors are chosen for their sealing qualities and material costs rather than their adhesive properties. This is because strength requirements are minimal, the bond areas are very large, and processing involves a series of manual adhesive application and assembly operations. Other important selection considerations are flexibility of the adhesive system for low-temperature impact loads, fast fixturing for limited-processing fixtures, and bead application from caulking cartridges in nonsag forms.

Construction adhesives may range in performance requirements from latex adhesives for bonding decorative wall panels to concrete or wood furring strips, to silicone adhesives for glazing and window mounting, to butyl elastomeric sealants/adhesives for bonding and sealing dissimilar substrates in low-stress applications, to contact cements or other solvent-base adhesives for bonding laminates.

Two applications that depart from normal construction applications are door bonding (residential, commercial, subway car, and truck cabs all use the same adhesive system) and solar panel bonding. Door construction has changed dramatically over the past decade from wood and wood-glue con-

struction to metal sheathed, foam-filled security doors with high insulative factors and durability.

Methacrylate adhesives are used to bond galvanized steel sheets together to a steel frame to form a unitized steel shell that is then filled with polyurethane foam (Fig. 21). This adhesive method uses surface activator and robotics (tracer) adhesive applicators, and greatly reduces labor costs while increasing productivity (Fig. 22). Adhesives have replaced spot welding, which was slow and required extensive surface finishing to hide the weld marks. Methacrylates are generally popular for steel sheet bonding applications because they require little, if any, surface preparation, even when bonding to oily rust-preventative coatings on steel.

Methacrylates offer a good compromise between strength, toughness, flexibility, shock resistance (cured properties), application ease (liquid properties), and fast fixturing, which requires 3 to 5 min at room temperature, 1 min with mild heat, and full cure within 1 h. As noted above, doors for subway cars, trucks, and homes and other steel-paneled construction products use this family of adhesive, with slight application technique and curing variations. Because many doors are painted with heat-cured products, methacrylate adhesives can be fully cured in minutes during the paint bake cycle after initial activator fixturing.

Solar panel bonding is another planar bonding application involving outdoor exposure to UV energy, temperature extremes of −40 to 60 °C (−40 to 140 °F), wind, rain, and hail. The concept of using solar-electric generation to produce electricity commercially, in addition to residential hot water and room heater applications, is now being fully evaluated. In time, it is estimated that up to 15% of domestic electricity will be generated by solar energy technology, which therefore represents a growth area for adhesives applications.

Solar panels for power generation can measure up to 3 m (10 ft) long and 1 m (3.5 ft) wide and be mounted in computer-driven arrays to constantly face the sun. Individual glass mirrors (~500 000) are bonded to metal panels that reflect the energy of the sun to a central collection point for steam generation. A combination of silicone adhesive/sealant and premixed, two-part methacrylate adhesives for fast fixturing and reinforcing are used in their construction. The use of mild heat for the methacrylates allows production cycles of 11 min for mirror bonding. Double sealants (silicone and methacrylates) are used because of the silver and copper paint on the mirror surfaces, which would corrode and peel if exposed to moisture.

Another solar power technology involves photovoltaic cell construction (direct conversion of light electricity), using methacrylate adhesives to bond an ABS plastic housing to the acrylic lenses to form a durable bond and watertight seal. The ability to withstand UV degradation and maintain an effective moisture seal is critical to the nature of the application and the choice of adhesive. Thermal cycling of the modules between -30 and $105\ °C$ (-18 and $220\ °F$) tests the flexibility and long-term durability of the bond on plastic substrates.

REFERENCES

1. P. Gwynne, "Adhesives. Bound for Boundless Growth," Special Report, Jan 1982
2. *Adhesives*, 4th ed., DATA Inc., 1986
3. D. Dunn and R. Batson, "Chemical Joining of Plastics," Training Program, Center for Professional Advancement Adhesives, 1989
4. D. Dunn and R. Batson, "Ultraviolet Curing Adhesives," Training Program, Center for Professional Advancement Adhesives, 1989
5. *Electronic Market Data Book*, Electronic Industries Association, 1989
6. R. Lilly, Advanced Adhesives, Assembly Engineering, May 1987, p 78-81
7. H. Kraus, "Adhesives for Microelectronics," Abelstik Labs
8. J. Coleman, Sticky Issues in Electronic Assembly, Assembly Engineering, Feb 1989, p 29-32
9. W. Hastie, Adhesives for Electronics, *Circuits Manuf.*, Jan 1981, p 52-62
10. K. Drain and S. Grant, "Adhesives. A Vital Element in Surface Mount Technology," Paper presented at SMART II Conference, 1984
11. "SMD Adhesives Application and Curing," Technical Brochure, Signetics Corporation, July 1986
12. Conductive Epoxy Is Tested for SMT Solder Replacement, *Electron. Packag. Prod.*, Feb 1985
13. F. Kulesza, Conductive Epoxy Solves Surface Mount Problems, *Electron. Prod.*, March 1984, p 83-88
14. E. Frauenglass, J. Moran, and R. Batson, New Adhesives Speed Heat Removal, *Circuits Manuf.*, Feb 1984
15. D. Francis, Plastic Process and Materials Selection, *MDDI*, Nov 1987, p 45-49

Electronic Packaging Applications for Adhesives and Sealants

Harry K. Charles, Jr., The Johns Hopkins University, Applied Physics Laboratory

THE RAPID ADVANCEMENT OF INTEGRATED CIRCUITS (ICs) and associated microelectronic technologies toward complex, high-density, high-speed devices has placed increasing demands on device protection, encapsulation, and packaging methods. As clock speeds exceed 100 MHz, and rise times drop below 1 ns, the need for properly terminated transmission lines that preserve pulse shape and minimize propagation delay is paramount. The increased number of inputs and outputs of chips associated with the exponential rise in chip device density (Ref 1) (Fig. 1) and the exponential decline in device feature size (Ref 2) (Fig. 2) for very large scale integrated circuits (VLSIC) have forced significant increases in packaging density and the need for high-conductivity and low-capacitance interconnects.

Because increased device density and high-speed operation usually imply greater power consumption, more attention must be paid to the overall thermal management and the reliability and mechanical integrity of electronic interconnection, encapsulation, protection, and packaging schemes. All these rapidly changing packaging parameters (Ref 3) have created a growing need for advanced materials for device interlayer dielectrics, top surface passivation and overcoating layers, die adhesives, encapsulants, package sealants, and circuit interconnections.

In general, the use of adhesives and sealants in the electronics industry is widespread and contributes directly not only to the manufacturability of the electronic device but also to its long-term operation and survivability. Device performance or lifetime can be adversely affected by a variety of environmental and handling parameters, including moisture, ionic contamination, electromagnetic radiation, energetic particles, extremes of temperature and pressure, chemicals, physical shock, electrostatic discharge, and chip handling and processing. Because the device processing and packaging operations are so critical to device performance and to the type of adhesives and sealants used, a brief review of both wafer

processing and die packaging are presented in this article, followed by an overview of the use of adhesives and sealants in board-level packaging.

Because of the myriad of materials used and the potential for adhesive and sealant use throughout most microelectronics operations, this article will focus on general classes of materials (for example, epoxies, silicones, metals, alloys, and so on) and their application in two specific areas: die (and substrate) attachment and electrical interconnection,

and die and board passivation and encapsulation. In general, all the materials to be described fall into two broad categories: inorganic and organic. Particularly in the areas of die interconnect and attachment, organic resins filled with inorganic materials (metals and high-thermal-conductivity ceramics) are quite common. Similarly, certain organic vehicles containing inorganic dielectrics are used as overcoats.

Inorganic materials include metals and metal alloys used for die attachment and

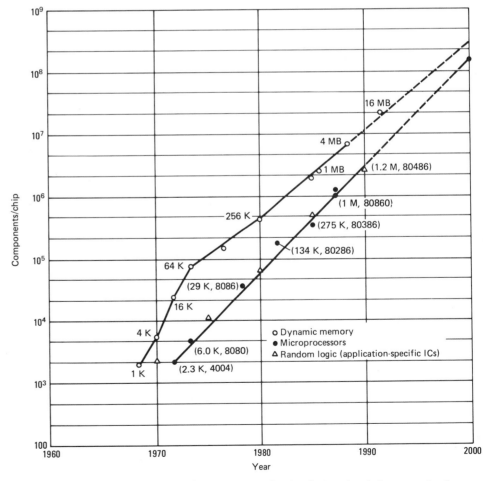

Fig. 1 Number of components per chip versus time, for the three major device categories (memory, microprocessors, and random logic)

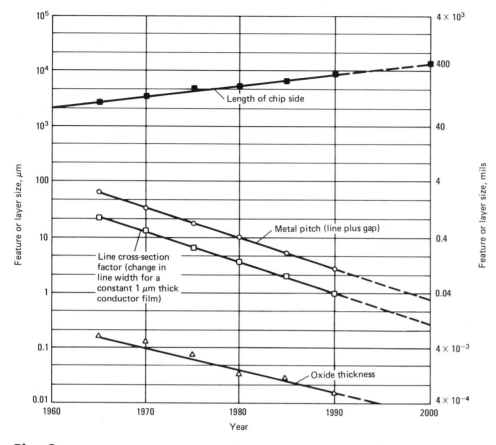

Fig. 2 Device and chip feature size versus time. Channel length (not shown) parallels line cross section.

electrical interconnection, and silicon oxides, silicon nitrides, metal oxides, glasses, and certain ceramics used for device passivation and encapsulation. The inorganic coatings have proved to be most useful for device passivation and hermetic caps on IC dies (Ref 4).

Organic materials include polyimides, epoxies, and thermoplastic and thermoset resins used for die attachment and interconnection, as well as epoxies, polyimides, silicones, urethanes, silicone-polyimides, parylenes, benzocyclobutene, block copolymers, sol-gels, and so on, which are used in a myriad of coating and encapsulation activities. Many of these materials and their compositions, properties, and uses are described in the last three sections in this article.

Wafer Processing

Wafers. Integrated circuits start with the growth of single-crystal silicon (Si) and/or gallium arsenide (GaAs). The crystals are grown from molten semiconductor materials (melting temperature of 1414 °C, or 2577 °F, for silicon and 1238 °C, or 2260 °F, for gallium arsenide) contained in a heated crucible by the Czochralski process (Ref 5).

In the Czochralski method (Fig. 3), the crucible is quartz (crystalline SiO_2) for silicon and graphite (carbon) for gallium arsenide. The crucibles are heated by radio frequency induction heating. Because arsenic is a volatile material, gallium arsenide decomposes with the loss of gaseous arsenic (thereby producing nonstoichiometric Ga_xAs_y, with $x > y$) unless an arsenic overpressure is maintained during growth. In addition, an inert liquid capping layer, such as boron trioxide (B_2O_3), can be used to cover the melt to prevent arsenic escape and melt-crucible interaction (Ref 6). The liquid capping process is known as liquid-encapsulated Czochralski (LEC).

Once the proper melt conditions have been achieved, a seed crystal is inserted into the melt and slowly withdrawn at a controlled rate (both pull and rotation). Crystal growth occurs by freezing or solidification at the interface between the solid crystalline seed and melt. Crystal growth proceeds by layer addition at the moving solid-liquid interface. The typical ingot (boule) growth rate is 10 μm/s (0.4 mil/s). Large, perfect, single-crystal ingots of silicon over 1 meter (3.3 ft) long and 250 mm (10 in.) in diameter have been grown.

After growing, surface finishing, and x-raying to determine ingot orientation (for proper device alignment and chip separation later), the ingot is cut with high-speed diamond saws into slices, or wafers, ap-

proximately 100 to 250 μm (4 to 10 mils) or more in thickness (depending on the semiconductor material and ingot thickness). The sawn wafers are then evaluated for dislocations, resistivity, and other properties. The accepted wafers are polished and etched to a mirrorlike surface on at least one side. Surface roughnesses less than 25 nm (1 μin.) are not uncommon in highly polished wafers. The polished wafers are then used to process integrated circuits.

A typical IC processing flow for a simple set of circuit elements is shown in Fig. 4. Photolithographic patterning and other processing operations have been treated in an article by Charles and Clatterbaugh (Ref 7). In the following subsections, further details of diffusion, metallization, and, of course, passivation (oxidation) are presented.

Diffusion. After the wafers are prepared, one of the first steps in IC processing is the growth of an oxide layer on the silicon. Controlled thicknesses of silicon dioxide (SiO_2) are grown by placing the wafers in a quartz tube furnace in an oxidizing ambient at temperatures of 900 to 1100 °C (1650 to 2010 °F). Oxygen diffuses through the growing oxide layer to reach the silicon surface ($Si + O_2 \rightarrow SiO_2$). The oxide layer contains 2.2×10^{22} molecules/cm³ (36×10^{22} molecules/in.³) of SiO_2 (with a density of 2.2 g/cm³) and consumes a thickness of silicon equal to approximately 0.45 times the thickness of the oxide layer.

The oxide layer can act as a barrier or as a source for the introduction of dopants (controlled impurities) into the silicon. A simple example of the SiO_2 masking property (or barrier) is shown in Fig. 5, where the dopant species (boron) is introduced by ion implantation. The energetic boron ions penetrate the exposed region of the silicon but are blocked by the areas covered by the SiO_2. See Table 1 for diffusion coefficients of various impurities in SiO_2, compared to those of silicon. The boron ions typically penetrate less than 0.1 μm (4 μin.) into the silicon, depending on the ion beam energy.

To distribute the impurities deeper into the silicon, high-temperature diffusion (Fick's laws, Ref 8) must be used. The silicon oxide is also a key element in impurity doping using gaseous diffusion. Once a window is opened in the protecting SiO_2 layer, as shown, for example, in Fig. 5, the wafer with a patterned oxide layer is placed in a high-temperature furnace with an atmosphere containing oxygen and a dopant gas that is in an inert carrier, such as argon or nitrogen. At high temperatures, the dopant gas decomposes and becomes incorporated in the growing oxide layer. Typical reactions for boron and phosphorus impurity species are shown in Fig. 6.

As the doped oxide layer grows on the regions of exposed silicon, the elemental impurities diffuse into the silicon according to conventional diffusion laws (Ref 8). Typ-

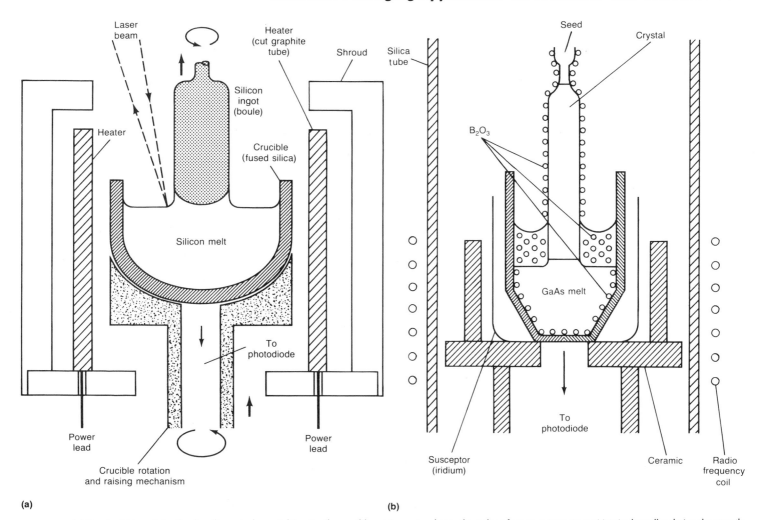

Fig. 3 (a) Czochralski crystal puller for silicon single-crystal ingots. The crucible raising system keeps the melt surface at a constant position in the puller during the growth process. The photodiode and the laser beam provide feedback for temperature control. (b) Liquid encapsulated Czochralski (LEC) puller for gallium arsenide. Entire unit is enclosed in pressure vessel at several times the atmospheric pressure of argon.

ical diffusion coefficients for various impurities in silicon (and gallium arsenide) are given in Table 1.

Metallization. Silicon devices are typically metallized or interconnected (both bonding pads and conductor traces) using aluminum metal because of its high electrical conductivity and its ability to form a protective oxide layer on its top surface. The metal makes contact with the silicon only in the openings in the protecting oxide layer, as shown in Fig. 7. Although aluminum is the dominant metallization used by most IC manufacturers, it has one significant problem. Silicon is soluble in aluminum (typically 0.5 to 1.0% in the 450 to 500 °C, or 840 to 930 °F, temperature range) and during postdeposition (usually physical vapor deposition, Ref 9, or sputtering, Ref 10) heat treatment, silicon diffuses into the aluminum, leaving holes or voids in the silicon, which are filled by aluminum metallization or "spikes." Spikes can be minimized by reducing processing times and rapidly increasing temperatures (Ref 11).

As device junctions became shallower with scaling in the very large scale integration generation, aluminum spikes were deep enough to destroy transistor operation. To reduce even minimal spiking, aluminum alloys with 1 to 5 at.% Si are routinely deposited. This material also has problems in that too much silicon in the aluminum can cause silicon to precipitate and form silicon crystallites on the pad surface during heat treatment. The crystallites are typically p-type (aluminum is a p-type dopant for silicon) and can form undesirable electrical behavior if they come in contact with underlying n-type silicon layers.

As interconnect trace dimensions shrink, the current densities carried by circuit traces increase. Once current densities reach 10^6 A/cm^2 (6.45×10^6 A/in.^{2}) the interconnects begin to fail by a process called electromigration (Ref 12). Pure aluminum interconnects will crack (Ref 13) along conductors carrying a current of 10^6 A/cm^2 (6.45×10^6 A/in.^{2}) for periods of less than 100 h at 220 °C (430 °F). The cracks and concurrent resistance increases are due to

the mass transport of conductor materials. Atoms move through grain boundaries, creating voids on one end and hillocks on the other. Electromigration is reduced by decreasing grain boundaries (that is, increasing grain size) and by reducing grain-boundary diffusion by the introduction of a boundary-pinning impurity such as copper. A typical aluminum alloy includes 4 wt% copper and 1 to 2 wt% Si. Such alloys can increase the electromigration resistance by factors of 50 to 100. The introduction of titanium can further improve the electromigration resistance of the aluminum-copper-silicon interconnect system (Ref 14).

Because of the above problems, the silicon-aluminum interface is basically unstable, requiring the development of more stable and uniform metallized contacts. The interface between the aluminum and silicon can be improved by using silicide contacts. These metal-silicon contacts are formed by a uniform (no pits) reaction between the silicon and the deposited metal layer during a thermal annealing process. Silicides make good electrical contacts, with known Schott-

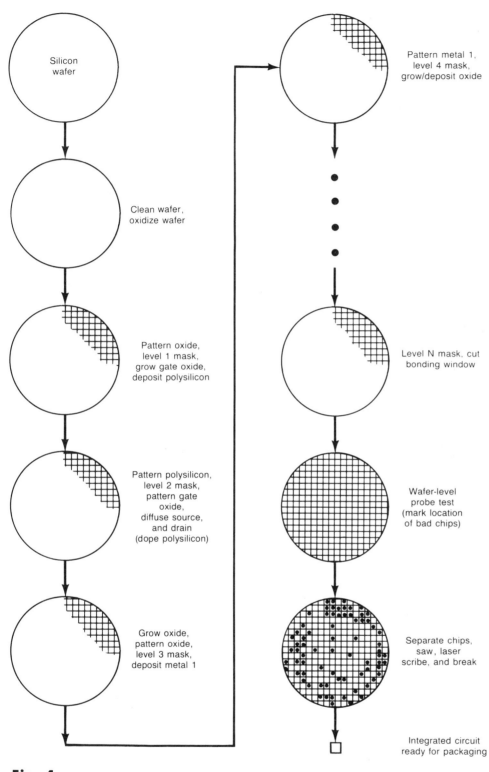

Fig. 4 Generic silicon IC process flow

[Figure labels, top to bottom, left column:]
Silicon wafer

Clean wafer, oxidize wafer

Pattern oxide, level 1 mask, grow gate oxide, deposit polysilicon

Pattern polysilicon, level 2 mask, pattern gate oxide, diffuse source, and drain (dope polysilicon)

Grow oxide, pattern oxide, level 3 mask, deposit metal 1

[Figure labels, right column:]
Pattern metal 1, level 4 mask, grow/deposit oxide

Level N mask, cut bonding window

Wafer-level probe test (mark location of bad chips)

Separate chips, saw, laser scribe, and break

Integrated circuit ready for packaging

ky barrier heights and resistivities as low as 15 $\mu\Omega \cdot$ cm (Ref 15). Titanium silicide (TiSi$_2$) is a particularly useful contact barrier layer. A typical TiSi$_2$ contact is shown in Fig. 8.

The metallization of gallium arsenide is a significantly more difficult problem. As in the case of silicon, metallizations on gallium arsenide can form either ohmic contacts or Schottky barriers. Because of the special properties of gallium arsenide, the forming of contacts is especially difficult. These properties include the fact that gallium arsenide decomposes into Ga and As$_2$ gas at about 590 °C (1095 °F) (Ref 16), the Schottky barrier height on n-type gallium arsenide

is large ($\sim$0.8 eV) and essentially independent of metallization; it is difficult to highly dope GaAs, which usually makes ohmic doping levels of 5×10^{18}/cm^3 (80×10^{18}/in.^{3}) impossible to reach and annealing at high temperatures (up to 850 °C, or 1560 °F) is needed to activate implanted dopants, necessitating capping or high arsenic overpressures to prevent gallium arsenide decomposition. Ohmic contacts are difficult to achieve, especially on n-type gallium arsenide. Much research has been done on GaAs contact metallizations, with AuGe$_2$ being the most stable GaAs contact material (Ref 17). Silicon has also been shown to be a useful material.

Passivation. In addition to the use of oxide and nitride layers as an impurity for diffusion barriers, they are also used as insulating layers between metallization levels (or lines) and as insulating protective covers. The layers are generally deposited by chemical vapor deposition (CVD) techniques in which silane (SiH$_4$) and oxygen react to form SiO$_2$:

$$SiH_4 + O_2 \rightarrow SiO_2 + 2H_2$$

As shown in the section "Inorganic Materials: Passivations and Encapsulants" in this article, the reactions can take place at 450 °C (840 °F) or below, depending on the gas mixtures and whether plasma enhancement is used. These low-temperature reactions allow the SiO$_2$ films to be deposited over aluminum metallization. An important property of deposited SiO$_2$ films is the stress in the film. Low stresses in the as-deposited films improve their stability under thermal cycling. The addition of phosphorus pentoxide (P$_2$O$_5$) while depositing the SiO$_2$ film can significantly reduce the built-in stress. Silicon dioxide doped with phosphorus is called phosphosilicate glass (PSG). PSG can be produced by the simultaneous pyrolysis of silane and phosphine in oxygen. Typical reactions are:

$$SiH_4 + O_2 \rightarrow SiO_2 + 2H_2$$

$$4PH_3 + 5O_2 \rightarrow 2P_2O_5 + 6H_2$$

Water may also be a by-product of the pyrolysis of phosphine. It appears that large amounts of P$_2$O$_5$ may be incorporated in SiO$_2$; however, SiO$_2$ films become increasingly hygroscopic with increasing P$_2$O$_5$ content. In passivation applications, PSG films typically contain less than 8 wt% phosphorus. The incorporation of P$_2$O$_5$ reduces the built-in stress of as-deposited silica films, which is 3×10^8 N/m^2 (tension). Experimentation has shown that nominally at approximately 20 wt% P$_2$O$_5$, the internal silica film stress is zero, which greatly improves film stability.

The coefficient of thermal expansion (CTE) of SiO$_2$ films with incorporated P$_2$O$_5$ increases with increasing P$_2$O$_5$ content. PSG films can be tailored to match the CTE

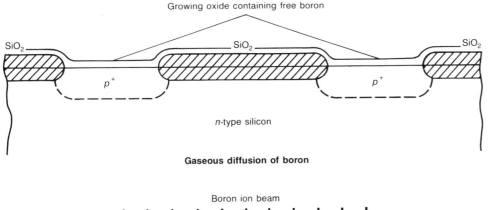

Gaseous diffusion of boron

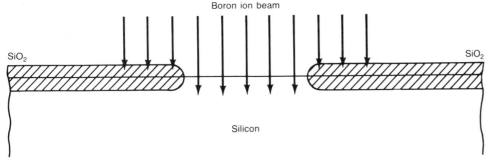

Ion implantation of boron

Fig. 5 Examples of the masking properties of silicon dioxide

of the underlying substrate, thus significantly improving passivation-layer stability under thermal cycling. The CTE of undoped SiO_2 can be increased from 6×10^{-7}/K to about 6×10^{-6}/K by the incorporation of approximately 22 wt% P_2O_5. The latter CTE closely matches that of gallium arsenide, thereby improving the capability of PSG to act as a stable passivation for gallium arsenide. PSG is also an extremely effective diffusion barrier to such elements as zinc and tin because of its dense microstructure (compared to SiO_2). It also stops sodium ion transport through oxide layers in metal-oxide semiconductor field-effect transistors (MOSFETs).

Silicon nitride can also be grown by chemical vapor deposition by reacting silane gas with ammonia, as shown in Fig. 9. Other reactions for the deposition of both SiO_2 and Si_3H_4 are given in the section "Inorganic Passivations."

Testing and Separation. Once fabricated, silicon or GaAs ICs are electrically tested at the wafer level using various probing techniques. Both conventional contact probes and noncontact beam (electron and laser) technologies can be used (Ref 18). After the testing and mapping of the good circuits or dies, the die area is separated from the wafer by sawing. High-speed (30 000 to 50 000 rpm) diamond-impregnated saw blades are used. Once separated, the good dies are placed in clean storage containers awaiting shipment or packaging. Details of device packaging are given below.

Device Packaging

The basic operations in device packaging involve placing the IC into a package (die attach), interconnecting the circuit to the packaging (interconnection bonding) either directly or through an intervening substrate (for example, hybrid, multichip module), and, finally, sealing or encapsulating the package. Of course, inspections, testing operations, and environmental screening are used to verify the performance of the product.

Die Attach. The electrical connection to most ICs is made to bonding pads on the top surface or face of the chip. To provide mechanical support (and sometimes a back-surface electrical and/or thermal interconnect), the chip or die is bonded to the substrate or package using eutectic die attach, alloy (solder) attach, and/or organic

Table 1 Diffusion coefficients for various impurities in silicon, silicon dioxide, and gallium arsenide, with $D = D_o e^{-E_A/K_B T}$

	Silicon D_o			Silicon dioxide D_o			Gallium arsenide D_o		
Element	cm²/s	in.²/s	E_A(eV)	cm²/s	in.²/s	E_A(eV)	cm²/s	in.²/s	E_A(eV)
Boron	0.76	0.12	3.46	7.23×10^{-6}	1.12×10^{-6}	2.38	...	...	...
Gallium	225	34.8	4.12	1.04×10^5	0.16×10^5	4.17	2.9×10^8	0.45×10^8	6.0
Phosphorus	3.85	0.60	3.66	5.73×10^{-5}	0.89×10^{-5}	2.30	...	...	...
Arsenic	0.066	0.01	3.44	67.25(a)	10.42	4.7	0.7	0.11	3.2
				3.7×10^{-2}(b)	0.57×10^{-2}	3.7			
Antimony	12.9	2.0	3.98	1.31×10^{16}	0.20×10^{16}	8.75	...	...	...
Aluminum	0.5	0.08	3.0	...	...	...	...	...	...
Carbon	1.9	0.29	3.1	...	...	...	...	...	...
Platinum	...	...	...	1.2×10^{-13}	0.19×10^{-13}	0.75	...	...	...
Gold	1.1×10^{-3}	1.7×10^{-4}	1.12	8.2×10^{-10}	1.3×10^{-10}	0.8	2.9	0.45	2.64
				1.52×10^{-7}	0.23×10^{-7}	2.14			
Silver	2×10^{-3}	3.1×10^{-4}	1.6	...	...	...	25	3.9	2.27
Chromium	0.01	1.5×10^{-3}	1.0	...	...	...	4.3×10^3	665	3.4
Copper	0.04	6×10^{-3}	1.0	...	...	...	0.03	4.7×10^{-3}	0.53
Tin	32	5.0	4.25	...	...	...	0.038	5.9×10^{-3}	2.7
Indium	16.5	2.6	3.9	...	...	...	(c)	(c)	...
Beryllium	...	...	...	...	...	...	7.3×10^{-6}	1.13×10^{-6}	1.2
Lithium	$0.9–2.3 \times 10^{-3}$	$1.4–3.6 \times 10^{-4}$	0.63–0.78	...	...	...	0.53	0.082	1.0
Oxygen	0.07	0.011	2.44	2.7×10^{-4}	0.42×10^{-4}	1.16	2×10^{-3}	3.1×10^{-4}	1.1
Silicon									
T > 1050 °C, or 1920 °F	1.5×10^3	233	5.02	...	...	...	1.5×10^{17}(d)	0.23×10^{17}	6.85
T < 900 °C, or 1650 °F	1.5	0.23	4.25	...	...	...	0.11(e)	0.017	2.5

(a) Ion implant Na. (b) Ion implant O_2. (c) $D = 7 \times 10^{-11}$ cm²/s (1.1×10^{-11} in.²/s) at 1000 °C (1830 °F). (d) Low concentration, with SiO_2 cap. (e) High concentration, with SiO_2 cap

$$4POCl_3 + 3O_2 \longrightarrow 2P_2O_5 + 6Cl_2$$

$$2P_2O_5 + 5Si \longrightarrow 5SiO_2 + 4P \text{ (diffuses)}$$

n–type (phosphorus)

$$4BBr_3 + 3O_2 \longrightarrow 2B_2O_3 + 12Br$$

$$2B_2O_3 + 3Si \longrightarrow 3SiO_2 + 4B \text{ (diffuses)}$$

p–type (boron)

Fig. 6 Typical reactions for boron and phosphorus impurity species

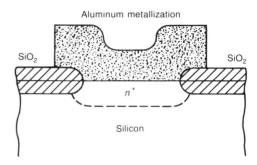

Fig. 7 Typical aluminum metal contact. Silicon contact is made only through openings in the oxide.

adhesive attach. By eutectic die attach, one typically means the gold-silicon eutectic (94 wt% Au, 6 wt% Si), as shown in the phase diagram in Fig. 10. The Au-Si eutectic melts at 370 °C (698 °F).

There are several ways to form the gold-silicon eutectic bond. The simplest is to place a silicon die into intimate contact with a gold-plated surface and then raise the temperature above 370 °C (698 °F). A simple scrubbing motion aids the formation. A gold metallization on the chip back also helps. The use of a thin sheet of the eutectic alloy (a preform) between the chip and the substrate (or package) metallization is a preferred technique of several manufacturers. Regardless of the technique, it is essential that sufficient gold be present to ensure that the molten alloy composition remains on the left branch (on the left, or gold-rich, side of the eutectic point) of the phase diagram as melting continues until all the gold is consumed or the temperature drops below 370 °C (698 °F).

Other metal alloys with either eutectic points or solidus-liquidus ranges have been used for die attach. These include the gold-tin eutectic (80 wt% Au, 20 wt% Sn, with a melting point of 280 °C, or 536 °F), the tin-gold eutectic (90 wt% Sn, 10 wt% Au, with a melting point of 217 °C, or 423 °F), tin-lead solders (for example, 63 wt% Sn, 37 wt% Pb, with a 183 °C, or 361 °F, eutectic point), and other lead alloys (such as lead-indium). Because they do not rely on the interdiffusion of gold with silicon (and the formation of the gold-silicon eutectic), either a thin sheet of the alloy or pretinning (in the case of tin-lead solders) must be used between the metallized back of the chip and

the metallized region of the substrate or package bottom. Alloy die bonds provide good mechanical strength, electrical conductivity, and low thermal impedance. Some alloys or solders require flux to aid in the reflow and prevent excessive oxidation. If flux is used, it must be removed, thereby introducing extra processing, cleaning, and/or inspection steps.

The die can also be bonded to the package or substrate using organic adhesives. Typical adhesives used are epoxy, polyimide, and other thermoset or thermoplastic materials. Organic adhesives are typically poor thermal and electrical conductors unless filled with thermally and/or electrically conductive particles. Gold and silver are the materials commonly used to provide electrical (and thermal) conductivity. Improved thermal performance is achieved by the introduction of thermally conductive particles, such as alumina (Al_2O_3), beryllia (BeO), and so on, without sacrificing electrical isolation.

A perceived disadvantage of the organic adhesive bond is the potential for outgassing, or the subsequent release of gas during periods of thermal stress. The outgas products (known and unknown contaminants) have the potential for causing adverse effects on device reliability and, hence, product lifetime. The poor thermal stability of organic adhesive (especially above the glass transition temperature), coupled with the outgassing of trapped solvents and reaction by-products, have produced significant concerns for VLSI die attach reliability.

Silver-filled, low-melting-point glass materials (~400 °C, or 750 °F) have emerged as a bonding alternative for VLSI die attachment. The use of 400 °C (750 °F) temperatures and oxidizing ambients (necessary for achieving proper adhesion reactions) has caused major concerns in processing. These materials also contain solvents and binders,

leading to the outgassing problems associated with organic adhesives.

Interconnection Bonding. Integrated circuit electrical interconnection or bonding is accomplished in three primary ways: wire bonding, tape automated bonding, and inverted-chip reflow, or flip-chip, technology. In wire bonding, a metal interconnect wire is placed in intimate contact with the chip bonding pad, the substrate package metallization, or the package pin. Under proper conditions of temperature, pressure, and perhaps ultrasonic energy, the metal wire and the underlying metallization are fused or welded to effect the electronic interconnect.

Wire bonding is divided into three categories, depending primarily on the mechanism used to create the weldment. The types of wire bonding are thermocompression bonding (ball-wedge), thermosonic bonding (ball-wedge), and ultrasonic bonding (wedge-wedge).

In thermocompression bonding, a gold wire that has been fed through a heated capillary has its end formed into a molten ball by an electronic arc or hydrogen torch. While still plastic, the ball on the end of the capillary is lowered onto the chip bonding pad, causing deformation of the gold ball (under pressure and temperature) and the fusing of the resultant nail-head-shape bond to the chip bonding pad (typically aluminum). The capillary is then raised, and the work is moved into position to form the second bond (tail or wedge bond), which is formed by the capillary edge deforming the gold wire under pressure and elevated temperature conditions. Once the second bond is formed, the wire is broken mechanically, and the cycle (flame off, first bond, second bond, wire separation) begins again.

In thermosonic bonding, the high capillary and substrate heats associated with thermocompression bonding are reduced by the introduction of ultrasonic energy. The bonding process is similar in most respects to thermocompression bonding, except when the wire and chip bonding pad or the wire and the second bond pad or metallization are in contact. In these two instances, ultrasonic energy is introduced to achieve

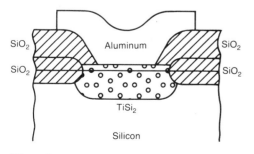

Fig. 8 Titanium silicide contact to silicon

$$3SiH_4 + 4NH_3 \xrightarrow[\text{Atmospheric pressure}]{700-900 \text{ °C}} Si_3N_4 + 12H_2$$

Fig. 9 Silicon nitride produced by chemical vapor deposition

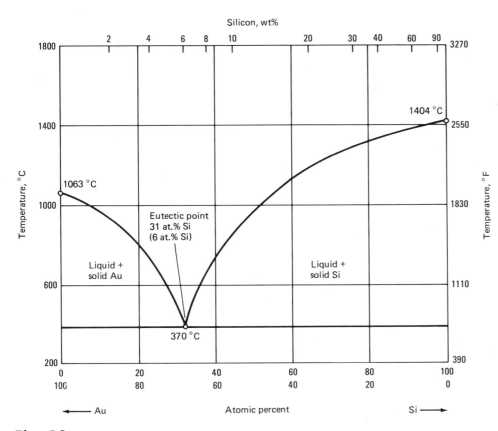

Fig. 10 Gold-silicon phase diagram. Source: Ref 8

Metal packages can have large numbers of input and output connections (>200), yet each must be brought out through an isolation glass bead. The lead is sealed to the bead, and the bead to the package by a glass-to-metal sealing process (Ref 21). Over 75% of all metal packages are welded, using parallel seam sealing, opposed electrode welding, or laser welding. The remaining metal packages are solder sealed using either gold-tin or conventional tin-lead alloys. Solder sealing can be performed in a conveyer furnace, with a seam sealer or a heated platen or cap (Ref 22).

There are many types of ceramic packages, including flat packages, ceramic dual-in-line packages (DIPs), chip carriers, grid arrays, and so on. In ceramic packages, a Kovar lead frame is attached to a glazed-alumina package base by temporarily softening the glazing or glass layer. After the die is placed in the package and wire bonded, the glazed-ceramic cap is put on and sealed to the assembly in a furnace or with a hot platen or cap sealer. The glasses used for sealing are high-lead content vitreous or devitrifying glasses with seal temperatures of 400 °C (750 °F) or more. Other ceramic package styles have a Kovar seal ring attached in addition to the lead frame attached to the ceramic body. After plating the lead frame and the seal ring with suitable metals, a metal lid is soldered or brazed to the seal ring. A ceramic seal frame is often used instead of the Kovar seal ring. A glazed ceramic lid is typically fused to the ceramic seal ring.

Encapsulation. The major chip encapsulation (packaging) techniques are cavity filling, saturation (impregnation), and coating (dip, surface, and conformal). The common cavity-filling processes are potting, casting, and molding.

In potting, an electronic component (single die, hybrid, and so forth) within a container is filled with a liquid resin that is then cured. The material, container, and circuit become an integral part of the final assembly. Typical resins used for potting include epoxies, silicones, and polyurethanes. Containers are usually made of metal and/or polymeric materials.

Casting is very similar to potting, except that the container (or outer casing) is removed after the cavity-filling material is cured. Typically, no heat or pressure is used in the process, although some vacuum might be used to allow the filling of remote recesses and to help outgas the polymeric resin.

Molding involves injecting a premelted polymeric material into a mold containing the electronic circuit or assembly, allowing the resin to set (harden), and then opening or releasing the encapsulated electronic part from the preshaped cavity or mold. The exact control of mold pressure, viscosity of the molding compound, cavity design, and

the high interface temperatures required for weld formation. The ultrasonic energy is introduced by vibrating or resonating the capillary now affixed to the end of an ultrasonic transducer horn. In thermosonic bonding, bond quality is determined by pressure, capillary temperature, and the amplitude and duration of the ultrasonic pulse. A typical ball-wedge bonding operation is shown in Fig. 11.

In ultrasonic, or wedge-wedge, bonding, the bonds are formed by capturing the wire against the impending bonding surface using a flat or wedge-shape tool. The bond is formed by applying pressure and an appropriate burst of ultrasonic energy (typically 20 to 60 kHz frequency). Ultrasonic bonding is accomplished primarily with aluminum wire, although gold and some other materials have been used (Ref 19). Typical steps in a wedge bonding process are shown in Fig. 12.

The tape automated bonding (TAB) process involves bonding ICs to patterned metal on multilayer tape (copper patterned film on polyimide carrier film) using thermocompression bonding techniques. Once attached to the carrier film, the IC can be tested, placed into a package, encapsulated, and environmentally stressed prior to excisement from the tape and attachment to the board or substrate by outer lead bonding. The TAB leads are planar beams. In multilayer tape structures with ground

planes, they can provide controlled impedance interconnects. Full details of the tape automated bonding process have been described in Ref 20.

The inverted-chip reflow, or flip-chip, process involves the formation of balls or bumps on the chip bonding pads (either solderable or nonsolderable bumps). Once "bumped," the chip is inverted over appropriate metallization pads on the substrate, and an interconnection is formed by solder reflow or thermocompression and/or ultrasonic techniques. The inverted interconnect can be extremely high density with excellent electrical performance and great amenability to automation. The inverted nature hides the device face, which causes concern regarding inspection. Also, welded-type interconnects may be difficult to repair. Details of this method are described in Ref 20.

Package Sealing. Packages fall into two generic types: hermetic and nonhermetic. In polymer-sealed or encapsulated packages (described in the following section, "Encapsulation"), moisture will penetrate in a relatively short time (hours), as shown in Fig. 13. Thus, the only true hermetic packages are made of metals, ceramics, and glasses. Metal packages fall into several types: the transistor outline (TO) (or round header) type, the butterfly (or flat) package, the platform package, and the monolithic (or bathtub) package. These various package styles are shown in Fig. 14.

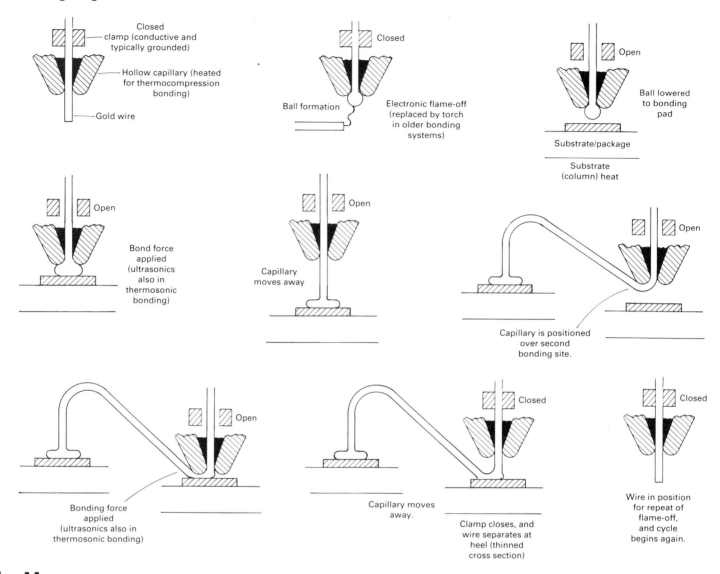

Fig. 11 Steps in the ball-wedge bonding process

filling mechanism(s) is critical to the success of the molding process. The size and shape of the dies, leads, and cavity determine internal stress buildup in the molded electronic part. Large dies and packages have been known to crack during soldering because of residual stresses induced during the molding process (Ref 23). Such stresses must be analyzed and reduced (by the proper choice of package shape, molding resin, lead frame design, and so on) in order to have a high-yield molding system. Finite-element techniques are particularly useful for this type of analysis (Ref 24).

Historic molding processes such as injection and pressure-type techniques are being replaced by platen transfer and reactive injection methods using new low-stress molding compounds to minimize the stresses associated with VLSI devices. Details of various molding processes are described in Ref 22.

Saturation and surface-coating techniques include impregnation, dipping, and conformal coating. Impregnation involves the application of a low-viscosity resin to the component, which already has a thin layer of the material bonded to the surface (perhaps by surface coating). This is typically used with a cavity-filling process or a conformal-coating technique. Dip coating is performed by dipping the component in the encapsulation resin, withdrawing the component, drying, and finally curing the encapsulant. Important parameters for producing effective dip-coated parts (that is, controlled-thickness encapsulation) include resin viscosity, withdrawal rate, and resin temperature. Conformal coating is accomplished by spinning or flow coating. The encapsulant rheology must be tailored, not only to ensure a dense coating with uniform flow, but also to accommodate circuit geometries, especially in hybrid applications (Ref 25).

Hermetic Versus Nonhermetic Packaging. Historically, in the microcircuit industry, the need for hermeticity (the sealing of electronic circuits from the ambient) was considered essential for reliability. This traditional view was originally based on the reliability testing of plastic and hermetically sealed transistors. Consequently, the circuits designed for use in high-reliability environments were packaged in either metal or ceramic chip packages with soldered, welded, or fused lids. Such circuits were typically expensive; suffered from low yields due to die attach, interconnect, and lid sealing; and would actually seal in corrosive or hazardous vapors from some of the adhesives or residual processing chemicals.

In particular, when some of the early epoxies were used for die attach, thermal aging would produce significant outgassing, and the sealed ambient would contain large amounts of water vapor and ammonia. These gases were correlated with electrical failures. Thus, in certain circumstances, the hermetic nature of the package actually accelerated failure.

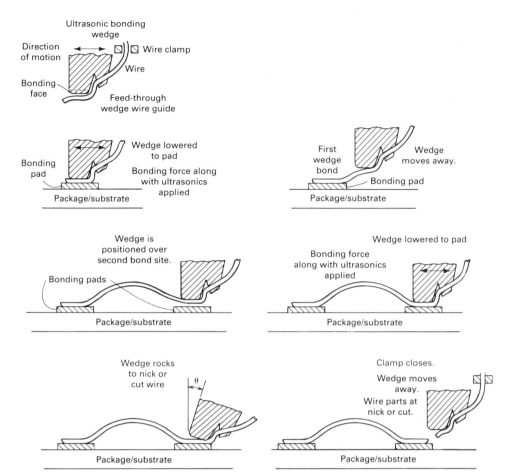

Fig. 12 Steps in the wedge-wedge ultrasonic bonding cycle

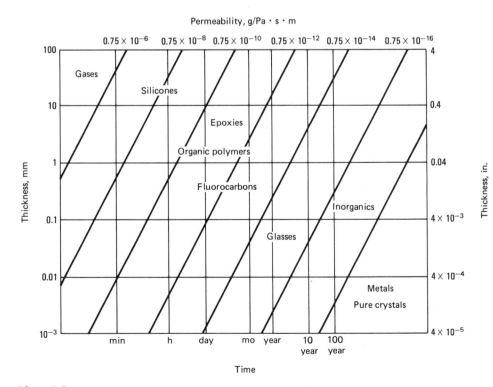

Fig. 13 Moisture penetration versus thickness for various materials. Source: Ref 3

Improved epoxies, metal attaches, package platings, and so on have produced highly reliable hermetic packages in recent times. These packages, although extremely reliable, are expensive. However, the substitution of low-melting-point glass seals for either lid welding or soldering can reduce the cost of hermetic packages by a factor of five (Ref 26).

Even lower costs can be achieved by replacing the hermetic package parts with low-cost organics, for example, epoxy seals, plastic package structure, or organic overcoating. Details of the organic overcoats, encapsulants, and package sealing materials are contained in the last two sections in this article. The use of organics that are definitely not hermetic (see Fig. 13) raises the issue of reliability because of the penetration of moisture.

Many studies have shown that appropriate inorganic capping (refer to the section "Inorganic Passivations"), organic junction coatings (such as certain silicones) and then plastic (epoxy) encapsulation, have produced extremely reliable IC packages (Ref 27). The organic junction coatings form excellent chemical bonds with the chip surface, precluding moisture interaction with the underlying IC metallization.

A two-phase organic packaging technique (epoxy encapsulation and silicone die overcoat) is usually necessary because the encapsulating epoxies contain leachable impurities and typically do not adhere well to the inorganic surfaces of silicon chips. Thus, when moisture ultimately penetrates through the epoxy to the IC surface, it will usually leach impurities from the epoxy, which in turn will cause corrosion at the IC surface. Silicones are very pure and chemically neutral and, despite their high moisture permeability, they bond effectively with the surface of the IC as shown in Fig. 15. The chemisorption of silicon into the IC passivations allows the reaction of free radicals (hydroxyl groups) on both surfaces to create oxygen bonds at the interface. This close bonding prevents the formation of galvanic corrosion cells, even though moisture freely penetrates through the bulk of the polymer. Corrosion can only occur after the chemical bond is disrupted, for example, by chemicals (acids, reactive solvents, and so on), by mechanical stress, and by disbonding due to trapped impurity layers.

Board-Level Packaging

The structure and the process of joining packages to higher-level assemblies, for example, cards, boards, or substrates, have a major influence on electronic system design, manufacturability, cost, and reliability. This choice of board interconnection methodology plays a primary role in determining board configuration and material selection. In addition, the choice of intercon-

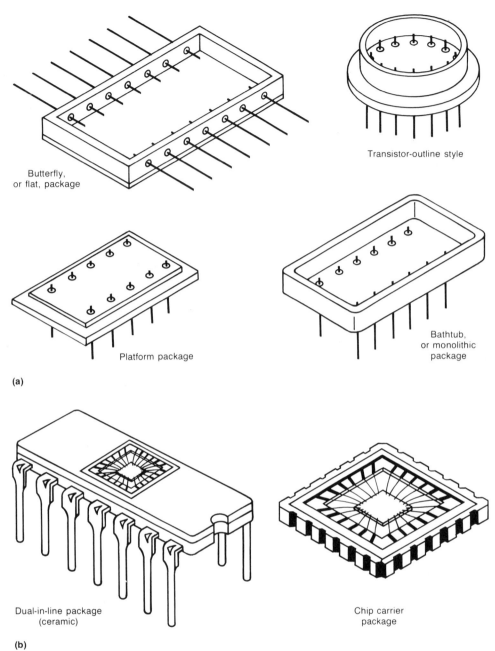

Butterfly,
or flat, package

Transistor-outline style

Platform package

Bathtub,
or monolithic
package

(a)

Dual-in-line package
(ceramic)

Chip carrier
package

(b)

Fig. 14 Common hermetic package styles. (a) Metal packages. (b) Ceramic packages (ceramic flat packages not shown)

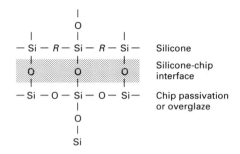

$$-Si - R - Si - R - Si-$$ Silicone

Silicone-chip interface

$$-Si - O - Si - O - Si-$$ Chip passivation or overglaze

Si

Fig. 15 Reactive bonding of silicone chip overcoat to the SiO_2 passivation on a silicon integrated circuit

nection method is influenced by the need for removability to replace failed parts or to change a part to achieve enhanced performance.

These criteria usually lead to a solder joint (or sometimes a socket) that can produce a high-performance electrical interconnect with sufficient reliability (in properly designed structures) and an appropriate degree of impermanence to allow device repair. Basic forms of packages and soldered interconnects are shown in Fig. 16. The major alternative to a solder joint for interconnecting the package to the board is a mechanical, pluggable (pin-socket) interconnection. The pin-socket interconnection allows easy package removal and is particularly suited for upgrade and repair in the field, as well as initial module testing and the like. Pin-socket interconnections typically provide inferior-performing interconnections compared to the more permanent solder joint. Other alternatives may include conductive adhesives or direct chip-on-board type (Ref 28) with wire-bonded interconnections, although repair in both these cases may be extremely difficult.

The increasing density of IC's has forced a significant increase in chip input/output (I/O) and a corresponding increase in wiring density on the mating circuit boards. The need for increased density has forced the development of greater packaging and board density than could be achieved by the standard DIP with its regular, wide lead spacing (1.27 or 2.54 mm, or 50 or 100 mil grid) column vias and through-hole mounting. The new packaging direction is surface (i.e. leadless) mounting on boards with staggered (blind) vias and no wasted real estate. The through-hole, mounted, pinned package now accounts for about 66% of the industrial/consumer electronic package markets, but it is projected that by 1995 this number will drop to less than 50%, as the transition to automated surface-mount assembly continues worldwide.

Through-Hole Mounting. As shown in Fig. 16, in through-hole mounting, the lead from the package is formed (shaped or bent, in the case of an axially leaded device) and then pushed through a hole in the printed wiring board (PWB). The hole is typically plated, and the lead is soldered into the hole, to effect both a mechanical and electrical interconnection. The mainstay through-hole package has been the DIP, which has been used for IC packaging since the early 1960s. In the DIP, the high-density chip I/O is fanned out over a larger area to meet the wiring density of the PWB.

The DIP typically has leads on 2.54 mm (0.1 inch) centers along two opposing edges of the package. It is plugged into a socket or soldered into a matching pattern of plate through holes in the PWB. DIPs have either plastic or ceramic bodies, which provide mechanical support and environmental protection for the IC chip. The plastic DIP is low cost, but suffers from limitations in hermeticity, as discussed above, and also in heat dissipation (Ref 29). DIPs come with up to 64 leads, but large DIPS suffer significantly in electrical performance over equivalent-lead-count packages of the surface-mount type such as chip carriers or quad flat packages. The longest lead to shortest lead length ratio on a large DIP is approximately 7 to 1, whereas in an equivalent chip carrier the same ratio is 1.5 to 1.

In a pin-type package, the primary alternative to the DIP is the pin grid array (PGA) in which the entire area of the package bottom is covered with pins. The PGA

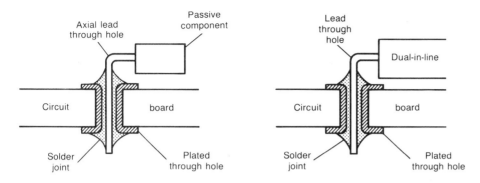

Pin/lead through-hole mounting

(a)

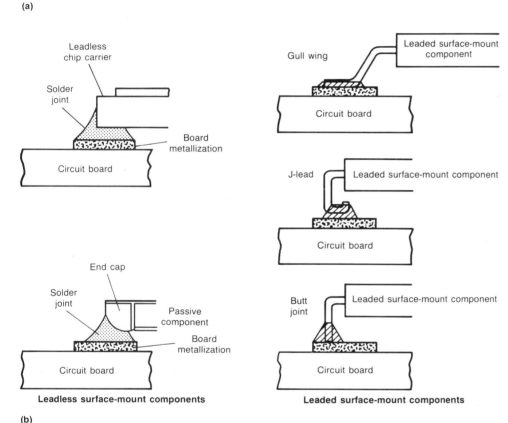

Leadless surface-mount components

Leaded surface-mount components

(b)

Fig. 16 Through-hole and surface-mount packaging technology. (a) Pin/lead through-hole mounting. (b) Leaded/leadless surface mounting

involves higher cost, but offers higher performance in all areas than the DIP, especially in I/O connections for a given package area (see Fig. 17). PGAs are available in both ceramic and plastic forms. The plastic PGAs outperform the ceramic versions in several important ways, in addition to cost. First, the lowered dielectric constant of the epoxy or polyimide organics allows higher-speed operation for a given geometry. Second, the improved conductivity of the pure copper metallization (over either thick film or cofired ceramic metallizations) minimizes electrical transmission loss in critical circuit applications. Despite the low cost and better electrical performance, reliability and nonhermeticity are still major plastic

PGA concerns. Thermal management in plastic PGAs is more difficult than in their ceramic counterparts (due to the lower thermal conductivity of the organics).

Surface Mounting Technology (SMT) provides many alternatives to pinned packages and through-hole mounting. There are many drivers for using SMT, including increasing density, reduced package size (and cost), improved board area utilization, better electrical performance, and improved repairability. There are two principal types of surface-mount packages: leadless packages, which include leadless chip carriers, pad grid arrays, and certain fundamental surface-mount elements such as chip capacitors, resistors, and inductors; and leaded

packages, including chip carriers (J-lead), quad flat packages, gull wing packages, and butt-joint packages.

In the leadless packages, solder forms the bridge between the metallized areas of the package and the board. The solder geometry and, hence, its performance (Ref 30) are determined by the package placement, solder volume, pad lengths, solder material, and the solder reflow processing during the joint operation. In the leaded packages, a compliant lead extends from the package and forms the interconnection of the package by means of a solder joint to the board. In either case, the solder joint forms both the electrical and the mechanical connection between the board and the package. As mentioned previously, SMT will continue to grow as automated assembly continues to increase. Through-hole mounting, however, will continue to be used. Furthermore, it is quite common to find both SMT and through-hole mounted components on the same board.

The major distinction between leaded and leadless SMT is the elastic compliance of the interconnecting lead in the leaded device between the package and the board. The elastic compliance becomes important under conditions of thermal and power cycling (Ref 31) of the package-board systems. Many analyses have been performed to estimate the reliability (cycles to failure) of both configurations. Results suggest certain design rules for the design and fabrication of reliable solder joints (Ref 30) but the process is far from complete. Solder joint reliability is the major concern in all SMT applications (Ref 32).

A circuit board may be considered a composite of organic and inorganic materials with both internal and external wiring that allows components to be mounted (mechanically supported) and electronically interconnected. For the sake of this discussion, thin film and thick film inorganic circuit board or substrate constructs are excluded. These are described by Charles and Clatterbaugh (Ref 7) and Borland (Ref 33). There are four basic printed board technologies: rigid boards, flexible boards, metal core boards, and injection-molded boards.

The materials used in the manufacture of printed boards (with the exception of metal core boards and injection-molded boards) are copper-clad laminates of organic dielectrics, which include phenolic-paper, epoxy-glass, polyimide glass, and others (to a much lesser degree). Polyimide is being used increasingly because of its high-temperature stability and excellent handling properties. Epoxy-glass and its variants are by far the most widely used printed board material today. The properties of epoxy-glass and polyimide materials are compared in Table 2. The basic definitions used in the description of printed boards are shown in Fig. 18.

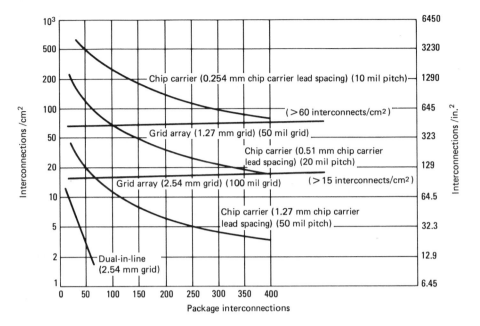

Fig. 17 Interconnection density versus number of package inputs/outputs for various package styles

Table 2 Basic properties of epoxy-glass and polyimide-glass circuit boards

Property	Epoxy-glass (FR-4)	Polyimide-glass
Glass transition temperature, °C (°F)	120–130 (250–265)	210–220 (410–430)
Surface resistivity ($\Omega \cdot$ cm)	10^{13}	10^{13}
Dielectric constant at 1 MHz	4.0–5.5	4.0–5.0
Dissipation factor at 1 MHz	0.02–0.03	0.01–0.015
Water absorption, 23 °C (73 °F), %	0.3	0.15–0.4
Coefficient of thermal expansion, 10^{-6}/K		
$x, y, T < T_g$	14–20	12–16
$z, T < T_g$	50–70	40

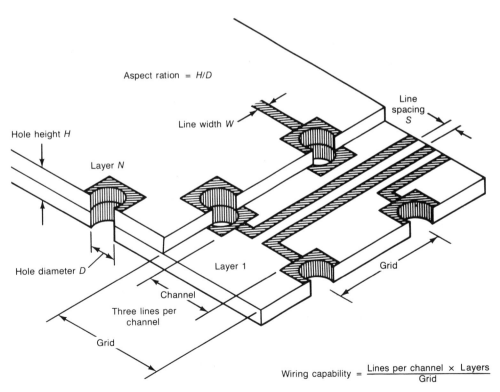

Fig. 18 Printed board schematic showing basic board pattern definitions

Printed boards range in complexity from single-sided to complex, multilayer boards. Examples of board cross sections are shown in Fig. 19. The interconnection structures shown can be extended farther and, in principle, to any number of layers (for example, boards of 40 to 50 layers have been demonstrated) (Ref 34). Typical cores (an insulation layer sandwiched between two patterned metal layers) are interleaved with the insulation layers (epoxy-glass prepreg) to build the board structure to the desired thickness, and then the whole sandwich is laminated. Drilling and plating connect the layers vertically with plated-through-holes. Buried vias can occur when the two circuit layers of a core are interconnected by a plated via (through hole) prior to lamination. Details of printed board fabrication have been described by Tummula (Ref 35).

As mentioned above, epoxy-glass laminates (of the FR-4 type, which is a fire-retardant, epoxy-glass cloth) (Ref 36), have received the most widespread use. Polyimide-glass is the second most important board technology. Properties of both are given in Table 2. The typical epoxy resin is the diglycidyl ether of 4,4'-bis(hydroxyphenyl) methane, which is referred to as a bisphenol A (BPA)-base system. Other resins and reinforcements used in printed boards are summarized in Table 3.

Fire retardancy is achieved by adding tetrabromobisphenol A to the BPA so that the resultant material has 15 to 20% bromine. The addition of 10% or more epoxy novolac resin can raise the glass transition temperature (T_g) of the board and improve board chemical resistance (Ref 37). The widely used curing agent is dicyandiamide (DICY). Because it has a low solubility in common organic solvents, a catalyst (a tertiary amine such as tetramethyl butane diamine, or TMBDA) is used to achieve good flow under lamination. Cured physical properties are good, including resiliency and a T_g between 120 and 135 °C (250 and 275 °F), depending on processing conditions. The disadvantage of DICY is its solubility, which may allow free DICY to be incorporated in the cured resin, possibly resulting in blistering, delamination, and increased moisture absorption.

Fiberglass is the major reinforcing material in epoxy-glass and polyimide-glass boards. Because printed board fiber-reinforced mats are woven of many fibers, a sizing is put on the fibers during the drawing process to prevent damage during the weaving operation. Once woven, the fibers are thermally cleaned (the sizing is burned off). Because thermally cleaned fibers do not form a good bond with epoxy resins, a coupling agent such as an organosilane is used. A good coupling agent is essential to the maintenance of the insulation resistance of the dielectric (Ref 38).

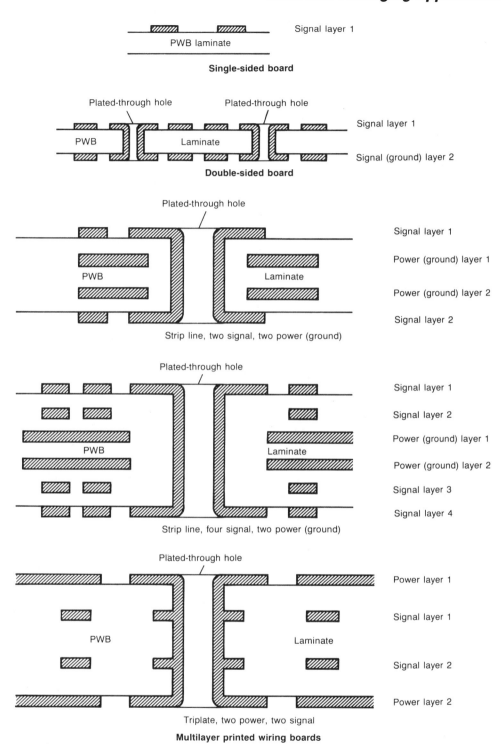

Single-sided board

Signal layer 1
PWB laminate

Double-sided board

Plated-through hole   Plated-through hole
Signal layer 1
PWB   Laminate
Signal (ground) layer 2

Plated-through hole
Signal layer 1
Power (ground) layer 1
PWB   Laminate
Power (ground) layer 2
Signal layer 2

Strip line, two signal, two power (ground)

Plated-through hole
Signal layer 1
Signal layer 2
Power (ground) layer 1
PWB   Laminate
Power (ground) layer 2
Signal layer 3
Signal layer 4

Strip line, four signal, two power (ground)

Plated-through hole
Power layer 1
Signal layer 1
PWB   Laminate
Signal layer 2
Power layer 2

Triplate, two power, two signal

Multilayer printed wiring boards

Fig. 19 Generic printed board cross sections

The copper conductors on printed boards are made by additive (plating) or subtractive (etching) techniques. The most common is the etching of copper foil bonded to the surface of the epoxy-glass board or core. The copper foil is produced by electrodeposition inside a rotating drum. The foil is stripped from the drum with the smooth side (drum side) placed upward and the rough side placed next to the laminate (res-

in) to promote adhesion. Foil made by the electrodeposition method is very pure and free from defects, which is essential for fine-line patterning.

Many polymers, copolymers, and polymer blends (as described in more detail in the last two sections of this article) have been developed over the years, but none has replaced the versatility of epoxies within the use temperature range of 125 to 150

°C (255 to 300 °F). Polyurethanes have excellent mechanical properties, but are limited thermally and are subject to hydrolysis, which affects dimensional stability in critical applications. Acrylics are typically used as adhesives for polyimides because of their versatility at high temperatures. Polyimide siloxane hybrids have been used to replace acrylics because of their superior high-temperature performance. Other materials, such as perfluoroethylene-propylene copolymer films, offer the promise of even higher performance.

Flexible circuits are typically one or two signal layers made on thin, flexible boards of polyimide or polyester (polyethylene terephthalate) film rather than rigid epoxy-glass fabric. Surface wiring is photolithographically patterned copper foil that has been laminated to the film dielectric with either epoxy or acrylic adhesives. Via holes are typically punched and plated using conventional printed board processes. By mounting rigid board elements to a flexible board or stiffening flexible boards in certain regions, the manufacturer can build a fold-up board system (folding in the nonrigid regions), which becomes a three-dimensional packaging scheme. This rigid-flex system can be assembled, soldered, and tested flat and then folded to fit the shape needed for the system application. Because of the flexible board interconnects between "boards," connectors and cables are eliminated. In certain applications, chips tend to be placed directly on the flexible cables and, in principle, can support finer lines because of their extremely smooth surfaces (compared to standard printed boards). Other materials, such as polytetrafluoroethylene (PTFE), are finding increased flex-circuit usage, especially for microwave applications.

Injection-molded board technology is a potential low-cost printed board manufacturing method. It differs from basic epoxy-glass technology in that resins containing fillers to improve thermal and/or mechanical properties are molded to the desired shape, even three-dimensional features. The resulting board is metallized by conventional methods or screen printing to form surface tracks and to place through holes. Other methods involve transfer molding of prefabricated three-dimensional film layers.

By using proper fillers and epoxy compounds or molded-in metal cores of aluminum, copper, or copper-Invar-copper, the CTE of the molded board can be tailored to match metal interconnections or silicon dies, permitting direct die bonding. It is projected that metal core molding will be especially useful in providing significant improvements in thermal management, especially if the insulation between the heat element and the metal core is thin.

Board overcoatings or encapsulations provide mechanical protection, moisture

Table 3 Printed board material systems

Material(a)	Glass transition temperature °C	°F	Dielectric constant at 1 MHz	Dissipation factor at 1 MHz	Coefficient of thermal expansion, 10^{-6}/K	Volume resistivity, $\Omega \cdot$ cm	Moisture absorption, %
Laminates							
Epoxy-fiberglass............	120–130	250–265	4.0–5.5	0.02–0.03	14–20 (x,y), 50–70(z)	10^{12}	0.1–0.3
Polyimide-fiberglass.........	210–220	410–430	4.0–5.0	0.01–0.015	12–16 (x,y), 40(z)	10^{14}	0.15–0.4
Epoxy-aramid..............	125	255	3.9	· · ·	6–8	10^{16}	0.85
Polyimide-aramid..........	250	480	4.0	· · ·	5–8	10^{12}	1.5
Epoxy-quartz..............	125	255	· · ·	· · ·	6–12	· · ·	· · ·
Polyimide-quartz..........	250	480	3.4	0.005	6–12 (x,y), 34(z)	· · ·	· · ·
PTFE-fiberglass	75	165	2.3	· · ·	20	10^{10}	1.10
Epoxy-Kevlar..............	· · ·	· · ·	· · ·	· · ·	6–10 (x,y), 90(z)	· · ·	· · ·
Polyimide-Kevlar...........	· · ·	· · ·	3.6	0.008	3–7 (x,y), 83(z)	· · ·	· · ·
PTFE-Kevlar..............	· · ·	· · ·	2.6	· · ·	· · ·	· · ·	· · ·
PTFE-μfiber	· · ·	· · ·	2.3	0.0008	· · ·	· · ·	· · ·
PTFE-fiberglass	· · ·	· · ·	2.3–2.5	0.0008–0.002	· · ·	· · ·	· · ·
Insulating polymers							
Epoxy (novolac)............	140–170	285–340	3.2	0.016	· · ·	10^{14}–10^{16}	1–6
Polyimide.................	>300, 250	570	2.9–4.2	0.002–0.006	3–50 (x,y), 3–45 (z)	10^{16}	0.5–3.5
PTFE	(19)(b)	480 (66)	2.1–2.8	0.0004	65 (x,y), 84(z)	· · ·	· · ·
Polyacrylate...............	· · ·	· · ·	3.0–3.8	· · ·	62.5	· · ·	· · ·
PEEK....................	143	290	3.0	· · ·	40	· · ·	· · ·
PBZT	250	480	3.0	· · ·	9 (x,y), 20(z)	· · ·	· · ·
Benzocyclobutene...........	250–350	480–660	2.6–2.7	0.0004–0.003	35–66	· · ·	0.02

(a) PTFE, polytetrafluoroethylene; PEEK, polyetheretherketone; PBZT, poly(p-phenylene benzobisthiazole). (b) Crystal structure change, triclinic-trigonal

and hostile gas protection, and dielectric shock protection and also can act as a solder mask or dam. The coating must be durable enough to provide scratch resistance and, when applied prior to soldering, must be able to withstand the soldering process. If repair is permitted, the coating must withstand the hand soldering or hot gas jets used in component rework. In addition to providing protection against moisture, the coating should protect against chlorine and sulfur.

Coatings can be applied by screen printing, roller or dip coating, evaporation (parylene), and the fluidized-bed process, the latter of which provides thicker coatings. Sometimes a silane pretreatment is necessary to achieve adhesion between the board and the coating.

Epoxy, polyimide, parylene, silicone, and glass-base systems have been used for circuit protection. Ultraviolet (UV)-curable acrylics can be easily applied and cured in a few seconds. Epoxies require temperatures of at least 70 °C (160 °F) and about 2 h to cure. Parylene is vacuum deposited at room temperature. Polyimide is suitable for performance at high temperatures

(>150 °C, or 300 °F) but requires curing at temperatures between 300 and 400 °C (570 and 750 °F). Glass-type protectants require curing above 400 °C (750 °F) for times in excess of 30 min.

Both polyimides and acrylics exist in solid-sheet form and can be laminated on flat boards using pressure-sensitive adhesives. Pressure-sensitive acrylic and silicone adhesives are available in various thicknesses. Silicones cost more than acrylics, but can be used at high temperatures. Solid films are shaped by punching or die cutting. Properties of various coating materials are listed in Table 4.

Inorganic Materials: Passivations and Encapsulants

Silicon dioxide (SiO_2), silicon nitride (Si_3N_4), and silicon-oxynitride are the commonly used inorganic passivations and/or encapsulants. They are typically deposited in thin layers (nominally 1 to 2 μm, or 0.04 to 0.08 mil) using sophisticated thermal, plasma, or radiation-assisted deposition techniques. The inorganic passivations are

excellent moisture and mobile-ion barriers, with silicon nitride being typically superior to silicon dioxide (Ref 39). The inorganic oxides and nitrides typically fall in the range of the glasses in terms of moisture diffusion, as shown in Fig. 13. The mobile-ion protection ability of silicon dioxide can be significantly improved by the addition of phosphorus (usually less than 6 wt%). Typical properties for both silicon nitride and silicon dioxide are shown in Table 5.

Other inorganic oxides used for passivation of both semiconductor and metal layers include: aluminum oxide (Al_2O_3), produced by the anodization of aluminum metal and various deposition techniques on nonaluminum substrates, including sputtering and radio frequency ion plating (Ref 40); tantalum oxide (Ta_2O_5), formed by the anodization of the metal or reactive sputtering from tantalum or Ta_2O_5 targets on nontantalum substrates (Ref 41); titanium oxide (TiO_2), again deposited by anodization or sputtering, depending on substrate materials (Ref 42) and so forth. Details of the deposition and potential applications of these materials are contained in the cited references and will not be considered further in this article.

Table 4 Typical properties of conformal overcoat materials

Material	Glass transition temperature °C	°F	Dielectric constant at 100 Hz	Dissipation factor at 100 Hz	Dielectric strength kV/mm	MV/in.	Volume resistivity, $\Omega \cdot$ cm	Moisture absorption, %
Epoxies	140–170	285–340	3.2	0.02–0.03	20–35	0.5–0.9	10^{14}	1–6
Polyimides	>300	570	2.9–4.2(a)	0.002–0.006	>240	6.1	>10^{16}	0.5–3.5
Silicones	−55 to 150(b)	−65 to 300	2.5–3.0	0.0016–0.002	20–45	0.5–1.1	10^{14}–10^{16}	· · ·
Parylenes...............	290–420(c)	555–790	2.6–3.0(a)	0.0006–0.013(a)	185–275	4.7–7.0	10^{16}–10^{17}	<0.1
Polyurethanes...........	−70 to 130	−95–265	3.5(d)	0.03–0.04	14–35	0.36–0.9	10^{12}–10^{15}	0.4–0.65
Acrylics	· · ·	· · ·	2.0	0.01	15–40	0.38–1.0	10^{13}–10^{14}	· · ·

(a) At 1 MHz. (b) Useful temperature range. (c) Melting point. (d) At 10 GHz

Table 5 Properties of inorganic passivations

| Properties | Thermal | Silicon dioxide | | | | Silicon nitride | |
		Plasma	SiH$_4$ + O$_2$	TEOS	SiCl$_2$H$_2$ + N$_2$O	LPCVD	PECVD
Deposition temperature °C (°F)........	800–1200 (1470–2190)	200 (390)	450 (840)	700 (1290)	900 (1650)	900 (1650)	300 (570)
Density, g/cm^3	2.27	2.3	2.1	2.2	2.2	2.8–3.1	2.5–2.8
Refractive index....................	1.46	1.47	1.44	1.46	1.46	2.0–2.1	2.0–2.1
Dielectric constant..................	3.8–3.9	. . .	. . .	. . .	. . .	. . .	. . .
Dielectric strength, kV/mm (MV/in.)...	500–1000 (12.7–25.4)	300–600 (7.6–15.2)	800 (20.3)	1000 (25.4)	1000 (25.4)	1000 (25.4)	600 (15.2)
Compressive stress at 25 °C (77 °F), MPa (ksi)	200–400 (29–58)	300 (43.5)(a)	300 (43.5)(b)	100 (14.5)	300 (43.5)	1000 (145)(b)	200 (29)(c)
Coefficient of thermal expansion, 10^{-6}/K.................	0.5	0.64–1.1	0.64	0.64	0.64	4	4–7
Bulk resistivity, Ω · cm	10^{14}–10^{16}	. . .	. . .	. . .	. . .	10^{15}–10^{17}	10^{15}

TEOS, tetraethylorthosilane + O$_2$. (a) Value for tensile stress is 300 MPa (43.5 ksi) as well. (b) Value represents tensile, rather than compressive stress. (c) Value for tensile stress is 500 MPa (72.5 ksi).

Deposition Methods. The thermal chemical vapor deposition process is by far the most widely used generic technique for the preparation of silicon dioxide, silicon nitride, silicon-oxynitride and, of course, polysilicon (Ref 43). Typical CVD reactions for the production of inorganic passivations, with their reaction temperatures, are listed in Table 6. There are two basic forms of CVD: reduced- or low-pressure CVD (that is, LPCVD) (Ref 44) and atmospheric pressure CVD (Ref 45).

In reduced-pressure CVD, the reactive gases are pumped through a heated reactor chamber, as shown in Fig. 20. The wafers are held vertically in a multiple wafer holder inside the reaction chamber. Because the passivation deposition rate is dependent upon the reactive gas concentration and the chamber pressure, there is a reactive gas concentration gradient within the reaction chamber (high at the beginning of the heated section and low at the end). To compensate for the falling gas concentration with distance, a multiple-zone furnace is used, with increasing furnace temperature (high CVD reaction rate) toward the gas exit (thus

essentially preserving a uniform deposition rate despite falling gas concentration). Typical temperatures range from 300 to 900 °C (570 to 1650 °F), with chamber pressures ranging from 25 to 250 Pa (0.004 to 0.04 psi). The gas flow rates are between 100 and 1000 cm^3/min (6 and 60 in.^{3}/min). This process produces high-quality uniform films in large-batch processed loads. However, most of the gases used in this process are toxic.

In atmospheric CVD, the furnace is typically operated in a continuous mode (conveyor belt style), as shown in Fig. 21. The reactive gas is fed from a multiple orifice manifold and flows uniformly on and around the wafers and through the open belt structure at atmospheric pressure. The process typically produces uniform, high-quality SiO$_2$ and Si$_3$N$_4$ in a high-throughput mode. The process consumes large quantities of the reactive gas. The furnace also forms particulates that must be removed by frequent chamber cleaning.

The thermal process, although producing high-quality inorganic films, requires relatively high deposition temperatures, nominally 600 to 900 °C (1110 to 1650 °F). In modern device technologies such as complementary metal-oxide semiconductor (CMOS), the aluminum metallization tends

to form "hillocks" (Ref 46) due to the intermetallic diffusion problems associated with a high-temperature process. Plasma-assisted CVD processes offer attractive low-temperature alternatives (≤400 °C, or 750 °F) for the formation of SiO$_2$, Si$_3$N$_4$, and silicon-oxynitride. Plasma-assisted CVD reactors exist in two primary types, parallel plate and hot wall.

The parallel-plate plasma-assisted CVD (Ref 47) is typically a cylindrical glass chamber with parallel aluminum plates acting as the top and bottom electrodes. The bottom electrode, which is grounded, serves as the substrate platen. It is typically heated from 100 to 400 °C (212 to 750 °F). A radio frequency voltage is applied to the top plate to create a flow discharge between the parallel plates. Reactive gases are introduced between the plates and then decompose to form silicon nitride and silicon oxide. The advantage of this process is its low deposition temperature, but it is a low-throughput-batch operating process.

The hot-wall plasma-assisted CVD process (Ref 48) uses a three-zone heated quartz reaction tube in which the wafers are stacked vertically, parallel to the gas flow. The reactive gas flows from end to end with a plasma discharge being struck from a top electrode to the aluminum or graphite sam-

Table 6 Reactions for the formation of oxides and nitrides and of films of silicon

| Material | Temperature | |
	°C	°F
Silicon dioxide		
SiH$_4$ + CO$_2$ + H$_2$	850–950	1560–1740
SiCl$_2$H$_2$ + N$_2$O.........	850–900	1560–1650
SiH$_4$ + N$_2$O	750–850	1380–1560
SiH$_4$ + NO	650–750	1200–1380
Si(OC$_2$H$_5$)$_4$	650–750	1200–1380
SiH$_4$ + O$_2$	400–450	750–840
SiH$_4$ + N$_2$O	250–350(a)	480–660
Silicon nitride		
SiH$_4$ + NH$_3$	700–900(b)	1290–1650
SiCl$_2$H$_2$ + NH$_3$	650–750(c)	1200–1380
SiH$_4$ + NH$_3$	200–350(a)	390–660
SiH$_4$ + N$_2$	200–350(a)	390–660
SiH$_4$ + N$_2$O	200–350(a)	390–660
SiH$_4$ + N$_2$O + NH$_3$	200–350(d)	390–660
Silicon		
SiH$_4$	600–650	1110–1200

(a) In addition to plasma. (b) At atmospheric pressure. (c) At reduced pressure. (d) In addition to plasma (silicon-oxynitride)

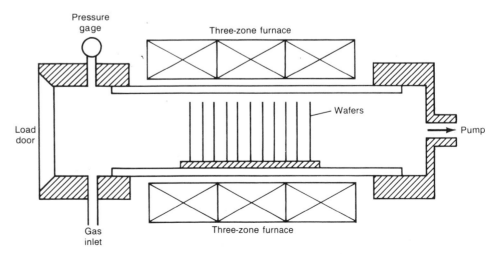

Fig. 20 Reduced-pressure CVD reactor (hot wall)

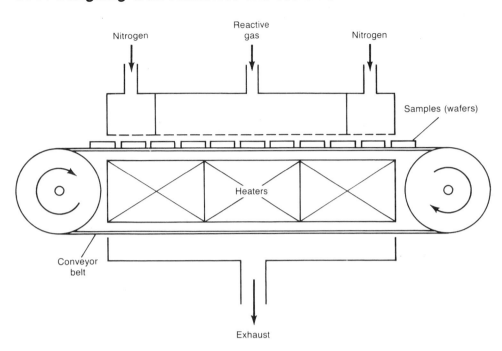

Fig. 21 Atmospheric pressure CVD reactor

ple holders. This is an extremely low-temperature deposition process (for example, 200 to 350 °C, or 390 to 660 °F), and the large three-zone tube allows greater throughput (despite manual loading and unloading) than the parallel-plate system. Again, like the parallel-plate system, the hot-wall plasma-assisted CVD is subject to particulate generation, which can damage the wafers and the reactor. As can be seen in Table 5, silicon nitride, silicon oxynitride, and silicon dioxide can be prepared by this process.

A third relatively new deposition technique, radiation-assisted or radiation-stimulated deposition, has been developed by Peters (Ref 49). In this technique, UV radiation is used to stimulate a mercury catalytic reaction with silane and ammonia (or hydrazine) to form Si_3N_4 (Ref 50). The reaction does not require heat, but relies on the absorption of UV photon energy by the reactants, leading to the dissociation of their chemical bonds.

In similar work, Erhlich (Ref 51), Boyer (Ref 52), and researchers at Lawrence Livermore Laboratories (Ref 53) have developed low-temperature laser-deposition processes. The lasers typically provide high-intensity deep-UV sources (nominally 200 nm, or 8 μin.) for the photodissociation of reactive gases (metallo-organics such as metal hybrids), which then deposit metal or dielectric layers on areas of the substrate under the laser beam. These layer pantography techniques have been described in Ref 54, and, because of the high laser cost and deposited-film purity, have not been fully commercialized.

The mercury-sensitized radiation stimulation by Peters, however, uses simple (low-cost) UV lamps and has been licensed commercially. A typical process for Si_3N_4 can be accomplished with minimal wafer heat (that is, 100 °C, or 212 °F). Film quality is not as good as with the CVD processes, because film layers are typically porous and contain trace amounts of mercury. Mercury is an extremely rapidly diffusing impurity in silicon, causing serious reliability concerns. Such mercury-containing passivation layers may be ideally suited as caps for mercury cadmium telluride (Hg-Cd-Te) radiation detectors and thermal sensors (Ref 55).

Organic Materials: Adhesives, Passivations, Sealants, and Encapsulants

Organic polymer sealants and encapsulants are divided into three broad categories: thermosetting materials, thermoplastic polymers, and elastomers. The thermoset materials cross link during their curing process and cannot be reversed to their original state (polymer) after curing. Examples of thermoset polymers include polyimides, epoxies, polyesters, butadiene-styrenes, silicon-modified polyimides, acrylates, and silicon-epoxies.

Thermoplastic materials, when subjected to high temperatures, melt and flow. When cooled, they solidify without cross linking. Thermoplastic processes are reversible, and these materials have many engineering applications. Typical thermoplastic materials include polyvinyl chloride, polystyrene, polyethylene, fluorocarbon polymers, acrylics, poly(p-xylylene), and perhaps some preimidized silicon-modified polyimides.

Elastomers are special thermosetting materials (that is, they cross link) and have high elongation ability. These materials consist of long, linear, flexible molecular chains that are joined by interval cross linking (Ref 56). Examples of elastomer materials include silicone rubber, silicone gel, natural rubbers, and polyurethanes.

Of all the materials mentioned above, the materials that have been shown to be most useful in electronic applications are epoxies, polyimides, fluorocarbon polymers, poly(p-xylylene), urethanes, and silicones (including silicone gel). Each of these will be treated in some detail in the following subsections.

Epoxies. Of all the polymeric materials available as potential electronic adhesives and encapsulants, epoxy-base resin systems are by far the most dominant. The basic properties of epoxy resins, as a class, provide great versatility and wide acceptance in electronics packaging applications. Such properties include high strength, good electrical insulation, easy cure mechanisms, chemical resistance, excellent adhesion, and minimum shrinkage. Epoxies are a diverse group of resins, but the epoxide group (or oxirane ring) is characteristic of all epoxy materials. The oxirane ring is shown in Fig. 22, along with the basic structures of a few common epoxies.

Epoxies can react with each other to form a rigid cross-linked structure or they can be cured (coreacted) with a wide variety of other materials. Three epoxy cure mechanisms (amine cured, anhydride cured, and novolac hardened or cured) are shown in Fig. 23.

The amine-cured systems (aliphatic, cycloaliphatic, and aromatic) offer fast cures with long shelf life, but poor moisture resistance and reduced electrical performance. Amine-cured systems typically find limited use in electronics applications. Anhydride-cured resins work well at high temperature, but their ester-type linkages are easily hydrolyzed which limits the application of these materials in humid environments.

The ether linkages formed during the action of epoxy resins with novolac hardeners, while not being as thermally stable as the linkages in anhydride-cured systems, have excellent moisture resistance. Novolac-hardened epoxies are widely used in the electronics industry. Other curing materials include catalytic Lewis acids and bases and latent curing agents such as DICY, as mentioned in the section "Inorganic Passivations."

In electronic applications, the three most commonly used epoxy resins are the diglycidyl ethers of bisphenol A and bisphenol F (yielding the phenolic novolacs and cresol novolacs, respectively) and the cycloaliphatic epoxides. The basic structures of these resins are shown in Fig. 24. Figure 25 presents a multifunctional resin, known

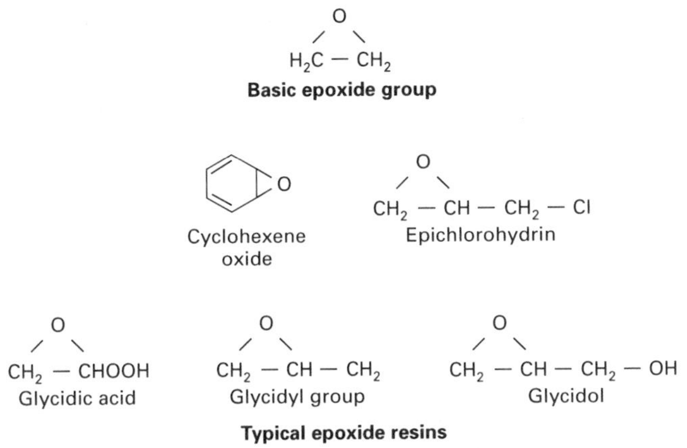

Fig. 22 Basic epoxide (oxirane) group plus several common epoxy resins

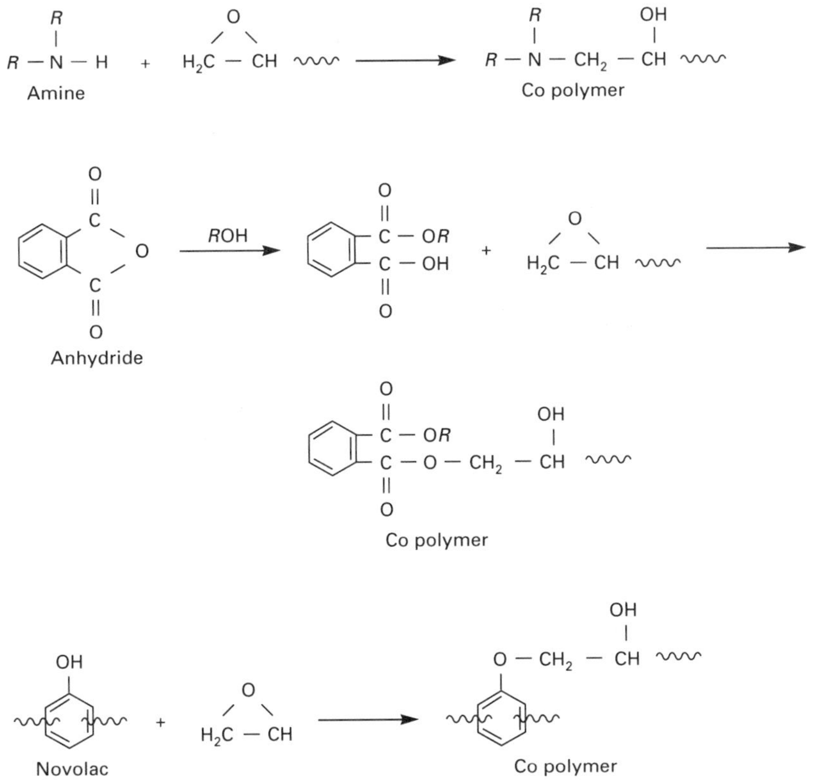

Fig. 23 Basic epoxy curing mechanisms

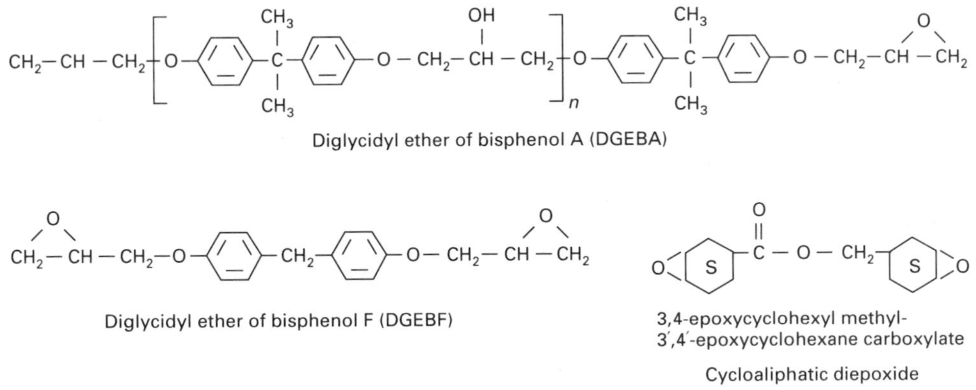

Fig. 24 Basic structure of diglycidyl ethers of bisphenol A and F and cycloaliphatic epoxide. S, saturated (has hydrogens everywhere instead of double bonds)

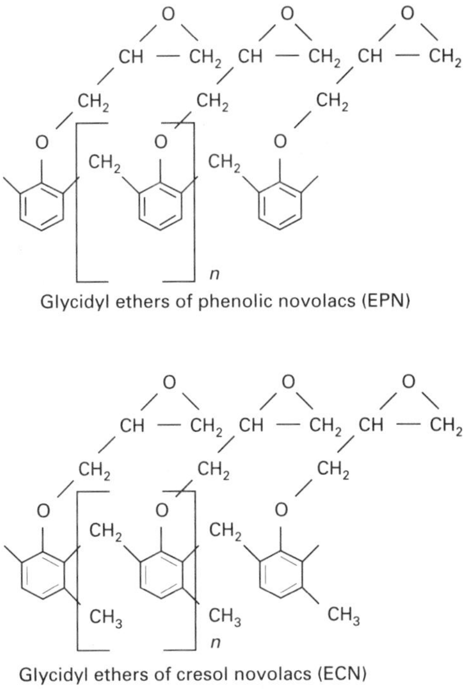

Glycidyl ethers of phenolic novolacs (EPN)

Glycidyl ethers of cresol novolacs (ECN)

Fig. 25 Basic structures of phenolic and cresol novolac epoxies

as epoxidized phenol novolac (EPN), whose cross-link density is independent of molecular weight. The resultant material has low CTE, low moisture absorption, and high second-order T_g. Epoxidized cresol novolac (ECN), also shown in Fig. 25, is the basis for most modern rigid encapsulants. The terms novolac epoxy or epoxy B are most commonly associated with ECN resin hardened with phenol-formaldehyde novolac.

In the actual reactions that form ECN and/or EPN, the novolac is first reacted with epichlorohydrin, with the epoxide group of the epichlorohydrin joining the active hydroxyls of the novolac. The epoxide is then re-formed with the evolution of HCl. In practice, the reactions never quite follow the ideal behavior because side reactions and incomplete dehydrohalogenation can yield unwanted species. The unwanted species typically reduce the amount of epoxide in the resin and reduce the cross-link density. The unwanted species also contain chlorine, which can be activated under prolonged exposure to heat and humidity. It should be pointed out that the cycloaliphatic epoxides (or peracid epoxides) are the only class of epoxies with no residual chlorine.

Modern epoxy encapsulants (and adhesives) are highly complex materials that can contain, in addition to the resin and hardener (curing agent), a catalyst, a filler system, lubricants (mold release), pigments, coupling agents, flame retardants, and so on. Table 7 lists typical epoxy casting resin systems and some of their physical proper-

Table 7 Representative epoxy casting resins

Cure type	Anhydride	Aliphatic amine	Aromatic amine	Catalytic	Dianhydride	High temperature anhydride	Novolac anhydride
Resin/curing agent	DGEBA/HHPA	DGEBA/DETA	DGEBA/MPDA	DGEBA/BF$_3$-MEA	DGEBA/PMDA	DGEBA/NMA	Novolac/NMA
Catalyst	BDMA	...	...	...	...	BDMA	BDMA
Cure cycle, °C (°F)	150 (300)	25 (77) + 200 (390)	150 (300)	120 (250) + 290 (390)	150 (300) + 200 (390)	200 (390)	90 (190) + 165 (330) + 200 (390)
Number of hours	4	24 + 2	2	2 + 2	4 + 14	3	2 + 4 + 16
Tensile strength at 25 °C (77 °F), MPa (ksi)	72 (10.4)	75 (10.9)	85 (12.3)	43 (6.2)	22 (3.2)	75 (10.9)	66 (9.5)
Elongation, %	5.6	6.3	5.1	3.0	...	2.7	3.2
Volume resistivity, $\Omega \cdot$ cm	4×10^{14}	2×10^{16}	1×10^{16}	...	...	2×10^{14}	10^{16}
Dielectric constant at 1 MHz (25–150 °C, or 77–300 °F)	3.2–3.6	3.3–4.2	3.8–4.6	3.2–3.3	3.3–3.6	2.9–3.3	3.2
Dissipation factor at 1 MHz (25–150 °C, or 77–300 °F)	0.013–0.02	0.034–0.033	0.038–0.015	0.024–0.012	0.022–0.013	0.021–0.13	0.016

Resins: DGEBA, diglycidyl ethers of bisphenol A. Curing agents: HHPA, hexahydrophthalic anhydride; DETA, diethylene triamine; MPDA, metaphenylene diamine; BF$_3$-MEA, boron trifluoride monoethylamine; PMDA, pyromellitic dianhydride; NMA, nadic methyl anhydride. Catalysts: BDMA, benzyldimethyl amine

ties. Figure 26 shows the effects of various fillers on the thermal expansion and conductivity of various epoxy systems.

Epoxy resins are used as adhesives because of their excellent bonding characteristics. Most commercial epoxy adhesives are proprietary formulas based on generic formulations of the resins and other components discussed above. Epoxy adhesives for electronics applications fall into three primary categories: insulating (or bonding) adhesives, electrically conductive adhesives, and thermally conductive adhesives. Bonding adhesives, widely used throughout the electronics industry, are available in thixotropic pastes or sheet (tape) form. These materials may contain fillers such as silica or alumina to increase their electrical resistivity and improve adhesion. Certain types of the tape or sheet materials contain fiberglass mats or meshes to enhance their stability and dielectric behavior.

Metal fillers such as silver, gold, or copper are added for electrical conductivity. To achieve low resistivity, filler loadings can be quite high (relative to the resin content); this reduces cured resin strength and adhesion. Bond strengths (in comparison to structural or bonding adhesives) can be reduced by factors of four to ten, depending on filler content.

For electrical insulation but high thermal conductivity, powder fillers such as aluminum nitride (AlN) and beryllia (BeO) are added to epoxies. If electrical conductivity can be permitted in addition to good thermal conductivity, then metal powders such as gold, silver, and copper can be used. Graph-ite has also been used in these applications. Strengths tend to be intermediate when compared to pure bonding adhesives.

Polyimides have many uses in electronics, including wire coatings, printed boards, and flexible circuits. They are also used as insulators or protective coatings for semiconductors and hybrid microcircuits. Polyimides are typically sold as polyamic acid precursors because the polyimides themselves (fully cured aromatic polyamic acids) are soluble only in strong polar solvents (acids or bases). To produce the classic polyimide, one typically starts with a dianhydride (pryomellitic dianhydride, or PMDA) and an aromatic diamine (oxydianiline, or ODA). The solvent is usually N-methyl pyrrolidone (NMP), although other solvents such as dimethylacetamide

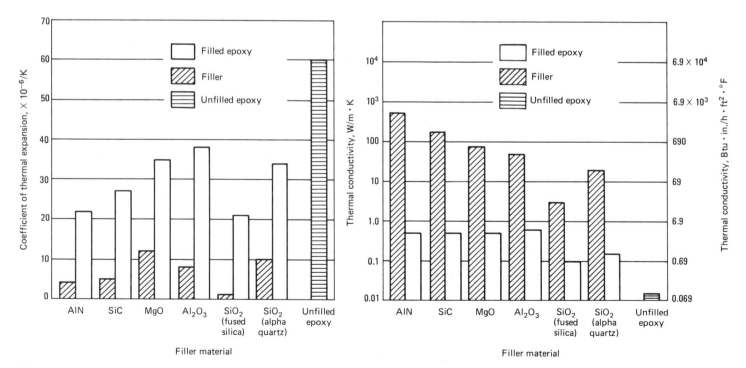

Fig. 26 Effect of epoxy fillers in casting resins on (a) coefficient of thermal expansion and (b) thermal conductivity

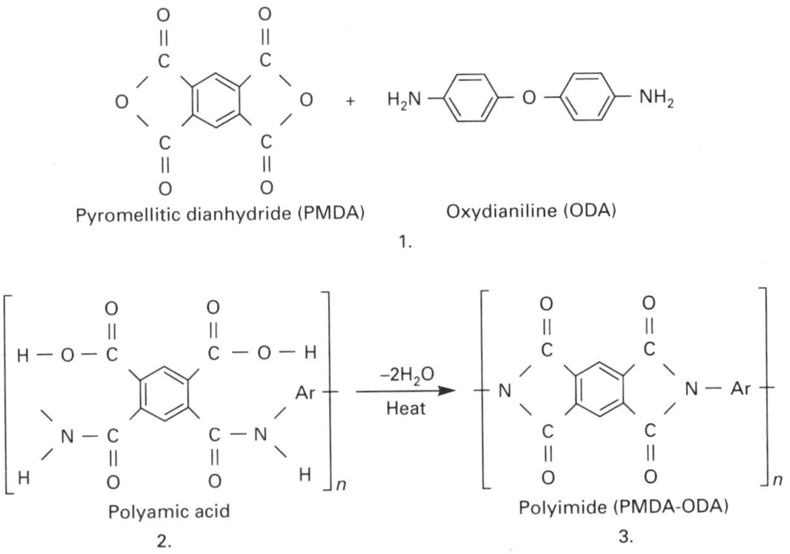

Fig. 27 Basic reaction for forming a cured polyimide film from the dianhydride-diamine combination. PMDA/ODA is a typical example.

(DMAC), dimethyl sulfoxide (DMSO), and dimethyl formamide (DMF) have been used. NMP is usually preferred because of its lower toxicity.

The reaction between PMDA and ODA produces a soluble polyamic acid, which, upon curing, has the desirable thermal, electrical, and mechanical properties. Curing agents could also be used, but this is not typical of the commercial solution-type polyamic acid coating materials. A typical polyimide thermal cure process is shown in Fig. 27.

There are many other aromatic dianhydrides that can be used, including hexafluoro diaphenyl dianhydride, oxydiphenylene dianhydride (ODPA), biphenyl dianhydride (BPDA), and 4,4-benzophenone dicarboxylic dianhydride (BTDA). Other aromatic diamines include 4,4'-diamino-diphenyl ether (DADPE), para-phenylene diamine (PPD), methylene dianiline (MDA), meta-phenylene diamine (MPD), and para-phenylene dianiline (PPDA).

Typically, polyamic acids and cured polyimide layers and coatings can be prepared with various combinations of the above anhydride-diamine combinations in reactions similar to that for PMDA/ODA. By varying the ingredients, the polymer properties can be changed over wide ranges. For example, PMDA-benzidine is a very rigid polyimide with a modulus of elasticity of approximately 11.8 GPa (1.7×10^6 psi), with a 2% elongation at break, whereas flexible PMDA-ODA can have an elastic modulus up to 3.4 GPa (0.50×10^6 psi), with a 100% elongation at break.

Photosensitive polyimides shorten processing time and eliminate a difficult polyimide etching process. The photosensitivity is built in by the addition of an unsaturated monomer such as hydroxyethyl-methacrylate (HEMA) to the polymer backbone. The

HEMA is either covalently or ionically bonded. All commercial photosensitive polyimides to date are negative working (that is, as in negative photoresist, that which is exposed is what remains). Details of the processing of conventional and photosensitive polyimide film layers have been described by Clatterbaugh and Charles (Ref 57). The differences in structure between conventional and photosensitive polyimide are shown in Fig. 28.

Basic mechanical, thermal, and electrical properties of some typical polyimides are given in Table 8. As mentioned above, polyimide properties may be tailored by blending appropriate anhydrides and amines. For example, by blending a high and a low CTE polyimide, one can achieve a desired CTE encapsulant that matches the substrate CTE and reduces the thermal stress associated with encapsulated devices under temperature cycling. The moisture affinity, high curing temperature, and high cost of polyimides are its only drawbacks. Advances in polyimide synthesis have produced materials with lower moisture absorption and improved long-term adhesion.

The newer siloxane-polyimide copolymers, which combine both the silicone and polyimide properties, offer great promise in electronics. A typical siloxane-polyimide is a block of copolymer of BTDA with methylene dianiline (MDA) and bis-gamma aminopropyltetramethyl disiloxane (GAPD). This block copolymer has higher solubility and lower thermal stability ($\leq$400 °C, or 750 °F) than PMDA-type polyimides. The siloxane-polyimide typically requires no adhesion promoter (for example, aminosilane/gamma-amino propyl silane) when used on silicon wafers. The basic BTDA, MDA, and GAPD structures are shown in Fig. 29.

Silicone coatings have long been recognized in the electronics industry for their performance (excellent electrical properties, elasticity, low moisture absorption, ionic impurity, thermal stability, and so on), ease of application, and rapidity of low-temperature cure. Silicone coatings are usually classified by cure mechanism (heat cure, room-temperature or vulcanizing cures, and UV cure). UV curing can occur in as little as 10 s. Silicones are available either as dispersions in solvents or solventless. They are typically impervious or very resistant to oxidation, electrical arcing, ozone, UV radiation, and fire. The ash/smoke from burning silicone is nontoxic, containing only silica, CO_2, and low levels of CO.

The low surface tension of silicones provides excellent water and moisture resistance. Although silicone is permeable by water vapor, as described in the section "Hermetic Versus Nonhermetic Packaging," the coated surface is protected, provided that the adhesion remains intact. The silicone prevents the formation of a continuous film of moisture at the interface, thereby preventing a conducting path. Silicone is very pure, with no mobile ions, and its low moisture absorption in general prevents surface corrosion. Silicones resist fungus, and some formulations have been show to be biocompatible (Ref 58).

The basic composition of all silicones is shown in Fig. 30. They exist in two broad categories, solvent borne (solvent dispersion of high molecular weight solids) and room-temperature vulcanizing (RTV) (either moisture cure or addition cure). The solvent-borne materials must be heat cured. First, the solvent must be evaporated at room temperature or at a somewhat elevated temperature. Once the solvent is evaporated, the resultant sticky material must be cured (cross-linked thermoset) with additional heat or by extended room-temperature cure. For example, DC 1-2577 consists of a high molecular weight silicone resin in toluene. This material can be either dipped, spray coated, or brushed. Once coated, the board is typically air dried for 30 min (accelerated drying would be at 80 °C, or 175 °F, for a shorter time) and then cured rapidly at temperatures above 100 °C (212 °F). Cured DC 1-2577 is a flexible film in layer thicknesses of less than 0.15 mm (6 mils). In thicker layers, this highly branched polysiloxane cures to an elastoplastic material with a Shore D hardness between 25 and 30. The low elastic modulus of this material makes it particularly useful as a nondamaging overcoat material.

There are two types of RTV materials: first, RTV-I, which cures upon exposure to molecular moisture/humidity and is typically a one-part system, and, second, RTV-II, a two-component formulation that cures at room temperature when the components

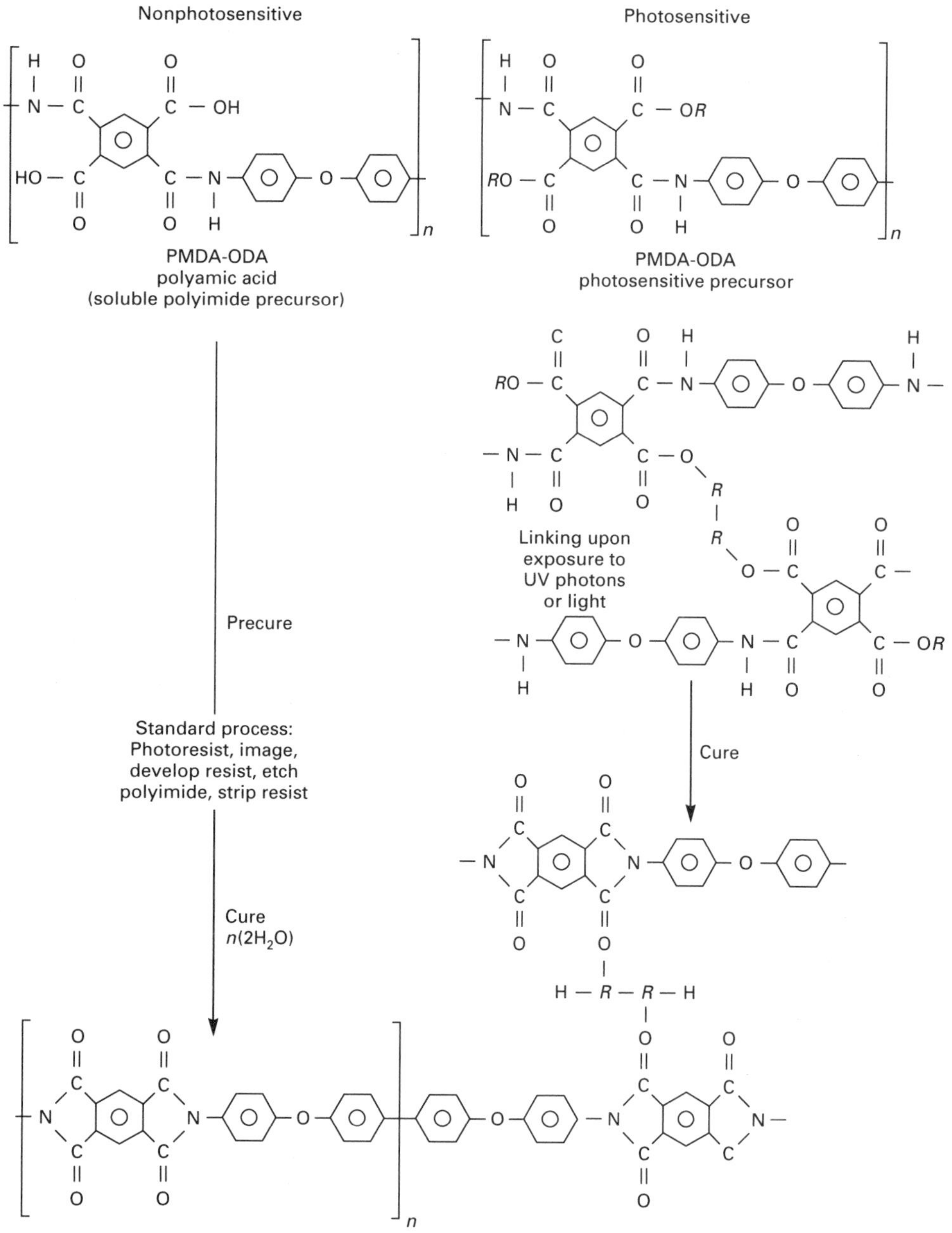

Fig. 28 Conventional and photosensitive polyimide chemical reactions. *R*, photoreactive group

are mixed. The moisture cure mechanism is shown in Fig. 31. If the product is kept away from moisture, it will remain viable for long periods. Once moisture (air) is introduced, curing will begin. Ideal conditions are 25 °C (77 °F), 50% relative humidity for 72 h. Moisture-cured RTVs give off cure by-products (*R*—OH), which can potentially corrode electronics if they are based on acetoxy cure mechanisms. Other mechanisms, such as ethoxy curing, will not corrode electronics.

A typical addition-cured mechanism is shown in Fig. 32. The cross-linking reaction occurs by the addition reaction of a hydride-functionalized cross linker across the vinyl group of the silicone polymer. The polymer is catalyzed by low levels of platinum metal complexes (for example, 20 to 30 ppm). Care must be taken not to poison (inhibit) the platinum catalyst by the incorporation of amines or sulfides. Because these silicons are typically low-viscosity fluids, no solvents are needed in processing. If thinning is needed, xylene or 1,1,1-trichloroethane can be added. Additional processing, however, will be required to evaporate the added solvent. One component of the RTV-II formulation is now available with inhibited catalysts that retard polymerization at room temperature. Thus, these materials must be cured at elevated temperatures (120 to 150 °C, or 250 to 300 °F, for up to 30 min).

UV-curable silicone polymers can be made by attaching radiation-curable organic groups to the silicone backbone and using UV photoinitiators rather than tin salts or platinum complexes for catalysts. The two most common UV-curable silicones involve the attachment of acrylate (or methacrylate) or epoxide groups. A typical acrylated silicone is shown in Fig. 33. This is a modified silicone structure in which organic groupings, rather than polysiloxane bonds, provide the cross linking. The addition of the acrylate groups lowers the thermal stability and somewhat compromises flexibility and electrical performance.

Table 8   Properties of typical polyimides

Property	PMDA/ODA	BTDA/ DADPE	BTDA/DADPE (photosensitive)	BPDA/PPD
Density, g/cm³	1.42	1.41	1.61	. . .
Stress, MPa (ksi)	22 (3.2)	36 (5.2)	64 (9.3)	9 (1.3)
Tensile strength, MPa (ksi)	103 (14.9)	133 (19.3)	133 (19.3)	343 (49.7)
Elongation, %	40	15	15	25
Elastic modulus, GPa (10⁶ psi)	1.4 (0.20)	2.4 (0.35)	. . .	8.3 (1.2)
Moisture uptake, %	2–3	2–3	1–3	0.5
Dielectric constant at 1 kHz, 50% relative humidity	3.5	3.3	3.3	2.9
Dissipation factor at 1 kHz	0.002	0.002	0.002	0.002
Volume resistivity, Ω · cm × 10¹⁶	>1	>1	>1	>1
Dielectric breakdown, MV/m (MV/in.)	>240 (6)	>240 (6)	>240 (6)	>240 (6)
Glass transition temperature, °C (°F)	>400 (750)	>320 (610)	>320 (610)	>400 (750)
Coefficient of thermal expansion, 10⁻⁶/K	20	40	40	3
Thermal conductivity, W/m · K (Btu · in./h · ft² · °F	0.155 (1.07)	0.147 (1.02)	0.147 (1.02)	. . .
Decomposition temperature, °C (°F)	580	560	550	620
Weight loss, in 2 h, air, at 500 °C (930 °F), %	3.6	2.9	0.5	1.0

Table 9 Typical property ranges of silicone coatings

Properties	Value
As supplied	
Viscosity, Pa · s	0.12–70
Flash point, closed cup, °C (°F)	5–100 (40–212)
Temperature range, useful, °C (°F)	−55 to +150 (−65 to 300)
Cure time, days(a)	1–7
Shelf life, months	6–12
Pot life, days	0.5–30
Cured	
Dielectric constant at 100 kHz and 25 °C (77 °F)	2.6–3.1
Dissipation factor at 100 kHz and 25 °C (77 °F)	0.0004–0.002
Dielectric strength, kV/mm (kV/in.)	20–50 (510–1270)
Coefficient of thermal expansion, 10^{-6}/K(b)	200–1000
Tensile strength, MPa (ksi)	1.7–6.2 (0.25–0.90)
Elongation, %	50–1200
Shore A hardness	20–70

(a) Depending on layer thickness. (b) At a range of 25 to 100 °C (77 to 212 °F)

Table 10 Properties of parylene

Property	Parylene N	Parylene C	Parylene D
Density, g/cm³	1.11	1.29	1.42
Refractive index (sodium O-line)	1.661	1.639	1.669
Melting point, °C (°F)	420 (790)	290 (555)	380 (715)
Coefficient of thermal expansion at 25 °C (77 °F), 10^{-6}/K	69	35	. . .
Thermal conductivity, W/m · K (Btu · in./h · ft² · °F)	0.12 (0.83)	0.082 (0.57)	. . .
Dielectric constant at 1 MHz	2.65	2.95	2.80
Dissipation factor at 1 MHz	0.0006	0.013	0.002
Dielectric strength, MV/m (MV/in.)(a)			
Pulse	275 (7.0)	220 (5.6)	215 (5.5)
Step	235 (6.0)	185 (4.5)	. . .
Volume resistivity at 23 °C (73 °F), 50% relative humidity, Ω · cm	1.4×10^{17}	8.8×10^{16}	2×10^{16}
Surface resistivity at 23 °C (73 °F), 50% relative humidity, Ω · cm	1×10^{13}	1×10^{14}	5×10^{16}
Water absorption, %	<0.1	<0.1	<0.1
Tensile strength, MPa (ksi)	45 (6.5)	70 (10.2)	75 (10.9)
Yield strength, MPa (ksi)	42 (6.1)	55 (8.0)	60 (8.7)
Elongation at break, %	30	200	10
Yield elongation, %	2.5	2.9	3

(a) Measured on 25 μm (1 mil) thick film

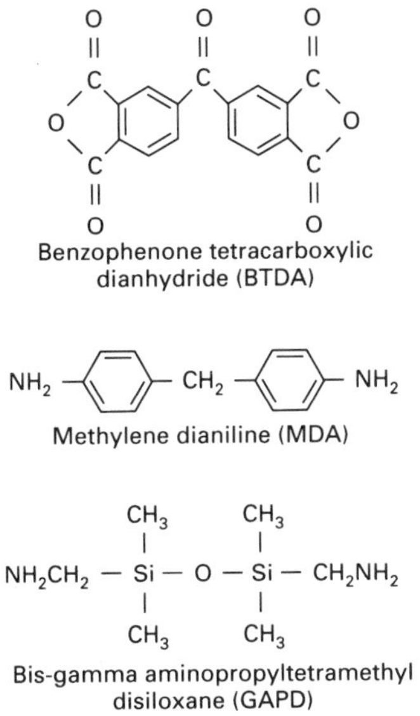

Benzophenone tetracarboxylic dianhydride (BTDA)

Methylene dianiline (MDA)

Bis-gamma aminopropyltetramethyl disiloxane (GAPD)

Fig. 29 Basic building blocks of the siloxane-modified polyimide systems

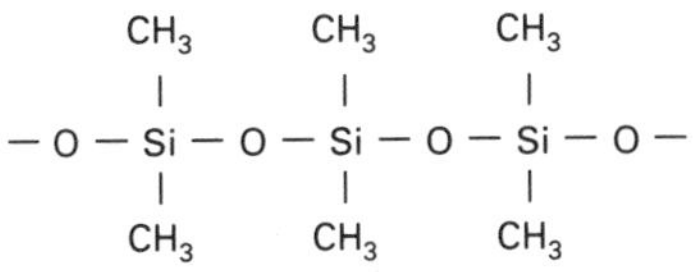

Fig. 30 Basic silicone structure

compromise in electrical and thermal performance is to be expected. Shadow curing (Ref 59), or the lack thereof, is also a cause for concern. A list of the basic properties of silicones is presented in Table 9.

Poly(*p*-Xylylene), or Parylene. To process parylene, the dimer di-*p*-xylylene, a [2.2]-paracyclophane, is thermally disassociated into a monomer (paraxylylene). Although the monomer is stable as a gas at low pressure it spontaneously polymerizes upon condensation to produce a coating of high molecular weight, linear poly-paraxylylene. The reaction process, also known as the Gorham process (Ref 60), is shown in Fig. 35. First a thermal reactor is used to vaporize the dimer (at 150 to 200 °C, or 300 to 390 °F, and ~135 Pa, or 0.02 psi, pressure) to the monomer, which serves as a stable intermediate. Deposition of the spontaneously polymerized monomer occurs at room temperature (25 °C, or 77 °F, and about 15 Pa, or 0.002 psi). The process was devel-

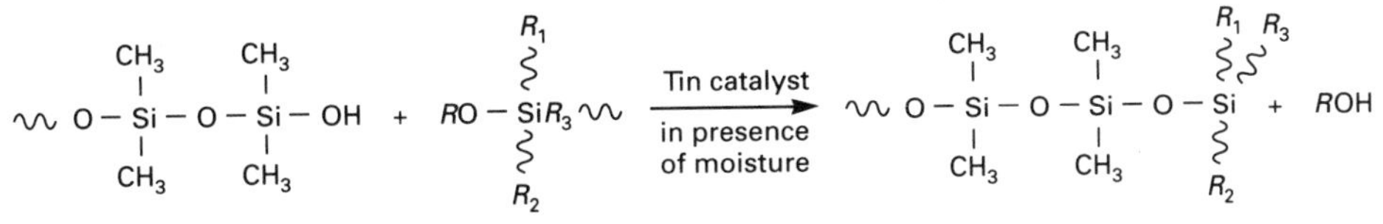

Epoxidized silicones provide an alternative method of producing UV-curable silicones. Epoxide groups are attached to the silicones (as shown in Fig. 34), and curing is produced by the photodecomposition with liquid diaryliodonium salt as a catalyst. This decomposition produces a strong acid (hexafluoro-antimonic acid) which reacts with the epoxy-functionalized silicone to form polyepoxide cross links. Again, some

Fig. 31 Silicon moisture cure mechanism. R_1, R_2, R_3 are radical by-products. *R*OH is a cure by-product.

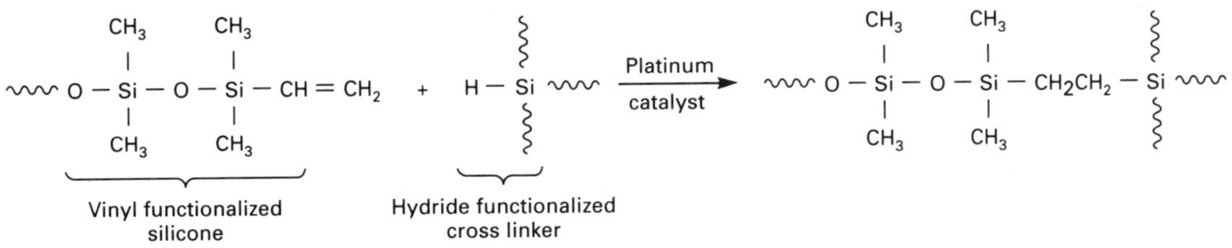

Vinyl functionalized silicone

Hydride functionalized cross linker

Fig. 32 Silicone addition cure mechanism

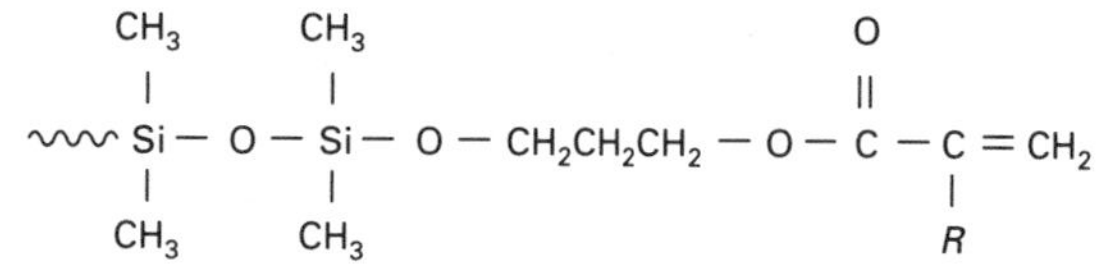

Fig. 33 Basic structure of acrylated silicons. R is either CH_3 or H.

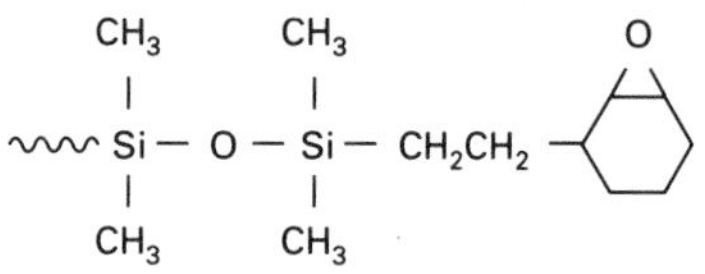

Fig. 34 Basic structure of photosensitive silicone (epoxidized siloxane). The cure consists of photo decomposition with a liquid diaryliodonium salt $(Ar_2I^+ \ SbF_6^-)$ catalyst.

oped by Union Carbide Corporation, and Parylene is a trademark of that company (Ref 61).

The room-temperature deposition makes the parylene process an attractive encapsulation process for all electronics. The deposit (nominally 25 to 50 μm, or 1 to 2 mils, in thickness) is an excellent conformal coat with superior step coverage and extremely low deposition stress. Films have excellent chemical resistance with no incorporated impurities (especially in the nonchlorinated form) and no solvents. The parylene coating is tough, scratch resistant, and, because it is a vacuum deposition technique, it is suit-

able for coating hard-to-reach areas around and under chips, beam leads, and so forth.

There are three commercially available substituted dimers and their resultant parylenes. Parylene N is an unsubstituted C-16 hydrocarbon, [2.2]-paracyclophane (II). Parylene C is an aromatic chlorination of the dimer with one chlorine atom per aromatic ring. Parylene D is an aromatic chlorination of the dimer with two chlorine atoms per aromatic ring. Properties of Parylenes N, C, and D are summarized in Table 10.

Urethanes or polyurethanes are available in one- or two-part systems that are solvent

based to achieve low viscosity for dipping, spraying, brushing, and so on. They are formed by reacting an isocyanate (R_1-NCO) with a hydroxyl (R_2-OH) compound to form a urethane group, as shown in Fig. 36. Urethanes can be cured by heat or with internal catalysts. In thermally cured systems, the reactive isocyanates are inhibited by a readily volatilized or easily decomposed compound until the system is heated. In catalytically cured systems, the part containing the isocyanate is mixed with a second part containing the catalyst. In some systems additional cross linking is achieved by reacting amines or hydroxyl groups with the isocyanate groups. Because the isocyanates are highly reactive with moisture, moisture must be avoided prior to use.

The advantages of polyurethanes are good adhesion, rapid cure, good flexibility at low temperatures, excellent chemical resistance, low dielectric constant, and good moisture resistance. Basic properties are listed in Table 11.

Fluoropolymers. The basic fluoropolymer structure is shown in Fig. 37. The fluorine content of the fluorodiol can be varied during synthesis by choosing the size of the perfluoroalkyl group. Examples are listed in Fig. 37. Fluoropolymer coatings are used in environments that require high coating reliability: water, oil, and silicone oil. They possess a very low dielectric constant, excellent weatherability, and low ion migration rate and they resist microbiological attack. Table 12 summarizes the commonly available fluoropolymers.

Properties of representative fluoropolymers are given in Table 13. Fluoropolymers display good adhesion with substrates such

Table 11 Typical properties of urethanes

Property	Value
Dielectric strength, kV/mm (kV/in.)	14–35 (350–890)
Dielectric constant at 100 Hz–10 GHz	5–3.5
Dissipation factor at 100 Hz–10 GHz	0.016–0.04
Volume resistivity, $\Omega \cdot cm$	10^{12}–10^{15}
Thermal conductivity, W/m · K (Btu · in./h · ft² · °F)	0.5×10^{-4} (3.45×10^{-4})
Arc resistance, s	180
Water absorption, %	0.4–0.65
Tensile strength, MPa (ksi)	7–35 (1.0–5.1)
Elongation, %	300–400
Shore A hardness	40–70
Shrinkage during cure, %	2
Heat distortion temperature, °C (°F)	10–65 (50–150)
Safe-use temperature, °C (°F)	90–130 (195–265)
Coefficient of thermal expansion, 10^{-6}/K	200–300

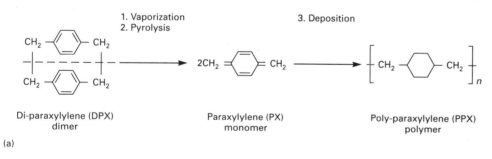

1. Vaporization
2. Pyrolysis

3. Deposition

Di-paraxylylene (DPX) dimer   Paraxylylene (PX) monomer   Poly-paraxylylene (PPX) polymer

(a)

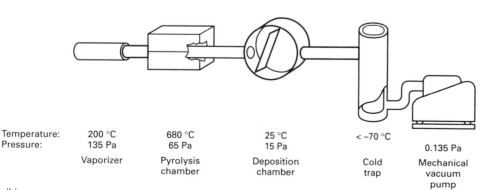

Temperature:	200 °C	680 °C	25 °C	< –70 °C	
Pressure:	135 Pa	65 Pa	15 Pa		0.135 Pa
	Vaporizer	Pyrolysis chamber	Deposition chamber	Cold trap	Mechanical vacuum pump

(b)

Fig. 35 Basic Gorham process for producing parylene. (a) Chemical reaction. (b) Schematic apparatus with approximate temperatures and pressures

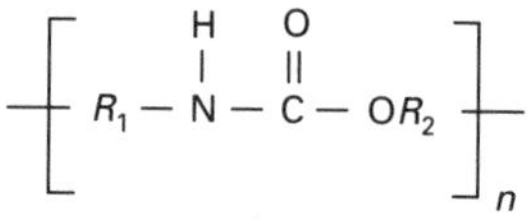

Fig. 36 Basic urethane chemical structure

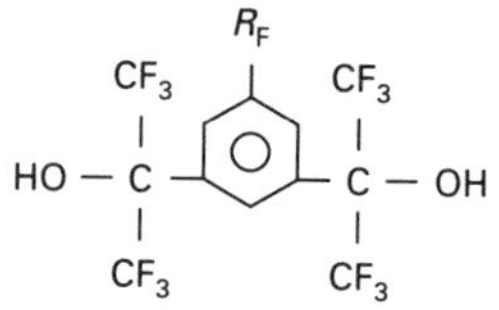

Fig. 37 Basic fluorodiol structure

Table 12   Basic fluoropolymer products

Fluoropolymer product	Fluoropolymer(a)	Description(b)
Fluorad FC-721	PFOM	MMA backbone with a $C_{17}F_{15}$ moiety
Fluorad FC-723	PFOM	MMA backbone with a C_7F_{15} moiety
Fluorad FC-725	Fluorinated terpolymer with 27 wt% fluorine	Acrylate backbone with fluoroaliphatic and aliphatic side chains

(a) PFOM, poly-1H, 1H-pentadecafluorooctyl methacrylate. (b) MMA, methylmethacrylate

Table 13   Properties of representative fluoropolymers

Property	FC-721	FC-725
As supplied		
Form....................................	Thin liquid (2% solids)	Viscous liquid (30% solids)
Appearance...........................	Clear, light-colored solution	Clear, light-yellow solution
Specific gravity at 25 °C (77 °F)	1.5	0.97
Flash point, °C (°F)	>95 (203)(a)	26 (79)(b)
Viscosity at 25 °C (77 °F), Pa · s.............	. . .	0.3
Dried film		
Appearance	. . .	Transparent, colored under ultraviolet light
Refractive index	1.36	. . .
Specific gravity...........................	. . .	1.3
Thickness, dip coated, μm (μin.)	0.05–0.075 (2–3)	. . .
Dielectric constant at 100 kHz	5.50	2.64
Dissipation factor at 100 kHz	0.012	0.016
Dielectric strength, MV/m (MV/in.).........	60 (1.5)	100 (2.5)
Volume resistivity, $\Omega \cdot$ cm..................	5.2×10^{10}	4.5×10^{15}

Note: FC-273 is composed of the same polymer as FC-721, but with 0.2 wt% solids. (a) Per Setaflash closed cup. (b) Per Tag closed cup

as copper, glass, epoxy, and polytetrafluoroethylene (PTFE) glass laminates. Under certain conditions, fluoropolymers have been reported to have superior moisture resistance, even to silicone (Ref 62). Typical fluoropolymers are unreactive, but most commercial fluoropolymer formulations are proprietary. Thermal degradation of poly-1H, 1H-pentadecafluorooctyl methacrylate (PFOM) is a two-stage process. The first stage begins at 140 °C (285 °F) and is complete at 200 °C (390 °F), while the second stage begins at 220 °C (430 °F) and is complete by 320 °C (610 °F). The degradation product is assumed to be the monomer from which the polymer was made (Ref 63). Other published data suggest thermal stability above 200 °C (390 °F).

Thermoplastics. The increased industry pressure for faster processing times, lower temperatures, lower pressures, and recycling are bringing thermoplastic materials in competition with traditional thermoset formulations. To date, thermoplastics have been proposed for both die attach and encapsulation applications. Preliminary reports on thermoplastic die attach adhesives have already appeared in the literature (Ref 64). In general, however, most of the discussion of thermoplastic use in electronic applications has centered on printed boards, where recently introduced thermoplastics match or exceed the properties of epoxy-glass laminate, as shown in Table 14.

As can be seen from Table 14, polysulfone (polyether sulfone) and polyether-imide have lower properties (dielectric constant, dissipation factor, and so on) than FR-4 and approach those of PTFE-glass board materials. PTFE board material is difficult to prepare for metallization, whereas engineered thermoplastics are relatively easy to process, making them potentially several times cheaper. Further, the thermoplastics are more stable with higher glass transition temperatures. For example, the T_gs of polysulfone and polyether-imide are 190 °C (375 °F) and 217 °C (425 °F), respectively, while the T_g of FR-4 is only 120 °C (250 °F).

Thermoplastics offer the capability to produce boards with three-dimensional features, such as tapered holes, chamfered edges, support ribs, bosses, holes, and slots. In addition to the five thermoplastics listed in Table 14, others that offer promise include formulations of reinforced nylon, polyesters (polybutylene terephthalate and polyethylene terephthalate), aromatic copolyester, polyamide-imide, other polyimide-silicones, polyarylate, polyphenylene sulfide, polyetheretherketone, and fluorocarbons.

The focus of electronics applications of thermoplastics to date has been in second- and third-level packaging, with the notable exception of the recent work on substrate adhesive bonding (Ref 64). Thermoplastics have not as yet been applied to encapsulation because of two primary problems, high viscosity and low adhesion.

Thermoplastics (depending on shear rate during molding) can have apparent viscosities between 1000 and 10 000 Pa · s. By shear thinning, melt viscosities can be reduced to 1000 to 3000 Pa · s for polyphenylene sulfide with resultant shear stresses from 0.14 to 0.35 MPa (0.02 to 0.05 ksi), which can cause wire sweep and actual wire crossovers (Ref 36). Even epoxies in transfer-molding operations with much lower viscosities have caused bond pad shearing. Thus, unless wire bonds of much greater strength can be developed, thermoplastics will not be used for the encapsulation of chip and wire circuits. In TAB technology, the corresponding shear stresses for beam/bump fracture are between 20 and 80 MPa (2.9 and 11.6 ksi) (for 0.1 × 0.13 mm, or 4 × 5 mils, contact area with shear forces between 0.25 and 0.50 N, or 0.06 and 0.11 lbf), which are significantly greater than the above-mentioned melt front viscosity stress.

Thermoplastics generally have poor adhesion (weak) to surfaces because of the lack of chemical reaction sites. Nevertheless, physical coupling by means of acid/base reactions or solubilization can occur. Physical bonding is much more likely to degrade under stress than is its chemical counterpart. For example, polyphenylene sulfide encapsulation on ceramic capacitors was reported to fail because of poor adhesion (Ref 65).

Chemical modifications of the basic structure have the potential for developing thermoplastics for encapsulation. For ex-

Table 14   Properties of thermoplastic materials

Material	Glass transition temperature °C	°F	Dielectric constant at 1 MHz	10 GHz	Dissipation factor at 1 MHz	10 GHz	Coefficient of thermal expansion, 10^{-6}/K	Volume resistivity, $\Omega \cdot$ cm	Moisture absorption, %
Polysulfone.............	190	375	3.0	3.0	0.004	0.006	. . .	. . .	. . .
Polyether-imide	217	425	3.1	3.4	0.007	0.003	56	. . .	. . .
Polyethylene...........	115	240	. . .	2.3	. . .	0.0005	200	10^{16}	<0.01
Polyimide-silicone	150–240	300–465	. . .	3.0	. . .	0.007	300–500	10^{16}	<1
Polyether sulfone........	230	445	3.5	. . .	0.006	. . .	. . .	. . .	. . .
Polyimide-glass	210	410	4.0–5.0	. . .	0.01–0.015	. . .	12–16 (x,y), 40(z)	10^{14}	0.15–0.4
Epoxy-glass	120–130	250–265	4.0–5.5	. . .	0.02–0.03	. . .	14–20 (x,y), 70(z)	10^{12}	0.1–0.3

ample, a reactive site "graft" can improve adhesion of the plastic matrix to electronic device surfaces. The copolymerization of the main thermoplastic resin chain with a low molecular weight monomer compatible with resin fillers or device surfaces would also improve adhesion. The making of polymer alloys could improve the resultant polymer properties over those of the constituent polymer elements. Under the proper conditions, an interconnected polymer network can be generated (Ref 66).

Interpenetrating polymer networks (IPNs) of two or more cross-linked polymers hold extreme promise (Ref 67) for improved encapsulants with minimum shrinkage, low polymerization exotherms, high density (less permeability), thermal stability (higher T_gs than constituent elements), and good wetting (low shear stresses and good adhesion). Typical systems might include epoxies and silicones, epoxies and polyurethanes, polyurethanes and polyesters, or silicones and nylon. Details of IPN polymers have been presented by Sperling (Ref 67).

REFERENCES

1. G. Moore, VLSI, What Does the Future Hold?, *Electron. Aust.*, Vol 42, 1980, p 14
2. S.M. Sze, *VLSI Technology*, McGraw-Hill, 1983
3. C.P. Wong, Chapter 3, Integrated Circuit Device Encapsulants, in *Polymers for Electronic Applications*, J.H. Lai, Ed., CRC Press, 1989, p 63
4. C.P. Wong, Integrated Circuit Encapsulants, in *Polymers in Electronics*, Vol 5, John Wiley & Sons, 1986, p 638
5. J. Czochralski, *Z. Phys. Chem.*, Vol 92, 1918, p 219
6. E.P.A. Metz, R.C. Miller, and R. Mazelsky, *J. Appl. Phys.*, Vol 33, 1962, p 2016
7. H.K. Charles, Jr. and G.V. Clatterbaugh, Thin-Film Hybrids, in *Packaging*, Vol 1, *Electronic Materials Handbook*, ASM INTERNATIONAL, 1989, p 313
8. A.B. Glaser and G.E. Subak-Sharpe, Diffusion, in *Integrated Circuit Engineering, Design, Fabrication and Applications*, Addison-Wesley, 1979, p 197
9. L. Holland, *Vacuum Deposition of Thin Films*, Chapman and Hall, 1966
10. B. Chapman, *Glow Discharge Processes*, John Wiley & Sons, 1980
11. C.S. Pai, E. Cabreros, S.S. Lau, T.E. Seidel, and I. Suni, *Appl. Phys. Lett.*, Vol 47 (No. 7), 1985, p 652
12. J.M. Poate, K.N. Tu, and J.W. Meyer, Ed., *Thin Films—Interdiffusion and Reactions*, John Wiley & Sons, 1987
13. A.J. Learn, *J. Electron. Mater.*, Vol 5, 1974, p 531

14. J.J. Steppan, J.A. Roth, L.C. Hall, D.A. Jeannotte, S.P. Carbone, A Review of Corrosion Failure Mechanisms During Accelerated Tests: Electrolytic Metal Migration, *J. Electrochem. Soc. Solid-State Sci. Technol.*, Jan 1987, p 175
15. N.G. Einspruch and G.B. Larrabee, Ed., *VLSI Electronics: Microstructure Science*, Vol 6, Academic Press, 1983
16. J.H. Pugh and R.S. Williams, *J. Mater. Res.*, Vol 1 (No. 2), 1986, p 343
17. R. Beyers, K.B. Kim, and R. Sinclair, *J. Appl. Phys.*, Vol 61 (No. 6), 1987, p 2195
18. E. Wolfgang, Future Trends in Electron Beam Testing, *Microelectron. Eng.*, Vol 7, 1987, p 435; R.K. Jain, Picosecond Optical Techniques Offer a New Dimension in Microelectronics Test, *Test Meas. World*, June 1984, p 40
19. J. Ling and C.E. Albright, The Influence of Atmospheric Contamination on Copper to Copper Ultrasonic Welding, *IEEE Proc. Electron. Compon.*, May 1984, p 209
20. H.K. Charles, Jr., Electrical Interconnection, in *Packaging*, Vol 1, *Electronic Materials Handbook*, ASM INTERNATIONAL, 1989, p 224
21. C.J. Leedecke, P.C. Baird, and K.D. Orphanides, Glass-to-Metal Seals, in *Packaging*, Vol 1, *Electronic Materials Handbook*, ASM INTERNATIONAL, 1989, p 455
22. O.M. Uy and R.C. Benson, Package Sealing and Passivation Coatings, in *Packaging*, Vol 1, *Electronic Materials Handbook*, ASM INTERNATIONAL, 1989, p 237
23. S.A. Gee, W.F. VandenBogert, V.R. Akylas, and R.T. Shelton, Strain Gauge Mapping of Die Surface Stresses, in *Proceedings of the 39th Electronic Components Conference*, Institute of Electrical and Electronics Engineers, 1989, p 343
24. H.K. Charles, Jr. and G.V. Clatterbaugh, "Solder Joint Reliability—Design Implications From Finite Element Modeling and Experimental Testing," Paper 89-WA/EP-22, presented at the ASME Annual Winter Meeting, American Society of Mechanical Engineers, Dec 1988
25. K. Moran, Ed., *Hybrid Microelectronic Technology*, Gordon & Breach, 1984
26. N. Sinnadurai, High Density Packaging of Chips and Subcircuits, in *Handbook of Microelectronics Packaging and Interconnection Technologies*, Electrochemical Publications Ltd., 1985, p 95
27. N. Sinnadurai, An Evaluation of Plastic Coatings for High Reliability Microcircuits, in *Proceedings of the Third European Hybrid Microelectronics Confer-*

ence, International Society for Hybrid Microelectronics, 1981, p 482
28. D.J. Bodendorf, K.T. Olsen, J.F. Trenko, and J.R. Ninnard, Active Silicon Chip Carrier, *IBM Tech. Disclosure Bull.*, Vol 15 (No. 2), July 1972, p 656
29. C.H. Mitchell and H. Berg, Thermal Studies of a Plastic Dual In-Line Package, *IEEE Trans. Compon. Hybrids Manuf. Technol.*, Vol CHMT-2 (No. 4), 1979
30. C.V. Clatterbaugh and H.K. Charles, Jr., Design Optimization and Reliability Testing of Surface Mount Solder Joints, *Int. J. Hybrid Microelectron.*, Vol 8, 1985, p 31
31. C.V. Clatterbaugh and H.K. Charles, Jr., Thermomechanical Behavior of Soldered Interconnects for Surface Mountings: A Comparison of Theory and Experiment, in *Proceedings of the Electronic Components Conference*, Institute of Electrical and Electronics Engineers, May 1985, p 60
32. W. Engelmaier, Thermo-Mechanical Effects, in *Packaging*, Vol 1, *Electronic Materials Handbook*, ASM INTERNATIONAL, 1989, p 740
33. W. Borland, Thick-Film Hybrids, in *Packaging*, Vol 1, *Electronic Materials Handbook*, ASM INTERNATIONAL, 1989, p 332
34. H. Nakahara, Complex Multilayer Boards Vie for Space in Japanese Computers, *Electron. Packag. Prod.*, Vol 27 (No. 2), Feb 1987, p 70
35. R.R. Tummala and E.J. Rymaszewski, Ed., *Microelectronic Packaging Handbook*, Van Nostrand Reinhold, 1989
36. H.W. Markstein, Laminates Support Technology Advances While Offering Alternate Choices, *Electron. Packag. Prod.*, Vol 24 (No. 6), June 1984, p 83
37. M. Schlack, Guide to High Performance Engineering Plastics, *Plast. World*, Vol 45 (No. 4), April 1987, p 30
38. M.K. Antoon, B.E. Zehner, and J.L. Koenig, Spectroscopic Determination of the In-Situ Composition of Epoxy Matrices in Glass-Fiber-Reinforced Components, *Polym. Compos.*, Vol 1 (No. 1), Jan 1980, p 24
39. C.P. Wong, Integrated Circuit Device Encapsulants, Chapter 3, in *Polymers for Electronic Applications*, J.H. Lai, Ed., CRC Press, 1989, p 69
40. S. Yamanaka, T. Ibara, T. Maeda, T. Takikawa, and H. Yoshino, Applications of Ceramic Thin Film Technology to Hybrid Microelectronics, in *Proceedings of the International Microelectronics Symposium*, International Society for Hybrid Microelectronics, Oct 1989, p 439
41. R.W. Berry, P.M. Hall, and M.T. Harris, *Thin Film Technology*, D. Van Nostrand, 1986, p 271
42. L.I. Maissel and R. Glang, Chapters 3

and 4, in *Handbook of Thin Films*, McGraw-Hill, 1970

43. A.C. Adams, Dielectric and Polysilicon Film Deposition, in *VLSI Technology*, S.M. Sze, Ed., McGraw-Hill, 1983, p 93

44. W. Kern and V.S. Ban, Chemical Vapor Deposition of Inorganic Thin Films, in *Thin Film Processes*, J.L. Vossen and W. Kern, Ed., Academic Press, 1978, p 257

45. W. Kern and G.L. Schnable, Low Pressure Chemical Vapor Deposition for Very Large-Scale Integration Processing—A Review, *IEEE Trans. Electron. Dev.*, Vol ED-26, 1979, p 647

46. J.M. Mayer and S.S. Lau, *Electronic Materials Sciences: For Integrated Circuits in Silicon and GaAs*, Macmillan, 1990, p 283

47. A.C. Adams and C.D. Capio, The Deposition of Silicon Dioxide Films at Reduced Pressure, *J. Electrochem. Soc.*, Vol 126, 1979, p 1042

48. W.A. Pliskin, Comparison of Properties of Dielectric Films Deposited by Various Methods, *J. Vac. Sci. Technol.*, Vol 14, 1977, p 1064

49. J.W. Peters, U.S. Patent 4,371,587, 1981

50. T.C. Hall and J.W. Peters, Photonitride Passivating Coating Enhances IC Reliability and Simplifies Fabrication, *Insul. Circuits*, Vol 27 (No. 1), 1981, p 22

51. D.J. Ehrlich, R.M. Osgood, Jr., and T.F. Deutsch, Laser Microphotochemistry for Use in Solid State Electronics, *IEEE J. Quantum Electron.*, Vol 16 (No. 11), 1980, p 1233

52. P.K. Boyer, G.A. Roche, and G.J. Collins, Laser Photodeposition of Silicon Oxides and Silicon Nitrides, *Electrochem. Soc. Extended Abstr.*, Vol 82-1, 1982, p 102

53. D.B. Tuckerman, D.J. Ashkenas, E. Schmidt, and C. Smith, "Die Attach and Interconnection Technology for Hybrid WSI," *1986 Laser Pantography Status Report UCAR-10195*, Lawrence Livermore Laboratories, 1986

54. H.K. Charles, Jr., VHSIC Packaging for Performance, *APL Tech. Rev.*, Vol 1 (No. 2), 1989, p 47

55. H. Okake, *Photochemistry of Small Molecules*, John Wiley & Sons, 1978, p 219

56. C. Hepburn, *Polyurethane Elastomers*, Applied Science Publishers, 1982

57. G.V. Clatterbaugh and H.K. Charles, Jr., The Application of Photosensitive Polyimide Dielectrics in Thin Film Multilayer Hybrid Circuit Structures, in *Proceedings of the 1988 International Microelectronics Symposium*, International Society for Hybrid Microelectronics, Oct 1988, p 320

58. C.R. McMillin, Current Topics in Biomedical Rubbers and Elastomers, *IEEE Eng. Med. Biol. Mag.*, Vol 8 (No. 2), June 1989, p 30

59. C.J. Kreider, Silicone-Base Coatings, in *Packaging*, Vol 1, *Electronic Materials Handbook*, ASM INTERNATIONAL, 1989, p 773

60. W.F. Gorham, U.S. Patent 3,342,754, 1967

61. W.F. Gorham and W.D. Neigisch, Parylene, in *Encyclopedia of Polymer Science and Technology*, Vol 15, John Wiley & Sons, 1971, p 98

62. A. Christou, J.R. Griffith, and B.R. Wilkins, Reliability of Hybrid Encapsulation Based on Fluorinated Polymeric Materials, *IEEE Trans. Electron. Dev.*, Vol ED-26 (No. 1), 1979, p 77

63. H.B. Tompkins, Degradation of Thin Films of a Fluoropolymer, *J. Appl. Polym. Sci.*, Vol 17, 1973, p 3795

64. A.A. Shores, Thermoplastic Films for Adhesive Bonding Hybrid Microcircuit Substrates, in *Proceedings of the 39th Electronic Components Conference*, Institute of Electrical and Electronics Engineers, 1989, p 891

65. T.Y. Su, Encapsulation of Ceramic Capacitors With Thermoplastic Materials, in *Proceedings of the 32nd Electronic Components Conference*, Institute of Electrical and Electronics Engineers, 1982, p 174

66. H.L. Frisch and E. Wasserman, Chemical Topology, *J. Am. Chem. Soc.*, Vol 83, 1961, p 3789

67. L.H. Sperling, *Interpenetrating Polymer Networks and Related Materials*, Plenum Publishing, 1981

Applications for Sealants

Kent Adams, I.A.M., Inc.

THE FIRST APPLICATIONS for sealants, which date back to early civilization, were related to construction and transportation. Boats and shelters were sealed to help keep water and wind away, and nature supplied reliable materials in abundance for these purposes. Although modern commercial applications have proliferated into a wide range of assembly applications, of which construction and transportation are primary, the basic function of sealants is still to keep out or seal in air, fluids, and contaminants. In both historical and modern settings, sealants also add strength to the assemblies of builders, help reduce noise, and provide a more visually pleasing result.

Sealant technology, however, has evolved appreciably from early materials. New polymers meet ever-increasing commercial demands. The aerospace, automotive, electrical/electronics, and construction markets represent excellent examples of current sealant applications.

This article provides an overview of sealant technology and application designs for aerospace application; construction, automotive, and electrical/electronics applications are reviewed in more depth. The diversity of sealant applications in these markets gives a sense of direction for today's sealant technology. The scope of materials and application methods has increased, enabling designers and assembly engineers to meet new performance requirements.

Sealant/coating companies are now included among design teams. Sealants do more than merely compensate for assembly parameters or provide safe seals, and they are an integral part of design and performance, as exemplified by the aerospace industry.

Aerospace Market

Developments in metals and composite materials to meet advanced design requirements have resulted in corresponding developments in sealant technology. Conversely, numerous new sealant applications have evolved. The strong creative relationship between designers, materials suppliers, and sealant suppliers has been a key element in the successful engineering of sealants for the aerospace market.

Sealant applications for this market include not only aerospace vehicles, but also the specialized ground support equipment. The diversity of sealant technology and the creativity in its use can be seen in fuel tank applications. Rubber-base sealants such as Buna-N have been used from the start. The changing requirements for fuel storage have led to specialized products, such as sealants for groove injection channels. Curing and self-repairing fuel tanks use sealants for rivet heads, faying surfaces, and fillet joints. Sealants are also used on inspection plates. The types of sealants that have met the demanding requirements for durability, chemical resistance, and environmental integrity are fluorosilicones and polysulfides (Fig. 1).

Reliability and environmental integrity are critical in pressurized-cabin and compartment applications. Specially designed mixing and dispensing equipment, such as that provided by Pyles Division of Sealed Power Corporation and Semco, Inc., has

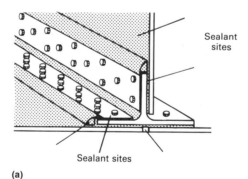

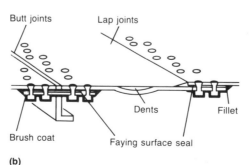

Fig. 1 Sealant sites. (a) Of integral fuel tank. (b) In structural applications

made the sealant easy to apply. Applications include fillets, faying surfaces, and rivets, and the sealants include room-temperature-vulcanizing (RTV) silicones, polyurethanes, polysulfides, fluorosilicones, solvent-base acrylics, and two-component rubber compounds. These materials provide environmental seals, corrosion protection, and structural integrity.

Selecting a sealant for a cabin, compartment, access door, fire wall, fuselage, or fuel tank application requires a careful, multistep analysis of the functional requirements, as well as physical descriptions of the bonded surfaces, including projected time required to apply the sealant. Each application usually requires two or three different sealants to complete the assembly properly. The first step is to establish environmental requirements such as temperature, pressure, and fluid exposure. Flammability and fire resistance properties are important in cabin applications. Second, structural performance criteria, including flexibility, adhesion, and shear strength, must be established. The third step is to determine sealant thickness and bonding areas. The fourth step is to establish working time, tack-free time, pot life, and cure speeds necessary to complete the assembly. Finally, application procedures and the sequence of the application of sealants must be determined. The exact equipment used to apply sealants must be anticipated.

In a pressurized cabin, cargo compartment, or fuselage application, a range of sealants is used on a variety of different surfaces. Faying surfaces (overlapping metal joints) are coated with sealant prior to assembly. Grooves are injected with a curing or noncuring sealant, depending on requirements. Before a fillet of sealant is applied to seams, the dimension of the fillet, which varies by application, must be carefully specified (Fig. 2). Rivet heads are sealed in most applications.

Cabin and galley floors are sealed around the perimeter and at joints. Fire walls are sealed with specialty flame-resistant products; expansion joints are sealed. The hole and conduit are sealed with a potting compound, whereas access doors are sealed using a gasketing-type product. A conductive sealant is sometimes required around

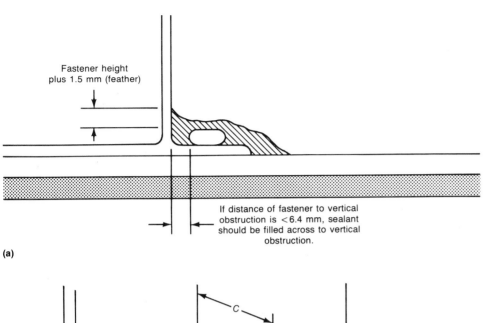

(a)

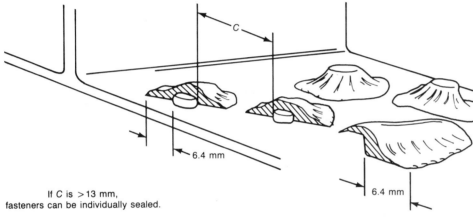

(b)

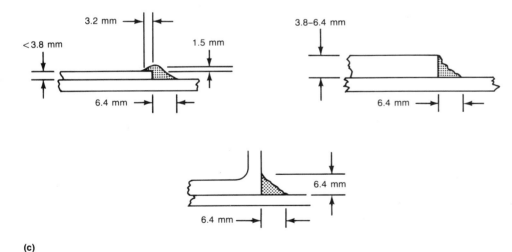

(c)

Fig. 2 Dimension specifications. (a) In the sealing of near-vertical obstruction. (b) In the sealing of individual fasteners. (c) Other configurations

veloped, as have specialty windshield and canopy sealants that meet the extreme demands of sealing and bonding. A special sealant has been developed for jet turbine blades.

Sealants with dielectric, chemical-resistance, and thermal-resistance properties are used on hydraulic lines and electrical connectors. Sealants are used to prevent bimetallic corrosion, which can occur in assemblies. Engines use specialty chemical gaskets to ensure reliability.

It can be seen from the foregoing that sealant technology has evolved to meet aerospace industry demands for lighter, more reliable, and more economical assemblies. As noted earlier, applications continue to expand as new designs evolve.

Service equipment for aerospace vehicles is a secondary market that has similar sealant product demands, but specialized applications. The service and repair companies require specialized application equipment for the sake of economy. New proportion-mixing cartridges, application guns, and unit dosage packaging have been developed. The same product quality and reliability can be obtained with this equipment as is possible at the original-equipment-manufacturing level, without the need for larger-volume equipment.

Construction Market

Sealants for the commercial construction market are designed to act as a barrier to water, air, atmospheric pollution, vibration, insects, dirt, and noise, while compensating for movement that occurs in the structure being sealed. This movement can be caused by many factors, including changes in temperature and moisture content, permanent loading or load transfer, and chemical changes such as sulfate attack and wind loads.

Concrete structures use three principal types of joints: expansion, contraction (control), and construction. Expansion joints relieve the compressive stresses of abutting structural units that may develop when concrete expands. Contraction joints are designed planes of weakness for regulating cracking that might occur because of an unavoidable decrease in concrete volume. Construction joints are used between adjacent pours of concrete. They also result from positioning of precast units. Joints in concrete structures may be horizontal, vertical, or inclined, depending on the nature of the structure. Some typical structures in which the aforementioned joints are used include precast and poured-in-place (including tilt-up) concrete, retaining walls, bridge abutments, hydraulic structures, tunnels, highways, culverts, bridges, and stadiums.

Other typical uses for sealants in construction include perimeter caulking of exterior and interior panels and frames; joints

doors to ensure electrical continuity, and corrosion-resistant sealants are used in galleys and heads. Aerodynamic sealants are used for smoothing, and thermally conductive sealants are used for heat transfer.

Aerospace requirements have spawned the development of special low-volatility sealants for cases in which vacuum exposure is encountered. Special ablative and thermal barrier sealants have also been de-

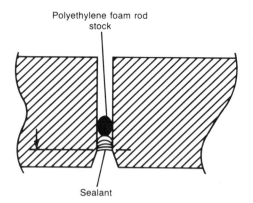

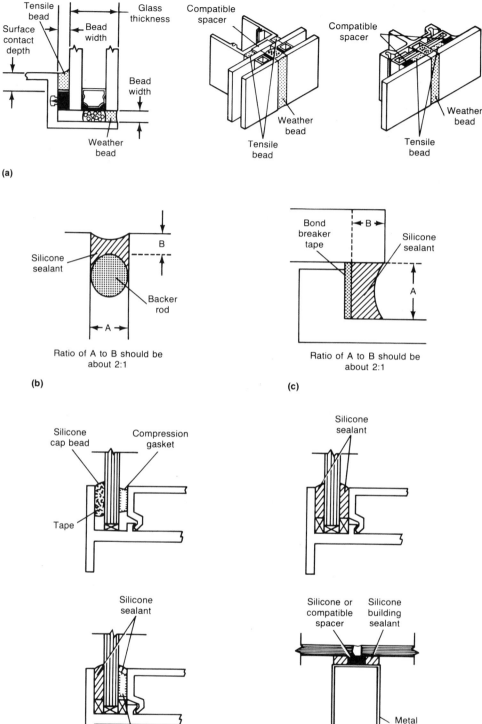

Fig. 3 Typical vertical joint

of the undersides of precast planks or beams; perimeter caulking of sanitary fixtures, flashing, skylights, reglets, pre-engineered metal buildings, masonry structures, copings, and electrical and mechanical openings; and exterior insulation systems.

Sealants find widespread use in construction glazing applications where they are used to install glass into frames or openings to ensure weather tightness. Face glazing is the method used to install glass into openings. In this method, a rabbeted sash is used, and a triangular bead of sealant is applied to hold the glass in place while acting as a watershed. In channel glazing, usually one fixed stop and one removable stop are used, with the rabbet forming the confines for the sealant.

With the advent of silicone sealants, innovative glazing systems known as structural glazing have developed. Structural glazing is a system of bonding glass to the structural framing of a building using silicone sealants. In this system, wind loads are transferred from the glass to the structural supports by the sealant.

The three most-used structural glazing systems are the total vision system and the two- and four-sided structural curtain wall systems. In the total vision system, large, monolithic glass is used in conjunction with glass mullions. Silicone sealants bond the glass components to one another and transfer the wind loads to the framing system. Two-sided structural glazing uses a glass-supported head and sill in a conventional glazing rabbet, using stops, gaskets, and so on; the two available sides are then bonded to a structural mullion with a silicone sealant. A four-sided structural glazing system uses no mechanical fasteners to hold the glass; instead, all four sides of the glass are bonded to the structural framing system with silicone sealant.

The renovation of existing structures has resulted in new requirements for sealant systems. Architects, contractors, materials suppliers, and sealant companies must preserve historical uniqueness while meeting the demands of modern office buildings.

Fig. 4 (a) Curtain wall construction. (b) Recommended joint design. (c) Bond breaker tape. (d) Typical glazing details

Polyurethane sealants can seal terra-cotta, brick, brownstone, marble, and cast iron without staining. Urethane sealants, available in a variety of architectural colors, allow close color matches to existing materials. New windows are installed using advanced glazing techniques and are sealed in place with RTV silicone sealants.

An example of the challenges posed for sealants in renovation is the Litz Brothers building in Philadelphia. The project involved 12 interconnected buildings built between 1859 and 1916. The goal was to restore the buildings to the appearance of the 1916 to 1920 period. Six different materials were used on the exterior facade. A urethane sealant system available in colors to match the building materials was selected for the exterior perimeter caulking. The same low-modulus sealant was selected for expansion joints.

Aluminum replacement windows were custom manufactured to replicate the original windows. These were sealed in place using a one-part urethane sealant selected in colors to match exterior materials. A close working relationship between the sealant manufacturer, architect, construction manager, window fabricator, and installation contractor made this project possible (Ref 1).

For the installation of new granite facades on existing structures, a urethane sealant is preferred. A polyethylene foam backing is inserted into the joint to control sealant depth. Sealant is then applied using an automatic or manual sealant gun. This creates a uniform bead of the proper cross section to give maximum performance. Two-component polyurethanes are popular in this application because of the faster cure times, which help reduce renovation costs (Fig. 3).

Glass curtain wall construction is a cost-effective method to help create new facades. One- and two-component silicone sealants have been developed to transfer loads precisely and deflect wind shear to the steel members. Numerous combinations of glazing sealants, gaskets, and insulation sealants have been developed to meet the specific needs of a design. The structural requirements should be calculated using the actual contact dimension of the sealant, ensuring proper strength. Deflection stress due to wind loading must be considered in sealant selection. Manufacturers have carefully developed formulas to assist in structural design (Fig. 4).

The trapezoidal loading theory is used to calculate stress on sealants used in glass curtain wall construction. The theory states that panels under wind loading will deflect stress in a tributary area fashion rather than remain flat. The contact dimension of the sealant is calculated by multiplying one-half the length of the shortest panel by the projected wind loads, divided by sealant design strength. (Industry standard design strength is 140 kPa, or 20 psi.) This theory holds true for two- or four-sided systems (Ref 2).

Curtain wall construction places strenuous demands on sealants used to provide an exterior weather seal. The sealant must withstand repeated expansions and compressions due to thermal effects. Because sealants must have both high strength and

Fig. 5 Two-sided butt-joint glazed skylight. Courtesy of General Electric Company, Silicone Products Division

high elastic modulus, high-strength silicones are preferred.

The increased use of atrium designs and skylights has created demands for high-performance sealants. The design considerations for skylights are similar to those for curtain walls. The demands vary significantly, and the design concepts used to solve the problems have been innovative.

One example of unique requirements is a large, multiuse construction in the southeastern United States, where visitors can walk from either one of two six-level parking decks to a hotel or an office building, or through the mall area, without once contacting environmental elements. Over a galleria is an 11 000 m² (117 000 ft²) skylight more than 430 m (1400 ft) long, with an apex 45 m (150 ft) from the floor. One of the largest skylights in the country, it was created using a two-sided butt-joint/glazing process (Fig. 5). In conventional glazing systems, pollutants gather on the glass and are washed down the glass with the rain, becoming trapped on the exterior horizontal stops. This causes the glass to become streaked and dirty. A butt-glazed system has no external traps, allowing the rain to run down freely and resulting in a cleaner glass.

The two-sided butt-glazed skylight not only maximizes aesthetic benefits, but also minimizes maintenance time and costs and the problems associated with thermal expansion. The skylight contractor on this project custom extruded aluminum supports for the skylight vaults and designed a complex system of slots in the central support beams to decrease movement due to

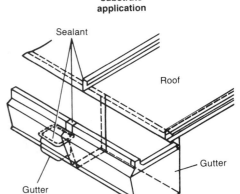

Typical metal substrate application

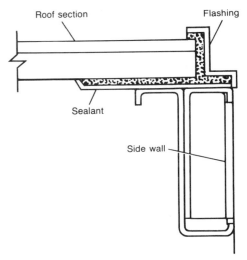

Typical metal building application

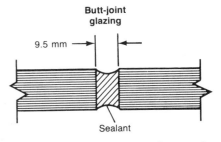

Butt-joint glazing

Fig. 6 Gutter, roof, and butt-joint glazing applications for sealants

temperature changes. The skylight was designed with laminated safety glass units. During its construction, a new technique was used, that is, the use of tension cables to reduce the deflection of the steel arch beams caused by the weight of the glass.

Along with skylights, roofs and gutters are a common service problem. The majority (90%) of roofing leaks occur at flashings, parapet walls, copings, and gutters. Anticipating these problems and applying butyl, silicone, urethane, or acrylic caulks can significantly reduce maintenance costs (Fig. 6).

The standing seams of metal roofs can be sealed using a unique sealant concept based on foamed hot-melt adhesives. The sealant is applied using a stationary heat gun at the final stage of the roll former. The final seam is made, and the hot-melt sealant expands to fill the seam completely. The benefits over conventional seams are increased reliability, improved insulation, and improved seals. The foamed sealant, which moves with the expansion and contraction of the roof, provides reliable performance.

Roof-mounted air conditioning units require sealants that provide environmental barriers and give insulation properties. Urethane foams are used to fill the gaps around duct work, pipes, and conduits. A secondary sealant is applied to exposed surfaces. Mounting bolts and footers are sealed using a urethane or RTV silicone sealant.

Concrete structures require proper sealing to ensure maximum service life, particularly parking structures and horizontal decks subjected to snow removal equipment and chemicals. The expansion joints are filled with a low elastic modulus sealant, a nonskid expansion joint cover is bolted in place, and the edges of the seal are filleted with a low-modulus sealant. The stress cracks are routed out and then filled with sealant. The drains and the perimeter of the structure are sealed. Pipes, conduits, and drains through concrete structures must be sealed to ensure proper service life. Urethanes and silicones are popular sealant choices for sealing concrete to a dissimilar material.

Plumbing sealants have changed with the increased use of plastic pipe. Specific liquid sealants have been developed for the expanded range of plastic pipe. Threaded plastic connectors employ special sealants that aid in assembly and help prevent leaks. The proper selection of plastic pipe sealant is critical because of the potential of stress cracking. Teflon tapes used on iron pipe are being replaced with liquid sealants that have Teflon additives. Chemical pipe sealants effectively prevent corrosion, reduce loosening tendencies, and help prevent leaks. Pipe sealants for metals have been designed to work on steel, copper, and cast iron. The use of chemical pipe sealants in plumbing applications is well specified by building codes.

Epoxy systems and polyester cloth overwraps are used in plumbing repairs. These systems have extremely good durability and are easy to apply. Solid, two-component epoxy putty is well suited for repair applications.

Sinks, tubs, and floor drains are sealed to prevent water damage. Acrylic caulks with silicone are economical and easy to use. RTV silicone caulks are used where more flexibility is required. The perimeter of

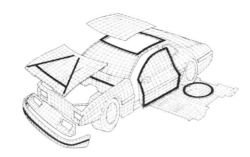

Fig. 7 Automotive body assembly applications for sealants

sinks and basin undersides are sealed with acrylic sealants. Joints between tile walls and flooring are sealed with silicone or acrylic sealants or with high-adhesion polyvinyl acetate sealants. Toilets have an integral seal, but secondary sealing to walls and floors are completed with caulks. Bathroom and kitchen sealants contain mildew-resistant additives. Many applications require white materials, but custom colors are gaining popularity.

Interior heating and air conditioning systems use sealants. Duct work is sealed in critical areas to reduce noise and increase efficiency. Vents are sealed to prevent the leakage of fumes. Sealants are used on the perimeters of furnaces to prevent vibration and noise.

The use of metal framing for interior walls has created new needs for sealants. Wiring conduits are sealed into the metal frames, where the wires are tacked into place and sealed to prevent damage. Communications and security system wiring are also sealed in place.

Each subcontractor at a job site uses special sealants designed for specific applications. As discussed in this section, the application areas are: concrete, foundation, exterior structure, windows, glazing, electrical, plumbing, roofing, communication systems, security systems, and heating, ventilation, and air conditioning (HVAC).

Automotive Market

Sealant technology and application methods have responded quickly to the changing demands of the automotive industry. Coordinated efforts between parts suppliers, sealant manufacturers, application/assembly equipment companies, and automotive designers have spawned innovative solutions to complex design problems. Materials development teams are working together on new engine designs, body designs, and electronic control and sensing systems.

Sealant applications fall into three very broad categories of engine, body, and specialty-component sealants. These three classes include more than a dozen different

Fig. 8 Windshield sealant application. Courtesy of Essex, Specialty Product Corporation

sealant families. The body assembly sealant may well have undergone the most dynamic changes in the 1980s and will continue to change and evolve throughout the 1990s.

The combined changes of new lightweight corrosion-resistance metals; new low-temperature-bake, clear, base-coat systems; increased use of plastics; and low-drag-coefficient designs have significantly altered sealant requirements in automotive body assembly. Basic applications, such as seam sealants, windshield sealants, and gas tank sealants, have expanded into a broader range of diverse sealant types and applications (Fig. 7).

Windshield sealants have changed significantly from those used with rubber channels to polysulfide and polyurethane direct-glaze systems. The new sealant/adhesive systems have been dictated by flush-mount glass installation, aerodynamic styling, and rollover safety standards. These sealant application procedures have become fully automated processes, using high-viscosity urethanes (Fig. 8).

Sealant technology has contributed to passenger safety. For example, in the case of windshield design, it has enabled the manufacture of windows that blow out and away from the passenger compartment.

The flush-mount window demanded for windshields has also been adopted for other fixed windows. The sealant not only bonds windows in place, but also provides environmental seals and structural strength. Epoxy and urethane sealant/adhesive systems have enabled new, lighter weight, easier-to-assemble designs.

The taillight systems have been integrated into aerodynamical body design. The expansive plastic lens covers that have been popular since 1986 are assembled without mechanical fasteners. The gasket material bonds the assembly into place and provides an environmental seal. These sealed units are faster to assemble without resorting to exterior mechanical fasteners, and help prevent corrosion.

"T" tops and sunroofs have adopted windshield and fixed-window sealant techniques for solving problems of leaks, wind noise, and flutter. Urethane, RTV silicone, and polysulfide sealants have primarily been applied manually because of low volumes. Some of the more custom-designed units are installed by specialty manufacturers.

The manufacture of lighter, more corrosion-resistant body and frame assemblies has been facilitated by new sealants. Many of these sealants are partially cured in the assembly process and fully cured in the paint bake cycle. Urethanes, hot melts, and vinyl plastisol products have been used for bonding and sealing. Sealants used in conjunction with adhesives create stronger, more reliable assemblies that distribute stress loads more evenly and assist in corrosion resistance.

Galvanized metal use has increased because of demands for rust-resistant vehicles. The spot welding of galvanized steels can reduce the rust resistance due to coating damage. The use of adhesive bonding with sealants has addressed this problem. Galvanized metals require more preparation prior to bonding to ensure long-term adhesion. The same types of preparation that are required for primer adhesion are used for top coat adhesion.

New solid hot-melt sealants, popular with European automotive manufacturers, are available in preformed shapes that are easily and quickly applied on fasteners and in seams, without the need for special equipment. The sealant melt flows out and cures during the paint baking process. Typical applications are in trunks and passenger compartments. These solid products are appropriate for lower-volume application or for cases in which the automated application of liquid products is not economical.

The bonding of rigid structural plastic body components to metal frames provides four functions in certain 1990 vans. Two-component epoxies and urethane products bond the panel in place, provide structural integrity, distribute forces evenly, and provide the environmental seal for the passenger compartment. New polymer technology has been combined with precise application methods to yield a strong, reliable sealant/adhesive assembly. This trend toward lighter, more efficient body designs is expected to increase during the 1990s. Thus, sealants used in automotive body assemblies are no longer represented solely by pitch-and-rubber coatings used to spot leaks and compensate for poor door fits. Sealants now assist in the assembly process and lend structural integrity to the automotive vehicle. Teams of designers, plastic suppliers, metal suppliers, paint manufacturers, and sealant and adhesive companies work together to provide reliable assemblies.

Engine sealants continue to meet the new challenges of tighter tolerances, precise torque tension values, and the sealing of new materials. Engine applications range from microporosity sealants to thread sealants to chemical gaskets. Fluids that leak from engines are not only quality issues, they are also classified as emissions by governmental standards.

Sealing microporosity has become increasingly important with the shift to new, lightweight metals. The applications include transmission cases, intake manifolds, transaxles, power steering housings, pump covers, and even engine blocks. The vacuum impregnation process is the most popular method of application. Parts are submerged in a tank of resin, and a vacuum is pulled to remove air from the metal pores. The vacuum is removed, and the liquid resin fills the voids in the metals. The resin then cures. Anaerobic polymers are used in this process, though alternative methods are available. Microporosity sealants have increased parts reliability and have helped reduce service costs for automotive manufacturers. In aluminum engine blocks, for example, the resin seals porosity in cooling systems and oil passages to ensure that fluids will not leak. It is projected that more than 20 million automotive castings made of aluminum and other lightweight metals will be sealed for microporosity in 1990 (Ref 3).

Chemical gasketing has facilitated the manufacture of more reliable, low-service-cost engines. RTV silicones and anaerobics are the dominant products used for chemical gasketing. The smaller, lighter, and higher-velocity (of rotation) engines require new and improved sealants that resist higher temperatures and fluids, adhere better, and are easier to apply.

Automatic application equipment has encouraged the growth of gasket sealant use. Tracing equipment designed for three-dimensional application can easily apply precise beads to complex parts. The equipment can be changed, with some ease, from one application to the next because sealant manufacturers have worked closely with equipment manufacturers to develop versatile systems. Sealant formulations have been adjusted to compensate for pump limitations. Sealant thixotropy is an important factor in obtaining a proper sealant bead. Because of the increasing range of gasket-type sealants, equipment is tuned to a particular sealant and manufacture (Fig. 9).

Pre-1980 engines used a limited selection of sealants. Post-1980 four-, six-, and eight-cylinder engines use a broader range of products. Turbocharged engines have placed additional demands on high-temperature-range, higher-flexibility sealants. The

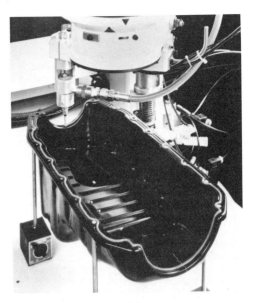

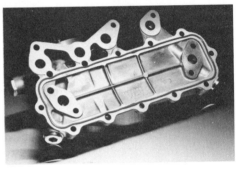

Fig. 9 Gasket sealant application. Courtesy of Three Bond Corporation

challenges continue with the new engine designs planned for the 1990s.

The fouling of oxygen sensors is an example of the problems faced by sealant manufacturers. The oxygen sensor adjusts the timing to ensure proper engine operation. This sensor, first used in California models, was experiencing fouling, partially because of volatiles from the sealants. A new generation of low-volatility RTV silicone gasket sealants was developed.

Demands for more precise fastening and tighter fits in engines of the late 1980s have led to new sealants. Because these engines run hotter and require the sealant to withstand more deflection, a new higher-temperature and higher-flexibility grade of sealants was developed. These sealants are low-volatility and low-corrosion types. Adhesion levels were increased to ensure "leak-free" performance.

Turbocharged engines pushed sealant requirements even further. New 370 °C (700 °F) temperature-resistant sealants were developed. They combine the low-volatility, low-corrosion, high-flexibility, and improved fluid-resistance properties of other chemical gaskets with new, higher heat ranges.

Fig. 10 Electronic surface-mount component to which RTV silicone gel can be applied for thermal-stress and environmental-damage protection. Courtesy of General Electric Company, Silicone Products Division

Typical applications for chemical gaskets are water pumps, timing chain covers, engine pans, valve covers, transaxles, transmission cases, dipstick tubes, thermostats, freeze plugs, exhaust gas recirculation valves, camshaft covers, split cases, and bearing caps. Many of these applications would not be possible with cut gaskets.

One automotive manufacturer reported a 0.08 km/L (0.2 mile/gal) increase after a precut gasket was replaced by a chemical gasket. On certain engines, the gasket on the "G" rotor oil pump was replaced by an anaerobic sealant. The tighter fix on the pump yielded a more efficient operation and resulted in increased mileage (Ref 4).

Other chemical sealants that aid in engine assembly, help prevent corrosion, and prevent the leakage of fluids and fumes are thread-fitting sealants. These sealants are used on air conditioning pumps and vacuum line fittings, as well as on the fittings used to mount sensors. On water pump assemblies, through bolts are coated with a thread sealant to help prevent leakage.

Automotive electrical systems have undergone the most radical changes of all automotive components. Computer-controlled engine systems and self-diagnostic electric circuits have totally changed automotive wiring. The number of car fuses alone has expanded from 14 to more than 70

in some newer models. These multiple circuits require connectors, wires, and sensors, which are held in place or protected by sealants.

New circuit devices are assembled using sensitive surface-mount components and other environmentally sensitive components. A dielectric RTV silicone gel can be used to protect these components. A low-viscosity, two-component sealant, it flows around the components, providing soft cushioning. Upon cure, it seals and protects against thermal and mechanical shock, external vibrations, moisture, and atmospheric contamination. This unique product has solved some of the problems encountered in the adaptation of microelectronics to the automotive market (Fig. 10) (Ref 5).

The precise placement and environmental protection of sensors is critical to the operation of an electrical system. The wires that connect sensors and circuits must be tacked in place, and the connectors sealed. New ultraviolet (UV)-curable RTV silicones have proved to be extremely effective because they can be applied as a liquid with precision applicators, and they cure in place upon exposure to UV light. They are noncorrosive, flexible, and conform easily to surfaces. In addition to bonding, these sealants provide thermal and mechanical shock protection for delicate components. Con-

ventional RTV silicones, moisture-cure urethanes, and epoxies are also used in these tacking and sealing applications.

Heating and air conditioning ducts also must be bonded and sealed in place. An adhesive/sealant not only prevents leakage, but also prevents noise caused by vibration. Double-face tapes, urethanes, butyls, water-base acrylics, and RTV silicones have been specified for this application.

A variety of seam sealants are used in gas tank applications. The new vans, light trucks, and specialty vehicles have created new demands for gas tank design. Sealant reliability is critical in the closed, no-emission tank designs. The traditional Buna-N sealant systems, as well as urethanes, polysulfides, and fluorosilicones are used. Sealant locations include tank level sensors, seams, fill pipes, and inspection plates.

Light truck and utility vehicle construction has used aircraft design concepts. Fewer rivets, welds, and mechanical fasteners are used in assembly. Bonding and seaming sealants have contributed in providing lighter-weight and lower-cost constructions. Sealants are used in channels to bond and seal inner and outer skins. Foamed sealants are used in standing seams. The lower-volume applications use manually operated air-powered caulking guns.

The modification of vans and light trucks into recreational vehicles is accomplished with the extensive use of sealants. The chassis is prepared using urethane sealants. The flooring, seats, and supports are bonded and sealed in place. The new roofs are sealed. Skylights, rails, and ladders are installed and sealed using urethanes and RTV silicones. New windows, splash guards, and mirrors are mounted and sealed (Fig. 11).

The changes in automotive design specifications and assembly continue to draw sealant manufacturers into a closer relationship with paint, metal, and plastic suppliers. Automated application procedures will continue to be a critical part of sealant design. The robotic applications of sealants are now common and will become a primary application method. Advanced electronics will be protected by specialty sealants.

Electrical and Electronics Market

Sealants have enjoyed the largest growth percentage in the electrical and electronics market. Sealants perform the traditional roles of keeping out contamination, moisture, and air and sealing in fluids or inert atmosphere. Sealants also provide thermal and mechanical shock protection and enhance dielectrical properties.

Electrical applications for sealants, which tend to be less demanding than electronic component protection applications, have several categories: protecting terminals, contacts, and relays; sealing fuse and junction boxes; and tacking and splicing wires and protective insulators. The national codes and testing laboratories have established well-defined standards for sealants. These standards should be reviewed before a sealant is selected.

Although the protection of terminals and wiring connections covers a broad range of applications, the basic requirements are to seal out moisture, prevent corrosion and loss of contacts, eliminate arcing, and help prevent short circuits. A popular sealant is a PVC liquid coating, which is generally solvent based and cures by evaporation. This product is available in a variety of colors to assist in color-coding connections.

A range of electrical motor connections are coated with sealants after assembly. Examples of these mechanically fastened connections are the starter motor for automobiles and battery connections for stationary power sources. The sealant is used to protect against corrosion, help prevent loosening, and protect contacts from accidential shorts.

Potting compounds are used in more demanding electrical contact assemblies. The connection assembly is filled with a two-component epoxy, or a urethane, silicone, or polyester material, totally encapsulating the contacts. Pumps that require water or fluid immersion also have contacts that are potted. Aviation applications that experience thermal shock or hostile environments use potting compounds, as well. The protection provided by potting compounds is far superior to that provided by terminal coatings.

Dielectric sealants are used to protect relays and contacts. The most common sealants are greases, which provide both protection and lubrication. Silicone sealants are occasionally used because of their excellent dielectrical and flexibility properties.

Junction boxes, fuse boxes, and breaker switches are also protected by sealants. Gasket-type sealants are ideal for these applications because they conform easily to irregular surfaces and can be cured to one side of the panel, allowing easy access. RTV silicone, acrylic, and urethane sealants are preferred. The demands on the sealant are generally moderate: sealing out air and moisture.

Component cover applications are more demanding. The covers protect connections and fuses on electrical devices that may be submerged in water or other fluids. These sealants can be exposed to high heat and mechanical stress. In extremely demanding applications, silicone and fluorosilicone polymers are used.

A range of flowable sealants is used to coat the inside of connection boxes, where total sealing is essential. These two-component sealants flow and coat the inside of

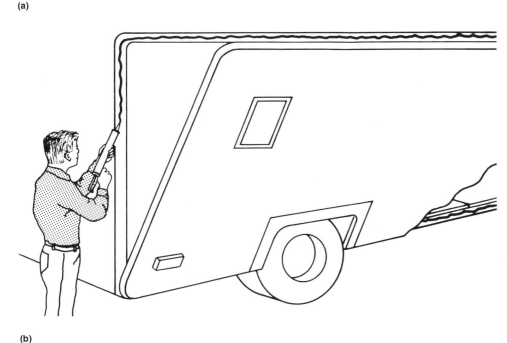

(a)

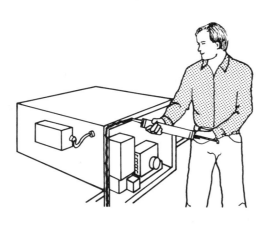

(b)

(c)

Fig. 11 Sealant applications in recreational vehicle conversion. (a) In chassis, sealing cockpit floors, cockpit steel, fire wall. (b) Shelling (bonding and sealing) sidewalls to floors, roof to sidewalls, rear to sidewalls. (c) Sealing reefer, hot-water-heater, and generator compartments

Fig. 12 Sealing of small components. Courtesy of Three Bond Corporation

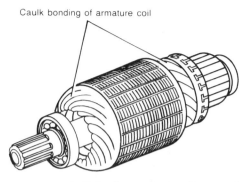

Caulk bonding of armature coil

Fig. 13 Caulk bonding application. Courtesy of Three Bond Corporation

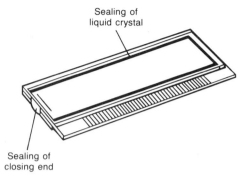

Sealing of liquid crystal

Sealing of closing end

Fig. 14 UV curing resin sealing application. Courtesy of Three Bond Corporation

compartments to prevent fluids or contaminants from entering.

Liquid rubber sealants are used to protect wire splices. These sealants coat the wires and are allowed to dry. Undersea cables are spliced using epoxy or polyester resins. These splices are totally encapsulated by the resin, forming a hard water-and-solvent-resistant shell. Low-thermal-conductivity sealants are used where intermittent exposure to heat is encountered. Sealants are usually thick plates, cured with large cross sections. Additives are included in the formulations to improve thermal-barrier properties.

High-voltage power transmission systems are subject to arcing and flashover. Electrical insulators are protected with dielectric sealants, generally in paste form, to improve performance. The switches, relays, and connectors are coated with a thixotropic sealant to allow flow into the recesses of the equipment.

The high-voltage circuits in both the automotive and aerospace industries present an added problem, that of dielectric breakdown. The silicones have proved to be very effective, especially where flexibility is important. Epoxies have dielectric resistance and are formulated for rigid applications.

In certain applications, transformers, coils, and relays require a thermally conductive encapsulating polymer. These electronic compounds are filled with conductive materials to help the sealant transmit heat properly. The formulations used are two-component epoxy urethanes and hybrid polymers.

Electronic components have sealant requirements similar to those of electrical applications. The application of the sealants tends to be automated because there are numerous components of the same design. A few examples are described below.

Component terminals are sealed with acrylics, urethanes, or silicones. The UV-curable silicones and acrylics are widely used when production rates are high, and they are ultimately lower in cost (Fig. 12).

Armature coils are sealed and bonded using an epoxy resin. The sealant protects the assembly from corrosion and contamination. The sealant must cure completely, yielding an assembly capable of temperature resistance up to 150 °C (300 °F) (Fig. 13).

Liquid crystal displays are potted and sealed using UV-curable resins. An epoxy resin exposed to UV light cures to a highly moisture-resistant and anticorrosive coating (Fig. 14).

The case that protects surface-mount assemblies is itself protected by dielectric gels. Upon cure, these gels provide thermal and mechanical stress protection. The gels can be mechanically applied and cured using UV light. In more demanding applications, the exterior of the case is coated with a conventional sealant.

Semiconductor components are coated with elastomeric sealants to provide passivation and protection. The one- and two-component elastomeric compounds are specifically formulated to meet the diverse needs of electronic circuitry. Products range from dielectric gels that relieve stress to particle barrier systems. Modifications in cure time, flow out, and environmental performance have been made for specific products to meet selected requirements.

Microporosity sealants are finding application in metal-to-plastic electronic components. Dual-in-line and single-in-line packages are coated with microporosity sealant, and increases in reliability have been reported (Ref 6).

Wire tacking on motor assemblies is accomplished using UV-curable resins. These resins have fast cure times and allow automated-assembly applications. Wire tacking can also be accomplished with conventional sealant systems.

Electrical and electronics applications for sealants continue to grow with the ever-increasing number of electrical devices and electronically managed systems. Sealants help ensure the reliability of these systems. They can be both electrically insulative and conductive, as well as thermally insulative and conductive. These sealants help manage environmental stress and provide the normal protection from air, moisture, and contaminants.

ACKNOWLEDGMENT

The following companies have provided information regarding specific applications for sealants that has been invaluable to the development of this article: Dow Corning Corporation, Essex Specialty Product Corporation, General Electric Company, Henkel Corporation, Loctite Corporation, Norton Corporation, Polymeric Systems Inc., Product Research and Chemical Corporation, Three Bond Corporation, Tremco Incorporated, Sika Corporation, and Uniroyal Incorporated.

Information provided by J.V. Morris of Polymeric Systems, Inc. is also appreciated.

REFERENCES

1. *Tremco Constr. News*, Vol S1 (No. 3), Fall 1988
2. O'Brien and Harschig, *Exterior Wall Systems: New Technology and Design*, American Society for Testing and Materials, 1988; and Dow Corning Corporation brochure
3. Press Release, Loctite Corporation, 1985
4. Press Release, Loctite Corporation, 1986
5. Press Release, General Electric Company, Silicone Products Division, March 1989
6. *Electron. Packag. Prod.*, June 1988

Environmental Effects

Chairman: K.L. DeVries, University of Utah
Co-Chairman: P.R. Borgmeier, University of Utah

Introduction

The use of adhesives to replace mechanical connectors and other joining methods has enjoyed rapid growth. A major reason for this is that the load is distributed over a larger area with the use of adhesives, thereby reducing stress concentrations. Nevertheless, adhesives introduce their own design problems. For one thing, stress distribution in adhesive bonds is often not well understood. An equally important problem is the role of environment. The first of these problems has been addressed in a number of papers and reviews (for example, Ref 1 to 5). Although considerable work has been done in the areas of stress analysis, adhesive fracture mechanics, and adhesive failure analysis, there is still much that is not known, and more research is needed. The subject of this Section, environmental effects on adhesives, has received less attention in the literature.

Environmental considerations are of much more than purely academic interest. For example, adhesives may have great advantages for applications in which reducing weight is a major consideration; these include uses in the aerospace, transportation, and electronics industries. Adhesives for such applications must be able to resist stresses while exposed to extreme environmental conditions.

Adhesives used in aircraft are exposed to extremes in temperature and humidity. Sitting on the runway, an airplane can have skin temperatures exceeding 50 °C (120 °F) at nearly 100% relative humidity; a few minutes later, the plane can be at a high altitude where the ambient temperature may be −50 °C (−60 °F). These low temperatures are accompanied by very low pressures, which can cause rapid moisture removal and lead to blistering and void formation in an adhesive. At the same time, the adhesives in the aircraft may experience high vibratory stresses and/or shock loadings. Furthermore, adhesives used in airframes may be exposed to jet fuel, cleaning solvents, deicing fluids, and other chemicals such as the agents used to melt ice on runways. Many aircraft have useful lives of many decades (for example, the B-52, still currently in use, was built in the 1950s and 1960s), and this must be considered along with environmental factors.

Adhesives used in space vehicles are exposed to greater temperature extremes (both hot and cold) and lower pressures. Electromagnetic wave and particle radiation, such as that present in space, can also dramatically affect polymers. Most adhesives are polymers and therefore are susceptible to chain scission, cross linking, and other radiation effects that can significantly affect properties such as modulus, strength, and interfacial bonding.

Even earthbound modes of transportation, such as automobiles, trucks, and buses, experience formidable environmental conditions. These vehicles also are exposed to extremes in temperature and humidity. Salts and other agents used to melt ice and clean highways can not only chemically interact with adhesives but also corrode the metal adherends. Rough roads and chuckholes expose adhesive joints to vibratory and shock stresses. These vehicles commonly encounter the shocks and extreme stresses of minor fender benders or more serious accidents. Because of an accident or for other reasons, an automobile might be sanded, stripped, chemically cleaned, and repainted, thereby exposing the adhesives in the vehicle to the solvents and chemicals associated with such operations. Finally, automobiles, trucks, and buses—and the adhesives used in them—are exposed to the environments of cities and highways, which include ozone, NO_x, SO_2, and other constituents of smog.

Adhesives have long been used in the construction and building industries. Their use continues to increase in this area because they offer improved quality as well as economic advantages. They are used in many different applications, including construction of large laminated beams, plywood, and veneer paneling; bonding of flooring materials; and cementing of cabinets and tabletops. The adhesives used for these applications may be exposed to various substances and conditions, including water and moisture; high and low temperatures (including freezing of absorbed moisture); solvents that may be present in stains and paint; cleaning agents and chemicals; insects, bacteria, mold, and other biological agents; as well as the deterioration associated with aging over several decades.

Adhesivelike materials are finding increasing use in other consumer products. Parts of household and office appliances are often adhesively bonded together. In many electrical and electronic applications, adhesives are used not only as bonding agents to hold component parts in place but also as sealants to protect electric components from potentially harmful atmospheres. Adhesives are also used to maintain vacuums or other desirable atmospheres around parts. Again, these seals are generally expected to maintain their integrity for the life of the unit, which very well may span several decades.

The listing of environmental requirements for adhesives applications could continue in greater detail; however, it is hoped that the above provides insight into many of the environmental considerations that can affect performance. Once in place, adhesives may be expected to function for very long periods of time. Because adhesives are not easily inspected, conditions of exposure, loading, intensity, or duration of loading cannot always be accurately anticipated.

The strength of an adhesive is perhaps its single most important property, but it is not an easy property to define, and experimental measurements of strength often yield ambiguous results. Even so, there is no dearth of experimental techniques for testing the strength of adhesives. The American Society for Testing and Materials (ASTM) and its counterparts in other countries have formalized and standardized a wide variety of tests for investigating adhesive properties. The tests compiled by ASTM are in *Adhesives*, Volume 15.06 of the *Annual Book of ASTM Standards*; some other volumes contain specialized tests for specific materials. Members of these organizations perform a valuable service by designing and publishing details of standardized methods for testing and reporting tests results, thereby facilitating comparisons of results from different laboratories. Standardized tests are available for measuring many different properties.

The most commonly used adhesive strength tests can be broadly classified into three categories: tensile tests, shear tests, and peel tests. Various tests in each of these categories can be quite different in terms of testing geometry, required testing apparatus, and so on. The measured strength of a given adhesive is generally very dependent on the details of test geometry, sample preparation, and testing parameters. Even the relative ranking between two different adhesives can depend on these parameters. For example, it is possible that an adhesive that appears much stronger than another adhesive in a lap-shear test may exhibit a much lower strength in a peel test.

Most standard tensile and lap-shear tests recommend or specify that test results be reported as force at failure divided by the bonded area. Some designers have used this average stress as a parameter for the design of practical bonded joints. This should be done only with extreme care. Several factors must be taken into consideration. The stresses in typical joints are very nonuniform, and the maximum stresses are markedly greater than the average values. The gross strength of a joint is generally dependent on adhesive thickness, adherend thickness, and other geometric factors. Surface conditions, surface treatment, curing conditions, and other factors also affect joint integrity.

The consequences of decreasing the bonded area of the overlap region in standard lap-shear specimens by 50% (that is, 50% of the overlap region is left unbonded) can serve as an example of the variation between average and maximum stress results. If the average stress failure criteria had much validity, it would be expected that the joint strength would be reduced by approximately half. Recently published test results for a high-strength epoxy adhesive and steel adherends (Ref 6) indicate that the reduction in joint strength was less than 25% for half-area edge debonds and less than 10% for half-area centered debonds.

Recently some success has been enjoyed in using fracture mechanics methods to apply the results from one test geometry to predict the performance of other joints (Ref 1, 2, 7-12). ASTM D 3433 describes a test geometry and methods for obtaining fracture mechanics parameters. Other test geometries have also been proposed and used as standard fracture mechanics test specimens.

A popular and graphically appealing approach to fracture mechanics (Ref 13, 14) views a joint as a system with two failure requirements. First, the stresses at the crack tip (in the materials making up the joint or in the interphase between the materials) must be sufficient to break bonds. Second, there must be an energy balance consistent with the first law. In this model, it is hypothesized that even if the stresses at the crack tip are very large (as they would be for a stressed sharp crack) a crack can grow only if sufficient strain energy is released from the stress field—as the crack grows—to account for the energy required to create the new crack or debond surface area. The specific value of this energy (measured in joules per square meter of crack area) has been given various names (for example, fracture energy, fracture toughness, and work of adhesion).

Fracture mechanics analysis of a joint indicates that its load-carrying capability depends on the joint geometry, the moduli of the adherends and the adhesive, the fracture energies of the materials or that of the adhesive bondline, and the size of inherent flaws (cracks in the joint). Consistent with such an approach, it would be expected that the best and most logical way to explore environmental effects on the strength of a joint would be to investigate environmental effects on these more basic properties rather than attempting to measure their impact on gross strength. Unfortunately, the state of the art has not yet developed to this stage, but it is a worthy long-range goal. In the interim, however, a satisfactory alternative exists for determining environmental effects on an adhesive joint. The tested joint must duplicate the desired geometry of the end-use joint, and all other factors must be matched as closely as possible while the joint is being exposed to the given environmental conditions. The articles that follow clearly illustrate many important environmental considerations and effects. The reader is cautioned, however, that the results are most applicable to joints that closely duplicate the test geometries. In other cases they provide some qualitative insight into anticipated effects.

Another caution is worth noting. It is not always sufficient to independently explore the effects of different agents. Different agents can interact in such a way as to reinforce and intensify their overall influence. Cohesive failure research on some polymer systems has shown that there are sometimes synergistic interactions between applied stresses and environmental agents such as ozone and NO_x. In these cases the combined effect of the two factors is larger than the sum of their independent effects. Such may also be the case for some adhesive systems.

The fundamental mechanism or mechanisms for failure have not been completely and unambiguously defined. In fact, several molecular or physical mechanisms may be involved for different materials or even in the same material (Ref 2, 15). The commonly proposed mechanisms include mechanical interlocking dispersive; primary bonding, for example, as with silane coupling agents (Ref 16), molecular diffusion and entanglement; and acid-base reactions. Again, it would be anticipated that a knowledge of which fundamental mechanisms are active would be useful in predicting the influence of various agents. Very sophisticated analytical instruments are being used to help define these roles, for example, electron spectroscopy for chemical analysis, Fourier transform spectroscopy, secondary ion mass spectroscopy, ion-scattering spectroscopy, Auger electron spectroscopy, scanning electron microscopy, and surface energy measurements; a comprehensive understanding of the role of the various failure mechanisms may be on the horizon. At present, these methods provide some useful insights into expected behavior.

The articles in this Section explore a myriad of environmental considerations for adhesive joints. Some of the issues addressed in these articles can be summarized as follows:

- "Thermal Effects on Adhesive Joints": Most adhesives are polymers that exhibit viscoelastic and other temperature- and time-dependent properties. Temperature might also affect cross linking, cause oxidation of the adhesive or adherends, cause brittleness, and so on

- "Moisture Effects on Adhesive Joints": Many polymers are relatively permeable to water; most absorb from a fraction of a percent to several percent. Moisture can influence polymer properties through hydrolysis, and it can act as a plasticizer and have other similar effects. The interface (interphase) may play an important role in moisture effects. Wicking is a well-known moisture effect; in some cases, water actually tends to displace the adhesive
- "Electrochemical and Corrosion Effects on Adhesive Joints": When adhesives are used with metals in a aqueous environment, electrochemical corrosion must be a consideration. Because many structures for use in aqueous environments have long desired lifetimes, time is an important parameter
- "Chemical Resistance of Structural Adhesive Bonds": Ozone, solvents, fuels, oils, and other agents are known to interact and affect the properties of the polymers typically used in adhesives. They may affect interactions in the interphase as well as adherend properties
- "Radiation and Vacuum Effects on Adhesive Materials in Bonded Joints": Adhesive joints and seals used in outer space are exposed to various types of radiation and to very high vacuum. Irradiation can change the tensile lap-shear strength and other properties of adhesive materials. It might be thought that vacuum would remove environmental agents that can be detrimental to the joint. However, vacuum can itself have deleterious and/or beneficial effects through outgassing, volatilization, removal of plasticizers and other low-molecular-weight components, and so on
- "Combined Temperature-Moisture-Mechanical Stress Effects on Adhesive Joints": These agents, which are treated separately in articles mentioned above, have various synergistic effects when combined
- "Weathering and Aging Effects on Adhesive Joints": It is not always easy to simulate actual service conditions. Weatherometers and accelerated-aging techniques are commonly used, but the reliability of the results obtained by these methods may be questioned
- "Durability Assessment and Life Prediction for Adhesive Joints": Several techniques in common use to explore these aspects of joints are discussed and evaluated
- "Environmental Considerations Unique to Sealants": Sealants are similar to adhesives in that both must tightly adhere to

another material. Sealant strength is usually a less-important consideration than its ability to form a coherent continuous boundary to prevent ingress or egress of water or other aggressive agents. How this ability might be influenced by the environments is an important consideration in sealant design

REFERENCES

1. G.P. Anderson, S.J. Bennett, and K.L. DeVries, *Analysis and Testing of Adhesive Bonds*, Academic Press, 1977
2. A.J. Kinloch, *Adhesion and Adhesives Science and Technology*, Chapman and Hall, 1987
3. R.D. Adams and W.C. Wake, *Structural Adhesive Joints in Engineering*, Elsevier, 1984
4. A.N. Gent, *Adhes. Age*, Vol 25 (No. 2), 1982, p 27-31
5. H.F. Brinson, J.P. Wightman, and T.C. Ward, Ed., *Adhesion Science Review*, The Virginia Tech Center for Adhesion Science, 1987
6. P.R. Borgmeier, K.L. DeVries, J.K. Strozier, and G.P. Anderson, "Evaluation of the Strength and Mode of Fracture of Standard Lap Shear Joints With and Without Initial Large Debonds," Paper presented at the Winter Annual Meeting of the American Society of Mechanical Engineers (San Francisco), Dec 1989
7. K.M. Liechti and C. Lin, *Structural Adhesives in Engineering*, Institute of Mechanical Engineering, 1986
8. S. Mostovoy and E.J. Ripling, *Adhesion Science and Technology*, Vol 9B, L.H. Lee, Ed., Plenum Publishing, 1975
9. W.G. Knauss, *J. Compos. Mater.*, Vol 5, 1971, p 176-192
10. G.G. Trantina, *J. Compos. Mater.*, Vol 6, 1972, p 371-385
11. W.S. Johnson and S. Mall, *A Fracture Mechanics Approach for Designing Adhesively Bonded Joints*, STP 876, American Society for Testing and Materials, 1985, p 189-199
12. A.S.D. Wang and F.W. Crossman, *J. Compos. Mater. Suppl.*, Vol 14 (No. 1), 1980, p 71-108
13. M.L. Williams, *J. Adhes.*, Vol 4, 1972, p 307-332
14. A.A. Griffith, in *Proceedings of the 1st International Congress of Applied Mechanics*, 1924, p 55-63
15. S.R. Hartshorn, Ed., *Structural Adhesives Chemistry and Technology*, Plenum Publishing, 1986
16. E.P. Plueddemann, *Silane Coupling Agents*, Plenum Publishing, 1982

Thermal Effects on Adhesive Joints

John Comyn, John Comyn Ltd. (United Kingdom)

AN ADHESIVELY BONDED JOINT has three basic elements: the adherends, the interface (or interphase), and the adhesive. The thermal properties of adherends are not the prime concern of this article; in fact, little is known about the effect of temperature on the interface. However, the first section of this article does describe interfacial effects in terms of six adhesion theories, interfacial tension, and differential thermal expansion. With metallic, glass, and ceramic adherends, it is the adhesive itself that is susceptible to increases in temperature, and such changes weaken the joint and generally lead to cohesive failure (that is, failure within the adhesive).

Polymers are the major components of all adhesives; therefore, they are the principal factors influencing any property of an adhesive. There is relatively little specific information in the literature on the thermal properties of adhesives. Therefore, this article uses information on the nearest equivalent polymer in many instances. For example, polymethyl methacrylate can be used as a model for reactive acrylic adhesives, polydimethylsiloxane for silicone adhesives, nylon for polyamide hot melts, and ethylene for ethylene-vinyl acetate (EVA) copolymer hot melts.

Interfacial Effects

Theories of Adhesion. There are six theories of adhesion: the weak boundary layer theory, diffusion theory, electrostatic theory, chemical bonding theory, mechanical interlocking theory, and physical adsorption theory. They all appear to be valid in some circumstances. These theories have been reviewed by Wake (Ref 1) and Kinloch (Ref 2).

The weak boundary layer theory proposes that clean surfaces are capable of forming strong adhesive bonds, but that they can be prevented from doing so by a layer of contaminant that is cohesively weak. Oils and greases on metals are classic examples of such contaminants. If the weak boundary layer is removed, then a strong adhesive bond can be obtained.

The diffusion theory mainly applies to samples of identical non-cross-linked rubbers in contact. In such a case the polymer

molecules interdiffuse, and the boundary eventually disappears.

The mechanical interlocking theory is based on the surface of the substrate being so uneven that the adhesive can enter its irregularities and then harden. The relevant consideration of thermal effects for this theory, as well as for the diffusion theory, is in the context of the bulk properties of the adhesive.

The electrostatic theory originated in the proposal that if two metals are placed in contact, electrons will be transferred from one to the other, forming an electrical double layer. This layer gives a force of attraction at the interface. The effects of heat on such an interface are difficult to speculate.

The chemical bonding theory is based on the idea that covalent or ionic bonds are formed across the interface. These are strong interactions that give strong adhesive joints. At elevated temperatures, these bonds might be broken in such a manner that bonds in the adhesives might be in the processes of depolymerization and oxidative degradation.

The physical adsorption theory considers that the adhesive and substrate are in intimate molecular contact, and that weak attractive van der Waals forces operate between them. Because they occur between any two molecules in contact, van der Waals forces contribute to all adhesive bonds. Although they are weak in comparison with chemical bonds, they are more than adequate to account for the strengths of the strongest adhesive joints.

Interfacial Tension. Wu (Ref 3) has shown that the variation with temperature of the surface tensions of polymers

$-(\delta\gamma/\delta T)$ is quite small and in the range of 0.05 to 0.08 mN/m · K. The term $-(\delta\gamma/\delta T)$ is in fact the surface entropy, and the small value is due to the limited number of conformations of chain molecules at the surface. Values of surface tension and of $-(\delta\gamma/\delta T)$ for a number of polymers are shown in Table 1.

Girifalco and Good (Ref 4) give Eq 1 for the interfacial tension between two phases, 1 and 2. Here, γ_1 and γ_2 are the surface tensions of the two phases, γ_{12} is the interfacial tension, and ϕ is an empirical parameter that is in the range of 0.5 to 1.2:

$$\gamma_{12} = \gamma_1 + \gamma_2 - 2\phi(\gamma_1\gamma_2)^{1/2} \qquad \text{(Eq 1)}$$

Differentiation of Eq 1 gives Eq 2:

$$\frac{\delta\gamma_{12}}{\delta T} = \frac{\delta\gamma_2}{\delta T} + \frac{\delta\gamma_2}{\delta T} - \phi\left[\left(\frac{\gamma_1}{\gamma_2}\right)^{1/2}\left(\frac{\delta\gamma_2}{\delta T}\right)\right. $$
$$\left. + \left(\frac{\gamma_2}{\gamma_1}\right)^{1/2}\left(\frac{\delta\gamma_1}{\delta T}\right)\right] \qquad \text{(Eq 2)}$$

Equation 1 is an approximation, but it has the virtue of being relatively simple. Fowkes (Ref 5) separates the interfacial term into dispersive and polar components to give an equation that has been widely used in the literature and that is a better approximation to reality. However, differentiation of Fowkes's equation would give a complicated expression, including a number of terms that have not been determined by experiment. Values of some interfacial tensions appear in Table 2, where it can be seen that interfacial tensions and therefore joint strengths decrease with temperature. The data in Table 2 are for polymer-to-polymer interfaces, but because polymer

Table 1 Dependence of surface tensions of polymers on temperature

| Polymer | Molar mass | Surface tension, γ ||||| Surface entropy, $-(\delta\gamma/\delta T)$, mN/m · K |
| | | At 20 °C (68 °F) || At 140 °C (285 °F) || At 180 °C (355 °F) || |
		mN/m	lbf/ft(a)	mN/m	lbf/ft(a)	mN/m	lbf/ft(a)	
Polystyrene	M_v = 44 000	40.7	2.79	32.1	2.20	29.2	2.00	0.072
PMMA	M_v = 3 000	41.1	2.82	32.0	2.19	28.9	1.98	0.076
PVA	M_w = 11 000	36.5	2.50	28.6	1.96	25.9	1.77	0.066
HDPE	M_w = 67 000	35.7	2.45	28.8	1.97	26.5	1.82	0.057
PDMS	0.06(b)	19.8	1.36	14.0	0.96	12.1	0.83	0.048
EVA (25% VA w/w)	M_n = 17 000	35.5	2.43	27.5	18.8	24.8	1.70	0.067

PMMA, polymethyl methacrylate; PVA, polyvinyl acetate; HDPE, high-density polyethylene; PDMS, polydimethylsiloxane; EVA, ethylene-vinyl acetate copolymer; M_v, viscosity average molar mass; M_w, weight average molar mass; M_n, number average molar mass. (a) All lbf/ft values should be multiplied by 10^{-3}. (b) Kinematic viscosity for PDMS given in m^2/s. Source: Ref 3

Table 2 Variation of interfacial tension with temperature

| | Interfacial tension, γ_{12} | | | | | | |
| | At 100 °C (212 °F) | | At 140 °C (285 °F) | | At 180 °C (355 °F) | | Interfacial entropy, |
Interface	mN/m	lbf/ft(a)	mN/m	lbf/ft(a)	mN/m	lbf/ft(a)	$-(\delta\gamma_{12}/\delta T)$, mN/m · K
HDPE/PS	6.7	0.459	5.9	0.404	5.1	0.349	0.020
HDPE/PVA	12.4	0.849	11.3	0.774	10.2	0.700	0.027
PS/PVA...........	3.9	0.267	3.7	0.253	3.5	0.240	0.0044
PS/PMMA..........	2.2	0.151	1.6	0.110	1.1	0.075	0.013
PMMA/P*n*BMA.....	2.4	0.164	2.0	0.137	1.5	0.103	0.012
PDMS/P*n*BMA	3.9	0.267	3.8	0.260	3.6	0.247	0.0037

HDPE, high-density polyethylene; PS, polystyrene; PVA, polyvinyl acetate; PMMA, polymethyl methacrylate; P*n*BMA, poly-*n*-butylmethacrylate; PDMS, polydimethylsiloxane. (a) All lbf/ft values should be multiplied by 10^{-3}. Source: Ref 3

Table 3 Coefficients of linear thermal expansion

Material	
Polymers, 10^{-4}/K^{-1}	
Low-density polyethylene	1.0
High-density polyethylene...........	1.3
Polymethyl methacrylate	$\approx$2.6 (below T_g)
	$\approx$5.3 (above T_g)
Polystyrene......................	6–8
Other materials, 10^{-6}/K	
Aluminum	29
Steel...........................	$\approx$11
Titanium	$\approx$ 9
Soda-glass......................	8.5
Wood (along grain)	3–5
Wood (across grain)	35–60

surface tensions decrease with temperature, it is likely that all adhesive-substrate interfacial tensions would weaken with increasing temperature.

Differential Thermal Expansion. One important and comprehensible interfacial phenomenon is that of differential thermal expansion between adhesive and adherend. Coefficients of linear expansion of polymers are greater than those of other materials by a factor of about 100; examples are given in Table 3. Values for epoxide adhesives, both above and below the glass transition temperature (T_g), are given in Table 4. Expansion coefficients in the leathery state are about 2.5 to 3.5 times higher than those in the glassy state. One way to reduce expansion coefficients of adhesives is by the incorporation of mineral fillers.

Thermal stresses may develop in adhesive joints because of differences in coefficients of thermal expansion (CTE). Consider the case of a high-temperature-curing structural adhesive, such as an epoxide. As the adhesive cures at a temperature of about 150 °C (300 °F), it is converted from being a liquid to a rubbery polymer; it will decrease slightly in volume, and this alone will be a source of stress. As the joint is cooled to the glass transition temperature, additional differential stresses will develop between adhesive and adherend, but the compliant adhesive will tend to relax. Once the glass transition temperature is passed, however, the adhesive becomes less compliant and stresses will build up in the joint. Thermal stresses will thus depend on the differences in expansion coefficients of adherend and adhesive and the amount of cooling below the glass transition temperature. Simply stated, the smaller the difference in coefficients, the smaller the stress buildup. The incorporation of

mineral fillers represents one way to reduce the CTEs of adhesives.

Thermal Transitions in Adhesives

The Glass Transition. All polymers have a glass transition temperature, which is a property of the amorphous phase. Below the T_g, they are relatively hard and rigid and are described as glasses; above the T_g, they are soft and flexible and are termed rubbery or leathery. Molecular motion is the underlying phenomenon. Below T_g, translational or rotational motion in the polymer backbone is greatly restricted. Once the T_g is exceeded, the onset of this motion results in a material that is compliant and may flow if it has not been cross linked.

Heat distortion temperature and softening point are closely related to T_g. For a given polymer, these three temperatures will be similar, but not usually identical, because of the use of different heating rates and the fact that the phenomena are pressure sensitive.

Both glassy and leathery polymers are used as adhesives. Examples of the former are epoxides, phenolics, and reactive acrylics; rubber-base adhesives are examples of the latter. However, because of the large changes in mechanical properties, it is unacceptable for the glass transition temperature to exist in the same region of temperature in which an adhesive bond has to operate.

Values of T_g for some adhesive polymers are given in Table 5. The glass transition temperature of an adhesive depends on its chemical composition. Any factor that tends to decrease molecular motion will have the effect of increasing T_g. Such factors include the presence of polar or inflex-

ible chemical groups. In contrast, large nonpolar chemical groups and the presence of plasticizing additives lower T_g. Plasticization of adhesives can occur when, for example, water is introduced into epoxides and tackifiers are introduced into pressure-sensitive adhesives.

Some of these effects are illustrated by the data in Table 5. For example, the T_g of the series of methacrylate polymers falls as the size of the nonpolar ester-alkyl group increases from having one carbon atom (polymethyl methacrylate) to eight carbon atoms (poly 2-ethylhexyl methacrylate). Polymethyl methacrylate forms the basis of structural acrylic adhesives, whereas polybutylmethacrylate and poly 2-ethylhexylmethacrylate are used, with tackifying additives, in pressure-sensitive adhesives. The diglycidyl ether of bisphenol A (Table 5) is widely used as a resin in epoxide adhesives.

The three principal groups of compounds that are used to cure such epoxide adhesives are aliphatic amines, aromatic amines, and acid anhydrides. Cure with aliphatic amines occurs at lower temperatures than

Table 5 Glass transition temperatures of selected polymers and adhesives

| | T_g | |
Polymer	°C	°F
Linear polymers		
Polymethyl methacrylate..............	105	220
Polyethyl methacrylate	65	150
Polypropyl methacrylate	35	95
Polybutyl methacrylate	20	70
Poly 2-ethylhexylmethacrylate........	−10	15
Polydimethylsiloxane	−127	−195
Styrene-butadiene rubber	−61	−78
Epoxide adhesives (resin-curing agent)		
DGEBA-di(1-aminopropyl-3-ethoxy) ether	67	153
DGEBA-triethylenetetramine..........	99	210
DGEBA-1,3-diaminobenzene	161	320
DGEBA-4,4'-diaminodiphenylmethane	119	246

Source: Ref 7, 8

Table 4 Coefficients of linear thermal expansion of cured epoxide adhesives, both above and below T_g

| | Above T_g 10^{-4}/K^{-1} | Below T_g 10^{-4}/K^{-1} | T_g °C | T_g °F |
Hardener				
BF$_3$-amine..................................	1.7	0.63	141	285
Diethylenetetramine	1.8	0.60	122	250
1,2-diaminobenzene	2.1	0.57	190	375
Methylene bis (1-chloro aniline).................	1.9	0.83	149	300
Aliphatic diamine	2.0	0.77	47	115

Note: Resin is diglycidyl ether of bisphenol A. Source: Ref 6

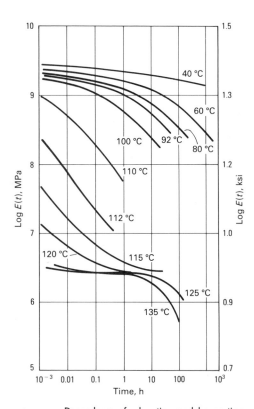

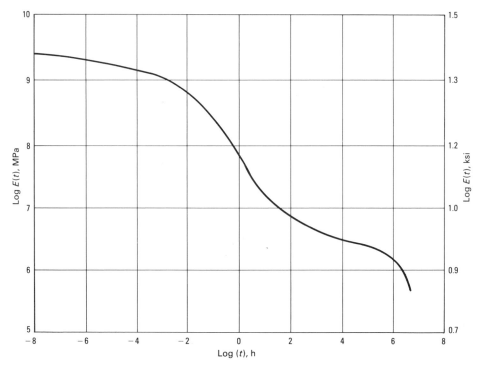

Fig. 1 Dependence of relaxation modulus on time for polymethyl methacrylate at temperatures between 40 and 135 °C (105 and 275 °F). Number average molar mass = 3 600 000. Source: Ref 9

Fig. 2 Dependence of relaxation modulus on time for polymethyl methacrylate at 110 °C (230 °F). Plot based on data in Fig. 1

that with aromatic amines and acid anhydrides. This temperature difference is reflected in the T_g of the cured adhesive. The di(1-aminopropyl-3-ethoxy)ether and triethylenetetramine in Table 5 are aliphatic amines, whereas 1,3-diaminobenzene and 4,4′-diaminodiphenylmethane are aromatic. However, one factor behind aromatic amines yielding epoxides that have a high T_g is the relative inflexibility of the hardener units.

Cured rubber-modified structural adhesives have a glassy matrix and a disperse phase of small rubber particles, which have a reinforcing action. Such materials have two T_gs, a high one associated with the matrix, and a lower one for the rubber particles. The T_g of the rubbery phase represents a lower temperature limit for these adhesives.

Melting of Semicrystalline Polymers. Most polymers used in adhesives are amorphous and therefore do not have a melting point. Exceptions to this are polyvinyl alcohol, which is used in remoistenable adhesives, and hot-melt adhesives based on ethylene copolymers and polyamides. The melting point of low-density polyethylene is about 117 °C (240 °F), and its degree of crystallinity will depend on its thermal history. For hot-melt adhesive formulations, vinyl acetate is used as a comonomer, and because this reduces the regularity of the

polymer chains, there is a large reduction in degree of crystallinity.

Polyamides have melting points that are too high for hot-melt adhesives; for example, nylon 6 melts at 225 °C (435 °F), nylon 66 at 264 °C (510 °F), and nylon 12 at 180 °C (355 °F). Their melt viscosities are also high for this particular application. However, copolymerization and terpolymerization of nylon monomers give products with more irregular structures and reduced levels of intermolecular hydrogen bonding, thereby yielding lower melting points.

Viscoelastic Properties of Adhesives. The term viscoelastic is used to describe the properties of polymers that show a combination of the viscous properties of a liquid and the elastic properties of a solid. This means that their properties are time- and temperature-dependent. If a constant strain is applied to a polymer at a fixed temperature, then the stress needed to maintain that strain will decrease (relax) with time. The time-dependent stress can be expressed as the stress modulus $E(t)$, also commonly called the relaxation modulus. The dependence of $E(t)$ on time for polymethyl methacrylate at a number of temperatures is shown in Fig. 1. Using the principle of time-temperature superposition, the time range of the data in Fig. 1 can be extended both to longer and shorter times at any chosen temperature. According to this principle, any locus can be slid horizontally along the log (time) axis until it superimposes and extends to its neighbor; thus, data at a temperature below the

reference temperature can be used to extend to shorter times, whereas data at higher temperatures can be used to reach longer times. This has been done on the lines in Fig. 1, keeping 110 °C (230 °F) as the reference temperature, and the result is shown in Fig. 2.

The result of this operation is a single locus of log $E(t)$ against log (time) at a single temperature (here, 110 °C, or 230 °F), which extends over a wider time range. A further consequence of time-temperature superposition is that plots of log $E(t)$ against temperature, at constant time, are similar in shape to those of $E(t)$ against log (time). To measure such data, a stress would be applied to a polymer, and the strain would be measured after a fixed time. If, for example, the fixed time were 10 s, then it is the 10-s modulus that would be measured.

Figure 3 shows the modulus-temperature plot for a linear polymer, which has four regions of viscoelastic behavior. At the lowest temperatures, the material is a high-modulus glass. As temperature increases, it passes through the glass transition region and its modulus falls to that of a leathery material. It holds a stable modulus over a modest region of temperature, but then enters a region of flow. The major difference between this case and that of a cross-linked polymer is that the cross links prevent flow from occurring.

Temperature Effects. The effect of temperature on the strength of some aluminum alloy joints bonded with epoxide adhesives

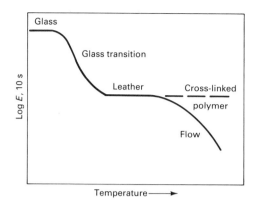

Fig. 3 Typical dependence on temperature of the 10-s modulus of a linear polymer

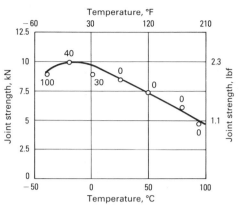

Fig. 4 Effect of temperature on the strength of aluminum lap joints with FM 1000 adhesive. Numbers corresponding to points represent percentage of interfacial failure. Source: Ref 10

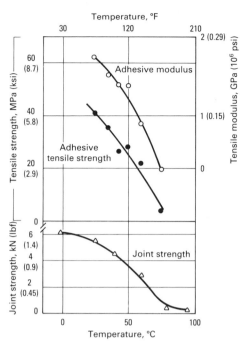

Fig. 5 Temperature dependence of strength of aluminum lap joint using epoxide adhesive compared to mechanical properties of the adhesive. Adhesive is DGEBA cured with di(1-aminopropyl-3-ethoxy)ether. Source: Ref 11

has been investigated by Brewis, Comyn, and Shalash (Ref 10, 11). In the case of the epoxide-polyamide adhesive FM 1000 (Cyanamid), modulus and tensile strength were measured on film samples of the cured adhesive over the temperature range of 16 to 85 °C (60 to 185 °F), and a large fall in modulus was observed at the glass transition temperature of 40 °C (105 °F). The strengths of single-lap joints were measured over the wider temperature range of −60 to 95 °C (−75 to 205 °F), and results are shown in Fig. 4. It can be seen that joints have maximum strength at subzero temperatures. As the temperature rises, joints become weaker and the mode of failure becomes 100% cohesive. Data on the variation of the mechanical properties of the adhesive, with varying temperatures, are shown in Table 6.

The mechanical properties of films of the cured epoxide adhesive based on DGEBA and di(1-aminopropyl-3-ethoxy)ether were measured over the temperature range of 25 to 75 °C (77 to 165 °F). Joint strengths were measured over the temperature range of 0 to 95 °C (30 to 205 °F). The T_g of the adhesive was 65 °C (150 °F). The data in Fig. 5 show that joint strengths are controlled by the mechanical properties of the adhesive, and that bulk properties dominate interfacial properties in controlling the strengths of these joints (that is, as the strength of the adhesive is reduced, so is that of the adhesive joint).

Thermal Degradation of Adhesives

Depolymerization. Two conditions must be satisfied for depolymerization to occur. The first is that the temperature must exceed the "ceiling" temperature of the polymer, and the second is that there must be a mechanism available for depolymerization. An example that illustrates this particularly well is poly-1-methylstyrene, which has a ceiling temperature of 61 °C (140 °F); once heated above that temperature, poly-1-methylstyrene is thermodynamically unstable. However, nothing usually happens until temperatures much higher than this are reached, at which point free radicals will be produced, resulting in homolytic fission of chemical bonds. The free radicals provide active centers for further depolymerization to occur.

The free energy of polymerization (ΔG) is related to the heat ΔH and entropy ΔS of polymerization by Eq 3; T is the absolute temperature:

$$\Delta G = \Delta H - T\Delta S \qquad \text{(Eq 3)}$$

For nearly all polymerizations, both ΔH and ΔS are negative, which means that there is a critical temperature, (known as the ceiling temperature, T_c) above which it is impossible for polymerization to occur. At the ceiling temperature, $\Delta G = 0$; therefore, T_c

is related to the heat and entropy of polymerization by Eq 4:

$$T_c = \Delta H/\Delta S \qquad \text{(Eq 4)}$$

Entropy of polymerization does not vary widely for polymers; it is usually in the range of 100 to 125 J/mol · K (25 to 30 Btu/lb · mol · °R) for liquid vinyl monomers. This is because in each case the dominant factor is the loss of transitional freedom that occurs when a monomer molecule is incorporated into a polymer chain. Table 7 gives some values for entropy of polymerization. When stating these thermodynamic parameters it is essential to indicate the state of the monomer and polymer. This is done by using two letters, the first of which refers to the monomer. States are indicated as g = gas, l = liquid, c = amorphous condensed phase, and c' = semicrystalline condensed phase.

In contrast, heats of polymerization do vary from monomer to monomer. The range

Table 6 Tensile mechanical properties of FM 1000 adhesive

| Temperature | | Yield stress | | Strength | | Elongation, | Modulus | | Failure |
°C	°F	MPa	ksi	MPa	ksi	%	MPa	ksi	
16	60	· · ·	· · ·	82.3 ± 1.0	11.9 ± 0.15	5.4 ± 0.2	2030 ± 26	294 ± 3.8	Brittle
25	77	· · ·	· · ·	72.6 ± 0.6	10.5 ± 0.09	5.0 ± 0.6	1877 ± 12	272 ± 1.7	Brittle
35	95	56.1 ± 1.2	8.1 ± 0.17	60.4 ± 1.2	8.8 ± 0.17	249 ± 6	1473 ± 32	214 ± 4.6	Ductile
40	105	13.4 ± 0.8	1.9 ± 0.12	60.3 ± 1.3	8.7 ± 0.19	272 ± 3	338 ± 16	49 ± 2.3	Ductile
50	120	· · ·	· · ·	48.0 ± 1.4	7.0 ± 0.20	275 ± 5	176 ± 7	26 ± 1.0	Rubbery
60	140	· · ·	· · ·	35.1 ± 0.5	5.1 ± 0.07	267 ± 1	4.1 ± 0.4	0.59 ± 0.06	Rubbery
70	160	· · ·	· · ·	35.5 ± 1.4	5.1 ± 0.20	293 ± 4	3.4 ± 0.1	0.49 ± 0.01	Rubbery
85	185	· · ·	· · ·	30.1 ± 3.1	4.4 ± 0.45	194 ± 12	5.6 ± 0.2	0.81 ± 0.03	Rubbery

Source: Ref 10

Table 7   Entropies of polymerization

Polymer	States	ΔS	
		J/mol · K	Btu/lb · mol · °R
Methyl methacrylate......	1c	−117	−28
1-methylstyrene.........	1c	−110	−26
Styrene	1c	−104	−25
Ethylene oxide...........	gc′	−174	−42
	1c	−86	−21
Propylene oxide.........	gc′	−189	−45
	1c′	−101	−24

Note: 1c′ values for the cyclic oxides have been calculated by subtracting 88 J/mol · K (21 Btu/lb · mol · °R) from gc′ values. This is the entropy of evaporation of unassociated liquids as given by Trouton's rule. Source: Ref 12

Table 8   Heats of polymerization

Polymer	States	ΔH	
		kJ/mol	Btu/lb · mol × 10⁻³
Methyl methacrylate.................	1c	−55.5	−23.6
1-methylstyrene....................	1c	−35 to −39	−14.9 to −16.9
Ethylene oxide.....................	gc′	−127.3 or −140	−54.1 or −59.5
Ethylene	gc	−101.5	−43.1
Isobutene	gc	−72	−30.6

Source: Ref 12

of values for liquid vinyl monomers is from 30 to 100 kJ/mol (28 to 94 Btu/mol). One factor that is particularly important in causing monomers to have different values is steric hindrance in the polymer chain. Table 8 gives some values for heat of polymerization. Changes in heat, rather than entropy of polymerization, cause polymers to have different ceiling temperatures. Some ceiling temperatures are given in Table 9.

When a reactive adhesive cures, the heat of polymerization will be evolved. If cure is slow, allowing the evolved heat to be transferred to the surroundings, then the process will be isothermal. Fast cure can lead to temperature rises and possibly to damage to the adhesive.

Oxidative Degradation. Oxygen in the atmosphere generally attacks polymers. In many cases, the rate of attack, even at ambient temperatures, necessitates the use of stabilizing additives. Examples of adhesives that normally contain antioxidants are hot melts, those based on natural and synthetic rubbers, and pressure-sensitive adhesives.

Although all polymers will degrade at elevated temperatures in the absence of air, the rate of degradation is generally increased when there is a supply of oxygen. In adhesive joints, the supply of oxygen will depend on the oxygen permeabilities of both adhesive and adherends, and on the distance that oxygen must diffuse in either of these. Metallic, ceramic, and glass adherends are impermeable to oxygen and therefore represent a complete barrier to it, but composites with polymeric matrices are permeable to atmospheric gases. However, the rate of oxygen diffusion into adhesives at elevated temperatures will probably be sufficiently high for impermeable adherends

Table 9   Ceiling temperatures

Polymer	States	T_c	
		°C	°F
Methyl methacrylate	gc	164	327
1-methylstyrene	1c	61	142
Nylon 6	1c′	277	530
Polydimethylsiloxane	1c	110	230

Source: Ref 12

to offer little protection against oxidative degradation.

Oxygen is a diradical and can participate in polymerization and degradation reactions by acting as an initiator. Oxygen is absorbed by the polymer to form a hydroperoxide (Fig. 6a) that subsequently decomposes to give radicals (Fig. 6b) that then initiate degradation.

Figure 7 shows the amount of oxygen absorbed by a typical polymer in air as the temperature is raised. The pure polymer has an induction period in which the hydroperoxide is formed, but this is absent in the polymer to which hydroperoxide groups have previously been added. In the presence of an antioxidant, there is a long induction period in which the stabilizer scavenges any free radicals. The end of the induction period marks consumption of the stabilizer. Oxidative degradation is a complex series of chain reactions. Karger-Kocsis, Senyei, and Hedvig (Ref 13) have demonstrated induction periods in the oxygen uptake of a number of hot-melt adhesives.

Different polymers have varying resistance to oxidative degradation. For some vinyl polymers in solution, stability decreases in the order shown in Fig. 8, representing the ease with which a hydrogen atom can be removed.

Solid polystyrene is much more stable than polystyrene in solution because at normal temperatures it is a glass (T_g = 100 °C, or 212 °F), and the rate of oxygen diffusion in polymeric glasses is generally less than it is in solutions or in leathery polymers. In the case of polybutadiene, stability is further reduced by the high chemical reactivity of the carbon-carbon double bond.

The products of polymer degradation can be a monomer, other volatiles, and a modified polymer. In the absence of oxygen, polymethyl methacrylate degrades to give a high yield of monomer, which is representative of a trend shown by other poly-

mers derived from 1,1-disubstituted ethylenes. Polymers from monosubstituted ethylenes tend to give off volatiles other than monomer. Polymers can be modified by cross linking or by the loss or reaction of side groups. Two examples of the latter, although not particularly relevant to adhesives, are the loss of hydrogen chloride by polyvinyl chloride to yield a polyene, and the carbonization of polyacrylonitrile.

The excellent resistance to oxidation possessed by high-temperature adhesives such as polyimides, polyquinoxalines, and polybenzamidazoles, is due to their high content of aromatic groups.

Polymers are also attacked by ultraviolet (UV) radiation. The combined effects of UV radiation and oxygen are particularly damaging. However, unless one adherend is transparent to UV rays, this is not a factor in the weakening of polymer adhesives.

Thermal Conductivity of Adhesives

Adhesives filled with metal powders, especially silver, are conductors of heat and electricity. To increase thermal conductivity alone, metal oxide fillers can be used. The most effective of these is beryllium oxide, which is both toxic and expensive; aluminum oxide is a practical alternative. Some values of thermal conductivity are collected in Table 10.

Kilik, Davies, and Darwish (Ref 15) have measured thermal conductivities of a toughened epoxide adhesive filled with copper and aluminum powders. The effect of the size and shape of particles is less significant than the amount of filler used.

Polymer — H + O₂ = Polymer — O — O — H

(a)

Polymer — O — O — H = Polymer — O· + ·OH

(b)

Fig. 6   Role of oxygen in polymerization and degradation reactions. (a) Oxygen absorption by polymer to form hydroperoxide. (b) Decomposition of hydroperoxide

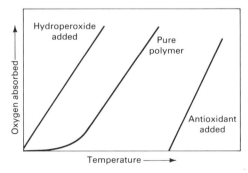

Fig. 7   Typical oxygen absorption of pure and modified polymers

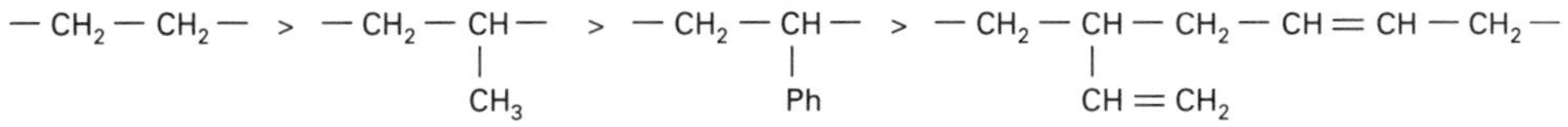

$$-CH_2-CH_2- \quad > \quad -CH_2-CH- \quad > \quad -CH_2-CH- \quad > \quad -CH_2-CH-CH_2-CH=CH-CH_2-$$

Polyethylene   Polypropylene   Polystyrene   Poly-1,2-butadiene

where the branches: CH_3 under polypropylene, Ph under polystyrene, $CH=CH_2$ under poly-1,2-butadiene.

Fig. 8 Decreasing order of stability of vinyl polymers in solution

Table 10 Thermal conductivities

Material	W/m · K	Btu · in./h · ft³ · °F
Silver	410	2830
Copper	370	2555
Beryllium oxide	220	1520
Aluminum oxide	34	235
Unfilled epoxide	0.17–0.26	1.17–1.80
Epoxide with 25% by weight of Al_2O_3	0.34–0.51	2.35–3.52
Epoxide with 50% by weight of Al_2O_3	0.51–0.68	3.52–4.69
Epoxide with 75% by weight of Al_2O_3	1.3–1.7	8.97–11.7
Epoxide with 50% by weight of Al	1.7–3.4	11.7–23.5
Best epoxide with Ag filler	1.7–6.8	23.5–46.9

Source: Ref 14

Temperature Limits of Adhesives

The temperatures at which different types of adhesives will survive in service for a period of 1 year will vary somewhat among particular materials, but the following general guidelines can be used. The upper limit for acrylates and cyanoacrylates is 80 °C (175 °F), whereas epoxides cured with polyamides have a service limit of 65 °C (150 °F). Epoxides cured with aliphatic amines have an upper limit of 100 °C (212 °F), whereas this limit is 150 °C (300 °F) for those cured with aromatic amines, and 175 °C (350 °F) for those cured with acid anhydrides. The service temperature is 200 °C (390 °F) for silicones and 260 °C (500 °F) for polyimides. Many of these adhesives can be exposed at higher temperatures for shorter periods without detrimental effects.

Rubbery adhesives and rubber-modified structural adhesives have lower temperature limits (the glass transition temperature). In many cases, the development of thermal mismatch stresses upon cooling will increasingly weaken joints.

REFERENCES

1. W.C. Wake, Theories of Adhesion and Uses of Adhesives: A Review, *Polymer*, Vol 19, 1978, p 291-308
2. A.J. Kinloch, The Science of Adhesion Part 1, Surface and Interfacial Aspects, *J. Mater. Sci.*, Vol 15, 1980, p 2141-2166
3. S. Wu, Interfacial and Surface Tensions of Polymers, *Rev. Macromol. Chem.*, Vol 11, 1974, p 1-73
4. L.A. Girifalco and R.J. Good, A Theory for the Estimation of Surface and Interfacial Energies: Part I, Derivation and Application to Interfacial Tension, *J. Phys. Chem.*, Vol 61, 1957, p 904-909
5. F.M. Fowkes, Attractive Forces at Interfaces, *Ind. Eng. Chem.*, Vol 56 (No. 12), 1964, p 40-52
6. D.H. Kaelble, J. Moacanin, and A. Gupta, Physical and Mechanical Properties of Cured Resins, in Chapter 6, *Epoxy Resins Chemistry and Technology*, 2nd ed., C.A. May, Ed., Marcel Dekker, 1988, p 603-651
7. P. Peyser, Glass Transition Temperatures of Polymers, in *Polymer Handbook*, 3rd ed., J. Brandrup and E.H. Immergut, Ed., John Wiley & Sons, 1989, p VI/209 to VI/277
8. D.M. Brewis, J. Comyn, R.J.A. Shalash, and J.L. Tegg, Interaction of Water With Some Epoxide Adhesives, *Polymer*, Vol 21, 1980, p 357-360
9. J.R. McLoughlin and A.V. Tobolsky, The Viscoelastic Behavior of Polymethyl Methacrylate, *J. Colloid Sci.*, Vol 7, 1952, p 555-568
10. D.M. Brewis, J. Comyn, and R.J.A. Shalash, The Effect of Moisture and Temperature on the Properties of an Epoxide-Polyamide Adhesive in Relation to its Performance in Single Lap Joints, *Int. J. Adhes. Adhes.*, Vol 2, 1982, p 215-222
11. D.M. Brewis, J. Comyn, and R.J.A. Shalash, The Effect of Water and Heat on the Properties of an Epoxide Adhesive in Relation to the Performance in Single Lap Joints, *Polymer*, Vol 24, 1983, p 67-70
12. W.K. Busfield, Heats and Entropies of Polymerization, Ceiling Temperatures, Equilibrium Monomer Concentrations; and Polymerizability of Heterocyclic Compounds, in *Polymer Handbook*, 3rd ed., J. Brandrup and E.H. Immergut, John Wiley & Sons, 1989, p II/295 to II/334
13. J. Karger-Kocsis, Z. Senyei, and P. Hedvig, Comparative Thermal and Thermomechanical Properties of Hot Melt Adhesives, *Int. J. Adhes. Adhes.*, Vol 1, 1980, p 17
14. J.C. Bolger, Conductive Adhesives, in Chapter 43, *Handbook of Adhesives*, 3rd ed., I. Skeist, Ed., Van Nostrand Reinhold, 1990, p 705-712
15. R. Kilik, R. Davies, and S.M.H. Darwish, Thermal Conductivity of Adhesive Filled With Metal Powders, *Int. J. Adhes. Adhes.*, Vol 9, 1989, p 219-224

Moisture Effects on Adhesive Joints

Nak-Ho Sung, Laboratory for Materials and Interfaces and Department of Chemical Engineering, Tufts University

ADHESIVE JOINTS for structural applications typically consist of metal (such as aluminum, titanium, or steel) or glass substrates bonded with organic adhesives. These substrates generally have high surface energies, and therefore the adhesives used tend to be polar in nature (Ref 1). High initial adhesion strength is easily achieved provided that the surfaces are free of contaminants and properly treated mechanically or chemically. The use of these joints is, however, severely limited because of rapid loss of joint strength when the joint is exposed to various working environments.

One of the most common degradating agents is water. Water is a highly polar molecule that is permeable to most polymers, and it is practically impossible to prevent water from migrating to the interface where a high-energy surface adherend is present. Although water plasticizes polymers, it is the interface regions where water is believed to reduce the strength. Long-term durability in a humid environment is one of the key issues in broadening the application of structural joints beyond the aerospace industry. The effect of a hostile environment on joint strength is shown in Fig. 1, which compares the loss of strength in metal-epoxy joints after exposure in a hot-wet environment with the excellent durability that can be achieved under dry conditions.

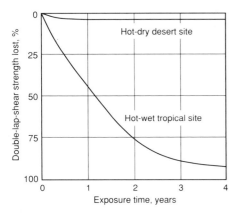

Fig. 1 Effect of outdoor weathering on the strength of aluminum alloy/epoxy-polyamide joints.
Source: Ref 2

Migration of Water to Adhesive Joints

When adhesive joints are exposed to wet environments, water molecules will migrate and be preferentially adsorbed into the interface region. This is because joint substrates, such as metal or metal oxides, have very high surface energies and water permeates through all organic adhesives. Water can enter either by diffusion through the bulk adhesive layer or by wicking along the adhesive-adherend interface. Ingression or wicking along the interface becomes important in a system where water may readily displace adhesives from the substrate, and the displacement is augmented by preexisting microcracks or debonded areas at the interface, which originate from poor wetting by the adhesives.

However, in a typical structural joint, such as epoxy-metal, many researchers (Ref 3-5) have found that water generally enters a joint system by diffusion through the epoxy rather than by passage along the interface.

Water Diffusion in Polymers. The water molecule is relatively small and in the liquid state is strongly associated through hydrogen bonding. The enthalpy of hydrogen bond formation is in the range of 14.2 to 27.6 kJ/mole (3.4 to 6.6 kcal/mole) (Ref 6); thus, strong localized interactions could develop between water molecule and polar groups in polymers. The rate of water diffusion in a polymer is typically "non-Fickian" such that the diffusion coefficient (D) is not a constant, but it may depend strongly on the concentration and also the sample thickness (Ref 7). The overall rate of water penetration increases because of a number of factors, such as increase in temperature, presence of internal and external stresses, and swelling stresses. A rigorous analysis on the kinetics of water diffusion in adhesive joints has been published recently in a series of papers (Ref 8) in which a comprehensive coupled model has been developed to relate the diffusion coefficient of water molecules in polymers to temperature, strain, and penetrant concentration.

Table 1 lists values of the permeability coefficient (P) and the diffusion constant D of water in some polymers. The permeability coefficient P is defined as the amount of vapor (cc STP) permeating 1 cm^2 of cross-sectional area of 1 cm thickness in 1 s with a pressure difference of 1 cmHg across the polymer. P is governed by the product of the diffusion coefficient (D) and the solubility (S) of water in the polymer, that is, $P \propto D \cdot S$. As shown in Table 1, both epoxy and phenolics, which are common adhesives for structural joints, show relatively low diffusion rates and thus are less susceptible to moisture attack (Ref 2).

For the purpose of estimating sorption kinetics in adhesive joints, however, it is reasonable to use a Fickian diffusion model. The rate of water diffusion into the laminated area (Fig. 2) of a joint may be analyzed by the following transient diffusion equation for a two-dimensional diffusion case (Ref 9):

$$\frac{\partial C}{\partial t} = D\left(\frac{\partial^2 C}{\partial x^2} + \frac{\partial^2 C}{\partial y^2}\right) \qquad \text{(Eq 1)}$$

with boundary conditions for a rectangular area given by $-a < x < a$ and $-b < y < b$, where a and b are half lengths of the laminated area. The solution to this equation is given by (Ref 10):

$$\frac{c(x,y)}{c(\infty,\infty)} = 1 - \phi(a,x) \cdot \phi(b,y) \qquad \text{(Eq 2)}$$

where $c(\infty,\infty)$ is the maximum absorption by the adhesive layer, and:

$$\phi(a,x) = \frac{4}{\pi} \sum_{N=0}^{N=\infty}$$

$$\left\{ \frac{(-1)^n}{2n+1} \exp\left[\frac{-(2n+1)^2 \cdot \pi^2 \cdot D \cdot t}{4a^2}\right] \right. $$
$$\left. \cdot \cos\frac{(2n+1) \cdot \pi \cdot x}{2a} \right\} \qquad \text{(Eq 3)}$$

The term $\Phi(b,y)$ has the same expansion. When monitoring the diffusion along the center line (x axis), Eq 2 simplifies to:

$$\frac{c(x,0)}{c(a,0)} = 1 - \phi(a,x) \cdot \phi(b,0) \qquad \text{(Eq 4)}$$

Equation 4 can be computed as a function of x for a given value of a. Figure 3 gives an example of such an analysis where water concentration in an epoxy is plotted as a

Table 1 Permeability coefficients (P) and diffusion constants (D) of water in various polymers

Polymer	Temperature °C	°F	$P \times 10^{-9}$	$D \times 10^{-9}$
Vinylidene chloride/acrylonitrile copolymer	25	75	1.66	0.32
Polyisobutylene.................................	30	85	7–22	...
Phenolic	25	75	166	0.2–10
Epoxy	25	75	10–40	2–8
Epoxy/tertiary amine	20	70	...	2.4
	40	105	...	6.5
	60	140	...	18.1
	90	195	...	60.7
Polyvinyl chloride	30	85	15	16
Polymethyl methacrylate	50	120	250	130
Polyethylene (low density).....................	25	75	9	230
Polystyrene	25	75	97	...
Polyvinyl acetate	40	105	600	150

Source: Ref 2

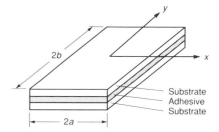

Fig. 2 Model adhesive laminate for bidirectional diffusion analysis

function of diffusion time and distance into the joint. This calculation is based on a diffusivity value of 18.1×10^{-9} cm²/s for the epoxy at 60 °C (140 °F) (Ref 2). The water front gradually advances toward the center, and the entire sample will eventually attain an equilibrium moisture content.

Liquid Water Versus Vapor—Sorption Isotherm. A wet environment may be either liquid water or a vapor of varying relative humidity (RH). At 100% RH, the chemical potential of the vapor is the same as that of the liquid; therefore, sorption, diffusion, and equilibrium moisture uptake will be the same as in a liquid water environment. At partial pressures of less than 100% RH, however, the equilibrium moisture uptake will be substantially lower than that in liquid water. Also, if diffusivity is concentration dependent such that the diffusion coefficient increases with concentration, a lower RH environment may slow the sorption kinetics as well.

Figure 4 shows a typical sorption isotherm at 25 °C (75 °F) for a DGEBA epoxy cured with diaminodiphenyl sulfone (DDS) at 167 °C (320 °F) for varying times (Ref 11). Equilibrium moisture uptake is also reduced

by the increase in degree of cross linking in the epoxy, as it is cured for a longer time.

Critical Water Concentration. It has been reported that a critical water concentration exists below which water-induced damage of the joint may not occur. For an epoxy system, it is estimated to be 1.35 wt% (Ref 2). Exposure of metal-epoxy joints to a humid environment (below 50% RH at 60 °C, or 140 °F, for example) will not cause any permanent damage to the joint strength (Ref 2). Any loss in the joint strength by the absorbed water can be restored upon redrying if the equilibrium moisture uptake was below the critical water concentration.

In a silane-primed joint of α-Al_2O_3/ polyethylene laminates, water immersion leads to a delamination that progresses from the laminate edges through the hydrolysis of the silane layers. Upon redrying the laminate, however, areas with less than a critical water concentration were restored fully, whereas the damage is irreversible in the areas where moisture uptake is greater than the critical concentration of water (Ref 9).

Strength Degradation and Failure Mode

Under dry conditions, failure of structural joints normally occurs by cohesive failure of the adhesive layer. Prolonged exposure to a wet environment, however, shifts the failure mode to adhesive failure through the polymer-substrate interface (Ref 2, 4, 12, 13). The loss of joint strength by water is therefore believed to be due to degradation of the interface rather than weakening of the bulk adhesive (Ref 14-16). In epoxy, for example, absorbed water plays the role of a plasticizer, thereby substantially reducing the tensile modulus and lowering the glass transition temperature (T_g). Upon dehydration, however, original values of the modulus and T_g are restored, suggesting that these effects are generally reversible (Ref 16).

In a similar study (Ref 14), changes of the strength of an epoxy matrix were compared with the shear strength loss of aluminum/ epoxy joints when exposed to vapors of water and ethanol. The results are shown in Fig. 5. The tensile strength of an epoxy adhesive was degraded much more by ethanol than by water vapor. However, in adhesive joints, shear strength was lost only in water vapor, whereas ethanol vapor had very little to no effect. These studies further support the idea that the loss in joint strength is primarily due to degradation of interfacial region through water-substrate interaction.

Mechanism of Strength Loss

Although a number of mechanisms have been proposed to account for the loss of interfacial strength, essentially two complimentary mechanisms are considered successful in explaining most of the observed phenomena. The following discussion explains their nature.

Displacement of Adhesive by Water. The stability of a metal oxide/polymer interface is examined in terms of the thermodynamic work of adhesion (Ref 2). The thermodynamic work of adhesion (W_A) is defined as the energy required to separate unit area of interface into the respective surfaces. The work of adhesion (Ref 17) can be expressed by:

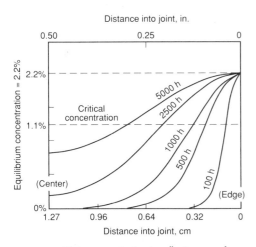

Fig. 3 Water concentration in adhesives as a function of distance into joint at a given time. Source: Ref 2

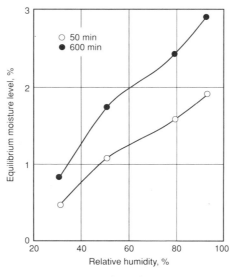

Fig. 4 Sorption isotherm for water in epoxy (DGEBA + DDS) cured at 160 °C (320 °F) for 50 mm (2 in.) and 600 min. Source: Ref 10

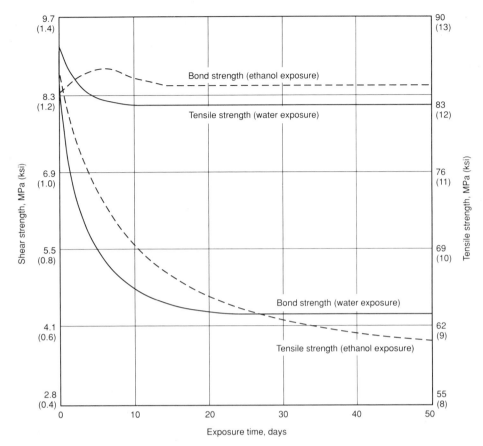

Fig. 5 Shear strength of aluminum epoxy joints and tensile strength of epoxy resin after exposure to water and ethanol at 90 °C. Source: Ref 14

$$W_A = \gamma_a + \gamma_b - \gamma_{ab} \qquad \text{(Eq 5)}$$

where γ_a and γ_b are the surface free energies of the polymer and metal oxide, respectively, and γ_{ab} is the interfacial free energy. In the presence of liquid, the work of adhesion (W_{Al}) becomes:

$$W_{Al} = \gamma_{al} + \gamma_{bl} - \gamma_{ab} \qquad \text{(Eq 6)}$$

A positive value of W_A or W_{Al} indicates that the interface is thermodynamically stable, whereas a negative value reflects the instability of the interface in which polymer and metal oxide layers dissociate spontaneously. The calculation of W_{Al} will thus predict the stability of the interface in a wet environment. The thermodynamic work of adhesion, W_A or W_{Al}, is calculated from the experimentally determined values of the surface free energies for solids and liquids.

Table 2 lists values for components of typical organic adhesive-metal oxide joint

systems. Using these values, the work of adhesion for a number of interfaces were calculated (Ref 2); the results are listed in Table 3 for an inert environment (W_A) and for water (W_{Al}). In a dry environment, interfaces show large positive values of work of adhesion indicating that they are stable. However, when water is present, the work of adhesion becomes negative. The change from a positive to a negative work of adhesion provides a driving force for the displacement of adhesive on the metal oxide or glass substrate by water. In contrast to metal-oxide/epoxy or glass-epoxy cases, a greater stability is found in carbon fiber-epoxy interface for which both W_A and W_{Al} remain positive.

The spontaneous dissociation described above would cause joint failure if the bond relied solely on these dispersive forces as is the case for a microscopically smooth adherend. The work of adhesion alone, however, cannot predict the stability of actual joints when interfacial adhesion may involve covalent bonding, interdiffusion, or mechanical interlocking mechanisms.

Hydration of Oxide Layers. Another way water can degrade the strength of adhesive joints is through hydration of the metal oxide layer at the interface. Common metal oxides, such as aluminum and iron oxides, undergo hydration. The resulting metal hydrates become gelatinous, and they act as a weak boundary layer because they exhibit very weak bonding to their base metals (Ref 23, 24).

In aluminum alloy (2024-T3)-epoxy joints, for example, the initial oxide produced on the aluminum substrate is usually amorphous Al_2O_3. Upon exposure to moisture, Al_2O_3 is converted to aluminum hydroxide with a chemical composition between that of boehmite ($Al_2O_3 \cdot H_2O$) and pseudoboehmite ($Al_2O_3 \cdot 2H_2O$). Failure surface analysis reveals that the hydroxide layer is normally attached to the adhesive side, suggesting that adhesion of the hydroxide to aluminum is very weak. Thus, once a hydroxide is formed it is separated easily from the substrate, causing failure of the joint. It is also suggested that the hydroxide attached to the adhesive layer is formed directly from the original aluminum oxide that was mechanically bonded to the aluminum (Ref 24).

As the crack opens up during its propagation, the freshly exposed aluminum metal surface will undergo further hydration through a corrosion reaction that can be described as:

$$2Al + 4H_2O \rightarrow 2AlOOH + 3H_2 \uparrow \qquad \text{(Eq 7)}$$

Figure 6 is the schematic representation of the model proposed by Venables and coworkers (Ref 25) to illustrate the mechanism of adhesion strength loss through the hydration of aluminum oxide discussed above.

Table 2 Dispersion and polar components of the surface free energies of typical joint components

Surface	γ^d mJ/m^2	γ^d ft · lbf/in.^{2}	γ^p mJ/m^2	γ^p ft · lbf/in.^{2}	γ mJ/m^2	γ ft · lbf/in.^{2}	Reference
Epoxy adhesive	41.2	19.6	5.0	2.4	46.2	22.0	5
Silica	78.0	37.1	209.0	99.5	287.0	136.6	18, 19
Aluminum oxide	100.0	47.5	538.0	256.0	638.0	303.6	20, 21
Water	22.0	10.5	50.2	23.9	72.2	34.4	22
Ferric oxide	107.0	50.9	1250.0	594.5	1357.0	645.7	5

Table 3 Work of adhesion values for various interfaces

Interface	Work of adhesion In inert medium mJ/m^2	Work of adhesion In inert medium ft · lbf/in.$^2 \times 10^{-6}$	Work of adhesion In water mJ/m^2	Work of adhesion In water ft · lbf/in.$^2 \times 10^{-6}$
Epoxy-ferric oxide	291	138	−255	−121
Epoxy-silica	178	85	−57	−27
Epoxy-aluminum oxide	232	110	−137	−65
Epoxy-carbon fiber	88–90	42–43	22–44	10–21

Source: Ref 2

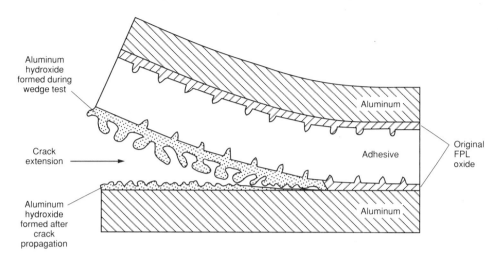

Fig. 6 Schematic drawing of the failure mechanism in an aluminum/polymer joint system during wedge testing in humid environment. The original oxide is converted to hydroxide, which adheres poorly to the aluminum substrate. Source: Ref 24

The stability of the oxide surface against hydration was found to vary significantly depending on the type of metal treatment employed (Ref 26, 27). For example, in the aluminum case, phosphoric acid anodization (PAA) (Ref 26) produces an oxide surface that outperforms the surface produced by the Forest Products Laboratory (FPL) (Ref 28) process in joint integrity as well as in long-term durability. In the FPL process, the surface is typically decreased and alkaline cleaned, followed by immersion in a solution containing $Na_2Cr_2O_7 \cdot 2H_2O$, H_2SO_4, and H_2O in a 1:10:30 ratio by weight for 15 to 30 min. In the PAA process, the surface is first treated by FPL process and then anodized in an aqueous solution containing 10% by weight of H_3PO_4 for about 25 min.

The better performance in the PAA-treated surface is attributed to the oxide morphology, which contains a thicker hexagonal cell structure with longer whiskerlike protrusions (10 nm × 100 nm, or 0.4 μin. × 4.0 μin.) than the FPL-treated surface, and this provides a polymer-oxide interface similar to the fiber-reinforced structure with a more effective mechanical interlocking (Ref 23).

Hydration studies on the PAA oxide reveal that improved durability is due to the presence of a monolayer of $AlPO_4$ that is adsorbed on the porous aluminum oxide. Hydration of the oxide is preceded by absorption of water in $AlPO_4$ and subsequent dissolution of phosphates. This prolongs the overall incubation time of oxide hydration and thus improves durability. This finding leads to the concept of applying a monolayer of hydration inhibitor to improve the joint durability (Ref 27).

Compared to aluminum oxide, titanium oxide, in general, is much more stable because hydration is extremely slow (almost to the point of not occurring), even at elevated temperatures (Ref 29). This oxide can undergo amorphous-to-crystalline TiO_2 (anatase) transformation under severe conditions without hydration. This process may be catalyzed by water, leading to joint strength degradation (Ref 30).

Improvement of Joint Durability

Wet strength durability of adhesive joints can be improved either by preventing water from reaching the interface in sufficient quantity (above the critical water concentration) or by improving the durability of the interface itself. Some approaches that have been taken by researchers are summarized below.

Increasing Barrier to Water Diffusion. The selection of a polymer having low values of permeability and diffusivity as an adhesive (Table 1) is an obvious approach to retard or reduce water migration into joints. However, it must be balanced with the other properties an adhesive should have (such as wettability and processibility, mechanical strength and toughness, and adhesion strength) and with cost. Another way to reduce permeability is to incorporate inert mineral or metallic fillers into the adhesive formulation (Ref 31) (assuming the water cannot easily debond the second phase). The inert second phase acts as a barrier to water diffusion, thereby effectively reducing the overall permeability. Another direct approach is to coat the exposed edges of the joint with sealants having very low moisture permeabilities. Such a layer must, however, be thick enough to be effective; this may become impractical from a processing point of view.

An approach has also been taken to chemically modify the adhesive to reduce the water permeation. For example, when an epoxy is highly fluorinated, the equilibrium moisture uptake is reduced by as much as 85% (Ref 32), thereby retarding water-induced degradation through the suppression of the local moisture levels below the critical water concentration.

Hydration Inhibition or Retardation. Because water-induced degradation in polymer-metal joints proceeds first by conversion of oxide to hydroxide, an approach was taken to retard or inhibit the hydration of oxides by applying a monolayer film of certain organic inhibitors (Ref 33). In aluminum-epoxy joints, for example, an FPL-treated aluminum oxide surface was treated with nitrilotris(methylene)phosphonic acid (NTMP) by dipping into an aqueous solution containing 10-300 ppm of NTMP; this led to a monolayer coverage of NTMP as determined by x-ray photoelectron spectroscopy (XPS) (Ref 33). The presence of this phosphonic acid on the oxide surface significantly improved the stability of the FPL-treated aluminum. Figure 7 shows the effect of NTMP treatment on the wedge-test crack length of aluminum-epoxy panels as a function of exposure time in 100% RH at 60 °C (140 °F). The data demonstrate that the performance of the FPL-treated surface is improved significantly—to a level equal to or better than that of the PAA-treated surface—by pretreatment in an inhibitor solution (2-10 ppm) (Ref 27).

The PAA-treated oxide is generally more stable against moisture than the FPL-treated oxide due to the adsorbed phosphate. However, it can be improved further by NTMP treatment (Ref 27). The application of an inhibitor monolayer, therefore, appears to be an important route for durability improvement.

Another approach in retarding oxide to hydroxide conversion is to create highly crystalline Al_2O_3, as opposed to the amorphous Al_2O_3 normally generated by FPL or PAA treatments. Crystalline Al_2O_3 is much more stable against moisture: It can survive boiling water immersion over a week, whereas amorphous Al_2O_3 undergoes hydration in a matter of minutes or hours (at 80 °C, or 175 °F, in water). At present, no method has been established for creating a crystalline oxide layer on aluminum substrates. However, the development of such a method would certainly offer an attractive alternative.

Application of Primer (Coupling Agents). A primer solution can be applied to the adherend surface prior to bonding with the adhesive to improve wet strength durability. One type of primer is a chemical coupling agent that can establish primary chemical bondings across the adhesive-adherend interface. The most common type is organosilanes, originally developed for the glass fiber/polymer system (Ref 34). Typical silanes have the structure $X_3Si(CH_2)_nY$, where n = 0 ~ 3, X is a hydrolyzable group

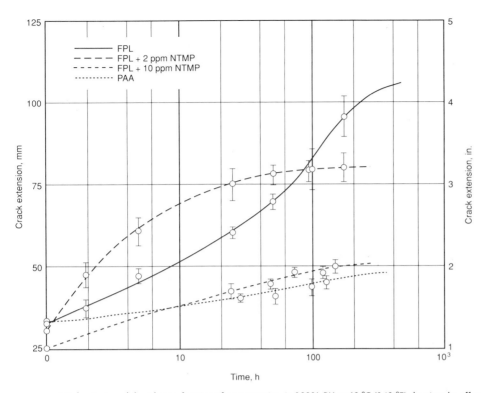

Fig. 7 Wedge-test crack length as a function of exposure time in 100% RH at 60 °C (140 °F) showing the effect of treating adherend by inhibitor solution (10 ppm of NTMP). Source: Ref 27

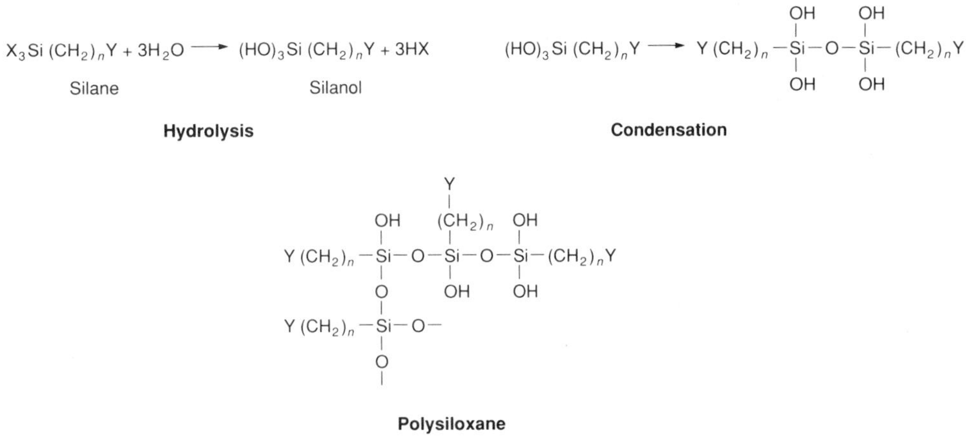

$$X_3Si(CH_2)_nY + 3H_2O \longrightarrow (HO)_3Si(CH_2)_nY + 3HX$$

Silane   Silanol

Hydrolysis

$$(HO)_3Si(CH_2)_nY \longrightarrow Y(CH_2)_n-\overset{\overset{\displaystyle OH}{|}}{Si}-O-\overset{\overset{\displaystyle OH}{|}}{Si}-(CH_2)_nY$$

Condensation

Polysiloxane

Fig. 8 Hydrolysis and condensation of silane applied from a dilute solution to form a polysiloxane network film on the adherend surface

$$Y(CH_2)_n-\overset{\overset{\displaystyle OH}{|}}{\underset{\underset{\displaystyle OH}{|}}{Si}}-OH + HO-Si\,\text{glass} \longrightarrow Y(CH_2)_n-\overset{\overset{\displaystyle OH}{|}}{Si}-O-Si\,\text{glass}$$

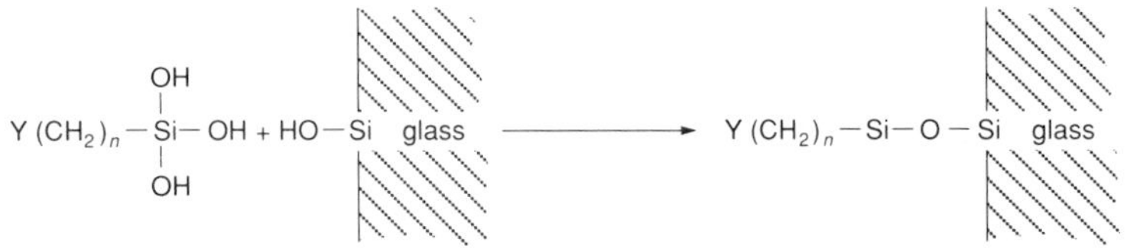

Fig. 9 Reaction of silanol with the surface hydroxyl group to establish Si—O—Si covalent bonding on the glass in a glass/polymer system

such as alkoxy (—OR), and Y is an organofunctional group capable of reacting with the polymer matrix. When applied from a dilute solution, silane undergoes hydrolysis and condensation to form a polysiloxane network film on the adherend surface, as shown in Fig. 8.

In a glass/polymer system, the silanol reacts with the surface hydroxyl group on the glass under certain conditions to establish Si—O—Si covalent bonding (Fig. 9); such chemical coupling is believed to enhance the interface stability in wet environments (Ref 35, 36).

Silanes have also been applied to metal-polymer systems, which resulted in improved durability (Ref 37-40). For example, applying γ-glycidoxylpropyltrimethoxy silane to a steel substrate led to a substantial improvement in the durability of steel-epoxy joints (Ref 37, 38). Likewise, γ-aminopropyltriethoxy silane (γ-APS) was successfully used in an aluminum-epoxy joint system (Ref 39, 40). It is generally recognized that the increased durability is directly related to the presence of oxirane bonds (metal—O—Si) that are formed through the reaction of hydroxyl groups on the metal oxide surface and silanol groups (Ref 37, 38). However, it must be pointed out that metal—O—Si or Si—O—Si bonds themselves are susceptible to hydrolysis, and the exact mechanism by which silanes improve the wet strength is still not clearly understood.

In silane-primed joints, fracture surface analysis by Auger and XPS reveals that failure is often by cohesive failure through the polysiloxane layer (Ref 41); that is, the silane primer improves joint durability by increasing the water resistance of the metal oxide-polysiloxane interface; however, the primer layer itself becomes the weakest part of the joint because it is susceptible to hydrolysis. Also, the hydrophilic nature of silane tends to accelerate water penetration into the interface region. This points to the importance of the structure and properties of the polysiloxane network developed on the substrate prior to bonding with adhesives.

Recent work (Ref 9, 40, 42) showed that both the structure and the thickness of polysiloxane are indeed important variables in determining the strength of primed joints. For example, γ-APS adsorbed from an aqueous solution and dried at room temperature is only partially cross linked. Such a layer is surprisingly detrimental to the wet strength of aluminum-epoxy joints, that is, the strength is worse than the unprimed case. Only when it is highly cross linked by extended dehydration is the wet strength stability drastically improved, as shown in Fig. 10 for the wet strength of aluminum/γ-APS/epoxy joint where cross linking is varied by dehydration time (Ref 40). Also, there appears to exist an optimum thickness of the silane layer that produces maximum strength (Ref 9).

Although silanes are the most common primers, other coupling agents have been used for metal-polymer joints with varying degrees of success; these include organotitanates (Ref 43), betaketone (Ref 13), and multifunctional mercaptoesters (Ref 4).

Phenolic-base primers are also found to provide a very good durability in metal-polymer joints (Ref 44). High-temperature cure conditions employed for phenolics are believed to be conducive to the formation of ether linkages between the oxide surface and phenolics, and the presence of such

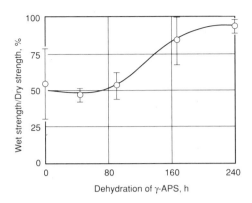

Fig. 10 Wet strength retention of Al/γ-APS epoxy joint as a function of dehydration cross linking of APS prior to joint formation (0.2 μm thick γ-APS, after 25 days of water immersion at 55 °C, or 130 °F)

bonding is considered to be responsible for the improved performance. Again, however, such linkage is highly susceptible to water attack, and thus the exact mechanism needs further explanation.

REFERENCES

1. W.C. Wake, *Adhesion and Formulation of Adhesives*, Applied Science Publishers, 1982
2. A.J. Kinloch, Interfacial Fracture Mechanical Aspects of Adhesion Bonded Joints—Review, *J. Adhes.*, Vol 10, 1979, p 193
3. A.W. Bethune, *SAMPE J.*, Vol 11 (No. 14), 1978, p 4
4. D.M. Brewis, J. Comyn, and J.L. Tegg, *Int. J. Adhes. Adhes.*, Vol 1, 1980, p 35
5. R.A. Glenhill and A.J. Kinloch, Environmental Failure of Structural Adhesive Joints, *J. Adhes.*, Vol 6, 1974, p 315
6. M. Gordon, C.S. Hope, L.D. Loan, and R.-J. Roe, *Proc. R. Soc.*, Vol A258, 1960, p 215
7. J. Crank and G.S. Park, Ed., *Diffusion in Polymers*, Academic Press, 1968
8. D.R. Lefebvre, D.A. Dillard, and T.C. Ward, A Model for the Diffusion of Moisture in Adhesive Joints, Part I: Equations Governing Diffusion, *J. Adhes.*, Vol 27, 1989, p 1; D.R. Lefebvre, D.A. Dillard, and M.F. Brinson, A Model for the Diffusion of Moisture in Adhesive Joints, Part II: Experimental, *J. Adhes.*, Vol 27, 1987, p 19; S. Roy, D.R. Lefebvre, D.A. Dillard, and J.N. Reddy, A Model for the Diffusion of Moisture in Adhesive Joints, Part III: Numerical Simulations, *J. Adhes.*, Vol 27, 1989, p 41
9. A. Kaul, N.-H. Sung, I. Chin, and C.S.P. Sung, Durability and Failure Analysis of Silane Treated α-Al₂O₃/Polyethylene Joint in Wet Environment, *Polym. Eng. Sci.*, Vol 24, 1984, p 493
10. H.S. Carslaw and J.C. Jaeger, *Conduction of Heat in Solids*, 2nd ed., Oxford University, 1959, p 173
11. J.B. Enns and J.K. Gillham, Effect of the Extent of Cure on the Modulus, Glass Transition, Water Absorption and Density of an Amine-Cured Epoxy, *J. Appl. Polym. Sci.*, Vol 28, 1983, p 2831
12. W. Brockman, Durability of Polymer Bonds, in *Adhesion Aspects of Polymeric Coatings*, K.L. Mittal, Ed., Plenum Press, 1982, p 265
13. A.J. DeNicola and J.P. Bell, Synthesis and Testing of β-Diketone Coupling Agents for Improved Durability of Epoxy Adhesion to Steel, in *Adhesion Aspects of Polymeric Coatings*, K.L. Mittal, Ed., Plenum Press, 1982, p 443
14. C. Kerr, N.C. MacDonald, and S. Orman, Effect of Hostile Environments on Adhesive Joints, *J. Appl. Chem.*, Vol 17, 1967, p 62
15. J.S. Noland, Some Factors for Achieving Environmental Resistance in 120 °C Structural Adhesives, *Polym. Sci. Technol.*, Vol 9A, 1975, p 413
16. R.I. Butt and J.L. Cotter, *J. Adhes.*, Vol 8, 1976, p 11
17. W.A. Zisman, Influence of Constitution on Adhesion, in *Handbook of Adhesives*, 2nd ed., I. Skeist, Ed., Van Nostrand, 1962, p 33
18. G.E. Boyd and H.K. Livingstone, *J. Amer. Chem. Soc.*, Vol 64, 1942, p 2373
19. L. Shartsis and S. Spinner, *J. Res. Nat. Bur. Stand.*, Vol 46, 1951, p 385
20. H. Schonhorn, *J. Phys. Chem.*, Vol 71, 1967, p 4578
21. J.J. Ramussen and R.P. Nelson, *J. Am. Ceram. Soc.*, Vol 54, 1971, p 3981
22. F.M. Fowkes, in *Treatise on Adhesion and Adhesives I*, R.L. Patrick, Ed., Marcel Dekker, 1967, p 325
23. J.D. Venables, D.K. McNamara, J.M. Chen, and T.S. Sun, *Appl. Surf. Sci.*, Vol 3, 1979, p 88
24. J.D. Venables, Review—Adhesion and Durability of Metal-Polymer Bonds, *J. Mater. Sci.*, Vol 19, 1984, p 2431
25. J.D. Venables, D.K. McNamara, J.M. Chen, B.M. Ditchek, T.I. Morgenthaler, T.S. Sun, and R.L. Hopping, *Proceedings of the 12th National SAMPE Technological Conference*, Society for the Advancement of Material and Process Engineering, Oct 1980, p 909
26. G.S. Kabayashi and D.J. Donnelly, Report DG-41517, The Boeing Company, Feb 1974
27. J.S. Ahearn, G.D. Davis, T.S. Sun, and J.D. Venables, Correlation of Surface Chemistry and Durability of Aluminum/Polymer Bonds, in *Adhesion Aspects of Polymeric Coatings*, K.L. Mittal, Ed., Plenum Press, 1983, p 288
28. H.W. Eichner and W.E. Schowalter, Report 1813, Forest Products Laboratory, 1950
29. B.M. Ditchek, K.R. Breen, T.S. Sun, and J.D. Venables, *Proceedings of the 12th SAMPE Technological Conference*, Society for the Advancement of Material and Process Engineering, Oct 1980, p 882
30. M. Natan and J.D. Venables, *J. Electrochem. Soc. Extended Abstracts*, Vol 83, 1983, p 379
31. *Handbook of Fillers and Reinforcements for Plastics*, H.S. Katz and J.V. Milewski, Ed., Van Nostrand, 1978
32. J.R. Griffith and J.D. Bultman, *Ind. Eng. Chem., Prod. Res. Dev.*, Vol 17 (No. 1), 1978, p 8
33. G.D. Davis, T.S. Sun, J.S. Ahearn, and J.D. Venables, *J. Mater. Sci.*, Vol 17, 1982, p 1807
34. *Composite Materials*, Vol 6, *Interfaces in Polymer Matrix Composites*, E.P. Plueddemann, Ed., Academic Press, 1974
35. A. Ahagon, A.N. Gent, and E.C. Hsu, Adhesion Science and Technology, in *Polymer Science and Technology*, Vol 9a, L.H. Lee, Ed., Plenum Press, 1975, p 281
36. E.P. Plueddemann, *Silanes*, Plenum Press, 1982
37. M. Gettings and A.J. Kinloch, *J. Mater. Sci.*, Vol 12, 1977, p 2511
38. M. Gettings and A.J. Kinloch, *Surf. Interface Anal.*, Vol 1, 1979, p 189
39. F.J. Boerio and C.A. Gosselin, Paper 2G, presented at 36th Annual Conference of Reinforced Plastics/Composites Institute, Society of the Plastics Industry, 1981
40. A. Kaul, N.-H. Sung, I.J. Chin, and C.S.P. Sung, Effect of Bulk Structure of Amino Silane Primer on the Strength and Durability of Al/Epoxy Joints, *Polym. Eng. Sci.*, Vol 26, 1986, p 768
41. M. Geetings, F.S. Baker, and A.J. Kinloch, Use of Auger and X-Ray Photoelectron Spectroscopy to Study the Locus of Failure of Structural Adhesive Joints, *J. Appl. Polym. Sci.*, Vol 21, 1977, p 2375
42. N.-H. Sung, A. Kaul, I. Chin, and C.S.P. Sung, Mechanism Studies of Adhesion Promotion by γ-Aminopropyltriethoxy Silane in α-Al₂O₃/Polyethylene Joint, *Polym. Eng. Sci.*, Vol 22, 1982, p 637
43. P.D. Calvert, R.R. Lalanandham, and D.R.M. Walton, in *Adhesion Aspects of Polymeric Coatings*, K.L. Mittal, Ed., Plenum Press, 1982, p 457
44. J.C. Bolger and A.S. Michaels, in *Interface Conversion for Polymer Coatings*, P. Weiss and G.D. Cheever, Ed., Elsevier, 1968, p 3

Electrochemical and Corrosion Effects on Adhesive Joints

Andrew Stevenson, Materials Engineering Research Laboratory Ltd. (England)

POLYMER-METAL adhesive joint durability in aqueous liquids has been a subject of concern to engineering end users for a long time. A lack of knowledge about mechanisms of joint failure in service and about reliable life prediction methods is still a considerable obstacle to the wider use of adhesive joints in critical engineering applications. However, important advances were made during the 1980s in understanding these areas, and adhesive joints are being used in some critical structural applications, particularly in aerospace and offshore engineering.

The role of electrochemical potentials in causing adhesive joint failure has only recently been studied scientifically (Ref 1). Studies of the role of electrochemical effects in the corrosion of metals are, on the other hand, well established. There has been some debate concerning which comes first: corrosion or adhesive bond failure. However, when the adhesive joint is immersed in an electrolyte and the adherend is electrically connected to become a cathode, the situation is clear: Corrosion of the metal is suppressed, but the rate of bond failure is enhanced. In this article, corrosion is considered to be an aftereffect rather than a cause of adhesive bond failure. However, an intimate relationship exists between the mechanisms of corrosion and cathodic bond failure; the latter has been considered by some authors to be a form of localized corrosion (Ref 2). The ability of cathodic potentials to cause bond failure appears to be common to all polymers and adhesive systems.

Analogous problems to those of adhesive joints have been reported for organic coatings (Ref 2, 3). Although the detailed mechanisms are different for coatings, the role of electrochemical potentials may be similar. If an adhesive system has such poor durability in water that it fails by other mechanisms before cathodic disbondment can occur, then the role of electrochemical effects is less clear. It is most clear for adhesive joints that otherwise have excellent durability in water. Elastomer-metal bonds formed under heat (for example, 150 °C, or 300 °F) and pressure (for example, 1380 kPa, or 200

psi) during vulcanization of the elastomer and in the presence of compatible bonding agents are generally considered to be the most durable adhesive joints in water-base environments. These bonds have been shown to be durable in seawater for several years with no evidence of any deterioration (Ref 1, 2). Corrosion is also inhibited where the elastomer is bonded to carbon steel.

This article is devoted predominantly, but not exclusively, to such highly durable adhesive joints. The resulting mechanisms are, however, of wider significance. Such adhesive joints are of considerable engineering importance because they have been used in critical structural components such as deep-water compliant oil platforms and subsea (thick-build) pipe coatings, where they are required to function for 20 or 30 years (Ref 4). The role of electrochemical potentials is particularly important because large numbers of sacrificial anodes are frequently used in subsea applications to inhibit corrosion of exposed steelwork. Delamination of a subsea structural bearing from cathodic bond failure could present a serious problem, even if it occurred gradually over a period of 10 years.

Exposure in Electrochemically Inert Conditions

In electrochemically inert conditions, elastomer-metal adhesive bonds have been found to be extremely durable, even after several years of total immersion in salt water. Elastomer-titanium adhesive joints have been immersed in seawater for periods exceeding 2½ years, with no measurable bond failure (Ref 1).

Diffusion of water into the polymer layer was studied in detail as a part of these experiments. One elastomer type (purified cis-polyisoprene) absorbed only 1% of water after 2½ years, whereas another (magnesia-cured polychloroprene) absorbed 35%. However, no bond failure occurred in either case when care was taken to maintain electrochemically inert conditions. This was done by ensuring that all fittings in the

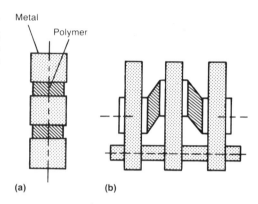

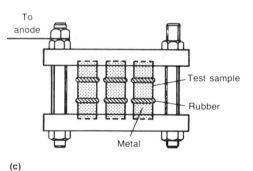

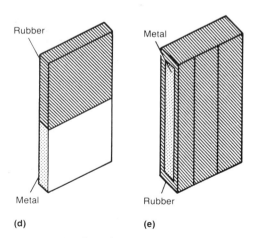

Fig. 1 Simple shear testpiece geometries. Testpieces are 25 mm (1 in.) in diameter with 10 mm (0.4 in.) layers of elastomer. (a) Unstrained. (b) Simple shear at 100% strain. (c) Compression at 25% strain. (d) Butt testpiece. (e) Peel testpiece

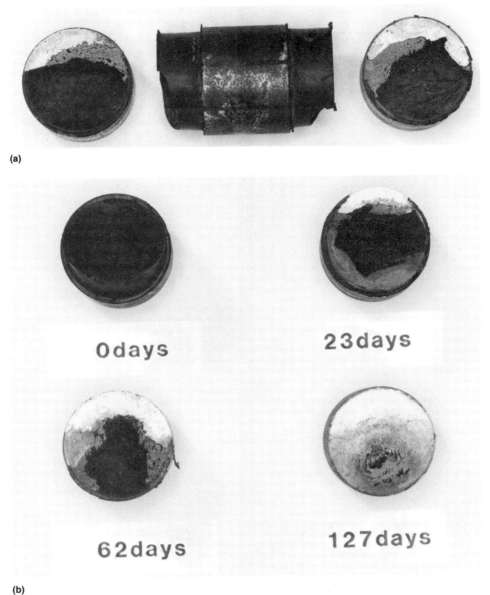

(a)

0days   23days

62days   127days

(b)

Fig. 4 Typical NR-carbon steel bond failure surfaces. (a) After 65 days synthetic seawater exposure at 3 °C (38 °F) with 1.05 V sacrificial anode. Upper part of interface exhibits blue deposit. (b) Progressive bond failure in synthetic seawater at 45 °C (104 °F) with 1.05 V sacrificial anode. Color of deposit is white.

Fig. 2 Partial NR-carbon steel bond failure after 1-year on-site deep-sea exposure. The bonds had no cathodic protection. Upper part of surface shows red-brown corrosion after bond failure.

environmental test tanks were of glass or titanium, thereby avoiding any electrochemical interactions between dissimilar metals. The water was circulated but not replenished. The same results were obtained at temperatures of 3, 23, and 40 °C (38, 73, and 104 °F). Figure 1 shows the type of testpiece geometries used for these experiments.

Similar tests were performed with three elastomer types, natural rubber (NR), acrylonitrile-butadiene rubber (NBR), and chloroprene rubber, using an industry standard adhesive primer system, Chemlock 205/220. The adhesive bonds were formed in each of these cases during vulcanization of the elastomer under heat (150 °C, or 300 °F) and pressure (1380 kPa, or 200 psi).

Similar tests have been performed with an adhesive system consisting of only a thin layer of epoxy resin. In these tests, failure was relatively rapid in salt water at 40 °C (104 °F). An unstrained testpiece geometry (see Fig. 1a) was used for the epoxy resin tests.

Similar results have also been obtained for carbon steel-elastomer systems (for ex-

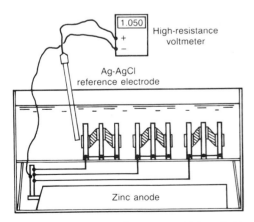

Fig. 3 Arrangement for tests at high cathodic potentials

ample, carbon steel with NR/C205-220) where the supply of oxygen was not constantly replenished. When it was replenished, for example, in 120 m (400 ft) deep-sea trials performed off the coast of northwestern Scotland with an undercurrent of oxygenated water, the results were different and both bond failure and metal corrosion were observed. The bond failure always appeared to precede the onset of metal corrosion (Fig. 2). In another laboratory experiment using the same polymer-adhesive-metal system, immersion in a solution with pH 12 for 6 months did not cause dry bond failure (Ref 5). This suggests that in electrically inert conditions, high-pH solutions alone are not enough to cause failure of at least this type of adhesive joint.

Effect of High Cathodic Potentials

The presence of high cathodic potentials has the effect of accelerating bond failure when some adhesive joints are exposed to salt water. A zinc sacrificial anode in combination with steel adherends is commonly used in cathodic protection engineering applications. Figure 3 shows a laboratory simulation of this system. The testpieces used were predominantly elastomer-steel of the type shown in Fig. 1(a), although earlier work also used the types shown in Fig. 1 (d and e). The simulation produced a measured cathodic potential of 1040 to 1070 mV at the testpiece surface (with reference to an Ag-AgCl reference electrode). Bond failure

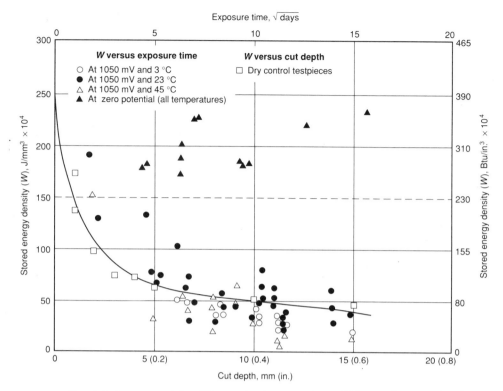

Fig. 5 Stored elastic energy at failure (W) versus bond exposure time at 1050 mV and 3, 23, and 45 °C (38, 73, and 104 °F), and at zero potential (all temperatures). W versus cut depth is plotted for dry control testpieces.

was rapid at all temperatures tested without any corrosion of the carbon steel. Figures 2 and 4 show typical failure surfaces from such tests (Ref 6).

Each failure surface consists of three regions. The first is a black region where failure occurs through the polymer. This region may be asymmetrically located because, in simple shear, bond failure initiates at the leading edge, where local shear or surface tensile strains are at a maximum. After bond failure, a gap may open between elastomer and metal in this region. The second region is adjacent to the region of polymer fracture; it is a gray area of bare metal where the failure locus is predominantly at the interface between the metal and its primer (Chemlock 205) coating. Small particles of fractured polymer may be left adhering to the steel in this region. This type of bond failure occurs only after a pull-to-failure test has been carried out. The third and outermost failure region exhibits a

powder deposit. This deposit was of a bluish color for tests at 3 °C (38 °F) (Fig. 4a) and a cream-white color for tests at 45 °C (113 °F) (Fig. 4b). The color was intermediate at 23 °C (73 °F).

Chemical analysis showed that the blue-white deposit contained predominantly magnesium, with lesser amounts of sulfur, chlorine, calcium, iron, and zinc, whereas the cream-white deposit was predominantly zinc, with lesser amounts of sulfur, calcium, magnesium, and chlorine. All of these elements were constituents of the seawater used in the experiments. These deposits formed on the steel surface after bond failure and increased in thickness until, after about 200 days of seawater exposure, they filled the gap opened between rubber and metal by bond failure and inhibited further bond failure. Similar deposits occurred on all exposed electrically connected steel surfaces, thus indicating that the cathodic protection provided by the sacrificial zinc an-

odes was working correctly. Salt deposition on the passivated steel surface (cathode) is part of the electrolytic process.

Figure 4(b) also shows how the area of bond failure in both the first and second regions increases with time of saltwater exposure, whereas the area of polymer failure (the third region) decreases. When determining any change in adhesive bond strength (interfacial failure energy), it is important to know if there was any change in the cohesive failure strength of the polymer. Clearly, if the polymer fracture strength decreased during a test, then the area of bond failure could be anomalously smaller than it would be if the strength had remained constant; such an occurrence would cause a misleading picture to be obtained.

To investigate this, the energy required to cause fracture was measured by integrating the force-deflection curve obtained from each shear pull-to-failure test. This energy was then used to derive a characteristic stored energy density (W) at fracture. In the absence of bond failure, W characterizes the critical fracture energy of the polymer and yields a value comparable with results from standard fracture or tear tests. With an area of partial bond failure, however, the fracture geometry is more complicated, although the same approach can be used.

It is sufficient for present purposes to describe the results empirically. Figure 5 shows the stored energy density plotted against the seawater exposure time for each testpiece with a cathodic potential of 1050 mV. The root exposure time correlates with the area of bond failure, the latter acting as a starter crack for subsequent cohesive polymer fracture.

As a control experiment, a set of new dry testpieces was cut at the bond to various depths, simulating the full range of observed bond failure depths. The geometry of bond failure was also simulated. These new unexposed testpieces failed at identical failure energies to those of the testpieces that had exhibited bond failure following seawater exposure for up to 2 years. This result clearly indicates that no significant decrease in polymer fracture strength took place during the tests; it also indicates that reductions in force to failure are only geometrical in nature. The failure penetration depths are thus entirely representative of changes in bond strength at the adhesive-metal interface.

The characteristic stages of the bond failure process are illustrated in Fig. 6, in which the area of bond failure penetration has been plotted against the square root of seawater exposure time. If diffusion processes are involved, a linear relation between penetration depth and the square root of time is expected from diffusion theory.

The first stage is a time lag observed before any bond failure can be measured (that is, <0.5 mm, or 20 mil, failure pene-

Table 1   Bond failure penetration depths at various imposed cathodic potentials during seawater exposure at 23 °C (73 °F)

Cathodic potential, mV	Induction period, days	Rate of bond failure, mm/$\sqrt{day}$	Maximum penetration depth mm	in.
1050	1	1.0	15	0.6
950	5	1.0	9	0.36
850	5	1.0	9	0.36
750	20	1.0	7	0.28
610	30	1.0	4	0.16
0	>1000	...	0	0

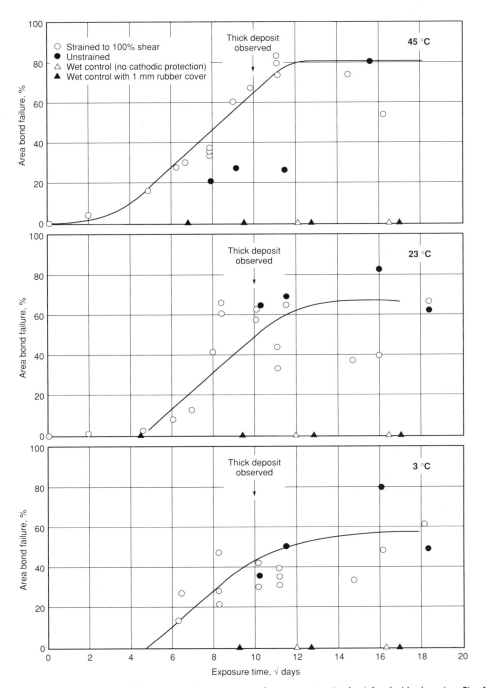

Fig. 6 Area of bond failure versus the square root of exposure time (in days) for double-shear (see Fig. 1c) NR-carbon steel testpieces with cathodic protection

penetration tests for a range of cathodic potentials in seawater at 23 °C (73 °F), including the zinc anode case as 1050 mV. This mechanism of bond failure requires exposure of the bond edge to the electrolyte. Figure 6 also includes results for testpieces (which had been encapsulated in a 1 mm thick cover of elastomer) when no bond failure at all was recorded.

Effect of Impressed Current

Impressed current methods may create lower potential differences, thereby reducing the rate of bond failure. The effect of potential differences lower than 1.05 V on carbon steel testpieces can be illustrated by describing a series of experiments that used a potentiostatic arrangement to supply a constant impressed current at a chosen potential difference. A series of platinum electrodes were installed and connected to a current source to provide constant cathodic potentials of 0.95, 0.85, 0.75 and 0.61 V at the bond edge of the carbon steel testpieces. The test method was otherwise the same as that described for the experiments with zinc anodes. Separate seawater tanks were used for each different cathodic potential. All tests with impressed currents were performed at 23 °C (73 °F).

Substantial delamination occurred after long periods of seawater exposure in all cases. No steel corrosion was observed with cathodic potentials of 950 or 850 mV, although some was observed after long periods with 750 mV, and more with 610 mV. No thick powder deposits were observed on exposed steel surfaces. Once again, all changes in modulus or energy at a break could be attributed to the geometrical changes caused by bond failure; that is, there were no reductions in the cohesive polymer fracture strength itself.

The induction period progressively increased as the level of imposed cathodic potential was reduced; it went from 1 day at 1050 mV to 30 days at 610 mV (Fig. 7). After the induction period, bond failure began to be observed and the failure penetration rates were the same for all potentials as far as could be ascertained within the limits of experimental error. This rate was 1.0 (±2) mm/(day)$^{1/2}$. After very long exposure periods (up to 3 years), the penetration depth approached an equilibrium value that increased from 4 mm (0.16 in.) at 610 mV to the maximum possible (≧12.5 mm, or 0.5 in.) at 1050 mV. These results are summarized in Table 1. Identical testpieces exposed to seawater in the same restraining jigs (Fig. 1b), but without electrical connection to a potentiostat or to a dissimilar metal (that is, 0 mV), showed no bond failure whatsoever—even after more than 3 years of seawater exposure. This interpretation is influenced by the well-known square root law of classical diffusion theory, whereby

tration depth). This induction period increases with decreasing temperature. The second stage is marked by an increasing area of bond failure into the testpiece; during this period, the increase in the bond failure area is an approximately linear function of the square root of exposure time. This suggests that liquid diffusion is a controlling factor; however, unlike the situation for organic coatings, only lateral diffusion along the interface is considered to be important, not vertical diffusion through the bulk polymer. A failure penetration rate can be defined for each test condition associat-

ed with each area; for example, it would be defined as the average of the maximum penetration depth for each of four bond surfaces in a testpiece of the type shown in Fig. 1(a).

Finally, there was an asymptote to an equilibrium maximum depth of bond failure penetration. For the testpiece design used, the maximum possible depth was 15 mm (0.6 in.) (assuming some asymmetry). In a number of cases, the equilibrium failure penetration depth was found to be considerably less than 15 mm (0.6 in.). Table 1 summarizes results of adhesive bond failure

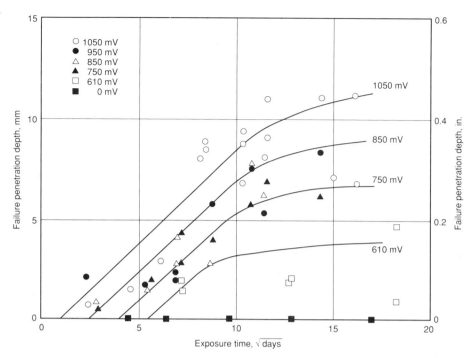

Fig. 7 Penetration depth of bond failure versus the square root of seawater exposure time for NR-carbon steel testpieces at 23 °C (73 °F)

electrochemical conditions and all temperatures. Again, there is a definite time lag, or induction period, before the observation of any bond failure. This time lag increases with decreasing temperature and decreasing cathodic potential. The dissimilar metal combination showing the largest cathodic potential is carbon steel-zinc, at 1050 mV. The exact values of the local potentials caused by the other metallic combinations could not be determined, but they are estimated to be less than 1 V. The measured rates of failure penetration are all 1.0 (±0.2) mm/(day)$^{1/2}$ within experimental error after the induction period for all temperatures and all electrochemical conditions. It is possible that the rate varied slightly with temperature, but the precision of the experiment did not permit this to be discriminated. There appeared to be an approach to a final equilibrium depth of maximum bond failure in all cases. This increased with increasing temperature and increasing cathodic potential. The results of this investigation are summarized in Table 2. The most striking differences are in the induction period, which increased with decreasing temperature or decreasing potential.

The results of this investigation confirm that electron flow caused by the potential difference between anode and cathode is the dominant effect. This phenomenon of cathodic disbondment can occur due to dissimilar metals being in contact, without a specially introduced sacrificial anode. These data indicate a need to widen the range of practical applications of adhesion for which electrochemical potentials are considered as factors that determine durability.

the rate of liquid ingress is expected to increase proportionally with the square root of time. This occurs after a time lag that is dependent on the electrochemical conditions. The similarity of diffusion rates is most likely explained and determined by polymer composition, which remains the same.

Effect of Dissimilar Metals in Contact

When placed in electrical contact in an electrolyte, dissimilar metals can create similar potential differences depending on, for example, their place in the galvanic series. Adherends of Inconel, two types of titanium, and stainless steel were investigated using the same adhesive system and test arrangement described earlier. The specimens were exposed to seawater at 3, 23, and 45 °C (38, 73, and 104 °F) in separate tanks, but all were in electrical contact with carbon steel. The polymer layers in the testpieces were strained to 100% shear in

the holding jigs shown in Fig. 1(b). These holding jigs were all fabricated from carbon steel.

The system was not electrochemically inert for the Inconel, stainless steel, or titanium testpieces because the possibility of electrochemical interactions between the dissimilar metals in contact existed throughout the seawater exposure period. The outer surfaces of the holding jigs were initially painted with a chlorinated rubber coating to prevent gross corrosion of the carbon steel. However, this coating lasted only about 1 year, after which some corrosion of the outer surfaces of the holding jigs and a layer of corrosion product on the floor of the test tank could be observed. Carbon steel testpieces were also mounted in the carbon steel holding jigs and electrically connected to a large block of zinc to create a known cathodic potential of 1050 mV.

The full set of results for this investigation is shown in Fig. 8, which is a plot of the failure penetration depth versus the square root of the seawater exposure period for all

Effect of Mechanical Strain

The asymmetric appearance of the region of bond failure in Fig. 2 and 4 suggests that mechanical strain on the elastomer layer may influence the bond failure penetration depth. A series of experiments were conducted using various elastomer strains. Carbon steel testpieces were exposed to seawater at 23 °C (73 °F) with a cathodic potential of 1050 mV and zero elastomer strain, 100% shear strain, or 25% compression strain (the last in the holding jigs shown in Fig. 1c).

Table 2 Bond failure penetration depths for various combinations of dissimilar metals during seawater exposure

Metal combination	Induction period, days, at 3 °C (38 °F)	23 °C (73 °F)	45 °C (104 °F)	Maximum penetration depth at 3 °C (38 °F) mm	in.	23 °C (73 °F) mm	in.	45 °C (104 °F) mm	in.
Carbon steel-zinc	5.8	1.1	0.8	10	0.40	11	0.44	15	0.60
Inconel-carbon steel	151	96	53	5	0.20	11	0.44	10	0.40
Stainless steel-carbon steel	151	96	53	5	0.20	10	0.40	7	0.28
Pure titanium-carbon steel	676	262	154	>2	>0.08	6	0.20	7	0.28
Titanium alloy-carbon steel	864	237	164	>1	>0.04	4	0.16	6	0.20
Inconel alone	>1000	>1000	>1000	0	0	0	0	0	0
Carbon steel alone	>1000	>1000	>1000	0	0	0	0	0	0

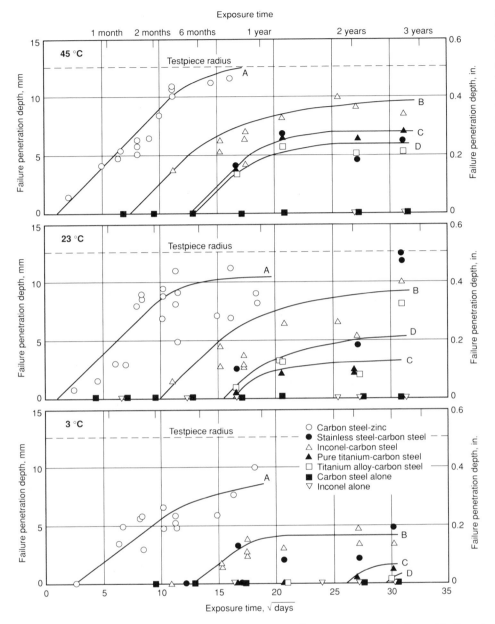

Fig. 8 Penetration depth of bond failure versus the square root of seawater exposure time for combinations of dissimilar metals at 3, 23, and 45 °C (38, 73, and 104 °F). Curve lines: A, carbon steel-zinc; B, Inconel-carbon steel; C, pure titanium-carbon steel; D, titanium alloy-carbon steel

Each of these strain conditions resulted in bond failure. Within experimental error, no significant difference existed in induction period, failure rate, or final bond failure penetration depth among any of the different elastomer strains applied during testpiece exposure. This suggests that elastomer strain has no effect on bond durability. Up to the depth studied, the presence of local shear strain does not accelerate bond failure, nor does the presence of a compressive strain retard it. Figure 9 illustrates the experimental evidence that supports these conclusions.

The effect of a uniform mechanical strain on the rate of bond failure should not be confused with the tendency of strain distri- bution to open up a crevice between elas- tomer and metal *after* bond failure has oc- curred. The conclusion reached from these data is limited to the proposition that a uniform mechanical strain field does not enhance this mechanism of bond failure.

Effect of Corrosion and Adhesive Joint Failure

Electrochemical potentials have been found to play a dominant role in limiting the durability of some types of adhesive joints. The corrosion behavior of these joints is the reverse of normal corrosion activity in that bond failure is more likely to occur when the adherend is a more noble metal that is

relatively protected from corrosion. In cor- rosion studies of metals, it has been found that it is not just the relative position of the metals in the galvanic series that determines corrosion activity, but also their polariza- tion behavior (Ref 4). Polarization effects can lead to a breakdown of the metal oxide structure, which later creates the possibility of, for example, pitting corrosion on stain- less steels in seawater. In the present case, it is possible that some change occurs to the metal oxide surface during the induction period before any bond failure is observed. Thereafter, the rate of failure can be deter- mined simply by a reaction rate influenced by the rate of lateral diffusion of electrolyte through the polymer layer at the interface. This model was derived from an empirical consideration of adhesion experiments. However, an analogous model has been considered to explain cathodic delamination of organic coatings on metals (Ref 3).

When two dissimilar metals are in electri- cal and electrolytic contact, the difference in electrochemical potential between them will cause electron flow. Corrosion of the more noble metal is usually decreased, and that of the more active metal is increased. In general, any electrochemical reaction can be divided into oxidation and reduction reactions, and there can be no net accumu- lation of electric charge during such reac- tion (Ref 4). Because of this, the evolution of hydrogen at the cathode of the more noble metal is likely. It is possible that the molecular hydrogen enters into an adverse reaction at the metal interface that destroys the bond integrity. Corrosion in natural environments is described in Ref 7.

Theoretical Models and Failure Mechanisms

A simple equivalent circuit can be used to characterize the components of the electri- cal system at the adhesive interface with the metal (Fig. 10). Water diffusion is expected to reach quasi-equilibrium in an element of polymer close to a free surface and at the interface, thus completing changes in ca- pacitance before interfacial effects of bond failure begin. Observations of metal corro- sion may correlate with changes in the resistance to charge transfer at the metal interface (R_i) that occur after the comple- tion of bond failure in any such local ele- ment.

Due to their relative positions in the elec- trochemical series (Fe, 0.61 V; Zn, 1.05 V), the steel adherends form cathodes, whereas the zinc forms anodes. Hydrogen ions mi- grate toward the cathode, together with avail- able metal ions (M^{2+}) from solution. Several reactions are possible at the interface:

$$2H^+ + 2e^- \rightarrow H_2 \qquad (Eq\ 1)$$

$$2H^+ + 2e^- + \tfrac{1}{2}O_2 \rightarrow H_2O \qquad (Eq\ 2)$$

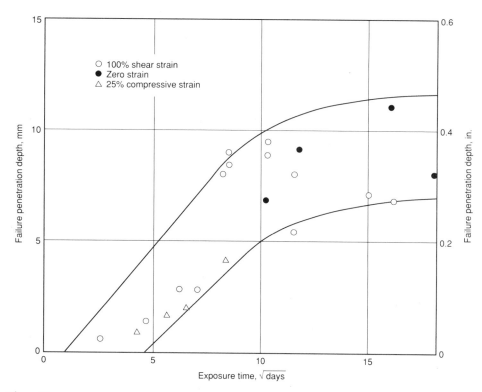

Fig. 9 Effect of elastomer strain on bond durability at 23 °C (73 °F) and 1050 mV

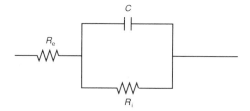

Fig. 10 Simple equivalent circuit. R_e, electrolyte solution resistance; C, double-layer capacitance at the interface; R_i, resistance to charge transfer at the metal interface

$$H_2O + \tfrac{1}{2}O_2 + 2e^- \rightarrow 2OH^- \qquad \text{(Eq 3)}$$

$$M_a O_b + 2bH^+ + ae^- \rightarrow aM^o + bH_2O \qquad \text{(Eq 4)}$$

$$M^{n+} + nOH^- \rightarrow M(OH)_n \qquad \text{(Eq 5)}$$

When the potential difference is high enough (for example, about 1 V), it is thought that hydrogen evolution (Eq 1) is the dominant reaction. If the metal surface is catalytically active, then cathodic disbondment can occur. There is some evidence that phosphated steel surfaces or zinc surfaces inactivated with small amounts of cobalt reduce the catalytic activity and the rates of disbondment (Ref 3). The availability of electrons for reaction (Eq 1) can be indicated by means of a potentiodynamic polarization curve. A polarization curve for steel in aerated NaCl is shown in Fig. 11. Hydrogen gas is not normally observed at ambient pressures, but there is evidence (from unpublished work) that it may accumulate at the interface under some circumstances.

The physics of the adhesive bond formed between each of the three interfaces (metal/primer-bonding agent-polymer) is not well understood. The extremely strong and durable nature of the bonds between vulcanized rubber and steel, even after prolonged exposure to moisture (in the absence of electrochemical reactions), makes it unlikely that they simply involve secondary intermolecular forces. On the other hand, these bonds do possess some features (for example, rapid dissolution in the presence of certain selected solvents) that make it unlikely that they are all primary chemical bonds. It is possible that some form of ionic bonding contributes to perhaps the normal secondary intermolecular bonding that is thought to occur for other adhesive systems. Such bonds may be relatively more susceptible to disruption by electron flow and electrochemical activity.

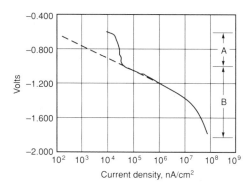

Fig. 11 Potentiodynamic polarization curve for steel in aerated NaCl of 0.5 M and 6.5 pH. Dashed line indicates extrapolation of the polarization curve representing the hydrogen evolution reaction. Dominant reactions: A, $H_2O + (\tfrac{1}{2})O_2 + 2e^- = 2OH^-$; B, $2H^+ + 2e^- = H_2$.

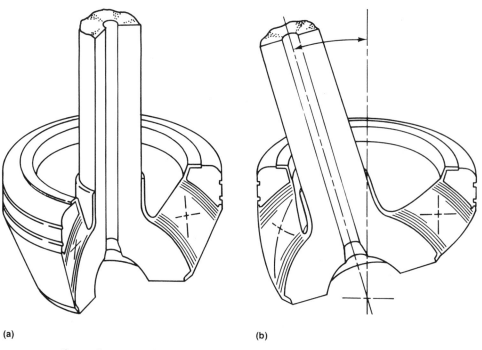

Fig. 12 Flexjoint for use on a deep-sea compliant oil platform. (a) No deflection. (b) 16° conical deflection. Groups of parallel lines indicate location of stainless steel reinforcing interleaves that are adhesively bonded to layers of structural elastomer.

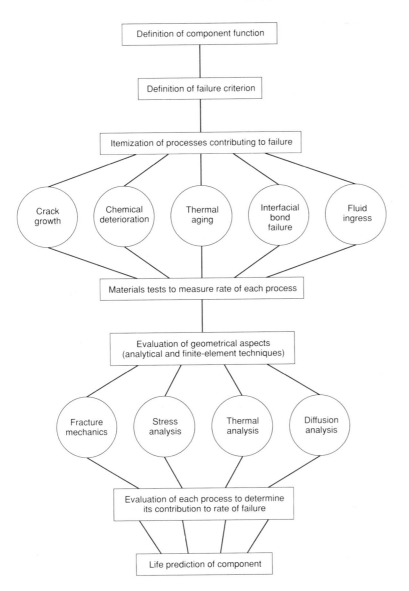

Fig. 13 Life prediction methodology for establishing the durability of adhesive joints in engineering components

The physical effect of H_2 gas evolution may cause bond failure. Once evolved, hydrogen can dissolve and increase the pH of the solution, at least locally. Although some types of adhesive joints (for example, epoxy-metal) fail after immersion in some high-pH solutions alone, others do not. The elastomer-steel system discussed in this article was tested in a sodium hydroxide solution of pH 12 at 23 °C (73 °F), but with no cathodic potentials (Ref 5). No bond failure was observed even after 6 months exposure; however, in seawater solution *with* cathodic potentials, bond failure was observed after 3 days. This provides clear evidence that a high-pH solution alone is not enough to cause cathodic disbondment for all adhesive systems.

The physical effects of hydrogen evolution are thus considered to be the primary cause of bond failure at high enough potential differences. Increases in solution pH are considered secondary, given the physics of the system. It is fair to say, however, that there is not universal agreement in the literature on the primary mechanism and where the chemistry of the materials alone has been considered. The view has been advanced that all effects are due to the development of local, high-pH solutions.

When the potential difference is low enough (for example, −0.6 to −0.9 V), the dominant reaction may be that in Eq 3 or even that in Eq 2, with oxygen reductions playing a dominant role. Changes may occur to the metal oxide layer present at the metal surface, as indicated by Eq 4 and 5. This may disrupt the adhesive bond at the interface, although evidence suggests that adsorption of oxidized species occurs after bond failure has occurred (Ref 8). There is evidence that when the reaction in Eq 3 is the dominant reaction and oxygen is excluded from the system, cathodic delamination is suppressed (Ref 3).

Careful studies have been made by Boerio *et al.* (Ref 9) of the composition of steel surfaces after cathodic disbondment using x-ray photoelectron spectroscopy (XPS). When steel surfaces were etched in phosphoric acid, failure was reported to occur more slowly than when the surfaces were polished, and XPS results showed evidence of phosphates. Low failure rates were also reported for grit-blasted surfaces. The authors concluded that one of two mechanisms was responsible for bond failure. Either local increases in pH caused degradation of the phenolic resin base of the primer, or the presence of soluble metal chlorides at the bond line (due to dehydrohalogenation of chlorinated rubber in the adhesive) caused large osmotic pressures at the interface.

Either mechanism implies some degradation of the primer or the adhesive layers between the vulcanized polymer and steel bulks. Because no similar effects were observed when testpieces were exposed to immersion in a large excess of high-pH solution (Ref 5), it is perhaps more likely that the presence of high-pH solution (and primer or adhesive degradation) is an aftereffect and not the primary cause of bond failure. Evolution of hydrogen at the steel substrate, either in atomic form or molecular accumulations, may be the primary mechanism. This hydrogen evolution may become relatively dispersed unless there is an area of weak bonding, where it may accumulate to form blisters beneath the adhesive or coating.

Based on the available evidence, it is difficult to say whether bond failure allows gas accumulations, which then may go into solution as $2H^+$ to decrease solution pH, or if H_2 evolution causes microblistering to physically disrupt the bonds. If tests are performed at high liquid pressures, the volume of accumulated gas may expand when the pressure is relieved, allowing positive detection of hydrogen gas. Another possible mechanism is associated with a greater role for ionic bonding as an adhesion mechanism. In this case electron flow itself may contribute to interfacial changes that cause bond failure.

Increasing Resistance to Cathodic Bond Failure in Adhesive Joint Applications

Cathodic disbondment is potentially a widespread mechanism of failure that can limit the durability of adhesive joints in many different applications. It is of particular importance where high cathodic potentials are used for corrosion protection of the reinforcement (steelwork) that is electrically connected to the polymer-metal interface. It may also be important where local cathodic potentials are set up by dissimilar metals in contact. Cathodic disbondment will only arise as a failure mechanism for those more durable types of bonds that do not first fail by other mechanisms. The resistance of an adhesives system to cathodic disbondment can be increased by ensuring that no bare metal-polymer interface is directly exposed to electrolyte and that the metal adherend surface is electrically insulated. This can be best arranged by design of the adhesive joint. The joint design should avoid exposed edges and specify a polymer coating over the metal that is sufficient to encapsulate it and provide a reasonable measure of electrical isolation.

Attempts have been made to enhance the chemistry of the adhesive system for increased resistance to cathodic disbondment (Ref 10). Some success was reported in reducing the rate of delamination. However, no modified chemical formulations of adhesive systems have been successful as yet in completely resisting any cathodic disbondment.

The principles involved with adhesive joint design can be illustrated by a case study of a structural application in which adhesive joints play a critical role. Figure 12 shows a composite flexjoint used as a critical structural bearing to provide flexibility in a deep-water compliant oil platform, the tension leg platform, developed for use in the North Sea. The flexjoint consists of nearly 20 alternating layers of stainless steel spherical interleaves and elastomer. The metal layers are indicated by the groups of parallel lines. Each stainless steel reinforcement is bonded to the elastomer (nitrile rubber) using an adhesive system. The edge of each metal is rounded and designed to be protected by an outer cover of elastomer with a thickness of at least 10 mm (0.4 in.).

Each lower flexjoint is immersed in seawater for the 20-year life of the structure, and each is a critical component in a primary load-bearing path. Therefore, adhesive joints of high durability are critically important. Careful adhesive joint design at the outer edges of the flexjoint ensures that no bare edges are directly exposed to a potential electrolyte (seawater). Extensive use is made of sacrificial anodes to protect steelwork elsewhere in the structure. Each steel interleaf is electrically isolated from the cathodically protected steelwork by the successive layers of polymer. Polymer is also extended far beyond each bond edge on both the shaft and the outer bulk metal. Although the design was not primarily for this purpose, it demonstrates the kind of design approach necessary to maximize resistance to cathodic disbondment. After 5 years of service immersed in seawater, the flexjoint has exhibited no obvious evidence of adhesive bond failure between the elastomer and the steel reinforcements (Ref 11).

If engineers are to have the confidence to design such critical components for long life, adhesive joint durability can be considered within a broader approach to component life prediction (Fig. 13). The rate of each failure mechanism needs to be characterized so that expected life can be calculated. Measuring rates at various temperatures allows extrapolation from higher rates at higher temperatures to lower rates at lower temperatures, as used in this article for seawater exposures at 3 °C (38 °F) and 45 °C (104 °F). Figure 13 shows one overall scheme for life assessments of polymer engineering components. Interfacial bond failure would usually be a life-limiting factor for any engineering component in which the adhesive joint performed a critical function.

REFERENCES

1. A. Stevenson, On the Durability of Rubber/Metal Bonds in Seawater, *Int. J. Adhes. Adhes.*, Vol 5 (No. 2), 1985, p 81-91
2. H. Leidheiser, W. Wang, and L. Igetofft, The Mechanism for the Cathodic Delamination of Organic Coatings From a Metal Surface, *Prog. Org. Coatings*, Vol II, 1983, p 19-40
3. W. Wang and H. Leidheiser, "A Model for the Quantitative Interpretation of Cathodic Delamination," Paper presented at the Pourbaix Symposium of the Electrochemical Society (New Orleans), Oct 1984
4. F. Sedillot and A. Stevenson, Laminated Rubber Articulation for the Deep Water Gravity Tower, *J. Energy Resour. Technol. (Trans. ASME)*, Vol 105, 1983, p 480
5. A. Stevenson, Materials Engineering Research Laboratory Ltd., unpublished research, 1986
6. A. Stevenson, The Effect of Electrochemical Potentials on the Durability of Rubber/Metal Bonds in Sea Water, *J. Adhes.*, Vol 21, 1987, p 313-327
7. R. Baboian and G. Haynes, *Corrosion in Natural Environments*, STP 558, American Society for Testing and Materials, 1974, p 171-184
8. S.E. Castle and J.F. Watts, *Corrosion Control by Organic Coatings*, National Association of Corrosion Engineers, 1981, p 78
9. F.J. Boerio, S.J. Hudak, M.A. Miller, and S.G. Hong, Cathodic Disbondment of Neoprene From Steel, in *Proceedings of the 10th Adhesion Society Conference*, L. Sharpe, Ed., Gordon & Breach, 1988, p 567
10. J. Bullock, R. Thomas, J. Thornton, and R. Rushing, "Durability of Rubber/Metal Bonds in Marine Environments," Paper presented at the 1988 Conference of the American Chemical Society Rubber Division (Dallas), April 1988
11. W. McIntosh and A. Stevenson, Operating Experience With Elastomer Composites on the Hutton Tension Leg Platform, in *Proceedings of the 9th International Conference on Offshore Mechanics and Arctic Engineering*, Vol III, Part A, 1990, p 127-138

Chemical Effects on Adhesive Joints

Theodore J. Reinhart, Systems Support Division, Materials Laboratory,
Wright Research and Development Center

THE SELECTION of an adhesive system for any given structural (load-carrying) application is basically made by empirical testing and evaluation efforts that prove that the system, that is, the adhesive, primer (if any), and the adherend surface preparation, as well as the bonding process perform satisfactorily in a given component geometry. The choice among single- and multicomponent adhesives, adhesive application methods, tooling and fixturing techniques, and curing process will be dictated by the performance level required from the bonded component. Adhesive selection, as well as adherend selection, are also dictated by the necessity to survive aggressive or even corrosive environments.

For example, rapid-curing single- and multicomponent adhesives can be formulated for robotic application and the development of bonding strength in short times. In many cases where moderate performance levels are required, these materials also display good resistance to aggressive environmental exposures. Structural film type adhesives normally require pressure and extended cure cycle times at temperatures of 120 °C (250 °F) or 175 °C (350 °F), depending on adhesive type and service conditions. Typically, these high-performance films provide the highest levels of bonded joint performance and the greatest resistance to environmental degradation.

This article describes chemical agents that an adhesive joint usually encounters; the chemical resistance of important adhesives types, as well as common adherends; methods of bonded joint protection; and test methods for chemical resistance. First, however, other factors that govern the adhesive selection process are summarized below. The basic criteria that should be considered are:

- The type of material to be bonded, whether the adherends are metal, ceramic, or of composite materials. Adherend physical properties, including size porosity and heat distortion temperatures, should also be considered. The prebond surface preparation utilized to prepare metallic surfaces for bonding must present a bondable, thermally and chemically stable, and highly adherent surface for the adhesive. When bonding dissimilar materials, selection of adhesives must take into consideration the varying bonding abilities that may exist with permeable and impermeable adherends. Analysis must consider stresses arising from different coefficients of thermal expansion. Problems of electrolytic corrosion and necessary electrical isolation must be carefully evaluated and solved

- All possible process or bonding procedures. The simplest process that will satisfy the design requirements will normally be the best one to select. All assembly steps, including surface preparations, adhesive mixing and applications, and priming and curing should be under rigid control

- All service conditions. Stress analysis must show that loads are sufficiently low in magnitude to prevent premature bond failure. Loading cycles (cyclic rates and continuous versus intermittent loading) are important. Considerations of environmental conditions are also fundamental in the selection of the adhesive and any bonded joint protection system that may be devised. Actual service conditions should be closely duplicated for adhesive evaluation purposes

- The economics of the systems and the selected process. Normally, the adhesive cost is minor compared with the cost of the component. Automated equipment usually provides lowest cost assembly where production quantities can support the investment necessary

Often-Encountered Chemical Agents

During service in an aerospace application, an adhesive bonded assembly is likely to come into contact with a relatively well-defined set of fluid materials. These include jet fuels (kerosene-base hydrocarbons); lubricating oils (mineral and synthetic ester base); numerous greases; glycol and silicone base heat exchange/cooling fluids; anti-icing formulations; cleaning agents; and, in the extreme case, phenolic, methylene chloride, and formic acid base paint-stripping formulations. Atmospheric moisture is also present in various amounts, depending on service locations.

Two fortunate circumstances work to protect the adhesive bonded joint from the intrusion of aggressive materials. First, most cross-linked polymer adhesives, such as epoxies, phenolics, polyurethanes, and acrylics, are highly resistant to many aggressive materials, at least at temperatures below their glass transition temperatures (T_g). Second, because the bond line is quite thin and is easily protected, the area of adhesive exposed to the aggressive agent is actually quite small.

Casual or intermittent exposure of a properly cured adhesive bond line to any of the aforementioned fluids at room temperature (Ref 1) has been shown to cause no degradation of the properties of the bonded joint. In some instances, exposure to materials such as glycols and silicones has resulted in a slight increase in the measured strength of the bond. However, areas receiving continuous exposure and exposures at elevated temperatures while under stress have resulted, in some instances, in the serious degradation of properties. Paint stripper formulations are the only materials with the potential for causing chemical degradation of the adhesive; those based on phenols, chlorinated phenols, and formic acid can cause serious reductions in polymer molecular weight. Although all of the other fluid materials, including water and the synthetic ester hydraulic fluids, act as plasticizers to one degree or another, long-term exposures have shown that no chemical degradation of the polymer results.

Because strength and stiffness are the primary performance requirements of the adhesive bonded joint, it is important to fully define and characterize the influence of any aggressive fluid in concert with the actual service loads and temperatures to be encountered. Therefore, only tests that combine the service loads (stresses), whether cyclic or steady, with temperature and the fluids that the bond line could potentially be exposed to, will accurately show what the bonded joint will do.

When bonded joint design criteria are being established, the above-mentioned fac-

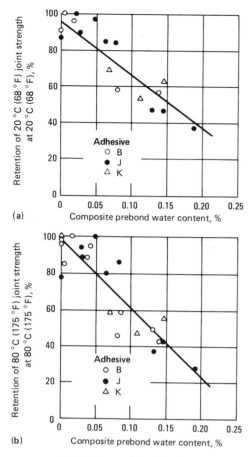

(a)

(b)

Fig. 1 Effect of composite prebond water content on joint strength retention of various adhesives. (a) At 20 °C (68 °F). (b) At 80 °C (175 °F)

tors must be adequately represented in the test program. Moisture absorbed into the bond line is perhaps the most serious factor that must be considered. Small amounts of moisture absorbed at the edges of an adhesive bond can improve joint strength in cases where a brittle adhesive is necessary.

Some adhesive formulations are more sensitive to property losses caused by absorbed water (see Fig. 1) than others. Testing is the only sure way to determine the performance of a given adhesive formulation.

Joint strengths at room temperature are not affected by moisture or other plasticizers. Severe reductions in hot-wet joint strength are always encountered at elevated temperatures and reach a maximum as the T_g of the polymer is approached. Prolonged heating at temperatures below the T_g will dry out polymer and restore bond joint mechanical properties. Too rapid a dryout (Ref 2 and 3) can cause polymer degradation and cracking. It appears, from the strength and failure mode data, that there are three mechanisms by which water in an adhesive joint can reduce joint strength. These are the formation of voids/cracks caused by too rapid heating or heating above the polymer T_g; plasticization of the adhesive, causing a severe reduction in polymer shear modulus, and desorption of the adhesive or primer from the adherend.

Contamination of the adherends prior to bonding (see Table 1) by foreign materials such as release agents (silicones or fluoro-

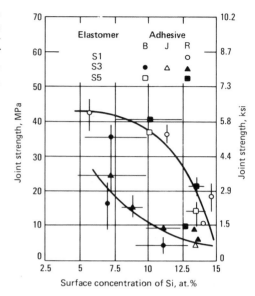

Fig. 2 Effect of surface concentration of silicon and silicone elastomer on strengths of composite joints bonded with different adhesives

carbons) or fuels and lubricants (hydrocarbons) by any means can result in disaster (see Fig. 2 to 4). Every effort should be expended to prevent adherend or adhesive contamination prior to processing the joint.

Removal of contaminants from nonporous surfaces can, in many cases, be easily accomplished by cleaning with waterbase detergent solutions and solvent vapor degreasing. Contaminated porous surfaces,

Table 1   The effect of composite surface contamination and pretreatment on joints

Release agent	Surface pretreatment	Contamination, at.%	Joint strength MPa	Joint strength ksi	Failure mode, % Composite	Failure mode, % Interface	Failure mode, % Adhesive
120 °C (250 °F) cured matrix resin							
Uncoated nylon peel ply	None	Si 2	42.3 ± 1.9	6.13 ± 0.28	53 ± 13	25 ± 11	22 ± 14
	Hand abrasion	Si 0	44.5 ± 0.8	6.45 ± 0.12	23 ± 5	60 ± 17	12 ± 8
	Grit blast	Si 1	44.8 ± 3.1	6.50 ± 0.45	73 ± 6	0	27 ± 6
PTFE-coated glass cloth	None	F 7, Si 1	46.1 ± 2.0	6.69 ± 0.29	65 ± 11	12 ± 11	22 ± 5
	Hand abrasion	F 3, Si 0	47.4 ± 0.8	6.87 ± 0.12	78 ± 16	1 ± 1	21 ± 16
	Grit blast	F 0, Si 3	47.8 ± 2.2	6.93 ± 0.32	76 ± 10	0	24 ± 10
PVF film	None	F 1, Si 0	50.2 ± 1.7	7.28 ± 0.25	21 ± 4	0	79 ± 4
	Hand abrasion	F 0, Si 0	48.4 ± 3.3	7.02 ± 0.48	41 ± 20	1 ± 2	58 ± 19
	Grit blast	F 0, Si 1	46.7 ± 2.3	6.77 ± 0.33	62 ± 5	14 ± 14	25 ± 12
Silicone liquid	None	Si 3	47.9 ± 2.1	6.95 ± 0.30	21 ± 8	3 ± 4	78 ± 11
	Hand abrasion	Si 1	50.9 ± 1.7	7.38 ± 0.25	75 ± 12	0	25 ± 12
	Grit blast	Si 1	48.5 ± 1.3	7.03 ± 0.19	79 ± 9	11 ± 10	9 ± 5
180 °C (355 °F) cured matrix resin							
Uncoated nylon peel ply	None	Si 1	31.6 ± 1.2	4.58 ± 0.17	16 ± 5	84 ± 5	0
	Hand abrasion	Si 0	27.5 ± 3.6	3.99 ± 0.52	11 ± 5	89 ± 5	0
	Grit blast	Si 1	43.7 ± 2.9	6.34 ± 0.42	92 ± 2	1 ± 2	2 ± 2
Silicone-coated nylon	None	Si 1	37.8 ± 1.2	5.48 ± 0.17	72 ± 15	0	28 ± 15
	Hand abrasion	Si 1	43.1 ± 1.2	6.25 ± 0.17	91 ± 5	0	9 ± 5
	Grit blast	Si 1	45.8 ± 2.3	6.64 ± 0.33	89 ± 4	0	11 ± 4
PVF film	None	F 21, Si 0	27.7 ± 2.4	4.02 ± 0.35	0	100	0
	Hand abrasion	F 8, Si 0	44.6 ± 1.7	6.47 ± 0.25	19 ± 11	81 ± 11	0
	Grit blast	F 0, Si 1	45.3 ± 2.4	6.57 ± 0.35	90 ± 10	3 ± 2	7 ± 10
Halocarbon	None	F 34, Si 0	11.4 ± 1.8	1.65 ± 0.26	4 ± 1	95 ± 2	1 ± 1
	Hand abrasion	F 27, Si 0	21.0 ± 2.6	3.05 ± 0.38	42 ± 17	57 ± 17	1 ± 1
	Grit blast	F 0, Si 1	41.5 ± 1.1	6.02 ± 0.16	91 ± 4	2 ± 4	7 ± 3
Silicone liquid	None	Si 0	47.4 ± 2.6	6.87 ± 0.38	60 ± 18	7 ± 6	33 ± 13
	Hand abrasion	Si 0	45.8 ± 2.8	6.64 ± 0.41	40 ± 8	57 ± 11	3 ± 4
	Grit blast	Si 1	46.7 ± 2.3	6.77 ± 0.33	80 ± 6	9 ± 8	13 ± 8

F, fluorine; Si, silicon; PTFE, polytetrafluoroethylene; PVF, polyvinyl formal

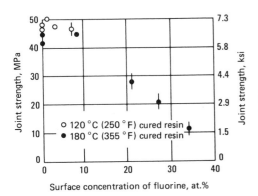

Fig. 3 Effect of composite surface contamination on strength of joints bonded with adhesive B

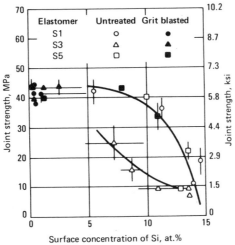

Fig. 4 Effect of surface pretreatment on the surface composition of carbon fiber reinforced plastic (CFRP) molded against silicone elastomers, and the strength of joints bonded with adhesive R

however, can be very difficult to clean, and in many instances removal of the surface via mechanical means may be required (Ref 4). The Section "Surface Considerations" in this Volume provides numerous articles on this area of concern.

Exposure of adhesive bonded joints to solid or semisolid lubricants normally results in no loss in joint mechanical properties. However, it has been shown that a number of reagents included in paint strippers and ester-base hydraulic fluids can cause serious losses in bonded joint mechanical properties if the exposure is more than just casual contact. Absorbed moisture will cause large reductions in joint properties, particularly at elevated temperatures.

Chemical Resistance of Adhesives by Chemical Type

There are roughly one dozen major family groups of adhesives of value to the engineer, as described in the Section "Adhesives Materials" in this Volume. These broad family groups include cyanoacrylates, epoxies, phenolics, polyurethanes, polyimides, and acrylics (Ref 5, 6), each of which is briefly described below in terms of chemical resistance properties.

Cyanoacrylate adhesives form the basis for moisture- and anaerobic-curing adhesives utilized as structural adhesives in many industrial applications. Formulations are available with a range of toughness values and very modest heat resistance. Normally, these materials are not highly resistant to degradation by chemicals, solvents, or moisture. Warm, moist conditions and moderate stress levels can cause very rapid degradation of joint properties.

Epoxy polymers, combined with one of a number of curing agents or hardeners, can be found in a large variety of adhesive formulations. Often they are formulated to meet the specific requirements of specialized applications. They are the primary adhesives utilized in structural applications in the aerospace industry because they are very strong and versatile. As a general rule,

epoxy adhesive formulations have high levels of mechanical properties, and depend on base resin and curing agent, possess excellent resistance to degradation by aggressive chemicals, solvents, and moisture. However, the rule of extensive applications testing and evaluation of any adhesive still applies to their use.

Phenolic adhesives are based on phenol-formaldehyde resins and are normally modified by elastomeric base materials to enhance toughness. Nitrile, vinyl, and other elastomers form excellent adhesives when combined with phenolics. Phenolic adhesives possess excellent resistance to degradation because of exposure to hot hydraulic fluids and are utilized to bond clutch facings and brake linings. These bonds must resist fluids, heat, and high stresses. Phenol-formaldehyde resins are condensation-curing materials and release water during the cure cycle. This makes it very difficult to obtain void-free bond lines without high curing pressures.

Polyurethane-base adhesives, like the epoxies, can be individually formulated to meet a wide variety of mechanical requirements. They are normally two-component formulations that cure at room or moderate temperatures. Although they have a greater degree of susceptibility to loss of properties because of moisture or solvent attack compared to some other structural adhesives, the polyurethanes are considered to be durable, load-bearing adhesives as long as environmental conditions are not too severe and the loads imposed are modest.

Polyimide polymers form the basis for a class of adhesives that have high heat resistance and excellent chemical and moisture degradation resistance. High curing temperatures, difficult processing, and brittleness are some of the drawbacks that are encountered in their use.

Toughened acrylic adhesives, in general, are materials modified by additions of elastomers and or thermoplastic ingredients. These modifiers can be physically dispersed particulates that are either physically separate or chemically attached to the adhesive polymer. Substantial increases in overall mechanical performance can be achieved by toughening. However, toughened materials, in some instances have shown increased sensitivity to moisture degradation and solvent attack.

A comprehensive test and evaluation program is indispensable in the determination of the chemical resistance of a given adhesive.

Chemical Resistance of Common Adherends

Aluminum alloy adherends must be carefully protected from exposure to galvanic couples, mild acids, bases, salt-laden atmospheres, and high humidity. High-strength aerospace alloys tend to be active chemically and require a well-designed and applied protection scheme.

Titanium alloys are normally highly resistant to acids, bases, and salt-laden atmosphere exposures. Care should be taken when titanium alloys are used in contact with anhydrous organic solvents. Rapid crack formation under load can occur under these conditions.

Steel alloys can be readily attacked by mild acids, and rust formation is quite rapid under high humidity or salt-laden atmosphere exposures.

Stainless steel alloys are normally highly resistant to acids, bases, and high humidity and salt-laden atmosphere exposure. They require little in the way of corrosion protection, even under the most aggressive of environments.

Fiber-reinforced plastic composites, depending on the matrix utilized, are normally highly resistant to degradation upon exposure to mild acids, bases, salts, and humidity. They do not rust or corrode and do not degrade unless temperatures above the polymer T_g are encountered.

Methods of Bonded Joint Protection

Bonded joints are commonly protected in two major ways. The first is by design configuration and the second by protective coatings and/or sealants. Correct design procedures include placement of bonded joint seams to prevent continual exposure to corrosive fluids due to pooling, as well as the selection and application of corrosion-inhibiting primers, organic coatings, and elastomeric sealants. Protective coatings and elastomeric sealants (Ref 5, 7) have been developed to a high degree of effectiveness and should be employed to protect adhesive bonded joint seams wherever necessary.

Protective coating technology is described in this article in terms of these categories: paint systems, water-displacing inhibitors, and elastomeric sealants. Metal-chemical conversion coatings are not discussed here. They may have some applicability to protecting the bond, but it is very limited.

Paint Systems

Paint systems (Ref 7) have been the most important method for protecting metals and other structural materials from degradation because of aggressive environmental exposure. The development of improved paint systems is continuing, and overall usage is increasing. The main reason for this is that paint systems, when properly selected and applied, generally offer better corrosion protection than other approaches.

Paint systems can also be formulated to meet many simultaneous requirements, including corrosion resistance, adhesion, flexibility, abrasion resistance, fluid resistance, air/bake dry, and easy application. The selection of the proper system is critical to performance and minimizes chances for failure in service.

Paint systems are combinations of several discrete layers, not all of which are paint layers. These layers are usually designated by specific names, such as primer and top coat. It is the total paint system that gives the required protection to the substructure. Each separate coating of the system serves one or more critical functions.

The most important function of the primer coat is to form a tie between the substrate and the top coat. A primer must be compatible with both and be capable of transferring loads between the two. In some instances, a single primer or top coat cannot provide the desired characteristics, and multiple coatings must be used. The promotion of primer-to-substrate adhesion is not covered in detail here, but it is noteworthy that surface preparation to achieve strong, durable adhesion is as critical when painting as it is in structural adhesive bonding.

Basically, paint systems protect substrate materials in two ways: They act as a barrier to the (aggressive) environment and as a reservoir for the required corrosion-inhibiting additives.

Failure of paint systems occurs when the system can no longer function to protect the substrate. Failure because of appearance degradation is also a possibility but is not considered further here. When paint failure occurs by degradation, the aggressive environment has direct access to the substrate. Causes of paint failure can include: abrasion or impact damage, cracking or crazing, loss of adhesion and peeling, blistering, osmosis, and underfilm corrosion. Different types of underfilm corrosion can occur as

the result of differences in metal, metal oxide, and coating materials. These reactions are not discussed here.

Types of Paint Systems. Primers of the etch/wash type are designed to replace pretreatments such as anodizing or chemical conversion coatings. These systems do not provide the high levels of durability of other pretreatments but are easier to apply. The resistance of these materials to hydraulic fluids and solvents such as paint strippers is low. This can be an advantage where facilitation of coating removal is a consideration. Etch/wash primers must always be followed with a second primer and a top coat for optimum protection.

Etch primers consist basically of a polyvinyl butyral resin, zinc tetroxychromate, and phosphoric acid in a water-base media. Chemical reaction upon drying causes formation of a resin-metal complex. This complex forms a base for subsequent coatings.

Alkyd primers, either air-dry or bake-on types, are low-cost, single-component coatings. They are durable to water exposure, but their hydraulic fluid and solvent resistance is poor. Experience has shown that alkyd primer performance does not match that provided by epoxy primers. As a result, alkyd primers are almost obsolete.

Epoxy primers currently are the mainstay of the aerospace industry. These primers have good resistance to water and solvents and generally give good corrosion protection to aluminum alloys. They adhere well to pretreated metal surfaces but are not suited for applications to untreated aluminum. Epoxy primers used in aerospace applications are normally two-component formulations that form a highly cross-linked polymeric film. Both amine and polyamide cure systems are available. Those based on amines have better acid and solvent resistance than those based on polyamides, but the polyamide-cured systems generally give the best overall performance in service, having better flexibility, adhesion, gloss retention, and acid resistance. Epoxy primers can be used with a variety of top coats to form a paint system.

Special primers have been developed for compatibility with polyurethane paint top coats. Those based on epoxy are the choice of the industry because they have been formulated to perform under the polyurethane top coats, which have relatively high moisture diffusion coefficients.

Polysulfide-base primers have been developed for applications where the structure is subject to large amounts of flexing. Aircraft wings, springs, and torsion bars are typical examples of flexible structures. Epoxy-base primers would crack under these service conditions.

Top coats of acrylic, epoxy, and polyurethane are described below. Nitrocellulose base paints are now obsolete as far as the aerospace community is concerned.

Acrylic top coats are regularly used by aerospace manufacturers and their durability is reasonable. For aircraft lacquers, methyl methacrylate is used. A high-molecular-weight polymer in a formulated solvent system forms the paint mixture. These paints have fair to poor resistance to fluids and solvents. Their main advantages are: good weatherability, rapid drying, one-component mix, and easily strippable. Some of the disadvantages are: low buildup per coat, low gloss unless polished, poor over etch primers, and high shrinkage. Service experience has shown that the effective life of these materials under aggressive conditions is about 2 years.

Epoxy top coats have found extensive applications in the commercial sector but are not the preferred system for use on aircraft. This situation is solely because of the tendency of epoxy top coats to chalk and yellow upon exposure to ultraviolet (UV) light. They generally give good substrate protection and can be used directly over etch or epoxy primers. Epoxies are generally two-component systems that have a limited pot life after mixing and cure very slowly below about 15 °C (60 °F). For exterior applications, the effective life of these materials is about 4 years.

Polyurethane top coats (Ref 7) are the most important class of materials presently in use on commercial and military aircraft. These coatings owe their popularity to a combination of good mechanical and chemical properties and excellent weathering (durability) properties. They possess good resistance to abrasion, water, organic solvents, and fluids. Polyurethane top coats can have an effective life of over 5 years under very aggressive service exposure conditions.

Polyurethanes are generally two-component materials or single-component systems requiring atmospheric moisture to cure. These materials are slow drying/curing, and single-coat thickness buildup is low. Because of their excellent solvent resistance, abrasion resistance, and adhesion, these top coats are comparatively difficult to strip.

Coating system selection for the protection of bonded joint seams, edges, and such is governed by the types of structure and the service conditions to be encountered. On aircraft, inner and outer surfaces above the cargo floor and wing exterior surfaces can be subjected to: rain (acid rain), salt, chloride ion, deicing solutions, glycols, hydraulic fluids, ester base, UV light, smoke/exhaust stack gases, and atmospheric contaminants. Under these conditions, epoxy primers and polyurethane top coats will provide the longest service lives. In the case of flexible wing aircraft, such as cargo aircraft, polysulfide primers are used. Adhesive bonded joints thus protected should be free from degradation due to external environments, as long as the coating film remains intact.

Wing interiors and the fuselage below the floor, where constant exposure to fuel, hydraulic fluids, or moisture is encountered, will probably require extra protection and sealant materials. The interior structure is subject to condensing moisture, and in some cases, high chloride ion concentrations. In these cases, the mixtures of fluids and water can pool or wet the insulation blankets in contact with the structure, thus creating conditions that will promote rapid bond degradation due to corrosion. In situations where these conditions are likely to exist, extra protection is obtained from multiple primer coats and the use of epoxy base top coats. Epoxy top coats are to be preferred here due to their lower moisture diffusion rates compared to polyurethanes. Good design and adequate draining can go a long way toward eliminating problems in these areas.

Water-Displacing Materials

A class of corrosion-preventive compounds called water-displacing corrosion inhibitors (WDCIs) (Ref 7) have a number of desirable properties. In addition to providing good corrosion inhibition, they displace water from both surfaces and crevices.

Upon evaporation of the solvent from the WDCI compound, a soft or hard wax coating is formed. The commercially available products are proprietary and compositions vary considerably.

The degree of protection of the adhesive bonded seam or joint edges depends not only on the corrosion inhibition, wettability, and water displacement of the WDCI, but on its penetrability, protective film formation, and compatibility with the materials it contacts, as well.

Corrosion inhibition is obtained through the use of organic inhibitor additives. To be effective, the WDCI, which is normally applied by light spray, must be able to spread on the surface and penetrate into crevices. The unique capability of the WDCI is its ability to creep under water drops or films attached to the surface or in crevices and displace them, thus eliminating the corrosive environment. Petroleum sulfonates and partially fluorinated carboxylic acids are generally utilized as water-displacing additives. Penetrability is obtained through the use of solvents that possess low surface tension. Protective films are normally formed by the use of high-molecular-weight hydrocarbons dissolved in the solvents. WDCIs are available for interior or exterior applications.

Very good results can be obtained when using WDCIs as a corrosion prevention measure on aircraft in conjunction with a paint system. Thus, the WDCI is a secondary line of defense. Its effectiveness is

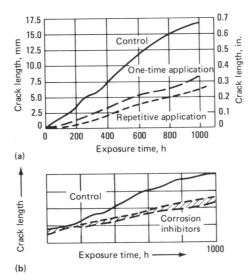

(a)

(b)

Fig. 5 WDCI application. (a) Difference between effect of one-time and repeat applications. (b) Effect on stress-corrosion cracking

shown in Fig. 5. A second application in 1 or 2 years may be necessary in bilge areas and other areas prone to corrosion. Reapplication requires only a general precleaning and removal of excess moisture. The WDCIs have prevented general corrosion and have been highly effective in reducing disbonding and bond line corrosion of adhesive bonded joints.

Application of WDCI is recommended by a number of aircraft manufacturers in areas where disbonding is being experienced in order to retard and arrest bond line corrosion (Ref 8). The use of WDCIs is also recommended in areas of good bonds so that disbonding can be prevented (Ref 9).

Laboratory testing has shown significant reductions in bond degradation and elimination of bond line corrosion as a result of WDCI application to lap joints exposed to highly aggressive, hot, humid environments. However, the use of WDCIs can have a negative influence on the static and fatigue properties of mechanical joints (Ref 9, 10). Generally, WDCIs do not have any adverse effect on polyurethanes, epoxies, elastomers, sealants, or any metals. However, tests should be performed to confirm this situation when using a specific WDCI.

Elastomeric Sealants

Sealants (Ref 5, 7) are formulations of elastomeric materials especially designed to seal off certain parts of aircraft structures, such as fuel tank or pressure cabin sealants. This article focuses on the use of elastomeric sealants to prevent corrosive liquids (generally moisture), fuels, hydraulic fluids, and solvents from penetrating an adhesive bonded structure.

A sealing compound is composed of a polymeric binder blended with fillers and additives to provide the desired engineering properties, for example, elongation, cohe-

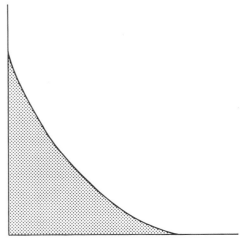

Fig. 6 Correctly feathered fillet

sive strength, viscosity, and adhesion. Sealants are normally designed to seal against fuel, water, or air penetration into or out of a given aircraft structure. Therefore, they require specific properties, such as fluid and solvent resistance, water resistance, and hydrolytic stability. Sealants may be of the curing (thermosetting) or noncuring (thermoplastic) variety.

To protect the edges of adhesive bonded structures, a sealant is normally applied as a fillet sealer or coating. Fillet seals are applied by means of a sealant gun and require extrudable sealant formulations. Sealant is normally brushed on the joint seam to form a base coat for the extruded fillet. After fillet application, a brushable sealant is used to overcoat the fillet (Ref 11).

Fillet geometry is critical to optimum performance. The edges of the fillet should be feathered (see Fig. 6) to prevent penetration of water or other fluids. Poor fillet configurations are depicted in Fig. 7, whereas various sealant geometries are shown in Fig. 8 and 9. The correct procedures for surface preparation are provided in Table 2.

Sealant Types. The main sealant varieties that are used in the majority of aerospace applications are: polysulfides, polyurethanes, and the silicones (as well as fluorosilicones). Other chemical families used in various applications are described in the articles constituting the Section "Sealant Materials" in this Volume.

Polysulfide sealants have excellent resistance to fuels, fluids, solvents, and moisture. They can survive continuous service temperatures of about 120 °C (250 °F) in air. The polysulfides have a long record of excellent service performance when properly applied. Polysulfides require carefully cleaned surfaces and the application of adhesion promoters in order to function properly.

Polyurethane sealants have given excellent performance in long-term exposure to water, fuels, and fluids. These materials

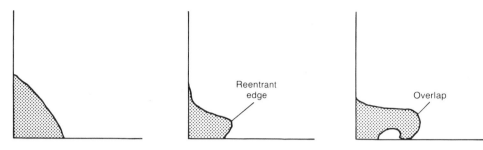

Fig. 7 Poor fillet configurations

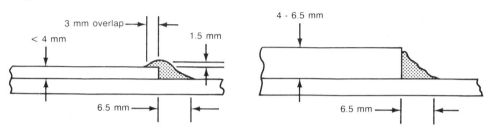

Fig. 8 Fillet sealing at overlaps

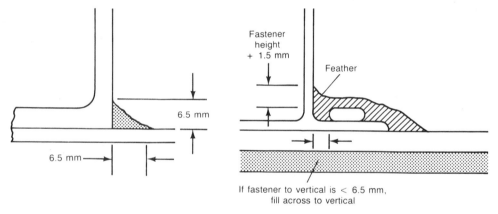

Fig. 9 Fillet sealing at vertical numbers

Sealant application techniques are:

● Before sealing, clean the surface. Wipe cleaners in and wipe them off (with clean rags—no lint)
● Do not use too much sealant; this becomes messy and too hard to clean
● Clean up any sealant mess as you go

Sealant types and uses are:

● *Sprayable*: Build thickness by multiple layers, not by heavy spraying
● *Brushable*: Apply to a thickness of approximately $1/16$ in. or less. Apply with short, stiff brush. Work into surface with small circular motion
● *Extrudable*: Hold sealant gun nearly perpendicular to the surface ($\approx 30°$ off) with nozzle tip pointing in the direction of movement. This minimizes air entrapment
● *Faying surface sealant*: Extrude a bead across the surface, then spread uniformly with roller or wide blade to thickness of approximately 250 to 380 μm (10 to 15 mils)
● *Note*: In using sealants containing volatile solvents, allow time for much of the solvent to evaporate before closing the surface

Chemical Resistance Test Methods

The testing (Ref 5) of structural adhesives in order to generate real engineering properties is still relatively new. Mechanical properties determination of adhesives is a complicated subject because of the complex interactions between adhesive and substrate. In addition, the relationships between adhesive stresses in a standard test specimen and a more complicated structural assembly are unknown. Yet, many continue to rely on standard tests to generate properties data. Standard and special tests are described in the Section "Testing and Analysis" in this Volume.

Although the American Society for Testing and Materials (ASTM) Committee D-14 on adhesives has developed a number of tests specific to adhesives (Table 3), there are no test methods designed specifically to evaluate the effects of aggressive chemicals on adhesive bonded joints. Any test to evaluate such properties, in either real or accelerated time, must combine stress, temperature, and the aggressive environment(s) that the service conditions represent.

Joint geometry can be an overriding factor in the performance of the bonded assembly. Stress distribution in the joint is one of the most significant factors complicating the relationship between mechanical properties of the adhesive and joint performance. The following variables must all be considered:

● Adhesive properties: elongation, strength, and modulus (shear)

must be carefully formulated for use in specific environments.

Silicone sealants provide high heat-resisting capabilities and excellent resistance to water exposure. However, fuel, hydraulic fluid, and solvent resistance are poor. Adhesion promoters must be used with these materials in order to obtain adequate adhesion.

Fluorosilicone sealants provide high heat resistance and excellent resistance to water, fuels, fluids, and solvents.

Table 2   Equipment and procedures for surface cleaning prior to sealant

Equipment	Procedure

New surfaces
Cheesecloth cleaning pads (many)
Alkaline cleaner
Organic cleaner
Squirt bottles for holding the cleaning liquid
Rubber gloves
 	
Old surfaces (after removing old sealant)
Stripper
Materials mentioned above
Implements for removing residual sealant from
  cracks and crevices
  Scrapers (wood and plastic)
  Brushes (wire and plastic)
  Probes, claws, picks
  Bendable wire with scraper blade
  Clock spring with scraper blade
  Dental mirror with ball joints
	1. Place the cleaning liquid in squirt bottle
	2. Squirt liquid onto cheesecloth pad. Scrub; turn and pad frequently
	3. Change pads frequently. Squirt cleaning liquid onto pad. Never dip pad into a container of cleaning liquid. (It contaminates the cleaning liquid)
	4. When surface appears clean, wet a fresh pad. Give final rinse
	5. With a new dry pad, wipe the surface dry; don't allow to evaporate
	Notes:
	Clean from the top downward
	Clean from the inside to the outside. This avoids recontaminating the surface
	If the sealant is not applied in 24 h, clean it again

Table 3 List of adhesives-applicable ASTM test methods

ASTM test method	Subject
D1002	Shear strength by tensile loading (lap-shear test)
D1876	T-peel test adhesion
D1780	Creep of metal-to-metal adhesives
D1781	Climbing drum peel test
D2182	Disk shear strength of adhesive joints by compression loading
D2294	Creep properties of adhesives by tensile shear loading
D2918	Durability of adhesive joints statically stressed in peel
D2919	Durability of adhesive joint statically stressed in shear by tension loading
D3166	Fatigue properties of single-lap shear specimen by tensile loading
D3167	Floating roller peel resistance

- Adhesive thickness
- Length of bonded joint
- Adherend properties: yield strength, thickness, and modulus

The following "rule of thumb" may be of help in selecting bonded joint geometries for test specimens:

- The failing load of lap joints increases as the width of the joint increases
- The failing load of lap joints does not increase proportionally with length
- Stronger adhesives and higher elongation adhesives normally produce stronger joints
- The stiffness of the adherends and adhesives influences the failure load of a joint. Stiffer adherends generally produce stronger joints
- Adhesive layer thickness can significantly influence adhesive bonded joint performance. Too thin or too thick bond lines can detract from performance

REFERENCES

1. B.M. Parker and R.M. Waghorne, Surface Pretreatment of Carbon Fibre Reinforced Composites for Adhesive Bonding, *Composites*, Vol 13 (No. 3), 1982, p 280-288
2. B.M. Parker, The Effect of Composite Prebond Moisture on Adhesive-Bonded CFRP-CFRP Joints, *Composites*, Vol 14 (No. 3), 1983, p 226-232
3. G.N. Sage and W.P. Tiu, The Effect of Glueline Voids and Inclusions on the Fatigue Strength of Bonded Joints in Composites, *Composites*, Vol 13 (No. 3), 1982, p 228-232
4. S.H. Myhre, J.D. Labor, and S.C. Aker, Moisture Problems in Advanced Composite Structural Repair, *Composites*, Vol 13 (No. 3), 1982, p 289-297
5. *Adhesives, Sealants, and Primers Desk-Top Data Bank, Adhesives*, 4th ed., Data, Inc.
6. *Adhesives in Modern Manufacturing*, American Society of Manufacturing Engineers, 1987
7. W.A.J. Mdonen, "Protective Coatings for Aircraft Structures," Report LR-413, Delft University of Technology, Dec 1983
8. K. O'Brien, "Corrosion Prevention Measures—Water Displacing Organic Inhibitors LPS-3, April 1972
9. V.A. Yeager, "LPS-3 Effects of Test Life of 737 Fuselage Cap Splices," April 1972
10. J. Schijve, Effect of Anti-Corrosion Penetrant on Fatigue of Riveted Joints, Technical Report NLR TR 77103U, University of Delft, 1977
11. R.E. Meyer, Polysulfide Sealants for Aerospace, *SAMPE J.*, March/April 1983, p 27-36

Radiation and Vacuum Effects on Adhesive Materials in Bonded Joints

Kristen T. Kern, Department of Physics, Old Dominion University
Sheila Ann T. Long and William G. Witte, Jr., NASA Langley Research Center
Wynford L. Harries, Department of Physics, Old Dominion University

UNDERSTANDING ENVIRONMENTAL EFFECTS on adhesive performance is important because adhesives are used in many diverse applications. In particular, adhesively bonded joints for aerospace structures are exposed to various types of radiation and thermal extremes (Ref 1). The effects of ultraviolet (UV) light, x-rays, gamma rays, neutrons, and protons on adhesives are briefly reviewed from the literature, and new results are presented here on the effects of exposure to electron radiation and thermal cycling between temperature extremes on the performance of six aerospace adhesives.

Review of Literature. Much work has been done to determine the effects of ionizing radiation, especially electrons, on epoxy resins, both neat and in composite materials (Ref 2-5). However, the effects of radiation on adhesives in bonded joints are not as well known. Early space applications required the investigation of the optical properties of adhesives for use in solar cells. The transmittance of several adhesives was shown to decrease after exposure to UV radiation (Ref 6, 7).

The mechanisms of degradation due to electron radiation of the epoxy resin formed by tetraglycidyl 4,4'-diaminodiphenyl methane cured with 4,4'-diaminodiphenyl sulfone (TGDDM/DDS) have been investigated (Ref 8, 9). Wilson *et al.* reported a decrease in the glass transition temperature (T_g) of TGDDM/DDS after exposure to 100 MGy of electron radiation (Ref 10).

Adhesives for Space Applications

A proposed design for space structures features tubular composite struts connected by aluminum fittings (Ref 11). The aluminum fittings would be adhesively bonded to the struts, as shown in Fig. 1. The adhesive is thus a critical structural component, and its performance in the space environment must be understood. The space environment includes various types of radiation and thermal exposures. In particular, a space structure in polar earth orbit may receive a dose of 3 MGy from electron radiation during a 30-year mission (Ref 12). Simultaneously, temperature extremes during each orbit are expected to be as high as 66 °C (150 °F) and as low as −101 °C (−150 °F).

Six commercial aerospace adhesives were evaluated in aluminum-to-composite joints exposed to a simulated space environment. Aluminum (2024-T3) was grit blasted, sulfuric acid anodized, and primed with American Cyanamid Company BR-127 primer before being bonded to T300/934 composite material. The composite was 12-ply material made of Fiberite T300 carbon fibers and 934 epoxy resin. The area of the composite to be bonded was grit blasted and cleaned. The section "Results of Adhesive Evaluation" in this article provides full details.

Methods of Evaluation

Analytical techniques were used to determine the elemental constituents of thin adhesive castings. Electron-dispersive analysis by x-ray (EDAX) and proton-induced x-ray emission (PIXE) were used for preliminary identification of heavier elements. The mass concentrations of each heavy element and certain lighter elements were then determined by elemental analysis and atomic spectroscopy.

To evaluate the tension lap-shear strength of the adhesives at extreme temperatures, the aluminum-to-composite single-lap shear (SLS) specimens, shown in Fig. 2, were

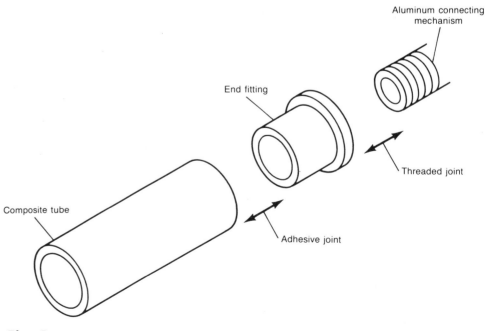

Fig. 1 Proposed strut tube connection

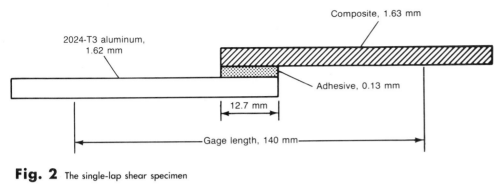

Fig. 2 The single-lap shear specimen

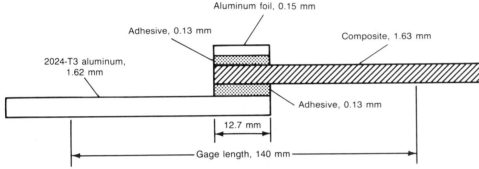

Fig. 3 The modified single-lap shear specimen

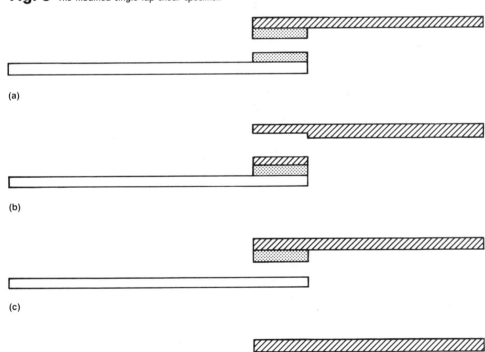

Fig. 4 Cohesive and adhesive types of failure. (a) Cohesive failure in the adhesive. (b) Cohesive failure in the composite. (c) Adhesive failure at the adhesive-aluminum interface. (d) Adhesive failure at the adhesive-composite interface

mechanical grips of an Instron universal mechanical testing machine with a heating/cooling chamber. The gage length was 140 mm (5.5 in.). Load was applied to the specimens parallel to their long axis. The crosshead speed was 1.27 mm/min (0.05 in./min). Specimen temperature was held at the test temperature for 15 min before application of tensile load. The tensile load was recorded as a function of displacement until specimen failure occurred. The shear strength of the bond was defined as the load at failure divided by the bond area.

The modified single-lap shear (MSLS) specimen, shown in Fig. 3, was used to evaluate the shear strength of each adhesive after exposure to environmental conditions. The MSLS specimen is an aluminum-to-composite SLS specimen with a 0.15 mm (0.006 in.) thick aluminum foil shield bonded to the composite adherend to simulate the protective cover on strut tubes (Ref 14). MSLS specimens were tested at room temperature with the same test parameters as those used for the SLS specimens tested at various temperatures.

Each SLS and MSLS specimen was examined optically to determine the mode of failure. Cohesive failure in the adhesive, shown in Fig. 4(a), was considered to occur when adhesive remained on both adherends. Cohesive failure in the composite (Fig. 4b) resulted in the fracture occurring in the bulk of the composite material, rather than in the adhesive. Adhesive failure (Fig. 4c and d), was considered to occur if the adhesive separated from either the aluminum or the composite adherend. Cohesive failure in the adhesive provides the best measure of the strength of the adhesive material. Adhesive failures and cohesive failures in the composite can only be used as lower bounds of adhesive strength.

Dynamic mechanical analysis (DMA) measures the resonant frequency and damping of a specimen as it is deformed by a constant-amplitude sinusoidal strain (Ref 15, 16). From the resonant frequency and damping values, dynamic storage and loss moduli values can be calculated. The temperature of the specimens may be varied to give the moduli as functions of temperature. Specimens with geometry similar to the overlap (bonded) region of the SLS specimens were used for the DMA. The DMA specimens, shown in Fig. 5, used composite identical to the composite used for the SLS specimens. DMA specimens were bonded as 110 × 130 mm (4.31 × 5.1 in.) plates and then cut into 12 specimens 50.8 mm (2.0 in.) long and 12.7 mm (0.5 in.) wide. A Du Pont model 982 DMA was used in this study. The samples were clamped at a 25.4 mm (1 in.) gage length. The resonant frequency, in hertz, and the damping signal, in millivolts, were recorded for temperatures from −120 °C (−189 °F) to 320 °C (608 °F). The temperature increased at a rate of 5 °C (9 °F) per

tested at −101 °C, 24 °C, and 66 °C (−150 °F, 75 °F, and 150 °F). Test procedures followed American Society for Testing and Materials (ASTM) Standard D 3163 (Ref 13). The materials were bonded in 120 mm (4.7 in.) wide plates, which were then cut into four 25.4 mm (1.0 in.) wide specimens. For testing, the specimens were held in the

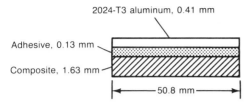

Fig. 5 The dynamic mechanical analysis (DMA) specimen

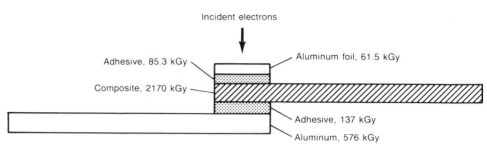

Fig. 6 Radiation dose distribution in modified single-lap shear specimens

minute. A 0.2 mm (0.008 in.) oscillation amplitude was maintained throughout the test. The dynamic loss modulus curve of each specimen reached a peak corresponding to the T_g of the adhesive and a second, higher-temperature peak corresponding to the T_g of the composite resin.

Simulated Space Environment

Tension lap-shear tests and DMA were performed on specimens that had been subjected to no exposure, to a 3 MGy dose of 1 MeV electron radiation, 1000 thermal cycles between −101 °C (−150 °F) and 66 °C (150 °F), and to a 3 MGy dose of radiation followed by 1000 thermal cycles. Irradiation was performed in accordance to ASTM standard D 1879, procedures B and C (Ref 17). Radiation exposures were made with a Radiation Dynamics, Inc. Dynamitron, under a vacuum, with the specimens mounted on a water-cooled exposure surface to prevent specimen heating. The dose rate was 0.3 MGy per hour for 10 h. Beam current was measured by a Faraday cup. Energy deposition per incident electron was calculated for each type of specimen, with each adhesive using a one-dimensional, multilayer electron transport program entitled TIGER (Ref 18). The duration of exposure for a given beam current and dose rate could then be calculated by the equation:

$$t = DaT\rho/IA \, (2.78 \times 10^{-10})$$

where t is the time of exposure in seconds, D is the desired dose in Gy, a is the area of the Faraday cup in cm^2, T is the thickness of the sample in cm, ρ is the average density of the sample in g/cm^3, I is the beam current in amperes, and A is the energy deposited per incident electron in keV.

The same equation could be solved to calculate the dose of radiation in each layer

of material if the value of A is taken to be the energy deposited in that layer. Representative dose calculations are shown in Fig. 6 and 7 for MSLS specimens and DMA specimens, respectively. The error in the transport code and in the dose calculation was approximately ±10%. Specimens were subjected to thermal cycling by moving them on a carriage between heat lamps and a liquid nitrogen manifold. The entire cycle was completed under a vacuum. A complete cycle lasted approximately 30 min. Specimens were stored in a vacuum at room temperature between exposure and testing.

Results of Adhesive Evaluation

Presented below are results of tension lap-shear tests and DMA on adhesively bonded specimens. Average shear strengths are presented with standard deviations. The averages include only the shear strength values of those specimens that failed cohesively in the adhesive. In the event that less than six specimens failed in this manner, the average of those specimens is reported and the number of values averaged is noted. For tests in which none of the specimens failed in the adhesive, the average of all specimens tested is presented with a ''greater than'' sign to indicate that the value is a lower limit of the strength of the adhesive. Glass transition temperature values, determined by DMA, are averages of two or three values and are reported with one-half the difference between the highest and lowest of the values as a measure of error.

EA 934NA, a gray, epoxy-based, two-part paste adhesive manufactured by The Dexter Corporation, Hysol Division, was examined using the above methods. The

manufacturer-suggested uses of EA 934NA are potting, filling, fairing, and shimming (Ref 19). This adhesive is certified MMM-A-132 and is reported to have superior shear strength at 177 °C (350 °F). The manufacturer product information indicated that the adhesive could be cured for 5 days at room temperature or for accelerated cures not exceeding 93 °C (200 °F). Specimens for this study were cured at 91 °C (196 °F) for 1 h. Elemental analysis and atomic spectroscopy of EA 934NA castings revealed mass concentrations, as shown in Table 1.

The results of tensile lap-shear tests on joints bonded with EA 934NA are presented in Tables 2 and 3. The average shear strength values in Table 2 demonstrate good strength retention at 66 °C (150 °F). The average shear strength values of specimens tested at 24 °C (75 °F) agree well with the shear strength of aluminum-to-aluminum SLS specimens provided by the manufacturer, also reported in Table 2. The shear strength values of MSLS specimens are shown in Table 3. Unexposed MSLS specimens had an average shear strength slightly lower than the aluminum-to-composite specimens tested at 24 °C (75 °F). The average shear strength of MSLS specimens

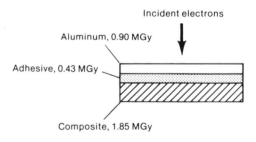

Fig. 7 Radiation dose in DMA specimens

Table 1 Major elemental constituents of EA 934NA adhesive

Element	Mass concentration, %
Carbon	44.84
Aluminum	24.50
Oxygen	12.37
Hydrogen	6.01
Nitrogen	5.15
Silicon	2.62
Other	4.49

Table 2 Effect of temperature on the shear strength of EA 934NA

Temperature		Shear strength			
		Aluminum-to-composite		Aluminum-to-aluminum	
°C	°F	MPa	ksi	MPa	ksi
−55	−67	...	...	19.3	2.80
24	75	21.5 ± 0.79(a)	3.12 ± 0.12(a)	21.4	3.10
66	150	20.0 ± 1.7(b)	2.90 ± 0.25(b)	...	...
82	180	...	...	12.4	1.80
149	300	...	...	7.58	1.10
260	500	...	...	3.10	0.45

(a) Average of three values. (b) Average of five values

Table 3 Effect of radiation on the shear strength of EA 934NA adhesive

Specimens	Shear strength	
	MPa	ksi
Unexposed	19.4 ± 1.4	2.81 ± 0.21
Irradiated	24.1 ± 1.7	3.50 ± 0.25
Thermally cycled	22.8 ± 1.5	3.31 ± 0.22
Irradiated and thermally cycled	20.9 ± 2.4	3.04 ± 0.34

Table 4 Effect of radiation on the glass transition temperature of EA 934NA

Specimens	Glass transition temperature	
	°C	°F
Unexposed	166 ± 10	330 ± 18
Irradiated	186 ± 3	367 ± 5
Thermally cycled	184 ± 2	363 ± 4
Irradiated and thermally cycled	188 ± 0	370 ± 0

Table 5 Major elemental constituents of EA 9309.3NA adhesive

Element	Mass concentration, %
Carbon	69.73
Oxygen	16.12
Hydrogen	7.56
Nitrogen	3.03
Silicon	0.51
Other	3.05

Table 6 Effect of temperature on the tensile lap-shear strength of EA 9309.3NA

Temperature		Shear strength			
		Aluminum-to-composite		Aluminum-to-aluminum	
°C	°F	MPa	ksi	MPa	ksi
−101	−150	16.0 ± 3.6	2.31 ± 0.52	...	...
−55	−67	...	...	27.6	4.00
24	75	29.2 ± 1.5	4.24 ± 0.22	29.0	4.20
66	150	>13.8 ± 3.3	>2.00 ± 0.47	...	...
82	180	...	...	6.90	1.00

Table 7 Effect of radiation on the shear strength of EA 9309.3NA adhesive

Specimens	Shear strength	
	MPa	ksi
Unexposed	>27.0 ± 2.4	>3.91 ± 0.35
Irradiated	>26.3 ± 2.7	>3.81 ± 0.38

Table 8 Effect of radiation on the glass transition temperature of EA 9309.3NA

Specimens	Glass transition temperature	
	°C	°F
Unexposed	131 ± 1	268 ± 1
Irradiated	131 ± 2	268 ± 4

Table 9 Major elemental constituents of EA 9320NA adhesive

Element	Mass concentration, %
Carbon	67.92
Oxygen	16.27
Hydrogen	7.78
Nitrogen	4.23
Silicon	1.38
Other	2.42

tested after each of the three exposure conditions was higher than that of the unexposed specimens. The largest increase occurred for the irradiated specimens, whereas specimens tested after exposure to both radiation and thermal cycling had an average shear strength that was less than that of the specimens exposed to thermal cycling alone.

Table 4 presents the T_gs of EA 934NA adhesive subjected to no exposure, to 3 MGy of electron radiation, to 1000 thermal cycles, and to 3 MGy of electron radiation followed by 1000 thermal cycles. The T_g determined by the peak in the DMA loss modulus of the irradiated adhesives was 20 ± 13 °C (37 ± 23 °F) higher than that of the unexposed adhesive. Similarly, the T_gs of the thermally cycled adhesive and of the irradiated and thermally cycled adhesive increased 18 ± 12 °C (33 ± 22 °F) and 22 ± 10 °C (40 ± 18 °F), respectively.

EA 9309.3NA is a purple-pink, epoxy, two-part paste adhesive manufactured by Dexter Hysol. This adhesive is reported to be resistant to humidity, water, and most fluids, while possessing high shear and peel strengths (Ref 20). The manufacturer cure specifications are 3 to 5 days at a temperature greater than 24 °C (75 °F), with accelerated cures allowed. Specimens for this study were cured for 1 h at 82 °C (180 °F). Table 5 contains elemental mass concentration data as determined by elemental analysis and atomic spectroscopy.

Residues of EA 9309.3NA adhesive on the MSLS specimens outside the bond were exposed to direct irradiation. The color of these residues changed from a purple-pink hue, prior to the exposure, to pale yellow, after the exposure.

In Table 6, the tensile lap-shear strength of aluminum-to-composite SLS specimens is compared with that of aluminum-to-aluminum SLS specimens tested by the manufacturer. At 24 °C (75 °F), the aluminum-to-composite specimens had a shear strength that agrees with the aluminum-to-aluminum result. The shear strength of specimens tested at −101 °C (−150 °F) was 45% lower than the shear strength of the specimens tested at 24 °C (75 °F). The adhesive in the specimens tested at 66 °C (150 °F) had a shear strength greater than 13.8 MPa (2.00 ksi).

The tensile lap-shear strengths of unexposed and exposed MSLS specimens made with EA 9309.3NA adhesive are shown in Table 7. The values obtained are lower bounds to the strength of the adhesive and are less than the shear strength of the aluminum-to-composite SLS specimens tested at 24 °C (75 °F).

The T_g of EA 9309.3NA, shown in Table 8, did not change for specimens analyzed after exposure to 3 MGy of electron radiation.

EA 9320NA is a pale-blue, epoxy, two-part paste adhesive manufactured by Dexter Hysol. EA 9320NA is reported to have good peel strength and elevated-temperature resistance (Ref 21). This adhesive meets the requirements of MMM-A-132 type 1, class 2. The cure cycle suggested by the manufacturer is 5 to 7 days at a temperature greater than 21 °C (71 °F), with accelerated cures at higher temperature being acceptable. Specimens for this study were cured for 1 h at 82 °C (183 °F). Elemental analysis and atomic spectroscopy of cured adhesive castings revealed the elemental constituents shown in Table 9.

Direct exposure to irradiation caused the adhesive residue on the MSLS specimens to

Table 10 Effect of temperature on the shear strength of EA 9320NA

Temperature		Shear strength			
		Aluminum-to-composite		Aluminum-to-aluminum	
°C	°F	MPa	ksi	MPa	ksi
−55	−67	...	...	23.4	3.40
24	75	32.4 ± 2.8	4.70 ± 0.41	31.7	4.60
66	150	22.3 ± 1.1	3.23 ± 0.16	...	...
82	180	...	...	10.3	1.50

Table 11 Effect of radiation on the shear strength of EA 9320NA adhesive

Specimens	Shear strength	
	MPa	ksi
Unexposed	>23.8 ± 2.7	>3.45 ± 0.38
Irradiated(a)	>16.3 ± 1.9	>2.37 ± 0.28
Thermally cycled(a)	>16.6 ± 5.0	>2.41 ± 0.72
Irradiated and thermally cycled(a)	>14.4 ± 1.5	>2.09 ± 0.22

(a) Average of three values

Table 12 Effect of radiation on the glass transition temperature of EA 9320NA

Specimens	Glass transition temperature °C	°F
Unexposed	134 ± 7	273 ± 13
Irradiated	133 ± 2	272 ± 3
Thermally cycled	132 ± 7	270 ± 13

Table 13 Major elemental constituents of EA 9334 adhesive

Element	Mass concentration, %
Carbon	67.80
Oxygen	15.84
Hydrogen	8.19
Nitrogen	7.45
Chlorine	0.28
Other	0.44

Table 14 Effect of temperature on the shear strength of EA 9334

Temperature °C	°F	Shear strength Aluminum-to-composite MPa	ksi	Aluminum-to-aluminum MPa	ksi
−101	−150	9.86 ± 0.19(a)	1.43 ± 0.03(a)	. . .	. . .
24	75	18.6 ± 1.9	2.70 ± 0.28	13.8	2.00
66	150	20.7 ± 1.2(a)	3.00 ± 0.17(a)	. . .	. . .
177	350	. . .	. . .	4.1	0.60

(a) Average of two values

change color. Prior to exposure, the adhesive residue was pale blue. After exposure to electron radiation, the adhesive residue had a rust-brown color.

Tables 10 and 11 contain the results of tension lap-shear tests on specimens bonded with EA 9320NA. The average shear strength of aluminum-to-composite specimens tested at 24 °C (75 °F) was slightly greater than that of aluminum-to-aluminum SLS specimens tested by the manufacturer. The average shear strength of specimens tested at 66 °C (150 °F) was 31% lower than that of specimens tested at 24 °C (75 °F). The shear strengths of MSLS specimens are lower limits of the strength of the adhesive.

The T_gs of unexposed and exposed EA 9320NA are shown in Table 12. The T_g of irradiated material was not significantly different from that of unexposed material or that of the thermally cycled material.

EA 9334 is an amber, epoxy, two-part paste adhesive produced by Dexter Hysol. This adhesive is resistant to UV light and is useful as a coating-bonding material (Ref 22). EA 9334 is reported to have excellent high-temperature strength. The manufacturer specifies a cure cycle of 5 to 7 days at a temperature greater than 21 °C (71 °F), with accelerated cures up to 93 °C (203 °F). Specimens in this study were cured for 1 h at 81 °C (181 °F). Mass concentrations determined by elemental analysis and atomic spectroscopy are presented in Table 13.

As shown in Table 14, the average tensile lap-shear strength of three aluminum-to-

composite SLS specimens tested at 24 °C (75 °F) was 36% greater than the value reported by the manufacturer. The tensile lap-shear strength of specimens tested at −101 °C (−150 °F) was 47% less than the 24 °C (75 °F) value. Specimens tested at 66 °C (150 °F) had an average tensile lap-shear strength that was greater than that of the specimens tested at 24 °C (75 °F), although the standard deviations overlapped.

Tensile lap-shear tests of unexposed and irradiated MSLS specimens resulted in the lower limit values presented in Table 15. The average tensile lap-shear strength of thermally cycled specimens, however, was less than that of the aluminum-to-composite SLS specimens tested at 24 °C (75 °F).

The T_g of irradiated EA 9334, shown in Table 16, was 5 ± 5 °C (10 ± 7 °F) greater than the T_g of unexposed adhesive.

EA 9396 is a clear, epoxy, two-part paste adhesive produced by Dexter Hysol. EA 9396 is reported to have excellent strength properties at temperatures between −55 °C (−67 °F) and 177 °C (350 °F) (Ref 23). The manufacturer suggests a cure of 5 to 7 days at a temperature above 21 °C (71 °F), with accelerated cures up to 30 min at 82 °C (183 °F). Specimens for this study were cured for 30 min at 80 °C (179 °F), with additional time at elevated temperatures during warmup and cool-down. Elemental mass concentrations for EA 9396 as determined by elemental analysis and atomic spectroscopy are presented in Table 17.

The lower limits of the shear strength of EA 9396 in aluminum-to-composite SLS specimens and in MSLS specimens are shown in Tables 18 and 19. The average tensile lap-shear strength of aluminum-to-composite SLS specimens tested at 24 °C (75 °F), shown in Table 18, are considerably less than the results reported by the manufacturer. However, the aluminum-to-composite specimen shear strengths at 24 °C (75 °F) and at 66 °C (150 °F) are lower bounds of adhesive strength. Similarly, the shear

strengths of MSLS specimens tested after exposure to each simulated environmental condition, presented in Table 19, are also lower limits to the strength of the adhesive.

The T_g of EA 9396 adhesive was not found to change significantly between unexposed and irradiated specimens, as shown in Table 20. The T_g of the thermally cycled adhesive was higher than that of the unexposed adhesive, although the margins of error overlapped. The T_g of the irradiated and thermally cycled adhesive was not significantly different from the T_g of the unexposed adhesive.

FM 300M is a modified epoxy adhesive film manufactured by the American Cyanamid Company. This adhesive is designed for bonding metal-to-metal and sandwich composite structures (Ref 24). FM 300M is reported to have excellent resistance to moisture and corrosion. The variation used in this study was a 150 g/m² (0.03 lb/ft²) formulation on a random mat polyester carrier. The film had a green color and was cured for 1 h at 175 °C (350 °F) in accordance with manufacturer instructions. The elemental constituents, as determined by elemental analysis and atomic spectroscopy, are shown in Table 21.

The color of adhesive residue on the MSLS specimens was green prior to exposure to electron radiation. After exposure, the residue had a darker green color.

The average shear strength of FM 300M in aluminum-to-composite SLS specimens is shown in Table 22. The aluminum-to-composite shear strength value at 24 °C (75 °F) was 22% less than the aluminum-to-aluminum SLS result reported by the manufacturer at 24 °C (75 °F), also shown in Table 22. The specimens tested at 66 °C (150 °F) had a slightly higher average tensile lap-shear strength than those tested at 24 °C (75 °F).

The tensile lap-shear strengths of MSLS specimens are shown in Table 23. The av-

Table 15 Effect of radiation on the shear strength of EA 9334 adhesive

Specimens	Shear strength MPa	ksi
Unexposed(a)	>14.5 ± 0.97	>2.10 ± 0.14
Irradiated(b)	>15.3 ± 0.39	>2.22 ± 0.06
Thermally cycled	15.7 ± 2.1	2.28 ± 0.31

(a) Average of four values. (b) Average of two values

Table 16 Effect of radiation on the glass transition temperature of EA 9334

Specimens	Glass transition temperature °C	°F
Unexposed	169 ± 1	335 ± 1
Irradiated	174 ± 4	345 ± 6

Table 17 Major elemental constituents of EA 9396 adhesive

Element	Mass concentration, %
Carbon	64.84
Oxygen	14.59
Nitrogen	10.46
Hydrogen	8.54
Chlorine	0.22
Other	1.35

Table 18 Effect of temperature on the shear strength of EA 9396

Temperature		Shear strength			
		Aluminum-to-composite		Aluminum-to-aluminum	
°C	°F	MPa	ksi	MPa	ksi
−55	−67	⋯	⋯	24.1	3.50
24	75	>12.7 ± 0.62	>1.84 ± 0.09	29.0	4.20
66	150	>13.6 ± 1.9(a)	>1.97 ± 0.28(a)	⋯	⋯
82	180	⋯	⋯	24.1	3.50
149	300	⋯	⋯	13.1	1.90
177	350	⋯	⋯	8.27	1.20

(a) Average of five values

Table 19 Effect of radiation on the shear strength of EA 9396 adhesive

Specimens	Shear strength	
	MPa	ksi
Unexposed	>14.4 ± 1.6	>2.09 ± 0.23
Irradiated	>12.8 ± 1.0	>1.86 ± 0.15
Thermally cycled(a)	>12.8 ± 0.44	>1.86 ± 0.06
Irradiated and thermally cycled	>13.5 ± 0.60	>1.96 ± 0.09

(a) Average of five values

Table 20 Effect of radiation on the glass transition temperature of EA 9396

Specimens	Glass transition temperature	
	°C	°F
Unexposed	205 ± 3	401 ± 5
Irradiated	205 ± 4	401 ± 7
Thermally cycled	213 ± 7	415 ± 13
Irradiated and thermally cycled	208 ± 2	406 ± 4

Table 21 Major elemental constituents of FM 300M adhesive

Element	Mass concentration, %
Carbon	59.69
Oxygen	15.00
Hydrogen	5.85
Nitrogen	3.89
Other	15.57

Table 22 Effect of temperature on the shear strength of FM 300M

Temperature		Shear strength			
		Aluminum-to-composite		Aluminum-to-aluminum	
°C	°F	MPa	ksi	MPa	ksi
24	75	23.2 ± 3.0	3.37 ± 0.43	29.8	4.33
66	150	24.1 ± 1.9	3.49 ± 0.27	⋯	⋯
120	250	⋯	⋯	23.2	3.36
149	300	⋯	⋯	15.9	2.31

erage tensile lap-shear strength of unexposed MSLS specimens was greater than that of the aluminum-to-composite SLS specimens tested at 24 °C (75 °F) and less than one standard deviation below the result obtained by the manufacturer. Irradiated specimens had an average shear strength 47% less than that of the unexposed specimens. Specimens that were thermally cycled before testing showed good strength

Table 23 Effect of radiation on the shear strength of FM 300M adhesive

Specimens	Shear strength	
	MPa	ksi
Unexposed(a)	27.5 ± 3.1	3.98 ± 0.45
Irradiated	14.7 ± 6.2	2.14 ± 0.89
Thermally cycled(b)	27.7 ± 0.82	4.02 ± 0.12
Irradiated and thermally cycled(c)	20.7 ± 3.6	3.00 ± 0.52

(a) Average of three values. (b) Average of two values. (c) Average of four values

Table 24 Effect of radiation on the glass transition temperature of FM 300M

Specimens	Glass transition temperature	
	°C	°F
Unexposed	182 ± 9	360 ± 15
Irradiated	198 ± 2	388 ± 4
Thermally cycled	198 ± 0	388 ± 0
Irradiated and thermally cycled	199 ± 1	390 ± 2

retention, having an average tensile lap-shear strength that was not significantly different from that of the unexposed specimens. However, specimens that were irradiated and then thermally cycled had an average tensile lap-shear strength 27% lower than that of the specimens subjected to thermal cycling alone.

The T_gs of unexposed and exposed FM 300M are shown in Table 24. The T_g of irradiated specimens was 16 ± 11 °C (28 ± 19 °F) greater than the T_g of unexposed adhesive. The T_g of the adhesive exposed to thermal cycling and that of the adhesive exposed to irradiation followed by thermal cycling increased by a similar amount.

Results Overview. A qualitative summary of the effects of electron radiation is presented in Table 25. The color of three adhesives was affected by the radiation, whereas three were unaffected. The data indicate increases in the shear strengths of two adhesives, whereas that of a third decreased. No conclusions could be drawn about the shear strengths of the other three adhesives. No changes occurred in the T_gs of three adhesives, whereas the data indicate increases in the T_gs of the other three adhesives.

The effect of thermal cycling on the shear strengths of four adhesives is not yet determined. Of the two remaining, one had an increase in shear strength while the other showed no effect. The T_gs of four adhesives increased due to thermal cycling while two showed no change in T_g.

REFERENCES

1. C.K. Purvis, Overview of Environmental Factors, in *NASA/SDIO Space Environmental Effects on Materials Workshop, Part 1*, NASA Conference Publication 3035, National Aeronautics and Space Administration, 1989
2. J.D. Memory, R.E. Fornes, and R.D. Gilbert, Radiation Effects on Graphite Fiber Reinforced Composites, *J. Reinf. Plast. Compos.*, Vol 7, Jan 1988, p 33-65
3. S. Egusa, Mechanism of Radiation-Induced Degradation in Mechanical Properties of Polymer Matrix Composites, *J. Mater. Sci.*, Vol 23, 1988, p 2753-2760
4. D. Evans, J.T. Morgan, R. Sheldon, and G.B. Stapleton, Post Irradiation Mechanical Properties of Epoxy Resin/Glass Composites, Science Research Council, Issue 15, 1970, p 2427
5. R.E. Bullock, Mechanical of a Boron-Reinforced Composite Material Radiation-Induced of Its Epoxy Matrix, *J. Compos. Mater.*, Vol 8, Jan 1974, p 97
6. L.B. Fogdall and S.S. Cannaday, "Space Radiation Effects of a Simulated Venus-Mercury Flyby on Solar Absorptance and Transmittance Properties of Solar Cells, Cover Glasses, Adhesives, and Kapton Film," Paper presented at the AIAA 6th Thermophysics Conference, Tullahoma, TN, American

Table 25 General effects of radiation on adhesives

Material	Color	Shear strength	Glass transition temperature
EA 934NA	No change	Increase	Increase
EA 9309.3NA	Purple-pink to pale yellow	Undetermined	No change
EA 9320NA	Gray-green to rust-brown	Undetermined	No change
EA 9334NA	No change	Increase	Increase
EA 9396	No change	Undetermined	No change
FM 300M	Green to dark green	Decrease	Increase

Institute of Aeronautics and Astronautics, April 1971

7. E. Anagnostou and A.E. Spakowski, "The Effect of Electrons, Protons and Ultraviolet Radiation on Plastic Materials," Paper presented at the Eighth Photovoltaic Specialists Conference, Seattle, Aug 1970

8. A. Gupta, D.R. Coulter, F.D. Tsay, and J. Moacanin, Mechanisms of Interactions of Energetic Electrons With Epoxy Resins, in *Proceedings of the Symposium on ESA Spacecraft Materials in a Space Environment*, June 1982

9. G.M. Kent, J.M. Memory, R.D. Gilbert, and R.E. Fornes, Variation in Radical Decay Rates in Epoxy as a Function of Crosslink Density, *J. Appl. Polym. Sci.*, Vol 28, 1983, p 3301-3307

10. T.W. Wilson, R.E. Fornes, R.D. Gilbert, and J.D. Memory, Effect of Ionizing Radiation on an Epoxy Structural Adhesive, in *Proceedings of the Symposium on Cross-Linked Polymers: Chemistry, Properties, and Applications*, April 1987

11. D.M. Mazenko, G.A. Jensen, and P.J. McCormick, Joint Technology for Graphite Epoxy Space Structures, *SAMPE J.*, May/June 1987, p 28-34

12. D.R. Tenney, Structural Materials for Space Applications, in *NASA/SDIO Space Environmental Effects on Materials Workshop, Part 1*, NASA Conference Publication 3035, National Aeronautics and Space Administration, 1989

13. "Standard Test Method for Permanganate Time of Acetone and Methanol," D 1363, *Annual Book of ASTM Standards*, American Society for Testing and Materials

14. H.W. Dursch and C.L. Hendricks, Protective Coatings for Composite Tubes in Space Applications, *SAMPE Q.*, Vol 19 (No. 1), Oct 1987

15. "982 DMA Dynamic Mechanical Analyzer Operators Manual," Rev. B, E.I. Du Pont de Nemours & Company, Inc., Sept 1983

16. B.E. Read and G.D. Dean, "The Determination of Dynamic Properties of Polymers and Composites," Adam Hilger Ltd., 1978

17. "Standard Practice for Exposure of Adhesive Specimens to High-Energy Radiation," D 1879, *Annual Book of ASTM Standards*, American Society for Testing and Materials

18. J.A. Halbleib, Sr. and W.H. Vandevender, TIGER: A One-Dimensional, Multilayer Electron/Photon Monte Carlo Transport Code, Sandia Laboratories, March 1984

19. "EA 934NA Product Information Sheet," Hysol Aerospace and Industrial Products Division, The Dexter Corporation, March 1987

20. "EA 9309.3NA Product Information Sheet," Hysol Aerospace and Industrial Products Division, The Dexter Corporation, March 1987

21. "EA 9320NA Product Information Sheet," Hysol Aerospace and Industrial Products Division, The Dexter Corporation, March 1987

22. "EA 9334 Product Information Sheet," Hysol Aerospace and Industrial Products Division, The Dexter Corporation, March 1987

23. "XEA 9396 Product Information Sheet," Hysol Aerospace and Industrial Products Division, The Dexter Corporation, Sept 1987

24. "FM 300 Adhesive Film Product Information Sheet," American Cyanamid Company, Jan 1987

Combined Temperature-Moisture-Mechanical Stress Effects on Adhesive Joints

Gary Good, Battelle Memorial Institute

RELIABILITY is one of the most important characteristics of an adhesive. In this context, reliability refers to the ability of an adhesive to function above its performance requirements for a specified length of time. Over the past several years, industry and the general public have continually expected improved reliability. In order to ascertain whether an adhesive will offer improved reliability, its service life must be estimated. Most uses invariably expose an adhesive to a multitude of environmental stresses. (In this article, the terms stresses and environmental stresses refer generically to all forces acting on an adhesive. Mechanical stresses will be specifically referred to as mechanical stresses to differentiate them from the generic forces.)

A commonsense approach involves exposing an adhesive to the major stresses and determining how these stresses affect its service life. As in all endeavors, prior planning of the work yields major dividends in time and cost. Therefore, this article will first discuss experimental design and statistical considerations, and then provide information on how actual physical considerations affect the accuracy of service life predictions.

Evaluation Parameters

During the adhesive development or selection process, most of the effort is spent on evaluating unaged adhesives. The goal of this effort is to obtain an adhesive that will meet or surpass the predetermined performance requirements. If an adhesive does not meet or surpass these performance requirements in its unaged state, it is not likely to meet or surpass them during its useful or service life. Almost nothing improves with age.

Compared with unaged testing, service life evaluations take much more time. To conduct service life evaluations, samples must be prepared, exposed to environmental stresses for extended lengths of time, and then tested. Thus, the primary differences between service life measurements and unaged measurements are the length of time the samples are exposed to the environmental stresses and the cost associated with the exposure. Normally, the cost of a service life evaluation is a fractional increase over that of determining the unaged properties. Therefore, cost is not as critical a factor as time; service life tests can take 10 or even 100 times longer than unaged tests. It is not feasible to run a set of service life tests for each set of unaged tests because such a practice would greatly increase the length of the selection or development process. Service life tests are usually run in the final stage of the development or selection process, if they are run at all.

In recent years, industry and the consumer public have demanded increased levels of reliability from adhesive products. In this sense, "reliability" refers to performance at or above the lifetime performance requirements of the product. No degradation of performance over the product lifetime is acceptable. In fact, when any property levels drop below the performance requirements, the product is in a failed state and its useful life is over.

During this same time period, adhesive technology has expanded at an ever-increasing rate, and newer and better adhesives have been introduced with dizzying frequency. Not too long ago it was quite difficult to get a new product introduced because a large segment of the potential market would take a wait-and-see position before buying. In today's environment, however, those who are waiting for the bugs to be worked out of new products often find themselves using inferior products based on outmoded technologies.

The demand for greater reliability and the rate at which new adhesives are being introduced place an even greater burden on service life evaluations of adhesives. Additional difficulties have been raised by the fact that substrates (the materials joined by adhesives) have been experiencing the same market and technological forces as adhesives. These forces have resulted in a predicament facing those who attempt to evaluate adhesive performance: An adhesive must be reliable over its lifetime even though it degrades throughout that lifetime and the time available for testing it is much less than its expected lifetime. Fortunately, the methodologies associated with evaluating the aged properties of materials have kept pace with the advances in adhesives and other materials.

Accelerated Service Life Testing

The only way to solve the predicament described above is to conduct accelerated service life testing, in which the levels of environmental stresses are raised above the stress levels that the adhesive would experience during a normal lifetime. Traditionally, this has been done by raising one stress at a time to a level much greater than that which would be encountered under normal operating conditions and observing the behavior of the adhesive. However, there is a severe drawback to this type of acceleration. This drawback can be illustrated by an analogy: If an egg is placed under a chicken and a sufficient length of time elapses, a chick will hatch; however, if this process is accelerated by placing the egg in boiling water, a drastically different product results. Some of the accelerated service life tests that have been administered in the past have been as ineffective in reaching their objectives as this example.

The problem, in scientific or engineering terms, is that in the attempt to accelerate the aging process, the stress is increased to such a level that the adhesive itself is in a raised-stress regime (RSR). The RSR produces behavior and failures that are thermodynamically impossible at normal levels of environmental stresses. Thus, the results of

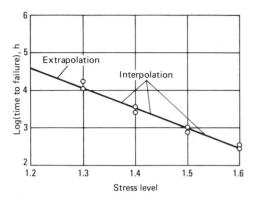

Fig. 1 Extrapolation over a single stress

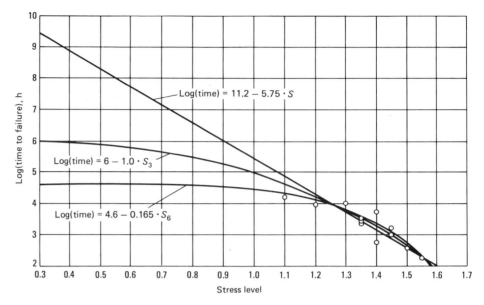

Fig. 2 Comparison of three models

RSR investigations have no direct bearing on the reliability of the adhesive. The most that can be inferred from RSR tests is that adhesives that perform better in the tests will perform better at normal levels of environmental stresses. Although this is often true, the results taken by themselves are qualitative.

Unless historical data are available that can be used to correlate accelerated results with field results, there can be no answers to questions such as: If an adhesive lasts 3 h longer than the control adhesive, how much more reliable is it than the control? If another adhesive lasts 2 days longer than the control, how much better is it than the other two adhesives? As previously noted, because of the rapid growth of adhesive technology, the adhesive may be obsolete by the time field data are obtained.

Predicting product service life is the goal of accelerated service life testing. To accomplish this, accelerated service life testing should remain within the normal-operating-stress regime (NOSR) while one or more stress levels are raised above normal operating levels. The stresses must be raised to accelerate the aging process, but they must not be raised to such a level that they force the system into an RSR.

Selecting the Number of Experiments

The validity of accelerated service life testing depends on the validity of extrapolating normal operating stress levels from raised stress levels. For this reason, it is important to consider the validity of extrapolating lower stress levels from higher stress levels (see Fig. 1). Errors of extrapolation come primarily from bias and variance. Bias errors occur from using an inadequate model; variance errors are the inherent errors associated with experimental data.

Variance errors can be reduced by increasing the number of data points. This is illustrated by Eq 1, which predicts the variance of service life estimated by a linear extrapolation (Ref 1, 2):

$$v(L_{noc}) = S^2 \left\{ (1/N) + (T_{noc} - T_{mean})^2 / \left[\sum_{i=1}^{N} (T_i - T_{mean})^2 \right] \right\}$$ (Eq 1)

where v is the estimated variance at normal operating conditions, L_{noc} is the service life at normal operating conditions, S is the estimated standard deviation of replicates, N is the total number of data points, T_{noc} is the normal operating stress level, T_{mean} is the mean stress over all stresses, and T_i is a single stress. Dividing both sides of Eq 1 by S^2 yields:

$$\beta = v(L_{noc})/S^2 = (1/N) + (T_{noc} - T_{mean})^2 / \left[\sum_{i=1}^{N} (T_i - T_{mean})^2 \right]$$ (Eq 2)

where β is the ratio of estimated variance to the estimated standard deviation of replicates.

This equation was further modified (Ref 1) by replacing the stress with dimensionless stress, which is given by:

$$\phi_j = (T_j - T_{mean})/(T_{max} - T_{min})^2]$$ (Eq 3)

where ϕ_j is the dimensionless stress j and T_j is the stress j. This yields:

$$\beta = (1/N) + \left[\phi_{noc}^2 / \left(\sum_{i=1}^{N} \phi_i^2 \right) \right]$$ (Eq 4)

and because:

$$1 = \phi_i^2$$ (Eq 5)

Eq 4 reduces to:

$$\beta = (1 + \phi_{noc})^2 / N$$ (Eq 6)

A reasonable criteria would be for the estimated variance of the predicted point to be equal to that of the replicate variance, or β must be equal to 1. Letting β equal 1, the number of experiments would be:

$$N = 1 + \phi_{noc}^2$$ (Eq 7)

The number of experiments run should be greater than or equal to N. For example, if the maximum level of the stress is 1.65, the minimum level of stress is 1.1, and the use level is 0.3, then by using Eq 3 and 7, these calculations result:

$$T_{mean} = (1.65 + 1.1)/2 = 1.375$$
$$\phi_{noc} = (0.3 - 1.375)/[0.5(1.65 - 1.1)]$$
$$= -1.075/0.275$$
$$= -3.91$$

$$N = 1 + (-3.91)^2$$
$$= 1 + 15.28$$
$$= 16.28$$

Thus, 17 or more experiments need to be run for the variance of the extrapolated prediction to be approximately the same as the replicate variance.

Selecting the Model

The bias error can be reduced by carefully selecting the model. An incorrect model can result in a major error in the extrapolated result (Ref 3). The predicted times to failure for the three models shown in Fig. 2 are 42 000, 920 000, and 3.25×10^9 h. Table 1 contains the data that were used to generate Fig. 1 and service lives extrapolated from that data. Increasing the number of experiments should not substantially improve the estimate of the service life for any of the models.

Table 1 Data for Fig. 1 and extrapolated service lives

Stress level	Life, h
1.10	15 992
1.20	9 111
1.30	11 241
1.30	10 212
1.30	11 264
1.35	3 171
1.35	2 754
1.35	3 567
1.40	592
1.40	5 710
1.40	5 274
1.45	1 634
1.45	993
1.50	399
1.50	412
1.55	162
1.55	189
1.65	20
1.65	25

Table 2 Five models and their statistics

Run	Time p	Time Transform	Stress p	Stress Transform	F_r	$100R^2$	F_{rof}
1	-1	$1/t$	-1	$1/S$	9	30	61
2	-1	$1/t$	1	S	14	20	51
3	1	t	-1	$1/S$	61	77	6
4	1	t	1	S	49	55	8
5	0	$Log(t)$	0	$Log(S)$	72	80	4

Note: p is the transformation parameter. Source: Ref 3

One of the most critical aspects of accelerated-aging testing is the selection of the model. The service life prediction depends on the model selected. A systematic approach was derived for determining the most appropriate model (Ref 3, 4). In this approach, a single functional form is selected, and then the model is evaluated over a family of transformations based on dependent and independent variables:

$$T = Z \cdot p \text{ for } p < 0 \text{ and } 0 < p \qquad \text{(Eq 8)}$$
$$T = F(Z) \text{ for } p = 0$$

where Z is the variable (independent or dependent) being transformed, p is the transformation parameter (usually between -3 and 3), and T is the transformed variable.

A family of five different regression analyses is given in Table 2. The $100R^2$ value and the significance of the regression F value (F_r) are used to determine which model yields the best results (Ref 2, 3, 5, 6). Table 2 also contains the $100R^2$ and F_r values for the respective models. Based on this information, the log functions are the most appropriate model.

After the model has been selected, the scale sensitivity must be determined (Ref 3). The scale sensitivity is given by:

$$K = F_{r,mean,high} - F_{r,mean,low} \qquad \text{(Eq 9)}$$

Table 3 Seven models and their statistics

Run	$p(S)$(a)	Term	F_r	$100R^2$	F_{lof}	Standard deviation
1	1	S	96	84	3.1	0.352
2	2	S_2	128	88	2.1	0.312
3	3	S_3	164	90	1.4	0.279
4	4	S_4	201	92	0.94	0.254
5	5	S_5	227	93	0.69	0.240
6	6	S_6	235	93	0.62	0.236
7	7	S_7	224	92	0.72	0.242
8	8	S_8	201	92	0.94	0.254

(a) p, transformation parameter; S, stress

where K is the scale sensitivity factor, $F_{r,mean,high}$ is the mean of F_r with p at its high level, and $F_{r,mean,low}$ is the mean of F_r with p at its low level. Using the data from Table 1, $K(t) = [(61 + 49)/2] - [(9 + 14)/2] = 43.5$ and $K(S) = [(49 + 14)/2] - [(61 + 9)/2] = -3.5$. It is obvious from the scale sensitivity factor that the time to failure is much more sensitive than the stress.

The next step involves selecting several values of $p(t)$ in the range of -3 to 3 using the log of the stress. Even though the log(t) log(S) functional form has the best $100R^2$ and F_r values, it also has an F lack of fit (F_{lof}) of 4.3, which indicates that the model can be improved. As stated earlier and shown in Fig. 2, a model must adequately represent the data. A need for further investigation is indicated by these results. Even though the stress is not as sensitive as the time to failure, the $100R^2$ and F_r values decrease slowly but steadily as p increases from 1 to 6. A p value of 6 yields the best model (Table 3); the F_{lof} is less than 1, indicating that the model is an adequate representation of the data.

Extreme care must be exercised when selecting a model. It must represent the data as accurately as possible to ensure that the extrapolations are the best possible estimates, given the available data. Even those models that have acceptable statistics can vary greatly in their extrapolated values. For the models in Fig. 2, the F_r values are greater than 100, the $100R^2$ values are greater than 90, and the F_{lof} values are less than 3; all three models would be viewed as statistically valid. Also, the data that were used did not have any replicates at the lower two stresses. It would have been better to have replicate stress levels, which certainly would have provided a greater confidence in the extrapolated predictions. The information was taken from actual experiments, for which conditions are seldom ideal. All in all, the best model is:

$$Log(t) = 4.6 - 0.165 \cdot S^6 \qquad \text{(Eq 10)}$$

where t is the time to failure and S is the stress.

The models represented by the last seven rows of data in Table 1 exhibited a different failure than those represented by the first twelve rows. In other words, the data in Table 1 are from two different stress regimes; as pointed out earlier, this is a potentially dangerous practice. The methodol-

ogy for selecting a model was applied to the data in the first twelve rows of Table 1, and the results are shown in Fig. 3. Using this approach, the best model is:

$$Log(t) = 4.3585 - 0.06312 \cdot S^8 \qquad \text{(Eq 11)}$$

where t is the time to failure and S is the stress. The other curve, which has stress raised to the fourth power, is included for comparison. As shown in Fig. 2, the normal operating stress is assumed to be 0.7. This level was back-calculated from Eq 3 and 7 with N equaling 12. In the previous example, N was 19.

The predicted value for the best model is 22 830 h. The two standard deviation boundaries are 6000 and 86 000 h. This is a rather large difference, but it is consistent with the data. The life measurements are very noisy, and the accuracy of the predictions suffers because of this inherent variability.

As information becomes available from the field or from longer-term tests, it should be used to verify the model. Field data cannot be used to correct the model because field data do not come from controlled experiments. Field data can, however, be compared to the model for consistency with the extrapolated predictions.

Multiple Stresses

As discussed in the section "Accelerated Service Life Testing" in this article, one of the most serious hazards associated with this testing is the possibility of inducing a raised stress regime. This is more likely when a single stress (as opposed to multiple stresses) is used to accelerate the aging process. In the example from the preceding section, if the testing had been performed

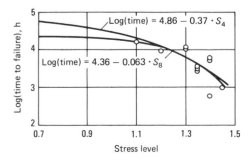

Fig. 3 Comparison of two models

Table 4 Experimental design for three stresses in four experiments

Experiment number	Dimensionless stresses		
	ϕ_1	ϕ_2	ϕ_3
1	−1	−1	1
2	−1	1	−1
3	1	−1	−1
4	1	1	1

Table 5 Comparison of β for single- and multiple-stress extrapolations

Total extrapolation distance	Single stress	Multiple stress
3	2.5	1
4	4.25	4.58
5	6.5	2.33
6	9.25	3.25
10	25.25	8.58

entirely within the RSR, the predicted life for a stress level of 1.30 would be 28 655 h. This compares with an observed service life of approximately 11 000 h and a predicted service life of 5750 h. The normal operating stress level of 1.3 was back-calculated from Eq 3 and 7, with N equal to 7. In the previous two examples, N was 12 and 19. If the expected life was projected to an operating stress level of 0.3, the extrapolated service life would be 23 016 × 10^9 h. This example indicates the importance of testing an adhesive within its NOSR.

Two or more stresses raised above their normal operating levels will accelerate aging in such a way that the adhesive remains with its NOSR and failure occurs within a reasonable length of time. Multiple stresses also more closely parallel the conditions in the end-use of the product. In the real world, an adhesive will experience thousands of different stresses. Most of these will have a minimum effect on the life of the adhesive. The few that are critical should be tested in multiple-stress accelerated-aging experiments.

Multiple-stress accelerated aging experiments have been shown to reduce the variance of service life predictions (Ref 1). For example, if four experiments are run as indicated in Table 4, then the ratio of estimated variance to replicate variance (Ref 2) is given by:

$$\beta = (1 + \phi_1^2 + \phi_2^2 + \phi_3^2)/N \qquad \text{(Eq 12)}$$

To compare Eq 6 with Eq 12, it should be assumed that the individual stresses are scaled in such a way that a unit increase in any dimensionless stress results in an equal reduction in failure time; that is, a decrease of one unit in ϕ_1 will decrease the failure time by the same amount as a one-unit decrease in ϕ_2 and a one-unit decrease in ϕ_3. Table 5 compares Eq 6 and Eq 12, given these assumptions; the data in Table 5 clearly demonstrate that multiple-stress service life testing can be used to reduce the variance of service life predictions.

Temperature, Moisture, and Mechanical Stress

The preceding sections dealt with the mathematical and statistical treatment of accelerated service life evaluations. Understanding these mathematical and statistical methodologies is as important as understanding the physical and chemical charac-

teristics of temperature, moisture, and mechanical stress in obtaining accurate service life predictions. The remaining discussion focuses on the chemical and physical nature of the effects of temperature, moisture, and mechanical stress. The molecular-free-volume concept and stress effects are discussed first to provide a basic conceptual approach to understanding the effects of these forces on adhesives. Then, temperature, moisture, and mechanical stress are each discussed individually, followed by an explanation of how temperature and mechanical stress interact to degrade the performance of adhesives.

Molecular Free Volume

The concept of molecular free volume (Ref 7) provides a basis for understanding diffusion and the degradation of adhesives. It is presented as a means of conceptually depicting the interaction of adhesives and water molecules, as well as the effect of temperature, moisture, and mechanical stress on adhesives at the molecular level.

An adhesive is actually a physical mixture of several ingredients. The main ingredient is the adhesive molecule or molecules. If the adhesive is a thermoset, the mixture will include one or more curatives, and it may include other materials with a number of different functions. Even though discrete physical domains can exist at the molecular level within an adhesive, there will also be empty domains or locations among the molecules. These domains are called the molecular free volume.

It may be helpful to picture a tangled wad of different colored strings, where each string represents different molecules in the adhesive. The places where no strings exist appear as voids within the wad. These voids correspond to the molecular free volume. For the remaining discussion, the adhesive mixture will be referred to simply as the adhesive and no distinction will be made among its various components.

A molecule of water can enter and exist in an adhesive only within the free volume. It should be clear that a water molecule cannot exist at the same location as one of the adhesive molecules. As more and more water molecules enter the adhesive, the free volume will change because the original free volume is being replaced by water mole-

cules. At some point—depending on the chemical structure of the adhesive, the amount of water within the adhesive, and the extent of interaction between the two—the adhesive volume will begin to increase. As they push their way into the adhesive, the water molecules actually force the adhesive molecules farther and farther apart. This phenomenon is referred to as swelling. The degree of swelling is used to calculate the total free volume; that is, given the volume of the adhesive in its original state, the volume increase due to swelling can be used to calculate the total free volume within the adhesive.

Recent studies using the positron annihilation spectroscopy free-volume microprobe have conceptually represented the free volume as two components: (1) the number of free-volume sites and (2) the average volume of these sites. This breakdown facilitates a greater understanding of diffusion and degradation. For example, in a chain scission process, the total free volume can remain relatively constant while the average number of sites decreases and the average site size increases. This breakdown also allows the development of conceptual mechanisms that involve changes in the distribution and the total number of free-volume sites.

Stress

By itself, stress does not limit the service life of adhesives. As with all materials, adhesives have ultimate tensile strengths and ultimate elongations. If these ultimate property levels are exceeded, catastrophic failure of the adhesive will result. As long as the environment produces mechanical stresses that are well within the ultimate tensile properties, the adhesive should not experience any degradation of performance.

Cyclic stress is potentially damaging because it can cause fatigue failures in the adhesive. Fatigue failures are more likely to be functions of thermal degradation than functions of direct mechanical degradation. Adhesive material can convert mechanical energy into thermal energy, thereby raising the temperature of the adhesive. Even this cyclic effect is minimal as long as the maximum cyclic stress does not approach the ultimate mechanical properties of the adhesive. Provided that the mechanical stresses are well within the ultimate properties of the adhesive, mechanical stress does not limit the service life of the adhesive.

Mechanical stress has been shown to have a tendency to increase the crystallinity of adhesives (Ref 7, 8). Crystallinity in adhesives is measured in localized areas where the molecules align themselves. Most adhesives have very little crystallinity. Crystallinity induced by mechanical stress decreases the permeability of an adhesive to a level below that of a nonstressed adhesive. However, this is generally a minor effect because the degree of stress-induced crystallinity is usually quite small.

Temperature

An increase in temperature will decrease the service life of an adhesive by increasing the rate of autodegradation. Most adhesives degrade by chain scission; that is, the polymer molecule is cut in two. Such degradation has the effect of increasing the average free-volume site size and decreasing the total number of free-volume sites. As this process continues, the ultimate mechanical properties of the adhesive increase due to the chemical degradation of the adhesive molecules. Eventually, the ultimate mechanical properties degrade to such a degree that they approach the mechanical stress fields that are present in the adhesive due to its normal operating conditions; at this point, the adhesive will exhibit a catastrophic failure.

Moisture

Moisture is the most damaging of the three environmental stresses. The degree of the threat posed by moisture depends on the chemical structure of the adhesive. The more polar the adhesive, the greater the potential that moisture will greatly reduce its service life. If the adhesive is nonpolar, then moisture will have a minimal effect because it is unable to permeate into the adhesive. It should be noted that even though the basic polymer molecule of an adhesive may itself be nonpolar, some adhesive mixtures contain hydrophilic ingredients that cause the adhesive to act as a polar adhesive. At low concentrations, moisture acts as a plasticizer and essentially softens the adhesive; this is a relatively benign effect that has a minimal negative impact on the adhesive. As the concentration of moisture increases and the free volume becomes packed with water molecules, the resulting internal stresses can actually rip the adhesive apart from within.

Acidic moisture can cause chain scission in adhesive molecules. As was noted in the discussion of temperature effects, chain scission tends to lower ultimate mechanical properties.

Temperature-Moisture

The interaction of temperature and moisture causes greater degradation of performance than can be attributed to either temperature or moisture by itself. There is, of course, the temperature effect in chain scis-

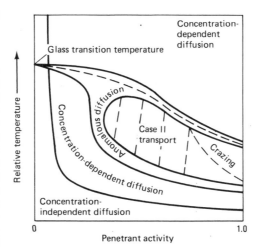

Fig. 4 Diffusion mechanism diagram. Source: Ref 9

sion reactions: The rate of reaction increases with increasing temperature above the normal operating temperatures.

The most critical temperature-moisture effect is the diffusion of the water molecules into the adhesive. The dependence of the diffusion mechanism on temperature is described in Ref 9 and 10 and shown in Fig. 4. At normal operating conditions, most adhesives exhibit a case II transport mechanism. This mechanism is relaxation driven, and thus is not dependent on the concentration of the moisture. The concentration-dependent transfer is often called Fickian diffusion. The rate of the diffusion within the case II region is less than that in the Fickian region. Because service life testing involves raising the temperature to accelerate the aging process, there is considerable danger that the system will pass through the anomalous diffusion region into a Fickian region. In the anomalous region, both concentration-dependent diffusion and relaxation-dependent diffusion take place. This region is normally very narrow and the system quickly passes into the concentration-dependent region.

The anomalous region follows the glass transition temperature. Thus, if the system passes through this temperature, the system is controlled by concentration-dependent diffusion. Because the system passes from a normal-operating-stress regime into an elevated-stress regime, the diffusion rate into the adhesive increases and life estimates decrease. The life measurements in the elevated-stress regime will result in shorter service life predictions than those obtained within the case II region.

Moisture and Mechanical Stress

Temperature, moisture, and mechanical stress interact to cause solvent crazing. Crazing is the formation of microcracks within the adhesive. Even though these cracks appear quite small, they are very large when compared with the molecular-level free volume. These cracks can act as channels for moisture, and they tend to weaken the adhesive. Failure by solvent crazing occurs by a brittle fracture mechanism (Ref 10, 11), and a moisture penetration zone always initiates the failure. As seen in Fig. 4, this crazing occurs below the glass transition temperature at high activities. This region can be shifted to lower activities by increasing the stress. Again, if the adhesive is unlikely to exhibit stress crazing in its normal-operating-stress region, acceleration into a region where stress crazing exists will yield inaccurate and low estimates of the service life.

REFERENCES

1. G.C. Derringer, Considerations in Single and Multiple Stress Accelerated Life Testing, *J. Qual. Technol.*, Vol 14 (No. 3), July 1982
2. N.R. Draper and H. Smith, *Applied Regression Analysis*, 2nd ed., John Wiley & Sons, 1966
3. G.C. Derringer, M.J. Cassady, and M.M. Epstein, A Reassessment of the Strength-Failure Time Regression Extrapolation for Estimating the Long-Term Strength of Plastic Pipe, in *Proceedings of the Seventh Plastic Fuel Gas Pipe Symposium*, Gas Research Institute, Vol 15, 1980
4. G.E.P. Box and D.R. Cox, *J. R. Stat. Soc.*, Vol B-26, 1964, p 211
5. G.A.F. Seber, *Linear Regression Analysis*, John Wiley & Sons, 1977
6. S. Chatterjee and B. Price, *Regression Analysis by Example*, John Wiley & Sons, 1977
7. D. Fox, M.M. Labes, and A. Weissberger, *Physics and Chemistry of the Organic Solid State*, John Wiley & Sons, 1965
8. S.W. Lasoski, Jr. and W.H. Cobbs, Jr., *J. Polym. Sci.*, Vol 36, 1959, p 21
9. H.B. Hopfenberg and H.L. Frisch, *Polym. Lett.*, Vol 7, 1969, p 405
10. D.W. Van Kerelen, Ed., *Properties of Polymers: Their Estimation and Correlation With Chemical Structure*, 2nd ed., Elsevier, 1976
11. B.J. MacNulty, *Br. Polym. J.*, Vol 6, 1974, p 39

Weathering and Aging Effects on Adhesive Joints*

H. Kollek, Fraunhofer Institute for Material Science

THE ENVIRONMENT can change the mechanical properties of adhesive joints. In most cases, a decrease in strength is observed. The environmental factor that has the most influence is the absorption of water in a humid climate. Water absorption weakens the adhesive to some degree and adhesion is often irreversibly changed. The time-dependent behavior of adhesive joints is a special problem in the prediction of durability. For this reason, testing methods must simulate within a short time period the same changes of properties that might occur over years in service. Many different test methods have been developed, and their reliability is discussed in this article and compared with long-time exposure.

Definition of an Environment

An environment can be defined as the sum of all factors acting on an adhesive joint; this means the surrounding media, temperature, radiation, and, in a wider sense, mechanical forces, too. These factors and internal conditions, such as residual stress in monomers in the adhesive or other characteristic stresses, superimpose each other. This network of influencing factors governs the time-dependent behavior of an adhesive joint. It is impossible to investigate all these influences at the same time. Therefore, the effect of a surrounding atmosphere will be discussed here. Climate can be described as the oxygen component from the air, water in vapor or liquid form, temperature, and ultraviolet radiation, although radiation is often not of great importance because much of it is reflected or absorbed by the parts being joined, thereby it affects just the edges of the bond line.

The two parameters water and temperature are normally combined in the term climate, which is described more precisely as a temperature value and the level of relative humidity. However, humidity is a value that depends strongly on temperature, although this dependency can be eliminated if the concen-tration of water vapor in an atmosphere is equal to the vapor pressure, a condition which would equate to a relative humidity of 100%. This, of course, is the maximum concentration of water that can be present in an atmosphere at a given temperature. Higher amounts of water cannot evaporate. In an open system where pressure is constant, the water vapor will displace the other gases, such as air, and the sum of the partial pressures will remain constant.

The relative humidity is the quotient of the actual vapor pressure and the maximum vapor pressure that could be present at that temperature. Figure 1 shows that the pressure of water vapor is dependent on temperature. All existing climates must lie under this curve; otherwise water condensation will occur. Figure 1 indicates that a climate of 50 °C (122 °F) and 95% relative humidity contains three times more water vapor than a climate of 30 °C (86 °F) and 95% relative humidity. Theoretically this higher water content should accelerate the aging mechanisms in an adhesive joint.

Adhesive joints exposed to an outdoor environment are less influenced by the climate than are those subjected to many severe aging tests. The average climate on earth rarely exceeds values of 30 °C (86 °F) and 95% relative humidity, although values this high can be experienced in some tropical or subtropical forests. Long-duration, higher temperatures do exist, but mostly in tandem with sunshine and therefore typically with a lower relative humidity. In the microclimate that directly surrounds the adhesive joint, differing values can occur. A closed box containing water can be heated by the sun to 100 °C (212 °F), with a relative humidity of 90% or more.

In addition to the influence of climate, the substances in the joint must be considered as well. If an adhesive joint is enclosed, a microclimate is present in which substances can be enriched. From this standpoint, the construction of a joint is important. In such a microclimate, which can be called a closed climate, aging mechanisms are often observed that are different from those in an

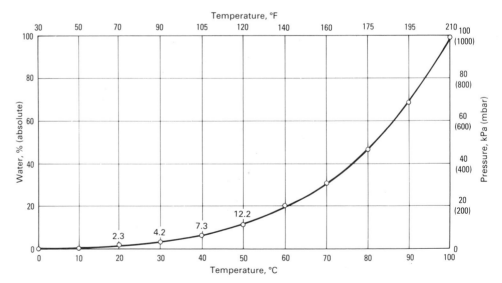

Fig. 1 Vapor pressure of water versus temperature

*Written and translated by H. Kollek from "Die Diffusion des Wasserdampfes der feuchten Luft in die Klebschichten von Metallklebungen," DFVLR-Report 79-06, 1979

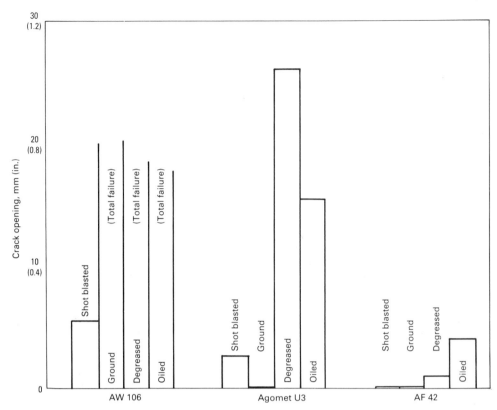

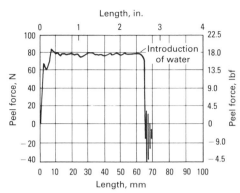

Fig. 3 Wet peel test of a steel joint. Source: Ref 1

Fig. 2 Crack growth in a wedge test of steel bonds, with various surface treatments: AW 106 is a cold-curing two-component epoxide adhesive, Agomet U3 is a cold-curing polyester adhesive, and AF 42 is a hot-curing epoxide adhesive. Source: Ref 1

open climate with respect to exchanges of gases and vapors. Such a climate can be verified in many commercial climate chambers. A closed climate is present in desiccators, for example.

As previously mentioned, oxygen is an important factor influencing the aging behavior of adhesive joints. It is especially responsible for corrosion in glue lines.

Short-Term Testing

Test specimens, such as those described in DIN 53 295 or ASTM D 1183, "Standard Test Methods for Resistance of Adhesives to Cyclic Laboratory Aging Conditions," are used to determine the durability of adhesive joints. The specimens used may have lap-shear, peel, or wedge configurations. They are usually stored in an environment of constant temperature and humidity. This practice does not always produce results that correspond to the experiences of failed bonds in service. Therefore, a large number of other tests have been developed for many different applications, such as the methods described in ASTM D 1581, "Standard Test Method for Bonding Permanency of Water- or Solvent-

Soluble Liquid Adhesives for Labeling Glass Bottles" and ASTM D 2918, "Standard Practice for Determining Durability of Adhesive Joints Stressed in Peel." Many other specialized tests are also used by industry. The results from all of these different tests cannot always be compared. For example, the automotive industry in Europe prefers test cycles that combine a hot and humid climate test, a deep temperature test, and a salt spray test. These tests can have durations ranging from several weeks to 1 year. However, they must still be considered short-term tests, compared to the estimated lifetime of a typical automobile joint, which is usually more than a decade.

Highly reliable short-term tests that provide immediate information on proposed durability are not well known, although the wedge test is one example. It is performed by joining two parts approximately 25 × 100 mm (1 × 4 in.) in size. A wedge is driven between the plates at one of the small edges of the bonded joint, thereby creating a crack that runs into the glue line. The specimen is then stored in a humid climate over several hours, after which the increase in crack length that occurs during storage is measured. Figure 2 presents examples of the results for such tests for steel joints using various adhesives and different surface treatments (Ref 1).

Another example of a short-term test is the wet peel test, for which a peel specimen is used. During peeling, a few drops of water containing some tenside are applied with a syringe at the crack tip. This can lead to an immediate adhesive failure, sometimes with a dramatic breakdown of peel strength, as shown in Fig. 3. Here, a peel test similar to that of ASTM D 903 was used, with a steel foil and a cold-curing epoxy adhesive. The surface treatment was grinding. As shown in Fig. 4, such a breakdown is not observed if the joints are made from chemically treated aluminum, according to aircraft specifications.

Storage in a constant climate is often performed over several months to evaluate the durability of adhesive joints. Figure 5

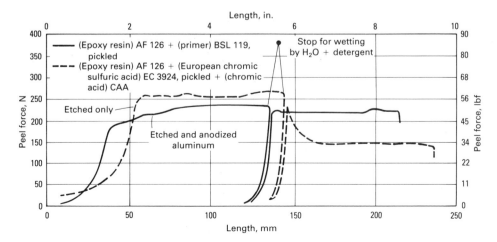

Fig. 4 Wet peel strength of aluminum joints with different surface treatments

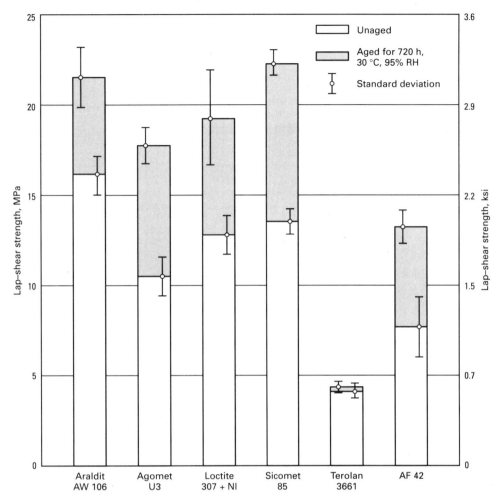

Fig. 5 Increase of lap-shear strength after 4 weeks storage in a humid climate. Specimen was stainless steel, shot-blasted, using different adhesives. Source: Ref 1

shows the results of a 4-week storage in a constant climate of 30 °C (86 °F) and 95% relative humidity, which show increases of lap-shear strength. Specimens of steel with different surface treatments were used. A problem in such experiments is the rate of water diffusion in the adhesive. At temperatures up to 40 °C (105 °F), water diffusion of 1 to 2 mm (0.04 to 0.08 in.) has been measured for epoxy resins after 1 month (Ref 1, 2). The results indicate that the saturation of the glue line with water or an equilibrium with the environment cannot be attained in a few weeks. This may cause problems when evaluating long-term behavior based on such short-term tests, as described in the article "Durability Assessment and Life Prediction for Adhesive Joints" in this Section of the Volume. It should also be noted that the velocity of diffusion depends on the temperature (Ref 3).

Long-Term Testing

A comparison of long-term and short-term test results was first given in 1966 by Eichhorn, as described in Ref 4 and 5. The long-term exposures in different climates and in water lasted over 4 years. Some mechanical strength values due to exposure in the forests of Surinam over 12 years are shown in Fig. 6 and 7 and described in Ref 6 to 9. It is noteworthy that even after several years, relatively fast decreases in strength and ultimate failure can still occur. Unfortunately, a failure analysis is not provided in the literature, but it can be assumed that corrosion is the main reason behind this failure.

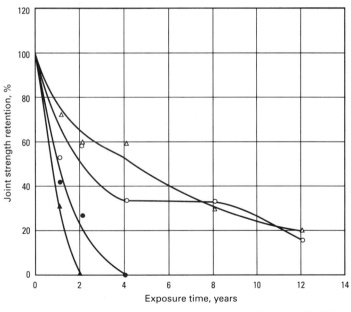

Fig. 6 Lap-shear strength of aluminum joints for up to 12 years with different surface treatments and alloys. Joints made from a cold curing epoxide. Source: Ref 5

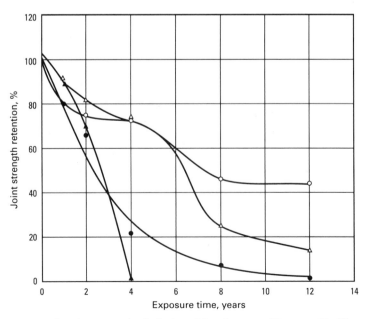

Fig. 7 Lap-shear strength of aluminum joints for up to 12 years with different surface treatments and alloys. Joints were made from a hot curing epoxide.

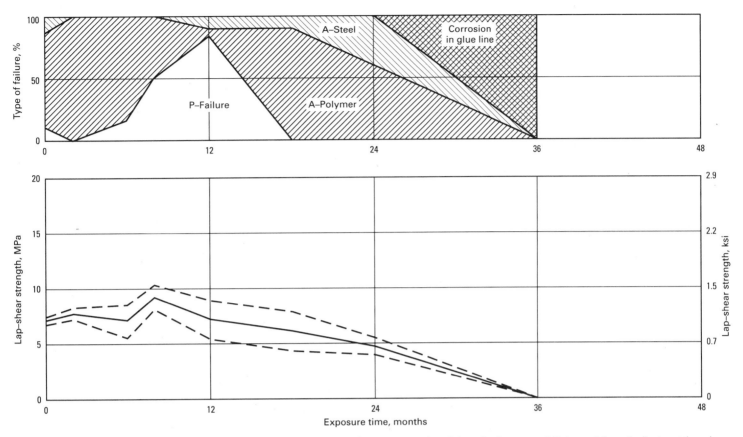

Fig. 8 Lap-shear strength and fracture surface representation of steel-polypropylene joints. P-Failure, failure of polymer part; A-Polymer, failure of adhesion at the polymer part; A-Steel, failure of adhesion at the steel part

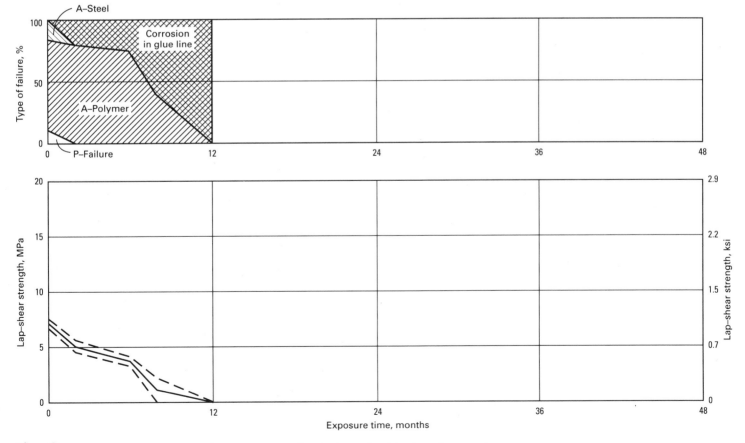

Fig. 9 Lap-shear strength and fracture surface representation of steel-polypropylene joints in a salt spray test

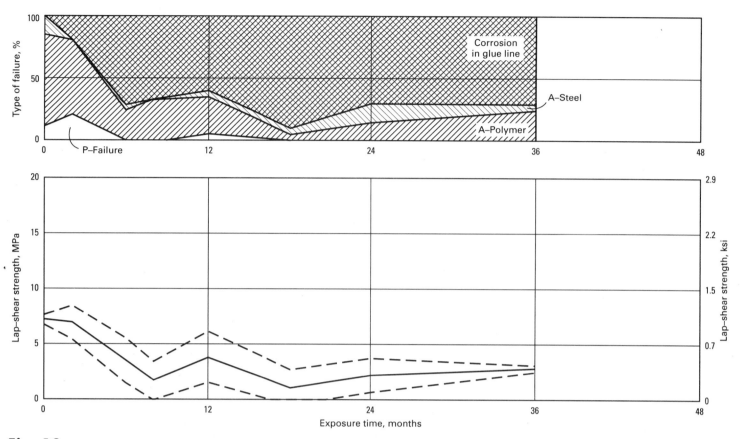

Fig. 10 Lap-shear strength and fracture surface representation of steel-polypropylene joints in an outdoor exposure

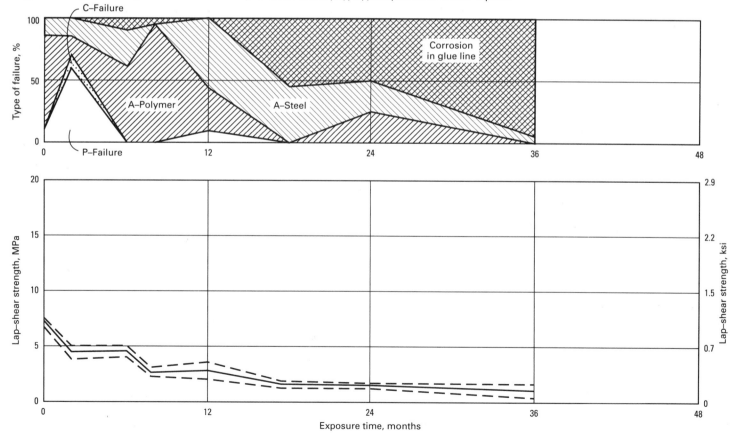

Fig. 11 Lap-shear strength and fracture surface representation of steel-polypropylene joints in a climate of 30 °C (86 °F) and 95% relative humidity. C-Failure, adhesive cohesion failure

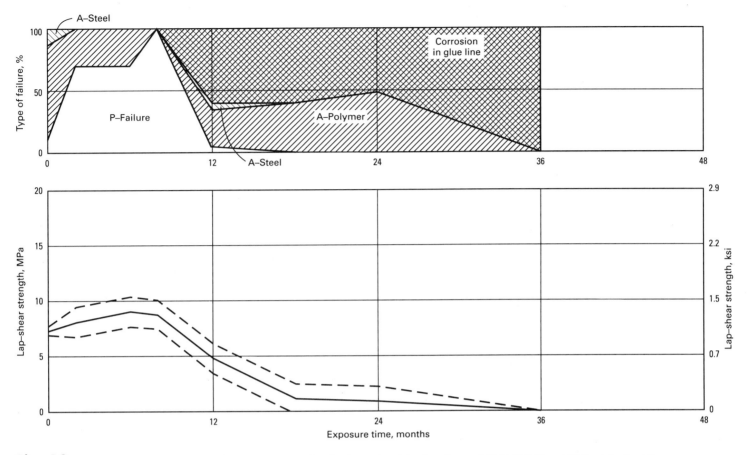

Fig. 12 Lap-shear strength and fracture surface representation of steel-polypropylene joints in a climate of 50 °C (120 °F) and 60% relative humidity

The time dependence of the mechanical properties of adhesive joints and the failure mechanisms are also described below and in Ref 10 for the case of steel-polypropylene joints. The adhesive was a cold-curing epoxy resin. All tests used a lap-shear specimen made of a polypropylene part that was 4 mm (0.16 in.) thick and an automotive steel part that was 1.0 mm (0.04 in.) thick. The lap-shear strengths and the different fracture types are identified in Fig. 8, in terms of percent of joining area. The lap-shear strength curves and standard deviations in Fig. 8 are plotted over the time of exposure in a climate of 30 °C (86 °F) and 60% relative humidity. The data are based on five specimens. Measurements were made at the beginning, and after 2, 6, 8, 12, 24, and 36 months.

The upper diagram in Fig. 8 illustrates the type of failure. The classifications explained in the figure caption are those given in DIN 54 456, the German specification for durability tests. The hatched areas were plotted using straight lines between the points of measurement. The start of corrosion is a nonpredictable statistic effect and is not visible before testing of the specimens. In some cases, this leads to a backwards-running corrosion in the plots.

The data in Fig. 8 show that in the first 2 months of exposure, lap-shear strength in-

creases. There are two reasons for this effect. First, a postcuring of the adhesive occurs and, second, the adhesive begins to weaken at the edges of the glue line because of the uptake of water, which lowers the stress maxima at these edges, resulting in higher strength. Lap-shear strength subsequently decreases and reaches a minimum after about 8 months. This is due to the changes of the water distribution in the glue line and weakening of the adhesive. The next step is an increase of strength at about 12 months, at which point the adhesive joint reaches an equilibrium with the environment over the entire joining area, and no further changes in water concentration in the adhesive take place. After 12 months, a slow decrease of strength can often be observed. This is due to long-term aging effects and ultimately governs the life span of the joint. After 24 months, rapid corrosion is observed, leading to rapid, total failure of the joint.

Figures 9 to 13 represent additional examples of this behavior. Figure 9 is based on exposure in a salt spray test. The rapid corrosion, which starts immediately, is the determining failure mechanism. Figure 10 gives the results of outdoor exposure in northern Germany, a maritime climate but without the direct influence of seawater. In this case, corrosion starts soon after exposure, but not as fast as in the salt spray test. Figures 11 and

12 are based on exposure to open climates of 30 °C (86 °F) and 95% relative humidity and 50 °C (120 °F) and 60% relative humidity, respectively. Results from a closed climate of 20 °C (68 °F) and only 60% relative humidity are given in Fig. 13.

These figures demonstrate the importance of corrosion as a failure mechanism. In some cases, it started after a year or more of exposure and then led to a rapid total failure within an additional few months. The start of corrosion could not be predicted in the experiments, but once it began, it always progressed quickly. To summarize these results, it can be said that a salt spray test or exposure in a closed climate without satisfactory aeration will cause more and faster corrosion than storage in an open climate.

Another problem is that corrosion can start in the glue line and often is not visible from the outside. This means that an adhesive joint can be destroyed by corrosion with hardly any evidence of corrosion on the outside of the joint.

Comparison of Short- and Long-Term Testing

A comparison of the reliability of different test methods for technical applications is always difficult and requires some expe-

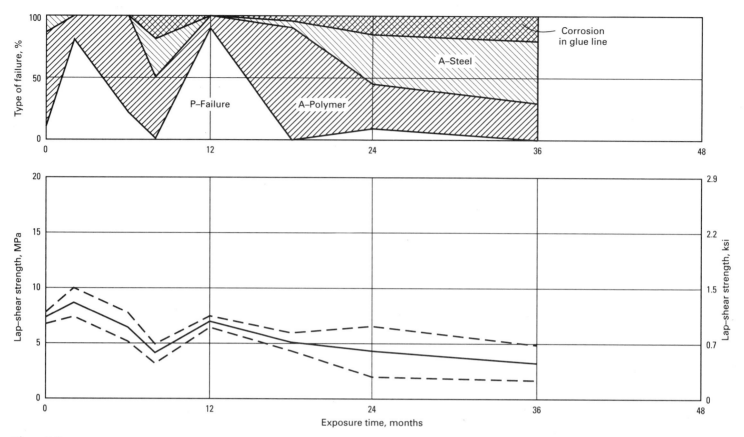

Fig. 13 Lap-shear strength and fracture surface representation of steel-polypropylene joints in a climate of 20 °C (68 °F) and 60% relative humidity

rience. The environmental conditions in which the adhesive must serve must be determined carefully. In particular, the possible appearance of corrosion must be considered, and appropriate techniques to prevent it must be applied.

The real short-term tests, such as the wet peel and the wedge tests, allow a comparative evaluation of the stability of adhesion, but they do not correspond directly to the long-term behavior of a joint because they represent tests for different adhesion levels. For example, the wet peel test requires a higher level of adhesion stability than does the wedge test.

Aging procedures over months or even years give more reliable results, but still require experience to establish correlations between these tests and real long-term behavior. Measuring mechanical properties, as well as observing failure mechanisms, is very useful in establishing correlations to long-term behavior.

REFERENCES

1. W. Brockmann and H. Kollek, *Metallkleben im Maschinenbau, FKM-Heft*, Vol 81, 1980
2. W. Althof, in "Die Diffusion des Wasserdampfes der feuchten Luft in die Klebschichten von Metallklebungen," DFVLR-Report 79-06, 1979
3. J. Comyn, Kinetics and Mechanisms of Environmental Attack, in *Durability of Structural Adhesives*, A.J. Kinloch, Ed., Applied Science Publishers, 1983
4. E. Eichhorn *et al.*, "Untersuchungen über das Alterungsverhalten, die Temperaturbeständigkeit und Zeitstandfestigkeit von Metallklebverbindungen mit und ohne Füllstoffzusätze," Forschungsberichte 1734, Landes Nordrhein-Westfalen, 1966
5. A. Matting, *Metallkleben*, Springer-Verlag, 1969
6. J.D. Minford, Durability of Aluminum Bonded Joints in Long Term Tropical Exposure, *Adhes. Adhes.*, Vol 2, 1982, p 25-31
7. J.D. Minford, Adhesives, in *Durability of Structural Adhesives*, A.J. Kinloch, Ed., Applied Science Publishers, 1983, p 135/214
8. J.D. Minford, Aluminium Adhesive Bond Permanence, in *Treatise on Adhesion and Adhesives*, Vol 5, R.L. Patrick, Ed., Marcel Dekker, 1981, p 45-138
9. J.D. Minford, Permanence of Adhesive-Bonded Aluminium Joints, in *Adhesives in Manufacturing*, G.L. Schneberger, Ed., Marcel Dekker, 1983
10. H. Kollek, Leistungsmöglichkeiten der Klebtechnik, in *Kleben von Kunststoff mit Metall*, W. Brockmann, L. Dorn, H. Käufer, Ed., Springer-Verlag, 1989

Durability Assessment and Life Prediction for Adhesive Joints

W. Brockmann, Kaiserslautern University, H. Kollek, Fraunhofer Institute for Material Science

LIFE PREDICTION for adhesive joints employing laboratory tests requiring only days, weeks, or months is a complex problem. The tests are carried out with the hope of simulating important effects that may happen to a joint during its lifetime. Much experience is needed to understand the test concepts, carry out the tests, and evaluate the results.

Initial strength values of bonded joints, such as shear or peel properties, are obtainable from the adhesive manufacturers or the literature. They are useful to compare different adhesives and to demonstrate the effect of parameters such as bond line thickness, overlap length or curing conditions, and, in some cases, the surface state, although these parameters may change the observed strength of the bonded joint. However, based on initial strength values alone, it is impossible to describe the real efficiency of the adhesive bonding technique for joining system assemblies if the assemblies must serve many years at elevated temperatures, especially under humid environmental conditions.

It is well known today that, except for degradation by concentrated acids or bases, x-rays or gamma-rays, some organic solvents of high polarity, or temperatures of more than 100 or 150 °C (212 or 300 °F), the most common and important factor influencing the long-term behavior of adhesively bonded joints is the presence of humidity or liquid water.

Bond line corrosion is extremely important in the case of bonded metal joints without protection against primary corrosion to water or humidity in the environment. Because of their small dimensions and high polarity, water molecules can diffuse into practically all organic materials, including cross-linked organic adhesive layers. In most cases, the diffusion process ends at a state of equilibrium, the level of which is related to the adhesive, and with only some exceptions, is reversible. Such a diffusion process cannot be avoided by coating the edges of the bonded area with paints or sealants, which only inhibits primary corrosion of the areas and slightly delays the diffusion process.

Water molecules between the adhesive polymer chains can initiate two important mechanisms that change the long-term properties of the bonded joints. The first of these is a slight swelling process leading to a decrease in tensile strength, and in most cases, to an increase in the plastic deformation properties of the polymer and therefore to higher peel resistance. Chemical reactions between the diffusing water and the organic molecules that compose the bulk adhesive are possible, but as far as it is known, they do not occur to any great extent and are not very significant in determining the lifetime of bonded joints if cross-linked adhesives are used.

Figure 1 depicts a typical example of the diffusion-dominated strength characteristic of bonded joints between steel and polycarbonate adherends as dependent on exposure time in a natural climate in northern Germany.

The second important mechanism is the capability of water molecules to diffuse into the interfacial boundary zone, destroying the adhesion between the adhesive and the metal surface. Degradation of the boundary layer of an adhesive may lead to the well-known interfacial failures considered from a macroscopic viewpoint, such as a total separation of the adhesive from the metal surface without changing the nature of the metal surface. The degradation process in the boundary layer which finally leads to failure, is not reversible and consequently is the most dangerous mechanism. In this boundary zone, the polymer may be affected by water by hydrolytic degradation. Such a mechanism will occur more readily if the polymer in these regions has not been cross linked fully.

Figure 2 represents a typical example of the adhesion-dominated strength characteristic of bonded joints between steel and polycarbonate adherends, as dependent on exposure time in an artificial humid climate of 20 °C (68 °F) and 70% relative humidity (RH). The decrease of strength after an

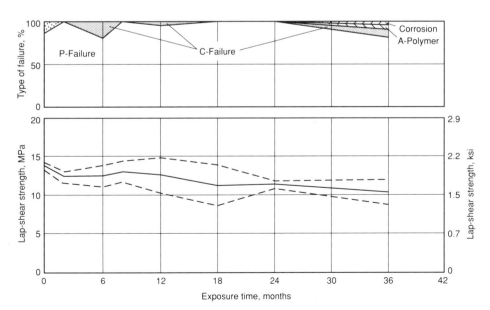

Fig. 1 Shear strength of bonded joints between steel and polycarbonate, as dependent on exposure time in natural climate in northern Germany. Adhesive: two-component polyurethane. P-Failure, failure of polymer part; C-Failure, failure of adhesive in cohesion; A-Polymer, failure of adhesion at the polymer part

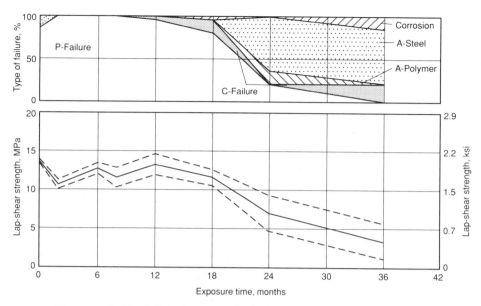

Fig. 2 Shear strength of bonded joints between steel and polycarbonate, as dependent on exposure time in humid climate of 20 °C (68 °F) and 70% RH. Adhesive: two-component polyurethane. A-steel, failure of adhesion at the steel part; A-Polymer, failure of adhesion at the polymer part; C-Failure, failure of adhesive in cohesion

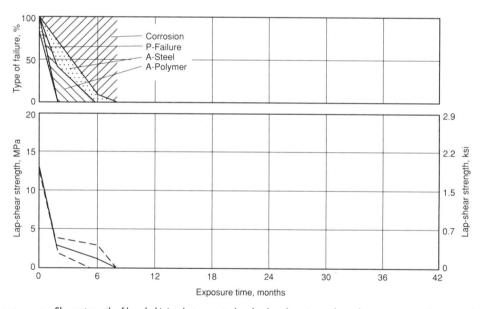

Fig. 3 Shear strength of bonded joints between steel and polycarbonate, as dependent on exposure time in a salt spray test (ASTM B 117). Adhesive: two-component polyurethane. A-steel, failure of adhesion at the steel part; A-Polymer, failure of adhesion at the polymer part; P-Failure, failure of polymer part

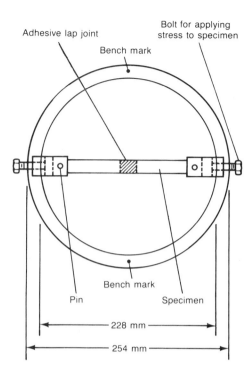

Fig. 4 Minford ring used for durability tests. Source: Ref 4

18-month exposure time is due to the increasing adhesional failure.

As far as is known, a third failure mechanism can occur in bonded joints with metal parts without outer corrosion protection. If humidity or water from the environment produces primary corrosion on the metal near the adhesive layer, rapid bond line corrosion into the joint is induced. Figure 3 depicts the bond line corrosion-dominated strength characteristic of bonded joints between steel and polycarbonate, as dependent on exposure time in a salt spray chamber (ASTM B 117).

It must be stated here that an exact interpretation of failure mechanisms when considering only strength values or strength/time curves is difficult and sometimes dangerous. More information about the real degradation mechanisms can be obtained by a visual inspection of the fracture surfaces of the destroyed joints. This is true for long-term tests and in particular, for short-term tests.

In addition to the durability test methods described below, a large number of different or merely modified tests are used. In most cases, particularly in the case of a macroscop-

ic analysis of the fracture surfaces of the bonded joints, the reason for failure can be tracked to the three principal degradation mechanisms discussed in this article, along with useful investigation methods mentioned.

Durability Test Techniques

Diffusion-Dominated Durability. One of the best laboratory techniques to evaluate diffusion-dominated strength behavior is the measurement of the creep of mechanically loaded shear specimens using optical or electromechanical devices, either after or during elevated-temperature and high-humidity exposure (Ref 1-3).

Different techniques have been developed to apply a load. One is the so-called Minford ring, which is shown in Fig. 4 (Ref 4). A lap-shear specimen is fixed in the ring, which acts as a spring. Because this ring is a lightweight device, many of them can be placed in a climate chamber without fear of overloading. The ring is relatively inexpensive because it can be made from tubes that are available in many diameters and wall thicknesses. Both diameter and thickness influence the elasticity of the tube and the maximum deformations and loads that can be applied to a Minford ring. The ring is very useful if the creep of the adhesive joint is not too high because creep will lower the load to a certain extent.

Mechanical load also can be applied by other systems that contain spiral springs. Figure 5 gives an example of such a device. The advantage is that joints with a high level of creep also can be investigated.

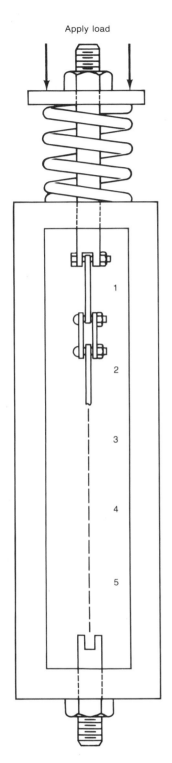

Apply load

Fig. 5 Spring-loaded fixture. Source: Ref 4

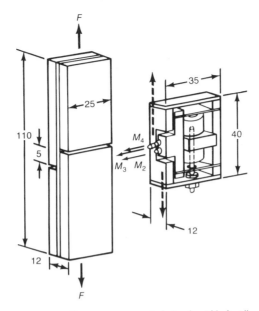

Fig. 6 Creep measurement device by Althof. All dimensions given in mm (Ref 1)

- Dimensions are small
- Deformations are measured only in the area of lap and only in the load direction; therefore, only a small part of adherend deformation is included in the total measured deformation
- The electrical measuring system records measured data on plotters or recorders
- It can be combined with electrical load devices
- Measuring begins at the start of the test and continues up to the fracture of the joint

Some results of such measurements are given in Fig. 7 for three different adhesives after exposure in various environments. It can be seen that the most change is the result of moisture and heat. The graduations of change are different for the three adhesives. For FM 1000 (one-component epoxy-nylon system), the shear modulus and proportionality limit are decreased considerably, the strength is decreased slowly, and the strain at fracture is increased. For AF 126-2 (one-component epoxy-nitrile system), the graduation of change is less than it is for FM 1000. For FM 73 (rubber-modified one-component epoxy system), the shear modulus is unchanged by the exposure; however, the proportionality limit is displaced to small shear stresses. The results were determined at ambient temperature and in some cases at 50 °C (120 °F) after the exposure. The passage of the shear stress-strain curve from linearity to a bent curve is displaced to smaller shear stresses as temperature is increased. Many such diagrams exist in the literature. The change of curves is also variable with time. An asymptotical curve can be found after about 25 weeks.

A second possible technique is to measure the creep properties of bonded joints

Both techniques can be used when measuring creep during exposure, using the specimen shown in Fig. 6. The relation of the maximum shear stress on the end of overlap to the mean shear stress computed by the finite element method is N = 1.135.

Figure 6 also shows the extensometer developed by Althof (Ref 1). Its advantages over other instruments are:

under load in humid environments under elevated temperatures. Typical creep curves, as dependent on exposure time in a humid climate of 40 °C (105 °F) and 95% RH of different hot curing adhesives under different shear loads, are plotted in Fig. 8. If single-lap joints in the form of standard specimens are used, then a measurement time of about 1000 h in a constant climate of 30 to 50 °C (85 to 120 °F) and 95% RH is usually sufficient. From such tests, cold-curing adhesives such as aminoamide-cured epoxide resins and second-generation acrylate adhesives showed a long-term strength of only 1 N/mm^2 (145 lbf/in.^{2}), which is not more than 5 or 10% of the initial shear strength. During the loading process in a humid environment, shear strains of tan γ = 0.4 occurred in the bond line. Under these small loads, the creep process came to an end after 100 to 400 h. Under higher loads, higher rates of creep were observed, and the creep processes did not stop. Practically all the specimens under higher loads were destroyed under loading times of 200 to 300 h. If the temperature of the environment was higher than 40 °C (105 °F), practically no long-term load capability was measurable.

Totally different properties are observed if hot-curing epoxide adhesives, such as rubber-modified epoxide systems, used in the aircraft industry, are tested over 1000 h in an environment of 50 °C (120 °F) and 95% RH. These adhesives are able to withstand shear stresses of 15 N/mm^2 (2175 lbf/in.^{2}) and more. This is equivalent to 50 or 60% of the initial strength. During this aging procedure, the creep processes came to an end after about 500 h, and the measured creep strains were in the range of tan γ = 0.5 to 1.0.

Only these more realistic test methods, which demonstrate the large differences in the load capability and durability of cold- and hot-curing adhesives, give clear indications of the actual load capability of a bonded joint. If such a joint has to serve within the temperature range of 30 to 50 °C (85 to 120 °F) over long times in the presence of humidity, but is designed so that the load is not higher than 5% of the initial strength in the case of cold-curing adhesives, and is not higher than 30 to 50% of the initial strength in the case of hot-curing adhesives, then aging problems and failures within the adhesives over long times should not be observed.

Adhesion-Dominated Durability. The reliable evaluation of the water resistance of adhesion using standard peel or shear specimens requires relatively long exposure times (6 to 12 months) in different humid environments. This can be demonstrated in the case of bonded aluminum joints (surface treatment of chromic-sulfuric acid etching), in which a two-component epoxide-aminoamide curing adhesive is used, the durability properties of which are relatively poor

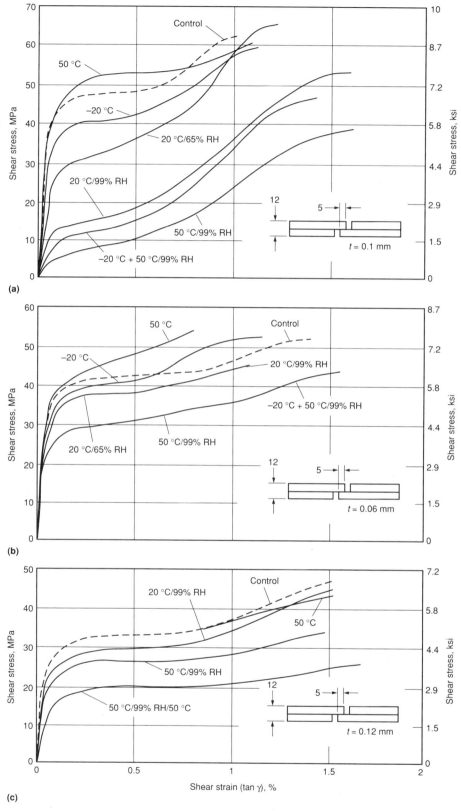

Fig. 7 Shear stress-strain diagrams of adhesives after 8-week exposure in various environments. (a) Adhesive FM 1000. (b) Adhesive AF 126-2. (c) Adhesive FM 73. RH, relative humidity; *t*, adhesive thickness. Source: Ref 1

becomes visible (Fig. 9, middle was tested unloaded, and bottom was tested under load). The shear strength reduction after this exposure time is approximately 50%. This weakening of the adhesion will, in the course of time, increase and can lead to a total destruction of the joint, as opposed to the swelling reaction in the adhesives. This is illustrated in Fig. 10, which shows the cracked surfaces of bonded steel joints that have been tested for their strength after 1 year of outdoor exposure in northern Germany. The failure areas of steel joints where the surfaces have been shotblasted (Fig. 10, top), and the failure areas where the steel surfaces have only been degreased (Fig. 10, bottom) are illustrated. A phenolic resin, which is very resistant, was employed to join the steel. In the case of the degreased surfaces, considerable adhesive failures and corrosion commencing from the edges could be established after just 1 year. However, in the case of the shotblasted surfaces, changes in the boundary layers could not be detected.

Wedge Test

All results concerning the durability of bonded steel joints reported so far involve tests that require expensive equipment, such as climate chambers and testing machines, as well as well-trained personnel to carry out the experiments. In most factories or production lines where bonded joints are assembled, it is usually not possible to carry out such complex experiments to control the quality of the surface pretreatment process or the bonded joints. However, a relatively simple short-time test method that has been developed is the wedge test (Ref 5), shown in Fig. 11.

The specimens used in the wedge test are cut from two metal sheets that have been bonded together. In the aircraft industry, aluminum sheets of 3.2 mm (0.13 in.) thickness are common, but in the experiments on the durability of adhesion of steel joints described in Ref 5, sheets 1.6 mm (0.064 in.) thick were used. After bonding, the metal plates were separated by driving a steel wedge of 4 mm (0.16 in.) thickness into one end of the adhesive layer. This wedge produces a crack in the unaged specimen that extends several millimeters down into the adhesive layer, which is easy to observe from the edges. The wedge between the plates produces high stresses at the tip of the initial crack in the adhesive layer. The specimens are then exposed to a hostile environment, such as a humid climate, for several hours. The dependence of the water stability of the adhesion during this aging process is illustrated by the crack propagating from the middle of the glue line into the boundary layer, and, in the case of unstable adhesion, at a relatively high velocity. In the case of stable adhesion, it propagates only within the adhesive and only over a

when compared, for example, with aircraft adhesives.

Standard lap-shear specimens in the initial strength test failed in a cohesive failure mode (Fig. 9, top). After a 1-year exposure time in an artificial changing climate (12 h at 30 °C, or 86 °F, and 95% RH; 12 h at 0 °C, or 32 °F), an increase of adhesional failure

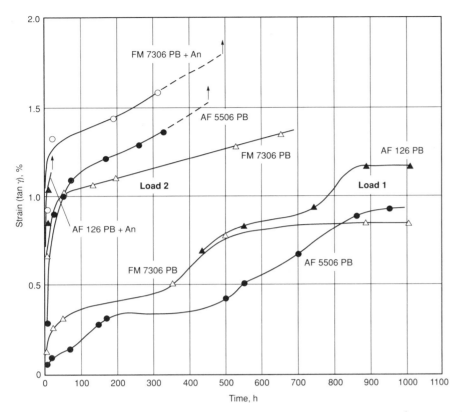

Fig. 8 Creep curves of bonded aluminum joints (standard lap joints) under load 1 of 15 N/mm² (2175 lbf/in.²), and load 2 of 22.5 N/mm² (3260 lbf/in.²) in humid climate of 40 °C (105 °F) and 95% RH

Fig. 9 Fracture surfaces of aluminum joints. Adhesive: epoxide-aminoamide system. Top, unaged; middle, after 1 year of changing climate without load; bottom, after 1 year of changing climate and load

small distance of between 0 and approximately 8 mm (0 and 0.32 in.).

The wedge test is a good tool for detecting relatively poor adhesion (not water resistant), which can be demonstrated by results obtained from steel bonds. The results are plotted in Fig. 12. In this investigation the thickness of the sheets used in the wedge test was only 1.6 mm (0.064 in.). The adhesives used were a polyaminoamide-epoxy resin, a styrene-peroxide cured acrylate, and a one-component epoxide resin. The mild steel substrates were pretreated using different procedures.

The crack lengths plotted in Fig. 12 were measured on the specimen sides after an exposure time of 48 h in an artificial humid

Fig. 10 Fracture surfaces of steel bonds after 1 year aging in a natural climate. Adhesive: phenolic resin. Top, sandblasted; bottom, degreased

climate of 30 °C (86 °F) and 95% RH. It may be readily seen from the crack elongation that the surface treatment has a major influence on the durability behavior. Compared with the results obtained from a classical aging test on single-lap joints (Fig. 12, 18), it is clear that the wedge test leads to results of the same general character. If it is not possible to measure the crack length before and after storing the specimens in a humid environment, it is sufficient in most cases to initiate the crack without measuring the crack length, expose the specimens in a humid environment for some hours, and then separate the joint by mechanical force. If the initial cohesive crack has not changed into an apparent interfacial mode during the exposure time, then the water stability of the adhesion is sufficient for most purposes.

One disadvantage of this test method is that the results are only of a qualitative nature. Therefore, it is not possible to distinguish quantitatively between different grades of interface stability.

A second disadvantage, a low detection level for only very poor adhesion, can be demonstrated by results obtained with 3.2 mm (0.13 in.) thick aluminum specimens (Ref 6).

Wedge specimens from 2024 clad and bare aluminum alloy were prepared with the adhesive-primer system of FM 73-BR 127. Surface preparation was made by chromic-sulfuric oxide acid (CSA) etching during various time periods. Except for etch-

ing time, this process is quite similar to the optimized Forest Products Laboratory (FPL) process (Ref 5), as shown in Table 1.

Times chosen for the CSA process were 1, 2, 3, 4, 5, 6, 7, 10, 15, 20, 25, and 30 min. After 10 min, the CSA process leads to micromorphology structures similar to FPL etching. Figure 13 shows some of these wedge specimens of 2024 clad alloy. After an etching time of more than 4 min, no adhesion failure will occur. In the case of 2024 bare aluminum alloy, there is no failure of adhesion even after shorter etching times (Fig. 14). For these specimens, aging time was 30 days at 40 °C (105 °F) and 95% RH, although normally an aging time of only 2 days is recommended (Ref 7). On the other hand, it is known that surfaces etched for less than 25 min may lead to metal bonds having lower durability. From this point of view, the wedge test seems only to be a test

Table 1 Comparison of CSA and FPL processes not specified, but comparable to CSA

Parameter	CSA	FPL
Na₂Cr₂O₇, kg/m³ (lb/ft³) ...	80 (5)	88 (5.5)
H₂SO₄, kg/m³ (lb/ft³)	280 (17.5)	305 (19)
Al, kg/m³ (lb/ft³)	3.0 (0.19)	9 (0.56)
Cu, kg/m³ (lb/ft³)..........	0.3 (0.02)	(a)
Temperature, °C (°F)	60 (140)	65 (150)
Etching time, min	30	10

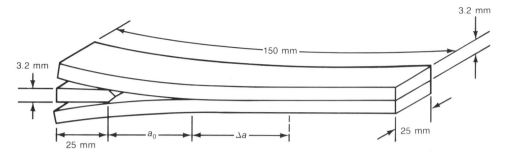

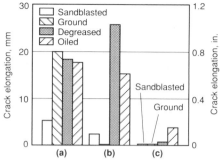

Fig. 11 The Boeing Company wedge test specimen configuration. Source: Ref 5

Fig. 12 Crack extension in mild steel wedge specimens and dependence on surface treatment. Adhesives: (a) two-part cured epoxy, (b) two-part acrylic, (c) one-part hot-cured epoxy. Aging process, 48 h in artificial climate of 30 °C (86 °F) and 95% RH. Mild steel specimens have tensile strength of 363 MPa (53 ksi).

of adhesional properties, even when a small level of adhesion durability is required.

Thus, it must be pointed out that the normal wedge test is not able to detect differences in the adhesional stability of metal bonds produced by using hot-curing adhesives with corrosion-inhibiting primers and chemical surface treatments such as chromic-sulfuric acid etching or anodizing of aluminum alloys. These treatments and adhesives systems are commonly used to produce reliable bonded aluminum joints with service lives of 40 000 h and more in civil aircraft structures.

Wet Peel Test

Higher sensitivity can be attained with the so-called wet peel test, of which the floating roller peel test is but one example (Ref 8, 9). The test commences in the dry state. Then, after about half of the sample length is peeled, the process is stopped and water that contains a detergent is applied to the crack opening. The detergent content of the water (about 0.5 to 1%) is intended to give a better wetting. Contrary to original assumptions, the type of detergent applied has no significant influence. During subsequent peeling in the presence of moisture that contains detergent, the cohesive failure that is usually found in dry joints changes almost spontaneously, either partially or completely, into so-called adhesive failure, which is connected to the corresponding

reduction of the resistance to peeling. Such a transformation, however, does not occur with very moisture-resistant joints.

The detection level of this wet peel test has to be considered high, compared with the wedge test. This can be demonstrated by wedge and wet peel test results obtained from aluminum (T 2024 clad) bonds produced with a corrosion-inhibiting primer and a one-component epoxide adhesive. The surfaces of aluminum parts were pretreated by three modifications of the chromic-acid-anodization (CAA) process (Ref 10). In Europe, the unsealed 40/50 V CAA is preferred, because the long-term stability of such pretreated Al-bonds is proved for up to 3 decades in service.

The anodizing is preceded by an alkaline degreasing, alkaline etch, and a CSA etch, which is similar to the optimized FPL etch. In this sequence of pretreatment steps, the need for the CSA process is unknown. Three principal operations are executed by the CSA etch:

- Forming a distinct surface morphology (preimpregnation before anodizing)
- Etching (removal of about 3 μm, or 0.12 mil, of Al)
- Dismutating or descaling (conversion of surface acidity, and if unclad material is used, removal of alloy elements)

To determine which of these functions was superfluous, testing that involved variation

of the pretreatment process was conducted (Table 2).

Wedge test results given in Table 3 show that no differences in the adhesion stability were observed. Based on peel results obtained in the wet state and the corresponding failure mode, Table 4 allows the conclusion that only alkaline degreasing without subsequent chromic acid anodizing before anodizing is an insufficient treatment system. With present-day aircraft adhesives, joints that show no impairment of the adhesive zone in the wet peel test can only be produced if the surfaces of the aluminum

Table 2 Variations of the pretreatment

Pretreatment 1	Alkaline degrease
	Alkaline etch
	(CAA)
Pretreatment 2	Alkaline degrease
	Alkaline etch
	HNO_3 dismutating
	(CAA)
Pretreatment 3 (production process)	Alkaline degrease
	Alkaline etch
	(CSA)
	(CAA)

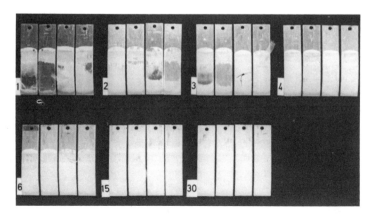

Fig. 13 Fracture surfaces of wedge specimens from 2024 T3 clad alloy. Adhesive: FM 3/BR 127. Numbers denote the various etching times in minutes.

Fig. 14 Fracture surfaces of wedge specimens from 2024 T3 bare alloy. Adhesive: FM 73/BR 127. Numbers denote the various etching times in minutes.

Table 3 Variation of the pretreatment: wedge test result

Pretreatment step	Initial crack length mm	in.	Crack extension mm	in.	Failure mode
1	32.6	1.28	5.6	2.2	Cohesive
2	32.3	1.27	4.3	1.7	Cohesive
3	35	1.38	6	2.4	Cohesive

Table 4 Variation of the pretreatment: results of the peel and salt spray test

Pretreatment step	Peel strength Dry N/25 mm	lbf/in.	Wet N/25 mm	lbf/in.	Failure mode Dry	Wet	Corroded area, % (after 90 days, salt spray test)
1	281	63	59	13	C	A	15–20
2	284	64	261	59	C	Cp	10–20
3 (standard)	275	62	245	55	C	C/Cp	10–25(a)

(a) Scatter band of numerous experiments. C, cohesive failure; A, adhesive failure; Cp, cohesive failure, mostly in the primer layer

are perfectly anodized. In all other cases (apart from a few exceptions), at least partial adhesional failure occurs. This is illustrated in Fig. 15, which shows failure zones in dry and wet peeled samples where, after phosphoric acid anodizing, the quality of the final rinsing water was varied. Only when using distilled water was there no adhesional failure, either in the wet or dry state, whereas the use of demineralized water and tap water led to partial adhesional failure.

By means of this wet peel test, it is possible to prove that failure mechanisms in the boundary layer zone are mainly initiated by diffusing water. This can be observed in a series of tests to evaluate the influence of the quantity of water available on the behavior in the boundary layer zone. In order to obtain meaningful results, the water concentration of the liquid introduced had to be reduced without changing the quantity of the liquid. Therefore, a liquid that would not diffuse into the polymer during the test

time had to be added to the water. Furthermore, its polarity had to be close to that of water. A liquid polymer, polyethylene glycol, with a mean molecular weight of 9600 (PAG 200), was chosen.

CSA-pretreated samples bonded with a corrosion-inhibiting primer-one-component epoxy adhesive system were investigated. In the wet peel test with water, these samples showed a clear decrease of peel resistance and an increase in the proportion of adhesional failure, compared to the dry peel test. In a wet peel test with pure PAG 200, there is a slight increase of peel resistance that can be attributed to the rheology of this more viscous liquid in the crack tip. With mixtures of PAG 200 and water, the peel resistance decreases as the amount of water increases until the peel resistance value reaches that of the normal wet peel test. Figure 16 presents the results of two sample series that differ slightly as to the thickness of the primer. It can be deduced from these results that the wet peel test reacts specifically toward water; the concentration of the available water is important.

In those zones where failure is identified macroscopically as being adhesional, transmission electron microscopic investigations show that the failure does not lie in the actual polymer-oxide boundary zone, but, on the contrary, in a polymer layer close to the

boundary zone. Investigations by Auger spectroscopy on the same sample prove that there is no aluminum on the primer side; however, on the oxide tips in all zones there are considerable concentrations of carbon. The fracture runs, therefore, in a polymer layer close to the interface, the state of which is influenced by the characteristics of the surface; in this case, anodized aluminum.

The results that can be obtained using the wet peel test as a short-term method to investigate the water stability of adhesion are so convincing it must be mentioned that this test and the wedge test are usable only in bonded joints with cross-linked adhesives. If the adhesion layer is not a cross-linked polymer (for example a hot melt), the polymeric material in the tip of the crack does not change its water-diffusion properties by deformation in such a drastic manner as is the case with cross-linked systems that allow the rapid transport of water from the mid layer of the glue line into the boundary zone, which produces adhesional failure.

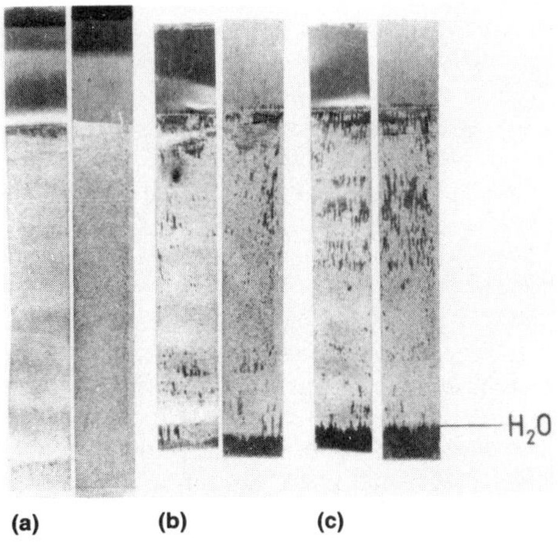

Corrosion-Dominated Durability

For adhesive-bonded metal joints, corrosive environments are a more serious problem than the influence of moisture. This can be demonstrated by fracture surfaces of mild steel specimens bonded with an epoxide-nitrile resin, originally developed for chemically treated aluminum alloys (Fig. 17). The lap-shear specimens were aged without protection of the steel parts in the natural climate of northern Germany.

In the case of shotblasted steel surfaces, small areas of interfacial failure and corrosion initiating from the edges are visible after 1 year. In the case of degreased-only steel surfaces after 30 and 60 days in the unloaded and loaded states, total delamination and subsequent corrosion of the metal surfaces had occurred. The delamination must be corrosion-initiated. If corrosion were to occur after debonding, it would be harder on grit-blasted surfaces than on sur-

Fig. 15 Cracked regions of specimens (2024 plates, PAA, EC 3924/AF 126 adhesive system) after the wet peel test (floating roller peel) in (a) distilled water, (b) demineralized water, and (c) tap water

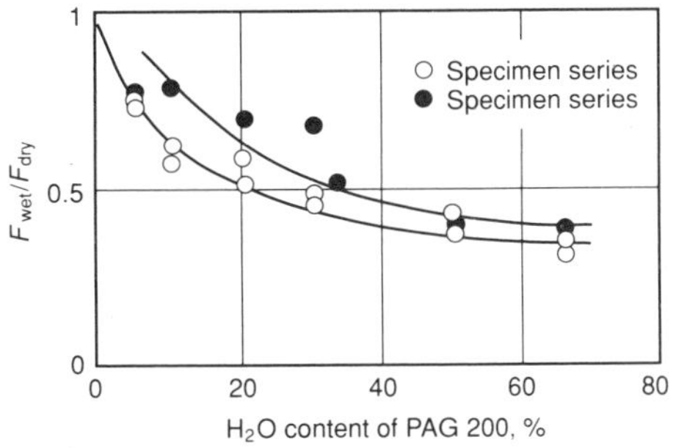

Fig. 16 Peel strength as a function of water content of liquid in wet peel test

Fig. 17 Fracture surfaces of steel joints after aging in a natural climate. Adhesive: epoxide-nitrile resin. Top, sandblasted; bottom, degreased

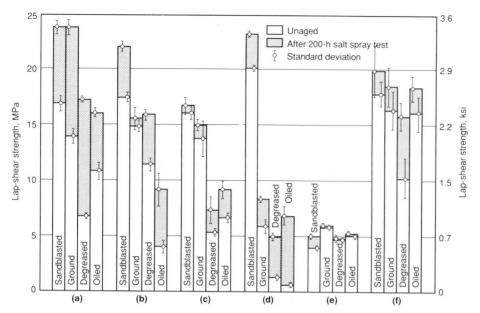

Fig. 18 Shear strength of mild steel joints. (a) Two-part cured epoxy. (b) Two-part acrylic. (c) Anaerobic acrylic. (d) Cyanoacrylate. (e) PVC plastisol. (f) One-part hot-cured epoxy

faces that were only degreased due to the higher electrochemical potential that exists after grit blasting. No connection between a cyclic loading of the specimens during the aging processes and the occurrence of interfacial failure was observed. If these surfaces are compared with those shown in Fig. 10, the results show that the type of adhesive, as well as the state of the metal surface, are important parameters for the corrosion stability of a bonded steel joint. For both adhesives, a mechanical roughening process of the metal surface by a shotblasting process appears to have a positive effect on the durability of the joints.

The easiest and fastest method to investigate such corrosion-dominated degradation in bonded metal joints is to store the bonded parts in a salt spray chamber with a continuous spraying of 5% NaCl solution in 95% RH at 35 °C (95 °F) (ASTM B 117). For the investigation of the corrosion resistance of bonded steel joints, dependent on the type of adhesive and different surface treatments, only several hundred hours of exposure in the salt spray chamber are needed, as shown in Fig. 18.

The influence of a particular surface treatment on the corrosion behavior of bonded steel joints is not equivalent with all different adhesives. Plotted in Fig. 18 are the initial and the residual strength values obtained with joints of mild steel employing different adhesives and different surface treatments of steel adherends. It can be seen that different adhesives show a very complex aging behavior. The degree of aging is dependent on the state of the surfaces, which were either shotblasted, ground by emery paper, degreased in acetone, or coated with a thin film of corrosion-inhibiting oil. From these results it may

be concluded that the durability of steel bonds produced with a hot-curing epoxide resin is better than that of specimens produced with cold-curing epoxide resins and other systems. As expected, cold-curing epoxide resins and cyanoacrylate adhesive are characterized by relatively low durability on untreated or oiled surfaces, while the plastisol adhesive shows very good durability on untreated or oily mild steel surfaces. These particular properties of the plastisol adhesives are the reason for their wide use in the automotive industry. In practice, the relatively low shear strength of the system is not a disadvantage.

The salt spray test is useful to evaluate the corrosion resistance of bonded aluminum joints used in aircraft. The type of degradation initiated in the salt spray test is absolutely the same that is observed occasionally in real structures, when the edges of the specimens are unprotected and rough (sawn). This method should be avoided without fail during production. The investigation results can only be compared when this edge treatment is the same, and service knowledge has indicated that the same rules of precedence are found concerning the stability behavior of different materials and process combinations. The strong influence proceeding from the edge also makes comparative interpretations more difficult.

To obtain clearly visible bond line corrosion in the case of very stable joints, it is necessary to age them for as long as 300 days in a salt spray chamber. Figure 19 shows typical examples of the course of bond line corrosion, demonstrating three effects. On one hand, bond line corrosion sometimes sets in point-by-point, starting from the edge of the joint, and sometimes it does not. This makes quantitative determination difficult be-

cause at certain times, large regions of the test specimen may not yet be corroded, while other regions can be totally delaminated. This can be seen in Fig. 19, which shows the respective failure zones of two specimens (left/right) produced in the same manner and tested at the same time. On the other hand, the figure also demonstrates the effect of surface treatment (a and b) and of the adhesive. In Fig. 19(b) and (c), the surfaces of the sheet metals had the same pretreatment and the same primer, but different adhesives were used. It is evident that the type of adhesive chosen can considerably influence the intensity of the bond line corrosion.

The fact that such relatively long aging times were necessary before bond line corrosion could clearly be observed caused difficulties in the development and optimization of adhesive systems, and a highly informative short-term test would therefore make work easier. So far, however, all experiments in this direction have failed. Configurations have been investigated by which the specimens were dipped in different corrosive liquids that also contained adhesive extract. An intensified salt spray test, containing acetic acid and copper ions in the spray solution (CASS test), seemed to be very promising. However, like the salt spray test, this test on CAA-treated surfaces only indicated whether bond line corrosion occurred at all. The variation of the results in terms of extent and reproducibility was so large that wrong conclusions could be drawn from the results. Thus, by repetition of comparative investigations, other rules of precedence concerning the quality of adhesive joints have been established.

After opening an adhesive joint that has been affected by bond line corrosion, the bond line can be identified by the following macroscopic characteristics, outlined in Fig. 20. The starting point for the degradation, as shown in Fig. 15, is always on the edge of the sheet metal where corrosion has set in. Heavy corrosion occurs in the bonded zone around this starting point. Next there is a largely corrosionless, adhesive-delaminated zone (bright crescent) between the polymer and the adherend, adjacent to which is a noncorroded, cohesive failure zone in the adhesive.

The sensitivity of the joints toward the effects of bond line corrosion clearly depends on the kind of metal pretreatment and the choice of adhesive system. In general, anodized surfaces are superior to etched surfaces with respect to their resistance to bond line corrosion. Furthermore, it could be established from such tests that an adhesive system can only be judged as a whole, that is, as primer and adhesive, and at the very best in connection with the chosen surface pretreatment process and base material.

It is remarkable that some phenolic systems, such as Redux 775, even under bond

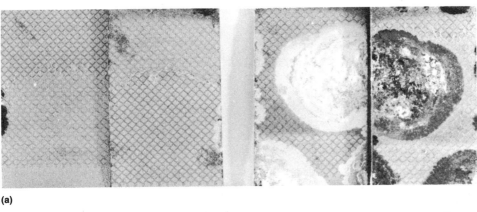

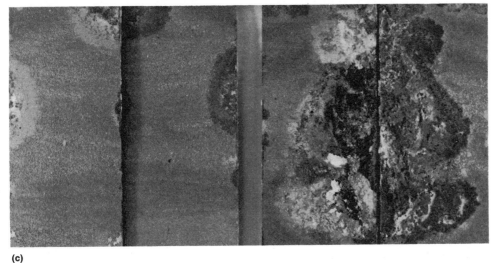

Fig. 19 Bond line corrosion in 2024 T3 clad alloy joints subject to 4480-h salt spray test. (a) CAA (The Boeing Company) surface treatment, EC 3924/AF 163-2K adhesive. (b) CAA (VFW) surface treatment, EC 3924/AF 163-2K adhesive. (c) CAA (VFW) surface treatment, EC 3924/AF 126-2 adhesive

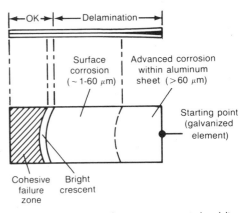

Fig. 20 Diagram showing macroscopic bond line corrosion

er, the bond line corrosion behavior is clearly dominated by the adhesive. It therefore must be concluded that diffusion of adhesive substances plays an important role (see the section "Durability Test Techniques" in this article).

In this connection it is interesting that the sensitivity to bond line corrosion seems to depend largely on the oxide layer thickness, at least in the case of anodizing processes. For a layer thickness of more than 2 μm (0.08 mil) on clad 2024 T3 alloy, no significant improvements can be established.

Further, it has been found in most cases that the cladding of aluminum alloys, such as 2024 and 7075, encourages bond line corrosion ("clad is bad"). With other bare, corrosion-resistant alloys (1100, 5056) no bond line corrosion could be initiated. The same applies to aluminum-lithium alloys; in the extreme CASS test, local and layer corrosion occurred, originating at the exposed adherend sides; yet the bond line remained absolutely free from delamination and corrosion.

REFERENCES

1. W. Althof and W. Brockmann, *Adhes. Age*, Vol 20 (No. 9), 1977, p 28
2. H.-D. Steffens and W. Brockmann, Aging Resistance of Light Metal Bonded Joints Using New Surface Treatment Methods, Library Translation 1662, Royal Aircraft Establishment, 1972
3. W. Brockmann, in *Durability of Structural Adhesives*, A.J. Kinloch, Ed., Applied Science Publishers, 1983, p 281
4. J.D. Minford, in *Durability of Structural Adhesives*, A.J. Kinloch, Ed., Applied Science Publishers, 1983, p 135
5. J.A. Marceau, Y. Moji, and J.C. McMillen, in *Proceedings of the National SAMPE Symposium*, Vol 21, Society for the Advancement of Material and Process Engineering, 1976, p 332
6. H. Kollek and W. Brockmann, in *Pro-*

line corrosion conditions, turn out to be extraordinarily resistant. This has been confirmed by the experiences of Fokker. According to this company, practically no cases of bond line corrosion have occurred in aircraft during total operating times of more than 20 years. Even the use of normal hot-curing industrial adhesives without aero-nautical primers often leads to joints that are less sensitive to bond line corrosion than those joints that use aeronautical adhesives. Interesting results have been achieved in tests in which epoxy resin adhesive and phenolic resin primer and vice versa were employed. Results indicated that although primary failure occurs in the primer-oxide boundary lay-

ceedings of 26th National SAMPE Symposium, Vol 26, Society for the Advancement of Material and Process Engineering, 1981, p 170

7. E.W. Thrall, Primary Bonded Structure Technology (PABST), Paper 6539, Douglas Aircraft Company

8. K.K. Knock and M.C. Locke, in Proceedings of the 13th National SAMPE Conference, Society for the Advancement of Material and Process Engineering, 1981, p 445

9. W. Brockmann, et al., Int. J. Adhes. Adhes., Vol 6 (No. 3), 1986, p 115

10. W. Marz, in Proceedings of 17th National SAMPE Conference, Society for the Advancement of Material and Process Engineering, 1985, p 1088

Environmental Considerations Unique to Sealants

Arthur M. Usmani, Boehringer Mannheim Corporation

SEALANTS are used to fill joints, gaps, and cavities between two or more substrates that may be similar or dissimilar. Discontinuities in structures are eliminated by using sealants to satisfy requirements related to economy, convenience, and function. Sealants primarily isolate an environment and serve as a barrier to the passage of gases, liquids, and solid particulates. Sealants also serve to maintain a pressure differential, for example, in pressurized aircraft cabins; reduce mechanical shock, vibration, and sound; protect electronic devices; and provide an electrical path, in the case of conductive sealants.

A sealant must not only fill the gap and create a barrier, but must be capable of being extended and compressed (movement stability) without losing its sealing effectiveness and adherence. In contrast, structural adhesives are used to bond materials that must sustain load transfer; therefore, thin formulations must possess much higher tensile and shear strength capabilities. Potting and encapsulating compositions are formulated for insulation and shock reduction characteristics. Some adhesives and potting materials are formulated from the same polymers that are used for sealants.

Economics. During the past 50 years, a plethora of sealants, notably high-performance sealants, have been developed to meet the high-technology needs of the aerospace, construction, and transportation segments of the economy. Curtain wall construction as a new architectural design method was introduced in the early 1950s, thus creating the need for sealants that adjust for foundation settling, temperature swings, and loads on the curtain wall panels because of winds. The cured sealant must possess and retain adhesion and movement

capability while resisting the elements, for example, temperature extremes, water, ultraviolet light, and environmental pollutants. Another example of increased sealant performance requirements is sealants that can perform in the harsh environment experienced in space applications.

Commercial and residential construction and repair applications account for the largest single use of sealants, about 75%. Automobiles, trucks, trains, aircraft, and other transportation vehicles represent another area that makes extensive use of sealants. The current sealant business, an important segment of the polymer industry, uses 225 $\times 10^9$ g (500 $\times 10^6$ lb) per year at a total cost of about \$700 million. Market growth is expected to be about 4% annually through the 1990s.

The distribution of one- and two-component sealants, grouped by market and by polymeric binder system, is shown in Tables 1 and 2, respectively.

Chemistry and Technology of Sealants

General Formulation Aspects (Ref 1). Selection of a polymer system to use in designing and formulating a sealant to match expected requirements is very important. In addition to a polymer system, a sealant formulation typically includes a plasticizer, fillers, and additives. Table 3 gives the functions of sealant components.

Polymer System. Sealants are basically highly filled (40 to 80 wt%) compositions based on a wide variety of polymers. The selection of the polymer system is most critical and is directly related to the end-use of the sealant. About 15 to 20 different families of polymers, alone or in combination as polyblends, are used to achieve the shelf stability, application characteristics, performance, and weatherability required for a specific end-use at a competitive cost. Commonly used polymer systems are polyurethanes, silicones, polysulfides, polymercaptans, thioethers, acrylics, butyls, polyvinyl acetates, chlorosulfonated polyethylenes, styrene tri-block copolymers (styrene-ethylene/butylene-styrene), asphaltics, and other bituminous and oleoresinous base materials. For severe environments, such as those encountered in space, cyanosilicones, fluoroalkylarylenesiloxanylene (FASIL), phosphonitrilic fluoroelastomer (PNF), and flexible polyimides have been suggested.

Plasticizers reduce the stiffness and modulus and thus improve the flexibility of the cured sealant. In general, plasticizers also improve the processibility and extrudability of the sealants. Plasticizers are available in a wide array, varying in their chemical structure, viscosity, surface properties, and molecular weight. The plasticizer must be compatible with the polymer system. Because of their inherent chemical nature and nonreactivity, plasticizers have no adhesion properties. In fact, the adhesion of a sealant

Table 1 Current annual U.S. consumption of sealants

Market segment	By wt%	By revenue, %
Construction	42	39
Transportation	15	17
Manufacturing	11	11
Consumer retail	32	33

Table 2 Sealant distribution categorized by polymeric binders

Type	Sealant performance	Volume 10^9 g	Volume 10^6 lb	Dollars, \$000
Polyurethane	High	21	46	70
Silicone	High	20	44	125
Polysulfide	High	19.5	43	68
Butyl	Medium	20	44	53
Acrylic	Medium	18	40	50
Polyvinyl acetate	Medium	16	36	18
Oleoresinous	Low	17	37	19
Asphaltic	Low	33	73	31
Preformed seal	...	25	55	70
Putty	Low	20	44	12
Hot melts	...	0.9	2	40
Other	...	10	22	80

Table 3 Functions of sealant components

Component	Functions
Polymer	Provides elastomeric and adhesive properties, movement capability
Plasticizer	Flexibilizes sealant, improves processing and extrudability, lowers cost
Filler	Reinforces and lowers cost
Additive(s)	Improves and enhances specific sealant features, such as durability, adhesion, shelf stability, and curing

Source: Ref 1

Table 4 Types of sealants

Type	Description
Hot-pour sealants	Melt to required temperature for proper handling and performance, specifically adhesion
Cold-pour, multicomponent	Cure by cross linking of polymers. Sealant base, curative, and color pack, if required, are thoroughly mix blended. Blended material can be poured into joints
Nonsag, noncured sealants	Apply by knife into joint; may require warming for proper application
Nonsag, one-component, chemically cured sealants	Cure by chemical reactions of polymers. Applied by knife, caulking cartridge, or bulk dispensing equipment
Nonsag, multicomponent, chemically cured sealants	Cure by chemical reactions of polymers. Sealant base, curative and color pack must be mix blended. Filled into cartridges and dispensed by automatic dispensers or applied with a knife
Heat-softened, nonsag sealants	Require heating for dispensing and application
Strip sealants, cold applied	Apply strip to the joining surface, remove backing paper, and place second joining surface for sealing
Strip sealants, hot applied	Cut, remove backing paper, and fit into slot to be sealed. Heat, making sure that heat-softened material has continuous contact with sides of forming surfaces to be sealed
Compression seals	Place preform material in compression between the forming surfaces

tends to decrease with increased levels of plasticizers.

Fillers provide reinforcement as well as reduce the overall cost of the sealants. Calcium carbonate and silicates are commonly used because of their availability, low cost, and compatibility with most polymer systems. Fillers must be carefully selected to avoid disastrous results in the sealant. Filler type, amount to be used, and particle size must also be considered. The incorporation of large amounts of small-particle-size fillers into the sealant formulation generally leads to an overall increase in modulus and decrease in flexibility. Use of a larger-particle-size filler generally results in a reduction in modulus and improvement in flexibility.

Certain thixotropic additives are used as fillers to decrease sag characteristics in sealants. Thixotropic materials resist flow until a force is applied. Sealants that do not contain a thixotropic agent may therefore flow out of the joint after application. Fibers, specifically cellulosics, also impart antisagging characteristics and are used in sealants.

Other additives are commonly used in small amounts to impart specific properties to the sealant, for example, adhesion and package stability. In the case of water-base latex sealants, protection against bacteria (biocides) and freeze-thaw cycles (ethylene glycol or propylene glycol) is required for package stability. Surfactants in latex sealants help to disperse the filler, aid in extrudability, and improve the wettability of the sealant to the applied substrate.

Adhesion promoters and coupling agents, for example, silanes and titanates, are commonly used to improve adhesion characteristics, whereas antioxidants and UV absorbers are used to improve resistance to UV light. Popular antioxidants are hindered phenols, whereas hindered amines and substituted benzotriazoles are frequently used as UV absorbers. Mildewcides and fungicides are added to the sealant formulation to resist fungal attack once the sealant has begun its service life.

Catalysts, for example, peroxides, tertiary amines, and tin compounds, are used to improve cure characteristics. The curing of the sealant by conversion into a hardened state is accomplished by chemical methods, for example, oxidation, vulcanization, polymerization, or by physical action, such as evaporation of the solvent.

Forms and Types. Sealants packaged in tubes or cartridges are applied by hand or power-activated caulking guns. In bulk form, one- or two-component sealants are sold in quantities from 0.004 to 0.190 m³ (1 to 50 gal). They are then applied by knife or filled into cartridges. Sealant materials with two or more components are mixed by powered drill mixers or meter mixing equipment. Finally, extruded tapes are available as preformed ribbons or profiles of butyl rubber or polyvinyl chloride, in either dense solid or expanded forms. These tapes may be composite structures, where a rubber core, a rubber rod spacer, or aluminum shims are embedded in the sealant. Commonly available types of sealants are briefly described in Table 4.

Performance. Sealants may be divided into four classes: low performance, medium performance, high performance, and advanced performance. Performance refers to movement capability, except in the case of advanced performance for aerospace applications, where severe environments of extremely high and low temperatures also require durability. Low-performance sealants do not usually cure and are commonly based on air oxidation of —C≡C— in fatty oil. These sealants "dry" or "set" by evaporation of solvent or water. Although they are inexpensive, they also have low flexibility and tend to crack when subjected to moderate joint movement.

Medium-performance sealants are generally solvent-base acrylics and butyl rubbers. These sealants, due to solvent loss, do produce joint shrinkage and are somewhat stringy during sealant application. Latex sealants based on vinyl acetate homo- and copolymers, vinyl acrylics, and acrylics have received much attention in recent years. These sealants can be painted, but shrinkage is a drawback. The joint movement capability of latex sealants was previously limited, but new technology has given rise to latex sealants that meet high-performance specifications.

Among the high-performance sealants, polyurethanes, polysulfides, and silicones are the most important (Ref 1). These three sealants are all curing types, and therefore produce very little shrinkage, resulting in excellent joint movement (Ref 1). Styrene block copolymers have recently joined this group of sealants, specifically for use in hot-melt applications. Acrylic latex high-performance sealants that are transparent and have good weatherability characteristics have also been developed (Ref 1).

There are presently few commercially available sealants that can sustain the extreme high and low temperatures required for aerospace applications. FASIL, developed by the Air Force Materials Laboratory, has a service temperature range of −54 to 260 °C (−66 to 500 °F). The phosphazenes, originally developed as a PNF, are now available from Ethyl Corporation with a service temperature range from −65 to 175 °C (−85 to 345 °F).

A comparison of sealant performance parameters is provided in Table 5.

Features of Sealant Types

The following descriptions represent a few chemical features of some important sealant types.

Oleoresinous. Vegetable oils, usually blown or heat-bodied linseed or soya oils, are

Table 5   Advantages and disadvantages of various categories of sealants

Sealant	Advantages	Disadvantages
Low performance (Ref 1)		
Oleoresinous	Low cost	Poor joint movement
	Easy application	Limited adhesion
		Cracks on aging
Medium performance (Ref 1)		
Acrylic, solvent-based	Good adhesion	Limited joint movement
	Good UV resistance	Odor
		Slow "setting"
Butyls, solvent-based	Low cost	Joint shrinkage
	Broad adhesion range	Flammability
		Stringy when applied
Latex, vinyl acrylic	Water cleanup	High joint shrinkage
	Good shelf stability	Limited water resistance
		Poor extrudability at low
		temperatures
High performance (Ref 1)		
Polyurethane	Excellent joint movement	Limited UV resistance
	Excellent adhesion range	Limited package stability
	No shrinkage	Prone to hydrolysis
		Slow cure (one-component)
Polysulfide	Excellent joint movement	Limited package stability
	Excellent adhesion range	Limited UV resistance
	No shrinkage	Bad sulfur-type odor
Silicone, acetoxy	Superior UV resistance	High cost
	Excellent joint movement	Poor adhesion to concrete
	Rapid curing	Cannot paint sealant surface
	No shrinkage	
	Excellent water resistance	
Silicone, modified	Excellent joint movement	Medium cost
	Good UV resistance	Limited adhesions to
	Good adhesion range	concrete and some plastic
	Paintable	substrates
	Odorless	
Styrene block copolymer	Excellent joint movement	Flammability
	Excellent UV resistance	High joint shrinkage
Advanced performance		
FASIL	Broad service temperature	
	range	
	Excellent adhesion to	
	titanium and aluminum	
	JP-4 fuel resistant	
	Excellent hydrolytic stability	
Phophazenes	Outstanding resistance to	
	fuel, oils, fluids, chemicals	
	Stable to O_2 and O_3	
	Low compression set	
	High modulus	
	Broad service temperature	
	range	
Cyanosilicones	Resistant to hot hydrocarbon	Withdrawn from market
	fuels	
	Broad operating temperature	
	range	
Flexible polyimide	High-temperature fuel	
	stability	
	Low-temperature flexibility	
	Excellent JPF resistance	
	Excellent adhesion to Ti and	
	Al	
	No stress corrosion to Ti	
	Excellent tensile properties	

used as binders for putties, elastic glazing compounds, and architectural caulking compounds.

Asphaltic materials come in solvent cut, emulsion, or hot-pour forms. Rubber is used as a reinforcement in high-solids formulations. Typically, an asphaltic sealant contains 50 to 60% asphalt blend, 20 to 30% ground rubber from tire scrap, and 20% cyclic hydrocarbon. The sealant is applied at 120 to 205 °C (250 to 400 °F).

Butyls. Polyisobutylene produced by cationic polymerization of isobutylene $H_2C{=}C{-}(CH_3)_2$ has been used as a sealant since the 1930s. Butyl rubber, a copolymer of isobutylene and 15 to 45% isoprene, can be easily cross linked by *p*-quinone dioxime and a peroxide. Both polyisobutylene and butyl rubber sealants usually contain fillers, for example, carbon black, zinc oxide, clays, silicas, and tackifiers such as pentaerythritol ester of rosin. These formulations contain solvents, for example, cyclohexane, the evaporation of which acts to harden the sealant.

Butyl rubber sealants are used for joint and panel sealing applications and for glazing. Tape butyl rubber sealants are pressed into cavities that are to be filled by sealants. Thermoplastic butyl hot-melt systems (100% solids) are used as insulation sealants for glass windows and doors.

Acrylics. Polymer solutions of acrylic esters have been used in sealants since the 1970s and are similar to high-solids (90%) acrylic coatings. A typical acrylic sealant contains about 40% pigmented polymer solids, with calcium carbonate, talc, and fumed silica as thickening and rheological agents. The formulation also includes pine oil as a dispersing and penetrating agent, ethylene glycol for activation of the fumed silica, and xylylene for viscosity control and ease of extrusion. Acrylic sealants are used for industrial sealing and glazing, concrete roofs, masonry panels, and metal-to-masonry joints, such as skylights.

Water-base (latex) acrylic sealants are used in residential construction and home repair. They have been rated as medium- to low-performance products because they shrink upon curing and have limited flexibility. Recent advances in technology by DAP Inc. (Dayton, OH) have resulted in products that meet commercial construction criteria for high performance. Latex acrylic sealants are easy to apply and simple to clean with water. They can be formulated for clarity. Their other features are low odor, fast tack-free time, excellent weatherability, paintability, and good adhesion. Acrylic sealants can be siliconized by inclusion of 1 to 10% silicone.

Polyvinyl Acetate Latex. Sealants based on homo- and copolymers of vinyl acetate have been available since the 1950s, mostly for residential applications such as bathtub caulking, wall tile joints, and wallboard joints. Ethylene vinyl acetate and acrylic vinyl emulsions are also used.

Polyurethanes are the most versatile polymer systems used to formulate sealants. Polyurethane sealants are available as one- or two-component compositions in cartridge or bulk form. Applications include insulated glass and automotive windshield sealants.

Polyurethanes are produced by the reaction of a diisocyanate with polyols to form a urethane linkage.

$$OCN{-}R{-}NC + HO{-}R{-}OH \longrightarrow$$

$$\underset{\displaystyle O}{\overset{\displaystyle H}{{-}R{-}\overset{\displaystyle |}{N}{-}\overset{\Vert}{C}{-}O{-}R{-}}}$$

Many aliphatic, as well as aromatic, diisocyanates are available. The selection of polyols is also very broad. Therefore, polyurethanes can be made with almost any desired combination of properties. Additional versatility is possible by the reaction of prepolymer diisocyanates with diamines to produce urea linkages. Prepolymers can also react with water to give urea linkages.

Table 6 One-component polysulfide sealant formulation

Component	Wt%
Polysulfide liquid polymer (LP-32)	50
Epoxidized soya oil	4
Pyrogenic silia	2
Calcium carbonate	5
Titanium dioxide, rutile	22
Hydrated lime	2
Synthetic zeolite	2
Calcium peroxide	4
Phthalate plasticizer	2
Gamma-aminopropyltriethoxysilane	1
Toluene	6
Total	**100**

Note: Manufactured and packed under anhydrous conditions

Table 7 Two-component polysulfide sealant formulation

Component	Pounds	Wt%
Base		
Polysulfide polymer, liquid	100	54.42
Phthalate plasticizer	20	10.89
Calcium carbonate, precipitated, surface treated	35	19.05
Titanium dioxide, rutile	10	5.44
Carbon black, furnace	5	2.72
Stearic acid	1	0.54
Organic-modified bentonite clay	3.0	1.64
Fumed silica	2.5	1.36
Phenolic resin	5	2.72
Gamma-aminopropyltriethoxysilane	0.15	0.08
Sulfur	0.1	0.05
Toluene	2	1.09
Subtotal, base	**183.75**	**100.00**
Curative		
Lead dioxide, technical	7.5	50.00
Stearic acid	0.75	5.00
Dibutyl phthalate	6.75	45.00
Subtotal, activator	**15.00**	**100.00**

Polyurethane prepolymers are made from glycols based on polyesters, polyethers, polythioethers, and polybutadiene backbones. The various backbones impart specific characteristics to the sealants. Polyester-base materials have good fuel and weather resistance, with excellent strength and temperature stability, but suffer from a lack of low-temperature flexibility and poor hydrolytic stability. Polyether derivatives from polypropylene glycol and polyoxytetramethylene glycol exhibit good water and weather resistance, good physical properties, and low-temperature flexibility. Polybutadiene products have limited low- and high-temperature properties. They swell excessively in hydrocarbon fuels, but their water resistance is excellent. Their physical properties are average, but their weatherability is poor due to unsaturation (oxidation of double bond).

Many formulations of two-component polyurethanes are available to suit most applications. Generally, the NCO/OH ratio is in the range of 1.05 to 1.10. One of the components is pigmented and filled (titanium dioxide, calcium carbonate, talc, silica), and the other component is a hydroxylated polymer that is also pigmented and contains catalyst, such as dibutyl tin dilaurate. These two components should be mixed thoroughly on the job. Pot life is generally 1 to 4 h, and full properties are achieved in 1 to 2 days.

In one-component formulations, isocyanate prepolymers with NCO equivalent weights of 1000 to 2000 are used. A typical sealant may contain 30 to 60% of pigmented prepolymer with titanium dioxide, calcium carbonate, and silica. Water must be excluded during the manufacture of the one-component sealant.

Polyurethanes adhere well to a wide range of substrates and have excellent abrasion resistance. Their main drawback is poor UV resistance. To improve weatherability, UV stabilizers are generally added.

Liquid polysulfide polymers have been used since the 1950s to make polysulfide sealants for construction, glazing, marine, and aircraft applications. These sealants are available in one- and two-component forms.

Liquid polysulfide is prepared by reacting an excess of aqueous sodium polysulfide with bis(2-chloroethyl) formal. Cross linking is usually introduced by adding small amounts of 1,2,3-trichloropropane. Heating to 100 °C (212 °F) produces chain lengths with —C—S—S—C— and —C—S—S—S—C—; the ends are capped by polysulfide terminals. Reaction with sodium sulfite reduces the polysulfides to disulfides and the terminal groups to thiols.

Mercaptan-terminated liquid polymers (HS—R—SH) are available from Morton Thiokol in a range from 1000 to 7000 in molecular weight. Cross-linking densities range from 0.5 to 2.0 mol%.

The liquid polysulfide polymers are usually cured by oxidizers, for example, metal oxides or peroxides. Examples are dichromate, manganese dioxide, and calcium peroxide. Curing is accelerated by sulfur and water and retarded by stearic acid.

Tables 6 and 7 give formulations for one- and two-component sealants. Polysulfide sealants have been successful in aircraft and construction applications due to their fuel and solvent resistance, good low-temperature performance, and good weatherability.

Recently, a new class of high-performance polysulfide polymer has been developed by reacting liquid polysulfide polymer with dithiols (HS—R—SH), as shown in Fig. 1. The modified polysulfide polymers show low viscosity and good compatibility with plasticizers and fillers. They are more chemical resistant than liquid polysulfides.

Silicone sealants were developed in the early 1940s. They have outstanding heat resistance, low-temperature flexibility, and good exterior durability. Two-component types have been developed for room-temperature vulcanization (RTV) and high-temperature vulcanization (HTV). One-component silicone sealants that are cured by exposure to moisture were introduced in the 1960s.

Silicone sealants are used in construction, industrial, and consumer end-uses. Outstanding properties include ease of application with a wide range of rheological properties, rapid cure, excellent elastomeric properties, good thermal stability, and good UV, ozone, and chemical resistance.

In terms of chemistry, the silicone repeating unit contains the silicon-oxygen backbone (Fig. 2). The major physical properties of the sealant depend on the R group and the curing agent used.

Silicones are classed into types by the curing agent used for vulcanization and the small molecule that splits off as a by-product. The most commonly used silicones are the acetoxy type (Fig. 3), which gives off acetic acid during sealant preparation and curing (Fig. 4). The undesirable odor of acetic acid and the marginal paintability of

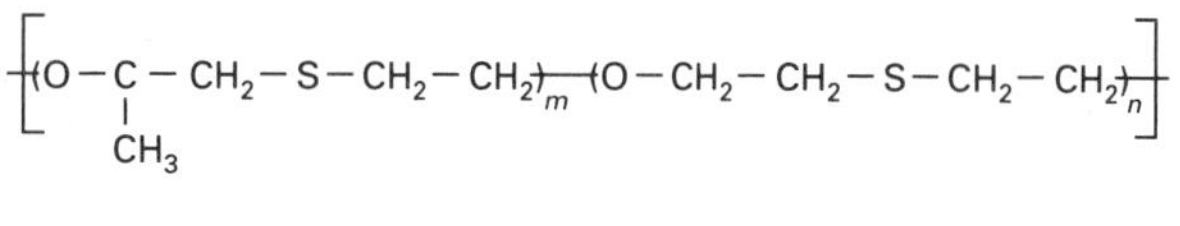

$$\left[\begin{array}{c} \text{(O} - \overset{\displaystyle |}{\underset{\displaystyle CH_3}{C}} - CH_2 - S - CH_2 - CH_2 \text{)}_m \text{(O} - CH_2 - CH_2 - S - CH_2 - CH_2 \text{)}_n \end{array} \right]$$

Fig. 1 Dithiol-modified liquid polysulfide polymer

$$HO - (Si - O)_m - \overset{\displaystyle \underset{\displaystyle R}{\overset{\displaystyle R}{|}}}{Si} - OH$$

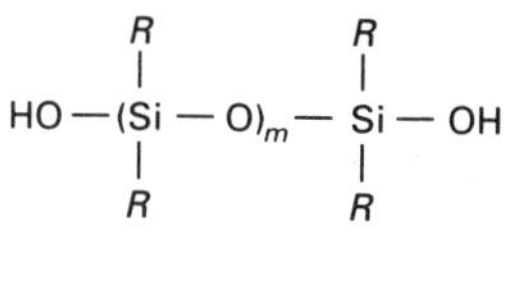

Fig. 2 Silicone sealant chemical structure

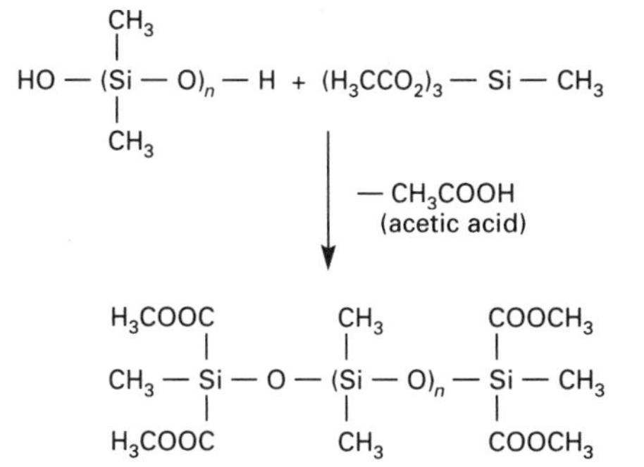

$$CH_3$$
$$HO-(Si-O)_n-H + (H_3CCO_2)_3-Si-CH_3$$
$$CH_3$$

$$\downarrow \quad -CH_3COOH \ (\text{acetic acid})$$

$$H_3COOC \quad CH_3 \quad COOCH_3$$
$$CH_3-Si-O-(Si-O)_n-Si-CH_3$$
$$H_3COOC \quad CH_3 \quad COOCH_3$$

Fig. 3 Sealant preparation reaction and resulting acetoxysiloxane prepolymer. Source: Ref 1

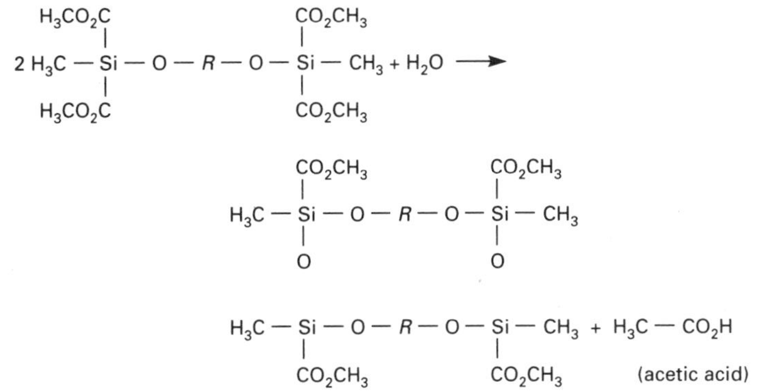

$$H_3CO_2C \quad CO_2CH_3$$
$$2\,H_3C-Si-O-R-O-Si-CH_3 + H_2O \longrightarrow$$
$$H_3CO_2C \quad CO_2CH_3$$

$$CO_2CH_3 \quad CO_2CH_3$$
$$H_3C-Si-O-R-O-Si-CH_3$$
$$O \quad O$$

$$H_3C-Si-O-R-O-Si-CH_3 + H_3C-CO_2H$$
$$CO_2CH_3 \quad CO_2CH_3 \quad (\text{acetic acid})$$

Fig. 4 Moisture-curing reaction of acetoxysiloxane prepolymer, where R is a repeating unit. Source: Ref 1

acetoxy silicone sealants are its major shortcomings. Other types of silicone sealants are based on curing agents, such as alkoxy or oxime silanes, that split off alcohol or oximes and are thus less noxious than the acetoxy type. Furthermore, these curing systems give better adhesion to concrete and metals.

The two-component silicone sealants are based on a hydroxyl-terminated polysiloxane and a cross linker such as ethyl orthosilicate. The curative component is a catalyst, for example, dibutyl tin dilaurate in a paste form. When the two components are mixed, curing takes place. The two-component silicone sealants find application in aircraft and device encapsulation.

Block Copolymers. The technology of styrene-butadiene rubber has been known for many years. Early copolymers were based on styrene and butadiene. Sealant formulations based on this copolymer weathered poorly due to unsaturation, thus making it undesirable for extensive outdoor application. Recent advances in styrene triblock copolymers have resulted in novel and better polymers, for example, styrene-butadiene-styrene (SBS), styrene-isoprene-styrene (SIS), and styrene-ethylene/butylene-styrene (SEBS). SEBS has excellent UV resistance if formulated with a suitable UV absorber.

Hybrids (Ref 1). A new area of increasing interest is the synthesis of "hybrids" to yield a high-performance sealant. In hybrids, strengths of two combined materials may be obtained while individual inherent weaknesses may be eliminated. A polyether polymer (typically used in polyurethane) terminated with silyl groups (related to silicones) is referred to as modified silicone. Sealants based on this polymer have better weathering characteristics than polyurethane sealants, but without the odor and unpaintability problems of acetoxy silicone sealants. In addition, adhesion, abrasion resistance, and low-temperature extrudability are also improved.

Anaerobic polymers cure only in the absence of atmospheric oxygen and in the presence of metal ions on the substrate. The polymers are esters of alkylene glycols of acrylic, or methacrylic acid formulated with a latent initiator (for example, cumene hydroperoxide), plasticizers, accelerators, and stabilizers. Anaerobic sealants are used to lock nuts and bolts, seal threaded pipes, and improve the fit of parts. Properties of anaerobics and the sealant types described thus far are given in Table 8. The sealant types described below are all considered advanced sealants.

Advanced Sealants. Cyanosilicones, fluoroalkylarylenesiloxanylene (FASIL), phosphonitrilic fluoroelastomers (PNF), flexible polyimide, tetrafluoroethylene oxide phenylquinoxaline elastomer (FEX), and perfluoroalkyl ethers have been suggested as advanced sealants for harsh environments in aerospace applications. Such advanced sealants have been used for fuel tank sealing, channel sealing, and aerodynamic smoothing. Their basic requirements are high service temperature, fault toler-

Table 8 Properties of sealants

Properties	Bituminous	Oleoresinous	Polybutene	Butyl	Styrene-butadiene	Acrylic solution	Acrylic emulsion	Polyvinyl acetate	Polysulfide	Polyurethane	Silicone
Nonvolatiles, wt%	70–90	96–99+	99+	74–99+	60–70	80–85	80–85	70–75	90–99+	94–99+	98+
Maximum joint movement, ±%	5	5	5–10	10–15	5–10	10–15	5–10	5	25	25–40	25–50
Recovery after joint movement(a)	Poor	Poor	NA(b)	Fair to good	Poor to fair	Fair	Fair	Poor to fair	Fair	Good	Excellent
Shrinkage, %	10–20	4	1	1–20	20–30	10–20	10–20	20–25	10	6	2
Life expectancy, exterior, year(c)	1–2	2–10	5–10	5–15	3–10	5–20	2–20	1–3	10–20	20+	30+
Cure type	Evaporation	Oxidation	Noncuring	Evaporation	Evaporation	Evaporation	Evaporation	Evaporation	Chemical cure	Chemical cure	Chemical cure
Practical service temperature range Maximum, °C (°F)	65 (150)	65 (150)	80 (180)	95 (200)	80 (180)	80 (180)	80 (180)	65 (150)	120 (250)	120 (250)	205 (400)
Minimum, °C (°F)	−18 (0)	−18 (0)	−40 (−40)	−30 (−20)	−25 (−10)	−18 (0)	−18 (0)	−18 (0)	−40 (−40)	−40 (−40)	−65 (−90)
Adhesion to common building materials, unprimed(d)	Good	Good	Good to excellent	Good	Fair	Excellent	Good	Fair to good	Good	Good	Fair to good
Components(e)	1	1	1, T	1, T	1	1	1	1	1, 2	1, 2	1, 2

(a) E, less than 10% compression set; G, 10–20%; F, 20–30%; P, <30%. (b) NA, not applicable. (c) Mild to severe condition. (d) Adhesion of all materials improved with suitable primers. (e) 1, one-part; 2, two-part; T, tape

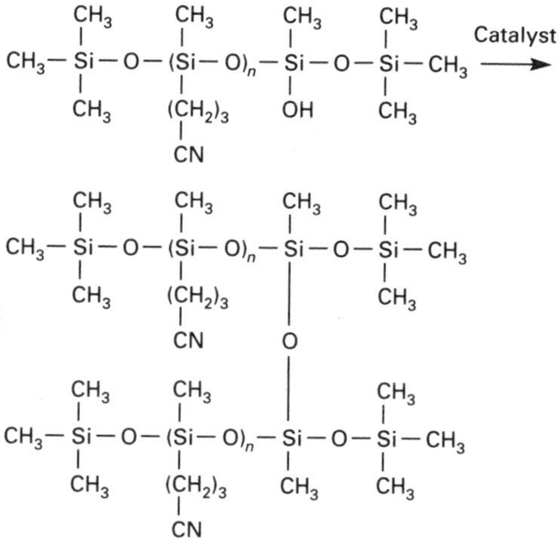

Fig. 5 Polycyanosiloxane polymer

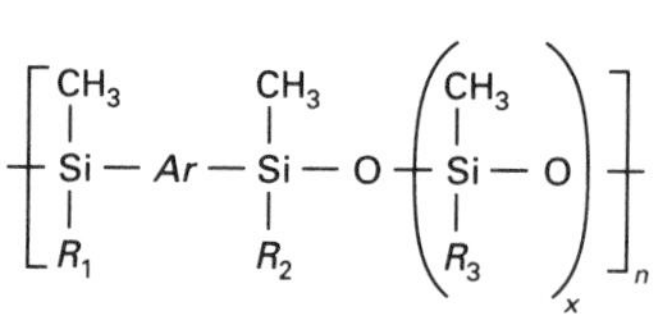

Fig. 6 FASIL chemical formula, where R_1, R_2, and R_3 are either methyl or 3,3,3-trifluoropropyl and $x = 0$, 1, or 2; and Ar is either m-phenylene or m-xylylene.

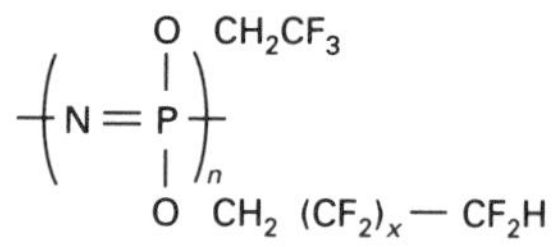

Fig. 7 Structure of PNF, where x can be 1, 3, or 5

ance, and good adhesion to metals and advanced composites.

Cyanosiloxanes are made by catalytic addition of an unsaturated nitrile to methyl hydrogen polysiloxanes. The one unreacted hydride (—Si—H) is hydrolyzed to —SiOH that can condense to give high molecular weight (Fig. 5). The dark brown, clear, 500 Pa · s product can undergo liquid-gel transformation at room temperature. This polymer is resistant to hot hydrocarbon fuels but is soluble in polar solvents.

One product has been used for environmental sealing of structural faying surfaces of the space shuttle orbital airframe, and another is used to seal integral fuel tanks of aircraft. These sealants are a noncuring type so that the channel can be resealed without a need for disassembling the structure.

FASIL sealant, as previously described, has a service range of −55 to 260 °C (−66 to 500 °F). Adhesion to titanium and aluminum are excellent. It is hydrolytically stable and resistant to JP-4 fuel. The general formula of FASIL is shown in Fig. 6.

PNF is prepared from a soluble chlorophosphazene precursor and a mixture of fluorinated alkoxides (65% trifluoroethoxide and 35% telomer fluoroalkoxides). The structure of PNF is shown in Fig. 7. PNF contains about 55% fluorine and has limited unsaturated sites for cross linking. Because of its high fluorine content, the sealant has excellent resistance to fuels, oils, hydraulic fluids, and many other chemicals. The service temperature ranges between −70 and 175 °C (−91 and 345 °F).

Because there are no weak linkages in its chemical structure, PNF is resistant to oxidation and ozone. It has low compression set, high modulus, good abrasion resistance, and outstanding flexural fatigue characteristics.

Flexible Polyimides. Monomers that can react to give a flexible polyimide sealant are shown in Fig. 8. The reaction product gives a sealant with up to 260 °C (500 °F) fuel stability, low-temperature flexibility to −45 °C (−50 °F), high adhesion and no stress corrosion on titanium substrates, solventless formulation, excellent tensile properties, and easy curing and processibility.

Solar Collector Sealants. In solar collector cells, the temperature can reach as high as 200 °C (390 °F). High humidity, ozone, and UV radiation are also prevalent. Sealing materials must be durable under such harsh conditions. A fluoroelastomer in the form of preformed seals or gaskets performs well. The elastomer is expensive and has a high compressive set at temperatures of 10 °C (50 °F) and lower, which presents a problem in colder climates. Butyl rubbers, if stabilized for UV radiation, are better than silicones for edge sealing and may be substituted for the high-cost fluoropolymers.

Elements of Degradation

Resistance to the elements in the environment often determines the lifetime of a sealant in an intended application. Useful life is influenced by degradative processes during polymer synthesis and sealant processing to service-life limits. Durability requirements vary. Stresses may be imposed by mechanical processes, heat, cold, radiation (UV, x-ray, nuclear), gases (oxygen, ozone, sulfur dioxide), humidity, solvents (jet fuels, oils, hydraulic fluids, water), corrosive chemicals, and bacterial challenges. Physical and mechanical properties change as the service life progresses because of induced changes, for example, chain scission, depolymerization, substitution reactions, elimination, and cross linking.

Bond scission, caused by many energy sources, is very important in polymer degradation. Bond scission, the combined effects of several elements, and the sensitivity of the partially degraded polymer to further and faster attack is shown in Fig. 9.

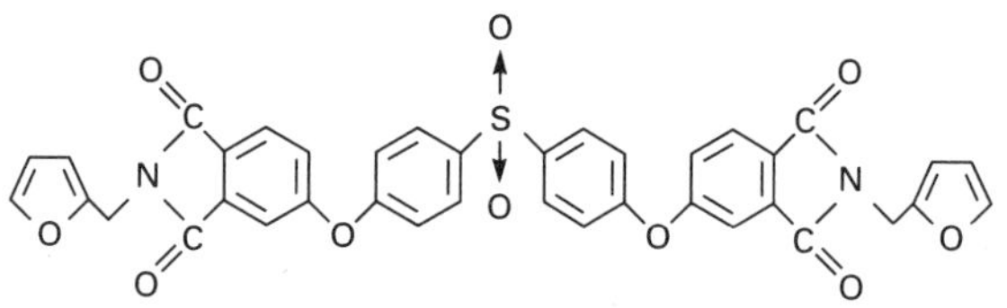

Bis (furfuryl) imide of bis [4-(3,4-dicarboxyphenoxy) phenyl] sulfone dianhydride

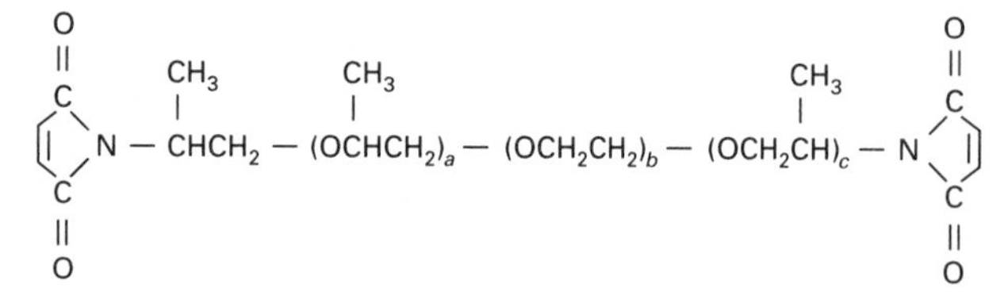

Bis (maleimide) of Jeffamine ED*

* Repeat units *a*, *b*, and *c* are integers yielding molecular weights of approximately 600, 900, and 2000

Fig. 8 Monomers for flexible polyimide sealant

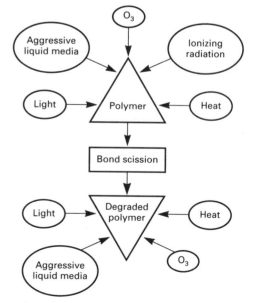

Fig. 9 Degradation sources. Aggressive liquid media include water, jet fuels, oils, hydraulic fluids, and chemicals

Accelerated testing usually involves thermal analysis and mechanical testing. However, oxidative and photo-oxidative degradation is studied using Fourier transform infrared spectroscopy, electron spin resonance, and chemiluminescence.

Predicting Sealant Performance. The outside envelope of a building has a cyclic movement in response to ever-changing temperature and humidity. This may consist of panels of various kinds with joints between. The elements of a building may expand or contract as one unit, upon changes in temperature and/or humidity. The sealant should be formulated to keep its shape as applied and harden to a viscoelastic rubberlike material. It also must be able to withstand expansion or contraction because the sealant will expand at low temperatures and contract at higher temperatures. Thus, demands on sealants are severe, specifically in colder climates.

Factors that regulate the behavior of sealants are stress, temperature, rate of deformation, humidity, air, light, substrate, water, and chemicals. Tensile tests, under controlled conditions, along with mathematical manipulation, can be used to predict a sealant's performance and durability. After the tensile testing method is defined as a characterization method, a weathering study on the sealant must be done to predict performance.

Water-Related Problems. The durability of sealants that are mostly synthetic materials are affected by environmental factors. In harsh environments, synthetic materials degrade and fail in their expected performance. The detrimental environmental elements are moisture (water), temperature, radiation, and chemicals.

Polymers are permeable to moisture or liquid water. Water attacks the sealant, the interface, or the substrate. Fluorine and silicon-containing polymers have low water permeability. Most sealants are polar in nature, and hence, the moisture can diffuse in bulk. Sorbed water can also cause changes. Polyurethane and cyanoacrylates are known to hydrolyze. Hydrolysis can lead to debonding. Construction sealants, for example, polysulfides, polyurethanes, and acrylics, are moisture sensitive. As an example, the major failure cause of an insulated glass unit is because of penetration of moisture as a result of poor workmanship. The condensed and trapped water can swell and hydrolyze the polysulfide sealant. Hydrolysis leads to sealant bond breaking that can result in cohesive failure. Physical swelling can induce stresses at the interface, causing bond separation.

REFERENCE

1. V.R. Foster, Polymers in Caulking and Sealant Materials, *J. Chem. Educ.*, Vol 64 (No. 10), Oct 1987, p 861-865

Manufacturing Use

Chairman: Sam C. Aker, Sam C. Aker, Inc.

Introduction

THE MANUFACTURING PROCESSES adhesively bonded structures and those for the application of sealants have enough areas of commonality that they can be considered together. Both begin with surface preparation and use many of the same basic materials and curing technologies. The application of a sealant might be compared to the preparation of a one-sided adhesive joint. The level of performance of the adhesive material in a structural adhesively bonded joint is probably higher than that of a sealant material in a sealing application, but the basic mechanisms of adhesion between the adherend or basic structure and a polymer are similar. In general, the adherend is passive and the adhesive or sealant is the active material in that the polymer must wet the substrate and, in some cases, change its state, that is, polymerize, cross link, cure, or react. Additional similarities exist in storage and control methods, chemical process controls, mixing and application equipment and facilities, and curing facilities.

Of course, many differences in the two systems may be pointed out. One of the most significant of these differences is that adhesive joining is usually employed as the basic fabrication technique in the beginning of a process, whereas sealing is most frequently done as one of the final operations in a manufacturing cycle. The significance of this difference is that the processes for the surface preparation of the unassembled details of an assembly may not be applicable to a semifinished assembly. For example, hot chemical solutions commonly used for adhesively bonded metal details would probably not be used to prepare an assembly containing riveted, bolted, or bonded joints for a sealing operation.

Joining the details of a structure by the use of adhesives is a common practice for many types of structures, and the reasons for the choice of adhesive bonding are many and varied. The transportation industry probably makes the widest use of adhesives; in this category, the aircraft industry features the most critical applications. The critical nature of these applications may be due to the use of aluminum as the principal aircraft structural material. Aluminum is more difficult to weld than steel, which is the structural base material of automobiles, trucks, and trailers.

Adhesive bonding of aluminum and reinforced plastic (composite) materials is a basic process in the aircraft industry. A whole series of practices and processes have developed that are somewhat peculiar to the aerospace industry. The adhesive-bonding facilities of most aerospace companies are similar to one another in many ways and are quite different from the types employed in the rest of the transportation industry. These aerospace facilities are mainly dedicated to structural bond processing using one-component adhesives purchased in film or sheet form; these adhesives are applied to chemically prepared substrates under environmentally controlled conditions. The resultant assemblies are usually cured in heated autoclaves under closely instrument-controlled temperature and pressure conditions. The tooling for these operations is generally tailored for each particular assembly, and production lots are small as compared with those of most industrial operations. With a few exceptions, this sector of adhesive technology is much the same in every aircraft factory.

While many aircraft structural bonding facilities have presses and other special curing fixtures, the autoclave is the most commonly used device for curing bonded assemblies. The traditional autoclave curing fixture consists of a metal structural base configured to the controlled surface of the bonded assembly, together with various arrangements of cauls and rubber or plastic curing bags, which are sealed to the base tool. These curing bags are used to apply the bonding pressure to the details of the assembly. Heat for curing the adhesive is supplied by heating the pressurizing medium, which, in turn, heats the tools and parts contained under the curing bag. This same arrangement has been adapted to lamination and bonding of composite materials for aircraft components.

The use of one-part film-type adhesives and autoclave curing systems is characteristic of aircraft-type structural bonding facilities. The use of either single- or multiple-component liquid or paste adhesives is fairly common in aircraft applications, but such adhesives are largely confined to secondary bonding or repair and rework activities.

After the materials from which the item is to be made have been chosen, the next step is the selection of the family of adhesives to be used. With these two elements defined, the process parameters applicable to the combination will largely define the manufacturing process. The facilities required for each phase of the bonding process can be simple or elaborate, depending on the number, size, and structural complexity of the product.

The most elementary requirement is that the facility must have suitable areas in which the operations can be performed. It is preferable in all cases to have a segregated area for application of the adhesive and for assembly of the detail parts of the structure to be bonded. Once the details are assembled and placed in the bonding tool or curing fixture, segregation of the operation from the rest of the manufacturing spaces becomes less important.

The choice of adhesive systems will usually dictate the curing procedure, and here again, the equipment may be simple or elaborate, depending on the job requirements. The equipment can be as simple as a clamp to hold the details in place while the adhesive cures at room temperature, or it can be as complicated as a special fixture designed to hold parts in configuration while providing heat and pressure; the latter might be used for producing large numbers of critical structural assemblies.

The most common beginning pieces of equipment for a bond shop for curing adhesives are an oven and a vacuum pump. The oven can supply heat for curing the adhesive, and the vacuum system can be used in combination with a vacuum bag and a bond form to hold the details in place and in contact during the cure. Autoclave bonding works in the same manner, but the autoclave provides higher pressures, better heat transfer, and better control of temperature and heat rate. The specific cure requirements for various combinations of base materials, adhesives, and part configurations are given in earlier articles in this Volume that deal with specific adhesives and materials.

Storing Adhesive and Sealant Materials

James Van Twisk, McCann Manufacturing Co.
Sam C. Aker, Sam C. Aker Incorporated

THE STORAGE and handling of adhesives and sealants are very important to the manufacturer who is preparing to purchase and store production quantities of adhesive systems. Adhesive and sealant materials have finite lives in storage. Once the materials are compounded, their useful lives are limited. Shelf lives can vary from a few days to more than 1 year, depending on the basic nature of the materials from which the adhesives are made and on the manner in which they are handled. Because of the limited-life nature of these materials, provisions for the storage and handling of inventories are required to ensure the existence of adequate supplies of material in good condition.

Information on the shelf life and storage requirements for any material is available from its manufacturer. Manufacturer recommendations should provide the basis for determining the storage facilities and requirements for the intended use.

The first section of this article discusses storage considerations for the constituent materials in typical adhesive and sealant formulations. The following section is a description of the effects of aging on stored adhesive and sealant materials. Next, the three common types of storage facilities (freezer, refrigerator, room-temperature containment) and the applicability of each are described; this section is followed by descriptions of inventory control mechanisms. The final two sections address general and specific handling considerations; the latter includes a discussion on handling hazardous materials.

Formulation Constituents

Typical formulations comprise a basic resin plus a hardener or curing agent, along with other components that modify the basic resin properties. The as-purchased material may be one-part, that is, it may contain all the component parts of the material, or the package may consist of two components, which are mixed at the point of use.

Generally, the two-component forms have much longer shelf lives or use times and are storable for longer times. One-part materials generally require storage at temperatures near −18 °C (0 °F) if they are to be kept for as long as 6 months.

For short-term storage, 5 °C (40 °F) will usually suffice. Ambient-temperature storage is possible for many two-part systems, but controlled storage at 5 °C (40 °F) is safer. Ambient- or room-temperature storage may be quite safe for many multicomponent formulations. However, because the term room temperature does not specify a definite temperature, it is safer to use refrigerated storage. With refrigeration, the temperature variation is smaller, and the temperature can be controlled and recorded for proof of proper storage conditions.

Many constituent materials of adhesive and sealant formulations are flammable. Some formulations, particularly those containing solvents, have low flash points. Most of the materials cure by chemical reaction initiated by the introduction or activation of catalyst and/or cross-linking agents. These polymerization reactions are accelerated by heat. The curing reaction, once initiated, can be endothermic (that is, heat is required to complete the reaction) or exothermic, and the mixture may not require further heating to complete the cure. In some cases the exothermic reaction is so vigorous that cooling is required to prevent too much heat buildup. It is necessary, therefore, to handle the materials carefully and to keep storage temperatures within those recommended by the manufacturer for the specific material involved. In addition, because some of these mixes can become very warm, the different containers used for two- or three-part materials should be stored separately to prevent accidental mixing of components in the event of spillage.

Effects of Aging

As-purchased adhesives and sealant materials are affected by temperature and hu-midity, and even by light. These effects are always detrimental, the rate of deterioration depends on the type of base resin and on the form of the material chosen.

Adhesives aging in an environment that hastens their deterioration will develop certain characteristics such as an increase in viscosity or, for films, a loss of tack and drapeability. Brittleness and cracking of the uncured films will start to occur. If the adhesive is subjected to elevated temperatures during its cure cycle, less flow is likely to occur during cure. Gel times of epoxy film adhesives become shorter. If high levels of humidity are also present, the materials can absorb moisture, which will affect their performance when cured.

Extended exposure of the uncured material to temperatures and humidities above those recommended by the manufacturer will reduce its cohesive and adhesive strength when cured. Surface effects, such as feathering and crazing, may be evident on the bonded substrates. All these factors contribute to reduced performance and premature failure of the adhesive bonded joint.

Figure 1 shows the effect of aging conditions on an epoxy film adhesive with a 120 °C (250 °F) recommended cure cycle. Samples of the adhesive were aged 90 days at 21 °C (70 °F) and 90 days at 32 °C (90 °F). Material aged at 21 °C (70 °F) was then placed in a 95 to 100% humidity chamber at 50 °C (120 °F) for 1 h. The tensile shear values show decreasing strength with increased exposure to temperature and humidity. Figure 2 shows the increase in viscosity of a paste adhesive as a function of time at room temperature. The viscosity change appears just before the expiration of the adhesive shelf life.

Storage Facilities

A production plant may have several room-temperature storage facilities. For example, there may be a hold area where material is received and samples are taken for quality testing, a main storage area, and a lab or department storage area.

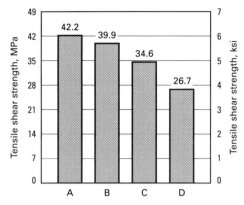

Fig. 1 Effect of aging conditions on epoxy film adhesive. A, fresh adhesive; B, cured after 90 days at 24 °C (75 °F); C, cured after 90 days at 32 °C (90 °F); D, cured after 1 h at condensing humidity

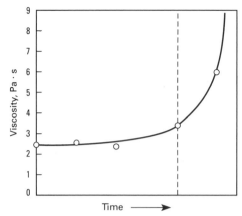

Fig. 2 Increase in viscosity of a paste adhesive as a function of time at room temperature

Upon receiving the materials at the facility, the user must place the containers in a dedicated area. This area must be suitable for handling flammable and hazardous materials. At this point, the material must be checked for compliance with packaging and handling requirements that are set down in the written procedures of the company. If the material requires inspection testing for acceptance, such tests will be made at this time, and the material will remain in a hold area until tests are completed.

Each material is then ready for storage in a segregated and dedicated area that meets the requirements for the specific material. Storing the material in a dedicated area will permit the material to be easily identified and controlled. Control of stored material involves keeping records of storage-life limits and maintaining proper inventories to meet production commitments; the section "Inventory Control Mechanisms" in this article provides more details.

An outside storage building is recommended if several drums of flammable materials are to be stored. The section "Han-

dling Considerations" in this article contains additional information about the storage of hazardous materials.

To properly store adhesive and sealant materials it is necessary to review manufacturer recommendations concerning proper temperature conditions for the specific product type. Three levels of storage temperature are common: room temperature (15 to 27 °C, or 60 to 80 °F), refrigerated (2 to 5 °C, or 35 to 40 °F), and frozen (at or below −18 °C, or 0 °F). Some products are specifically designed to have long storage lives in exposed outdoor environments.

The size of the storage area will depend on certain factors. Primary considerations are the production unit use requirements, the predicted production rate, storage life requirements, and procurement lead time. Other considerations, such as minimum order size or discounts for volume buying, can also influence the amount of space required. Described below are the types of facilities based on temperature requirements.

The freezer storage facility provides the greatest degree of flexibility for storage of limited-life materials. The −18 °C (0 °F)

temperature is low enough to prevent aging of the stored material for periods of 6 months to 1 year in most cases. Many standard freezer storage units are available in the industry in a range of sizes.

Storage units must be large enough to accommodate the volume required for storing specific size containers while providing full access to all items so that "first in, first out" inventory control is possible. The freezers can be built-in units or self-contained portable units; the portable units are the most flexible in operation. If an appreciable amount of space is needed, it may be better to use multiple storage units in case of unit failures. All freezer storage units should be equipped with a temperature-recording device to ensure that proper storage conditions have been maintained.

A central storage freezer for a manufacturing plant can be as large as 75 to 110 m^2 (800 to 1200 ft^2). The actual amount of total storage space may be proportional to the amount of adhesives used over a 3- to 6-month time period. If the freezer is equipped with two doors, different methods for storage and retrieval of materials can be employed. In one method (Fig. 3), the adhesives are received into one end of the freezer and removed for use from the other. This creates the periodic emptying and refilling of storage areas or racks along the sides of the freezer. Another way to use a two-door freezer is to allow personnel from different departments to use each end of the freezer.

The maintenance of the freezer storage units must include regular defrosting, either manual or automatic. In addition, periodic inventories (usually every month) are required to ensure that the stored materials are within the time limits recommended by the product manufacturer and the specific documents of the user specifications. Access to these freezers should be restricted to the department(s) specifically assigned to the use and handling of these materials.

The heat removal capacity of the unit must be sufficient to maintain temperatures near −18 °C (0 °F) in the freezer during daily use. Insulation factors of the walls, floor, and ceiling of the freezer must be taken into account. Also, the amount of traffic in the freezer and the amount of time the freezer doors are open affect the ability of the unit to maintain the proper temperature: A cooling system may be capable of reducing the freezer temperature to −18 °C (0 °F), but it may not be effective when the doors are open for a significant length of time.

Refrigerated storage units are useful for storing materials that:

- Do not require the very low temperature of frozen storage
- Cannot tolerate subfreezing temperatures because of their composition
- Will be used within a few weeks or months

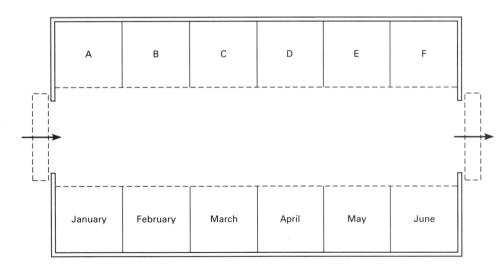

Fig. 3 Floor plan for freezer storage of adhesives

Product Code #:	
Date Received:	Shelf Life:
Expiration Date:	Recertification Date:
Vendor:	
Storage Condition:	
Disposition:	
Out Time Record Required (Yes/No)?:	
Out Time Record (If Required):	

Fig. 4 Example of a shelf-life identification tag

● Have been cut to size and are waiting to be used (film details, for example)

Aqueous solutions and dispersions, for example, must be stored in refrigerators rather than freezers because freezing can damage their emulsions. In addition, many two-part systems with limited room-temperature shelf lives store well in refrigeration temperatures.

An advantage of this type of storage is that it allows materials to warm up to room temperature in less time than is required for a frozen product. This is an important time saver when using materials such as partial rolls or precut details of sheet materials that have to be held for a short time before being removed from storage to be reused.

Refrigeration units set at 2 to 5 °C (35 to 40 °F) present less of a problem with defrosting than do the −18 °C (0 °F) freezer units. Like the freezer units, refrigeration units should be equipped with temperature recorders and should receive scheduled periodic maintenance. Shipping by refrigerated truck is the most common method of transporting standard industrial amounts of temperature-sensitive adhesives.

Ambient- or room-temperature storage is acceptable for short-term storage of one-part systems and for the storage of many two-part systems when the base and catalyst are separate and neither is affected by temperature. The problem with ambient- or room-temperature areas is that it is difficult to account for limited-life materials in such areas because they are less segregated from materials that do not have life limitations. This distinction is much clearer with the closed-box environment of refrigeration or freezer storage.

Inventory Control Mechanisms

Record Keeping. When an adhesive is received, a certification or product data sheet should accompany the product. This article guarantees or otherwise states the product shelf life from the date of manufac-

ture, ship date, or the date of receipt by the buyer. A standard policy is to attach a label to the container acknowledging the date of receipt, the date (if any) when quality testing of the material is to be completed, and the number of days that the material has been standing at room temperature after it has begun to be used. A sticker indicating the shelf life (in days) that is available at room temperature for a given material should also be attached to the package of adhesive. If −18 °C (0 °F) storage is not required, then the daily record keeping on the tag is not necessary. An example of this tag is shown in Fig. 4.

Inventory/Purchasing. In order to have fresh material on hand when it is needed for production, an inventory control system should be established. This system supplies data to a program that translates the amount of raw materials necessary for finished goods. Such a process needs a current inventory status on all the adhesive materials being used at the manufacturing facility. Next, all sales orders for finished products are accumulated. Each finished product requires a given amount of certain raw materials. The system collates the total amount of each type of raw material needed for all products (plus a 10 to 25% safety margin or current scrap rate).

The system may be able to project how much of each adhesive will be needed within 1 month, 2 months, 6 months, or up to 1 year. In this way the buyer is informed when to purchase material. The program needs to be both flexible enough to incorporate last-minute changes in orders and timely enough to permit quality assurance testing, if necessary, before the material can be used.

Handling Considerations

Adhesives must be stored in sealed containers for any period of time longer than a day or two at room temperature. Otherwise, moisture absorption can be a problem. It is also important to reseal bags of adhesive films before they are refrigerated or frozen.

A material is most vulnerable to moisture collection when it is removed from cold storage and the package or container is not properly sealed. Moisture condenses on the adhesive and quickly reduces its mechanical capabilities.

Safety Precautions

Safety is another consideration when handling resins, hardeners, films, solvents, and other materials. Cotton, leather, or rubber gloves should be worn to protect the hands from repeated contact with the materials. There is not much acute danger with many of these systems, but repeated contact over long periods of time can sensitize the skin and produce unpleasant reactions such as itchiness, redness, swelling, and blisters.

Solvent-base sprays should be used only in well-ventilated areas; in many cases, the operator should wear a fume mask. Goggles are also required in the event of splashing or spilling of liquid adhesives. A user should always request a material safety data sheet from the supplier or manufacturer of an adhesive and read it before use. There may be certain handling and storage conditions that were not previously known. Copies of these sheets should be kept on file in the using area and in the department that has responsibility for the control of stored materials. This applies to the separate components as well as to the uncured mix.

Hazardous Materials

The safe storage of adhesives containing flammable, corrosive, or hazardous substances is of major importance. Many organic solvents used in liquid adhesives are flammable.

There are two storage area possibilities for these materials: storage in a small building or shed separate from the main building, or storage in the main plant. An outside storage building is recommended if several drums of flammable materials are to be stored. The structure should be built of fire-resistant materials and equipped with a sprinkler system. In the event of a fire, the sprinkler unit will serve to cool the other drums. Dry chemical, foam, or CO_2 fire extinguishers must be available nearby to put out a fire of flammable solvents.

For storage of just a few drums of flammable materials, insulated metal storage cabinets are available that will protect the drums by closing their doors automatically in the event of a fire. It is best to provide a ground for the drums by hooking a wire from the drum to a suitable ground, so that a potentially explosive spark will not occur. The most appropriate internal location for a flammable storage area is a room with brick or masonry walls and fire doors that open outward.

Flammables. Some combustible materials have low flash points. These hazardous materials need to be stored in metal cabinet enclosures that provide both isolation and protection. In addition, the areas or enclosures can be equipped with sprinklers for fire suppression.

Corrosives. Some adhesive products contain corrosive compounds that require special containers and need to be separated from other materials to avoid interaction. In two-part adhesives, some amine hardeners are considered corrosives, as are anhydride hardeners. Plastic high-density polyethylene containers or containers with an inside phenolic coating are used most often with these materials. Drums of corrosives and other hardeners should be stored separately from adhesive film and resins because a spill or leakage of these products could produce an uncontrolled polymerization reaction, producing excessive heat and possibly toxic fumes.

SELECTED REFERENCES

- J.T. Rice, Adhesive Selection and Screening Testing, in *Handbook of Adhesives*, 3rd ed., I. Skeist, Ed., Van Nostrand Rheinhold, 1990, p 94
- Standard Test Method for Storage Life of Adhesives by Consistency and Bond Strength, D 1337, *Annual Book of ASTM Standards*, American Society for Testing and Materials

Metering and Mixing Equipment

William Devlin, Products Research and Chemicals Corporation

HIGH-PERFORMANCE multicomponent materials are utilized throughout industry to improve design, aesthetics, and assembly methods; increase joint strength; reduce weight; and protect from environmental conditions. Before adopting a course of action that requires a multicomponent adhesive or sealant, careful consideration must be given to the methods and equipment used for metering and mixing the materials.

To assure the quality and reproducibility of an end product, the metering and mixing system must be compatible with the handling properties of the material and the manufacturing/environmental conditions encountered in the work area. This article describes the various equipment alternatives and offers guidelines for a logical and systematic study of the application, as well as performance parameters that should be considered.

In addition to the prime advantage of higher performance levels, a multicomponent adhesive or sealant system offers significant advantages through its ability to provide a faster cure or set. This allows the bonded part or assembly to be handled, moved, packaged, shipped, used, or stressed sooner than slower-curing one-part materials would allow. Fast-reacting materials mixed at or near the point of application reduce or eliminate the need for heat-curing equipment, such as autoclave or batch ovens, and bulky fixtures or jigs. Multicomponent systems also permit in-the-field use of high-performance bonding or sealing materials, providing a significant advantage in construction, repair, equipment installation, and rework applications.

Although the advantages of these materials are numerous, and may promise productivity and quality gains, the user must be concerned with the accurate metering, mixing, and dispensing of the material. Accordingly, these questions must be addressed:

- Will the components be proportioned accurately and consistently?
- Will the material be mixed thoroughly?
- Have all the necessary health and safety issues related to handling of the separate and mixed components been thoroughly considered?

- How will the unused material be handled? Is it hazardous to handle or dispose of?
- Does the system require volatile solvents for purging mixers, for example?
- Has the cost of waste been computed?
- Are current production personnel qualified to perform the metering/mixing functions or operate the equipment?

Methods and Systems

The methods described below and/or their corresponding delivery systems include manual weighing and mixing; proportioned kits; preweighed, packaged kits; and bulk meter and mix systems.

Manual Weighing and Mixing. The ancient Egyptians developed a method to combine various natural materials into a resinous conglomeration of materials to conduct processes germane to their era, such as mummification. Modern industry still has instances where manual weighing and mixing are required.

Many materials can be successfully hand-proportioned and mixed, particularly those for which the ratio is simple to determine (such as 1:1 by volume) and the viscosities of the individual components are similar and low enough to facilitate hand mixing with an appropriate tool. However, manual weighing, volumetric portioning, and mixing of sophisticated, high-performance materials can be risky, whether they are silicone, urethane, polysulfide, acrylic, polyester, or epoxy. Problem areas include air incorporated by hand mixing, which requires degassing to prevent voids in cured materials; the difficulty in judging fluid volumes, as opposed to weight; and "off ratios" that result from operator errors, such as pouring the contents of one container into another, leaving material stuck to the side of the emptied one, or from "adjusting" the ratio to compensate for an environmental or production condition. An example of this problem would be the addition of "a little extra" catalyst or hardener to speed cure when ambient temperatures drop or when production requires more rapid curing. This also causes wasted mate-

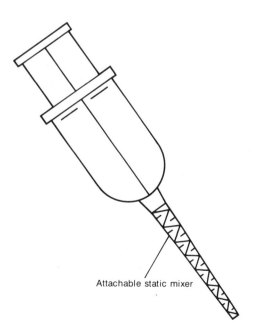

Fig. 1 Side-by-side (dual) syringe (50 ml size)

Attachable static mixer

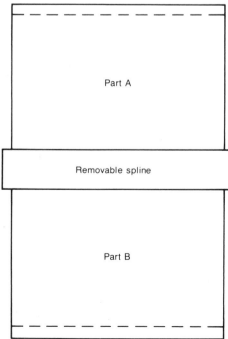

Fig. 2 Divider bag or hinge pack

Part A

Removable spline

Part B

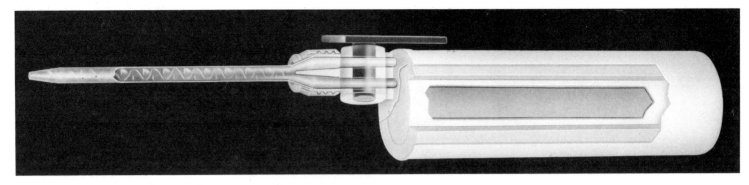

Fig. 3 Coaxial cartridge, static mix system. Source: Liquid Control Corporation

rial because an inadequate quantity of catalyst is left for the remaining quantity of sealant. Off ratios or improper mixing adversely affects the performance of a bonded assembly.

Proportioned Kits. Most material manufacturers supply products in containers designed to facilitate the combining and mixing of their contents. In other words, part A and part B are packaged separately and supplied as a kit, and, normally, one of the containers has enough empty space to accommodate the other component and a little extra room for mixing the two (or three)

components without spilling. This is the case whether the product comes in 55 g (2 oz) tube kits; pint, quart, or gallon kits; or other configurations.

In theory, the metering is done at the factory, and the only thing the operator must do is follow the mixing instructions. In practice, these undesirable scenarios can result from the use of these kits:

- The operator buys gallon kits but never uses more than a few ounces at a time. This involves manual metering and mixing, with all of its uncertainties, plus the

added problems of potential contamination and compromising of shelf life by repeated opening and closing of the containers

- Part A is dumped into the part B container, but the first container is not totally emptied (due to such things as residue on side walls and under the lip of the can). This results in an off-ratio condition that, depending on the material's sensitivity to ratio, can be a major problem
- Part A and/or B requires premixing or agitation to ensure filler content dispersion in one or both components, but the

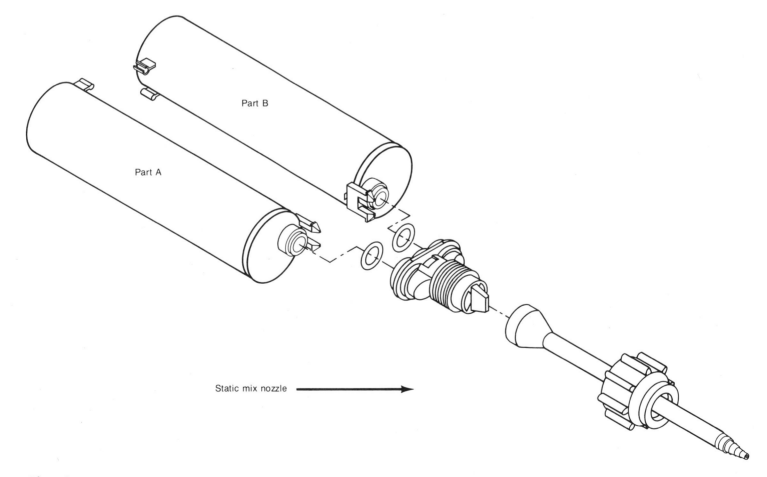

Fig. 4 Side-by-side, static mix system. Courtesy of Semco

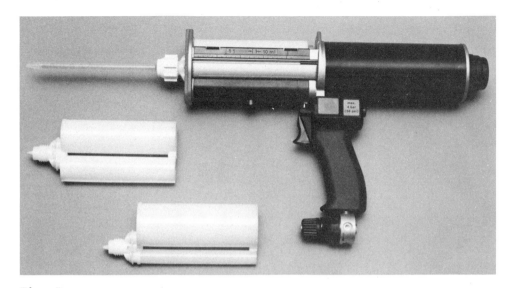

Fig. 5 Side-by-side cartridge and pneumatic dispenser. Courtesy of Semco

operator fails to carry out this procedure properly prior to blending or mixing the two components. This can result in an "out of specification" condition in the final mixed and cured material
- Parts A and B are properly combined, but the manufacturer mixing instructions are not followed exactly. It can be difficult to

determine, strictly by visual observation, when some materials are adequately mixed. This is particularly true if the color of parts A and B are the same or similar or if the ratio of A to B is much different than 1:1. Some materials will not perform as expected when they are overmixed. Loss of application life, wetting

ability, and changes in handling properties are common overmixing problems. Of course, undermixing is never acceptable
- The adhesive or sealant manufacturer may supply kits in matched sets, which means that only those components that are matched by the manufacturer can be mixed. If other components are used, there is a risk of mismatching base and accelerator. Often, the negative results are not immediately obvious because the material cures adequately. However, adhesion systems or exposure resistance capabilities may be thoroughly compromised by such a mismatch

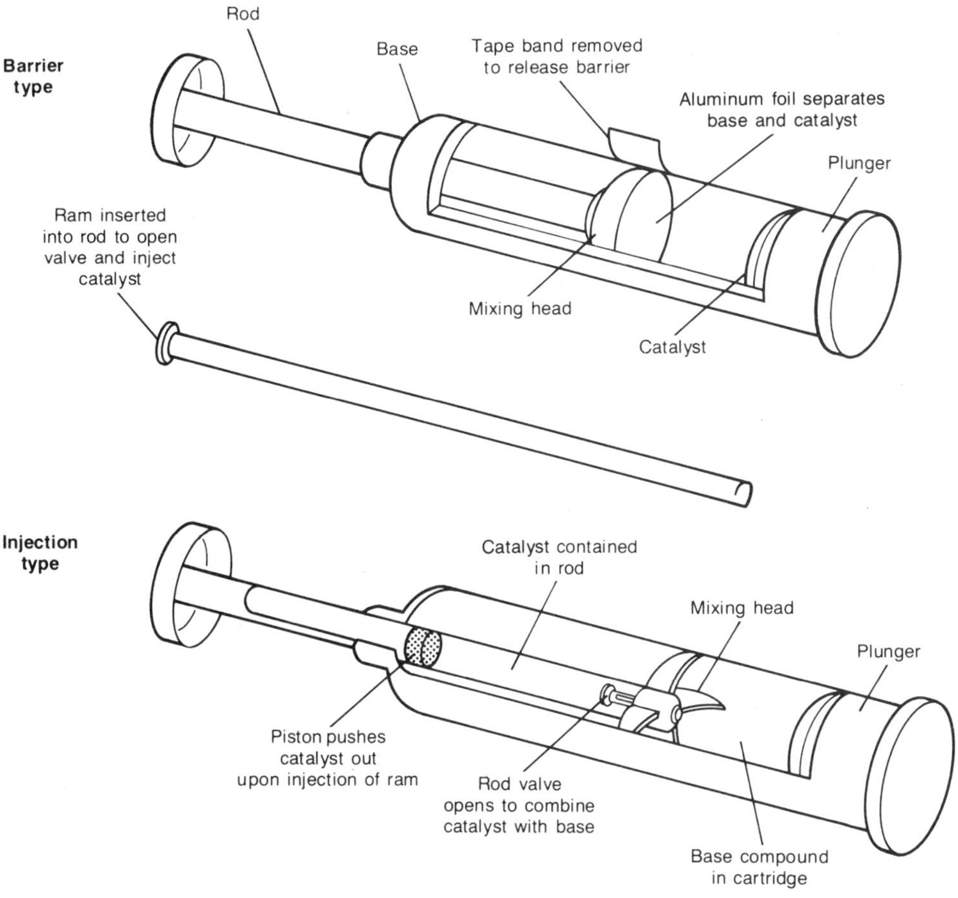

Preweighed, Packaged Kits. A variety of practical solutions are available to users of multicomponent materials when the application requires a limited amount of material (for example, less than 570 g, or 20 oz, per use). Specially designed packages—from simple mechanical devices to self-contained disposable, storing, metering, mixing, and application systems—are described below.

Side-by-side syringes (Fig. 1) are designed for low-to-medium-viscosity (self-leveling) adhesives in 1:1 and 2:1 ratios. This package features a dual plunger that empties the parallel syringe barrels simultaneously. The only task remaining is to ensure proper mixing. Side-by-side syringes can be equipped with attachable static mix tubes to mix material as it is dispensed.

Divider bags (Fig. 2) contain parts A and B in the same plastic sleeve but separate them with removable "spline." The operator can remove the spline and knead the materials together within the plastic pouch. After kneading, he can cut open the pouch and squeeze out the adhesive. With some materials, this mixing method may not provide adequate mixing because of wide variation in viscosity and flow characteristics of the two components, as well as variation in operator handling.

Coaxial cartridge, static mix systems (Fig. 3) use two individual cylinders, one within the other, to separate the reactive materials until used. An extrusion gun is used to simultaneously push the pistons and deliver properly ratioed material into a "static mixer" that also acts as an application nozzle or dispensing tip. This system allows intermittent use, by virtue of its disposable/replaceable static mixer, and is capable of handling several different ratios (for example, 1:1, 2:1, and 10:1). The maximum volume available with these systems is usually 380×10^{-6} m^3 (380 mL).

Side-by-side, static mix systems (Fig. 4) package components in separate cartridges. An extrusion gun (mechanical or pneumatic) extrudes the materials through a static, or motionless, mixer that also serves as an application orifice (Fig. 5). The static mixer

Barrier type

Rod
Base
Tape band removed to release barrier
Aluminum foil separates base and catalyst
Plunger
Ram inserted into rod to open valve and inject catalyst
Mixing head
Catalyst

Injection type

Catalyst contained in rod
Mixing head
Plunger
Piston pushes catalyst out upon injection of ram
Rod valve opens to combine catalyst with base
Base compound in cartridge

Fig. 6 Barrier- and injection-style kits. Source: Techcon Systems Inc.

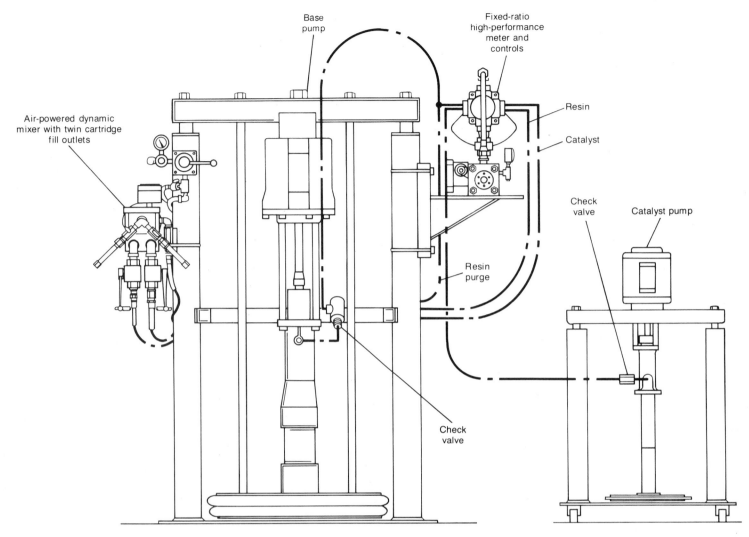

Fig. 7 Fixed-ratio meter/mix machine. Source: Pyles

is disposable, allowing intermittent use of the package. Cartridge sizes range from 75 to 750×10^{-6} m³ (75 to 750 mL), and ratios range from 1:1 to 10:1.

Barrier- and injection-style kits (Fig. 6) can accommodate any ratio of material. Materials are preweighed and ready for mixing at the point of application. The entire content of the kit is mixed within the package. This eliminates operator exposure to the chemicals, which is also the case in the systems described above. After mixing, a nozzle (many options are available for specific applications) is attached and material is dispensed manually or with a gun (mechanical or pneumatic). The entire package is disposable. No intermittent use of this package is possible. This system is based on dynamic mixing, rather than static or motionless mixing. Mixing energy, either manual or mechanical, is required. The barrier kit is appropriate for compounds with a volumetric ratio or more than 10 parts per 100, and the injection kit is appropriate for ratios of less than 10 parts per 100. In the

barrier kit, catalyst and base are separated by an aluminum barrier formed over the mixing head, whereas the injection kit contains the catalyst inside a valved mixing rod and the base inside the cartridge.

The principal advantages of preweighed, packaged kits are that they:

- Ensure correct ratio
- Minimize exposure to unreacted chemicals
- Provide a mixing system (although the material manufacturer should be contacted to determine if adequate mix can be accomplished in the particular kit under consideration)
- Can be sized to the job
- Facilitate the field use of high-performance materials
- Can be convenient in a production environment where bulk meter/mix/dispense equipment is either not an acceptable alternative or is available but not working

Bulk Meter/Mix Equipment. Typically, production requirements dictate whether an

investment will be made to automate the metering, mixing, and dispensing steps. The amount of material being used may require a more efficient handling system, the production process may have enough repeatability to standardize such variables as shot size, or product performance may require a level of consistency that only a machine can provide.

Metering and Mixing Equipment

Many alternatives are available in pumping, metering, and mixing apparatus. However, the permutations derived when mixing and matching these alternatives can be confusing. The more common types of equipment for each operation are described below. Bear in mind that the right tool for a specific job is generally determined by two factors:

- The handling characteristics of the selected adhesive or sealant material (such as

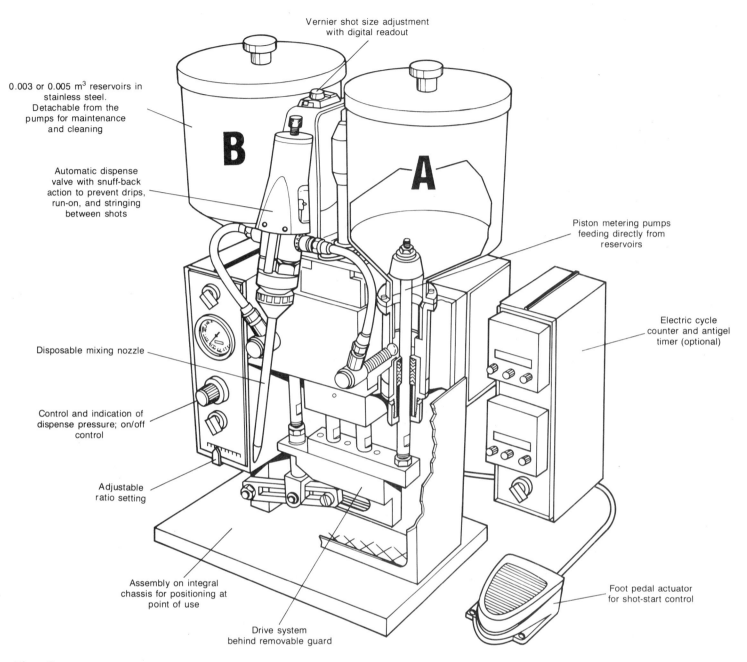

Vernier shot size adjustment with digital readout

0.003 or 0.005 m³ reservoirs in stainless steel. Detachable from the pumps for maintenance and cleaning

Automatic dispense valve with snuff-back action to prevent drips, run-on, and stringing between shots

Piston metering pumps feeding directly from reservoirs

Electric cycle counter and antigel timer (optional)

Disposable mixing nozzle

Control and indication of dispense pressure; on/off control

Adjustable ratio setting

Assembly on integral chassis for positioning at point of use

Foot pedal actuator for shot-start control

Drive system behind removable guard

Fig. 8 Variable-ratio meter/mix machine. Source: Liquid Control Corporation

viscosity, flow rate, temperature range, amount and type of fillers and/or solvents, pot life, cure rate, gel time, degree of exotherm, thixotropy, and mixability)

• Production requirements and environmental conditions (such as units per hour, line speed, assembly time, ambient temperature, and humidity)

Material Transfer Pumps. With these pumps, the adhesive or sealant is often unloaded or emptied directly from its shipping container into the metering section of the machine. In some cases, the transfer pump also acts as the metering mechanism and delivers material directly to the mixing point. Again, the type of adhesive or sealant

selected dictates which transfer system is most appropriate. The choices are:

• Gear pumps
• Positive displacement pumps
• Reciprocating pumps
• Reciprocating pumps with follower plates to keep pumps primed and containers sealed
• Pressure vessel pumps
• Gravity-fed pumps
• Auger-fed pumps
• Extrusion/ram fed pumps

Metering Systems. The handling characteristics of the adhesive or sealant determines the most appropriate metering mechanism. More sophisticated systems

employ a combination of devices to assure an accurate and consistent ratio. For example, in the system depicted in Fig. 7, a carefully monitored supply pump pressure maintains sufficient meter filling to a positive-displacement, two-way meter. In addition, back pressure on the outboard sides of the meter is controlled by pressure compensation valves and hose diameters, to assure accurate delivery to the mixing chamber.

When low-viscosity materials are used, gear pumps can supply consistent metered amounts of the two components. In some systems, a common belt or chain drives separately sized pulleys attached to the gear motor shafts.

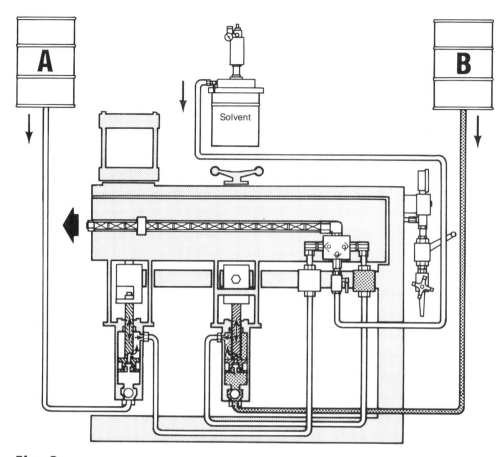

Fig. 9 Operation of solvent-purge meter/mix machine. Source: Sealant Equipment and Engineering Inc.

Another common metering device, a variable ratio meter, delivers a consistent amount of the base compound. However, the amount of catalyst or hardener can be varied by shifting the position of the pivot arm on the catalyst "meter." This type of equipment offers the flexibility needed when materials or material ratios are changed periodically (Fig. 8).

Regardless which metering system is used, it is important to make a ratio check to determine if the correct amounts of parts A and B are being delivered by the system.

Mixing Systems. Innovation and imagination have been important elements in engineering various approaches for combining separate adhesive or sealant components. Some materials mix quite readily, whereas others require a considerable amount of energy (shear), temperature, pressure, and/or time to initiate the chemical reaction. The selection of mixing systems, therefore, is very important.

Increasing awareness of costs associated with material waste and hazardous waste disposal have had a major impact on mixing technology. Because the specific characteristics of adhesives materials are important to the selection process, the amount of mixing necessary to achieve the desired performance properties must be considered. Also to be considered is the most efficient way of dealing with the material left over in the mixer (and associated equipment) when the application is completed. Common mixing systems are described below.

Mechanical Mix. In this approach, rotating blades shear the materials, assuring optimum blending. Depending on the material and time in the mixing chamber, cooling jackets are often employed to extend the application, or pot, life of the material because the friction of the mixer and the exotherm of the adhesive or sealant can cause substantial heat buildup that typically shortens both handling and curing time.

When shut down, the mixer can be solvent purged, base purged, or frozen. During solvent purge, a cleaning solvent is introduced into the mixing chamber, flushing materials out of the mixer so that no cured material remains in the chamber (Fig. 9). Problems include handling and disposal of the solvent and waste sludge as well as the

quantity and cost of solvent required to adequately clean the mixer.

During base purge, the base component only is pumped into the mixing chamber, displacing the mixed material and filling the chamber with a nonreactive compound. When ready for next use, mixed material is pumped in and displaces the base material, which is then discarded.

In the quick-freeze process, which is commonly used in the aerospace industry, mixers fill special, plastic cartridges that are quick-frozen (at -75 °C, or -100 °F, or lower) and stored at low temperature (-40 °C or °F) until needed. After the mixing operation, the entire mixing assembly is placed in a freezer where the low temperatures inhibit the cure of the material. When ready to resume, the mixer is thawed (approximately ½ h at room temperature) and freshly mixed material is used to flush out the mixing chamber.

Static Mix. Sometimes referred to as "motionless mixing," this versatile mixing system depends on flow of the components through a mix tube that contains baffles or chambers that combine, separate, and recombine the materials repeatedly. Mixing tube size varies with material type, back pressure, viscosity, and so forth. (Static mixing tubes are depicted in Fig. 3 to 5.) Cleanup can be accomplished by a solvent flush, solvent/air flush, or base purge, or by using disposable elements and nozzles.

Durable stainless steel varieties can be "burned-out" to remove cured residue (material and mixer suppliers should be consulted before this alternative is attempted). Static mixers can also be quick-frozen to inhibit curing of the material and then flushed with mixed material when restarting the equipment.

Impingement Mix. This style of mixing is limited to low-viscosity (typically less than 0.1 Pa · s) materials and those that can be heated to lower their viscosity. High-pressure streams of each component collide or impinge on each other in a small chamber and are immediately dispensed. The mixing chamber is usually less than 5×10^{-6} m^3 (0.3 in.^{3}) in volume and is purged by a close-fitting piston that evacuates the chamber without the use of cleaning solvents. Viscosity, pressure, and temperature must all be closely monitored in such systems.

SELECTED REFERENCES

- J. Drake, Impact of Encapsulation Compounds on the Selection of Dispensing Equipment, *Electr. Manuf.*, Nov 1989
- M. Petterborg, Barrier/Injection Kits Reduce Costs and Production Time, *Adhes. Age*, July 1988

Dispensing and Application Equipment for Adhesives and Sealants

James E. De Vries, Automotive Business Group, Nordson Corporation

ADHESIVES AND SEALANTS are being used for product assembly in today's global industries as a means of reducing manufacturing costs and improving product quality. As a result of these opportunities, a growing number of bulk-dispensed adhesive and sealant applications are being implemented by various manufacturers.

The successful implementation of production processes to apply these materials is dependent on the clear definition of process objectives, an understanding of material properties and their relationship to dispensing system performance, and the careful selection of system components, in light of the objectives and equipment limitations. The available types of dispensing and application equipment for bulk-dispensed materials are surveyed and reviewed, as are the material properties that determine equipment specification.

The level of dispensing system automation ranges from simple, manual systems to large, complex robotic systems. The level of automation used presents trade-offs that must be considered relative to the objectives of the process. The advantages and limitations of manual, mechanized, and automated dispensing and application equipment for bulk-dispensed adhesive and sealant materials will be discussed.

Equipment technology has also evolved to handle new adhesive and sealant materials. General dispensing system improvements relative to accuracy, repeatability, and reliability have been realized. Equipment features such as precise temperature control, system diagnostics, and process monitoring capabilities have made adhesives and sealants applications increasingly cost effective.

Classification of Materials

Adhesives and sealants can be classified various ways because hundreds of basic materials are available in an almost infinite number of formulations. For the purpose of specifying dispensing equipment, adhesives and sealants can be classified as follows:

Heat-processed adhesives and sealants

- Hot melts
- Warm melts
  Weld-through sealers
  Weldable/expandable sealers
  Single-part epoxies

Room-temperature-applied adhesives and sealants

- Plastisols
- Silicones (room-temperature vulcanizing, RTV)
- Single-part epoxies
- Urethanes

Note that there is also a broad category of two-component materials, which are not covered here.

This classification is important in terms of dispensing equipment because it determines whether the equipment needs to be heated. Many materials require heating to reduce the viscosity of the material to a pumpable value. Other materials require heating to cut through the die oils on steel components to provide acceptable adhesion. Equipment capable of melting or reducing material viscosity by heating is more sophisticated and costly than equipment to apply materials at room temperature because of the electrical equipment required to heat and maintain temperature control of all the system components that contact the material during dispensing.

Dispensing System Components

The four elements common to all dispensing systems are pumps or pumping system, header system, dispensing valve or gun, and system controls. Careful selection of each element is required to ensure the intended performance of the material dispensing system. The basic dispensing system elements and factors regarding their selection are described below.

Pumping System

The heart of any dispensing system is the pump or pumps that extract the adhesive or sealant from the shipping container and increase the material pressure to a level capable of generating the required material flow rate through the dispensing gun. If the material to be dispensed needs to be melted or the viscosity reduced by heating, a heated pump or bulk melter is used. Bulk unloaders are unheated pumps used for materials to be dispensed at or near room temperature. Typical container sizes include 0.02 m³ (5 gal) pails (Fig. 1a), 0.2 m³ (55 gal) drums (Fig. 1b), and 1.1 m³ (300 gal) totes (Fig. 2). Pumps are available to dispense directly from these container sizes.

The container size should be selected based on the material consumption rate, the shelf life of the material, available container storage space, and container disposal costs.

Replacing an empty container with a full container typically takes 15 to 30 min. The container selected should require no more than 1 or 2 changes per 8 h shift. Conversely, the container of material should be consumed before the shelf life of the material at the application temperature and pressure is exceeded.

Totes of 1.1 m³ (300 gal) in volume are used in applications where the material consumption rate is such that the use of smaller containers is not practical because of container storage requirements and the frequency of container changes.

Environmental concerns have forced users of bulk-dispensed adhesives and sealants to consider the disposal of the shipping containers. Normally, 0.02 m³ (5 gal) pails and 0.2 m³ (55 gal) drums made of steel or fiber are not reused.

The 1.1 m³ (300 gal) totes, which are made of steel, have an epoxy liner, which makes them reusable. They fit two abreast

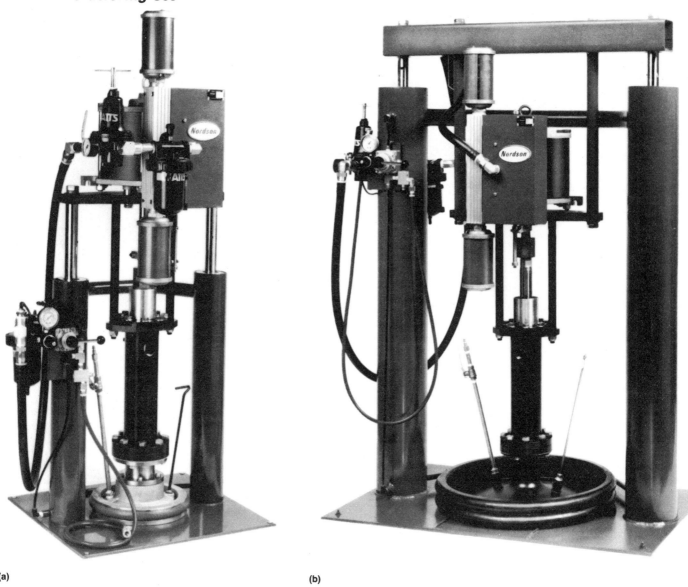

(a) (b)

Fig. 1 Bulk unloaders. (a) 0.02 m³ (5 gal). (b) 0.2 m³ (55 gal)

on a standard truck bed and the empty totes are returned to the material supplier, where they are cleaned and refilled with new material. The use of 1.1 m³ (300 gal) totes minimizes container storage requirements, reduces labor costs because of fewer container changes, and eliminates container disposal costs.

Regardless of container size, all bulk melters and bulk unloaders are functionally similar. A pump, which is connected to a follower plate or platen, is mounted in a frame assembly or ram (Fig. 3, 4). The ram uses air cylinders to lower the pump and platen assembly into the container of material, forcing the material into the pump inlet.

The amount of downforce required depends on the viscosity of the material and the size of the container. Higher-viscosity materials require larger cylinders to generate sufficient downforce to force material into the pump. Rams of 0.02 m³ (5 gal) have

one or two cylinders, 0.2 m³ (55 gal) rams have two cylinders, and 1.1 m³ (300 gal) rams have four cylinders for the required stability and alignment with the container. In rare applications, hydraulic cylinders are used to generate extremely high downforce. However, standard shipping containers cannot withstand the high pressures that hydraulic cylinders are capable of generating and special containers or container reinforcements are required.

The frame assembly also includes the pneumatic controls for raising and lowering the ram, positioning and removing the container, and controlling the pump. In addition, bulk melters have an electrical enclosure attached to the frame assembly that incorporates temperature-control hardware and motor controls if a gear pump is used (see Fig. 5).

When a follower plate or platen is lowered into the material, it seals against the

inside diameter of the container. The platen seals must be compatible with the material to be dispensed and are available in several materials, including nitrile rubber, silicone rubber, or polytetrafluoroethylene (PTFE)-encapsulated silicone rubber. The PTFE-encapsulated seals are the more expensive and difficult to install but offer the greatest chemical resistance.

Material bleed and container removal or blow-off mechanisms are also built into the platen assembly. The material bleed valve is actuated to release air that is trapped under the platen when a new container of material is inserted. The release of all air is critical to prevent air pockets from being introduced into the system. The container blow-off mechanism is a pressure-set air supply that is connected to the platen when an empty container is to be removed. Air is introduced under the platen, which breaks the vacuum between the platen and the contain-

Fig. 2 1.1 m³ (300 gal) bulk unloader

er and allows the platen to be raised out of the container with the ram.

Bulk melter platens are heated to melt or reduce the viscosity of material to be fed into the pump. The melt rate must be equal to or greater than the pump rate or the pump will cavitate and the flow of material from the pump will be interrupted. Different platen configurations are available, depending on the required melt rate. High melt rate platens have 75 mm (3 in.) fins (Fig. 6a) for maximum heat transfer surface area, and lower melt rate platens have 13 mm (0.5 in.) pins (Fig. 6b) that also minimize the amount of material left in the container. Gear pump bulk melters have the pump mounted in close proximity to the heated platen so that the pump is heated by conduction. The

platen and pump are both heated on piston pump bulk melters because of the additional mass of the piston pump.

Different types of pumps can be used in either bulk melters or bulk unloaders, depending on the application and material requirements. The three most common types of pumps for bulk-dispensed adhesives and sealants are screw, gear, and reciprocating piston pumps.

Screw pumps are output-limited relative to the pump size and create a significant amount of heat because of the pumping action of the screw. As a result, screw pumps are reserved for difficult-to-melt materials and materials that are not sensitive to shear heating in low flow rate applications.

Gear and reciprocating piston pumps are positive-displacement pumps that dispense a specific volume of material per revolution or stroke of the pump. A gear pump has an electric motor that drives a gear set, which relies on intimate metal-to-metal contact of the meshing gear teeth to seal the pump cavities and create the pumping action (see Fig. 7). Gear pumps provide the steady, uninterrupted material flow that is required for most automatic or robotic applications. Unfortunately, gear pumps typically are limited to materials that are less than 200 Pa · s in viscosity. The downward force of the ram on the platen is the only force used to feed material into the pump. The viscosity of many adhesives and sealants is greater than 200 Pa · s and will cause gear pump cavitation because of the inability of the ram force to provide a sufficient supply of material to the pump inlet.

The abrasivity of the material can be the determining factor in the decision to use a gear or piston pump. Material abrasivity is due in large part to the content of filler materials. Fillers are added to material formulations to reduce cost, build viscosity, and improve weldability, among other reasons. Filler materials include calcium carbonate, calcium silicate, clays, chalk, metallics, and glass spheres. The action of the abrasive fillers on the sliding metal surfaces of a gear pump can rapidly wear the gear set, causing premature failure. Some materials have a relatively high content of petroleum oils and/or waxes, which act as lubricants that counteract some of the material abrasivity. When in doubt, the material should be tested for lubricity prior to pump specification.

Reciprocating piston pumps have a plunger to displace material from upper and lower chambers in the pump that are alternately opened and closed with internal check valves (see Fig. 8). An air motor drives the plunger in both vertical directions, so that material is dispensed on both the upstroke and the downstroke. Because piston pumps do not rely on the contact of meshing gears to create the pumping action, they are more resistant to abrasive materials. As a result, most materials with high filler content should be pumped with piston pumps.

The material forced into a piston pump is assisted by a shovel at the bottom of the plunger, which reaches into the material on the downstroke and shovels material into the lower pump chamber on the upstroke. This shovel assist, in addition to the downward force of the ram, enables piston pumps to pump many high-viscosity materials that gear pumps cannot handle.

Piston pumps can use ball checks for low- to medium-viscosity materials (up to 100 Pa · s) or flat/chop checks for medium- to ultra-high-viscosity materials (50 to 10 000 Pa · s). The limiting factors of a ball check piston pump are the small passages through

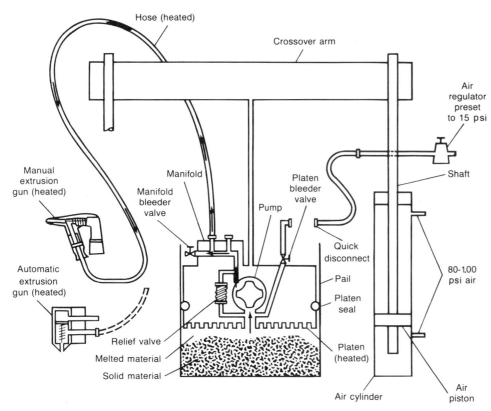

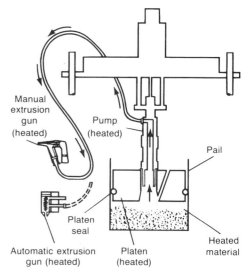

Fig. 3 Schematic diagram of 0.02 m³ (5 gal) gear pump bulk melter

Fig. 4 Schematic diagram of 0.02 m³ (5 gal) piston pump bulk melter

the pump and the inability of the ball check to seat and unseat properly with high-viscosity materials.

Piston pumps can generate higher system pressures than gear pumps because of the high diametral ratio of the air motor piston to that of the piston in the pump. Some piston pumps can theoretically generate system pressures in excess of 45 MPa (6.5 ksi).

The disadvantage of a piston pump is that the reciprocation of the plunger causes system pressure fluctuations. When the plunger reaches the top or bottom of the stroke and changes direction, the pump output pressure momentarily drops, resulting in material flow rate variations at the dispensing gun. To compensate for pressure fluctuations, ancillary devices such as mastic regulators, accumulators, or metering pumps are required to provide a constant material output rate.

To specify the size of any pump, the required flow rate and pressure must be determined. The flow rate is determined by dividing the required volume of material by the dispense time. For example, if a 0.64 mm (¼ in.) diam circular bead of material 610 mm (24 in.) long must be dispensed in 5 s, the flow rate would be calculated by first determining the cross-sectional area of the bead: $[\pi(0.25)^2]/4 = 30$ mm² (0.049 in.²). Then, multiply the cross-sectional area by the length of the bead to get the total volume of material required: 30 mm² (0.049

in.²) × 610 mm (24 in.) = 20 × 10⁻⁶ m³ (1.18 in.³). Finally, divide the total volume of material required by the dispense time: 20 × 10⁻⁶ m³ (1.18 in.³)/5 s = 4 × 10⁻⁶ m³/s (0.25 in.³/s). This is a relatively low flow rate because the volume of material is small and the application rate is slow. Spray applications typically require more material, and robotic application rates of 510 mm/s (20 in./s) are not uncommon.

The required output pressure from the pump is a function of the pressure drop of all system components and the required flow rate. The equipment supplier should be able to supply pressure drop data on all system components and make a recommendation as to the pump capacity.

Gear pumps are limited by the maximum rev/min of the motor and its torque capacity. If the pressure to generate the required flow rate exceeds the torque capacity of the motor, the motor will go into current limit and the desired flow rate will not be achieved. Piston pumps provide material on demand because the pump remains pressurized and only strokes when the dispensing gun is opened. Piston pumps are limited by the maximum cycle rate of the pump and maximum rated output pressure. The theoretical output pressure is determined by multiplying the supply air pressure by the pump ratio. Piston pump ratios of 24:1, 48:1, and 65:1 are common.

For applications where a continuous flow of material is required, a pumping system

with automatic changeover is used to eliminate downtime caused by container changes (Fig. 9). This type of system is made up of two pumps, each with a hose that connects to a common changeover manifold that supplies material to the header system.

Header System

The header, or delivery, system is the section of the dispensing system that delivers the material from the pumping system to the dispensing guns.

The header system configuration depends on the relative location of the pumping system to the application points, material to be dispensed, and the required material delivery rate. Figure 10 shows a simple header system for dispensing multiple droplets of adhesive.

The header system is usually composed of rigid pipes and/or flexible hoses and can range in complexity from a single hose delivering material to a dispensing gun, to systems with over 90 m (300 ft) of pipe and hose supplying multiple dispensing guns. Hose diameters range from 6.4 to 38 mm (0.25 to 1.5 in.), whereas pipe diameters range from 13 to >75 mm (0.5 to >3 in.).

Other header system components include valves to isolate the pumping system or sections of the header system, valves to drain and clean the system, and material pressure regulators to balance the flow of material to each dispensing gun in multiple-gun systems.

The sections of header pipe can be fabricated with flanges that bolt together for easy installation and maintenance, or the sections can be welded together on site. Plant equipment specifications may dictate welded pipe for certain system pressures. A drawback of welded pipe is that the entire header must be replaced or pipe sections

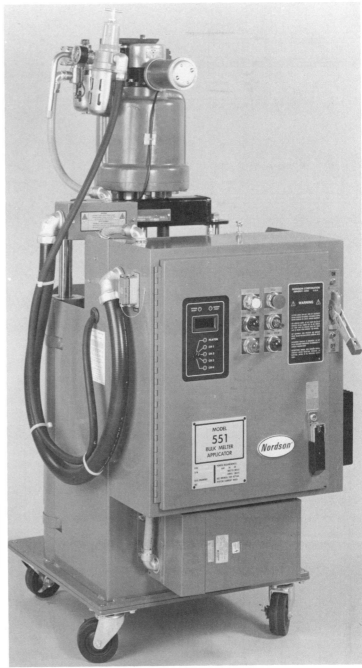

Fig. 5 Bulk melters. (a) 0.2 m³ (55 gal). (b) 0.02 m³ (5 gal)

cut out, cleaned, and rewelded if material builds up on the inside diameter of the pipe.

Of the many different material properties that affect the dispensing system design, viscosity has the biggest impact on the header system. Viscosity is the measure of the resistance to flow of the material and is defined as the ratio of shear stress to shear rate.

The apparent viscosity of the adhesive or sealant, when flowing in a specific system configuration, is a function of the material rheology. Rheology deals with the behavior of the adhesive or sealant material flowing in the dispensing system.

The viscosity of Newtonian materials remains constant, regardless of shear rate. The viscosity of non-Newtonian materials is dependent on the shear rate, and Bingham plastic, dilatant, or pseudoplastic behavior is demonstrated. The viscosity of Bingham plastic materials is initially high but reduces as shear rate increases. The viscosity of a dilatant material increases as the shear rate is increased. The viscosity of a pseudoplastic, or shear thinning, material decreases as the shear rate increases (see Fig. 11).

In some material compositions, there can be a unique interaction between the material components. The components in the material can form a network that has a higher viscosity than the unnetworked components. When the material is subjected to shear stress, the network will break down and the viscosity will decrease to the unnetworked viscosity. The longer the material is subjected to the shear stress, the lower the viscosity will become, until the unnet-

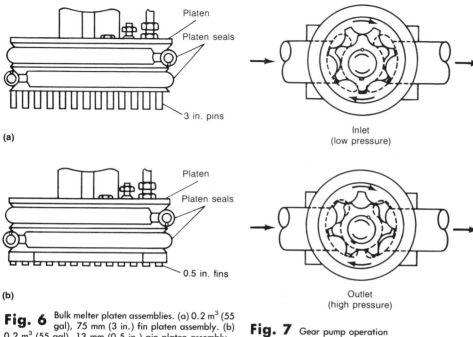

(a)

(b)

Fig. 6 Bulk melter platen assemblies. (a) 0.2 m³ (55 gal), 75 mm (3 in.) fin platen assembly. (b) 0.2 m³ (55 gal), 13 mm (0.5 in.) pin platen assembly

Fig. 7 Gear pump operation

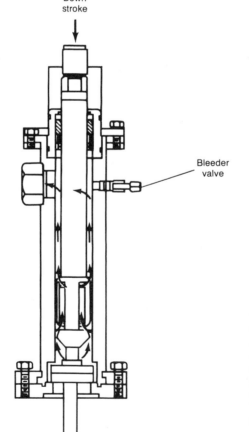

Fig. 8 Piston pump operation

worked viscosity is reached. A material that displays this type of rheological behavior is called thixotropic (see Fig. 12).

Shear thinning and thixotropic behavior are relatively common material characteristics that result in lower apparent viscosity. If the resulting flow rates at different dispensing system pressures are plotted for different rheological types of materials, the system implications of viscosity are clearly evident (see Fig. 13).

The relationship between the material viscosity and the size of the flow passages in the dispensing system is critical. For a Newtonian material, the flow resistance of a component in the dispensing system is a function of the length of the flow passage divided by the diameter to the fourth power. As a result, relatively minor changes in the configuration of system components can have a major impact on the flow resistance of the system. Dispensing systems with long, complex header systems and/or components with small flow passages will require higher pump pressures to generate the required material flow rate. High-viscosity materials also require more system pressure to generate a given flow rate. System components, including hoses, pipes, manifolds, and guns, must be designed to withstand the high pressure. As a result, high-pressure system components are larger and more expensive. Furthermore, booster pumps may be required within the header system to generate the required pressure at the dispensing gun.

Heated hoses and pipes are used for hot melts and other materials that are applied at temperatures above ambient. The hoses and pipes are wound with heater elements, along with a temperature-sensing element that measures the temperature over a length of the hose or pipe (Fig. 14). Sensing temperature over a long length is preferable to sensing at a single point, which will generate erroneous readings if there are hot or cold spots in the system resulting from the location of the hose or pipe.

The temperature-sensing element provides feedback to the controller on the bulk melter in small systems, or the system control console in large systems. Electrical connectors are used to connect the various heated system components to the system controls. In large systems, it is highly recommended to divide the header system into individual electrical components that are each connected to the system controller. This simplifies installation and troubleshooting.

The temperature control of individual hoses and sections of pipe provides superior temperature control accuracy as well as the flexibility to set components at different temperatures. Also, individually heated hoses and pipes are much easier to replace than large sections of piping that are wound with heater tape.

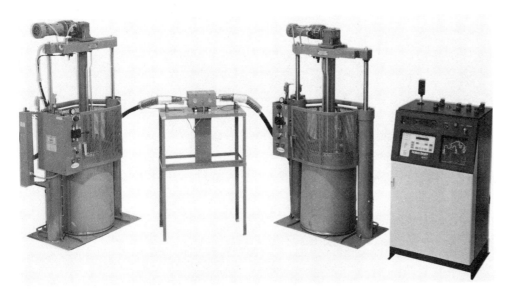

Fig. 9 Heated pumping system with automatic changeover

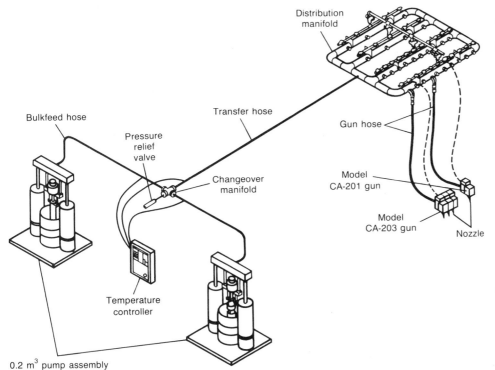

Fig. 10 Typical droplet dispensing system configuration

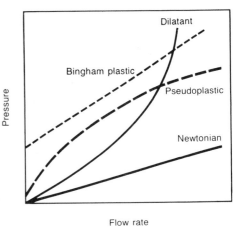

Fig. 11 Viscosity versus shear rate

Fig. 12 Viscosity versus time

Fig. 13 Pressure versus flow rate

Some low-viscosity materials can be pumped without heating. However, if the viscosity is very temperature sensitive, temperature conditioning is required to provide constant material output. In this case, the material is pumped out of an unheated bulk unloader, and a heated header system is used to stabilize the material temperature regardless of ambient conditions. Variable-orifice dispensing guns can also help compensate for variations in viscosity caused by temperature. If the material is very sensitive to shear heating or if the system will be subjected to extremely high ambient temperatures, a heat exchanger may be required to heat and cool the header system.

Dispensing Gun

The dispensing gun, or valve, controls the flow of material onto the substrate. The dispensing gun can be manually operated or actuated automatically. It can be a simple on/off valve or a sophisticated variable-orifice dispensing gun, depending on the requirements of the application (see Fig. 15-18).

The proliferation of robotic dispensing applications and the demand for quality improvement has driven the development of dispensing guns that are capable of placing the exact amount of material in the exact location, part after part. The exact amount of material must be maintained regardless of robot velocity and material pressure fluctua-

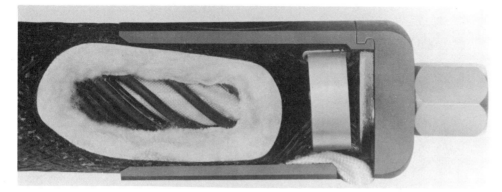

Fig. 14 Heated hose construction

tions. In response, a variable-orifice gun that varies material output in proportion to robot velocity has been developed. Using a signal from the robot controller to convey changes in robot velocity, the gun orifice size is increased or decreased. Fluctuations in system pressure due to the pump are detected by a pressure transducer in the gun and fed back to the system control loop, resulting in an adjustment to the gun orifice size.

Variable-orifice guns can also be used in conjunction with flow feedback to compensate for material viscosity changes resulting from ambient-temperature changes or batch-to-batch material variations.

There are a variety of application techniques and material deposition methods that can be used in manual or automatic applications depending on the volume of material to be dispensed, the location, and the desired shape of the material deposition (Fig. 19). Each technique is described below.

Extrude. The material is extruded out of the nozzle in one of many bead cross sections, including round, semicircular, or triangular.

Flow Brush. The material is extruded into the center of a flow brush. This allows the material to be spread out or brushed into seams.

Droplet. A droplet of material is dispensed simultaneously out of multiple automatic guns that are mounted in a manifold assembly.

Spray techniques can be airless, air-assisted, or swirl based.

Airless. The material is dispensed at high pressure through a nozzle that is designed to disperse the material into a specific spray pattern.

Air-assist is similar to airless, except air is directed at the material stream as it exits the nozzle to contain the width of the spray pattern.

Swirl. Air impinging on the material stream as it exits the nozzle creates cohesive material fibers that are rotated into a swirl pattern.

System Controls

System controls range from the pneumatics on a single, unheated piston pump to sophisticated control consoles (Fig. 20) used in large, heated dispensing systems. In a simple system with a single bulk unloader, all the controls for the pump and

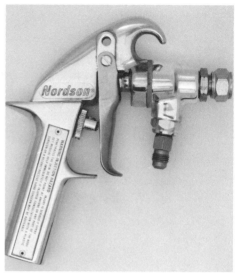

Fig. 15 Unheated manual spray gun

ram are mounted on the bulk unloader. Stand-alone bulk melters also have the electrical controls for heating the pump and plate, and a limited number of hoses and guns.

Larger systems in high-production applications require automatic changeover from one bulk unloader or bulk melter when one container is empty, so that the empty container can be replaced without interrupting the flow of material.

In an unheated system, the automatic changeover is performed pneumatically and the controls are mounted on the bulk unloaders or on a separate control panel. Heated systems have the automatic changeover function integrated into the system control console.

Although a heated dispensing system is used here for illustration, the sequence of operation is equally relevant for unheated dispensing systems. In a typical heated sealer application (see Fig. 21), bulk melter A is the on-line pump. The on-line bulk melter is the pump that is currently heating the material to application temperature and pumping the material to the changeover manifold at the required temperature, pressure, and flow rate. Bulk melter B is the standby pump, which is set at a standby or reduced temperature to minimize the amount of time the material is at application temperature while ensuring that minimal time is required to reach full application temperature when commanded from the system controller. When bulk melter A is almost empty, the system controller will heat up bulk melter B to full application temperature.

When bulk melter A is actually empty, the system controller will stop the pump on unit A, reduce the temperature of unit A to standby temperature, and start the pump on

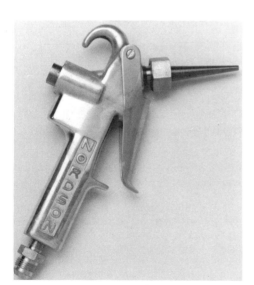

Fig. 16 Unheated manual extrusion gun

Fig. 17 Unheated automatic gun

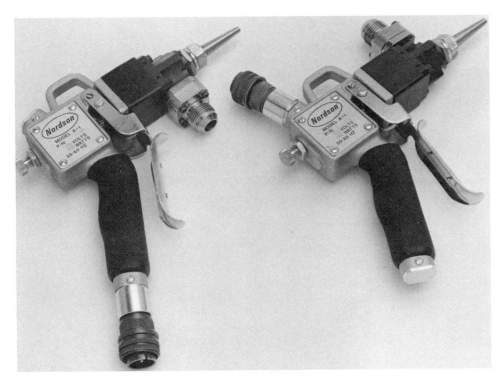

Fig. 18 Heated manual extrusion guns

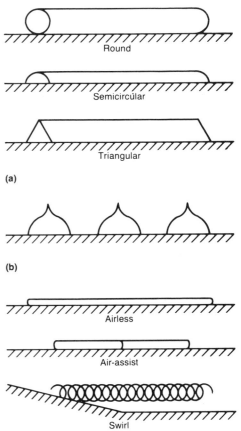

Fig. 19 Application techniques. (a) Extruded beads. (b) Droplets. (c) Spray

Fig. 20 Heated system control console

unit B. In this way a continuous supply of material is provided.

In a heated system the controller also monitors and controls the temperature of all system components, including pipes, hoses, and guns. The controller should have individual temperature controls for all system components for maximum flexibility.

The viscosity of most materials decreases as the material is heated. However, there is typically a temperature after which some materials will degrade (for example, hot-melt adhesives), or the viscosity will increase because of fusion of the material components (for example, plastisols).

Some materials are designed to react at specified temperatures. One-part reactive epoxies and weldable/expandable sealants are two examples of materials that are designed to react when exposed to temperatures between 150 and 175 °C (300 and 350 °F). Both materials are applied at elevated temperatures to reduce the viscosity, but it is critical that the material is not exposed to a temperature for a long enough period to cause a reaction.

For heated materials, especially ones that are thermally reactive, accurate temperature control is required with minimal steady-state error and temperature overshoot. The steady-state error in a temperature control system is the difference between the set point or desired temperature and the actual temperature after the system has reached equilibrium. Temperature overshoot is the maximum temperature that the system reaches during heat-up before the system settles to steady state.

Over-temperature protection to detect when a set point is inadvertently increased above a specified maximum temperature or a runaway temperature fault is critical to prevent accidental material reaction or degradation.

Other process variables, such as material pressures and flow rates, can also be monitored by the system controller. Control systems that incorporate statistical process control (SPC) are available that log all pertinent process data, including time and date, flow rate, number of jobs, and amount of material dispensed per job. This data can be transferred via diskette or RS232 port to a personal computer for viewing and statistical analysis. The article "Statistical Process Control" in this Volume provides numerous details.

System diagnostics pinpoint open or shortened heaters or temperature sensors in the system for easy troubleshooting and maintenance. More sophisticated controllers allow the user to define faults as major or minor so that appropriate corrective action can be taken.

Automation interface features allow the dispensing system controller to communicate with a cell or plant controller when the system is ready to dispense material or when there is fault that requires supervisory action by the robot or cell controller.

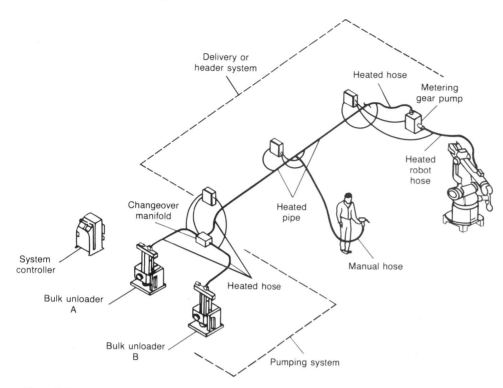

Fig. 21 Typical dispensing system configuration with robotic and manual application points

Fig. 22 Robotic spray application

Dispensing System Automation

Today's dispensing systems range from simple manual systems to complex robotic systems. However, the demand for lower manufacturing costs, improved product quality, and environmental issues have created a trend toward robotic application of adhesives and sealants.

There are three general levels of dispensing system automation:

• Manual
  Hand tools (no automation)
  Manually operated automation
• Fixed/hard automation
• Robotic

Manually operated dispensing equipment requires the lowest initial investment. The systems are simple, usually consisting of the pumping system, a hose, and a dispensing gun, although manual systems can be large if there are multiple application points.

Manual systems are best suited to applications that do not require complex orientation of the dispensing gun relative to the substrate. Even though the human hand is a very sophisticated tool, it is very difficult for the operator to perform complex dispensing paths repeatedly for extended periods.

Manually operated automation is used to address the accuracy and fatigue issues of more complex manual dispensing applications. Fixtures are used to guide the operator along the dispensing path in such a way that the material is accurately deposited on

every part. This type of system has its place, although the application is still manual. As such, the complexity of the dispensing path is limited, and complex fixtures (three-dimensional, for example) will become less accurate as wear occurs, resulting in quality and maintenance problems.

Fixed or hard automation, as defined here, is equipment designed and built to perform a specific dispensing operation. No manual operation is required, with the possible exceptions of inserting the part into the fixture and/or starting the operation, although these tasks are typically performed automatically in high-production applications. Another definition of fixed or hard automation is an automated system that does not use programmable, multiaxis robots.

The major limitation of fixed or hard automation is the lack of flexibility because the system will perform only the dispensing operation for which it was designed. The complexity of the dispensing path has limitations that are defined by the availability of standard automation components. Dedicated equipment can be designed to perform virtually any task; even so, the cost quickly reaches a point where a multiaxis robot makes the most economic sense. Fixed or hard automation does eliminate most operator error and can be a cost-effective way to perform relatively complex, well-defined applications.

Robotic dispensing systems are the most flexible for dispensing applications. Complex dispensing applications that are not possible with other levels of automation can

be performed with multiaxis robotics (Fig. 22).

The robot can also be reprogrammed if the geometry of the part changes or if the process requirements change. An example of a common process change would be the requirement for an additional bead of adhesive to be deposited on the component. Fixed automation would require mechanical modifications to the automation or complete redesign. A robotic system can usually accommodate a process change with reprogramming, provided the change does not exceed the physical limits of the robot. A change in part geometry can also be handled with reprogramming. However, changes in part geometry can necessitate fixture changes regardless of the type of automation.

The Future of Dispensing Systems

Trends in manufacturing methods and productivity goals in the plant dictate changes in equipment capabilities. Factors such as production volume requirements and increased emphasis on process control will require increasingly sophisticated equipment with capabilities for process surveillance and self-diagnostics.

In order to reduce costs, productivity goals have translated into faster cycle times. To meet faster cycle times, the entire dispensing system must be capable of higher material output rates.

Dispensing equipment design and development is directed toward meeting ever-changing production needs that result from competitive pressures and advances in material technology. Future development efforts will be directed toward the design of lower-maintenance equipment capable of handling new and unique materials, as well as continued emphasis on addressing the production requirements of increased material output, consistent material placement, and improved process control.

Adhesive Bonding Preparation, Application, and Tooling

Hans J. Borstell and Valerie Wheeler, Grumman Aircraft Systems

ADHESIVE BONDING is an ancient technology that typically is used to make three types of joints:

- *Insert bond*: One adherend is bonded to another in either a cavity or a very small local area. Many fasteners are bonded in place to prevent loosening after installation
- *Doubler bond*: A second layer of structural material is bonded locally onto a larger sheet in an area of high stress, such as the edge of an automobile hood
- *Panel bond*: Two face sheets and a core are joined to produce a light stiff panel, such as an aircraft cabin floor, or an attractive panel from a less-attractive core, such as a kitchen countertop

These adhesive joints are chosen because they can be used to distribute applied loads over larger surface areas than spot welds and mechanical fasteners. This load distribution enables adhesive joints to avoid the penalties of stress concentrations introduced by these other types of joints; stress concentrations can lead to structural failure or heavy designs.

The basic steps in the adhesive bonding process are:

- Collection of all the parts in the bonded assembly, which are then stored as kits
- Verification of the fit to glue line tolerances
- Cleaning of the parts to promote good adhesion
- Application of the adhesive
- Mating of the parts and adhesive to form the assembly
- Application of a force to produce a good fit
- Application of force (concurrent with application of heat to the adhesive to promote a chemical reaction, if needed)
- Inspection of the bonded assembly

The versatility of the process is evidenced by the broad spectrum of assemblies that can be produced: The same steps used to produce a kitchen countertop covered with decorative laminate can be used to produce a highly structural flight-safety-critical floor panel for a jetliner.

Bonding Preparations

Bonding processes are so versatile that the selection of the proper process and adhesive can be a complicated task. This section addresses the decisions and preparations that must be made during the prebonding period.

Selection. The key questions that must be answered before process and adhesive selection are:

- What are the operating conditions of the assembly?
- What are the pertinent economic factors such as perceived value and competitive considerations?
- How good does the assembly really have to be?

The following production requirements must be met to ensure adequate bond performance:

- A suitable adhesive must be selected, and the design must allow for a bond area that is large enough to ensure that the bond will have sufficient strength to withstand the applied stress during service
- Separations, entrapped volatiles, and contamination in the bond line must be kept to a minimum by the selection of suitable bonding process parameters and equipment
- Suitable tooling and alignment techniques must be selected so that the components of the assembly are bonded in the correct configuration

In general, bond performance objectives are met by using methods and equipment that are appropriate for the particular step of the bond preparation process. For example, the substrate and decorative laminate for a countertop probably are prepared for bonding by blowing off dust and loose dirt. However, structural aircraft-grade bonded titanium panels require a different preparation process. Because titanium forms a tenacious oxide to which most organic substances do not bond, a timed sequence of grit blasting and chemical dips is required. The effectiveness of the cleaning usually

must be verified by the fabrication and testing of process control coupons. Obviously, the higher value of the titanium-skinned panel as compared with that of the countertop justifies the higher cost of the preparation process for the aircraft part.

Technical data and advice provided by the adhesive supplier can be useful when making process and adhesive decisions. A review of pertinent industry or government specifications and standards is also useful. Instituting a quality control network that frequently monitors the various aspects of the operation can also help ensure success. Such a network can track and confirm that the adhesive and process are within specifications.

Prekitting of Adherends. Many adhesives have a limited working life at room temperature, and adherends, especially metals, are contaminated by exposure to the environment. Thus, it is normal practice to kit the adherends so that application of the adhesive and buildup of the bonded assemblies can proceed without interruption. The kitting sequence is determined by the product and production rate. The following are examples of kitting practices for particular products:

- *Hollow core doors*: The frames are stacked, and the face sheets are stored on a dolly
- *Low-volume aircraft panels*: Kit parts are kept in a sealed carton or plastic bag
- *Electronic insert potting*: Circuit boards are placed on a conveyor belt while components are loaded into the hopper of an insertion device

Bond Line Thickness Control. Controlling the thickness of the adhesive glue line is a critical factor in bond strength. This control can be obtained by matching the quantity of available adhesive to the size of the gap between the mating surfaces under bonding conditions. Higher applied loads during bonding tend to reduce glue line thickness. A slight overfill is usually desired to ensure that the gap is totally filled.

The precise glue line thickness is determined by the bond strength required per unit area and the cohesive strength and strain capability of the adhesive. The glue line thickness can influence the failure behavior of the bond. If cohesive failure (that is, failure by separation within the adhesive mass) occurs, then the glue line thickness must be reduced to promote adhesive failure (that is, failure by disbonding from the adherend). Conversely, if all the adhesive is squeezed out of a local area due to a high spot in one of the adherends, a disbond will result.

Fortunately, most bonds do not require optimum strength and can tolerate some local disbonds. For these bonds, an attempt should be made to produce faying parts to a tolerance of 0.8 mm (0.03 in.), and enough adhesive should be applied to fill the gap. The use of excess adhesive adds weight and unnecessary material cost; it also increases cleanup costs. However, these negative factors must be tolerated in order to avoid disbonds during service.

Some of the most critical structural bond applications are found in highly loaded aircraft structures. For applications of this kind, the adhesive used is in the form of a calendered film with a thin fabric layer. The fabric maintains the bond line thickness by preventing contact between the adherends. Voids are not permitted. In the most common case, the bond line can vary from 50 to 230 μm (2 to 9 mils). Extra adhesive can be used to handle up to 510 μm (20 mil) gaps. Larger gaps must be accommodated by reworking the metal parts; tolerances are achieved by reworking of the various details so that they fit as required.

Prefit Evaluation. A prefit-checking fixture is often used in high-precision bonding. This fixture simulates the bond by locating the various parts in the exact relationship to one another as they will appear in the actual bonded assembly (see Fig. 1). For high-value assemblies, the glue line thickness is simulated by placing a vinyl plastic film or the actual adhesive encased in plastic film in the bond lines. The assembly is then subjected to the heat and pressure normally used to chemically react the epoxy adhesive to form the bond. The parts are disassembled, and the vinyl film or cured adhesive is then visually or dimensionally evaluated to see what corrections are required. These corrections include sanding the parts to provide more clearance, reforming metal parts to close the gaps, or applying additional adhesive (within the permissible limits) to particular locations in the bond.

Verification of bond line thickness may not be required for all applications. However, the technique can be used to validate the fit of the mating parts prior to the start of production or to determine why large voids are produced in repetitive parts. When using paste adhesives, the adhesive thickness can be simulated by encasing the adhesive in plastic film (as mentioned above) or alu-

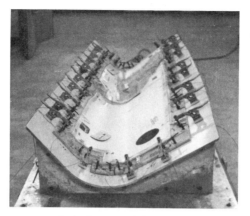

Fig. 1 Prefit fixture used for a complex-contour aircraft door

minum foil. Once the fit of mating parts has been evaluated, the necessary corrections can be made. For cases in which the component parts can be dimensionally corrected, it is much more efficient to make the correction than risk having to scrap the parts or having them fail in service. For example, after the frames for hollow core doors are assembled, they often are planed for smoothness on both faces. The face sheets can be bonded to the planed frames with reasonable confidence that void-free bonds will be produced. The use of glass fibers or fabric to prevent excess resin from being squeezed from the joint also increases the reliability of the bonded joint.

Surface Treatment Prior to Adhesive Application. A major cause of bond failure is contamination of the surface of the adherend. The contaminant can be dust, grease, oil, oxides (on metallic surfaces), or particles of the adherend. These contaminants prevent the adhesive from wetting the surface of the adherend to produce a good bond. Once again, the bond strength requirements determine how much contamination can be tolerated. In the case of dry, previously sanded wood products, removal of the dust by an airstream or brush may be adequate.

Researchers have long sought to develop adhesives that yield acceptable but not necessarily optimum bond strengths on slightly soiled metal substrates. Some new adhesives have been developed that appear to meet this requirement. These adhesives have sparked interest in the automotive industry as possible replacements for, or supplements to, fasteners or spot welds for bonding metal or plastic panels.

In the aircraft industry, optimum bond strengths are desired to reduce the size and weight of joints. Structural integrity concerns and the need to avoid in-service failures mandate chemical preparation of adherend surfaces. Typical procedures for several different adherends are given in Table 1. Aluminum parts are placed in baskets or racks and cleaned by immersion in

Table 1 Surface preparation techniques for various adherends

Adherend	Process
Wood	Sand to remove contamination, blow off dust
Paper	Eliminate nonbonding coatings during paper manufacture
Steel	Grit blast and vapor degrease
Aluminum	Immerse in nonetch alkali detergent, rinse; immerse in sodium dichromate/sulfuric acid solution, rinse, dry; apply primer to protect surface
Titanium	Wipe with detergent to remove soil, dry hone, rinse; immerse in proprietary oxidizing solution, rinse, dry; apply primer to protect the surface
Composites	Remove peel ply to expose a chemically clean surface or mechanically abrade and blow off dust

large tanks. Similar processes can be used in other industries; however, the overall tendency is to keep pretreatment at a minimum to hold down processing costs. For example, fiberglass panels are often bonded in the as-molded condition in commercial work, whereas the bonding surfaces of composite aircraft parts are always prepared by mechanical abrasion or by the removal of a strippable ply that had been added to the surface to prevent contamination.

Adhesive Application

The application of adhesives and the accuracy of the application have a major impact on cost and quality. Enough adhesive must be applied to form an acceptable joint, but any excess represents wasted material. The labor and equipment costs should also be kept to a minimum. The method of application is a function of the physical form of the adhesive. The choice of method is also influenced by the volume and sophistication of the work.

The most commonly used adhesives are supplied as liquids, pastes, or prefabricated films. The liquid and paste systems may be supplied as one-part or two-part systems. The two-part systems must be mixed before use and thus require scales and a mixer. A paint shaker to agitate low-viscosity adhesives is also desirable.

The storage facility plays a vital role in a production bonding operation. Many one-part adhesives and film adhesives must be stored at temperatures below ambient. Adhesives containing organic solvents must be segregated to reduce the risk of fire, and others containing polymers and solvents must be shaken periodically to prevent settling or gelation.

One factor that must be considered in adhesive application is the time interval available between adhesive preparation and final assembly of the adherend. This factor,

Fig. 2 Robot used for spraying adhesive

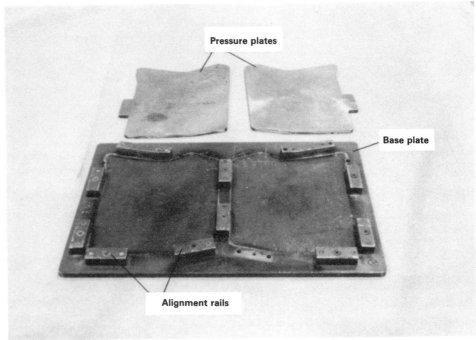

Fig. 3 Press bonding fixture

which is referred to as open time or working life, must be matched to the production rate. Obviously, materials that are ready to bond quickly are needed for high-rate applications such as those found in the automotive and appliance industries.

The following are factors that influence the open time of particular adhesive systems:

- To prevent solvent entrapment in the joint, solvated adhesives often must dry to a tack-free condition before the adherends are mated
- Two-part systems that cure by chemical reaction often have a limited working life before they become too viscous to apply
- Water-based adhesives often take a very long time to lose enough moisture to prevent squeeze out under moderate pressure

Application of liquid adhesives can be accomplished using brushes, rollers, manual sprays, or robotically controlled sprays. The robot (Fig. 2) can apply tightly controlled quantities of adhesive to specific areas. Solvated two-part systems are sprayed using equipment with two pumps; preset quantities of each component are pumped through the spray head where they are blended into a single stream. Of course, many plants use several different application systems simultaneously for their various job shop requirements.

Application of paste adhesives can be accomplished by brush, by spreading with a grooved tool, or by extrusion from cartridges or sealed containers using compressed air. For the latter, the combination of the orifice diameter and the applied pressure controls the size of the bead applied to the work. Robots can move the application head in a constant path at a repetitive surface speed to enhance the accuracy of bead placement and size. The use of robots to apply paste adhesive is analogous to their use to locate spot welds. In the automotive industry, several vans with plastic skins bonded to a steel structural frame are as-

sembled using paste adhesives applied by robots.

Applications of Film Adhesives. Film adhesives are costly and thus are used mainly in aircraft applications. They consist of an epoxy, bismaleimide, or polyimide resin film and a fabric carrier. The fabric guarantees a glue line because it prevents contact of the adherends. These adhesives are manually cut to size, usually with knives, and placed in the glue lines.

New Application Techniques. A developing technology involves the use of dispensing heads to precisely place drops of anaerobic adhesives in the desired locations. Typical applications include prepositioning of gaskets during the assembly of small engines, preattachment of lug nuts on wheels during automobile assembly, and bonding of components to circuit boards.

Another developing technology is the use of prefabricated shapes of epoxy adhesive made from resins that react and move only at elevated temperatures. These doughnut-shaped preforms can be placed on a circuit board by an automated system before the component to be bonded is installed.

Tooling

Adherends must be in a specified relationship to one another during bond formation. Slippage of one of the constituents of the bonded assembly will result in a need for costly reworking, or the entire assembly might be scrapped. When a paste or liquid adhesive is used, it is usually helpful to have a load applied to the joint to deform the adhesive to fill the glue line. In applications

such as aircraft parts, higher pressures are used to force the adherends to fit to within small tolerances. The adhesives used in high-pressure bonding typically do not have a high flow; they contain a thin scrim fabric to ensure that a bond line is maintained. Fixtures can be used to maintain part alignment, or fixtureless concepts can be used in which some other method provides the required alignment.

Fixtures. Bonding fixtures represent an investment in tooling and have a significant impact on cost. Every time they are used, they must be loaded with parts, unloaded, and maintained. Fixtures can be used for both high- and low-rate production. A bonding fixture contains several basic elements:

- A tooling surface matching the contour of the bonded panel (face sheet)
- A series of jig points, side rails, and pressure blocks as needed
- A support structure to prevent warpage of the face sheet for contoured tools
- A dolly or removable wheels to permit moving the fixture if needed

Press Bonding Fixtures. Many flat panels are bonded in press bonding fixtures. The face sheet of the fixture is flat, and the alignment rails are usually thinner than the thickness of the structural panel. For very thin panels, the rails may be higher than the panel to ensure that alignment is maintained. In this type of fixture, a pressure plate is used to take up the space between the panel and the top of the rails. The pressure plate fits within the alignment rails (see Fig. 3). Often, press bonding tools are made to standard thicknesses so that sever-

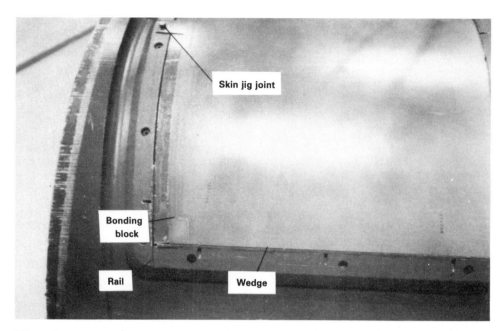

Fig. 4 Detail of a pressure bag bonding fixture

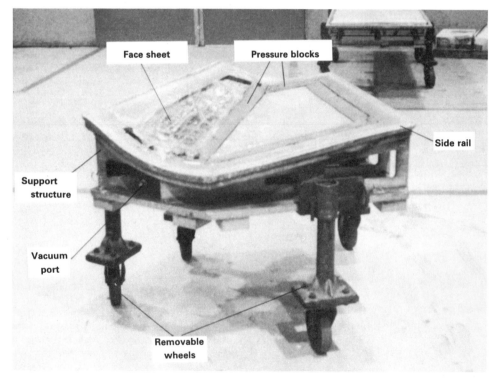

Fig. 5 Autoclave bonding fixture

al tools can be placed in the press cavity at the same time.

Pressure Bag Fixtures. Parts with significant contour must be bonded using a pressure bag process, because the thickness of the tooling needed for the parallel flat surfaces of press bonding fixtures is excessive for these parts. Also, the extra thickness causes problems with thermal uniformity, tool cost, and handling.

In the pressure bag process, a plastic-film bag is attached to the face sheet of the tool and sealed in place. The air under the bag is evacuated by a vacuum part to produce a pressure of 95 kPa (14 psi). The heat required to chemically react the adhesive is provided by an oven. If higher pressure is needed, the tool is placed in a heated pressure vessel called an autoclave. The additional pressure required to obtain a good

bond is obtained by pressurized air or an inert gas. As the pressure is applied, the vacuum under the bag is vented to prevent damage to the honeycomb assembly from entrapped vacuum.

Additional vacuum ports must be provided to evacuate the vacuum bag. Typically, two holes are drilled in the face sheet and tapped with pipe thread. A pipe sealed with Teflon tape is mounted in each hole. Quick-disconnect fittings and hoses are used to connect the tool to the vacuum systems in the oven or autoclave. One port is used to apply vacuum and the second is connected to a recorder that monitors the pressure under the bag.

The rails are pinned to the face sheet so that they stay in place despite the side loads produced by the applied pressure. If the rails were to move, damage to the panel would result.

Panels with straight edges can be bonded using only the edge rails. However, many panels have one skin that projects beyond the core and second skin. Usually, the central core is enclosed by Z-shape members that are bonded to both the larger external skin and the smaller internal skin. The bonding fixture for these panels is configured to place the larger skin next to the face sheet of the tool. The rails surround the larger skin. The space between the rail and the vertical edge of the Z-shape member is filled with removable pressure blocks. Wedges are placed between the side rails and pressure blocks to force the blocks inward to ensure good contact with the Z-shape member. Blocks are also required when a space is needed between the skins for attachment of a fitting (see Fig. 4).

Pressure bag fixtures (Fig. 5) are more complex than press bonding fixtures. These fixtures are fitted with a base to support the tooling surface in the proper contour. The base must contain provisions for moving the tools, such as removable wheels or hoist points for lifting the tool on the dolly or platens for positioning the tool in the autoclave (Fig. 6) or an oven. Flat tables are often used because they can accept tools of

Fig. 6 Flat and contoured bonding fixtures mounted on a platen

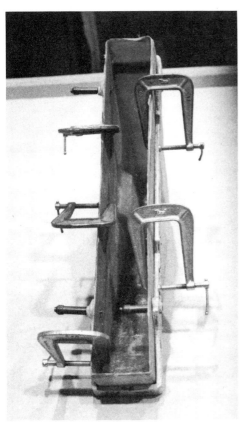

Fig. 7 Fiberglass assembly bonded using clamps and temporary fasteners

Fig. 8 Honeycomb splicing fixture using screw jacks to apply load

Fig. 9 Flat-panel bonding press. Courtesy of M.C. Gill Corporation

any configuration, provided that there are several flat areas in the tool base. To prevent warpage, the base must be configured to maximize airflow under the face sheet.

Fixture Design. Several secondary rules apply to the design of a successful bonding fixture:

- The tooling surface must be precisely machined to the desired contour so that the skin fits well into the cavity produced by the face sheet and the rails
- The face sheet must be free of nicks and gouges that could damage the skin
- The side rails must be easily detachable from the face sheet; this can be accomplished by using pins or screws to attach the rails
- The bonding fixture should be coated with Teflon to prevent adhesion before the tool is used for the first time
- Mold release or fluorocarbon film must be applied before each use to prevent sticking of the blocks to the adhesive squeezeout

Including tapped, threaded holes in the bonding blocks to permit the use of a T wrench will simplify block removal. Also, jig points, locating blocks, or precisely located alignment pins must be provided for each major solid detail located within the honeycomb. These devices ensure that the parts are always in the same location so that

the prefit operation performed before the panel is disassembled for cleaning will be valid for the panel when it is reassembled with adhesive after cleaning.

The coefficient of thermal expansion of the face sheet of the fixture and that of the panel skins should match so that the relative position of the tooling and skins is maintained during elevated-temperature curing of the adhesive. This is especially important for a sandwich structure. Both bonds of such a structure should be formed by chemical reaction within a narrow temperature range. As soon as one skin is bonded to the core, the relative positions of these bond constituents are locked in. As the assembly is heated further, the skins and core continue to expand thermally until the second skin is bonded. Because the second skin has expanded more than the first skin prior to bond formation, the panel has thermally induced warpage when used at ambient. The solution to this problem is to use a fixture of minimum mass and provide a uniform flow of heated air above and below the tool surface. Typically, aluminum panels are bonded in aluminum fixtures and composite panels are bonded in composite or steel fixtures. However, steel fixtures can be used to replace composite fixtures because the coefficients of thermal expansion of the two materials are closely matched; electroformed nickel tools are the next best choice.

Because the fixtures for large panels are heavy, handling equipment is required in a production bonding area. Cranes or forklifts are needed, and the tools must be fitted with the proper hoist points. Because the tools must be moved from the clean room (where the adhesive is applied to the individual parts) to the curing equipment, dollies or removable wheels must be provided. Some fabricators use permanently attached steel wheels to move the tools. If large platens are used, steel tracks in the floor and steel wheels on the platen are needed to move the plate in and out of the curing equipment.

Fixtureless Bonding. Alternatives to the use of fixtures are desirable for both high- and low-rate production. One method involves using the fully assembled part as the fixture, as in the case of some metallic automotive panels. Once the adhesive is applied to the edge of the internal skin and the external skin is folded over, the skins are locked into position and cannot move relative to each other. Another example is the various forms of insert bonding used to assemble electronic connectors and circuit boards. The bonding of O-rings and gaskets and the attachment of labels also use the workpiece as a bonding fixture.

Another fixtureless method for maintaining alignment of the adherends during bonding involves drilling precision alignment holes in the workpiece, applying the adhesive, and installing temporary fasteners. The fasteners are coated with mold release to prevent them from bonding to the assembly. The fasteners also apply force to the joint to produce adhesive flow. A similar concept is used to mate the decks and hulls of small fiberglass boats. After bonding, the alignment holes are often enlarged to receive structural fasteners.

Pressure Applicators. A common approach for applying bonding pressure involves using clamping devices, such as C-clamps or woodworker's clamps, to hold the adherends in place. A fiberglass panel bonded with C-clamps is shown in Fig. 7.

Flat panels covered with decorative laminate are often bonded in U-shape clamps, which consist of a plate fitted with threaded rod stock at each end. Another plate (fitted with holes to accept the rods) and nuts are used to provide the mechanical load. The workpiece is located between the rods, which provide the alignment needed to locate the decorative covering relative to the edge of the panel core.

Spring-loaded clamps that are similar to old-fashioned laundry clamps are also used for certain bonding applications because they can be installed and removed rapidly. One patented type of clamp has one movable surface and one fixed surface; the adjustable clamps are permanently attached to a base, and the movable jaws open and close as required. In high-rate production, pneumatically actuated clamps are used to accelerate the loading of parts or the removal of the assembly from the fixture.

Mechanical screw jacks can also be used to apply load and align adherends. For example, blocks of honeycomb material are often spliced in the vertical direction with a foaming adhesive that expands to fill the space between mating blocks. The operation is done in a fixture that defines the horizontal surface of the assembly. Pressure plates, strong backs, and screw jacks are used to push the core blocks toward the horizontal surface as the adhesive is cured in an oven (see Fig. 8).

Curing Equipment. Various types of curing equipment are used for adhesives requiring heat for polymerization. Conventional ovens and full or partial vacuum are used for lightweight assemblies with thin skins. A variation of this concept is used in the automotive industry: Heated tunnels concurrently cure adhesives and exterior paint in automobile bodies. Multicavity platen presses (Fig. 9) are used to bond flat panels under heat and pressure. In the aircraft industry, most panels have a complex geometry and are bonded by the pressure bag technique. The bond may occur in an oven or an autoclave.

Adhesives and Sealants Curing and Cure Controls

Todd Taricco and Mark Moulding, Thermal Equipment Corporation

THE PROCESSING of structural adhesives by thermal curing and related techniques is the focus of this article. The information presented applies also to those sealants that require cure of this type. Autoclave process control requirements and their computerization are also described in detail.

The basic requirements for the curing of thermoset and thermoplastic adhesives are, first, to provide the proper thermal cycle and the proper positioning and, second, to provide a clamping force for uniform compaction of the resin system (usually accomplished using a vacuum bag and predefined tool surface). There are actually many devices and methods for accomplishing these requirements; those that are of major importance are described below.

Ovens

Convection Oven/Autoclave. The curing of adhesives in an oven/autoclave is probably the most widely used and economical method of processing adhesives. This method uses forced gas convection (typically, air) to heat the component being processed to a preset temperature profile. The atmosphere in the oven/autoclave is circulated by means of a fan-driven system of either a single or multiple configuration; hence, the term convection is appropriate. The autoclave is primarily a convection oven that operates at elevated internal pressures (Fig. 1).

Fig. 1 Autoclave system

It is important to select the proper air velocity in the oven/autoclave work space to provide adequate heat transfer to the part load (tools and adhesive) and to allow uniform heat distribution to the load configuration. With regard to the direction of air flow, optimal air flow is usually parallel to the largest surface area of the part. This provides uniform heat transfer through the tool surface to the part. A uniform part heat-up rate provides maximum cure control efficiency.

The air velocity is primarily a compromise between the attempt to optimize heat transfer (higher velocity) and the prevention of damage to the part-holding fixtures (tools) or vacuum bags (used to apply clamping force). A good solution to the velocity problem is a variable-speed fan-driven system. This will accommodate ovens used to process a wide variety of parts over a wide temperature range, as well as autoclaves that use wide ranges of pressure.

There are many heating methods used in ovens. The method selected depends on the parts being processed, the size of the equipment, utility availability, and cost constraints. Electric heat is common for small ovens and is quite controllable, although it has some operating cost disadvantages. When using direct-fuel firing, the products of combustion enter the work space. These combustion products contain CO_2, CO, NO_x, and water vapor, which can be harmful to components or bagging films used in the oven. Indirect gas and/or fuel firing is the best of both worlds, because it allows low costs per unit and no combustion by-products are emitted in the work space. The major drawback of this heating method is its high capital equipment cost due to the materials used in the heat exchanger system. Steam-heated ovens are very efficient if steam is available and processing temperatures are relatively low. In all cases, controllable proportional cooling should be provided to the oven or autoclave to allow a controlled cool-down rate, which helps prevent induced stresses in the component as a result of a rapid decrease in part temperature.

Radiant ovens are normally used in semicontinuous or continuous manufacturing processes and on components with uniform cross sections and emissivities (for example, the continuous forming of adhesively bonded films).

A derivation of the radiant oven is the vacuum-assisted radiant oven. With this oven, the entire work space is evacuated, which serves to remove volatiles and water from the adhesive prior to curing. A drawback of this method is that it requires mechanical clamping because vacuum bagging is not effective as a result of the lack of any pressure differential. This problem can be avoided by placing the adhesive under the vacuum bag with a slight differential relative to the oven (that is, 98 kPa, or 29 in. Hg, in the bag and 75 kPa, or 22 in. Hg, in the oven), thereby preventing the trapping of volatiles used for fragile components (for example, the bonding of solar cells). Radiant heat transfer features can be included in convection ovens and in autoclaves for supplemental heat.

Microwave ovens and microwave-assisted ovens are used in the processing of several large composite structures, but are rarely used in adhesive curing, with the exception of some continuous processes. The problems of microwave design are numerous, and extreme caution must be used because of the spurious radiation and sparking produced with metal parts. Several manufacturers have had success with microwave-enhanced autoclaves and convection ovens.

Autoclave Processing System

In an autoclave, the pressure ranges appropriate for adhesives bonding typically vary between 105 and 1050 kPa (15 and 150 psi). The amount of pressure used is a function of the ability of the part structure and tool to withstand the pressure; the type of adhesive system being used; part geometry; and the required amount of ply compaction.

The component to be placed in the autoclave is first vacuum bagged. The vacuum bag provides a membrane across which the pressure differential is generated, providing the clamping force on the component. When this clamping force is applied to the adhesive, the properties of the part are improved by reducing porosity, voids, ply delamination, and so forth. Autoclaves are the primary tool used in the aerospace structures industry because they provide all of the functions needed for an optimal cure of the typical adhesives and resin systems used.

Co-curing is another processing function conducted in an autoclave, in which fiber-reinforced preimpregnated materials (prepregs) are cured simultaneously with adhesive films and core materials. Co-curing eliminates the multiple autoclave steps of skin manufacture and subsequent assembly bonding. However, it is necessary to maintain a uniform or adequate hydrostatic pressure on the prepreg material. Improper pressure can result in porosity, improper resin content, and improper consolidation. Many of these problems can be addressed by the correct selection of prepregs and cores for co-curing. The key factor in this selection is the assumption that the adhesives and prepregs are fluid and that uniform pressure will be maintained at the prepreg-to-adhesive core interface. This is why solid-type cores (that is, foams) perform better than does large open-cell honeycomb and why net resin low-pressure prepregs outperform high-bleed, high-pressure systems.

Another innovative method of co-curing has been demonstrated in which the selected adhesive and core are able to withstand 175 °C (350 °F) and 700 kPa (100 psi) conditions. The adhesive is selected to cure at 120 °C (250 °F) while maintaining strength at 175 °C (350 °F). The prepreg is selected to stage at 120 °C (250 °F) at low pressure (that is, 240 kPa, or 35 psi) and has a 175 °C (350 °F), 690 kPa (100 psi) final cure. The cure begins by recouping to 120 °C (250 °F) at 240 kPa (35 psi), thereby curing the core to the adhesive. Then, the pressure is raised to 690 kPa (100 psi), and the now-rigid adhesive distributes the load between the core cells, providing a pseudo-uniform 690 kPa (100 psi) pressure on the prepreg. The cure continues to 175 °C (350 °F), where the prepreg is cured. This process may be feasible for many parts with process variations and modifications.

A thermoplastic material that simulates a film adhesive can be used to simulate the assembly of a structure and to check the fit of the skin and core bond lines by examining the impression left in the thermoplastic film. This process is one method of checking ply and detail part compaction within the structure. When using this product, care should be taken to prevent core damage, due to excessive or uneven vacuum and/or pressure.

Bagless Autoclave Processing. Many adhesives, both thermoset and thermoplastic, respond well to autoclave processing without vacuum bagging. This process is used on fairly high-viscosity materials, such as those used on thin film-to-metal applications. These materials can be cured on rolls in bulk form. The process functions by applying autoclave pressure directly to the adhesive which raises its pressure without infiltrating the adhesive. This increase in pressure reduces the void size and inhibits the generation of water and volatile-produced voids by raising their boiling points to a point where the adhesive viscosity allows the fluids to be held in solution.

Mechanical Restraint Methods. As previously discussed, the vacuum bag is an effective method for providing clamping force on a component up to approximately 98 kPa (29 in. Hg), depending on vacuum supply and altitude in conventional oven cures and autoclave cures. This vacuum can be harmful to many adhesives, because just as increasing the hydrostatic pressure raises the boiling point of materials, lowering the pressure decreases the boiling points and can cause severe porosity in adhesives. This is why, when an autoclave is used, many adhesives require that the vacuum level be returned to atmospheric pressure or higher after internal pressurization of the autoclave has reached an acceptable level for that resin system.

Press curing is another method of providing mechanical clamping force to a component. Presses are very cost effective when dealing with high production rates and relatively fast-curing adhesives. When evaluating a process or component, the uniform application of force across the bonding area and normal to the bonding surface provides optimal results (part strength). Structures and components with flat or shallow contours and constant thicknesses are easily pressed. As part geometry becomes more complex and core materials are added, the problems to be addressed by the tool designer increase dramatically. Internal pressurized bags, mechanical pressure mechanisms, and the regulation of pressure rates are examples of the issues.

Press Tooling

Matched metal tooling is a viable option for assembling a structure when production quantities can be used to amortize the cost sufficiently, as in the case of helicopter blades. The tool designer has several design obstacles to address. First and foremost is the necessity to maintain proper clamping force on all the composite components with the tool closed, at temperature, thereby overcoming the problems of uneven thermal expansion of the materials and tooling.

One way to circumvent the problem of maintaining uniform pressure in matched

Fig. 2 Clam-shell autoclave with flexible diaphragm

metal tooling is by the use of intensifiers (either pneumatically inflated ones or ductile rubbers) to distribute the force or create pressure due to thermal expansion. The expansion of certain silicone rubbers is fairly predictable and is capable of exerting large forces. These pressure intensification methods are required in many parts because of problems that arise when a part is being assembled and adhesives being used flow slightly, reducing their thickness. Therefore, if the tool is closed cold, the part could be damaged or there may not be adequate clamping force at that temperature.

Also, when curing a part, it may be necessary to heat the part with very low press force until the adhesives soften adequately to allow the tool to be closed without part damage. It may also be necessary for the platens of the press to float out of parallel, with one end of the tool closing before the other, because adhesives in the part heat at different rates with various viscosities.

Single-Sided Tooling. There is a family of pressed and horizontally split autoclaves that are fitted with soft rubber pads or inflatable diaphragms to provide uniform pressure on a component while it is being heated. This feature can also be integrated into a press tool by means of an upper cavity and a flexible diaphragm (Fig. 2).

In this process, the tooling is single sided and is much like an autoclave tool with an integral base. These presses perform well on parts that have shallow contours or that can withstand some lateral force from the elongated bladder or from the rubber pad.

Roll Consolidations. Many fast-curing adhesives or thermoplastics can be processed by heated roll consolidation. This process brings the materials and adhesives in contact under a single- or multiple-roll arrangement. This method is common in the manufacture of many film and reinforced materials. Applications are also seen in continuous forming and bonding, that is, a metal-composite-metal continuous section to be used as a stiffener or structural member.

The Thermex process is a relatively new patented process that is applicable for curing under high pressures (2 to 70 MPa, or 0.3 to 10 ksi) and at temperatures ranging from ambient to 815 °C (1500 °F.) The process uses a powdered silicone rubber that is pressurized while integrally heating the tooling. This process can be accomplished in a Thermex vessel or in dedicated tool configurations.

Repair Using Adhesives. The issue of structure repair, both in the field and in the shop, is receiving more and more attention as the number of composite parts, especially in the aircraft industry, is increasing.

Metal bonded repair can best be accomplished by remanufacturing the part or placing the damaged skin in the same autoclave used originally to manufacture the part. This also applies to composite parts where large areas are being replaced. For smaller repairs, portable flexible heating pads have been developed to allow a localized repair. Typically, the damaged area is machined with a hand-held router. The damaged core is replaced with a new section of core. A film adhesive is placed between the old and new core, and a surface patch is prepared. These surface patches should be "B-staged," or "preautoclaved," or pressed because the portable equipment can typically provide only vacuum pressure. The patch and a layer of film adhesive are then placed over the prepared core, and the flexible heater and vacuum force are applied. Various manufacturers can provide detailed procedures for this type of repair upon request.

There are several areas where caution is advised when attempting a repair. First and foremost, an engineering analysis must be made on the part to evaluate structural loads, repair strength, and the testing/acceptance criteria. Also, because a component may have moisture or water in it, heating of any kind may damage other areas of the part. Vacuum desiccation is an effective means of removing water when the part is heated slightly in a vacuum environment. This can be accomplished in a vacuum bag having breathers or in a vacuum oven. Many attachments or details may have been added to the part after the original cure and, when heated, may cause differential thermal expansion and subsequent local damage around fasteners. The part may have been exposed to oils or fluids that may compromise the bonded repair. These are some of the issues that must be addressed before attempting any repair. The section below describes the computer controls and methodology applicable to the thermal processing of adhesives, with an emphasis on adhesives used in metal-bonded and composite structures.

Computer-Controlled Cure Systems

As the use of composite materials increases, greater demands are being placed on the production process for increased part quality and production efficiency. The computer is being successfully applied to satisfy this need.

Evolution of Control Methods. Originally, autoclave cures of composites required the abilities of a skilled operator. Because these vessels often had no automatic means of controlling temperature, pressure, or vacuum to a set point, an operator performed this function manually. Simple thermostatic control was soon added to the temperature process, although frequently a manual vent and pressure valve continued to be used for pressure control. Occasionally, a cam-type programmer was used to adjust the temperature set point automatically through a rudimentary cure cycle.

As solid-state digital set point controllers became cheaper and more common on the factory floor, autoclave control systems became more sophisticated. The set point controllers frequently had ramp-soak programmers built in, and controlled part vacuum headers were sometimes added. Microprocessor-based controllers that incorporated two or more control loops and a step programmer in a single package were frequently used.

The introduction of data logging to monitor and record part temperatures during the cure process was the first step toward the true closed-loop control of part curing. This was accomplished by using either multi-trace chart recorders or integrated data loggers. These units provided a permanent record of the actual part temperatures during the cure cycle and were used to create more accurate cure definitions and prove cure compliance to engineering specifications without wasting excessive time.

At first, the data loggers were used only to record part data for later analysis and quality control. Soon, however, some combination of the part temperature readings was used to influence the temperature control loop. This was frequently accomplished by a blended cascade control configuration, or some other construction that was affected by the part readings.

As the combination algorithms became more complex, the next logical step was the use of computers to integrate the data acquisition, vessel control, and operator interface portions of the control system into a single work station. Typical configurations would include a dedicated computer with an attached data acquisition unit and some form of mass storage (usually a hard disk). Vessel control would usually be accomplished by separate set point controllers receiving a remote set point from the computer, but would sometimes be controlled directly by the computer system.

Especially in the early stages of computerization, a single computer system would be used to control more than one vessel. In fact, some installations had as many as six

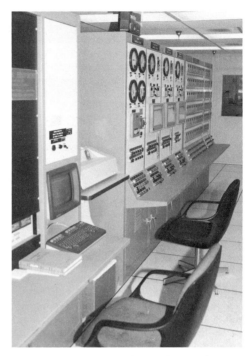

Fig. 3 Computer control system for autoclave curing

vessels under the control of a single computer. There are significant disadvantages to this structure, especially in view of the current low cost of computing resources. In particular, failure of the computer system affected all vessels controlled by that system, and could cause the loss of many parts. In fact, many installations now have two completely independent computer systems for each vessel, in a "hot backup" configuration. Often, the cost of a single lost part can more than justify the additional hardware required.

The key concept in current state-of-the-art computer-controlled cure systems is the closed-loop feedback that allows the temperature profile of the part, rather than that of the autoclave or other vessel, to be controlled (Fig. 3). In order to implement this sort of closed-loop control, the computer system must have a reliable method of acquiring temperature data from the parts in process.

Temperature Sensor

Almost without exception, the temperature sensor of choice is the thermocouple. This is a very simple device, consisting only of a junction of two dissimilar metals. Usually, insulated wires of the two metals are bared and twisted at one end to form a probe. When a temperature differential exists between the junction and the other end of the wires, a voltage (known as the Seebeck effect) may be measured across the junction. This voltage is proportional (although not linearly) to the temperature differential.

The thermocouple owes its popularity to several characteristics. It is compact, very inexpensive, and quite rugged. It can easily be manufactured on site, with simple tools, and its thermal mass is low enough to permit quick response to temperature changes. Offsetting these advantages, however, are a few complications in the area of data acquisition.

The Seebeck effect produces a very low voltage change, ranging from a maximum of about 25 μV/°C to as low as 3 or 4 μV/°C. Because of these very low signal levels, the data acquisition system must be quite sensitive and must have a facility for integrating or averaging data to reduce errors due to the inevitable noise. (This problem is eased somewhat by the low source impedance of a thermocouple.) In addition, the temperature at the voltmeter must be known precisely because the potential is generated by the temperature differential at both ends of the wires. Finally, because of the very nonlinear relationship between voltage and temperature, the acquisition system must have a method (usually a polynomial approximation performed in software) to convert the readings to a usable temperature value.

Additional problems arise in practical application. One of the most common problems is the loss of continuity by the thermocouple ("going open"). This can occur at the plug in the autoclave wall, at the junction, or anywhere along the length of the wire. The better thermocouples are welded at the junction to reduce this type of failure. A more insidious failure results from junction contamination, which can cause a thermocouple to read incorrectly and often erratically. Work hardening of the wire can cause measurable errors, although they are not usually significant errors, in a production environment.

The data acquisition portion of a computer control system must be capable of recognizing and effectively handling thermocouple failures. Because the cure process is closed loop, incorrect feedback from the parts will cause an erroneous and often noncompliant cure. Worse, because of the compensation afforded by the system, the data logs will show that the parts were apparently correctly cured.

Thermocouple Failure Detection. The most common thermocouple failure is loss of continuity. This can be discovered in several ways. The easiest, but least effective, way is simply from evidence of a high resistance between one of the legs of the thermocouple and a relatively high (5 to 12 V) voltage source. If the thermocouple "goes open," the voltmeter will receive an input that is orders of magnitude greater than any reasonable reading, and it can take appropriate action. This is called "up-scale burnout protection" because it simulates an extremely high reading in failure mode ("down-scale burnout" also exists). Unfor-

tunately, this will not expose thermocouples that are erratic or partially open (a common failure mode), and it also has the negative effect of biasing readings of even normal thermocouples slightly toward the high side of the actual temperature.

A far superior method is explicit resistance measurement of the thermocouple, performed as a separate operation from the temperature measurement. This can best be performed by passing a known current (1 mA is common) through the junction and measuring the resultant voltage. Ohm's law allows the accurate calculation of the junction resistance from this voltage. Because this measurement is performed separately from the temperature measurement, the temperature reading is not affected, and much more critical evaluations may be made of a thermocouple.

Another effective test of thermocouple validity is the rate of change of its measured temperature. Especially when the thermocouple is in intimate contact with a part or tool, rates of temperature change in excess of 25 °C/min (45 °F/min) are virtually impossible. The control program can monitor each thermocouple for this condition, and handle it appropriately.

Temperature Control

One of the prime advantages of computer control of the curing process is the ability to control with accuracy the temperature profile of the parts being cured. Several parameters affect the curing process during its various stages; these are discussed in this section of the article.

A typical resin cure consists of a rate-controlled temperature ramp up to the cure temperature, followed by a soak of predetermined duration. Of these two portions, the ramp presents the greater number of complications to a computer control algorithm.

Of primary importance in most production settings is the accurate control of the temperature rise rate of the resin system. The complex changes in material properties over this portion of the cure greatly influence the final structure of the part being made. For example, an excessive temperature rise rate can cause the adhesive to begin its glass transition before adequate flow has occurred, while too slow a heat-up rate can cause resin depletion when the resin viscosity remains low for too long a period. Production time efficiencies are a significant consideration as well.

Because of the varying thermal masses of parts (and even portions of the same part) in a production setting and the finite limit of heat transfer from the pressurizing medium into the parts, it is theoretically impossible to cause all parts to heat at the same rate. Therefore, the control system must make decisions on which part readings to use for its rate control feedback. Because there is

at least some tolerance permitted in the rise rate of the parts, the algorithm has some leeway in which to work. From a production standpoint, the most efficient method is to control the "lightest" (highest rate) part to the maximum allowable rate. This will ensure that no parts exceed the allowable maximum and that the heavier parts will heat up as fast as possible. An alarm function should be implemented to indicate whether any parts are not meeting at least the minimum allowable rate.

Because the process of resin cure is exothermic, the parts actually undergo some self-heating during the cure. This causes the effective thermal mass of the parts to change dynamically. For this reason, a single part may not (in fact, probably will not) retain the highest rate during the ramp phase. In addition, it is important to note that the hottest part is not necessarily the part that heats the fastest. The control system must be able to handle these factors in a correct and dynamic manner.

This exothermic process can also develop into a runaway condition, with the part self-heating at an excessive rate. This is especially true of heavy, thick parts, where a large quantity of resin exists in bulk. The control system must recognize this condition and respond to it appropriately. An alarm should be given, and the heat rate may be reduced. In a production environment, however, it is important that other parts in the load not be sacrificed to save the one that is running away. The decision to abort or continue the cure may be made manually, or the control algorithm may have some logic to handle this situation.

In addition to temperature rate, several other parameters must be carefully controlled in a practical composite cure system. The internal temperatures of various portions of the same part must not be allowed to deviate beyond a specified maximum. If this deviation is allowed to become too great, portions of the part may undergo a transition to the glass phase before others, causing warpage of the part, latent internal stress, or microfractures. To prevent this from happening, the control system should be able to reduce the heat-up rate as necessary, so that temperatures within the affected part can equalize. Obviously, the system must continue to monitor for parts failing to meet the minimum required temperature rate.

Equally critical is control of the temperature differential between the part and the pressurizing medium surrounding it. If this differential is allowed to become too great, the finite thermal conductivity of the part will cause a significant temperature drop between the surface of the part and its inside bulk. This can cause "skin curing," with effects similar to those mentioned above. Because thermocouples placed on the surface of the part will not, by themselves, detect this condition, it can sometimes be beneficial to embed a thermocouple in a waste area of the part, or between the part and the tool. As noted above, the control algorithm must reduce the temperature rise rate of the system to correct this condition.

Although the actual temperature of the pressurizing medium itself should not be of primary concern to the control algorithm (except as described above), there are still real-world limits that must be imposed. Of course, the safety limitation of the autoclave or other vessel must never be compromised; this should be protected by hardwired controls as well. More commonly, the part bagging materials, thermocouple plugs, and composite tools will define hard constraints on the maximum temperature that should be applied during a cure. This too, should be a programmable parameter of the system.

Because the ramp portion of a cure cycle is so heavily concerned with rate control, the absolute accuracy of temperature control is not overly critical. However, the soak or dwell portion of a cure is primarily concerned with upper and lower temperature limits. Although the actual temperature used is not critical, it must be accurately controlled. The amount of cure that a resin system experiences is proportional to the integral of its temperature exposure over time; therefore, both must be known to prove the completeness of the cure of a part. Typical front-to-back accuracies for a quality system are on the order of ±1 °C (±2 °F).

Pressure Control

During the course of a composite system cure, pressure control is also critical. By its nature, pressure is considerably easier to control from a mechanical standpoint than is temperature. The requirements of the control algorithm are also less complex, but there are still several areas of key importance.

Gas. The first and most fundamental decision to be made in the selection of equipment to perform composite curing is the pressurizing medium. Most autoclaves are currently pressurized with nitrogen because it is relatively inexpensive, can be stored compactly in liquid form, and provides an inert atmosphere for the parts. Only ordinary air is less expensive or more convenient, but it poses significant practical problems. Because air is about 20% oxygen, it supports combustion, and, at elevated temperatures and pressures, spontaneous flash fires frequently occur. Although these are rarely dangerous (being contained inside the pressure vessel), a fire will destroy the parts involved and require minor to major repair to the autoclave.

Carbon dioxide is also sometimes used. Because it is much heavier than nitrogen, it allows better heat transfer to the parts, permitting more uniform temperatures and, therefore, higher heating rates. It can be stored under pressure in liquid form indefinitely (unlike nitrogen, which continuously boils away). Offsetting these advantages is its much greater cost, both direct and indirect. Because of its greater mass, the motor that operates the circulating fan in the autoclave must be larger, and the fan design must be different from the design for air or nitrogen use. Also, as carbon dioxide under pressure is allowed to expand, it becomes so cold that some of it undergoes a transition directly to a solid phase, that is, dry ice. Particles of dry ice are quite abrasive and cause significant erosion of the valves and other components of the pressurizing system.

Because of the great amount of stored energy, all gaseous pressurizing systems can be extremely dangerous if proper design is not used throughout the vessel and its control system. In addition, regular maintenance and rigorous enforcement of safety procedures are mandatory.

Water. Instead of gas, water is sometimes used as the pressurizing medium. It has the advantage of much greater heat transfer and is typically used at much higher pressures (7 MPa, or 1 ksi, operating pressures are common). The mechanical design of a hydroclave is somewhat more complicated than that of an autoclave, and there is more chance of part contamination because any bag leak will allow water to enter the part. The problem of phase change into steam limits practical maximum operating temperatures to about 205 °C (400 °F). All the safety considerations of autoclaves also apply to hydroclaves; with the latter, the energy is stored not as compressed gas, but as superheated liquid water.

Mechanical Press. Another popular method for applying pressure to the structure being cured is use of a mechanical press. This allows rapid pressure cycling, requires no pressurizing medium, and is quite safe. However, it is frequently difficult to apply uniform pressure to the part, and tool design becomes much more critical. Original and on-going mechanical costs are substantially higher, and the number of parts that can be cured simultaneously is limited.

Other. In an effort to address several of these issues, work is being done on the development of a vessel that uses a flexible, powdered, solid, pressurizing medium. Heat is applied directly to the parts by a heat blanket, and therefore the medium itself remains cool. Because it is nearly incompressible, no stored energy is present, which greatly enhances the safety of the system.

The control algorithm must sequence the application of pressure in concert with the thermal cure cycle. The application of pressure should ideally take place after the viscosity of the adhesive drops and must be completed by the beginning of the glass transition phase. The application of pressure is usually based on part temperature in order to synchronize this method accurately.

It is also important to control the rate of pressure application. Pressurization that is too rapid can physically jar or move the parts, especially in complex structures. More important, pressures must be given a chance to equalize within the part in order to prevent the hydrostatic flow of the resin or the formation of voids. The accuracy of the rate of pressure application is especially critical in parts that contain a core structure which could be damaged.

Vacuum Control

In order for any pressure to be applied to the parts, a pressure differential between the part and the vessel sides of the bag must exist. Although simply venting the part to the atmosphere would accomplish this goal, the application of vacuum has several other benefits. During initial transportation and loading of the part lay-up into the vessel, the vacuum provides mechanical stabilization for the parts. This setup also provides a convenient method for verifying the integrity of the vacuum bag before the cure commences.

During the early stages of the cure, the resin system releases several gaseous by-products. These can come simply from trapped moisture, but are more frequently the result of solvents inherent in the composition of the prepreg material. If not removed before significant curing has progressed, these trapped volatiles can lead to porosity of the part, voids, or ply separations. By applying vacuum during this time, the vapor pressure of the outgassing is substantially reduced, and the contaminating volatiles can be drawn out before harm occurs. This outgassing can occur to a lesser degree throughout the cure cycle.

The application of vacuum also allows ongoing verification of the integrity of the bag. To accomplish this, additional hoses are typically run to the parts being cured. Originally monitored by gages, the hoses are now typically connected to vacuum voltage transducers, which can then be monitored by the computer control system. In both cases, steps must be taken to ensure adequate protection of the monitoring devices in the event of a bag break. Under that condition, the monitoring system may be pressurized to the full internal pressure of the vessel. Mechanical protection is usually used for gages, whereas the careful selection of transducers allows both ruggedness and accuracy in electronic monitoring. The fully automatic calibration of these transducers is available in the more sophisticated systems.

To reduce the number of lines run to the part, the vacuum is sometimes measured at the vacuum source for each part. Although much more convenient, this arrangement does not provide as accurate a measurement of the actual part vacuum. A bag break may not cause sufficient flow to lower the vacuum at the source, but because of the limited flow capacity of the breather material, the part may have areas that do not receive adequate vacuum.

As with the pressurizing medium, careful control of the rate of vacuum application or removal is necessary, especially in the case of core parts. Trapped pressure between laminations can cause voids or destruction of the core material. Ideally, pressure is applied to the parts at the same time that vacuum is removed, so that the hydrostatic pressure on the part remains constant during the transition. This is a feature of more sophisticated computer control systems.

A recent innovation in the use of part vacuum involves headers that can be pressurized. The pressurization of the vacuum supply must be kept below the current pressure of the autoclave in order to prevent rupture of the bagging material. There are several ramifications of this arrangement.

Because of the relatively small internal volume of the vacuum supply system, it is possible to change the pressure very rapidly. By increasing both the vessel and header pressures simultaneously, the differential pressure on the part can be kept low. By suddenly venting the header, or even applying vacuum, very rapid pressure changes may be made to the part. This can be useful in the processing of thermoplastic (rather than thermosetting) materials.

Another benefit of this system is that the absolute pressure on the part can be kept high to ensure efficient heat transfer, without applying excessive differential pressure to the part. This can allow the use of a higher heating rate, with commensurate production efficiency. However, it is possible for the elevated pressures to become trapped inside the part structure, resulting in voids or delamination during depressurization, or even latent failure. Also, the pressurizing medium can actually dissolve at elevated pressures within some resin systems, resulting in porosity or internal stresses.

Reporting Functions

A major strength of computerized control systems is the capability to analyze and create reports of acquired data. This capability in a composite cure control system allows reports with a degree of detail that was never available with simpler systems (data loggers). This reporting system can be used to inspect the cure of the parts for quality and conformance to specification, to monitor production efficiencies, and to perform research and analysis.

Foremost among the data that are reported are the actual temperature and pressure readings of the part sensors. The autoclave process values are also normally included in this group. To ease the task of quality inspection and analysis, the readings are usually grouped by part and presented in a tabular or graphic format. Some systems also can produce reports that show only the exceptions to the specified cure cycle, which can be very useful for speeding quality inspection. In addition to generating the part readings, some systems maintain a log of events that occur during the cure. These events can include, for example, the exact time that pressure is applied or soak timing begins, any part alarms that the control system can detect, and any operator interaction that occurs. These features are helpful in tracking down the source of problems that occur during a cure.

An important feature of some systems is an on-line or off-line method of interactively analyzing the acquired part data. A comprehensive package permits the interactive manipulation of part reading and the subsequent graphic display or plotting of the results. It may also be possible to perform statistical analysis on the data from several past runs.

To meet official record-keeping requirements and to facilitate the tracking of production, some means of permanently archiving the data from past runs is necessary. The simplest solution is to save the reports generated from each run. This can result in a very large volume of paper, unless microfiche is used. A superior method involves the storage of the cure data on machine-readable media. The use of floppy disks, for example, results in a huge space savings over the use of paper, with the added advantage that the data can be retrieved in the future for further analysis, if necessary. Other types of media, such as tape cartridges or optical disks, provide even more capacity, with added data integrity, as well.

The latest development in data archiving is the use of a separate host computer system, which acquires data from all the machines in the "cure cell" for permanent archiving. This centralizes all record-keeping functions and provides the power of a typically larger computer system for archive search and retrieval. A cure cell management computer system can provide numerous other benefits as well, as explained below.

Future Computer Control Techniques

The state of the art in computer control systems is constantly advancing, as higher-performance resin systems and more wide-ranging applications are devised. Currently, most development is centered on two issues: the more accurate determination of degree of cure of the resin, and more efficient cure cycle control algorithms.

Thermocouple temperature measurements do not report the actual degree of cure of the parts to which they are attached. Currently, this can only be calculated from the thermal exposure of the parts and the rheology of the resin system being used. Other sensor types that report this parameter directly are being explored. Among the most promising are viscosity sensors and dielectric devices.

Viscosity sensors attempt to measure the viscosity of the part under test, which is, after all, the exact parameter of interest. Methods receiving the most attention at this writing involve the transmissivity, refraction, or attenuation of an ultrasonic signal passed through the part. Substantial difficulties remain in the development of a practical method, primarily because of the adverse environment inside an autoclave and the difficulty in achieving reliable sensor coupling to the part.

Electromagnetic signals have also been shown to be influenced by the state of cure of the resin, but the amount of signal change that can be affected by most methods is so small that this method is not yet of practical use for production purposes. An exception is the measurement of the dielectric loss of the resin under cure. If the resin is used as the dielectric for a capacitor, the loss factor of that capacitor provides a very clear indication of the state of resin cure. Solid-state sensors that combine the capacitive element and signal conditioning circuitry on a single die not only exist, but are able to survive normal production cure temperatures. Unfortunately, they are relatively expensive and must be embedded in the resin, which is not practical for many types of production parts.

An alternative dielectric sensor method uses an external measurement system, with relatively inexpensive sensors placed inside the autoclave only. Several sensor designs have been tried, but none has won industry-wide acceptance.

Most present-day computer control systems use a variation of a formula for specifying a cure cycle definition. Typically, this formula variation takes the form of a list of events and times or, in more sophisticated systems, a series of segments that control the various aspects of the cure.

More flexible methods of specifying cure cycle definitions are being explored. The most promising ones are language based and are divided into two categories. The first is an algorithmic language, which allows the user, in terms of variables, to use conditional statements and procedures in order to define the desired cure cycle. More esoteric research uses a rule-based language, in which the user merely specifies the desired goals and constraints of the cure cycle, and the control system applies these rules to complete the goals in the most efficient way. The algorithmic approach is slightly more robust and easier to program, whereas the rule-based technique has the potential for somewhat more efficient cures. The automatic incorporation of the resin rheology into the cure cycle is possible with both methods.

Automation and Robotics for Adhesives and Sealants Use*

Herb Turner, Automotive Business Group, Nordson Corporation

THE DISPENSING OF ADHESIVES and sealants (hot, warm, or cold) has been established as a proven market for the application of robotics. Robots are being used across industries for applying adhesive and sealants to increase quality and to reduce labor and material costs. The automotive segment has led the way for the application of robotics. This article will deal with that segment and some of the criteria associated with automating particular applications, emphasizing the relationships between the application objectives and the adhesive/sealant dispensing equipment. Successful applications take into consideration the capabilities and limitations of the adhesive/sealant materials, the dispensing equipment, the tooling and fixtures, and the robot.

Selection of the robot and the dispensing equipment are of primary importance when automating sealant and adhesive application processes. Those installations that have taken into account all aspects of the application requirements have been successful.

Robots are used to dispense adhesives and sealants throughout the automotive industry today. Automation requirements will be examined for windshield bonding to the vehicle, door manufacturing, passenger compartment interior seam sealing, and the application of sealants during the buildup of the vehicle body. These account for the vast majority of robotic sealant and adhesive dispensing applications in the automotive industry.

Dispensing Equipment for Robotic Applications

The three elements of a dispensing system are the pumping system, the delivery or header system, and the dispensing gun or valve. Careful selection of each element is required to deliver the desired bead at the required flow rate. Lack of performance in any one of the three elements will limit the benefits of the automated dispense system.

Pumping System

The pumping system is typically made up of two bulk unloaders, two feed hoses delivering the pumped material to a common changeover manifold, and a system controller. Two bulk unloaders (A and B) are normally selected in a robotic system to provide a continuous flow of material. The bulk unloaders could be provided with either piston or gear pumps, depending on application and material requirements.

A heated dispensing system is used here for illustration; however, many of the system variables are equally relevant for cold dispense systems. In a typical heated sealer application (Fig. 1), bulk unloader A would be the on-line unloader. The on-line unloader is the unit currently heating the material to application temperature and pumping the material to the changeover manifold at the required temperature, pressure, and flow. Bulk unloader B is the standby unloader and will usually be at a setback (reduced) temperature to minimize the amount of time the sealer is at application temperature while ensuring that minimum time is allotted to come up to full application temperature when commanded from the system controller.

The control logic for the two bulk unloaders is as follows:

- Bulk unloader A provides material as needed by the dispensing guns while bulk

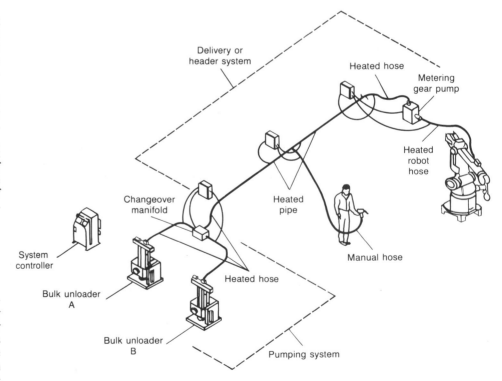

Fig. 1 Typical dispensing system configuration with robotic and manual application points

*Adapted from ''Robotic Dispensing of Sealants and Adhesives,'' in *Handbook of Adhesives*, Third Edition, Irving Skeist, Ed., Van Nostrand Reinhold, 1990, p 743-756. With permission. The last section, ''Developing a Robotic System,'' is reprinted with permission from *Adhesive Age*, April 1983

unloader B waits in a standby (reduced temperature) mode until needed

- When bulk unloader A reaches a predetermined level known as *low drum*, a limit switch sends a signal to the system controller, which in turn increases the setpoint temperature of bulk unloader B from the setback temperature to the required application temperature (usually the same as A)
- When bulk unloader A reaches *empty drum*, another limit switch sends a signal to the system controller indicating that bulk unloader A is out of material. The system controller automatically then switches bulk unloader B to *on-line* status and signals manufacturing personnel that a changeover has occurred and that a new drum of material is required. Bulk unloader A will remain at application temperature to facilitate easy removal of the drum follower plate. After the follower plate has been raised and the empty drum removed and replaced by the new drum, the follower plate is then reinserted followed by purging of air from beneath the follower plate (total time takes about 3 to 8 min depending on training and skill). A *ready* button on unloader A is then activated by the operator to place drum A into the standby mode by the system controller

Bulk unloaders come in three common sizes: 20 L, 200 L, and 1150 L (5, 55, and 300 gal). Selection of the appropriate size is application dependent and requires several considerations. First, acceptable time intervals between drum changes should be determined. The volume of the bead deposition per application and the number of applications per hour will dictate the amount of time between drum changes. Some materials have a very long shelf life and therefore provide more flexibility in container size. Other considerations are economies of larger drums and accepted plant practice for material handling. For materials having shorter shelf life or where there is low daily material usage, a smaller drum size is optimal. For example, in the process of building vehicle doors, approximately 34 cm³ (2.1 in.³) of material are required in the assembly process. If 120 doors per hour are manufactured to meet production requirements, then 4.13 L (252 in.³, or 1.09 gal) per hour will be needed for the pumping system. In this case the proper bulk unloader/material container size would be 200 L (55 gal) and not 20 L (5 gal) 1150 L (300 gal). A 1150 L (55 gal) drum of material allows for 6.3 days per bulk unloader of production, as compared to 4.6 h for a 20 L (5 gal) drum or 34.4 days for a 1150 L (300 gal) container. Sized correctly and with tandem bulk unloaders, scheduled drum changes can be made instead of emergency/reactionary drum changes. The tandem un-

loaders also provide backup for each other in the event of failure, or during system maintenance, and they output a continuous bead even during the automatic changeover sequence.

System design and controls monitor material usage and other system conditions to ensure continuous production. Another feature of a well-designed automatic control system is diagnostics, particularly in the area of pinpointing the location of failed heaters or temperature sensors. A system that monitors all temperature fluctuations throughout the distribution path, providing immediate feedback, has advantages relative to inferred sensing controls. An example of inferred sensing would be a controller that was designed to monitor the minimum set-point temperature. For inferred sensing, an alarm will sound when the actual running temperature reaches the minimum set-point temperature and infers from the low temperature that a heater has failed rather than a controller that was designed to immediately sense a failed heater. Because of the inferred temperature sensing mechanism, a time delay of 20 min may elapse before a system failure is detected. A control system that pinpoints heater loss and location within the system immediately prevents lost production and simplifies system troubleshooting. Immediate feedback also signals maintenance before critical downtime occurs.

Another system control feature often required for robotic systems is an automation interface card that allows signals to be sent back and forth between the robot, the cell controller, and the system controller. Signals such as SYSTEM READY indicate that there are no major malfunctions and that the system has reached application temperature. Another valuable system control feature in a robotic dispense system is simplified gun purge capability. In the event that skilled personnel who know how to facilitate a purge signal from the robot cannot be located, an easily located button on the system controller for this purpose is important. All of these features are needed to minimize downtime, which is often the yardstick by which robotic systems are measured.

Header System

The delivery or header system usually consists of flexible hoses and rigid pipes. The design of this system must be coordinated with the plant layout, the material to be dispensed, and the instantaneous material delivery rate required. The two primary factors are material and delivery rate. Each material to be dispensed (adhesive or sealant) requires a unique and specific pumping pressure based on header size (hose and pipe inner diameter) and delivery rate. Calculations should be made to estimate system pressure drop for a proposed system configuration to determine whether suffi-

cient system pressure availability exists. Tests should also be conducted to verify system design. These calculations and tests should take into account maximum robot velocity and bead size to be dispensed to determine instantaneous delivery rate. Available cycle time is *not sufficient* information for calculating instantaneous delivery rate requirements. Available dispense time and maximum robot velocity are also critical considerations for proper selection of delivery system components.

Other considerations for both heated and unheated delivery/header systems include modularity for system configuration or redesign, backup temperature sensors, hose and pipe sizing, ease of maintenance, and routing of hoses on the robot with regard to its movements during robot cycles and for robot maintenance.

Dispensing Gun

Over the years many attempts have been made at designing a robotic dispensing valve. The progression from each design has been evolutionary rather than revolutionary in the attempt to design dispensing equipment that has the same level of performance as the robot. Many of the dispense valve designs have fallen short of meeting the same performance criteria that robots have met in the area of speed of response. Recently Nordson Corporation released a variable-orifice dispensing gun (patent pending) that has met the design criterion that all systems have tried to obtain—speed of response equal to or greater than the robot. The Nordson system (Fig. 2) has the ability to adjust material flow rate as the robot adjusts velocity to achieve uniform bead deposition.

Bead control, or management of bead deposition, is driven by both the manufacturing process and the rheology of the materials to be dispensed. Process requirements may dictate a short cycle time to dispense the material; hence varying velocities and accelerations are required to meet path accuracy and cycle time constraints. As the robot varies its velocity to meet these sometimes opposing constraints, management of the deposition becomes more important in dispensing uniform beads. As materials develop, the amount dispensed and the cost play an important role in the dispensing equipment selection. A material dispensed in too small or large a quantity may negatively impact part quality as well as material usage.

The key to uniform material deposition using robotics is the interface between the equipment and robot. The robot sends and receives electrical signals from the dispensing gun in a continual feedback system. The signals sent from the robot to the dispensing system cover basic functions, such as gun on/off, while the dispensing system offers diagnostic signals that indicate fault condi-

Fig. 2 Robotic application of hem flange adhesive; Nordson Corporation system

tions and provide troubleshooting assistance. These signals are important because they aid in minimizing manpower for checking quality (verifying bead deposition), system maintenance, and downtime.

Typical Methods of Bead Management

Alternative methods of bead management include the pump and gun combination, the shot metering system, and the electropneumatic control of gun orifice size relative to changing robot velocity and to hydraulic and pneumatic pressure variations. The relative merits of each method will be discussed.

The simplest method for robotic dispensing is the pump and gun combination. The pump is used to unload material from bulk containers and create hydraulic pressure in a distribution system. The gun is used to turn the flow on and off while at the same time providing a fixed flow resistance to the material. The greater the flow resistance, and consequent pressure drop, the greater pump pressure required to increase flow or bead size. There are many drawbacks in this type of system. Variations in pump pressure output during the robot cycle result in variations in flow (bead size), which can result in bead deposition that does not meet process specification. Variations in robot velocity combined with no bead control other than gun on/off can also result in inconsistent bead size. As a result of changes in robot velocity, excess material is often applied to ensure that the minimum bead size is always applied.

Single pump and gun dispensing equipment of this type requires constant robot velocity to provide a constant and consistent bead. However, to achieve the objectives for cycle times, productivity goals, and quality needed to meet today's standards, robot velocities are varied to maintain path accuracy, especially in cornering. The equipment configurations and limitations discussed are often frustrating to the manufacturing/process engineer who is responsible for product quality and productivity.

Programmable transfer pumps, better known as shot pumps, are another equipment variation available for robotic dispensing of adhesives and sealants (Fig. 3). In shot pump systems, the dispensing gun has a fixed needle-and-seat arrangement, providing a constant orifice diameter.

The shot pump may be air, hydraulic, or electrically controlled. The shot pump controller accomplishes bead management by processing an output control signal from the robot and varying the shot pump output proportional to the robot signal. This robot output control signal is either a constant DC voltage or a variable DC voltage that is proportional to robot tool center point (TCP) velocity. This technique provides greater control over the preceding method of constant pressure and fixed orifice. However, there are several constraints affecting the application of shot pump dispense systems.

The first constraint is floor space requirements for the shot pump and shot pump controller. The second constraint is the distance between the shot pump and the gun tip. Typically, the shot pump is one hose length away from the dispensing gun tip, causing response delays in the required fluid output variance. To compensate for the delayed change in material flow at the gun tip, robot manufacturers have tried both prepressurizing the system and programming the robot in anticipation of material output. The resulting system hysteresis from pressurizing and uncontrolled pressure decay often provides poor overall bead management.

As mentioned earlier, it is possible to model the dispensing systems performance characteristics and to compensate for them in robot programming. However, due to the complexity of the dispensing configuration, overall bead management is typically not optimized. Reported system response time to changes in the robot signal varies from 200 milliseconds to over 1 second. Today's robots are known to update and respond in as little as 30 milliseconds. With a robot velocity of 510 mm/s (20 in./s), a 200 millisecond response would result in an incorrect bead size for up to 100 mm (4 in.) while the metered output lags behind the robot signal.

The remaining two issues associated with shot pumps, limited dispense output volume

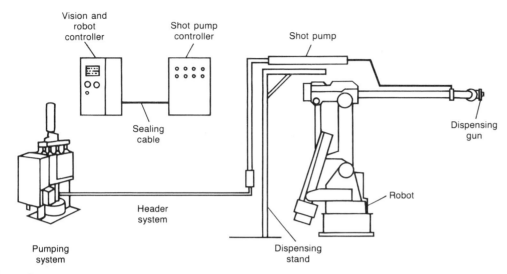

Fig. 3 Typical system configuration for a shot meter installation

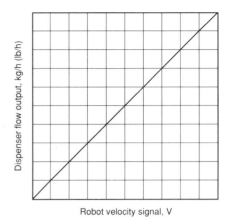

Fig. 4 Linear relationship between robot input command voltage and flow

and a condition known as "packing out," pertain directly to the pump itself. These two issues are interrelated. A properly sized shot pump will evacuate all material in the cylinder during the dispense cycle. As a result, the system is inflexible to meet increases in dispensing requirements. Between the time the dispensing equipment was first specified and the actual installation date, the material dispensing requirements are likely to have changed because of automobile design changes or to meet the manufacturing/tooling capabilities. If the shot cylinder is oversized relative to the specification, then a given amount of material will not be evacuated for each application. The material remaining in the cylinder is then subjected to repeated high pressure applications, potentially causing fillers to be compacted due to squeeze-out of the carrier (for example, plastisol). This compacting, also known as packing out, results in a solid plug in the cylinder that cycles back and forth for each application. Problems occur when bits and pieces of this solid plug break off and cause nozzle clogs during dispense. Once a nozzle clog occurs, the shot pump will usually continue to pressurize the system until the weakest system component ruptures to relieve the overpressure condition.

Another variation of the programmable transfer pump involves the use of an electric servomotor-driven gear pump. These pumps have the same deficiencies associated with shot pumps. Response lags caused by gear pump inertia, as well as the distance between the pump and the dispensing gun, contribute to limited control and poor response over the length of the bead. In addition, floor space is consumed by the control panel required for receiving signals from the robot and sending control signals to the servomotor controller and ensuing servomotor/gear pump combination.

Pump wear is another problem associated with gear metering systems since tolerances must be held tight to minimize pump leakage across the side plates from high upstream pressure. In addition, many of the materials dispensed have a high content of abrasive fillers, further contributing to excessive pump wear.

Advancements in Dispensing Technology

To meet the challenge of today's robotic applications, a new approach to automated dispensing has been developed involving electropneumatic control of the gun orifice size as a function of robot velocity. The system is able to respond with equal or greater speed than the robot-to-input command signals. The system is also designed to handle a wide range of materials and features a linear relationship between robot input command voltage and flow (Fig. 4) for accurate bead proportioning. Additional product features include near-zero floor space, ease of maintenance, repairability, and modular components for compatibility of parts across many models of guns. Finally, the gun design incorporates dowel pins and piloted fits to ensure accuracy and repeatability from gun to gun for successful automation.

For accurate metering and flow adjustment, and to ensure fast response time, adjustment of flow occurs close to the gun nozzle. To accomplish this, pressurized material flows into the gun and across a variable orifice. The variable orifice is comprised of a solid carbide needle and seat for maximum wear resistance. Downstream of the needle and seat, but before the nozzle, a pressure transducer provides pressure feedback to the dispenser's closed loop control (Fig. 5). A nozzle pressure control is used to

compensate for variations in both hydraulic and pneumatic system pressures.

Figure 6 is a graphical presentation of the control loop illustrating the operating sequence. The diagram shows the relationship between the robot signal, pneumatic servo, hydraulic/pneumatic pressure supplies, pressure feedback, and nozzle orifice. As the diagram illustrates, a change in the incoming robot signal causes a change in the controlled nozzle pressure and subsequently adjusts gun orifice size to provide accurately controlled material flow.

Four operator-adjustable controls are used to ensure ease of use and maximum controllability. The first control is a purge function. With a simple switch closure from the robot, the system is placed into a purge mode that is fully controllable via the system control settings. Increasing the purge value causes an increase in the flow, while decreasing the purge setting decreases the flow. During the purge cycle the robot controls the duration of gun on time by a timed switch closure. The second control setting controls bead size during the dispense cy-

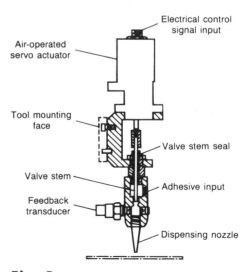

Fig. 5 Variable-orifice dispensing gun

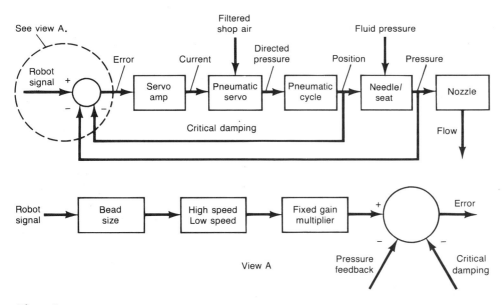

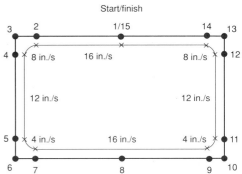

Fig. 7 Test path with varying velocities

Fig. 6 Analog control loop for variable-orifice dispensing gun

cle. The bead size control function and incoming robot signal are multiplied together to form a proportional control. To ensure the linear relationship between input signal and flow, two additional controls are used. The two additional controls, referred to as *high speed* and *low speed*, compensate for material properties such as shear thinning. These controls transform material nonlinearities into a linear function over a wide range of flow.

Figures 7 and 8 display the speed of response by the system during the course of the sample path. The path (Fig. 7) has speed variations ranging from 100 mm/s to 405 mm/s (4 in./s to 16 in./s). The inverted pressure transducer signal and the accompanying robot command signal are illustrated in the response curve in Fig. 8, which also shows how the system compensates for pump wink. Pump wink occurs when the piston changes direction and causes a dip in the system supply pressure. Variations in pump supply pressure such as pump wink are minimized by the electropneumatically controlled dispense gun to provide uniform material flow.

Applications

Dispensing of adhesives and sealants is accomplished robotically in many indus-

tries. The following robotic dispensing applications of adhesive and sealant are taken from the automotive industry, where the majority of robotic dispensing applications are used.

Interior Seam Sealing

This application involves the sealing of body joints necessary for the unibody construction of automobiles. These seals are critical because they seal the passenger compartment from moisture, dust, and wind noise. Any one of these quality defects resulting from improper seam sealing can lead to costly warranty repairs for the automotive manufacturer. For interior seam sealing the bead applied has a ribbonlike appearance that is generally 20 to 40 mm (0.75 to 1.50 in.) wide and 1.3 to 3.2 mm (0.050 to 0.125 in.) thick. In order to achieve this ribbonlike deposition, an air mix airless paint tip is used. The air mix, which uses various combinations of horn and face air, produces fan or conical spray patterns. The conical spray pattern is predominantly used because little or no orientation of the spray pattern is required relative to the substrate. The net result is more time to dispense with less time used to orient the robot wrist to achieve a directional spray pattern. Inadequate management of the bead as the robot varies its velocity results in nonuniform material deposition.

Figure 9 shows the typical seams to be sealed on an automotive chassis. This sealing process is normally found in the paint shop in the plant. The sealant material is applied immediately, prior to topcoat paint processes. The ensuing paint ovens then cure both the paint and sealant.

Typical equipment configuration for robotic seam sealing includes high-response variable-orifice guns, recirculating pumping systems, and material temperature conditioning elements.

The materials applied for interior seam sealing are predominantly *plastisol sealants*. These plastisol materials may be complex rheological systems designed to promote interactions between the suspended component particles, yielding unique pump-

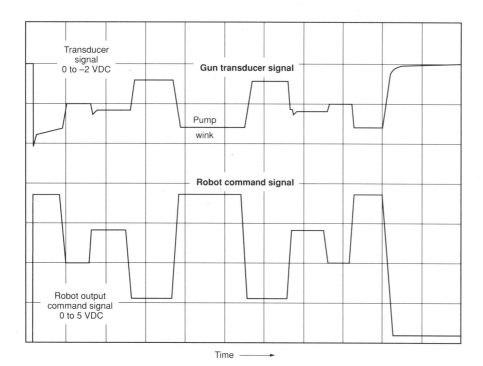

Fig. 8 Variable-orifice gun response to input command signal from robot

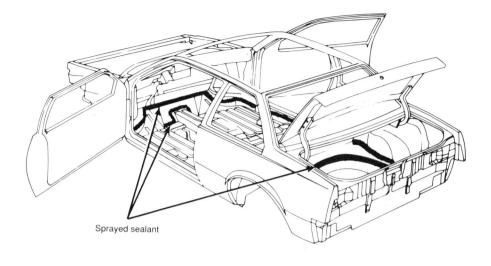

Sprayed sealant

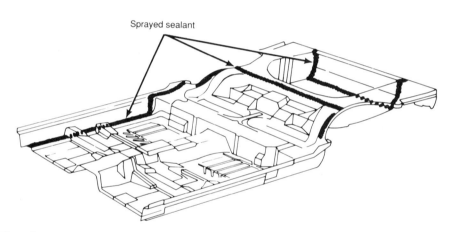

Sprayed sealant

Fig. 9 Typical seams to be sealed robotically on an automobile chassis

ing characteristics. These interactions form a network that raises the inherent viscosity of the material. However, when the material is sheared by pumps, fittings, hoses, and pipes, the network is broken down and the viscosity returns to a lower stabilized value. The material is said to be thixotropic when it displays this type of shear-thinning rheological behavior. Shear-thinning material behavior is desirable, as it allows the material to be sprayed without regard to orientation of the vehicle body without running, dripping, or sagging. Material recirculation throughout the dispense system prior to daily start-up of production allows stabilization of material viscosity to ensure consistent spray patterns and ribbon depositions, critical to the success of the robotic installation.

Another material property that can influence spray pattern and ribbon deposition is temperature-induced viscosity variation. In general, as the material temperature is raised (usually due to changes in ambient air temperature), the material viscosity will

also decrease. To minimize the impact of temperature changes on the material viscosity, the material can be temperature conditioned. There are two strategies for material temperature conditioning. The first technique is to water jacket the header system to provide heating and cooling for viscosity stabilization. Traditionally the temperature has been set at 21 °C (70 °F) to stabilize material viscosity. The major drawbacks for this type of system are cost, reliability, and maintenance. Because of the water lines, heat exchangers, pumps, and installation complexity, the cost for such a system is often high. Reliability and maintenance of the system also play an important role, especially if the pumps or heat exchangers fail, allowing inconsistent material temperature. Since the material is no longer temperature conditioned, variations in spray pattern or ribbon deposition result.

The second method of temperature conditioning uses heaters and sensors throughout the distribution path. This system temperature conditions the material by applying

heat in the header system equivalent to the maximum ambient temperature. The system utilizes a modular piping network outfitted with zoned temperature controls (similar to Fig. 1) that allow the materials temperature to be raised gradually to the desired application temperature. The two basic pieces of information required to apply this system correctly are the temperature-viscosity curve for the material and the maximum recommended material application temperature. Since the material temperature is equal to ambient temperature while in the bulk drum, during storage and shipping, raising the material temperature so it is slightly below the maximum ambient temperature minimizes viscosity variations. Stabilized material viscosity provides uniform spray depositions on the automobile seam. Material temperature conditioning equipment of this type should include a header system with machine wound heater tape on the pipes for uniform heating and temperature control. A control system is required to accurately measure and control the temperature to ±0.6 °C (±1 °F). Maximum uptime is achieved with modular/zoned temperature controls. Should one of the zones deviate from the temperature set point, the controller can automatically turn that zone off while continuing to run the remainder of the system with adequate temperature conditioning. Figure 10 shows a robotic spray sealer application gun reaching across the inside of a vehicle applying a ribbon of material to the automobile seams.

Adhesive Bonding of Automotive Doors

As competition for automotive market share continues to intensify and manufacturers extend warranties for corrosion protection, adhesive bonding of components rather than spot welding has become the required assembly technique. Many of today's automobiles are constructed with two-sided galvanized sheet metal. When the galvanized metal is spot welded, the galvanization is burned through, leaving an area vulnerable to corrosion. In addition, spot welding in certain areas of the automobile is labor intensive and may require time to repair and touch up for customer acceptance.

This application covers adhesive bonding of the hem flange for automotive doors. For hem flange bonding the door is made up of an inner skin and an outer skin. The inner skin mounts the lock and window mechanisms as well as the various trim pieces, hinges, and crash bar, while the outer skin provides design contour and accepts the color coat of paint.

The process of building door assemblies requires the application of a structural adhesive to the outer door panel. The next step in the assembly of doors is the joining of the inner and outer panels. Finally, a

Fig. 10 Robotic seam sealing. Arrows point to seam.

hemming die is used to turn a small section of the outer panel over the inner panel to form a hem. Accurate bead placement and uniform material deposition are critical to achieving a structural bond. These requirements, combined with high production speeds, make hem flange bonding well suited for robotic automation. Figure 11 depicts

the entire door assembly line process complete with hemming die and induction cure fixture. The induction cure fixture applies localized heating that causes the material to cure to a green strength sufficient to hold the door in position during shipping and handling until the final cure in the assembly plant paint ovens.

Figure 12 depicts the process of joining inner and outer panels. When the hem is formed by the dies, the bead of material is pressed out to form a thin film of adhesive between the two door panels. If too little material is applied, a film bond of inadequate strength is formed between the two panels. On the reverse side, if too much material is applied to the outer panel, excess adhesive will be forced out of the hemmed joint (known as squeeze-out) during the hemming operation, causing contamination of the die tooling. As this adhesive begins to build up on the die, it often transfers to the door assembly and results in expensive rework after final paint has been applied. The other drawback to squeeze-out is in the area of increased maintenance for the hemming dies. Over time, the material builds up on the die and hardens, ultimately requiring a rework to maintain die specification.

The best solution to meet hem flange application requirements is to use a robot to position the dispensing gun around the programmed path and to utilize a dispensing system with sufficient response to control the size of the bead adequately regardless of velocity changes. It is not uncommon for

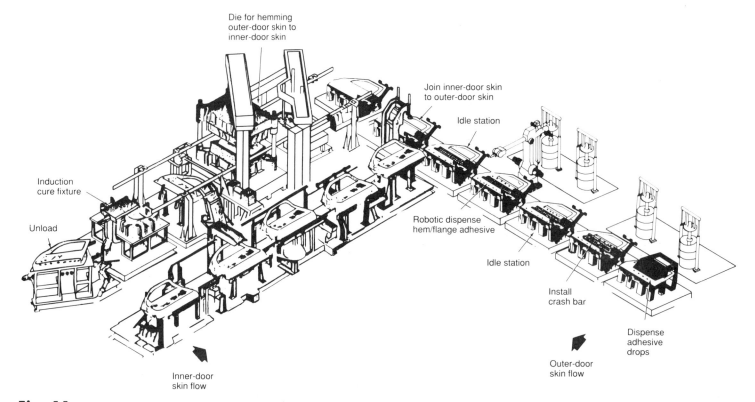

Fig. 11 Typical door assembly line at a stamping plant or in the body shop of a final assembly plant

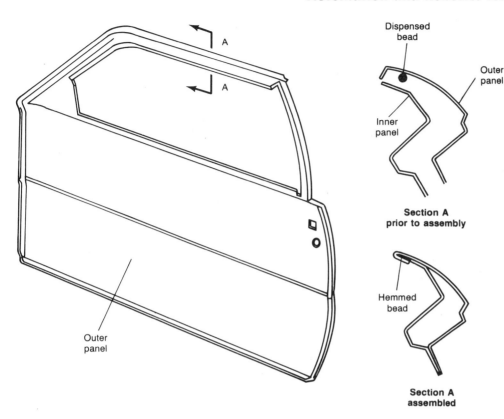

Fig. 12 Adhesive bonding of an automobile door hem flange

doors to have a 3 mm (0.120 in.) bead that is 2.5 m (100 in.) long robotically dispensed in 4 s. Generally, two robots are used, with one robot applying adhesive around the window frame area while the second robot applies adhesive around the perimeter of the lower half of the door. Systems are often set up to ensure maximum production by having the two robots back each other up. Should one of the robots have a failure of some type, the second robot would run in a mode known as *degrade* and execute both robot programs in an effort to maintain production. The upper robot velocity limit for dispensing material has traditionally been 760 mm/s (30 in./s), with 510 mm/s (20 in./s) being the average velocity for dispensing material. Material temperature conditioning is also appropriate for hem flange bonding. For this application, the material is heated to about 28 °C (82 °F) to stabilize the viscosity as well as to ensure that the material has enough heat to wet out the oily metal and provide good adhesion to avoid movement of the bead until it has been hemmed in the manufacturing process.

Windshield Bonding and Body Shop Robotic Sealing

Figures 13 and 14 depict possible robot system configurations for windshield bonding and body shop sealing. All of the appli-

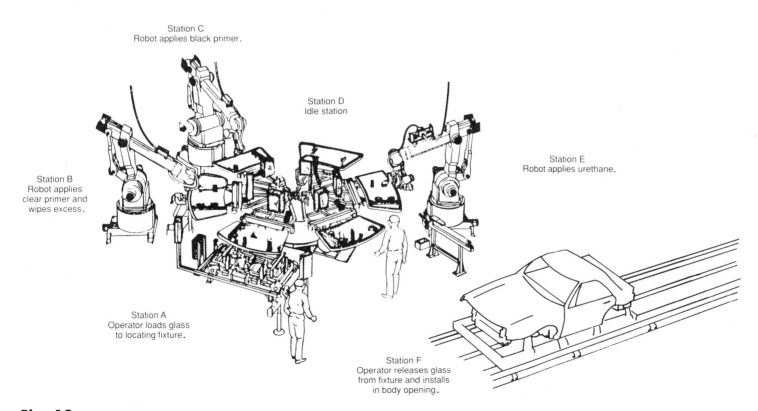

Fig. 13 Trimline windshield primer and urethane robotic application

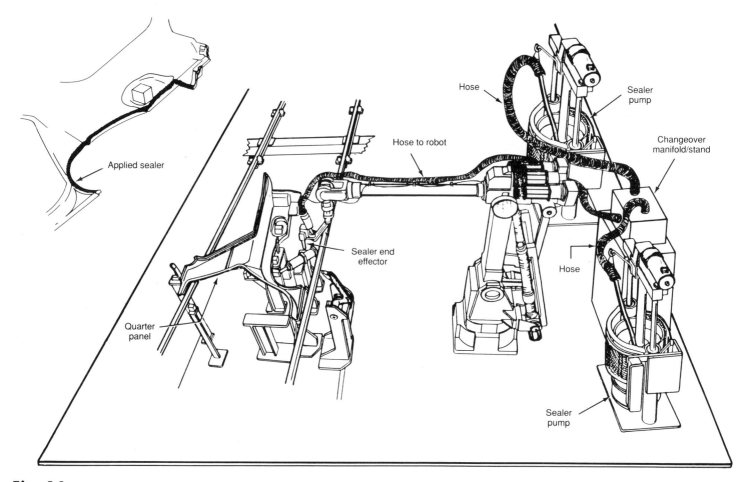

Fig. 14 Robotic sealing in the body shop

cations mentioned have similar considerations for evaluation of robotic applications. The common considerations for robotizing these and other adhesive and sealant applications include required bead profile (round, flat, or triangular); instantaneous flow or delivery rate; robot speed, accuracy, and repeatability; and workspace of the robot.

Windshield bonding often requires the dispensing of a triangular bead at 255 to 455 mm/s (10 to 18 in./s). The dispensing of a triangular bead that has a base dimension of 10 mm (0.400 in.) and a height of 12 mm (0.470 in.) requires that the gun nozzle be constantly oriented about the glass to obtain the desired bead profile.

The material traditionally used in windshield bonding has been a single-component moisture-cure *urethane*. These urethanes are often high in viscosity resulting in high application pressures to meet the robotic dispense rate.

Robotic sealing in the body shop, like seam sealing and door assembly, is performed to seal out dust and moisture while protecting weld locations from corrosion due to the galvanization being burned off. The bead dispensed is nominally a 4.0 to 5.1 mm (0.160 to 0.200 in.) diameter, at velocities that approach 510 mm/s (20 in./s). The

robotic systems used in this system are similar to Fig. 1.

The materials used in body shop sealing vary in application temperature. Traditionally, cold (or ambient) materials are applied during sheet metal body construction. In some instances, cold body shop sealers have caused manufacturing difficulties in ensuing production operations, for example, during the phosphate process. In those assembly areas where material washout has caused quality or process problems affecting the paint finish or phosphate washer, the best solution has been to dispense a warm or hot applied weldable sealant. These materials are able to bite through the oils on the sheet metal to ensure good adhesion.

Developing a Robotic System

When developing a robotic system for product assembly using adhesives, sealants, or gaskets, the procedures below may be useful.

Set Up a Team

- Establish a project team with representatives who have strong managerial, organizational, and technical skills. It is im-

portant to establish a competent team because the equipment includes sophisticated electronics, mechanics, application, and interface equipment
- If a consultant is considered, choose one who has experience with robotics and is familiar with your manufacturing process

Select Proper Tasks

- Start simple. Do not become involved in something overly complex, particularly if this is your first experience with robots
- To begin, identify tasks that will not require major changes in the manufacturing process
- Define a task that requires only one type of application equipment
- Define material requirements for the task
- If the application requires only one or two axes of movement, then some other form of automation may be better. However, batches of simple operations can be combined at some centralized location where robots can provide the flexibility to handle a variety of operations, each of which would otherwise require its own equipment design
- Product variety for one robot is limited by its memory size. Check for both the

number of positions and programs that the robot can store
- Be sure that the part can be accurately positioned in the robot's workspace. It is generally less expensive to design new fixtures than to modify existing fixtures
- Construct a work flow chart

Define Objectives

Quality

- Define requirements for the accuracy, uniformity, and consistency of the bead of the hot melt (adhesive or sealant)
- Discuss the equipment quality record. Identify the expected time interval during which the system will operate reliably and the service requirements
- Identify who will have responsibility for installation and service of the system
- Identify which standards are applicable to the system

Productivity

- Establish cycle time goals for each application
- Define the minimum amount and frequency of downtime for the system

- Develop backup procedures

Cost Savings

- Calculate material and labor savings
- Identify expenses for the equipment, manufacturing analysis, and relocation of equipment
- Investigate tax credits

Safety

- Define systems and precautions that are necessary

Training

- Consider impact to workforce and what retraining will be involved
- Identify who will need training and on which parts of the system
- Determine how and where the training will be given
- Recognize the benefits of the exposure to automation for other manufacturing operations

Future Requirements

- Discuss how needs and processes may change in the future and how the flexibil-

ity of the robotic system can contribute to those changes

Design the System

- Involve materials, application equipment, and robot vendors early in the design process
- Understand the interrelationship of each piece of equipment and identify one vendor who will coordinate these interrelationships
- Ask for trials with your product and materials
- Identify how part positioning and part identification requirements will be designed and controlled
- Determine how much the present facility will have to be redesigned

ACKNOWLEDGMENT

Special thanks to Sharon Dodson, Market Analyst, Automotive Business Group, Nordson Corporation, for taking the time to help edit and critique this article.

SECTION 9

Nondestructive Inspection and Quality Control

Chairman: Yoseph Bar-Cohen, Douglas Aircraft Company, McDonnell Douglas Corporation

Introduction

THE QUALITY OF ADHESIVE-BONDED JOINTS is determined by a large number of interrelated elements, including the constituents and the process of making them adhere. Process control is needed at every stage, beginning with the moment the adhesive and adherends are obtained. Surface treatment is a fairly well established chemical process, whereas the bonding of the adherends depends largely on the qualifications of the operator. The actual curing process requires accurate process monitoring of temperature, time, and pressure while the components are in the autoclave or press. The quality of the bond depends also on the application method, making the bonded joint history very important. As part of the quality assurance process, it is necessary to review the joint history by examining all pertinent documents to ensure that each step of the process has met the requirements.

Once an assembly has been bonded, its inspection is limited to checking the dimensions of the product and determining whether or not a cohesion has been formed. From this stage, nondestructive evaluation (NDE) techniques are needed. In the simplest form of NDE, unbonding can be detected by tapping the structure and listening to changes in the sound characteristics. An unbonded area will respond with a lower damping of the sound and a sound pitch (frequency) that decreases with the increase in defect size. Although this method is fast, simple, and inexpensive, it is very limited and unreliable and can only be used to detect gross unbonds. The increase in the application of bonding to flaw-sensitive structures and to complex configurations prompts the need for highly reliable NDE methods to safeguard bond performance.

The assurance of the performance of a bond requires a method that is capable of detecting and characterizing unbonds and determining the adhesive properties. The ultimate need is to be able to measure, nondestructively, the strength of the bond.

To reach this goal, parameters that might be sensitive to the strength were sought by many investigators (Ref 1-3). Some limited success was observed when the variables of the materials were carefully controlled and a direct relation existed between the average strength and a single property of the adhesive material, such as thickness. However, the practical application of any NDE concept to determine the strength of a bond is not expected to be reliable. Unfortunately, strength is not a physical parameter, but a structural one, and it is an indication of the highest level of stresses that the weakest spot of a specific structure is able to carry. While it might be possible to estimate the average strength of a bonded system, this measure is meaningless for the performance of the actual structure. This is because the bond will fail when the stress exceeds the material's weakest spot strength rather than the average strength, which might be substantially higher.

The strength of an adhesive bond depends on two elements: cohesion and adhesion. Cohesion is associated with the strength of the bond between the various molecules of the adhesive layer. Cohesive strength is determined by the type of adhesive, its elastic properties, and its thickness. Limited information can be obtained about these parameters when using NDE methods.

Adhesion is related to the strength of the bond of the adhesive and the adherends. The quality of the adhesion is critical to the performance of the adhesive as a bond between the components of an assembly. The interface layer is a fraction of a micrometer thick, making it very difficult to characterize nondestructively. A weakness of this layer, which can be caused by poor surface preparation, is not detectable by any existing NDE method.

As a result of the above limitations in determining bond strength, NDE methods are mostly used to detect and characterize unbonds. Generally, the most sensitive and reliable NDE methods for bonded joints are based on ultrasonics, because of the sensitivity of ultrasonic devices to physical continuity over interfaces. This affects the characteristic response of the wave propagating through or reflected from the bonded region and the interfaces. Unfortunately, ultrasonics cannot provide a complete answer to all aspects of NDE because it is plagued by edge effects, need for couplant, strong attenuation in the bonded system, and sensitivity to the properties of the bonded media. Other methods, such as holography and thermal wave imaging, are used in cases where ultrasonics is limited. In addition, computers are incorporated to improve the evaluation process by providing a higher signal to noise and better detectability of defects.

This Section reviews the quality assurance techniques employed from the preparation for bonding to production and in-service use. The individual articles in this Section provide an understanding of the basis of the various NDE and quality assurance techniques as they relate to raw materials, the bonding process, and end-product bonded joints (both metallic and composite joints). The capability of the various NDE techniques is described to help the reader understand the reasoning behind the use of one method or another and the expected results from each of the methods. The engineer will be able to acquire the background for a common language with the NDE personnel and to have a knowledge of what to expect.

REFERENCES

1. J.L. Rose and P.A. Meyer, Ultrasonic Procedures for Predicting Adhesive Bond Strength, *Mater. Eval.*, Vol 31 (No. 6), 1973, p 109
2. P.L. Flynn, Cohesive Bond Strength Prediction for Adhesive Joints, *J. Test. Eval.*, Vol 7 (No. 3), May 1979, p 168-171
3. G.A. Alers, P.L. Flynn, and M.J. Buckley, Ultrasonic Techniques for Measuring the Strength of Adhesive Bonds, *Mater. Eval.*, Vol 44 (No. 4), April 1977, p 77-84

Raw Materials Quality Control

William D. Brown and Gerald K. McKeegan, Dexter Adhesives & Structural Materials Division

QUALITY CONTROL to assure batch-to-batch consistency of adhesive products has been recognized as a necessity since the earliest conception of adhesive bonded aircraft. Because the first aircraft structures were constructed primarily of wood, and because the adhesive was required to sustain shear loads only, it is easy to understand why the pioneers of the aircraft industry selected a wooden lap-shear specimen for their measure of quality control. The adhesive products were required to attain either a minimum shear value or cause failure of the wooden adherends. With the natural inconsistencies that occur in wood, quality control testing of these early structural adhesives must frequently have been a frustrating experience.

The applications of adhesives in aircraft construction have become more sophisticated through the years, as has adhesive technology. As new adhesive chemistry, forms, and uses have been developed, the need for equally sophisticated quality control of the adhesive has become of paramount importance to both the adhesive manufacturer and the adhesive user.

Early aircraft adhesives were of the liquid/paste form, where two materials were mixed together and applied to the adherends. The adherends were assembled (usually by clamping them together), and the mixed resin was allowed to polymerize to a solid, thus bonding the adherends together. Over the intervening years, this basic technology of adhesive bonding, while requiring more sophisticated techniques, has changed little in concept. A significant portion of the total aerospace adhesive market is still represented by two-part adhesives. Usually supplied in kit form, consisting of a resin component and a curing agent component, they require the thorough mixing together of specified quantities of each part prior to use. Polymerization is usually effected over a period of time at ambient temperature, accelerated by application of heat to the bond line.

It was not until the late 1940s and early 1950s that adhesives in the form of calendered films found their way into aerospace applications. In manufacturing calendered films, the supplier premixes all the components required to effect polymerization. The adhesive is then calendered into film at a thickness specified by the customer. Film adhesives usually are supplied with a woven or mat carrier, but there is an increasing demand for unsupported films to aid reticulation techniques (where the adhesive is applied only to the edges of honeycomb cell walls).

Recognizing that the premixing and calendering operation brings together all the components needed to promote polymerization, it is easy to understand why the vast majority of manufactured films also require storage at low temperatures to inhibit this polymerization and thereby prolong their useful life. Films manufactured this way are usually referred to as having been supplied in the B-stage condition. Once returned to ambient temperature, the polymerization process accelerates, resulting in a finite time period in which to complete the final assembly and cure operation.

The same time constraints apply to paste adhesives once the resin and curing agent components have been mixed. In either case—films or pastes—it is essential to assemble the adherends quickly, before the polymerization process advances to the point where poor flow may interfere with the development of expected bonding properties. Pressure must be maintained on the adherends, at the bond line, until the adhesive is fully cured.

As stated above, polymerization of the adhesives is usually accelerated by raising the temperature of the bond line in the assembly. This application of heat also enhances the flow and wettability characteristics of the adhesive as it passes from a film or paste form, through a reduced viscosity stage, and finally to the desired fully polymerized, cured state (Fig. 1).

The heat required for curing is usually supplied by an oven, press, or autoclave. In the two latter methods, pressure can be more uniformly applied. When pressure is not uniform, the adhesive attempts to flow from high-pressure to low-pressure areas during the cure cycle, usually resulting in varying thin and thick spots in the cured bond line. One method for achieving uniform pressure in an oven is to use a vacuum bag around the bonded assembly.

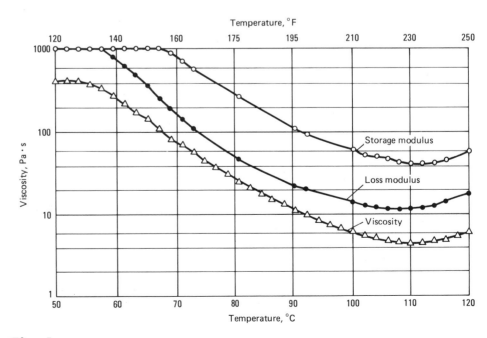

Fig. 1 Adhesive rheometrics curve

Product Acquisition Approach

The quantity and/or quality of the raw materials entering either the film or paste manufacturing processes, together with the variations in bonding and curing parameters described above, can have a significant effect on the quality of the final bonded assembly.

Typically, the quality of the raw materials purchased for use by the adhesive supplier are controlled by his own internal specifications. Likewise, control of material quantities, times, temperatures, mixing, and other processing variables are controlled by the supplier's manufacturing documents. After manufacture, the finished goods are invariably tested to the physical and mechanical requirements of the supplier's internal specification. In addition, they are usually tested to the requirements of a material specification referenced on the customer's purchase order. A certification is then issued with the product, showing that the material met the required properties of the specification at the time it was shipped.

Recognizing that the material supplier has a mandate to test the physical and mechanical properties of his product prior to shipping, serious consideration needs to be given as to what level of additional testing is justified by the consumer. Almost without exception, it is the current accepted practice in the aerospace industry that incoming adhesive materials are again subjected to the identical battery of tests as those already performed and certified by the supplier. The only possible rationale for this extreme level of redundant testing is that the consumer has little faith in the capabilities or integrity of the supplier, or something adverse and beyond the control of the supplier or the customer could happen to the product while it is being shipped.

Consider the latter point first. The only circumstances that could affect the integrity of properly compounded adhesive materials are mechanical damage or an adverse heat or moisture history that is sufficient to affect the degree of polymerization. Evidence of mechanical damage would be part of the normal receiving inspection procedures, regardless of the product. Heat history while the product is in transit can easily be monitored by inexpensive and reusable temperature recorders.

Consideration of the first point, whether redundant testing is conducted by the consumer because of a lack of faith in the capabilities and integrity of his supplier, is an issue that is best addressed from the supplier's viewpoint. An approved and qualified supplier of a product might wonder why any consumer with doubts about his capability or integrity would continue to be a customer; why the consumer is compelled to repeat all the testing done by the supplier, and why the consumer can justify imposing these same conditions on the supplier. A somewhat more radical viewpoint is that many suppliers might prefer to negotiate a lower price for their products if they could be relieved of the onerous burden of certification testing and assured that the customer will perform this testing regardless. A further consideration of the issue is that it is inconceivable that any reputable supplier would intentionally provide an inferior product that would fail to perform as specified and risk either having his name removed from the list of approved suppliers or the possibility of being associated with a serious aircraft accident.

With these considerations in mind, it is appropriate to seriously evaluate what tests are required, and when. A well-balanced approach to this issue could assure high quality, but at minimum cost to both the supplier and consumer.

Product Test Phases

The variety of test procedures that must currently be performed on an adhesive system, from design through production stages, generally can be categorized as design allowable, qualification, certification, or receiving inspection (acceptance) tests. Design allowables and qualification tests are aimed at establishing that the initially supplied product, and any products subsequently added to a material specification, are fully capable of performing in all anticipated conditions for which the specification was designed. Certification and acceptance tests are intended to measure the continuing reliability of the adhesive product as supplied. For the purposes of this article, certification and receiving inspection tests are both considered to be raw material quality control tests.

Design allowables are those tests performed to support the currently planned, and probable future, applications of the adhesive system. Typically, this type of testing is conducted to fully characterize and understand the mechanical properties of the cured adhesive and will, as a minimum, encompass tests simulating the anticipated environmental extremes of the proposed end-use item. At this early stage of evaluation, the design allowables testing program usually includes a variety of adherend materials, both metallic and composite, to support end-use applications. The effects of long-term environmental conditions, such as heat, moisture, UV exposure, or fluids would also be evaluated in this phase.

The full characterization of the adhesive provides the designer with data representing the capabilities of the adhesive system under a variety of design and environmental conditions. Design allowables testing is usually performed on a minimum of three batches obtained from the potential supplier and compounded from different batches of raw material on production—as opposed to laboratory—equipment. This three-batch requirement is aimed at providing both the designer and the materials engineer with information relative to possible minor variations that can occur in the compounding of organic materials.

After completion of the design allowables program, the resulting data are usually incorporated into a material specification. The purpose of a material specification is to delineate the requirements that must be met before a product can be added to a qualified products list (QPL). Establishing these requirements, and conducting certification and acceptance tests, are intended to assure the continuing reliability of further purchases.

Qualification test requirements that are written into a material specification normally follow a pattern that is similar to, but somewhat reduced from, those tests performed in the design allowables phase. Whereas a design allowables program usually includes a variety of adherends, overlap shear lengths (L/t), and honeycomb cell sizes, qualification tests are normally standardized to include one adherend, a typical L/t ratio of 8 for shear specimens, and a single honeycomb cell size. At this stage of development, the materials engineer is primarily interested in demonstrating that alternative products that meet the specification requirements will display physical and mechanical properties that are at least equivalent to those of the originally selected product. For this reason, the three-batch requirement discussed in the design allowables section above would normally be enforced, as would determination of the effects of long-term environmental conditions.

Certification. After being approved/qualified as a potential source for the adhesive system defined by the material specification, the supplier is required to certify that each manufactured batch of this product meets the physical and mechanical properties of that specification. Compared to the properties evaluated in the design allowables or qualification programs, this phase can, and should, represent a much reduced testing program. Because it is obviously impractical to impose long-term environmental evaluations on products with a finite shelf life, these products should be excluded from certification requirements.

Acceptance. After the products have been evaluated and found to conform to the requirements of the material specification, a written certification to that effect is provided to the customer by the supplier when shipment is made. Almost without exception, it is the current practice of aerospace customers to repeat *all* the tests performed for certification by the supplier, on each shipment received. These tests are usually

Table 1 Typical quality control certification tests

Test type	Temperature range		Method
	°C	°F	
Shear			
Single-overlap .... −55 to 290		−67 to 550	ASTM D 1002
			MMM-A-132A
Blister .... −55 to 150		−67 to 300	ASTM D 3165
Beam flexure 25		75	MIL-A-25463
Tube.......................... −55 to 175		−67 to 350	Customer defined
Flatwire tension................... 25 to 175		75 to 350	ASTM C 297
Peel			
Metal-to-metal climbing drum ..... −55 to 80		−67 to 180	ASTM D 1781
Floating roller .... −55 to 150		−67 to 300	ASTM D 3167
T-peel .... −55 to 80		−67 to 180	ASTM D 1876

referred to as acceptance or incoming receiving tests.

Raw Material Quality Control. Once it has been established that the product form and composition will give the desired performance, it becomes critical to assure that each batch of the adhesive will provide continued, uniform performance. Ultimately, this assurance is derived from controlling the adhesive manufacturing process. The raw materials and the procedures used in making each batch must be consistent with those used to make the batches subjected to the design allowables and qualification processes.

Because an adhesives user is unable to control the manufacturing process used by the adhesives suppliers, there is a natural tendency for the user to employ some or all of the tests designed for qualification as tests to accomplish certification and acceptance of each shipment of adhesive. Although this methodology provides a reasonable guarantee that unacceptable adhesive will be prevented from reaching the production floor, it is not necessarily the cheapest or most effective means of assuring batch-to-batch consistency. Because qualification tests are often unique to a specific adhesive application, their use in the acceptance testing process can entail an expensive variety of test materials, cure cycles, and specimen configurations when the adhesive can be applicable for a variety of end uses.

However, if a single test matrix is defined, along with standardized materials and procedures, and the same tests are performed on all adhesives, it has been found that each product gives a characteristic result in that matrix. That characteristic result is repeatable, batch-to-batch, if the adhesive composition and manufacturing procedures are consistent with those used in the qualification process. It is this consistency, that is, quality, that is the object of certification and acceptance testing.

Test Techniques

The vast array of mechanical testing currently conducted under the aegis of "quality control" includes a majority of the tests shown in Table 1. For generic purposes,

these tests can be grouped as shear, flatwise tension, or peel tests, each of which is described below in terms of its own merit. In addition, full descriptions of these and other tests are provided in the Section "Testing and Analysis" in this Volume.

Shear. True shear conditions exist only in the double-overlap shear configuration assembled according to ASTM D 3528, "Test Method for Strength Properties of Double Lap Shear Adhesive Joints by Tension Loading." For economic and practical reasons, this "ideal" specimen is not used for quality control purposes. Rather, the single-overlap shear specimen is often substituted. This specimen is described in MMM-A-132A, "Adhesives, Heat Resistant, Airframe Structural, Metal-to-Metal" and ASTM D 1002, "Test Method for Strength Properties of Adhesives in Shear by Tension Loading (Metal-to-Metal)." Although the ultimate shear strength of an adhesive system tested in this manner can be severely impacted by variations in the modulus and yield strength of the adherends, under controlled conditions the single-overlap shear specimen is a useful tool for quality control purposes. Both the single-lap and double-lap configurations are shown in Fig. 2.

There are additional nuances in lap-shear testing, such as the varying grip lengths required by MMM-A-132 and ASTM D 1002, but these have not been shown to significantly affect test data. However, rate of application of load to the specimen has been shown to have a major effect on the ultimate load and needs to be defined in the material specification.

Flatwise tension testing according to ASTM C 297, "Method for Tension Test of Flat Sandwich Construction in Flatwise Plane," is invariably used to evaluate the bond strength between skin details and a core material, usually honeycomb. More in theory than in practice, this is an ideal method of assessing the capabilities of the adhesive to form the fillets that are necessary to achieve a satisfactory bond to the cell walls of honeycomb core. If the test specimen shown in Fig. 3 is fabricated by bonding it to the tension blocks with a room-temperature curing adhesive, this test

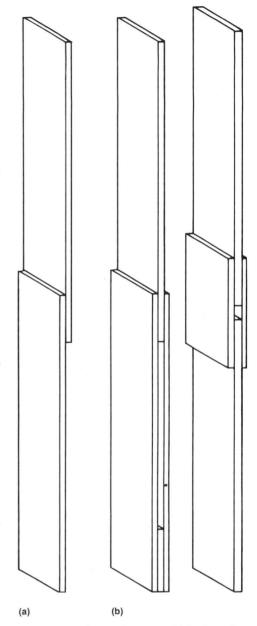

Fig. 2 Lap-shear configurations. (a) Single-overlap shear. (b) Double-overlap shear

would be more meaningful than it is with the current widely used practice of assembly with a heat-cured adhesive. The extensive use of heat-cured film adhesives for the bonding of specimens to test blocks has come about primarily for three reasons: ease of application, reduced cure time, and the fact that film adhesives usually have a higher glass transition point than their paste cousins.

It is this widespread use of heat-cured adhesives for the block bonding operation that undermines the validity of the flatwise tension test for quality control purposes. Assuming a case where the supplier had inadvertently omitted a portion of the curing agent, it is not hard to understand that a

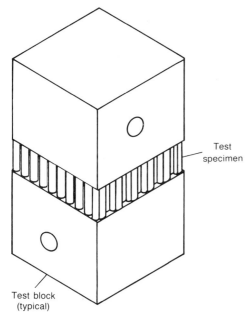

Fig. 3 ASTM C 297 flatwise tension test assembly

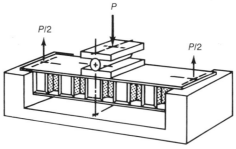

Fig. 4 Honeycomb sandwich edge strength test

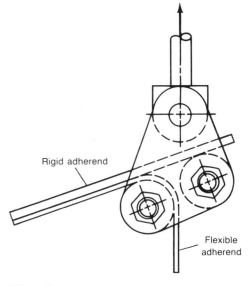

Fig. 5 ASTM D 3167 roller drum peel test

significant degree of undercure could occur, which would be compensated for when the specimen receives a second cure during the block bonding operation.

Peel. A wide variety of peel testing techniques exists for both metal-to-metal and metal-to-honeycomb applications. Theoretically, the peel test is a measure of the toughness of an adhesive, which is considered by many to be a desirable property. It is virtually impossible to formulate tough adhesives that also exhibit good high-temperature thermal aging properties.

Many applications can be cited for the successful use of extremely brittle adhesives (with peel properties so low as to be virtually unmeasurable) on air-wetted surfaces under extremely adverse environmental conditions. Careful analysis of failure modes where skins are peeled away from the substrates always indicates that the initial failure mode occurred by either shear or flatwise tension failure. Hence, peel is a secondary mode of failure. From a quality control point of view, peel testing can also be considered an unreliable test because products with a latent degree of undercure will, to some extent, exhibit higher peel values than are typical for the fully cured product.

Alternative Techniques

The shortcomings of many current and historical evaluation concepts have already been described. This section identifies potential and practical alternate procedures in terms of the two principal objectives of quality control testing: to assure batch-to-batch consistency and to assure that the product suffered no adverse conditions during shipment.

Consistency Controls. As discussed above, the single-overlap shear test, conducted under sufficient controls, is adequate to determine that an adhesive is fully capable of polymerization during the specified cure cycle. Although lap-shear testing is capable of identifying discrepancies in the formulation of a batch, potential variations are best identified by Fourier transform infrared (FTIR) and high-performance liquid chromatography (HPLC) techniques, where the resulting graphs can be compared against a standard. FTIR and HPLC characterizations are also able to indicate variations in the heat history (staging) of the product. These variations, while not necessarily detectable by lap-shear testing, could severely impact the capability of the adhesive to form adequate fillets during a honeycomb core bonding operation.

The ability of the adhesive system to flow and form the desired fillets to cell walls is best evaluated using a honeycomb specimen tested in the flatwise tension configuration described in ASTM C 297. A future alternative to this test may be the so-called honeycomb sandwich edge strength test (Ref 1), shown in Fig. 4. Both of these honeycomb structure tests are suited primarily to film adhesives, rather than to pastes.

Toughened adhesives, which are intended to yield high peel strength, should be evaluated using the roller drum peel test shown in Fig. 5 and described in ASTM D 3167, "Test Method for Floating Roller Peel Resistance of Adhesives." This test is excellent for testing both film and paste adhesives.

Adhesive primers consist of an adhesive chemistry system that usually includes corrosion inhibitors and a dye, dissolved or dispersed in a volatile solvent. Adhesive primers are tested in conjunction with a film or paste adhesive, using the same performance tests described in this article. Primers are usually sold in 1 gal or 5 gal pails, which are filled from a larger batch. It is important to assure that the solvent/adhesive ratio is uniform throughout all containers. Therefore, an additional test for primers is to check the volatile material content of several randomly chosen containers using the procedure described in ASTM D 2369, "Test Method for Volatile Content of Coatings."

There are several foaming adhesives on the market, designed primarily for edge trimming or splicing of honeycomb core structures. The commonly used tube shear test shown in Fig. 6 is ideal for these materials.

Testing Considerations

Test Temperatures. Equally as important as the type of testing selected for quality control purposes is the temperature at which these tests are performed. It is normal for mechanical properties, such as lap-shear strength, of an organic material to deteriorate as the temperature rises. At some point, usually very close to the glass transition temperature (T_g) of the cured adhesive, the rate of this deterioration increases rapidly, as shown in Fig. 7. Therefore, an elevated-temperature lap-shear or flatwise tension test may be appropriate as a certification or acceptance test, but the temperature chosen for such tests should be 10 to 15 °C (18 to 25 °F) below the T_g.

On the other hand, peel tests performed at or near the T_g can be equally misleading in that the more "brittle" adhesive systems often exhibit a markedly higher degree of toughness as they approach this temperature, and can thus produce indications of uncharacteristically high peel values. In general, elevated-temperature peel tests for certification or acceptance should be avoided.

Materials/Procedures/Control. Control of the materials and procedures used for certification and acceptance testing, with emphasis on standardization, will assure repeatable results and minimal testing costs.

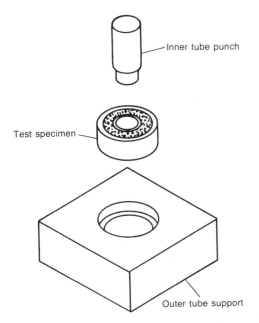

Fig. 6 Tube shear test

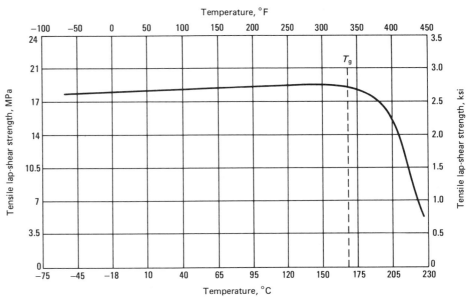

Fig. 7 Typical tensile lap-shear profile for 175 °C (350 °F) curing epoxy film adhesive

Although a variety of substrate materials are used in the aerospace industry, it is neither necessary nor practical to conduct certification and acceptance testing on every type of substrate. Rather, verification that an adhesive is compatible with all intended substrates should be done during the design allowables and qualification phases. Again, the purpose of certification and acceptance testing is to determine only that the product is equivalent to that which was qualified.

Aluminum is the most widely used substrate for aerospace adhesive testing. Many alloys and tempers are available, but the most common one used is 2024-T3, with either the bare alloy surface or a pure aluminum-clad surface. Because the same adhesive will give different results on each of these surfaces, users should pick one surface for all testing to assure consistent results.

The surface of aluminum test panels must be prepared for optimum bonding. Surface preparation procedures vary widely, from sanding or grit blasting, to acid etching or electroanodizing. The objective is to achieve a surface that is absolutely free of contaminants and that presents a uniformly roughened surface. The ideal roughness is found at the level of a few hundred Å. Therefore, surface treatments that achieve this level are best. A standard surface treatment acid etching of the surface is recommended, followed by phosphoric acid anodizing in accordance with ASTM D 3933, "Practice for Preparation of Aluminum Surfaces for Structural Adhesives Bonding (Phosphoric Acid Anodizing)." After being treated, the surface should not be exposed to contaminants from fingers, cotton gloves, or solvents.

Honeycomb core also comes in a variety of alloys, cell sizes, thicknesses, and densities. A standard configuration is 6.4 mm (¼ in.) cell, 5052 alloy, 0.1 mm (0.004 in.) wall gage, 13 mm (½ in.) high, which meets the requirements of specification MIL-C-7438F, "Core Material, Aluminum, for Sandwich Construction." This will provide adequate evaluation of fillet formation, as well as bond strength, at a modest cost. Aluminum core should be cleaned by vapor degreasing and air drying, but further surface preparation is normally not required for testing purposes. The Section "Surface Considerations" in this Volume contains articles that cover surface analysis techniques, the use of primers and coupling agents, and the surface preparation of a variety of substrates.

Some paste adhesives may require bond line thickness control. One common method for doing this is the addition of 0.5% by weight 0.13 or 0.25 mm (5 or 10 mil) glass beads to the mixed adhesive. The optimum bond line thickness depends on the test mode (for example, peel or lap-shear). Many adhesives include bond line control materials in their formulations, which precludes additional efforts. The supplier's technical data sheet should address this issue and define the desired thickness for a given test configuration. Use of wires in the bond line should be avoided because this will disrupt the surface energy matrix, adversely affecting test results.

Shipping Controls. Although the probability of adverse thermal history occurring on domestic shipments is low, it is nonetheless real and even more possible when international transportation is necessary. As mentioned above, inexpensive temperature recorders can be included with the ship-

ment. Where doubt exists, the degree of polymerization also can be evaluated by FTIR and HPLC techniques, which are then compared to similar data from the supplier.

Finally, all adhesives should be subjected to physical inspection for damage caused by improper handling during transit. Such damage is usually not the result of poor product quality, but sometimes can be prevented by better packaging or carrier selection.

Acceptance Alternatives

Even with standardization of test methods and materials, acceptance testing of adhesives is an expensive and time-consuming process. Waiting for acceptance testing invariably delays production processes and ties up valuable storage space (as well as capital) for idle inventory.

Two alternatives are available to reduce or eliminate much of the cost of acceptance testing. Both alternatives depend on the relationship between the user and the adhesive supplier. As stated earlier, the supplier has a mandate to test the product against the customer specifications, as well as his own internal specifications. If the user and supplier have a relationship based on trust and confidence, then the supplier's data ought to be adequate evidence of the quality of the product.

With that concept in mind, one alternative approach is the source inspection of the adhesive. The supplier notifies the user when the adhesive has been manufactured and is ready for testing, and a representative from the user's quality or inspection department goes to the supplier's facility to observe the testing. The source inspector certifies the test reports and approves the adhesive for immediate use on arrival at the user's facility.

A second alternative is to annually evaluate the supplier's testing facilities and procedures, and certify their lab as a qualified testing facility. Once certified, the supplier's lab would send test reports and certifications, either before or along with each adhesive shipment. Those documents would represent the basis for the user's acceptance of the product. Again, the result is that the adhesive is immediately available for use when it arrives at the user's facility.

With either of these alternatives, a "comfort factor" can be added by periodic audit testing of randomly selected batches. This will assure the continuing reliability of the supplier's data.

REFERENCE

1. R.B. Krieger, Jr., A Test for Allowable Stress on Face Bonds to Aluminum Honeycomb Core in Hostile Fluid Environments, in *Proceedings of the 31st International SAMPE Symposium*, Society for the Advancement of Material and Process Engineering, 1986

Processing Quality Control

Richard W. Roberts, General Dynamics Convair Division

THE USE OF POLYMERIC adhesives and sealants in a structural application requires a great deal of confidence in the capabilities of the materials. These capabilities include not only the structural strength or integrity of the material but more importantly its ability to provide repeatable and reproducible results. There are numerous different kinds of sealants and adhesives. Their types and chemistry are explained in the Sections "Adhesives Materials" and "Sealant Materials" in this Volume. However, because of the chemically reactive nature of these materials, they all share several common requirements for quality production use.

This article takes the approach that quality is not a characteristic that can be added by inspection, but rather must be built in by using the best possible manufacturing practices and standards. Traditional inspection-based quality (reactive) is compared to proactive ways of planning quality from within.

The processing of the material is key in attaining consistent results. There are several factors within the processing cycle that directly or indirectly affect the final outcome of the adhesive or sealant application. These factors include: adhesive manufacturing batch-to-batch consistency, material age-life history, moisture exposure, application (adherend) geometry, surface preparation, handling and exposure to contaminants, tooling, lay-up technique, and curing (time, temperature, vacuum, and pressure). This article will deal primarily with the common requirements of these systems and how each of the above factors contributes to, and can be controlled to achieve, optimum results. Generally, the example used for this article is an epoxy film adhesive. However, the rules apply to most materials and only exceptions will be noted. A typical adhesive bonding operation follows the process flow depicted in Fig. 1.

Material Manufacturing Consistency

The adhesives manufacturing industry has worked for many years to provide the manufacturing industry with consistent materials. The adhesives and sealants available today offer a wide range of properties and environmental exposure possibilities. Each formulation of a material is unique, in terms of constituents and their quantities, in order to provide these properties. This causes the opportunity for some variation in the properties of different batches of material.

Much work has been done in the characterization of materials to identify the property effects caused by the chemical changes from batch-to-batch variation and the means to identify it. The use of high-pressure liquid chromatography (HPLC) and other techniques provides the means to analyze the chemical content of the materials, but is just barely yielding significant correlations to final physical properties. Some work documented by R.J. Hinrichs in the *Proceedings of the SAMPE 34th International Symposium* shows a direct relationship of the batch-to-batch variations of adhesives constituents to the wettability and rheology of the adhesive, and shows how this can be compensated for in the cure cycle. This will be discussed more in the Section "Curing" in this article.

Industry has historically responded to this problem by testing all lots of incoming material prior to use. The tests are often curtailed or reduced on the basis of statistical process control (SPC) after a proven track record is established. Because these tests are expensive for the material purchaser, and thereby add to the real cost of the material, many material suppliers are implementing SPC on their own processes to maintain a competitive edge. The theory of material supplier certification is gaining support within the industry as a means to improve quality and cost issues associated with adhesives.

Material Age-Life History

Because all sealants and adhesives are chemically reactive species, they are sensitive to age-life history. Age-life history is a term that is used within this article to describe the storage conditions of an adhesive system prior to full cure. This includes temperature and length of exposure time for material shipping, receiving, kitting, cutting, and handling prior to full cure. The degree to which age-life history affects the properties of the final-cured material varies greatly. Generally, the higher the final strength of the material, the more sensitive it is to age-life history. Also, the more critical the application of a material is, the more controlled the material must be in terms of age-life. The entire age-life cycle of a polymeric material is shown in Fig. 2.

The term shelf life is often applied to describe age-life history. The use of shelf life is merely an indicator of the actual state of the adhesive resin. The shelf life of a household two-part epoxy may be several years at room temperature. However, the shelf life of a one-part film adhesive is usually six months to a year at depressed temperatures, −18 to 4 °C (0 to 40 °F). The

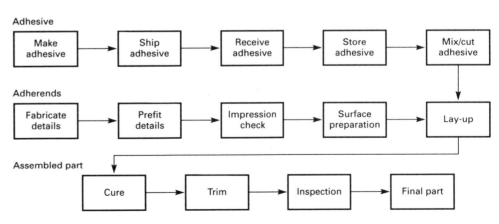

Fig. 1 Typical process flow for adhesive bonding

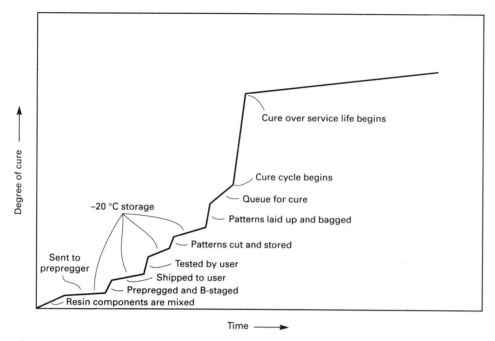

Fig. 2 Material age-life representation. Source: Applied Polymer Technology Inc.

radical difference in these shelf lives can be explained in two ways. First, the household epoxy is a two-part material, where one part is commonly called the hardener (catalyst) and the other, the resin. In two-part systems, the components are not mixed until the adhesive is about to be used. In a one-part system (such as the film adhesive), the resin and catalyst are mixed at the time of adhesive filming. Therefore, they have the ability to react at all times. The use of cold storage takes advantage of the generally exponential dependence of time-temperature exposure and reaction speed. This is typically explained by an Arrhenius-type equation, as shown in Eq 1.

$$k = Ae^{-(E_a/RT)} \qquad \text{(Eq 1)}$$

where k is the reaction rate, A is a constant, E_a is the activation energy, R is the gas constant, and T is the temperature in absolute units. The material continues to react, but at a greatly reduced rate. Notably, the two-part resin will also self-polymerize, given adequate time.

If one were to believe that shelf life is a good measure of age-life history, then one might wonder why so many materials are requalified after their shelf life expires. Actually, shelf life is an easily measurable parameter to quantify the ability of the resin to react. The number is usually determined by the development of test data over a period of time. As the resin ages, it loses properties by a reasonably predictable behavior. The shelf life is determined to be the period of time that, assuming the resin ages predictably at the specified storage conditions, the resin will continue to yield results within the design allowables at the time of full cure. Essentially, this means that if correctly processed within its shelf life, the resin will produce good parts.

The actual life of a material is difficult to quantify. The actual time-temperature exposure that a material experiences is difficult to both accurately measure and assume its accuracy. These are the "open loops" of the process, which often drive manufacturers to impose the use of test coupons to verify material acceptability at the time of cure. In an effort to remove these reasons for expensive tests, several manufacturers (material and users) are experimenting with systems to actually measure temperature and time exposure independently of any operator intervention. The combined use of certified material, within its shelf life, and the automatic measurement of age-life history, are major steps in the assurance of product quality.

Moisture Exposure

A condition that is very closely related to material age-life history is the absorption of moisture by polymeric materials. Moisture is a significant chemical inhibitor in epoxy resin systems and can cause voids and porosity in epoxy and all other resin types. Interestingly, moisture is required by silicone-base one-part sealants and adhesives in order to cure. This is more fully covered in the curing section "Curing" in this article. The measurement of the moisture content of a material sample prior to cure is the subject of much research at present. Because no easy, nondestructive way exists to measure and remove excess moisture from the resin system prior to lay-up and cure,

the industry has adopted the approach of using a controlled environment in order to control moisture absorption.

This technique uses a clean-room-type atmosphere with controlled temperature and humidity in order to minimize the moisture absorbed by the adhesive resin. The moisture absorption rate varies somewhat by type of material but generally is consistently affected by the temperature and humidity of its surroundings. Generally, the humidity within the clean room is maintained at less than 50% relative humidity (RH) and the temperature between 18 and 24 °C (65 and 75 °F). Some companies have tighter or looser restrictions. This temperature range provides both the operators and the material with a reasonable environment.

Some moisture is helpful in epoxies because it plays a large part in the tackiness and thereby the handleability of the material during lay-up. For example, southern California occasionally experiences periods of extreme dryness in the atmosphere, that is, 8 to 20% RH. During these times it is often necessary to restrict complex lay-ups simply because the adhesives have insufficient tack for the lay-up process. The reasons for this will be explained in the section "Lay-Up Technique" in this article.

One of the most important ways to maintain low moisture content on adhesives that are frozen is to ensure that the adhesive is sealed from atmospheric moisture during the thawing cycle. Opening the adhesive just once while frozen and allowing moisture to condense on the adhesive will allow more moisture to diffuse to the adhesive than would occur in normal handling exposure to a controlled clean room for several days.

Application Geometry

Adhesives have very distinct limits to the thickness of bond lines that will provide acceptable properties after cure. A graph of bond line versus bond strength has the appearance shown in Fig. 3. Additionally, in applications where honeycomb is used between face sheets, the adhesive will not adequately bond to the honeycomb unless direct contact is made. The adhesive attaches to the honeycomb by wicking up the cells of the honeycomb to provide an adequate fillet and bond.

The verification of bond line thickness is often done using a technique called verifilm. The verifilm material is sandwiched within the adherends, similar to the way the adhesive would be, and is then cured. The verifilm can be one of several types of materials, including polyvinyl chloride (PVC) film, the actual adhesive sandwiched within clear release film, or a specifically formulated substitute adhesive material that does not adhere to the adherends.

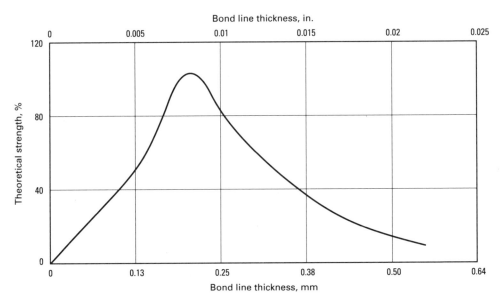

Fig. 3 Bond line thickness versus strength for a typical adhesive

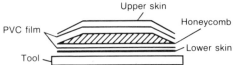

Fig. 4 Typical polyvinyl chloride verifilm lay-up sequence

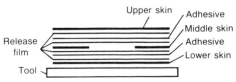

Fig. 5 Typical adhesive film impression check lay-up

The verifilm process usually occurs at a time called prefit in the process shown in Fig. 1. The prefit operation and its relation to the tooling are very important. Good tooling is essential to correct bond line achievement. This will be more completely explained in the section "Tooling Verification." The details (adherends) are placed in the tool as they would normally for final cure. However, they receive no special surface treatment for bonding, for example, Forest Products Laboratory (FPL) etch or phosphoric acid anodize. They are generally solvent cleaned to remove grease and metal pieces, but are not otherwise prepared for bonding. During prefit, it is essential that the adherends be checked for proper fit on the tool, tooling tabs, holes, and pins, which must fit without binding. The parts must be able to move against each other at the time of pressure application. Any binding of parts in the tool will cause uneven pressure distribution, and the bond line thickness will vary considerably.

The verifilm is placed between the adherends in place of the adhesive. The PVC-type film is used exclusively for honeycomb applications. It does not flow adequately to give precise measurements of metal-to-metal bonds. Typical lay-ups are shown in Fig. 4 and 5. Generally, the lay-ups cannot be successfully mixed because of the difference in cure temperature. However, the sandwiched adhesive will work on honeycomb core with some risk to the core, because the core cuts the release film.

The lay-up is completed by bagging the part using the same methods that will be used on the final part, and the part is cured. The cure for verifilm is generally shorter than it is for the final cure. The PVC film is heated to about 80 to 105 °C (180 to 220 °F) and held for 5 to 10 min, pressure is released

(if possible), and the part cooled. Standard adhesive and verifilm adhesive are generally heated to their respective standard cure temperatures and held for a much shorter time, such as 10 to 20 min, instead of 60 to 120 min. This allows the adhesive to flow and gel, but does not allow the adhesive to complete cross linking to build maximum strength, something that is not necessary when verifilm is used.

The parts are then debagged and disassembled. PVC film will clearly show an imprint from the honeycomb, if contact was made during the cure. A heavy imprint shows good contact, and a light imprint, marginal contact. Localized areas where the imprint cuts through the PVC generally means that the core is too tall in those areas and should be trimmed. If the PVC is cut through in most areas of the imprint, the material was heated too much or held at temperature too long, or the cure pressure was unreasonably high. Future "cures" of the verifilm must be revised to reduce the appropriate parameter to alleviate this problem. If the pressure is reduced on verifilm, it must also be reduced on the final cure.

Adhesive film layers are removed from the lay-up, and after removal of the release material, are measured with a micrometer to determine thickness. Areas of light and dark are easily discernible, but not necessarily quantitative. Careful measurement of the adhesive is required until the adhesive is well characterized as to its coloring. Any areas where the adhesive is squeezed out from the scrim cloth (in a supported film adhesive) indicates a much too thin bond line and must be corrected prior to bond. The impression check is often repeated several times until an acceptable impression check is obtained.

The verifilm inspection of bond-line thickness is a very difficult and expensive procedure. Essentially, the part is layed-up twice in the fabrication cycle. In addition, the verifilm adhesive will nearly always leak onto the adherends and has to be cleaned off prior to surface preparation for bonding. Additionally, the release films can inhibit the flow of adhesives as they would the typical flow during a regular cure cycle, thereby introducing uncertainty into the results of the process.

In summary, although the verifilm process yields some very important information about the lay-up of the part, its "value added" can be suspect. Some companies in the aerospace industry use verifilm and destructive testing of first articles and sampled parts in an effort to reduce the cost of fabrication by certification. This process is more fully described in the section on tooling.

Surface Preparation

Surface preparation of metal bonded parts is a very important step in the bonding process. It can consist of any of the following methods or a combination of them: solvent wipe, abrasion (sandpaper, Scotchbrite, or sandblasting), chemical processing, and application of adhesive primers.

The most simple surface preparation method is usually a type of abrasive treatment followed by a solvent cleaning. This provides adequate strength bonds for noncritical joints. This method is most often used with sealants and paste adhesives. In this case, the failure of the bond is not usually critical to the successful completion of the system's function. The problems of this method are that the preparation is very dependent on the skill and care of the operator. The final preparation is not inspectable by any means other than destructive testing.

Structural bonding with film-type adhesives usually uses a form of chemical surface treatment followed by an adhesive priming operation. This type of treatment is very rigorous in its ability to provide a good bonding surface for the adhesive. The chemical processing methods provide a microscopically rough surface for the interface of adhesive molecules. These surfaces are formed by several different processes, such as FPL etch (a sodium dichromate, chromic acid etch), phosphoric anodize, and chromic anodize. All of these processes form good bonding surfaces. The FPL etch and phosphoric acid anodize treatments are superior to the chromic acid anodize in terms of bond strength, but are less corrosion resistant.

The phosphoric acid anodize is remarkable in that it is inspectable after completion. The inspection involves the examination of the finished surface with a polarized filter at an oblique angle. Rotation of the polarizing filter shows a change of the incoming light. The actual technique requires some practice and demonstration, but this technique is important in that it allows the quality of the bonding surface to be evaluated prior to bond.

Chromic acid anodize is useable, but because of the chemical seal coats that are generally placed on the anodized surface, bond strengths are not optimum. Also, the anodize process must have the chemical concentrations in the solutions controlled to very precise levels. Generally, standard chromic acid anodize for prepainting treatment is not adequate enough to attain a good bonding surface.

The application of adhesive primer to the prepared surface is performed for a few reasons. The primer is generally the adhesive resin, dissolved in a solvent. By applying the primer on the prepared surface, it performs two functions. First, it seals the bonding surface from the atmosphere and generally will extend the period of time that the bonding surface can be stored prior to final bonding. Second, it gives the adhesive film a surface that is easily wetted during the cure. It is extremely important that the adhesive primer be applied per the manufacturer instructions. The primer layers must be the correct thickness and must be applied wet. If the primer is applied dry (that is, the solvent has flashed off before the primer is on the metal surface), the too-thick primer will inhibit the bonding of the adhesive to the metal substrate and the required adhesive strength will not be obtained. Primer application technicians must be well trained. Occasional random checks of primer thickness and test coupons can be used to assure continued priming operation quality.

Chemical processing quality is checked by the use of the wedge crack extension test. In this test, two 150 × 150 mm (6 × 6 in.) adherends that are 4.5 mm (⅛ in.) thick are processed by the chemical method of choice. The panels are then primed, laid up, cured, and cut into five 25 × 150 mm (1 × 6

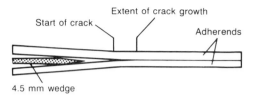

Fig. 6 Wedge crack test specimen

in.) panels. A 4.5 mm (⅛ in.) wedge is driven into one end of each specimen at the bond line. The wedge causes the adhesive to crack. The end of the crack is marked and the specimens are exposed to moisture in a humidity chamber. After a period of time the specimens are removed and the growth of the crack is marked and measured. Excessive growth (greater than 13 mm, or ½ in.) indicates that the process does not provide adequate bond preparation, and is not in control. This is a very rigorous test. A drawing depicting a completed wedge crack test specimen is shown in Fig. 6.

The wedge crack test is an excellent method to certify and continually verify the condition of the surface preparation. The wedge crack test should be run about once a week, along with chemical tests to verify concentrations and surface preparation quality. By doing this on a scheduled basis, the variability of the process can be kept under control and continually verified. This promotes a system that alleviates the need for test coupons to be run with every part.

Handling and Exposure to Contaminants

Adhesives and adherends are easily contaminated by many different sources. Bare hands and fingerprints can cause enough contamination to cause delaminations and disbonds. The oil in the hair and on the hands of lay-up operators is the single biggest source of contamination. Operators must never handle adhesive or details with bare hands after processing, or allow their hair to contact the same surfaces. These factors can be controlled by the use of white gloves, lab coats, and hair nets in the lay-up areas.

The introduction of silicone adhesives, sealants, or contaminants into a metal bond clean-room environment is a dangerous proposition that must be avoided. The lingering effects of airborne and touchborne silicon can be detrimental to bond integrity for years. Air motors, gasoline-powered carts, or even unfiltered air nozzles can introduce oil contamination, which is also very difficult to detect and remove. No such equipment must be allowed in the lay-up environment.

The clean-room air must be filtered to removal large airborne contaminants such

as smoke and dust. The actual level of filtration is dependent on the sensitivity of the adhesive systems. No smoking or drinking should be allowed in the clean-room facility because these activities produce significant risk for the maintenance of a clean bonding environment. The cleanliness of the facility must be monitored by periodic inspections and daily cleaning and damp mopping, in order to ensure that the lay-up environment will not be the cause of bonding problems.

Tooling Verification

Tooling is unquestionably one of the most important facets in the production of bonded assemblies. The tool has many functions. First, it must hold all of the details in the correct relationship for the adhesive bonding operation. Second, the tool must hold the parts in the right location without impairing the transfer of correct bonding pressure to the bond lines and/or supply the required bonding pressure. Third, the tool must provide the final contour of the part while performing and not inhibiting the above two functions. Fourth, the tool must allow the part to heat to cure temperature in a reasonable time. Fifth, it must provide a means to transition the pressure application over areas of sharp change in contour. Lastly, it must be sized to accommodate the even heating of the part to the cure temperature.

The design of the tool must simultaneously address all of the above issues and be sturdy and durable enough for multiple cure cycles without changing contour and provide the means for interfacing to the cure device (oven or autoclave). The tool must be proofed prior to its release to production. This can be accomplished simultaneously with the first-article impression checks and destructive test.

Unless the part to be bonded contains very expensive components, few parts are to be made, or the adherends are very heavy, a destructive test of a metal bonded assembly is the best way to ensure correct tool configuration. The process should follow the steps of Fig. 1, exactly, for standard production parts. The impression check is performed and used to map out any potential problem areas in the part. Correction of obvious deficiencies in the tool should be performed at this point, and the impression check repeated to ensure that adequate corrections were made. Upon a successful impression check, the parts are processed, laid-up, and cured. On particularly difficult to disassemble parts, the processing cycle can be skipped and a tool surface release agent can be applied to the adherends to aid in disassembly. This should be avoided if possible because it can change the surface tension forces that affect the way the adhesive resin flows during cure.

Fig. 7 Destructive test skin peeler

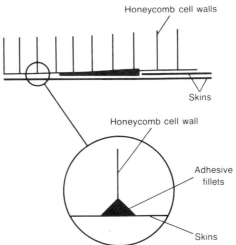

Fig. 8 Filleting and adhesive bridging of honeycomb core

The ability of the tool to meet heat-up rates and the part temperature gradient must also be checked during the first-article proofing. For this phase of the proofing, the part is extensively instrumented with thermocouples for the proofing run. Thermocouple data are collected and judgments are made as to the ability of the tool to provide correct heating. The tool must be designed to minimize the part temperature gradient to about 11 °C (20 °F) over the entire part. The reasons for this are described in the section "Curing."

After cure, the part is removed from the tool, measured for finished part thickness and conformance to contour requirements (this sometimes requires completion of the trimming of the part and test installation at its next assembly), nondestructively tested by ultrasonic C-scan or a similar technique, and then disassembled. The disassembly of a thin-skinned laminate can be eased by the construction of a "peeler," shown in Fig. 7. This looks much like an old sardine can opener, with the end being able to be opened in order to remove the skin. The strength of these adhesives is readily apparent during the act of disassembly. It will require considerable effort to make the adhesive fail in its typically lowest-value property, that of peel. After the first skin is peeled, the bonding surface is examined for the presence of voids, porosity, and bond line thickness. The results of the C-scan and thickness measurements are correlated to the actual condition of the part. This also helps in the validation of the nondestructive test methods in order to provide confidence that those techniques can determine actual flaw size and location.

Honeycomb areas are examined for good filleting of the adhesive at each cell wall and for the proper filling of any bridge areas, as shown in Fig. 8. After the successful completion of all layers, the tool can be certified as being capable of providing acceptable parts. If the test is not successful, the regions requiring rework are generally very easy to identify. Often, rework can be conducted without the requirement to repeat the destructive test, as long as the rework can be verified by other means, such as dimensional verification.

On first examination, this process appears to be expensive. However, this single nonrecurring expense can preclude the need for expensive recurring impression checking all parts. It also can provide the basis for accepting the parts based on only a nondestructive test and inexpensive dimensional checks.

Lay-Up Technique

The most important aspect of the actual lay-up is the training of the lay-up technicians. Because manual lay-up is the least controllable portion of the process, training and certification of the technicians are essential. The actual lay-up technique is relatively straightforward, but requires practice.

The important parameters to stress in the education of the lay-up technicians are the methods of adhesive cutting and adhesive application. The application of any adhesive requires that the adhesive be applied without entrapped air, which will cause voids. All entrapped air can rarely be bled out of the bond lines. The adhesive must be applied and smoothed to ensure that no air is entrapped. The tack of the adhesive is important because it allows the adhesive to be applied smoothly to the adherend and not peel off. Air bubbles under the adhesive layer will cause air bubbles and voids in the final part. All bubbles must be worked out

by puncturing the adhesive or by working the bubble to an edge. The adhesive must be applied to one side at a time. When the adherends are assembled, similar procedures must be followed to minimize entrapped air between the layers of metal.

There is a tendency by lay-up personnel to cut the adhesive to the shape of the part using a razor knife and using the actual part as a template. This is good in one respect, in that the adhesive can be cut to the exact size required. However, care must be taken to ensure that the metal is not scored or scratched by the knife. This includes scraping off the adhesive primer on the side of the part.

Curing

Curing is a complex subject that is examined here from the perspective of the different types of adhesives and sealants. First, the elevated-temperature curing of single-part or filmed adhesives is covered, providing insight not only into the film adhesives but also into many factors that affect all other types of adhesives. Next, two-part adhesives and sealants, and, finally, one-part adhesives and sealants, are covered, but only in terms of their exceptions.

Cure of Film Adhesives

Time and Temperature. The curing of the adhesive is the time during which the adhesive develops the majority of its desired properties. Correct cure is essential to the production of acceptable parts. The cure cycle can be performed in a variety of cure devices, including autoclaves, ovens, and specially designed cure tools. There are three very important facets to cure cycles: time, temperature, and pressure. Time and temperature are the components that allow the adhesive to develop the strength characteristics it requires by chemical reactions, and pressure is responsible in many ways for the actual physical configuration of the bond line.

Time and temperature control the kinetics of the chemical reaction, leading to the polymerization of the resin system. In Eq 1, it was shown that the reaction rate is exponentially dependent on the temperature in which the reaction is occurring. The higher the temperature, the faster the reaction. The time component is important to the reaction because it determines how long the reaction is allowed to progress on a given reaction path. Time is also important in the determination of the rheological curve that the resin follows during cure. The components are completely dependent on each other in terms of producing acceptable parts.

In the production environment, it is desirable to shorten the cycle as much as possible in order to gain efficiency on the cure device. This philosophy has some hid-

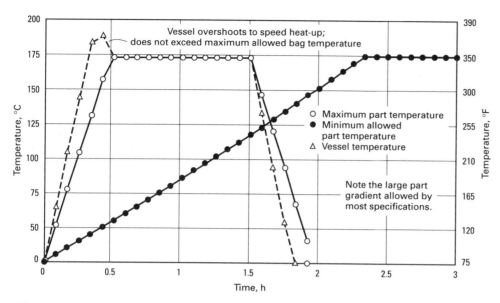

Fig. 9 Typical cure cycle for adhesives

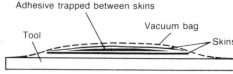

Fig. 10 "Trapping off" in a metal bond cure

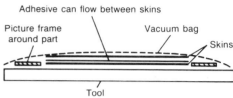

Fig. 11 Tooling concept to alleviate trapping off

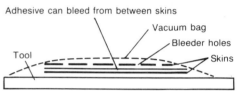

Fig. 12 Bleeder holes in adherends to allow adhesive to flow

den pitfalls in that the adhesive may cure to a given acceptable strength in a number of different heat rates, but the ability of the adhesive to wet out and attach itself to the adherend is very dependent on the period of time that the adhesive is allowed to flow at or around its low-viscosity point. Whereas a good batch of adhesive is often able to perform over a wide range of heat-up rates, marginal batches may not be as tolerant to the variations in heat-up rates.

This is a factor in why representative test coupons are not necessarily representative of the actual production part. The configuration of the test coupon and its fixtures rarely allows the test coupon bond line to experience the exact cure profile that the actual part experiences. This leads to scatter in the comparison of test coupon data to that of the actual parts. There are numerous incidences where the test coupon failed test, but, in an effort to use the part, a small piece of the actual part is destructively tested to check quality and easily passes the test. The review of the cure will often reveal that the test coupon either heated much faster or much slower than the actual part. This is by no means the only reason that test coupons fail, but there is little doubt that it contributes significantly.

Time and temperature are relatively easy to track during the cure cycle. There are several control systems that control the application of time and temperature on the parts. It is very important that the control system actively use the cure device ambient temperature to control the temperature of the part, not just the ambient temperature. The actual temperature of the cure device is not really significant, with one exception. The temperature of the ambient air in an oven or autoclave must not exceed the temperature limitation of the bagging materials. A chart showing this is depicted in

Fig. 9. The highest temperature line during the heat up is the ambient temperature of the cure device. The other lines represent the minimum and maximum part temperatures.

The gradient of the part temperature within the same part and among all parts during cure is very important. It is useful to remember Eq 1, where the rate of reaction is dependent on part temperature. If the part experiences any significant part temperature gradient at the time of gelation, large differences in properties can result. This is because where a resin experiences different temperatures, the part will differentially cure and induce stresses in the bond line and, possibly, differences in the bond strength in the various areas of the part.

Thermocouples are used extensively in the curing of adhesives. They provide real-time feedback to the controller on the temperature condition within the part. It is important that the thermocouples be correctly placed in order to accurately represent the temperature of the part. One method of doing this is to install a great number of thermocouples on the first-article destructive test part (see the section "Tooling Verification"). These multiple thermocouples are used to establish the highest and lowest temperature regions of the part. Thereafter, only the two thermocouples that represent these areas are installed for normal cures. This reduces the need to install multiple thermocouples and provides consistency in the data provided during cures. This method allows the cure methodology to be certified, and cure cycle conformance can be verified on a very small thermocouple sample.

Pressure. There are several myths concerning the use of pressure during metal bonding that must be corrected. First, the foremost reason for using pressure in the curing of adhesives is to provide bond line

thickness control by forcing the adherends next to one another at the time of cure. A small effect of pressure is to control the formation of voids and porosity in the adhesive during cure. The myth concerning pressure use is that more pressure is better. This is wrong, especially in metal bonding, where adherends are not porous. Figure 10 shows a cross section of a metal bond joint that exhibits a phenomenon called trap off. The application of pressure to this bond causes the edges of the top layer to squeeze off against the lower layers and prevents excess resin or volatiles from escaping. This causes uneven bond lines and entrapped air at the bond line. Tooling can be made to accommodate this type of problem by employing transition bars and pressure intensifiers. This concept is depicted in Fig. 11.

Another method of dealing with the problem of resin flow and air entrapment is the concept of bleed holes on the adherends, which allow the resin and air to escape from between the adherends into the core or bleeder area. This is shown in Fig. 12.

Honeycomb core necessitates additional considerations for pressure application. Two problems can arise in the use of core in a metal bond structure. One is core markoff and the other is core crush, but both of these problems are caused by the application of pressure during the bonding cycle. Markoff is the imprinting of the core onto the surface of the part. This is undesirable for aerodynamic and aesthetic reasons. Markoff can be minimized by using minimum pressure during the cure, thicker face sheets, or by using a caul sheet between the face sheet and the bag. Core crushing is also

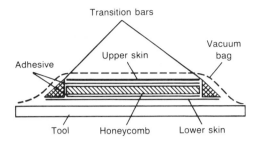

Fig. 13 Tooling utilizing transition bars

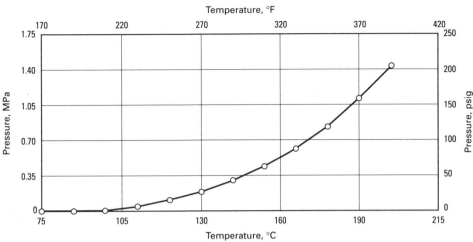

Fig. 14 Steam pressure versus pressure table

caused by excessive pressure during cure or by the uneven application of pressure to the surface of the part. Correct design of transition bars can alleviate this problem, as shown in Fig. 13. Another way that core gets crushed during cure is by the venting of the vacuum that is used during vacuum bagging. When the air tries to reenter the areas that previously held a vacuum, the sides collapse and the core is crushed. This also occurs when a vacuum bag breaks during the cure cycle. Usually, this effect occurs only with very lightweight honeycomb core. However, the effect is seen on all types of bag breaks. The vacuum problem can be addressed by limiting the vacuum used to bag parts for cure. The effect is minimized by reducing the applied vacuum to 17 to 34 kPa (5 to 10 in. Hg) instead of 68 to 100 kPa (20 to 30 in. Hg).

Vacuum Control. As noted above, vacuum can have a profound effect on the physical condition of the final part. However, it can also be used to manipulate resin chemistry in order to optimize the characteristics of the final part.

Vacuum is very important in the process of bonded structures because it allows the part to be bagged and tested for integrity at ambient pressure. Vacuum also has many detrimental effects to the curing resin, the most profound of which is the reduction of the ambient pressure surrounding the resin. This reduction allows the formation of volatiles that might otherwise be in solution in the liquid resin, which causes voids and porosity in the bond lines. This is the major reason that no adhesively bonded part should be cured under vacuum. Autoclave-cured parts must be at least vented to atmospheric pressure. Internal bag pressurization is described below. Parts cured in ovens must derive their pressure by mechanical means, rather than from vacuum.

The application of pressure to the entire structure will have the effect of transferring pressure into the resin itself, and is known as the fluid hydrostatic pressure (FHP) of the resin. This FHP has the ability to suppress the evolution of volatiles during the curing of the resin. The problem is that the applied external pressure is very inefficiently (roughly 10%) transferred to the resin as fluid hydrostatic pressure.

The alternative to producing FHP from external pressure is to induce pressure by introducing it (pressure) inside the bag via the "vacuum" lines attached to the bag. The technique is limited to a pressure that is less than the pressure applied to the exterior of the bag. The effect of this pressure is to suppress the formation of volatiles and to redissolve gases that have evolved into the resin. Additionally, air bubbles are reduced in size proportionally to the amount of pressure applied, thereby reducing some of the dependence of the process on lay-up technique.

The suppression of volatiles follows the curve of the steam tables very closely. The reason for this is clear. Generally, the volatile with the highest vapor pressure is water. If water vapor evolution can be suppressed, then all other volatiles have been suppressed. The stream chart is shown in Fig. 14.

Curing offers one of the best opportunities to incorporate certified control. The use of computer-controlled cure devices and the storage of data collected from carefully selected locations on the part represent a way to assure that the part has experienced the correct cure cycle. Additionally, the cure cycle can be manipulated in untraditional ways in order to assure that part quality is optimized.

Cure of Two-Part Adhesives and Sealants

Two-part adhesives and sealants present an additional uncertainty to the control of the process, that of mixing the adhesive. All of the same guidelines described in this article apply for surface preparation and use, but care must be exercised in the determination of the correct proportions and the thorough mixing of the adhesive.

Two-part adhesives most often have a pot-life requirement after mixing. It is essential that the adhesive be applied to the adherends and pressure applied before the expiration of the pot life. The adhesive may remain workable after the expiration of its pot life, but it must not be used because the ability of the adhesive to cure to the proper strength begins to degrade rapidly after that time. The pot life often necessitates an inspection operation to verify that the adhesive is applied in the correct time. However, the most recent approach is to certify the operator and rely on his craftsmanship to ensure the operation is carried out correctly.

Nearly all two-part adhesives have room-temperature and elevated-temperature cures. These can generally be used interchangeably, as both have been verified to produce the same results. The elevated-temperature cures are generally considerably shorter and therefore may provide dramatic improvements in flow time. Production rates and facility requirements will dictate the most appropriate cure method.

Correct cure and mixing can be easily verified for two-part adhesives. Generally, extra adhesive is mixed and a small sample is allowed to cure, along with the actual part. The sample is examined visually, and, possibly, by hardness to verify that the adhesive is correctly cured. This represents a simple non-intrusive test for adhesive quality.

Curing of Single-Part Sealants

Single-part sealants comprise both silicone and nonsilicone types, which are used in areas where either an extremely long pot life is desired, such as faying surface sealing, or extreme physical properties are needed. These sealants cure at room temperature or elevated temperature and generally cure in relationship to the thickness of the bond line and the distance to air.

Many single-part sealants (such as silicones) cure by a reaction with the moisture in air. Therefore, the rate of cure is directly in relation to the ability of the moisture to diffuse through the bond line to react with the material. These materials do not generally have accelerated cure cycles.

One-part sealants and adhesives are very difficult to verify as to final cure and application effectiveness. Representative samples are difficult and expensive to fabricate and test. Samples left in a cup to cure alongside the actual parts are not representative of the application and therefore suspect. Unfortunately, these materials are generally only verified by full-scale *in situ* testing. An example of this is the pressure test of an aircraft fuselage after assembly. These tests are slow and require a very long time between application and test. This results in a very slow evaluation of quality, which leads to very slow feedback as to part quality.

Quality Assurance

The assurance of adhesive and sealant quality can take two general forms. The first is to use part-by-part verification test coupons, which follow the part through the process. The second is to use process certification, with periodic recertification and random sampling of parts by nondestructive evaluation. The cost of each method must be weighed for each project that is going to be in production. If a part is very expensive or may only be built a few times, then part-by-part verification may be the appropriate choice. However, for any part that will be produced in more than a few units, process certification can achieve significant cost savings. The fabrication and testing of individual test coupons can be a very expensive process. In the current defense and commercial environments, cost savings measures that do not compromise quality constitute the correct path to profitable production programs.

End Product Nondestructive Evaluation of Adhesive-Bonded Metal Joints*

Donald J. Hagemaier, Douglas Aircraft Company, McDonnell Douglas Corporation

ADHESIVE-BONDED JOINTS are extensively used in aircraft components and assemblies where structural integrity is critical. Figure 1 shows how adhesives are used to join and reinforce materials, whereas Fig. 2 and 3 show typical adhesive-bonded joints used in aircraft structural assemblies. However, the structural components are not limited to aircraft applications, but can be translated to commercial and consumer product applications as well. In terms of production cost, ability to accommodate manufacturing tolerances and component complexity, facility and tooling requirements, reliability, and repairability, adhesive bonding is very competitive when compared to both co-cure and mechanical fastener joining methods.

Variables such as voids (lack of bond), inclusions, and variations in glue line thickness are present in adhesive-bonded joints and affect joint strength. This article addresses the problem of how to inspect bonded assemblies so that all discrepancies are identified. Once the flaws are located, engineering judgment and analysis can determine whether adequate strength exists or whether the parts need to be reworked or scrapped. When very large and costly assemblies have been bonded, the importance of good nondestructive inspections can be fully appreciated.

This article describes several techniques and presents drawbacks and limitations where they exist. Because no single method or procedure can locate every flaw, it is important for personnel to be aware of the available techniques and their limitations. Bonded assemblies are inspected immediately after the adhesive cure cycle, after the parts have been cleaned and removed from the manufacturing jigs or fixtures. The next important time for inspection is after the assemblies have been in use and the condition of the parts is known. It is necessary to

determine whether new delaminations have occurred or whether any metal cracks have developed. Inspection in the field requires that the instruments be easily transportable and be usable in confined places.

The terms that are commonly used describe what is to be done. For example, nondestructive testing (NDT) comprises the testing principles methodology, nondestructive inspection (NDI) involves inspection to meet an established specification or procedure, and nondestructive evaluation (NDE) is the actual examination of materials, components, and assemblies to define and classify anomalies or discontinuities in terms of size, shape, type, and location. This article addresses the three concepts as they apply to adhesive-bonded joints within structures.

Inspection reference standards must be developed for each instrument, and the standards must be made available to and used by inspectors to standardize their equipment and to permit them to detect the anomalies that indicate a flaw. Inspectors must also be familiar with the internal details of the assembly being inspected so that

they can distinguish between flaws and legitimate structural details, for example, they must be aware of the number of layers of metal that were bonded at any one spot.

Description of Defects

A wide variety of flaws or discontinuities can occur in adhesive-bonded structures. Table 1 lists possible generic flaw types and their producing mechanisms. Metal-to-metal voids or unbonds are the most frequently occurring rejectable flaws, as indicated in Table 2.

Interface defects are the result of errors made during the pretreatment cycle of the adherends prior to the actual bonding process. In practice, pretreatment flaws are reduced by careful process control and by adherence to specification requirements and inspection before proceeding with the bond cycle. Controls generally include the water-break test and measurement of the anodic layer and primer thickness. Interface defects can be caused by improper or inadequate degreasing, deoxidizing, anodizing,

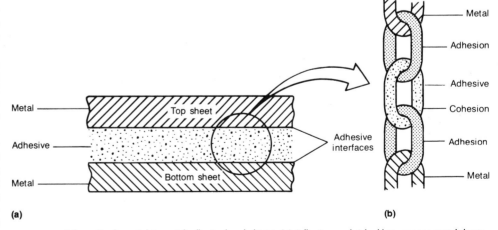

Fig. 1 Schematic of a metal-to-metal adhesive-bonded joint. (a) Adhesive sandwiched between two metal sheets. (b) Analogy of adhesive-bonded joint components to individual links in a chain

*Adapted with permission from *Adhesive Bonding of Aluminum Alloys*, Marcel Dekker, 1985, p 337-423

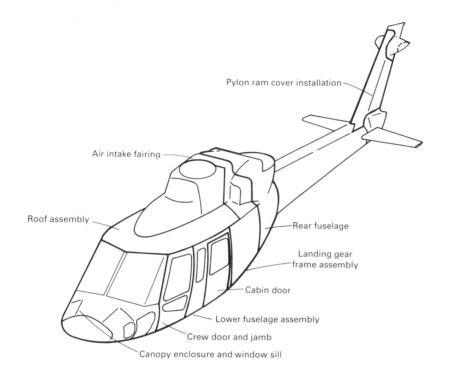

(a)

Pylon ram cover installation

Air intake fairing

Roof assembly

Rear fuselage

Landing gear frame assembly

Cabin door

Lower fuselage assembly

Crew door and jamb

Canopy enclosure and window sill

■ Fiber-reinforced plastic structure
▨ Fiber-reinforced aluminum honeycomb
▥ Aluminum honeycomb
▩ Metal-to-metal interface
▤ Titanium-faced honeycomb

(b)

Fig. 2 Typical applications of adhesive-bonded joints in aircraft. (a) Helicopter components. (b) Lockheed C-5A transport plane, with various types of honeycomb sandwich structures totaling 2230 m² (24 000 ft²) in area

or drying, by damage to the anodizing layer, or by excessive primer thickness.

Interface defects are generally not detectable by state-of-the-art NDT methods.

Therefore, test specimens are processed along with production parts and sent to the laboratory for evaluation. Applicable wedge crack specimens, lap-shear specimens, or honeycomb flatwise tension specimens are fabricated and tested to determine whether the process meets specification requirements before the bonding cycle starts.

Considerable effort at Fokker in the Netherlands led to the important discovery that the ideal oxide configuration for adhesion on aluminum alloys can be detected by inspection with an electron microscope at suitable magnification (Ref 2-4). Because in order to inspect with the electron microscope a piece of the structure must be removed, the electron microscope became a useful tool for adhesion quality control. Another physical parameter that was used as a basis for the NDT of surfaces for the ability to bond was contact potential, which is measured by a proprietary method developed by Fokker that uses what is known as a contamination tester (Ref 3, 5). This instrument is based on Kelvin's dynamic-condenser method but avoids the disturbances usually associated with it. There is sufficient evidence that the contamination tester is able to detect nondestructively the absence of the optimum oxide configuration arising from incomplete anodizing and/or subsequent contamination (Ref 4).

More recently, Couchman *et al.* (Ref 6) at General Dynamics Corporation developed an adhesive bond strength classifier algorithm that can be used to build an adhesive bond strength tester. Lap-shear specimens were fabricated using Reliabond 398 adhesive. The test specimens include:

● A control set with optimum bond strength
● An undercured set
● A weak bond produced by an unetched surface
● A thin-bond adhesive that was cured without a carrier

The weakest bond was observed to fail at 725 kPa (105 psi), while the strongest held to 15.7 MPa (2.27 ksi). Tabulated results showed the following:

● *Undercured set*: 725 to 6410 kPa (105 to 930 psi)
● *Unetched surface set*: 5.79 to 7.58 MPa (0.840 to 1.10 ksi)
● *Control set*: 13.4 to 14.8 MPa (1.94 to 2.15 ksi)
● *Thin-adhesive set*: 14.1 to 15.6 MPa (2.05 to 2.27 ksi)

The accept/reject value was set at 13.1 MPa (1.90 ksi), and all specimens were classified correctly. The important factor is that an interface defect (unetched surface), which results in poor adhesion, was detected.

Defects within the cured adhesive layer can be one or more of the following:

● Undercured or overcured adhesive

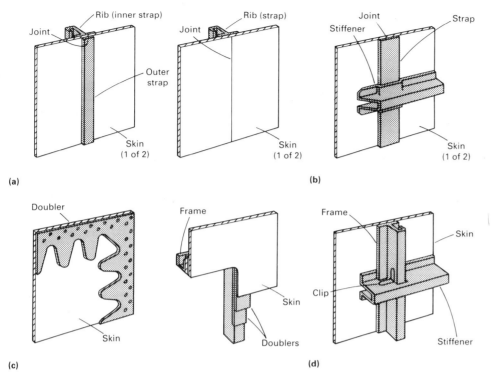

Fig. 3 Some typical adhesive-bonded joints used to join components in structural assemblies. (a) Skin splices. (b) Stiffener runout. (c) Bonded doublers. (d) Shear clip

Table 1 Generic flaw types and flaw-producing mechanisms

Flaw-producing mechanism	Generic flaw type				
	Metal-to-metal	Metal-to-core	Core	Surface	Adhesive
Disbonds, internal	X				
Disbonds, part edge	X	X			
Disbonds, high core	X				
Porosity	X	X	X		X
Unremoved protective release film from adhesive	X	X			
Foam adhesive in film adhesive bond line	X	X			
Cut adhesive	X	X			X
Adhesive gaps	X	X			
Missing adhesive	X	X	X		
Weak bonds	X	X	X		
Extra layers of film adhesive	X	X			
Foreign objects	X	X	X		X
Double drilled or irregular holes	X			X	
Disbonds, low core		X			
Void or gap, chemical milled land		X			
Void or gap, doublers		X			
Missing fillets		X			
Void, closure-to-core		X			
High-density inclusions (chips, and so on)	X	X			X
Voids, foam joint		X	X		
Disbond, shear ties	X	X			
Lack of sealant at fasteners		X			
Thick foam adhesive		X			
Broken fasteners	X	X			
Crushed core			X		
Wrinkled core			X		
Condensed core			X		
Distorted core			X		
Blown core			X		
Node bond separation			X		
Missing core (short core)		X	X		
Cut core		X	X		
Water in core			X		
Cracks				X	
Scratches				X	
Blisters				X	
Protrusions				X	
Indentations (dents/dings)				X	
Wrinkles				X	
Pits				X	

Source: Ref 1

- Thick adhesive resulting in porosity or voids due to improper bonding pressure or part fit-up
- Frothy fillets and porous adhesive caused by a too-fast heat-up rate
- Loss of long-term durability due to excessive moisture in the adhesive prior to curing

In normal cases, the curing time is very easily controlled. The curing temperature and temperature rate are controlled by proper positioning of the thermocouples on the panel and by regulating the heat-up rate.

Thick glue lines occur in a bonded assembly because of the inadequate mating of the facing sheets or blocked fixing rivets, and they result mostly in porosity and voids. However, a thick glue line made with added layers of adhesive is usually free of porosity. Porosity has a significant effect on the strength of the adhesive, with higher porosity related to a greater loss in strength and a void condition resulting in no strength. These defects occur quite frequently (Table 2). Porosity can also be caused by the inability of volatiles to escape from the joint, especially in large-area bond lines. Excessive moisture in the adhesive prior to curing can be prevented by controlling the humidity of the lay-up room. The entrapped moisture, after curing, cannot be detected by NDT methods unless it results in porosity.

Other defects that occur during fabrication can include:

- No adhesive film
- Protective film left on adhesive
- Foreign objects (inclusions)

In practice, these conditions must be prevented by process control and by training the personnel engaged in the bonding operation. The first two conditions occur infrequently. Shavings, chips, wires, and so on, can result in porosity or voids. Honeycomb core assemblies have been found with all types of foreign material.

Metal-to-Metal Defects

Voids. A void is any area that should contain, but does not contain, adhesive.

Table 2 Frequency of rejectable flaws in adhesive-bonded assemblies

Defect	Number of defects	Percentage of total
Metal-to-metal voids and disbonds	378	74
Skin-to-core voids and disbonds	19	3
Gap in core-to-closure bond	9	2
Lack of foaming adhesive or voids in foaming adhesive	22	4
Difference in core density	6	2
Lack of fillets	1	1
Crushed or missing core	32	6
Short core	40	8
Total	**507**	**100**

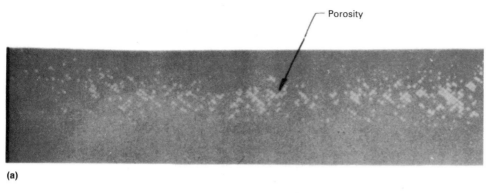

Porosity

(a)

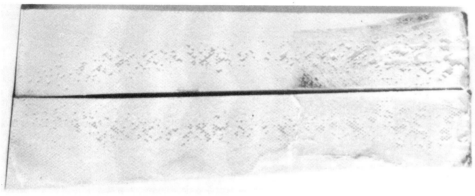

(b)

Fig. 4 (a) Neutron radiograph and (b) visual confirmation of porosity in an adhesive-bonded joint

Voids are found in a variety of shapes and sizes and are usually at random locations within the bond line. Voids are generally surrounded by porosity if caused by a thick bond line and may be surrounded by solid adhesive if caused by entrapped gas from volatiles.

Unbonds or Disbonds. Areas where the adhesive attaches to only one adherend are termed unbonds. Unbonds can be caused by inadequate surface preparation, contamination, or improperly applied pressure. Because both adherends are not bonded, the condition is similar to a void and has no strength. Unbonds or disbonds are generally detectable by ultrasonic or sonic methods.

Porosity. Many adhesive bond lines have some degree of porosity, which may be either dispersed or localized. The frequency and/or severity of porosity is random from one assembly to the next. Porosity is defined as a group of small voids clustered together or in lines. The neutron radiograph shown in Fig. 4(a) confirms the presence of porosity in the bond lines visible to the eye in Fig. 4(b). Porosity is detected equally well using conventional radiography or neutron radiography (Fig. 5). Scattered linear and dendritic porosity is usually found in adhesives supported with a matte carrier. Linear porosity generally occurs near the outer edge of a bonded assembly and in many cases forms a porous frame around a bonded lami-

nate. Porosity is usually caused by trapped volatiles and is also associated with thick (single-layer adhesive) bond lines that did not have sufficient pressure applied during the cure cycle. The reduced bond strength in these porous areas is directly related to its density frequency and/or severity.

Porous or Frothy Fillets. This condition results from a too-high heat-up rate during curing. The volatiles are driven out of the adhesive too rapidly, causing bubbling and a porous bond line, which is distinguished by the frothy fillets. This defect is visually detectable and should also be seen in the test specimens processed within the production parts.

Lack of Fillets. Visual inspection of a bonded laminate can reveal areas where the adhesive did not form a fillet along the edge of the bonded adherends or sheets. In long, narrow joints, a lack of fillet on both sides generally indicates a complete void. This defect is considered serious because the high stresses near the edges of a bond joint can cause a cracked adhesive layer due to shear or peel forces. A feeler gage can be used to determine the depth of the defect into the joint. If the gap is too tight for a feeler gage, ultrasonic or radiographic techniques can be used to determine the depth of the edge void.

Fractured or Gouged Fillets. These defects are detected visually. Cracked fillets are usually caused by dropping or flexing

the bond assembly. Gouges are usually made with tools such as drills or by impact with a sharp object. Fractured and gouged bond lines are considered serious for the reasons stated above for lack of fillets.

Adhesive Flash. Unless precautions are taken, adhesive will flow out of the joint and form fillets as well as additional adhesive flow on mating surfaces. Although the condition is not classified as a defect, it is considered unacceptable if it interferes with ultrasonic inspection at the edges of the bonded joint where stresses are highest.

Burned Adhesive. The adhesive may be burned during drilling operations or when bonded assemblies are cut with a band saw. The burned adhesive is essentially overcured, causing it to become brittle and to separate from the adherend. Also, the cohesive strength of the burned adhesive is drastically reduced. Figure 6 shows burned adhesive around hole 9 caused by improper drilling, as well as bond delamination along the edge of the panel adjacent to holes 16 through 20 caused by band sawing. Improper drill speed or feed, coupled with improper cooling, can cause these types of defects. The burned adhesive around hole 9 in Fig. 6 is detectable by ultrasonic C-scan recording methods (Fig. 7).

Adherend Defects

Adherend defects can be detected visually and do not include the processing procedures.

Fractures (Cracks). Cracks in the adherend, whatever the cause, are not acceptable.

Double-Drilled or Irregular Holes. Some bonded assemblies may contain fastener holes. When holes are drilled more than once, have irregular shapes, or are formed with improper tools, they are considered defects. The load-carrying capacity of the fasteners is unevenly distributed to the adherends, resulting in high local stresses that may cause fracture during service.

Dents, Dings, and Wrinkles. These defects are serious only when extensive in nature, as defined by applicable acceptance criteria or specifications. They are most detrimental close to, or at, a bond joint. Dents are usually caused by impact with blunt tools or other objects and are usually rounded depressions. Dings result from impact with sharp objects or when an assembly is bumped at the edge. Dents or dings may cause bond line or adherend fractures. Wrinkles are bands of distorted adherends and are usually unimportant.

Scratches and Gouges. A scratch is a long, narrow mark in the adherend caused by a sharp object. Deep scratches are usually unacceptable because they can create a stress raiser, which may generate a metal crack during service. On the other hand, gouges are blunt linear indentations in the adherend surface. Deep gouges, like scratches, are generally unacceptable.

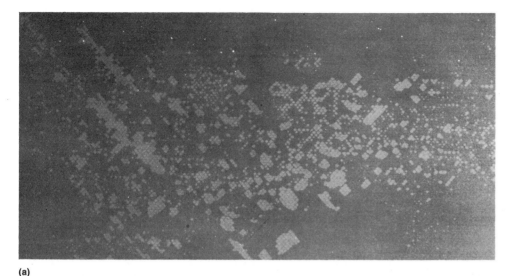

(a)

(b)

Fig. 5 Porosity in an adhesive-bonded joint. (a) X-ray radiograph. (b) Neutron radiograph

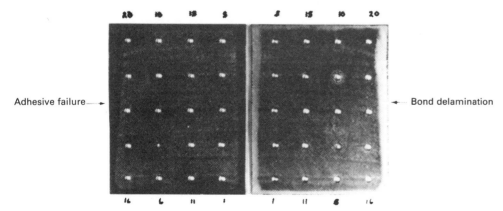

Adhesive failure → ← Bond delamination

Fig. 6 Examples of burned adhesive

Honeycomb Sandwich Defects

The most prominent defects found or generated in honeycomb sandwich assemblies are summarized in Table 1. Adhesion and/or cohesion defects may also occur in bonded honeycomb sandwich assemblies. The metal-to-metal closure areas for honeycomb panels may exhibit the types of defects discussed in the preceding section. In addition, sandwich assemblies can have defects in the honeycomb core, between the core and skins, between core and closure, at chemically milled steps, and in core splices. These bond areas are shown in Fig. 8 for a typical honeycomb assembly.

Water in Core Cells. Upon completion of the bonded assembly, some manufacturers perform a hot-water leak test to determine whether the assembly is leakproof. If the assembly emits bubbles during the leak test, the area is marked and subsequently repaired. To ensure that all bonded areas are inspected and that no water remains trapped in the assembly, it is then radiographed. This is important because water can turn into ice during operational service and rupture the cells, or it can initiate corrosion on the skin or core. Water in the core can be detected radiographically when the cells are filled to at least 10% of the core height. Also, x-ray detection sensitivity is dependent on the sandwich skin thickness and radiographic technique. An additional problem is the ability to determine whether the suspect area has excessive adhesive, filler, or water. Water images usually have the same film density from cell to cell or for a group of cells, whereas adhesive or filler images may vary in film density within the cells or show indications of porosity. A radiographic positive print of moisture in honeycomb is shown in Fig. 9.

Crushed Core. A crushed honeycomb core may be associated with a dent in the skin or may be caused by excessive bonding pressure on thick core sections. Crushing of the core greatly diminishes its ability to support the facing sheets. Figure 10 shows an x-ray positive print of crushed core. Generally, crushed core is most easily detected with angled x-ray exposures. Crushed core is defined as localized buckling of the cell walls at either face sheet, when associated with the halo effect on a radiograph. On the other hand, for wrinkled core, the cell walls are slightly buckled or corrugated. Radiographically, the condition appears as parallel lines in the cell walls. A wrinkled core is generally acceptable.

Condensed core occurs when the edge of the core is compressed laterally. Lateral compression may result from bumping the edge of the core during handling or lay-up, or slippage of detailed parts during bonding. The condition occurs most often near honeycomb edge closures. Figure 11 shows a positive radiograph of various degrees of condensed core.

Node separation results when the foil ribbons are separated at their connecting points or nodes, as shown schematically in Fig. 12 and in the photograph in Fig. 13. Node separation usually occurs during core fabrication. It may also result from pressure buildup in cells as a result of vacuum bag leaks or failure, which allows the pressurizing gas to enter the assembly and core cells.

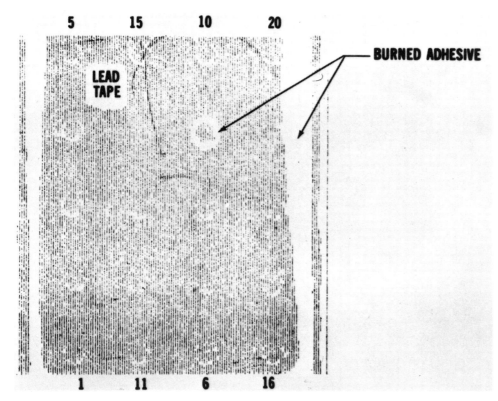

Fig. 7 Ultrasonic C-scan recording of right side of Fig. 6

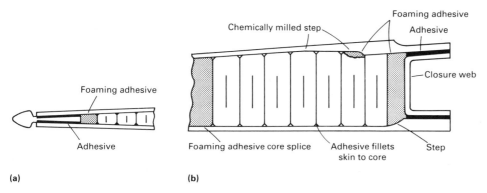

Fig. 8 Typical configuration of bonded honeycomb assembly. (a) Trailing edge. (b) Leading edge

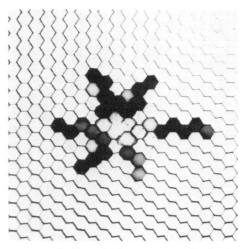

Fig. 9 Positive print from x-ray negative showing water intrusion into honeycomb cells

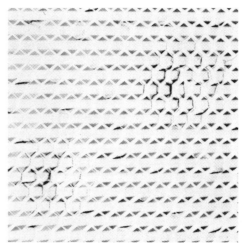

Fig. 10 Positive print from x-ray negative showing crushed honeycomb core

Blown core occurs as a result of a vacuum bag leak or because of a sudden change in pressure during the bonding cycle. The pressure change produces a side loading on the cell walls that can either distort the cell walls or break the node bonds. Radiographically, this is indicated as:

- Single-cell damage, usually appearing as round or elliptical cell walls with partial node separation
- Multicell damage, usually appearing as a curved wave front of core ribbons that are compressed together

The blown core condition is most likely to occur at the edge of the assembly in an area close to the external surface, where the greatest effect of sudden change in pressure occurs. This condition is most prevalent whenever there are leak paths, such as gaps in the closure ribs to accommodate fasteners, or chemically milled steps in the skin where the core may not fit properly. When associated with skin-to-core unbonds, the condition is detectable by pulse echo and through transmission ultrasonic techniques. The condition is readily detectable by radiography when the x-ray beam centerline is parallel to the core cell walls, as shown in Fig. 14.

Voids in Foam Adhesive Joints. Defects in core-to-closure, core-to-core splice, core-to-trailing edge fitting, and chemically milled steps (Fig. 8) in foam adhesive joints can result from certain conditions:

- The foam adhesive can slump or fall, leaving a void between the core and the face skins. This particular condition is most readily detected by ultrasonics
- The core edge dimension may be cut undersize, and as a result the foam does not expand uniformly to fill the gap between the closure and the core
- The foaming adhesive can fail to expand and surround the core tangs, as illustrated in Fig. 15

Protective Film Left on Adhesive. Protective films are usually given a bright color so that they can be seen and removed from the adhesive before bonding. If they remain on the adhesive, the adhesive is prevented from contacting one of the adherends. This condition is very difficult to detect by ultrasonic techniques and x-ray radiography, with which it would appear as a mottled condition. It generally produces porosity, especially at the perimeter, which aids in its detection.

Metal-to-Metal Defects. The most prominent defects are voids and porosity. Disbonds can occur under particular conditions:

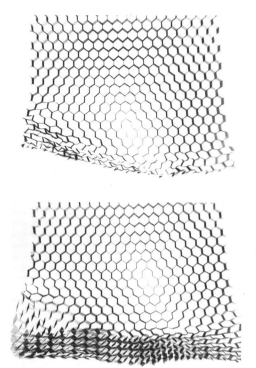

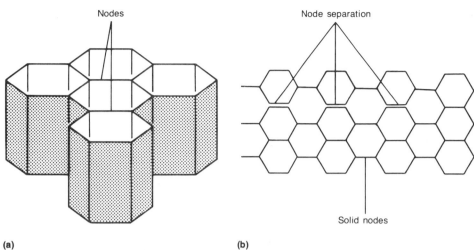

Fig. 11 Two x-ray positive radiographs showing various degrees of condensed core

Fig. 12 Examples of honeycomb core separation. (a) Joined and solid nodes. (b) Node separation

- A core slightly higher than a closure member
- Lack of applied pressure from tooling
- Entrapped volatiles in the bond joint prior to cure
- Excessive moisture in the adhesive prior to cure
- Contaminated honeycomb core

Voids and porosity are detectable by low-kilovoltage (15 to 50 kV) x-ray techniques using a beryllium-window x-ray tube, by thermal neutron radiography, and by ultrasonic C-scan techniques employing small-diameter or focused search units operating at 5 to 10 MHz. If the flaw is the result of insufficient pressure, the adhesive will be porous, as shown in Fig. 4 and 5. The lower the kilovoltage and/or the thicker or denser the adhesive, the higher the resolution of the flaw image. In general, the flaw size detectable by radiography is smaller than that detected by ultrasonics. Also, some adhesives, such as AF 55 and FM 400, are x-ray opaque, resulting in a much higher contrast between voids, porosity, and solid

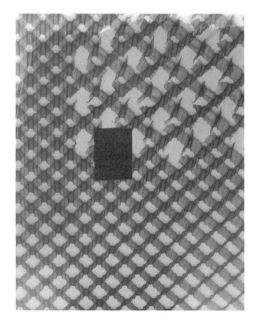

Fig. 13 Radiograph of node separation

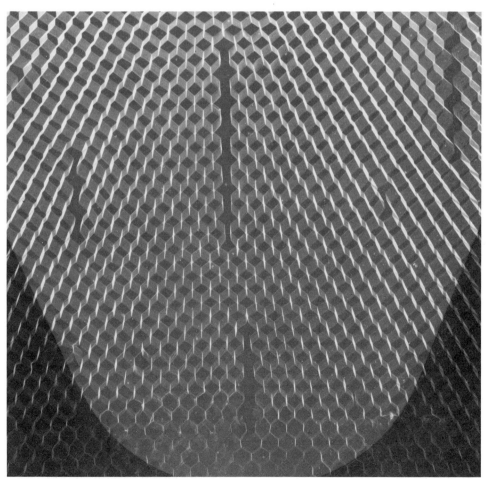

Fig. 14 Radiographic illustration of a blown core

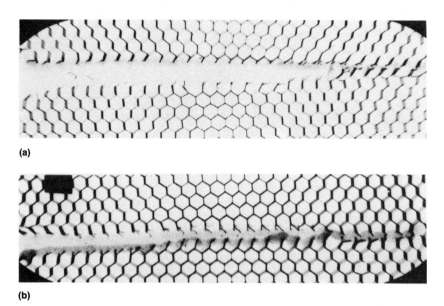

(a)

(b)

Fig. 15 Positive print from x-ray radiograph of foaming adhesive at core splice. (a) Lack of or unacceptable foaming at core splice. (b) Acceptable foaming at core splice

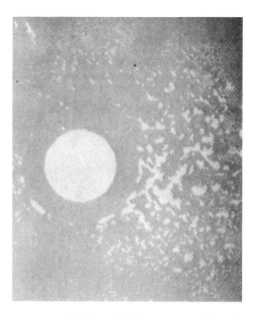

Fig. 16 Positive print from neutron radiograph showing void and porosity in an adhesive bond line. Two 13 mm (½ in.) thick aluminum adherends bonded with EA 9628 adhesive are shown; AF 55 adhesive bond would have yielded similar results with x-ray radiography.

adhesive (Fig. 16). Disbonds (separation between adhesive and adherend) are best detected by ultrasonic techniques.

One of the most important types of voids in honeycomb assemblies is the leakage-type void, which is oriented normal to the metal-to-metal bond line and penetrates to the core. Moisture can penetrate such a void during operational service and cause corrosion or ice damage to the core. This type of void is shown in Fig. 17(a) and (b).

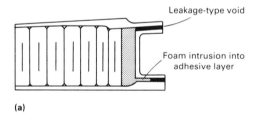

(a)

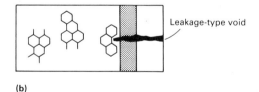

(b)

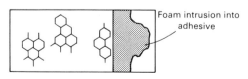

(c)

Fig. 17 Schematics of leakage-type void and foam intrusion in metal-to-metal joints. (a) Leakage-type void and foam intrusion into adhesive layer due to excessive gap between extrusion and skins. (b) Leakage-type void. (c) Foam intrusion into adhesive

Foam Intrusion. Another type of defect that can cause problems in radiography is foam intrusion into a metal-to-metal joint near a closure. This type of flaw is shown in Fig. 17(a) and (c). Ultrasonically, it will appear bonded; therefore, ultrasonics cannot confirm the radiographic findings. Because this may or may not be a defect, it must be determined by engineering analysis.

Skin-to-Core Voids at Edges of Chemically Milled Steps or Doublers. This condition occurs when the adhesive fails to bridge the gap at the edges of chemically milled or laminated steps or doublers (Fig. 8). This is detected radiographically as a dark line or an elliptically shaped dark image. Ultrasonically, it will appear as a linear void along the chemically milled step position.

Missing Fillets. As pressure is applied during the bonding cycle, adhesive fillets are formed at the edges of each honeycomb cell. Fillets will not be formed if pressure is not maintained. If the adhesive is x-ray opaque, this condition is readily detected by directing the radiation at an angle of approximately 30° with respect to the centerline of the core or closure web. If the adhesive is not x-ray opaque, then ultrasonic, eddy sonic, and tap tests can be used to locate the area having unbonded cells. Ultrasonic C-scan can be used to record skin-to-adhesive and adhesive-to-core voids. Adhesive-to-core voids are more difficult to detect than skin-to-adhesive voids.

Short core can exist if the core edges are cut shorter than the assembly into which it will be placed and bonded. It is detectable by x-ray radiography and is evident as core edges that do not tie into the closure via the foaming adhesive.

Foreign Objects. Honeycomb assemblies may contain foreign objects as a result of poor fabrication practices. Nuts, rivets, small fasteners, metal chips, and similar items can be left in the honeycomb cells before bonding or as a result of drilling operations for fastener installations near hinge points. These objects are easily detected by x-ray radiography. They are usually not detrimental if they can be potted in place with a room-temperature adhesive.

Repair Defects

All the flaws or defects defined previously for metal-to-metal or honeycomb sandwich assemblies can occur during salvage or repair of these components. Repairs must be of good quality in order to maintain the reliability of the bonded structure during continued service.

In-Service Defects

Most defects or flaws caused during service originate from impact damage, corrosion, or poor fabrication.

Impact Damage. Many bonded assemblies are made from thin materials and are susceptible to damage by impact. Damage can be caused by small projectiles, work stands, dropped tools, personnel walking on no-step assemblies, stones or other debris thrown by aircraft wheels, or similar damage. Impact imposes strain on the adhesive, causing it to crack or separate from the adherends. Impact can cause crushed honeycomb core, resulting in loss of strength. The crushed core can resonate during service and slowly degrade the adhesive by fatigue until it debonds from the adherend or until cracks occur adjacent to the skin-

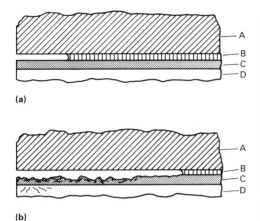

Fig. 18 Schematics illustrating the causes of adhesive delamination for a metal adherend. (a) Results of moisture entry in the unstable oxide. (b) Corrosion of alcladding and base aluminum. A, adhesive primer system; B, oxide; C, alcladding; D, base aluminum

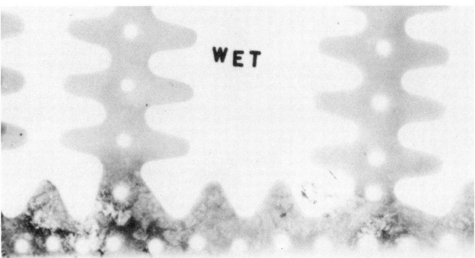

Fig. 19 Positive print from x-ray radiograph showing interface corrosion of an adhesive-bonded aluminum laminate

to-core fillets. Fortunately, impact damage generally leaves a mark in the surface of the part. These surface marks pinpoint or indicate possible subsurface damage, which can then be evaluated by NDT inspections.

Corrosion can be found in all structural concepts. However, a good example of bonding excellence is the honeycomb acoustic panels for the DC-10 and 747 engine inlets. These applications are quite demanding because of the combination of sonic and ambient environment being introduced into the perforated sandwich structure. Stable oxide surface preparation is an essential part of the bond foundation. Improper surface preparation can result in an unstable oxide layer, which may allow the entry of moisture, delamination, and/or crevice corrosion (Fig. 18).

Moisture entry to the core of a honeycomb assembly eventually leads to corrosion of the core. This occurs when moisture moves along the bond line to the individual cells. Moisture moves more rapidly in an assembly if the core is perforated. Fortunately, this moisture problem is detectable by NDT inspection methods. Water is detectable by x-ray radiography or by acoustic emission testing with a hot-air gun or heat lamp to cause boiling or cavitation of the water (Ref 7-9). Core corrosion and subsequent core-to-skin delamination is detectable by a variety of NDT methods, such as x-ray, contact ultrasonic ringing, sonic bond testers, eddy sonic, tap test, and acoustic emission with a heat source.

Figure 19 shows a section from a commercial aircraft wing trailing edge that had moisture entry at the leading and trailing edges, resulting in bond line corrosion. The corrosion in this panel was detected by x-ray radiography, bond testers, tap test, neutron radiography (which looked much like Fig. 19), and acoustic emission using a heat source.

Poor Fabrication (Flaws or Weak Bonds). In-service failures can occur from weak bonds (adhesion failures) caused by poor surface preparation, unstable oxide failure, and corrosion of alcladding and base aluminum alloy (Fig. 18). Many tests show that the alclad on 7000-series aluminum sheet is very susceptible to corrosion attacks in the bond line and must be thoroughly tested before consideration for bonding operations. The 2000-series alclad aluminum alloys are not as susceptible to this condition. Both series of alloys with no alclad surface but with good surface preparation, adhesive, and corrosion-inhibiting primer will have no corrosion problems.

Another manufacturing condition that leads to adhesion failures is caused by improper cleaning of the honeycomb core prior to bonding. This condition results if glycol (which is often used to support the core when it is machined) is not completely removed prior to bonding. Incomplete removal of glycol will cause a weak bond to exist, and in-service stresses may cause a skin-to-core delamination. This condition is controlled by adding a fluorescent tracer to the glycol. After the core is machined and cleaned, it is inspected, using an ultraviolet (black) light, for any residual glycol prior to bonding.

Applications and Limitations of NDT to Bonded Joints

A variety of NDT methods are available for inspection. Only the methods applicable to the inspection of bonded structures will be discussed in this section. All the methods or techniques presented can be used in fabrication inspections, while only a limited number are applicable to on-aircraft or in-service inspection. The methods described below have proved to be the most success-

ful in detecting flaws in bonded laminates and honeycomb assemblies. The Section "Methods of Nondestructive Evaluation" in *Nondestructive Evaluation and Quality Control*, Volume 17, *Metals Handbook*, ASM INTERNATIONAL, contains articles that describe in detail many of the techniques identified below, and others as well.

Visual Inspection

All metal details must be inspected to ensure conformity with design. Before large assemblies are ready for bonding, the details are assembled in the bonding jig as though bonding were to occur. In place of the adhesive, a sheet of Verifilm is used. This material has nearly the same flow characteristics as the adhesive, but it is prevented from bonding by release film so that it will not stick to the details. The whole assembly is placed in an autoclave or press just as though it were being bonded. After the heat-pressure cycle, the parts are disassembled, and the Verifilm is inspected visually to determine whether the pressure marks are uniform throughout. A uniform marking gives good assurance of proper pressure at the bond lines. All areas showing no pressure must be inspected and parts must be modified to obtain proper fit and pressure during the cure cycle. A poor showing on the Verifilm is cause to rerun the check.

When the details go through an anodize cycle, a visual check can be made for phosphoric acid anodize but not for other types of anodize. For visual verification of phosphoric anodize, the inspector looks through a polarizing filter, at an angle of approximately 10° from parallel to the surface of the anodized part. The surface of the panel is well lighted by a fluorescent white light tube, and the inspector rotates the polarizing filter while observing the anodized sur-

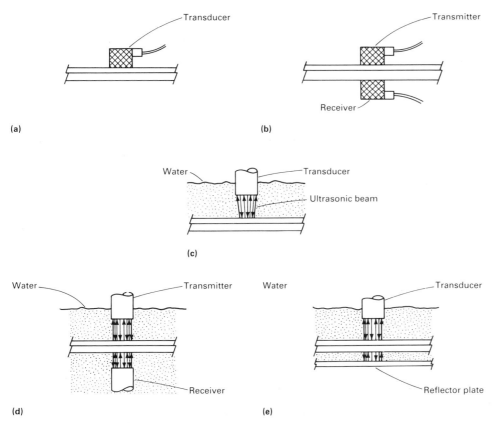

Fig. 20 Ultrasonic inspection techniques. (a) Contact pulse echo with a search unit combining a transmitter and receiver. (b) Contact through transmission. Transmitting search unit on top and receiving search unit on bottom. (c) Immersion pulse echo with search unit (transmitter/receiver) and part inspected under water. (d) Immersion through transmission with both search units (transmitter and receiver) and part under water. (e) Immersion reflector plate. Same as (c) but search unit requires a reflector plate below the part being inspected

face. If the panel has been properly anodized, the inspector will see a change in hues equivalent to the colors of the rainbow.

From the opened-wedge-crack specimen, the inspector looks at the failed surface to determine the type of failure. Areas of poor adhesion occur where the adhesive has separated from the substrate. This condition is manifested by variations in color and texture. A cohesive failure occurs through the adhesive and is uniform in color and texture.

After an assembly is bonded, it is inspected visually for scratches, gouges, dings, dents, and buckles. The adhesive flash and fillets can be inspected for cracks, voids, and unbonds at the edge. Feeler gages can then be used to determine the depth of the unbond at the edge. The in-service visual inspection of bonded joints can reveal cracked metal or adhesive fillets, delamination or debonding due to water intrusion or corrosion, impact or foreign object damage, and blisters, dents, and other mechanical damage.

Ultrasonic Inspection

A number of different types of ultrasonic inspection techniques using pulsed ultrasound waves at 2.25 to 10 MHz can be applied to bonded structures. Following are brief descriptions of various techniques.

Contact Pulse Echo. In this technique, the ultrasonic beam is transmitted and received by a single search unit placed on one surface of the part (Fig. 20a). The sound is transmitted through the part, and reflections are obtained from voids at the bond line. If the bond joint is of good quality, the sound will pass through the joint and will be reflected from the opposite face, or back side. Bonded aircraft structure is usually composed of thin skins, which result in multiple back reflections on the cathode ray tube (CRT) screen of the pulser/receiver. The appearance of multiple reflections on the CRT has prompted some inspectors to term this the ringing technique. When a void is present, the reflection pattern changes on the CRT and no ringing is seen. For bonded parts having skins of different thickness, inspection should be conducted from the thin-skin side.

Contact through transmission (Fig. 20b) is useful for inspecting flat honeycomb panels and metal-to-metal joints. Special search-unit holding devices have been fabricated so that the test can be performed by one inspector. Such a device is used for inspecting the metal-to-metal closures of a bonded honeycomb panel. Longer and wider-spaced holders have been fabricated from tubing. The holding tool should be custom designed for the assembly being inspected. To perform the test, liquid couplant must be applied to both sides of the assembly.

Immersion Method. For this method, the assembly must be immersed in a tank of water, or water squirters must be used to act as a couplant for the ultrasonic beam. There are three fundamental techniques: pulse echo (Fig. 20c), through transmission (Fig. 20d), and reflector plate (Fig. 20e). Typical CRT displays for bonded and unbonded samples using these three immersion techniques are shown in Fig. 21. These techniques are applicable to bonded laminates and honeycomb structures. The choice of technique is partially based on the thickness and configuration of the bonded assembly as well as physical size.

Bonded structures are generally inspected by the immersion method using a C-scan recorder, which is an electrical device that accepts signals from the pulser-receiver and prints out a plan-view record of the part on a wet or dry facsimile paper recording (Fig.

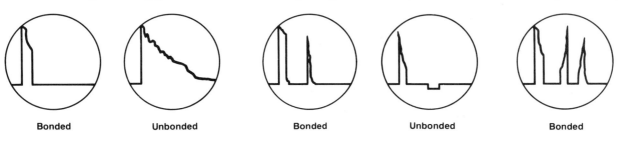

(a)   Bonded   Unbonded   (b)   Bonded   Unbonded   (c)   Bonded   Unbonded

Fig. 21 Typical CRT C-scan displays obtained for both bonded and unbonded structures using three pulsed ultrasound techniques. (a) Pulse echo. (b) Through transmission. (c) Reflector plate

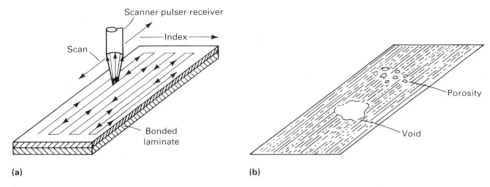

Fig. 22 Schematics of immersion ultrasonic C-scan (a) scanner motion and (b) plan-view record of C-scan recording on facsimile paper

22). To obtain the recording, the bonded panel is placed under water in a tank, and the ultrasonic search unit is automatically scanned over the part. The ultrasonic signals for bond-unbond conditions are determined from built-in defect reference standards. The signals are displayed on the pulser-receiver CRT screen, and the signal amplitude is used to operate the recording alarm after setting the electronic gate around the signal of interest (Fig. 23). Generally, high-amplitude signals will record, but low-amplitude signals will not record. The recording level is adjustable to select the desired signal amplitude. Figure 23 shows a typical ultrasonic C-scan recording system.

Ultrasonic immersion techniques, employing a C-scan recorder, are used extensively by aircraft manufacturers to inspect adhesive-bonded assemblies. The C-scan recordings provide a permanent record with information on the size, orientation, and location of defects in bonded assemblies. The C-scan systems are designed to inspect particular parts and therefore vary considerably in size and configuration. Computer-operated controls have been incorporated into some systems to control the scanning motions of the search units and to change instrument gain at changes of thickness in the assembly.

The through transmission and reflector plate techniques are easier to perform than the pulse-echo technique and are useful for producing C-scan recordings of test specimens and flat laminates. Special equipment is required for large panels, squirters for honeycomb panels that cannot be immersed, and contour followers for contoured parts.

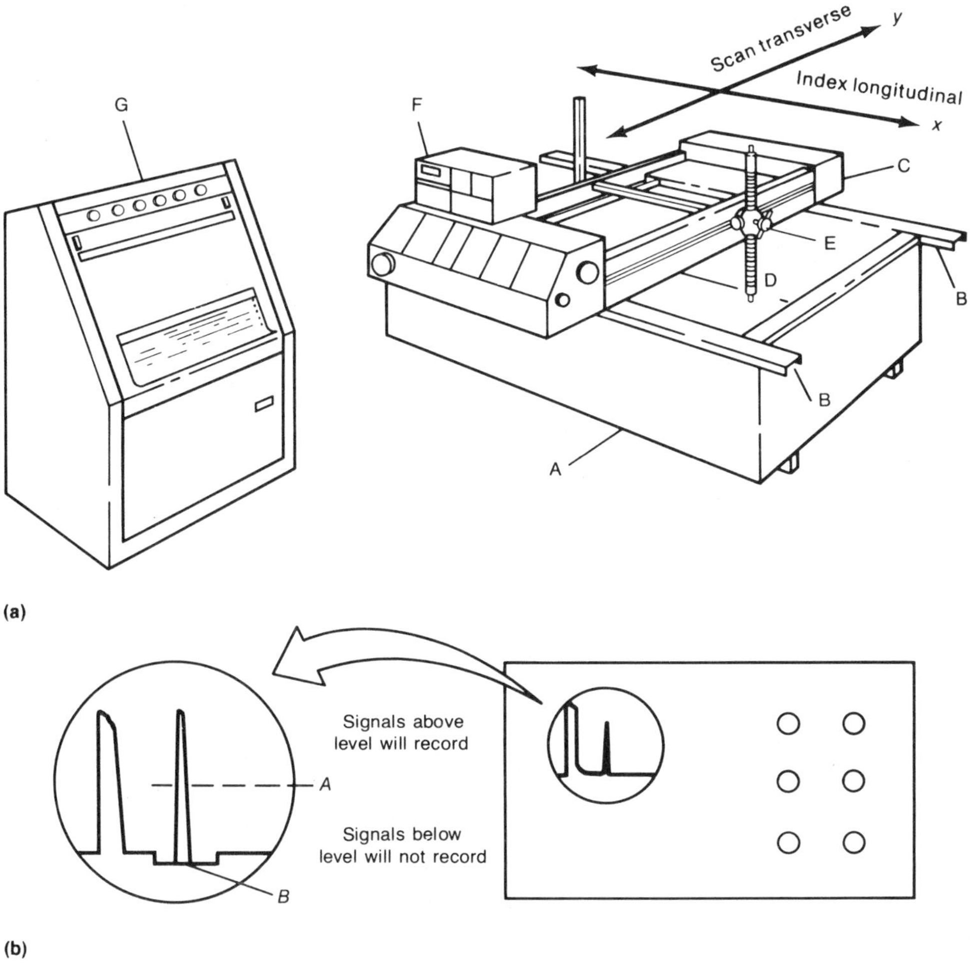

Ultrasonic Inspection Limitations. The ultrasonic method suffers from destructive wave interference. Adhesive thickness interference effects are most notable at the pinch-off zone near the edges of laminates bonded with adhesive using a mat (nonwoven) carrier. Metal thickness interference effects (Fig. 24) are usually obtained from any laminate having a bonded tapered doubler. The problem with the interference effects is that they make the parts appear unbonded. The phenomenon is complex and will require further study before its effects can be eliminated.

Either ultrasonic bond test methods or neutron radiography is used to determine the acceptability of bonded assemblies yielding C-scan results showing interference effects (see the section "Ultrasonic Bond Testers" in this article). Figure 25 shows a through transmission C-scan of a bonded honeycomb assembly. The lead tape (generally used for radiography) attenuates the ultrasonic beam and is therefore used to simulate voids. It also is used as an index or orientation marker to relate the location of flaws in the part to the C-scan recording. In the contact pulse-echo ultrasonic inspection of bonded laminates, inspection from the thin adherend side produces the best results.

Ultrasonic Bond Testers

A wide variety of ultrasonic bond testers have been developed over the past 20 years. As a consequence, this mode of bond inspection probably requires the most study because of the number of instruments available and the claims of the manufacturers. Various independent studies have been con-

Fig. 23 Ultrasonic C-scan immersion system. (a) The system consists of a large water tank (A) with side rails (B), which support the movable bridge (C). The search unit (D) is held at a proper distance above the part by the scanner arm (E), moves across the part, and is then indexed along the longitudinal axis. The visual pulser-receiver unit (F) is mounted on the bridge. A permanent record is made by the C-scan recorder (G). (b) Close-up of pulser-receiver on the right and a close-up of the CRT presentation on the left. A recording level (A) is set on the instrument. All signals above will record on the C-scan machine, and signals below will not show. Item B is an automatic recording gate associated with the mechanics of record making.

REFLECTOR PLATE C-SCAN AT 5 MHZ

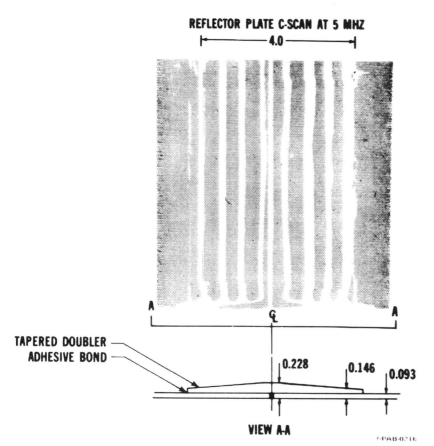

VIEW A-A

TAPERED DOUBLER
ADHESIVE BOND

0.228   0.146   0.093

Fig. 24 Reflector plate C-scan at 5 MHz of ultrasonic destructive interference effects due to varying adherend thickness

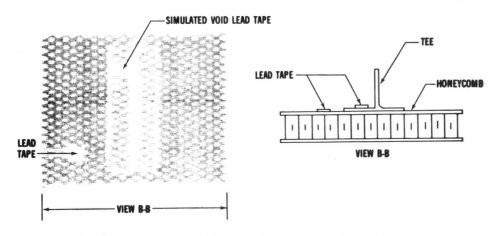

SIMULATED VOID LEAD TAPE

LEAD TAPE

VIEW B-B

TEE

LEAD TAPE

HONEYCOMB

VIEW B-B

Fig. 25 Through transmission C-scan of adhesive-bonded honeycomb. (a) C-scan of section B-B at 5 MHz using lead tape to simulate void. (b) Cross-sectional view B-B of bonded honeycomb showing position of lead tape

ducted in an attempt to clarify the situation (Ref 2, 3, 10-20). The results of these studies revealed that all the instruments are capable, in varying degrees, of determining the quality of the bond. None of the instruments is capable of establishing the adhesive quality of a bond that is defective as a result of:

- Poor surface preparation of the substrate
- Insufficient cure temperature
- Contamination of the adhesive or substrate prior to bonding

The conclusions indicate that ultrasonic bond test instruments are the most reliable and sensitive for detecting voids, porosity, thick adhesive, and corrosion at the bond line, and that they can be used to inspect metal-to-metal and honeycomb structures.

Instruments such as the Coinda Scope, Stub Meter (Ref 13, 14), Sonic Resonator

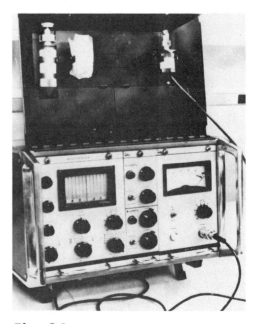

Fig. 26 Fokker bond tester

(Ref 11), and Arvin Acoustic-Impact Tester (Ref 15) have not found general acceptance and are no longer on the market. Following is a short description of the ultrasonic bond testers currently being used to inspect adhesive-bonded assemblies.

The Fokker Bond Tester (Fig. 26) is based on the analysis of the dynamic characteristics of the mass-spring and dashpot system formed by the combination of the bonded assembly with a piezoelectric transducer having known mass and resonance characteristics. Changes in the viscoelastic properties of the adhesive layer are detected as variations in the resonance frequency and impedance of the system. The calibrated body acts as a transducer that can be driven at different frequencies. The dimensions of the transducer are chosen in relation to the total thickness of the metal sheet to be tested and to the required mode of resonance. The response of the total system, as shown by the impedance curve over the swept frequency band, is displayed on a CRT display (A scale) (Fig. 27), and the peak-to-peak amplitude of the curve is shown by a microammeter (B scale).

For inspecting metal-to-metal bonded joints, the probe containing the calibrated body is placed on top of a piece of sheet material with the same thickness as the upper sheet material of the bonded laminate. The central frequency of the oscillator is selected such that the lowest point of the impedance curve is in the center of the A scale. Simultaneously, the B scale is adjusted to 100 (Fig. 27). Calibration of the instrument on a nonbonded sheet ensures that, in all cases of a complete void, the peak position will return to the center of the A scale and that the B scale reading will be 100 (Fig. 28). The next calibration places the probe

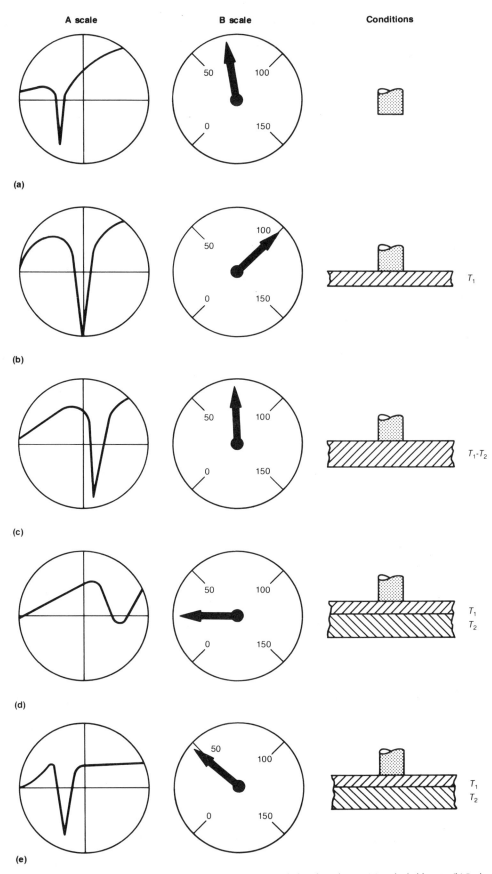

A scale	B scale	Conditions

(a)

(b)

(c)

(d)

(e)

Fig. 27 Several typical Fokker bond tester readings for specific bond conditions. (a) Probe held in air. (b) Probe is on top of the upper adherend (no bonds) and is a calibration for unbond condition. (c) Probe is placed on a single piece of metal as thick as the combined parts being bonded and calibrated for ultimate quality. (d) Probe is placed on bonded joint with good-quality bond (no voids); high-strength bond. (e) Probe is placed on bonded joint with porosity; low-strength bond

on a piece of metal sheet equivalent in thickness to all the metal sheets in the subject bonded laminate. The peak obtained in this test is the resonance frequency to be expected of an ideally bonded laminate (Fig. 27 and 28).

The quality of cohesion from any test can be accurately determined by comparing the instrument reading with established correlation curves. In practice, the acceptance limits are based on the load or stress requirements of the adhesive for each joint. The accuracy of the prediction of quality depends primarily on knowing the manufacturing variables and the accuracy of the nondestructive and destructive tests conducted in accordance with MIL-STD-860 (Ref 21). The destructive test for metal-to-metal joints uses the lap-shear specimen and, for honeycomb panels, uses the tension specimen for bond strength correlations. Comparative tests by several investigators indicate the Fokker bond tester to be more reliable in quantitatively measuring bond strength as related to voids and porosity in the joint (Ref 2, 3, 5, 10, 16, 18, 22, 23). For the inspection of honeycomb, calibration is accomplished in the same manner except that the micrometer (B scale) on the instrument is used. The degree of quality is reflected on the B scale. Low-quality bonds will give a high B reading, and good-quality bonds will give a low B reading.

The Fokker bond tester has been successfully applied to a wide variety of bonded sandwich assemblies and overlap-type joints of various adherends, adhesive materials, and configurations. The method has proved suitable for joints having reasonably rigid adherends, including metallic and nonmetallic materials. Highly elastomeric or porous adherends attenuate ultrasonic response. The method is most sensitive to the properties of the adhesive and is particularly sensitive to voids, porosity, and incomplete wetting (unbond). Fokker tests readily detect:

- Voids in either adhesive or nonmetallic adherend materials
- Cracks and delaminations in adherends
- Unbonded, flawed, ruptured, unspliced, or crushed honeycomb core

Data indicate the test to be capable of measuring bond degradation caused by such factors as moisture, salt spray, corrosion, heat aging, weathering, and fatigue.

NDT-210 Bond Tester (Fig. 29). Sound waves from a resonant transducer are transmitted into, and received from, the bonded structure. Unbonds or voids in the structure alter the sound beam characteristics, which in turn affect the electroacoustic behavior (impedance) of the transducer. A frequency range extending from less than 100 Hz to more than 6 MHz is provided for maximum performance. The unit automatically adjusts to the resonant frequency of the probe.

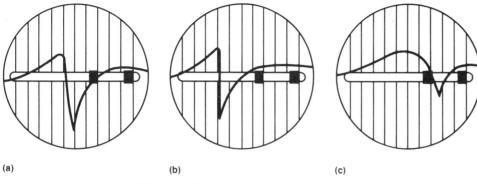

(a)   (b)   (c)

Fig. 28 Typical displays of Fokker bond tester A scale for various qualities of adhesive-bonded joints. (a) Central frequency, no strength. (b) Higher frequency, low strength. (c) Lower frequency, high strength

A precision automatic gain control oscillator maintains a constant voltage at all frequencies, eliminating the need for manual adjustment. Test response is displayed on a 114 mm (4½ in.) wide meter. The operator can set an adjustable alarm to trigger when the response exceeds a preset level.

Metallic, nonmetallic, or joints that are a combination thereof can be inspected for voids and unbonds. The joints can be adhesively bonded, brazed, or diffusion bonded.

The Shurtronics Mark I harmonic bond tester (Fig. 30), now identified as the Advanced Bond Evaluator, is a portable, low-frequency, eddy sonic instrument that uses a single transducer for the inspection of thin metal laminations and metal-to-honeycomb bonded structures. The instrument utilizes a coil that electromagnetically vibrates the metal face sheet. This coil induces a pulsed eddy current flow in a conductive material by an oscillating electromagnetic field in the probe. The alternating eddy current flow produces an accompanying alternating magnetic field in the part. The attraction of the magnetic fields causes acoustomechanical vibrations in the part. When a structural variation is encountered, the change in

acoustic response is detected by a broadband receiving microphone located inside the eddy current probe.

An ultrasonic oscillator generates a 14 to 15 kHz electric signal in the probe coil, creating an oscillating electromagnetic field. The resulting acoustomechanical vibration in the testpiece is detected by a microphone with a bandwidth of 28 to 30 kHz. The microphone signal is filtered, amplified, and displayed on a microammeter. The instrument is calibrated to read just above zero for a condition of good bond and to read full scale for unbonds 13 mm (½ in.) in diameter or larger. This method does not require a liquid couplant on the part surface. It is useful for the quick detection of unbonds in large-area bonded metal laminates, metal-to-nonmetal laminates, or honeycomb structures. It is highly effective for detecting skin-to-core unbonds or voids in acoustic-honeycomb panels utilizing perforated facing sheets. The sensitivity of the instrument decreases rapidly for skin thickness

Fig. 29 NDT-210 bond tester

over 2.0 mm (0.080 in.) and near the edge of the part.

The Advanced Bond Evaluator (ABE) (Fig. 31) has the following improvements:

- *1% alarm accuracy*: Vernier control ensures test repeatability with zero hysteresis
- *Phase comparator*: Detects defects in nonmetallic composite materials and provides additional information on metallic structures

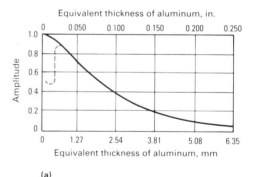

(a)

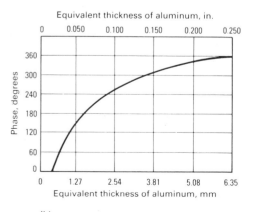

(b)

Fig. 32 Variation of (a) amplitude and (b) phase as detected with a Sondicator

Fig. 30 Shurtronics harmonic bond tester

Fig. 31 Advanced Bond Evaluator bond tester. Courtesy of M. Gehlen, Uniwest/ Shurtronics Corporation

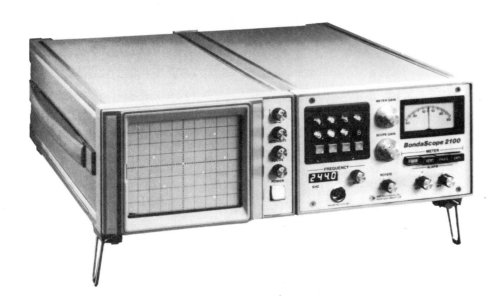

(a)

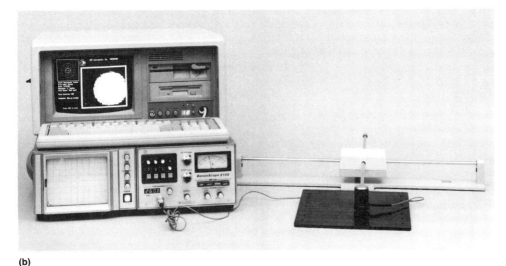

(b)

Fig. 33 Bondascope 2100. (a) Front view of unit showing CRT readout of bond line depths. (b) Unit used as a component in a Portascan bond testing system. Courtesy of R.J. Botsco, NDT Instruments

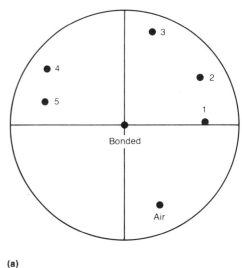

(a)

(b)

Fig. 34 Bondascope ultrasonic impedance plane presentation for multilayer laminar. (a) Orientation of Bondascope display of the unbonds shown in (b). (b) Multilayer bonded laminate with unbonds located at 1, 2, and so on

- *Portable operation*: Has self-contained batteries and built-in recharger
- *Modular construction*: Has plug-in circuit boards for fast field servicing and maintenance

The Sondicator is a pulsed transmit/receive ultrasonic portable instrument that is capable of operating in a very low (25 to 50 kHz) acoustic frequency range. The instrument operates within this range at a selected single frequency obtained by manual tuning for best instrument performance. The model S-2 contains two meters and associated electronic circuitry, along with a test probe containing two Teflon-tipped transducers mounted approximately 19 mm (¾ in.) apart. One transducer imparts vibration at 25 kHz into the surface. Vibrations travel laterally through the material to the other transducer. The second transducer detects the amplitude and phase relationship of the vibrating surface, directing the associated signals to the two meters of the test equipment. If the probe encounters a delaminated or unbonded area, it will in effect be introducing vibration into an area that is thinner than the bonded area. The amplitude of vibration will increase, and the phase of vibration will shift, as the thinner section vibrates more vigorously. The needles on the two meters will move toward each other.

The instrument is primarily used to detect delaminations of the bond or laminar-type voids in metal and nonmetal structures. The inspection can be performed from one face of the structure or by through transmission. It requires no liquid couplant. Multiple bond lines and part edges will reduce the sensitivity of the instrument. Because of the directionality of the sound from the transmitter to the receiver, the part must be scanned uniformly. The unit is capable of detecting defects 25 mm (1 in.) in diameter and larger in most materials. Metallic and nonmetallic materials can be inspected without changing probes. Audible and visible alarms are activated by received signals. The instrument can detect internal delaminations and voids in bonded wood, metal, plastic, hard rubber, honeycomb structures, Styrofoam, and composites. It can measure skin or face sheet thickness in composite structures as a change in amplitude or phase (Fig. 32).

Bondascope 2100. The Bondascope (Fig. 33) is an advanced microprocessor-based device that operates on the principle of ultrasonic impedance analysis. This technique allows the total ultrasonic impedance vector for the material to be monitored as a flying dot on a scope display. With the center of the scope screen as the origin (reference), impedance phase changes are displayed circumferentially, while impedance amplitude changes are displayed radially. Thus, the position of the dot on the scope immediately reveals the phase and amplitude of the impedance of the material (Fig. 33a). Therefore, each defect at a different bond line possesses a characteristic dot position on the scope display.

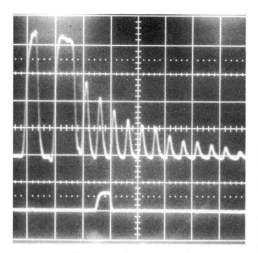

Fig. 35 Novascope CRT display of a 0.15 mm (0.006 in.) steel sheet. Upper trace: A-scan echo pattern. Lower trace: Thickness gate reading obtained from two successive echoes. Courtesy of R.T. Anderson, NDT Instruments

With the addition of a 640-KB random access memory (RAM) portable computer having a 229 mm (9 in.) built-in color monitor, a high-speed plug-in data acquisition card (DAC), a software package, and a 1.4 to 1.8 kg (3 to 4 lb) scanner incorporating a ball-bearing gimbal probe holder, the Bondascope 2100 becomes an integral part of the PortaScan (portable ultrasonic color scan imaging) system that can be used for both C-scan imaging and set-up flying dot (impedance plane) imaging (Fig. 33b).

Figure 34 illustrates the typical Bondascope response to unbonds at different bond lines (depths) in a multilayered adhesive-bonded laminate. The instrument was calibrated so that the dot was in the center of the screen (origin) when the probe was placed on a well-bonded section of the laminate. The numbered dots represent the signals obtained when the probe was placed over regions having unbonds located at different respective bond line depths in the laminate. The dot labeled "air" is the signature obtained when the probe was off the sample. Thus, improper ultrasonic coupling into the material is quickly recognized.

A meter on the Bondascope, by means of selectable push buttons, can display the signal amplitude, phase, or vertical resolved component of the amplitude. Through the use of a phase-rotator control, the signal response can be positioned so that nonrelevant signals are suppressed from the meter/alarm readings.

A keyboard push-button matrix allows the operator to digitally program calibration-sample reference dots on the scope display to aid in interpreting signals obtained during actual testing. This type of storage display not only facilitates operation and interpretation but also eliminates the confusing permanent streaks that would occur if a conventional storage scope were used to display the flying dot.

The Bondascope is suitable for inspecting metallic and nonmetallic bonded structures (multilayered laminates or honeycomb). Graphite-epoxy and other fiber-matrix composites can also be tested with this device.

The NovaScope is a sophisticated high-resolution, pulsed ultrasonic thickness gage with a digital readout and scope display for qualifying the echo pattern. It is primarily intended for either manual contact probe or focused immersion applications for which the conventional digital-only readout gages are not suitable. This is the case for structures having complex, difficult-to-test shapes or whenever the product is in motion. The instrument has a variety of controls for optimizing the test response to the desired ultrasonic echo periods.

Figure 35 shows the resolving power of the instrument for gaging thin materials. In this case, the material is a steel sheet only 1.5 mm (0.006 in.) thick. The upper trace is the A-scan echo pattern, while the lower trace shows the thickness gate adjusted between two successive echoes from the thickness of 1.5 mm (0.006 in.). This instrument is useful for testing thin composite laminates for delaminations, for detecting corrosion in aircraft skins, and for bond line thickness gaging.

Ultrasonic and Bond Tester Method Sensitivity. The defect detection sensitivity of ultrasonic techniques with respect to defect size and location is highly dependent on the changing conditions of part complexity, number of bond lines, operator experience, and so

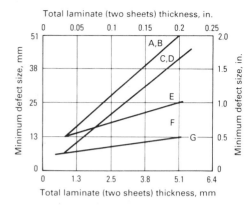

Fig. 36 Relative sensitivity of instruments used to inspect bonded laminates of increasing total thicknesses. A, Sondicator, contact method; B, Shurtronics harmonic bond tester; C, Fokker bond tester; D, NDT-210 bond tester; E, Sondicator, through transmission method; F, ultrasonic pulse echo; G, ultrasonic through transmission

on. Most instruments detect major defects most of the time, but in some cases special techniques and skills are needed to conduct a reliable inspection. Whenever possible, bonded structures should be inspected from opposite surfaces to detect small defects. In the ultrasonic bond test methods, different-size search units or probes are available. The smallest flaw that can be readily detected is of the order of one-half the diameter of the search unit or probe element.

Test Frequency. The higher the test frequency, the smaller the flaw that can be detected. Ultrasonic testing at 5 or 2.25 MHz will detect smaller flaws than can be

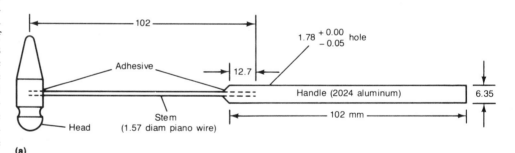

(a)

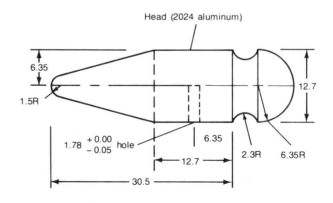

(b)

Fig. 37 Construction details for an inspection tap hammer. (a) Complete assembly. (b) Tap hammer head. Liquid/paste adhesive can be used if desired. The hole in the handle/head can be reduced to provide an interference fit and to preclude the need for the adhesive. Dimensions given in millimeters

detected with the Sondicator (25 to 50 kHz) and the harmonic bond tester (15 kHz). With multiple bond lines in a structure, smaller flaws become progressively more difficult to detect from the surface through succeeding bond lines. This does not apply to ultrasonic through transmission or reflector plate testing, in which the small-diameter sound beam passes completely through the part. Figure 36 shows the relative defect sensitivity of ultrasonic and bond tester techniques.

X-ray Radiography

Radiography is a very effective method of inspection that allows a view into the interior of bonded honeycomb structures. The radiographic technique provides the advantage of a permanent film record. On the other hand, it is relatively expensive, and special precautions must be taken because of the potential radiation hazard. With the radiographic method, inspection must be conducted by trained personnel. This method utilizes a source of x-rays to detect discontinuities or defects through differential densities or x-ray absorption in the material. Variations in density over the part are recorded as various degrees of exposure on the film. Because the method records changes in total density through the thickness, it is not preferred for detecting defects (such as delaminations) that are in a plane normal to the x-ray beam.

Some adhesives (such as AF 55 and FM 400) are x-ray opaque, allowing voids and porosity to be detected in metal-to-metal bond joints (Fig. 5a). This is extremely advantageous, especially for complex-geometry joints, which are difficult to inspect during fabrication. The x-ray inspection should be performed at low kilovoltage (25 to 75 kV) for maximum contrast. A beryllium-window x-ray tube should always be used when radiographing adhesive-bonded structures. To hasten exposures, medium-speed, fine-grain film should be used. Selection of the film cassette should be given special consideration because some cassette materials produce an image on the film at low kilovoltages.

Neutron Radiography

Neutron radiography is very similar to x-ray or γ-ray radiography in that both depend on variations in attenuation to achieve object contrast on the resultant radiograph. However, significant differences exist in the effectiveness of the two methods, especially when certain combinations of elements are examined. The mass absorption coefficients of the different elements for x-rays assume a near-linear increase with atomic number, while the coefficients for thermal neutrons show no such proportionality. The attenuation of x-rays is largely determined by the density of the material being examined. Thus, thicker and/

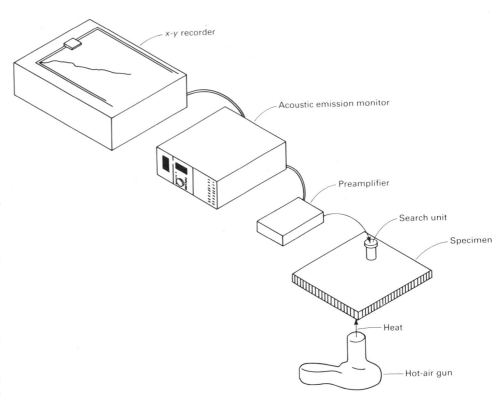

Fig. 38 Acoustic emission detection of active corrosion in adhesive-bonded structures

or denser materials appear more opaque. The absorption for thermal neutrons is a function of both the scattering and capture probabilities for each element. The density or thickness of a material or component is less important in determining whether it will be transparent or opaque to the passage of neutrons. For example, x-rays will not pass through lead easily, yet they will readily penetrate hydrocarbons. In contrast, neutrons will penetrate lead and are readily absorbed by the hydrogen atoms in an adhesive or hydrocarbon material.

Neutron Sources. Neutrons are produced from accelerator, radioisotope, or reactor sources. Neutrons, like x-rays, can be produced over an enormous energy range with large differences in attenuation at the various energy levels. The major efforts in neutron radiography have been performed using thermal neutrons because the best detectors exist in this energy regime. Neutron sources generally produce γ-rays of moderate intensity, so that a neutron detector sensitive to γ-ray radiation has a γ-ray image superimposed on the neutron image.

Direct Versus Transfer Method. The most widely used imaging method is the use of conventional x-ray film with converter screens. The rate of radioactive emission of the converter screen divides the photographic imaging into prompt emission or delayed emission. The prompt-emission converters (gadolinium

or rhodium) require that the film be present during neutron exposure. This process is referred to as the direct method. The delayed-emission converters (indium or dysprosium) allow for activation of the converter and transfer of the induced image to the film after neutron exposure. This technique is referred to as the transfer method. In the original film, the neutron opaque areas appear lighter than the surrounding material. Image contrast can be increased by making a contact positive print from the film negative. When this is done, the neutron opaque area will be darker than the surrounding material.

Evaluation of Adhesive-Bonded Structures. Hydrogenous (adhesive) materials inspected by neutron radiography can be delineated from other elements in many cases where x-ray radiography is inadequate. However, the neutron radiography inspection method does not appear to be cost effective for the routine inspection of adhesive-bonded structures. It is extremely useful for evaluating the quality of built-in defect reference standards or for failure analysis. If the adhesive is not x-ray opaque, neutron radiography can be used to detect voids and porosity. The hydrogen atoms in the adhesive absorb thermal neutrons, rendering it opaque.

Tap Test

Perhaps the simplest inspection method for ensuring that a bond exists between the

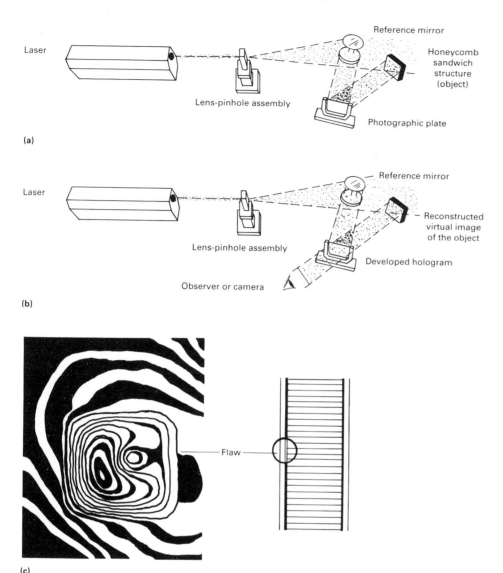

Fig. 39 Holographic recording and viewing. (a) Setup for recording hologram. (b) Setup for real-time viewing of hologram. (c) Example of real-time hologram of a flaw (void in adhesive) in a honeycomb core assembly. The part was heated to give thermal stressing.

honeycomb and the facing sheet is that of coin tapping. An unbond is readily apparent by a change in the tone or frequency of sound produced when an adhesive-bonded structure is tapped with a coin or rod, compared to the sound produced for a bonded area. Coins such as silver dollars or half-dollars are used for this test. For standardized production testing, a 13 mm (½ in.) diam solid nylon or aluminum rod, 102 mm (4 in.) in length, with the testing end smoothly rounded to a 6.4 mm (¼ in.) radius, is used. Another version of a tap tester is the aluminum hammer shown in Fig. 37.

Tap testing with a coin or a small aluminum rod or hammer is useful for locating large voids or disbonds of the order of 38 mm (1½ in.) in diameter or larger in metal-to-metal or thin facing-sheet honeycomb

assemblies. Tap testing is limited to the detection of upper facing sheet to adhesive disbonds or voids. It will not detect voids or disbonds at second, third, or deeper adherends. The test method is subjective and may yield wide variations in test results. On thin face sheets, the coin tap will produce undesirable small indentations. In such cases, an instrument such as the Shurtronics harmonic bond tester can be used effectively.

Acoustic Emission

In certain cases, acoustic emission techniques are more effective than x-ray or conventional ultrasonic methods in detecting internal metal corrosion and moisture-degraded adhesive in bonded panels. The principle is based on the detection of sound or stress wave signals created by material

undergoing some physical or mechanical transformation. Regarding the detection of bond line corrosion, the acoustic signals apparently arise from cavitation or the boiling of moisture within the joint. If the corroded joint is dried out before the acoustic emission test is performed, no acoustic response is obtained.

The equipment consists basically of an amplifier and a piezoelectric sensor with a resonant frequency of about 200 kHz. The emission level is recorded on a chart or counter. Other equipment includes a signal processor, a search unit, a 50 dB preamplifier, an x-y chart recorder or counter, and a hot-air gun (Fig. 38). Simple heating methods employing a hot-air gun or heat lamp are used to increase emissions from active corrosion sights or to boil moisture. Heating can be done from the search-unit side.

The panel to be tested is heated to about 65 °C (150 °F) by holding the hot-air gun within 50 to 75 mm (2 to 3 in.) of the surface of the panel for about 15 to 30 s. Immediately after heating, the transducer or search unit is placed a short distance from the heated spot. The transducer is held in position for 15 to 30 s to obtain a complete record of any emission in the heated area. The inspection is conducted on a 152 mm (6 in.) grid. An important consideration during the test is the manner in which the transducer is held against the part surface. Because movement of the transducer can produce appreciable noise, care must be taken in its placement and holding.

Corrosion has been detected in a number of adhesive-bonded honeycomb structures (Ref 7-9). A direct inspection cost savings of more than 75% over comparative x-ray inspection has been achieved.

Special NDT Methods

Holographic Interferometry. Defects such as core-to-skin and core-splice voids, delaminations, crushed core, and bond line corrosion can usually be detected by holographic nondestructive test techniques (Ref 3, 24). A hologram of the test specimen is first recorded by means of laser light reflected from its surface and superimposed on a mutually coherent reference beam in the plane of a high-resolution photographic emulsion. The hologram provides a complete record of all visual information about the entire illuminated surface of the specimen, including the phase and the amplitude of the reflected wave front.

The specimen is then stressed in one of several possible ways, including heat, pressure change, evacuation, and acoustic vibration. A surface displacement differential of only a few microinches between a defective and a good-quality bond is adequate to produce a holographic image. Differential surface displacements, caused by subsurface anomalies, are observed or recorded

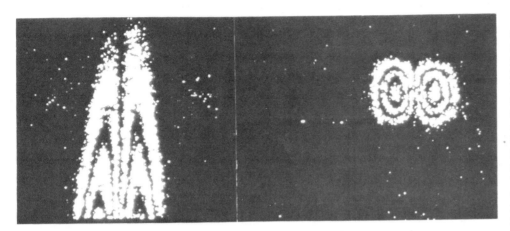

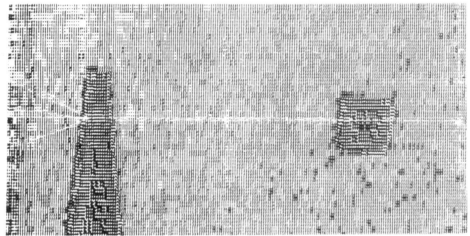

Fig. 40 Real-time shearography images (top) of graphite-foam core sandwich panel with an edge pull-out defect and a Teflon insert at the foam-to-core bond line. Bottom image is an ultrasonic C-scan recording of the part (not quite to the same scale). Test time for the shearography image of the defects was 2 s. Courtesy of Laser Technology, Inc.

with one or more of the following three holographic techniques:

- *Real time*: The stress-illuminated specimen is viewed through its developed hologram (made in a different state of stress)
- *Time lapse*: Two holographic exposures are made on the same plate, with the specimen in two states of stress, and then reconstructed
- *Time average*: The hologram is recorded during many cycles of sinusoidal vibration of the specimen

A fluid-gate photographic plate holder allows the in-place development of holograms and enhances the speed of the operation (Ref 24). Because of the acute sensitivity of holographic interferometry to small disturbances, it is necessary to reduce spurious and unwanted acoustic noise and temperature gradients within the environment of the testing system. Floor vibrations arising from heavy traffic, the loading/unloading of vehicles, and heavy-duty machinery generate noise that must be minimized.

Systems for testing sandwich structures generally use helium-neon gas lasers, which normally deliver between 60 and 80 mW of power at 632.8 nm (6328 Å). This power is sufficient to examine at least 0.19 m^2 (2 ft^2) of surface area, and the deep red color of the light is close enough to the sensitivity peak of the eye for good fringe-to-background contrast.

A typical real-time hologram showing a skin-to-core void revealed by thermal stressing is illustrated in Fig. 39. In large measure, the basic limitation of holography is related to the stressing techniques utilized. Holography cannot be used without a surface manifestation of a defect during stressing. If the material thickness precludes detection of a resolvable fraction of the fringe spacing, then the technique is ineffective. It is not useful for inspecting complex laminates, because of the inability to stress the void areas. Thermal stressing of aluminum is unsatisfactory, because of its high thermal conductivity (heat is transferred laterally rather than vertically through the joint). Holography is satisfactory for inspecting honeycomb structures, but few voids exist at the skin-to-core interface. It is limited in locating voids at honeycomb closures or multilaminate metal-to-metal areas. Numerous flaws in the closure areas of bonded honeycomb assemblies (voids and porosity) that were detected by ultrasonics and x-ray methods were not detected by holographic interferometry.

At Fokker-VFW, a holographic installation for testing bonds was developed and put into use for production inspection (Ref 3). The installation uses a single hologram for components up to 6 m (20 ft) in length. It has adjustable optics for optimum sensitivity and component scanning. The defect area can be magnified and the results presented on a highly sensitive video monitor that records the information. The bonded components can be deformed by either vibrations or thermal effects.

Interference holography shows much more than just bond defects, and training is needed for personnel to learn to distinguish areas that are defective from those that are merely not ideal (Ref 2, 3, 21). Core-splice location and quality can be inspected, and core machining anomalies are visible. Also, the thermal method is the most practical deformation method for sandwich structures, especially for graphite-epoxy skins bonded to aluminum core. Extensive studies have shown that thermal deformation is difficult to use for metal-to-metal structures (Ref 21). These difficulties led workers at Fokker to concentrate on vibration testing because of the relation between resonance properties and cohesion quality. Much time will be needed to develop a universal system of quality testing based on interference holography. However, a specific interference holography system has been developed for the inspection of truck and aircraft tires (Ref 25). A wide variety of tire defects have been detected and categorized for both new and retread tires and have been related to durability.

Shearography offers the sensitivity of holographic interferometry to skin-to-core unbonds as well as to deep unbonds in both honeycomb and foam sandwich structures.

Table 3 Structure, flaw type, and shearography stressing method

Structure	Flaw	Test method
Composite honeycomb	Crushed core	Vacuum
Composite honeycomb	Unbond	Thermal, vacuum
Multiple bond line honeycomb	Unbond	Vacuum
Graphite panels	Impact damage	Thermal, vacuum
Graphite panels	Voids, porosity	Vacuum
Rubber bonded to aluminum	Unbond	Vacuum
Composites	Moisture	Microwave excitation
Bearing plating	Unbond	Vacuum

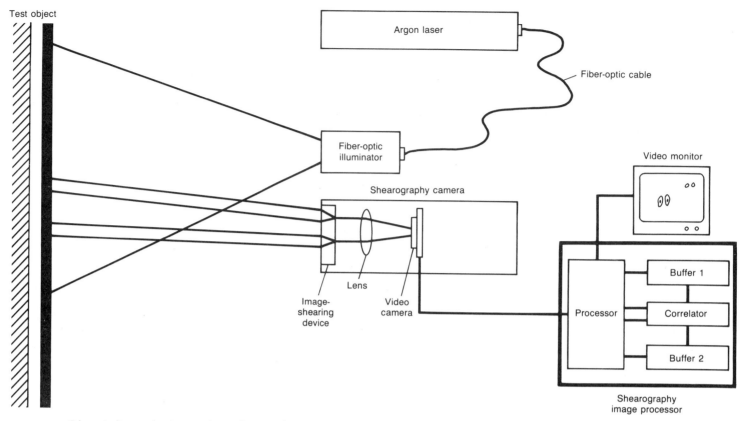

Fig. 41   Schematic diagram showing a production shearography system. Part is illuminated with a laser through a fiber-optic cable. Electronic shearography camera captures interferogram and stores reference image in a video fram buffer. Real-time correlator produces a real-time video image of strain pattern when test part is stressed.

Unlike holographic interferometry, shearography is real-time, all electronic, and insensitive to environmental vibration and ambient light. As with holography, an appropriate stressing technique is of primary importance in order to allow the subsurface flaw to induce a strain concentration on the surface of the test part. This may be achieved using the stressing techniques in Table 3. For composite structures, the primary stressing techniques are thermal loading for delamination and impact damage and pressure reduction loading for unbonds. In the late 1980s shearography was applied with great success to honeycomb structures in the manufacture of advanced aircraft (Ref 26). In many cases, the NDT capability exceeds that for ultrasonic C-scan with an inspection throughput that is 100 to 1000 times greater. Figure 40 depicts a shearography image of a graphite-foam core sandwich panel.

Electronic shearography was invented by Y.Y. Hung at Oakland University in 1984 (Ref 27) and licensed exclusively to Laser Technology, Inc. Shearography uses a laser to illuminate an area on the test part for inspection. A typical shearography system of the type used for aircraft manufacturing is shown in Fig. 41. The laser beam is introduced to the test part through a fiber-optic cable. In this way, only the small 0.9 kg (2 lb) shearography camera and the laser

illuminator are mounted on the end of a gantry for manipulation over the test part. The output is an image-processed video display in color that can show both the size and shape of subsurface defects and a qualitative measure of depth.

Shearography has been used to inspect multiple bond line honeycomb and can in many cases see unbonds at many levels within the panel. In helicopter blades, sheared core, crushed core, unbonds, and corrosion have been easily detected with a pressure reduction technique (Fig. 42). In shop applications, a large test chamber with an air blower is used to reduce the air pressure around the part. On-vehicle applications can utilize a portable system available from Laser Technology, Inc. (Ref 28). This system has a lightweight inspection head that can pull a vacuum on the test part and inspect 0.20 m² (2 ft²) at a time in as little as 3 to 5 s (Fig. 43).

Shearography has been used on large rockets such as the Atlas Centaur launch vehicle (Ref 29) to inspect foam-to-tank bond lines. A portable vacuum window was used to stress the foam and reveal all unbonds using a portable shearography system. Other applications include skin-to-foam unbonds and foam-to-foam bond lines. The successful detection of corrosion and unbonds in F-111 flaps, debonding of space shuttle orbiter thermal protection tiles, un-

bonds in rubber coatings on aluminum, and cracks in rivet holes in an aluminum fuselage structure have been reported by the system manufacturer.

Infrared or Thermal Inspection. The thermal NDT of adhesive-bonded structures has been performed by a variety of techniques, including the following:

- Infrared radiometer testing (Ref 22, 30)
- Thermochromic or thermoluminescent coatings (Ref 22, 31)
- Liquid crystals (cholesteric) (Ref 32, 33)

For infrared radiometer testing, the detector employs a moving heat source and records variations in heat absorption or emission while scanning the part surface. The scanners, or detectors, used in infrared testing are called radiometers. The radiometer generates an electrical signal exactly proportional to the incident radiant flux. Because scanning and temperature sensing are performed without contact, the observed surface is not disturbed or modified in any way. The thermal pattern from the part under test can be observed on a CRT, storage (memory) tube, or x-y recorder.

Tests are performed by heating the sample with a visible-light heat source (quartz lamp or hot-air gun) and then observing the surface heating effects with the radiometer. Heat is applied to both bonded and unbonded areas. The bonded areas will con-

Fig. 42 A production shearography system for aircraft and missile parts using pressure reduction stressing. Courtesy of Laser Technology, Inc.

duct more heat than the unbonded areas because of good thermal conduction to the honeycomb structure. The test panel is first coated with flat black paint to ensure uniform surface emissivity. The sample is then scanned, as shown in Fig. 44. Graphs are obtained by connecting the horizontal movement of the sample to one axis of an x-y recorder and the radiometer output to the other axis. Typical line scan and area scan test results are shown in Fig. 44(b) and 44(d), respectively. Similar results are obtained with the AGA Thermovision (in the pitch or catch or through transmission modes) (Ref 22). Good results are generally obtained when testing graphite-fiber or bo-

ron-fiber composite face sheets bonded to aluminum core. Honeycomb structures fabricated exclusively from aluminum (skin and core) or aluminum skin bonded to plastic core is difficult to inspect by infrared methods because of the lateral heat flow in the aluminum face sheets.

Thermochromic or Thermoluminescent Coatings. The use of an ultraviolet-sensitive coating containing a thermoluminescent phosphor that emits light under excitation by ultraviolet radiation (black light) permits the direct visual detection of disbonds as dark regions in an otherwise bright (fluorescent) surface (Ref 22, 31). The coating is sprayed on and dried to a 75

to 130 μm (3 to 5 mil) thick plastic film. Defects appear as darkened areas when the panel is heated to 60 °C (140 °F) and viewed under ultraviolet light (Fig. 45). The defective regions can then be marked on the plastic film with a felt-tip pen. The rate of fluorescent reversal and retention depends on the thermal conductivity of the underlying structure.

Thermochromic paints consist of a mixture of temperature-indicating materials that change color when certain temperatures are reached. Thirty-six materials covering the temperature range of 40 to 1628 °C (104 to 2962 °F) are available, with the low-temperature materials being used for bond inspection. The low-temperature paint changes from light green to vivid blue upon reaching 40 °C (104 °F). The color change will last (in the defect area) for about 15 min, depending on the relative humidity. The dark areas can be made to revert to the original color by applying moisture in the form of steam. The tests can be repeated a number of times without destroying the properties of the paint. The paints are easy to apply and remove.

Liquid crystals are a mixture of cholesteric compounds that change color when their temperature changes as little as 0.84 °C (1.5 °F), and they always attain the same color at a given temperature for a specific crystal composition. After suitable surface preparation or cleaning, a thin coating of liquid crystals is applied by spray or brush to the test object surface. When the object is correctly heated (relatively low temperatures) with a heat lamp or hot-air gun, the defects are indicated by differences in color (Ref 32, 33). Unfortunately, the color continues to change through a specific color band as it is heated and cooled. Therefore, the defects must be marked on the surface of the part as they appear and disappear. Photographs can be taken for a record of the test results after specific defects are located. The test can be repeated a number of times without destroying the liquid crystals. In some cases, black paint is required under the liquid crystal coating to obtain uniform emissivity and color contrast in the defect areas. The theory and results obtained using liquid crystals are virtually the same as those obtained with thermoluminescent and thermochromic coatings.

Bondscan Thermography Inspection System. The Bondscan is a portable instrument used to perform nondestructive testing of honeycomb and composite structures in production or in the field. It consists of a high-resolution infrared Vidicon television camera, video monitor, built-in hot-air blower, and a video frame memory with a built-in video printer (Fig. 46). This piece of equipment is very effective for finding unbonds in composite-skinned honeycomb, entrapped water, and impact damage in graphite-epoxy structures (Fig. 47).

Fig. 43 A portable electronic shearography system. Courtesy of Laser Technology, Inc.

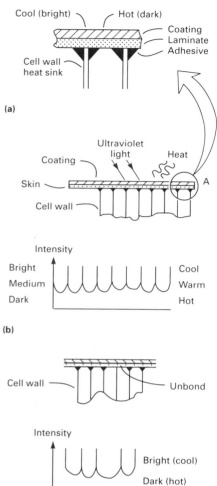

(a)

(b)

(c)

Fig. 45 Thermoluminescent coating technique on boron composite aluminum honeycomb flap assembly. (a) Detail of honeycomb structure. (b) Plot of intensity versus location for bonded joint at locations shown. (c) Plot of intensity versus location for unbonded joint at location shown

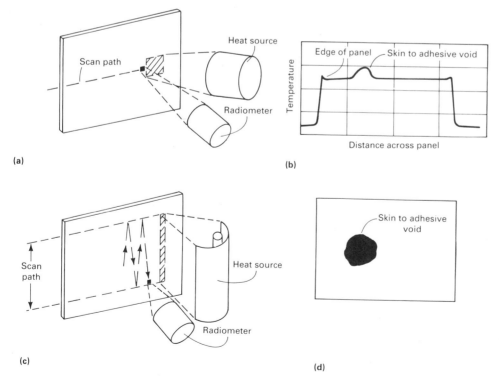

(a)

(b)

(c)

(d)

Fig. 44 Schematics and readouts illustrating infrared radiometer tests. (a) Arrangement for line scan test. (b) Example of readout and method of void detection. (c) Arrangement for area scan test. (d) Example of readout and method of void detection

Fig. 46 The Bondscan contains a long-wavelength infrared television camera, heater, and video memory in a small, lightweight package for NDT of composites using thermography NDT. Courtesy of Laser Technology, Inc.

Fig. 47 Indication of an unbond in a honeycomb aircraft control surface. The trailing edge can be seen at the bottom. Courtesy of Laser Technology, Inc.

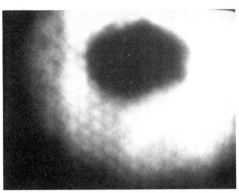

Fig. 48 Indication of entrapped water in a helicopter blade made of honeycomb. Detection time was less than 5 s. Courtesy of Laser Technology, Inc.

All thermography NDT relies on the differential coefficient of thermal conductivity at the site of a flaw. Defects that produce no change in thermal conductivity are undetectable with this method. Also, if the thermal conductivity of the material is comparable to that for aluminum (230 W/m · K, or 1600 Btu · in./h · ft^2 · °F), the time difference between the heating of the part and the diffusion of the thermal wave into the surrounding material is so short that the flaws are not detected. Most composite structures have coefficients of thermal conductivity in the range of 2 to 0.2 W/m ·

K (14 to 1.4 Btu · in./h · ft^2 · °F), which is the range for which the product was designed. As the Bondscan is moved from left to right, the heater projects a fan of warm air onto the test part. Impact-damaged areas in composite structures in which the matrix of the structure is fractured produce local thermal conductivities from 2 to 20 times greater than is the case in undamaged material. As a result, these areas remain warmer after the surface temperature is raised several degrees Celsius by the Bondscan heater. The operator then scans the heated area with the Bondscan

camera and sees the warmer, defective areas as white on a dark background.

Conversely, water may be identified as dark areas in honeycomb. The material in contact with the water remains dark against a light background because of the greater thermal mass (Fig. 48).

The Bondscan is lightweight and uses a long-wavelength infrared camera to observe defects. It has been used both in hangar and on air-aircraft applications with excellent results for impact damage, unbonds, and delamination in honeycomb and composite panels.

Leak Test. A hot-water leak test is generally conducted on all bonded honeycomb assemblies immediately after fabrication or repair. The test is performed by immersing the part in a shallow tank of water heated to about 65 °C (150 °F). The heat causes the entrapped air to expand. If there are any leakage paths, bubbles will be generated at the leakage site. The panel is monitored visually for escaping bubbles, and the leakage site is marked on the panel for subsequent sealing. A part that leaks after fabrication will generally develop problems in service. After the leak test, the assembly is radiographed for water that may have become entrapped during the leak test.

Acoustical holography provides a way to observe the interior properties of composite laminates or adhesive-bonded joints. This acoustical technique employs ultrasound to obtain three-dimensional information on the internal structure of the test specimen. The acoustical hologram is the converter, or recorder, which allows acoustic information to be visualized much as film is the converter or recorder for light. The acoustic imagers employ different holographic recording techniques, depending on the specific application of the instrument (Ref 34).

Liquid-Surface Acoustical Holography. In this method, the liquid surface acts as a dynamic film for momentary storage of the hologram while it is converted to a visual image through the use of coherent laser light. This technique (Fig. 49) allows real-time acoustical imaging, supplying the operator with an instantaneous view of all internal structures. Because the part must be moved through the fixed acoustic beam, part size is a limiting factor in this technique.

Scanning acoustical holography employs a scanning technique to construct the hologram. The hologram is then recorded on either transparency film or a storage oscilloscope. This technique provides a permanent record of the holographic information, allowing the operator to observe the reconstructed image at a later time. By illuminating either the liquid-surface hologram or the transparency film with laser light, the operator can immediately observe all interior properties of the test sample. Figure 50

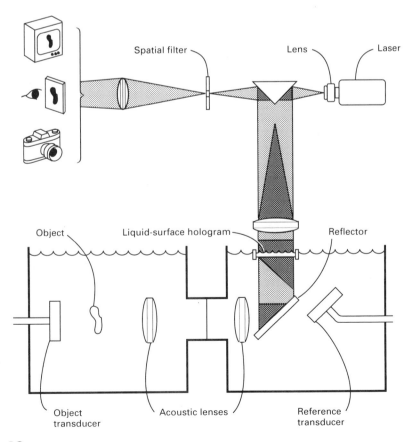

Spatial filter · Lens · Laser

Object · Liquid-surface hologram · Reflector

Object transducer · Acoustic lenses · Reference transducer

Fig. 49 Schematic of a liquid-surface imaging system

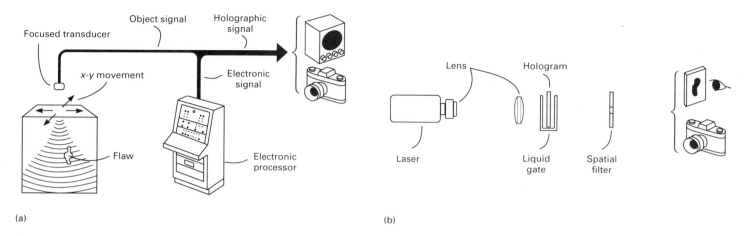

Fig. 50 Schematic of (a) pulse-echo acoustical holography system and (b) holographic reconstruction unit

Table 4 Correlation of NDT results for built-in defects in laminate panels

| | Radiography | | Ultrasonic bond testers | | | | Ultrasonic inspection(b) | | | | | | |
	Low-kilovolt x-ray	Neutron radiography	Fokker bond tester	Sondicator	Harmonic bond tester	210 sonic bond tester	Contact pulse echo	Contact through transmission	Immersion C-scan pulse echo	Immersion C-scan through transmission	Immersion C-scan reflector plate	Coin tap test	Remarks
Laminate defects													
1. Void	(c)	D	D	D	D	D	D	D	D(d)	D	D	PD(e)	c,d,e,f
2. Void (C-14 repair)	(c)	D	D	D	D	D	D	D	D(d)	D	D	PD(e)	d,e
3. Void (9309 repair)	(c)	D	D	D	D	D	D	D	D(d)	D	D	PD(e)	d,e
4. Lack of bond (skin to adhesive)	(c)	ND	D	D	D	D	D	D	D(d)	D	D	PD(e)	d,e
5. Manufacturer sheet (FM123-41)	(c)	PD(g)	D	D	ND	PD	ND	PD	D(d)	D(g)	D(g)	ND	d,g
6. Thick adhesive (1, 2, 3 ply)	(c)	ND	D	ND	ND	D	ND	ND	ND(e)	PD(d)	D(d)	ND	d,e
7. Porous adhesive	(c)	D	D	ND	ND	D	D	D	D	D	D	ND	c
8. Burned adhesive	(c)	ND	D	ND	ND	D	D	D	PD	D	D	ND	h
9. Corroded joint	D	D	D	PD	PD	D	D	D	PD	D	D	PD	i

(a) ND, not detected; PD, partial detection; D, detected. (b) Contact surface wave was tried but did not detect any built-in defects. (c) Panels were made using FM 73, which is not x-ray opaque. With x-ray opaque adhesive, defects 1, 2, 3, 5, 6, 7 are detected. (d) Method suffers from ultrasonic wave interference effects caused by tapered metal doubles or variations in adhesive thickness. (e) MIL-C-88286 (white) external topcoat and PRI432G (green) plus MIL-C-83019 (clear) bilge topcoat dampened the pulse-echo response. (f) Minimum detectable size approximately equal to size of probe being used. (g) Manufacturer separator sheet not detectable but developed porosity and an edge unbond during cure cycle, which was detectable. (h) Caused by drilling holes or band sawing bonded joints. (i) Moisture in bond joint (Armco 252 adhesive, Forest Products Laboratory etch)

Table 5 Correlation of NDT results for built-in defects in honeycomb structures

| | Radiography | | Ultrasonic bond testers | | | | Ultrasonic inspection | | | | | | |
	Low-kilovolt x-ray	Neutron radiography	Fokker bond tester	Sondicator	Harmonic bond tester	210 sonic bond tester	Contact pulse echo	Contact through transmission	Contact shear wave	Immersion C-scan pulse echo	Immersion C-scan through transmission	Coin tap test	Remarks
Honeycomb defects													
1. Void (adhesive to skin)	ND(b)	D	D	D	D	D	D	D	ND	D	D	D	Replacement(b) standard
2. Void (adhesive to skin) repair with C-14	···	···	···	···	···	···	···	···	···	···	···	···	No void improperly made
3. Void (adhesive to skin) repair with 9309	···	···	···	···	···	···	···	···	···	···	···	···	No void improperly made
4. Void (adhesive to core)	ND	D	D	···	D	D	ND	D	ND	D	D	D	Replacement standard
5. Water intrusion	D	D	ND	ND	ND	ND	ND	D	ND	ND	D	D	···
6. Crushed core (after bonding)	D(c)	PD	PD(d)	ND	PD(d)	ND	ND	PD(d)	ND	PD(d)	PD(d)	D	(c, d)
7. Manufacturer separator sheet (skin to adhesive)	ND(e)	D	ND	ND	PD	ND	PD	D	ND	D	D	PD	(e)
8. Manufacturer separator sheet (adhesive to core)	ND(e)	D	ND	ND	ND	ND	D	ND	ND	D	PD	D	(e)
9. Void (foam to closure)	D	D	D	D	D	D	D	D	ND	D	D	D	···
10. Inadequate tie-in of foam to core	D	D	D	D	D	D	ND	D	ND	D	ND	D	···
11. Inadequate depth of foam at closure	D	D	D	D	ND	PD(f)	ND	ND	ND	ND	ND	PD	···
12. Chemically milled step void	ND(e)	D	ND	ND	ND	ND	ND	ND	ND	ND	ND	ND	(e)

(a) ND, not detected; PD, partial detection; D, detected. (b) Panels were made using FM 73, which is not x-ray opaque. (c) The 0.13 and 0.25 mm (0.005 and 0.010 in.) crushed core detected by straight and better by angle shot. (d) Detects 0.25 mm (0.010 in.) crush core. (e) Has been detected by x-ray when adhesive was x-ray opaque (FM 400). (f) Discloses defect at a very high sensitivity

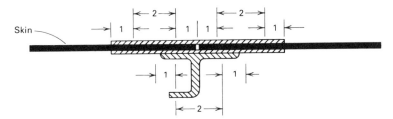

Fig. 51 Bonded joint divided into quality zones based on design shear stresses. The example shown is a longitudinal skin splice with an internal longeron. Quality grade 1 is within 13 mm (½ in.) of the edge of any bonded member, and grade 2 is the bond between any two grade 1 bonds.

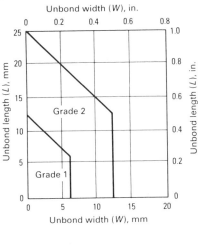

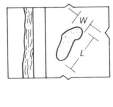

Fig. 52 Definition of acceptance grades for voids or unbonds

schematically illustrates a scanning holographic system. The capability of acoustical holography to reveal flaws in bonded structures is covered in Ref 35. One drawback to the use of scanning acoustical holography is the high cost of the equipment. In addition, considerable time is needed to reconstruct the various holograms of the entire part.

Selection of NDT Method

A universally accepted single test method for evaluating bonded structures has not yet been developed. Therefore, a combination of test methods is required for complete and reliable inspection. Generally, the selection of a test method is based on:

- Part configuration and materials of construction
- Types and sizes of flaws to be detected
- Accessibility to the inspection area
- Availability of equipment/personnel

The application of NDT methods in the production cycle for honeycomb sandwich panels generally includes:

- Material property tests
- Surface preparation checks (wedge crack, T-peel, and so on)
- Verifilm (prebond) tooling check
- Visual inspection
- Hot-water leak check
- Radiographic check for water and other internal discontinuities
- Ultrasonic C-scan for high-resolution inspection of skin-to-core bond
- Manual inspection of doublers, close-out members, and so on, using acoustic methods such as pulse-echo ultrasonic, Shurtronic harmonic bond tester, or Fokker bond tester

Built-in defects were produced in typical laminate and honeycomb specimens. These specimens were evaluated by state-of-the art NDT methods to determine which methods reliably detected the different defects. The results of this investigation are given in Table 4 for adhesive-bonded laminates and in Table 5 for adhesive-bonded honeycomb structures (Ref 6). The NDT methods described in this section are used to test sandwich structures having aluminum or titanium skins and aluminum core. Most of the methods can be effectively applied to structures with metal facing sheets and non-

metallic core and structures with graphite fiber, boron fiber, and fiberglass face sheets bonded to aluminum core. Only a limited number of the methods are applicable for structures having both core and facing sheets made from nonmetallic materials.

NDT in the Product Cycle

Nondestructive testing serves not only as a basis for process control, but also to establish quality control standards. The acceptance or rejection of a bonded assembly is directly related to the quality level desired by the designer. The quality level, in turn, is based on the importance of the part or component in terms of performance and safety. The frequency/severity flaw criteria and sensitivity of the test should be based on the desired quality level. To avoid unwarranted inspection costs, it is advisable to divide the bonded joint into zones based on stress levels or criticality of part function. Bonded joints are generally designed to withstand shear loads. Stress analysis of bonded joints reveals that the outer edges of a bonded joint will be subjected to higher stress levels than the center of the joint. Therefore, higher quality may be desired at the edges, and parts should be zoned accordingly. The zoned area should be dimensioned so that the inspector can define the two different zones prior to inspection. Figure 51 shows an example of quality zoning for bonded laminate skin splice. A definition of inspection grade numbers versus allowable void sizes is illustrated in Fig. 52.

For multiple bond line joints, it may not be possible to inspect to a high quality at the edge of all adherends, because of the loss of test sensitivity at each successive bond line below the surface. If the edges of a bond joint are stepped or staggered, high-quality inspection is possible at the edges of all bond lines. The width of the step should be large enough to accommodate the test instrument probe. Joint edges that are not stepped should be inspected from opposite sides.

Acceptance/Rejection Criteria

Before inspection can be performed on a bonded structure, accept/reject criteria must be established by engineering. These criteria are usually in the form of frequency

(number of flaws per unit area) and/or severity (maximum allowable size flaw data). To avoid unnecessary rejections, the engineering function must not specify unrealistic or overly conservative criteria. The frequency and severity criteria should be based on sound engineering judgment as related to calculated levels. It should be noted that flaws do not grow under cycle loads if good, durable adhesives are used and if generous overlaps exist.

Military Guidelines. For military contracts, the guidelines or requirements for preparing such criteria are specified in MIL-A-83377 (Ref 36). Specific acceptance criteria for adhesive-bonded aluminum honeycomb structures are specified in MIL-A-83376 (Ref 37). Specification MIL-A-83377 states that the contractor shall prescribe nondestructive tests and a complete NDT process specification when required by MIL-I-6780 in evaluating the quality of adhesive-bonded structures (Ref 38). The NDT process specification must be submitted for approval by the government prior to bonded assembly production. The contractor is required to specify the type and size of the defects that are acceptable. The size of the defect must be consistent with the capability of the NDT method to be used, realizing that NDT methods and instrumentation may have different capabilities. Application of the various NDT methods is limited by material types and design configuration. For aluminum honeycomb sand-

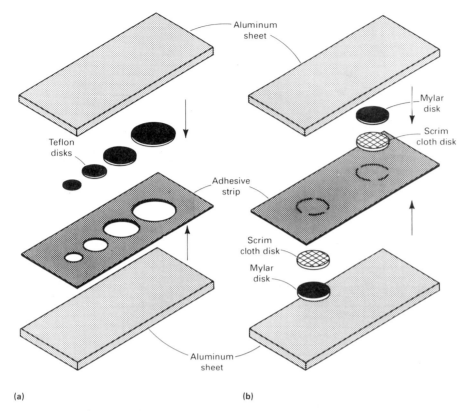

Fig. 53 Two methods of fabricating unbond reference standards. (a) Method 1 uses two pieces of aluminum sheet that have been anodized and primed and then bonded together with a piece of adhesive that has various-size cutouts filled with 0.13 mm (0.005 in.) thick Teflon disks. (b) Method 2 uses metal, as in method 1, but the voids are produced by placing a disk sandwich of Mylar and scrim cloth of desired size on the uncured adhesive. The assembly is then bonded.

wich, the allowables of MIL-A-83376 can be used, if determined by the contractor to be acceptable minimums for the particular application.

Specification MIL-A-83377 also requires that the NDT of adhesive-bonded structures be performed only by qualified NDT personnel. Personnel qualification requirements for NDT are specified in MIL-STD-410 (Ref 39).

Records and Specifications

It is recommended that four basic types of inspection records be prepared and maintained. They are specifications, an inspection log, a detailed written test procedure, and a rejection or nonconformance record.

Specifications. Test method specifications should be obtained for each type of production inspection instrument to be used. Each specification should specify controls for the particular method and should define reference standard configurations. In addition, an acceptance criteria specification should be prepared.

Inspection Log. Production inspection personnel should maintain a daily inspection log. This log should contain the part number, part dash number, part serial number, fabrication outline serial number, inspector name or stamp, date and method of inspection, and a remarks column to indi-

cate pass or reject. The rejection tag number can be entered, along with a description of the flaw, and its exact location.

Detailed Written Test Procedure. Nondestructive testing inspections should follow a detailed written procedure for each component tested. The written procedure must comply with the test method specification requirements and, as a minimum, must include:

- A sketch of the part or configuration showing in detail the part thickness, alloy material and temper, and the adhesive system used
- Manufacturer and model number of instrumentation to be used
- Model or designation number of probe to be used
- Any recording, alarm, or monitoring equipment and test fixtures or inspection aids that are to be used
- Applicable reference standard(s) serial number and design description
- Type of fluid couplant to be used
- Instrument calibration procedure, if required
- Testing plan. The part surface from which the test will be performed, the testing technique(s) for each joint, and the maximum probe index or shift per scan should be described

- Acceptance limits for each joint of the assembly
- Reference point or identifying mark, which is placed on each part prior to inspection; this mark is used for orientation when reviewing inspection results
- Outline of the areas where multiple layers of adhesive have been used
- Identification of section area changes not visible to the inspector
- Verification of the test procedure on a production part prior to approval

Rejection or Nonconformance Record. All nonconforming flaws should be described on a rejection tag. The inspection agency should outline on the part surface the location and shape of unacceptable bond line flaws. If it is necessary to mark the part permanently, the inspector must remove the couplant and wipe with solvent the area to be marked, and then reapply the marks. Next, the marks are covered with a very light coat of clear epoxy enamel. Marking the problem areas in this way facilitates their location after the part has been in service.

Reference Standards

Reference test standards for calibrating and standardizing the bond inspection instruments are essential. Even with adequate reference standards, inspection often produces conflicting information or a deviation of results. In these cases, more than one inspection method should be used in order to improve the accuracy of the data and to define the deviating condition. Under ideal conditions, the test standard with built-in defects of selected sizes should closely duplicate the structure adherends to be inspected. The laminate skin should be of the same thickness, and the honeycomb core should be of the same size and density. Other variations to be encountered in the test specimen, such as tapered core, chemically milled skins and doublers, and so on, should be incorporated into the test standard.

Except for built-in defects, the test standard for adhesive-bonded structures should be fabricated in the same manner as the production assembly. The void or defect should be introduced in the same bond line as that to be inspected in the structure. The standard can be a series of simple test specimens composed of details identical to the several areas of the assembly to be inspected and having a bond of good quality but with controlled or known defect locations and size.

Attempts to produce voids or controlled understrength bonds by the local application of grease or other foreign materials have been found to be ineffective. All adhesives and primers vary and may not respond to the following suggested methods of standard preparation. Therefore, the finished

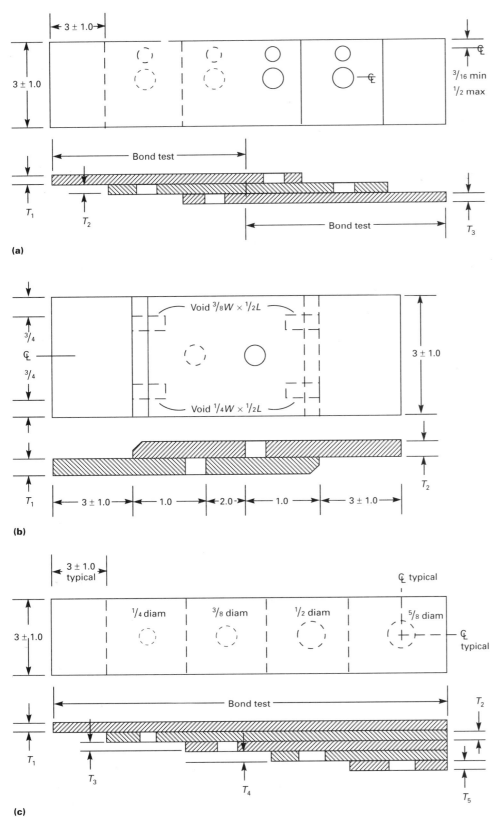

Fig. 54 Additional methods for fabricating unbond reference standards. (a) Method 3 uses three flat, unchamfered aluminum sheets of different thicknesses. This standard can be inspected from either side. Hole diameters can vary from 6.4 to 13 mm (¼ to ½ in.) and are free of adhesive. (b) Method 4 uses two chamfered aluminum sheets of different thicknesses with drilled holes 9.5 or 13 mm (⅜ or ½ in.) in diameter (free of adhesive) and edge voids made by inserting Teflon shims of the size shown. Shims must be removed after bonding. This standard can be inspected from either side. (c) Method 5 is a flat step standard using a buildup of various sheet thicknesses that matches the structure to be inspected. There should be no adhesive at the base of the holes. This standard should be inspected from the top side only. Dimensions given in inches

standards should be evaluated by various NDT methods prior to the validation of their use as an inspection standard. Neutron radiography is an excellent method for verification if voids, porosity, or both have been generated in the standard. The standard must then be inspected by the test instrument chosen for the particular part to confirm its ability to locate the flaw.

Metal-to-Metal Reference Standards. Various methods have been developed for producing simulated voids, porosity, or unbonds in metal-to-metal bonded laminates. Numerous ways to produce reference standards for metal-to-metal bonded joints or assemblies are described in Ref 10. Because of the variety of such standards, it is not possible to describe all of them in this article, and it is recommended that the reader consult the references for additional information.

Figure 53 shows two methods for producing standards for simulated skin-to-adhesive unbonds. The method 1 (Fig. 53a) or 2 (Fig. 53b) standards are useful for all NDT application methods, provided that the Teflon or Mylar inserts do not bond to the adherends.

Voids, Porosity, and Unbond Standards. Voids or unbonds of specified sizes are difficult to produce because free-flowing adhesive may fill the anticipated void or unbond area. One researcher was successful in producing natural flaws in the adhesive layer of metal-to-metal bond joints (Ref 1). By trial-and-error methods, defect conditions such as a void produced by removing a section of adhesive, a void produced by inserting a 0.64 mm (0.025 in.) diam wire between adherends, and porosity caused by inserting a 0.38 mm (0.015 in.) diam wire between adherends prior to bonding can be produced at specific locations in a bonded joint and with the desired sizes to be detected in an x-ray radiograph.

The methods 3, 4, and 5 standards (Fig. 54a, b, and c) are useful for manual bond test methods (such as Fokker, harmonic, and 210 bond test applications) and for pulse-echo ultrasonics. The holes at the edge and center of the method 3 standard are required if pinch-off of adhesive occurs near the edge, with thicker adhesive near the center of the standard. For maximum sensitivity, a bond joint should be inspected from both sides, and methods 3 and 4 standards are used. If the joint can be inspected from only one side, the method 5 standard is used. Holes are drilled after the part is bonded.

Standards containing porosity can be made by the wire insertion method or by bonding a large-area panel 0.6 × 0.6 m (2 × 2 ft, minimum) at low pressure (275 kPa, or 40 psi, maximum). In bonding with a mat (nonwoven) carrier, the adhesive pinches off at the outer edge, producing a picture frame of gross porosity close to the edge. Scattered porosity and some small voids are

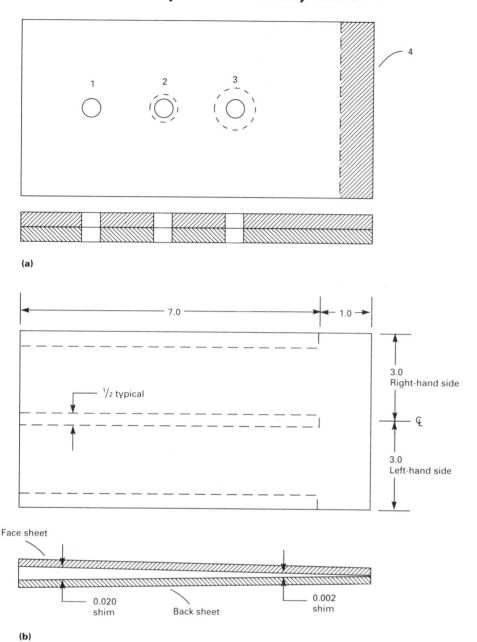

Fig. 55 Additional methods for fabricating unbond reference standards. (a) Method 6 presents a burned-adhesive standard in which feature 1 is a good-quality hole, 2 is a slightly burned adhesive, 3 is a grossly burned adhesive, and 4 is a grossly burned edge caused by abusive band sawing. (b) Method 7 becomes a variable-quality standard by inserting tapered shims between face and back sheet before bonding. Three 13 mm (½ in.) wide shims are shown in place. Tapered shims cause increasing porosity going from right to left. Half of the face sheet can be removed after inspection to reveal the condition of the adhesive. Dimensions given in inches

produced throughout the rest of the panel. After ultrasonic C-scan or radiographic inspection, representative specimens or small standards can be removed from the panel. The methods for producing voids and porosity are useful for creating defects at specific locations in slow-cycle fatigue test specimens. This type of specimen is used to study the effects of defects under load, cycle rate, or environment.

Burned-Adhesive Standards. Method 6 (Fig. 55a) is used to fabricate a standard representing burned adhesive caused by improper drilling or sawing of bonded panels. If improper drills, excessive drill speed and feed, or no coolant is used, burning of the adhesive adjacent to the hole may occur.

Variable-Quality Standards. Method 7 (Fig. 55b) is a way to fabricate a variable-quality standard. Through the use of tapered shims, good quality is obtained at the thin end and poor quality is obtained at the thick end. Figure 56 shows a positive reproduction made from an x-ray negative of AF 55 adhesive in a tapered standard. For visual correlation, only the right-hand upper skin need be removed from the panel, thus revealing the variable-adhesive quality. The left-hand side is kept intact to determine variations in NDT instrument response to different adhesive quality conditions.

Calibration of Bond Testers. When the size of the unbond or void is not of primary concern or when calibrating instruments to a bond/no-bond response, the following standards prove adequate. These standards are especially useful for resonant-type instruments such as the Fokker, harmonic, Sondicator, and 210 bond testers. They relate well to the bonded part when a 120 to 175 °C (250 to 350 °F) curing epoxy or epoxy-phenolic resin system, which does not attenuate the ultrasonic energy, is used.

Basically, the standards are made by cutting 38 mm (1.5 in.) diam holes in a 13 mm (½ in.) thick plywood panel and then bonding 75 × 75 mm (3 × 3 in.) aluminum sheets, using room-temperature-curing adhesives, so that the center of the sheet is centered over the holes in the plywood. The thickness of the aluminum sheets is chosen to match the thickness of the aluminum adherends of the test part. The number of sheets is governed by the number of adherends in the test part. When the aluminum sheet is bonded to a wooden base, it does not resonate freely and reacts similarly to unbonds in the part being inspected.

Honeycomb Reference Standards. Method 1 or 2 (Fig. 53) is used to create a void or unbond reference standard for honeycomb. For honeycomb standards, the unbonds are created between the adhesive-to-skin and adhesive-to-core interfaces. Skin-to-adhesive voids or unbonds are usually easier to detect than adhesive-to-core unbonds. Another way to prepare a void or unbond standard for a honeycomb panel is illustrated in Fig. 57. This standard is useful when attempting to detect unbonds on both sides of the panel if inspection can be performed from only one side.

Honeycomb crushed core standards are made by inserting 25 mm (1 in.) diam, 0.13 and 0.25 mm (0.005 and 0.010 in.) thick metal disks between the skin and core and then applying pressure. The disks are removed and the panel is bonded. The 0.13 mm (0.005 in.) crushing is very subtle and is difficult to detect radiographically. The 0.25 mm (0.010 in.) crushing is easily detectable.

Water intrusion standards are produced by hot-bonding one skin to the core, adding various quantities of water into the cells, and then cold-bonding the second facing sheet in place.

Core-splice standards are made by cutting the edges of the core with a taper and then applying foaming adhesive to one side of the splice before bonding. Voids in the foaming adhesive are detectable by low-kilovoltage radiography. Core-to-closure voids are made by eliminating foaming adhesive at small areas along the joint.

Separator sheet standards are produced by bonding-in a desired size of the material

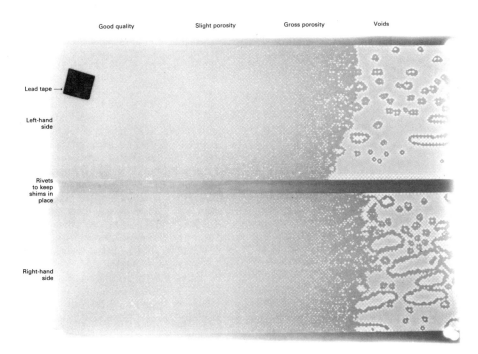

Fig. 56 Positive print from x-ray negative of tapered-shim variable-quality standard shown as method 7 in Fig. 55

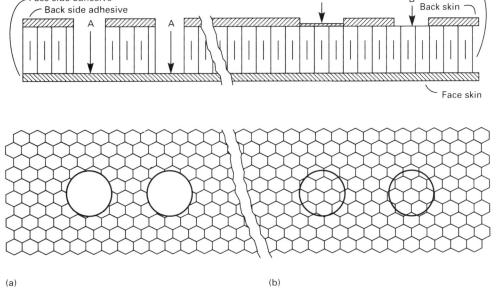

Fig. 57 Construction details for making a honeycomb standard for voids and unbonds. (a) Back skin and core cut out to desired diameter. Cutting through face sheet adhesive shows skin-to-adhesive unbond; cutting to the adhesive shows adhesive-to-core unbond. (b) Cutting through back skin first to the adhesive to show skin-to-adhesive unbond and then through the adhesive to show adhesive-to-core unbond. Teflon disks can be used to help separate adhesive from skins at cutouts. All tests are to be made from the face skin side.

at the skin-to-adhesive and adhesive-to core interfaces.

A blown-core standard can be made by mechanically damaging the core prior to bonding.

Corroded-core or core-to-adhesive corrosion standards can be made by bonding only one skin to the core (open-face honeycomb) and then alternately spraying the honeycomb side in a salt spray system. The degree of corrosion can be monitored visually. If it is desired to close up the panel, the second facing sheet can be cold-bonded to the other side of the core.

Substitute Standards. When the ideal standards are not available, other standards of construction similar to that of the test part can be used. However, honeycomb standards should not be used for metal-to-metal laminate structure, and vice versa.

Inspection Without Standards. When standards are not available, a known undamaged area can be used as the standard. Repaired areas are inspected by comparing them to an unrepaired area. The instrument reading may change because of structural changes resulting from the repair. Repeated inspection scans must be conducted under various instrument settings, and the inspection should be verified by other methods and/or instruments. A knowledge of the part configuration (for example, number of bond lines, adhesive thickness, skin thickness, and so on) is essential for interpreting the test results.

Evaluation and Correlation of Inspection Results

The experience, confidence, and effectiveness of NDI technicians and engineering personnel are gained by correlating NDI results with destructive testing. Comparisons can be made between the NDI results and the actual size, shape, location, and type of defect. Mistakes will occur in the early stages of the bonding program, but if accurate records, photographs, and sketches are kept, they can serve as effective training devices. Sections of defective parts should be kept for training purposes. All these items add to the confidence of the inspector in evaluating and accurately reporting defects. They also add to the engineer's confidence in the inspector and in the data reported by the inspector.

Procedure

Before new adhesive-bonded assemblies are fabricated on a mass production basis, the first assembly is evaluated by numerous methods. Inspections are performed and NDT results are evaluated to detect any variations in the part that are indicative of discontinuities. The part is then cut up, and special precautions are taken to section through the discontinuities or to separate the joint at the discontinuity to identify and verify the NDI results. To determine which NDT method(s) detected the discontinuity, the NDI technicians use different color markings—for example, blue for x-ray, green for ultrasonic, red for a bond tester, and so on. These markings can be affixed to the surface of the part or placed on a transparent overlay of the part. When the part is sectioned, the results will identify which NDT method(s) correctly identified the location, size, shape, and type of discontinuity. Although it is expensive, cutting the adhesive-bonded assemblies is sometimes the only way to determine the nature of the discontinuity or to verify its existence. All inspection results should be carefully documented for future reference.

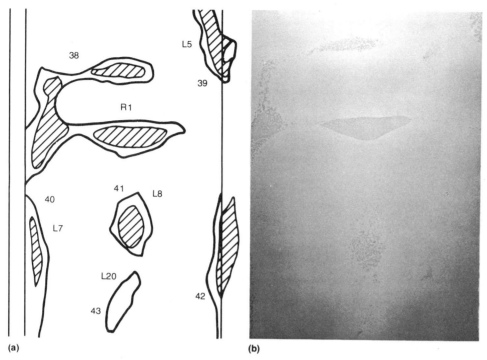

Fig. 58 Correlation of NDI results. The same specimen was inspected by (a) Fokker bond tester and (b) neutron radiography. See text for discussion.

Once a highly reliable inspection technique is established, the remaining defective assemblies can be used for developing repair procedures. The original NDI techniques can then be used for determining whether the repairs are effective. In some cases, defective parts are consistently produced, and the material, production process, or design must be changed to correct the condition. This condition is usually evident when the same type of defect occurs in the same location in all manufactured parts.

Reliable feedback from manufacturing personnel is beneficial to the NDI technician. This information can include such items as a too-high or too-low oven temperature, incorrect heat-up rate, leaks in the vacuum bag, additional layers of adhesive added at a particular spot in the assembly, and so on. With this information, the NDI technician is better prepared to inspect a particular group of parts.

Correlations of Destructive and NDI Results

Following is a typical example of the correlation of NDI results and destructive testing. In this example, a metal-to-metal joint containing a bonded doubler and longerons and frames was rejected for numerous voids and porosity. The initial inspection was performed using the Fokker bond tester, and the defective areas were marked on the surface of the part, as shown in Fig. 58(a). The inspection results were verified using the 210 bond tester. The part was then sent out for thermal neutron radiography.

There was excellent correlation between the Fokker bond test results and the neutrograph (Fig. 58b).

The panel was then ultrasonically C-scanned using the through transmission technique. Again, the correlation was excellent, as shown in Fig. 59. A plastic overlay was made to record the Fokker bond tester results.

The part was then subjected to chemical milling to dissolve the aluminum adherends, leaving the adhesive layer intact. The adhesive thickness was measured and recorded. The destructive-to-nondestructive correlation was again excellent and revealed that the adhesive thickness in and around the defects was greater (0.33 to 0.46 mm, or 0.013 to 0.018 in.) than in the nondefective areas (0.18 to 0.28 mm, or 0.007 to 0.011 in.). In addition to the correlations noted above, lap-shear specimens could have been made through the different defect conditions, and variations in bond strength could have been determined.

Because of the variety of defects that can occur in honeycomb assemblies, defects detected by one test method should be verified by an additional but similar test method. Time and facilities permitting, the assembly should also be x-rayed in the defective area.

Ultrasonic Wave Interference Effects

In the ultrasonic inspection of adhesive-bonded joints, two conditions may cause ultrasonic wave interference (Fig. 24) effects:

- When inspecting bond lines under tapered metal doublers
- When inspecting close to the edge of a bonded laminate and the adhesive is thin due to pinch-off (Ref 10)

Ultrasonic theory predicts that constructive interference (increased amplitude) occurs when the transmitted and reflective waves are in phase with one another (Ref 40, 41). Conversely, destructive interference (decreased amplitude) occurs when the transmitted and reflected waves are 180° out of phase with one another. Destructive wave interference causes a drop in signal amplitude, and the C-scan recorder will not record in the area of interference. An inexperienced operator might misinterpret these areas as voids or unbonds. However, interference effects (caused by tapered adherends) are uniform across the part, resulting in bands of light and dark on the C-scan recording (Ref 10). When destructive interference effects are shown in C-scan recordings of bonded splices having tapered adherends, the bond joints are given another inspection by a different method, such as the Fokker bond tester.

Destructive wave interference effects can also be seen in C-scan recordings of adhesive-bonded laminates made with an adhesive having a mate carrier. It is less evident in laminates bonded with adhesive having a woven carrier. The destructive interference effects are most evident near the outer edges of the assembly where the adhesive is pinched off to thicknesses of 0.08 mm (0.003 in.) or less. The interference effects are frequently dependent and therefore change position if the test frequency is altered (Ref 40, 41). Less wave interference is experienced with through transmission testing than with pulse-echo testing at similar frequencies. Bonded laminates exhibiting destructive wave interference effects require inspection by another test method in the area in question to verify bond quality.

Correlation of Bond Strength and Fokker Bond Tester Readings

The ability of the Fokker bond tester to measure bond strength is completely dependent on the accuracy of the correlation curves, which plot Fokker readings versus either lap-shear strength (for metal-to-metal joints) or flatwise tensile strength (for honeycomb joints). For metal-to-metal joints, the correlation is made between the Fokker A-scale readings and tensile lap-shear strength. For honeycomb structure, the correlation is made between Fokker B-scale readings and flatwise tensile strength.

Flatwise tensile specimens are prepared in accordance with MIL-STD-401 (Ref 42), except that the adherends and core should simulate those of the bonded test assembly being tested. Fabrication procedures should

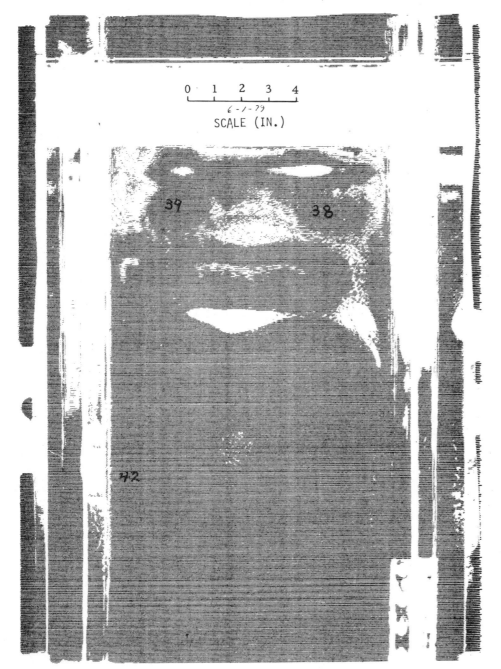

Fig. 59 Ultrasonic C-scan recording of panel in Fig. 58 showing voids and porosity. Through transmission method was used with low-gain (31 dB).

Effect of Porosity and Voids on Bond Testing. A relationship can be established between Fokker bond tester A-scale frequency shift and lap-shear strength as related to porosity. Tests have shown that the percentage of porosity increases with adhesive thickness (single layer) (Ref 5, 10). Porosity first appears at an adhesive thickness of 0.25 mm (0.010 in.) and becomes almost completely void at a thickness of 0.51 mm (0.020 in.).

The first step is to produce high-quality lap joints (25.4 mm, or 1.0 in., wide by 13 mm, or ½ in., overlap) containing 100, 80, and 50% width bond lines, as shown in Fig. 62. The test specimens are approximately 178 to 203 mm (7 to 8 in.) long, and a minimum of six specimens should be made for each percentage bond and thickness, that is, 1.0 mm/1.0 mm (0.040 in./0.040 in.) (in general, the t_1 thickness is equal to the t_2 thickness), and so on. Typical average lap-shear values for 2024-T3 and FM 73 adhesive are shown in Fig. 63.

The second step is to make specimens with increasing thickness of the single adhesive layer. This will produce specimens of variable quality resulting from porosity. To obtain the quality variation, spacer shims are used, as shown in Fig. 64. The thickness of the spacers is increased in steps (0.10 mm, or 0.004 in., per step) from 0.10 to 0.51 mm (0.004 in. to 0.020 in.), resulting in a highly porous bond joint. The lap-shear specimens are numbered and neutron radiographed. The specimens are then Fokker bond tested, and the A-scale readings are recorded for each specimen. The lap-shear coupons are cut from each panel, and the adhesive thickness (for each specimen) is accurately measured. The adhesive will generally be found to be slightly thicker than the associated shim. The lap-shear test is performed in accordance with Federal Test Method Standard No. 175, method 1033, or Federal Specification MMM-A-132. The neutron radiographs indicate areas of nonuniformity of porosity, and results from these nonuniform areas should be discarded.

The third step is to prepare shimmed specimens using additional layers of adhesive. This will produce good-quality specimens of thick adhesive. The panels are neutron radiographed, Fokker bond tested, cut into specimens, and lap-shear tested, with the data subsequently evaluated. Plots of lap-shear strength versus adhesive thickness for the single porous layer and multiple nonporous layers of FM 73 are shown in Fig. 65.

The fourth step is to plot lap-shear strength versus Fokker A-scale readings for each adherend thickness (Fig. 66). If the thickness of the adherends is not equal, the back (unprobed) thickness is plotted. Unequal-thickness adherend specimens are tested from both sides, and two quality

be identical to those used in preparing the assembly to be tested. Specimens are prepared with various degrees of bond quality by the introduction of porosity in the cured adhesive. The degree of porosity increases with the thickness of the adhesive layer. Figure 60 illustrates the procedure for honeycomb-to-skin joints.

Nondestructive testing of the panels is performed in accordance with procedures outlined in MIL-STD-860 (Ref 23) or as specified in the Fokker bond tester training manual. Instrument readings should be taken within predetermined areas of the test

specimen. Each reading should be recorded and correlated with the probe location on the specimen. The specimens are then destructively tested, and the tensile strength values are recorded. Flatwise tensile strength versus average ultrasonic readings for each specimen should be plotted, and a representative curve drawn through the plotted points. Two correlation curves should be drawn for specimens with dissimilar thickness adherends. A typical correlation curve for Fokker B-scale readings versus flatwise tensile strength is shown in Fig. 61.

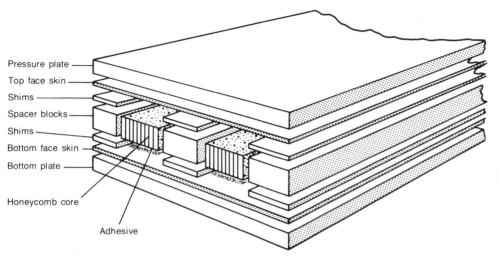

Fig. 60 Honeycomb sandwich test panel

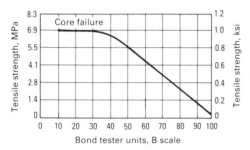

Fig. 61 Typical Fokker B-scale correlation curve versus honeycomb flatwise tensile strength. Cell structure: 0.64 mm (0.025 in.) face skin, 3.18 mm (0.125 in.) cell size, and 0.025 mm (0.001 in.) foil thickness. Probe, 3414

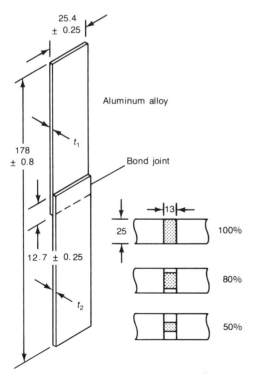

Fig. 62 Lap-shear test coupon. Dimensions given in millimeters

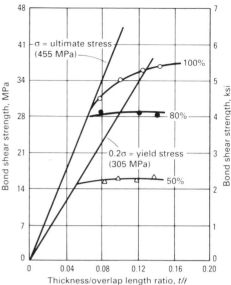

Fig. 63 Variation in lap-shear strength with t/ℓ ratio and bond length using aluminum alloy 2024-T3 and adhesive FM 73. t_1/t_2 (mm): 1.0/1.0, 1.3/1.3, 1.6/1.6, and 1.8/1.8

curves are established. With reference to Fig. 66 and 67, the bond strength can be determined for 100, 80, and 50% specimens having various thickness or t/ℓ ratios. By evaluating all the data, various classes for single-layer (porous) and multiple-layer (nonporous) joints can be specified:

- *Class A*: Greater than 80%
- *Class A/B*: Greater than 65%
- *Class B*: Greater than 50%
- *Class C*: Greater than 25%

A-Scale Frequency Shift Versus Bond Quality. The large scatter in results for the 25% quality level specimens can be attributed to the extensive porosity in these specimens. This does not pose a problem to the inspector because quality levels below 50% are usually not acceptable. The actual relationship between the A-scale frequency shift (as seen by the inspector) and the bond quality (class) is illustrated in Fig. 67.

Cohesive Bond Strength Acceptance Limits. Finally, the acceptance limits are incorporated into a table for use by the inspector. Table 6 lists the acceptance limits for single bond line joints with FM 73 adhesive and specific-thickness adherends. Establishing Fokker quality diagrams for mul-

tiple bond line joints is possible but complicated. Fokker-VFW should be contacted to obtain the recommended procedures for developing such quality diagrams.

ACKNOWLEDGMENT

The sections "Shearography" and "Bondscan Thermography Inspection System" in this article were furnished by John W. Newman, Laser Technology, Inc.

REFERENCES

1. M.T. Clark, "Definition and Non-Destructive Detection of Critical Adhesive Bond-Line Flaws," AFML-TR-78-108, U.S. Air Force Materials Laboratory, 1978
2. R.J. Schliekelmann, Non-Destructive Testing of Adhesive Bonded Metal-to-Metal Joints, *Non-Destr. Test.*, April 1972
3. R.J. Schliekelmann, Non-Destructive

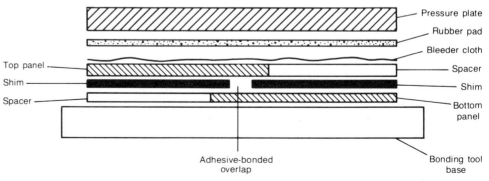

Fig. 64 Metal-to-metal lap-shear test panel

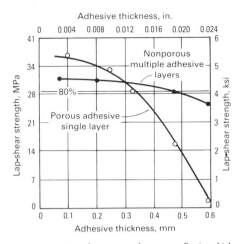

Fig. 65 Lap-shear strength versus adhesive thickness for porous and nonporous adhesives. Adherends, 1.6 mm/1.6 mm; adhesive, FM 73 with mat carrier; primer, BR127

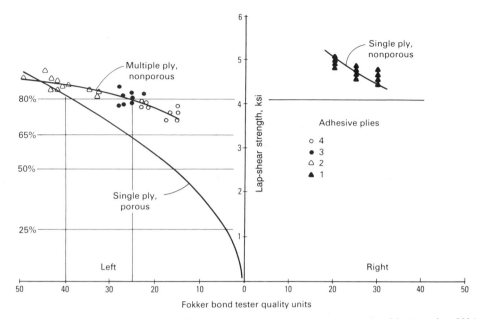

Fig. 66 Lap-shear strength versus Fokker bond tester quality units. Bond tester, model 70; probe, 3814; adherends, 1.6 mm/1.6 mm; adhesive, FM 73; primer, BR127

Testing of Bonded Joints—Recent Developments in Testing Systems, *Non-Destr. Test.*, April 1975

4. P. Bijlmer and R.J. Schliekelmann, The Relation of Surface Condition After Pretreatment to Bondability of Aluminum Alloys, *SAMPE Q.*, Oct 1973

5. K.J. Rienks, ''The Resonance/Impedance and the Volta Potential Methods for the Non-destructive Testing of Bonded Joints,'' Paper presented at the Eighth World Conference on Nondestructive Testing, Cannes, France, 1976

6. J.C. Couchman, B.G.W. Yee, and F.M. Chang, Adhesive Bond Strength Classifier, *Mater. Eval.*, Vol 37 (No. 5), April 1979

7. J. Rodgers and S. Moore, *Applications of Acoustic Emission to Sandwich Structures*, Acoustic Emission Technology Corporation, 1980

8. J. Rodgers and S. Moore, ''The Use of Acoustic Emission for Detection of Active Corrosion and Degraded Adhesive Bonding in Aircraft Structures,'' Sacramento Air Logistics Center (SM/ALC/MMET), McClellan Air Force Base, 1980

9. The Sign of a Good Panel Is Silence, *Aviat. Eng. Maint.*, Vol 3 (No. 4), April 1979

10. D. Hagemaier and R. Fassbender, Nondestructive Testing of Adhesive Bonded Structures, *SAMPE Q.*, Vol 9 (No. 4), 1978

11. R. Botsco, The Eddy-Sonic Test Method, *Mater. Eval.*, Vol 26 (No. 2), 1968

12. J.R. Kraska and H.W. Kamm, ''Evaluation of Sonic Methods for Inspecting Adhesive Bonded Honeycomb Structures,'' AFML-TR-69-283, U.S. Air Force Materials Laboratory, 1970

13. N.B. Miller and V.H. Boruff, ''Evaluation of Ultrasonic Test Devices for Inspection of Adhesive Bonds,'' Final

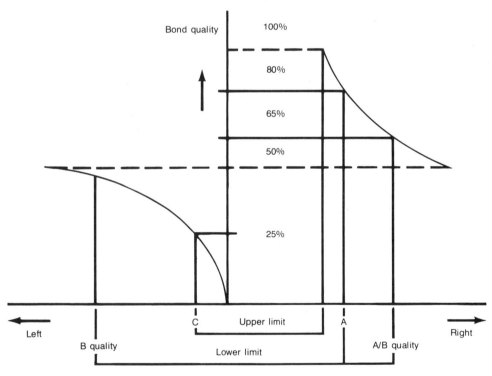

Fig. 67 Establishment of acceptance limits (sheet to sheet)

Table 6 Fokker cohesive bond strength acceptance limits for single bond line using FM 73 adhesive

Lower sheet thickness		Model 70 probe	Upper limit(a)	Lower limits(b)			
mm	in.			Class A(c) 80%	Class A/B(c) 65%	Class B(c) 50%	Class C(c) 25%
1.0	0.040	3814	>R8	>L54	>L40	>L25	>L10
1.3	0.050	3814	>R10	>L48	>L36	>L20	>L9
1.6	0.063	3814	>R12	>L42	>L30	>L18	>L8
1.8	0.071	3814	>R16	>L40	>L28	>L18	>L7
2.1	0.081	3814	>R18	>L38	>L25	>L15	>L6

(a) R, right-hand side. (b) Skin/stringer and skin/frame bonds shall be class minimum; doubler/skin bonds shall be class A/B minimum. Additional adhesive layers shall not yield readings less than the lower limits by the following factors: 2 layers (−15); 3 layers (−25); 4 layers (−30). (c) L, left-hand side

Report ER-1911-12, Martin Marietta Corporation, 1962

14. J. Arnold, "Development of Non-Destructive Tests for Structural Adhesive Bonds," WADC-TR-54-231, Wright Air Development Center, 1957

15. R. Schroeer, et al., The Acoustic Impact Technique—A Versatile Tool for Non-Destructive Evaluation of Aerospace Structures and Components, Mater. Eval., Vol 28 (No. 2), 1970

16. H.M. Gonzales and C.V. Cagle, Nondestructive Testing of Adhesive Bonded Joints, in Adhesion, STP 360, American Society for Testing and Materials, 1964

17. D.F. Smith and C.V. Cagle, Ultrasonic Testing of Adhesive Bonds Using the Fokker Bond Tester, Mater. Eval., Vol 24, July 1966

18. H.M. Gonzales and R.P. Merschell, Ultrasonic Inspection of Saturn S-II Tank Insulation Bonds, Mater. Eval., Vol 24, July 1966

19. R.J. Botsco, Nondestructive Testing of Composite Structures With the Sonic Resonator, Mater. Eval., Vol 24, Nov 1966

20. R. Newschafer, Assuring Saturn Quality Through Non-Destructive Testing, Mater. Eval., Vol 27 (No. 7), 1969

21. "Fokker Ultrasonic Adhesive Bond Test," MIL-STD-860, 1978

22. R.J. Schliekelmann, Non-Destructive Testing of Adhesive Bonded Joints, in Bonded Joints and Preparation for Bonding, AGARD-NATO Lecture Series 102, Advisory Group for Aerospace Research and Development, 1979

23. D.J. Hagemaier, Bonded Joints and Non-Destructive Testing: Bonded Honeycomb Structures, 2, Non-Destr. Test., Feb 1972

24. D. Wells, NDT of Sandwich Structures by Holographic Interferometry, Mater. Eval., Vol 27 (No. 11), 1969

25. R.M. Grant, "Conventional and Bead to Bead Holographic Non-Destructive Testing of Aircraft Tires," Paper presented at the 1979 Assembly and Test Area Nondestructive Testing Forum, Seattle, WA, Sept 1979

26. C.F. Lewis, What's Wrong With This Picture?, Mater. Eng., Jan 1989, p 49-52

27. Y.Y. Hung, Shearography: A New Optical Method for Strain Measurement and Nondestructive Testing, Opt. Eng., May/June 1982, p 391-395

28. J. Newman, Advanced Shearography NDT, in Proceedings of the Conference on Nondestructive Testing, American Society for Nondestructive Testing, Oct 1989

29. D. Burleigh and J. Engel, NDT of an Adhesively Bonded Fixed Foam Insulation for Atlas Centaur Cryogenic Fuel Tanks, in Proceedings of the Conference on Quantitative NDT, University of Iowa, July 1989

30. P.R. Vettito, A Thermal I.R. Inspection Technique for Bond Flaw Inspection, J. Appl. Polym. Sci., Appl. Polym. Symp. No. 3, 1966

31. C. Searles, Thermal Image Inspection of Adhesive Bonded Structures, in Proceedings of the Symposium on the NDT of Welds and Materials Joining, 1968

32. W. Woodmansee and H. Southworth, Detection of Material Discontinuities With Liquid Crystals, Mater. Eval., Vol 26 (No. 8), 1968

33. S. Brown, Cholesteric Crystals for Non-Destructive Testing, Mater. Eval., Vol 26 (No. 8), 1968

34. D.H. Collins, Acoustical Holographic Scanning Techniques for Imaging Flaws in Reactor Pressure Vessels, in Proceedings of the Ninth Symposium on NDE, San Antonio, April 1973

35. A Sample of Acoustical Holographic Imaging Tests, Holosonics Inc.

36. "Adhesive Bonding (Structural) for Aerospace Systems, Requirements for," MIL-A-83377

37. "Adhesive Bonded Aluminum Honeycomb Sandwich Structure, Acceptance Criteria," MIL-A-83376

38. "Inspection Requirements, Nondestructive: For Aircraft Materials and Parts," MIL-I-6870 (ASG)

39. "Nondestructive Testing Personnel Qualification and Certification," MIL-STD-410

40. B.G. Martin, et al., Interference Effects in Using the Ultrasonic Pulse-Echo Technique on Adhesive Bonded Metal Panels, Mater. Eval., Vol 37 (No. 5), 1979

41. P.L. Flynn, Cohesive Bond Strength Prediction for Adhesive Joints, J. Test. Eval., Vol 7 (No. 3), 1979

42. "Sandwich Construction and Core Materials; General Test Methods," MIL-STD-401

End Product Nondestructive Evaluation of Adhesive-Bonded Composite Joints

Yoseph Bar-Cohen, Douglas Aircraft Company, McDonnell Douglas Corporation
Ajit K. Mal, Mechanical, Nuclear and Aerospace Engineering Department,
University of California, Los Angeles

RECENTLY DEVELOPED nondestructive evaluation (NDE) methods for inspecting bonded composite joints in aircraft structures are the focus of this article. This type of joint is used increasingly in new aircraft, replacing mechanically fastened metallic composite assemblies. For example, bonded joints are now used in helicopter rotor blades, aircraft wings, fuselages, and rudders. A typical structure consists of either laminates bonded to a core in sandwich form (during or after curing) or a laminate bonded to another laminate.

Bonded composite joints are substantially more difficult to inspect than are metal joints because the composite and the bond must be examined simultaneously, and both are problematic areas for nondestructive evaluation. Some NDE methods that are used for metallic structures are also applicable to bonded composite joints. However, the special nature of composites, namely their anisotropic, layered, nonhomogeneous structure, introduces certain limitations to the use of these methods. The methods that are applicable to composite bonded joints include ultrasonics, acoustic emission, radiography, holography, and thermography, each of which is described in this article. A brief review of the state of the art is also provided. A useful reference work as regards metallic structures is *Nondestructive Evaluation and Quality Control*, Volume 17 of the 9th Edition of *Metals Handbook*, ASM Internaitonal, 1989, as well as the article "End Product Nondestructive Evaluation of Adhesive-Bonded Metal Joints" in this Volume.

Although the simplest, fastest, and most common NDE technique is tap testing, in which the inspector taps the test structure with a hammer or coin and listens to the sound characteristics, it is very limited in scope and can only be used to detect gross unbonds. Cawley and Adams made a sys-tematic study of the tapping method by examining the force input by the tapper to the test structure (Ref 1). They found that an unbond changes the characteristics of the force input. Further, the impact duration increases and the peak force decreases as the defect becomes larger. Even though Cawley and Adams substantially improved the reliability and detectability of the technique, it is still limited to thin layers (up to about 2 mm, or 0.08 in. thick), and the smallest detectable unbond is about 10 mm (0.4 in.) in diameter in typical laboratory test specimens.

The integrity and durability of a bonded joint can be ensured only by using methods that detect and characterize unbonds and determine the properties of the adhesive as well. The ultimate need is to be able to measure the strength of the bond using nondestructive means. To attain this capability, parameters that might be sensitive to strength have been sought by many investigators (see, for example, Rose and Meyer, Ref 2). Limited success was achieved when the material variables were carefully controlled and when there was a direct relationship between the average strength and a single property of the adhesive material, such as thickness. However, in practice it is difficult to correlate NDE data with the strength of a bond. This is because strength is a structural rather than a physical parameter and, in fact, is an indication of the highest stresses that the weakest spot of a specific structure is able to carry. Generally, there is no NDE method that can search systematically through a structure to identify all the weak points and determine which is the weakest. Further, bonding is particularly sensitive to interface characteristics, which cannot yet be determined by nondestructive means. Thus, even though it may be possible to estimate the average strength of a bonded system, this measure may not be very useful. The bond will simply fail when the stress exceeds the material strength in its weakest spot rather than the average strength, which may be substantially higher.

The mechanical strength of an adhesive bond depends on cohesion and adhesion. Cohesion is associated with the bond between the various molecules of the adhesive layer. Cohesive strength is determined by the type of adhesive, its elastic properties, and its thickness. Nondestructive evaluation can provide limited information about these parameters. Adhesion, on the other hand, is related to the bond between the adhesive and the adherend. The quality of the adhesion is critical to the performance of the adhesive as a bond between the components of an assembly. Because the interface layer is often a fraction of a micrometer thick, it is difficult to characterize it by nondestructive techniques. Weakness in this layer, which can be caused by poor surface preparation, is not detectable by NDE. Because of these limitations in determining strength, NDE methods are normally used to detect and characterize unbonds rather than bond strength.

Ultrasonic NDE

Ultrasonics is one of the principal means of nondestructively examining composite joints. Various wave modes are applied, including longitudinal, shear, and Lamb waves. The parameters analyzed include the amplitude, time-of-flight, and response frequency. Most ultrasonic methods are based on normal incidence of an acoustic or ultrasonic beam. The most important of these methods are described below.

Impedance. Instruments that measure by acoustic impedance are widely used for the NDE of both metallic and composite joints. This method is based on establishing a standing wave in the test material (Ref 3). To limit the effect of attenuation, frequen-

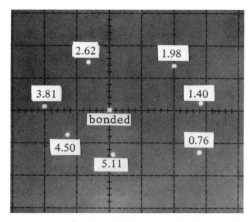

Fig. 1 Indication of delamination depth using the impedance technique (Bondascope) for a graphite-epoxy 10.2 mm (0.41 in.) thick laminate with 10 mm (0.40 in.) diam flaws

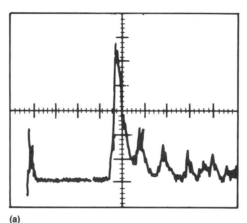

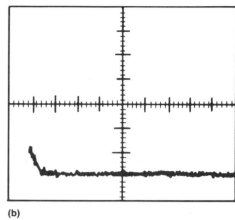

(a)

(b)

Fig. 2 Through-transmission response of a sandwich of graphite-epoxy skins and a metallic honeycomb core. (a) Bonded. (b) Unbonded

cies well below the megahertz range are used. The change of phase or frequency and amplitude are determined by measuring their loading effect on a piezoelectric transducer.

The change in the acoustic impedance of the test specimen because of unbonding also changes the electric impedance of the transducer that is loaded by the material. The change in the load depends on the distance traveled in the material, or on its effective thickness, causing an appropriate change in the acoustic impedance. The electrical impedance of the transducer and its resonant frequency for bonded materials are used for reference in searching for unbonds.

The use of impedance as an NDE technique for detecting unbonds is based on the difference in the characteristic response between bonded and unbonded areas. Two instruments that use this method, the Bondascope 2100 and the Fokker bond tester, are widely used. An example of the response of the Bondascope 2100 to unbonds and delaminations at different depths is shown in Fig. 1.

Ultrasonic Pulse-Echo and Through-Transmission Methods (Ref 4, 5). A wave traveling from one material to another will be partially reflected and partially transmitted through the interface. The amplitude of each of the two generated components of the wave is determined by the degree of acoustic impedance mismatch between the

two materials. The larger the impedance mismatch, the higher the reflection coefficient. The two components of the wave are employed in the pulse-echo and through-transmission NDE techniques. They are used with short ultrasonic pulses and are operated at higher frequencies (0.5 to 10 MHz) than the impedance technique. To increase the detectability of defects in a through-transmission test, a reflector plate can be used to reflect the transmitted signal back to the transducer. Thus, a single transducer is used, and the effect of the attenuation is increased by doubling the wave path.

Examples of the response when using through-transmission and pulse-echo techniques are shown in Fig. 2 and 3, respectively. The through-transmission technique is involved in evaluating the amplitude of the signal transmitted through the composite structure. Unbonds or delaminations cause a reduction or reflection of the transmitted signal, which depends on the defect size relative to the transducer diameter. The difference between bonded and unbonded areas is obvious from the appearance or disappearance of the signal after traveling through the part. This technique is widely used because it is fast and inspects the entire sectional thickness in one scan. However, it is more difficult to apply the technique in field conditions because of the need

for access to both surfaces of the material, the required alignment of the two transducers during inspection, and the need to maintain efficient ultrasonic coupling.

The pulse-echo technique, on the other hand, provides detailed information, including depth information, about the bonded areas. It is possible to determine whether the unbond is above or below the adhesive layer or whether it is indeed an unbond or a delamination. When a delamination is detected, its depth can be determined with the relatively high accuracy of ±0.1 mm (±4 mils) in typical specimens (Ref 5). The detected defects are accepted or rejected by comparing their response to a reference defect with a size defined by the material specifications. It is common to define levels of quality that depend on the defect size by the letters A, B, and C, where A identifies the highest quality as used for primary structures and C identifies the lowest (Ref 4).

To deal with the complexity associated with the inspection of bonded composite joints, computerized C-scan systems are increasingly employed. Such systems provide detailed information about the depth distribution of delaminations using time-of-flight data. The information is presented in colors with a ''look-up'' table indicating the depth range (see Fig. 4). Once the ultrasonic data are acquired, a user can process the

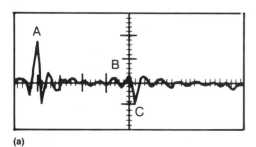

(a)

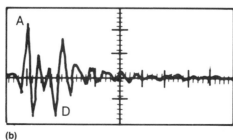

(b)

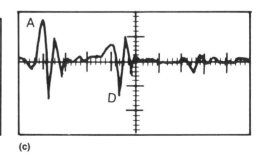

(c)

Fig. 3 Pulse-echo indication from a sandwich specimen where graphite-epoxy [0,90]$_{2s}$ skin and glass-epoxy [0] are bonded with FM 73 adhesive to an aluminum honeycomb. (a) Bonded sample. (b) Delamination between fourth and fifth layers of the graphite-epoxy skin. (c) Unbond between the glass-epoxy and the adhesive layer. A, front surface reflection; B, reflection from the glass/epoxy layer; C, reflection from the adhesive back surface; D, delamination

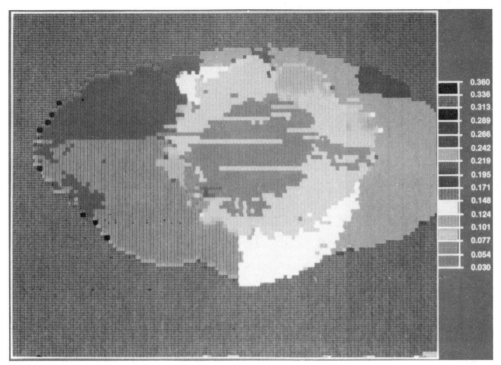

Fig. 4 C-scan image of impact damage in a graphite-epoxy laminate using zoom capability. Depth of damage given in inches

Ultrasonic Spectroscopy. The signals measured by pulse-echo or through-transmission techniques are examined in the time domain, using a broadband transducer. The received signal can also be analyzed in the frequency domain by means of a frequency analyzer, which continuously displays the Fourier transform. The advantage of this spectroscopy technique is its ability to reveal frequency-dependent features that cannot be easily identified in time-domain signals. Further, the signal can be processed with various enhancement techniques that improve the detectability of defects. Examples of such processes include filtering, convolution, and correlation. Ultrasonic spectroscopy has been investigated for use as an NDE tool, mostly during the 1960s and the 1970s (see, for example, F.H. Chang *et al.*, Ref 6). An example of the response from a graphite-epoxy laminate with and without delamination is shown in Fig. 7. As indicated, there is an increase in the frequency interval between the minima on the reflection coefficient with an increase in depth.

Leaky Lamb Waves. All the techniques described earlier are based on a normal incidence of the ultrasonic waves to the specimen. The leaky Lamb wave (LLW) technique is associated with an angular insonification of the tested joint. When a transducer insonifies a test part at an oblique angle, the wave is refracted, as well as mode converted, to induce guided waves that can be very useful in NDE. These waves, when excited, propagate along the plate and are strongly affected by the prop-

data and determine a wide variety of details using various image manipulation features. As shown in Fig. 4, the image of impact damage to a graphite-epoxy skin can be "zoomed" so that the user can observe it in more detail. Further, the rotation of three-dimensional displays can be used to present the depth distribution of defects, such as impact damage, in visual form (see Fig. 5). Computers offer programming and control capabilities that allow contours to be followed. Such systems are increasingly applied to the inspection of large structures, such as wings, after production. A typical view of a contour-following system is shown in Fig. 6.

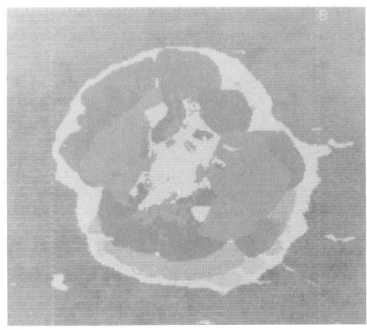

(a)

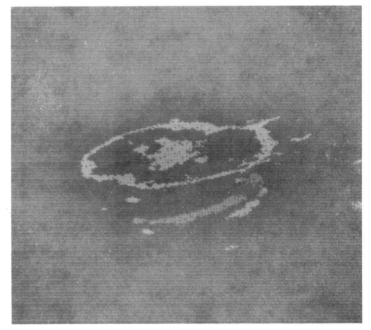

(b)

Fig. 5 (a) C-scan image of impact damage in a graphite-epoxy laminate using three-dimensional rotation. (b) Side view obtained by rotation of data

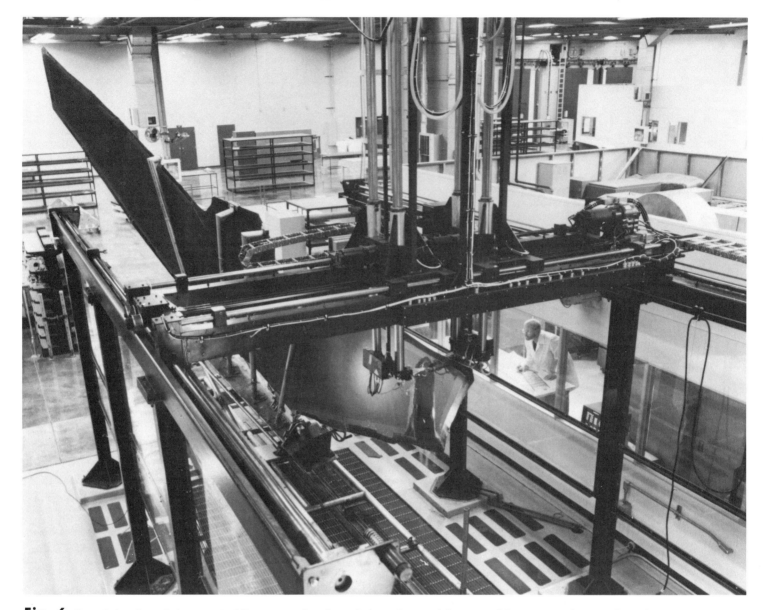

Fig. 6 General view of a typical computerized C-scan system that allows a high-speed contour-following capability. Courtesy of McDonnell Douglas Corporation

erties of the laminae, the adhesive, and the interfaces (Ref 7).

The LLW setup is based on two transducers immersed in water. For a fixed angle of insonification, the acoustic waves are mode converted to induce Lamb waves at certain frequencies, resulting in leakage of radiation into the fluid. When a leaky wave is introduced, the field of the specularly reflected wave is distorted. The specular part of the reflected wave and the leaky wave interfere, a phase cancellation occurs, and two components are generated with a null between them (Ref 8). A schematic description of the leaky Lamb wave experiment is shown in Fig. 8.

LLW phenomena have two characteristics that make them useful for the NDE of bonded joints. First, the phase cancellation in the null zone of the LLW field is very sensitive to any

change in the interface conditions. The presence or absence of a bond, as well as the change in adhesive properties, significantly alters the LLW response. Further, two types of stress, compression and shear, are encountered simultaneously when a Lamb wave travels in a plate, in contrast to the single compression-type stress of conventional NDE methods. Because the two types of stress are affected differently by different material and defect parameters of the interface, the LLW technique can provide better diagnostics of interfacial zones than can conventional NDE techniques.

An example of the applications of the LLW phenomena for the NDE of a bonded composite joint is shown in Fig. 9. A precured sandwich of AS4/3501-6 $[0,90]_{2s}$ skins is bonded to a core of a 12.7 mm (0.50 in.) high, 3.2 mm (0.13 in.) wide cell of honey-

comb. Unbonds between the laminate and the honeycomb were simulated using polytetrafluoroethylene wafers with diameters of 25.4, 19.1, 12.7, and 6.4 mm (1, 0.75, 0.50, and 0.25 in.) (Ref 9). The sample was insonified at 15° and the LLW modes were measured. A C-scan system was connected to the LLW setup, and the amplitude was recorded as a function of location. The test was conducted at 5.31 MHz, which represents one of the LLW modes for the unbonded skin as a layer. The test result shown in Fig. 9 clearly identifies the unbonds caused by the generation of the mode, which, in turn, creates a null that is detected by the receiver.

LLW is particularly useful for the NDE of defects in laminates. A graphite-epoxy unidirectional plate with delaminations, porosity, and ply gaps was tested using pulses

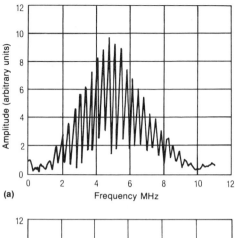

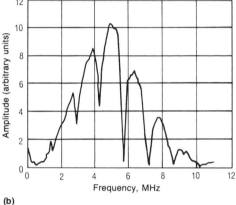

Fig. 7 Frequency-domain response from a sound and defective graphite-epoxy laminate [0,90]$_{2s}$ using 5 MHz transmitter. (a) Defect-free area. (b) 25 mm (1 in.) diam delamination between sixth and seventh layer

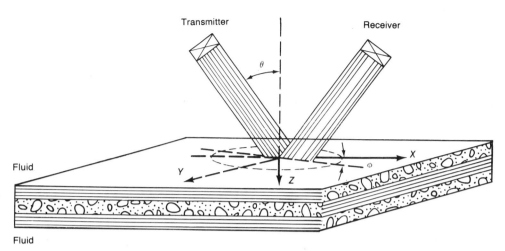

Fig. 8 Schematic diagram of the leaky Lamb wave field

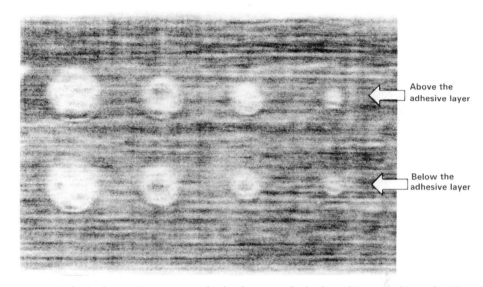

Fig. 9 Leaky Lamb wave C-scan image of unbonds in a sandwich of graphite-epoxy skins and a Nomex honeycomb core

at a 20° angle of incidence along the fibers. The LLW setup was incorporated into a computerized C-scan system where the phase response was coded as a color look-up table. Results have shown that each of the defects is clearly identified, and the porosity, which was simulated by 40 μm (1.6 mil) diam microballoons is easily indicated (see Fig. 10).

Other NDE Methods

The methods described below are based on various physical phenomena associated with the travel of certain types of energy through the test structure. This traveling energy in the form of a wave interferes with the material interfaces and discontinuities. Detectors are used to transform the interacted wave into a form that can be visualized on a film or a computer monitor. Some of the methods are used with metallic bonded joints; their principles are discussed in the article "End Product Nondestructive Evaluation of Adhesive-Bonded Metal Joints" in this Section of the Volume. These methods have been attractive to the NDE community because of their ability to cover large areas of the bonded joints at high speed without the need for a cumbersome coupling material. However, their applicability has limitations, which are also described below.

Acoustic Emission. Thermal or mechanical stresses lead to the initiation or growth of defects, which, in turn, release elastic waves. These waves, which are in the form of ultrasonic pulses, can be received by piezoelectric transducers. This method, known as the acoustic emission technique, is used mostly in research and development applications. Averbukh and Gradinar (Ref 10) reported that the level of acoustic emission as a function of the load depends on the quality of the bond.

Radiographic techniques employ short-wavelength electromagnetic radiation to determine the quality of parts. Penetrating x-rays or gamma rays are transmitted through the tested parts. The amount of the energy absorbed is monitored by a film or a real-time detector (for example, electro-optical devices, or a fluorescent screen). Variations in overall thickness or density of the tested material along the radiation path cause a change in the detector response.

Radiography is widely used to detect porosity, inclusions, and damage to the core of sandwich structures. This method is used to detect damaged honeycomb, contaminations, and debris, as well as the absence of the adhesive bond (Ref 11). Radiography cannot detect unbonds directly because the separation of surfaces is not sufficient to have any effect on the degree of absorption of x-rays or gamma rays (the unbonds are oriented perpendicular to the beam, making detection very difficult). Enhancement techniques were suggested in recent years to aid the visualization of defects by allowing fluid with a high-radiation absorption capability to penetrate the defect cavity. This method requires that the defect have an opening to the surface to allow the capillary action to take place.

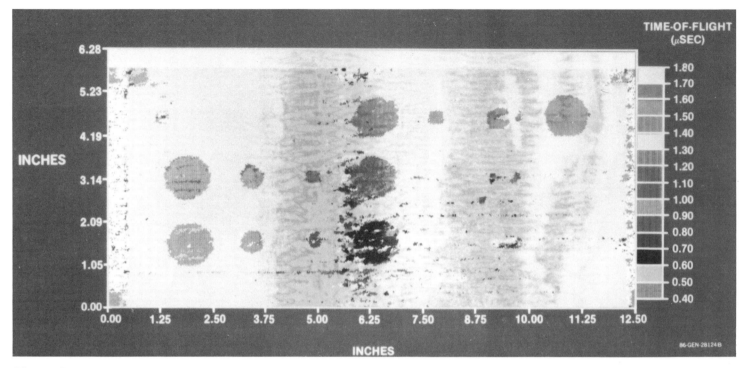

Fig. 10 Pulse LLW C-scan imaging of a $[0]_{24}$ graphite-epoxy laminate. Color was assigned to the different depths.

Holography is an optical technique that employs coherent light to generate interference between the reflection from a tested object and a reference reflector. The recorded interference pattern hologram serves as a reference image. To detect unbonds or delaminations, the part is strained, and a new hologram is superimposed over the reference image. When the combined hologram is viewed in the reconstruction process, a three-dimensional image of the part is seen with a pattern of interference fringes superimposed on it. The latter usually represents out-of-plane deformation contours (Ref 12). Because unbonds or delaminations will lead to a change in the displacement field, a deviation in the fringe pattern will be observed at the defect area (see Fig. 11).

Holography is a very sensitive method capable of indicating displacement levels of fractions of a wavelength of laser light (a few microinches). Various methods of image enhancement have been developed in recent years, and methods of eliminating fringe patterns that are not related to defects are being increasingly applied. Unfortunately, holography is extremely sensitive to vibrations, and therefore requires special considerations for field applications.

In recent years many efforts have been made to overcome the sensitivity of holography to ambient vibration, which prohibits its application in field conditions. One approach that has been successful is to use shearography, which can be applied either with films or electronically in real time. Shearography is a method of recording the strain with an image-shearing device to provide two overlapping and laterally sheared images to a recording medium (video camera or film). The image yields a fringe pattern that is related to the loci of surface strain. If the two interferograms represent different strain states of the test object, the difference in strain patterns is displayed. Flaws that lead to a perturbation of the surface strain pattern are easily detected by shearography.

Portable systems are now available for the inspection of airframe structures under field conditions typically in a shop or depot. This type of system is shown in operation in Fig. 12, while the typical response from

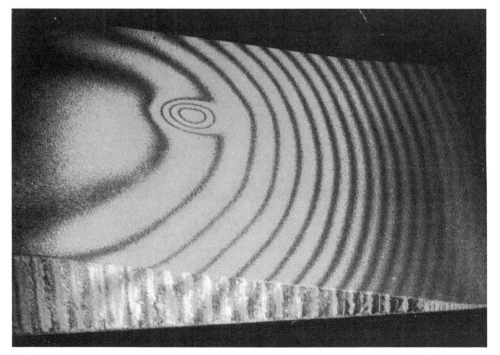

Fig. 11 Holographic indication of an unbond in a helicopter rotor blade. Source: Ref 12

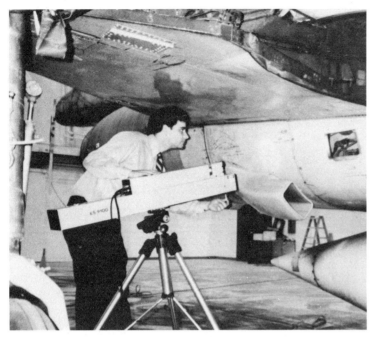

Fig. 12 A shearographic system in operation under shop conditions. Courtesy of Lockheed Corporation

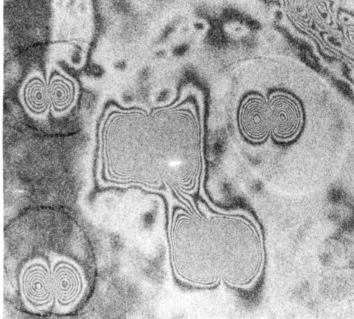

Fig. 13 Typical view of a response from unbonds using shearography. Courtesy of Laser Technology, Inc.

unbonds is shown in Fig. 13. An unbond looks like a double bull's-eye because the technique renders an image of the displacement gradient rather than its absolute value. Shearography is a very sensitive method and can indicate strain that results from a 0.3 μm (12 mil) deformation. Further, it can cover areas as large as 9 m² (100 ft²) in an hour.

Thermography. The interaction of thermal energy while traveling through a test structure is monitored using infrared or heat sensors. In the 1950s and 1960s, liquid pastes, known as liquid crystals, were commonly used to indicate the variation in heat conductivity through the test part. In recent years, there has been a significant shift toward reliance on infrared detectors. This resulted from the tremendous improvement in sensitivity and imaging capability of real-time thermographic systems that use infrared sensors. Thermographic systems display the variation in conductivity through the bonded part, where an unbond is a poor conductor. Images are displayed in colors that represent different temperature ranges. A deviation in color from a reference value indicates the presence of an unbond (Ref 13, 14).

The monitored thermal energy can be the result of either an external heat source or an internal temperature change (Ref 15). The latter, which is also known as the passive technique, tends to be produced as a result of friction. For example, under cyclic loading, crack surfaces will rub against each other to excite heat (Ref 16). Because the passive technique is associated with low levels of temperature change, transient monitoring techniques are applied to indicate the temperature variations. When a heat source is used

to induce thermal gradients in the test material, the method is called active thermography. The heat source pattern is convolved with the image, making interpretation of the results difficult.

In an effort to make thermography a standard NDE technique, many investigators are seeking methods of exciting uniform heat patterns or extracting the image of the heat source. The major problem concerning the reliability of the thermal technique is its inability to distinguish between a bonded joint and the condition in which two laminates are in close contact. Moreover, the sensitivity of thermography decreases significantly with an increase in the overall thickness of the bonded joint. These limitations have been the main obstacles to the use of thermography, even though this technique has been available for many years.

State of the Art

State-of-the-art NDE technology does not provide any physical parameter that can be directly correlated with bond strength. NDE techniques are more capable of detecting unbonds at the adhesive layer, at its interface with the adherends, or between the layers of the skins. Because adhesive bond strength strongly depends on the quality of the surface preparation, an NDE method for determining the quality of the surface preparation, such as the presence of contamination, is essential (Ref 17).

Ultrasonic NDE methods are capable of providing information about adhesive bond properties, but not all relevant parameters in the detected signals are being used at

present. An increase in the signal acquisition speed, an improvement in signal processing techniques, and an increase in the size and speed of access to computer memory are expected to extend the capability of the ultrasonic technique. Such improvement should allow several parameters to be captured while a bonded area is being scanned and should disclose features that enable an assessment of the skin and the adhesive bond quality.

Presently, pulse-echo and through-transmission methods are the most widely used NDE techniques for bonding in a production environment, whereas impedance resonance and pulse-echo methods are used in field conditions. An increase in the use of the LLW method is expected because of its superiority in providing quantitative information. Theoretical analysis of wave propagation in anisotropic, multilayered media has led to a better understanding of wave behavior (Ref 18, 19). Such a development should allow LLW modes to be predicted for bonded and unbonded laminates, which are presently too complex to interpret for multilayered laminates.

Shearography and thermography can cover large areas at a single field of view without the need for a couplant. Progress in overcoming limitations of these techniques makes them increasingly attractive for the field inspection of large structures.

REFERENCES

1. P. Cawley and R.D. Adams, The Mechanics of the Coin-Tap Method of Nondestructive Testing, *J. Sound Vib.*, to be published

2. J.L. Rose and P.A. Meyer, Ultrasonic Procedures for Predicting Adhesive Bond Strength, *Mater. Eval.*, Vol 31 (No. 6), 1973, p 109

3. R.J. Botsco and R.T. Anderson, Ultrasonic Impedance Plane Analysis of Aerospace Laminates, *Adhes. Age.*, June 1984, p 22-25

4. D.J. Hagemaier, "NDT of Adhesive Bonded Structure," Paper 6652 presented to the Air Transport Association of America, Nondestructive Testing Forum, Douglas Aircraft Company, 13-15 Sept 1977

5. Y. Bar-Cohen, U. Arnon, and M. Meron, Defect Detection and Characterization in Composite Sandwich Structures by Ultrasonics, *SAMPE J.*, Vol 14 (No. 1), 1978, p 4-8

6. F.H. Chang *et al.*, Principles and Applications of Ultrasonic Spectroscopy in NDE of Adhesive Bonds, *IEEE Trans. Son. Ultrason.*, Vol SU-23 (No. 5), 1976, p 334-338

7. A.K. Mal, C.-C. Yin, and Y. Bar-Cohen, The Influence of Material Dissipation and Imperfect Bonding on Acoustic Wave Reflection From Layered Solids, in *Review of Progress in Quantitative NDE*, Vol 7A, D.O. Thompson and D.E. Chimenti, Ed., Plenum Press, 1988, p 927-934

8. A. Schoch, Sound Transmission in Plates, *Acustica*, Vol 2 (No. 1), 1952, p 1-17

9. Y. Bar-Cohen and D.E. Chimenti, NDE of Composite Laminates by Leaky Lamb Waves, in *Review of Progress in Quantitative NDE*, Vol 5B, D.O. Thompson and D.E. Chimenti, Ed., Plenum Press, 1986, p 1199-1206

10. I.I. Averbukh and V.V. Gradinar, Inspecting the Strength Characteristics of Composite Materials, *Sov. J. Nondestr. Test.*, Vol 9 (No. 3), 1973, p 261-264

11. R.R. Teagle, Recent Advances in Mechanical Impedance Analysis Instrumentation for the Evaluation of Adhesive Bonded and Composite Structures, in *Proceedings of the 3rd European Conference on NDT*, 18 Oct 1984, p 139-162

12. T. Reed, Holographic NDT Looks to the 90s, *Photonics Spectra*, Nov 1989, p 139-142

13. P.V. Mclaughlin, E.V. McAssey, and R.C. Deitrich, Nondestructive Examination of Fiber Composite Structures by Thermal Field Techniques, *NDT Int.*, Vol 13, 1980, p 56-62

14. W.N. Reynolds and G.M. Wells, Video-Compatible Thermography, *Br. J. Non-Destr. Test.*, Vol 26, 1984, p 40-43

15. C.C. Guyott, H.P. Cawley, and R.D. Adams, The Nondestructive Testing of Adhesively Bonded Structures: A Review, *J. Adhes.*, Vol 20 (No. 2), 1986, p 129-159

16. S.S. Russell and E.G. Henneke, Dynamic Effects During Vibrothermographic NDE of Composites, *NDT Int.*, Vol 17, 1984, p 19-25

17. L.J. Hart-Smith, "Adhesive Layer Thickness and Porosity Criteria for Bonded Joints," Report AFWAL-TR-82-4172, Air Force Wright Aeronautical Laboratories, 1982

18. A.K. Mal, Wave Propagation in Layered Composite Laminates Under Periodic Surface Loads, *Wave Motion*, Vol 10, 1988, p 257-266

19. D.E. Chimenti and A.H. Nayfeh, Experimental Ultrasonic Reflection and Guided Waves Propagation in Fibrous Composite Laminates, in *Proceedings of the Joint ASME and SES Applied Mechanics and Engineering Sciences Conference*, 20-22 June 1988, A.K. Mal and T.C.T. Ting, Ed., AMD-Vol 90, p 29-38

Statistical Process Control

Thomas C. Bingham, Boeing Commercial Airplane Company

STATISTICAL PROCESS CONTROL (SPC), pioneered by Walter Shewhart in the late 1920s, is the discipline of measuring and controlling variation of process output. With the use of SPC, the natural variation of a process can be distinguished from abnormal fluctuations. The variation that is characteristic to the process, known as common cause, behaves randomly and can be modeled using statistical theory. A nonrandom impact, usually induced by some assignable source, is called a special cause. When special causes of variation have been eliminated, the process is said to be in a state of statistical control. It is then meaningful to measure the process capability or process spread and compare it to the requirements levied by engineering specifications.

Statistical process control methodology was well understood before the outbreak of World War II and was used in many industrial applications. However, the devastation of the war left the manufacturers in the United States operating without international competition, and U.S. industries developed a false sense of manufacturing superiority during the decade following the war. This outlook fed attitudes of arrogance and complacency with respect to product quality and customer satisfaction. By the 1970s, the customer base was hungry for products that consistently met its needs. Eastern manufacturers had been preparing for such an opportunity and seized it. Many entire industries in the West were unable or unwilling to make the internal changes necessary to meet the challenge. Ironically, the tools this new competition used were the same ones that the West had developed and then forgotten only a few decades earlier.

Western manufacturers are now relearning SPC and embracing what is called the total quality concept, which requires quality to be the responsibility of everyone. This holds true at all levels of the organization: engineering, planning, tooling, manufacturing, accounting, or any other function. It rejects any system that depends on a separate inspection group to ensure the acceptability of a product. The total quality concept focuses on minimizing the variation of process output in order to maximize product quality.

Unfortunately, most Western manufacturing firms still depend on end-item inspection that is separate from the manufacturing process to protect the consumer from defects. This procedure is driven by the assumptions that the manufacturing system is unable to consistently create acceptable products and that good parts must be sorted out from the bad. Operating under these assumptions, 100% of the products need to be checked for conformance upon completion.

The cost of this activity led manufacturers to employ sampling plans that dramatically reduced the amount of inspection while controlling the average outgoing quality, which is measured by the percentage of defective products. Agreements between suppliers and customers set acceptable quality levels (AQL), which specify a certain maximum (on the average) of defective products that the customer would accept.

This movement has been devastating for three reasons. First, it promotes quality complacency because, once the AQL is achieved, there is no incentive to reduce product variability any further. Second, contrary to its intent, it actually increases product costs by trading inspection costs for the far greater costs of in-service poor quality. Third, it stifles an attitude of quality ownership on the part of the producing organization, because that organization depends on an independent entity to control quality.

It has been shown that for any stable production system, the optimum amount of end-item inspection is either 100% or 0%. The only factors that determine which is appropriate for a particular system are the downstream cost of escape, the defective rate, and the cost of inspection. If the defective rate times the cost of the escape exceeds the cost of inspection, then 100% inspection is the cheapest solution; otherwise, no inspection is warranted (Ref 1). Of course, such a formula assumes that the inspection activity is error-free. This has proved to be a very poor assumption because 15 to 20% inspection error is not uncommon, and some complex products have much poorer results.

The impact of poor quality is, at best, very difficult to measure. Tolerance stacking and poor service life can result from within-tolerance variations. Customer satisfaction and market share can be measured, but correlating them to product consistency is a nearly impossible task.

One of the few documented product variation studies that had an impact on customer satisfaction was conducted by a division of a major automotive manufacturer in the mid-1980s. The difference between customer satisfaction with supposedly identical transmissions produced inside and outside the United States was shocking. Although the significant components of the U.S.-produced transmission were well within the engineering specification limits, the foreign transmissions built to the same specifications used just slightly over one-quarter of the tolerance region. This difference in quality was observable to customers, leading to an extensive study that was later videotaped.

This article focuses on statistical tools designed to expose sources of variability. These tools are intended to be used by the process owner, not by the product inspector. They will enable the owner to identify and eliminate recurring impacts to the process. By using these statistical methods as early warning mechanisms, the owner can correct a degrading process before it results in defective products. The absence of these defects will cause a chain of improvements in the downstream tasks that depend on proper input. These tools should enable management to develop ever-improving systems that not only meet but exceed customer expectations.

Process Capability and Product Quality

Although many attempts have been made to define quality, functional definitions depend on the specific product in question. However, one common denominator has been identified that destroys quality across the board: variation. All product characteristics must consistently achieve their respective targets. Variation from the target results in a degradation of quality; this is true even if variation is constrained within specification limits. Any deviation from a nominal dimension results in some econom-

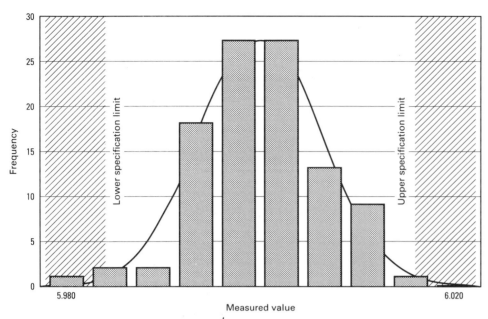

Fig. 1 Normal distribution histogram compared to the engineering specification. Process is centered, but incapable of meeting the tolerance region. Source: Tom Tosch, Boeing Computer Services

ic loss, either for the producer or the customer. The greater the deviation, the greater the loss. Conversely, reducing the variation reduces the loss.

The most common tool used to display variation is the histogram, which is an accumulated history of the measurements displayed in a single time-independent graph. The measurements are pooled together without respect to the order in which they were taken. The entire range of values is broken up into a fixed number of evenly spaced subintervals, or classes. Each measurement is examined and placed into the subinterval that bounds its value. The number of measurements in each class is counted. A bar is then drawn that has a length proportional to the frequency within a given class. This is repeated for each class, completing the histogram, as shown in Fig. 1.

The histogram provides a picture of variability that is easy to interpret. It displays three major characteristics: the process center, the range of distribution, and the shape of distribution. Processes that have one specific target and that are subject only to common-cause variation usually have a normal distribution, which is indicated by a bell-shape pattern on the histogram (Fig. 1).

Although this is not universally true, it occurs commonly enough that the assumption forms a useful base in measuring variability. Normal distributions are characterized by two parameters: the mean (μ) and the standard deviation (σ). Neither parameter can be determined exactly; both must be estimated from measurements for any given process.

Because the mean and the standard deviation must be estimated, the process output must be sampled. This requires obtaining a number of independent measurements that are representative of the normal behavior of the process output. Even under such conditions, the estimates of both the mean and the standard deviation will themselves be subject to variability. The term "estimate" is appropriate because the parameters can never be known exactly. However, the data can be formulated into statistics that provide the closest possible measures. All such statistics improve as more observations are taken, but the rate at which they improve is not always commensurate with the cost of obtaining the additional data. As a result, samples must be obtained in such a way as to optimize both economic and statistical considerations. The goal is to provide ac-

ceptable yet thrifty measures of the parameters.

In order to measure variability, the standard deviation must be estimated. The best statistic available, assuming normally distributed data, is called the sample standard deviation. It is defined as:

$$S = \sqrt{\frac{1}{n-1} \sum_{i=1}^{n} (x_i - \bar{x})^2}$$

where n is the number of measurements taken in the sample, x_i is the ith measurement in the sample and $\bar{x}$ is the average of the sample (also called the sample average). The region starting at 3 standard deviations below the average and ending at 3 standard deviations above the average comprises nearly all the natural variability of the process output. This spread of 6 standard deviations is called the process capability, and it accounts for all but 0.3% of the normal distribution. A simple comparison can be made between the process specifications and the process capability in the form of a ratio: C_p = (upper specification limit − lower specification limit)/6σ. Assuming that the process is both normally distributed and centered, the data in Table 1 can be used to predict the percentage of the process that will fall outside of the specifications.

The C_p, which is often used to measure what is called the process potential, compares the process capability (that is, the process spread) to the tolerance region. If the process is centered between the upper and lower specification limits, then the reciprocal of C_p is simply the proportion of the tolerance that the process consumes. If, however, the process is operating nearer to one specification than the other, or if only one specification exists, then different indices are needed to compare the actual performance of the process relative to the specification limits. The following are standard in the industry: C_{pl} = ($\bar{x}$ − lower specification limit)/3; C_{pu} = (upper specification limit − $\bar{x}$)/3; and C_{pk} = min(C_{pu}, C_{pl}).

The lower process capability ratio (C_{pl}) assumes that only a lower specification exists. Like C_p, it increases as the variability decreases and vice versa. Unlike C_p, it decreases as the process average gets closer to the lower specification limit. This means that the index is sensitive to the closeness of the distribution to the specification limit. "Closeness" is meaningful only if it is relative to the variability that is measured by the standard deviation. Therefore, C_{pl} is a good indicator of the risk of defective output when only a lower specification exists. The upper process capability ratio (C_{pu} behaves identically to C_{pl}, except it assumes that only an upper specification limit is present.

If both upper and lower specifications exist, then the two-sided process capability ratio (C_{pk}) is determined as the smaller

Table 1   Parts per million out of tolerance

C_p	Fallout	C_p	Fallout	C_p	Fallout
0.30	368 120	1.05	1632	1.45	13.6
0.40	230 139	1.10	966	1.50	6.8
0.50	133 614	1.15	560	1.60	1.59
0.60	71 860	1.20	318	1.70	0.341
0.70	35 728	1.25	176	1.80	0.068
0.80	16 395	1.30	96	1.90	0.016
0.90	6 933	1.35	51	2.00	0.002
1.00	2 699	1.40	26	3.00	0.000

Note: Process is centered within specification limits; measurements are normally distributed.

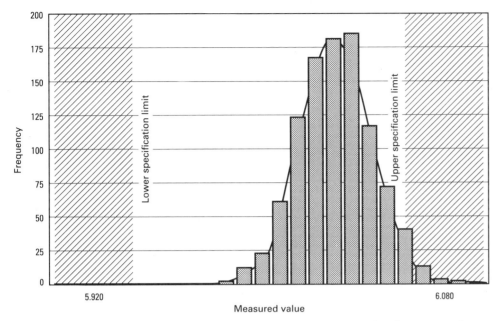

Fig. 2 Histogram for a process with a 30% shift of the process mean relative to the tolerance range. Source: Tom Tosch, Boeing Computer Services

(worse) of C_{pl} and C_{pu}. The C_{pk} is sensitive to whichever specification limit has the greatest risk of being violated. Figure 2 is a histogram for a process with a process potential that is clearly able to achieve the tolerance band but has poor performance due to a lack of centering. In order to predict process fallout, a revised table that makes use of both C_p and C_{pk} must be constructed. The two indices working together can be used to predict process fallout for both specifications at once.

A process that has a C_{pk} of at least 1 is often called just barely capable, or simply capable. This term may be a bit confusing because the capability of a process is measured purely as the process spread, that is, 6 standard deviations. A capable process is one with a process capability, or process spread, that fits within the tolerance region. In other words, it is capable of achieving the specifications. Therefore, it is possible for a single process to be capable of meeting some specifications but not others.

Table 2 provides the estimated process fallout based on the normal distribution. There is no assumption that the process is centered. The process fallout includes out-of-tolerance conditions for both specification limits. These process capability measures should never be used to advertise process fallout to the level of accuracy indicated on the table. Both C_p and C_{pk} can only be estimated from finite collections of data. They are therefore subject to natural variability themselves; if they were recomputed on a different collection of data from the same process, the new results would be somewhat different.

Also, Tables 1 and 2 are based very heavily on the normality assumption, which is usually a reasonable approximation but never a perfect fit. Therefore, the fallout tables should be considered approximate for any real application. Nevertheless, the use of capability ratios forces the measurement of the process output. The value of this over merely tracking defectives cannot be overstated. Tracking defective products provides information only after the trouble has arrived, whereas studying measurements provides information on trends that could result in future defective products. This information enables the use of preventative measures instead of reactive measures only.

Increasing the number of measurements always increases the confidence we can have in both C_p and C_{pk}. This fact raises the question: How much is enough? To answer this, some basic parameters must be established:

- The minimum acceptable C_{pk}
- The acceptable risk of incorrectly crediting a process as meeting or exceeding the minimum C_{pk}
- A predefined safety margin above the minimum acceptable C_{pk}

The most common criteria for selecting the minimum acceptable C_{pk} is the fallout table, assuming a centered process (see Table 1).

Table 2   Process fallout assuming normal distribution

C_p	Parts per million defective for a $C_p - C_{pk}$ of								
	0.00	0.05	0.10	0.15	0.20	0.25	0.30	0.35	0.40
0.30	368 120	373 486	389 323	414 863	448 896	489 854	535 930	585 206	635 776
0.40	230 139	235 367	250 867	276 099	310 183	351 943	399 953	452 607	508 198
0.50	133 614	137 979	151 000	172 447	201 925	238 852	282 451	331 741	385 556
0.60	71 861	75 060	84 672	100 732	123 267	152 245	187 527	228 813	275 603
0.70	35 729	37 813	44 128	54 858	70 274	90 694	116 420	147 675	184 544
0.80	16 395	17 611	21 331	27 774	37 280	50 288	67 291	88 788	115 229
0.85	10 772	11 665	14 410	19 214	26 404	36 414	49 752	66 966	88 596
0.90	6 934	7 572	9 547	13 041	18 348	25 868	36 089	49 560	66 855
0.95	4 372	4 817	6 202	8 681	12 505	18 024	25 676	35 978	49 497
1.00	2 700	3 002	3 950	5 666	8 357	12 313	17 913	25 614	35 944
1.05	1 633	1 833	2 466	3 626	5 475	8 246	12 250	17 878	25 595
1.10	967	1 097	1 509	2 274	3 515	5 412	8 211	12 231	17 868
1.15	561	643	905	1 398	2 212	3 480	5 393	8 201	12 226
1.20	318	369	532	842	1 363	2 193	3 470	5 388	8 198
1.25	177	207	306	497	823	1 353	2 188	3 468	5 387
1.30	96	114	172	287	487	818	1 351	2 186	3 467
1.35	51	61	95	163	282	484	817	1 350	2 186
1.40	27	32	51	90	160	281	484	816	1 350
1.45	14	17	27	49	89	159	280	483	816
1.50	7	8	14	26	48	88	159	280	483
1.60	2	2	4	7	13	26	48	88	159
1.70	0	0	1	2	3	7	13	26	48
1.80	0	0	0	0	1	2	3	7	13
1.90	0	0	0	0	0	0	1	2	3
2.00	0	0	0	0	0	0	0	0	1

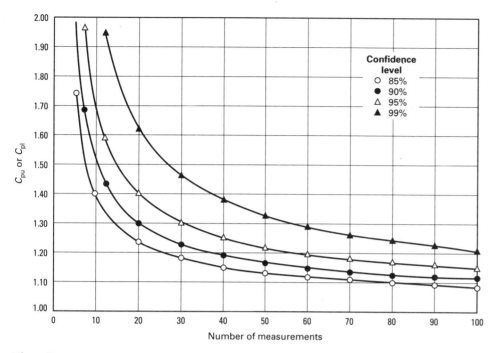

Fig. 3 Minimum computed process capability ensuring a true capability greater than 1.0

The acceptable risk is purely a business decision. It represents the frequency in which a company is willing to accept a process as meeting or exceeding the acceptable C_{pk}, when in fact the process falls short of that standard. To obtain any confidence that the true, but unknown C_{pk} exceeds the minimum acceptable C_{pk}, the estimated C_{pk} must actually exceed the target by a certain interval. For highly capable systems, the process can be certified with very few measurements, provided that the gap between the actual C_{pk} and the acceptable C_{pk} is large. Marginal systems, on the other hand, will require much more data to be able to demonstrate process acceptability. This gap, or safety margin, must be specified to determine the number of measurements needed.

These three criteria are business decisions, not statistical decisions. For example, suppose a company wishes to be 85% sure that the true C_{pk} is at least 1.0. In other words, the company has required the process to be just barely capable as a minimum. The company is willing to take a 15% risk that the data used to estimate the C_{pk} is overly optimistic, simply because a limited

Table 3   Minimum observed process capability required to ensure a true process capability that exceeds the specified value (90% assurance for one-sided capability index)

Sample size	90% probability that C_{pl} (or C_{pu}) exceeds										
	1.000	1.100	1.200	1.300	1.400	1.500	1.600	1.700	1.800	1.900	2.000
250	1.068	1.174	1.279	1.385	1.491	1.597	1.703	1.809	1.915	2.021	2.127
225	1.072	1.178	1.284	1.390	1.496	1.602	1.709	1.815	1.921	2.028	2.134
200	1.077	1.183	1.289	1.396	1.502	1.610	1.716	1.823	1.930	2.036	2.143
175	1.082	1.189	1.296	1.403	1.511	1.617	1.725	1.832	1.939	2.046	2.154
150	1.089	1.197	1.304	1.412	1.520	1.628	1.735	1.843	1.952	2.059	2.167
140	1.092	1.201	1.309	1.416	1.525	1.633	1.741	1.849	1.958	2.066	2.174
130	1.097	1.205	1.313	1.422	1.530	1.638	1.747	1.855	1.964	2.073	2.182
120	1.101	1.210	1.318	1.427	1.536	1.645	1.753	1.863	1.972	2.080	2.189
110	1.107	1.215	1.324	1.434	1.543	1.652	1.762	1.871	1.980	2.089	2.199
100	1.112	1.222	1.331	1.441	1.551	1.660	1.770	1.880	1.990	2.100	2.210
90	1.119	1.229	1.340	1.449	1.560	1.670	1.780	1.891	2.002	2.112	2.223
80	1.127	1.238	1.349	1.460	1.571	1.682	1.793	1.904	2.016	2.127	2.239
70	1.137	1.249	1.361	1.473	1.584	1.697	1.809	1.921	2.033	2.145	2.257
60	1.151	1.263	1.376	1.489	1.602	1.715	1.828	1.942	2.055	2.168	2.282
50	1.167	1.282	1.395	1.510	1.625	1.740	1.855	1.969	2.084	2.199	2.314
48	1.171	1.286	1.401	1.516	1.631	1.746	1.861	1.976	2.092	2.207	2.322
46	1.176	1.291	1.406	1.521	1.637	1.752	1.868	1.984	2.099	2.215	2.330
44	1.180	1.297	1.412	1.528	1.643	1.759	1.875	1.991	2.107	2.223	2.340
42	1.186	1.301	1.417	1.534	1.650	1.766	1.883	2.000	2.116	2.233	2.349
40	1.192	1.308	1.424	1.541	1.658	1.774	1.892	2.008	2.126	2.243	2.360
38	1.198	1.314	1.431	1.548	1.666	1.784	1.901	2.019	2.136	2.254	2.372
36	1.205	1.321	1.439	1.557	1.675	1.793	1.911	2.029	2.148	2.266	2.385
34	1.212	1.329	1.448	1.567	1.685	1.804	1.923	2.042	2.161	2.280	2.399
32	1.220	1.338	1.457	1.577	1.696	1.815	1.935	2.055	2.175	2.294	2.414
30	1.229	1.348	1.469	1.588	1.708	1.829	1.949	2.070	2.190	2.311	2.431
28	1.239	1.360	1.481	1.601	1.723	1.844	1.965	2.087	2.208	2.329	2.451
26	1.251	1.372	1.494	1.616	1.738	1.861	1.983	2.106	2.228	2.351	2.473
24	1.264	1.387	1.510	1.633	1.757	1.880	2.004	2.127	2.252	2.375	2.499
22	1.280	1.404	1.529	1.654	1.779	1.904	2.029	2.154	2.279	2.404	2.529
20	1.299	1.424	1.551	1.677	1.804	1.930	2.057	2.184	2.311	2.438	2.565
18	1.321	1.449	1.577	1.705	1.835	1.963	2.092	2.221	2.350	2.479	2.608
16	1.349	1.479	1.610	1.741	1.872	2.004	2.136	2.267	2.399	2.530	2.662
14	1.384	1.518	1.652	1.786	1.921	2.055	2.191	2.326	2.461	2.596	2.731
12	1.432	1.570	1.708	1.847	1.986	2.125	2.265	2.404	2.544	2.683	2.823
10	1.500	1.643	1.788	1.933	2.079	2.225	2.371	2.516	2.662	2.809	2.954
9	1.546	1.694	1.843	1.993	2.143	2.293	2.442	2.593	2.743	2.894	3.045
8	1.605	1.760	1.914	2.069	2.225	2.380	2.536	2.692	2.848	3.005	3.161
7	1.685	1.847	2.010	2.172	2.336	2.498	2.662	2.826	2.990	3.154	3.318
6	1.800	1.972	2.147	2.321	2.495	2.668	2.844	3.018	3.193	3.368	3.544
5	1.979	2.170	2.362	2.552	2.744	2.937	3.129	3.321	3.513	3.706	3.900

Source: Fritz Scholz, Boeing Computer Services

Table 4 Minimum observed process capability required to ensure a true process capability that exceeds the specified value (95% assurance for one-sided capability index)

Sample size	95% probability that C_{pl} (or C_{pu}) exceeds										
	1.000	1.100	1.200	1.300	1.400	1.500	1.600	1.700	1.800	1.900	2.000
250	1.088	1.196	1.303	1.410	1.518	1.626	1.733	1.841	1.948	2.057	2.165
225	1.093	1.201	1.309	1.416	1.525	1.633	1.741	1.849	1.958	2.066	2.174
200	1.099	1.208	1.316	1.425	1.533	1.641	1.750	1.859	1.967	2.077	2.186
175	1.107	1.215	1.325	1.434	1.543	1.653	1.762	1.871	1.980	2.090	2.200
150	1.116	1.226	1.336	1.446	1.556	1.666	1.777	1.887	1.997	2.107	2.218
140	1.121	1.231	1.341	1.452	1.562	1.673	1.783	1.893	2.004	2.116	2.226
130	1.126	1.237	1.347	1.458	1.569	1.680	1.791	1.902	2.013	2.124	2.236
120	1.132	1.243	1.354	1.466	1.577	1.689	1.800	1.911	2.023	2.135	2.247
110	1.138	1.250	1.362	1.474	1.586	1.698	1.810	1.922	2.035	2.147	2.260
100	1.146	1.258	1.371	1.483	1.597	1.709	1.822	1.935	2.048	2.161	2.274
90	1.155	1.268	1.382	1.495	1.608	1.722	1.835	1.949	2.063	2.177	2.291
80	1.166	1.280	1.394	1.509	1.623	1.738	1.852	1.967	2.082	2.197	2.311
70	1.179	1.294	1.410	1.526	1.641	1.757	1.873	1.989	2.105	2.221	2.337
60	1.197	1.313	1.430	1.547	1.665	1.782	1.899	2.016	2.134	2.252	2.369
50	1.220	1.338	1.457	1.576	1.695	1.814	1.935	2.054	2.173	2.293	2.413
48	1.225	1.344	1.464	1.583	1.702	1.823	1.943	2.063	2.183	2.304	2.423
46	1.231	1.351	1.471	1.590	1.711	1.831	1.952	2.072	2.193	2.313	2.435
44	1.237	1.358	1.478	1.599	1.720	1.841	1.961	2.083	2.204	2.326	2.447
42	1.245	1.365	1.487	1.607	1.729	1.851	1.972	2.094	2.216	2.338	2.460
40	1.252	1.373	1.495	1.617	1.739	1.862	1.984	2.106	2.229	2.352	2.474
38	1.260	1.383	1.505	1.628	1.751	1.874	1.996	2.120	2.243	2.367	2.490
36	1.269	1.393	1.516	1.640	1.763	1.886	2.011	2.135	2.259	2.383	2.507
34	1.279	1.404	1.527	1.652	1.776	1.901	2.026	2.151	2.276	2.401	2.526
32	1.290	1.415	1.541	1.666	1.791	1.917	2.043	2.170	2.295	2.422	2.548
30	1.303	1.429	1.556	1.682	1.809	1.935	2.062	2.190	2.317	2.444	2.572
28	1.317	1.444	1.572	1.700	1.828	1.956	2.085	2.213	2.341	2.469	2.599
26	1.333	1.462	1.591	1.719	1.849	1.979	2.109	2.239	2.369	2.499	2.629
24	1.352	1.482	1.613	1.743	1.875	2.007	2.138	2.270	2.401	2.533	2.664
22	1.373	1.505	1.638	1.771	1.904	2.037	2.171	2.305	2.439	2.572	2.707
20	1.399	1.534	1.669	1.804	1.939	2.075	2.211	2.347	2.484	2.619	2.755
18	1.430	1.568	1.706	1.844	1.982	2.121	2.260	2.399	2.537	2.676	2.815
16	1.469	1.610	1.752	1.893	2.036	2.177	2.320	2.463	2.606	2.749	2.891
14	1.520	1.665	1.812	1.958	2.104	2.251	2.399	2.546	2.694	2.841	2.988
12	1.588	1.739	1.892	2.045	2.198	2.351	2.505	2.658	2.812	2.966	3.120
10	1.686	1.847	2.008	2.171	2.333	2.495	2.659	2.821	2.985	3.148	3.311
9	1.754	1.922	2.090	2.258	2.427	2.597	2.766	2.936	3.105	3.275	3.444
8	1.844	2.019	2.195	2.372	2.550	2.727	2.905	3.082	3.261	3.439	3.618
7	1.965	2.152	2.340	2.528	2.717	2.908	3.096	3.287	3.475	3.666	3.857
6	2.143	2.347	2.552	2.757	2.963	3.171	3.377	3.583	3.792	3.998	4.206
5	2.430	2.663	2.895	3.130	3.364	3.598	3.833	4.067	4.302	4.540	4.774

pool has been collected. Suppose also that the company is willing to define a gap of 0.20 between the required C_{pk} and the estimated value. This means they will require the C_{pk} estimated from the data to be at least 1.20 before they believe that the true but unknown C_{pk} is at least 1.0. The bottom curve of Fig. 3 can then be used to determine the number of measurements that must be collected.

As seen from the curve in Fig. 3, 26 measurements are needed to satisfy the requirements for this example. If, however, management is willing to widen the minimum gap to 0.25, then only 18 measurements are needed. On the other hand, if only a 10% risk is permitted, then the same gaps produce sample sizes of 60 and 26, respectively. Table 3 provides the numerical values of the curve with 10% risk for various target C_{pk} values. Table 4 provides the same type of information as Table 3, but with the risk decreased to 5%.

To be academically honest, these curves were developed assuming that either C_{pl} or C_{pu} was used rather than C_{pk}. Individually, the variability of the C_{pl} and C_{pu} can be described using noncentral t distributions. However, C_{pk}, which is the minimum of the two, cannot be so easily described. Yet, for all practical purposes, the risk analysis of C_{pk} can be approximated through the confidence table and graph for C_{pl}.

There may be some concern about this approximation. However, the concern is based upon whether one is taking the perspective of the consumer or the producer. Traditionally, the producer wishes to demonstrate quality no worse than it actually is. The consumer, on the other hand, depends on quality assurance methods that minimize the risk of crediting quality better than it is. Under the modern quality philosophy, the supplier is supposed to adopt the customer's point of view; therefore, statistics are considered acceptable if they are conservative in favor of the consumer.

Figures 4 and 5 illustrate the conservative nature of these risk estimates under two scenarios. The first scenario (Fig. 4) comes from a process that is normally distributed with a mean of 0 and a standard deviation of 1. The upper specification limit has been established at 3, and the lower has been set at 3. Therefore, the process is centered, with a true $C_{pk} = C_p = 1$. In this first scenario, a computer generates 1000 samples of data to simulate long-term production. Each sample contains exactly 50 independent measurements. Both C_{pk} and C_{pl} are estimated for each of the 1000 samples. In this case, the true C_{pk} and C_{pl} are both known to be 1 because the data are deliberately created that way. However, calculated C_{pk} and C_{pl} may vary from sample to sample. Figure 4 shows the variability of both statistics in the form of a histogram. The shaded area is the region beyond a 90% threshold value taken from the first column of Table 3 for a sample size of 50.

There is a fairly wide distribution of both the C_{pl} and the C_{pk} values in Fig. 4. The C_{pu} values are not shown only because they have the same underlying distribution of C_{pl} and therefore are not needed in the discussion. In this scenario, 8.5% of the C_{pl} values lie above the threshold value. This is reasonable considering that a 90% confidence factor was designed into Table 3. These C_{pl} values indicate that a process on the brink of being inadequate for the requirements will fail to achieve the threshold value for C_{pl} most of the time. Occasionally, the estimated C_{pl} will surpass the threshold, giving an overly optimistic measure of the process acceptability.

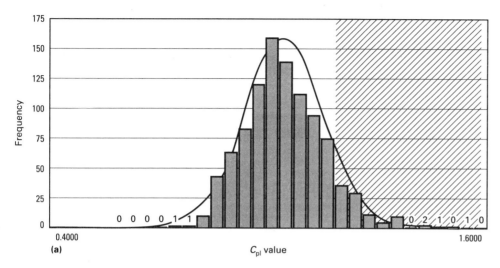

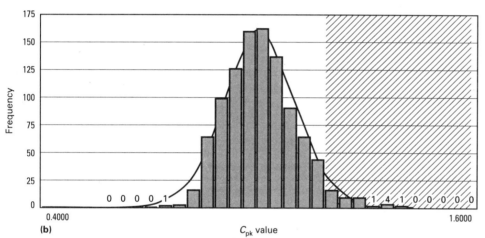

Fig. 4 Histograms for (a) C_{pl} and (b) C_{pk} values taken from 1000 samples of 50 measurements. Normal distribution: mean = 0; standard deviation = 1; upper specification limit = 3; lower specification limit = −3. Source: Tom Tosch, Boeing Computer Services

The C_{pk} escapes above the threshold less frequently than C_{pl}. This is due to the conservative nature of Tables 3 and 4. Figure 5 shows another process that also has a true C_{pk} = 1.0, but that is no longer centered. The standard deviation has dropped to 0.95, but the mean has shifted toward the lower specification by 0.15 units. This shift is actually very small and represents a very slight change in the variability. Using the computer simulator under these assumptions and displaying the data in the same form as Fig. 4 shows that C_{pk} and C_{pl} become very close. Thus, whatever conservatism exists in the approximation of using C_{pl} risk tables disappears very rapidly with processes that are even slightly off-center.

Any process capability analysis is at best an approximation because it depends on knowledge of the underlying probability distribution. All the tables in this article assume normality.

Testing Normality

Properly constructed histograms divide the range of the observed data into equally spaced intervals to provide a pictorial representation of the frequency distribution. This representation allows the user to compare the shape of the histogram to the appropriate normal curve. Such a comparison can serve as a somewhat subjective test of normality. However, the number of intervals that are selected affects the positioning of the class limits for each cell. Randomness within the sample can cause noticeable differences in the histogram if the class limits are changed. Changing the class limits is all too easy with the widespread availability of computer software, and a user might be tempted to adjust the number of intervals (and class limits) to more closely achieve a preconceived shape.

As a rule of thumb, the square root of the number of observations is used to determine the number of subintervals. However, there is nothing magic about that number. If a number of values cluster near a class boundary, then even slight changes in the number of subintervals can change the shape of the graph. It is difficult to check the normality assumption under these circumstances.

The use of a probability plot reduces, but does not eliminate, the subjectivity of normality testing. A probability plot sorts the measurements and assigns them integer ranks (Fig. 6). These ranks are converted to an approximately linear percentage scale. The percentage ranks are plotted against the actual values themselves, but the ranks are not plotted on a linear scale. The scale is tight in the center and is expanded as it moves toward both extremities. The scale is expanded in such a manner that normally distributed data will plot in an approximately straight line. The plot provides a graphic means of inspecting the normality assumption without having to deal with an arbitrary interval size. If a plot substantially deviates from a straight line, then nonnormal variation is present.

As can be seen in Fig. 6, the extreme ends of the probability plot tend to lose the linear nature seen in the central region. This is fairly common even in perfectly normal distributions. If, however, sweeping curves or erratic jumps are present within the central part of the graph, then the normality of the data should be questioned. If it is determined that the data are nonnormal, this does not invalidate findings obtained with on-line monitoring tools, which are fairly forgiving to violations in the normality assumption. (These tools are described in the remaining section of this article.) Fallout tables, however, depend very heavily on the normality assumption.

Monitoring Process Output Over Time

Histograms and probability plots provide excellent pictures of process variability. However, they fail to display changes in the process over time. The simplest time-dependent graph displays are described in this section.

A run chart (Fig. 7) plots each measurement as a single point, one at a time in the order in which they are received. The value of each measurement is plotted against its sample number. It is usually assumed that the measurements are collected at regular intervals.

If the engineering nominal dimension and tolerances are known, they are often plotted as horizontal lines on the graph. These lines provide a quick visual check of the acceptability of the output. The average of the plotted points is often computed; if plotted, it is called the centerline. The centerline can be compared to the nominal dimension to check for bias, that is, the degree to which a process is off-center.

Attempting to adjust a process based on a single measurement or very few measurements is always a poor strategy. If only a few measurements are obtained, then the degree of variability is unknown. A single point can easily be far off the mark but also

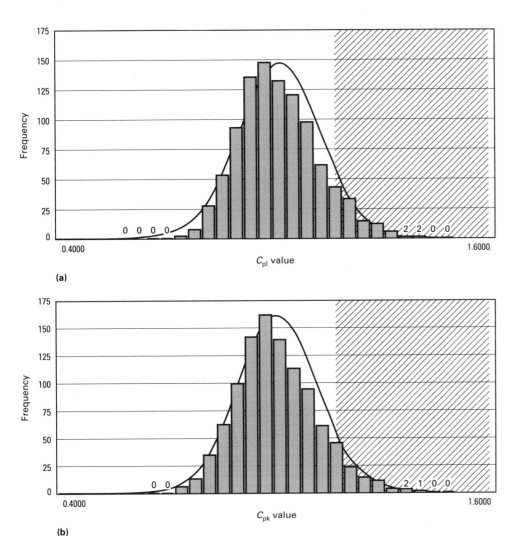

(a)

(b)

Fig. 5 Histograms for (a) C_{pl} and (b) C_{pk} taken from 1000 samples of 50 measurements. Normal distribution: mean = −0.15; standard deviation = 0.95; upper specification limit = 3; lower specification limit = −3. Source: Tom Tosch, Boeing Computer Services

well within the normal operating range of a perfectly well-targeted process. Making adjustments based on single values will substantially increase the variation. By comparing the engineering nominal dimension

to the process average, the operator can accurately target the output without falling into the trap of overcontrol.

As helpful as run charts can be, they do have some serious limitations. Because

each point plotted represents exactly one measurement, the process variability cannot be determined without combining measurements taken at different time periods. This can be remedied simply by obtaining several measurements at any one point in time. Even two or three measurements will provide a basis for measuring process consistency.

A run chart can be generalized to take advantage of these additional measurements. If a homogeneous collection of measurable quantities is available at regular intervals and the variation of this group is representative of the process capability, then a fixed sample taken from the group is called a subgroup. The number of measurements per sample is called the sample size. It is very important for each measurement to be independent of every other measurement. For example, measurements of the diameters of two holes created by two separate drilling operations in the same type of metal (assuming negligible tool wear) are independent. However, measurements of the same hole using two different orientations are not.

Because the measurements within the sample are all taken at the same time (or as close to the same time as possible), they should all be plotted at the same horizontal location. The array of plotted points for each sample forms a tier. The tiers are accentuated by connecting all the points in the sample with a vertical line. In this manner, the variation within the sample is nicely measured by the length of the line, which is called the range (see Fig. 8).

As with the run chart and the histogram, the tier chart can have specification limits plotted as horizontal lines on the graph. For a capable process, none of the tiers should fall outside of this tolerance band, except on very rare occasions. Certainly, the less tolerance used, the greater the ability of the process to achieve the specification limits.

Process Control

The tier chart contains all the information necessary to adequately monitor the output of a process, but it does not separate special and common causes of variation. Using the tier chart alone will not enable the user to determine if the process is being impacted from a force that is not part of its normal operation. Producers often mistakenly believe that the specification limits can be used to accomplish this. Unfortunately, the tier chart (like all traditional quality control systems) reacts only if finished products fall outside of tolerance. Using this approach, producers are able to respond only after defective products are made because the tier chart gives no indication of the source of the problem that led to the rejected part.

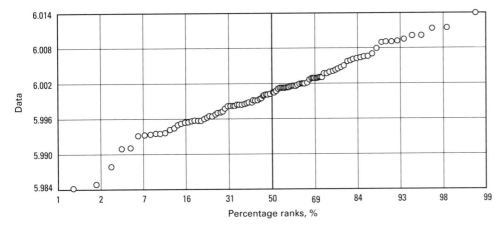

Fig. 6 Probability plot for normally distributed data. Source: Tom Tosch, Boeing Computer Services

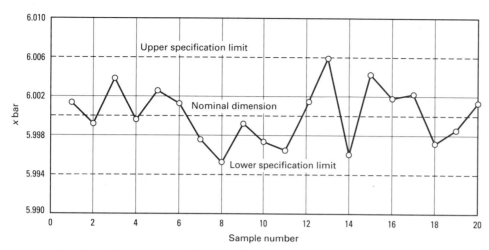

Fig. 7 Run chart with engineering specifications and nominal dimension. Source: Tom Tosch, Boeing Computer Services

As an alternative, the same data collected on the tier chart can be used to form what are called control charts. Control charts can provide information about the causes of variation without being influenced by the specification limits. Statistically based control limits are established that bound the natural variation intrinsic to the process itself. The control limits are wide enough to accommodate almost all variation that is built into the nature of the process, that is, common-cause variation. Any points that fall outside the control limits are considered to be influenced by special causes of variation that are not characteristic of normal operation. This conclusion is based on the remoteness of the probability that the process would produce such extreme output under normal operating conditions.

Figure 9 shows a control chart that monitors the activity of both the process average and the process variability. In general, the actual measurements are not directly plotted on the control chart. The few observations displayed in a given tier are combined into a single statistic that best characterizes a particular source of variation. The two most common sources of interest are the central tendency of the process and the variability. These are generally monitored using sample averages and ranges.

The average is obtained simply by adding all measurements within a given sample (or tier) and dividing that figure by the number of terms used in the sum. The range is obtained by subtracting the smallest value from the largest within the sample. This corresponds to the length of the line segment connecting the points in the tier. In this way, an average and range can be obtained for each sample (tier). These averages and ranges are then plotted against time or sequence on their own separate charts, forming the basis for what are called the $\bar{x}$ chart and R chart. The $\bar{x}$ chart is useful for determining how well the process is staying on track over time. The R chart monitors changes in the consistency of the output.

A completed control chart must include a centerline and upper and lower control limits. The centerline is simply a horizontal line that marks the average of the plotted points. The standard deviation of the plotted points is estimated using special formulas appropriate for the statistic being plotted. The upper control limit is constructed by drawing a line 3 standard deviations above the centerline; the lower control limit is established by drawing another line 3 standard deviations below the centerline. These control limits are just barely wide enough to encompass the common-cause variability; special causes fall outside of these lines. The control chart is vitally important to statistical process control because it provides a mechanism for distinguishing between variation that is characteristic to the process (common-cause variation) and that which is influenced by a specific source.

In order to take advantage of this powerful tool, three very important principles must be applied. First, the control charts must be maintained by the operator. Second, the operator must keep a record of any changes in the operation or the environment surrounding it. Third, a corrective action system must be in place that reacts to out-of-control conditions to correct processes experiencing special causes of variation.

These principles assume that the operator is most knowledgeable about the operation

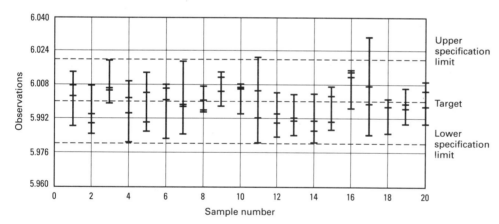

Fig. 8 Tier chart with a sample size of 4. Source: Tom Tosch, Boeing Computer Services

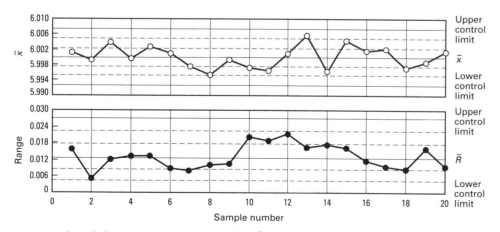

Fig. 9 Control chart (control region only) showing $\bar{x}$ and R data. Source: Tom Tosch, Boeing Computer Services

Table 5   Constants used in the calculation of control limits for measurement data

Sample	Range method				S method		
	A2	D3	D4	d2	A3	B3	B4
2............	1.881	0.000	3.269	1.128	2.659	0.000	3.267
3............	1.023	0.000	2.574	1.693	1.954	0.000	2.568
4............	0.729	0.000	2.282	2.059	1.628	0.000	2.266
5............	0.577	0.000	2.114	2.326	1.427	0.000	2.089
6............	0.483	0.000	2.004	2.534	1.287	0.030	1.970
7............	0.419	0.076	1.924	2.704	1.182	0.118	1.882
8............	0.373	0.136	1.864	2.847	1.099	0.185	1.815
9............	0.337	0.184	1.816	2.970	1.032	0.239	1.761
10...........	0.308	0.223	1.777	3.078	0.975	0.284	1.716
11...........	0.285	0.256	1.744	3.173	0.927	0.321	1.679
12...........	0.266	0.284	1.716	3.258	0.886	0.354	1.646
13...........	0.249	0.307	1.693	3.336	0.850	0.382	1.618
14...........	0.235	0.328	1.672	3.407	0.817	0.406	1.594
15...........	0.223	0.347	1.653	3.472	0.789	0.428	1.572
16...........	0.212	0.363	1.637	3.532	0.763	0.448	1.552
17...........	0.203	0.378	1.622	3.588	0.739	0.466	1.534
18...........	0.194	0.391	1.609	3.640	0.718	0.482	1.518
19...........	0.187	0.403	1.597	3.689	0.698	0.497	1.503
20...........	0.180	0.415	1.585	3.736	0.680	0.510	1.490
21...........	0.173	0.425	1.575	3.778	0.663	0.523	1.477
22...........	0.167	0.435	1.565	3.820	0.647	0.534	1.466
23...........	0.162	0.443	1.557	3.858	0.633	0.545	1.455
24...........	0.157	0.452	1.548	3.896	0.619	0.555	1.445
25...........	0.153	0.460	1.540	3.931	0.606	0.566	1.434

and therefore should be the "owner" of the quality he produces. The operator is best able to detect a problem, can take corrective action the quickest, and has the best insight into problem sources. By correcting problems before allowing the output to move to the next production step, the process owner can avoid the cost of expending additional time and effort on a defective product. Most existing quality control systems do not involve the operator in such a way and thus result in unnecessarily high amounts of scrap generation, product reworking, and out-of-sequence production. By keeping a process log, the operator has a basis for correlating changes in the work environment with real, nonrandom changes in the process.

Control charts warn the operator when the process has truly changed. There is almost no risk that a process will trigger an out-of-control condition unless a clearly significant impact has occurred. If the operator leaves the process alone while it is in statistical control, then operator overcontrol can be avoided. Avoiding overcontrol is very important because adjusting a process as a reaction to common-cause variation results in an increase in the total variability. On the other hand, correcting a process that has experienced a special cause of variation results in bringing it back to its normal operation. Control charts provide an early-warning mechanism that can lead toward self-correcting defect prevention systems; such systems can potentially eliminate out-of-tolerance conditions altogether.

Mechanics of Control Limit Calculation. Shewhart variable control charts estimate the standard deviation only on short-term variation, which is illustrated by the length of the tiers on the tier chart. The longer the tiers, the more inconsistent the process. The shorter the tiers, the more capable the process.

If the tiers are small but jump or migrate around the chart, then either special causes of variation are at work that will impact the process average, or other large sources of variability are at work in between the samples. In the latter case, more sophisticated methods of accounting for all sources of variation may be needed. In the former case, it is important that these special causes are removed so that they do not influence the control limit calculations. These special causes artificially widen the separation between the limits, thereby hiding special causes rather than highlighting them, and impeding fruitful investigation and elimination of process problems.

The short-term variation represented by the ranges (lengths of the tiers) indicates what the process is capable of at any given time. Therefore, the average of the ranges is used as the basis for determining process variability. In order to establish these control limits, the process must be monitored over a sufficient period of time to gain an accurate measure of the variability. This period is called the control region. A group of 15 to 20 samples is generally enough to establish stable control limits and centerlines.

Conversion factors established by Shewhart facilitate the calculation of control limits using the average range. Table 5 lists conversion constants that can be used to construct the control limits for $\bar{x}$ charts and R charts. The constants depend only on the sample size as indicated. Table 6 sum-

marizes the formulas that use the constants for variable data. These two tables provide a very simple mechanism for calculating the limits. The first two charts in Table 6 are the most commonly used because of their simplicity and versatility.

The centerline of the $\bar{x}$ chart (called CL_x) is computed by averaging the sample averages. The result is called the grand average and is equivalent to a computation of the average of all the individual measurements. Next, the average of all the ranges ($\bar{R}$) is computed. The control limits for the $\bar{x}$ chart, denoted by UCL_x for the upper control limit and LCL_x for the lower control limit, can be calculated by using the formulas in the table.

The calculations for the R (range) chart are even simpler and make use of the same average range as above. The control limits are denoted as UCL_r and LCL_r. The centerline is denoted as CL_r. It is very important that the R chart be in statistical control before the average range is used to represent the process variation. Out-of-control ranges indicate assignable (nonrandom) causes that are not characteristic of the process. They inflate the average range, widening the control limits for both charts.

Because out-of-control points must be removed from the calculations before the control limits can be trusted, a potentially iterative procedure results. This procedure involves calculating limits, removing out-of-control points, and recalculating the limits. This apparently circular approach seems, on the surface, to run the risk of whittling away all the data as the control limits draw closer together during each iteration. In practice, however, the calculations stabilize quickly, provided that at least 15 to 20 time periods survive the first reduction.

Once the control limits are established, they are extended beyond the control region in anticipation of new measurements. The new averages and ranges are then plotted on the same chart. However, the control limits and centerlines are not computed. The new data are compared to the established control limits to determine if and when the process changes.

If a new sample average falls outside of either control limit, then there is strong evidence that the process average has changed. Similarly, if the range leaves the control band, then a true change has occurred that affects the capability of the process. With both charts, this indication of change may be the result of an external shock to the process or it may signal a developing trend. Whatever its source, the operator must investigate the change to avoid recurring problems and to reverse degradations developing in the system.

It should always be remembered that out-of-control conditions do not indicate

Table 6 Control limit formulas for variable data

Chart	Purpose	Plot statistic	Centerline(a)	Lower	Upper
Sample size (n), 3–5					
R	Monitor change in the process capability based on within-sample variation	$r_i = \max(x_{i1}, x_{i2}, \ldots, x_{in}) - \min(x_{i1}, x_{i2}, \ldots, x_{in})$	$\bar{R} = \dfrac{r_1 + r_2 + \ldots r_m}{m}$	$D_3\bar{R}$	$D_4\bar{R}$
$\bar{x}_r$	Monitor change in the average based on within-sample variation	$\bar{x}_i = \dfrac{x_{i1} + x_{i2} + \ldots + x_{in}}{n}$	$\bar{\bar{x}} = \dfrac{\overline{x_1} + \overline{x_2} + \ldots + \overline{x_m}}{m}$	$x - A_2\bar{R}$	$x + A_2\bar{R}$
Sample size (n), ≥10					
S	Monitor change in the process capability based on within-sample variation	$S_i \sqrt{\dfrac{(x_{i1} - \overline{x_i})^2 + \ldots + (x_{in} - \overline{x_i})^2}{n-1}}$	$\bar{x} = \dfrac{S_1 + S_2 + \ldots S_m}{m}$	$B_3\bar{S}$	$B_4\bar{S}$
$\bar{x}_s$	Monitor change in the average based on within-sample variation	$\overline{x_i} = \dfrac{x_{i1} + x_{i2} + \ldots + x_{in}}{n}$	$\bar{\bar{x}} = \dfrac{\overline{x_1} + \overline{x_2} + \ldots + \overline{x_m}}{m}$	$\bar{\bar{x}} - A_3\bar{S}$	$\bar{\bar{x}} + A_3\bar{S}$
Sample size (n), ≥1					
MR	Monitor change in process capability based on adjacent average differences	$r_i = \|(\bar{x}_i - \bar{x}_{i-1})\|\ (i > 1)$	$\overline{MR} = \dfrac{r_2 + r_3 + \ldots r_m}{m-1}$	0.0	$1.88\overline{MR}$
$\bar{x}_{mr}$	Monitor change in the average based on short-term variation between samples	$x_i = \dfrac{x_{i1} + x_{i2} + \ldots + x_{in}}{n}$ $(i \geq 1)$	$\bar{\bar{x}} = \dfrac{\overline{x_1} + \overline{x_2} + \ldots + \overline{x_m}}{m}$	$\bar{\bar{x}} - \dfrac{\overline{MR}}{1.128}$	$\bar{\bar{x}} + \dfrac{\overline{MR}}{1.128}$

(a) m, number of samples in control region (generally 15 to 20)

good or bad. They simply indicate change and give some insight into where the change is occurring. Some change is for the good. Reductions in variability can show up as out-of-control conditions, and these reductions are desirable. However, systems rarely change for the better on their own. If management is testing intentional changes in the process, then control charts can be an excellent tool to verify the impact of the change.

Figure 10 shows five new averages and ranges plotted on the chart from Fig. 9. The averages are out of control, indicating a process shift. The last two ranges are also out of control, indicating an increase in variability. As soon as one point falls outside of the control limits, investigation of its cause should commence. It is important to note that these results make no statement about the acceptability of the product. Specifications play no role here. If the process capability is sufficiently high, it is possible that no defective parts have been made. If such is the case, then the control chart has served as an early warning mechanism that allows the operator to detect a developing problem and take action to prevent the introduction of defective items into the system.

Control limits should never be modified unless it can be shown that the process has truly improved. If such is the case, a new higher standard is in order and a new control region must be established. If the process has degraded, then it must be fixed. The process should be brought back to the original quality level; control limits should not be adjusted to accommodate a poorer standard.

An improvement can be made in the $\bar{x}$ and R chart by replacing the sample range with the sample standard deviation (S). The range only makes use of the extreme points, but the standard deviation makes use of all the data. The construction of the charts is almost identical to that of the $\bar{x}$ and R charts. S is computed for each sample and plotted on its own chart. S is computed as the average of these samples. The formulas for $\bar{x}$ and S are contained in the second group of Table 5.

It should be noted that $\bar{S}$ and $\bar{R}$ both measure short-term variability because both the ranges and standard deviations are computed for each sample individually. However, S is a better measure of variability than R, but the improvement is very slight for samples in the typical range of 3 to 5. Differences do not become very noticeable until the sample size is approximately 10. Because of this, and because the range is more understandable to the majority of people, the R charts are generally preferred over the S charts. However, in the rare instances when large sample sizes are desirable, the S chart should be used because of the decreasing ability of the range to measure variability.

Control Charts for Individual Measurements

In some situations it is impossible to define a rational subgroup from which to sample because measurements are available on an individual basis only. Several circumstances can cause this condition, including the following:

- Destructive testing is required to obtain the measurement
- Measurements are expensive
- A chemical process that changes slowly is being studied (multiple measurements in a short period of time do not reflect process changes)
- Measurements are being taken from a very slow production process that produces only one measurable output at a time

If only one measurement is available per subgroup, then the average becomes the measurement itself. However, under this scenario, a range or a standard deviation is impossible to compute. In such cases, a moving range is employed. A moving range makes use of each adjacent measurement and subtracts the larger from the smaller. This results in one less range than there are measurements. The average range is still computed in the usual manner.

The control limits can still be computed, despite the fact that successive ranges share one measurement. The appropriate formulas are contained in the last group of Table 5. This procedure stretches the Shewhart philosophy somewhat because there is no longer a subgroup structure that represents the short-term variability of the process;

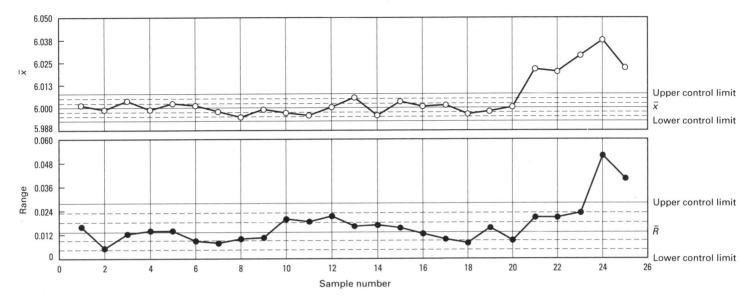

Fig. 10 Control chart from Fig. 9 plotting new data using extended control limits. Source: Tom Tosch, Boeing Computer Services

however, the moving range is a workable alternative.

In some situations, the moving-range model is still desirable even though multiple measurements are possible for a subgroup. This occurs when the variability within the sample does not adequately represent the process capability. Chemical processes typically fall into this category because the variability of multiple measurements of a single solution taken at one time corresponds only to measurement or laboratory error, not process variation. Some time needs to pass before the process itself has had a chance to change. In such cases, the sample average can be used as if it were a single measurement.

Other Control Charts for Measurements

Since the early pioneering work of Shewhart, other methods have been devised that improve sensitivity to changes in the process. These charts permit the aggregation of measurements over time and plot the aggregated results. This is a major change from the Shewhart approach; Shewhart methods plot statistics that are a function only of the samples themselves (or nearly so in the case of the moving range).

One such chart is the cumulative sum (CuSum) chart, which accumulates the differences of the sample means from the process target. These cumulative sums, rather than the averages, are then plotted. This approach increases the sensitivity of the chart to small shifts in the mean; however, the chart loses some ability to detect large process spikes because it treats recent measurements on an equal par with older ones. This disadvantage can be compensat-

ed for by applying scaling factors to each sample that decrease with the age of the sample. Such an approach has been developed and is called the geometric moving average (GMA) chart. The GMA chart plots the following statistic:

$$z(i) = r*x(i) + (1 - r)*z(i - 1)$$

where $z(i)$ is the ith point plotted, r is a scalar selected in advance, and $x(i)$ is the sample average for the ith time period. By using this construction, influence of older measurements decreases geometrically, which accounts for its name.

Although the GMA chart will not be covered in detail in this article, it is useful to compare it to the $\bar{x}$ chart using computer-generated data simulating a normal distribution. Figure 11 shows both charts using samples 1 through 20 to form the control region. In this region, only common-cause variation is present, and the variation follows a normal distribution. Starting at sample 21, the process average is shifted upward by one-third of a standard deviation. This is held until sample 31, where another one-third σ shift occurs, for a total of a two-thirds shift from the original control region.

There is a delay in detecting a change in both charts. The $\bar{x}$ chart, however, gives little illumination as to when the shift occurred. Even worse, it appears that only occasional shocks are occurring to the system as indicated by (1) the out-of-control condition at sample 26 and (2) the violation of a company rule at sample 40. The GMA chart, on the other hand, very clearly demonstrates a sustained shift and points to sample 20 or 21 as the likely point at which the shift occurred. This information can be gathered from the apparent trend in the plotted curve.

In practice, it is risky to depend very heavily on trends when using charts based on a cumulative statistic. However, if an out-of-control condition occurs, then a process shift is a very likely cause for the trend. Reference 2 contains detailed descriptions of charts based on cumulative statistics.

Attribute Control Charts

Frequently, a quantitative measure for quality, which is required for variable control charts, cannot be determined. A part with a complex geometry can be compared to a template to determine a proper match; the part either matches or it doesn't. Plated parts can be tested for proper adhesion by peeling a tape of known adhesive strength from the surface and checking for plate removal. Any plate removed constitutes a failure. Scratches, nicks, blemishes, or other imperfections can be counted to determine the quality of a surface. The frequency or percentage of occurrence of these attributes can also be plotted on a control chart. The centerline and control limits can be computed in a manner similar to their computation for the variable charts.

For a percentage defective chart, or p chart, a sample of n items is inspected and each is classified as being either acceptable or defective. The number of defective items is divided by the number inspected. This fraction is then plotted (generally on a percentage scale). Sampling is continued over evenly spaced time periods, and the resulting percentages of defective items are plotted for each time period.

Just as with the variable control charts, the purpose of the attribute charts is to detect process change. If any given per-

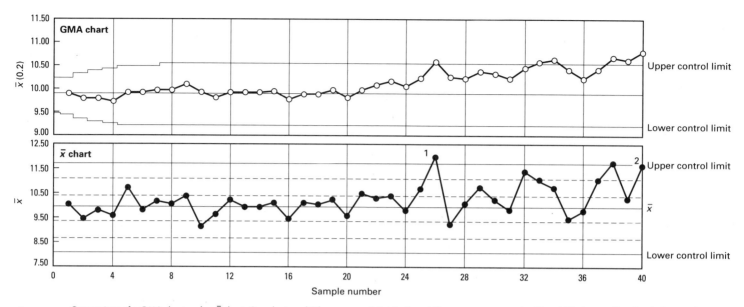

Fig. 11 Comparison of a GMA chart and an $\bar{x}$ chart. Sample size of 3 from a normal distribution; shifts can occur at samples 21 and 31. Source: Tom Tosch, Boeing Computer Services

centage falls outside the control limits, then a distinct change has been detected in the process quality. On the surface, an upward trend would appear to be bad and a downward trend would appear to be good; however, this is not necessarily the case. A change in the percentage of defective items can be due either to a change in the process quality or a change in the detection ability. If unchecked, the latter can be more devastating than the former.

As with all the Shewhart control charts, the centerline is the expected average of the plotted points, the upper control limit is 3 standard deviations above the centerline, and the lower control limit is 3 standard deviations below the centerline. Table 7 contains formulas for the most commonly used attribute control charts.

If the sample size (n_i) is constant across all time periods, then the centerline is truly the average of the plotted percentages. If the sample size changes, then the centerline is not the average in the conventional sense, but the formula in the table still applies. The control limits themselves will fluctuate inversely to the square root of the sample size because the degree to which the percentages vary depends on the number of tests that went into their construction.

To avoid fluctuating control limits, some p chart applications use an average sample size as an approximation of the exact control limits. If the sample sizes do not vary dramatically, this is a reasonable approximation.

The p chart should never be used to track out-of-tolerance conditions deter-

mined by comparing quantitative measurements to specification limits. If such measurements are available, variable charts such as the $\bar{x}$ and R charts should be employed. Attribute charts do not react until defects are introduced into a process. Variable charts react whenever the process changes, regardless of whether or not out-of-tolerance conditions have occurred. Variable charts therefore can provide an early warning of coming trouble, whereas attribute charts cannot. Furthermore, p charts require a large sample size, generally at least 30. Variable charts typically require 3 to 5 measurements per sample; or as few as 1.

Figure 12 illustrates the relative detection abilities of an $\bar{x}$ chart and a p chart for a process surpassing the specification requirements. Control charts using quantita-

Table 7 Control limit formulas for attribute data

Chart	Purpose	Sample size (n)	Plot statistic(a)	Centerline(b)	Control limits Lower(c)	Control limits Upper
p	Monitor change in the proportion defective. Can be used if the sample size varies	≥ 30	$p_i = \dfrac{d_i}{n_i}$	$\bar{p} = \dfrac{d_1 + d_2 + \ldots d_m}{n_1 + n_2 + \ldots n_m}$	$\bar{p} - 3\sqrt{\dfrac{\bar{p}(1-\bar{p})}{n_i}}$	$\bar{p} + 3\sqrt{\dfrac{\bar{p}(1-\bar{p})}{n_i}}$
np	Monitor change in the number defective. Requires constant sample size	≥ 30	d_i (p_i must still be computed as in the p chart)	$\overline{np}$	$n\bar{p} - 3\sqrt{n\bar{p}(1-\bar{p})}$	$n\bar{p} + 3\sqrt{n\bar{p}(1-\bar{p})}$
c	Monitor change in the number of nonconformities on a single part or unit	1	c_i	$\bar{c} = \dfrac{c_1 + c_2 + \ldots c_m}{m}$	$\bar{c} - 3\sqrt{\bar{c}}$	$\bar{c} + 3\sqrt{\bar{c}}$
u	Monitor change in the number of nonconformities per unit. Can be used if the sample size varies	≥ 1	$u_i = \dfrac{c_i}{n_i}$	$\bar{u} = \dfrac{c_1 + c_2 + \ldots c_m}{n_1 + n_2 + \ldots n_m}$	$\bar{u} - 3\sqrt{\dfrac{\bar{u}}{n_i}}$	$\bar{u} + 3\sqrt{\dfrac{\bar{u}}{n_i}}$

(a) d_i = number of defective units; c_i = number of nonconformities. (b) m = number of samples in control region (generally 15 to 20). (c) Result cannot be <0.

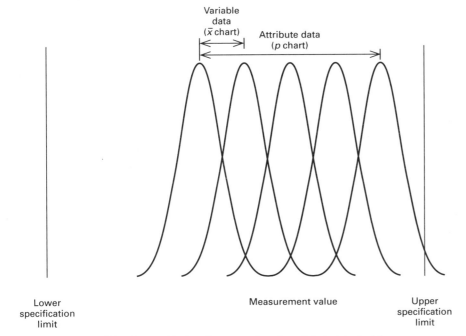

Variable data ($\bar{x}$ chart)

Attribute data (p chart)

Lower specification limit

Measurement value

Upper specification limit

Fig. 12 Detection ability of a variable data chart versus that of an attribute data chart for a process surpassing specification requirements

tive measurements ($\bar{x}$ charts) detect process changes without reference to specification limits; charts that are based on tracking defective products (p charts) react only to products that have exceeded the tolerance band. If the process shifts slightly, the $\bar{x}$ chart will go out of control. By its very nature, the p chart cannot react until defects are actually generated.

On a minor modification to the p chart called the np chart, the operator plots the number of defective items rather than the percentage. The sample size must still be known; in fact, it cannot be changed throughout the control chart. Table 7 contains the appropriate control limit formulas for the np chart.

The sensitivity of attribute charts can by improved by broadening the concept of a defective unit. Some flaws or other nonconformities may not actually render a unit rejectable, but they may still be undesirable. For example, military standards limit the number of pits that can be found on a test panel of anodized sheet metal undergoing salt spray tests; however, the panel will be considered acceptable if the number does not exceed a fixed threshold value. The pits can be considered nonconformities even though they do not result in rejection of the part.

Data of this type must be handled differently than that treated in a p chart. Many nonconformities can exist on any given unit, but p charts have no mechanism for dealing with them. With a p chart, a unit is either defective or acceptable; there is not gradation of acceptability. In a more particularized system, the quality of a unit is a function of the number of nonconformities. The natural variability of data of this type is best described by a Poisson distribution. The control limit calculations are very simple because the standard deviation is the square root of the mean. When nonconformities are plotted for one unit at a time (that is, the sample size equals 1), the resulting chart is called a c chart. The c chart formulas are given in Table 7.

A modification to the c chart is appropriate if nonconformities are very rare and sample sizes greater than 1 are possible. If n_i is the sample size for time period i, then $u_i = d_i/n_i$ (defects per unit) is plotted for time period i. This is called a u chart and is defined in Table 7.

Application of Statistical Process Control

It is not feasible to apply control charts and process capability studies to anything and everything that are measurable. Therefore, any company attempting to apply SPC must focus its efforts on characteristics that are key to the quality of the finished products. The quality of key characteristics is generally a function of a few key process parameters. For example, the strength of titanium-cadmium-plated steel parts is directly affected by the speed at which hydrogen is removed from the plated part. A key characteristic for this part is the strength; however, because the strength can be measured directly only with a destructive test, the process measurement is the rate of hydrogen removal.

Subjective tools are available to help teams identify and document key process characteristics. Adaptations of traditional failure modes and effects analysis can be helpful. A class of tools that can quantify the degree to which many process variables affect product characteristics is called design and analysis of experiments.

Designed experiments are an immensely powerful tool for isolating factors that not only improve process output but also enable the output to be stable against unavoidable fluctuations in nuisance factors. Process factors that are found to be insignificant are useful because they may allow the relaxation of specification limits. All of these benefits can usually be achieved by running a sequence of a few relatively small experiments; each sequence typically contains between 8 and 32 tests. Each experiment is capable of analyzing the effects of many input factors on one or more output variables. This topic is too broad to be covered in this article, but Ref 3 and 4 provide a good foundation for industrial experimentation.

After the key process parameters that drive the quality of the specified key characteristics are isolated, control charts should then be employed on each. Specifications should be derived for each of the process characteristics that eliminate the possibility of out-of-tolerance conditions in the product key characteristic. The analysis of a designed experiment is an excellent way to determine these limits. By controlling and achieving sufficient capability on each key process, routine inspection on the end characteristic becomes unnecessary. By isolating key processes, a company increases its leverage on quality. Furthermore, these upstream processes are generally far more repeatable than the individual products that are impacted by these processes.

Statistical process control is a simple yet powerful asset to quality control. It focuses on defect prevention at the process level rather than defect detection for the finished product. With the advent of just-in-time production systems, smaller and smaller lot sizes are being required for many industries. By focusing on the process, the quality of combined sets of small lots can be ensured as a whole, thereby eliminating the need for separate quality control systems for each individual lot.

If used in the proper total-quality environment, SPC becomes a tool of the producing organization; thus, the burden of quality is placed on the process owner, who has the most timely and direct influence on the output. Through the application of SPC, built-in waste can be transformed into profit, a leverage item worth many times its value in equivalent sales. Investment in SPC generally provides several full returns within the first year of application.

REFERENCES

1. W.E. Deming, *Out of the Crisis,*, Center for Advanced Engineering Studies, Massachusetts Institute of Technology, 1986
2. D.C. Montgomery, *Introduction to Statistical Quality Control*, John Wiley & Sons, 1985
3. J.S. Hunter *et al.*, Design and Analysis of Experiments, in *Juran's Quality Control Handbook*, 4th ed., J.M. Juran and F.M. Gryna, Ed., McGraw-Hill, 1988, p 26.1-26.81
4. D.C. Montgomery, *Design and Analysis of Experiments*, John Wiley & Sons, 1984

SELECTED REFERENCES

● C.A. Bicking and F.M. Gryna, Jr., Process Control by Statistical Methods, in *Quality Control Handbook*, 3rd ed., J.M. Juran, Ed., McGraw-Hill, 1974
● *Statistical Quality Control*, Western Electric/Delmar Printing, 1983

Bonded Structure Repair

Chairman: Ray E. Horton, Boeing Military Airplane Company

Introduction

THE REPAIR OF BONDED METAL AND ADVANCED COMPOSITE structural components is the subject of this Section. Its articles have been especially selected to cover key repair aspects. These include prerepair preparation of the damaged region, suggested repair materials, configuration of the repair, and strength analysis. The procedures address repairs made either at a depot or under less sophisticated field conditions. The information is based largely on methods developed by aircraft manufacturers and users. The procedures should, however, have considerable generic application. Although the article on bonded metal repair is specifically for aluminum, with some modification many of the principles will apply to other metals. In the case of titanium, for example, PasaJell 107 rather than PasaJell 105 might be used for the surface preparation, and, if appropriate, a higher-temperature adhesive could be used for bonding.

The application of adhesive bonding for aluminum has been of two types, metal laminates and sandwich construction. Metal laminate construction has commonly been used for such applications as built-up edge bands and fuselage tear straps. Sandwich construction has typically been used for secondary structures such as wing and empennage trailing edges. The latter are usually in the form of thin skins and honeycomb core. Many of the failures associated with bonded laminate construction have been delaminations, either primarily caused by damage, or in older aircraft, caused by the use of earlier evolution surface treatments and adhesive systems. The latter problem has largely been remedied on later built airplanes with the adoption of such items as phosphoric anodized surface preparation, corrosion resistant primers, and improved high-temperature curing adhesives. In cases where repairs are required but facilities such as acid cleaning tanks and autoclaves do not exist, bond quality is still a concern. Such is the case at field bases where facilities are limited, aircraft turnaround time is critical, and many repairs must be made without removing the parts from the airplane. The articles presented provide procedures for these types of repairs, which are expedient yet minimize compromises in repair quality.

Considerable coverage is also given to the repair of aluminum skin honeycomb core sandwich. Because of its light construction and typically vulnerable location on the aircraft, it frequently needs repair. Often moisture enters into the core, commonly at an edge detail or damage site, causing core corrosion or possibly imbalance of a control surface. In any event, repair is required. Consequently, the information noted above regarding surface preparation and the use of proper adhesives will apply. Additionally, procedures are given for the removal of moisture from the core prior to repair closure. Even a small amount of remaining moisture will cause high pressure in the core during an elevated-temperature cure, causing the core to collapse or "blow." Thus, a previously bad condition can become considerably worse.

Understandably, the major portion of the articles is directed to the repair of advanced composites. As with bonded repairs on metal, the proper treatment of the surfaces to be bonded is critical. The process may be more difficult, beginning with paint removal to obtain a bondable surface. Chemical paint removers may damage the composite, and mechanical abrasion is both tedious and time consuming. Plastic media or grit blasting may provide the best immediate solution, although disposal of the resulting residue can present a problem. Considerable development work is currently being done in this area. This ranges from the use of protective coatings to act as stripping barriers, to the development of robot laser systems with automatic surface sensors to control stripping depth.

Again, in contrast to metal, composites tend to absorb moisture or contaminating fluids such as fuel or hydraulic oil. If a good-quality bonded repair is to be made, these must be removed by drying and/or the use of cleaning solvents before bonding is accomplished. The long drying period required may not be permissible; thus repair materials with minimum surface sensitivity to contamination should be selected.

Quality field repairs are a particular challenge. Time and facilities are probably limited. It is usually expedient to use lower-temperature-curing adhesives, or resins and pressure, at best, may be applied by vacuum bagging. (Note: Some resin systems become porous and lose strength when cured under full vacuum pressure. Consequently, in some cases partial vacuum rather than full vacuum may be preferred.) Where applicable, it is suggested that a wet lay-up procedure or a built-up patch of prepreg be considered. Because most wet lay-up resins are two-part systems and are catalyzed during application, they have a long storage life. The shelf life of many prepreg systems can also be very good if they are prestaged prior to storage.

Relative to resin storage, considerable work is being done on the development of methods using thermoplastics for repair. Proponents cite excellent material storage life and rapid application using heat sources such as induction bonding and ultrasonics. The main hurdle seems to be the high temperatures required to melt the resins. Currently the most promising application seems to be for repairing the high-temperature thermoplastic composite systems.

Coverage in the articles is provided for both thin and thick laminates. The latter provide the greater challenge because of the long bonded joint overlaps required to achieve high load transfer. The discussions consider structure that may be designed to sustain a high strain level, a capability which must be restored. Requirements for a flush outer surface and possibly only one-side access are also considered, as is the length of splice plates and the tapering necessary to prevent peeling at joint edges and premature failure.

Several levels of repairs are discussed. Those with the best strength and environmental durability result from careful surface preparation, proper design, proper materials, and elevated-temperature/adequate-pressure cures. Such repairs can usually be made at the field level, as well as at a depot.

Repair of Aluminum Aircraft Structures*

M.H. Kuperman, United Airlines Maintenance Operations Center
R.E. Horton, The Boeing Company

ADHESIVE-BONDED aluminum aircraft structures require repairs when mechanical damage occurs on the manufacturing or flight line or because of in-service problems. Most manufacturers elect to substitute a new part from their production process for the damaged unit and either recycle or scrap the old part. The preference of airline operators is usually to install a new or rebuilt structure and remove the old, except in the case of cosmetic damage. Mechanical damage, however, actually represents a small fraction of the overall problem necessitating bonded structure repair, compared to in-service needs. Thus, fixing in-service problems, involving alloys 2024 and 7075, is the focus of this article. The techniques, surface preparations, materials, and processes for the repair of an in-service bonded structure are also applicable to manufacturer-based units. The end result of the repair process is to restore the structural integrity of the bonded part and, in some cases, to increase the service period beyond the initial design life.

Repair Considerations

Alternatives

The in-service repair of adhesive-bonded aluminum aircraft structure has, as its primary concern, the safety and dependability of the aircraft. Within this limitation is an optimal time frame for performing repair work, consistent with aircraft scheduling requirements. The aircraft operator receives from the aircraft manufacturer a structural repair manual, which conservatively estimates the damage tolerance of the adhesive-bonded structures. This tolerance is a function of the design, weight, performance criteria, and service environment of the aircraft. As a rule, balanced flight control surfaces are more critical than other units and require more stringent controls. Guided by the design tolerance, an operator might elect (time permitting) to repair the damage on the aircraft in a permanent man-

ner. If the damaged surface is interchangeable, the operator might elect to replace the structure with a spare unit, again, time permitting. The damaged surface is then repaired in the aircraft operator's shop or other repair station. A third option is to perform a temporary repair, again within the constraint envelope of damage tolerance. A temporary repair is life limited, requiring inspection at suitable frequencies for deterioration and replacement by a permanent repair at a scheduled aircraft maintenance opportunity. Hence, the determination to perform either a temporary or a permanent repair depends primarily on the extent of damage but also on aircraft scheduling requirements.

Damage and Assessment

The types of in-service damage that aluminum aircraft adhesive-bonded structures most often experience are:

- Dents, usually defined as skin depressions in which the adhesive bond is left intact (undelaminated)
- Scratches and gouges, defined as surface damage not resulting in material removal (scratches) and damage resulting in material removal (gouges), which is considered the same as a crack
- Cracks, which are any break, usually somewhat linear in nature, in the surface of a honeycomb panel
- Punctures, defined as damage resulting from the penetration of the skin surface
- Delamination, which is an adhesive-type failure between either the core and the adhesive (if a honeycomb panel) or the skin and the adhesive

With delaminations, if skins are primed before bonding, the separation may be between the adhesive and the primer or between the primer and the skin. In any case, most delaminations result in visible skin deformations. If not, they may be detected by tapping with a blunt, metallic object. A dull thud indicates a delaminated area,

whereas a crisp ring indicates a sound bond. In some cases, particularly in multiple-layer bonded structures, delamination cannot be detected by simple tapping techniques. More sophisticated methods, such as ultrasonic bond testers or similar instruments, are required. Service experience with the part in question dictates the type of inspection requirement.

Water Contamination. Honeycomb panel skin damage by punctures, cracks, sealant deterioration, or adhesive delamination may allow water or other aircraft fluid to enter. Water contamination cannot be tolerated in honeycomb panels for many reasons, the more important ones being, first, that water freezes in the panel, with the resulting expansion causing core damage and delamination; second, that the weight of the panel increases, causing an imbalance in flight control surfaces; and third, that the presence of moisture in the metal honeycomb core causes accelerated corrosion. Moisture can also enter a honeycomb panel by absorption when a fillet seal becomes damaged, or it may be absorbed by or diffused through the adhesive. X-ray inspection can be used to detect the presence of water and is a common nondestructive inspection technique. Weight-history methods are also used on parts that can easily be removed and replaced.

Environmental deterioration is usually defined as failure of the adhesive bond due to interaction between the adhesive, primer, metal surface, or honeycomb and the aircraft environment. Aluminum adhesive-bonded structures on aircraft in commercial service rarely fail in a cohesive (within the adhesive) manner. Except for mechanical damage, most failures start from an interface with the aircraft environment. This interface may be a tooling hole in which the sealant has failed, allowing moisture to penetrate the honeycomb. It could be the edge of a part (leading/trailing), a rivet or fastener hole through the structure, a cracked honeycomb potting compound, or edge close-out material. On U.S. commercial transport aircraft fabricated before 1970, interface

*Adapted from *Adhesive Bonding of Aluminum Alloys*, Marcel Dekker, 1985, p 425–493

failures result in deteriorated or missing honeycomb and debonded skins, with the debond mechanism in the skins being alclad dissolution. Honeycomb, previously perforated to allow ease of bonded-structure fabrication, was changed to nonperforated material. This reduced the rate of moisture propagation through the core. This honeycomb was later given a corrosion-inhibiting treatment, and the primary failure mode became skin-to-core disbond. The Boeing Company postulates the environmental attack of the bonded surface to be:

- Moisture penetrates through the adhesive by diffusion and reaches the aluminum oxide-bonded surface
- A weak hydrated oxide forms and then fails under stress. This action forms a crevice
- Corrosion continues within the crevice, which further accelerates the adhesive disbond and destroys the surface

New developments in the United States to ensure the environmental stability of the bonding surface consist of anodizing in a solution of phosphoric acid. Chromic acid anodizing has been used overseas for years as a method of obtaining an environmentally durable bonding surface. Surface treatments for a variety of adherends are described in the Section "Surface Considerations" in this Volume. However, the following discussion on surface preparation considers this topic from a repair standpoint.

Surface Preparation

Surface preparation is discussed below in relation to service life, qualitative procedure comparison, and nontank procedures.

Relationship to Service Life

The importance of properly preparing the aluminum surface prior to bonding cannot be overemphasized. This is especially true if the bonds are to be made with a room-temperature or 120 °C (250 °F) curing adhesive system.

Satisfactory service life as a function of bond quality specifically related to surface preparation has been demonstrated conclusively by tests on several high-life aircraft (Ref 1). Evaluations were made on bonded structures where no problems were evident, and comparisons were made of locations adjacent to areas where delaminations had occurred. Although the quality of the areas was indistinguishable when lap-shear or portable shear test procedures were employed, the difference was readily apparent when specimens were loaded in cleavage in a hot (60 °C, or 140 °F), high-humidity environment by use of the wedge test (Fig. 1). The measured rate of crack growth for the specimens was high when test coupons were taken adjacent to the failed areas, but typically low for a satisfactorily performing

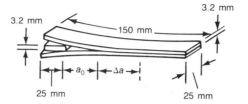

Fig. 1 Wedge test specimen configuration

structure. Additionally, the former specimens typically failed adhesively, leaving one of the metal surfaces shiny and bare, while the specimens from the satisfactory areas failed cohesively through the center of the bond line.

The results were important for two reasons. First, they indicated that the weak link was typically the bond of the adhesive to the metal, not the internal strength of the adhesive itself. Second, the evaluations established the wedge test as the much-sought-after short-time test that could meaningfully indicate whether a particular surface preparation/adhesive combination could be expected to have long-time service durability.

Those working at a repair facility might use the wedge test to rate their current metal cleaning and bonding procedures. If a typical 120 °C (250 °F) curing adhesive is used, the initial crack opening should typically be 30 to 38 mm (1.2 to 1.5 in.). Growth of the crack after 1 h of exposure to 50 °C (120 °F) and 100% relative humidity should not exceed 5 mm (0.2 in.). A 175 °C (350 °F) curing adhesive usually has a longer initial crack opening. However, its subsequent crack growth rate should essentially be zero. More important, when the specimens are split apart, the failure should be cohesive and should show no evidence of separation from the base metal.

A less quantitative but more dramatic demonstration of surface preparation quality can be attained by the use of the peel specimen detailed in Fig. 2. A constant but moderate force is exerted to pull the bonded pieces apart, while a drop of water (or coffee) is placed in the crack at the start of the bond. Typically, a bond with poor or marginal surface preparation will unzip with minimal effort, leaving a shiny metal surface. The bond to a well-prepared surface will remain tough, and failure will be through the center of the adhesive.

Another method of illustrating surface preparation quality is to press a piece of 3M 250 masking tape on the cleaned surface. On a well-prepared surface, such as one prepared by phosphoric acid hand anodizing, the adhesive will strip from the tape as it is removed.

Qualitative Procedure Comparison

A fairly comprehensive program designed to evaluate various quality levels of surface

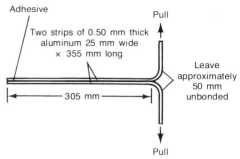

Fig. 2 Metal-to-metal peel test specimen

preparation was conducted by Boeing under contract to the U.S. Air Force (Ref 2). In this program, a 120 °C (250 °F) curing epoxy adhesive (AF 126) was used with EC 2320 primer (3M Company) to bond wedge specimens, which were cleaned by a variety of commonly used methods. These included solvents, such as Freon (DuPont), methyl ethyl ketone (MEK), and trichloroethylene; hand-applied acids, such as sulfuric-dichromate (Forest Products Laboratory, or FPL, etch), 2% hydrofluoric acid, and PasaJell 105 (Semco Sales and Service Company); and acid tank processes, including the conventional and an optimized sodium dichromate/sulfuric acid etch and phosphoric acid anodizing, the latter being a Boeing proprietary process designated BAC 5555. Additionally, the solvent procedures were evaluated with and without a supplemental conversion coating treatment. The coating used was Alodine 1200 (Amchem Products). The results are shown in Fig. 3. Comparative results can also be seen by noting the differences in the surfaces of failed specimens in Fig. 4.

The results indicated that hand cleaning with solvents was an inadequate surface preparation method, at least for the AF 126/EC 2320 system. Solvent cleaning with a conversion coating prior to bonding gave considerable improvement. Similar results were obtained in Boeing in-house tests (Table 1). In these tests, methods 6 and 7, in which Alodine 1200 was not used, gave much more rapid crack growth. In Fig. 3 and 4, progressively more improvement can be seen with the use of the hand-applied acid procedures. Evidence of adhesive failure almost totally disappeared when tank cleaning was used.

It should be mentioned that results depend on the particular adhesive system used. The AF 126/EC 2320 system has a comparatively wide tolerance to surface preparation quality. However, most room-temperature-curing adhesives have a narrow tolerance. This is seen in Fig. 5, where wedge specimen failure surfaces are shown for four typical room-temperature-curing systems. Surfaces were tank cleaned using the optimized FPL etch procedure. The results can be compared with equivalent conditions for AF 126, shown in Fig. 4(d).

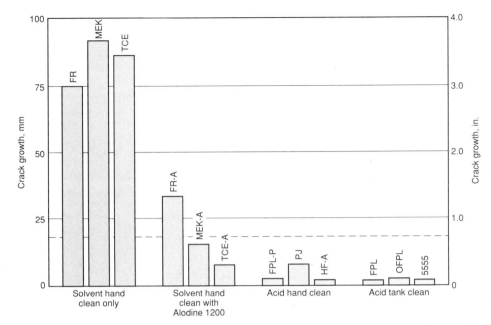

Fig. 3 Wedge test data showing comparative quality of various surface preparation methods. For AF 126/EC 2320 adhesive; 2024-T3 bare aluminum, 3.2 mm (0.125 in.) thick; 1 h at 50 °C (120 °F) and 100% relative humidity. FR, freon; MEK, methyl ethyl ketone; TCE, trichloroethylene; A, chromate conversion coating (Alodine 1200); FPL, Forest Products Laboratory etch; FPL-P, FPL paste; PJ, PasaJell 105; HF, 2% hydrofluoric acid; OFPL, optimized FPL etch; 5555, phosphoric acid anodized

The advantages of the precise control that tank equipment has over solution chemistry, temperature, immersion time, and access to rapid and complete rinsing cannot be equaled by hand procedures. Wherever practical, tank-type surface preparation facilities should be used.

Among the tank procedures, phosphoric acid anodizing has proved to be superior for use with 120 °C (250 °F) curing epoxy systems (Ref 2-4). Its use, where possible, is also recommended for room-temperature-curing adhesives. The 175 °C (350 °F) curing adhesives, at least those evaluated in Ref 5, are seemingly less discriminating and perform equally well with the optimized FPL etch, chromic acid, or phosphoric acid anodizing procedures.

At locations where adequate surface preparation facilities are not available, consideration should be given to stocking prepared skins, or sheet stock. This material can be cleaned using good-quality tank procedures. The skin is primed, and a nylon peel ply is bonded to the surface with an approved adhesive film. The prepared skin is then stored until needed. At that time the suitable patch or splice plate configuration can be cut, the nylon peel ply stripped from the bond surface, and the skin piece bonded in place. The bond surface on the part to be repaired must still be cleaned, but the total amount of cleaning is minimized. The remaining surface can be cleaned using one of the nontank surface preparations discussed below in the section "Nontank Procedures."

Evaluation of the Phosphoric Acid Nontank Anodizing Procedure

In many instances, tank cleaning is neither practical nor possible. For example, although repair details such as patch plates can be immersed in an acid tank, typically the part to be repaired cannot. At repair locations where cleaning or surface preparation tanks are not available, a nontank or hand cleaning procedure must be used.

The wide quality range of several nontank methods has been discussed. Understandably, the quality of the various methods is almost directly inverse to the ease of application. It should be noted, however, that even the best and, hence, most time-consuming methods involve only a minor portion of the repair task, and their use is strongly recommended.

A phosphoric acid hand anodizing procedure has been developed that is decidedly superior for typically sensitive room-temperature-curing and 120 °C (250 °F) curing adhesives. The method is not overly difficult. Gauze is saturated in a paste form of the acid, which is applied to the surface. This is overlaid with a stainless steel screen. Anodizing is accomplished quickly with low voltage (10 min at 6 V dc). After being rinsed and dried, the area is ready for bonding. Because handling freshly anodized surfaces results in poor bond strength, care must be taken prior to adhesive application.

The phosphoric acid nontank anodizing method has been thoroughly evaluated and optimized (Ref 6). The prime advantage is that the durability of the bonds, as indicated by the wedge test, is closely comparable to bond durability achieved when the best tank cleaning procedure is used. This can readily be seen in Table 2, which compares anodizing with three other nontank methods and two tank methods.

Another convincing comparison is shown in Fig. 6. The tank method in Fig. 6(a) and the nontank methods in Fig. 6(b) and (c) gave comparable results until water was introduced. Then the peel strength of the bond prepared with PasaJell 105 dropped to nearly zero, while the strength of the phosphoric acid nontank anodized specimen remained high (Fig. 6b).

Morphology studies also indicate that nontank phosphoric acid anodizing should perform comparably to the tank process. The surface oxide growths are quite similar, as seen by the two comparative scanning electron microscope (SEM) photographs in Fig. 7.

The practicality of using the nontank anodizing procedure to repair real structures has been thoroughly demonstrated. Numerous aircraft parts were received from the U.S. Air Force and Navy and satisfactorily repaired using this process in the Air Force/Boeing program (Ref 6). The anodizing procedure used to repair damage to an AWACS wing is shown in Fig. 8. The process was also demonstrated in making large bonded repairs to fuselage structures in a program supportive of the McDonnell Douglas/Air Force Primary Adhesively Bonded Struc-

Table 1 Effect of conversion coating (Alodine 1200) on bond stability

Surface preparation method(a)	Salt spray wedge test, days to failure	50 °C (120 °F), 100% relative humidity wedge test, days to failure	60 °C (140 °F) water wedge test, crack growth	
			mm	in.
1. 2% HF, Alodine 1200, dried 1 h	85	85	6.4	0.25
2. 2% HF, Alodine 1200, dried 4 h	79	85	6.4	0.25
3. Alodine 1200 only, dried 1 h	72	77	9.7	0.38
4. 2% HF, Alodine 1200, baked 60 min at 71 °C (160 °F)	114	92	3.0	0.12
5. 2% HF, Alodine 1200, dried 1 h, plus DeSoto 513–707 primer	No failure	No failure	9.6	0.38
6. Same as method 1, except no Alodine 1200	7	10	1.9	0.75
7. Same as method 5, except no Alodine 1200	...	...	Failure in 5 s	

(a) 2024-T3 nonclad aluminum, AF 126 adhesive, EC 2320 primer except as noted. HF, hydrofluoric acid. Source: The Boeing Company

(a)

(b)

(c)

(d)

Fig. 5 Typical wedge specimen adhesive failure surfaces for four room-temperature-curing adhesives. Optimized FPL etch. Courtesy of The Boeing Company

Fig. 4 Various specimen failures after 1 h exposure to 50 °C (120 °F)/100% relative humidity, AF 126 adhesive. (a) Adhesive failure of solvent hand-cleaned wedge specimens. (b) Failure surfaces of solvent plus Alodine 1200 hand-cleaned wedge specimens. (c) Failure surfaces of acid hand-cleaned wedge specimens. (d) Cohesive failure of tank-cleaned wedge specimens. Courtesy of The Boeing Company

tures Technology (PABST) program (Ref 4). Figure 9 shows surface anodizing of simulated damage to the skin and frame being accomplished in this program. After repair, specimens were removed for lap-shear and wedge testing. The results were satisfactory. It was learned, however, that when large, complex areas are to be anodized, special care must be taken to ensure proper contact of the anodizing components.

Nontank Procedures

Surface preparation procedures using a tank, such as the optimized sodium dichromate/sulfuric acid etch, phosphoric acid anodizing, and chromic acid anodizing are described in Ref 7. Their use is recommended when facilities are available.

Before describing individual cleaning procedures, it should be noted that many of the materials used for this purpose are hazardous; they may be toxic, flammable, or corrosive. Care must be exercised in their handling. Regulatory safety rules should be observed, including:

- Keep repair area well ventilated
- Minimize breathing of cleaning solvent fumes
- Wear rubber gloves, goggles, and protective clothing when handling acid or caustic cleaning solutions. If skin or eyes contact these solutions, immediately flush the area with water and seek medical attention
- Protect surrounding areas by masking and do not allow acid or caustic solutions to enter cracks or crevices
- Do not allow acids to contact the surface of high-strength steels
- Remove all cleaning solutions by thorough rinsing
- Keep solvents and cleaners in safety containers
- Use explosive-proof electrical equipment

Nontank precleaning and cleaning procedures are listed below in order of preference relative to bonding with room-temperature-curing or 120 °C (250 °F) curing adhesives. The procedures and their steps are summarized in Table 3. PasaJell 105 appears equal to nontank anodizing for the 175 °C (350 °F) curing systems and is preferred because of the ease of application. Precautions must be taken, however, because of its extreme corrosivity.

Precleaning. Prior to repair, undamaged areas, crevices, and fasteners should be protected from acid contamination by

Table 2 Comparison of various tank and nontank surface preparation methods using the wedge test

Alloy(a)	Surface preparation(b)	Adhesive	Primer	Initial crack length mm	Initial crack length in.	Crack growth At 1 h mm	Crack growth At 1 h in.	Crack growth At 4 h mm	Crack growth At 4 h in.	Crack growth At 30 days mm	Crack growth At 30 days in.
2024-T3B	TCE-A	EA 9601	EA 9261	40.1	1.58	Failure		. . .		. . .	
2024-T3B	PJ	EA 9601	EA 9261	40.9	1.61	6.07	2.39	Failure		. . .	
2024-T3B	HF-A	EA 9601	EA 9261	40.1	1.58	4.29	1.69	Failure		. . .	
2024-T3B	OFPL	EA 9601	EA 9261	38.6	1.52	0.33	0.13	0.33	0.13	1.09	0.43
2024-T3B	5555	EA 9601	EA 9261	39.6	1.56	0.20	0.08	0.33	0.13	1.24	0.18
7075-T6C	NTA	EA 9601	EA 9261	38.6	1.52	0.33	0.13	0.33	0.13	0.56	0.22

Note: All tests at 60 °C (140 °F) 100% relative humidity. (a) B, nonclad; C, alclad. (b) TCE-A, trichloroethylene plus Alodine; PJ, PasaJell 105; HF-A, 2% hydrofluoric acid plus Alodine; OFPL, optimized FPL etch (tank); 5555, phosphoric acid anodize (tank); NTA, phosphoric acid anodize (nontank)

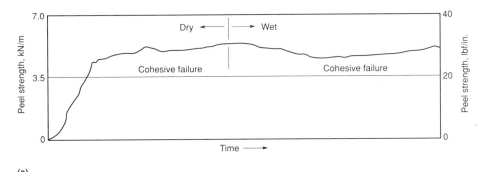

(a)

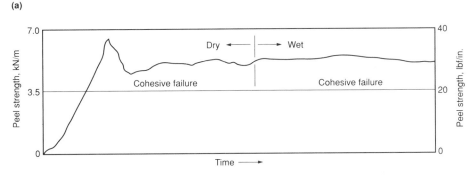

(b)

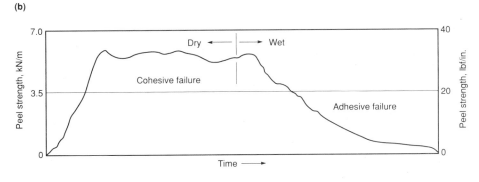

(c)

Fig. 6 Comparative dry-wet peel test results using three surface preparation procedures, FM-73, and 7075-T6 nonclad. (a) Optimized FPL etch. (b) Phosphoric acid, nontank anodized. (c) PasaJell 105. Courtesy of The Boeing Company

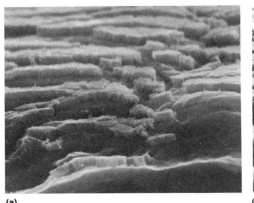

(a)

(b)

Fig. 7 Scanning electron photomicrographs of an oxide layer formed by phosphoric acid anodizing. (a) Using tank process. (b) Using nontank process. Courtesy of The Boeing Company

Fig. 8 Anodizing to install a patch on an aircraft lower-wing surface. Courtesy of The Boeing Company

Fig. 9 Anodizing the surface to be bonded in the repair of a large fuselage panel. Courtesy of The Boeing Company

masking them off with suitable tape and plastic film. Then, the following precleaning procedure should precede each of the subsequently described surface preparation methods.

1. Wipe the surfaces surrounding the repair area with solvent to remove oil, dirt, and grease
2. Remove the organic finish with solvents and mechanical aids or with approved

strippers. The former method is preferred. If stripper is used, it must be prevented from entering existing bond lines because it may have a deleterious effect on the bond

Phosphoric Acid Nontank Anodizing. This procedure requires the use of acid (safety notes are listed above). Steps include:

1. Solvent wipe with MEK, trichloroethane, or equivalent
2. Roughen the surface with a nonwoven abrasive, such as nylon abrasive pads or equivalent
3. Dry wipe with clean gauze to remove dust and debris
4. Apply a uniform coat of gelled 12% phosphoric acid to the aluminum surface. Gelled phosphoric acid compound can be obtained from Products Research & Chemical Corporation (PR 50) or made by adding a fumed silica powder to 12% phosphoric acid until thickened

Table 3 Nontank surface preparation procedures for aluminum

Method	Steps
Nontank phosphoric acid anodizing	Solvent wipe and abrade
	Apply gelled acid, gauze, and screen
	Use substrate as anode and screen as cathode and apply 6 V dc for 10 min
	Remove screen and gauze; rinse
	Dry; check for color
PasaJell 105	Solvent wipe and abrade
	Apply PasaJell 105 for 10–15 min
	Rinse
	Dry
2% hydrofluoric acid and chromate conversion coating	Solvent wipe and abrade
	Apply 2% HF, wipe after 10–30 s
	Rinse; apply chromate conversion coating within 15 min
	Dry
Abrasion, no acid etch	Solvent wipe and abrade
	Remove dust and debris; wipe clean

5. Place three or four layers of gauze over the coating; apply another coat of gelled phosphoric acid to saturate the gauze completely
6. Embed a piece of stainless steel screen over the coated gauze. Apply another coating of the gelled phosphoric acid. The stainless screen should be in continuous contact with the gauze but must not contact any part of the aluminum surface being anodized
7. Connect the screen as the cathode (−) and the aluminum as the anode (+). Setting up the screen as the cathode (−) and the aluminum as the anode (+) should be checked before proceeding
8. Apply a dc potential of 6 V for 10 min. A rectifier may be used to supply the voltage and current during anodizing. The current density should be in the range of 11 to 75 A/m² (1 to 7 A/ft²). In an emergency, a fresh or fully charged dry or wet cell battery may be used to anodize small areas
9. At the end of the anodizing time, open the circuit and remove the screen and gauze
10. Moisten clean gauze with water. Lightly wipe off the gelled acid with the moistened gauze without delay. The rinse delay time is limited to less than 5 min. Do not rub the anodized surface. Immersion or spray rinsing should be used if possible
11. Air dry a minimum of 30 min at room temperature or forced-air oven dry at 60 to 70 °C (140 to 160 °F)
12. Check the quality of the prepared sur-

face. A properly anodized surface will show an interference color when viewed through a polarizing filter rotated 90° at a low angle of incidence to fluorescent light or daylight
13. If no color change is observed, repeat steps 4 through 12. Machined or abraded surfaces are sometimes difficult to inspect for color. Rotation of the polarizing filter is required because some pale shades of yellow or green are so close to white that without a color-change inspection, they might be considered "no color." This would falsely indicate the absence of anodic coating

Caution: Do not touch the dried, anodized surface. Do not apply tape to the surface.

PasaJell 105 Method. This procedure requires the use of acid (safety notes are listed in the section "Precleaning"):

1. Solvent clean with MEK, trichloroethane, or other approved solvent
2. Abrade with a nylon abrasive pad or 400-grit aluminum oxide abrasive paper
3. Dry wipe with clean gauze pads
4. Apply PasaJell 105 to the surface with a spatula, brush, or gauze
5. Leave PasaJell 105 on the surface for 10 to 15 min
6. Lightly wipe off the PasaJell, but do not wipe it onto adjacent areas or crevices or allow it to enter honeycomb core
7. Dampen a clean gauze pad with clean water and lightly wipe the treated area. Repeat as necessary and check with litmus paper to be sure that all traces of acid have been removed. Do not rub the surface
8. Allow the area to air dry before applying primer or bonding. Do not handle freshly etched surface

Two Percent Hydrofluoric (HF) Acid Method. This procedure requires the use of acid (safety notes are listed in the section "Precleaning"):

1. Solvent clean with MEK, trichloroethane, or other approved solvent
2. To remove contaminants, abrade the surface thoroughly with nylon abrasive pads (MIL-A-9962 type A) or 400-grit aluminum oxide abrasive paper
3. Dry wipe with clean gauze to remove abrasive residue
4. Moisten (do not saturate) clean gauze with 2% hydrofluoric acid solution. Wipe the abraded area briskly with the pad. Treat this procedure as though the HF were a solvent
5. After the acid has worked for 10 to 30 s, dampen a clean gauze pad with clean water and wipe the acid-treated areas. Do not drag gauze through nontreated areas onto treated areas
6. Within 15 min, apply chromate conversion coating. Instructions for the appli-

cation of the conversion coating are given below

Solvent Wipe and Abrade. This method should be used only on areas internal to the honeycomb panel or in areas where contamination of adjacent structure could not be prevented if acid were used.

1. Solvent clean with MEK, trichloroethane, or other approved solvent
2. Abrade with a nylon abrasive pad (MIL-A-9962 type A) or 400-grit aluminum oxide abrasive paper
3. Gauze wipe or vacuum to remove dust. Solvent wipe again
4. Following the conventional solvent wipe and abrade procedure (that is, steps 1 through 3), it is recommended that a chromate conversion coating be applied to the surface

Chromate Conversion Coating. It has been demonstrated that the application of a chromate conversion coating as a final step to some surface preparation procedures significantly improves bond durability. For this reason, inclusion of the conversion coating (MIL-C-5541, class 1A) is recommended if these surface preparation methods are to be used.

The conversion coating also has a second use. When cleaning in preparation for the repair bond, it may be necessary to bare some of the peripheral surface area. Treating these surfaces with the chromate coating provides an excellent paint base and acts as a corrosion inhibitor. Before applying the coating, these surfaces should be cleaned as previously described.

The procedure for applying a chromate conversion is described below. Although the procedure described pertains specifically to the application of a common conversion coating, Alodine 1200, this is not intended to indicate a preference. If another coating is used, the recommended procedure of the manufacturer should be followed.

1. Apply Alodine 1200 evenly and liberally with a fiber or nylon brush, clean cheesecloth, or cellulose sponge
2. Allow the solution to remain for 3 to 4 min to form a bronze or golden-brown coating. Keep the treated area from drying by gently blotting with cheesecloth moistened with Alodine 1200. Do not allow the alodine surface to become powdery
3. Rinse with clean water by gently contacting the treated surface with wet (not saturated), clean cheesecloth. Contact (blot) for 1 to 2 min and repeat. *Caution:* Exercise care when rinsing and contacting the treated surface to avoid scratching or removing the freshly formed coating

4. Gently contact the surface with clean, dry cheesecloth to absorb excess liquid. Repeat as necessary
5. Check crevices with blue litmus paper for possible acid contamination. If litmus paper turns red, acid is still present, and steps 3 to 5 should be repeated
6. Air dry at room temperature or use hot air (60 °C, or 140 °F, maximum)

Caution: Cloths or other materials used to apply the conversion coating should be rinsed out thoroughly in water after use and deposited in an approved safety container. To dispose of chromate conversion and rinse solutions, follow local safety and health regulations.

Repair Materials

Selecting the proper repair materials involves many details and presents many alternatives. Because the intent is usually to restore the structure to its predamaged condition, consideration should be given to the use of the original materials. However, if examination proves that the original material has been deficient or if its use is not compatible with the repair situation, a more suitable replacement must be substituted. This is especially true when a part is being rebuilt. A light-density core in a damage-prone area, for example, usually can be replaced with one of heavier density if the balance of the part is not critical.

If the aircraft has been in a fleet for some time, the manufacturer is a good source for an improved-material recommendation. The manufacturer is constantly evaluating new products, and it is impossible to keep the repair manual up to date. An inquiry will also alert the manufacturer to a problem area that requires attention.

When selecting repair materials, consideration must also be given to the structural criticality of the damaged area, the time that is available to make the repair, and the availability of tools and facilities at the repair location. In many cases, temporary repairs can be made quickly to secondary structures, using fiberglass-resin patches and a heat lamp to accelerate the cure. However, if the damage is significant, the part should be replaced where practical. The old part can be sent to the shop for repairs of a more permanent nature.

Metal Sheet Materials. Attention to permanent, good-quality repairs usually requires that metal repair parts be replaced. The manufacturer repair manual is a source of material identification. The engineering staff should be consulted when it is necessary to make substitutions. In general, for small repairs, substitutions are allowable when the replacement is a stronger material. For example, 7075-T6 sheet may be substituted for 2024-T3, or 5056 aluminum honeycomb core for 5052. The substitution may be reversible if the consideration is one of stiffness or bond line corrosion resistance and not strength. In the latter case, a strength substitution becomes permissible by increasing the gage or density of the material. Such substitutions allow maximum utility of material inventories. If larger repairs or rebuilding are to be accomplished, however, original materials should generally be used. Selection of the initial material involves the consideration of many properties, for example, toughness and corrosion resistance, in addition to strength and stiffness. Therefore, general substitutions are not recommended without a detailed analysis of the problem.

When design information is required on aluminum facing materials, reference should be made to MIL-HDBK-5 (Ref 8) or specification QQ-A-250 (Ref 9). Where possible, bonds should be made to nonclad faces. The use of fiberglass-resin composites to accomplish large aluminum repairs is usually not recommended, primarily because of the low modulus of elasticity of the material. However, the relative ease and speed with which these repairs can be made often make them useful where stiffness is not a critical criterion and stresses are low. Fiberglass-resin repair procedures are especially adaptable to areas that have compound curvatures or for honeycomb edge close-outs. Either wet lay-up or preimpregnated glass layers (prepregs) can be used. Typical material data are given in military specification MIL-C-9084 (Ref 10). Mechanical strength and design information is given in MIL-HDBK-17 (Ref 11).

Honeycomb Core. Aluminum core is available in a wide range of standard cell sizes and densities. The replacement core should match the core cell size and density of the original. The core ribbon should be oriented in the same direction. If the original core size or density is not available, use of the next higher core density or next smaller cell size is permissible. Corrosion-resistant core should be used as a replacement in all repairs regardless of the originally specified aluminum core type. This honeycomb should be nonperforated. Aluminum 5056 alloy core may be substituted for 5052 core. Aluminum core information is given in MIL-C-7438 (Ref 12).

Nonmetallic cores, such as those that are glass-fabric reinforced or fibrous-nylon based, can also be used for repairs. When substituted for aluminum core, they should equal the aluminum in strength and stiffness. Information on fiberglass core is given in MIL-C-8073 (Ref 13). Fibrous-nylon base (that is, Nomex) core data are given in Aerospace Material Specification (AMS) 3711 (Ref 14).

Adhesives and Sealants. Mechanical property data for adhesive systems are available from several sources, such as the adhesive supplier and the airframe manufacturer who uses the particular material. General information can also be obtained from military specifications such as MIL-A-25463 (Ref 15) and MMM-A-132 (Ref 16). Care must be taken in using these data to allow for any subsequent effects of environmental degradation or strength loss of the material at elevated temperatures. It should also be noted that these data typically apply to 13 mm (0.5 in.) lap-length splices. Strength values, particularly of the 175 °C (350 °F) curing adhesives, may show drastic reductions for longer laps.

It is very important that the processing procedures used to determine the design strengths duplicate those that are used later for the repair. Many of the commonly used adhesives were tested with cleaning and bonding methods representative of repair procedures in U.S. Air Force contracts F33615-73-C-5171 and F33615-76-C-3137. These data are reported in Ref 1 and 6. Typical strength plots for various bonded joint overlap lengths are given for several 120 and 175 °C (250 and 350 °F) curing adhesives.

Data for several room-temperature-curing adhesives tested with 13 mm (0.50 in.) lap slices are given in Ref 1. Two other excellent sources for test data are Ref 2 and 3, which have information that is valuable for adhesive comparison purposes, based on lap lengths of 13 mm (0.50 in.). Values for other lap lengths, if required, will have to be determined. Because the data in Ref 1 to 3 and 6 are typical values, it is recommended that they be reduced to a minimum of two-thirds for design utility.

The use of primers is recommended to improve durability and to minimize contamination before adhesive cure. Both noncorrosion-inhibiting and corrosion-inhibiting primers are available and may be applied by spray or wipe-on techniques. The use of corrosion-inhibiting primers is especially recommended for the 120 °C (250 °F) curing adhesives when proper equipment and necessary personnel skills are available for application. Bonds incorporating corrosion-inhibiting adhesive primers (CIAPs) have considerably increased durability over non-CIAP bonds when the primer is applied correctly to a well-prepared surface. A disadvantage is that the CIAP must be baked prior to cure of the primary bond. However, this disadvantage is offset by the long storage life of aluminum pieces with the CIAP coating if properly protected from sunlight and dirt. This allows the stocking of precleaned and primed sheet material at a repair facility. The material can then be cut to size as required and merely solvent wiped prior to bonding.

The application of corrosion-inhibiting primers is best suited to a large repair station. The surface must be well prepared by tank cleaning or hand anodizing, and

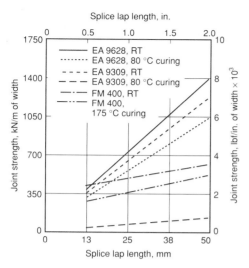

Fig. 10 Comparative effect of lap strength and temperature on the strength of three classes of adhesives. Cyclic conditions: 60 °C (140 °F) at 100% relative humidity; 15 min at maximum load, 5 min at 0 load, 3 cycles/h. RT, room temperature

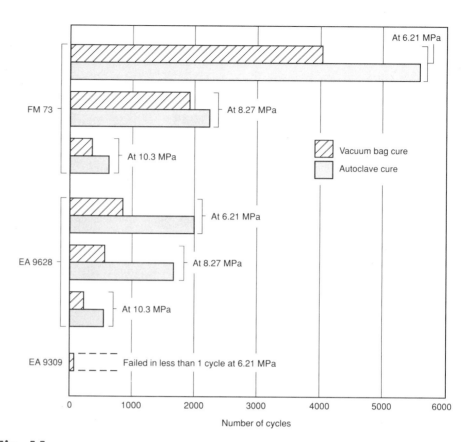

Fig. 11 Comparison of cyclic load-environmental exposure data for three adhesive systems

the application and thickness must be carefully controlled. The use of noncorrosion-inhibiting primers is recommended when conditions are less than optimal and when the additional time required to bake cure CIAP primer is not permissible. The use of primer with 175 °C (350 °F) curing adhesive is recommended, but is not considered as important as with 120 °C (250 °F) systems. In general, the 175 °C (350 °F) systems are in themselves more environmentally stable and more tolerant of surface preparation.

The adhesives used for permanent repairs should be of the same class as those used in the original structure. Typically, a 120 °C (250 °F) curing epoxy is used for subsonic aircraft where temperatures may reach a maximum of 80 °C (180 °F), such as on the ground in desert areas. If higher temperatures are encountered, such as around the auxiliary power unit (APU), adhesives with higher-temperature capability, usually curing at 175 °C (350 °F), may be required. These systems are also typically required for supersonic aircraft.

When operating temperatures permit, 120 °C (250 °F) systems are preferred because they are typically stronger, tougher, and easier to process. By contrast, a 175 °C (350 °F) system has lower strength and is more applicable to honeycomb core/face sheet bonds. The strength-versus-lap length curve is typically quite flat (Fig. 10), and little strength is gained by lengthening a splice. In many cases, consideration must be given to supplementing the bond strength by using mechanical fasteners.

Room-temperature-curing adhesives have application for temporary repairs to lightly loaded structures. However, their use is not recommended for permanent repairs or for temporary repairs in critical areas.

Even the best-candidate adhesives carefully screened for evaluation in Ref 1 gave poor wedge test results and had limited strength at the moderate upper surface temperature ranges (Fig. 5). Results of stressed load tests performed on two 120 °C (250 °F) curing adhesives and a candidate room-temperature-curing system are shown in Fig. 11. The exposure temperature at 60 °C (140 °F) was not high, yet the room-temperature-curing system failed quickly under sustained load. The use of an autoclave is compared to vacuum bagging for 120 °C (250 °F) systems in Fig. 11. Other comparative data of interest are shown in Table 4.

Once the adhesive cures, sealing the assembly becomes important. The adhesive flash and fastener holes through bond lines should be protected, as should all bare edges exposed by trimming, drilling, or machining, or those scuffed through the surface finish. Two suitable sealants are polysulfides governed by MIL-S-8802 (Ref 17) and corrosion-inhibiting polysulfides governed by MIL-S-81733 (Ref 18). Silicone sealants are usually used for sealing applications in excess of 150 °C (300 °F). Painting the entire structure with a filiform corrosion-resistant polyurethane primer and enamel system will greatly increase the environmental durability of the 120 °C (250 °F) cured bond.

Type of Repair

From the perspective of facilities and techniques, repairs can be grouped in five categories, described in the remaining sections of the article. First is small-area line maintenance work, where fiberglass overlay patches, aluminum skin patches, aluminum honeycomb plug patches, or riveted aluminum sealed with polysulfide sealant will suffice. Second is life-limited repairs using rivets and room-temperature-curing adhesives, necessary for aircraft dispatch but requiring short-time inspection intervals and replacement at the first available maintenance opportunity. Third are partial repairs using a heating blanket and vacuum bag technique either for localized areas or when the chance of damaging the original bond with more extensive heat application is a factor. Perforated metal repair skins are used in some instances. The fourth category consists of partial repairs using the autoclave to replace honeycomb core and as much as one side of the damaged structure. If surface preparation techniques are less than optimal, the parts are monitored for deterioration and rebuilt as time permits. The final category is fully repaired or rebuilt bonded structures in which the old spar, end ribs, and so on, are used as the basis for building a reconditioned part. All surfaces are given optimal preparation prior to bond-

Table 4 Lap-shear and peel test results

Adhesive	Cure and standard deviation	13 mm (0.5 in.) lap shear								Metal-to-metal peel					
		At room temperature		At 80 °C (180 °F)		After 14-day salt spray		Stressed 60 days at 4.1 MPa (0.6 ksi)(a)		At room temperature		At −55 °C (−67 °F)		After 14 days at 120 °C (250 °F), 100% relative humidity	
		MPa	ksi	MPa	ksi	MPa	ksi	MPa	ksi	N·m/m	lbf·in./in.	N·m/m	lbf·in./in.	N·m/m	lbf·in./in.
EA 9628	Autoclave(b)	31.7	4.60	20.0	2.90	31.8	4.606	31.8	4.610	272	61.1	51.6	11.6	279	62.7
	Standard deviation	92	92	38	38	154	154	66	66	5.6	5.6	1.0	1.0	3.9	3.9
	Vacuum(c)	27.6	4.01	15.7	2.27	28.0	4.06	29.6	4.29	...	...	...	...	...	...
	Standard deviation	93	93	72	72	38	38	64	64	...	...	...	...	...	...
FM 73	Autoclave(b)	27.5	3.99	18.1	2.62	27.6	4.01	27.5	3.99	251	56.4	70.3	15.8	252	56.7
	Standard deviation	87	87	88	88	92	92	125	125	5.8	5.8	5.5	5.5	2.5	2.5
	Vacuum(c)	26.1	3.78	16.5	2.39	27.0	3.92	26.3	3.82	...	...	...	...	...	...
	Standard deviation	110	110	153	153	114	114	348	348	...	...	...	...	...	...
EA 9309	Vacuum(d)	24.1	3.49	4.0	0.58	23.9	3.46	Failure in 5, 11, 14, and 18 days		248	55.8	20.0	4.5	195	43.9
	Standard deviation	442	442	34	34	593	593			5.9	5.9	1.2	1.2	3.9	3.9

(a) Exposure at 50 °C (120 °F) and 100% relative humidity. (b) At 240 kPa (35 psi), 90 min. (c) At 85 to 95 kPa (25 to 28 in. Hg), 95 °C (200 °F), 2 h. (d) At 85 to 95 kPa (25 to 28 in. Hg), room temperature

ing, and corrosion-inhibiting honeycomb, corrosion-inhibiting adhesive primer, film adhesive, and new aluminum skins are used.

Line Maintenance Repairs

Small-area line maintenance repairs can be subdivided into a number of classifications, described below.

Surface Dent Repair. As a rule of thumb, dent filling is never allowed where bond separation could allow the filler material to be ingested into an engine or accessory intake opening. Typical requirements for the extent of allowable dent filling (excluding balanced panels, in which weight additions are minimized without rebalancing) are:

- The skin must not be cracked
- The depth of any gouges or scratches must not exceed 20% of the skin gage
- The maximum dent depth must not exceed 3.2 mm (⅛ in.)
- The size of a single dented area must not exceed 0.016 m² (25 in.²), and all dented areas must not exceed 20% of the total skin area on one surface of the part
- No delaminations are allowed in the dented area and around the periphery. Nondestructive testing must be used to determine whether delamination exists
- Dent filling is required only on aircraft external surfaces
- Minor dents of less than 25 mm (1 in.) diam or less than a 0.90 mm (0.035 in.) depth do not require repair if they are smooth and free of punctures, cracks, or scratches

The usual surface preparation for a dent repair is to sand or chemically strip off the paint and primer and then hand wipe with a grease-free solvent. The area is then lightly and evenly abraded with a nylon web containing aluminum oxide particles or an abrasive cloth. The residue is then wiped with a solvent such as methyl ethyl ketone on a lint-free, grease-free cloth. This is followed by a dry-cloth pickup. Cleaning is continued until the cloth shows no residue. It is recognized that this cleaning technique does not yield an environmentally durable bond, but line maintenance schedule opportunities preclude more sophisticated methods. Immediately upon completion of abrasive cleaning, the dent is filled with a room-temperature-curing epoxy adhesive. The adhesive may be rigid (epoxy polyamine) or flexible (epoxy polyamide), depending on whether the surface to be filled is rigid or flexible. Subsurface air bubbles are eliminated during application, and the adhesive is formed to the surrounding surfaces using a parting film cover during the working life of the adhesive. Heat can be used to accelerate adhesive cure. After curing, rough spots are sanded, and the repair is covered with a chromate-containing primer and a finish coat of polyurethane paint. This provides a certain amount of environmental durability.

Puncture and gouge repairs up to 25 mm (1 in.) in diameter use room-temperature-curing adhesive and serrated rivets. These repairs use fast-curing epoxy adhesives with the ability to withstand temperature excursions of −55 to 70 °C (−65 to 160 °F) without adhesion problems. Epoxy polyamine formulations have been found to be satisfactory. First, the damage should be assessed and it should be ascertained that the adjacent area is not delaminated. The puncture or gouge should be cleaned by trimming or routing/grinding, and the repair diameter determined. The damaged area is flushed with a nonflammable solvent such as 1,1,1-trichloroethane and dried using a hair dryer. Then, depending on the repair diameter, the steps depicted in Fig. 12 should be followed. A serrated rivet and/or washer should be cleaned with solvent and installed through the use of a rigid, room-temperature-curing epoxy adhesive. The repair should be flush with the panel surface. Repairs exposed to temperatures higher than 70 °C (160 °F) can be plotted with a room-temperature-cure and/or a heat-post-cure system to provide elevated-temperature strength. The repair is covered with a chromated primer and compatible polyurethane paint top coat.

Skin Damage or Delamination, With No Honeycomb Damage. Typical wet lay-up repair is used. Damage to honeycomb panels can be removed by first sanding off all paint and primer in the damaged area. A minimum-size cutout in the skin is then required to remove the damage. This cutout can be circular or oblong. The edges of the resulting hole should be burr-free and approximately 90° to the surface. Any entrapped water can then be removed by heating the open core cells, with care being taken not to heat sealed parts of the panel above 80 °C (180 °F) (for 120 °C, or 250 °F, cured original structure).

Adjacent parts of the panel should also be checked for water entrapment. In nonperforated honeycomb panels, each cell must be ventilated to remove water. Rather than drill a hole in the skin that covers each cell, time can be saved by removing the skin, drying the affected area, and replacing the skin. After the cutout is completed, the dust and chip residue should be vacuumed. The skin is then cleaned as described above for dent repair, and the honeycomb is flushed with nonflammable solvent. After drying, one aircraft company specifies priming with a Buna-N type of coating (MIL-S-4383) (Ref 19) followed by impregnating 120 and/or 181 (1581/7781)-type fiberglass with an epoxy polyamide resin system. This system, although slow to cure, resists hydraulic fluid and yields a structural repair. Additional environmental resistance is obtained by covering the repair with polyurethane paint over a compatible primer.

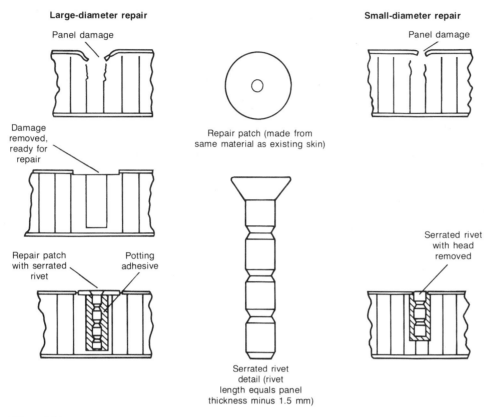

Fig. 12 Rivet plug repair for punctures and gouges

fiberglass layers required. Each ply must be 13 mm (½ in.) larger in diameter or edge distance than the ply below. The final ply must overlap the edge of the repair by at least 38 mm (1.5 in.) (see Fig. 13)

3. Mix resin and catalyst. The weight of the mixture should approximate the weight of the dry fiberglass cloth
4. Apply a thin coat of resin to the repair area (over the primed surface)
5. Place the first fiberglass ply in position and work out the entrapped air. Stipple with a brush until the resin comes to the surface. Apply another layer of mixed resin and the next ply. Repeat this operation until all plies have been applied. Fair out the resin about 13 mm (½ in.) beyond the last ply to provide environmental protection for the primer. Complete the repair within the pot life, or work life, of the resin
6. Cover the repair with a parting film and squeegee to work out excess air, being careful not to remove too much resin and cause a resin-starved laminate. Wipe off resin outside the parting film area
7. Allow the repair to cure. Vacuum may be used to increase the strength of the repair. Typically, vacuum-bagged laminates are approximately 20% stronger than those cured under contact pressure (see Fig. 14). Vacuum can also be used to hold the repair in position during cure, especially on under-wing or vertical lay-ups. Heat from lamps, blanket, or hot air can be used to accelerate curing. Do not exceed a temperature that is 30 to 40 °C (50 to 70 °F) below the original cure temperature of the panel

The steps in the repair procedure are:

1. Apply a thin coating of primer and allow to dry before the resin overcoat is applied
2. The gage of aluminum skin requiring repair determines the number of fiber-

glass layers, or plies, to be used. Thickness to 0.50 mm (0.020 in.) requires two plies of 1581-style cloth, to 0.81 mm (0.032 in.) requires three plies, and to 1.0 mm (0.040 in.) requires five plies. If 120-style cloth is desired, double the number of plies. Cut the number of

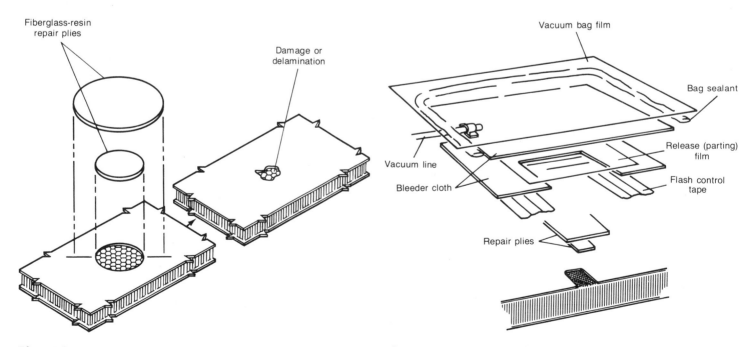

Fig. 13 Fiberglass-resin repair technique (no core damage)

Fig. 14 Application of pressure (vacuum) during cure for skin ply replacement

8. After the cure, remove the parting film, lightly sand the surface, and cover with a polyurethane primer and paint system

A technique called the preimpregnated cloth method is recommended by most manufacturers as an alternative to conventional wet lay-up techniques. Its usefulness is shown when working on vertical or overhead positions. The steps of the procedure are:

1. Precut the desired number of fiberglass plies to the required sizes
2. Cut some parting film (hard material) at least 75 mm (3 in.) larger than the largest ply
3. Mix the required amount of resin (approximately equal in weight to that of the fiberglass cloth)
4. Tape one piece of parting film to a flat surface. Spread a layer of mixed resin over the film
5. Place one or two fiberglass plies on top of the parting film and resin
6. Lay the other piece of parting film on top of the fiberglass plies and work the mixed resin into the cloth using a squeegee or roller. Be sure to wet the entire area of fiberglass
7. Peel back the top release film and cover the resin-cloth mix with another layer of resin. Add one or two dry fiberglass plies on top of the existing plies
8. Replace the parting film and work the mixed resin into the fiberglass cloth
9. Continue these steps until the required number of plies are used
10. Cut the preimpregnated patch to the required outside size and apply a thin coating of resin to the previously prepared repair area
11. Remove the inside layer of parting film and apply the patch over the damaged area. Remove and replace the remaining parting film with a larger piece and apply it over the repair
12. Allow the patch to cure and finish the repair by covering with the appropriate primer and paint system

Skin and Honeycomb Core Damage Requiring Wet Lay-Up, Potted Core Technique. As in the previous skin damage repair sequence, the damage must be cut out, and the area prepared for bonding. All loose and damaged honeycomb is removed, but any well-bonded material is left intact (see Fig. 15). Solvent is used to flush the area and allowed to evaporate before potting compound is added. Potting compounds are selected for their low density, stiffness, and compatibility with the honeycomb panel service temperatures. These repairs are usually small because the potting compound adds weight (and mass) to the panel, which could be a problem in vibration areas and on balanced assemblies.

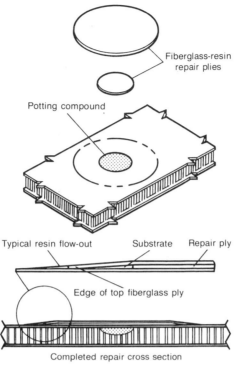

Fig. 15 Potted core repair with fiberglass-resin overlay

The mixed potting compound is applied until it fills the damaged area (Fig. 15). The potting compound is allowed to cure, with or without the accelerating effect of heat, as required. The surface is sanded smooth, and the wet lay-up resin-fiberglass repair is applied to one surface, as discussed previously.

Skin and Honeycomb Core Damage Requiring Wet Lay-Up, Honeycomb Core Replacement. Figure 16 depicts a repair cross section. The damaged honeycomb should be cut out flush with the inside skin, with care being taken not to damage the surface. A router and hole saw with stop is usually used for this operation. The damaged core is cut around the perimeter and then routed from the inside adhesive surface. Before cleaning, a core plug is made from the same material as the original honeycomb panel (Fig. 16). Then, the core plug, core cutout area, and area to be repaired are cleaned, as described previously. The honeycomb core is potted in using potting compound material on the sides and bottom of the insert. After cure, the wet lay-up repair is accomplished, assuming that the core was cut and potted to the correct thickness. To increase the repair strength, vacuum pressure can be used during the cure of the potted honeycomb and fiberglass overlay (Fig. 14 and 17).

Skin and Core Damage, Both Surfaces, Requiring Aerodynamically Flush Patch. Access to repair must be available from both sides. The damaged skin and honey-

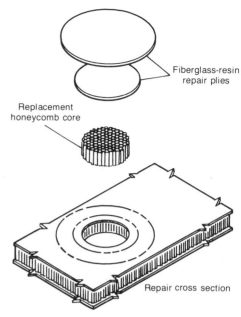

Fig. 16 Honeycomb core plug repair with fiberglass-resin overlay

comb are cut out to gain access to the panel from the interior or internal skin surface (see Fig. 18 for repair cutout configuration). All areas to be bonded are cleaned, and then a piece of parting film is taped over the hole in the external skin. This is followed by the lay-up of the required number of resin-fiberglass plies, based on skin thickness, to provide for the skin repair. The first ply must mate with the hole in the skin, and the last ply should just fit inside the honeycomb core cutout area (Fig. 19). Using a potting compound with a curing agent that is compatible with the wet lay-up resin, the core plug is inserted, and the external skin repair and potting compound are allowed to cure. The core plug is milled flush with the internal skin when the potting and fiberglass lay-up have cured, and the residue is flushed out with solvent. After the solvent has evaporated, the next step is to lay up the required number of internal skin plies. The repair is finished after the resin has cured.

If access is available from only one side of the panel, this type of repair can still be accomplished. The repair would start with the internal skin side, and the cutout configuration would be the same as shown in Fig. 19 (substitute internal skin for external skin). The wet lay-up resin-fiberglass patch should be precured and bonded to the inner skin. After a light sanding and cleaning, the repair would then proceed as described previously, but the external surface patch would not be flush aerodynamically.

Skin and Core Damage, Both Surfaces, With a Nonaerodynamic Patch. Access to repair is required from both sides. First, the entire damaged area is removed and the affected area of the panel is cleaned, as in previous procedures. Either side is repaired

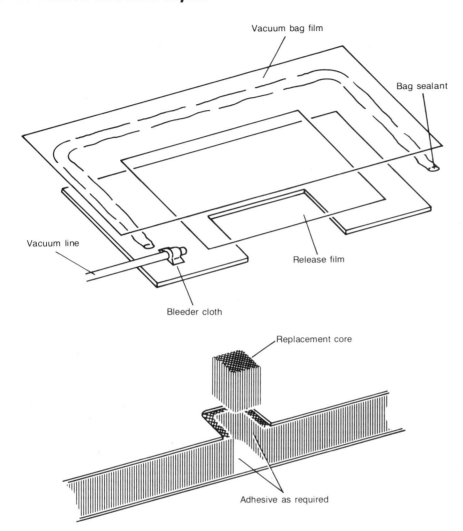

Fig. 17 Application of pressure (vacuum) during cure for core replacement

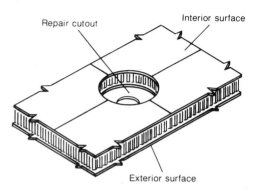

Fig. 18 Repair cutout configuration for aerodynamic flush patch (surfaces reversed for clarity)

thin-gage aluminum wraparound skin that can be bonded over the trailing edge with a MIL-S-8802-type sealant. If special environmental protection is desired, an edge coating of silicone or a similar product is applied.

Major damage to the trailing edge is repaired by the method shown in Fig. 22. Obvious areas of delamination should be cut out. Depending on the size of the damage, the area is then cleaned and filled with potting compound or with a new piece of honeycomb, which is cut to fit, placed, and potted. A fiberglass-resin wet lay-up repair (the number of plies depending on the skin thickness) to both surfaces completes the operation. Alternatively, if vacuum pressure is available to hold a trailing-edge wraparound doubler in position, the repaired area can be covered with a thin-gage

using a fiberglass-resin lay-up with the number of plies calculated from the original skin thickness (Fig. 20). After curing, the core plug is inserted, and the potting compound is applied around the perimeter and to the new skin. As soon as the potting compound has cured, the honeycomb is milled flush with the skin, and an additional fiberglass-resin lay-up is applied. The panel is finished as required after curing.

Trailing-Edge Damage Repair. Minor damage to the trailing edge of wedge-shape structure can be repaired using a dent filler. Any scratches or irregular areas should be dressed out with a file or similar device. If there is no delamination between the skins and honeycomb, the surface is then thoroughly cleaned and filled in with a dent-filling compound (Fig. 21). If the repair is visually objectionable, it is covered with a

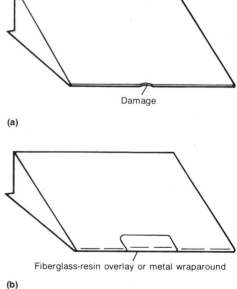

(a)

(b)

Fig. 21 Minor trailing-edge damage repair. (a) Damage is dressed out and filled. (b) Covered with fiberglass-resin overlay or metal wraparound

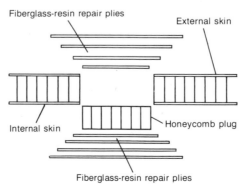

Fig. 19 Aerodynamically flush honeycomb plug repair with fiberglass-resin overlay

Fig. 20 Nonaerodynamic honeycomb core plug repair with fiberglass-resin overlay

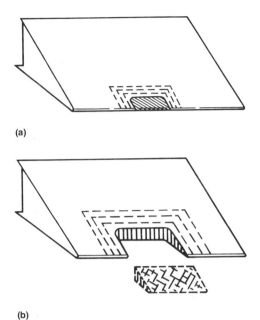

(a)

(b)

Fig. 22 Major trailing-edge damage repair. (a) Cut out damage is filled with potting compound, covered with fiberglass-resin or metal overlay. (b) Same as for (a) except honeycomb core replacement used for weight savings

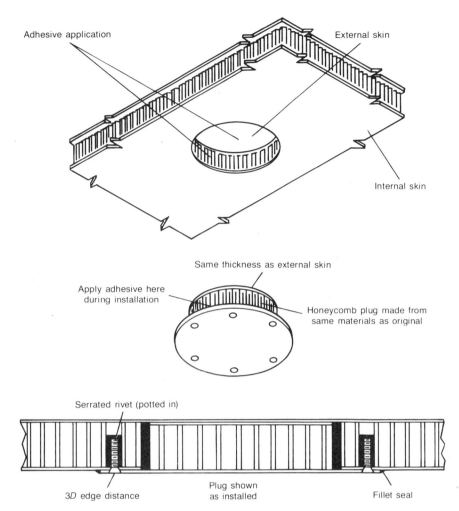

Fig. 23 Honeycomb core-aluminum skin plug repair

aluminum sheet bonded with MIL-S-8802 sealant.

Skin and Core Damage, With a Prefabricated Nonaerodynamic Plug. This type of repair can be accomplished quickly if repair inserts are available. Plugs of various diameters can be prefabricated using the same skin construction, honeycomb core, and adhesive as used in the original part. Standard repair cutout sizes will then interface with these prefabricated plugs (Fig. 23). Because the skin thickness of the repair plug will be equivalent to that of the original panel, it is important to determine in advance whether the repair will be accomplished from an internal or external surface. If damage occurs through both skins, an insert is bonded to the repair plug, which fills in the skin contour. This configuration is similar to the fiberglass wet lay-up repair insert. In any event, the damage must be dressed out, and the area to be repaired thoroughly cleaned. Prior to this, the location of the holes for serrated-rivet installation must be determined, and both the repair plug and the panel must be line drilled. Using an L-shape tool, the honeycomb inside the panel hole is opened to provide some space for the rivet potting compound. Then, as shown in Fig. 23, the repair plug is glued into the cutout area with a room-temperature-curing or elevated-temperature-curing structural epoxy adhesive. If a high-temperature adhesive is used, the cure temperature should be kept about 30 °C (50 °F) below the original cure temperature of the panel, and the panel and repair should be kept under pressure while curing. At the same time

that the plug is being installed and cured, the serrated rivets should be installed as described previously. After the repair cures, the edges are filleted with an environmental sealant, and the plug and panel are covered with an environmentally durable primer and paint system. This reduces the incidence of corrosion in the repair area. The repair may be circular or oblong, with size limitations to be determined by the aircraft manufacturer.

Skin and Core Damage, With a Prefabricated, Aerodynamically Flush Plug. This repair, similar to the plug configuration described previously, starts with removal of the damaged skin and honeycomb within allowable size limitations. The repair patch plug should be manufactured from the same materials as used in the original panel, with both skins the thickness of the damaged surface. After cleaning, the patch is bonded into the damaged panel cutout, as described previously. The depth of the cutout must be controlled to mate with the thickness of the repair plug (Fig. 24). After the repair, the patch plug must fit flush with the existing skin. To complete the repair, the perimeter is sealed with an environmental sealant

such as MIL-S-8802 and painted with polyurethane paint and a compatible primer.

Life-Limited Repairs

Life-limited repairs can be completed in 1 or 2 h if preplanning and heat curing are accomplished. These repairs must be inspected periodically for delamination, corrosion, or other bond degradation. Evidence of such degradation requires replacing the repair, replacing the assembly (time permitting), or accomplishing a permanent autoclave-type repair. Maintaining the environmental seal is the secret of a long-lived interim or life-limited repair. This seal should be replaced if it deteriorates because the repair is sensitive to moisture. Life-limited repairs such as those described in this section should be confined to single skin areas of honeycomb panels, and the location of the repair should not interfere with sliding, moving, or stationary adjacent surfaces.

To accomplish this repair, several thicknesses of perforated 2024-T3 sheet coated with corrosion-inhibiting adhesive primer

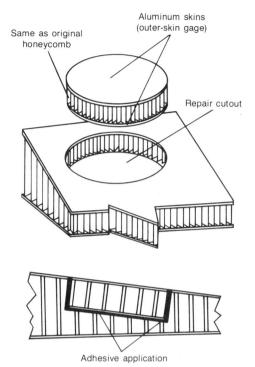

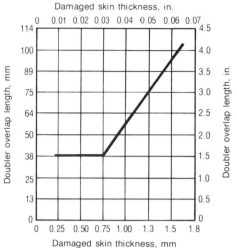

Fig. 25 Overlap distance for aluminum doublers. When overlap exceeds 65 mm (2.5 in.), perforated doubler or mating face sheet is used.

Fig. 24 Aerodynamically flush honeycomb core-aluminum skin plug repair

should be available. Repair doublers are perforated with 3.2 mm (⅛ in.) holes on 50 mm (2 in.) centers and contain a minimum corner radius of 13 mm (½ in.). The preprimed doublers are packaged so that only a solvent wipe is required before bonding. An original thickness of 0.30 to 0.50 mm (0.012 to 0.020 in.) requires a 0.65 mm (0.025 in.) doubler, a thickness of 0.55 to 0.10 mm (0.021 to 0.040 in.) requires a 1.3 mm (0.050 in.) doubler, and a thickness of 1.0 to 1.6 mm (0.040 to 0.063 in.) requires a 2.0 mm (0.080 in.) doubler. Double overlap distances as a function of damaged skin thickness are shown in Fig. 25.

First, the damage should be evaluated and the damaged skin thickness measured. The proper preprimed, perforated repair

doubler should be selected and cut to size. Assuming that only one skin and the honeycomb are damaged, the repair steps are (Fig. 26):

1. Stop drill the skin cracks or tears and, within the perimeter of the repair doubler, perforate the dented skins with a pattern similar to that on the repair doubler. Be careful not to penetrate the undamaged skin
2. Mark the location of the repair doubler on the damaged surface. Drill holes through the doubler and skin on 38 mm (1½ in.) spacings, 13 mm (½ in.) inside the outer perimeter, to fit repair rivets. Remove any protective finish in the doubler overlap area on the damaged surface
3. Vacuum any residue from the damage; clean the area with abrasive and solvent wipe
4. Solvent wipe the repair doubler with clean solvent and a purified cheesecloth
5. Mix the repair adhesive and force or inject the mixed material into the cracks and holes. A lightweight fast-curing ep-

oxy is recommended. Fill the core and level the adhesive to the surface, coat the repair doubler, and place it in position on the damaged surface
6. Blind rivet the doubler to the damaged surface within the pot life of the adhesive
7. Apply a vacuum bag over the repair, using a parting film to contact the adhesive, and heat cure the patch. Maintain a repair cure temperature at least 30 °C (50 °F) below the original cure temperature of the panel
8. After the cure, remove the vacuum bag materials and determine that the repair contains no delamination. Cover with metallized polyester tape for an environmental seal

For damage to both skins, repair is from the outside surface (Fig. 27). If the inside skin is cracked but not punctured, it should be stop drilled and an aluminum doubler bonded over the stop drill hole. If the skin is punctured, the damage (minimum size) should be cut out and the skin thickness measured. A nonperforated doubler should be prepared to fit inside the inner skin. After deburring all edges, the doubler and skin are drilled for rivet installation, as described previously. Surfaces are cleaned for bonding, and the adhesive is mixed and applied to the inner skin cutout and the doubler area. Rivets are applied within the adhesive pot life. The doubler inside skin is cleaned immediately, and the single-skin repair described previously can proceed. If honeycomb of the proper density and thickness is available, it should be used in place of the potting adhesive to reduce the weight of the repair. The ribbon of the replacement core plug should be oriented in the same direction as the original core.

Partial Repairs

Partial repairs are accomplished using a heating blanket and vacuum bag technique and a 120 or 175 °C (250 or 350 °F) curing structural film adhesive that has been tested to ensure that it will retain its strength and density after a vacuum pressure/heat exposure. The maximum allowable amount of applied vacuum during cure requires investigation to ensure that no adhesive film constituents are removed under heating conditions.

The optimal heat-up rate is another parameter that must be determined in order to minimize adhesive porosity. Because portions of the vacuum pressure heat-cured bond have a tendency to become slightly porous, proper sealing and seal maintenance are required for long-lasting repairs. For some adhesive systems, the use of partial, rather than full, vacuum pressure results in less bond porosity. Supplier recommendations should be consulted. All perforations in the repair doubler, all adhesive flash, all edges exposed by machining or

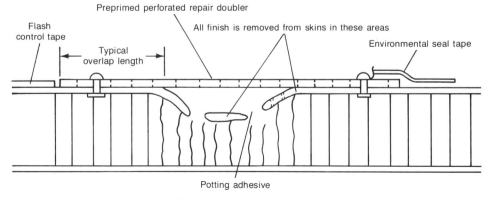

Fig. 26 Fast heat-curing interim (life-limited) repair to aluminum honeycomb structures. Damage limited to outside skin and honeycomb

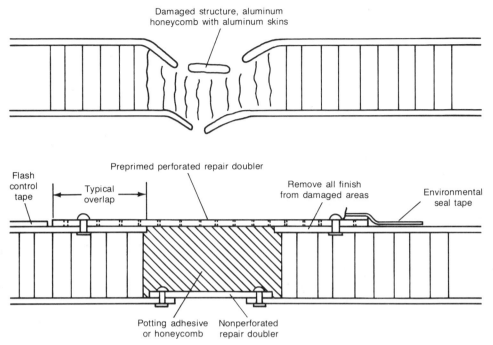

Fig. 27 Fast heat-curing interim (life-limited) repair to aluminum honeycomb structure with damage to both skins. Repair accomplished from outside

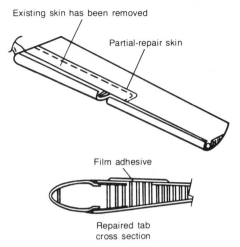

Fig. 28 Honeycomb tab. Typical partial repair configuration

trimming, and any fasteners through the bond should be sealed with polysulfide sealant and/or fluid-resisting primer and polyurethane paint. A typical partial repair configuration is shown in Fig. 28.

Two kinds of perforation patterns are used for vacuum cure repairs, the choice depending on the size of the repair. Areas of metal-to-metal overlap for repairs up to 0.016 m² (25 in.²) require perforating the inner-skin surface to allow air to escape during the bonding cycle. Holes 3.2 mm (⅛ in.) in diameter on 38 mm (1.5 in.) centers are drilled in the overlap area. At least two staggered rows of holes are used (see Fig. 25 for the proper doubler overlap distance as a function of thickness). All holes should be a minimum of 13 mm (½ in.) from the edge of the overlap. Partial repairs greater than 0.016 m² ((25 in.²) should be manufactured from perforated metal. Otherwise, the edges of the repair tend to pinch off when heat and vacuum are applied, and the repair becomes oval shaped, resulting in poor bond strength in the center of the repair. A perforation pattern of 3.2 mm (⅛ in.) holes on 25 mm (1 in.) straight centers, 0.002/mm² (1.5/in.²) is adequate and can be purchased commercially. As mentioned previously, different-size patches can be premanufactured by processing with a nonsilicated alkaline cleaner, sodium dichromate/sulfuric acid etch, phosphoric acid anodizing, and then, for 120 °C (250 °F) curing adhesives, a corrosion-inhibiting adhesive primer.

Silicone heating blankets are fairly effective as a means for curing film adhesive. However, heat can also be obtained from cartridge heaters encased in a form-fitting aluminum block, radiant heat from heat lamps, a hot-air blower, gas heater, and so on, if an autoclave is unavailable. A temperature controller and thermocouples should be used to ensure that the structure does not overheat and that the requirements for heat-up rate and maximum cure temperature are maintained.

An important requirement for vacuum pressure and heat-cured repairs is that all parts must be prefitted and formed to ensure that the surfaces have good contact over the entire bonded area. All surfaces and edges must be free of burrs and other surface imperfections. This allows adhesive to flow out of each perforation during cure, which indicates the presence of the proper pressure on the repair. At present, vacuum-heat repairs are limited to a maximum of 0.18 m² (280 in.²). A silicone heating blanket with a power density of 7.8 kW/m² (5 W/in.²) is used. If a vacuum pump is unavailable, a vacuum aspirator that uses shop air pressure (Fig. 29) can be commercially purchased.

One way to accomplish a vacuum bag-heating blanket repair is to use a blanket that is its own vacuum bag. Such blankets are expensive and are usually limited to use on smaller repairs. The steps for this type of repair are given below, but other styles of heating blankets and assembled vacuum bags (such as those shown in Fig. 14 and 17) can be used.

1. Assess the damage and clean the area as described previously. A typical puncture is shown in Fig. 30. Figure 31 shows the same panel after cleanup

2. Surface preparation on the damaged structure could consist of abrasion and solvent cleaning, followed by acid paste etching, being careful not to trap acid during application or rinse. Aluminum foil tape is usually adequate for this purpose. The repair patch can be cleaned similarly or can be given an environmentally durable treatment, described previously. Any finish in the damaged area is removed prior to final cleaning

3. Potting is applied and smoothed out prior to adhesive application. A heat-setting expansion-type material that cures at the same temperature as the film adhesive can be used (see Fig. 32)

4. Adhesive is applied to the repair patch (Fig. 33), and the patch is secured in position with metal bond flash control tape (Fig. 34). This type of tape strips clean (without a residue) from the metal surface after cure. Because vacuum pressure is insufficient to form thicker skin contours, any joggle or overlapping repair must be preformed, and a layer of adhesive approximately equal to the thickness of the existing skin must be placed in the joggle location. This prevents void formation after adhesive curing with resulting "oil-canning" in service

5. The vacuum heating blanket and instrumentation needed to pressurize and cure the repair are shown in Fig. 29. The instrumentation consists of a temperature indicator and controller, thermocouple connections, isolation transformer with ground fault indicator, vacuum aspirator, and vacuum gage with associated hoses. The blanket itself contains a built-in thermocouple, a composite rubber-metal caul plate (to prevent mark-off), and a bleed pattern, which eliminates the need for a separate bleeder cloth. The thermocouple is located adjacent to the repair area and is connected

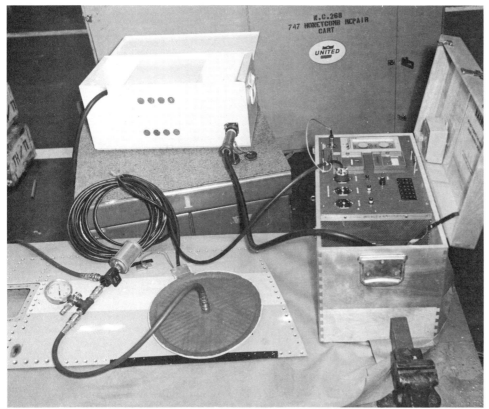

Fig. 29 Vacuum heating blanket and instrumentation for curing the repair. Courtesy of United Airlines

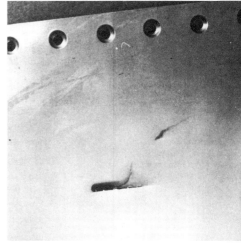

Fig. 30 Punctured honeycomb surface. Courtesy of United Airlines

to a jack plug, as shown. A piece of release film is placed over the repair. The vacuum aspirator is connected to a 585 kPa (85 psi) shop air line, and a vacuum hose is connected to the heating blanket. The control thermocouple within the blanket is located adjacent to the caul plate, and the heating blanket is placed over the repair area. Hand pressure on the edge seal is required to obtain a vacuum. Independent thermocouples are used to determine the exact temperature of the repair, because the control thermocouple is located near the vacuum bag heaters and may be affected by them. The blanket is connected to the electrical supply box, which is then attached to the ground fault indicator and isolation transformer. The isolation transformer is then connected to line voltage

6. Because the blanket is under vacuum, the next step is to raise the repair temperature incrementally to the curing range of the adhesive. The temperature controller is used for this purpose, and increments usually do not exceed 1.7 °C/min (3 °F/min). The repair is then held at cure temperature for the required time

7. Upon cure completion, power is disconnected and vacuum is held on the blanket until the temperature falls below 50 °C (125 °F). Vacuum is then disconnected, the release film is re-

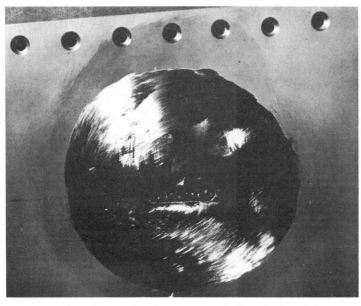

Fig. 31 Same honeycomb surface as in Fig. 30, cleaned and ready for repair. Courtesy of United Airlines

Fig. 32 Potting/stabilizing the honeycomb with a foaming adhesive. Courtesy of United Airlines

Fig. 33 Film adhesive applied to the repair doubler. Courtesy of United Airlines

moved from the repair, and any excess resin is peeled from the silicone blanket. The completed repair is shown in Fig. 35

As mentioned previously, a separate, flat heating blanket (commercially available) placed inside a vacuum bag (Fig. 14) can be used in place of an integral vacuum bag-heating blanket. Placement depends on the surface contour and desired heating rate. Because more heat must be used to penetrate the vacuum bag and vacuum insulating layer with the blanket outside the bag, heat-up rates are slower. However, it is not good practice to force the heating blanket against external fittings or over tight radii using vacuum pressure. This will damage the blanket and may cause an electrical malfunction.

Autoclave Repair Techniques

The autoclave is used for repair purposes when the amount of damage to an adhesive-bonded part is extensive enough to require partial or complete rebuilding. After years of flight service, many bonded structures contain a significant amount of corrosion, and partial repairs are no longer justified. The autoclave, with a supporting tool backing up the structure to be repaired, is used to provide the pressure-temperature environment necessary for adhesive curing. Alternatively, mechanical damage of the aircraft during production necessitates a permanent repair (Ref 6). If a honeycomb or metal bonded structure can be removed from the aircraft, it is processed using an autoclave, as described below. If autoclave support tools are unavailable, evacuated plastic bags filled with dry sand are sometimes used. Small particles or balls (shot) are also used for the same purpose when placed within a frame. If repairs must be accomplished on the aircraft, pressure is sometimes applied by means of removable fasteners secured through holes in perforated doublers and the original structure. Springs are used to exert curing pressure. (See Ref 6 for further details.) Autoclave cure techniques, which are described in Ref 7, will not be discussed here except to show how they affect the repair process. Similarly, there will be no discussion of surface treatments prior to bonding. See the Section "Surface Considerations" in this Volume for detailed descriptions.

The repair of adhesive-bonded structures is accomplished in a sequence of basic steps:

1. Prepare the aluminum surface
2. Prime the skins, spar, end ribs, and other metal details, excluding honeycomb
3. Heat bake the primer to drive off solvents and, in some cases, cure the film
4. Verify the correct primer thickness
5. Apply film adhesive to skin details, locating spar, end ribs, and honeycomb as required; apply core splice and spar bond adhesives. Assemble test specimens, if required
6. Vacuum bag the assemblies
7. Autoclave cure the assemblies
8. Repeat steps 5 through 7 if more than one cure cycle is needed, shaving honeycomb and adding more details as required
9. Accomplish quality control inspection of cured structures and test specimens
10. Seal and paint assembly

Surface preparation varies, depending on the need of the service. The optimal treatment is applied to details requiring full skin, partial repairs, and rebuilt structures. This treatment consists of hot-alkaline cleaning, hot-acid etching, and phosphoric acid anodizing, followed by the application of corrosion-inhibiting primer. Partial skin, spar, rib, and honeycomb repairs for which limited life is anticipated need only be alkaline cleaned and acid etched. Priming with a corrosion-inhibiting material is optional. Details that cannot be immersed in cleaning and etching tanks are etched with acid paste or anodized with nontank phosphoric acid. If the old adhesive is securely bonded, with no evidence of corrosion or other deterioration, sanding and feathering to the surrounding area is required. The adhesive is then solvent wiped until clean. Aluminum honeycomb core is liquid/vapor degreased or washed in clean solvent.

Fig. 34 Repair doubler secured in position with flash control tape. Courtesy of United Airlines

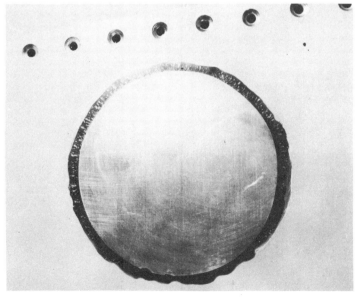

Fig. 35 Completed repair, ready for final priming and painting. Courtesy of United Airlines

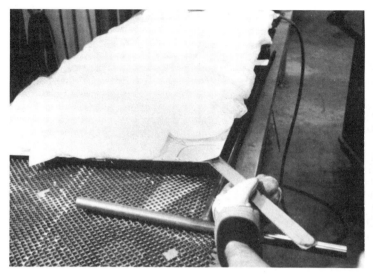

Fig. 36 Removal of the upper skin. Courtesy of United Airlines

Fig. 37 Upper skin removal with "can opener." Courtesy of United Airlines

Other pertinent observations that relate to adhesive application and cure as they apply to repair processes are:

- Avoid the buildup of several layers of foaming film adhesives (core splice, spar bond) in areas where their heat of cure will not be distributed by metal details. The expanding material can also crush honeycomb if the fit is too tight
- Autoclave cure pressure should, in part, be determined by the condition of the old core when accomplishing partial repairs. Lower the cure pressure if the honeycomb is deteriorated
- Particularly when curing 120 °C (250 °F) adhesive-bonded assemblies with more than one cure cycle, the length of the first cycle may be reduced to 30 to 35 min because a later heat cycle will fully cure the assembly

The steps required to rebuild a typical aircraft spoiler are defined below and are based on the philosophy of repairing parts using the latest environmentally durable processes for adhesive-bonded surfaces. The goal is to salvage as much of the old structure as possible. This necessitates the selection of a point in the service life of the part at which the spar, end ribs, and fittings can be saved. Beyond this point in time, corrosion will dictate the purchase of new structures. Experience determines the optimal repair condition without sacrifice of service life. To rebuild a spoiler:

1. Repair begins by indicating the damaged or delaminated areas and deciding whether a partial repair or total rebuilding is needed
2. Dry ice is used to help peel off the upper skin. A chisel starts the operation (Fig. 36)
3. A "can opener" and some hand labor complete the skin peeling process (Fig. 37)
4. Areas of clad dissolution and subsequent disbond become evident (Fig. 38)

5. A new upper skin is then outlined and sheared
6. The old, corroded structure is cut away
7. After an immersion stripping process that removes finishes, sealant, adhesive, and adhesive foam, the spar and end ribs, together with skins and doublers, are racked and immersed in hot alkaline cleaning solution
8. After spray rinsing, the rack is immersed in a hot sodium dichromate/sulfuric acid etch solution
9. The rack is then spray rinsed in demineralized water after exposure to hot acid
10. Parts and rack are then immersed in a phosphoric acid anodizing solution
11. Spray rinsing follows the anodizing operation
12. The rack is then air dried
13. Any 120 °C (250 °F) curing adhesive requires the application of corrosion-inhibiting adhesive primer to the freshly anodized surface; other systems may

Fig. 38 Alclad dissolution and disbond. Courtesy of United Airlines

Fig. 39 Assembly of spar to skin with film adhesive. Courtesy of United Airlines

Fig. 40 End rib and clip replacement. Courtesy of United Airlines

Fig. 41 Placement of corrosion-inhibiting honeycomb. Courtesy of United Airlines

require a compatible primer. These coatings protect the anodized surface from damage

14. The corrosion-inhibiting primer is then cured. Other primers may require only partial cure and solvent removal
15. The part details are disassembled from the handling rack, and adhesive lay-up begins with the spar-to-upper skin assembly (Fig. 39). The adhesive has already been inspected
16. End ribs and clips are added (Fig. 40)
17. Corrosion-inhibiting honeycomb is located inside the rib-skin-spar assembly and trimmed to fit (Fig. 41)
18. Release film is placed over the uncured assembly and test specimens (Fig. 42). Bleeder cloth is added, and a vacuum bag is superimposed on the subassembly
19. The vacuum-bagged spoiler assembly is then readied for the first-stage autoclave cure
20. After the adhesive cure cycle, the vacuum bag and release materials are removed and the honeycomb is contoured to final shape
21. The lower honeycomb surface skins and doublers are covered with adhesive and positioned over the contoured honeycomb
22. Doublers are fixed in position with flash control tape
23. Release film is placed over the assembly, bleeder cloth is added, and another vacuum bag is fabricated
24. The surface undergoes a second autoclave cure
25. The vacuum bag is removed, and some fittings and seals are added. (The final assembly is shown in Fig. 43)
26. The spoiler is covered with a polyurethane filiform corrosion-resisting primer and paint system

Similar processes apply to other aircraft parts, such as flaps, aileron tabs, rudder tabs, and flap vanes, in fact, any bonded assembly.

REFERENCES

1. J.A. Marceau and Y. Moji, A Wedge Test for Evaluating Adhesive Bond Surface Durability, *Adhes. Age*, Oct 1977
2. J.A. Marceau and J.C. McMillan, Exploratory Development on Durability of Adhesive Bonded Joints, AFML-TR-76-172, U.S. Air Force Materials Laboratory, Oct 1976
3. J.C. McMillan, J.T. Quinlivan, and R.A. Davis, Phosphoric Acid Anodizing Aluminum for Structural Adhesive Bonding, *SAMPE Q.*, April 1976
4. "Primary Adhesively Bonded Structures Technology (PABST)," Technical Bulletin 2, U.S. Air Force Contract F 33615-75-C-3016, May 1975
5. M.C. Locke, R.E. Horton, and J.E. McCarty, Anodize Optimization and Adhesive Evaluations for Repair Applications, AFML-TR-78-104, U.S. Air Force Materials Laboratory, July 1978

Fig. 42 Release film placement during vacuum bagging. Courtesy of United Airlines

Fig. 43 Final assembly. Courtesy of United Airlines

6. R.E. Horton, W.M. Scardino, and H. Croop, "Adhesive Bonded Aerospace Structures Standardized Repair Handbook," Contract AF 33615-73-C-5171, U.S. Air Force Flight Dynamics Laboratory, 1973

7. E.W. Thrall and R.W. Shannon, Ed., *Adhesive Bonding of Aluminum Alloys*, Marcel Dekker, 1985, p 21-74

8. *Metallic Materials and Elements for Aerospace Vehicle Structures*, MIL-HDBK-5, *Military Standardization Handbook*, U.S. Department of Defense, May 1989

9. "Aluminum and Aluminum Alloy Plate and Sheet; General Specification for," QQ-A-250

10. "Cloth, Glass, Finished, for Resin Laminates," MIL-C-9084, Military Specification

11. *Polymer Matrix Composites*, Vol 1, MIL-HDBK-17, *Military Standardization Handbook*, U.S. Department of Defense, June 1989

12. "Core Material, Aluminum, for Sandwich Construction," MIL-C-7438, Military Specification

13. "Core Material, Plastic Honeycomb, Laminated Glass Fabric Base, for Aircraft Structural and Electronic Applications," MIL-C-8073, Military Specification

14. "Core, Honeycomb, Fibrous Aramid Base, Phenolic Coated," AMS 3711, Aerospace Material Specification

15. "Adhesive, Film Form, Metallic Structural Sandwich Construction," MIL-A-25463, Military Specification

16. "Adhesives, Heat Resistant, Airframe Structural, Metal-to-Metal," MMM-A-132, Military Specification

17. "Sealing Compound, Temperature-Resistant, Integral Fuel Tanks and Fuel Cell Cavities, High Adhesion," MIL-S-8802, Military Specification

18. "Sealing and Coating Compound, Corrosion Inhibitive," MIL-S-81733, Military Specification

19. "Sealing Compound, Topcoat, Fuel Tank, Buna-N," MIL-S-4383, Military Specification

Repair Concepts for Advanced Composite Structures*

S.H. Myhre, Northrop Corporation
C.E. Beck, Air Force Flight Dynamics Laboratory

THE REPAIRABILITY of advanced composite structures has improved considerably since the late 1970s. Although new techniques will continually be devised for special situations, the basic issue has been resolved: Graphite-epoxy is a repairable structural material.

Two aspects of repairability, high-strength restoration and low cost, are depicted in Fig. 1. Of course, aircraft downtime and performance degradation should be evaluated and included in cost considerations. Strength restoration of 80% of the tension ultimate allowable (F_{tu}) is generally considered sufficient to cover all but the most unusual cases. The compression ultimate allowable (F_{cu}) may be higher than tension but is rarely used as the critical design allowable except where F_{cu} is degraded, as in the case of the hot-wet environment.

Early efforts to repair advanced composite materials generally resorted to an external-patch concept. Alternate layers of titanium foil and film adhesive have been used to make a concentric, stepped patch (Ref 1). Such a repair may be adequate in particular cases, but it suffers from the high shear and peel stresses at the edge of the hole. The external protrusion increases, and the joint efficiency decreases, as the thickness of the laminate and the area of the repair become larger.

By contrast, the scarf joint originally seemed particularly well adapted to advanced composite materials, as it was to wood, the original composite material. (A scarf joint is made by cutting away segments of two adherends at similar angles and bonding the adherends with the cut areas fitted together.) However, attempts to use this repair joint did not meet with outstanding success. A boron-epoxy 27-ply multidirectional laminate joined to an equal thickness (3.5 mm, or 0.135 in.) of 6Al-4V-Ti alloy through a 3° scarf with MB 329 adhesive developed an ultimate load of about 54% of the laminate allowable strength (Ref 2). A repair joint that develops 80% or more of the laminate allowable is better suited for the permanent repair of composite primary aircraft components.

Although these meager results and other attempts were not encouraging, the scarf joint with a co-cured patch that was essentially flush with the outer mold line looked promising if it could be understood and optimized. The U.S. Air Force funded the Large Area Composite Structure Repair (LACOSR) program to answer in a systematic way the question of what constitutes a safe, reliable, and operationally suitable repair for large-area damage to composite structures (Ref 3, 4). "Large area" was defined as an area that is essentially unlimited with respect to the applicability of the repair techniques in question. This program has evaluated the environmental effects of oil, fuel, paint stripper, and absorbed moisture; the service conditions of tension and compression loads, prior fatigue history, and service temperature; processing variables such as vacuum bag curing versus autoclave curing, shop machining versus portable-equipment dressing of the damaged area, and precured versus co-cured adherends; and, finally, a group of inherent features, including the laminate material thickness, stacking order, structural form, accessibility, and protrusion limitations.

The work described in this article has demonstrated that graphite-epoxy structures designed in a variety of forms for a wide range of load intensities can be repaired effectively to their original strength. Repair concepts have been validated through the required environmental and load condition tests. Repairs in flat panels have been successfully demonstrated. Procedures used have proved to be not only structurally sound, but also relatively easy to implement.

Basic Considerations

Closed Options. Those inherent features of a repair that represent closed options to a repair team and that were determined somewhat arbitrarily for the LACOSR program should be considered first. All parent laminates discussed here were made from AS/3501-5 graphite-epoxy tape. A 16-ply graphite laminate $[(0/\pm45/90)_2]_s$ was chosen as being representative of a large class of laminates that carry both normal loads and shear. The number of plies was doubled or reduced by half in some cases, and the stacking order was later changed so that this generic laminate was considered from many aspects. The ultimate tension and compression fiber-failure allowables are given in Table 1. Minimum damage size is defined as a 100 mm (4 in.) diam hole that completely penetrates the panel.

Two structural forms were considered: a flat laminate panel and a flat honeycomb sandwich panel. Eight-ply laminates $(\pm45/0/90)_s$ were used on the faces so that the sandwich represented a relatively lightly loaded structure. Although curved panels were not repaired, the need to address the problems of curvature was not ignored. The approach described here is applicable to curved panels and shapes.

The operational environment and load conditions imposed on the repair joint concepts included temperatures between −55 and 130 °C (−65 and 265 °F), as well as tension and compression, static and postfatigue, and wet and dry conditions. The term dry refers to the repair assembly. All of the parent laminates were soaked in 80 °C (180 °F) deionized water until the laminate had absorbed 1% moisture by weight. "Wet" repairs were subsequently moisture condi-

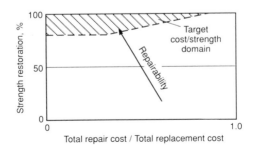

Fig. 1 Repairability

*Reprinted in edited form with permission from *Journal of Aircraft*, Volume 16 (No. 10), October 1979, p 720

Table 1   AS 3501-5 graphite-epoxy 16-ply parent laminate allowables

| Temperature(a) | | Tension allowable | | | Compression allowable | | |
| °C | °F | Loading | | | Loading | | |
		kN/m	10³ lbf/in.	Strain, %	kN/m	10³ lbf/in.	Strain, %
−55	−65	1085	6.21	· · ·	1705	9.74	· · ·
RT	RT	1200	6.86	1.1	1330	7.60	1.2
130	265	1006	5.75	· · ·	825	4.71	· · ·

Note: Cured at 690 kPa (100 psi) and 175 °C (350 °F) for 2 h; laminate stacking [(±45/0/90)₂]ₛ; water content, 1.0%; B basis tension allowables from Ref 5, 90% of average of 20 tests; compression allowables, 80% of average of 3 tests. (a) RT, room temperature

tioned at 80 °C (180 °F) and 95% relative humidity for 30 days.

The postfatigue test is a determination of residual strength after a two-lifetime history of fatigue loading using the load spectrum developed for the F-5E lower inboard wing skin. The exceedance curve (Fig. 2) represents 1000 h of flying with compressive loads deleted. Three specimens of a given repair were run with the maximum fatigue load at 41.5% of parent laminate allowable, and another three at a maximum fatigue load of 22.9% of parent allowable. The residual strength reported here is the minimum of these two results.

Design Options. With the repair job thus defined, the available options could be considered, such as whether to bolt or bond, the adhesive and patch materials to use, the type of patch and its arrangement, and curing details. Based on the proposition that graphite-epoxy is inherently a bonded structure and the bonding to it would not be difficult, the first decision was to bond.

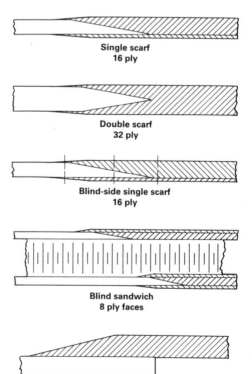

Single scarf
16 ply

Double scarf
32 ply

Blind-side single scarf
16 ply

Blind sandwich
8 ply faces

External patch
16 ply

Fig. 3 Repair joint concepts

The patch material chosen was AS/3501-6 graphite-epoxy cured at 105 kPa (15 psi) vacuum bag pressure and 175 °C (350 °F) (heated at 1.7 °C, or 3 °F/min) for 2 h and cooled to 65 °C (150 °F) before the release of vacuum pressure. The tensile strength of the patch laminate was about 84% of the parent allowable because the patch was vacuum cured while the parent laminate was cured at a higher pressure in an autoclave. The patch ply stack-up arrangement was generally the same as the parent laminate, with the addition of a few plies as necessary in the primary loading directions. Co-cured patches were used to eliminate close-tolerance fit problems in the scarf joint.

The adhesive FM 400 (0.34 kg/m², or 0.07 psf film) was selected to meet the 130 °C (265 °F) service temperature requirement. Rail shear tests on a seven-ply laminate of the adhesive cured at 105 kPa (15 psi) and 175 °C (350 °F) for 1 h resulted in room-temperature shear strengths of 45.8 MPa (6.64 ksi) dry and 29.2 MPa (4.24 ksi) wet (immersed in 80 °C, or 180 °F, water for 7 days). Meaningful shear properties are most difficult to obtain. The cure conditions, adhesive thickness, moisture content, and test methods all influence the results. The properties and the strength values to be used can also be dictated by the degree of sophistication of the analysis to which the data are applied.

Repair Concepts

The five repair joint concepts shown in Fig. 3 were developed for generic laminate repair. Each of these will be discussed in detail.

Bonded-Scarf Joint Flush Repair. A co-cured scarf joint of graphite-epoxy composite material capable of restoring nearly the full parent allowable strength was developed. Several preliminary tests were made to verify specific points indicated by the analysis, as described in the Appendix of this article.

The first point concerned the scarf taper ratio: length to thickness. Increasing the taper ratio improves the strength but not in proportion to the increase in length. A taper of 2.5 mm (0.10 in.) per ply (18/1) was selected, although this was much lower than had been commonly thought necessary for wood or advanced composites. Joining the 16-ply generic parent laminate

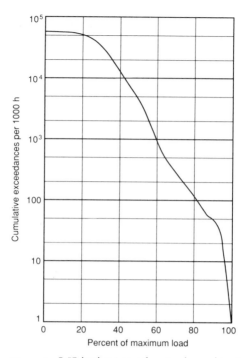

Fig. 2 F-5E load spectrum for wing lower skin at buttock plane, 0.75 m, or 29.5 in.

to a 16-ply patch with the same [(0/±45/90)₂]ₛ stacking order as shown in Fig. 4 resulted in shear failure in the joint at 51% joint efficiency.

The second point concerned the ends of the longest 0° patch ply. The high shear stress and peeling action known to exist at the end of the ply had to be resolved. Special ply termination tests were run on the 16-ply generic parent laminate with a 0° ply bonded to each surface on each end but terminating near the center of the specimen. Tension load was applied. The control specimen with the terminating plies cut off straight and perpendicular to the longitudinal axis failed at 77% of the parent allowable by the peeling away of the extra ply. Of the specimens with several kinds of serrations cut into the end of the terminating plies (Table 2), all failed nearly 100% of the parent allowable and failed in laminate tension. No peeling occurred on any of nine

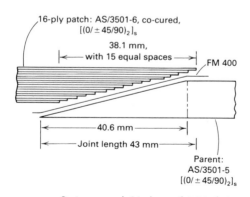

16-ply patch: AS/3501-6, co-cured, [(0/±45/90)₂]ₛ

38.1 mm with 15 equal spaces

FM 400

40.6 mm

Joint length 43 mm

Parent: AS/3501-5 [(0/±45/90)₂]ₛ

Fig. 4 Basic co-cured 16-ply scarf joint design. Tension joint efficiency was 51%.

Table 2 Results of ply termination tests

Ply end cut		Tension ultimate strength MPa	ksi	Mode of failure	Percent of parent allowable
	Control	36.3	5.26	Peel	77
5.1 mm (0.20 in.)	2.5 mm (0.10 in.)	45.7	6.63	Tension	97
10 mm (0.40 in.)	2.5 mm (0.10 in.)	48.7	7.06	Tension	103
3.2 mm (0.125 in.)	6.4 mm (0.25 in.)	47.6	6.90	Tension	100

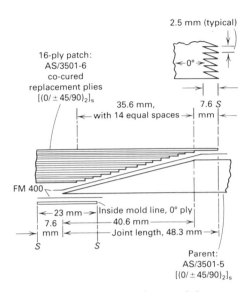

Fig. 5 Refined scarf joint design, including serrations. Tension joint efficiency was 70%. *S*, serrate end of ply

specimens having a total of 36 serrated ply ends. A pair of commercial pinking shears cuts a satisfactory serration 3.2 mm (⅛ in.) deep.

The third point concerned the end of the parent laminate. It should be noted that shop personnel must be instructed to scarf the laminate all the way to the inner surface. The edge may look a bit ragged, and they may be inclined to trim it off. However, if this inner-surface ply is a 0° ply (that is, aligned with the load) it should be scarfed more than the others.

In order for this inner-surface 0° ply to be unloaded through two surfaces rather than one, a small serrated patch ply was applied, as shown in Fig. 5. Also, the outer-surface patch ply was extended slightly and serrated, as shown. By adding this small amount of material and serrating the ply ends, joint efficiency was raised to 70% and the mode of failure became patch laminate tension.

Having resolved the details of the splice, the repair itself could now be considered. The generic parent laminate was assumed to be loaded in shear as well as in longitudinal tension and compression. Therefore, one ply was added in each of three directions (0, +45, and −45) on both sides of the laminate (Fig. 6). This joint developed 97% joint efficiency (in room-temperature, dry, tension conditions) and consistently failed in parent laminate tension. This is a very satisfactory repair concept.

An earlier version involved essentially the same splice, but the six added plies did not continue to the end of the specimen. These specimens failed in repair laminate tension with joint efficiencies of 83% at room-temperature, dry conditions, 80% at −55 °C (−65 °F), dry conditions, and 87% at room-temperature, dry conditions and with prior fatigue history. Tests on these latter conditions were not repeated with the improved design. Rather, the results obtained were scaled by the ratio of 97:83, giving an estimated joint efficiency of 93% at −55 °C (−65 °F) and 101% with prior fatigue history, as shown in Fig. 6. The same specimen was tested at 130 °C (265 °F), where failure was parent laminate tension at 110% of the 130 °C (265 °F) tension allowable.

The stacking order of the generic parent laminate was changed to $[(\pm45/0/90)_2]_s$ because this was thought to be more typical of evolving aircraft practice. The patch was changed accordingly to keep the replacement plies in the same order, but otherwise the repair remained essentially unchanged from the design of Fig. 6. The test results, completed with this stacking order, are shown in Fig. 6 for tension and compression, dry and wet, at −55 °C (−65 °F), room temperature, and 130 °C (265 °F). Apparently, the structural efficiency of the single-scarf joint is not significantly degraded by moisture, by cold or hot temperature conditions, or by prior fatigue history.

Double-Scarf Joint Flush Repair. The double-scarf joint permits a shorter joint length and the removal of less parent material than a single-scarf joint. Using the same 18:1 taper ratio, the design for a 32-ply generic laminate repair was patterned after the single-scarf joint. Three plies (±45/0) were added to both surfaces, and the two longest 0° plies on either side were serrated.

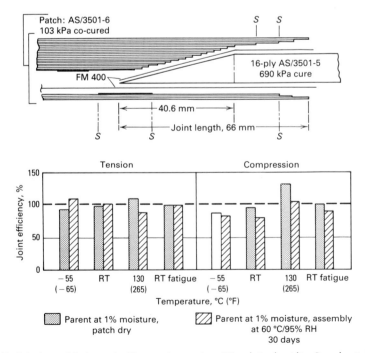

Fig. 6 Final single-scarf flush repair. RT, room temperature; RH, relative humidity. Bars denoted by RT fatigue represent specimens tested at room temperature after the fatigue history described in text.

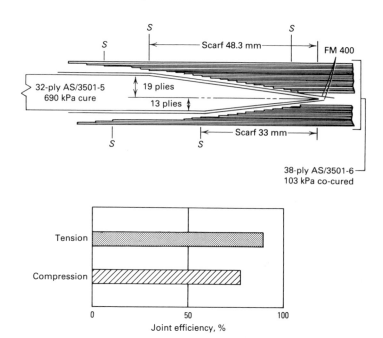

Fig. 7 Double-scarf flush repair

Tension and compression specimens and the results of room-temperature, dry tests are shown in Fig. 7. The results satisfactorily demonstrate the validity of this approach, although additional patch material appears to be required for full strength restoration.

Blind-Side Bonded-Scarf Repair. The difficulties involved in repairing a panel with access from one side only are obvious; the solutions are not. The substantial improvement (37%) achieved by even a small patch bonded to the inside surface (Fig. 5, compared to Fig. 4) could not be ignored. The design presented here evolved out of numerous considerations, not the least of which were how to locate the inside patch and how to effect a good bond.

A precured AS/3501-6 laminate $(\pm45/0)_s$ was chosen for the inside patch for the following reasons: no new bonding materials or logistics were involved; tapering the patch was made easy by cutting the plies to the proper size and serrating the edge where necessary; and forming the patch to a compound curvature, if necessary, could be done without expensive tools, using the outer surface of the part to make a mold. The laminate stacking order was symmetric and continued to be even as internal plies were terminated.

With this type of repair, locating the internal patch may require some ingenuity. With access limited to the external surface, the patch must be installed through the damage cutout. A long, narrow hole offers the easy solution of rotating the panel in place; however, bending the thin panel to insert the patch through a circular hole could be considered. The six-ply panel can be elastically bent to a 50 mm (2 in.) radius of curvature without damage. Once in place, the panel is held against the blind-side (inner) surface of the panel while holes are drilled through both pieces from the outside and Clecos (blind-hole clamps) are installed.

The splice detail in Fig. 8 shows the blind-side panel bonded to the inner surface. The drill template remains in place during this bonding cycle to provide support to the Clecos beyond the edge of the parent laminate and through the scarfed area. The holes must be filled after removal of the Clecos to prevent vacuum leaks during the secondary co-curing of the flush patch. The co-cured part of the repair, consisting of $(\pm45/0)$ plies plus replacement plies, is bonded in the second cure with vacuum pressure, while bagged on the outside surface only.

Results of the blind-side repair tests at various temperatures, moisture, and load conditions are given in Fig. 8. The lowest joint efficiency shown is 84%, and, in general, the results indicate that a blind-side repair can be made every bit as good as one with access to both sides if the blind-side patch can be adequately bonded using techniques like those demonstrated here.

Blind-Side Sandwich Repair. The approach to honeycomb sandwich repair taken here is a scarf joint on both inner and outer eight-ply faces (Fig. 9). The inner face is repaired first, following much the same procedure as that used for a blind-side monolithic laminate just discussed. The blind-side patch is a five-ply symmetric precured laminate that is Cleco bonded with FM 400 adhesive to the blind-side surface of the inner face sheet. The open-side patch consists of two parts: The replacement plies are co-cured against the scarf surface, and the extra open-side plies are a five-ply symmetric precured laminate. The entire patch is bonded in place with FM 400 adhesive.

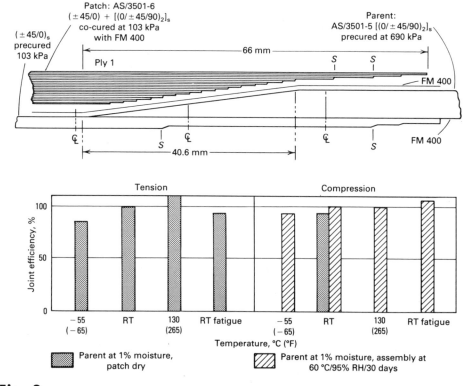

Fig. 8 Blind-side single-scarf flush repair. C$_L$, 2.49 mm (0.098 in.) diam Cleco hole

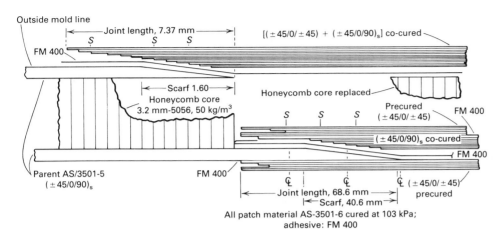

Table 3 External patch concepts

FM 400 bonded joints (width = 25 mm, or 1 in.)	Failure mode	Joint efficiency, %
Control	Peel and shear	0.52
One 3.2 mm (1/8 in.) rivet	Shear and patch net tension	0.73
One rivet and glass fabric	Shear and rivet head pull-through	0.78

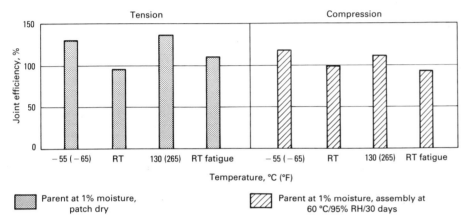

Fig. 9 Blind-side sandwich repair. C_L = 0.18 mm (0.070 in.) diam Cleco hole

Pressure for co-curing the patch is applied by installing the replacement core as a temporary plug and bagging over this core to make the vacuum seal against the outside surface. After this bond cycle, the core is removed for inspection of the inner face repair and then reinserted to bond the core to the inner skin and make the foam-bond core splice. The core is then sanded flush with the inside surface of the outer face. The patch for the outer face, consisting of replacement plies against the scarf and a five-ply supplement outside the original mold line, is entirely co-cured. Because no repair material can be placed against the inner surface of the outer face, the scarf in Fig. 9 was made at 5.1 mm (0.20 in.) per ply instead of using the slope of the single-scarf laminates. All 0° ply ends were serrated to alleviate the higher shear stresses in the bond line at these terminations.

The complete test results on the blind-side sandwich repairs are given in Fig. 9. The lower of the tension and compression allowables was used to determine joint efficiency because both faces are loaded equally in opposite senses. The results show essentially full allowable strength restoration in every case tested. Multiple failure modes were common, indicating that the repair was well matched to the parent laminate capability, where failures had generally occurred.

Bonded External Patch Repair. The no-scarf external-patch option offers some advantage of simplicity. The damaged area is removed, leaving a straight hole. A graphite patch is precured to a surface representing the outer mold line and then bonded to the damaged surface. Such a patch, being eccentric to the skin, must be designed for the bending induced by this eccentricity. The bending strains are surprisingly high. Additional plies required for the bending increase the eccentricity, however; the problem does converge.

The more important problem is the peeling and high shear stress on the bond line at the edge of the hole. This peeling problem was attacked empirically because it is not amenable to analysis. Six configurations were investigated, five of which involved some simple treatment at the edge of the hole. Three configurations and the test results are given in Table 3. Small blind rivets at 25 mm (1.0 in.) spacing made a significant improvement by resisting the peel force. The glass fabric and FM 400 adhesive filling the undercut made another improvement by relieving the shear-stress peak. Observing the specimen as the load is increased reveals no indication of resistance to shear by the rivets until after the bond has failed.

To stabilize the specimen for compression loading, the repaired laminate was assembled into a honeycomb sandwich beam. This specimen form was then used for all subsequent tests in tension as well as compression. The beam assembly prevented local rotation of the laminate and any subsequent peeling. This raised the joint efficiency from 78% to 93% for room-temperature, dry, tension conditions. Complete results of these four-point beam tests of the external-patch repair are shown in Fig. 10. With the change in specimen form, the mode and location of failure also changed from peeling at the edge of the parent laminate to a failure at the edge of the patch. Special treatment was also given at the edge of the patch by increasing the step length and serrating the end of the first three 0° plies. Having these three 0° plies on the surface now appears to be detrimental to the repair in this area. The compression failures consistently appeared at the end of the second longest ply of the patch. The effect of the joggling of the longest ply over the end of the second longest ply will be seen more dramatically in the larger panels discussed below.

Repairs of Panels With 100 mm (4 in.) Diam Holes. With the validation of several repair joint concepts by means of the uniaxial testing just discussed, the extension of these concepts to three-dimensional, real-world applications required substantiation. To some extent, this was accomplished for the scarf joint flush repair and the external-patch repair using the 16-ply 305 × 1220 mm (12 × 48 in.) panels of Fig. 11. Each panel was loaded as a four-point beam. Both repairs are direct applications of the joints previously evaluated.

Four panels with the scarf joint flush repair were constructed. Three were scarfed with preliminary router cuts and finished with a sanding disk. The fourth was prepared with a milling machine with the intent of determining the sensitivity of repair strength to the method of making the scarf. This panel and two

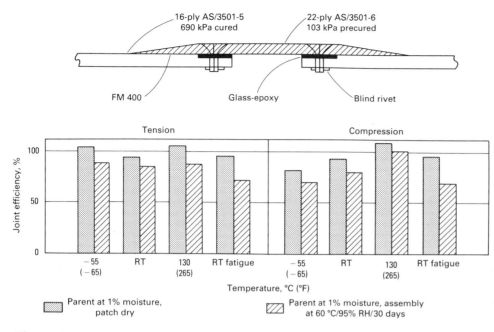

Fig. 10 External-patch joint

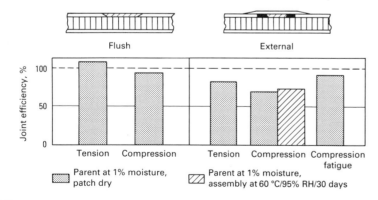

Fig. 11 Holed-panel repairs

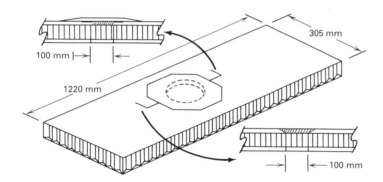

Fig. 12 Holed-panel results

Ten panels, each with a 100 mm (4 in.) diam hole, were repaired with an external patch designed for longitudinal loading only. The results shown in Fig. 12 are seen to be lower than the corresponding results on the scarf joint flush repair and are also considerably lower than the corresponding results on 25 mm (1 in.) wide beams. No sign of incipient failure was evident in any of the tension panels. However, those panels with the repair in compression showed considerable incipient failure beginning at 25% of the parent ultimate compression allowable.

Failure began where the surface ply (0°) joggled over the next longest ply. While this joggle was only a 0.14 mm (0.0055 in.) step, it nevertheless had a significant effect on the ability of the ply to resist failure. Once a local failure occurs, it grows and is joined by others as load is increased, until the entire patch is peeled off. Concern over fatigue loading appears to be justified where the maximum compression loads are high enough to cause such incipient failure. However, in this test spectrum compression fatigue loading to a maximum of 37% of the parent compression allowable resulted in no incipient failure. After fatigue loading, incipient failure began during residual-strength testing at 45% of the parent compression allowable. It appears that a stress-relieving mechanism takes place under fatigue loads, thereby suppressing the incipient failure.

Demonstration Panels

The remaining tasks of the LACOSR program included a series of tests on a partially damaged 50-ply graphite laminate and a 64-ply hybrid laminate. A box beam, large enough for 0.48 × 1.5 m (19 × 60 in.) uniformly loaded test panels, was used to test the large-scale demonstration components. Five different test panels were constructed, representing low- and high-load components, sandwich and plate construction, damage on aircraft and off, partial damage, and through damage in areas up to 0.06 m² (100 in.²). These five panels demonstrated failure strengths of 122 to 155% of the design ultimate load of the virgin panel. A summary discussion of this concluding portion of the program may be found in Ref 6.

Absorbed moisture in the parent laminate has been found to be one of the most significant problems in the repair of advanced composite structures that are more than 30 plies thick. Bonded repairs, cured at elevated temperature, may result in a blistered laminate or deteriorated adhesive caused by the absorbed moisture. A technique has been developed (Ref 7) to evaluate the moisture content of a laminate in service. Reliable, high-strength bonded repairs can be achieved if proper moisture control and material selection are exercised.

others were loaded with the repair in tension, resulting in 109% joint efficiency. All failed in parent laminate tension. No improvement was gained by the milling machine scarfing. The one panel tested with the scarf repair in compression achieved 92% of the parent laminate ultimate compression allowable, failing in patch compression and peel. No indication of early incipient failure was evident in any of the scarf joint flush repairs. In Fig. 12 these results are compared with results from the panels repaired with an external patch.

Appendix: Load-Transfer Analysis

An elementary analysis based on the assumption that the load on each adherend is proportional to the relative adherend stiffness is found to be entirely satisfactory when applied at each ply of the adherends. Although this procedure does not account for shear lag in the adhesive, it does account for the heterogeneous nature of the laminate and gives a satisfactory description of the shear distribution along the entire splice.

The procedure is shown in Fig. 13 for a flush scarf joint of identical 16-ply $[(0/\pm45/90)_2]_s$ laminate adherends. The stiffness distribution on both laminates is given on the small bar chart at upper left. The values are a function of the laminate ply distribution and the lamina elastic properties. For the laminate shown where the load is aligned with the 0° plies, 65% of the load is carried by the 0° plies, 15% by each of the ±45° plies, and 5% by the 90° plies. By assigning

the value of unity to the thickness per ply, the relative tensile stiffness (Et) values of each adherend are determined by accumulating the relative E values at each station as thickness is increased by one ply per station increment. Knowing the (relative) Et of each adherend, the total Et is found, and the relative load on one adherend is determined from the assumption that the load on each adherend is proportional to the adherend Et.

The difference in the load fraction at two adjacent stations is the average relative shear (RS) loading for the increment:

$$(RS)_{i,i-1} = \left(\frac{L}{T}\right)_i - \left(\frac{L}{T}\right)_{i-1}$$

where L_i = left Et of station i and T_i = the total Et of station i. The relative shear diagram (Fig. 13) gives some insight into the load-transfer mechanics. No claim of preci-

sion is made because shear lag is neglected, but the general nonuniformity is rationally determined. The higher shears exist on those steps where the 0° ply of the thinner adherend is bonded.

The strength of the splice is inversely proportional to the maximum relative shear:

$$P_{ult} = F_{su} \left(\frac{\Delta x}{MRS}\right)$$

where F_{su} = adhesive ultimate shear strength; Δx = station increment, and MRS = maximum relative shear.

Adhesive shear strengths are uniquely related to the kind of test performed. The difficulty arises when applying such data to design. The joint analysis has roughly determined the shear distribution on the bond line using station increments corresponding to a thickness change of one ply. Shear strength determined by the rail shear test on FM 400 adhesive has been found to relate adequately to this analysis.

The significant design goal is to reduce the maximum relative shear. The intent of this simple analysis is to draw attention quickly to such problem areas, thereby making innovative remedial action possible.

The scarf joint in Fig. 13 is flush on both surfaces and is therefore 5.1 mm (0.2 in.) shorter than the length of the scarf. The effect of increasing the overlap of the two adherends is to reduce the maximum relative shear, as shown in Fig. 14.

Another approach to reducing the maximum relative shear is to add 0° plies to both surfaces over the splice and have them extend a short distance beyond, as shown in Fig. 15(a). Both added plies should terminate at the same station. The maximum relative shear of 0.1226 at the end of the added ply results from assuming, again, a 2.5 mm (0.1 in.) increment to transfer the load. The next largest relative shear is 0.1156 in Fig. 15(a). If the basic lap is increased to 43 mm (1.7 in.), together with the added plies, as in Fig. 15(b), the MRS = 0.1226 remains unchanged, but the next largest relative shear is reduced to 0.0958. To reduce the maximum value, the extended lap length is increased to 53 mm (2.1 in.) and the ends of the added plies are serrated, as shown in Fig. 15(c). Serrating can be done easily with pinking shears. With 2.5

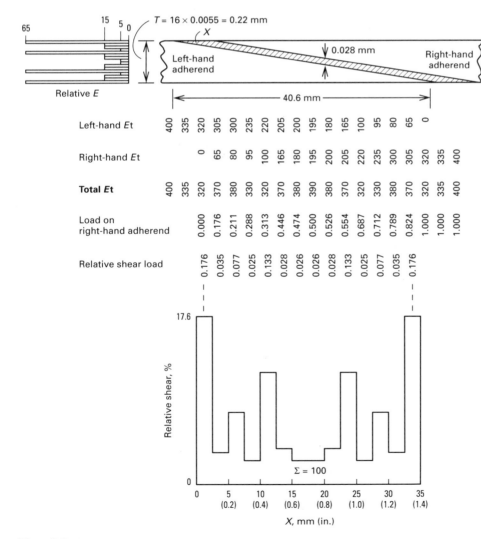

Fig. 13 Shear distribution analysis for scarf joint with identical $[(0/\pm45/90)_2]_s$ adherends

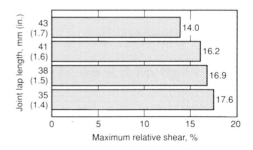

Fig. 14 Effect of lap length on MRS

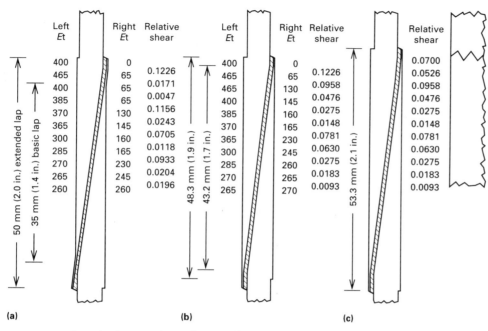

Fig. 15 Shear distribution analysis with added 0° plies. Joint (c) has an extended lap with serrated ply end (detail at right).

mm (0.1 in.) deep serrations and again assuming 2.5 mm (0.1 in.) to load any fiber, the relative shear at the longest ply end is reduced by nearly half. The maximum relative shear then becomes 0.0958, as seen in Fig. 15(c). The precise effect of the serration is beside the point here, if in fact the relative shear at the ply end is reduced to below the 0.0958 level, which now defines the shear strength of the splice. Tests have indeed shown that serrating the ends of the long plies substantially improves the strength of a splice.

The effect of hot-wet conditions should be considered carefully because several changes occur. The matrix-dominated elastic properties (E_2 and G_{12}) are greatly reduced at 130 °C (265 °F) wet conditions. The resulting relative E values for the (0/±45/90)

laminate at 130 °C (265 °F) wet are (73/13/13/1), respectively (more heterogeneous than for room-temperature, dry conditions). Shear analysis for the joint of Fig. 15(c) using these values results in a maximum relative shear of 0.1022. This 7% increase in the *MRS* weakens the joint. In addition, the thermochemical effects reduce the strength of the adhesive.

If a shear flow is to be transferred across the joint, the analysis is similar. The essential difference is that lamina shear stiffnesses rather than extensional stiffnesses must be used. A pair of plies at ±45 is about six times as stiff in shear as a 0/90° pair. The relative G values are arranged in the stacking order, and the Gt of each adherend is determined. The splice of Fig. 15(c) ana-

lyzed for shear flow resulted in a maximum relative shear in two areas. As before, this fraction of shear flow q is transferred in the increment length Δx. Hence:

$$q_{ult} = F_{su}\left(\frac{\Delta x}{MRS}\right)$$

ACKNOWLEDGMENT

The results presented are from work sponsored by the Air Force Flight Dynamics Laboratory under Contract F33615-76-3017, "Large Area Composite Structure Repair."

REFERENCES

1. "Composite Development Program—Graphite/Epoxy Repair Concepts," MDC AC3715, Vol 4, McDonnell Aircraft Company, March 1976
2. G. Lubin *et al.*, "Repair Technology for Boron/Epoxy Composites," AFML-TR-71-270, Grumman Aerospace Corporation, Feb 1972
3. A.L. Scow *et al.*, "Large Area Composite Structure Repair," First Interim Report, AFFDL-TR-77-5, Northrop Corporation, Jan 1977
4. R.W. Kiger and S.H. Myhre, "Large Area Composite Structure Repair," Second Interim Report, AFFDL-TR-77-121, Northrop Corporation, Nov 1977
5. R.M. Verette and J.D. Labor, "Structural Criteria for Advanced Composites," AFFDL-TR-76-142, Vol 1, Northrop Corporation, March 1977
6. S.H. Myhre and J.D. Labor, Repair of Advanced Composite Structures, *J. of Aircraft*, Vol 8 (No. 7), July 1981, p 546
7. S.H. Myhre, J.D. Labor, and S.C. Aker, Moisture Problems in Advanced Composite Structural Repair, *Composites*, Vol 13 (No. 3), July 1982, p 289

Repair of Advanced Composite Commercial Aircraft Structures

Jerry W. Deaton, Materials Division, NASA Langley Research Center

IN THE EARLY 1970s, a systematic program for the design, fabrication, testing, and flight-service evaluation of numerous advanced composite components was initiated by the National Aeronautics and Space Administration (NASA). The necessity for repair of composite structures was demonstrated in the NASA flight-service programs (Ref 1 and 2). The objective of the flight-service program was to assess the long-term durability of advanced composites in various flight environments and to provide the necessary data to encourage aircraft manufacturers and operators to make production commitments for the use of composite structures on commercial aircraft. Major emphasis was placed on the evaluation of advanced composites on commercial transport aircraft because of their high utilization rates, exposure to worldwide environmental conditions, and systematic maintenance procedures.

The flight-service program was expanded in 1979 when the U.S. Army and the NASA Langley Research Center initiated joint programs to evaluate composite components on military and commercial helicopters. In general, helicopters accumulate fewer flight hours than do transport aircraft; however, the flight environments and fatigue loading are sometimes more severe for the helicopter components. A summary of NASA's composite structures flight service through July 1989 is given in Table 1. Additional details of NASA's flight-service program are given in Ref 1 and 2. In 16 years, a total of 350 components have been put into flight service and have amassed over 5 million flight hours. Inspection and maintenance results are given in Table 2. Typically, flight-service components on commercial aircraft are inspected every 12 or 13 mo, and components on military aircraft are inspected at 6- or 12-mo intervals by engineering personnel from the component manufacturing company. All components are inspected visually, and some are also inspected with ultrasonics.

The use of advanced composites on more complicated structural components may require additional inspection methods, such as x-ray or eddy current (see Ref 3 and 4). Data from Table 2 indicate that some composite components from over half of the flight-service programs have experienced either damage or deterioration in the flight-service environment. The types of damages experienced include foreign object impact, bird strike, hailstones, moisture intrusion and corrosion, disbonds, lightning strike, and damage caused by ground handling equipment. Also given in Table 2 are the pertinent references (Ref 5-24) for the design, fabrication, and flight-service evaluation for each component. Documented repair procedures are also included in the references where applicable.

The increased use of advanced composite structures on aircraft is anticipated. Current military aircraft routinely use composite structures, and future aircraft will certainly expand composite use. Although applications of composite structures in commercial transports is not aggressive, secondary structures are now in production for various aircraft. Development of advanced repair techniques must be commensurate with the increased use of composite structures and the anticipated wider range of material systems so that airline operators can maintain the components.

Repair Development

The repair of structures fabricated from composite materials requires knowledge from several disciplines. Inspection techniques must be available to identify defective or damaged areas, and analysis methods must be able to determine if located damage is detrimental to structural performance. Efficient repair processes are necessary, and tests must be performed to demonstrate that repaired composite structure can be put back into service for the remainder of the design life. Research at NASA Langley (Ref 25-35) was augmented through grants to universities (Ref 36-43) and contracts to aerospace companies (Ref 43-52) to investigate the repair problem. It should be noted that the majority of references cited are dated in the late 1970s and early 1980s and utilized state-of-the-art material systems and adhesives for that time period. Higher-performance material systems and adhesives have been developed in recent years that are applicable to primary aircraft structures. The repair processes discussed in the references should be adaptable to many of the new material systems and adhesives. The repair research emphasized graphite-epoxy (Gr-Ep) composites for commercial transport applications, and, to

Table 1 NASA composite structures flight service summary (as of July 1989)

Aircraft component	Total components(a)	Start of flight service	Cumulative flight hours	
			Highest time on one aircraft	Total
L-1011 fairing panels	18 (15)	Jan 1973	47 650	709 230
737 spoiler	108 (34)	July 1973	42 000	2 593 740
C-130 center wing box	2 (2)	Oct 1974	9 650	19 220
DC-10 aft pylon skin	3 (2)	Aug 1975	40 360	97 220
DC-10 upper aft rudder	15(b) (10)	April 1976	50 300	473 820
727 elevator	10 (8)	March 1980	34 300	284 400
L-1011 aileron	8 (8)	March 1982	25 930	203 040
737 horizontal stabilizer	10 (10)	March 1984	14 540	139 970
DC-10 vertical stabilizer	1 (1)	Jan 1987	11 070	11 070
S-76 tail rotors and horizontal stabilizer	14 (0)	Feb 1979	5 860	53 150
206L fairing, doors, and vertical fin	160 (51)	March 1981	9 800	430 000
CH-53 cargo ramp skin	1 (1)	May 1981	1 700	1 700
Total	350 (142)			5 016 560

(a) Numbers in parentheses indicate number of components still in service. (b) Five more rudders to be installed

Table 2 Summary of composite-component inspection and maintenance results

Aircraft component	Inspection interval	Inspection method	Status	Reference
L-1011 fairing panels	12 mo	Visual	Minor impact damage, surface cracks, fiber fraying, and hole elongation	5, 6
737 spoiler	12 mo	Visual, ultrasonic	Moisture intrusion and corrosion, impact damage caused by hailstones, bird strike, and ground handling equipment	7, 8
C-130 center wing box	6 mo	Visual, ultrasonic	No defects after 15 years of service	9, 10, 11
DC-10 aft pylon skin	12 mo	Visual	Surface corrosion and cracks	12
DC-10 upper aft rudder	3, 12 mo	Visual, ultrasonic	Seven incidents, minor disbonds, rib damage caused by ground handling, and skin damage due to lightning strike	13, 14
727 elevator	13 mo	Visual	Two damaged by lightning strike and two damaged during ground handling	15
L-1011 aileron	12 mo	Visual	No damage	16
737 horizontal stabilizer	12 mo	Visual, ultrasonic	Two de-icer impacts, one damaged by engine fragment	17, 18
DC-10 vertical stabilizer	12 mo	Visual	No damage	19
S-76 tail rotors and horizontal stabilizer	100 h surface damage, 1000 h, or 12 mo	Visual	Small disbonds in one torque box of stabilizer; no defects in tail rotor spars	20, 21
206L fairing, doors, and vertical fin	1200 h or 12 mo	Visual	Lightning strike, underdesigned metal hinges, and skin-to-core disbonds	22, 23
CH-53 cargo ramp skin	12 mo	Visual	No damage	24

a lesser extent, graphite-polyimide (Gr-PI) composites for higher-temperature applications. Three areas of research have been selected from the sponsored programs to illustrate repair development accomplishments.

Scarf Repairs

Scarf joint repair methods were developed and verified experimentally (Ref 39). Figure 1 gives results for three repair configurations evaluated. At the top of the figure is a cross section of a scarf repair to a damaged laminate. The damaged area was cleaned by a machining operation. A machined precured patch is then adhesively bonded in the cleaned-out area, and a doubler of the same material may or may not be added. The data shown are for a graphite-epoxy AS/3501-6 laminate having a ply orientation of $(0_2/\pm45/90/\pm45/0_2)s$. The first bar shows a 50% reduction in tensile strength for a specimen that has a 13 mm (0.5 in.) slot. The next bar indicates that a plain-scarf repair for a severed laminate restores 65% of the tensile strength. The next bar indicates that the addition of a uniform doubler does not increase the tensile capability over the plain scarf. This is due to the existence of a large peeling stress at the end of the doubler, which initiates failure. However, if the doubler is tapered as indicated in the last bar, then 90% of the laminate strength can be developed. The data shown are for a scarf angle of 6°; however, a scarf angle of 3° has been shown to be more efficient but much more difficult to machine. Additional results are presented in Ref 39.

Graphite-Epoxy Composite Materials

In the late 1970s, the Lockheed-California Company was contracted to study repair

techniques for graphite-epoxy structures for commercial transport applications (Ref 51). The research program was performed in four phases. Phase 1 consisted of three tasks: Task A, survey of theoretical and experimental work on composite defect sensitivity; Task B, survey of airline damage experience and airline repair and maintenance procedures; and Task C, preparation of a matrix defining and categorizing flaws and damage.

Phase 2 consisted of a single task to develop and verify repair methods suitable for use in airline maintenance operations. Repair methods selected for evaluation

were based on results from the Phase 1 surveys on defect sensitivity of composites and on airline damage experience and maintenance/repair capabilities. The surveys indicated a need for repair methods adaptable to both maintenance base and line station (overnight) operations. The surveys also indicated a need for a wide choice of repair methods based on simplicity and ease of processing versus structural efficiency. The results indicated the need for both flush and external patch repairs, and the need for structural bonded (elevated-temperature), room-temperature bonded, and bolted repairs.

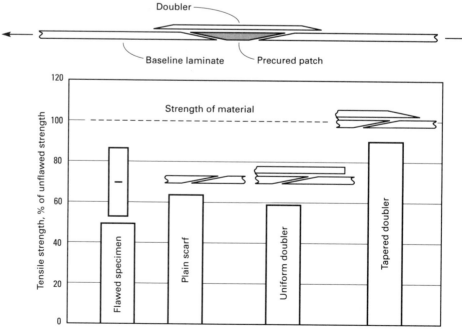

Fig. 1 Strength of repaired AS/3501-6 laminates using precured patch

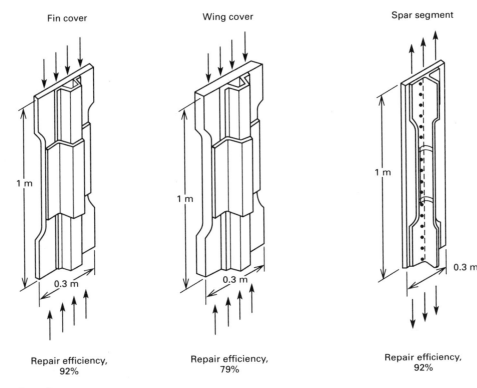

Fig. 2 Repair of Gr-Ep subelement specimens

Repairs were accomplished using laminates fabricated with T300/5208 graphite-epoxy representative of materials used in the NASA flight-service programs. Repair laminates included a 50-ply lay-up representative of a highly loaded structure, such as a wing cover, and a 16-ply lay-up representative of a lightly loaded structure, such as a stabilizer or control surface. Repair methods were evaluated using a sandwich beam specimen incorporating the repaired laminates on one face of the sandwich. A few tabbed laminate repair specimens were included for comparison. All repairs were performed after moisture conditioning of the parent laminates to simulate typical in-service condition of composite parts. The repair specimens were tested in static tension and compression at room temperature, −54 °C (−67 °F) and 82 °C (180 °F). Some fatigue tests were also run on both the 16-ply and 50-ply specimens using fatigue spectra representative of a vertical stabilizer and wing cover, respectively.

The test results (Ref 51) indicated that flush graphite repairs with tapered bond lines, which incorporated structural grade adhesives and prepregs with ply-to-ply orientation and overlap, provided the greatest structural efficiency and strength recovery and restored design strength for both lightly loaded and highly loaded structures. These repairs are the most complex and expensive and are limited to use by properly equipped maintenance base operations. External graphite patch repairs that incorporate structural grade

systems are less complex and less expensive and provide an adequate restoration of design strength for lightly loaded components. Vacuum cure pressure proved to be satisfactory for both precured bonded and cure-in-place graphite repairs using structural grade adhesive and prepreg systems. Room-temperature bonded, wet lay-up, and bolted repairs with blind fasteners are well adapted to the limitations of line station operation but provide limited strength recovery. This limited recovery is adequate, however, for many lightly loaded components.

Phase 3 of the Lockheed contract also consisted of a single task—to demonstrate repairability and repair quality on structural subelements. This activity involved simulated damage and repair of three subelement specimens: a lightly loaded hat-stiffened skin cover (L-1011 composite vertical fin cover), a highly loaded hat-stiffened skin cover (wing design), and an L-1011 vertical fin spar segment representative of a substructure for a lightly loaded component (see Fig. 2). These subelement specimens consisted of a single stiffener element for the two hat-stiffened covers and a cap and partial web section for the spar segment. The damage consisted of a complete cut through the stiffener element and adjacent skin for the two hat-stiffened cover subelements and through the cap and web segment of the L-1011 vertical fin spar specimen. The damage was then repaired using the following concepts developed and evaluated in the Phase 2 coupon tests:

- *Fin cover.* An external cure-in-place Gr-Ep patch skin repair and a precured bonded graphite hat splice
- *Wing cover.* A flush cure-in-place Gr-Ep patch skin repair and a precured bonded Gr-Ep hat splice
- *Fin spar.* A bolted repair with aluminum splice plates mechanically attached to the cap and web sections

The two repaired cover segments were tested to failure in compression, along with undamaged control specimens. The repaired spar segment was tested to failure in tension, also with an undamaged control specimen. The control and repaired specimens were conditioned to 1% moisture content prior to repair, and the bonded repairs were moisture conditioned prior to test.

The test results on the control specimens correlated with predicted unflawed strengths, and the three repaired specimens achieved from 79 to 92% of unflawed strength, which in all cases was well above design strength levels. Thus, Gr-Ep patch repairs, either precured bonded or cured-in-place, appear to be satisfactory for the repair of both lightly loaded and highly loaded parts. These repairs can be satisfactorily accomplished using vacuum pressure, although there may be some limitations on use of vacuum cure-in-place Gr-Ep patches for very high load levels.

Phase 4 of this same contract consisted of a single task identified as the repairability and repair quality of large-area repairs on structural components. This activity utilized the full-scale ground test article (GTA) of the L-1011 advanced composite vertical fin (ACVF) developed under another NASA contract.

Large-area damage, representative of lightning strike damage, was inflicted on the ACVF cover and adjacent hat stiffeners (see Fig. 3) following two lifetimes of fatigue cycling. A bonded external precured Gr-Ep patch was applied to the skin, followed by mechanical attachment of the disbonded stiffeners to the repaired skin. After repair, the fin GTA was subjected to one additional lifetime of fatigue cycling, followed by loading to ultimate strength and failure.

The fin GTA failed at 120% of design ultimate load. The repair patch remained intact and unaffected by the fatigue cycling and testing to failure. Strain measurements indicated no effect of the patch on far-field strains in the fin component.

The tests verified that the patch met and exceeded the design requirements of restoring the capability of the damaged area to withstand design loads for the fin. The results also validated the concept of repairing lightly loaded components such as the vertical fin using vacuum cured, external bonded patches. This approach is relatively simple to accomplish in any type of repair

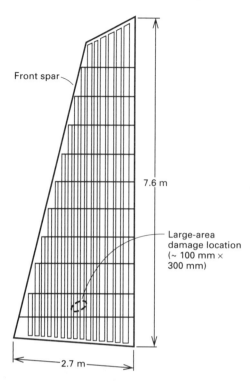

Front spar

7.6 m

Large-area
damage location
(~ 100 mm ×
300 mm)

2.7 m

Fig. 3 Large-area damage location on L-1011 vertical fin ground test article

Fig. 4 Graphite-polyimide repair demonstration specimens

situation likely to be encountered in commercial aircraft service. Complete details of all four phases of this program are presented in Ref 51.

Graphite-Polyimide Composite Materials

In 1980, Rockwell International was contracted to develop repair techniques for Celion 6000/LARC-160 Gr-PI composite material for higher-temperature applications (Ref 52). The types of flaws that were repaired are representative of manufacturing defects, in-service damage, and in-service deterioration. Repair bonding methods were developed on coupon specimens using bonding approaches similar to those reported in Ref 30 and 33. Repairs were made on damaged specimens configured as flat laminates with either 25 mm (1.0 in.) or 51 mm (2.0 in.) diam holes, honeycomb sandwich panels, and hat-stiffened skin-stringer panels (see Fig. 4). Techniques developed under this program utilized autoclave-bonded repairs for each of the three specimen configurations. Non-autoclave repair processes need to be developed for Gr-PI materials to be considered for application to primary aircraft structures.

Flat-Laminate Repair. Several flush and doubler repair design concepts were implemented on 152 × 305 × 2.3 mm (6.0 × 12.0 × 0.091 in.) flat 16-ply laminates with a lay-up of $(0/\pm45/90)_{2s}$ using a LARC-160 prepreg resin cocure process developed on coupon specimens. Ultrasonic inspection indicated that void-free cocure bonds and patches were achieved for both 3° and 6° scarf angle flush repairs. However, cocure repaired laminates with external doublers indicated some void areas. Laminates were

(a)

(b)

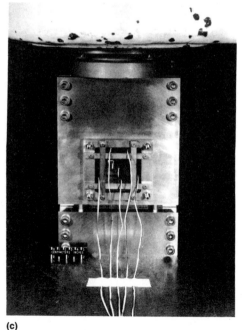

(c)

Fig. 5 Thin laminate compression test fixture. (a) Fixture with a 51 × 51 mm (2 × 2 in.) unsupported insert. (b) Thin laminate compression specimen placed in fixture. (c) Compression fixture in test machine

Fig. 6 Graphite-polyimide sandwich panel prepared for repair with prepreg preforms

tested in compression at room temperature using a special test fixture developed by NASA Langley and shown in Fig. 5. Elevated-temperature tests were not performed because of test fixture constraints. A flush repair with a 3° scarf angle gave the best results, restoring nearly 80% of the undamaged laminate strength. An improved cocure repair design, which added serrated patch plies, was fabricated but could not be tested in the special test fixture. The improved design was predicted to return over 90% of the original laminate strength.

Honeycomb Sandwich Repair. Two types of honeycomb sandwich panel configurations were used for repair development and demonstration. The first type was a panel designed to carry a compression load of 2.1 MN/m (12 000 lbf/in.). The second type was designed to carry a compression load of 0.53 MN/m (3000 lbf/in.). The demonstration of honeycomb sandwich repair involved testing three types of specimens: undamaged control, damaged with a 51 mm (2.0 in.) diam hole, and repaired specimens.

The same processes developed for the repair of flat laminates were directly extended to the repair of honeycomb sandwich panels. The repair consisted of a precured solid laminate plug bonded into a recess in the core, as shown in Fig. 6. The plug provided the base for cocuring the patch plies on the face sheet laminate and prevented dimpling. The patch had a 4.5° scarf with three external serrated octagonal patch plies added. A completed repair is shown in Fig. 7.

Specimens were tested in compression at room temperature and 316 °C (600 °F). Repair efficiency of the heavily loaded specimens at room temperature ranged between 76 and 91% of undamaged control specimen strength, and 75% of undamaged control specimen strength at 316 °C (600 °F). Elevated-temperature test results indicated significant reductions in honeycomb sandwich compressive strength: undamaged strength, 63% of room-temperature value; damaged strength, 56% of room-temperature damaged value; and repaired strength, 53% of repaired room-temperature value. Similar results were obtained for the repair of lightly loaded honeycomb sandwich specimens.

Hat-Stiffened Skin-Stringer Repair. Two types of hat-stiffened skin-stringer panel configurations were used for repair development and demonstration. One configuration was designed to carry 1.05 MN/m (6000 lbf/in.) in compression, whereas the other configuration was designed to carry 0.33 MN/m (1900 lbf/in.) in compression. The specimens were 152 mm (6.0 in.) wide and 305 mm (12.0 in.) long. Hat-stringer repairs were developed on elements with a simulated 51 mm (2.0 in.) damage length in the hat assembly. A variation of the cocured flush scarf angle patch with external serrated edged doubler was implemented in repair of hat-stringer parts. All repairs had an excellent aesthetic appearance, with a smooth transition between the repair area and parent material, as shown in Fig. 8. The

Fig. 7 Completed repair of Gr-PI sandwich panel

serrated edges terminating the external doubler plies showed good definition and all layers were well compacted. Ultrasonic inspection showed low voids in the web and flange areas. However, the cap repair area showed some voids in the tapered bond section. Control and repaired specimens were tested in compression at room temperature and 316 °C (600 °F).

Repair efficiency of the heavily loaded repaired specimens ranged between 81 and 94% at room temperature, and 70 and 95% at elevated temperature. All specimens failed just outside of the repair area. Compression strength of the undamaged specimens tested at 316 °C (600 °F) was 77% of the undamaged control specimen tested at room temperature. Similar tests were performed on the lighter loaded skin-stringer specimens. All failures also occurred outside of the repair area. Failure of the undamaged control specimen tested at room temperature achieved only 80% of the target design strength. The elevated-temperature (316 °C, or 600 °F) strength was 94% of the room-temperature value. The elevated-temperature strength was expected to be about 75% of the room-temperature strength, indicating that the room-temperature specimen had a premature failure and the analysis was not in error. The reason for the premature failure was not determined. Because of the lower than expected undamaged control specimen strength, repair efficiencies are not given. Complete details of this program are presented in Ref 52.

Repair Durability

The long-term durability of repairs is being addressed in a 10-year outdoor exposure program at NASA Langley (Ref 35). Four repair methods are being evaluated, which

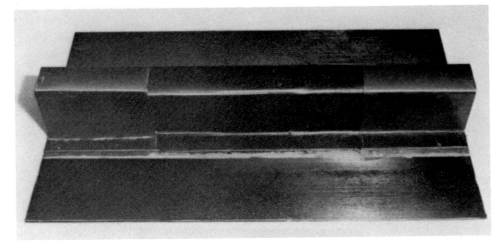

Fig. 8 Completed repair of Gr-PI hat-stiffened skin-stringer panel

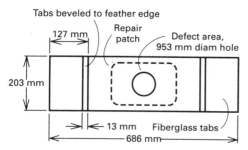

Fig. 9 Tabbed laminate specimen

Table 3 Exposure and test plan for repaired graphite-epoxy laminates

| Static strength | Outdoor exposure condition | | | | Test date |
| | No load, years | Constant load, years | No load + Fatigue(a) | | |
			Years	Lifetimes	
Baseline	0	0	0	2.5	5/83
Residual strength	1	1	1	0.25(b)	6/84
	3	3	3	0.75	6/86
	5	5	5	1.25	6/88
	7	7	7	1.75	6/90
	10	10	10	2.50	6/93

(a) Fatigue loads applied indoors using closed-loop electrohydraulic testing system. (b) 0.25 lifetime of fatigue applied each year

include: externally bolted aluminum-plus adhesive; precured, bonded external Gr-Ep; cure-in-place external Gr-Ep; and cure-in-place flush Gr-Ep. All four repair concepts being evaluated utilized structural grade adhesives and/or prepreg material systems having a 177 °C (350 °F) cure temperature. The repair methods are listed in increasing order of complexity and expense of performing. The aluminum patch repair utilized pressure from the fasteners during adhesive cure, whereas the other three repair methods used vacuum pressure during cure. Repairs were accomplished using laminates fabricated from T300/5208 Gr-Ep. A complete description of how each of the four repairs was achieved is given in Ref 35.

Tabbed Laminate Specimen

A sketch of an exposure specimen is shown in Fig. 9. Each repair specimen is 686 mm (27 in.) long and 203 mm (8 in.) wide. Fiberglass tabs 3.6 mm (0.14 in.) thick were bonded to each end on both surfaces of the specimen. Each tab was beveled to a feather edge extending 13 mm (0.5 in.) from the edge adjacent to the test section. The tabs were bonded to the specimen with Metlbond 329 (M-329) supported film adhesive in a single operation.

The damage to be repaired consisted of a 953 mm (3.75 in.) diam hole located in the center of the test section. This specimen provides a realistic geometric representation of an actual repair and permits the testing of a damaged, unrepaired specimen. All repair specimens were prepared for bonding by sanding lightly with 180-grit abrasive paper and wiping clean with a cloth soaked with methyl ethyl ketone (MEK). All laminates to be repaired were conditioned to a 1% moisture level prior to repair installation. Also, all specimens received an aircraft finish coat of polyurethane paint before being exposed to the outdoor environment.

Exposure and Test Plan

The exposure and test plan for repaired Gr-Ep specimens is shown in Table 3. A total of 108 specimens are included in the test program. The test program includes 18 undamaged control specimens, 18 damaged unrepaired specimens, and 18 repaired

specimens for each of the four repair concepts being evaluated. One specimen type is included for each outdoor exposure condition and duration shown. The baseline static strength control specimens were the same size (203 mm, or 8 in.) wide by 686 mm, or 27 in. long) as the repair specimens. However, the control specimens for determination of residual static strength after exposure were 25 mm (1 in.) by 305 mm (12 in.) composite tension specimens (ASTM D 3039, Ref 53).

The baseline specimens identified as fatigue specimens were exposed to 2.5 lifetimes of the fully reversed fatigue loading of the L-1011 fin spectrum given in Ref 51 for similar 16-ply Gr-Ep specimens. Briefly, the L-1011 fin spectrum is defined for a 36 000 flight lifetime and consists of blocks of climb, cruise, and descent flight spectra typically experienced by the L-1011 aircraft and applied in a random manner. Maximum

loads in the spectra were set to cause maximum tensile strains of 0.2%, which represents limit load design strains for the L-1011 fin skin. The residual static strength fatigue specimens received 0.25 lifetime of fatigue (9000 flights) each year, as indicated in Table 3. Baseline static strength tests were performed in May 1983, residual static strength tests after 1 year of outdoor exposure in June 1984, 3-year outdoor exposure residual strength tests in June 1986, and 5-year outdoor exposure residual strength tests in June 1988. Additional residual static strength tests are planned after 7 and 10 years of outdoor exposure.

Outdoor Exposure Test Setup

Figure 10 shows the outdoor exposure test setup for the control and the repaired specimens. The rack of specimens shown in Fig. 10(a) is associated with the no-load condition. Another rack of specimens (not

Fig. 10 Outdoor exposure test setup. (a) Unstressed specimens. (b) Sustained stress control specimens. (c) Sustained stress repaired specimens

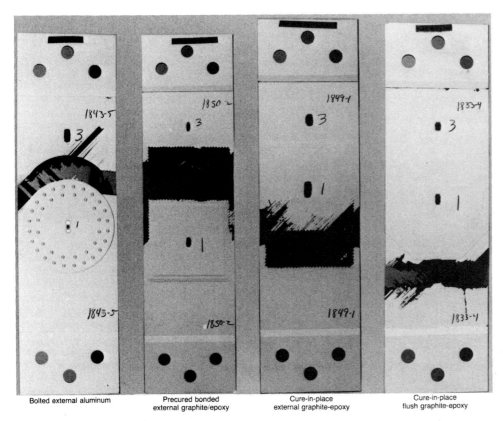

| Bolted external aluminum | Precured bonded external graphite/epoxy | Cure-in-place external graphite-epoxy | Cure-in-place flush graphite-epoxy |

Fig. 11 Failures of baseline repaired graphite-epoxy laminates

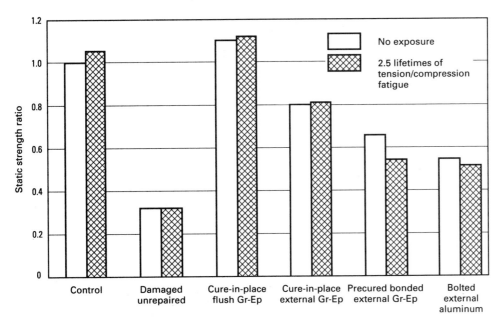

Fig. 12 Relative baseline strengths of repaired graphite-epoxy laminates

load corresponding to limit load strain (0.2%) of the L-1011 vertical fin skin, which is equivalent to 22% of ultimate load for the laminate under consideration. The loading frames also faced south and the repaired specimens were at an angle of approximately 45° to the horizontal. The smaller sustained load control specimens were exposed in a horizontal plane because of interference between the clamping fixtures and the loading frame structure.

Baseline Results

Photographs of the failure of the baseline repaired Gr-Ep laminates are shown in Fig. 11. Failure of the bolted external aluminum repair specimen initiated through the fastener holes. In Ref 51, it was observed that for similar external metal doubler repairs, residual strength was low because of premature pull-out of the blind fasteners. It has been demonstrated, however, that significant strength recovery can be provided by a metal patch repair by using standard titanium screws and plate nuts (Ref 54 and 55). The failure of the precured bonded external Gr-Ep repair (Fig. 11) was due to disbonding of the repair patch. For the cure-in-place external Gr-Ep repair, a similar failure was obtained, along with fiber splitting in the outermost ply of the repair. The saw-tooth appearance along the edges of these repairs is caused by the cutting of the repair materials with pinking shears. The cure-in-place flush Gr-Ep repair failed in tension outside the repair area.

Figure 12 shows the relative baseline strengths for each of the four repair concepts for no outdoor exposure and after 2.5 lifetimes of fully reversed fatigue cycling. The precured bonded external Gr-Ep repair is the only repair that indicates a loss (approximately 19%) in residual tensile strength because of the fatigue cycling. However, it should be noted that only one specimen has been tested at each condition, and this may be within the scatter band for this type of test.

The cure-in-place flush Gr-Ep repair exceeded the control specimen strength by approximately 10% and is the only repair that failed outside the repair area. The cure-in-place external Gr-Ep repair was next in bond efficiency at approximately 79% of control specimen strength. The precured bonded external Gr-Ep and the bolted external aluminum repairs were about 67% and 55% effective, respectively, in restoring control specimen strength.

Exposure Results

Figure 13 shows photographs of the repaired Gr-Ep laminates that were tested to failure after 5 years of outdoor exposure. Failures of the bolted external aluminum, precured bonded external Gr-Ep, and the cure-in-place external Gr-Ep repairs were similar to those of the baseline repaired

shown) is associated with the no-load plus fatigue exposure condition. Each rack was positioned so that the specimens were oriented longitudinally, facing south, at an angle of approximately 45° to the horizontal. The sustained load test setup for the

control and repaired specimens are shown in Fig. 10(b) and (c), respectively. Load was applied to the specimens by a screw-driven jack and the load cell was aligned along the horizontal axis of the specimens. All sustained load specimens were stressed to a

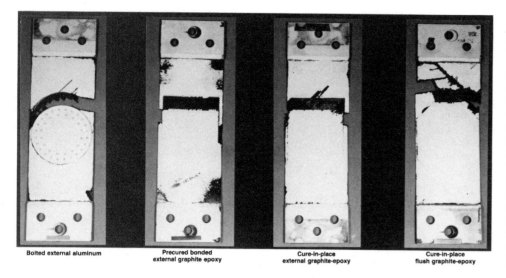

Bolted external aluminum   Precured bonded external graphite epoxy   Cure-in-place external graphite-epoxy   Cure-in-place flush graphite-epoxy

Fig. 13 Failures of repaired graphite-epoxy laminates after 5 years of outdoor exposure

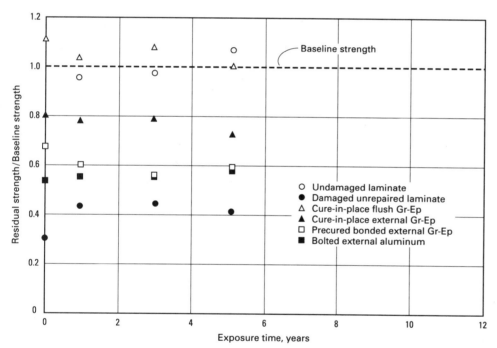

Fig. 14 Residual tensile strength of repaired graphite-epoxy T300/5208 after outdoor exposure

Ep repair specimen failed in tension through the repair, whereas the failure of the baseline specimen failed outside the repair area.

Figure 15 shows the residual tensile strength of repaired Gr-Ep T300/5208 after sustained stress outdoor exposure. All specimens were subjected to a sustained stress of 22% of ultimate for the undamaged control specimen, which corresponds to a strain level of 0.2%. Again, the data shown on the ordinate are repeated from Fig. 12. The 1-, 3-, and 5-year data shown for the nonrepaired specimens are, again, from the 25 mm (1.0 in.) by 305 mm (12.0 in.) specimen size. Comparison of the sustained stress exposure data for each of the four repair concepts indicates that residual tensile strength is essentially unchanged after 5 years of sustained-stress outdoor exposure.

Failure modes after 5 years of outdoor exposure and 1.25 lifetimes of fatigue are similar to those of the baseline repaired specimens shown in Fig. 11, except that the cure-in-place flush Gr-Ep repaired specimen failed in tension through the repair.

The effects of combined outdoor exposure and fatigue cycling on the residual tensile strength of repaired Gr-Ep T300/5208 are shown in Fig. 16. The data shown (flagged symbols) on the ordinate are repeated from Fig. 12 for the specimens subjected to 2.5 lifetimes of fully reversed fatigue cycling previously described. The 1-, 3-, and 5-year data shown for the nonrepaired specimens are, again, from the 25 mm (1.0 in.) by 305 mm (12.0 in.) specimens. The cure-in-place external Gr-Ep repair is the only repair concept of those being evaluated that shows a loss in residual tensile strength after 5 years of outdoor exposure plus 1.25 lifetimes of fatigue (16% loss of strength). The other three repair concepts remain essentially unchanged after 5 years of outdoor exposure plus 1.25 lifetimes of fatigue.

Program Synopses

Efficient repair techniques for damaged graphite-epoxy composites have been developed utilizing structural grade adhesive and prepreg materials cured with vacuum pressure. These repair methods have been demonstrated and evaluated on coupon, subelement, and full-sized composite components applicable to secondary graphite-epoxy structure for commercial transport aircraft.

The technology base developed for repair of graphite-epoxy structures was extended to repair technique development for graphite-polyimide structures for higher-temperature applications. Repair methods have been developed and demonstrated on coupon specimens, flat laminate, honeycomb sandwich, and hat-stiffened skin-stringer compression panels fabricated from graphite-polyimide material. Repair efficiencies from room-temperature tests of repaired graphite-polyimide specimens were, in gen-

specimens previously shown in Fig. 11. The cure-in-place flush Gr-Ep repair failure occurred through the repair area with severe fiber splitting in all plies of the Gr-Ep patch.

The residual tensile strengths of repaired Gr-Ep T300/5208 specimens after outdoor exposure are shown in Fig. 14. The data shown on the ordinate (no exposure) are repeated from Fig. 12. The 1-year, 3-year, and 5-year exposure data shown for the nonrepaired specimens are from the 25 mm (1.0 in.) by 305 mm (12.0 in.) specimen size and may exhibit some scale effect. Comparison of the exposure data for each of the four repair concepts indicates that only the

precured bonded external Gr-Ep repair experienced a loss in residual tensile strength due to outdoor exposure. A 10% loss is indicated after 1 and 5 years of exposure and a loss of about 18% is indicated after 3 years of outdoor exposure. The residual tensile strength of the other three repair methods is essentially unchanged after 5 years of outdoor exposure. Again, it should be noted that each data point represents a single test.

Failure modes after 5 years of sustained stress outdoor exposure are similar to the baseline repaired specimens shown in Fig. 11, except that the cure-in-place flush Gr-

eral, above 75%. However, test results reported at 316 °C (600 °F) indicated significant reductions in compression strength.

The long-term durability results of four repair methods for graphite-epoxy specimens exposed outdoors at the Langley Research Center have been presented. Exposure included unstressed, constant sustained stress, and unstressed conditions, plus a yearly imposition of 0.25 lifetime of fatigue. Test results indicated that the residual strengths for all repairs evaluated are not significantly affected after 2.5 lifetimes of fatigue, 5 years of outdoor exposure, 5 years of sustained stress outdoor exposure, or 5 years of outdoor exposure plus 1.25 lifetimes of fatigue.

It should be noted that the repair programs discussed in this article utilized material systems that were considered to be state of the art in the late 1970s and early 1980s, specifically, the use of an untoughened epoxy resin matrix. Since that time, many changes have occurred in material forms, innovative structural concepts, and in the use of toughened resins that could influence the repair of composite structures in the future.

REFERENCES

1. H.B. Dexter and A.J. Chapman, NASA Service Experience With Composite Components, in *Proceedings of the 12th National SAMPE Technical Conference*, Society for the Advancement of Material and Process Engineering, Oct 1980, p 77-99
2. H.B. Dexter, "Long-Term Environmental Effects and Flight Service Evaluation of Composite Materials," Report TM 89067, National Aeronautics and Space Administration, Jan 1987
3. M.L. Phelps, "Assessment of State-of-the-Art of In-Service Inspection Methods for Graphite/Epoxy Composite Structures on Commercial Transport Aircraft," Report CR-158969, National Aeronautics and Space Administration, Jan 1979
4. M.L. Phelps, "In-Service Inspection Methods for Graphite/Epoxy Structures on Commercial Transport Aircraft," Report CR-165746, National Aeronautics and Space Administration, Nov 1981
5. J.H. Wooley, D.R. Paschal, and E.R. Crilly, "Flight Service Evaluation of PRD-49/Epoxy Composite Panels in Wide-Bodied Commercial Transport Aircraft," Report CR-112250, National Aeronautics and Space Administration, March 1973
6. R.H. Stone, "Flight Service Evaluation of Kevlar-49/Epoxy Composite Panels in Wide-Bodied Commercial Transport Aircraft," Tenth and Final Annual Flight Service Report, Report CR-172344, National Aeronautics and Space Administration, June 1984
7. R.L. Stoecklin, "A Study of the Effects of Long-Term Ground and Flight Environment Exposure on the Behavior of Graphite/Epoxy Spoilers—Manufacturing and Test," Report CR-132682, National Aeronautics and Space Administration, June 1975
8. R.L. Coggeshall, "737 Graphite Composite Flight Spoiler Flight Service Evaluation," Report CR-172600, National Aeronautics and Space Administration, May 1985
9. W.E. Harvill, J.J. Duhig, and B.R. Spencer, "Program for Establishing Long-Time Flight Service Performance of Composite Materials in the Center

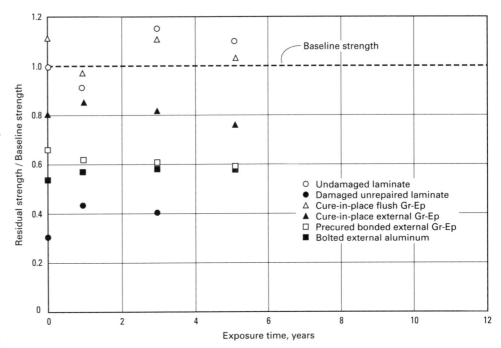

Fig. 15 Residual tensile strength of repaired graphite-epoxy T300/5208 after outdoor exposure stressed at 22% of ultimate

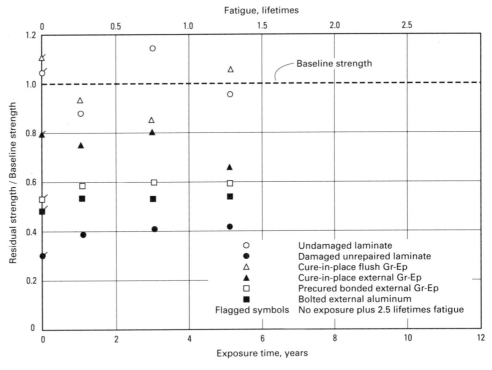

Fig. 16 Residual tensile strength of repaired graphite-epoxy T300/5208 after outdoor exposure and fatigue

Wing Structure of C-130 Aircraft: Phase II, Detailed Design," Report CR-112272, National Aeronautics and Space Administration, April 1973

10. W.E. Harvill and A.O. Kays, "Program for Establishing Long-Time Flight Service Performance of Composite Materials in the Center Wing Structure of C-130 Aircraft: Phase III, Fabrication," Report CR-132495, National Aeronautics and Space Administration, Sept 1974

11. W.E. Harvill and J.A. Kizer, "Program for Establishing Long-Time Flight Service Performance of Composite Materials in the Center Wing and Structure of C-130 Aircraft: Phase IV, Ground/Flight Acceptance Tests," Report CR-145043, National Aeronautics and Space Administration, Sept 1976

12. B.R. Fox, "Flight Service Evaluation of an Advanced Metal Matrix Aircraft Structural Component," Flight Service Final Report, Report J-3827, McDonnell Douglas Corporation Aug 1985

13. G.M. Lehman, "Flight-Service Program for Advanced Composite Rudders on Transport Aircraft," Sixth Annual Summary Report, Report J-6574, McDonnell Douglas Corporation Aug 1982

14. B.R. Fox, "Flight Service Program for Advanced Composite Rudders on Transport Aircraft," Ninth Annual Summary Report, MDC Report J-3871, Sept 1985

15. D.V. Chovil *et al.*, "Advanced Composite Elevator for Boeing 727 Aircraft," Vol 2, Final Report, Report CR-15958, National Aeronautics and Space Administration, Nov 1980

16. C.F. Griffin and E.G. Dunning, "Development of an Advanced Composite Aileron for the L-1011 Transport Aircraft," Report CR-3517, National Aeronautics and Space Administration, Feb 1982

17. J.E. McCarty and D.R. Wilson, Advanced Composite Stabilizer for Boeing 737 Aircraft, in *Proceedings of the Sixth Conference on Fibrous Composites in Structural Design*, Report AMMRC MS 83-2, Army Materials and Mechanics Research Center, Nov 1983

18. R.L. Coggeshall, "Boeing/NASA Composite Components Flight Service Evaluation," Report CR-181898, National Aeronautics and Space Administration, Nov 1989

19. C.O. Stephens, Advanced Composite Vertical Stabilizer for DC-10 Transport Aircraft, Report CR-172780, Contract NAS1-14869, Douglas Aircraft Company, 22 Jan 1979

20. M.J. Rich and D.W. Lowry, "Flight Service Evaluation of Composite Helicopter Components," First Annual Report, Report CR-165952, National Aeronautics and Space Administration,

June 1982

21. M.J. Rich and D.W. Lowry, "Flight Service Evaluation of Composite Helicopter Components," Second Annual Report (May 1982-Sept 1983), Report CR-172562, National Aeronautics and Space Administration, April 1985

22. H. Zinberg, "Flight Service Evaluation of Composite Components on the Bell Model 206L: Design, Fabrication, and Testing," Report CR-166002, National Aeronautics and Space Administration, Nov 1982

23. H. Zinberg, "Flight Service Evaluation of Composite Components on the Bell Model 206L," First Annual Flight Service Report, Report CR-172296, National Aeronautics and Space Administration, March 1984

24. D.W. Lowry and M.J. Rich, "Design Fabrication, Installation, and Flight Service Evaluation of a Composite Cargo Ramp Skin on a Model CH-53 Helicopter," Report CR-172126, National Aeronautics and Space Administration, April 1983

25. J.W. Deaton, "Repair Technology in Composite Materials," Paper presented at the Navy-sponsored Composite Material Maintenance/Repair Workshop, San Diego, 26-28 Sept 1978

26. J.W. Deaton, "Composites for Advanced Space Transportation Systems (CASTS) Repair Technology," Report TM-80038, National Aeronautics and Space Administration, March 1979, p 95-97

27. J.W. Deaton, "NASA's Repair Technology Program in Composite Materials," Paper presented at the International Workshop on Defense Applications of Advanced Repair Technology for Metal and Composite Structures, Naval Research Laboratory, Washington, D.C., 22-24 July 1981

28. J.W. Deaton, "A Repair Technology Program at NASA on Composite Materials," Report TM-84505, National Aeronautics and Space Administration, Aug 1982

29. J.W. Deaton, "A Process for Repairing Graphite/Epoxy Structures," Paper presented at the Sixth Conference on Fibrous Composites in Structural Design, New Orleans, 24-27 Jan 1983

30. J.W. Deaton and N.A. Mosso, "Preliminary Evaluation of Large Area Bonding Processes for Repair of Graphite/Polyimide Composites," Paper presented at the 28th National SAMPE Symposium and Exhibition, Anaheim, CA, Society for the Advancement of Material and Process Engineering, 12-24 April 1983

31. B.A. Stein and W.T. Hodges, Rapid Adhesive Bonding Concepts for Specimens and Panel Fabrication and Field Repair, in *Proceedings of the 16th Na-*

tional Conference, Society for the Advancement of Material and Process Engineering, 1984, p 103-118

32. J.W. Deaton, "Residual Strength of Repaired Graphite/Epoxy Laminates," Paper presented at the Tri-Service Composites Repair Technology Workshop, Dayton, OH, 30 Oct-1 Nov 1984

33. H. Brown, Ed., *Composite Repairs SAMPE Monograph No. 1*, Society for the Advancement of Material and Process Engineering, 1985

34. J.W. Deaton, "Residual Strength of Repaired Graphite/Epoxy Laminates After 3 Years of Outdoor Exposure," Paper presented at Second DoD/NASA Composites Repair Technology Workshop, San Diego, 4-6 Nov 1986

35. J.W. Deaton, "Residual Strength of Repaired Graphite/Epoxy Laminates After 5 Years of Outdoor Exposure," NASA CP-3087, Part 2, 1990, p 439-454

36. R.B. Pipes and R.L. Daugherty, "Damage Repair Technology for Composite Materials," Report 77N70820, or NASA CR-149110, National Aeronautics and Space Administration

37. R.B. Pipes and R.L. Daugherty, "Damage Repair Technology for Composite Materials," Report 77N81104, or NASA CR-153995, National Aeronautics and Space Administration

38. R.L. Daugherty and R.B. Pipes, "Damage Repair Technology for Composite Materials," Report 78N70369, or NASA CR-155367, National Aeronautics and Space Administration

39. D.W. Adkins and R.B. Pipes, Planar Scarf Joints in Composite Repair, in *Proceedings of the Second International Conference on Composite Materials*, Metallurgical Society of the American Institute of Mining, Metallurgical, and Petroleum Engineers, 16-20 April 1978, p 1200-1210

40. R.L. Daugherty and S.A. Howard, Effects of Debonds on the Strength of Composite Plates, in *Proceedings of the Second International Conference on Composite Materials*, Metallurgical Society of the American Institute of Mining, Metallurgical, and Petroleum Engineers, 16-20 April 1978, p 1224-1236

41. J.D. Webster, "Flaw Criticality of Circular Disbond Defects in Compressive Laminates," Report 81N32195, or NASA CR-164830, National Aeronautics and Space Administration

42. R.B. Pipes and D.W. Adkins, "Strength and Mechanics of Bonded Scarf Joints for Repair of Composite Materials," Report 83N15359, or NASA CR-169708, National Aeronautics and Space Administration

43. D.A. O'Brien, "Finite Element Stress Analysis of Idealized Composite Damage Zones," M.S. thesis, Virginia Polytechnic Institute and State University,

Jan 1978

44. R.H. Stone, "Composite Repair Procedures for Commercial Transport Applications," Paper presented at the SME Fourth Structural Composite Conference, Los Angeles, Society of Manufacturing Engineers, 6-8 June 1978

45. R.H. Stone, "Composite Repair Procedures for Commercial Transport Applications," Paper presented at the Composite Material Maintenance/Repair Workshop, San Diego, CA, 26-28 Sept 1978

46. R.H. Stone, "Activities and Plans for Composite Repair," Paper presented at Maintenance and Technology Workshop on Shipboard Repair Composite Structural Materials, Salt Lake City, UT, 9-11 Sept 1980

47. R.H. Stone, "Development of Repair Procedures for Graphite/Epoxy Structures on Commercial Transports," Paper presented at the 12th National Technical Conference, Seattle, 6-8 Oct 1980

48. R.H. Stone, "NASA/DoD Sponsored Maintenance and Repair Technology," Paper presented at Composite Primary Aircraft Structures Maintenance and Repair Workshop, San Francisco, 2-3 March 1981

49. J.F. Knauss and R.H. Stone, "Demonstration of Repairability and Repair Quality on Graphite/Epoxy Structural Subelements," Paper presented at the SAMPE 27th National Symposium and Exhibition, Society for the Advancement of Material and Process Engineering, May 1982

50. J.S. Jones, "Celion/LaRC-160 Graphite/Polyimide Composite Processing Techniques and Properties," Paper presented at the SAMPE 14th National Technical Conference, Society for the Advancement of Material and Process Engineering, Oct 1982

51. R.H. Stone, "Repair Techniques for Graphite/Epoxy Structures for Commercial Transport Applications,"

Report CR-159056, National Aeronautics and Space Administration, Jan 1983

52. J.S. Jones and S.R. Graves, "Repair Techniques for Celion/LaRC-160 Graphite/Polyimide Composite Structures," Report CR-3794, National Aeronautics and Space Administration, 1984

53. "Standard Test Method for Tensile Properties of Fiber-Resin Composites," *Annual Book of ASTM Standards*, American Society for Testing and Materials

54. J.D. Labor, "Large Area Composite Structure Repair," Reports AFFDL-TR-77-5, AFFDL-TR-121, AFFDL-TR-78-83, Northrop Corporation

55. C.F. Griffin, L.D. Fogg, and E.G. Dunning, "Advanced Composite Aileron for L-1011 Transport Aircraft Design and Analysis," Report CR-165635, National Aeronautics and Space Administration, 1981

Surface Preparation of Composites for Adhesive-Bonded Repair*

THE IMPORTANCE of proper surface preparation for a reliable repair cannot be overstated. This has been proved with experimental data in which contamination associated with several standard lay-up and cleaning procedures were accurately measured (Ref 1). In-house applications at Douglas Aircraft Company have supported the data and conclusions of Ref 1, which are integrated into this article.

Because composite repair techniques vary, surface preparation depends on the requirements for repairing a particular composite. This article first describes an accurate test used to determine the effectiveness of preparation methods already applied to a surface. Next, the individual steps in the preparation process are described.

Water-Break Test

Verifying the adequacy of surface preparation is extremely important. This can be accomplished using the standard water-break test, which has been used for decades in metal bonding applications and is universally accepted as the final step in surface preparation for metal structures. This test, used by Douglas Aircraft Company for more than 20 years in the preparation of plastic surfaces for painting or filling, has been found to be very accurate. Addition-

ally, this test should be in fibrous composite adhesive bonding specifications and is usually in the repair manuals of aircraft manufacturers.

The word fibrous in the term fibrous composites refers to glass fibers, which have been used for decades in fillets and fairings, as well as the lighter aramid and carbon (or graphite) fibers that are used in conjunction with organic resin matrices, usually epoxy, phenolic, or polyester.

The equipment for the water-break tests is extremely simple, consisting of a plastic squirt bottle that contains deionized water and an electric forced-air dryer, which is used to dry the surface afterwards. A clean recirculating air oven could also be used to dry large parts that are not attached to the aircraft in question. However, parts made using the lost-wax process, such as radomes, must not be heated because embedded wax will migrate to the surface and prevent adhesion.

Figure 1 shows that water forms a thin, continuous coating and wets the entire area of a properly prepared composite surface, whereas it collects in discrete beads and is repelled by an unprepared or improperly prepared surface. This simple test is very discriminating and easily interpreted. Figure 2 shows the surface water being removed by the dryer, which is held at a

sufficient distance from the composite to avoid scorching the surface. An air temperature of about 120 °C (250 °F) at the surface is optimum and can easily be measured with a thermometer. The rate of drying, which is of the order of 645 mm^2/s (1 in.^{2}/s), is simply monitored by watching the progressive evaporation of the water (see Fig. 2).

The reason for using deionized, rather than tap, water is shown in Fig. 3. When solids, such as salt, are dissolved in tap water, they become concentrated in the final drops of water during the process of evaporation and leave beach marks (stains). Wiping off all excess water with a clean cloth before drying by heat diminishes this problem. The application of the least amount of water possible is preferable because a thick film of water can bridge a small local area of contamination and conceal a defect such as a fingerprint (see the four photographs of the same surface in Fig. 4). Any excess water should be removed by shaking, if possible, or by waiting for some of the water to evaporate to expose any covered-over contamination. An alertness to such a situation can be developed by practicing with both deliberately contaminated surfaces and properly cleaned ones.

The same phenomenon occurs with inadequate sanding after the removal of a peel ply. Therefore, the assessment should not

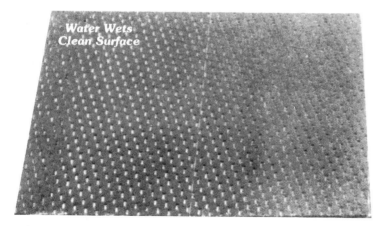

Fig. 1 Comparative water-break test results. Left: Proper preparation. Right: Improper preparation. Courtesy of Canadair

*Adapted, with permission, from Surface Preparation of Fibrous Composites for Adhesive Bonding or Painting, by L.J. Hart-Smith, R.W. Ochsner, and R.L. Radecky, Douglas Aircraft Company, in *Douglas Service*, First Quarter, 1984

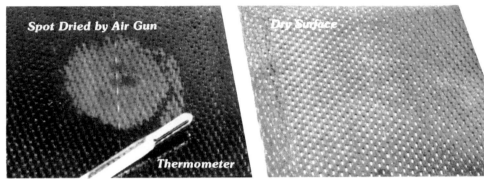

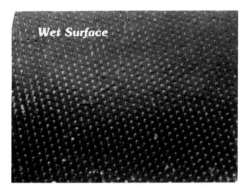

Fig. 2 Progressive monitoring of surface water evaporation. Courtesy of Canadair

be based on a quick inspection when a thick surface layer of water is present. Contaminated areas are revealed as the water dries or is forcibly evaporated. A properly cleaned surface dries steadily and continuously, whereas contaminated areas "dry" instantly, once the water film that had covered them becomes sufficiently thin. Once a part has passed the water-break test and has dried out, it should be handled only with clean white cotton gloves, and only outside the area to be bonded, and then bonded or primed promptly to prevent contamination.

Pumice/Abrasive Cleaner Scrubbing

If the condition of the surface to be bonded or painted is unknown, or if the surface is known to have been contaminated by being cured against a laminating tool coated with a spray-on or rub-on release agent, the first operation is to remove any gross contamination with a strong solvent, such as 1,1,1-trichloroethane or methyl ethyl ketone. Next, the surface should be scrubbed thoroughly with a water-base abrasive paste, using Scotchbrite pads or

Fig. 3 Strain left by rinse with tap water. Courtesy of Canadair

their equivalent. Pumice has been found to work very well as an abrasive.

This cleaning operation should be sufficient to produce a bondable surface, as indicated by a water-break-free surface during scrubbing. However, if the contamination is too deep, grit blasting, which is described below, will be needed.

The prime purpose of the scrubbing operation is to remove the contaminated surface layer of resin from a smooth laminate. Similarly, simply removing a peel ply that was part of the laminate lay-up does not usually produce a good bondable surface; ply removal should be followed by wet sanding, as discussed in the section "Peel Plies" in this article. The pumice paste should be rinsed off with water after the scrubbing has been completed. The method for verifying the adequacy of the rinse is natural or forced-air drying, after which any residue will be evident in the form of white powder. The final rinse, with deionized water, can be used as the water-break test.

Grit Blasting

A light grit blast, using aluminum oxide grit, is the best surface preparation known. It is a technique that should be mastered because, in addition to being effective in surface preparation, it is the only reliable treatment for recleaning a surface later in the life of a laminate (long after any initial peel ply has been removed). The superiority of grit blasting has been confirmed by a series of tests conducted at a number of aerospace companies. Bonded joint test results have shown that after grit blasting, bonding strengths are optimal and scatter is minimal. Grit blasting must be used with great care because excessive pressure can damage the surface plies of both aramid and graphite.

The art of grit blasting is easy to learn, provided the machine being used is not excessively powerful. Very little material should be removed (13 to 25 μm, or 0.5 to 1 mil, maximum), and the prepared surface should have a characteristic dull, or matte, finish, which is easily recognized.

A typical grit blasting machine has a ring vacuum collector that recovers the grit. Although a skirt of bristles restricts the spread of debris, the operation creates sufficient dust that it should not be performed in laminate lay-up areas.

Recirculation of the grit is not recommended. As Ref 1 states (in relation to the grit blasting of a surface cured against a caul plate with a baked-on release agent), "Even three passes of the grit blast gun succeeded in removing only two-thirds of the contamination (and in contaminating the grit)."

Although the precise details for using grit blasting need to be established for the specific equipment of each operator, a useful grit is No. 280 dry alumina grit at 140 kPa (20 psi) pressure. Even so, there is minute but detectable damage to the surface layer of the composite laminate.

Figures 5 to 7 show progressively higher magnifications of the same carbon-epoxy laminate and reveal individual damaged fibers within 5 μm (0.2 mil) of the surface. The damage extends only for a very small fraction (one hundredth) of the outermost layer, which was 0.13 mm (5 mils) thick in the case shown. Excessive blast pressure or inadequate gun motion can erode a hole completely through thick composite laminates in less than 1 min. Practice is essential. It has been shown that, in the absence of a suitable low-power grit blaster, a very thorough job of hand sanding will suffice.

It has also been shown that the use of a primer enhances the cohesive bond strength of adhesive to grit-blasted composite surfaces (Ref 2). A primer is probably better able to flow and wet this surface than a low-flow adhesive.

Peel Plies

Common practice on flat or single-curve panels is to include a peel ply (usually a woven nylon fabric) on the inside and outside of each lay-up. (This is rather difficult to do on the outside of a compound-curve skin, on which a smooth exterior finish is required, because of wrinkles or overlaps in the peel ply.) Most peel plies are coated with a release agent to ensure that their removal does not

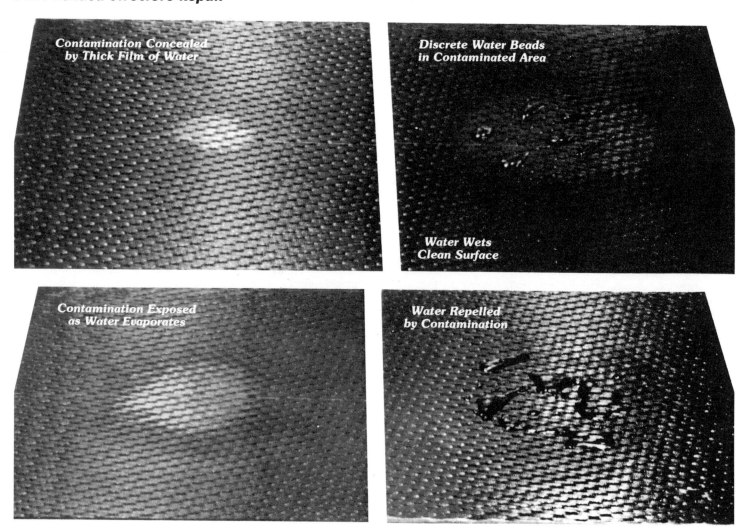

Fig. 4 Contamination concealed by thick water film and revealed by evaporation. Courtesy of Canadair

damage the underlying plies. Upon their removal, peel plies take with them any release agent that has transferred to the lay-up mold surface. Even those peel plies that do not transfer any release agent leave a very smooth surface, making adherence of adhesive or paint difficult. This can be verified easily by a microscopic examination of the laminate surface. As Fig. 8 shows, there is a clear imprint of the texture of the removed peel ply. Very little resin has actually been pulled away; the resin that remains shows up

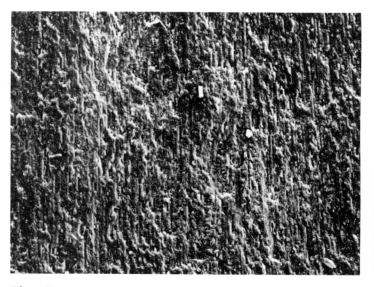

Fig. 5 Grit-blasted carbon-epoxy laminate surface. 100×. Courtesy of Canadair

Fig. 6 Grit-blasted carbon-epoxy laminate surface. 1400×. Courtesy of Canadair

Fig. 7 Grit-blasted carbon-epoxy laminate surface. 2000×. Courtesy of Canadair

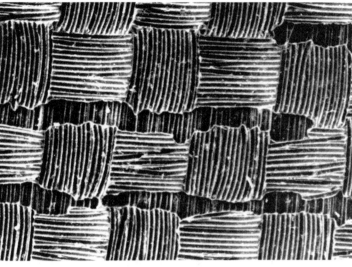

Fig. 8 Imprint on laminate surface from removal of peel ply. 70×. Courtesy of Canadair

as slightly darker areas with exposed fibers, running vertically, in the underlying laminate. An examination of the underside of the peel ply reveals little resin.

Some organizations use peel plies as part of a two-step procedure. The peel ply serves to protect the surface against gross contamination and, as mentioned above, its removal (first step) takes any transferred release agent away from the lay-up mold surface. The second step is to sand this "clean" surface enough to roughen the *bottoms* of all the depressions left by the weave in the peel ply. Therefore, in order to sand the surface, it is necessary to remove all high spots in the textured surface. Another important reason for sanding is to preclude the entrapment of an air bubble in each depression, which would reduce the effective bond area by 40%.

Grit blasting can be substituted for the sanding operation, particularly if a large area is involved. The important consideration is that the simple removal of a peel ply does not constitute adequate surface preparation for either adhesive bonding or painting. A water-break test, conducted just after the removal of a fine-weave ply that has not been coated with a release agent, will usually indicate surface acceptability because of the very small distances between the fractured high spots and the slick low spots. This is not surprising, in view of the demonstration in Fig. 4, in which a thick layer of water bridges at least 13 mm (½ in.) of contamination. The need for either sanding or grit blasting after removing a peel ply has been demonstrated by test results, which show consistently stronger joints than those attained without the extra abrading.

Figure 9 shows (under low magnification) the surfaces created by the simple removal of the peel ply (top), the minor improvements due to insufficient sanding (center),

and the distinctly matte finish associated with grit blasting (bottom).

Figure 10 shows the same results at higher magnifications. If wet sanding is performed properly, adequate surface preparation will be indicated by a water-break-free surface.

Solvent Wipe Cleaning of Surface Contamination

It has been standard practice to prepare composites for adhesive bonding by a technique known as scuff-sand and solvent wipe. As will be described later, this is usually a grossly inadequate treatment. Solvents, as well as release-coated peel plies, should be regarded with the utmost caution. The problem with using solvents to clean fibrous composite laminates can be illustrated by comparing Fig. 1 (left) and Fig. 11. The specimen in Fig. 1, which had been cleaned properly, was subsequently wiped with reagent-grade methyl ethyl ketone that was allowed to dry by evaporation. The subsequent water-break test, on the specimen in Fig. 11, shows water being repelled by the ring that was left as the solvent evaporated. Actually, most of the nonwet area had not been contaminated. With shop-grade solvent or with gross surface contamination, the solvent would have left a uniformly bad surface, and a water-break test would have looked like the specimen shown at the right in Fig. 1.

Nevertheless, solvents can be used to clean surfaces rather than contaminate them. The difference is entirely in the technique. Allowing the solvent to evaporate off the surface results in the spreading of any contamination in a thin film over the entire area to be bonded. The correct technique is a two-handed operation: One hand contains the solvent-soaked cloth, and the other hand contains a clean cloth that is used to

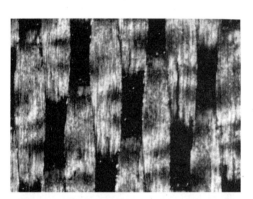

Fig. 9 Laminate surfaces. Top: After removal of peel ply only. Center: After peel ply removal and sanding. Bottom: After peel ply removal and grit blasting. Courtesy of Canadair

Fig. 10 Laminate surfaces shown in Fig. 9, but magnified at 25× (top and center) and 30× (bottom). Courtesy of Canadair

wipe up the solvent before it has time to dry. Only a small area can be done at a time. The process is repeated until the cleaning cloth ceases to pick up dirt.

Isopropyl alchohol is preferred for the final cleaning operation. It is more miscible in water than are the stronger solvents and can be removed more completely by rinsing in water. In order to master this technique, it should be practiced on sample pieces of laminate, using the water-break test to verify the adequacy of the treatment.

Drying Laminates That Contain Absorbed Moisture

The hot-bonded repairs of fibrous composite laminates that have been in service for a few months are complicated greatly by even an equilibrium moisture content of less than 1% in glass-epoxy or carbon-epoxy laminates or 2% in aramid-epoxy laminates. In an extensive repair test program conducted for the U.S. Air Force, and described in the article "Repair Concepts for Advanced Composites Structures" in this Volume and in Ref 3 of this article, the laminates to be repaired were deliberately preconditioned and not dried prior to elevated-temperature bond repairs. Typical consequences were a 50% loss in the strength of a thin scarf joint with a one-in-twenty slope and the inability to keep even a single 0.13 mm (0.005 in.) tape layer bonded under fatigue loads unless the edges had been serrated with pinking shears.

These documented indications of the problems associated with absorbed moisture in laminates being repaired by hot bonding or by co-curing of the patch are typical of the numerous unpublicized problems observed by many organizations. If there is sufficient absorbed moisture in a · cured laminate, the rapid heat-up rate associated with adhesive bonding or with a co-cured repair can cause delaminations from the pressure of the steam being generated.

It is most important to dry laminates thoroughly and slowly before repairing them at elevated temperatures. Any water that migrates into the uncured resin or adhesive during the cure cycle has the same adverse effect on mechanical properties as does moisture absorbed by prepregs (composite material preimpregnated with resin) or adhesive films exposed to humid environments. The water-break test should not contribute to this moisture problem because the water applied during the test contacts

Fig. 11 Surface contamination left by the evaporation of cleaning solvent. Courtesy of Canadair

the surface for such a short time that it is only necessary to dry the surface afterward; no moisture will migrate to the interior. To place this in perspective, laminates 6.3 mm (0.25 in.) thick need 24 h of drying at 135 °C (275 °F) to enable interior moisture to migrate out of the laminate; thus, surface moisture will not migrate far into the laminate in a few minutes at room temperature.

There is a special need for caution when drying out fibrous composite panels containing honeycomb or foam cores. Any moisture that has accumulated in these cells could be converted to steam, if heated excessively, and explode the whole panel. Such panels should be dried at a temperature no greater than 70 °C (160 °F) unless the entire panel is bagged and pressurized externally during the drying cycle.

REFERENCES

1. B.M. Parker and R.M. Waghorne, Surface Pretreatment of Carbon Fibre-Reinforced Composites for Adhesive Bonding, *Composites*, Vol XIII, July 1982
2. R.L. Radecky, "Adhesive Bonding of Precured Graphite/Epoxy Laminates," Report LR-10737, Materials and Process Engineering Laboratory, Douglas Aircraft Company, Nov 1982
3. J.D. Labor and S.H. Myhre, "Large Area Composite Structure Repair," Technical Report AFFDL-TR-79-3040, Air Force Flight Dynamics Laboratory, March 1979

Surface Contamination Considerations

Donald F. Sekits, Boeing Military Airplanes, Parts, Materials, and Processes Technology

SURFACE CONTAMINATION must be addressed whenever bonding or adhesion is required in the manufacture or repair of a structural component. Surfaces are generally contaminated as a result of the service environment. However, residual contaminants from the original part manufacture must also be considered. Compatibility with the substrate to be repaired, as well as with all repair materials, must be evaluated and allowable limits established. Removal of contaminants to acceptable limits could be conducted as a specific step in the repair process or could be accomplished during damage removal, surface preparation, or even during the bonding process.

Although the focus of this article is the role of contaminants in the bonded repair process, the influence of surface contamination on a component's service history is also examined. Bonded metals and composites, in both laminate and sandwich structures, are briefly discussed. Fuller descriptions of the surface preparation of a variety of substrates are presented in the Section "Surface Considerations" in this Volume.

Contamination

The principal elements that contribute to the requirement to perform bonded structural repairs, other than the part exceeding its design limits, include marginal integrity as a result of nondetectable defects created during part manufacture, finish system failure, and both maintenance-induced and foreign object damage. When there is damage, exposure of the damaged region to the service environment usually accelerates the component's degradation. In fact, the environment is usually the source of surface contamination.

During the manufacture of adhesively bonded assemblies, such contaminants as mold releases, parting films, and employee carelessness (transferring fingerprints, hand lotions, food, and other contaminants to parts) have led to the production of inferior components. Surface contaminants will also affect the finish system adhesion if they are not properly removed. Two elements of finish systems should be considered in terms of their ability to provide contaminant protection. First, some systems incorporate a so-called wash primer to facilitate the removal of the finish system itself resulting in less than optimal adhesion. Second, paint films applied to exterior structures are not impenetrable moisture barriers. Therefore, finishes do not fully protect the structure from environmental contaminants.

Foreign object or maintenance-induced damage obviously creates a path for the intrusion of contaminants and usually results in degradation either of the exposed surfaces or of the interior of sandwich structures. On aluminum surfaces, loosely adhered oxide corrosion products are formed. For both aluminum and composite sandwich structures, the freeze/thaw or expansion/contraction cycle inherent in flight envelopes causes delamination and rapid increase in the size of the damaged area.

A contaminant itself can be defined as any substance or surface condition that is detrimental to a structural adhesive bond. Consequently, the contamination must be removed or treated prior to or during the bonding process. Aerospace experience shows that the principal surface contaminants on parts that have been damaged in service include moisture, hydraulic fluid, and fuel. Moisture is an electrolyte for the galvanic corrosion of metals. It can be wicked into sandwich structures and absorbed by composite materials. Hydraulic fluid usually is found on movable control surfaces because of spills or leaks. Fuel exposure is generally limited to aircraft structural areas that are sealed in order to provide fuel storage.

Metallic Substrates

Aluminum is the prevalent metallic material used in subsonic aircraft structures. Both bonded metal-to-metal and metal sandwich structures have been in service since the 1950s. Besides external factors, corrosion is the principal reason for structural degradation of these parts, and, in most cases, is initiated by inadequate prebond surface preparation. As the oxide forms, it occupies more space than the base metal, thereby delaminating the bonded structure. In the repair process, the first step is to remove the close-proximity finishes and sealants. The structure must then be completely dried. This is particularly important with sandwich configurations. Next, all corrosion products must be removed. Corroded honeycomb core is usually obvious because it is often reduced to a powder. Because of its thinness, corroded honeycomb must be removed and replaced. The nonporous nature of metals allows the use of standard solvent cleaning as a means of removing organic soils.

The most important consideration in repairing metallic structure is the prevention of subsequent corrosion. This is best accomplished by phosphoric acid anodizing, followed by the application of a corrosion-inhibiting adhesive primer. The phosphoric acid anodize produces a very thin oxide that is stable and provides excellent primer adhesion. The oxide by itself does not offer corrosion resistance, but in conjunction with the primer, provides an order of magnitude better protection than sealed acid anodized oxides. The best surface preparation procedures available need to be employed when producing bond lines that are exposed to the environment.

Composite Substrates

For the repair of polymeric composite materials, surface contamination considerations primarily involve the matrices. Because the fibers that reinforce the matrices—whether they are carbon, glass, or aramid fibers—are typically impregnated with the resin matrix, their surfaces usually are not subject to contamination. The matrices vary significantly in chemical composition and processing history and therefore have different affinities for contaminants. For this discussion, the matrices are categorized as thermosets and thermoplastics. Although thermosetting epoxy matrices are primarily addressed, the information is generally applicable to bismaleimides and thermoset polyimides as well, unless noted otherwise.

Thermosets. The absorption of moisture by epoxy-base composites under service conditions has a severe impact on the ability to successfully repair damaged structures. Conventional repair technology commonly requires cure temperatures equivalent to those originally used in part manufacture (120 to 180 °C, or 250 to 355 °F). Unless moisture is completely removed, additional damage often results during the repair cure cycles. Similarly, hydraulic fluid and jet fuel must also be removed from surfaces before repair bonding is performed.

Figure 1 shows the moisture absorption of 16-ply carbon fiber reinforced epoxy panels under ambient laboratory conditions and exposed to condensing humidity at 71 °C (160 °F). The 1.3% weight gain shown for the panel exposed to high humidity approaches the reported saturation values for epoxy-base composites. It also corresponds to moisture contents measured on aircraft parts that have been in service for long periods of time in tropical environments.

However, from a contamination consideration, a significant loss of material properties occurs in less than 24 h under ambient conditions. Lap-shear data (Ref 1) show that when using a standard repair adhesive cured under vacuum pressure and at 180 °C (355 °F), there is a 10% statistically significant reduction in shear strength unless the panels are thoroughly dried (24 h at 95 °C, or 200 °F) before repair. From a practical point of view this means that if a part is damaged during the mechanical assembly process of its initial manufacture, it should be dried before attempting a bonded adhesive repair. This is especially true for a part that has been autoclave cured and nondestructively tested with water-coupled through-transmission ultrasonics (industry standard). The allowable level of contami-

nation is directly related to the moisture sensitivity of the repair adhesive system.

The presence of moisture will affect the adhesive bond in sandwich structures similarly, but there is also a risk of debonding of the honeycomb core nodes and skin/core delamination in areas other than those being repaired. Therefore, as an example, it is standard practice at the Air Force Air Logistics Centers to bake all sandwich parts that require repair for a minimum of 48 h at 95 °C (205 °F).

In the case of hydraulic fluids, Fig. 2 shows a weight loss for 16-ply epoxy laminated with an initial residual water content of 0.12% when soaked in hot (65 °C, or 150 °F) MIL-H-83282 fluid. The weight loss may be caused by a replacement of residual moisture. In order to test this theory, additional panels were dried prior to the soak in hot hydraulic fluid and in fact did gain 0.03% by weight after 24 h. However, based on lap-shear results, exposure of carbon-epoxy laminates to 65 °C (150 °F) hydraulic fluid does not appear to have any detrimental effect on subsequent hot bonded repairs as long as the surface is lightly abraded and solvent wiped prior to adhesive application.

Exposure of laminates to jet fuel (JP-4) at room temperature showed a weight loss similar to that of the hydraulic fluid soaked panels (Fig. 3). When the panels were dried prior to exposure, a slight weight gain resulted within 24 h. The standard abrade and solvent wipe preparation were not sufficient to prepare fuel-soaked panels for bond repairs that required cures at the same temperature as the original part. It appears that, like moisture, complete fuel (JP-4) contaminant removal is required prior to adhesive application.

A less extensive but similar study was conducted with 20-ply carbon-bismaleimide (BMI) panels. The moisture absorption as determined by weight gain is shown in Fig.

4. For the BMI, more than half of the moisture was picked up within 48 h of exposure to condensing humidity at 71 °C (160 °F). After 49 days at ambient conditions the weight gain was 0.51%. The best results for lap-shear tests based on the same adhesive as was used for the epoxies and cured under vacuum at 180 °C (355 °F), were for the panels held at ambient conditions. Saturating the panels per Fig. 4 or completely drying them resulted in significantly lower lap-shear values of almost 15% less for the dried panels and up to 25% less for the moisture-saturated panels.

After a 7-day exposure to hot hydraulic fluid and fuel, dried BMI panels showed weight gains of 0.05% and 0.11%, respectively. After 1 day, the hydraulic fluid panel had absorbed 0.05%, but the fuel-soaked panel had only gained 0.04%. As with the epoxies, bonded repair mechanical properties were not affected by the hydraulic fluid but were significantly reduced by only minute quantities of fuel absorbed in a relatively short time. Two types of failure modes were noted in the lap-shear specimens. For the nonexposed and hydraulic fluid exposed panels, the failure was interlaminar within the composite, whereas for the moisture-and-fuel exposed panels, the failure was generally within the adhesive, similar to the failure of epoxy composites.

Thermoplastics. For this discussion, experience with the thermoplastic polyetheretherketone (PEEK)-base composite is described. This carbon-PEEK composite has been extensively characterized by the aerospace industry and is one of the most inert of the thermoplastic matrices. The data presented are for a 20-ply laminate. The absorption values for the various contaminants are tabulated in Table 1.

The carbon-PEEK laminates absorb much less moisture than the thermoset materials but have a greater propensity for both hy-

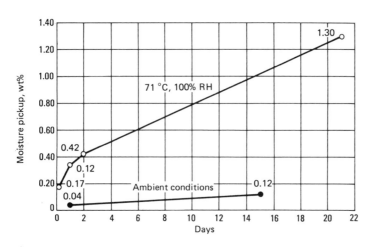

Fig. 1 Moisture absorption of 16-ply carbon-epoxy laminates

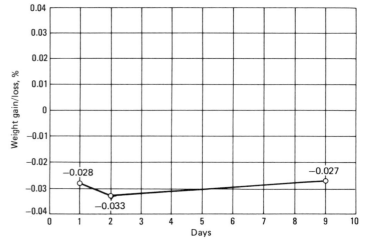

Fig. 2 Hydraulic fluid absorption of 16-ply carbon-epoxy laminates

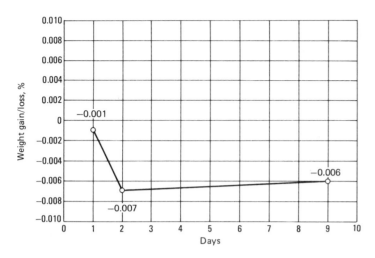

Fig. 3 Fuel (JP-4) absorption of 16-ply carbon-epoxy laminates

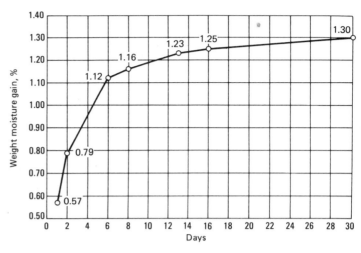

Fig. 4 Moisture absorption of 20-ply carbon-bismaleimide laminate

Table 1 Carbon-PEEK composite fluid absorption rates

Exposure time, days	Moisture absorption, wt%(a)	Hydraulic fluid absorption, wt%(b)	Fuel (JP-4) absorption, wt%(c)
1	0.03	0.06	0.17
7	0.10	0.13	0.34
27	0.20	...	...

(a) At 71 °C (160 °F), 100% relative humidity. (b) At 65 °C (150 °F). (c) At ambient temperature

draulic fluid and fuel. The much higher processing temperatures required by thermoplastics cause additional contamination problems. The transfer or decomposition of shop aides used during high-temperature processing act as sources of this surface contamination. These aides include vacuum bagging films and sealants, mold releases, and polytetrafluoroethylene-coated breather films.

In the case of PEEK-base composites, the inertness of the surface appears to mask the effects of all of the potential contaminants discussed above. The standard surface preparation procedures used for thermosets are generally not adequate. However, those that have been found acceptable for surface preparation, such as plasma treatment or etching with solutions used to activate polytetrafluoroethylene surfaces, also do an excellent job of removing surface contaminants. This points out that the most important considerations are to maximize the surface activity and adhesive compatibility. Contaminants must be recognized but not necessarily treated separately.

Many of the other thermoplastics, such as the polysulfones and polyamide-imides, fall between thermosets and PEEK in terms of surface contamination considerations. They are all processed at high temperatures and therefore are susceptible to contamination by the manufacturing aides used in processing, but may not require the severe surface preparation of PEEK-base composites.

REFERENCE

1. D.F. Sekits and M.A. Walker, "Repair of Contaminated Graphite Composite Structures," Paper presented at the 34th International SAMPE Symposium and Exhibition, Closed session, Society for the Advancement of Material and Process Engineering, 8-11 May 1989

SELECTED REFERENCE

● S. Emerick, D.F. Sekits, and M.A. Walker, "Advanced Depot Repair Processes," Air Force Contract F33615-85-C-5052, April 1990

Metric Conversion Guide

This Section is intended as a guide for expressing weights and measures in the Système International d'Unités (SI). The purpose of SI units, developed and maintained by the General Conference of Weights and Measures, is to provide a basis for world-wide standardization of units and measure. For more information on metric conversions, the reader should consult the following references:

- "Standard for Metric Practice," E 380, *Annual Book of ASTM Standards*, Vol 14.02, 1988, American Society for Testing and Materials, 1916 Race Street, Philadelphia, PA 19103
- "Metric Practice," ANSI/IEEE 268–1982, American National Standards Institute, 1430 Broadway, New York, NY 10018

- *Metric Practice Guide—Units and Conversion Factors for the Steel Industry*, 1978, American Iron and Steel Institute, 1133 15th Street NW, Suite 300, Washington, DC 20005
- *The International System of Units*, SP 330, 1986, National Bureau of Standards. Order from Superintendent of Documents, U.S. Government Printing Office, Washington, DC 20402-9325
- *Metric Editorial Guide*, 4th ed. (revised), 1985, American National Metric Council, 1010 Vermont Avenue NW, Suite 320, Washington, DC 20005–4960
- *ASME Orientation and Guide for Use of SI (Metric) Units*, ASME Guide SI 1, 9th ed., 1982, The American Society of Mechanical Engineers, 345 East 47th Street, New York, NY 10017

Base, supplementary, and derived SI units

Measure	Unit	Symbol	Measure	Unit	Symbol
Base units			Entropy	joule per kelvin	J/K
			Force	newton	N
Amount of substance	mole	mol	Frequency	hertz	Hz
Electric current	ampere	A	Heat capacity	joule per kelvin	J/K
Length	meter	m	Heat flux density	watt per square meter	W/m^2
Luminous intensity	candela	cd	Illuminance	lux	lx
Mass	kilogram	kg	Inductance	henry	H
Thermodynamic temperature	kelvin	K	Irradiance	watt per square meter	W/m^2
Time	second	s	Luminance	candela per square meter	cd/m^2
			Luminous flux	lumen	lm
Supplementary units			Magnetic field strength	ampere per meter	A/m
			Magnetic flux	weber	Wb
Plane angle	radian	rad	Magnetic flux density	tesla	T
Solid angle	steradian	sr	Molar energy	joule per mole	J/mol
			Molar entropy	joule per mole kelvin	J/mol · K
Derived units			Molar heat capacity	joule per mole kelvin	J/mol · K
Absorbed dose	gray	Gy	Moment of force	newton meter	N · m
Acceleration	meter per second squared	m/s^2	Permeability	henry per meter	H/m
Activity (of radionuclides)	becquerel	Bq	Permittivity	farad per meter	F/m
Angular acceleration	radian per second squared	rad/s^2	Power, radiant flux	watt	W
Angular velocity	radian per second	rad/s	Pressure, stress	pascal	Pa
Area	square meter	m^2	Quantity of electricity,		
Capacitance	farad	F	electric charge	coulomb	C
Concentration (of amount of			Radiance	watt per square meter steradian	W/m^2 · sr
substance)	mole per cubic meter	mol/m^3	Radiant intensity	watt per steradian	W/sr
Conductance	siemens	S	Specific heat capacity	joule per kilogram kelvin	J/kg · K
Current density	ampere per square meter	A/m^2	Specific energy	joule per kilogram	J/kg
Density, mass	kilogram per cubic meter	kg/m^3	Specific entropy	joule per kilogram kelvin	J/kg · K
Electric charge density	coulomb per cubic meter	C/m^3	Specific volume	cubic meter per kilogram	m^3/kg
Electric field strength	volt per meter	V/m	Surface tension	newton per meter	N/m
Electric flux density	coulomb per square meter	C/m^2	Thermal conductivity	watt per meter kelvin	W/m · K
Electric potential, potential			Velocity	meter per second	m/s
difference, electromotive force	volt	V	Viscosity, dynamic	pascal second	Pa · s
Electric resistance	ohm	Ω	Viscosity, kinematic	square meter per second	m^2/s
Energy, work, quantity of heat	joule	J	Volume	cubic meter	m^3
Energy density	joule per cubic meter	J/m^3	Wavenumber	1 per meter	1/m

850 / Metric Conversion Guide

Conversion factors

Area

To convert from	to	multiply by
in.^{2}	mm^2	6.451 600 E + 02
in.^{2}	cm^2	6.451 600 E + 00
in.^{2}	m^2	6.451 600 E − 04
ft^2	m^2	9.290 304 E − 02

Bending moment or torque

To convert from	to	multiply by
lbf · in.	N · m	1.129 848 E − 01
lbf · ft	N · m	1.355 818 E + 00
kgf · m	N · m	9.806 650 E + 00
ozf · in.	N · m	7.061 552 E − 03

Bending moment or torque per unit length

To convert from	to	multiply by
lbf · in./in.	N · m/m	4.448 222 E + 00
lbf · ft/in.	N · m/m	5.337 866 E + 01

Current density

To convert from	to	multiply by
A/in.^{2}	A/cm^2	1.550 003 E − 01
A/in.^{2}	A/mm^2	1.550 003 E − 03
A/ft^2	A/m^2	1.076 400 E + 01

Electric field strength

To convert from	to	multiply by
V/mil	kV/m	3.937 008 E + 01

Electricity and magnetism

To convert from	to	multiply by
gauss	T	1.000 000 E − 04
maxwell	μWb	1.000 000 E − 02
mho	S	1.000 000 E + 00
oersted	A/m	7.957 700 E + 01
Ω · cm	Ω · m	1.000 000 E − 02
Ω circular-mil/ft	μΩ · m	1.662 426 E − 03

Energy (impact, other)

To convert from	to	multiply by
ft · lbf	J	1.355 818 E + 00
Btu(a)	J	1.054 350 E + 03
Btu(b)	J	1.055 056 E + 03
cal(a)	J	4.184 000 E + 00
cal(b)	J	4.186 800 E + 00
kW · h	J	3.600 000 E + 06
W · h	J	3.600 000 E + 03

Flow rate

To convert from	to	multiply by
ft^3/h	L/min	4.719 475 E − 01
ft^3/min	L/min	2.831 000 E + 01
gal/h	L/min	6.309 020 E − 02
gal/min	L/min	3.785 412 E + 00

Force

To convert from	to	multiply by
lbf	N	4.448 222 E + 00
kip (1000 lbf)	N	4.448 222 E + 03
tonf	kN	8.896 443 E + 00
kgf	N	9.806 650 E + 00

Force per unit length

To convert from	to	multiply by
lbf/ft	N/m	1.459 390 E + 01
lbf/in.	N/m	1.751 268 E + 02
kip/in.	N/m	1.751 268 E + 05

Fracture toughness

To convert from	to	multiply by
ksi $\sqrt{\text{in.}}$	MPa $\sqrt{\text{m}}$	1.098 800 E + 00

Heat content

To convert from	to	multiply by
Btu/ft^3 (volume)	kJ/m^3	3.725 895 E + 01
Btu/lb (mass)	kJ/kg	2.326 000 E + 00
cal/g	kJ/kg	4.186 800 E + 00

Heat flow intensity

To convert from	to	multiply by
Btu/ft^2 · h	W/m^2	3.154 591 E + 00
cal/cm^2 · h	W/m^2	1.163 000 E + 01

Heat input

To convert from	to	multiply by
J/in.	J/m	3.937 008 E + 01
kJ/in.	kJ/m	3.937 008 E + 01

Impact strength

To convert from	to	multiply by
ft · lbf/ft	J/m	4.448 222 E + 00
ft · lbf/ft^2	J/m^2	1.459 002 E + 01
ft · lbf/in.	J/m	5.337 866 E + 01
ft · lbf/in.^{2}	J/m^2	2.102 043 E + 03

Length

To convert from	to	multiply by
Å	nm	1.000 000 E − 01
μin.	μm	2.540 000 E − 02
mil	μm	2.540 000 E + 01
in.	mm	2.540 000 E + 01
in.	cm	2.540 000 E + 00
ft	m	3.048 000 E − 01
yd	m	9.144 000 E − 01
mile	km	1.609 300 E + 00

Length per unit mass

To convert from	to	multiply by
in./lb	m/kg	5.599 740 E − 02
yd/lb	m/kg	2.015 907 E + 00

Mass

To convert from	to	multiply by
oz	kg	2.834 952 E − 02
lb	kg	4.535 924 E − 01
ton (short, 2000 lb)	kg	9.071 847 E + 02
ton (short, 2000 lb)	kg × 10^3(c)	9.071 847 E − 01
ton (long, 2240 lb)	kg	1.016 047 E + 03

Mass per unit area

To convert from	to	multiply by
oz/in.^{2}	kg/m^2	4.395 000 E + 01
oz/ft^2	kg/m^2	3.051 517 E − 01
oz/yd^2	kg/m^2	3.390 575 E − 02
lb/ft^2	kg/m^2	4.882 428 E + 00

Mass per unit length

To convert from	to	multiply by
lb/ft	kg/m	1.488 164 E + 00
lb/in.	kg/m	1.785 797 E + 01

Mass per unit time

To convert from	to	multiply by
lb/h	kg/s	1.259 979 E − 04
lb/min	kg/s	7.559 873 E − 03
lb/s	kg/s	4.535 924 E − 01

Mass per unit volume (includes density)

To convert from	to	multiply by
lb/ft^3	g/cm^3	1.601 846 E − 02
lb/ft^3	kg/m^3	1.601 846 E + 01
lb/in.^{3}	g/cm^3	2.767 990 E + 01
lb/in.^{3}	kg/m^3	2.767 990 E + 04
oz/in.^{3}	kg/m^3	1.729 994 E + 03

Power

To convert from	to	multiply by
Btu/s	W	1.055 056 E + 03
Btu/min	W	1.758 426 E + 01
Btu/h	W	2.930 711 E − 01
erg/s	W	1.000 000 E − 07
ft · lbf/s	W	1.355 818 E + 00
ft · lbf/min	W	2.259 697 E − 02
ft · lbf/h	W	3.766 161 E − 04
hp (550 ft · lbf/s)	kW	7.456 999 E − 01
hp (electric)	kW	7.460 000 E − 01

Power density

To convert from	to	multiply by
W/in.^{2}	W/m^2	1.550 003 E + 03

Pressure (fluid)

To convert from	to	multiply by
atm (standard)	Pa	1.013 250 E + 05
bar	Pa	1.000 000 E + 05
in. Hg (32 °F)	Pa	3.386 380 E + 03
in. Hg (60 °F)	Pa	3.376 850 E + 03
lbf/in.^{2} (psi)	Pa	6.894 757 E + 03
torr (mm Hg, 0 °C)	Pa	1.333 220 E + 02

Specific area

To convert from	to	multiply by
ft^2/lb	m^2/kg	2.048 161 E − 01

Specific energy

To convert from	to	multiply by
cal/g	J/g	4.186 800 E + 00
Btu/lb	kJ/kg	2.326 000 E + 00

Specific heat capacity

To convert from	to	multiply by
Btu/lb · °F	J/kg · K	4.186 800 E + 03
cal/g · °C	J/kg · K	4.186 800 E + 03

Stress (force per unit area)

To convert from	to	multiply by
tonf/in.^{2} (tsi)	MPa	1.378 951 E + 01
kgf/mm^2	MPa	9.806 650 E + 00
ksi	MPa	6.894 757 E + 00
lbf/in.^{2} (psi)	MPa	6.894 757 E − 03
MN/m^2	MPa	1.000 000 E + 00

Temperature

To convert from	to	multiply by
°F	°C	5/9 · (°F − 32)
°R	K	5/9
°F	K	5/9 · (°F + 459.67)
°C	K	°C + 273.15

Temperature interval

To convert from	to	multiply by
°F	°C	5/9

Thermal conductivity

To convert from	to	multiply by
Btu · in./s · ft^2 · °F	W/m · K	5.192 204 E + 02
Btu/ft · h · °F	W/m · K	1.730 735 E + 00
Btu · in./h · ft^2 · °F	W/m · K	1.442 279 E − 01
cal/cm · s · °C	W/m · K	4.184 000 E + 02

Thermal expansion

To convert from	to	multiply by
μin./in. · °C	10^{-6}/K	1.000 000 E + 00
μin./in. · °F	10^{-6}/K	1.800 000 E + 00

Velocity

To convert from	to	multiply by
ft/h	m/s	8.466 667 E − 05
ft/min	m/s	5.080 000 E − 03
ft/s	m/s	3.048 000 E − 01
in./s	m/s	2.540 000 E − 02
km/h	m/s	2.777 778 E − 01
mph	km/h	1.609 344 E + 00

Viscosity (dynamic and kinematic)

To convert from	to	multiply by
poise (P)	Pa · s	1.000 000 E − 01
cP	Pa · s	1.000 000 E − 03
lbf · s/in.^{2}	Pa · s	6.894 757 E + 03
ft^2/s	m^2/s	9.290 304 E − 02
ft^2/h	mm^2/s	2.580 064 E + 01
in.^{2}/s	mm^2/s	6.451 600 E + 02

Volume

To convert from	to	multiply by
in.^{3}	m^3	1.638 706 E − 05
ft^3	m^3	2.831 685 E − 02
fluid oz	m^3	2.957 353 E − 05
gal (U.S. liquid)	m^3	3.785 412 E − 03

Volume per unit time

To convert from	to	multiply by
ft^3/min	m^3/s	4.719 474 E − 04
ft^3/s	m^3/s	2.831 685 E − 02
in.^{3}/min	m^3/s	2.731 177 E − 07

Wavelength

To convert from	to	multiply by
Å	nm	1.000 000 E − 01

(a) Thermochemical. (b) International Table. (c) kg × 10^3 = 1 metric ton

SI prefixes—names and symbols

Exponential expression	Multiplication factor	Prefix	Symbol
10^{18}	1 000 000 000 000 000 000	exa	E
10^{15}	1 000 000 000 000 000	peta	P
10^{12}	1 000 000 000 000	tera	T
10^{9}	1 000 000 000	giga	G
10^{6}	1 000 000	mega	M
10^{3}	1 000	kilo	k
10^{2}	100	hecto(a)	h
10^{1}	10	deka(a)	da
10^{0}	1	BASE UNIT	
10^{-1}	0.1	deci(a)	d
10^{-2}	0.01	centi(a)	c
10^{-3}	0.001	milli	m
10^{-6}	0.000 001	micro	μ
10^{-9}	0.000 000 001	nano	n
10^{-12}	0.000 000 000 001	pico	p
10^{-15}	0.000 000 000 000 001	femto	f
10^{-18}	0.000 000 000 000 000 001	atto	a

(a) Nonpreferred. Prefixes should be selected in steps of 10^3 so that the resultant number before the prefix is between 0.1 and 1000. These prefixes should not be used for units of linear measurement, but may be used for higher order units. For example, the linear measurement, decimeter, is nonpreferred, but square decimeter is acceptable.

Abbreviations and Symbols

a   crack length
A   ampere
A   area; ratio of the alternating stress amplitude to the mean stress
Å   angstrom
AAPS   N-2-aminoethyl-3-aminopropyl trimethoxy silane
ac   alternating current
AES   Auger electron spectroscopy
AMS   Aerospace Material Specification
ANSI   American National Standards Institute
AP   alkaline peroxide
APA   3-aminophenylacetylene
APB   1,3-bis(3-aminophenoxy) benzene
APU   auxiliary power unit
AQL   acceptable quality levels
at.%   atomic percent
bal   balance
BMA   butadiene-co-maleic anhydride
BNIs   bisnadimides
BPO   benzoyl peroxide
BTDE   benzophenone tetracarboxylic acid dimethyl ester
CAA   chromic acid anodization
CIAP   corrosion-inhibiting adhesive primer
CLS   cracked lap-shear (specimens)
cpm   cycles per minute
cps   cycles per second
CSA   chromic-sulfuric acid; Canadian Standards Association
CTBN   carboxy-terminated butadiene acrylonitrile
CTE   coefficient of thermal expansion
CTSWZ   crack tip (stress) whitening zones
d   an operator used in mathematical expressions involving a derivative (denotes rate of change)
d   depth; diameter
dc   direct current
DCB   double-cantilever beam
DDS   diaminodiphenyl sulfone
DETA   diethylene triamine
DH   acrylic adhesives Du Pont Hypalon
diam   diameter
DMAC   dimethyl acetamide
DMF   dimethyl formamide
DMSO   dimethyl sulfoxide
DSC   dynamic scanning calorimetry
DTA   differential thermal analysis
e   natural log base, 2.71828
E   modulus of elasticity (Young's modulus)
EAA   poly(ethylene-co-acrylic acid)
EB   electron beam
ECN   epoxidized cresol novolac

EDAX   electron dispersive analysis by x-ray
EDS   energy dispersive spectroscopy
EELS   electron energy loss spectroscopy
EEW   epoxide equivalent weight
ENF   end-notched flexure
EPN   epoxided phenol novolac
Eq   equation
ESCA   electron spectroscopy for chemical analysis
ESD   electron stimulated desorption
et al.   and others
EXAFS   extended x-ray absorption fine structure
F   force
FAB   fast ion bombardment
FD   finite difference
FE   finite element
Fig.   figure
FIM   field ion microscopy
FMVSS   Federal Motor Vehicle Safety Standards
FPL-P   Forest Products Laboratory paste
ft   foot
FTIR   Fourier transform infrared
g   gram
G   shear modulus
G'   storage modulus
G"   loss modulus
gal   gallon
GPa   gigapascal
GSA-FSS   General Services Administration, Federal Supply Service
h   hour
H   height
hexa   hexamethylenetetramine
HFP   hexafluoropropylene
HI   high impact
HLB   hydrophile-lipophile balance
Hz   hertz
IAES   ion-induced Auger electron spectroscopy
ID   inside diameter
IIR   isoprene-isobutylene rubber
ILMMS   independently loaded mixed-mode specimen
IML   inner mold line
in.   inch
INS   ion neutralization spectroscopy
IPN   interpenetrating polymer networks
IR   infrared (radiation)
ISO   International Organization for Standardization
ISS   ion scattering spectroscopy
J   joule
k   notch sensitivity factor

K   Kelvin
K   coefficient of thermal conductivity; bulk modulus of elasticity
K_c plane-stress fracture toughness
K_I stress-intensity factor
K_{Ic} plane-strain fracture toughness; mode I critical stress-intensity factor
K_{Iscc} threshold stress intensity for stress-corrosion cracking
kg   kilogram
km   kilometer
kPa   kilopascal
ksi   kips (1000 lbf) per in.^{2}
kV   kilovolt
L   liter
L   length
lb   pound
lbf   pound-force
LEED   low-energy electron diffraction
LEFM   linear elastic fracture mechanics
LFM   linear fracture mechanics
ln   natural logarithm (base *e*)
log   common logarithm (base 10)
m   meter
MAA   methacrylic acid
MDA   methylene dianiline
MDI   diphenylmethane-4,4'-diisocyanate
meq   milliequivalent
Mg   megagram
mg   milligram
min   minute; minimum
MJ   megajoule
mL   milliliter
mm   millimeter
MMA   methyl methacrylate
MMF   mixed-mode flexure
mol%   mole percent
MPa   megapascal
MSLS   modified single-lap shear
MW   molecular weight
MVEMA   methyl vinyl ether-co-maleic anhydride
N   Newton
N   fatigue life (number of cycles)
NASA   National Aeronautics and Space Administration
NBR   acrylonitrile-butadiene rubber
NBS   N-bromosuccinimide
NDE   nondestructive evaluation
NDI   nondestructive inspection
NDT   nondestructive testing
NE   nadic ester
NIST   National Institute of Science and Technology
nm   nanometer
NMP   N-methyl-2-pyrrolidone
No.   number

NOSR normal operating stress regime
NPFC Naval Publications and Forms Center
NPT National pipe thread
NR natural rubber; neutron radiography
NTMP nitrilotris (methylene) phosphoric acid
NVH noise, vibration, and harshness
OD outside diameter
OML outer mold line
OSHA Occupational Safety and Health Administration
oz ounce
p page
P applied force; pressure
Pa pascal
PAA phosphoric acid anodization
PDE partial differential equations
PDMS polydimethylsiloxane
PF phosphate-fluoride (etch)
PIB polyisobutylene
PIXE proton-induced x-ray emission
PMMA polymethyl methacrylate
PMR polymerization from monomeric reactants
ppm parts per million
PPTA poly-*p*-phenylene terephthalamide
PSA pressure-sensitive adhesive
psi pounds per square inch
QPL qualified products list
R radius
RA reduction of area
RBS Rutherford backscattering spectroscopy
Ref reference
rem remainder
RFGD radio-frequency glow discharge
RGA residual gas analysis
RH relative humidity
RHEED reflection high-energy electron diffraction
rms root mean square
rpm revolutions per minute
RSR raised-stress regime
RTV room-temperature-vulcanizing
s second
SAA surface-activated acrylic
SAE Society of Automotive Engineers
SALI surface analysis with laser ionization
SAM scanning Auger microprobe
SEM scanning electron microscopy
SEN single-edge notched
SEXAFS surface extended x-ray absorption fine structure
SHA sodium hydroxide anodization

SIMS secondary ion mass spectrometry
SIS styrene-isoprene-styrene
SLS single-lap shear
SMPP sintered metal powder process
S-N stress-number of cycles (fatigue)
SNMS secondary neutral mass spectrometry
sp gr specific gravity
STEM scanning transmission electron microscope
STM scanning tunnelling microscopy
STP standard temperature and pressure
t thickness; time
T temperature
TBA Torsional braid analysis
TCA trichloroethane
TCE trichloroethylene
TDCB tapered double-cantilever beam
TDI toluene diisocyanate
TDR time-domain reflectometry
TEM transmission electron microscopy; transverse electromagnetic
TFE tetrafluoroethylene
T_g glass transition temperature
TGDDM/DDS tetraglycidyl 4,4'-diamino-diphenyl methane (cured with) 4,4'-diamino-diphenyl sulfone
THB temperature-humidity-bias (test)
T_m melt temperature
TTT time-temperature-transformation
TTU through-transmission ultrasonics
TWA time-weighted average
UHV ultrahigh vacuum
UPS ultraviolet photoelectron spectroscopy
UV ultraviolet
V volt
VF_2 vinylidene fluoride
vol volume
vol% volume percent
WDCI water-displacing corrosion inhibitors
WDS wavelength dispersive spectroscopy
WLF Williams-Landel-Ferry
wt% weight percent
XAES x-ray-induced Auger electron spectroscopy
XANES x-ray absorption near-edge spectroscopy
XES x-ray emission spectroscopy
XPS x-ray photoelectron spectroscopy
yr year

Z impedance; atomic number; standard normal distribution
° angular measure; degree
°C degree Celsius (centigrade)
°F degree Fahrenheit
$\rightarrow, \leftarrow, \leftrightarrow$ direction of reaction
÷ divided by
= equals
≅ approximately equals
≠ not equal to
≡ identical with
> greater than
≫ much greater than
≧ greater than or equal to
∞ infinity
∝ is proportional to; varies as
∫ integral of
< less than
≪ much less than
≦ less than or equal to
± maximum deviation
− minus; negative ion charge
× diameters (magnification); multiplied by
· multiplied by
/ per
% percent
+ plus; positive ion charge
√ square root of
~ approximately; similar to
∂ partial derivative
α thermal diffusivity; angle of incidence
β angle of refraction
γ shear strain
δ skin depth
Δ change in quantity; increment; range; phase shift
ε normal strain; emissivity; dielectric coefficient
η viscosity
θ Curie temperature
μ linear attenuation coefficient; magnetic permeability; the mean (or average) of a distribution
μin. microinch
μm micron (micrometer)
ν Poisson's ratio; frequency
π pi (3.141592)
ρ density; resistivity
σ stress; standard deviation; electrical conductivity
Σ summation of
τ shear stress
Φ angle of refraction
ω circular frequency (angular velocity)
Ω ohm

Greek Alphabet

A, α	alpha	I, ι	iota	P, ρ	rho
B, β	beta	K, κ	kappa	Σ, σ	sigma
Γ, γ	gamma	Λ, λ	lambda	T, τ	tau
Δ, δ	delta	M, μ	mu	Υ, υ	upsilon
E, ε	epsilon	N, ν	nu	Φ, φ	phi
Z, ζ	zeta	Ξ, ξ	xi	X, χ	chi
H, η	eta	O, o	omicron	ψ, ψ	psi
Θ, θ	theta	Π, π	pi	Ω, ω	omega

Index